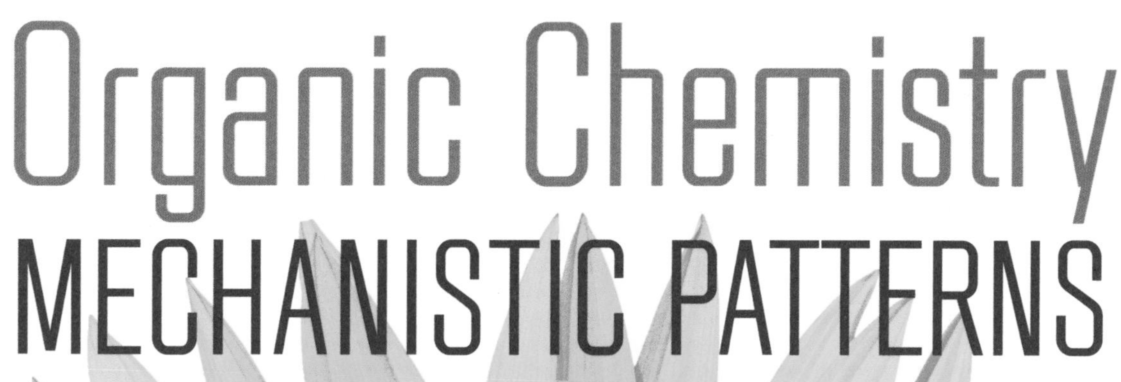
Organic Chemistry
MECHANISTIC PATTERNS

ORGANIC CHEMWARE

Organic ChemWare for use with *Organic Chemistry: Mechanistic Patterns* is a comprehensive collection of learning objects to aid in the teaching and learning of organic chemistry at the postsecondary level. Designed for both individual study and classroom projection, Organic ChemWare empowers students while redefining the lecture experience. It bridges the gap between the static imagery of textbooks and the dynamic world of organic chemistry.

Organic ChemWare includes more than 180 interactive, web-based multimedia simulations with an emphasis on:

- Lewis structures
- curved arrow notation
- reaction mechanisms
- orbital interactions
- conformational analysis
- stereochemistry
- ^{1}H- and ^{13}C-NMR

In the default "Study Mode," all animations (and orbital depictions, if applicable) are accompanied by informative text vignettes, pausing the animations and describing key points and reaction details. Toggling to "Presenter Mode" hides all text vignettes and zooms the animation to promote classroom focus while reducing cognitive load.

All animated mechanisms are depicted in dash/wedge bond line notation; the kinematic effect of bond motion helps students to perceive and understand the three-dimensionality of organic structures inferred by the notation and to "think tetrahedral."

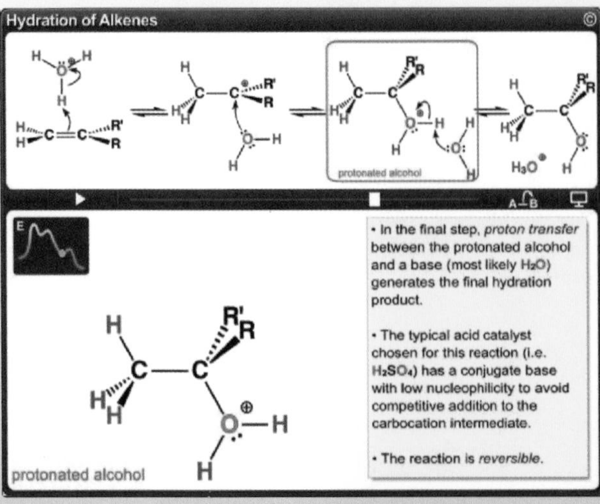

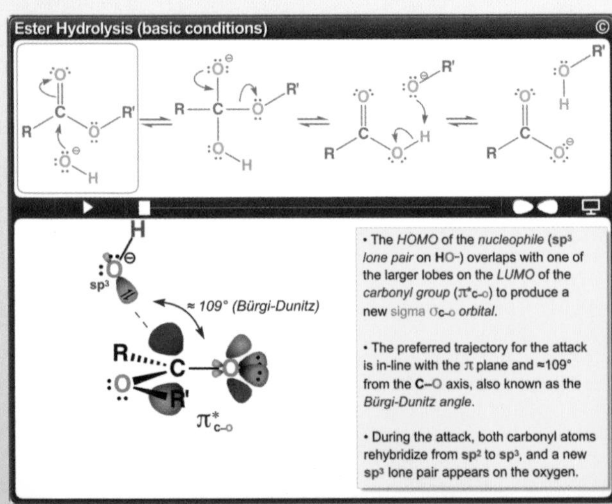

Organic ChemWare is included with every purchase of a new text.

ORGANIC CHEMISTRY

Mechanistic Patterns

William Ogilvie
University of Ottawa

Nathan Ackroyd
Mount Royal University

C. Scott Browning
University of Toronto

Ghislain Deslongchamps
University of New Brunswick

Felix Lee
The University of Western Ontario

Effie Sauer
University of Toronto Scarborough

NELSON

NELSON

Organic Chemistry

by William Ogilvie, Nathan Ackroyd, C. Scott Browning, Ghislain Deslongchamps, Felix Lee, Effie Sauer

Senior Publisher, Digital and Print Content:
Paul Fam

Marketing Manager:
Terry Fedorkiw

Technical Reviewers:
Philip Dutton, Barb Morra

Content Development Manager:
Katherine Goodes

Photo and Permissions Researcher:
Kristiina Paul

Production Project Manager:
Lila Campbell

Production Service:
Cenveo Publisher Services

Copy Editor:
Wendy Yano

Proofreader:
A. Malik Basha

Indexer:
BIM Creatives LLC

Design Director:
Ken Phipps

Managing Designer:
Pamela Johnston

Interior Design:
Cathy Mayer

Cover Design:
Courtney Hellam

Cover and Mechanistic Re-View Image:
Yuliyan Velchev/Shutterstock.com

Organic ChemWare Icon:
Ghislain Deslongchamps

Art Coordinator:
Suzanne Peden

Illustrators:
Crowle Art Group, Cenveo Publisher Services

Compositor:
Cenveo Publisher Services

Printed and bound in Canada
1 2 3 4 20 19 18 17

For more information contact Nelson Education Ltd., 1120 Birchmount Road, Toronto, Ontario, M1K 5G4. Or you can visit our Internet site at nelson.com

Library and Archives Canada Cataloguing in Publication Data

Ogilvie, William Walter, author
Organic chemistry: mechanistic patterns / William Ogilvie (University of Ottawa), Nathan Ackroyd (Mount Royal University), Felix Lee (The University of Western Ontario), Scott Browning (University of Toronto), Ghislain Deslongchamps (University of New Brunswick), Effie Sauer (University of Toronto).

Includes bibliographical references and index.
ISBN 978-0-17-650026-9 (hardcover)

1. Chemistry, Organic—Textbooks. I. Ackroyd, Nathan, author II. Title.

QD251.3.O45 2017 547
C2016-907181-2

ISBN-13: 978-0-17-650026-9
ISBN-10: 0-17-650026-X

BRIEF CONTENTS

CONTENTS

ABOUT THE AUTHORS

Courtesy of Pamela Trudeau

William Ogilvie, PhD, is an Associate Professor in the Department of Chemistry at the University of Ottawa. He was an NSERC 1967 Scholar who received his PhD from the University of Ottawa in 1989. Following this, he was an NSERC postdoctoral fellow at the University of Pennsylvania and at the Scripps Research Institute. In 1990, he joined Boehringer-Ingelheim Pharmaceuticals (then BioMega) in Montreal working as a research scientist and spent 11 years in the industry before moving to the University of Ottawa. His teaching focus has been organic and medicinal chemistry, and he has also taught large science classes for non-scientists. He was awarded the Excellence in Education Prize by the University of Ottawa in 2006.

Courtesy of Nathan Ackroyd

Nathan Ackroyd, PhD, is an Associate Professor of Chemistry and faculty member at Mount Royal University in Calgary. He has always been interested in how the world works as it does. Trying to find detailed answers to broad questions led him to an early interest in chemistry and physics. After earning a Bachelor of Science in Chemistry from Brigham Young University, he moved to the University of Illinois where he focused on the organic synthesis of imaging agents to simplify the diagnosis of breast tumours. In addition to Organic Chemistry, Dr. Ackroyd teaches Biochemical Pharmacology and Drug Discovery for fourth-year biology students. Through these courses, he hopes to increase students' understanding of how the chemicals we are made of interact with the chemicals we use every day.

Courtesy of Scott Browning

C. Scott Browning, PhD, is an Associate Professor, Teaching Stream, in the Department of Chemistry at the University of Toronto. After finishing his doctorate, Dr. Browning completed a postdoctoral term as a JST Fellow at the National Institute of Bioscience in Japan, developing novel, platinum-based, anti-cancer prototypes. He is interested in chemistry education, public scientific literacy, and the use of information technology in the teaching and learning of postsecondary science. His research pursuits include molecular modelling as both a teaching and research tool, focusing on small molecules in reactions of chemical and biological interest.

Rob Blanchard Photo

Ghislain Deslongchamps, PhD, is Professor and Chair of Chemistry at the University of New Brunswick. Upon joining the department, he quickly established a name for himself in the research field of molecular recognition. His research interests currently include organocatalysis, computer-assisted molecular design, and visualization in chemical education. He has always showed a strong commitment to teaching and how technology can help students learn more effectively. He has been recognized by *Maclean's* magazine as one of UNB's top professors. Developing new computer-based visualization techniques for chemical education since 2000, he is the creator of Organic Chemistry Flashware and Organic ChemWare published by Nelson. Dr. Deslongchamps is a past director of the SHAD program at UNB, Canada's top summer enrichment program, which empowers exceptional high school students.

Courtesy of Felix Lee

Felix Lee, PhD, is an Assistant Professor in the Department of Chemistry at The University of Western Ontario. Dr. Lee is a two-time recipient of Western University's Award of Excellence in Undergraduate Teaching, awarded by the University Students' Council, The Bank of Nova Scotia, and the UWO Alumni Association. He is also a recipient of a Marilyn Robinson Award for Excellence in Teaching. As one student describes, "He has not only turned my most hated subject into my favourite; he has inspired me to do well in subsequent courses and life events." According to another professor, "He is obviously recognized as an excellent teacher, and now he is helping the faculty by being a teacher's teacher." Dr. Lee has extensively been involved in the restructuring of first-year chemistry at The University of Western Ontario, and he is currently a co-director of the new Western Integrated Science program.

Courtesy of Effie Sauer

Effie Sauer, PhD, is an Associate Professor, Teaching Stream, in the Department of Physical and Environmental Sciences at the University of Toronto Scarborough. With the department since 2009, she has taught a variety of courses including general, organic, and green chemistry. In 2012, Dr. Sauer was honoured to be named one of UTSC's "Professors of the Year" by the student-run newspaper, *The Underground*. More recently, she was awarded the UTSC Faculty Teaching Award (2013). Prior to her appointment at UTSC, Dr. Sauer completed her PhD at the University of Ottawa (2007), followed by a postdoctoral fellowship at Yale University.

This group of authors has applied a "special teams" approach to the development of this text. Each author has contributed in a focused way to different aspects of the book to ensure consistency throughout. By taking on separate tasks in writing the book, they have focused on each person's strength in making the project the best it could be.

FOREWORD

Courtesy of Molly Shoichet/ Photographer Brigitte Lacombe

Organic chemistry permeates all parts of our everyday lives, from the soap we use to clean dishes, to the pharmaceutical drugs we take for our ailments, to the polymers used in clothing. Organic chemistry is also used to design new drugs—such as antibody–drug conjugates that are being used to more effectively treat cancer—and to create materials that can more effectively capture the sun's energy for a clean, environmentally friendly source of power. With organic chemistry, we can design molecules to overcome current challenges, resulting in a better future.

While organic chemistry can be daunting if students think about it as a large list of reactions that have to be memorized, it can be super exciting and straightforward when considered from a mechanistic perspective—that is, understanding how and why reactions occur. This textbook approaches organic chemistry from a mechanistic perspective while at the same time giving students some practical touch points in the "Why It Matters" section of every chapter.

I particularly like this approach to teaching organic chemistry. By teaching students how and why reactions occur, they can begin to appreciate when they will occur. This is particularly satisfying for students and can be complemented with practical laboratory experiments and creative critical-thinking projects. The latter are most useful for any future studies involving independent research or creative problem solving.

Molly S. Shoichet, PhD, NAE, O. Ont.
University Professor and Tier 1 Canada Research Chair
Department of Chemical Engineering & Applied Chemistry
University of Toronto

PREFACE

Organic chemistry is a science that has existed for less than 200 years. The traditional way to teach this discipline is based on the laboratory technology for identifying organic substances that existed in the eighteenth century, in which chemical tests that detected the presence of particular functional groups were used to identify molecular structure. Because of the importance of these chemical tests, it was natural that classroom instruction would focus on the functional groups that were the targets of these tests. Although successful, this approach required extensive rote memorization without understanding. Deep understanding of the discipline therefore required a long time and considerable experience to acquire.

In the 1930s, the idea of understanding reactivity by considering the movements of electrons, rather than just atoms, was pioneered. This mechanistic method of analyzing reactivity is a more general and powerful way of thinking about organic chemistry, making it possible to describe *why* a reaction occurred, and to explain many concepts that had previously been derived from empirical measurement. But...

Today, textbooks and courses are still organized around the functional group concept.

Today, textbooks and courses are still organized around the functional group concept. Mechanisms are taught today, but typically in the context of the older functional group way of studying the discipline. Because chemists learn the discipline according to functionality, they tend to teach the subject the way they have been taught—grouping by molecular structure. It is difficult to move beyond this traditional way of thinking about organic chemistry. We, as educators, tend to fall back into old patterns, and utilize the functional-group-centred approach.

For example, ozonolysis is often taught as part of alkene reactivity, presenting a complex cycloaddition to students who are still trying to master the concepts of nucleophile and electrophile. Texts often compound this challenge by presenting "magic" reactions where no mechanistic insight is provided. In the case of ozonolysis, a reducing agent is often shown to magically transform the ozonide into two carbonyl components, with no understanding of how the process operates.

A mechanistic method is—in principle—more general, easier to understand, and provides a better way to achieve a deep understanding of chemical reactivity. But a mechanistic method requires a mechanistic approach. A curriculum must be organized around reactivity, not structure.

But a mechanistic method requires a mechanistic approach. A curriculum must be organized around reactivity, not structure.

Organizing a curriculum around chemical reactivity rather than structure has many advantages. Chemical reactions are often more difficult to understand than molecular shapes and patterns. Therefore, organizing a curriculum around reactivity breaks down the hardest problem into manageable chunks. Recognizing patterns of electron flow between seemingly different reactions can allow a chemist to predict how a chemical will react, even if they have never seen a particular reaction before. Visualizing reactivity as a collection of patterns in electron movement is a more powerful and systematic way of approaching learning in organic chemistry. It still requires memorization, but because this is directly linked to reaction patterns, a deeper understanding of the discipline is possible. This lowers student workload and gives more structure to the discipline. For example, many students are currently taught elimination reactions, and are later shown the oxidation of alcohols and aldehydes. Because two different terms are used, students do not realize that these reactions follow the same reactivity pattern. Therefore, they simply memorize them. If they understand eliminations, they can understand oxidation if the mechanistic similarities are pointed out.

Visualizing reactivity as a collection of patterns in electron movement is a more powerful and systematic way of approaching learning in organic chemistry.

The mechanistic method requires a shift in philosophy in organic instruction. The functional group approach arranges lessons around structure. A mechanistic view of organic chemistry arranges lessons around patterns of electron movement and considers functional groups as participants in these movements. Study a reaction, and then consider the functional groups that can carry out the transformation.

In writing this book, we have taken great care to establish a progression of *reactivity*, from simple to complex. Functional groups are introduced as necessary, while focusing on the reaction at hand rather than on the various things each functional group does. This provides the student with a set of tools they can use and understand, rather than just having a list of reactions to memorize.

At each stage, we have placed an emphasis on understanding the underlying principles of each reaction. Care has been taken to point out many details that are usually glossed over in other mechanistic descriptions.

Pedagogy for the Mechanistic Approach

Throughout the chapters, assorted pedagogy promotes student learning and engagement based on the mechanistic approach.

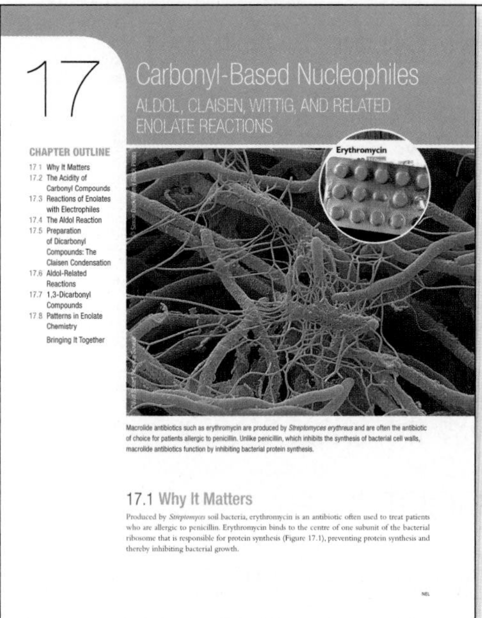

17

Carbonyl-Based Nucleophiles

ALDOL, CLAISEN, WITTIG, AND RELATED ENOLATE REACTIONS

CHAPTER OUTLINE

17.1 Why It Matters
17.2 The Acidity of Carbonyl Compounds
17.3 Reactions of Enolates with Electrophiles
17.4 The Aldol Reaction
17.5 Preparation of Dicarbonyl Compounds: The Claisen Condensation
17.6 Aldol-Related Reactions
17.7 1,3-Dicarbonyl Compounds
17.8 Patterns in Enolate Chemistry
Bringing It Together

Macrolide antibiotics such as erythromycin are produced by *Streptomyces erythreus* and are often the antibiotic of choice for patients allergic to penicillin. Unlike penicillin, which inhibits the synthesis of bacterial cell walls, macrolide antibiotics function by inhibiting bacterial protein synthesis.

17.1 Why It Matters

Produced by *Streptomyces* soil bacteria, erythromycin is an antibiotic often used to treat patients who are allergic to penicillin. Erythromycin binds to the centre of one subunit of the bacterial ribosome that is responsible for protein synthesis (Figure 17.1), preventing protein synthesis and thereby inhibiting bacterial growth.

NEL

Why It Matters begins each chapter and provides an introduction to the relevancy of the material about to be covered.

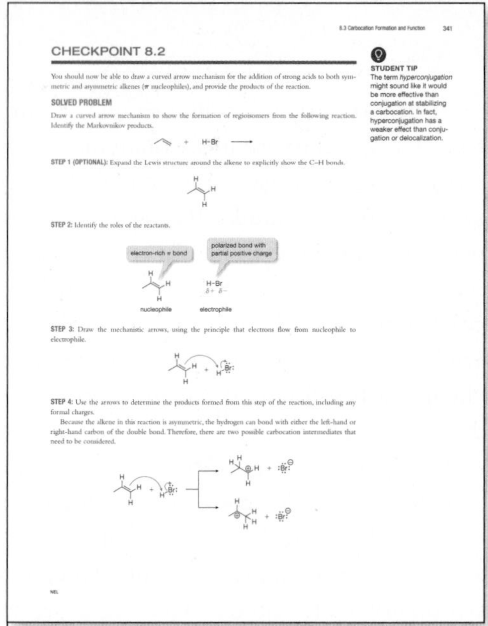

8.3 Carbocation Formation and Function 341

CHECKPOINT 8.2

You should now be able to draw a curved arrow mechanism for the addition of strong acids to both symmetric and asymmetric alkenes (π nucleophiles), and provide the products of the reaction.

SOLVED PROBLEM

Draw a curved arrow mechanism to show the formation of regioisomers from the following reaction. Identify the Markovnikov products.

STEP 1 (OPTIONAL): Expand the Lewis structure around the alkene to explicitly show the C–H bonds.

STEP 2: Identify the roles of the reactants.

STEP 3: Draw the mechanistic arrows, using the principle that electrons flow from nucleophile to electrophile.

STEP 4: Use the arrows to determine the products formed from this step of the reaction, including any formal charges.

Because the alkene in this reaction is asymmetric, the hydrogen can bond with either the left-hand or right-hand carbon of the double bond. Therefore, there are two possible carbocation intermediates that need to be considered.

STUDENT TIP

The term *hyperconjugation* might sound like it would be more effective than conjugation at stabilizing a carbocation. In fact, hyperconjugation has a weaker effect than conjugation or delocalization.

NEL

Each chapter features several **Checkpoints**, which follow the description of key material in the text. These inform the student explicitly about what they should now be able to do or understand, illustrated with a solved problem. Related exercises are included along with a problem that integrates several ideas.

Student Tips identify shortcuts or common mistakes that students make.

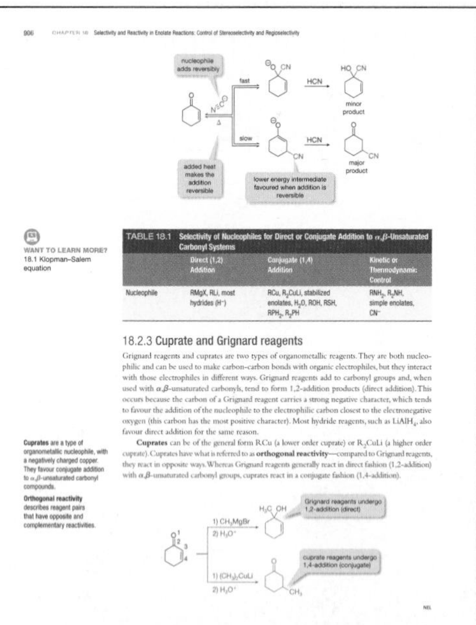

906 CHAPTER 18 Selectivity and Reactivity in Enolate Reactions: Control of Stereoselectivity and Regioselectivity

WANT TO LEARN MORE?
18.1 Klopman–Salem equation

TABLE 18.1 Selectivity of Nucleophiles for Direct or Conjugate Addition to α,β-Unsaturated Carbonyl Systems

	Direct (1,2) Addition	Conjugate (1,4) Addition	Kinetic or Thermodynamic Control
Nucleophile	RMgX, RLi, most hydrides (H^-)	RCu, R_2CuLi, stabilized enolates, H_2O, ROH, RSH, RPH_2, R_2PH	RNH_2, R_2NH, simple enolates, CN^-

18.2.3 Cuprate and Grignard reagents

Grignard reagents and cuprates are two types of organometallic reagents. They are both nucleophilic and can be used to make carbon-carbon bonds with organic electrophiles, but they interact with those electrophiles in different ways. Grignard reagents add to carbonyl groups and, when used with α,β-unsaturated carbonyls, tend to form 1,2-addition products (direct addition). This occurs because the carbon of a Grignard reagent carries a strong negative character, which tends to favour the addition of the nucleophile to the electrophilic carbon closest to the electronegative oxygen (this carbon has the most positive character). Most hydride reagents, such as $LiAlH_4$, also favour direct addition for the same reason.

Cuprates can be of the general form RCu (a lower order cuprate) or R_2CuLi (a higher order cuprate). Cuprates have what is referred to as **orthogonal reactivity**—compared to Grignard reagents, they react in opposite ways. Whereas Grignard reagents generally react in direct fashion (1,2-addition) with α,β-unsaturated carbonyl groups, cuprates react in a conjugate fashion (1,4-addition).

Cuprates are a type of organometallic nucleophile, with a negatively charged copper. They favour conjugate addition to α,β-unsaturated carbonyl compounds.

Orthogonal reactivity describes reagent pairs that have opposite and complementary reactivities.

NEL

Want to Learn More provides content on the Student Companion Website that describes a topic in more detail. These illustrate a reaction or concept beyond the scope of the text, but which may be of interest to advanced students or to those who use the book as a reference.

Organic ChemWare, an extensive collection of more than 180 interactive animations on the Student Companion Website, has been integrated throughout the text. With an emphasis on Lewis structures, electron flow, resonance, reaction mechanisms, and orbital interactions, the animations support and reinforce the mechanistic philosophy of the text, making a direct connection between the static imagery of the text and the dynamic reaction processes they represent.

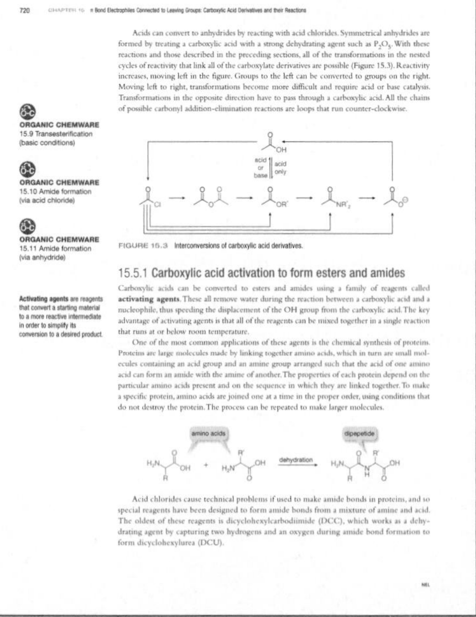

720 CHAPTER 15 π Bond Electrophiles Connected to Leaving Groups: Carboxylic Acid Derivatives and their Reactions

Acids can convert to anhydrides by reacting with acid chlorides. Symmetrical anhydrides are formed by treating a carboxylic acid with a strong dehydrating agent such as P_2O_5. With these reactions and those described in the preceding sections, all of the transformations in the nested cycles of reactivity that link all of the carboxylate derivatives are possible (Figure 15.3). Reactivity increases, moving left in the figure. Groups to the left can be converted to groups on the right. Moving left to right, transformations become more difficult and require acid or base catalysis. Transformations in the opposite direction have to pass through a carboxylic acid. All the chains of possible carbonyl addition-elimination reactions are loops that run counter-clockwise.

ORGANIC CHEMWARE
15.9 Transesterification (basic conditions)

ORGANIC CHEMWARE
15.10 Amide formation (via acid chloride)

ORGANIC CHEMWARE
15.11 Amide formation (via anhydride)

FIGURE 15.3 Interconversions of carboxylic acid derivatives.

15.5.1 Carboxylic acid activation to form esters and amides

Carboxylic acids can be converted to esters and amides using a family of reagents called **activating agents**. These all remove water during the reaction between a carboxylic acid and a nucleophile, thus speeding the displacement of the OH group from the carboxylic acid. The key advantage of activating agents is that all of the reagents can be mixed together in a single reaction that runs at or below room temperature.

One of the most common applications of these agents is the chemical synthesis of proteins. Proteins are large molecules made by linking together amino acids, which in turn are small molecules containing an acid group and an amine group arranged such that the acid of one amino acid can form an amide with the amine of another. The properties of each protein depend on the particular amino acids present and on the sequence in which they are linked together. To make a specific protein, amino acids are joined one at a time in the proper order, using conditions that do not destroy the protein. The process can be repeated to make larger molecules.

Activating agents are reagents that convert a starting material to a more reactive intermediate in order to simplify its conversion to a desired product.

Acid chlorides cause technical problems if used to make amide bonds in proteins, and so special reagents have been designed to form amide bonds from a mixture of amine and acid. The oldest of these reagents is dicyclohexylcarbodiimide (DCC), which works as a dehydrating agent by capturing two hydrogens and an oxygen during amide bond formation to form dicyclohexylurea (DCU).

NEL

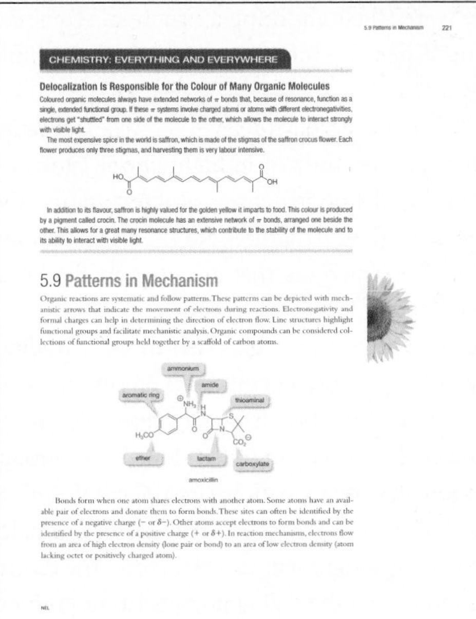

Chemistry: Everything and Everywhere boxes describe applications or stories related to the material in the text. The topics have been chosen to be recent subjects that will interest university students.

The **Patterns in...** sections tie together the concepts shown in the chapter in a visual way. Reaction mechanisms are shown in a "stacked" format so that the underlying patterns are easily visible. Reactions and structures are aligned to highlight repeating electron flows or controlling elements, with some text to describe the key reactivity patterns.

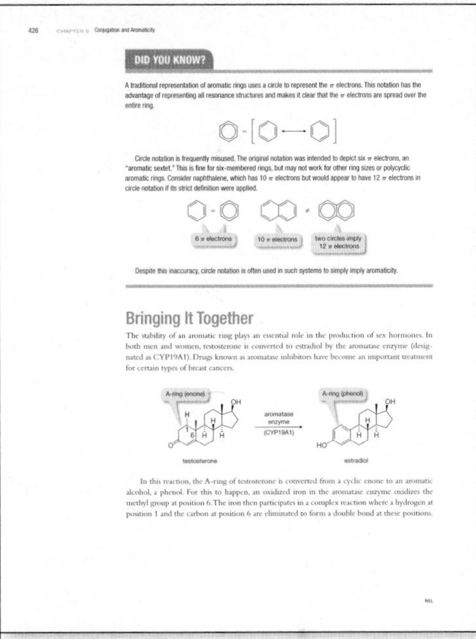

Did You Know? boxes provide extra detail about chemical reactivity. These are optional sections that give a deeper explanation of concepts or provide information beyond the scope of the text.

You Can Now lists the skills that each student should have acquired by reading the text and completing the questions and exercises.

A Mechanistic Re-View is a list of the reactions (with mechanisms) that were described in each chapter.

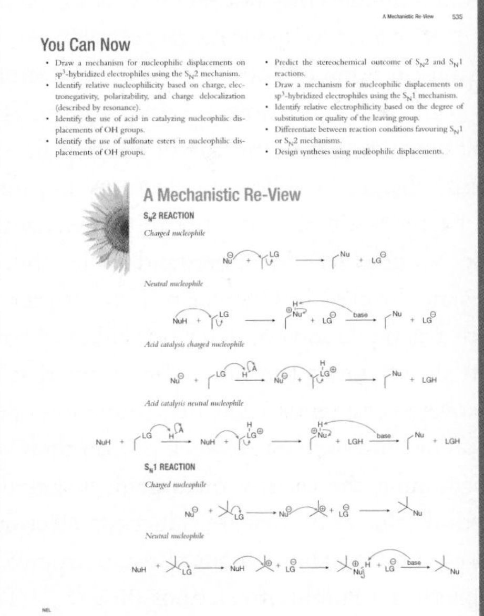

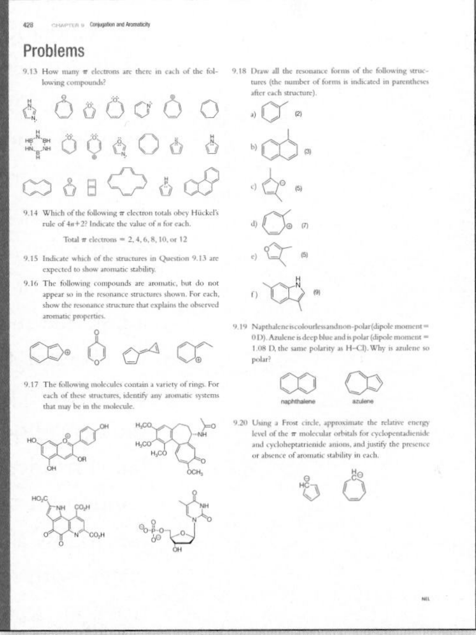

Problems, including MCAT Style Problems and Challenge Problems, are included at the end of each chapter.

Organization

A key part of this approach is a careful reorganization of the overall organic curriculum, progressing from simple reactions to complex ones.* We were all taught using a structural sequence and have a tendency to fall back into familiar patterns. When teaching your course, try to think of increasing complexity of reaction, not structure.

The first two chapters of this book are intended to be partial reviews, as many organic students have taken introductory chemistry in their first year of study or in high school. One key element of Chapter 1 is the use of Lewis structures and bond-line structures, and techniques for manipulating these to understand chemical reactivity. Bond-line structures are used throughout the textbook for two main reasons: they are, after all, the structures that are used in the "real world" and they are easier to understand because they contain less visual clutter.

Chapter 2 describes nomenclature and molecular properties and is intended to be a reading assignment or review. Organic nomenclature is taught in high school chemistry, as are the roles of intermolecular forces. Organic functional groups are described in this chapter in terms of group properties rather than bulk properties of simple molecules containing that functional group. The philosophy is that most organic molecules contain more than one functional group, and therefore it is more important to look at the contribution of the groups to reactivity, rather than, for example, what simple aldehydes smell like. Using this chapter as a reading assignment also recognizes the reality that, in 2017, computers have greatly diminished the importance of the skill of nomenclature, both by providing automated ways of naming (ChemDraw/ChemDoodle) and searching (SciFinder).

Chapters 3 and 4 are traditional chapters covering alkane structure, conformation, and stereochemistry. Although considerable detail is presented, not all the material needs to be covered in lectures. Much of this can serve as a general reference. It is anticipated that the first three weeks of instruction using this text will cover Chapters 1, 3, and 4 (with Chapter 2 as a reading assignment).

Chapter 5 covers the basics of the curved arrow notation and mechanisms as a *tool* to understand reactivity. Although students may not yet know any organic reactions, they can apply the principles introduced in this chapter to deduce even complex electron flows. Many complex reactions are shown in this chapter. It is important to remember that students do not need to know anything about the reactions at this point; reactions are simply given as a way to practise using the curved arrow notation. Basics including the direction of electron flow are described, along with methods of determining formal charges by following electrons and using mechanistic arrows. Resonance is discussed in this chapter as a tool to practise using mechanistic arrows. Since only π bonds are involved, students do not need to fully understand nucleophiles or electrophiles at this stage.

Acids and bases are covered in Chapter 6. This chapter serves as an important foundation for many subsequent reactions, and we describe acids and bases in some detail, although we do assume that students already know the basics. One topic that has been explicitly introduced, and which is not often covered elsewhere, is the determination of the relative acidity of charged acids, a task that many students find difficult to work out on their own.

Rather than beginning the section on organic reactivity with the traditional chapter on S_N1/S_N2-type reactions, this book first introduces π electrophiles and π nucleophiles. Indeed, the conventional way of introducing chemical reactivity involves the simplest functional group (alkyl halides) but presents a family of reactions ($S_N1/S_N2/E1/E2$ rearrangements) that form a continuum of competing reactivities. Based on electron flow, these reactions look simple but in fact follow a very complex network of reactivity patterns. To avoid the high cognitive load associated with this traditional organization, we introduce chemical reactivity using π bonds. These reactions follow simpler patterns (adding to carbonyls or proceeding through the most stable carbocation) and are the reason for the grouping of Chapters 7 through 10.

* Flynn, A.B. and W.W. Ogilvie. "Mechanisms before reactions: A mechanistic approach to the organic chemistry curriculum based on patterns of electron flow," *Journal of Chemical Education*, 2015, *92*, 803–810.

First this format presents biological *reactions* in order of increasing complexity. In this way biological subjects can each be used as examples when new reactions or concepts are introduced. This approach provides the opportunity to explain what is happening in more chemical terms and at a level of detail that goes beyond other texts.

Secondly, the reactions that happen in living things are fundamentally the same ones that happen in laboratory flasks. The electron flows are the same, and the roles of the various reagents are the same. By mixing the biological content with the "regular" content the idea is reinforced that there is nothing "magical" about biological reactions, they just happen in enzyme active sites rather than in free solution.

Instructor Resources

The **Nelson Education Teaching Advantage (NETA)** program delivers research-based instructor resources that promote student engagement and higher-order thinking to enable the success of Canadian students and educators. Visit Nelson's **Inspired Instruction** website at nelson.com/inspired to find out more about NETA.

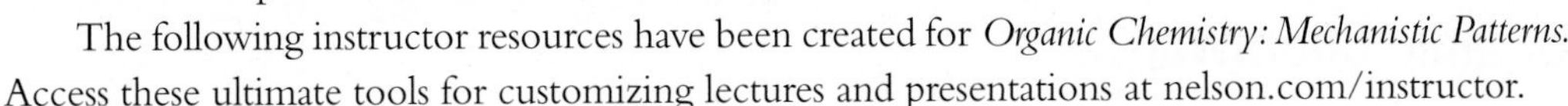

The following instructor resources have been created for *Organic Chemistry: Mechanistic Patterns.* Access these ultimate tools for customizing lectures and presentations at nelson.com/instructor.

NETA Test Bank

This resource was written by Anthony Chibba, Trent University. It includes 1000 multiple-choice questions written according to NETA guidelines for effective construction and development of higher-order questions. Also included are 500 true/false, 200 short-answer, and 200 fill-in-the-blank questions.

The NETA Test Bank is available in a new, cloud-based platform. **Nelson Testing Powered by Cognero®** is a secure online testing system that allows instructors to author, edit, and manage test bank content from anywhere Internet access is available. No special installations or downloads are needed, and the desktop-inspired interface, with its drop-down menus and familiar, intuitive tools, allows instructors to create and manage tests with ease. Multiple test versions can be created in an instant, and content can be imported or exported into other systems. Tests can be delivered from a learning management system, the classroom, or wherever an instructor chooses. Nelson Testing Powered by Cognero for *Organic Chemistry: Mechanistic Patterns* can be accessed through nelson.com/instructor.

Instructor's Solutions Manual

This manual, prepared by Neil Dryden, University of British Columbia, and Nathan Ackroyd, Mount Royal University, has been independently checked for accuracy by Philip Dutton, University of Windsor. It contains complete solutions to all in-text and end-of-chapter problems, the Checkpoint Practice and Integrate the Skill problems, and the Challenge Problems.

NETA PowerPoint®

Microsoft® PowerPoint® lecture slides for every chapter have been created by Mark Vaughan, Quest University. There is an average of 50 to 60 slides per chapter, many featuring key figures, tables, and photographs from *Organic Chemistry: Mechanistic Patterns*. NETA principles of clear design and engaging content have been incorporated throughout, making it simple for instructors to customize the deck for their courses.

Image Library

This resource consists of digital copies of figures, short tables, and photographs used in the book. Instructors may use these jpegs to customize the NETA PowerPoint or create their own PowerPoint presentations.

TurningPoint® Slides

TurningPoint® classroom response software has been customized for *Organic Chemistry: Mechanistic Patterns* by Mark Vaughan, Quest University. Instructors can author, deliver, show, access, and grade, all in PowerPoint, with no toggling back and forth between screens. With JoinIn, instructors are no longer tied to their computers. Instead, instructors can walk about the classroom and lecture at the same time, showing slides and collecting and displaying responses with ease. Anyone who can use PowerPoint can also use JoinIn on TurningPoint.

Student Ancillaries

Organic ChemWare

Organic ChemWare for use with *Organic Chemistry: Mechanistic Patterns* makes even the most complex concepts easily understood. Open your eyes to the dynamic, molecular world of organic chemistry through a comprehensive collection of more than 180 interactive animations and simulations designed to help you visualize chemical structures and organic reaction mechanisms. *Organic ChemWare* ties back to the key concepts presented in the text to make sure that you gain a thorough understanding of organic chemistry.

Follow the simple instructions to access *Organic ChemWare* using the Printed Access Card included with each new copy of this text. Once you have accessed the site, use the search bar to easily search for the key terms provided in the margin of the text. In just seconds, you will find interactive simulations that will bring the text concepts to life.

Standalone versions of *Organic ChemWare* are also available via NELSONbrain.com. The standalone version includes an additional 50 learning objects, covering advanced topics and reactions.

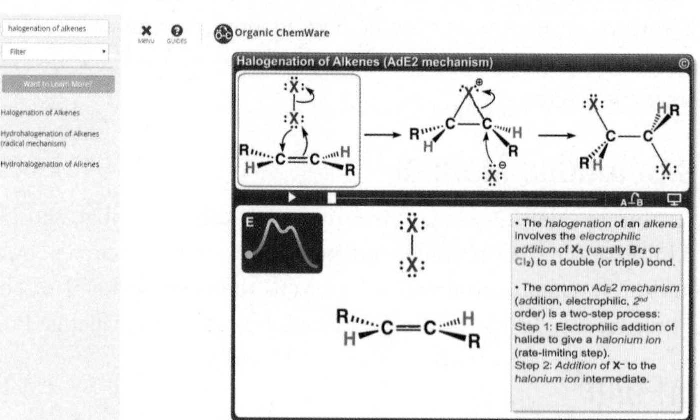

Student Solutions Manual

The *Student Solutions Manual* contains detailed solutions to all odd-numbered Checkpoint and end-of-chapter Problems, and MCAT Style Problems, as well as the solutions to all Challenge Problems in each chapter. Solutions match problem-solving strategies used in the text. Prepared by Neil Dryden, University of British Columbia, and Nathan Ackroyd, Mount Royal University, the solutions have been also technically checked to ensure accuracy.

Acknowledgments

Courtesy of Alison Flynn

Alison Flynn, University of Ottawa, was an initial collaborator and contributed significantly to the development of the curriculum and to the philosophy of mechanistic organization. She also made key contributions to the design of Checkpoints, You Can Now, and solutions. Professor Flynn's research is focused on how students learn organic chemistry, and on how they understand concepts such as synthesis and mechanism. Her "break it down" approach to teaching the subject has heavily influenced many of the pedagogic elements in the text.

The authors greatly appreciate the work and suggestions of Tyra Montgomery Hessel, University of Houston, at the onset of this project. The authors are also indebted to the substantive editors, David Peebles and Carolyn Jongeward, for their suggestions and comments, ensuring the overall consistency in voice, tone, and style of writing. As well, the technical checks by Philip Dutton, University of Windsor, and Barb Morra, University of Toronto, were much appreciated!

Nathan Ackroyd would specifically like to thank students in the Winter 2011 class of Chemistry 2101 for providing valuable feedback and suggestions regarding early drafts of Chapter 13, "Structure Determination I: Nuclear Magnetic Resonance Spectroscopy."

The authors also wish to thank the following instructors who provided thoughtful comments and guidance throughout the writing of this text via the review process:

Athar Ata, University of Winnipeg
Yuri Bolshan, University of Ontario Institute of Technology
John Carran, Queen's University
Anthony Chibba, McMaster University
Fran Cozens, Dalhousie University
Shadi Dalili, University of Toronto Scarborough
Philip Dutton, University of Windsor
Nola Etkin, University of Prince Edward Island
Robert Hudson, The University of Western Ontario
Philip Hultin, University of Manitoba
Ian Hunt, University of Calgary
Norm Hunter, University of Manitoba
Anne Johnson, Ryerson University
Uwe Kreis, Simon Fraser University
Larry Lee, Camosun College
Jennifer Love, University of British Columbia
Stephen MacNeil, Wilfrid Laurier University
Susan Morante, Mount Royal University
Barb Morra, University of Toronto
Arturo Orellana, York University
Stanislaw Skonieczny, University of Toronto
Jackie Stewart, University of British Columbia
Paul Zelisko, Brock University

1 Carbon and Its Compounds

All life on Earth, no matter how big or small, is based on the element carbon. But why?

CHAPTER OUTLINE

1.1 Why It Matters

The molecules upon which life is based are composed mostly of carbon: the lipids that make up our cellular membranes, the DNA responsible for cellular reproduction, the reactants and products of our biological processes, as well as the enzymes that catalyze them. Organic chemistry—the chemistry of carbon—seeks to understand the structures and reactivities of molecules that contain carbon, including our biomolecules.

biomolecules contain mostly carbon

$H_2N{-}CH(COOH){-}CH_2{-}CH_2{-}C(=O){-}NH_2$

most organic compounds also contain hydrogen

A quick inspection of a periodic table shows there are almost 100 naturally occurring chemical elements on Earth. Among these, carbon represents a very small percentage of the total number of atoms. On average, only about 16 out of every 10 000 atoms on Earth are carbon.

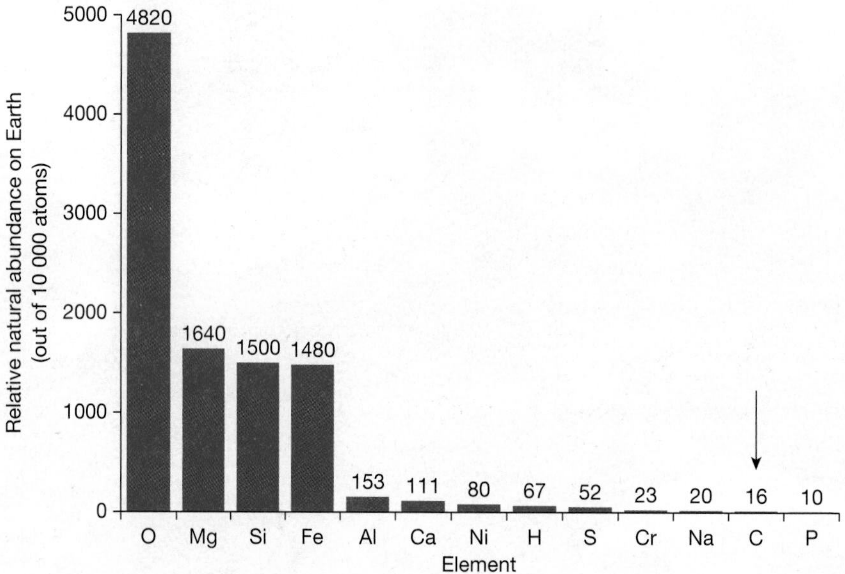

Although carbon is a rare element, it is by far the most abundant one among the chemicals that make up living things. Why is carbon, above all other elements, so important to life? The answer lies in carbon's unique properties. It can bond to itself and form long chains, rings, and complex molecules. This allows carbon to form three-dimensional structures and react in a "modular" way, making life possible.

This chapter reviews the basics of bonding in organic molecules and also introduces techniques used to represent the way that atoms are connected in molecules.

1.2 Organic Molecules from the Inside Out I: The Modelling of Atoms

An **atomic orbital** (AO) maps out, point by point in the volume of space surrounding the nucleus, the likelihood (probability) of finding its electron at each point. It is a map of probability. Atomic orbitals are often represented as a surface within which the electron(s) may be found 95 percent of the time.

Every atom has a set of atomic orbitals that describes the relative probabilities of finding electrons about the atom. Each electron is said to "fill" or "occupy" an **atomic orbital** that describes its distribution in space around the nucleus. Since every atom has more atomic orbitals than electrons, the remaining orbitals that are not occupied by electrons are "empty."

The five most important atomic orbitals of organic chemistry are plotted in Figures 1.1 and 1.2. Each atomic orbital is labelled 1s, 2s, or 2p according to its distinctive characteristics of size and shape. Each orbital describes a different distribution of the probabilities of finding an electron in the space about the nucleus. Both the 1s and 2s atomic orbitals are spherical, but the 2s orbital is larger than the 1s orbital. This means there is a greater likelihood of finding a 2s electron at larger distances from the nucleus than a 1s electron.

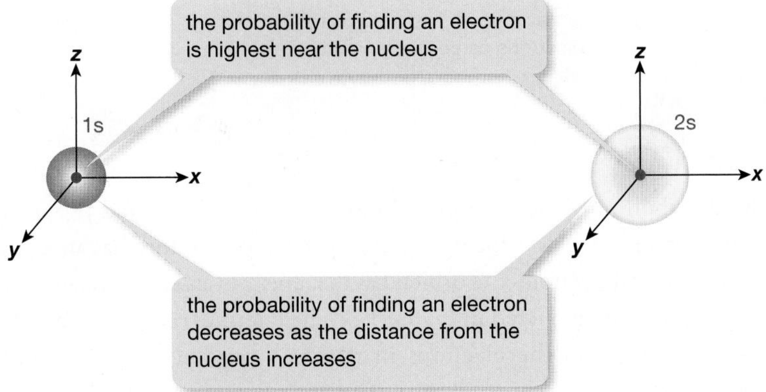

FIGURE 1.1 Plots of the 1s and 2s atomic orbitals. The intensity of orbital colour at any point in space reflects the likelihood of finding the electron at that point.

In contrast, the three 2p orbitals point in specific directions; they lie perpendicular to the others and are labelled p_x, p_y, and p_z (Figure 1.2). Each 2p orbital has two lobes that point in opposite directions away from the nucleus. The different colours of the lobes of the 2p orbitals depict the phase of the orbitals (positive or negative). Phase is a mathematical description of the **wavefunction** of the electrons in the orbital. Whether a given phase is positive or negative is not important. What matters is whether phases match or not with the orbitals of neighbouring atoms. Regardless of whether a lobe is one phase or the other, their shapes and intensity of colour are identical: there is an exactly equal likelihood of finding a 2p electron at the same point in either lobe.

A **wavefunction** is a mathematical description of a particular quantum state of an electron or other particle.

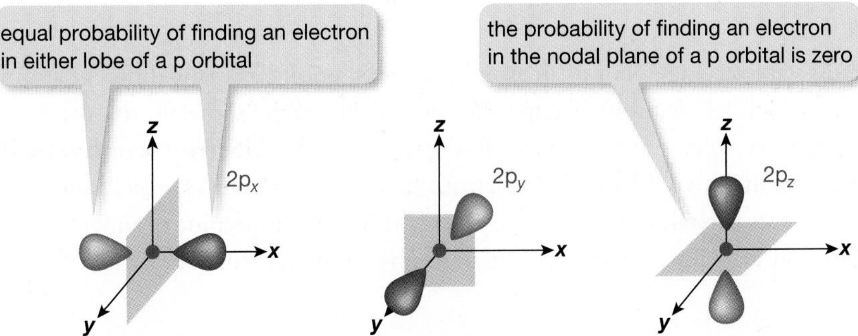

FIGURE 1.2 Plots of the three 2p atomic orbitals. Each of the three 2p orbitals lies perpendicular to the other two. The colour of the orbital lobe reflects its phase (positive or negative).

The region in space where a 2p orbital changes phase is a nodal plane: the place where the value of the orbital is exactly zero and, as a result, the probability of finding the electron in that plane is exactly zero.

An orbital can be occupied by zero, one, or two electrons. If two electrons occupy the same orbital, they must be spin-paired—that is, they have opposite spins. This is often shown in orbital diagrams by using small arrows to denote electrons.

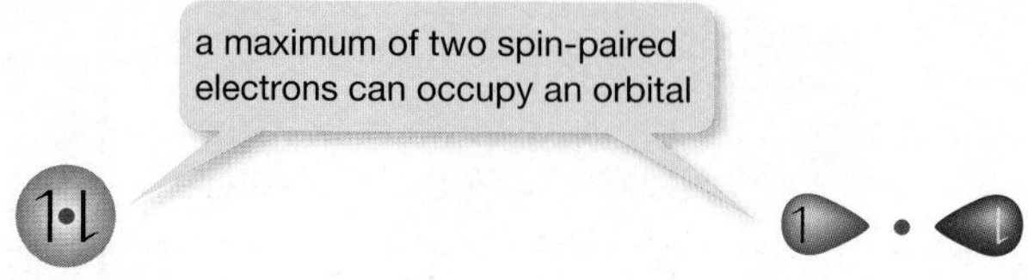

The specific distribution of electron probabilities in the space around the nucleus fixes the energy of the orbital at a particular value. Every atomic orbital therefore has an energy associated with it, and any electron that occupies the orbital has that energy value. The relative energies of the 1s, 2s, and three 2p orbitals are depicted in Figure 1.3. The energies of the orbitals are **quantized**, which means each orbital (and the electrons in it) has a particular fixed quantity of energy. The three 2p atomic orbitals have precisely the same energy and are referred to as **degenerate atomic orbitals**.

Quantized refers to the particular fixed value of the energy of an atomic orbital.

Degenerate atomic orbitals are any set of orbitals that have the same energy value.

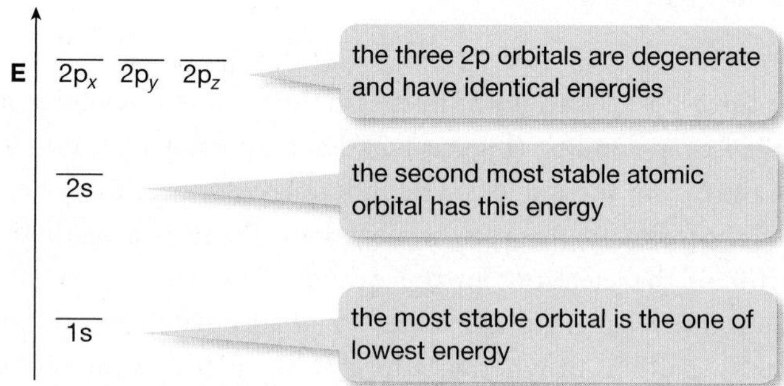

FIGURE 1.3 The relative energies of the 1s, 2s, and three 2p atomic orbitals. The most stable orbitals are those of lowest energy. Actual energy values of the orbitals are not given. Orbitals of higher energy than the 2p orbitals are not shown.

The most stable arrangement of an atom's electrons among its many different atomic orbitals is the one in which its electrons occupy the most stable orbitals, that is, the orbitals of lowest energy. For the six electrons of carbon, this **ground-state electron configuration** is the arrangement in which two spin-paired electrons fill the most stable 1s orbital, another two similarly occupy the 2s orbital, and its last two electrons (of the same spin) occupy two of the three degenerate 2p orbitals. This electron configuration is often written as $1s^2 2s^2 2p^2$ (Figure 1.4).

The **ground-state electron configuration** is the one of lowest energy: that is, the most stable one. All other arrangements, which are necessarily of higher energy, are called *excited states*.

E
2p$_x$ 2p$_y$ 2p$_z$
2s
four valence orbitals (2s, 2p$_x$, 2p$_y$, 2p$_z$) and four valence electrons
1s
two spin-paired electrons in the 1s atomic orbital

FIGURE 1.4 Ground-state electron configuration of a carbon atom. Actual energy values of the five orbitals are not given. Orbitals of higher energy than the 2p orbitals are not shown. The ground-state electron configuration for carbon is $1s^2 2s^2 2p^2$.

For carbon, nitrogen, oxygen, and fluorine atoms, their two 1s electrons are so stable (low in energy), that they do not participate in bonding to other atoms. Instead, the four less stable 2s and 2p orbitals of C, N, O, and F, and the electrons that occupy them, are involved in bonding in organic molecules. These less stable orbitals are known as **valence orbitals**, and the electrons that occupy them are called **valence electrons**.

An atom's **valence orbitals** are the occupied orbitals of highest energy (and any accompanying empty orbitals of similar high energy).

Valence electrons occupy valence orbitals.

CHECKPOINT 1.1

You can now draw the electronic configuration for atoms and identify their valence electrons and orbitals.

ORGANIC CHEMWARE
1.1 Electron configuration

SOLVED PROBLEM

(a) Draw the atomic orbital energy diagram for a nitrogen atom. (b) From this diagram, write its ground-state electron configuration. (c) How many valence electrons does nitrogen have? (d) How many valence orbitals does nitrogen have? (e) Draw the shape of each valence orbital of nitrogen.

STEP 1: Nitrogen has seven protons and therefore seven electrons. Placing its seven electrons in a diagram of atomic energy levels yields the following:

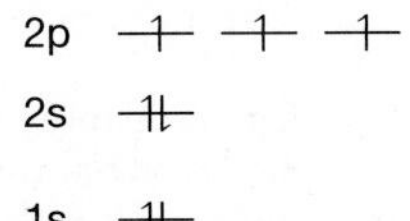

STEP 2: The ground-state electron configuration of nitrogen is therefore $1s^22s^22p^3$.

STEP 3: The valence orbitals of nitrogen are its 2s and its three 2p orbitals (higher shell orbitals). From the diagram, note there are a total of five valence electrons among nitrogen's four valence orbitals.

STEP 4: The shapes of the four valence orbitals of nitrogen are shown here.

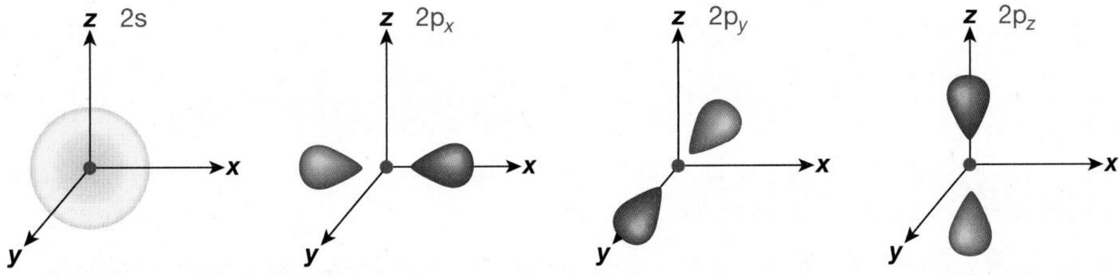

PRACTICE PROBLEM

1.1 Draw the electronic configuration of the following atoms.
 a) S
 b) Cl
 c) Na^+

1.3 Organic Molecules from the Inside Out II: Bonding

Organic molecules are composed mostly of carbon, hydrogen, nitrogen, and oxygen, and occasionally with phosphorus, sulfur, or halogens; but many other elements are also possible. Atoms behave very differently in organic molecules than they do by themselves. Their behaviour in a molecule is dominated by their interactions with neighbouring atoms. These interactions hold the organic molecule together and also impart to the compound its chemical and physical character.

Electrostatic attraction is the attraction of opposite charges to each other.

Ionic bonds result from the transfer of electrons from one atom to another, which creates opposite charges that are attracted to each other.

A **covalent bond** is the energetically favourable sharing of two electrons; this holds atoms in close proximity to each other.

The force holding atoms together in all molecules is **electrostatic attraction**. The bonds holding atoms together are the result of the attraction between positive and negative charges. There are two types of bonds that hold atoms together: **ionic bonds** and **covalent bonds**. Ionic bonds occur when electrons are transferred from one atom to another. This creates opposite charges on the two atoms involved, which holds the atoms together. Ionic bonds normally occur when the electronegativity difference between the two atoms is very different. The resulting compounds are called *salts*.

$$NH_4^{\oplus}\ Cl^{\ominus} \qquad CH_3CO_2^{\ominus}\ Na^{\oplus}$$

Covalent bonds are the result of sharing electrons between atoms, and are by far the most common bond type for organic molecules. Normally, each atom in the bond contributes one electron, forming an electron pair (with opposite spins) that is shared between the two atoms. Again, it is charge that holds the atoms together. Since both nuclei (positive) are attracted to the two bonding electrons (negative), the electrons between the nuclei hold them together. This is a more stable arrangement for an electron than remaining exclusively in the electric field of its own atom.

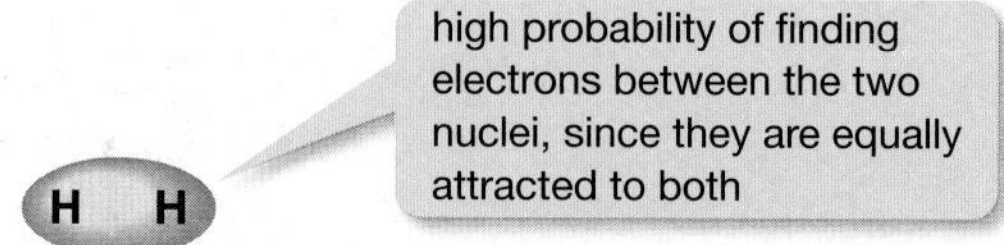
H H
high probability of finding electrons between the two nuclei, since they are equally attracted to both

Atoms form bonds using valence electrons from their valence shell, the outermost layer of electrons in an atom. When forming bonds, atoms tend to follow the octet rule: the total number of electrons in their valence shell is eight. This rule is strictly followed for first-row elements (B, C, N, O, F). Elements in the lower rows also follow the octet rule, but can occasionally exceed an octet of electrons.

1.4 Organic Molecules Represented as Lewis Structures

The **Lewis structure** depicts the bonding in a molecule. A line between the participating atoms represents the two shared valence electrons of each covalent bond. Non-bonded electrons are represented by dots.

Non-bonded (lone pair) electrons reside on one atom, occupying space around that atom.

Lewis structures effectively represent the way atoms are connected in organic molecules. In a **Lewis structure** each bond between atoms is represented by a line, and **non-bonded** (lone pair) electrons are shown as dots. When they are present, non-bonded electrons are arranged as distinct pairs around the elemental symbol of the atom they reside on.

bonds are shown as lines between the elemental symbols of the atoms

```
    H H   H
    | | ..|
  H-C-C-O-C-H
    | | ..|
    H H   H
```

non-bonded electrons are shown as dots arranged in pairs around the atom they are associated with

Carbon, because it has four valence electrons, can make four bonds. Each valence electron can potentially share an orbital with one electron from another atom, thereby making four bonds. Oxygen, because it has six valence electrons, typically forms only two bonds because the other four electrons must remain paired.

In some compounds, atoms are connected by more than one bond. In these cases each bond is represented by a separate line. Such double and triple bonds are important in organic structures, particularly because they form a key site for organic reactions.

```
 H  H    :O: H
 |  |  ..  ||  |
H-C-C-O-C-C-C≡N:
 |  |  ..      |
 H  H          H
```

Some atoms are connected by double or triple bonds. Each line represents a pair of electrons.

Some atoms in Lewis structures carry a **formal charge** (FC). Formal charge is a bookkeeping-type method of tracking charged atoms in a structure. It is based on the number of valence electrons that the charged atoms bring to the molecule. It compares the number of shared (bonded) and non-shared (non-bonded) electrons to the valence number of the atom. The formal charge is calculated by subtracting the number of bonds and non-bonded electrons from the group number (column of the periodic table it appears in) of the atom.

The **formal charge** (FC) of an atom describes a deficit or excess of electrons based on formally comparing the number of electrons an atom shares in a Lewis structure with the number of electrons it should have to be electrically neutral. A Lewis structure is not complete without formal charges.

Some atoms carry formal charges. The formal charge is calculated by subtracting the number of bonds and non-bonded electrons from the group number of the atom.

Oxygen is in group 6.
It is surrounded by 3 bonds and 2 non-bonded electrons.
FC = (group #) − (# of bonds) − (# of non-bonded electrons)
FC = 6 − 3 − 2 = + 1

```
           H
           |
 H  H     ⊕O: H
 |  |  ..  ||  |
H-C-C-O-C-C-H
 |  |  ..      |
 H  H          H
```

1.4.1 Formal charge method of drawing Lewis structures

The following procedure can be used to draw Lewis structures. This general method is especially useful for constructing unfamiliar structures and functional groups. (See Chapter 2 for more information on functional groups.)

1. Count the total valence electrons in the structure based on the group number of each of the atoms in the molecule.

$CH_3CO_2CH_3$

3 carbons (group 4)	$3 \times 4 = 12$
6 hydrogens (group 1)	$6 \times 1 = 6$
2 oxygens (group 6)	$2 \times 6 = 12$
	30 valence electrons

ORGANIC CHEMWARE

1.2 Lewis structure: Ethane

1.3 Lewis structure: Ethene

1.4 Lewis structure: Ethyne

1.5 Lewis structure: Ethyl carbanion

1.6 Lewis structure: Ethyl carbocation

1.7 Lewis structure: Bromoethane

1.8 Lewis structure: Methanol

1.9 Lewis structure: Methyl acetate

1.10 Lewis structure: Methoxide anion

1.11 Lewis structure: Methylamine

1.12 Lewis structure: Methylammonium cation

1.13 Lewis structure: Ethanitrile

2. For charged molecules or groups, add one electron for each negative charge, and subtract one for each positive charge.

$CH_3CO_2^{\ominus}$

2 carbons (group 4)	$2 \times 4 = 8$
3 hydrogens (group 1)	$3 \times 1 = 3$
2 oxygens (group 6)	$2 \times 6 = 12$
1 negative charge	$1 \times 1 = 1$
	24 valence electrons

$CH_3CH_2NH_3^{\oplus}$

2 carbons (group 4)	$2 \times 4 = 8$
8 hydrogens (group 1)	$8 \times 1 = 8$
1 nitrogen (group 5)	$1 \times 5 = 5$
1 positive charge	$1 \times 1 = -1$
	20 valence electrons

3. Draw the connected atoms in the structure using single bonds only (do not exceed four bonds for first-row elements).

$CH_3CO_2CH_3$

3 carbons (group 4)	$3 \times 4 = 12$
6 hydrogens (group 1)	$6 \times 1 = 6$
2 oxygens (group 6)	$2 \times 6 = 12$
	30 valence electrons

```
  H O   H
  | |   |
H-C-C-O-C-H
  |     |
  H     H
```

4. Count the total number of bonds drawn. Each bond has two electrons. Multiply by 2 to get the total number of bonding electrons. Subtract this number from the total number of electrons to get the number of non-bonded electrons.

10 bonds shown (20 electrons)

$CH_3CO_2CH_3$

```
  H O   H
  | |   |
H-C-C-O-C-H
  |     |
  H     H
```

(30 total valence electrons) − (20 bonded electrons) = (10 non-bonded electrons)

5. Add these non-bonded electrons to the structure, starting with the most electronegative atoms. Continue until the octets are filled, and then move to the next atom until all electrons have been distributed.

Add the non-bonded electrons to the structure (most electronegative atoms first). Continue until octets are filled.

$CH_3CO_2CH_3$

```
    ..
  H:O:  H
  | |   |
H-C-C-O-C-H
  |  ¨  |
  H     H
```

6. Calculate formal charges using the formula FC = (group number) − (number of bonds) − (number of non-bonded electrons). Any carbons with exactly four bonds will not carry a charge, and their formal charge does not need to be calculated.

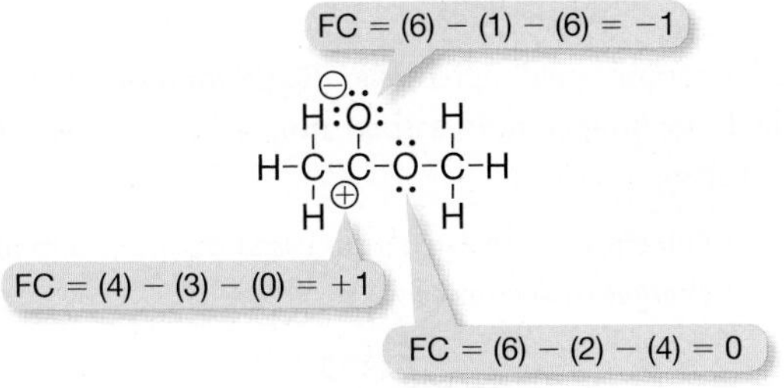

FC = (6) − (1) − (6) = −1

FC = (4) − (3) − (0) = +1

FC = (6) − (2) − (4) = 0

7. Use electron pairs from negative atoms to make extra bonds with *adjacent* positive atoms that don't have filled octets. To check the structure, recalculate the formal charges on any of these atoms. Try to arrive at a structure with the fewest number of charges possible.

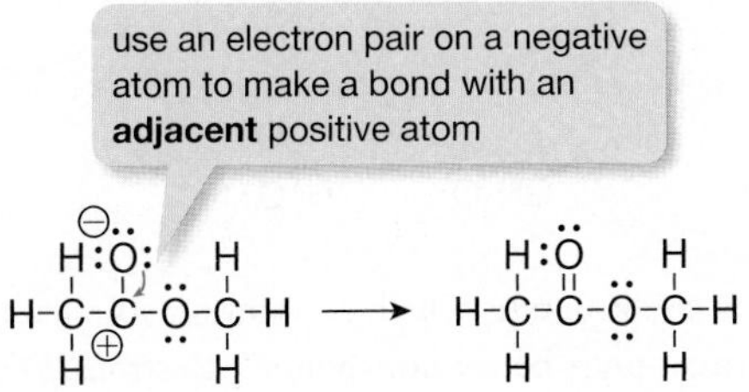

use an electron pair on a negative atom to make a bond with an **adjacent** positive atom

1.4.2 Exceptions to the octet rule

Atoms in the first row of the periodic table do not exceed an octet of valence electrons when bonding (octet rule) and do not form structures in which there are more than four groups of electrons surrounding them. However, organic materials commonly involve structures in which some first-row atoms have an incomplete octet; that is, they are surrounded by less than eight electrons. Such structures are often unstable and contribute to reactivity.

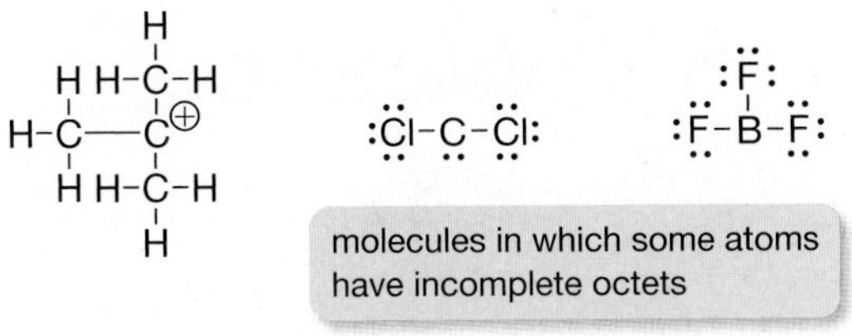

molecules in which some atoms have incomplete octets

Elements in lower rows of the periodic table can form structures in which they are surrounded by more than eight electrons. Sulfur and phosphorous are two such elements that form these structures. The method described earlier can be used to arrive at structures such as these.

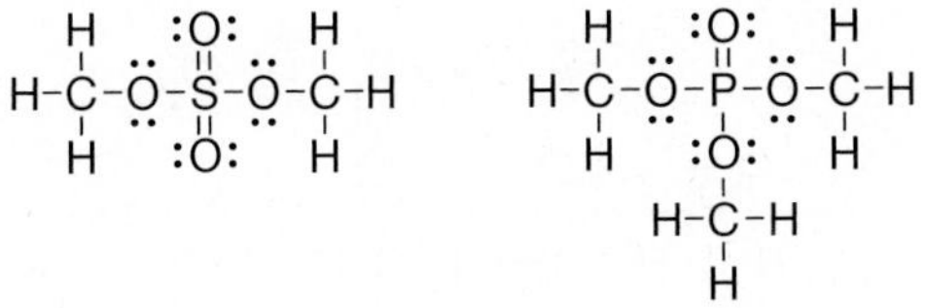

CHECKPOINT 1.2

You can now draw Lewis structures for simple organic molecules.

SOLVED PROBLEM

Formaldehyde, CH_2O, is an important building block in the creation of more complex organic molecules. Both H atoms of formaldehyde are bonded to its carbon atom. Draw its Lewis structure in which all atoms fill their valence orbitals by sharing electrons.

STEP 1: Count the total valence electrons in the structure based on the group number of each atom in the molecule. There are no overall charges to account for.

	CH_2O
1 carbon (group 4)	$1 \times 4 = 4$
2 hydrogens (group 1)	$2 \times 1 = 2$
1 oxygen (group 6)	$1 \times 6 = 6$
	12 valence electrons

STEP 2: Draw the connected atoms in the structure using single bonds only. Remember both hydrogens are connected to the carbon.

CH_2O O | H-C-H

STEP 3: There are three bonds in the structure, which account for six valence electrons. Based on the total number of electrons available, there must be six non-bonded electrons. Distribute these on the structure, starting with the most electronegative atom (oxygen). The resulting structure is as follows:

:Ö: | H-C-H

STEP 4: Formal charges can now be calculated using the formula FC = (group number) − (number of bonds) − (number of non-bonded electrons). For the oxygen [(group 6) − (1 bond) − (6 non-bonded electrons)], this gives a formal charge of −1. For the carbon [(group 4) − (3 bonds) − (0 non bonded electrons)] this gives a formal charge of +1.

FC = (6) − (1) − (6) = −1

⊖:Ö: | H-C-H ⊕

FC = (4) − (3) − (0) = +1

STEP 5: Use electron pairs from negative atoms to make double and triple bonds with adjacent positive atoms that have incomplete octets. To check your result, recalculate formal charges for the atoms involved.

FC = (6) − (2) − (4) = 0

⊖:Ö: | H-C-H ⊕ ⟶ :O: ‖ H-C-H

FC = (4) − (4) − (0) = 0

PRACTICE PROBLEM

1.2 Draw the Lewis structure of the following molecules. Identify the bonding and non-bonding (lone) pairs of electrons of the molecules.

a) $CH_3CH_2NH_2$
b) $CH_3S(O)CH_3$
c) CH_3CH_2CN
d) $(CH_3)_2CHO^{\ominus}$
e) $(CH_3)_4N^{\oplus}$
f) $HSO_3^{\ominus}$ (hydrogen is connected to oxygen only)
g) $HSO_3^{\oplus}$ (hydrogen is connected to oxygen only)

ORGANIC CHEMWARE
1.14 Lewis structure: Ethanal

1.5 Covalent Bonding: Overlap of Valence Atomic Orbitals

Most bonds in organic molecules are covalent. This type of bond results from the overlap of atomic orbitals between atoms to form new orbitals of electrons surrounding both nuclei. Atomic orbitals may overlap and share electrons in two ways: either head-on or side-by-side. The head-on overlap forms a **σ bond** (pronounced *sigma* bond), in which the overlap takes place along the axis connecting the two nuclei. In this case, because the orbitals point directly at each other, there is a high probability that the bonding electrons will be found in the region between the nuclei. The result is an effective, direct sharing of the two valence electrons between the two nuclei, and this renders most σ bonds quite strong (Figure 1.5).

A **σ bond** (pronounced *sigma* bond) is a covalent bond in which the direct line through the nuclei presents the highest probability of finding the shared electrons.

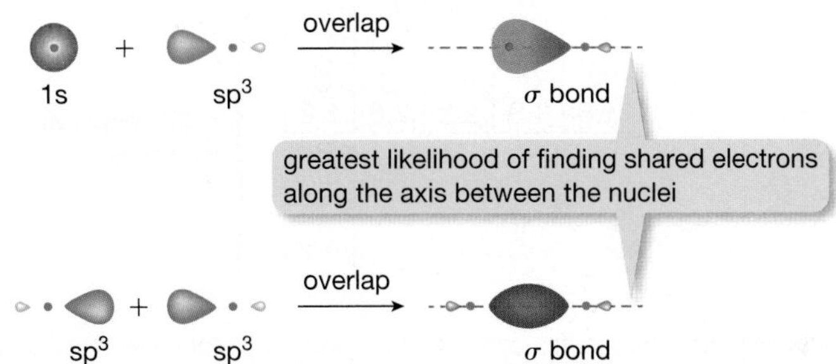

FIGURE 1.5 σ bonds form by the head-on overlap of two atomic orbitals. Top: Overlap between a 1s orbital and an sp^3 orbital (see also Section 1.7.1). Bottom: Overlap between two sp^3 orbitals. The nuclei are shown as green dots.

ORGANIC CHEMWARE
1.15 Molecular orbitals: C–H σ bonds (sp^3 + 1s)

ORGANIC CHEMWARE
1.16 Molecular orbitals: C–C σ bonds

In contrast to head-on overlap, side-by-side overlap forms a **π bond** (pronounced *pi* bond), in which the orbitals are oriented perpendicular to the axis through the nuclei. See Figure 1.6, for example, where the two 2p orbitals are oriented perpendicular to the axis through the nuclei. The result is that the greatest likelihood of finding the shared electrons of a π bond lies equally on each side of this axis. This equal probability above and beneath the line together constitutes *one* π bond. By contrast to the σ bond, there is zero probability of finding the bonding electron pair along the axis through the nuclei. Because the p orbitals that contribute to the π bond are not pointing directly at each other, the overlap achieved in a π bond is generally less than that of a σ bond, and this results in a poorer sharing of electrons. Therefore, the π bond is a weaker covalent bond than the σ bond.

A **π bond** (pronounced *pi* bond) is a covalent bond in which the highest probability of finding the shared electrons occurs equally above and below the line through the nuclei.

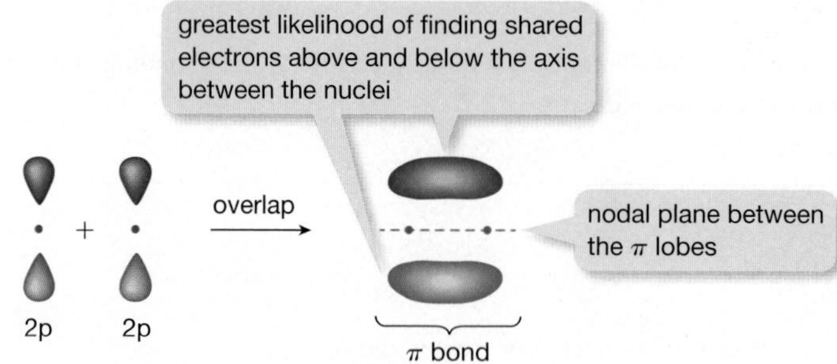

FIGURE 1.6 A π bond from the overlap of two 2p orbitals that lie perpendicular to the line through the nuclei. The nodal plane of the orbital contains the nuclei. There is zero likelihood of finding π electrons in this nodal plane.

ORGANIC CHEMWARE
1.17 Molecular orbitals: C=C π bonds

ORGANIC CHEMWARE
1.18 Molecular orbitals: σ bond types

1.5.1 Unequal sharing of electrons: Electron-rich and electron-deficient atoms

In molecules such as H_2, the two electrons of the covalent bond are equally shared between the nuclei of the two hydrogen atoms. This is because the two atoms are the same (both H atoms), and each pulls equally on the shared electrons. In the covalent bonds of organic molecules, the two electrons may not be equally shared because the two atoms of the bond are different. The difference in **electronegativity** of the two atoms that share a bond accounts for the fact that atoms of different elements vary in their ability to draw electrons toward themselves.

The **electronegativity** of an atom is its ability to pull electrons toward itself from the surrounding atoms to which it is bonded. The greater the electronegativity, the greater is the ability of the atom to draw electrons from its neighbours.

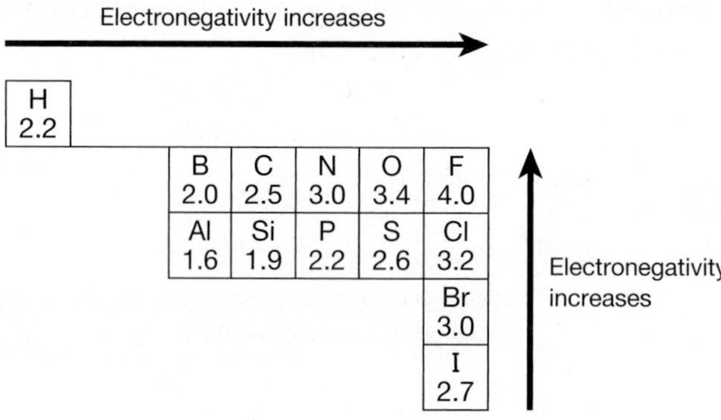

FIGURE 1.7 Electronegativity values for some of the more important elements of organic chemistry. Electronegativity increases left to right and bottom to top.

The electronegativity values for some of the more important elements of organic chemistry are presented in the partial periodic table of Figure 1.7. A comparison of the electronegativities of two atoms that share a bond provides an approximate measure of the difference in electron sharing between them, and this measure is an indication of the bond's polarity. For example, the O–H bond in CH_3OH is polar. The oxygen of this bond is more electronegative than the hydrogen. This means the oxygen pulls more strongly on the shared electrons of the O–H bond, creating a substantial **bond dipole**. Across the O–H bond, there is an excess of negative charge at one end (the oxygen atom) and an equal excess of positive charge at the other end (the hydrogen atom). This is typically indicated as $\delta+$ at the electron-deficient atom and $\delta-$ at the electron-rich atom. Alternatively, the difference in charge may be depicted as a dipole arrow (denoted with a + across the arrow) pointing from the electron-deficient atom toward the electron-rich atom.

A **bond dipole** is a dipole created across a chemical bond. It is the result of differences in electronegativity between the nuclei involved.

The + is placed on the side of the arrow near the electron-deficient atom. Depicting the charge difference with an arrow offers the advantage that the relative magnitudes of bond dipoles can be expressed by the lengths of the arrows.

STUDENT TIP
A bond that has a substantial dipole due to the unequal sharing of electrons is referred to as a **polar covalent bond.** Polar bonds are a source of reactivity in many organic molecules.

A **polar covalent bond** is a covalent bond in which there is a significant difference in electronegativity between the atoms involved.

Oxygen is much more electronegative than hydrogen as seen in this electron density map. There is an unequal sharing of electrons between them, making the oxygen slightly negative and the hydrogen slightly positive. In this depiction, the size of the orbital lobe is used to depict the likelihood of finding electrons around each atom in the bond.

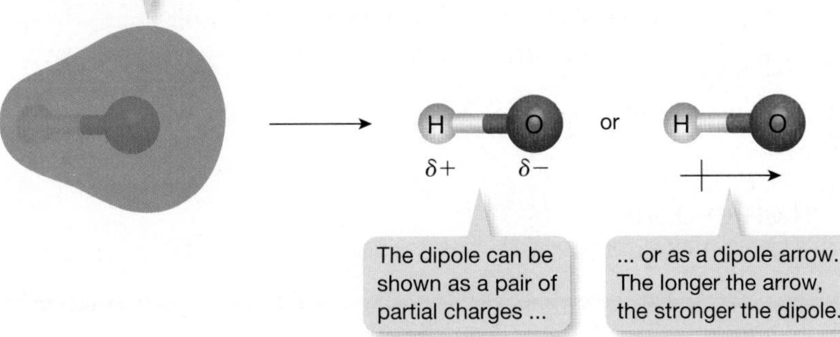
H
O
or
H
O
δ+
δ−
The dipole can be shown as a pair of partial charges ...
... or as a dipole arrow. The longer the arrow, the stronger the dipole.

The C–H bonds of organic molecules are generally considered to be weakly polar or non-polar (not polar at all). This is due to the very small difference in electronegativity between carbon and hydrogen.

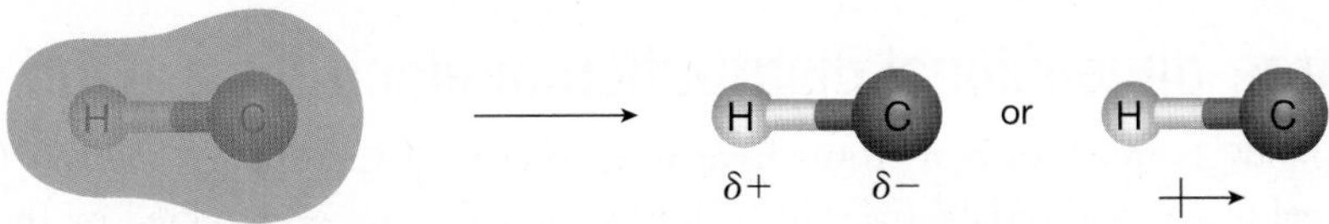
H
C
H
C
or
H
C
δ+
δ−

Whether polar or non-polar, covalent bonds are the force holding every organic molecule together; sharing of electrons between its atoms arises from overlap of their valence orbitals. This sharing of electrons between atoms to fill their valence orbitals provides a basis for predicting the structures, physical properties (Chapter 2), and reactivity of organic molecules.

CHECKPOINT 1.3

You can now recognize the Lewis structure as a simple representation of the covalent bonding in a molecule. You can also recognize these covalent bonds as a sharing of two valence electrons between atoms (overlap of atomic orbitals on the participating atoms).

SOLVED PROBLEM

(a) How many electrons are there in each covalent bond between carbon and nitrogen in CH_3CN? (b) In which direction does the dipole of the C–N bond lie? (c) What does the direction of the bond dipole tell you about this C–N bond?

STEP 1: Draw the Lewis structure of CH_3CN.

H
H–C–C≡N:
H

STEP 2: The bond between the carbon and nitrogen is a triple bond consisting of three distinct pairs of electrons.

STEP 3: Nitrogen is to the right of carbon in the periodic table. This means that nitrogen must be more electronegative than carbon. The electrons in the carbon-nitrogen triple bond tend to occupy the space around the nitrogen more than they occupy the space around carbon. This makes nitrogen $\delta-$ and carbon $\delta+$.

H $\delta+$
H-C-C≡N:
H $\delta-$

PRACTICE PROBLEM

1.3 Draw the Lewis structure of the following. Identify any dipoles that may be present.

a) $(CH_3)_3CCl$
b) $CH_3C(O)CH_3$
c) $CH_3CH_2CH_2CHOHCH_3$
d) $HOCH_2CH_2CH_2CH_2CHO$

1.6 The Shapes of Atoms in Organic Molecules

The Lewis structure of an organic molecule can be used to predict a molecule's structural features, including the geometries adopted by each atom in a compound.

1.6.1 Three-dimensional distribution of electrons around atoms

When covalently bonded, carbon atoms have three possible structural geometries (or shapes): linear, trigonal planar, and tetrahedral. These three geometries are exemplified by the following simple molecules:

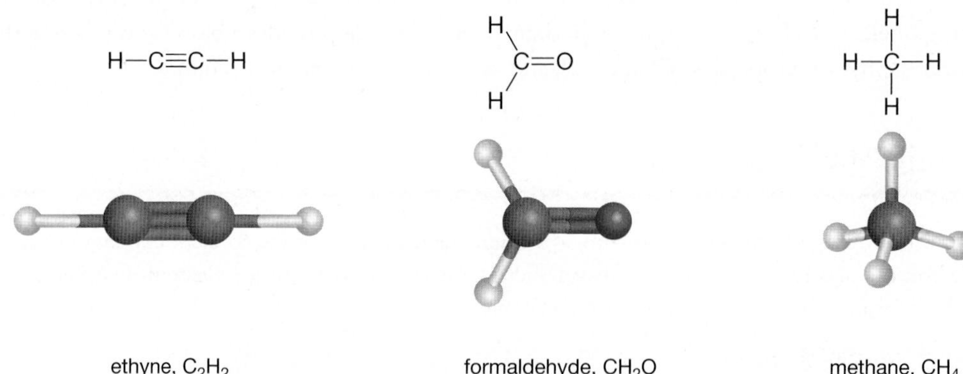

ethyne, C_2H_2 formaldehyde, CH_2O methane, CH_4

1. *Tetrahedral geometry.* In methane (CH_4), the carbon atom exhibits a tetrahedral geometry. The four hydrogen atoms are equally displaced in a pyramid arrangement around the central carbon atom. A tetrahedral geometry is characterized by bond angles of roughly 109° between the atoms.

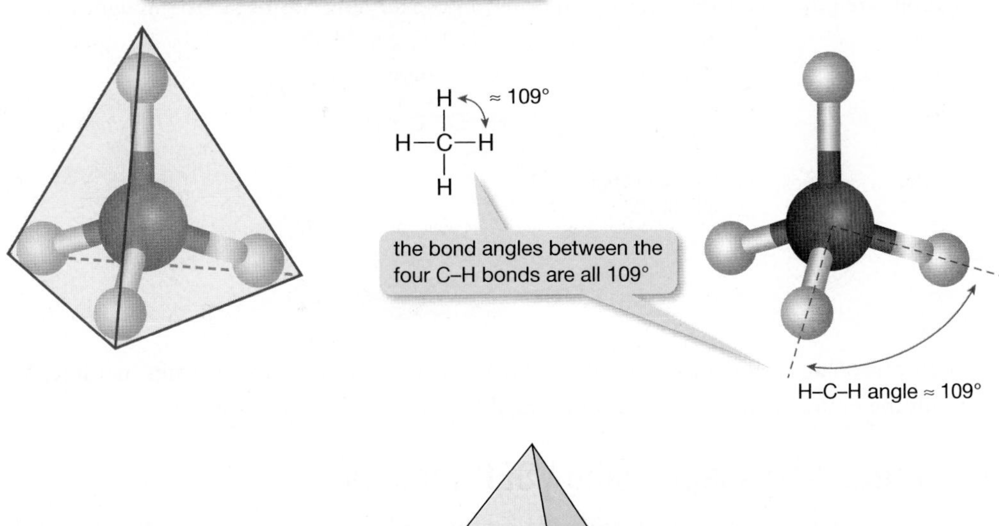
the carbon is at the centre of a tetrahedron defined by the four outer hydrogens
≈ 109°
the bond angles between the four C–H bonds are all 109°
H–C–H angle ≈ 109°

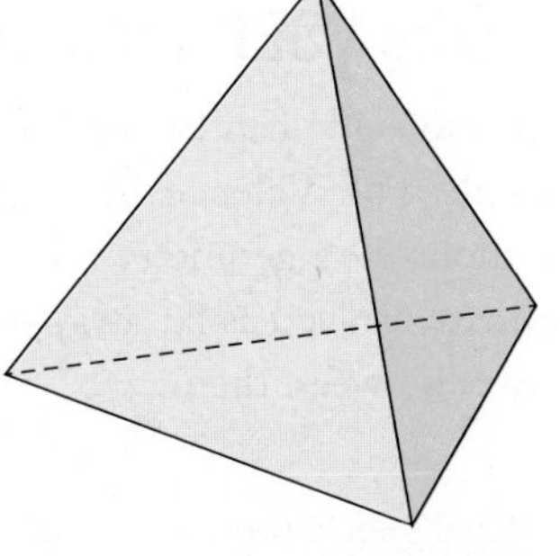

STUDENT TIP

The tetrahedron is an important shape in organic chemistry. A perfect tetrahedron is a pyramid-shaped polyhedron consisting of four equilateral triangles. Every triangle shares each of its sides with one of the other triangles. Three triangles meet at each vertex.

Unlike the linear and trigonal planar geometries (which are flat shapes), the tetrahedron is three-dimensional and cannot be well-represented using standard Lewis notation. Instead, a system of special wedge bonds is used to show the three-dimensional structure around a tetrahedral atom (Section 1.6.2).

2. *Trigonal planar geometry.* The carbon atom of formaldehyde (CH_2O) has a trigonal planar geometry. The carbon is surrounded by three atoms (two hydrogens and one oxygen), which all lie in the same plane as the carbon. The bond angles of any three atoms are approximately 120°. In the following diagram, the trigonal plane of the carbon atom is shown in gold.

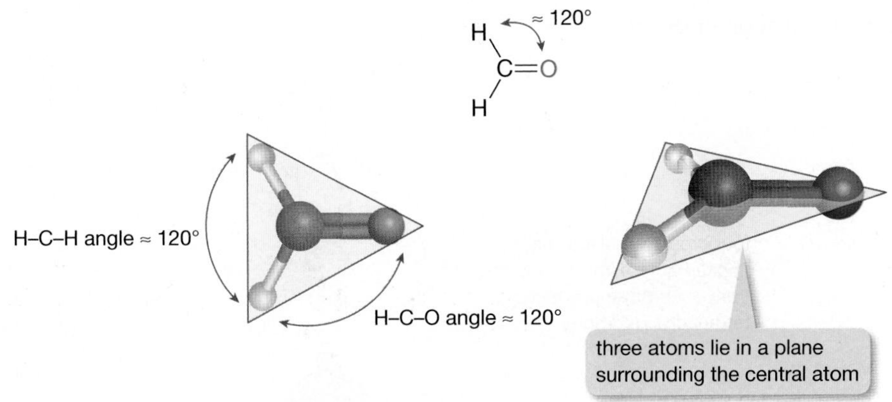
≈ 120°
H–C–H angle ≈ 120°
H–C–O angle ≈ 120°
three atoms lie in a plane surrounding the central atom

3. *Linear geometry.* It has been experimentally determined that the carbon atoms in ethyne (C_2H_2) exhibit linear geometry. The two atoms bonded to each carbon (hydrogen and carbon) are positioned 180° from each other, and so all three atoms lie in a straight line.

180°

H—C≡C—H

the bond angles are all 180°

H-C-C angle ≈ 180°

Both carbon atoms in ethyne adopt this linear structural feature: three atoms in a line. As a result the entire ethyne molecule is linear; all four atoms lie on a single line.

1.6.2 Predicting shape using VSEPR theory

The Lewis structure of any organic molecule can be used to predict the geometry around the non-hydrogen atoms in that molecule. The **V**alence **S**hell **E**lectron **P**air **R**epulsion (VSEPR) theory explains this aspect of the molecule's geometry. This concept states that the groups of valence electrons around each atom are arrayed as far away from each other as possible so as to minimize electron-electron repulsions between them.

Repulsions between groups of valence electrons around an atom dictate its geometry.

H-C-H (with H above C and :Ö-H below C)

In the Lewis structure of methane, there are four groups of electrons around its carbon atom: all of them are C–H bonding pairs. Positioning four groups of electrons as far away from each other as possible is achieved only by a tetrahedral arrangement. The carbon atom of methane adopts this geometry to minimize the repulsions among its four bonding pairs; this is consistent with experimental observations.

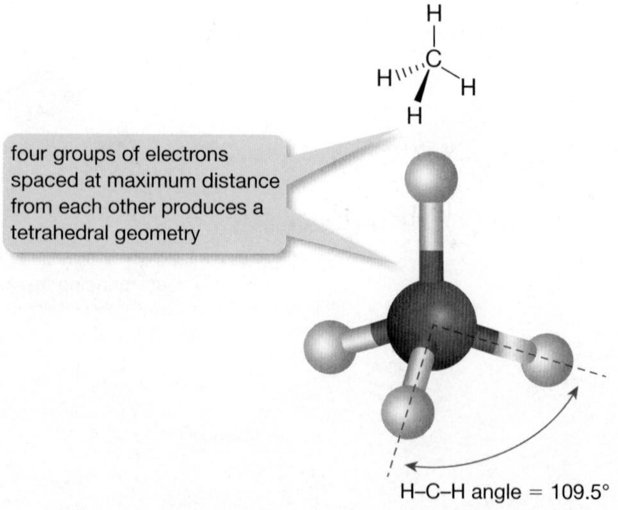
H
H
C
H
H
four groups of electrons spaced at maximum distance from each other produces a tetrahedral geometry
H-C-H angle = 109.5°

To describe the three-dimensional geometry of tetrahedral bonds in molecules, a system of special wedge bonds has been developed. Bonds that are drawn as simple lines lie in the plane of the page. Bonds that point toward the reader, out of the page, are shown using dark wedges. Bonds that point away from the reader, into the page, are drawn using hashed wedges. The use of the wedged bond and hashed bond represents the three-dimensional shape at a tetrahedral carbon atom.

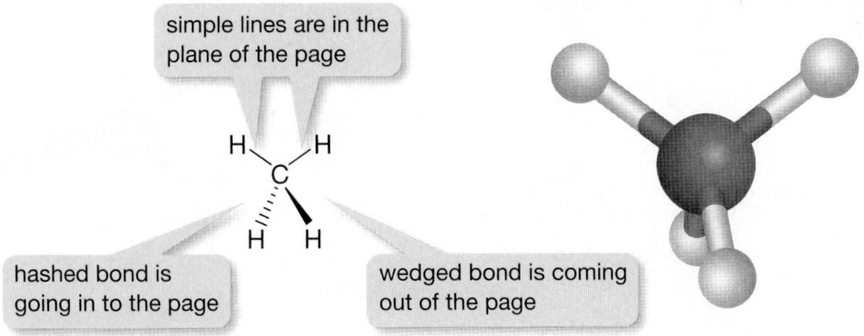

To properly capture the tetrahedral shape, two of the bonds to a particular atom should always be drawn as a "V" shape in the plane of the page using simple lines. The hashed bond and wedged bonds (one each per atom) are always drawn next to each other and in the wider space around the carbon atom; this positions the carbon at the centre of the tetrahedron.

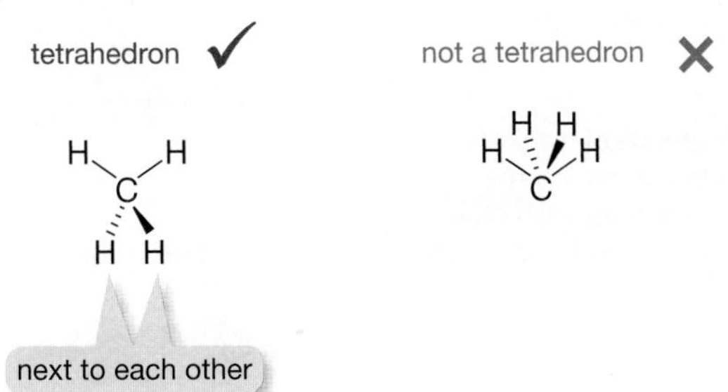

STUDENT TIP

Hashed bonds are not the same as dashed bonds, so be careful when drawing them. A hashed wedge denotes a three-dimensional shape; a dashed bond is a partial bond.

hashed bond shows three-dimensional orientation

C''''H

dashed bond is a partly broken bond

C---H

The tetrahedral arrangement around atoms can be shown in any orientation, but two bonds should always be drawn in the plane as single lines: one bond drawn above (dark wedge), and one bond drawn below (hashed wedge).

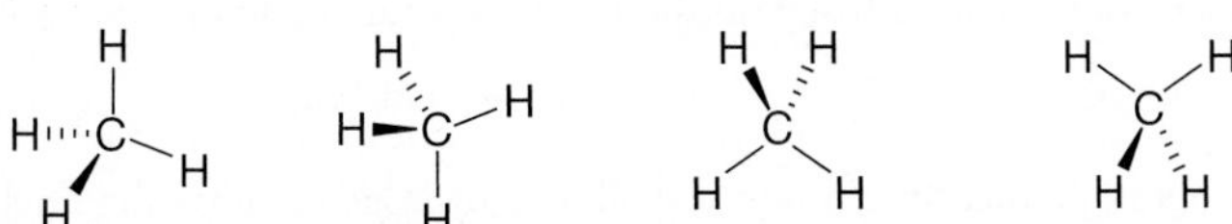

The observed geometry at the carbon atom of formaldehyde may similarly be predicted from its Lewis structure. The carbon atom has three groups of valence electrons around it: two C–H bonds and one C=O double bond. Positioning three groups of electrons as far away from each other as possible requires that they occupy a plane around the central atom at an angle of

120° to each other. The carbon atom of formaldehyde therefore adopts a trigonal planar geometry of atoms in order to minimize the repulsions between its three groups of electrons. As predicted by the VSEPR concept, the trigonal planar geometry is experimentally observed at the carbon atom of formaldehyde.

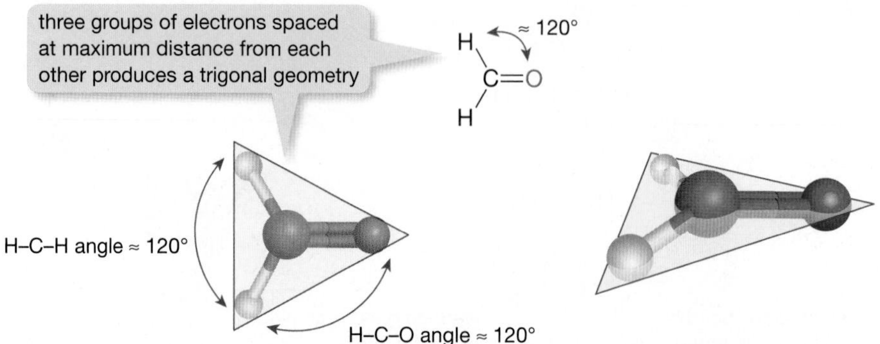
three groups of electrons spaced at maximum distance from each other produces a trigonal geometry
≈ 120°
H
C=O
H
H-C-H angle ≈ 120°
H-C-O angle ≈ 120°

In the Lewis structure of ethyne, both carbon atoms have two groups of valence electrons around them: a C–H bond and a C≡C triple bond. To get as far away from each other as possible, these two groups of electrons must be positioned on opposite sides of the carbon atom with an angle of 180° between them. The result is a linear geometry around each of the carbon atoms in ethyne, which is experimentally observed.

180°

H—C≡C—H

two groups of electrons spaced at maximum distance from each other produces a linear geometry

H-C-C angle ≈ 180°

CHECKPOINT 1.4

Applying the concept of repulsions among groups of valence electrons around an atom, you should now be able to identify the shape of each atom based on the Lewis structure of an organic molecule.

SOLVED PROBLEM

Use the VSEPR concept to predict the geometry at all non-hydrogen atoms of methyl acetate.

```
     H
     |1
   H-C-H  H
     |  3 |5
  2:O-C-C-H
     4||  |
      :O: H
```

There are five non-hydrogen atoms in the structure.

Carbon 1 shows four groups of electrons around it, so a tetrahedral arrangement of the four groups is predicted.

Oxygen 2 shows two bonding groups and two non-bonding (lone pair) groups, so a tetrahedral arrangement of the four groups is predicted.

Carbon 3 shows three bonding groups and zero non-bonding (lone pair) groups, so a trigonal arrangement of the three groups is predicted.

Oxygen 4 shows one bonding group and two non-bonding (lone pair) groups, so a trigonal arrangement of the four groups is predicted.

Carbon 5 shows four groups of electrons around it, so a tetrahedral arrangement of the four groups is predicted.

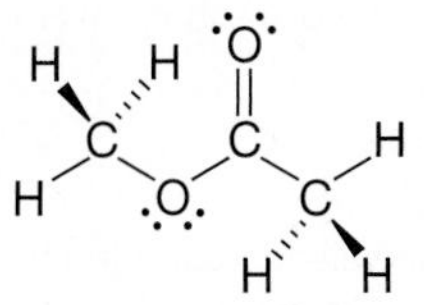

PRACTICE PROBLEM

1.4 Use the VSEPR concept to predict the geometry at each non-hydrogen atom in the following Lewis structures.

a)

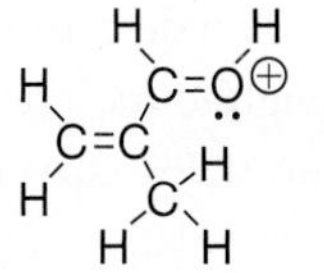

b)

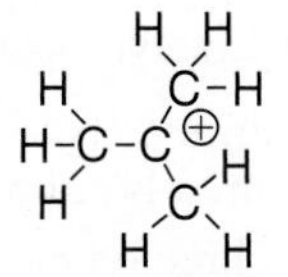

c) $CH_3CH_2\overset{\oplus}{C}NH$

INTEGRATE THE SKILL

1.5 To the following Lewis structure, add lone pairs of electrons based on the formal charges shown. Then use the VSEPR concept to predict the electron pair geometry at each non-hydrogen atom.

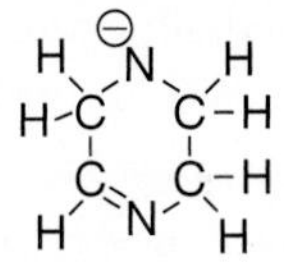

1.7 The Valence Bond Approach to Electron Sharing

As discussed in Section 1.5, a covalent bond arises from the overlap of atomic orbitals between two atoms and involves a sharing of two spin-paired electrons within that space. A Lewis structure provides a simple picture of which atoms are sharing electrons, but it does not explain the nature of orbital overlap between them.

There are two commonly used theories of bonding: the valence bond (VB) approach and the molecular orbital (MO) approach. Both of these model the bonding in organic molecules using valence orbitals.

The valence bond model extends the idea that a bond involves the sharing of two spin-paired electrons through the overlap of atomic orbitals between atoms; specifically, each atom

A **localized bond** involves the sharing of two electrons by means of the overlap of two atomic orbitals on two adjacent atoms. It is confined to the region between the two atoms.

contributes one orbital and one electron to form the bond. Such bonds are known as **localized bonds**; they are confined to the region between the two participating atoms. The better the orbital overlap, the better the sharing of electrons and the stronger the bond.

Lewis structures provide a good starting point for describing valence bonding. Each line in the Lewis structure represents a localized bond that involves valence orbital overlap between the two participating atoms.

In valence bond terms, each line in a Lewis structure represents a localized bond in which the probability of finding the electron pair is confined to the regions between the atoms.

However, the three types of bonding geometries introduced in Section 1.6 determine which valence orbitals the atoms use to establish covalent bonds with their neighbours. The valence bond approach uses the concept of hybrid orbitals to describe and explain the observed geometries in organic molecules.

Orbital hybridization is not a physical process; rather it is a mathematical combination of wavefunctions to produce new descriptions of the observed bonding. It involves calculations based on the overlap and mixing of an atom's valence orbitals to model new atomic orbitals for bonding. These new atomic orbitals, called **hybrid orbitals**, form the required bonding geometry because they have the appropriate mix of an atom's 2s and 2p orbitals. Three types of hybrid orbitals—sp^3, sp^2 and sp—are described in the following sections. They correspond to the three types of bonding geometries.

Hybrid orbitals are atomic orbitals that form a bonding geometry by a suitable mixing of the 2s and 2p orbitals of the atom.

1.7.1 Tetrahedral geometries require sp^3 hybrid orbitals

The VSEPR concept predicts that the carbon atom of methane has a tetrahedral geometry—that is, four groups of bonding electrons positioned tetrahedrally around the carbon atom. According to the valence bond model, the tetrahedral arrangement of C–H bonds results from the mixing of the four atomic orbitals arranged tetrahedrally around the carbon with the valence orbitals of the four hydrogen atoms. However, the 2s and 2p orbitals of carbon cannot form a tetrahedral distribution of bonds because they are not arranged tetrahedrally; the 2s orbital is non-directional (sphere), and the three 2p orbitals are arrayed at 90° to each other. The requirement for a tetrahedral distribution of bonds is resolved by the generation of hybrid orbitals. Four tetrahedrally distributed atomic orbitals of carbon can be generated by hybridizing one 2s orbital and *three* 2p orbitals on the same atom.

A carbon with a tetrahedral arrangement of four groups requires ...

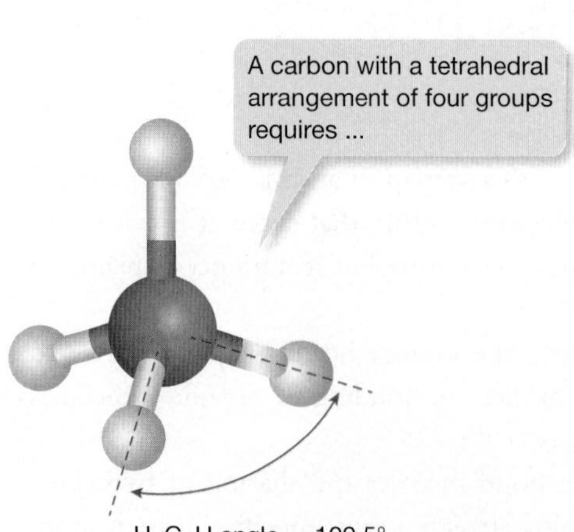

H–C–H angle = 109.5°

... a tetrahedral arrangement of its orbitals: four sp^3 hybrid atomic orbitals.

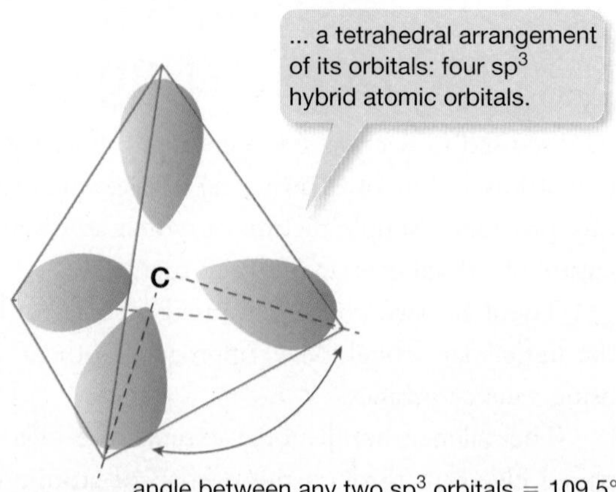

angle between any two sp^3 orbitals = 109.5°

This results in a set of four hybrid orbitals arranged tetrahedrally around the carbon, as shown in the following diagram. Because each hybrid is composed of one part 2s and three parts 2p, each of the four hybrid orbitals are designated as an sp^3 orbital. Each of these consists of a large lobe and a small lobe of different phases (depicted with light and dark shades of green), separated by a node (depicted by an area where the colour changes). Only the large lobe is important for overlap with the orbitals of the surrounding atoms.

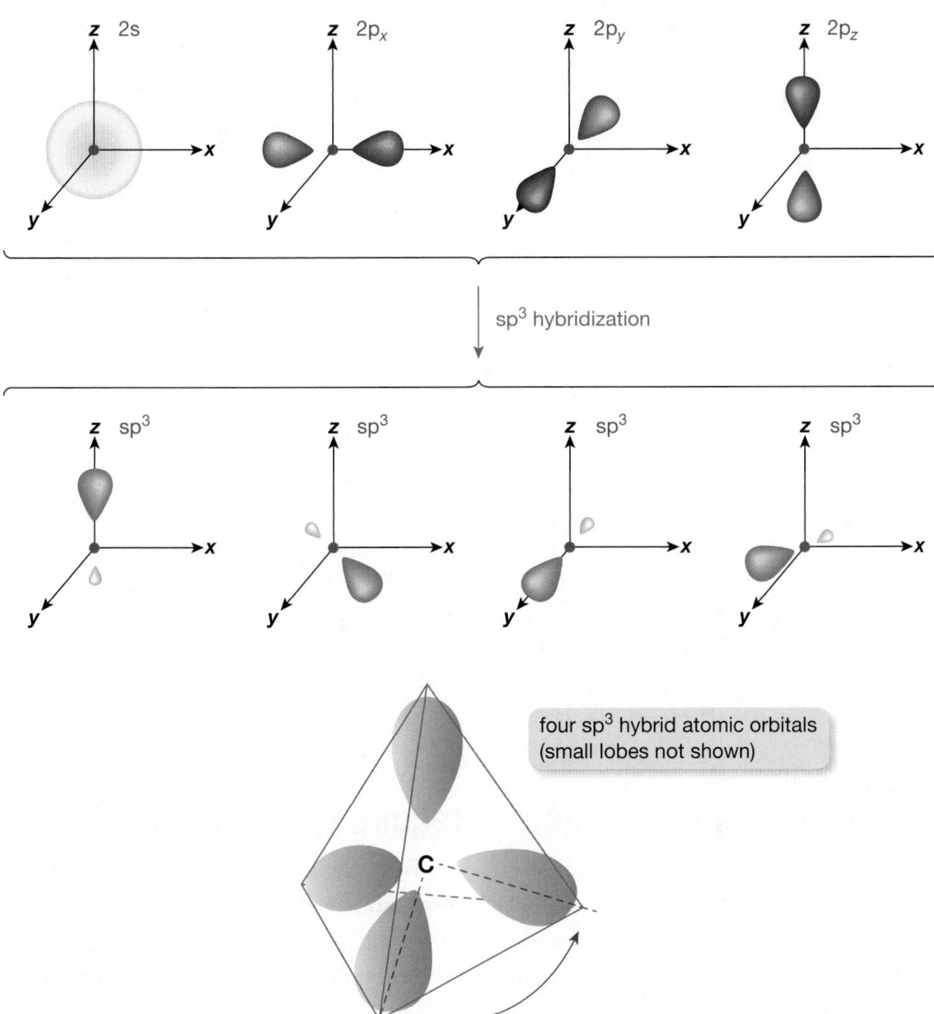

Methane and ammonia provide two examples of the tetrahedral distribution of electrons involving sp^3 hybrid orbitals. In methane, the bonding consists of four localized σ bonds, each obtained from the head-on overlap of a 1s orbital on a hydrogen with the carbon sp^3 orbital that points toward it. Each line between C and H in the Lewis structure of methane then represents such a localized interaction.

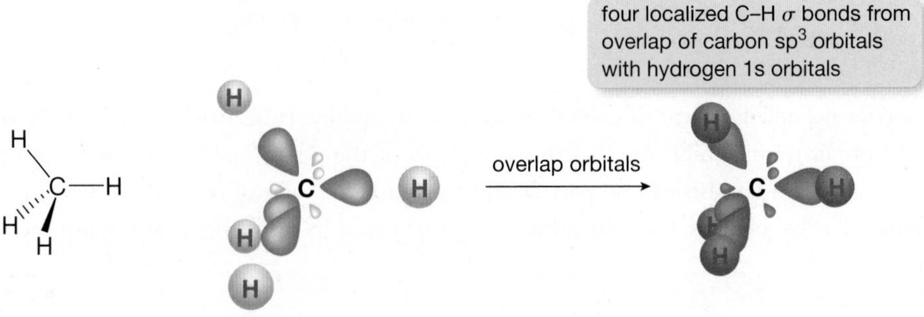

In ammonia, the tetrahedral arrangement of bonding and the lone pairs result from three localized N–H σ bonds formed from the head-on overlap of sp^3 orbitals on nitrogen with the 1s orbitals of three hydrogen atoms. The fourth sp^3 hybrid orbital on the nitrogen contains the atom's lone pair of electrons.

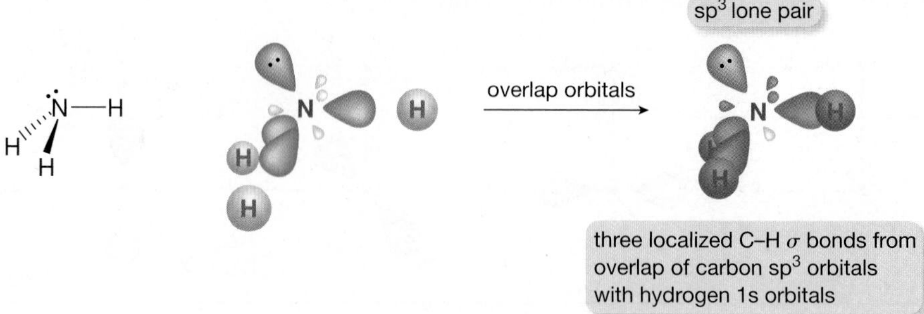

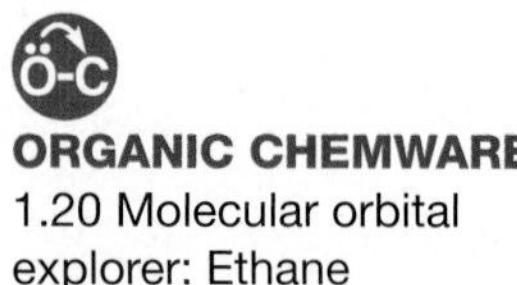

ORGANIC CHEMWARE
1.20 Molecular orbital explorer: Ethane

ORGANIC CHEMWARE
1.21 Molecular orbital explorer: Methanol

When sp^3-hybridized atoms are linked to form chains, the hybridization at each atom creates a zig-zag pattern of tetrahedrons, which gives a very different representation from the linear one implied by Lewis structures.

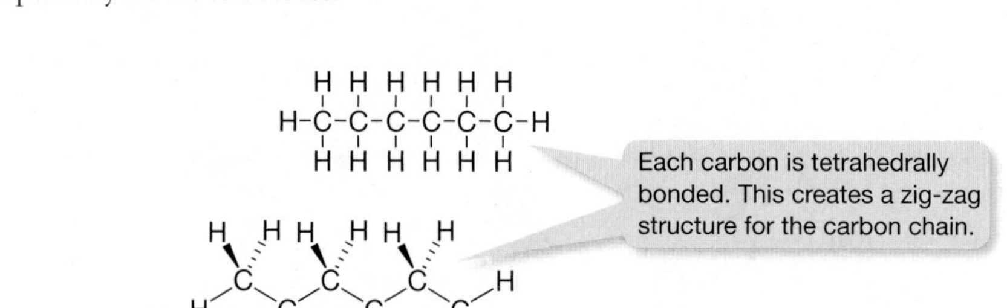

Each carbon is tetrahedrally bonded. This creates a zig-zag structure for the carbon chain.

1.7.2 Trigonal planar geometries require sp^2 hybrid orbitals

The VSEPR concept predicts that the carbon atom of formaldehyde has a trigonal planar geometry—that is, three groups of electrons positioned 120° apart from each other in the same plane as the carbon atom. Valence bonding at such a carbon requires a trigonal planar arrangement of orbitals around the carbon so as to establish overlap with the orbitals of the surrounding hydrogen and oxygen atoms.

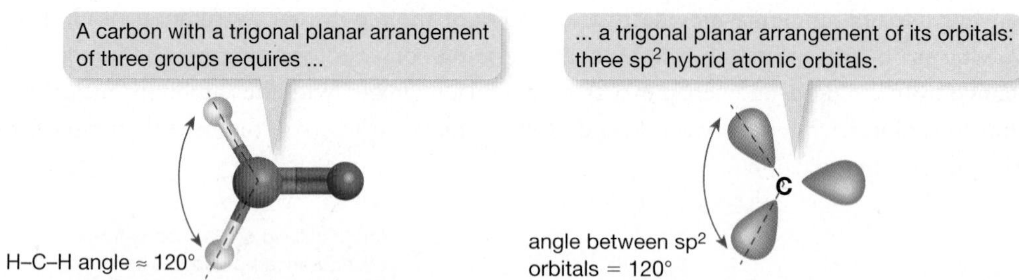

This trigonal arrangement of orbitals is attained by sp^2 hybridization of the valence orbitals of the carbon atom, in which the 2s orbital and *two* of the 2p orbitals of the atom are mixed. Because each hybrid consists of one part 2s and two parts 2p, each of these three hybrid orbitals are designated as sp^2 orbitals. One 2p orbital is not involved in the hybridization and so remains unchanged.

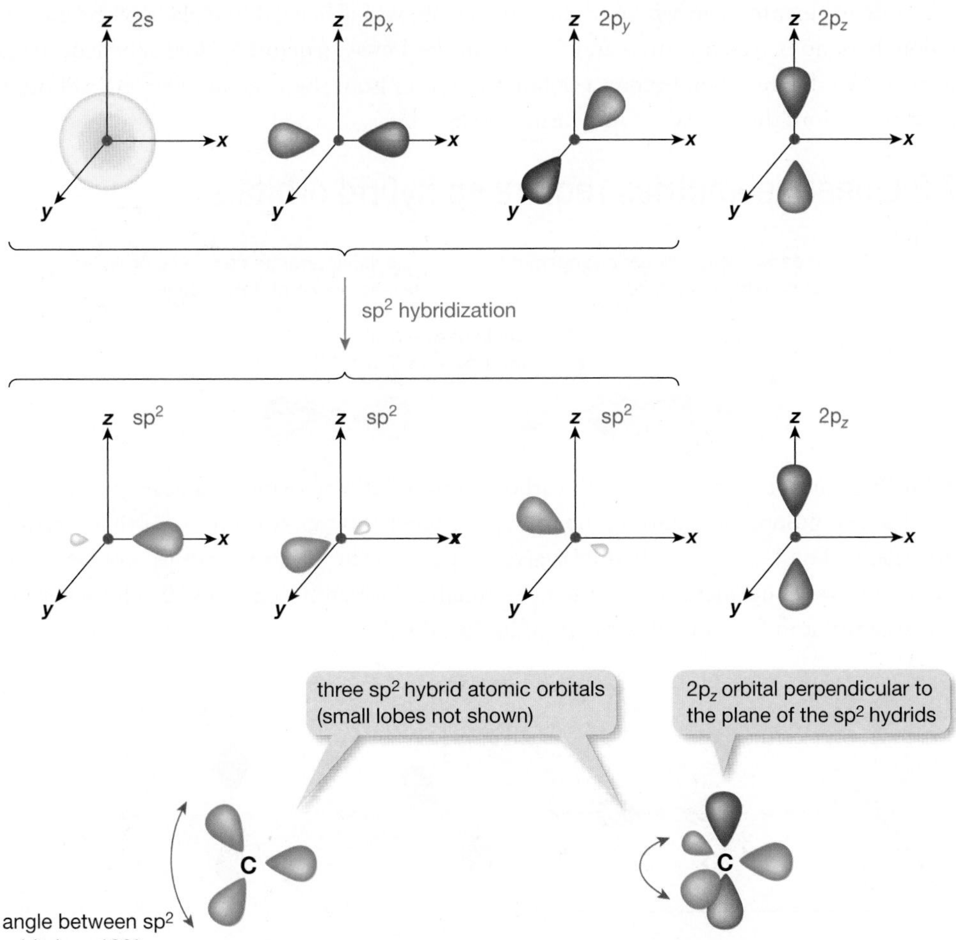

ORGANIC CHEMWARE
1.22 Hybridization: sp^2

The VSEPR concept also predicts a trigonal planar geometry for the oxygen atom of formaldehyde, which indicates the oxygen is also sp^2 hybridized. Two of the sp^2 hybrid orbitals of carbon are positioned to form σ bonds from head-on overlap with the 1s atomic orbitals of the surrounding hydrogen atoms; the third hybrid orbital forms a σ bond through head-on overlap with one of the sp^2 hybrid orbitals of the oxygen atom. In this molecular structure, the remaining two sp^2 hybrid orbitals of oxygen contain its two lone pairs of electrons.

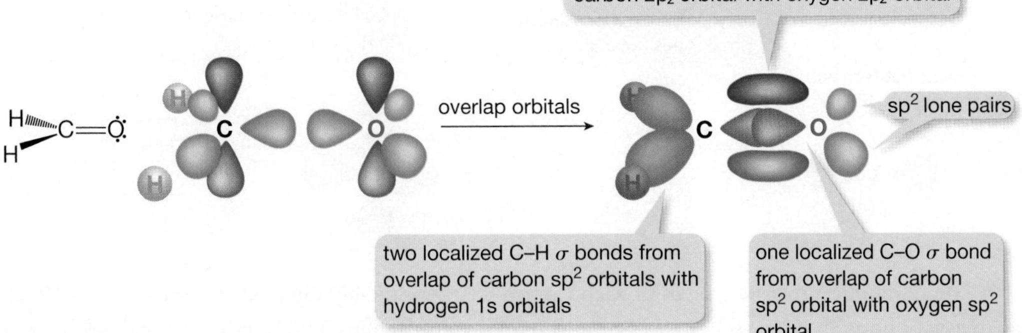

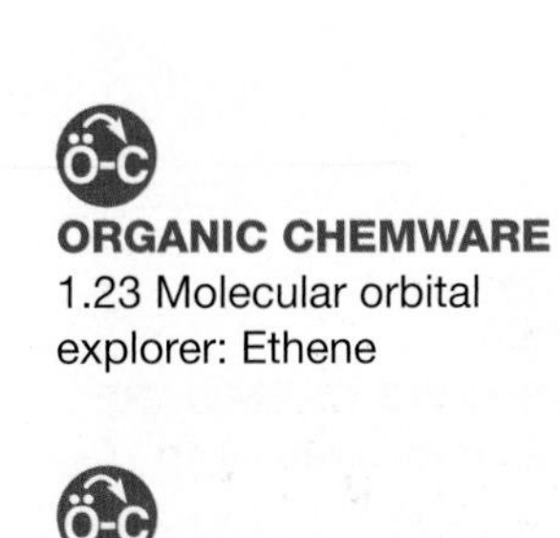

ORGANIC CHEMWARE
1.23 Molecular orbital explorer: Ethene

ORGANIC CHEMWARE
1.24 Molecular orbital explorer: Formaldehyde

Both of the sp^2-hybridized carbon and oxygen atoms carry an unhybridized 2p orbital that lies perpendicular to the plane of the three sp^2 hybrid orbitals. These orbitals overlap in a side-by-side manner to form a π bond that projects above *and* beneath the plane of the molecule. The double bond between carbon and oxygen in the Lewis structure of formaldehyde actually consists of two different bond types: a σ bond resulting from the overlap of sp^2 hybrid orbitals, and a π bond from the overlap of unhybridized 2p orbitals.

1.7.3 Linear geometries require sp hybrid orbitals

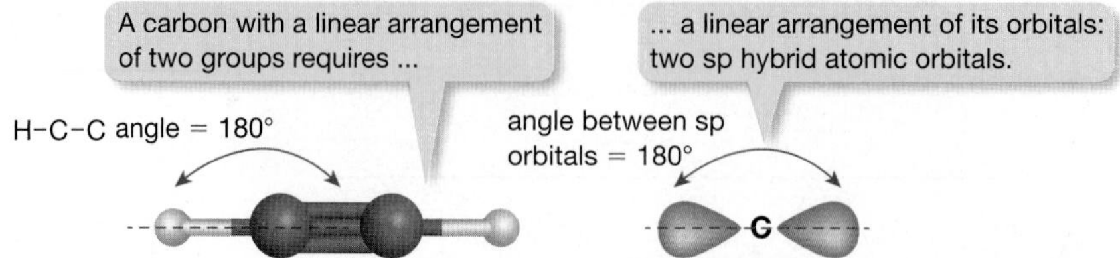

The VSEPR concept predicts that both carbon atoms of ethyne exhibit a linear geometry. Each carbon has two groups of electrons around it, positioned 180° apart from each other. According to the valence bond approach, the linear arrangement of two electron groups around an atom requires a linear arrangement of two valence orbitals at that atom. This can be achieved only by hybridizing the atom's 2s orbital with *one* of its 2p orbitals.

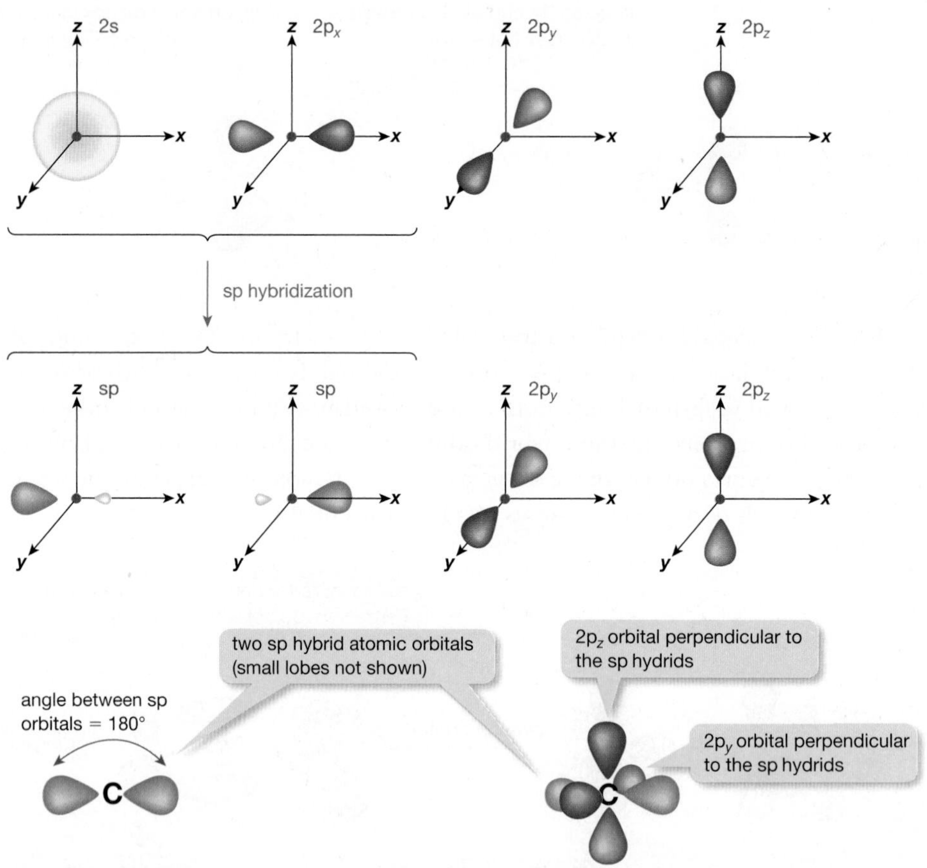

ORGANIC CHEMWARE
1.25 Hybridization: sp

The resulting sp hybrid orbitals of each carbon are then suitably directed for head-on overlap to form a σ bond with each other as shown in the following diagram. The bonds between carbon and hydrogen are constructed by overlap between an sp orbital on the carbon and the 1s orbital on each hydrogen. The two 2p orbitals that remain unhybridized on each carbon overlap side by side to establish two separate π bonds that lie perpendicular to each other. A triple bond in valence bond terms consists of a σ bond and two orthogonal π bonds between two sp-hybridized atoms.

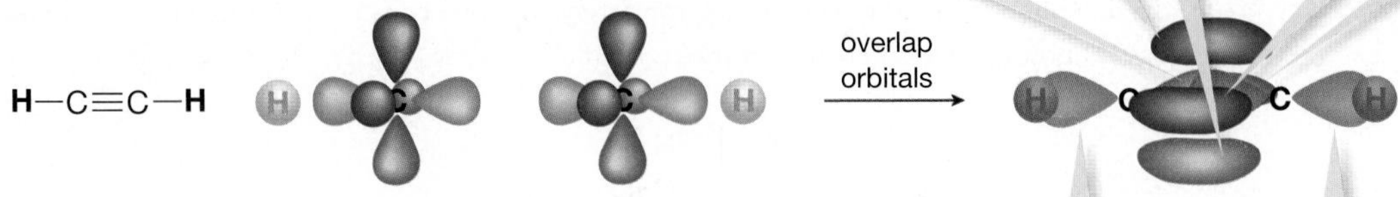

CHECKPOINT 1.5

You have learned about the valence bond model: what a localized covalent bond is and how it arises from the overlap of an atomic orbital on each of the two participating atoms.

ORGANIC CHEMWARE
1.26 Molecular orbital explorer: Ethyne

SOLVED PROBLEM

Prenol occurs naturally in citrus fruits and berries and has a fruity aroma. (a) Predict the geometry at each non-hydrogen atom in the Lewis structure of prenol shown here. (b) Use the geometries you determined in part (a) to establish the hybridization at each of these atoms.

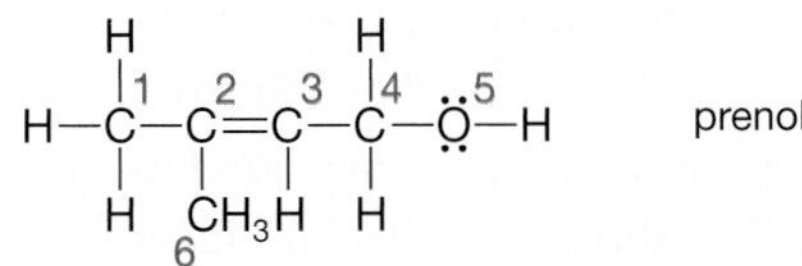

Atom 1: The carbon atom has four groups of electrons around it. The VSEPR concept tells us that the four groups (three hydrogens and one carbon) will be arranged tetrahedrally around the carbon atom. This carbon must be sp^3 hybridized to provide orbitals in a tetrahedral arrangement.

Atom 2: The carbon atom is surrounded by three groups of electrons: one double bond and two single bonds. The VSEPR model predicts the carbon atom will be trigonal planar in shape. This carbon atom must be sp^2 hybridized to obtain orbitals pointing in the correct directions of 120° away from each other.

Atom 3: This carbon atom is sp^2 hybridized. See the explanation for atom 2.

Atom 4: This carbon atom is sp^3 hybridized. See the explanation for atom 1.

Atom 5: The oxygen atom is surrounded by four groups of electrons: two single bonds and two non-bonding (lone) pairs. Applying the VSEPR concept, the four groups will adopt a tetrahedral arrangement about the oxygen atom. According to the VB approach, the valence orbitals of the oxygen atom will be sp^3 hybridized; each of the lone pairs of oxygen resides in a sp^3 hybrid orbital.

Atom 6: This C atom is sp^3 hybridized. See the explanation for atom 1.

STUDENT TIP
Notice that the σ bonds of all observed geometries require hybrid orbitals—either sp^3, sp^2 or sp hybrids—so that they point in the direction of the surrounding atoms. In contrast, π bonds arise from overlap between unhybridized p orbitals. Hybrid orbitals are not involved in π bonding.

STUDENT TIP
The sum of the superscripts in hybrid orbital designations is equal to the number of groups of electrons that surround the hybridized atom.

PRACTICE PROBLEM

1.6 a) Use the VSEPR concept to determine the electron pair geometry at each of the non-hydrogen atoms in the following Lewis structure of 2,5-dihydrofuran.

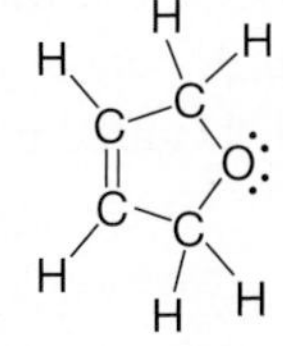

b) Use the geometries from part (a) to establish the required hybridization at each non-hydrogen atom in the VB approach to modelling the bonding in this molecule.

INTEGRATE THE SKILL

1.7 a) To the following Lewis structure, appropriately add lone pairs of electrons and formal charges based upon the overall charge on the molecule.

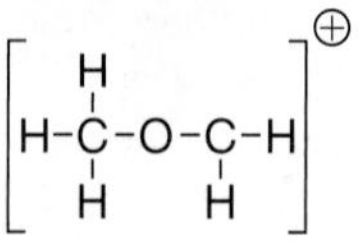

$$\left[\begin{array}{c} \text{H} \\ | \\ \text{H-C-O-C-H} \\ | \quad\quad | \\ \text{H} \quad\quad \text{H} \end{array}\right]^{\oplus}$$

b) Use the VSEPR concept to predict the electron pair geometry at each non-hydrogen atom.

c) Use the geometries from part (b) to predict the required hybridization at each non-hydrogen atom.

d) This molecule can be represented by another Lewis structure containing a double bond. Draw this structure and answer the questions in parts (b) and (c) for this new structure.

1.8 Resonance Forms: Molecules Represented by More than One Lewis Structure

Resonance is a *tool* used to describe the delocalization of π electrons in a molecule.

The Lewis structure is a simple representation of the bonding in an organic molecule. However, some bonding situations—specifically those that involve π bonds—are not well described by single Lewis structures. Because of the special way that π orbitals interact, it is sometimes impossible to represent the bonding by just one Lewis structure. In these situations, the concept of **resonance** is used to describe how the atoms are connected.

Consider the positively charged allyl cation. The VSEPR concept predicts that the geometry at each carbon atom is trigonal planar, and the valence bond model suggests that all three trigonal planar atoms are sp^2 hybridized. The three sp^2 hybrid orbitals of each carbon form σ bonds with their three neighbours—by overlap with the sp^2 orbital of an adjacent carbon or by overlap with the 1s orbital of an adjacent hydrogen atom. The π bond results from the sharing of a pair of electrons arising from the overlap of unhybridized 2p orbitals of the carbon on the left and the carbon in the middle. The unhybridized p orbital of the carbon on the right is apparently empty, consistent with the atom's formal charge of +1.

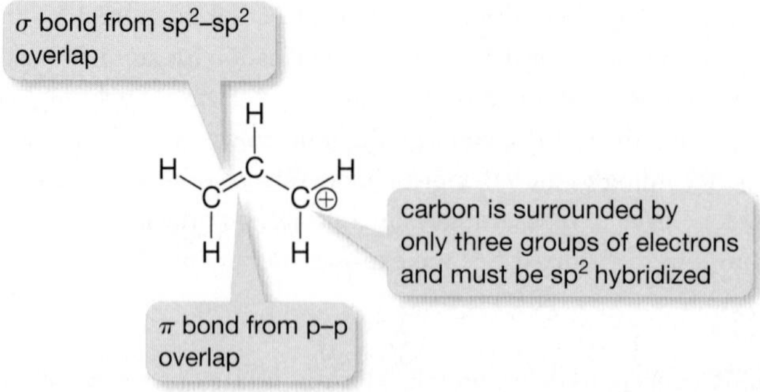

However, this single Lewis structure does not provide a complete description of the electron distribution and bonding in the allyl cation. A view of the allyl cation from the side (Figure 1.8) shows the potential overlap of all three unhybridized p orbitals.

p orbital overlap is the same between all three carbons

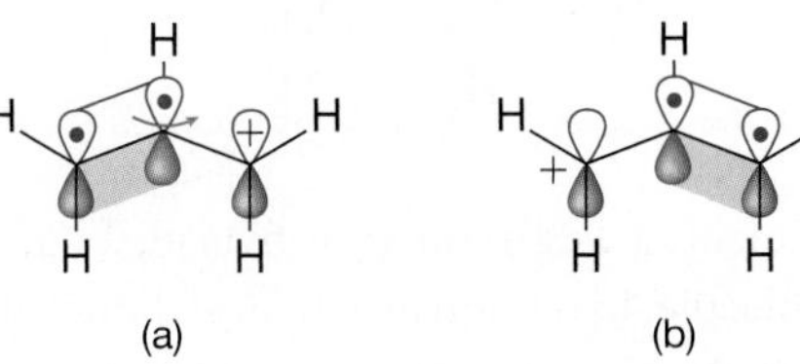

(c)

FIGURE 1.8 Side-view drawing of the allyl cation showing its unhybridized p orbitals: (a) sharing of two π electrons to the left; (b) sharing two π electrons to the right; (c) sharing of two π electrons over all three carbons.

The empty p orbital on the right in Figure 1.8a is coplanar with and overlaps with the p orbital on the middle carbon. The filled p orbital on the left also overlaps with the p orbital on the middle carbon, and does so to the same extent as the empty p orbital on the right. Through this overlap, the two electrons of the π bond can "spill over" to the right side (red arrow in Figure 1.8a), so they are now shared between the carbon in the middle and the carbon on the right. This provides an equivalent structure (shown in Figure 1.8b). The two π electrons have an equal likelihood of being on the right as in Figure 1.8b as they do of being on the left as in Figure 1.8a. This suggests that *both* bonding descriptions must somehow contribute to the π bonding of the molecule, and that the two electrons are shared by all three carbons. Sharing two electrons between more than two atoms is called **delocalization**, and the electrons are called **delocalized electrons**.

Delocalization occurs when π electrons are shared by more than two atoms.

Delocalized electrons are not associated with a single atom or bond. Instead, they are shared among several atoms; they are delocalized.

A **resonance form** is one of a set of related Lewis structures used to describe bonding situations in which π electrons are not confined to individual atoms or bonds. Each resonance form provides a bonding description based upon a different location of the molecule's π electrons.

The two bonding descriptions of Figure 1.8 can be represented by two corresponding Lewis structures. These structures are **resonance forms**: Lewis structures of a molecule that differ only in the distribution of π electrons among *overlapping p orbitals* on adjacent atoms. They are denoted by the double-ended arrow shown here and by placing brackets around all of the structures. In resonance forms, the hybridization of the atoms and the σ bonds between them remain unchanged.

double-ended arrows are used to indicate and separate resonance forms

Neither of the two resonance forms of the allyl cation on its own properly describes the π bonding in the molecule. Although each resonance form suggests that one carbon-carbon bond is a double bond and the other is a single bond, experimental data shows that the two carbon-carbon bonds of the allyl cation are identical. In fact, the C–C bond length is in between that of a single and a double bond. Consideration of both resonance forms *together* reveals a more complete picture of the π bonding in the allyl cation: two π electrons are equally shared (i.e., delocalized) among three carbon atoms. Taken together, the two resonance forms indicate that the two carbon-carbon bonds are the same—a full σ bond plus a partial π bond. The allyl cation is a planar (flat) molecule; all eight atoms are in the plane of the page.

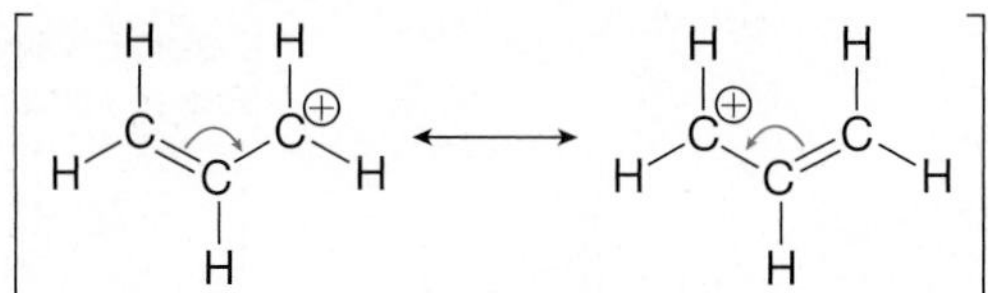

ORGANIC CHEMWARE
1.27 Resonance: Allyl cation

Resonance structures are sometimes represented as a resonance hybrid, a single structure that represents all the resonance forms for a molecule. In a resonance hybrid, bonds that occur in some forms but not in others are shown with a dashed line. Charges that occur in some forms but not in others are shown by the symbol δ (delta); this indicates the labelled atoms carry a partial charge (part of a total charge is found on that atom).

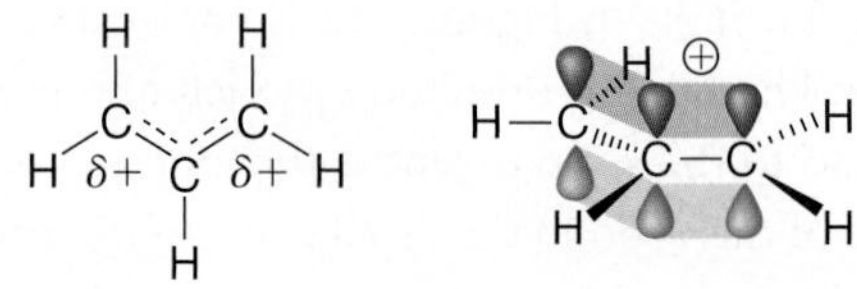

resonance hybrid

STUDENT TIP
Molecules do not interchange back and forth between resonance forms. Rather, the actual structure is a composite of all the contributing resonance forms.

Resonance forms are normally more useful to work with than resonance hybrids. Resonance forms provide information about chemical reactivity and structure in a manner that is easier to use than that in the hybrid representation. It is very important to remember that resonance is *not* a physical process. Rather, it is a *tool* used to compensate for the limitations of Lewis structures with respect to depicting the delocalization of electrons in certain molecules.

Quantum mechanical calculations can be used to visualize the delocalization of electrons in molecules, which is what resonance forms try to depict. These calculations produce molecular orbitals—that is, orbitals that span the whole molecule and show how electrons can delocalize over large areas (Figure 1.9).

STUDENT TIP
Resonance forms are very useful for predicting chemical reactivity as well as structure.

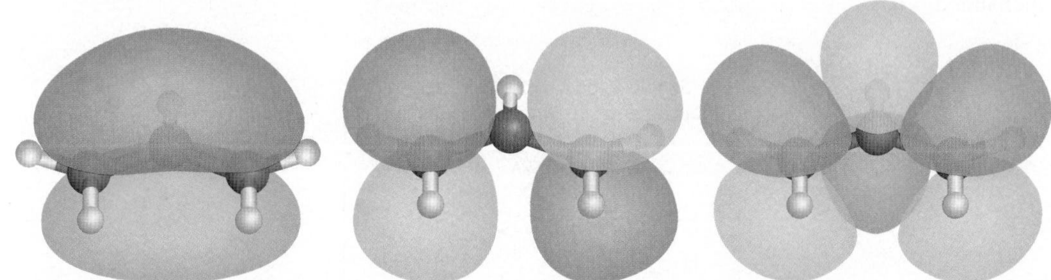

FIGURE 1.9 The complete π orbital system of the allyl cation that corresponds to the resonance among the three carbon atoms. The different colours represent the different phases of the molecular orbital lobes. The orbital on the left is the lowest energy and contains the pair of π electrons in the ground state of the molecule. Notice how the orbitals extend over *all three* carbon atoms. See Chapters 6 and 20 for further descriptions of resonance and molecular orbitals.

Consider the case of the allyl anion. Once again, a single Lewis structure is inadequate to represent the structure and reactivity of this molecule. In fact, there are two equivalent structures for this molecule.

two equivalent Lewis structures

Although VSEPR theory assumes that the carbanion centre is sp^3 hybridized, experiments reveal that the geometry of the allyl anion is planar, just like its carbocation counterpart. Likewise, the two C–C bond lengths are identical! Assuming that all three carbon atoms are sp^2 hybridized, including the carbon that bears the formal charge in the individual Lewis structures, resonance can be rationalized by experimental observations. This presents three collinear p orbitals on three adjacent carbons, which allows delocalization of the four p electrons over these atoms.

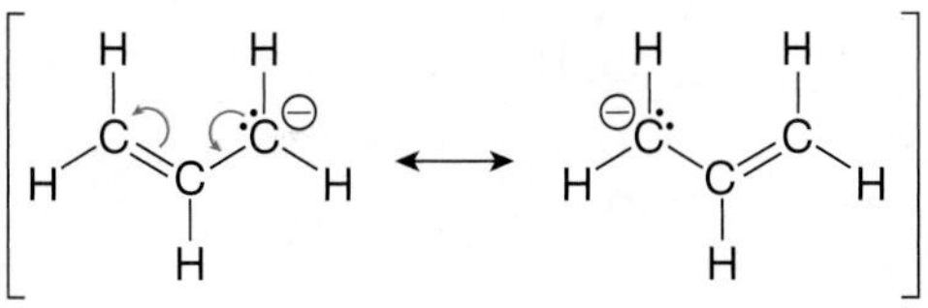

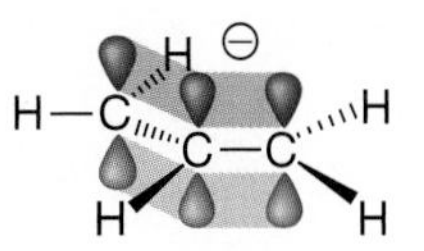

In general, Lewis structures of molecules that have a non-bonded pair of electrons (i.e., a lone pair) next to a π bond are indicative of resonance. The gain in molecular stability imparted by resonance easily compensates for the change in hybridization of the carbanion centre from sp^3 to sp^2 to produce a proper alignment of its p orbitals.

1.8.1 Ways to draw simple resonance forms from Lewis structures

Two general methods are used to draw contributing resonance forms from Lewis structures. The first method uses the formal charge method of drawing Lewis structures (Section 1.4.1) to obtain a structure in which all the bonds are single and the formal charges have been assigned. From this, all the possible resonance forms of the structure can be drawn. The second method starts with a resonance structure and "breaks" π bonds to obtain formally charged atoms that can be used to construct "new" π bonds.

When using the first method for drawing resonance forms, add the non-bonding electrons to the structure so that it has opposite charges alternating (+, −, +, etc.) along the chain of atoms involved. For example, to construct the resonance forms of the allyl cation, use the formal charge procedure to arrive at a structure in which there are no multiple bonds, and the electrons are distributed to give an alternating arrangement of − and + charges.

$$[CH_2CHCH_2]^{\oplus}$$

3 carbons (group 4)	3 × 4 =	12
5 hydrogens (group 1)	5 × 1 =	5
1 positive charge	1 × −1 =	−1
		16 valence electrons

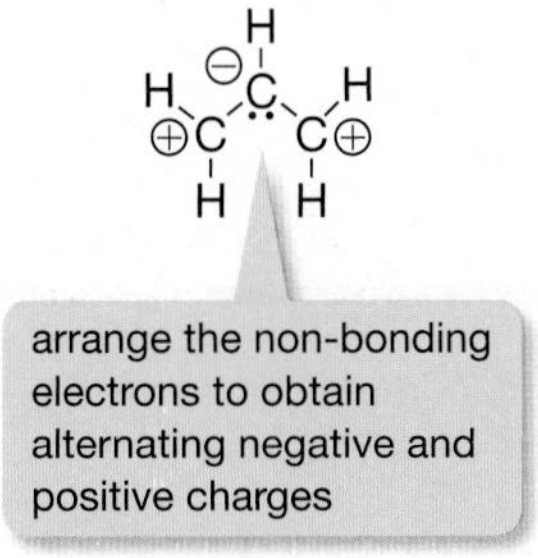

arrange the non-bonding electrons to obtain alternating negative and positive charges

Next, construct double and triple bonds in the typical way. Note that in this example there are two ways to do this, and each produces a resonance form. Consider all the ways that charges can be combined to form multiple bonds to construct all the resonance forms for a molecule.

Electrons can be combined with the positive carbon on both the left and the right. Each leads to a different resonance form.

When using the second method for drawing resonance structures, start with a given Lewis structure and "break" bonds to generate a structure with the opposite charges alternating. An effective way to do this is to move electrons toward positive charges and away from negative ones. Once you have the alternating-charge structure, create double and triple bonds as described for the first method of drawing resonance structures.

Move electrons from a π bond onto one of the atoms of the bond. Move electrons toward positive charges and away from negative ones so as to obtain alternating negative and positive formal charges.

More advanced methods of constructing and analyzing resonance structures are described in Chapter 5.

It is important to remember that resonance does not involve the molecule interchanging back and forth between resonance forms. Resonance is not some kind of equilibrium, but is a tool in which a blend of *all* of the resonance forms *at the same time* provides a *picture* of the π bonding in a molecule. Resonance forms are used because they are easier to work with than resonance hybrids or calculated orbitals.

CHECKPOINT 1.6

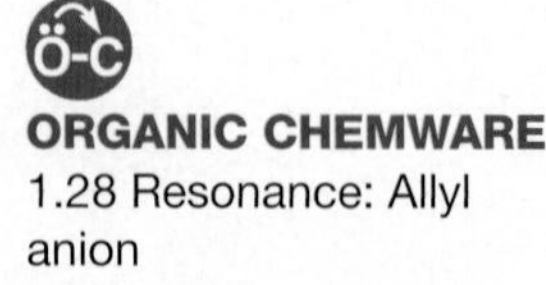

ORGANIC CHEMWARE
1.28 Resonance: Allyl anion

You can now draw resonance forms from a simple Lewis structure.

SOLVED PROBLEM

Given the following Lewis structure for the allyl anion, draw the second resonance form for this structure.

METHOD A

STEP 1: Redraw the structure from scratch (except for the π bond) using the formal charge method. Carefully distribute the non-bonding electrons to obtain alternating negative and positive charges.

3 carbons (group 4)	$3 \times 4 = 12$
5 hydrogens (group 1)	$5 \times 1 = 5$
1 negative charge	$1 \times 1 = 1$
	18 valence electrons

STEP 2: Combine electrons from negatively charged atoms with adjacent positive atoms to form π bonds. There are two ways to do this, and each produces a resonance form.

ORGANIC CHEMWARE
1.29 Resonance: Carboxylate anion

ORGANIC CHEMWARE
1.30 Resonance: Enolate anion

METHOD B

Break any π bonds to form alternating positive and negative charges. Move electrons toward positive charges and away from negative ones.

Break π bonds to make alternating formal charges. Move electrons away from negative charges and toward positive charges.

ORGANIC CHEMWARE
1.31 Resonance: Pentadienyl cation

ORGANIC CHEMWARE
1.32 Resonance: Pentadienyl anion

PRACTICE PROBLEM

1.8 Draw the second resonance form of the following molecules. Where appropriate, include formal charges in your resonance form (three resonance forms are possible for part d).

a) $H_3C\text{–}C(=O)\text{–}CH_3$

b) $H_3C\text{–}NO_2$

c) $H_3C\text{–}C(=\overset{\oplus}{O}H)\text{–}CH_3$

d) $H_3C\text{–}C(=O)\text{–}NH_2$

ORGANIC CHEMWARE
1.33 Resonance: Carbonyl group

ORGANIC CHEMWARE
1.34 Resonance: Imine

ORGANIC CHEMWARE
1.35 Resonance: Nitrile

INTEGRATE THE SKILL

1.9 The following molecule has four significant resonance forms. Use the formal charge method to draw each of them.

$H_2N\text{–}C(=\overset{\oplus}{N}H_2)\text{–}NH_2$

A **molecular orbital** (MO) maps out, point by point in space, the probabilities of finding electrons in the volume of space around the nuclei of a molecule. Molecular orbitals are often represented as a shape that depicts the probability of finding an electron within that volume of space 95 percent of the time.

A **σ molecular orbital** (σ MO) has its amplitude concentrated along the axis between the nuclei of the molecule.

A **bonding molecular orbital** is a molecular orbital with a high electron density in the volume of space between the atoms of the molecule. Electrons in bonding molecular orbitals stabilize a molecule.

1.9 Molecular Orbital Approach to Electron Sharing

The valence bonding approach described in Section 1.7 is one method of modelling the bonding in organic molecules. The molecular orbital (MO) approach is a second method, which extends the idea of bonding to the entire molecule. This approach involves mixing a set of atomic orbitals to obtain a set of **molecular orbitals** that describe the likelihood of finding electrons in the space around the molecule.

Two of the molecular orbitals of carbon monoxide, CO, are shown in Figure 1.10. The distinct shape of each orbital describes the probability of finding electrons around the carbon monoxide molecule. This shape can be interpreted as an enclosure within which the electrons can be found 95 percent of the time. Both orbitals describe a different probability distribution, but they both have amplitude along the line running through the two nuclei. For this reason, these orbitals are known as **σ molecular orbitals** (σ MO).

The σ molecular orbital on the bottom of Figure 1.10 shows continuous electron density in the region between the carbon and oxygen atoms. Orbitals that provide this "zone" of electrons between atoms are known as **bonding molecular orbitals**. Electrons in these orbitals stabilize the molecule and hold the atoms together.

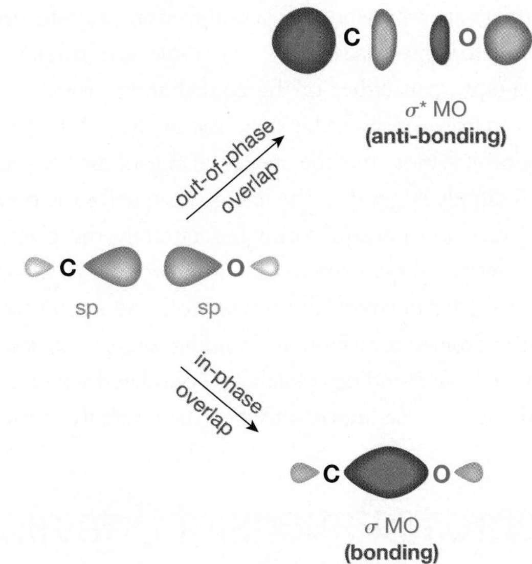

FIGURE 1.10 Two of the several molecular orbitals of carbon monoxide: a σ molecular orbital and a σ* molecular orbital.

In contrast, the orbital on the top of Figure 1.10 has a region between the two atoms where the probability of finding electrons is zero. This region between the carbon and oxygen atoms represents a destabilizing interaction. It is an **anti-bonding molecular orbital**, written as σ★, where the asterisk "★" designates an anti-bonding orbital. Electrons in a σ★ molecular orbital force the atoms apart.

Molecular orbitals describe bonding over the entire molecule. However, the focus of organic chemistry is typically the orbitals involved in chemical reactions or particular structures. If a reaction requires the breaking of a carbon-bromine bond, for example, only the orbitals involved between the carbon and bromine are analyzed, not those of the entire molecule.

Molecular orbitals can be thought of as a combination of the atomic orbitals of the molecule's atoms. For example, the two molecular orbitals of carbon monoxide associated with the σ bond can be considered a combination of the sp orbital on its carbon atom and the sp orbital on its oxygen atom. A σ molecular orbital arises from the in-phase overlap of these orbitals, and a σ★ orbital results from the out-of-phase overlap of the atomic orbitals.

ORGANIC CHEMWARE
1.36 Molecular orbitals: C–O σ orbitals

An **anti-bonding molecular orbital** (σ*) has nodes between the adjacent atoms of the molecule. Electrons in anti-bonding orbitals destabilize the molecule and force atoms apart.

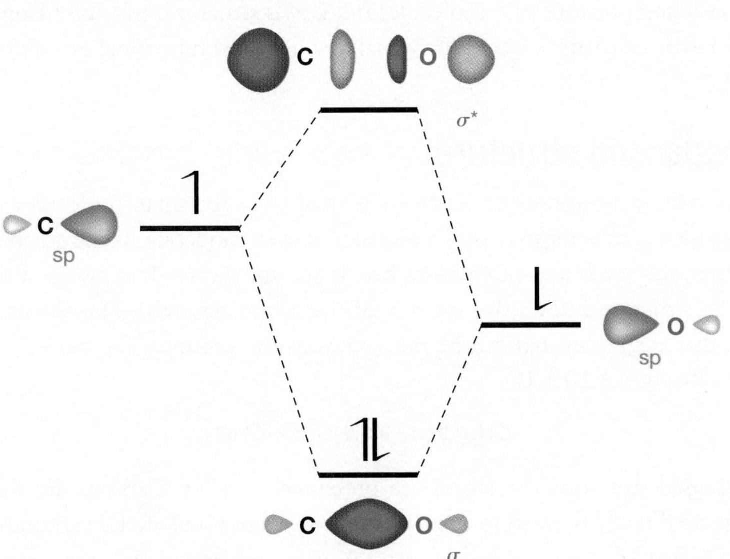

Although molecular orbitals are represented as a combination of atomic orbitals, the total number of orbitals is preserved. Two atomic orbitals make two molecular orbitals. In-phase overlap generates orbitals with lower energy than either of the contributing orbitals, and produces a bonding orbital between the atoms (electron density in between the nuclei). Out-of-phase overlap results in a higher-energy anti-bonding orbital. Note that the energy change of the anti-bonding orbital relative to the starting atomic orbitals is slightly larger than the energy change for the bonding orbital.

To represent the combination of atomic orbitals accurately, the placing of electrons in new orbitals requires filling the lowest energy orbitals first. This produces a favourable energy situation for the molecule because electrons between the nuclei hold the nuclei together.

Only the filled molecular orbitals contribute to bonding, and so *only the occupied molecular orbitals contribute to the bonding in the molecule.* Bonding orbitals are considered when analyzing the structure of molecules. Anti-bonding orbitals become important when the reactivity of molecules is analyzed.

1.10 Other Representations of Organic Molecules

So far in this chapter, the molecules have been drawn as Lewis structures. A Lewis structure uses atomic symbols to show all atoms, lines to display the molecule's bonding electron pairs, and dots to represent non-bonding electrons. The dots are properly arrayed in pairs (lone pairs) around the atomic symbol of the atom they are associated with.

```
                    H
                    |
   H  H  H  H  H   :O: H        :O:
   |  |  |  |  |    |  |         ||
H -C -C -C -C -C  - C -C = C  -  C -H
   |  |  |  |  |    |      |
   H  H  H  H  H    H      H
```

This method of depicting a detailed structure has drawbacks. First, it is time consuming to create them. Second, Lewis structures explicitly present so much information that they become cluttered, making it difficult to identify key features that contribute to structure and reactivity. Alternatively, to facilitate the writing and interpreting of chemical structures, several shorthand styles can be used to represent organic molecules. The most common of these are condensed structures and line structures.

Sometimes the different shorthand drawing styles can be combined in one structure, depending on the feature of interest in the molecule. For example, condensed formulas are used for non-reacting portions of a molecule, and Lewis structures may be helpful for studying reactivity. The Lewis structure is especially valuable when new functional groups or reactions are encountered.

1.10.1 Condensed structure

Typically, a condensed structure (or formula) is used only for small molecules or portions of molecules because it is difficult to depict molecules of even moderate size using this style. In condensed structures, the solid lines of covalent bonds are not shown. It is assumed that the person interpreting the structure knows the valence of the atoms involved. The valence—number of connections—that each atom has can be derived from the group of the periodic table that the atom resides in (Section 1.10.1.1).

$$CH_3(CH_2)_4CHOHCHCHCHO$$

In a condensed structure, the atoms are presented in a list. Carbons are followed by the hydrogens that are directly bonded to them. If the hydrogens listed after a carbon represent fewer than four bonds (maximum number for carbon), it is understood that the next atom listed is bonded to that carbon.

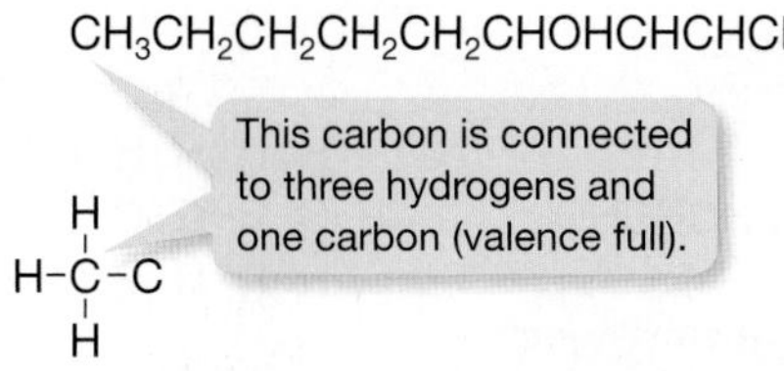

$CH_3CH_2CH_2CH_2CH_2CHOHCHCHCHO$

This carbon is connected to three hydrogens and one carbon (valence full).

H–C(H)(H)–C

If a structure cannot account for four connected atoms there must be multiple bond connections.

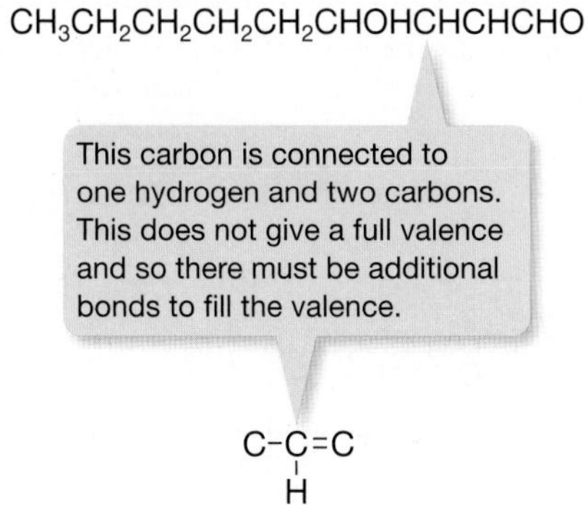

$CH_3CH_2CH_2CH_2CH_2CHOHCHCHCHO$

This carbon is connected to one hydrogen and two carbons. This does not give a full valence and so there must be additional bonds to fill the valence.

C–C(H)=C

If multi-atom groups are attached to a carbon, those groups may be listed in parentheses, including any necessary subscripts.

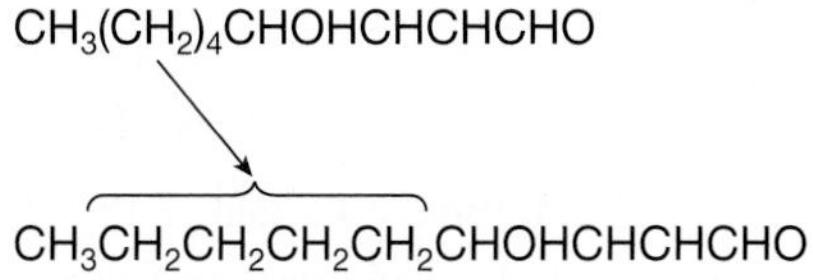

$CH_3(CH_2)_4CHOHCHCHCHO$

$CH_3CH_2CH_2CH_2CH_2CHOHCHCHCHO$

Some groups are very common and have condensed abbreviations that must be recognized.

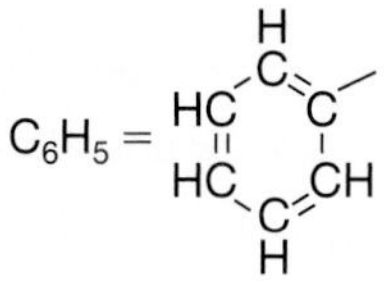

C_6H_5 =

1.10.1.1 The HONC rule

To interpret condensed structures properly, it is useful to remember the appropriate number of bonds for each atom.

H	O	N	C	element
1	2	3	4	"normal" number of bonds

Hydrogen can accommodate only one pair of bonding electrons. Each of the other elements in organic molecules typically accommodates four electron pairs. Based on the number of electrons "supplied" with each atom, this leads to a "normal" number of bonds for each element. Oxygen forms two bonds (and carries two lone pairs). Nitrogen forms three bonds (and carries one lone pair). When oxygen or nitrogen atoms carry less than the normal number of bonds, they usually have an extra lone pair and are negative. When these atoms form more than this number of bonds, they effectively over-share some electrons and become positive. Carbon atoms that carry three bonds can be negative or positive, depending on whether or not they have a lone pair.

carbon normally forms four bonds

$CH_3(CH_2)_4CHOHCHCHCHO$

oxygen normally forms two bonds

The position of elements in the periodic table can also be used as a guide to find the number of bonds an atom "normally" forms (atoms in lower periods have the ability to form more bonds than the group number indicates).

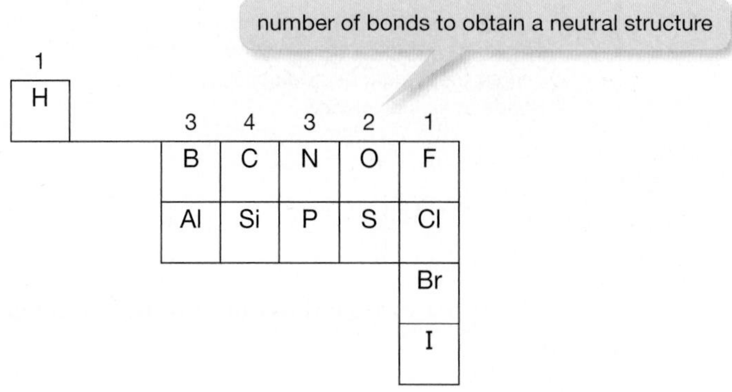

1.10.2 Line structure

Line (or bond-line) structures are used extensively to depict the shape of a molecule, effectively reducing the clutter of a Lewis structure. Line structures are valuable because they can be created quickly, and also they emphasize the shapes and functional (reactive) parts of molecules. The following conventions apply when drawing and interpreting line structures.

1. All bonds between atoms (other than hydrogen) are drawn as solid lines in a zig-zag pattern.
2. The atomic symbols of the carbon atoms are not shown; rather, each vertex and line terminus represents a carbon atom, and carbons are understood to be present at each bond transition.

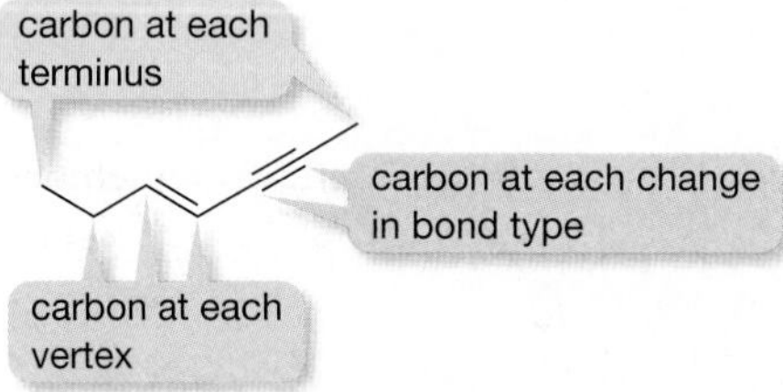

3. The elemental symbols of heteroatoms (any element other than carbon or hydrogen) are shown, but lone pairs of electrons are not shown.

4. Hydrogens are not shown unless the hydrogen is connected to an explicitly drawn atom. When hydrogens are shown, the bonds to them are not; rather the hydrogens are listed after the symbol of the atom to which they are connected.

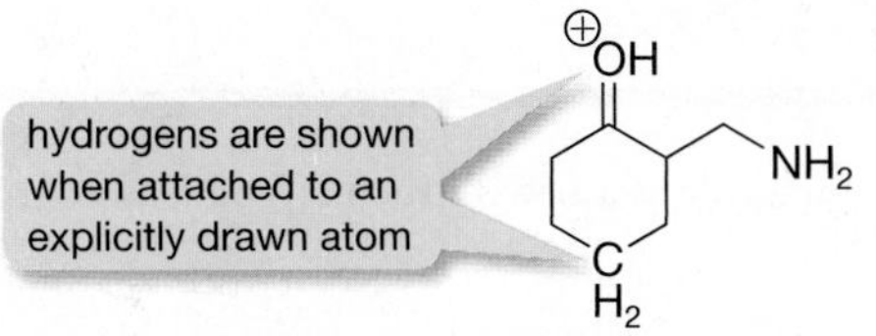

5. The number of hydrogens connected to each carbon is implied by the fact that the number of groups of electrons at each carbon must total four. For example, if three bonds are shown to a carbon, the fourth bond must be to a hydrogen that is not shown.

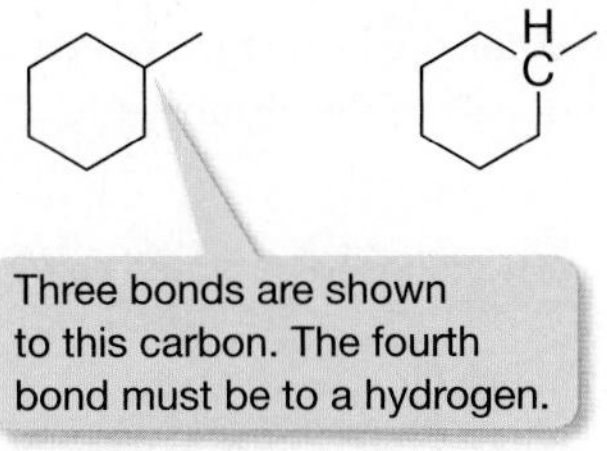

H
C
Three bonds are shown to this carbon. The fourth bond must be to a hydrogen.

6. While lone pairs are not usually included on line structures, it is important to remember they are there; in fact, it is a good habit to add them to line structures to aid in problem solving.

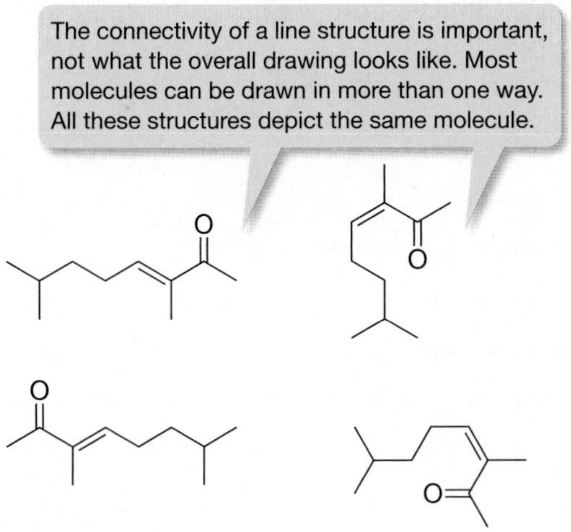

The connectivity of a line structure is important, not what the overall drawing looks like. Most molecules can be drawn in more than one way. All these structures depict the same molecule.
O
O
O
O

1.10.2.1 Wedged and hashed bonds in line structures

The following two structures show the same molecule represented by two different drawings. When drawn correctly, the wedged bonds and hashed bonds at one of the carbons convey the tetrahedral geometry at that carbon atom.

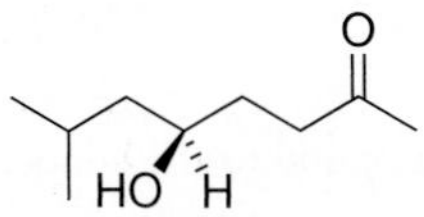

In the second drawing, the carbon-hydrogen bonds can be removed without a loss of information. Hydrogen atoms connected to carbons are not typically included in line structures. The bonds that remain—two in the plane of the page and one projecting outward—indicate that the carbon-hydrogen bond not represented must be projecting back into the page.

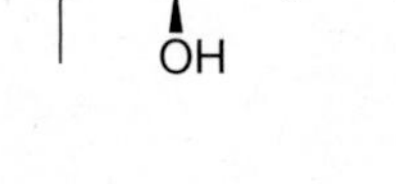

ORGANIC CHEMWARE
1.37 Line-angle structure: Clovene

ORGANIC CHEMWARE
1.38 Line-angle structure: Cholesterol

ORGANIC CHEMWARE
1.39 Line-angle structure: Codeine

CHECKPOINT 1.7

You can now draw an organic molecule as a Lewis structure, condensed structure, or a line structure.

SOLVED PROBLEM

Draw the line structure for the molecule $CH_3CH_2NHCH_2C(CH_3)_2(CH_2)_2Cl$. Use wedged bonds and hashed bonds at the carbon atom indicated in blue.

STEP 1: Examine the condensed structure proceeding from left to right to identify the longest chain. Note the atomic sequence C–C–N–C–C–C–C–Cl has eight heavy (non-hydrogen) atoms. Two of these are carbons from the $(CH_2)_2$ grouping in the condensed structure. Note that the two CH_3 groups in parentheses, bonded to the blue carbon atom, are not part of the chain.

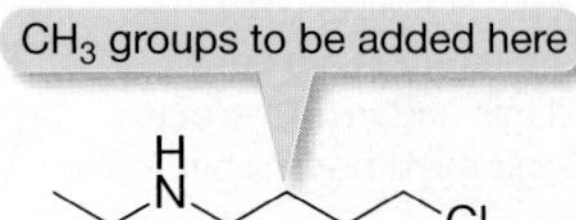

STEP 2: Add the two CH_3 groups to the line structure using a dashed bond and a wedged bond; make sure they are placed on the correct carbon atom.

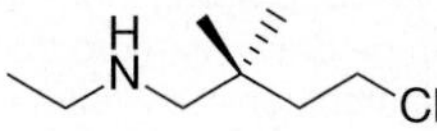

PRACTICE PROBLEM

1.10 a) Draw the line structures of the following molecules.

i)

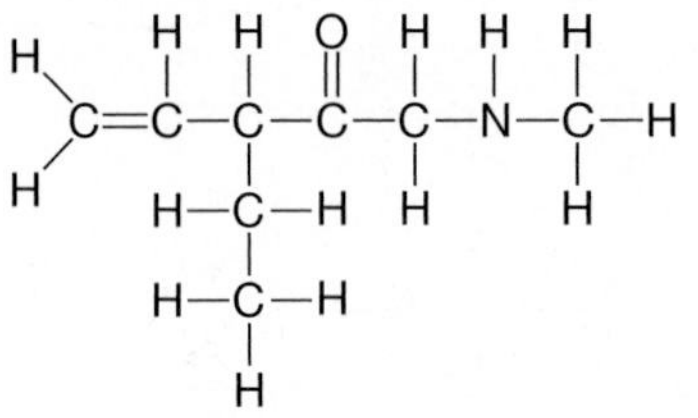

ii)

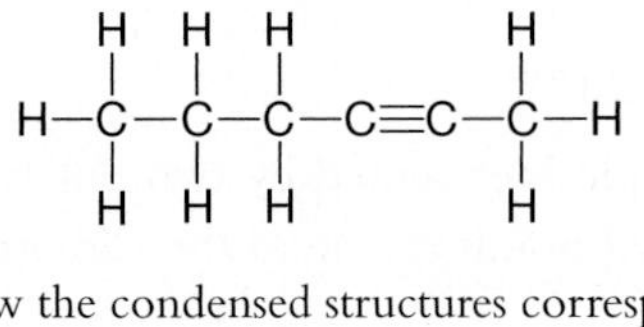

b) Draw the condensed structures corresponding to the compounds shown in part (a).

c) Draw Lewis structures, including lone pairs of electrons, of the following condensed structures.

i) $CH_3CH_2COCH_3$

ii) $(CH_3)_2CCHCH(OH)CH_3$

d) Draw the line structures that correspond to the condensed structures shown in part (c).

INTEGRATE THE SKILL

1.11 Draw the Lewis structure of the following molecule, including lone pairs of electrons.

$$[CH_3(CH_2)_3CHOHCH_2CCC(OH)CH_3]^{\oplus}$$

CHEMISTRY: EVERYTHING AND EVERYWHERE

Graphene: A Material of the Future?

Carbon occurs in nature in a surprising variety of structures. One of these elemental forms, diamond, consists of an extended, regularly repeating arrangement of sp^3-hybridized carbon atoms. Their tetrahedral shape yields an extended network of strong covalent carbon-carbon single bonds that extends three-dimensionally from one end of the diamond to the other. A diamond is essentially a single (giant) molecule, whose interlocking array of bonds gives it extreme hardness and mechanical strength.

123dartist/Shutterstock.com

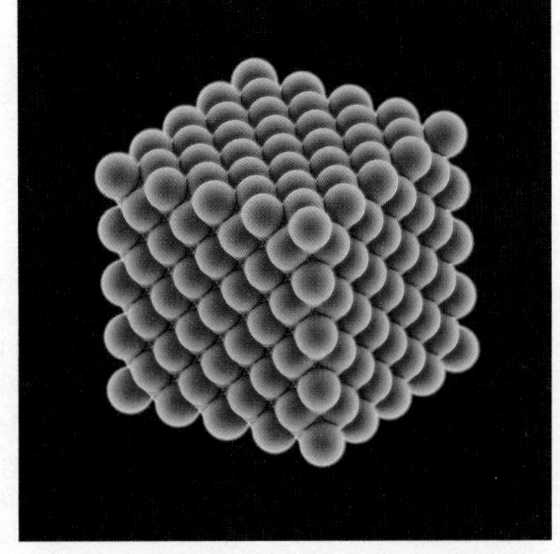

Laguna Design/Science Source

A particularly interesting form of elemental carbon is graphene. All the carbon atoms of graphene are sp^2 hybridized. The trigonal planar shape at each carbon atom yields a "chicken wire" network of repeating hexagons of carbon that extends in two dimensions, rather than across three dimensions as in diamond.

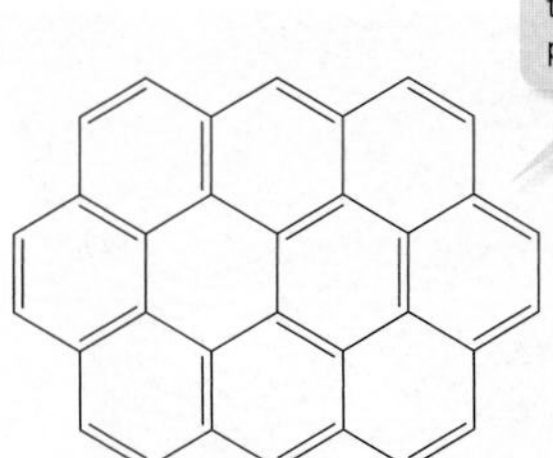
trigonal planar shape at each carbon produces a two-dimensional sheet

a (very!) small sheet of graphene

This sheet of carbon atoms could consist of vast numbers of carbon atoms, yet it is only one atom thick! The carbon atoms of a graphene sheet can be seen in the following image produced by a scanning tunnelling microscope.

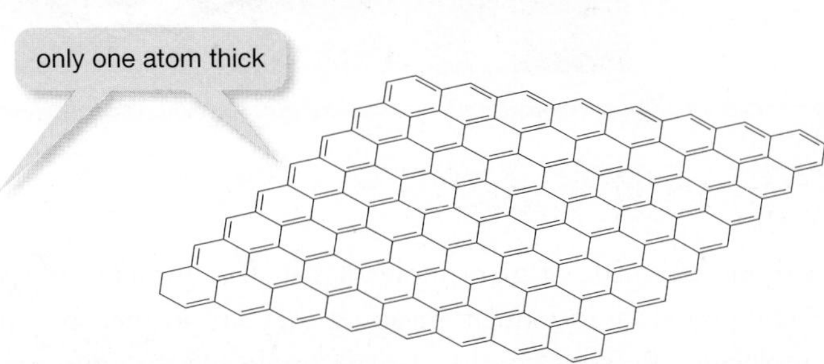
only one atom thick

SOURCE: A. T. N'Diaye, J. Coraux, T. N. Plasa, C. Busse, T. Michely, "Structure of epitaxial graphene on Ir(111)," *New J. Phys.* 10, 043033 (2008) ("IOP select"-Artikel, "Best of 2008")

In addition to the σ bonding between the sp^2-hybridized carbon atoms, the molecular bonding model of graphene describes delocalized π bonding across the entire sheet. The following figure shows one of the many bonding π molecular orbitals of a small graphene sheet. Note the absence of any nodes in this particular bonding

Continued

orbital, which describes the extensive sharing of two π electrons among all the carbon atoms of the sheet. This delocalized sharing arises from in-phase overlap of the 2p orbitals on each sp^2-hybridized carbon atom. Like diamond, a graphene sheet—such as ones already made a few centimetres long—is a single, giant molecule.

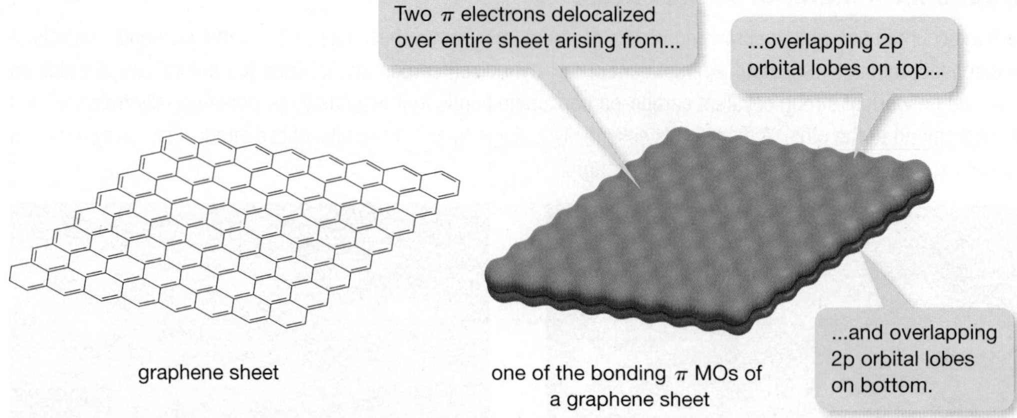
Two π electrons delocalized over entire sheet arising from...
...overlapping 2p orbital lobes on top...
...and overlapping 2p orbital lobes on bottom.
graphene sheet
one of the bonding π MOs of a graphene sheet

The planarity and bonding of the graphene sheet imparts some remarkable and potentially useful properties. Since carbon is one of the lightest elements of the periodic table, graphene is, by weight, the strongest material currently known—more than 200 times stronger than steel on a kilogram-for-kilogram basis! One square metre of graphene sheet—only one atom thick—would weigh less than 0.001 grams, about the same as a human hair 15 centimetres long.

In addition to strength, graphene has remarkable bonds that allow it to conduct electricity better than most metals and make it highly impermeable to gases and liquids. Its bonds also allow sheets of graphene to be twisted or stretched much more than almost any other substance.

Carbon nanotubes, another interesting form of the element, can be thought of as graphene sheets rolled into the shape of a tube. Due to their extraordinary electrical and mechanical properties, scientists are currently investigating carbon nanotubes for potential applications. Because of their strength, some are suggesting that nanotubes may serve as microscopic cables. Although their diameters are only about 1×10^{-9} metres, the longest nanotube produced to date (a single molecule) is over half a metre in length!

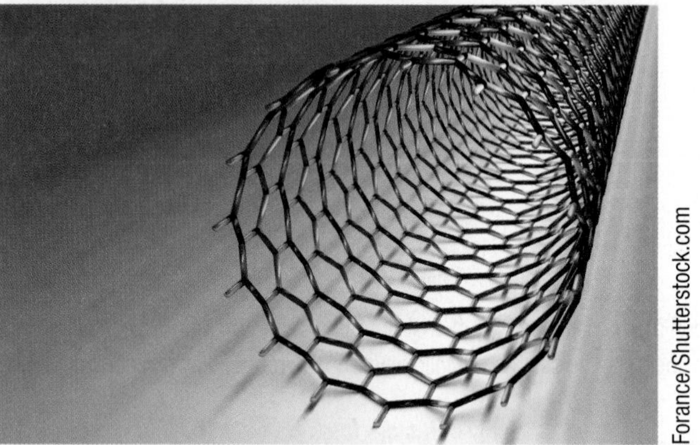

Forance/Shutterstock.com

Bringing It Together

Living things are made of giant molecules such as proteins, polysaccharides, and nucleic acids. These compounds are modular, meaning they are formed by linking together long chains of small molecules. Proteins consist of many molecules of amino acid, polysaccharides consist of many sugars, and nucleic acids consist of nucleotide bases.

The small building-block molecules are simpler than the giant complex molecules, and they are mostly made of carbon. These small structures result from the ability of carbon to bond with itself and form extended chains and rings. This ability was dramatically illustrated by the experiments of Stanley Miller in 1952.

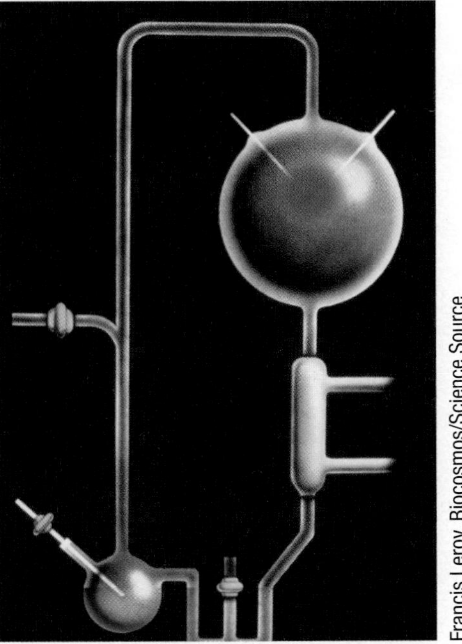
Francis Leroy, Biocosmos/Science Source

To simulate the primordial atmosphere of Earth, Miller constructed a synthetic mixture of gases—water, ammonia, methane, and hydrogen. He then subjected this mixture to an electric discharge to simulate lightning. After several days he found a gooey pink material in the flask. By analyzing this "primordial ooze," he determined it was composed of amino acids identical to those found in modern living things.

More recently, researchers have simulated the primordial ocean near undersea vents. The mineral-rich water, when heated (like it is beside an undersea volcano), contains substances that are also rapidly converted to the molecules of life.

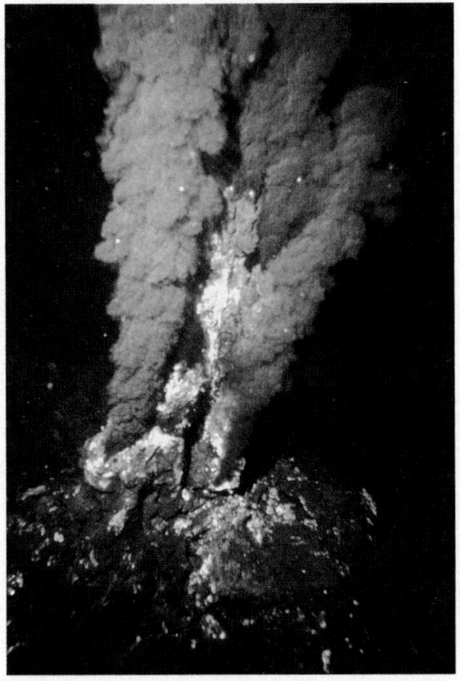
Science Source

The building blocks of life assemble themselves under the right conditions (which are surprisingly common). This self-assembly occurs because carbon readily bonds with itself, a unique property among elements.

You Can Now

- Identify bond types in organic molecules as being σ bonds or π bonds.
- Predict the geometry of atoms in molecules using VSEPR theory.
- Predict the hybridization of atoms in molecules using the valence bond model.
- Predict the bonding in molecules using a molecular orbital description.
- Draw Lewis structures for molecules using the formal charge method.
- Draw condensed structures of organic molecules.
- Draw organic molecules as line structures.
- Construct simple resonance forms using the formal charge method.
- Construct resonance hybrids from resonance forms.

Problems

1.12 Organic compounds containing boron (atomic number 5) are becoming increasingly important in the creation of new and important organic molecules in the laboratory.

a) Draw the atomic orbital energy diagram for a neutral boron atom. Be sure to show its electrons in the diagram.

b) From this diagram, write its ground-state electron configuration.

c) Identify the valence electrons of boron in the diagram. How many valence orbitals does a boron atom have?

1.13 a) Explain why the 1s orbital of a hydrogen atom is its valence orbital, but its 2s orbital is not.

b) Explain why the 1s orbital of a carbon atom is not one of its valence orbitals.

1.14 Silicon, phosphorus, sulfur, and chlorine are important third-row elements in organic chemistry. The orbital energy diagram of an atom that includes its 3s, 3p, and 3d valence orbitals is shown here.

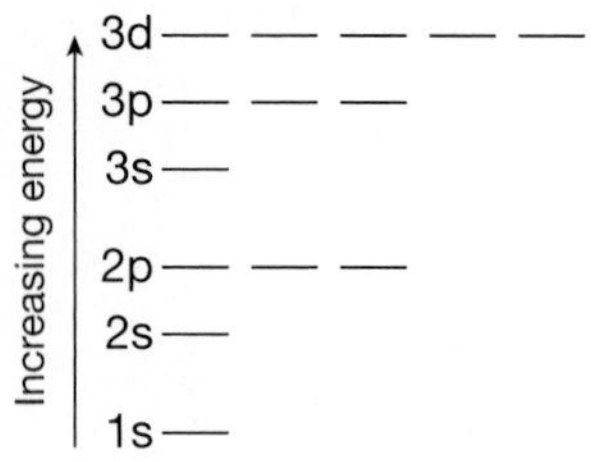

a) Identify the degenerate sets of atomic orbitals in the diagram.

b) Which orbitals are of lower energy: the 3p or the 3d?

c) Knowing the shape of a 1s and 2s atomic orbital, what shape do you predict for the 3s orbital?

d) Based on the shape of 2p orbitals, what shape do you predict for the 3p orbitals?

e) A silicon atom has a total of 14 electrons and is located directly beneath carbon in the periodic table. Use the atomic energy level diagram for carbon shown in Figure 1.4 as a basis for determining the ground-state electron configuration of silicon. Place the 14 electrons of silicon in the correct positions on the right-hand side of the diagram.

f) If the ground-state electron configuration of carbon can be written as $1s^22s^22p^2$, write out the ground-state electron configuration of silicon based on your work in part (e).

g) Based upon your work in part (e), how many valence electrons does a silicon atom have? Use your answer to draw the Lewis dot diagram of a silicon atom. How does this compare with the Lewis dot diagram of carbon?

1.15 Draw the following molecules as Lewis structures.

a) CH_3NH_2
b) $CH_3CHCHCH_2CH_3$
c) C_2H_2
d) CH_3CH_2CHO
e) $CH_3CH_2OH_2^{\oplus}$
f) $(CH_3CH_2)_3N$
g) CH_3CN
h) $CH_3CH(OH)CH_3$
i) CH_3NCO
j) $(CH_3)_3C^{\oplus}$
k) $CH_3CH_2O^{\ominus}$
l) $[CH_3C(OH)CH_3]^{\oplus}$ (two structures)

1.16 Methylamine, CH_5N, has a C–N single bond. Draw the Lewis structure of methylamine, filling the valence atomic orbitals of all atoms. Identify the bonding and non-bonding (lone) electron pairs of the molecule.

1.17 Draw the complete Lewis structure of a molecule with the following characteristics (parts a–d), showing all atoms, lone pairs of electrons, and formal charges. Make sure that your assigned formal charges sum to the overall charge on the molecule.

a) a molecule of formula C_2H_5N having no formal charge on any atom
b) a cation of formula $C_2H_8N^+$
c) an anion of formula $C_2F_3O^-$ having a C=O double bond
d) two neutral molecules of formula C_2H_3N, both having a C≡N triple bond (Hint: one has no formal charges, whereas the other has two.)

1.18 Draw complete Lewis structures for the following condensed structures.

a) $(CH_3)_2CHCH_2NH_2$
b) $HO(CH_2)_2CH{=}C(CH_2CH_3)_2$
c) $Cl_2CHCH_2CONHCH_3$
d) $NH(CH_2CN)_2$

1.19 In the following molecules, assign (non-zero) formal charges to those atoms that have them.

a) 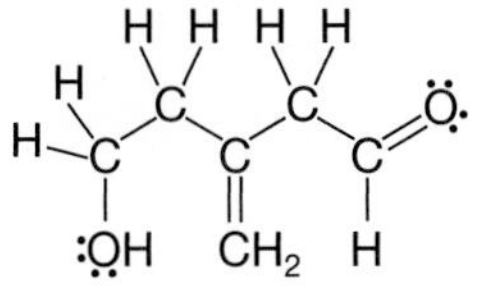 H H H H / H / H–C–C(=CH₂)–C–C(H)=Ö: / :ÖH

b) H–C(H)=C(H)–C̈(CH₃)–CH₃

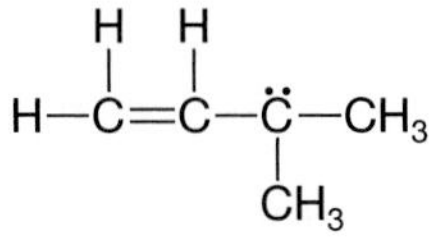

c) (H₃C)(CH₃)N=C(H₃C)(CH₃)

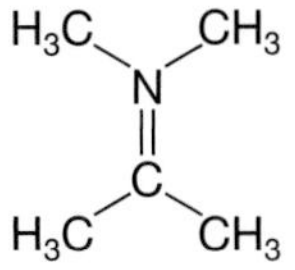

d) 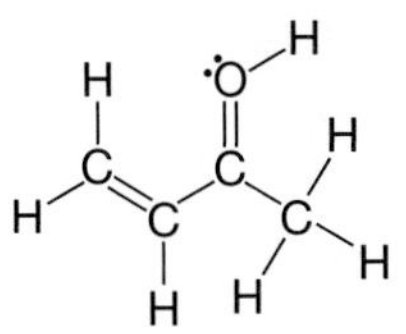 e) 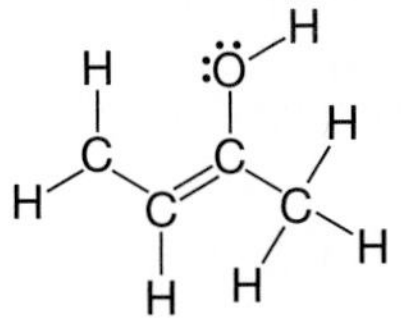 f)

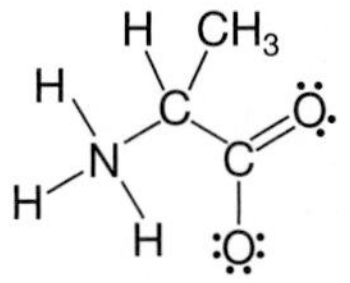

1.20 Add the appropriate lone pairs of electrons to each atom in the following Lewis structures.

a) 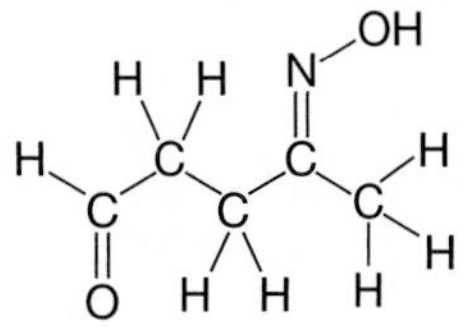 b) 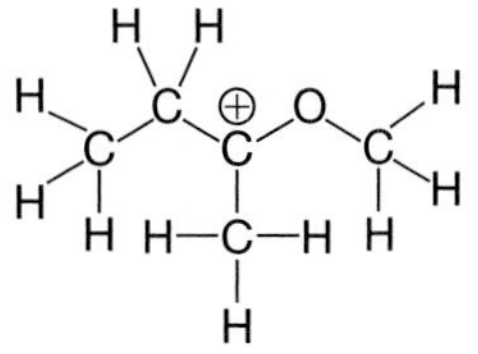 c) 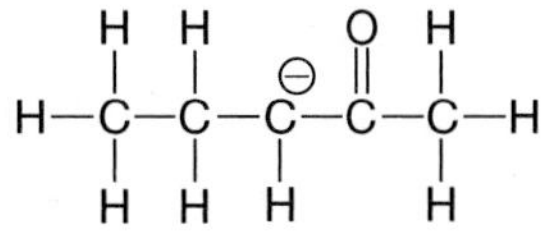 d)

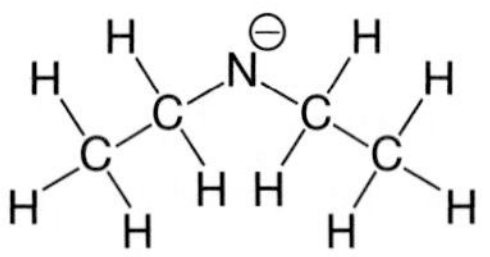

1.21 Use the electronegativities provided in Figure 1.7 as a basis for answering the following questions.

a) Which is more polar: a N–H bond or a B–H bond? What is the important distinction between them?
b) Redraw the following molecules and place dipole arrows to indicate the direction of the dipole at each of the bonds highlighted in blue.

i) 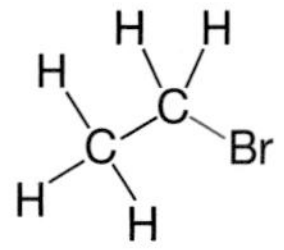 ii) 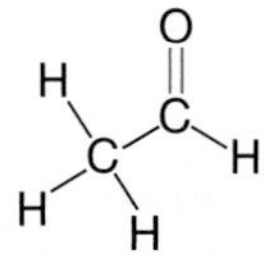 iii) 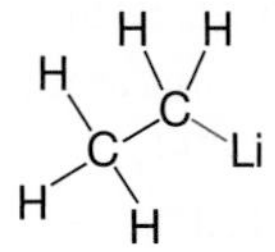 iv)

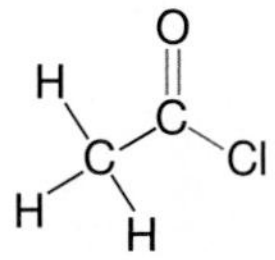

c) For each molecule shown in part (b), identify which carbon atom you would you expect to be the most electron deficient.

1.22 Use the VSEPR concept to predict the electron pair geometry at each non-hydrogen atom in the following molecules.

a) 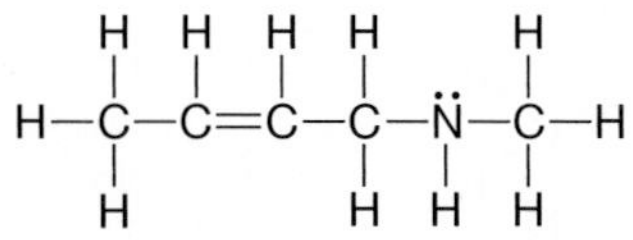 H H H H H
H—C—C=C—C—N̈—C—H
H H H H

b) H H
H C Ö
C C
⊕N C—H
H C H
H H

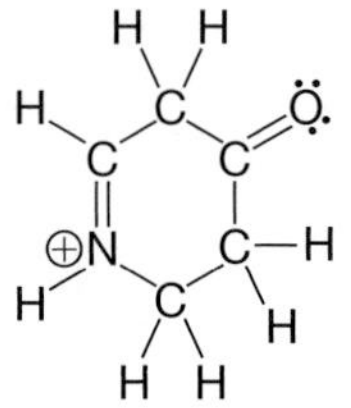

c) O (ring)—≡—F

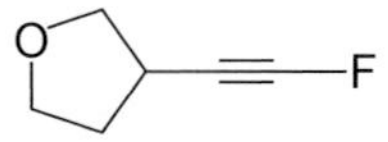

d) 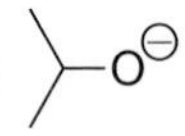)—O⊖

1.23 For the molecules of Question 1.22, employ the VB model of bonding and use the geometries predicted to assign a hybridization for each non-hydrogen atom.

1.24 Draw the Lewis structure of each of the following, showing the geometry around each atom with the proper bond notation. (Hint: draw the longest chain of atoms in the structure in a zig-zag style in the plane of the paper.)

a) $CH_3CH_2CH_2OH_2^{\oplus}$
b) $CH_3CH_2C(O)CH_2CHCH_2$
c) $(CH_3CH_2)_2NH$
d) $(CH_3CH_2C(CH_3)_2$
e) 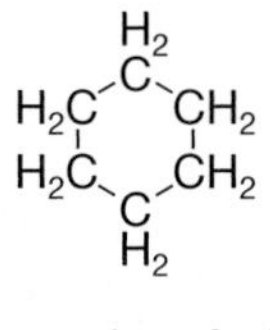 H_2C C H_2 CH_2
H_2C CH_2
C H_2

1.25 For each of the following structures, indicate the hybridization of each non-hydrogen atom, predict the geometry of the electrons pairs around the atom, and predict the geometry of the atoms around each atom.

a) $HOCH_2CHCHC(O)CH_2CH_3$
b) $CH_3CH(CH_3)CCCH_2CH_2CN$
c) 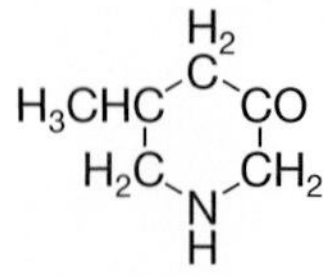 H_2 C
H_3CHC CO
H_2C CH_2
N
H
d) 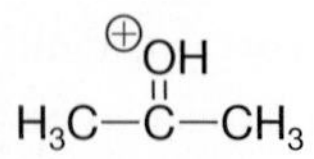 ⊕OH
‖
H_3C—C—CH_3
e) $(CH_3)_3C^{\oplus}$

1.26 Draw the resonance forms for each of the following molecules.

a) $[CH_3CHCHCHCHCH_3]$ (three forms)
b) $CH_3C(NH)CH_3$ (two forms)
c) 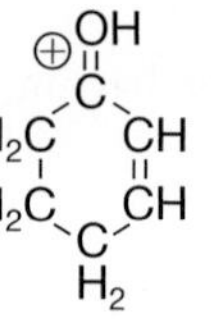 ⊕OH
‖
C
H_2C CH
H_2C CH
C
H_2
(two other forms)
d) 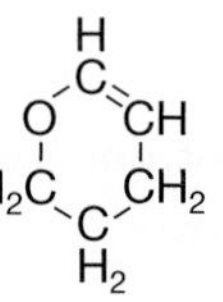 H
C
O CH
H_2C CH_2
C
H_2
(one other form)

1.27 Draw line structures for each of the following Lewis structures.

a) 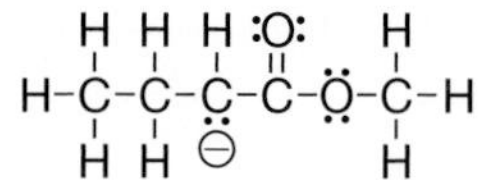 H H H :O: H
H-C-C-C-C-Ö-C-H
H H ⊖ H

b) H
⊕ C H
:O H
H C H
C C
C C
H C H
:Br: H

c) H
H-C-H
H H ⊖ H
H-C-C-N̈-C-C-H
H H H
H-C-H
H

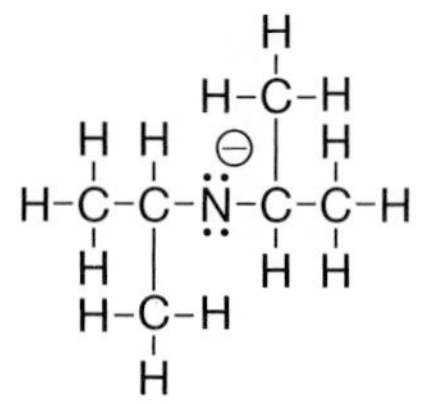

d) 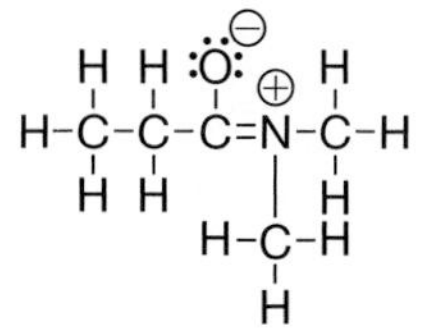 H H :Ö:⊖ ⊕ H
H-C-C-C=N-C-H
H H H
H-C-H
H

e) 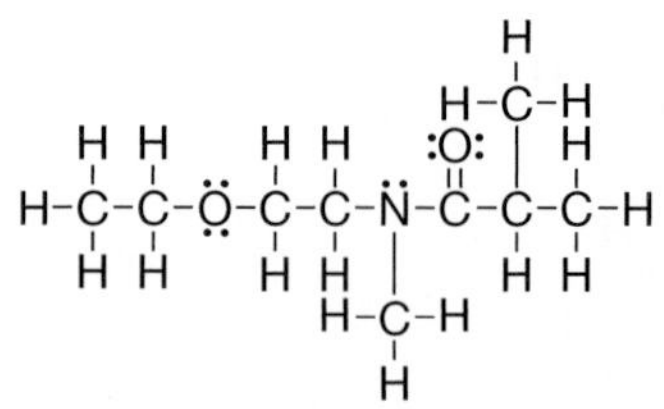 H
H-C-H
H H H H :O: H
H-C-C-Ö-C-C-N̈-C-C-C-H
H H H H H H
H-C-H
H

f) 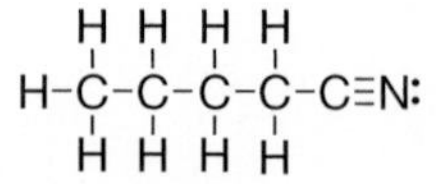 H H H H
H-C-C-C-C-C≡N:
H H H H

g) 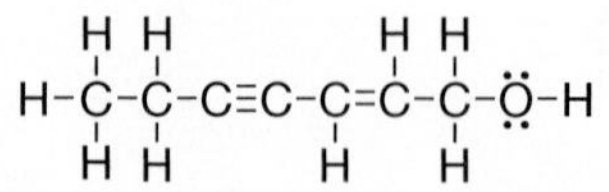 H H H H
H-C-C-C≡C-C=C-C-Ö-H
H H H H

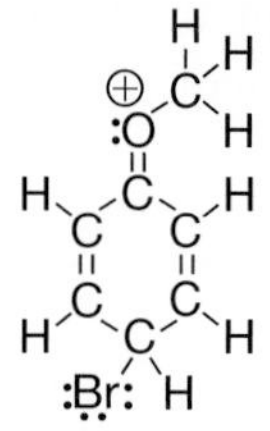

1.28 Draw the complete Lewis structure for each of the following line structures.

a) 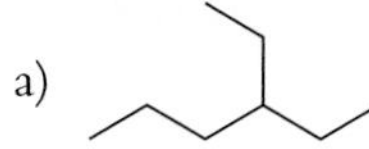

b)

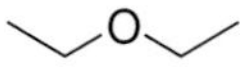

c)

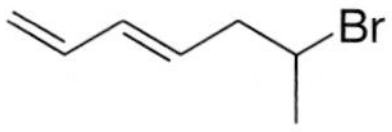

d)

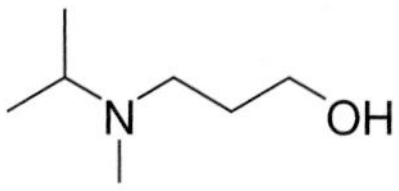

e)

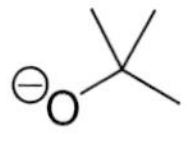

f)

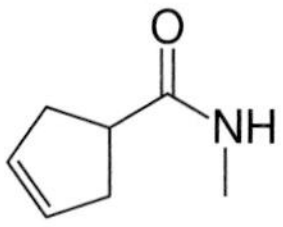

g)

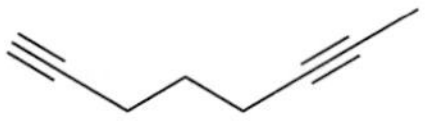

h)

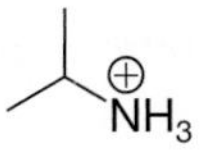

i)

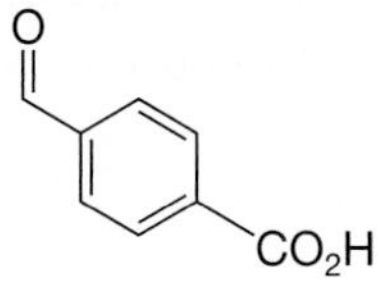

j)

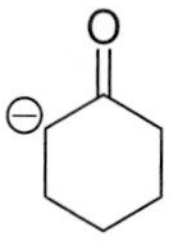

1.29 Convert the following to Lewis structures.

a)

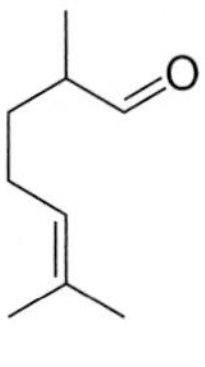

b)

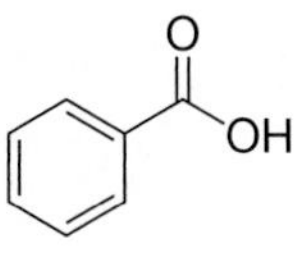

c)

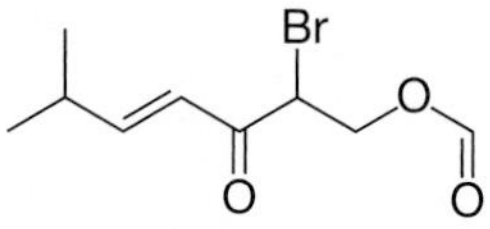

d)

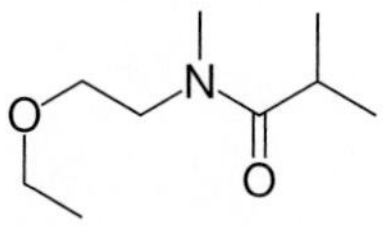

e)

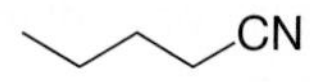

f)

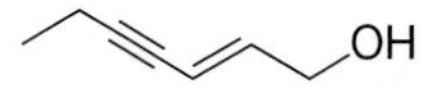

1.30 Draw all of the molecules in Question 1.29 as condensed structures.

1.31 Convert the following to Lewis structures.

a)

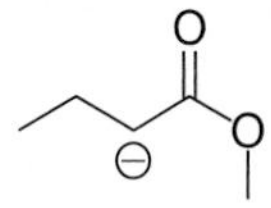

b) 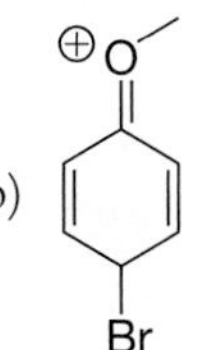

c) 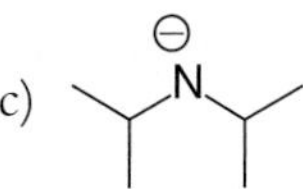

d) 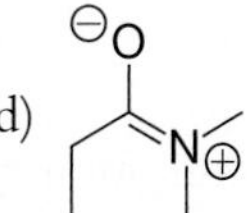

1.32 Draw all of the molecules in Question 1.31 as condensed structures.

1.33 Draw line structures for the following condensed structures.

a) $CH_3CH_2CH(CH_3)CH_2CO_2H$
b) $CH_3(CH_2)_2N(CH_2CH_3)_2$
c) $CH_2{=}CHOCH_2CH(CH_3)_2$
d) $CH_3CH_2CO(CH_2)_2CHO$

1.34 Draw line structures for the following molecules. Use wedged bonds and dashed bonds at any atom with tetrahedral geometry.

a) $NH_2CH_2\mathbf{C}(CH_3)_2OH$
b) $BrCH_2CH_2\mathbf{C}H(CH_2CH_2Br)CH_2CH_2Br$
c) $(CH_3)_2\mathbf{C}H(CH_2)_3OCH_3$
d) $CH_3CH_2COCH_2\mathbf{C}(OH)(CH_3)CH_2COCH_2CH_3$

1.35 Consider the following molecule.

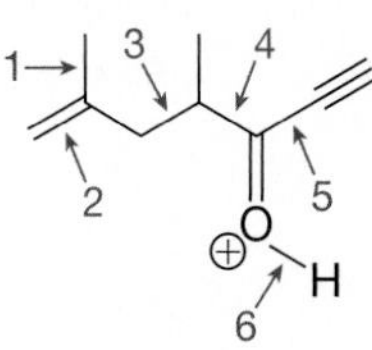

a) Predict the geometry at each non-hydrogen atom.
b) The localized bond labelled as "1" arises from the overlap of an sp^2 hybrid orbital on one carbon atom with an sp^3 hybrid orbital on the other.

Based upon the geometries you determined for part (a), assign hybridizations to each non-hydrogen atom and then use them to describe each of the labelled localized bonds, 2 through 6, in terms of the atomic orbital overlap between the two participating atoms.

1.36 In formamide, CH_3NO, the N, O, and one H atom are bound to the carbon atom, while the other two H atoms are bonded to the N atom.

a) Draw a complete Lewis structure of formamide, in which all valence atomic orbitals are filled and no atom bears a formal charge.

b) Formamide can be drawn using two other possible Lewis structures that are resonance forms of the molecule. Draw these Lewis structures.

1.37 a) Boron lies one position to the left of carbon in the periodic table. Based on this information, how many valence electrons does boron have?

b) Borane, BH_3, is an important reagent in organic chemistry. Draw its Lewis structure and predict its shape.

c) Based on its shape, assign a hybridization to the boron atom of borane. What is the total number of electrons in the valence orbitals of the boron atom due to sharing with its three neighbours?

1.38 What is the fundamental difference between a bonding π molecular orbital and an anti-bonding π molecular orbital in terms of the π bonding in a molecule?

MCAT STYLE PROBLEMS

1.39 What is the geometry and hybridization of the atom indicated by the blue arrow?

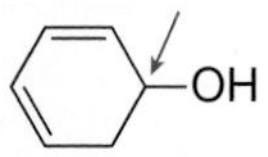

OH

a) tetrahedral and sp^3 hybridized
b) trigonal planar and sp^2 hybridized
c) trigonal pyramidal and sp^2 hybridized
d) trigonal planar and sp^3 hybridized

1.40 Which of the following structures carries an overall molecular charge?

a) $CH_3CH_2CHCHCH_2COOH$
b) $CH_3CH_2CHCHCH_2OH$
c) $CH_3CH_2CHCHCHO$
d) $CH_3CH_2CHCHCHOH$

CHALLENGE PROBLEM

1.41 The following molecule represents an important group that acts as an electron pair donor in many types of chemical reactions. There are three significant resonance forms of this structure (including the one depicted). Draw these forms and use them to predict the sites on the molecule that can act as electron pair donors.

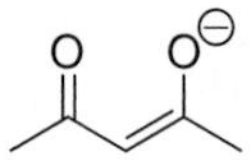

O O⊖

Anatomy of an Organic Molecule

2

CHAPTER OUTLINE

Cholesterol, which is found in food and also made in our bodies, has both hydrophobic and hydrophilic regions due to its chemical composition and structure.

2.1 Why It Matters

Cholesterol is one of the most important molecules in animal cells. While it can be obtained from foods such as eggs, meats, and dairy products, most cholesterol is actually manufactured by our bodies. Cholesterol is an essential component of cell membranes, where it contributes to structural integrity. It is also the starting compound in the production of several hormones.

cholesterol

The line structure of cholesterol shows this molecule is largely composed of σ bonds between sp^3-hybridized carbon atoms; there is also an OH group and a double bond. The structure of an organic molecule determines its function, which raises the question: what aspects of cholesterol's structure are associated with each of its functions? To facilitate discussion of the structure of organic molecules, in this chapter we establish a vocabulary of names and features that help us identify organic molecules, their structural elements, and some of the properties that arise from them.

2.2 Structural Features of Molecules

The structure of an organic molecule—its particular array of carbon, hydrogen, and other atoms in three-dimensional space—is the source of its chemical behaviour and physical properties. Principally, the structure of a compound dictates its melting point, its reactivity toward other chemical species, its response to ultraviolet light, and its benefit to an organism.

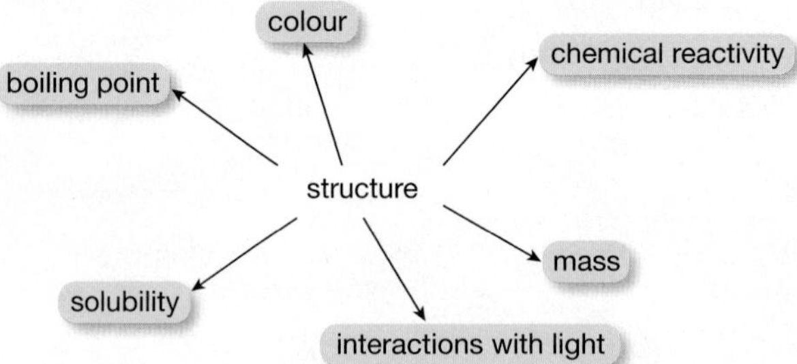

Molecules that share structural features also share properties, and the closer their structural similarity, the greater their similarity in properties and reactivity. For example, ethylamine and methylamine are both gases at room temperature, behave similarly in their reaction with acids and other organic molecules, and have very similar odours.

ethylamine methylamine

A line structure diagram shows the key features of an organic molecule and emphasizes the particular parts that contribute to its reactivity and physical properties. The key structural features of organic molecules include the hydrocarbon, the functional groups, and any substituents. The **hydrocarbon** is the portion of the molecule that consists of carbon and hydrogen only. This region, depicted by lines or zig-zags, indicates the overall shape of the molecule.

A **hydrocarbon** is a compound consisting of only carbon and hydrogen atoms.

functional group

OH

hydrocarbon

HO

hydrocarbon

OH

R

Each functional group usually reacts in the same ways in all molecules it is found in. It is common to simplify structures by using the abbreviation R to highlight the reactions of the functional group.

A **functional group** is an atom or group of atoms that, because of its structure, exhibits its own distinct pattern of chemical reactivity.

Alkanes are hydrocarbons that have only single bonds between carbon atoms. All the carbons are sp^3 hybridized.

A **substituent** is a particular atom or group of atoms that replace a hydrogen atom in an organic molecule.

The hydrocarbon portion of most organic compounds does not typically participate in chemical reactions. Rather, the chemical reactions that molecules undergo happen primarily through the **functional groups** on the molecule. Functional groups, the second key feature of molecular structure, contain reactive bonds and non-bonded electrons. The electronegativity difference between their atoms is an important element of chemical reactivity (Chapter 5), as is the bonding pattern between their atoms.

Hydrocarbons frequently occur as atoms connected in a continuous chain, and such hydrocarbons are called *linear hydrocarbons*. Hydrocarbons in which all the carbons are sp^3 hybridized are called **alkanes**. Linear alkane hydrocarbons can be transformed into branched structures when certain hydrogen atoms are replaced with other atoms or groups of atoms. Such groups are called **substituents**, the third key feature of molecular structure.

STUDENT TIP

The term *linear alkane* refers to a continuous chain of carbon atoms, but its carbon atoms do not actually lie in a line. Each carbon atom is tetrahedral and has bond angles of about 109°; therefore, the molecule has an alternating zig-zag structure. As such, the arrangement of carbon atoms drawn in a line structure is sometimes referred to as zig-zag.

alkanes—all carbons are sp^3 hybridized

linear alkane

longest linear chain

substituent

branched alkane

Carbon-hydrogen and carbon-carbon σ bonds constitute the hydrocarbon parts of a molecule. These bonds are strong—having good overlap between atomic orbitals—and difficult to break. Such regions are therefore generally chemically unreactive and function primarily as a scaffold that holds the functional parts of the molecule (functional groups) together.

STUDENT TIP

A number of line structures and Lewis structures can be drawn to represent almost any organic molecule. The way atoms are connected determines molecular structure, not the way they are drawn.

same molecule

A **saturated** molecule has no π bonds or rings and therefore has the maximum amount of hydrogen that its atoms can accommodate.

Molecules and parts of molecules that do not contain a ring and have only single bonds connecting the atoms are known as **saturated** molecules. Such structures have the maximum number of hydrogens that is allowed by the valency of the other atoms in the structure. Saturated hydrocarbons have the general empirical formula C_nH_{2n+2}.

An **unsaturated** molecule has at least one π bond or a ring to which hydrogen atoms may be added.

A **cyclic** molecule has at least one ring of atoms. Acyclic molecules have no rings.

Organic molecules that have either a ring or double or triple bonds are known as **unsaturated** molecules. The π bonds in structures with double or triple bonds can potentially be reacted with H_2, which adds one hydrogen atom to each of the atoms connected by each π bond. **Cyclic** molecules have one or more rings of atoms. These compounds are also unsaturated because two of their bonds—between the heavier atoms and hydrogens—are replaced by a direct connection between the heavier atoms.

NH_2	NH_2	HN
C_2H_7N saturated	C_2H_5N unsaturated	C_2H_5N unsaturated

2.2.1 Types of functional groups

Most organic molecules contain functional groups—that is, groups of one or more atoms whose structure confers a particular pattern of reactivity. There are many functional groups in organic chemistry, and each has a different structure. Functional groups tend to have predictable reactivities due to the patterns of connection among their atoms. Consequently, particular functional groups react in the same way in all compounds where they occur. A functional group reacts similarly whether it is attached to a carbon, a hydrogen, or a larger group. Occasionally, sections of a molecule that do not have a functional group are represented in a line structure diagram with the alkane portion abbreviated as R; in this case, R refers to *remainder* or *residue*.

= R

NH_2 = R NH_2

STUDENT TIP
The abbreviation R can be used to represent *anything* on a molecule, not just hydrocarbons. R could even represent a single hydrogen atom.

Functional groups can be identified by two main features: π bonds and heteroatoms.

2.2.1.1 π bonds

There are four types of hydrocarbon functional groups:

1. alkanes (carbon-carbon single bonds, carbons are sp^3 hybridized)
2. alkenes (carbon-carbon double bonds, carbons are sp^2 hybridized)
3. alkynes (carbon-carbon triple bonds, carbons are sp hybridized)
4. aromatics (special ring structures with alternating patterns of single and double bonds—described in Chapter 10)

Although alkanes are usually not very reactive, the other hydrocarbon types are functional. Alkenes, alkynes, and aromatics are composed partly of π bonds. These bonds are weaker and more reactive than hydrocarbon σ bonds, and therefore structures that have π electrons (π bonds) will be functional. In line structure diagrams, π bonds always appear as parallel bond lines, as shown in the following examples.

alkene
alkyne
π bonds are functional groups
alkyne
aromatic
alkene
many molecules contain more than one functional group

2.2.1.2 **Heteroatoms**

A second type of functional group contains heteroatoms: any atoms other than carbon or hydrogen. In organic molecules, typical heteroatoms have one or more non-bonded pairs of electrons (lone pairs) that can participate in reactions. Because the electronegativity of such atoms differs from that of carbon, the electron density around a heteroatom is not the same as around the carbon atom to which it is bonded; this feature tends to induce reactions. A π bond directly connected or adjacent to a heteroatom should be considered as part of a single functional group involving the bond and the heteroatom.

some common functional groups containing heteroatoms

OH hydroxyl
NH_2 amine
NO_2 nitro

amide
nitrile
anhydride

An organic molecule can be regarded as a hydrocarbon framework with attached functional groups that have various patterns of reactivity. Most organic molecules contain more than one functional group and exhibit behaviours that can be attributed to all of those groups. By understanding *how* particular reactivity patterns are tied to a molecule's structure, predictions can be made about the kinds of reactions that any organic molecule will undergo. This knowledge can also be used to create new organic molecules designed to have specific properties.

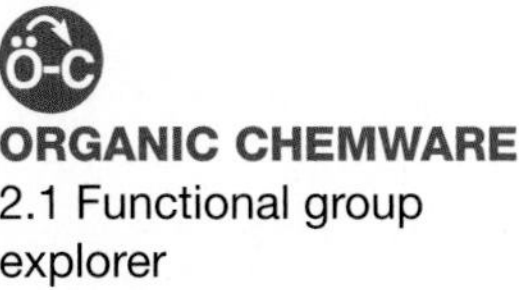

TABLE 2.1 Common Heteroatom-Containing Functional Groups

Functional group structure	Name	Functional group structure	Name
R–OH	Alcohol	RC(=O)OH	Carboxylic acid
R–O–R	Ether	RC(=O)OR	Ester
RC(=O)R	Ketone	RC(=O)NR$_2$ RC(=O)NHR RC(=O)NH$_2$	Amide
RC(=O)H	Aldehyde	R–X X = F, Cl, Br, I	Halide
R_3N R_2NH RNH_2	Amine	$R_2C(OR)_2$	Acetal
$R_2C(OH)_2$	Hydrate	$R_2C(OH)(OR)$	Hemiacetal
R–NO$_2$	Nitro group	R–C≡N	Nitrile

CHECKPOINT 2.1

You should now be able to identify the structural features of a molecule.

SOLVED PROBLEM

Identify the functional groups in the following molecule, including the hydrocarbon regions.

$$BrCH_2CH(CH_3)COCH_2CH_2N(CH_2CH_3)_2$$

STEP 1: Draw the structure in line form to better see how its atoms are connected.

Structure:

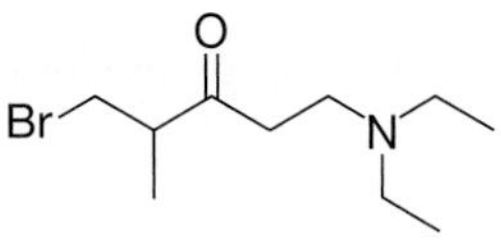

STEP 2: Identify the hydrocarbon portions of the molecule. These are the regions with only carbon and hydrogen atoms.

STEP 3: Identify the functional groups involving heteroatoms or π bonds (or both) using the information in Table 2.1.

PRACTICE PROBLEM

2.1 Identify functional groups in the following molecules.

a)

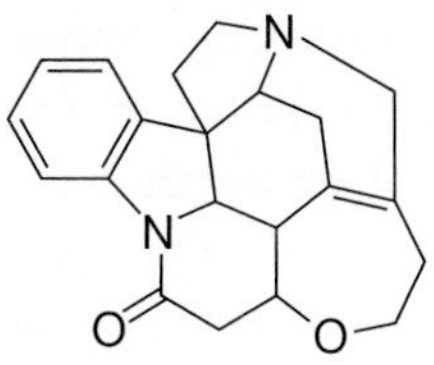

b)

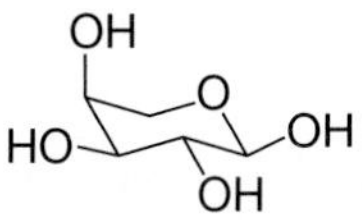

c)

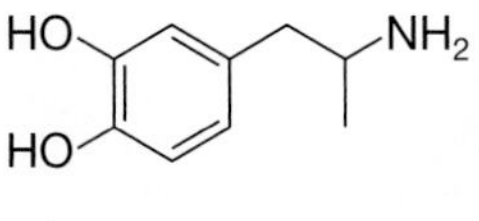

d)

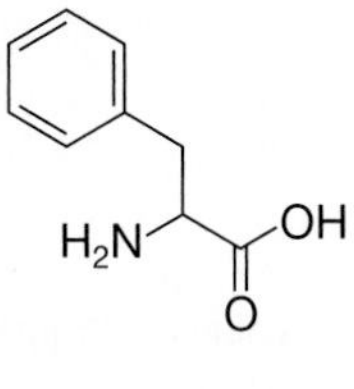

INTEGRATE THE SKILL

2.2 Describe the type of bonds involved (σ, π) in each of the functional groups in the molecules in Question 2.1.

2.3 Functional Groups and Intermolecular Forces

Intermolecular forces are different kinds of weak attractive forces that molecules exert on each other when they are in close proximity.

The physical properties of a substance are determined by the distribution of electrons about the constituent molecules—that is, the location in each molecule where the electrons are more likely and less likely to be. The distribution of electrons generates **intermolecular forces** *between* organic molecules. Such forces are much weaker than those of covalent bonds *within* organic molecules. However, organic molecules that are close to each other—such as in solids, liquids, or solutions—exhibit attractive forces toward each other that are collectively strong enough to impart important properties, including its melting point, boiling point, and its solubility.

Organic molecules interact with each other by means of four types of intermolecular forces:

1. electrostatics
2. dipole-dipole interactions
3. hydrogen bonding
4. London forces (dispersion forces)

The relative importance of each of these as an attractive force between molecules depends primarily on the structure of the participating molecules. In addition, the strength of intermolecular forces depends on the contributions of all the functional groups in the molecule, which in turn determines the overall behaviour of the molecule.

Intermolecular forces result from charge interactions; opposite charges attract and like charges repel. The distribution of electrons in a molecule produces areas of high electron density (negative-charge areas) and areas of low electron density (positive-charge areas). This distribution can be calculated by quantum mechanical computer modelling and depicted in a molecule's **electrostatic potential map** (Figure 2.1). This type of map uses a colour spectrum to represent the relative likelihood of finding electrons at any point around a molecule. The colours range from red, depicting regions that have an excess of electrons (δ^-), to blue, showing regions that are electron deficient (δ^+).

An **electrostatic potential map** is a plot of the forces on a point charge measured at a fixed distance from a molecule. It is often interpreted as regions of negative and positive charge on a molecule.

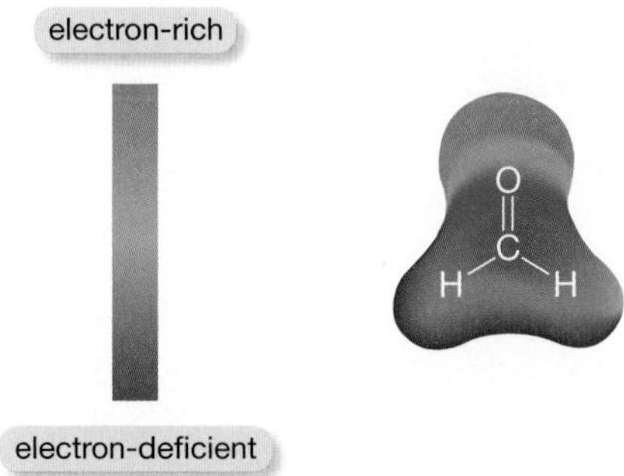

FIGURE 2.1 Calculated electrostatic potential of formaldehyde. Red illustrates regions with an excess of electron density, and blue denotes regions with electron deficiency.

The positive regions of one molecule are attracted to the negative regions of neighbouring molecules, and vice versa. Such intermolecular forces cause molecules to be attracted to other molecules of the same compound and also to other molecules that have similar types of dipoles. The strength of such attractions depends on the number and size of the charges involved. Larger charges or more interactions bind molecules together tightly, whereas small charges or fewer interactions lead to weaker intermolecular forces.

2.3.1 Electrostatic interactions

One type of intermolecular force between organic molecules is **electrostatics**. Electrostatic interactions take place where organic functional groups have a full formal charge, creating strong attractive forces between molecules. These interactions represent the strongest type of intermolecular force and result in ionic structures. The creation of charge in organic structures often involves acid-base chemistry (Chapter 6), although permanently charged groups are also possible.

Electrostatics are attractive forces that result from a full formal charge.

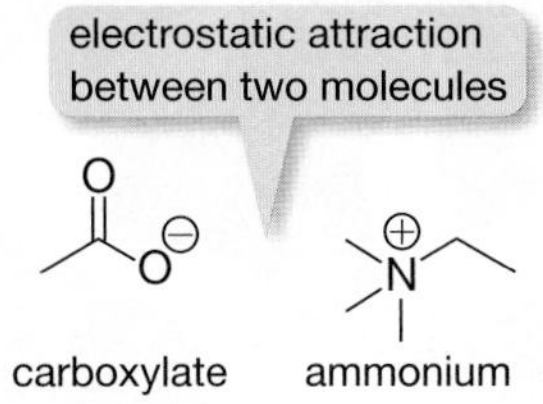

electrostatic attraction between two molecules

carboxylate ammonium

2.3.2 Dipole-dipole interactions

The second type of intermolecular force between organic molecules is **dipole-dipole interaction**. This type of interaction results from attractive forces between the poles of the functional groups on one molecule and the opposite poles of the groups on the neighbouring molecules. Some functional groups have no charge but carry a permanent dipole due to the electronegativity difference between the atoms in the group. For example, carbonyl groups (C=O) have a dipole in which the more electronegative oxygen carries a negative pole and the less electronegative carbon carries a positive pole (as previously shown for formaldehyde). This dipole structure is represented in line structure diagrams either by a dipole arrow or by indicating partial positive and negative charges beside the appropriate atoms.

A **dipole-dipole interaction** is an attractive force between the negative end of a permanent dipole in a molecule and the positive end of a permanent dipole in a neighbouring molecule (or vice versa).

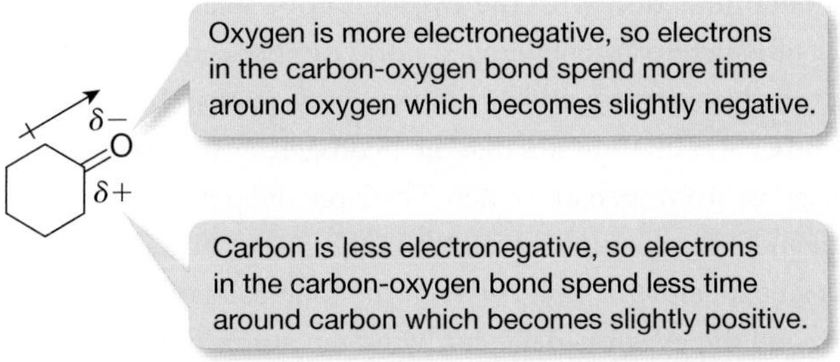

$\delta-$ O $\delta+$

Oxygen is more electronegative, so electrons in the carbon-oxygen bond spend more time around oxygen which becomes slightly negative.

Carbon is less electronegative, so electrons in the carbon-oxygen bond spend less time around carbon which becomes slightly positive.

When located close to each other, functional groups that have permanent dipoles engage in dipole-dipole interactions (Figure 2.2). Dipoles are possible whenever there is a significant electronegativity difference between atoms in a functional group, and the compounds resulting from such interactions are said to be **polar**. Dipole-dipole interactions occur in many functional groups that contain electronegative atoms—for example, ethers, halides, or any group with a carbonyl (C=O).

A **polar** molecule has a net overall dipole.

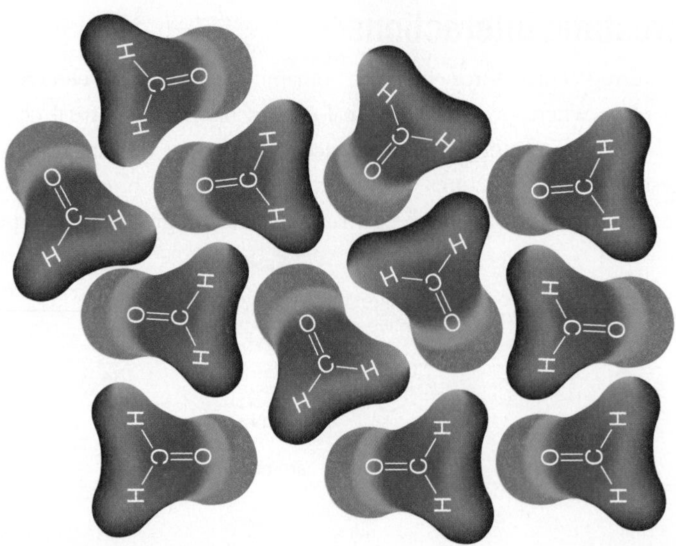

FIGURE 2.2 Dipole-dipole interactions. When near to each other, polar molecules such as formaldehyde tend to align so that the negative end of their permanent dipoles are in close proximity to the positive end of the dipole of their neighbours, and vice versa.

2.3.3 Hydrogen bonding

Hydrogen bonding (H-bonding) is an attractive force between a lone pair of a nitrogen or oxygen atom in a group and a hydrogen atom that is covalently bonded to a N or O atom in a neighbouring group.

A third type of intermolecular force is **hydrogen bonding** (H-bonding). This is a special kind of dipole-dipole interaction that is possible for groups with very electronegative atoms bonded to hydrogen atoms. This interaction is very important in the case of OH and NH groups. The electronegativity difference between the heteroatoms (oxygen or nitrogen) and the hydrogen is large and generates a very strong dipole. This dipole involves the lone pairs of the heteroatom forming a dipole-dipole interaction with the hydrogen of a nearby similar group.

The large dipole of the heteroatom-hydrogen bond can be seen in the electrostatic map of water (see the diagram later in this section), which shows a very intense blue colour (region of positive charge) around the hydrogens. This indicates that the hydrogen atoms are very electron deficient when bound to an electronegative atom such as oxygen. The large difference in the distribution of electrons between the hydrogen and oxygen atoms within the OH groups of water enables hydrogen bonding between them. This is an explicit attraction between the lone pairs of the electron-rich oxygen atom of one water molecule and the electron-deficient hydrogen atoms of surrounding water molecules.

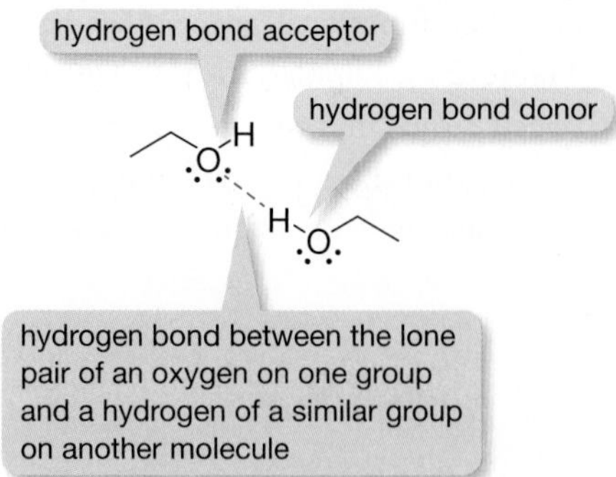

Because of the high electronegativity difference required for hydrogen bonds, usually only the nitrogen and oxygen atoms of organic functional groups are able to engage in hydrogen bonding. Similarly, the hydrogen atom is usually sufficiently electron deficient to hydrogen bond only when it is connected to a nitrogen or oxygen atom.

When two functional groups participate in hydrogen bonding, the group that provides the hydrogen atom is called the **hydrogen bond donor**. The group that provides a lone pair of electrons to form the hydrogen bond is called the **hydrogen bond acceptor**. Organic functional groups that can act as hydrogen bond donors also have electronegative atoms that allow them to act as hydrogen bond acceptors. Because of the greater dipole involved in these intermolecular forces, hydrogen bonding is generally stronger than dipole-dipole interaction.

Hydrogen bond donors are functional groups in which oxygen or nitrogen atoms are connected to hydrogens and can participate in hydrogen bonding.

Hydrogen bond acceptors are functional groups in which oxygen or nitrogen atoms have lone electron pairs.

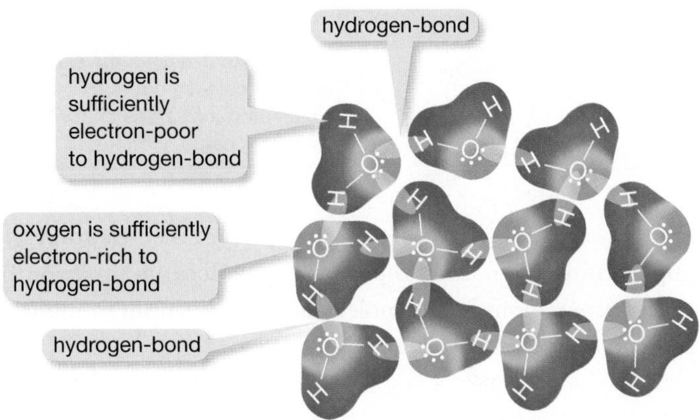

Functional groups in which oxygens or nitrogens are not bound to hydrogen can also act as hydrogen bond acceptors. Such hydrogen bond acceptors can only hydrogen bond with functional groups that act as hydrogen bond donors. This type of interaction is especially important in the solvation of organic molecules in water and other **polar protic solvents**.

Polar protic solvents are those solvents capable of acting as hydrogen bond donors.

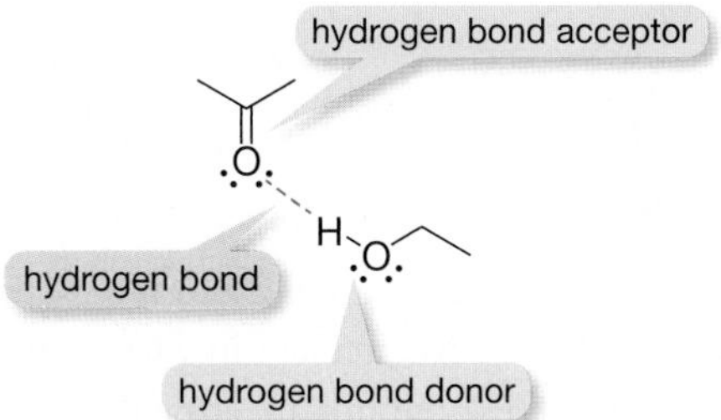

2.3.4 London forces

A fourth type of intermolecular force results from small temporary dipoles called **London forces** or **dispersion forces**. The hydrocarbon portion of an organic molecule does not have a permanent dipole. The electronegativity difference between hydrogen and carbon is very small, and the hydrogens surrounding the carbon atoms effectively cover the entire surface of each carbon atom. Since hydrocarbons do not have any kind of permanent dipole, they are **non-polar**. The electrostatic map of methane shows a uniform distribution of electrons with no obvious positive or negative regions.

London forces (dispersion forces) are attractive interactions that exist between *all* molecules in close proximity to each other, regardless of whether or not they engage in other intermolecular interactions. They are the result of small temporary dipoles induced in each molecule by the other.

A **non-polar** molecule has no overall dipole, or a very small one.

STUDENT TIP
Keep in mind that all molecules—polar or non-polar—establish London forces with their neighbours. Polar molecules can additionally engage in dipole-dipole or H-bonding interactions with each other, but non-polar molecules cannot.

Such non-polar hydrocarbons are weakly attracted to each other when they are in close proximity and thereby generate the small, temporary dipoles known as London (or dispersion) forces.

The electrons surrounding the atoms of hydrocarbons are constantly in motion. This produces temporary situations in which, at any instant, portions of the hydrocarbon are slightly negative and other regions are slightly positive.

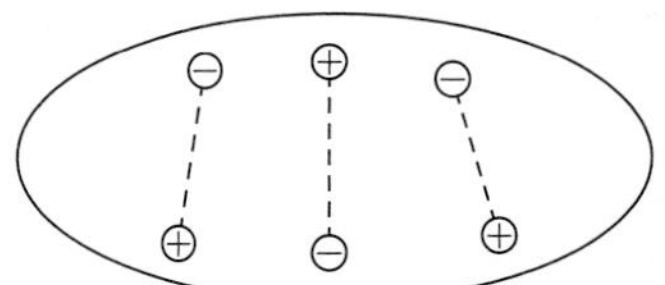

At any instant in time, the electrons in a molecule are not evenly distributed, which produces areas of positive and negative charge.

When another hydrocarbon is nearby, the negative parts of the first group repel the electrons in the second group, creating an induced dipole that produces an attraction between the two molecules. Similarly, positive regions in one molecule attract electrons from the second molecule, creating other small dipole attractions.

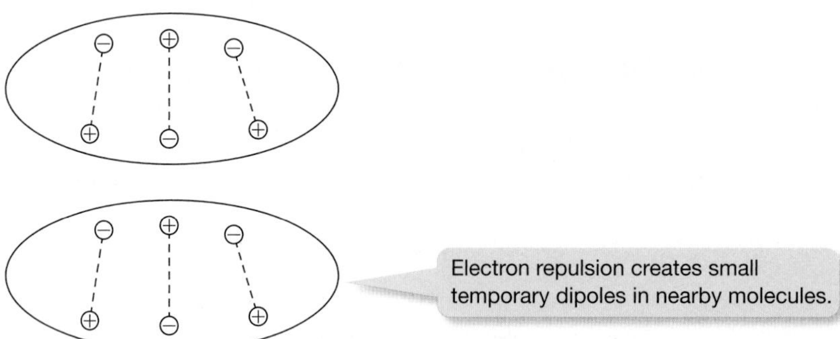

Electron repulsion creates small temporary dipoles in nearby molecules.

Such induced dipole-dipole interactions are small and temporary; in the next instant, these diploes will change. As electrons continue to circulate in each group, a new pattern of negative and positive regions is generated, and these regions are attracted to the induced opposite dipoles on the nearby molecule. The net effect of London forces is to attract the two groups together.

London forces are generally the weakest of the intermolecular forces. The strength of these interactions usually depends on the amount of surface area contact between the groups (or molecules) involved. As hydrocarbons become longer, the attraction between them increases. Branched hydrocarbon structures tend to reduce the number of possible close-contact areas between such groups and thereby limit the attractive forces.

The four intermolecular forces—electrostatics, hydrogen bonding, dipole-dipole interactions, and London forces—contribute to the overall attraction of one molecule to another. Electrostatics usually result in the strongest interactions. Hydrogen bonding is normally stronger than dipole-dipole interactions, and London forces are usually the weakest of the intermolecular forces.

CHECKPOINT 2.2

You should now be able to identify the types of intermolecular forces involved in various molecular structures.

SOLVED PROBLEM

Identify the functional groups and the possible types of intermolecular forces in the following molecule.

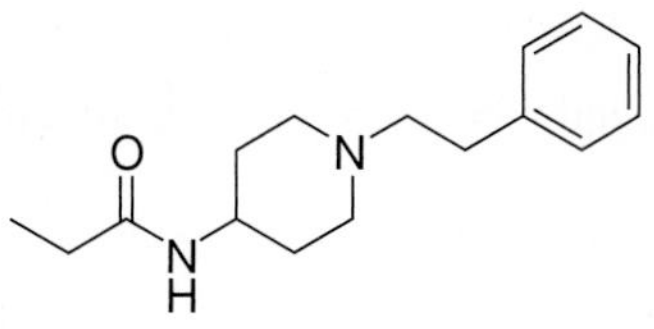

STEP 1: Identify the hydrocarbon portions of the molecule. These are regions of a molecule composed of carbon and hydrogen only. On line structure diagrams, regions with no elemental symbols are hydrocarbons.

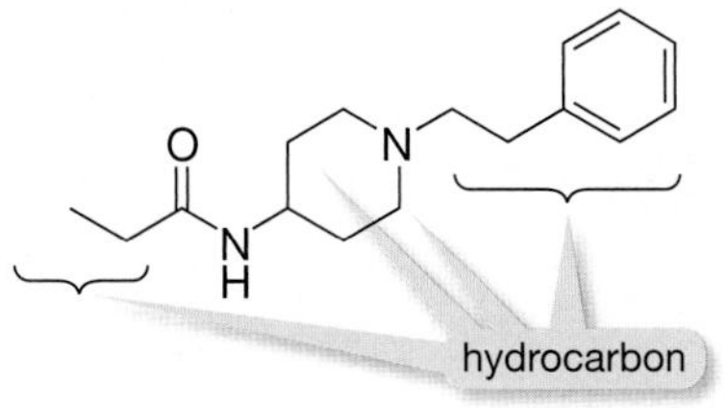

STEP 2: Functional groups consist of π bonds, heteroatoms, or combinations of both. Table 2.1 lists the names of common functional groups.

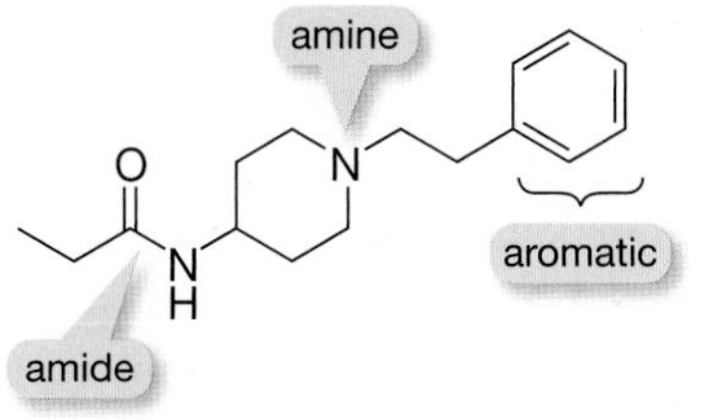

STEP 3: Hydrocarbons generally produce only London forces. There are no charged groups on this molecule, so electrostatics are not possible. The amine can create a dipole with neighbouring carbons and, because it has a lone pair of electrons, can act as a hydrogen bond acceptor. The amide is capable of dipole-dipole interactions and can act as a hydrogen bond donor (NH) and hydrogen bond acceptor (oxygen, nitrogen).

PRACTICE PROBLEM

2.3 Identify the functional groups and the possible types of intermolecular forces in the following molecules.

a)

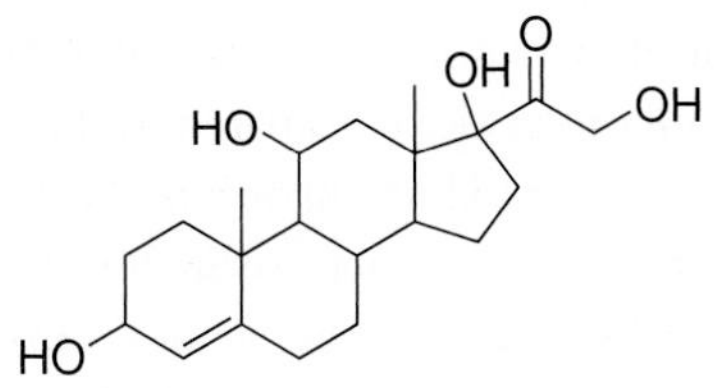

b)

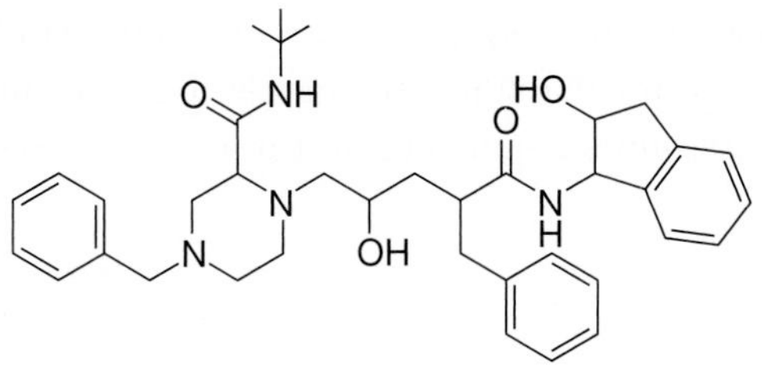

INTEGRATE THE SKILL

2.4 Identify the functional groups, and show the hybridization of the atoms in each group.

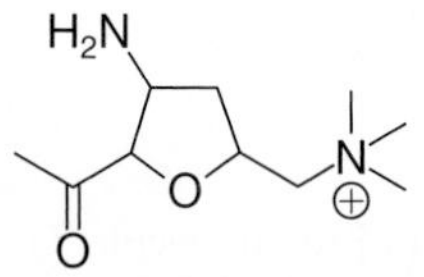

2.4 Relation between Intermolecular Forces, Molecular Structure, and Physical Properties

The structure of an organic molecule determines what intermolecular forces the molecule is capable of, and so a molecule's structure controls its physical properties. Structure dictates the distribution of electrons in the molecule, which in turn determines the types and magnitudes of forces that act between molecules. A critical examination of structure can lead to predictions of molecular properties and behaviour.

All the intermolecular forces generated by the molecules of a substance are responsible for the physical properties of that substance. Different aspects of molecular structure, including size and shape and presence of functional groups, determine the type and strength of intermolecular forces occurring in particular substances, and these result in unique physical properties, such as melting and boiling points and solubility.

2.4.1 Boiling point and melting point

The greater the intermolecular forces, the more difficult it is to separate individual molecules from each other, and therefore more energy is required to do so. The melting point or boiling point of a substance provides a measure of the amount of energy required to separate molecules. Substances with strong intermolecular forces require more energy to break apart the molecules, so they have higher boiling points and melting points.

The size and shape of molecules affects the amount of intermolecular forces and thereby the physical properties of substances. For example, long-chain molecules (compared to shorter-chain molecules) have more surface area contact with one another, which increases the amount of London forces between molecules and results in a corresponding increase in the boiling and melting points of substances composed of these molecules. Branched molecules (compared to chain molecules) have less surface area available for contact, which results in substances with slightly lower melting or boiling points. Molecules that contain rings are more rigid than open-chain compounds, and this improves molecular packing (the molecules fit together easily). This increases intermolecular forces, and the temperatures required to melt or boil such compounds increases.

	Melting point (°C)	Boiling point (°C)
(butane)	−134	−1
longer chains give more surface area contacts (octane)	−57	126
branching lowers surface area available for London forces (2,2,4-trimethylpentane)	−107	99
ring structures pack together well, increasing intermolecular interactions (cyclooctane)	14	149

Among the types of intermolecular forces, dipole-dipole interactions are stronger than dispersion forces. Therefore, molecules with functional groups that carry permanent dipoles have a strong influence on molecular properties. The greater the polarity of the functional group, the stronger is the force of attraction between the molecules. Hydrogen bonding in particular provides a very strong intermolecular attraction.

	Boiling point (°C)
(pentane)	36
O (3-pentanone)	101
OH (3-pentanol)	115

Electrostatics, the strongest intermolecular force, has a corresponding large effect on physical properties, particularly a large influence on melting point and boiling point. Organic molecules with permanent charges form ionic interactions with counter ions. This significantly increases the boiling and melting points of particular substances.

	Melting point (°C)	Boiling point (°C)
O, H (acetaldehyde)	−125	21
OH (ethanol)	−114	78
O, OH (acetic acid)	16.2	117
O, $O^{\ominus}$ $Na^{\oplus}$ (sodium acetate)	>300	N/A

The number of functional groups on a molecule also influences the physical properties of compounds. The more functional groups there are, the more polar interactions are possible, which tends to result in substances with higher melting and boiling points. The exact values (boiling point, for example) of physical properties cannot always be predicted, but predictions can be made about the relative magnitudes of these properties. As the complexity of molecular structure increases so does the complexity of the intermolecular forces involved. Interactions due to intermolecular forces do not contribute equally or in the same way in different molecules, which makes the prediction of physical properties such as boiling point difficult in some cases.

Functional groups that carry dipoles are usually more polar than hydrocarbons. The degree of polarity depends not only on the amount of dipole character in the functional group but also on the orientation of the dipole. For example, the four polar bonds in tetrachloromethane (CCl_4) are distributed in such a way that the individual dipoles cancel each other out to make the molecule non-polar.

2.4.2 Solubility: Like dissolves like

A **solvent** is a liquid medium in which compounds may be dissolved. It may be water or an organic substance.

The **solute** is the material that is dissolved in the solvent.

The intermolecular forces a molecule is capable of control other types of physical properties, such as their solubility in different solvents. Solubility is governed by the type and strength of the intermolecular forces between molecules of a **solvent** and molecules of a **solute**, the material that is dissolved. Whether in a living cell or the lab, the majority of organic reactions take place in a liquid solvent. Solvents are also used to purify or sometimes transport chemical substances. The solvent may be water, or it may be an organic substance that is a liquid at room temperature. Typical organic solvents are small molecules that may differ considerably in their polarity; however, most solvents are much less polar than water.

Organic solvents may be classified into three families (Figure 2.3) according to the type of intermolecular forces they can create.

1. Polar protic solvents, usually the most polar of the solvents, can act as hydrogen bond donors and are often miscible with water. In fact, water is a polar protic solvent.
2. Polar aprotic solvents have strong dipoles, and most can act as hydrogen bond acceptors, which makes many of them highly water-soluble.
3. Non-polar solvents have their molecules held together primarily by London forces.

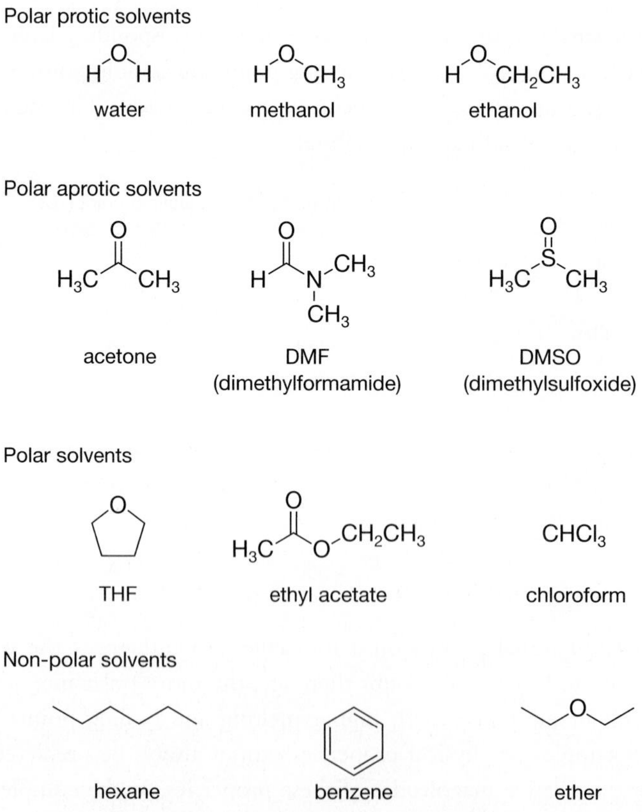

FIGURE 2.3 Commonly encountered solvents and their polarity classification.

The primary role of a solvent is to dissolve a solute. The process involves molecules of the solvent surrounding molecules of the solute forming a uniform substance. For this to take place, the solvent and the solute must favourably interact with each other, specifically by means of one or more of the four types of intermolecular forces described in Section 2.3. The type of intermolecular force interaction established between the solvent and the solute depends on the structure of their molecules. Typically, molecules are most soluble when the types of intermolecular forces are well matched between solute and solvent: *like dissolves like*.

The overall properties of a solvent are often described in terms of the polarity of the solvent—that is, the degree to which the intermolecular forces depend on permanent dipoles. Hydrocarbon solvents have no permanent dipoles and are considered non-polar. At the opposite end of the polarity scale is water, which is completely hydrogen bonded and is very polar. Most organic solvents have a permanent dipole, which gives some degree of polarity. However, these molecules also have a non-polar hydrocarbon portion, and so organic solvents tend to be much less polar than water. The polarity of most organic solvents falls between the polarity of water and that of hydrocarbons such as hexanes.

DID YOU KNOW?

Many organic molecules that have a permanent dipole are often considered non-polar. Their dipole is so small that any dipole-dipole interactions do not significantly contribute to the molecule's intermolecular forces. For example, the common organic solvent diethyl ether does have a small permanent dipole but is generally considered a non-polar solvent. In many cases, there is a clear difference in the polarity of molecules (for example, CH_3OH is much more polar than CH_4); in other situations, the overall polarity difference between molecules can be very small.

In general, a very polar solvent does not dissolve non-polar molecules. This is because the solvent molecules maintain their (stronger) dipole-dipole attractions with each other rather than break those attractions and surround a non-polar molecule with which they can establish only (weaker) London forces. By contrast, polar solvents such as acetone dissolve polar organic molecules because a favourable dipole-dipole interaction (and perhaps H-bonding) can be established between them. For similar reasons, non-polar solvents dissolve non-polar organic molecules in which the predominant intermolecular force of attraction between the solvent and solute molecules is London forces.

2.4.3 Hydrophilicity and hydrophobicity

Water, being highly polar, presents a special situation for organic compounds. Water solubility is important for laboratory purifications and for certain organic reactions, and predicting the water solubility of organic compounds is essential for many situations. Water is also the solvent that makes life possible, and is necessary for foods, pharmaceuticals, and other products.

Although some organic compounds dissolve in water, the vast majority do not. Very non-polar compounds such as hydrocarbons do not dissolve in water, and these molecules are often referred to as **hydrophobic**. In this situation, the attractive force between polar water and the temporary dipole in a non-polar molecule is too weak to replace the strong hydrogen bonds between water molecules. Consequently, the non-polar molecules are excluded from water. Small organic molecules (usually those with less than five carbons) that contain polar functional groups do dissolve in water. However, the more carbon atoms there are in an organic molecule, the less water-soluble the molecule becomes, to the point that certain molecules are effectively insoluble in water.

A **hydrophobic** ("water-fearing") compound establishes only weak London forces with surrounding water molecules; therefore, water molecules maintain hydrogen bonds with each other, rather than forming new ones with the solute.

In water, the hydrocarbon parts of larger molecules tend to aggregate together rather than interact with water, which leads to insolubility in water.
H_2O
OH
HO
H_2O
H_2O
OH
HO
H_2O
H_2O
OH
HO
H_2O

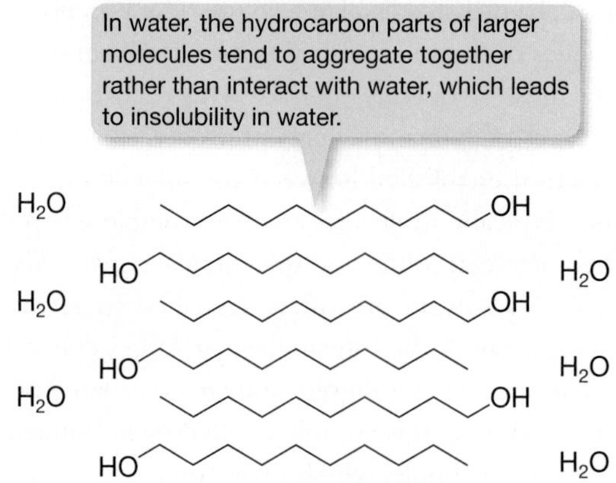

A **hydrophilic** ("water-loving") molecule or group of atoms is polar enough to form favourable intermolecular interactions, including hydrogen bonding, with water.

Water solubility increases as the number of polar groups on a molecule increases. Such molecules are said to be **hydrophilic**. In addition to London forces, hydrophilic molecules can establish dipole-dipole interactions and hydrogen bonds with the surrounding water molecules. In particular, water solubility greatly increases when the polar groups are capable of hydrogen bonding.

Water solubility (g/L)
O
H
5
OH OH
HO
O
OH OH H
909
adding polar functional groups to molecules increases the water solubility of the compound

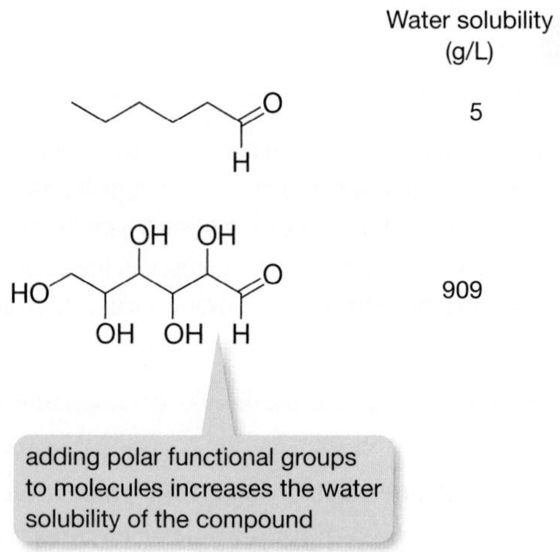

The degree to which an organic compound dissolves in water—its water solubility—is governed by the relative proportion of its hydrophilic and hydrophobic regions. Molecules that have a smaller proportion of polar groups relative to their hydrophobic region are less soluble in water. Water solubilites increase as the polarity of the functional groups increases, and as the polarity of the molecule increases in relation to the molecule's size.

Water solubility (g/L)
OH
miscible
OH
5.6
increasing the size of the non-polar portion of the molecule relative to the polar part tends to reduce water solubility
O
OH
10.8
more polar functional groups lead to higher water solubility

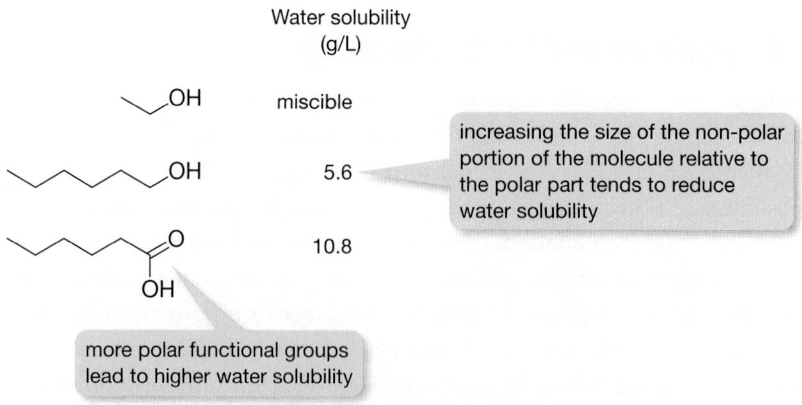

Organic molecules with branching in a hydrocarbon chain have less surface area than unbranched long-chain molecules. The branching structure makes it easier for surrounding water molecules to hydrogen bond with each other. As a result, a branched organic molecule such as *tert*-butyl alcohol is more soluble in water than its linear analog, containing the same number of carbon atoms.

	Water solubility (g/L)
(1-butanol) OH	7.7
(*tert*-butyl alcohol) OH	miscible

Water H-bonds around branched molecules better than around unbranched molecules, which tends to increase the solubility of such compounds in water.

Charged molecules can interact with water by electrostatics and often show significant water solubility. In general, most organic molecules are soluble in organic solvents, but relatively few are highly soluble in water. Many organic molecules that carry a net charge can dissolve to a significant degree in water.

	Water solubility (g/L)
(pentanoic acid) OH, O	11
(sodium pentanoate) ONa, O	428

the sodium salt is very water-soluble

CHEMISTRY: EVERYTHING AND EVERYWHERE

Your Own Personal Carboxylic Acid Signature

Maridav/Shutterstock.com

Continued

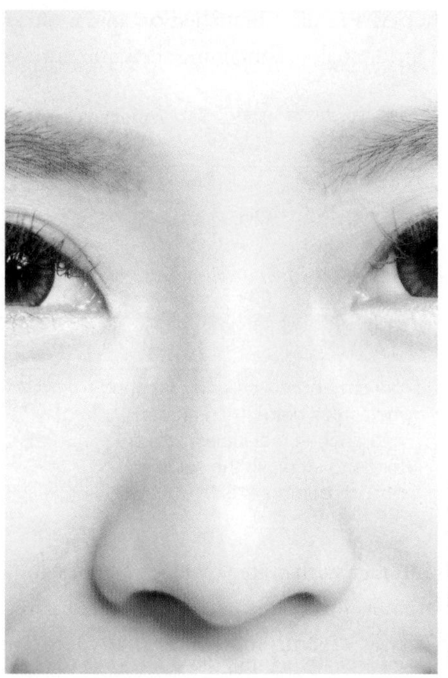

aslysun/Shutterstock.com

When the study of chemistry was in its infancy, one of the ways to identify and classify newly discovered compounds was by their odour. The scientists of the time would smell or even taste these substances! For obvious reasons, this practice is no longer acceptable, but the odour of a substance is a property of the molecule. Smell is the result of molecules binding to receptor molecules in our nose. As molecular structures change so does the way in which molecules bind to individual receptors; thus different compounds have different odours.

Human body odour derives from a complex mixture of dozens of organic compounds. The substances excreted from our sweat glands are actually quite odourless. However, the bacteria on our skin metabolize these substances into the odoriferous ones. Many of these metabolic products are carboxylic acids, some of which are shown here. Two in particular, shown in blue, are the major components in sweat responsible for the unpleasant odour associated with physical activity.

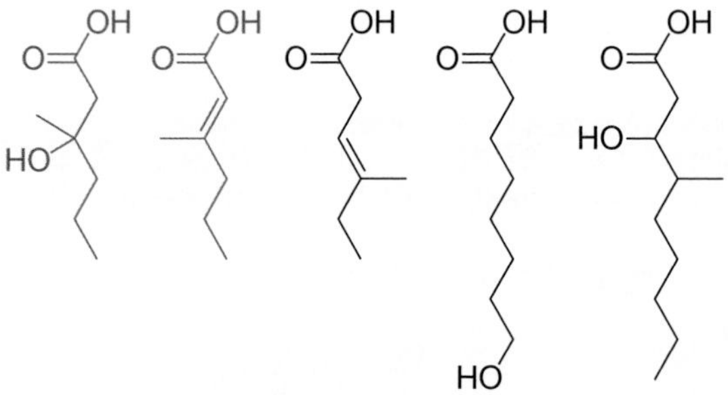

The relative amounts of the compounds in sweat differ from person to person. This gives each of us an aroma profile that is almost as unique as our fingerprints. Our carboxylic acid "signature" makes us readily identifiable to dogs and other animals, including humans: mothers can identify their children by their unique body odour.

Note that the presence of the carboxylic acid and hydroxyl functional groups makes the compounds soluble in sweat, which is mostly water. This also means that the odoriferous compounds in sweat are easily washed away in the shower after exercise. Showering not only washes away these compounds but also controls the populations of bacteria that produce them!

CHECKPOINT 2.3

You should now be able to (1) identify the intermolecular forces that exist between molecules based on their structure; and (2) predict differences in properties, such as boiling point, that have their basis in intermolecular forces arising from differences in structure between molecules.

SOLVED PROBLEM

The boiling point of $CH_3(CH_2)_6CO_2H$ is 240 °C. Would you expect the boiling point of $HOCH_2(CH_2)_6CO_2H$ to be higher or lower than this? Provide an explanation for your choice.

SOLUTION: The –OH group in $HOCH_2(CH_2)_6CO_2H$ provides an additional site of favourable dipole-dipole and H-bonding interactions with neighbouring $HOCH_2(CH_2)_6CO_2H$ molecules within the pure liquid. So the intermolecular forces between molecules of $HOCH_2(CH_2)_6CO_2H$ are greater than the intermolecular forces between molecules of $CH_3(CH_2)_6CO_2H$. More heat is therefore required to separate molecules of $HOCH_2(CH_2)_6CO_2H$. This means it has the higher boiling point.

PRACTICE PROBLEM

2.5 Do you expect the solubility of $HOCH_2(CH_2)_6CO_2H$ in water to be greater or less than the solubility of $CH_3(CH_2)_6CO_2H$? Provide an explanation for your choice. Do you expect the solubility of $HOCH_2(CH_2)_6CO_2H$ in hexanes to be greater or less than the solubility of $CH_3(CH_2)_6CO_2H$?

INTEGRATE THE SKILL

2.6 Write the Lewis structure of ethyl acetate, $CH_3CO_2CH_2CH_3$, and draw the bond dipoles of each bond in which the atoms have an electronegativity difference of 0.5 or higher. Indicate its hydrophobic and hydrophilic regions.

2.5 Naming Organic Molecules

A systematic approach to the naming of organic molecules has been established to enable clear communication within the scientific community. Known as the IUPAC[1] system, this set of rules generates unambiguous names for organic molecules based on their structure.

Knowledge of the basic rules and conventions regarding the naming of organic structures makes it possible to do the following:

1. Derive the correct name of an organic compound from its structure.
2. Draw the structure of a molecule based on its name.

The second is perhaps the more important skill, as it is necessary when searching for information in the scientific literature.

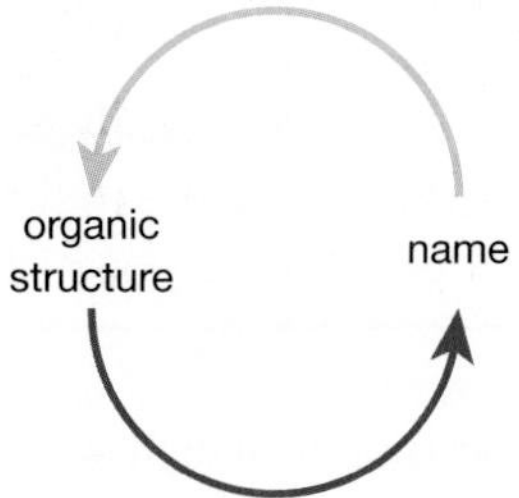

1. IUPAC stands for International Union of Pure and Applied Chemistry, a scientific non-governmental organization that serves to advance the chemical sciences.

In this section, the focus is on the skill of naming a structure. Section 2.5.4 introduces the process of generating the structure of an organic compound from its name.

2.5.1 The basics of naming structures

The name of an organic compound is composed of up to three segments: a root name, a required suffix to that root name, and a prefix—which may or may not be needed depending on the molecule.

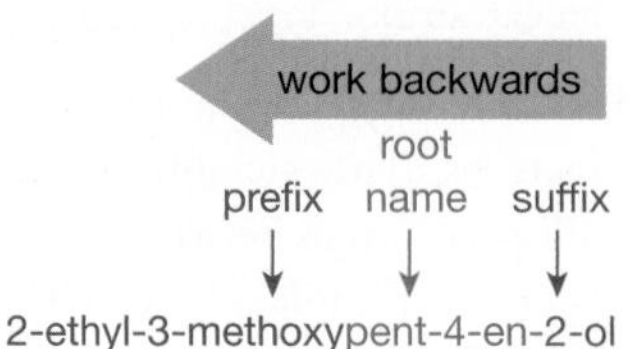

Usually it is best to name compounds by working backwards. Start by identifying the appropriate suffix, which is based on the functional groups present. For example, alkanes are identified by the suffix *ane*. Then identify the root name (the *parent* chain), which is the longest chain in the molecule that is directly attached to the functional group of the suffix. The root name depends on the number of carbons in that chain. Table 2.2 lists the root names of linear hydrocarbons up to 20 carbons long. For example, a five-carbon chain would have a root name of *pent*. Finally, identify the prefix, which is based on the type and placement of substituents in branched hydrocarbons. For example, 2,4,5-trimethyl is a prefix in front of the root name. The following sections describe how to identify root names and prefixes for simple hydrocarbons, branched hydrocarbons, and cyclic alkanes.

2.5.1.1 Simple alkanes

The root name describes the longest chain—sometimes called the *parent* chain—in the molecule that is attached to the suffix group. The root name is identified according to the number of carbons in this chain. Root names for carbon chains up to 20 carbons long are listed in Table 2.2.

TABLE 2.2 Root Names of Carbon Chains

Number of Carbons	Root Name	Number of Carbons	Root Name
1	meth	11	undec
2	eth	12	dodec
3	prop	13	tridec
4	but	14	tetradec
5	pent	15	pentadec
6	hex	16	hexadec
7	hept	17	heptadec
8	oct	18	octadec
9	non	19	nonadec
10	dec	20	eicos

The compound shown is a hydrocarbon in which all the carbons are sp^3 hybridized. Such a compound is called an alkane, and the suffix *ane* applies. There are eight carbons in this chain, and so the root name is *oct*, which when combined with the suffix gives the name *octane*.

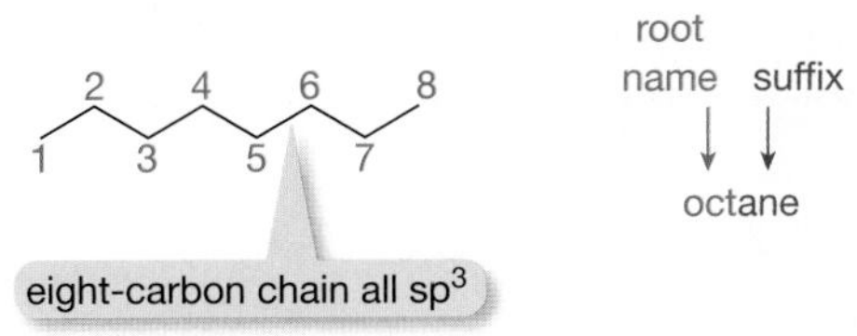

2.5.1.2 Branched hydrocarbons

Some chains contain carbons that are connected to more than two other carbons. When this happens the structure of the molecule becomes branched. To name such structures, first identify the longest carbon chain (parent chain) to determine the root name of the compound. The suffix added to this root name identifies the functional group.

Each branch is also known as a substituent. A branch is named by adding the suffix *yl* to the root name that corresponds to the number of carbons in the branch. The branch is further named in terms of the location of the branch on the parent chain—that is, the number of the carbon to which the substituent is attached on the parent chain. Parent chains are numbered starting at the end of the chain closest to the first branch point.

The location number of the branch is placed, separated by a hyphen, before the name of the branch. Together, the position and name of the branch become the prefix directly in front of the root name of the branched hydrocarbon.

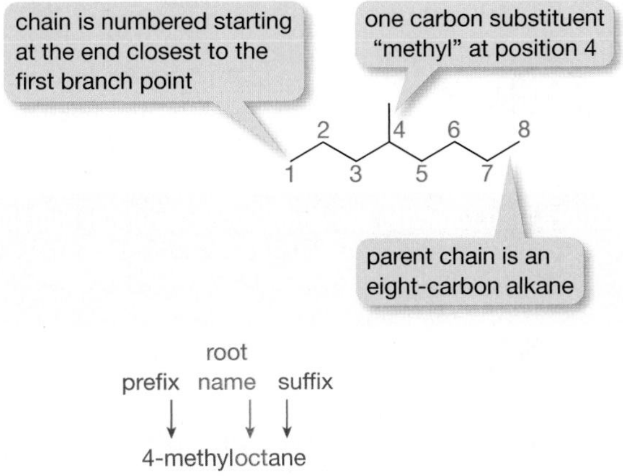

When a molecule has more than two substituents (branches), these are listed in the prefix in alphabetical order. Note that numbering on the parent chain is based on proximity to the first branch point, not on alphabetical order.

The compound depicted in the following example shows the longest chain has 10 sp^3-hybridized carbons; so *decane* is the root name and suffix for the parent chain. The molecule has two substituents. One substituent has two carbons and is therefore an ethyl group (root name *eth* indicates two carbons; the suffix *yl* indicates a substituent). The second substituent is a methyl group (root name *meth* and suffix *yl*). To name the location of the substituent, the parent chain is numbered starting at the end closest to a branch point. Note that substituent type is not involved in this aspect of the naming.

To determine the full name of the compound, the prefix is added to the root name. The prefix includes the substituent names listed in alphabetical order, and the numbers are always separated from letters by hyphens. The prefix and root name are not separated.

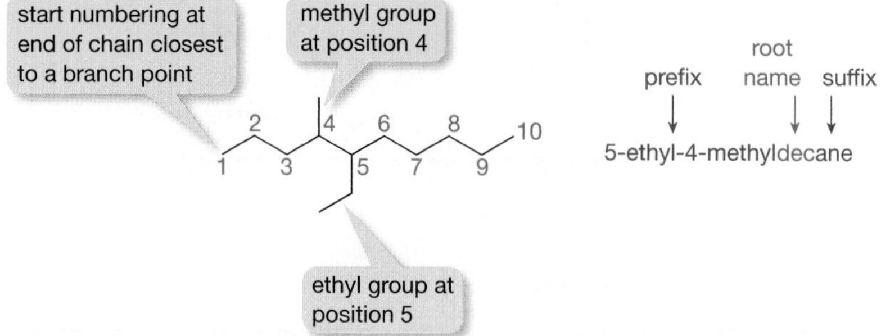

For naming molecules with two or more identical groups, a descriptor such as di, tri, or tetra is added to the branch prefix to indicate the number of identical branches (see Table 2.3). The positions of each branch are listed at the beginning of the prefix describing the branch type, separated by commas: for example, 2,4,5-trimethyl. Note that the prefixes such as di, tri, and tetra are not considered when alphabetizing the substituents for naming.

In the following molecule, there are three methyl groups on a heptane chain. Rather than using a prefix of 2-methyl-4-methyl-5-methyl, the more compact 2,4,5-trimethyl prefix is added to the root name to obtain the overall name.

three methyl substituents

1 2 3 4 5 6 7

root
prefix name suffix
2,4,5-trimethylheptane

TABLE 2.3 Descriptors that Indicate the Number of Identical Substituents

Number of identical substituents	Descriptor	Number of identical substituents	Descriptor
2	di	11	undeca
3	tri	12	dodeca
4	tetra	13	trideca
5	penta	14	tetradeca
6	hexa	15	pentadeca
7	hepta	16	hexadeca
8	octa	17	heptadeca
9	nona	18	octadeca
10	deca	19	nonadeca
		20	eicosa

2.5.1.3 Complex branches

In some organic molecules, the branches are complex because they are also branched. These branched substituents are named according to the same rules and conventions of other organic molecules, except for two small differences:

1. The suffix *yl* is used to indicate a substituent.
2. Numbering begins at the carbon that attaches the substituent group to the parent chain.

The full name of the branched substituent (including its prefix) is placed within parentheses: for example, (1-methylpropyl). A number and hyphen are placed in front of the opening parenthesis to indicate the position of the branched group along the main chain: for example, 4-(1-ethylpropyl) indicates that the complex branch (1-ethylpropyl) is found at position 4 of the main chain. The complex branch consists of a propyl group with a methyl attached at position 1.

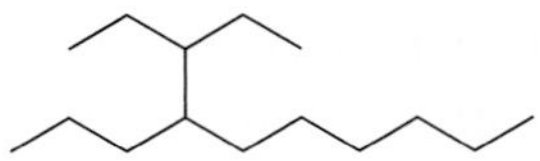

In this example, the longest carbon chain has 10 carbons, which gives a root name and overall suffix of decane. This parent chain is numbered starting at the end closest to the first branch point: the location of a complex substituent on carbon 4 of the parent chain. To name this branch, first identify and number the longest linear chain in the substituent, *starting at the point at which it attaches to the parent chain*. The longest such chain in this group has three carbons, which gives the root name of propyl. Since a two-carbon substituent is attached to carbon 1 of this group, the full name of the entire substituent is (1-ethylpropyl). Parentheses are used to indicate that this is the full name of a branched substituent, and the number preceding the parenthetical prefix indicates the location of the branched substituent on the molecule's parent chain.

The substituent name(s), listed in alphabetical order, form the prefix of the full structure. Note that a space or hyphen is *not* placed after the closing parenthesis of the last group listed.

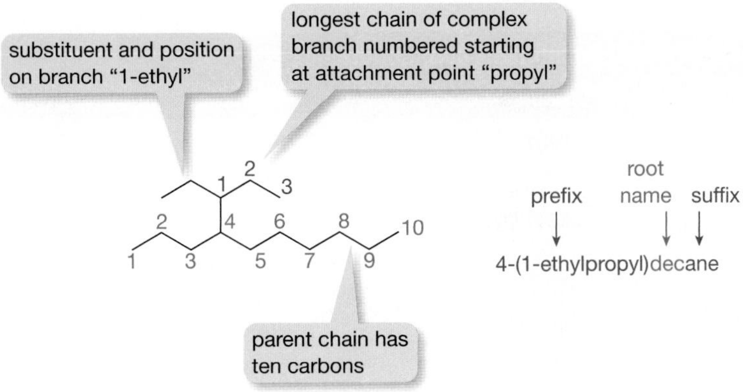

substituent and position on branch "1-ethyl"

longest chain of complex branch numbered starting at attachment point "propyl"

parent chain has ten carbons

prefix | root name | suffix

4-(1-ethylpropyl)decane

2.5.1.4 Cyclic alkanes

To name a cyclic compound, or a compound that contains a ring, the prefix *cyclo* is added before the part of the name that describes the cyclic chain.

cyclohexane

This compound contains a ring consisting of six carbons. Because this is an alkane with six carbons in the parent chain, the compound is a hexane. To indicate that the six carbons of the chain are in a ring, the prefix *cyclo* is added; so, the compound is named cyclohexane.

The following two examples show compounds that consist of a ring and a chain. In the first example, the ring forms the largest chain of the molecule's structure, so the ring is identified as the parent chain and provides the root name for the molecule. There is only one substituent attached to the ring, so its location is considered to be at carbon 1 of the ring. In this case of only one substituent located on carbon 1, stating the position number is optional. Therefore, the molecule is named either propylcyclohexane or 1-propylcyclohexane.

The second compound has a ring containing five carbons and a straight chain containing nine carbons. The parent chain in this case is the linear portion of the molecular structure because it has more carbons than the ring. This molecule is named either cyclopentylnonane or 1-cyclopentylnonane; since the cyclopentyl is attached to the first carbon of the nonane, the number is optional.

2.5.1.5 Common names

Several substituents are encountered so frequently that they are typically referred to by common names, sometimes called *trivial names*. These names are a holdover from those used more than a hundred years ago, before systematic nomenclature was invented. Some of the more common ones are listed in Table 2.4.

TABLE 2.4 Selected Substituent Common Names

Structure	Name	Structure	Name
	vinyl		allyl
	isopropyl		*tert*-butyl
	phenyl		

CHECKPOINT 2.4

You should now be able to name alkanes.

SOLVED PROBLEM

Name the following structure.

STEP 1: Identify the functional group and the parent chain. All the carbons in this structure are sp^3, and so it is an alkane. The longest chain (parent chain) has nine carbons.

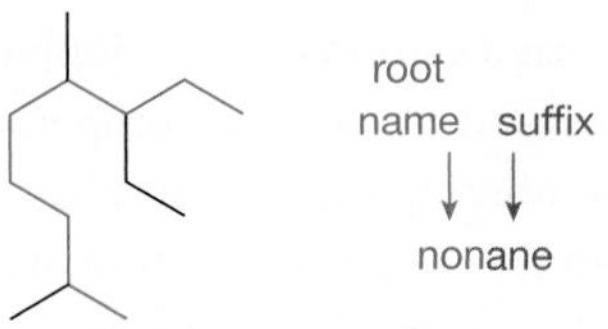

root name suffix

nonane

STEP 2: Identify any substituents. There are two one-carbon groups and a two-carbon group. The parent chain is numbered starting at the end closest to the first branch point. Since there are two methyl groups, use the descriptor di.

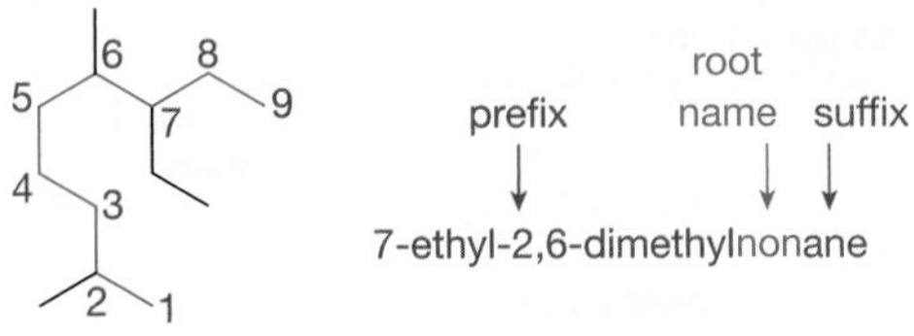

prefix root name suffix

7-ethyl-2,6-dimethylnonane

PRACTICE PROBLEM

2.7 Provide names for the following molecules.

a) $CH_3CH_2CH_2CH_2CH_3$

b)

c)

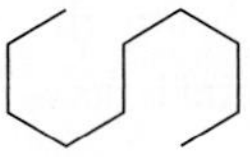

d)

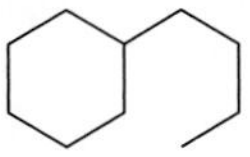

e)

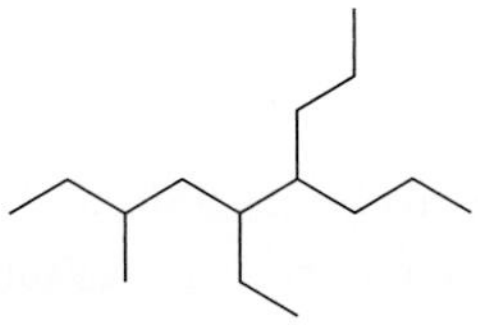

f)

g)

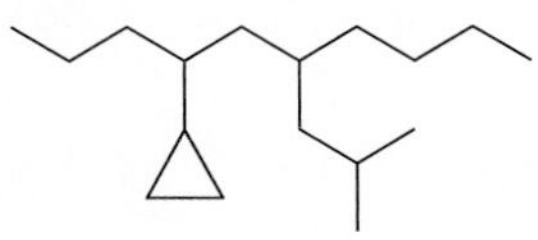

INTEGRATE THE SKILL

2.8 Isooctane is an important hydrocarbon added to gasoline to improve its combustion properties. The structure of isooctane is as follows:

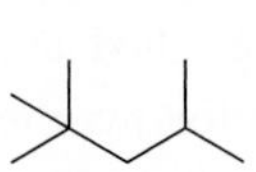

a) Give the proper IUPAC name for this compound.

b) The boiling point of isooctane is 99 °C. The boiling point of octane is 126 °C. Account for this difference.

2.5.2 Naming alkenes and alkynes

Hydrocarbons that have π bonds require a particular suffix to identify their functional group. The suffix *ene* indicates a molecule has a carbon-carbon double bond, so a molecule with at least one carbon-carbon double bond is called an *alkene*. To properly name an alkene, first indicate the position of the double bond by numbering the carbon chain starting at the end closest to the double bond, and then insert the number of the *first* carbon of the double bond between the root name and the suffix. When the double bond is located at position 1, it is not necessary to include a position number for the bond. The following compound is pent-2-ene.

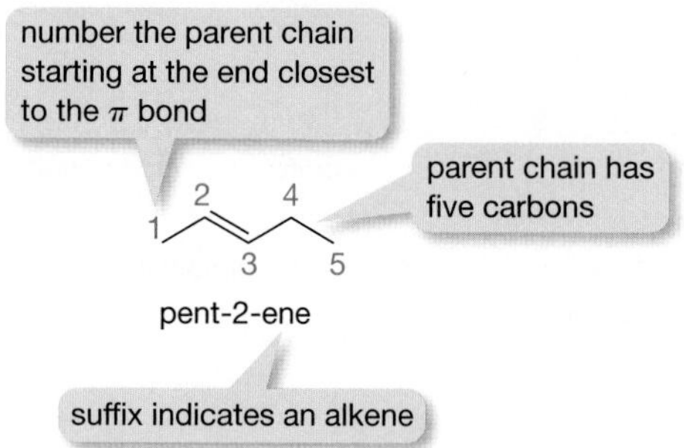

Compounds that have more than one double bond require a descriptor that indicates the number of double bonds in the molecule. These descriptors are the same as the ones used to indicate the number of identical substituents (di, tri, tetra, etc.). The descriptor is added just before the *ene* of the suffix. The positions of the double bonds are given as a list. The following compound is deca-2,6-diene.

STUDENT TIP

Double bonds can adopt one of two geometries. These are described using a special naming system, which is illustrated in Chapter 4.

For compounds that have double bonds located in a ring, numbering begins at one carbon of the double bond, and proceeds in the direction of the double bond. When the molecule has substituents, the numbers proceed across the double bond in the direction toward the nearest substituent. The following compound is 3-methylcyclohexene (implied that the double bond is at position 1).

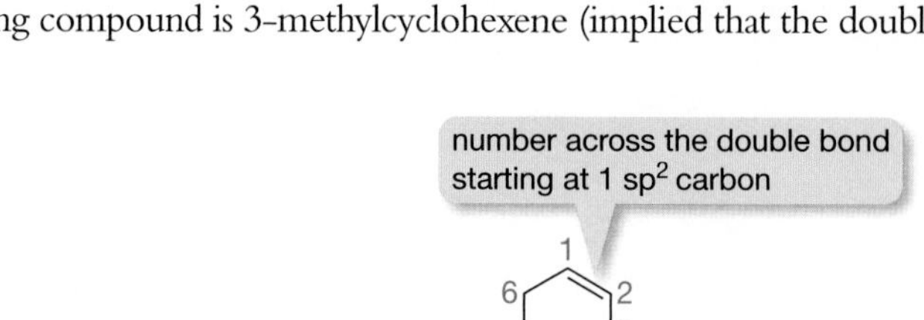

Compounds that have triple bonds are called *alkynes*. The suffix *yne* indicates the functional group is an alkyne. The naming of alkynes follows the same process as the naming of alkenes.

Priorities for suffixes are established by the oxidation state of the group. Common functional groups and their priorities are listed in Table 2.5. If there are both double and triple bonds in a molecule, the alkyne has higher priority; therefore, it is listed last, and chain numbering is based on the alkyne. Such a molecule is an *–enyne*.

CHECKPOINT 2.5

You should now be able to name alkenes and alkynes.

SOLVED PROBLEM

Provide a name for the following structure.

STEP 1: Identify the functional groups present to establish the proper suffix and parent chain. There are two double bonds (alkenes) and one triple bond (alkyne). The alkyne has higher priority, and so the suffix is *yne*. The parent chain must contain this functional group.

alkyne is the highest priority functional group—suffix is *yne*

number the parent chain starting at the end closest to the suffix functional group

parent chain is the longest chain that contains the suffix functional group

STEP 2: There are two alkene groups in the parent chain, which has nine carbons. The alkenes are located at positions 4 and 7 (first carbon of the group). This gives the root name of *non-4,7-dien*. Combine this with the suffix for *non-4,7-dienyne* (because the alkyne occurs at position 1, no number is necessary).

STEP 3: There are two substituents: methyl groups at positions 4 and 8. This gives a prefix of *4,8-dimethyl*. The molecule is 4,8-dimethylnon-4,7-dienyne.

PRACTICE PROBLEM

2.9 Name the following molecules.

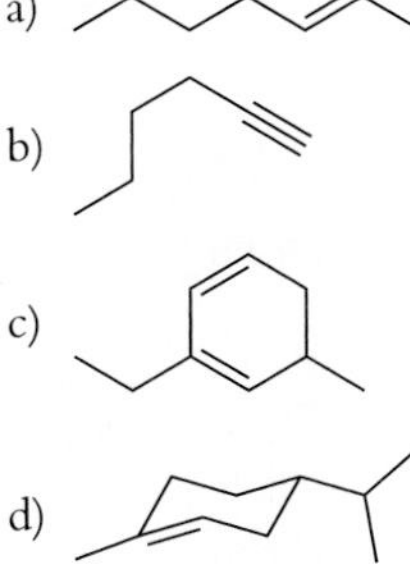

INTEGRATE THE SKILL

2.10 Draw the structure of 4-methyl-5,5-dipropyldodec-3-ene.

2.5.3 Functional groups

Functional groups other than akenes or alkynes are distinguished either by a prefix or suffix. If a molecule has only one such functional group, the suffix in its name corresponds to the functional group, and the parent chain is the one to which the group is attached. Numbering on the parent chain starts at the end closest to the suffix group. When the parent chain contains an alkane, alkene, or alkyne, these groups become part of the root name, and the trailing *e* is removed from the name of the carbon structure.

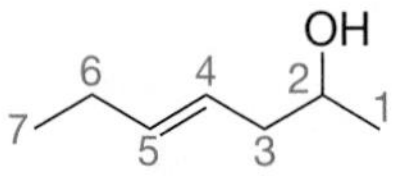

root ↓ suffix ↓

hept-4-en-2-ol

In the case of carboxylic acids and aldehydes, the carbon of the group becomes carbon 1 of the parent chain, and numbering takes place starting with this atom. Because the functional group—specified by the suffix—is attached at carbon 1, numbering the suffix location is optional.

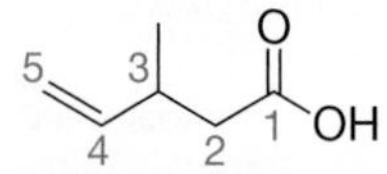

3-methylpent-4-enoic acid

TABLE 2.5 Common Functional Groups Described by Suffixes and Prefixes

Structure	Functional group (priority)	Suffix	Prefix
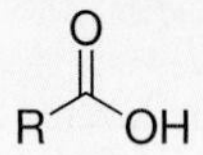 RC(O)OH or RCO_2H or RCOOH	Carboxylic acid	oic acid	
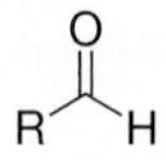 RC(O)H or RCHO	Aldehyde	al	oxo
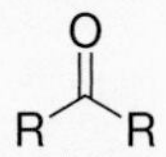 RC(O)R or RC(O)R or RCOR	Ketone	one	oxo
R–OH	Alcohol	ol	hydroxy
$R–NH_2$ R_2NH R_3N	Amine	amine	amino
R–F R–Cl R–Br R–I	Halide		fluoro chloro bromo iodo

For ketones, the carbon that is attached to the oxygen is indicated in the name. The position of this ketone group on the chain is identified by the number placed between the root name and the suffix, in the same manner as for alkenes and alkynes. Numbering of the chain begins at the end closest to the group indicated by the suffix (ketone). For other functional groups as well, the position to which the group is attached is indicated by a number between the root name and the suffix. When the group is attached at position 1, this position number is optional.

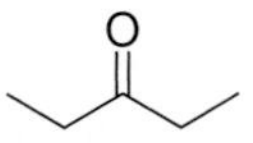

pentan-3-one

When a molecule has more than one functional group, the functional group with the highest priority is used to establish the suffix. Other groups are named as substituents using the prefixes listed in Table 2.5.

CHECKPOINT 2.6

You should now be able to name simple organic molecules.

SOLVED PROBLEM

Determine the name of the organic molecule shown here.

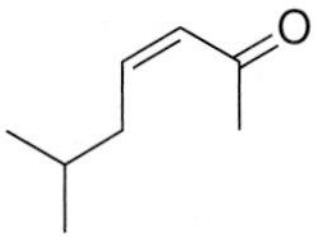

STEP 1: Suffix: Inspect the molecule and identify the functional group of highest priority. This is a ketone (priority 3, Table 2.5), so the name of the molecule ends in *-one*.

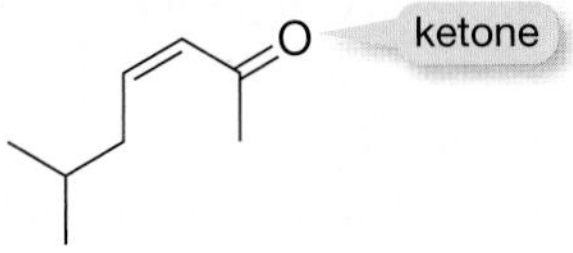

ketone

STEP 2: Root name: The root name is based on the parent chain that contains the suffix group. This is the linear chain of seven carbon atoms shown here in red, so the root name begins with *hept*. The parent chain is numbered starting at the end closest to the ketone. Therefore, this molecule is a *2-one*.

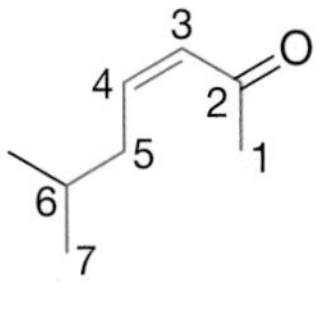

With the numbering in place, the second part of the root name can be determined based on the locations of any double or triple bonds along the parent chain. There is one double bond that begins at position 3 (*3-en*). The completed root name is therefore *hept-3-en-*, and adding the *2-one* suffix provides *hept-3-en-2-one*.

STEP 3: Prefix: All other substituents must be named as prefixes. There is only one methyl substituent at position 6, which gives a prefix of 6-methyl. Adding the prefix to the root name and suffix provides the complete name: 6-methylhept-3-en-2-one.

PRACTICE PROBLEM

2.11 Determine the name of the following organic molecules.

a)

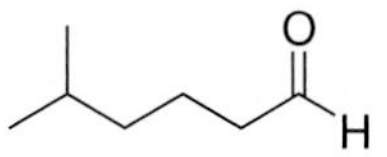

b)

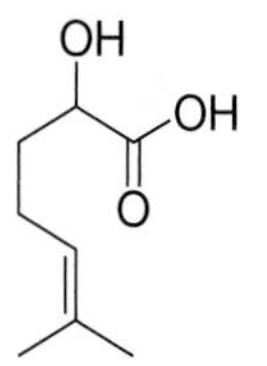

c)

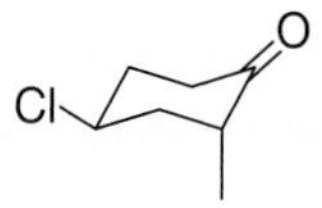

d)

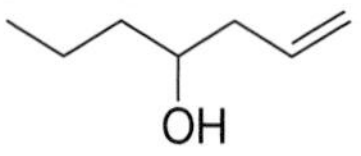

INTEGRATE THE SKILL

2.12 Determine the systematic name of the following organic molecule and paste the name into your favourite online search engine to find its common name and one of its uses.

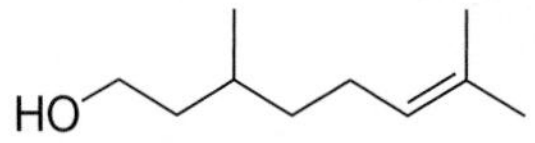

2.5.4 Generating structures from names

Typically, scientific literature, especially the older literature, uses compound names. To find further information about chemical compounds, it is very important to be able to convert names to structures. The following four steps provide a systematic way to draw a molecular structure based on the name of the compound. First, examine the root name. The root name provides the number of carbons in the structure, and these can be quickly drawn and numbered. Second, examine the suffix, which indicates the highest priority functional group, and add these at the correct position to the parent chain. If no number is provided with the suffix, the group is attached at position 1. Third, examine whether the root name includes alkenes or alkynes, and add these to the drawing of the structure. Fourth, examine the prefix of the overall name to determine if there are any substituents, and add these to the drawing.

The following steps describe how to generate the structure of 3-ethyl-4-methylpent-4-en-2-ol based on its name.

1. In the root name pent-4-en, the pent indicates a five-carbon chain with a double bond at position 4.

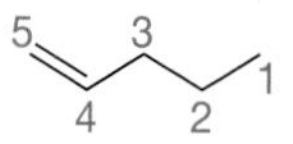

2. The suffix 2-ol indicates an OH at position 2 of the chain.

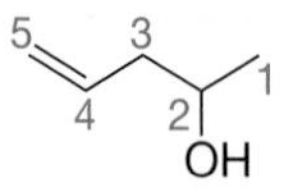

3. The prefix shows the presence of two substituents, an ethyl group at position 3 and a methyl group at position 4.

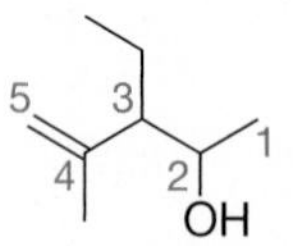

CHECKPOINT 2.7

You should now be able to convert a name to a structure.

SOLVED PROBLEM

Draw the line structure of 5-butyl-8-fluoro-7-hydroxy-6,6,9-trimethyldecan-3-one.

STEP 1: Identify the prefix (if any), the root name, and suffix. The suffix is easiest to identify since it is short and often separated from the rest of the name by the number that identifies its location. The suffix is *-3-one.*

STEP 2: Root name: *decan*. The parent chain *dec* consists of ten carbon atoms. The *an* indicates that the parent chain contains single bonds only. The ten-carbon atom chain can be drawn, its atoms numbered, and the ketone placed at position 3 to obtain the structure of decan-3-one:

O
1 3 5 7 9
2 4 6 8 10

STEP 3: Prefix: 5-butyl-8-fluoro-7-hydroxy-6,6,9-trimethyl. There is a total of six substituents in the prefix.

- 5-butyl. The molecule has a four-carbon alkyl group, $CH_3CH_2CH_2CH_2–$, connected to position 5 of the parent chain.
- 8-fluoro. The molecule has a fluorine atom, F–, connected to position 8 of the parent chain.
- 7-hydroxy. The molecule has an alcohol functional group, –OH, connected to position 7 of the parent chain.
- 6,6,9-trimethyl. There are three methyl groups in the molecule: two at position 6 and one at position 9.

O OH
F

To further simplify the drawing of a molecule, it is common practice to represent small alkyl groups using the appropriate two-letter abbreviation: Me for methyl, Et for ethyl, etc. (Table 2.6). The molecule could be equally drawn as shown in which two abbreviations are used.

O OH
F

O Bu OH
Et
F

TABLE 2.6 Common Abbreviations for Small Substituents

Structure	Substituent name	Abbreviation
CH_3	Methyl	Me
CH_3CH_2	Ethyl	Et
$CH_3CH_2CH_2$	Propyl	Pr
$(CH_3)_2CH$	Isopropyl (1-methylethyl)	*i*-Pr
$CH_3CH_2CH_2CH_2$	Butyl	Bu

PRACTICE PROBLEM

2.13 Draw the line structure of the following.

5,5-difluoro-1-(4-methylheptan-2-yl)cyclohexa-1,3-diene

Bringing It Together

The important structural features of cholesterol can now be examined. The four rings contain mainly carbon-carbon single bonds, as does the 1,5-dimethylhexyl "tail" of the molecule. To some extent, cholesterol may be viewed as a relatively unreactive framework of cycloalkane and alkane that contains two functional groups: the alcohol and the alkene positioned in close proximity. Cholesterol easily embeds itself in the hydrophobic environment of the cell membrane because its structure is primarily a non-polar hydrocarbon, with only the polar alcohol group providing any opportunity for hydrogen bonding. The four rings of cholesterol are fused together—that is, each ring shares a carbon-carbon single bond with its neighbouring rings. Locking the four rings together in this manner gives cholesterol a very rigid shape. The chemically inert cycloalkyl and alkyl regions lower the reactivity of cholesterol, which helps in its role as a structural component of the membrane.

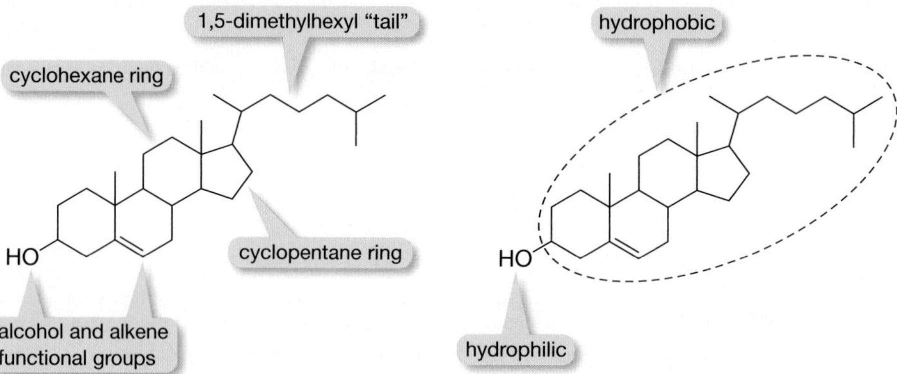

ORGANIC CHEMWARE
2.2 Line-angle structure: Cholesterol

The two functional groups are a primary source of reactivity in cholesterol. For example, cholesterol is not stored in cells in its native form. Instead, it is stored as an ester through a reaction at its alcohol functional group. Notice how its ester is even more hydrophobic than cholesterol itself!

HO → $CH_3(CH_2)_{15}$ O O

The common name *cholesterol* is used in all scientific circles with good reason. It is not easy to refer to cholesterol by its systematic IUPAC name: 10,13-dimethyl-17-(6-methylheptan-2-yl)-2, 3,4,7,8,9,11,12,14,15,16,17-dodecahydro-1*H*-cyclopenta[*a*]phenanthren-3-ol. Using this name would not facilitate scientific communication. The naming of cholesterol is well beyond the scope of this textbook, but try to spot its prefix, root name, and suffix.

You Can Now

- Identify the structural features of a molecule: its substituents; its functional groups, including π bonds and heteroatoms; and its hydrophobic and hydrophilic groups.
- Identify whether an organic molecule is saturated or unsaturated.
- Identify and name the functional groups in an organic molecule.
- Recognize the London forces and the dipole-dipole, hydrogen bonding, and electrostatic interactions that can exist between molecules; and also understand how the magnitude of these forces determines the physical properties of a compound such as its boiling point and solubility.
- Write the systematic (IUPAC) name of an organic compound, given the structure of the organic molecule.
- Draw the line-bond structure of an organic compound, given its systematic (IUPAC) name.

Problems

2.14 Identify the functional groups in each of the following molecules.

a)

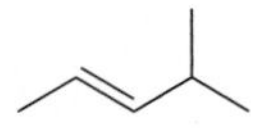

b)

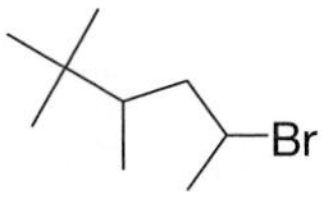

c)

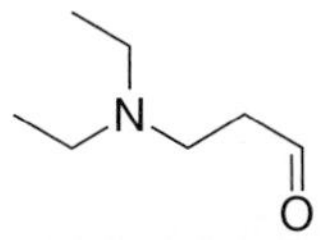

d)

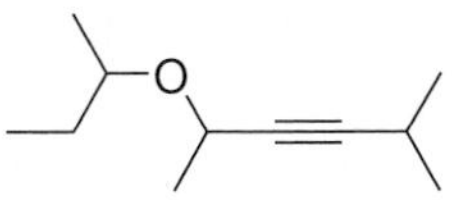

e)

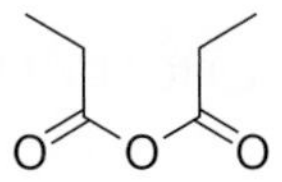

f)

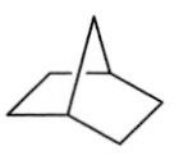

g)

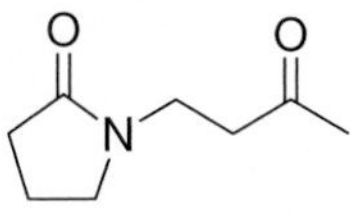

h)

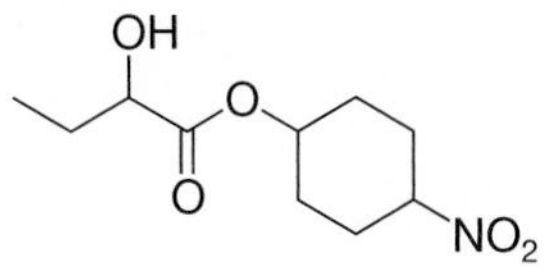

2.15 Identify each of the molecules in Question 2.14 as saturated or unsaturated.

2.16 Draw the line structure of the following molecules and identify their functional groups.

a) $ClCH_2CH(CH_3)COCH(CH_3)_2$

b) $CH_3CH(OH)CH_2CHO$

c) $(CH_3)_2C{=}CHCH_2NHCO(CH_2)_2CH_3$

d) $HC{\equiv}CCH_2OCH_2COOC_2H_5$

2.17 Provide a reason for the difference between the electrostatic potential map of methane and that of CF_4 shown here.

2.18 Identify the negative, positive, and non-polar regions of the following molecule.

2.19 The boiling point of $CH_3(CH_2)_4CO_2H$ is 206 °C. Do you expect the boiling point of 2-ethylhexanoic acid to be higher or lower than this? Provide an explanation for your choice.

2.20 Place the following compounds in decreasing order of their *expected* boiling points (that is, from highest boiling point to lowest boiling point). Briefly justify your choices by identifying the types of intermolecular forces possible.

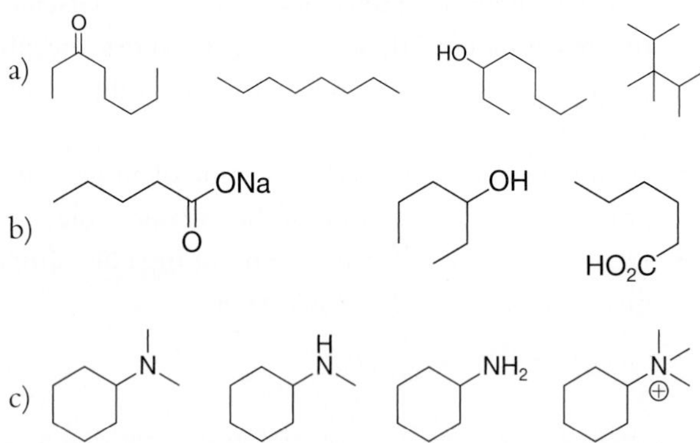

2.21 Place the following molecules in the expected order of increasing solubility in water.

OH   N H   OH

OH   N H   OH

2.22 Identify the hydrophobic and hydrophilic regions of the following molecules. Which of them are likely to dissolve in water? Which are likely to be soluble in organic solvents?

a) HO, HO, OH, OH, O, OH

b) O, $N(CH_3)_2$

c) HS, N, O, O, $O^{\ominus}$

d) H_2N, S, NH_2, O, O

2.23 Name the following molecules.

a)

b) $CH_3CH_2CH_2CH_2CH_2CH_2CH_2CH_3$

c)

d) H-C-C-C-C-H (each C bearing two H)

2.24 Name the following branched hydrocarbons.

a)

b) $CH_3CH_2CCH_3CH_2CH_2CH(CH_2CH_3)$-$CH_2CH_2CH_3$

c)

d)

2.25 Name the following branched hydrocarbons.

a)

b) $CH_3CH_2CCH_3CH_3CH_2CH(CHCH_3CH_3)$-$CH_2CH_2CH_3$

c)

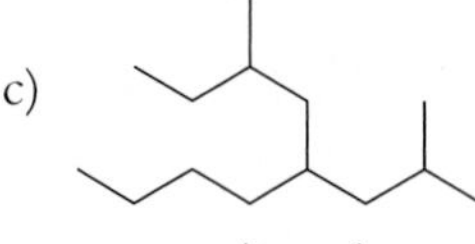

d)

2.26 Name the following unsaturated hydrocarbons.

a) 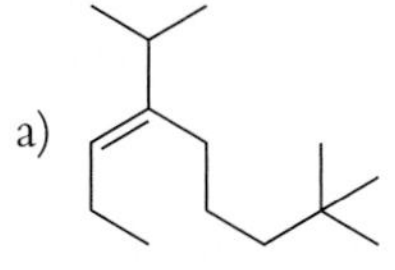

b)

c)

d)

2.27 Consider the molecule shown here and answer the following.

a) Identify the functional group of highest priority by putting a square around it on your molecule.
b) Identify the parent chain of the molecule, and put a circle around each carbon atom of the parent chain on your molecule.
c) Identify, by number, the carbon atom where the highest priority functional group resides, and use this to number the atoms of the parent chain appropriately.
d) Write the combined root name and suffix for the molecule.
e) Identify the substituents on the parent chain by putting a triangle around them on your molecule.
f) Name each of the substituents of part (e) and identify their location by carbon number.
g) Use numeric descriptors (di, tri, etc.) for multiple substituents having the same structure, and arrange them alphabetically.
h) Write the full name of the molecule.

2.28 The following molecules have names that consist only of the root name and suffix. Determine the systematic (IUPAC) names of the following compounds.

a)

b)

c)

2.29 Determine the systematic names of the following.

a)

b)

c)

d)

e)

2.30 For each of the following descriptions, draw the structure of an organic molecule that has the desired characteristics. More than one structure may be possible.
a) A cycloalkane of molecular formula $C_{13}H_{25}Cl$ that has exactly three substituents: a chlorine atom, a *sec*-butyl group, and an isopropyl group.
b) A saturated molecule of molecular formula $C_3H_8O_2$ that is a hydrogen bond acceptor but not a hydrogen bond donor.
c) A dicarboxylic acid of molecular formula $C_6H_{10}O_4$ possessing two methyl groups.

2.31 Provide an example in which a substituent is not a functional group and one in which a functional group is not a substituent.

2.32 Write the systematic name of the following molecules.

a)

b)

c) 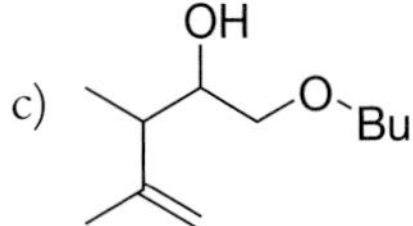

2.33 Glucose can adopt either a linear or cyclic structure. A simplified representation of its linear form is shown here. Provide the systematic name of this linear form of glucose.

2.34 Provide the systematic names of the following molecules:

a) tBu—◇—OH

b)

2.35 a) Identify the prefix, root name, and suffix in each of the following systematic names.

i) 2-methylpropan-1-al
ii) 2,3-dichlorocyclopent-1-ene
iii) heptan-3-ol
iv) 5,6-diethyl-7-hydroxyoct-1-yn-3-one

b) Identify the functional group of highest priority in each of the names in part (a).

c) For each of the names in part (a), determine the length of the parent chain and whether it contains any double or triple bonds. If so, report their location by atom number.

d) List any additional substituents in the names of part (a), and also their locations.

e) Draw the structure of the molecules of part (a).

2.36 For each of the systematic names that follow, draw the hydrocarbon.

a) heptane
b) 2,6-dimethyl-4-propylheptane
c) 4-ethylhept-2-yne
d) 3,3-diisopropylcyclohepta-1,4-diene

2.37 Draw the molecules for each of the following IUPAC names.

a) butanoic acid
b) 4-propoxybutanoic acid
c) 3-methylbut-2-enoic acid
d) 2-ethylpent-3-ynoic acid

2.38 Draw the line structure of the following molecules.

a) propyl 3-oxopentanoate
b) 1-methoxy-4,5-dimethylhex-5-en-3-ol
c) 1-(4-hydroxyphenyl)propan-1-one
d) 3,6-dimethylcyclohex-4-ene-1,2-diol
e) 4-(2-chloro-4-methylpentan-3-yl)-3-methyloctanal

MCAT STYLE PROBLEMS

2.39 Which of the following names is the systematic (IUPAC) name of the molecule shown here?

O

a) 2-oxo-propylbenzene
b) 1-cyclohexylpropan-2-one
c) 1-cyclohexylacetaldehyde
d) 2-oxo-propylcyclohexane

2.40 Place the following molecules in order of increasing solubility in water.

$H_3\overset{+}{N}$ Cl^- A; CO_2H B; CO_2CH_3 C; CO_2CH_3 D

a) D < C < B < A
b) C < B < D < A
c) C < D < A < B
d) C < D < B < A

2.41 Identify which of the following line structures does NOT represent 2,4-dichloro-5-ethyloctane.

a) Cl Cl

b)

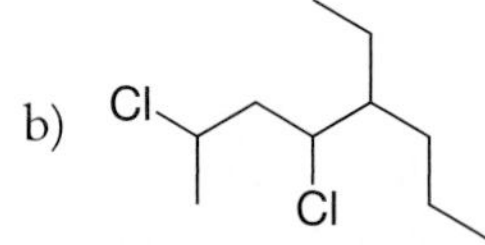

c)

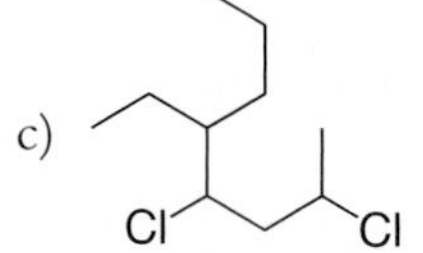

d) Cl Cl

e)

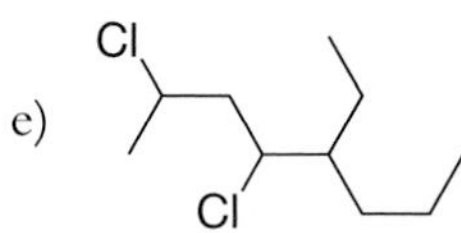

2.42 Which of the following is the electrostatic potential map of CH_2FOH?

a)

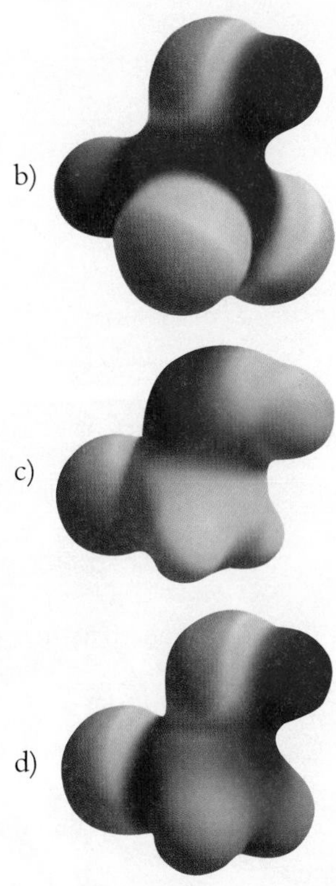

CHALLENGE PROBLEMS

2.43 The electrostatic potential map of the indicated molecule shows that one of the alkene carbon atoms is more electron deficient than the other.

a) Determine the systematic name of the molecule.
b) Use resonance structures to explain the electron distribution in the electrostatic potential map.

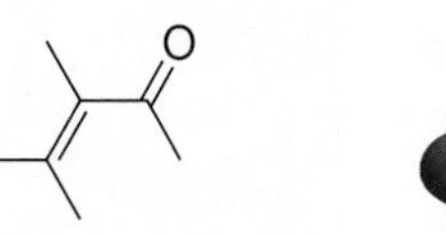

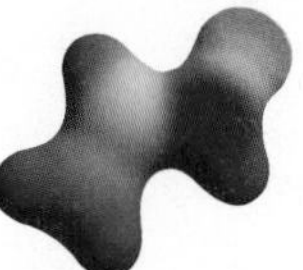

2.44 By treating each successive set of parentheses as a separate name, draw the structure of the following molecules.
a) 4-(1-(1-chloroethoxy)propan-2-yl)octane
b) 3-(5-(2-hydroxyethyl)-6-methylhept-5-en-3-yl) cyclohexan-1-one

2.45 Write the systematic (IUPAC) name of the molecule shown here.

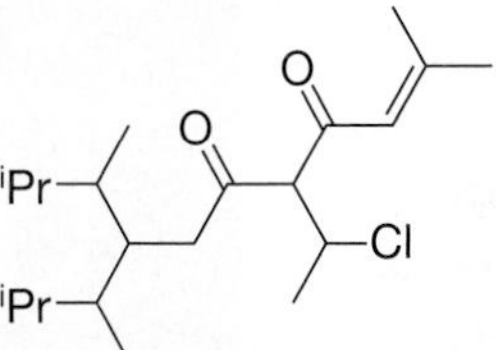

3 Molecules in Motion: Conformations by Rotations

CHAPTER OUTLINE

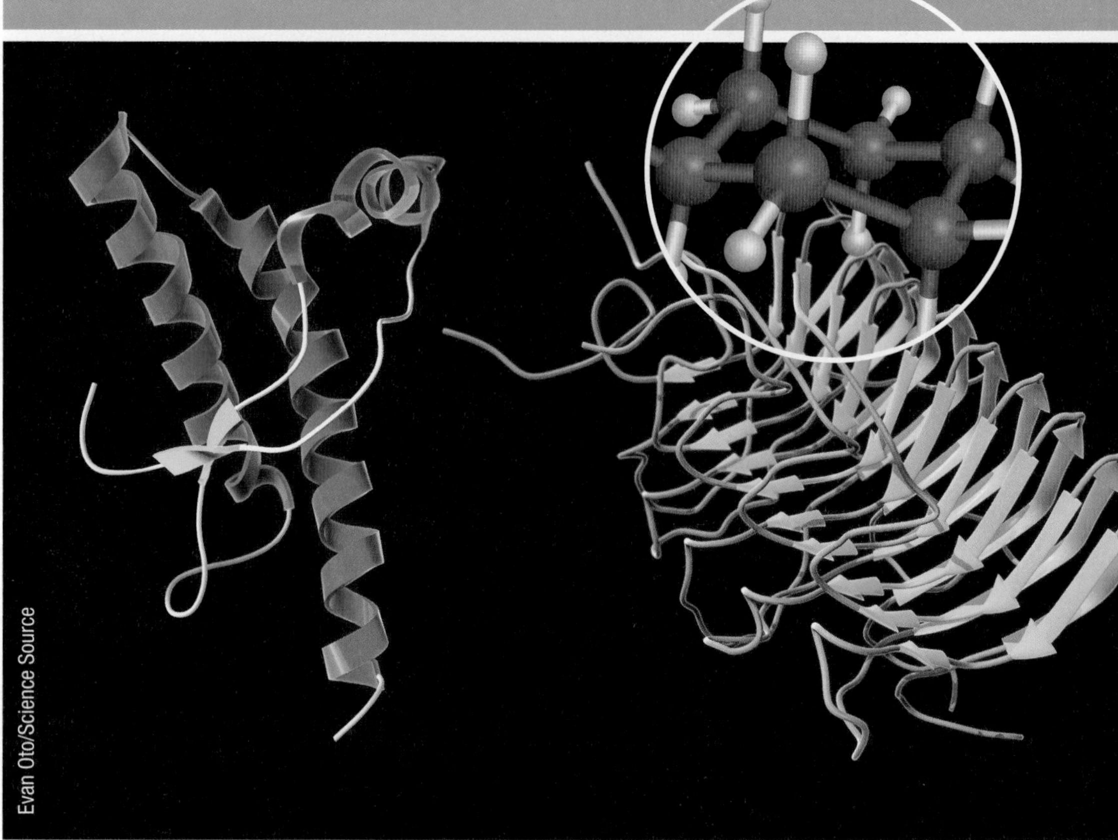

Cellular prion protein (PrP^{C}), when normally folded (left), has a biological function in the cells that protect the nerves in the brain. However, when the protein is incorrectly folded (in an incorrect conformation—right), it causes prion diseases such as Creutzfeldt–Jakob ("mad cow") disease.

3.1 Why It Matters

The enzyme lysozyme is an important component of the human immune system. Present in white blood cells, tears, saliva, mucus, and breast milk, lysozyme destroys harmful bacteria by breaking down their cell walls.

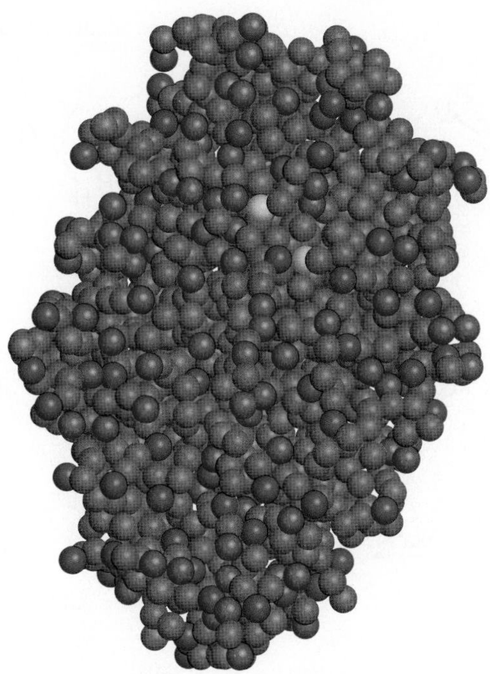

Bacterial cell walls are largely composed of long linear chains of two alternating carbohydrate rings, denoted A and B in Figure 3.1. Each chain is at least several hundred units long. Lysozyme cleaves the chains at a bond between a B ring and a linking substituent oxygen atom; one of these bonds is marked with a blue arrow in the figure.

A B A B A B

FIGURE 3.1 The alternating A-B structure of the carbohydrate chain in a bacterial cell wall. The R group of the B rings has the structure $-OCH(CH_3)COOH$.

How does lysozyme break this specific bond? As you will see in this chapter, rotations about bonds within the carbohydrate ring play a critical role in enabling lysozyme to cleave the chains in the cell walls.

3.2 Rotation about Single Bonds

At room temperature, organic molecules are constantly in motion. Not only do the molecules tumble past each other, the atoms within each molecule move relative to each other. In particular, atoms rotate about the single bonds of the molecule.

The single bond between two carbon atoms in a molecule is the result of overlap between an sp^3 hybrid orbital on each atom, as shown for a molecule of ethane.

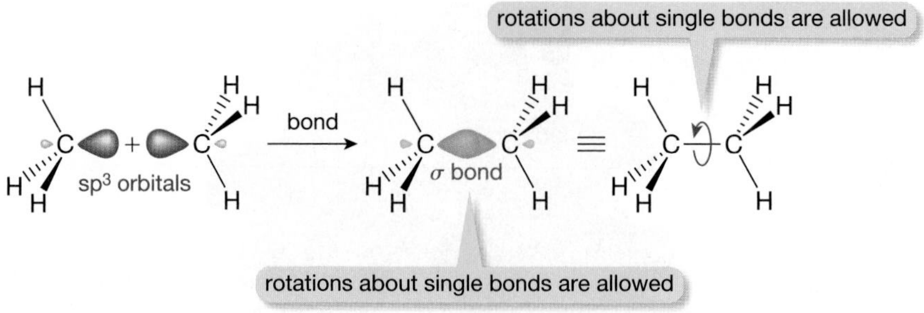

Because the sp^3 orbitals are symmetric about the internuclear axis, the rotation of one methyl group about the C–C σ bond does not change the overlap of these orbitals. So the σ bond remains intact.

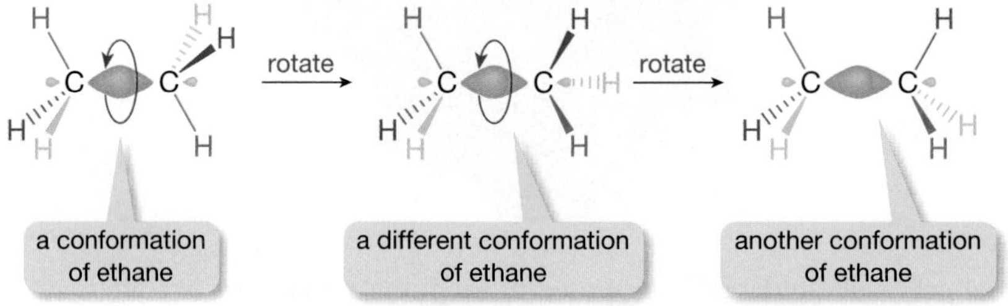

Rotation about most single bonds is a relatively low-energy process, which occurs readily at room temperature. When such rotations take place, the atoms of one part of the molecule continually change position relative to the rest of the molecule. The different spatial arrangements produced by rotation about single bonds are called **conformations**. The terms *conformers* or *rotamers* refer to the different shapes a molecule can adopt by rotations about its single bonds.

A **conformation** is a particular arrangement of atoms in a molecule resulting from rotation about the single bonds of the molecule. Molecules that exist in different conformations are called *conformers* or *rotamers*.

The conformation around a single bond can be clearly represented by a Newman projection. Such projections show the arrangement of the atoms as they would appear when viewed from one end of the axis through the single bond. The following diagram shows a Newman projection for ethane.

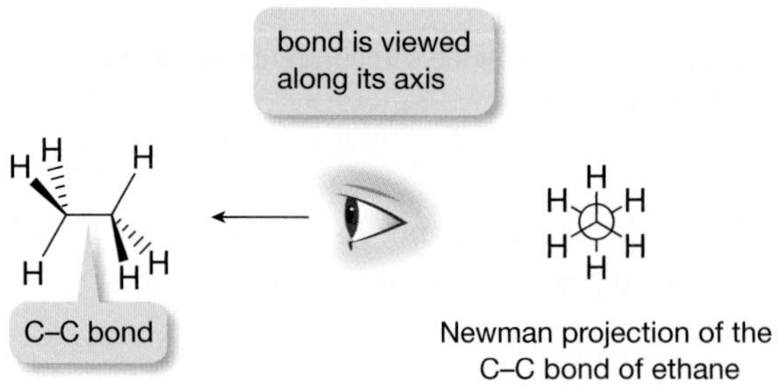

The carbon closest to the viewer is depicted as a point at which the three bonds to its substituents meet; these atoms are shown in blue in the following diagram. The carbon at the other end of the C–C bond (further from the viewer and highlighted in red) is drawn as a circle; the three substituents on this carbon atom are depicted as being connected to the *outside* of the circle.

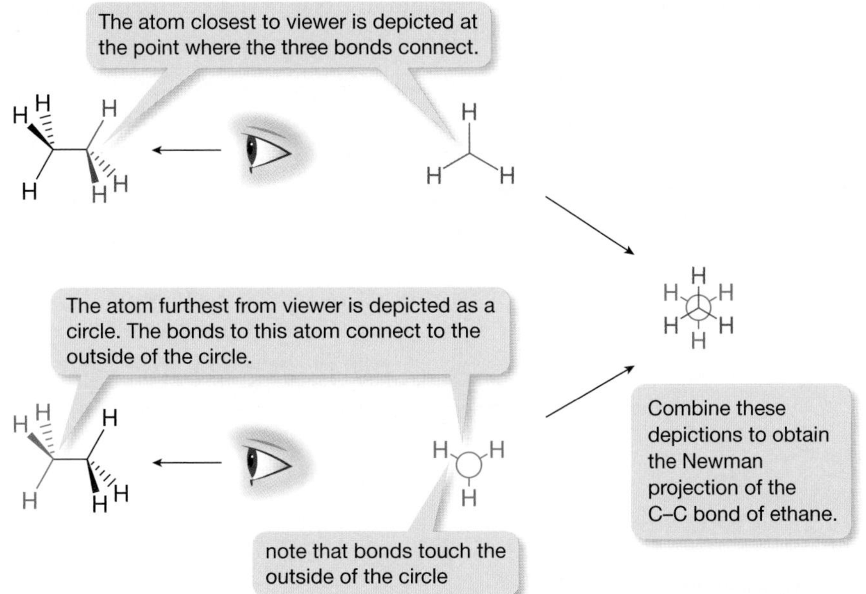

STUDENT TIP

Darling molecular models provide a special two-part sphere that attaches to atoms with sp^3 and sp^2 orbitals. You can use this sphere to represent the rear atom for a Newman projection.

ORGANIC CHEMWARE

3.1 Newman projection: Ethane

ORGANIC CHEMWARE

3.2 Newman projection: Butane

ORGANIC CHEMWARE

3.3 Newman projection: 1,2-Difluoroethane

Since each infinitesimal increment in rotation about a single bond produces a different conformation, even a simple molecule has an infinite number of possible conformations. Rotation about a particular bond is described by the **torsion** or **dihedral angle**, which is the angle between one plane defined by the single bond and a substituent on one of the atoms in the bond and a second plane defined by the single bond and a substituent on the other atom of the bond. Thus, the torsion angle is specified by four atoms: the two atoms of the bond about which rotation takes place plus one substituent on each of the bond atoms, as shown. For ethane, these four atoms are the two carbon atoms plus two hydrogen atoms, one bonded to each carbon. In the corresponding Newman projection, this angle is displayed directly.

The **torsion** or **dihedral angle** is the angle between the bonds to a substituent on each of the atoms of the single bond that constitutes the axis of rotation.

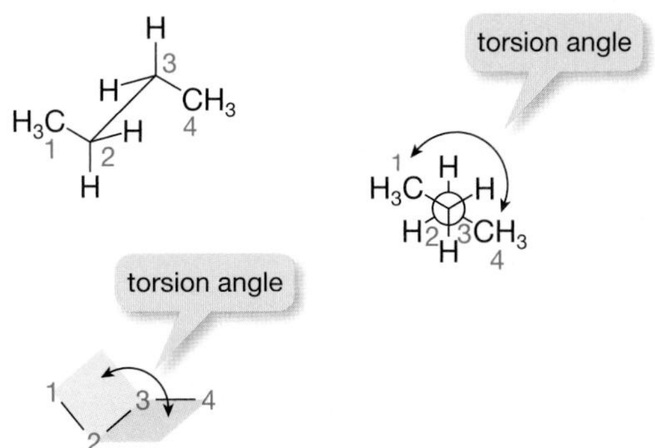

In the first conformation in the following diagram, the coloured C–H bonds that define the torsion angle of the ethane molecule are perfectly aligned, giving a torsion angle of 0°. All three C–H bonds to the front carbon atom are aligned with the three C–H bonds to the back atom, thus *eclipsing* the bonds to the back atom. In this **eclipsed conformation**, the three bonds of the front atom are close enough to the three bonds at the back that the electron pairs of these aligned bonds repel each other. This repulsion makes the eclipsed conformation of ethane the least stable—that is, highest energy—conformation. The repulsion generates **torsional strain**: a force that drives rotations away from eclipsed conformations. Rotation about the C–C bond of ethane in an eclipsed conformation moves the front and rear atoms further apart, reducing the

An **eclipsed conformation** has the substituents on the front atom aligned with those on the back atom.

Torsional strain is a strain within a molecule that arises from repulsions between electrons in bonds on adjacent atoms.

A **staggered conformation** has substituents on adjacent atoms as far away from each other as possible.

STUDENT TIP
The bonds in an eclipsed conformation are normally drawn as close to one another as possible, without overlapping, so that all the bonds and substituents can be seen.

STUDENT TIP
When depicting Newman projections of bonds involving heteroatoms, be sure to properly indicate those atoms. Keep in mind that most heteroatoms also carry lone pairs, and these should be shown on a Newman projection.

torsional strain. This torsional strain is at a minimum in a **staggered conformation**, where the three bonds of the atoms at the front are positioned precisely midway among the three bonds at the back, making a torsion angle of 60°.

eclipsed conformation: greatest torsional strain → high energy

H-C-C-H torsion angle = 0°

bonds on adjacent C atoms *closest* → *greatest repulsion* between C–H bonds (greatest torsional strain)

rotate C–C bond

staggered conformation: least torsional strain → low energy

H-C-C-H torsion angle = 60°

bonds on adjacent C atoms *furthest apart* → *least repulsion* between C–H bonds (least torsional strain)

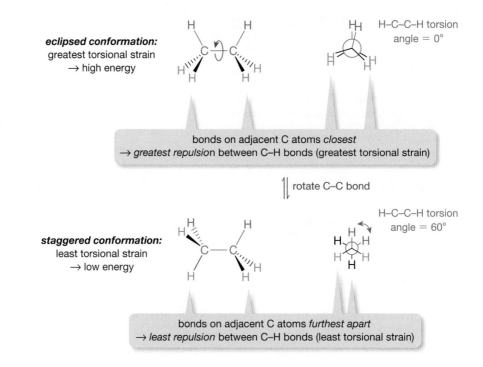

CHECKPOINT 3.1

You should now be able to convert line drawings into Newman projections for both staggered and eclipsed conformations.

SOLVED PROBLEM

For the following structure, draw a Newman projection as seen when viewed from the right and looking down the indicated bond (the bond between the atoms marked with an asterisk). Determine whether the conformation is staggered or eclipsed.

Cl OH
Br F

STEP 1: Determine which atom will be in front of the projection and which will be behind.

The question asks to view the molecule from the right, so the carbon with the OH group attached will be in front and the carbon with the bromine will be behind.

back carbon
Cl OH
Br F
front carbon

STEP 2: Draw an open circle for the carbon atom in the back and fill in its three substituents.

When viewed from the right-hand side, the ethyl group is pointing straight up, the bromine atom is pointing to the left, and the fluorine atom is pointing to the right.

Note: If you're having trouble with this step, or any of the other steps, try building a model of the original structure and physically rotating it so that you are looking down the bond needed for the Newman projection.

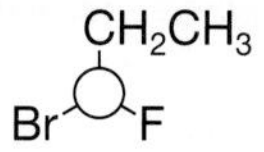

CH_2CH_3

Br F

The three bonds are drawn such that they just touch the outside of the circle representing the back atom, but they do not go through the circle.

STEP 3: Add the front carbon atom with its three substituents.

When viewed from the right-hand side, the methyl group is pointing straight down, the chlorine atom is pointing to the left, and the OH group is pointing to the right. Filling in the substituents on the front carbon gives the following:

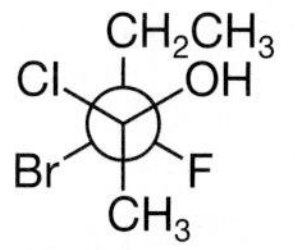

CH_2CH_3

Cl OH

Br F

CH_3

STEP 4: Determine whether the conformation is staggered or eclipsed.

The groups attached to the front and back carbons are staggered as can be seen from the 60° torsional angle between them (e.g., the Cl–C–C–Br bond has a 60° torsional angle), so the conformation is staggered. If the conformation were eclipsed, all of the bonds would be overlapping.

PRACTICE PROBLEM

3.1 Draw the corresponding Newman projections for each of the following structures, and classify each as either staggered or eclipsed. View each structure from the right-hand side, looking down the indicated bonds.

a) 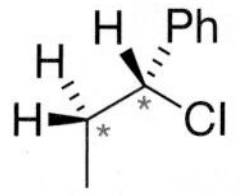 H H Ph H Cl

b) 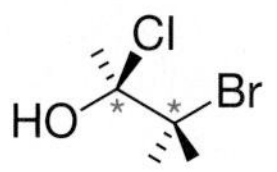 Cl HO Br

c) 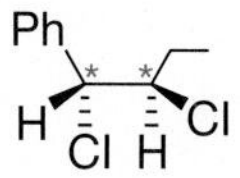 Ph H Cl Cl H

d) 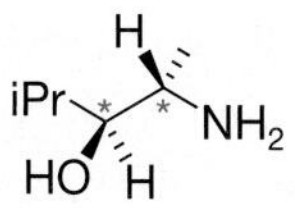 H iPr NH_2 HO H

INTEGRATE THE SKILL

3.2 a) Draw Newman projections of both the staggered and eclipsed conformations of 2,2,2-trifluoroethan-1-ol.

b) Convert the following Newman projection into a line-bond drawing without changing the conformation of the molecule.

Br
HO CH_3
H CH_3
Br

Rotation about the C–C bond to a torsion angle of 120° (Figure 3.2) raises the energy of the molecule because torsional strain increases as the molecule moves into another eclipsed conformation. The eclipsed conformations at 0° and 120° have identically eclipsed C–H bonds, and hence identical torsional strains and identical energies—2.9 kcal/mol (12 kJ/mol) greater than that of the staggered conformation at 60°.

ORGANIC CHEMWARE
3.4 Conformational analysis: Ethane

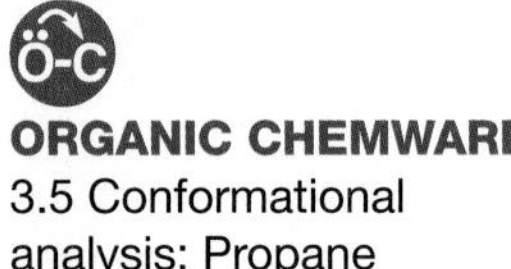

ORGANIC CHEMWARE
3.5 Conformational analysis: Propane

ORGANIC CHEMWARE
3.6 Conformational analysis: Butane

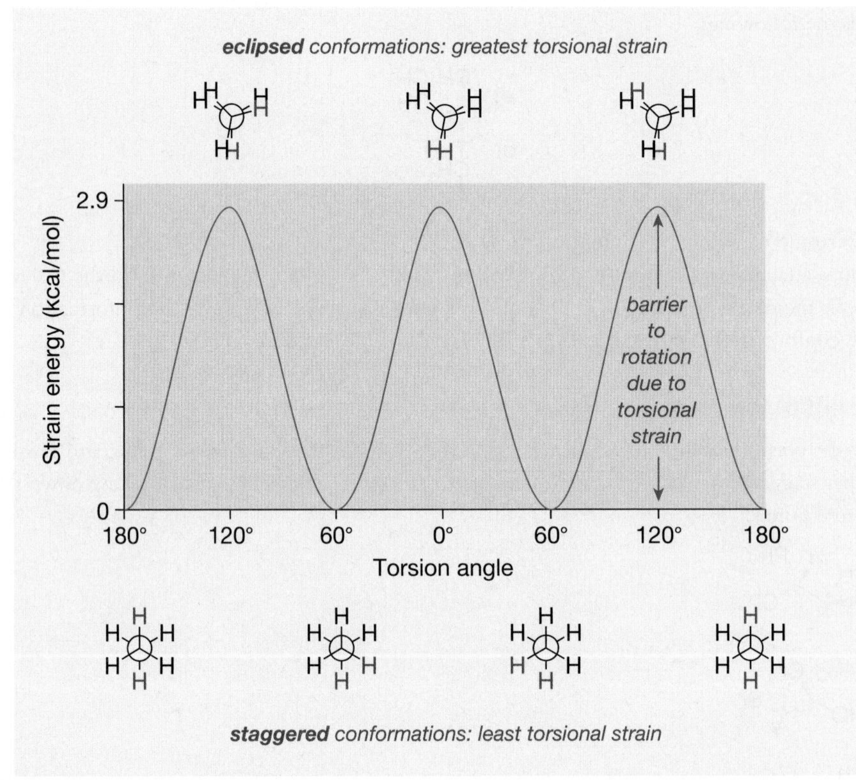

FIGURE 3.2 The energy levels of ethane molecules vary periodically with torsion angle.

Figure 3.2 shows that the energy of an ethane molecule varies periodically as the torsion angle increases from 0 to 180° (moving right from 0°). Energy minima at 60°, 180°, and −60° correspond to the three most stable staggered conformations of ethane, and maxima at 0°, 120°, and −120° correspond to the three least stable eclipsed conformations. Between these extremes are an infinite number of other conformations with intermediate energy levels. The exploration of the geometries and resulting energies of molecular conformations is called **conformational analysis**.

Conformational analysis is the study of the geometries and resulting energies of the conformations of a molecule.

Torsional strain creates energy barriers, which molecules must overcome in order to rotate from one energy minimum to another (see Figure 3.2). At room temperature, most of the molecules in a sample of ethane do not have enough energy to reach the eclipsed conformation at the top of the energy barrier. Instead, the vast majority of the molecules are in conformations close to the energy minima, with equal numbers having torsion angles of about 60°, 180°, or −60°. These three stable conformations are all conformers of ethane. At any given moment a small fraction of the ethane molecules in the sample have torsion angles that deviate substantially from these energy minima, and a few have torsion angles close to one of the three eclipsed conformations.

CHEMISTRY: EVERYTHING AND EVERYWHERE

Ethane in the Solar System

NASA/JPL-Caltech/University of Arizona/University of Idaho

Titan, a moon of Saturn, has a surface temperature of 94 K (−179 °C). This moon has lakes that consist largely of liquid ethane. At 94 K, there is little heat that ethane can use for rotation to higher energy conformations. On Titan only about one in four million ethane molecules has enough energy to cross the 2.9 kcal/mol energy barrier and rotate to a different conformation. Almost all of the ethane molecules have conformations close to the three energy minima, and constantly undergo small rotations around the C–C bonds, usually of only a few degrees.

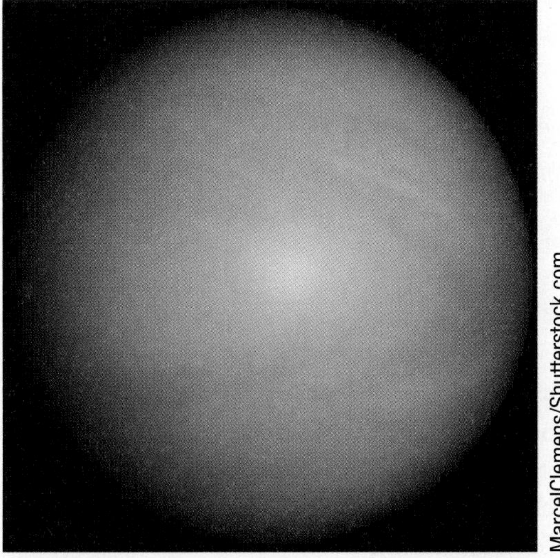

MarcelClemens/Shutterstock.com

The mean surface temperature on Venus is a scalding 735 K (462 °C): hot enough to melt lead. So Venusian ethane is a gas rather than a liquid. At any given moment about one in eight of these ethane molecules has enough energy to move from one energy minimum to the next. The constant rotations for ethane on Venus are, on average, much larger than those for ethane on Titan.

3.3 Steric Strain

Steric strain is a repulsive force within a molecule that arises from the interpenetration of the electron clouds of atoms that are close to each other but not directly bonded.

In addition to torsional strain, the vast majority of organic molecules experience some **steric strain**. This strain occurs when atoms that are not directly bonded to one another are close enough that their electron clouds interpenetrate. The resulting repulsion forces the atoms away from each other if they are free to move. Often the bonds to the atoms limit their movement, so the repulsive force strains these bonds. This steric strain raises the energy level of the conformation, making it less stable.

atoms are pushed apart by repulsions that arise when their electron clouds interpenetrate

Ethane has negligible steric strain since the electron clouds of its hydrogen atoms have virtually no overlap. However, the larger substituents in molecules such as 1,2-dichloroethane can cause significant steric strain.

The **antiperiplanar** or **anti-conformation** of a molecule $CH_2X–CH_2Y$ has a torsion angle of 180° between substituents X and Y.

The four atoms that most conveniently define the torsion angle in 1,2-dichloroethane are the two carbon atoms about which rotation takes place and the two chlorine atoms. The staggered conformation that has a Cl–C–C–Cl torsion angle of 180° is called the **antiperiplanar** or **anti-conformation**. Since the substituent bonds are staggered, the anti-conformation has no torsional strain. The steric strain in this conformation is also negligible because the separation between each chlorine atom and the other substituents is large enough to effectively eliminate repulsion between their electron clouds. Consequently, the anti-conformation is the lowest energy conformation of 1,2-dichloroethane.

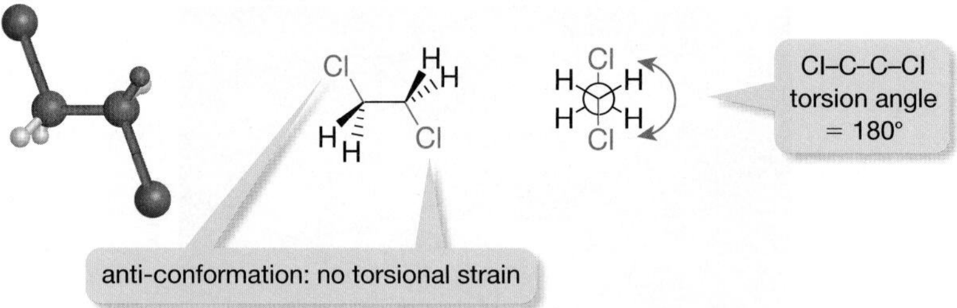

The **gauche conformations** of a molecule $CH_2X–CH_2Y$ have a torsion angle of either 60° or 300° between substituents X and Y.

The two other staggered conformations have torsion angles of 60° and 300° and are called **gauche conformations**. The chlorine atoms in these conformations are close enough to each other that repulsions between their electron clouds strain the molecule. This steric strain makes the energy of both gauche conformations about 0.6 kcal/mol (2.5 kJ/mol) greater than that of the anti-conformation (see Figure 3.3).

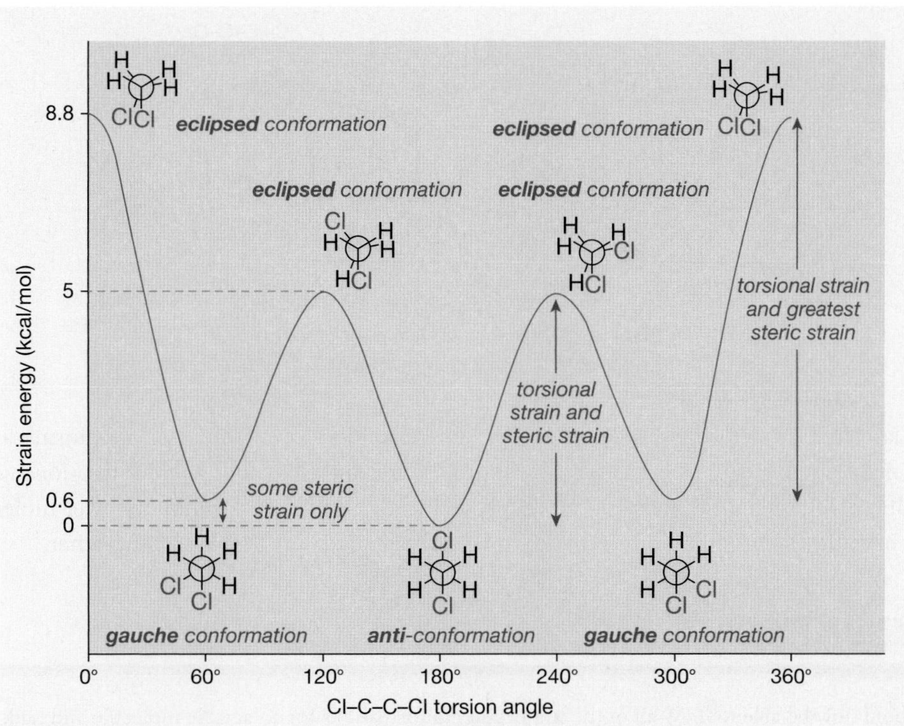

FIGURE 3.3 Strain energy of conformations of 1,2-dichloroethane.

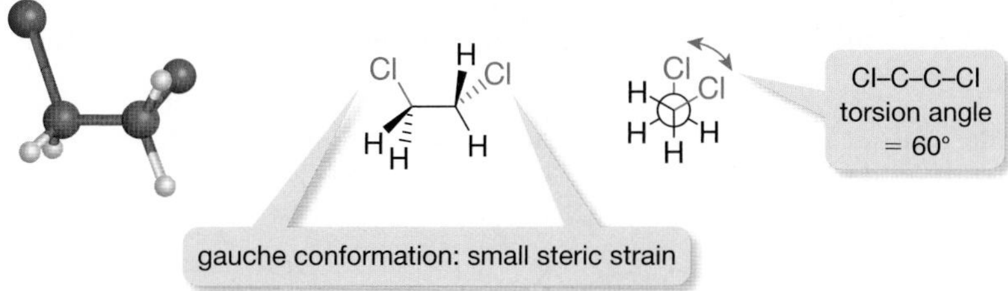

Similarly, the three eclipsed conformations of 1,2-dichloroethane do not all have the same energy. These eclipsed conformations experience similar torsional strains, but the conformation at 0° has the greatest steric strain because the large chlorine atoms are closest to each other (larger atoms and groups increase the amount of strain). As a result, the energy of this conformation, sometimes called a *totally eclipsed conformation*, is 8.8 kcal/mol (38 kJ/mol) greater than that of the anti-conformation.

Cl–C–C–Cl
torsion angle
= 0°

eclipsed conformation: greatest torsional strain and steric strain

The other two eclipsed conformations, at 120° and 240°, experience less steric strain because each chlorine atom eclipses a smaller hydrogen atom on the adjacent carbon. Both these conformations are 5 kcal/mol (22 kJ/mol) less stable than the anti-conformation. At room temperature fewer than 1 out of 1000 molecules of 1,2-dichloroethane are in either of these two eclipsed conformations.

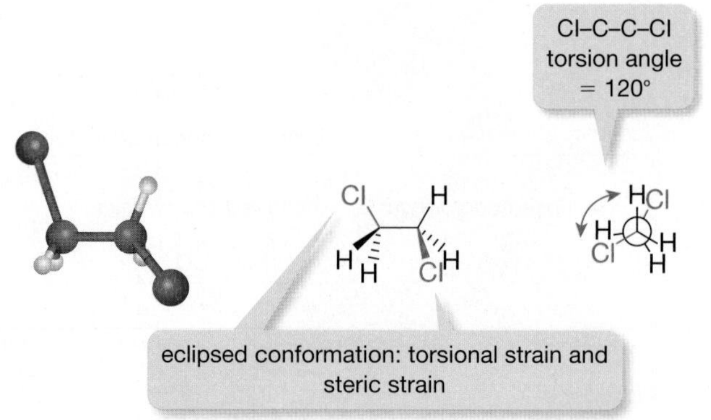

If larger substituents, such as isopropyl groups or phenyl rings, replace the chlorine atoms in 1,2-dichloroethane, the steric strain would be greater, and the eclipsed conformations would have much higher energies than those of 1,2-dichloroethane. Correspondingly, fewer molecules would have the energy required to rotate from one staggered conformation to another.

CHECKPOINT 3.2

You should now be able to draw all of the important conformations for an acyclic molecule and rank their relative energies based on the amount of torsional and/or steric strain that occurs.

SOLVED PROBLEM

Use Newman projections to draw all possible staggered and eclipsed conformations of 1-bromo-2-chloroethane, and rank them according to their relative energies. Label the antiperiplanar and gauche conformations.

STEP 1: Draw the structure in line-bond form, including all hydrogen atoms bonded to the backbone.

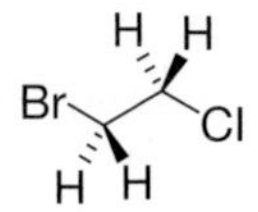

STEP 2: Draw a Newman projection of your line drawing following the steps in Checkpoint 3.1 as a guide.

Note that there are many possible correct answers, depending on how you drew your line drawing in Step 1 and the viewing angle you choose for your Newman projection. We chose to view the molecule from the right-hand side, giving the Newman projection shown here.

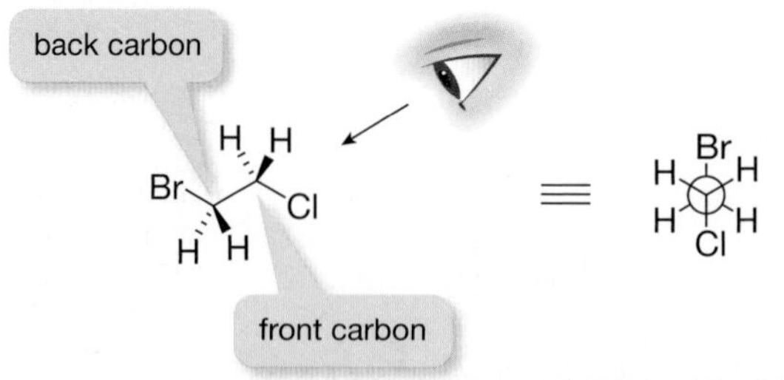

STEP 3: Keeping the back carbon and its substituents in the same position, rotate the front carbon by 60° in the clockwise direction.

Rotating the front carbon 60° puts the front bonds directly in line with the bonds attached to the back carbon. For clarity, the Newman projection of this eclipsed conformation has been offset slightly so that the atom labels in the front and back can both be seen.

atoms offset for clarity; meant to represent eclipsed atoms

rotate 60°

STEP 4: Repeat the previous step four more times until you have a total of six conformations.

By continuing to rotate the front carbon, there will be alternating staggered and eclipsed conformations until, eventually, no more new conformations can be generated. The six possible conformations for this molecule are as follows:

(original)

STEP 5: Identify the antiperiplanar conformation.

The antiperiplanar conformation places the largest atoms/groups on the front and back carbons 180° from each other. In the following molecule, shown in the antiperiplanar conformation, the bromine and chlorine atoms are kept as far apart as possible with a Br–C–C–Cl torsional angle of 180°.

STEP 6: Identify the gauche conformation(s).

There are two gauche conformations with Br–C–C–Cl torsional angles of 60° and 300°. In these conformations, the bonds are staggered; however, the Br and Cl atoms are close enough to each other to cause steric strain.

PRACTICE PROBLEM

3.3 a) Draw Newman projections for the staggered and eclipsed conformations of butane, looking down the C-2–C-3 bond. Rank them according to their relative energies.

b) For each of the following compounds, draw Newman projections of the most stable conformation(s) and the least stable conformation(s).

i) 2-methylpentane (along the C-2–C-3 bond with C-2 in front)

ii) bromopentane (along the C-1–C-2 bond with C-1 in front)

iii) N,N,O-trimethylhydroxylamine (along the O–N bond with O in front). Be sure to include the lone pairs.

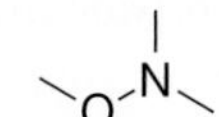

N,N,O-trimethylhydroxylamine

INTEGRATE THE SKILL

3.4 a) Use the energy diagram in Figure 3.3 to answer the following questions about 1,2-dichloroethane.

i) What is the height of the energy barrier to convert a gauche conformation into its anti-conformation?

ii) What is the height of the energy barrier to convert an anti-conformation into its gauche conformation?

iii) What is the height of the energy barrier to convert a gauche conformation with a torsional angle of 60° into a gauche conformation with a 300° torsional angle?

b) Plot a graph of strain energy as a function of the torsion angle defined by the four atoms of 1,2-dichloroethane shown here in blue. How does this graph differ from the one in Figure 3.3 where the Cl–C–C–Cl torsion angle is used?

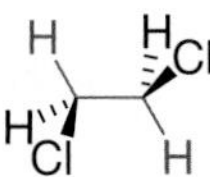

3.4 Strains in Cyclic Molecules

STUDENT TIP
Use a molecular model to help you visualize the conformations of ring structures.

Many cyclic or ring structures consisting of atoms connected by single bonds are found in nature. Such structures are often drawn as simple polygons. Although these polygon line drawings make cyclic molecules appear planar, the atoms of most single-bonded ring structures do not all lie in same plane.

square — cyclobutane, oxetane

hexagon — cyclohexane, piperidine

3.4.1 Conformations of three-membered rings

Three-membered rings, such as cyclopropane, are always planar because the three-ring atoms define a plane. Three-membered rings are highly strained and, therefore, high in energy. Consequently, such rings are much more reactive than corresponding open-chain structures. For example, more than 36 kcal (150 kJ) is required to convert a mole of propane into cyclopropane.

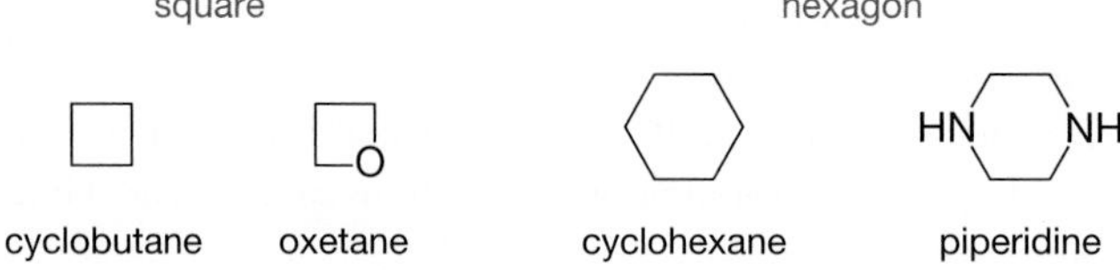

Angle strain is a strain that arises from bond angles that do not permit maximum orbital overlap between the atoms of a molecule.

Most of the energy needed to convert propane into cyclopropane comes from the potential energy that exists in strained bonds in the cyclic molecule. **Angle strain** arises when the atoms of a ring have bond angles that deviate from the normal bond angles. Because the carbon atoms of propane are sp^3 hybridized, they achieve maximum atomic orbital overlap with their neighbours at tetrahedral bond angles of about 109.5°. Since the triangular geometry of cyclopropane requires C–C–C bond angles of 60°, maximum overlap of sp^3 orbitals cannot occur. Instead, each carbon atom has "bent" bonds with reduced orbital overlap to the two adjacent carbon atoms. Energy is required to replace the tetrahedral C–C bonds in propane with the weaker bonds in the unfavourable triangular configuration of cyclopropane.

In addition to angle strain, three-membered rings also possess significant torsional strain. All of the C–H bonds of cyclopropane are eclipsed. As the following diagram shows, each C–C bond has a pair of eclipsed C–H bonds above the plane of the ring and another pair below the plane.

Despite the strain present in three-membered rings, they do occur in nature. The following compound, found in the bacterium *Streptomyces sp.*, contains six cyclopropane rings!

3.4.2 Conformations of four-membered rings

Rings with four or more single-bonded atoms have non-planar conformations. Planar cyclobutane—a square of carbon atoms—has 90° C–C–C bond angles and all the C–H bonds eclipsed. A limited amount of rotation is possible in a four-membered ring, so the angle and torsional strain pushes cyclobutane into a non-planar butterfly conformation.

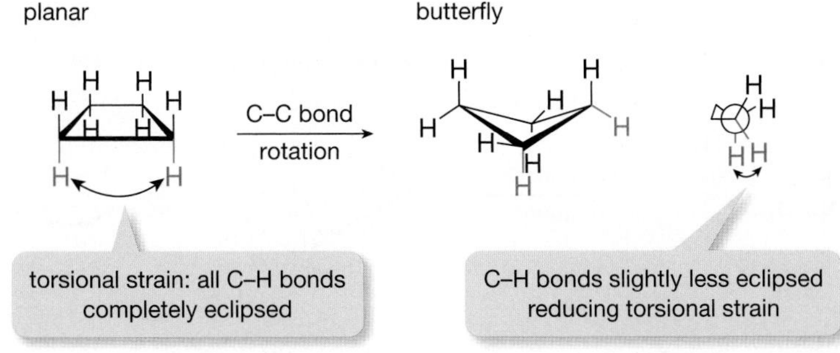

The butterfly conformation of cyclobutane has C–C–C bond angles of 88°. Therefore, this conformation has considerable angle strain, though much less than cyclopropane. The butterfly conformation is an energy minimum because the decrease in torsion strain is greater than the slight increase in angle strain.

Like acyclic compounds, cyclic molecules constantly rotate about their single bonds and often convert from one conformation to another. Cyclobutane has two butterfly conformations of equal energy, one with its wings flipped up and the other with its wings down. In converting from one conformation into the other—a process known as **ring inversion**—the cyclobutane molecule passes through the planar conformation of cyclobutane.

Ring inversion is the conversion of a cyclic molecule from one conformation to another by rotation about the single bonds of the ring.

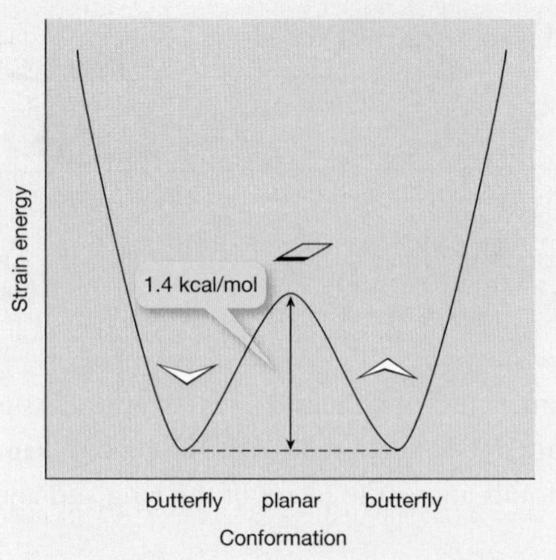

Like the energy barrier to rotation in ethane, the barrier to ring inversion is low: about 1.4 kcal/mol (6 kJ/mol). At room temperature, molecules of cyclobutane rapidly flutter their wings, readily changing from one conformation to the other. Since the torsional and angle strains are identical in both butterfly conformations, these conformations have the same energy and are therefore present in precisely equal amounts.

rapid ring inversion

3.4.3 Conformations of five-membered rings

For five-membered rings, the planar conformation is a regular pentagon. Little angle strain exists because the internal angle of a regular pentagon is 108°, very close to the 109.5° angle required for maximum orbital overlap. However, the planar structure has significant torsional strain because all the C–H bonds are eclipsed.

planar

envelope

C–C bond rotation

torsional strain from 10 totally eclipsed C–H bonds

108° ≈ 109° little angle strain

reduced torsional strain

The atoms in five-membered rings can rotate further about the single bonds than the atoms in four-membered rings, so five-membered rings have conformations that reduce the torsional strain by a greater amount. For cyclopentane, the most stable conformation is the **envelope conformation**, which has one carbon atom of the ring out of the plane of the other four ring atoms, giving a structure that resembles an envelope with its flap open. Though some torsional strain still exists in the planar part of the envelope conformation, the almost staggered arrangement of C–H bonds around the carbon atom of the flap reduces torsional strain, as shown in the previous Newman projection. The energy of the envelope conformation of cyclopentane is about 4.8 kcal/mol (20 kJ/mol) lower than that of the planar form. Cyclopentane has 10 envelope conformations, all with the same energy. The molecules flutter between these conformations, with each carbon atom taking a turn as a flap that points either up or down. So there is no "normal" conformation of cyclopentane.

The **envelope conformation** of a five-membered ring has one ring atom that is displaced from the plane of the other four ring atoms.

CHECKPOINT 3.3

You should now be able to use angle and torsional strain to explain the preferred conformations of cyclic molecules and compare their energies to those of less favourable conformations.

SOLVED PROBLEM

Which molecule do you expect to have greater strain in its C–C bonds: cyclopropane or hexabromocyclopropane? Explain your answer.

STEP 1: Draw the two structures, including any hydrogen atoms attached to the carbon backbone.

STEP 2: Compare the angle strain, if any, present in each molecule.

The carbon atoms in both structures are sp^3 hybridized and therefore prefer to adopt angles of 109.5°. In both structures, the internal C–C–C angles have been reduced to ~60°, resulting in significant angle strain. Since the difference between the optimal bond angles and the actual bond angles is the same in both structures, the amount of angle strain is also similar.

STEP 3: Compare the torsional strain, if any, present in each molecule.

In both molecules, the atoms attached to the ring are eclipsed, resulting in significant torsional strain. However, in cyclopropane, the eclipsed atoms are small hydrogen atoms. In contrast, hexabromocyclopropane has eclipsing bromine atoms that are much larger and will therefore result in more significant torsional strain. Overall, the second molecule, hexabromocyclopropane, is expected to have greater strain in its C–C bonds.

PRACTICE PROBLEM

3.5 For each pair of structures, predict any differences expected between the amounts of angle and/or torsional strain.

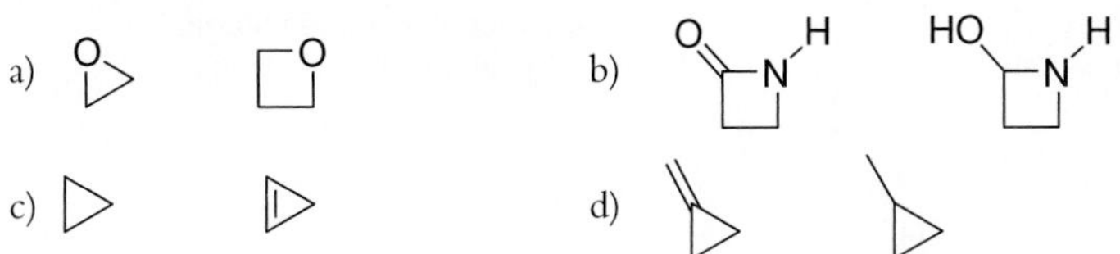

INTEGRATE THE SKILL

3.6 Based on what you have learned so far in this chapter, answer the following.

a) How many eclipsed interactions would you expect if cyclohexane existed in a planar conformation?

b) Would you expect any significant angle strain to exist in planar cyclohexane?

3.5 Conformations of Six-Membered Rings

The **chair conformation** of a six-membered ring has a zig-zag configuration with the ring atoms alternately above and beneath the equatorial plane.

The amount of rotation possible about each bond of a six-membered, single-bonded ring is greater than that for smaller rings. Consequently, six-membered rings can deviate farther from planarity to form a conformation that is virtually free of torsional and angle strain. In this **chair conformation**, the ring atoms are displaced alternately above and below the equatorial plane of the molecule. Note that the equatorial plane of a ring compound is the same as the plane of the planar conformation.

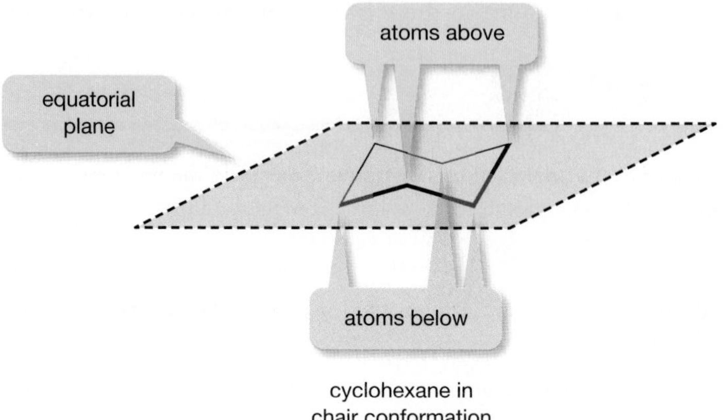

cyclohexane in chair conformation

Axial substituents of a six-membered ring are displaced in directions perpendicular to the equatorial plane of the ring.

Equatorial substituents are displaced in directions nearly parallel to the equatorial plane.

Six of the twelve hydrogen atoms of cyclohexane are displaced axially; that is, the bonds to them are oriented perpendicular to the equatorial plane and hence parallel to an axis through the centre of the ring. These **axial** hydrogens are displaced alternately above and below the equatorial plane (they are coloured yellow in the following diagram). The remaining six hydrogens (coloured red) are termed **equatorial** because their positions lie close to the equatorial plane of the ring. Each carbon atom in the ring is bonded to one axial hydrogen and one equatorial hydrogen.

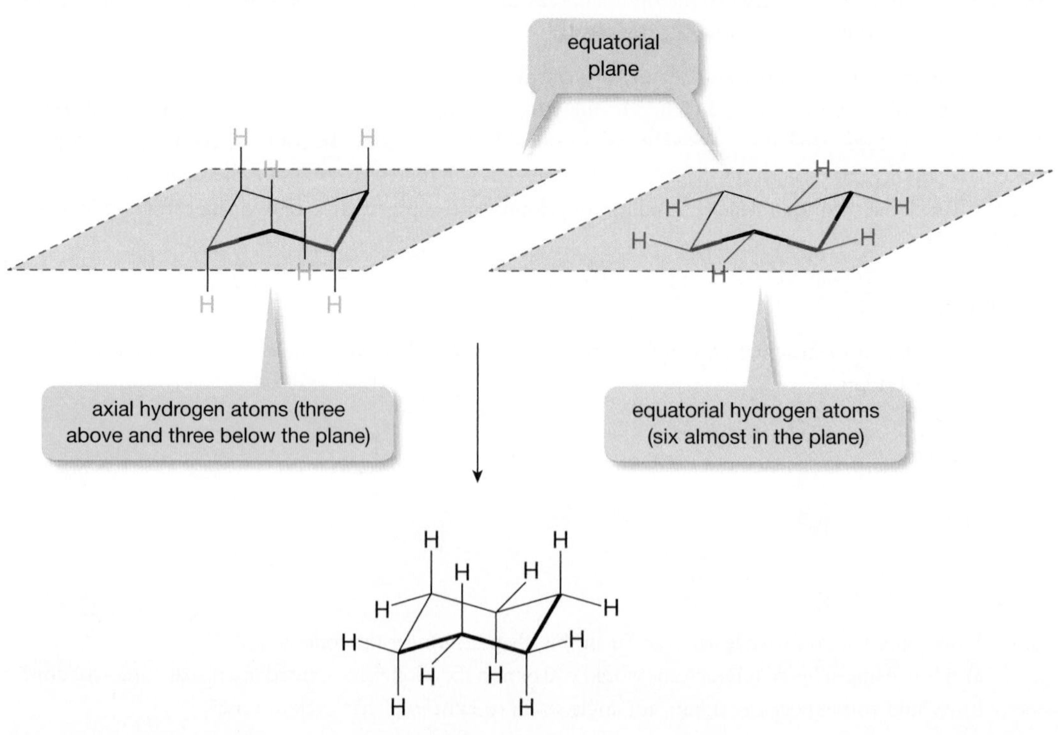

Thus, the chair conformation of cyclohexane is a puckered ring that has no torsional strain. As shown by the following Newman projection, this conformation has a staggered orientation of hydrogens about all of its carbon-carbon bonds. Additionally, there is virtually no angle strain because the puckering allows bond angles that are very close to 109° between the carbon atoms in the ring. The resulting stability of the chair conformation of six-membered rings makes it a common and important structure in nature. The stability of this shape can impart a critical rigidity to organic molecules, such as the cell wall compound in Figure 3.1.

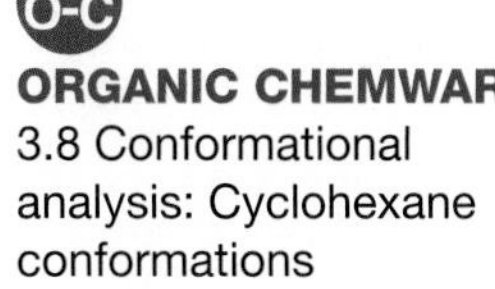

ORGANIC CHEMWARE
3.7 Conformational analysis: Cyclohexane—chair

ORGANIC CHEMWARE
3.8 Conformational analysis: Cyclohexane conformations

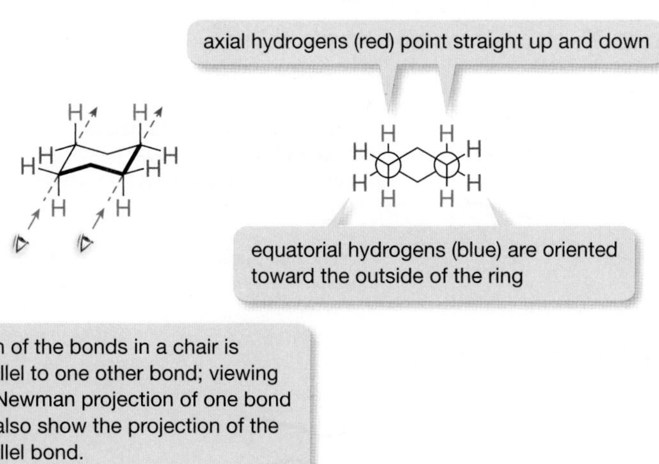

axial hydrogens (red) point straight up and down
equatorial hydrogens (blue) are oriented toward the outside of the ring
Each of the bonds in a chair is parallel to one other bond; viewing the Newman projection of one bond will also show the projection of the parallel bond.

Note that the axial hydrogens (red) point straight up and down, whereas the equatorial hydrogens (blue) are oriented toward the sides of the ring.

3.5.1 Drawing chair conformations

It is important to be able to draw chair conformations properly. This section provides an example of how to draw a chair conformation—specifically, an approach for drawing the chair conformations of cyclohexane. To better illustrate the process, the bonds drawn parallel to each other are presented in the same colour. All lines representing bonds in the ring should be the same length since the actual bonds are all the same size.

Part 1: Ring framework

Step 1: Draw a shallow "V".

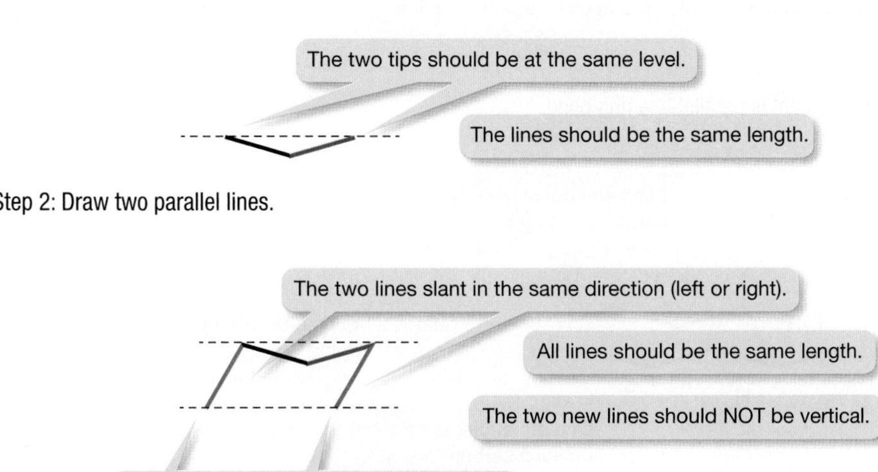

The two tips should be at the same level.

Step 2: Draw two parallel lines.

The two tips should be at the same level.

Note: There are two chair conformations. In one conformation, the two red lines slant from right to left. In the other chair, the two red lines slant left to right.

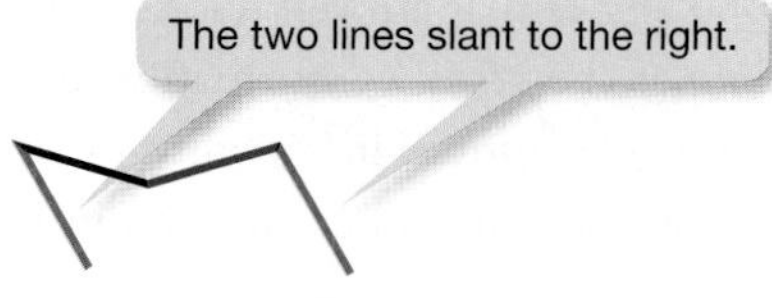

Step 3: Draw a second "V"; this one is inverted relative to the first.

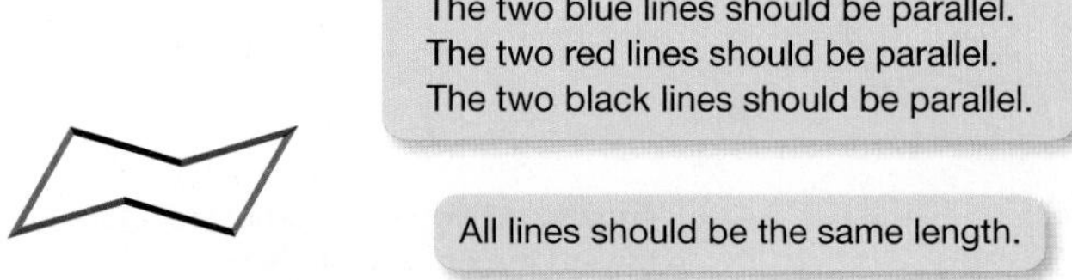

The two blue lines should be parallel.
The two red lines should be parallel.
The two black lines should be parallel.
All lines should be the same length.

Note: The second chair is completed in the same way.

Part 2: Drawing substituents

Axial substituents should be vertical.

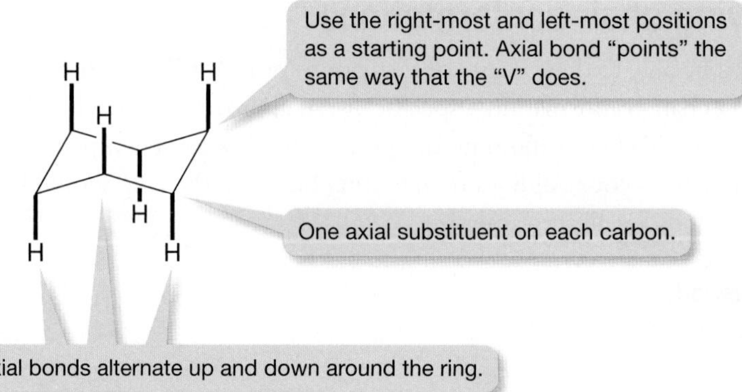

Each *equatorial* substituent is parallel to a ring bond.

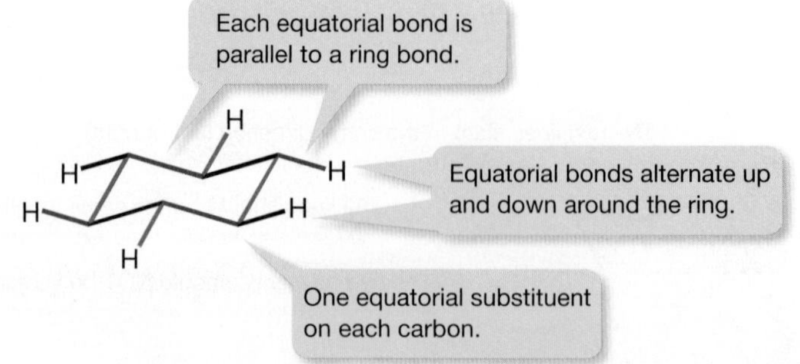

Each carbon has one axial and one equatorial position.

Two bonds should have gaps to show that they lie behind other bonds.

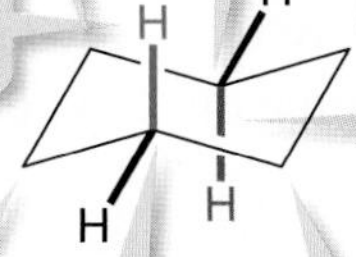
The top three carbons are assumed to lie away from the viewer.
This axial bond lies in front of the ring bond.
This axial bond lies behind the ring bond.
The bottom three carbons are assumed to lie toward the viewer.

Avoiding common errors

1. There should be no horizontal or vertical lines in the ring structure.

incorrect

correct

Ensure that you draw the first "V" carefully.

tips touch the same horizontal level

2. The axial bonds should point in the same direction as the "V" to which they are connected.

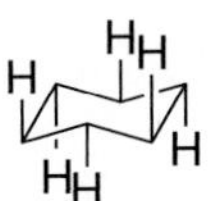
incorrect

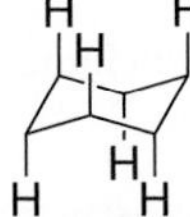
correct

Use each "V" as a guide. Start at one clear location and alternate around the ring.

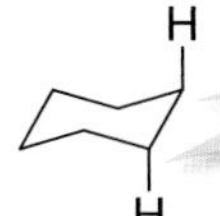
axial bond points in same direction as the "V" connected to it

3. The equatorial bonds should be parallel to the ring bonds.

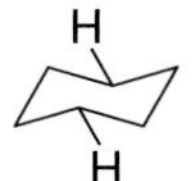
incorrect

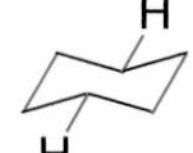
correct

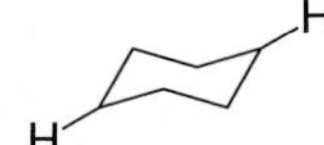
incorrect

correct

Make sure each equatorial bond is parallel to a ring bond.

4. The bottom three carbons are assumed to lie toward the viewer.

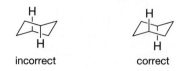

Practising drawing chairs is the best way to learn the proper perspective.

When cyclohexane is drawn as a hexagonal line structure, the molecule is viewed from above. However, when cyclohexane is drawn in a chair conformation, the molecule is viewed from the side, so some of the ring carbon atoms will be closer to the viewer than others. Changing from a top view to a side view brings the two carbon atoms at the bottom of the hexagon closer to the viewer and the two carbon atoms at the top of the hexagon further from the viewer. This positioning is occasionally depicted by representing the C–C bonds with wedges projected toward the viewer.

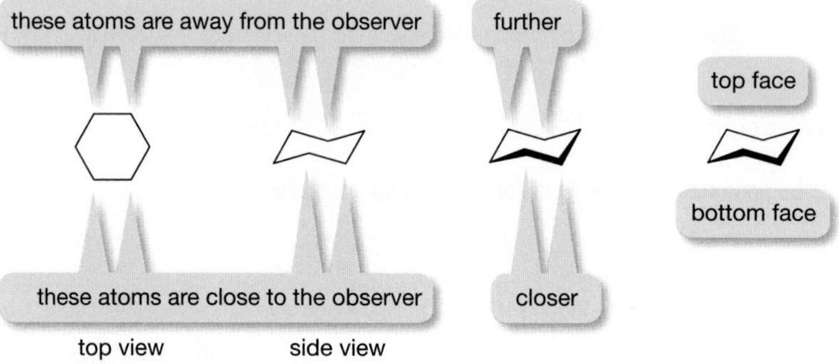

CHECKPOINT 3.4

You should now be able to draw chair conformations of cyclohexane-based structures.

SOLVED PROBLEM

Draw a chair conformation of 1,3-dichlorocyclohexane where one chlorine is axial and the other is equatorial.

STEP 1: Draw the backbone of a chair structure following the guidelines in Section 3.5.1.

When the ring is drawn as a series of parallel lines as shown in Section 3.5.1, the structure shown in parentheses is formed. Removing the coloured and bolded lines gives the desired chair template.

STEP 2: Add an axial chlorine atom to one of the carbon atoms and designate it as carbon 1.

We chose to make carbon 1 the right-most carbon atom; however, this choice is arbitrary. Since this carbon atom is pointing up in the ring structure, its axial bond must be drawn pointing straight up.

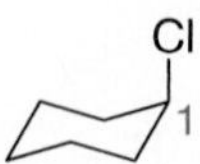

STEP 3: Add an equatorial chlorine atom at carbon 3.

Counting three carbons over from carbon 1, we label carbon 3 (note that you could also have gone the other direction and placed carbon 3 in the back). An equatorial bond at this position needs to be drawn parallel to the bonds highlighted in blue.

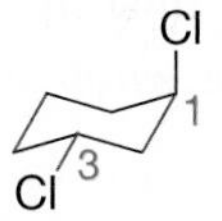

The final structure is as follows:

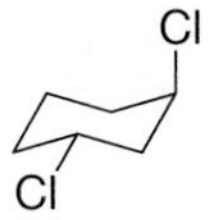

PRACTICE PROBLEM

3.7 a) Draw chair structures for each of the following:

i) 1,1,3,3-tetramethylcyclohexane

ii) 4-bromo-1,1-dimethylcyclohexane (place Br in an equatorial position)

iii)

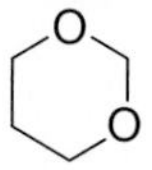

b) Identify the mistake in each of the following chair structures. Draw the correct structure.

i)

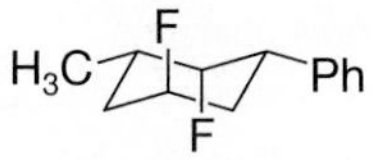

ii)

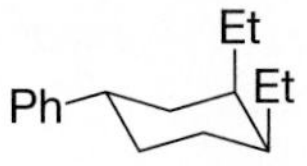

iii)

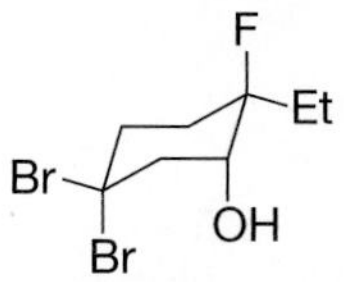

INTEGRATE THE SKILL

3.8 a) Draw a Newman projection of the following chair structure, looking down the indicated bonds. Be sure to include the hydrogen atoms that have not been explicitly shown.

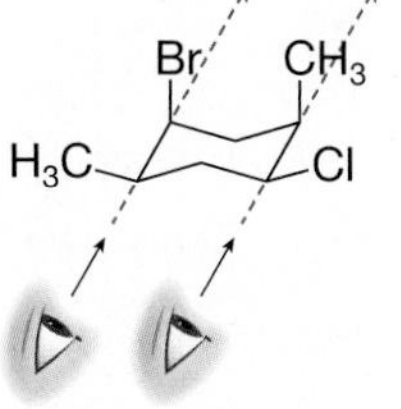

b) Identify two differences between an envelope conformation of cyclopentane and a chair conformation of cyclohexane.

3.6 Six-Membered Rings Flip Their Chairs

The six carbon-carbon bonds of cyclohexane are constantly rotating, causing the structure to wiggle. The molecules spend most of their time in a chair conformation or in conformations close to it. Occasionally, the rotations are large enough to flip the chair into other conformations, including a different chair.

side bonds slant to the right

chair flip

side bonds slant to the left

A ring inversion from one chair to another also flips each axial substituent to an equatorial position and each equatorial substituent to an axial position.

chair flip

axial hydrogens rotate to equatorial positions

chair flip

equatorial hydrogens rotate to axial positions

The process of converting from one chair conformation to another involves three other conformations that have different energy maxima and minima: the half-chair, the twist boat, and the boat. The **half-chair** conformation sits at an energy maximum, approximately 10.3 kcal/mol (43 kJ/mol) greater than the chair conformation. The twist-boat conformation is an energy minimum attained as the bonds in the half-chair continue to rotate. Further rotations flip the twist boat into a boat conformation, then into a different twist boat, followed by a different half-chair, and then finally the other chair.

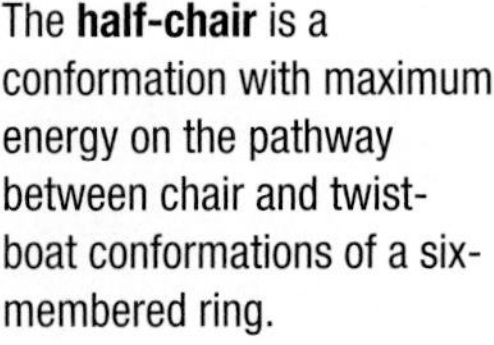

The **half-chair** is a conformation with maximum energy on the pathway between chair and twist-boat conformations of a six-membered ring.

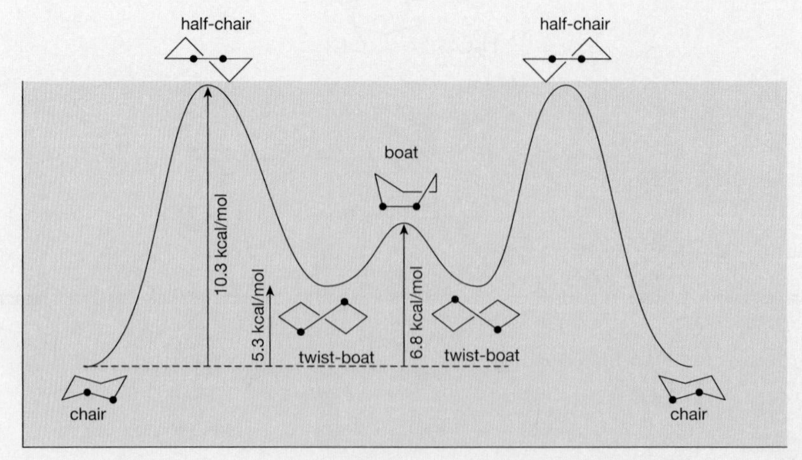

The rate at which ring inversion occurs depends on how frequently stable chair conformations can overcome the barrier to inversion by attaining the high-energy conformation of the half-chairs. The 10.3 kcal/mol (43 kJ/mol) energy difference between the chair and half-chair conformations of cyclohexane is a relatively low barrier, so ring inversion of this compound occurs frequently at room temperature; in fact, each molecule flips between conformations thousands of times a second! While attaining the half-chair conformation is necessary for each cyclohexane molecule to move from one of its chair conformations to the other, the molecules are in one of the chair conformations 99 percent of the time.

3.7 Six-Membered Rings with Substituents

3.7.1 Monosubstituted six-membered rings

A six-membered ring bearing one substituent, such as methylcyclohexane, exists as a mixture of two rapidly interconverting chair conformations. The two chair conformations of cyclohexane have the same geometry and therefore the same energy, but those of methylcyclohexane do not. In one chair conformation of methylcyclohexane, the methyl group is positioned equatorially; in the other chair conformation the methyl group has an axial position. The conformation with an equatorial methyl group has almost no steric strain as the methyl group is well separated from the hydrogen atoms of the ring. However, the conformation with an axial methyl group has significant steric strain, as the methyl group is close to the two axial hydrogens on the same face of the ring. Each one of these two **1,3-diaxial interactions** raises the energy of the conformer by 0.9 kcal/mol (3.8 kJ/mol). So the axially substituted chair conformation with an axially displaced CH_3 group is 1.8 kcal/mol (7.6 kJ/mol) less stable than the equatorially substituted conformation. Note that 1,3-diaxial interactions between two hydrogens are negligible because the electron clouds of the hydrogens are so small.

1,3-diaxial interactions are steric strains that arise from interpenetration of the electron clouds of ring substituents separated by two carbon atoms.

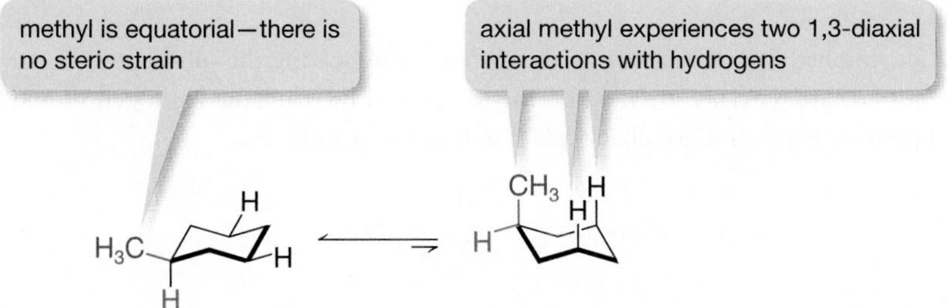
methyl is equatorial—there is no steric strain
axial methyl experiences two 1,3-diaxial interactions with hydrogens
H
H₃C
H
H
CH₃
H
H
H

STUDENT TIP

The numbering of 1,3-diaxial interactions arises from numbering the ring atom bearing the substituent as atom 1. The adjacent carbon atom is atom 2, and the next carbon atom in the ring is atom 3. Thus, the diaxial interactions are between the substituent on atom 1 and the two hydrogen atoms (or other substituents) on atom 3.

At room temperature, more than 95 percent of the molecules of methylcyclohexane adopt the conformation with an equatorial methyl group. This ratio is indicated qualitatively by the equilibrium arrows of unequal length.

The steric strain from 1,3-diaxial interactions depends on the size and nature of the ring substituents. Greater steric strain increases the ratio of the equatorial conformation to the axial conformation. For example, the steric strain arising from an axial *tert*-butyl group, $C(CH_3)_3$, is large enough that less than 0.03 percent of *t*-butylcyclohexane molecules exist in the axial chair conformation at room temperature. Measurements of the proportion of equatorial conformations can provide an estimate of the relative size of the substituents (Table 3.1).

TABLE 3.1 Relative Populations of Equatorial and Axial Conformations in Some Monosubstituted Cyclohexanes at 298 K

The A-value is the difference in free energy between the higher energy conformation (axial) and the lower energy conformation (equatorial).

Substituent	% Equatorial	% Axial	A-Value (kcal/mol)
–Cl	67.4	32.6	0.6
$–OCH_3$	73.4	26.6	0.55
$–CH_3$	94.6	5.4	1.8
$–C_6H_5$	99.4	0.6	3.0
$–C(CH_3)_3$	99.96	0.04	3.0

ORGANIC CHEMWARE
3.9 Conformational analysis: Monosubstituted cyclohexanes

CHECKPOINT 3.5

You should now be able to draw both possible chair conformations for cyclohexane-based structures with one or no substituents. When substituents are present, you should also be able predict the lowest-energy chair conformation based on the steric interactions present.

SOLVED PROBLEM

Draw the chair conformation that results from the inversion of the following molecule, and determine which chair is the most stable.

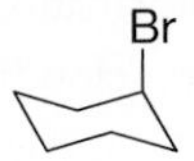

Br

STEP 1: Draw the implied hydrogen atom on the carbon atom bearing the substituent.

With the bromine atom axially oriented on the top face of the ring, the hydrogen should be drawn in the equatorial position beneath it, parallel to the bonds in red as indicated.

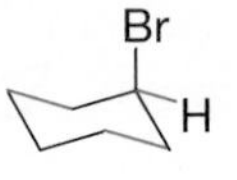

Br
H

STEP 2: Draw the backbone of the chair formed after ring inversion, following the steps described in Section 3.5.1.

Drawing the ring as a series of parallel lines as shown in Section 3.5.1 gives the structure in parentheses. Removing the bold coloured lines gives the backbone.

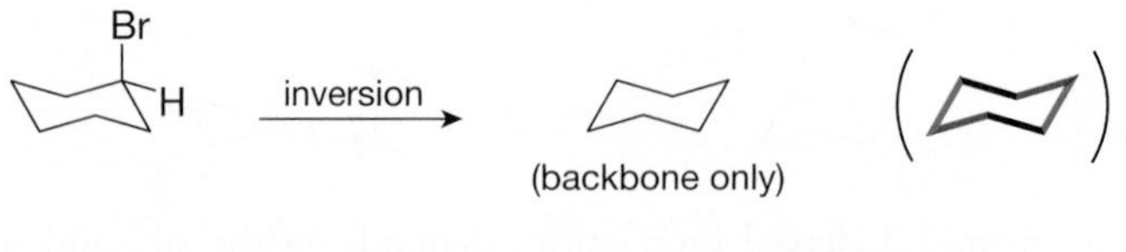

Br
H
inversion
(backbone only)

STEP 3: Locate the atom bearing the bromine atom in the new chair structure.

The left-most and right-most atoms in a chair do not change position during inversion, so they can be used as reference points to locate the other atoms in the ring. The right-most atom has been marked in the following structures. In the original structure, the carbon with the bromine attached is one position counterclockwise of the marked carbon. Likewise, the carbon bearing the bromine atom in the inverted chair will be one position counterclockwise of the right-most atom, as indicated.

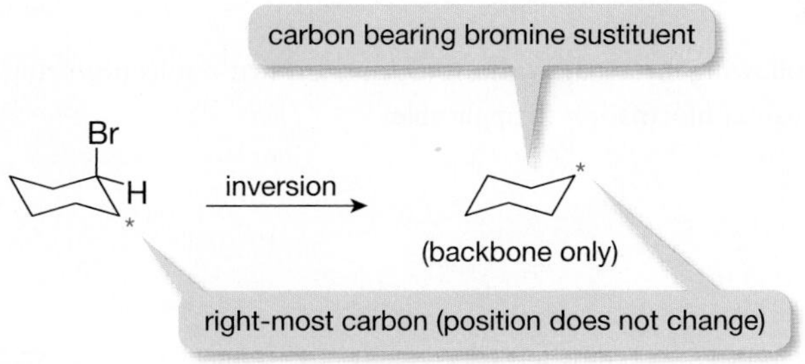

Note that the position of the carbon bearing the bromine atom appears to "move." This is due to an illusion created by perspective in the chair structure. To avoid confusion, use simple labels to locate the "fixed" positions (such as the left-most and right-most positions), and relate the other substituents to these positions.

STEP 4: Add the bromine and hydrogen atom to the carbon bearing the bromine substituent.

The carbon atom in the new ring with the bromine has its axial bond pointing straight down. Its equatorial bond will be pointing in a direction that it is parallel to the indicated bonds in red. Although the orientation of the bonds has changed, the relative positions of the bromine and hydrogen atoms do not. Since the bromine is on the top face in the original structure, it remains on the top face in the inverted structure, placing it in the equatorial position. Likewise, the hydrogen atoms remain on the bottom face in both structures.

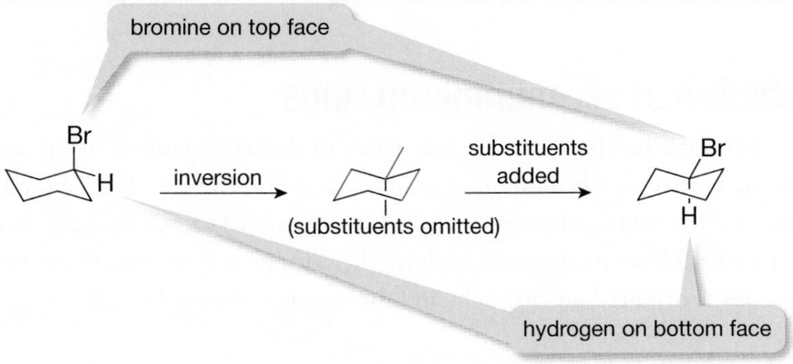

STEP 5 (OPTIONAL): Redraw the inverted structure omitting the hydrogen atom.

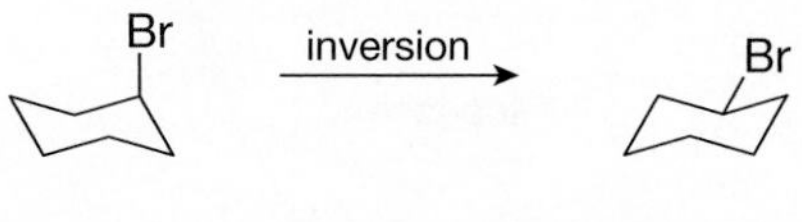

STEP 6: Evaluate the steric interactions in each chair structure to determine their relative stability.

When substituents are in an axial position, they experience 1,3-diaxial interactions with the other axial atoms/groups across the ring. Therefore, the original structure, with its axial bromine atom, is the least stable conformation, and the inverted chair is more stable.

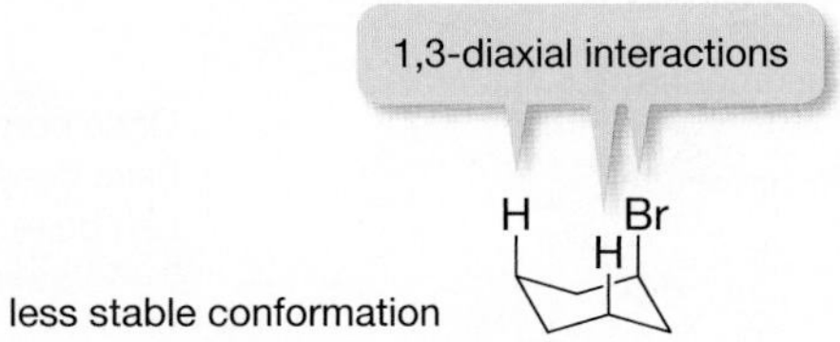

PRACTICE PROBLEM

3.9 For each of the following chairs, draw the conformation that results from ring inversion, and select the most stable chair conformation (if applicable).

a) 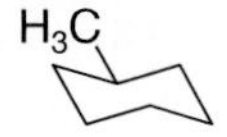 H_3C

b) 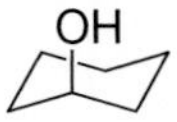 OH

c) 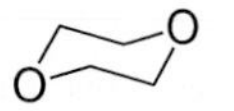 O O

d) 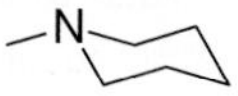 —N

INTEGRATE THE SKILL

3.10 a) Draw Newman projections for the two possible chair conformations of methylcyclohexane. Set up the projection so that you are looking down the C-1–C-2 and C-4–C-5 bonds.

b) The two possible chair conformations of ethynylcyclohexane exist in nearly equal quantities at 298 K. By contrast, ethylcyclohexane exists primarily in the equatorial conformation. Suggest a reason for this difference.

3.7.2 Disubstituted six-membered rings

The ***cis* isomer** of a cycloalkane has its two substituents on the same face of the ring.

The ***trans* isomer** of a cycloalkane has one substituent bonded to each face of the ring.

1,4-Dimethycyclohexane has two isomers that differ in the arrangement of the methyl groups on the ring. The ***cis* isomer** has both methyl groups bonded to the *same* face of the ring, whereas the ***trans* isomer** has the two methyl groups bonded to *opposite* faces of the ring. In the top view of the *cis* isomer in the following diagram, both methyl groups project toward the viewer. One of these methyl groups is oriented equatorially, and the other is oriented axially.

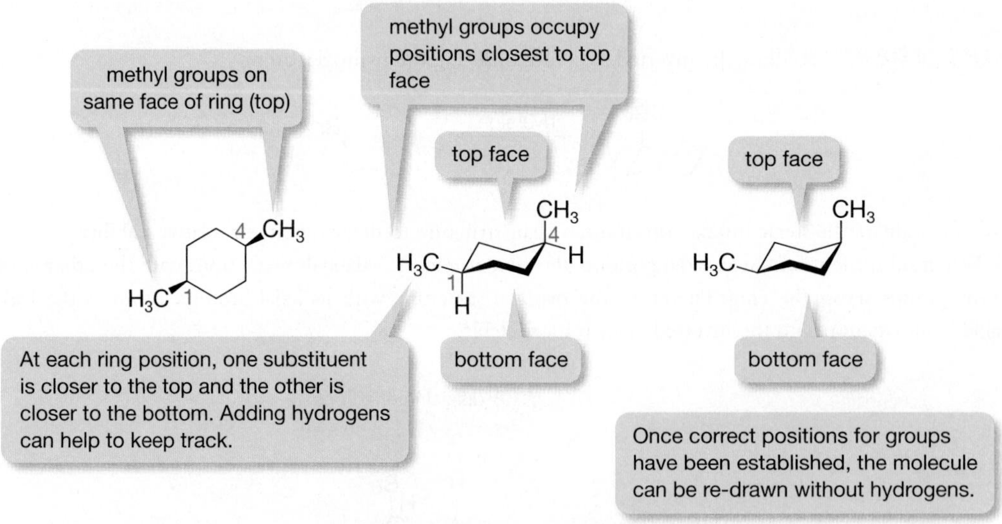

Ring inversion of a chair conformation of the *cis* isomer moves the equatorial methyl group to an axial position and the axial methyl group to an equatorial position. So both chair conformations of *cis*-1,4-dimethylcyclohexane have one equatorial methyl group and one axial methyl group. The conformations have identical 1,3-diaxial interactions and therefore have equal energies.

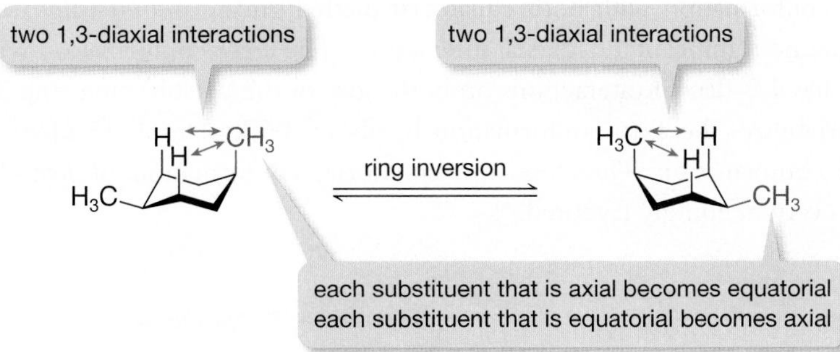

In the following diagram, the top view of the *trans* isomer of 1,4-dimethylcyclohexane shows that one methyl group projects toward the viewer and the other projects away. This configuration gives chair conformations in which the two methyl groups are either in both equatorial or both axial positions.

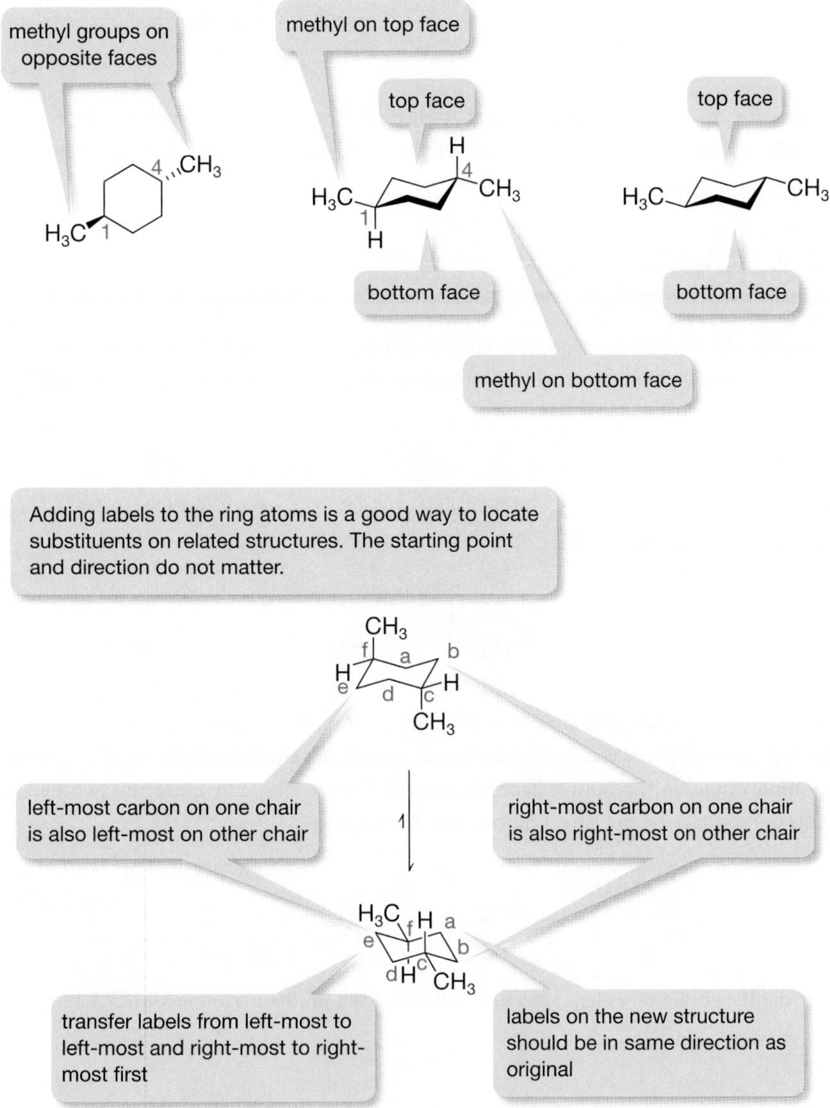

STUDENT TIP

The diagrams that show the zig-zag structure of the two chair conformations are side views from slightly different angles. This difference in viewpoint sometimes causes an illusion that substituents move around the ring when the chairs flip. To ensure that you place substituents correctly when drawing chairs, remember that the right-most and left-most carbons remain in those positions when chairs flip. Labelling the ring atoms is also helpful. Label the right-most and left-most atoms first in both chairs, then go around both rings in the same direction to label the remaining ring atoms (i.e., either clockwise in both or counter-clockwise in both). Use the labels to locate substituents.

The chair conformation with both equatorial methyl groups has virtually no steric strain because there are no significant 1,3-diaxial interactions. The chair conformation with both axial methyl groups has 1,3-diaxial interactions on both sides of the cyclohexane ring. The resulting steric strain destabilizes the axial conformation by about 3.6 kcal/mol (15 kJ/mol) relative to the equatorial conformation. Therefore, the equatorial conformation of *trans*-1,4-dimethylcyclohexane is overwhelmingly favoured.

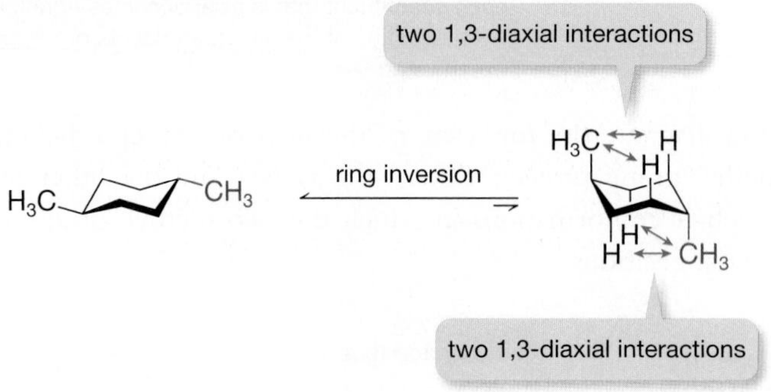

two 1,3-diaxial interactions

H_3C CH_3 ring inversion H_3C H H H H CH_3

two 1,3-diaxial interactions

Note that the *cis* and *trans* isomers of 1,4-dimethylcyclohexanehave the same connectivity; they differ only in the *orientation* of the bonds to the methyl groups. *Cis* and *trans* isomers are examples of *stereoisomers*: a class of isomers described in Chapter 4.

CHECKPOINT 3.6

You should now be able to draw both possible chair conformations for cyclohexane-based structures bearing any number of substituents. You should also be able to predict which chair conformation will be the lowest in energy.

SOLVED PROBLEM

Draw the two chair conformations of *cis*-1,3-dimethylcyclohexane from the hexagonal top view shown here, and compare their relative stabilities.

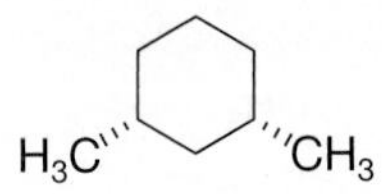

H_3C CH_3

STEP 1: Draw the implied hydrogen atoms on the carbon atoms bearing substituents.

From the top view of the given molecule, the *cis* isomer has both methyl groups projecting into the page, so they must reside on the bottom face of the cyclohexane ring. Therefore, the two hydrogen atoms must reside on the top face of the ring, pointing out of the page.

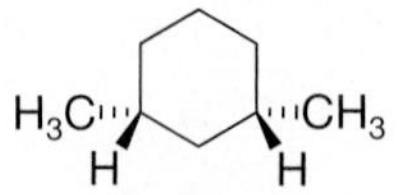

H_3C CH_3 H H

STEP 2: Draw the backbone of a chair structure following the guidelines in Section 3.5.1.

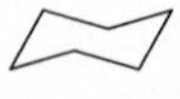

STEP 3: Assign carbons 1 and 3, and draw in the axial and equatorial bonds at these positions.

Number the positions in the "hexagon" structure. The starting points and direction do not matter, but it is convenient to start at one methyl and number towards the other. Transfer the numbers to the chair. The starting point does not matter, but the direction of numbering (counterclockwise in this example) should be the same. Draw the equatorial and axial bonds at any positions needing substituents.

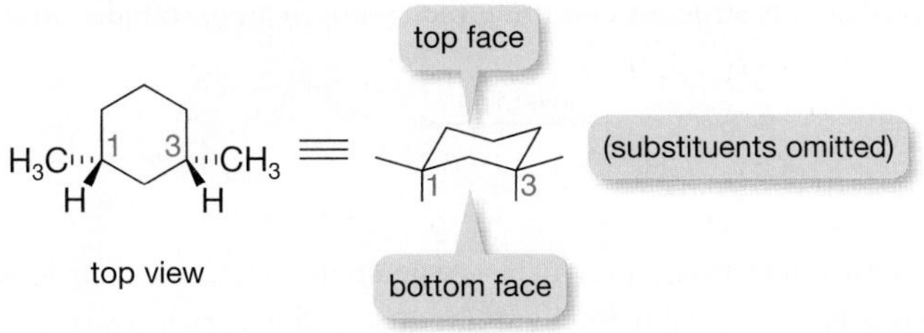

STEP 4: Add the hydrogen atoms and the methyl groups, making sure to maintain their positions relative to the provided top view of the molecule.

In the top view of the molecule, the hydrogen atoms are on the top face of the ring, and the methyl groups are on the bottom face. This relationship needs to be maintained when drawing the side view (chair structure). Therefore, the methyl groups will be in the axial positions pointing downward, whereas the hydrogen atoms will occupy the equatorial positions toward the top of the structure.

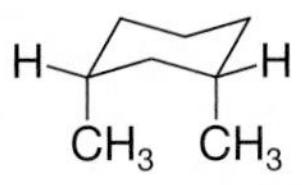

STEP 5: Following the steps described in Section 3.5.1, draw the backbone of the chair that forms after inversion of the ring.

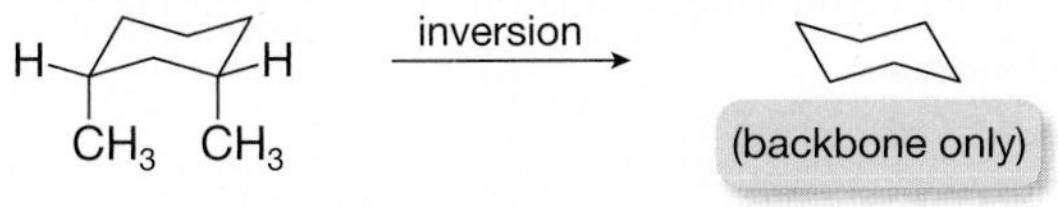

STEP 6: Locate the carbon atoms bearing substituents in the new chair structure and draw in their axial and equatorial bonds.

Using the right-most carbon as a point of reference, the carbon atoms bearing methyl groups will be found at the positions shown (labelled carbons 1 and 3). Since these atoms are pointing up, their axial bonds will also point up, and the equatorial groups will point to the side with a downward angle.

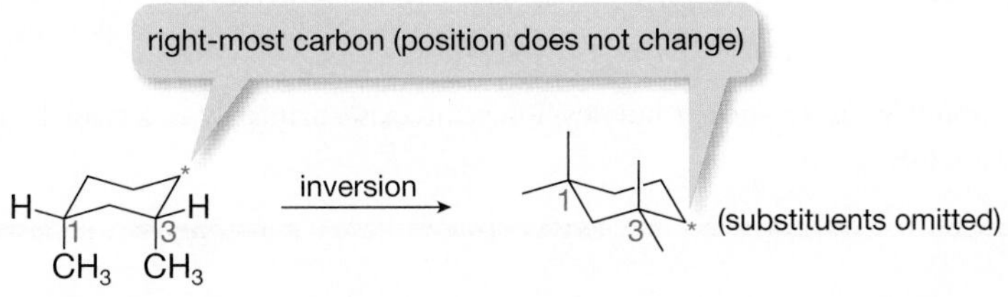

STEP 7: Add the hydrogen atoms and the methyl groups, making sure to maintain their relative positions.

Since the hydrogen atoms are pointing toward the top face in the original chair structure, they should be placed in the axial position in the inverted structure so that they remain on the top face of the ring. Likewise, the methyl groups are pointing downward to the bottom face, so they are placed in the equatorial positions.

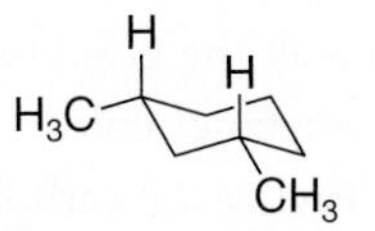

STEP 8 (OPTIONAL): Redraw the structures omitting the hydrogen atoms bonded to the ring.

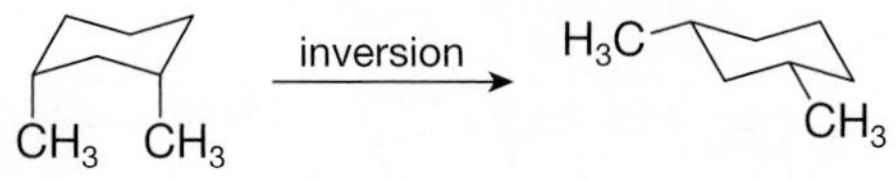

STEP 9: Evaluate the steric interactions in each chair structure to determine their relative stability.

The first chair drawn has two axial methyl groups; these will experience significant 1,3-diaxial interactions with each other across the ring, as well as with the other axial hydrogen on the bottom face of the molecule. Therefore, the second chair drawn, with the methyl groups in equatorial positions, is the more stable conformation.

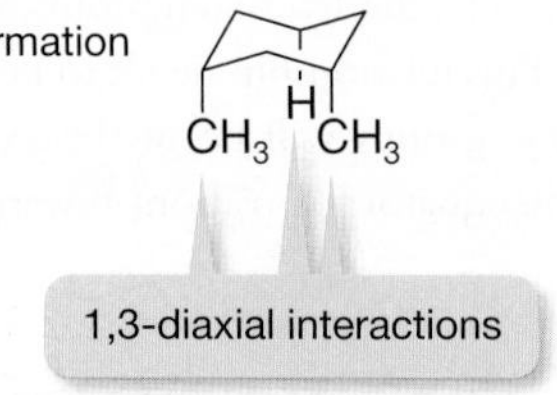

PRACTICE PROBLEM

3.11 For each of the following molecules, draw the two possible chair conformations and predict which one will be favoured under equilibrium conditions. Explain your choices.

a) *trans*-3-bromocyclohexanol

b) *cis*-1-chloro-2-fluorocyclohexane

c)

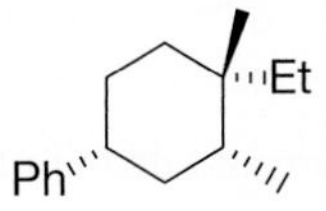

d)

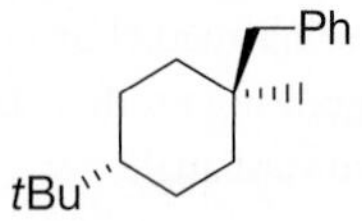

INTEGRATE THE SKILL

3.12 Suggest a reason why *cis*-1,4-di-*tert*-butylcyclohexane exists primarily in a twist-boat form rather than a chair conformation.

Bringing It Together

As mentioned in Section 3.1, the enzyme lysozyme breaks down the carbohydrate chains of bacterial cell walls. Each carbohydrate ring favours the chair conformation, as shown in Figure 3.4. Lysozyme begins its attack by attaching itself to a six-ring A-B-A-B-A-B segment of the carbohydrate chain. In doing so, the lysozyme forces some of the B rings into the less-stable twist-boat conformation. Now the lysozyme can break the bond between the B ring and the oxygen atom that links it to the adjacent A ring. The enzyme replaces this C–O bond with a bond to an OH group from a molecule of water, thus cleaving the carbohydrate chain.

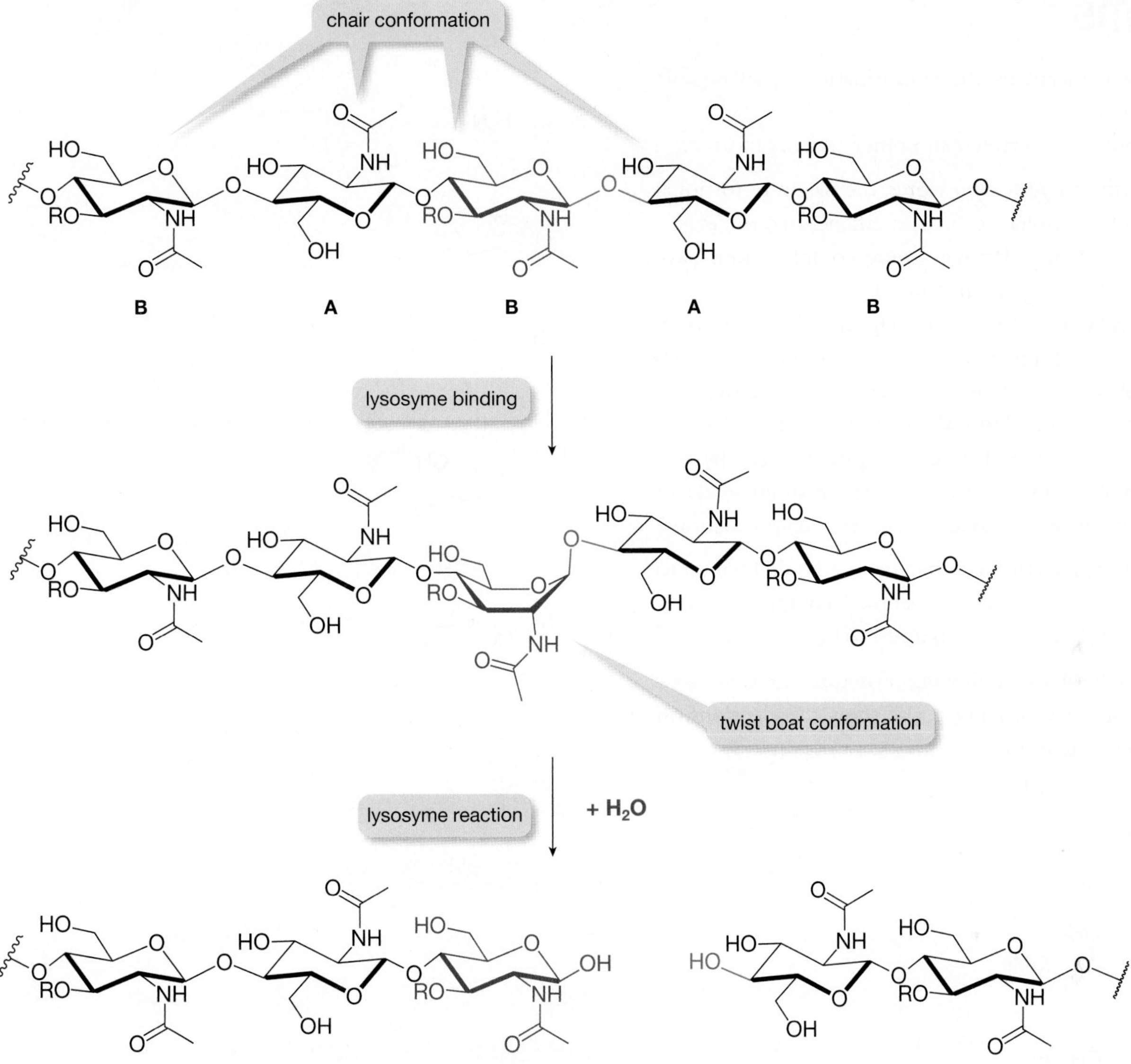

FIGURE 3.4 Cleavage of the carbohydrate chain by lysozyme (R = $-OCH(CH_3)COOH$).

You Can Now

- Distinguish between torsional, steric, and angle strain, and use these concepts to predict the low-energy conformations of small molecules.
- Convert between line drawings and Newman projections for acyclic molecules.
- Draw the various conformations generated through rotation of the single bonds in acyclic molecules, and rank the conformations according to the amount of torsional and/or steric strain that occurs.
- Distinguish between staggered, eclipsed, antiperiplanar, and gauche conformations.
- Rationalize the preferred conformations of small cyclic molecules in terms of the torsional and steric strain present in their unstable planar forms.
- Draw the two possible chair conformations of cyclohexane-based structures.
- Rank the relative stability of the two chair conformations in cyclohexane-based structures that have substituents, based on the presence or absence of 1,3-diaxial interactions.
- Convert between two-dimensional hexagonal drawings of cyclohexane-based structures and their corresponding chair structures, including those with multiple substituents that can exist as either *cis* or *trans* isomers.

Problems

3.13 a) What is meant by the conformation of an organic molecule?

b) Differentiate between an isomer and a conformer.

3.14 a) Describe, in your own words, the origin of torsional strain in a molecule such as ethane. In your explanation, account for why some conformations have higher torsional strain than others.

b) Differentiate between the origins of torsional strain and steric strain in an organic molecule. Does torsional strain exist only between the substituents of adjacent atoms? How about steric strain?

c) Although torsional strain exists in ethane, why is it generally considered that no steric strain exists in either of its staggered or even eclipsed conformations?

3.15 A friend suggests that Lewis structures could be used to describe the staggered and eclipsed conformations of ethane. Do you agree or disagree? Why?

3.16 Draw each of the following structures in *both* chair forms. For parts (b) and (c), also show the electrons on the ring heteroatom.

a) 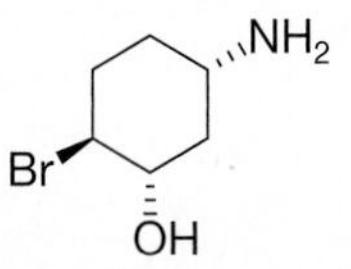 NH_2, Br, OH

b) 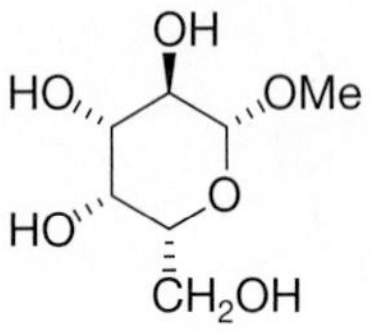 OH, HO, OMe, O, HO, CH_2OH

c) 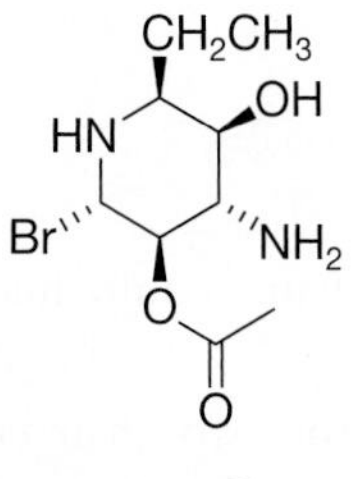 CH_2CH_3, OH, HN, Br, NH_2, O, O

d) 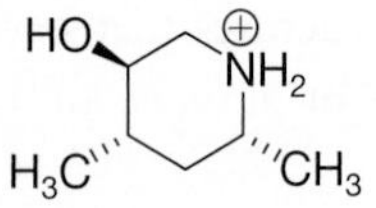 HO, $\overset{+}{N}H_2$, H_3C, CH_3

3.17 Draw each of the following using line notation (hexagon) indicating stereochemistry.

a) 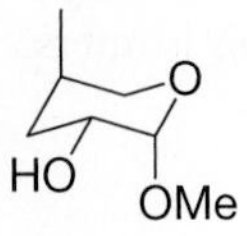 O, HO, OMe

b) 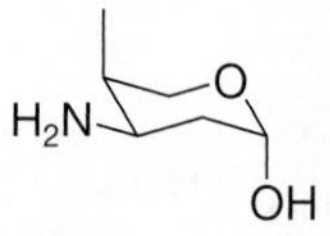 H_2N, O, OH

c) 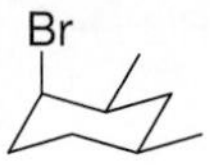 Br

d) 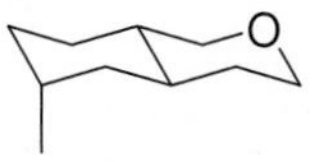 O

3.18 Draw the other chair conformation of the following.

a) 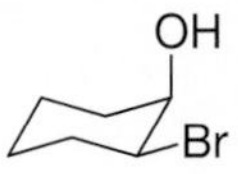 OH, Br

b) 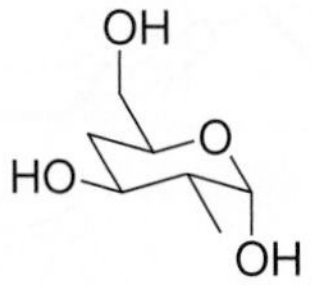 OH, O, HO, OH

c)

d) 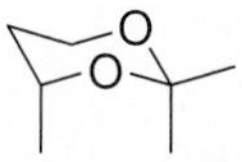 O, O

e) 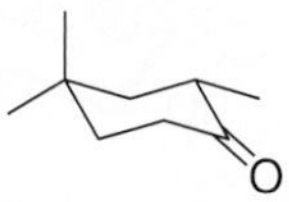 O

3.19 a) Sighting along its C–N bond, draw the Newman projection of the conformation of dimethylamine given in perspective below. (Note the presence of the lone pair of electrons on the nitrogen.)

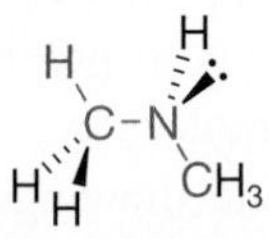

H, H, C–N, H, H, CH_3

b) Qualitatively plot a graph of strain energy vs. torsion angle for a full 360° rotation about this C–N bond. Define the torsion angle by the two C atoms, one N atom, and one H atom of dimethylamine as shown above in blue.

c) How many of the conformations of dimethylamine have the same energy?

d) How many of the barriers to rotation in dimethylamine have the same height?

e) Do you expect these barriers to rotation to be greater, lesser, or the same height as those of methylamine (CH_3NH_2)? Why?

3.20 a) Sighting along its C-1–C-2 bond, draw in perspective the highest energy conformation and lowest energy conformation of 1-chloropropane and identify the strains present in each.

b) If the torsion angle in 1-chloropropane is defined by its Cl atom and three C atoms, what are the torsion angles of the two conformations in part (a)?

3.21 a) Qualitatively plot a graph of strain energy vs. torsion angle for a full 360° rotation about the C-2–C-3 bond in 2-methylbutane. Define the torsion angle by the three C atoms and one H atom shown below in blue.

b) Draw the three staggered conformations of 2-methylbutane in Newman projection looking down the C-2–C-3 bond from C-2 to C-3.

c) From your graph, identify by their torsion angle which of these three conformations have the same energy. Identify the sources of strain in these conformations.

d) Draw the most stable *eclipsed* conformation of 2-methylbutane in Newman projection. How many conformations have this energy and what are their torsion angles?

3.22 It is not possible for an acyclic alkane to have *cis* and *trans* isomers as cycloalkanes do. Explain why this is so.

3.23 a) Draw perspective representations of the two conformations of *cis*-1,3-dimethylcyclobutane. Which of these two conformations do you expect to be the most stable? Provide an explanation in terms of strain within the molecule.

b) Similarly, draw the two conformations of *trans*-1,3-dimethylcyclobutane. Which of these two conformations do you expect to be the one adopted most frequently in a large sample of *trans*-1,3-dimethylcyclobutane molecules? Provide an explanation in terms of strain within the molecule.

3.24 Does the phrase "inversion by rotation" describe the ring-flipping process in cyclohexane? Comment both favourably and unfavourably on the accuracy of the phrase.

3.25 Part A of the following diagram depicts methylcyclohexane, and the part of the chair conformation in red shows one of the two 1,3-diaxial interactions between the axial methyl group and an axial hydrogen atom. This interaction creates 0.9 kcal/mol of steric strain in this conformation.

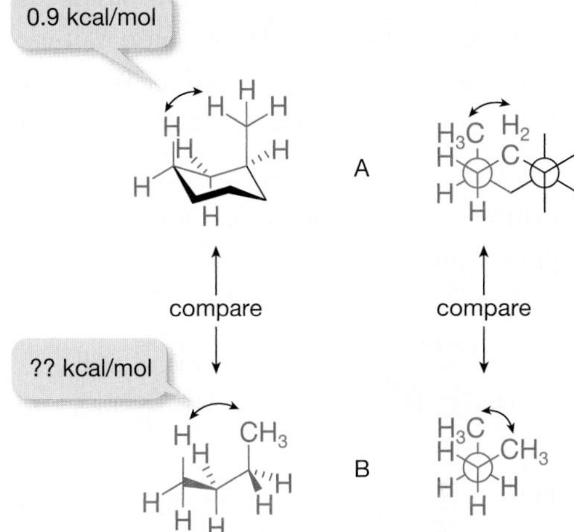

a) Compare the structure of the red part of the methylcyclohexane in both figures of part A with the gauche conformation of butane shown in the two figures of part B, and predict the amount of steric strain present in the gauche conformation of butane.

b) How much more steric strain exists in methylcyclohexane than in the gauche conformation of butane?

3.26 a) Draw both chair conformations of *trans*-1,2-dichlorocyclohexane shown below.

b) Describe the positions of the chlorine atoms in each conformation as axial or equatorial.

c) How do these positions of the chlorine atoms compare to their positions in the two chair conformations of *trans*-1,3-dichlorocyclohexane?

3.27 a) Draw *cis*-4-methylcyclohexanol and *trans*-4-methylcyclohexanol. Identify which stereoisomer is *cis* and which is *trans*.

b) Draw one chair conformation of each molecule, and determine the position of the hydroxyl group in each isomer as equatorial or axial.

c) Draw the other chair conformation of each molecule, and determine the position of the hydroxyl group in each isomer as equatorial or axial.

d) While the two 1,3-diaxial interactions of an axial methyl group contribute a steric strain energy of 1.8 kcal/mol (7.6 kJ/mol), the two 1,3-diaxial interactions of the smaller hydroxyl group contribute about 1.0 kcal/mol (4.2 kJ/mol). Using this information, calculate the relative energies of each of the conformations in parts (b) and (c). Which chair conformation will each isomer "prefer" to reside in?

3.28 a) Draw the perspective representation of the conformation of the two molecules shown here in Newman projection:

b) Draw the chair structures of the cyclohexane conformations shown here in Newman projection.

c) For each of the four molecules above, draw their Newman projections obtained by sighting down the same C–C bonds as described above but viewed from the other direction.

3.29 The following diagram shows a Newman projection of cyclohexane in which the red "back atoms" are carbons 4 (left) and 6 (right). Copy the Newman projection and draw in all substituents of the most stable conformation of *cis*-1,3-diphenylcyclohexane.

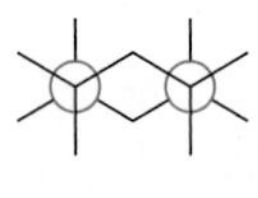

3.30 Explain why 1,2-diaminocyclopentane has *cis* and *trans* isomers but 1,1-dimethylcyclopentane does not.

3.31 While the torsion angle of the anti-conformation of 1,2-dichloroethane is indeed precisely 180°, the torsion angles of its two gauche conformations are not precisely 60° and 300° but actually lie closer to 67° and 293°. Draw one of the gauche conformations of 1,2-dichloroethane in Newman projection and provide an explanation for this in terms of the sources of steric strain present.

3.32 Draw the following perspective representations in Newman projection, sighting the molecule along the bond(s) indicated by the arrow(s).

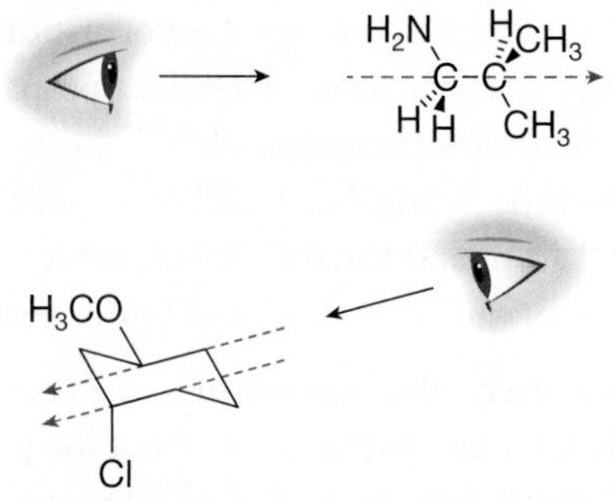

3.33 α-Glucopyranose, one of the cyclic isomers of glucose, is shown here. Draw its chair conformations.

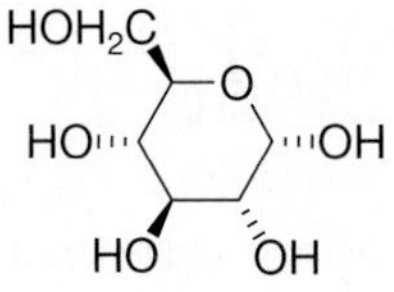

3.34 a) Ring A in the bacterial carbohydrate chain of Figure 3.1 is known as *N*-acetylglucosamine. Redraw the structure of *N*-acetylglucosamine shown below, adding the missing hydrogens to the carbon atoms of the ring in the appropriate axial or equatorial positions.

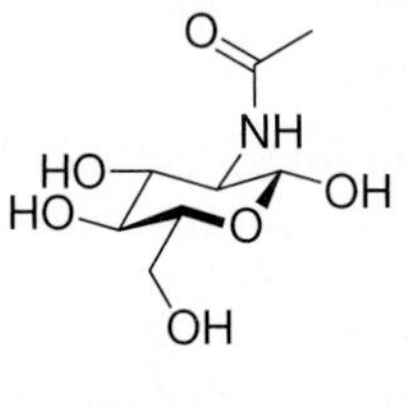

b) Draw the structure of the chair conformation of *N*-acetylglucosamine that results from its ring inversion, including axial and equatorial hydrogen atoms.

c) Which of its two chair conformations do you expect is the most stable? Provide a reason for your choice.

d) On the planar line structure shown here, draw the substituents of *N*-acetylglucosamine as lines and wedges.

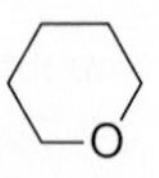

3.35 a) Draw the chair conformation of *trans*-1-*tert*-butyl-3-methylcyclohexane in which the methyl substituent is positioned axially.

b) Draw the chair conformation obtained by ring inversion of the conformation in part (a).

c) Would the conformation in part (a) have a higher, lower, or the same strain energy as the conformation in part (b)? Provide a rationale for your choice.

d) Draw the more stable chair conformation in Newman projection, sighting along the C-1–C-6 and C-3–C-4 bond.

3.36 As discussed in Section 3.7, the difference in energy between the chair conformations of *trans*-1,4-dimethylcyclohexane is about 15 kJ/mol. The difference in energy between the chair conformations of *trans*-1,2-dimethylcyclohexane is only about 11 kJ/mol.

a) Draw the two chair conformations of *trans*-1,2-dimethylcyclohexane and the two chair conformations of *trans*-1,4-dimethylcyclohexane.

b) Identify and compare the sources of strain in the less stable conformation of each compound.

c) Draw the more stable conformations of *trans*-1,2-dimethylcyclohexane and *trans*-1,4-dimethylcyclohexane in Newman projection, sighting down C-1–C-2 and C-5–C-4 bond. Compare the sources of strain present in each.

d) Provide an explanation as to why the difference in energy between the two conformations of *trans*-1,2-dimethylcyclohexane is about 4 kJ/mol less than the difference in energy between the two conformations of *trans*-1,4-dimethylcyclohexane.

e) *trans*-1,2-Dibromocyclohexane is an atypical disubstituted cyclohexane in that, unlike *trans*-1,2-dimethylcyclohexane, its diequatorial chair conformation is *less* stable than its diaxial conformation. Use your knowledge of bond dipoles and your answer to part (d) to offer a reason as to why this is the case.

3.37 1,2,4,5-Tetramethylcyclohexane exists as five (and only five!) different isomers depending upon which face of the cyclohexane ring each of the four methyls is bonded. In the arrangement shown here, three of the four methyl groups are on the top face of the ring. Draw the other possible isomers of 1, 2, 4, 5-tetramethylcyclohexane.

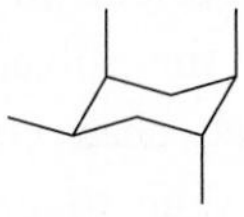

3.38 a) Draw *trans*-1-chloro-3-methylcyclohexane in planar projection.

b) Draw its two chair conformations and identify all sources of strain present in each.

c) Which chair conformation would equilibrium favour?

3.39 a) Draw the two conformations of 1-methylphosphetane oxide, shown here.

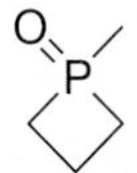

b) Do you expect the energy of these two conformations to be equal? Explain why or why not.

3.40 In terms of overlap between atomic orbitals, explain why azetane (left) has less angle strain than aziridine (right).

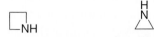

3.41 2-Methylcyclopropanamine (left) is a cyclic constitutional isomer of 2-methylazetidine (right). There are at least eight other cyclic constitutional isomers of 2-methylazetidine. Draw six of them. Hint: remember that *cis* and *trans* isomers are stereoisomers, not constitutional isomers.

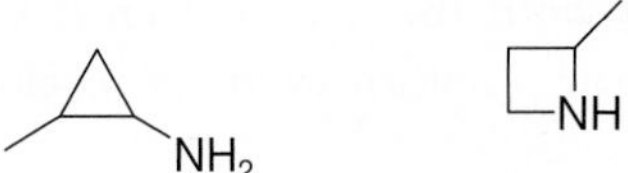

3.42 The five- and six-membered rings of *monosaccharides* or simple sugars are often represented in *Haworth projection*, as the following shows for one isomer of galactose, a sugar found in milk and other dairy products. The Haworth projection permits the viewer to readily identify the spatial arrangement of the hydroxyl groups about the ring of the sugar. A friend suggests that Haworth projections are not physically realistic descriptions of the actual structure of a six-membered ring.

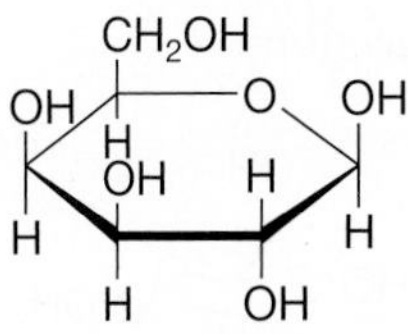

a) Identify two things that are physically inaccurate about the Haworth projection.

b) Draw the most stable chair conformation of the galactose molecule that is depicted in Haworth projection above.

3.43 Draw a qualitative energy diagram for the inversion of a twist-boat conformation of cyclohexane. How many chair and half-chair conformations lie along the pathway of twist-boat inversion?

3.44 *trans*-1,3-Di-*tert*-butylcyclohexane is one of the few cyclohexane compounds in which the twist-boat form is believed to be more stable that the chair form. However, the *cis* isomer exists overwhelmingly in chair form. By drawing the chair conformations of each, explain this difference in behaviour.

3.45 Draw line structures of cyclopropane and cyclopropene. Do you expect cyclopropene to have greater, lesser, or the same amount of angle strain as cyclopropane? Give a reason for your choice.

3.46 Like six-membered rings, larger rings (of 7–10 atoms) exist as interconverting conformers having staggered or near-staggered orientations around the ring to minimize strain. The conformations adopted by these rings are often described in terms of chair and boat segments. For example, cyclooctane is thought to have several conformations, the most stable of which is the boat-chair conformation.

a) Identify both the boat and chair segments of the boat-chair conformation of cyclooctane shown above.
b) For this conformation, identify the location of any eclipsed arrangements of atoms that would cause torsional strain. (Hint: Build a model of cyclooctane and view it in Newman projection down its carbon-carbon bonds.)

3.47 A derivative of the monosaccharide known as altrose is shown here. Draw this compound in planar projection (a) as viewed from above the molecule; and (b) as viewed from beneath the molecule. Pay particular attention to the positions of atoms 1 and 4 by numbering them in your planar projections.

3.48 Decalin, $C_{10}H_{18}$, is a hydrocarbon consisting of two cyclohexane rings that are *fused* because they share a ring C–C bond. Decalin exists in the following two forms. One is the *cis* isomer, in which the two blue carbon atoms of the ring on the right represent *cis* substituents of the ring on the left. The other is the *trans* isomer, in which the two blue carbon atoms of the ring on the right are *trans* substituents of the ring on the left.

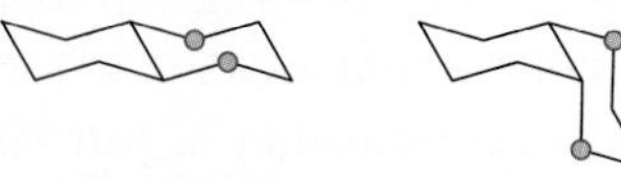

a) Which of the above isomers is the *cis* isomer and which is the *trans*?
b) Draw each isomer as line structures in planar projection, paying particular attention to differentiating between the isomers in their *cis–trans* relationship.
c) By identifying sources of steric strain in each isomer, predict which isomer is most stable.
d) One of the two isomers of decalin cannot undergo ring inversion. Which one? Provide a reason for your choice. If necessary, build a molecular model of each isomer to assist you in answering this question.

3.49 a) The aptly named tetrahedrane, shown here, is a highly strained hydrocarbon that has never been discovered or made in the lab. While it has the appearance of four interconnected cyclopropane rings, its total strain energy, calculated to be about 130 kcal/mol (550 kJ/mol), is much more than four times that of cyclopropane. Offer an explanation in terms of angle strain as to why this is so.

b) Another highly strained alkane, cubane, having a total strain energy of almost 170 kcal/mol (700 kJ/mol), has been synthesized in the lab. Can you think of a reason why the synthesis of tetrahedrane is proving to be more challenging than that of cubane, even though cubane has the higher total strain energy?

3.50 a) Sighting along the C-1–C-2 (and C-5–C-4) bond, draw the two chair conformations of *trans*-1,2-dibromocyclohexane in Newman projection and identify the sources of strain in each chair conformation.
b) When *trans*-1,2-dibromocyclohexane is dissolved in a variety of organic solvents, it is experimentally observed that the relative amounts of its two chair conformations present in solution depend upon the polarity of the solvent. Consider the dipole moments of each chair conformation of *trans*-1,2-dibromocyclohexane to determine which conformation would be favoured in a non-polar solvent such as CCl_4.

MCAT STYLE PROBLEMS

3.51 Which of the following statements (i)–(iii) is/are true?

i) The 1,3-diaxial interactions present in methylcyclohexane are a source of torsional strain in the molecule.

ii) All of the strain in cyclopropane is angle strain.

iii) Several conformations of a molecule may have exactly the same energy.

a) Only statements (i) and (ii) are true.

b) Only statement (ii) is true.

c) Only statements (i) and (iii) are true.

d) Only statement (iii) is true.

3.52 Which of the following molecules will have the greatest difference in energy between its chair conformations?

a) 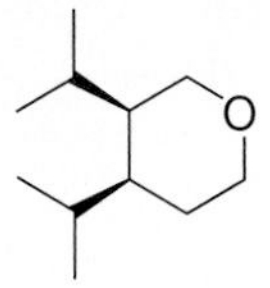 O

b) 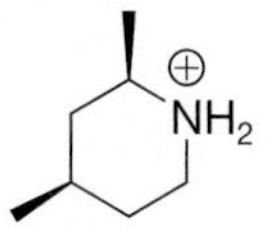 $\overset{\oplus}{N}H_2$

c) 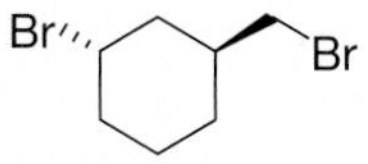 Br Br

d) 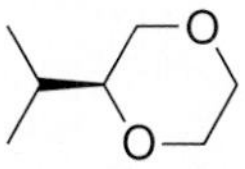 O O

3.53 Which of the following statements (i)–(iii) is/are true?

i) The highest energy conformation attained during the ring inversion of cyclohexane is the twist-boat conformation.

ii) Each of the conformations of an organic molecule represents a different constitutional isomer of the molecule.

iii) Ring inversion converts *cis* isomers into *trans* isomers and vice versa.

a) Only statements (i) and (ii) are true.

b) Only statement (i) is true.

c) Only statement (ii) is true.

d) None of the statements (i)–(iii) is true.

3.54 Which of the statements below correctly describes the chair conformations of *trans*-1,3-diisopropylcyclohexane.

a) The two chair conformations are equal in energy.

b) The higher energy chair conformation contains two axial isopropyl groups.

c) The higher energy chair conformation contains two equatorial isopropyl groups.

d) The lower energy chair conformation contains two equatorial isopropyl groups.

CHALLENGE PROBLEMS

3.55 In humans and other vertebrates, *gamma*-aminobutyric acid (GABA) acts as a neurotransmitter by regulating the flow of chloride ions (Cl^-) into the cells of the brain and central nervous system. To better understand the relationship between its biochemical role and its conformations, a number of compounds have been studied that are structurally very similar to GABA but conformationally more rigid.

$H_3N(CH_2)_3CO_2^{\ominus}$

GABA

a) One of these is the compound (+)-CAMP, which has the structure shown here.

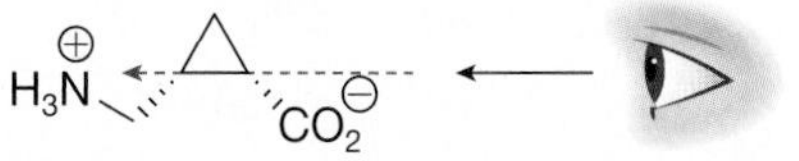

$\overset{\oplus}{H_3N}$ $CO_2^{\ominus}$

Draw the Newman projection of GABA in the conformation adopted by (+)-CAMP as viewed by sighting down the bond indicated by the arrow. Do you think that GABA is likely to adopt this conformation? Provide a reason for your answer.

b) Another structural analogue of GABA is piperidinium carboxylate. In the following line structure of piperidinium carboxylate, GABA can be "seen" in blue.

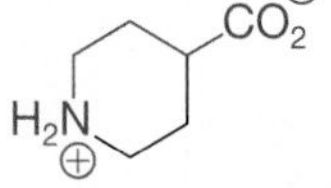

$CO_2^{\ominus}$ $H_2\underset{\oplus}{N}$

Draw piperidinium carboxylate in its most stable chair conformation. If GABA adopted this conformation, would this be a different conformation for GABA than that which you drew for GABA in part (a)?

c) It has been suggested that muscimol, a compound isolated from mushrooms of the genus *Amanita*, might also serve as a GABA analogue. Draw the line structure of GABA in the conformation in which its four carbon atoms and ammonium group are positioned as described by the four carbon atoms and ammonium group of muscimol shown here. Describe this conformation of GABA by identifying whether each of the four bonds between its two functional groups lies in a staggered or eclipsed conformation. Is this a different conformation for GABA than the one you drew for GABA in either part (a) or part (b)?

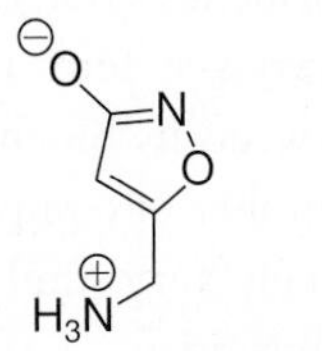

3.56 The following scheme shows the formation of a new sp^3 centre from an sp^2 one. Using the Newman projections provided, predict the structure of the major product and briefly justify your choice.

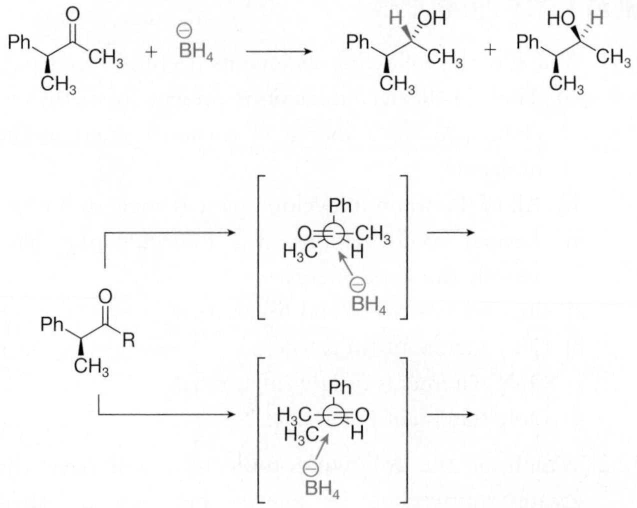

Stereochemistry
THREE-DIMENSIONAL STRUCTURE IN MOLECULES

4

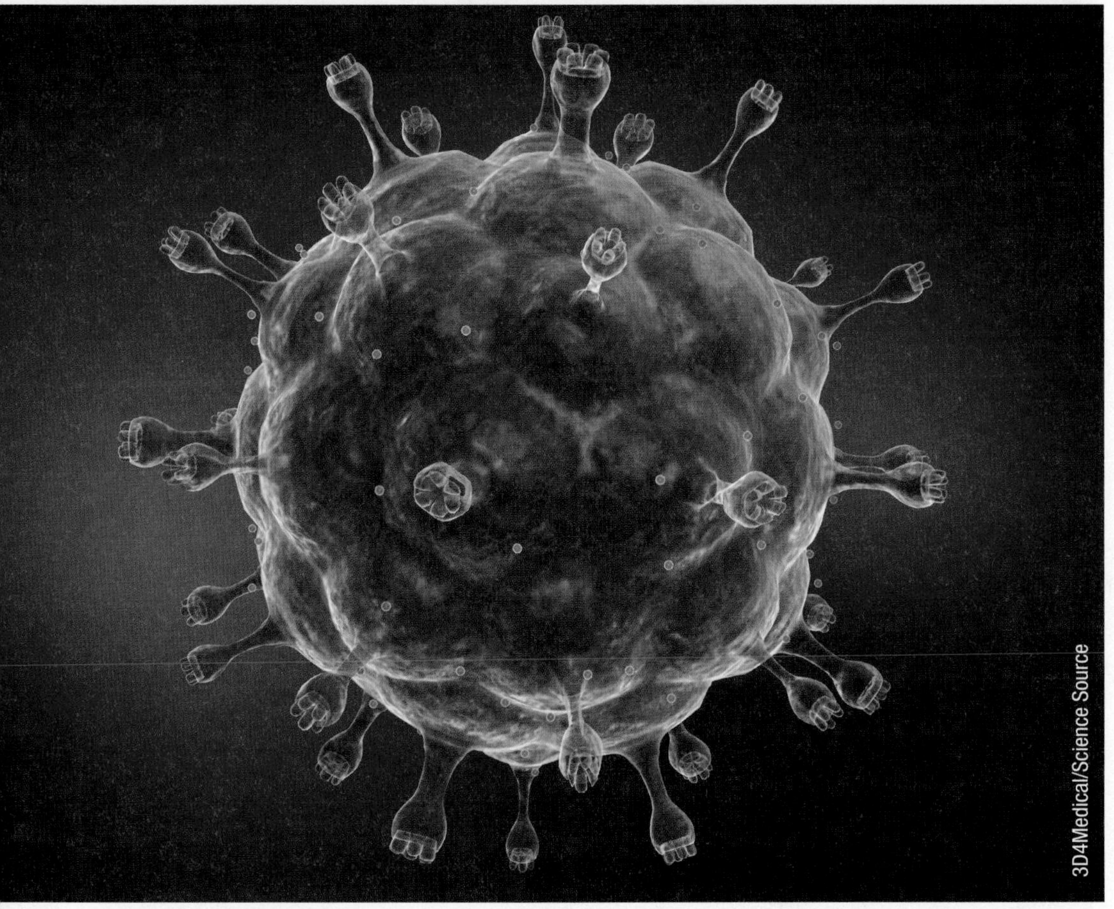

How did a drug invented by a chemist at McGill University turn an HIV infection from a death sentence into a manageable disease?

CHAPTER OUTLINE

4.1 Why It Matters

In 1981, a fatal new disease emerged in California. The later stages of the infection were characterized by rare cancers and infections by unusual pathogens. In 1982, this condition was named AIDS: acquired immune deficiency syndrome. By 1986, researchers had determined that AIDs is caused by the human immunodeficiency virus (HIV), a previously unknown retrovirus that destroys the immune system of people infected with it.

The first drug approved to treat HIV infections was AZT, a failed anti-cancer drug. This drug is structurally similar to thymidine, one of the building blocks of DNA. When replicating its DNA, HIV uses some AZT rather than thymidine, thus producing defective DNA that is not functional. Unfortunately, the side effects of AZT are severe because this drug also interferes with human DNA production.

AZT thymidine

In 1988, a new drug, 3TC (also called lamivudine), was developed by researchers in Montreal. The effectiveness of 3TC is comparable to that of AZT, but the side effects are much less dangerous. The new medication has two molecular forms: 3TC itself, and its mirror image, termed *ent*-3TC. The *ent*-3TC form is highly toxic because its structure is so similar to the DNA building block deoxycytidine that it interferes with human production of DNA.

The enzymes that HIV uses to make DNA differ from those the human body uses. The virus reacts the same way with both forms—3TC and *ent*-3TC—so both forms block viral DNA synthesis. However, 3TC is relatively safe for humans because the human machinery that makes DNA does not react with molecules that have shapes similar to the mirror image of deoxycytidine.

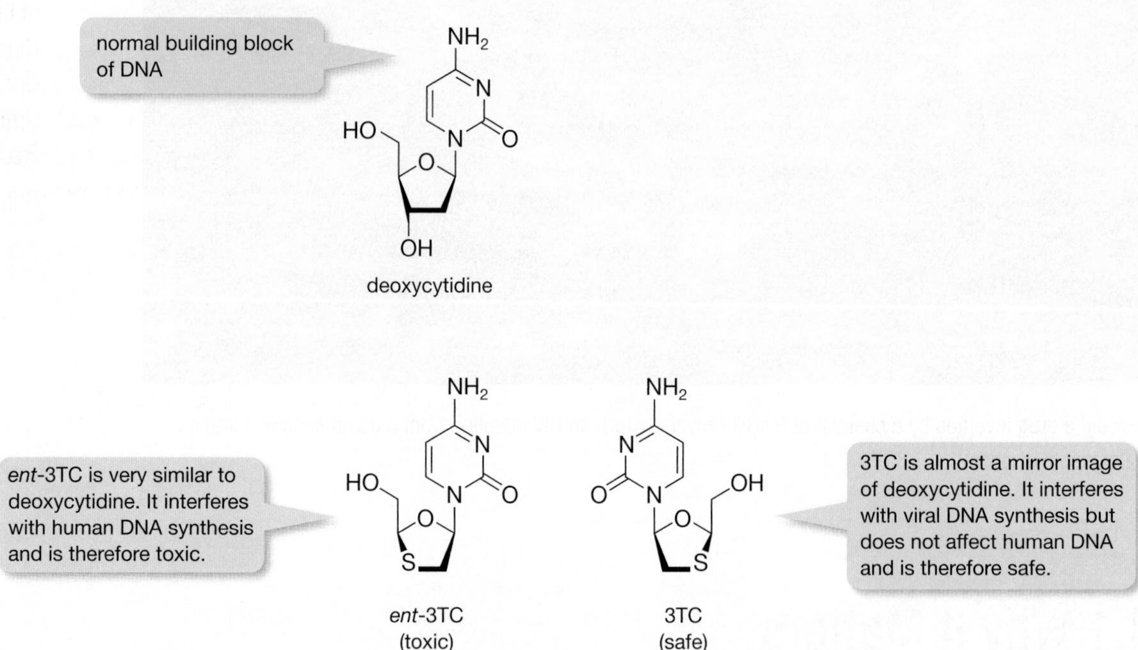

Molecules that are mirror images of one another, such as 3TC and *ent*-3TC, have the same atoms and the same connectivity, but they differ in the spatial arrangement of the atoms and consequently in their reactions in the body. Many of the molecules in living things have forms with different three-dimensional arrangements, so these forms affect many processes in everyday life. Stereochemistry, the subject of this chapter, concerns the study of the different spatial arrangements of atoms that form the structure of molecules, including the terms used to describe the different arrangements and the ways to represent them in structural formulas.

Dr. Bernard Belleau, inventor of 3TC.

4.2 Constitutional Isomers and Stereoisomers

The way that atoms are connected in a molecule determines the properties and chemical behaviour of that molecule. A given set of atoms can usually be connected in more than one way, thus producing two or more different molecules. Different molecules formed from the same set of atoms are called **isomers**. There are several types of isomers. For example, the atoms in compounds with the formula $C_4H_{10}O$ can be connected in seven ways. Each of the compounds is a **constitutional isomer** of the others, and has distinct physical and chemical properties.

Isomers are molecules that have the same atoms, but their atoms are connected in different ways.

Constitutional isomers are molecules that have the same chemical formula, but their atoms are connected in different ways.

Seven constitutional isomers of $C_4H_{10}O$

ORGANIC CHEMWARE
4.1 Isomers

Many molecules that have three-dimensional structures can have **stereoisomers**. The atoms in stereoisomers are bonded in the same way, but are arranged in space differently. The different arrangements of bonds that give rise to these structures are called **configurations**. The configuration of atoms in three dimensions is usually represented on paper with solid wedge bonds (indicating that substituents project out of the page) and hashed bonds (indicating that substituents project into the page).

Stereoisomers have the same atoms connected in the same sequence, but they differ in the three-dimensional arrangement of those atoms.

The **configuration** of a molecule is the three-dimensional arrangement of the bonds that connect the atoms. The configuration of a molecule is permanent.

plain bonds are in the plane of the paper
hash bonds project behind the plane of the paper
OH
OH
wedge bonds project upward from the plane of the paper

Enantiomers are non-superposable mirror images.

For example, 2-butanol can be formed in two stereoisomers that are mirror images of each other. These mirror images are different molecules; they cannot be superposed on each other. Mirror-image structures that cannot be superposed on each other are called **enantiomers**. So the mirror-image forms of 2-butanol are enantiomers of each other. Rotating enantiomers does not make their structures match. For example, as shown in the following diagram, rotating the 2-butanol molecule on the right 180° clockwise to place the OH on the other side of the chain does not produce the structure on the left. A second rotation can make the carbon backbones of the enantiomers match, but leaves the OH groups projecting in opposite directions. So the two structures still do not match.

STUDENT TIP
When comparing enantiomers, use the term *superposable*, not super*im*posable. Superimposed objects are simply placed one on top of the other. For example, a chair can be superimposed on a table. *Superposed* objects are placed over each other such that that all parts coincide exactly, so the objects must be identical. Thus, a chair cannot be superposed on a table.

2-butanol can exist as non-superposable mirror images
mirror plane
OH
OH
enantiomers cannot be superposed on one another
keep same
rotate 180°
OH
OH
rotate 180°
OH
these are still enantiomers
keep same
OH

Chiral objects are not superposable on their mirror images.

Chirality is the ability of objects to exist as non-superposable mirror images of each other.

Enantiomers are **chiral** molecules (Figure 4.1). **Chirality** is the ability to exist as non-superposable mirror images.

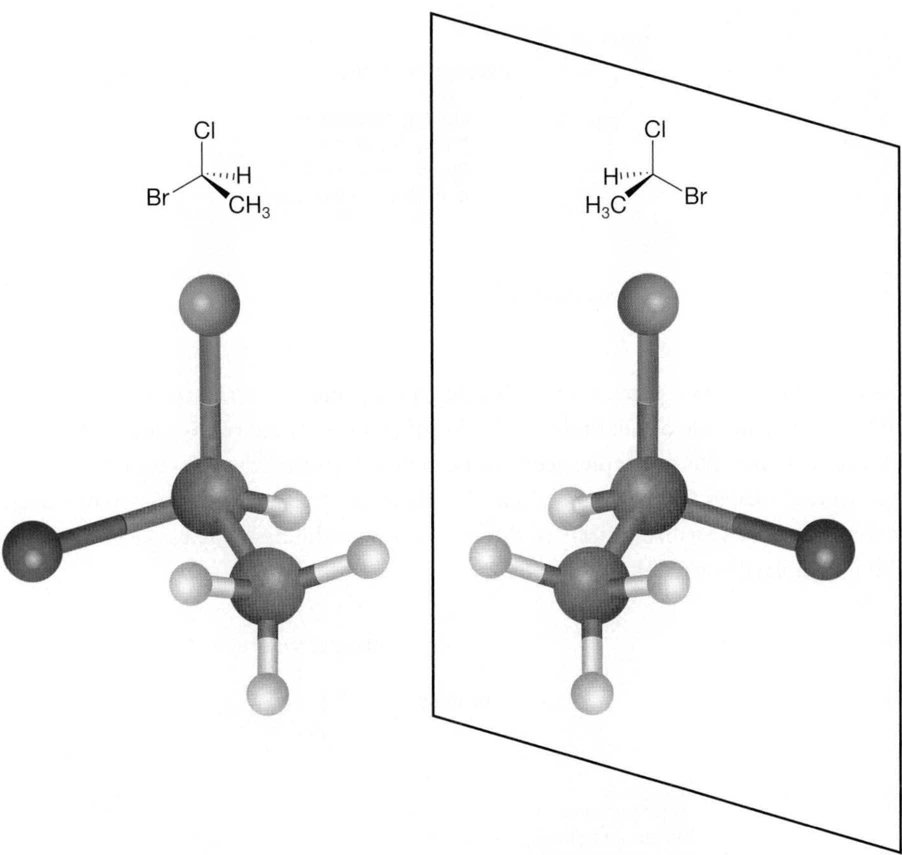

FIGURE 4.1 Mirror-image, three-dimensional structures of bromochloroethane.

It is always possible to draw a mirror image of a molecule, but the molecule is chiral only if the mirror images are non-superposable. For example, 1-butanol can be drawn as a pair of mirror images, but these mirror-image structures can be superposed as shown in the following diagram. So the two drawings are simply different views of the same molecule; they are *not* enantiomers. A compound that does not have an enantiomer is said to be **achiral**.

Achiral objects are superposable on their mirror images.

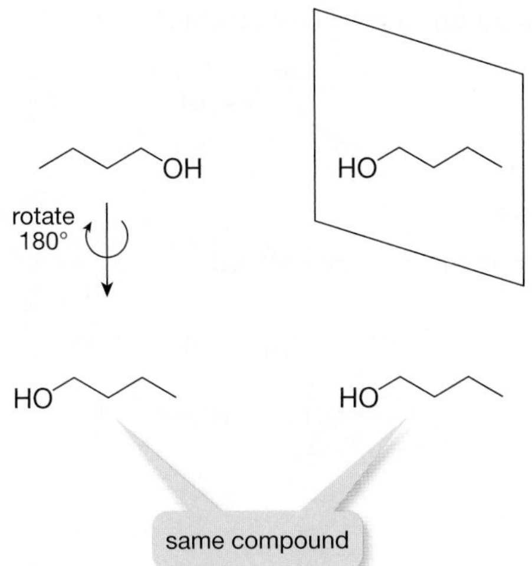

One way to determine whether an object is chiral or achiral is to look for a plane of symmetry. If a plane of symmetry exists, the object is achiral. A drawback to this method, especially in the case of acyclic molecules, is that it relies on the way a molecule is drawn.

STUDENT TIP
The most reliable way to determine if a molecule is chiral is to build a molecular model of the molecule and its mirror image and then try to superpose the models by moving or re-orienting them. You can rotate parts of the structure around single bonds as long as no bonds are broken.

STUDENT TIP
If a plane of symmetry exists in a molecule, that molecule cannot be chiral. The plane of symmetry may only be apparent for certain conformations. It may be necessary to rotate the single bonds in an achiral molecule to find the plane of symmetry.

bond up

plane of symmetry

zig-zag notation may create the illusion that the two sides of a molecule are different

OH

OH

bond down

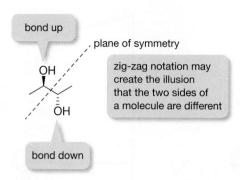

The preceding diagram depicts the molecule with a plane of symmetry. But due to the way it is drawn, the left-hand side of the molecule looks different from the right-hand side. To properly evaluate the structure, the molecule needs to be redrawn so that the left- and right-hand sides appear as mirror images. When this is done the plane of symmetry becomes more noticeable. Unless you can easily visualize objects in three dimensions, the best way to check for chirality is by building a molecular model.

OH OH

rotate around the single bond

re-draw

plane of symmetry

HO OH

CHECKPOINT 4.1

You should now be able to determine whether a molecule is chiral or achiral based on whether its mirror image is superposable onto the original structure.

SOLVED PROBLEM

Determine whether the following structure is chiral or achiral.

OH

APPROACH A: WITHOUT MODELS

STEP 1: Draw the mirror image of the structure provided.

mirror image

OH HO

STEP 2: Attempt to superpose the mirror image onto the original structure.

For two structures to be superposable, they must be identical. Begin by trying to transform the mirror image into the original structure by rotating about different axes.

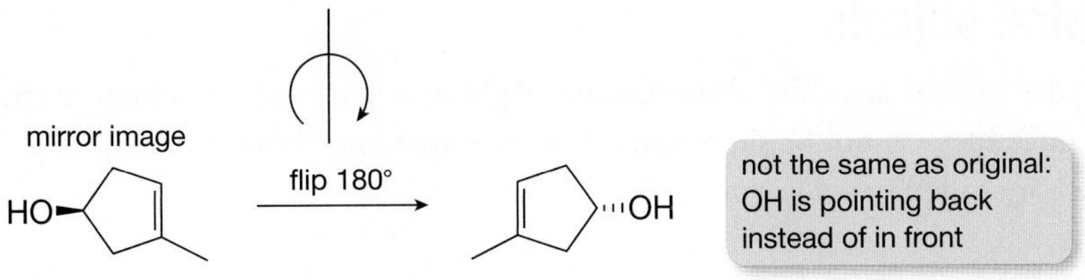

mirror image

flip 180°

not the same as original: OH is pointing back instead of in front

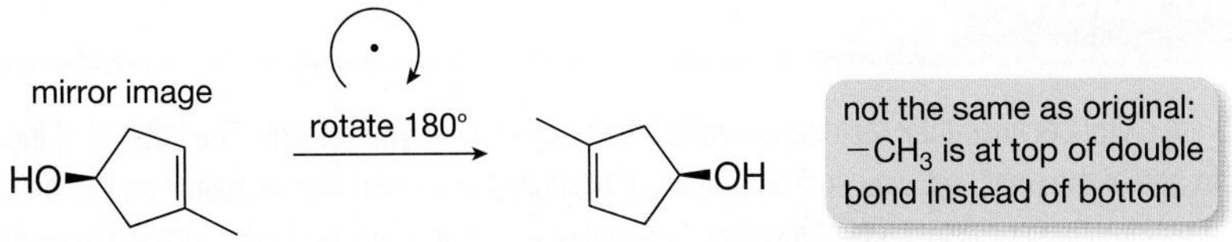

mirror image

rotate 180°

not the same as original: $-CH_3$ is at top of double bond instead of bottom

Rotating or flipping the mirror image fails to produce the original structure, so the two structures cannot be superposed.

STEP 3: Classify the molecule as either chiral or achiral.

The structure is chiral since the mirror image is not superposable onto the original structure.

APPROACH B: USING MODELS

STEP 1: Build a model of the original structure.

STEP 2: Build a model of the structure's mirror image. You might find it easier to first draw the mirror image, as shown in Approach A, Step 2, and then build a model of your drawing.

STEP 3: Attempt to superpose the two models onto each other. If you have built your two models correctly, you should find that it is impossible to superpose them.

STEP 4: Classify the molecule as either chiral or achiral.

Since the two structures are not superposable, the molecule is chiral.

PRACTICE PROBLEM

4.1 Use models to determine which of the following molecules are chiral.

a)

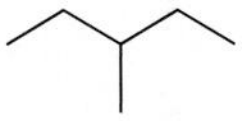

b)

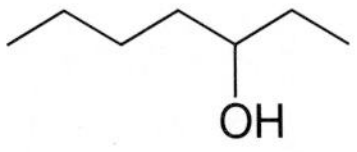

c)

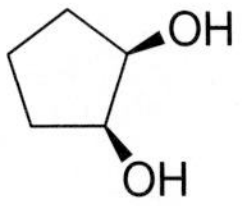

d)

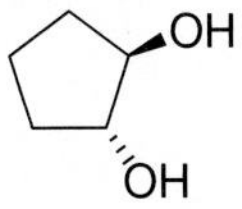

INTEGRATE THE SKILL

4.2 Determine which of the following compounds are chiral.

a) propan-2-ol

b) 1-bromo-1-chloroethane

c) 1,3-dimethylcyclopent-3-en-1-ol

4.2.1 Chiral objects

Many everyday objects are chiral. For example, right and left hands are chiral mirror images of each other since they cannot be superposed (try it). Pairs of non-molecular chiral objects such as hands or feet are called *enantiomorphs*.

DID YOU KNOW?

Internal organs are chiral because their distribution within the body is non-symmetrical. The left half of the human body is not a mirror image of the right half. Viewed from the front, typically the human liver is mainly on the left side of the body, and the large intestine coils clockwise. However, in about one person in 10 000 the internal organs are arranged in a way that is the mirror image of the normal layout of human organs (Figure 4.2). This condition, called *situs inversus,* is not harmful, can be detected only by a medical examination, and has nothing to do with being left-handed.

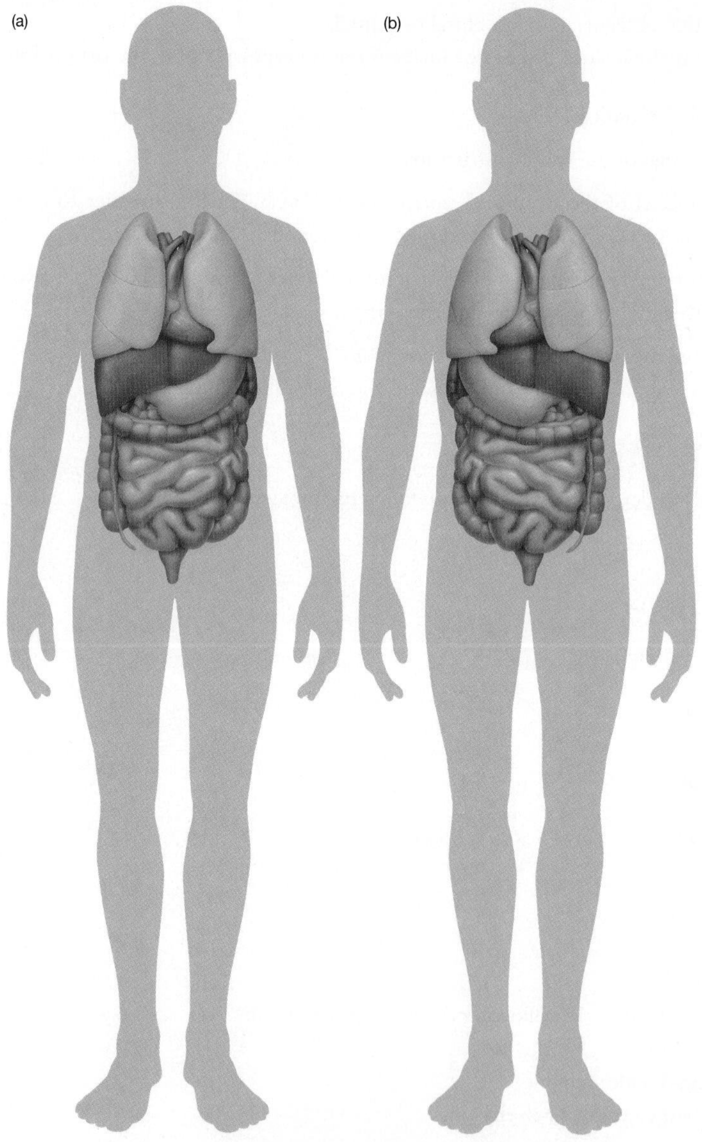

FIGURE 4.2 (a) Normal layout of internal organs in the human body; and (b) layout of internal organs in *situs inversus* humans.

CHECKPOINT 4.2

You should now be able to classify both objects and molecules as chiral or achiral based on whether a plane of symmetry occurs.

SOLVED PROBLEM

Classify each of the following as either chiral or achiral based on whether or not a plane of symmetry can be found.

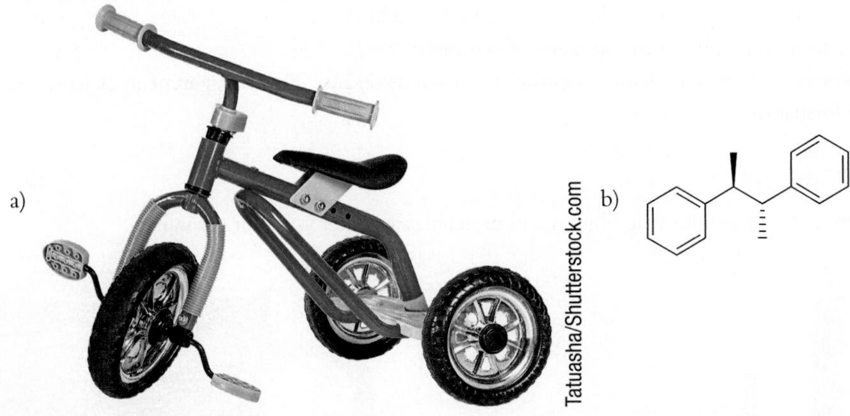

Tatuasha/Shutterstock.com

STEP 1: Look for a plane of symmetry in the object or molecule using the original view or conformation.

a) No plane of symmetry exists for the tricycle in its current view.

b) No plane of symmetry exists for the molecule in its current conformation.

STEP 2: Consider alternative views or conformations of the object or molecule to see whether a plane of symmetry can be found.

a) The tricycle viewed head on appears to have a plane of symmetry down the centre line, but closer inspection of the pedals shows that one is up while the other is down. Therefore, there is no plane of symmetry.

Tatuasha/Shutterstock.com

b) Rotating the original structure of the molecule 180° about the carbon-carbon single bond gives a new conformer with a plane of symmetry.

rotate C–C bond

plane of symmetry

STEP 3: Classify the object or molecule as either chiral or achiral.

a) The tricycle is chiral since no plane of symmetry exists.

b) The molecule is achiral since a plane of symmetry exists. This is apparent in at least one of its conformations.

PRACTICE PROBLEM

4.3 Classify each of the following objects and molecules as either chiral or achiral.

INTEGRATE THE SKILL

4.4 Prove that the following molecules are achiral by redrawing each molecule in a conformation that clearly shows a plane of symmetry.

a)

b)

c)

4.3 Chirality Centres

Molecules can be chiral only if they have three-dimensional shapes, which require atomic geometries that distribute bonds in three dimensions. Consequently, most chiral molecules possess at least one sp^3-hybridized atom.

Atoms that are connected to four different groups can give rise to chirality in molecules. These atoms are called **chirality centres, asymmetric centres, stereogenic centres**, or **stereocentres**. Carbon atoms are chirality centres more commonly than any other element, but many atom types are capable of being chirality centres. A sulfur atom can be a chirality centre when it is part of a sulfoxide functional group—that is, when the sulfur atom is connected to oxygen, a lone pair of electrons, and two *different* carbon groups.

Chirality centres are atoms that are connected to four different groups. These atoms are also called **asymmetric centres, stereogenic centres**, or **stereocentres**.

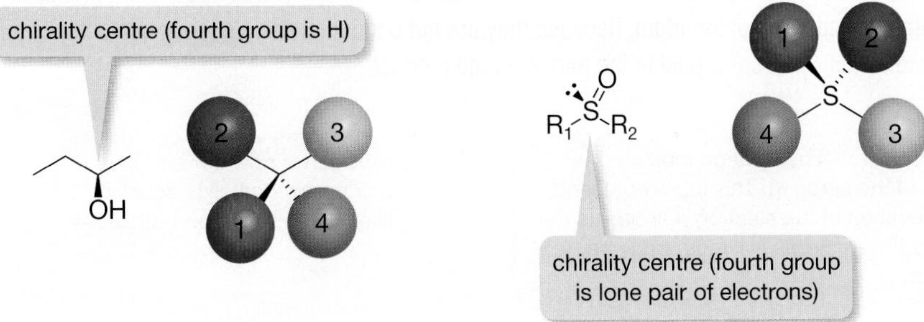

In a quaternary ammonium ion, nitrogen is connected to four groups. If the four groups are all different, the nitrogen atom is a chirality centre.

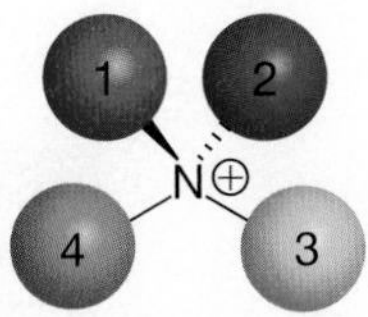

Nitrogen is most commonly bonded to three groups, but has a "hidden" fourth substituent: a lone pair of electrons. In principle, such nitrogens are stereogenic because they are surrounded by four different groups; the lone pair is counted as a group. However, the lone pair of electrons on most sp^3 nitrogens can rapidly "flip" from one side of the atom to the other because the sp^3 orbital has lobes on both sides of the atom. This process, known as nitrogen inversion, makes these mirror-image forms impossible to separate. Such atoms are *not* considered to be chirality centres unless other factors in the molecule lock the configuration of the atom in place. For example, both nitrogen atoms in Tröger's base are chirality centres because the rigid ring stops the nitrogen atoms from inverting (Figure 4.3). Consequently, the enantiomers of Tröger's base are stable and can be isolated.

Most nitrogen atoms are not considered to be chirality centres because the sp^3 orbital containing the lone pair permits a rapid equilibrium between mirror images, and so separate enantiomers cannot be isolated.

The rigid structure of Tröger's base does not allow the nitrogen atoms to invert, and so each nitrogen atom in this molecule is a chirality centre.

FIGURE 4.3 Tröger's base. Both nitrogens are chirality centres in this molecule.

DID YOU KNOW?

Nitrogen inversion is possible because a lone pair of electrons in an sp^3 orbital has a small probability of being found on the opposite side of the atom. Because they are not bonded, these electrons can actually tunnel to the other side to establish an sp^3 orbital in the mirror-image product.

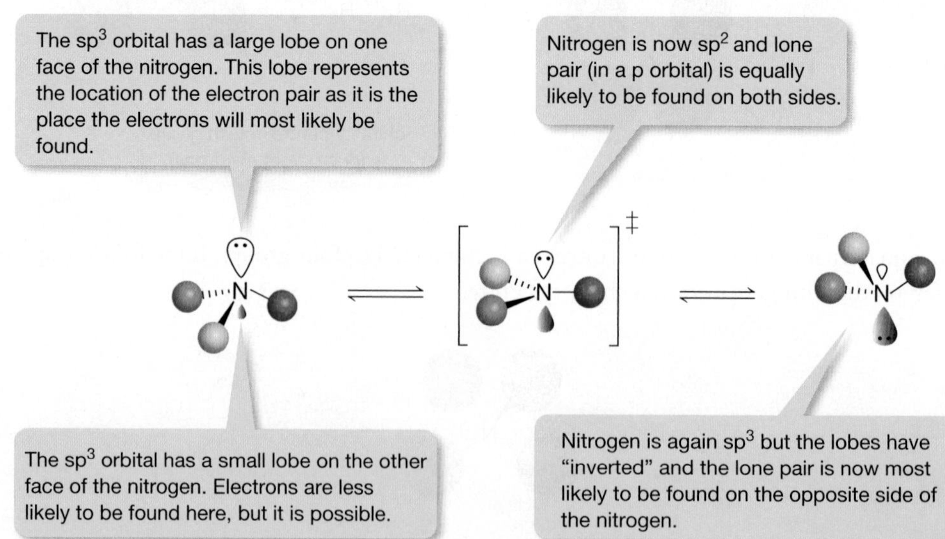

The sp³ orbital has a large lobe on one face of the nitrogen. This lobe represents the location of the electron pair as it is the place the electrons will most likely be found.
Nitrogen is now sp² and lone pair (in a p orbital) is equally likely to be found on both sides.
N
N
N
‡
The sp³ orbital has a small lobe on the other face of the nitrogen. Electrons are less likely to be found here, but it is possible.
Nitrogen is again sp³ but the lobes have "inverted" and the lone pair is now most likely to be found on the opposite side of the nitrogen.

A connected atom cannot easily move from one side of a nitrogen to the other. Because the nucleus involved in the bond is positively charged, the negatively charged electrons spend most of their time on the face that is toward the atom, and this prevents inversion from happening.

N
The atom involved in the bond prevents the electrons in the bond from straying too far. The nitrogen atom cannot invert.

STUDENT TIP

The presence of a chirality centre is a useful clue when you want to determine whether a molecule is chiral. Molecules with a single chirality centre are always chiral; however, molecules with more than one chiral centre may or may not be chiral, as you will see in Section 4.6.

CHECKPOINT 4.3

You should now be able to identify the chirality centres in a molecule.

SOLVED PROBLEM

Tetracycline is a commonly prescribed antibiotic. Determine the number of chirality centres in the following structure of tetracycline.

STEP 1: Identify all of the sp^3-hybridized carbon atoms in the molecule.

There are six sp^3-hybridized carbon atoms, indicated by the asterisks in the following structure.

* sp^3 carbons

STEP 2: Determine which of the sp^3-hybridized carbon atoms have four unique atoms or groups bonded to them. These are chirality centres. Of the six sp^3-hybridized carbon atoms, five have four unique atoms or groups attached. (Hint: It might be helpful to explicitly draw any missing hydrogen atoms before proceeding with this step.)

* chirality centres

STEP 3: Check for any non-carbon-based chirality centres, such as conformationally rigid nitrogen atoms or sulfoxides.

There are no other chirality centres in the molecule.

STEP 4: Count the total number of chirality centres identified.

There are five chirality centres in tetracycline, all carbon-based.

PRACTICE PROBLEM

4.5 Locate all of the chirality centres in the following molecules.

a) HO O H N O O H₂N O

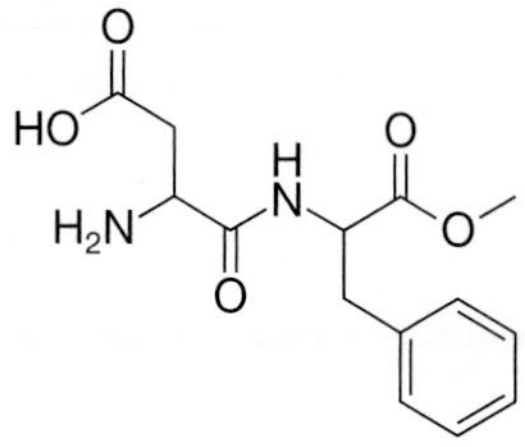

b) O

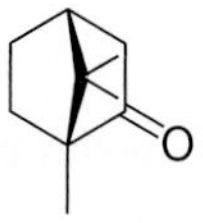

c) OH HO

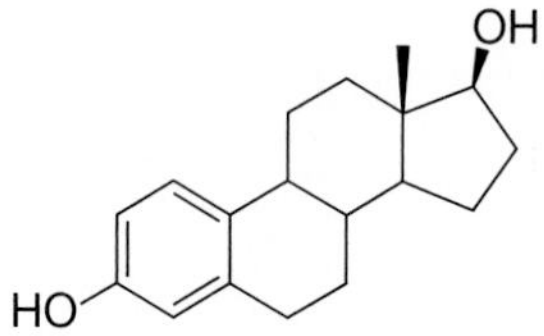

d) OH O O S NH₂

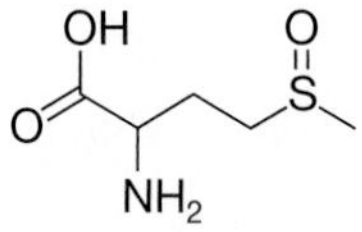

INTEGRATE THE SKILL

4.6 There are 14 constitutional isomers of $C_5H_{12}O$. Draw all 14 structures, and determine which of them contain at least one chirality centre.

Some molecules are chiral even though none of their atoms is a chirality centre. These molecules normally contain some kind of structural twist that imparts chirality. Allenes and some alkenes have this kind of asymmetric structure. Although the allene shown has no sp^3-hybridized atoms holding four different substituents, the π orbitals on the central carbon atom are at right angles to each other (see Chapter 1), giving the molecule a three-dimensional shape that resembles a stretched tetrahedron. (This structure is easiest to see in a model.) This shape renders the mirror images non-superposable, so the molecule is chiral.

chiral allene

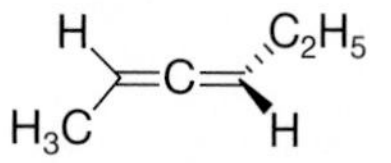

H C_2H_5 C H_3C H

The π bonds in this molecule are 90° to each other. This locks a twist into the molecule and the two mirror images cannot superpose.

Some molecules are chiral because they have features (called *conformational restrictions*) that prevent full rotation of some bonds. Materials containing such restrictions can have *conformations* that are non-superposable mirror images. The resulting enantiomers do not interconvert easily and can be separated. Such enantiomers are called **atropisomers**.

Atropisomers are stereoisomers that result from hindered rotation around a single bond.

OH
OH
HO
HO

Binol exists as non-superposable mirror images because the single bond connecting the rings cannot rotate through 360°.

Viewing the molecule along the single bond from this direction (top view) shows the butterfly shape.

Ring hydrogens collide, preventing full rotation. The molecule exists in a chiral butterfly shape.

H
H
OH
OH

OH
OH
top view

An example of such a material is binol. All the carbon atoms in this compound are sp^2 hybridized, and there are two flat-ring systems joined by a single bond. In principle, this single bond can rotate, but rotation to produce the mirror image is blocked by interference between the two hydrogen atoms that are explicitly drawn in the lower left structure above. As a result, the molecule has propeller-shaped atropisomers in which the two flat double rings are perpendicular to each other, as shown in the top view of the molecule. The atropisomers are distinct molecules that can be separated and isolated.

Some cyclic chiral molecules, called *helicenes*, do not have rotatable bonds or stereogenic atoms; instead, they are helical (corkscrew-shaped). These helical structures curve either clockwise or counter-clockwise. The resulting atropisomers cannot easily interconvert because the top and bottom rings overlap.

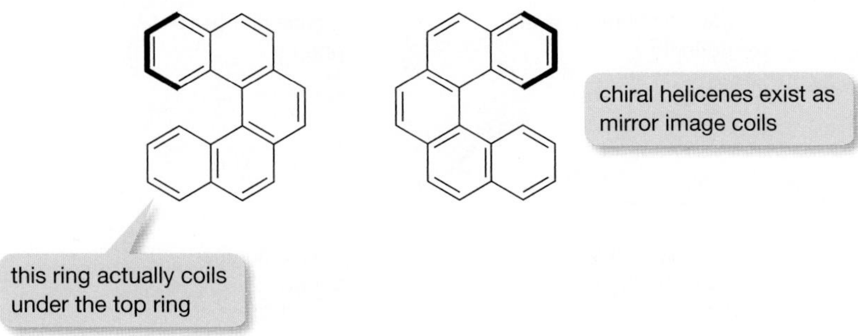
chiral helicenes exist as mirror image coils
this ring actually coils under the top ring

4.4 Cahn-Ingold-Prelog Nomenclature

The **absolute configuration** of a chirality centre describes the exact three-dimensional arrangement of the atoms around a chirality centre.

The Cahn-Ingold-Prelog system is used to describe the **absolute configuration** of a chirality centre. When using this system, each chirality centre in a molecule is separately assigned a configuration. The first step in the process is to identify each chirality centre. Stereochemical notation (hash or wedge bonds) in a molecular drawing often indicates that an atom *may* be a chirality centre. The second step is to assign a priority to each of the substituents based on the atomic numbers of the atoms *directly* attached to the centre. The third step is to determine the configuration of the centre by rotating the molecule according to the orientation of the assigned priorities.

4.4.1 Cahn-Ingold-Prelog priorities and configurations

Each of the substituents attached to a chirality centre is assigned a *priority* based on the atomic numbers of the atoms *directly* attached to the centre. The atom with the highest atomic number is assigned the highest priority; the atom with the second highest atomic number is given the second priority, and so on, until all four substituents have been assigned a priority. Electron pairs have an atomic number of 0.

For example, the following diagram shows the carbon bearing the bromine is connected to four different substituents (CH_3, Br, OCH_3, H); therefore, it is a chirality centre. Bromine has the highest atomic number, so it has first priority. Oxygen has an atomic number of 8 and has second priority, and carbon (atomic number 6) is next, with third priority. Only the atoms that are directly attached to the chirality centre are considered when assigning priority. The additional atoms on the oxygen and methyl carbon are not considered unless there is a "tie"; that is, the atoms bonded directly to the centre of chirality are the same element (see Section 4.4.2). The fourth atom that is connected to the chirality centre is a hydrogen, which has the lowest atomic number and therefore the lowest (fourth) priority.

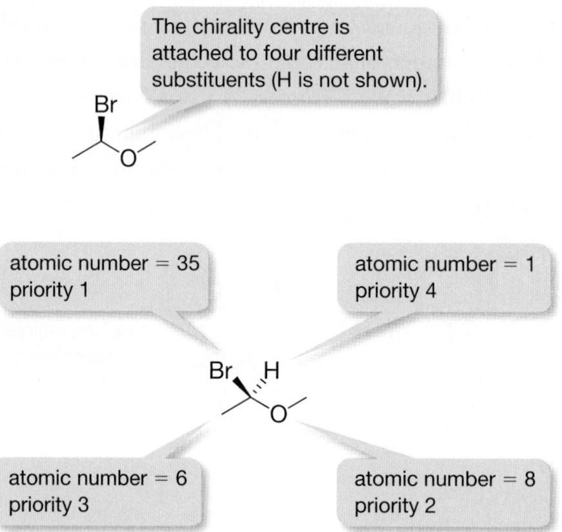

After the priorities have been assigned, the configuration of the chirality centre is determined by rotating the molecule to place the lowest (fourth) priority group facing away from the viewer (in the diagram, the fourth priority group already projects back into the page). The orientation of the first three priority groups (1, 2, 3) determines the configuration of the centre. If the sequence proceeds clockwise, the chirality centre is designated *R* (*rectus*, Latin for "right"). If the priority sequence is counter-clockwise, the centre is *S* (*sinister*, Latin for "left"). In the following diagram, the priority sequence of the substituents is clockwise; therefore, the configuration of the stereocentre is *R*.

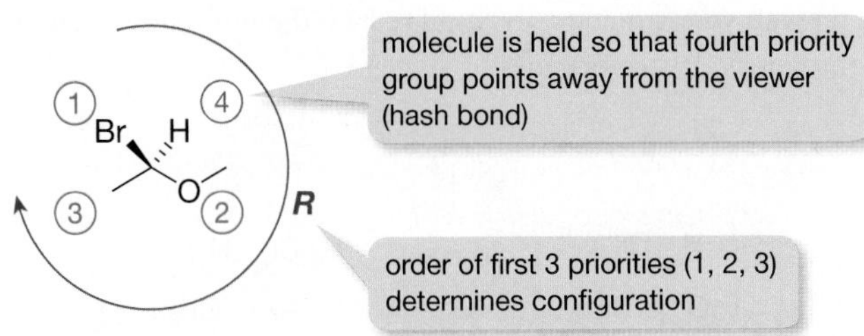

ORGANIC CHEMWARE
4.2 Cahn-Ingold-Prelog sequence rules

STUDENT TIP
By using a model, you can reliably rotate a molecule to place the lowest priority group in back. Faster manipulations can be done on paper using the techniques described in Section 4.5.1.

ORGANIC CHEMWARE
4.3 *R*/*S* configuration

CHECKPOINT 4.4

You should now be able to determine the absolute configuration of a chirality centre with four different groups.

SOLVED PROBLEM

Halothane is a commonly used anesthetic. Both the *R* and *S* enantiomers of halothane are effective. Determine which enantiomer is shown below.

Cl
F_3C Br

STEP 1: Locate the chirality centre in the molecule. There are two sp^3-hybridized carbon atoms in this molecule. However, only one of them is attached to four different groups: the one marked with the asterisk. (The dashed wedge used for the carbon-chlorine bond is a good hint that this carbon might be a chirality centre.)

Cl
F_3C * Br

STEP 2: If a hydrogen atom has been omitted from the chirality centre, redraw the structure to show the stereochemistry of the hydrogen atom. For tetrahedral geometry, the implied hydrogen must be drawn coming out of the plane of the page.

H Cl
F_3C Br

STEP 3: Assign priorities to each atom bonded to the chirality centre based on the atomic number of the bonded atom.

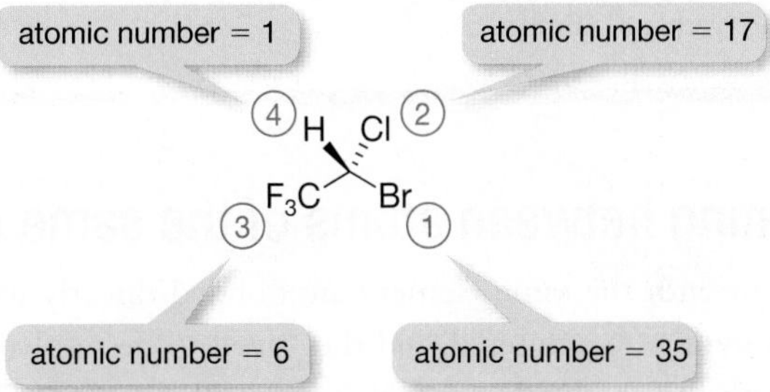

STEP 4: Rotate the structure so that the lowest priority group is pointing back into the page.

You can do this either with a model or on paper. To use a model, build the structure exactly as drawn above and then hold it so that the lowest priority group (the hydrogen atom) is pointing away from you.

Alternatively, imagine flipping your structure 180° and redraw the molecule. (Section 4.5 describes some additional techniques for redrawing stereocentres.)

flip 180°

STEP 5: With the lowest priority group pointing away from you, trace a pathway through the first three priority groups, and assign the absolute configuration of the stereocentre.

The pathway traced is clockwise; therefore, this is the *R* enantiomer of halothane.

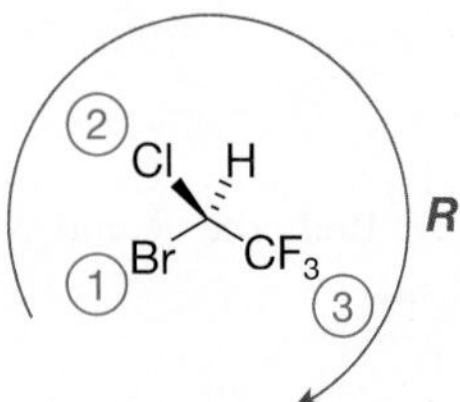

PRACTICE PROBLEM

4.7 Determine the absolute configuration of each chirality centre in the following molecules.

a)

b)

c)

d)

INTEGRATE THE SKILL

4.8 Draw the *R* enantiomer of each of the following compounds.

a) 1-iodoprop-2-en-1-ol

b) 1-fluoro-1-methoxyethan-1-amine

4.4.2 Differentiating between atoms of the same element

When substituent groups with the same element are bonded directly to the chirality centre, a tie occurs. Ties are frequently encountered, and they need to be resolved in order to differentiate between atoms of the same element and assign them priorities. This is accomplished by examining the atoms immediately adjacent to the ones directly attached to the chirality centre. The adjacent atoms are compared by starting with the ones with the highest atomic numbers and then working down the chain until the *first* difference is encountered. *Stop at the first point of difference.* Note that adjacent atoms are considered *only* when two or more directly bonded atoms are tied.

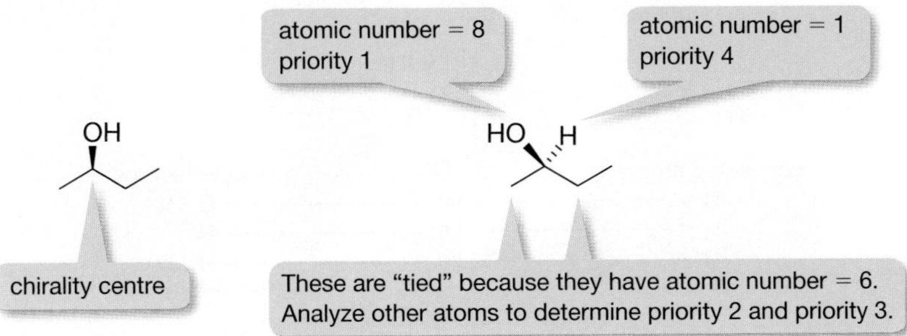

For example, the chirality centre of the molecule 2-butanol—shown in the following diagram—bears four different groups (CH_3, CH_3CH_2, OH, H). Of the four directly attached atoms, oxygen has the highest priority and hydrogen the lowest. However, the two carbon atoms attached to the stereocentre are tied. To resolve the tie, the atoms directly attached to these two carbons need to be considered. The carbon on the left is connected to three hydrogens; the carbon on the right is connected to a carbon and two hydrogens. Since carbon has a higher atomic number than hydrogen, the group on the right (ethyl) has a higher priority than the group on the left (methyl). This resolves the tie, makes it possible to assign priorities, and determines the overall configuration is *R*.

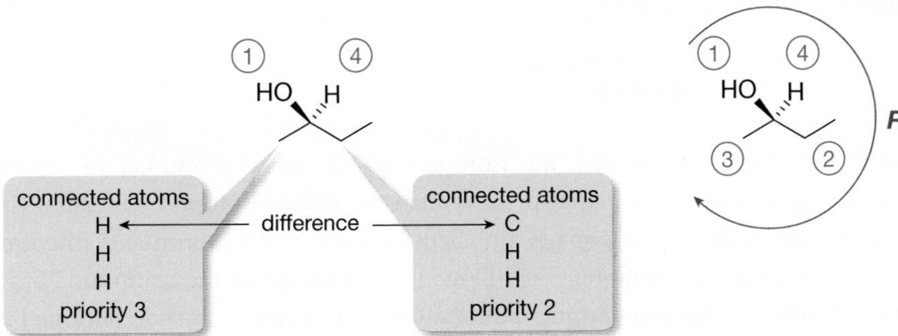

To look for more ties, compare all the immediately adjacent atoms before advancing further along a chain. The cyclic compound in the following diagram has a stereogenic carbon connected to an OH group. Of the substituents on this carbon, oxygen has the highest atomic number and hydrogen the lowest, and the two carbon atoms connected on either side are tied.

atomic number = 1
priority 4
atomic number = 8
priority 1
H OH
atomic number = 6
atomic number = 6
O
tie

The atoms directly connected to these two carbons are also tied.

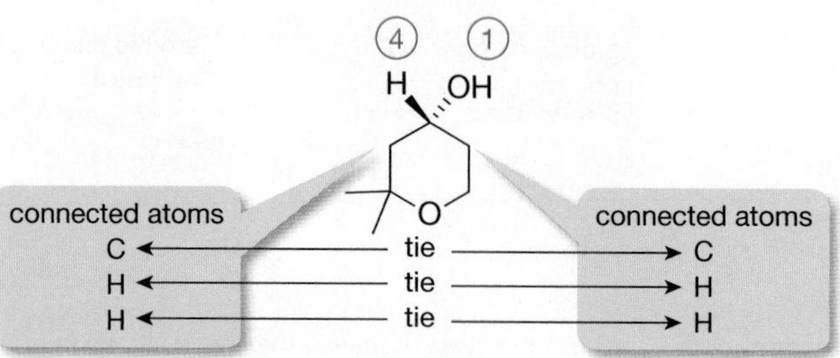

To resolve the tie, consider the next atom in the chain on each side. Again, compare the atoms with the highest atomic numbers first. Since the comparison in this case shows a difference, the substituent groups can be assigned priorities. Based on these priorities, the configuration of the centre is *R*.

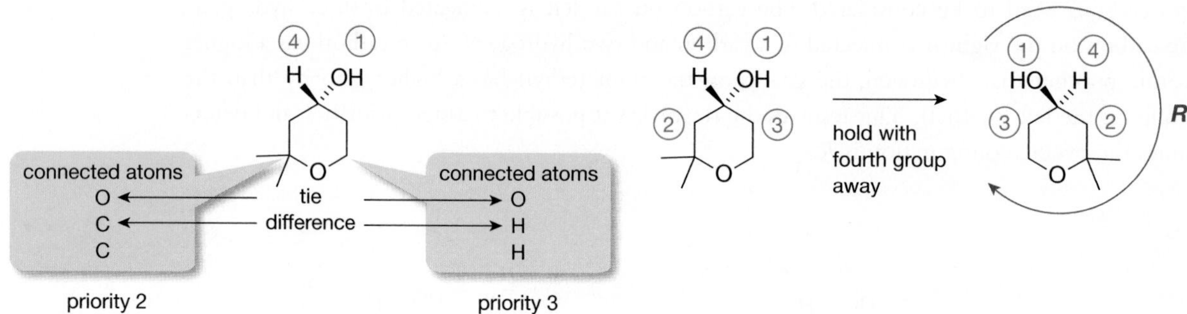

When isotopes are encountered, the isotope with the greater number of neutrons (and hence greater atomic mass) has a higher priority than the lighter isotope. Deuterium (symbol D or ^{2}H) is an isotope of hydrogen that has a nucleus consisting of a proton and a neutron. Since deuterium is not the same as hydrogen, the following diagram shows the compound has a stereocentre. At this centre, the oxygen (atomic number 8) has first priority; carbon (atomic number 6) has second priority; the deuterium atom (atomic number 1, *mass* 2) has third priority; and the hydrogen atom (atomic number 1, mass 1) has fourth priority. The configuration of this centre is therefore *R*.

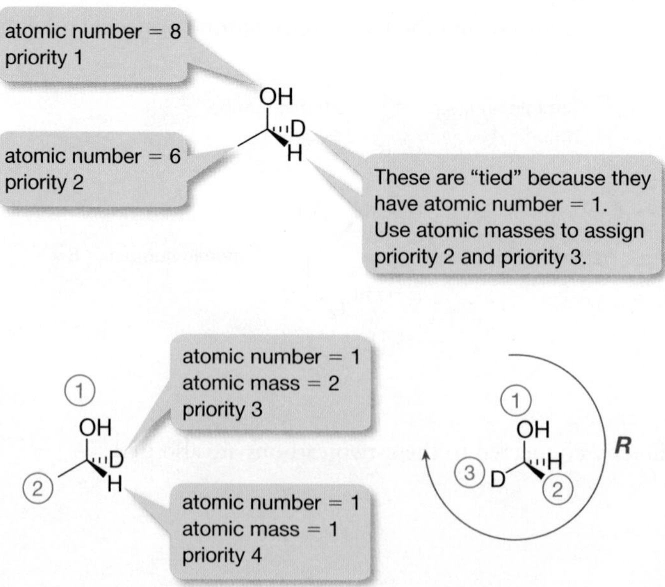

4.4.3 Groups with multiple bonds

The Cahn-Ingold-Prelog system treats double or triple bonds as if they were multiple single bonds by using a procedure known as the "phantom atom rule." To assign priorities to substituents on atoms with double or triple bonds, redraw the substituents, duplicating the atoms that are attached by π bonds to create singly bonded groups and atoms. Note that the resulting diagram is only an analytic tool for assigning priorities; it includes phantom atoms that do not actually exist, so it does not represent the real structure of the molecule.

For example, in the following diagram, the carbon in the imine group is doubly bonded to a nitrogen, so the π bond on this carbon is replaced by a single bond to a phantom nitrogen atom. Similarly, the nitrogen is connected by the π bond to the carbon, so a phantom single-bonded carbon is added to the nitrogen. Notice that only the π-bonded atoms are added as phantoms, not entire groups.

This atom is connected to a nitrogen by the π bond, so duplicate a phantom nitrogen atom.
H
N
CH3
H
N
N
CH3
C
This atom is connected to a carbon by the π bond, so duplicate a phantom carbon atom.

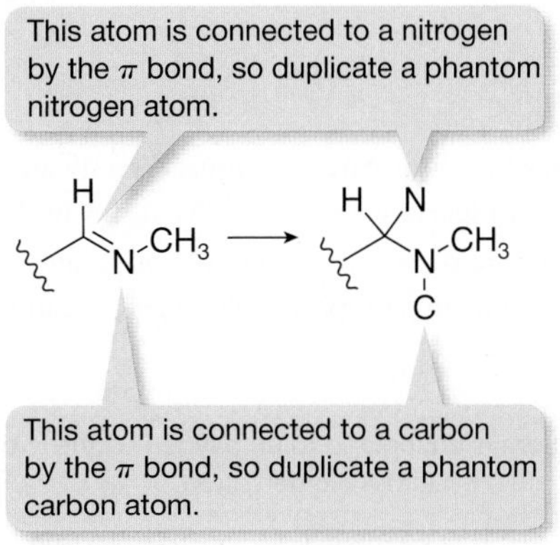

CHECKPOINT 4.5

You should now be able to determine the absolute configuration of a chirality centre with four different groups.

SOLVED PROBLEM

Determine the configuration of the following aldehyde.

OH
O
H
O
O

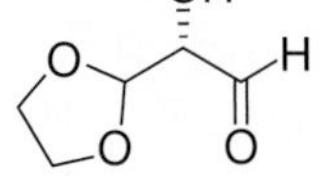

STEP 1: Locate the chirality centre in the molecule. The central carbon in the molecule is a stereocentre since it carries four different groups.

STEP 2: If a hydrogen atom has been omitted from the chirality centre, redraw the structure to show the stereochemistry of the hydrogen atom. For tetrahedral geometry, the implied hydrogen must be drawn coming out of the plane of the page.

HO
H
O
H
O
O

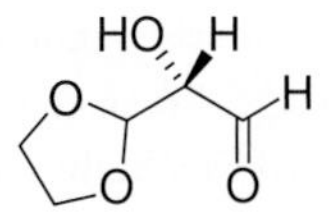

STEP 3: Begin assigning priorities to each group attached to the chirality centre based on the atomic numbers of the directly bonded atoms.

The OH group has the highest priority (atomic number for oxygen is 8) and the hydrogen has the lowest priority (atomic number 1). The two carbons are tied for second and third priority (atomic number 6).

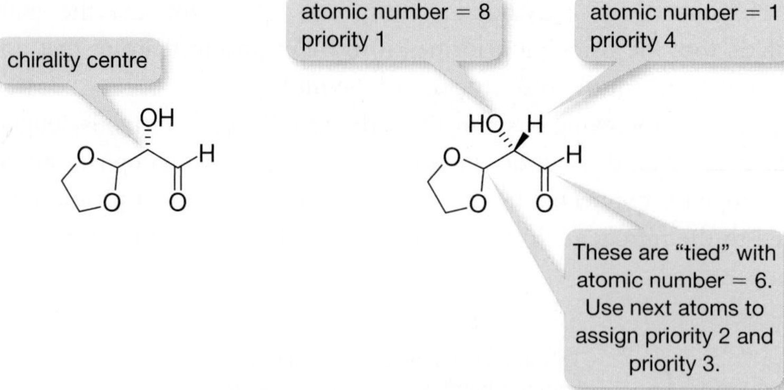

STEP 4: Attempt to resolve any ties by comparing the atoms directly attached to the tied atoms. If further ties are encountered, continue the comparison with the next atoms in the chain. If double or triple bonds are part of the comparison, they should be redrawn using phantom atoms. Since one of the tied carbons is part of a double bond, it cannot be directly compared. The C—O double bond needs to be redrawn with phantom atoms replacing the π bonds.

This atom is connected to an oxygen by the π bond; duplicate a phantom oxygen atom.

becomes:

This atom is connected to a carbon by the π bond; duplicate a phantom carbon atom.

Each of the carbons is now connected to two oxygens and a hydrogen, so the carbons are still tied. A further branch must be examined.

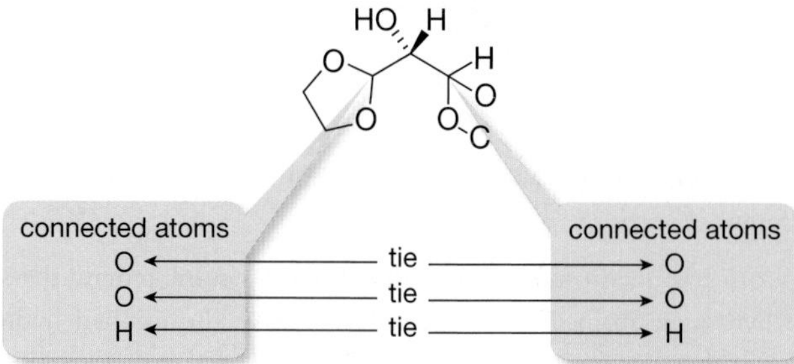

In the chain on the left side, the next atom with the highest atomic number is an oxygen. On the right, the next atom with the highest atomic number is also an oxygen. The oxygen indicated in the following diagram is compared first because it is connected to a higher priority chain than the other oxygen is.

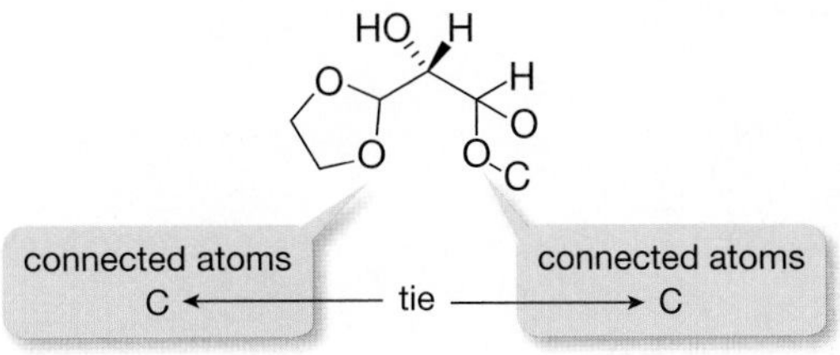

connected atoms C ← tie → C connected atoms

Once again, the connected atoms are tied. All the atoms in the branch have been examined, so the next branch is now considered. This comparison does show a difference, which finally resolves the tie and allows the priorities to be assigned.

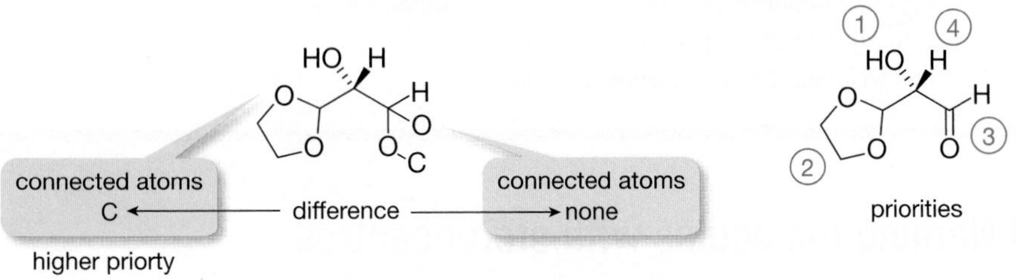

connected atoms C ← difference → none connected atoms

higher priorty

priorities

STEP 5: Rotate the structure so that the lowest priority group is pointing back into the page.

With the current view, the lowest priority group (H) is pointing out of the page. The structure needs to be viewed from the other side. You can either build a model of the structure and physically rotate it so that the H is pointing away from you, or you can redraw the structure, flipping it 180° as shown.

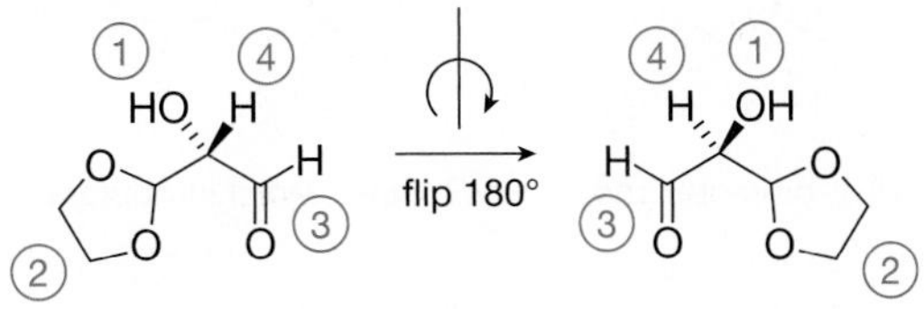

flip 180°

STEP 6: With the lowest priority group pointing away from you, trace a pathway through the first three priority groups, and assign the absolute configuration of the stereocentre. The resulting first three substituent priorities are arranged clockwise, giving the chirality centre an *R* configuration.

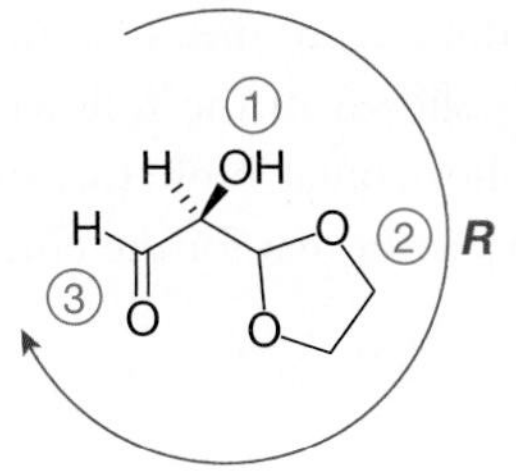

PRACTICE PROBLEM

4.9 Determine the absolute configuration of each chirality centre in the following molecules.

a)

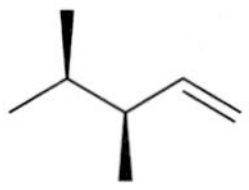

b)

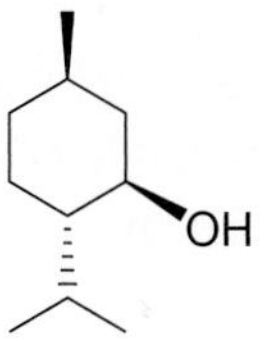

c) HO OH OH OH

d) H N O CF3

INTEGRATE THE SKILL

4.10 Draw the *S* enantiomer of each of the following compounds.
   a) 3-(*tert*-butyl)-3-isopropylpent-1-en-4-yne
   b) 1-formylcyclohex-2-ene-1-carboxylic acid

4.4.4 Naming molecules with stereocentres

If a molecule contains only one stereocentre, the configuration is indicated by an italic *R* or *S* in parentheses at the start of the name of the compound. When the molecule has more than one chirality centre, the atom number indicating location precedes each configuration symbol.

(*R*)-2-bromobutane (2*R*, 3*S*)-2,3-dichlorobutane

4.5 Drawing Enantiomers

Several drawing techniques are used to represent, compare, and classify the structures of enantiomers.

The fundamental way to draw the enantiomer of a molecule is to use a mirror plane to reflect the molecule. For example, as shown in the following diagram of 2-bromobutane, the mirror plane can be either vertical or horizontal. Note that the configuration of *each* stereocentre in the mirror image is always the opposite of that for the corresponding stereocentre in the original molecule.

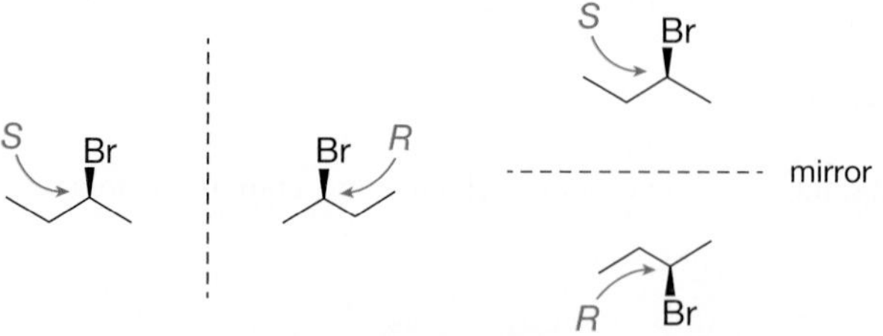

A more common method for drawing the opposite enantiomer is to *invert* the chirality of the centre by changing a hashed bond to a wedge or a wedge to a hashed bond. When using this method, it is important to preserve the orientation of the backbone of the structure.

S Br invert R Br

backbone zig-zags remain the same

Using this latter inversion process generally makes changes in configuration readily identifiable. The process interchanges two of the substituents attached to the chirality centre (H and Br). Exchanging *any two* groups on a chirality centre changes that chirality centre to the opposite configuration. So exchanging the bromine and the CH_3 group of (*S*)-2-bromobutane also produces the *R* enantiomer.

enantiomers

S Br switch any two groups (e.g., CH_3 and Br) R Br

Two sequential exchanges on a chirality centre can be used to generate a new representation of that centre. Exchanging any two groups on a chirality centre inverts that centre. Making a second exchange of any two groups reverts the centre back to its original configuration. So making two exchanges with different pairs of groups produces a different view of the original centre.

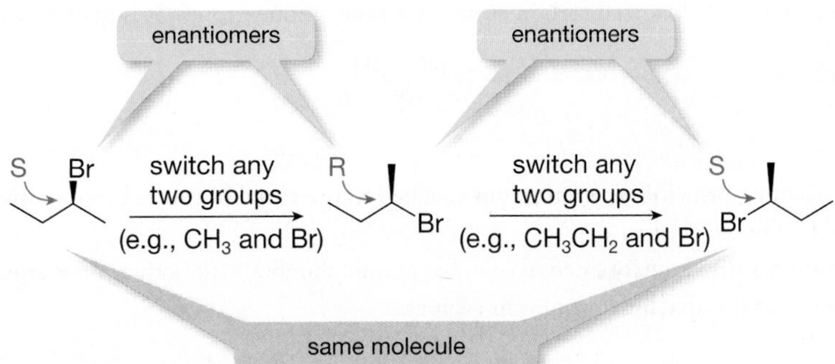

STUDENT TIP
Atoms in line structures are assumed to carry full octets. It may be helpful to draw the hydrogens on chirality centres before manipulating the structures.

4.5.1 Use of double exchanges in assigning configuration

To understand the double-exchange process for assigning configuration to a chirality centre, consider the molecule in the following diagram. In this case, to classify the configuration of the chirality centre, the molecule must be drawn such that the fourth priority group occupies the hashed-bond position. The first step involves exchanging the fourth priority and second priority groups, but this move also produces the enantiomer of the original molecule. The second step involves exchanging any two substituents *other than the fourth priority group*; this produces a view of the starting molecule with the fourth priority group at the back. This view makes it possible to determine the configuration of the chirality centre.

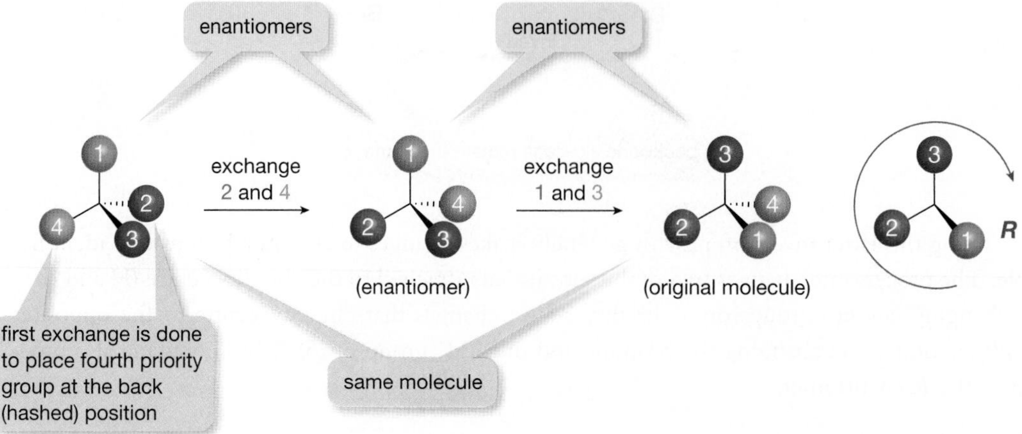

CHECKPOINT 4.6

You should now be able to use inversions and double inversions to draw enantiomers and alternative views of molecules, respectively.

SOLVED PROBLEM

Determine the absolute configuration of the following molecule *without using molecular models.*

STEP 1: Locate the chirality centre in the molecule.

There is only one sp^3-hybridized carbon atom with four unique groups. It is marked with an asterisk.

STEP 2: Begin assigning priorities to each group attached to the chirality centre based on the atomic numbers of the directly bonded atoms.

The OH group has first priority since oxygen has atomic number 8; the other three groups are tied and have a carbon atom as the attachment point in each case.

STEP 3: Attempt to resolve any ties by comparing the atoms directly attached to the tied atoms. If further ties are encountered, continue comparing the next atoms in the chain. If double or triple bonds are part of the comparison, redraw them using the phantom atom rules.

The nitrile group (CN) contains a triple bond, so a direct comparison of the tied carbon atoms is not possible until the CN group is redrawn using the phantom atom rules.

Using the artificial structure, you can now compare the three tied carbon atoms and assign priorities.

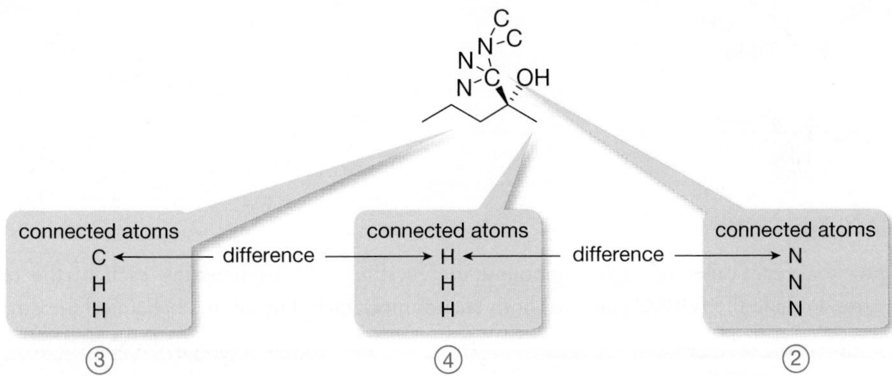

STEP 4: Rotate the structure so that the lowest priority group is pointing back into the page.

Use a double inversion to redraw the molecule in a new orientation. To get the lowest priority group (methyl) pointing back into the page, switch the methyl and OH groups, thus generating the enantiomer of the original structure. To get back to the original structure, switch any other two groups (not the methyl group since it needs to stay pointing back behind the page). The resulting structure is a conformer of the original structure.

switch 1 and 4 → switch 1 and 2 →

enantiomer of original structure

original structure, redrawn as new conformer

STEP 5: With the lowest priority group pointing away from you, trace a pathway through the first three priority groups and assign the absolute configuration of the stereocentre. The resulting first three substituent priorities are arranged clockwise, giving the chirality centre an *R* configuration.

PRACTICE PROBLEM

4.11 Determine the absolute configuration of each chirality centre in the following molecules *without using molecular models.*

a)

b)

c)

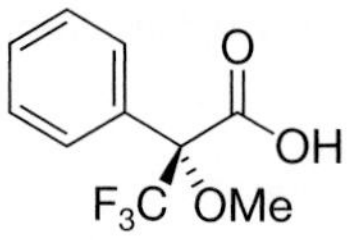

d)

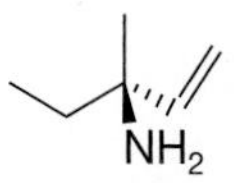

INTEGRATE THE SKILL

4.12 Draw the enantiomer of each compound in Question 4.11 by inverting each of the chirality centres. Provide the IUPAC name for both enantiomers, including all stereochemical information.

4.6 Diastereomers

Diastereomers are stereoisomers that are not enantiomers.

Molecules often contain more than one chirality centre; most biological molecules contain many. Since each chirality centre can have either *R* or *S* configuration, molecules with multiple centres have many stereoisomers, some of which are not mirror images of each other. Stereoisomers that are not enantiomers are called **diastereomers**.

In the following diagram, the amino alcohol shown on the left has two chirality centres. The carbon connected to the OH group has an *R* configuration, as does the carbon bearing the NH_2 group. This three-dimensional molecule has a non-superposable mirror image. In the mirror image, the configurations of *all* of the stereocentres are inverted relative to those in the original structure. Inverting *all* of the stereocentres in a molecule creates the mirror image of the molecule. Once the configurations of the centres in one isomer are known, the configurations of the chirality centres in its enantiomer and diastereomers are readily assigned since a configuration can only be *R* or *S*.

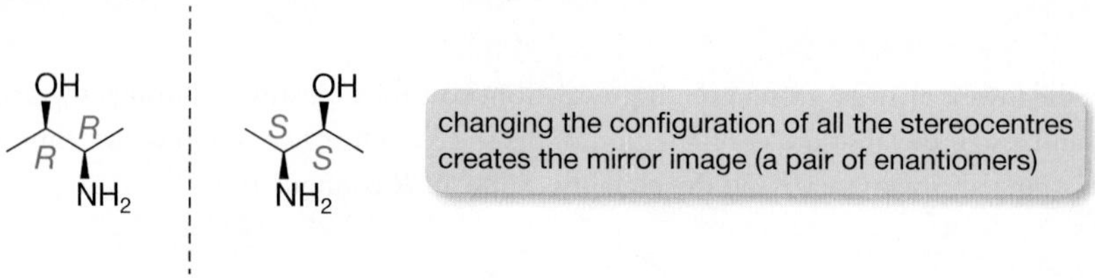

Changing the configuration of *some* of the stereocentres produces a diastereomer. In the following example, the stereocentre connected to the oxygen has been inverted relative to the configuration of the (*R*, *R*) stereoisomer, whereas the chirality centre bearing the nitrogen is unchanged. This new molecule is a stereoisomer of both enantiomers of the original molecule, but is not a mirror image of either of them.

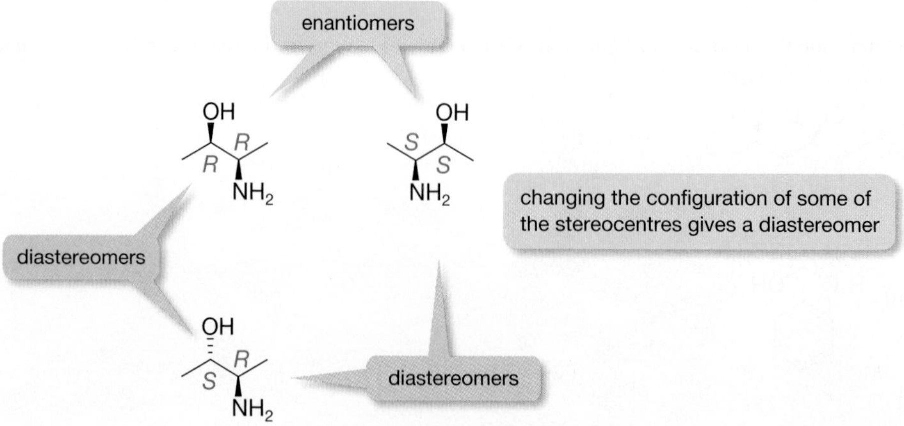

The *R* configuration of the nitrogen is unaltered between the (*R*, *R*) stereoisomer and its diastereomer. In the diastereomer, the configuration of the stereocentre with the OH group is inverted, and therefore must be *S*.

The mirror image of the compound on the lower left is non-superposable, so this mirror image and the compound on the lower left are enantiomers. A comparison of the configurations of the chirality centres shows that the compound on the lower right differs from both enantiomers in the top row, and so it is a diastereomer of each.

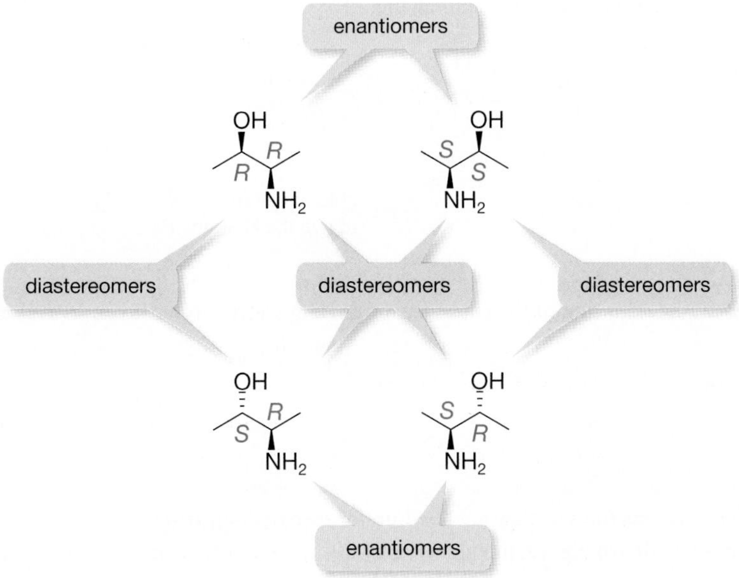

CHECKPOINT 4.7

You should now be able to distinguish between stereoisomers that are enantiomers and those that are diastereomers.

SOLVED PROBLEM

Determine whether the following stereoisomers are enantiomers or diastereomers.

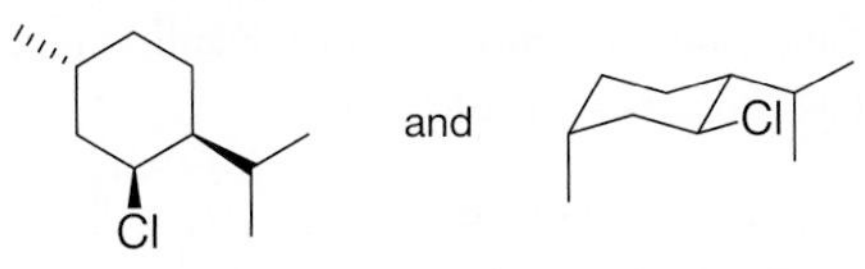

APPROACH A: COMPARE STRUCTURES

STEP 1: Since the structures are shown in different formats, you first need to redraw one of the structures so that both are shown in the same format. For example, redraw the chair conformation as a zig-zag line drawing. (Alternatively, you could convert the zig-zag structure into a chair representation.)

Start by drawing in the missing hydrogen atoms on the chirality centres of isomer B:

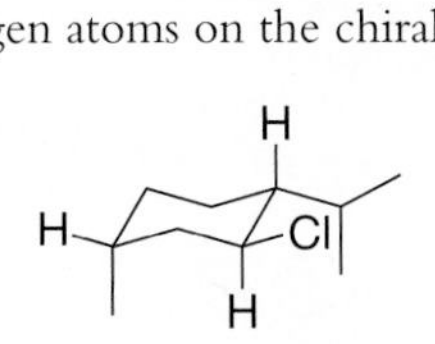

To convert the zig-zag line drawing, imagine viewing the structure from above. Since the isopropyl and methyl groups point down relative to their hydrogen atoms, they point back into the plane of the page in the zig-zag drawing.

isopropyl group is below the H atom

methyl group is below the H atom

At the middle chirality centre, the chlorine atom is above the hydrogen atom and therefore points in front of the plane of the page in the zig-zag drawing.

chlorine atom is above the H atom

Thus the zig-zag line drawing of the chair structure when viewed from the top is as shown as follows:

STEP 2: Re-orient the structures so they are lined up for ease of comparison.

Rotating the newly drawn zig-zag line drawing by 60° gives an orientation matching that of isomer A.

rotate 60° clockwise

STEP 3: Attempt to superpose the two structures onto each other.

With the two structures now in the same form and drawn with the same orientation, it is possible to see that they cannot be superposed. While the methyl and chloro groups match up, one isopropyl group points up and the other points down. Since some stereocentres match and some do not, the structures cannot be enantiomers. Therefore, these stereoisomers are diastereomers.

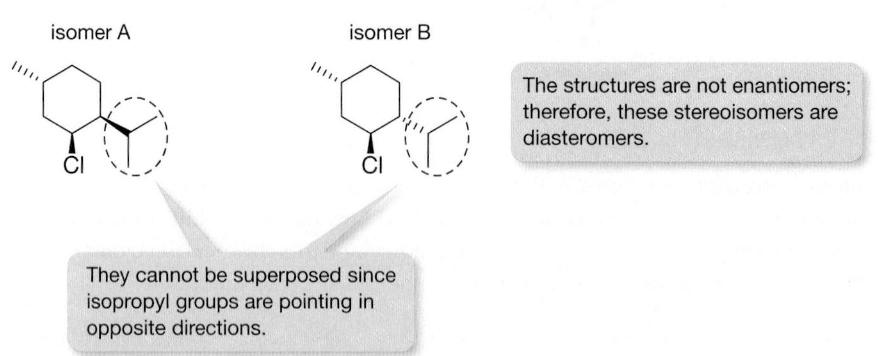

APPROACH B: COMPARE ABSOLUTE CONFIGURATIONS

STEP 1: Assign absolute configurations to all chirality centres in the two stereoisomers. Once you have assigned the configurations in isomer A, you can use the configurations already assigned for isomer A as a starting point to assign the centres in isomer B. Only the chirality centre with the isopropyl group is different; therefore, it must have the opposite configuration in the two structures.

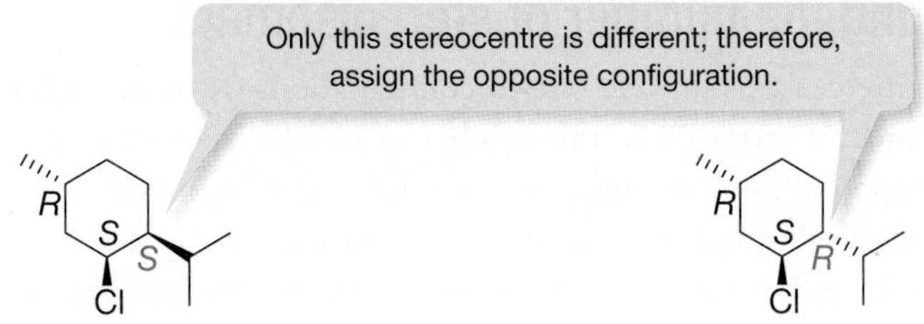

Only this stereocentre is different; therefore, assign the opposite configuration.

isomer A

isomer B

STEP 2: Compare the absolute configurations at each chirality centre.

If all of the chirality centres in two stereoisomers have the opposite configurations, then the two isomers are *likely* enantiomers. (An important exception is discussed in Section 4.7.) If some of the absolute configurations are the same while others are different, then the two isomers are diastereomers.

In this example, two of the stereocentres have the same absolute configuration, and one stereocentre has a different configuration. So isomers A and B are diastereomers, not enantiomers.

PRACTICE PROBLEM

4.13 Classify each pair of stereoisomers as either enantiomers or diastereomers.

a) 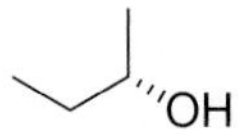 and (*R*)-2-butanol

b) 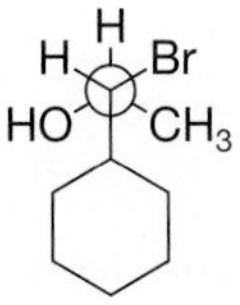 and

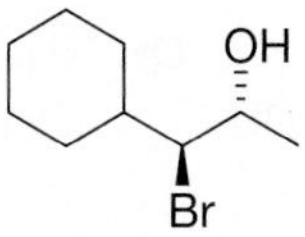

c) 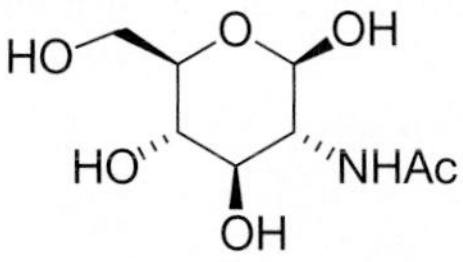 and

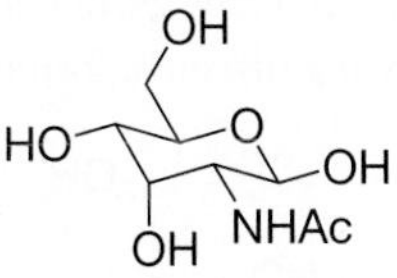

d) 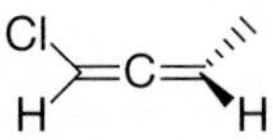 and

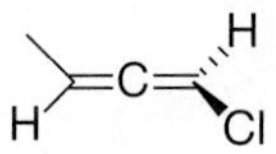

INTEGRATE THE SKILL

4.14 Classify each pair of structures as either enantiomers, diastereomers, constitutional isomers, or identical compounds.

a) 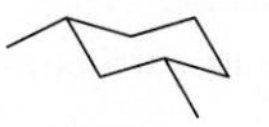 and

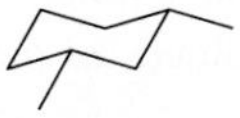

b) 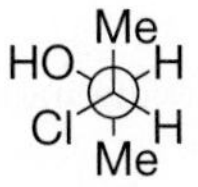 and

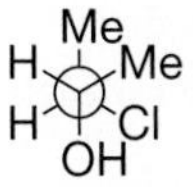

c) 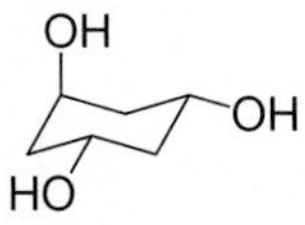 and

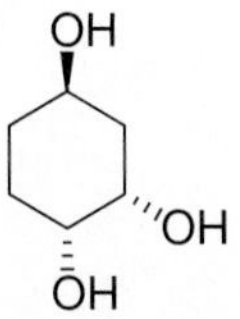

4.6.1 Estimating the number of stereoisomers

Since a chirality centre can exist in one of two configurations (*R* or *S*), each chirality centre in a molecule contributes a factor of 2 to the number of possible stereoisomers for that molecule. Consequently, a compound that contains n stereocentres can exist in as many as 2^n stereoisomers. This value is the *maximum* number of stereoisomers that are possible for a given compound. For some compounds, certain combinations of configurations produce identical structures, thus reducing the actual number of stereoisomers.

CHECKPOINT 4.8

ORGANIC CHEMWARE
4.4 Isomers

You should now be able to calculate the maximum number of stereoisomers possible for a structure containing chirality centres.

SOLVED PROBLEM

A monosaccharide (sugar) with six carbons is called a hexose. Many hexoses adopt the following cyclic structure. How many stereoisomers are possible for this type of sugar?

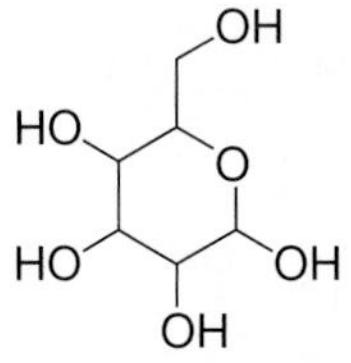

STEP 1: Count the number of chirality centres in the molecule. To identify chirality centres, first look for any sp^3-hybridized carbon atoms that have four different groups attached. In this case, there are five stereocentres; these are marked with asterisks in the following structural formula.

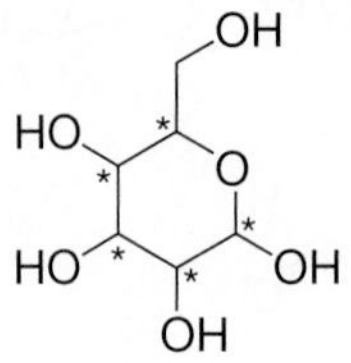

Next, double-check to ensure there are no non-carbon-based chirality centres, such as conformationally rigid nitrogen atoms or sulfoxides. In this case, there are none.

STEP 2: Calculate the maximum number of stereoisomers possible using the formula 2^n.

With five stereocentres, there are $2^5 = 32$ possible stereoisomers for this cyclic sugar structure.

PRACTICE PROBLEM

4.15 Determine the maximum number of stereoisomers possible for each of the following molecules.

a) 2,3,5,6-tetrahydroxy-2-methylheptanoic acid

b) 1,3,5-trimethylcyclopent-1-ene

c)

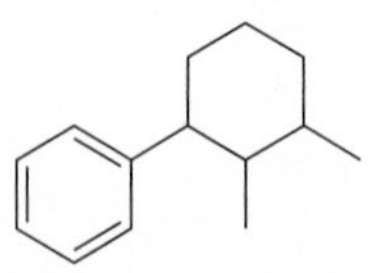

d)

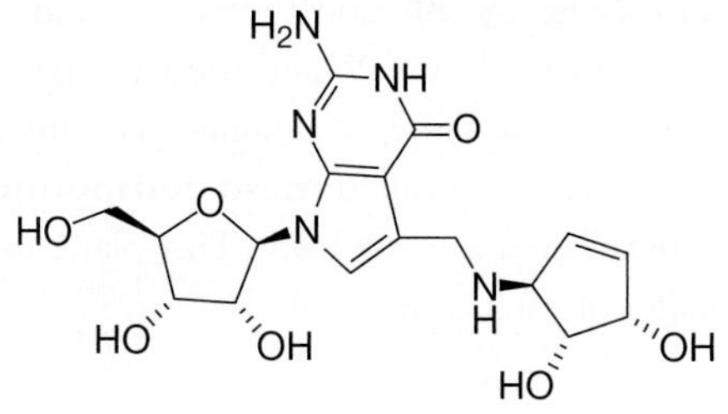

INTEGRATE THE SKILL

4.16 For each of the molecules shown in Question 4.15, draw the stereoisomer that has an *R* configuration at each chirality centre.

4.7 Meso Compounds

Molecules that contain only one chirality centre are always chiral and have non-superposable mirror images. However, molecules that contain more than one chirality centre are not always chiral. For example, *trans*-1,2-dibromocyclopentane has an enantiomer. All of the stereocentres in the mirror image of the *trans* isomer have the opposite configuration to that of the corresponding centre in the original molecule, and the two structures are not superposable.

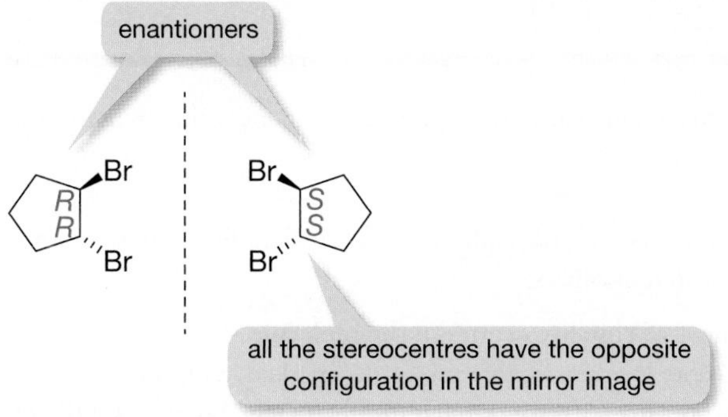

enantiomers

all the stereocentres have the opposite configuration in the mirror image

The 2^n formula indicates that a maximum of four stereoisomers are possible ($n = 2$) for 1,2-dibromocyclopentane. Inverting the lower stereocentre in each of the compounds in the top row gives two new structures that are mirror images of each other, but are not enantiomers.

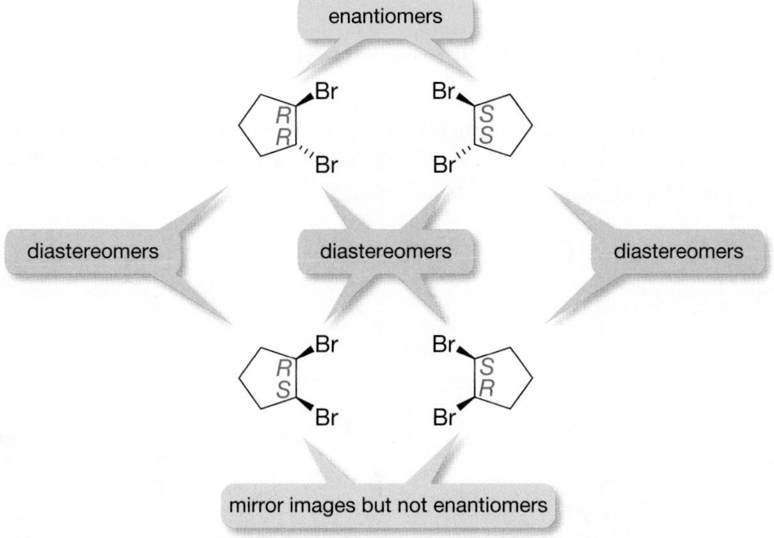

enantiomers

diastereomers

diastereomers

diastereomers

mirror images but not enantiomers

Meso compounds contain more than one stereocentre and have superposable mirror images.

Rotating the left-hand molecule counter-clockwise by 90° and the right-hand molecule clockwise by 90° places the structures in the orientations shown. When the two structures are drawn this way, it is possible to see that they are exactly the same. Molecules that contain more than one stereocentre and have superposable mirror images are called **meso compounds**. Meso compounds contain a plane of symmetry, sometimes called a *mirror plane*. This plane bisects the molecule such that the two halves are mirror images of each other.

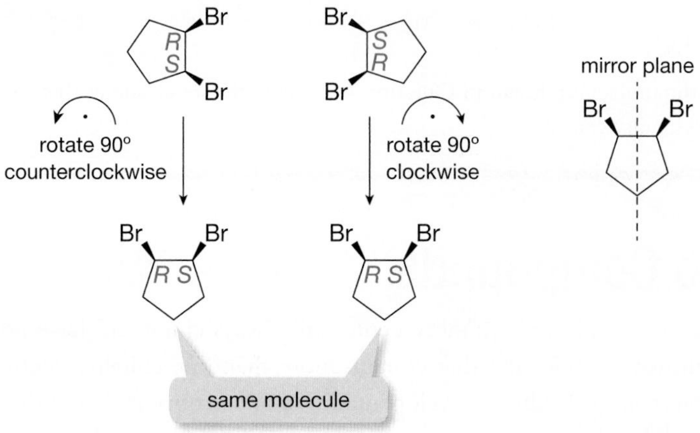

rotate 90° counterclockwise

rotate 90° clockwise

mirror plane

same molecule

CHECKPOINT 4.9

You should now be able to recognize meso compounds and identify their relationship to other stereoisomers.

SOLVED PROBLEM

Draw all possible stereoisomers of 3,4,5-trifluoroheptane. Identify any meso compounds, and classify their relationships to the other stereoisomers.

STEP 1: Count the number of chirality centres in the molecule.

There are three chirality centres in this molecule, marked with asterisks in the following structure. Although the central carbon atom has two similar groups attached to it ($-CHFCH_3$), these groups contain chirality centres and, therefore, could either be the same or be different.

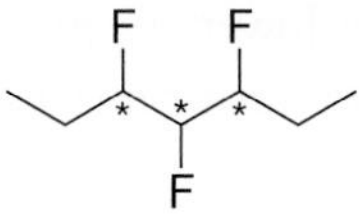

STEP 2: Calculate the maximum number of stereoisomers possible: $2^3 = 8$. There are eight possible stereoisomers for this compound.

STEP 3: Draw all possible stereoisomers. There are eight possible isomers, which means there will be four diastereomers, each with a mirror image. Construct the first four isomers by systematically changing one stereocentre at a time. Then, draw the mirror image of each.

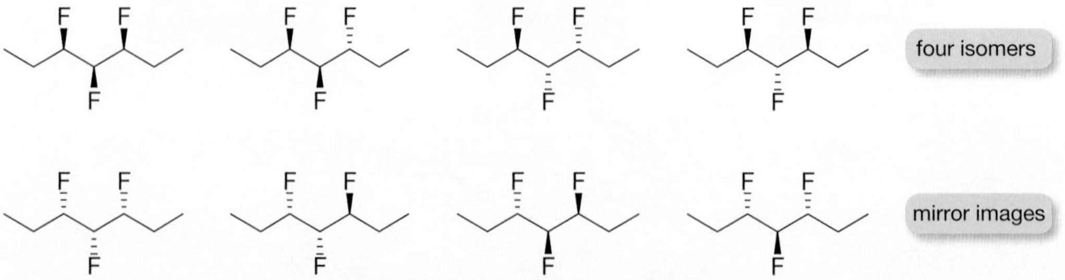

four isomers

mirror images

STEP 4: Determine whether any of the structures are meso compounds by checking for planes of symmetry. The first and fourth structures in the top row each have a plane of symmetry. Since they also both have chirality centres, these compounds are meso compounds.

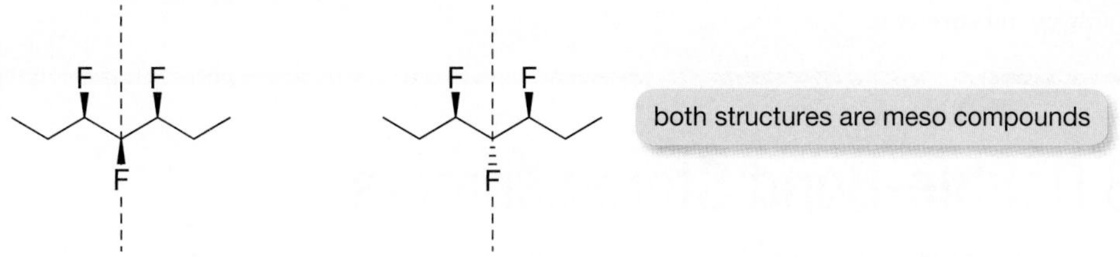

both structures are meso compounds

The second and third structures in the top row are two representations of the same compound. Because the left and right portions of each molecule are the same, the centre carbon is not a chirality centre in this molecule.

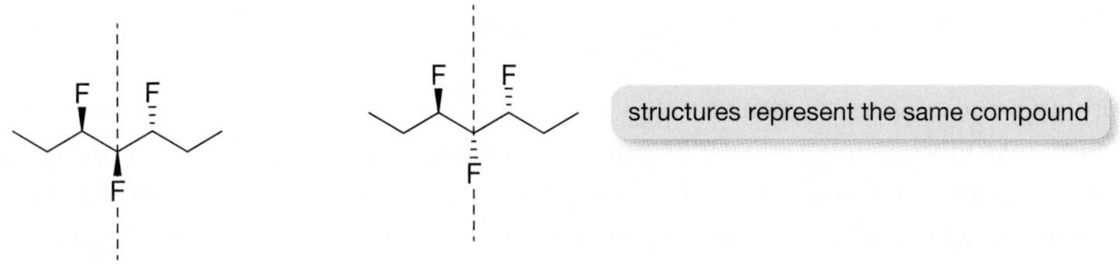

structures represent the same compound

STEP 5: Determine the relationships between each meso compound and the other stereoisomers.

To be enantiomers, structures must be non-superposable mirror images of each other. Neither meso compound is the mirror image of any other isomer, nor of each other. Therefore, both meso compounds must be diastereomers of each other, and diastereomers of the other isomer. The mirror image of the remaining isomer represents its enantiomer and is a diastereomer of the other molecules.

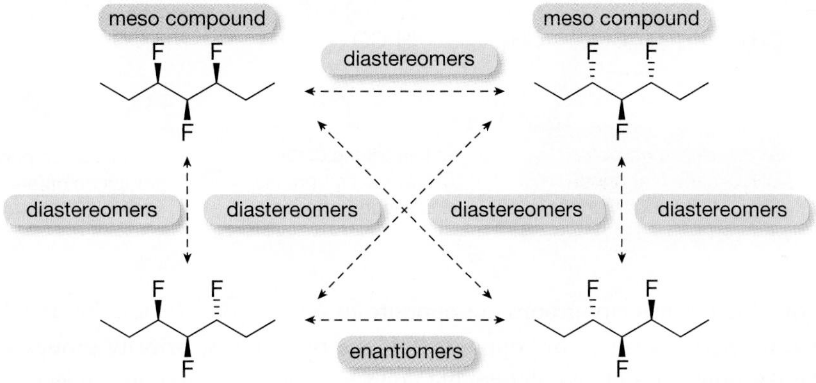

meso compound
meso compound
diastereomers
diastereomers
diastereomers
diastereomers
diastereomers
enantiomers

PRACTICE PROBLEM

4.17 Classify each of the following compounds as chiral or achiral and as meso or non-meso compounds.

a) 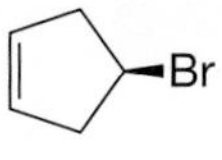 Br

b)

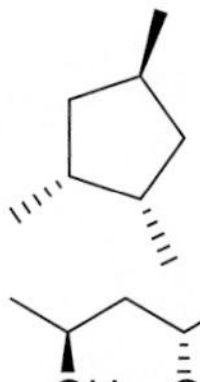

c) 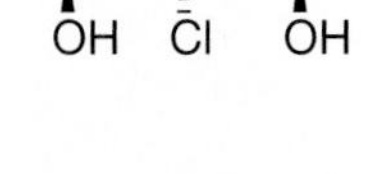 OH Cl OH

d)

e) 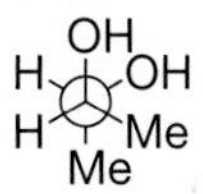 OH, H, OH, H, Me, Me

f) 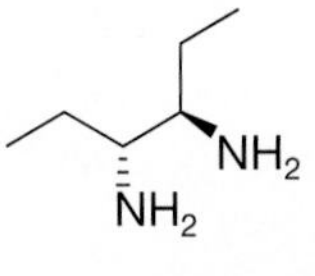 NH_2, NH_2

INTEGRATE THE SKILL

4.18 When asked to draw and name the stereoisomers of pentane-2,4-diol, you might mistakenly list both (2*R*,4*S*)-pentane-2,4-diol and (2*S*,4*R*)-pentane-2,4-diol. Explain what is wrong with this answer and correct it.

4.8 Double-Bond Stereoisomers

The arrangement of groups around a double bond produces stereoisomers. For each double bond, two geometries are possible, which are often specified by using the *cis* and *trans* nomenclature.

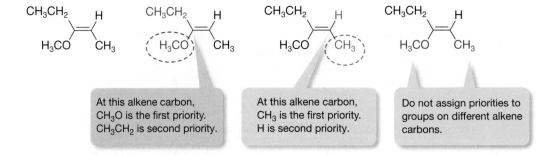

cis *trans*

Although the *cis/trans* nomenclature is useful, it cannot describe double bonds that have more than two substituents. For this reason, the configuration of double bonds are specified using the Cahn-Ingold-Prelog system. The two substituents on *each* of the sp^2-hybridized carbon atoms of the double bond are assigned a first or second priority using the Cahn-Ingold-Prelog procedure. In the following example, the CH_3O group on the left carbon of the double bond has first priority and the CH_3CH_2 substituent has second priority. Similarly, the CH_3 group on the right carbon of the double bond has first priority, and the hydrogen substituent has second priority. Note that the substituents on one carbon atom of the double bond are *not* compared to the substituents on the other carbon atom in the bond.

CH_3CH_2 H H_3CO CH_3

At this alkene carbon, CH_3O is the first priority. CH_3CH_2 is second priority.

At this alkene carbon, CH_3 is the first priority. H is second priority.

Do not assign priorities to groups on different alkene carbons.

STUDENT TIP
To help you remember that *E* stands for "opposite" while Z stands for "together," use the trick that both E and O (for *opposite*) are vowels, whereas Z and T (for *together*) are consonants.

If the two highest priority groups are opposite to one another (*trans*), the double bond is designated *E* (*entgegen*, German for "opposite"). If the two highest priority groups are on the same side of the double bond (*cis*), the double bond has a *Z* configuration (*zusammen*, German for "together"). In the preceding diagram, the alkene has *Z* configuration. Like the *R* and *S* designations, *E* and *Z* are written in italic and placed in parentheses at the start of the name of the compound. If more than one designator is required, the atom number for the location of each double bond is inserted before each *E*/*Z* designation.

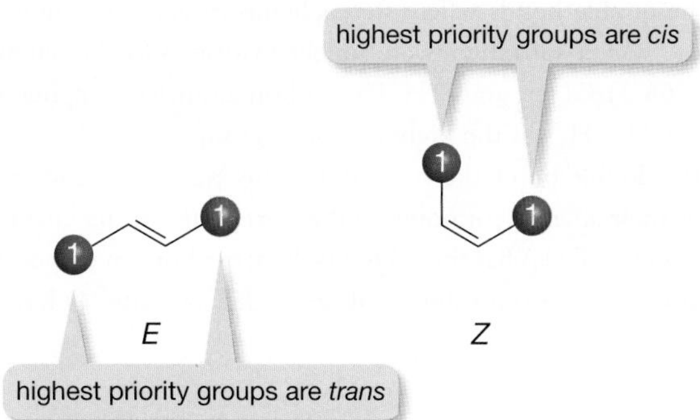
highest priority groups are cis
1
1
1
1
E
Z
highest priority groups are trans

ORGANIC CHEMWARE
4.5 *E/Z* notation

STUDENT TIP
Cis/trans nomenclature can be used to describe only the stereochemistry on certain double bonds (those with only a single substituent on either side), whereas *E/Z* nomenclature can be used for any double bonds with stereochemistry. Therefore, when naming compounds, it is best to use *E/Z* nomenclature to describe the geometry of a bond.

CHECKPOINT 4.10

You should now be able to name structures with double bonds by using either *cis/trans* nomenclature or *E/Z* nomenclature, as appropriate.

SOLVED PROBLEM

Name the following compound, including any relevant stereochemical information.

OH

STEP 1: Determine the IUPAC name for the compound, ignoring the stereochemistry of the double bond for now. The longest carbon chain has six carbon atoms and is numbered starting from the OH group (highest priority group). There are two alkenes, positioned at carbons 2 and 4. Thus, the base name is hexa-2,4-dien-1-ol. After adding the two methyl groups attached to carbons 2 and 5, the IUPAC name—without stereochemistry—becomes 2,5-dimethylhexa-2,4-dien-1-ol.

4 5 6 3 2 1 OH

STEP 2: Determine which double bonds, if any, can produce stereoisomers. To give rise to stereoisomers, a double bond needs to have two unique groups bonded to each end of the double bond. The left-most double bond has two identical methyl groups attached to one of its carbons, so this double bond cannot produce stereoisomers. The middle double bond, however, has unique groups bonded to both ends of the double bond. Therefore, stereoisomers are possible, and appropriate stereochemical information for this double bond must be included in the compound name.

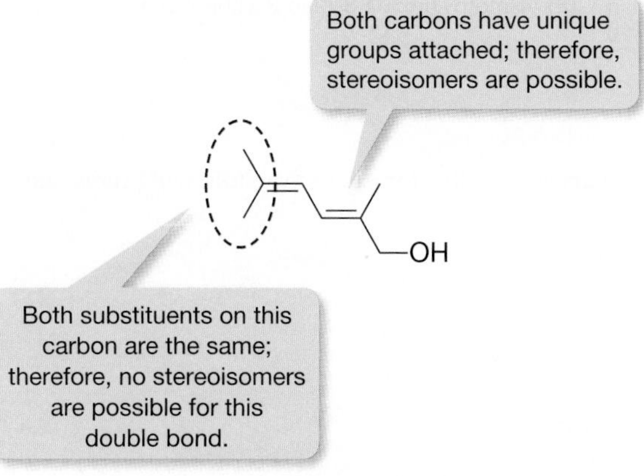
Both carbons have unique groups attached; therefore, stereoisomers are possible.
OH
Both substituents on this carbon are the same; therefore, no stereoisomers are possible for this double bond.

STEP 3: To account for the double bonds where stereochemistry exists, examine each end of the double bond in turn to determine which substituents are higher priority. On the left side of the double bond, the two substituents are $-CCH(CH_3)_2$ and $-H$. The carbon atom has a higher atomic number than the hydrogen atom, so the $-CCH(CH_3)_2$ is the higher priority group.

On the right side of the double bond, the two substituents are $-CH_3$ and $-CH_2OH$. Both substituents have a carbon atom as their attachment point, so the next atoms in the chain need to be compared to resolve the tie. The top carbon ($-CH_3$) has three Hs attached; the bottom carbon ($-CH_2OH$) has two Hs and an O attached. Therefore, the bottom substituent has higher priority (O has a higher atomic number than H).

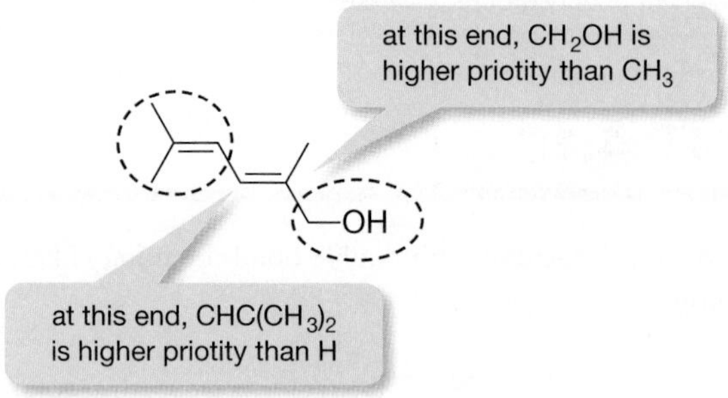

at this end, CH_2OH is higher priotity than CH_3

OH

at this end, $CHC(CH_3)_2$ is higher priotity than H

STEP 4: Designate each double bond as either *E* or *Z* based on the location of the two higher priority groups. The two higher priority groups are on opposite sides of the double bond (one is the top substituent; the other is the bottom substituent). Therefore, the double-bond geometry is *E*.

STEP 5: Incorporate the *E*/*Z* designation into the compound name. Since there is only one double bond with stereochemistry in this molecule, the designator *E*, put in parentheses, goes at the front of the name. The compound's name is (*E*)-2,5-dimethylhexa-2,4-dien-1-ol.

PRACTICE PROBLEM

4.19 Each of the following alkenes has a mistake in its name. Identify the mistakes and provide the correct names.

a) 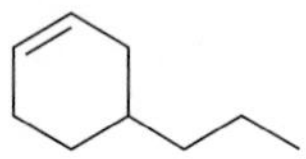 (*E*)-4-propylcyclohex-1-ene

b) 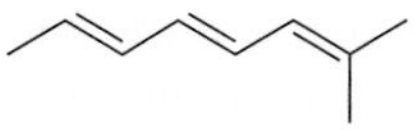 (2*E*,4*E*,6*E*)-2-methylocta-2,4,6-triene

c) 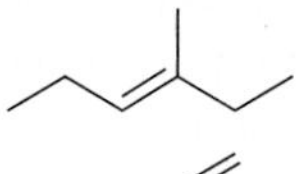 *trans*-3-methylhex-3-ene

d) 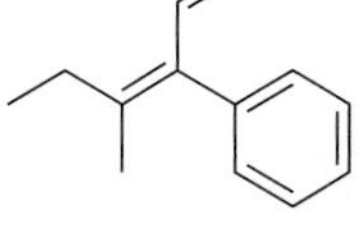 (*Z*)-(4-methylhexa-1,3-dien-3-yl)benzene

INTEGRATE THE SKILL

4.20 How many stereoisomers are possible for each of the following molecules?

a) 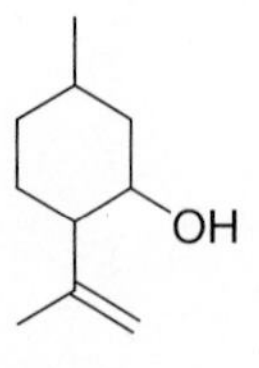 OH

b)

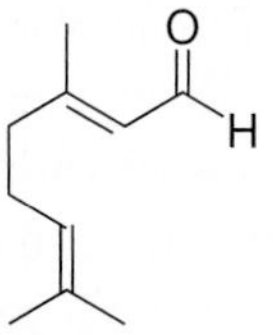

c)

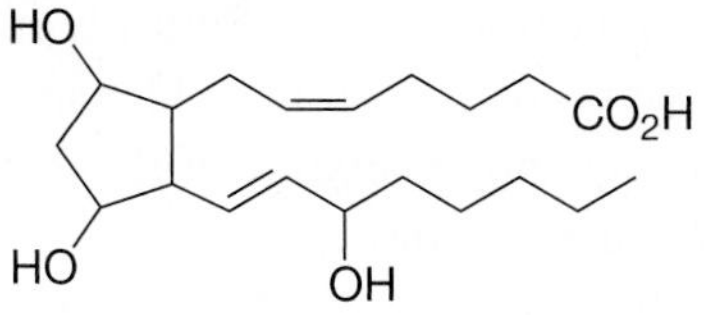

4.9 Physical Properties of Enantiomers and Diastereomers

The diastereomers of a compound have distinct molecular structures, so they differ from one another in their physical and chemical properties, such as melting points, boiling points, physical states, and solubility. Diastereomers also differ in terms of their chemical reactivity.

By contrast, pure enantiomers of a compound have exactly the same melting point, boiling point, and all other physical properties except one. Enantiomers are completely indistinguishable from each other except under two conditions: (1) when they interact with another chiral material (described in this section); or (2) when they interact with polarized light (Section 4.10).

When interacting with another chiral material, the two enantiomers of a substance react in different ways and at different rates because these interactions form diastereomers that do have different properties. For example, enantiomers may have different solubilities in a chiral solvent, or one enantiomer may react faster with another chiral molecule than the opposite enantiomer does.

Because mirror-image molecules cannot interact with each other in the same way that identical molecules can, mixtures of the enantiomers of a compound often have physical properties that differ from those of the pure *R* or *S* enantiomer (see Table 4.1).

TABLE 4.1 Physical Properties of Enantiomers and Mixtures of Enantiomers

Compound	Melting point
(*S*)-Alanine 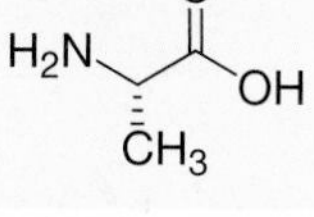	314.5 °C
(*R*)-Alanine 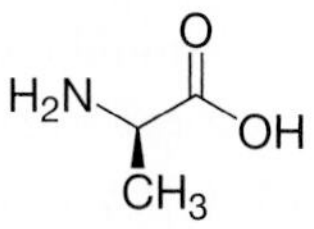	314.5 °C
(*RS*)-Alanine (50%*R* + 50%*S*)	289 °C

To understand such differences in properties, consider the behaviour of some macroscopic chiral objects. Hands are chiral objects. A right hand and a left hand interact with an achiral object such as a ball in exactly the same way, but they differ in how they interact with a chiral object such as a glove or scissors. Right hands and left hands do not interact with each other in the same way. Shake hands with another person, both using a right hand. Then, with your right hand and the other's left hand, shake hands again. The interaction changes. Similarly, molecules of one enantiomer pack together with molecules of the same enantiomer in a different way than they do with molecules of the opposite enantiomer. Consequently, a mixture of *R* and *S* alanine has a different melting point than pure (*S*)-alanine or (*R*)-alanine does.

4.10 Optical Rotation

Optical rotation is rotation of plane-polarized light by a substance or mixture.

The only physical measurement that differentiates enantiomers is **optical rotation**, which involves the interaction of enantiomers with plane-polarized light. Light has both wave and particle properties. As light propagates, its photons (particles) oscillate sinusoidally, producing an electric field and a magnetic field that are perpendicular to each other and to the direction of propagation. Viewed head-on, the electric field of a single photon appears to oscillate back and forth along a single plane perpendicular to direction of the photon. In Figure 4.4, this plane (represented in two dimensions as a line) happens to be directed straight up and down.

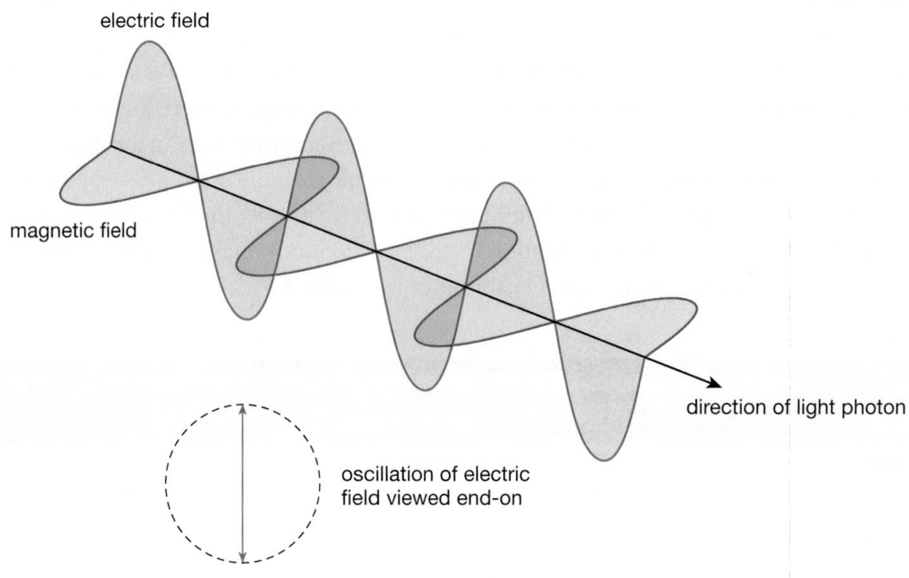

FIGURE 4.4 Oscillation of the electrical and magnetic fields of light.

Ordinary light contains a huge number of photons, which oscillate in all planes around the direction of propagation. Viewed end-on, a beam of light has photons oscillating at every angle perpendicular to the direction of travel. Polarizing filters pass only light that is oscillating in a particular plane. Shining ordinary light through such a filter produces plane-polarized light, in which all of the oscillations are occurring in a single plane (Figure 4.5).

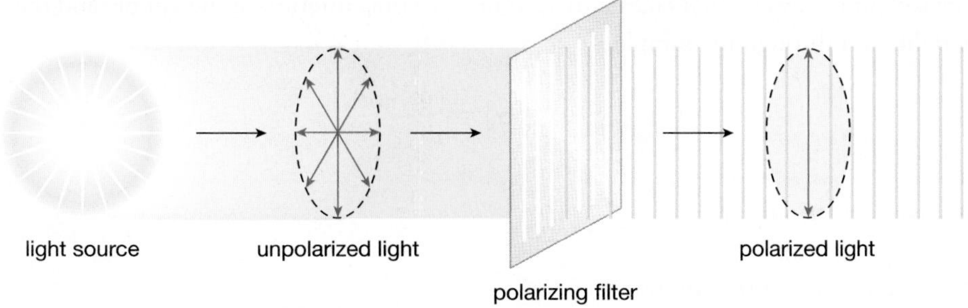

FIGURE 4.5 Using a polarizing filter to create polarized light.

When plane-polarized light is passed through a sample of a chiral substance, the plane of polarization rotates as the light passes through the sample. The light emerges with the oscillation plane at a different angle. Materials that interact with plane-polarized light in this way are called **optically active**, and only chiral materials have this property. The angle that the light is rotated is measured with an instrument called a polarimeter (Figure 4.6).

Optically active molecules rotate plane polarized light.

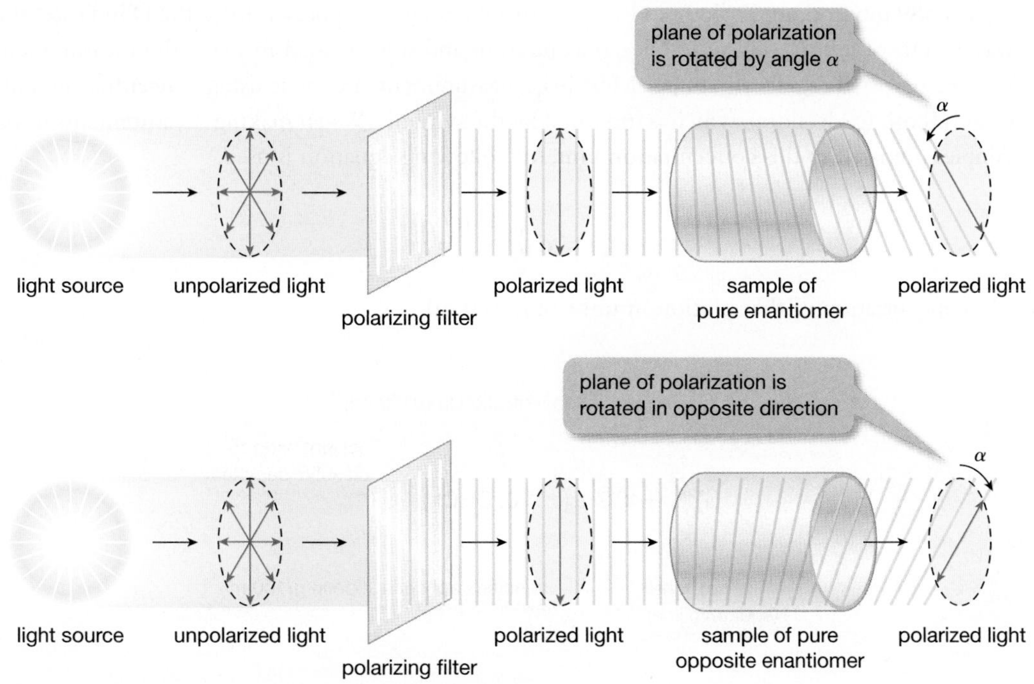

FIGURE 4.6 Schematic representation of the operation of a polarimeter.

STUDENT TIP
The sign of the optical rotation for a sample does *not* indicate the absolute configuration of chirality centres in the compound.

The two enantiomers of a substance rotate plane-polarized light in opposite directions. The angle that a given sample rotates plane-polarized light is called the **optical rotation**, α, of the sample. Clockwise rotation is defined as positive, whereas counter-clockwise rotation is negative. Compounds that rotate plane-polarized light clockwise ($\alpha > 0$) are called *dextrorotary* (*d* or +), and those that rotate light counter-clockwise ($\alpha < 0$) are called *levorotary* (*l* or −).

The **optical rotation** (α) is the angle that a sample rotates plane-polarized light.

The **specific rotation** ($[\alpha]$) is the angle that a sample rotates plane-polarized light after correction for sample size and concentration.

The magnitude of optical rotation depends on the number of molecules that the light encounters, so samples are compared by calculating the **specific rotation, $[\alpha]$**, of each sample.

Specific rotation is a value that takes into account the concentration of the sample and the length of the light path through the sample.

$$[\alpha] = \frac{\alpha}{c \cdot l}$$

where
$[\alpha]$ = specific rotation
α = optical rotation
c = concentration of the solution in g/mL
l = length of cell in dm

When $[\alpha]$ is reported, the wavelength of light used and the temperature of the measurement are normally included. The following notation is used for this:

$$[\alpha]_{\lambda}^{T}$$

where
T = temperature
λ = wavelength of light used in nm

Modern polarimeters are equipped with sodium lamps that provide monochromatic yellow light at 589 nm (sodium D-line), and so most specific rotations appear as $[\alpha]_D$ (the D indicates the wavelength of light is 589 nm). The concentration and solvent used to make the measurement must be reported together with $[\alpha]$. Most $[\alpha]_D$ measurements are made using concentration units of g/100 mL (g/decilitre), which is indicated by the symbol *c*. When making measurements using samples prepared in this concentration unit, the following equation is used.

$$[\alpha]_{\lambda}^{T} = \frac{100\alpha}{c \cdot l}$$

c = concentration of the solution in units of g/100 mL

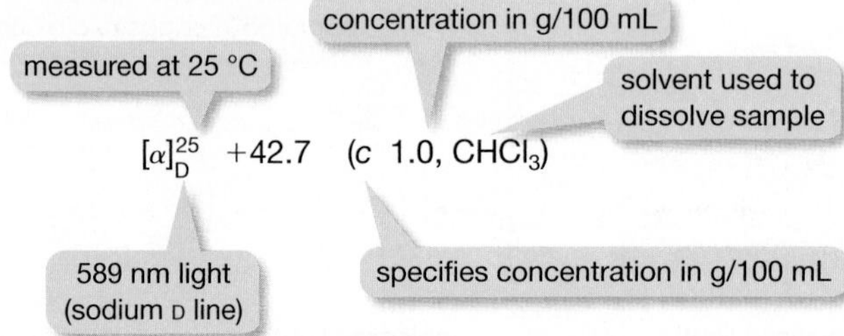

The enantiomers of a compound have a specific rotation of the same magnitude but an opposite sign. For example, (+)-alanine has an $[\alpha]_D$ of +14.5 (*c* 2.0, 1M HCl), whereas the levorotary enantiomer (−)-alanine has an $[\alpha]_D$ of −14.5 (*c* 2.0, 1M HCl).

CHECKPOINT 4.11

You should now be able to calculate the specific rotation of a sample from given polarimetry data.

SOLVED PROBLEM

A 20.0 mg sample of glyceraldehyde ($C_3H_6O_3$) was dissolved in 2.0 mL of $CHCl_3$. When this solution was placed in a sample cell 1.0 cm long and then analyzed in a polarimeter using light from the D-line of a sodium lamp, the observed rotation was 0.043° counter-clockwise. Calculate the specific rotation of the sample.

STEP 1: Decide which formula to use. Since the most commonly used unit of concentration is grams per 100 millilitres, you can use the following formula:

$$[\alpha]_\lambda^T = \frac{100\alpha}{c \cdot l}$$

STEP 2: Calculate the concentration of the sample. Remember that the *unit* is g/100 mL (g/decilitre).

amounts from the problem

convert the units of mL to units of (100 mL): decilitres

$$c = \frac{20.0\ \text{mg}}{2.0\ \text{mL}} \times \frac{1\ \text{g}}{1000\ \text{mg}} \times \frac{100\ \text{mL}}{1 \times (100\ \text{mL})}$$

convert mg to g

$$c = 1.0\ \text{g}/100\ \text{mL}$$

STEP 3: The length of the sample cell needs to be converted into decimetres:

$$l = 1.0\ \text{cm} \times \frac{1\ \text{dm}}{10\ \text{cm}}$$

$$l = 0.10\ \text{dm}$$

STEP 4: Calculate the specific rotation, and report the answer using proper notation. The rotation was counter-clockwise and is therefore negative. Filling in the known values gives the specific rotation as follows:

$$[\alpha]_D = \frac{100 \times -0.043°}{1.0\ (\text{g}/100\ \text{mL}) \times 0.10\ \text{dm}}$$

$$[\alpha]_D = -43°$$

To report with proper notation, you need to include the concentration (units of g/100 mL are represented using the symbol *c*) and the solvent used to prepare the sample. If a temperature was specified for the measurement, it needs to be added as a superscript to the right of the [α] symbol. Since no temperature was specified for this problem, assume a standard temperature of 20 °C, but omit this temperature from the answer. The final answer is as follows:

$$[\alpha]_D = -43°\ (c,\ 1.0,\ CHCl_3)$$

PRACTICE PROBLEM

4.21 Baclofen, a commonly prescribed muscle relaxant, can exist as either the *S* or the *R* enantiomer. The *R* enantiomer, shown below as its HCl salt, is significantly more potent than the *S* enantiomer. Answer the following questions related to the optical properties of (*R*)-baclofen.

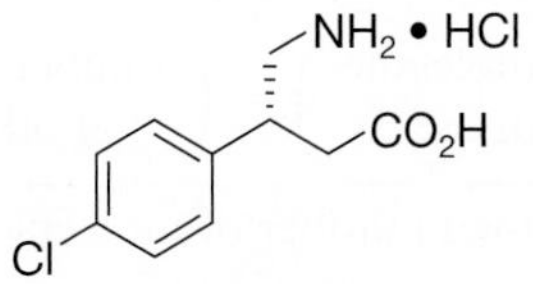

$NH_2 \bullet HCl$

CO_2H

Cl

a) A 100.0 mg sample of (*R*)-baclofen is dissolved in 50.0 mL of water. The solution is placed in a 1.0 dm polarimeter tube and analyzed using the 589 nm D-line of sodium at 25 °C. The observed rotation is −0.004°. Determine the specific rotation for (*R*)-baclofen.

b) What is the specific rotation of (*S*)-baclofen under the same conditions?

c) At what concentration would the observed rotation of (*R*)-baclofen be −0.08°?

d) What would be the observed rotation if 30.0 mg of (*R*)-baclofen were dissolved in 10.0 mL of water?

INTEGRATE THE SKILL

4.22 A chemist is asked to determine the specific rotations of all stereoisomers of cyclopentane-1,2,3-triol. The chemist needs only one measurement to determine all the specific rotations. Explain why.

4.11 Optical Purity

In **optically pure** samples, all the molecules are the same substance with the same absolute configuration.

A **racemic mixture (racemate)** is a mixture of equal amounts of both enantiomers of a compound.

The **optical purity** of a sample is the ratio of the specific rotation of a sample to the specific rotation of a pure enantiomer of the same compound.

The **enantiomeric excess** (ee) of a sample is the ratio of the amount of one isomer to the total amounts of all isomers in the sample.

Samples that contain only one enantiomer are said to be **optically pure**. An optically pure sample produces a maximum value of $[\alpha]$ for the compound.

Many samples of chiral materials exist as mixtures of enantiomers. In these mixtures, the molecules of the (+) isomer rotate plane-polarized light clockwise, whereas the molecules of the (−) isomer rotate the plane-polarized light counter-clockwise. If there are equal amounts of (+) and (−) molecules in the sample, no rotation is observed because positive rotation due to the (+) isomer is offset by the negative rotation of the (−) isomer. A sample containing a 50:50 ratio of enantiomer isomers is called a **racemic mixture**, or a **racemate**. Since racemic mixtures give an $[\alpha]_D$ of 0, they are optically inactive.

A solution that contains different amounts of each enantiomer is optically active, but the less concentrated isomer cancels the effect of an equal quantity of the other isomer. Thus, the observed rotation corresponds to the difference between the amounts of the enantiomers. The specific rotation for the mixture can be used to calculate the relative amounts of each enantiomer, which is often expressed in terms of the **optical purity** or **enantiomeric excess** (ee) of the solution. Optical purity and enantiomeric excess are numerically equivalent but are each calculated using different equations.

$$\text{optical purity} = \frac{|\text{observed}[\alpha]|}{|[\alpha]\text{ of pure enantiomer}|} \times 100\%$$

Optical purity is normally reported as a positive value. The isomer that is in excess is specified.

Suppose the rotation of a pure enantiomer of a substance is +10°. If a measurement of a mixture containing both enantiomers of that substance gives a specific rotation of −7°, the optical purity of the sample is as follows:

$$\text{optical purity} = \frac{|-7°|}{|+10°|} \times 100\% = 70\% \text{ purity of } (-) \text{ isomer}$$

The sign of the rotation observed for a sample indicates which enantiomer is in excess. In this case, the sample must have more of the (−) isomer than the (+) isomer to give a negative value for the observed $[\alpha]$. On a molecular level, the %ee reflects the populations of both isomers in the sample:

$$\%\text{ee} = \frac{\begin{pmatrix}\text{number of molecules}\\ \text{of isomer in excess}\end{pmatrix} - \begin{pmatrix}\text{number of molecules}\\ \text{of other isomer}\end{pmatrix}}{\text{total number of molecules}} \times 100\%$$

Optical purity and %ee are numerically equivalent (optical purity = %ee). For a sample that is 70% optically pure (−), the remaining 30% is optically inactive. That inactive portion contains equal amounts of the two enantiomers. Thus 15% of the sample is (+) isomer and 85% is (−) isomer (70% excess + half of the inactive portion).

CHECKPOINT 4.12

You should now be able to relate the specific rotation of a sample to its optical purity.

SOLVED PROBLEM

The specific rotation of the *S* form of naproxen is $-20.0°$. A drug maker is selling naproxen on the Internet. To save money, the company makes a mixture of the beneficial *R* form and the inactive *S* form. Health Canada measures a sample of this "counterfeit" naproxen and finds a specific rotation of $+5°$. Determine the relative amounts of the *R* and *S* isomers in the mixture.

STEP 1: Determine which enantiomer is in excess.

The specific rotation of the *S* form is negative, and the observed optical rotation is positive; therefore, the *R* enantiomer is in excess.

STEP 2: Determine which equation to use based on the data provided.

Because the problem gives information about optical rotations, use the following equation:

$$\text{optical purity} = \frac{|\text{observed}[\alpha]|}{|[\alpha] \text{ of pure enantiomer}|} \times 100\%$$

STEP 3: Calculate the optical purity (%ee = optical purity). The observed optical rotation is $+5°$, whereas the pure *S* enantiomer has a specific rotation of $-20°$. Substituting these values into the equation gives the following:

$$\text{optical purity} = \%\text{ee} \; \frac{|+5|}{|-20|} \times 100\% = 25\%$$

Therefore, the mixture being sold online has a 25% excess of the *R* enantiomer.

STEP 4: Use the calculated %ee to determine the percentage of each enantiomer present in the mixture.

An optical purity of 25% means that 75% of the material is racemic (a 1:1 mixture of the two enantiomers), and 25% is pure (*R*)-naproxen. So the mixture is 37.5% (*S*)-naproxen (half of the 75% of the mixture that is racemic) and 62.5% is (*R*)-naproxen (half of the 75% of the mixture that is racemic, plus the 25% that is pure (*R*)-naproxen).

PRACTICE PROBLEM

4.23 An enantiomerically enriched sample of 2-butanol has an observed optical rotation of $+6.7°$. Enantiomerically pure (*S*)-(+)-2-butanol has a specific rotation of $+13.5°$. Use this information to answer the following questions.

a) What is the optical purity of the mixture?
b) What is the ratio of the *S* and *R* enantiomers in the mixture?
c) If the ratio of enantiomers were adjusted to 9:1, what would be the optical purity of the sample?
d) What ratio of enantiomers is needed to obtain an optical purity of 99%?

INTEGRATE THE SKILL

4.24 A chemist has been asked to determine the specific rotation of a chiral compound. However, no enantiopure samples (a sample with only a single enantiomer) are available. Instead, the chemist decides to measure the optical rotation of a sample with an optical purity of 50%. At a concentration of 0.4 g per mL of water, and using a 0.5 dm sample tube, the polarimeter gives a reading of $+20.0°$ using the sodium D-line. Calculate the specific rotation for the enantiopure form of this compound.

4.12 Fischer Projections

Fischer projections are a method for making two-dimensional representations of three-dimensional molecular structures. Used primarily for carbohydrates, these projections are an efficient way to draw molecules that contain many stereocentres. In a Fischer projection, each chirality centre is drawn as a cross. Vertical lines depict bonds angling *into* the page, and horizontal lines depict bonds projecting *up* and *out* of the plane of the paper. The following shows zig-zag bond-line diagrams and Fischer projections for glyceraldehyde and arabinose.

zig-zag notation | Fischer projection

chirality centre (indicated with an asterisk)

glyceraldehyde

arabinose

When molecules contain more than one chirality centre, the backbone chain bearing the chirality centres is usually drawn vertically as a "stack" of chirality centres. If the backbone is drawn horizontally, the convention that horizontal bonds project above the plane of the paper still applies.

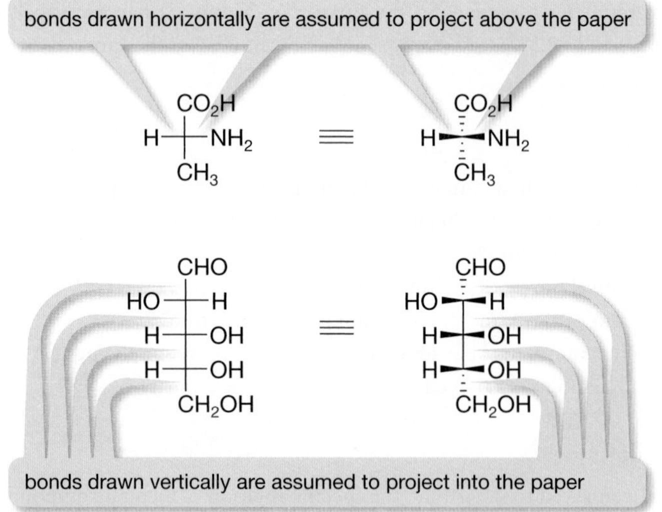

When viewed from the side, the backbone of a Fischer projection forms an arch with the substituents pointing up above it. Except for smaller molecules with a single chirality centre, this conformation results in many eclipsed atoms and therefore does not represent a conformation that most molecules would ever adopt.

when viewed from the left side

becomes

becomes

high energy conformation with many eclipsed atoms

When the backbone of a molecule contains atoms that are not chirality centres, these atoms do not need to be fully depicted in Fischer projections.

Backbone atoms that are not chirality centres do not need to be fully depicted in Fischer projections.

CHECKPOINT 4.13

You should now be able to convert between zig-zag bond-line drawings and Fischer projections.

SOLVED PROBLEM

Convert the following Fischer projection into a zig-zag line drawing.

STEP 1: Redraw the Fischer projection, adding the stereochemistry. Draw the horizontal bonds so that they are coming out of the plane of the page; draw the top-most and bottom-most vertical bonds so that they are pointing back behind the plane of the page. The vertical bonds in the middle can be left as they are.

STEP 2: Draw a side view of the backbone of the Fischer projection, omitting the substituents on the chirality centres for now. When drawing the side view of a Fischer projection, you can imagine looking from either the left or right side. It does not matter which you choose; however, it is a good idea to pick one approach and use it consistently. The following diagram shows the side view seen from the left. Viewed from the left, the CO_2H group at the top of the projection ends up on the left side, whereas the bottom $-CH_3$ group ends up on the right side. Since there are four chirality centres in the Fischer projection, draw four chirality centres in the side-view structure, omitting the actual substituents for now. To help keep track of the chirality centres, number them as shown.

when viewed from the left side

becomes

side view of backbone (without substituents)

top of Fischer projection points to the left

bottom of Fischer projection points to the right

STEP 3: Fill in the substituents on the chirality centres. When you view a Fischer projection from the left side, the horizontal bonds to the left point toward you (out of the plane of the page), and the horizontal bonds to the right point away from you (in behind the plane of the page). Now fill in the substituents on your side-view structure based on whether they were pointing left or right in the original Fischer projection. Use numbers to make sure you put the correct substituents on each chirality centre.

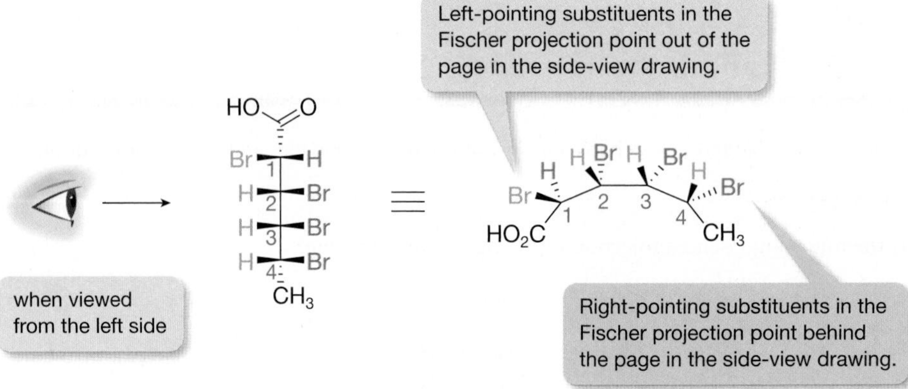

STEP 4: Convert the side-view structure into a proper zig-zag line structure by rotating every other chirality centre in the side-view structure by 180°. Since there are four chirality centres in the backbone of this molecule, you need to rotate the second and fourth chirality centres to generate the desired zig-zag backbone. Leaving all other atoms and groups in place, redraw the second and fourth chirality centres so that they are below the rest of the backbone. Also redraw the substituents attached to these carbons: the substituents should be pointing downward instead of upward, and the substituents will have switched places (i.e., those pointing behind the page are now in front and vice versa). If you left all the remaining atoms in place, your final structure, while correct, will likely look a bit awkward. You can redraw it with proper bond lengths and angles to get a tidier structure.

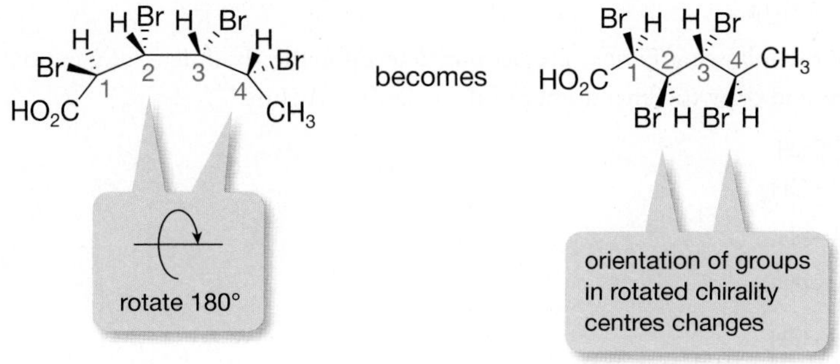

H Br H Br H Br 2 3 4 Br 1 HO$_2$C CH$_3$ becomes HO$_2$C 1 2 3 4 CH$_3$ Br H H Br Br H Br H

rotate 180°

orientation of groups in rotated chirality centres changes

PRACTICE PROBLEM

4.25 For each of the following Fischer projections, draw the corresponding zig-zag line drawing, and for each of the zig-zag line drawings, draw the corresponding Fischer projection.

a) 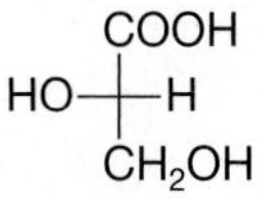 COOH, HO—H, CH$_2$OH

b) 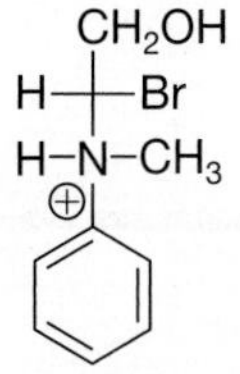 CH$_2$OH, H—Br, H–N$^{\oplus}$–CH$_3$, phenyl

c) 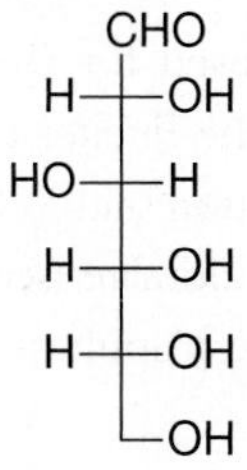 CHO, H—OH, HO—H, H—OH, H—OH, OH

d) 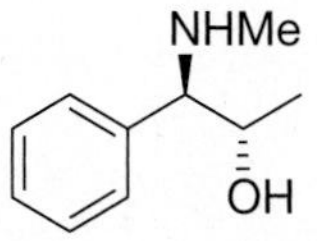 NHMe, OH

e) 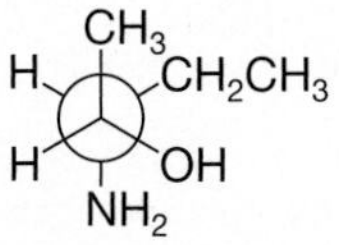 CH$_3$, H, CH$_2$CH$_3$, H, OH, NH$_2$

f) 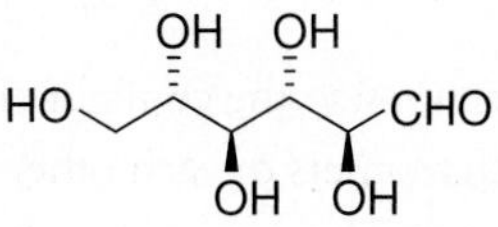 HO, OH, OH, CHO, OH, OH

INTEGRATE THE SKILL

4.26 For each of the following Fischer projections, determine the absolute configuration of each chirality centre, and draw the enantiomer of the original structure.

a)

CO_2H
H—CH_3
H_3C—H
HO—CH_3
H—OH
CH_2CH_3

b)

CH_2OH
H—Br
H—Cl
CH_3

c)

O O
H—

4.12.1 D and L nomenclature

Fischer projections form the basis of a nomenclature system commonly used for the enantiomers of biomolecules such as carbohydrates and amino acids. In this system, the Fischer projection of the molecule is drawn such that the functional group with the highest oxidation state (Chapter 7) is uppermost. For sugars, the designation D applies if the OH group of the penultimate (second-last) carbon points to the right. If this OH group faces to the left, the designation L applies.

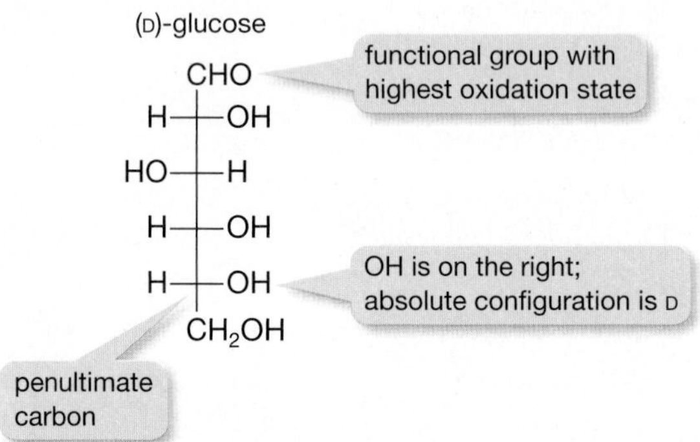

This system names enantiomers rather than individual chirality centres, so the configuration D or L applies to the *entire molecule*. Glucose, mannose, and galactose are diastereomers of each other, and each of these sugars can exists as either a (D) enantiomer or an (L) enantiomer. The names in each case refer to the specific diastereomer, and also specify a pattern of stereochemistry for the OH groups on carbons 2, 3, 4 and 5 of each sugar. Normally only the (D) enantiomer of these sugars are found in living things; (L) sugars are less common in nature.

diastereomers

(D)-glucose	(L)-glucose	(D)-mannose	(D)-galactose
CHO	CHO	CHO	CHO
H–C–OH	HO–C–H	HO–C–H	H–C–OH
HO–C–H	H–C–OH	HO–C–H	HO–C–H
H–C–OH	HO–C–H	H–C–OH	HO–C–H
H–C–OH	HO–C–H	H–C–OH	H–C–OH
CH_2OH	CH_2OH	CH_2OH	CH_2OH

enantiomers

For amino acids, the designations (D) or (L) refer to the orientation (left or right) of the amino (NH_2) group when the molecule is drawn in a Fischer projection with the group having the highest oxidation state (COOH) at the top. The amino acids in your body all have an (L) configuration. The (D) enantiomers of amino acids do occur naturally, but are very rare.

The designations (D) and (L) do not indicate the optical activity of a molecule (*d* or *l*). For example, some (L) enantiomers are dextrorotary (*d*).

(L)-alanine	(D)-phenylalanine	(L)-threonine
CO_2H	CO_2H	CO_2H
H_2N–C–H	H–C–NH_2	H_2N–C–H
CH_3	CH_2–(phenyl)	H–C–OH
		CH_3

CHECKPOINT 4.14

You should now be able to apply the D/L nomenclature system to biological molecules including sugars and amino acids.

SOLVED PROBLEM

Unlike most sugars, the (L) enantiomer of arabinose is more frequently found in nature than its (D) enantiomer. Draw Fischer projections for all possible stereoisomers of (L)-arabinose and label each as either L or D.

(L)-arabinose
CHO
H–C–OH
HO–C–H
HO–C–H
CH_2OH

STEP 1: Determine the maximum number of stereoisomers possible.

There are three chirality centres in (L)-arabinose, marked with an asterisk in the following structure. Therefore, there are $2^3 = 8$ stereoisomers possible.

CHO
H—*—OH
HO—*—H
HO—*—H
CH_2OH

STEP 2: Draw all eight possible stereoisomers.

Systematically interchanging the position of the OH group at each chirality centre results in the following eight possible stereoisomers:

CHO	CHO	CHO	CHO
HO—H	H—OH	HO—H	H—OH
HO—H	HO—H	H—OH	H—OH
HO—H	HO—H	HO—H	HO—H
CH_2OH	CH_2OH	CH_2OH	CH_2OH

CHO	CHO	CHO	CHO
HO—H	H—OH	HO—H	H—OH
HO—H	HO—H	H—OH	H—OH
H—OH	H—OH	H—OH	H—OH
CH_2OH	CH_2OH	CH_2OH	CH_2OH

STEP 3: Designate each stereoisomer as either D or L based on the stereochemistry of the penultimate carbon. The penultimate carbon is second from the bottom. Examine the OH group attached to it. If the OH points left, the structure is an (L) enantiomer; if it points right, the structure is a (D) enantiomer.

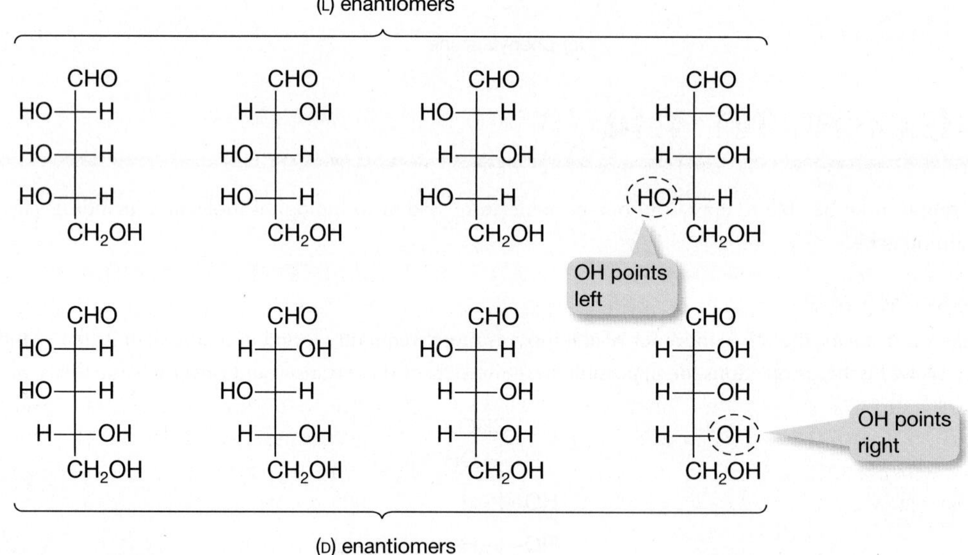

PRACTICE PROBLEM

4.27 The following diagram shows the structure of (D)-fructose. Draw all possible stereoisomers of (D)-fructose, and label each isomer as either D or L. Classify the relationships between (D)-fructose and each of the other isomers.

(D)-fructose

INTEGRATING THE SKILL

4.28 Convert each of the following amino acids to a Fischer projection and determine whether they are (L) or (D) enantiomers.

a) aspartic acid

b) cysteine

c) histidine

CHEMISTRY: EVERYTHING AND EVERYWHERE

Discovery of Enantiomers

The first observation of optical activity of organic substances was made by Jean-Baptiste Biotin in 1815, but at the time no one knew why some substances rotated plane-polarized light and others did not. In 1848, Louis Pasteur made a puzzling discovery about the optical activity of sodium ammonium tartrate. Tartaric acid is present in grapes, and tartrates (salts of this acid) commonly precipitate out of wine as it ages. These tartrates were known to cause positive optical rotation. However, Pasteur found that a solution made with sodium ammonium tartrate that he had purchased from a chemical company did not rotate plane-polarized light. This batch was racemic because it had been synthesized rather than isolated from wine by-products. Pasteur then crystallized the tartrate and examined the crystals under a microscope. He noticed that the sample was composed of two types of crystals that were mirror images of each other (Figure 4.7).

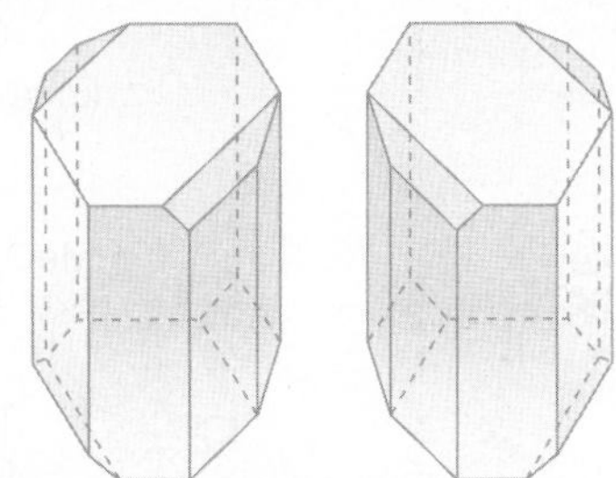

FIGURE 4.7 Mirror-image crystals of sodium ammonium tartrate.
SOURCE: Based on George B. Kauffman, Robin D. Myers, "The resolution of racemic acid: A classic stereochemical experiment for the undergraduate laboratory," *J. Chem. Educ.*, 1975, 52 (12), p 777.

Continued

STUDENT TIP
Chiral resolution is the process of separating mixtures of enantiomers into pure samples of enantiomers.

Pasteur painstakingly separated the mirror-image crystals with tweezers, and measured the optical activity of solutions of each type. One solution gave the same $[\alpha]_D$ as sodium ammonium tartrate derived from wine; the other solution rotated plane-polarized light by an equal amount in the opposite direction. Years later, Pasteur realized that this observation was a consequence of the molecular structure of the materials, and he correctly deduced that the molecules were non-superposable mirror images. Pasteur's remarkable insight led to the discovery of the tetrahedral geometry for carbon.

Pasteur had performed the first separation of enantiomers, a process called *chiral resolution*. He was fortunate to be using sodium ammonium tartrate as it is one of the very few compounds that will crystallize from a racemic solution into single enantiomer crystals.

Bringing It Together

Most of the molecules in your body are chiral. As a result, the enantiomers of a chiral drug can have different effects. Since manufacturing drugs as single enantiomers is often difficult and expensive, many drugs are made and sold as racemic mixtures when the differences in how the enantiomers react in the body are not significant. However, the isomers of some drugs can have drastically different effects on the body.

Cocaine, an illegal narcotic, has a total of eight stereoisomers: four diastereomers plus the four corresponding enantiomers. Only (*l*)-cocaine, the form found in coca leaves, has narcotic effects.

(*l*)-cocaine
found in coca leaves

(*d*)-cocaine
enantiomer of cocaine

(*l*)-pseudococaine

(*d*)-pseudococaine

(*l*)-allococaine

(*d*)-allococaine

(*l*)-allopseudococaine

(*d*)-allopseudococaine

diastereoisomers of cocaine

diastereoisomers of cocaine

Several studies estimate that cocaine has total sales greater than that for any other drug except cannabis. The original wording of narcotics laws in the United States applied only to cocaine derived from coca leaves. Over the years, US defence attorneys have argued that the presence of any of the other isomers indicates that the cocaine is not extracted from coca leaves, and therefore is technically legal. In 1984, one drug dealer in the United States was acquitted because of this argument. To protect against this defence, some jurisdictions have since revised their narcotics laws to include stereoisomers, because a mixture containing (*l*)-cocaine still has narcotic effects.

You Can Now

- Distinguish between conformational isomers and different types of stereoisomers (enantiomers and diastereomers).
- Classify molecules and objects as either chiral or achiral.
- Locate the chirality centres in a molecule.
- Assign an absolute configuration to a chirality centre.
- Name compounds with stereochemistry—including the absolute configuration of any stereocentres as well as the geometry of any stereoisomeric double bonds.
- Calculate the maximum number of stereoisomers possible for a given compound.
- Identify meso compounds and relate their structures to other stereoisomers.
- Calculate specific rotations for chiral compounds.
- Use optical rotation data to evaluate the optical purity of a sample.
- Convert between zig-zag line drawings and Fischer projections.
- Distinguish between the D/L nomenclature system and the Cahn-Ingold-Prelog naming system.
- Describe processes by which enantiomers can be separated.

Problems

4.29 Classify each of the following molecules as either chiral or achiral.

a)

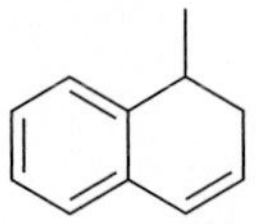

b)

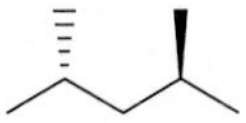

c) 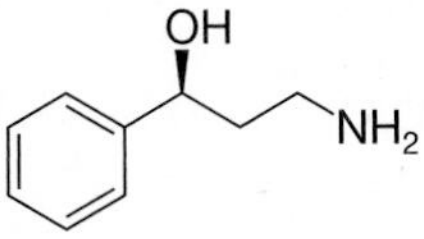 OH NH$_2$

d)

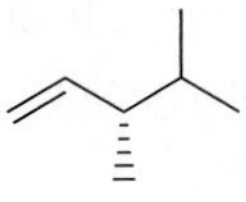

e)

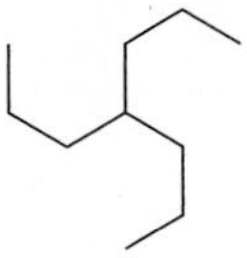

f) 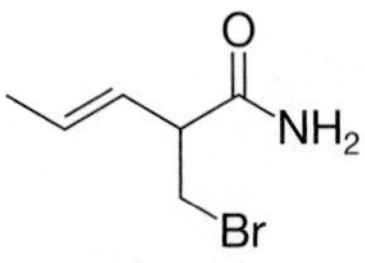 O NH$_2$ Br

g) 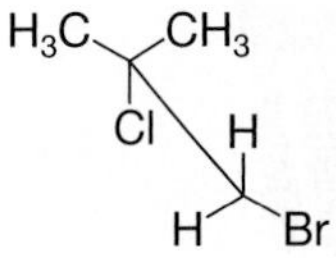 H_3C CH_3 Cl H H Br

h) Br Br

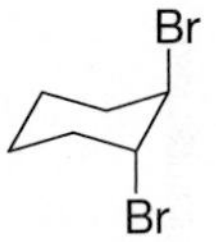

i)

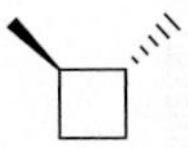

4.30 Classify each of the following objects as either chiral or achiral.

a)

n7atal7i/Shutterstock.com

4.31 The following three compounds are commonly prescribed drugs. For each drug, identify its chirality centres by marking them with an asterisk (*).

a) ritalin

b) fexofenadine (Allegra)

c) penicillin

4.32 Determine whether the structures in each of the following pairs are enantiomers, diastereomers, identical molecules, or constitutional isomers.

a) and

b) and

c) and

d) 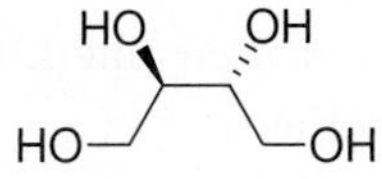 and

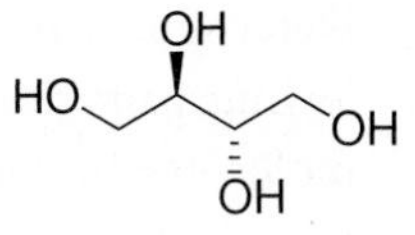

e)  and

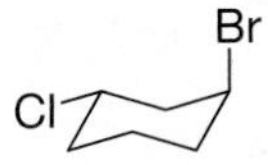

f) 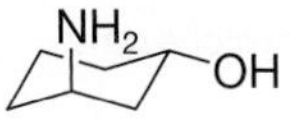 and

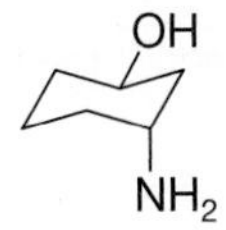

g) 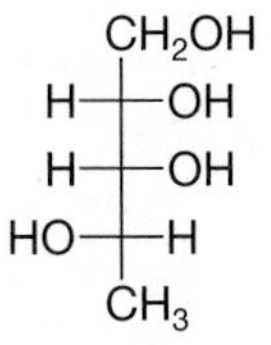 and

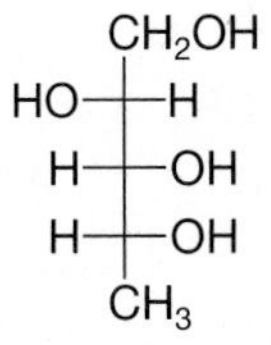

h) 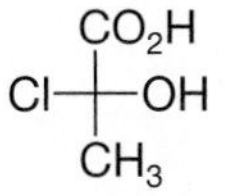 and

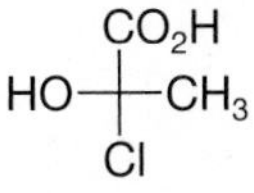

i)

j)

4.33 Assign the absolute configuration to each chirality centre in the following molecules.

a)

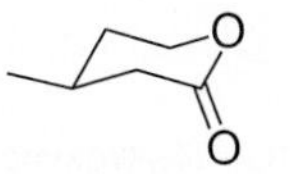

b)

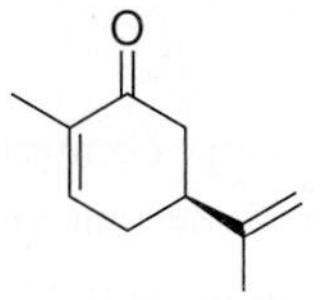

c)

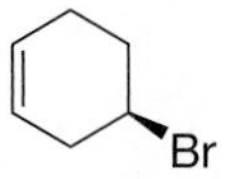

d)

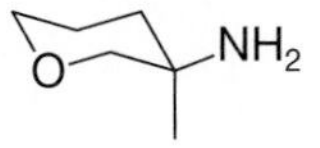

e)

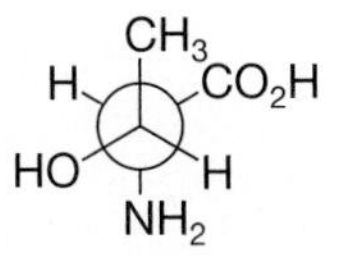

f)

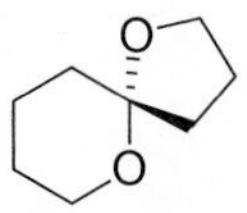

g)

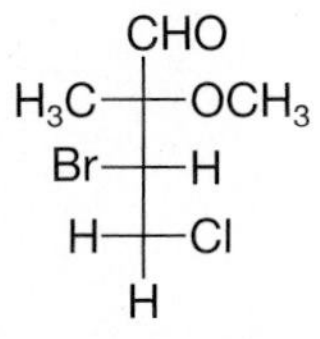

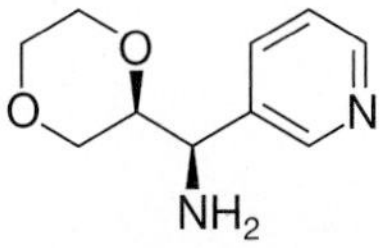

h)

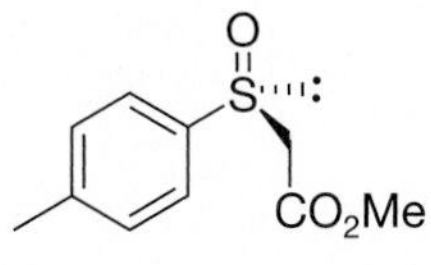

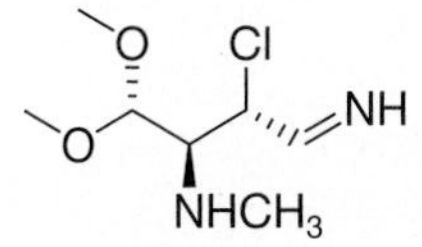

4.34 Provide the IUPAC name for each of the following compounds.

a)

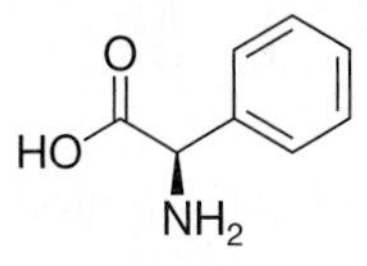

b)

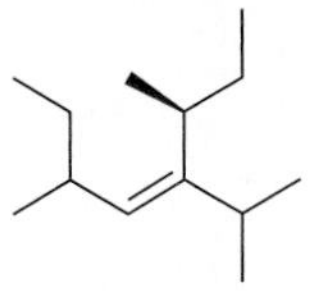

c)

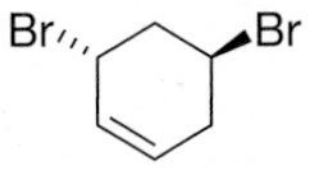

d)

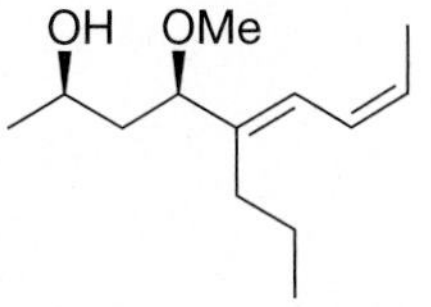

4.35 A sample of sucrose (table sugar) weighing 2.6 g is dissolved in 10.0 mL of pure water. Using a 1.0 cm sample tube and light from the D-line of sodium, the sample is found to rotate plane-polarized light +1.73°. Determine the specific rotation of sucrose.

4.36 The specific rotation of (*S*)-phenylalanine is −35.2°. A sample of phenylalanine has an observed optical rotation of −32°.

a) Find the %ee of the sample.

b) What is the predominant isomer in the mixture?

c) Find the ratio of isomers in the mixture.

4.37 Convert each of the following structures to a Fischer projection.

a)

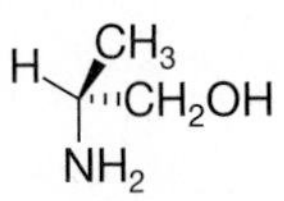

b)

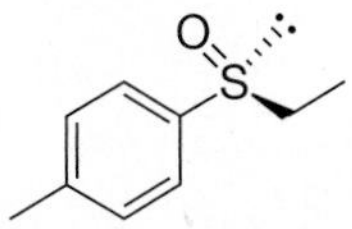

c)

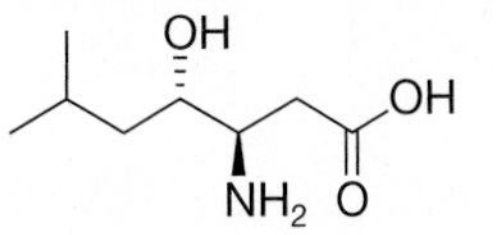

d)

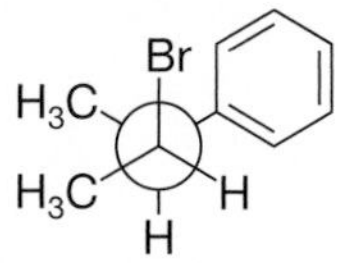

e)

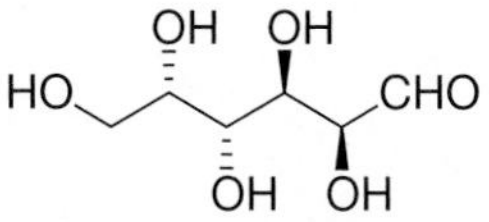

f)

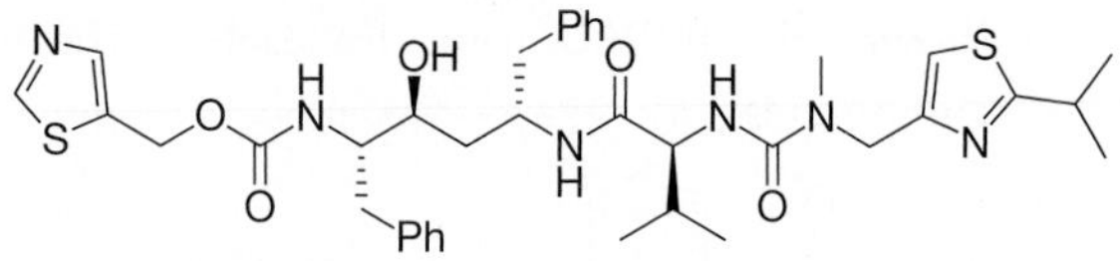

4.38 Convert each of the following Fischer projections to a zig-zag line drawing.

a)

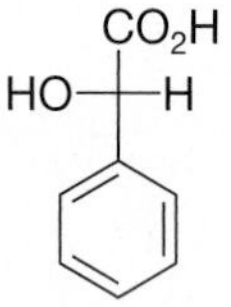

b)

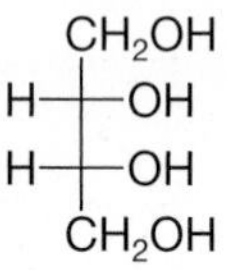

c)

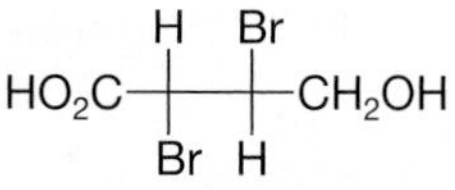

d)

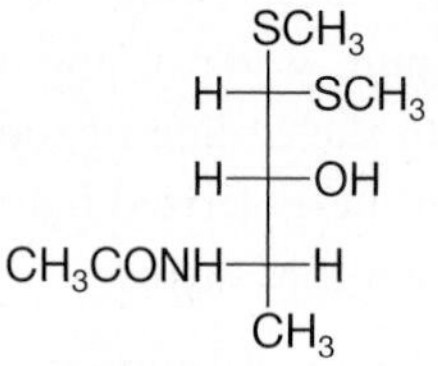

e)

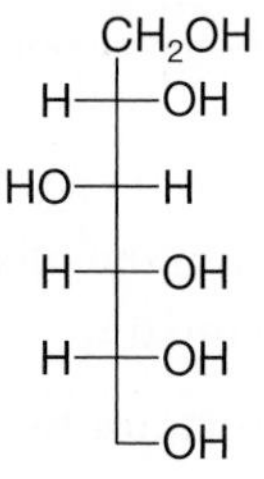

f)

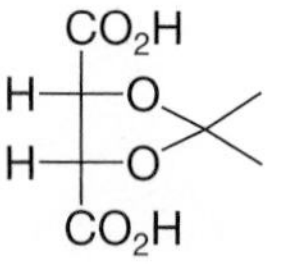

4.39 Determine whether the structures in each of the following pairs are enantiomers, diastereomers, identical molecules, or constitutional isomers.

a) 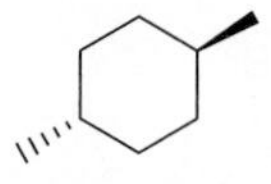 and

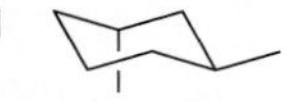

b)  and *cis*-1,3-dimethylcyclohexane

c) 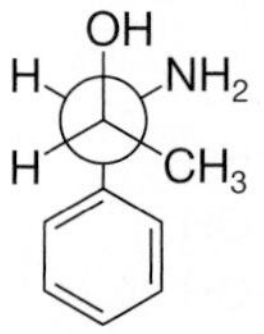 and

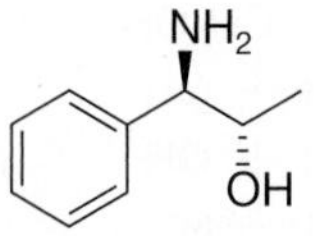

d) 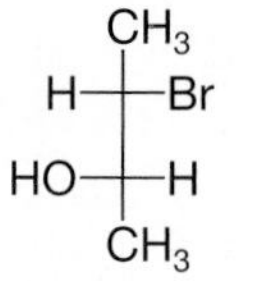 and

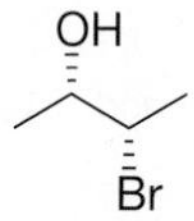

e) 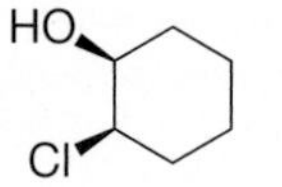 and

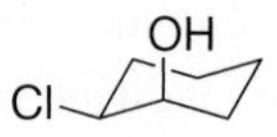

f) 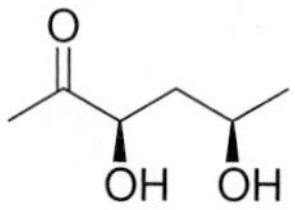 and (3*R*,5*S*)-3,5-dihydroxyhexan-2-one

4.40 Determine which of the following compounds are chiral and which are achiral. For those that are achiral, indicate whether they are also a meso compound.

a)

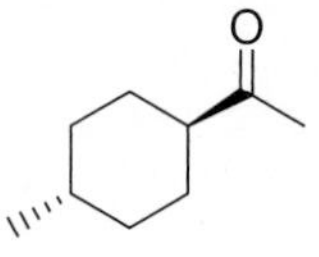

b)

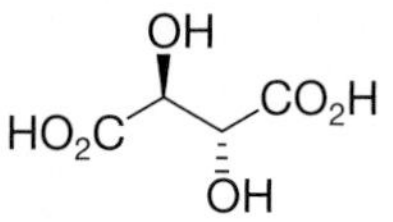

c)

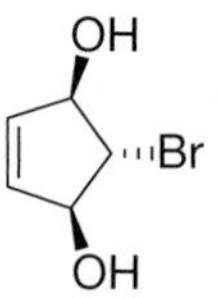

d)

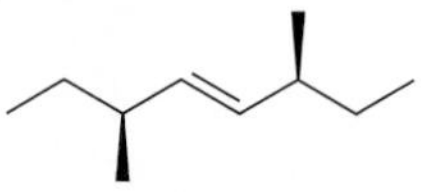

e)

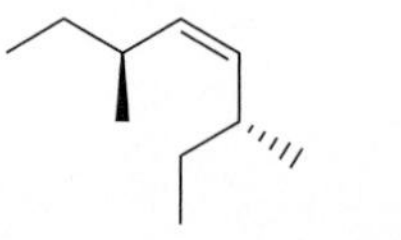

f)

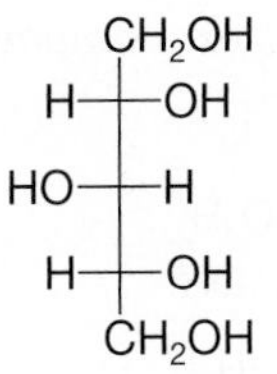

g)

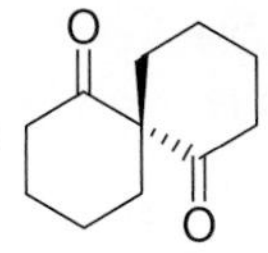

h)

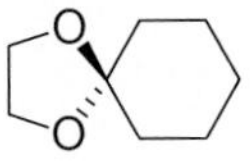

4.41 Calculate the maximum number of stereoisomers that could exist for each of the following structures.

a)

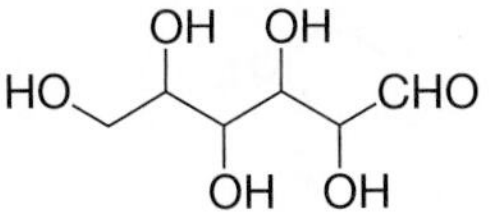

b)

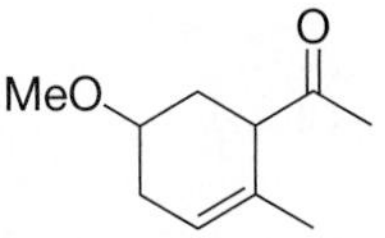

c)

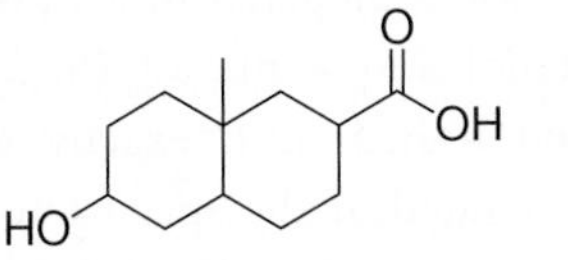

d)

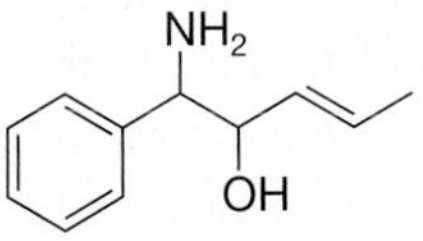

4.42 The following compounds can exist as several different stereoisomers. Draw all possible stereoisomers and provide the IUPAC name for each one.

a) 2-chloro-4,5-dimethylhexan-2-ol
b) 1,3-dimethylcyclohexane
c) 3,6-dimethyloct-3-ene
d) 1,3-dimethylcyclopent-3-en-1-ol

4.43 Which of the following structures correspond to (2*R*,4*S*)-5-hydroxy-2,4-dimethylpentanoic acid?

a)

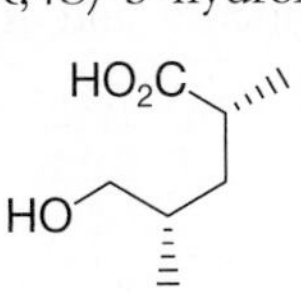

b)

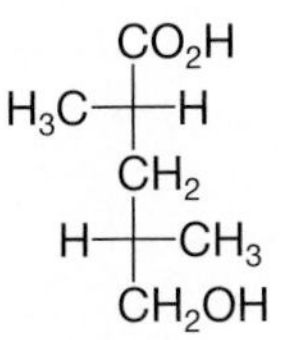

c)

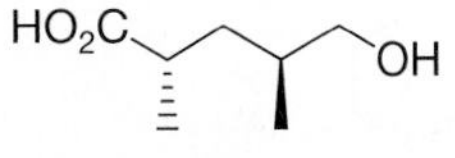

d)

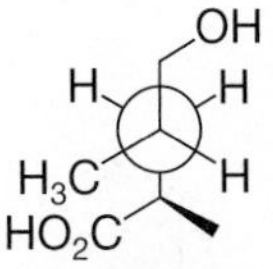

4.44 Calculate the %ee and optical rotation of a mixture containing 6.3 g of D-(+)-glucose and 2.2 g of L-(−)-glucose. The specific rotation of pure L-(−)-glucose is −53°.

4.45 What is the optical purity of a solution that contains 1 mmol of D-tartaric acid and 5 mmol of L-tartaric acid?

4.46 A sample of a pure enantiomer gives a rotation that is exactly 180°. Describe an experiment that would allow you to determine whether the rotation was +180° or −180°.

4.47 Peptides are formed when amino acids join together in a condensation reaction. For example, the dipeptide L-alanyl-D-phenylalanine is formed when L-alanine condenses with D-phenylalanine, as shown here.

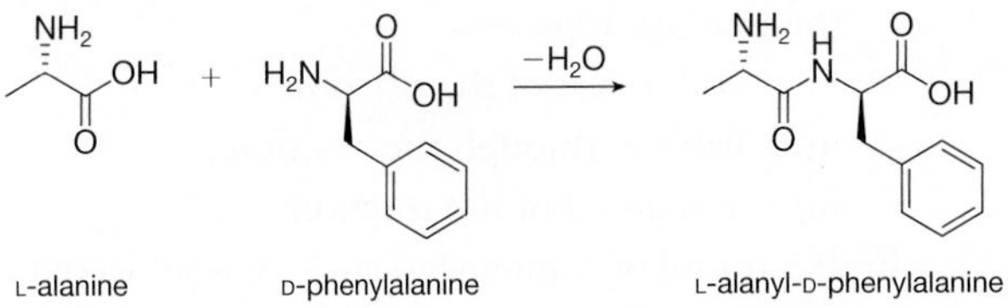

L-alanine + D-phenylalanine $\xrightarrow{-H_2O}$ L-alanyl-D-phenylalanine

The following structure is a tripeptide made up of cysteine, leucine, and serine, drawn as a Fischer projection. Determine whether the amino acid residues occur in their D or L form.

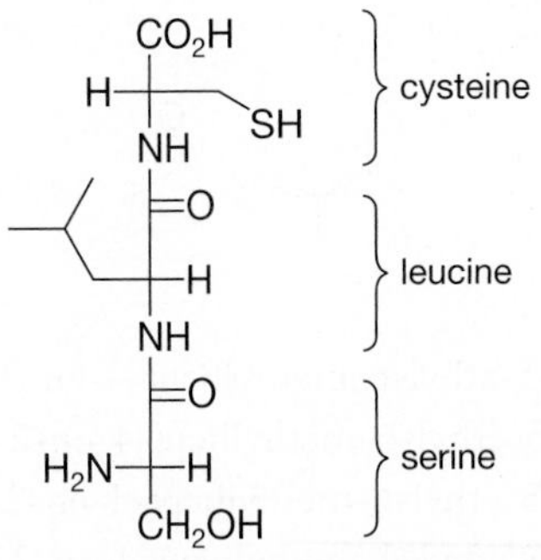

cysteine

leucine

serine

4.48 Which of the following pairs of compounds do you expect would be easily separated by physical methods such as distillation or recrystallization?

a) (1*R*,3*S*)-1,3-diphenylpropane-1,3-diol and (1*S*,3*S*)-1,3-diphenylpropane-1,3-diol

b) 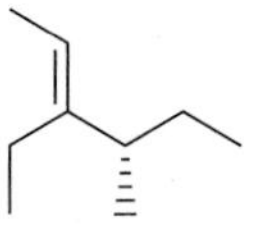 and

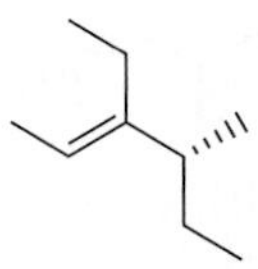

c) 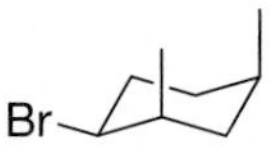 and

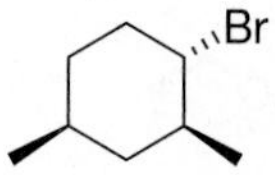

4.49 Draw all the isomers of $C_4H_6Br_2$ that contain a cyclobutane ring, then answer the questions that follow.

a) How many constitutional isomers are there among the structures drawn?
b) Identify any pairs of enantiomers.
c) Identify any pairs of diastereomers.
d) Identify any achiral compounds.
e) Identify any meso compounds.

4.50 A racemic mixture of 1-phenylethan-1-ol is readily separated by a kinetic resolution using a chiral catalyst. The *S* enantiomer reacts preferentially with the catalyst and is transformed into a ketone (acetophenone), whereas the *R* enantiomer remains unreacted.

a) What is the expected enantiomeric excess (%ee) of the 1-phenylethan-1-ol at the following points of the resolution process:
   i) at the start of the reaction
   ii) halfway through the reaction
   iii) at the end of the reaction
b) If 5 mmol of 1-phenylethan-1-ol is subjected to this kinetic resolution, what is the maximum amount of (*R*)-1-phenylethan-1-ol that can be isolated?

MCAT STYLE PROBLEMS

4.51 Select the correct IUPAC name for the following compound:

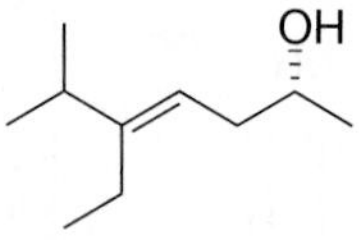

a) (*R*,*Z*)-5-ethyl-6-methylhept-4-en-2-ol
b) (*S*,*Z*)-5-ethyl-6-methylhept-4-en-2-ol
c) (*R*,*E*)-5-ethyl-6-methylhept-4-en-2-ol
d) (*S*,*E*)-5-ethyl-6-methylhept-4-en-2-ol

4.52 Which of the following pairs of compounds are enantiomers?

a) 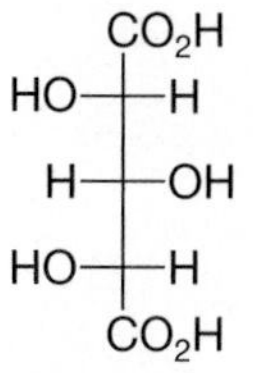 and

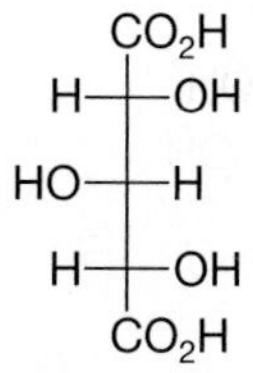

b) 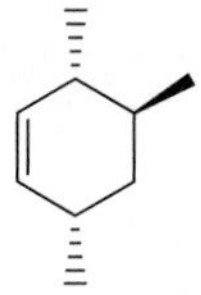 and

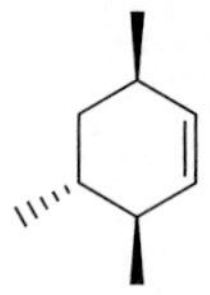

c) 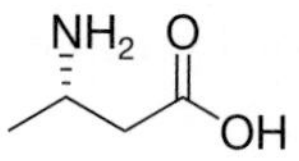 and

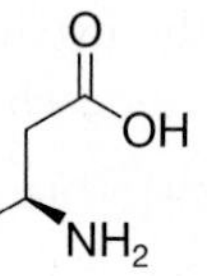

d) 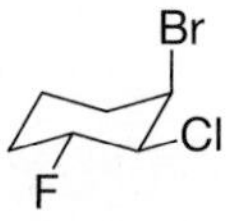 and

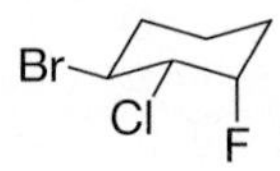

4.53 A mixture of 2-butanol is prepared in a beaker using 6 mL of (*R*)-2-butanol and 4 mL of (*S*)-2-butanol. The mixture is then boiled until exactly 4 mL of the solution has evaporated. If the specific rotation of (*R*)-2-butanol is −13.5°, which of the following statements is true of the solution left in the beaker?

a) It will rotate plane-polarized light by −13.5°.
b) It will rotate plane-polarized light by +13.5°.
c) It will rotate plane-polarized light by +4.5°.
d) It will rotate plane-polarized light by −2.7°.

4.54 Glucose and gulose are both examples of six-carbon sugars known as aldohexoses. Which of the following statements about (D)-glucose and (D)-gulose is false?

a) They both rotate light in the clockwise direction.
b) They are diastereomers of each other.
c) They both have an *R* configuration at the penultimate carbon.
d) Their enantiomers are (L)-glucose and (L)-gulose, respectively.

CHALLENGE PROBLEMS

4.55 For each of the following classes of molecules, draw the smallest possible compound that could exist with at least one chirality centre. Note that there may be more than one possible answer for some parts.

a) acyclic alkane (no rings)

b) cycloalkane

c) cycloalkene

4.56 A sample of *trans*-cyclooctene is found to rotate plane-polarized light, whereas a sample of *trans*-cyclodecene does not. Explain this observation. (Hint: You may wish to use molecular models.)

rotates plane-polarized light

does not rotate plane-polarized light

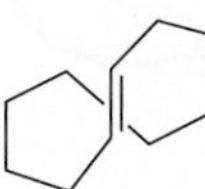

5 Organic Reaction Mechanism

USING CURVED ARROWS TO ANALYZE REACTION MECHANISMS

CHAPTER OUTLINE

Protein molecules in food are large and cannot be absorbed from the digestive tract. How are they broken down into smaller molecules?

5.1 Why It Matters

When you digest food, your digestive tract breaks complex biomolecules down into smaller compounds that the gut can absorb. Your body produces enzymes (protein-based catalysts) that greatly accelerate the chemical reactions that are involved in food digestion. Pepsin (Figure 5.1), an enzyme produced in the stomach, carries out the digestion of proteins by selectively breaking amide bonds in the proteins. This reaction is an example of *hydrolysis*, a reaction with water that breaks bonds in a compound. Hydrolysis of an amide bond produces a carboxylic acid and an amine.

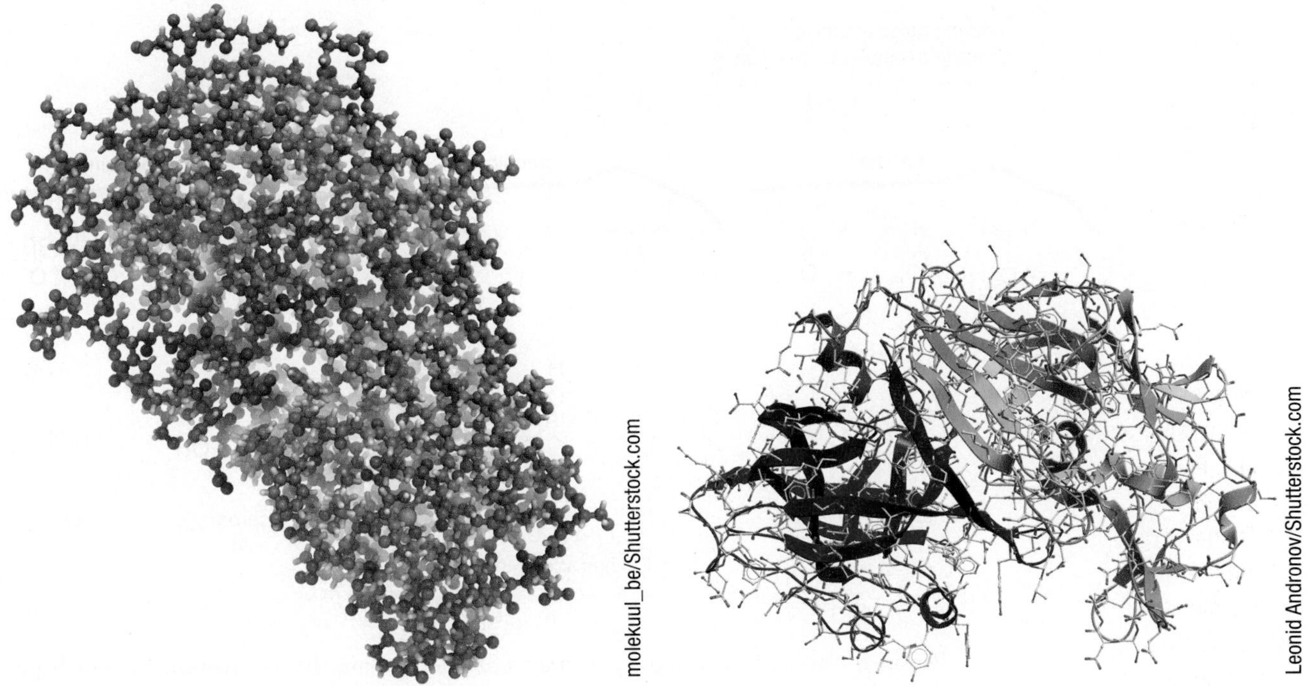

molekuul_be/Shutterstock.com
Leonid Andronov/Shutterstock.com

FIGURE 5.1 Space-filling (left) and ribbon (right) representations of pepsin.

$$\text{amide} + H_2O \longrightarrow \text{carboxylic acid} + \text{amine}$$

Without the enzyme pepsin, this reaction requires very harsh conditions because amide bonds are very resistant to hydrolysis: at pH 7 and 25 °C, a typical amide bond has a **half-life** of 40 000 years! To hydrolyze amide bonds, chemists must heat them in a concentrated aqueous acid, such as HCl, for many hours. With pepsin, the same reaction occurs in just minutes at 37 °C in the stomach.

Half-life is the time required for half of a given amount of substance to be consumed in a reaction.

Pepsin has both an acid (RCOOH) and a base ($RCOO^-$) at the exact location where the reaction with the amide happens. These two groups are perfectly oriented to simultaneously interact with an amide group and a water molecule. Flows of electrons are controlled, and acid–base reactions are guided to exactly the right location to make the reaction work. The motions of the electrons as the reaction is proceeding constitute the mechanism of the reaction. Organic molecules react by the movement of electrons, which cause the bonds between atoms to break and form.

enzyme holds atoms in the right place to catalyze a reaction
pepsin
pepsin
pepsin
amide
H_2O has added to amide
carboxylic acid
amine

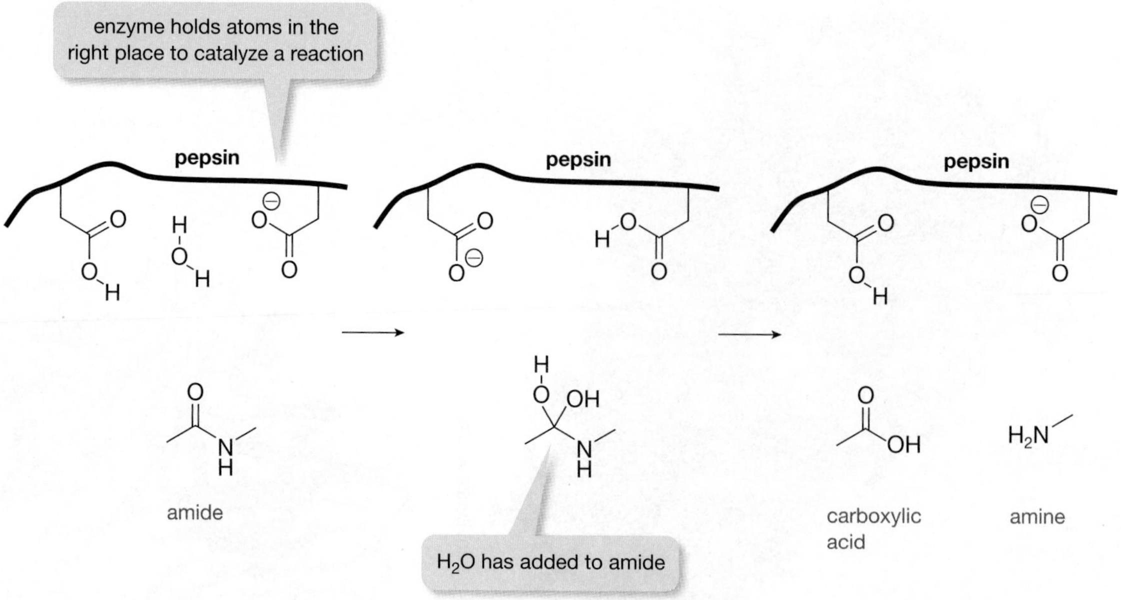

Pepsin is classified as an *aspartyl protease* enzyme because the mechanism by which pepsin catalyzes a hydrolysis reaction involves functional groups found in aspartic acid (an amino acid within the enzyme). Another example of an aspartyl protease is *HIV protease* (Figure 5.2), an enzyme involved in the life cycle of HIV, the virus that causes AIDS. A class of HIV-AIDS drugs known as *HIV protease inhibitors* work by selectively blocking HIV protease, but not other aspartyl enzymes. The discovery of these HIV protease inhibitors was made possible by first understanding the reaction mechanisms of such enzymes.

This chapter describes the curved arrow notation that is used both to represent the movement of the electrons that occurs as bonds form and break during organic reactions and also to predict reaction mechanisms and products. This chapter also explains the concept of *resonance*, which is an important tool for analyzing structure and reaction mechanisms. Finally, it discusses how resonance structures can be used to depict the sharing (delocalization) of π electrons between several atoms.

HIV protease inhibitor

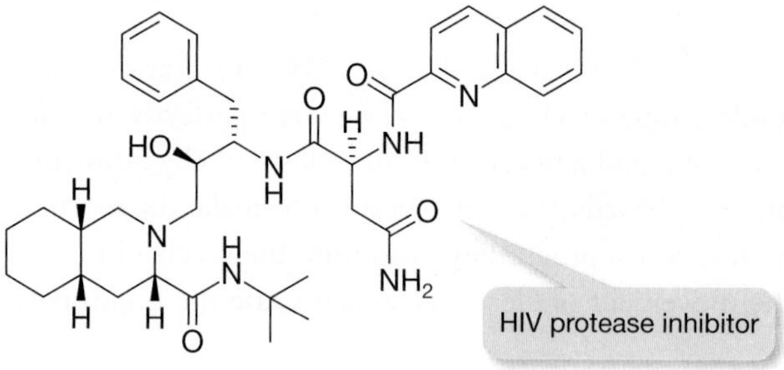

saquinavir

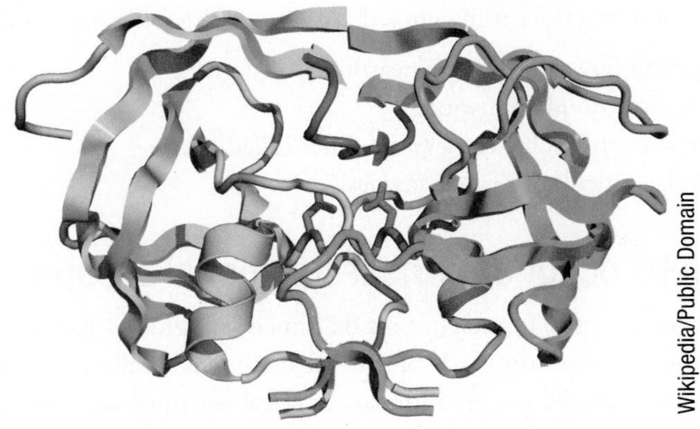

FIGURE 5.2 Computer representation of saquinavir (tube structure) in the active site of HIV protease (shown as purple and blue surfaces).

5.2 Organic Reaction Mechanisms

In reactions, bonds between atoms in the reagent molecules break, and new bonds form between other atoms to make the product molecules. Bonds are pairs of electrons, and the key to understanding chemical reactivity is the movement of these electrons as the bonds are formed or broken.

Consider the following reaction equation of iodoethane with potassium hydroxide. The chemical equation includes all the reagents used and the products formed in the reaction. However, it does not provide any information about the mechanism of the reaction; that is, the equation does not describe *how* the reaction proceeds.

$$\text{I} + \text{KOH} \longrightarrow \text{OH} + \text{KI}$$

Curved arrow notation depicts the flow of electrons during a reaction. This notation focuses on the *valence electrons* and how their overall movements result in the formation and breaking of bonds. The notation is a powerful tool for tracking the process of complex reactions. The following diagram uses curved arrow notation (coloured arrows) to represent the electron flows in the reaction of iodoethane with potassium hydroxide.

new oxygen-carbon bond forms

new oxygen-carbon bond formed

$$\text{I:} + \text{K}^{\oplus} + {}^{\ominus}\text{:O-H} \longrightarrow \text{OH} + \text{K}^{\oplus} + \text{:I:}^{\ominus}$$

carbon-iodine bond breaks

iodide takes the electrons from the broken bond

During this reaction, the hydroxide oxygen donates a pair of electrons to a carbon of iodoethane (red arrow), forming a new bond. At the *same time*, the bond between the iodine and the carbon breaks (blue arrow). The overall result is a substitution reaction that produces a molecule of ethanol and an iodide ion. The curved arrows indicate which electrons are involved in forming and breaking bonds, and show the sequence of events for multi-step reactions. In this way, these arrows describe the overall reaction mechanism.

To analyze organic reactions using curved arrows, use the following points as a guide:

1. Note that electron flows occur at or near functional groups.
2. Use arrows to show bond formation.
3. Use arrows to show bond breaking (sometimes called *cleaving*).
4. Use arrows to keep track of formal charges.

5.2.1 Reactions occur at or near functional groups

When looking at organic structures, focus on the functional groups. Functional groups are sites on molecules where reactions happen. Molecules that are structurally very different but have the same functional groups usually undergo similar chemical reactions involving those groups. Most organic molecules have several functional groups (Chapter 2). The following describes some general features to look for when identifying functional groups.

STUDENT TIP
Functional groups are easier to see when bond-line structures are used.

5.2.1.1 π bonds

The π bonds in double or triple bonds are weaker and more reactive than σ bonds. In line-structure diagrams, π bonds always appear as parallel bond lines, and they react in predictable ways.

5.2.1.2 Heteroatoms

In organic molecules, heteroatoms (any atom besides C or H) often possess one or more non-bonded pairs of electrons (*lone pairs*) that can participate in reactions. Since the electronegativities of such atoms differ from that of carbon, the electron density around a heteroatom is not the same as around the carbon atom it is bonded to, a feature that tends to induce reactions. A π bond directly connected to a heteroatom should be considered as part of a single functional group involving the bond and the heteroatom.

ORGANIC CHEMWARE
5.1 Functional group explorer

STUDENT TIP
Electrons, not charges, are involved in mechanisms. When determining a reaction mechanism, make sure to start with a pair of electrons and end at an atom that can accept electrons.

5.2.1.3 Formal charges

The presence of a formal charge on a molecule raises its energy state relative to its uncharged form, often facilitating a chemical reaction. (The calculation of formal charges is described in Chapter 1.) Negative charges on atoms are typically associated with non-bonded electron pairs, which can be shared with other atoms to form bonds. Positive formal charges indicate sites that *may* accept electrons from another atom. If such a positive charge exists on an atom that has an incomplete octet, that atom can accept electrons from other atoms to form bonds. If the positively charged atom has a complete octet, it usually accepts only a pair of electrons if one of its bonds also breaks.

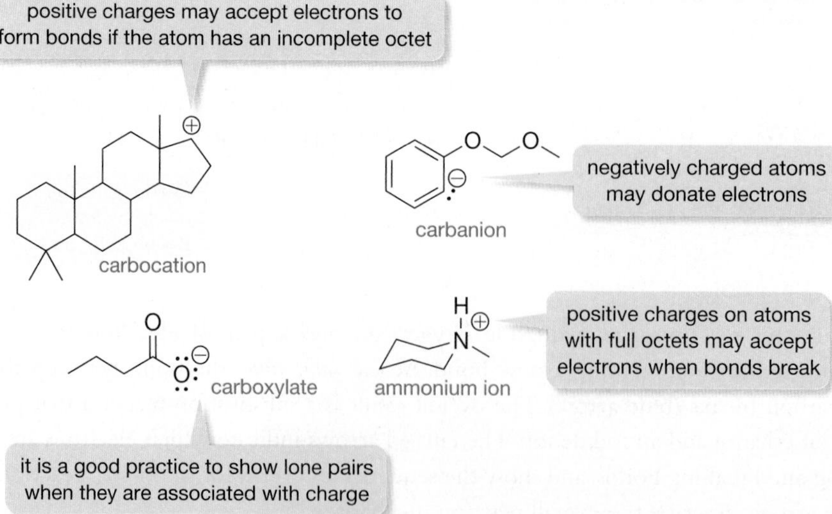

5.2.1.4 Unpaired electrons

An unpaired electron is often referred to as a *radical* (see Chapter 19). Since an unpaired electron usually indicates a prime site for reactivity, line structures should always include any *unpaired* electrons.

H H OMe radical

O radical Cl CN Cl CN O ⊖

ORGANIC CHEMWARE
5.2 Lewis structures: Ethyl carbocation

ORGANIC CHEMWARE
5.3 Lewis structures: Ethyl carbanion

ORGANIC CHEMWARE
5.4 Lewis structures: Ethanoate anion

ORGANIC CHEMWARE
5.5 Lewis structures: Methylammonium cation

5.2.1.5 Hydrocarbons

The hydrocarbon portion of most organic compounds usually does not participate in chemical reactions, and can be simply considered as a scaffold that bears the functional groups. For simplicity, when the hydrocarbon portion is not involved in a reaction mechanism, this portion is often omitted from the reaction diagram or represented by the letter R (for or "residue").

scaffold OH hydroxyl

OH hydroxyl scaffold

Both of these molecules behave like

scaffold R–OH hydroxyl

A hydroxyl group is like the OH group of water, except it is bonded to a carbon atom.

H–OH
R–OH

STUDENT TIP
The abbreviation R can be used to represent *anything* on a molecule, not just hydrocarbons. R could even represent a single hydrogen atom.

5.2.2 Expanding Lewis structures and drawing conventions

Electron pairs are normally not shown in line structures in order to make drawings less cluttered and to highlight the overall structures more clearly. To analyze the mechanism of a reaction and predict the products, it is important to consider the electron pairs in the functional group. Most of the heteroatoms in organic compounds are surrounded by a full octet of electrons, counting both lone pairs and bonds. This provides a quick way to establish how many lone pairs there are on a given atom. Note, however, that atoms below the second period (row) of the periodic table can accommodate more than eight valence electrons and some elements, such as boron, may have incomplete octets.

Non-bonded electrons are implied in bond-line structures.

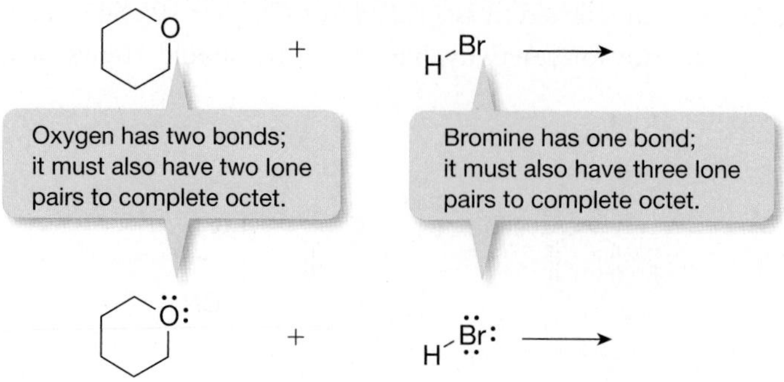

Oxygen has two bonds; it must also have two lone pairs to complete octet.

Bromine has one bond; it must also have three lone pairs to complete octet.

When drawing a reaction mechanism, the source of electrons should be explicitly drawn. Bonds and charges are already included in line drawings, but lone pairs may need to be added if they are acting as an electron source. Strictly speaking, only the lone pairs on atoms that share their electrons need to be shown; however, it can be useful to expand the Lewis structures of the entire functional group, especially when dealing with an unfamiliar reaction. Also, by showing the lone pairs on both the reactants and the products, it is easier to keep track of any changes to formal charges that take place during the reaction.

Hydrogen atoms are also typically omitted from line drawings; however, when they are directly involved in a reaction mechanism, they need to be explicitly shown. Once one hydrogen atom is drawn, it is good practice to include any others that are also bonded to the same atom.

As an example of how these drawing conventions are put into practice, consider the following reaction. The overall reaction has been drawn with typical line-bond drawings: the lone pairs have been omitted, and only hydrogens attached to the heteroatom are shown.

Overall reaction:

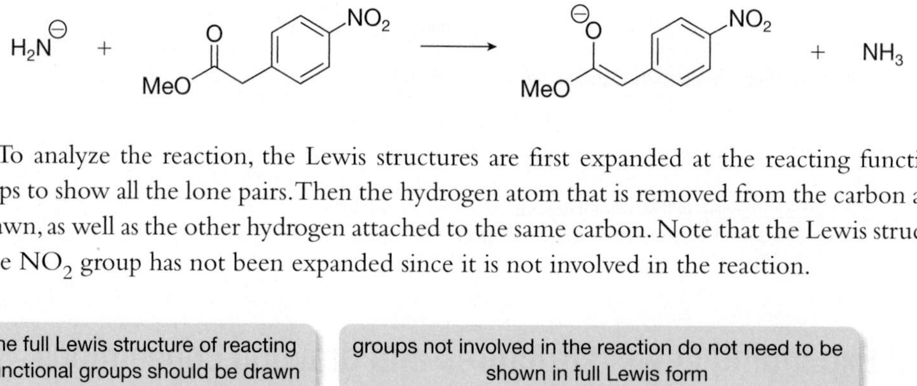

To analyze the reaction, the Lewis structures are first expanded at the reacting functional groups to show all the lone pairs. Then the hydrogen atom that is removed from the carbon atom is drawn, as well as the other hydrogen attached to the same carbon. Note that the Lewis structure of the NO_2 group has not been expanded since it is not involved in the reaction.

the full Lewis structure of reacting functional groups should be drawn

groups not involved in the reaction do not need to be shown in full Lewis form

all hydrogens on a reacting site should be shown

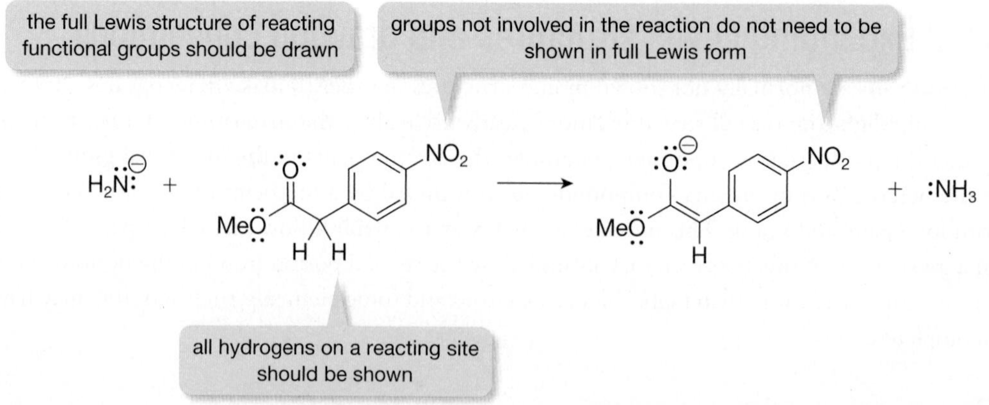

CHECKPOINT 5.1

You should now be able to identify the reactive sites in a molecule and expand their Lewis structures to show the reacting electrons.

SOLVED PROBLEM

Identify the functional groups in the following drug used to control blood pressure. Expand the Lewis structure of each functional group.

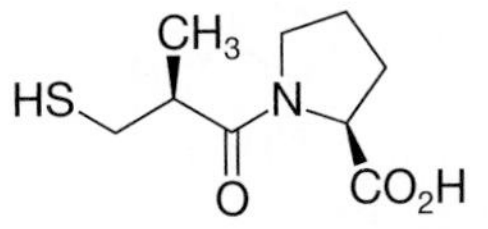

STEP 1: Functional groups can be identified through heteroatoms or π bonds. Start by identifying these in the line structure. Heteroatoms must be explicity drawn, which provides a rapid way of identifying them. Table 2.1 (page 52) lists some common functional groups. If you see a functional group that is not listed in the table, try searching online using your favourite search engine.

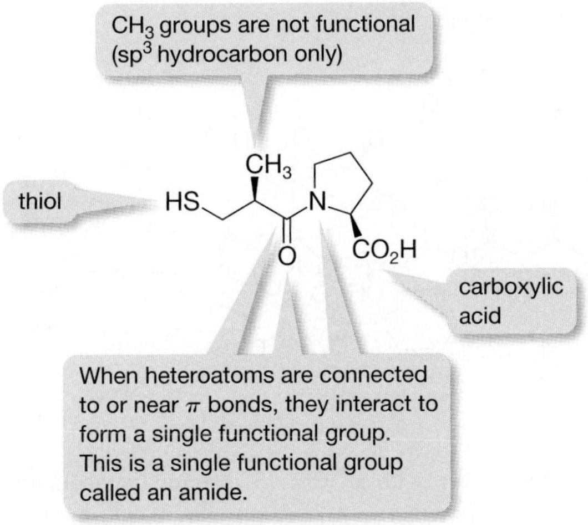

CH$_3$ groups are not functional (sp^3 hydrocarbon only)

thiol

carboxylic acid

When heteroatoms are connected to or near π bonds, they interact to form a single functional group. This is a single functional group called an amide.

STEP 2: Use the formal charge method to expand the Lewis structure of the functional group. To do this, replace the attachment with R (supplies one electron).

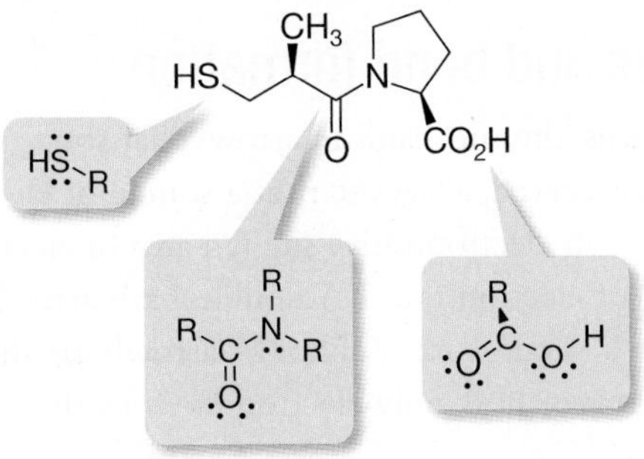

STEP 3: Once the Lewis structures are determined using the R method, redraw the complete structure.

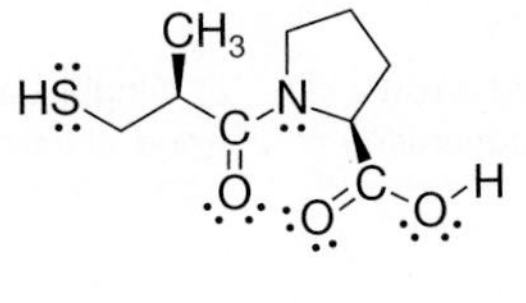

PRACTICE PROBLEM

5.1 Below are three anti-cancer drugs currently in use. For each drug, identify the functional groups and expand their Lewis structures to show the lone pairs.

a) Doxorubicin

b) Mitomycin C

c) Paclitaxel

INTEGRATE THE SKILL

5.2 Draw an example of a molecule with at least 15 carbons that contains the following functional groups: an aromatic ring, a carboxylic acid, an amide, an alkene, an alkyne, a chiral alcohol with *S* configuration, and a chiral amine with *R* configuration. Show all lone pairs in the functional groups.

STUDENT TIP
The terms electropositive, low electron density, positive charge, and partial positive all indicate an area of electron deficiency. The terms electronegative, high electron density, negative charge, and partial negative all indicate an area of excess electrons.

5.2.3 Curved arrows and bond formation

To depict reaction mechanisms, the mechanistic arrows that show electron movement need to be included. These arrows are curved, always start at a source of electrons (a bond, lone pair, or unpaired electron), and end at an electropositive site (an area of electron deficiency). The arrowhead points in the direction of electron travel. A double-barb arrowhead depicts the movement of an electron pair, whereas a single-barb (fishhook) arrowhead shows the movement of one electron. Note that curved arrows show only electron flow, *not* the movement of atoms.

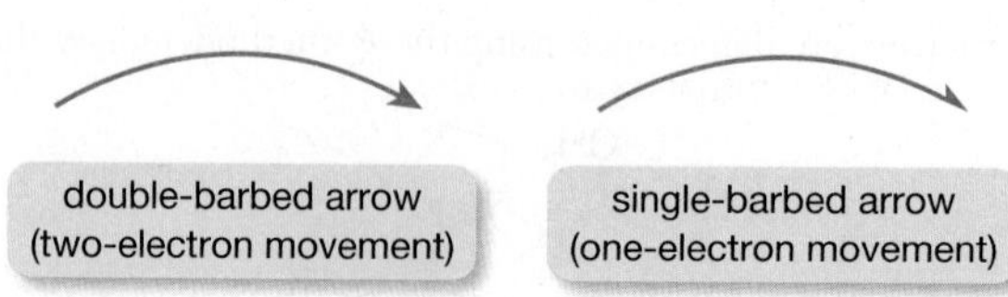

The curved arrows in Figure 5.3 depict electron flow as bonds are formed. In each of the three examples, the following are conserved during the reactions: (1) the number of electrons, (2) the number of atoms, and (3) the total charge.

STUDENT TIP
IMPORTANT! Curved arrows start at electrons and point to an atom or space between atoms. A common error is to use arrows to show atom movement, especially in acid–base reactions (H^+). Remember that curved arrows show electron movement.

a) $H_3C–\ddot{O}:^{\ominus}$ (alkoxide) + $Mg^{\oplus}–Br$ ⟶ $H_3C–\ddot{O}–Mg–Br$

b) alkoxide + BH_3 ⟶ $:\ddot{O}–BH_3^{\ominus}$

c)

FIGURE 5.3 Examples of bond formation using mechanistic arrows.

In example (a) of Figure 5.3, a double-barb curved arrow starts at a pair of electrons on the oxygen and ends at the positively charged magnesium. This arrow indicates that a lone pair on the oxygen atom of the alkoxide is shared with (donated to) the magnesium atom to make a new O–Mg bond. The oxygen started with three non-bonded electron pairs but finished with two non-bonded pairs of electrons. (These two pairs were not involved in the reaction, so they remained on the oxygen.)

Example (b) shows a negatively charged alkoxide oxygen giving electrons to a neutral boron atom to form a new bond. The oxygen loses electron density and becomes neutral, while the boron atom gains electrons and becomes negative.

In example (c), electrons from a negatively charged oxygen form a new bond with a positively charged carbon elsewhere on the same molecule. Both of the reacting atoms end up with neutral charge in the product.

Mechanistic arrows are curved. The number of curves and their overall shape is not significant.

STUDENT TIP
Curved arrows can have more than one shape. Because of the complexity of some molecules, it is sometimes necessary to "bend" the arrows more than once to keep the drawings clear.

Mechanistic arrows are used to show the flow of electrons to form bonds between the atoms at either end of the arrow. The parts of the molecules with no mechanistic arrows are unchanged. Changes in charges resulting from electron movement can be determined by formal charge calculations, or by applying the concept of "donating" or "accepting" electrons. Section 5.3 describes such changes in formal charges in more detail.

ORGANIC CHEMWARE
5.6 Curved arrow notation: Break/form X–Y σ bond

5.2.4 Curved arrows and bond breaking

When an electron flow causes a bond to break, the curved mechanistic arrow starts in the *centre* of the bond that breaks. This arrow points to the atom that will receive the electrons from the bond, usually the more electronegative atom in the bond.

ORGANIC CHEMWARE
5.7 Curved arrow notation: Break/form H–X σ bond

STUDENT TIP

When bonds are broken in organic reactions, the electrons generally flow to the more electronegative atom or best electron attractor.

two barbs indicates a two-electron movement

arrow starts in centre of C–Br bond and points toward the more electronegative atom of the bond (bromine)

atom that the arrow points to receives the electrons when the bond breaks

ORGANIC CHEMWARE

5.8 Curved arrow notation: Break/form X–Y^+ σ bond

ORGANIC CHEMWARE

5.9 Curved arrow notation: Break/form H–X^+ σ bond

Breaking a carbon–bromine bond in an alkyl bromide produces formal charges. The mechanistic arrow showing electron flow for this process starts at the centre of the line depicting the bond. The arrow points from the bond toward the bromine, indicating that the bromine will gain the electrons from the bond and become negative, whereas the carbon will lose electrons and become positive.

When bonds involving positive charges break, the electrons move toward the positive charge (a positively charged atom is a very strong electron attractor). In the following example, the oxygen atom starts out positively charged and then gains electrons when the bond breaks. The charge on the oxygen becomes zero. The carbon loses electron density and acquires a positive charge.

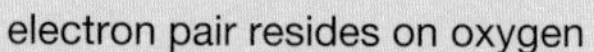

arrow starts in centre of C–O bond and points toward the strongest electron-attracting atom (oxygen)

5.2.5 Combining bond formation with bond breaking

Reaction mechanisms often involve the simultaneous formation and breaking of bonds. In the following example, the first arrow shows the formation of a bond between the oxygen of the alkoxide and the hydrogen of HCl, and the second curved arrow shows the bond between the hydrogen and the chlorine atom breaking. The arrows indicate the location of bonds, electrons, and charges at the end of the reaction.

bond forms between oxygen and hydrogen

bond between hydrogen and chlorine breaks, and the bonding pair moves to chlorine (most electronegative)

ORGANIC CHEMWARE

5.10 Acid–base chemistry: Proton transfer (B^- + H–A)

ORGANIC CHEMWARE

5.11 Acid–base chemistry: Proton transfer (B + H–A)

ORGANIC CHEMWARE

5.12 Acid–base chemistry: Proton transfer (B + H–A^+)

The next example shows a reaction involving two neutral molecules. Although the atoms are different from those in the previous example, the flow of electrons is essentially the same. The sulfur of the sulfide donates a lone pair of electrons to the carbon of the alkyl bromide, forming a bond with the carbon atom. The carbon from the alkyl bromide simultaneously gives away a pair of electrons. These electrons flow to the bromine atom (which is more electronegative than carbon), thus breaking the C–Br bond. This results in a sulfur carrying a positive charge and a bromide ion with a negative charge.

bond between carbon and bromine breaks, and the bonding pair moves to bromine
bond forms between sulfur and carbon
to avoid exceeding the octet on carbon, a bond must break as another is formed

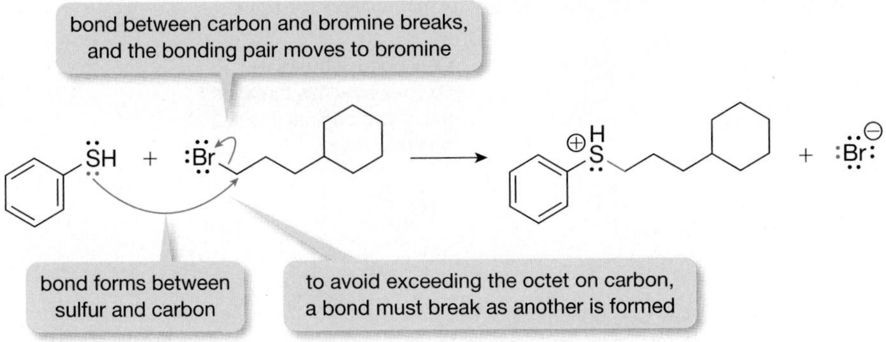

In both of the preceding examples, bonds are breaking at the same time as new bonds are forming. This is often necessary to avoid over-filling the valance shells of the reacting atoms.

CHECKPOINT 5.2

You should now be able to determine the product of a reaction given its starting materials and the reaction mechanism in curved arrow notation.

SOLVED PROBLEM

Draw the product(s) of the following reaction.

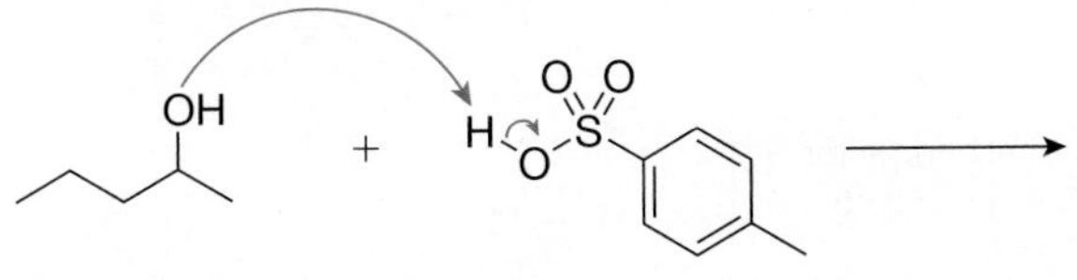

STEP 1: Expand the Lewis structure of the starting materials near the reacting centres.

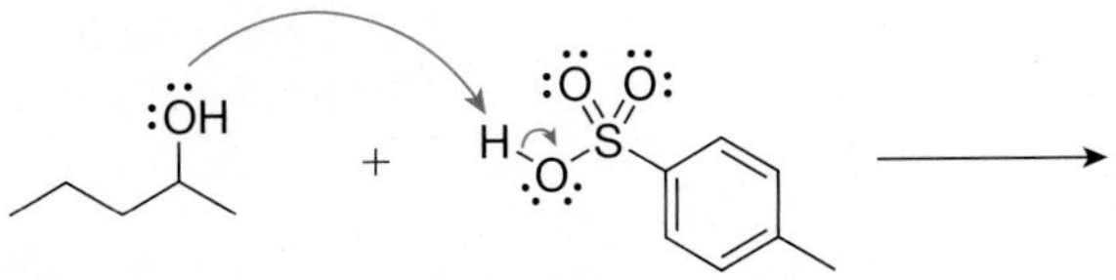

STEP 2: Use the curved arrows to determine which bond(s) will break and which new bond(s) will form. You might find it helpful to colour-code your arrows and the involved atoms and bonds.

Arrows that start at an atom or lone pair are bond-forming arrows, whereas arrows that start in the middle of a bond are bond-breaking ones. The first arrow, red in the following diagram, starts at the lone pair on oxygen and ends at the hydrogen atom, indicating that a new bond is forming between the oxygen and hydrogen atoms. The second arrow, in blue, starts in the middle of the H–O bond and ends at the oxygen atom, indicating that the H–O bond is breaking during this reaction step and the electrons from that bond are moving to the oxygen atom.

Red arrow
New bond forms between oxygen's lone pair and hydrogen.
Blue arrow
H–O bond breaks and the oxygen takes the electrons.

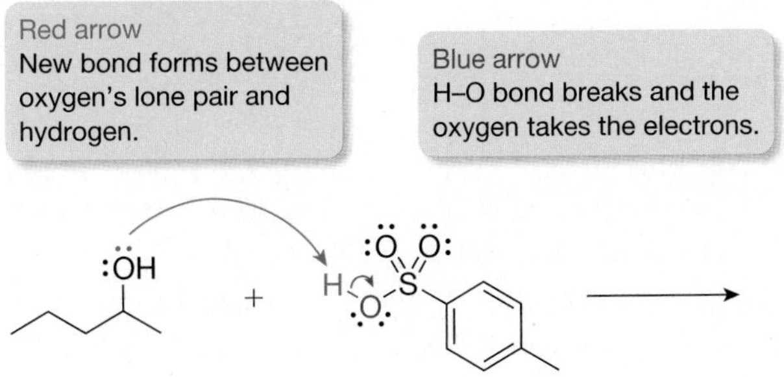

STEP 3: Draw the resulting products, omitting any formal charges for now.

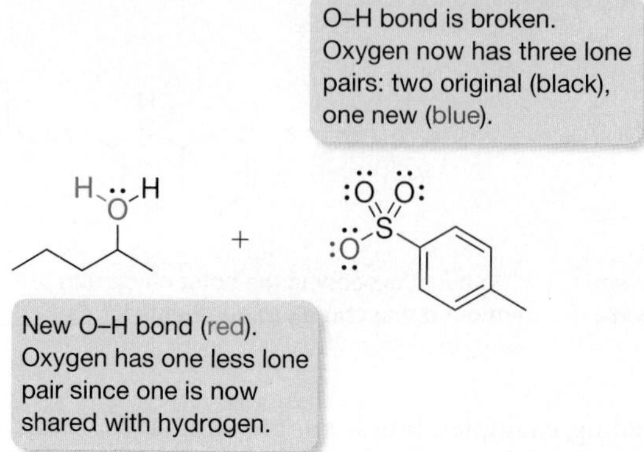
O–H bond is broken.
Oxygen now has three lone pairs: two original (black), one new (blue).
New O–H bond (red).
Oxygen has one less lone pair since one is now shared with hydrogen.

STEP 4: Calculate the formal charges for any atoms directly involved in the reaction and add them in the final structures.

There are three atoms directly involved in the reaction: two oxygen atoms and one hydrogen atom. Apply the formula for calculating the formal charge to each of the three atoms:

FC = (number of electrons in atom's neutral valence) − (number of bonds) − (number of non-bonding electrons)

Formal charge for O
FC = 6 − 3 − 2
= +1

Formal charge for H
FC = 1 − 1 − 0
= 0

Formal charge for O
FC = 6 − 1 − 6
= −1

STEP 5 (OPTIONAL): Redraw the final products as line drawings without all of the extra lone pairs.

While including the lone pairs is useful when examining a reaction, they can be omitted from final answers to avoid cluttering the structures. However, formal charges must always be included.

PRACTICE PROBLEM

5.3 Use the curved arrows to predict the product(s) of the following reactions.

a) 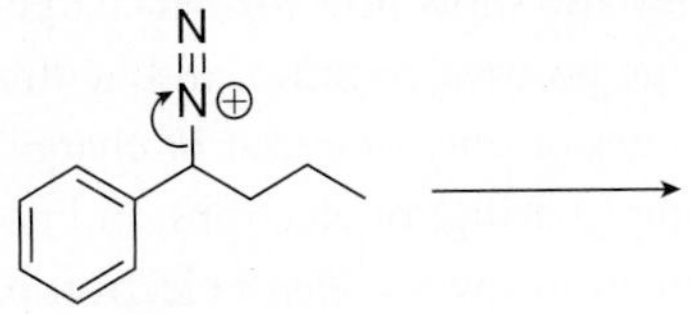 N ≡ N⊕ →

b) 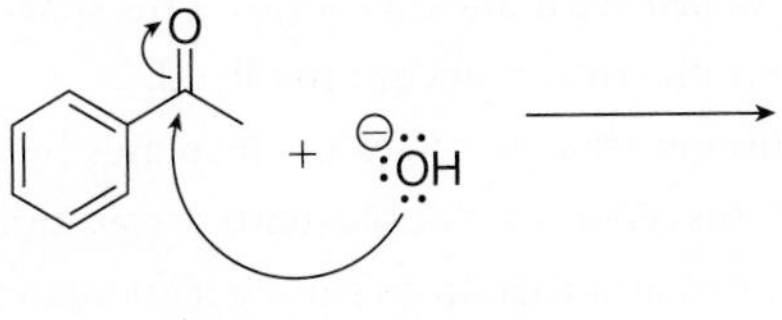 O + ⊖:OH →

c) 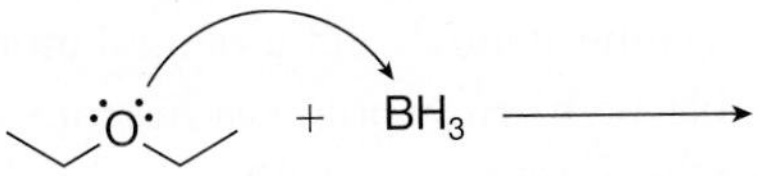 :O: + BH_3 →

d) 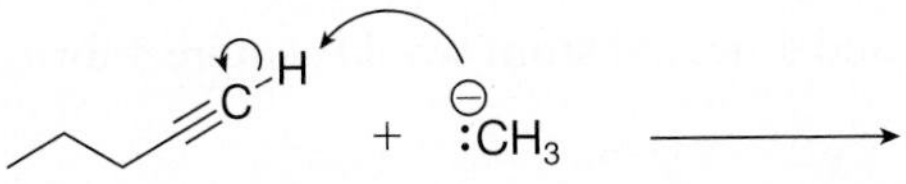 C–H + ⊖:CH_3 →

e) 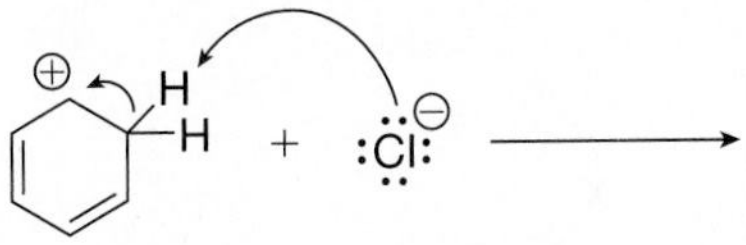 ⊕ H –H + :Cl:⊖ →

f) 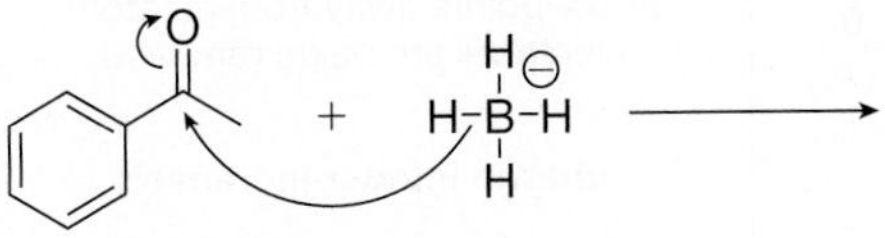 O + H–B⊖–H (H, H) →

INTEGRATE THE SKILL

5.4 Identify the error in each of the following mechanisms. Redraw each mechanism to give the products shown.

a) 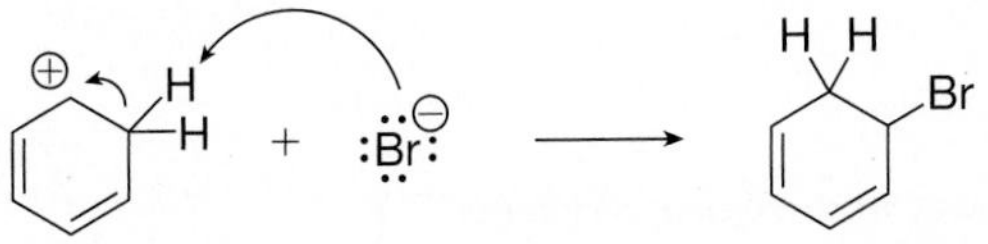 ⊕ H –H + :Br:⊖ → H H Br

b) 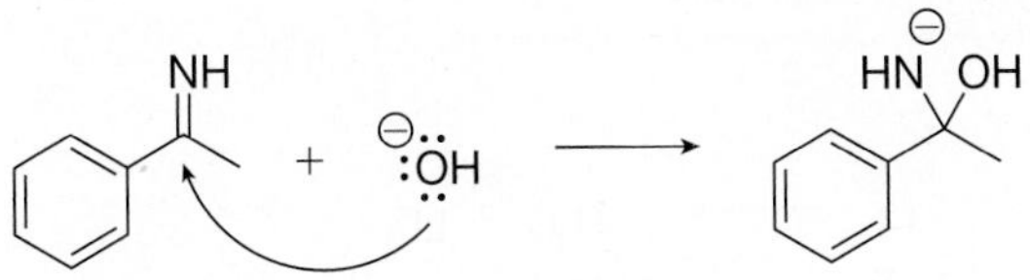 NH + ⊖:OH → ⊖ HN OH

c) 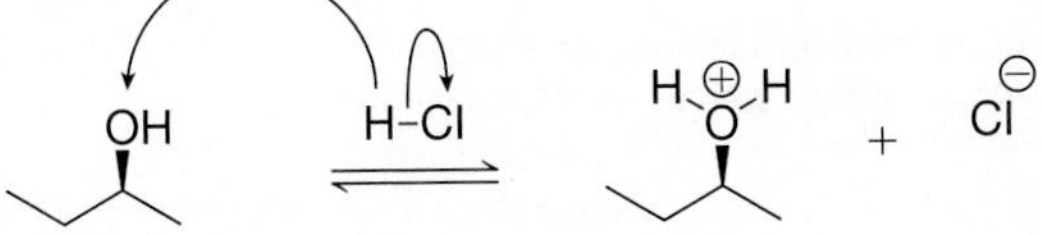 OH H–Cl ⇌ H⊕H O + Cl⊖

5.3 Curved Arrows and Formal Charges

Because mechanistic arrows show electron movement, they also show how formal charges change during reactions. The flow of a *pair* of electrons can change positive, negative, and neutral formal charges by *one unit*. Moving two electrons results in changes of only one unit of charge because electron movements involve bond formation and breaking (sharing) of electrons, and not simply the transfer of a pair from one atom to another. Thus, an atom gives a *shared* electron pair away, and when electron pairs are accepted they become *shared* with another atom. This sharing is the reason formal charges only change by one unit when electron pairs are involved.

The formal charge can be treated as a simple integer that increases or decreases by one unit depending on the direction a mechanistic arrow points. When a double-barb arrow points *away* from an atom, electrons are moving away from the atom and that atom's integer increases by one. Thus, an atom with a formal charge of −1 would become neutral, and a neutral atom would acquire a formal charge of +1. Similarly, when a double-barb arrow points *to* an atom, electrons are moving toward the atom, and that atom's integer decreases by one unit. Thus, an atom with a formal charge of +1 would become neutral, and a neutral atom would acquire a formal charge of −1.

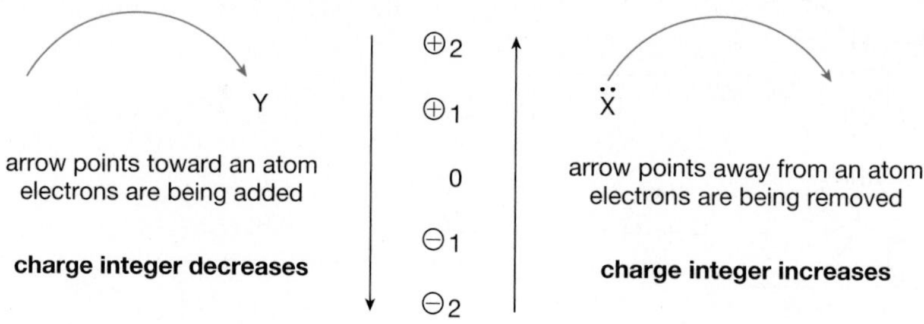

In the following example, the curved arrow points away from the oxygen, so the formal charge integer on this atom increases from −1 to 0. Since the arrow points toward the lithium atom, the formal charge integer on this atom decreases from +1 to 0. In this process the total net charge is conserved.

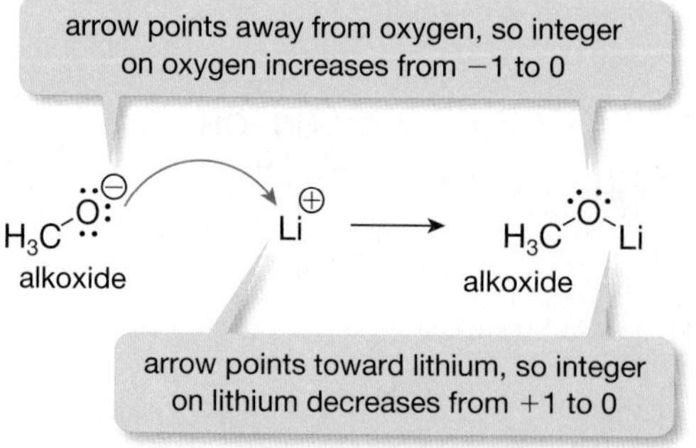

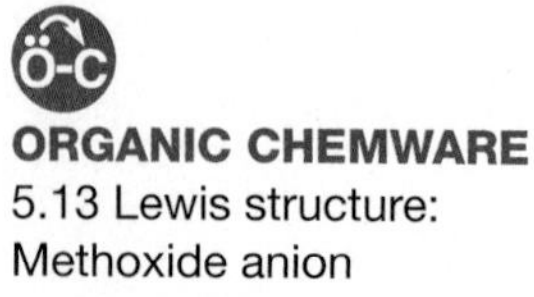

ORGANIC CHEMWARE
5.13 Lewis structure: Methoxide anion

When the σ bond in *tert*-butyl bromide breaks, the electrons are drawn toward the more electronegative atom in the bond, so the mechanistic arrow points toward the bromine atom. The formal charge integer on this atom therefore decreases from 0 to −1. Since the arrow points away from the carbon, the formal charge integer on the carbon increases from 0 to +1.

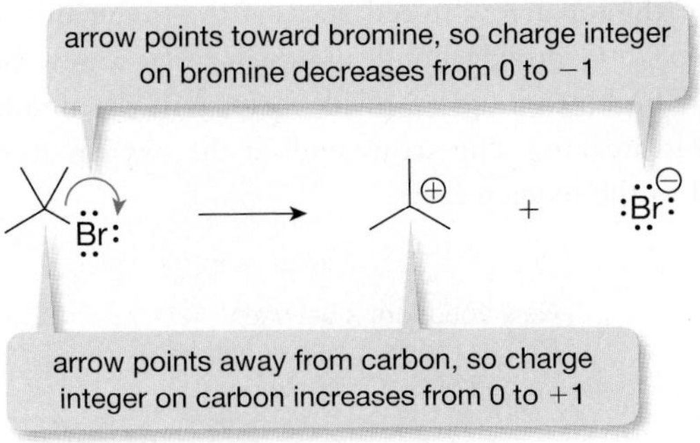

arrow points toward bromine, so charge integer on bromine decreases from 0 to −1

arrow points away from carbon, so charge integer on carbon increases from 0 to +1

When an atom both gains and loses a pair of electrons, the formal charge on the atom does *not* change because the distribution of electrons (bonds and lone pairs) on the atom does not change. For example, in the following reaction diagram, a mechanistic arrow points toward a carbon atom, and another arrow points away from the same atom. Therefore, the charge on this carbon does not change during the reaction.

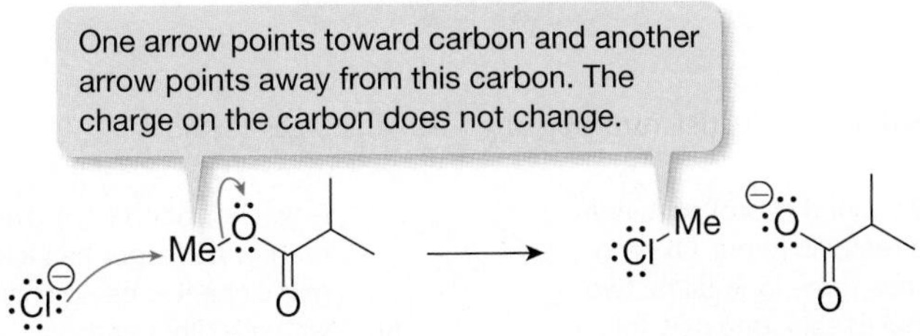

One arrow points toward carbon and another arrow points away from this carbon. The charge on the carbon does not change.

CHECKPOINT 5.3

You should now be able to use curved arrows to quickly determine any changes in formal charges that take place during a reaction.

SOLVED PROBLEM

Draw the products of the following reaction, using the curved arrows to assign any formal charges.

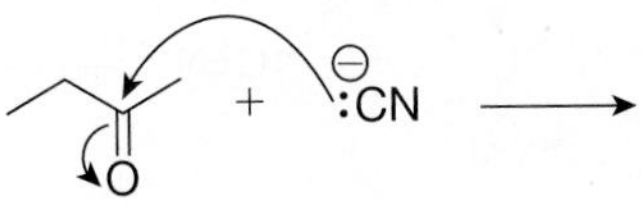

STEP 1: Expand the Lewis structure of the starting materials near the reacting centres.

Note that there is some flexibility here in terms of how much you want to expand the Lewis structures. For example, once you are more comfortable with common reaction mechanisms, you might not need to explicitly show the C atom of the carbonyl group as has been done here.

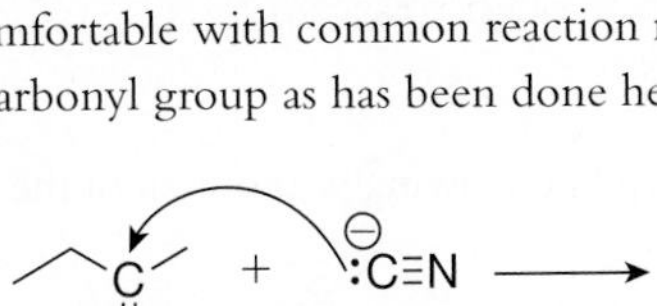

STEP 2: Use the curved arrows to determine which bond(s) will break and which new bond(s) will form. You might find it helpful to colour-code your arrows and the involved atoms and bonds.

In the following figure, the top arrow, in red, starts at the lone pair on the carbon of CN^- and ends at the carbon of the C=O double bond, indicating that a new bond is forming between these two carbon atoms. The bottom arrow, in blue, starts in the middle of the C=O π bond, indicating that this bond is breaking. The arrow ends at the oxygen atom, so the π electrons are moving from the π bond to the oxygen atom.

Red arrow
New bond forms between CN^- carbon's lone pair and carbon of C=O double bond.

Blue arrow
π bond of C=O double bond breaks and electrons move to the oxygen.

STEP 3: Draw the resulting product(s), omitting any formal charges for now.

C=O π bond is broken; single bond remains (blue). Oxygen now has three lone pairs: two original (black), one new (blue).

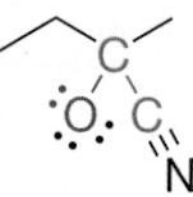

New C–C bond (red). This carbon no longer has a lone pair since it is being shared with the other carbon.

STEP 4: Determine the formal charges for any atoms directly involved in the reaction, using the curved arrows to guide you.

One arrow points toward carbon and another points away; the formal charge on this atom does not change.

Arrow points away from carbon, so it will increase its charge integer by +1; formal charge in product is 0.

Arrow points toward oxygen, so it will reduce its charge integer by 1; formal charge in product is −1.

STEP 5: Redraw the final product as a line drawing without all of the extra lone pairs.

PRACTICE PROBLEM

5.5 The products of the following reaction steps have already been drawn; however, they are missing formal charges. Use the provided curved arrows to complete the structures of the products.

a) Ö: + BF_3 ⟶ O BF3

b) Br: ⟶ + Br

c) ⊕ H Ö H ⟶ H O H

d) HÖ ⊕ÖH H ⟶ OH O H

e) Ö: Ö N⊕ Ö ⟶ O N O O

INTEGRATE THE SKILL

5.6 Below are two multi-step reaction mechanisms. Lone pairs have been drawn on the starting materials; however, the formal charges are missing. The rest of the structures are missing both lone pairs and formal charges. Calculate the missing formal charges in the reactants, and then use the given curved arrows to fill in both the formal charges and any necessary lone pairs for the rest of the mechanism.

a) :Ö Ö: ÖEt ⟶ O OEt O ⟶ O O + OEt

b) Ö: HÖCH_3 ⟶ O O CH_3 H HOCH_3 ⟶ O O CH_3 + H O CH_3 H

5.4 Intramolecular Reactions

When separate molecules react, the process is called an **intermolecular reaction**. The functional groups on one molecule interact with the functional groups on another molecule, resulting in the formation of products, as in the following example.

Intermolecular reactions occur between two or more molecules.

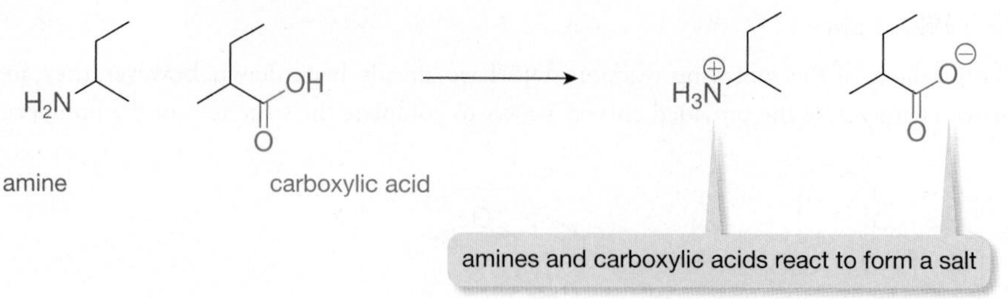

amine carboxylic acid

amines and carboxylic acids react to form a salt

Intramolecular reactions occur within a single molecule.

Reactions that occur *within* a molecule are called **intramolecular reactions.** These reactions can occur when a molecule contains two or more functional groups that are arranged such that they can interact. The flows of electrons are the same as those in intermolecular reactions. Intramolecular reactions can be very fast because the functional groups are positioned close to each other.

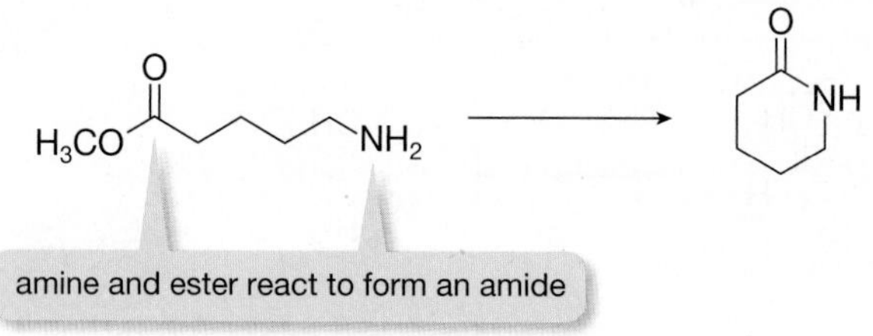

amine and ester react to form an amide

CHECKPOINT 5.4

You should now be able to distinguish between intermolecular and intramolecular reactions.

SOLVED PROBLEM

Decide whether each step in the following reaction is intra- or intermolecular. Add mechanistic arrows to show the flow of electrons.

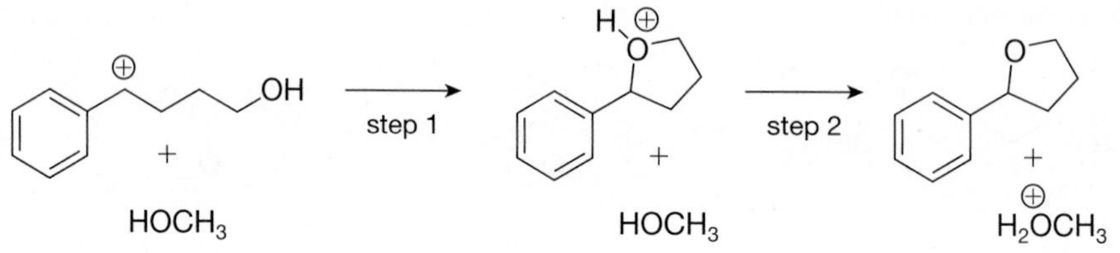

step 1 step 2

$HOCH_3$ $HOCH_3$ $H_2\overset{\oplus}{O}CH_3$

STEP 1: Identify the molecules involved in each step of the reaction to determine whether the reactions are intermolecular or intramolecular.

In the first step of the process, the methanol is unchanged, so it must not be involved in the reaction. Hence, step 1 of the reaction involves only the other molecule: the one bearing the positively charged carbon (carbocation) and the hydroxyl group. Therefore, step 1 is an intramolecular process.

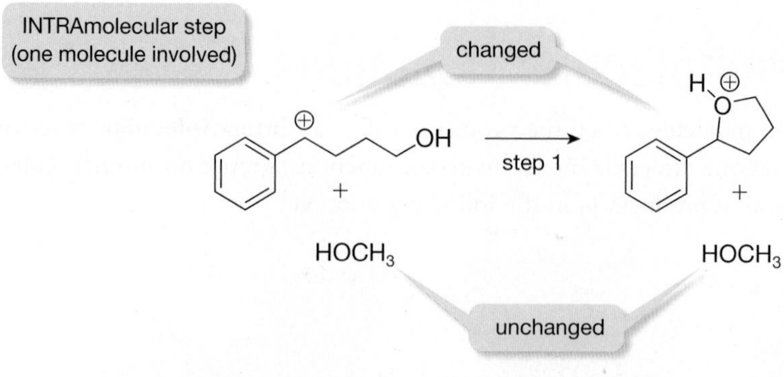

INTRAmolecular step (one molecule involved)

changed

step 1

$HOCH_3$ $HOCH_3$

unchanged

In the second step, a proton is transferred from the oxonium to the methanol, and both of the molecules change. Therefore, step 2 is an intermolecular reaction.

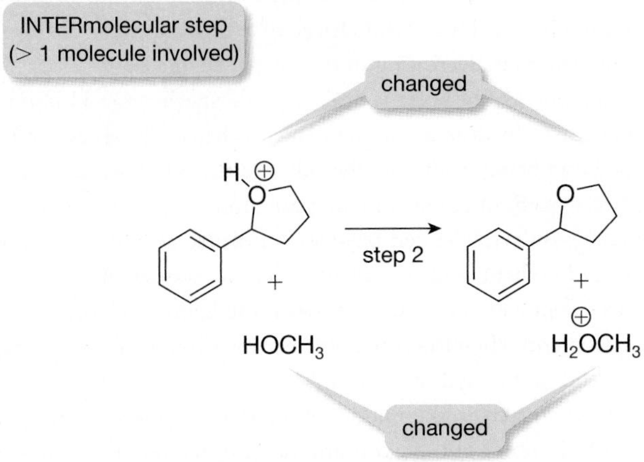

INTERmolecular step (> 1 molecule involved)

changed

step 2

$HOCH_3$ + ... → ... + $H_2OCH_3^{\oplus}$

changed

STEP 2: Expand the Lewis structure around the parts of the molecules that are involved in the reaction steps.

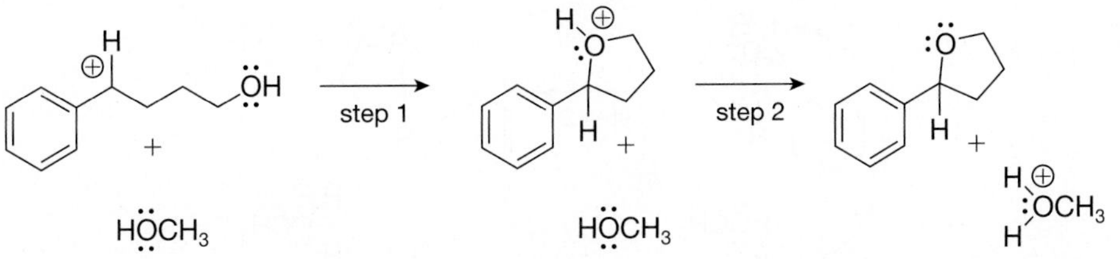

step 1 step 2

STEP 3: Determine which bonds were made and which were broken during each step of the reaction.

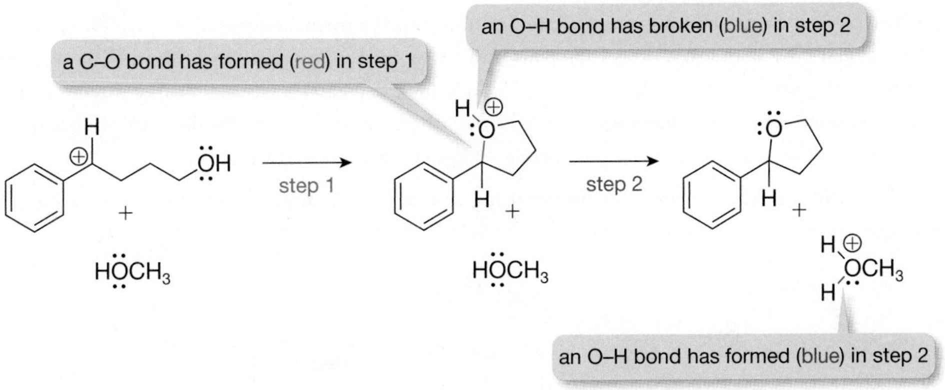

a C–O bond has formed (red) in step 1

an O–H bond has broken (blue) in step 2

step 1 step 2

an O–H bond has formed (blue) in step 2

STEP 4: Add curved arrows that can account for the broken and formed bonds, as well as any changes to formal charges and lone pairs.

In step 1 of the reaction, a new C−O bond forms, so there will be a curved arrow connecting these two atoms. Remember that curved arrows always start at an electron source and move toward an area of electron deficiency. For this reaction, start the arrow at one of oxygen's lone pairs (electron source) and end it at the positively charged carbon—an area of electron deficiency since it has only six valence electrons.

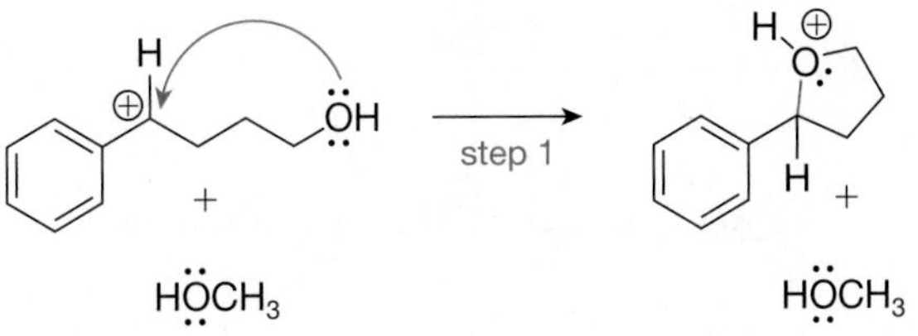

step 1

To verify that this arrow is correct, confirm that it fits with any changes to lone pairs and formal charges. After the reaction, the oxygen atom has one less lone pair and a formal charge of +1. This fits with the oxygen now sharing a lone pair with the carbon atom, and having an arrow pointing away from it, increasing its formal charge by +1. The formal charge of the carbon atom changes from +1 to 0, which fits with an arrow pointing toward it, decreasing in the formal charge by 1.

In step 2 of the reaction, one O−H bond is broken while another O−H bond is formed. In effect, a hydrogen transfers from one oxygen atom to another. The mechanism is represented by two curved arrows: one corresponding to the bond being made, and the other to the bond being broken.

The bond-forming arrow needs to connect the oxygen atom on a methanol molecule to the hydrogen atom on the oxonium ion (positively charged oxygen). To decide at which atom the arrow should start, compare the two options and evaluate which is a better electron source. In this case, the methanol oxygen atom has lone pairs (an excellent electron source), whereas the hydrogen atom is next to a positive charge (making it very electron deficient). Therefore, the curved arrow should start at one of the two lone pairs on the methanol oxygen and end on the hydrogen atom.

Bond-breaking arrows always start in the middle of the bond being broken, so the second arrow must start in the centre of the O−H bond of the oxonium ion. Just like bond-forming arrows, bond-breaking arrows also end at an electron-deficient site (or electronegative atom). The positive charge on oxygen is a very strong electron attractor, so the arrow should point toward the oxygen atom.

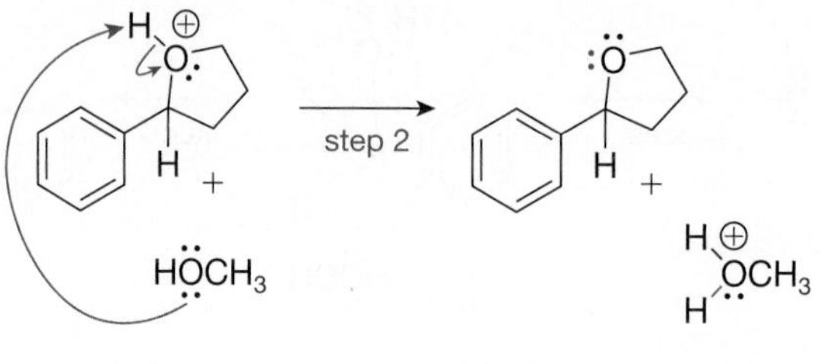

step 2

$HOCH_3$

$H_2OCH_3^+$

To check that the arrows are correct, confirm that they account for any changes to lone pairs and formal charges. The oxygen on methanol has lost a lone pair and gained a formal charge of +1, which makes sense since this oxygen now shares one of its lone pairs and has an arrow pointing away from it. The oxygen on the oxonium ion gains a lone pair during the reaction, which decreases its formal charge integer from +1 to 0. This makes sense because this oxygen gained a pair of electrons from the breaking O−H bond, and it has an arrow pointing toward it, causing its formal charge integer to decrease by 1.

STEP 5 (OPTIONAL): Redraw your final answer using line drawings, omitting any unnecessary lone pairs and hydrogen atoms.

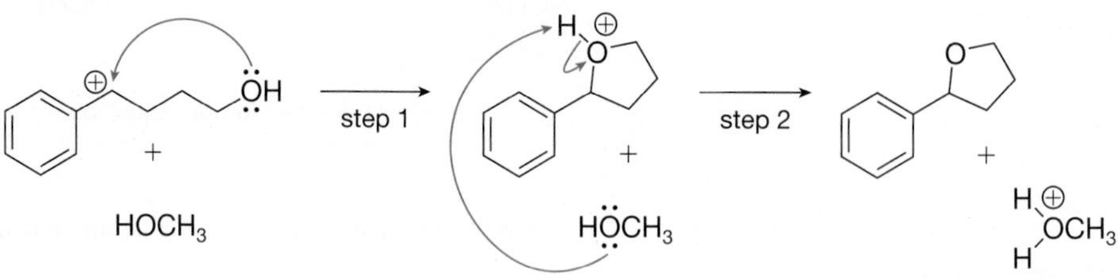

step 1 step 2

$HOCH_3$ $HOCH_3$ $H_2OCH_3^+$

PRACTICE PROBLEM

5.7 Classify each of the following reactions as intra- or intermolecular. Add mechanistic arrows to show the flow of electrons in each case.

a)  + HCl ⟶ + Cl^-

b) 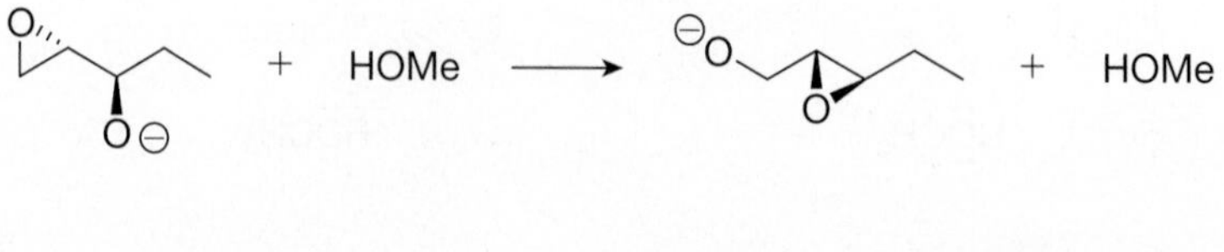 + HOMe ⟶ + HOMe

c) 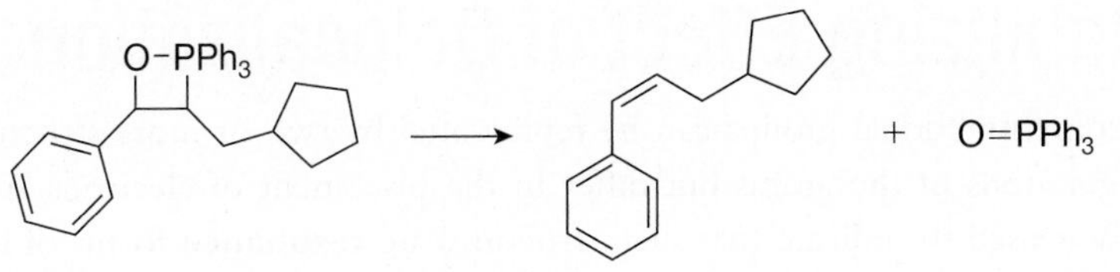 + O=PPh$_3$

d)

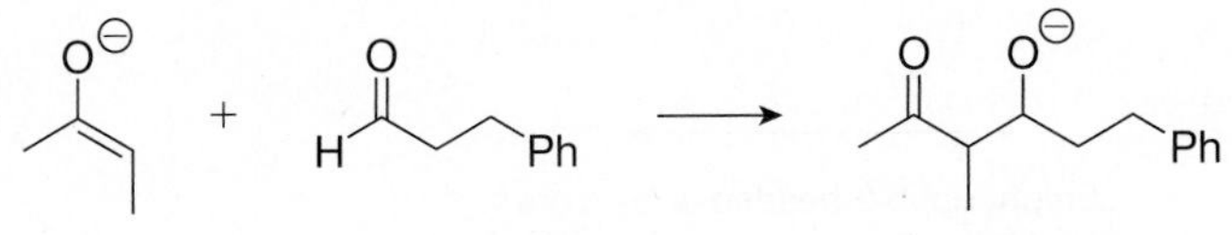

INTEGRATE THE SKILL

5.8 For each of the following reactions:
   i) Classify the reaction as either inter- or intramolecular.
   ii) Draw the expected product.
   iii) Provide the hybridization and geometry of every carbon atom directly involved in the reaction, both in the starting materials and the final product(s).
   iv) Locate any chirality centres formed during the reaction.

a)

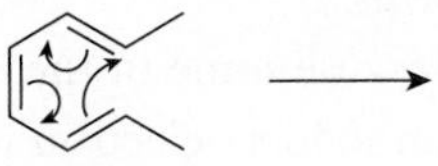

b)

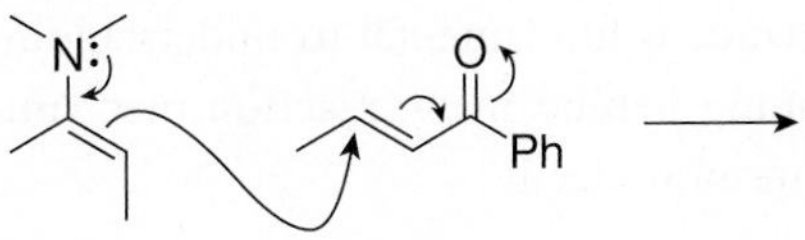

c)

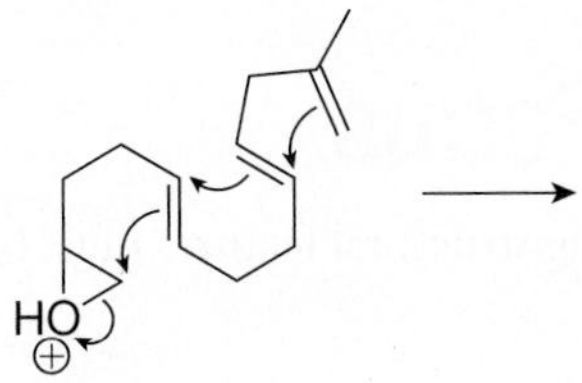

5.4.1 Representing the movement of π bond electrons

Some reactions occur through the movement of π electrons only. In these cases, it is common to show bond formation with an arrow that points to the *space between atoms*. For π electron movement, you can use this positioning interchangeably with an arrow that points toward an atom.

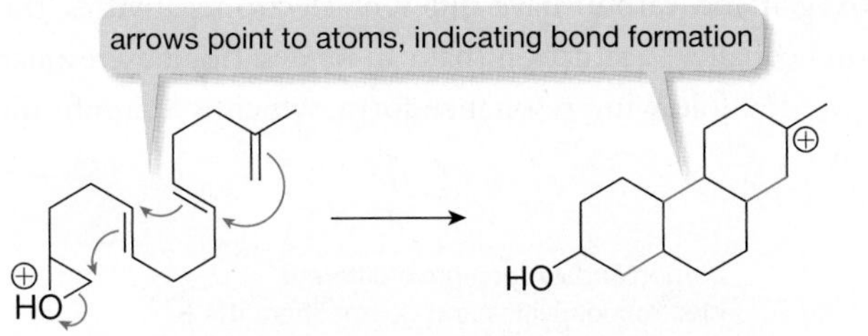

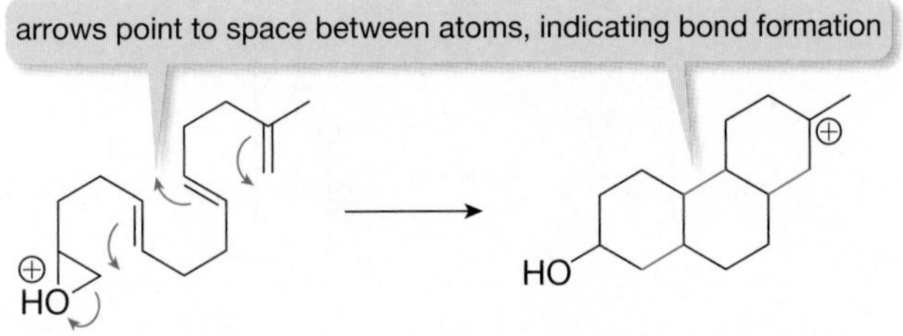

arrows point to atoms, indicating bond formation

arrows point to space between atoms, indicating bond formation

5.5 The Stabilizing Effect of Delocalization

Resonance is a method of describing the structure and properties of a molecule that cannot be drawn as a single Lewis structure.

In Lewis diagrams, some functional groups can be represented by two or more structures that have identical configurations of the atoms but differ in the placement of electrons. A straight double-headed arrow is used to indicate that such structures are **resonance** forms of the same functional group.

⟷

straight double-headed arrows are used to separate resonance forms

The actual structure of a group undergoing resonance is a *blend of all of the resonance forms at the same time*. The molecule does not "flip" back and forth between the resonance forms, which is why resonance is indicated by a special double-ended arrow instead of an equilibrium arrow. The individual resonance forms can be used as tools to describe the structure and reactivity of a molecule, but they are not real structures. The actual molecule is best represented by a resonance hybrid, or computer model, that is a composite of the resonance forms.

Functional groups that participate in resonance gain stability because some of the electrons in the structure are shared between several atoms. This delocalization reduces electron repulsion and increases bond strength. The concept of resonance is fundamental to understanding a large portion of organic chemistry, including the reasoning behind many reaction mechanisms. The rest of this chapter describes resonance structures in more detail.

5.6 Constructing Resonance Forms

For a molecule to have resonance, at least one of the following structural features must be present in the molecule:

1. a π bond made up of atoms with different electronegativities
2. a π bond directly beside at least one of the following features:
   a) paired or unpaired electrons
   b) atoms with incomplete octets
   c) other π bonds
   d) charged atoms lacking octets or carrying lone pairs
3. an atom with an incomplete octet adjacent to an atom with a pair of non-bonding electrons

Carbonyl groups are composed of a carbon and an oxygen connected together by a double bond. Because the oxygen and carbon have different electronegativities, they do not share the electrons in the π bond equally—a situation that can be described by resonance. Breaking the π bond of the group gives the following resonance form, which is a significant contributor to the structure of the group.

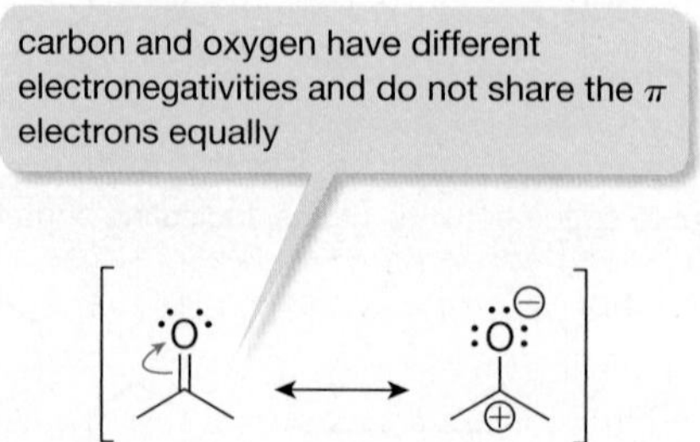

carbon and oxygen have different electronegativities and do not share the π electrons equally

An amide has a π bond involving two elements with different electronegativities, so resonance is possible for this functional group. A nitrogen with a lone pair is beside the π bond, and so this lone pair may participate in the same resonance.

π bond made up of two different elements

lone pair adjacent to the π bond

amide

Curved arrows can be used to draw the other resonance forms of the functional group. When no formal charges are present, a good place to begin is by considering resonance forms in which the π bond is broken. Consider the example in following diagram. In the C=O bond of the amide (structure A), the oxygen is more electronegative than the carbon, and will have a stronger attraction for electrons. The π bond of the double bond can be broken, with the electrons flowing to the oxygen, as depicted by the first curved arrow. The result is structure B, in which a lone pair is located on a neutral nitrogen atom *adjacent* to a positive carbon that lacks an octet. This configuration allows a double bond between the nitrogen and carbon atoms to form by donating the lone pair from nitrogen to the positively charged carbon. This flow of electrons corresponds to the second curved arrow and produces structure C. It is not possible to draw any further structures (no other π bonds, charges, or lone pairs), so these three structures are the only possible resonance forms of the amide.

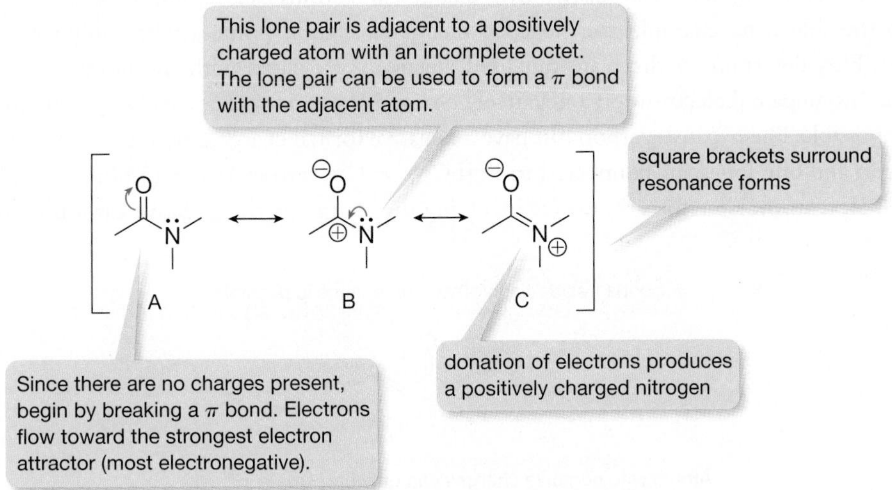

ORGANIC CHEMWARE
5.14 Resonance: Amide

An α,β-unsaturated ketone contains two adjacent double bonds, an arrangement that makes resonance possible. Since no formal charges are present, bonds must be broken to make resonance forms. Which of the π bonds should be broken first? In which direction should the electrons from the bonds flow? Just as in the previous example, these questions can be addressed by considering electronegativity. Heteroatoms usually have different electronegativity than carbon, and provide an excellent starting point to analyze resonance. Therefore, begin with the π bond involving the oxygen.

Resonance form B can be constructed by moving the π electrons of the carbon-oxygen bond onto the more electronegative oxygen. The positive charge in this structure now determines the direction electrons flow to break the adjacent carbon–carbon π bond. Positive charges are strong electron attractors, and so the electrons in the π bond will flow toward this charge, producing resonance form C. Once the last π bond has broken, there are no more π electrons, and so the analysis is complete. It is important to work with one bond at a time to avoid missing possible resonance structures.

The heteroatom is a good starting point because it has a different electronegativity than carbon. This can be used to determine the proper direction to break the bond.

The mechanistic arrow points away from this carbon; the charge on the carbon therefore increases from 0 to +1.

A ⟷ B ⟷ C

This π bond involves two carbon atoms. There is no difference in electronegativity between them and so breaking this bond produces insignificant resonance structures.

The positive charge determines the direction of π bond breaking. The electrons in the π bond will be attracted to the positive charge.

ORGANIC CHEMWARE
5.15 Resonance: Propenal (acrolein)

ORGANIC CHEMWARE
5.16 Resonance: 2-Butenal (crotonaldehyde)

STUDENT TIP
Not all negatively charged atoms carry lone pairs. Negatively charged atoms without lone pairs do not participate in resonance.

To draw resonance structures for molecules that have an existing formal charge, use the charge as the starting point for resonance analysis. A positive charge attracts electrons and tends to "pull" adjacent lone pairs and π bond electrons toward it. Negative charges repel electrons and "push" them into adjacent π bonds, positive charges, and atoms lacking full octets.

In the following example, the negative charge is a good point for beginning resonance analysis. First determine if there are unpaired electrons associated with the negative charge. If there are no unpaired electrons on a negatively charged atom, resonance involving that atom will not be possible. The negative carbon can have a negative formal charge only if the carbon has one hydrogen and one lone pair connected to it: [FC = −1 = (group IV) − (3 bonds) − (2 non-bonded electrons)]. Therefore, resonance involving the charge on the negative carbon is possible.

π bonds beside each other—resonance is possible

Atoms with negative charges can only participate in resonance if they carry a lone pair, so first determine if lone pairs are present using formal charge calculation.

The carbon must have a lone pair to carry a formal charge of −1: FC = −1 = (group IV) − (3 bonds) − (2 non-bonded electrons)

Once the possibility of resonance has been determined, start by moving the electrons associated with the negative charge toward the adjacent carbon carrying a π bond. This π bond must break at the same time so that the carbon receiving the electrons from the negative carbon does not exceed its octet. The electron pair from the breaking π bond flows to the left carbon in the bond (the carbon farther from the negative charge), giving resonance form B. In this form, the negative charge is adjacent to the C=O π bond, so a further electron movement is possible. This movement produces resonance structure C. Since electrons cannot move any farther along the molecule, the three structures shown are the only resonance forms for this molecule.

STUDENT TIP

Do *not* break single bonds when constructing resonance forms; only π bonds (in double and triple bonds) may be broken.

CHECKPOINT 5.5

You should now be able to generate a complete set of resonance forms for a given compound.

SOLVED PROBLEM

Determine which of the structures below have resonance. For those that do, draw all possible resonance forms, using curved arrows to show how each structure is generated.

STEP 1: Expand the Lewis structure of the atoms near π bonds, heteroatoms, and charges.

Start by adding in any missing lone pairs that are next to π bonds or atoms with charges. When first dealing with resonance structures, it is a good idea to add any implied hydrogens on or near these features, since explicitly showing these hydrogens can help you keep track of the formal charges in the resonance structures.

STEP 2: For those structures with π bonds, see if the atoms directly adjacent to the π bonds have any of the features that make resonance possible.

The adjacent atoms on structure II have two of the features that make resonance possible, so this structure will have resonance. Structure III, however, has none of the possible features on the atoms adjacent to its π bond and therefore does not meet this criterion for resonance.

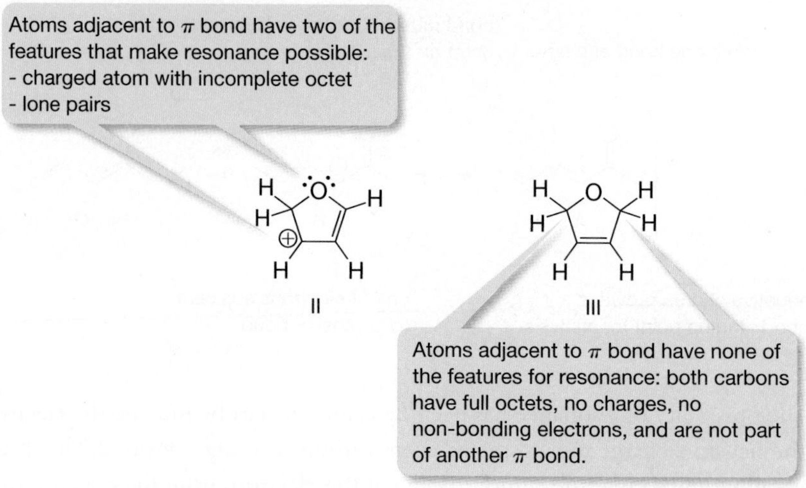

Atoms adjacent to π bond have two of the features that make resonance possible:
- charged atom with incomplete octet
- lone pairs
II
III
Atoms adjacent to π bond have none of the features for resonance: both carbons have full octets, no charges, no non-bonding electrons, and are not part of another π bond.

STEP 3: For those structures still under consideration, look for π bonds between atoms with different electronegativities.

Although structure III has a π bond, it is a C=C double bond, so the two atoms in the bond have identical electronegativities. Therefore, this bond does not meet this criterion for resonance.

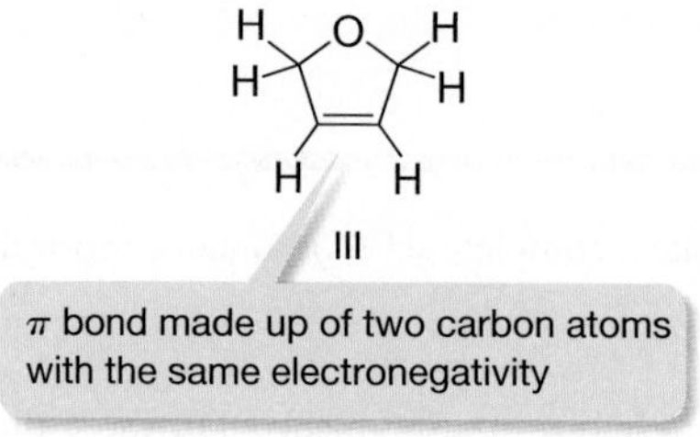

STEP 4: For those structures still under consideration, look for any non-bonding electrons next to atoms lacking a complete octet.

Structures I and IV have atoms with an incomplete octet: in both cases, a positively charged carbon (carbocation). In structure I, neither of the atoms next to the carbocation have any non-bonding electrons to share. Therefore, structure I fails to meet this criterion for resonance. Structure IV, on the other hand, has an oxygen atom with lone pairs next to its carbocation. These lone pairs could share their electrons, so structure IV has resonance.

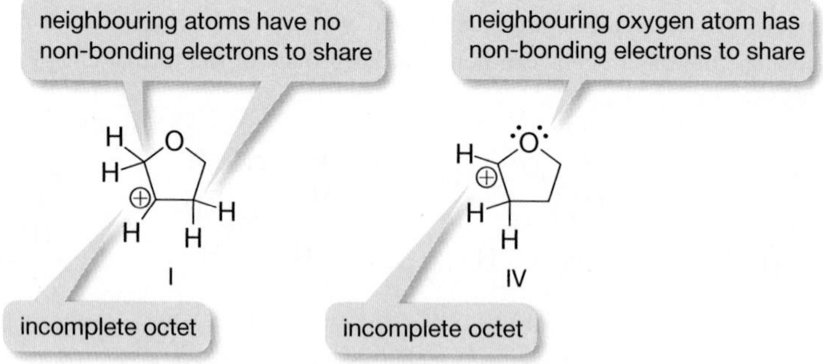

Therefore, of the four structures, structures II and IV can participate in resonance.

STEP 5: To generate new resonance contributors for structures with resonance, look for any charges or non-bonding electrons next to a π bond.

Structure II has a positively charged carbocation directly adjacent to the π bond. This carbocation can receive electrons from the π bond since the carbon atom lacks a full octet. A curved arrow shows the flow of electrons from the π bond to the carbocation, and generates resonance form B.

Structure II

A
B
Carbocation lacks a full octet; the electrons from the adjacent π bond move over to be shared with the carbocation.

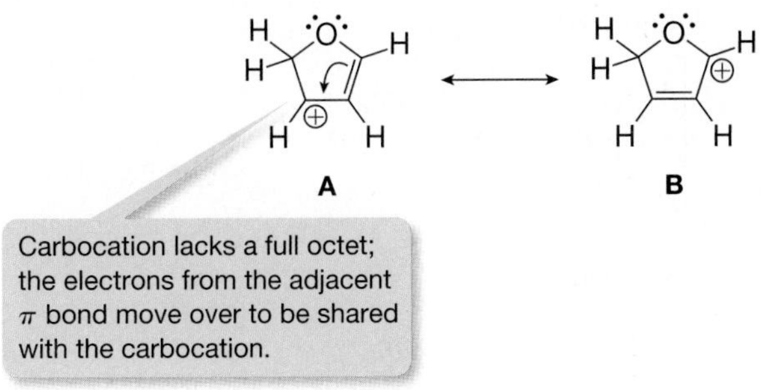

Resonance form B has a new carbocation with an incomplete octet. The two lone pairs on the neighbouring oxygen atom can be shared with the carbocation. Drawing a curved arrow from the lone pair on oxygen to the positive charge generates the third and final resonance structure, C, with a new C=O bond and a positive charge on the oxygen.

Structure II

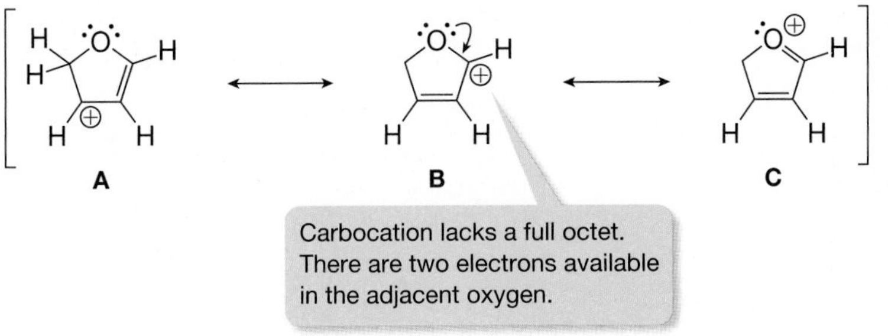

A
B
C
Carbocation lacks a full octet. There are two electrons available in the adjacent oxygen.

Structure IV has no π bond with adjacent charges or lone pairs.

STEP 6: If there are no charges or lone pairs next to π bonds, look for π bonds that can be broken by pushing electrons to a more electronegative atom.

Structure IV has no π bonds.

STEP 7: If there are no π bonds, look for non-bonding electrons that can be shared with atoms lacking a complete octet.

The lone pairs on the oxygen atom of structure IV are right next to the positively charged carbocation. Since this carbocation lacks a complete octet, it can accept one of the lone pairs from oxygen to make a new C=O double bond. The curved arrow starts at the lone pair and ends in the middle of the C−O bond (it could also start at the lone pair and point to the carbon). This electron flow creates a new C=O π bond and results in a neutral carbon and a positively charged oxygen.

lone pair on oxygen shared with carbocation
Structure IV:

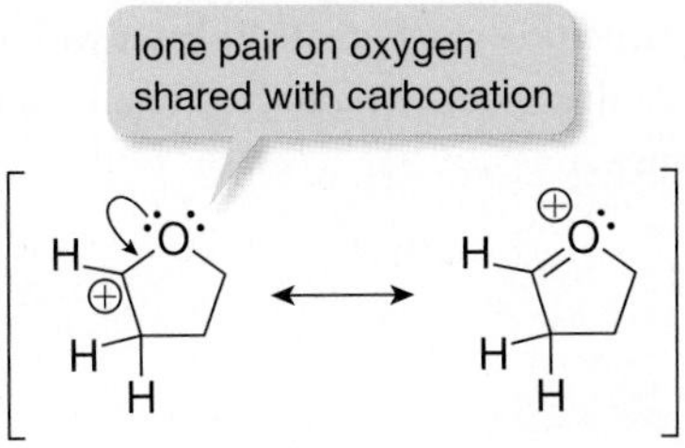

STEP 8: Double-check your work.

When drawing resonance structures, it is easy to involve too many bonds and electrons at once, and thus accidentally skip over a possible structure. Make sure you worked only one bond at a time. Also check that you considered all the nearby atoms and electrons that could possibly participate in resonance, and that you included all the formal charges on each structure.

STEP 9: Redraw your final answer using line drawings, omitting any unnecessary lone pairs and hydrogen atoms.

Structure II

Structure IV

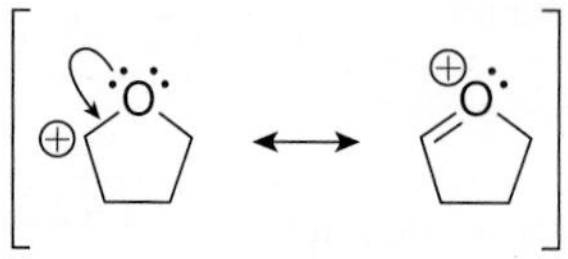

PRACTICE PROBLEM

5.9 Determine which of the following structures can participate in resonance. For the ones that can, draw all possible resonance forms.

a)

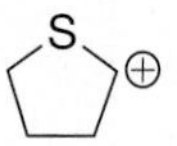

b)

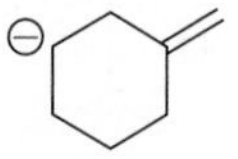

c)

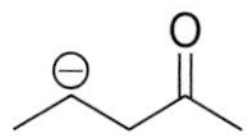

d)

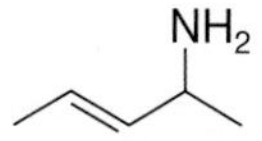

e)

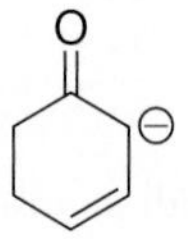

f)

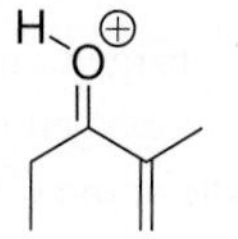

INTEGRATE THE SKILL

5.10 Each of the following structures can potentially react with positively charged molecules. Identify the site(s) on each molecule that would be attracted to an atom with a positive charge, using resonance forms to justify your answer.

a)

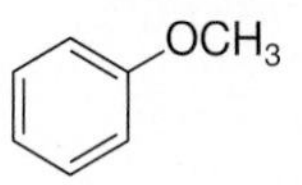

b)

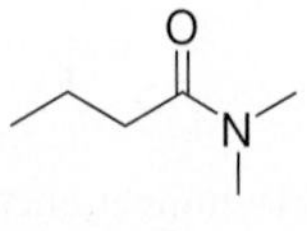

c)

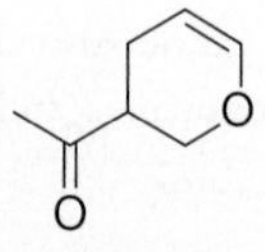

5.7 Evaluating Resonance Form Contributions

When constructing resonance hybrids from resonance forms, it is important to recognize that the relative contributions of resonance forms are not always equal. Forms with favourable electron distributions make larger contributions to the hybrid structure than forms with less favourable electron distributions. In drawings of chemical structures and reactions, functional groups are normally represented by the "best" (highest contributing) resonance forms. The relative "quality" of resonance forms also provides a guide to the most likely sites of reactivity in a given functional group.

The following guidelines are used to assess the quality of individual resonance forms. These guidelines are listed in order of decreasing importance.

In general, contributing resonance forms have the following characteristics:

1. the most atoms with full octets
2. the fewest number of formal charges
3. a) negative formal charges located on the most electronegative atoms
   b) positive formal charges located on the most electropositive atoms
4. a) like charges separated by the maximum distance possible
   b) opposite charges as close together as possible

In the following example, both resonance forms of the compound have full octets on all atoms and a minimum number of formal charges. These structures therefore make equal contributions to the resonance hybrid structure.

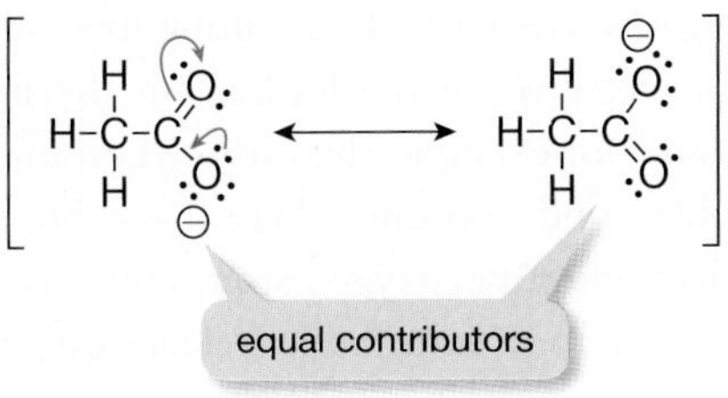

In the next example, the major resonance contributor of the amide is structure A, in which all atoms have octets and no atoms are charged. Structure B has two charges and an atom with less than eight valence electrons, so this form makes only a minor contribution to the overall structure. Structure C makes an intermediate contribution since it has two charged atoms, but all the atoms have full octets.

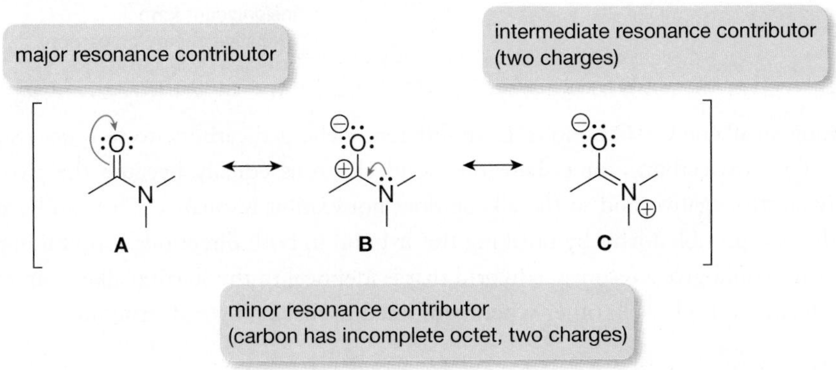

An amide group is therefore drawn as structure A, since this is the major resonance contributor. The *reactivity* of this functional group can be predicted by examining the other resonance forms. In this case, the reactivity of the group derives from form C (intermediate), and to a lesser extent form B (minor) (see Chapters 7 and 15 for an explanation of this reactivity).

5.7.1 Significant and insignificant structures

Some resonance forms make no contribution to the functional group structure. Such insignificant structures often have more than two formal charges on a functional group or connected system. A common example of this is a carboxylate, the conjugate base of a carboxylic acid. Starting with either of the two significant resonance forms, breaking the oxygen-carbon double bond produces a resonance form with three formal charges. This is an insignificant form that does not contribute to the structure or the reactivity of the group.

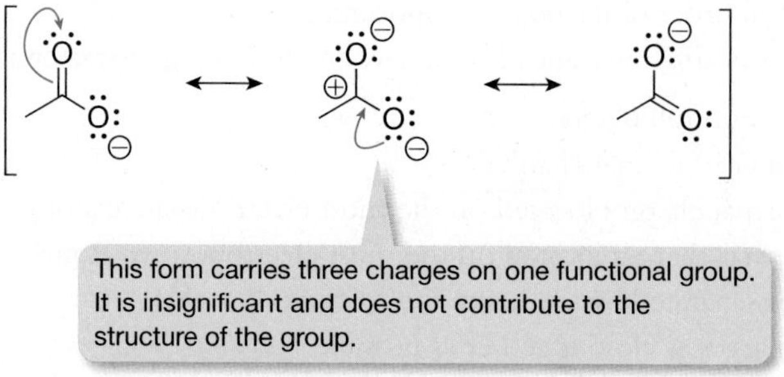

This form carries three charges on one functional group. It is insignificant and does not contribute to the structure of the group.

Insignificant resonance forms also occur when electrons from a breaking π bond flow to an atom with inappropriate electronegativity. In general, when drawing resonance structures, π bonds may be broken if the double bond involves two different elements, and if the electrons transfer to the more electronegative element. The resulting structure will have adjacent negative and positive charges, with the negative charge located on the more electronegative atom of the pair. In the following diagram, for example, the carbonyl group has two significant resonance forms: one with a C=O double bond, and one where the π bond is broken to put a negative formal charge on the more electronegative oxygen atom (blue arrows). Breaking the π bond in the other direction pushes the electrons onto the less electronegative carbon atom, resulting in an insignificant resonance form.

significant form (negative charge on oxygen)

significant form (no charges)

insignificant form (negative charge on carbon)

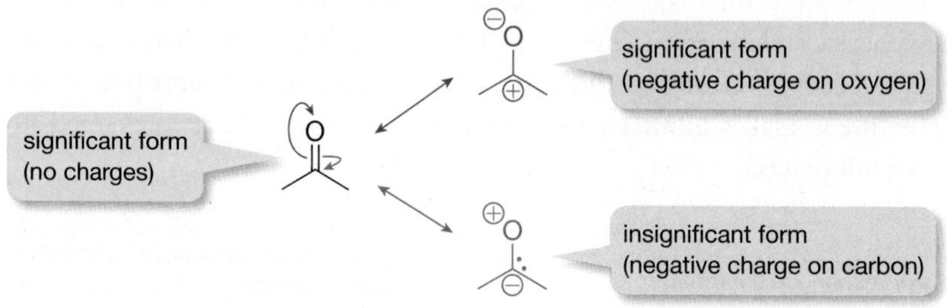

Breaking an alkene C=C bond to form differently charged carbon atoms is *not* a significant possibility. The two carbon atoms share the bond electrons equally because the two carbons are equally electronegative, and so the alkene does not exhibit resonance. This can be shown in drawing the two possible forms by breaking the π bond in both directions. Combining the resonance forms would give a resonance hybrid that is identical to the starting alkene. In effect, the resonance forms "cancel" each other when combined to make a hybrid structure.

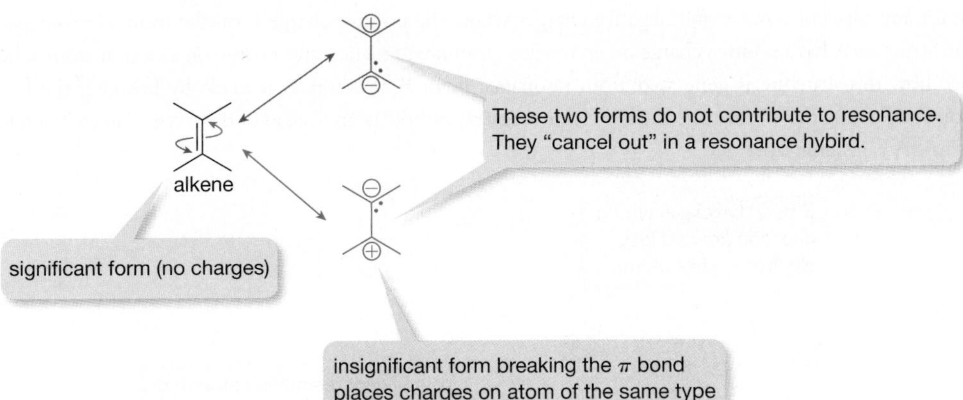
These two forms do not contribute to resonance. They "cancel out" in a resonance hybird.
alkene
significant form (no charges)
insignificant form breaking the π bond places charges on atom of the same type

CHECKPOINT 5.6

ORGANIC CHEMWARE
5.17 Resonance: Carbonyl group

ORGANIC CHEMWARE
5.18 Resonance: Carbonyl group (protonated)

You should now be able to rank the quality of forms contributing to a resonance hybrid and identify any insignificant resonance forms.

SOLVED PROBLEM

Examine the proposed set of resonance contributors below. Identify any insignificant resonance structures, and explain why they should not be included. For the remaining structures, rank them according to their contribution to the overall resonance hybrid (1 = contributes the most).

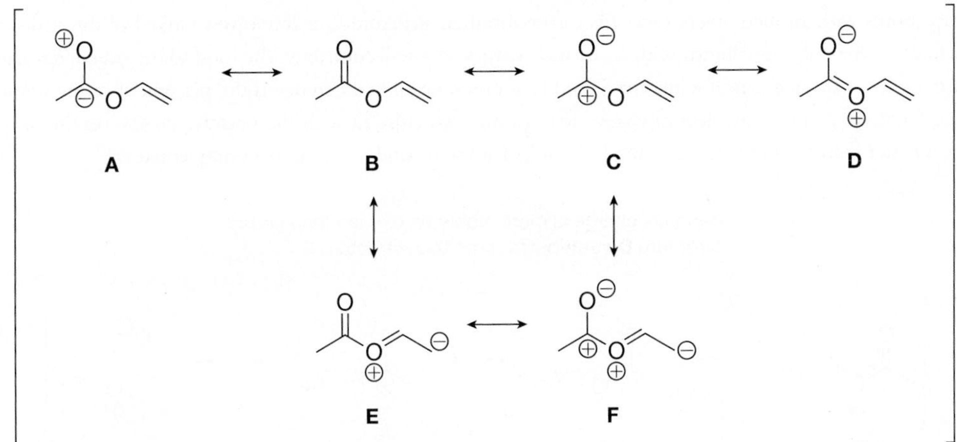
A
B
C
D
E
F

STEP 1: Look for any structures with more than two formal charges.

Structure F has four formal charges, whereas all the others have either two or none. Therefore, structure F can be considered an insignificant resonance structure.

too many formal charges; insignificant contribution

F

STEP 2: Look for any structures that are the result of breaking a π bond and moving the electrons away from the more electronegative atom.

Without the curved arrows showing the generation of each resonance structure, it can be difficult to spot these types of insignificant structures. You can either draw all the missing curved arrows (good practice!) or look for any adjacent positive and negative charges where the positive charge is on the more electronegative atom. Structure A has a positive charge on an oxygen atom next to a negative charge on a carbon atom. Check to see how this structure is generated: from resonance form B, structure A is made by breaking the C=O π bond and giving the electrons to the less electronegative carbon atom instead of the oxygen atom. Therefore, structure A is insignificant.

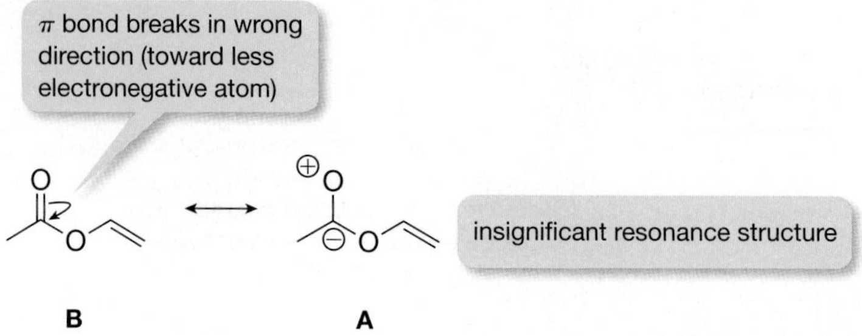

π bond breaks in wrong direction (toward less electronegative atom)

insignificant resonance structure

B A

STEP 3: Look for any structures that result from breaking a π bond made of two atoms with identical electronegativities.

Again, this can be difficult to spot without the curved arrows. Instead, you can look for structures that have positive and negative charges on adjacent atoms of the same element (e.g., two side-by-side carbon atoms, one with a positive charge and one with a negative charge). None of the resonance contributors shown have this feature.

STEP 4: Rank the significant structures based on—and in this order of decreasing contribution—the presence of complete octets, the number of formal charges, the placement of charges on appropriate atoms, and the separation of charges.

Of the six structures provided, only B, C, D, and E are significant structures. Structure C is the only one that has any atoms with an incomplete octet (its carbocation), so structure C is the lowest ranked of the structures. Structure B is the only contributor with no formal charges, so it will contribute the most to the overall resonance hybrid. Finally, a comparison of structures E and D shows their only difference is the placement of the negative charge. Since oxygen is more electronegative than carbon, structure D, with the negative charge on the oxygen atom, is better than E. Therefore, structure D is ranked as the second most-contributing structure.

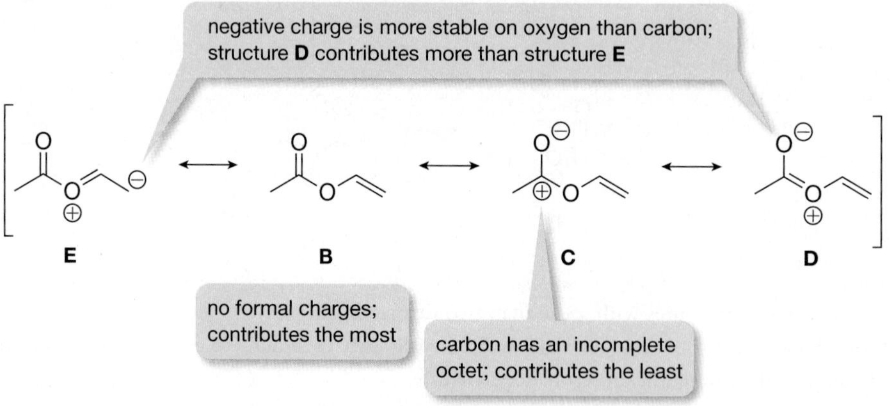

negative charge is more stable on oxygen than carbon; structure **D** contributes more than structure **E**

E B C D

no formal charges; contributes the most

carbon has an incomplete octet; contributes the least

The final ranking is as follows:

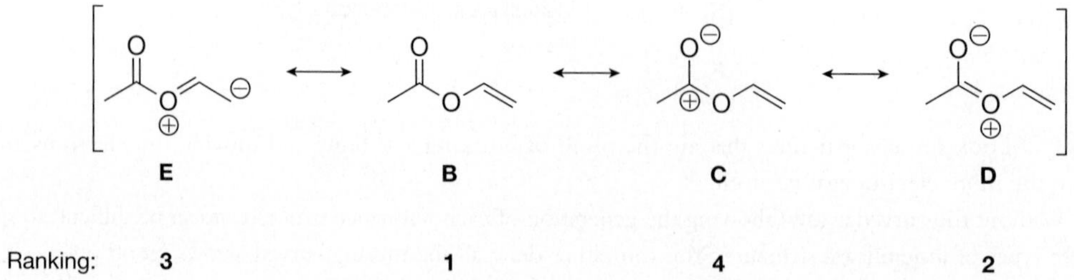

E B C D

Ranking: 3 1 4 2

PRACTICE PROBLEM

5.11 Rank the resonance structures of each set in order of decreasing quality. Some structures might have equal quality. Justify your answers.

a) 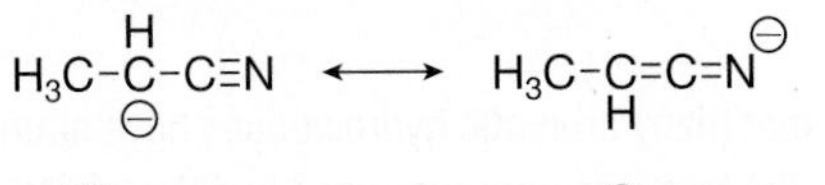 $H_3C-\overset{H}{\underset{\ominus}{C}}-C\equiv N \longleftrightarrow H_3C-\underset{H}{C}=C=N^{\ominus}$

b)

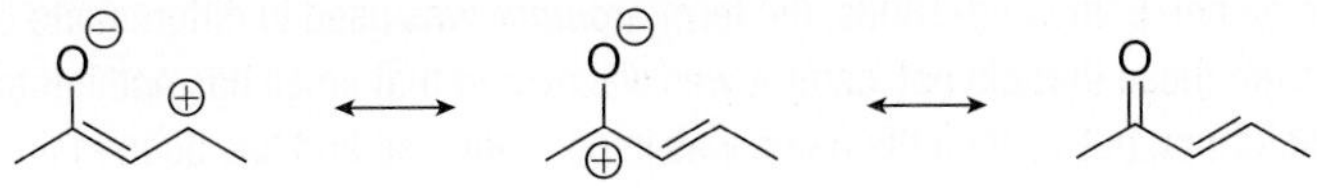

c)

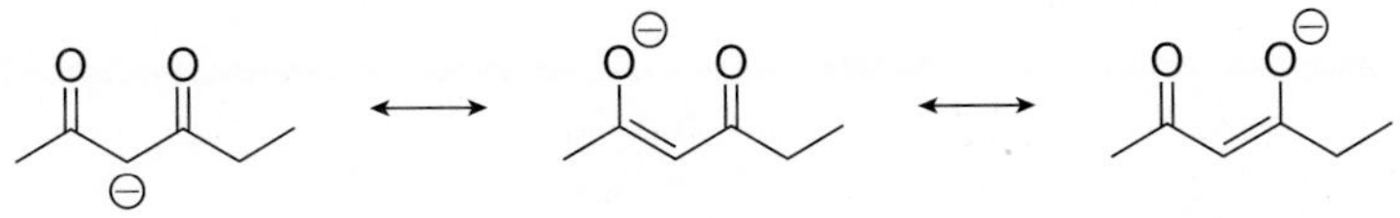

d)

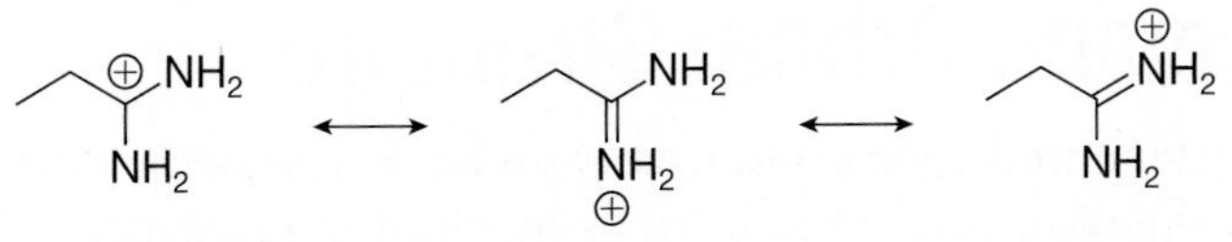

e)

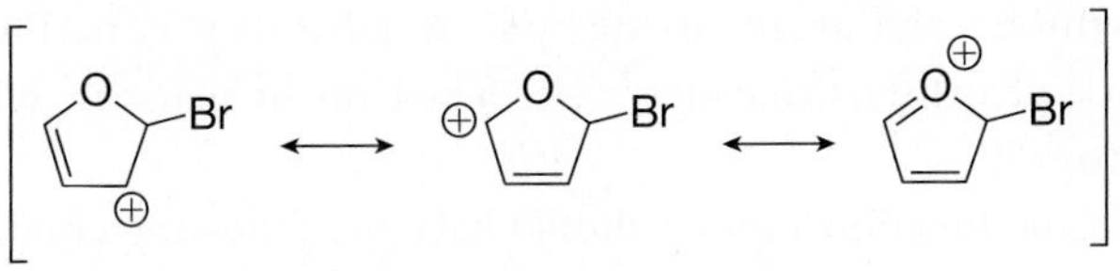

INTEGRATE THE SKILL

5.12 The following diagram shows the first step in a multi-step reaction. The structure formed after the first step is a hybrid of the resonance structures shown. Add curved arrows to show the mechanism for the first step of the reaction (from the starting materials to the first resonance structure drawn) and for the formation of each resonance form. Then rank the resonance structures according to their contribution to the resonance hybrid (1 = contributes the most). Note that formal charges are missing from the resonance structures and need to be added.

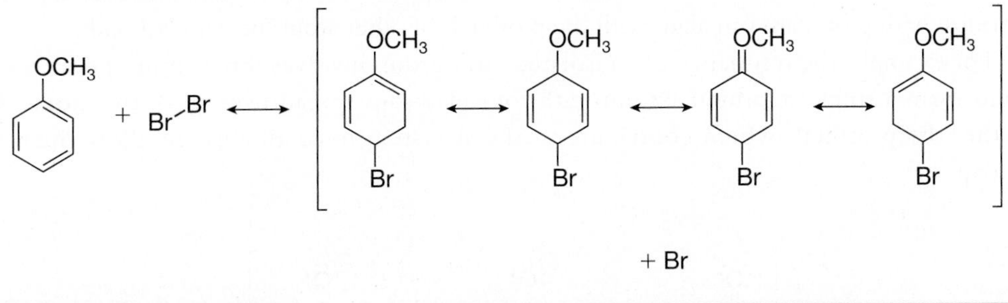

+ Br

DID YOU KNOW?

Many organic compounds exhibit a special kind of resonance called *aromaticity*. In these structures, there is a continuous band of π electrons that circulates in a ring structure that creates a very strong connection between the atoms that make up the ring. The bonding in these systems can be described by drawing resonance forms that involve the movement of π electrons in a circular fashion. Following the guidelines for the construction of mechanistic arrows, it is possible to construct resonance forms all the way around these rings, and arrive back at the starting structure. These compounds are described in detail in Chapter 9.

Continued

The name *aromaticity* is derived from the observation that many aromatic hydrocarbons have strong smells, whereas aliphatic hydrocarbons do not. In the mid-1800s, the term *aromatic* was used to differentiate between hydrocarbons that had a smell from those that did not. Later it was discovered that smell has nothing to do with whether a compound is aromatic or not, but by then the name was in common use and has been maintained to this day.

5.8 Resonance and Orbital Structure

The actual structure of a group undergoing resonance can be represented by drawing all the p orbitals involved in the resonance. The electrons involved in resonance then form a *single* orbital structure in which each atom of the functional group contributes one orbital. The result is a network of orbitals that describes a region in space in which the resonating electrons may be found.

For resonance to be possible, the functional group should have the following conditions:

1. a π bond made up of atoms with different electronegativities
2. a π bond directly beside at least one of the following features:
   a) paired or unpaired electrons
   b) atoms with incomplete octets
   c) other π bonds
   d) charged atoms lacking octets or carrying lone pairs
3. an atom with an incomplete octet adjacent to an atom with a pair of non-bonding electrons

All of these conditions create a situation in which atoms with p orbitals (or those that can hybridize with p orbitals) can align with the p orbitals of other atoms or with π bonds.

For example, the resonance of an isolated ester group involves three atoms. These three atoms form a single π orbital system with four electrons spread over all of the atoms. To do this, the p orbitals which contribute to the π system must all align to allow them to overlap.

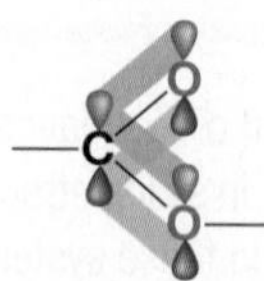

This requires that each atom involved in the resonance has the correct hybridization. Because resonance involves π bonds, the atoms involved must be either sp^2 or sp hybridized (to provide a p orbital that can participate in π bonding). To determine the hybridization involved, consider the hybridization of an atom in all the forms. Although the actual hybridization is a "blend," it is usually considered to be the one with the most s character (highest contribution from an s orbital).

CHEMISTRY: EVERYTHING AND EVERYWHERE

Delocalization Is Responsible for the Colour of Many Organic Molecules

Coloured organic molecules always have extended networks of π bonds that, because of resonance, function as a single, extended functional group. If these π systems involve charged atoms or atoms with different electronegativities, electrons get "shuttled" from one side of the molecule to the other, which allows the molecule to interact strongly with visible light.

The most expensive spice in the world is saffron, which is made of the stigmas of the saffron crocus flower. Each flower produces only three stigmas, and harvesting them is very labour intensive.

HO O O OH

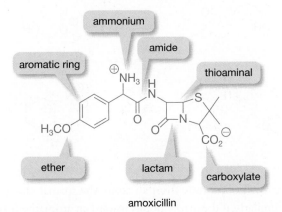

In addition to its flavour, saffron is highly valued for the golden yellow it imparts to food. This colour is produced by a pigment called crocin. The crocin molecule has an extensive network of π bonds, arranged one beside the other. This allows for a great many resonance structures, which contribute to the stability of the molecule and to its ability to interact with visible light.

5.9 Patterns in Mechanism

Organic reactions are systematic and follow patterns. These patterns can be depicted with mechanistic arrows that indicate the movement of electrons during reactions. Electronegativity and formal charges can help in determining the direction of electron flow. Line structures highlight functional groups and facilitate mechanistic analysis. Organic compounds can be considered collections of functional groups held together by a scaffold of carbon atoms.

ammonium
amide
aromatic ring
thioaminal
$\oplus NH_3$ H N S
H_3CO O O N $CO_2^{\ominus}$
ether
lactam
carboxylate

amoxicillin

Bonds form when one atom shares electrons with another atom. Some atoms have an available pair of electrons and donate them to form bonds. These sites can often be identified by the presence of a negative charge (− or $\delta-$). Other atoms accept electrons to form bonds and can be identified by the presence of a positive charge (+ or $\delta+$). In reaction mechanisms, electrons flow from an area of high electron density (lone pair or bond) to an area of low electron density (atom lacking octet or positively charged atom).

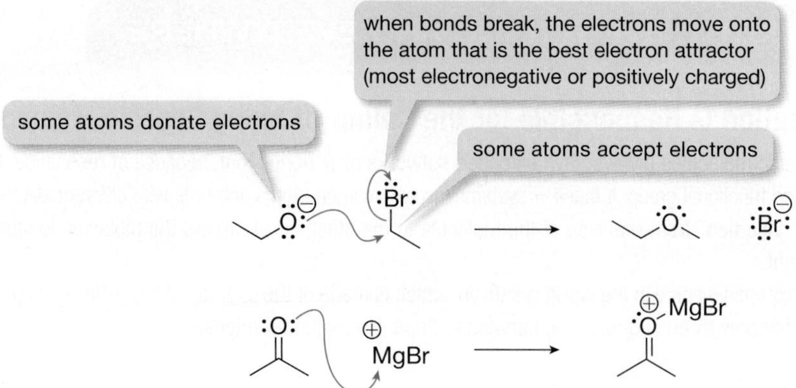

When a bond is broken, the bonding electrons tend to move toward the atom in the bond that is the strongest electron attractor. Often this is the more electronegative atom, but positive charges also attract electrons very strongly. When a bond involving a positive charge breaks, its electrons generally flow toward a positive charge.

when a bond breaks, the bonding electrons move to the more electronegative element

π bonds involving positive charges break toward the positively charged atom

If a double-barb arrow points away from an atom, the formal charge integer on the atom increases by one unit. Similarly, if the arrow points toward an atom, the formal charge integer on the atom decreases by 1. If an atom has one arrow pointing toward it and another arrow pointing away from it, the charge on that atom does not change.

source of electrons in a reaction increases its charge integer

destination of electrons in a reaction reduces its charge integer

In each step of a reaction mechanism, the number of electrons, the number of atoms, and the total net charge are conserved.

5.10 Patterns in Resonance

Of all the factors that stabilize organic molecules, delocalization (described by resonance) is the most profound. The overall patterns of electron movement are systematic and predictable; the same patterns apply to both neutral molecules and charged species. For example, electrons in a π bond are attracted to the atom that attracts electrons most strongly, usually either a positively charged atom or the most electronegative atom nearby.

isolated double bonds:

resonance is possible if the atoms are different types
resonance is not possible if the atoms are the same

positive charge on an sp^2 atom will lead to resonance

When a π bond is adjacent to a charge, a lone pair, another π bond, or an atom with an incomplete octet, it will be possible to construct resonance forms. Look for these features in molecules, and work systematically with mechanistic arrows to fully analyze resonance structures. Working with one bond at a time is an effective technique for drawing all possible resonance forms.

double bonds beside charges and non-bonded electrons

positive charge lacking octet adjacent to a π bond will lead to resonace

negative charge with lone pair adjacent to a π bond will lead to resonace

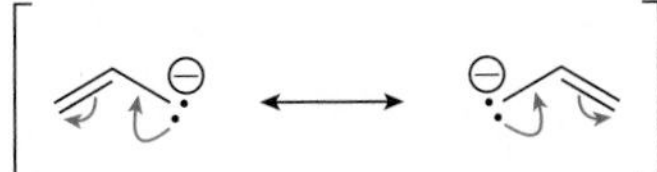

lone pair adjacent to a π bond will lead to resonance

ORGANIC CHEMWARE
5.19 Resonance: Carbonyl group

ORGANIC CHEMWARE
5.20 Resonance: Carbonyl group (protonated)

ORGANIC CHEMWARE
5.21 Resonance: Allyl cation

ORGANIC CHEMWARE
5.22 Resonance: Allyl anion

ORGANIC CHEMWARE
5.23 Resonance: Enol

ORGANIC CHEMWARE
5.24 Resonance: 1,3-Butadiene

CHECKPOINT 5.7

You should now be able to identify resonance patterns in a wide variety of organic compounds.

SOLVED PROBLEM

Here are four compounds in which resonance is possible.

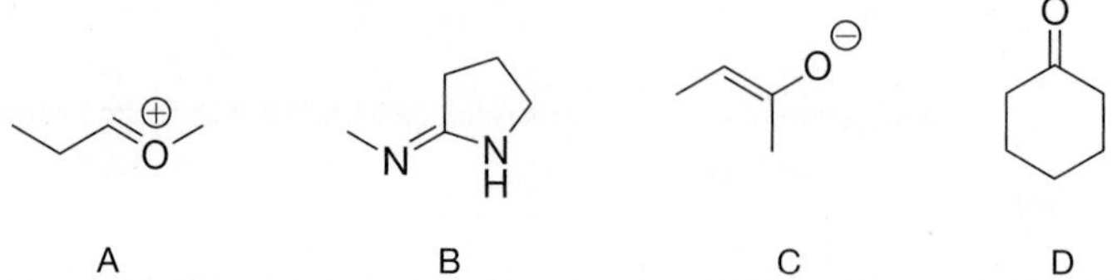

a) Group the molecules together by the type of resonance pattern they exhibit.

b) Draw the resonance structures for each molecule. Use curved arrows to show the similarities between the resonance patterns in the grouped structures.

STEP 1: Expand the Lewis structure of the atoms near π bonds, heteroatoms, and charges.

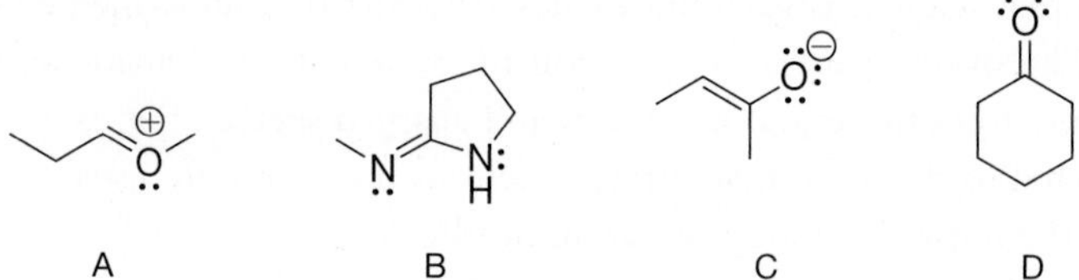

STEP 2: Determine the structural features of each molecule that give rise to resonance.

Compounds A and D both have an isolated π bond that is not adjacent to any non-bonding electrons, charges, or atoms with incomplete octets. The π bonds are, however, made up of two different elements with different electronegativities and can therefore participate in resonance. Compounds B and C each have a π bond with an adjacent heteroatom bearing lone pairs. These lone pairs can participate in resonance with the π bond.

A B C D

STEP 3: Redraw the structures so that the matching electron systems are drawn in the same orientation.

A D | B C

STEP 4: Draw resonance structures, using curved arrows to depict the movement of electrons.

In structures A and D, there is only an isolated π bond with which to work. The π bond needs to break and electrons will flow toward the more electronegative oxygen.

In structures B and C, there are lone pairs adjacent to the π bond. These lone pairs are the starting point for resonance and push into the existing π bond, forming a new π bond with the adjacent carbon atom. Since this carbon atom already has a full octet, the original π bond on the carbon must break, sending these electrons to the left-most atom (N or C) as a new lone pair of electrons.

PRACTICE PROBLEM

5.13 The following seven compounds can be described using resonance.

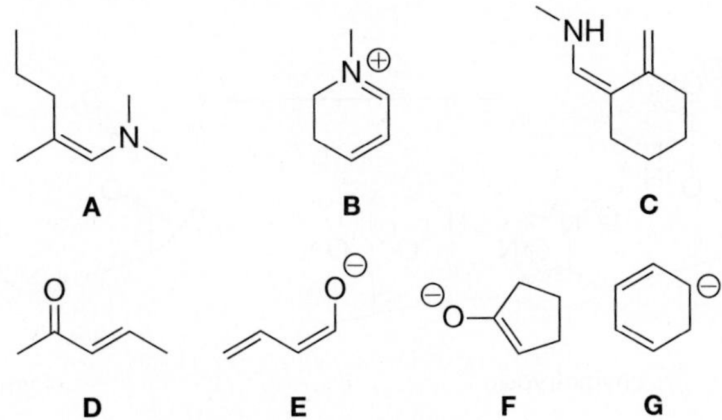

a) Group molecules together by the type of resonance pattern they exhibit.

b) Draw the resonance structures for each molecule. Use curved arrows to show the similarities between the resonance patterns in the grouped structures.

INTEGRATE THE SKILL

5.14 For each of the following molecules, draw two additional molecules that would have similar resonance patterns.

a)

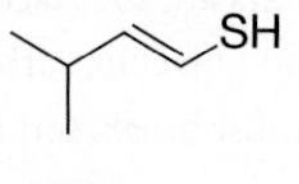

b)

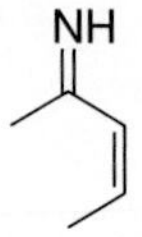

Bringing It Together

The digestion of a protein involves the catalyzed hydrolysis of amide bonds to generate smaller protein fragments that can be absorbed by the body. As mentioned at the beginning of this chapter, the same hydrolysis without an enzyme catalyst requires very harsh conditions because amide bonds are very stable.

Chymotrypsin is a digestive enzyme produced in the pancreas. Like pepsin, this enzyme aids digestion of proteins by selectively catalyzing the hydrolysis of amide bonds. Although the two enzymes act with different mechanisms, they enable the same hydrolysis reaction:

The reaction mechanism with chymotrypsin is more elaborate than with pepsin, but essentially chymotrypsin acts like a giant molecule of KOH. The active site of this enzyme has special functional groups that modify the acidity of an alcohol through resonance and delocalization. During the reaction, chymotrypsin enables the generation of a special nucleophile that acts like a super version of OH^-.

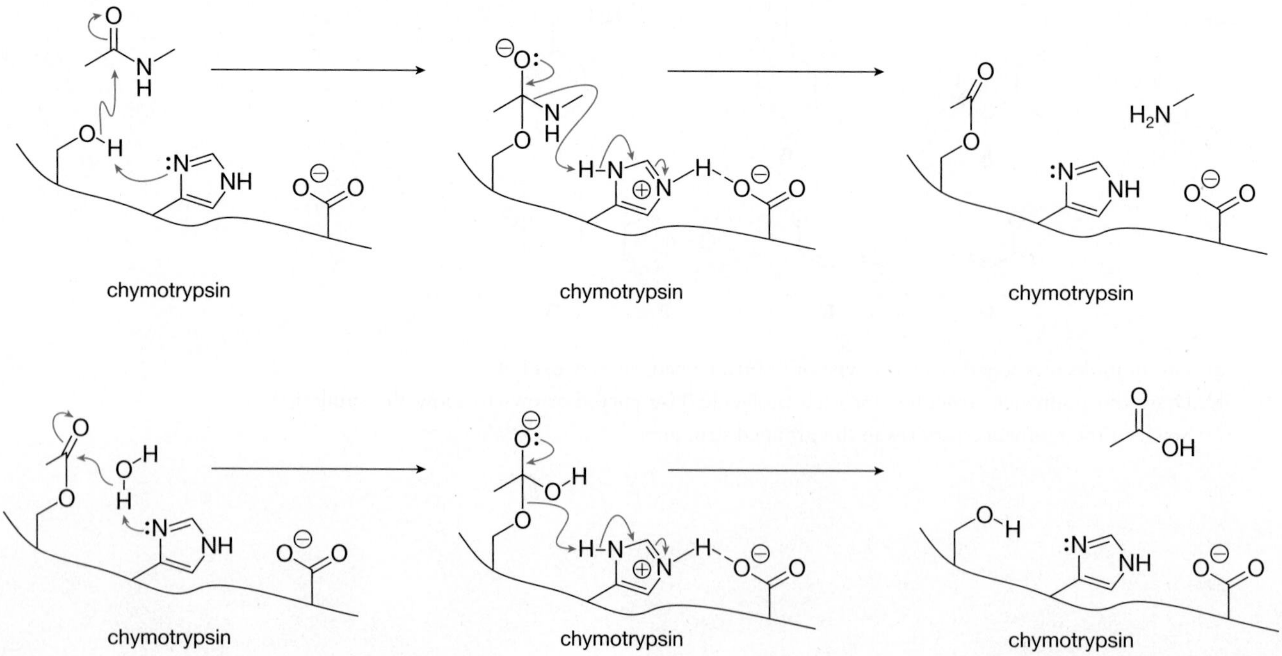

chymotrypsin

chymotrypsin

chymotrypsin

chymotrypsin

chymotrypsin

chymotrypsin

The enzyme has two functional groups that act like a strong base by sharing a proton (H^+), and thereby distribute a positive charge over a relatively large region. Thus, this enzyme controls both the reactivity and the order of proton exchanges during the reaction. These processes are critical in organic reactions and in biochemical processes.

You Can Now

- Identify functional groups within larger structures.
- Interpret curved arrow notation for both bond-breaking and bond-forming steps.
- Predict reaction products based on a reaction mechanism drawn with curved arrow notation.
- Provide a one-step mechanism using curved arrow notation to explain the formation of a product from its starting reactants.
- Predict any changes in formal charges that may accompany individual steps of a reaction mechanism drawn with curved arrow notation.
- Distinguish between intermolecular and intramolecular reactions.
- Use curved arrows to draw all possible resonance structures for a given molecule.
- Rank resonance contributors by their relative contributions to the resonance hybrid.
- Distinguish between significant and insignificant resonance contributors.
- Explain the origin of resonance in terms of the orbitals involved.
- Identify resonance patterns within a wide variety of organic compounds.

Problems

5.15 Calicheamicin γ_1 is a powerful anti-tumour antibiotic. Its structure is shown below. Find an example of each of the following functional groups in this structure.

a) alcohol
b) alkene
c) alkyne
d) amine
e) aromatic
f) ether
g) halide
h) ketone

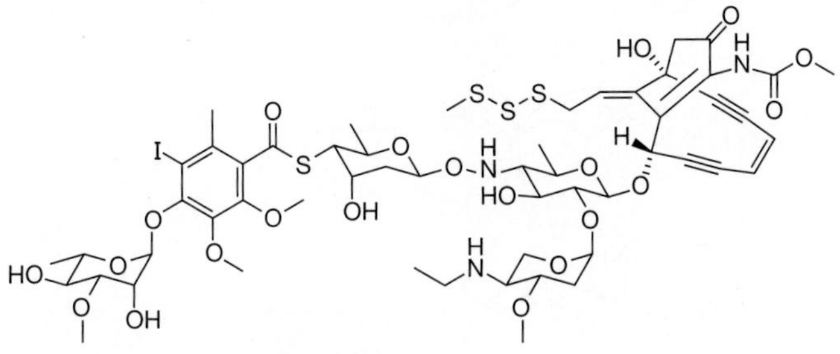

calicheamicin γ_1

5.16 Expand the Lewis structures of the following compounds to show the missing lone pairs. For atoms bearing a formal charge, also show any implied hydrogens.

a)

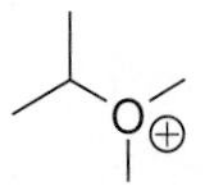

b)

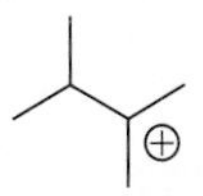

c)

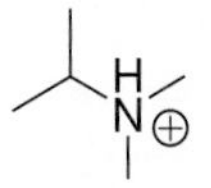

d)

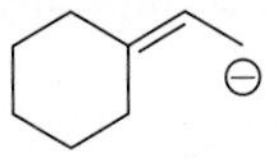

e)

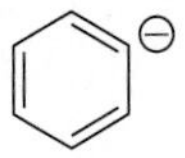

f)

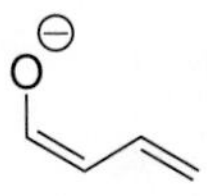

g)

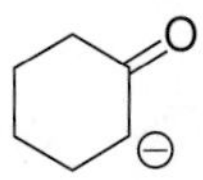

h)

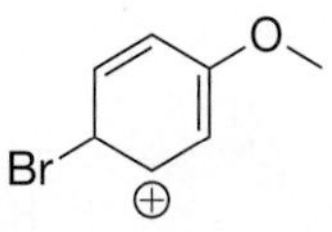

5.17 Lone pairs are involved in the reaction of BCX4430, an experimental drug for the Ebola virus. Draw all the lone pairs present in the molecule of BCX4430, shown here.

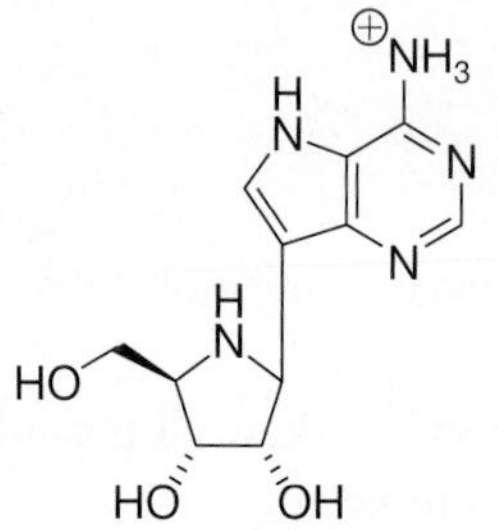

BCX4430
Ebola drug

5.18 Add curved arrows to explain the formation of each product.

a)

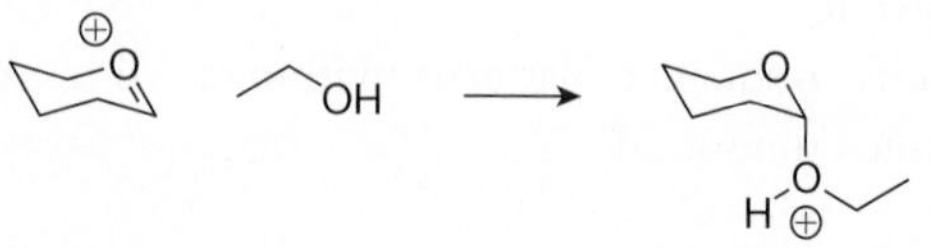

b)

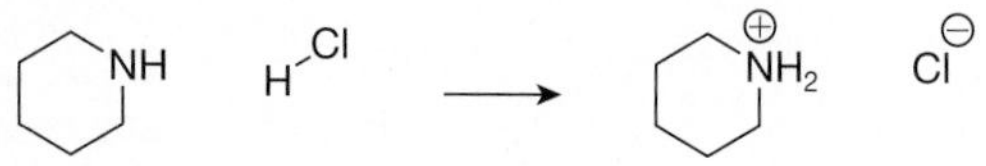

c)

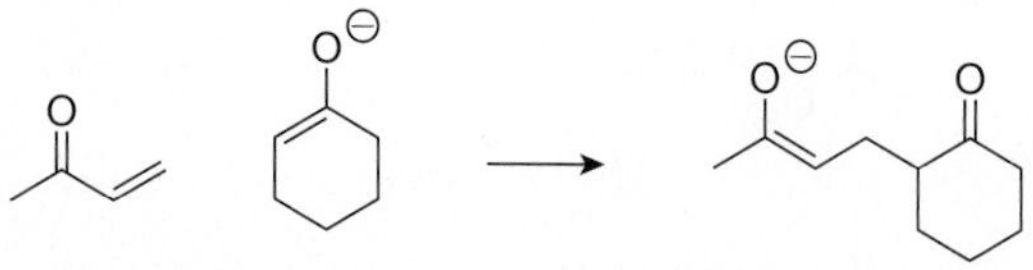

d)

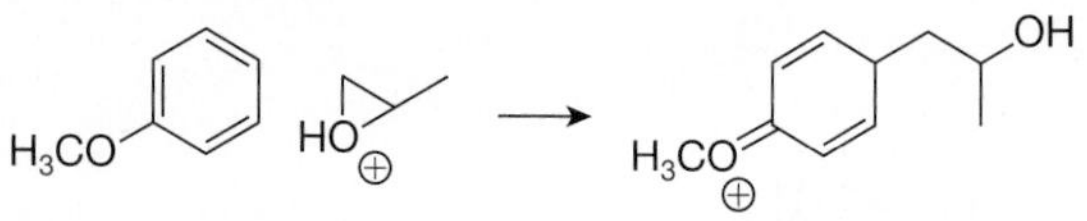

5.19 Use the given curved arrows to determine any missing formal charges in these reactions.

a)

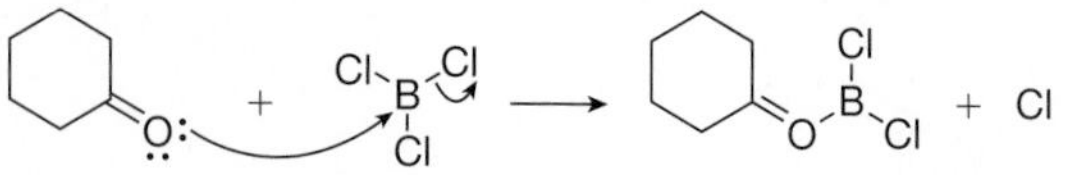

b)

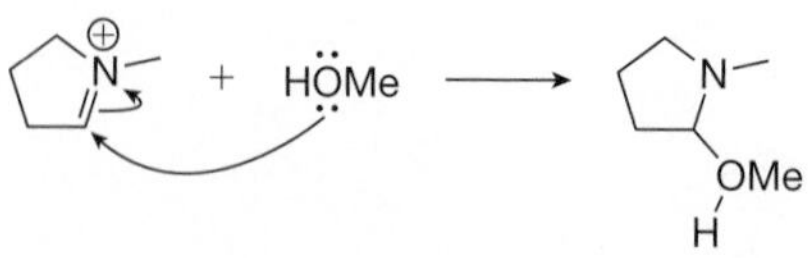

c)

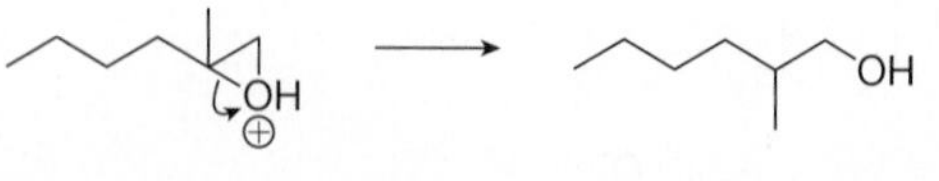

d)

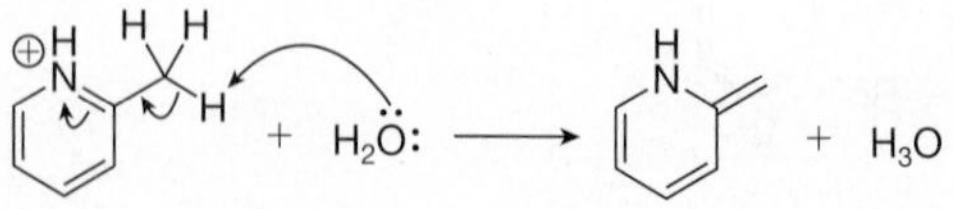

e)

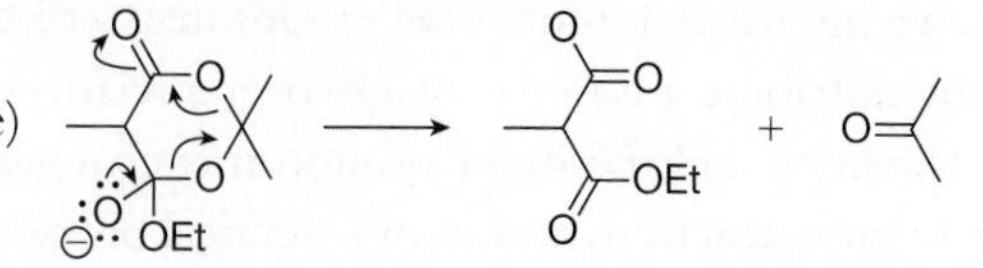

5.20 Classify each of the reactions in Question 5.19 as either intermolecular or intramolecular.

5.21 Draw the products of the following reactions, including any formal charges.

a)

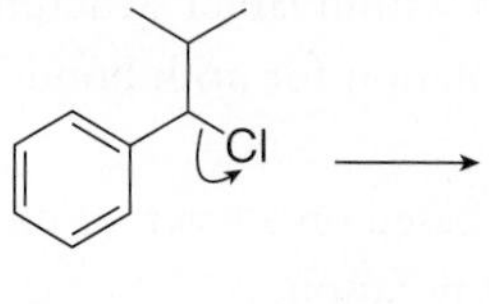

b)

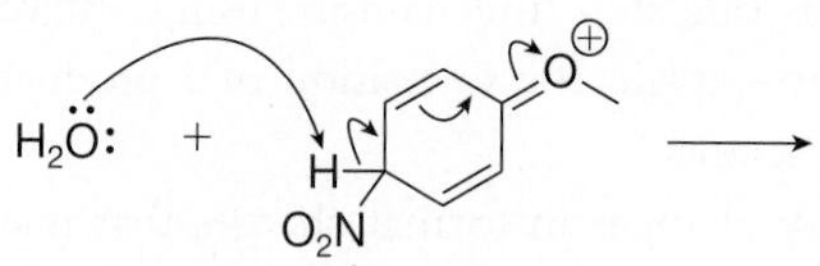

c)

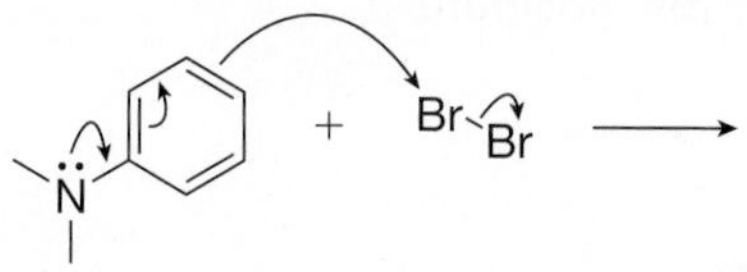

d)

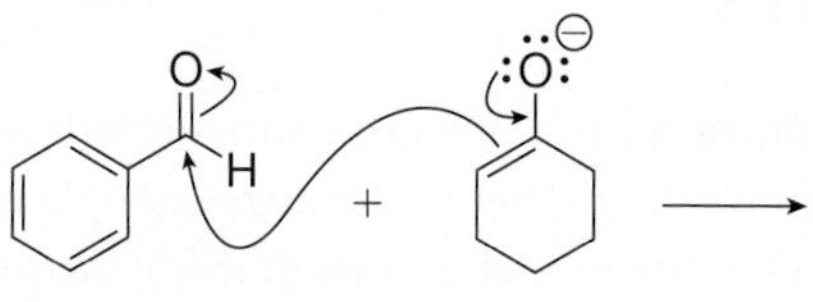

e)

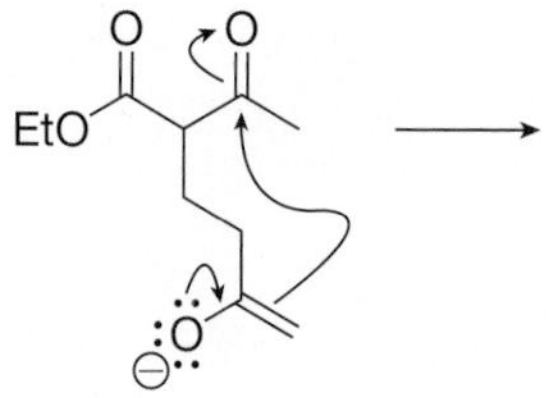

f)

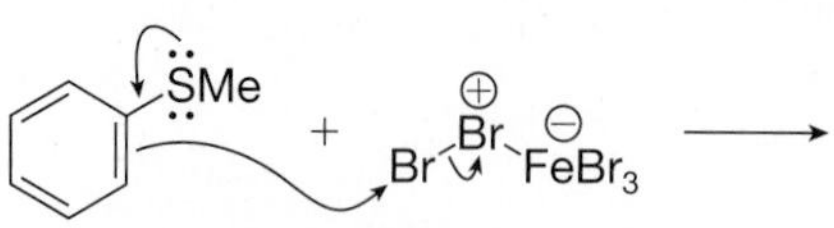

g)

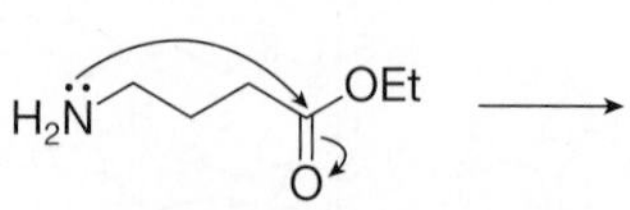

h)

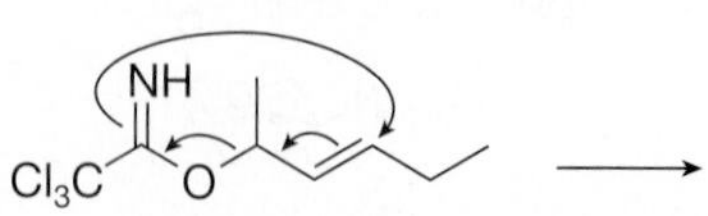

5.22 For each of the following reactions, add mechanistic arrows to show the flow of electrons. Where appropriate, include lone pairs or hydrogens.

a) 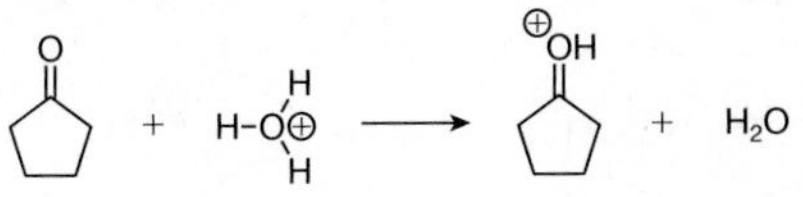 + H-O⊕(H)(H) ⟶ ⊕OH + H_2O

b) + ⟶ + H-N⊕

c) + H-Cl ⟶ ⊕CH₃ O-H + HCl

d) + CH_2N_2 ⟶ + ⊕CH_3N_2

e) H_3CO + O=N⊕=O ⟶ H_3CO⊕ NO_2

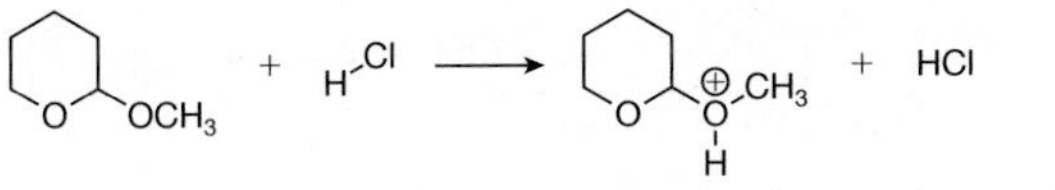

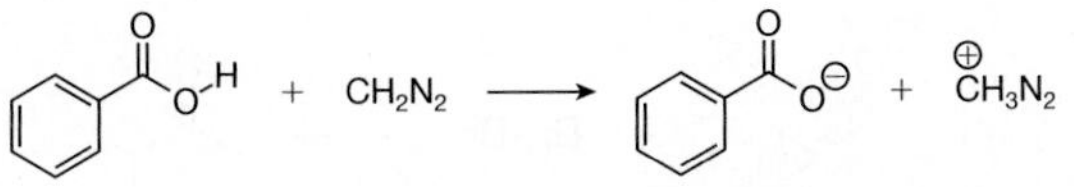

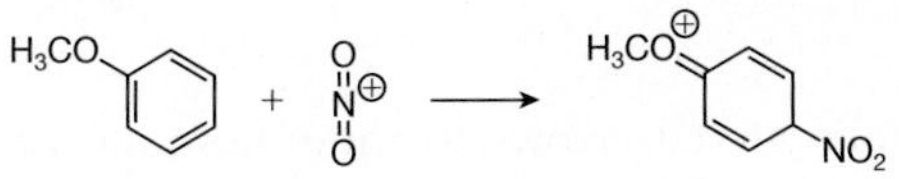

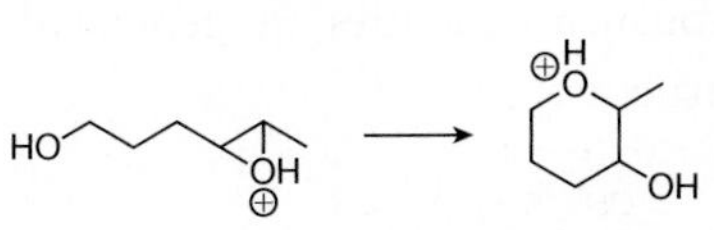

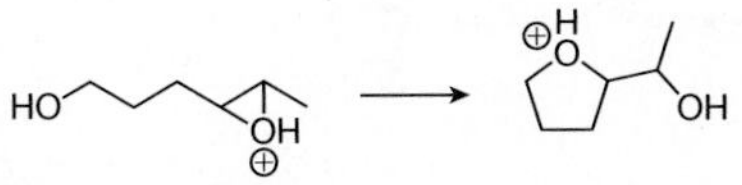

f) HO OH⊕ ⟶ ⊕H O OH

g) HO OH⊕ ⟶ ⊕H O OH

h) ⊕H O OH ⟶ HO ⊕OH

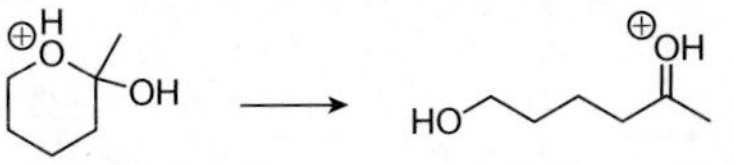

i) 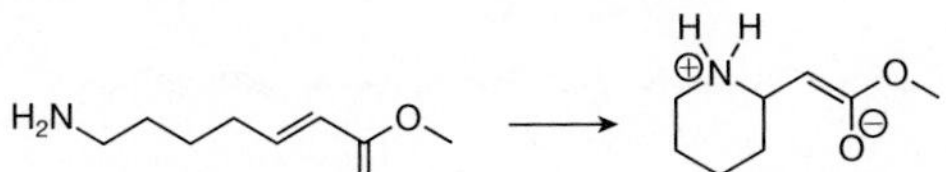 H_2N O O ⟶ H H ⊕N O O⊖

5.23 For each of the following reactions, add mechanistic arrows to show the flow of electrons. Where appropriate, include lone pairs, as well as any formal charges missing from the products.

a) 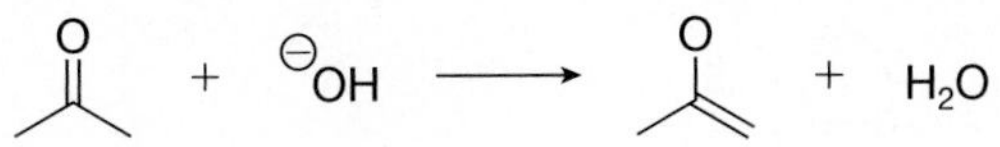 O + ⊖OH ⟶ O + H_2O

b) 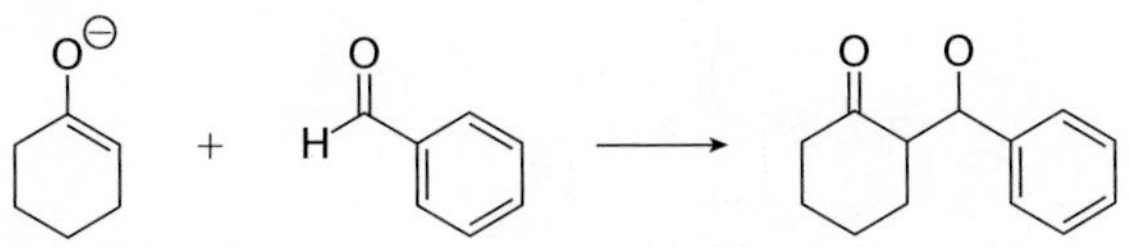 O⊖ + O H ⟶ O O

c) 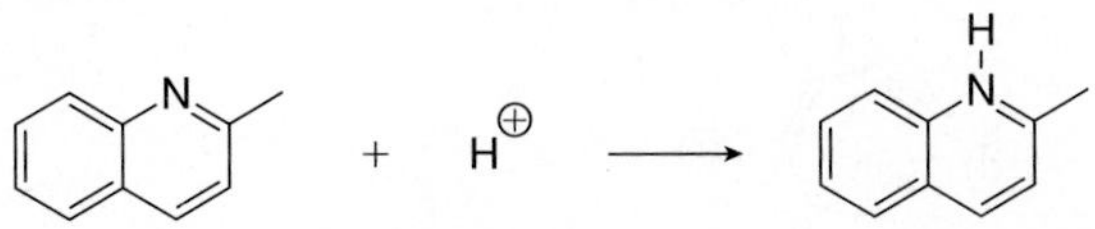 N + H⊕ ⟶ H N

d) 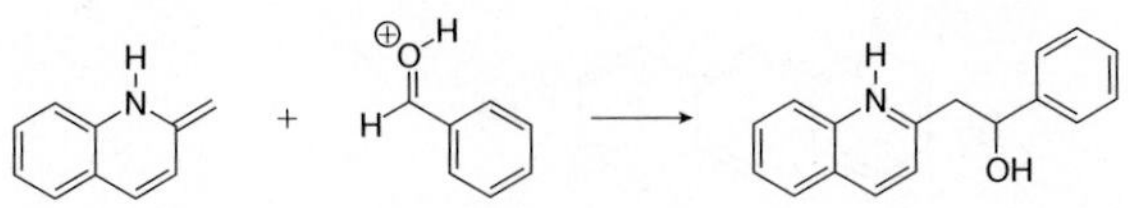 H N + ⊕O-H H ⟶ H N OH

5.24 For each of the following reactions, add mechanistic arrows to show the flow of electrons. Where appropriate, include lone pairs, as well as any formal charges missing from the reactants.

a) 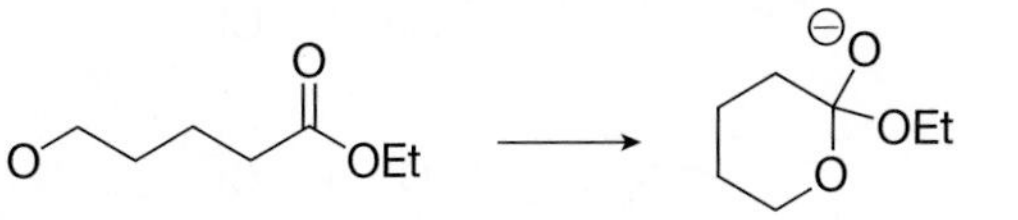 O O OEt ⟶ ⊖O OEt O

b) 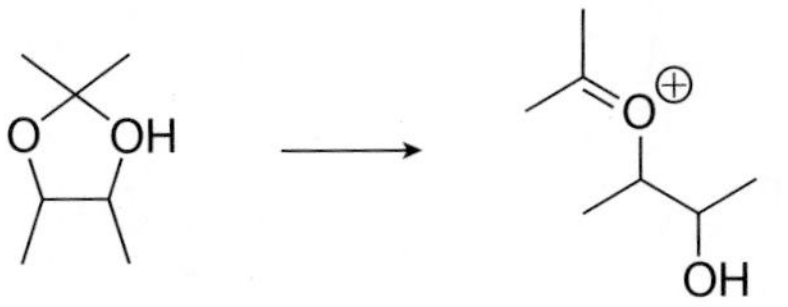 O O OH ⟶ O⊕ OH

c) 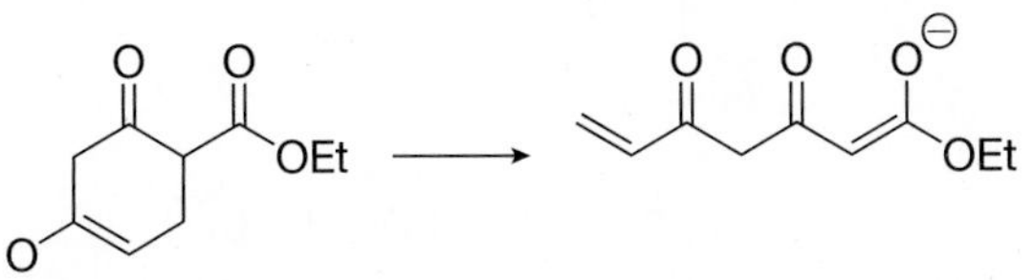 O O O OEt ⟶ O O O⊖ OEt

d) 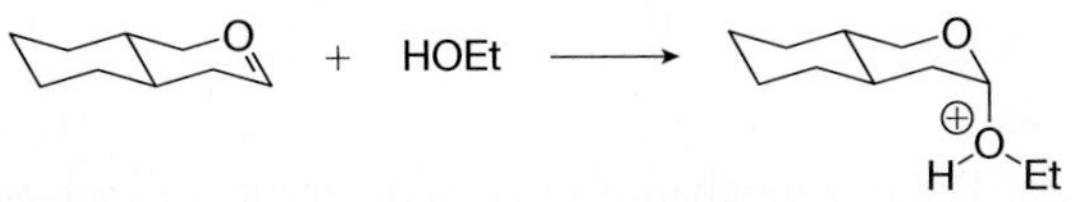 O + HOEt ⟶ O ⊕O H Et

e) 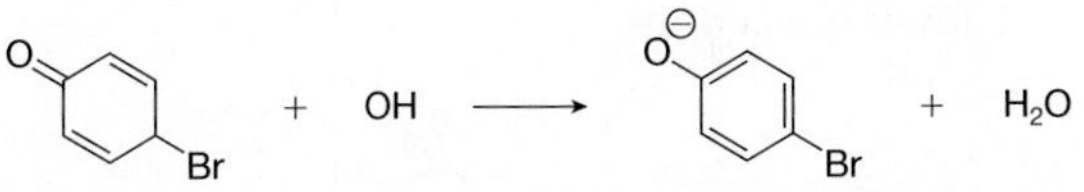 O Br + OH ⟶ O⊖ Br + H_2O

5.25 Consider the following four-step reaction. For each step of the mechanism:

a) Draw the missing curved arrows, as well any necessary lone pairs.

b) Classify the step as either intramolecular or intermolecular.

step 1

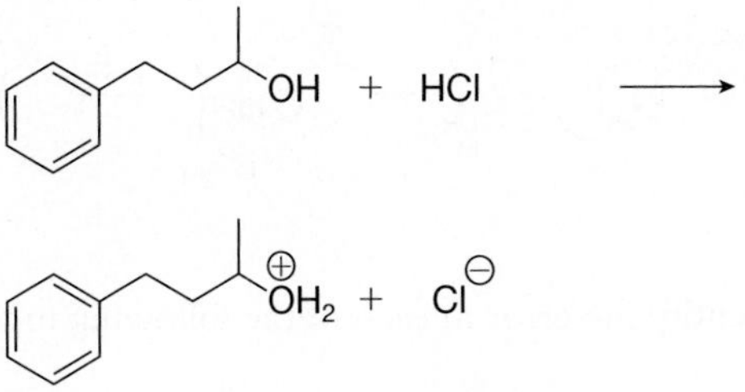

OH + HCl ⟶

⊕OH_2 + Cl⊖

step 2

⊕OH_2 + Cl⊖ ⟶

⊕ + H_2O + Cl⊖

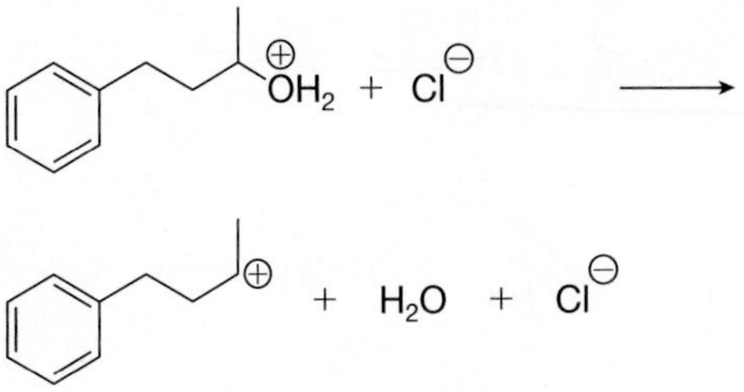

step 3

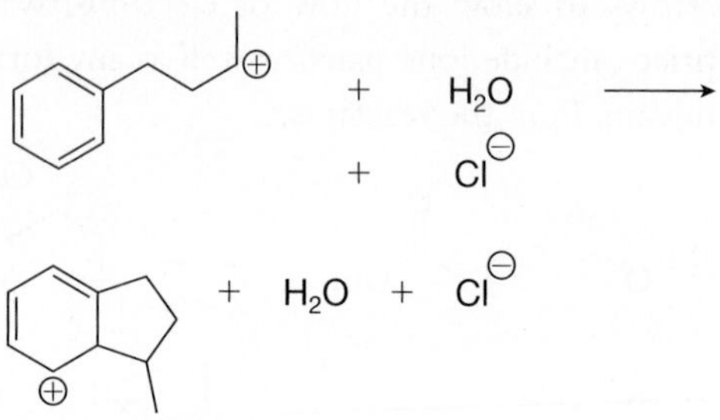

+ H_2O → + Cl^-

+ H_2O + Cl^-

step 4

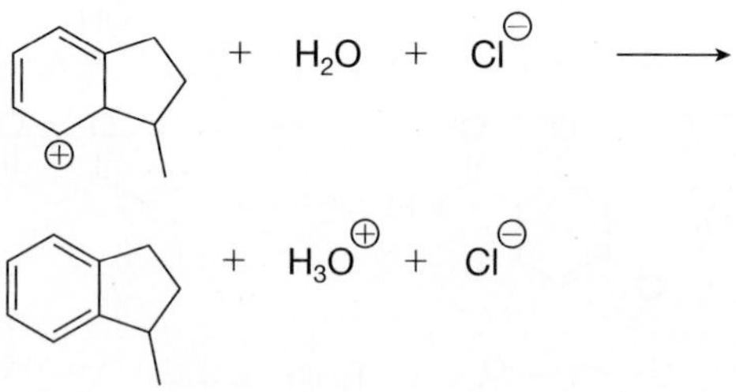

+ H_2O + Cl^- →

+ H_3O^+ + Cl^-

5.26 Write a mechanism that is the *reverse* of each of the following reactions:

a)

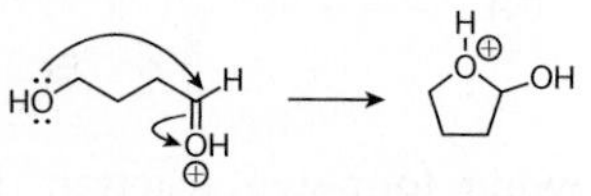

b)

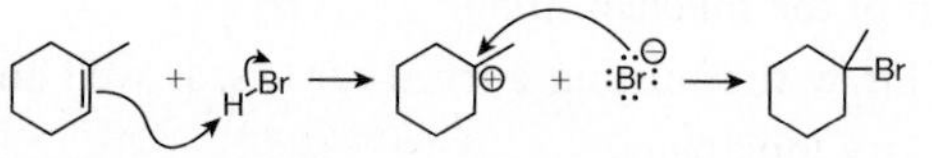

c)

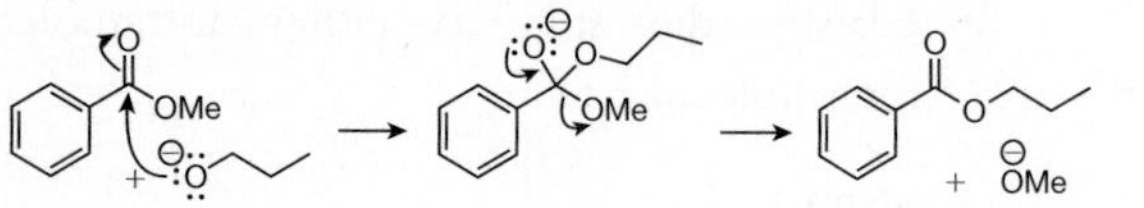

d)

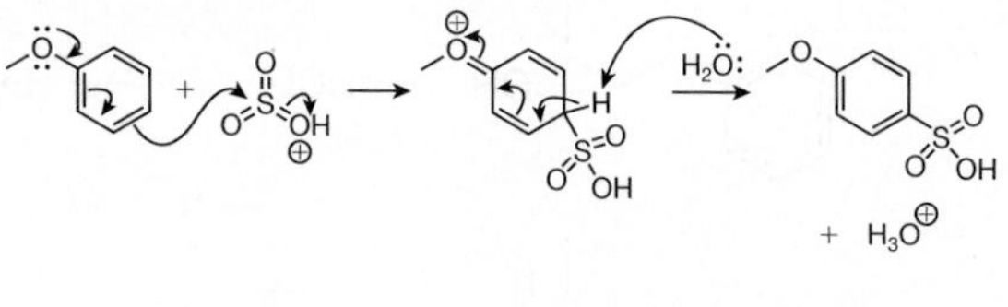

5.27 Identify the error in each of the following mechanisms.

a)

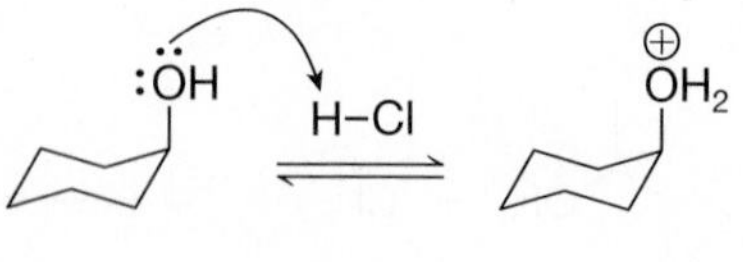

b)

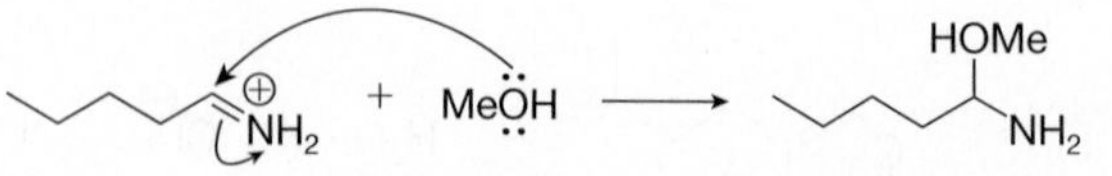

c)

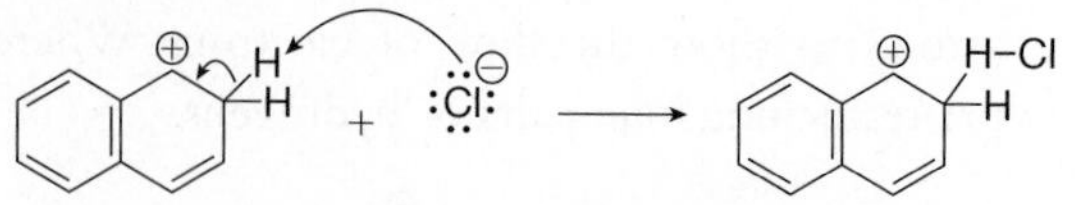

d)

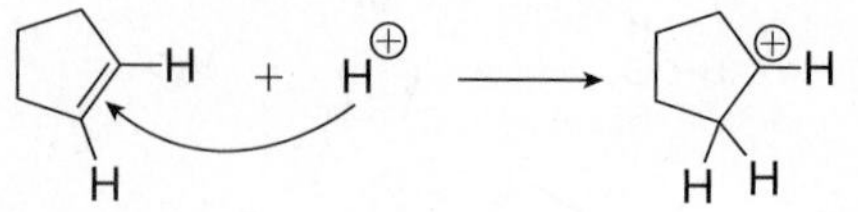

e)

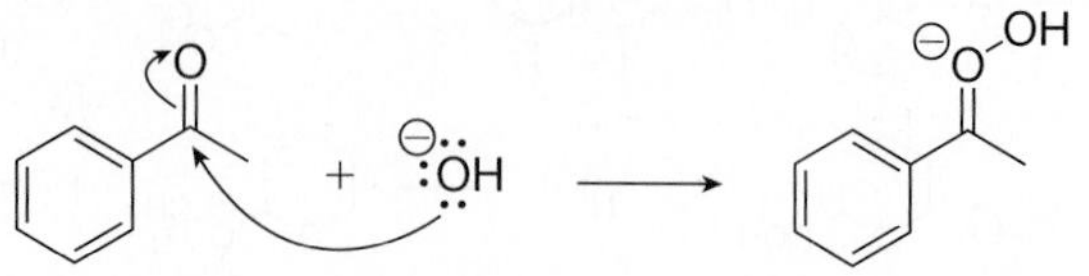

f)

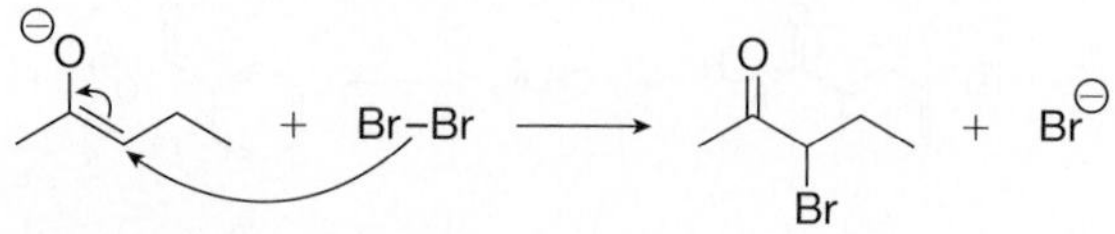

5.28 Provide curved arrows to show how the each of the following resonance structures are generated from the proceeding structure.

a)

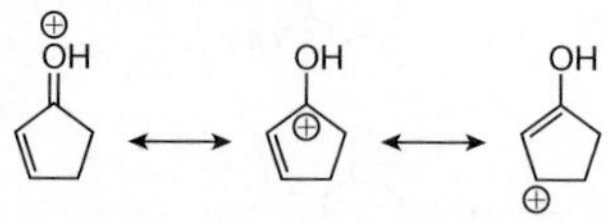

b)

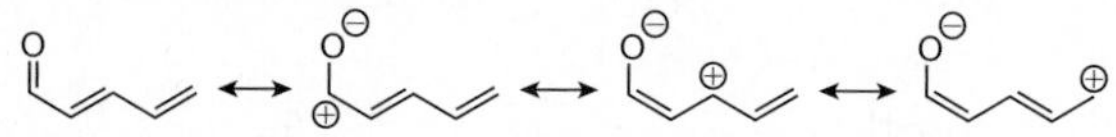

c)

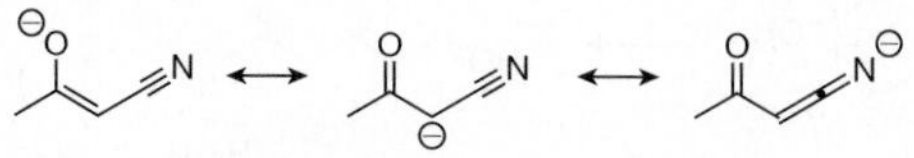

d)

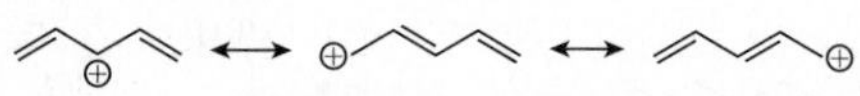

5.29 Draw all the significant resonance structures for each of the following molecules. The numbers in parentheses show the total number of contributors for that structure.

a)  (2)

b)  (5)

c) 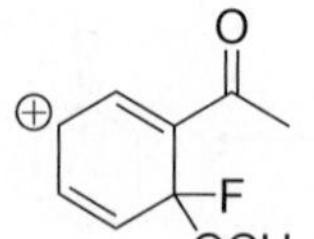 (4)

d)  (7)

e) 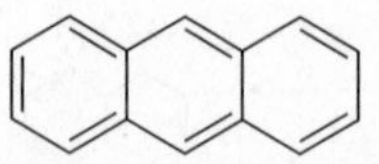 (4)

5.30 Draw all the significant resonance structures for each of the following molecules.

a)

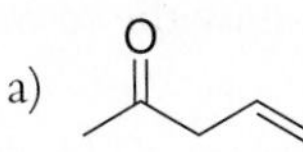

b)

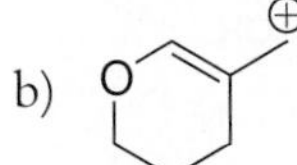

c)

d)

e)

f)

g)

5.31 Draw all the significant resonance structures for each of the following molecules and rank them according to their contribution to the overall resonance hybrid (1 = contributes the most). If two structures contribute equally, rank them the same.

a)

b) CH_2N_2

c)

d)

e)

f)

g)

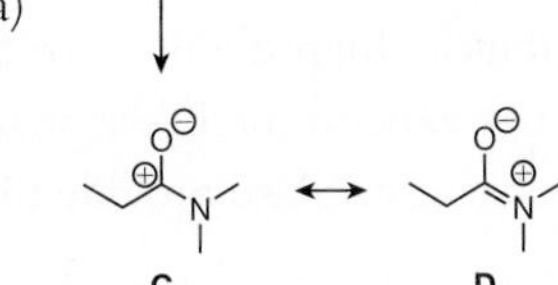

5.32 Provide curved arrows to show the generation of the proposed resonance contributors, starting from structure A. For each set of structures, identify any insignificant contributors and explain your selections.

a)

b)

c)

d)

e)

5.33 For each pair of ions, determine which is more stable and justify your answer.

a) 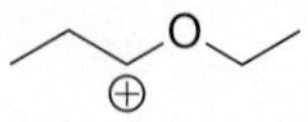 or

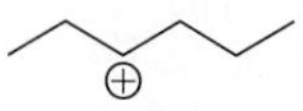

b) 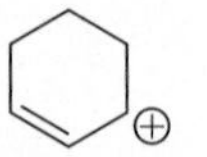 or

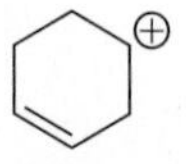

c) $\overset{\ominus}{C}H_2CH_3$ or $\overset{\ominus}{C}H_2COCH_3$

d) 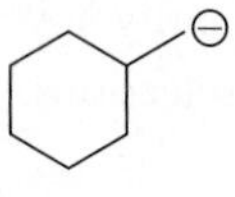 or

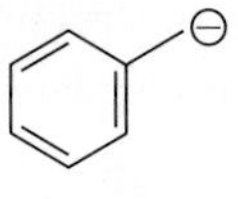

5.34 Consider a test question that asks you to use curved arrows to generate the significant resonance structures for the following series of compounds and to label the most significant contributor. Identify the errors that would occur if you do not expand the Lewis structures or double-check the mechanisms. Also provide the correct answers.

a) 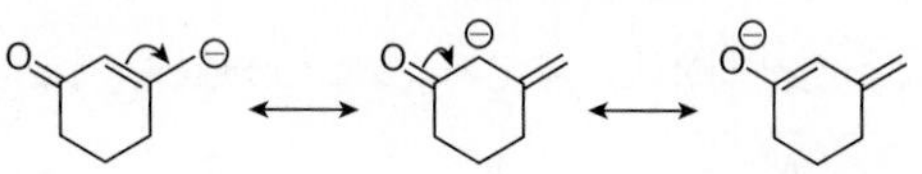 most significant contributor

b) 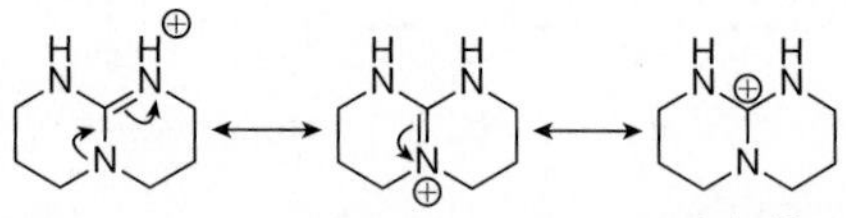 most significant contributor

c) 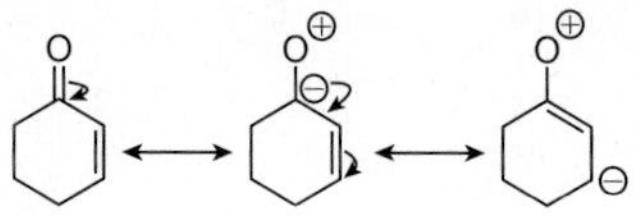 most significant contributor

d) 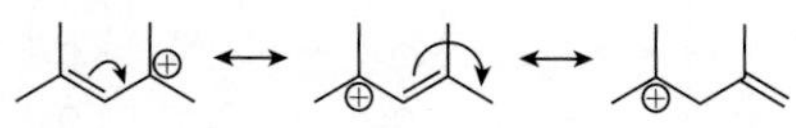 all contribute equally

e) 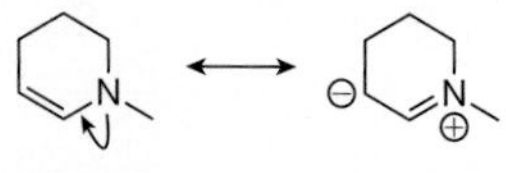 most significant contributor

f) 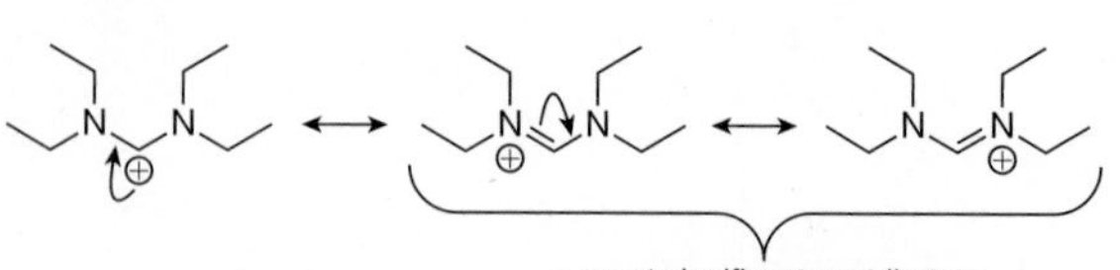 most significant contributors (contribute equally)

5.35 Group the following seven structures according to the type of resonance patterns they exhibit. Justify your selection by using curved arrows to draw the resonance contributors for each molecule.

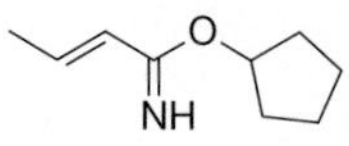

A 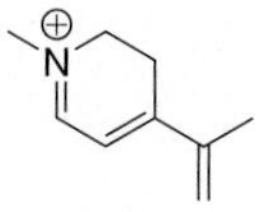 B 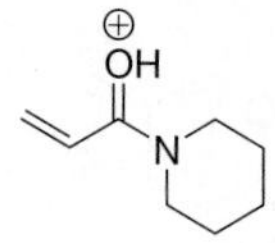 C 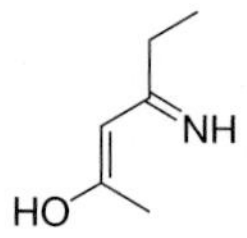 D 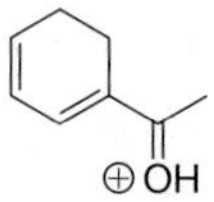 E 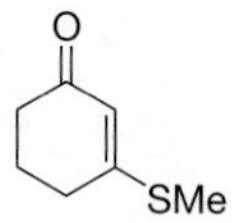 F 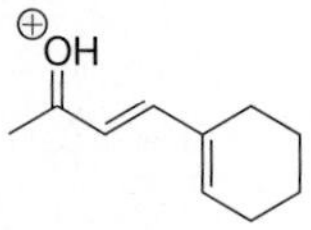 G

5.36 Use resonance structures to predict which sites on the following molecules would react with an electron donor. (Hint: Look for atoms lacking octets.)

a)

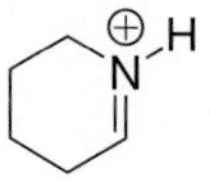

b)

c)

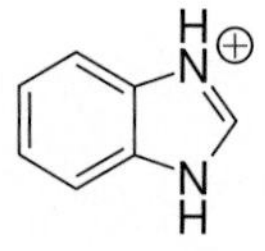

d)

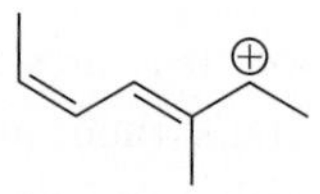

e)

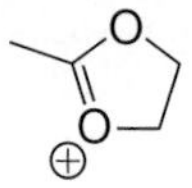

f)

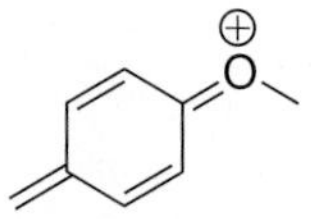

5.37 Use resonance structures to predict which sites on the following molecules would react with an electron acceptor. (Hint: Look for lone pairs.)

a)  CN

b)

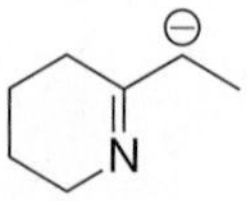

c) 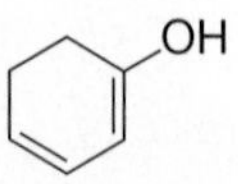 OH

d)

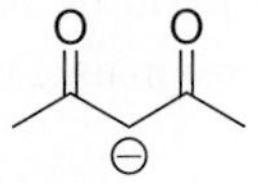

e)

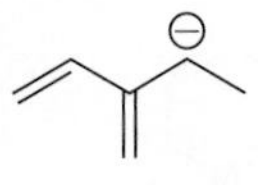

f) 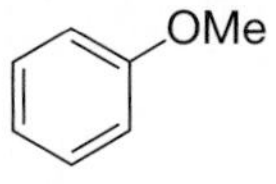 OMe

5.38 Use resonance structures to predict the sites on each molecule that can react with (i) electron-rich species; and (ii) electron-poor species.

a)

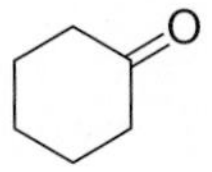

b)

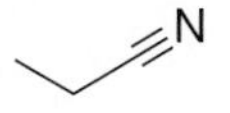

c)

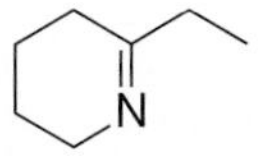

d)

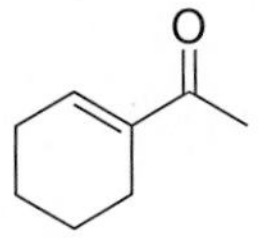

e)

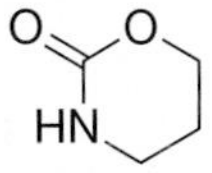

f)

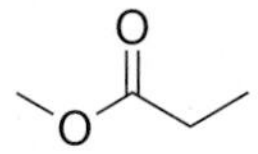

g)

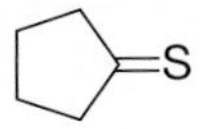

MCAT STYLE PROBLEMS

5.39 Paxil, shown here, is a commonly prescribed anti-depressant. How many non-bonding electrons does each molecule of Paxil have?

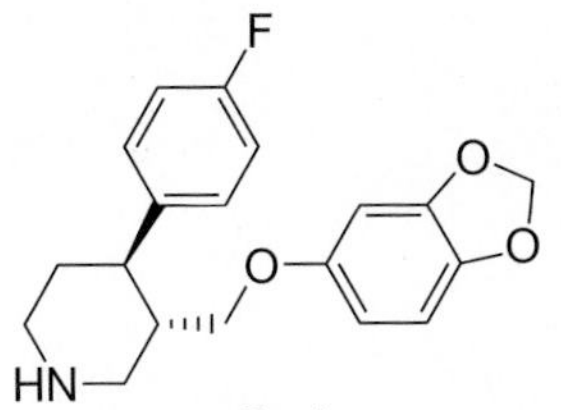

Paxil

a) 5
b) 10
c) 14
d) 20

5.40 Which of the following mechanisms does *not* contain an error?

a)

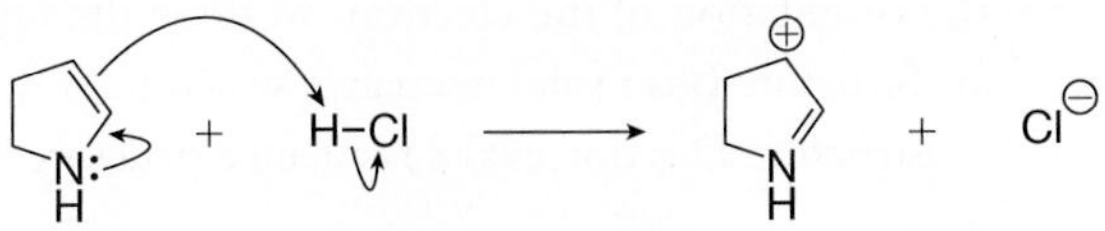

b)

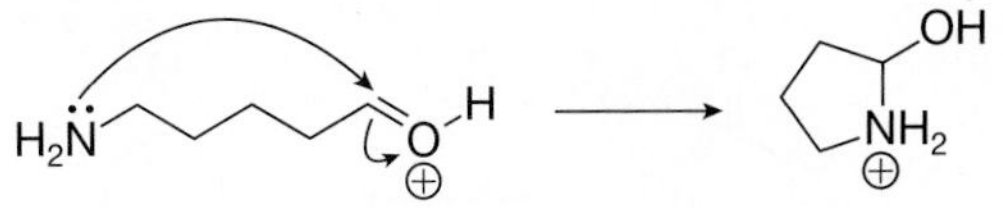

c)

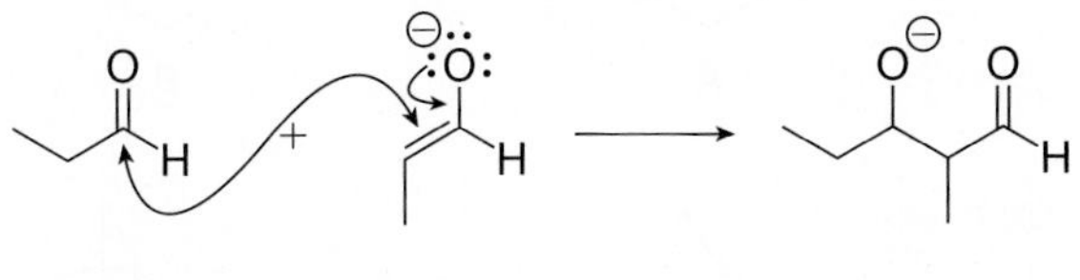

d)

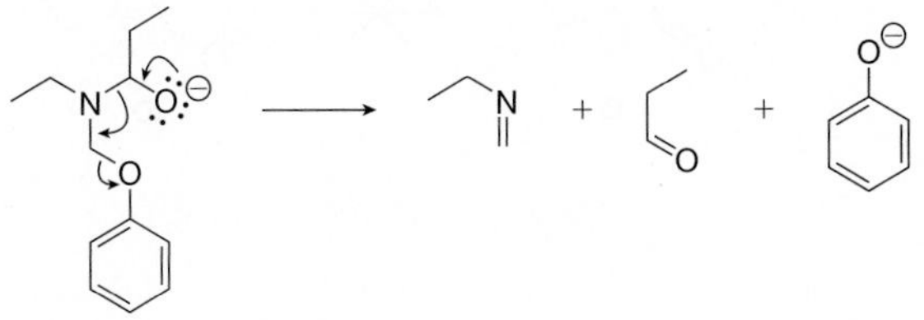

5.41 How many total resonance structures does the following molecule have?

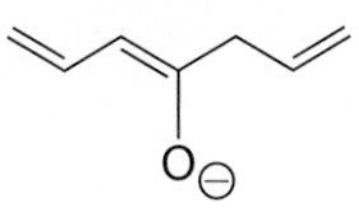

a) 2
b) 3
c) 4
d) 5

5.42 Which of the following is *not* a significant resonance structure for aniline, shown here?

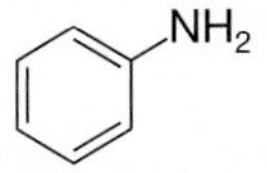

a)

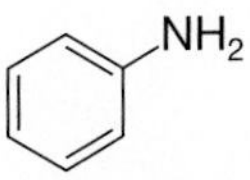

b)

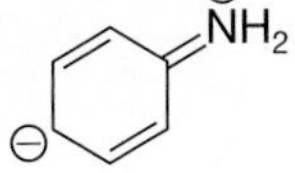

c)

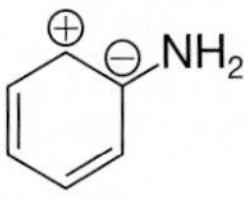

d)

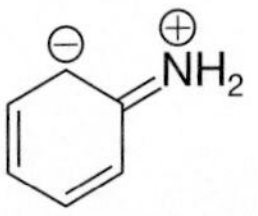

CHALLENGE PROBLEM

5.43 Explain the following statements. (Hint: Think about the orientation of the electrons in three dimensions.)

a) Structure B is a valid resonance structure of A; however, structure D is not a valid resonance structure of C.

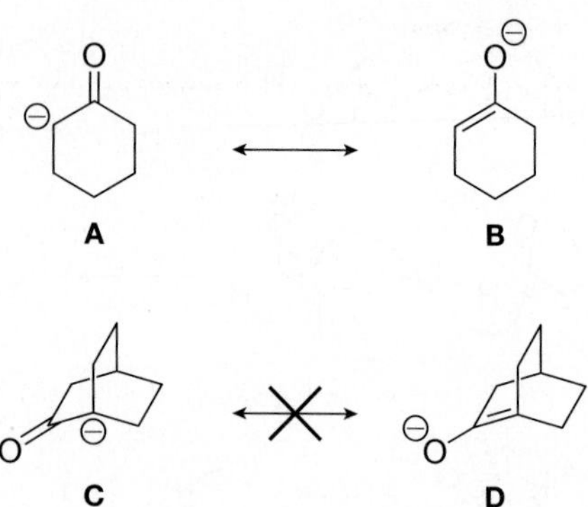

b) Although the nitrogen atom in pyridine has a lone pair, it does not participate in resonance with the neighbouring π bonds.

pyridine

Acids and Bases

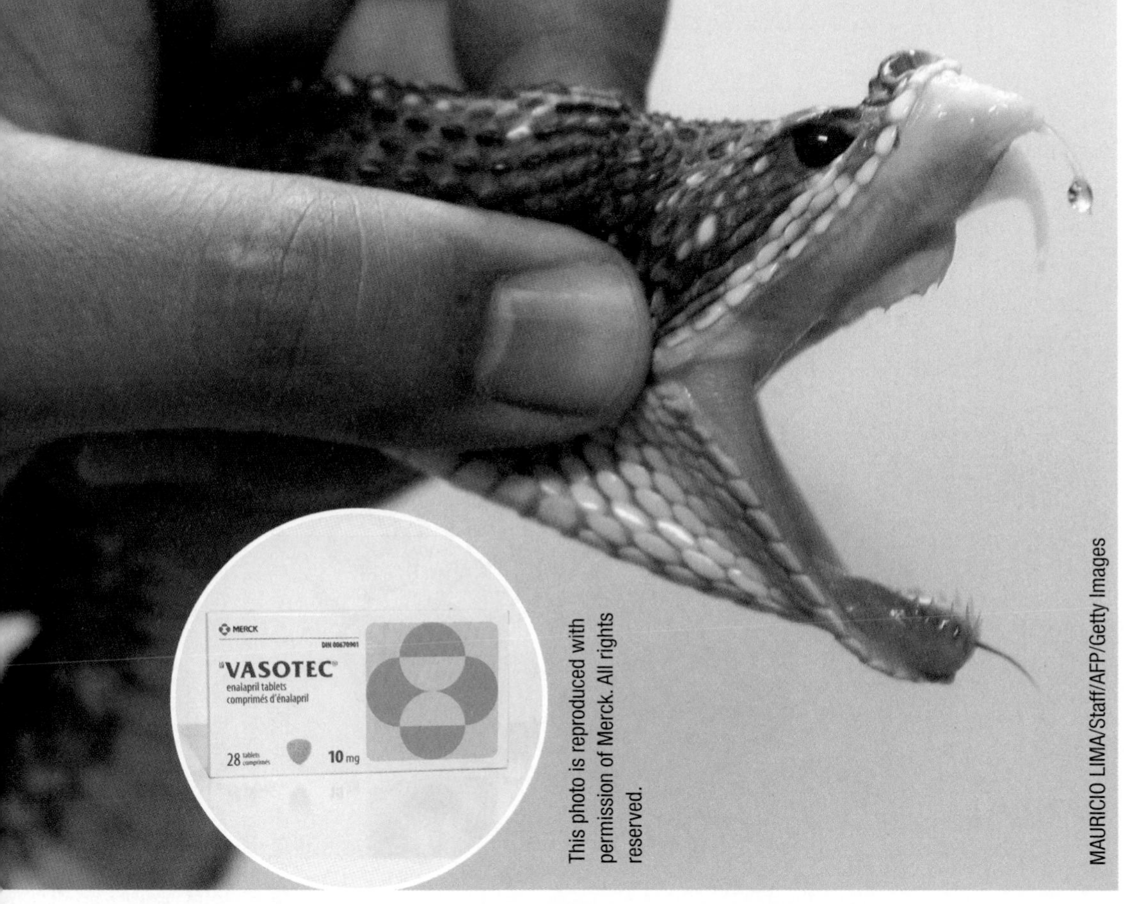

What do a Brazilian pit viper and a bottle of Vasotec (enalapril) have in common?

CHAPTER OUTLINE

6.1 Why It Matters

High blood pressure can substantially increase your risk of heart attack or stroke. One of the ways your body regulates blood pressure involves an enzyme called angiotensin converting enzyme (ACE). This enzyme hydrolyzes angiotensin I, a peptide found in the blood, to make a compound called angiotensin II (when a peptide is hydrolyzed, water reacts with an amide functional group to make an amine and a carboxylic acid). Among other functions, angiotensin II is a potent vasoconstrictor, a compound that causes the walls of blood vessels to contract, leading to high blood pressure.

peptide containing 10 amino acids

Angiotensin I Asp-Arg-Val-Tyr-Ile-His-Pro-Phe-His-Leu

ACE + H_2O ↓ ACE selectively hydrolyzes this amide bond

Angiotensin II Asp-Arg-Val-Tyr-Ile-His-Pro-Phe

peptide containing 8 amino acids

ACE belongs to a family of enzymes called zinc proteases. The hydrolysis mechanism involves a Zn^{2+} ion that functions as a Lewis acid when bound to particular functional groups within the enzyme. In this environment the Zn^{2+} ion accepts electrons from an amide functional group on angiotensin I.

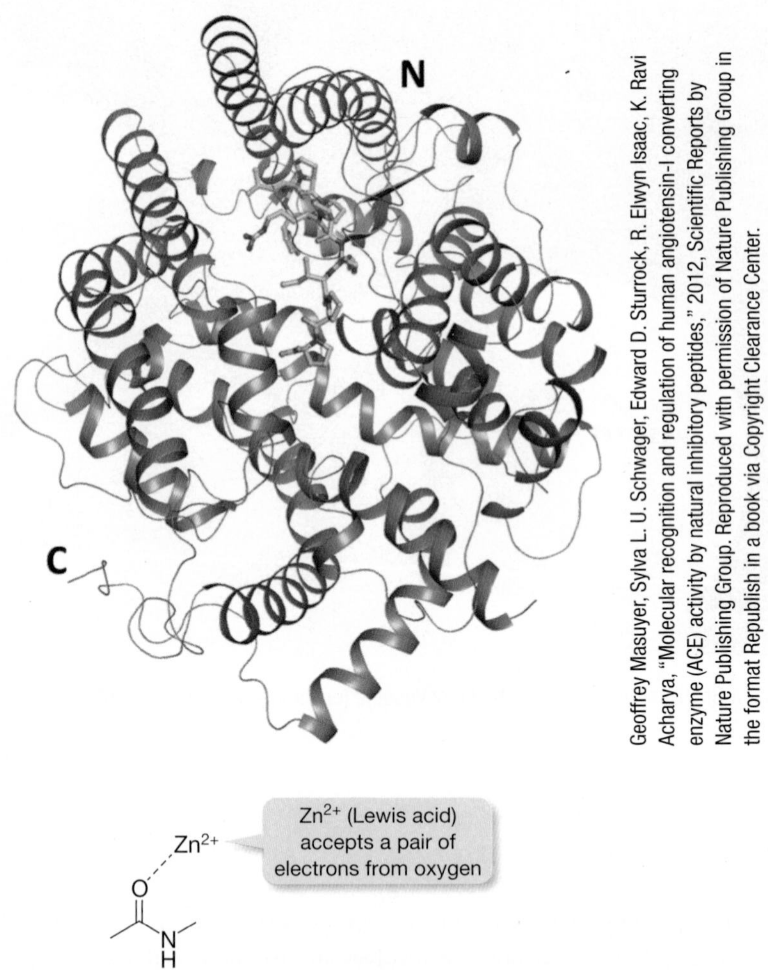

Geoffrey Masuyer, Sylva L. U. Schwager, Edward D. Sturrock, R. Elwyn Isaac, K. Ravi Acharya, "Molecular recognition and regulation of human angiotensin-I converting enzyme (ACE) activity by natural inhibitory peptides," 2012, Scientific Reports by Nature Publishing Group. Reproduced with permission of Nature Publishing Group in the format Republish in a book via Copyright Clearance Center.

Zn^{2+}

O

N

H

Zn^{2+} (Lewis acid) accepts a pair of electrons from oxygen

In the 1970s, scientists discovered teprotide in the venom of the Brazilian pit viper. Teprotide binds to the Zn^{2+} ion in ACE (teprotide is a Lewis base), preventing the enzyme from catalyzing the production of angiotensin II.

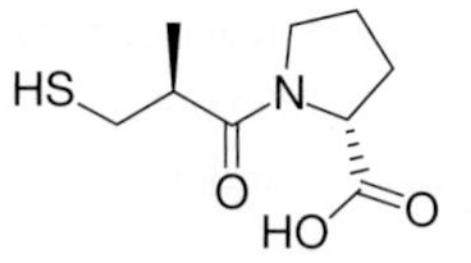

captopril

Teprotide was not very effective, and it degrades in the gastrointestinal tract, so this compound could not be used as a drug. However, by studying how teprotide blocks ACE from functioning, chemists were able to develop medications that were more effective and "drug-like" than teprotide. Captopril was the first of these synthetic ACE inhibitors. Other commonly prescribed ACE inhibitors were soon discovered including enalapril, lisinopril, and ramipril, which are all structurally related. All of these discoveries were possible because chemists could identify the best bases to react with the zinc Lewis acid in ACE.

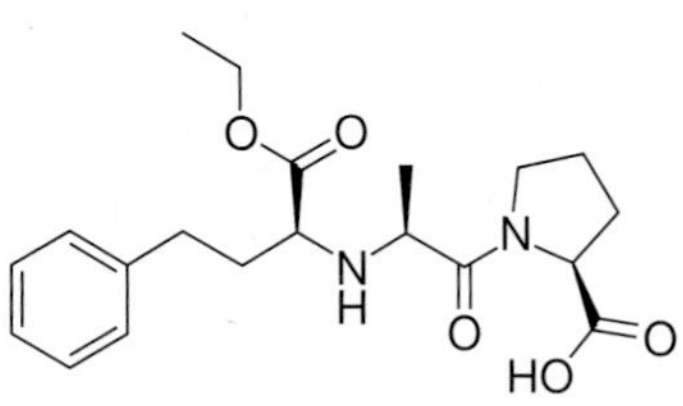

enalapril

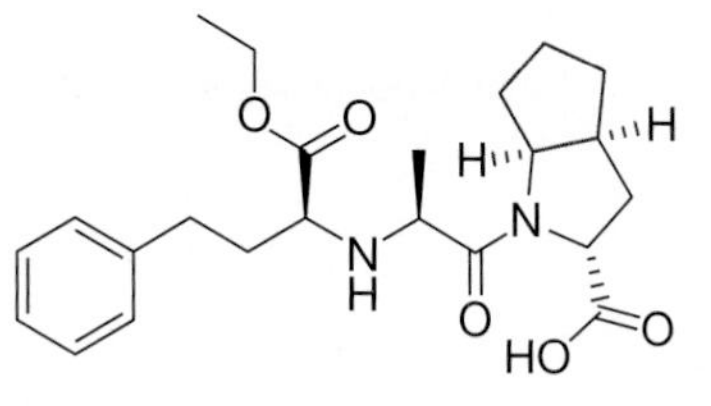

ramapril

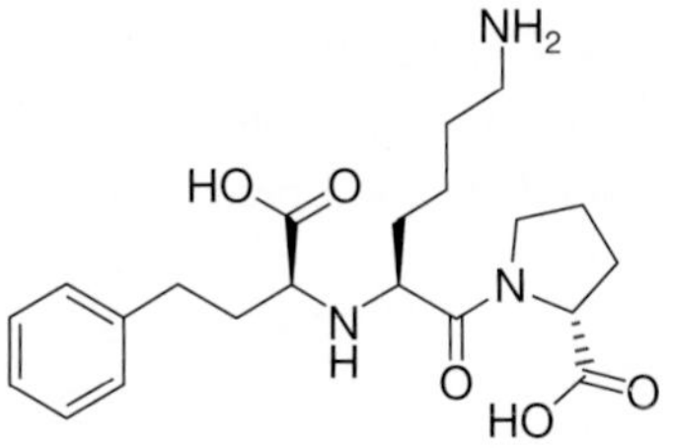

lisinopril

6.2 Electron Movements in Brønsted Acid–Base Reactions

According to the Brønsted theory, acids are proton (H^+) donors and bases are proton acceptors. Most acid–base reactions involve Brønsted acids and bases. For this reason, Brønsted acids and bases are normally termed *acids* or *bases* without specifying "Brønsted."

The behaviour of a typical Brønsted acid can be represented as an equilibrium dissociation of the acid (HA) into a proton (H^+) and conjugate base (A^-).

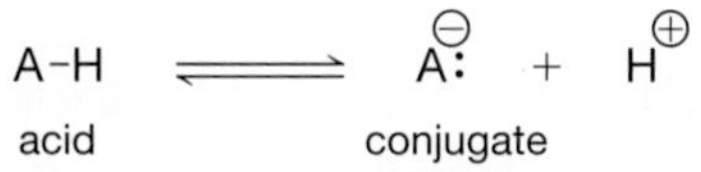

$$\underset{\text{acid}}{\text{A–H}} \rightleftharpoons \underset{\text{conjugate base}}{\text{A}:^{\ominus}} + \text{H}^{\oplus}$$

It is important to remember that H^+ does not exist by itself in chemical reactions. When acids dissociate, the H^+ is always attached to a base molecule. In water, H^+ is normally carried by a water molecule in the form of a hydronium ion, H_3O^+. Most organic reactions are performed in organic solvents that contain little or no water. In such solvents, H^+ ultimately ends up on the strongest base in a mixture. However, H^+ often temporarily "rides" on a weaker base during reactions.

In this textbook, generalized acid–base reactions include an unspecified base (B or B^-). When reactions are performed in dry (no water) organic solvents, this generalized base could be a solvent molecule, a reactant or product, or an added base; it is not always possible to know which is participating. Mechanisms shown in this book use the most likely base or a generalized base.

Simplified dissociation of an acid

electron pair moves toward the atom carrying the hydrogen

$$H{-}A \rightleftharpoons H^{\oplus} + :A^{\ominus}$$

What happens in water

$$H_2\ddot{O} + H{-}A \rightleftharpoons H_3O^{\oplus} + :A^{\ominus}$$

What happens in organic solvents

$$\ddot{B}^{\ominus} + H{-}A \rightleftharpoons HB + :A^{\ominus}$$

generalized base

Some texts represent H^+ by itself, rather than showing it bonded to a carrier molecule. Although this shortcut simplifies reaction equations, it is important to remember the carrier molecules are present in the reaction. It is good practice to always include a base, even if it is generalized.

CHECKPOINT 6.1

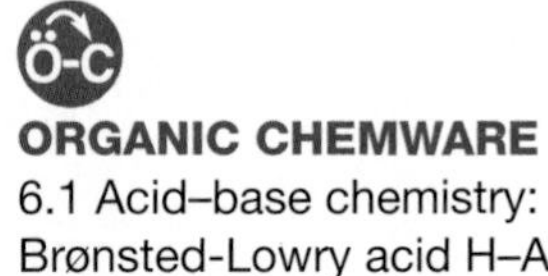

ORGANIC CHEMWARE
6.1 Acid–base chemistry: Brønsted-Lowry acid H–A

You should now be able to draw the mechanism of an acid–base reaction.

SOLVED PROBLEM

Draw the mechanism and products for the following acid–base reaction:

$$C_6H_5OH + H^{\ominus}$$

STEP 1: Expand the Lewis structure to show bonds and lone pairs in the functional groups.

$$C_6H_5{-}\ddot{O}{-}H + :H^{\ominus}$$

STEP 2: Determine which species will react as the base and which will react as the acid.

A base must have an electron pair available for sharing with the acid's proton (H^+). To find the base in the reaction, look for any available lone pairs—especially ones associated with a negative charge. In this reaction, the lone pair on the hydride (H^-), with its negative charge, makes this molecule the base.

An acid must have a hydrogen atom with a partial positive charge on it. Look for hydrogens attached to more electronegative atoms, especially ones with a positive charge or with nearby electron-withdrawing groups (see Section 6.4). In this reaction, there is a hydrogen atom bonded to an electronegative oxygen atom. The electronegativity difference between these atoms creates a polarized bond with a partial positive charge on the hydrogen atom. Therefore, this molecule is the acid.

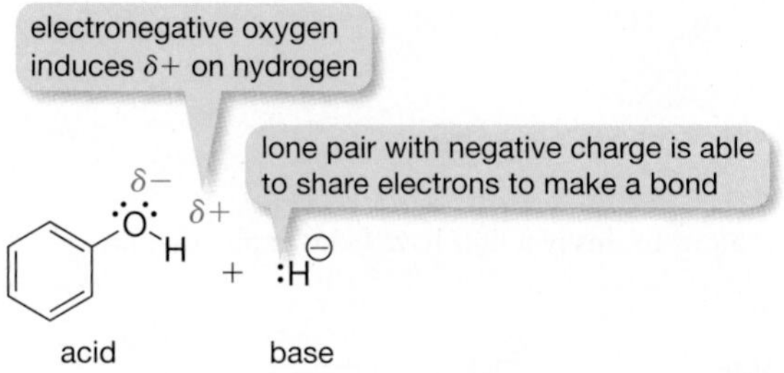

electronegative oxygen induces δ+ on hydrogen

lone pair with negative charge is able to share electrons to make a bond

acid base

STEP 3: Draw in the curved arrows, using the principle that negative charges are attracted to positive ones.

The lone pair on the hydride has a negative charge and acts as the electron source, so the curved arrows start here. Arrows move toward sites of electron deficiency, so the arrow ends at the hydrogen atom of the acid. (Note: Since a base is always electron rich and an acid is always electron poor, the curved arrows in an acid–base reaction always begin at the base and end at the acid). Since the hydrogen atom on the acid can't have more than two electrons in its valence shell, a second arrow must be drawn to correspond with the breaking of the O–H bond. This arrow starts in the middle of the bond and ends at the more electronegative oxygen atom.

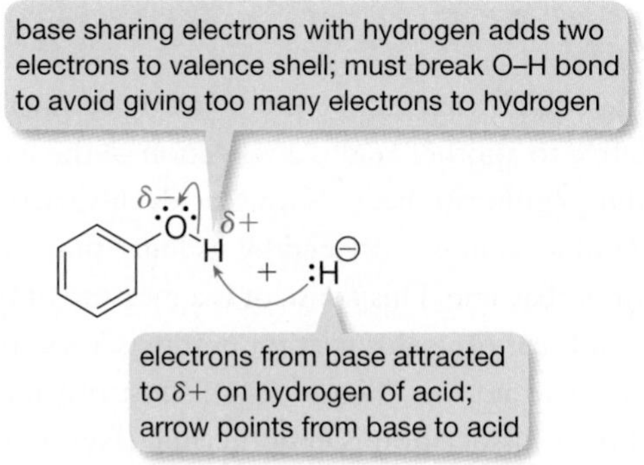

base sharing electrons with hydrogen adds two electrons to valence shell; must break O–H bond to avoid giving too many electrons to hydrogen

electrons from base attracted to δ+ on hydrogen of acid; arrow points from base to acid

STEP 4: Draw the products of the reaction, using the curved arrows to guide you. Don't forget the formal charges.

A new H–H bond forms while the O–H bond breaks, giving its electrons to the oxygen atom. The formal charge integer on the hydride increases by one unit since an arrow is pointing away from it, whereas the formal charge integer on oxygen decreases by one unit since an arrow is pointing to it. The resulting products, with their formal charges, are as follows:

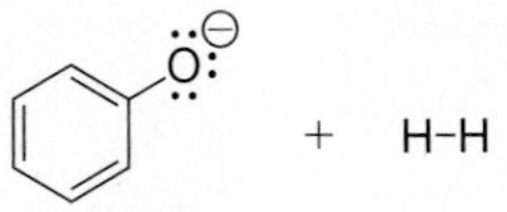

+ H-H

STEP 5 (OPTIONAL): Redraw the final reaction mechanism, omitting any unnecessary lone pairs and explicitly drawn bonds.

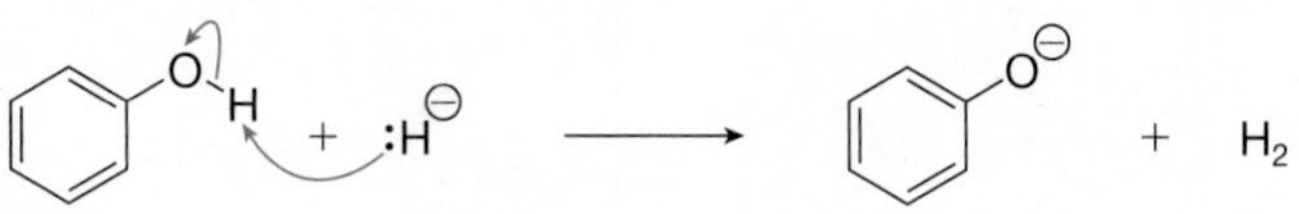

+ H_2

PRACTICE PROBLEM

6.1 Draw the curved arrow mechanisms for the following acid–base reactions and predict the products.

a) 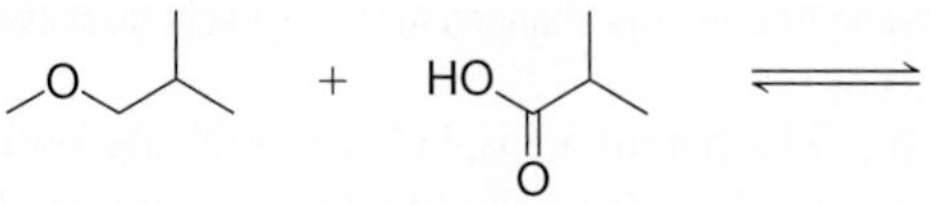 + HO

b) 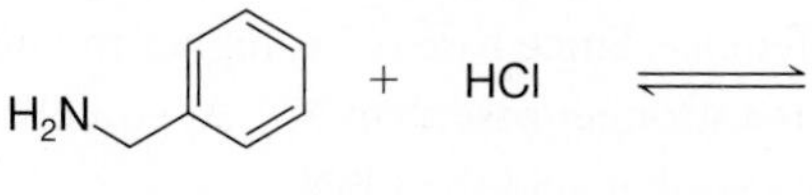 H_2N + HCl

c)

d) Note: the acidic protons in this reaction have been explicitly drawn.

INTEGRATE THE SKILL

6.2 For each of the acid–base reactions in Question 6.1, draw a curved arrow mechanism for the reverse reaction.

6.3 Free Energy and Acid Strength

The position of an acid–base equilibrium can be predicted by comparing acid or base strengths. The strength of an acid relative to another acid is a reflection of the extent of the dissociation of each into their corresponding conjugate bases. Strong acids dissociate much more readily than weak acids. The dissociation of an acid is described by its logarithmic dissociation constant, pK_a, which indicates the strength of that acid. This constant is a measure of the difference between the free energy (G) of the acid and that of its dissociation products. Very strong acids have a negative $\Delta G°$, and their dissociation is exothermic (Figure 6.1a). Most organic acids are weak and have positive values of $\Delta G°$, and their dissociation is endothermic (Figure 6.1b).

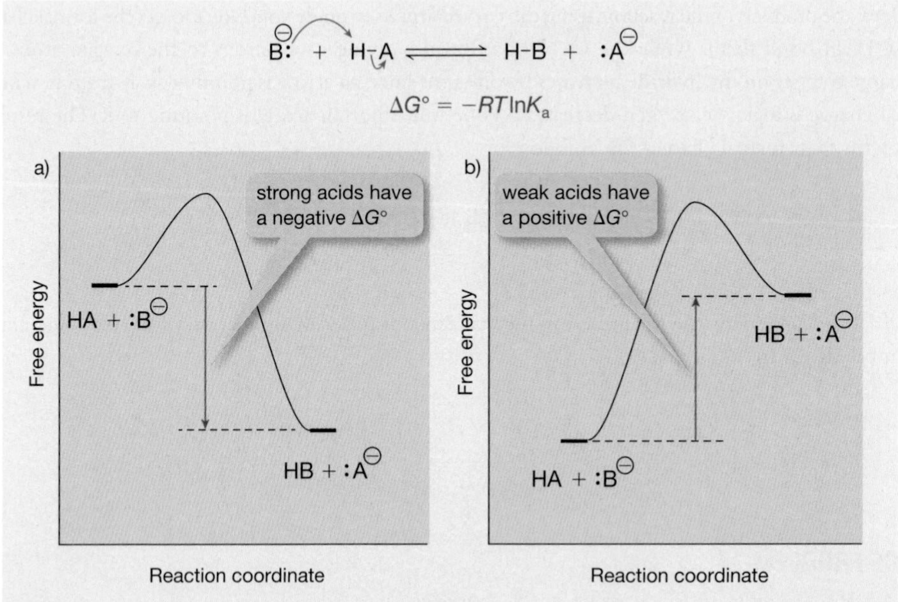

FIGURE 6.1 Reaction coordinate diagrams showing free-energy changes for typical acid dissociations.

Two weak acids are compared in Figure 6.2. The parent acids, HA and HX, are both neutral molecules, so the difference in free energy between them is negligible. Their conjugate bases A^- and X^- are charged and have a greater energy difference. Since base A^- is higher in energy than base X^-, A^- is less stable than X^-. Therefore, A^- is a stronger base than X^-. At equilibrium, HA is less dissociated than HX. In other words, HA is a weaker acid than HX.

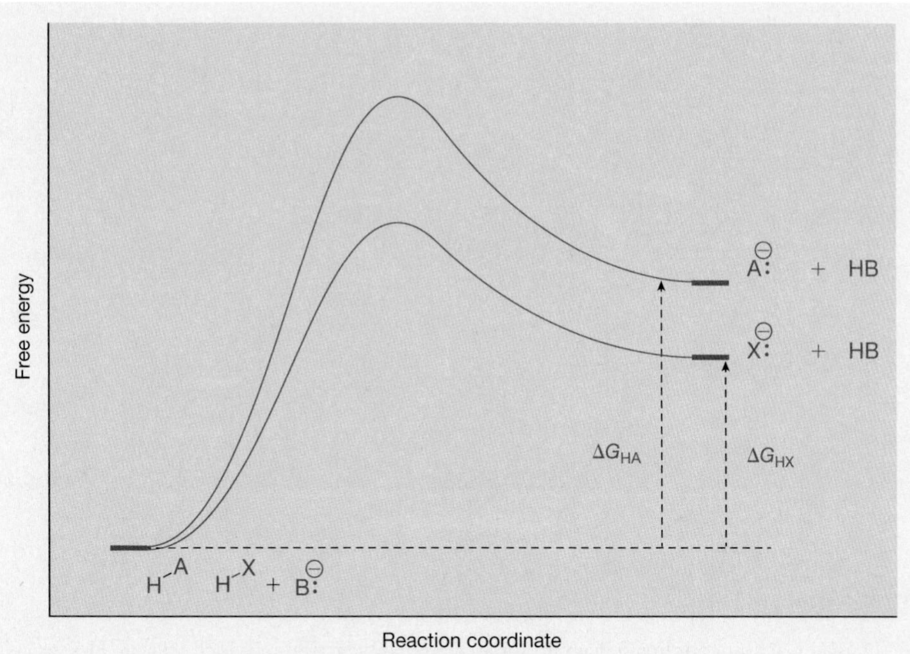

FIGURE 6.2 Free-energy changes for the dissociation of two hypothetical acids.

The relative strength of two acids (or two bases) in an organic reaction determines the order of the mechanistic steps during that reaction. There are both qualitative and quantitative methods for determining the strength of acids and bases.

CHECKPOINT 6.2

You should now be able to interpret and draw a reaction coordinate diagram to compare two acid–base reactions.

SOLVED PROBLEM

Draw a reaction coordinate diagram to demonstrate an exothermic acid–base reaction in which an acid, HA, reacts with a base, Y^-.

STEP 1: Draw the overall reaction.

$$\text{H–A} + :\text{Y}^{\ominus} \rightleftharpoons {}^{\ominus}\text{A}: + \text{H–Y}$$

STEP 2: Label the acid, base, conjugate base, and conjugate acid.

$$\underset{\text{acid}}{\text{H–A}} + \underset{\text{base}}{:\text{Y}^{\ominus}} \rightleftharpoons \underset{\text{conjugate base}}{{}^{\ominus}\text{A}:} + \underset{\text{conjugate acid}}{\text{H–Y}}$$

STEP 3: Identify the most stable base.

The reaction is exothermic, so the equilibrium lies toward the products. This means that A^-, the conjugate base, must be more stable than Y^-, the added base.

STEP 4: Draw the corresponding reaction coordinate diagram with the energy of the products lower than that of the reactants.

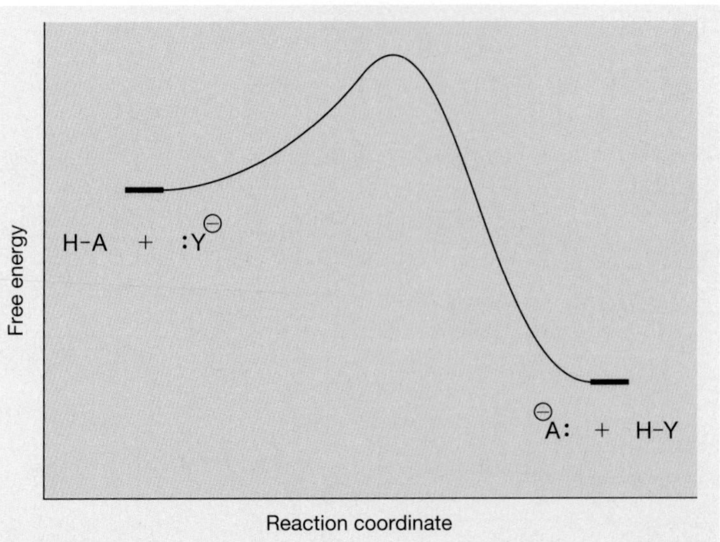

PRACTICE PROBLEM

6.3 a) Draw a reaction coordinate diagram for an acid–base reaction in which an acid, HA, reacts with Y^-. The equilibrium favours the reactants.

b) Draw a reaction coordinate diagram to demonstrate an acid–base reaction in which an acid, HA, reacts with Y^-. The equilibrium does not favour the starting materials or products.

c) Draw a reaction coordinate diagram for an acid–base reaction in which an acid, HA, reacts with Y^-. The conjugate base is less stable than the base.

d) Draw a reaction coordinate diagram for an acid–base reaction in which an acid, HA, reacts with Y^-. The conjugate base is more stable than the base.

INTEGRATE THE SKILL

6.4 Hydrofluoric acid (HF) is a weak acid that only partially dissociates in water. In contrast, hydrobromic acid (HBr) is a strong acid that completely dissociates in water.

a) Write the overall reaction equation for the dissociation of each acid in water.

b) Draw a curved arrow mechanism for each of the reactions drawn in part (a).

c) Which of the reactions from part (a) is best represented by the following reaction coordinate diagram?

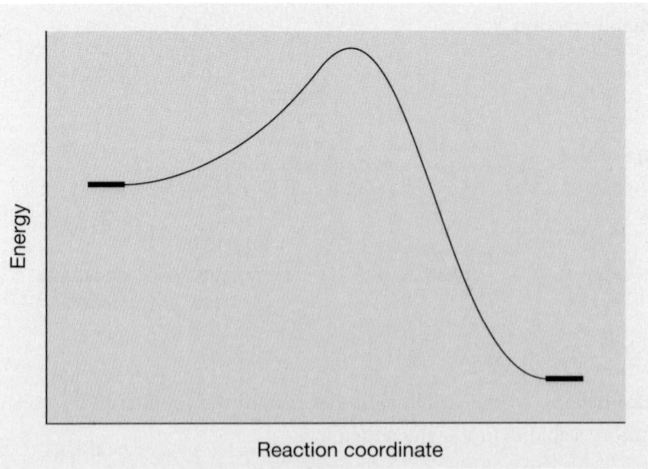

6.4 Qualitative Estimates of Relative Acidities

Acids always have a conjugate base, and either the acid or its conjugate base carries a charge. Energy differences between charged molecules are normally larger than those between neutral molecules, so the free energies of the charged species in dissociation expressions largely determine the strengths of acids.

Most organic acids are neutral. When the acids are neutral, it is easier to compare the conjugate bases, which bear a negative charge. The strength of a base is related to its ability to accommodate (i.e., stabilize) negative charge. When a charge is stabilized, the energy of the charged species is reduced. Bases that stabilize negative charges readily are weak (with a less positive $\Delta G°$), whereas bases that do not stabilize negative charges as much are stronger (with a more positive $\Delta G°$). The relative ability of a base to stabilize negative charge can usually be estimated using one of five factors that contribute to charge stabilization: electronegativity, atomic size, induction, hybridization, and delocalization (resonance).

6.4.1 Electronegativity

The electronegativity of the atom carrying the negative charge has a large impact on conjugate base stability. Electronegativity increases from left to right across the periodic table, so bases in which the atom carrying the negative charge lies farthest to the right stabilize negative charges the most. Consequently, such conjugate bases have lower energy, and their parent acids dissociate more, making these acids relatively strong.

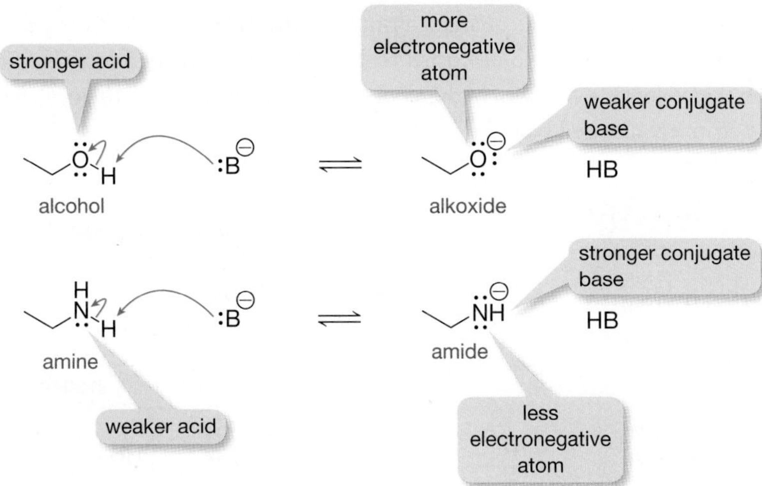

The conjugate bases of an alcohol and an amine are an alkoxide and an **amide**, respectively. Oxygen lies to the right of nitrogen on the periodic table and is therefore more electronegative. So, oxygen has a higher affinity for electrons, and a negative charge on oxygen is more stabilized than one on nitrogen. Consequently, the alkoxide is a weaker base than the amide. Because the alkoxide is the weaker base, the corresponding parent acid, the alcohol, is the stronger acid.

The term **amide** can refer to more than one functional group: a negatively charged nitrogen or a more complex structure (see Chapter 15).

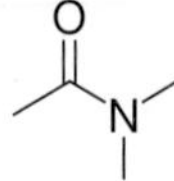

6.4.2 Atomic size

Large atoms can disperse a charge over larger volumes than small atoms can. Because electrons repel each other, spreading them out lowers their free energy. As a result, a negative charge on a large atom is more stable than one on a smaller atom. Thus, conjugate bases with the negative charge carried by a large atom are relatively weak, and their parent acids are relatively strong.

STUDENT TIP

When comparing atoms in the same *row* of the periodic table, always consider electronegativity.

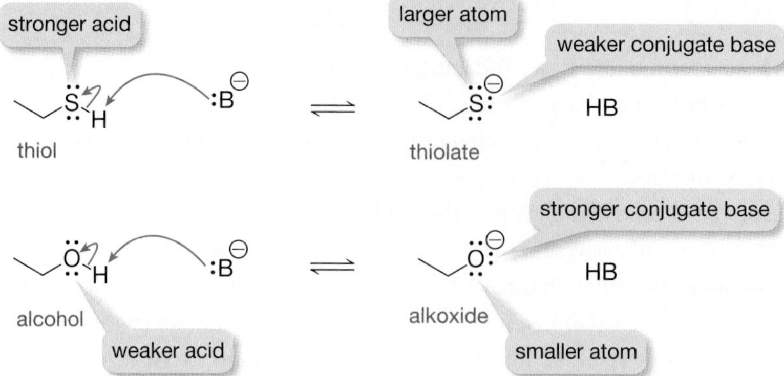

STUDENT TIP
When comparing atoms in the same *column* of the periodic table, always consider atomic size.

For example, the conjugate bases of a thiol and an alcohol are a thiolate and an alkoxide, respectively. Sulfur is below oxygen in the same column (group) of the periodic table, and is therefore larger. Its larger size stabilizes the thiolate relative to the corresponding alkoxide due to greater delocalization of the negative charge. Since the thiolate is more stable than the alkoxide, it is the weaker conjugate base, and the thiol is a stronger acid than the alcohol.

6.4.3 Induction

Induction is the removal of electron density from an atom by a strongly electronegative atom nearby.

Conjugate bases with their negative charge dispersed over several atoms generally have lower free energy than similar bases with charge localized on a single atom. The presence of electron-attracting groups (usually called electron-withdrawing groups) near a negative charge can spread out the electrons associated with the charged atom by **induction**. The electronegativity of atoms near the negative charge causes the electrons to be pulled toward the electronegative atom(s). This increases the volume over which these electrons are distributed (delocalizes them) and stabilizes the negatively charged base.

weaker acid
stronger conjugate base
H_3C alcohol :B ⇌ H_3C alkoxide HB
weaker conjugate base
F_3C trifluoroalcohol :B ⇌ F_3C trifluoroalkoxide HB
stronger acid
inductively withdrawing fluorines pull electron density away from oxygen

An oxygen atom carries the charge in the conjugate bases for both ethanol and trifluoroethanol. Ethanol has three hydrogen atoms at the end of its carbon chain, while trifluoroethanol has three fluorine atoms, the most electronegative element. The fluorine in the alkoxide of trifluoroethanol pulls electron density away from the negatively charged oxygen, spreading the electron density and reducing the effective charge on oxygen. This inductive effect is often shown as a charge dipole.

Induction makes the conjugate base of trifluoroethanol more stable than the conjugate base of ethanol. As a result, the trifluoroalkoxide is the weaker base, and trifluoroethanol is a stronger acid than ethanol.

Inductive effects occur through bonds. The strength of inductive effects depends on the electronegativity of the atoms involved, not the size of those atoms. Thus, CF_3CH_2OH is more acidic than CCl_3CH_2OH because the electronegativity of fluorine is stronger than that of chlorine.

STUDENT TIP

Induction occurs through σ bonds, while delocalization (described by resonance) involves π bonds.

weaker acid

stronger conjugate base

Cl_3C–O–H + :B⁻ ⇌ Cl_3C–O⁻ HB

weaker conjugate base

F_3C–O–H + :B⁻ ⇌ F_3C–O⁻ HB

stronger acid

fluorine is more electronegative and produces stronger inductive effect

The strength of induction depends on how many inductively withdrawing atoms are present and on how close these atoms are to the negative charge. For example, trichloracetate is more stabilized by induction than chloracetate is, because three chlorines are able to "pull" electrons more strongly than one. Therefore, trichloracetate is the weaker base, and trichloroacetic acid is stronger than chloroacetic acid.

stronger acid

more inductively withdrawing atoms (groups)

Cl_3C–C(=O)–O–H + :B⁻ ⇌ Cl_3C–C(=O)–O⁻ HB

$ClCH_2$–C(=O)–O–H + :B⁻ ⇌ $ClCH_2$–C(=O)–O⁻ HB

6.4.4 Hybridization

Atoms in organic molecules normally have their electrons in one of the three possible hybrid combinations of s and p orbitals: sp^3, sp^2, or sp. Since s orbitals are lower in energy than p orbitals, hybrid orbitals with more s character are lower in energy. For example, an sp orbital, which is a hybrid of an s orbital with one p orbital, has lower energy than an sp^3 orbital, which has contributions from one s and three p orbitals. Unpaired electrons are more stable in sp orbitals than in sp^3 orbitals. A lower energy orbital makes a negatively charged conjugate base more stable and hence weaker.

stronger acid
negatively charged carbon is sp
weaker conjugate base
:B⊖
HB
stronger conjugate base
:B⊖
$\ddot{C}H_2$
HB
weaker acid
negatively charged carbon is sp^3

Hydrogens bonded to sp-hybridized carbons dissociate more readily (they are easier to remove) than those bonded to sp^2-hybridized carbons. Similarly, hydrogens on sp^3-hybridized carbons are the most difficult to remove, and can be easily taken from only certain locations on molecules. This pattern also applies to functional groups involving heteroatoms.

6.4.5 Charge delocalization

Charge delocalization (described by resonance) can be a strong stabilizing influence for organic materials because it reduces electron repulsion. For example, the conjugate bases formed from an alcohol and a carboxylic acid (an alkoxide and a carboxylate, respectively) both have negative charges on an oxygen atom, so atomic size or electronegativity will not cause any differences in stability.

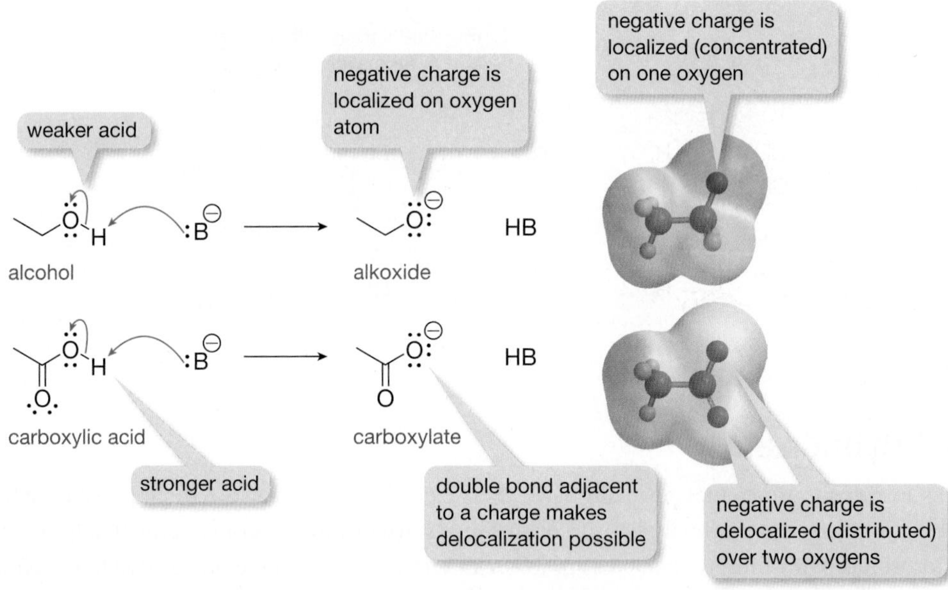

ORGANIC CHEMWARE
6.2 Resonance: Carboxylate anion

However, the alkoxide has a localized negative charge that resides on a single oxygen atom, while the negatively charged oxygen in the carboxylate is located next to a double bond that makes delocalization (depicted using two resonance forms) possible. Delocalizing the electrons over a larger volume (more atoms) lowers the energy of the carboxylate relative to the alkoxide, making the carboxylic acid the stronger acid.

negatively charged oxygen donates an electron pair to form a π bond with carbon

π bond breaks to avoid exceeding the octet rule at the carbon atom

Many organic reactions involve removing a hydrogen atom that is bonded to a carbon atom, leaving the electron pair from the bond on the carbon. Organic molecules in such reactions *act* like acids because a proton is being removed from them. However, hydrogen ions are removed from carbon atoms only when a *very* strong base is present. The *behaviour* of these molecules in acid–base reactions can be predicted by considering the compounds as Brønsted acids.

The two reactions shown illustrate a process called enolate formation, which is described in more detail in Chapter 17. In both reactions, protons are removed from carbons that are *adjacent* to C=O π bonds. The protons that are removed (transferred) are drawn in order to show a mechanism, just as electron pairs are drawn when they participate in reactions. Both molecules act as acids, and the stronger one (the molecule that the H^+ is more easily removed from) can be predicted by considering the resonance forms of the conjugate bases.

weaker acid

ketone

HB +

negative charge delocalized over two atoms

carbanion

enolate

1,3-diketone

HB +

carbanion

enolate

enolate

stronger acid

negative charge delocalized over three atoms

The negative charge on the ketone carbanion is delocalized as described by the enolate resonance structure. The 1,3-diketone carbanion has two enolate resonance forms, which delocalize the negative charge to both of its oxygen atoms. This greater delocalization makes the free energy of the 1,3-diketone carbanion lower than that of the ketone carbanion. The 1,3-diketone carbanion is the weaker base, and therefore the 1,3-diketone is the stronger acid (its hydrogen is more easily removed).

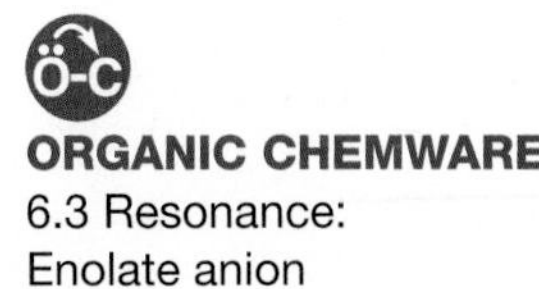

STUDENT TIP

Delocalization is usually the dominant factor when comparing the relative stability of charged species.

CHECKPOINT 6.3

You should now be able to identify the most acidic proton in a molecule.

SOLVED PROBLEM

Identify the most acidic proton in the following structure:

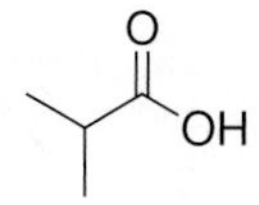

STEP 1: Expand the structure to show all the different *types* of hydrogen atoms and their bonds.

Note that the two CH_3 groups are equivalent, so only one of them needs to be expanded to show the hydrogens.

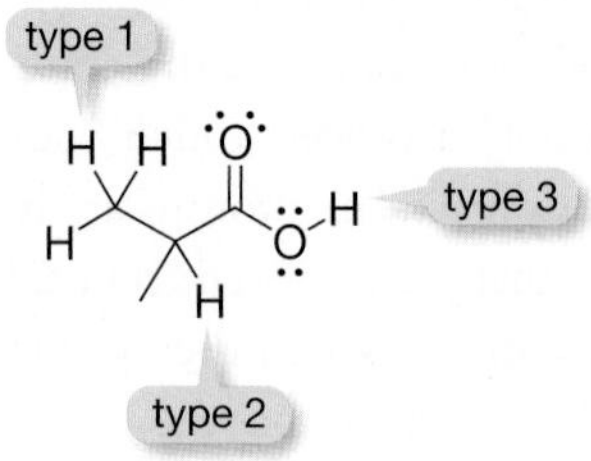

type 1

type 3

type 2

STEP 2: Draw all of the conjugate bases (CB) that correspond to deprotonating one of each type of hydrogen atom.

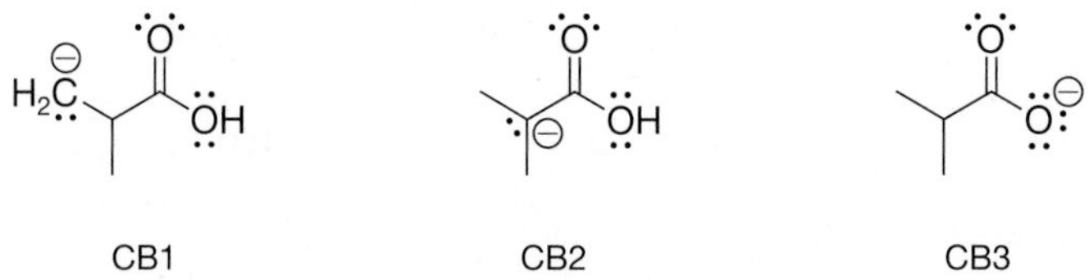

CB1 CB2 CB3

STEP 3: Compare the stability of the different conjugate bases.

CB1 and CB2 both have the negative charge on a carbon atom; however, CB2 has the possibility for delocalization, while CB1 does not. Since delocalization is a strongly stabilizing force, CB2 is more stable than CB1.

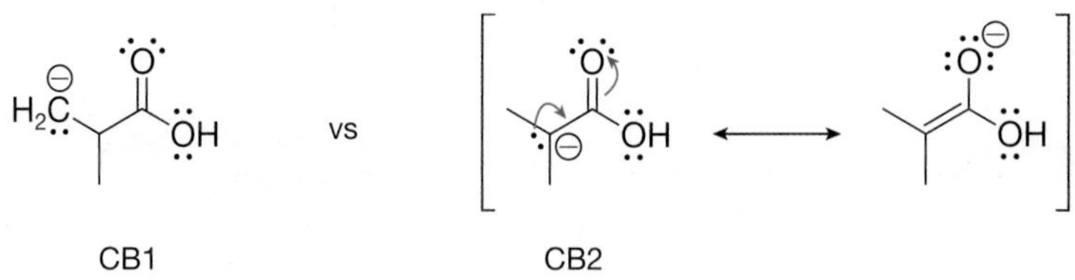

CB1 vs CB2

CB3 also has the possibility for delocalization (described by resonance) and needs to be compared with CB2. Both have two important resonance structures; however, in CB2, only one of the structures places the negative charge on the more electronegative oxygen atom, while in CB3, both resonance contributors have the negative charge on an oxygen. Since oxygen and carbon are in the same *row* of the periodic table, the difference in their electronegativities is the dominant stability factor. Oxygen is more electronegative than carbon, so the oxygen atom can better stabilize the negative charge than carbon atoms. This means CB3 is more stable than CB2.

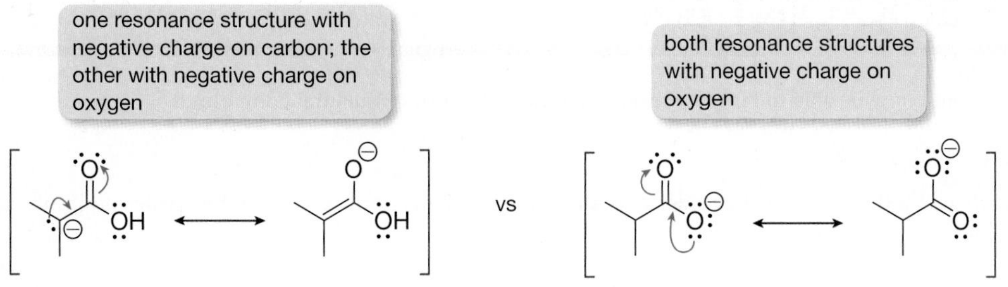

one resonance structure with negative charge on carbon; the other with negative charge on oxygen

both resonance structures with negative charge on oxygen

CB2 vs CB3

STEP 4: Use the relative stabilities of the conjugate bases to determine which set of protons is the most acidic.

CB3 is the most stable base, and therefore, the weakest. Recall the principle that the stronger the acid, the weaker its conjugate base (and vice versa). Since CB3 is the weakest base, the type 3 proton must be the most acidic (the most easily removed).

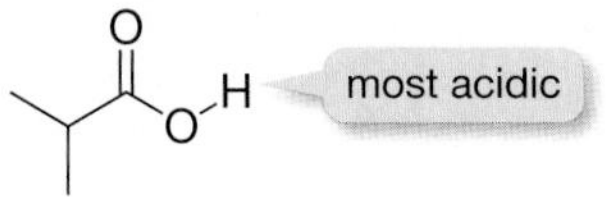

most acidic

PRACTICE PROBLEM

6.5 Determine the most acidic proton in each of the following compounds. Consider only the hydrogens that are explicitly drawn.

a)

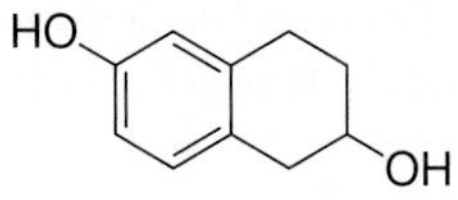

b)

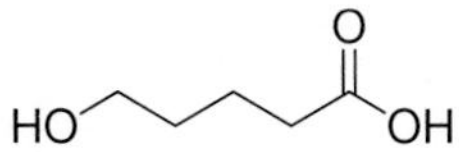

c)

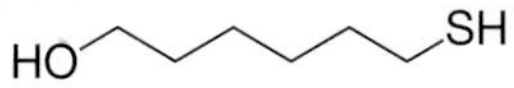

d)

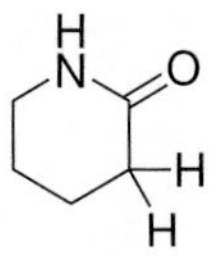

INTEGRATE THE SKILL

6.6 For each of the following compounds, identify the most acidic proton and then draw a mechanism for its reaction with a strong base, B^-.

a)

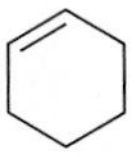

b)

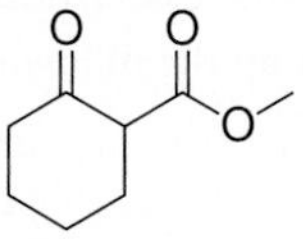

c)

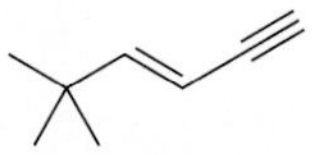

CHECKPOINT 6.4

You should now be able to rank the relative acidities of a series of neutral compounds.

SOLVED PROBLEM

Which acid is the strongest? Consider only the explicitly drawn protons in each molecule and justify your choice.

SH OH

acid 1 acid 2

STEP 1: Draw each conjugate base.

S⊖ O⊖

CB1 CB2

STEP 2: Compare the relative stabilities of the conjugate bases.

The sulfur and oxygen atoms are in the same *column* of the periodic table, so you need to consider the relative sizes of these atoms. The sulfur atom is larger than the oxygen atom and therefore disperses and stabilizes the negative charge more. Consequently, CB1 is more stable than CB2.

STEP 3: Use the relative stabilities of the conjugate bases to determine which compound is the most acidic.

Because CB1 is more stable and therefore a weaker base than CB2, acid 1 is stronger than acid 2.

PRACTICE PROBLEM

6.7 Rank each set of compounds in order of increasing acidity.

a) 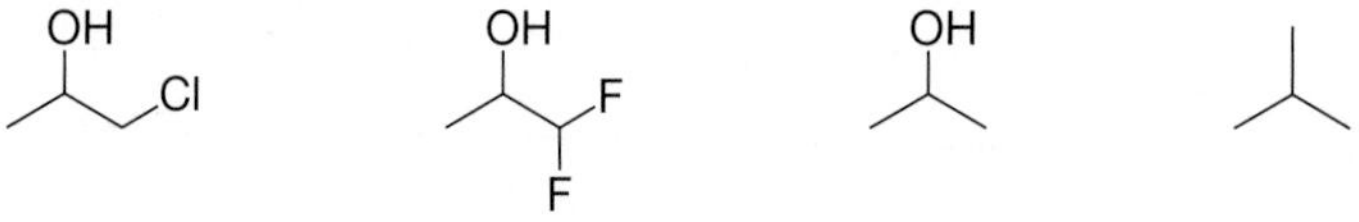 OH Cl OH F F OH

b) 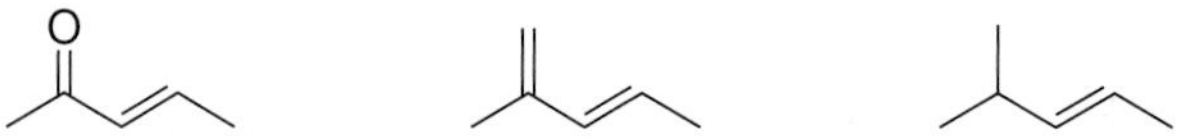 O

c) 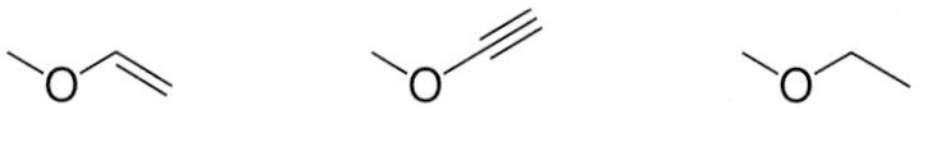 O O O

d) 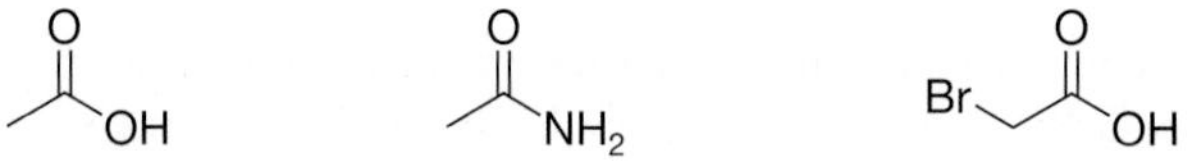 O OH O NH_2 Br O OH

INTEGRATE THE SKILL

6.8 The reaction coordinate diagram below shows the free-energy changes for the dissociation of three different acids, A, B, and C. Use the relative energies of the curves to match up the following structures with the acids in the diagram.

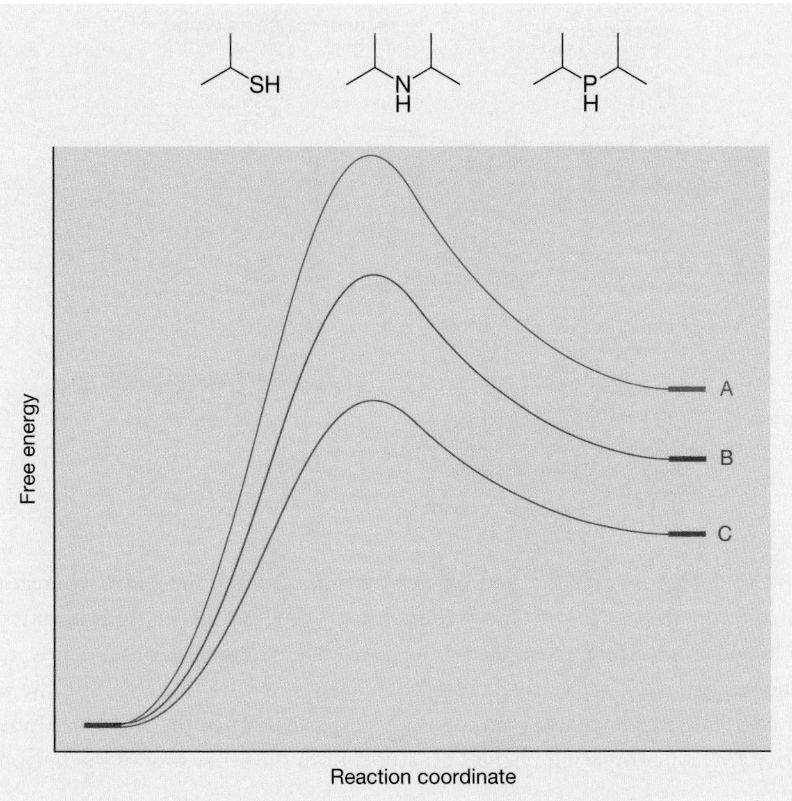

6.5 Relative Acidities of Positively Charged Acids

Acid–base reactions in organic chemistry sometimes involve positively charged acids that produce neutral conjugate bases. For example, a protonated alcohol oxonium can act as a proton donor (acid), forming a neutrally charged alcohol as the conjugate base.

$$H_3C\text{–}CH_2\text{–}\overset{\oplus}{O}H_2 \quad :B^{\ominus} \rightleftharpoons H_3C\text{–}CH_2\text{–}\ddot{O}H \quad HB$$

alcohol

As discussed in Section 6.4, charged species usually account for most of the difference in energy between the sides of a dissociation equation. The relative strengths of positively charged acids (and the relative strengths of their conjugate bases) can be estimated by evaluating the acids themselves using the same principles as apply for charged bases: electronegativity, atomic size, induction, hybridization, and resonance.

6.5.1 Electronegativity

Electronegative atoms destabilize positive charges, and so charged acids in which the element carrying the positive charge appears to the right side of the periodic table are less stable than acids in which the charged atom lies to the left of the periodic table. Acids with positive charge on more electronegative atoms are more acidic, and their conjugate bases are correspondingly less basic.

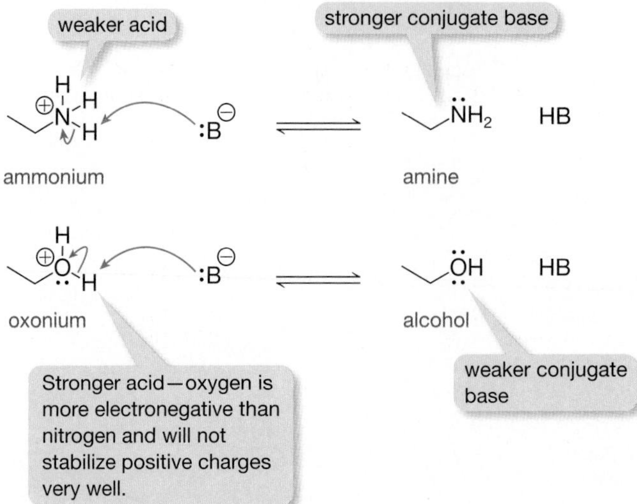

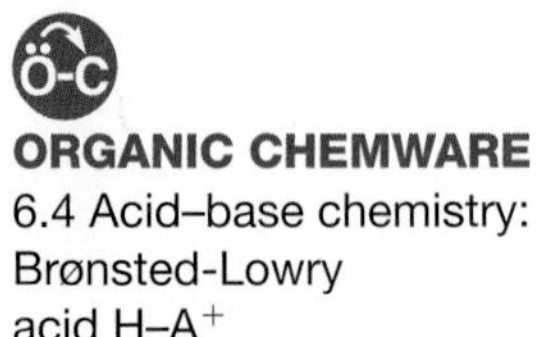

ORGANIC CHEMWARE
6.4 Acid–base chemistry: Brønsted-Lowry acid H–A+

The positive charge is carried on an oxygen atom in the oxonium ion, whereas the charge is carried on a nitrogen atom in the ammonium ion. Oxygen lies to the right of nitrogen on the periodic table and is therefore more electronegative than nitrogen. Consequently, the positive charge on the oxygen is less stable than the positive charge on the nitrogen, and the oxonium is the stronger acid. This means that the amine is a stronger base than the alcohol. Always consider electronegativity when comparing charges on atoms from the same row of the periodic table.

6.5.2 Atomic size

A positive or negative charge on a large atom is dispersed over a larger volume than it is on smaller atom. Consequently, positive charges on large atoms are more stable than charges on small atoms.

stronger acid

weaker conjugate base

oxonium :B ⇌ OH HB

alcohol

stronger conjugate base

thionium :B ⇌ SH HB

thiol

Weaker acid—sulfur is larger than oxygen and can delocalize charge over a larger volume.

For example, sulfur is lower than oxygen in the periodic table, which indicates that a sulfur atom is larger than an oxygen atom. Therefore, thionium is more stable than oxonium, and it is the weaker acid. Correspondingly, the conjugate base of the thionium is a stronger base than the alcohol. Always consider atomic size when comparing charges on atoms from the same column of the periodic table.

6.5.3 Induction

Nearby electronegative atoms *increase* the effective positive charge on an atom. They do this by withdrawing extra electron density from the positively charged atom, making the atom even more positive. This effect makes a positively charged molecule *less* stable and therefore a stronger acid.

weaker acid

stronger conjugate base

H_3C–$O^{+}H_2$ + :B^{-} ⇌ H_3C–OH HB

ethyloxonium

alcohol

weaker conjugate base

F_3C–$O^{+}H_2$ + :B^{-} ⇌ F_3C–OH HB

trifluoroethyloxonium

trifluoroalcohol

Stronger acid—inductively withdrawing fluorine atoms remove electron density from oxygen, making the positive charge even larger.

Trifluoroethyloxonium has three fluorine atoms close to the positively charged oxygen. The fluorine tends to pull electrons away from this oxygen, significantly increasing the positive charge on that oxygen. The increased charge makes the trifluoroethanol oxonium less stable, and hence more acidic, than the ethanol oxonium. The induced charge also makes trifluoroethanol a weaker base than ethanol.

Alkyl groups beside a positive charge will stabilize the positive charge. They do so by donating some electrons to the charged atom, thus lowering the effective charge on the atom. The positive charge is inductive (strongly electron withdrawing), and pulls electron density from the σ bonds of the alkyl group through a process called hyperconjugation (explained in detail in Chapter 8).

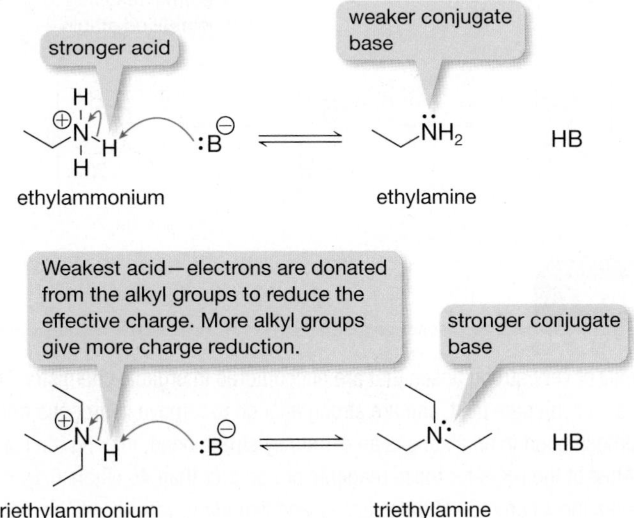

6.5.4 Charge delocalization

As with negatively charged molecules, delocalization can stabilize positively charged organic compounds by spreading the charge over a larger volume.

In the example shown, the positive charges on both oxonium ions are on sp^2-hybridized oxygen atoms. Therefore, the differences in strength between these two organic acids cannot be caused by atomic size, electronegativity, or hybridization. Instead, the difference in acidity is a consequence of the amount of delocalization possible for each acid.

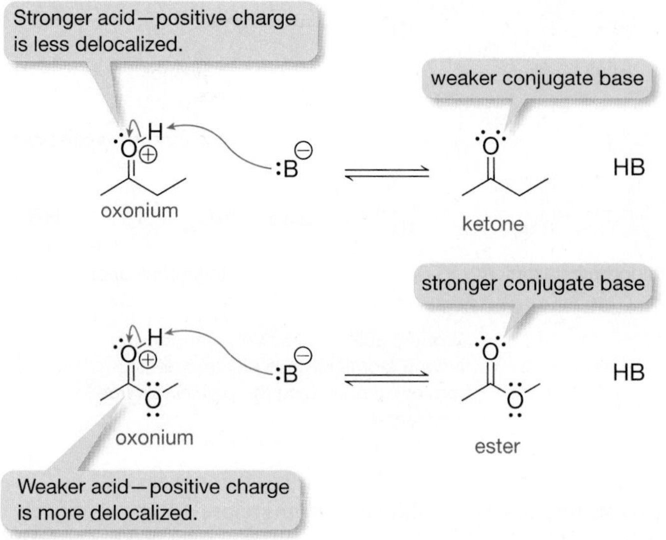

Stronger acid—positive charge is less delocalized.

weaker conjugate base

oxonium :B ketone HB

stronger conjugate base

oxonium :B ester HB

Weaker acid—positive charge is more delocalized.

The figure below shows that while such stabilization is possible for both oxonium ions, the oxonium from the ester has its charge spread out over three atoms rather than two, resulting in a lower energy relative to the ketone oxonium ion. Since the ester oxonium has the lower energy, it is the weaker acid.

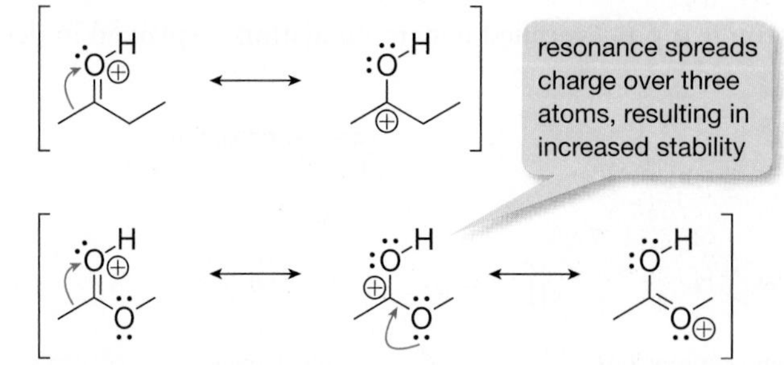

resonance spreads charge over three atoms, resulting in increased stability

DID YOU KNOW?

Alkyllithiums are a family of very strong bases that are encountered in organic chemistry. These bases have the corresponding alkane as a conjugate acid, and are strong enough to remove hydrogens from almost any molecule. The carbon-hydrogen bond in an alkane is an extremely strong bond, making the corresponding lithium bases very powerful. Most of the pK_a's for these reagents are greater than 45 (Table 6.1). *n*-BuLi (or BuLi—pronounced BULEE) is the most common of these bases, and like many other alkyllithiums, is commercially available.

The strength of these molecules as bases is high enough that they react violently and very exothermically with water. The amount of heat produced is so great that many alkyllithiums are pyrrophoric: they spontaneously ignite when exposed to moist air. Reactions involving alkyllithiums—in fact most organic reactions carried out today—are performed in special apparatus using inert atmospheres (nitrogen or argon) that cannot react with such reagents. Handling these compounds requires special equipment, training, and care.

TABLE 6.1 Some Common Alkyllithium Bases and the pK_a's of Their Corresponding Conjugate Acids (Alkanes)

Structure	Name	Abbreviation	pK_a
$CH_3CH_2CH_2CH_2Li$	*n*-butyllithium	*n*-BuLi or BuLi	50
$CH_3CH_2CHLiCH_3$	*s*-butyllithium	*s*-BuLi	51
$(CH_3)_3CLi$	*t*-butyllithium	*t*-BuLi	53
CH_3Li	methyllithium	MeLi	48

CHECKPOINT 6.5

You should now be able to identify the most acidic and the most basic compounds from a series of both neutral and charged structures, as well as the most acidic protons and the most basic sites on a given molecule.

SOLVED PROBLEM

Identify the most basic atom in this lactone (cyclic ester).

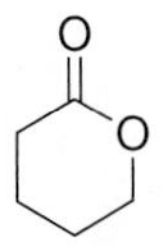

STEP 1: Expand the Lewis structure of the functional groups to show lone pairs.

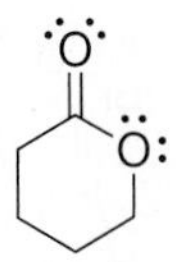

STEP 2: Identify the basic atoms in the structure.

The lone pairs on the oxygen atoms can act as potential basic sites.

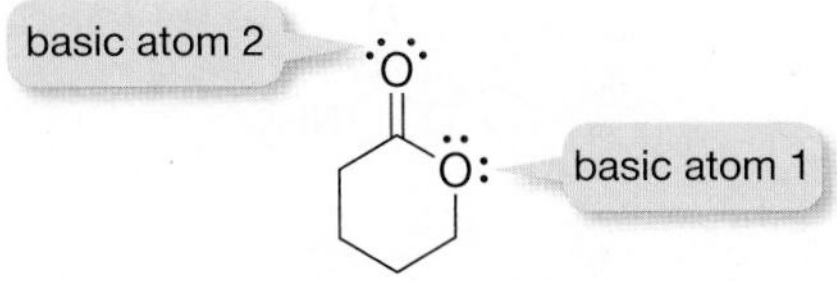

basic atom 2

basic atom 1

STEP 3: Draw the conjugate acids (CA) that would be produced from the protonation of each basic site.

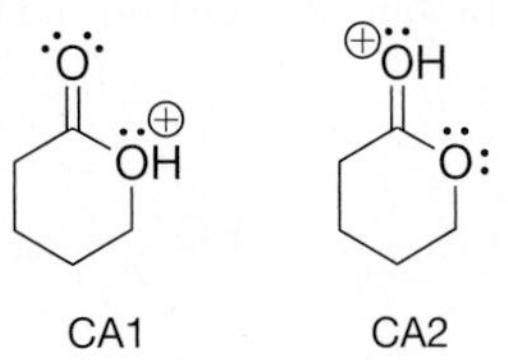

CA1 CA2

STEP 4: Compare the relative stabilities of the conjugate acids.

In both CA1 and CA2, the positive charge resides on an oxygen atom, so there is no difference in atom size or electronegativity. However, CA2 can be stabilized by delocalization (described by resonance), whereas CA1 cannot. Therefore, CA2 is the more stable conjugate acid.

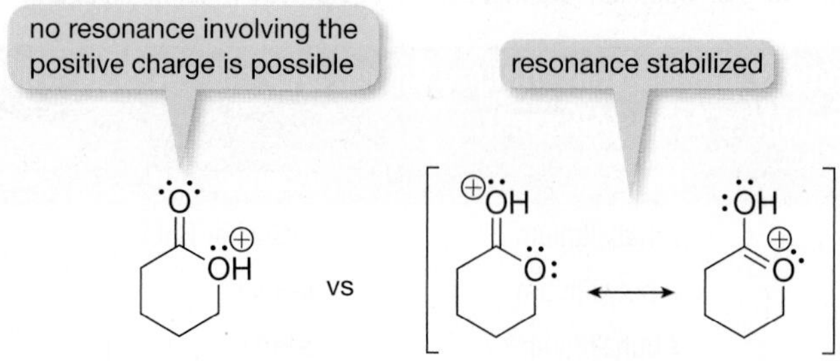

no resonance involving the positive charge is possible

resonance stabilized

CA1 vs CA2

STEP 5: Use the relative stabilities of the conjugate acids to determine which atom is the most basic.

The stronger the base, the weaker its conjugate acid (and vice versa). Since CA2 is the more stable and therefore weaker acid, basic atom 2 must be more basic than basic atom 1. Consequently, when the lactone reacts with an acid, the carbonyl oxygen is protonated to give CA2 as the product.

PRACTICE PROBLEM

6.9 a) Which of these compounds is the most acidic?

$H_2N^{\oplus}$ ... N vs HN ... $NH^{\oplus}$

b) Rank these compounds in order of increasing basicity.

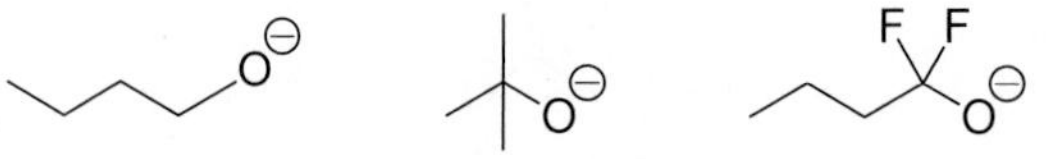

A B C

c) Identify the most acidic proton in the following compound.

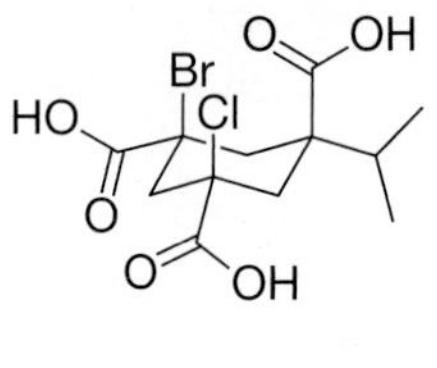

d) Identify the most basic atom in the following compound.

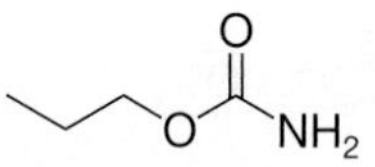

INTEGRATE THE SKILL

6.10 Predict the most likely acid–base reaction to take place between the following two compounds. Draw a curved arrow mechanism to show the reaction and the products.

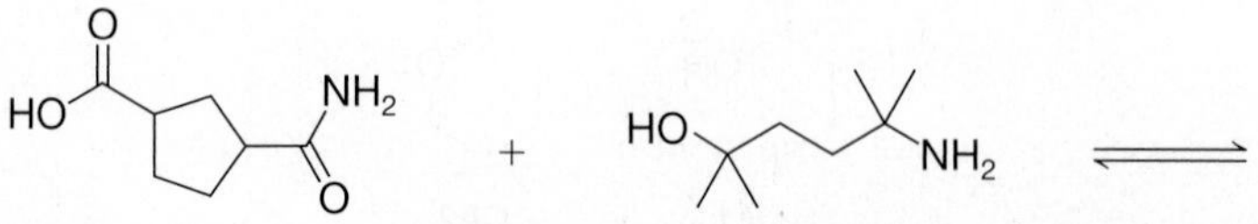

6.6 Quantitative Acidity Measurements

Acid strength has been measured accurately for many substances and is commonly expressed in terms of the logarithmic acid dissociation constant pK_a. A lower pK_a indicates a stronger acid, a higher pK_a a weaker acid. Figure 6.3 lists pK_a values for some common organic functional groups. To estimate the pK_a for other functional groups, use the values for the functional group that is the most similar to the structure being examined.

Acid	Conjugate base	pK_a of acid
$RC(=\overset{\oplus}{O}H)OH$	$RC(=O)OH$	−6
HI H_2SO_4 HBr HCl	$I^{\ominus}$ $HSO_4^{\ominus}$ $Br^{\ominus}$ $Cl^{\ominus}$	< −4
$R\overset{\oplus}{O}H_2$	ROH	−2 to −4
$H_3O^{\oplus}$	H_2O	−1.7
$RC(=O)OH$	$RC(=O)O^{\ominus}$	2 to 5
$Ar\overset{\oplus}{N}H_3$	$ArNH_2$	3 to 5
$R\overset{\oplus}{N}H_3$	RNH_2	9 to 13
RSH	$RS^{\ominus}$	9 to 12
H_2O	$HO^{\ominus}$	15.7
ROH	$RO^{\ominus}$	15 to 17
RNH_2	$RNH^{\ominus}$	~ 35

Increasing acid strength (upward); Increasing base strength (downward)

FIGURE 6.3 Approximate pK_a ranges for common organic functional groups.

6.6.1 Using pK_a to measure base strength

The strength of a base can be quantified using the pK_a of its conjugate acid. (The logarithmic base dissociation constant pK_b is not used in organic analysis.) The strongest bases are those for which the conjugate acids have the *highest* pK_a values. It is important to write out the pK_a expression properly, so that the correct acid is identified. The acid should always appear as a reactant, and the conjugate base as a product.

STUDENT TIP

Remember that pK_a is an *indirect* measure of the strength of a base; pK_a actually indicates the strength of the conjugate acid.

when evaluating the strength of a base, write the base on the right side of a pK_a expression

$$\text{HBase} \rightleftharpoons \text{Base}^{\ominus} \quad \text{H}^{\oplus}$$

$$\text{HBase}^{\oplus} \rightleftharpoons \text{Base} \quad \text{H}^{\oplus}$$

look for the conjugate acid in the pK_a table

Check the dissociation equation carefully to determine which compound is the base and what its conjugate acid is. For example, when ammonia, NH_3, is used as a *base*, its conjugate acid is NH_4^+, not NH_2^-.

when NH_3 acts as a base, use the pK_a of the corresponding conjugate acid, $NH_4^{\oplus}$

$$:\text{B}^{\ominus} + \text{NH}_4^{\oplus} \rightleftharpoons :\text{NH}_3 + \text{HB}$$

$$\text{p}K_a(\text{NH}_4^{\oplus}) = 9.42$$

Sodium and potassium amides, $NaNH_2$ and KNH_2, are strong organic bases. The strength of the base NH_2^- is determined by the pK_a of its conjugate acid NH_3, which behaves as a very weak acid to form NH_2^-.

the base strength of $NH_2^{\ominus}$ is described with the pK_a of the corresponding conjugate acid, NH_3

$$:\text{B}^{\ominus} + :\text{NH}_3 \rightleftharpoons :\ddot{\text{N}}\text{H}_2^{\ominus} + \text{HB}$$

$$\text{p}K_a(\text{NH}_3) = 41$$

STUDENT TIP

When evaluating the strength of a base, write the base on the right-hand side of a pK_a expression to help identify the corresponding conjugate acid.

The behaviour of NH_3 in these two acid–base reactions is quite different, as are the pK_a values of the acids in these reactions. It is critically important to identify the role of compounds and their conjugates in such reactions.

6.6.2 Finding pK_a values in a table

Tables provide the pK_a of hydrogens when attached to different types of functional groups. Once the appropriate acid has been identified by writing the pK_a equation, examine the functional group carefully. In particular, any π bonds, electron pairs, or open octets that can participate in resonance will strongly influence the acidity of the group. The functionalities in pK_a tables are organized this way. Look for the acid in the table that *most closely* resembles the group you are concerned with (it is rare to find the exact match).

CHEMISTRY: EVERYTHING AND EVERYWHERE

Drugs Must Be Lipophilic and Hydrophilic

Pharmaceutical companies face significant challenges during drug development. Perhaps the most difficult of these is getting the drug into the human body. The fluids in the body such as blood and digestive juice are mostly water, and so drugs need to be water-soluble (hydrophilic). However, biological membranes are very hydrophobic, and to enter cells drugs must be able to pass through this hydrophobic environment. This means that drugs must somehow dissolve in chemically opposite environments (hydrophobic and hydrophilic).

A good way to achieve this is through acid–base chemistry. Most drugs (>95%) are either weak acids or weak bases. At physiological pH (7.4), they are therefore charged and soluble in water. As long as the pK_a of the drug is not too far away from 7, the normal acid–base equilibrium ensures there are small amounts of neutral-form drug in solution. In neutral form, most drugs are non-polar enough to cross biological membranes. So the acid–base equilibrium provides a way for drugs to effectively be both charged and neutral at the same time.

$$\text{Drug–C(=O)OH} + H_2O \xrightleftharpoons{\text{pH 7.4}} \text{Drug–C(=O)O}^{\ominus} + H_3O^{\oplus}$$

soluble in biological membranes

soluble in water

6.7 Predicting Acid–Base Equilibria

6.7.1 Quantitative prediction of equilibria

The behaviour of organic acids and bases when mixed often determines the order in which bonds are broken and formed. The key to predicting the results of acid–base reactions is determining the equilibrium point, which indicates whether the equilibrium will favour the reactants or the products. Equilibrium reactions tend to shift toward the more stable materials. Therefore acid–base equilibria favour weaker acids and bases; that is, at equilibrium, the concentration of the weaker acid and base is greater than that of the stronger ones.

pK_a values can be used to predict whether an acid–base equilibrium will favour the reactants or the products. Identify the acids and compare their pK_a values. Since higher pK_a values correspond to weaker acids, the equilibrium will shift toward the acid with the higher pK_a.

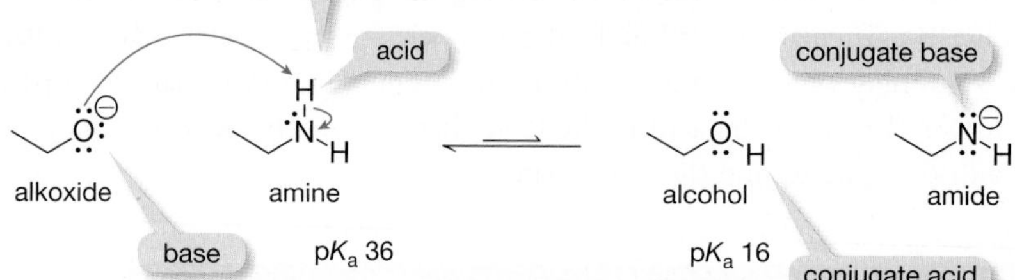

ORGANIC CHEMWARE
6.5 Acid–base chemistry: Proton transfer (B^- + H–A)

ORGANIC CHEMWARE
6.6 Acid–base chemistry: Proton transfer (B + H–A)

ORGANIC CHEMWARE
6.7 Acid–base chemistry: Proton transfer (B + $H–A^+$)

ORGANIC CHEMWARE
6.8 Henderson–Hasselbalch equation

To predict what will happen when an amine is mixed with an alkoxide, write the full dissociation equation for the possible reaction. Identify the two acids (acidic functional groups), and estimate their pK_a values by finding the structures that most closely resemble them in the pK_a table. In this example, the acids are an alcohol and an amine. The approximate pK_a for the amine is 36, while that for the alcohol is 16. The equilibrium will favour (shift toward) the reactants on the left since the acid with the higher pK_a is there. In other words, the concentrations of the compounds on the left side of the equation will be significantly greater than those of the compounds on the right side.

The difference in pK_a values indicates how *far* the equilibrium shifts. In this example, the difference is about 20 units, which means the amine is 20 *orders of magnitude* weaker of an acid than the alcohol is. This difference is so great that the equilibrium is shifted *completely* to the left side of the equation, indicating that the amine and the alkoxide will not react with each other in the manner written.

CHECKPOINT 6.6

You should now be able to predict the position of an equilibrium using pK_a values.

SOLVED PROBLEM

Use pK_a values to predict whether the following equilibrium will favour the reactants or products.

$\overset{\oplus}{O}H_2$ + NH_2 ⇌ OH + $\overset{\oplus}{N}H_3$

STEP 1: Label the acid, base, conjugate base, and conjugate acid.

$\overset{\oplus}{O}H_2$ + NH_2 ⇌ OH + $\overset{\oplus}{N}H_3$

acid base conjugate base conjugate acid

STEP 2: Use the pK_a table to estimate the pK_a values for the acid and the conjugate acid by finding the functional groups most similar to them.

$\overset{\oplus}{O}H_2$ + NH_2 ⇌ OH + $\overset{\oplus}{N}H_3$

acid base conjugate base conjugate acid

p$K_a \approx -3$ p$K_a \approx 11$

STEP 3: Estimate the equilibrium point based on the relative pK_a values.

The higher the pK_a, the weaker the acid. Acid–base equilibria lie to the side of the weaker species (in other words, the stronger species reacts), so this acid–base equilibrium favours the products.

PRACTICE PROBLEM

6.11 Use pK_a values to predict whether the following equilibria will favour the reactants or products.

a)

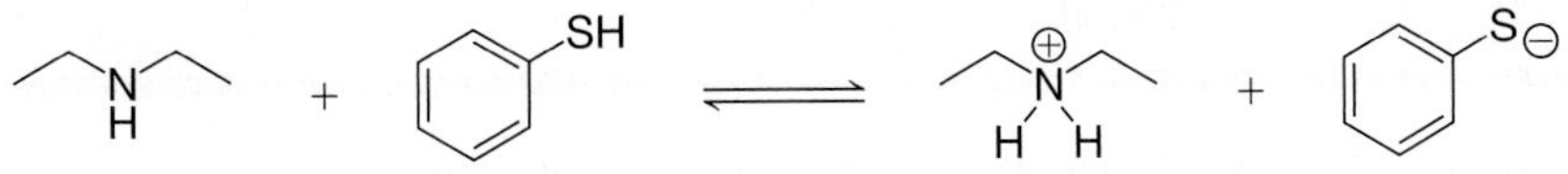

b)

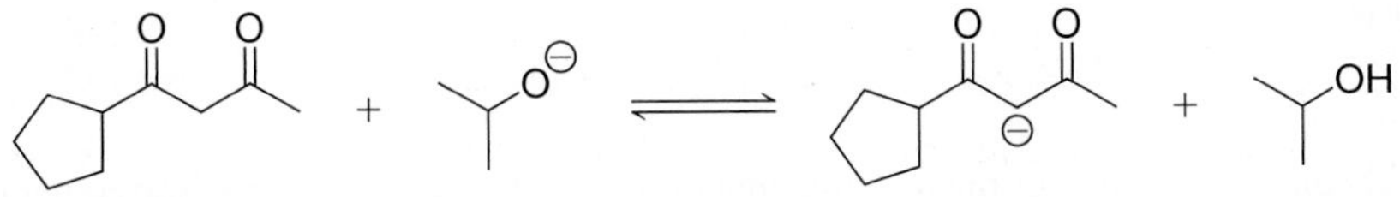

c)

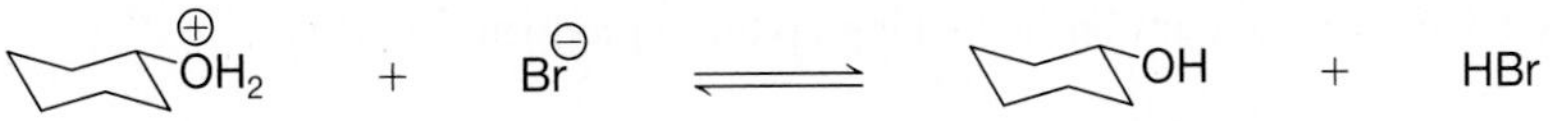

d)

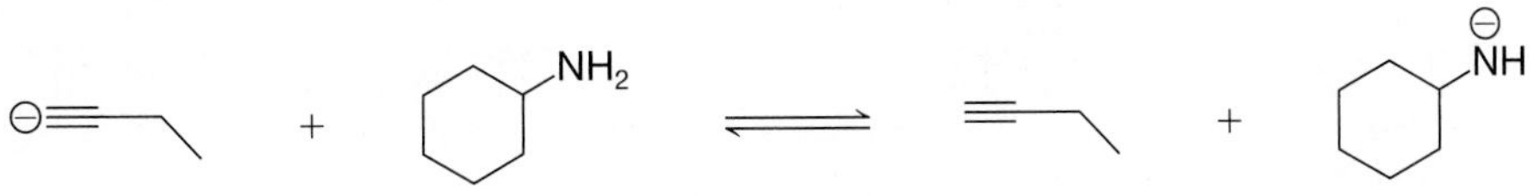

INTEGRATE THE SKILL

6.12 a) Use pK_a values to rank the acidity of the indicated protons.

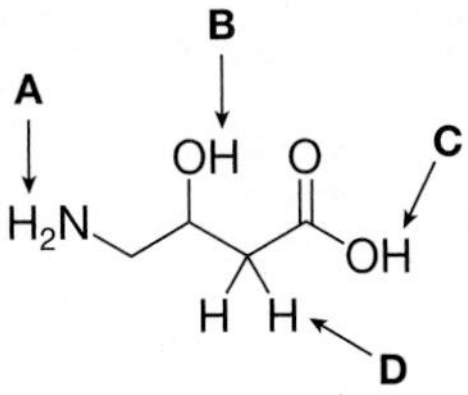

A B C D

b) Use pK_a values to rank the basicity of the indicated sites.

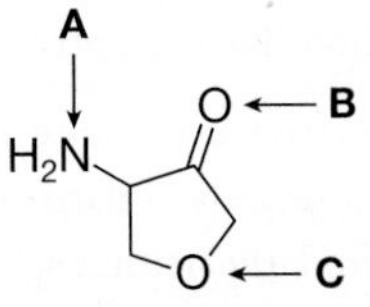

A B C

6.7.2 Qualitative prediction of equilibria

Since equilibria favour weaker acids and bases, information about the *relative* strength of acids and bases can be used to make qualitative predictions about acid–base equilibria. Comparing the stabilities of the *charged molecules* is the best way to obtain an accurate prediction.

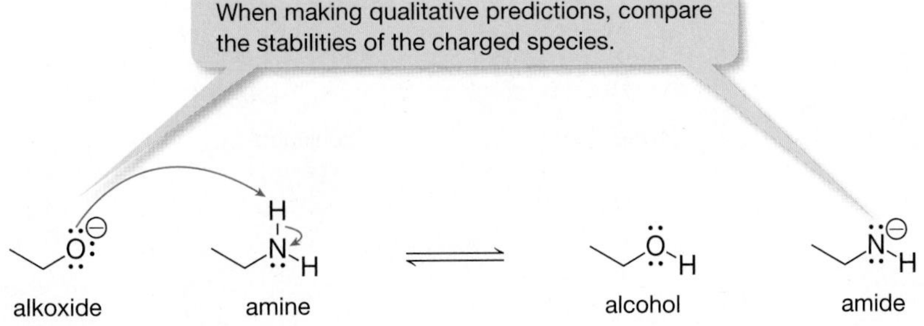
When making qualitative predictions, compare the stabilities of the charged species.
alkoxide
amine
alcohol
amide

The charged components in this example are the alkoxide and amide, which have negative charges on oxygen and nitrogen atoms, respectively. Oxygen is to the right on the periodic table and is more electronegative than nitrogen, so the negative charge is more stable on the oxygen than on the nitrogen, making the alkoxide the weaker base. This shifts the equilibrium toward the reactants.

CHECKPOINT 6.7

You should now be able to predict the position of an acid–base equilibrium both with and without the use of pK_a data.

SOLVED PROBLEM

Determine whether the products or the reactants are favoured by the following acid–base equilibrium, and justify your answer. Confirm your answer by looking up the appropriate pK_a data.

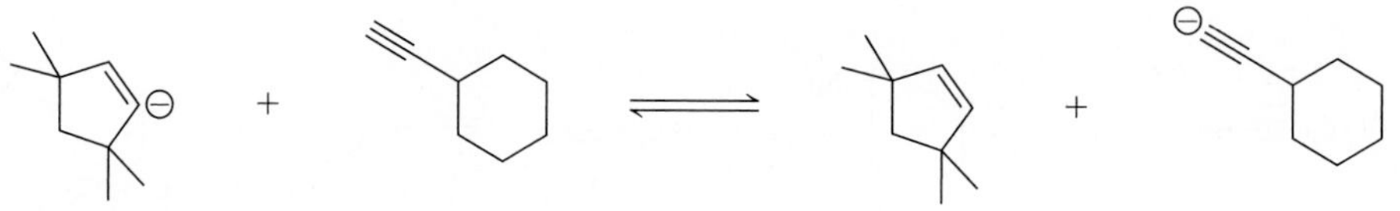

STEP 1: Label the acid, base, conjugate acid, and conjugate base.

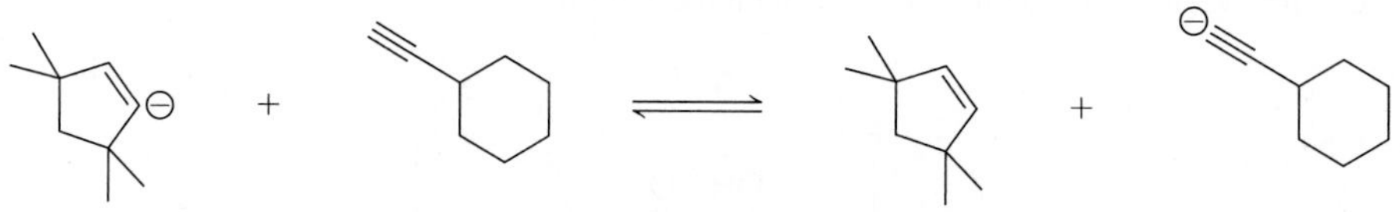

base acid conjugate acid conjugate base

STEP 2: Determine whether to compare the stability of the acid and conjugate acid or that of the base and conjugate base.

It is easier to compare the relative stability of charged species, so for this reaction, you would compare the base and the conjugate base since they both have negative charges.

STEP 3: Evaluate the relative stabilities of the base and conjugate base.

Both the base and conjugate base have a negative charge on a carbon atom, so atomic size and electronegativity will not be deciding factors. Similarly, neither species has any resonance stabilization, nor any nearby electron-donating or -withdrawing groups. The only remaining factor is hybridization. The base has its negative charge in an sp^2-hybridized orbital while the conjugate base has its charge in an sp-hybridized orbital. With greater s character, the sp orbital is better at stabilizing the negative charge, making the conjugate base more stable than the base.

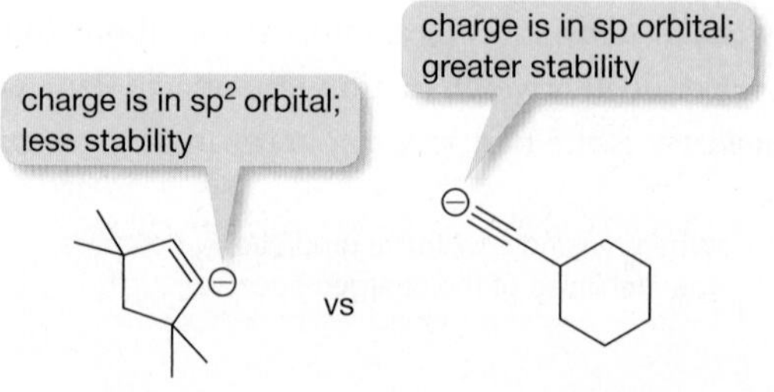

charge is in sp^2 orbital; less stability

charge is in sp orbital; greater stability

base vs conjugate base

STEP 4: Estimate the equilibrium point based on the relative stabilities of the base and conjugate base.

Since the conjugate base is more stable, it is also the weaker base. Acid–base equilibria favour formation of the weaker species (in other words, the stronger species reacts), so the products of this reaction will be favoured.

STEP 5: Use the pK_a table to estimate the pK_a values for the acid and the conjugate acid.

Although the base and conjugate base were used for the qualitative comparison, pK_a values are specific to acids; therefore, the conjugate acids must be used for a quantitative comparison. The pK_a of the acid is 25, whereas the pK_a of the conjugate acid is 44.

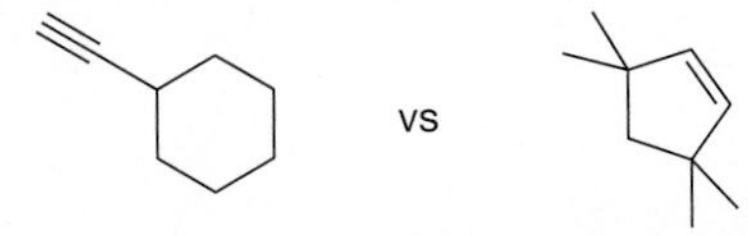

vs

acid

p$K_a \approx 25$

conjugate acid

p$K_a \approx 44$

STEP 6: Estimate the equilibrium point based on the relative pK_a values, and compare this answer with the one obtained by qualitative comparison.

The higher the pK_a, the weaker the acid. Acid–base equilibria lie to the side of the weaker species, so this acid–base equilibrium favours the products. This answer is in agreement with that found by qualitative comparison (as it should be!).

PRACTICE PROBLEM

6.13 Determine whether the products or the reactants will be favoured by each of the following acid–base equilibria. Confirm your answers by looking up the appropriate pK_a data.

a) OH + $O^{\ominus}$ ⇌ $O^{\ominus}$ + OH

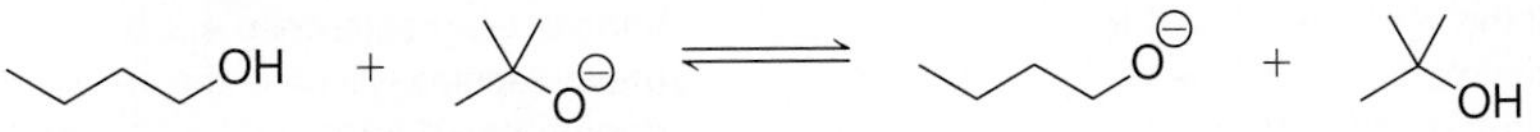

b) NH_2 / OMe + $^{\oplus}NH_3$ / NO_2 ⇌ $^{\oplus}NH_3$ / OMe + NH_2 / NO_2

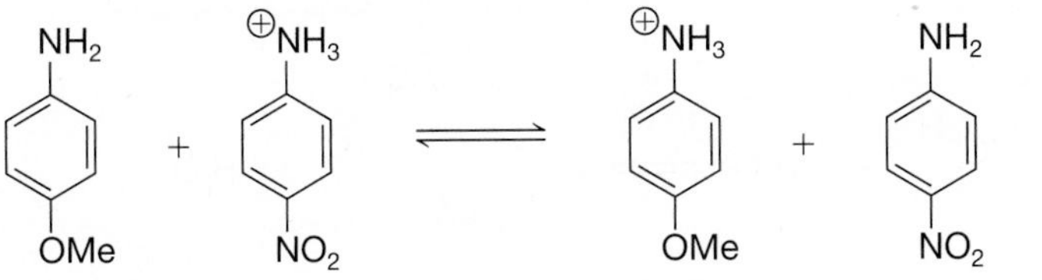

c) ⊖ + ⇌ + ⊖

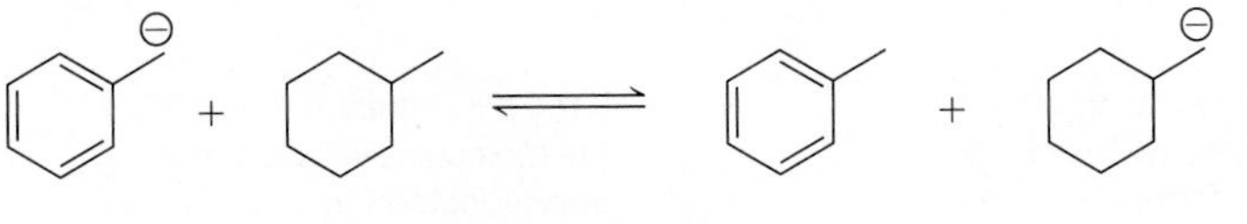

d) $\overset{\oplus}{N}H_3$ + O / O ⇌ NH_2 + $^{\oplus}OH$ / O

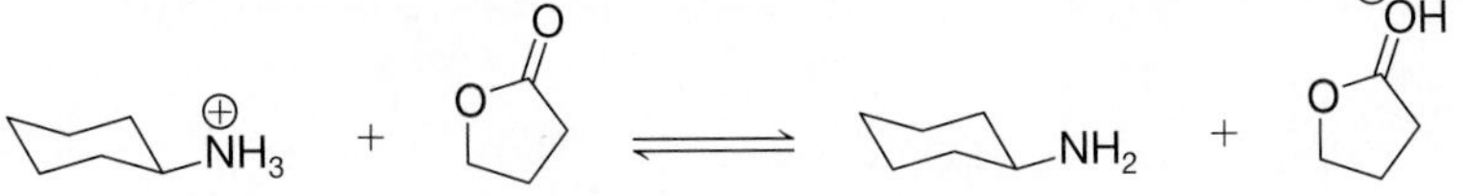

INTEGRATE THE SKILL

6.14 The following is a reaction coordinate diagram for the acid–base reaction between dimethylmalonate and diphenylamide (conjugate base of diphenylamine). Use the reaction coordinate diagram to determine which acid has the lower pK_a: dimethylmalonate or diphenylamine.

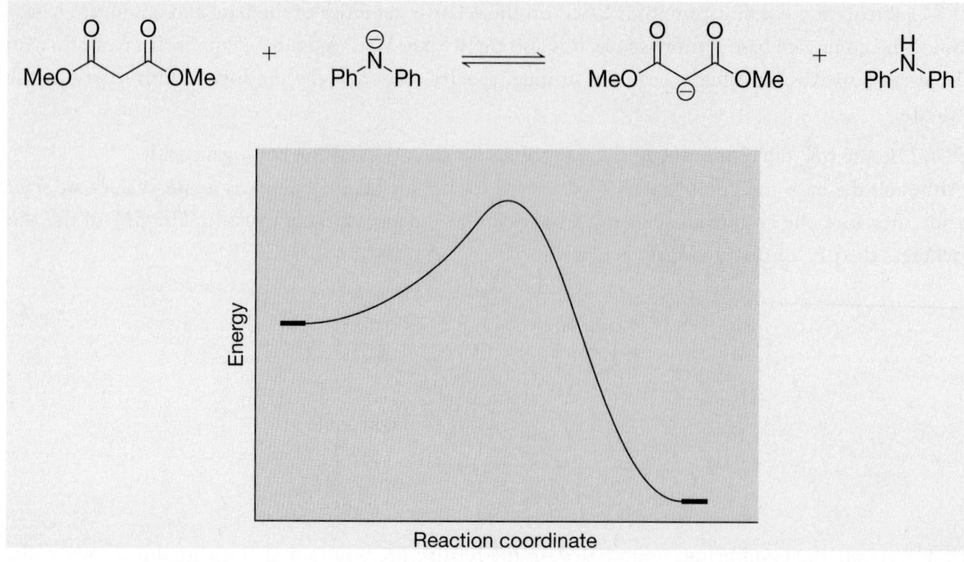

6.7.3 Protonation states at various pHs

The pK_a of an acid can be used to predict the protonation state at various pHs. Acids are primarily in their protonated form when the pH is less than the pK_a, and they deprotonate when the pH is greater than the pK_a.

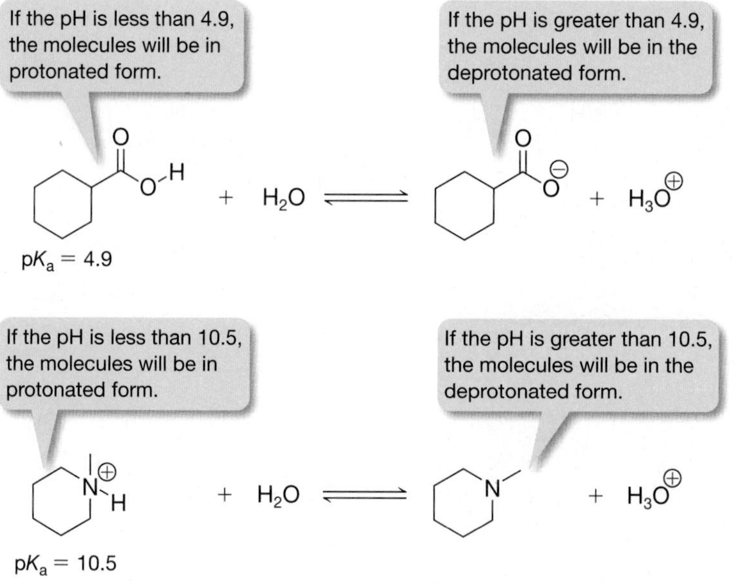

If the pH is less than 4.9, the molecules will be in protonated form.
If the pH is greater than 4.9, the molecules will be in the deprotonated form.
+ H2O
+ H3O
pKa = 4.9
If the pH is less than 10.5, the molecules will be in protonated form.
If the pH is greater than 10.5, the molecules will be in the deprotonated form.
+ H2O
+ H3O
pKa = 10.5

STUDENT TIP
The terms *protonated form* and *deprotonated form* refer to their position on an acid dissociation equation (reactants or products), not to whether a molecule is charged or not.

If the pH is *very* close to the pK_a value, a mixture of protonated and unprotonated states will be observed.

6.8 Lewis Acids in Organic Reactions

The **Lewis acid** theory classifies acids as electron pair acceptors and bases as electron pair donors. Lewis acids function just like protons (H^+) do. Metals are commonly used as Lewis acids, and these react with the strongest bases present, which can be identified using the guidelines provided in this chapter.

A **Lewis acid** is an electron pair acceptor. A Lewis base is an electron pair donor.

Brønsted acid

Lewis acid

boron has an unfilled octet

The metal in a Lewis acid must have empty orbitals to be able to accommodate an electron pair. Some Lewis acids expel leaving groups as part of the reaction pathway (Chapter 8). Some common Lewis acids include BF_3, BBr_3, $AlCl_3$, $FeCl_3$, $ZnCl_2$, $TiCl_4$ and $Hg(OAc)_2$.

6.9 Patterns in Acids and Bases

Trends in the periodic table can be used to quickly estimate the relative strengths of acids and bases. This information predicts whether a given acid–base reaction will proceed, and, if so, which compounds the equilibrium will favour. When comparing acids and bases, draw complete balanced equations, and compare the *charged components*. Energy differences are generally much greater for charged molecules than for neutral ones. This fact simplifies the process for determining the relative strengths of acids and bases.

If the charges are carried on heteroatoms within the same row of the periodic table, use electronegativity to determine relative stability. More electronegative elements stabilize negative charges, while positive charges will be *less* stable if found on these atoms.

If the charges are carried on heteroatoms within the same column of the periodic table, use size to determine relative stability. Large atoms stabilize both positive and negative charges by distributing them over a larger volume. Therefore, atoms lower in the periodic table will give more stable charged species than atoms in rows above.

The presence of electronegative atoms close to a charge can influence the stability of that charge through induction. Electronegative atoms withdraw electrons. This effect tends to stabilize a nearby negative charge by distributing it over a larger volume. Similarly, nearby electronegative atoms will destabilize positive charges. The magnitude of induction effects depends on the number of electronegative atoms present, the electronegativity of these atoms, and their proximity to the charged site.

Alkyl groups are electron donating (Chapter 8), and will stabilize adjacent positive charges. These groups also destabilize adjacent negative charges, but the effect is much weaker.

Of all the stabilizing factors, charge or electron delocalization (described by resonance) is the most profound. This effect strongly influences many processes by delocalizing electrons over several atoms. This effect is often possible when π bonds or lone pairs are adjacent to a charge.

pK_a values are used in organic chemistry to describe the strength of both acids and bases. This latter usage can be confusing, as it requires some knowledge of the role of the component being described.

To identify the pK_a of a base, use an acid dissociation equation and write the base on the right-hand side (disassociated side) of the equation. This can help to identify the proper conjugate acid to use to find its pK_a.

Bringing It Together

Omeprazole was the first of a family of drugs called *proton pump inhibitors*. These drugs are commonly prescribed to control excess stomach acid because they are more effective than most other anti-acid medications (also known as antacids) and have very few side effects. Proton pump inhibitors are largely unreactive except in the presence of a strong acid. Thus, these anti-acid drugs react only where they are needed, greatly reducing their side effects.

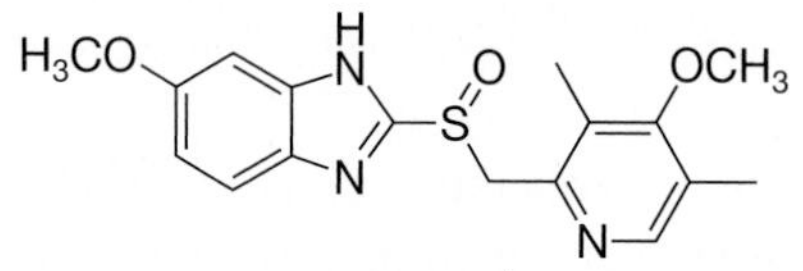

omeprazole

When omeprazole molecules enter canaliculi (the pores of the stomach where digestive acid is produced), the strongly acidic environment initiates a cascade of chemical reactions that change the structure and reactivity of the drug. The new molecule reacts to block the catalyzing action of the enzyme that the stomach uses to make HCl, thus reducing the amount of stomach acid. The following figure shows how HCl activates omepezole, letting it stop acid production while remaining inactive elsewhere in the body.

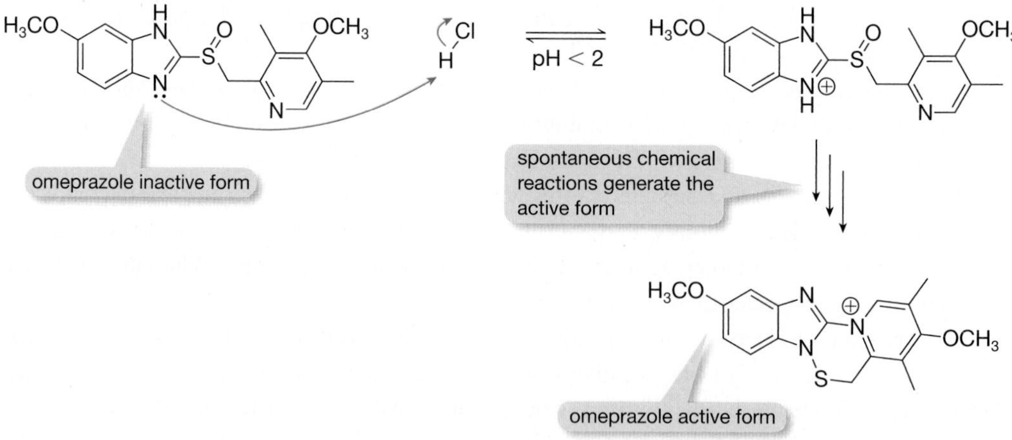

You Can Now

- Draw a curved arrow reaction mechanism for an acid–base reaction.
- Draw and interpret reaction coordinate diagrams for both weak and strong acids.
- Use reaction coordinate diagrams to compare relative acid strengths.
- Rank the relative acidities of various acids and justify the ranking in terms of structural features.
- Rank the relative basicity of various bases and justify the ranking in terms of structural features.
- Identify the most acidic proton on a molecule.
- Identify the most basic site on a molecule.
- Use pK_a data to compare the relative strengths of both acids and their conjugate bases.
- Use pK_a data to determine the direction of an acid–base equilibrium.
- Make qualitative predictions about the direction of an acid–base equilibrium by comparing the structures of either the acid and conjugate acid, or the base and conjugate base.

Problems

6.15 Predict the products of the following acid–base reactions, and draw a curved arrow mechanism showing their formation.

a) HBr + (diphenyl ether) ⇌

b) (tert-butoxide) + (N,N-dimethylurea derivative, NH_2) ⇌

c) (imidazolium, N–H, N⊕–H) + (cyclohexenol with OH) ⇌

d) HO(butanol) + HO–S(=O)(=O)–OH ⇌

e) 1) HA + B^- ⇌ A^- + HB
pK_a = 7 pK_a = 3

2) HA + B^- ⇌ A^- + HB
pK_a = −5 pK_a = 10

3) HA + B^- ⇌ A^- + HB
pK_a = 8 pK_a = −1

6.16 Aniline and cyclohexylamine are both very weak acids; however, aniline is slightly more acidic than cyclohexylamine. Using this information, draw a reaction coordinate diagram that shows the dissociation of both conjugate acids with a generic base, B^-, and their relative acidities.

aniline cyclohexylamine

6.17 Below are three acid–base equations and a reaction coordinate diagram. Use the provided pK_a data to determine which equation(s) could be represented by the reaction coordinate diagram.

1) HA + B^- ⇌ A^- + HB
pK_a = 7 pK_a = 3

2) HA + B^- ⇌ A^- + HB
pK_a = −5 pK_a = 10

3) HA + B^- ⇌ A^- + HB
pK_a = 8 pK_a = −1

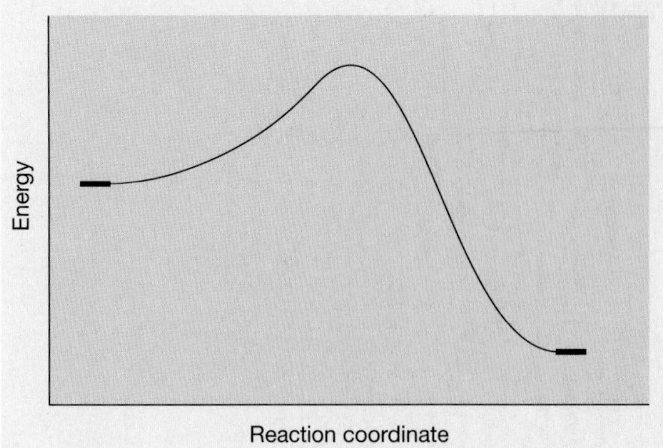

6.18 For each of the following pairs of compounds, predict which one will be the strongest acid. Justify your choice in terms of the compound's structure.

a) 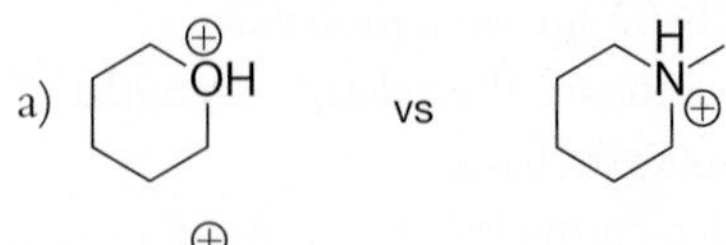

b)

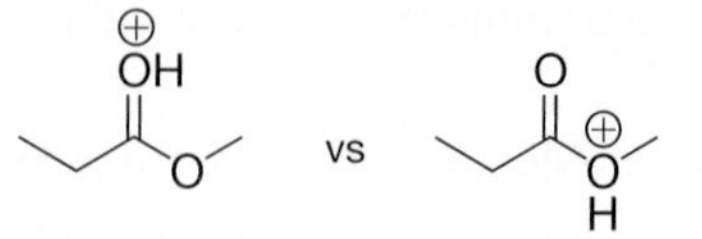

c) CBr_3CO_2H vs CH_3CO_2H

d) 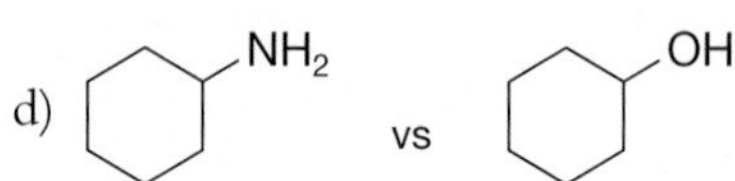

e) 

f) 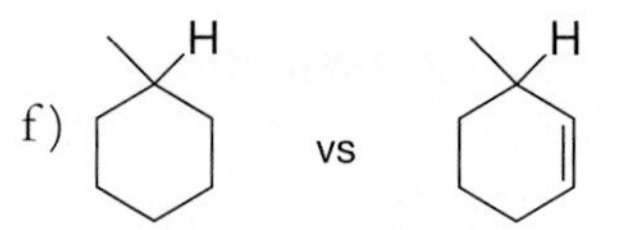

g) 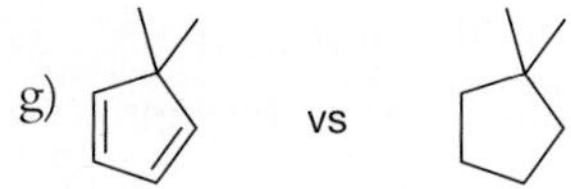

h) 

6.19 For each of the following pairs of compounds, predict which one will be the strongest base. Justify your choice in terms of the compound's structure.

a) 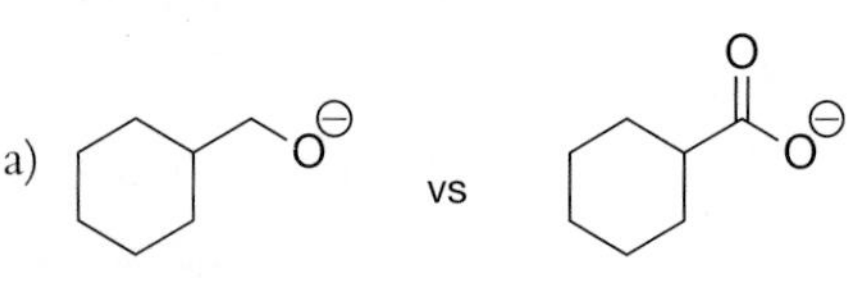

b) 

c)

d)

e)

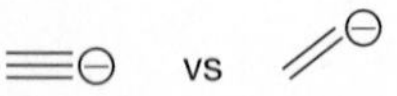

f)

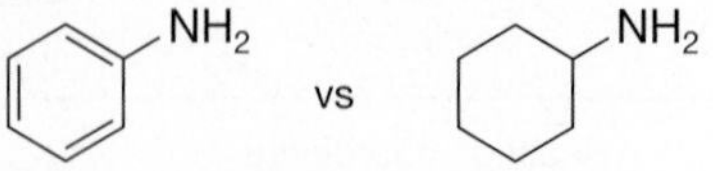

g)

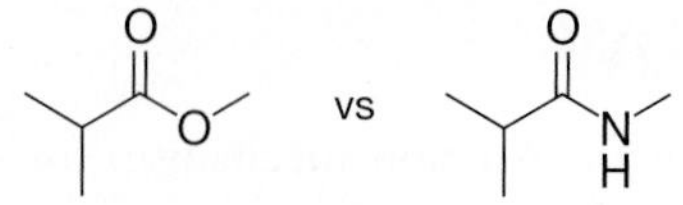

h)

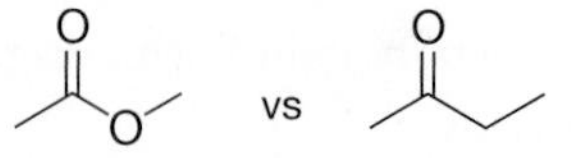

6.20 For each of the following sets of compounds, rank them in order of increasing acidity.

a)

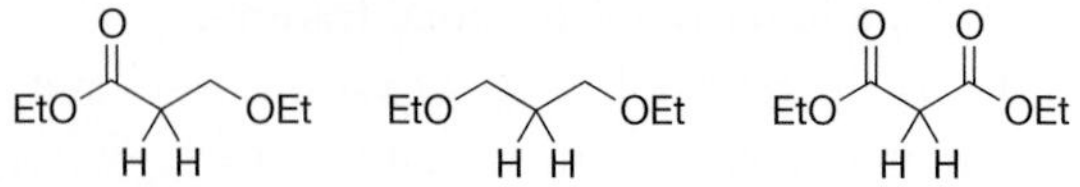

b)

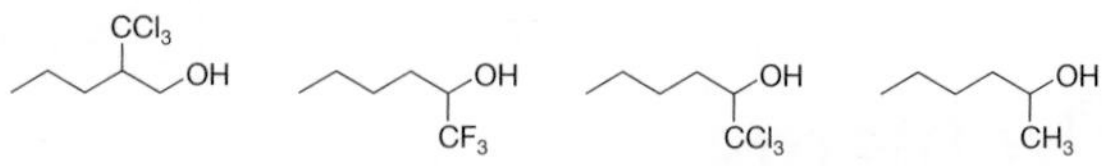

c)

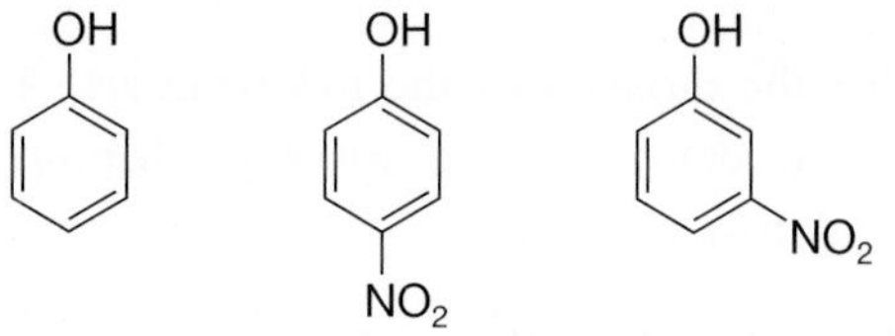

d)

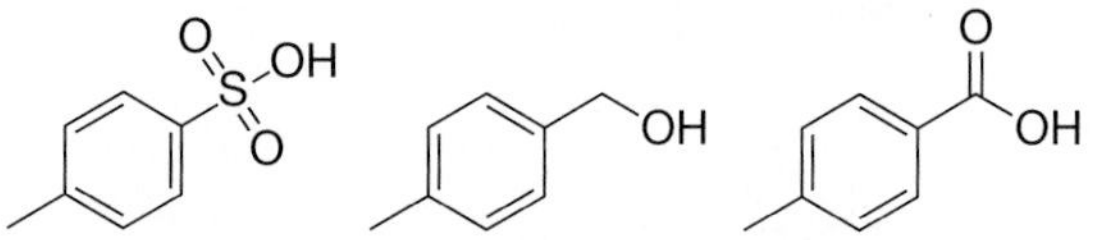

e)

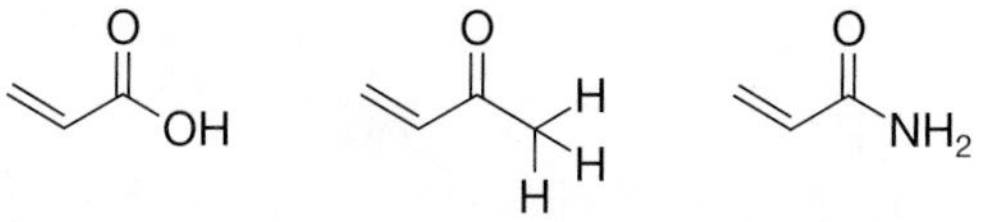

f)

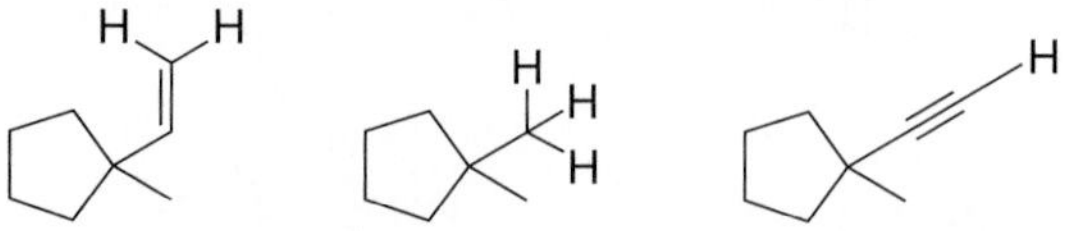

6.21 For each of the following sets of compounds, rank them in order of increasing basicity.

a)

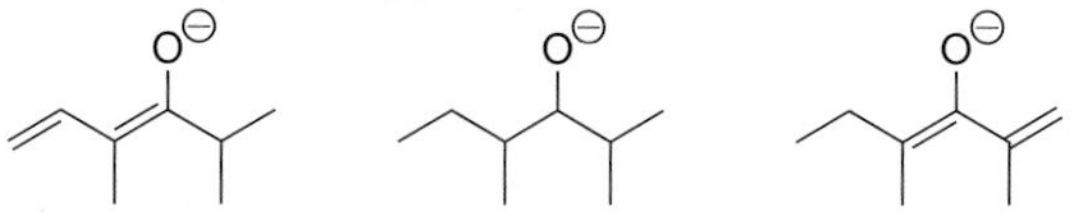

b)

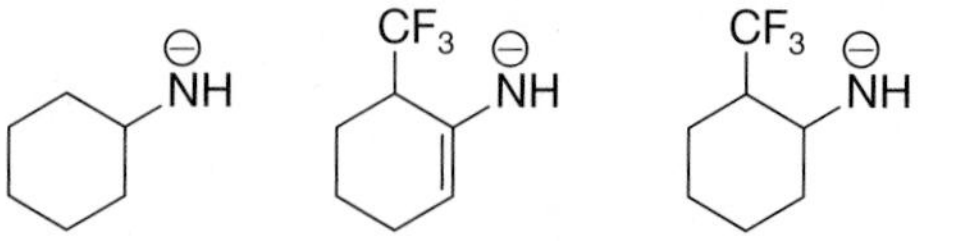

c)

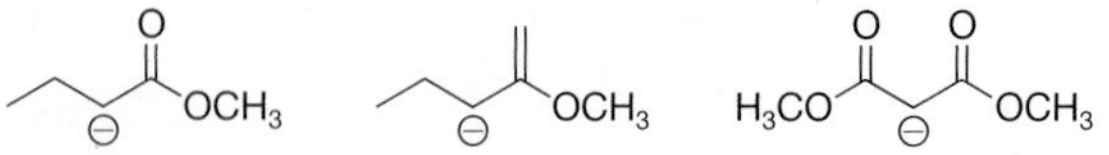

d)

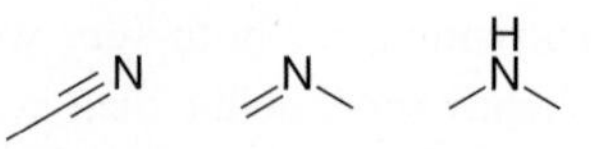

e) 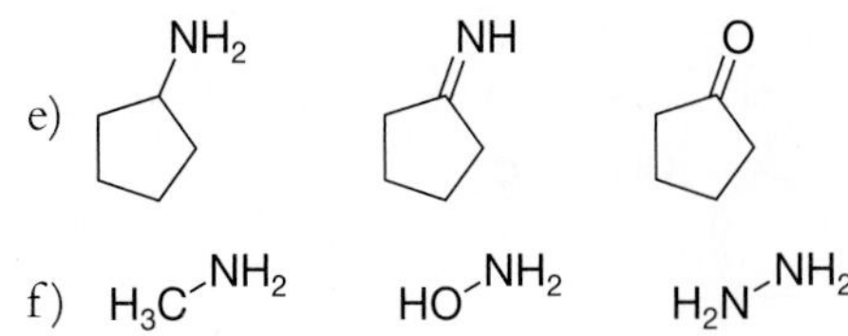

f) $H_3C{-}NH_2$ $\quad$ $HO{-}NH_2$ $\quad$ $H_2N{-}NH_2$

6.22 Predict the products of the following acid–base reactions, and draw a curved arrow mechanism showing their formation. Note that there are multiple acidic and/or basic sites in each reaction.

a) N, NH_2 + O, OH ⇌

b) O + $^{\ominus}NH_2$ ⇌

c) O, NH_2 + $\overset{\ominus}{N}$ ⇌

d) O, S, O, H, O + :O:, NH ⇌

e) $H_2\overset{\oplus}{O}$ + H_2N, OH ⇌

6.23 Predict the direction of the equilibrium using the appropriate pK_a values.

a) O, O + $H_3\overset{\oplus}{N}$ ⇌ O, $\overset{\oplus}{O}H$ + H_2N

b) $O^{\ominus}$, CH_3 + OH, NO_2 ⇌ OH, CH_3 + $O^{\ominus}$, NO_2

c) H, N + $\overset{\ominus}{N}$, N ⇌ $\overset{\ominus}{N}$ + H, N, N

d) O O, $\ominus$ + O O, O ⇌ O O + O O, O, $\ominus$

e) + $\ominus$ ⇌ $\ominus$ +

f) $\overset{\oplus}{O}H_2$ + O ⇌ OH + $\overset{\oplus}{O}H$

6.24 Draw the products of the following acid–base reactions and predict whether the reactants or products are favoured by the equilibrium based on the appropriate pK_a data.

a) HO, OH + N

b) OH + $\overset{\ominus}{N}$

c) O, H + $O^{\ominus}$

d) + $NaNH_2$

e) NH_2, O + NH, N

f) SH + $^{\ominus}O$, O

6.25 Draw the products of the following acid–base reactions, and predict whether the reactants or products are favoured by the equilibrium based on the structural features of the relevant species.

a) $O^{\ominus}$ + OH

b) O, OH + $^{\ominus}OH$

c) $\overset{\oplus}{O}$–H, OH + O, O–H

d) O, OH + OH

e) $\overset{\ominus}{N}$ + H, P

f) CF_3OH + $CF_3CH_2O^{\ominus}$

g) $CH_3CO_2^{\ominus}$ + CF_3CO_2H

h) CCl_3OH + $CF_3O^{\ominus}$

6.26 Predict the most likely site of protonation (addition of H^+) in the following compounds. Justify your selection in terms of the compound's structure.

a) 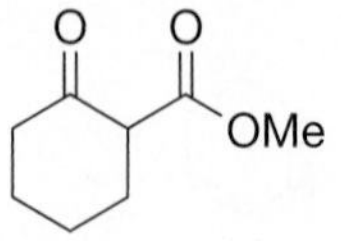 OMe

b) 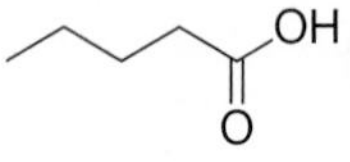 OH

c)

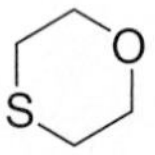

d)

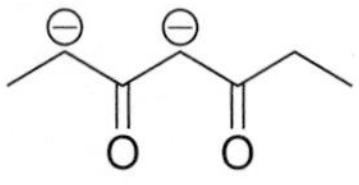

e)

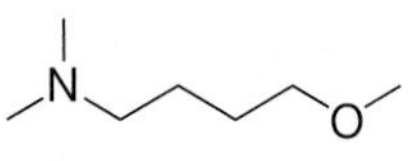

f)

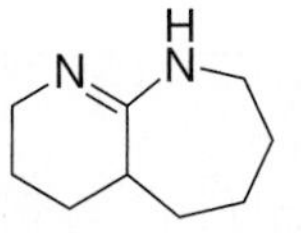

6.27 Rank the relative acidity of the indicated protons in order of increasing acidity.

a)

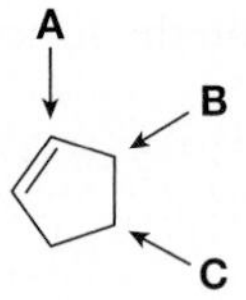

b)

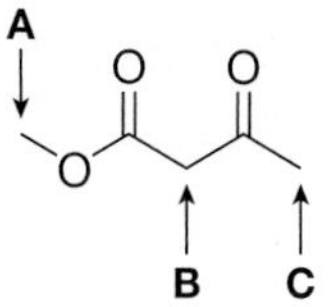

c) 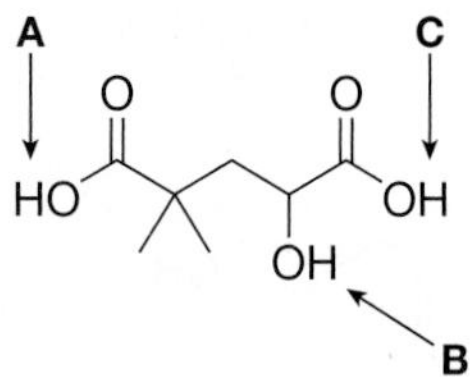 HO OH OH

d)

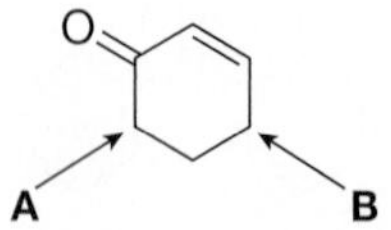

e) 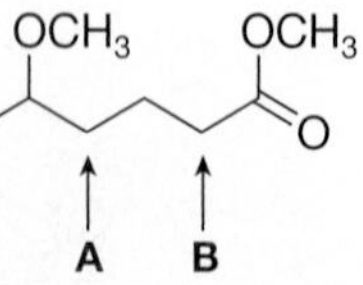 OCH_3 OCH_3

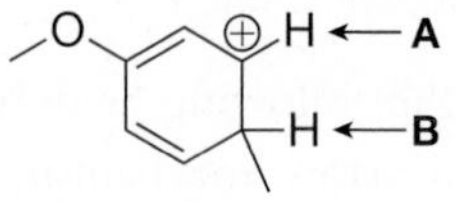

f) 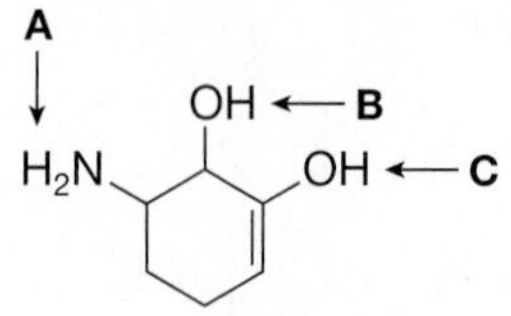 H_2N OH ← B OH ← C

g) H ← A H ← B

h) NH ← C

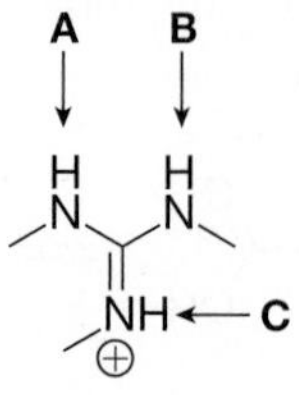

6.28 Use pK_a data to determine which of the following acids would react with OH^- to yield a reaction that would favour the products.

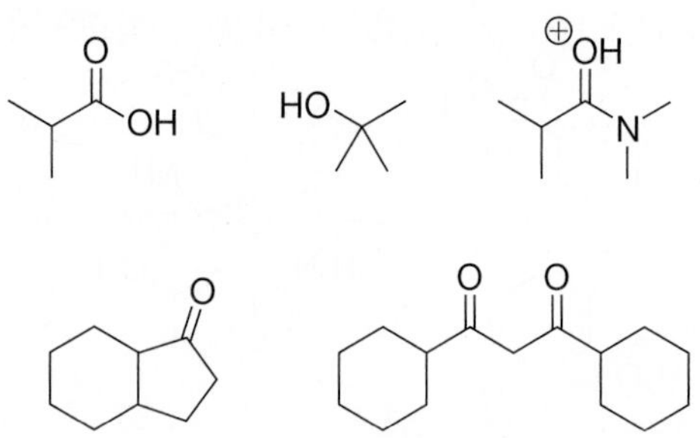

6.29 Use pK_a data to determine which of the following bases would react with acetic acid (CH_3CO_2H) to yield a reaction that would favour the products.

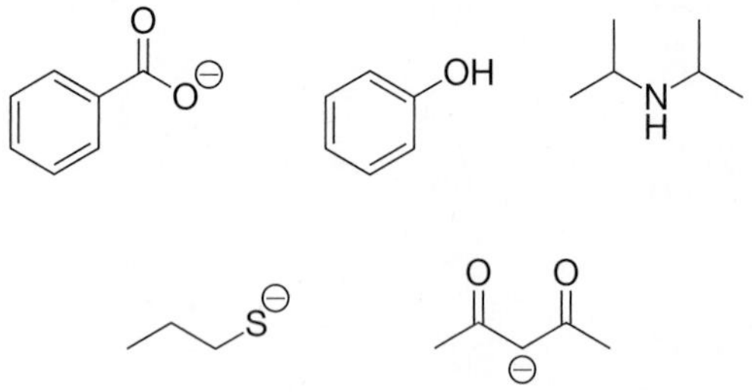

6.30 Compare the pairs of equilibria and decide which one will have the greater equilibrium constant (i.e., which one will favour the products more).

a)

b)

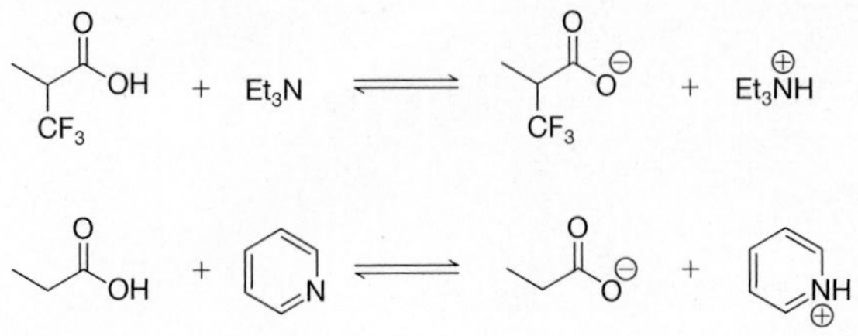

c)

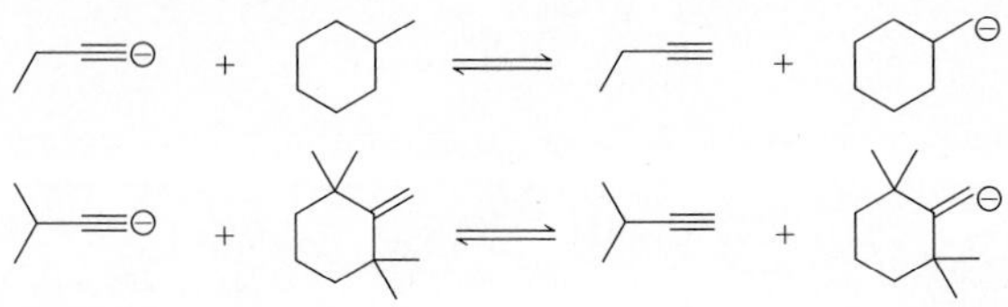

d)

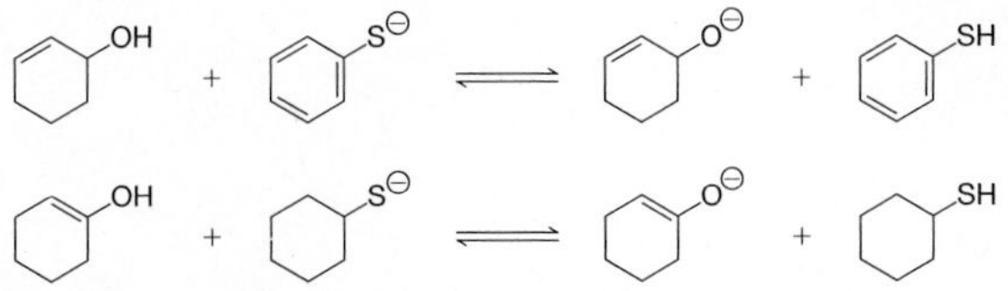

6.31 Atropine is a naturally occurring poison found in several species of plants belonging to the *Solanaceae* family. When used in carefully controlled doses, it is an effective muscle relaxant with applications in medicine.

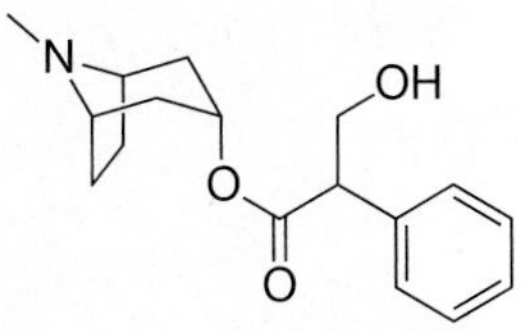

a) Identify and rank the basic sites.
b) Identify and rank the acidic sites (consider any proton with a pK_a less than 30).

MCAT STYLE PROBLEMS

6.32 When the anti-depressant citalopram is taken orally, it reaches the acidic environment of the stomach where an acid–base reaction takes place. Which of the following equations best represents the reaction?

a)

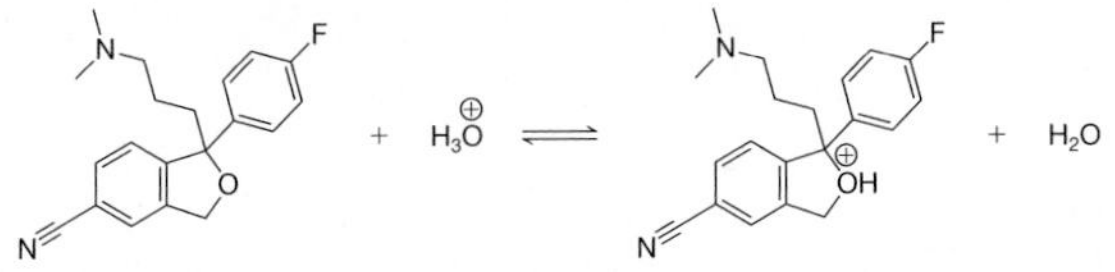

b)

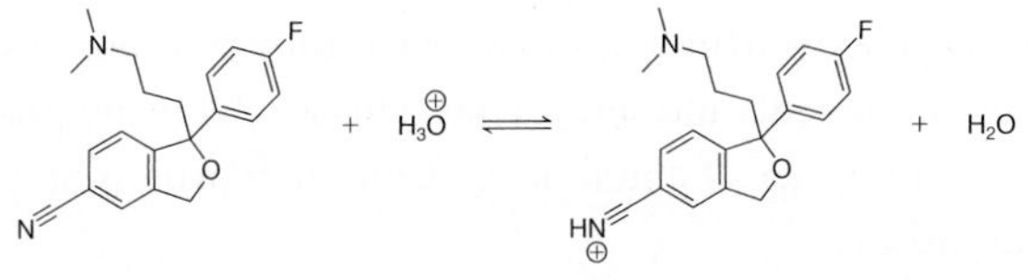

c)

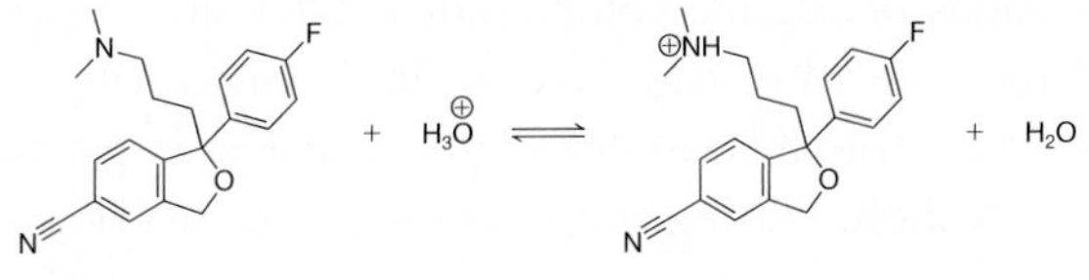

d)

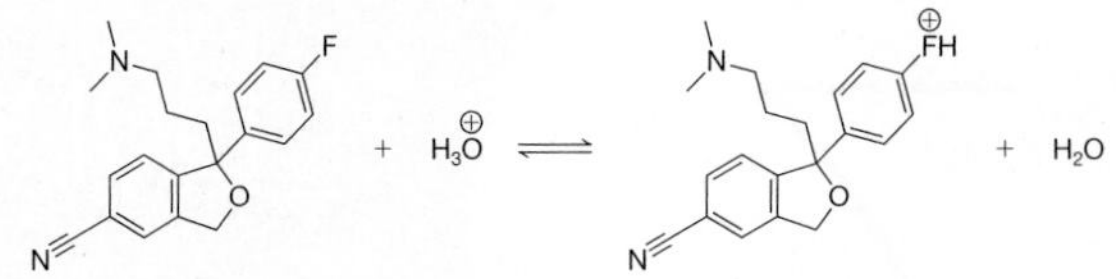

6.33 The active ingredient in an antacid tablet is calcium carbonate. Which of the following statements is true about the reaction that takes place between an antacid and stomach acid?

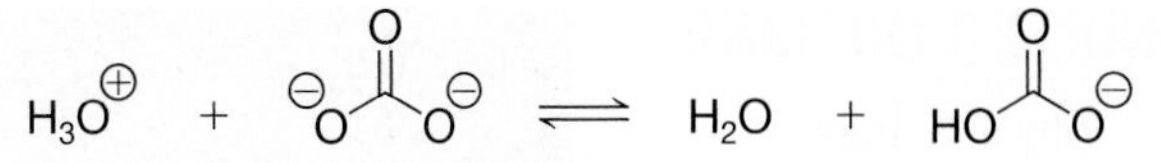

$$H_3O^{\oplus} + {}^{\ominus}O{-}C(=O){-}O^{\ominus} \rightleftharpoons H_2O + HO{-}C(=O){-}O^{\ominus}$$

a) The equilibrium favours the products.
b) Water is the conjugate acid.
c) Bicarbonate is the conjugate base.
d) Hydronium ion is a weak acid.

6.34 Amino acids have both acidic and basic sites. Therefore, they can undergo an acid–base reaction with themselves to form a zwitterion—a neutral molecule with both a negative and a positive charge. Which is the most likely zwitterion to form from the amino acid histidine?

a)

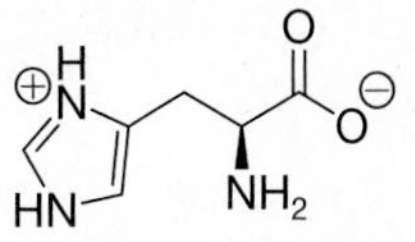

b)

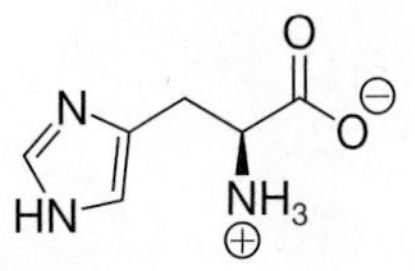

c)

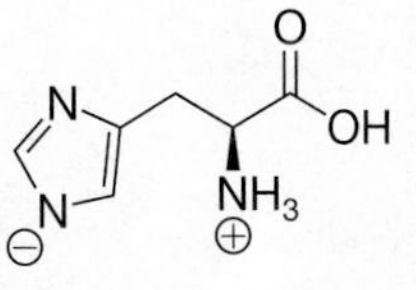

d)

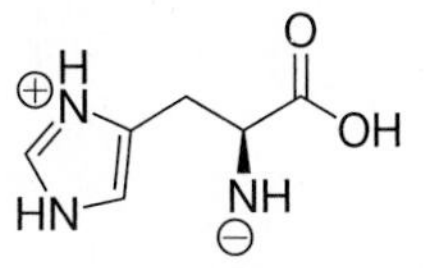

CHALLENGE PROBLEM

6.35 Compounds A and B below have the same pK_a value. In contrast, compound C has a much lower pK_a than compound D. Explain these observations.

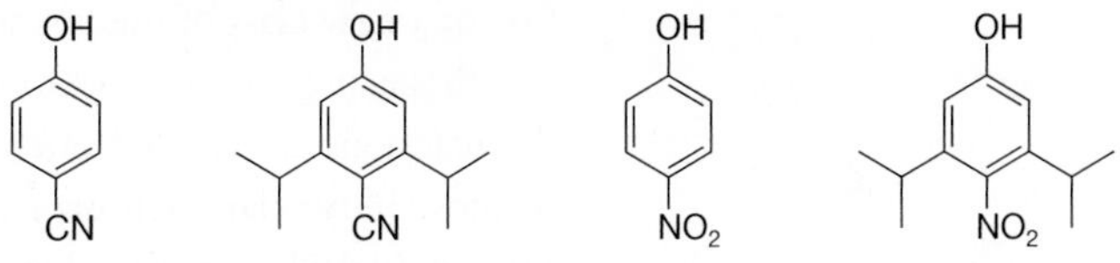

A B C D

7 π Bonds as Electrophiles

REACTIONS OF CARBONYLS AND RELATED FUNCTIONAL GROUPS

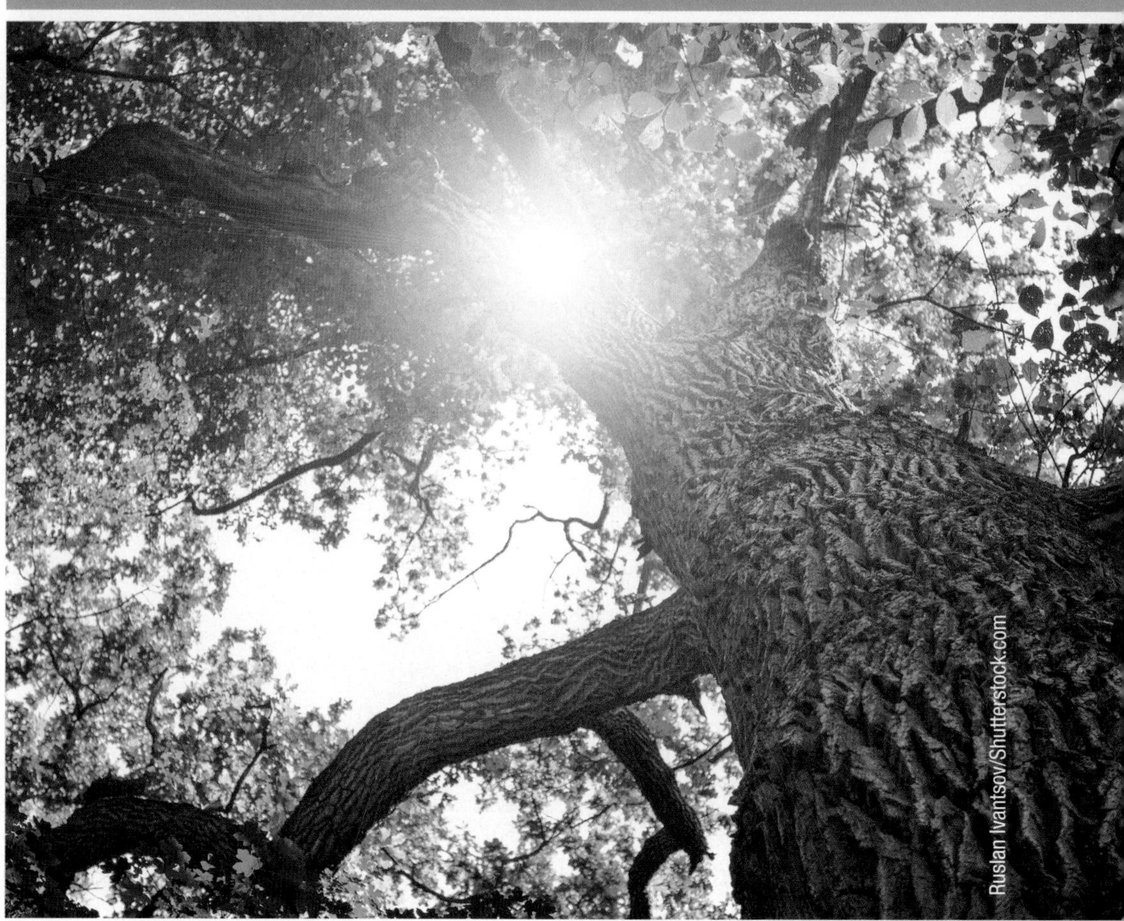

CHAPTER OUTLINE

7.1 Why It Matters

The molecules that make up living things exist mainly as single enantiomers. Consequently, the enantiomers of chiral drugs have different benefits and side effects. This is why many pharmaceutical companies manufacture a single enantiomer of chiral drugs. One such drug is Singulair, the first of a new class of medications for asthma.

Preparing single-enantiomer versions of organic compounds is difficult, especially when the reactions are carried out on a large scale. Most reagents used in manufacturing are achiral, so most industrial chemistry produces racemic mixtures of products. For example, the achiral reagent $NaBH_4$ converts ketones to alcohols. During the reaction, an sp^2-hybridized carbon atom is transformed into an sp^3-hybridized atom, creating a chirality centre and producing a racemic mixture. This process is an inefficient way to make a chiral drug because the portion of the resulting material that has the incorrect stereochemistry must be removed, which is often difficult and wastes half the material.

enantiomer needed for Singulair

Cl N O OMe O → NaBH₄ → racemic mixture (OH, O OMe)

The mechanism of this reaction involves the addition of a nucleophile (electron pair donor) to an electrophile (electron pair acceptor) to make a chemical bond. Knowledge of this reaction mechanism was the key to finding a version of the reaction used to manufacture the drug that could generate more of one enantiomer than the other.

This chapter describes mechanisms and products of nucleophilic addition to the electrophilic π bonds in carbonyl groups and related structures. It also discusses the function of reagents such as Grignards and organolithiums, and the use of acid and base catalysts to accelerate nucleophilic additions to carbonyls and other groups that contain electrophilic π bonds.

7.2 Carbonyls and Related Functional Groups Contain Electrophilic π Bonds

In organic chemistry, a key functional group is the carbonyl group, which consists of a carbon connected to an oxygen by a double bond (C=O). **Carbonyls** are a component of many larger functional groups, and their structure and unique electronic properties determines their reactivity.

A **carbonyl** is a functional group that consists of a carbon and oxygen connected by a double bond.

7.2.1 Orbital structure of the carbonyl group

The reactivity of the carbonyl group is a result of its structure and electronic properties. The simplest carbonyl-containing compound is formaldehyde. Both the carbon and oxygen are sp^2 hybridized, which places the four atoms and the two oxygen lone pairs in a single plane. All the atoms are connected by σ bonds, with the oxygen and carbon also connected by a π bond (shown in blue in Figure 7.1). This π bond lies perpendicular to the plane defined by the atoms. The π bond is the weakest bond in the carbonyl, and the geometry of the group makes the bond accessible to other molecules; these factors determine the reactivity of the carbonyl group.

ORGANIC CHEMWARE
7.1 Lewis structure: Ethanal

C and O are sp^2 hybridized

π bond is perpendicular to the plane of the carbonyl

FIGURE 7.1 Planar geometry of the carbonyl group of formaldehyde showing π orbitals (blue) perpendicular to the plane of the atoms and lone pairs (green).

Because oxygen is more electronegative than carbon, the carbonyl group has a permanent dipole moment, which is implied by the resonance form in which the oxygen carries a negative charge and the carbon bears a positive charge. This resonance form accounts for the polarity of the carbonyl group: the carbon has some positive character, and the oxygen has some negative character.

ORGANIC CHEMWARE
7.2 Resonance: Carbonyl group

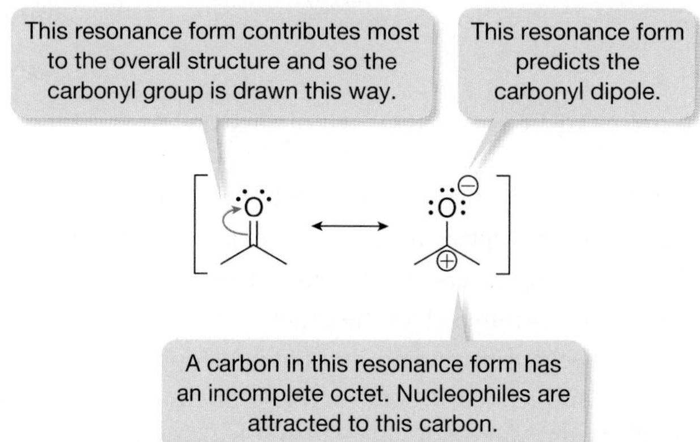

An **electrophile** is a group or atom that accepts electrons.

A **nucleophile** is a group or atom that donates or shares electrons.

The partial positive charge on the carbon of the carbonyl accounts for the reactivity pattern of the group. The carbon is an **electrophile**, which is an atom or group that is attracted to electrons and tends to accept electrons to make bonds. Electrophiles attract **nucleophiles**, which are atoms or groups that donate electrons to form bonds. The charged resonance form of ketones suggests an electrophilic reaction site—that is, the positive charge that resides on the carbon, which has an incomplete octet. The negatively charged oxygen has a full octet in all the resonance forms of the carbonyl, so it cannot act as an electrophile. Resonance predicts reactivity in reactions involving π electrons because both involve π electron movements.

ORGANIC CHEMWARE
7.3 Nucleophiles/ Electrophiles

To draw the charged resonance form shown above, the π bond of the carbonyl was "broken." The concept of an electrophilic π bond revolves around this electron movement. When the carbonyl acts as an electrophile, the π bond will break at the *same time*, moving its electrons toward the more electronegative oxygen.

7.2.2 Aldehydes and ketones

An **aldehyde** is a functional group that consists of a carbonyl connected to at least one hydrogen.

In **aldehydes**, the carbonyl is directly connected to at least one hydrogen atom. Aldehydes are often written using the shorthand formula RCHO, where R can be a carbon chain or a hydrogen.

Throughout this book, general structures may be shown with methyl groups instead of R groups.

aldehydes R–CHO (CHO), H–CHO (R = H), CH_3–CHO

A carbonyl directly connected to two carbon groups is called a **ketone**. The two-carbon chains (R groups) may be the same or may be different. If the groups are different, the notations R and R′ or R_1 and R_2 may be used. In this chapter, the simplest ketone, acetone (which has two methyl groups), is used to illustrate reactions with ketones.

A **ketone** is a functional group that consists of a carbonyl connected to two carbon groups.

Typically, aldehydes react more readily than comparable ketones with reagents that add a group to the carbonyl. However, the reactions with aldehydes and ketones have essentially the same mechanisms.

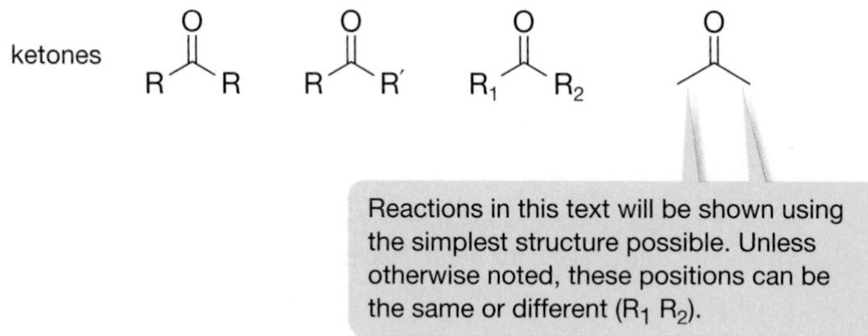

ketones
R R
R R′
R1 R2
Reactions in this text will be shown using the simplest structure possible. Unless otherwise noted, these positions can be the same or different (R1 R2).

7.2.3 Other functional groups with carbon-heteroatom π bonds

At first glance, the following three functional groups—imine, thiocarbonyl, and nitrile—may look different, but they all contain structures similar to a carbonyl group. Because of their structural similarity, these functional groups have reaction patterns similar to those of aldehydes and ketones.

ORGANIC CHEMWARE
7.4 Lewis structure: Ethanimine

ORGANIC CHEMWARE
7.5 Lewis structure: Ethanitrile

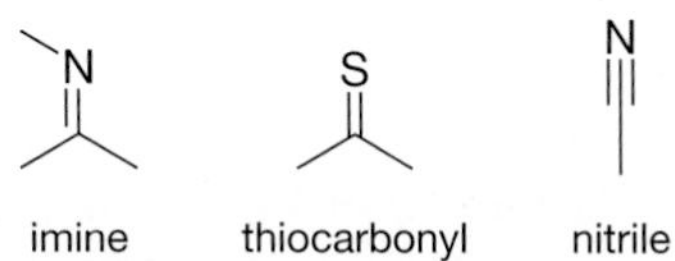

The **imine** functional group is a component of many natural compounds. The imine double bond connects a carbon and a more electronegative nitrogen. The reactivity of this group can be predicted by constructing its resonance forms. "Breaking" the π bond gives a resonance form in which the carbon has a positive formal charge (and an open octet) and the nitrogen has a negative charge. This charge distribution has the same pattern as a carbonyl group, indicating that the imine reacts much like a carbonyl.

An **imine** is a functional group that consists of a carbon connected to a nitrogen by a double bond.

Oxygen is surrounded by full octet in each resonance form and cannot accept electrons from nucleophiles.

Carbon in one resonance form does not have a full octet. Nucleophiles will add here.

Nitrogen is surrounded by full octet in each resonance form and cannot accept electrons from nucleophiles.

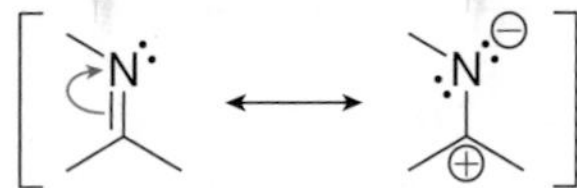

Carbon in one resonance form does not have a full octet. Nucleophiles will add here.

ORGANIC CHEMWARE
7.6 Resonance: Imine

A **thiocarbonyl** is a functional group that consists of a carbon connected to a sulfur by a double bond.

A **nitrile** is a functional group that consists of a carbon connected to a nitrogen by a triple bond.

A similar analysis shows that **thiocarbonyl** and **nitrile** groups also react like a carbonyl group.

Similar resonance analysis can be used to predict the reactivity of **oxocarbenium** and **iminium** ions, both of which have positive charges. An oxocarbenium ion contains a positively charged oxygen connected by a double bond to a carbon. The charged oxygen is much more electronegative than the carbon, so electrons from the π bond flow to the oxygen when this bond breaks. The singly bonded resonance form of this group shows that an oxocarbenium ion attracts nucleophiles to the carbon atom. An iminium ion, which contains a positively charge nitrogen, behaves the same way.

ORGANIC CHEMWARE
7.7 Resonance: Nitrile

An **oxocarbenium** ion contains a positively charged oxygen connected by a double bond to a carbon.

An **iminium** ion contains a positively charged nitrogen connected by a double bond to a carbon.

Heteroatom is surrounded by full octet in each resonance form. Nucleophiles cannot add here.

oxocarbenium ion

iminium ion

Carbon in one resonance form of each group does not have a full octet. Nucleophiles can add here.

All the resonance forms for both ions have positively charged atoms, but the carbon atom in each group is the only site that will react with nucleophiles. Nucleophiles cannot donate electron pairs to the heteroatoms in these groups because these atoms have full octets in all of the resonance forms. The carbons have a full octet in the resonance forms on the left, but are surrounded by only six valence electrons in the other forms. Therefore, the carbons are able to accept electrons. Even though the oxocarbenium ion and the iminium ion differ from a carbonyl in terms of charge and appearance, all these functional groups behave similarly because they have similar patterns of electron distribution and movement.

ORGANIC CHEMWARE
7.8 Resonance: Oxonium ion

ORGANIC CHEMWARE
7.9 Resonance: Carbonyl group (protonated)

STUDENT TIP
In reactions involving π electrons, reactivity can often be predicted by using resonance forms.

7.3 Nucleophilic Additions to Electrophilic π Bonds in Carbonyls and Other Groups

Nucleophiles form new bonds by sharing a pair of electrons with an electrophilic site. In a reaction with a carbonyl, the nucleophile donates two electrons to the carbonyl carbon, as shown in the following diagram by the mechanistic arrow pointing from a pair of electrons on the nucleophile (represented by Nu) to the electrophilic carbon atom. This sharing of electrons between the nucleophile (Nu) and the carbonyl carbon produces a new covalent bond. Since this new bond would result in the carbonyl carbon having more than eight valence electrons, the π bond between the central carbon and the oxygen of the carbonyl breaks; this is the weakest bond to the carbon. A second reaction arrow shows the electrons from the π bond flowing to the oxygen atom. Both arrows are shown on the same structure because the two electron flows happen at the same time.

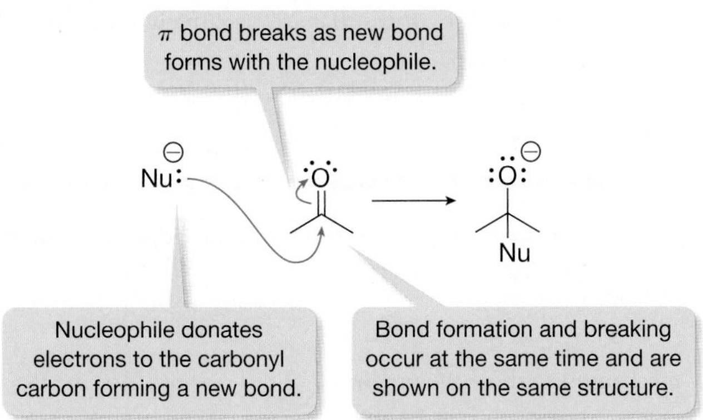

This reaction alters the hybridization of the central carbon of the carbonyl from sp^2 to sp^3, and the flat carbonyl group becomes a tetrahedral alkoxide. Having gained an extra pair of electrons, the oxygen of the product carries a negative charge. If a source of protons exists (for example, a protic solvent such as water or an alcohol), the alkoxide will react with this proton source to give the corresponding neutral alcohol. If no such protons are available, an additional step is needed, in which water and an acid are added to the reaction mixture to neutralize the charged alkoxide. The mechanism of the neutralization is the same with either method. At this point, the oxygen of the alkoxide becomes a nucleophile and donates a pair of electrons to the H^+ provided by the acid. A new O–H bond forms, generating the alcohol product. This stage of the reaction is an acid–base process, but it can also be viewed as a nucleophile–electrophile reaction.

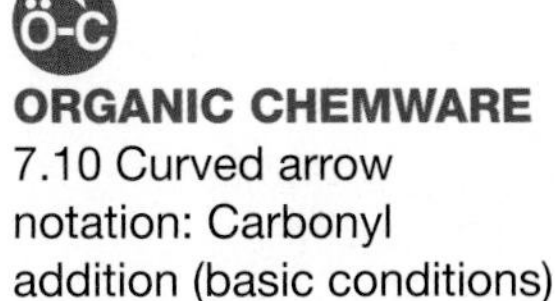

ORGANIC CHEMWARE
7.10 Curved arrow notation: Carbonyl addition (basic conditions)

Acid and water form H_3O^+, which is the species involved in the reaction. Many organic chemists use the shortcut H^+ to represent acid.

alkoxide → alcohol + H_2O

alkoxide → alcohol

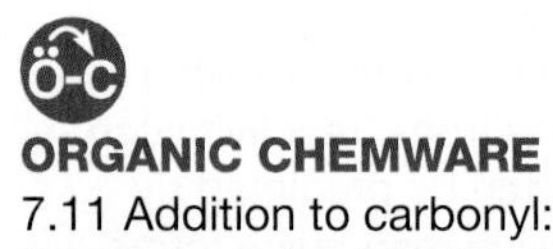
ORGANIC CHEMWARE
7.11 Addition to carbonyl: General mechanism (basic conditions)

7.3.1 Use of resonance forms in nucleophilic reaction mechanisms

In the reactions between a nucleophile and a carbonyl group, the electron flows of bond formation and π bond breaking are shown together in one step. In reactions where delocalization is possible, mechanisms should be drawn using the best resonance contributors of the molecules.

Resonance forms can be used as a *guide* to reactivity and electron flow in reactions involving π electrons because the patterns of electron movement in the reaction and in the drawing of resonance forms are the same. Nucleophiles add to those atoms with incomplete octets in at least one resonance form. Electrophiles add to atoms that are negatively charged and carry at least one lone pair in at least one resonance form.

CHECKPOINT 7.1

You should now be able to draw a curved arrow mechanism for nucleophilic addition to a carbonyl compound or related structure and predict the products.

SOLVED PROBLEM

Draw a curved arrow mechanism to show the product(s) of this reaction. Assume that water is both a reagent and the solvent.

$+ \; H_2O$

STEP 1: Identify the electrophile and nucleophile.

The structure on the left is positively charged, so it can attract electrons toward it and act as an electrophile. The oxygen of the water molecule has unpaired electrons available for sharing, so the water can act as a nucleophile.

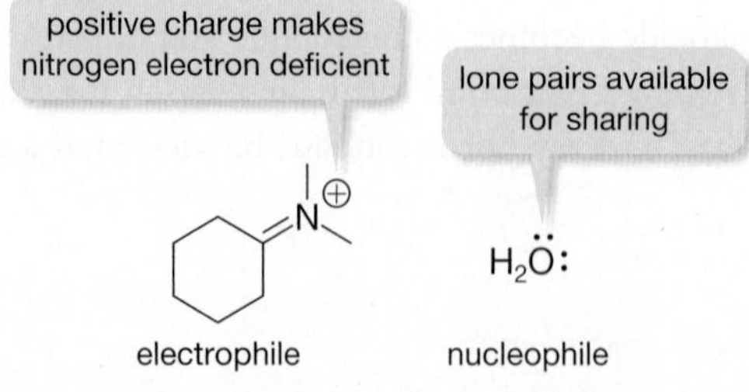

STEP 2: If necessary, use resonance forms to identify the reactive site in the electrophile (or nucleophile).

There are two resonance contributors for the electrophile. While both have a positive charge, only the second one has a positively charged atom with an incomplete octet. Therefore, the carbon acts as the reactive site, attracting electrons from the nucleophile. Note that although the contributor on the right is used to determine the reactivity of this functional group, the mechanism should be drawn with the best resonance contributor, which is the structure on the left in this case.

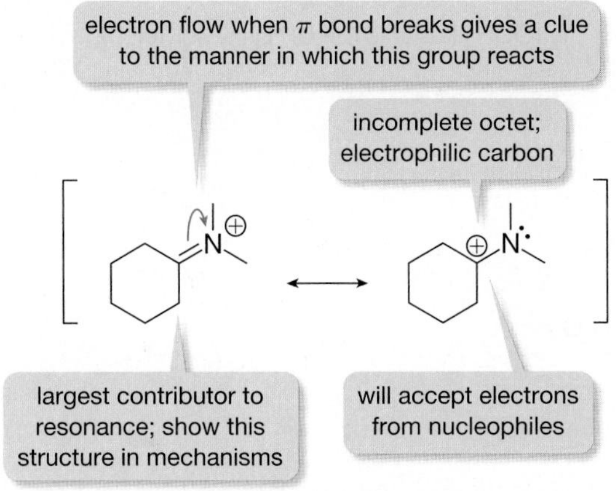

electron flow when π bond breaks gives a clue to the manner in which this group reacts

incomplete octet; electrophilic carbon

largest contributor to resonance; show this structure in mechanisms

will accept electrons from nucleophiles

STEP 3: Draw in the mechanistic arrows, using the principle that electrons flow from nucleophile to electrophile.

Electrons flow from one of the lone pairs on the oxygen to the electrophilic carbon on the electrophile. Adding electrons to this carbon gives the atom too many electrons, so a second arrow needs to be drawn to show the breaking of the C=N π bond, with the electrons flowing to the more electronegative nitrogen atom.

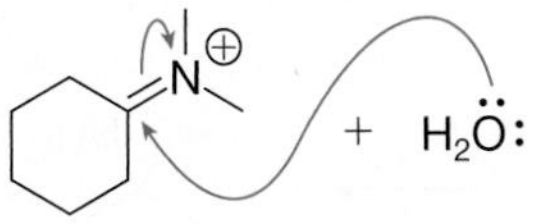

$+ \ H_2\ddot{O}:$

STEP 4: Use the arrows to determine the products of the reaction, including any formal charges.

A new bond has formed between the oxygen of water and the carbon of the iminium ion. At the same time, the C=N π bond has broken. The formal charge integer on oxygen changes from 0 to +1 (electrons moving away from the atom), and the formal charge integer on nitrogen changes from +1 to 0 (electrons moving toward the atom).

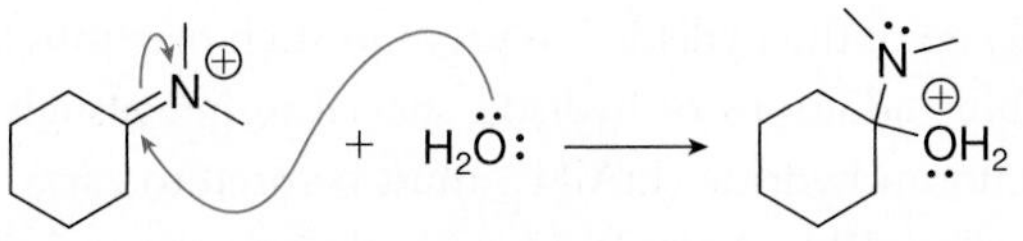

$+ \ H_2\ddot{O}: \longrightarrow$ $OH_2^{\oplus}$

STEP 5: Use a molecule of solvent to create a neutral product via an acid–base reaction.

The product of the first reaction step has a positive charge on oxygen. Since water is also the solvent, the excess water can act as a weak base to remove a proton from the product.

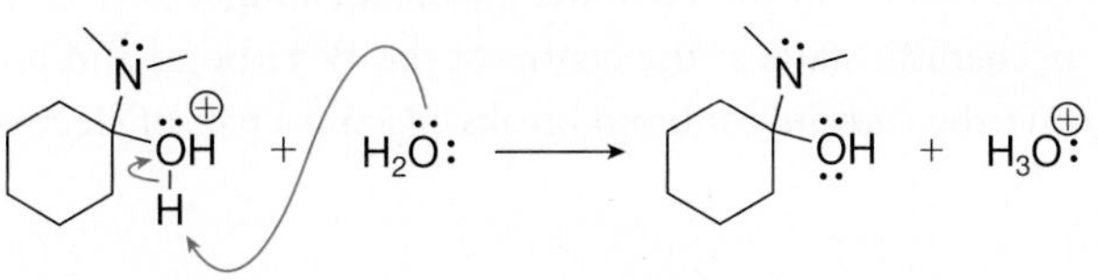

$+ \ H_2\ddot{O}: \longrightarrow$ $\ddot{O}H + H_3O^{\oplus}$

PRACTICE PROBLEM

7.1 Use curved arrows to show the mechanisms and predict the product(s) of the following. Unless otherwise indicated, assume that water is the solvent.

a) 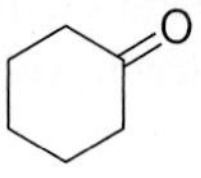 + $Cl^{\ominus}$

b) 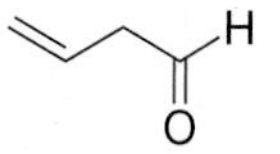 + $NaOCH_3$ (assume CH_3OH is the solvent)

c) 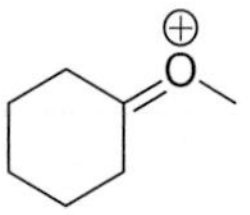 + $^{\ominus}OH$

d) 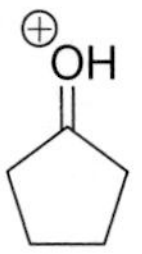 + H_2O

INTEGRATE THE SKILL

7.2 The electrophile is missing from each of the following reactions.

i) Use the products and the given nucleophiles to determine the structure of the missing electrophiles. Assume that water is the solvent for all these reactions.

ii) Write a mechanism for product formation.

iii) Write a mechanism for the *reverse* of each reaction.

a) electrophile A + $Br^{\ominus}$ ⟶ (HO Br)

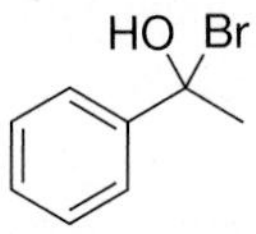

b) electrophile B + $^{\ominus}OH$ ⟶ (HO NH$_2$)

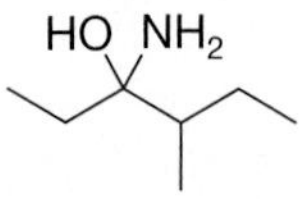

7.3.2 Addition of hydride nucleophiles to carbonyl groups

In principle, the simplest nucleophile that can be added to a carbonyl group is the hydride ion (H^-). However, the small size of the hydride ion prevents such reactions when they involve covalent bonds. For nucleophilic additions of hydride, special reagents such as sodium borohydride ($NaBH_4$) or lithium aluminum hydride ($LiAlH_4$) must be used to carry the hydride.

STUDENT TIP

Why not use NaH? Hydride reagents such as NaH or KH are very strong bases and are commonly used in acid–base reactions. Hydride by itself (H^-) is a very poor nucleophile.

In the borohydride anion (BH_4^-), the B–H σ bonds are polarized. Although boron carries a *formal* charge of −1, hydrogen is actually more electronegative, and so the electrons in the structure tend to reside near the hydrogens. For this reason, BH_4^- can act like a hydride nucleophile. As shown in the following diagram, when this happens, a B–H bond breaks, and its electron pair flows to the electrophilic carbon in the carbonyl group, forming a new C–H bond. The curved arrow showing this mechanism starts at the centre of the B–H bond and points to the carbonyl carbon. At the same time, the carbonyl π bond breaks, placing a pair of electrons onto the oxygen.

B–H bond acts as a nucleophile, donating an electron pair to the carbonyl carbon

π bond breaks as new bond forms with the nucleophile

EtOH

carbonyl

alkoxide

alcohol

square brackets are placed around transient intermediates that exist for short periods of time

alkoxide captures a proton from a molecule of solvent

Reactions involving $NaBH_4$ are normally done in alcohol solvents such as ethanol (EtOH), because these solvents actually participate in the reaction. The negatively charged oxygen of the product alkoxide removes a hydrogen from a molecule of solvent, producing the final product alcohol. The overall reaction happens in two steps, which happen one after the other to form the product. In the mechanism shown, the ethanol solvent is written over the reaction arrow (see Section 7.4).

STUDENT TIP

Note that the hydrogen atom shown in the alkoxide is omitted in the product alcohol. Hydrogen atoms bonded to carbons are normally omitted from line structures unless the hydrogen is needed to show a reaction mechanism.

The borane (BH_3) that is formed as a by-product of the reaction has a boron atom surrounded by only six electrons. The borane combines with a solvent alkoxide to form a compound with a tetravalent boron atom that has a formal charge of -1. The hydrogens bonded to the boron in this alkoxyborohydride product can transfer to three other carbonyl groups. So one molecule of $NaBH_4$ can, in principle, add H^- to four carbonyl groups—this is common for large-scale reactions. Typically, the excess $NaBH_4$ (i.e., more than 0.25 molar equivalent to the carbonyl reactant) is used to convert all the carbonyl groups in a sample into alcohol products.

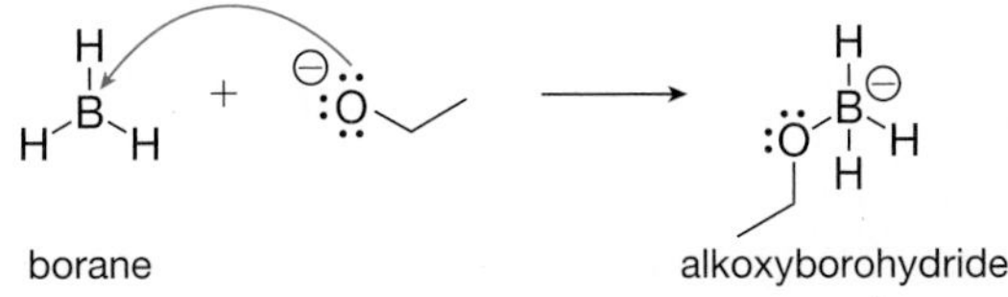

ORGANIC CHEMWARE

7.12 $NaBH_4$ reduction of aldehydes/ketones

Lithium aluminum hydride ($LiAlH_4$) is also commonly used to add hydride to a carbonyl group. The reaction mechanism is similar to that involving $NaBH_4$, but $LiAlH_4$ is much more reactive than $NaBH_4$. The overall process actually involves two distinct chemical reactions. $LiAlH_4$ reacts violently with water, and so reactions with this reagent must be performed only in anhydrous (dry) aprotic solvents, such as ether (Et_2O) or tetrahydrofuran [$(CH_2)_4O$, often abbreviated as THF].

square brackets are placed around transient intermediates that exist only for short periods of time in a reaction

THF

carbonyl

alkoxide

alane

alkoxyaluminum hydride

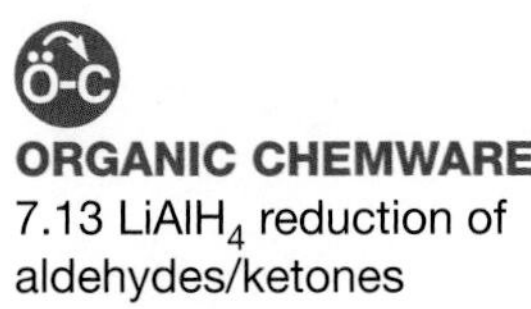

ORGANIC CHEMWARE
7.13 $LiAlH_4$ reduction of aldehydes/ketones

Because $LiAlH_4$ is used in aprotic solvents—which do not contain acidic hydrogen atoms—the process by which $LiAlH_4$ adds hydride to a carbonyl involves a second transformation. The initial alane by-product (AlH_3) has an aluminum atom surrounded by only six electrons. This electrophilic aluminum reacts with the intermediate alkoxide to form a compound with a new oxygen-aluminum bond in which the aluminum is tetravalent and has a formal charge of −1. The three hydrogens bonded to the aluminum in this alkoxyaluminum hydride product can transfer to three other carbonyl groups. Like sodium borohydride, one molecule of $LiAlH_4$ can, in principle, add H^- to four molecules containing carbonyl. The actual amount of $LiAlH_4$ used depends on the scale of the reactions. In large-scale reactions, a 4:1 ratio of carbonyl to $LiAlH_4$ is used. On small scale, it is common to use excess $LiAlH_4$.

NaOH or HCl, H_2O → alcohol

Once the conversion of the carbonyl has been completed, water and a strong acid or base are normally added to convert the aluminum-containing intermediates into alcohols. The overall process from carbonyl to alcohol therefore consists of two sequential reactions. These two reaction steps must not be combined because the reagents $LiAlH_4$ and H_2O are dangerously incompatible.

CHECKPOINT 7.2

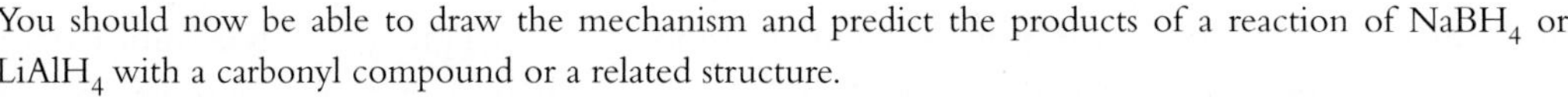

STUDENT TIP
Some mechanisms involve more than one step, which are usually shown using sequential arrows. Some processes involve more than one reaction. In these cases, each reaction is fully completed before the other is started.

You should now be able to draw the mechanism and predict the products of a reaction of $NaBH_4$ or $LiAlH_4$ with a carbonyl compound or a related structure.

SOLVED PROBLEM

Draw the mechanism and product of this reaction, given that the solvent is ethanol.

+ $NaBH_4$

STEP 1: Identify the electrophile and nucleophile.

The negative charge of the borohydride ion (BH_4^-) suggests that this reactant has electrons available for sharing and hence can act as the nucleophile. The sodium cation is a spectator ion, which is not involved in the reaction. Although the carbonyl of the aldehyde does not have a positive charge to attract electrons directly, it does have a polarized C=O bond, which makes the group electrophilic.

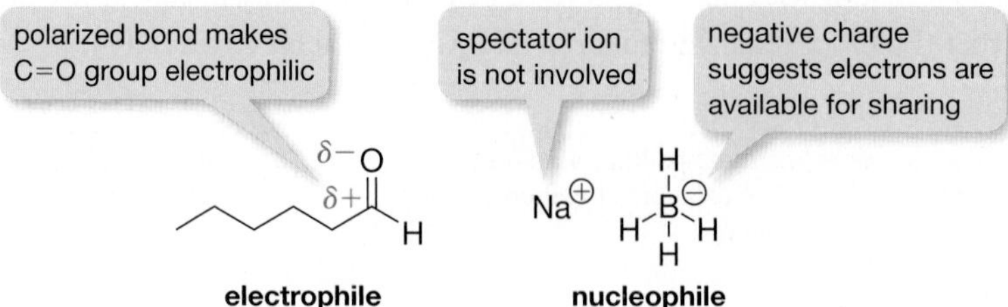

STEP 2: If necessary, use resonance forms to identify the reactive site in the electrophile.

There are two resonance contributors for the aldehyde, one of which places a positive charge on the carbon atom with an incomplete octet. The oxygen has a complete octet in both forms. Therefore, the carbon atom of the carbonyl group acts as the electrophilic site, attracting electrons from the nucleophile.

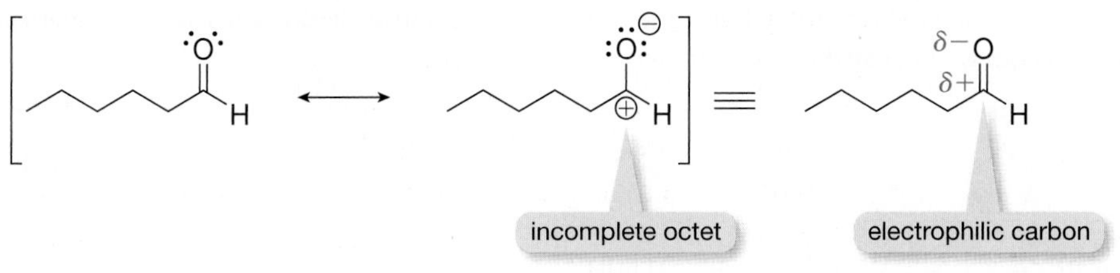

incomplete octet

electrophilic carbon

(Once you are familiar with the more common functional groups, you may not need to draw the resonance structures to determine the reactive site).

STEP 3: Draw the mechanistic arrows, using the principle that electrons flow from nucleophile to electrophile.

Electrons will flow from one of the B–H bonds to the carbon atom of the carbonyl group, forming a C–H bond. Adding electrons to this carbon gives it too many electrons, so a second arrow is needed to show the C=O π bond breaking, with the electrons flowing to the more electronegative oxygen atom.

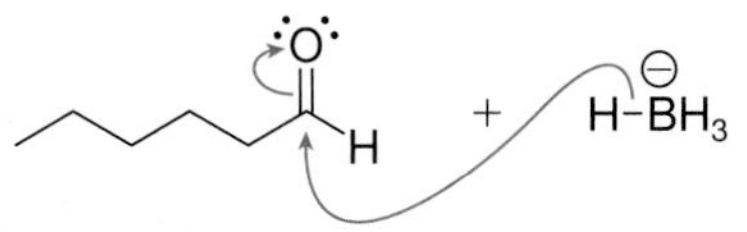

STEP 4: Use the arrows to determine the products of the reaction, including any formal charges.

A B–H bond has broken, and a new C–H bond has formed between the hydrogen and the carbon of the carbonyl group. At the same time, the C=O π bond has broken. The formal charge on boron changes from -1 to 0 (electrons moving away from the atom), and the formal charge on oxygen changes from 0 to -1 (electrons moving toward the atom).

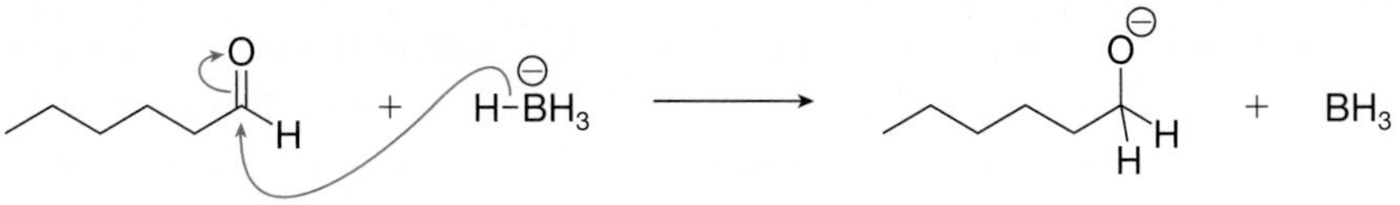

STEP 5: Use a molecule of solvent to create a neutral product via an acid–base reaction.

The product of the reaction has a negative charge. A molecule of ethanol solvent can act as a weak acid to protonate the alkoxide intermediate.

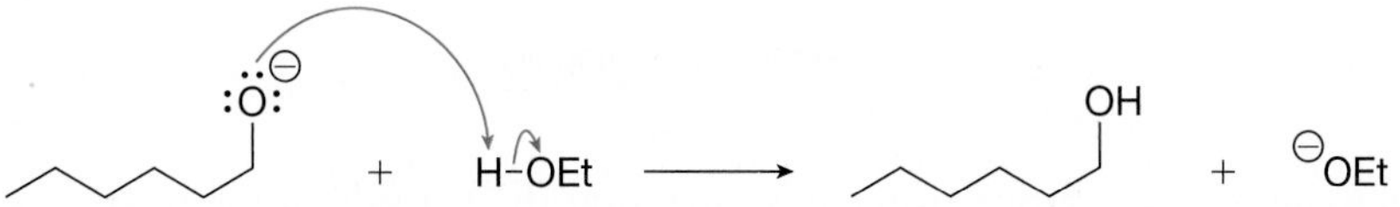

The EtO^- by-product will react with the BH_3 as described above, but the formation of this by-product does not affect the formation of the alcohol product.

PRACTICE PROBLEM

7.3 a) Draw the mechanisms and predict the products of the following reactions, given that ethanol is the solvent. D is the elemental symbol of deuterium, an isotope of hydrogen.

i) 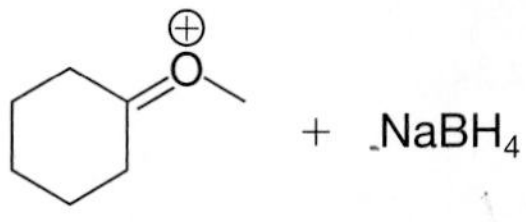 + $NaBH_4$

ii) 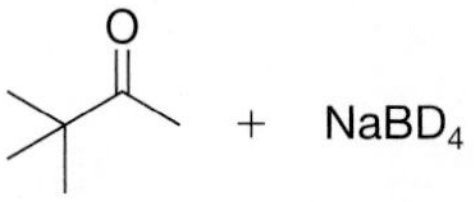 + $NaBD_4$

b) Predict the products of the following reactions. Assume that diethyl ether is the solvent and that aqueous acid is added in a second step.

i)  O + $LiAlH_4$

ii)  Ph O N + $LiAlH_4$

INTEGRATE THE SKILL

7.4 Predict the products that form when the following compounds are treated with an excess of $LiAlH_4$. Assume the reactions are carried out in tetrahydrofuran with aqueous acid added once the initial reaction is complete.

a) 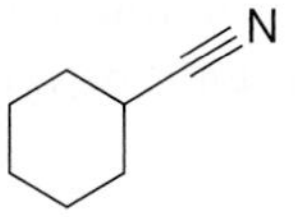 N

b) 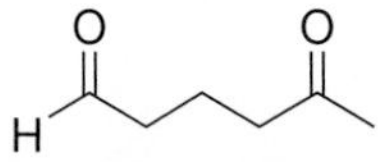 O O H

7.4 Over-the-Arrow Notation

Chemical reactions are formally written as balanced equations that show all the reactants on the left side of the reaction arrow and all the products on the right side. However, a more compact notation—called over-the-arrow—is often used, in which some of the information about reactants and products is written either over or under the reaction arrow. This method highlights the organic substrate (reactant) and product of a reaction and the transformations that take place.

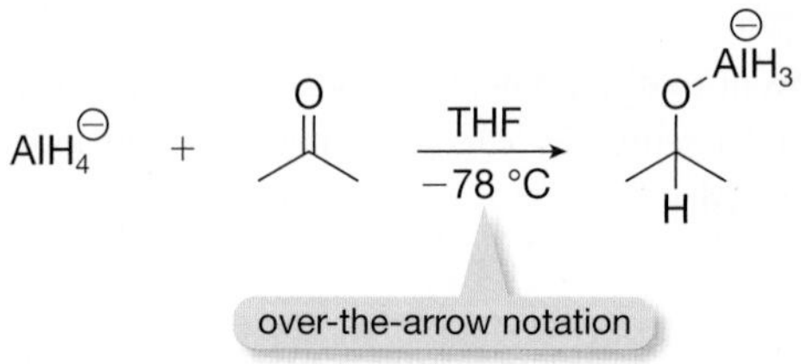

$AlH_4^{\ominus}$ + O —THF, −78 °C→ $O^{\ominus}$–AlH_3, H

over-the-arrow notation

Typically, the organic starting material is shown on the left side of the reaction arrow, and the final organic product is shown on the right. The other reagent(s) and, when necessary, the solvent(s) are written over-the-arrow. Also when necessary, information such as temperatures or pressures may be added.

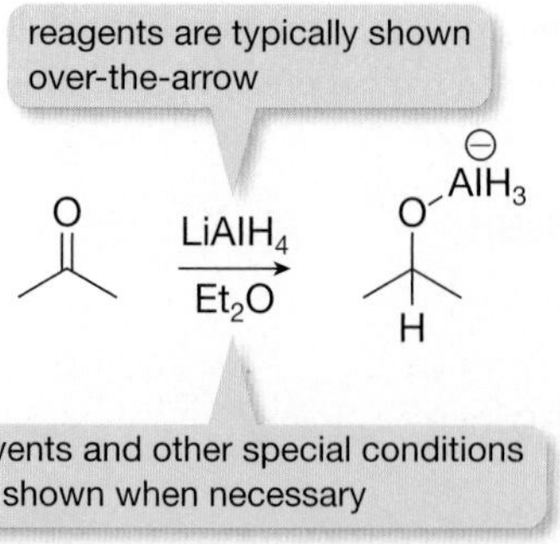

reagents are typically shown over-the-arrow

O —$LiAlH_4$, Et_2O→ $O^{\ominus}$–AlH_3, H

solvents and other special conditions are shown when necessary

Whether something is written above or below the reaction arrow has no special significance. The term *over-the-arrow* refers to both.

$$\xrightarrow[Et_2O]{LiAlH_4}$$

$$\xrightarrow[LiAlH_4]{Et_2O}$$

7.4.1 Writing sequential reactions

STUDENT TIP
Sequential reactions involve separate laboratory procedures. Each reaction may have its own series of mechanistic steps.

Some chemical transformations are actually a set of reactions that must be performed in sequence because the reactions require incompatible reagents or different conditions. For example, Section 7.3.2 described the two-stage conversion of a carbonyl compound into an alcohol using $LiAlH_4$. This process can be written with two separate reaction equations, as shown in the following diagram.

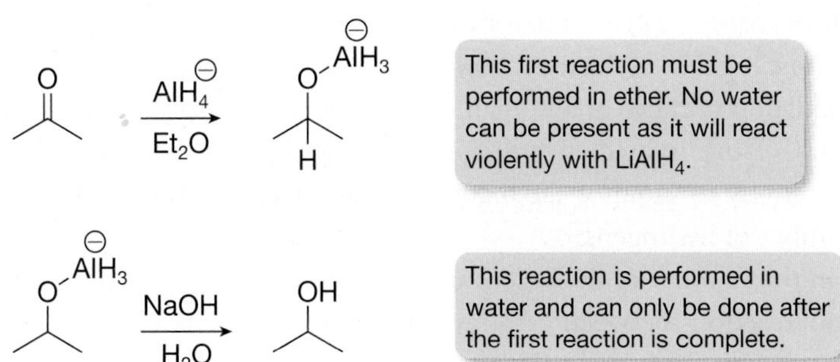

O
AlH4
Et2O
AlH3
O
H
This first reaction must be performed in ether. No water can be present as it will react violently with LiAlH4.
AlH3
O
NaOH
H2O
OH
This reaction is performed in water and can only be done after the first reaction is complete.

There are two other, more concise options for presenting the same information. The first is to show the second reaction as a continuation of the first. Since a new reagent and solvent are added, it is understood that the second reaction is performed after the first one is complete.

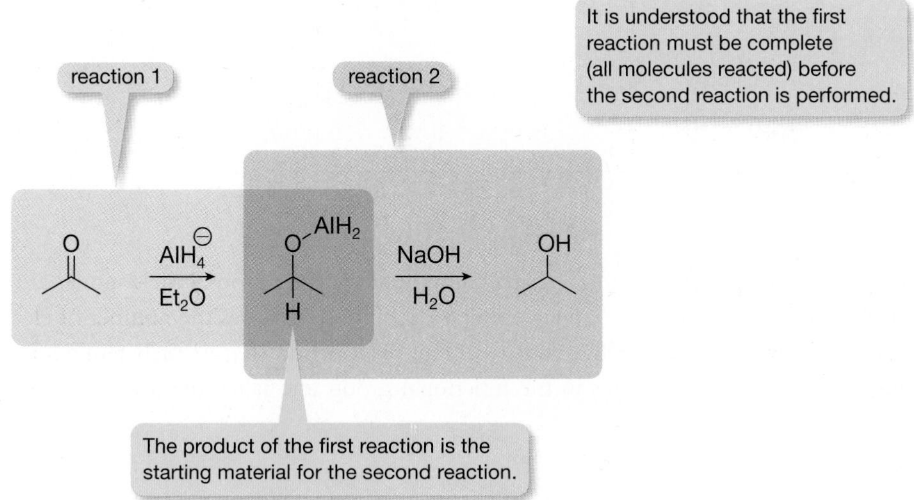

It is understood that the first reaction must be complete (all molecules reacted) before the second reaction is performed.
reaction 1
reaction 2
O
AlH4
Et2O
AlH2
O
H
NaOH
H2O
OH
The product of the first reaction is the starting material for the second reaction.

The second, even more compact option is a notation that uses a single reaction arrow pointing from the first starting material to the *last* product in the reaction sequence, without showing the intermediate product. The reagents for both transformations are listed over the reaction arrow and are numbered to indicate which reaction uses each set of reagents. Each number indicates a *separate* sequential reaction. In the following example, this notation shows that the first reaction uses $LiAlH_4$ as a reagent in ether, and the second reaction uses NaOH in water. It is understood that the product of the first reaction is the starting material for the second.

starting material
reaction 1
1) $LiAlH_4$, Et_2O
2) NaOH, H_2O
reaction 2
final product of all transformations

7.4.2 Organic oxidations and reductions

Organic compounds change oxidation states and undergo redox processes just as inorganic materials do. Because most organic molecules carry more than one functional group, it is impractical to evaluate the oxidation number of an organic molecule. It is usually more important to know the oxidation states of the individual functional groups than the overall oxidation state of a molecule. A practical way to monitor the oxidation state of a molecule or functional group is to determine the number of hydrogens.

An increase in the number of hydrogens in a molecule generally involves a reduction reaction. Therefore, the reactions between carbonyl groups and reagents such as $LiAlH_4$ or $NaBH_4$ are reductions. The shorthand notation [H] is sometimes used to indicate that a reduction process is occurring. Reactions in which the number of hydrogens decreases are considered to be oxidation reactions, and may be labelled with the symbol [O].

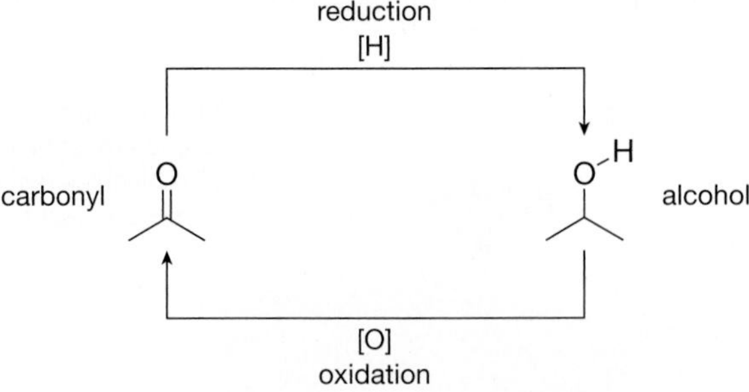

The basic definition of reduction is that a compound gains electrons. Nucleophiles are electron pair donors, so the addition of a nucleophile to a π bond increases the number of electrons in that functional group and thereby reduces it. The protonation step of such reductions does not change the number of electrons in the functional group, and is neither a reduction nor an oxidation.

reduction adds electrons to a functional group

reduction

hydride nucleophile adds a pair of electrons to the carbonyl

The second step does not change the oxidation state of the group. The total number of electrons does not change.

Oxidation reactions occur when π bonds are formed by the loss of a pair of electrons from the molecule, thereby oxidizing the functional group. Leaving groups will be discussed later in Section 7.8 and the details of oxidation reactions will be described in Chapter 12.

The second step is an oxidation that removes electrons from a functional group.

attach leaving group to the oxygen

oxidation

base

leaving group (LG) removes electron pair from alcohol

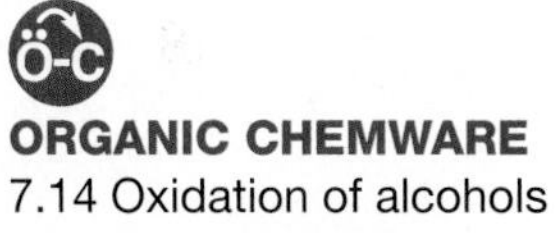
ORGANIC CHEMWARE
7.14 Oxidation of alcohols

Alkanes have the lowest oxidation state that organic molecules can attain. The introduction of heteroatoms increases the oxidation state of a molecule, as does the removal of hydrogen atoms to make π bonds.

alkane — alcohol — aldehyde — carboxylic acid

increasing oxidation state of carbon atom

Note that the reaction of $NaBH_4$ with a carbonyl is a reduction reaction, but it is also a nucleophilic reaction. It is important to use these terms properly to avoid confusion, but the reaction mechanism is the same, regardless of how the reaction is labelled.

CHECKPOINT 7.3

You should now be able to distinguish between oxidation and reduction reactions involving organic compounds.

SOLVED PROBLEM

Determine whether the following reaction is an oxidation or a reduction.

STEP 1: Count the number of hydrogen atoms before and after the reaction.

There are six hydrogen atoms in the starting compound, and eight hydrogen atoms in the product.

STEP 2: Count the number of heteroatoms before and after the reaction.

There are two oxygen atoms in both the reactant and product.

STEP 3: Determine whether the reaction is oxidation or reduction.

An increase in hydrogen atoms corresponds to reduction, whereas an increase in heteroatoms corresponds to oxidation. In this case, only the number of hydrogen atoms increased, so the overall reaction is a reduction.

PRACTICE PROBLEM

7.5 Classify each of the following transformations as an oxidation, a reduction, or neither.

a)

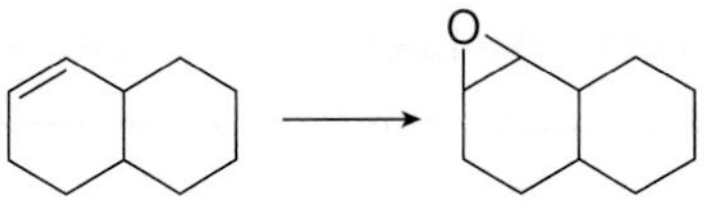

b)  OHC CHO → H O OH

c) 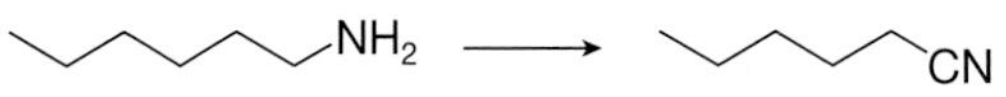 NH_2 → CN

d) 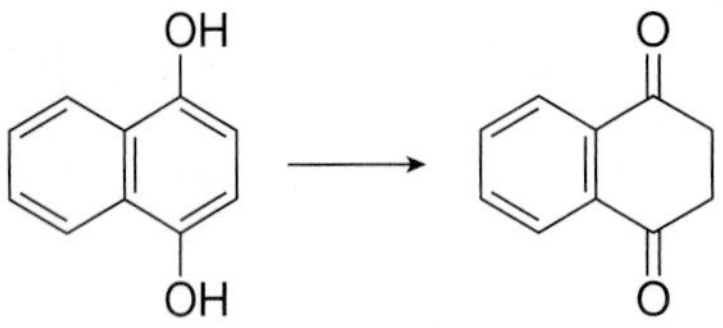 OH OH → O O

INTEGRATE THE SKILL

7.6 Determine whether the following compounds are oxidation or reduction products of phenol.

a) benzene
b) benzene-1,3-diol
c) cyclohexa-2,5-dien-1-ol
d) cyclohexanol

7.5 Addition of Organometallic Compounds to Electrophilic π Bonds

Organometallics are a large family of organic compounds in which carbon is bonded to a metal. Such compounds can participate in reactions that form carbon-carbon bonds in a way that joins multi-carbon pieces together. This is a very valuable reaction type because forming carbon-carbon bonds is key to the creation of new organic materials. Since metals are usually less electronegative than carbon, carbon-metal bonds are often polarized such that the carbon is partially negative and the metal is partially positive. The carbon in such bonds is nucleophilic (electron pair donor). Some organometallic reagents in common use include Grignards, organolithiums, and various acetylides.

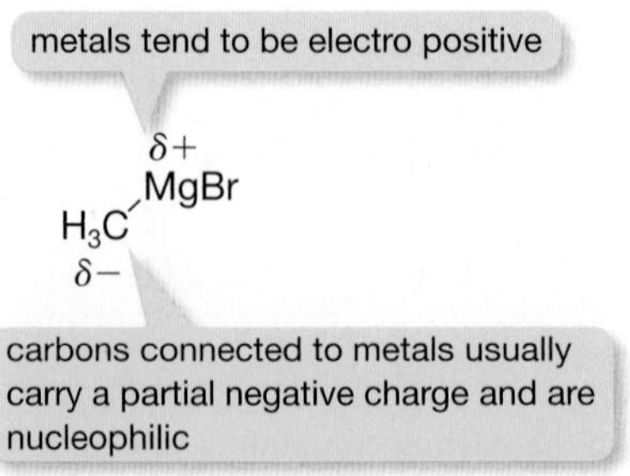

metals tend to be electro positive

$\delta+$ MgBr
H_3C $\delta-$

carbons connected to metals usually carry a partial negative charge and are nucleophilic

Many organometallics react violently with water, and some are even pyrophoric; that is, they ignite spontaneously on contact with the moisture in the air. These compounds are normally prepared and stored in solutions and are not isolated in pure form. The glassware, solvents, and all reagents must be anhydrous (free of water, completely dry) and unreactive atmospheres such as N_2 or Ar are used. Handling these materials requires special equipment and advanced training.

7.5.1 Addition of Grignard reagents

An important reagent in organic chemistry is the **Grignard reagent** (or simply Grignard), which contains a carbon bonded to a magnesium atom. It is used to introduce isotopic labels into organic materials, which makes it possible to determine the mechanism of certain reactions (see Section 7.5.2), and it is particularly useful for the formation of carbon-carbon bonds (see Section 7.5.3).

A **Grignard reagent** contains a carbon bonded to a magnesium atom.

7.5.1.1 Formation and properties of Grignard reagents

A Grignard reagent is formed by exposing an **ethereal solution** of an organic halide to magnesium metal. The solvent in an ethereal solution is typically ether or THF. In the following reaction equation, the solvent is written over-the-arrow. Note that solvents are included in the equation when they are important to a reaction.

An **ethereal solution** is one in which the solvent is an ether such as Et_2O or THF.

the solvent is important in this reaction because it interacts with the Grignard reagent and stabilizes it

Br + Mg $\xrightarrow{Et_2O}$ MgBr

organic halide

Grignard reagent

Br Mg Et O Et O Et Et

Grignards can be formed from organic chlorides, bromides, or iodides. Organomagnesium fluorides are very rare. Grignards are always made in ethereal solvents because they are not stable in other solvents. Magnesium metal is a very strong Lewis acid, which interacts with the lone pairs on the oxygen of the ether solvent. This interaction forms complexes with the ether molecules, stabilizing the Grignard reagent. For convenience, mechanisms involving Grignards are normally written without the coordinating solvent molecules; however, the solvent is often provided using over-the-arrow notation.

Because the magnesium atom is very electropositive, Grignard reagents have a highly polarized carbon-magnesium bond and behave as if there is a full negative charge on the carbon atom. Such reagents are called **carbanions**, very strong bases that have the corresponding alkane as the conjugate acid. The pK_a of a simple alkane is more than 45, making carbanions extremely reactive bases. Acid–base reactions involving Grignards are fast and *highly* exothermic. A Grignard reagent rapidly removes a proton from any functional group that contains a hydrogen with a pK_a lower than that of the Grignard's conjugate acid (the parent alkane). Susceptible functional groups are usually those that have hydrogens bonded to heteroatoms (for example, OH, NH, or SH). However, Grignards may remove a hydrogen from a carbon if the pK_a of that hydrogen is low enough.

A **carbanion** is a negatively charged carbon.

The carbon-magnesium bond is covalent, but this molecule behaves as if the carbon were negatively charged.
δ− δ+
H₂C—MgBr
H₃C
behaves as
H₂C:
H₃C
MgBr
carbanion
(negatively charged carbon)

Because acid–base reactions tend to be faster than most other processes, Grignards "prefer" to undergo acid–base reactions if possible. Therefore, Grignards can be prepared only when functional groups containing acidic (removable) hydrogens are not present. The reaction of Grignards with water is extremely rapid and may be violent. Care must be taken when working with Grignards to ensure that all solvents, reagents, and glassware are completely dry. Even the water in ambient air can be a problem. Grignards are usually prepared and used in dried air or preferably under an inert gas such as N_2 or Ar.

The reaction of a Grignard with water involves an acid–base reaction. As with $NaBH_4$, a bond in the Grignard acts a nucleophile. This bond donates electrons to a hydrogen in water, forming a new bond between the carbon of the Grignard and the hydrogen. The bond between the hydrogen and the oxygen of the water breaks, returning two electrons to the more electronegative oxygen atom. These electron flows produce an alkane, a magnesium cation, and a hydroxide anion. The latter two combine to form complex magnesium salts.

product is the alkane corresponding to the Grignard reagent
hydroxide and magnesium ions combine to form magnesium salts
MgBr
Grignard
H OH
H
alkane
:OH + MgBr
MgBr(OH)
Grignards are very strong bases and rapidly remove acidic hydrogens

The reaction of a Grignard with other OH, NH, or SH groups is exactly the same: the Grignard removes a H^+, which gives the corresponding negatively charged functional group.

7.5.2 Utility of the acid–base reactivity of Grignards

The acid–base reactivity of Grignards converts a carbon-halogen bond into a carbon-hydrogen bond. This process is commonly used to introduce isotopic labels into organic materials. If D_2O (heavy water) is used in place of H_2O, the hydrogen isotope deuterium (D or ^2H) will be placed at a known location on the target molecule. The location of the deuterium marker in the products of subsequent reactions helps researchers determine the mechanism of these reactions.

Br
+ Mg
Et₂O
MgBr
D₂O
D

7.5.3 Formation of carbon-carbon bonds with Grignard reagents

The most useful Grignard reactions are those that create carbon-carbon bonds. Because the carbon of the Grignard carries a partial negative charge, it is strongly nucleophilic and readily bonds to electrophilic carbons in various functional groups by using the carbon-magnesium bond as the source of electrons. For example, a Grignard bonds to the carbon of a carbonyl group, producing an alkoxide. This alkoxide then forms a bond with the positively charged magnesium ion, giving a magnesium alkoxide.

A magnesium alkoxide forms. The oxygen-magnesium bond is covalent.

BrMg — Grignard; carbonyl; Et_2O; MgBr; new bond made; magnesium alkoxide

Once the addition of the Grignard to the carbonyl group has been completed, the magnesium alkoxide is treated with a solution of dilute acid in water. This process breaks the oxygen-magnesium bond, producing an alcohol. This latter reaction is termed **hydrolysis**, because water (hydro) is breaking (lysing) a bond: the oxygen-magnesium bond in this example. Any acid may be used for this purpose, but aqueous NH_4Cl or HCl are most commonly employed.

Hydrolysis is a reaction with water that decomposes a functional group into other components.

acid and water are used to hydrolyze the oxygen-magnesium bond

O–MgBr; H_2O; NH_4Cl; OH; magnesium alkoxide

DID YOU KNOW?

Hydrolysis reactions also involve the flow of electrons, which can be described using arrow notation. Mechanisms involving acids usually start with a transfer of H^+ onto a molecule to create a positive charge. This charge is a very strong electron attractor and pulls electrons toward it. During hydrolysis, this positive charge "pulls" a molecule of water to the metal.

Continued

basic oxygen on product is protonated by acid

positive charge is a very strong electron attractor

by-product can transfer H^+ to another molecule of water to continue the cycle

The carbon-magnesium bond of a Grignard acts as the initial nucleophile. Although this bond is not ionic, considering the Grignard reagent as a pair of ions can help in predict the reagent's reactivity. The carbanion is the negative component of this pair and therefore the nucleophilic electron pair donor.

Grignard behaves like a carbanion

break bond by moving electrons onto more electronegative atom

CHECKPOINT 7.4

You should now be able to draw the mechanism and predict the products of a reaction of a Grignard reagent with a carbonyl compound or related structure.

SOLVED PROBLEM

Draw the mechanism and predict the products for each step of this reaction.

$$\xrightarrow[Et_2O]{EtMgBr} \qquad \xrightarrow[H_2O]{NH_4Cl}$$

STEP 1: Identify the electrophile and nucleophile in the first step of the reaction.

Ethylmagnesium bromide is a Grignard reagent. The highly electropositive magnesium atom creates a polarized C–Mg bond, making the carbon atom of the bond highly nucleophilic. The aldehyde, on the other hand, has a polarized carbonyl group, which creates an electrophilic carbon with a partial positive charge. (This charge distribution can be verified by drawing the resonance structures.)

nucleophilic carbon

electrophilic carbon

STEP 2: Add the mechanistic arrows, using the principle that electrons flow from nucleophile to electrophile.

Electrons flow from the carbon–magnesium bond to the carbon atom of the carbonyl group. However, adding electrons to the carbonyl carbon would overfill its valence octet, so the C=O π bond breaks, and a second arrow is needed to show the corresponding flow of electrons to the more electronegative oxygen atom.

O H + MgBr

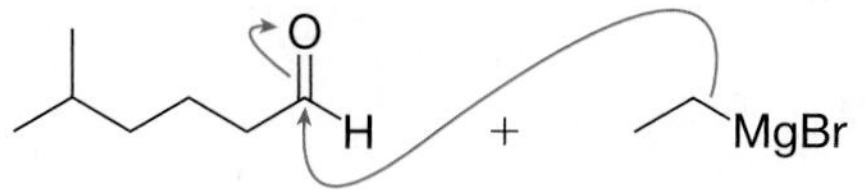

STEP 3: Use the arrows to determine the initial products of the reaction, including any formal charges.

The carbon–magnesium bond has broken to form a new carbon–carbon bond between the carbon of the ethyl group and the carbon of the carbonyl. At the same time, the C=O π bond has broken and electrons moved to the more electronegative oxygen atom. The formal charge on magnesium changes from 0 to +1 (electrons moving away from the atom), and the formal charge on oxygen changes from 0 to −1 (electrons moving toward the atom).

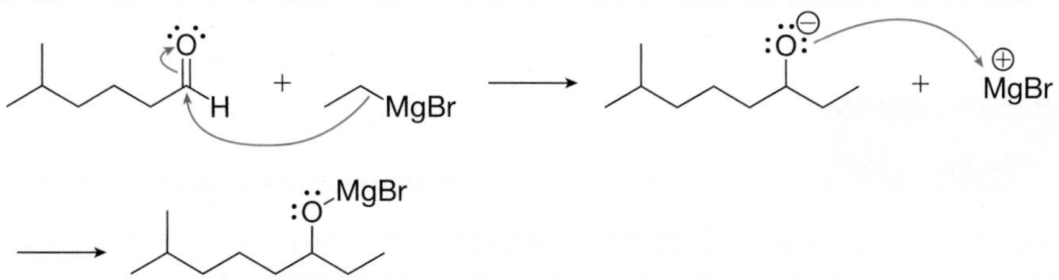

:O: H + MgBr → :O:⊖ + ⊕MgBr

→ :O–MgBr

A second set of arrows shows the negatively charged oxygen bonding with the positively charged magnesium ion. For simplicity, this step is sometimes omitted when drawing the reaction mechanism.

STEP 4: The aqueous acid added in the second step produces the neutral final product.

Ammonium chloride in water creates hydronium ions, which protonate the alkoxide intermediate to give a neutral alcohol and a molecule of water.

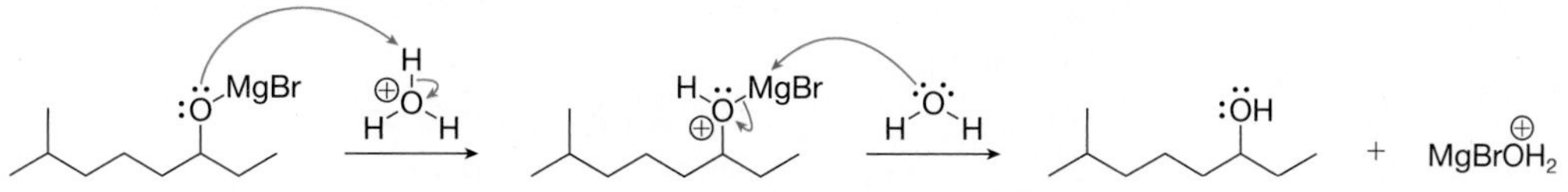

:O–MgBr → (H_3O^+) → H–O⊕–MgBr → (H_2O) → :ÖH + ⊕MgBrOH₂

PRACTICE PROBLEM

7.7 Draw a mechanism and predict the products for each step of the following reactions.

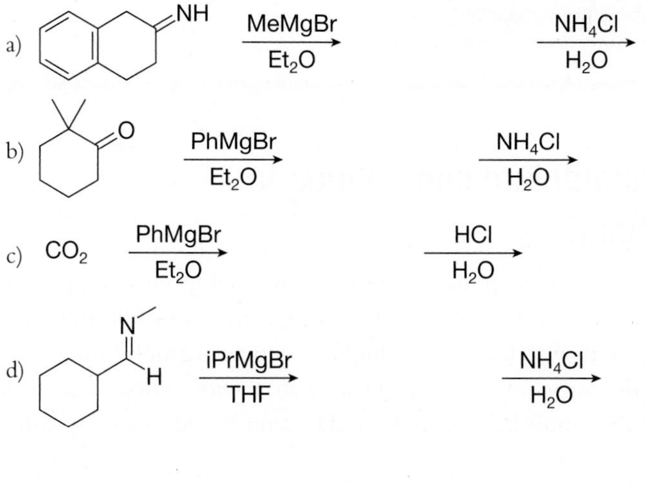

a) NH; MeMgBr / Et_2O → ; NH_4Cl / H_2O →

b) O; PhMgBr / Et_2O → ; NH_4Cl / H_2O →

c) CO_2; PhMgBr / Et_2O → ; HCl / H_2O →

d) N, H; iPrMgBr / THF → ; NH_4Cl / H_2O →

INTEGRATE THE SKILL

7.8 A chemist attempts to produce three different Grignard reagents to react with benzaldehyde (PhCHO). Determine which Grignard reagents will form successfully, and draw the product(s) of their reactions with benzaldehyde, assuming an aqueous work-up at the end of the reaction. If the Grignard formation is not possible, explain why.

a) Br (cyclohexyl bromide) + Mg $\xrightarrow{\text{EtOH}}$ MgBr (cyclohexyl)

b) Br (bromobenzene) + Mg $\xrightarrow{\text{Et}_2\text{O}}$ MgBr (phenyl)

c) Br (bromo-methylphenol, OH) + Mg $\xrightarrow{\text{Et}_2\text{O}}$ MgBr (OH)

DID YOU KNOW?

Although the exact mechanism by which Grignards add to carbonyls is not fully known, it is clear that the process involves an interaction between the magnesium of a Grignard and the oxygen of a carbonyl group. The actual reaction most likely involves two molecules of Grignard: one to activate the carbonyl by acting as a Lewis acid, and the other to serve as the nucleophile and form a bond with the carbon of the carbonyl group. The reaction is thought to have a circular flow of electrons, as shown in the following diagram.

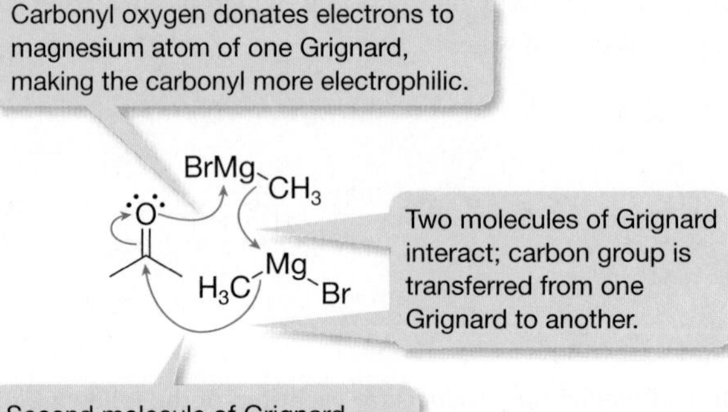
Carbonyl oxygen donates electrons to magnesium atom of one Grignard, making the carbonyl more electrophilic.
BrMg
CH3
O
Mg
H3C
Br
Two molecules of Grignard interact; carbon group is transferred from one Grignard to another.
Second molecule of Grignard delivers carbon group to carbonyl.

7.5.4 Addition of organolithium compounds to electrophilic π bonds

Organolithium compounds undergo nucleophilic addition to carbonyl groups in a process that is similar to Grignard reactions with carbonyl groups. Like Grignards, organolithiums are very strong bases that react violently with water. Although highly reactive, organolithiums are stable compounds if protected from the atmosphere. They are used in a wide range of reactions, so they are sold commercially, but they must be prepared, handled, and stored in strictly anhydrous conditions.

The preparation of organolithium reagents involves exposing the appropriate alkyl halide to lithium metal in ether, as shown in the following reaction equation. Because lithium atoms have only one valance electron, two lithium atoms are required for each molecule of organohalide.

Br + 2 Li —(Et_2O)→ Li + LiBr

During nucleophilic addition to carbonyl groups, the transferred group is the carbon-containing group of organolithiums. The lithium-carbon bond provides the pair of electrons for making the new carbon-carbon bond, as the following diagram shows.

Once all of the organolithium has reacted with the carbonyl, the resulting alkoxide salt is treated with water and acid to produce the corresponding alcohol. As with Grignards, weak acids such as NH_4Cl are normally used for this process.

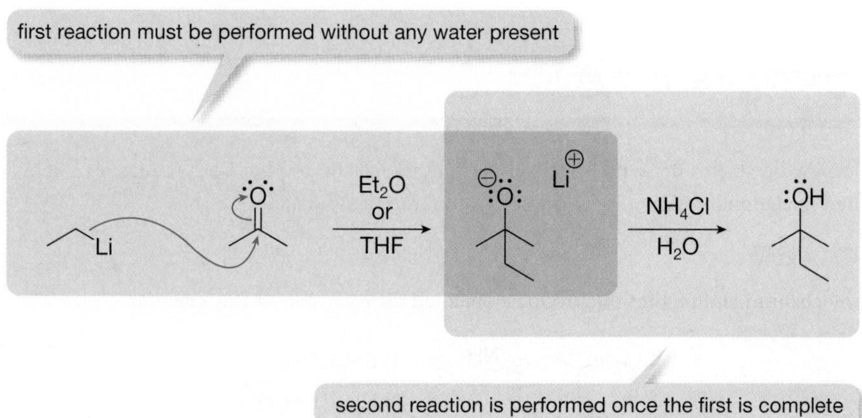

7.5.5 Addition of acetylides to electrophilic π bonds

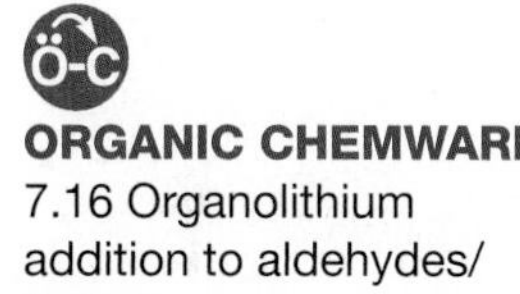
ORGANIC CHEMWARE
7.16 Organolithium addition to aldehydes/ketones

The hydrogen atoms of alkynes (acetylenes) are much more acidic than those in most other hydrocarbons, and are easily removed with strong bases. Acetylides (carbanions prepared from acetylenes) are very nucleophilic. Acetylene nucleophiles are usually made by deprotonating with a strong base such as organolithiums, Grignards, or sodium amide ($NaNH_2$, pK_a ~ 35). Acetylenic hydrogens have a pK_a of approximately 22.

H_3C—≡—H (stronger acid, pK_a 22) + $:NH_2^-$ (less stable base) —(NH_3)→ H_3C—≡—$C:^-$ (more stable base) + $:NH_3$ (weaker acid, pK_a 35)

Acetylides readily add to carbonyl groups in a process similar to that for other nucleophiles. Because water and carbanions are incompatible, the reaction involving water and acid is performed after the nucleophilic addition is complete. This reaction converts the alkoxide produced by the first reaction into the corresponding alcohol.

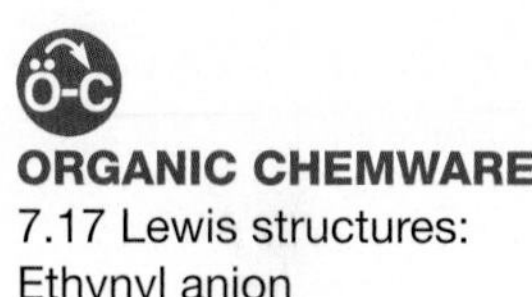
ORGANIC CHEMWARE
7.17 Lewis structures: Ethynyl anion

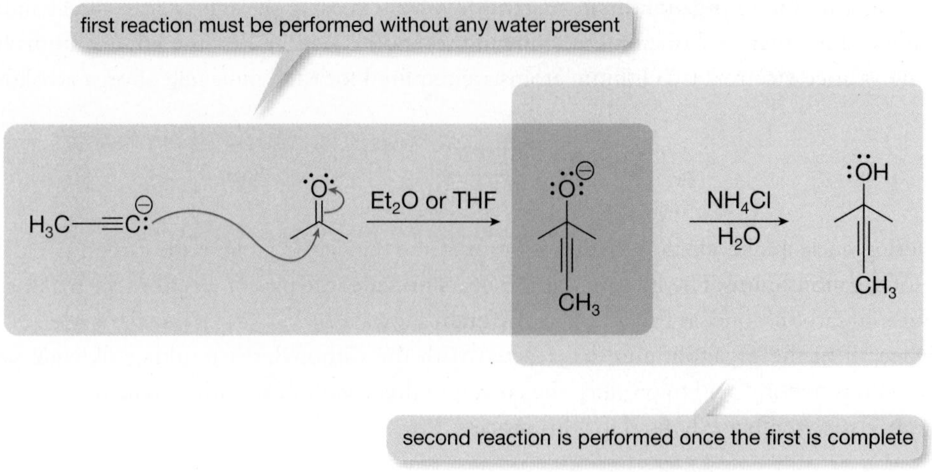

first reaction must be performed without any water present

$H_3C—≡C:^{\ominus}$ Et$_2$O or THF → NH$_4$Cl / H$_2$O → :ÖH, CH$_3$

second reaction is performed once the first is complete

CHECKPOINT 7.5

You should now be able to draw the mechanisms and predict the products of the reactions of a variety of carbon-based nucleophiles with a carbonyl compound or related structure.

SOLVED PROBLEM

Draw the mechanism and predict the products obtained for each step of this reaction.

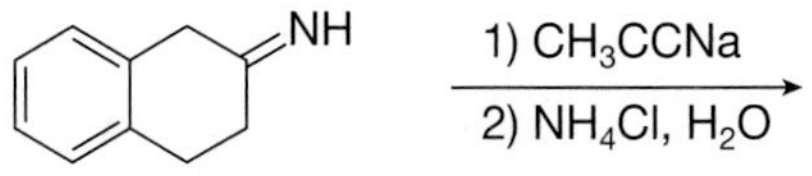

NH 1) CH_3CCNa 2) NH_4Cl, H_2O

STEP 1: Identify the electrophile and nucleophile in the first step of the reaction.

The reagent CH_3CCNa has a nucleophilic negatively charged carbon atom with a lone pair. The starting material, on the other hand, has an imine functional group with a polarized C=N bond, which makes the carbon atom a potential electrophile.

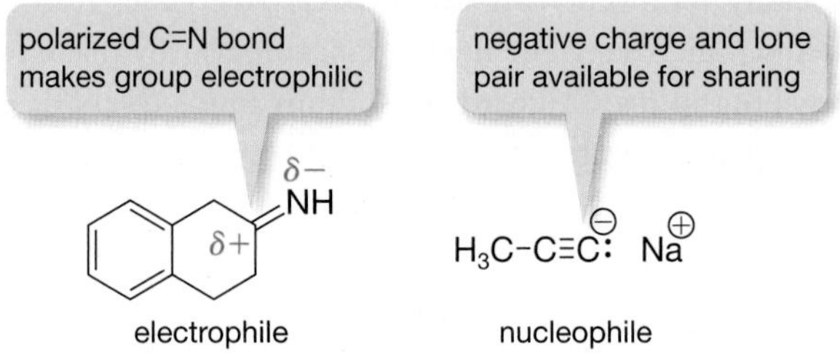

polarized C=N bond makes group electrophilic

negative charge and lone pair available for sharing

δ− NH δ+

electrophile

H_3C-C≡C: Na$^{\oplus}$

nucleophile

STEP 2: If necessary, use resonance to identify the reactive site in the electrophile.

There are two resonance contributors for the imine, one of which places a positive charge on the carbon atom with an incomplete octet. Therefore, the carbon atom of the imine group acts as the reactive site, attracting electrons from the nucleophile.

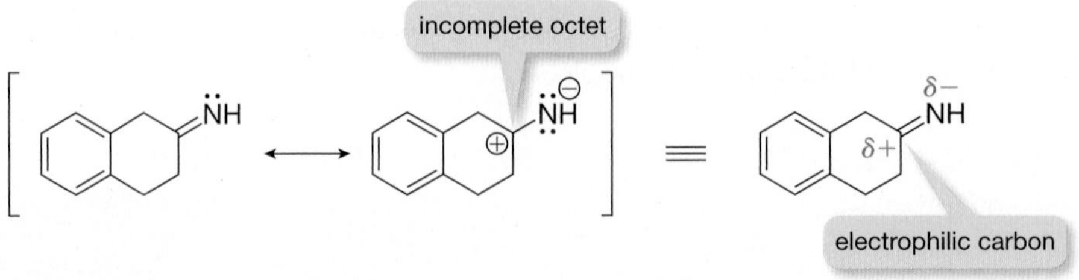

incomplete octet

NH ⟷ NH ≡ δ− NH δ+

electrophilic carbon

STEP 3: Draw in the mechanistic arrows, using the principle that electrons flow from nucleophile to electrophile.

Electrons will flow from the lone pair on the negatively charged carbon atom to the carbon of the imine group. Since this flow would give the imine carbon more than eight electrons, the C=N π bond breaks, and a second mechanistic arrow is needed to show the flow of electrons to the more electronegative nitrogen atom.

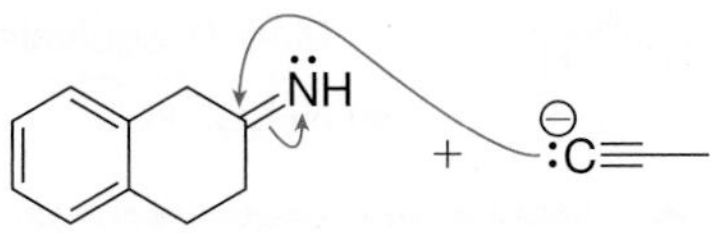

STEP 4: Use arrows to determine the initial products of the reaction, including any formal charges.

A new C–C bond has formed between the carbon of the alkyne and the carbon of the imine. At the same time, the C=N π bond has broken. The formal charge on the alkyne carbon changes from −1 to 0 (electrons moving away from the atom), and the formal charge on nitrogen changes from 0 to −1 (electrons moving toward the atom).

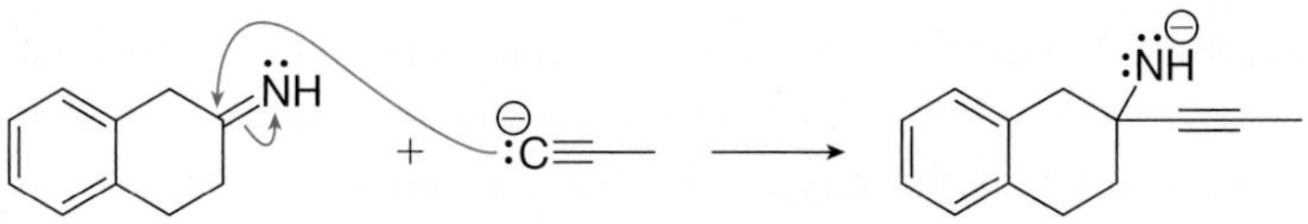

STEP 5: Use the aqueous acid added in the second step to produce the neutral final product.

Ammonium chloride in water creates hydronium ions, which protonate the amide intermediate to give a neutral amine and a molecule of water.

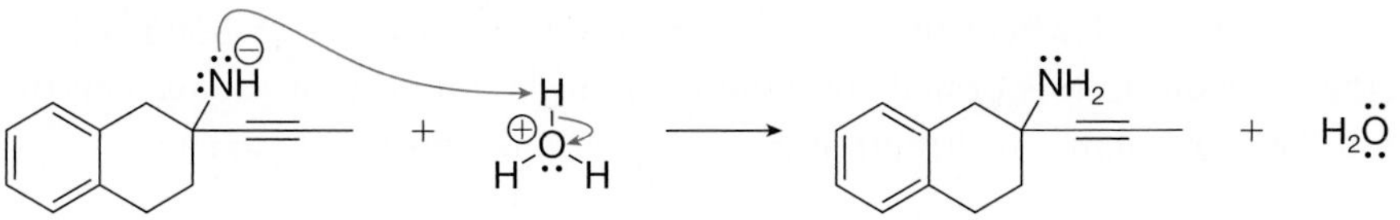

PRACTICE PROBLEM

7.9 Draw mechanisms and predict the products for the following reactions.

a) 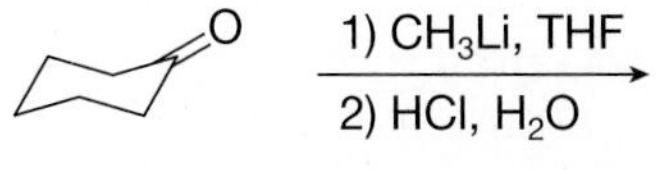 1) CH_3Li, THF; 2) HCl, H_2O

b) 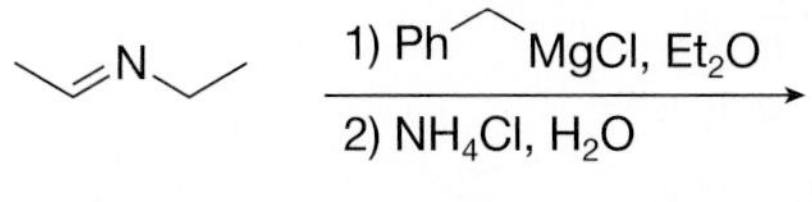 1) Ph MgCl, Et_2O; 2) NH_4Cl, H_2O

c) 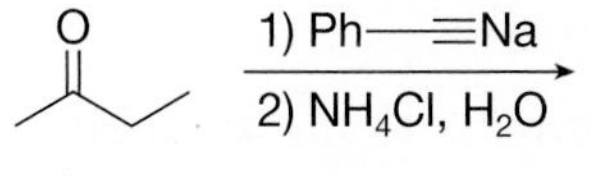 1) Ph—≡Na; 2) NH_4Cl, H_2O

d) 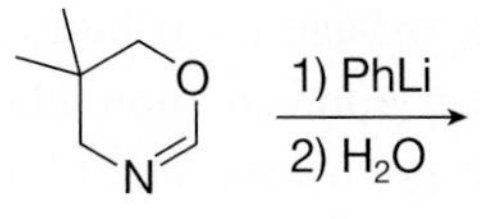 1) PhLi; 2) H_2O

INTEGRATE THE SKILL

7.10 The following compound has two electrophilic sites that could potentially react with an organolithium reagent such as methyllithium. Determine which site is more electrophilic, and draw the product of the favoured reaction. Assume only 1 mole of methyllithium is added for every mole of reactant (1 equivalent).

1) MeLi (1 equivalent), THF
2) NH_4Cl, H_2O

7.6 Using Orbitals to Analyze Reactions

So far in this chapter, reactivity has been analyzed using resonance structures and electronegativity. This method is effective and is used extensively to understand organic processes. However, examining the orbital interactions can give a deeper understanding of reaction pathways.

Since the oxygen atom of the carbonyl group is more electronegative than the carbon atom, the energy of the atomic orbitals of the oxygen atom is lower than that of the carbon atom. So the overlap between the orbitals of the carbon and the oxygen is not symmetrical. The sp^2-hybridized orbital of the oxygen is closer in energy to the σ molecular orbital than the sp^2-hybridized orbital of the carbon is. Consequently, the oxygen atom contributes more to the formation of this σ orbital, making this σ bond asymmetric. In Figure 7.2, the asymmetry is represented by the shape of the lobes drawn for the σ molecular orbital: the lobes are drawn larger near the oxygen. The size of the lobes corresponds to a factor called the *orbital coefficient* and indicates the relative probability of finding electrons. Electrons are most likely to be found where the orbitals are largest. Because the σ bond is larger near oxygen, the electrons in this orbital are more likely to be found near the oxygen than near the carbon. This probability distribution gives the carbon a positive character.

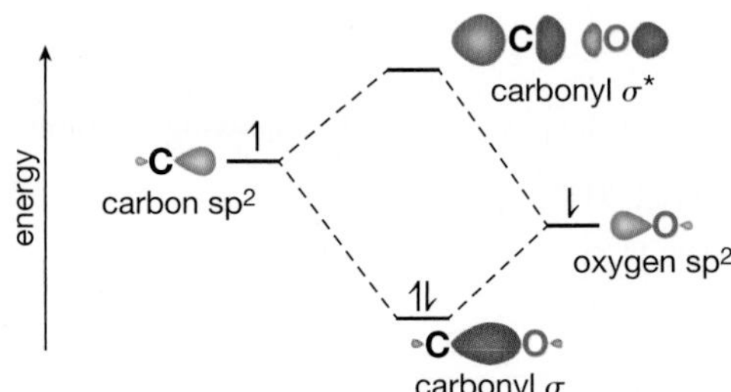

FIGURE 7.2 Construction of the σ orbital of a carbonyl group.

ORGANIC CHEMWARE
7.18 Molecular orbitals: C–O σ orbitals

Similarly, the atomic p orbital of the oxygen that contributes to the π bond is closer in energy to the π orbital than the p orbital of the carbon is. So the oxygen contributes more to this bond than does the carbon. Consequently, the electrons in the π bond are more likely to be found near the oxygen than near the carbon (Figure 7.3). To indicate this asymmetric distribution, the lobes of the π orbital near the oxygen are larger than those near the carbon.

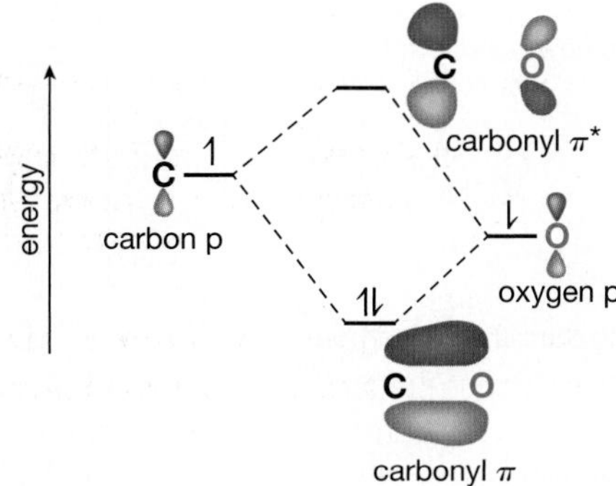

FIGURE 7.3 Construction of the π orbital of a carbonyl group.

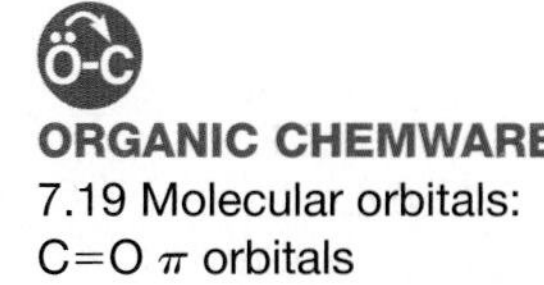

ORGANIC CHEMWARE
7.19 Molecular orbitals: C=O π orbitals

In order to accept electrons from an incoming nucleophile, an electrophile must have an empty molecular orbital. The nucleophile reacts with the lowest unoccupied molecular orbital (LUMO) of the electrophile to form a new bond. The LUMO of the carbonyl group is the $\pi^\star$ orbital (Figure 7.4).

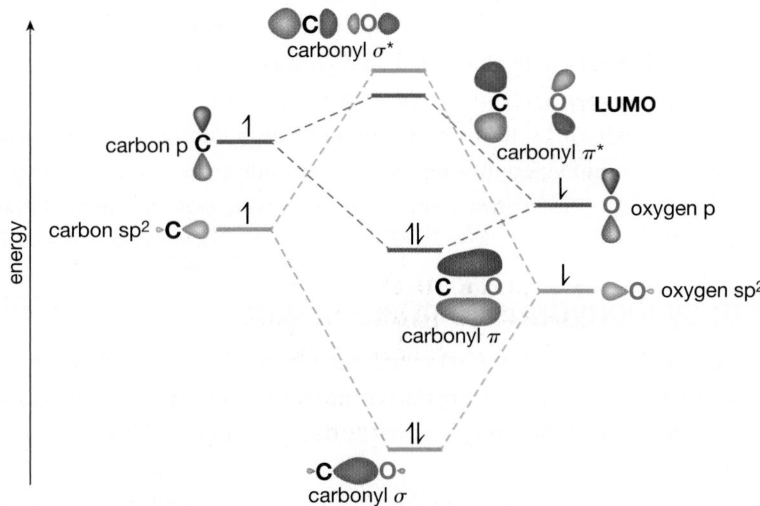

FIGURE 7.4 Complete orbital structure of a carbonyl group.

The carbonyl $\pi^\star$ orbital is closer in energy to the atomic p orbital of the carbon atom than to the p orbital of the oxygen atom. Therefore the carbon contributes more than the oxygen to this orbital, which is represented by larger orbital lobes near the carbon. A nucleophile approaching the carbonyl is more likely to interact with the larger lobes (larger coefficients) of the $\pi^\star$ LUMO. Because the larger lobes are located on the carbon, the nucleophile readily bonds to the carbon, but not to the oxygen.

7.7 Formation of Cyanohydrins from Carbonyls

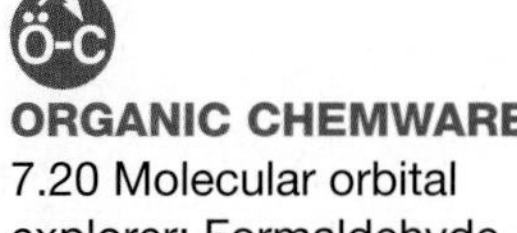

ORGANIC CHEMWARE
7.20 Molecular orbital explorer: Formaldehyde

In the nucleophilic cyanide ion, the carbon and nitrogen both have full octets with lone pairs. The carbon bears a formal charge of -1, and is the nucleophilic reaction site of the ion. Reactions between cyanide and carbonyl-containing compounds are normally performed by slowly adding acid to a mixture of sodium or potassium cyanide and the carbonyl compound. Adding acid slowly ensures that there is always plenty of free cyanide ion available to add to the carbonyl. The mechanism of this reaction is shown here.

The overall scheme of this reaction is:

ORGANIC CHEMWARE
7.21 Cyanohydrin reaction

A **cyanohydrin** is an sp^3-hybridized carbon connected to a hydroxyl group and a nitrile.

The reaction sequence and flow of electrons are much like those of the reactions involving $NaBH_4$ and $LiAlH_4$. The carbon atom of cyanide donates a pair of electrons to the carbonyl carbon while the π bond breaks, transferring a lone pair to oxygen. This step produces an intermediate alkoxide, which reacts with the added acid to form a **cyanohydrin.** Thus, the overall process adds the components of HCN *across the double bond* of the carbonyl.

The acid is added slowly in order to maintain a pH high enough to ensure a fast reaction. However, HCN is not a good nucleophile because the cyanide carbon has no lone pair to donate when H^+ is bonded to CN^-. So for this addition to work well, NaCN plus acid is added, rather than HCN alone.

7.7.1 Use of cyanohydrins to make sugars

The German chemist and Nobel Prize winner Emil Fischer (1852–1919) was one of the first chemists to make a naturally occurring organic compound from simpler materials. By 1894, he had proved the structures of all the sugars known at the time, and established many of the principles of stereochemistry.

A key part of this work was Fischer's synthesis of glucose from arabinose in 1890. He treated the five-carbon sugar arabinose (isolated from beets) with NaCN and acid to produce a cyanohydrin. Fischer then carried out a series of reactions that converted the nitrile into an aldehyde group, thus synthesizing the six-carbon sugars glucose and mannose. After reading this chapter, you should be able to draw mechanisms for all transformations in this synthesis except the final reduction step (which is described in Chapter 19).

CHECKPOINT 7.6

You should now be able to draw the mechanism and predict the products of a reaction of a cyanide ion with a carbonyl compound or related structure.

SOLVED PROBLEM

Draw the mechanism and predict the products of the following reaction.

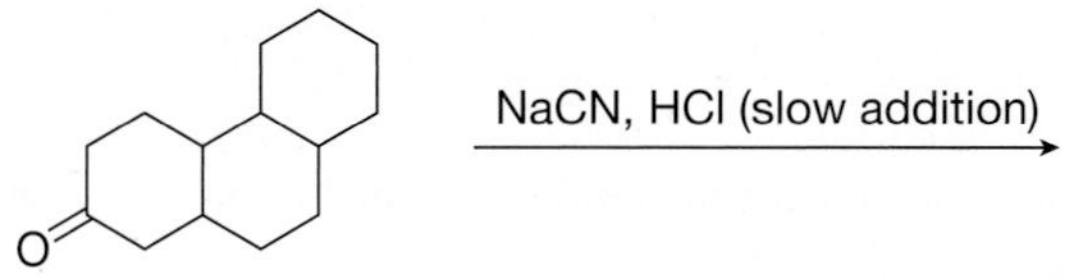

NaCN, HCl (slow addition)

STEP 1: Identify the electrophile and nucleophile in the first step of the reaction.

The starting material has a ketone functional group with a polarized C=O bond. The partial positive charge on the carbon of the carbonyl group makes this site an electrophile (which you can verify by drawing for the resonance structures). The NaCN reagent, on the other hand, has a negatively charged carbon atom with a lone pair available for sharing. This carbon atom will act as a nucleophile.

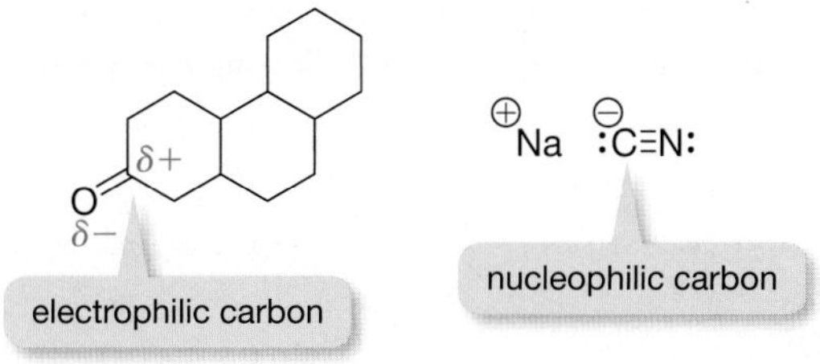

$\delta+$ $\delta-$ electrophilic carbon

$^{\oplus}$Na $^{\ominus}$:C≡N: nucleophilic carbon

STEP 2: Draw in the mechanistic arrows.

Electrons will flow from the lone pair on the negatively charged carbon atom to the carbon of the carbonyl group. Adding electrons to the carbonyl carbon would give it more than eight electrons, so the C=O π bond breaks, and a second arrow is needed to show the electrons flowing to the more electronegative oxygen atom.

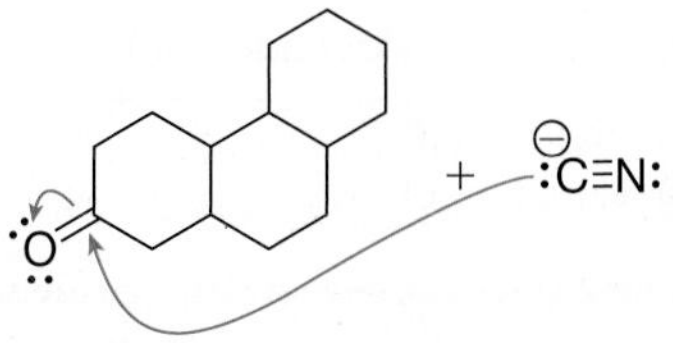

+ $^{\ominus}$:C≡N:

STEP 3: Use the arrows to determine the initial products of the reaction, including any formal charges.

A new carbon-carbon bond has formed between the cyanide's carbon atom and the carbon of the carbonyl. At the same time, the C=O π bond has broken. The formal charge on the nitrile carbon changes from -1 to 0 (electrons moving away from the atom) and the formal charge on oxygen changes from 0 to -1 (electrons moving toward the atom).

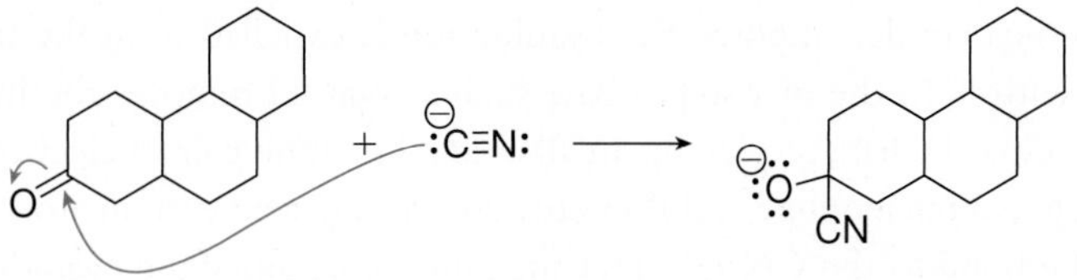

+ $^{\ominus}$:C≡N: ⟶ CN

STEP 4: Determine the neutral final product.

The alkoxide ion reacts with a molecule of HCl to give a neutral alcohol.

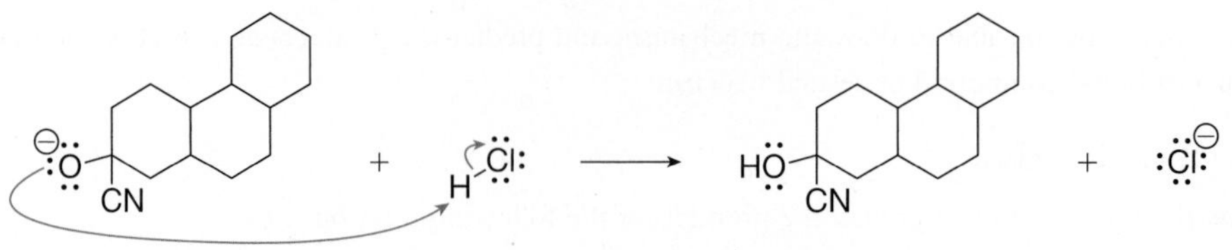

PRACTICE PROBLEM

7.11 a) Draw the mechanism and predict the products formed when each of the following reacts with potassium cyanide plus a slow addition of HCl.

i)

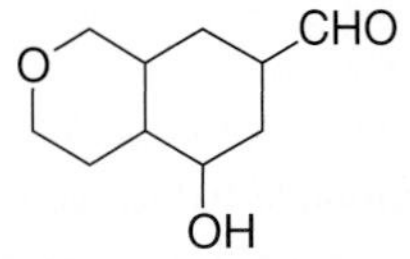

ii)

b) Predict the products formed when each of the following reacts with potassium cyanide.

i)

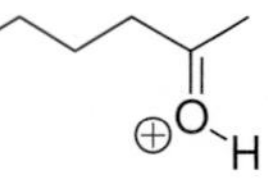

ii)

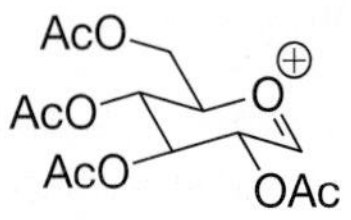

INTEGRATE THE SKILL

7.12 If you try to carry out the following sequence of reactions you will find that you can't isolate the desired amino alcohol at the end of the process. Explain why this synthesis does not work.

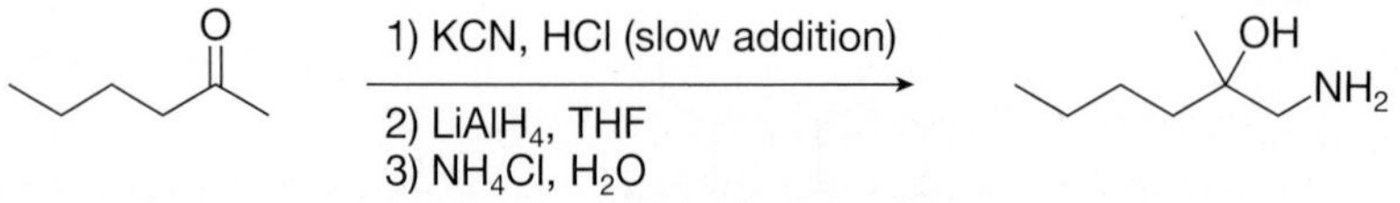

1) KCN, HCl (slow addition)
2) $LiAlH_4$, THF
3) NH_4Cl, H_2O

7.7.2 Reversing cyanohydrin formation

Exposing a cyanohydrin to basic conditions regenerates the carbonyl group while expelling the cyanide ion CN^-. Cyanohydrins must be handled carefully because the cyanide ion produced in the process can be dangerous if proper precautions are not taken (hydrogen cyanide is toxic). The mechanism of this transformation is simply the reverse of cyanide addition.

When a cyanohydrin decomposes, the cyanide ion is expelled from the molecule, taking a pair of electrons with it. In the first step, a base such as NaOH removes the hydrogen from the OH group of the cyanohydrin, producing an alkoxide. The lone pair of electrons on the oxygen of the alkoxide acts as a nucleophile and flows toward the adjacent carbon atom to form a new π bond, breaking the bond to the CN group at the same time. Since the cyanide ion carries away a pair of electrons it is known as a leaving group (see Section 7.8).

lone pair on oxygen acts as a nucleophile to form a π bond and expel cyanide

cyanohydrin

carbonyl

ORGANIC CHEMWARE
7.22 Curved arrow notation: Collapse of tetrahedral intermediate (basic conditions)

It is common practice to show this transformation in a single step. This is because the alkoxide is unstable, and it is converted to a carbonyl very quickly. So drawings of this mechanism often omit the alkoxide and instead show the π bond forming by a flow of the electrons toward the carbon.

ORGANIC CHEMWARE
7.23 Curved arrow notation: Collapse of tetrahedral intermediate (acidic conditions)

CHEMISTRY: EVERYTHING AND EVERYWHERE

Cyanide and Snake-Oil Medicine

Cyanohydrins are surprisingly common in nature. Some plants produce toxic or unpleasant-tasting cyanohydrins and related compounds to protect themselves from being eaten. Many plants store cyanide in the form of cyanogenic glycosides, in which the OH group of the cyanohydrin is converted to an OR group, making the cyanide more difficult to release. Cyanogenic glycosides are found in the seeds of many common fruits, including apples, apricots, cherries, and peaches. Bitter almonds contain amygdalin, which in the 1950s became the inspiration for a notorious medical fraud.

Cyanide is toxic to human cells and therefore toxic to cancer cells (which are out-of-control human cells). Laetrile has structural similarity to amygdalin, but releases cyanide much more readily. Laetrile, fraudulently promoted as a cancer cure by Ernst T. Krebs, was sold extensively in the 1950s and 1960s. Laetrile was not just ineffective as a cancer treatment, it hastened the deaths of many cancer sufferers. Krebs claimed that laetrile was a vitamin, which he called vitamin B17, to mislead patients and legislators into believing that laetrile was safe and effective. His disinformation was so successful that it was more than 20 years before his treatments were made illegal. Some cancer patients desperate for a cure, or searching for a "natural" way to treat their cancer, still seek laetrile treatments from quack practitioners in countries where it has not been banned.

amygdalin

laetrile

7.8 Leaving Groups

As described in Section 7.7.2, when a cyanohydrin decomposes, the cyanide ion takes away a pair of electrons when it is expelled from the molecule. This group is called a **leaving group** because it can stabilize a negative charge and leave with a pair of electrons when a bond breaks. Leaving groups tend to be electronegative atoms, or contain functional groups in which charge can be

Leaving groups remove electron pairs from functional groups.

well distributed by delocalization. Good leaving groups are very weak bases, so the ability of a functional group to act as leaving group can be estimated by considering the behaviour of the group in acid–base reactions. For example, the conjugate bases of the most commonly encountered strong acids all make excellent leaving groups (Table 7.1).

TABLE 7.1 Some Common Leaving Groups and Their Conjugate Acids

Strong Acid	Conjugate Base (weak base, excellent leaving group)
HI	I^-
HBr	Br^-
HCl	Cl^-
H_2SO_4	HSO_4^-
H_3O^+	H_2O

Good leaving groups stabilize negative charges well. The conjugate bases of strong acids make excellent leaving groups.

Br^- is a leaving group. It leaves with a pair of electrons.

The methods for estimating acid and base strengths (Chapter 6) can be applied to predict the ability of other potential leaving groups to receive and stabilize a negative charge during a reaction.

ORGANIC CHEMWARE
7.24 Leaving groups

CHECKPOINT 7.7

You should now be able to draw the mechanism and predict the products for a reaction that is the reverse of a nucleophilic addition to a carbonyl or related compound.

SOLVED PROBLEM

When the following compound is treated with a base, it quickly decomposes by expelling a leaving group. Draw the mechanism for the reaction and predict the final products formed.

OH OEt t-BuO$^-$

STEP 1: Identify the acid and the base.

The lone pairs on the alkoxide base (t-BuO$^-$) are associated with a negative charge, making them a strong base. The alcohol on the starting material is weakly acidic.

STEP 2: Draw the curved arrows for the acid–base reaction and the expected products.

OH OEt t-BuO$^-$ O$^-$ OEt + t-BuOH

STEP 3: Look for any nucleophiles and leaving groups attached to the same atom.

The lone pairs on the negatively charged oxygen atom make this atom nucleophilic. Two bonds over is an ethoxy group, which can potentially act as a leaving group. Although not the conjugate base of a strong acid, ethoxide (EtO^-) is able to stabilize a negative charge on its electronegative oxygen atom. Thus, ethoxide is a possible leaving group.

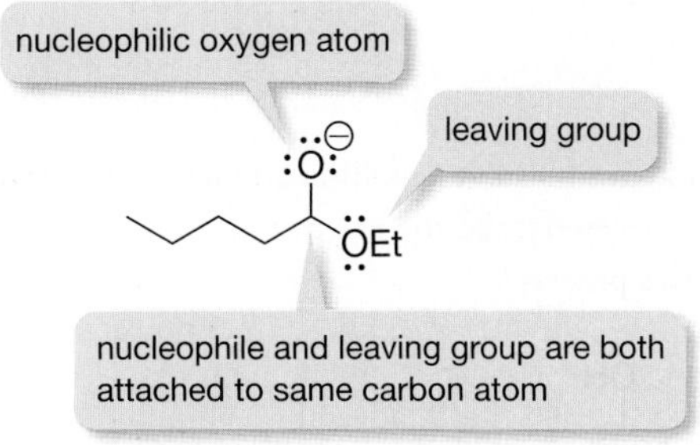

nucleophilic oxygen atom

leaving group

nucleophile and leaving group are both attached to same carbon atom

STEP 4: Draw in the curved arrows corresponding to the expulsion of the leaving group.

The arrows should begin at the source of electrons: a lone pair on the negatively charged oxygen. The electrons move to the carbon atom, creating a new C=O π bond. Since these electrons would overfill the valence shell of the carbon atom, the carbon-oxygen bond breaks. So you also need to draw a curved arrow to show the electrons from this σ bond flowing to the oxygen atom of the ethoxide leaving group.

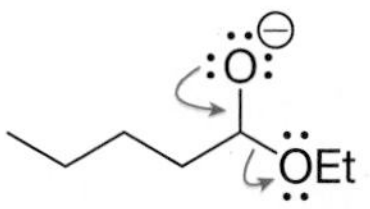

STEP 5: Use the arrows to determine the initial products of the reaction, including any formal charges.

A new C=O π bond has formed while a carbon-oxygen σ bond breaks. The formal charge on the alkoxide oxygen changes from -1 to 0, and the formal charge on the oxygen of the leaving group changes from 0 to -1.

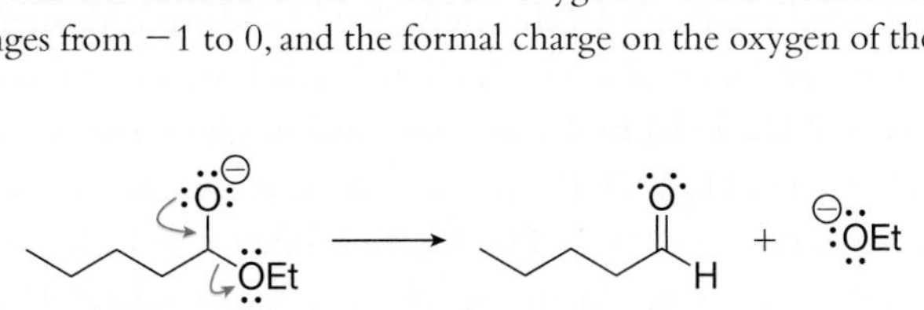

PRACTICE PROBLEM

7.13 a) The following structures are unstable and will quickly expel a leaving group. Draw the mechanisms and the final products for these processes.

i)

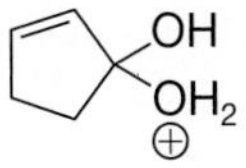

ii)

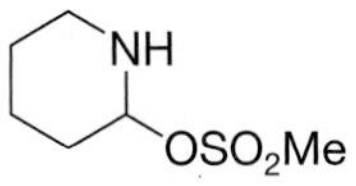

iii)

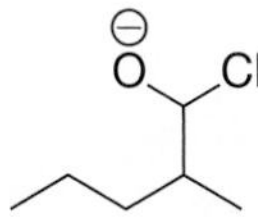

b) Predict the products of the following reactions.

i) HO CN, NaOH →

ii) OH, OH, NaOH →

INTEGRATE THE SKILL

7.14 Cyanohydrins can be made from bisulfite addition adducts. The advantages of this process are that the addition of acid is not necessary, and the reactions can be performed in water. Suggest mechanisms for all the steps in this process.

NaHSO$_3$ → $^{\ominus}O_3S$ OH, Na$^{\oplus}$ → NaCN → [H O] → NC OH + Na$_2$SO$_3$

7.9 Catalysis of Addition Reactions to Electrophilic π Bonds

A catalyst, often an acid or a base, can be used to accelerate a nucleophilic addition to a carbonyl group. The overall processes with acid and base catalysts are similar, but the order in which the hydrogen and nucleophile are introduced differs.

A **hydrate** is a functional group formed by the addition of water to a carbonyl group.

Water is not a very good nucleophile, and only slowly adds to carbonyl groups and related structures. However, in the presence of a base such as NaOH, water adds to the carbonyl of an aldehyde or ketone very quickly, and a **hydrate** is the final product. This occurs because OH^- is a much better nucleophile than H_2O. OH^- has three lone pairs and a negative charge, making it a much better electron donor than H_2O. The reaction begins with the nucleophilic addition of a hydroxide ion to the carbonyl, producing an alkoxide intermediate. The cation of the base is a spectator ion, so it is not shown in the reaction equation. The alkoxide quickly removes a hydrogen from a nearby molecule of water to produce the neutral hydrate and regenerate a molecule of hydroxide. Although this hydroxide molecule is not the one that began the sequence, there is no overall consumption of hydroxide. Thus, the base molecule catalyzes the addition process. Typically, only trace amounts of base are required.

water as an acid to protonate the intermediate
HO:$^{\ominus}$
:O:
NaOH
H_2O
:O:$^{\ominus}$
H O H
OH
carbonyl
:OH
:OH
hydrate
+ HO:$^{\ominus}$
hydroxide nucleophile adds to carbonyl as the π bond breaks
OH^- is a catalyst; the total amount present does not change during the reaction.

With an acid catalyst such as HCl, the addition process begins with protonation of the carbonyl oxygen. The lone pairs on the oxygen form a bond to the hydrogen of the acid, producing an oxonium intermediate (the spectator anion of the acid is not shown in the reaction equation). The oxonium is a very strong electrophile and readily accepts electrons from a nucleophile (water in this reaction). The oxygen atom of water has lone pairs of electrons that form a bond to the carbon atom of the oxonium, breaking the π bond. The resulting compound carries a positive charge. A water molecule, acting as a base, then removes a proton, producing the neutral final product. The acid is a catalyst since a hydronium is consumed in the initial step, but another is generated at the end. Many organic catalysts are consumed initially because they participate directly in the reaction's electron flow like other reagents, but are regenerated at the end of the reaction.

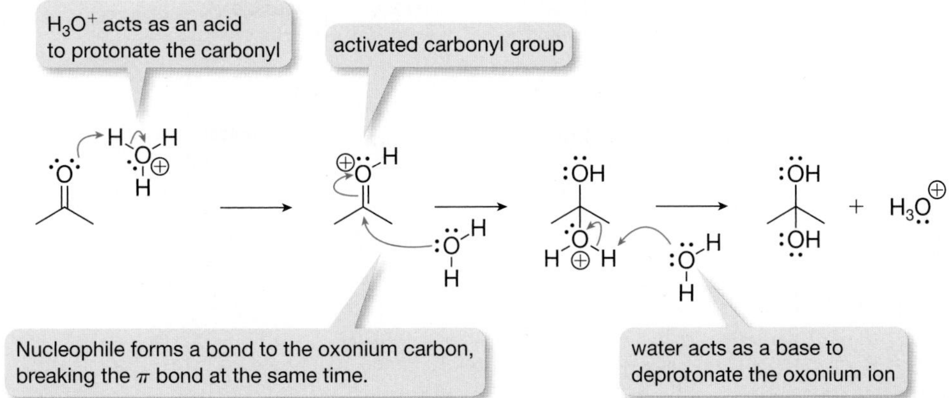

The reactions with both acid and base catalysts have a step involving the addition of a nucleophile to the carbonyl (shown in a box in the following diagram). The difference between the catalytic sequences is simply the order of protonation and deprotonation. Basic conditions accelerate many reactions by removing protons from the nucleophile, making it more able to donate electrons to the electrophile. Similarly, acids can accelerate reactions by making the electrophile more able to accept electrons from the nucleophile.

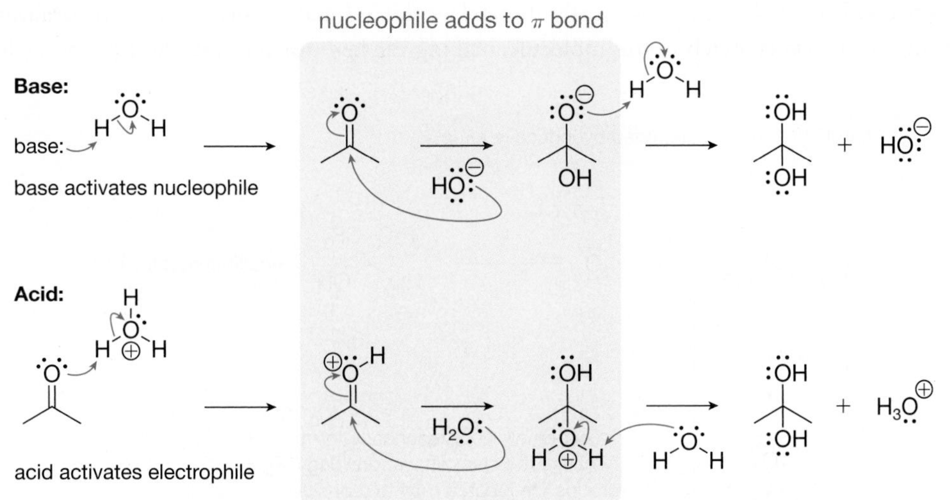

ORGANIC CHEMWARE
7.25 Hydration of aldehydes/ketones (basic conditions)

STUDENT TIP
When an acid such as HCl is used in water, the acid disassociates to form H_3O^+. Similarly, in other protic solvents such as alcohols, the solvents form H_2OR^+. The solvent normally acts as a base, not the anion of the acid.

ORGANIC CHEMWARE
7.26 Curved arrow notation: Carbonyl addition (acidic conditions)

ORGANIC CHEMWARE
7.27 Addition to carbonyl: General mechanism (acidic conditions)

ORGANIC CHEMWARE
7.28 Hydration of aldehydes/ketones (acidic conditions)

STUDENT TIP
Remember that mechanistic arrows show the movement of electrons, not atoms. A common error when writing mechanisms for hydrogen removal in acid–base reactions is to draw the mechanistic arrow pointing toward the hydrogen to indicate the movement of the atom.

7.9.1 Aldehydes and ketones in equilibrium with hydrates

The reaction between water and carbonyl groups is a very fast process. Any aldehyde or ketone placed in water can completely exchange oxygen atoms with the water in minutes (or seconds for the more reactive compounds). Hydrate formation is reversible, so the reaction is an equilibrium process. The position of the equilibrium is controlled by the relative energies—hence, stabilities—of the ketone (carbonyl) and hydrate forms of the molecule. For most aldehydes and ketones, the carbonyl form is much more stable than the hydrate form due to delocalization in the carbonyl form. The charged resonance form of a ketone is a minor contributor, but it usually does contribute to the stability of the molecule in the carbonyl form. For acetone, the ratio of ketone form to hydrate form is approximately 1000 : 1 in water.

resonance forms of the carbonyl

no resonance possible

+ H_2O ⇌ HO OH

equilibrium ratio 1000 : 1

Other factors can affect the position of the hydration equilibrium. For example, aldehydes hydrate more readily than ketones. The addition of water to an aldehyde carbonyl generates less steric hindrance than the addition to a ketone because the aldehyde has a single hydrogen atom at the location where the corresponding ketone has a carbon group or chain.

For some carbonyl-containing compounds, the hydrate is the more stable form; such molecules often exist primarily as hydrates. For example, the equilibrium for hexafluoroacetone favours the hydrate form by a ratio of approximately 1 : 10^6. The resonance form of hexafluoroacetone has a positive charge next to two very strong electron-withdrawing CF_3 groups. The inductive effects of these groups greatly destabilize this resonance form, and so delocalization does not happen very much in this molecule, making the keto form less stable than the hydrate.

carbonyl form is *destabilized* by inductive effects

F_3C CF_3 + H_2O ⇌ F_3C CF_3 HO OH

equilibrium ratio 1 : 10^6

an extremely poor resonance form because the electron withdrawing CF_3 groups are located next to a positive charge

CHECKPOINT 7.8

You should now be able to draw the mechanism and predict the products of the reaction of water (or alcohols) with a carbonyl compound under either acidic or basic conditions.

SOLVED PROBLEM

Draw the mechanism and predict the product(s) of the following reaction.

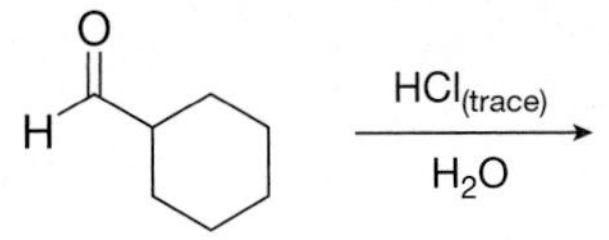

$HCl_{(trace)}$, H_2O

STEP 1: Identify the electrophile and nucleophile.

The starting reactant has an aldehyde functional group with a polarized C=O bond. The partial positive charge on the carbon of the carbonyl group makes this site electrophilic (as you can verify by drawing the resonance structures). Water mixed with a trace amount of HCl produces hydronium ions and chloride ions in water. Although both chloride ions and water molecules can act as nucleophiles (they both have lone pairs available for sharing), there are so many more water molecules than chloride ions that water is the most likely nucleophile in this reaction.

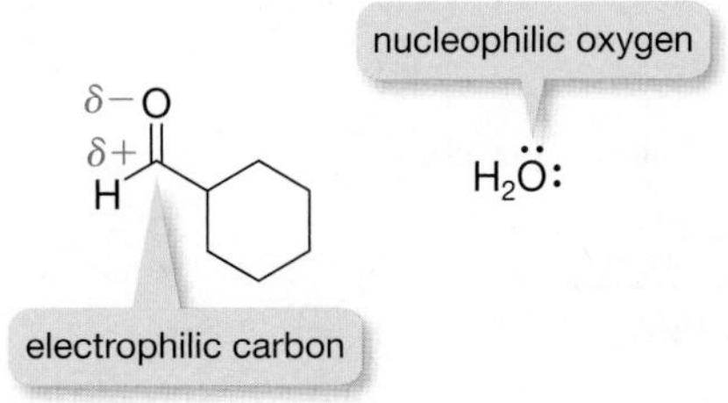

nucleophilic oxygen

electrophilic carbon

STEP 2: Determine whether an acid or a base is available to catalyze the reaction.

HCl is an acid. In water, it will form hydronium ions (H_3O^+), which can catalyze the reaction.

STEP 3: Use the acid to activate the carbonyl.

Acids catalyze carbonyl additions by protonating the carbonyl oxygen, making the group more electrophilic.

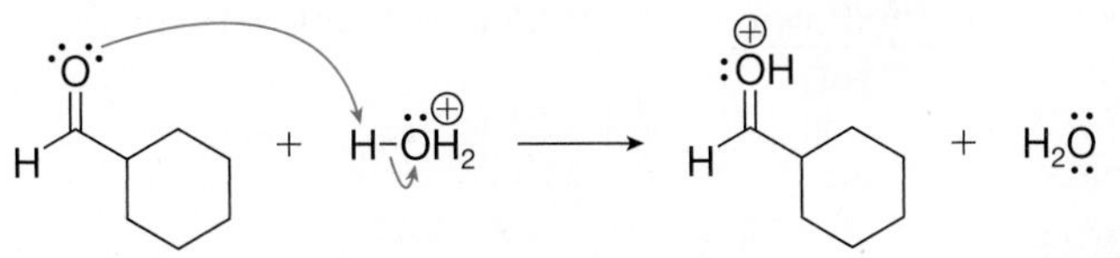

STEP 4: Draw in the mechanistic arrows.

The first arrow should begin at the lone pairs on the oxygen atom of the water and point to the carbon atom of the carbonyl. Since adding electrons to this carbon would exceed its valence octet, the C=O π bond breaks. So, the second arrow points from the C=O π bond to the positive charge on the oxygen atom.

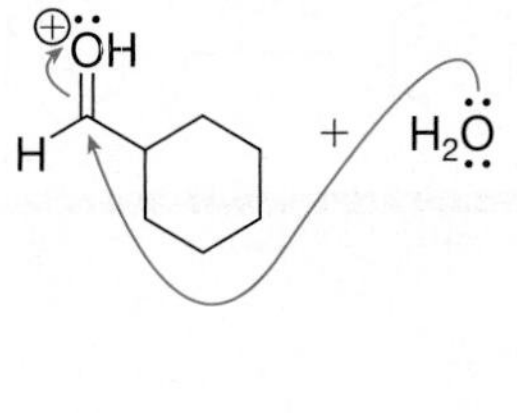

STEP 5: Use the arrows to determine the products of the reaction, including any formal charges.

A new C–O bond has formed between the oxygen of water and the carbon of the oxonium ion, while the C=O π bond has broken. The formal charge on the oxygen from water changes from 0 to +1, and the formal charge on the oxonium oxygen changes from +1 to 0.

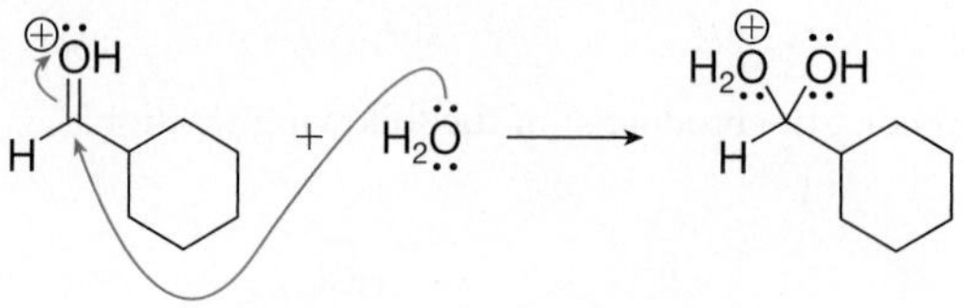

STEP 6: Use a molecule of solvent to produce the neutral final product.

A molecule of water acts as a weak base and deprotonates the intermediate product to give a neutral hydrate.

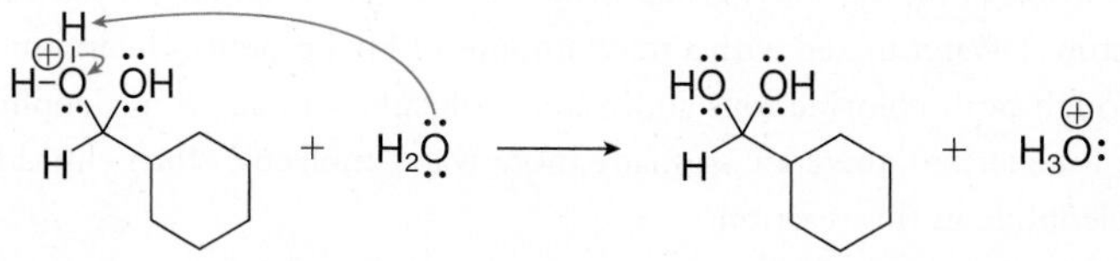

PRACTICE PROBLEM

7.15 Predict the products of the following reactions.

a) 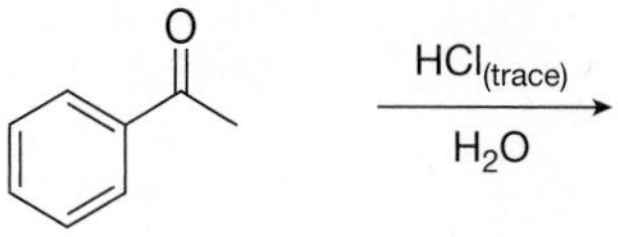 $\xrightarrow[H_2O]{HCl_{(trace)}}$

b) 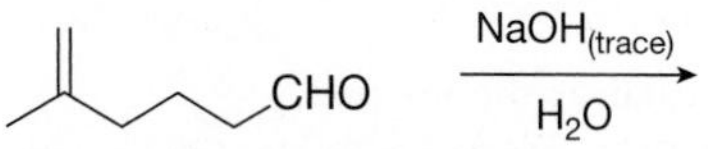 CHO $\xrightarrow[H_2O]{NaOH_{(trace)}}$

c) 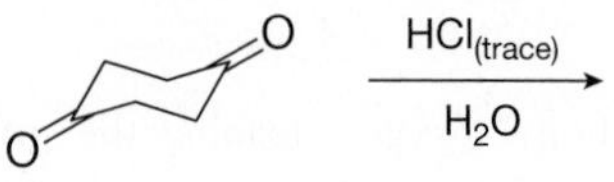 $\xrightarrow[H_2O]{HCl_{(trace)}}$

d) 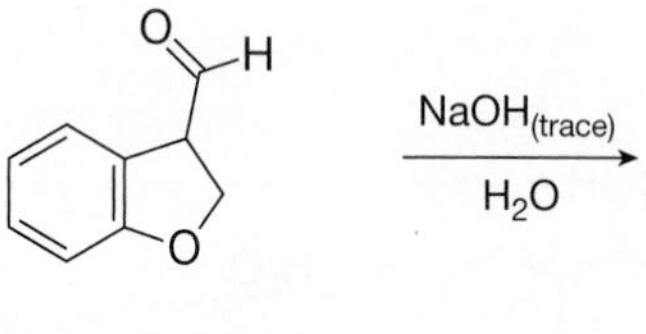 $\xrightarrow[H_2O]{NaOH_{(trace)}}$

INTEGRATE THE SKILL

7.16 When ketones are mixed with water, they very quickly undergo an "exchange" reaction with water. This process can be demonstrated by dissolving a ketone in water that is isotopically labelled with ^{18}O. Using the following structure, show how this process occurs, first with acid catalysis, then with base catalysis. (Hint: Hydrate formation is a reversible process.)

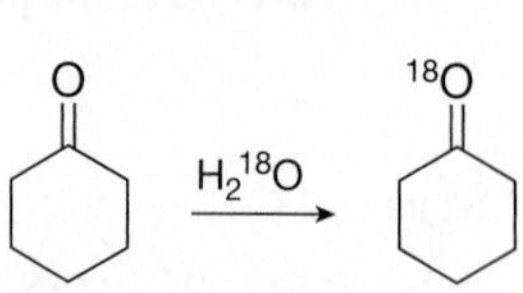

$\xrightarrow{H_2{}^{18}O}$

7.9.2 Hemiacetals and intramolecular reactions of carbonyl compounds

An acid can catalyze the addition of alcohol across a carbonyl group to form a **hemiacetal**, as shown in the following reaction equation. This reaction is essentially the same as a hydration, except that the nucleophile that adds across the carbonyl group is an alcohol, ROH, rather than water, HOH.

A **hemiacetal** is an sp^3-hybridized carbon connected to an OH group and to an OR group.

hemiacetal

A hydroxyaldehyde has the functional groups of both an aldehyde and an alcohol. When exposed to a small amount of acid, a hydroxyaldehyde can undergo an intramolecular reaction with the same mechanism as the addition reaction above. Although the transformation of the hydroxyaldehyde may appear more complex, both reactions are nucleophilic additions to a carbonyl group. The carbon chain of hydroxyaldehyde holds an electrophilic carbonyl and a nucleophilic hydroxyl in close proximity, favouring an intramolecular reaction. Once the roles of the functional groups are identified, mechanistic analysis is used to predict the formation of the cyclic product.

hemiacetal

Intermolecular and intramolecular forms of this reaction are essentially the same. The only difference is that the two reacting groups are linked by a carbon chain in the intramolecular version.

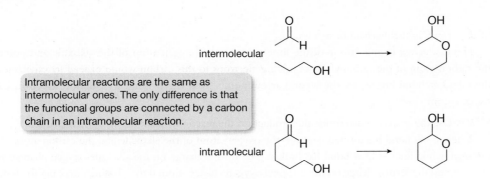

intermolecular

Intramolecular reactions are the same as intermolecular ones. The only difference is that the functional groups are connected by a carbon chain in an intramolecular reaction.

intramolecular

ORGANIC CHEMWARE
7.29 Intramolecular hemiacetal formation

ORGANIC CHEMWARE
7.30 Acetal and hemiacetal formation

Intramolecular reactions tend to be very fast when the reacting groups are close enough to form rings. In principle, any size of ring is possible; however, the formation of five- and six-membered rings is particularly favourable because the reacting groups are relatively close together in the original acyclic molecule and the geometry by which the groups can approach gives relatively low activation energies.

CHECKPOINT 7.9

You should now be able to draw the mechanism and predict the products for the reaction of an alcohol with a carbonyl compound under either acidic or basic conditions.

SOLVED PROBLEM

Use a mechanism to predict the product of the following reaction.

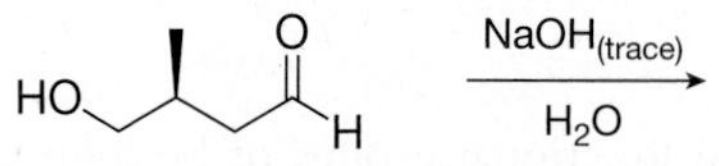

HO ... H (O) $\xrightarrow[H_2O]{NaOH_{(trace)}}$

STEP 1: Identify the electrophile and nucleophile.

The electrophile will be the carbonyl group with its partial positive charge on the carbon atom. There are, however, three possible nucleophiles: hydroxide, water, and the alcohol on the left side of the aldehyde. Hydroxide is present only in trace amounts, so it is likely a catalyst for the reaction. Since intramolecular reactions are faster than intermolecular reactions, the alcohol, which is directly connected to the electrophile, is more likely than water to act as the nucleophile in this reaction.

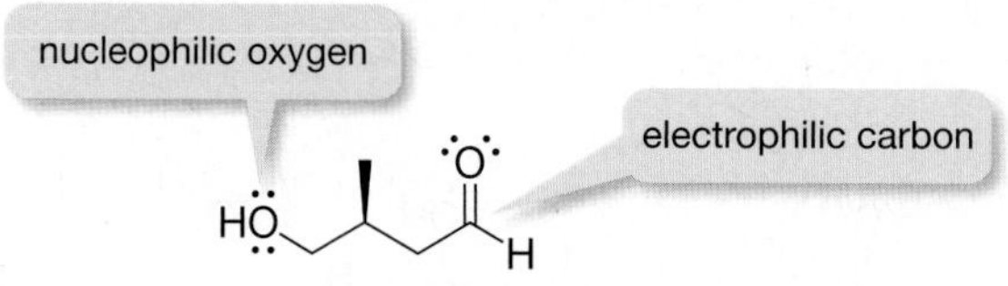

STEP 2: Determine whether an acid or a base is available to catalyze the reaction.

The base NaOH is present.

STEP 3: Use the base to activate the nucleophile by deprotonation.

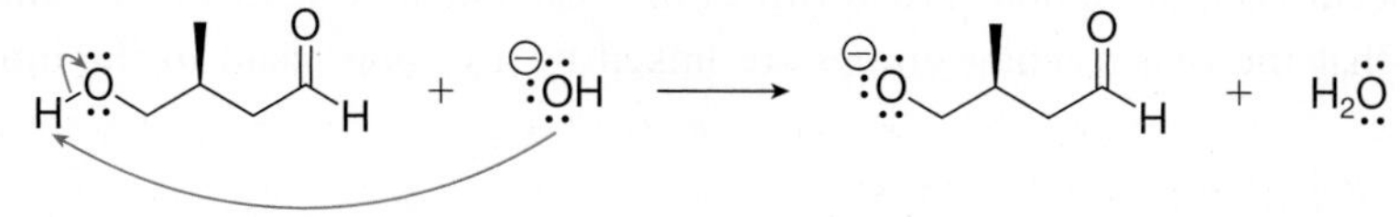

STEP 4: Draw in the mechanistic arrows.

The first arrow begins at one of the lone pairs on the oxygen atom of the alkoxide and points to the carbon atom of the carbonyl. Since adding electrons to this carbon would exceed its valence octet, the C=O π bond breaks. So, the second arrow points from the π bond to the more electronegative oxygen atom.

STEP 5: Use the arrows to determine the products of the reaction, including any formal charges.

A new C–O bond has formed between the oxygen atom of the alkoxide and the carbon atom of the carbonyl, while the C=O π bond has broken. The formal charge on the alkoxide oxygen changes from −1 to 0, and the formal charge on the carbonyl oxygen changes from 0 to −1. When drawing the product of an intramolecular reaction, numbering the atoms can help you keep track of the locations of the new bonds and any substituents on the ring. In this example, the new C–O bond forms between atoms 1 and 5, indicating that a 5-membered ring forms and the methyl substituent is located at carbon 3. (Note: When there are stereocentres on reacting molecules, there is no change to their configuration unless they are directly involved in the reaction.)

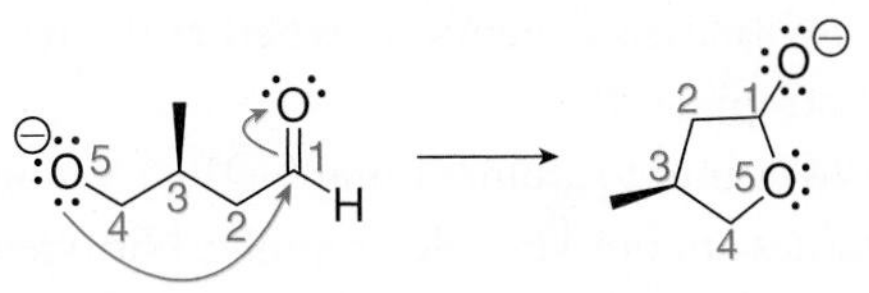

STEP 6: Use a molecule of solvent to produce the neutral final product.

A molecule of water acts as a weak acid and protonates the alkoxide intermediate. This reaction yields a neutral hemiacetal and regenerates the hydroxide catalyst.

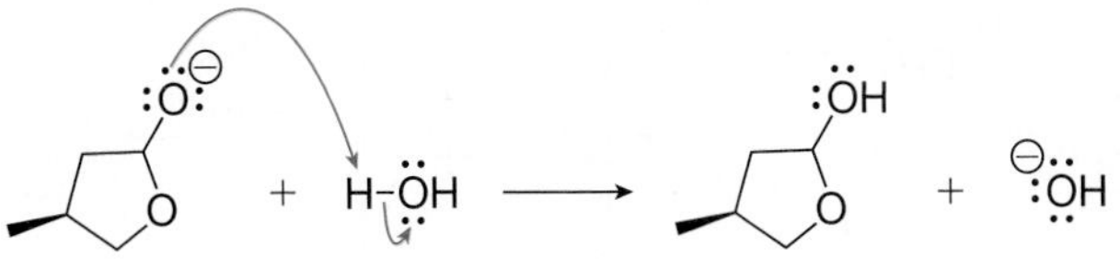

PRACTICE PROBLEM

7.17 Draw the mechanism and predict the products of the following reactions.

a) 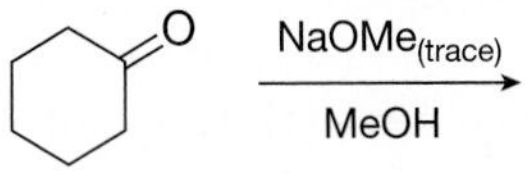 $\xrightarrow[\text{MeOH}]{\text{NaOMe}_{(\text{trace})}}$

b) 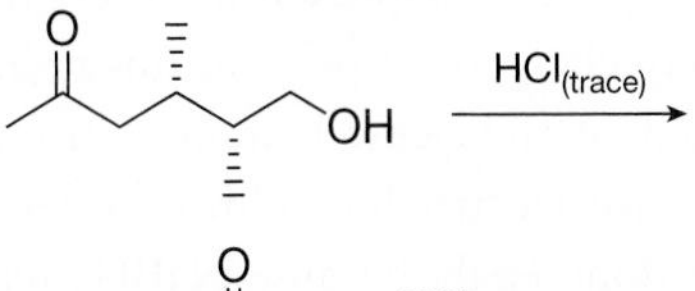 $\xrightarrow{\text{HCl}_{(\text{trace})}}$

c) 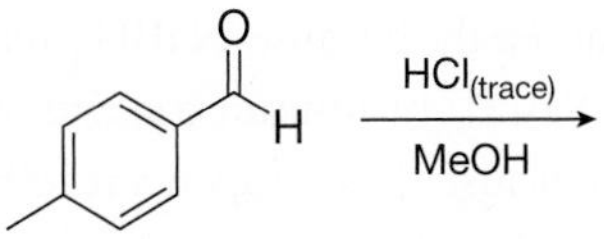 $\xrightarrow[\text{MeOH}]{\text{HCl}_{(\text{trace})}}$

d) 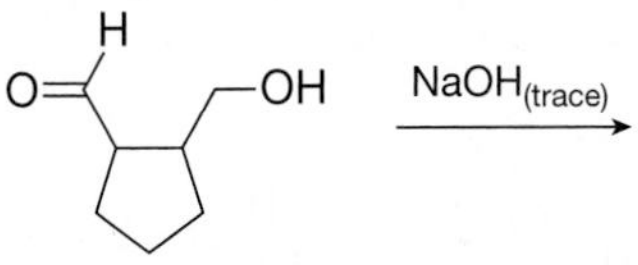 $\xrightarrow{\text{NaOH}_{(\text{trace})}}$

INTEGRATE THE SKILL

7.18 Although the following compound has both a ketone and an alcohol, it does not undergo an intramolecular reaction when treated with a catalytic amount of acid. Explain why this reaction does not take place, and predict the product(s) formed instead.

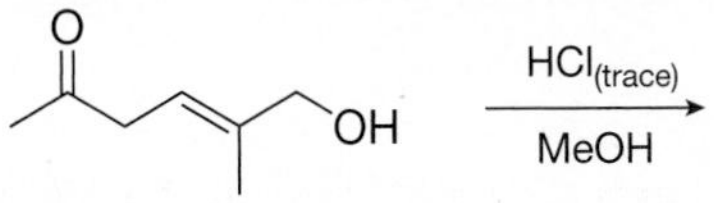

$\xrightarrow[\text{MeOH}]{\text{HCl}_{(\text{trace})}}$

7.9.3 Using catalysts to accelerate reductions

An acid or base can be used as a catalyst to accelerate a wide range of organic reactions. The selection of an acid or base as a catalyst normally depends on the reactivity of the functional groups involved and knowledge of the reaction mechanism. In general, *bases* accelerate a reaction by increasing the reactivity of a neutral *nucleophile* by converting it into its anionic conjugate base. This conjugate base donates ("pushes") electrons more readily than the neutral nucleophile. *Acid* catalysis activates a neutral *electrophile* by converting it to a positively charged material, which accepts ("pulls") electrons more readily.

STUDENT TIP
Catalyzed reaction mechanisms should be drawn to show either acid or base catalysis—never both, since the two would neutralize each other.

The choice of acid or base catalysis depends on whether the reaction requires the activation of the nucleophile or the electrophile.

Consider the reduction of imines to amines using $NaBH_4$. Without a catalyst, this reaction is usually very slow because imines are not very electrophilic. Nitrogen is less electronegative than oxygen, so the polarity of the carbon-nitrogen bond is not as strong as that of the corresponding carbon-oxygen bond. Furthermore, direct addition of hydride to the imine carbon produces a negative charge on the nitrogen, which is not very stable (amine $pK_a > 35$). Furthermore, $NaBH_4$ is not a strong nucleophile, so the reaction is slow.

EtOH (solvent)

The pK_a of the amine NH is approximately 35, meaning that N^- is very unstable and difficult to form.

The reduction of an imine can be accelerated in either of two ways: (1) the use of a stronger nucleophile, such as $LiAlH_4$; and (2) acid catalysis. This second approach involves treating an imine with acid to form an iminium ion. The positively charged nitrogen in this iminium ion is much more electron withdrawing than the nitrogen of the neutral imine. Thus, protonation of the imine by acid produces a much more reactive electrophile. However, reducing an imine with $NaBH_4$ in the presence of acid cannot be done easily because $NaBH_4$ reacts with acid. In acidic conditions, sodium cyanoborohydride ($NaCNBH_3$) or lithium triacetoxyborohydride ($LiBH(OAc)_3$) are commonly used to reduce imines because these reagents readily donate hydride but do not react with mild acids. The nucleophilic BH_3CN^- ion reacts with the activated iminium, producing the amine directly.

$H{-}BH_2CN$

acetic acid is a common solvent for imine reductions

7.10 Stereochemistry of Nucleophilic Additions to π Bonds

During the addition of nucleophiles to carbonyls and related compounds, the hybridization of the electrophilic carbon changes from sp^2 to sp^3. This change produces stereogenic centres if this carbon atom has four different substituents. Therefore, two product stereoisomers may be possible.

Consider a reaction in which groups A and B on the carbonyl reagent differ from each other and from the incoming nucleophile. Nucleophiles can approach the flat sp^2-hybridized carbonyl from above or below the plane of the carbonyl group. Approach from above produces one stereoisomer, and approach from below produces the other. If the carbonyl group is symmetric about the plane, the nucleophile is as likely to approach from above as from below, producing an equal mixture of the two possible isomers.

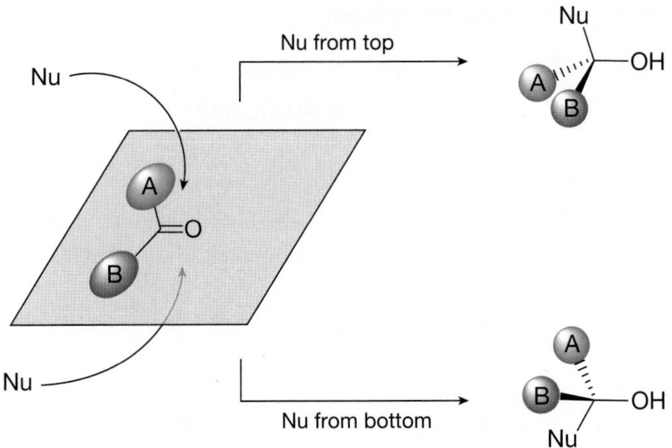

Sometimes the carbonyl compound is not completely symmetric with respect to the plane of the atoms in the carbonyl functional group. This asymmetry can make the reaction stereoselective. In the following example, the methyl group beside the carbonyl hinders the approach of a nucleophile to the bottom face of the carbonyl. Since only a little hydride addition occurs on this blocked face, the reaction produces largely the *cis* isomer.

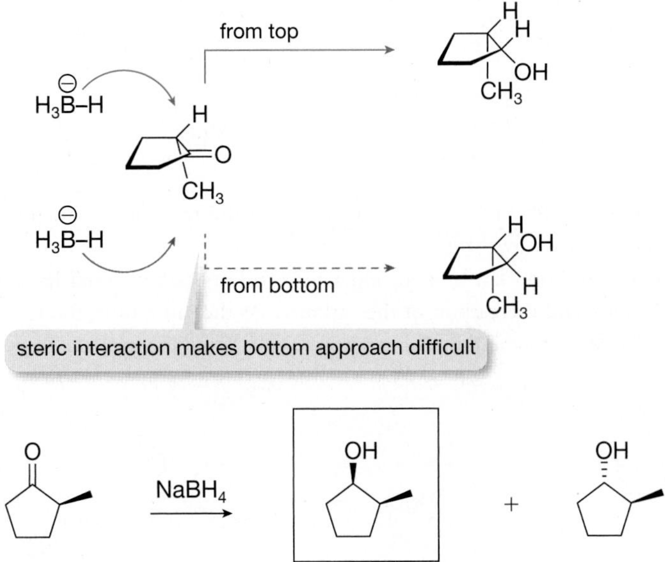

CHECKPOINT 7.10

You should now be able to predict the stereochemistry of the products formed when nucleophiles are added to carbonyl compounds and related structures.

SOLVED PROBLEM

Predict the product(s) of the following reaction, including any stereochemistry. If two or more stereoisomers can be made, indicate which, if any, the reaction will favour.

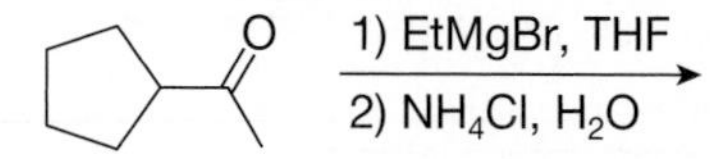

1) EtMgBr, THF
2) NH_4Cl, H_2O

STEP 1: Identify the electrophile and nucleophile.

Ethylmagnesium bromide is a Grignard reagent with a highly nucleophilic carbon atom. The ketone, on the other hand, has a polarized C=O bond with an electrophilic carbon atom.

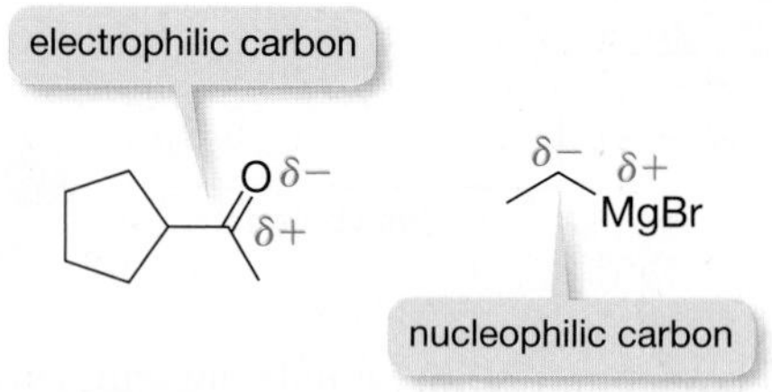

electrophilic carbon

$\delta-$ $\delta+$ $\delta-$ $\delta+$ MgBr

nucleophilic carbon

STEP 2: Draw in the mechanistic arrows.

Electrons flow from the carbon-magnesium bond to the carbon atom of the carbonyl group. To accommodate these electrons, the C=O π bond breaks. So the mechanism has two arrows: one pointing from the carbon-magnesium bond to the carbon of the carbonyl, and other pointing from the C=O π bond to the more electronegative oxygen atom.

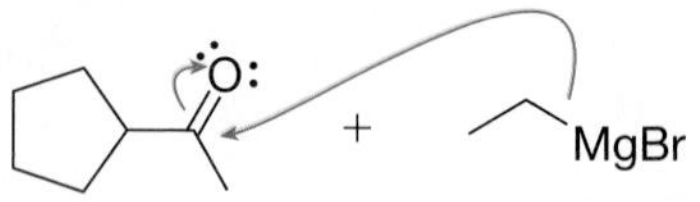

+ MgBr

STEP 3: Use the arrows to determine the initial products of the reaction, including any formal charges; ignore any stereochemistry, for now.

The carbon-magnesium bond has broken, and a new carbon-carbon bond has formed between the carbon of the ethyl group and the carbon of the carbonyl. At the same time, the C=O π bond has also broken. The formal charge on magnesium changes from 0 to +1, and the formal charge on oxygen changes from 0 to −1.

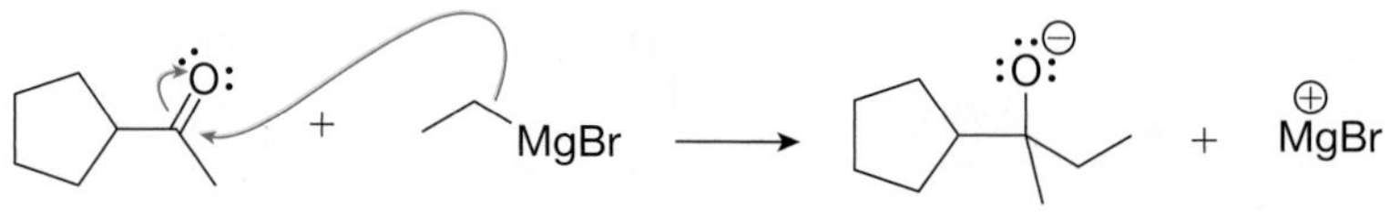

+ MgBr ⟶ + $\overset{\oplus}{\text{MgBr}}$

STEP 4: Use the aqueous acid added in the second step to produce the neutral final product.

Ammonium chloride in water creates hydronium ions, which protonate the alkoxide intermediate to give a neutral alcohol and a molecule of water.

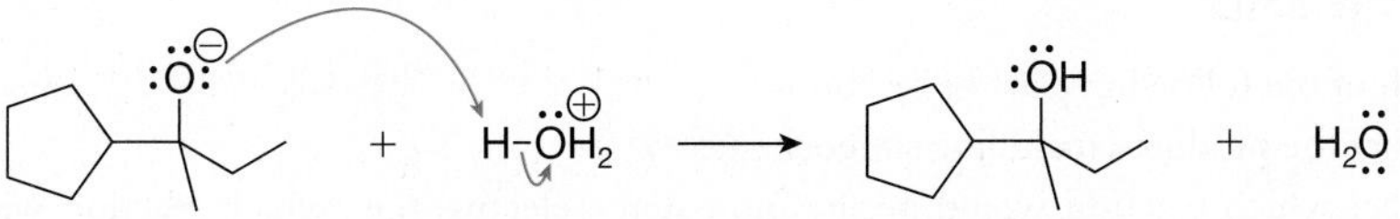

$$\text{alkoxide} + H\text{–}\ddot{O}H_2^{\oplus} \longrightarrow \text{alcohol} + H_2\ddot{O}$$

STEP 5: Determine whether any new stereocentres have been created during the reaction.

The product has one new chirality centre, which is marked with an asterisk in following structure.

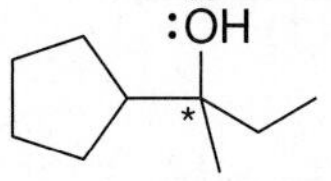

STEP 6: Draw all possible stereoisomers of the product.

Since the Grignard reagent can attack the carbonyl from either face, the new chirality centre has either the *R* or *S* configuration. Therefore, there are two possible stereoisomers of the product.

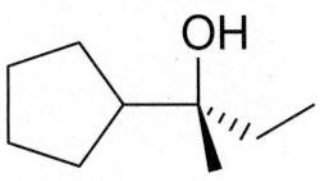

***R* enantiomer** 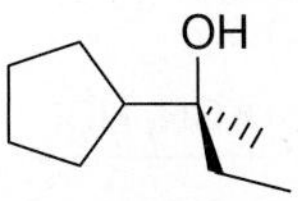 **S enantiomer**

STEP 7: Determine whether any stereocentres on the reactants would favour the formation of one stereoisomer over the other.

Since there are no chirality centres on either of the starting reactants, the two enantiomers will be formed in equal quantities.

PRACTICE PROBLEM

7.19 Predict the product(s) of each of the following reactions, including any stereochemistry. If two or more stereoisomers are possible, indicate which, if any, the reaction will favour.

a) 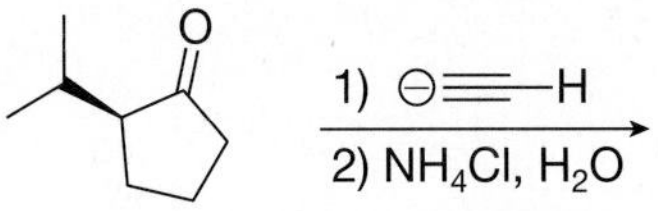 1) $\ominus$≡–H; 2) NH_4Cl, H_2O

b) 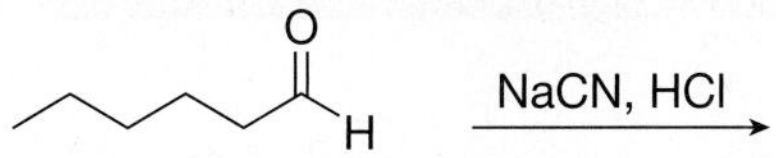 NaCN, HCl

c) 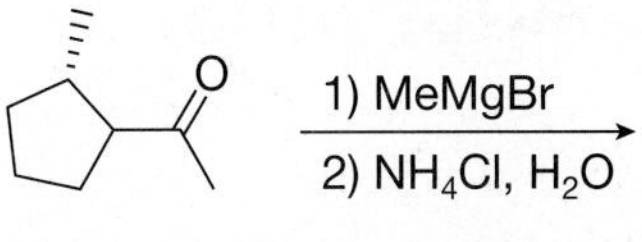 1) MeMgBr; 2) NH_4Cl, H_2O

d) 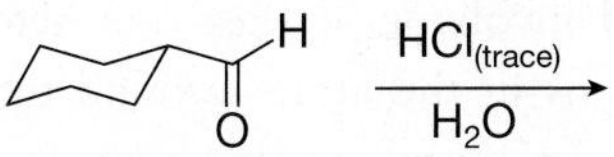 $HCl_{(trace)}$, H_2O

e) 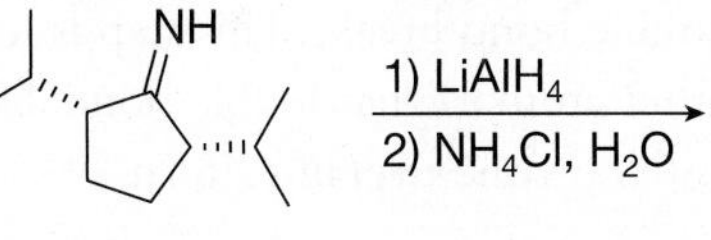 1) $LiAlH_4$; 2) NH_4Cl, H_2O

INTEGRATE THE SKILL

7.20 For each of the following pairs of reactions:

i) Predict the products, including stereochemistry.

ii) Predict which reaction would be the most stereoselective (i.e., which reaction will favour one isomer over the other to the greatest extent).

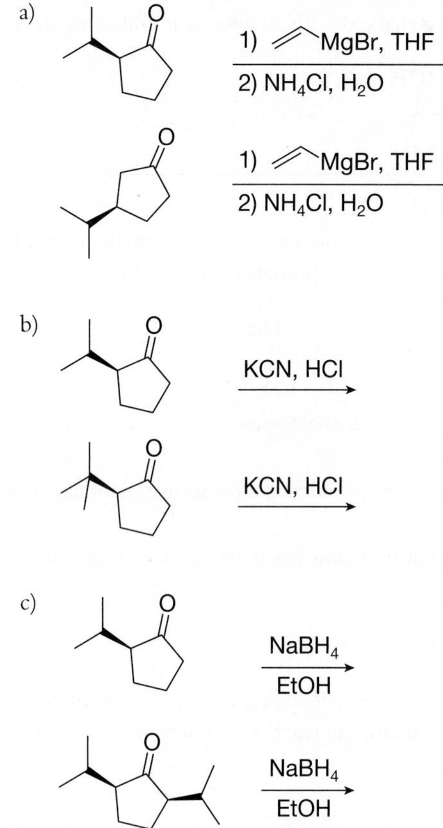

7.11 Patterns in Nucleophilic Additions to π Bonds

Nucleophilic additions to electrophilic π bonds (π bond involving at least one heteroatom) follow patterns that result from the distribution of electrons in the atoms around the double bond. The key reaction pathway involves the addition of a nucleophile to the most electrophilic atom of the double bond while the π bond of the double bond breaks. This step is followed by the addition of an electrophile (usually H^+) to the other atom of the double bond. Details vary from reaction to reaction, but the electron flows follow the same overall pattern.

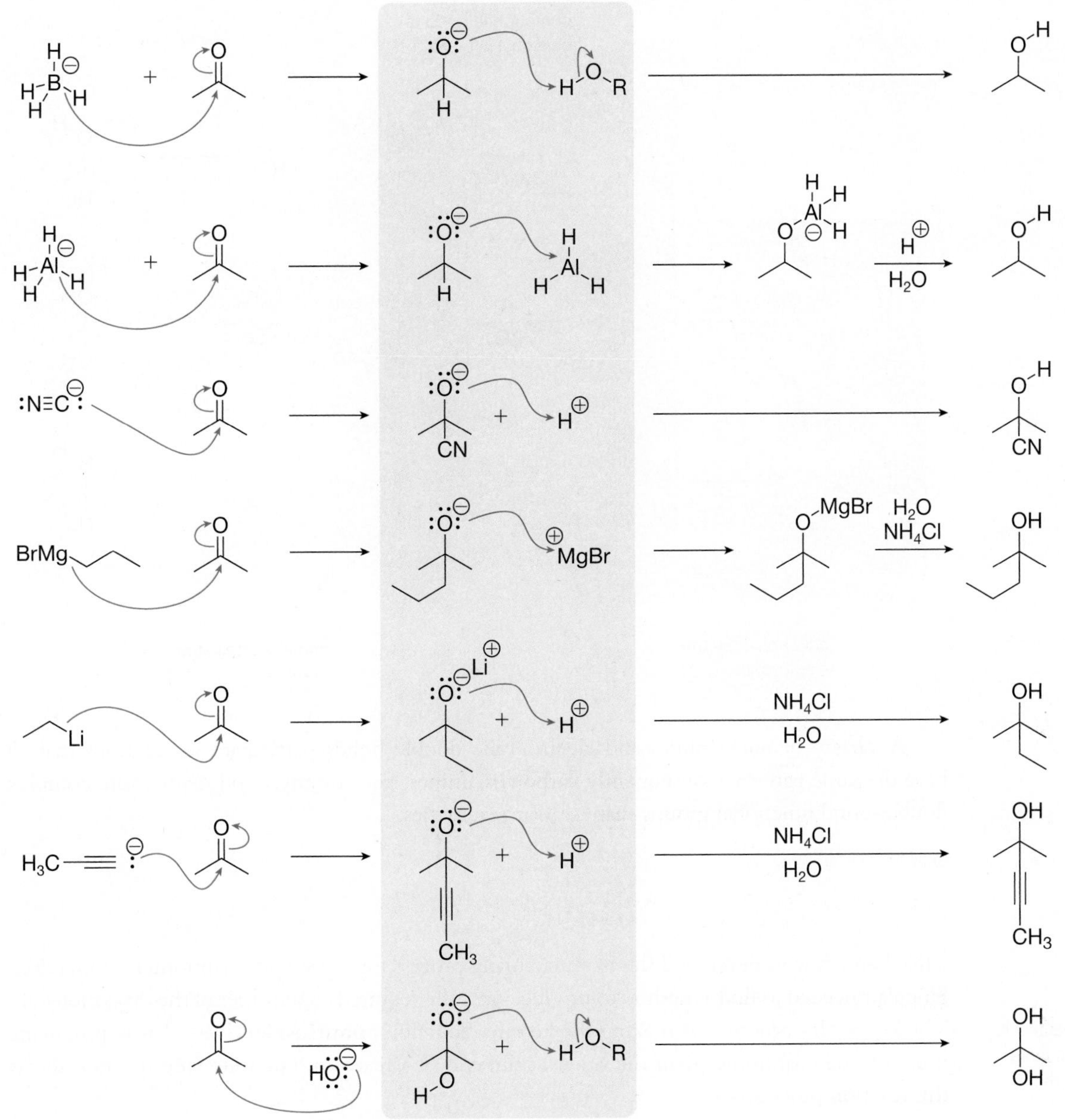

In acid conditions, the same processes occur, but the order of the protonation steps is reversed. Basic lone pairs on the most nucleophilic atom of the double bond first react with an acid. This reaction activates the electrophilic group, enabling it to accept electrons more readily. The nucleophile adds to the electrophilic carbon, in the process breaking the π bond in the same manner as in a base-catalyzed process. The result is often a positively charged intermediate that becomes neutral by giving up a proton in a later step. The bond between the hydrogen and the positively charged atom is broken by a flow of electrons onto the positively charged atom since it is very electronegative. This flow neutralizes the positive charge on that atom.

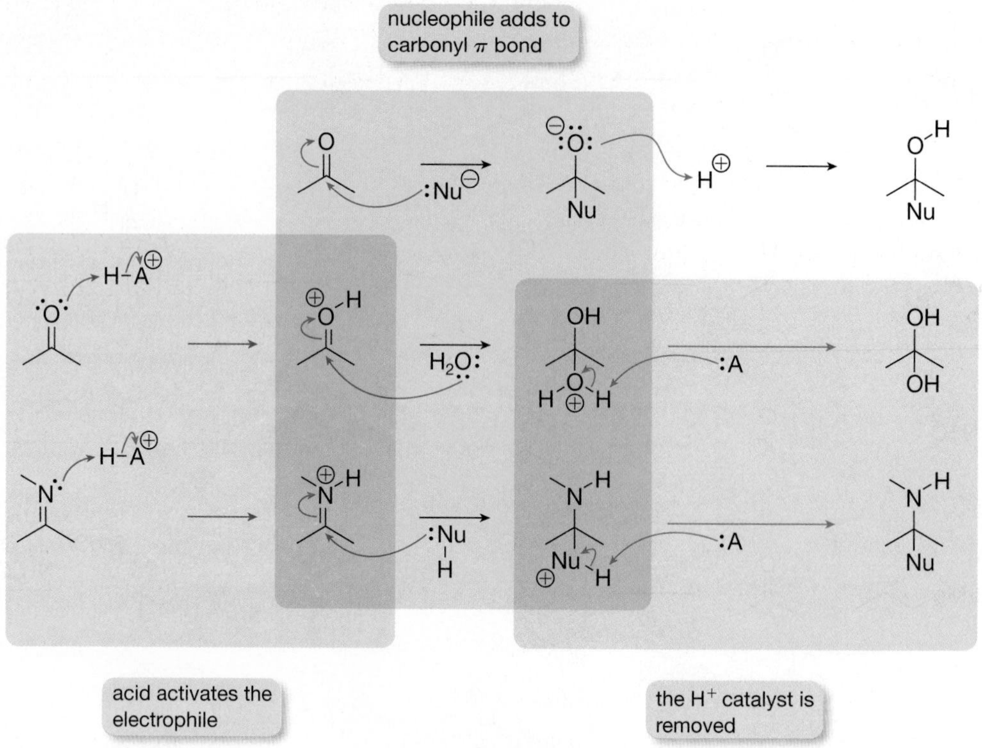

nucleophile adds to carbonyl π bond

acid activates the electrophile

the H^+ catalyst is removed

A variety of nucleophiles and electrophilic double bonds participate in reactions that all have the same pattern. Consequently, carbonyls, imines, thiocarbonyls, and many more complex double-bond functional groups share similar reactivities.

Bringing It Together

The chemists who developed the manufacturing process for the single-enantiomer asthma drug Singulair needed to find a method to produce only the required enantiomer of the drug molecule. The key to this process was making the ketone reduction **enantioselective**—that is, producing more of one enantiomer than the other at the end of a reaction. The following diagram shows this reaction process.

An **enantioselective** reaction is a reaction that produces more of one enantiomer than the other.

$NaBH_4$

racemic mixture

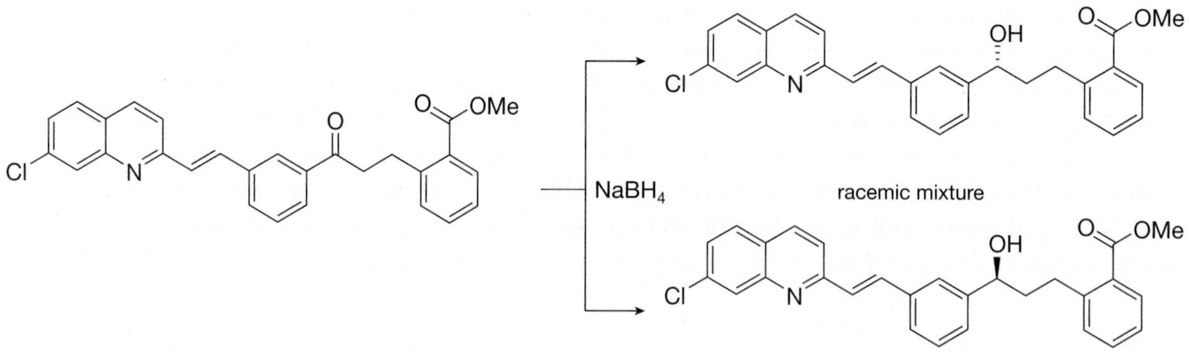

Certain reagents can be enantioselective because they react with only one face of an asymmetric ketone. As described in Section 7.10, this selective reactivity produces only one enantiomer. Recently, drug manufacturers have begun using enzymes and microorganisms to make chemical transformations enantioselective. Since all enzymes are chiral, they can interact selectively with achiral materials to form single enantiomers. The transition states leading to each enantiomer are diastereomers that have different energies. The pathway with the lowest energy prevails, so the reaction favours the resulting enantiomer.

One family of enzymes used to perform chiral reactions industrially is the ketoreductases, which transform ketones into alcohols. The mechanism catalyzed by these enzymes is very similar to the mechanism of $NaBH_4$ reduction. The active site of these enzymes is a chiral environment that allows the hydride to add to only one face of the ketone. Consequently, the product is a single enantiomer. The enzymes act with a co-factor called NADPH, which supplies a hydride to the ketone, while tyrosine (an amino acid) in the active site supplies a proton. NADPH is the biological equivalent of $NaBH_4$. The overall process adds a hydrogen atom to each side of the ketone double bond, forming an alcohol.

Cl, N, O, O, OMe — ketoreductase / NADPH → Cl, N, OH, O, OMe

one enantiomer

Such reactions facilitate the manufacture of modern drugs and advanced materials.

You Can Now

- Use the orbital structure and resonance forms of carbonyls and related structures to predict and explain their reactivity with nucleophiles.
- Draw the mechanism and predict the products for the reaction of carbonyl compounds and related structures with nucleophiles. Such nucleophiles include the following:
  - hydride equivalents (such as $NaBH_4$)
  - Grignard reagents
  - organolithiums
  - acetylides
  - cyanides
  - water and alcohols
- Determine whether the addition of either aqueous acid or aqueous base is needed at the end of a reaction with carbonyls and related structures.
- Interpret over-the-arrow notation for single- and multi-step reactions.
- Distinguish between oxidation and reduction reactions involving organic compounds.
- Identify a good leaving group.
- Draw mechanisms and predict the products for reactions that are the reverse of a nucleophilic addition to a carbonyl or related compound.
- Differentiate between acid- and base-catalyzed reactions, and draw mechanisms for both.
- Recognize structures with both a nucleophile and carbonyl (or related functional group), and draw the products of their intramolecular reaction.
- Predict the stereochemistry of the products formed when nucleophiles add to carbonyl compounds and related structures.

A Mechanistic Re-View

Addition of hydride to π bonds

Addition of Grignard, alkyl lithium, or acetylides to π bonds

Addition of cyanide ion to π bonds

Addition of water or alcohols to π bonds

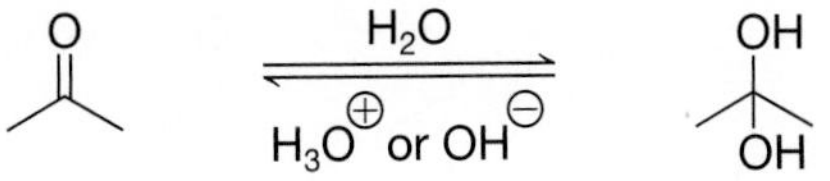

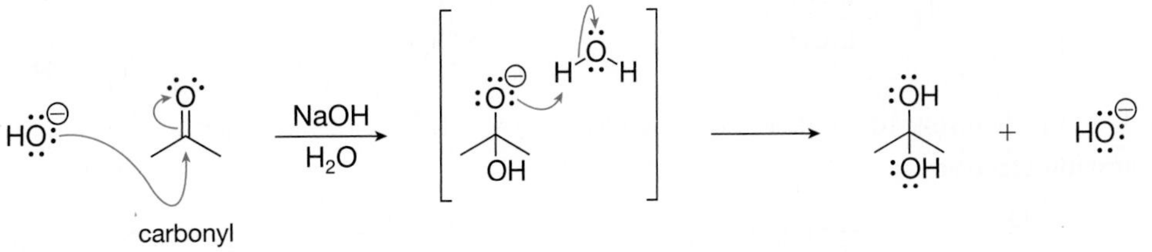

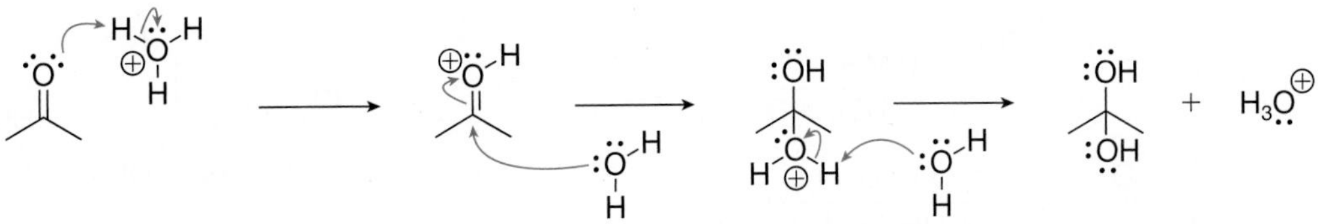

H_2O ; $H_3O^{\oplus}$ or $OH^{\ominus}$

NaOH, H_2O

carbonyl

Problems

7.21 Use resonance structures to predict the most electrophilic site(s) on the following molecules. Do not consider hydrogen atoms.

a)

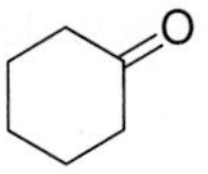

b)  CHO

c) 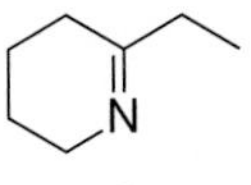 N

d) 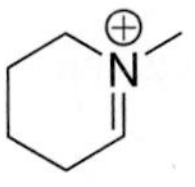 N

e) 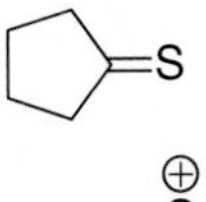 S

f) 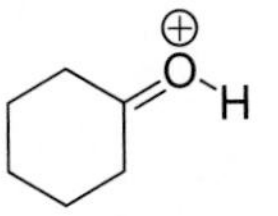 O H

g) O

h) N

i) 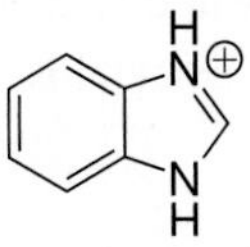 H N N H

7.22 Classify each of the reactions in the following multistep sequence as an oxidation, a reduction, or neither.

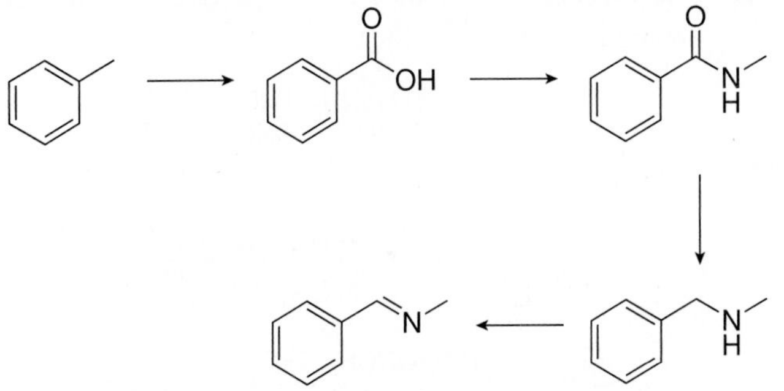

7.23 Use mechanisms to predict the products of the following reactions.

a) 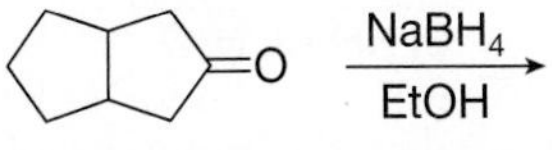 $NaBH_4$, EtOH

b) 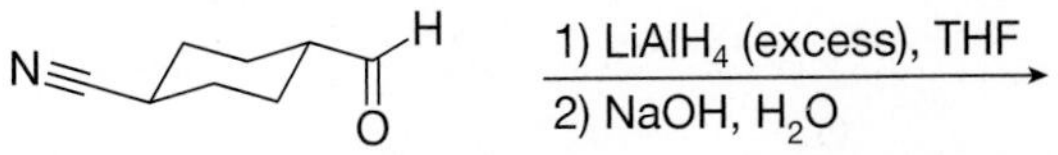 1) $LiAlH_4$ (excess), THF 2) NaOH, H_2O

c) 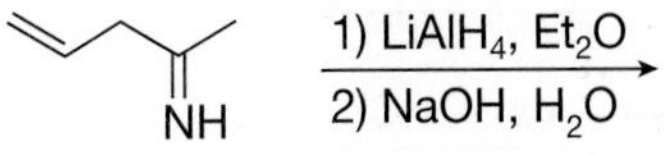 1) $LiAlH_4$, Et_2O 2) NaOH, H_2O

d) 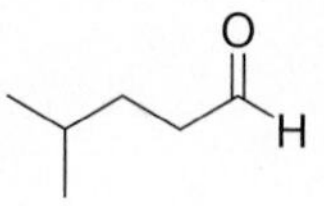 NaBH$_4$, EtOH

e) NaBD$_4$, EtOH

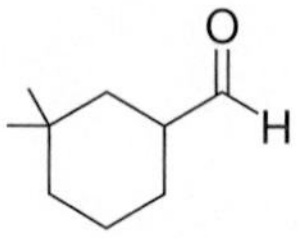

7.24 Use mechanisms to predict the products of the following reactions.

a) 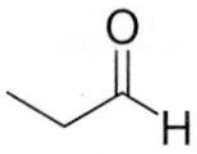 1) EtMgBr, THF 2) NH_4Cl, H_2O

b) 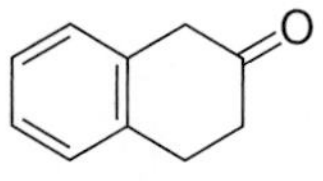 1) MeLi, THF 2) HCl, H_2O

c) 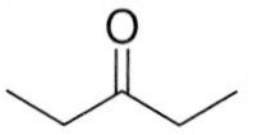 1) *n*-BuLi, Et_2O 2) NH_4Cl, H_2O

d) 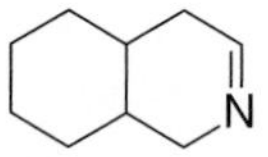 1) PhMgBr, Et_2O 2) HCl, H_2O

e) 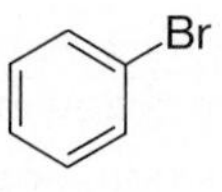 Br; 1) Mg, Et_2O 2) $(CH_3)_2CO$ 3) NH_4Cl, H_2O

7.25 Use mechanisms to predict the products of the following reactions.

a) 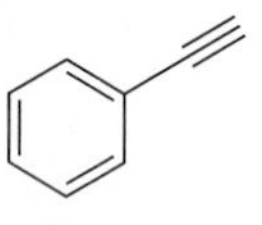 1) $NaNH_2$, THF 2) CH_3CH_2CHO 3) NH_4Cl, H_2O

b) 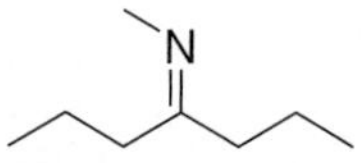 1) NaCCH, Et_2O 2) NH_4Cl, H_2O

c) 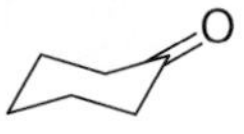 1) $NaCCCH_3$, THF 2) NH_4Cl, H_2O

d) 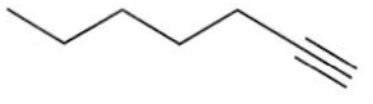 1) NaH, THF 2) 2-hexanone 3) NH_4Cl, H_2O

7.26 Use mechanisms to predict the products of the following reactions.

a) 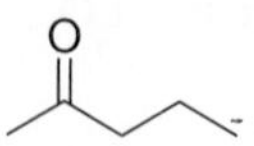 NaCN, HCl

b) 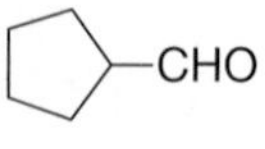 CHO; KCN, HCl

c) 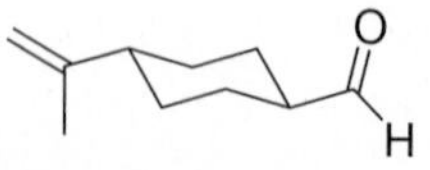 NaCN, HCl

d) 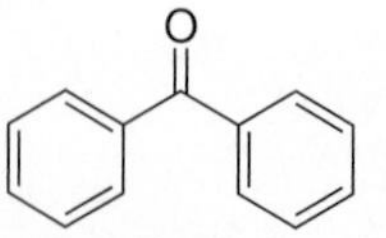 KCN, HCl

7.27 Use mechanisms to predict the products of the following reactions.

a) 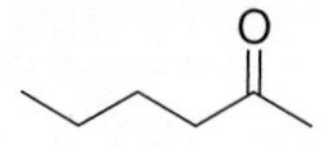 $HCl_{(trace)}$, MeOH

b) 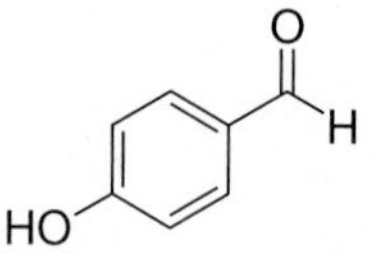 HO; $HCl_{(trace)}$, EtOH

c) 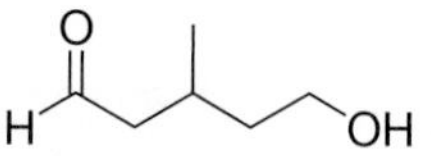 OH; $HCl_{(trace)}$

d) 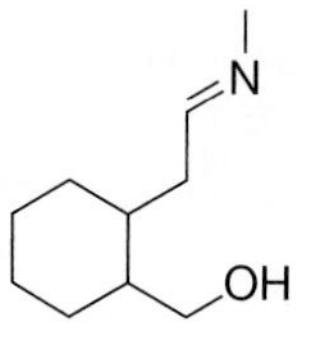 N; OH; $HCl_{(trace)}$

7.28 Predict the products formed when cyclopentanone is treated with each of the following. Write the mechanism for each reaction.

a) $NaBH_4$ in EtOH
b) NaCN and HCl
c) EtSH and a trace of HCl
d) PhMgBr in THF, then NH_4Cl in H_2O
e) propylamine and NaOH

7.29 Predict the products formed when benzaldehyde is treated with each of the following.

a) $NaBD_4$ in EtOH
b) KCN and HCl
c) *i*-PrOH and a trace of HCl
d) NaOMe in MeOH
e) *n*-BuLi in THF, then HCl in H_2O

7.30 Predict the products formed when the given compound is treated with each of the following.

N

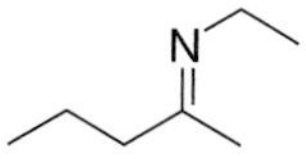

a) $LiAlH_4$ in THF, then HCl in H_2O
b) CH_3CH_2CCNa in Et_2O, then NH_4Cl in H_2O
c) EtMgBr in THF, then NH_4Cl in H_2O
d) NaOH in H_2O
e) PhLi in Et_2O, then HCl in H_2O

7.31 When the following molecule is treated with a catalytic amount of acid or base, a cyclic hemiacetal is formed.

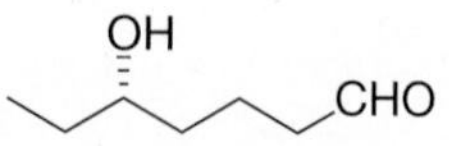

OH CHO

a) Use mechanistic arrows to predict the structure of the product and show how it is formed.
b) Draw all possible stereoisomers of the product using line-bond drawings.
c) Determine the absolute configurations of all the stereocentres in part (b).
d) Draw all possible stereoisomers of the product in both chair forms.
e) Write mechanisms for the reverse reaction (ring opening) using acid and base catalysis.

7.32 Provide a set of reaction conditions that could be used to carry out the following transformations.

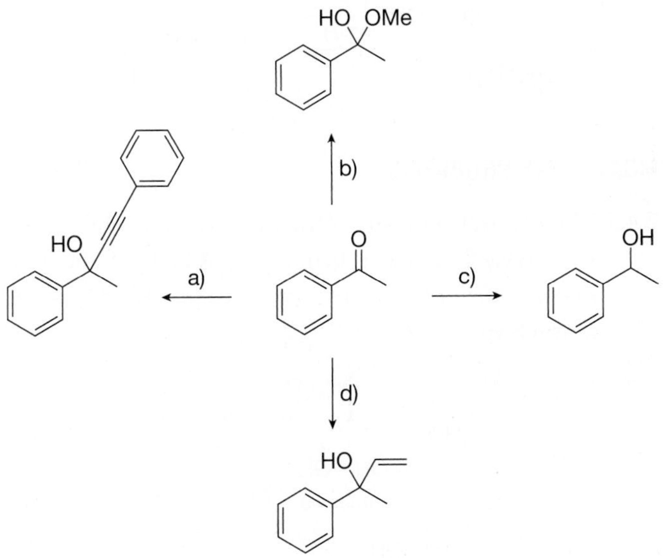

a) b) c) d)

7.33 Predict the products of the following reactions, including all possible stereoisomers. Indicate which stereoisomer, if any, the reaction will favour.

a) 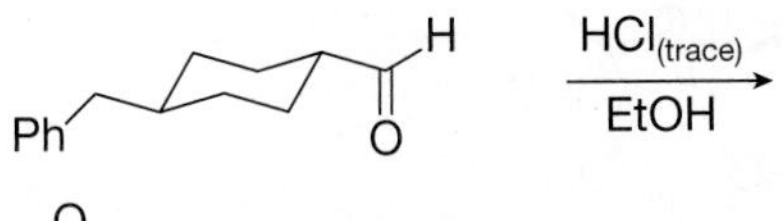 $HCl_{(trace)}$, EtOH

b) 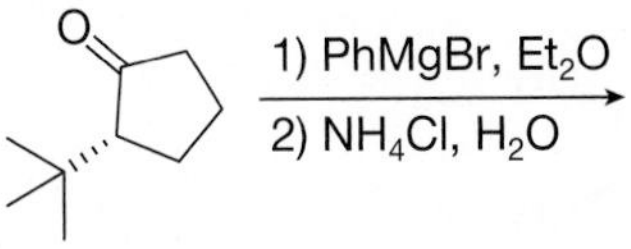 1) PhMgBr, Et_2O 2) NH_4Cl, H_2O

c) 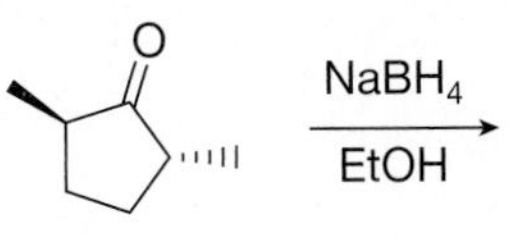 $NaBH_4$, EtOH

d) 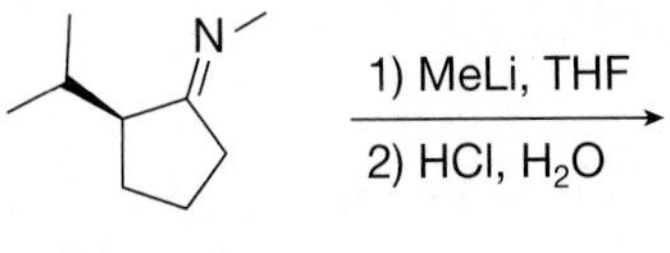 1) MeLi, THF 2) HCl, H_2O

e) 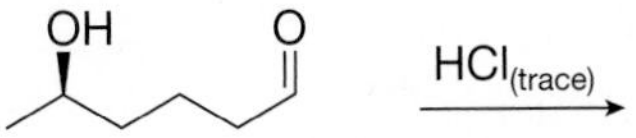 $HCl_{(trace)}$

7.34 The following structures yield carbonyl compounds when treated with aqueous base. Draw a mechanism for each transformation, and determine the aldehyde or ketone that results.

a)

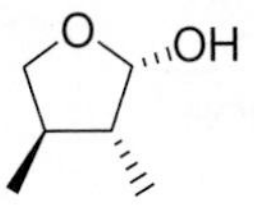

b)

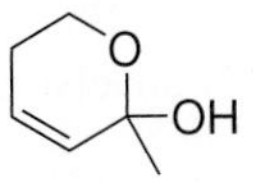

c)

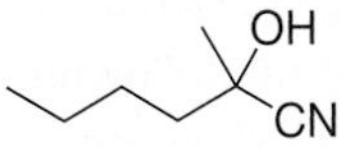

7.35 Write a detailed mechanism to explain the following reaction.

NaCN

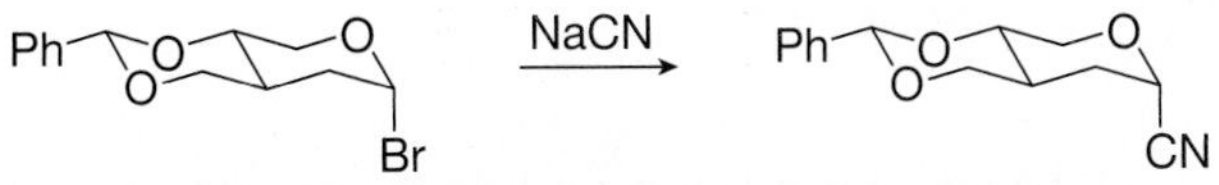

7.36 Depending on the nature of substituent X, the structure below might expel a leaving group X^- to form the following ketone.

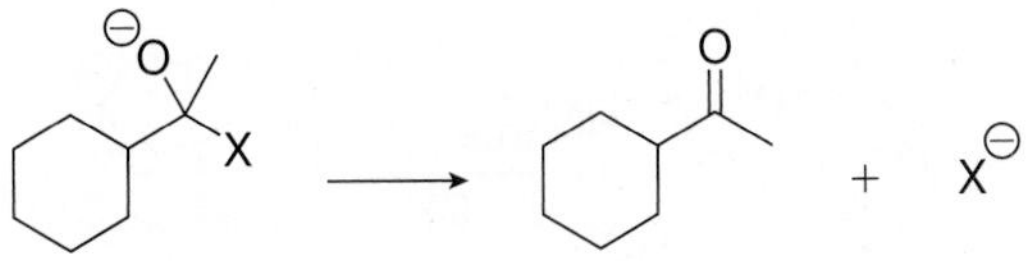

For which of the following substituents would you expect the above reaction to take place? Justify your answers.

a) CH_2CH_3
b) H
c) Br
d) Cl
e) OH
f) $OCOCH_3$
g) NH_2

7.37 In water, glucose exists as a mixture of isomers with the following structures.

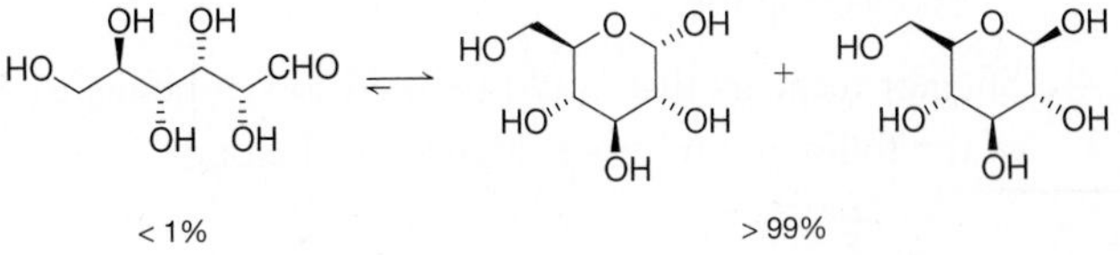

< 1% > 99%

Use mechanistic arrows to show how each of these materials forms in the presence of the following.

a) a catalytic amount of acid
b) a catalytic amount of base

7.38 When dissolved in ethanol with a small amount of acid, cyclohexanone is converted into a hemiacetal, which then transforms into an acetal.

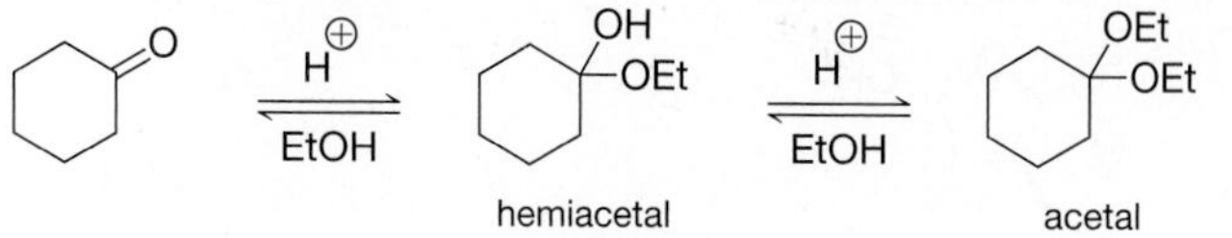

hemiacetal acetal

a) Draw the mechanism for the conversion of the ketone into the hemiacetal.

b) The first step in converting the hemiacetal into the acetal involves formation of the following oxonium ion. Draw the mechanism to show how this ion is made from the hemiacetal.

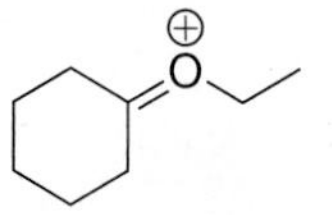

c) Propose a mechanism for how the oxonium structure in part (b) is converted into the final acetal product.

d) Use the mechanistic steps from parts (a), (b), and (c) to propose a mechanism by which the following acetal is formed from cyclohexanone.

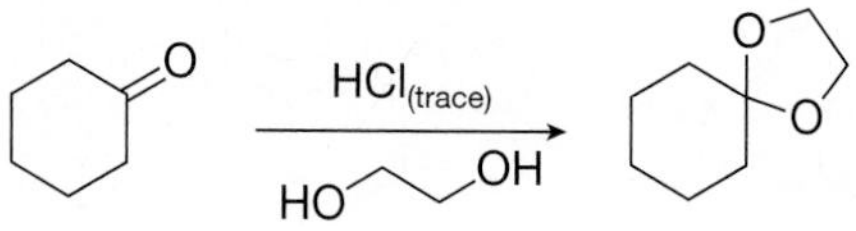

7.39 a) Draw an energy diagram representing the molecular orbitals of the following compounds.

b) Identify the LUMO of each.

c) Based on the contribution of each atom to the LUMO, predict the most likely site of reactivation with a nucleophile.

d) Based on the contribution of each atom to the HOMO, predict the most likely site of reaction with an electrophile.

7.40 Suggest reagents that could be used to synthesize each of the following from an aldehyde or ketone.

a) 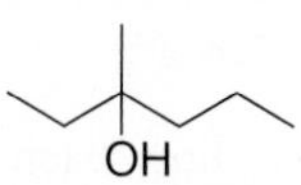 (3 methods)

b) 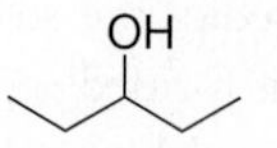 (2 methods)

c)

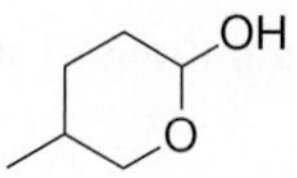

7.41 Propose a detailed mechanism to explain the following reaction.

a)

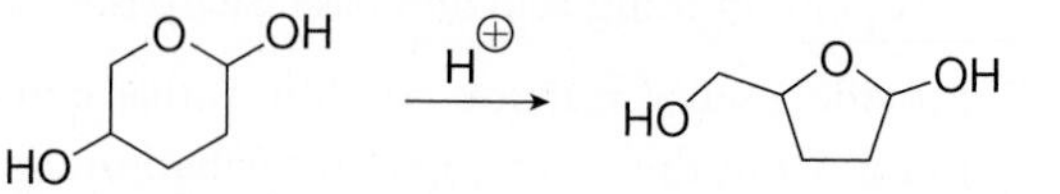

b)

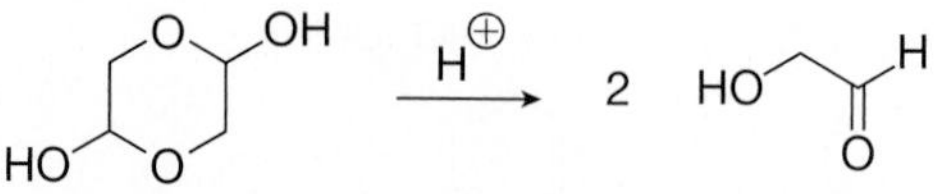

MCAT STYLE PROBLEMS

7.42 The open-chain form of fructose exists in equilibrium with several different hemiketals. Which of the following structures does *not* represent a possible hemiketal of fructose?

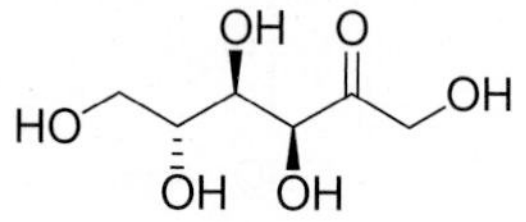

fructose

a)

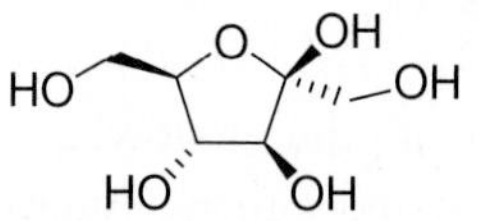

b)

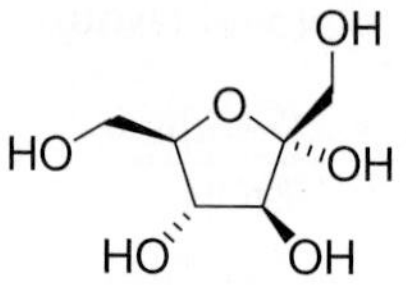

c)

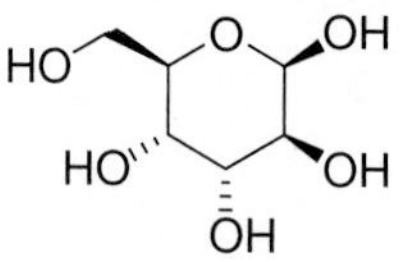

d)

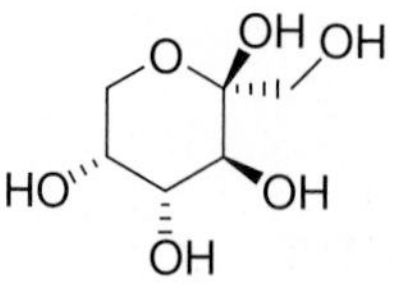

7.43 Which of the following is true of the reaction between glucose and NaCN in the presence of HCl?

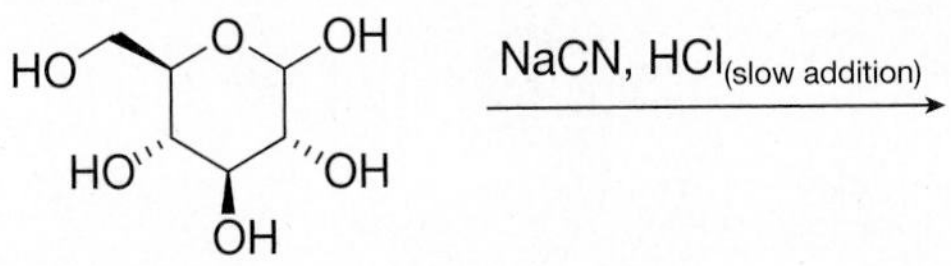

NaCN, $HCl_{(slow\ addition)}$

a) The products of the reaction are

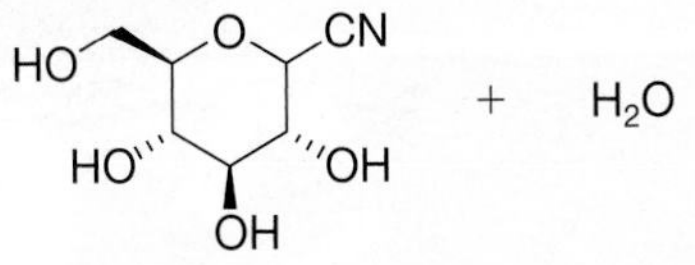

+ H_2O

b) The product of the reaction is

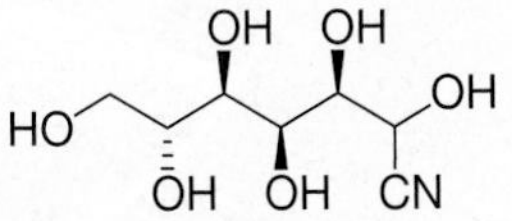

c) No reaction is possible because there is no reactive functional group on the cyclic form of glucose.

d) No reaction is possible because the HCl reacts with the hemiacetal before the cyanide ion is able to attack.

7.44 NADPH is a hydride donor used throughout biological systems to carry out reduction reactions. Which of the following reactions is likely to involve NADPH?

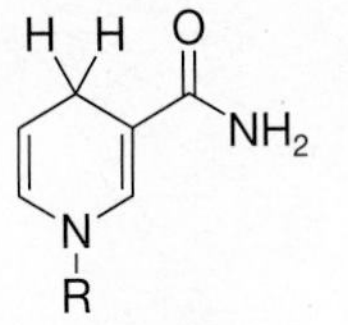

NADPH

a) Conversion of (*R*)-glycerol phosphate to dihydroxyacetone phosphate:

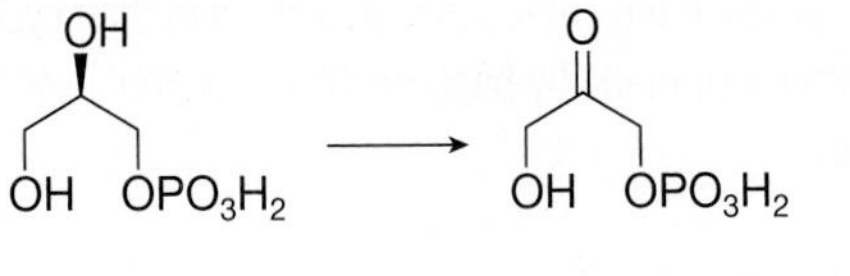

b) The final step in the biosynthesis of proline:

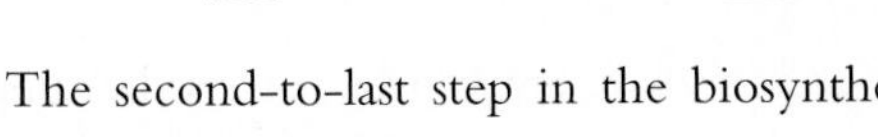

c) The second-to-last step in the biosynthesis of the amino acid histidine:

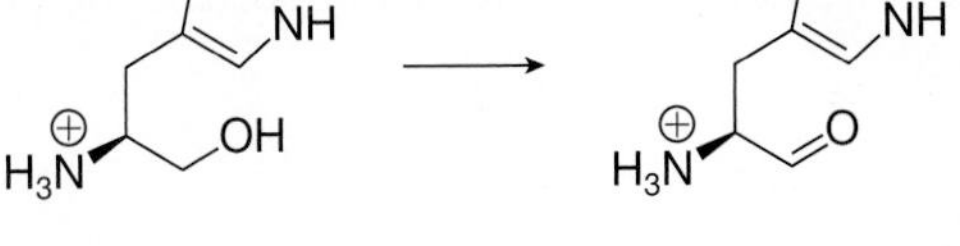

d) The reaction of glucose 6-phosphate catalyzed by glucose 6-phosphate dehydrogenase:

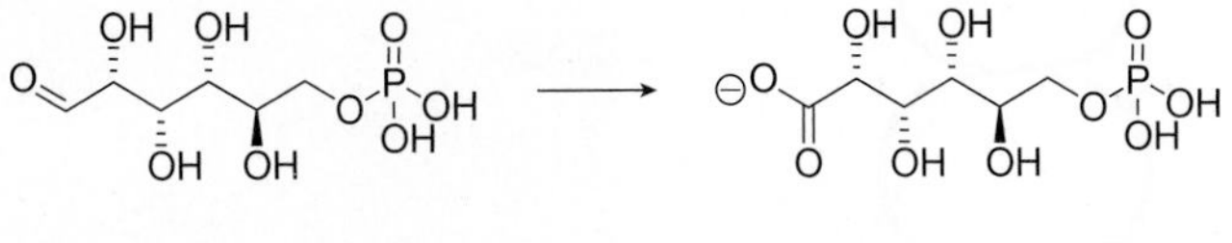

CHALLENGE PROBLEM

7.45 Lynestrenol is a synthetic hormone used to treat disorders related to progesterone deficiencies. A synthesis of lynestrenol from nortestosterone is shown here.

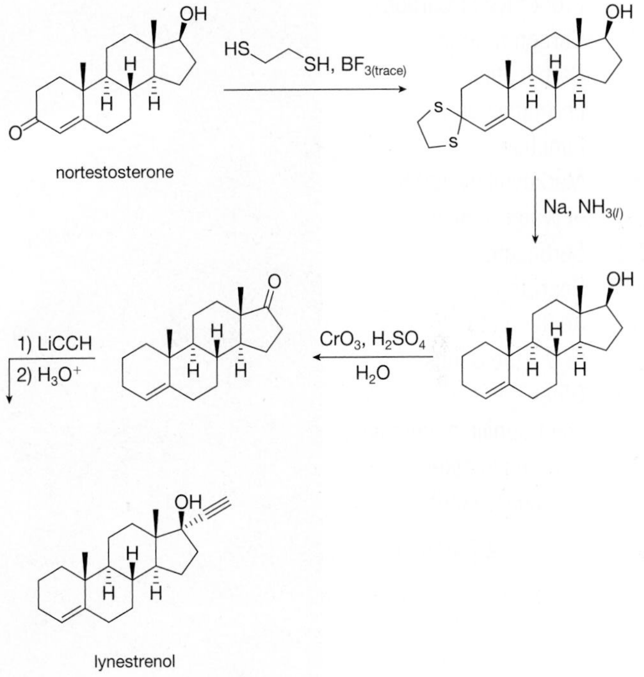

nortestosterone

HS⁀SH, $BF_{3(trace)}$

Na, $NH_{3(l)}$

CrO_3, H_2SO_4, H_2O

1) LiCCH
2) H_3O^+

lynestrenol

a) After reviewing Question 7.38, propose a detailed mechanism for the first step of the above synthesis. (Hint: BF_3 is a Lewis acid, which can catalyze reactions just like a Brønsted acid; the Lewis acid simply replaces H^+ in the mechanism.)

b) Identify both a reduction step and an oxidation step in the synthesis.

c) Draw the mechanism for the final conversion in the reaction, and explain the stereochemistry of product.

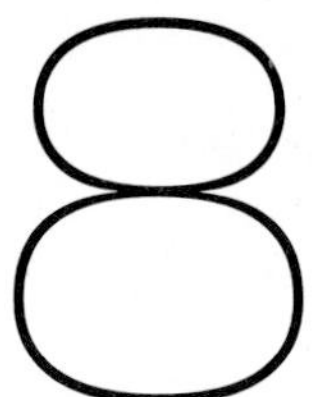

π Bonds as Nucleophiles
REACTIONS OF ALKENES, ALKYNES, DIENES, AND ENOLS

CHAPTER OUTLINE

What do sharks and bacteria have in common? Both produce the compound squalene and then cyclize it. However, the end products made by sharks differ from those made by bacteria. Shark liver oil used to be the primary source of squalene for commercial purposes.

8.1 Why It Matters

Eukaryotic organisms, which include plants and animals, are capable of synthesizing *steroids*. Steroids are *tetracyclic* compounds, which have a basic structure consisting of four fused rings, denoted A, B, C, and D in the following structure. Testosterone and estradiol, vital hormones that control secondary sex characteristics, are both steroids. Cholesterol, an important component of cell membranes in all animals, is also a steroid.

tetracyclic structural scaffold of steroids

testosterone

estradiol

cholesterol

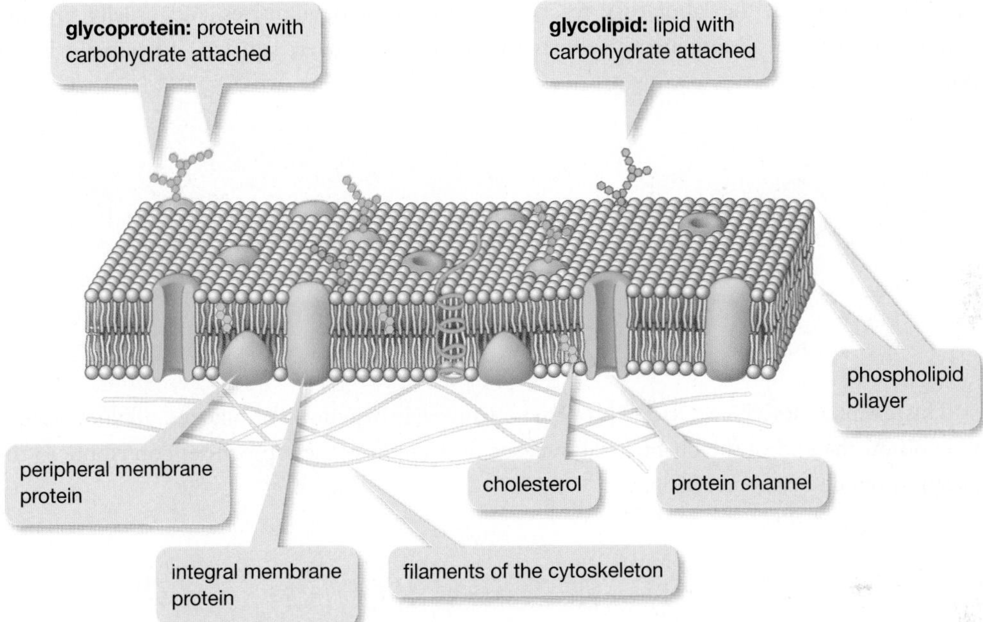

SOURCE: OpenStax College, Openstax Concepts of Biology. OpenStax CNX. http://cnx.org/contents/b3c1e1d2-839c-42b0-a314-e119a8aafbdd@8.39. Download for free at http://cnx.org/contents/b3c1e1d2-839c-42b0-a314-e119a8aafbdd@8.39.

ORGANIC CHEMWARE
8.1 Line-angle structure: Cholesterol

Prokaryotic (single-celled) organisms, such as bacteria, are unable to synthesize cholesterol and other steroids. Instead, they produce *hopanoids*, which are *pentacyclic* (five-ringed) compounds. Following are the structures of *bacteriohopanetetrol*, an abundant hopanoid in bacterial membranes, and its precursor *hopene*.

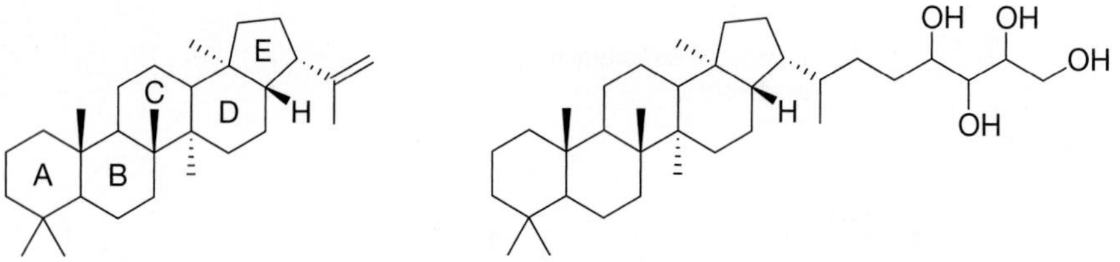

hopene

bacteriohopanetetrol

The acylic compound *squalene* is a precursor for both hopanoids and steroids. In living organisms, the reactions that produce these compounds each involve a different *enzyme*—proteins that catalyze biological reactions. An enzyme known as *squalene-hopene cyclase* catalyzes a key step in the production of hopanoids from squalene. In this complex step, the reaction creates five rings and nine stereocentres, with the necessary new covalent bonds and the proper stereochemistry (see the following diagram). The individual processes involved in the reaction can be performed in the laboratory, but the laboratory reactions require heat or additives, and usually produce mixtures of products. Consequently, efforts to chemically convert squalene to hopene in the laboratory have failed.

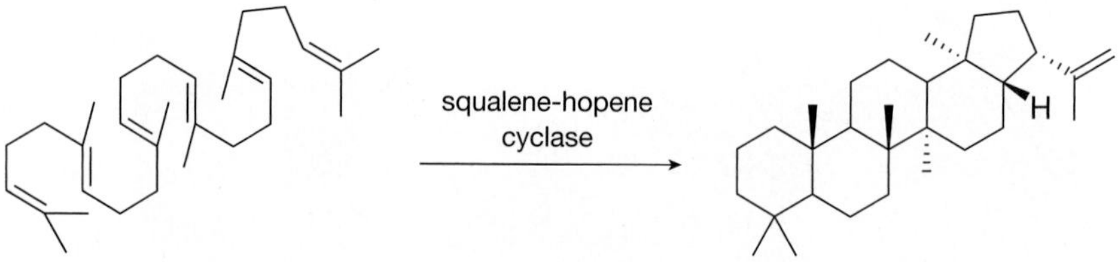

squalene-hopene cyclase

squalene

hopene

This chapter describes various processes involved in reactions of alkenes, alkynes, dienes, and enols. At the end of the chapter you will see in detail how these processes contribute to the squalene-hopene cyclase reaction.

8.2 Properties of Carbon-Carbon π Bonds

Carbon atoms that are connected by double or triple bonds constitute functional groups called alkenes or alkynes. These groups consist of carbon atoms connected by a σ bond and one or more π bonds. The multiple bonds connecting the carbons create an area of high electron density, making such groups good electron donors. The reactivity of these groups derives from the π bonds, because these π bonds are excellent nucleophiles.

8.2.1 Orbital structure of alkenes

The orbital construction of the alkene functional group is similar to that of carbonyls (Chapter 7). Both of the carbon atoms connected by the double bond are sp^2 hybridized; they are held together by one σ bond and one π bond. The π bond lies both above and below the plane defined by the two carbon atoms of the double bond and the four atoms connected to them by single bonds. These six atoms are shown in red in Figure 8.1. Note that the "top" and the "bottom" regions of the π bond together constitute one bond.

all red atoms lie in the same plane

FIGURE 8.1 Three-dimensional electronic structure of the alkene π bond, shown for 2-methyl-2-pentene.

Unlike a carbonyl group, the carbon-carbon double bond of most alkenes is non-polar because the two carbon atoms in the double bond have the same electronegativity. The electrostatic potential map in Figure 8.2 illustrates that the carbon atoms of ethene have equal and symmetric electron density.

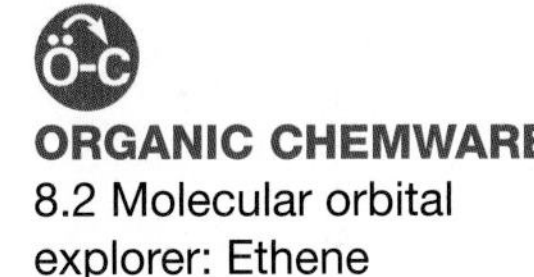

ORGANIC CHEMWARE
8.2 Molecular orbital explorer: Ethene

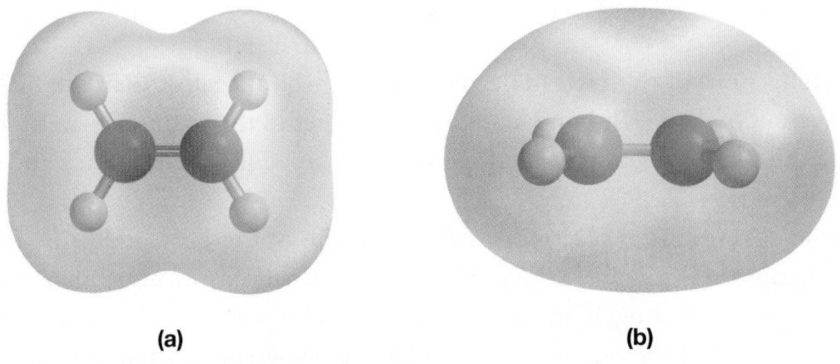

FIGURE 8.2 Electrostatic potential map of ethene. (a) top view; (b) side view. Red areas indicate high electron density; blue areas indicate low electron density.

Closer examination of the electrostatic potential map in Figure 8.2b reveals a region of high electron density (red) above the plane of the alkene functional group and an identical region below the plane, consistent with the location of the π bond. Alkene functional groups are *electron rich*, with the electrons of the π bond located on the two "faces" of the group. Electron-rich functional groups are able to donate electrons, which means they are nucleophilic. Alkenes usually act as nucleophiles and interact with electrophilic reagents.

STUDENT TIP
The two electron-rich faces of an alkene are attracted to electrophiles. Alkenes usually react as nucleophiles.

8.2.2 Electrophilic addition reactions

The most common reactions of alkenes involve the addition of atoms or groups (X and Y in the following diagram) to the carbon atoms in the double bond. During these reactions, the electrons of the π bond redistribute to form bonds to X and Y, leaving the two carbon atoms connected by a single bond in the reaction product.

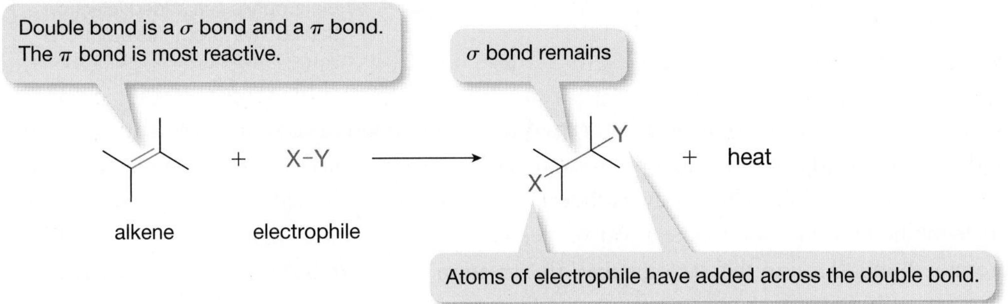

Because π bonds are higher in energy than σ bonds, they are more reactive than σ bonds. Reactions of alkenes normally involve the π bond only; the σ bond that is part of the double bond remains intact. When the π bond of an alkene reacts, the bond breaks, and two new σ bonds are created, one to each of the carbon atoms of the initial double bond. Since the two new σ bonds have a total energy less than that of the π bond of the alkene, addition reactions of this type are exothermic.

8.2.3 Addition of hydrogen halides to alkenes

The addition of hydrogen halides (HF, HCl, HBr, or HI) to an alkene follows this general equation.

2
1
+ HBr ⟶
Br
1
2
H
H
H and Br have added across the double bond

alkene electrophile

The bromine atom of the HBr molecule is more electronegative than the hydrogen atom, and so the electrons of the H–Br bond tend to reside nearer to the bromine than to the hydrogen, making the hydrogen slightly positive. This positively charged hydrogen attracts the nucleophilic π bond of the alkene.

A **carbocation** is a species containing a carbon atom with a +1 formal charge.

The alkene donates the π electrons to the hydrogen of the HBr molecule to form a new C–H bond. Because the valence orbital of hydrogen can hold no more than two electrons, the bond between the hydrogen and the bromine breaks, with the electrons moving toward the bromine, the most electronegative atom. This process forms a **carbocation**, an intermediate product in which one of the carbons that made up the π bond is now connected to the hydrogen from the HBr, while the other carbon now carries a formal charge of +1. Because of their charged and electron-deficient nature, carbocations have high energy and are very reactive.

the π electrons are used to form a σ bond between one carbon of the alkene and the hydrogen of the electrophile

addition of hydrogen forms a carbocation

bromide leaving group

H
H
+
δ+
H
δ−
:Br:
⟶
H H
⊕
H
+
:Br:⊖

alkene acts as nucleophile

electrons flow onto the most electronegative atom

the positive charge is adjacent to the carbon on which the electrophile has added (electrons flowed away from this atom)

A **reaction intermediate** is a chemical species formed in one step of a chemical reaction but consumed in a subsequent step.

The carbocation is a **reaction intermediate**. There is also a second intermediate: the bromide ion that acted as a leaving group from the HBr molecule. This ion has excess electrons and is therefore a nucleophile, whereas the carbocation is an electrophile with an incomplete octet. The nucleophilic bromide can donate a pair of electrons to the carbocation and form a new carbon-bromine bond. Because this step involves the creation of one σ bond, it is thermodynamically very favourable (exothermic).

carbocation is an electrophile

bromide is a nucleophile

alkyl halide

two electrons from the nucleophile are used to form a bond with the carbocation

The overall result of this two-step process is to add a hydrogen to one end of the double bond while adding a bromine to the other end. So the reaction is said to add HBr *across the double bond.*

bromine acts as leaving group in first step

alkyl halide

alkene acts as nucleophile in first step

carbocation is an electrophile in second step

bromide is a nucleophile in second step

ORGANIC CHEMWARE
8.3 Curved arrow notation: Electrophilic addition to alkene

ORGANIC CHEMWARE
8.4 Hydrohalogenation of alkenes

CHECKPOINT 8.1

You should now be able to draw a curved arrow mechanism for the addition of strong acids to symmetric alkenes (π nucleophiles) and provide the products of the reaction.

SOLVED PROBLEM

Draw a curved arrow mechanism and predict the products for the addition of hydrochloric acid to cyclopentene.

+ H-Cl ⟶

STEP 1 (OPTIONAL): Expand the Lewis structure around the alkene to explicitly show the C–H bonds.

Drawing the implied hydrogen atoms can help you keep track of where the new hydrogen atom adds and where the formal charges go. As you get comfortable with the reactions of alkenes, you'll be able to omit this step.

STEP 2: Identify the roles of the reactants.

The π bond in cyclopentene is electron rich and can act as a nucleophile. The hydrochloric acid is highly polarized due to the electronegativity difference between the atoms. The resulting partial positive charge on the hydrogen atom makes this species able to act as an electrophile.

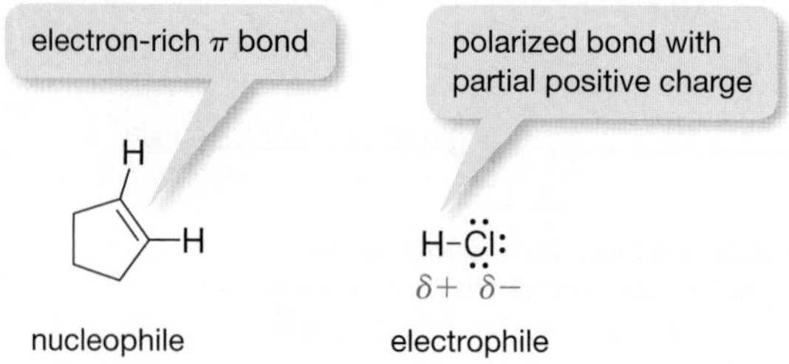

electron-rich π bond

polarized bond with partial positive charge

H–Cl: $\delta+$ $\delta-$

nucleophile electrophile

STEP 3: Draw in the mechanistic arrows, using the principle that electrons flow from nucleophile to electrophile.

The first arrow starts in the centre of the nucleophilic π bond. These electrons are donated to the hydrogen atom of the acid. In order to avoid giving too many electrons to hydrogen, a second arrow is drawn corresponding to the breaking of the H–Cl bond. The electrons from this bond move toward the more electronegative chlorine atom.

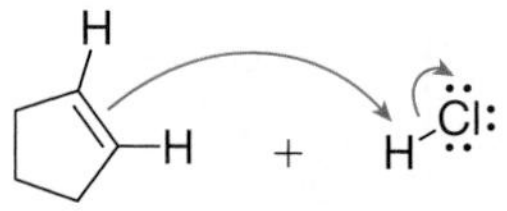

STEP 4: Use the arrows to determine the products formed from this step of the reaction, including any formal charges.

A new bond has been formed between one of the alkene carbons and the hydrogen from the acid. Since the alkene is symmetrical, it does not matter which carbon bonds with the hydrogen. In this case, we chose the top carbon. At the same time, the H–Cl σ bond has broken. The formal charge on the bottom carbon of the alkene changes from 0 to +1 (electrons moving away from the atom), and the formal charge on chlorine changes from 0 to −1 (electrons moving toward the atom).

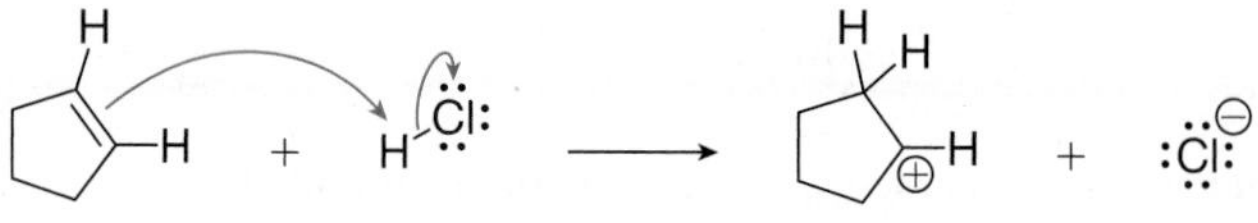

STEP 5: Repeat steps 2–4 until you obtain the final, neutral product. The products of the first step include a positively charged carbocation (an electrophile) and a negatively charged chloride ion (nucleophile). After the chloride ion shares its electrons with the carbocation, the final product is formed.

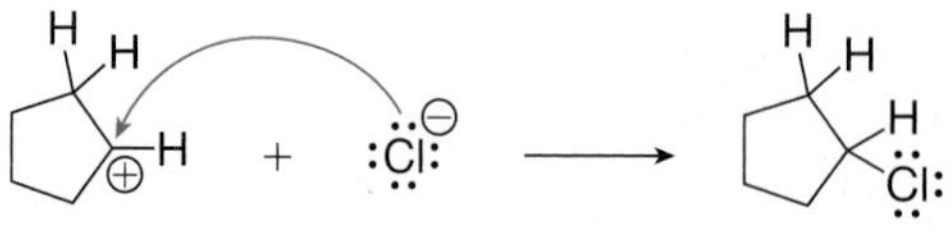

STEP 6 (OPTIONAL): Redraw the final product and mechanism as a line drawing without the extra hydrogens or lone pairs.

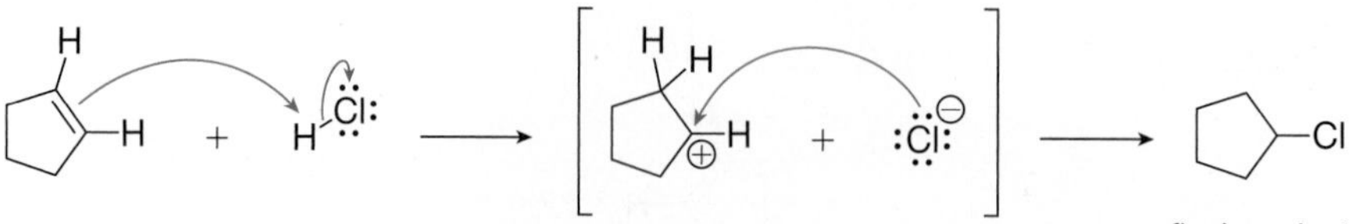

final product

PRACTICE PROBLEM

8.1 Draw a curved arrow mechanism and predict the products for the following additions of strong acids to alkenes.

a) + HCl

b) + HBr

c) + H_2SO_4

d) + CF_3CO_2H

INTEGRATE THE SKILL

8.2 Draw the mechanisms to show all the possible cations that result from addition of HCl to the alkenes below. Draw the resonance structures of each one, if applicable.

a) 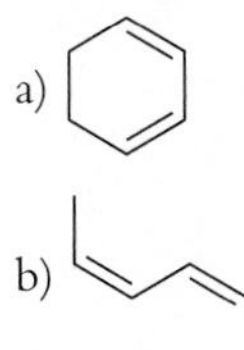

b)

8.3 Carbocation Formation and Function

The formation of a carbocation controls every aspect of electrophilic addition reactions. The rate at which carbocations are formed controls the reaction rate, the *locations* to which the atoms of the electrophile become attached to the alkene, and the relative *amounts* of each product. The *nature* of the carbocation can even influence the stereochemistry of the atoms in the products. All of these aspects are influenced by the energy of the carbocations—in particular, the energy of the transition states that produce them.

8.3.1 Energy changes control reactivity

The energy diagram for the addition of HBr to an alkene is shown in Figure 8.3. The first step in the overall reaction has the highest energy transition state. Since this step has the largest activation energy, it controls the overall rate of the reaction. The second step, the nucleophilic addition of the conjugate base to the carbocation, has much lower activation energy than the first step, and so it does not influence the overall rate of the process. In a chemical reaction, the step that has the highest activation barrier is called the **rate-determining step**. The rate of this reaction is governed by the rate at which the high-energy carbocation is formed.

The **rate-determining step** of a multi-step reaction mechanism is the step with the highest activation energy. Because it is the slowest step, it controls the overall rate of the reaction.

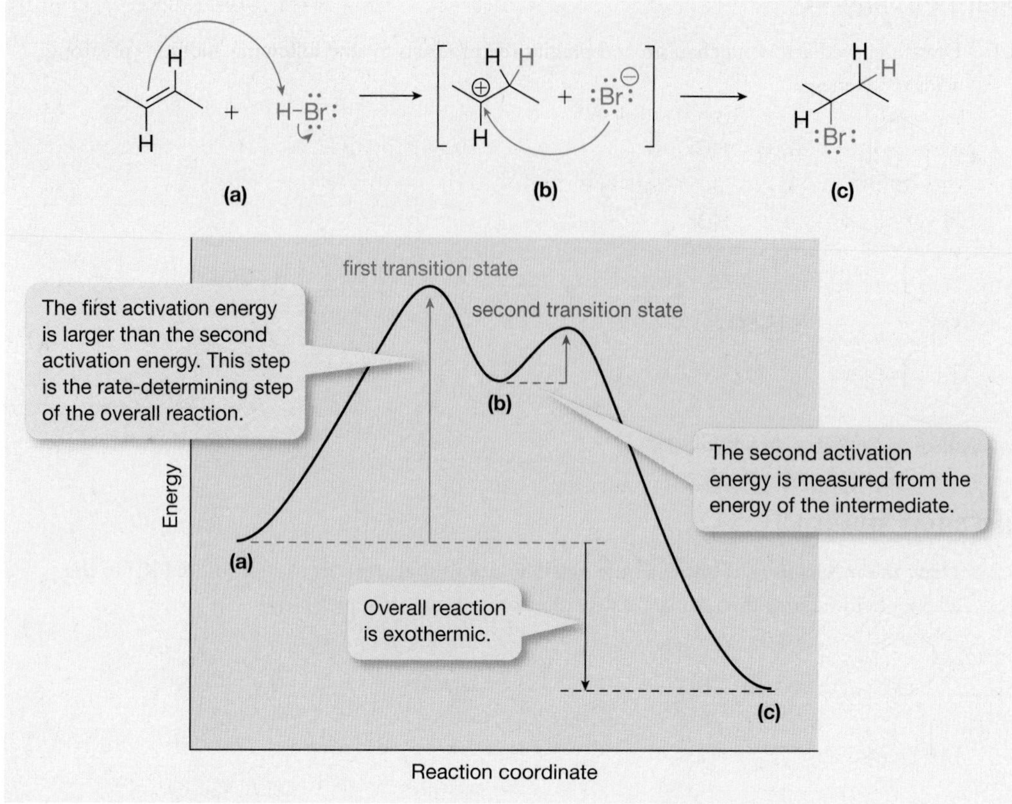
(a)
(b)
(c)
first transition state
second transition state
The first activation energy is larger than the second activation energy. This step is the rate-determining step of the overall reaction.
The second activation energy is measured from the energy of the intermediate.
Overall reaction is exothermic.
Energy
Reaction coordinate

FIGURE 8.3 Typical energy diagram for the electrophilic addition reactions of alkenes.

8.3.2 Regioselectivity in electrophilic addition reactions

Regioisomers are constitutional isomers that are formed from a chemical reaction.

When an unsymmetrical alkene reacts with a haloacid, the hydrogen atom in the acid can bond to *either* of the carbon atoms of the double bond. If the hydrogen adds to C-1, the resulting positive charge will reside on C-2, which in this case is connected to three other carbon atoms. If the hydrogen adds to C-2, the positive charge will reside on C-1. When halide ions react with these two carbocations, the resulting products are **regioisomers**: constitutional isomers formed when a reaction takes place at different locations (regions) of a molecule. In the addition of hydrogen halides to alkenes, the possible regioisomers differ in terms of the locations where the halogen and hydrogen atoms are added to the alkene.

hydrogen adds to C-2
less substituted carbocation forms slowly
slow
fast
the two possible constitutional isomers are a pair of regioisomers
more substituted carbocation is formed faster
hydrogen adds to C-1
major product

In this process, the more highly substituted carbocation, shown in the bottom pathway is much more stable than the carbocation on the top pathway. The more stable carbocation is formed much faster than the other carbocation, and so the 2-chloro product (bottom pathway) forms much faster than the 1-chloro product. Therefore, *more* of the 2-chloro product will be produced in this reaction. Reactions that produce more of one regioisomer than the other are said to be **regioselective**. The reasons for this selectivity are discussed later in this section.

Regioselective reactions favour the formation of one regioisomer over another.

8.3.3 Markovnikov's rule

In 1870, long before the mechanism was known, the Russian chemist Vladimir Markovnikov described the results of experiments involving the addition of HBr to propene, an asymmetric alkene. He found that the products contained a greater proportion of one regioisomer than the other.

the halide (nucleophile) has added to the site of the most stable carbocation (more substituted carbon)

propene + H–Br → 2-bromopropane + 1-bromopropane

Markovnikov product

Markovnikov described a method, known as *Markovnikov's rule*, for predicting the regioselectivity of reactions between an asymmetric alkene and a hydrogen halide. The product favoured in this reaction is known as the *Markovnikov product*, or the product of *Markovnikov addition*.

The *modern interpretation* of Markovnikov's rule is that the major product is the one formed from the most stable carbocation (often the most substituted carbocation). This version is the most useful and general interpretation of Markovnikov's rule.

STUDENT TIP

To determine the major product of an electrophilic addition reaction of an alkene, look for the most stable carbocation.

8.3.4 Stability of carbocations

Carbocations are high-energy species because of the formal positive charge on an atom with an incomplete octet. Any structural feature that reduces the effective positive charge on the carbocation makes it more stable and reduces the $\Delta G^{\ddagger}$ of formation. As a result, carbocations that belong to different structural classes have different stabilities and form at different rates.

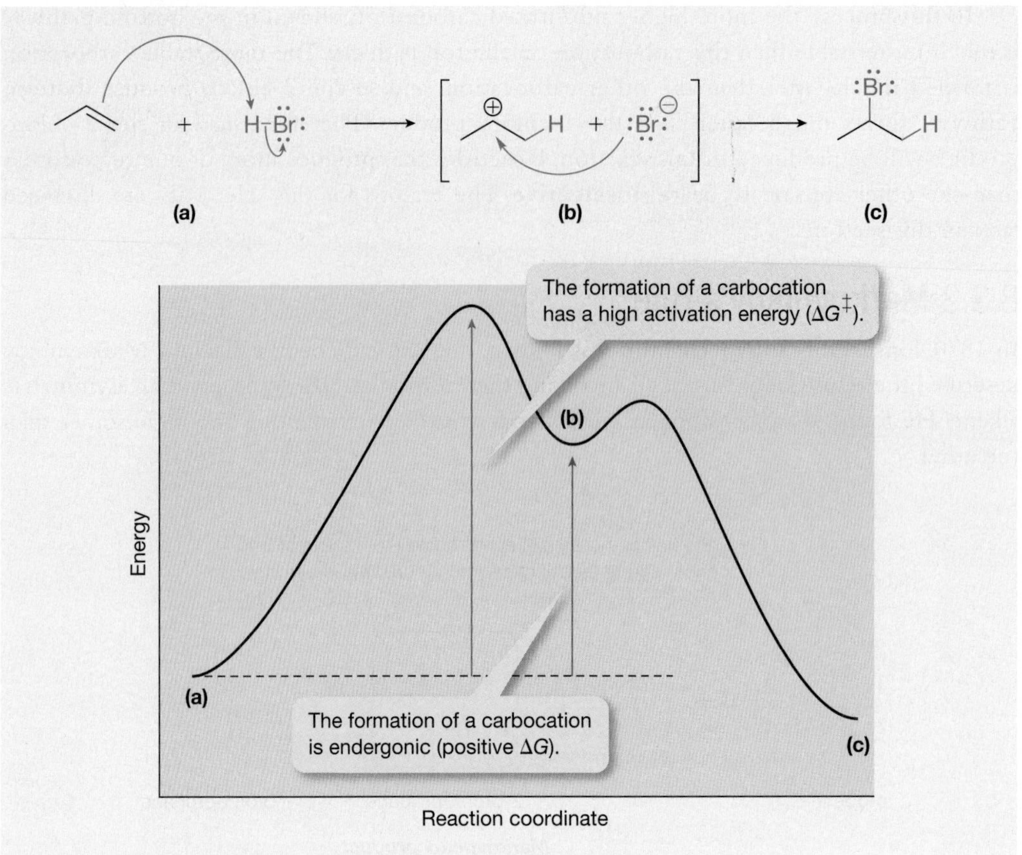

The structural classification of a carbocation depends on the number of *alkyl* carbons that are *directly* bonded to the carbon bearing the positive formal charge (see Table 8.1). If no alkyl carbons are connected, the carbocation is called a **methyl carbocation**. If there is only one carbon bonded to the charged carbon, the ion is a **primary carbocation** (the symbol 1° is sometimes used to denote primary). Two or three alkyl carbons directly connected to a carbocation give a **secondary** or a **tertiary carbocation**, respectively. Carbocations are classified only by the number of alkyl carbons directly connected to the charged carbon atom, and not by *what* the alkyl groups are. In general, the more substituents a carbocation has, the more stable it will be.

Methyl carbocations are not bonded to any alkyl groups. **Primary, secondary**, and **tertiary carbocations** are respectively bonded to one, two, and three alkyl groups.

STUDENT TIP
Carbocations that have more alkyl groups bonded directly to them are more stable.

TABLE 8.1 Relative Stabilities of Carbocations

Carbocation type	Symbol	Structure	Stability
Methyl	Me	$CH_3^{\oplus}$	Least stable
Primary	1°	$CH_3CH_2^{\oplus}$	↓
Secondary	2°		
Tertiary	3°		Most stable

Although all carbocations have a formal charge of +1, most carbocations carry an *effective* charge less than +1. *How much less* depends on the number of alkyl groups that are bonded to the carbocation. Because alkyl substituents are **electron-donating groups** that reduce the amount of positive charge on the cationic atom, the more groups attached, the lower the effective positive charge on the carbocation. This effect is shown by in the electrostatic potential maps in Figure 8.4. The amount of positive charge (blue) on each carbocation clearly decreases as the number of groups attached to the carbocation increases (the *overall* charge on each molecule is +1). Reducing the effective positive charge on a carbocation lowers its energy and increases its thermodynamic stability (the charge is spread over a larger volume).

Electron-donating groups are atoms or groups that donate electron density.

STUDENT TIP
More and less stable are *relative* terms for carbocations. Even tertiary carbocations are not actually stable; they are just less *unstable* than secondary carbocations.

primary carbocation ($CH_3C^{\oplus}H_2$)

secondary carbocation ($CH_3C^{\oplus}HCH_3$)

tertiary carbocation ($CH_3C^{\oplus}(CH_3)CH_3$)

FIGURE 8.4 Electrostatic potential maps of substituted carbocations.

8.3.5 Stabilization of carbocations by hyperconjugation

Alkyl groups cannot donate electrons through conjugation. Instead, alkyl substituents stabilize carbocations by donating electrons to the carbocation through a process called **hyperconjugation**, which is electron donation from a σ bond. Substituents on the positively charged carbon can rotate about their bonds. During such rotation, the other σ bonds in the substituents temporarily eclipse the empty p orbital of the carbocation and donate a small amount of electron density into the p orbital. This donation, called hyperconjugation, lowers the positive charge on the carbocation by a very small amount. When more alkyl groups are bonded to a carbocation, more hyperconjugation occurs, the localized positive charge on the carbocation is reduced, and the stability of the carbocation increases. Hyperconjugation occurs from the C–C or C–H bonds *adjacent* to the carbocation. By contrast, the carbon-hydrogen bonds directly connected to a carbocation are perpendicular to the p orbital. They cannot eclipse the empty p orbital and therefore cannot stabilize the carbocation.

Hyperconjugation is the interaction of the empty p orbital of the carbocation with filled σ bonds on adjacent carbon atoms.

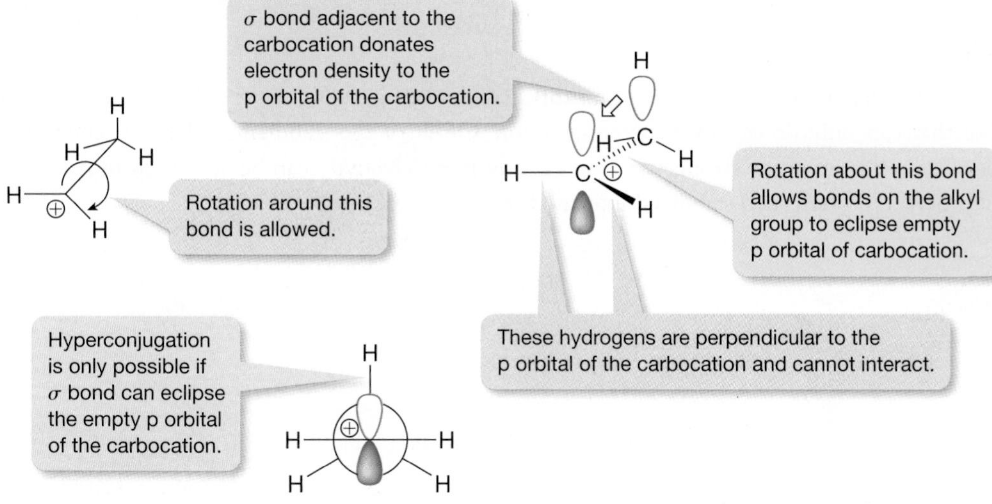

8.3.6 Stabilization of carbocations by charge delocalization

Any reduction or dispersion of the positive charge on a carbocation increases its stability. When resonance is possible, charge becomes delocalized, which stabilizes molecules. Most π bonds adjacent to carbocations stabilize those carbocations by delocalizing the charge over several atoms. This often increases the number of sites that can react in subsequent transformations. This delocalization and reactivity can be analyzed by examining the possible resonance structures of the carbocation.

Carbocation is stabilized by delocalization with π bond.

Nu

Nu

Nu

Nu

Two reaction sites are available for nucleophiles to react with. Reaction sites can be predicted from the resonance forms.

STUDENT TIP
Both hyperconjugation and charge delocalization stabilize carbocations by spreading the positive charge over multiple atoms.

Adjacent atoms with one or more non-bonding pairs of electrons also participate in delocalizing the positive charge of carbocations. This effect is very common with nitrogen and oxygen. Groups that stabilize carbocations this way are said to be *electron donating*.

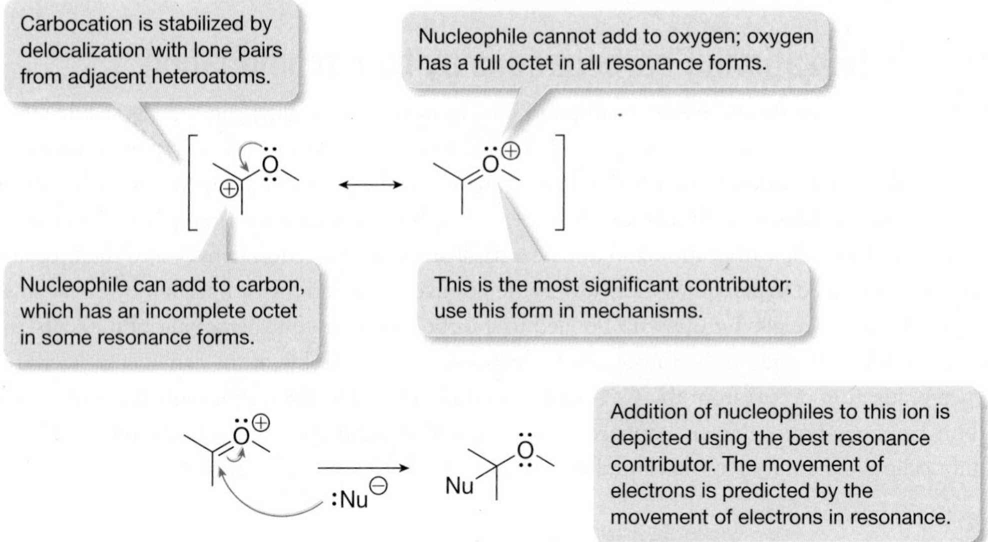

Carbocation is stabilized by delocalization with lone pairs from adjacent heteroatoms.
Nucleophile cannot add to oxygen; oxygen has a full octet in all resonance forms.
Nucleophile can add to carbon, which has an incomplete octet in some resonance forms.
This is the most significant contributor; use this form in mechanisms.
:Nu
Nu
Addition of nucleophiles to this ion is depicted using the best resonance contributor. The movement of electrons is predicted by the movement of electrons in resonance.

Groups that are electron donating in this way tend to have a greater stabilizing effect than those that can only donate by hyperconjugation. Of all of the stabilizing effects in chemistry, delocalization is often the strongest. Consequently, regioselectivity can be very high in reactions that proceed through such stabilized intermediates.

CHECKPOINT 8.2

You should now be able to draw a curved arrow mechanism for the addition of strong acids to both symmetric and asymmetric alkenes (π nucleophiles), and provide the products of the reaction.

SOLVED PROBLEM

Draw a curved arrow mechanism to show the formation of regioisomers from the following reaction. Identify the Markovnikov products.

+ H-Br ⟶

STEP 1 (OPTIONAL): Expand the Lewis structure around the alkene to explicitly show the C–H bonds.

STEP 2: Identify the roles of the reactants.

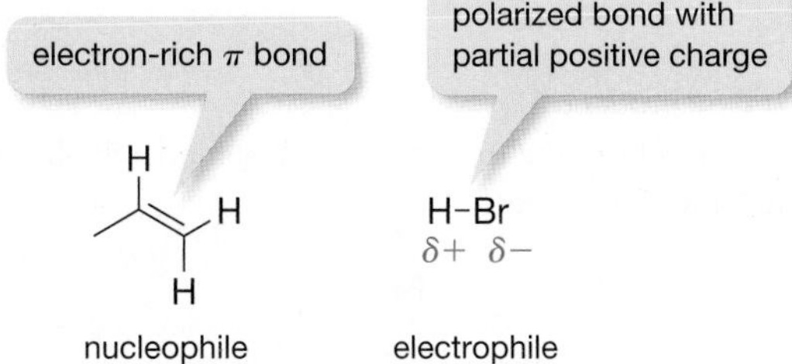

STEP 3: Draw the mechanistic arrows, using the principle that electrons flow from nucleophile to electrophile.

STEP 4: Use the arrows to determine the products formed from this step of the reaction, including any formal charges.

Because the alkene in this reaction is asymmetric, the hydrogen can bond with either the left-hand or right-hand carbon of the double bond. Therefore, there are two possible carbocation intermediates that need to be considered.

STUDENT TIP

The term *hyperconjugation* might sound like it would be more effective than conjugation at stabilizing a carbocation. In fact, hyperconjugation has a weaker effect than conjugation or delocalization.

STEP 5: Repeat steps 2–4 until you obtain the final, neutral product.

Path **(a)**

Path **(b)**

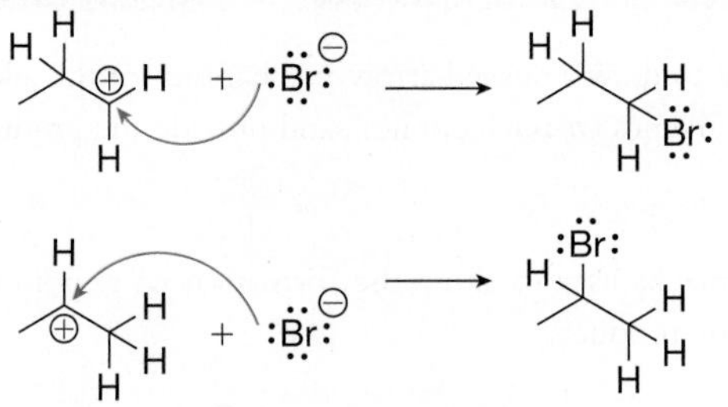

STEP 6 (OPTIONAL): Redraw the final products as line drawings without the extra hydrogens.

Path **(a)**

Path **(b)**

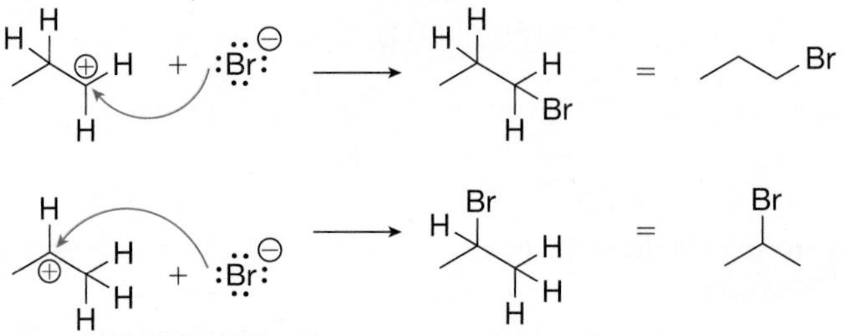

STEP 7: Identify the Markovnikov product.

The Markovnikov product arises from the most stable carbocation, which is the secondary one. So the Markovnikov product here is 2-bromopropane.

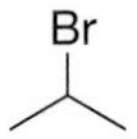

PRACTICE PROBLEM

8.3 Draw mechanisms for the formation of all possible products of the addition of HCl to each of the following compounds. If more than one regioisomer is possible, indicate the major product (the Markovnikov product).

a)

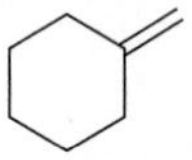

b)

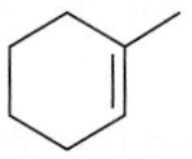

c)

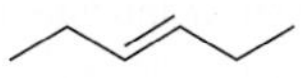

d)

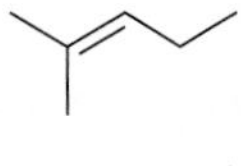

e)

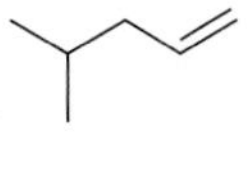

INTEGRATE THE SKILL

8.4 Draw the mechanisms for the formation of all products from the reactions of HBr with each double bond in the following molecules. Identify the carbocation that will form most rapidly, and predict the major products.

a)

b)

c)

8.3.7 Applying the Hammond postulate

The carbocation that forms fastest leads to the major product of the reaction. The rate at which such ions form depends on the activation energies for their formation. These, in turn, depend on the energy of the transition states leading to each one. Transition states are often difficult to study without special molecular modelling calculations. Since carbocations have a simpler structure and are easier to visualize, estimating rates of formation using carbocation stability is a useful approximation. The **Hammond postulate** uses the energy of an intermediate or product to estimate the energy of nearby transition states, which simplifies the analysis of regioselectivity.

Consider the three reaction coordinate diagrams shown in Figure 8.5. In part (a), the transition state is closer in energy to the reactants than to the products. This type of transition state is called an **early transition state**, and it resembles the reactants. In part (b), the transition state is closer in energy to the products than it is to the reactants. This type of transition state is called a **late transition state**, and it tends to resemble the products. In part (c), the transition state is not close to either the reactants or the products, and resembles neither.

The **Hammond postulate** states that the structure of a transition state resembles the species nearest to it in free energy.

In an **early transition state**, the structure of the transition state most resembles the structure of the starting materials.

In a **late transition state**, the structure of the transition state most resembles the structure of the products of that step.

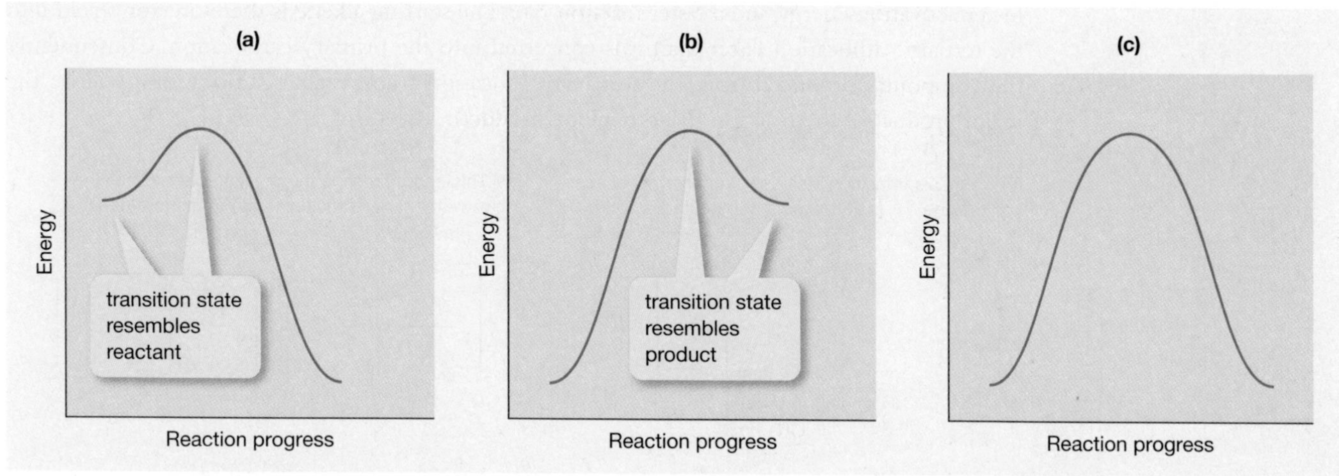

FIGURE 8.5 Reaction progress diagrams for (a) an early transition state, (b) a late transition state, and (c) a transition state that is neither early nor late.

When π bonds act as nucleophiles, they produce carbocations as intermediates. The reaction coordinate curve for the formation of these high-energy species indicates a late transition state, meaning that the first transition state resembles the carbocation (Figure 8.6). The energy of the transition state can therefore be explained by examining the structure of the carbocations that are formed as intermediates in the process.

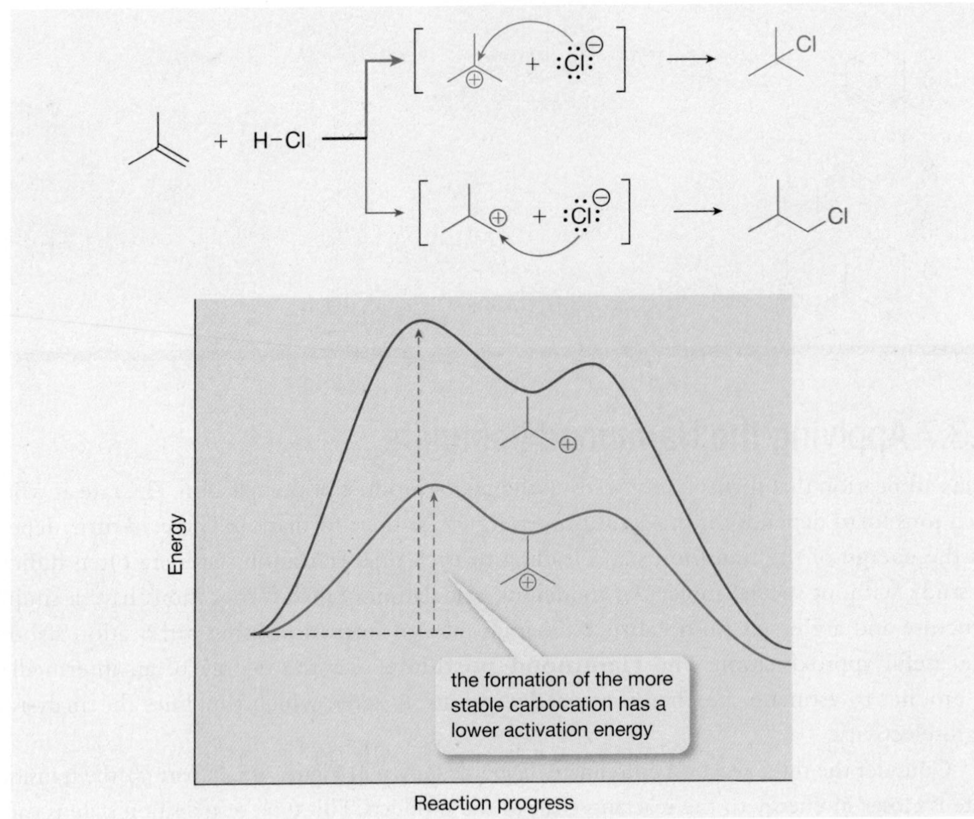

FIGURE 8.6 A reaction coordinate diagram comparing the formation of two carbocations of different stability.

In Figure 8.6, the tertiary carbocation has lower energy, and hence more stability, than the primary carbocation. Since the tertiary carbocation has lowest energy, the Hammond postulate indicates that the transition state leading to the tertiary carbocation is a lower energy transition state than the one that produces the primary carbocation. A lower energy transition state has lower activation energy and a faster reaction rate. The starting alkene is therefore converted into the tertiary carbocation faster than it is converted into the primary carbocation. Consequently, the compound produced from the tertiary carbocation (more stable carbocation), will be the major product of the reaction (Markovnikov product).

This transition state is lower energy because the $\delta+$ is at the more substituted carbon.

This carbocation is the most stable and has the lowest energy transition state for its formation.

fast

slow

transition states

carbocation intermediates

products

Transition states resemble the carbocations. Factors that stabilize + charges stabilize $\delta+$ charges.

CHECKPOINT 8.3

You should now be able to predict the major product(s) of a reaction between an alkene and a strong acid and rationalize its formation based on the mechanism.

SOLVED PROBLEM

The following reactants produce a single product. Write a mechanism to predict which product is formed, and explain why there are no other products.

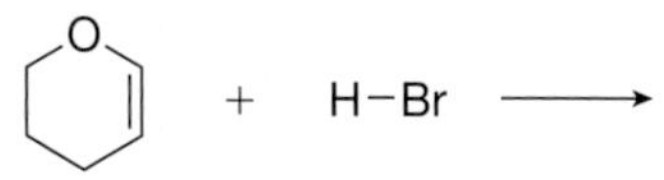

+ H–Br ⟶

STEP 1: Using steps 1–6 of Checkpoint 8.2 as a guide draw the mechanism for the formation of *both* possible products. Include any significant resonance structures of the carbocation intermediates.

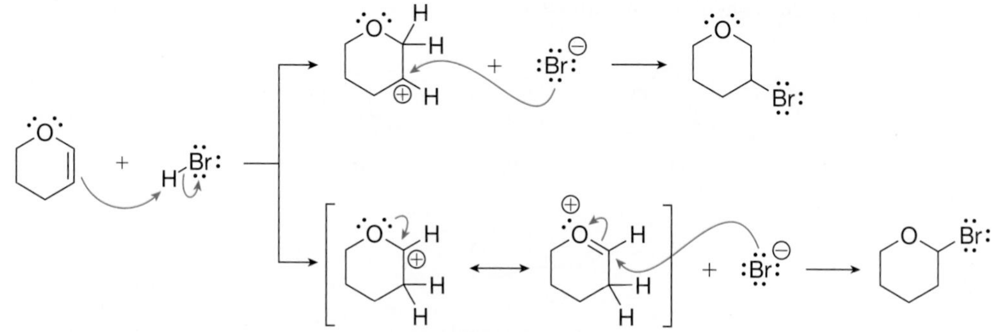

STEP 2: Determine which carbocation intermediate is more stable.

Both carbocations are secondary, but the bottom carbocation is stabilized by a lone pair of electrons from the oxygen, whereas the top carbocation is not. Therefore, the bottom carbocation is more stable.

STEP 3: Use the relative stability of the carbocations to predict which product will be the major product formed.

According to the Hammond postulate, formation of the more stable carbocation has the lower energy transition state, and thus forms faster, leading to the major product. Therefore, the bottom product will be the major product.

slow

fast

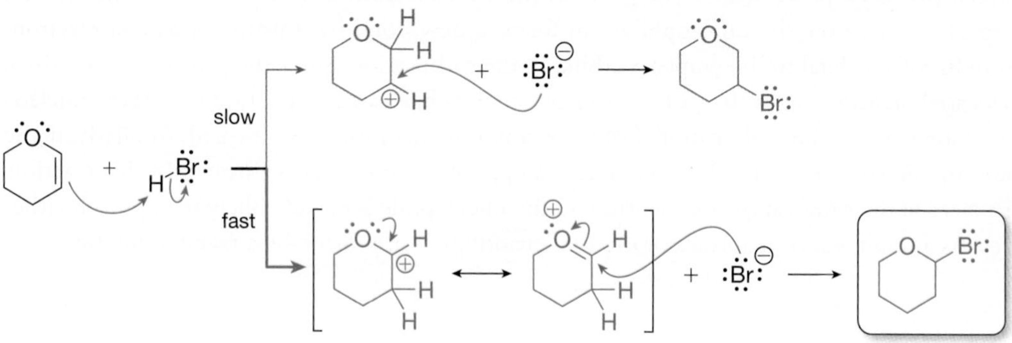

PRACTICE PROBLEM

8.5 Draw a curved arrow mechanism to show the formation of the major product formed when the following alkenes are reacted with HBr.

a)

b)

c)

d)

INTEGRATE THE SKILL

8.6 Which of the following alkenes (**a**, **b**, or **c**) can be expected to give the alkyl chloride shown here as the major product when reacted with HCl?

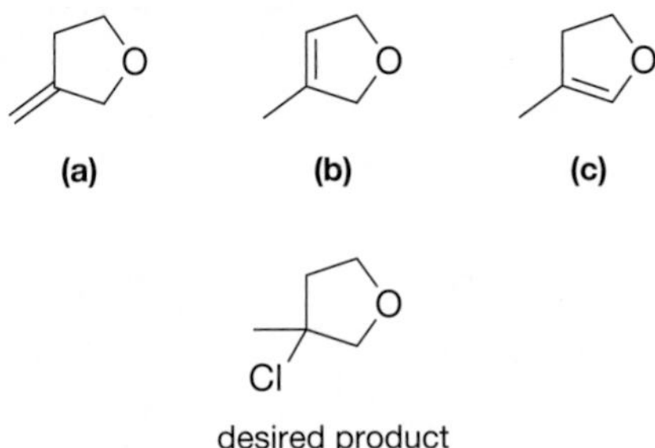

8.3.8 Reactions with carbocations are not stereoselective

The addition of a reagent across a π bond converts two sp^2 hybridized carbons into two sp^3 hybridized carbons. This addition may create new stereocentres in a molecule, which can lead to the formation of stereoisomeric products. During the addition to π bonds, the formation of the carbocation is regioselective because the most stable carbocation is produced preferentially.

Carbocations are sp^2 hybridized and therefore planar. The charged carbon has an empty p orbital that is perpendicular to the plane of the atoms directly connected to it. In the second step of the reaction, the nucleophilic ion forms a new bond by donating a pair of electrons from its filled orbital to the empty p orbital of the carbocation. Since the carbocation p orbital is located on *both* faces of the carbocation, the nucleophile can attack either face. If the nucleophile approaches the carbocation from the top, one enantiomer is formed. Similarly, if the nucleophile approaches the bottom face, the opposite enantiomer is formed. In this reaction, the faces of the carbocation are identical, so the nucleophile is equally likely to approach either face. As a result, equal quantities of the two enantiomers are formed—a racemic mixture.

The carbocation is sp^2 hybridized and flat. It has an empty p orbital that occupies both faces.

nucleophile adds to top of carbocation

pair of enantiomers

nucleophile adds to empty p orbital

nucleophile adds to bottom of carbocation

In hydrohalogenation reactions, the face on which the proton is added to the alkene has no influence on the direction from which the halide later approaches the carbocation. So the new carbon-hydrogen and carbon-halide bonds can be formed on the same face or the opposite faces of the alkene, which gives a mixture of stereoisomers. Hydrohalogenation reactions are usually *not* **stereoselective**; that is, they do not favour one stereoisomer over the other and therefore produce mixtures of stereoisomers.

Stereoselective reactions are those that *favour* the formation of one stereoisomer, though the other stereoisomer may also be produced as a minor product. These reactions tend to involve stepwise mechanisms.

8.4 Markovnikov Addition of Water to Alkenes

In the presence of a strong acid catalyst, such as H_3PO_4, alkenes react with water to form alcohols. The reaction has strong similarities to the addition of haloacids to alkenes, and the major regioisomer of the reaction is consistent with Markovnikov's rule: the major product arises from the most stable carbocation.

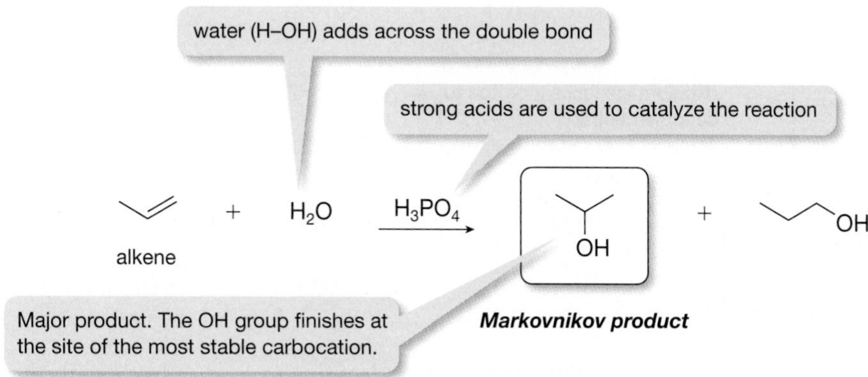
water (H–OH) adds across the double bond
strong acids are used to catalyze the reaction
+
H_2O
H_3PO_4
OH
+
OH
alkene
Major product. The OH group finishes at the site of the most stable carbocation.
Markovnikov product

Although HBr and HCl are strong enough acids to protonate π bonds, water is a very weak acid ($pK_a = 15.7$) and cannot protonate π bonds. The addition of water across a π bond therefore requires the presence of a strong acid catalyst such as sulfuric acid (H_2SO_4) or phosphoric acid (H_3PO_4). These acids react with water to form hydronium ions (H_3O^+), which are strong acids ($pK_a = -1.7$) that can react with π bonds.

$$H_2O + H_2SO_4 \rightleftharpoons H_3O^{\oplus} + HSO_4^{\ominus}$$

$pK_a = -5$ (stronger acid) $\qquad$ $pK_a = -1.7$ (weaker acid)

Because the reaction is performed under strongly acidic conditions, virtually no OH^- exists in the reaction mixture. The nucleophile that attacks the carbocation intermediate is therefore H_2O (the leaving group during carbocation formation). Since water is neutral, it is a weak nucleophile; however, the non-bonding electrons on its oxygen atom are sufficiently attracted to the electron-deficient carbocation that a reaction takes place.

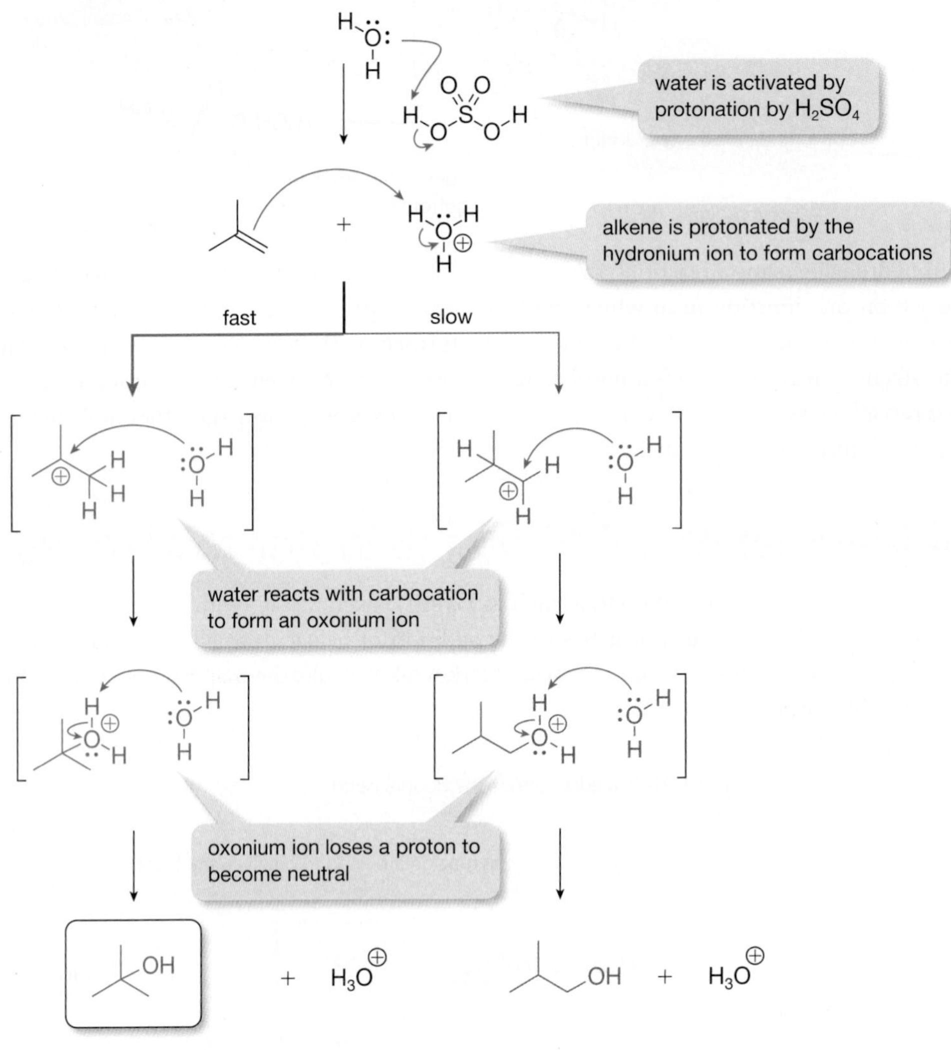

Markovnikov product

ORGANIC CHEMWARE
8.5 Hydration of alkenes

When water acts as a nucleophile and donates electrons to form a bond with the carbocation, the oxygen atom becomes positively charged. The resulting product is protonated and charged, and readily deprotonated by a conjugate base. In the last step of the process an H^+ is removed, which generates a neutral product. The addition of water to a double bond therefore requires four steps. The middle two steps involving the alkene are the same as in the HBr process. Using water as an electrophile adds an initial activation step to protonate the water and a final deprotonation to generate the neutral product.

Same electron movements
:Br:
alkene is protonated
nucleophile reacts with carbocation
OSO_3H
water is activated by protonation
product loses proton to become neutral

8.4.1 Addition of alcohols by acid catalysis

Structurally, water (H–OH) and alcohols (R–OH) differ only by the replacement of one hydrogen atom with an alkyl group. Because water and alcohols both contain a hydroxyl (OH) group, their reactions with many reagents are similar. More importantly, the reaction mechanisms are the same. The addition of alcohols to alkenes results in the formation of **ethers**, and the reaction obeys Markovnikov's rule.

Ethers are chemical compounds with two alkyl groups bonded to one oxygen atom.

acid catalyst
+ ROH
H_2SO_4
OR
+ OR
alkene
alcohol
Markovnikov product
the OR group has added to the site of the most stable carbocation

CHECKPOINT 8.4

You should now be able to draw the mechanism and products for the reactions of alkenes with water or alcohols in the presence of strong acids. Where relevant, you should also be able to predict the regiochemistry and stereochemistry of the major product(s).

SOLVED PROBLEM

Draw the curved arrow mechanism for the formation of the major product(s) of the following reaction.

OH
H_3PO_4

STEP 1: Identify the reacting nucleophile and electrophile.

The polarized O–H bonds of H_3PO_4 allow it to react as an electrophile. The starting material, on the other hand, has two potential nucleophilic sites: the electron-rich π bond and the lone pairs on the alcohol's oxygen atom. In principle, either nucleophile could react with the H_3PO_4, but the protonation of the alcohol is reversible and doesn't lead to any further reactions, so the alkene is the nucleophile of interest in this reaction.

STEP 2: Draw in the mechanistic arrows, using the principle that electrons flow from nucleophile to electrophile.

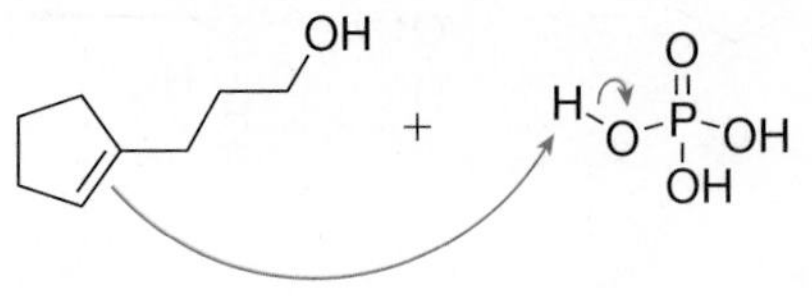

STEP 3: Use the arrows to determine the products formed from this step of the reaction, including any formal charges.

Because the alkene in this reaction is asymmetric, the hydrogen can bond with either the left-hand or right-hand carbon of the double bond. However, the major product will result from the more stable carbocation, in this case, the tertiary carbocation.

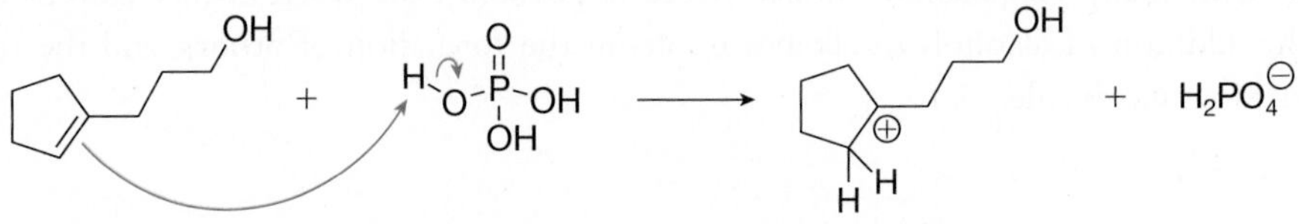

STEP 4: Repeat steps 1–3 until you obtain the final, neutral product.

After the first step of the reaction, there are two possible nucleophiles available to react with the carbocation: the conjugate base of the acid, $H_2PO_4^-$, and the alcohol attached to the same molecule. Intramolecular reactions are generally much faster than intermolecular reactions, so the alcohol is the reacting nucleophile. For an intramolecular reaction, numbering the atoms will help you keep track of the size of the ring, where the new bonds form, and where any substituents are located.

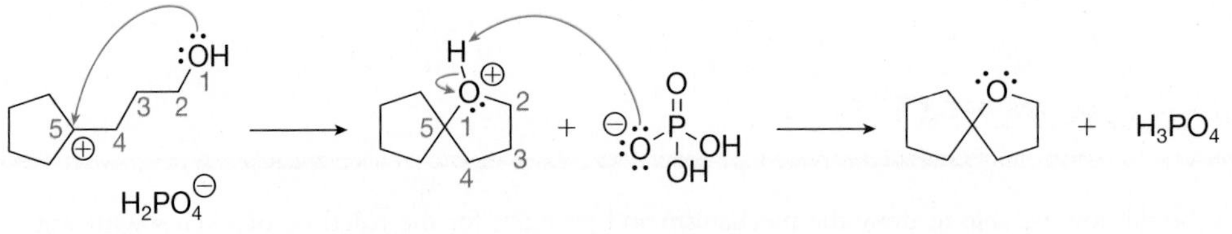

Therefore, the overall reaction is as follows:

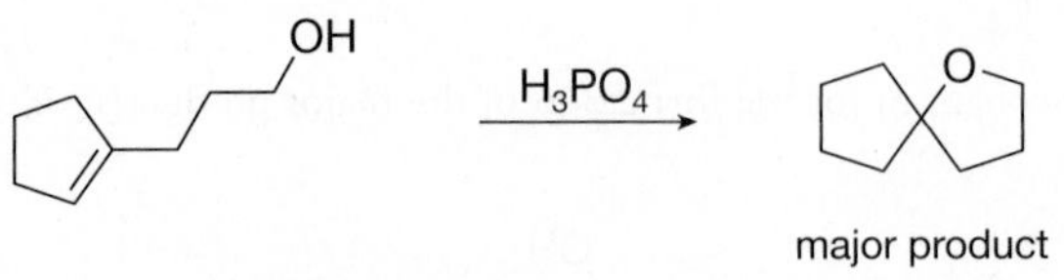

major product

PRACTICE PROBLEM

8.7 Draw the mechanism for the formation of the major products of the following reactions.

a) 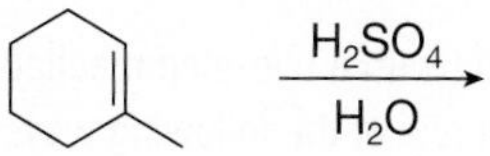 $\xrightarrow[H_2O]{H_2SO_4}$

b) 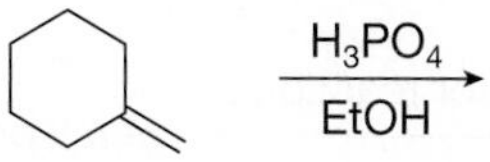 $\xrightarrow[EtOH]{H_3PO_4}$

c) 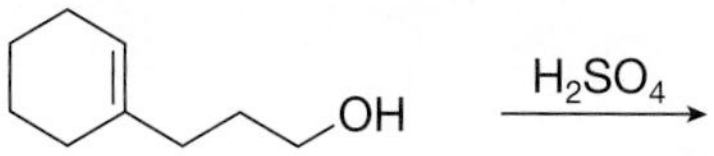 OH $\xrightarrow{H_2SO_4}$

d) 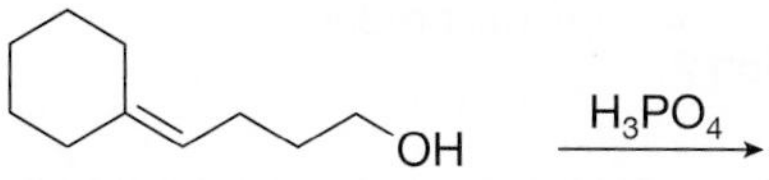 OH $\xrightarrow{H_3PO_4}$

e) 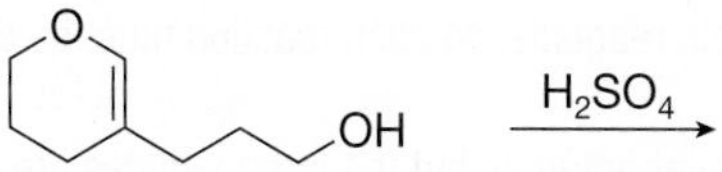 O ... OH $\xrightarrow{H_2SO_4}$

INTEGRATE THE SKILL

8.8 Sketch an energy diagram for the full reaction sequence of the addition of water to an alkene. Assume the products are lower in energy than the reactants.

8.4.2 Addition of water by oxymercuration-demercuration

The addition of water or alcohols to alkenes can sometimes require very harsh conditions. If the intermediate carbocation is not very stable, high temperatures or pressures must be used to initiate a reaction. If the structure of a molecule is complex, such conditions encourage undesired side reactions or decomposition. In addition, acid-catalyzed hydration reactions are reversible (see Chapter 12), and such equilibrium reactions can yield very little product if the alkene is quite stable.

Oxymercuration-demercuration is a much less problematic method for adding water or alcohols to the π bond of an alkene in a Markovnikov manner. This method actually involves two separate, sequential reactions: (1) oxymercuration using $Hg(OAc)_2$ ("Ac" is an abbreviation for acetyl, $C(O)CH_3$), and (2) reduction using sodium borohydride, $NaBH_4$.

Oxymercuration-demercuration is a Markovnikov reaction that adds water to an alkene using a mercury compound.

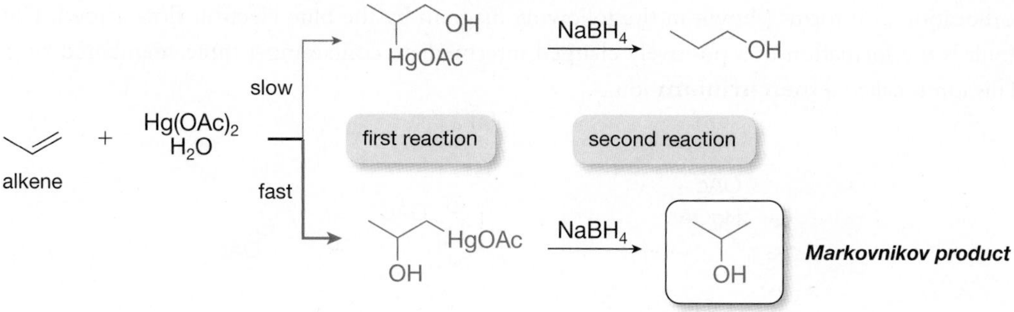

alkene + $Hg(OAc)_2$, H_2O

slow → OH, HgOAc $\xrightarrow{NaBH_4}$ OH

fast → HgOAc, OH $\xrightarrow{NaBH_4}$ OH — *Markovnikov product*

first reaction

second reaction

DID YOU KNOW?

Performing two separate reactions is not the same as performing one reaction with a two-step mechanism. An experimental method consisting of two separate reactions is typically written one of the following ways:

reactant $\xrightarrow{\text{reagents for reaction 1}}$ intermediate product $\xrightarrow{\text{reagents for reaction 2}}$ overall product

reactant $\xrightarrow[\text{2) reagents for reaction 2}]{\text{1) reagents for reaction 1}}$ overall product

In these sequences, the first reaction is performed and the intermediate product is isolated, then the second step is performed. The two reactions usually employ incompatible reagents, so each reaction must be done completely separately.

Reactions that have more than one step are written with similar notation, but the intermediates are usually not isolated. These intermediates normally exist for only a very short time within the reaction vessel.

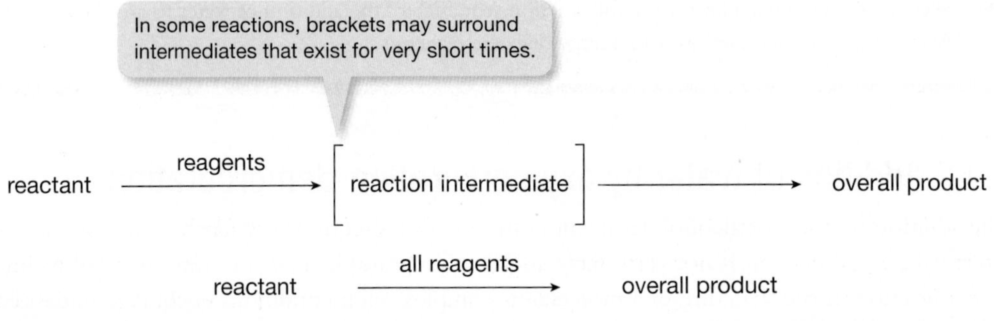
In some reactions, brackets may surround intermediates that exist for very short times.
reactant
reagents
reaction intermediate
overall product
reactant
all reagents
overall product

Although this method requires two separate reactions, it has several advantages over acid-catalyzed hydrations because the mechanism does not involve a carbocation intermediate. In the first step of the process, the nucleophilic π bond donates its pair of electrons to the electrophilic mercury atom, breaking the bond to one of the acetate groups in the process (red arrows). This flow of electrons is very similar to the flow when HBr reacts with a π bond.

Unlike hydrogen, the mercury atom contains non-bonded electrons that react with the carbocation as it forms (shown in the following diagram by the blue electron-flow arrow). The result is the formation of a positively charged intermediate containing a three-membered ring. This ion is called a **mercurinium** ion.

A **mercurinium** ion is a three-membered ring containing two carbon atoms and a positively charged mercury ion.

OAc
:HgOAc
alkene
⊕ OAc
Hg
mercurinium
+
⊖
OAc

DID YOU KNOW?

The formation of the mercurinium can be visualized as follows. Although this mechanism is not completely correct (the mercurinium actually forms in one step), it can be helpful for visualizing the flow of electrons from the mercury to the carbon.

Alkene reacts with $Hg(OAc)_2$ electron flow similar to reaction with HBr.

Hg has unpaired electrons (H does not). Lone pair forms bond with close-by carbocation.

OAc :HgOAc → ⊕ HgOAc → ⊕ HgOAc

OAc :HgOAc → ⊕ HgOAc

Because the ion forms in one step (all bonds form at the same time), the mechanism should show the ion formed directly.

Mercurinium ions are stabilized, with three resonance forms contributing to the overall structure. These forms are not equivalent and make unequal contributions to the actual structure. The greatest contributor is the structure that places the positive charge on the mercury atom because this large atom accommodates positive charge best, and because both carbon atoms have full octets in this structure. The other two resonance structures contain carbocations, which have incomplete carbon octets. One of these contributes very little to resonance because the charge is located on a very unfavourable primary position. The other resonance form contains a tertiary carbocation and so contributes more to the structure of the ion.

The result of this is that more positive character is exhibited by the tertiary carbon than the primary carbon, which favours a reaction with water at the tertiary site.

major contributing resonance form

⊕ HgOAc H H ⟷ ⊕ OAc Hg H H ⟷ HgOAc H ⊕ H

Contributing resonance form has tertiary carbocation. In the actual ion there is significant positive charge at this location.

minor resonance form has primary carbocation

The oxygen of the water nucleophile tends to react at the tertiary site because this location has the greatest positive charge resulting in a Markovnikov product. As with other alkene additions, the regioselectivity is controlled by the stabilization of a positive charge, which preferentially forms on the most substituted carbon of the double bond.

fast slow

positive charge develops on tertiary carbon

positive charge develops on primary carbon

This product forms faster. The transition state for the addition of water to the most substituted carbon atom provides the lowest energy pathway.

The concluding step of the mechanism of the oxymercuration is the deprotonation of the intermediate oxonium to form an organomercury alcohol. Since acetate is the strongest base present, it acts as the base for this process.

Depending on whether water or an alcohol is used in an oxymercuration reaction, the product is either an organomercury alcohol or an organomercury ether. This product is partly purified and then reacted with $NaBH_4$ to give the final product. This reaction is known as a **demercuration** reaction because it removes mercury from the compound.

Demercuration is a reaction that replaces the mercury in a C–Hg bond with a hydrogen atom.

$NaBH_4$ / EtOH

CHECKPOINT 8.5

WANT TO LEARN MORE?
8.1 Removal of mercury using $NaBH_4$

You should now be able to draw the mechanism for the oxymercuration reaction, predict the possible regioisomeric products formed, and identify the major product(s).

SOLVED PROBLEM

Draw the curved arrow mechanism and all regioisomeric products of the oxymercuration of the following alkene. Identify the major product(s).

$Hg(OAc)_2$
MeOH

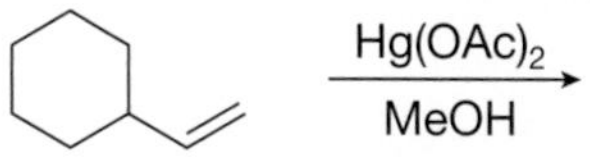

STEP 1: Identify the reacting nucleophile and electrophile.

The starting material has an electron-rich π bond that can act as the nucleophile, whereas the mercury atom in $Hg(OAc)_2$ is electrophilic.

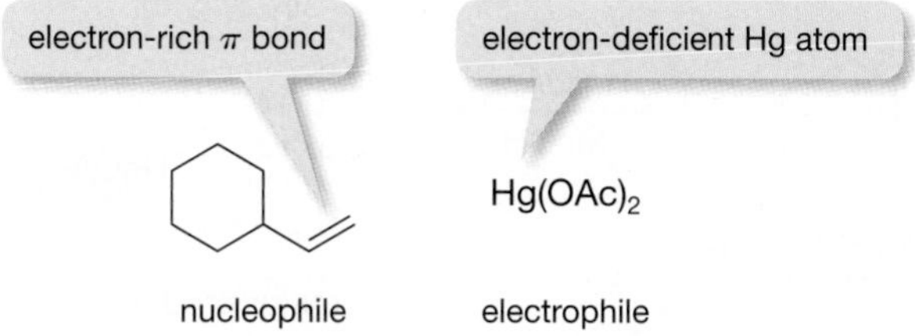

electron-rich π bond
electron-deficient Hg atom
$Hg(OAc)_2$
nucleophile
electrophile

STEP 2: Draw in the mechanistic arrows, using the principle that electrons flow from nucleophile to electrophile to guide you, as well as the sample mechanism in the text. Use the arrows to determine the products formed from this step of the reaction, including any formal charges.

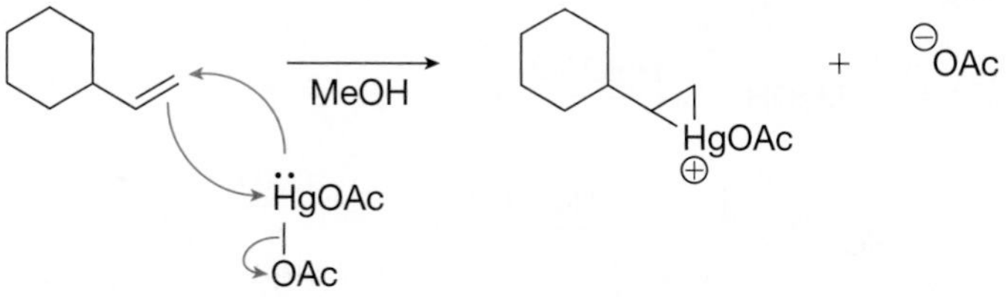

MeOH
HgOAc
OAc
HgOAc
⊕
+ ⊖OAc

STEP 3: Repeat steps 1 and 2 until you obtain the final, neutral products.

The product of the first step has a positive charge and will react with the nucleophilic lone pairs on the solvent, methanol. Since the mercurinium ion is asymmetric, the nucleophile can attack at either carbon. Both intermediates must be considered:

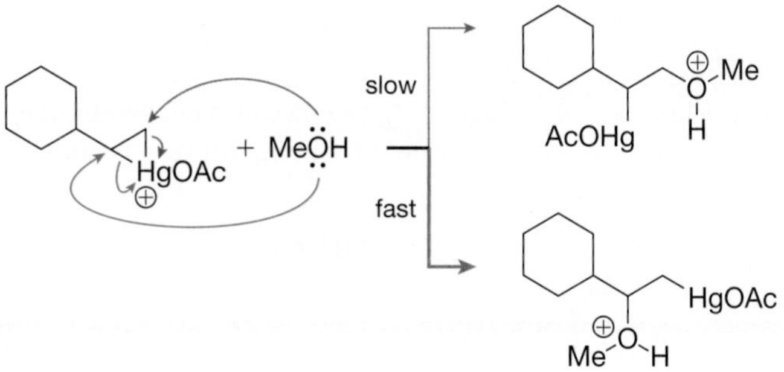

HgOAc
⊕
+ MeOH
slow
fast
AcOHg
⊕O–Me
H
HgOAc
⊕O
Me H

Finally, a molecule of acetate is used to neutralize the positive charges and create neutral products.

Path **(a)** AcOHg → AcOHg + HOAc

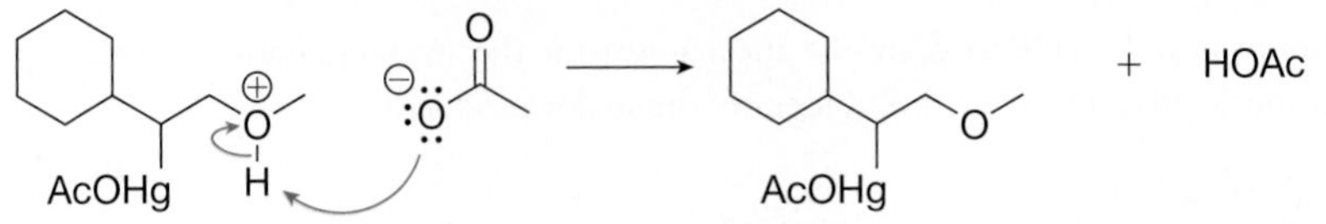

Path **(b)** HgOAc → HgOAc + HOAc

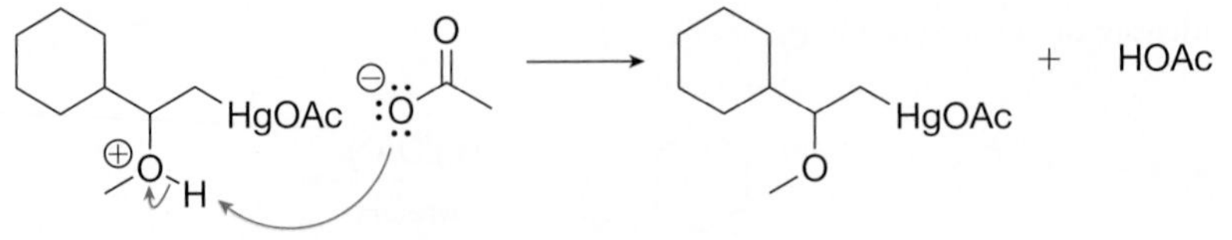

STEP 4: Determine which is the major product.

The major product will arise from reaction of the nucleophile at the carbon with the greater partial positive charge; that is, the one with the more substituted carbon. Therefore, the bottom product, from pathway B, will be the major one. The overall reaction is as follows:

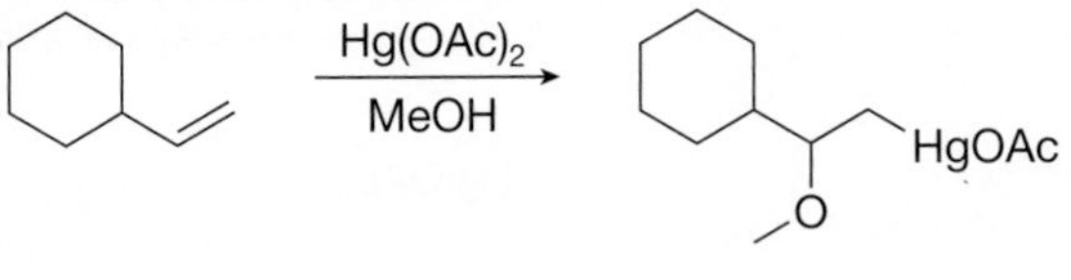

$Hg(OAc)_2$, MeOH → HgOAc

major product

PRACTICE PROBLEM

8.9 Draw the mechanism and regioisomeric products of the following oxymercuration reactions (you may ignore stereoisomers). If more than one regioisomer is formed, predict which, if any, will be the major product.

a)  + MeOH $\xrightarrow{Hg(OAc)_2}$

b) 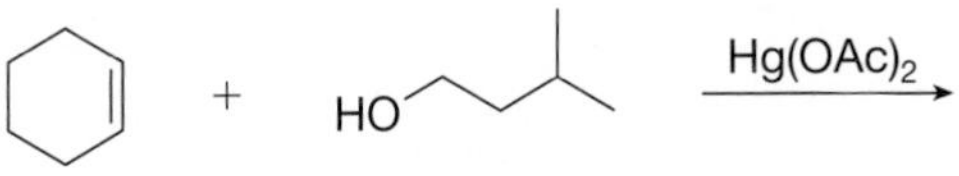 + HO $\xrightarrow{Hg(OAc)_2}$

c) 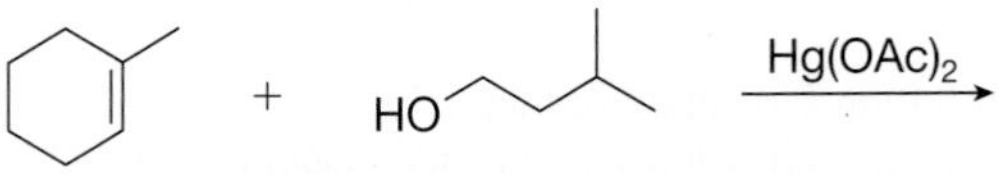 + HO $\xrightarrow{Hg(OAc)_2}$

d) 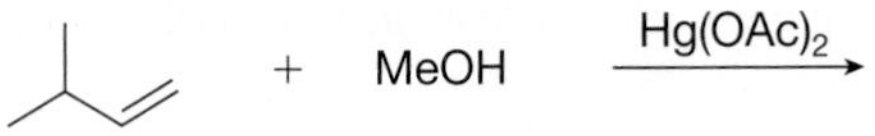 + MeOH $\xrightarrow{Hg(OAc)_2}$

INTEGRATE THE SKILL

8.10 When the alkene below reacts with $Hg(OAc)_2$, two mercurinium ions form in equal amounts. Draw both mercurinium ions and identify the relationship between them.

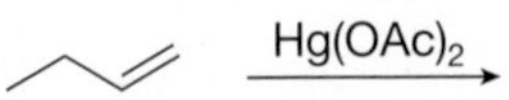

$\xrightarrow{Hg(OAc)_2}$

CHEMISTRY: EVERYTHING AND EVERYWHERE

Oxymercuration: Cooking Up Ecstasy in Clandestine Labs

In 1912, scientists searching for compounds that mimic the action of adrenaline, the body's fight-or-flight chemical, discovered 3,4-methylenedioxy-*N*-methylamphetamine (MDMA). It was a mild hallucinogen, and produced temporary feelings of openness, euphoria, and empathy, but it could not be used to treat a disease, so it was never commercialized.

MDMA was used as a recreational drug to a limited extent in 1960s, but became a much more widespread party drug in the1980s. Commonly known as "ecstasy," MDMA was still legal to produce and possess at that time. Because ecstasy can have dangerous side effects (including high body temperature, increased heart rate, and reduced judgment), the drug became illegal in most countries.

One of the ways to make MDMA involves an oxymercuration reaction. Unfortunately, this reaction produces mercury-containing impurities, which can be difficult to remove from the final product. Illegal drug manufacturers are not overly concerned with safety or purity, and so many ecstasy pills sold on the street contain significant amounts of mercury and other heavy metals, which are powerful neurotoxins—they destroy brain and nerve tissue. Contaminants in contraband ecstasy have caused severe illness and death.

ORGANIC CHEMWARE
8.6 Oxymercuration-demercuration

Safrole $\xrightarrow[H_2O]{Hg(OAc)_2}$ (OH, HgOAc) $\xrightarrow{NaBH_4}$ (OH) $\xrightarrow{steps}$ MDMA ($NHCH_3$)

8.5 Carbocation Rearrangements

A carbocation is a high-energy, unstable intermediate that reacts quickly with nucleophiles and bases. The presence of a carbocation can complicate reactions because it provides an opportunity for competing reactions to occur. This is especially true when reactions are heated. For example, carbocations might undergo elimination (Chapter 12) or add to an aromatic ring (Chapter 10). Even when nucleophilic addition is the preferred reaction pathway, a **carbocation rearrangement** may occur before the addition of the nucleophile.

A **carbocation rearrangement** changes the structure of the carbocation, often through a hydride or alkyl shift, to provide a more stable intermediate.

Hydride shifts involve the movement of a hydrogen, with its two electrons, from the carbon adjacent to the carbocation. When the bond moves, the existing carbocation becomes neutralized. The adjacent carbon is left short of electrons, creating a new carbocation.

[1,2-hydride shift] hydrogen moves with a pair of electrons (CH bond) from adjacent carbon

H–Br

tertiary carbocation more stable than secondary

Carbocations are more stable when they have more substituents because of the influence of hyperconjugation, but hyperconjugation can lead to rearrangements. As hyperconjugation takes place, the carbon-hydrogen σ bond can shift so that the more stable carbocation remains. Thus, hydride shifts occur due to an excess of shared electrons between a C–H bond and an adjacent empty p orbital.

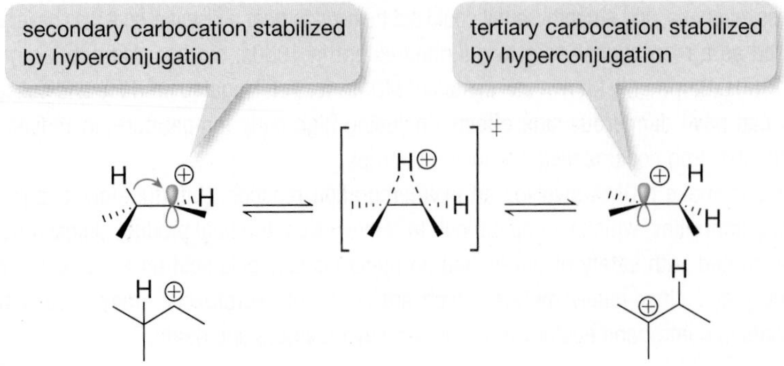

secondary carbocation stabilized by hyperconjugation

tertiary carbocation stabilized by hyperconjugation

ORGANIC CHEMWARE
8.7 Carbocation rearrangement: 1,2-Hydride shift

Rearrangements happen only when a carbocation with greater stability can form. A carbocation rearrangement forms a secondary carbocation from a primary, or a tertiary carbocation forms from a secondary. Allylic or benzylic cations are stabilized by conjugation, and can form from rearrangements even when the degree of substitution does not increase.

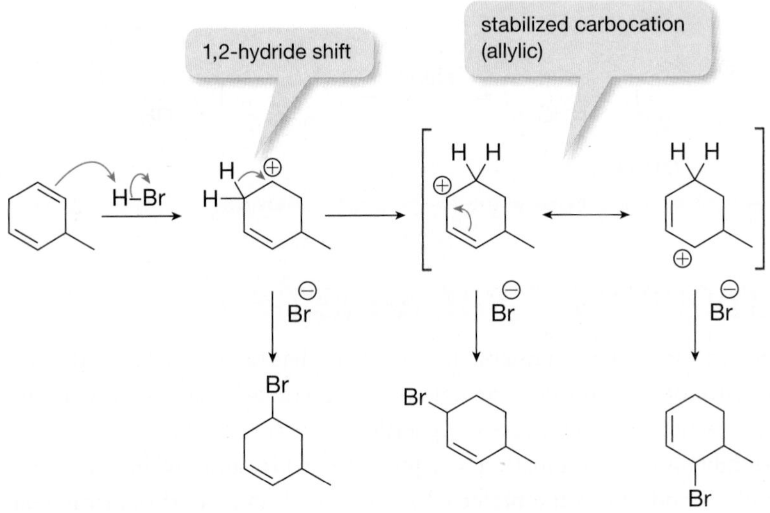

1,2-hydride shift

stabilized carbocation (allylic)

When there is no adjacent hydride, an alkyl group, such as a methyl, can also migrate.

8.5.1 Using lone pairs and hydrogens in mechanisms

Drawing full Lewis structures (electrons) and hydrogens on reacting functional groups is an excellent way to track electron movement and charge in chemical reactions. As long as electron movement rules are respected, it is not necessary to draw full Lewis structures for all functional groups; however, it is a good idea to use Lewis structures with unfamiliar functional groups. Similarly, not all hydrogens need be shown (except when they participate in reactions).

In the remaining chapters, mechanisms are depicted without explicitly showing hydrogens and lone pairs (unless necessary). Whenever you are unsure about what is happening in a reaction, draw full Lewis structures and hydrogens to clarify electron movement.

8.6 Addition of Halogens to Double Bonds

Alkenes react very rapidly with the halogens F_2, Cl_2, Br_2, and I_2. Because fluorine is highly electronegative, F_2 gas can react explosively with alkenes. The addition of halogens to double bonds involves two steps. In the first step a molecule of halogen adds to a double bond, forming a halonium ion. This ion is then opened by the halide leaving group of the first step to give the overall addition product.

The bond in a diatomic halogen molecule is easily broken by nucleophiles once a dipole is induced on the halogen. When a halogen molecule approaches orthogonally to the electron-rich π bond of an alkene, the electrons on the halogen atom closest to the π bond experience electron-electron repulsion, which induces a small dipole. The halogen atom closest to the alkene develops a $\delta+$ charge. The other halogen atom develops a $\delta-$ charge and therefore acts as a leaving group. For the following compound, this leaving group is a chlorine atom.

Cl δ−
Cl δ+
dipole induced onto Cl_2 molecule as it approaches π bond
Electron repulsion from the filled π bond forces electrons in the Cl–Cl bond onto the chlorine furthest from the π bond producing a dipole.

The π bond donates electrons to the partially positive halogen atom to form a carbon-halogen bond, and the halogen-halogen bond breaks to avoid exceeding the valency of the halogen atom. Like the mercury atom of an oxymercuration reaction, halogen atoms have unpaired electrons, which can make a bond with the carbocation as it forms. This **concerted** flow of electrons produces a cyclic **halonium ion**, in which both carbon halogen bonds form on the same face of the double bond, while displacing a halide ion as a leaving group.

In a **concerted** reaction, all bond-forming and bond-breaking events happen in the same step.

A **halonium ion** is a three-membered ring that contains a halogen with a +1 formal charge. If the halogen is chlorine, a chloronium is formed. Bromine produces a bromonium, and iodine gives an iodonium.

Lone pair on the halogen atom forms bond to carbon
:Cl–Cl
alkene
Cl ⊕
Cl ⊖
halonium ion
(chloronium)
Cl
Cl
chloride is a nucleophile to open the halonium ion

The second step of the mechanism of halogen addition is analogous to the second step of oxymercuration. A halide ion now becomes a nucleophile and opens the halonium ion ring. As with the mercurinium ion (and for the same reasons), this halide adds to the most substituted carbon atom of the halonium ion. Mechanistically this step is a Markovnikov addition. However, the atoms added to the alkene are identical, so the product is not called a Markovnikov product.

When the halide nucleophile approaches the carbon atom of the chloronium ion, it approaches the face *opposite* to that of the positive halogen for two reasons. First, the large halogen atom blocks the approach of the other atom from that side of the ion (a steric effect). Second, and more significantly, the nucleophile must overlap its filled orbital with an empty orbital on

STUDENT TIP
In cyclic halonium ions and cyclic mercurinium ions, a nucleophile attacks the *opposite* face of the ring from the Hg^+ or X^+ atom.

the electrophile. In this reaction, the empty orbital on the halonium ion is a σ^* orbital of the carbon-halogen bond. To overlap this anti-bonding orbital, the nucleophile must approach the electrophile from the side of the ring that is opposite to the halogen atom. The net result is the **anti-addition** of the two halogen atoms *relative to the original alkene.*

In **anti-addition**, the two new atoms or groups are added to opposite faces of a double bond.

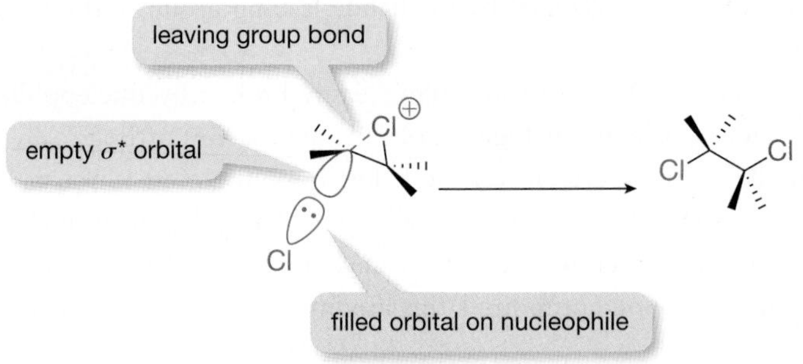

leaving group bond

empty σ^* orbital

filled orbital on nucleophile

ORGANIC CHEMWARE
8.8 Halogenation of alkenes

Anti-addition is most noticeable in the reaction of a halogen with a cyclic alkene. For example, the reaction between cyclohexene and Br_2 rapidly produces a bromonium ion, in which both C–Br bonds are formed on the same face of the double bond. Bromide opens this ion by approaching anti to a C–Br bond of the ion. As a result, the bromine atoms in the product show *trans* stereochemistry.

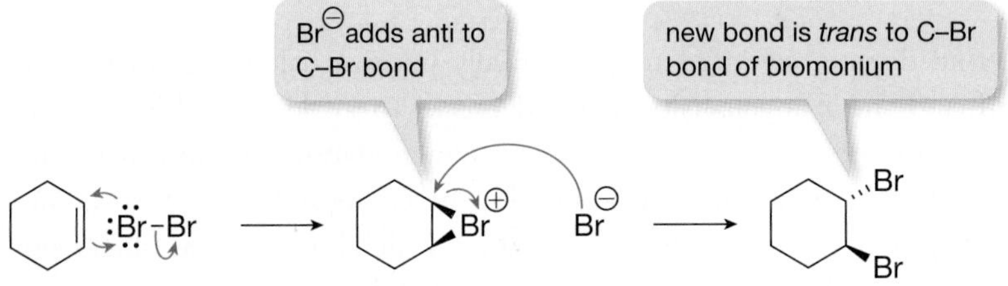

$Br^{\ominus}$ adds anti to C–Br bond

new bond is *trans* to C–Br bond of bromonium

8.6.1 Addition of hypohalous acids

Various reagents can open halonium ions in a Markovnikov process. For example, the addition of a hypohalous acid, such as hypochlorus acid (HOCl), forms addition products called **halohydrins**. These products form when an alkene is combined with a molecular halogen (e.g., Cl_2) in the presence of water or an alcohol. Halohydrin formation is a Markovnikov reaction because the OH group in the product is bonded to the most substituted carbon of the original alkene.

Halohydrins contain a halogen bonded to one carbon atom and a hydroxyl group bonded to an adjacent carbon atom.

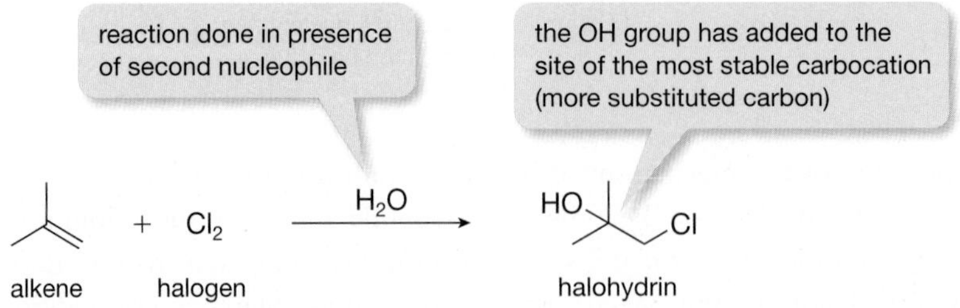

reaction done in presence of second nucleophile

the OH group has added to the site of the most stable carbocation (more substituted carbon)

alkene + Cl_2 (halogen) $\xrightarrow{H_2O}$ halohydrin

Like the formation of alkyl dihalides, formation of a halohydrin starts with a halogen molecule reacting with the alkene to produce a halonium ion. Water competes with chloride to open the halonium ion. The resulting charged intermediate quickly loses H^+ to form a halohydrin.

oxonium ion loses proton to become neutral
chloronium ion forms
water opens chloronium ion at location that best stabilizes carbocation (most substituted)

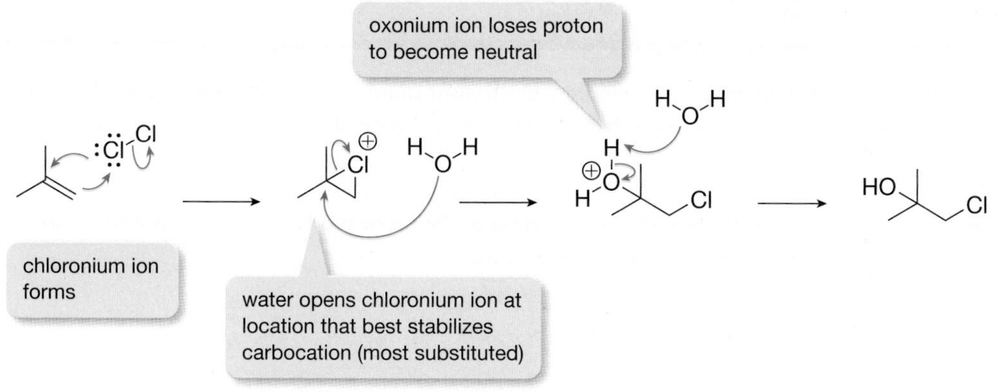

Halohydrin formation is regioselective. In the transition state for the ring opening, the water reacts with the carbon that has the most positive charge. This positive charge usually "prefers" to reside on the most substituted carbon. As a result, the major product is the regioisomer in which the OH group is connected to the more substituted carbon atom of the double bond.

The reaction is also stereoselective because the water must add to the opposite face of the halonium ion to the carbon-halogen bond. Thus, *anti*-addition products are formed, that is, the OH and halogen groups bond to opposite faces of the original alkene.

Both of these aspects are illustrated in the following reaction diagram. The halogen initially reacts with the double bond to generate a halonium ion. Notice that both carbon-halogen bonds form on the same face of the double bond (both up or both down). Nucleophilic water then attacks and opens the halonium ion, both regioselectively and stereoselectively.

H_2O attacks most substituted carbon
water removes $H^{\oplus}$ to form neutral product
Cl_2
H_2O
H_2O
alkene
halonium ion
halohydrin
anti addition of two groups to a cyclic alkene results in the two new groups being *trans* to each other

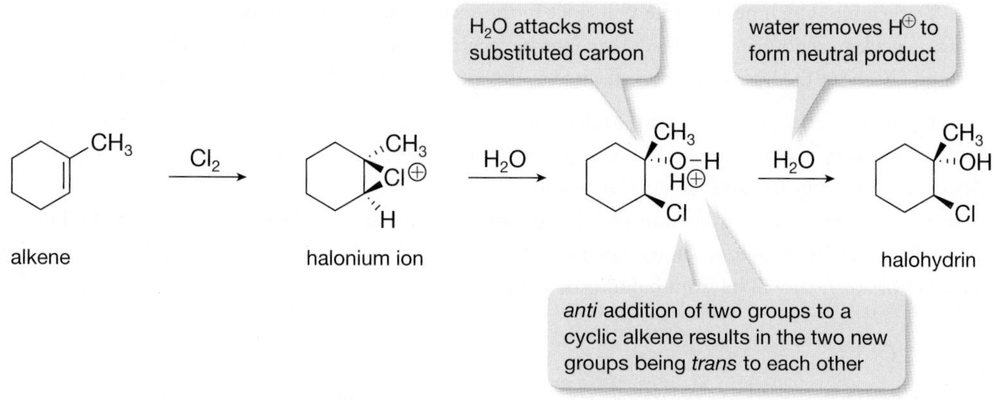

DID YOU KNOW?

The choice of drawing both carbon-chlorine bonds above the plane of the four atoms of the halonium structure is arbitrary. The starting alkene group is flat, and the reacting halogen molecules can approach either face of this structure. In a large collection of molecules, half of the molecules add to the top face and half add to the bottom. So the fact that the product is racemic is understood when writing this type of reaction. A single stereoisomer is shown in the drawing to illustrate the **relative stereochemistry** between the new stereocentres in the molecule. The actual product contains a 50:50 mixture of enantiomers.

Relative stereochemistry refers to spatial positions of two substituents *relative* to each other. The terms *cis* and *trans* refer to the relative stereochemistry of two substituents.

ORGANIC CHEMWARE
8.9 Halohydrin formation

CHECKPOINT 8.6

You should now be able to draw a curved arrow mechanism and predict the major products of a reaction between an alkene and a halogen-based electrophile, in the presence of competing nucleophiles.

SOLVED PROBLEM

Draw the curved arrow mechanism for the formation of the major product resulting from the following reaction. Include any relative stereochemistry in the product.

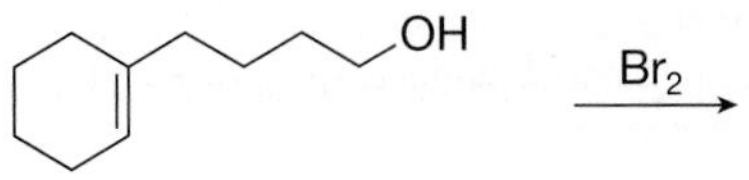

OH $\xrightarrow{Br_2}$

STEP 1: Using steps 1 and 2 of Checkpoint 8.5 as a guide, draw the mechanism and products for the first step of the reaction. Recall that electrophiles can add to either face of an alkene. Here, we chose to draw the C–Br bonds pointing out of the plane of the page, but this choice is arbitrary. In reality, an equal amount of each enantiomer is formed.

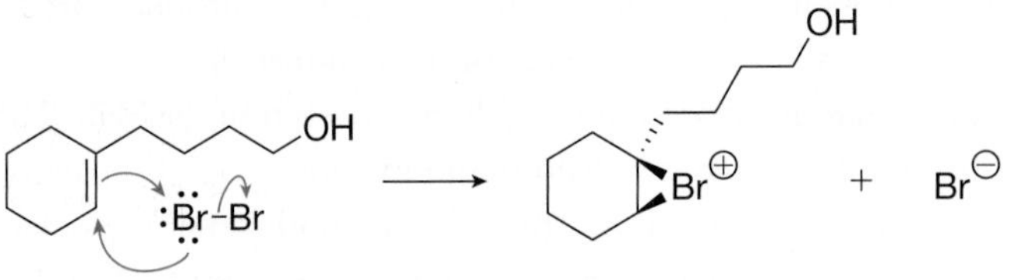

OH :Br–Br ⟶ OH Br⊕ + Br⊖

STEP 2: Determine which nucleophile will react in the second step of the reaction.

Intramolecular reactions proceed faster than intermolecular reactions for the formation of small- to medium-sized rings. Therefore, the hydroxyl group attached to the intermediate will react faster than the bromide ion.

STEP 3: Continue drawing the reaction mechanism and products, paying attention to both regiochemistry and stereochemistry. Numbering or labelling the atoms between reacting functional groups is a good way to determine final ring size and connection points.

The major product will result from the alcohol attacking the more substituted carbon of the bromonium ion. Since the bromonium ion has been drawn coming out of the plane of the page, the anti approach of the alcohol's oxygen will be from behind the plane of the page.

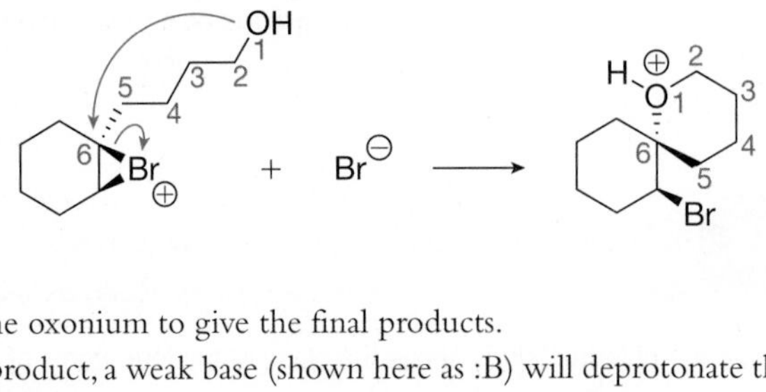

OH 1 2 3 4 5 6 Br⊕ + Br⊖ ⟶ H O⊕ 1 2 3 4 5 6 Br

STEP 4: Deprotonate the oxonium to give the final products.

To create a neutral product, a weak base (shown here as :B) will deprotonate the oxonium ion. The base could be a molecule of solvent (not specified in the question) or another molecule of the starting material.

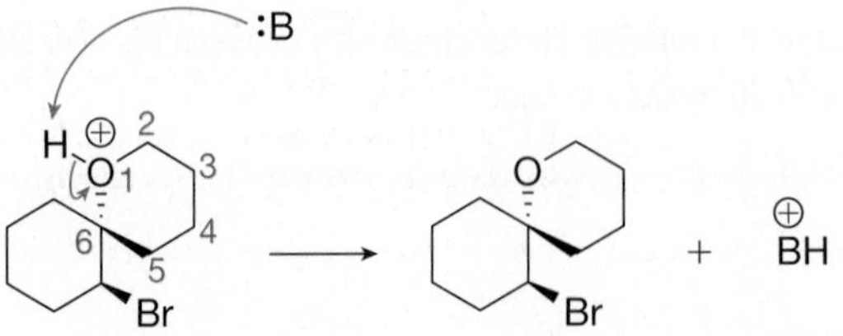

:B H O⊕ 1 2 3 4 5 6 Br ⟶ O Br + BH⊕

Therefore, the overall reaction is as follows:

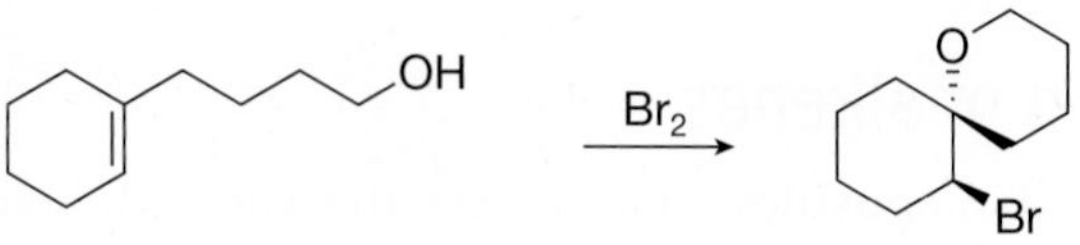

OH $\xrightarrow{Br_2}$ O Br

PRACTICE PROBLEM

8.11 Draw the major organic product for each of the following reactions. Include any relative stereochemistry in the products.

a) 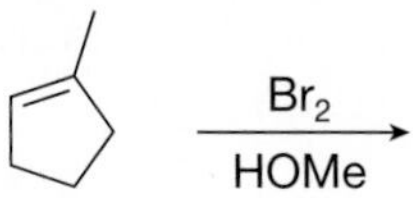 $\xrightarrow[\text{HOMe}]{Br_2}$

b) 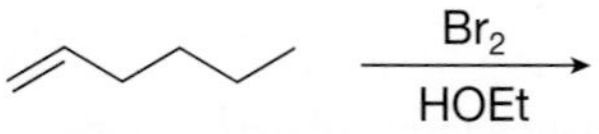 $\xrightarrow[\text{HOEt}]{Br_2}$

c) 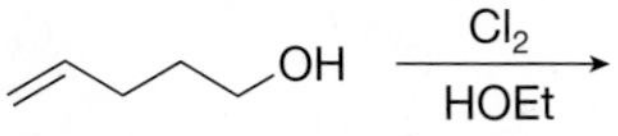 OH $\xrightarrow[\text{HOEt}]{Cl_2}$

d) 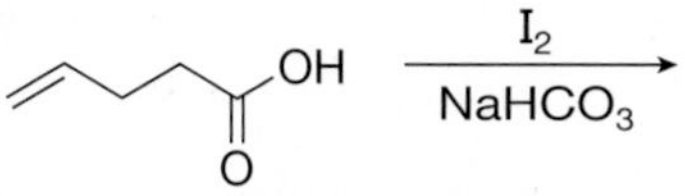 OH O $\xrightarrow[\text{NaHCO}_3]{I_2}$

e)  $\xrightarrow{\text{ICl}}$

f) 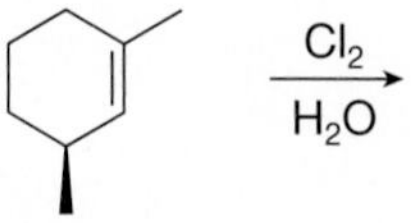 $\xrightarrow[H_2O]{Cl_2}$

INTEGRATE THE SKILL

8.12 Draw the alkenes and reagents that will react to give the products shown here.

a) 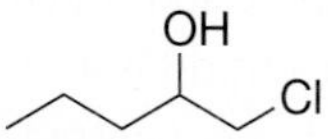 OH Cl

b) 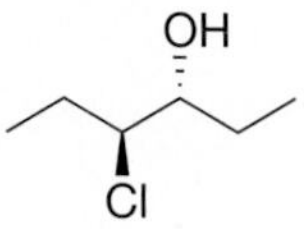 OH Cl

c) 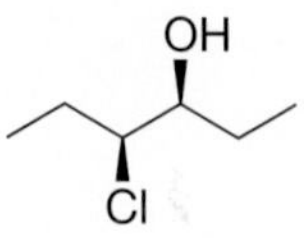 OH Cl

d) 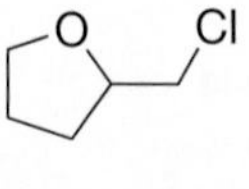 O Cl

8.7 Other Types of Electrophilic Additions

8.7.1 Epoxidation of alkenes

An **epoxide** is a cyclic ether with a ring consisting of one oxygen and two carbon atoms.

Alkenes can be converted to **epoxides** using peroxycarboxylic acids (sometimes called peroxyacids or peracids). The general structure of these acids is RC(O)OOH (sometimes written as RCO_3H). This epoxidation reaction bonds the same oxygen atom to both carbon atoms of the alkene double bond. Commonly used peroxyacids include monomagnesium peroxyperpthalate (MMPP) and *meta*-chloroperoxybenzoic acid (mCPBA) (Figure 8.7).

mCPBA

alkene

epoxide

a peroxycarboxylic acid

monomagnesium peroxyperphthalate (MMPP)

m-chloroperoxybenzoic acid (mCPBA)

Resonance forms of peroxyacids

this oxygen is NOT involved in resonance

the adjacent positive charge makes this oxygen susceptible to attack by a nucleophile

FIGURE 8.7 Structures of selected peroxycarboxylic acids.

The mechanism of the epoxidation of an alkene involves the same principles as the other reactions described in this chapter. These electron flows follow the same pattern as those of halogen addition and oxymercuration (red arrows in the following example). The reaction mechanism also involves rearrangement of the hydrogen on the leaving group. These additional electron flows (shown by black arrows) increase the complexity of the mechanism, but do not fundamentally alter what is happening. Epoxidation, like halonium ion formation, is a concerted reaction, so both carbon-oxygen bonds form on the same face of the alkene.

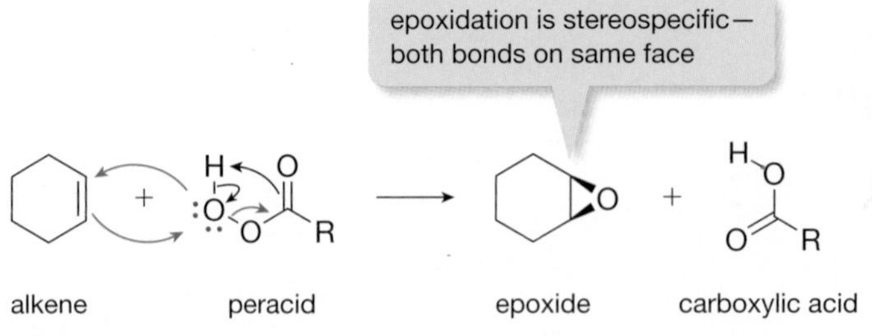

ORGANIC CHEMWARE
8.10 Epoxidation (with peracid)

Epoxides are useful compounds that can be opened with nucleophiles to create new bonds. Such reactions are highly regioselective and stereoselective. In the presence of acid catalysts, epoxides react with nucleophiles such as water or alcohols. The epoxide oxygen carries electron pairs that are protonated by the acid in the solution. The resulting positively charged oxonium intermediate is similar to the cyclic ions (mercurinium, halonium) described earlier in this chapter.

The reactions of oxonium rings are regioselective because the ion reacts with nucleophiles such that the nucleophile opens the protonated epoxide ring at the location of the most stable carbocation (most substituted position), similar to the reactions with bromonium and mercurinium ions. When water is the nucleophile, the result is a charged intermediate that loses a H^+ to a molecule of solvent to give a neutral product.

nucleophile reacts at carbon best able to stabilize a positive charge

nucleophile and leaving group anti

final product deprotonated to form neutral material

nucleophile reacts at carbon best able to stabilize a positive charge (most substituted carbon)

trans-diol

Oxonium reactions are also *trans* stereoselective. As with addition of halogens to alkenes, stereoselectivity arises due to the antiperiplanar approach of the nucleophile, dictated by the geometry of the σ^* LUMO of the protonated epoxide.

The regiochemistry of epoxide opening changes in basic conditions. Basic conditions activate the nucleophile rather than the electrophile. Because the epoxide does not have a positive charge in these reactions, there is no electronic preference to react at the site of carbocation stabilization. Instead, regiochemistry in these reactions is controlled by sterics, which favour the least crowded sites. Under basic conditions, epoxides react at the least substituted positions because the nucleophile can most easily approach these locations.

ORGANIC CHEMWARE
8.11 Epoxide opening (acidic conditions)

ORGANIC CHEMWARE
8.12 Epoxide opening (basic conditions)

Acid conditions:

positively charged rings are opened at the site of the most stable carbocation (most substituted carbon)

Base conditions:

no positive charge

neutral rings are opened at the least substituted carbon

Like the other reactions involving ring opening in this chapter, the approach of the nucleophile leads to the selective production of *trans* isomers.

OCH_3 adds antiperiplanar to original C–O bond, resulting in *trans* product

CHECKPOINT 8.7

You should now be able draw a mechanism for and predict the products formed when an epoxide is reacted with a nucleophile under either acidic or basic conditions.

SOLVED PROBLEM

Draw the curved arrow mechanism and predict the major product of the following reaction:

1) NaBr

2) NH_4Cl, H_2O

STEP 1: Identify the roles of the reactants.

The C–O bonds of the epoxide are polarized, making them electrophilic at the carbon atoms. The bromide of NaBr has lone pairs associated with a negative charge, making it nucleophilic.

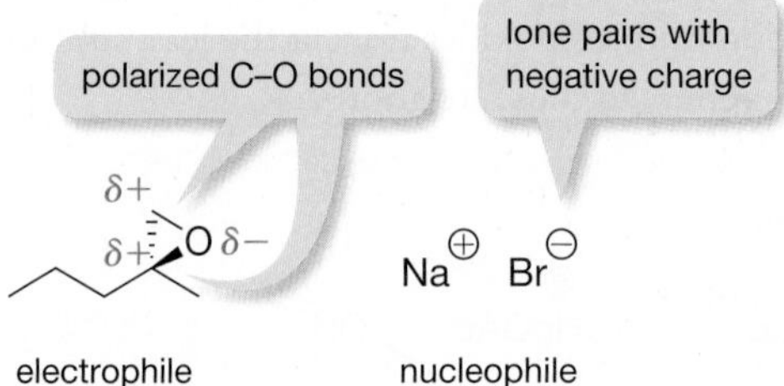

STEP 2: Draw in the mechanistic arrows, applying the principle that electrons flow from nucleophile to electrophile.

Under neutral or basic conditions, the nucleophile adds to the least substituted carbon, where steric interactions are minimized.

STEP 3: Use the arrows to determine the products formed from this step of the reaction, including any formal charges and relative stereochemistry.

STEP 4: Use the aqueous acid added in the second step to neutralize the final product. In water, acids form H_3O^+, which protonates the product.

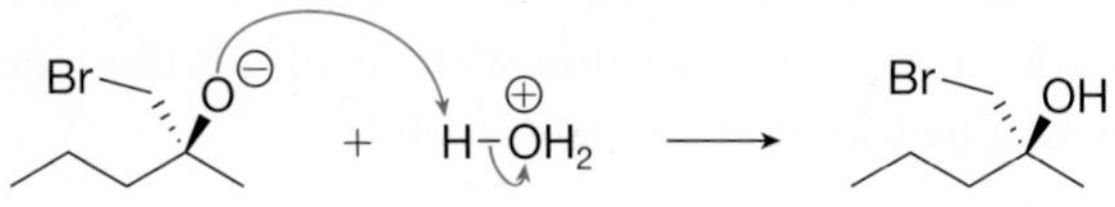

Br O⊖ + H–$\overset{\oplus}{O}H_2$ ⟶ Br OH

PRACTICE PROBLEM

8.13 For each of the following reactions, draw the curved arrow mechanism corresponding to the formation of the major product(s).

a) 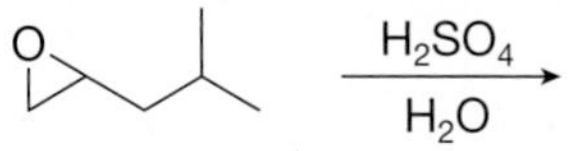 $\xrightarrow[H_2O]{H_2SO_4}$

b) 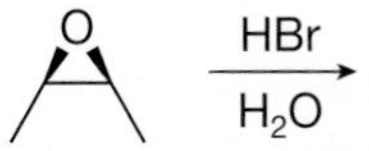 $\xrightarrow[H_2O]{HBr}$

c) 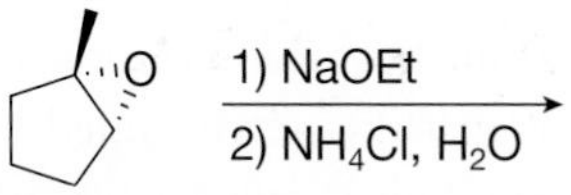 $\xrightarrow[2)\ NH_4Cl,\ H_2O]{1)\ NaOEt}$

d) 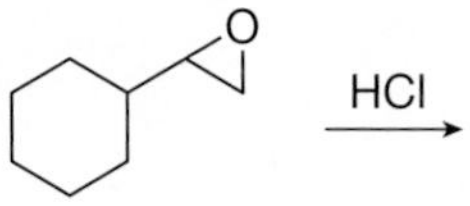 $\xrightarrow{HCl}$

INTEGRATE THE SKILL

8.14 Suggest two different methods that could be used to carry out the following transformation:

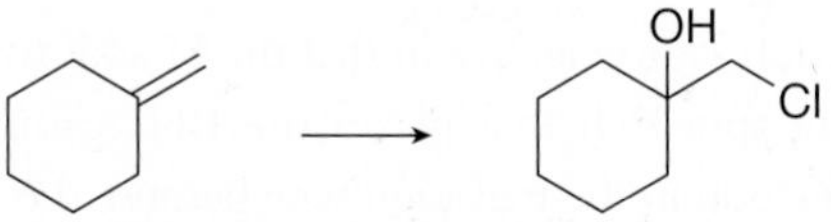

⟶ OH Cl

8.7.2 Hydroboration: An anti-Markovnikov addition

The addition of water to a π bond by acid catalysis or by an oxymercuration reaction is a Markovnikov reaction in which the major product has the hydroxyl group attached to the site of greatest carbocation stability (usually the more substituted carbon).However, alkenes can be converted to alcohols using reaction conditions that favour **anti-Markovnikov products**, which have the OH group on the other carbon of the double bond (usually less substituted). Reactions that produce such regioselectivity are called **anti-Markovnikov reactions**.

An **anti-Markovnikov product** places the nucleophilic atom on the carbon of the alkene that is the site of the less stable carbocation. These products arise from Markovnikov mechanisms in which a $\delta+$ charge is found on the site that best stabilizes it.

Anti-Markovnikov reactions are those that produce anti-Markovnikov products.

+ H_2O

OH — Markovnikov product

site of most stable carbocation

OH — anti-Markovnikov product

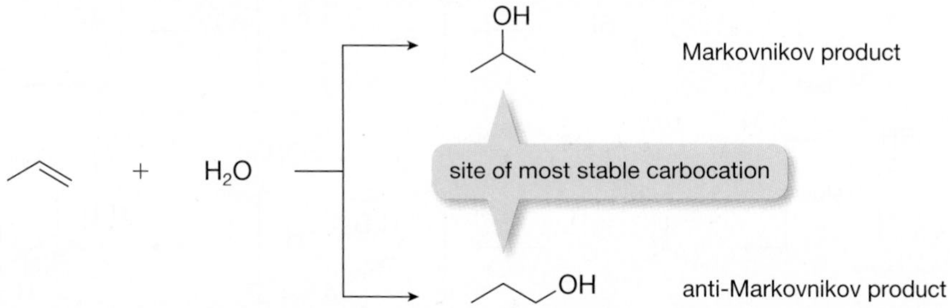

The anti-Markovnikov addition of water to an alkene is accomplished with borane (BH_3). The boron atom in BH_3 does not have a full octet, so this sp^2 hybridized atom is electron deficient and thus highly electrophilic. Since the electronegativity of hydrogen is slightly higher than that of boron, the boron-hydrogen bond of borane is polarized, with the boron atom carrying a partial positive charge and the hydrogen atom being partly negative. Consequently, when BH_3 undergoes an electrophilic addition reaction with an alkene, the boron acts as the electrophile, and the hydrogen (as a hydride) acts as a nucleophile.

δ− δ+ H—B H δ−
H δ−
Hydrogen is more electronegative than boron, so the electrons in the B–H bonds reside mostly near hydrogen. This extra electron density makes the hydrogens nucleophilic.
Boron has an incomplete octet and is very electrophilic.

The hydroboration of an alkene produces an alkylborane intermediate. This process occurs in a single concerted step that adds both the H and BH_2 groups across the double bond. Because all of the bond-forming events are occurring together, the new carbon-boron and carbon-hydrogen bonds form on the same face of the double bond. The result is that the newly added hydrogen and boron are *syn* to each other with respect to the carbon chain. This is called a ***syn* addition**.

In ***syn* addition**, the two new atoms or groups are added to the same face of an alkene.

Two factors make the reaction regioselective in that the H adds to the most substituted carbon atom of the π bond. First, the approach that places the BH_2 group near the least substituted carbon atom has less steric interaction in the transition state because the BH_2 portion of the molecule is larger than the hydrogen atom. Second, the hydroboration reaction is asynchronous: the C–H bond forms slowly, while the π bond breaks quickly. At the transition state, the carbon atom that forms a new carbon-hydrogen bond has carbocation character because the bond to the hydrogen atom is only partly formed. This partial positive charge is best stabilized if located on the *most substituted carbon* of the double bond. Thus, the same controlling element—the location of a stabilized positive charge—is at work in this reaction as in the other electrophilic additions to a double bond.

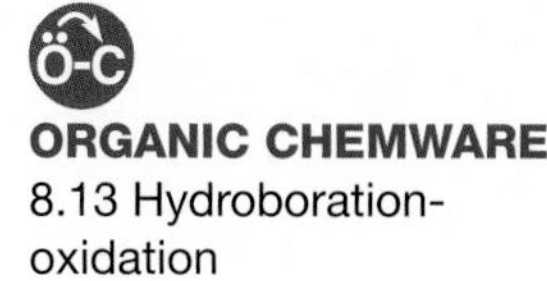
ORGANIC CHEMWARE
8.13 Hydroboration-oxidation

Since the alkylborane (RBH_2) product has an electrophilic boron atom and two B–H bonds, it can undergo two additional hydroboration reactions with alkene molecules. These further additions proceed by the same mechanism, with the regioselectivity increasing with each successive addition due to the steric factor of the extra R groups. The final product of the hydroboration reaction is a trialkylborane.

$H\text{-}B(H)CH_2CH(CH_3)_2$ → $H\text{-}B(CH_2CH(CH_3)_2)_2$ → $(H_3C)_2HCH_2C\text{-}B(CH_2CH(CH_3)_2)_2$

the large alkyl groups orient the boron atom toward the least substituted carbon atom of the alkene

For many laboratory processes, BH_3 is used in a 1:1 ratio with the alkene, so the reaction stops when the alkylborane is formed. When large-scale reactions are done, a 3:1 ratio of alkene to borane is used in order to economize (borane is expensive, and using excess reagents increases the amount of waste to be removed later), and trialkylboranes are formed.

After the hydroboration step is complete, a mixture of hydrogen peroxide (an oxidizing agent) and sodium hydroxide is added to the reaction mixture to convert the alkylborane into the product alcohol. The mechanism for this reaction is not discussed in this chapter, but it is similar to a Baeyer-Villager rearrangement, described in Chapter 20. An important feature of the process is that the carbon-boron bond is replaced by a carbon-oxygen bond with *retention of configuration*. This produces a product in which the added H and OH groups are *syn* to each other (a product of *syn* addition).

BH_2 $\xrightarrow[NaOH,\ H_2O]{H_2O_2}$ OH

oxygen replaces boron group with retention of stereochemistry

8.7.3 Stereochemistry in hydroboration reactions

Hydroboration is a *stereospecific* reaction, forming only one diastereomer. Because the transition state for hydroboration involves a ring of concerted electron flow, both the hydrogen and boron must add to the same face of the double bond. If the H and BH_2 both add to the top face of the double bond, they will force the methyl group to the back of the alkylborane molecule as shown in the following diagram. (Note that BH_3 is equally likely to add to the top or bottom face of the double bond, so the product is a racemic mixture. The stereochemistry shown is intended to depict relative configuration.) The peroxide step replaces the BH_2 group with an OH group. The configuration at the carbon to which the OH bonds does not change; this gives an alcohol product with the same relative stereochemistry as the alkylborane molecule. Redrawing the molecule as a line structure shows a *trans* configuration between the newly formed stereocentres.

STUDENT TIP
Even though hydroboration leads to the anti-Markovnikov product, the reaction itself still follows the Markovnikov rule in that it proceeds through the most stable cation ($\delta+$).

H and BH_2 have added *syn* to each other

BH_3 → H, BH_2 → H_2O_2, NaOH → H, OH → redraw → OH

stereocentre with BH_2 and OH groups maintains same configuration

CHECKPOINT 8.8

You should now be able to draw the mechanism and products for the hydroboration-oxidation of alkenes and predict the regiochemistry and stereochemistry of the products.

SOLVED PROBLEM

Draw the curved arrow mechanism for the first step of the following reaction and predict the major product.

1) BH_3
2) H_2O_2, NaOH

STEP 1: Identify the roles of the reactants.

The electron-rich alkene acts as a nucleophile, and the electron-deficient boron atom makes a good electrophile.

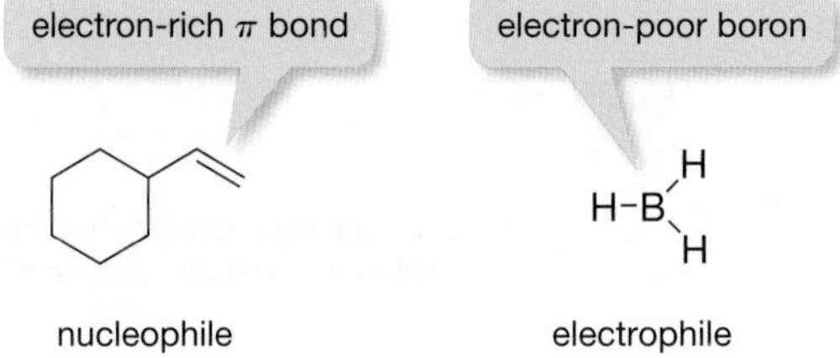

STEP 2: Align the reactants so the δ+ and δ– charges face each other.

partial positive charge is more stable on more substituted carbon

δ–
H
δ+
BH_2
δ+
δ–

STEP 3: Draw in the mechanistic arrows, applying the principle that electrons flow from nucleophile to electrophile and using the sample mechanism shown in this section.

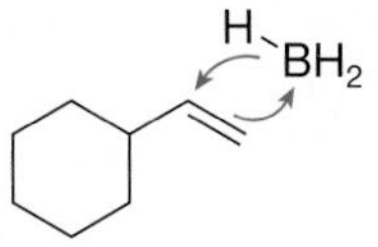

STEP 4: Use the arrows to determine the products formed from this step of the reaction, including any formal charges and relative stereochemistry.

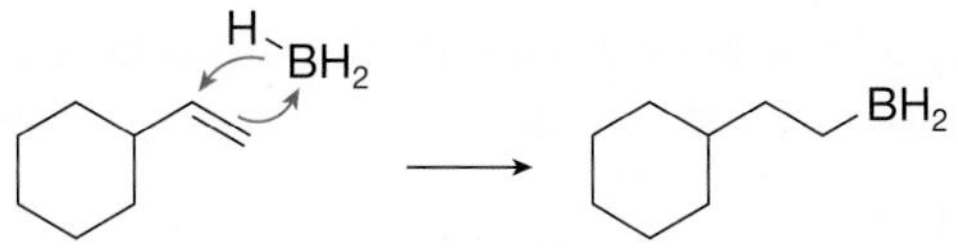

STEP 5: As an alternative method, draw both possible pathways. The pathway that proceeds through the lowest energy transition state (sterics and δ^+ at most substituted position) gives the major product.

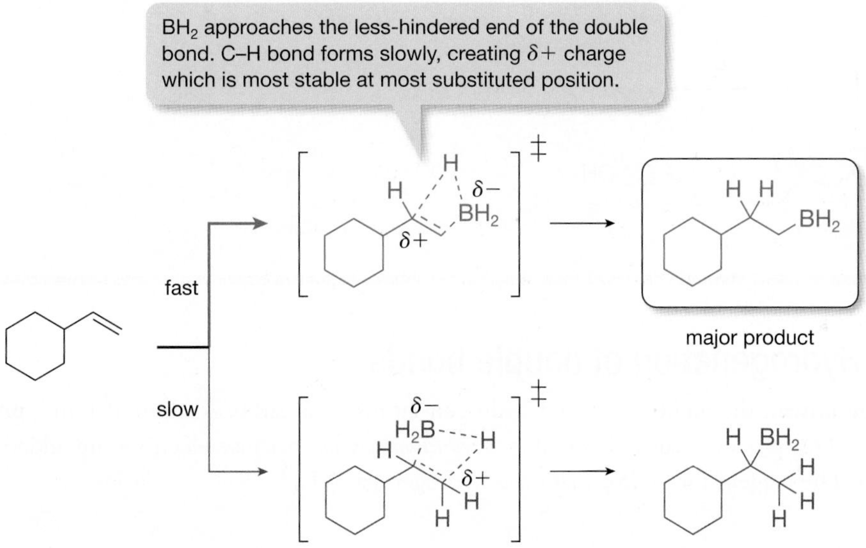

BH$_2$ approaches the less-hindered end of the double bond. C–H bond forms slowly, creating $\delta+$ charge which is most stable at most substituted position.

fast

slow

major product

STEP 6: Complete the reaction by replacing the $-BH_2$ with $-OH$ while preserving the configuration of any stereocentres that are present.

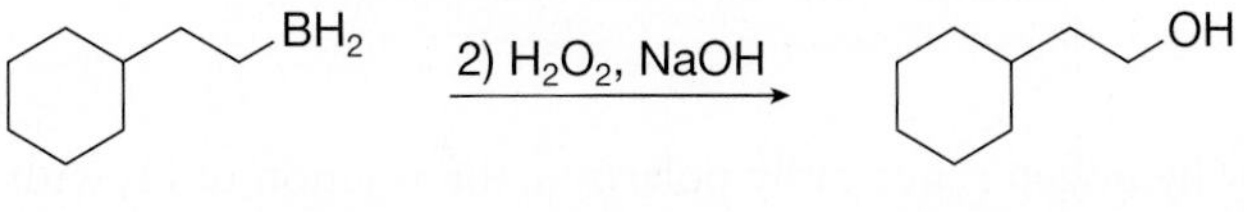

2) H_2O_2, NaOH

PRACTICE PROBLEM

8.15 Draw the mechanism of the first step of the hydroboration-oxidation of the following alkenes, and predict the major products formed after the oxidation step.

a)

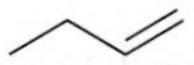

b)

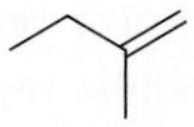

c)

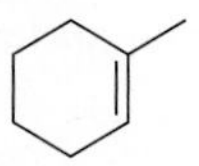

d)

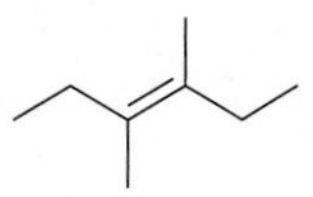

e)

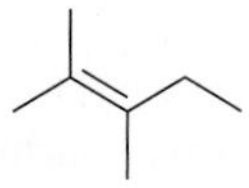

INTEGRATE THE SKILL

8.16 Determine which, if any, of these terms describe the first step of the following reactions: regioselective, regiospecific, stereoselective, and stereospecific.

a) 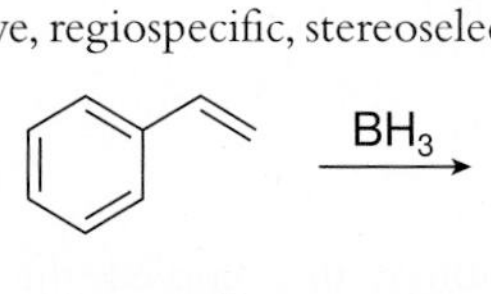 $\xrightarrow{BH_3}$

b) 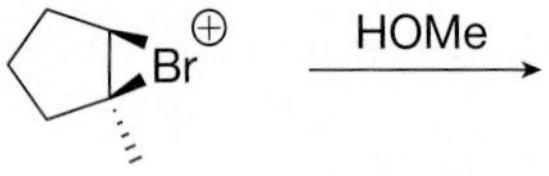 $\xrightarrow{HOMe}$

c) 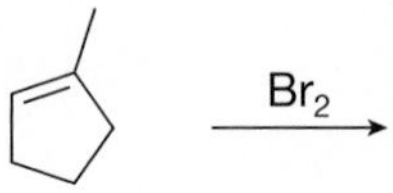 $\xrightarrow{Br_2}$

d) 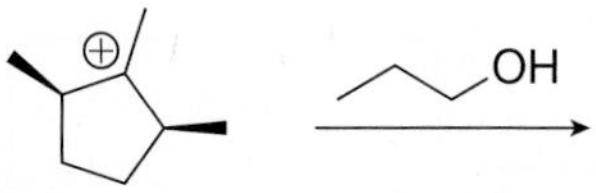 $\xrightarrow{\text{propanol (OH)}}$

8.7.4 Hydrogenation of double bonds

Hydrogenation is the addition of H_2 across a π bond.

Hydrogenation, the addition of two hydrogen atoms to a carbon-carbon π bond, produces an alkane. This process is also classified as a *reduction reaction* because electrons are added to the molecule. The reagent for hydrogenation is hydrogen gas (H_2).

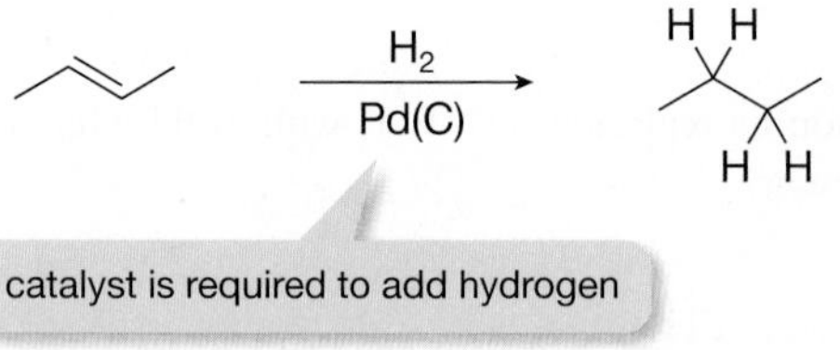

$\xrightarrow[Pd(C)]{H_2}$

catalyst is required to add hydrogen

Since molecular hydrogen is not easily polarized, the reaction of H_2 with a π bond requires a catalyst, typically a transition metal such as platinum (Pt) or palladium (Pd). These catalysts are commercially available as finely ground metal deposited on carbon powder. They are often referred to as "platinum on carbon," or "platinum on charcoal" and abbreviated as Pt/C (similarly for palladium). Since these catalysts are not soluble, hydrogenation reactions occur on the surface of the metal.

Pt and Pd have a high affinity for hydrogen and readily break the H–H bond, creating two new metal-hydrogen bonds on the metal surface. One face of the carbon-carbon π bond then interacts with the surface of the catalyst, attaching one carbon to the metal, while one of the hydrogen atoms adds to other carbon of the π bond. A carbon–hydrogen bond then replaces the metal–carbon bond very quickly, so the two hydrogen atoms are transferred to the same face of the alkene. Thus, hydrogenation is a stereospecific reaction.

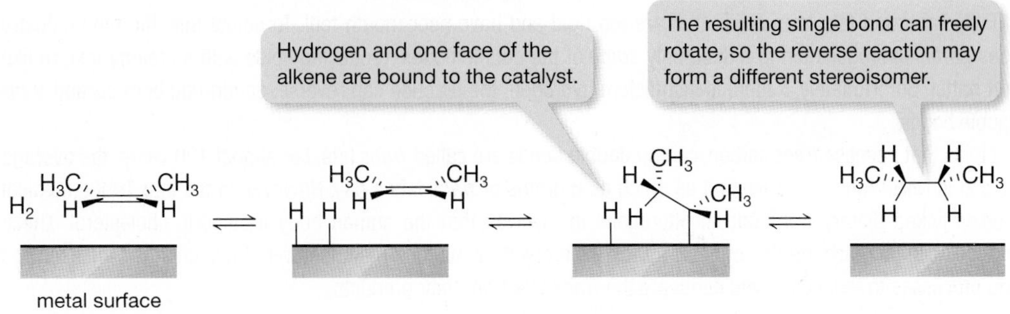

Hydrogen and one face of the alkene are bound to the catalyst.

The resulting single bond can freely rotate, so the reverse reaction may form a different stereoisomer.

metal surface

The steps involved in hydrogenation are reversible, and so an excess of hydrogen must be present to drive the equilibrium toward the formation of alkane. With insufficient hydrogen, not only is some of the alkene not hydrogenated, but some of the product alkane may *dehydrogenate*, producing an alkene that may be an isomer of the starting alkene.

ORGANIC CHEMWARE
8.14 Catalytic hydrogenation

CHEMISTRY: EVERYTHING AND EVERYWHERE

Trans Fats

Fats and oils are composed of lipids: molecules consisting of a unit of glycerol chemically linked with up to three fatty acids. Fats are found in animals and tend to be solid or semi-solid. The fatty acids in fats are either fully saturated (no carbon-carbon double bonds), or contain only one carbon-carbon double bond, which usually has a *cis* configuration.

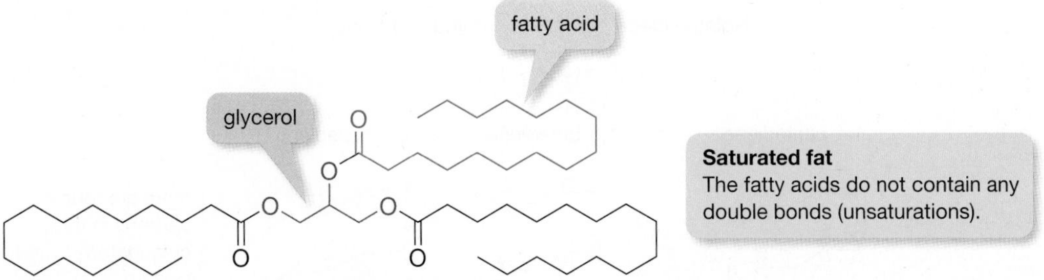

fatty acid

glycerol

Saturated fat
The fatty acids do not contain any double bonds (unsaturations).

Oils are found in plants and tend to be liquids. The fatty acids in oils are usually polyunsaturated (several carbon-carbon double bonds), and the double bonds generally have *cis* stereochemistry. The presence of these *cis* bonds prevents the oil molecules from fitting together to form solids, and so these compounds are usually liquids at room temperature.

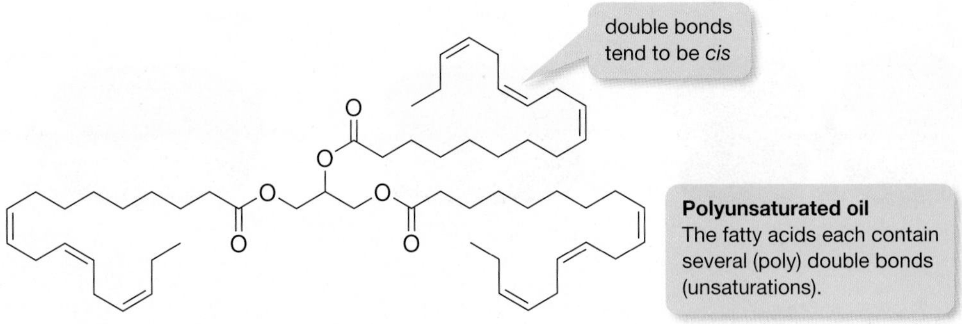

double bonds tend to be *cis*

Polyunsaturated oil
The fatty acids each contain several (poly) double bonds (unsaturations).

Continued

Because flavour molecules easily dissolve in fat, the food industry adds fat to processed food as a way of improving flavour. Oils are usually cheaper than fats, so oils are used whenever possible. However, people prefer eating creamy solids rather than oily substances because the texture or "mouth feel" of the semi-solid fat is more pleasant. Hydrogenation can improve the texture of oils by converting rigid double bonds into flexible single bonds, allowing the molecules to fit together and thereby making creamy solids.

Fully saturated fats form solids that are too rigid and have poor mouth feel. To adjust this, the food industry uses *partial* hydrogenation to reduce only *some* of the double bonds, producing lipids with a creamy texture like soft butter. Unfortunately, if there is insufficient hydrogen, the reaction can reverse, generating both *cis* and *trans* double bonds.

Lipids that possess *trans* carbon-carbon double bonds are called *trans* fats. For almost 100 years, the average person in North America consumed as much as 8 grams of *trans* fats a day. However, in the late 1990s medical studies linked dietary *trans* fats to alterations in the way that the human body transports cholesterol. These changes can increase the risk of heart attack. As a result, many food manufacturers have changed their recipes and processes to reduce or even eliminate the *trans* fats from their products.

8.7.5 Diene reactions

Dienes are compounds containing two alkenes.

In **isolated dienes**, the two alkenes are separated by more than one single bond.

In **conjugated dienes**, the two alkenes are separated by exactly one single bond.

Compounds containing two carbon-carbon double bonds are known as **dienes**. The two common types of dienes are **isolated dienes**, which have at least two single bonds separating the carbon-carbon double bonds, and **conjugated dienes**, which have only one single bond between the two carbon-carbon double bonds. The compounds 1,4-pentadiene and 1,3-butadiene are examples of an isolated and a conjugated diene, respectively.

alkenes separated by at least one sp^3 atom are isolated
adjacent alkenes (no sp^3 atoms in between) are conjugated
isolated diene
conjugated diene

1,4-pentadiene
1,3-butadiene
2,4-hexadiene
1
2
3
4
there are four carbons in this conjugated system
sp^3 carbon separates alkenes
isolated diene
adjacent alkenes
conjugated dienes

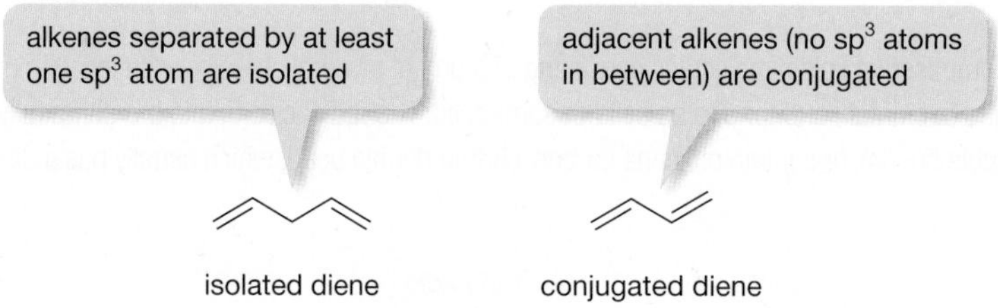

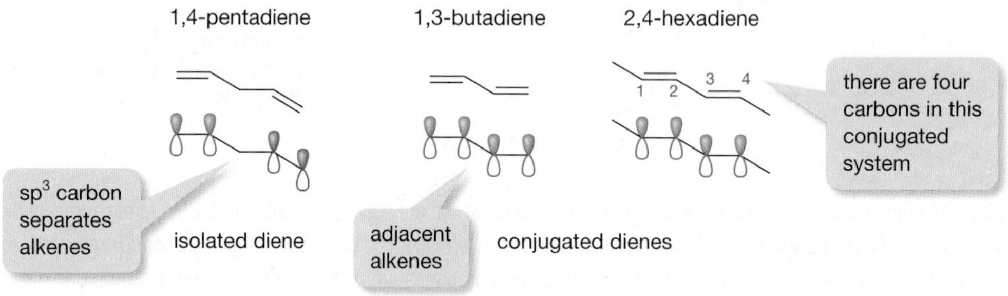

The π bond in each alkene is formed by the overlap of two adjacent p orbitals. In 1,4-pentadiene, the two π bonds are separated (isolated) by an sp^3 hybridized carbon atom; as a result, the two π bonds cannot interact with each other. In 1,3-butadiene, there are no sp^3 atoms between the two π bonds, so the π bonds can interact to form a molecular orbital system that spans all four carbons. These four carbons are referred to as the **conjugated system** of the diene. Only the carbon atoms involved in the interacting π bonds are part of the conjugated system; for example, carbons 1 and 6 of 2,4-hexadiene are not part of a conjugated system.

A **conjugated system** is a region of a molecule with interacting π orbitals. The electrons in a conjugated system are delocalized over all of the atoms involved.

8.7.6 Markovnikov addition of electrophiles to dienes

8.7.6.1 Addition to isolated dienes

Since the two π bonds do not interact with each other, each of the π bonds in an isolated diene reacts independently. Generally, the double bond that gives rise to a more stable carbocation upon the addition of an electrophile will react first because this provides the lowest energy pathway. If excess electrophile is used, the other double bonds in the molecule will react—in the order of their electron-donating ability (the double bond giving rise to the most stable carbocation is usually best able to donate electrons).

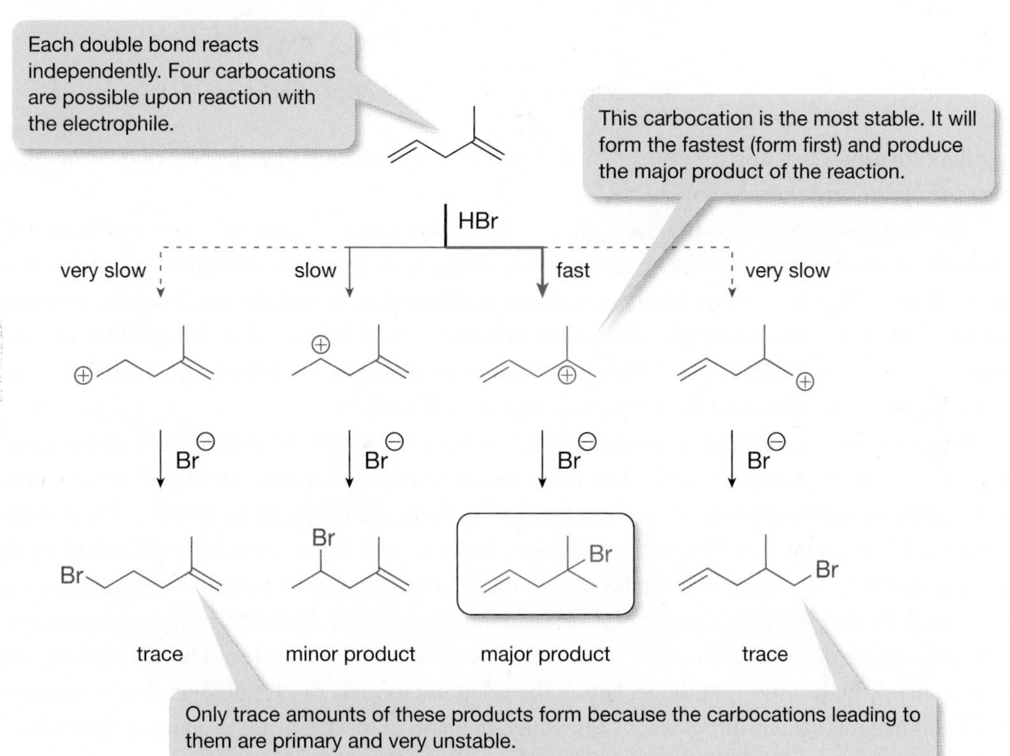

Each double bond reacts independently. Four carbocations are possible upon reaction with the electrophile.
This carbocation is the most stable. It will form the fastest (form first) and produce the major product of the reaction.
HBr
very slow
slow
fast
very slow
Br
Br
Br
Br
Br
Br
Br
Br
trace
minor product
major product
trace
Only trace amounts of these products form because the carbocations leading to them are primary and very unstable.

8.7.6.2 Addition to conjugated dienes

Conjugated dienes behave as if the two double bonds are one functional group. The carbocations that form are delocalized and often produce a mixture of constitutional isomers, whose structures can be predicted by analyzing the carbocations' resonance contributors. Two isomers are usually produced: a 1,2-addition product and a 1,4-addition product. The numbers refer to the carbon atoms of the conjugated system to which the electrophile is added. In the 1,4-addition product, a double bond is located between carbons 2 and 3 of the original conjugated system.

In the first step of the reaction mechanism, one double bond forms a bond to the electrophile, forming a stabilized carbocation. The contributions of the resonance forms predict the structures of the products that can form. Nucleophilic attack on C-2 by bromide results in the 1,2-addition product, whereas nucleophilic attack on C-4 gives the 1,4-addition product.

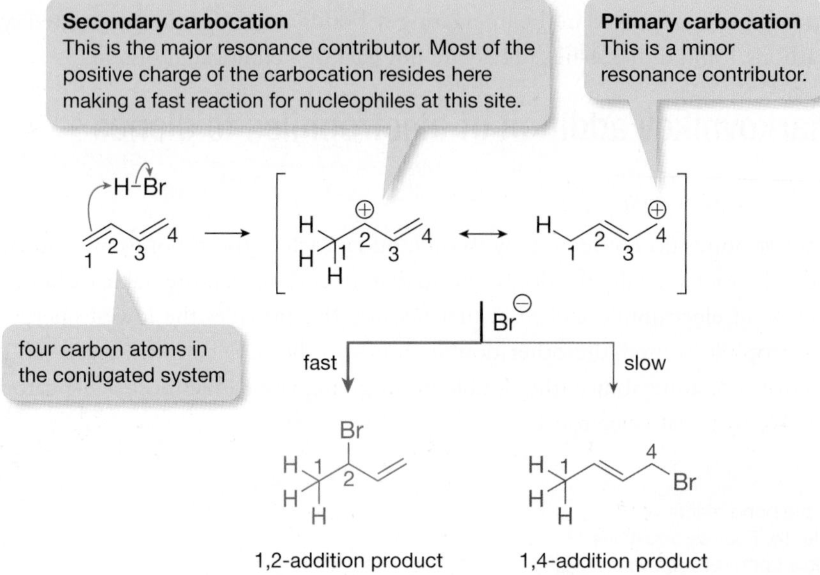

The final product distribution of these kinds of reactions is controlled by two competing factors: kinetics and thermodynamics.

The resonance form with a secondary carbocation makes a greater contribution to the structure of the carbocation because the positive charge is on the more substituted position. This means that C-2 has the greatest positive character and therefore attracts the nucleophilic bromide the strongest. The result is that the addition of bromide to C-2 is faster than the addition of bromide to C-4. Accordingly, the 1,2-addition product is known as the **kinetic product**. Kinetic products are usually obtained from reactions that are irreversible.

The **kinetic product** is the product that is formed the fastest.

If the reaction is performed at a higher temperature, the 1,4-addition product is favoured because the reaction now becomes reversible. The intermediate carbocation is delocalized, and at the higher temperature the carbon-bromine bond can break to re-form the carbocation, which can now form either the 1,2 or 1,4 product. The carbon-bromine bond in the 1,4 product is more difficult to break than the one in 1,2 product. This can be seen in the reaction coordinate diagram below. The activation energy for the formation of a carbocation from the more stable 1,4 product is larger than the activation energy for the formation of the same carbocation from the 1,2 product. The 1,4 product will react more slowly than the 1,2 product and, as the reaction proceeds, the 1,4 product slowly accumulates. Eventually, an equilibrium is reached in which the ratio of the 1,2 to 1,4 products is determined by difference between the free energies of the two products. Equilibrium reactions in which the free energies of the products determine the relative amounts formed are said to be under thermodynamic control, and the major products are called the **thermodynamic products**.

The **thermodynamic product** is the product that is the most stable.

The two competing factors, kinetics and thermodynamics, are shown in the reaction coordinate diagram and summarized as follows:

- The 1,2-addition product forms faster because the activation energy for this product is lower due to the large $\delta+$ on C-2. At low temperatures, the reaction is not reversible, and the 1,2-addition product is the major product.
- The 1,4-addition product forms slower but is the more stable product. At higher temperatures, the reaction becomes reversible, and the 1,4-addition product accumulates, becoming the major product.

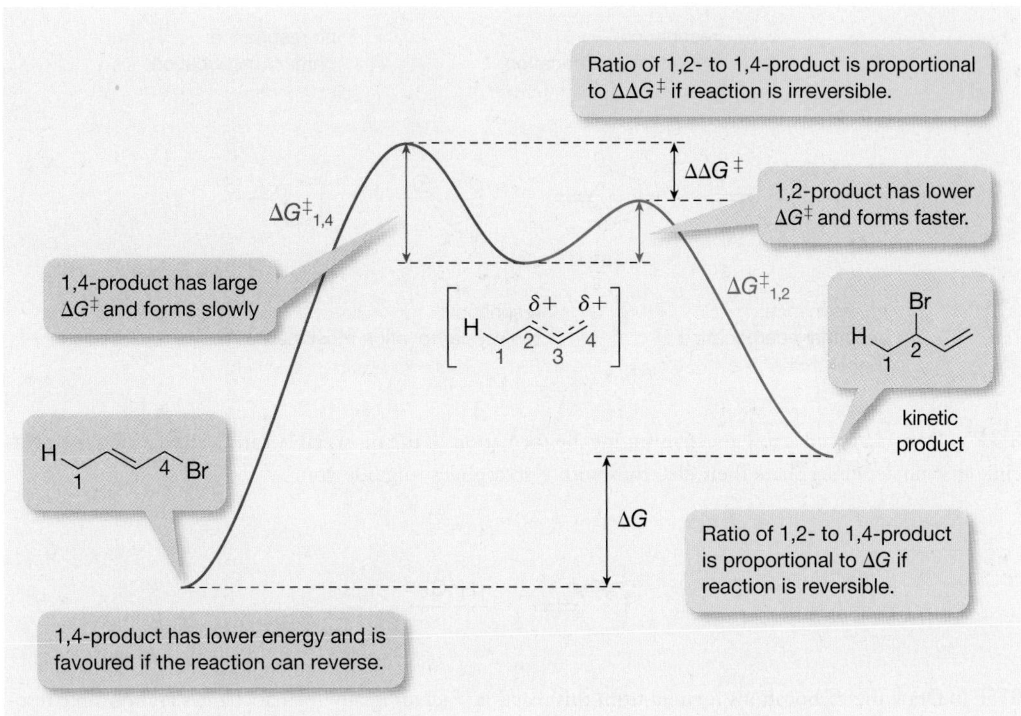

CHECKPOINT 8.9

You should now be able to draw the mechanism and predict the products for electrophilic additions to both isolated and conjugated dienes. In the case of conjugated dienes, you should also be able to distinguish between the kinetic and thermodynamic products.

SOLVED PROBLEM

Draw a curved arrow mechanism for the formation of the major product(s) from the reaction below. If relevant, identify the kinetic and thermodynamic products.

$\xrightarrow{HCl}$

STEP 1: Identify the roles of the reactants.

The electron-rich diene acts as the nucleophile, and the HCl, with its polarized H–Cl bond, serves as the electrophile.

STEP 2: Determine whether the double bonds are isolated or conjugated.

Since the double bonds are separated by a single σ bond, they are conjugated.

STEP 3: Consider the possible carbocation intermediates and determine which would be the most stable.

There are four possible carbocations that could form from the diene. Two of these are delocalized and therefore more stable than the others. Between these two, the one with the greater degree of substitution is the most stable.

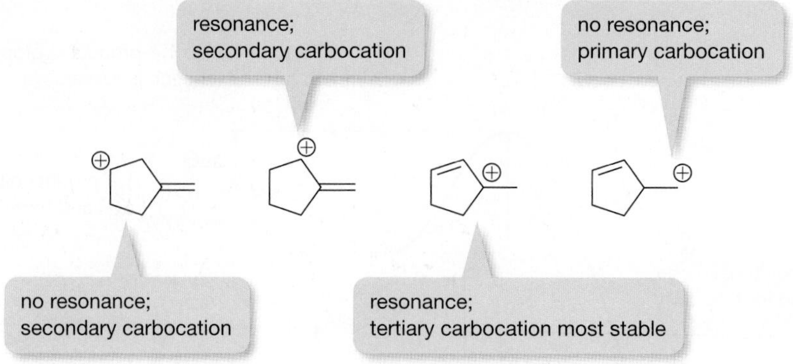

STEP 4: Draw a curved arrow mechanism for the formation of the most stable carbocation, using the principle that nucleophiles share their electrons with electrophiles to guide you.

H–Cl

STEP 5: Draw the carbocations formed from this reaction, including any formal charges and possible resonance structures.

H–Cl + $Cl^{\ominus}$

STEP 6: Draw curved arrow mechanisms and the resulting products for the reaction between the chloride ion nucleophile and each resonance structure of the carbocation.

Path **(a)** + $Cl^{\ominus}$ ⟶ Cl

Path **(b)** + $Cl^{\ominus}$ ⟶ Cl

STEP 7: Determine which is the thermodynamic product and which is the kinetic product.

The thermodynamic product is the one with the stronger C–Cl bond (i.e., the one where the reverse reaction forms a less stable carbocation). Therefore, the bottom reaction pathway (path b) in the previous step yields the thermodynamic product, and the top pathway (path a) gives the kinetic product.

STEP 8: Determine which is the major product.

Without more information about the reaction conditions, it is not possible to know whether the kinetic or the thermodynamic product would be favoured. At low temperature the kinetic product is favoured, at high temperature the thermodynamic product is favoured.

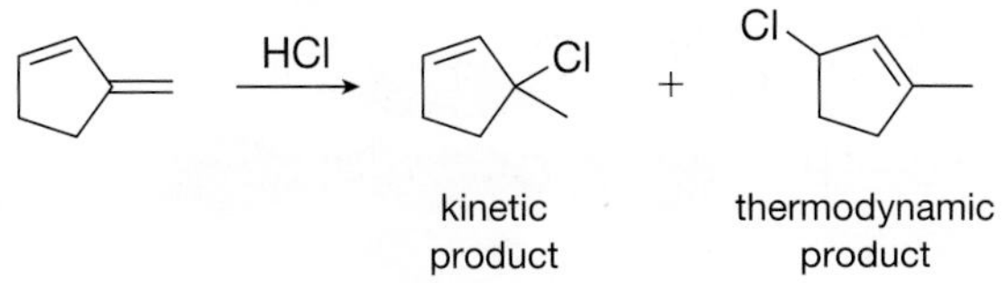

PRACTICE PROBLEM

8.17 For each of the following reactions, draw a mechanism showing the formation of the major products. Assume that the reactions are under thermodynamic control (i.e., the thermodynamic products are favoured) and that only one equivalent of the reagents are present (i.e., 1 mole of reagent for every mole of starting material).

a) 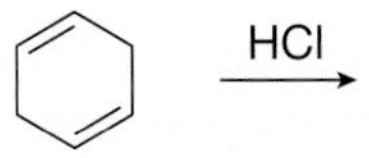

b) 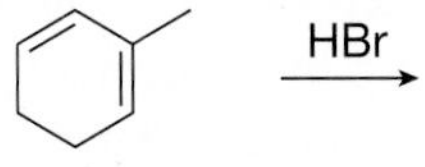

c) 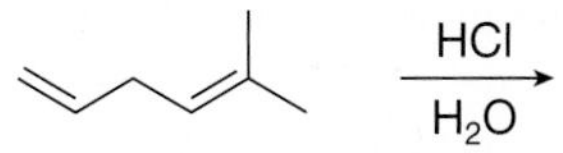

d) 

INTEGRATE THE SKILL

8.18 Consider the following reaction coordinate diagram, in which section (c) corresponds to the starting material.

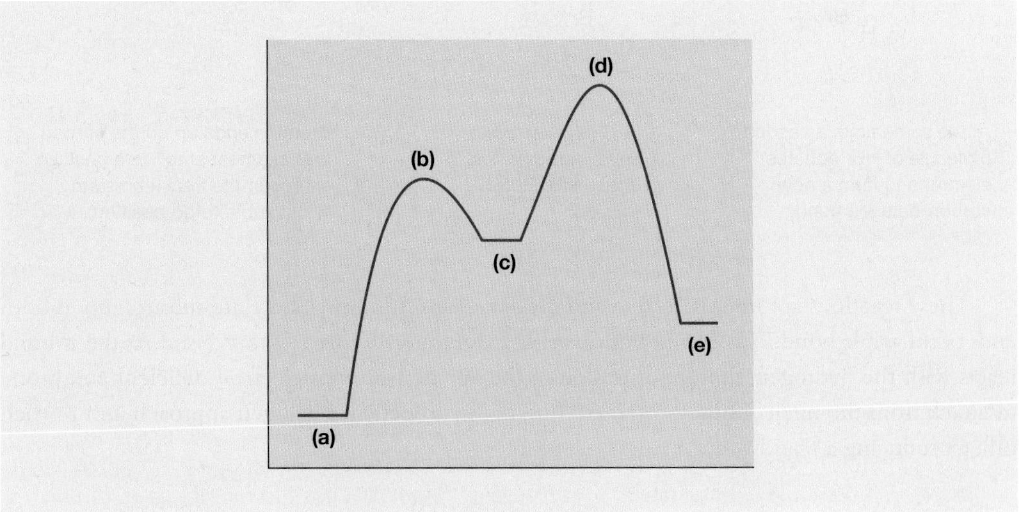

a) Label the axes.
b) Which section represents the kinetic product?
c) Which section represents the thermodynamic product?
d) Which section(s) represent(s) the transition state structure(s)?

8.7.7 Alkyne reactions

The triple bond in an alkyne consists of a σ bond with two π bonds that are orthogonal to each other (Figure 8.8). The π bonds in an alkyne react just like the π bonds in an alkene, so the chemistry of alkynes is very similar to that of alkenes.

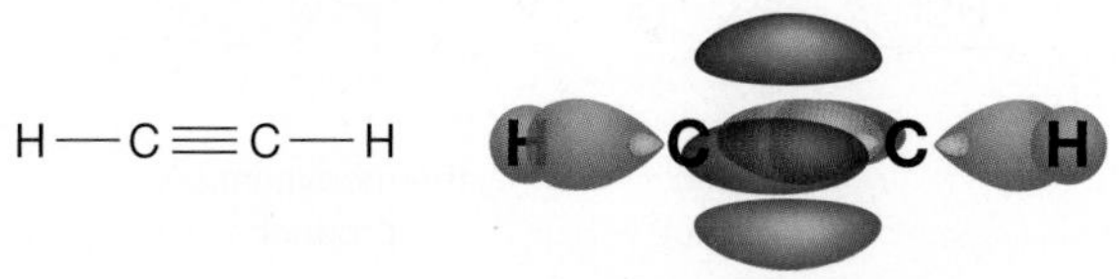

H—C≡C—H

FIGURE 8.8 Three-dimensional structure of an alkyne triple bond.

8.7.8 Markovnikov addition of electrophiles to alkynes

Vinyl halides are compounds that contain a halogen bonded directly to the carbon atom of a double bond.

Halide acids such as HBr undergo Markovnikov addition to alkynes to form **vinyl halides** (alkenyl halides). The regiochemistry in these reactions is controlled by the location of a developing positive charge in the transition state of the reaction. Alkynes react so quickly that the reaction does not produce discrete carbocations. The electron-rich triple bond begins the reaction by removing a hydrogen atom from a molecule of acid. As this hydrogen-carbon bond begins to form, a positive charge develops on the other carbon, which reacts with the halogen of a second molecule of HBr. The same factors that stabilize carbocations stabilize this $\delta+$ charge. Thus, the developing positive charge exerts Markovnikov control, and the major product has the bromine bonded to the more substituted carbon of the triple bond.

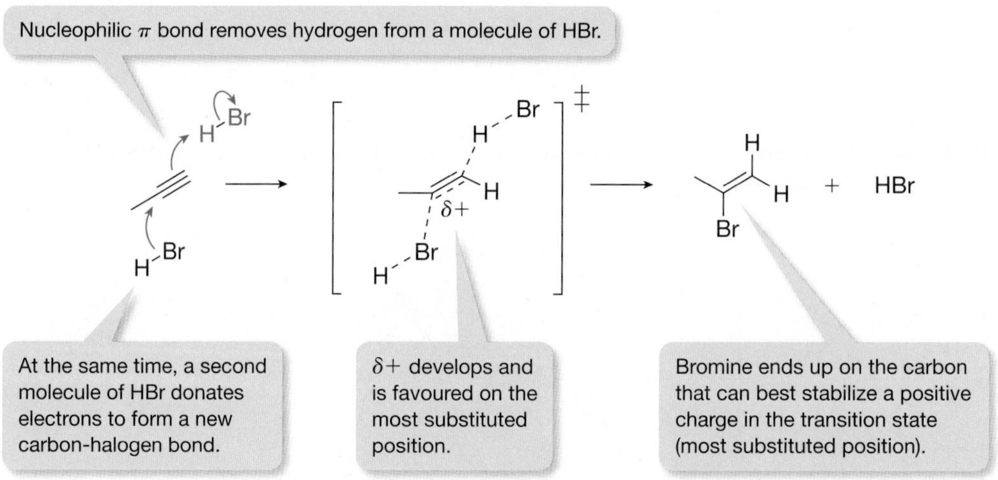

Nucleophilic π bond removes hydrogen from a molecule of HBr.

At the same time, a second molecule of HBr donates electrons to form a new carbon-halogen bond.

$\delta+$ develops and is favoured on the most substituted position.

Bromine ends up on the carbon that can best stabilize a positive charge in the transition state (most substituted position).

These reactions are stereoselective and produce *trans* isomers if there are substituents at both ends of the triple bond. This configuration arises from the structure of the π bond. As the π bond reacts with the hydrogen, the opposite side of the alkyne becomes electron deficient and prone to attack from the nucleophilic bromide. Thus, the hydrogen and halogen approach anti to each other, producing a *trans* isomer.

π orbital of double bond acts as nucleophile

Halide adds to π^* orbital of the double bond. As electrons enter this bond repulsion forces the electrons in the π bond onto the far side of the other carbon.

H and Br finish *trans* to each other

The vinyl halide product that results from this process contains a π bond that can undergo a further reaction with acid. This reaction is also under Markovnikov control, and preferentially produces a **geminal** dihalide when sufficient HBr is present.

Geminal describes two groups located on the same carbon.

nucleophilic π bond

Br is added to site of most stable carbocation (most substituted)

geminal dihalide

Water can be added to alkynes to generate enols (alk**en**e + alcoh**ol** functional group) using an acid catalyst in the same manner as in the hydration of alkenes. Adding an alcohol instead of water produces an enol ether. The addition of water follows Markovnikov's rule, placing the OH group on the location resulting from the most stable carbocation (more substituted).

Water adds across one π bond. OH ends up on the carbon that is best able to stabilize a carbocation (most substituted position).

The acidic conditions cause the enol formed to convert very quickly into the corresponding ketone. The mechanism of this process involves a Markovnikov-type protonation of the double bond producing a carbocation beside the oxygen. This location is favoured because the oxygen has lone pairs that can stabilize the carbocation. One resonance form is an oxonium ion, which is the major contributor to the hybrid structure (all the atoms have octets). This contribution is so strong that the mechanism for enol protonation is usually shown as leading directly to this oxonium. Removal of a proton from the charged oxygen by the solvent then gives the ketone.

resonance stabilized carbocation
this resonance form makes the largest contribution (all atoms have octets)
enol
ketone

Mechanism is drawn using most significant resonance form:

ORGANIC CHEMWARE
8.15 Tautomerization of ketones (acidic conditions)

Additions of water to alkynes can be done with other reagents. A combination of $Hg(OAc)_2$ and H_2SO_4 is frequently used. This reaction proceeds much like the oxymercuration of an alkene; however, the product is obtained directly because the mercury atom in the product rapidly exchanges with protons from the acid (the $NaBH_4$ reaction is not needed). The mechanism of this step is similar to the conversion of an enol to a ketone. The proton (H^+) adds to the double bond, and then the mercury group leaves as a positive ion. The resulting enol then converts rapidly to a ketone.

$Hg(OAc)_2$ carries out Markovnikov addition to one of the π bonds of the alkene
$Hg(OAc)_2$
H_2SO_4
H_2O
HgOAc
OH
acid catalyzes the removal of mercury and the formation of a ketone

Mechanism for mercury removal:

mercury is removed by a process that essentially is the reverse of protonation
π bond is protonated in Markovnikov fashion. Product ion is drawn using most significant resonance form.
π bond protonated in Markovnikov fashion
proton is removed to make the ketone

CHECKPOINT 8.10

You should now be able to recognize an enol and draw the mechanism of enol tautomerization, especially in the context of alkyne reactions with electrophiles.

SOLVED PROBLEM

Which of the following pairs of molecules represents an enol ketone pair? Draw a mechanism to show the conversion of the enol into its ketone form under acidic conditions.

(a)

(b)

(c)

STEP 1: Determine which pairs of structures contain an enol and its ketone form.

An enol has an alcohol directly attached to an alkene. Its corresponding ketone form differs only in the position of the alcohol's proton and the π electrons. Of the three pairings, only pair (b) meets these criteria. In pair (a), the alcohol is not an enol since the alcohol is not directly attached to the alkene. In pair (c), the oxygen atom of the ketone is in a different position. Therefore, it is not the correct ketone form of the given enol.

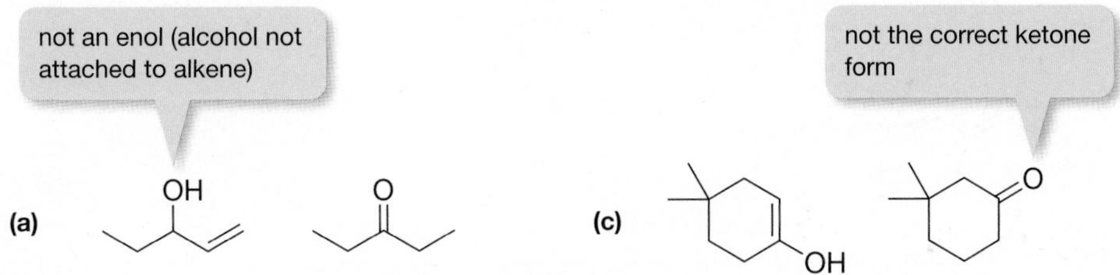

STEP 2: Redraw the enol from pair (b) with some aqueous acid present as a reactant.

STEP 3: Identify the roles of the reactants.

The enol is electron rich at both the lone pairs on oxygen and the π bond. The enol is the nucleophile. The hydronium ion, with its positive charge, acts as the electrophile.

STEP 4: Draw in the mechanistic arrows, applying the principle that electrons flow from nucleophiles to electrophiles and using the sample mechanism provided in this section.

STEP 5: Use the arrows to determine the products formed from this step of the reaction, including any formal charges.

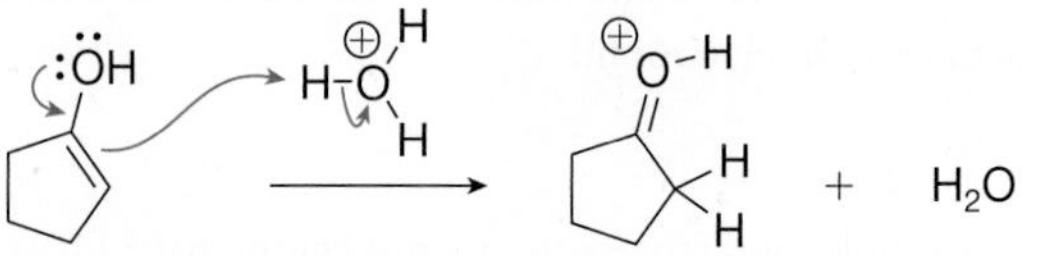

STEP 6: Repeat steps 3 to 5 until the neutral ketone product is reached.

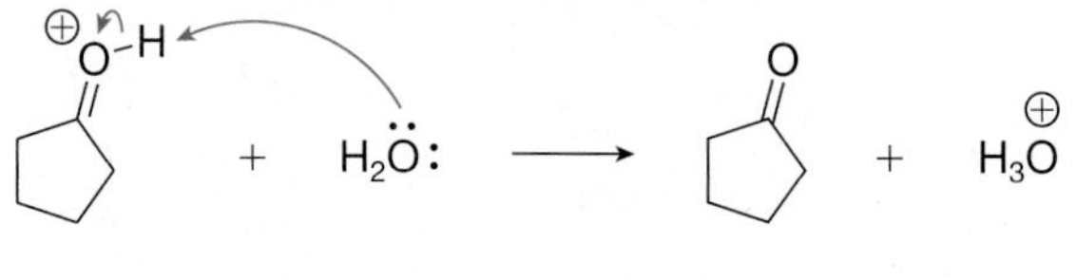

PRACTICE PROBLEM

8.19 Below are several incomplete enol-ketone pairs. Provide the missing structure from each pair and draw a curved arrow mechanism for their tautomerization under acidic conditions.

a)

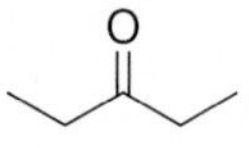

b)

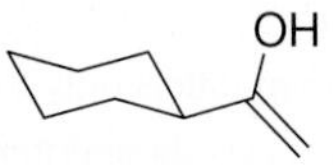

c)

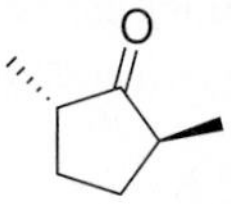

d)

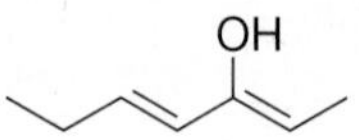

INTEGRATE THE SKILL

8.20 Propose a mechanism for Question 8.19, parts (a) and (b), under basic conditions.

8.7.9 Anti-Markovnikov additions to alkynes

Anti-Markovnikov addition of water to alkynes can be done with boranes, as for additions to alkenes. However, BH_3 is not used with alkynes because it is so reactive that it would hydroborate the vinyl borane intermediate. Sterically hindered disubstituted boranes are less reactive, and using them ensures that the reaction stops after a single borane addition. After the borane addition is complete, treatment with H_2O_2 and a base converts the vinyl borane into an enol, which rapidly transforms to the corresponding aldehyde.

borane carries out anti-Markovnikov addition to one of the π bonds of the alkyne

enol tautomerizes to carbonyl form

Cy_2BH; BCy_2; $\xrightarrow[NaOH]{H_2O_2}$; [enol OH]; aldehyde

large groups on borane prevent hydroboration of second π bond

once hydroboration reaction is complete, H_2O_2 is used to introduce oxygen

8.7.10 Hydrogenation of alkynes

As with alkenes, metal catalysts are needed to make H_2 reactive enough to interact with the nucleophilic π bonds of alkynes. If either Pt/C or Pd/C is used as a catalyst, the result of an alkyne reduction will be the corresponding alkane. The reaction initially generates an alkene, which is rapidly hydrogenated to produce the alkane.

$\xrightarrow[Pd/C]{H_2}$

Hydrogenation of an alkyne with a **Lindlar catalyst** can be used to prepare alkenes. Lindlar catalysts are "poisoned" forms of palladium, which reduce the activity of the Pd metal such that the reaction stops at the alkene stage. These catalysts add hydrogen to one face of the alkyne and therefore produce the *cis* isomer of the alkene exclusively.

A **Lindlar catalyst** is a palladium catalyst, deactivated with lead acetate and quinoline. Because it is less reactive, it only partially reduces alkynes, stopping at the *cis*-alkene product.

$\xrightarrow[Lindlar]{H_2}$

cis bond is formed preferentially

8.8 Patterns in Alkene Addition Reactions

The additions of electrophiles to nucleophilic π bonds are systematic. These reactions follow patterns controlled by the stabilization of the intermediate carbocations and $\delta+$ charges. Carbocations and partial positive charges are stabilized by adjacent electron-donating alkyl groups and conjugation.

In these reactions, the π bond acts as a nucleophile and reacts with an electrophilic atom. This often requires the departure of a leaving group from the electrophilic atom to avoid exceeding its valence when it is attacked by the π bond. Each reaction follows a pathway that leads to the formation of the most stable carbocation (Markovnikov's rule): the most stable carbocation is the one that is most highly substituted or best able to distribute charge. This carbocation is then attacked by an external nucleophile to generate the final product. Reactions with initial activation using acid catalysts have an extra step in which the H^+ is removed at the end of the process to generate a neutral compound.

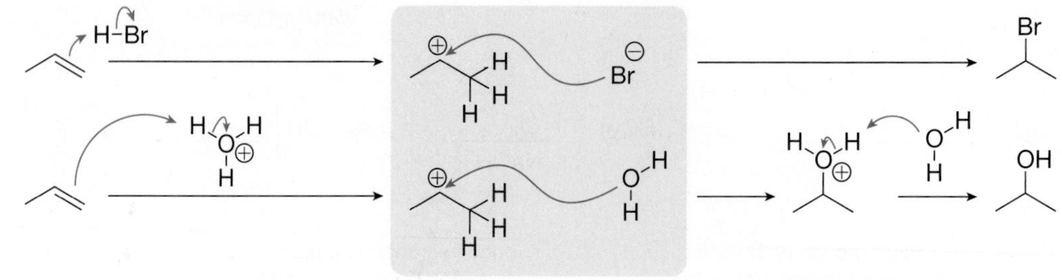

Some reactions proceed via cyclic intermediates because the electrophile added in the first step stabilizes the adjacent carbocation by electron pair donation. When the cyclic intermediates are opened, $\delta+$ charges are generated at the reaction site. The location of positive charge is the controlling element for reactions forming Markovnikov products.

Epoxide formation and opening fit this general pattern. The electron flow for epoxide formation is similar to the electron flow for bromonium ion formation. However, epoxidation has extra electron movement that reorganizes the bonds in the leaving group. Opening an epoxide in the presence of acid forms a three-membered cyclic ion. Similar to a bromonium ion, this ring is preferentially opened at the most substituted position, as this site is best able to stabilize the partial positive charge on that ion.

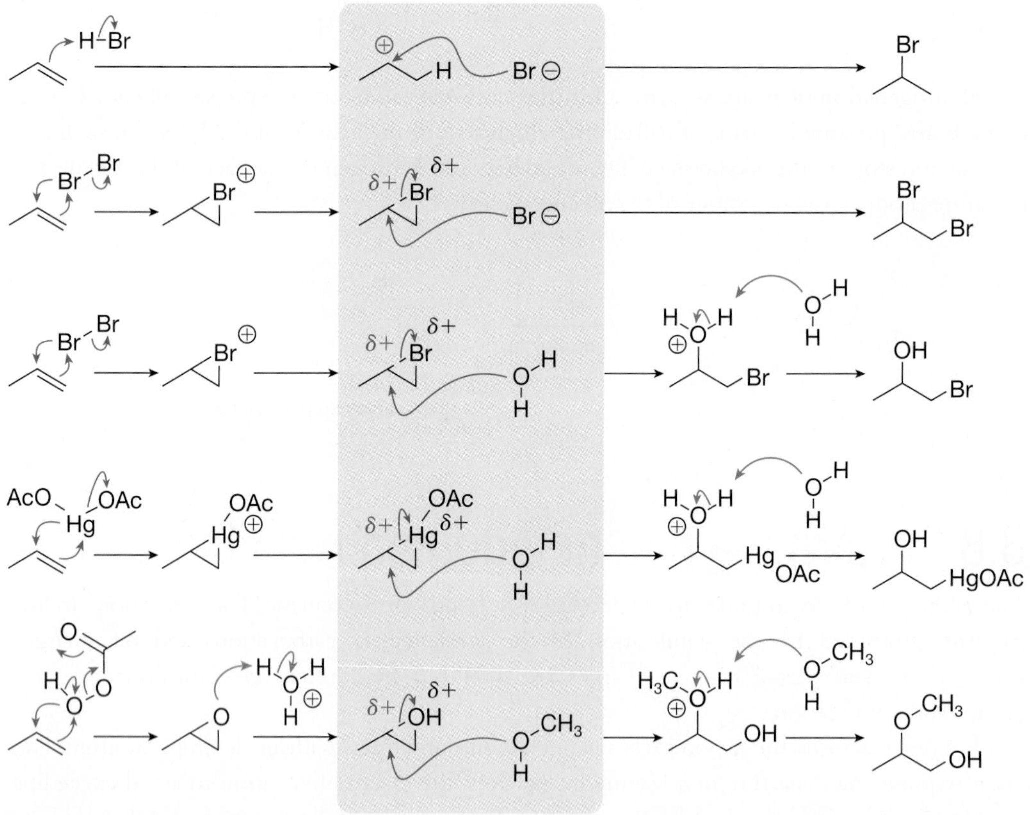

Some of the reactions shown above are stereoselective due to the presence of a ring in the reaction pathway. In these examples, the ring is opened by a nucleophile that must approach from the face that is anti to the departing atom.

Alkynes react in similar ways, but do not usually form full carbocations. Instead $\delta+$ charges develop in the transition states. The molecules react by pathways that proceed through intermediates and transition states that involve the most stable carbocations or $\delta+$ charges. Enols form oxonium ions, which are stabilized carbocations. These mechanisms are typically drawn using the oxonium ion directly; however, the same guiding principle is operating.

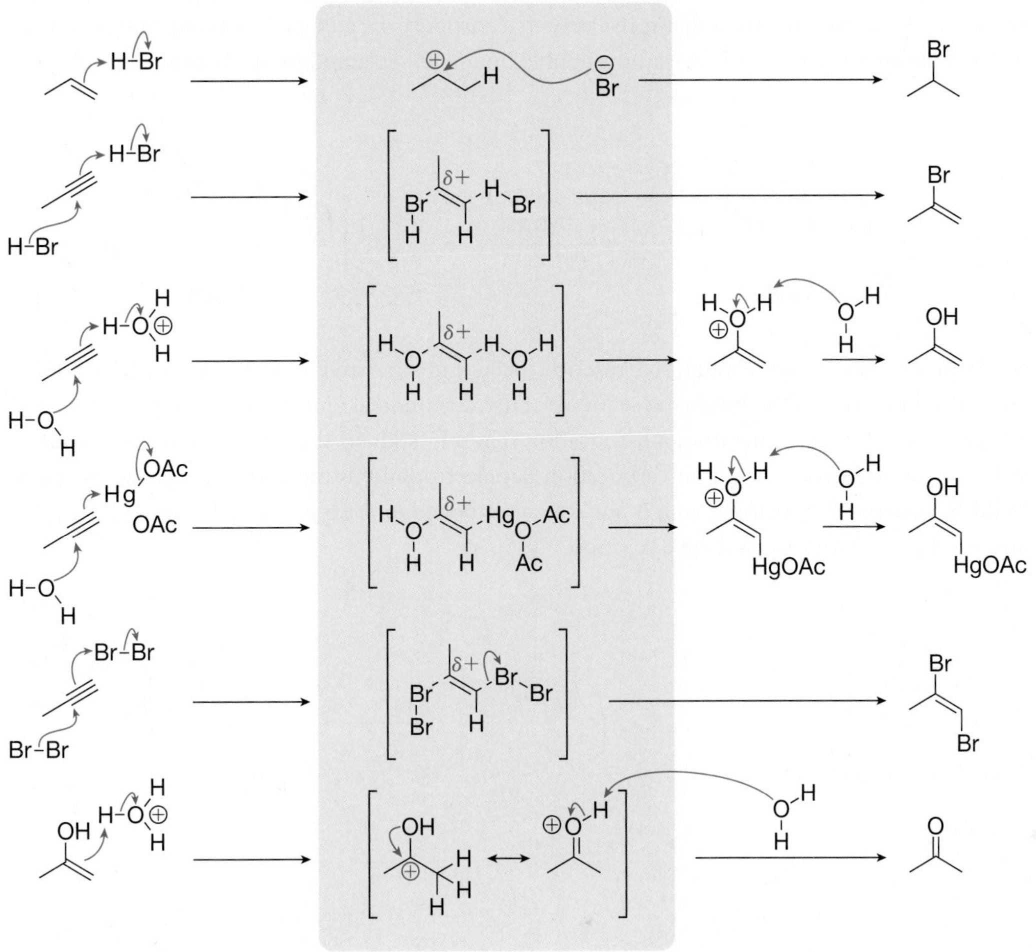

The involvement of a concerted cyclic transition state during the anti-Markovnikov hydroboration of alkenes ensures that the hydrogen and boron add to the same face of the double bond. The regiochemistry of hydroboration is controlled by the same principle as other reactions in this chapter. In the transition state of hydroboration, a $\delta+$ charge develops on one of the carbons of the alkene (or alkyne). The molecule will react via the pathway that best stabilizes this partial charge.

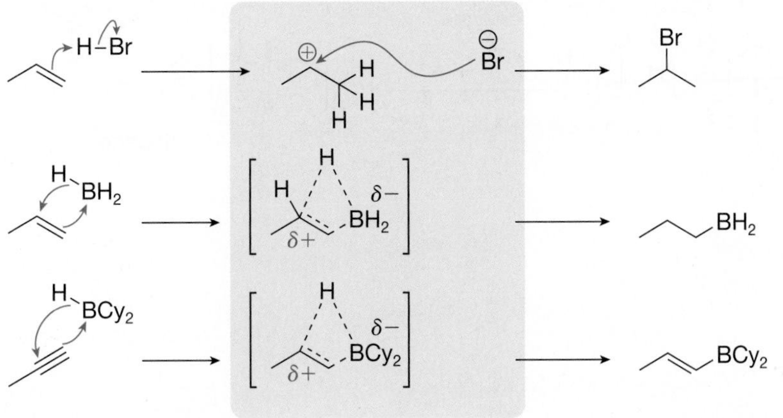

Bringing It Together

The conversion of squalene to hopene by the enzyme squalene-hopene cyclase can be analyzed in terms of electrophiles and nucleophiles. Because the reaction is enzyme-catalyzed, hopene is the *only* product. Enzymes provide a special environment that controls regiochemistry and stereochemistry in the reactions they catalyze. In this enzyme-catalyzed reaction, the enzyme orients each electrophile such that it reacts with a particular face of each π bond. Squalene is an isolated polyene with six carbon-carbon double bonds, all separated by sp^3 hybridized carbons.

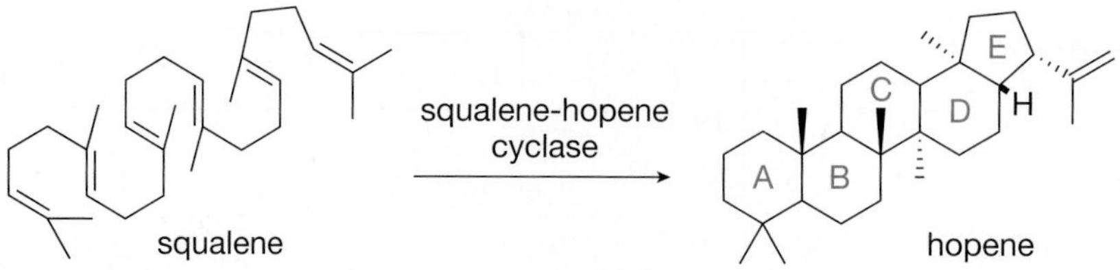

squalene
squalene-hopene cyclase
hopene

Although the enzyme-catalyzed reaction occurs in one step, it is easier to understand if considered in terms of the following sequence. First, an amino acid in the enzyme protonates the alkene on the far left, generating a carbocation. The H^+ adds to the alkene such that the most stable carbocation is formed. This carbocation (an electrophile) is then attacked by the closest π bond (a nucleophile), forming ring A and a new carbocation. Enzymes are able to provide acids and bases as necessary to catalyze a reaction.

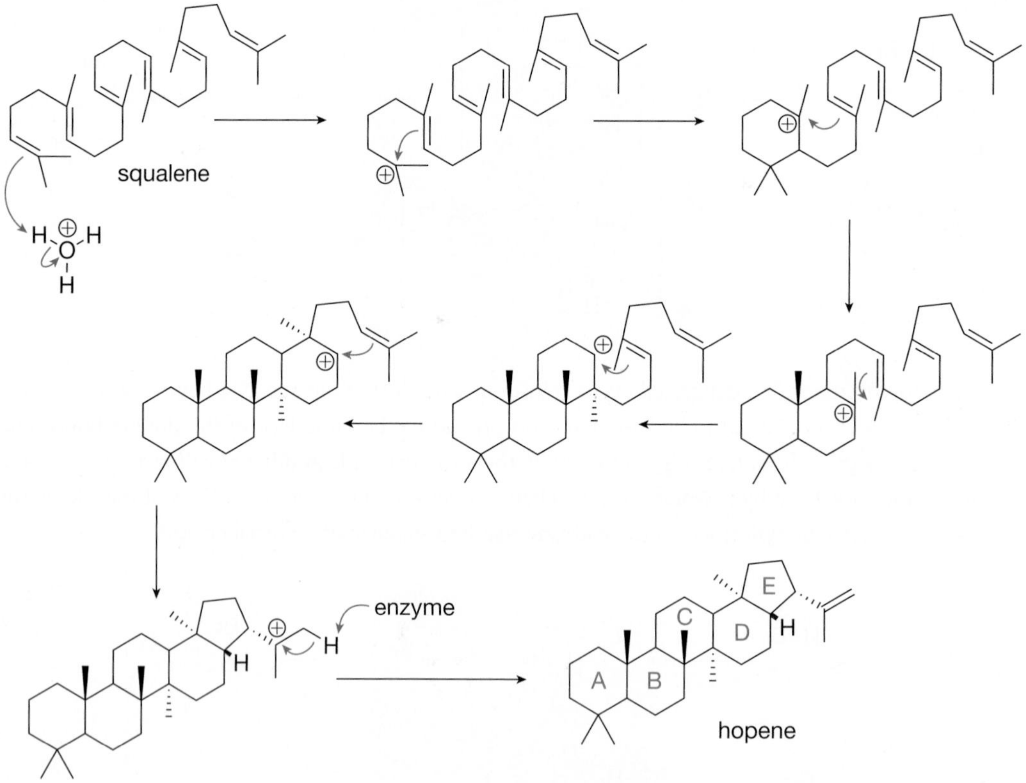

squalene
enzyme
hopene

Similarly, the new carbocation reacts with the closest π bond, forming another carbocation. This process repeats until all five rings have formed. Notice that formation of rings C and D involve secondary carbocations instead of the more stable tertiary carbocations. The secondary carbocations are not normally favoured, but the enzyme stabilizes the positive charge at secondary positions by placing electron-donating atoms nearby.

Finally, a basic amino acid in the enzyme removes a hydrogen ion from the carbocation to form an alkene; this step is simply the reverse of the addition of a proton to a π bond. The proton is selectively removed from one of the methyl groups to make a terminal alkene.

Even though the five fused rings of hopene are drawn as polygons, they are not planar. The four cyclohexane rings are in chair conformations. To be hopene, all the stereocentres must have a specific configuration. The conformation of hopene makes it a relatively flat molecule, facilitating its role in the cell membranes of bacteria. To ensure that hopene is the only product, squalene-hopene cyclase orients moveable groups in the squalene into a specific three-dimensional shape prior to the reaction. In particular, the nucleophilic π bonds are placed such that the electrophilic carbocations add only to the faces of the π bonds that result in chair conformation of each ring. During the reaction, various π bonds react such that the electrophile and nucleophile they interact with are anti to each other. This creates the best orbital overlap and results in the production of hopene.

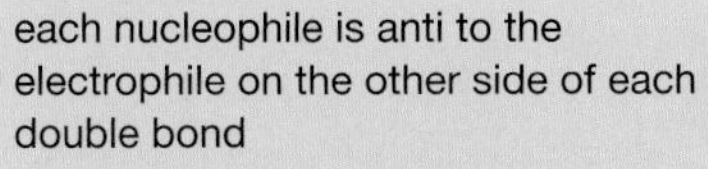

each nucleophile is anti to the electrophile on the other side of each double bond

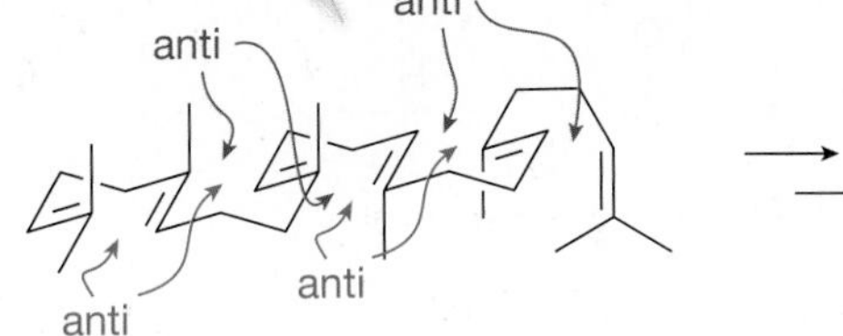

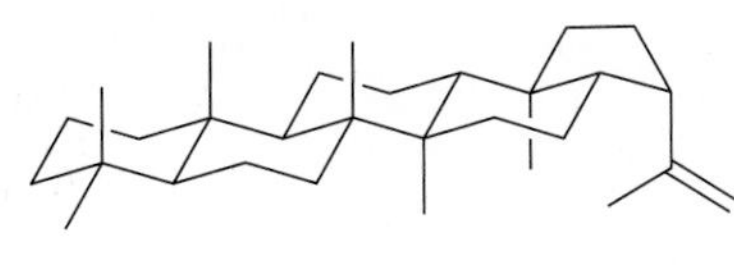

You Can Now

- Draw the mechanism and products for the reactions of alkenes with various electrophiles:
  - strong acids
  - $Hg(OAc)_2$
  - H_2O and HOR
  - X_2
  - X_2 with other nucleophiles present
  - peracids
  - BH_3
  - H_2
- Determine the major organic product of those reactions by
  - identifying symmetric versus asymmetric alkenes
  - drawing reaction coordinate diagrams
  - drawing the orbitals involved in each reaction
  - applying the Hammond postulate
  - assessing the relative degree of substitution and stability of carbocations
  - identifying regioselective and regiospecific reactions
  - identifying stereoselective and stereospecific reactions
  - assessing kinetic versus thermodynamic products
- Draw the mechanism and products for the reactions of alkynes with numerous electrophiles—reactions that are analogous to those of alkenes but include a tautomerization step when an enol is formed.
- Identify the major organic products of alkyne reactions.

A Mechanistic Re-View

MARKOVNIKOV-TYPE ADDITIONS

Addition of haloacids to π bonds

HBr

H Br

H H H H Br Br

Addition of water to π bonds

H_2O
H_3PO_4

H O

Oxymercuration of π bonds

$Hg(OAc)_2$
H_2O

OH HgOAc

$NaBH_4$

OH

AcO Hg OAc HgOAc H O H HgOAc HgOAc

Addition of halogens to π bonds

Cl_2

Cl Cl

Cl Cl Cl Cl Cl Cl

Addition of halogens to π bonds in presence of a second nucleophile

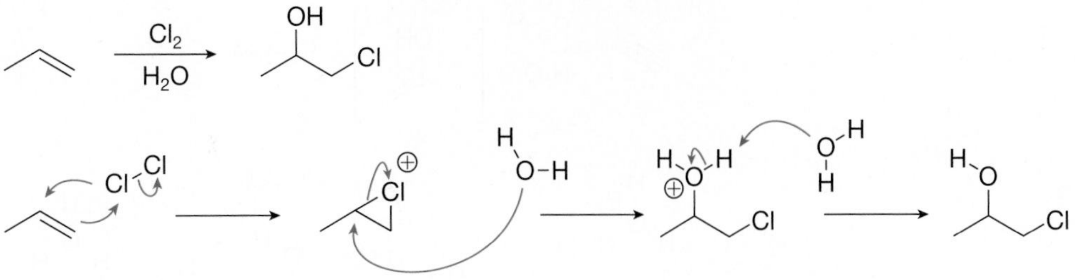

Cl_2, H_2O

OH, Cl

Cl, Cl, Cl, H, O–H, H, O, H, Cl, O, H, H, H, O, Cl

Epoxidation of π bonds

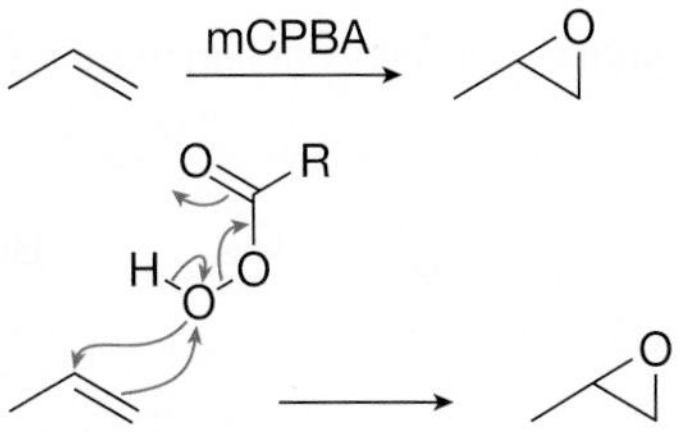

mCPBA

O

O, R, H, O, O, O

O

Opening of epoxides

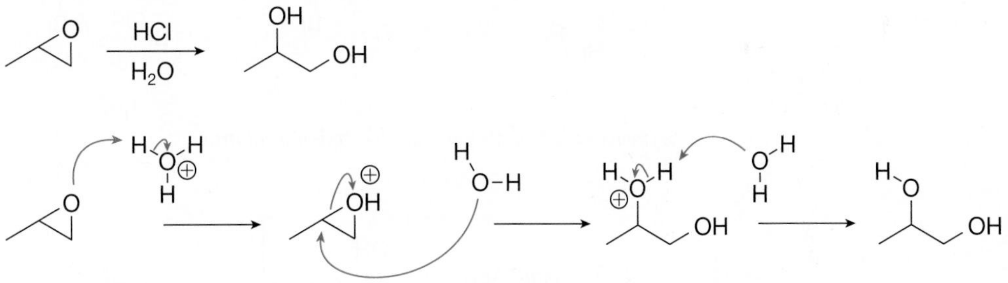

O, HCl, H_2O, OH, OH

H, O, H, H, O, OH, H, O–H, H, O, H, OH, O, H, H, H, O, OH

Addition of haloacids to alkynes, double π bond addition

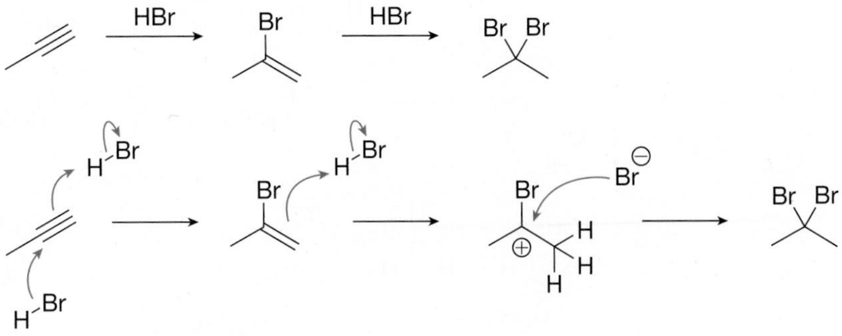

HBr, Br, HBr, Br Br

H, Br, Br, H, Br, Br, Br, H, H, H, Br Br

H, Br

Addition of water to alkynes, double π bond addition

$H_3O^{\oplus}$
H_2O
OH
O

H
H O H
H O H
H O H
H O H
H
H
H O
H O H
H
⊕ O H
O H
H
H
H
H
O

Addition of halogens to alkynes, double π bond addition

Br_2
Br
Br
Br_2
Br Br
Br
Br

Br Br
Br Br
Br
Br
Br
Br
Br
Br Br
Br
Br Br
Br
Br
Br Br
Br
Br

Oxymercuration of alkynes, double π bond addition

$Hg(OAc)_2$
H_2SO_4
H_2O
OH
H
HgOAc
O

OAc
Hg
OAc
H-O
H
H O H
O H
H
H
HgOAc
OH
HgOAc

OH
H
HgOAc
H O H
H
OH
H
H
HgOAc
H
O
H
OH
H
O
H H
⊕ O H
H
O
H
H
H
H
O

ANTI-MARKOVNIKOV ADDITIONS

Hydroboration of π bonds

1) BH_3 2) H_2O_2, NaOH, H_2O → OH

H–BH_2 → BH_2 ; H_2O_2, NaOH, H_2O → OH

Problems

8.21 When cyclohexene acts as a nucleophile and donates a pair of electrons to an electrophile, a cyclohexyl carbocation is formed. Explain why a *single* positive charge is formed on a carbon when a *pair* of electrons has been donated by the alkene.

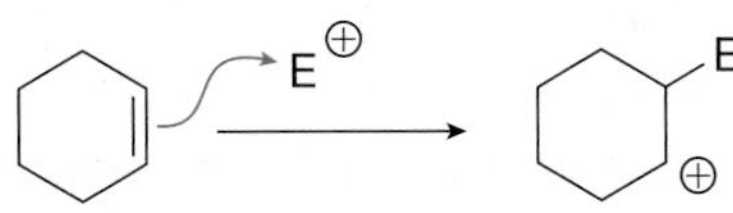

E⊕
E
⊕

8.22 Predict the major product(s) of the following reactions and give a mechanism to account for its formation.

a) HBr →

b) HI →

c) 1-methycyclohexene + HCl

d) HBr →

e) HO — H_3PO_4 →

8.23 Refer to the energy diagram shown. If the *reverse* reaction were to occur, would the reaction be thermodynamically favourable or unfavourable? Which step would determine the reaction rate?

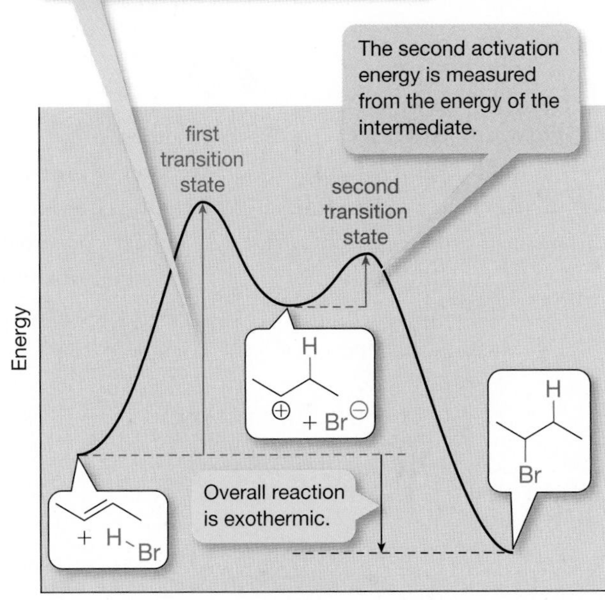

The first activation energy is larger than the second activation energy. This step is the rate-determining step of the overall reaction.
The second activation energy is measured from the energy of the intermediate.
first transition state
second transition state
Energy
H
⊕ + Br⊖
H
Br
Overall reaction is exothermic.
+ H–Br
Reaction coordinate

8.24 When 1,3-butadiene reacts with HBr, a mixture of 3-bromo-1-butene and 1-bromo-2-butene is formed. Propose a mechanism to account for this observation.

8.25 Propose a mechanism to account for the following reaction:

H_3PO_4 / H_2O → OH +

8.26 For each of the following pairs of carbocations, determine whether one carbocation is more stable than the other. Explain any difference in stability, and explain why the more stable carbocation would be formed faster than the other carbocation via the protonation of an alkene.

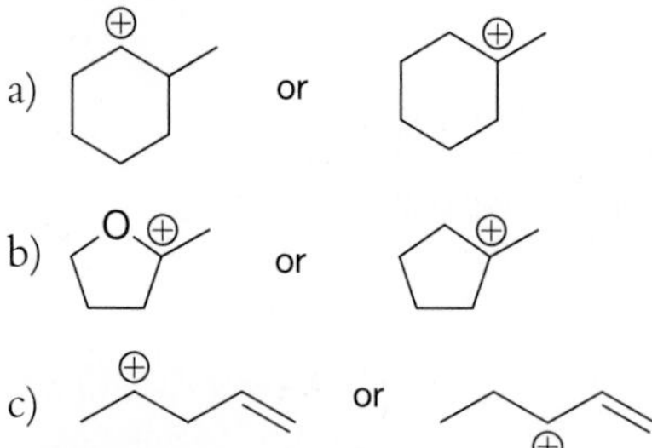

c) or

8.27 Draw curved arrow mechanisms for the formation of all possible carbocations from the reaction of HBr with the following alkenes. Classify each carbocation intermediate as either methyl, primary, secondary, or tertiary. For alkenes that form more than one carbocation, predict which carbocation is the most stable. Identify the Markovnikov products that result.

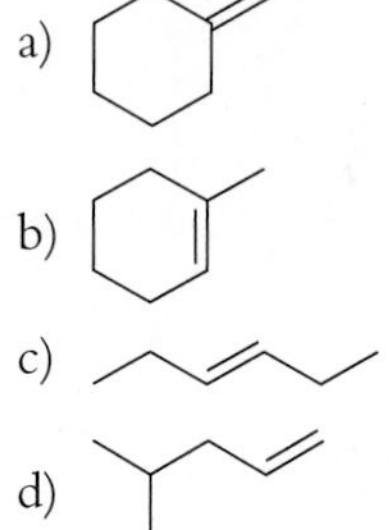

c)

d)

8.28 When HBr adds to compound (a), two Markovnikov products are formed. However, when HBr adds to compound (b), three Markovnikov products are formed. Explain.

a) or b)

8.29 a) Write mechanisms for the addition of HBr and the acid-catalyzed additions of H_2O and CH_3CH_2OH to cyclohexene.

b) In each step of each mechanism, label the nucleophiles and the electrophiles.

c) Why is an acid catalyst necessary for the addition of water and alcohols to alkenes?

d) Does the addition of a thiol, such as CH_3CH_2SH, to an alkene require an acid catalyst? Explain your answer.

8.30 Both the acid-catalyzed hydration and the oxymercuration-demercuration of an alkene are Markovnikov reactions. What are some advantages of the oxymercuration-demercuration reaction?

8.31 Explain why the addition of a nucleophile to a three-membered ring usually proceeds with *anti*-stereochemistry.

8.32 Which of the following reactions of alkenes are stereoselective? Which are stereospecific? Explain your reasoning.

a) acid-catalyzed addition of water
b) addition of HOCl
c) addition of Br_2
d) hydroboration-oxidation
e) hydrogenation
f) epoxidation

8.33 Identify the alkenes and the reagents that could be used to produce each of the following compounds in the highest yield.

a) OH

(racemic)

b) OH

(racemic)

c) OH

d) Cl

(racemic)

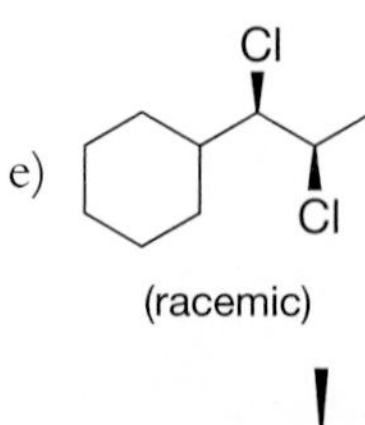

(racemic)

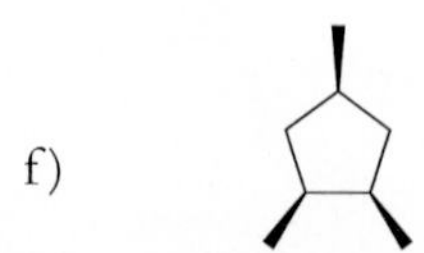

mixture of diastereomers

8.34 The following compound can be formed by the acid-catalyzed intramolecular cyclization of a starting material that contains both an alcohol and an alkene. What could the starting material be? Write a mechanism for the transformation.

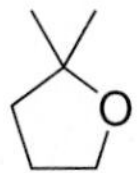

8.35 Draw the major product(s) formed in the following reaction sequences.

a) 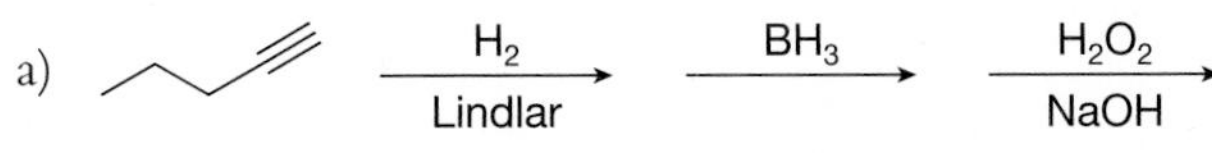

b) 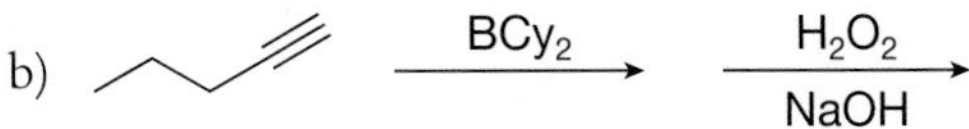

c) $\xrightarrow{H_3O^{\oplus}}$ $\xrightarrow{NaBH_4}$

8.36 Label the orbitals involved in each of the following reaction steps:

a) 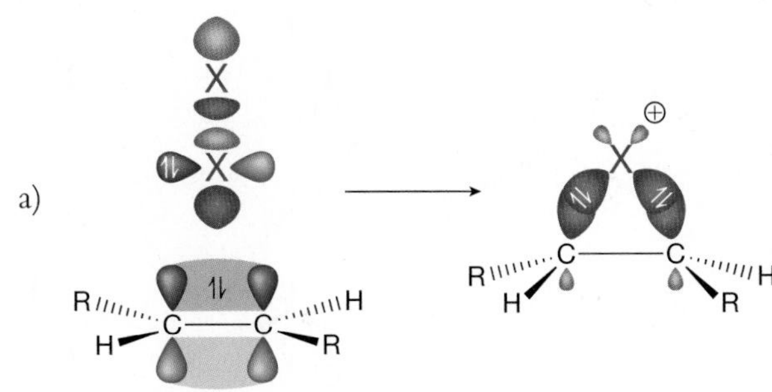

b) 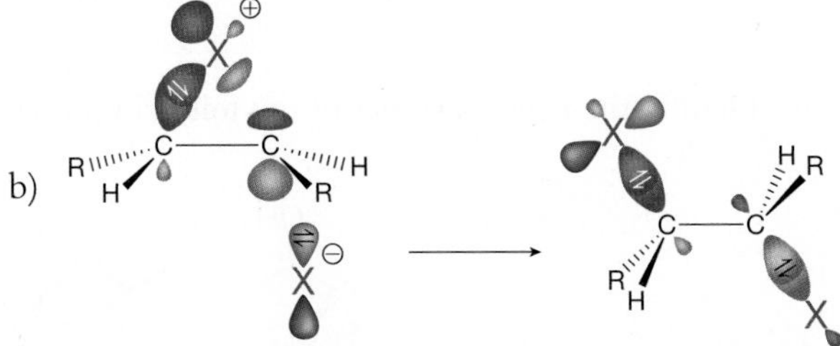

8.37 Show how you could prepare each of the following compounds, starting with methylenecyclohexane.

a) OH

b) O

c) OH Br

d) Br OH

e) OH

f) O

g) Cl

h)

8.38 Propose mechanisms to account for the following transformations. Explain the regioselectivity in each case.

a) $\xrightarrow{HBr}$ Br +

b) $\xrightarrow{HCl(excess)}$ Cl Cl

c) OH $\xrightarrow[H_2O]{H_3PO_4}$ [OH OH] → OH O

d) $\xrightarrow[2)\ H_2O_2,\ NaOH]{1)\ BHCy_2}$

[OH] → H O

8.39 *N*-Bromosuccinimide (shown here) is a source of electrophilic bromine (Br^+). When 1-methylcyclohexene is added to a mixture of *N*-bromosuccinimide (a source of electrophilic bromine Br^+) and NaF in CH_3CN a product is produced that has the molecular formula of $C_7H_{12}BrF$.

O N O
Br

a) Draw a mechanism for the reaction and predict the regiochemistry of the addition.

b) Draw one of the enantiomers of the product in the chair conformation.

c) Draw the same enantiomer in the other possible chair conformation.

d) What are the configurations of the stereocentres in the product shown in parts (b) and (c)?

8.40 Show how you could synthesize each of the following molecules from an alkyne of your choice.

a) a racemic mixture of (2*R*,3*R*)-2,3-dibromobutane and (2*S*,3*S*)-2,3-dibromobutane

b) 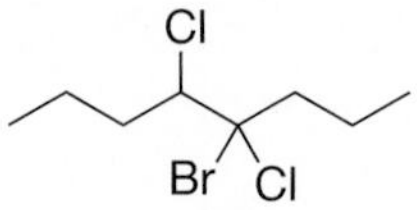 Cl, Br, Cl

c)

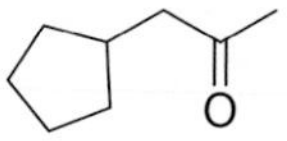

d) 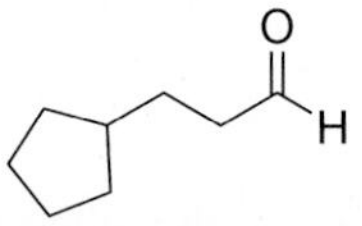 O, H

8.41 Give a mechanism to account for the following observation. (Hint: Remember that ketones are in equilibrium with their corresponding enols.)

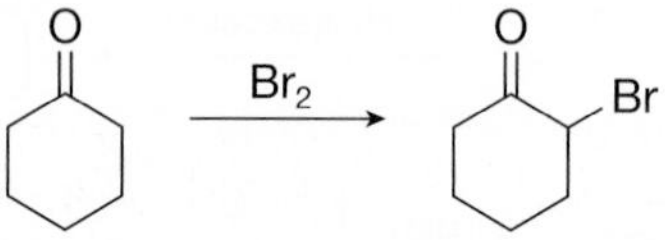

O, Br_2, O, Br

8.42 Nerol is converted into a mixture of terpineol and terpinene when it is heated with acid. Draw a mechanism to explain the formation of these products in this way. The related compound geraniol can be converted to the same two products using similar conditions. Draw a mechanism to explain this reaction, paying special attention to the stereochemistry of the starting alkene.

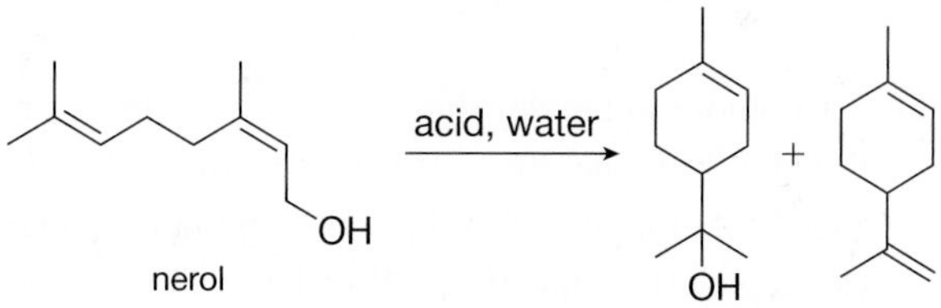

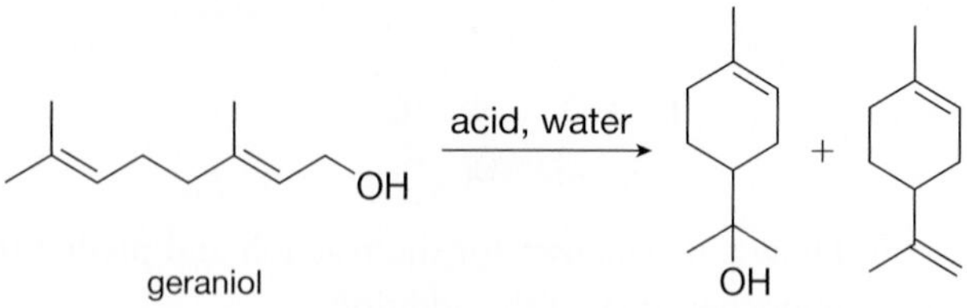

nerol, OH, acid, water, OH, +

geraniol, OH, acid, water, OH, +

MCAT STYLE PROBLEMS

8.43 Consider the mechanism for the addition of Br_2 to 1-butene in ethanol. Which step determines the reaction rate?

a) first
b) second
c) third
d) fourth

8.44 In the following structure, what is the hybridization of the carbon bearing the positive charge?

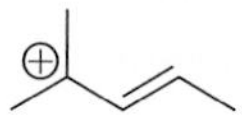

⊕

a) p
b) sp
c) sp^2
d) sp^3

8.45 Identify the major product of the following reaction:

Cl_2

A: Cl, Cl — B: Cl, Cl — C: Cl, Cl — D: Cl, Cl

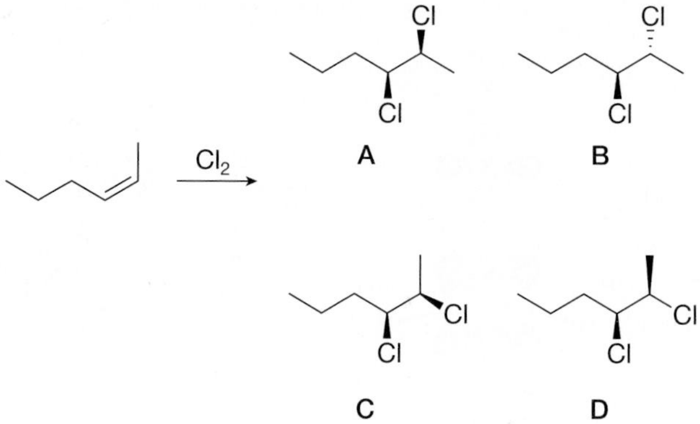

8.46 Identify the major product of the following reaction:

Cl_2, H_2O

A: OH, Cl — B: OH, Cl — C: Cl, OH — D: Cl, OH

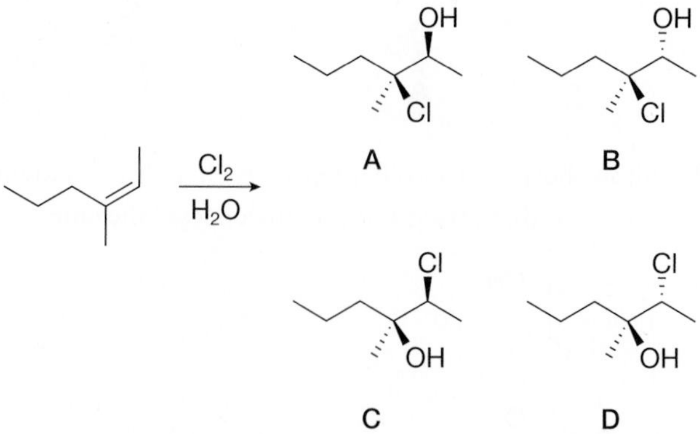

CHALLENGE PROBLEM

8.47 Terpineol can be converted to cineole (found naturally in the spice cardamom) by exposing it to a dilute solution of acid. Draw a mechanism to explain this conversion (it may be helpful to build a model).

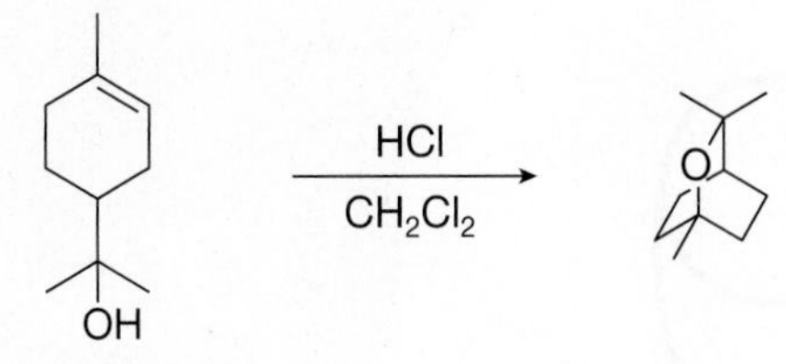

OH

terpineol

HCl
CH_2Cl_2

O

cineole

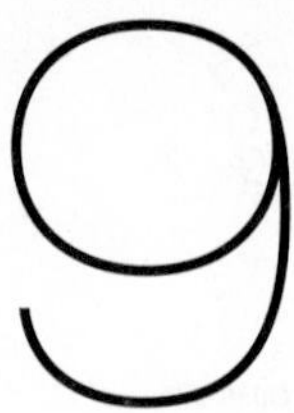

Conjugation and Aromaticity

CHAPTER OUTLINE

Andre Ringuette/Contributor/Getty Images

Modern professional hockey sticks are made from a number of high-tech composite materials, such as carbon fibre, that give each stick the proper amount of flex.

9.1 Why It Matters

Every hockey stick has a flex rating that corresponds to the number of pounds of force needed to bend the stick by one inch. The flex rating is important because the correct shooting technique for both slap shots and wrist shots requires bending the stick. Professional hockey players use sticks that have the proper flex rating based on their body weight and playing style. The range of flex possible for a wooden stick is limited compared to that of composite sticks containing carbon fibres.

The flex rating for a composite stick is tuned by controlling the arrangement of the carbon fibres. Longer carbon fibres aligned along the length of the shaft increase the flexibility. Short fibres aligned perpendicular to the shaft maximize the stiffness. The intermediate lengths of carbon fibres oriented at an acute angle provide intermediate stiffness. The durability of a stick is a function of the number of layers of composite, so the stiffness and durability of a stick is controlled by incorporating either multiple layers of flexible composite or fewer layers of less flexible composite.

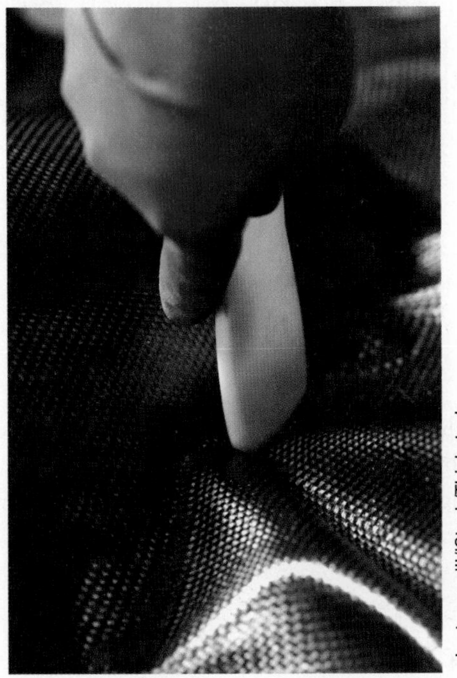

robertomorelli/iStock/Thinkstock

Carbon fibre is made by heating polyacrylonitrile in stages. The first stage causes the nitrile groups to react with each other and form a series of rings connected by imine groups.

C≡N C≡N C≡N Nu⊖ → C≡N C≡N C–Nu =N⊖ → C≡N C=N C–Nu N⊖ N → etc.

Oxygen is then introduced to convert the cyclic imines to pyridine rings. This latter reaction happens because pyridine rings have a special electronic property called *aromaticity*, which brings extra stability to the structure (see Section 9.3).

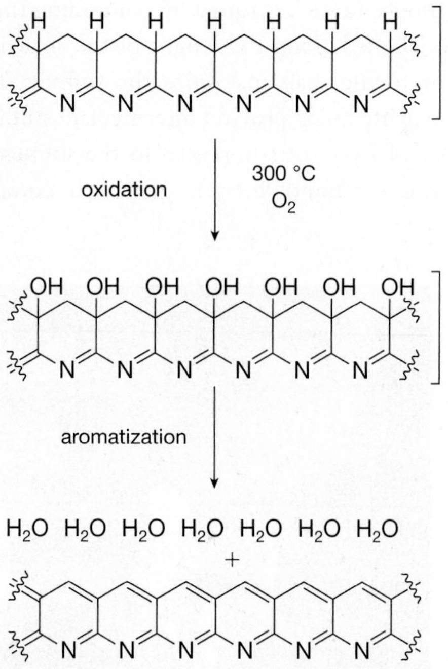

H H H H H H H
N N N N N N
oxidation
300 °C
O_2
OH OH OH OH OH OH OH
N N N N N N
aromatization
H_2O H_2O H_2O H_2O H_2O H_2O H_2O
+
N N N N N N

Heating the polypyridine chain to 2000 °C breaks the rings and drives off the majority of the nitrogen atoms, thereby forming a graphite sheet that is made up of aromatic all-carbon rings—an extremely strong structure. The extra stability of an aromatic ring gives the carbon fibre its strength and stiffness.

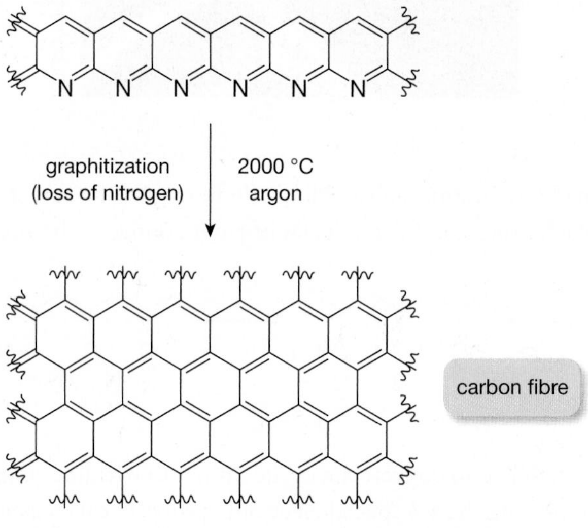

N N N N N N
graphitization
(loss of nitrogen)
2000 °C
argon
carbon fibre

This chapter describes the molecular properties of conjugated π systems and the special properties associated with aromaticity. It also explains how to recognize and predict whether particular compounds are aromatic, anti-aromatic, or non-aromatic based on specific criteria.

9.2 Molecular Orbital Review: Conjugated Systems

When a molecule has an alternating system of double and single bonds, the π bonds are said to be conjugated. In these systems, every atom is sp^2 hybridized (or sp hybridized), with a p orbital that can participate in π bonding. In a planar molecular geometry, all the p orbitals that make up these π bonds align with each other and can be conjugated; this is a form of electronic delocalization. The representation of conjugated systems as alternating single or double bonds does not accurately represent their actual bonding properties. The p orbitals in these carbons not only overlap to form the π bonds of the double bonds but also overlap across the single bonds to provide some π-bond character to the latter. For example, 1,3-butadiene is a conjugated molecule.

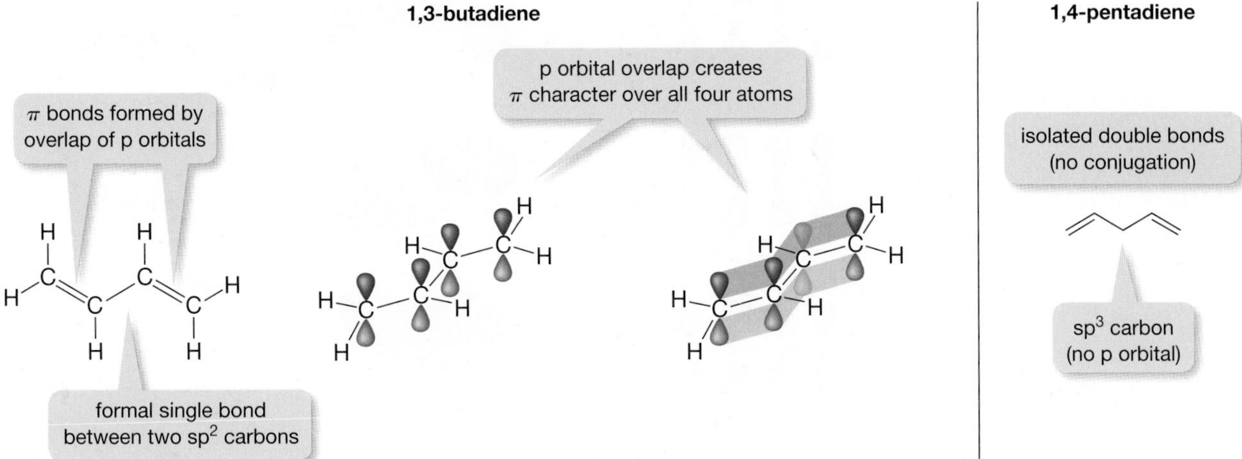

By comparison, 1,4-pentadiene is not a conjugated molecule. The central carbon is sp^3 hybridized and has no p orbital, so there can be no conjugation between the two isolated π bonds.

In general, no individual Lewis structure can account for conjugation. The bonding in conjugated systems can *sometimes* be described using resonance structures (Section 1.8). However, there are times when the molecular properties are not accurately predicted using the resonance hybrid formed from the combination of the different resonance contributors. For example, when describing acyclic all-carbon systems such as 1,3-butadiene, the resonance forms with charges are not significant contributors to the structure. Since it is the charged forms that allow for π-bonding between C-2 and C-3, the resonance model does not explain the strength of the π-bonding that extends along the entire length of the π system.

not significant contributors to the resonance hybrid

ORGANIC CHEMWARE
9.1 Resonance: 1,3-Butadiene

Systems such as this are more accurately described by consideration of the molecular orbitals (MOs) involved in the π-bonding (Section 1.9). According to molecular orbital theory, when the four atomic p orbitals of 1,3-butadiene overlap, four MOs of varying energy and symmetry are produced, depending on the relative phases of the p orbitals (Figure 9.1). The four valence electrons of this π system occupy the two lowest energy MOs: namely, (Ψ_1) and (Ψ_2). (See Chapter 20 for a detailed account of MOs.) The highest energy occupied orbital (in this case, Ψ_2) is referred to as the **highest occupied molecular orbital (HOMO)**. Similarly, the lowest energy unoccupied orbital is known as the **lowest unoccupied molecular orbital (LUMO)**.

The **highest occupied molecular orbital (HOMO)** is the highest energy orbital that contains electrons.

The **lowest unoccupied molecular orbital (LUMO)** is the lowest energy orbital that does not contain any electrons.

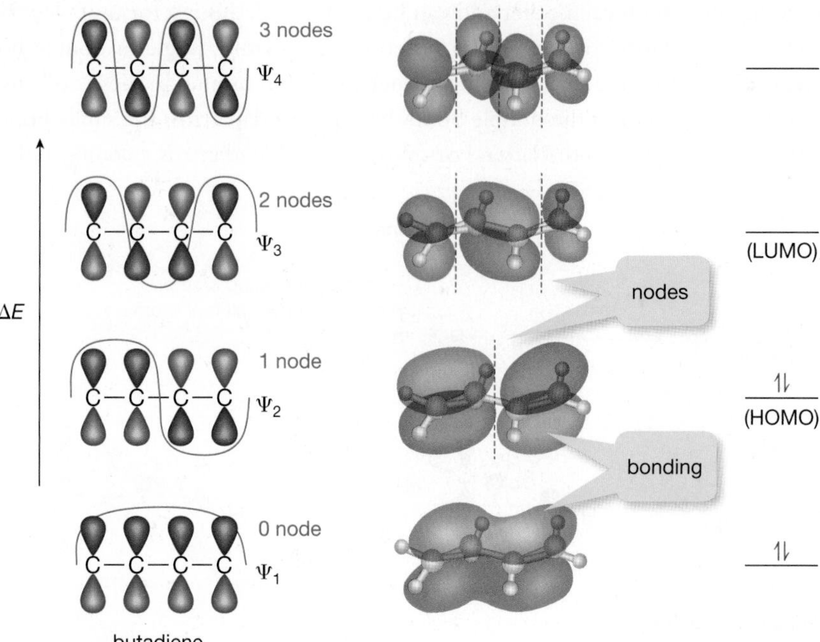

FIGURE 9.1 π-molecular orbitals of 1,3-butadiene.

The lowest energy π orbital (Ψ_1) has the highest symmetry and spreads electron density over the length of the system, which indicates conjugation over all the atoms of the functional group. The second MO (Ψ_2) has one orbital node and is more representative of two isolated π bonds. Overall, the MO picture supports the conjugation model, with the central bond having some degree of double-bond character.

Both valence bond theory (Section 1.7) and molecular orbital theory are models that were invented to explain and predict the physical phenomena observed with molecules. They are not real physical phenomena themselves. An actual atom is tetrahedral or trigonal planar; it is not physically sp^3 or sp^2 hybridized. Even though chemists often talk about hybridization as if it were a real phenomenon, hybridization is only a mathematical concept.

9.2.1 Bond rotation

The extent of π character between atoms in a conjugated system affects various properties, including bond lengths and the ability of certain bonds to rotate. In terms of bond rotation, single bonds typically rotate freely at room temperature. However, the "single" bond of 1,3-butadiene does not rotate easily because of its π-bond character. The rotation of the C-2–C-3 bond of butane has a rotational barrier of 3.62 kcal/mol (15.15 kJ/mol), whereas the corresponding bond of 1,3-butadiene has a rotational barrier of 5.93 kcal/mol (24.81 kJ/mol). The higher barrier to rotation means that the different forms require heat to equilibrate.

Δ

s-trans *s-cis*

There are only two conformations in which all the p orbitals remain parallel to each other and in conjugation. They are known as *s-cis* and *s-trans* conformations: the *cis* and *trans* isomerism about single bonds (*s* stands for sigma). The *s-trans* is the lowest energy conformation and cannot easily undergo rotation of its central single bond; otherwise, the conjugation would be disrupted. The *s-cis* conformation is also conjugated but higher in energy due to steric hindrance.

Heteroatoms with lone pairs that are adjacent to a π bond are typically planar, with the lone pair electrons in a p orbital. This allows them to participate in conjugation with the π bond, extending the length of the conjugated system. A classic example of this principle is the amide bond. The nitrogen of an amide is planar, and the C–N bond does not rotate at either room or body temperature. This is part of the reason that proteins have well defined and consistent secondary structures.

9.2.2 Bond lengths

The MO model for 1,3-butadiene suggests that the central bond has some double-bond character and, as a result, should be somewhat shorter than a normal single C–C σ bond; this is shown experimentally to be true. Likewise, the two formal double bonds of 1,3-butadiene are slightly longer than expected as they have some single-bond character.

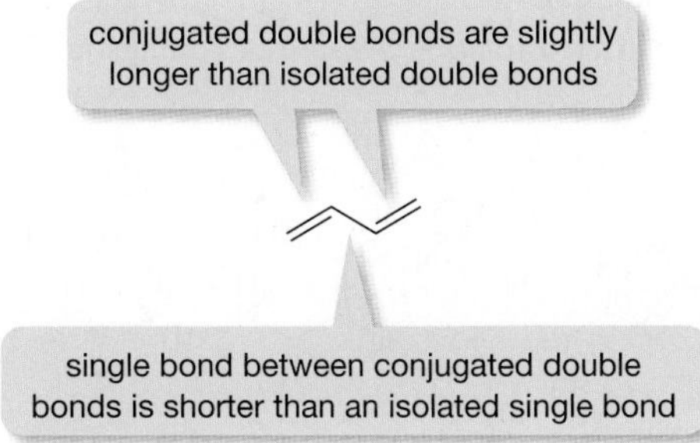
conjugated double bonds are slightly longer than isolated double bonds
single bond between conjugated double bonds is shorter than an isolated single bond

9.2.3 Heat of hydrogenation

Molecular stability generally increases when conjugation increases. For example, based on their relative heats of formation (ΔH_f), the conjugated 1,3-pentadiene requires 1.96 kcal/mol (8.21 kJ/mol) less energy to form than the non-conjugated 1,4-pentadiene.

conjugated

ΔH_f = 23.0 kcal/mol (96.12 kJ/mol)

(more stable)

non-conjugated

ΔH_f = 24.9 kcal/mol (104.33 kJ/mol)

(less stable)

Comparing relative heats of formation is one way to determine the stability of a compound. Another way is to measure the **heat of hydrogenation**—that is, the amount of energy released when an alkene is saturated to an alkane by a catalytic hydrogenation reaction.

The **heat of hydrogenation** is the amount of heat released when an alkene (or alkyne) is completely saturated by a catalytic hydrogenation reaction. It can be used as an indicator of molecular stability.

For example, the catalytic hydrogenation of 1,3-cyclohexadiene, compared to that of its 1,4-isomer in the presence of H_2 and a catalyst, reveals that saturating the conjugated diene is less exothermic (ΔH = 0.26 kcal/mol or 1.1 kJ/mol). Thus, the conjugated 1,3-cyclohexadiene is more stable than the non-conjugated 1,4-cyclohexadiene. Likewise, these two compounds can be subjected to thermodynamic equilibration in the presence of a catalyst, resulting in an 87 : 13 ratio of 1,3 to 1,4 isomers at room temperature, again favouring the conjugated isomer.

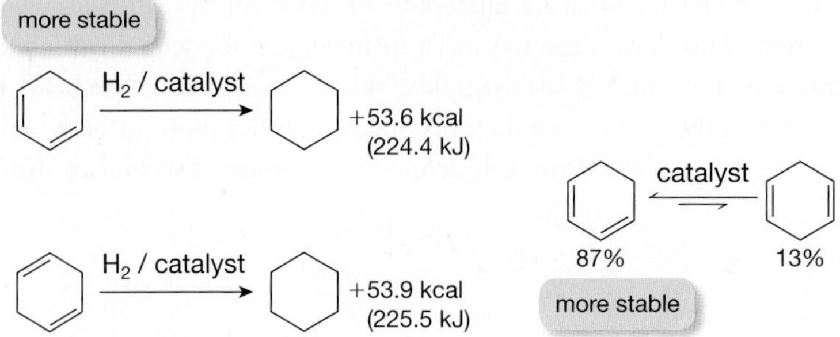

CHECKPOINT 9.1

You should now be able to predict molecular properties of conjugated π systems, such as bond rotation, length, and heat of hydrogenation.

SOLVED PROBLEM

A key factor in drug design is molecular flexibility. According to Veber's rules, a good drug structure should have fewer than 10 rotatable bonds. Consider raltegravir, an anti-HIV drug: excluding bonds to methyl groups, how many bonds have restricted rotation because of conjugation and how many bonds are fully rotatable?

raltegravir

STEP 1: Identify conjugated sections of the molecule that have alternating double and single bonds.

STEP 2: Include adjacent heteroatoms with lone pair electrons, which are in p orbitals, conjugated to the double bonds.

STEP 3: Exclude bonds to single atoms such as hydrogen and halides, and methyl groups.

STEP 4: Exclude all ring systems, because rings prevent full-bond rotation.

This example has no alkyl rings.

STEP 5: All remaining bonds are fully rotatable.

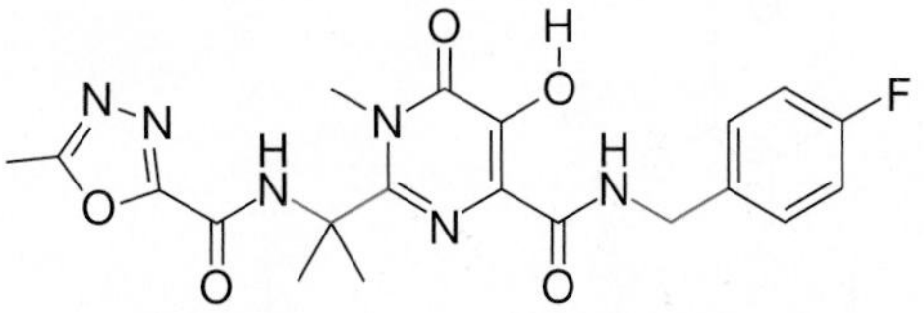

STEP 6: Consider bonds with restricted rotation, where rotation happens at elevated temperatures.

Veber's rule applies to drug-like molecules and describes their behaviours at body temperature. At 37 °C, most of the non-cyclic conjugated bonds can rotate. An exception to this rule is the amide bond, which has a stable configuration even at body temperature. In this example, an additional three bonds have restricted rotation that would count toward Veber's rule.

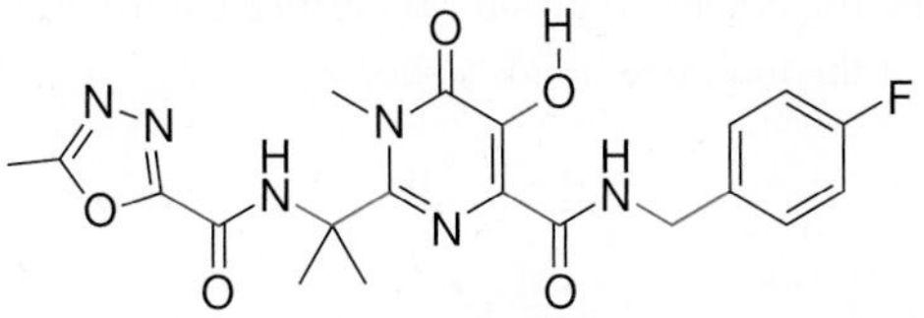

Raltegravir has a count of seven rotatable bonds, according to Veber's rule.

PRACTICE PROBLEM

9.1 a) How many rotatable bonds, excluding methyl groups, do each of the following drugs have?

i)

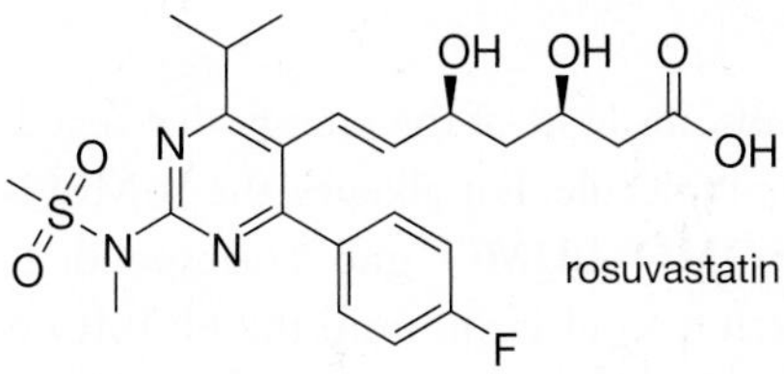

rosuvastatin

ii)

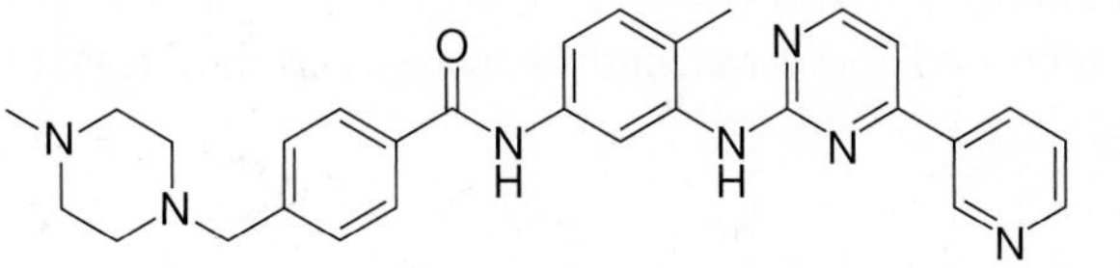

imatinib

iii)

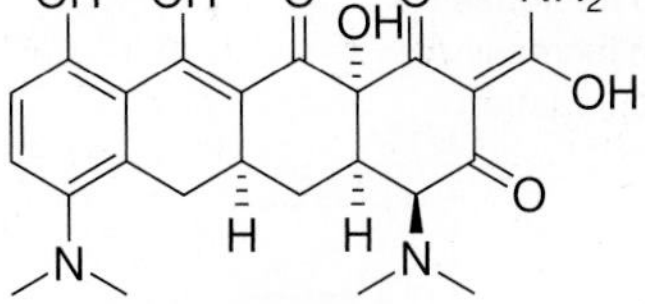

minocycline

b) Identify the shortest C–C "single" bond in each of the following compounds.

i)

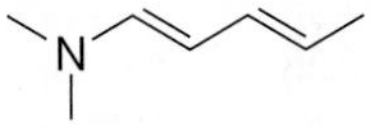

ii)

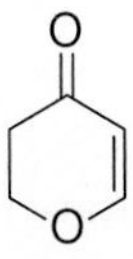

iii)

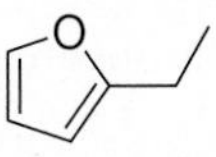

c) Which compound in each of the following pairs has the lowest heat of hydrogenation?

i) or

ii) or

iii) or

INTEGRATE THE SKILL

9.2 The properties of many conjugated systems are determined by the relative contribution of the resonance contributors. A fully conjugated polycyclic compound, naphthalene, has four different C–C bond lengths. Draw the other two possible resonance contributors of naphthalene, and use them to predict which of the indicated bonds is shortest.

9.2.4 UV-visible absorption

The **HOMO–LUMO gap** is the energy difference between the highest occupied molecular orbital (HOMO) and the lowest unoccupied molecular orbital (LUMO).

Whenever a molecule is exposed to light, it absorbs light of the energy that matches the energy gap between the HOMO and LUMO of the molecule. For alkenes, the π MO is the HOMO, and the $\pi^\star$ MO is the LUMO, and the **HOMO–LUMO gap** corresponds to light in the ultraviolet range. Absorption of such light excites an electron from the HOMO to the LUMO. According to Equation 9.1, energy (E) is inversely proportional to the wavelength (λ), so as the extent of conjugation is increased, the wavelength increases, and the HOMO–LUMO gap becomes smaller.

$$E = \frac{hc}{\lambda} \qquad (9.1)$$

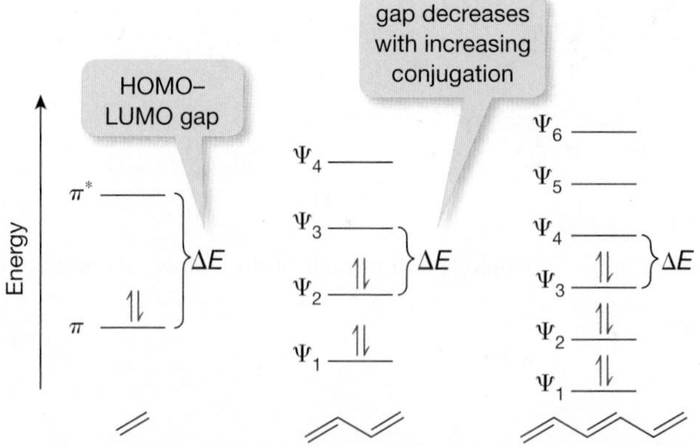

A molecule in the **ground state** is at its lowest energy.

A molecule in an **excited state** has absorbed energy. In the case of an electronic excited state, one of the lower energy electrons has been promoted to a higher energy molecular orbital.

Absorption of light promotes the molecule from the electronic **ground state** to an **excited state**. For ethene, the excitation of an electron from the π to the $\pi^\star$ orbital ($\pi \rightarrow \pi^\star$) is the only possible excitation. For 1,3-butadiene and 1,3,5-hexatriene, multiple transitions are possible, such as ($\Psi_1 \rightarrow \Psi_3$) or ($\Psi_1 \rightarrow \Psi_5$). Compounds with conjugated π systems that contain atoms with lone pairs such as oxygen or nitrogen also have possible excitations of non-bonding electrons, though this ($n \rightarrow \pi^\star$) absorption is generally weak.

excited electronic state
Energy
Ψ_4
Ψ_3
Ψ_2
Ψ_1
$h\nu$ absorption
indicates excited state
ground electronic state

Multiple electronic transitions are possible, the exact energy of which is affected by the vibrations of the molecule. Because of the large number of possible transitions for even very simple molecules, multiple broad peaks can be seen in a UV spectrum. The most strongly absorbed (maximum) wavelength is typically referred to as the λ_{max}. For example, the absorption spectra of both vitamins B6 and B12 shows multiple absorptions.

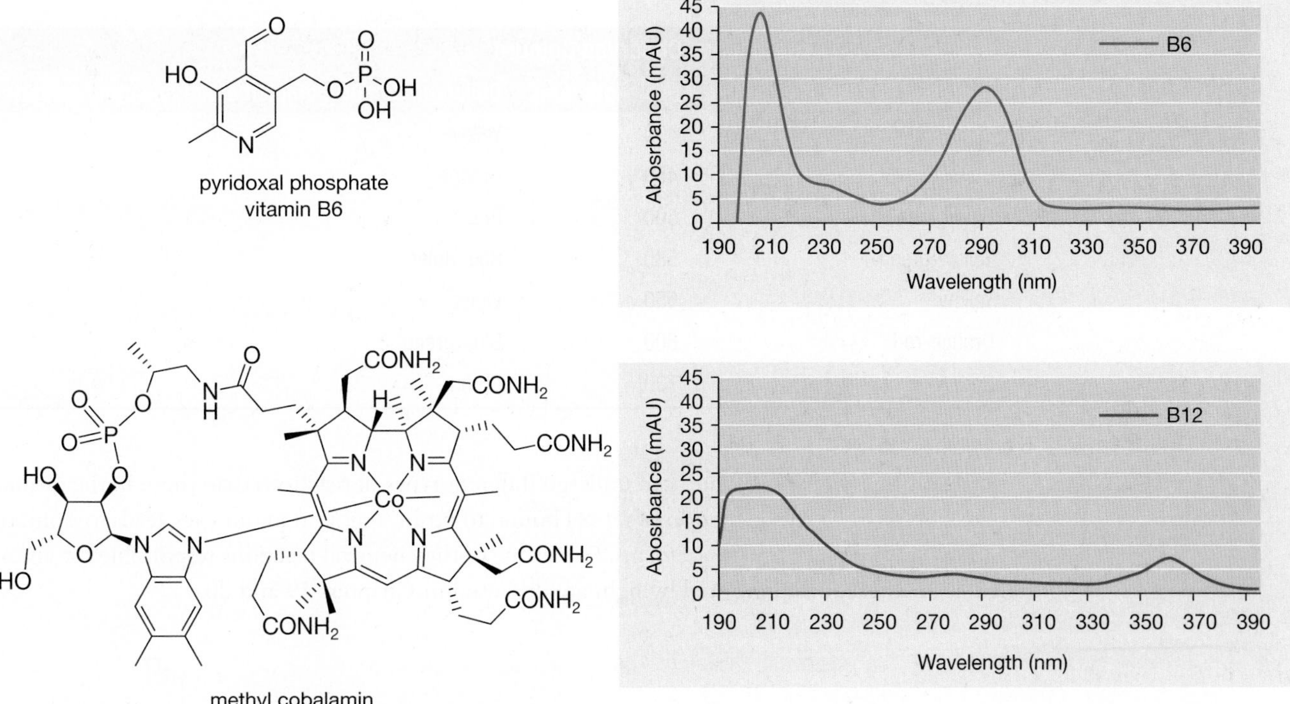

SOURCE: Based on data from Siew Hua Gan, Munvar Miya Shaik, "Rapid resolution liquid chromatography method development and validation for simultaneous determination of homocysteine, vitamins B6, B9, and B12 in human serum," *Indian Journal of Pharmacology*, Vol. 45, No. 2, March–April, 2013, pp. 159–167

Any increase in conjugation in a polyene π system increases the λ_{max} by about 40 nm for each double bond added (longer λ = smaller ΔE).

ethene
λ_{max} = 175 nm

1,3-butadiene
λ_{max} = 217 nm

1,3,5-hexatriene
λ_{max} = 258 nm

When conjugation increases sufficiently, the HOMO–LUMO gap becomes small enough that the absorbed light moves from the ultraviolet region into the visible region. When this happens, chemical compounds appear coloured. For example, β-carotene, a pigment found in carrots, absorbs light with a λ_{max} of 465 nm and is orange in colour.

β-carotene
λ_{max} = 465 nm

Pigments have colour because the wavelengths that aren't absorbed are reflected into the eyes of observers. The light absorbed by β-carotene at 465 nm is blue-violet. When this colour is removed from the full spectrum of visible light, the reflected light appears orange, giving carrots their colour. The colours observed, based on absorbed light, are listed in Table 9.1.

TABLE 9.1 Absorbed vs. Observed Colours

Absorbed Colour	Absorbed Wavelength (nm)	Observed Colour (complement of absorbed colour)
Violet	400	Yellow
Blue	450	Orange
Blue-green	500	Red
Yellow-green	530	Red-violet
Yellow	550	Violet
Orange-red	600	Blue-green
Red	700	Green

Molecules in excited states can undergo different types of reactions than those in the ground state. For example, the excitation of retinal bound to opsin, a protein in our eyes, leads to isomerization between the *cis* and *trans* forms. This is one of the chemical reactions responsible for vision. Additional reactions promoted by light are discussed in Chapters 19 and 20.

O

all *trans*-retinal
λ_{max} = 440 nm

O

11-*cis*-retinal
λ_{max} = 330 nm

CHECKPOINT 9.2

You should now be able to predict the relative HOMO–LUMO gap and the λ_{max} of conjugated systems. You should also be able to predict both the absorbed colour and reflected colour of a compound based on the absorbed λ_{max}.

SOLVED PROBLEM

Which of the following compounds is expected to have the longer λ_{max} of absorption?

O or OH

STEP 1: For each compound, identify each atom that has a p orbital contributing to the conjugated system.

O or OH

8 atoms in the conjugated system

6 atoms in the conjugated system

STEP 2: For comparable conjugated systems, the conjugated system that includes more atoms will have the smaller HOMO–LUMO gap.

O or OH

STEP 3: Because the HOMO–LUMO energy gap is inversely proportional to the wavelength of excitation, the compound with the smaller HOMO–LUMO gap also has the longer λ_{max}.

O or OH

expected λ_{max} = 313 nm

expected λ_{max} = 257 nm

PRACTICE PROBLEM

9.3 a) In each of the following pairs, which compound has the longer λ_{max}?

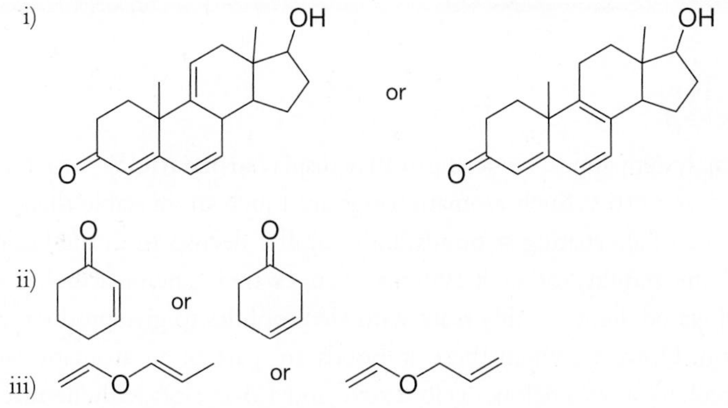

b) What is the expected observed colour of the following compounds? Refer to Table 9.1.

i)

lycopene
λ_{max} = 471 nm

ii)

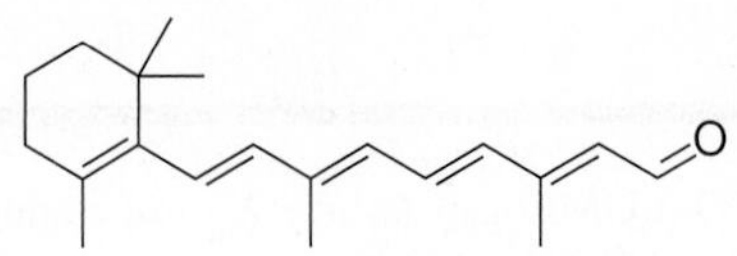

all-*trans* retinal
λ_{max} = 440 nm

iii)

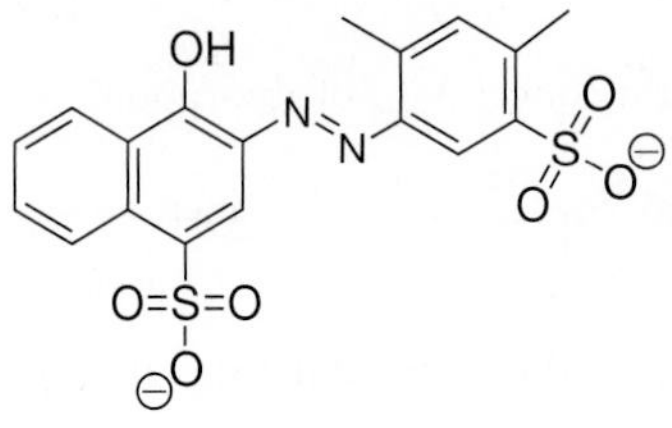

C.I. 14 700
λ_{max} = 502 nm

INTEGRATE THE SKILL

9.4 Absorption and excitation are often followed by relaxation back to the ground state and the emission of a photon of light. The process, called fluorescence, releases a photon with slightly longer wavelength than the absorbed photon. This wavelength difference, known as Stokes' shift, is the result of the excited state losing energy through other processes *before* the photon is emitted.

What colour do you expect quinine to appear under visible light? What colour do you expect quinine to release when excited with UV light of 230 nm?

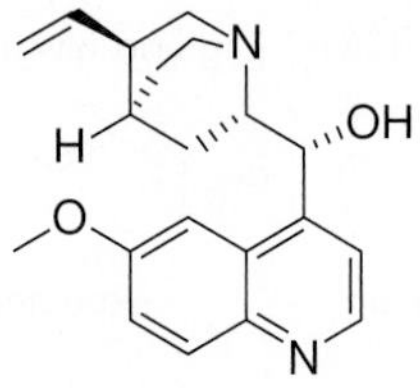

quinine
λ_{max} = 230 nm

9.3 Aromaticity

Aromaticity is a property of certain fully conjugated rings that confers special stability and other chemical properties.

When a fully conjugated π system appears in a ring, it may display **aromaticity**, which can drastically change its chemical properties. Such aromatic rings are much more stable than expected based solely on the presence of alternating π bonds. This stability derives from the bonds being arranged in a circle, and the stability of such systems changes their chemical behaviour. For example, alkenes and conjugated dienes readily react with electrophiles to give products resulting from electrophilic addition. However, when these π bonds are part of an aromatic ring, they react very differently, if at all. As shown below, cyclohexene and 1,3-cyclohexadiene react with a bromine molecule (Br_2) to give addition products, whereas benzene, their aromatic counterpart, does not react under the same conditions. Only in the presence of bromine and a certain catalyst does benzene react, yielding a substitution product instead of an addition. The reactions of aromatic rings are discussed further in Chapter 10.

In general, aromatic rings do not undergo addition reactions because the products would no longer be aromatic and, as described in Section 9.3.1, the energetic cost of disrupting the aromaticity is simply too high.

9.3.1 Aromatic stability

The energy released by the hydrogenation of 1,3,5-hexatriene is more than 79.3 kcal/mol (332 kJ/mol). By comparison, the hydrogenation of benzene (i.e., 1,3,5-*cyclo*hexatriene) is only 49.7 kcal/mol (208 kJ/mol). This indicates that the cyclic triene structure of benzene is much more stable than that of hexatriene, even though the two appear to have the same bonding pattern. The reason for this greater stability in benzene is that its π electrons can delocalize in a circular fashion.

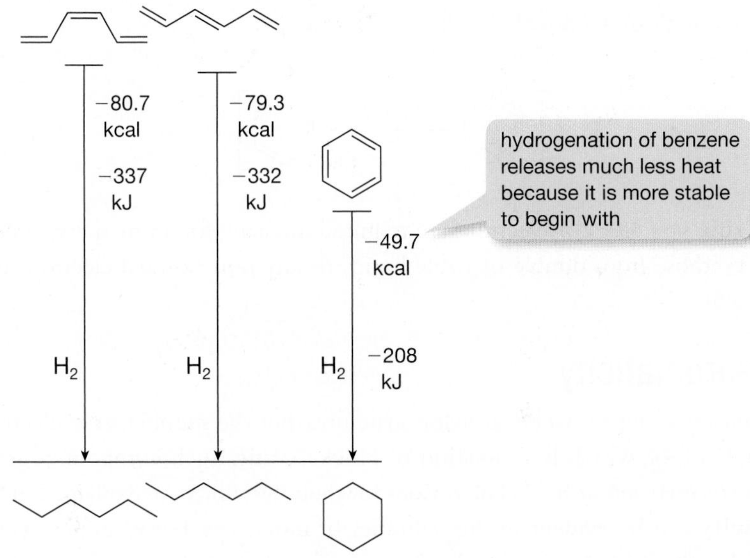

9.3.2 Criteria for aromaticity

For a structure to be aromatic, it must generally meet the following four requirements:

1. be cyclic
2. have a p orbital on all participating ring atoms (i.e., each ring atom is sp^2 or sp hybridized)
3. be planar
4. have $4n+2$ π delocalized electrons in the ring, where n is an integer value (Hückel's rule)

Benzene and pyridine, two common aromatic compounds, meet these requirements. Each atom in their rings has a p orbital (they are all sp^2 hybridized), the rings are planar, and six π electrons make up the network of π bonds (i.e., $[4 \times 1] + 2 = 6$). This allows the π electrons to circulate in the cyclic structure.

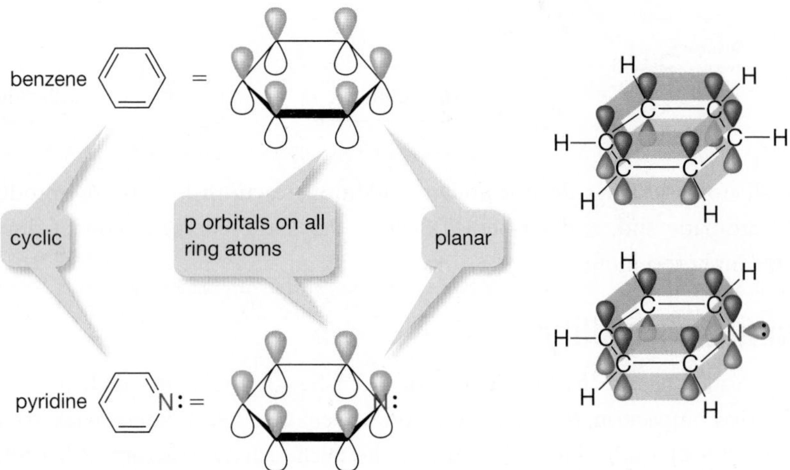

In fact, aromatic compounds can be represented by a series of resonance structures interchanged by curved arrows that depict electronic delocalization around the ring. The resonance picture accounts for the observation that all six C–C bonds in benzene are of the same length (in between that of a single and a double bond).

Hückel's rule states that the number of π electrons that participate in an aromatic ring structure must be a multiple of $(4n+2)$ where $n = 0, 1, 2, 3....$

Hückel's rule says $4n+2$ π electrons are required in a ring for aromaticity to occur. The π electrons may be those from double or triple bonds or any non-bonded electrons in p orbitals (see Section 9.3.5).

9.3.3 Anti-aromaticity

Anti-aromaticity is the severe instability incurred from having $4n$ π electrons in a delocalized ring. Most often, anti-aromaticity prevents products from having a planar conformation or from forming.

Some compounds may appear to be aromatic structures, but the number of π electrons in their ring is a multiple of $4n$, which is a violation of Hückel's rule. Such compounds are extremely unstable and are termed *anti-aromatic*. The reasons for their instability are explained in Section 9.4. **Anti-aromaticity** can be evident in the stabilities of molecules. For example, cyclobutadiene can be prepared in the lab but exists for about five seconds before decomposing.

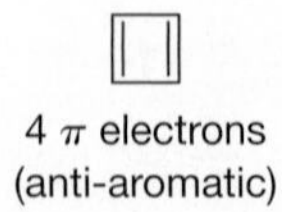

4 π electrons
(anti-aromatic)

9.3.4 Non-aromatic compounds

Other conjugated molecules are neither aromatic nor anti-aromatic. These compounds usually have at least one sp^3 atom as part of the ring, or they may exist in non-planar conformations. For example, cycloheptatriene is non-aromatic because one of its ring atoms is sp^3 hybridized, which disrupts the electron delocalization over the whole ring. Conjugation (and aromaticity) occurs only between properly aligned p orbitals, and these are available only on sp- and sp^2-hybridized atoms.

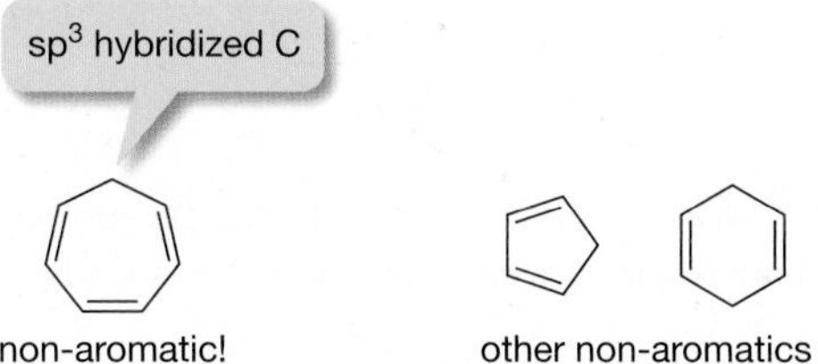

A different example of a non-aromatic molecule is cyclooctatetraene. At first glance, this compound may appear to be anti-aromatic (eight π electrons in a ring), and it would be if it were planar. However, bond rotation can prevent it from becoming planar, resulting in a boat-like conformation where the p orbitals cannot all be aligned with each other. Because the molecule avoids planarity, it cannot become anti-aromatic. In fact, cyclooctatetraene reacts as if it were four isolated alkene bonds.

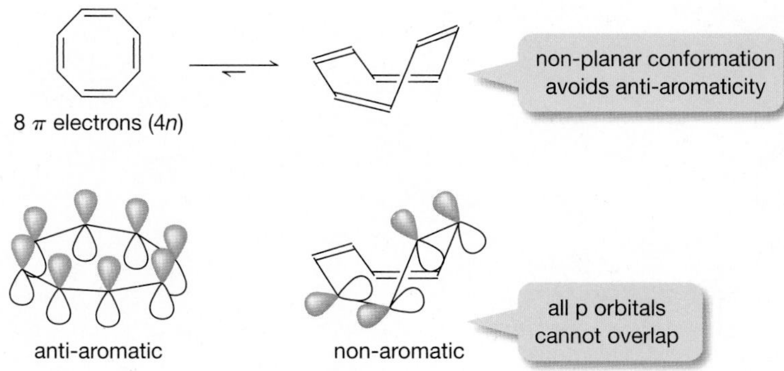

ORGANIC CHEMWARE
9.2 Resonance: Benzene

ORGANIC CHEMWARE
9.3 Resonance: Ortho-xylene

STUDENT TIP
MO Theory can be very complicated, and requires computer-based calculations to be done properly. On a less formal basis, organic chemists often use a combination of valence bond theory—to describe the σ-bond framework on a molecule—and an approximation of the molecular orbital theory—to describe any conjugated π-bond system.

CHECKPOINT 9.3

You should now be able to recognize and predict whether a neutral carbon ring molecule is aromatic, anti-aromatic, or non-aromatic, based on criteria including Hückel's rule.

SOLVED PROBLEM

Is the following compound aromatic, anti-aromatic, or non-aromatic?

cyclotetradecaheptaene

STEP 1: Determine if the compound is a fully conjugated ring.

Each carbon atom is part of a double bond, making each sp^2 hybridized, with a p orbital free to participate in a π–MO system.

STEP 2: Determine if the compound is planar.

There does not appear to be any significant strain that would prevent this molecule from being planar. The bond angles are close to what is expected for sp^2-hybridized carbons, and steric strain between the hydrogens pointing into the ring is minimal.

STEP 3: Count the number of π electrons to determine whether the molecule satisfies Hückel's rule.

Seven double bonds contribute fourteen π electrons: $4n + 2 = 14$, when $n = 3$. This molecule satisfies Hückel's rule for aromaticity.

PRACTICE PROBLEM

9.5 Assuming planarity, determine whether the following are aromatic, anti-aromatic, or non-aromatic.

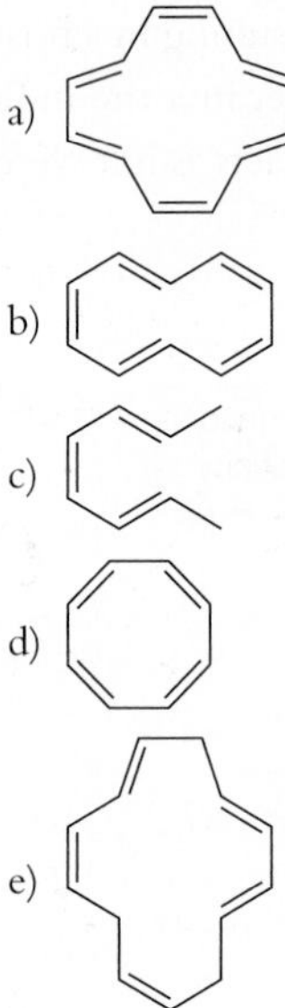

INTEGRATE THE SKILL

9.6 Pyrene may be considered an aromatic 14-membered ring. Alternatively, by counting the electrons in a different way, it may be considered a six-membered aromatic ring within the larger structure. Whether the actual electron flow follows one ring or another depends upon which provides the greatest aromatic stabilization.

Draw a resonance structure that maximizes aromatic stability by showing the presence of *two* six-membered aromatic rings within the larger structure.

9.3.5 Contribution of electron pairs to aromaticity

Some aromatic compounds incorporate heteroatoms as part of their rings. These heteroatoms may have lone pairs of electrons that may or may not contribute to aromaticity. Before applying Hückel's rule, it is necessary to determine if these unpaired electrons are in a p orbital that can participate in aromatic conjugation.

Pyridine is a heterocycle that contains a nitrogen in an aromatic ring. The nitrogen atom is sp^2 hybridized, and its p orbital participates in the ring aromaticity. The lone pair of electrons occupies an sp^2 orbital, which is orthogonal to the π system and therefore cannot participate in the ring conjugation. This sp^2 lone pair *cannot* be counted as part of the $4n+2$ criterion for aromaticity, so that pyridine is aromatic (3 double bonds = 6 π electrons).

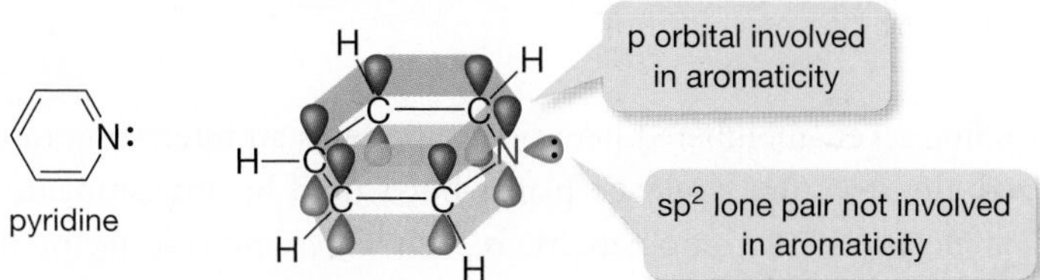

Pyrrole is another aromatic heterocycle. Its Lewis structure does not immediately suggest aromaticity because the ring does not show an alternating pattern of double bonds and single bonds. However, if the ring nitrogen is sp^2 hybridized, its lone pair must occupy a p orbital, aligned with the other p orbitals in the ring, and can therefore participate in the aromaticity. With two double bonds and one nitrogen lone pair, Hückel's rule is obeyed. In fact, atoms in molecules adopt whatever geometry results in the greatest stability. In the case of pyrrole, a planar nitrogen with a p orbital allows for aromatic stability that would be missing if the nitrogen were tetrahedral.

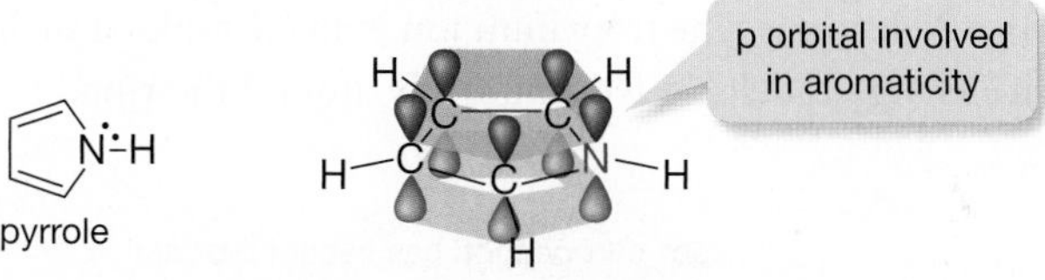

When furan is drawn as a Lewis structure, the oxygen carries two lone pairs, and aromaticity is not evident. If the oxygen is sp^3 hybridized, neither lone pair can participate in aromaticity, and so this molecule would be non-aromatic. However, if the oxygen is sp^2 hybridized, one lone pair occupies a p orbital and participates in aromatic conjugation, while the other electron pair is in an sp^2 orbital. This slight rehybridization of the oxygen results in the furan molecule becoming aromatic (i.e., four π electrons and two electrons from the oxygen p orbital).

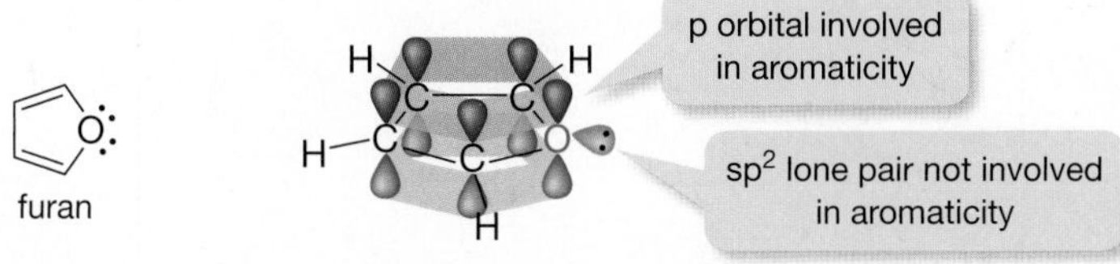

As with other aromatic compounds, the aromaticity of furan can be described as a series of resonance structures. Notice how the resonance is "circular." Working one bond at a time, electrons can delocalize around the entire ring, with a final electron flow regenerating the original structure.

STUDENT TIP

Two lone pairs on an atom cannot both be part of the same π system. Only one can be counted to satisfy Hückel's rule.

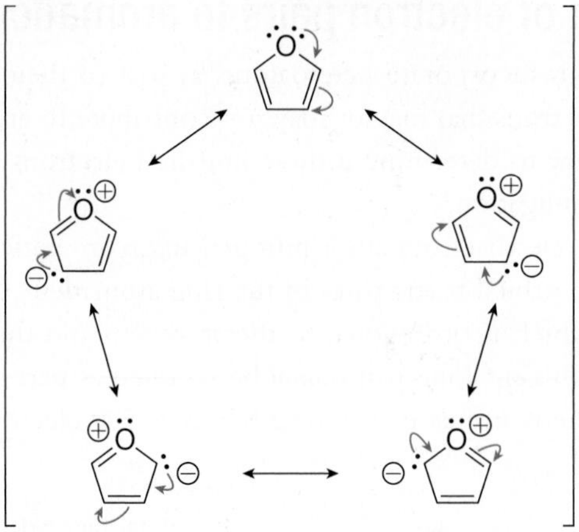

The corresponding seven-membered heterocycle presents an interesting case. If the oxygen were sp^2 hybridized, the structure could be planar but would be anti-aromatic (3 π bonds + 1 lone pair = 8 π electrons). Anti-aromaticity is such a destabilizing factor that the oxygen remains essentially sp^3 hybridized, and the molecule adopts a curved, non-aromatic conformation.

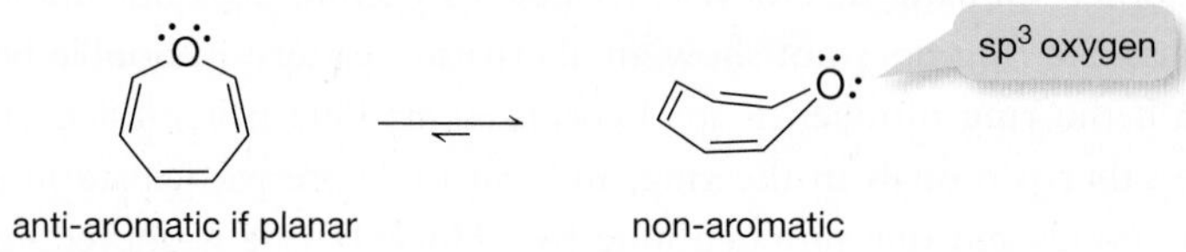

sp^3 oxygen

anti-aromatic if planar

non-aromatic

9.3.6 Aromatic ions

Anions and cations can also be aromatic. For instance, carbocations are sp or sp^2 hybridized and carry an empty p orbital. Their p orbital can participate in aromaticity if it is part of a ring of p orbitals delocalizing $4n+2$ electrons. The tropylium ion is an example of such an aromatic cation; notice how the positive charge is delocalized uniformly around the ring.

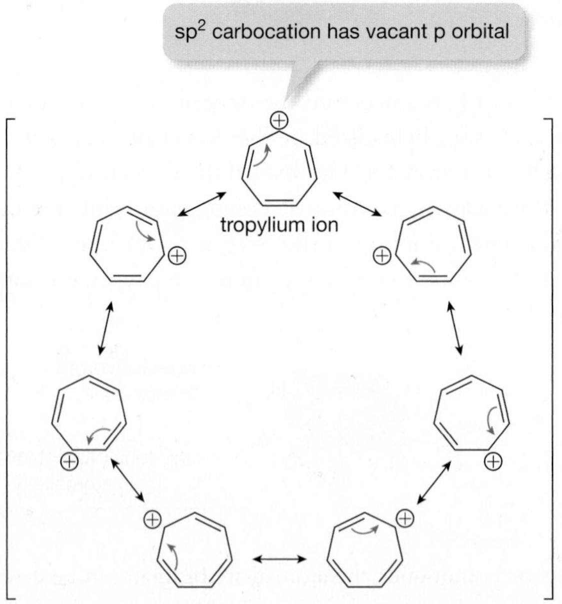

sp^2 carbocation has vacant p orbital

tropylium ion

Anions can also be aromatic if the electron pair occupies a p orbital. The cyclopentadienyl anion carries six π electrons delocalized over all five sp^2-hybridized carbons; it is an aromatic carbanion. Because of the aromatic stability of the cyclopentadienyl anion, neutral cyclopentadiene is approximately 10^{30}–10^{45} times more acidic than cyclopentane.

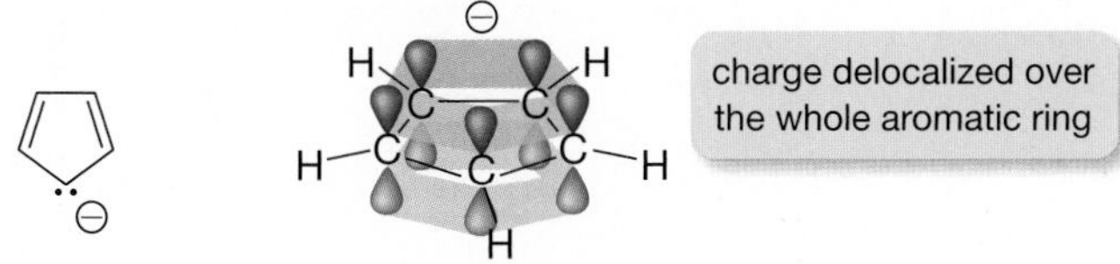

CHECKPOINT 9.4

You should now be able to recognize and predict whether heterocycles and ionic compounds are aromatic, anti-aromatic, or non-aromatic, based on criteria including Hückel's rule.

SOLVED PROBLEM

Is the following compound aromatic, anti-aromatic, or non-aromatic?

STEP 1: Determine if the compound is a fully conjugated ring.

Each carbon atom, as well as the nitrogen, is part of a double bond, making each sp^2 hybridized, with a p orbital free to participate in a π–MO system. Because it is adjacent to π bonds, the sulfur may also be sp^2 hybridized, providing a p orbital for conjugation.

STEP 2 (IF REQUIRED): Determine if the compound is planar.

There does not appear to be any significant strain that would prevent this molecule from being planar. The bond angles are close to what is expected for sp^2 atoms, and there are no steric interactions across the ring.

STEP 3: Identify any electrons not in double bonds that will contribute to the π system. Lone pairs such as those in carbanions and on heteroatoms may conjugate to the π system.

Only one set of electrons from any atom may be included.

If this atom becomes sp^2 hybridized, one lone pair must occupy an sp^2 orbital so only one pair can participate in aromaticity.

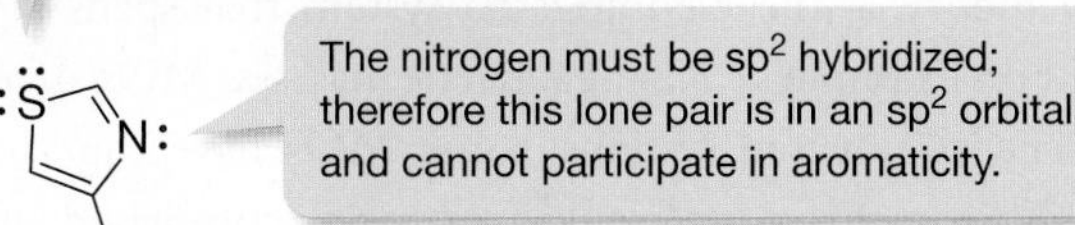

The nitrogen must be sp² hybridized; therefore this lone pair is in an sp² orbital and cannot participate in aromaticity.

STEP 4: Count the number of π electrons to determine whether the molecule satisfies Hückel's rule.

Two double bonds contribute four π electrons. One lone pair gives a total of $6 = 4n + 2$, when $n = 1$. This molecule satisfies Hückel's rule for aromaticity if the sulfur is sp^2 hybridized.

PRACTICE PROBLEM

9.7 Assuming planarity, determine whether the following are aromatic, anti-aromatic, or non-aromatic.

a)

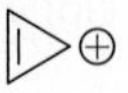

b)

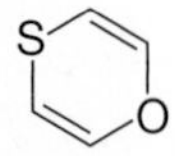

c)

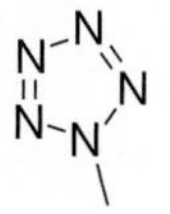

d)

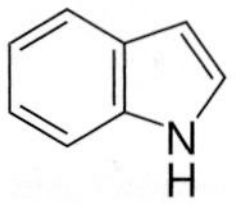

e)

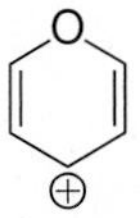

f)

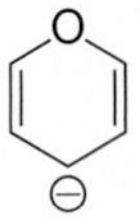

INTEGRATE THE SKILL

9.8 Draw resonance structures that clearly show aromaticity within each ring of caffeine.

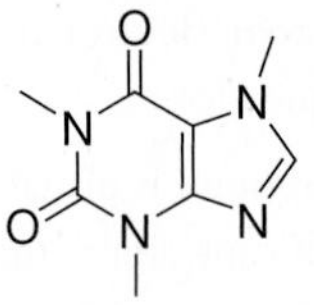

caffeine

9.4 Molecular Orbital Analysis of Aromatic Rings

Degenerate describes multiple orbitals that have the same energy and the same number of nodes.

The basis for Hückel's rule lies in the energies of the MOs that arise for cyclic π systems. According to MO theory, a linear conjugated system that spans *n* p orbitals forms *n* MOs of different energies. For example, 1,3,5-hexatriene has six MOs of different energies (Ψ_1–Ψ_6). In contrast, for cyclic conjugated systems, some MOs have the same energy and are referred to as **degenerate**. The aromatic molecule benzene is comprised of six π MOs, but two pairs are degenerate: the Ψ_2 and Ψ_3 MOs have identical energies, and each has one nodal plane; the orbitals Ψ_4 and Ψ_5 have two nodal planes and are also degenerate.

Energy

Ψ_6 Ψ_5 Ψ_4 Ψ_3 Ψ_2 Ψ_1

anti-bonding

bonding

top view

Ψ_1 Ψ_2 Ψ_3 Ψ_4 Ψ_5 Ψ_6

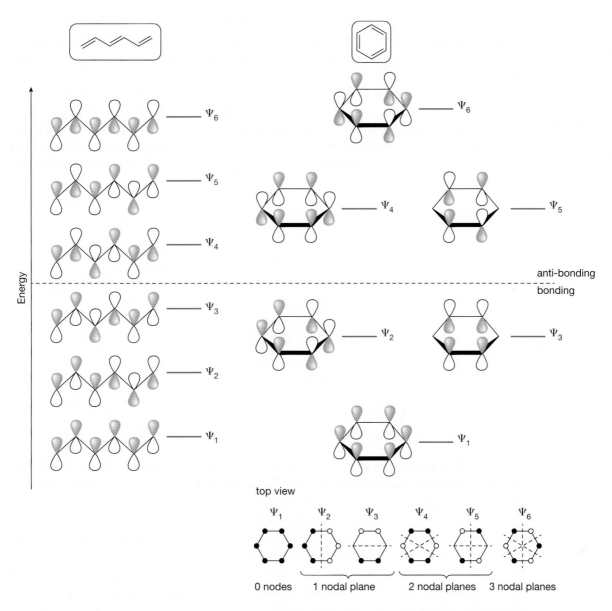

0 nodes | 1 nodal plane | 2 nodal planes | 3 nodal planes

The ground-state configuration of benzene places its six p electrons (i.e., $4n+2$) in the lowest energy MOs (Ψ_1–Ψ_3), which are all bonding and doubly occupied, producing a very stable structure that is aromatic.

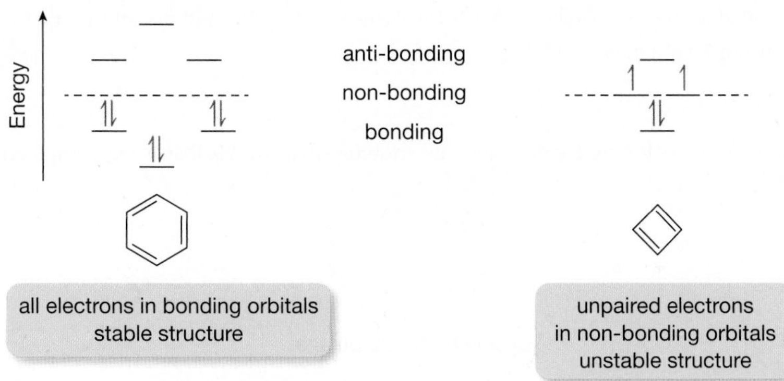

By comparison, the ground-state configuration of cyclobutadiene requires two non-bonding orbitals to be singly occupied. This produces a very unstable structure that would be anti-aromatic. In order to avoid anti-aromaticity, cyclobutadiene forms as a rectangle instead of a square. This prevents the formation of an anti-aromatic di-radical.

Frost circles are a mnemonic device for predicting the relative energy of the molecular orbitals of an aromatic system.

Simple MO patterns for conjugated ring structures can be constructed using **Frost circles**, which can help predict the relative energy of MOs in an aromatic system. To form a Frost circle, draw the compound to be analyzed as a polygon with one vertex pointing down within a circle connecting all the vertices. The relative location of each vertex corresponds to the relative energies (and degeneracies) of the π–MOs for that ring. The centre of the ring corresponds to the non-bonding energy level.

E
anti-bonding
non-bonding
bonding
vertex down

The information from a Frost circle can predict the presence or absence of aromaticity. For example, a cyclopentadienide anion has only five π orbitals, but six π electrons still completely fill the bonding orbitals, giving the ring its aromatic character.

Cyclobutadiene, on the other hand, has only four π electrons. The Frost circle analysis shows two electrons in non-bonding orbitals, which would make this system anti-aromatic instead of aromatic because the distribution of π electrons includes orbitals that do not contribute to bonding. Instead of existing as a high-energy, anti-aromatic compound, most structures predicted to be anti-aromatic avoid anti-aromaticity by adopting a conformation that does not permit the overlap of p orbitals over the entire ring system.

CHECKPOINT 9.5

You should now be able to predict relative MOs for conjugated cyclic systems, and predict aromaticity or anti-aromaticity using Frost circles.

SOLVED PROBLEM

Justify the aromaticity of cyclotetradecaheptaene by showing that the HOMOs are completely filled.

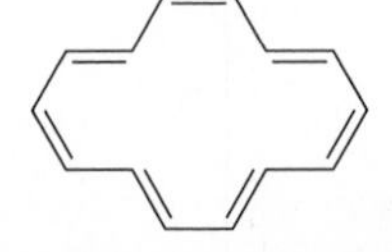

cyclotetradecaheptaene

STEP 1: Draw a Frost circle, with the ring to be considered pointing down, within a larger circle. To have the point of each carbon intercept the rim of the circle, the configurations of the double bonds must be changed to all-*cis*.

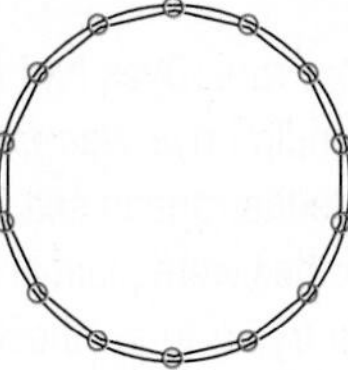

STEP 2: At each vertex of the ring and the circle, place a MO energy level.

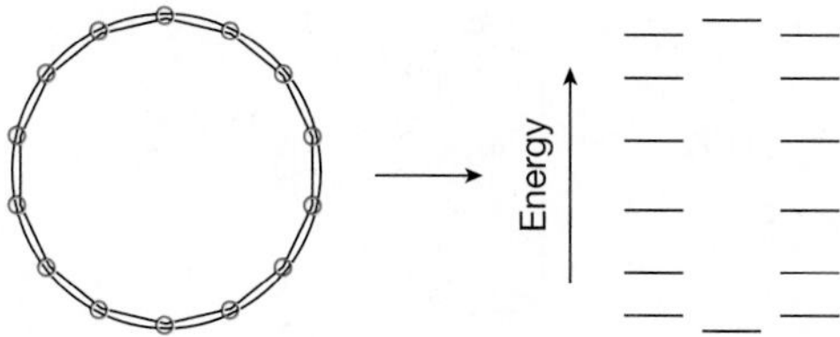

STEP 3: Starting at the lowest energy level, fill in the appropriate number of π electrons.

STEP 4: Determine if the HOMO is completely filled or half-filled. Completely filled indicates aromaticity. Half-filled indicates anti-aromaticity.

This molecule is predicted to show aromaticity.

PRACTICE PROBLEM

9.9 Predict the molecular orbital energy diagram for the following structures, and justify their aromaticity or anti-aromaticity.

a)

b) 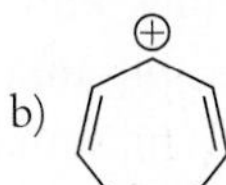

INTEGRATE THE SKILL

9.10 A cyclopentadienyl MO system is a flat, fully conjugated ring system. Show the MO energy diagram with the appropriate number of electrons for the cyclopentadienyl cation, anion, and radical. Use the energy diagram to predict the relative stability of the three related compounds.

CHEMISTRY: EVERYTHING AND EVERYWHERE

Aromatic Dyes

Before the mid-1800s, brightly dyed fabrics were rare. Dyes had to be extracted from natural sources and were only available in small amounts. For example, indigo dye was extracted from a variety of tropical plants, while Tyrian purple was obtained from a species of Mediterranean sea snails! These dyes were so expensive that only the wealthy could afford them; most people's clothes were plain and available in only dull colours, with dyes made from native plants. In 1856, William Perkin was trying to synthesize quinine using aniline, a chemical extracted from coal tar. He obtained instead a bright purple compound, which he found would colour fabric. Perkin patented this dye under the name of aniline purple, and it later became known as Perkin's mauve, the first artificial dye.

mauveine A

The raw material for producing Perkin's dye was coal tar, making his dye inexpensive and available in large quantities. So for the first time, the average person could afford to own brightly coloured clothes! Perkin's discovery launched the synthetic dye industry, which later founded the major pharmaceutical companies of the twentieth century. Indeed, advances in the preparation of aromatic dye compounds led to a variety of new synthetic methods for manufacturing useful products, including pharmaceuticals. While there are now dyes and drugs that are not aromatic, the early compounds that launched these industries are aromatic. Without aromaticity, the dye, paint, drug, and chemical industries would not have developed as we know them today.

9.5 Aromatic Hydrocarbon Rings

Benzene and benzene analogs such as pyridine are common both in nature and in synthetic compounds, and they are simple representatives of a vast number of other aromatic structures.

9.5.1 Annulenes

Annulene is a general term for any cyclic, fully conjugated system.

Benzene is the most common aromatic ring in a class of fully conjugated rings known as **annulenes**. Annulenes have an even number of carbons in the ring and are named [*n*]-annulene, where [*n*] specifies the ring size. According to this system, benzene would be named [6]-annulene.

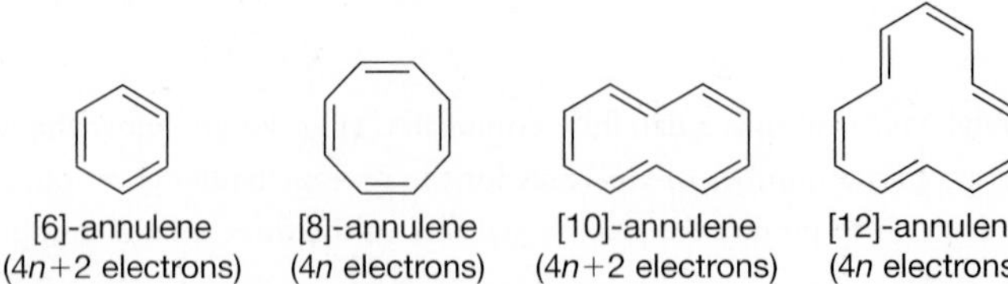

Annulenes are monocyclic and fully conjugated, but are not necessarily aromatic. Those that violate Hückel's rule are usually non-aromatic because they adopt non-planar conformations that avoid anti-aromaticity. If an annulene is large enough and obeys Hückel's rule, the ring can become planar, rendering the molecule aromatic. Some ring sizes cannot be planar because of steric interactions or bond angles and cannot be aromatic even if they satisfy the other criteria.

In the case of [10]-annulene, the predicted aromaticity is prevented because planarity is impossible despite its $4n + 2 = 10$ π electrons. The all-*cis* isomer cannot be planar because of the strained bond angles that such a geometry would impose. The preferred isomer has two *trans* double bonds but cannot be planar because of the severe transannular repulsion of two opposing hydrogens. Neither of these conformations is aromatic.

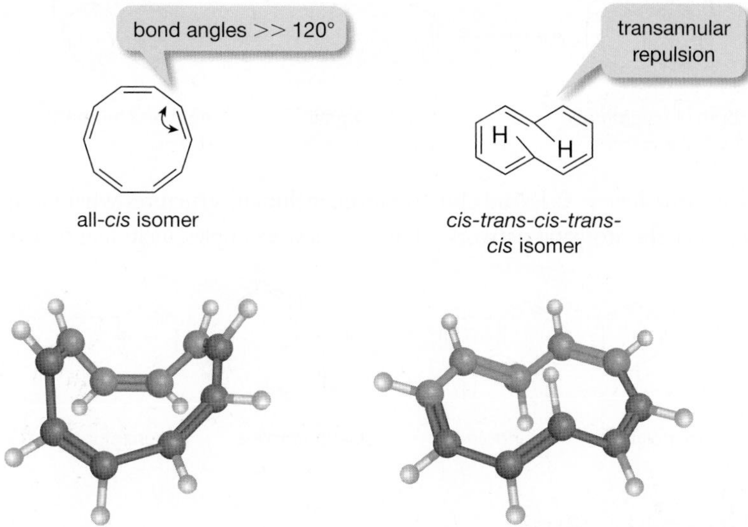

The first annulene larger than benzene that displays some aromaticity is [14]-annulene, though it has some ring strain. Its near-planar structure allows for enough conjugation for [14]-annulene to show some aromatic stability. The [18]-annulene is aromatic and strain free.

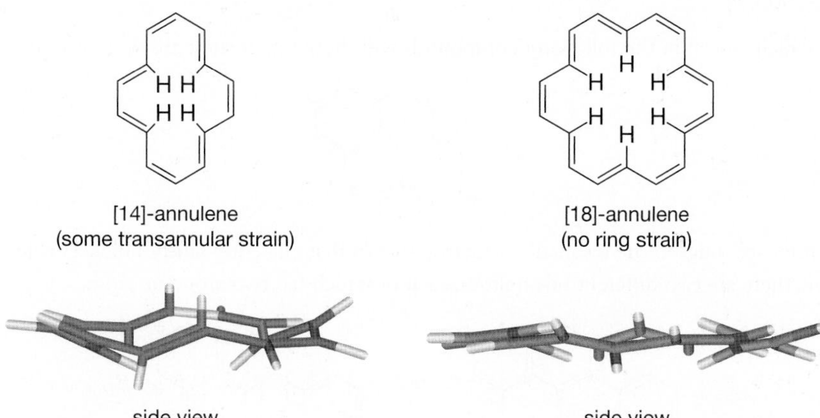

9.5.2 Polycyclic aromatic rings

Aromatic rings can be fused, producing polycyclic aromatic compounds such as naphthalene, anthracene, and phenanthrene. These compounds are typically referred to by trivial names.

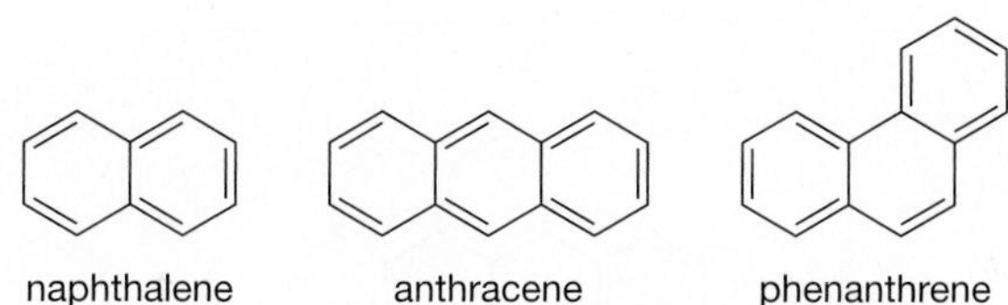

Each of these compounds satisfy Hückel's rule and are planar. Various resonance structures can be drawn to illustrate their aromaticity. For example, naphthalene has three major resonance contributors, two of which have one aromatic ring and one where either ring is aromatic. Because the actual structure is the composite of every resonance contributor, naphthalene shows somewhat less aromatic stabilization than two benzene rings.

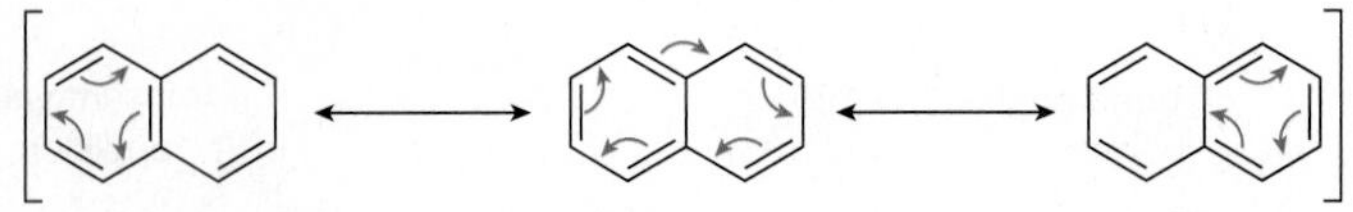

both rings aromatic

left ring aromatic

right ring aromatic

Polycyclic aromatic heterocycles are also common, including structures where lone pairs on the heteroatom are part of the aromatic network. Here are a few examples, including their trivial names:

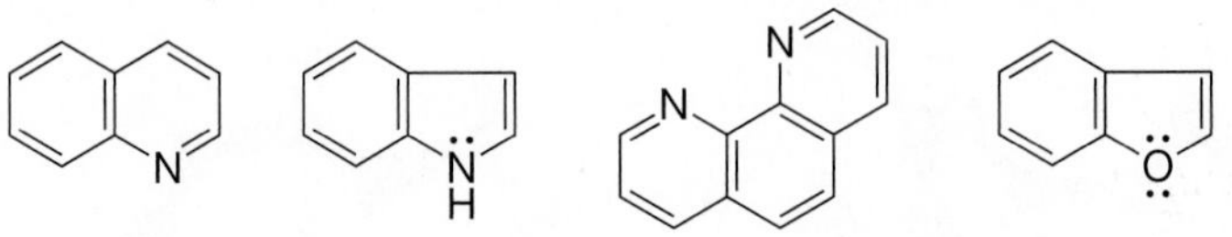

quinoline

indole

phenanthroline

benzofuran

CHECKPOINT 9.6

You should now be able to justify the relative aromaticity of polycyclic and macrocyclic conjugated systems, based on planarity and resonance contributors.

SOLVED PROBLEM

Determine which ring(s) in the following compounds will show the greatest aromatic stability.

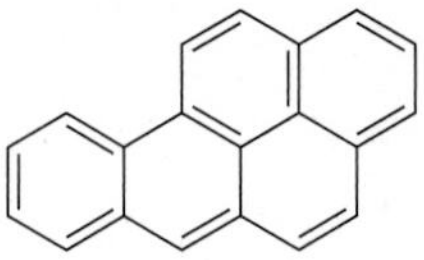

STEP 1: Identify any rings in the resonance structure shown that currently satisfy Hückel's rule.

As shown, there are two different possibilities, each of which has two aromatic rings.

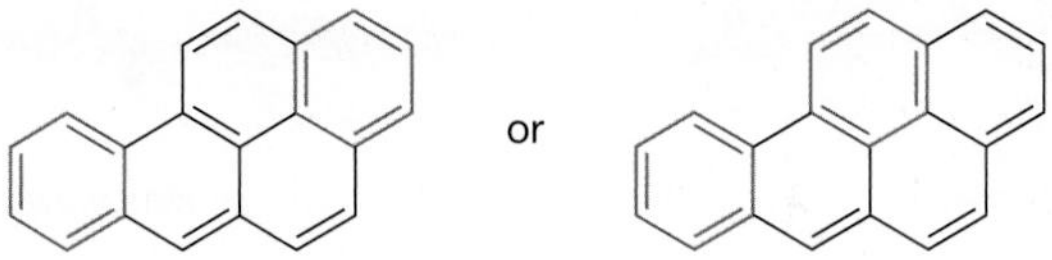

or

STEP 2: Draw resonance structures to consider other aromatic possibilities.

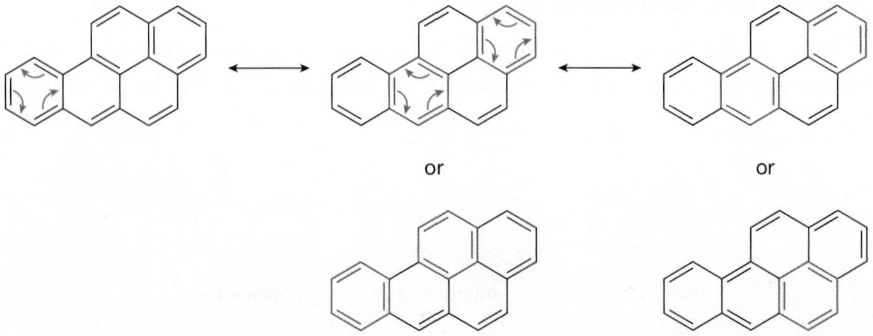

or

or

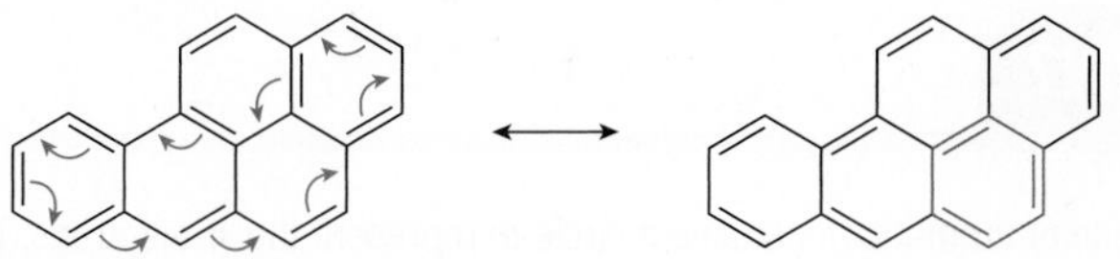

STEP 3: Evaluate and compare the relative stability of the different resonance contributors. Resonance contributors with fewer fully aromatic benzene rings within the larger structure are less stable and contribute less to the overall structure.

In this case, the resonance contributors with the least aromatic stabilization are those with the same benzene ring, suggesting that it is the least aromatic of the rings. Therefore, it will be the most likely to react as an alkene.

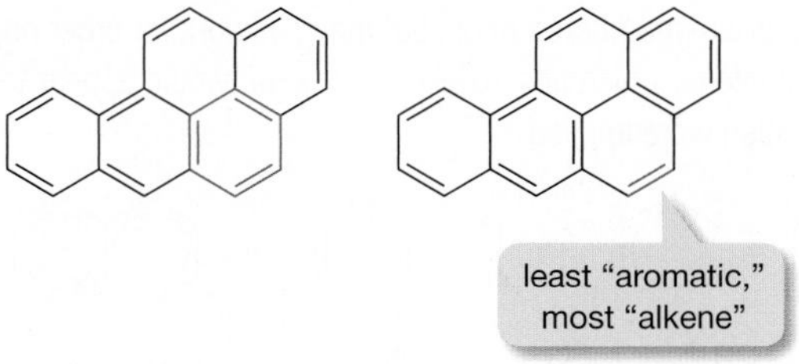

least "aromatic," most "alkene"

PRACTICE PROBLEM

9.11 a) Identify segments of the following structures that display lower aromatic stability.

i)

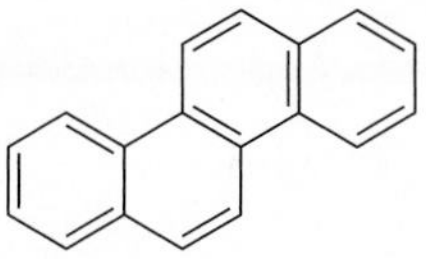

ii)

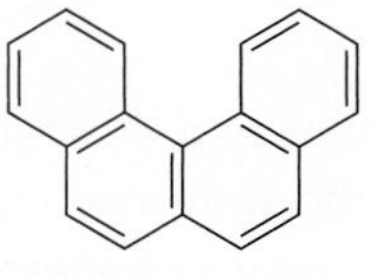

iii)

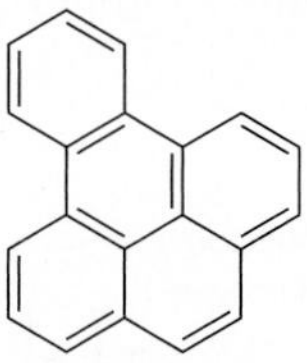

b) Are the following expected to be aromatic?

i) [18]-annulene

ii) [22]-annulene

INTEGRATE THE SKILL

9.12 The compound coronene is composed of six fused benzene rings. It can be written as any of 20 different resonance structures, with the possibility of many different aromatic systems. One common representation shows three fully aromatic benzene rings within the larger structure. Draw a resonance structure to justify the aromaticity of each of the rings, as well as a resonance structure with two aromatic rings, one along the outside of the ring and the other inside it.

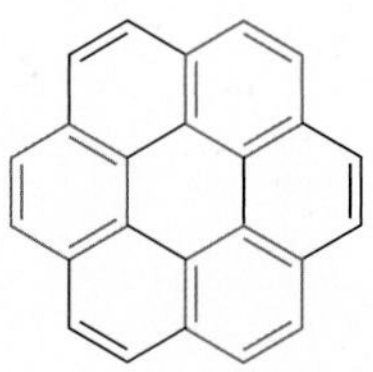

DID YOU KNOW?

A traditional representation of aromatic rings uses a circle to represent the π electrons. This notation has the advantage of representing all resonance structures and makes it clear that the π electrons are spread over the entire ring.

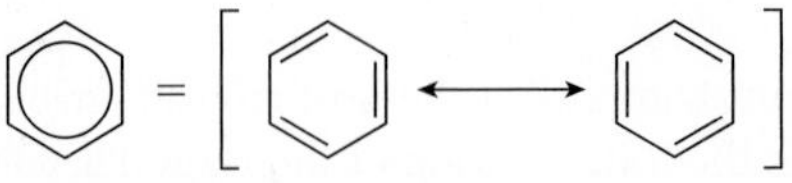

Circle notation is frequently misused. The original notation was intended to depict six π electrons, an "aromatic sextet." This is fine for six-membered rings, but may not work for other ring sizes or polycyclic aromatic rings. Consider naphthalene, which has 10 π electrons but would appear to have 12 π electrons in circle notation if its strict definition were applied.

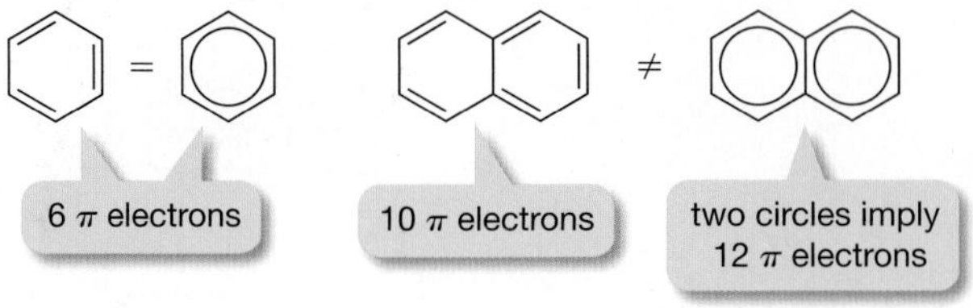

Despite this inaccuracy, circle notation is often used in such systems to simply imply aromaticity.

Bringing It Together

The stability of an aromatic ring plays an essential role in the production of sex hormones. In both men and women, testosterone is converted to estradiol by the aromatase enzyme (designated as CYP19A1). Drugs known as aromatase inhibitors have become an important treatment for certain types of breast cancers.

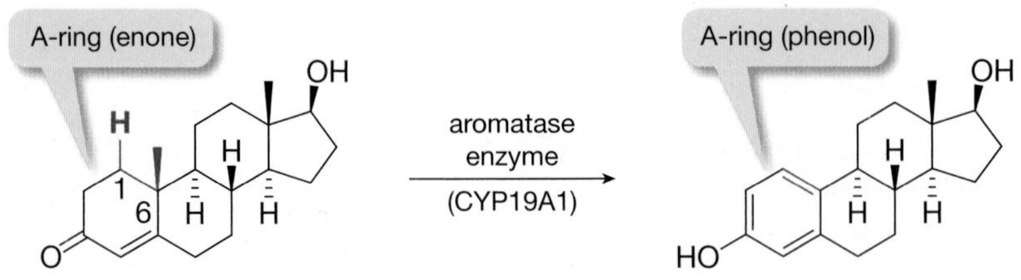

In this reaction, the A-ring of testosterone is converted from a cyclic enone to an aromatic alcohol, a phenol. For this to happen, an oxidized iron in the aromatase enzyme oxidizes the methyl group at position 6. The iron then participates in a complex reaction where a hydrogen at position 1 and the carbon at position 6 are eliminated to form a double bond at these positions.

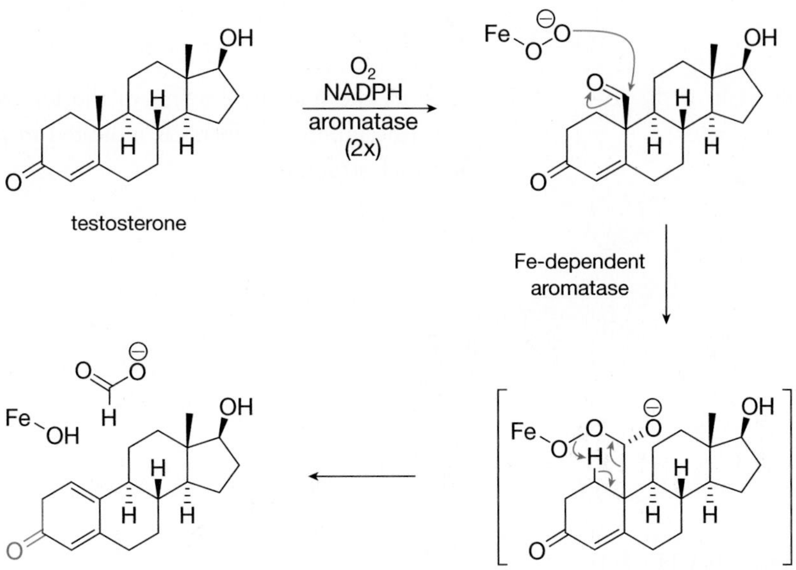

The initial product is a ketone with two ring double bonds, and it equilibrates very rapidly to its enol form. Although enols are usually much less stable than their corresponding keto-tautomer, in this case the enol tautomer is formed quasi-exclusively because the ring becomes aromatic.

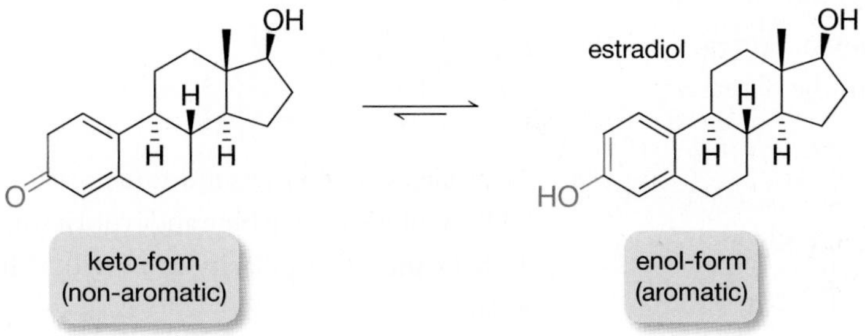

You Can Now

- Predict the relative energy levels, with phases and nodes, of π–molecular orbitals.
- Predict molecular properties of conjugated π systems, such as bond rotation, length, and heat of hydrogenation.
- Predict the relative HOMO–LUMO gap and the λ_{max} of conjugated systems.
- Predict both the absorbed and reflected colour of a compound, based on the absorbed λ_{max}.
- Recognize and predict whether a neutral carbon ring molecule is aromatic, anti-aromatic, or non-aromatic, based on criteria including Hückel's rule.
- Recognize and predict whether heterocycles and ionic compounds are aromatic, anti-aromatic, or non-aromatic, based on criteria including Hückel's rule.
- Predict relative MOs for conjugated cyclic systems, and justify aromaticity or anti-aromaticity, using Frost circles.
- Justify the relative aromaticity of polycyclic and macrocyclic conjugated systems, based on planarity and resonance contributors.

Problems

9.13 How many π electrons are there in each of the following compounds?

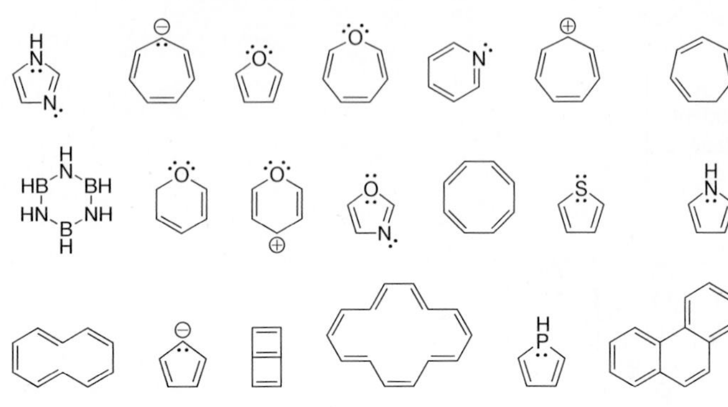

9.14 Which of the following π electron totals obey Hückel's rule of $4n+2$? Indicate the value of n for each.

Total π electrons = 2, 4, 6, 8, 10, or 12

9.15 Indicate which of the structures in Question 9.13 are expected to show aromatic stability.

9.16 The following compounds are aromatic, but do not appear so in the resonance structures shown. For each, show the resonance structure that explains the observed aromatic properties.

9.17 The following molecules contain a variety of rings. For each of these structures, identify any aromatic systems that may be in the molecule.

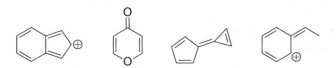

9.18 Draw all the resonance forms of the following structures (the number of forms is indicated in parentheses after each structure).

a) (2)

b) (3)

c) (5)

d) (7)

e) (5)

f) (9)

9.19 Napthalene is colourless and non-polar (dipole moment = 0 D). Azulene is deep blue and is polar (dipole moment = 1.08 D, the same polarity as H–Cl). Why is azulene so polar?

naphthalene azulene

9.20 Using a Frost circle, approximate the relative energy level of the π molecular orbitals for cyclopentadienide and cycloheptatrienide anions, and justify the presence or absence of aromatic stability in each.

9.21 Imidazole acts as a base, forming a conjugate acid with a pK_a of 7.05. Which nitrogen atom is protonated in the following acid–base reaction?

HN N HCl [HN N • HCl]

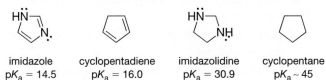

imidazole imidazole hydrochloride

9.22 The pK_a of cyclopentane is much higher than the pK_a of imidazolidine, showing that the electronegativity of nitrogen has a significant effect. The pK_a of imidazole is only slightly higher than the pK_a of cyclopentadiene. Why is cyclopentadiene so much more acidic than might be expected, based on the acidity of imidazole?

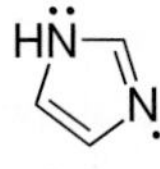

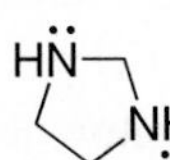

imidazole	cyclopentadiene	imidazolidine	cyclopentane
pK_a = 14.5	pK_a = 16.0	pK_a = 30.9	pK_a ~ 45

9.23 Generally, a reaction that forms an enol immediately tautomerizes to the ketone form, as in the conversion of cyclohexenol to cyclohexanone. However, such a reaction favours the enol form of pyridine-2-ol over the keto form, pyridine-2-one. Explain why this is so.

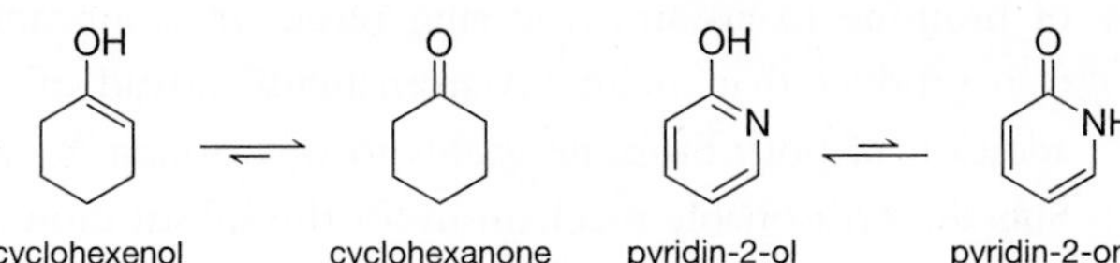

9.24 Cyclooctatetraene is not planar. What is the preferred conformation of the cyclooctatetraene dianion?

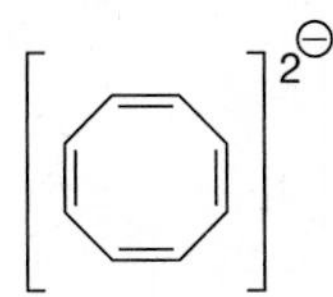

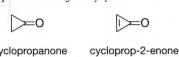

9.25 Which ketone is more polar: cyclopropanone or cycloprop-2-enone? Justify your answer.

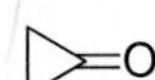

O

cyclopropanone cycloprop-2-enone

9.26 Under reducing conditions, 1,2,3-tri(9*H*-fluroen-9-ylidine)cyclopropane accepts two electrons. Draw the most stable resonance structure of the dianionic product.

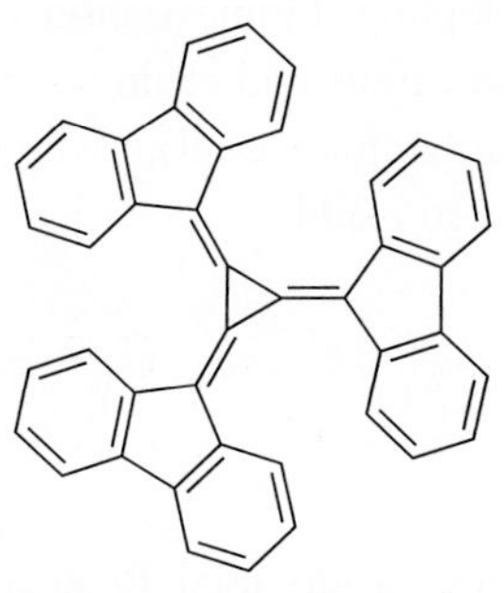

1,2,3-tri(9*H*-fluoren-9-ylidene)cyclopropane

9.27 DNA bases are all aromatic. Where necessary, draw resonance structures of each that clearly show their aromatic character.

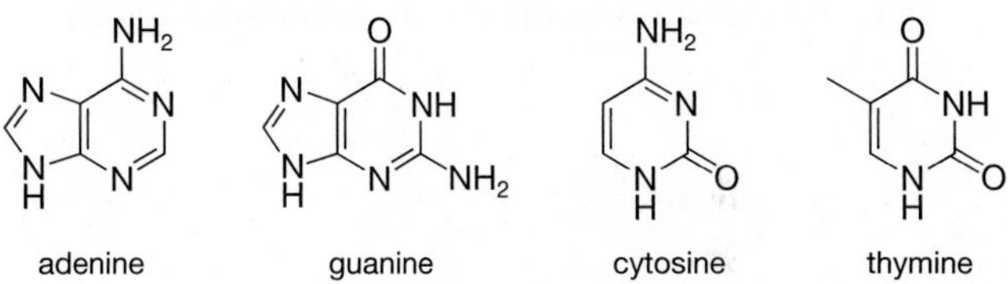

9.28 Deprotonation of either 1- or 5-methylcyclopenta-1,3-diene forms the same anion. Explain how this can be, and draw the structure of the product.

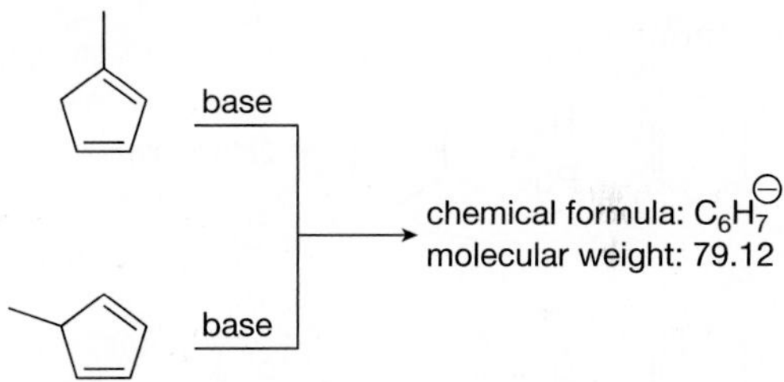

9.29 One of the following compounds is much more acidic than the others. Indicate which one this is, and briefly explain the unusual acidic nature of the compound.

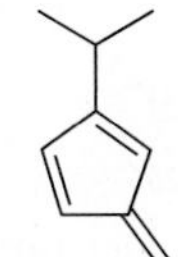

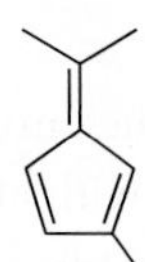

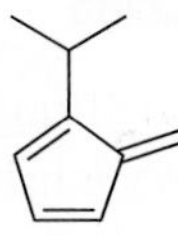

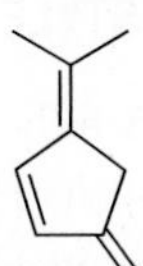

9.30 One of the most difficult problems in making synthetic proteins is maintaining stereochemistry when amino acids are connected together. The difficulty arises because amino acids tend to epimerize when "activated" for coupling. Epimerization can happen when the end of the amino acid chain forms the cyclic intermediate, as shown here. Explain why the ring structure can epimerize so readily.

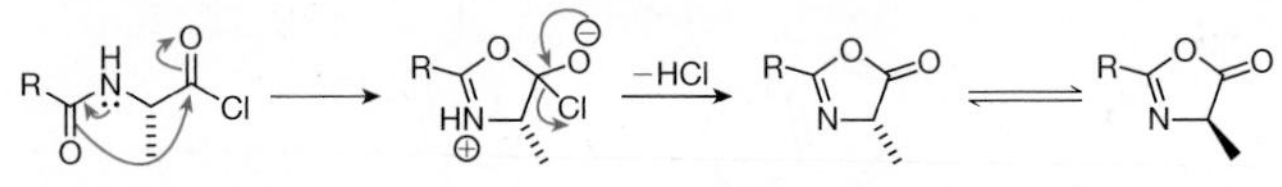

9.31 One protecting group used to keep an amine from acting as a nucleophile is the fluorenylmethyloxycarbonyl (fmoc) group. When it is time to remove fmoc, treatment with hydroxide removes an acidic proton. This causes the molecule to fragment, freeing the amine. Suggest a reasonable mechanism for the deprotection step, and explain why the proton removed is acidic.

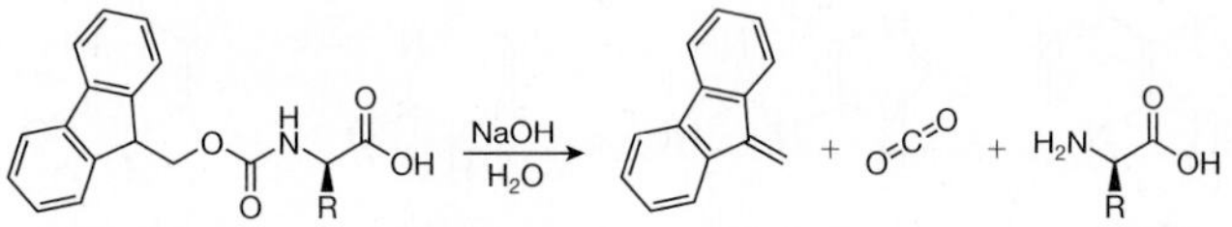

9.32 The hydrogenation of phenanthrene produces 135 kcal/mol (565 kJ/mol) of energy. The hydrogenation of cyclohexene produces 28.6 kcal/mol (119.7 kJ/mol). What is the aromatic stabilization energy of phenanthrene?

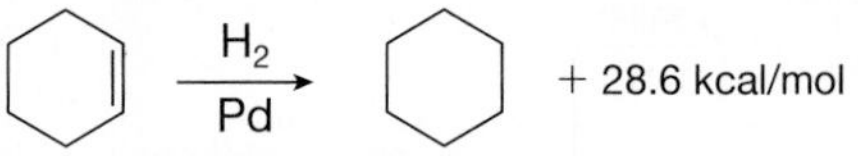

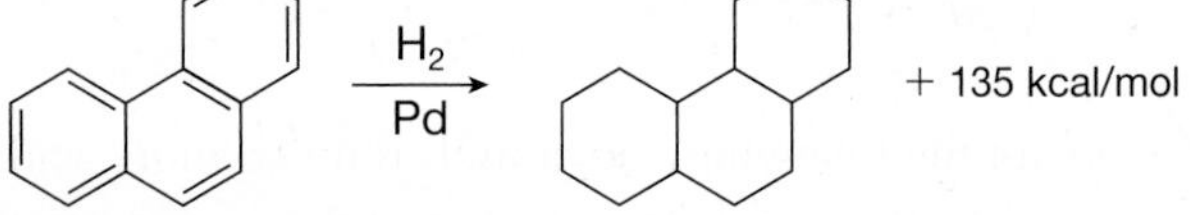

MCAT STYLE PROBLEMS

9.33 A fully conjugated ring becomes aromatic if
a) its HOMOs are half-filled with electrons.
b) its HOMOs are completely filled with electrons.
c) its LUMOs are half-filled with electrons.
d) its LUMOs are completely filled with electrons.

9.34 Which of the following structures are *not* aromatic?

a)

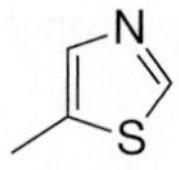

b)

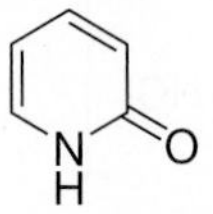

c)

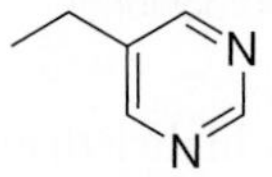

d)

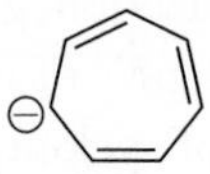

9.35 Cyclooctatetraene is
a) not planar, because it disrupts conjugation and avoids anti-aromaticity.
b) planar, because it ensures conjugation and allows anti-aromaticity.
c) not planar, because it disrupts conjugation and avoids aromaticity.
d) planar, because it ensures conjugation and allows aromaticity.

CHALLENGE PROBLEM

9.36 As described in Section 9.3, an attempted addition of bromine to an aromatic ring results in a substitution product that maintains aromaticity instead of an addition of both bromine atoms to one double bond. Suggest a reasonable mechanism for the substitution of a hydrogen atom by a bromine atom shown here.

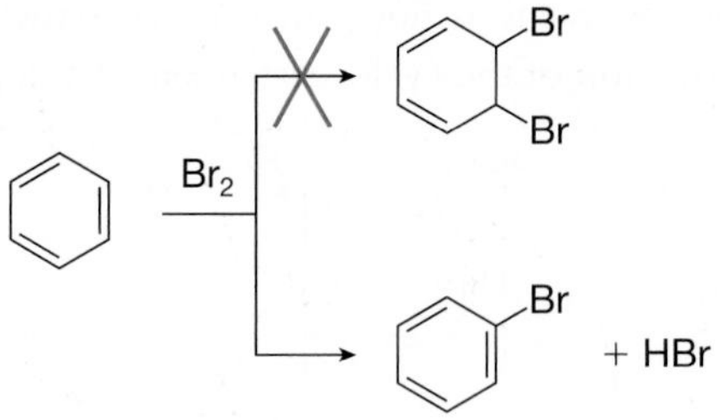

Synthesis Using Aromatic Materials

ELECTROPHILIC AROMATIC SUBSTITUTION AND DIRECTED *ORTHO* METALATION

10

CHAPTER OUTLINE

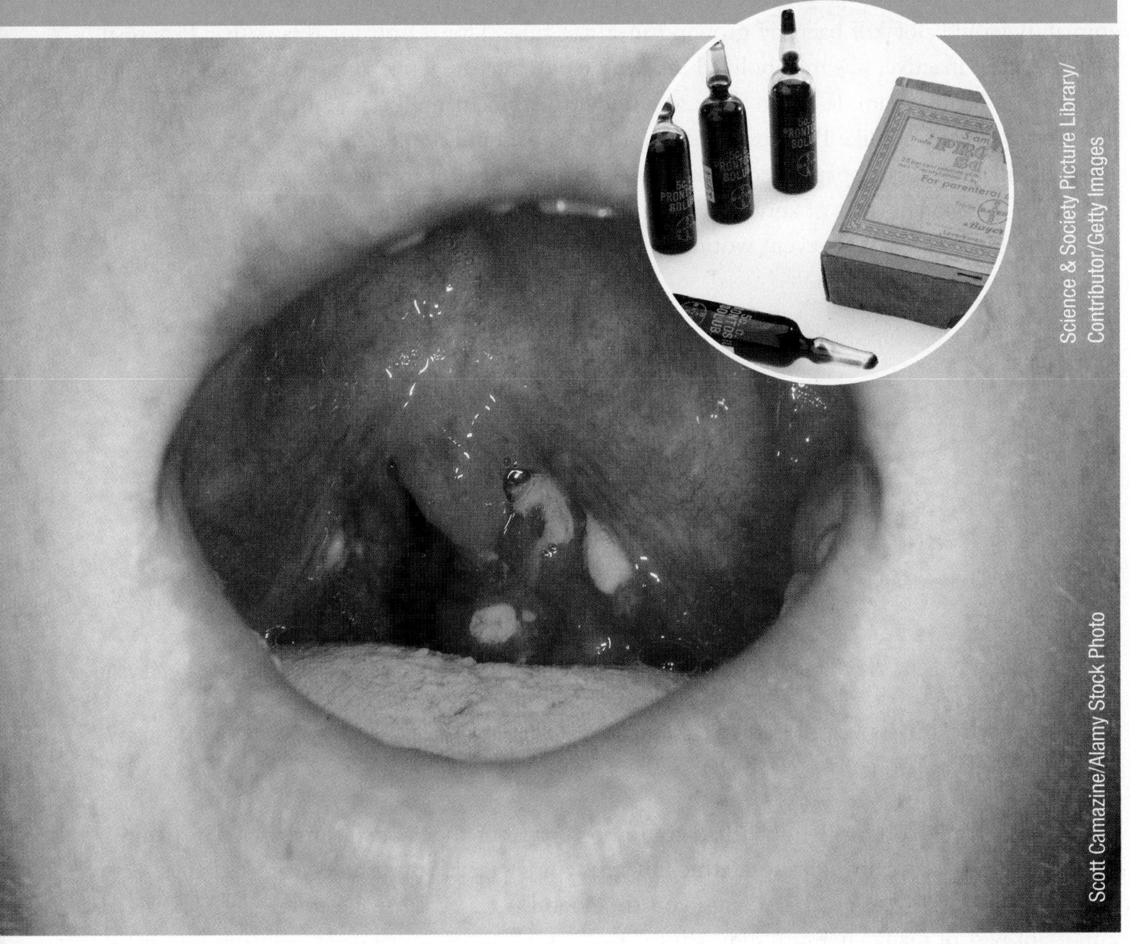

What do *Streptococcus* bacteria, the bacteria that causes *strep throat*, and a fabric dye have in common?

10.1 Why It Matters

In 1932, the first general antibiotic was discovered as a result of experiments done with Prontosil, a red dye developed by the German company I.G. Farben. Gerhard Domagk found that injecting Prontosil into mice protected them against bacterial infection. Later that year, Domagk's daughter became infected with *streptococcal* bacteria. She did not respond to other treatments and was near death when Domagk injected her with some Prontosil that he brought home from the laboratory. She made a full recovery as a result of the drug, and doctors began using Prontosil to treat other infections.

Prontosil → Prontalbin (sulfanilamide)

Later, scientists discovered that Prontosil was effective only when administered to an infected animal. It would not kill bacteria growing in a test tube. The reason for this is that Prontosil, which is itself inactive, was metabolized in living tissue into an active antibiotic called sulfanilamide. In 1935, sulfanilamide began to be sold under the trade name Prontalbin. It was a commercial success because, unlike Prontosil, it did not turn a patient's skin red!

Sulfanilamide is the parent compound of a larger class of aromatic drugs known as sulfa drugs, the first general class of antibiotics. The drug was issued to soldiers as part of their medical rations in World War II to prevent wound infections (Figure 10.1).

sulfanilamide

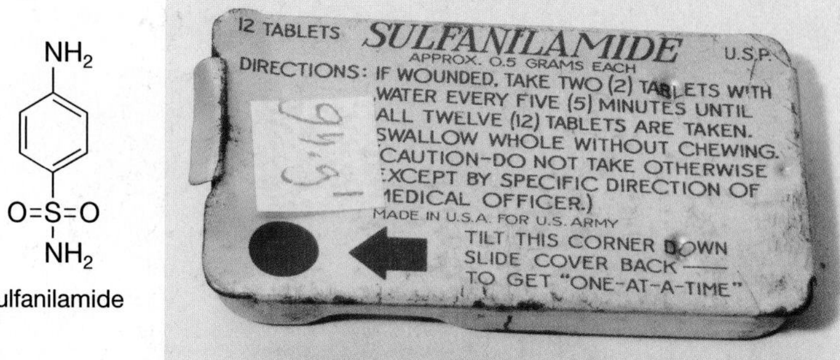

FIGURE 10.1 Early World War II medical rations for American soldiers included sulfanilamide to prevent wound infections.

SOURCE: KingaNBM. This file is licensed under the Creative Commons Attribution-Share Alike 4.0 International license: https://creativecommons.org/licenses/by-sa/4.0/deed.en

The use of sulfa drugs is much less common today, but they are still important in treating urinary infections and for fighting methicillin-resistant bacteria. Substituted aromatic rings form the core structure of all of these drugs, and the chemistry used to make them has become an important part of many chemical industries.

This chapter focuses on reactions that add electrophiles to aromatic rings, including halogenation, nitration, and sulfonation of aromatic rings. The chapter describes how aromatic substituents function as directing groups—*ortho*/*para* directing or *meta* directing—in activating or deactivating the ring toward electrophilic aromatic substitution, and also the use of directed *ortho* metalation as an alternative to electrophilic aromatic substitution.

Sulfanilamide Analog Sulfa Drugs

sulfafurazole   sulfalene   sulfadimethoxine

10.2 π Bonds Acting as Nucleophiles

As described in Chapter 8, the π bond of an alkene can act as a nucleophile in the presence of an electrophile (E^+), thereby generating a carbocation intermediate that further reacts to yield a final product.

For example, in the bromination of an alkene, the π bond attacks bromine to form a bromonium ion intermediate, which then leads to the formation of an addition product.

However, when benzene is mixed with bromine, no reaction happens. Although benzene contains π electrons, the stability imparted by the aromatic delocalization greatly reduces its reactivity toward typical electrophiles. Specifically, the activation energy for the formation of a carbocation intermediate is simply too high because the reaction disrupts the aromaticity.

Nevertheless, benzene can be made to react with particularly reactive electrophiles, such as bromine in the presence of a catalyst ($FeBr_3$). The catalyst increases the electrophilicity of Br_2 such that a reaction becomes possible. The overall reaction is also different: the product is the result of a substitution reaction rather than an addition reaction because the latter regenerates the aromaticity of the ring. This reaction, where hydrogen on an aromatic ring is replaced by an electrophile, is known as an **electrophilic aromatic substitution**: S_EAr is short for **S**ubstitution **E**lectrophilic **Ar**omatic.

Electrophilic aromatic substitution is a reaction where hydrogen on an aromatic ring is replaced by an electrophile.

10.3 Electrophilic Aromatic Substitution

Electrophilic aromatic substitution (S_EAr) reactions proceed by a general two-step mechanism: (1) addition of an electrophile (Chapter 8), and (2) elimination (explained in detail in Chapter 12). The combination of addition and elimination steps leads to a net substitution reaction.

first step: a pair of π electrons from the ring forms a bond to the electrophile
second step: base removes a hydrogen and restores the ring aromaticity
E
H
E
:base
E
aromatic ring
arenium ion
aromatic ring
non-aromatic

An **arenium ion** is a 1,3-cyclohexadienyl cation, the common intermediate in electrophilic aromatic substitution reactions.

In the first step of the mechanism, a pair of π electrons on the benzene ring forms a bond to the electrophile. The resulting carbocation intermediate is called an **arenium ion**, which is stabilized by conjugation. It is, however, no longer aromatic and very unstable compared to benzene.

arenium ion is stabilized by conjugation but is not aromatic

The second step of an aromatic substitution reaction involves the removal of a hydrogen atom by a base to restore the ring aromaticity, resulting in an overall substitution reaction. The molecule regains much stability by restoring the aromaticity; this, rather than addition of a nucleophile to the arenium ion, is the main driving force behind the formation of the substitution product.

Nu
Nu
addition product is not aromatic
substitution product is aromatic
base
nucleophile is aromatic
arenium ion is not aromatic and is not very stable

ORGANIC CHEMWARE
10.1 Electrophilic aromatic substitution: General mechanism

The general mechanism for all electrophilic aromatic substitution reactions is the same: attack of the electrophile by benzene followed by loss of a proton to restore aromaticity.

From the energetic standpoint, the first step of the reaction (i.e., the formation of the arenium ion) has a high activation energy barrier ($\Delta G^{\ddagger}_1$) because ring aromaticity is disrupted, and this step is highly endergonic. By comparison, the second step has a low activation energy barrier ($\Delta G^{\ddagger}_2$) because it restores the ring aromaticity, and is highly exergonic. Consequently, the first step limits the overall rate of the reaction; it is the rate-determining step of the electrophilic aromatic substitution.

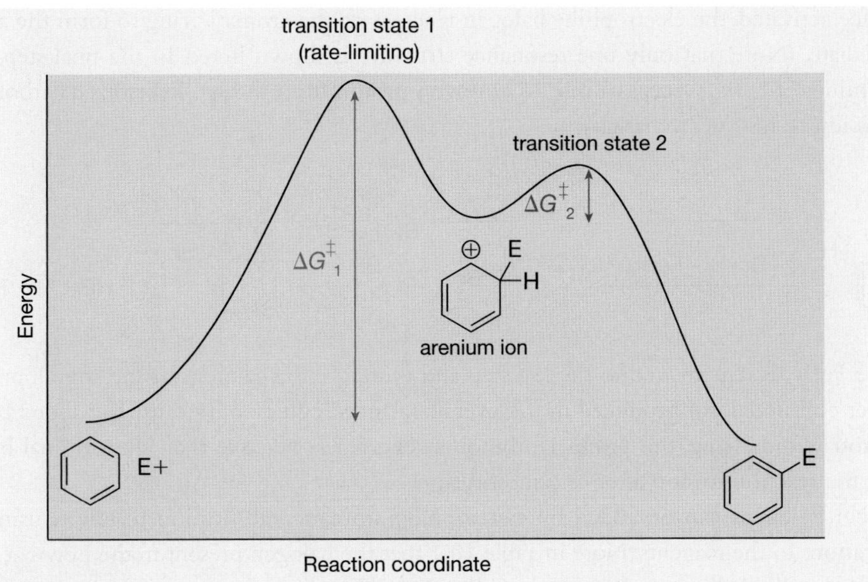

10.4 Types of Electrophiles Used in Electrophilic Aromatic Substitution

Although many electrophilic aromatic substitution reactions are known, they are mechanistically identical and tend to differ only in the way that a strong electrophile is generated. In this way, a variety of functional groups can be substituted onto aromatic rings, including halogens, nitro groups, sulfonic acids, alkyl, and acyl groups.

10.4.1 Halogenation

Aromatic rings can be halogenated with bromine, chlorine, or iodine using a Lewis acid catalyst to activate the halogen for reaction. (F_2 is reactive enough that no catalyst is needed, though this reaction is not a practical lab procedure.) The overall process is as follows:

Br_2, Cl_2, I_2

Ar–H $\xrightarrow[\text{Lewis acid catalyst}]{X_2}$ Ar–X + HX

To form the active electrophile in the halogenation reaction, the X_2 must first form a Lewis acid–base complex where a lone pair from X_2 is donated into an empty orbital of the Lewis acid. This renders one of the halogen atoms very electrophilic and reactive toward aromatic substitution. For example, Br_2 is activated by $FeBr_3$ through the donation of electrons from one bromine to form a strong leaving group.

very electrophilic Br

$Br{-}Br: + FeBr_3 \longrightarrow Br{-}\overset{\oplus}{Br}{-}\overset{\ominus}{Fe}Br_3$

leaving group

lone pair on Br forms a coordination bond with an empty orbital of Fe

Once activated, the electrophilic halogen reacts with the aromatic ring to form the arenium intermediate. (Note that only one resonance structure is shown here.) In the final step, one of the bromines of $FeBr_4^-$ acts as a base to remove a proton from the sp^3-hybridized carbon of the arenium ion to restore aromaticity.

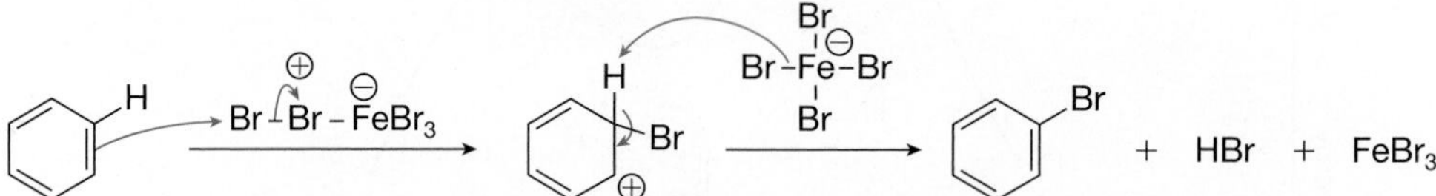

STUDENT TIP

Lewis acid–base theory defines acids as electron pair acceptors, instead of hydrogen ion donors. All acids can be defined this way; however, the term *Lewis acid* commonly refers to acids that are electron pair acceptors but are *not* hydrogen ion donors. Metals such as iron, aluminum, and boron are common Lewis acids.

The $FeBr_3$ is regenerated in the last step and therefore is a catalyst for the overall process; it is neither consumed nor produced in the overall balanced equation. Because the added bromine is electron withdrawing, the product bromobenzene is less reactive than the original benzene, causing the reaction to stop after one substitution.

Other halogens can be added by electrophilic aromatic substitution reactions using small modifications to the reagents. Note in Table 10.1 that the halogen present in the Lewis acid most often matches the halogen being substituted in the ring.

ORGANIC CHEMWARE

10.2 Electrophilic aromatic substitution: Halogenation

TABLE 10.1 Reagents and Catalysts Used for S_EAr Halogenation Reactions

Halogen Added	Electrophile	Lewis Acid
F	F_2	None
Cl	Cl_2	$FeCl_3$
Br	Br_2	$FeBr_3$
I	I_2	$CuCl_2$

CHECKPOINT 10.1

You should now be able to draw a mechanism for the halogenation of aromatic rings and draw the expected products.

SOLVED PROBLEM

Draw a mechanism for the following reaction and show the final product.

$$\xrightarrow[FeCl_3]{Cl_2}$$

STEP 1: Determine the roles of the reactants and the type of reaction taking place.

The starting material contains an aromatic ring that is weakly nucleophilic in the presence of highly reactive electrophiles. The molecular chlorine is not electrophilic enough to react with the benzene ring by itself, but in the presence of the Lewis acid $FeCl_3$, the chlorine is activated, making it capable of reacting in an electrophilic aromatic substitution.

STEP 2: Draw a mechanism for the formation of the activated electrophile. A lone pair on one of the chlorine atoms donates its electrons to an empty orbital on the $FeCl_3$ catalyst.

$$Cl{-}\ddot{\underset{..}{Cl}}{:} + FeCl_3 \longrightarrow Cl{-}\overset{\oplus}{Cl}{-}\overset{\ominus}{Fe}Cl_3$$

STEP 3: Draw a mechanism for the formation of the arenium ion. The electrons from one of the π bonds in the benzene ring shares its electrons with the electrophilic chlorine atom, forming a new C–Cl σ bond while simultaneously breaking the C–Cl σ bond in the electrophile. Note that although there are four positions on the ring where the chlorine could be added, the molecule is symmetrical and all four positions are in fact equivalent.

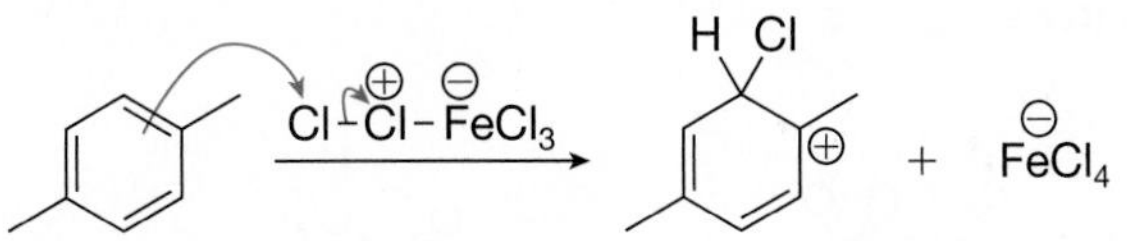

STEP 4: Draw a mechanism for the final step of the reaction. A chloride ion from $FeCl_4^-$ removes the highly acidic proton from the sp^3-hybridized carbon of the arenium ion. The electrons from the C–H σ bond flow toward the positive charge, re-forming the C=C π bond and regenerating the aromaticity of the benzene ring.

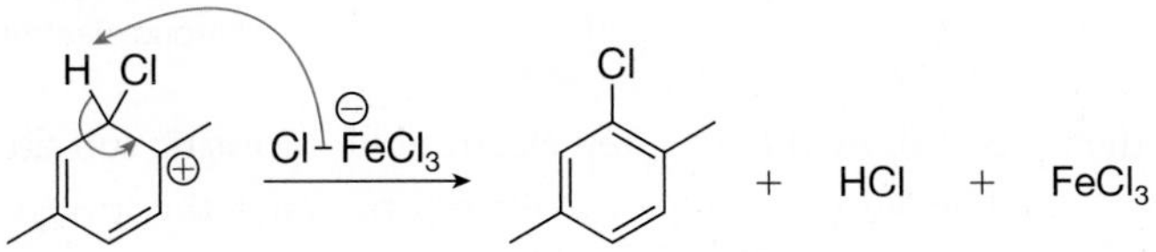

PRACTICE PROBLEM

10.1 a) Draw a mechanism for each of the following transformations.

i) 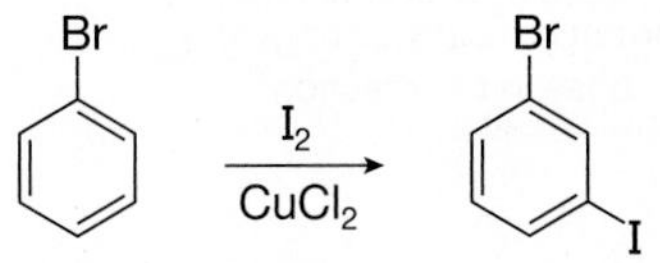 I_2, $CuCl_2$

ii) 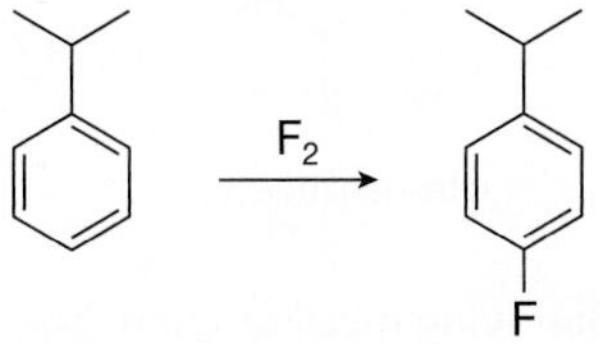 F_2

b) Predict the products of the following reactions:

i) 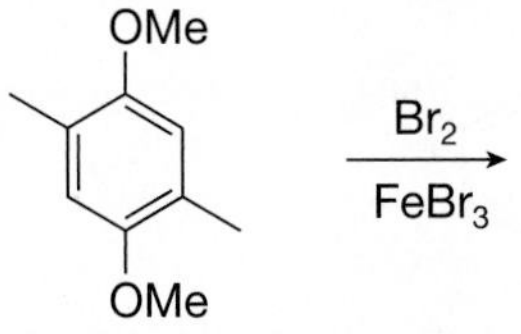 Br_2, $FeBr_3$

ii) 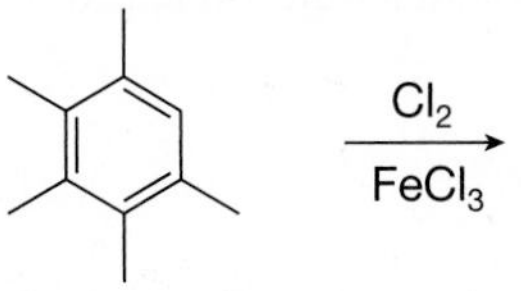 Cl_2, $FeCl_3$

INTEGRATE THE SKILL

10.2 a) Suggest a reason why bromination of benzene does not work in the presence of water.

b) Why is a Lewis acid catalyst required for chlorination but not for fluorination?

10.4.2 Nitration

In the **nitration** reaction, a hydrogen atom is replaced with a nitro group, NO_2, via electrophilic aromatic substitution. The electrophile in this reaction is the nitronium ion (NO_2^+).

Nitration is the substitution of a NO_2 group in place of a hydrogen atom on an aromatic ring.

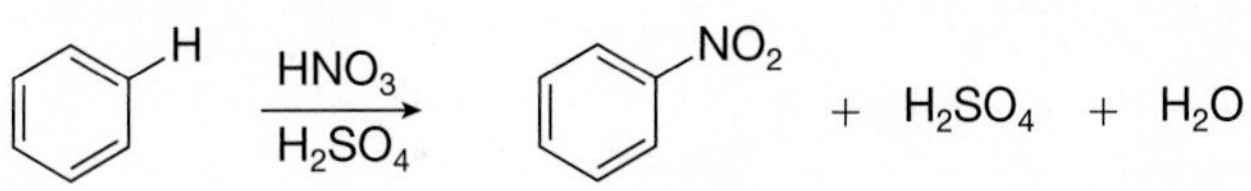

HNO_3, H_2SO_4 → $+ H_2SO_4 + H_2O$

The nitronium ion is generated by dehydration of nitric acid (HNO_3) with sulfuric acid (H_2SO_4). Sulfuric acid is a stronger acid (pK_a = −3) than nitric acid (pK_a = −1.4) and can readily protonate the latter in an acid–base reaction. The resulting HNO_3H^+ then decomposes to form NO_2^+ and water.

nitric acid (HNO_3) + sulfuric acid (H_2SO_4) ⇌ ... − H_2O → O=N=O nitronium ion (strong electrophile)

The nitration then proceeds by the two-step electrophilic aromatic substitution mechanism, as shown in Section 10.3. The last step can be carried out by water, the strongest base present in this highly acidic reaction medium, and results in the formation of nitrobenzene.

nitronium ion (electrophile)

water generated by NO_2^+ formation is the strongest base in the reaction

nitrobenzene + H_3O^+

ORGANIC CHEMWARE
10.3 Electrophilic aromatic substitution: Nitration

Nitro groups can be reduced to amine groups using a dissolving metal reaction. Iron or tin are the most commonly used together with a strong acid, such as HCl (see Chapter 19 for related reactions).

Fe or Sn / HCl

This second step is an important part of a key method of attaching NH_2 groups to aromatic rings.

10.4.3 Sulfonation

Sulfonation is the substitution of an SO_3H group in place of a hydrogen atom on an aromatic ring.

Arenesulfonic acids can be formed by heating aromatic compounds with concentrated H_2SO_4 or an anhydrous mixture of H_2SO_4 and SO_3, called fuming sulfuric acid. In this **sulfonation** reaction, a hydrogen on the aromatic ring is replaced by a sulfonic acid group, SO_3H.

benzene → (SO_3, H_2SO_4) benzenesulfonic acid + H_2SO_4

The active electrophile is likely SO_3H^+, which is produced either by dehydration of H_2SO_4 or by protonation of SO_3 by H_2SO_4.

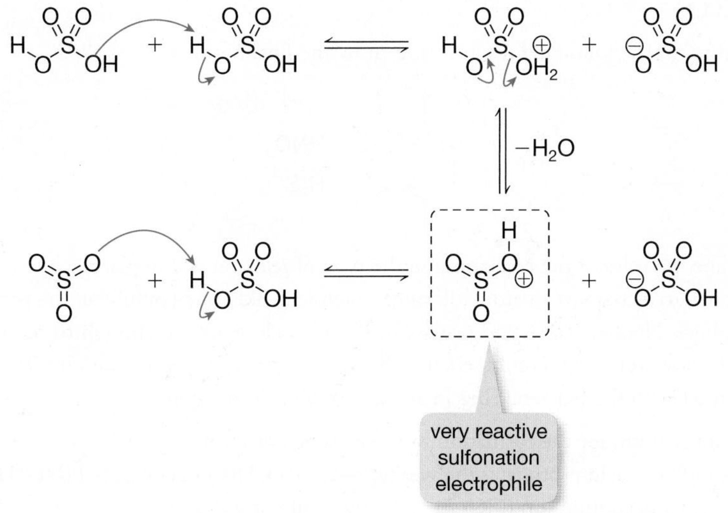

very reactive sulfonation electrophile

Once the reactive SO_3H^+ electrophile is formed, the aromatic substitution reaction proceeds: first by the addition of SO_3H^+, and then by the elimination of a proton adjacent to the carbocation (using the strongest base present; H_2O if sulfuric acid was used or HSO_4^- in the case of fuming sulfuring acid) to generate the arylsulfonic acid product.

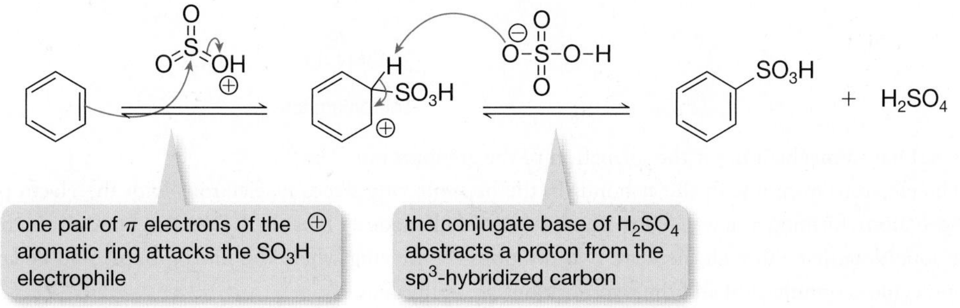

one pair of π electrons of the aromatic ring attacks the SO_3H electrophile

the conjugate base of H_2SO_4 abstracts a proton from the sp^3-hybridized carbon

ORGANIC CHEMWARE
10.4 Electrophilic aromatic substitution: Sulfonation

The sulfonation can be reversed using a strong acid, such as H_2SO_4, in water. The mechanism of this process is the exact reverse of the sulfonation reaction. First, the ring is protonated to form an arenium ion. Second, the ring loses SO_3H^+ to restore aromaticity. Water present in the mixture can react with SO_3H^+ to form H_2SO_4, removing SO_3H^+ from the reaction and driving the equilibrium to the right.

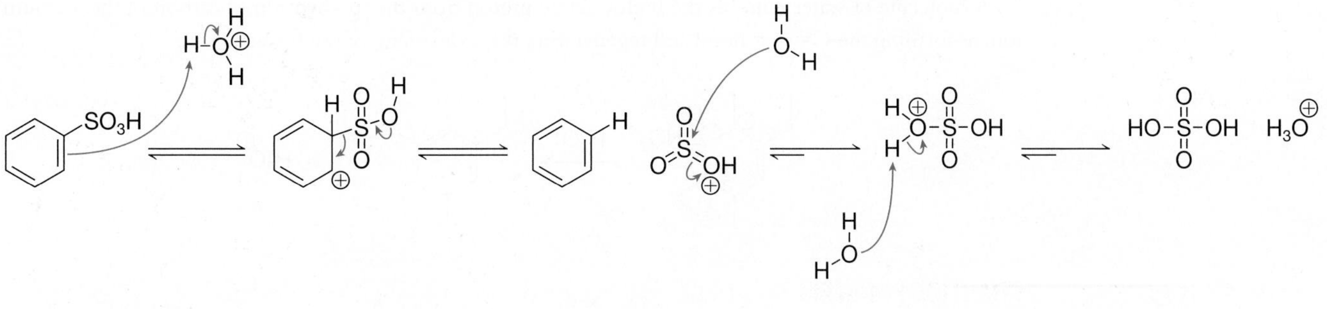

CHECKPOINT 10.2

You should now be able to draw a mechanism for the nitration and sulfonation of aromatic rings and draw the expected products. You should also be able to draw the products formed when nitrated aromatics are treated with a dissolving metal reduction.

SOLVED PROBLEM

Draw a mechanism for the following reaction and show the final product.

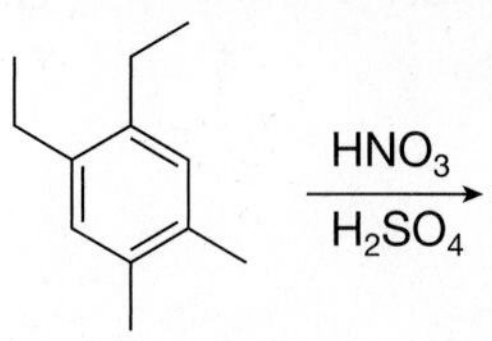

$\xrightarrow[H_2SO_4]{HNO_3}$

STEP 1: Determine the roles of the reactants and the type of reaction taking place.

The starting material contains an aromatic ring, which is weakly nucleophilic in the presence of highly reactive electrophiles. Neither nitric acid nor sulfuric acid is electrophilic enough to react with the benzene ring directly; however, when combined, they form highly reactive nitronium ions. These electrophilic nitronium ions react with the benzene ring in an electrophilic aromatic substitution.

STEP 2: Draw a mechanism for the formation of the activated electrophile.

A molecule of sulfuric acid protonates the weaker nitric acid. The intermediate HNO_3H^+ formed then decomposes to form a nitronium ion (NO_2^+) and a molecule of water.

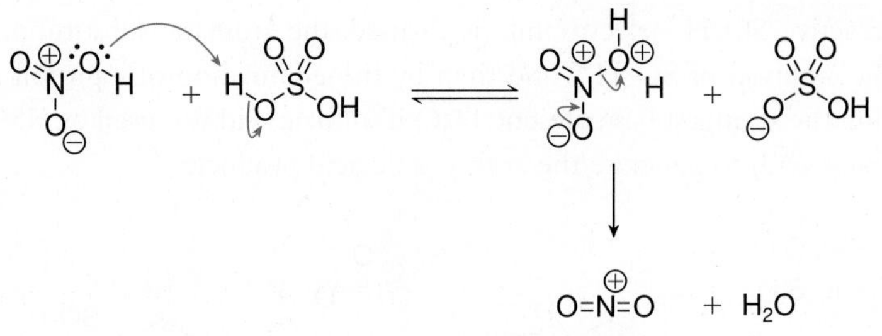

$O=\overset{\oplus}{N}=O$ + H_2O

nitronium ion

STEP 3: Draw a mechanism for the formation of the arenium ion.

The electrons from one of the π bonds in the benzene ring shares its electrons with the electrophilic nitrogen atom, forming a new C–N σ bond, while simultaneously breaking a N–O π bond in the nitronium ion. Note that although there are two positions on the ring where the nitro group could be added, the molecule is symmetrical and the two positions are equivalent.

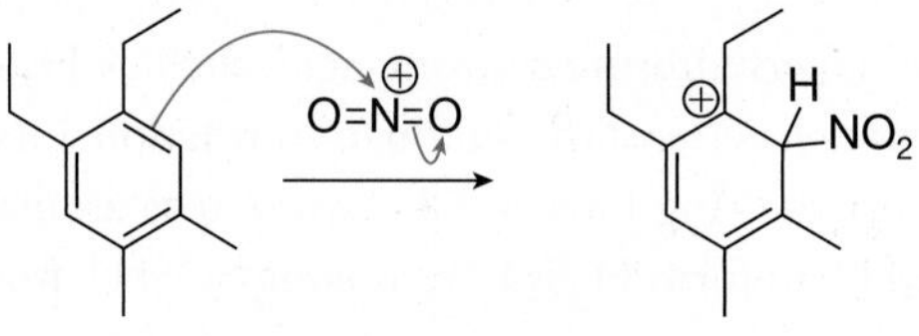

STEP 4: Draw a mechanism for the final step of the reaction.

A molecule of water removes the highly acidic proton from the sp^3-hybridized carbon of the arenium ion, re-forming the C=C π bond and regenerating the aromaticity of the benzene ring.

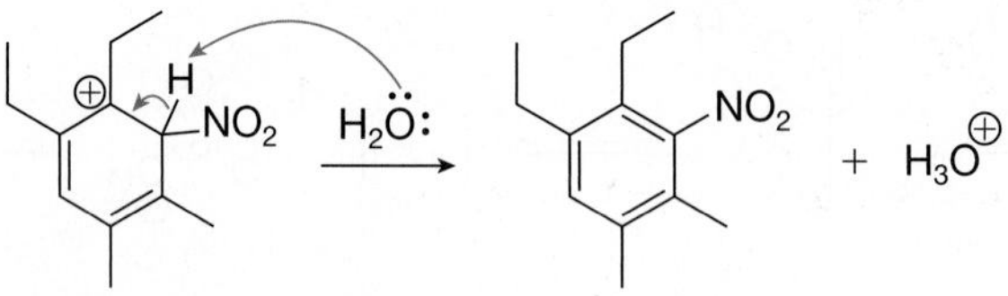

$+ H_3O^{\oplus}$

PRACTICE PROBLEM

10.3 a) Draw a mechanism for each of the following transformations.

i)

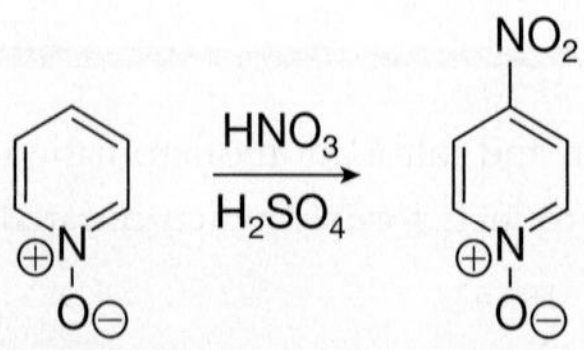

$\xrightarrow[H_2SO_4]{HNO_3}$

ii)

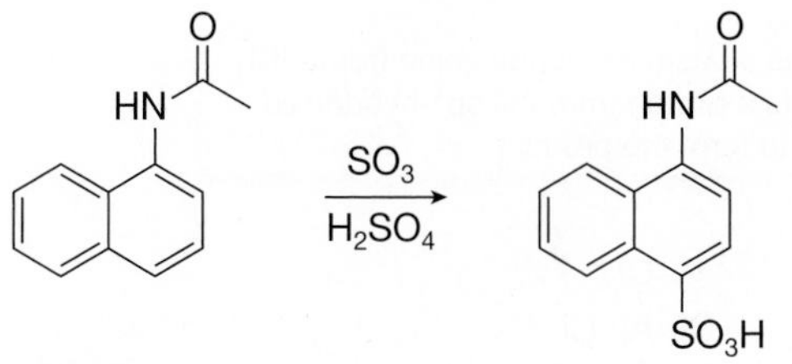

$$\xrightarrow[H_2SO_4]{SO_3}$$

b) Predict the products of the following reactions.

i)

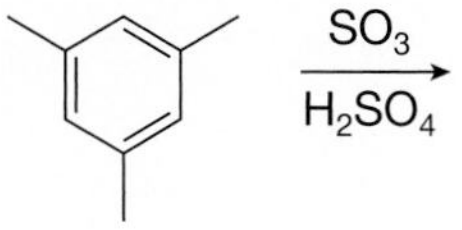

$$\xrightarrow[H_2SO_4]{SO_3}$$

ii)

$$\xrightarrow[2)\ Fe,\ HCl]{1)\ HNO_3,\ H_2SO_4}$$

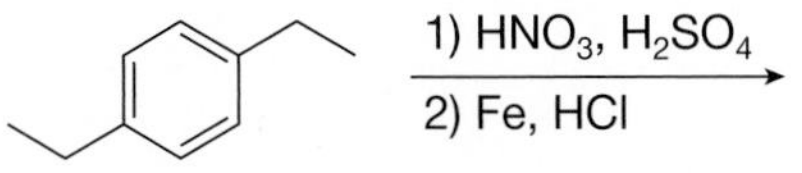

INTEGRATE THE SKILL

10.4 Draw reaction coordinate diagrams for the sulfonation of benzene and the desulfonation of benzenesulfonic acid. How are these energy diagrams related?

10.4.4 Friedel–Crafts alkylation

Friedel–Crafts alkylation is a reaction that substitutes an alkyl group in place of a hydrogen atom on an aromatic ring using an alkyl halide and a Lewis acid.

The **Friedel–Crafts alkylation**, a type of electrophilic aromatic substitution, is a procedure for adding alkyl groups to aromatic rings using an alkyl halide and a Lewis acid catalyst.

$$\xrightarrow[AlCl_3]{RCl} \quad + \quad HCl$$

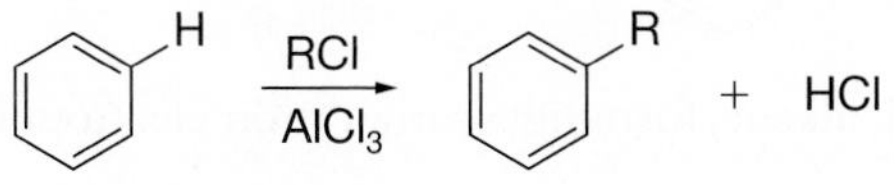

The Lewis acid converts the alkyl halide to a carbocation, a very strong electrophile that can then react with an aromatic ring. For example, *t*-butyl chloride and aluminum trichloride ($AlCl_3$) combine to form a reactive *t*-butyl cation, the active electrophile in a Friedel–Crafts alkylation. The vacant orbital on the aluminum of $AlCl_3$ accepts a lone pair of electrons from the chlorine atom of the alkyl chloride, just as $FeCl_3$ did in halogenation reactions (Section 10.4.1). This forms an active complex with an excellent leaving group, which forms a carbocation and aluminum tetrachloride.

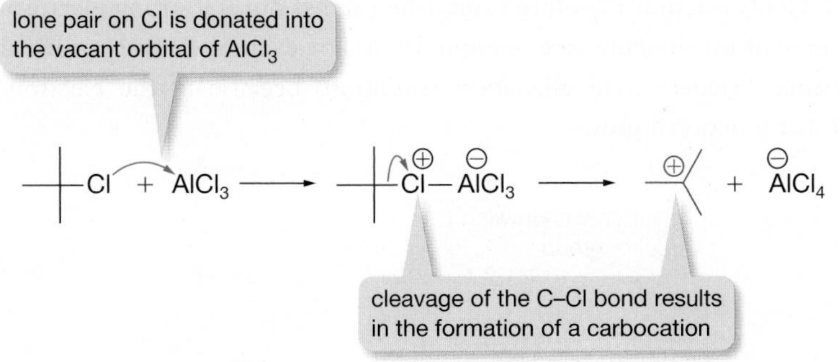

lone pair on Cl is donated into the vacant orbital of $AlCl_3$

$$-Cl + AlCl_3 \longrightarrow -\overset{\oplus}{Cl}-\overset{\ominus}{AlCl_3} \longrightarrow \oplus + \overset{\ominus}{AlCl_4}$$

cleavage of the C–Cl bond results in the formation of a carbocation

The electrophilic carbocation is then attacked by the aromatic ring following the S_EAr mechanism.

acting as a base, a chloride anion from $AlCl_4^{\ominus}$ abstracts a proton from the sp^3-hybridized carbon to form the product

+ HCl + $AlCl_3$

arenium ion intermediate

STUDENT TIP
Friedel–Crafts reactions are among the few ways of forming a C–C bond with an aromatic ring. The other method is to use a phenyl Grignard reagent with a carbonyl compound (Section 7.5.1).

ORGANIC CHEMWARE
10.5 Electrophilic aromatic substitution: Friedel–Crafts alkylation

Once again $AlCl_3$ is a catalyst and is not part of the balanced equation for the reaction. $FeCl_3$, the more common catalyst for halogenations, can also be used as a catalyst in Friedel–Crafts reactions. The alkyl halides are usually tertiary ($R_3C–Cl$) or secondary ($R_2CH–Cl$) because of the stability of their corresponding carbocations. As will be described in Section 10.4.5, the use of primary alkyl halides often leads to rearranged side-products.

An alternative (non-Friedel–Crafts) way to alkylate a benzene ring involves the use of an alkene and a mineral acid catalyst such as HF or H_2SO_4. (Note the counter ion must be non-nucleophilic so it will not react with the carbocation intermediate.)

HF

The acid protonates the alkene, forming a carbocation electrophile that then reacts with the aromatic reactant.

alkene + H–F ⟶ carbocation + $F^{\ominus}$

10.4.5 Limitations of the Friedel–Crafts alkylation

Three main limitations of the Friedel–Crafts reaction must be considered:

1. Carbocations are not reactive enough to couple with weakly nucleophilic aromatic rings. A Friedel–Crafts reaction therefore cannot be carried out if a strong electron-withdrawing group is present on the ring (see Section 10.6). For example, nitrobenzene does not react under normal Friedel–Crafts alkylation conditions because of the electron-withdrawing nature of that functional group.

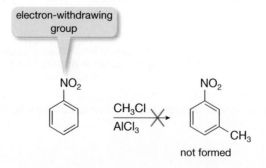

electron-withdrawing group

NO_2 —(CH_3Cl, $AlCl_3$)✕→ NO_2, CH_3

not formed

Likewise, anilines (aminobenzene derivatives) do not react either, although they bear an electron-donating group that should, in principle, facilitate the Friedel–Crafts alkylation. Under the reaction conditions, the amino group of the aniline (Lewis base) can form a bond to the Lewis acid. The resulting positive charge on nitrogen converts it to an electron-withdrawing group, which deactivates the molecule toward Friedel–Crafts alkylation.

NH_2 — CH_3Cl, $AlCl_3$ ⟶ (crossed) NH_2, CH_3

not formed

{ NH_2 + $AlCl_3$ ⟶ H, $AlCl_3$, N, H }

electron-withdrawing group

2. Because alkyl groups are electron donating, the product of a Friedel–Crafts alkylation is usually more nucleophilic, hence more reactive, than the starting reactant. For this reason, over-alkylation is common unless special measures are taken to ensure that the starting material is properly consumed—for example, using an excess of the aromatic reactant.

H — RCl, $AlCl_3$ ⟶ R + R, R + R, R

3. Carbocations can rearrange and lead to product mixtures. For example, the reaction of benzene with 1-chloropropane and $AlCl_3$ gives a mixture of products, with the desired *n*-propyl product forming as a minor product.

H — Cl, $AlCl_3$ ⟶ + + HCl

major minor

This is due to the rearrangement of the carbocation intermediate prior to the Friedel–Crafts alkylation step. The rearrangement can happen either by migration of an alkyl group ("alkyl shift") or a hydrogen ("hydride shift") next to the carbocation. During such rearrangements, the two bond electrons migrate along with the rearranging group.

alkyl shift **hydrogen shift**

R migrates with the two bond electrons

H migrates with the two bond electrons

R ⇌ R H H ⇌ H H

secondary carbocation tertiary carbocation primary carbocation secondary carbocation

The overall result is the formation of a more stable carbocation, driven by thermodynamic equilibration. The stability of a primary carbocation is *so* low that it is unlikely that the primary carbocation actually forms.

With an electrophile such as 1-chloropropane, the typical Friedel–Crafts alkylation mechanism cannot form the expected 1° carbocation, which is too unstable. That equilibrates rapidly to the more stable 2° carbocation via a hydride shift. Both carbocations can then react with the aromatic reactant to produce two different alkylation products. The reaction proceeds via two competing mechanisms that result in a mixture of products.

In the first, a hydride shift occurs at the same time as the leaving group departs, and a secondary carbocation forms.

Carbocation pathway

In the second mechanism, the leaving group is displaced by the attacking benzene in a concerted mechanism.

Concerted displacement pathway

In either case, a primary electrophile, which does not form primary carbocations, does not reliably add a primary alkyl chain to an aromatic ring.

Because of the potential for poly-alkylation and carbocation rearrangements, Friedel–Crafts alkylations cannot be reliably used to add primary alkyl groups to aromatic rings. These problems can be solved by a Friedel–Crafts variation that *acylates* rather than alkylates the ring.

STUDENT TIP

There are three limitations to the Friedel–Crafts alkylation reaction:

1. They do not work with electron-poor rings.
2. Addition of electron-donating alkyl groups increases reactivity and can lead to poly-alkylation.
3. Rearrangements prevent substitution using primary alkyl groups.

CHECKPOINT 10.3

You should now be able to draw a mechanism for the Friedel–Crafts alkylation of aromatic rings and draw the expected products.

SOLVED PROBLEM

Predict the major product of the following reaction and draw a mechanism to show its formation.

STEP 1: Determine the roles of the reactants and the type of reaction taking place.

The starting material contains an aromatic ring (a weak nucleophile) and an electrophilic alkyl bromide. The alkyl bromide is not reactive enough to react with the benzene ring. However, the $AlCl_3$ Lewis acid can activate the alkyl bromide to create a highly electrophilic carbocation. Under these conditions, we can expect a Friedel–Crafts alkylation to take place, a type of electrophilic aromatic substitution.

STEP 2: Draw a curved arrow mechanism for the formation of the carbocation electrophile.

A lone pair from bromine shares its electrons with an empty orbital in the $AlCl_3$ Lewis acid. The resulting complex expels a leaving group, leaving behind a secondary carbocation.

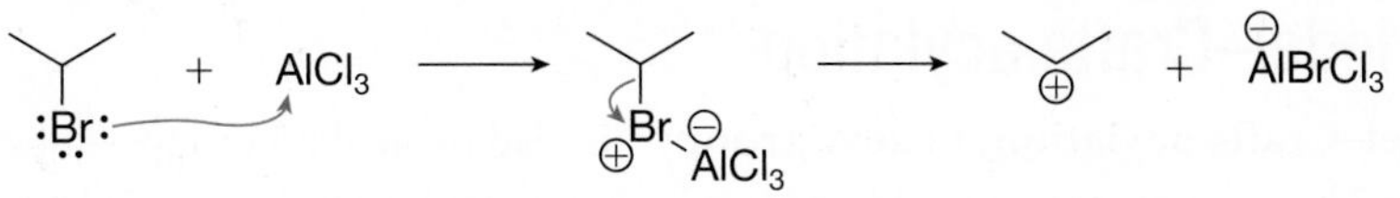

:Br: + $AlCl_3$ → $Br^{\oplus}$–$AlCl_3^{\ominus}$ → ⊕ + $AlBrCl_3^{\ominus}$

STEP 3: Check to see whether the carbocation can rearrange to a more stable structure. If it can, draw a mechanism for the rearranged structure.

This is the most stable carbocation available for this structure; no rearrangements are possible.

STEP 4: Draw a mechanism for the electrophilic aromatic substitution reaction. Use the steps in Checkpoints 10.1 and 10.2 to guide you if needed.

The electrons from one of the π bonds in the benzene ring shares its electrons with the empty p orbital of the electrophilic carbocation, forming a new C–C σ bond. Note that although there are three positions on the ring where the carbocation could react, the molecule is symmetrical and the three positions are equivalent. A chloride ion from $AlCl_3Br^-$ removes the highly acidic proton from the sp^3-hybridized carbon of the arenium ion, re-forming the C=C π bond and regenerating the aromaticity of the benzene ring.

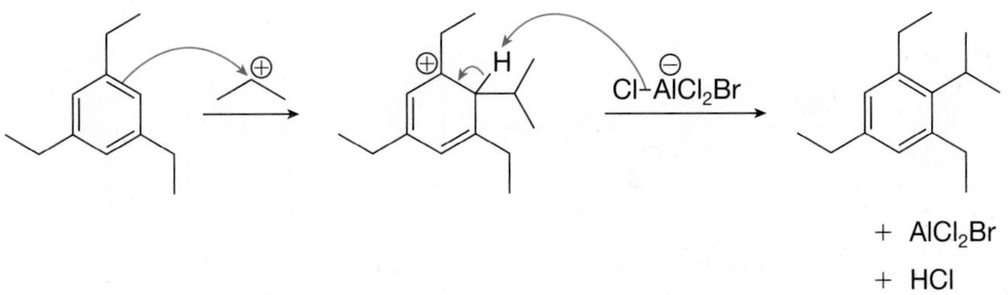

H Cl–$AlCl_2Br^{\ominus}$ →

+ $AlCl_2Br$

+ HCl

PRACTICE PROBLEM

10.5 Draw the expected product when benzene is reacted with the following alkyl halides in the presence of $AlCl_3$ catalyst.

a) ethylchloride
b) (1-bromoethyl)benzene
c) 2-chloroheptane
d) (bromomethyl)benzene

INTEGRATE THE SKILL

10.6 a) The following Friedel–Crafts alkylations do not proceed as shown. Use a mechanism to show why not, and predict the actual product obtained.

i) + Cl $\xrightarrow{AlCl_3}$

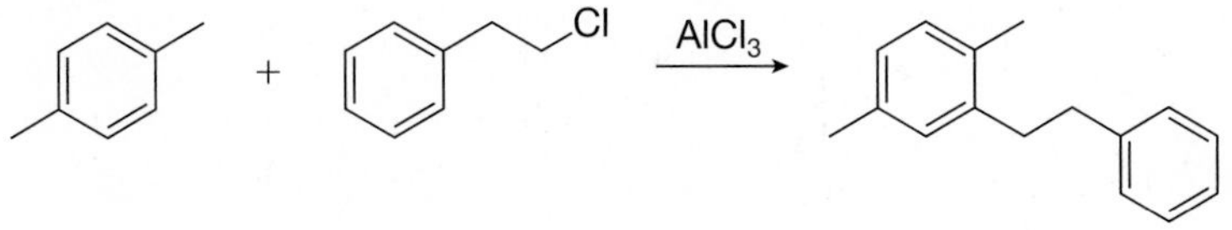

ii) + Cl $\xrightarrow{AlCl_3}$

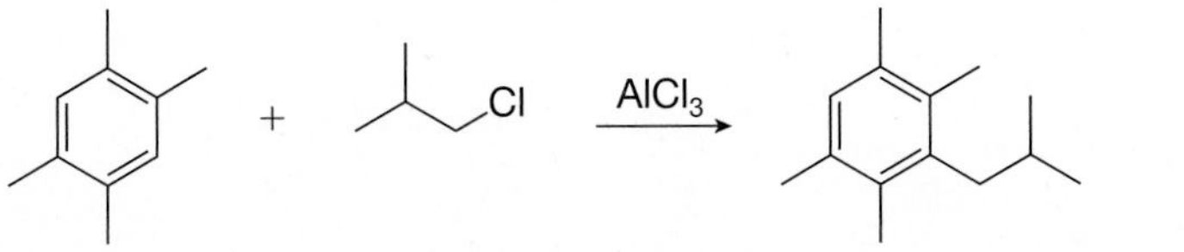

b) Suggest a reasonable mechanism for the following reaction.

HCl, $AlCl_3$

10.4.6 Friedel–Crafts acylation

Friedel–Crafts acylation, a variation of the Friedel–Crafts alkylation, substitutes an acyl group in the place of a hydrogen atom on an aromatic ring.

An **acyl group** consists of a carbonyl connected to an alkyl group.

In the **Friedel–Crafts acylation**, an **acyl group**—instead of an alkyl group—substitutes on an aromatic ring. The reagents are an acid halide and a Lewis acid catalyst such as $AlCl_3$.

$AlCl_3$ + HCl

The first steps of the process form an active electrophile called an *acylium ion*. First, a lone pair of electrons on the chlorine of the acid chloride is donated into the vacant orbital on the Lewis acid to improve the leaving group. This intermediate then decomposes to form a stabilized acylium ion that has alternate resonance structure. This is the actual electrophile that reacts with the aromatic ring. The process then follows the same mechanism as the Friedel–Crafts alkylation.

$AlCl_3$ + $AlCl_4^-$

acylium ion

loss of the activated leaving group forms an acylium ion

major resonance form (complete octets)

Compared to its alkylation counterpart, Friedel–Crafts acylation is a much more reliable reaction for making carbon-carbon bonds with aromatic groups. Because carbonyl groups are electron withdrawing, the addition of one acyl group to the aromatic ring *lessens* its reactivity toward further acylation, which prevents the addition of a second or third acyl substituent. Likewise, the acylium ion is not subject to rearrangement because it is stabilized by conjugation.

The Friedel–Crafts acylation, when paired with a reduction step, can produce the overall equivalent of the alkylation reaction, without the side-reactions. Common reduction reactions such as the Wolff–Kischner (NH_2NH_2/KOH, Chapter 16) or Clemmenson (Zn/HCl, Chapter 19) reductions can convert the carbonyl group of the acylation product to an alkyl group.

ORGANIC CHEMWARE
10.6 Electrophilic aromatic substitution: Friedel–Crafts acylation

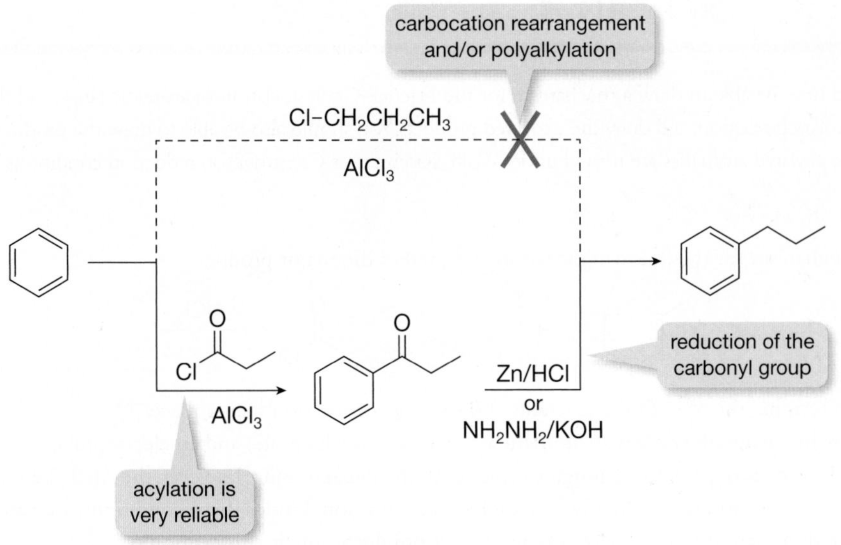

carbocation rearrangement and/or polyalkylation

$Cl-CH_2CH_2CH_3$

$AlCl_3$

reduction of the carbonyl group

Zn/HCl or NH_2NH_2/KOH

$AlCl_3$

acylation is very reliable

Friedel–Crafts acylations can also be carried out with an acid anhydride instead of an acid chloride.

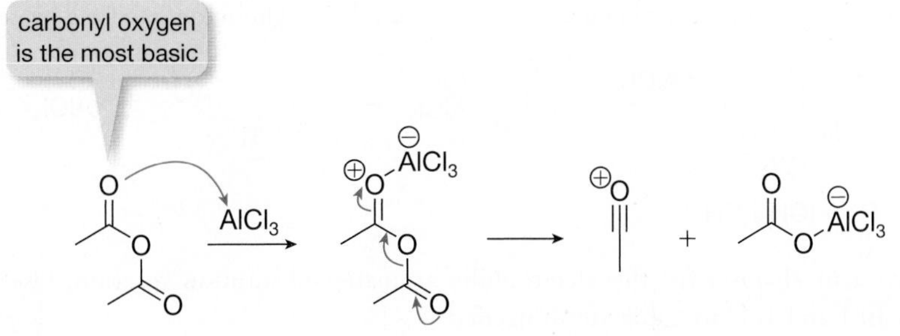

carbonyl oxygen is the most basic

$AlCl_3$

One limitation of the Friedel–Crafts acylation is that it cannot be used to form aldehydes because the required formyl chloride is not a stable compound.

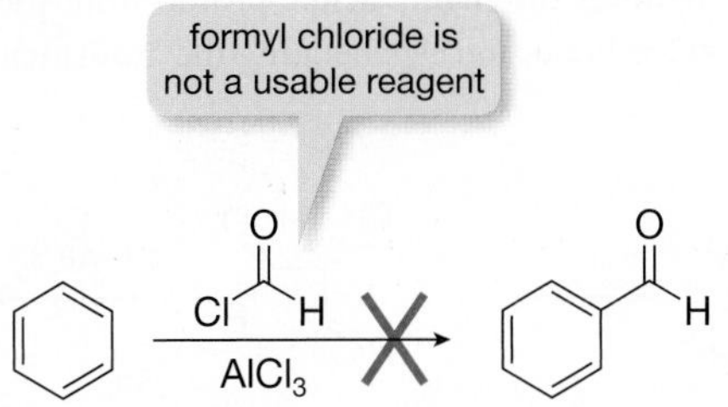

formyl chloride is not a usable reagent

$AlCl_3$

A variation on the Friedel–Crafts reaction called the Gatterman–Koch reaction uses carbon monoxide and mineral acid to make aromatic aldehydes.

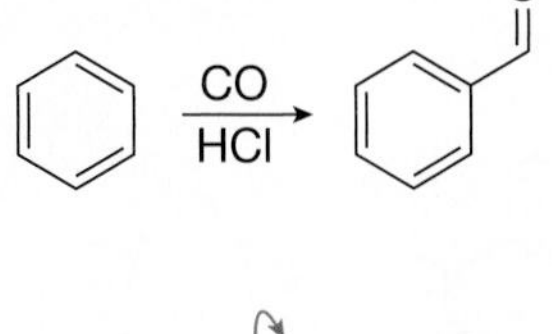

CO HCl

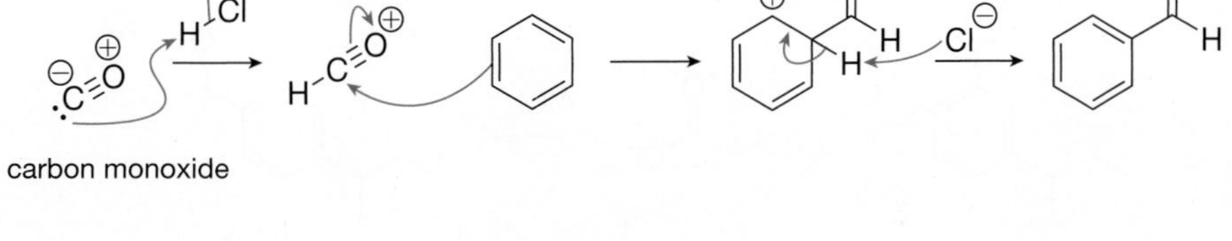

carbon monoxide

CHECKPOINT 10.4

You should now be able to draw a mechanism for the Friedel–Crafts acylation of aromatic rings, and the related Gatterman–Koch reaction, and draw the expected products. You should also be able to draw the products formed when these acylated aromatics are treated under Wolff–Kischner or Clemmenson reduction conditions.

SOLVED PROBLEM

Draw a mechanism for the following reaction and predict the major product.

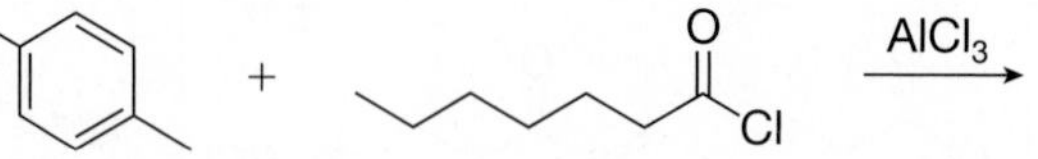

$\xrightarrow{AlCl_3}$

STEP 1: Determine the roles of the reactants and the type of reaction taking place.

The starting materials contain an aromatic ring (a weak nucleophile) and an electrophilic acid chloride. The acid chloride is not reactive enough to react with the benzene ring; however, the $AlCl_3$ Lewis acid can activate this reagent to create a highly electrophilic acylium ion. Under these conditions, we can expect a Friedel–Crafts acylation to take place, a type of electrophilic aromatic substitution.

STEP 2: Draw a mechanism for the formation of the acylium ion electrophile.

A lone pair from chlorine shares its electrons with an empty orbital on the $AlCl_3$ Lewis acid. The resulting complex then expels a leaving group, leaving behind an acylium ion stabilized by conjugation.

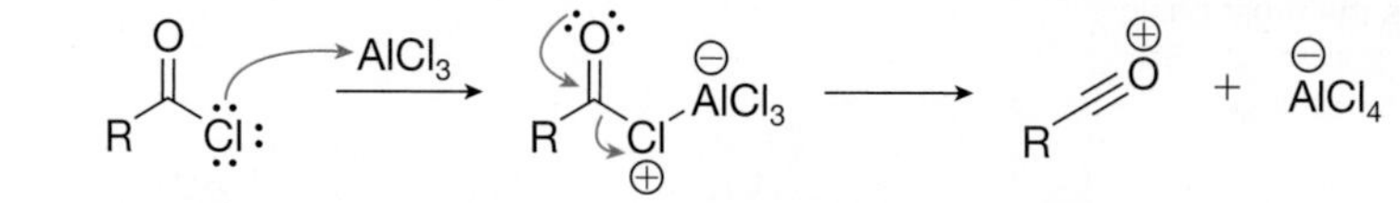

R=$-(CH_2)_5CH_3$

STEP 3: Draw a mechanism for the electrophilic aromatic substitution reaction. Use the steps in Checkpoints 10.1 and 10.2 to guide you if needed.

The electrons from one of the π bonds in the benzene ring are shared with the electron-deficient acylium ion, forming a new C–C σ bond. Note that although there are four positions on the ring where the acylium ion could react, the molecule is symmetrical, and all four positions are in fact equivalent. A chloride ion from $AlCl_4^-$ then removes the highly acidic proton from the sp^3-hybridized carbon of the arenium ion, re-forming the C=C π bond, and regenerating the aromaticity of the benzene ring.

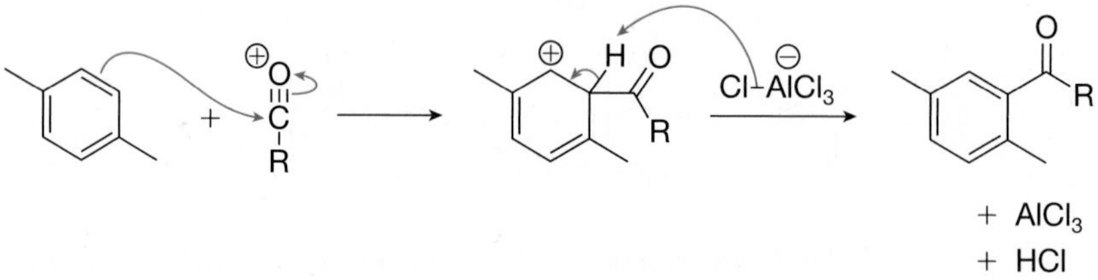

$+ AlCl_3$
$+ HCl$

PRACTICE PROBLEM

10.7 a) Draw a mechanism for each of the following transformations. Include all possible resonance structures for the arenium ion intermediates.

i) 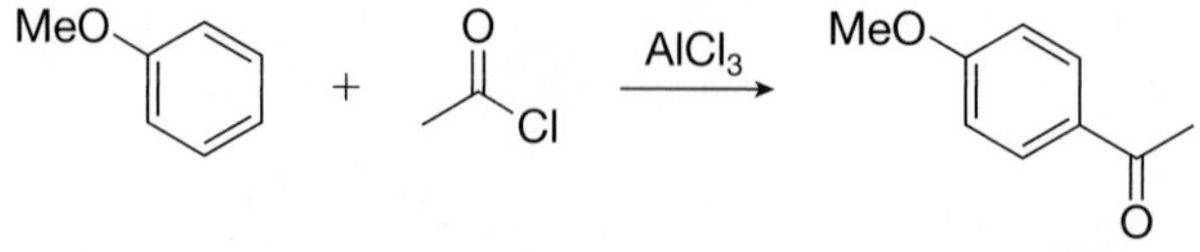 $\xrightarrow{AlCl_3}$

ii) 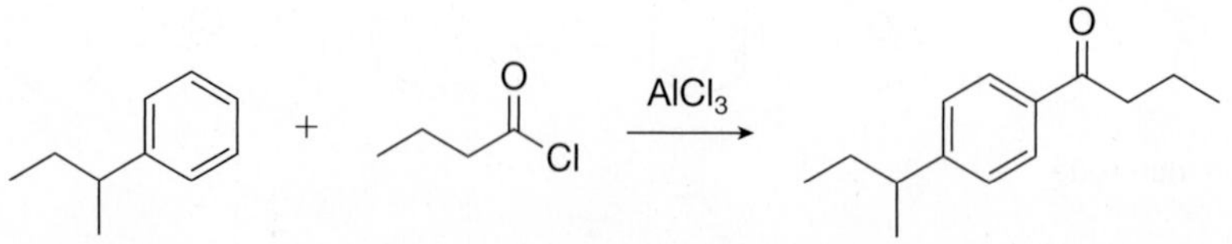 $\xrightarrow{AlCl_3}$

b) Predict the products of the following reactions.

i) 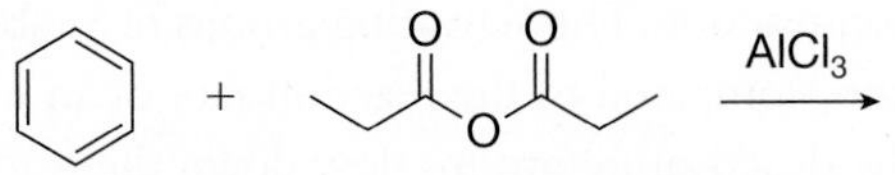 benzene + propanoic anhydride $\xrightarrow{AlCl_3}$

ii) 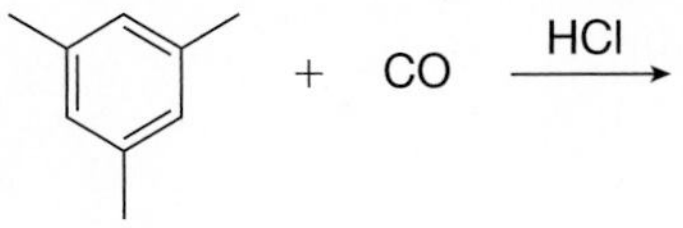 1,3,5-trimethylbenzene + CO $\xrightarrow{HCl}$

INTEGRATE THE SKILL

10.8 In Checkpoint 10.3, there were two examples of Friedel–Crafts alkylations that failed to give the desired product (Question 10.6a). Show how a Friedel–Crafts acylation could be used to successfully make the desired products.

10.5 Aromatic Nomenclature and Multiple Substituents

Some special nomenclature is used for aromatic rings that contain substituents. Many aromatic compounds have "trivial" names that do not follow the IUPAC system of nomenclature but are widely used nevertheless. Here are a few common trivial names.

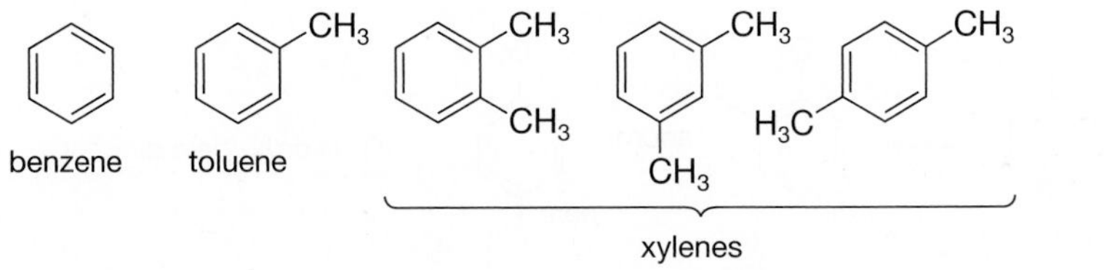

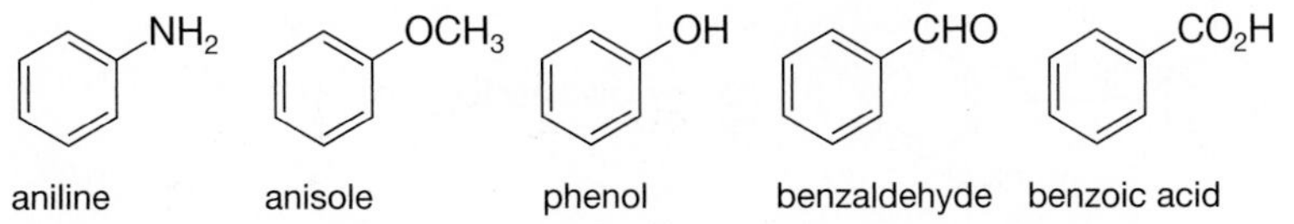

The relative positions of groups on an aromatic ring can be referred to by numbers or by Greek words. Groups located next to each other (1,2) are said to be *ortho*, groups located in a 1,3 manner are *meta*, and groups located in a 1,4 manner are *para*.

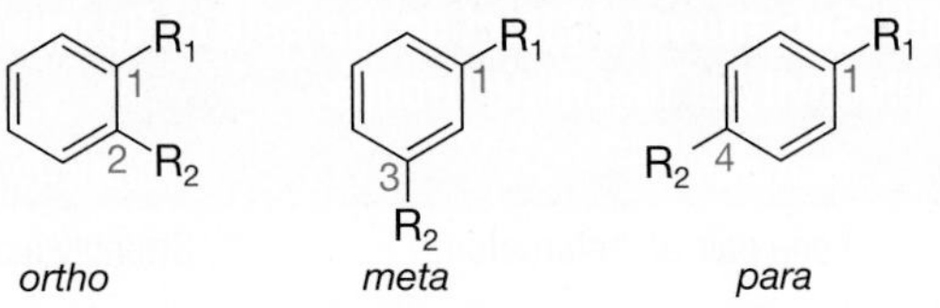

10.6 Directing Groups in Electrophilic Aromatic Substitution

When an aromatic compound already has one or more substituents, the outcome of an electrophilic aromatic substitution reaction is affected by those substituents. In particular, substituents can affect the *rate of the reaction* as well as the *regioselectivity* of the electrophilic aromatic substitution. These functional groups can be categorized based on two properties:

A substituent is **activating** if it increases the rate of reaction. In the context of S_EAr reactions, electron-donating groups are generally activating.

A substituent is **deactivating** if it decreases the rate of reaction. In the context of S_EAr reactions, electron-withdrawing groups are deactivating.

1. **Activating** or **deactivating** refers to the property of a substituent that influences the rate of the electrophilic aromatic substitution reaction. The activating groups of a substituted aromatic compound accelerate the reaction, compared to the reaction rate of an unsubstituted aromatic compound (i.e., benzene). The deactivating groups slow down the reaction relative to that of benzene.

EAS → reference reaction

EAS →
- if reaction is faster than for benzene: "G" is **activating group**
- if reaction is slower than for benzene: "G" is **deactivating group**

In general, electron-donating substituents on the aromatic ring are activating groups because they stabilize the arenium ion intermediate and also the transition state of the rate-limiting step of the overall reaction (see Section 10.3). Likewise, electron-withdrawing substituents on the aromatic ring are deactivating groups because they destabilize areniums and the transition states leading to them.

2. Directing refers to substituents that influence the regiochemistry of the electrophilic aromatic substitution reaction. ***Ortho/para* directors** favour the formation of *ortho* and *para* regioisomers, whereas ***meta* directors** favour the production of the *meta* isomer.

***Ortho/para* directing groups** are substituents that favour substitution at the *ortho* and *para* positions, relative to the directing group.

***Meta* directing groups** are substituents that favour substitution at the *meta* position, relative to the directing group.

EAS → *ortho* and/or *para* — "G" is *ortho/para* director

EAS → *meta* — "G" is *meta* director

Individual substituents on an aromatic ring tend to have an effect on both the reactivity and the regioselectivity of electrophilic aromatic substitutions, as summarized in Table 10.2. Conjugation and inductive effects are used to rationalize these observations.

TABLE 10.2 Aromatic Substituents' Effect on the Rate of Reaction and the Regioselectivity of Incoming Groups

Substituent	Features	Reactivity	Direction
NH_2, NHR, NR_2, OH, O^-	Lone pair on heteroatom	Strongly activating	*Ortho*/*para*
NHCOR, OCOR	Electron-delocalized lone pair on heteroatom	Moderately activating	*Ortho*/*para*
Alkyl, aryl	Hyperconjugation or weak electron delocalization	Weakly activating	*Ortho*/*para*
F, Cl, Br, I	Lone pair on electronegative halogen atom	Deactivating	*Ortho*/*para*
COH(R), CO_2H(R), CONHR, CN	Polar π bond conjugated to ring	Moderately deactivating	*Meta*
NO_2, NR_3^+, CX_3	Strong inductive electron-withdrawing group	Strongly deactivating	*Meta*

10.6.1 *Ortho*/*para* directing groups

Electron-donating groups tend to activate the electrophilic aromatic substitution, and they also direct the regioselectivity toward the preferred formation of *ortho* and/or *para* substitution products. The resonance structures of the arenium ion intermediates help explain the regioselectivity.

10.6.1.1 Strong *ortho/para* directors

All of these functional groups are directly linked to the aromatic ring via a heteroatom that bears at least one lone pair that can delocalize into the ring.

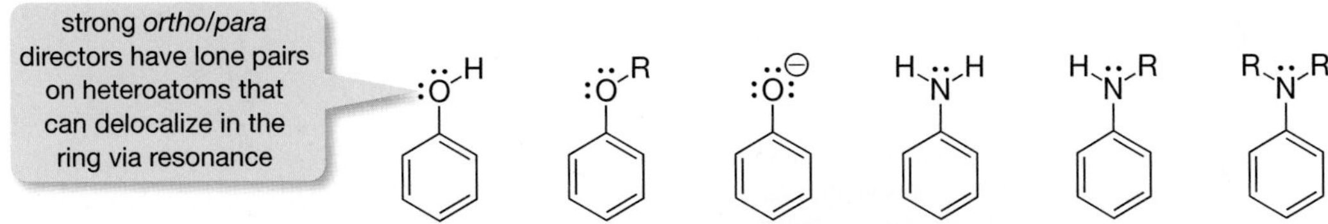

strong ortho/para directors have lone pairs on heteroatoms that can delocalize in the ring via resonance

According to the Hammond postulate (Section 8.3.7), an endothermic reaction preferentially goes through the transition state that leads to the more stable intermediate. The first step in the S_EAr mechanism is endothermic and rate limiting (it controls the overall rate of the reaction). The lowest energy arenium ion intermediate forms fastest, making it appear that molecules "choose" the lowest energy pathway available to them. This allows the functional groups on an aromatic ring to control the regioselectivity of the product by influencing the energy (stability) of the arenium ion intermediate.

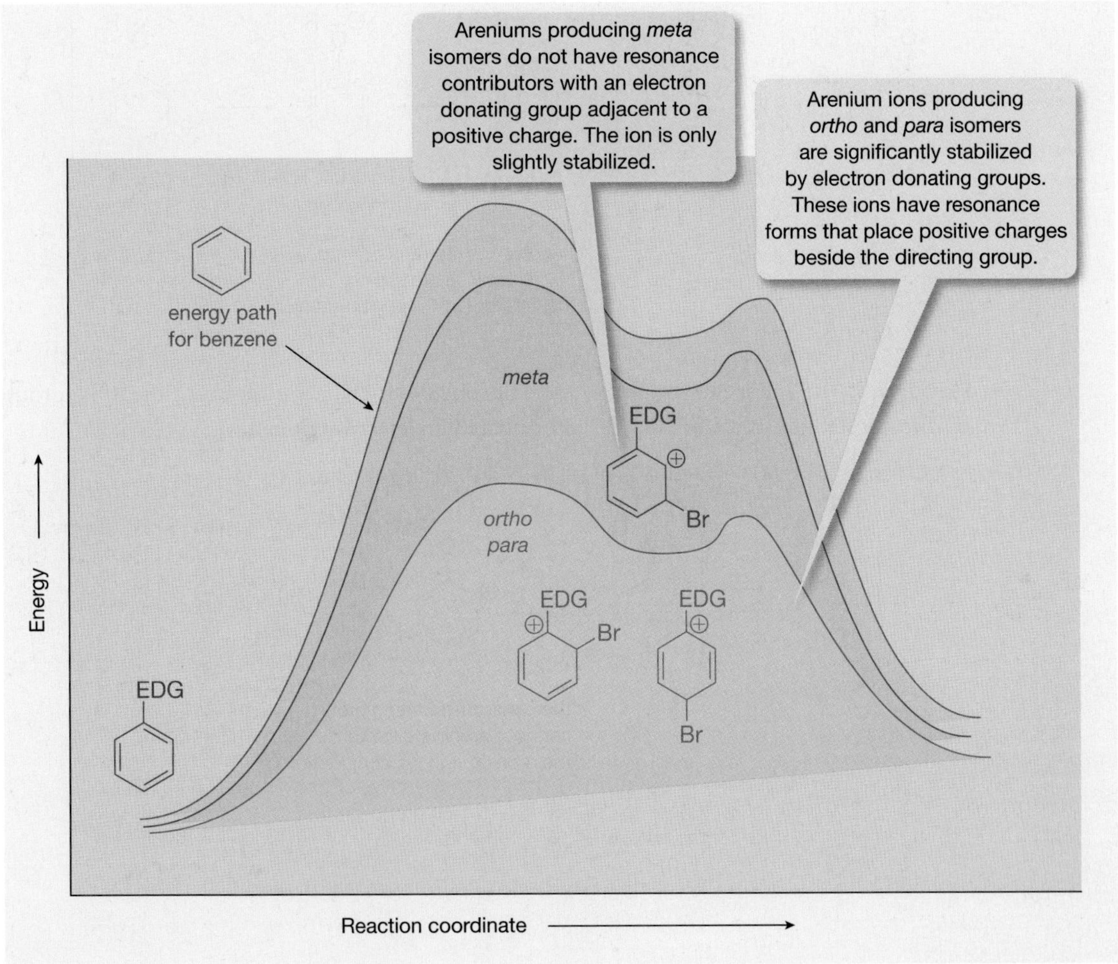

Consider the electrophilic aromatic substitution of an aryl ether at either the *ortho*, *para*, or *meta* position, and focus on the electron delocalization in the arenium ion intermediates.

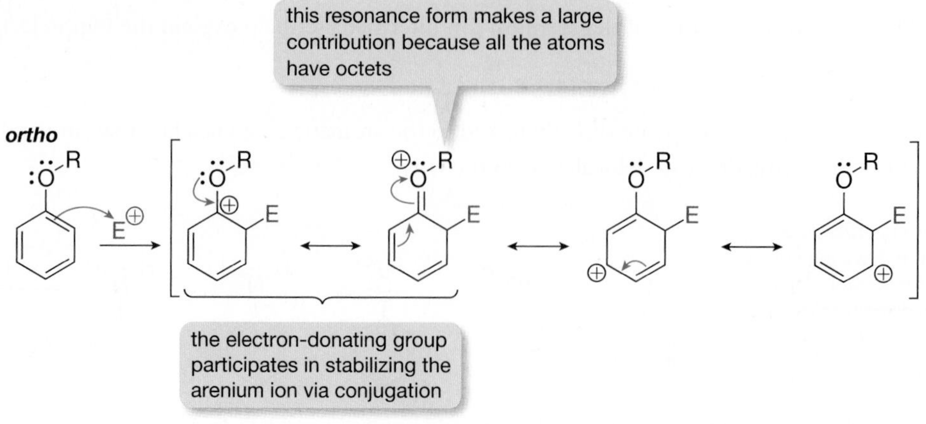

this resonance form makes a large contribution because all the atoms have octets

ortho

the electron-donating group participates in stabilizing the arenium ion via conjugation

If the initial electrophilic addition occurs at the *ortho* position, the OR group can participate in stabilization of the arenium ion intermediate by donating electrons. This lowers the activation energy of this rate-limiting step of the electrophilic substitution reaction. A similar situation arises if the electrophilic addition occurs at the *para* position; the OR group can also participate in electron delocalization with the arenium ion intermediate.

para

the electron-donating group participates in stabilizing the arenium ion via conjugation

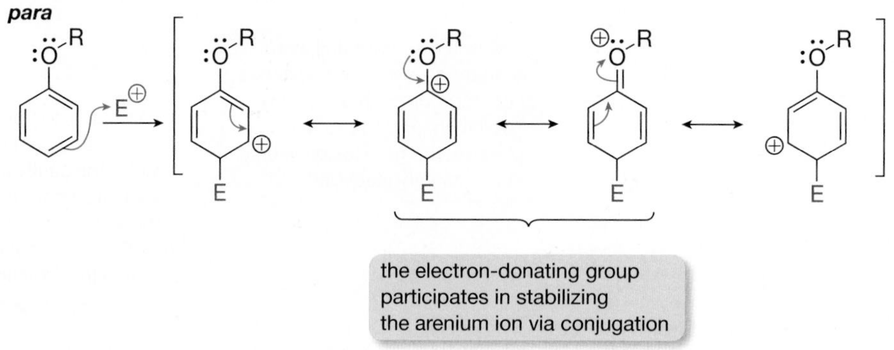

However, if the initial electrophilic addition occurs at the *meta* position, the OR group cannot directly stabilize the arenium ion intermediate by electron donation.

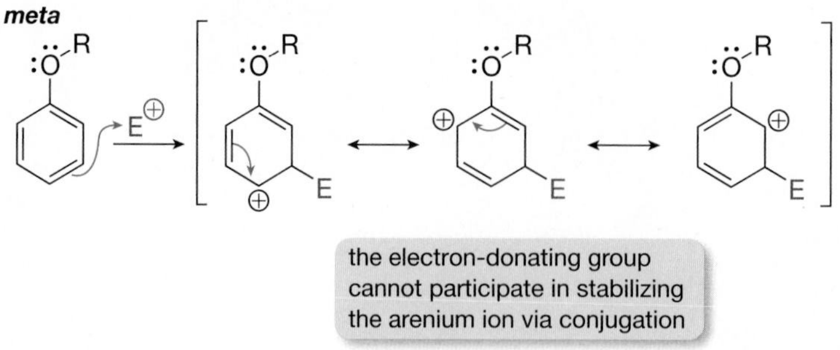

meta

the electron-donating group cannot participate in stabilizing the arenium ion via conjugation

This accounts for the overall regioselectivity of the electrophilic aromatic substitution when such strong *ortho/para* directors exist in the reactant. Addition *ortho* or *para* to the directing group significantly lowers the activation energy for the formation of the arenium ion, whereas *meta* addition does not result in a meaningful change in activation energy. Frequently, there is also a weak preference of *para* over *ortho* substitution because addition to the *ortho* position produces steric interference between the directing and the incoming group. Several factors can influence this regioselectivity, including steric effects, hydrogen bonding, and other parameters.

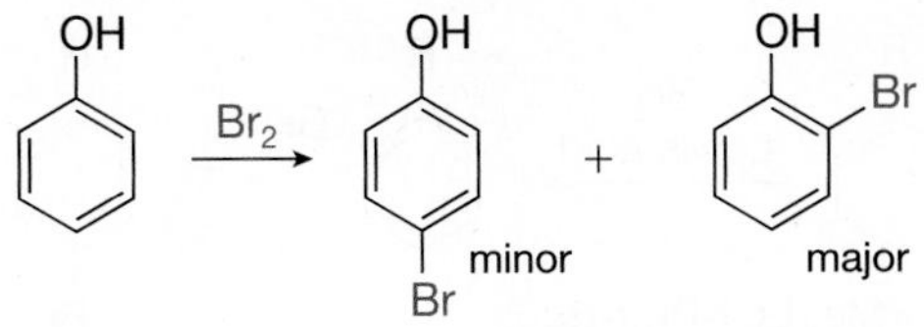

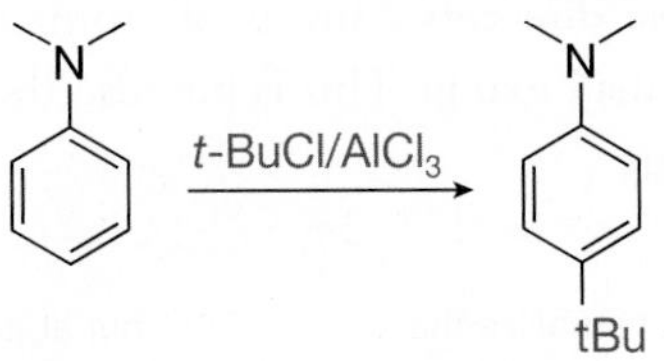

10.6.1.2 Moderate *ortho/para* directors

Some functional groups are also directly linked to the aromatic ring via a heteroatom that bears at least one lone pair. However, the lone pair is already delocalized into the functional group itself. This makes the groups weaker stabilizers of the arenium ion due to cross-conjugation.

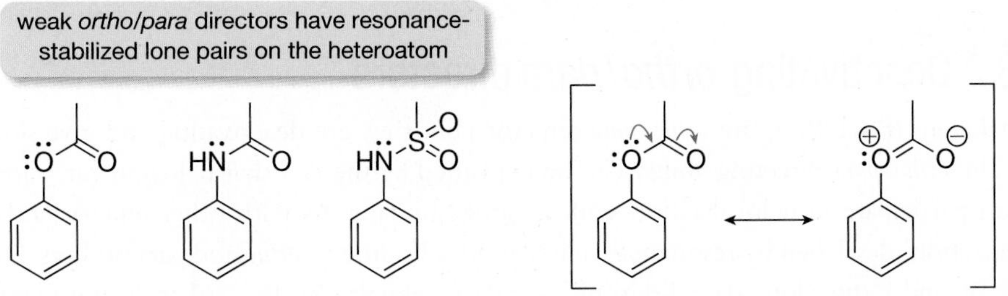

Consider the initial addition of an electrophile at the *ortho* position of a phenyl ester (Ar–OCOR). The ester group can still participate in the electron delocalization of the arenium ion intermediate, but much less strongly than an OH or OR substituent, due to cross-conjugation. The net effect is a moderate *ortho/para* director.

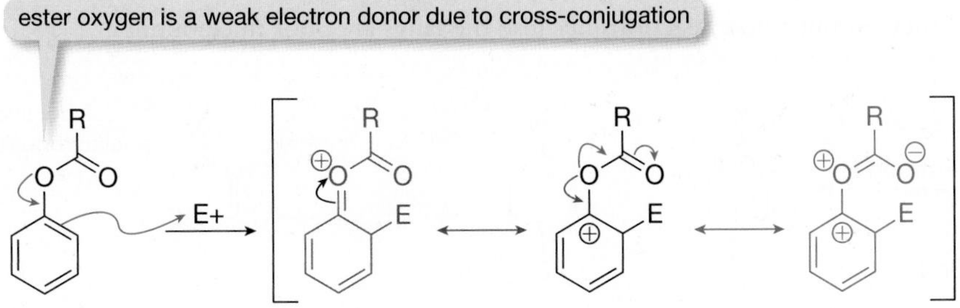

10.6.1.3 Weak *ortho/para* directors

There are two types of weak *ortho/para* directors. The first type consists of alkyl groups. These groups have weak electron-donating character due to hyperconjugation. Alkyl groups stabilize the positive charge of the arenium ion intermediate, but the degree of stabilization is limited. A branched alkyl group at the attachment point to the ring is *slightly* more electron donating than a straight chain. This is because the electron-donating ability of C–C bonds is greater than C–H bonds.

Br_2
Lewis acid
R=Me, Et, *i*-Pr, *t*-Bu...

The second type of weak *ortho/para* directors consists of aromatic rings. These functional groups can donate electrons only to a small extent. This is because their electron donation disrupts the aromaticity of the directing ring.

aromatic ring can stabilize the arenium ion by conjugation...

...but at the expense of its own aromaticity

10.6.2 Deactivating *ortho/para* directors

The halogens (F, Cl, Br, I) are *ortho/para* directors, but they are deactivating and give slow reactions. The *ortho/para*-directing ability can be explained by the fact that halogens carry lone pairs that can participate in delocalization with an aromatic ring. As with other *ortho/para* directors, delocalization (described by resonance) stabilizes ions leading to *ortho* and *para* isomers. However, for Cl, Br, and I, their lone pairs lie in high-order p orbitals (3p, 4p…) that do not overlap well with the 2p orbitals of the ring carbon. Although this reduces the resonance contribution, the effect is still strong enough to give a preference for *ortho* and *para* isomers.

At the same time, halogens are very electronegative and withdraw electrons from aromatic rings by the inductive effect. This tends to slow the overall rates of the reactions because the resulting aromatic rings are less nucleophilic (less-effective electron donors). Halogen substituents are therefore *ortho/para* directors because the corresponding arenium ions are resonance stabilized, but they produce slow reactions because the rings are poor nucleophiles.

strong $E^{\oplus}$
X=F, Cl, Br, I

resonance effect (electron-donating)
inductive effect (electron-withdrawing)

10.6.3 *Meta* directing groups

Electron-withdrawing groups typically deactivate an aromatic ring toward electrophilic aromatic substitution. Furthermore, these groups have an important effect on the regioselectivity of the reaction and predominantly lead to the formation of the *meta* product. Resonance structures of the arenium ion intermediates for the *ortho*, *meta*, and *para* additions can account for the regioselectivity.

EWG, $E^{\oplus}$, **ortho**

electron-withdrawing group directly attached to carbocation

EWG, $E^{\oplus}$, **para**

electron-withdrawing group directly attached to carbocation

EWG, $E^{\oplus}$, **meta**

carbocation is further away from electron-withdrawing group

For both *ortho* and *para* additions, one resonance structure of their corresponding arenium ion intermediate shows direct attachment of the electron-withdrawing group (EWG) to a carbocation. This destabilizes their overall arenium ion, compared to the one resulting from *meta* addition, where the positive charge is always further away from the EWG. The *meta* pathway becomes the lowest energy path, and is favoured overall.

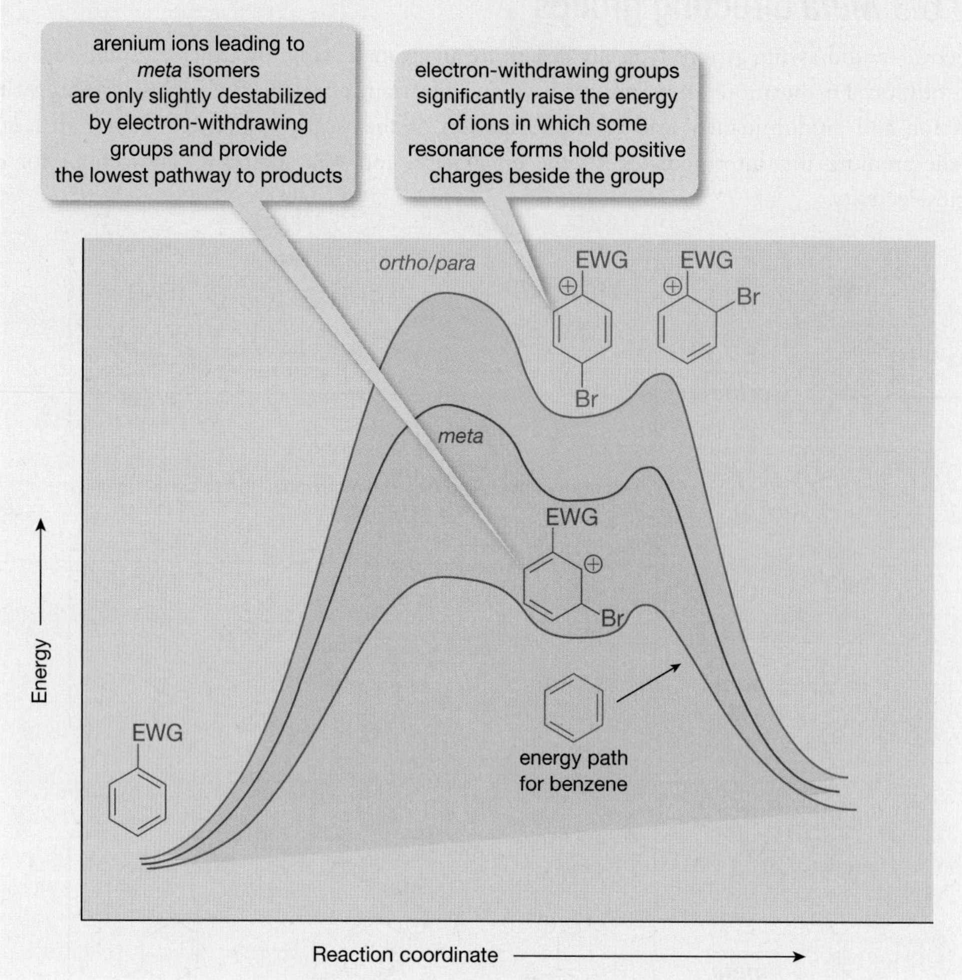

There are two types of *meta* directing groups.

10.6.3.1 Moderately deactivating *meta* directors

These groups contain polar π bonds that are conjugated to the aromatic ring and can withdraw electrons via electron delocalization. These π bonds connect to electronegative atoms, which are electron withdrawing and deactivate the aromatic ring toward electrophilic aromatic substitution. These groups destabilize the arenium ions that lead to the *ortho* and *para* regioisomers, and so *meta* products are formed preferentially.

X = H, R, OH, OR, NH_2, NHR

polar π bonds deactivate via conjugation

10.6.3.2 Strongly deactivating *meta* directors

Ring substituents that are electron withdrawing due to strong inductive effects are strongly deactivating and direct *meta* substitution. Nitro groups are particularly strong directors due to both inductive and π-delocalization effects. Groups that feature positive formal charges adjacent to the rings are particularly strong.

X=H, R

inductive effect

inductive effect and π delocalization

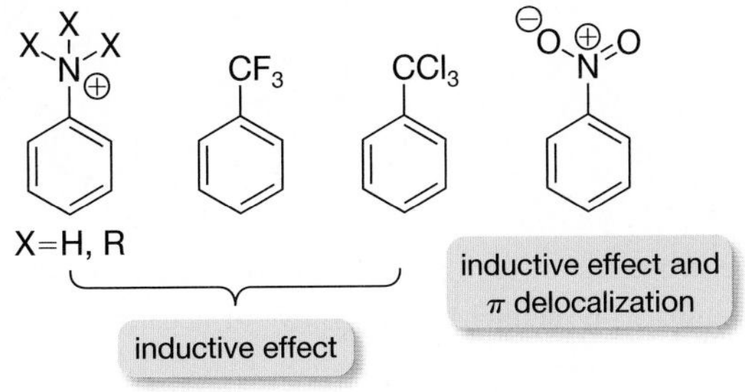

CHECKPOINT 10.5

You should now be able to classify aromatic substituents as either *ortho/para* directing or *meta* directing, and determine whether they will activate or deactivate the ring toward electrophilic aromatic substitution. You should also be able to use this information to predict the products of electrophilic aromatic substitution on monosubstituted benzenes.

SOLVED PROBLEM

Draw a mechanism for the following reaction and predict the major product.

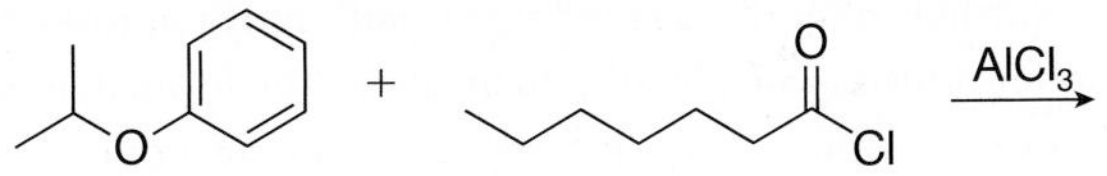

$AlCl_3$

STEP 1: Determine the nature of the substituent on the aromatic ring.

The alkoxy group on the aromatic ring contains is an electron-donating group, which increases the nucleophilicity of the ring. Activation of the acid chloride leads to a Friedel–Crafts acylation.

STEP 2: Draw a mechanism for the formation of the acylium ion electrophile. A lone pair from chlorine shares its electrons with an empty orbital on the $AlCl_3$ Lewis acid. The resulting complex then expels a leaving group, leaving behind a stabilized acylium ion.

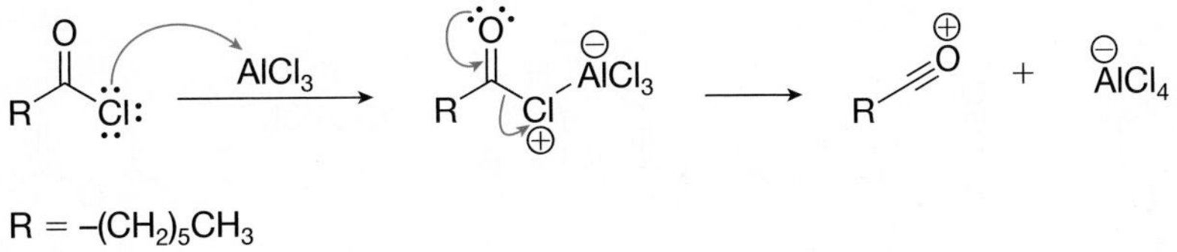

$R = -(CH_2)_5CH_3$

STEP 3: Draw a mechanism for the addition step, producing all possible cyclohexadienyl cation intermediates. Use the steps in Checkpoints 10.1 and 10.2 to guide you if needed.

The electrons from one of the π bonds in the benzene ring are shared with the electron-deficient acylium ion, forming a new C–C σ bond. In this example, the alkoxy group blocks addition to the *ortho* position. The *meta* and *para* positions are possible options. This addition produces a cation at either the *meta* or *para* position of the ring.

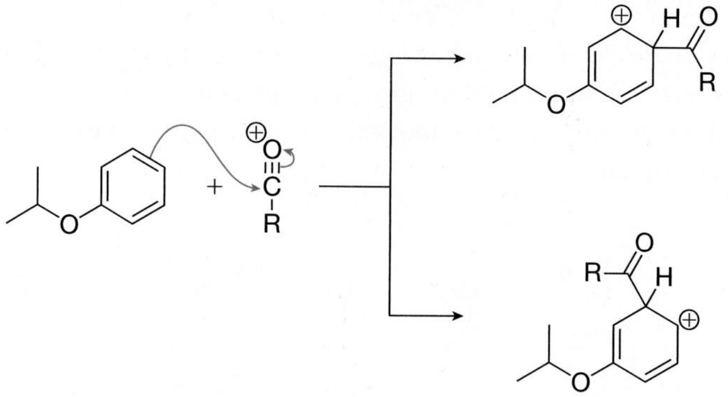

STEP 4: Draw the resonance structures for each possible intermediate, and identify any resonance contributors that have a significant impact on stability, either positive or negative.

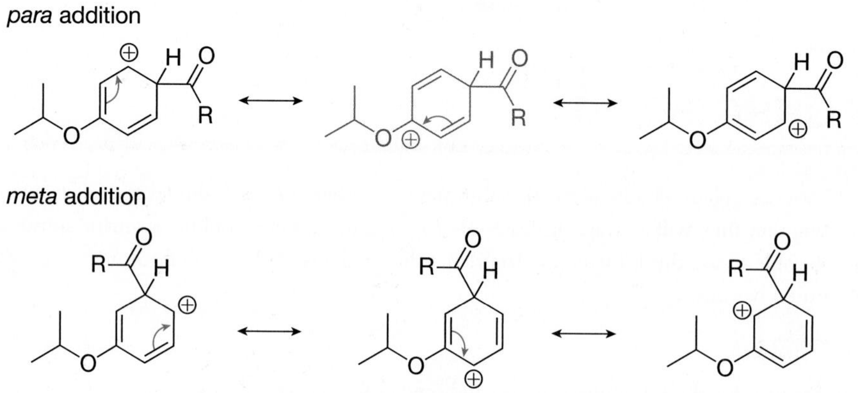

para addition

meta addition

One of the resonance contributors for the *para* addition significantly improves stability. Note that placing the cation next to the electron-donating alkoxy substituent allows the formation of an additional resonance structure. This structure is a very significant contributor because all the atoms carry octets.

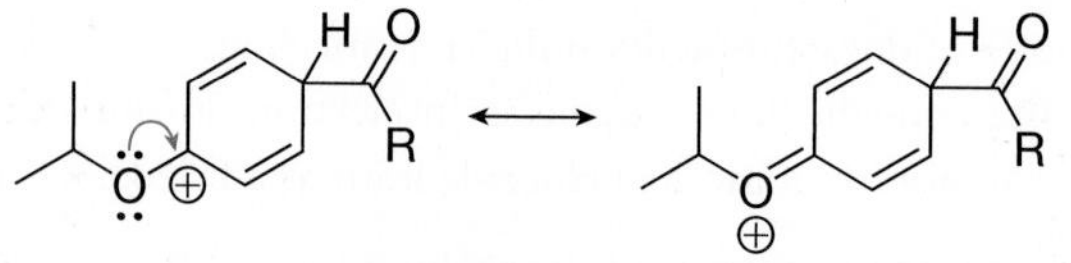

STEP 5: Using the more stable intermediate, complete the mechanism by adding the elimination step to form the preferred product.

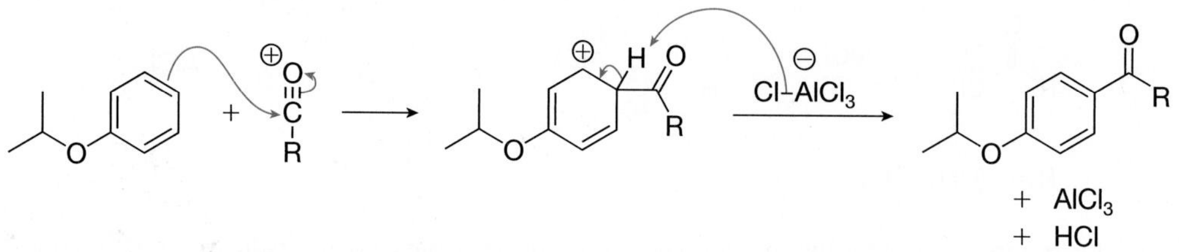

$+ AlCl_3$
$+ HCl$

PRACTICE PROBLEM

10.9 For each of the following, predict the expected major product with the correct regiochemistry.

a) 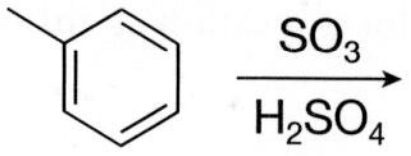 $\xrightarrow[H_2SO_4]{SO_3}$

b) 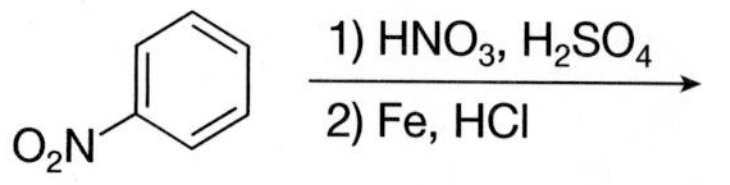 O_2N — 1) HNO_3, H_2SO_4; 2) Fe, HCl

c) 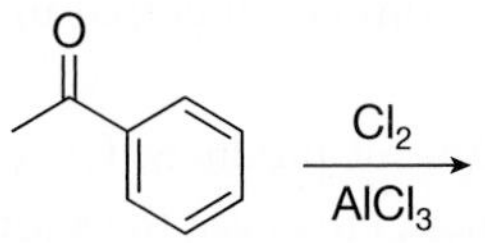

d) 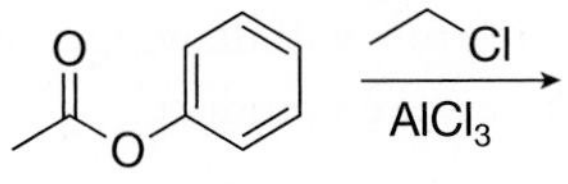

e) 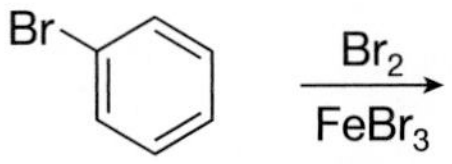

f) 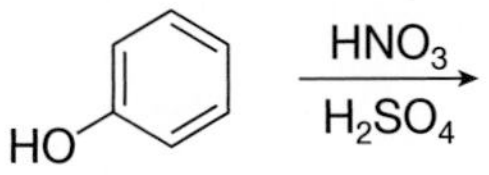

INTEGRATE THE SKILL

10.10 The alkylation of a hydroxylated biphenyl with a t-butyl group produces one major product. Predict the major product, and justify the regioselectivity by drawing resonance structures of the possible intermediates.

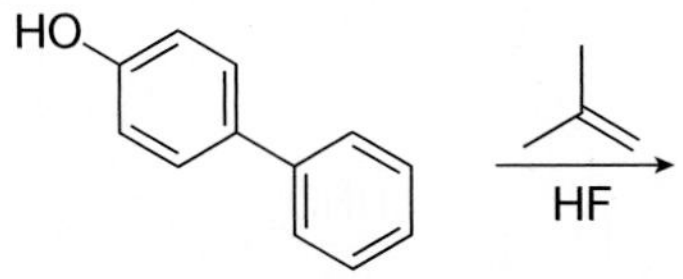

10.6.4 Modifying reactivity in electrophilic aromatic substitutions

Anilines (Ar–NR_2), phenols (Ar–OH), and anisoles (Ar–OMe) are so strongly activating that it can be difficult to stop electrophilic reactions from producing multiple substitutions. For example, the bromination of aniline produces complex mixtures of the mono-, di-, and tri-bromo products. Reactivity can be modified to favour monosubstitution by protecting the amino group as an amide (moderate *ortho/para* director) prior to carrying out the electrophilic aromatic substitution.

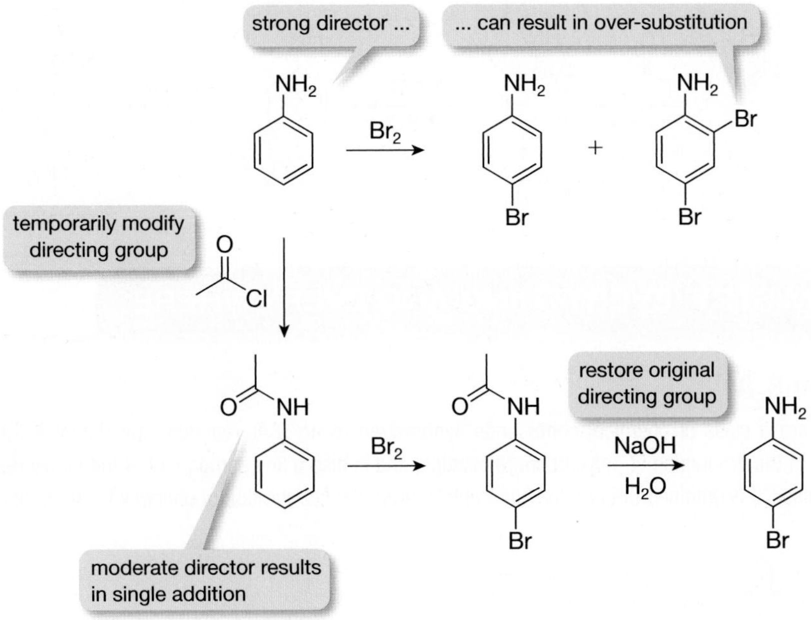

The mechanisms of the amide formation and hydrolysis reactions are discussed in detail in Chapter 15.

The nitration of aniline also produces an unexpected mixture of products. Generally, the electron-donating NH_2 group gives a mixture of *ortho*/*para* products. However, the strong acidic reaction conditions protonate the amine, converting it to an electron-withdrawing ammonium ion. This results in an unexpectedly large proportion of *meta*-substituted product.

expected major products

NH_2 $\xrightarrow[H_2SO_4]{HNO_3}$ NH_2 / NO_2 + NH_2 / NO_2 + NH_2 / NO_2

3–5% 35–50% 50–60%

meta product!

ortho/*para* director NH_2 $\underset{H_2SO_4}{\overset{HNO_3}{\rightleftharpoons}}$ $^{\oplus}NH_3$ *meta* director

Because amides are much less basic than amines, they do not convert to *meta*-directing groups in the presence of strong acid. The strong acidic reaction can be modified by using an amide-substituted aromatic ring, which greatly reduces the amount of *meta* substitution.

amide is a moderate *ortho*/*para* director and is not protonated by the reagent

NH_2 $\xrightarrow[\text{pyridine}]{Cl\text{, O}}$ HN, O $\xrightarrow[H_2SO_4]{HNO_3}$ HN, O / NO_2 $\xrightarrow{NaOH}$ NH_2 / NO_2

CHEMISTRY: EVERYTHING AND EVERYWHERE

BPA from S_EAr

Bisphenols are a class of chemical compounds synthesized using S_EAr reactions. Bisphenol A (BPA) is commonly used in the production of polycarbonate plastics, and is also a key component of the epoxy resins used to make fibreglass. It is obtained through the electrophilic aromatic substitution of phenol with acetone under acidic conditions.

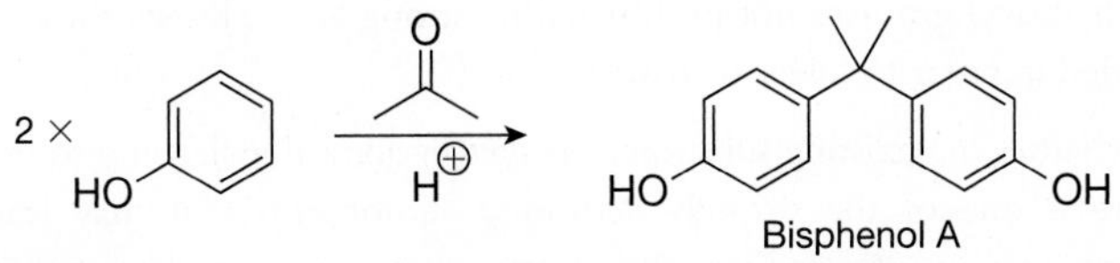

2 × HO H⊕ HO OH
Bisphenol A
(BPA)

The variety of possible bisphenols results from the variety of carbonyl compounds that will undergo this reaction. For example, the combination of phenol and acetone results in BPA, whereas combining acetaldehyde with phenol produces bisphenol E.

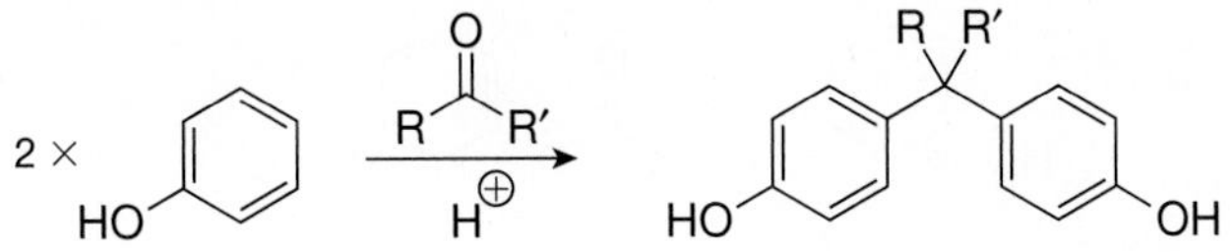

2 × HO R R′ H⊕ R R′ HO OH

R	R′	Bisphenol
CH_3	CH_3	Bisphenol A
H	CH_3	Bisphenol E

The mechanism involves two consecutive electrophilic aromatic substitutions. Because phenol is a strong *ortho*/*para* director, the reaction is relatively fast. Thousands of tons of various bisphenols are produced each year for manufacturing plastic products.

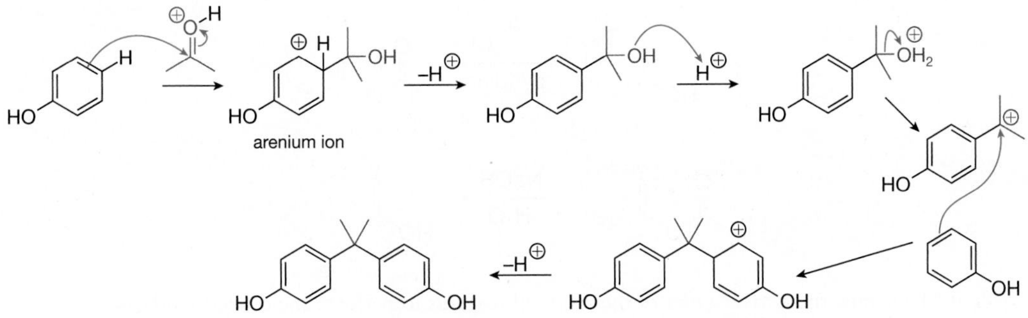

HO H ⊕O H arenium ion -H⊕ OH H⊕ OH_2 ⊕ HO OH arenium ion

CHECKPOINT 10.6

You should now be able to propose appropriate methods to modify the reactivity of strongly activating substituents to help control electrophilic aromatic substitution reactions.

SOLVED PROBLEM

Design a synthetic scheme to efficiently convert phenol to *p*-nitrophenol.

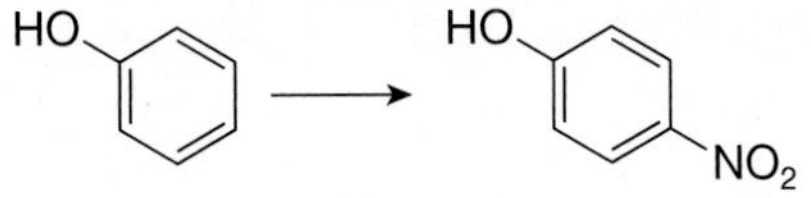

HO → HO NO_2

STEP 1: Identify the added substituent and the reagents required to add it.

The new substituent is a nitro group, which must be added using HNO_3 with H_2SO_4.

STEP 2: Determine whether the existing substituent will be altered by the required reaction conditions, leading to unexpected regioselectivity.

Unlike amines, a hydroxyl group is not protonated by strong acid. This synthesis does not require the hydroxyl to be modified in order to lower its basicity.

STEP 3: Determine whether the existing substituent is *strongly* activating, leading to polysubstitution.

A hydroxyl group is one of the strongly activating substituents that may lead to poly-nitration. Converting the hydroxyl to an ester reduces the electron donation into the ring, thereby reducing the ring's nucleophilicity.

STEP 4: Select appropriate conditions to modify the existing substituent and to reverse the modification.

Alcohol hydroxyl groups can be converted to esters using an acid chloride and pyridine. Esters can be converted to alcohols using NaOH in water. These reactions are discussed in detail in Chapter 15.

STEP 5: Determine the correct order of the required reagents to perform the desired synthesis.

The alcohol needs to be converted to an ester before the nitration step. Once the nitration has been completed, the ester must be converted back to the original hydroxyl substituent.

PRACTICE PROBLEM

10.11 Determine the reagents required to perform the following syntheses.

a)

b)

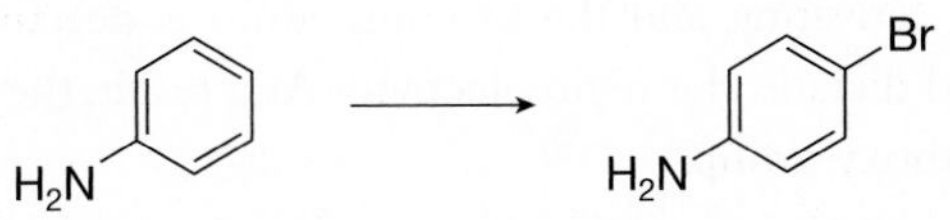

H_2N → H_2N, Br

c)

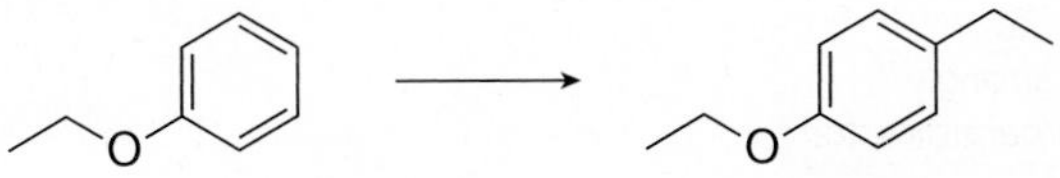

O → O

d)

O, O → HO, Cl

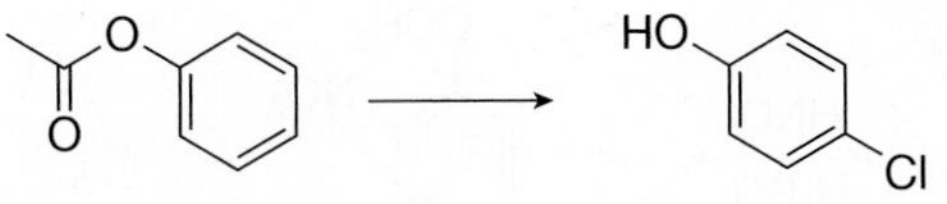

INTEGRATE THE SKILL

10.12 Select the reagents required to perform the following synthesis, using analogs of the esterification reagents used previously. Why does the nitro group add to the ring on the left instead of the ring on the right?

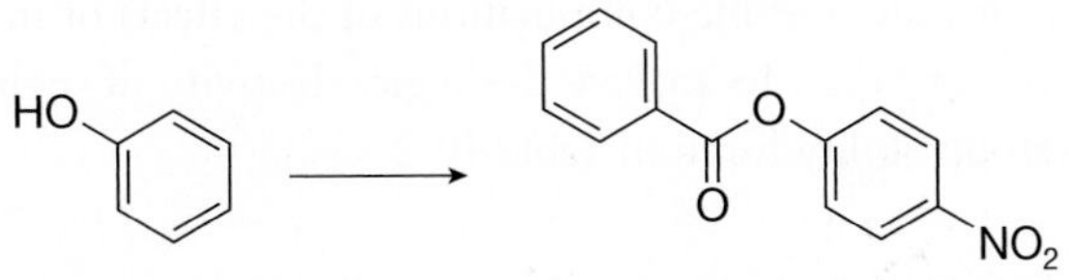

HO → O, O, NO_2

10.6.5 Strength of activation on polysubstituted benzenes

When multiple substituents are present on a ring, the collective effects of the directing groups must be considered. Specifically, the group that is most activating directs the location of the incoming electrophile in the major product, regardless of the directing nature of the other substituents. If none of the groups is activating (i.e., all *meta* directors), the regioselectivity is controlled by the group that is most strongly *deactivating* because the pathway is governed by the intermediate that is less destabilized.

Ideally, the directing properties of each group reinforce each other such that only one product is obtained. For example, in *p*-nitroanisole, the OCH_3 and NO_2 groups both direct toward the same position. This results in a single product being formed very selectively.

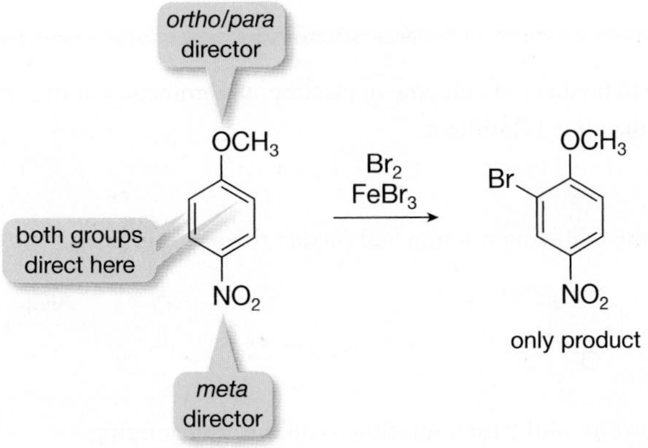

ortho/para director

OCH_3

both groups direct here

NO_2

meta director

Br_2, $FeBr_3$ →

Br, OCH_3, NO_2

only product

However, if groups on an aromatic ring have conflicting directing properties, the more strongly activating group normally dictates the regioselectivity. In the nitration of *p*-bromoanisole, the methoxy and bromo substituents are located in such a way that they direct to conflicting positions on the ring. Because the methoxy group is activating and the bromine atom is deactivating, the methoxy group exerts more influence and dictates the regioselectivity. As a result, the incoming nitro group is introduced *ortho* to the methoxy group.

OCH_3 is a strongly activating *ortho/para* director

OCH_3 / Br $\xrightarrow[H_2SO_4]{HNO_3}$ OCH_3, NO_2 / Br

Br is a moderately deactivating *ortho/para* director

In reactions of this sort, the possible combinations of the effects of multiple substituents are too many to consider in every case. To analyze the regioselectivity of such reactions, consult the hierarchy of directing-group ability listed in Table 10.2.

NO_2 is a strongly deactivating *meta* director

NO_2 $\xrightarrow[H_2SO_4]{HNO_3}$ NO_2, NO_2

acyl is a moderately deactivating *meta* director

electrophile adds *meta* to the stronger directing group

CHECKPOINT 10.7

You should now be able to predict the outcome of electrophilic aromatic substitution reactions on aromatic compounds with more than one substituent.

SOLVED PROBLEM

Draw a mechanism for the following reaction and predict the major product.

+ O Cl $\xrightarrow{AlCl_3}$

STEP 1: Determine the nature of the each substituent on the aromatic ring.

Both the alkoxy and alkyl groups on the aromatic ring are electron donating. Both substituents are *ortho/para* directing.

STEP 2: Determine whether the positions of each directing group reinforces or opposes each other's selectivity.

In this example, each group directs to the *ortho* and *para* positions. Since the position *para* to each group is already occupied by the other, only the *ortho* positions need be considered. Since the carbons *ortho* to each group are different, the direction from each group opposes that of the other.

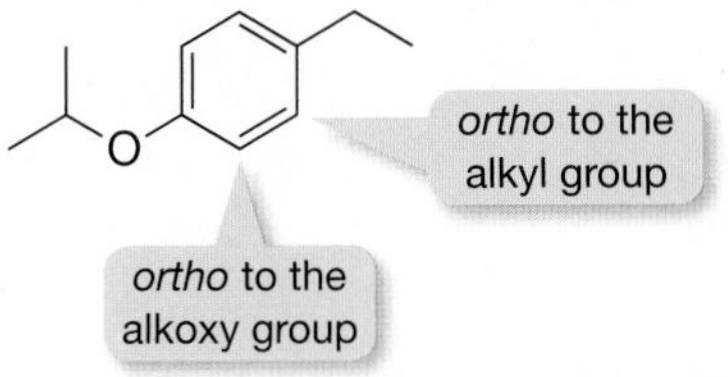

ortho to the alkyl group

ortho to the alkoxy group

STEP 3: In the case of competition, determine which group is *more* electron donating or activating. Consult Table 10.2 if needed.

Because alkoxy groups are strongly electron donating, the positions *ortho* to the alkoxy group are more activated than the positions *ortho* to the alkyl group, which is weakly activated. This makes the substitution *ortho* to the alkoxy group favoured.

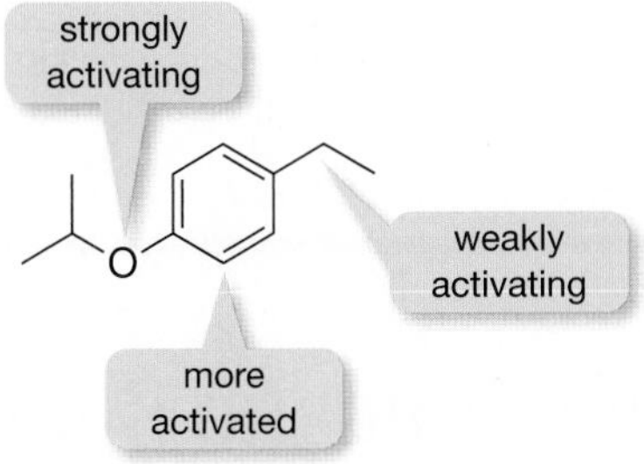

strongly activating

weakly activating

more activated

STEP 4: Complete the reaction at the more activated position.

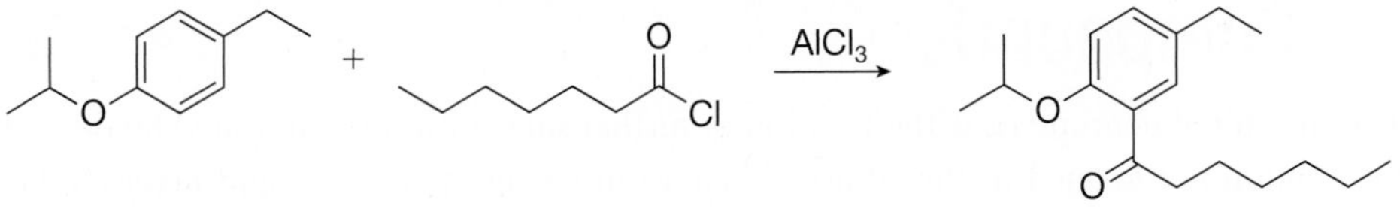

$\xrightarrow{AlCl_3}$

PRACTICE PROBLEM

10.13 For each of the following, predict the expected major product with the correct regiochemistry.

a) 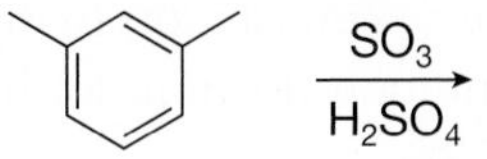 $\xrightarrow[H_2SO_4]{SO_3}$

b) 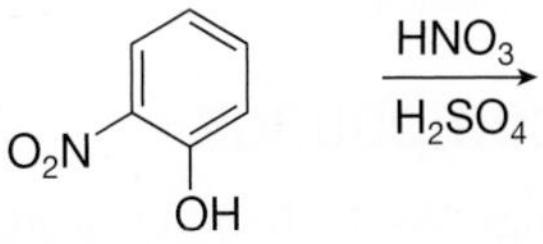 O_2N OH $\xrightarrow[H_2SO_4]{HNO_3}$

c) 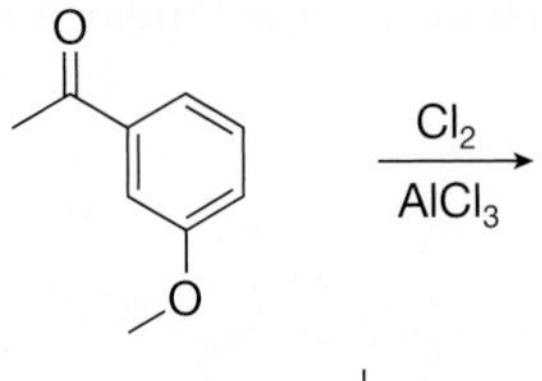 $\xrightarrow[AlCl_3]{Cl_2}$

d) 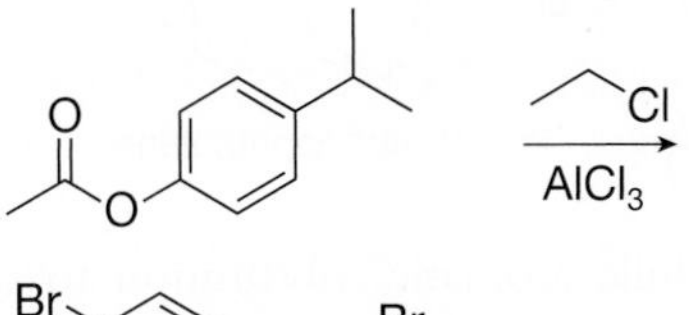 $\xrightarrow[AlCl_3]{\text{Cl}}$

e) 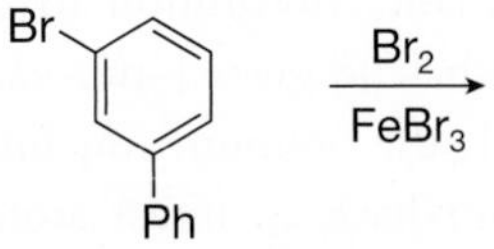 Br Ph $\xrightarrow[FeBr_3]{Br_2}$

f) 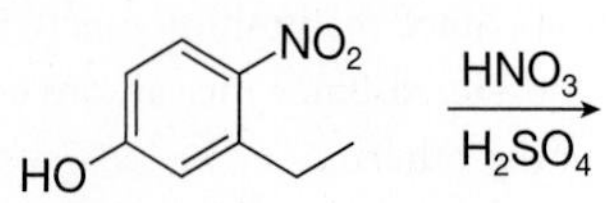 [structure: HO-substituted benzene with NO_2 and ethyl groups] $\xrightarrow[H_2SO_4]{HNO_3}$

g) 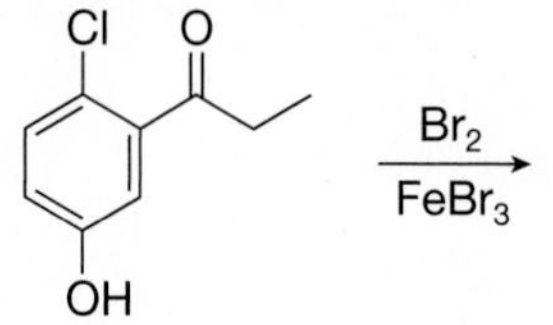 [structure: benzene with Cl, propanoyl (C=O) and OH groups] $\xrightarrow[FeBr_3]{Br_2}$

INTEGRATE THE SKILL

10.14 The alkylation of the following biphenyl with a *t*-butyl group produces one major product. Predict the major product, and justify the regioselectivity by drawing resonance structures of the possible intermediates.

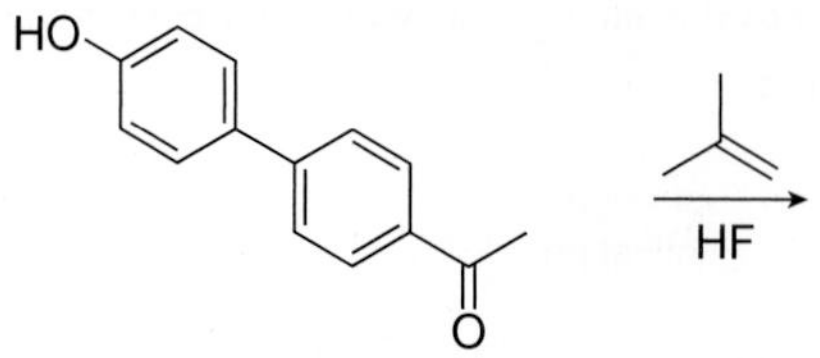

[structure: HO-biphenyl bearing acetyl group] $\xrightarrow{HF}$ (with isobutylene)

10.7 Electrophilic Aromatic Substitution of Polycyclic and Heterocyclic Aromatic Compounds

With a substituted benzene ring, the location of further substitution is controlled by the stability of the arenium ion formed in the addition step. Even though polycyclic and heterocyclic aromatic rings look different than benzene, their substitutions follow the same mechanism, and the regioselectivity is determined by the same factors.

Multiple aromatic rings and heteroatoms have stabilizing/destabilizing effects similar to the substituents in Table 10.2. The number and stability of the resonance contributors to the cationic intermediate of addition determine the regioselectivity of substitution reactions for both polycyclic aromatics (Section 10.7.1) and heterocycles (Section 10.7.2).

10.7.1 Reactivity of polycyclic aromatic compounds

Polycyclic aromatic compounds contain fused benzene rings. As with other aromatic compounds, these compounds undergo electrophilic attack in which the regioselectivity is controlled by the stability of the arenium ion intermediates.

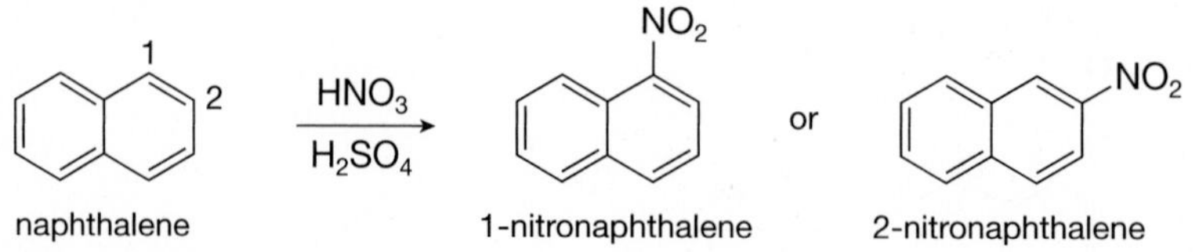

naphthalene (positions 1, 2) $\xrightarrow[H_2SO_4]{HNO_3}$ 1-nitronaphthalene (NO_2) or 2-nitronaphthalene (NO_2)

Naphthalene has two possible sites for electrophilic aromatic substitution to occur (the other sites are equivalent by symmetry). The nitration of naphthalene gives 1-nitronaphthalene as the major product. If the NO_2^+ electrophile adds to C-1, the arenium ion intermediate can be described by seven resonance structures, of which four have an intact aromatic ring.

In general, the greater the stability and the number of resonance contributors to the hybrid structure, the more stable the hybrid is. If the NO_2^+ electrophile adds to C-2, the arenium ion intermediate can be described by only six resonance structures, of which only two have an intact aromatic ring, making substitution at C-2 the higher-energy, disfavoured pathway.

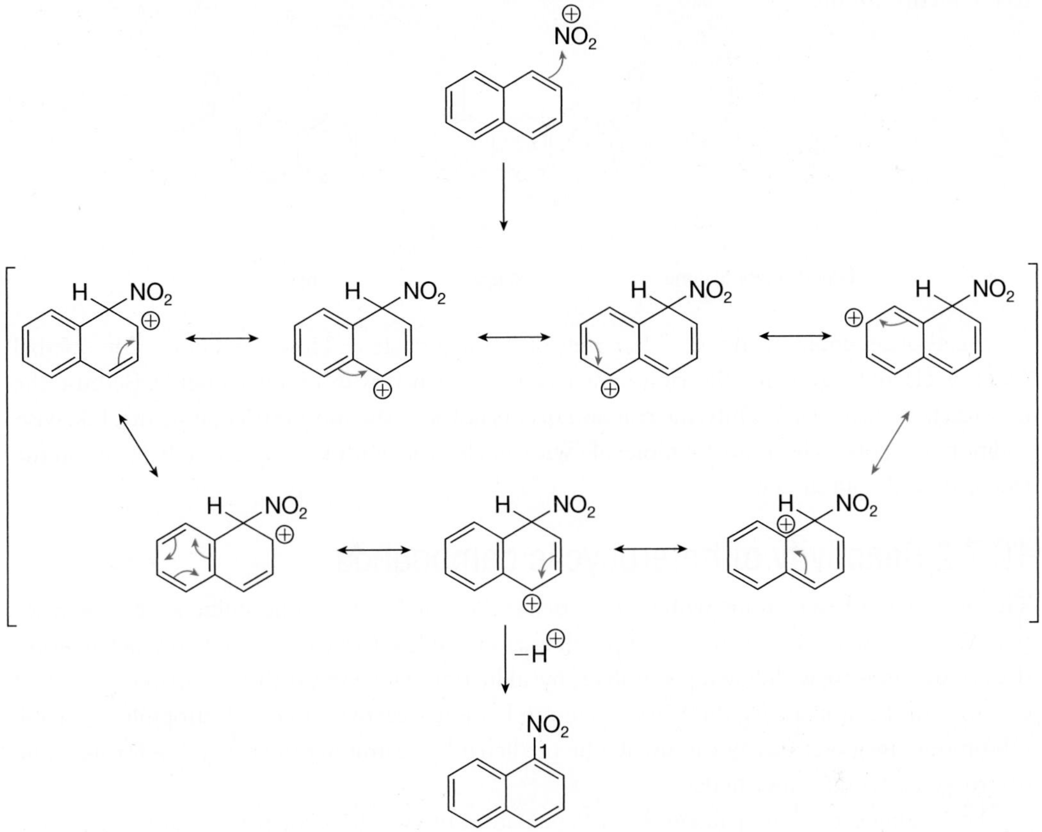

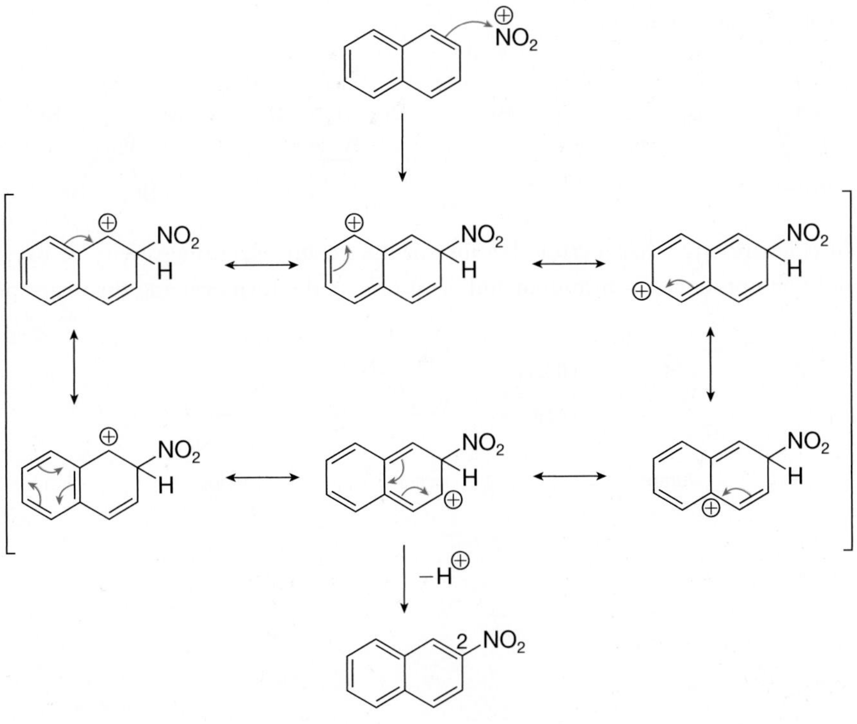

10.7.1.1 Reactivity of substituted polyaromatics

When substituted polyaromatics undergo electrophilic aromatic substitution reactions, the usual guidelines for directing groups apply. For example, the reaction of 1-methylnaphthalene with acetyl chloride in the presence of $AlCl_3$ provides a major product in which the new group is *para* to the methyl group.

O
Cl
$AlCl_3$
+
O
O
1-methylnaphthalene major minor

Acetylation could occur on either ring of the naphthalene. However, because the methyl group is electron donating, the ring to which it is attached is more nucleophilic. Because the ring reacts with an electrophile, the reaction occurs fastest at the most nucleophilic site. Likewise, addition to a polycyclic aromatic molecule with an electron-withdrawing group happens on the ring *without* the substituent.

10.7.2 Reactivity of heterocyclic compounds

The presence of heteroatoms within aromatic systems can have a strong influence on the reactivity and regioselectivity of the corresponding aromatic heterocycles. Second-row heteroatoms (O, N) are electron-withdrawing elements, by induction. However, if they contribute a pair of electrons to the aromaticity, the heterocycle can be quite reactive toward electrophilic aromatic substitution. Regioselectivity can usually be predicted by examining the resonance forms of the heterocyclic arenium intermediates.

For example, the lone pair on the nitrogen atom of pyrrole participates in aromaticity and makes the ring an excellent nucleophile. Pyrroles are so reactive that catalysts are less important, and over-reactions are common.

H N
Br_2
H N Br
+
Br H N Br
+
Br H N Br
Br Br
pyrrole

Furan is less reactive than pyrrole because of the greater electronegativity of oxygen. The nitration of furan produces 2-nitrofuran and 3-nitrofuran, the former being the major product.

O
HNO_3
H_2SO_4
O NO_2 2
+
O 3 NO_2
furan major minor

The regioselectivity can be explained by examining the relative stabilities of the two arenium ions. In the following example, the arenium ion obtained from nitration at position 2 has three main resonance forms; the most significant contributor (all atoms have octets) is conjugated. By comparison, the arenium ion obtained from nitration at position 3 has only two resonance forms; the more significant contributor is cross-conjugated, which prevents the formation of a third resonance form. As with other S_EAr reactions, the more stable ion is formed faster and determines the major product.

fast
HNO_3
H_2SO_4
slow
NO_2
conjugated arenium ion
cross-conjugated arenium ion

Fused heterocycles such as indole react preferentially on the heteroatom-containing ring because this one is the more nucleophilic of the two rings (six electrons over five atoms). In addition, the regioselectivity can be explained in terms of the pathway that forms an arenium ion without breaking the benzene aromaticity.

indole
$E^{\oplus}$
$-H^{\oplus}$
aromaticity preserved
aromaticity weakened

Pyridine is an electron-poor aromatic ring because the electrophilic nitrogen atom destabilizes the positive charge in the addition intermediate. Unlike the nitrogen in pyrrole and indole, the lone pair electrons in pyridine cannot participate in conjugation with the π system. This means that the effects of electronegativity are most important, and the pyridine nitrogen acts as a deactivating, *meta*-directing group.

nitrogen *meta* to ⊕ charge results in lower-energy pathway

pyridine

major

electronegative nitrogen destabilizes ⊕ charge

CHECKPOINT 10.8

You should now be able to predict the outcome of electrophilic aromatic substitution reactions on polycyclic and heterocyclic aromatic compounds.

SOLVED PROBLEM

The nitration of anthracene yields one of three possible products, at positions 1, 2, or 9. Which is the expected major product?

HNO_3, H_2SO_4

STEP 1: Determine the cationic intermediate produced by the addition step at each of the possible positions.

STEP 2: Determine which intermediate is more stable by drawing and evaluating all the resonance structures for each possibility.

In this example, each possible intermediate yields seven resonance structures. The addition to C-9 provides each structure with a full aromatic ring.

Addition to 9

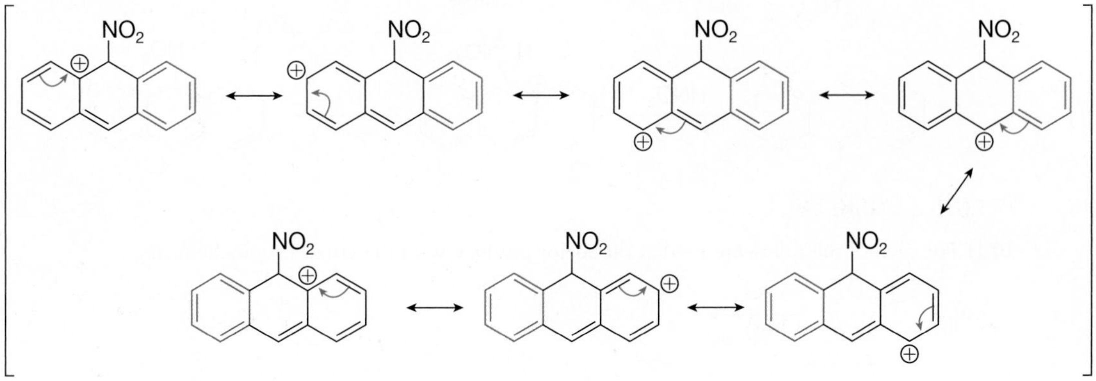

NO_2 NO_2 NO_2 NO_2

NO_2 NO_2 NO_2

Addition to C-1 gives seven resonance structures, only three of which contain a fully aromatic ring.

Addition to 1

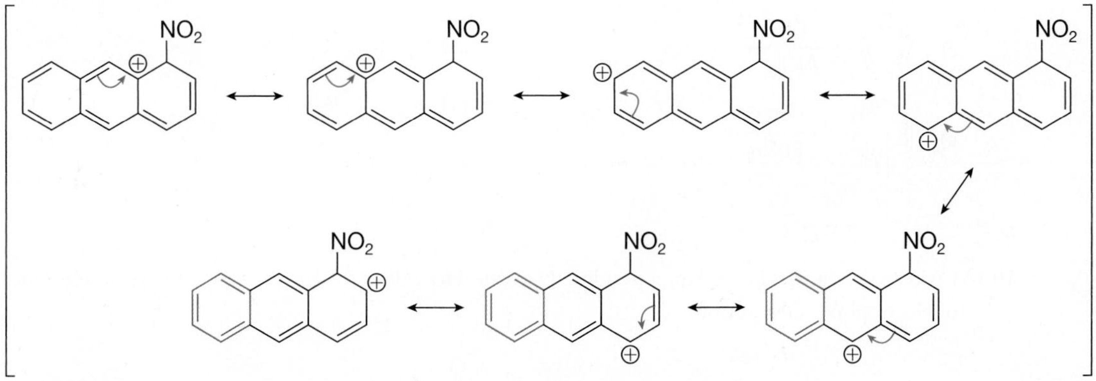

NO_2 NO_2 NO_2 NO_2

NO_2 NO_2 NO_2

Addition to C-2 yields only two resonance structures with a fully aromatic ring.

Addition to 2

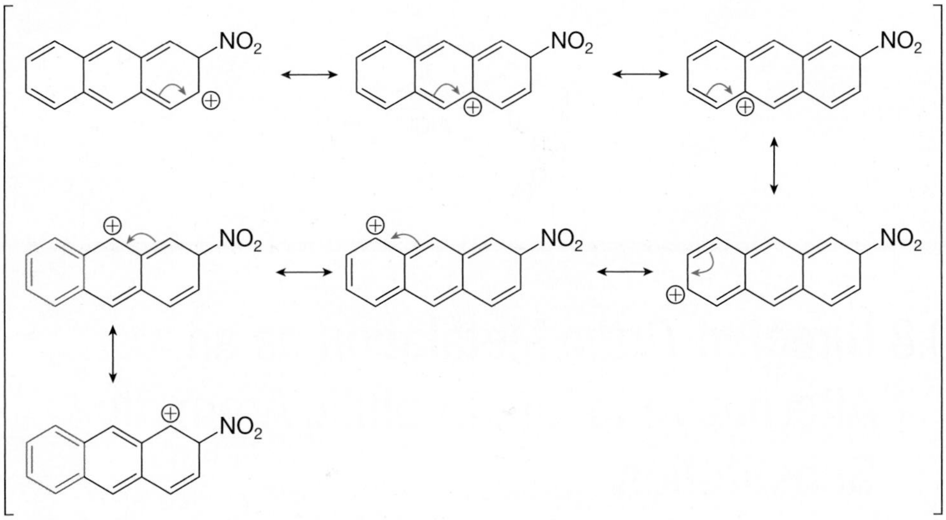

NO_2 NO_2 NO_2

NO_2 NO_2 NO_2

NO_2

STEP 3: Complete the reaction by doing the elimination step, using the most stable intermediate. The expected major product is 9-nitroanthracene.

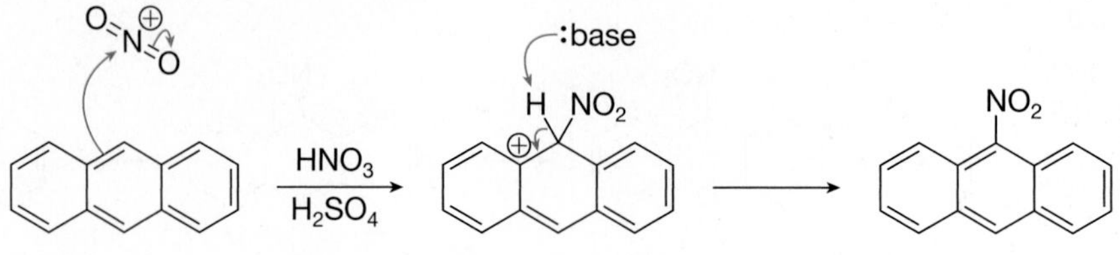

HNO_3 / H_2SO_4 → :base → NO_2

PRACTICE PROBLEM

10.15 For each of the following predict the major product, with the correct regiochemistry.

a) 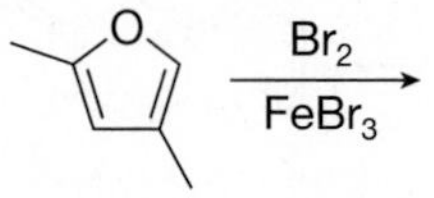 Br_2 / $FeBr_3$

b) 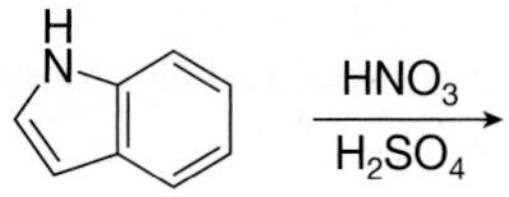 HNO_3 / H_2SO_4

c) 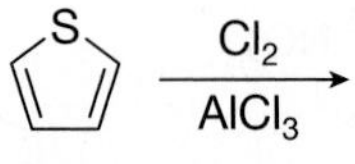 Cl_2 / $AlCl_3$

d) 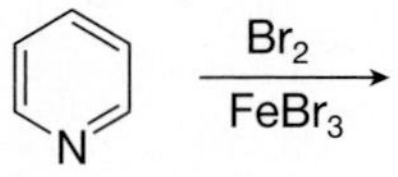 Br_2 / $FeBr_3$

INTEGRATE THE SKILL

10.16 Oxidation of a pyridine forms a pyridine N-oxide. How does oxidation change the regioselectivity of electrophilic substitution?

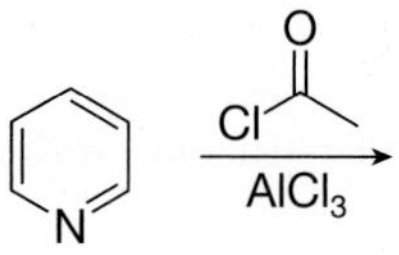

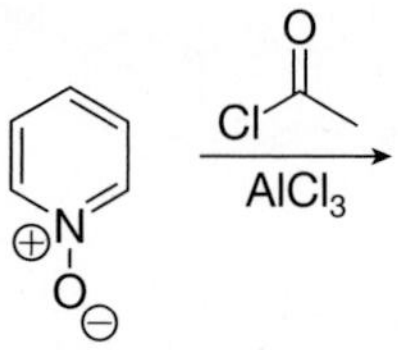

$AlCl_3$

$AlCl_3$

10.8 Directed *Ortho* Metalation as an Alternative to Electrophilic Aromatic Substitution

Directed *ortho* metalation (DOM) is an aromatic substitution reaction that first deprotonates the position *ortho* to a directed metalation group and then reacts with an electrophile.

Electrophilic aromatic substitution is a two-step process: (1) the electrophile adds to the ring, and (2) deprotonation restores the ring aromaticity. In **directed *ortho* metalation** (DOM), the same overall substitution reaction is achieved by reversing the order of these mechanistic steps.

Electrophilic aromatic substitution

add electrophile

remove hydrogen

$E^{\oplus}$

:base

Directed *ortho* metalation

remove hydrogen

add electrophile

DMG

R–Li

$E^{\oplus}$

+RH

$+Li^{\oplus}$

More specifically, the DOM reaction substitutes a functional group *ortho* to a **directed metalation group** (DMG). The first step of the reaction involves the deprotonation of a non-acidic Ar–H bond and requires the use of a very strong base such as an alkyllithium reagent (usually BuLi) at low temperature. The resulting aryllithium intermediate is highly reactive, is partially stabilized by the DMG, and reacts readily in the presence of an electrophile to complete the overall substitution.

Directed metalation group (DMG) is a substituent that favours deprotonation at the adjacent *ortho* position for a directed *ortho* metalation reaction.

Directed metalation groups must have a lone pair that can coordinate with lithium to direct the approach of an alkyllithium base to deprotonate the *ortho* position of the ring. The resulting aryllithium intermediate forms a complex with the *ortho* DMG, making it partially stabilized but retaining a strong carbanion character. Once formed, the aryllithium intermediate can react with a variety of electrophiles, including aldehydes, ketones, epoxides, and alkyl halides. Tetramethylethylenediamine (TMEDA, $Me_2NCH_2CH_2NMe_2$) is sometimes added to enhance the reactivity of the alkyllithium base. TMEDA complexes with the lithium, which increases the reactivity of the butyl group.

lone pair on DMG complexes the alkyllithium base

stabilized ArLi intermediate reacts with electrophile

BuLi

TMEDA

+ BuH

electrophile

deprotonation *ortho* to the DMG

aldehydes, ketones, epoxides, alkyl halides

10.8.1 Common directed metalation groups

A wide variety of directed metalation groups control deprotonation at the *ortho* position. There are a variety of electron-withdrawing DMGs that bear a polar π bond; these DMGs would be *meta* directing in a traditional electrophilic aromatic substitution. Many of these carbonyl-based DMGs can be introduced by using Friedel–Crafts acylation.

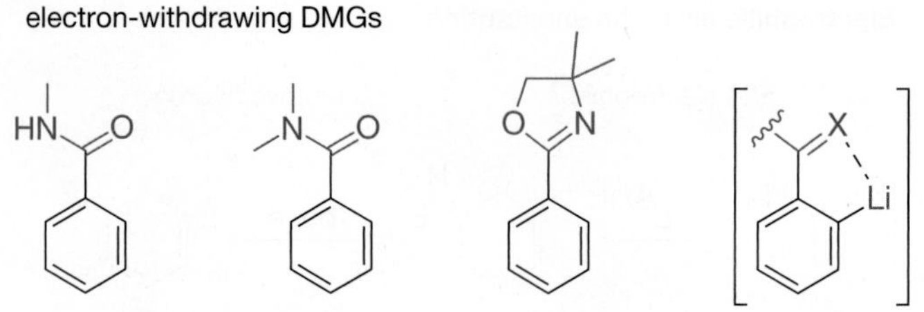

electron-withdrawing DMGs

Below are some electron-donating DMGs that bear a ring heteroatom; they are often derived from a phenol or aniline. With these groups, only *ortho* substitution occurs, rather than a mixture of *ortho* and *para* regioisomers as occurs in a traditional electrophilic aromatic substitution.

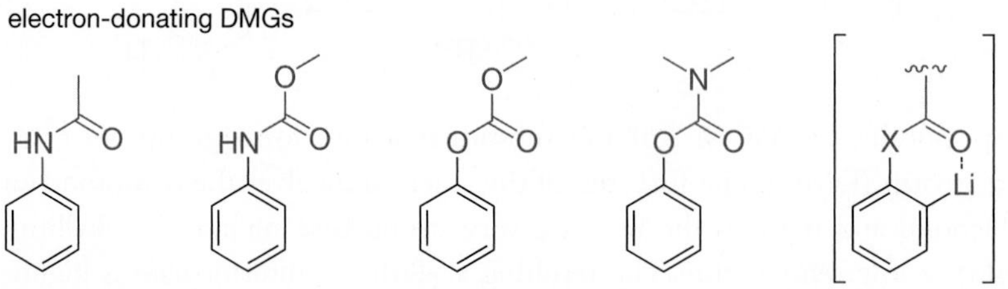

electron-donating DMGs

Once an aryllithium intermediate has formed, a variety of electrophiles can be added, including alkyl halides, carbonyl compounds, and epoxides. This process gives rise to a wide range of substitution products.

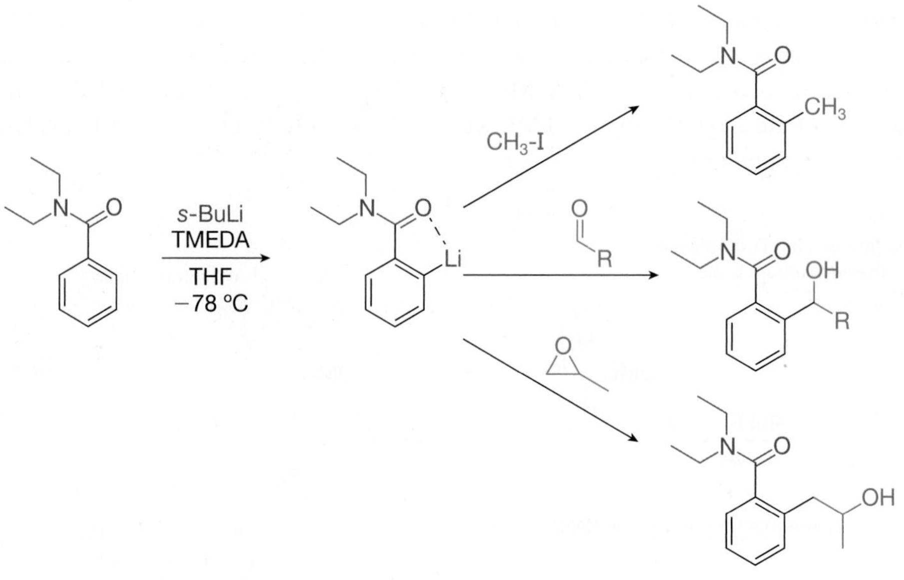

s-BuLi
TMEDA
THF
−78 °C

CH_3-I

CHECKPOINT 10.9

You should now be able to draw a mechanism for the directed *ortho* metalation of substituted aromatic compounds and draw the expected products.

SOLVED PROBLEM

What is the expected major product of the following reactions?

STEP 1: Identify the directed metalation group and coordinate the lithium base to it.

STEP 2: Draw mechanism arrows to deprotonate the adjacent *ortho* position to form an anion. The anion will coordinate to the lithium cation.

STEP 3: Use the aromatic anion as a nucleophile to attack the electrophile.

STEP 4: Add acid to neutralize the basic conditions.

PRACTICE PROBLEM

10.17 a) For each of the following, predict the expected major product.

i)

ii)

iii) 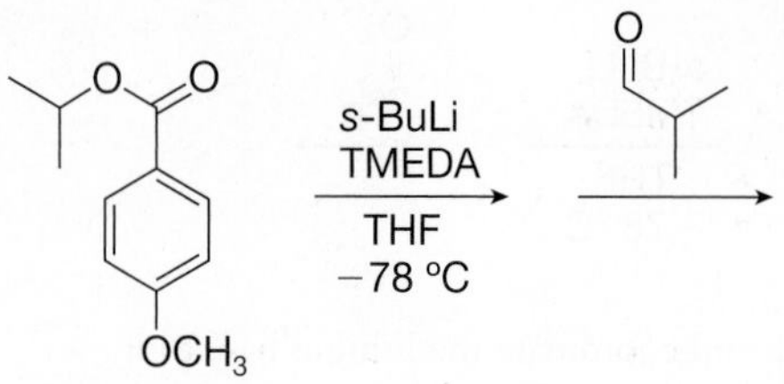 O O s-BuLi TMEDA THF −78 °C O OCH_3

b) For each of the following, identify the required reagents to form the desired products by means of directed metalation.

i) 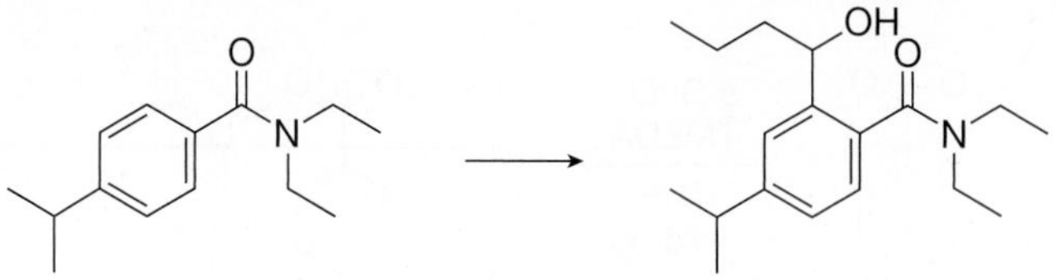 O N → OH O N

ii) 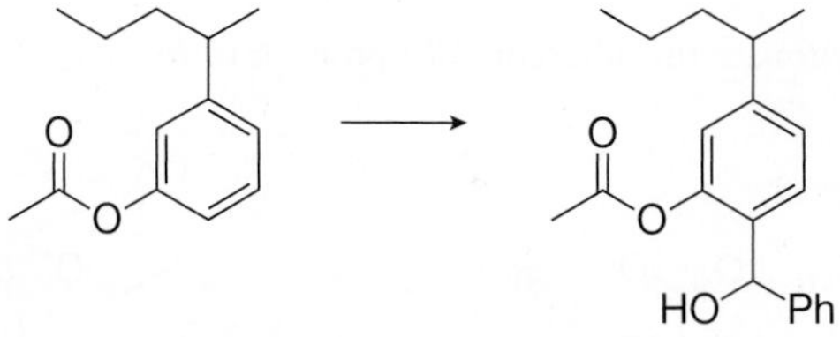 O O → O O HO Ph

iii) 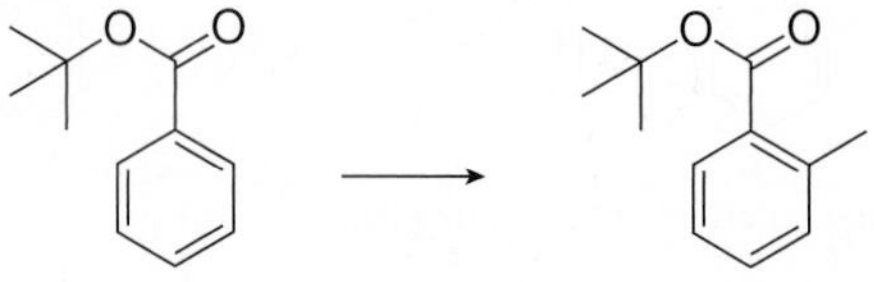 O O → O O

INTEGRATE THE SKILL

10.18 Reactions described in this chapter can be used to selectively methylate a ring either *ortho* or *para* to the ester. Identify the required reagents to synthesize the following compounds.

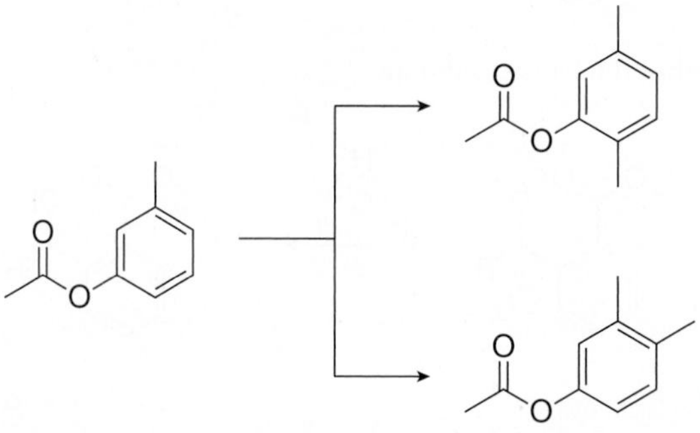

O O → O O ; O O

10.9 Retrosynthetic Analysis in Aromatic Synthesis

Synthesis is the process of making complex molecules from simpler ones.

Chemistry provides the means to create substances that would otherwise not exist. During the twentieth century, many new materials such as plastics, composites, pharmaceuticals, paints, coatings, and alloys were engineered to have specialized properties. These new substances are not usually assembled one atom at a time. Instead, they are produced by reacting different molecules to combine in a specific and controlled manner in a process called **synthesis**. Starting with a given reactant molecule, successive reactions are carried out to modify the molecule. Different chemical reactions are required for each modification, and so synthesis occurs in a stepwise fashion. Because synthetic sequences can be very long and include dozens of chemical reactions in succession, designing a workable synthesis is one of the most challenging tasks in organic chemistry.

An approach commonly used to design a synthesis is called **retrosynthesis** or *retrosynthetic analysis*. The process involves starting from the desired end-product and working backwards to devise suitable transformations. Working backwards from a desired target molecule breaks a complex problem into simpler ones. Retrosynthetic analysis uses a series of **disconnections**—imaginary processes of breaking bonds to visualize how the target molecule can be assembled. A special double arrow, called a *retrosynthetic* or *disconnection arrow*, is used to indicate disconnections. As well, a wiggly line can be added to indicate where a bond is cut in the disconnection.

Retrosynthesis is a technique of planning chemical synthesis in which a target molecule is analyzed in terms of what it can be made from.

A **disconnection** is a retrosynthetic step, an imaginary "reverse" reaction.

The following illustrates a simple retrosynthetic analysis for the synthesis of aniline. Read from left to right to see how the end-product can be obtained from nitrobenzene by a functional group transformation (a reduction in this case) and that nitrobenzene itself can be prepared by the nitration of benzene. Each retrosynthetic arrow corresponds to a possible chemical reaction. Specific reagents are not normally shown over these arrows. This visual representation is particularly useful for planning long multi-step syntheses.

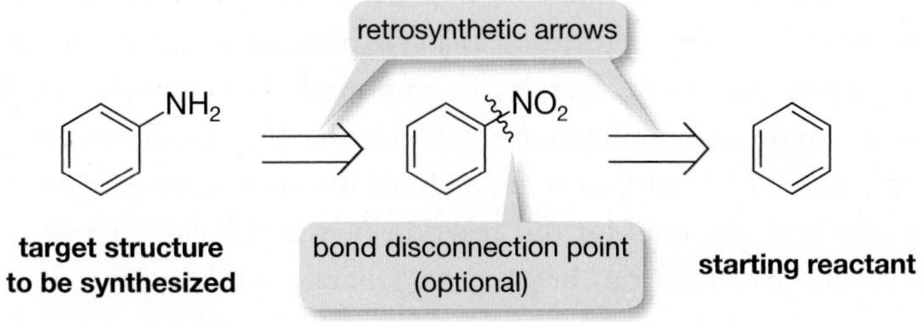

Reactions performed in the wrong sequence can create problems because certain functional groups may interfere or interact in the wrong way. Each disconnection can be evaluated to see if the forward reaction is actually workable, and this information can be used to determine the proper sequence of events.

For example, the two-step synthesis of *p*-nitro isopropylbenzene can be planned by using two different disconnection pathways. One pathway suggests a nitration as the final step; the other suggests a Friedel–Crafts reaction as the final step. Which of these disconnection pathways is the best to use can be determined by considering the forward reaction that would result from each step.

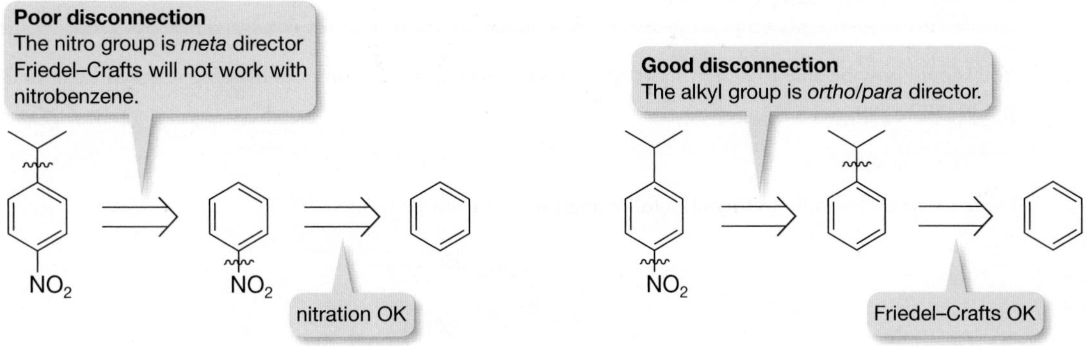

In the retrosynthetic analysis on the left, a nitration is followed by Friedel–Crafts alkylation. However, disconnecting the alkyl group first in the retrosynthetic route is a poor choice. This is because Friedel–Crafts alkylation does not work with a strongly deactivated ring such as nitrobenzene. Even if the reaction did work, the nitro group is a *meta* director, and so alkylation would be directed to an undesired position. In the retrosynthetic analysis on the right, Friedel–Crafts alkylation occurs first, followed by nitration. Disconnecting the nitro group first in the retrosynthetic route is a better option because the isopropyl group is a weak *ortho*/*para* director and will likely direct nitration to produce the correct regioisomer. Using the better retrosynthetic plan, the actual synthesis can be performed.

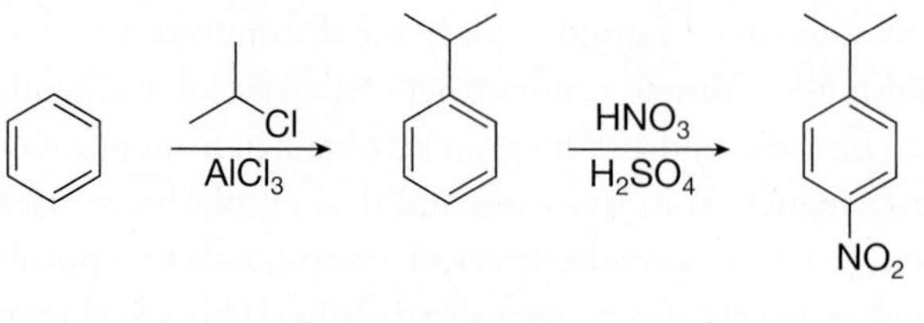

Cl, $AlCl_3$ → ; HNO_3, H_2SO_4 → ; NO_2

10.9.1 Using synthons in synthesis

A **synthon** is a fragment resulting from a disconnection that shows the general reactivity (nucleophile/electrophile) of the fragment.

A particular reaction can be carried out with a range of different reagents. In the design of a synthesis, these alternatives can be represented as a **synthon**, an imaginary component that captures the overall reactivity pattern of a series of compounds. For example, the following shows a retrosynthesis of *p*-amino propiophenone in which the first disconnection corresponds to a Friedel–Crafts acylation. The synthon is an acylium ion that represents the reactivity of the acylation electrophiles typically used for this reaction (i.e., acyl halides, esters, acid anhydrides). Mechanistic arrows and principles can be used to generate synthons, which can then suggest alternative reagent choices for particular reactions. The use of synthons also makes it possible to employ mechanistic arrows in a disconnection analysis.

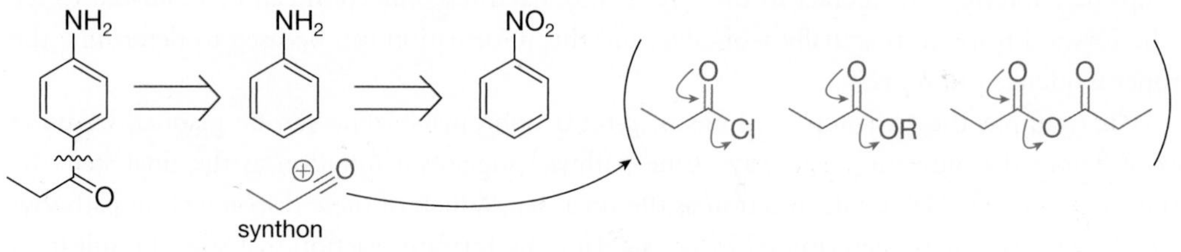

NH_2 ⇒ NH_2 ⇒ NO_2 ; O, Cl ; O, OR ; O, O, O ; O ; ⊕O ; synthon

CHECKPOINT 10.10

You should now be able to design short syntheses of small aromatic compounds using retrosynthetic analysis to guide you.

SOLVED PROBLEM

Design a synthetic route to form the following product from benzene.

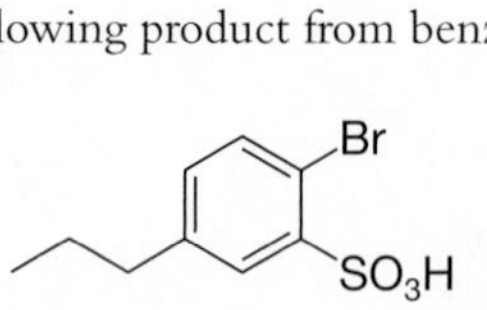

Br ; SO_3H

STEP 1: Identify the reactions needed to add each of the new substituents.

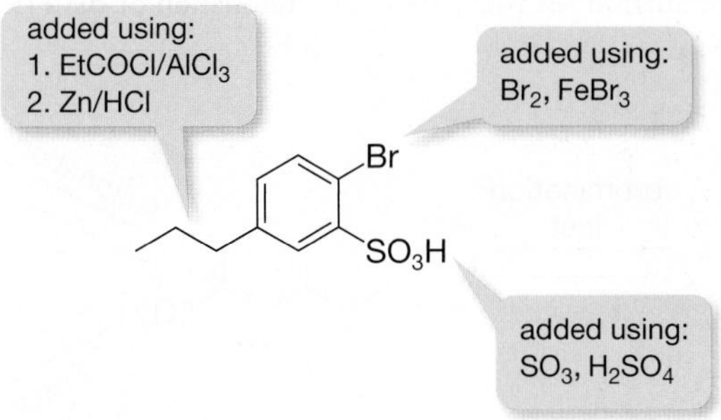

added using:
1. $EtCOCl/AlCl_3$
2. Zn/HCl

added using:
Br_2, $FeBr_3$

added using:
SO_3, H_2SO_4

STEP 2: Identify the directing properties of each substituent.

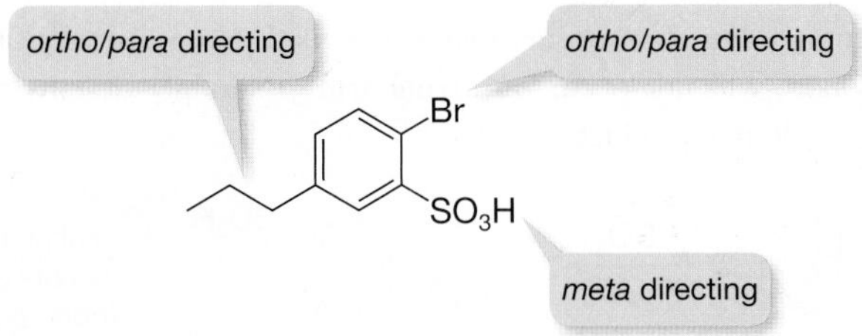

ortho/para directing

ortho/para directing

meta directing

STEP 3: Consider what would be needed if each possible reaction were the last one to be performed.

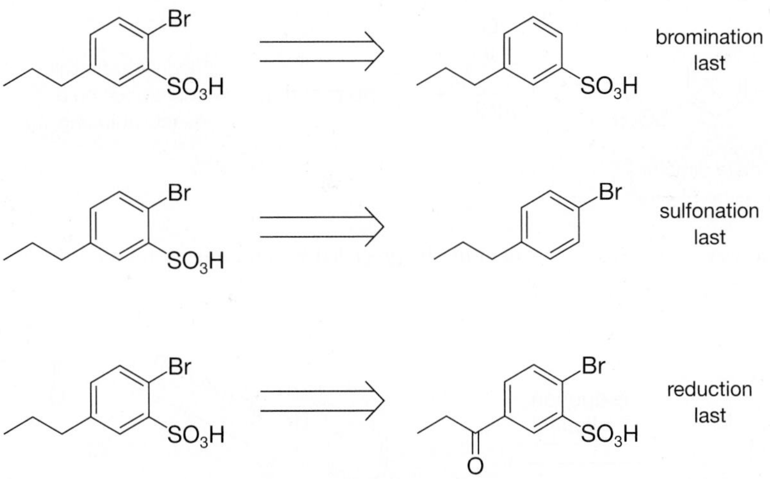

bromination last

sulfonation last

reduction last

STEP 4: In each case, consider whether the forward reaction would work as desired.

The required sulfonation would occur at the wrong position, so this route need not be considered.

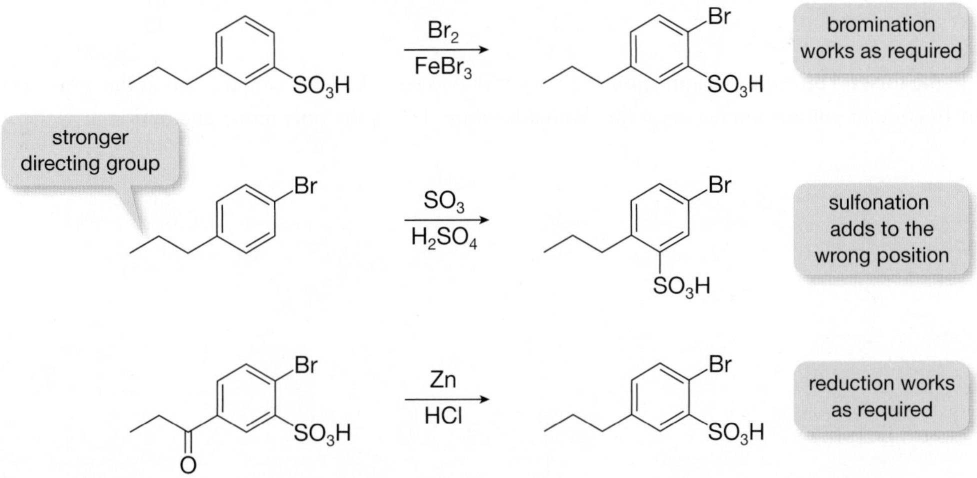

Br_2 / $FeBr_3$ — bromination works as required

stronger directing group

SO_3 / H_2SO_4 — sulfonation adds to the wrong position

Zn / HCl — reduction works as required

STEP 5: For the remaining options, repeat Steps 3 and 4 until the desired starting material has been reached. Two options for the bromination last route are either sulfonation or the Friedel–Crafts reaction.

Br
SO_3H
bromination last
SO_3H
sulfonation
Friedel–Crafts
SO_3H

In the forward direction, sulfonation occurs *para* to the alkyl group instead of the required *meta*. The other option, Friedel–Crafts, will not occur on a ring with a strongly deactivating sulfonic acid. This pathway cannot lead to the desired product.

SO_3
H_2SO_4
SO_3H
sulfonation occurs at the wrong position
ortho/para director

O
Cl
$AlCl_3$
SO_3H
no reaction
Friedel–Crafts will not happen on a deactivated ring
meta director

The last pathway of the reduction can either be preceded by a bromination or a sulfonation.

Br
SO_3H
reduction last
Br
SO_3H
O
bromination
SO_3H
O
sulfonation
Br
O

In the forward direction, bromination of a ring with two *meta* directors would occur at the wrong position. In contrast, sulfonation occurs at the desired location. This is the only remaining option.

bromination occurs at the wrong position

meta director

sulfonation occurs as required

meta director

ortho/para director

Repeating Steps 2 and 3 indicate that sulfonation must be preceded by Friedel–Crafts acylation and bromination, in that order. In the forward direction, bromination with the *meta* director would result in the wrong isomer. However, acylation of the bromobenzene favours the desired *para* position.

acylation

bromination

STEP 6: Put the reactions in the correct sequence in the forward direction.

PRACTICE PROBLEM

10.19 For each of the following, and starting with benzene, design a synthetic route to form each of the following compounds with the correct regioselectivity.

a) O, Br, NO_2

b) Cl, Br

c) Cl, Br, O_2N

INTEGRATE THE SKILL

10.20 This chapter focuses on reactions that add electrophiles to aromatic rings. Reactions that add only electrophiles to aromatic rings form only a limited range of products. Other reactions, discussed in Chapter 15, add a nucleophile to an aromatic ring and form another range of products. Explain why the following product cannot be synthesized from benzene, using only the reactions described so far.

Cl, C≡N, Cl

10.10 Patterns in Electrophilic Aromatic Substitution Reactions

Electrophilic aromatic substitution reactions all follow the same basic pattern of reactivity: the addition of an electrophile to form an arenium ion, followed by the elimination of a hydrogen to restore the aromaticity. Reactivity and regioselectivity are controlled by the stability of the arenium ions.

$E^{\oplus}$ → H, E, ⊕ :base → E

Because of the stability of the starting aromatic rings, most electrophiles require activation by a catalyst, which may add a few steps to the mechanistic sequence. However, the general electron flow for the S_EAr substitutions is preserved. What changes is the method of activating the electrophile. For halogenations and Friedel–Crafts reactions, the electrophile is activated through the formation of a Lewis acid complex. For nitration and sulfonation, the electrophile is the result of protonation with a Brønsted acid, which leads to the loss of water.

The regioselectivity of S_EAr reactions that involve substituted aromatic rings is controlled by the nature of the substituent. As a general rule, electron-donating groups are activating and stabilize the intermediates, leading to *ortho* and *para* substitution products. Electron-withdrawing groups are deactivating and destabilize any resonance forms with adjacent positive charges, making *ortho*/*para* pathways very unfavourable. Because the *meta* pathway is less affected than *ortho* and *para* directions, this pathway provides the lowest energy route to products, and the *meta* isomer dominates.

electron-donating groups *ortho* or *para* to the incoming electrophile stabilize these ions and favour *ortho*/*para* products

groups *meta* to the incoming electrophile do not strongly change the energy of these ions

electron-withdrawing groups *ortho* or *para* to the incoming electrophile destabilize these ions and favour *meta* products

Directed *ortho* metalation reverses the steps for substitution: the ring is first deprotonated, and then the electrophile is added. The ring must have a directed metalation group (DMG), which forms a complex with an alkyl lithium base and facilitates removal of the *ortho* hydrogen. This results in substitution *only* at the *ortho* position.

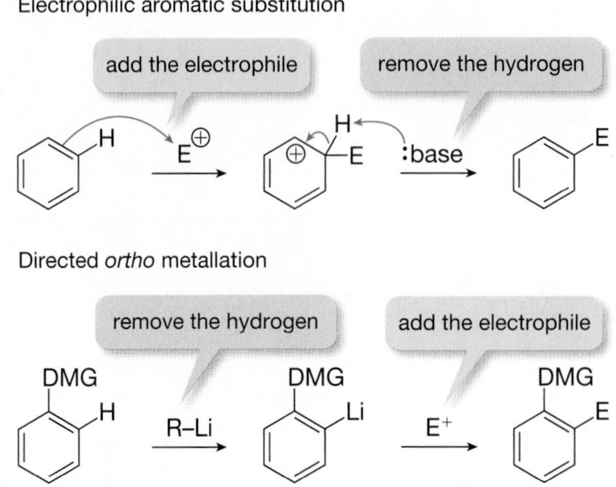

Bringing It Together

Before 1890, most medicines were ineffective or harmful. Medicines that actually worked were in limited supply and usually too expensive for the average person to afford. This situation changed in the late 1800s when synthetic drugs became available. These substances were aromatic compounds synthesized from abundant petroleum products such as oil and coal. The low manufacturing costs associated with these materials made the new synthetic drugs available to many people, resulting in a dramatic improvement in health care and life expectancy.

A large proportion of modern drugs are aromatic compounds, not because of their properties as drugs but because of two factors that kept manufacturing costs low for the pharmaceutical industry during the twentieth century. First, aromatic compounds are relatively easy to manufacture in large quantities (using many of the reactions shown in this chapter). Second, the source of most organic chemicals is petroleum (coal and later oil), which contains large amounts of aromatic molecules. Sulfanilamide can be made from benzene (readily available from oil), using two S_EAr reactions to add the main components. Nitration of benzene gives nitrobenzene, which is easily reduced to form aniline (also available from coal in large amounts). Acetylation of this material (Chapter 15) produces acetanilide, the amide group that is an *ortho/para* director. Sulfonation of acetanilide gives the *para* isomer selectively due to steric interference from the amide. Some small modifications to the functional groups produce sulfanilamide.

Sulphanilamide was the first antibiotic to be a commercial success, and it was widely available from the 1930s onward. One of the keys to its impact as a drug was its low manufacturing cost, which made this drug available to many who could not otherwise afford effective medication.

You Can Now

- Draw a mechanism for the halogenation of aromatic rings and draw the expected products.
- Draw a mechanism for the nitration and sulfonation of aromatic rings and draw the expected products.
- Draw the products formed when nitrated aromatics are treated with a dissolving metal reduction.
- Draw a mechanism for the Friedel–Crafts alkylation of aromatic rings and draw the expected products.
- Draw a mechanism for the Friedel–Crafts acylation of aromatic rings, and the related Gatterman–Koch reaction, and draw the expected products.
- Draw the products formed when acylated aromatics are treated under Wolff–Kischner or Clemmenson reduction conditions.
- Classify aromatic substituents as either *ortho/para* directing or *meta* directing, and determine whether they will activate or deactivate the ring toward electrophilic aromatic substitution.
- Use directing groups to predict the products of electrophilic aromatic substitution on monosubstituted and polysubstituted benzenes.
- Propose appropriate methods to modify the reactivity of strongly activating substituents to help control electrophilic aromatic substitution reactions.
- Predict the outcome of electrophilic aromatic substitution reactions on polycyclic and heterocyclic aromatic compounds.
- Draw a mechanism for the directed *ortho* metalation of substituted aromatic compounds and draw the expected products.
- Design short syntheses of small aromatic compounds using retrosynthetic analysis.

A Mechanistic Re-View

Electrophilic Aromatic Substitution (S_EAr)

$$\text{benzene} \xrightarrow{E^{\oplus}} \text{arenium ion (H, E, }\oplus) \xrightarrow{:\text{base}} \text{Ar–E}$$

Bromination

$$Br-Br + FeBr_3 \longrightarrow Br-\overset{\oplus}{Br}-\overset{\ominus}{Fe}Br_3$$

$$\text{Ar–H} + Br-\overset{\oplus}{Br}-\overset{\ominus}{Fe}Br_3 \longrightarrow \text{arenium ion (H, Br, }\oplus) \xrightarrow{Br-\overset{\ominus}{Fe}Br_3} \text{Ar–Br} + HBr + FeBr_3$$

Nitration

$$HNO_3 + H_2SO_4 \rightleftharpoons H_2NO_3^{\oplus} + HSO_4^{\ominus}$$

$$H_2NO_3^{\oplus} \longrightarrow O{=}\overset{\oplus}{N}{=}O$$

$$\text{Ar–H} + O{=}\overset{\oplus}{N}{=}O \longrightarrow \text{arenium ion (H, }NO_2\text{, }\oplus) \xrightarrow{H_2O} \text{Ar–}NO_2 + H_3O^{\oplus}$$

Sulfonation

$$H_2SO_4 + H_2SO_4 \rightleftharpoons H_3SO_4^{\oplus} \text{ (}OH_2^{\oplus}\text{)} + HSO_4^{\ominus}$$

$$\Big\Updownarrow -H_2O$$

$$SO_3 + H_2SO_4 \rightleftharpoons HSO_3^{\oplus} + H_2SO_4$$

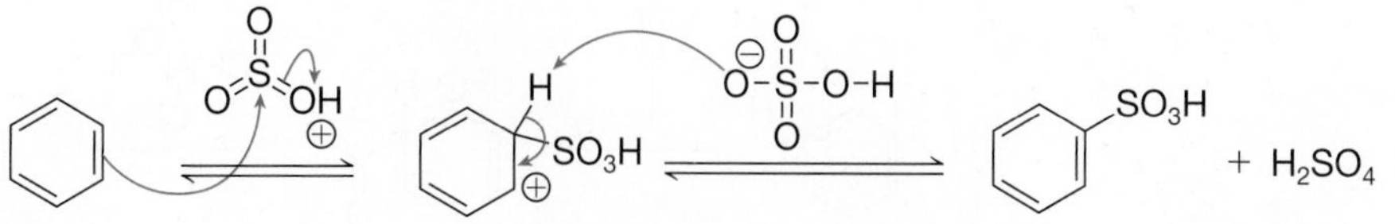

Desulfonation

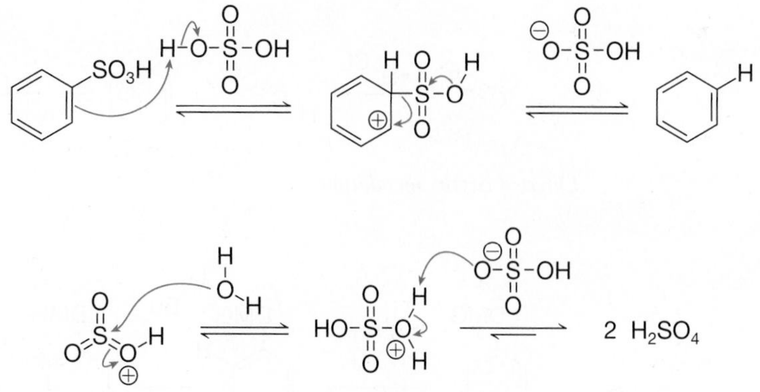

Friedel–Crafts alkylation

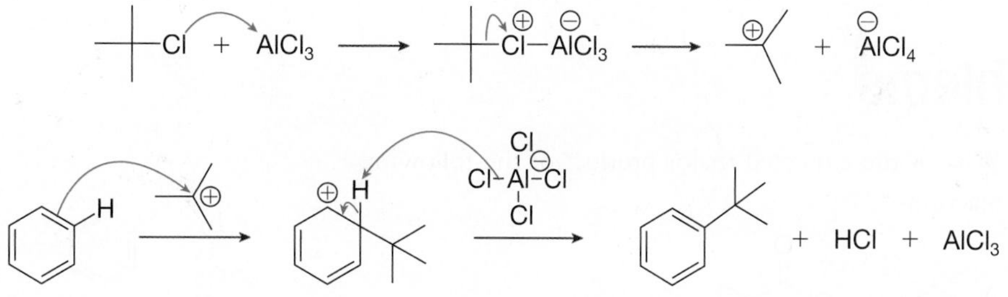

Carbocation rearrangements

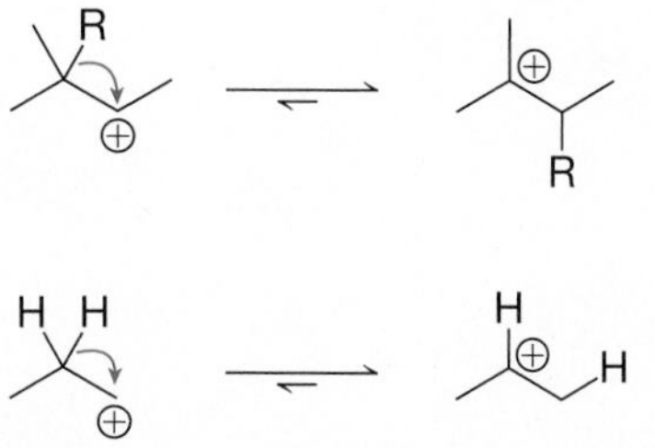

Friedel–Crafts acylation

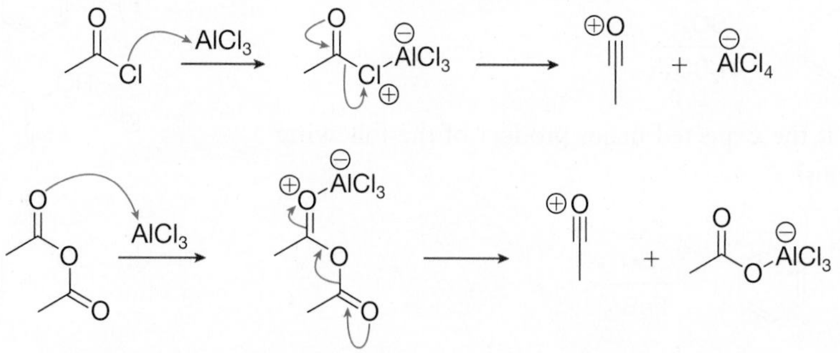

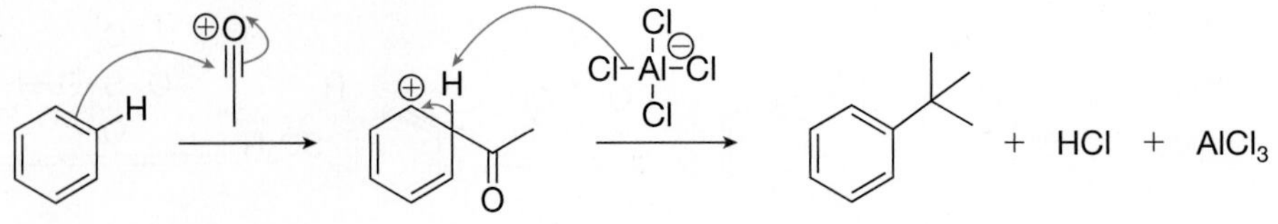

+ HCl + $AlCl_3$

Gatterman–Koch formylation

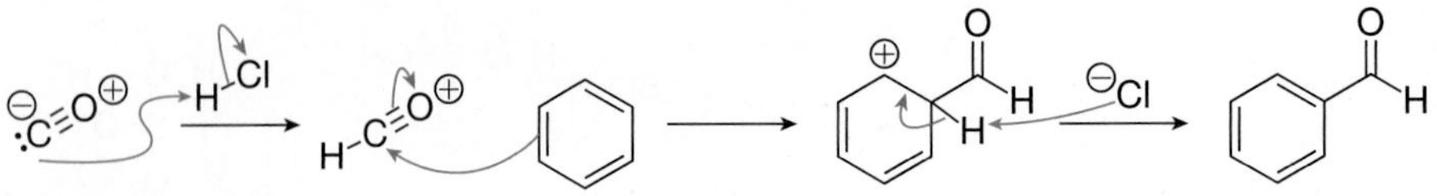

Directed ortho *metalation*

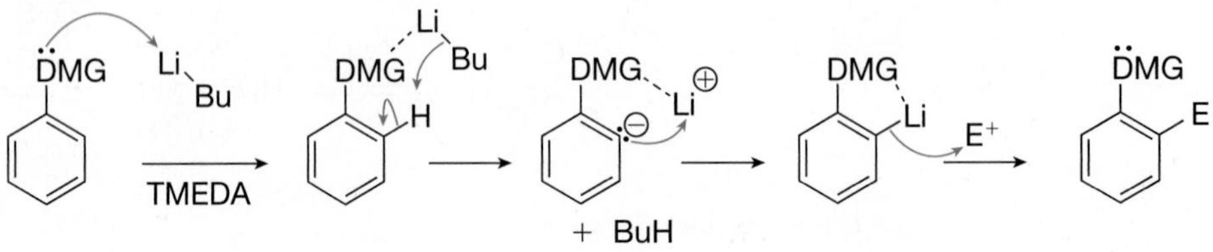

DMG Li Bu TMEDA + BuH E^+ E

Problems

10.21 What is the expected major product of the following reactions?

a) 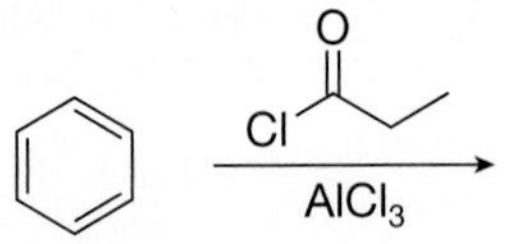 Cl (propanoyl chloride), $AlCl_3$

b) 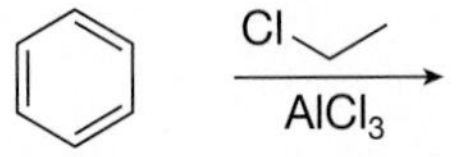 Cl (ethyl chloride), $AlCl_3$

c) 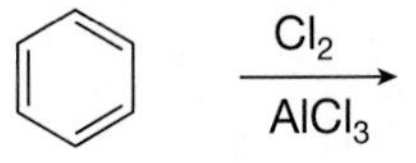 Cl_2, $AlCl_3$

d) 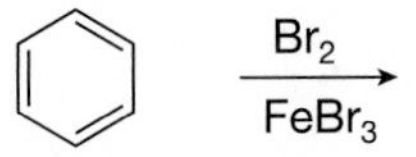 Br_2, $FeBr_3$

e) 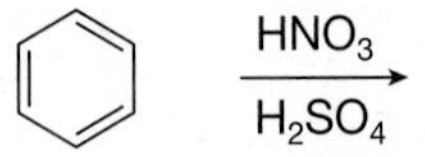 HNO_3, H_2SO_4

f) 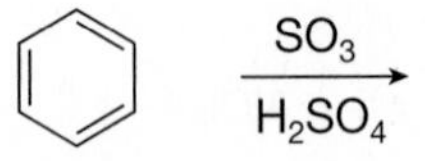 SO_3, H_2SO_4

10.22 What is the expected major product of the following reactions?

a) 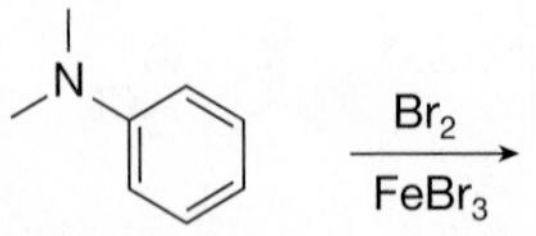 Br_2, $FeBr_3$

b) 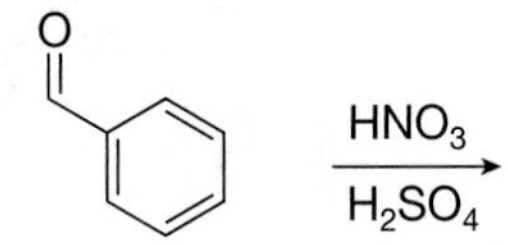 HNO_3, H_2SO_4

c) SO_3, H_2SO_4

d) Cl (propanoyl chloride), $AlCl_3$

e) H_3CO — Cl (propanoyl chloride), $AlCl_3$

f) Cl (ethyl chloride), $AlCl_3$

g) HO — Cl_2, $AlCl_3$

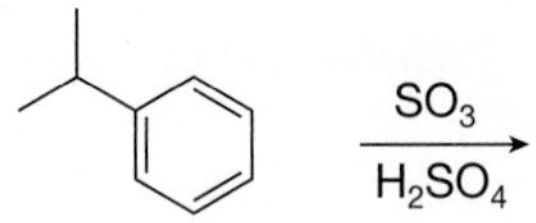

h) Br_2, $FeBr_3$

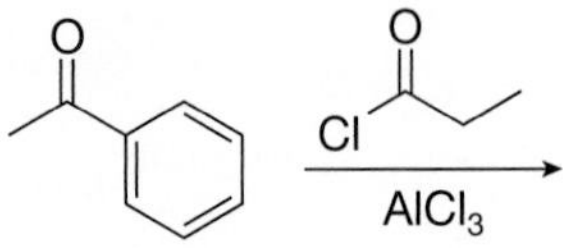

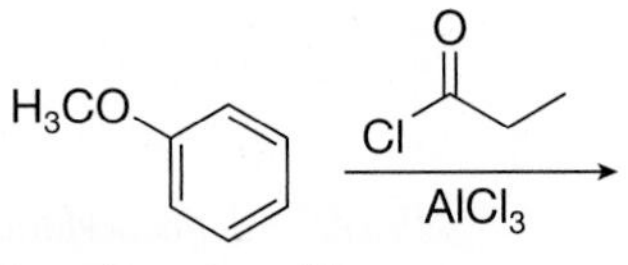

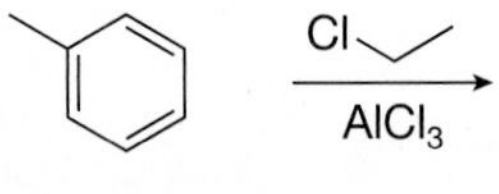

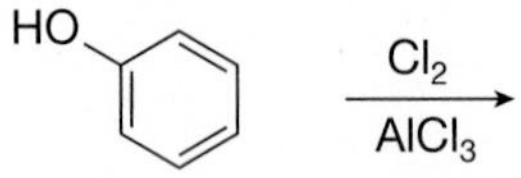

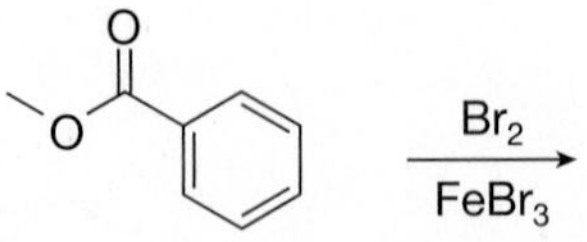

i) 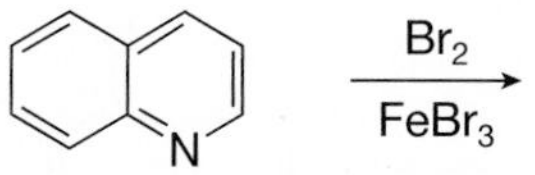 Br$_2$ / FeBr$_3$ (N)

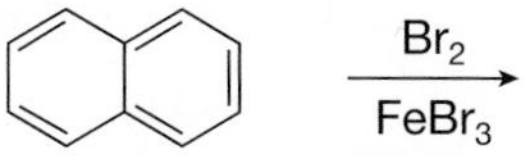

j) HO, O — SO$_3$ / H_2SO_4

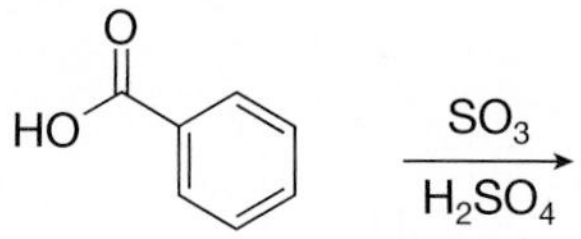

k) O_2N — Cl / $AlCl_3$

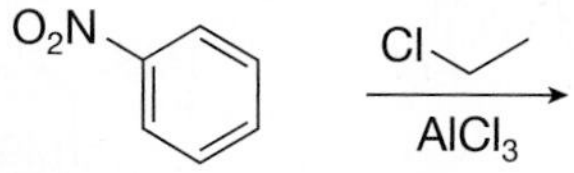

l) N≡C — Cl_2 / $AlCl_3$

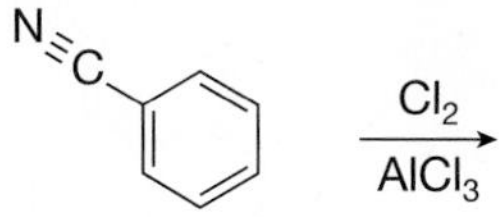

m) O, O — HNO_3 / H_2SO_4

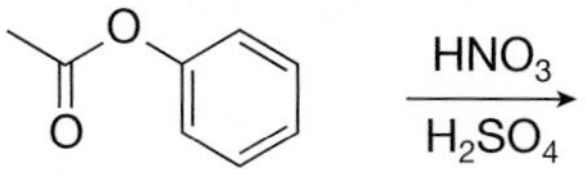

n) Br_2 / $FeBr_3$

o) N H — HNO_3 / H_2SO_4

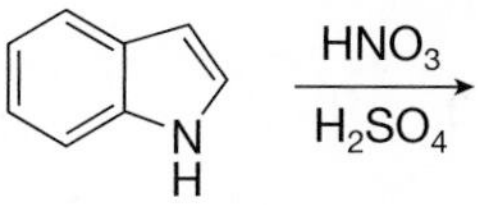

p) 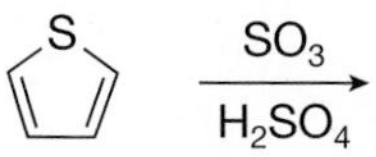 S — SO_3 / H_2SO_4

10.23 What is the expected major product of the following reactions?

a) 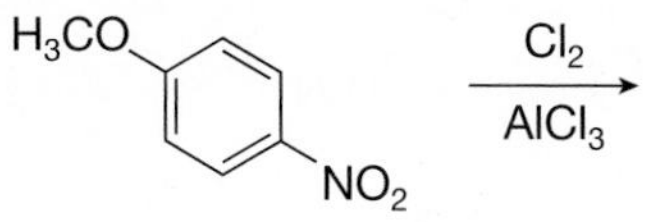 H_3CO, NO_2 — Cl_2 / $AlCl_3$

b) 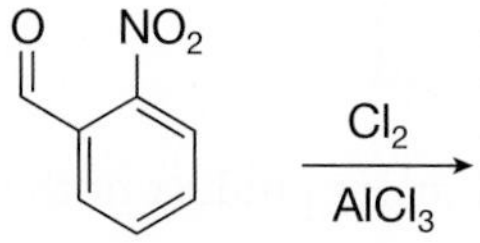 O, NO_2 — Cl_2 / $AlCl_3$

c) 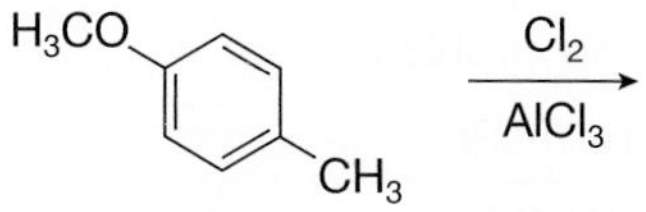 H_3CO, CH_3 — Cl_2 / $AlCl_3$

d) 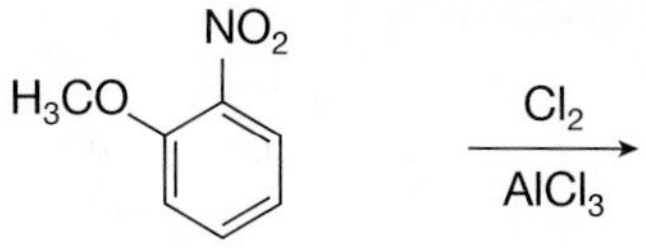 NO_2, H_3CO — Cl_2 / $AlCl_3$

e) 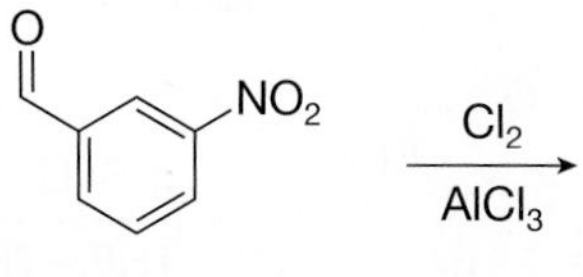 O, NO_2 — Cl_2 / $AlCl_3$

f) 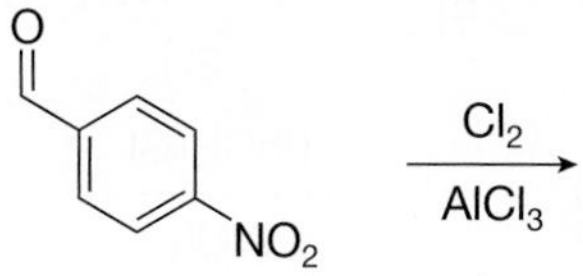 O, NO_2 — Cl_2 / $AlCl_3$

g) H_3CO, CH_3 — Cl_2 / $AlCl_3$

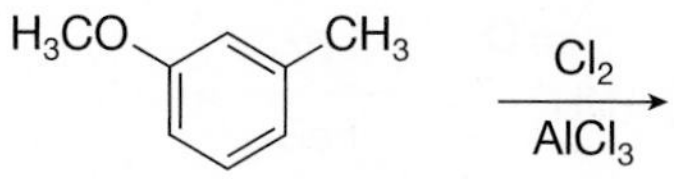

h) H_3CO, NO_2 — Cl_2 / $AlCl_3$

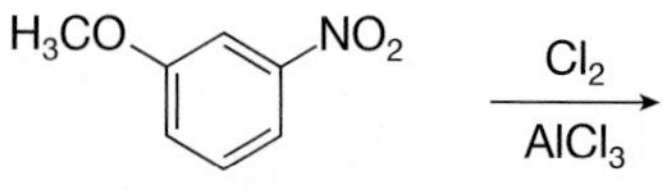

i) 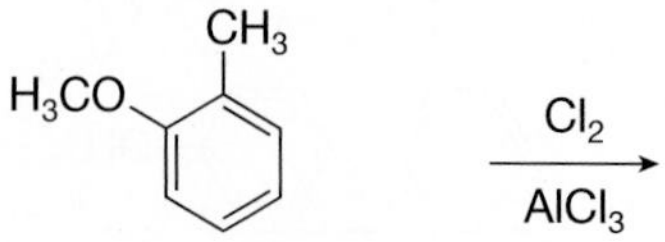 CH_3, H_3CO — Cl_2 / $AlCl_3$

j) 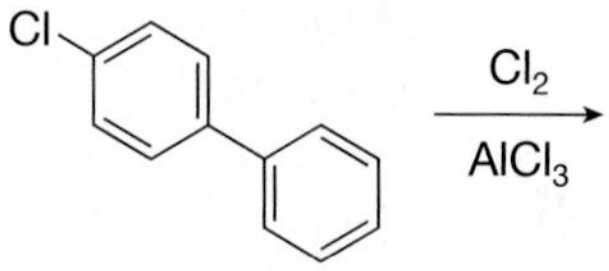 Cl — Cl_2 / $AlCl_3$

k) 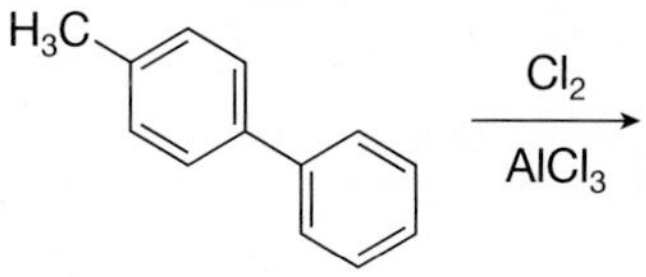 H_3C — Cl_2 / $AlCl_3$

l) 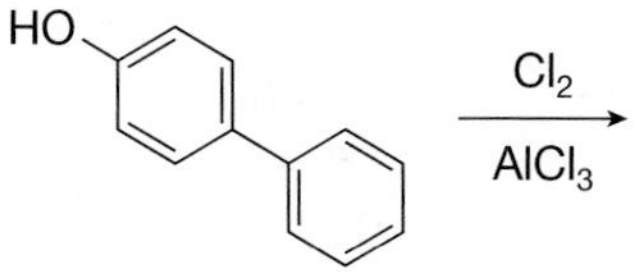 HO — Cl_2 / $AlCl_3$

10.24 What is the expected major product of the following reactions?

a) 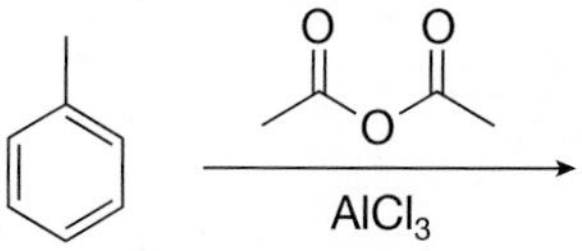 O, O, O — $AlCl_3$

b) 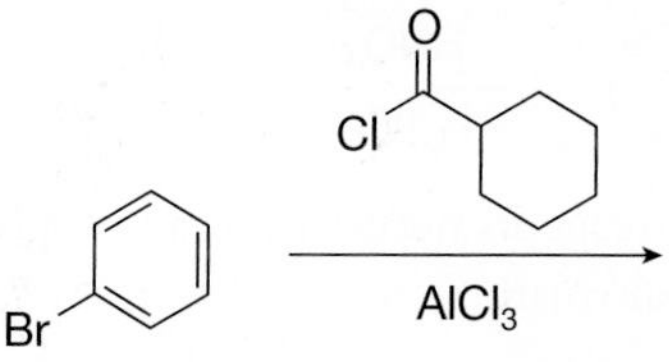 O, Cl, Br — $AlCl_3$

c) 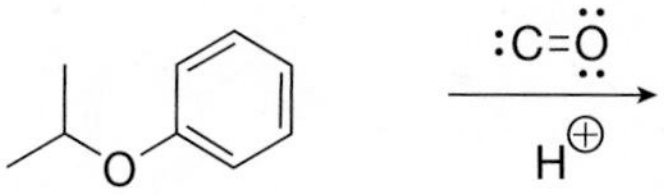 O — :C=Ö: / $H^{\oplus}$

d) S — Cl_2 / $AlCl_3$

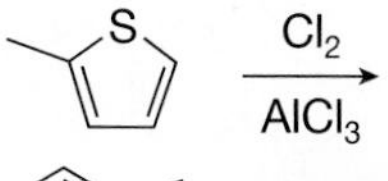

e) 1) HNO_3 / H_2SO_4 2) Zn/HCl

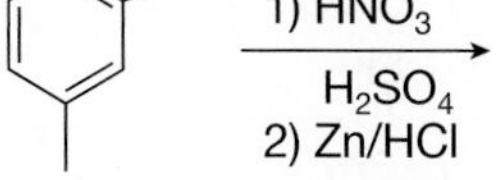

f) NO_2 — Cl_2 / $AlCl_3$

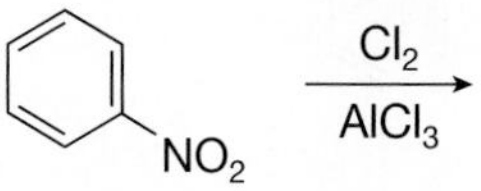

g) Cl — $AlCl_3$

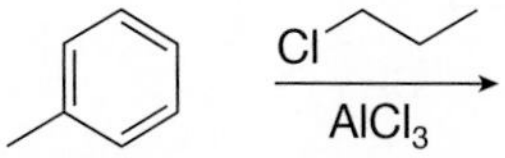

h) H N — Cl / $AlCl_3$

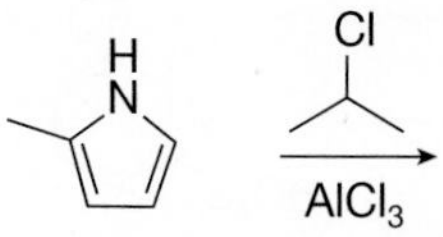

i) 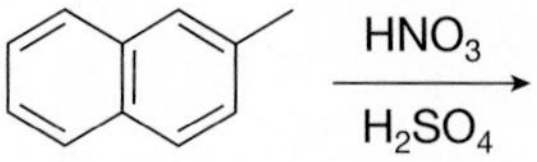 HNO_3 / H_2SO_4

j) 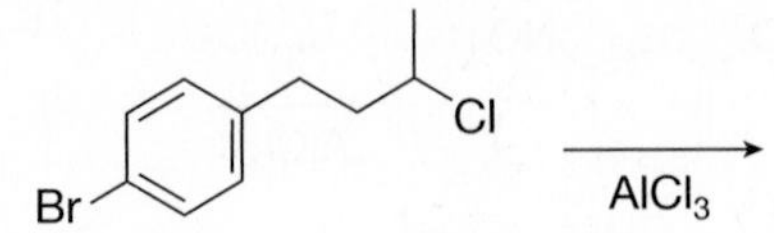 Br, Cl; $AlCl_3$

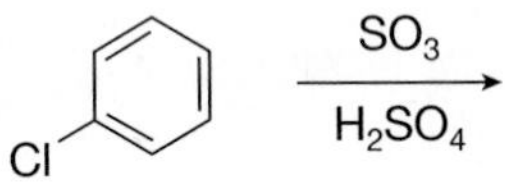

k) 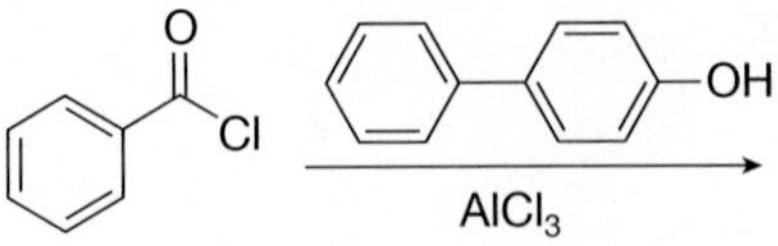 Cl, O, OH; $AlCl_3$

l) 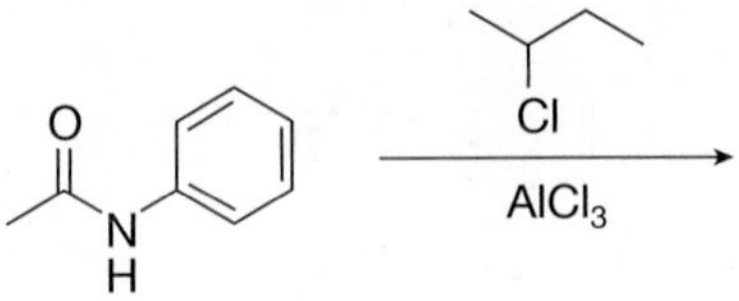 O, N, H, Cl; $AlCl_3$

m) Cl; SO_3 / H_2SO_4

n) 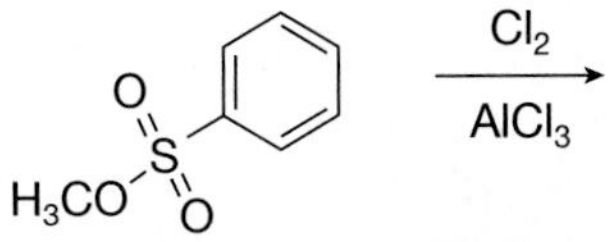 H_3CO, S, O, O; Cl_2 / $AlCl_3$

o) O, Cl; $AlCl_3$

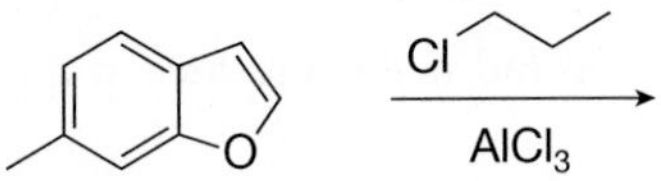

p) 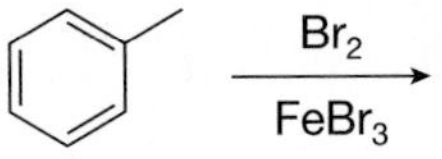 Br_2 / $FeBr_3$

q) 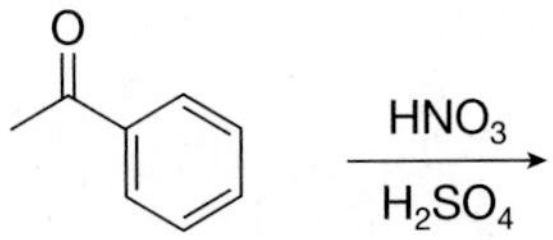 O; HNO_3 / H_2SO_4

10.25 Provide the reagents necessary to accomplish the following transformations.

a) 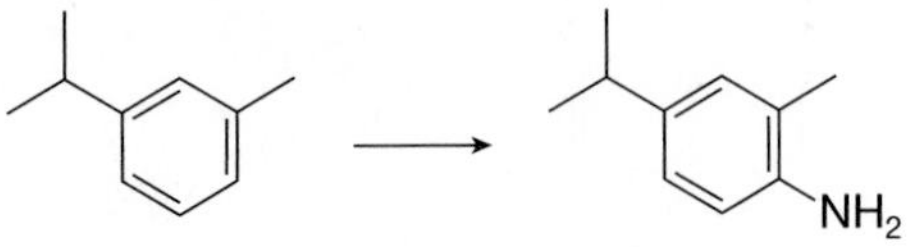 NH_2

b) 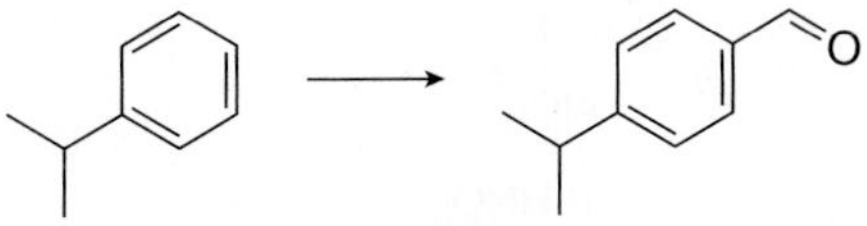 O

c) 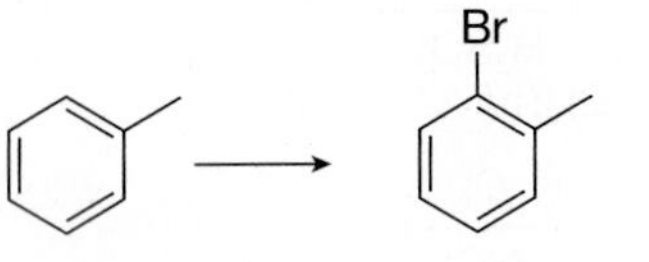 Br

d) 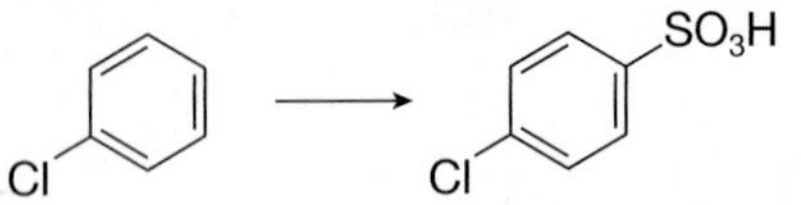 Cl → Cl, SO_3H

e) 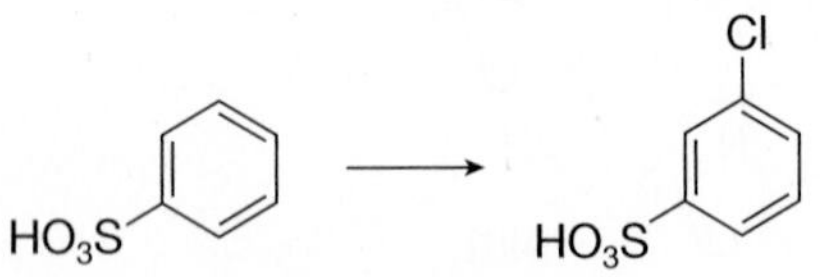 HO_3S → HO_3S, Cl

f) 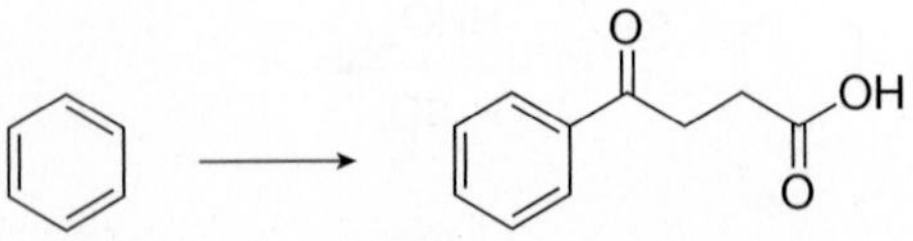 O, OH, O

g) 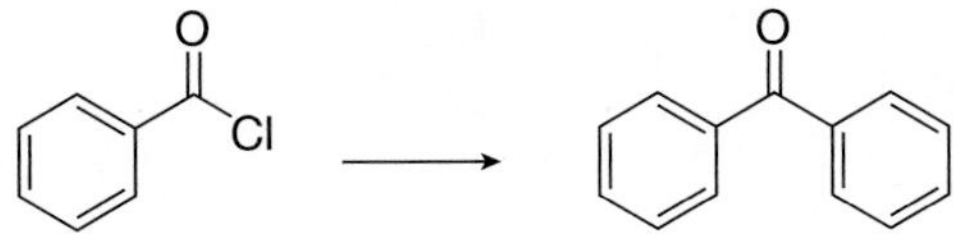 O, Cl → O

h) 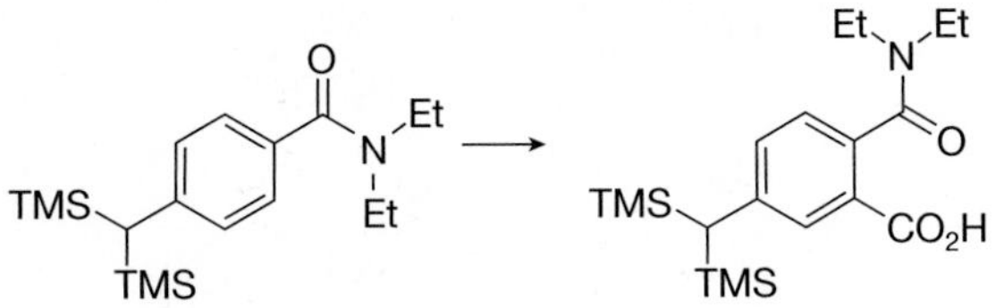 TMS, TMS, O, N, Et, Et → TMS, TMS, Et, N, Et, O, CO_2H

i) 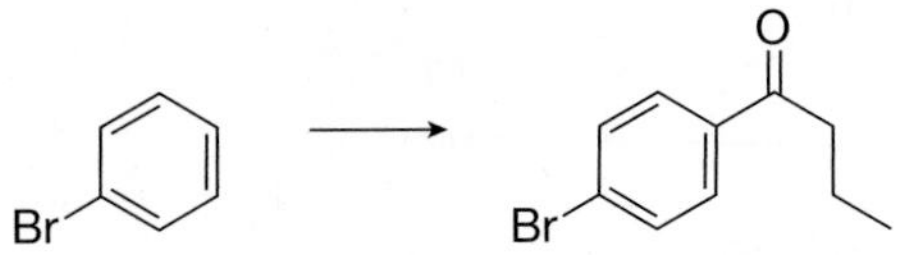 Br → Br, O

j) 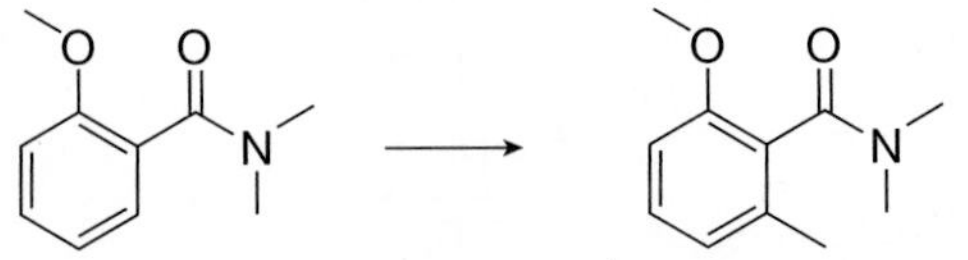 O, O, N → O, O, N

k)

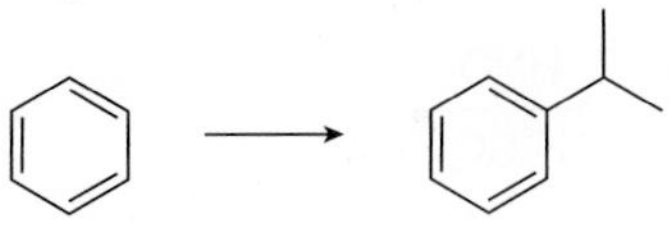

l) 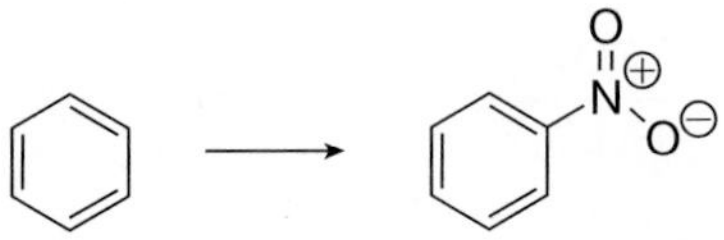 O, N⊕, O⊖

m) 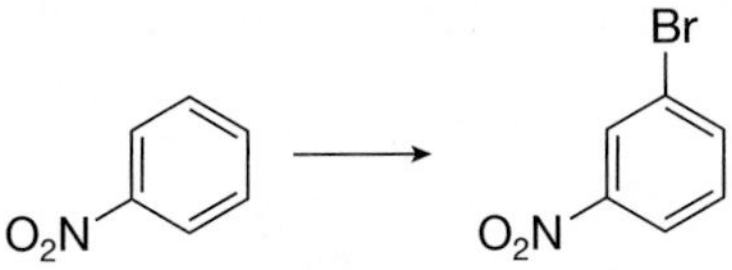 O_2N → O_2N, Br

n)

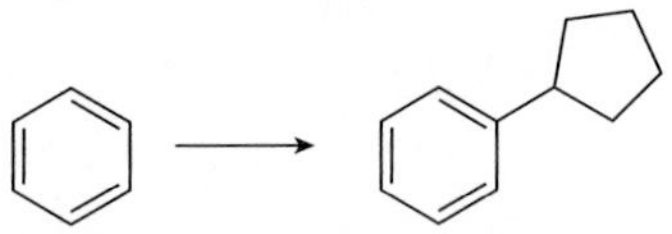

10.26 What is the expected major product of the following reactions? Include proper regio- and stereochemical information, if applicable.

a) 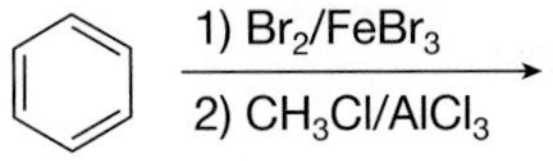 1) $Br_2/FeBr_3$ 2) $CH_3Cl/AlCl_3$

b) 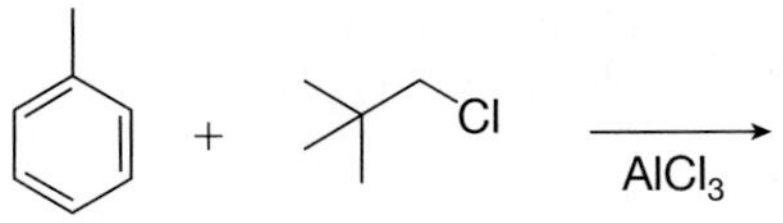 + Cl; $AlCl_3$

c) 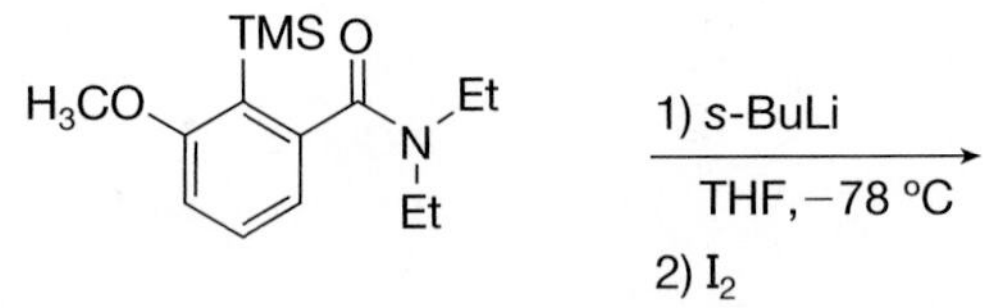 H_3CO, TMS, O, N, Et, Et; 1) *s*-BuLi, THF, −78 °C 2) I_2

d) 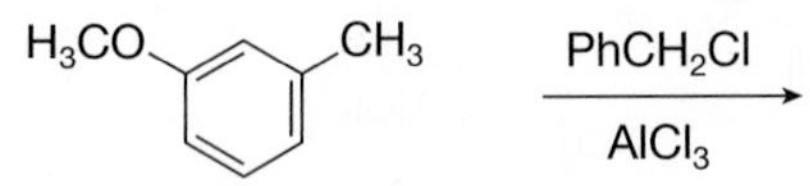 H_3CO, CH_3; $PhCH_2Cl$ / $AlCl_3$

e) 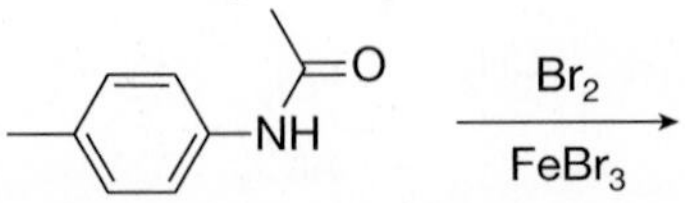 NH, O; Br_2 / $FeBr_3$

f) 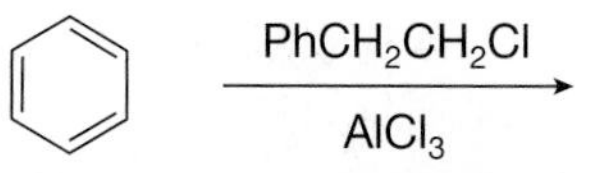 $\xrightarrow[\text{AlCl}_3]{\text{PhCH}_2\text{CH}_2\text{Cl}}$

g)  HO_2C–C6H4–C6H4–OH $\xrightarrow[\text{AlCl}_3]{\text{Cl}_2}$

h) 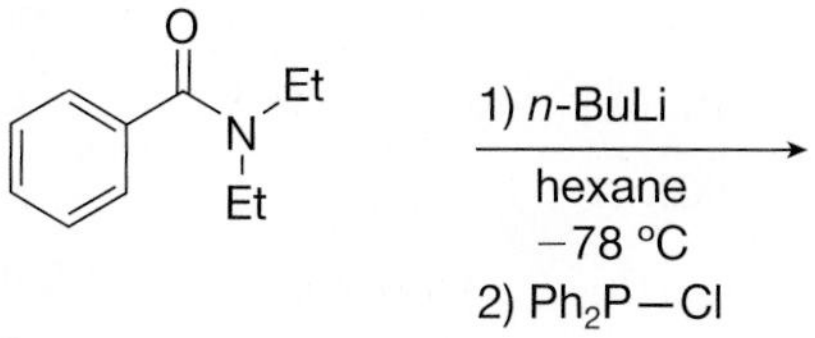 Et N Et $\xrightarrow[\text{hexane, } -78\ °\text{C; 2) Ph}_2\text{P—Cl}]{\text{1) } n\text{-BuLi}}$

10.27 Provide the reagents necessary to accomplish the following transformations. In each case, more than one step may be required.

a)

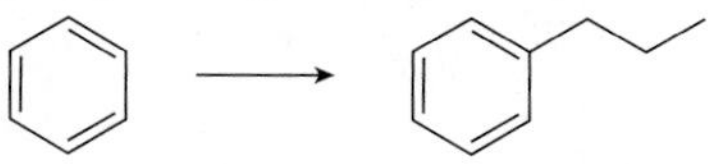

b) 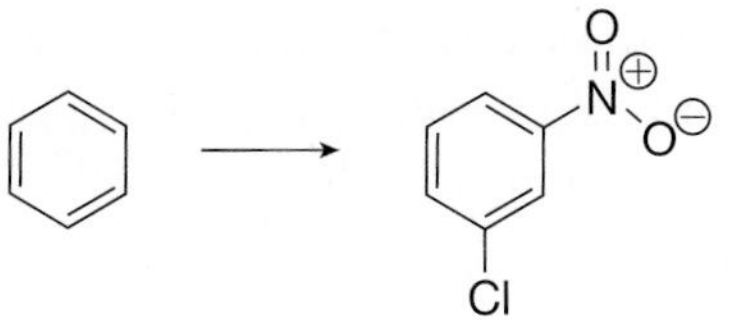 Cl

c) 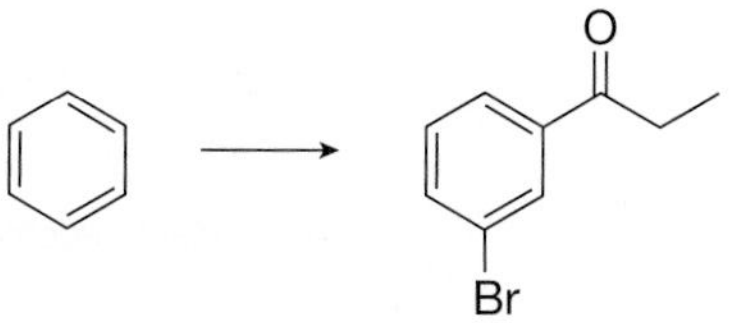 Br

d) 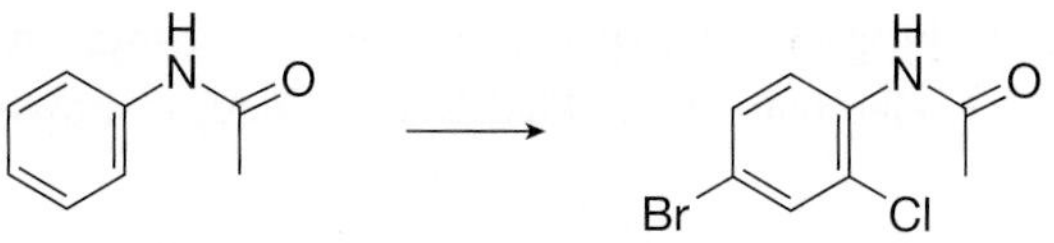 Br Cl

e)

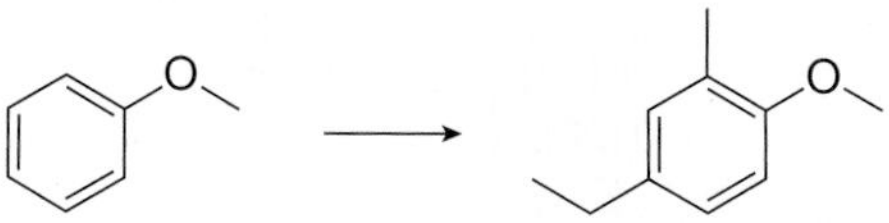

f) 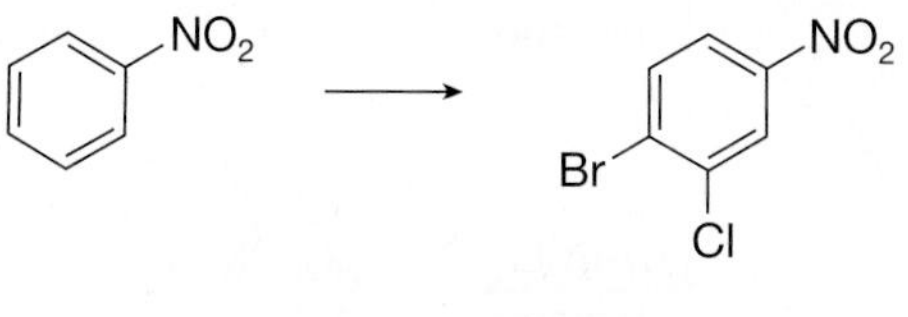 NO_2 → NO_2 Br Cl

10.28 Starting from benzene, synthesize the following molecules. You may use any other organic or inorganic reagents or solvents as needed.

a) 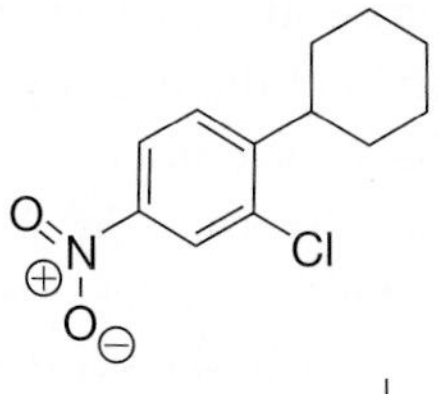 Cl

b) 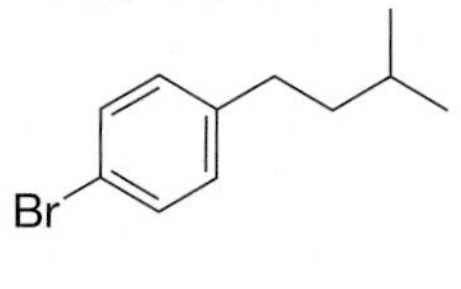 Br

c) 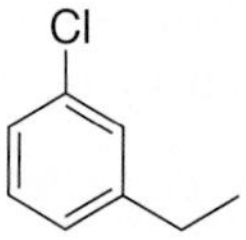 Cl

d)  NH_2 Br

e) 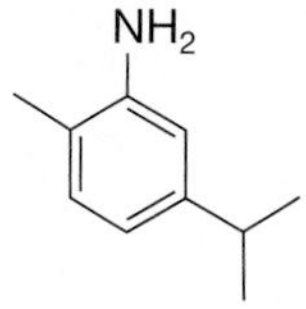 NH_2

f) 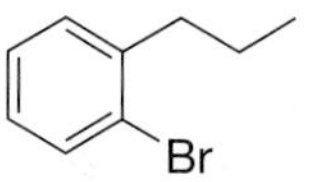 Br

g) 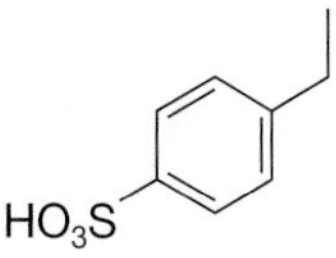 HO_3S

10.29 Benzene was reacted with benzoyl chloride and aluminum trichloride to give compound A ($C_{13}H_{10}O$). Compound A was reacted with *p*-methoxyphenylmagnesium chloride to give Compound B ($C_{20}H_{18}O_2$). What is the structure of each compound?

10.30 In the following reaction of nitrobenzene, two potential products are shown. One is correct and the other is not. Identify the correct product and explain why this intermediate is preferred.

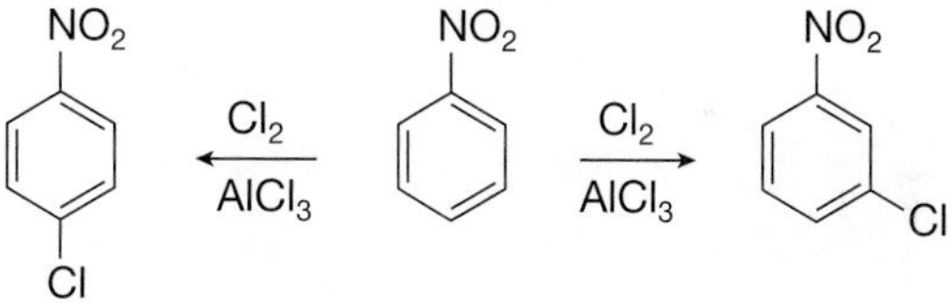

NO_2 Cl $\xleftarrow[\text{AlCl}_3]{\text{Cl}_2}$ NO_2 $\xrightarrow[\text{AlCl}_3]{\text{Cl}_2}$ NO_2 Cl

10.31 Which of the following products is expected as the major product? Justify your answer using the resonance structures of the relevant intermediates.

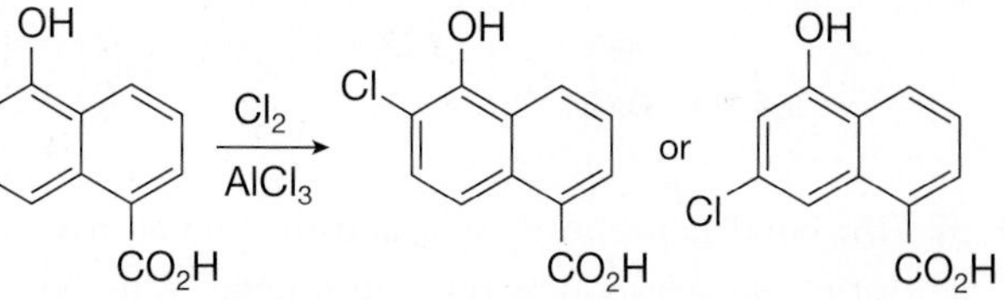

OH CO_2H $\xrightarrow[\text{AlCl}_3]{\text{Cl}_2}$ OH Cl CO_2H or OH Cl CO_2H

10.32 What is the expected product of the following bromination of benzeneselenol? Justify your answer by drawing the most stable resonance structure of the reaction intermediate.

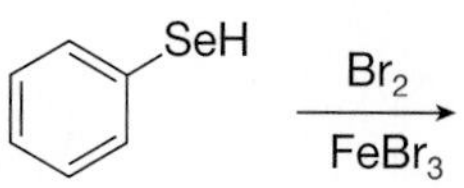

SeH $\xrightarrow[\text{FeBr}_3]{\text{Br}_2}$

10.33 Suggest a reasonable mechanism that explains the formation of the product of the following reactions.

a) $H^{\oplus}$

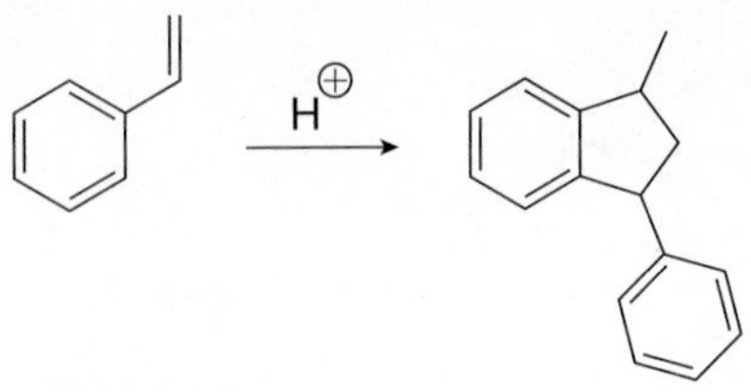

b) $AlCl_3$

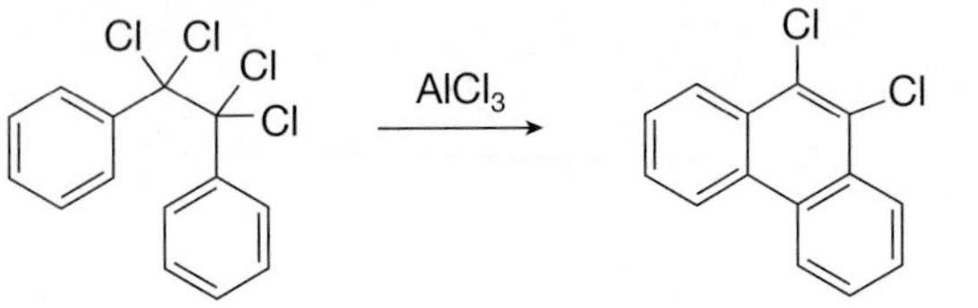

c) $H^{\oplus}$

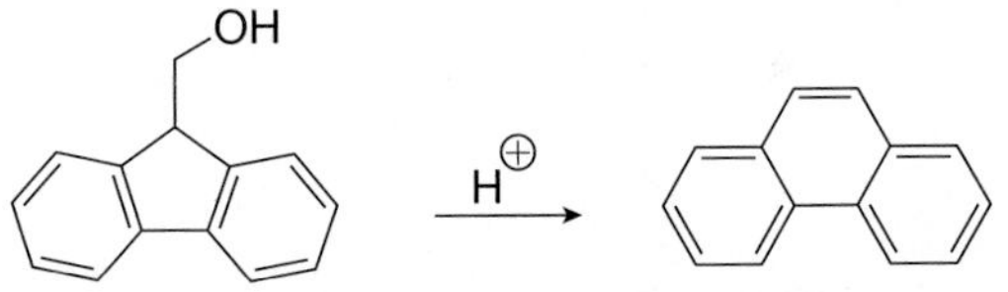

10.34 Propose a reasonable mechanism for the ring formation shown below.

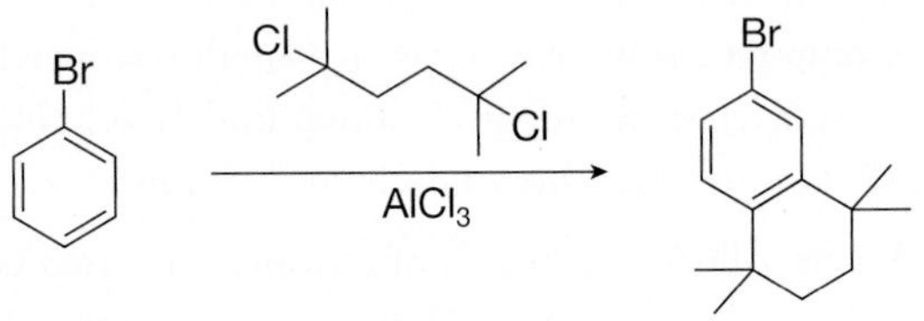

Br, Cl, Cl, $AlCl_3$, Br

10.35 Propose a reasonable mechanism for the nitrosation (addition of a nitroso substituent) of benzene, shown below.

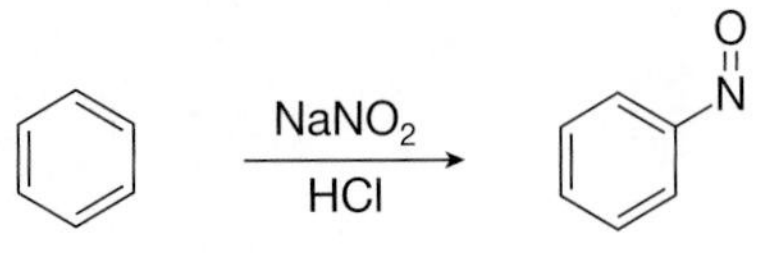

$NaNO_2$ / HCl

10.36 Propose a reasonable mechanism for the following reaction.

Br_2 / $AlCl_3$, Br, +

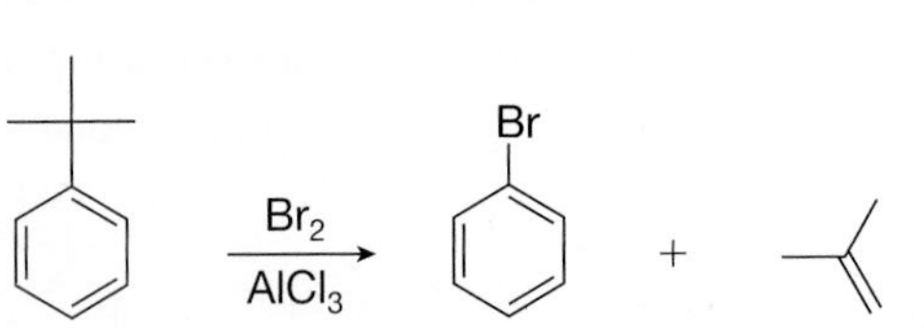

10.37 *Tert*-butyl-benzene, anisole, and toluene all have activating, *ortho*/*para*-directing substituents. Why does the product distribution (*ortho*:*para* ratio) vary the way it does in the reactions shown below?

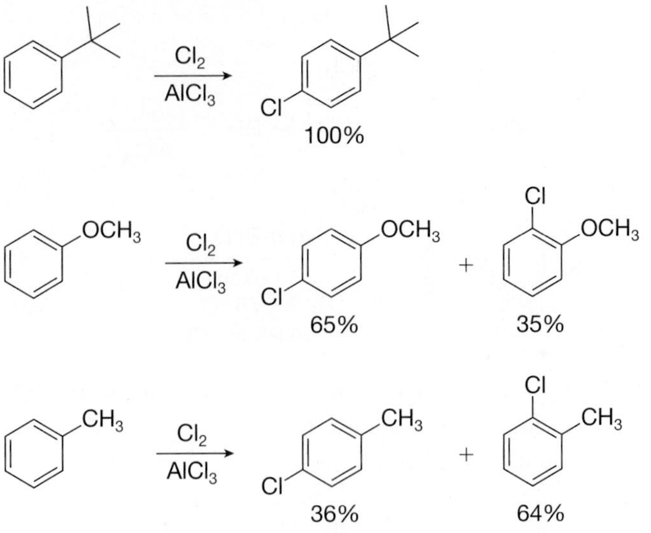

Cl_2 / $AlCl_3$ → Cl 100%

OCH_3 → Cl_2 / $AlCl_3$ → Cl, OCH_3 65% + Cl, OCH_3 35%

CH_3 → Cl_2 / $AlCl_3$ → Cl, CH_3 36% + Cl, CH_3 64%

10.38 *N,N*-dimethylaniline is very reactive toward electrophiles. The inclusion of an *ortho*-methyl greatly reduces reactivity, even though methyl is an activating group. Why is this?

N — very reactive; N — much less reactive

10.39 Propose an explanation for the change in lithiation selectivity when TMEDA is added, as shown below.

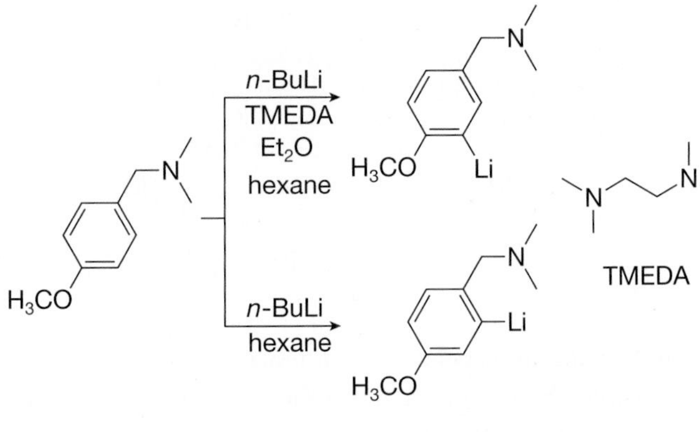

H_3CO, N; *n*-BuLi, TMEDA, Et_2O, hexane → H_3CO, N, Li; *n*-BuLi, hexane → H_3CO, N, Li; TMEDA

10.40 Trinitrotoluene is an important explosive used extensively in mining, industry, and the military. It is made from toluene by three successive nitrations, each one more difficult than the last. Explain why the nitration reactions become more difficult as the synthesis proceeds. What is the role of SO_3 in the last step?

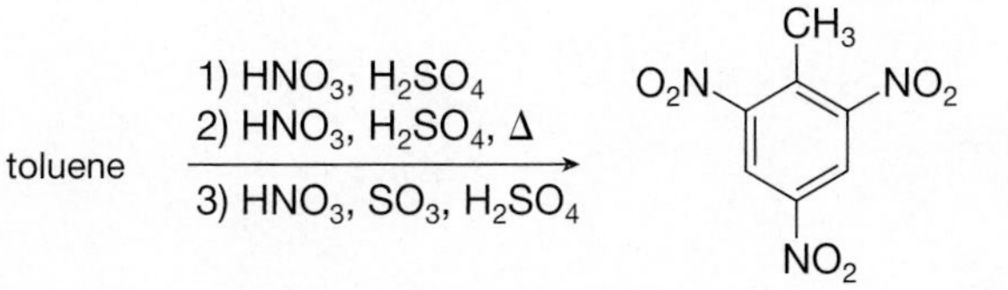

toluene → 1) HNO_3, H_2SO_4 2) HNO_3, H_2SO_4, Δ 3) HNO_3, SO_3, H_2SO_4 → (CH_3, O_2N, NO_2, NO_2)

10.41 Picric acid was one of the first shock-resistant explosives to be developed, and was used extensively in military high explosives until after World War I. It cannot be made directly from phenol because phenol is too reactive and forms polymer tars. Instead, two sulfonate groups are added, and this material is then nitrated. Explain the reason why this method works, and provide a mechanism for the process.

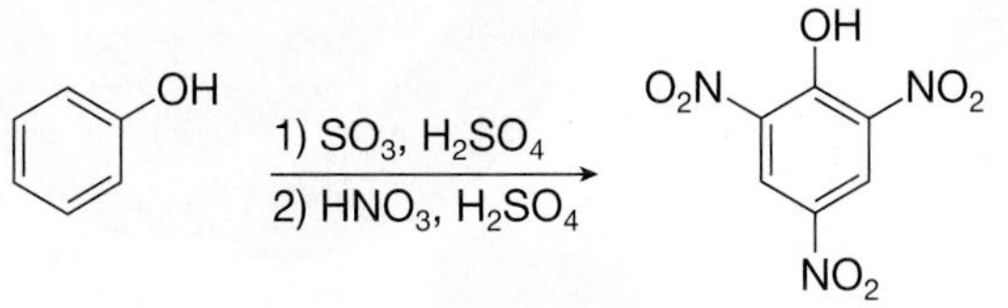

OH → 1) SO_3, H_2SO_4 2) HNO_3, H_2SO_4 → (OH, O_2N, NO_2, NO_2)

MCAT STYLE PROBLEMS

10.42 What is the expected major product of the following reaction?

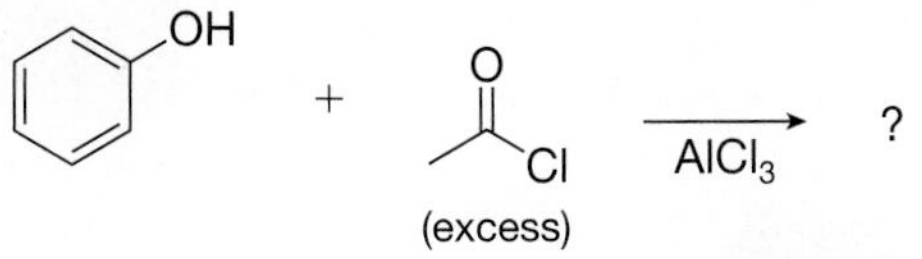

OH + (O, Cl) (excess) → $AlCl_3$ → ?

a) 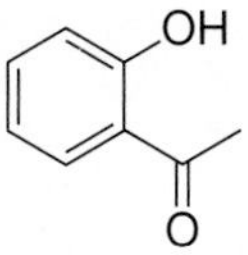 OH, O

b) 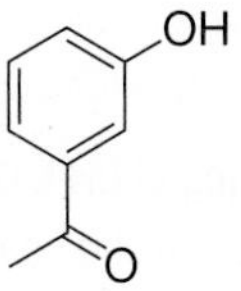 OH, O

c) 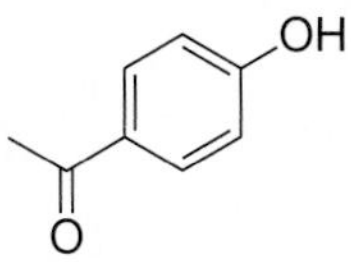 OH, O

d) 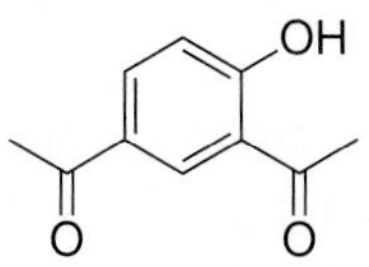 OH, O, O

10.43 Why is the Friedel–Crafts alkylation prone to form over-alkylated products?

a) The addition of electron-donating groups makes the cationic intermediate more stable.
b) The addition of electron-withdrawing groups makes the cationic intermediate more stable.
c) The addition of electron-donating groups makes the cationic intermediate less stable.
d) The addition of electron-withdrawing groups makes the cationic intermediate less stable.

10.44 Which of the following is *not* a proper resonance contributor to the intermediate formed from the addition of nitronium ion to phenol?

a) 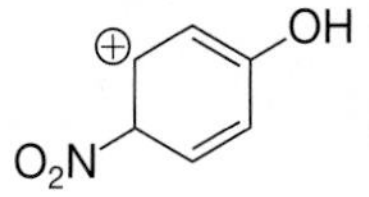 ⊕, OH, O_2N

b) 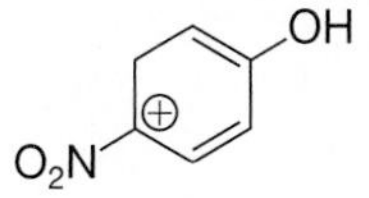 OH, ⊕, O_2N

c) 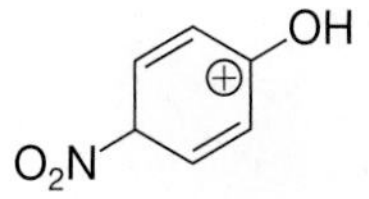 OH, ⊕, O_2N

d) 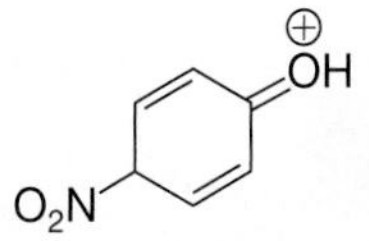 ⊕ OH, O_2N

CHALLENGE PROBLEMS

10.45 Salicylic acid is used in many medications. It occurs naturally in meadowsweet flowers, but it is more practical to make it from petroleum using the Kolbe–Schmidt reaction. Provide a mechanism for this process.

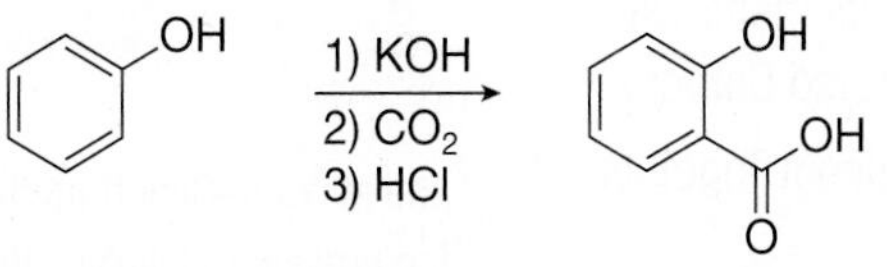

OH → 1) KOH 2) CO_2 3) HCl → (OH, OH, O)

10.46 The Fries rearrangement can be used to make a variety of substituted aromatic materials, such as the example shown below. Propose a mechanism to explain the Fries rearrangement. (Hint: Acids react preferentially with the strongest base, but can also react with weaker ones.)

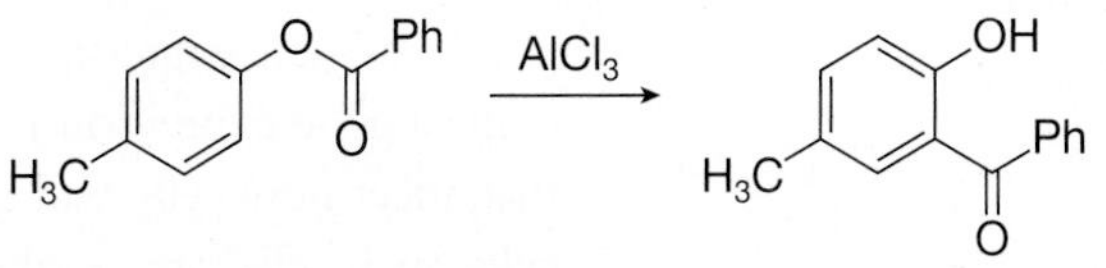

H_3C, O, Ph, O → $AlCl_3$ → (H_3C, OH, Ph, O)

11 Displacement Reactions on Saturated Carbons

S_N1 AND S_N2 SUBSTITUTION REACTIONS

CHAPTER OUTLINE

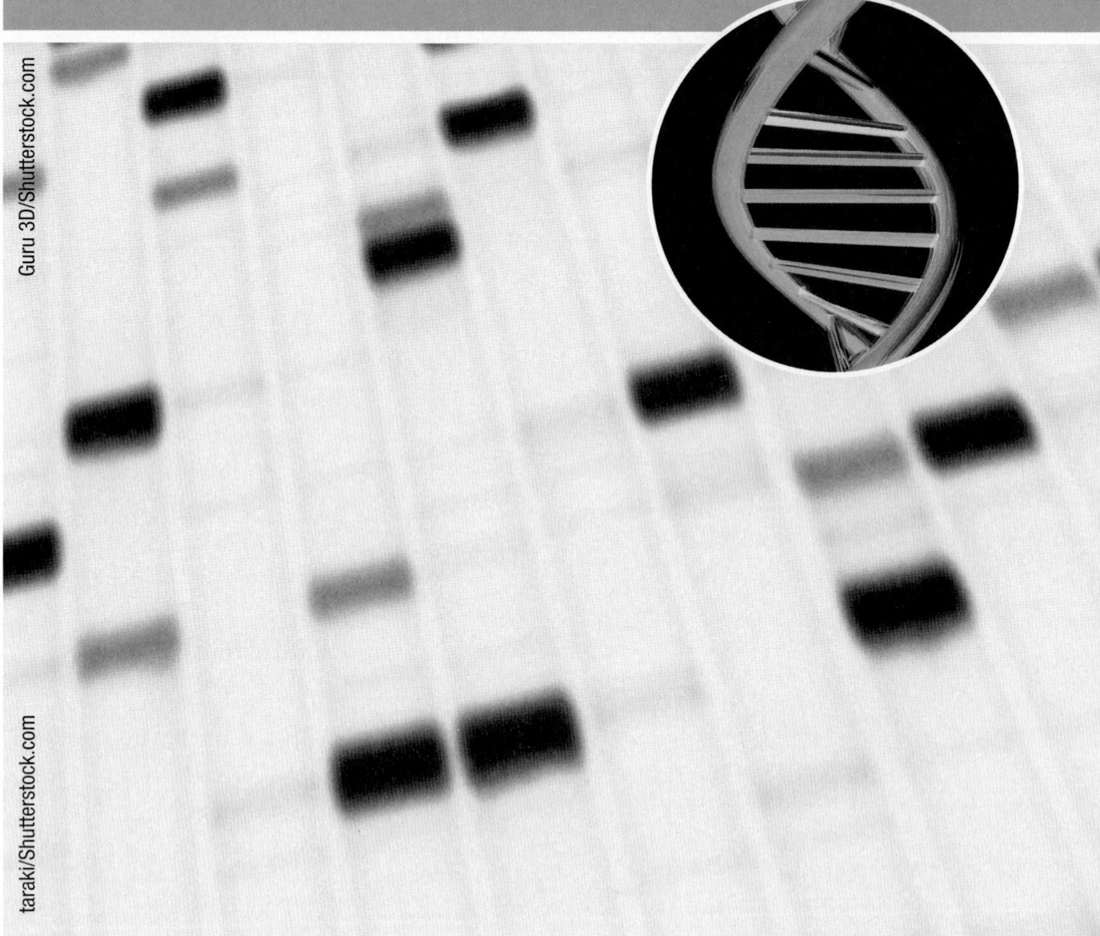

Guru 3D/Shutterstock.com

taraki/Shutterstock.com

The Maxam–Gilbert method was one of the first widely used procedures for the sequencing of DNA. One step of the procedure involves the treatment of DNA with dimethylsulfate, which methylates guanine and adenine and subsequently causes strand cleavage.

11.1 Why It Matters

Cells can select the genes they use in response to changes in their environment. This response, known as gene expression, allows cells to survive in many different conditions. One way that cells control gene expression is through epigenetics, which involves small chemical changes in DNA that affect how cells read genes. An important type of epigenetic control occurs in human stem cells. Such cells carry all the genetic information to transform into any type of cell. Stem cells are important in growth and cell-damage repair and have significant potential for the treatment of many degenerative diseases. Through a series of epigenetic changes, certain genes are "switched on," while other genes are "switched off," permanently changing the stem cell into a specialized cell. This process enables the human body to construct many different types of cells from a single cell prototype.

One of the mechanisms whereby DNA is modified for epigenetic control involves the enzymatic addition of methyl groups to some of its bases. This process of methylation involves a reaction called nucleophilic displacement, in which an enzyme adds a methyl group to certain DNA bases, such as adenosine, by displacement of a methyl group from another molecule.

adenosine (DNA) + *S*-adenosylmethionine (SAM) → (enzyme) *N*6-methyladenosine (modified DNA) + *S*-adenosylhomocysteine (SAH)

The methyl group that is added to DNA comes from *S*-adenosylmethionine (SAM), an enzyme co-substrate. SAM has an electrophilic methyl group that can react with adenosine, a nucleophilic site on DNA, producing *N*6-methyladenosine and *S*-adenosylhomocysteine (SAH). This displacement reaction proceeds by what is known as the S_N2 mechanism.

This chapter describes basic components of nucleophilic displacement reactions and, specifically, the mechanisms and stereochemistry of S_N1 and S_N2 reactions. A comparison of S_N1 and S_N2 displacements leads to ways to predict and to plan these types of reactions.

11.2 Displacement Reactions of Alkyl Halides

When one component in a molecule is replaced by another component, a **displacement reaction** occurs. When one of the components is a nucleophile, the reaction is called a *nucleophilic displacement*.

In a **displacement reaction**, one component of a molecule is replaced by another. These reactions are sometimes called replacements or substitutions. When the new component is provided by a nucleophile, the term *nucleophilic displacement* is used.

Alkyl halides contain a polarized carbon-halogen bond in which the bond electrons are drawn toward the halogen due to its greater electronegativity. This creates a partial negative charge on the halogen and, more importantly, a partial positive charge on the carbon. Areas of positive charge can direct where and how molecules react with nucleophiles. When nucleophiles react with alkyl halides, they attack (donate electrons to) the electrophilic carbon, displacing the halogen atom. In other words, the nucleophile replaces the halogen on the alkyl group, and this is called a *nucleophilic substitution reaction*.

bond breaks, moving the electrons onto the more electronegative atom

nucleophile has replaced the halogen on the carbon

$^{\ominus}OH + H_3C^{\delta+}–Br^{\delta-} \longrightarrow H_3C–OH + ^{\ominus}Br$

nucleophile reacts at electrophilic carbon (δ+)

When a nucleophile such as OH^- reacts with an alkyl halide such as CH_3Br, it forms a new bond to a carbon while displacing the bromide in a nucleophilic substitution reaction.

The reaction begins with the nucleophile donating a pair of electrons to the electrophilic carbon of the alkyl halide. During the reaction, the bond between the carbon and halogen breaks to avoid exceeding the valence of the carbon. As with other mechanisms, the bond breaks such that the electrons move onto the more electronegative atom. The result is a new compound in which the nucleophile has replaced the halogen.

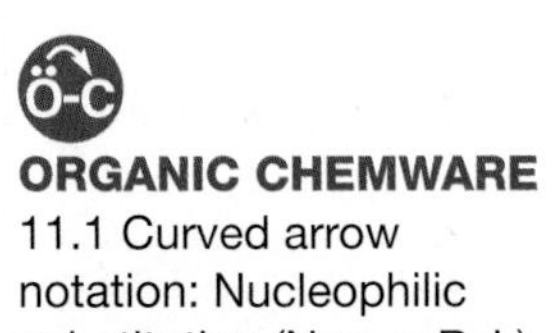

ORGANIC CHEMWARE
11.1 Curved arrow notation: Nucleophilic substitution (Nu^- + R–L)

Several key terms refer to the components of these reactions: *nucleophile*, *electrophile, leaving group*, *α-carbon*, and also *substrate* and *product.*

ORGANIC CHEMWARE
11.2 Curved arrow notation: Nucleophilic substitution (Nu + $R–L^+$)

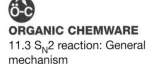

ORGANIC CHEMWARE
11.3 S_N2 reaction: General mechanism

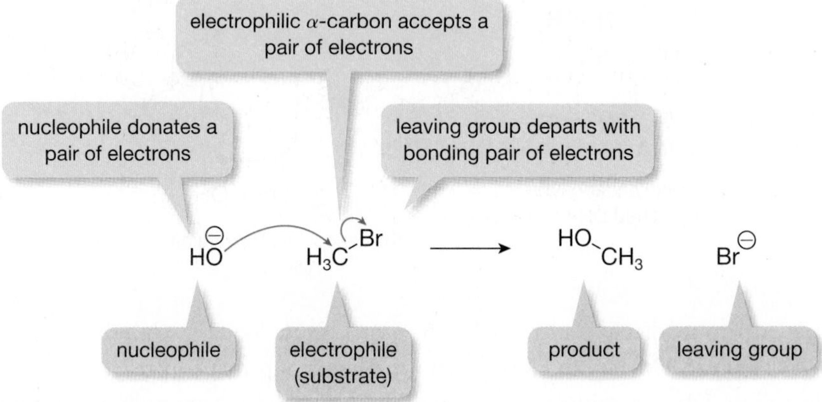

ORGANIC CHEMWARE
11.4 S_N2 reaction: $CH_3Cl + OH^-$

An **α-carbon** is the site of a chemical reaction during a nucleophilic displacement.

A **substrate** is the main organic molecule of interest in a chemical reaction.

The nucleophile donates a pair of electrons to the electrophile. The electrophile accepts electrons from the nucleophile to make a new bond. The carbon that accepts the new bond is called the **α-carbon**. The group that departs from the electrophilic α-carbon during the reaction, taking with it a pair of electrons, is the leaving group—the more electronegative group that takes the two electrons as the bond breaks.

In organic chemistry, the terms **substrate** and *product* are also used. The substrate is the main reactant and usually refers to the most complex molecule or the main molecule of interest. Both nucleophiles and electrophiles can be substrates, depending on the circumstance. The product is the molecule resulting from the chemical transformation of the substrate.

11.3 S_N2 Displacements

The following section describes S_N2 displacements in terms of reaction rates, mechanisms, and stereochemistry. Two key factors that influence reaction rates are also discussed: the nature of the nucleophile and the structure of the electrophile.

11.3.1 Reaction rates of S_N2 displacements

The displacement of a leaving group from an sp^3 atom by a nucleophile can occur in one or several steps. The mechanistic arrows used to describe the steps in a reaction sequence must reflect the actual sequence of events. Because molecules cannot be directly observed, indirect methods are used to assess the actual molecular interactions. The most powerful indirect method involves the determination of reaction rates. For example, the following expression describes the rate of the reaction between a nucleophile, OH^-, and an electrophile, CH_3Br.

$$\text{rate} = k\,[OH^{\ominus}][CH_3Br]$$

the rate of an S_N2 reaction depends on the concentration of both the nucleophile and the electrophile

Experiments show that the rate of the reaction depends on the concentration of both the nucleophile and the electrophile. Consequently, this reaction is designated S_N2, which stands for substitution nucleophilic **bimolecular**. Because the concentrations of both the nucleophile and the electrophile affect the reaction rate, both must be involved in the rate-determining (slowest) step of the process. With this information, the structure of the reaction's transition state can be deduced, as shown in the following reaction coordinate diagram. Indeed, the only way this type of reaction can occur is if the bond between the nucleophile and α-carbon forms at the same time as the bond between the α-carbon and the leaving group breaks. Such reactions where all the bond formation and bond breaking happens in the same step are called **concerted reactions**.

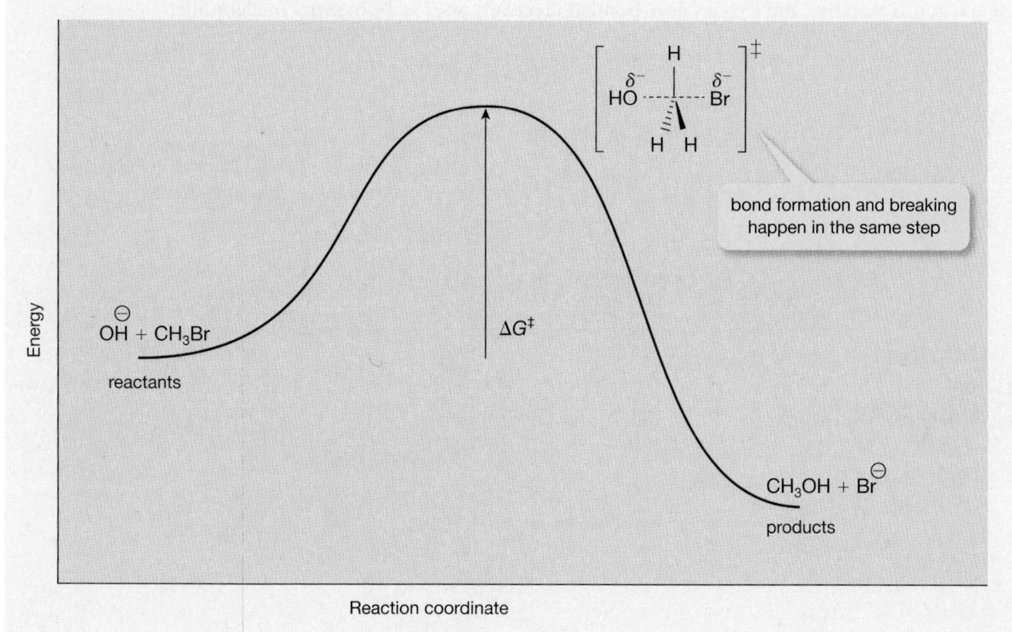

STUDENT TIP

Counter ions are present in mixtures that contain charged molecules. For example, OH^- has a counter ion (such as K^+ or Na^+). These counter ions are normally spectator ions in nucleophilic displacements but can occasionally be important in some reactions. To reduce clutter in diagrams we don't usually include them in most organic mechanisms.

If you prefer to draw the spectator ions, use ionic notation to help you visualize the reacting partners (K^+OH^- and not KOH).

The rate of a **bimolecular** reaction depends on the concentration of two reactants because two molecules are involved in the rate-determining step.

A **concerted reaction** is one in which multiple steps or processes occur together and essentially at the same time.

The α-carbon becomes trigonal bipyramidal in this transition state. Essentially the carbon becomes "flat," with the nucleophile and leaving group oriented on opposite sides of the α-carbon. Such transition state structures can be very useful for making predictions about the outcome of reactions.

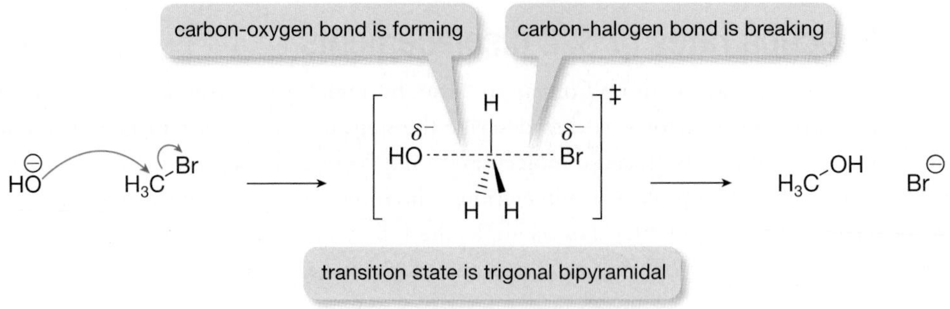

carbon-oxygen bond is forming

carbon-halogen bond is breaking

transition state is trigonal bipyramidal

CHECKPOINT 11.1

You should now be able to identify the basic components of a nucleophilic displacement and use mechanistic arrows to identify products.

SOLVED PROBLEM

Predict the product of the following reaction.

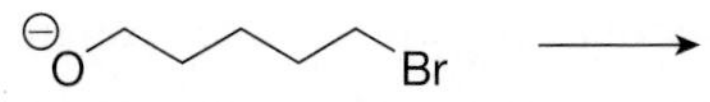

STEP 1 (OPTIONAL): Expand the Lewis structure around the reacting functional groups.

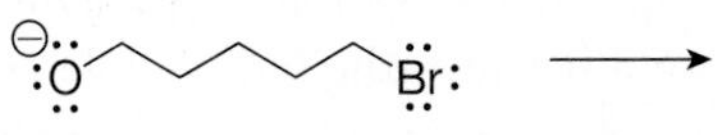

STEP 2: Identify the role of each functional group. First, the presence of a halide is a strong clue because it is normally an excellent leaving group. Therefore, the carbon attached to the halogen must be the α-carbon. The oxygen is negative and carries non-bonded electrons, so it is a potential nucleophile.

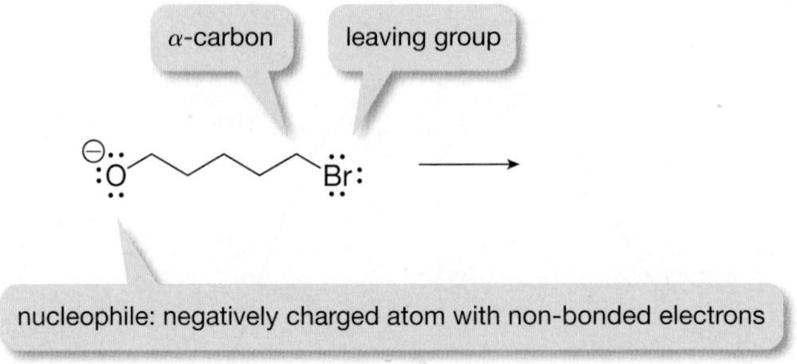

α-carbon

leaving group

nucleophile: negatively charged atom with non-bonded electrons

STEP 3: Use mechanistic arrows to predict the product. Note that for molecules that contain more than one functional group, it is often helpful to number the carbon chain.

nucleophile donates electrons to the α-carbon
leaving group
numbering the chain shows that a six-membered ring will form (new bond between O-1 and C-6)

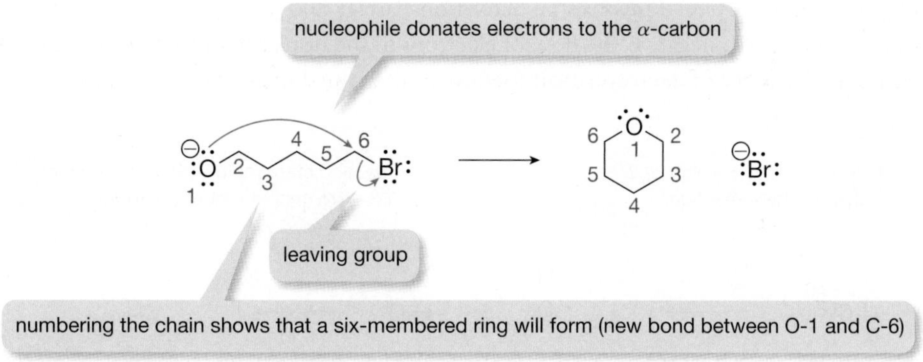

PRACTICE PROBLEM

11.1 Draw mechanisms and predict the products of the following structures.

a)

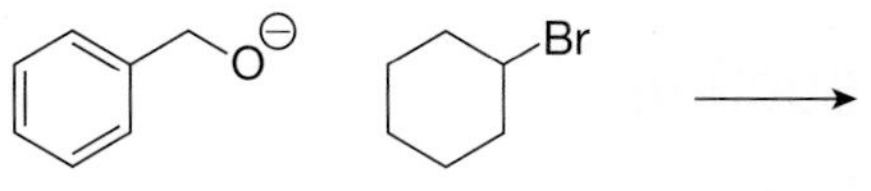

b)

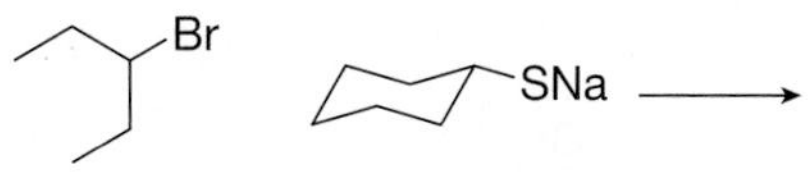

c)

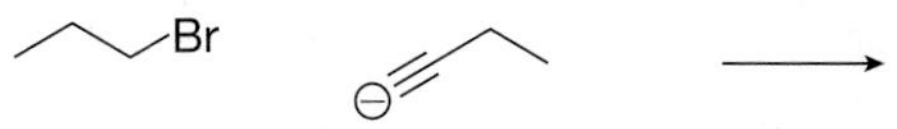

INTEGRATE THE SKILL

11.2 Use arrow notation to propose a mechanism and final product for the following reaction.

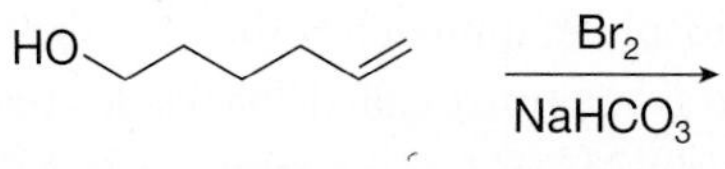

11.3.2 Mechanisms for S_N2 displacements

Depending on whether the nucleophile is negatively charged or neutral, mechanistic details for S_N2 displacement reactions can vary. For example, the S_N2 reaction using a negatively charged nucleophile such as CH_3O^- takes place in a single step, directly producing a neutral product. Note that sometimes the reaction may be preceded by a deprotonation step to form the anionic nucleophile.

reaction happens in one step
$CH_3O^- + H_3C{-}Br \longrightarrow CH_3OCH_3 + Br^-$
from deprotonation of CH_3OH

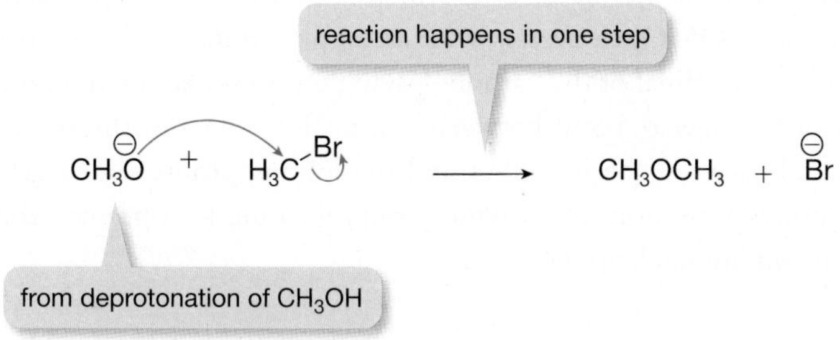

With neutral nucleophiles, the S_N2 reaction initially produces a reaction intermediate that has a positive charge. This is immediately followed by a rapid deprotonation by a base to produce the final, uncharged product. Note that the rate-determining step is still the S_N2 displacement step.

Each step in such a sequence involves an activation barrier and a transition state. Whether the nucleophile is neutral or negative, the flow of electrons in the displacement step is the same. What changes is just the order of deprotonation (before or after the displacement).

first step: rate-determining S_N2 displacement reaction

second step: rapid acid–base reaction to generate neutral product

$H_2O + H_3C{-}Br \xrightarrow{-Br^{\ominus}} H_3C{-}\overset{\oplus}{O}H_2 + H_2O \longrightarrow H_3C{-}O{-}H + H_3O^{\oplus}$

strongest base in this reaction is H_2O

11.3.3 Stereochemistry of S_N2 reactions

The S_N2 reaction inverts the configuration of the α-carbon.

Br (S) —— $HS^{\ominus}$ / DMF ——> SH (R)

nucleophile inverts the configuration at the α-carbon during S_N2 reactions

For this to occur, the nucleophile approaches the α-carbon from the opposite side of the leaving group by in-line fashion (sometimes called "backside attack"). As the nucleophile attacks the electrophilic α-carbon and the leaving group starts to leave, the geometry of the α-carbon changes. At the transition state of the reaction, the α-carbon adopts a trigonal bipyramidal geometry. By the end of the substitution reaction, the α-carbon regains a tetrahedral geometry, but with an opposite orientation. This is reminiscent of the inversion of an umbrella under high wind, and is referred to as a Walden inversion. In the above example, Walden inversion is evident because substitution of the *(S)*-2-bromobutane inverts the configuration of the stereogenic α-carbon to generate the *(R)*-thiol product.

ORGANIC CHEMWARE
11.5 S_N2 reaction: 2-chlorobutane + OH^-

Two factors account for the nucleophile's approach to the α-carbon. First, the frontside approach is sterically impeded by the presence of the departing leaving group, so the nucleophile cannot approach *syn* to the leaving group. Second, and more importantly, the backside approach allows for optimal overlap of the interacting orbitals. Specifically, the approaching nucleophile supplies a filled orbital (HOMO) to overlap with an empty orbital on the electrophile (LUMO), namely the $\sigma^\star$ molecular orbital of the carbon–leaving group single bond. This interaction leads to the formation of the new σ bond between the nucleophile and the α-carbon and to the breaking of the bond between the α-carbon and the leaving group. The larger lobe of the $\sigma^\star$ molecular orbital points away from the leaving group, allowing for optimal orbital overlap with the orbital of the incoming nucleophile.

$Nu^{\ominus}$ $sp^3{}_{Nu}$ (HOMO) + R, H, H, C–L σ^*_{C-L} (LUMO) $\longrightarrow$ $[Nu^{\delta-} \cdots C(R)(H)(H) \cdots L^{\delta-}]^{\ddagger}$ $\longrightarrow$ Nu–C(R)(H)(H) + $L^{\ominus}$

CHECKPOINT 11.2

You should now be able to predict the stereochemical outcome of an S_N2 reaction.

SOLVED PROBLEM

What isomer must be used to obtain the indicated product in an S_N2 reaction?

Br NaSH → SH

STEP 1: Identify the components of the reaction.

α-carbon
nucleophile
Br
$HS^{\ominus}$
SH
leaving group

STEP 2: The S_N2 reaction proceeds with inversion of the configuration. Therefore, the stereochemistry of the leaving group must be opposite to that of the product.

configuration at reacting site must be opposite to that of the product
Br
$HS^{\ominus}$
SH

STEP 3 (OPTIONAL): Once the nucleophile is identified, the product can be redrawn to highlight the stereochemistry of the bond being formed.

Br
$HS^{\ominus}$
SH
redraw the product to highlight the stereochemistry of the bond being formed

With the product redrawn in this way, the product stereochemistry may be more easily highlighted.

configuration at reacting site must be opposite to that of the product
Br
$HS^{\ominus}$
SH

PRACTICE PROBLEM

11.3 Use mechanistic arrows to determine the products of the following S_N2 reactions.

a) [cyclohexane with Br] $\xrightarrow{\text{NaOH}}$

b) Fischer projection: CH_3 top; H–C–OMe; H–C–Br; CF_3 bottom $\xrightarrow{\text{NaSH}}$

c) Newman projection: CH_3, H, H (front); H, Br, CH_3 (back) $\xrightarrow{CH_3CO_2^-}$

INTEGRATE THE SKILL

11.4 Even though the starting material for the following S_N2 reaction is enantiomerically pure, the *ee* of the product is only 20 percent.

[2-chlorobutane] $\xrightarrow{\text{NaI}}$

a) Assuming the *product* was enantiomerically pure, what is the expected product?
b) Provide an explanation for the observation that the ee of this product is only 20 percent.
c) What is the expected major isomer and why?

11.3.4 Nucleophilicity in S_N2 displacement reactions

Nucleophilicity is an expression of how fast (how easily) a nucleophile is able to react (donate electrons).

Several factors influence the rate of an S_N2 displacement; the most significant is the ability of the nucleophile to donate electrons. This is sometimes called the strength of the nucleophile—that is, its **nucleophilicity**. Nucleophilicity is a kinetic property related to the activation energy of a reaction. An important guide to understanding the strength of the nucleophile is the principle that nucleophiles are electron pair donors. The rate of S_N2 displacement depends on the ability of the nucleophile to donate electrons to the electrophilic α-carbon to form the new covalent bond. Strong (good) nucleophiles react quickly; weak (poor) nucleophiles react slowly. Although there are trends to nucleophilic behaviour, the way a nucleophile reacts depends on reaction conditions. For example, a good nucleophile in one reaction may perform poorly in other reactions.

The following sections describe—in order of importance—the factors that contribute to nucleophilicity. These include the presence of negative charge on the nucleophile, the electronegativity and size of the nucleophilic atom, the possibility of electron delocalization, and the amount of steric hindrance around the nucleophilic atom.

11.3.4.1 Effect of negative charge

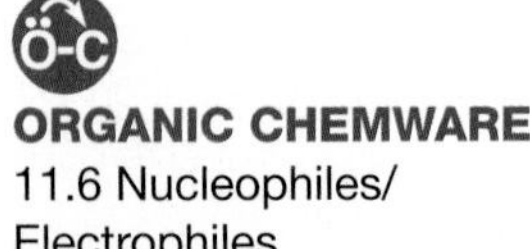

Negatively charged bases have more available electrons compared to their conjugate acids, and so they are better electron donors. So some bases are also strong nucleophiles. During a reaction that involves a negatively charged molecule, the energy of the charged nucleophile is higher than that of its neutral acid counterpart. This effect tends to lower the activation energy of nucleophilic displacement and result in a faster reaction.

formation of a charged intermediate has a large activation energy

$$H_2O + H_3C{-}Br \xrightarrow[H_2O]{\text{slower}} H_3C{-}\overset{\oplus}{O}H{-}H + H{-}O{-}H \longrightarrow H_3C{-}O{-}H + H_3O^{\oplus}$$

$$HO^{\ominus} + H_3C{-}Br \xrightarrow[H_2O]{\text{faster}} CH_3OH + Br^{\ominus}$$

Thus, OH^- is a much better nucleophile than H_2O. The negative ion more easily donates electrons to the α-carbon, and a fast reaction ensues. This is true for other base–conjugate acid pairs, such as CH_3O^- and CH_3OH or $CH_3CO_2^-$ and CH_3CO_2H.

STUDENT TIP

Do not confuse basicity with nucleophilicity. These two properties are not necessarily correlated. Basicity is a thermodynamic property that relates the energy of a base to its conjugate acid. Nucleophilicity is a kinetic property that relates to the activation energy of a reaction. In some circumstances, nucleophilicity increases as basicity increases; in other circumstances, nucleophilicity decreases as basicity increases.

11.3.4.2 Effect of electronegativity

Atoms with higher electronegativity are less nucleophilic because they are less able to share their valence electrons to make a bond. Electronegativity of the elements increases from left to right in the periodic table. For example, primary amines are more nucleophilic than the corresponding alcohols. This is because nitrogen is less electronegative than oxygen, so is more capable of donating electrons to form a new bond during the course of an S_N2 reaction. As a result, CH_3NH_2 reacts faster than CH_3OH under identical reaction conditions.

nitrogen is less electronegative and is a better electron donor

$$CH_3NH_2 + H_3C{-}Br \xrightarrow{\text{faster}} H_3C{-}\overset{\oplus}{N}H_2{-}CH_3 + CH_3NH_2 \longrightarrow H_3C{-}NH{-}CH_3 + CH_3\overset{\oplus}{N}H_3$$

$$CH_3OH + H_3C{-}Br \xrightarrow{\text{slower}} H_3C{-}\overset{\oplus}{O}H{-}CH_3 + CH_3OH \longrightarrow H_3C{-}O{-}CH_3 + CH_3\overset{\oplus}{O}H_2$$

oxygen is more electronegative and is a weaker electron donor

11.3.4.3 Effect of atomic size

Large atoms do not hold their valence electrons as tightly as small atoms do. This is because the nucleus has less effective attraction for electrons as the number of electrons increases and their distance from the nucleus becomes greater. Such atoms are **polarizable**; their electron clouds can be more easily distorted by nearby charges, and in particular, the electrons can be attracted to nearby positive charges to make bonds. This effect is seen moving down the periodic table; as size increases, so does polarizability and nucleophilicity. For instance, thiols are more nucleophilic than alcohols because sulfur is lower in the periodic table. Being larger and more polarizable, sulfur can donate electrons more easily than oxygen and is a more nucleophilic atom.

Polarizable atoms easily form dipoles. These atoms have large, diffuse electron clouds, which are easily disturbed by nearby charges.

11.3.4.4 Effect of electron delocalization

Delocalization in nucleophiles usually impairs their reactivity. Because delocalized electrons are shared over several atoms, they are not readily donated to form a bond. Delocalization also lowers the energy of the nucleophile, thereby stabilizing it and increasing the activation energy of a nucleophilic displacement. For example, hydroxide (OH^-) and acetate ($CH_3CO_2^-$) both contain negatively charged oxygen atoms, the nucleophilic atoms of each group. The oxygen lone pairs in hydroxide are localized on a single atom, whereas the oxygen lone pairs on acetate are delocalized over the atoms of the carboxylate group, reducing the electron density on each oxygen atom. Consequently, the acetate is a weaker nucleophile than hydroxide.

11.3.4.5 Steric congestion near the nucleophilic atom

The ability of a nucleophile to react is impaired by steric congestion due to groups or large atoms near the nucleophilic atom. As the nucleophile approaches the electrophilic α-carbon, steric repulsion with neighbouring groups interferes with the nucleophilic attack and slows the S_N2 reaction.

CHECKPOINT 11.3

You should now be able to identify the factors that control the relative nucleophilicity of functional groups.

SOLVED PROBLEM

What is the structure of the major product of the following S_N2 displacement?

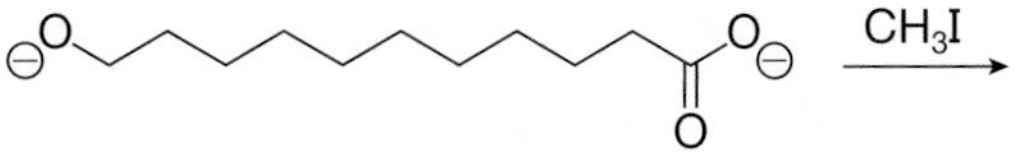

CH_3I

STEP 1: The rate of an S_N2 reaction depends on the relative ability of the nucleophiles to donate electrons. Better electron donors react faster. The structure shown involves competition between the two nucleophiles. The major product will be the one arising from the reaction with the best electron donor.

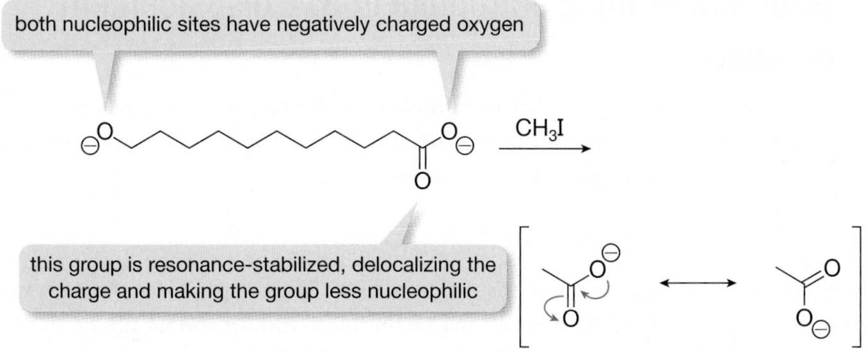

both nucleophilic sites have negatively charged oxygen

CH_3I

this group is resonance-stabilized, delocalizing the charge and making the group less nucleophilic

STEP 2: Identify the product arising from reaction at the most nucleophilic site.

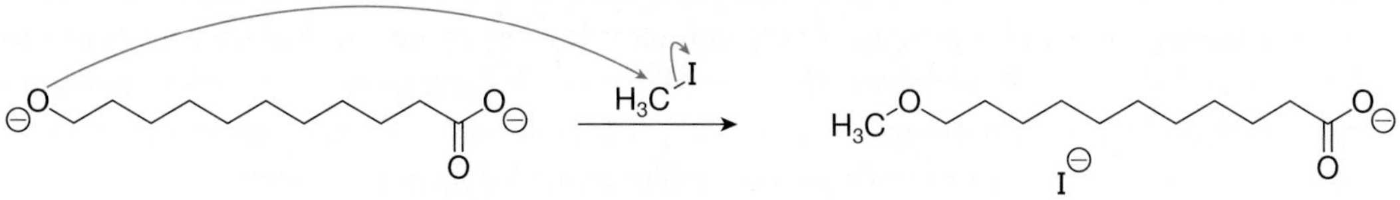

PRACTICE PROBLEM

11.5 Identify the better nucleophile among the following pairs.

a) 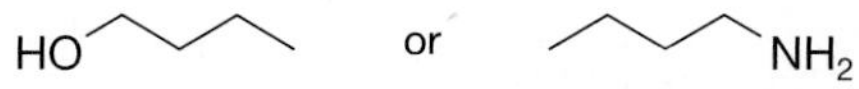 HO⌒⌒ or ⌒⌒NH_2

b) 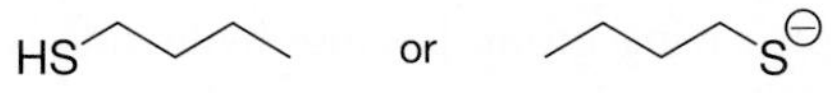 HS⌒⌒ or ⌒⌒$S^{\ominus}$

c) 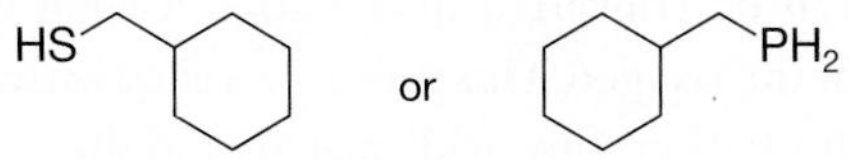 HS–CH₂–cyclohexane or cyclohexane–CH₂–PH_2

d) 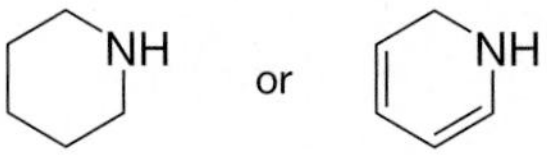 piperidine (NH) or tetrahydropyridine (NH)

INTEGRATE THE SKILL

11.6 The following pair of compounds produces different products depending on the base and order of mixing. Predict the product in each case, and provide a mechanism to explain your answer.

a) HO–cyclohexyl–NH_2 + benzyl bromide (Br) $\xrightarrow{K_2CO_3}$

b) HO–cyclohexyl–NH_2 $\xrightarrow[\text{2) benzyl bromide (Br)}]{\text{1) NaH}}$

11.3.5 Structure of the electrophile in S_N2 displacement reactions

The structure of the electrophile also has an influence on the rate of an S_N2 reaction. In the case of the electrophile, however, only two factors need to be considered: the ability of the leaving group to leave and the substitution pattern at the α-carbon.

11.3.5.1 Halogens as leaving groups

The leaving group in an S_N2 reaction has two functions. First, it polarizes the carbon–leaving group bond, thereby increasing the positive charge on the α-carbon and its attraction to nucleophiles. This factor should be considered when comparing leaving groups across the same row of the periodic table. Second, the leaving group departs with a pair of electrons. To understand this function, basicity guidelines are helpful because the same factors that make a weak base also make a good leaving group. Halogens are a very common leaving group for displacement reactions because the halides are all weak bases (F^- is usually a poor leaving group, however). Other atom types, such as oxygen, form stronger bases and may require special reaction conditions to function as leaving groups. The order of leaving group ability of the halogens is as follows:

$$I^- > Br^- > Cl^-$$

11.3.5.2 Ways to convert OH groups into good leaving groups

The OH^- group (or OR^- group) is a very poor leaving group because hydroxide is a strong base. However, an OH group (or OR group) can be converted into a good leaving group by chemical modifications that reduce the basicity of the oxygen. This process is called activating the OH as a leaving group. There are two main methods to do this: acid catalysis and the conversion to sulfonate esters.

reaction cannot proceed because OH^- is a very poor leaving group (strong base)

Br^- + H_3C–OH $\not\rightarrow$ H_3C–Br + ^-OH

Acid catalysis

The first method to activate OH as a leaving group is the process of acid catalysis. When a strong acid is added to an organic molecule containing an OH (or OR) group, the oxygen is basic enough to react with the acid. This has the effect of converting the OH group into an OH_2^+ (or ORH^+) group, which is a very good leaving group, departing as H_2O (or ROH).

protonation forms a good leaving group

$H_3C{-}OH$ + $H{-}Br$ ⟶ $H_3C{-}OH_2^{\oplus}$ $Br^{\ominus}$ ⟶ $H_3C{-}Br$ H_2O

hydroxyl group (base) reacts with HBr (acid)

water is a weak base and a very good leaving group

Strong acids such as HCl, HBr, and HI are good sources not only of acid but also of halide nucleophiles. When other nucleophiles are involved, H_2SO_4 or H_3PO_4 are normally used because their corresponding counter ions are non-nucleophilic. In these reactions, the nucleophile reacts in its neutral form, generating a charged intermediate that is neutralized by a second acid–base process.

The neutral form of the nucleophile must be involved when acidic conditions are being used (unless the conjugate base of the nucleophile is *very* weak). This is because negatively charged conjugate bases (such as OH^- or OR^-) undergo acid–base reactions very quickly. Consequently, very little (if any) of the negative forms are present in acidic mixtures.

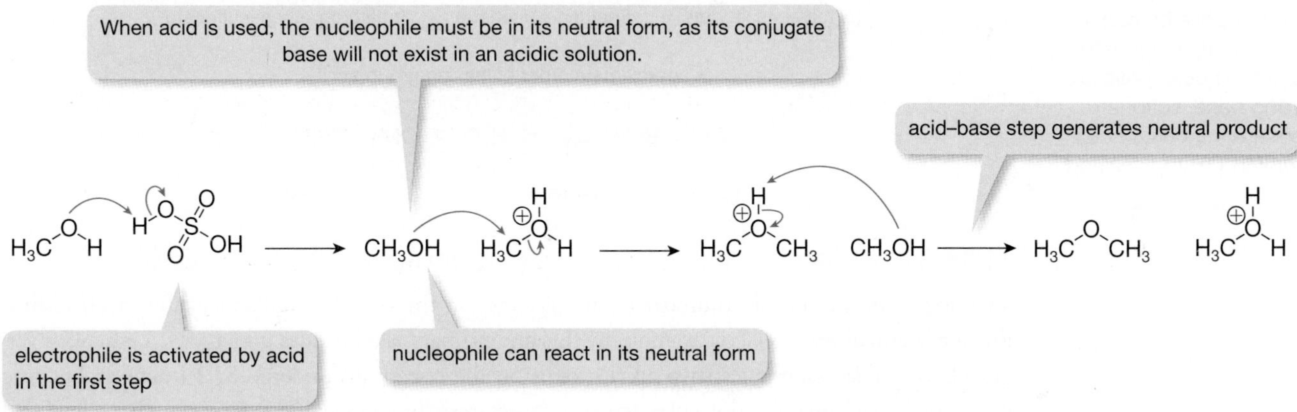

A general principle of catalysis in chemical reactions is that a base accelerates reactions by activating the nucleophile, whereas an acid accelerates reactions by activating the electrophile. Conversion of a neutral nucleophile into its conjugate base generates a charged molecule that is better able to give away (push) electrons. Conversion of a neutral electrophile into its conjugate acid generates a charged molecule that is better able to accept (pull) electrons. Converting an OH group into an OH_2^+ group increases its ability to pull electrons as it leaves.

Sulfonate esters

The second and more common method to convert OH into a good leaving group is to convert it into a sulfonate ester (this method cannot be used with OR groups). This process attaches a strong electron-withdrawing group onto the oxygen, which significantly reduces its basicity.

sulfonate ester

OH + tosyl chloride (TsCl) —pyridine→ alkyl tosylate (ROTs)

Sulfonate esters make good leaving groups because the leaving group, the sulfonate anion, is the conjugate base of a strong acid. Thus, it is a weak base (similar to the conjugate base of sulfuric acid).

$Br^{\ominus}$ + tosylate leaving group (OTs) → Br + tosylate anion ($TsO^{\ominus}$)

STUDENT TIP

Acid catalysis can be used to displace hydroxyl groups only when the desired nucleophile is a weak base. When the nucleophile is a stronger base, such as cyanide, acid catalysts deactivate the nucleophile by converting it to its conjugate acid. If the nucleophile is a strong base, sulfonate esters are the best choice to convert OH to a leaving group.

Among the many sulfonate esters available, three are most commonly used:

tosylate (OTs) (*p*-toluenesulfonate) | mesylate (OMs) (methanesulfonate) | triflate (OTf) (trifluoromethanesulfonate)

approximate order of leaving group ability

Cl < Br < I, OTs < H_2O < OMs < OTf

good — better — best

11.3.5.3 Electrophile substitution patterns near the α-carbon

The degree of substitution (number of substituents) on the α-carbon influences the mechanism of the displacement reaction. For example, methyl and primary alkyl halides undergo S_N2 reactions very quickly. Secondary substrates (two alkyl groups on the α-carbon) undergo S_N2 reactions, but react more slowly because their molecular structure causes steric hindrance that interferes with the approach of the nucleophile. Tertiary substrates (three alkyl groups on the α-carbon) do not react by S_N2 pathways; for example, in a tertiary α-carbon, the backside approach is completely blocked. Tertiary alkyl substrates can, however, react readily with nucleophiles to give substitution products; but the reaction must proceed by a different mechanism: S_N1.

Compound	α-Carbon structure	Symbol	Relative speed of S_N2
CH_3Br	methyl	Me	fast
CH_3CH_2Br	primary	1°	↓
$(CH_3)_2CHBr$	secondary	2°	slow
$(CH_3)_3CBr$	tertiary	3°	does not undergo S_N2

CHECKPOINT 11.4

You should now be able to predict the outcome of an S_N2 reaction.

SOLVED PROBLEM

Suggest a method to carry out the following reaction.

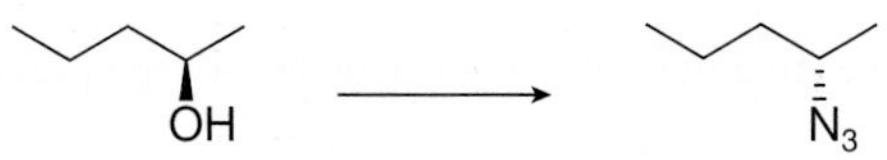

STEP 1: Evaluate the difference between the starting material and the product to identify the type of reaction required. In this example, the heteroatom exchanges from the starting material to the product, and the geometry of the configuration is inverted; this suggests an S_N2 reaction.

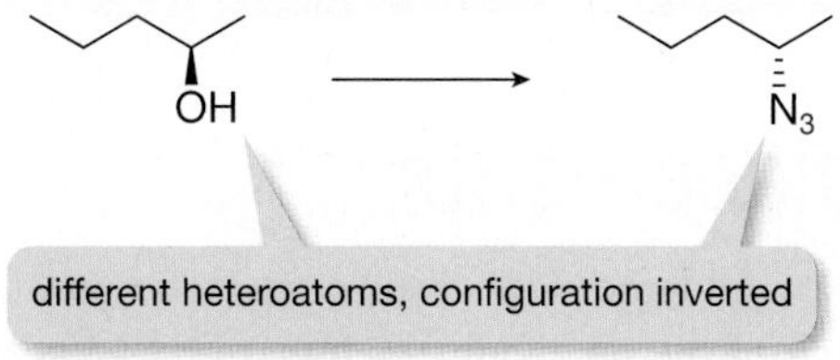

STEP 2: Identify the α-carbon. To do this, select the carbon whose configuration has been inverted. The OH group is replaced by N_3, suggesting the OH is a leaving group. However, OH^- groups are poor leaving groups and must be modified to make the reaction viable.

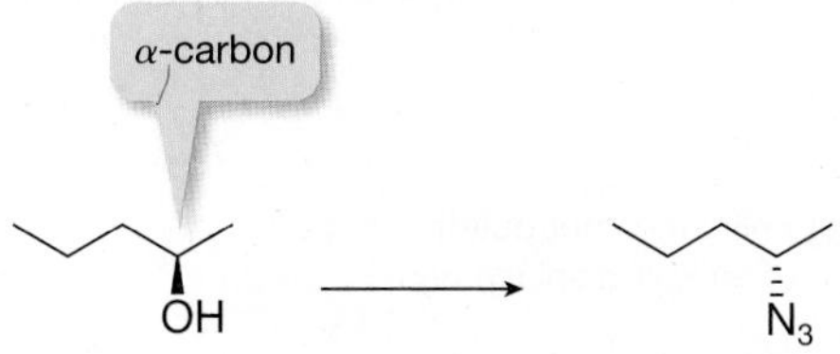

STEP 3: There are two ways to activate an OH group as a leaving group. Acid is one choice, but it is incompatible with the desired nucleophile (NaN_3), which is not a weak base. Therefore, a tosylate should be used.

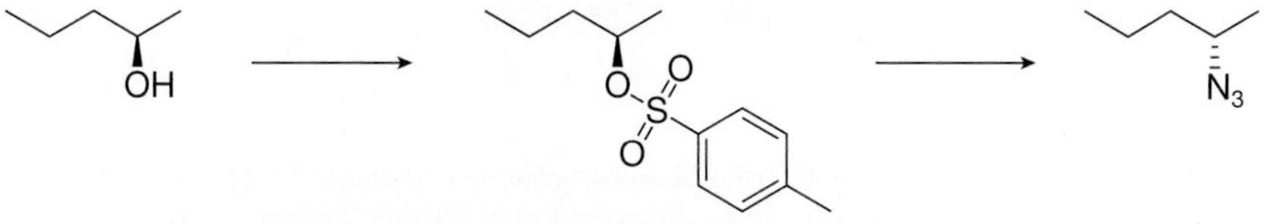

STEP 4: With the leaving group prepared, the reaction can proceed.

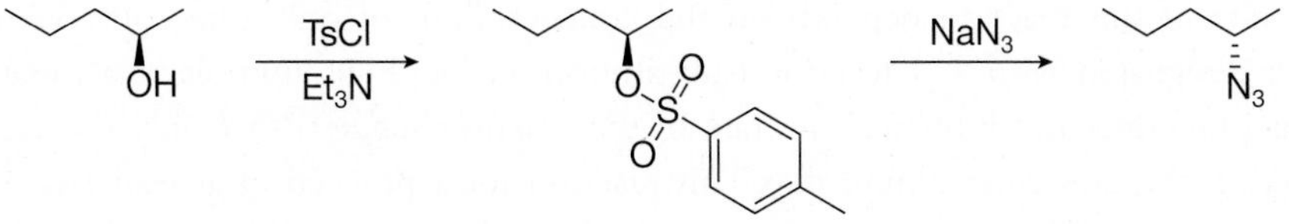

PRACTICE PROBLEM

11.7 Provide the missing starting material, product, or reagent as indicated.

a)

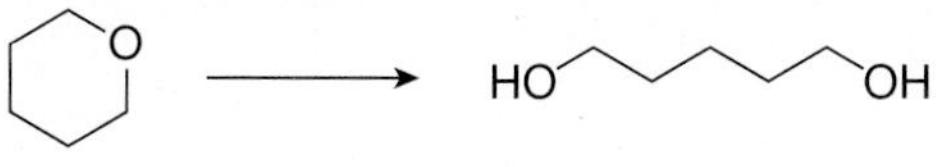

b) 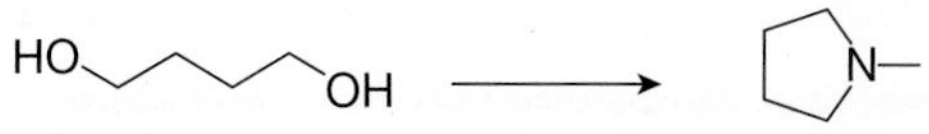 HO~~~OH ⟶ N—

c) 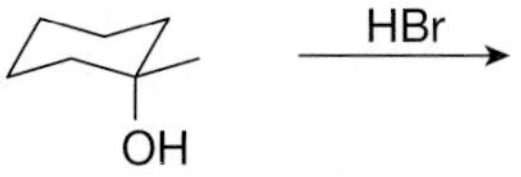 OH —HBr→

INTEGRATE THE SKILL

11.8 Suggest a method of converting the following aldehyde into the product shown.

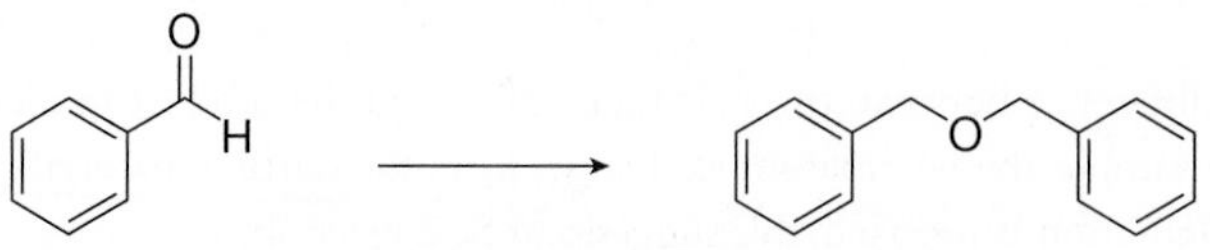

11.4 S_N1 Displacements

Tertiary alkyl halides cannot undergo S_N2 displacement because of severe steric hindrance blocking a backside approach of the nucleophile. However, displacement can occur through an S_N1 mechanism.

11.4.1 Reaction rate of S_N1 mechanism

The rate of reaction between a nucleophile and a tertiary substrate is described by the following expression:

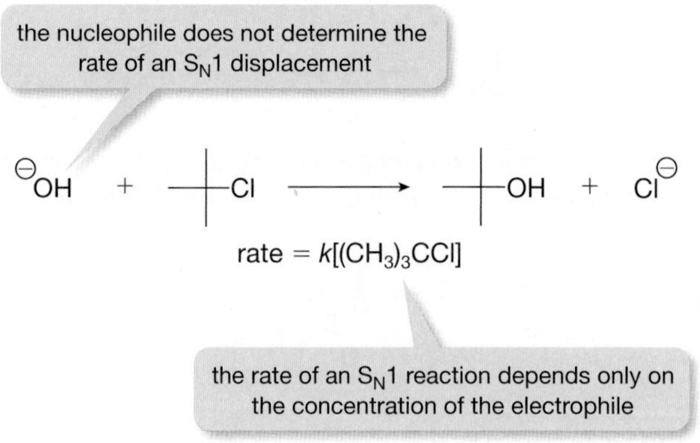

the nucleophile does not determine the rate of an S_N1 displacement

$$^{\ominus}OH + (CH_3)_3C\text{–}Cl \longrightarrow (CH_3)_3C\text{–}OH + Cl^{\ominus}$$

$$\text{rate} = k[(CH_3)_3CCl]$$

the rate of an S_N1 reaction depends only on the concentration of the electrophile

The rate of this reaction depends on the concentration of only one component, so the reaction is designated as an S_N1 reaction (substitution nucleophilic unimolecular). The fact that the nucleophile does not influence the rate of the reaction suggests that only the electrophile is involved in the rate-determining step. This reaction must proceed in at least two steps: one involves the electrophile, and one involves the nucleophile. This can happen only if the leaving group leaves in the first and rate-determining step and the nucleophile adds in the second step. When the leaving group departs with a pair of electrons, a carbocation is created. This provides a site for the nucleophile to add and complete the substitution reaction.

S_N1 reactions can involve several steps. Each step has a transition state associated with it, and each has a corresponding energy barrier.

ORGANIC CHEMWARE
11.7 S_N1 reaction: General mechanism

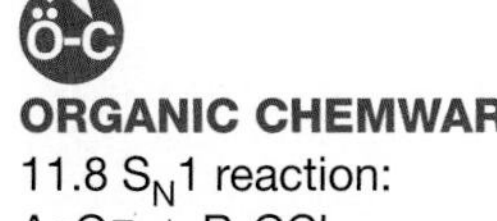
ORGANIC CHEMWARE
11.8 S_N1 reaction: $AcO^- + R_3CCl$

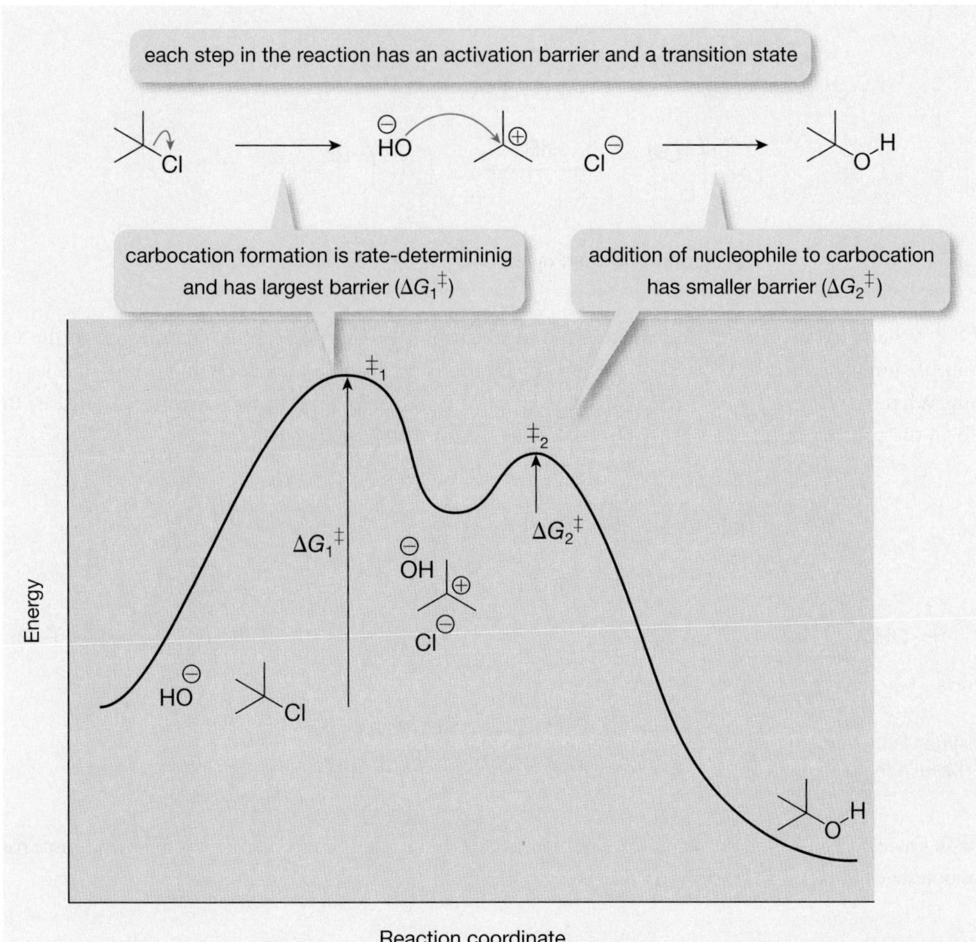

The S_N1 reaction is heavily dependent on the ability of the α-carbon to stabilize a positive charge. If the carbocation is too difficult to form, the S_N1 pathway becomes very slow. Then either an S_N2 reaction occurs or no reaction will take place. The activation energy for the formation of the carbocation controls the rate of the reaction, so the reaction can be analyzed by considering the transition state for this process. In practice, this is more commonly done by examining the carbocation directly (Hammond postulate). The more stable the carbocation, the more likely it will be involved in an S_N1 reaction. In general, tertiary substrates always displace by means of S_N1 mechanisms; secondary substrates may displace by either S_N1 or S_N2; and primary substrates, which rarely form carbocations, are limited to S_N2 displacements (the positive charge of primary cations adjacent to an aromatic ring can become delocalized by the ring).

CHECKPOINT 11.5

You should now be able to draw a mechanism for an S_N1 reaction.

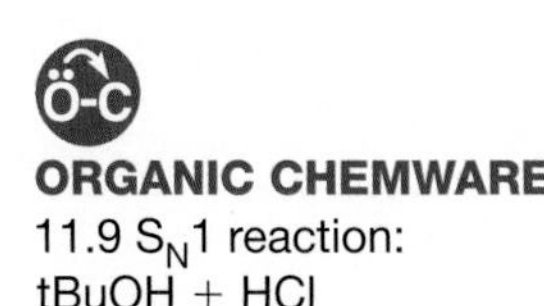
ORGANIC CHEMWARE
11.9 S_N1 reaction: tBuOH + HCl

SOLVED PROBLEM

Provide a mechanism for the following reaction, and draw the corresponding reaction coordinate diagram.

–OH —HBr→ –Br

STEP 1: Identify the difference between the starting material and the product. The Br has replaced the OH, which suggests a displacement. The OH appears to be the leaving group, which provides information about the α-carbon. This site is tertiary, so the reaction would be S_N1.

STEP 2: When acid is present, the first step in a mechanism is often protonation. The oxygen of the OH group has lone pairs that can act as a base. Protonation creates an oxonium ion, which is a good leaving group. When the leaving group leaves, a tertiary carbocation forms, and it can react with the leftover Br^- to form the product indicated.

STEP 3: Once you know the mechanism, draw a reaction coordinate diagram. Each step of the mechanism has a transition state. Formation of the carbocation is rate determining and has the highest-energy transition state.

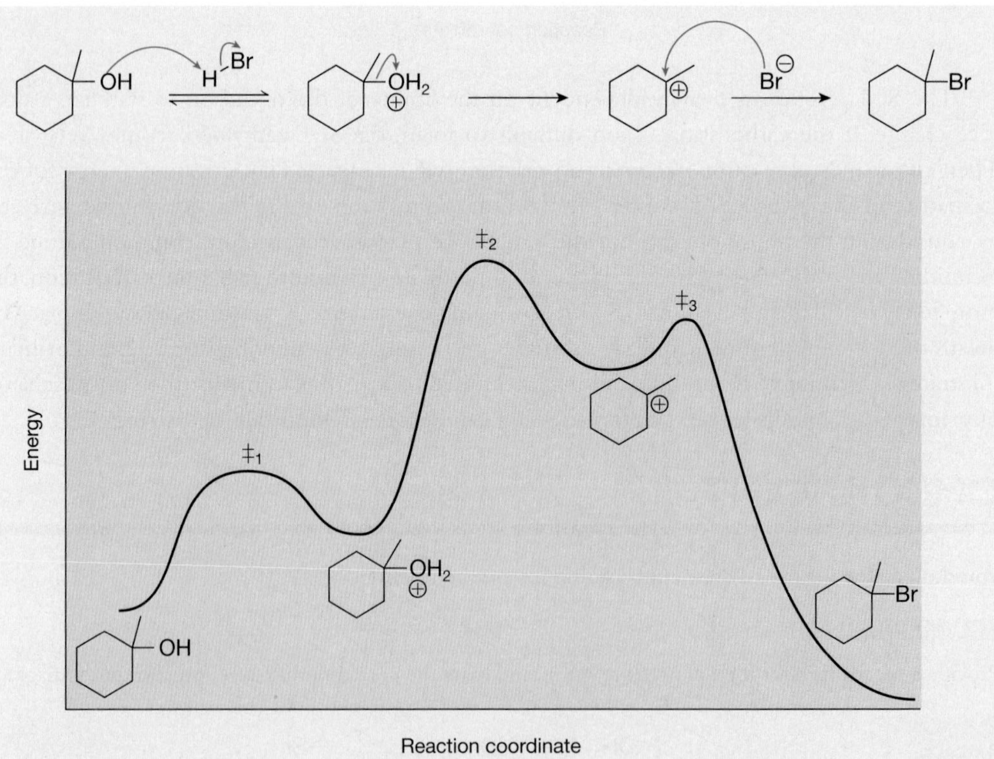

PRACTICE PROBLEM

11.9 Predict the products of the following reactions, and show the reaction coordinate diagram corresponding to each.

a) 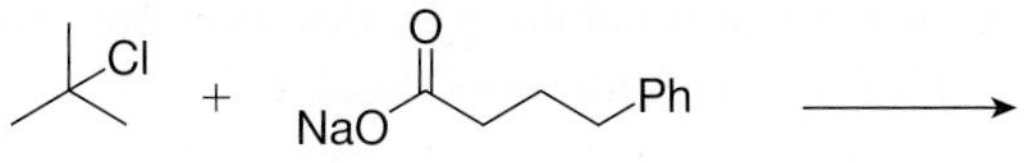 Cl + NaO(C=O)...Ph →

b) 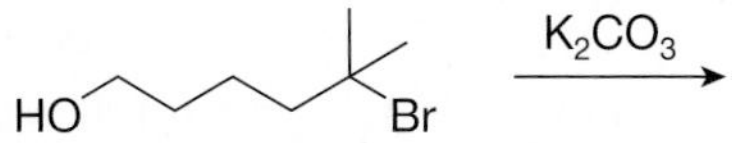 HO...Br $\xrightarrow{K_2CO_3}$

c) Br, CH$_3$, H, CH$_3$ $\xrightarrow{NaSH}$

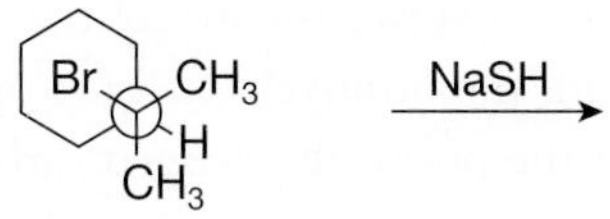

INTEGRATE THE SKILL

11.10 Carbon-based groups can also donate electrons by induction. Using this information, rank the following bases in order of increasing base strength. Why do the alkyl groups not donate electrons by hyperconjugation in this example?

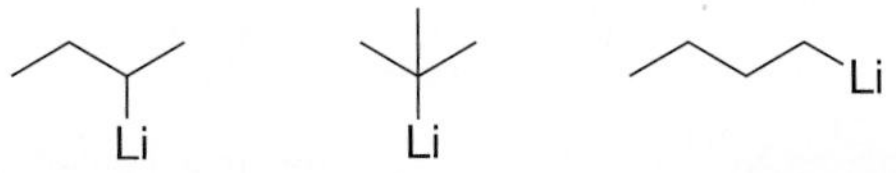

Li Li Li

11.4.2 Carbocations stabilized by charge delocalization

Carbocations can be stabilized by delocalizing their charge over several atoms. Substrates that form such stabilized cations tend to undergo S_N1 reactions, and may form more than one product.

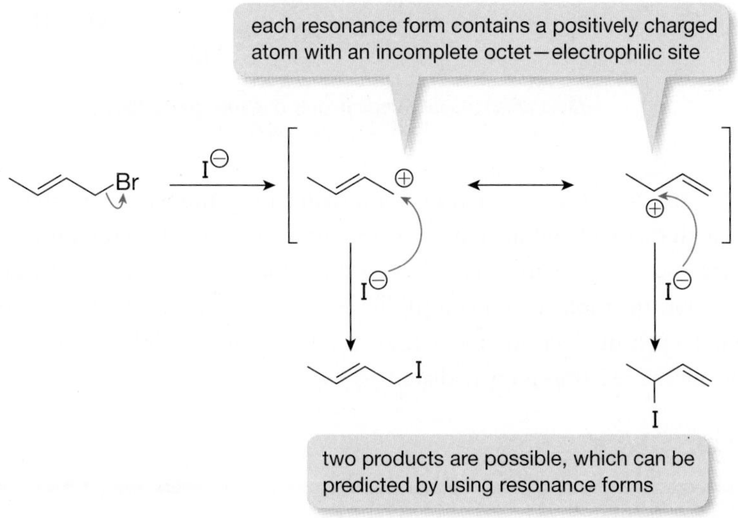

each resonance form contains a positively charged atom with an incomplete octet—electrophilic site

Br $\xrightarrow{I^{\ominus}}$ [⊕ ⟷ ⊕] ; $I^{\ominus}$; $I^{\ominus}$; I ; I

two products are possible, which can be predicted by using resonance forms

When the S_N1 pathway becomes viable, electrophiles react very quickly. In many cases, the reaction results in a single product. In other cases, especially when the resonance forms of the carbocation intermediates have similar contributions, the reaction results in mixtures of products. To determine whether the result is a single product or a mixture of products, the resonance forms need to be examined to see if they are similar or very different. When the two resonance forms are very different in terms of their contribution, the major product can change depending on the temperature of the reaction (see Chapter 7 for more detail).

Electrophiles that can form delocalized carbocations also make excellent substrates for S_N2 reactions. During these reactions the bonds form and break in a single step, but the reaction is asynchronous. The carbon–leaving group bond breaks much faster than the carbon-nucleophile bond forms. This creates a partial positive charge that is stabilized by delocalization (which can be described by resonance). See Sections 11.4.5 and 11.5 for more detail.

11.4.2.1 Nucleophilic reactions of oxonium ions

A very common family of S_N1 reactions involves the oxonium ion (Chapter 16). The following example shows the substitution of α-halo ethers. The first step involves the loss of the chloride to generate a stabilized carbocation. The resonance form with the positively charged oxygen is a much stronger resonance contributor than the form with the positively charged carbon. The second resonance form therefore contributes the most to the structure of this ion, and such ions are usually depicted as the oxonium form.

O Cl → [O ⊕ ↔ ⊕O] → O Nu

Nu

This resonance form has an incomplete octet on carbon. Nucleophiles add to that position.

This resonance form (oxonium ion) has full octets.

The mechanism is usually written using the best resonance form as it most closely resembles how the molecules actually react.

O Cl → ⊕O → O Nu

Nu

mechanism usually depicts only one resonance form

Of the two resonance forms, one has a carbon with an incomplete octet, which indicates the site of reaction with the nucleophile. This S_N1 reaction is typically shown with the intermediate oxonium ion because it depicts most closely how the molecules actually react (the "best" resonance forms resemble the molecules involved). To remember this, consider that the oxygen in the starting halide has a pair of electrons and acts as an internal nucleophile to push out the leaving group and generate the reaction intermediate.

CHECKPOINT 11.6

You should now be able to identify the product of an S_N1 displacement involving resonance.

SOLVED PROBLEM

What is the structure of the major product of the following reaction? (Hint: Silver interacts strongly with halides.)

O Cl → HO / $Ag_2(CO_3)$

STEP 1: Identify the reacting partners. The substrate has a halogen, which suggests a site for a nucleophilic displacement. The α-carbon is secondary, and is attached to an atom (oxygen) with lone pairs that could stabilize a carbocation. The question provides a clue that Ag^+ strongly interacts (bonds) with halogens. This suggests an S_N1 pathway in which the silver acts as a Lewis acid.

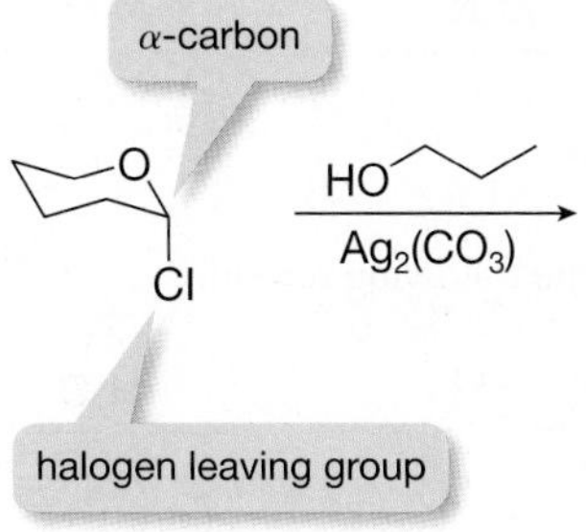

α-carbon

HO, $Ag_2(CO_3)$

Cl

halogen leaving group

STEP 2: Reactions involving Lewis acids are very similar to those involving Brønsted acids (metal$^+$ instead of H^+). Reaction between Ag+ and Cl generates a very good leaving group.

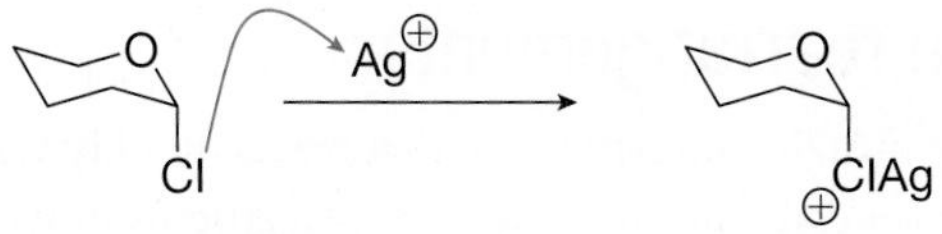

STEP 3: The next step of the reaction must be the loss of a leaving group (basic conditions). This generates a stabilized carbocation. Analysis of the resonance forms shows that the site of reactivity must be carbon (because it is the only atom with an incomplete octet), and that the oxonium must be the major resonance contributor (because all its atoms have octets).

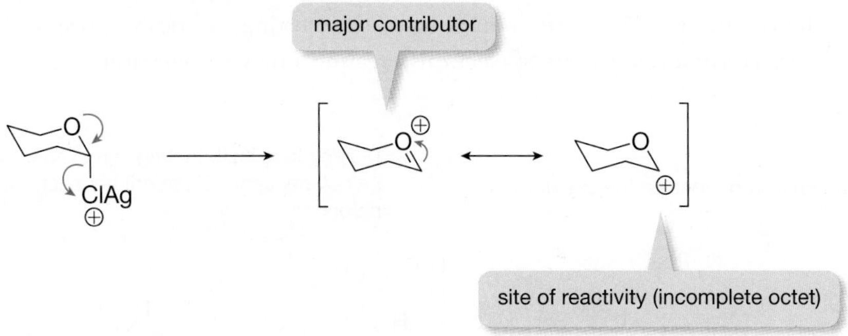

major contributor

site of reactivity (incomplete octet)

STEP 4: The nucleophile now reacts, most likely in the neutral form because the base that is present (CO_3^{-2}) is not strong enough to fully deprotonate the alcohol. This reaction forms an oxonium intermediate that quickly disappears with addition of the base.

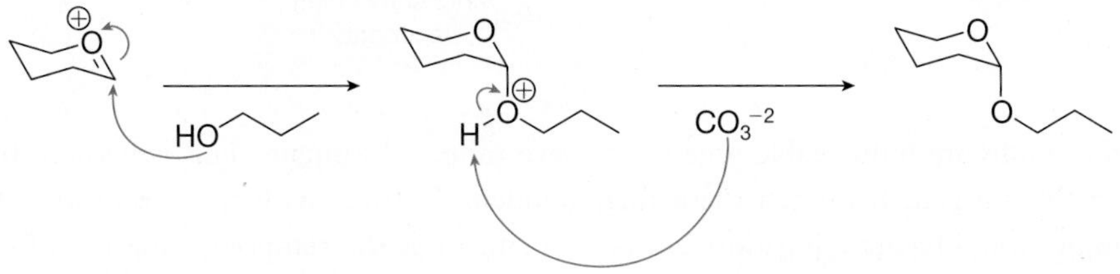

PRACTICE PROBLEM

11.11 What is the structure of the major product in each of the following reactions?

a) 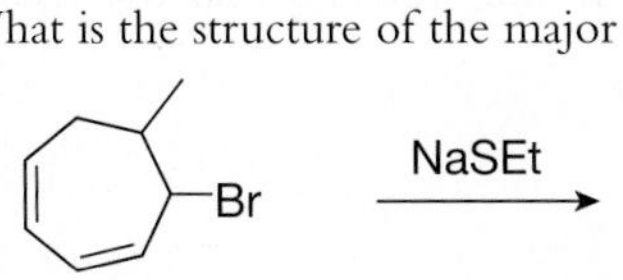 NaSEt

b) [cyclohexanol] 1) NaH; 2) [benzyl bromide] →

OH

Br

c) HO ... O ... Cl $\xrightarrow{Et_3N}$

INTEGRATE THE SKILL

11.12 What is the major product of the following reaction?

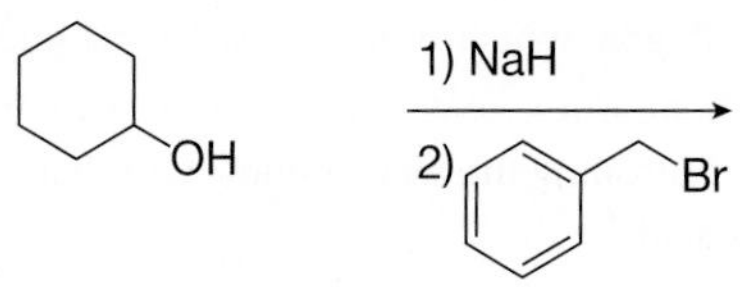

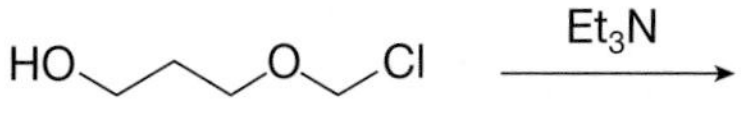

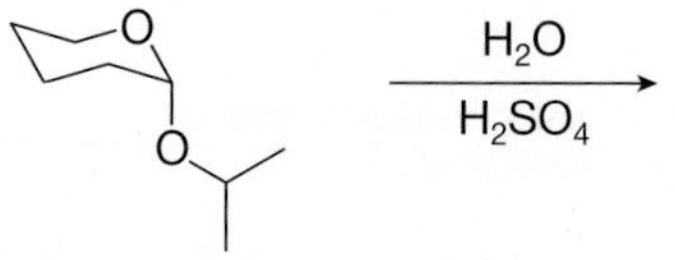

$\xrightarrow[H_2SO_4]{H_2O}$

11.4.3 Carbocation rearrangements

A carbocation is a relatively unstable intermediate that reacts quickly with nucleophiles and bases. The presence of a carbocation can complicate S_N1 displacements however because it provides an opportunity for competing reactions to occur. This is especially true when reactions are heated. For example, carbocations might undergo elimination (Chapter 12) or add to an aromatic ring (Chapter 10). Even when substitution is the preferred reaction pathway, a carbocation rearrangement may occur before the addition of the nucleophile.

Hydride shifts involve the movement of a hydrogen, with its two electrons, from the carbon adjacent to the carbocation. When the bond moves, the existing carbocation becomes neutralized. The adjacent carbon is left short of electrons, creating a new carbocation.

strong acid protonates the oxygen
[1,2]-hydride shift: hydrogen moves with a pair of electrons (CH bond) from adjacent carbon
OH
H–Br
$^{\oplus}OH_2$
H_2O
H
2
1
H H
2
1
tertiary carbocation more stable than secondary

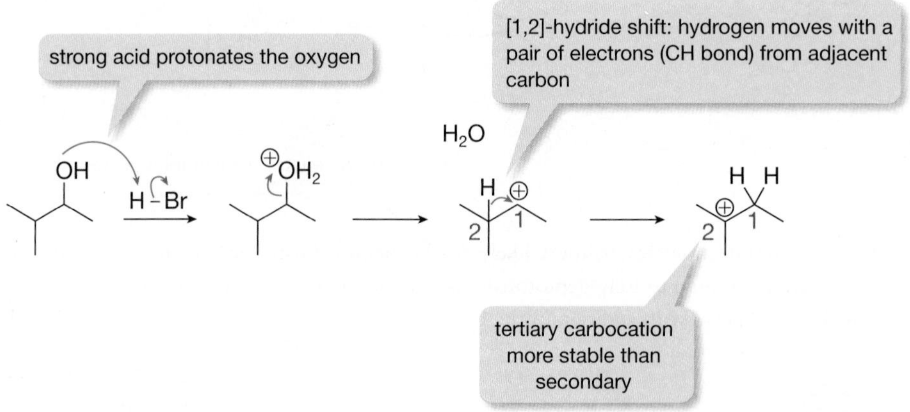

Carbocations are more stable when they have more substituents. This is because the C–H bonds on the adjacent hydrogen share their bonded electrons with the carbocation through hyperconjugation. Hyperconjugation can be thought of as the temporary sharing of electrons in σ bonds with the adjacent empty p orbitals of the carbocation. Hydride shifts occur due to the overlap of orbitals between a C-H σ bond and an adjacent empty p orbital. Rather than a simple temporary sharing, in a rearrangement the C-H σ bond shifts so that the more stable carbocation remains.

secondary carbocation stabilized by hyperconjugation

tertiary carbocation stabilized by hyperconjugation

Rearrangements happen only when a carbocation with greater stability can form. Note that a carbocation rearrangement forms a secondary carbocation from a primary, and a tertiary carbocation forms from a secondary. Allylic or benzylic cations are stabilized by charge delocalization and can form from rearrangements, even when the degree of substitution does not increase.

ORGANIC CHEMWARE
11.10 Carbocation rearrangement: 1,2-Hydride shift

[1,2]-hydride shift

delocalized carbocation (allylic)

When there is no adjacent hydride, an alkyl group such as a methyl can also migrate.

[1,2]-alkyl shift

secondary carbocation

tertiary carbocation (more stable)

ORGANIC CHEMWARE
11.11 Carbocation rearrangement: 1,2-Alkyl shift

CHECKPOINT 11.7

You should now be able to identify simple carbocation rearrangements.

SOLVED PROBLEM

What are all of the possible substitution products of the following reaction?

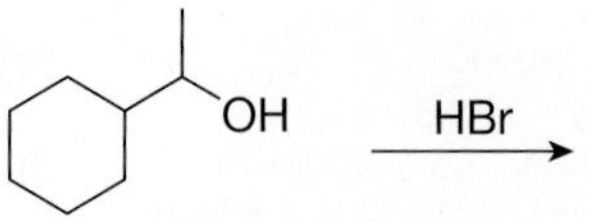

OH HBr

STEP 1: The OH in the substrate is a base and reacts with HBr, a strong acid. This reaction makes an oxonium ion, which is a good potential leaving group.

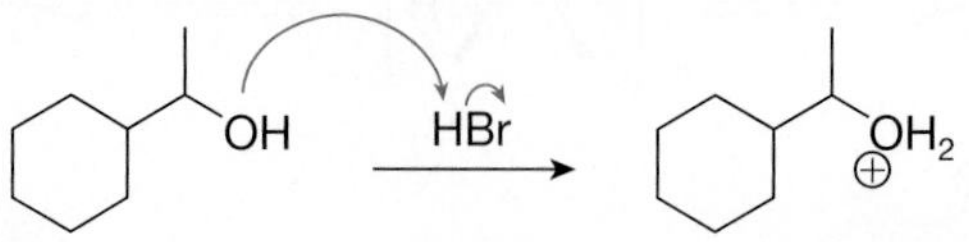

OH HBr $\overset{+}{O}H_2$

STEP 2: When the leaving group leaves, it forms a secondary carbocation with an adjacent carbon bearing a C–H bond that can migrate. There are two adjacent C–H bonds. Migration of the tertiary hydrogen gives a tertiary carbocation, whereas migration of the primary hydrogen gives a highly unfavourable primary carbocation. The tertiary carbocation is more stable than the original secondary ion, and so the reaction favours the rearrangement that forms the tertiary carbocation.

primary carbocation

secondary carbocation

tertiary carbocation

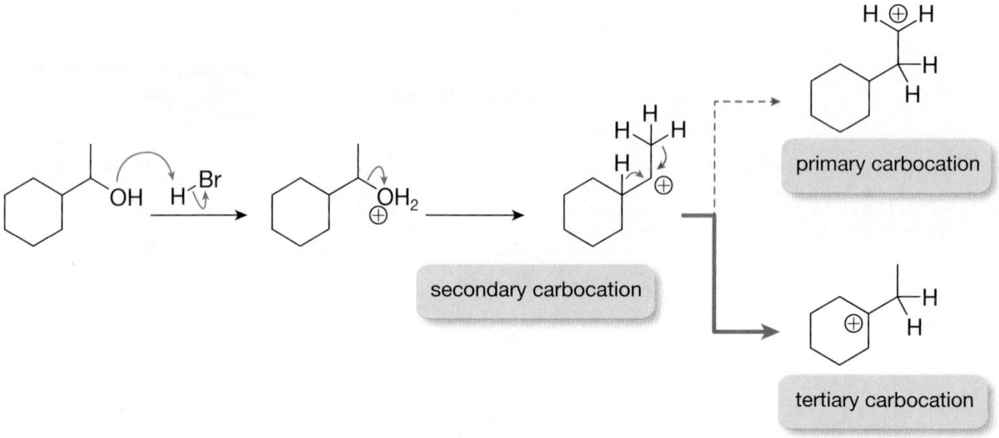

STEP 3: All three ions can react with the bromide. The primary ion is very unfavourable and results in only trace amounts of product (if any). The tertiary carbocation is more stable and provides the major product of the reaction.

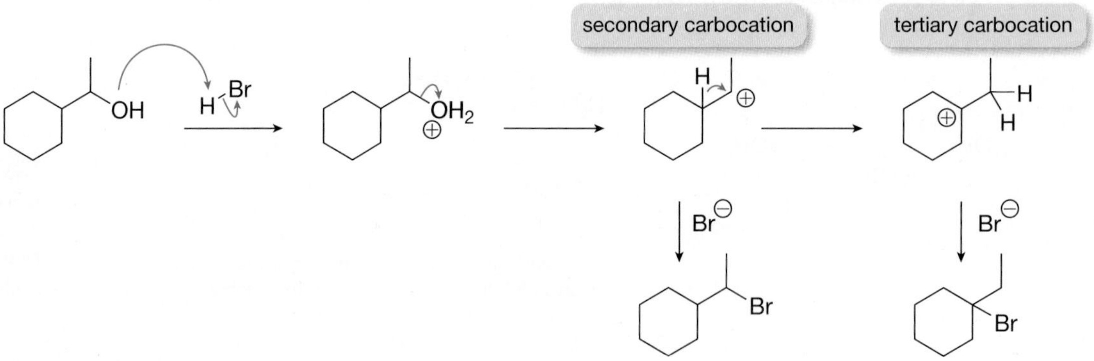

secondary carbocation

tertiary carbocation

Br^- Br^-

Br Br

PRACTICE PROBLEM

11.13 Propose a mechanism to explain each of the following observations.

a) 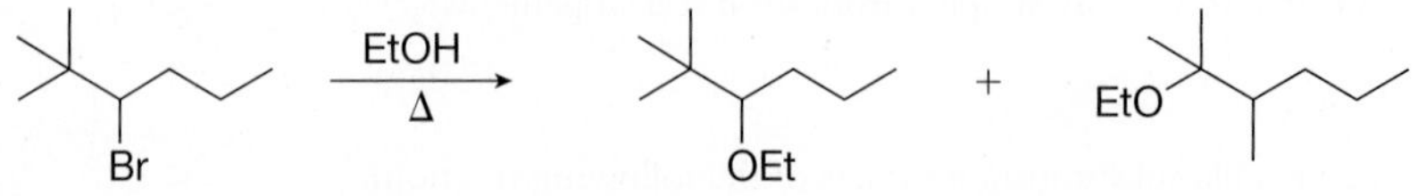 Br $\xrightarrow[\Delta]{EtOH}$ OEt + EtO

b) Cl $\xrightarrow[\Delta]{CH_3OH}$ OCH_3 + OCH_3

c) Br $\xrightarrow[\Delta]{EtOH}$ OEt + OEt

INTEGRATE THE SKILL

11.14 The reaction shown here is an example of a pinacol rearrangement. Use mechanistic arrows to explain the structure of the product.

OH OH $\xrightarrow{H_2SO_4}$ O

11.4.4 Effect of the leaving group

Because the leaving group is involved in the rate-determining step of the S_N1 reaction, its nature has a large influence on the reaction rate. In general, good leaving groups such as the sulfonate groups (OTs, OMs, OTf) accelerate S_N1 reactions.

approximate order of leaving group ability

Cl < Br < I, OTs < H_2O < OMs < OTf

good — better — best

11.4.5 Stereochemistry of S_N1 reactions

The stereochemical outcomes of S_N1 and S_N2 reactions are different because S_N1 reactions involve carbocations. Carbocations are sp^2 hybridized and flat, and so nucleophiles are free to approach either side of these ions (Chapter 7). If the α-carbon is connected to three different groups, the reaction can generate product mixtures with both configurations. Reaction conditions that yield equal amounts of both isomers produce a racemic mixture of products, even when the starting electrophile is enantiomerically pure.

nucleophile approaches from top face
Nu
CH3
H
carbocation is flat
H
CH3
Br
H
CH3
Nu
nucleophile can react from either face, resulting in a mixture of enantiomers
H
CH3
Nu
enantiomerically pure starting material
nucleophile approaches from bottom face

The carbocation that is formed during an S_N1 reaction is sp^2-hybridized, forming three σ bonds to its neighbours. This carbon is therefore flat and has an empty p orbital, the LUMO, which is perpendicular to the plane of the σ bonds. The incoming nucleophile provides a pair of electrons that reside in a HOMO.

To form the new bond, the HOMO of the nucleophile must overlap with the carbocation LUMO. The LUMO has an equal probability of accepting electrons on either face of the carbocation (the p orbital has equally sized lobes on both faces of the sp^2-hybridized carbon). This means that the nucleophile has an equal chance of forming a bond to either face of the carbocation.

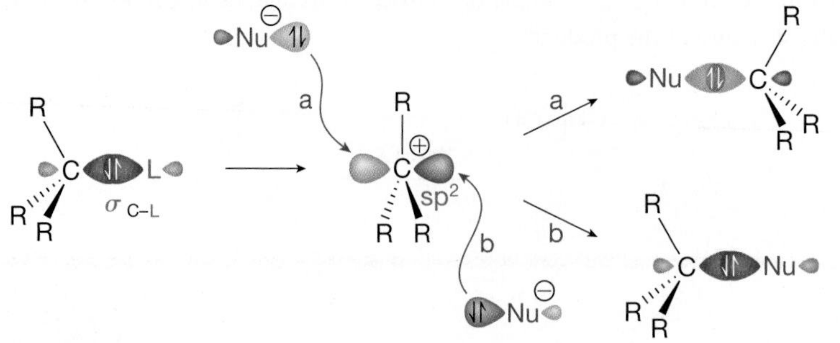

DID YOU KNOW?

It is usually not possible to carry out nucleophilic displacements on sp^2-hybridized carbons using S_N1 or S_N2 conditions. This is because the carbon leaving group bonds are stronger for sp^2-hybridized carbons than they are for sp^3-hybridized carbons. Also, the sp^2-hybridized carbon has greater s character, resulting in a lower orbital energy, which means that these carbons do not stabilize carbocations well.

Displacement of leaving groups on sp^2-hybridized carbons, such as those on alkenes or aromatic groups, requires special reagents. In these cases, palladium catalysts are used to carry out a cross-coupling reaction. These reactions have the appearance of a nucleophilic displacement; however, the mechanism of these reactions is not straightforward.

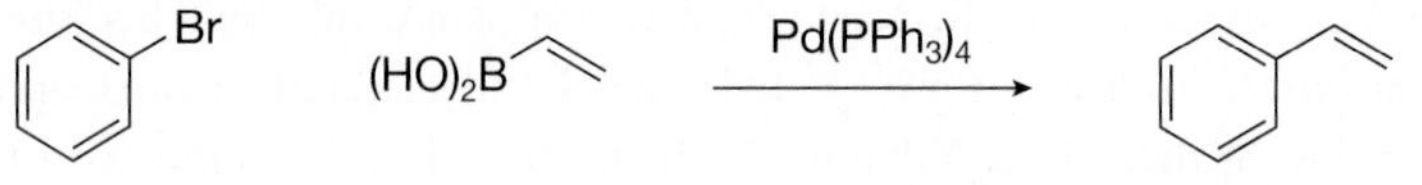

11.5 S_N1 and S_N2 as a Reactivity Continuum

To this point, the differences between S_N2 and S_N1 reactions have been emphasized in terms of their particular reactions rates, mechanisms, and stereochemistry. However, even though S_N2 and S_N1 reactions can be clearly distinguished as one-step or two-step processes, these pathways represent extremes of a reactivity continuum, and so many nucleophilic displacements display characteristics of both. Consequently, a particular displacement reaction mechanism may be more S_N2-like or more S_N1-like, without being purely one or the other.

The following examples illustrate S_N1 and S_N2 displacements as a reactivity continuum. These examples focus on the reactivity continuum in terms of sensitivity to inductive effects, degree of product mixture, and ability to form stabilized carbocations.

1. Many S_N2 substrates are sensitive to inductive effects (Chapter 6), suggesting that the α-carbon carries a small positive charge in the transition state. These reactions are concerted S_N2 reactions because no intermediate is formed, but the carbon–leaving group bond breaks faster than the nucleophile-carbon bond forms, creating a $\delta+$ on the α-carbon. The reaction

is S_N2 but involves a partial carbocation, suggestive of some S_N1 character. For example, the substrate with a CF_2 group beside the α-carbon induces a large charge on the α-carbon in the transition state, raising the energy of this species and slowing the reaction relative to the corresponding substrate substituted only with hydrogens.

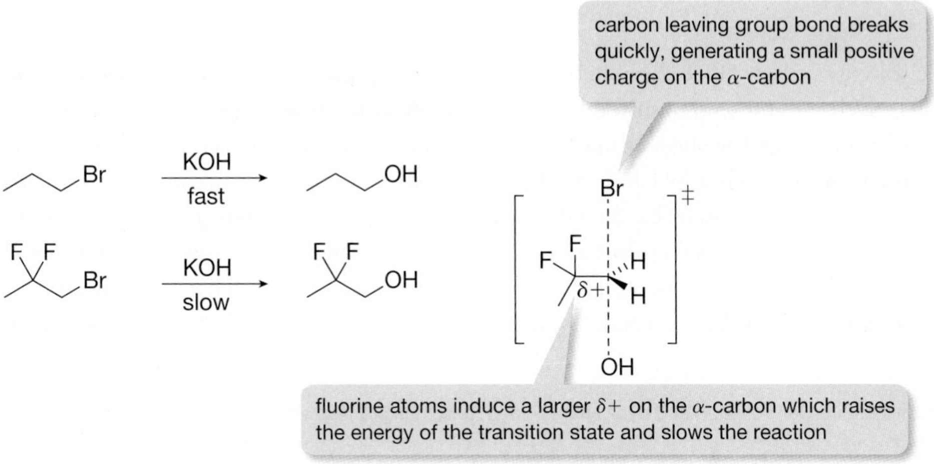

2. S_N1 reactions that produce completely racemic products from enantiomerically pure starting materials are relatively rare; it is much more common to find partial racemization. In these reactions, the major product has inverted stereochemistry relative to the starting material. This occurs because the nucleophile forms a bond to the carbocation before the leaving group completely diffuses away. As a result, the leaving group partially blocks one side of the carbocation, and the nucleophile most likely approaches the carbocation from the opposite side of the departing leaving group.

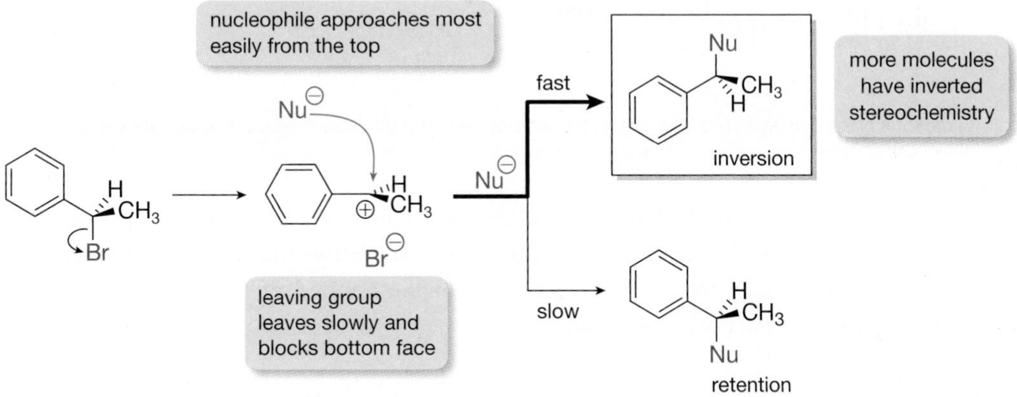

3. Electrophiles that can form stabilized carbocations can react by either S_N1 or S_N2 pathways. In each case, carbocations or partial carbocations are involved. These types of substrates can undergo nucleophilic displacements very quickly.

11.5.1 Different solvent effects

The effects of solvents on nucleophilicity differ depending on whether the reaction is an S_N1 or S_N2 displacement. In general, protic solvents stabilize carbocations very well, and these solvents favour S_N1 reactions. Polar aprotic solvents are especially good for S_N2 reactions because they enhance nucleophilicity.

As described in Section 11.3.4.2, the electronegativity of atoms increases from left to right across the periodic table, and nucleophilicity decreases accordingly. Atomic size increases moving down a column in the periodic table, and this affects nucleophilicity in a different way. The effect of atomic size on nucleophilicity is influenced by the solvent used in a reaction. For example, when protic solvents are used, nucleophilicity increases in relation to increasing atomic size, and when polar aprotic solvents are used, nucleophilicity decreases in relation to increasing atomic size (this latter effect is relatively small).

11.5.1.1 Protic solvents

Protic solvents such as H_2O, alcohols, or amines are very polar and capable of extensive hydrogen bonding. Such solvents tend to surround anions and solvate them very effectively through hydrogen bonding. Small anions have high charge densities and form strong hydrogen bonds. In protic solvents, these anions are less nucleophilic because the strong hydrogen bond network inhibits their ability to donate electrons. Large anions, on the other hand, have low charge density and so do not hydrogen bond as well. Because of this they do not interact as strongly with protic solvents and are better able to donate electrons. Such anions are stronger nucleophiles in protic solvents. Therefore, when protic solvents are used, nucleophilicity increases with increasing atomic size moving down the periodic table.

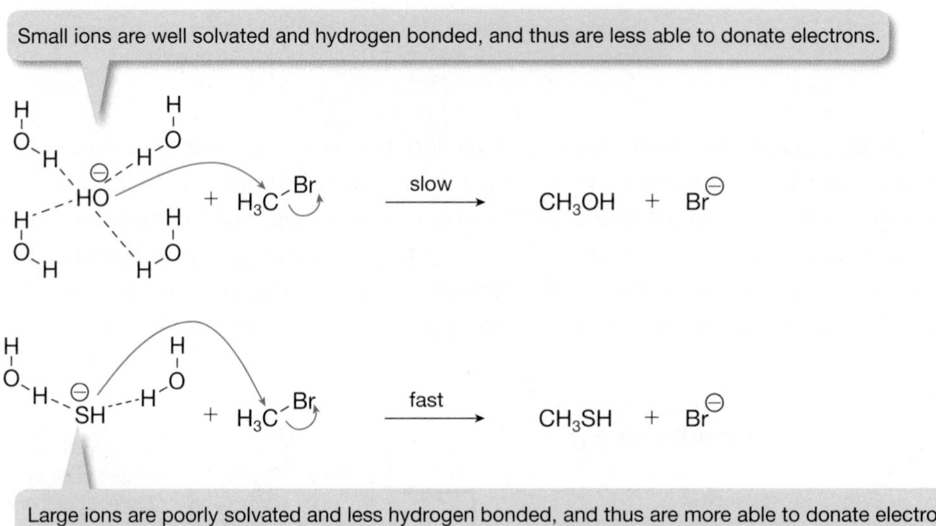

Protic solvents accelerate S_N1 reactions in two ways. First, the donation of a pair of electrons from the solvent to the carbocation stabilizes the carbocation and accelerates the reaction. Second, hydrogen bonding from the solvent assists in the removal of the leaving group. S_N1 reactions happen very quickly in protic solvents.

hydrogen bonding with the solvent helps the leaving group to leave

electron donation from the solvent stabilizes the carbocation

11.5.1.2 Polar aprotic solvents

Polar aprotic solvents do not have a hydrogen bond donating group, so they do not solvate anions well, and this affects nucleophilic behaviour. Poorly solvated, negatively charged nucleophiles become much better electron donors, and so S_N2 reactions can occur thousands of times faster in aprotic solvents. Polar aprotic solvents can also solvate metal cations very well due to their Lewis base character. This tends to create "naked" anions that are not impeded by their counter cations. Most modern S_N2 reactions are carried out in polar aprotic solvents such as DMF, DMSO, acetone, acetonitrile, nitromethane, or THF.

DMF DMSO acetone CH_3CN acetonitrile CH_3NO_2 nitromethane THF

These solvents provide dipole effects only, and are less effective in solvating carbocations. The use of these solvents slows the rate of reactions of S_N1 reactions considerably relative to the rate of the same reaction in protic solvents. On the other hand, dipolar aprotic solvents are highly beneficial in S_N2 reactions because they enhance nucleophilicity.

Non-polar solvents are rarely used for substitution reactions because they do not solvate charge well. Charged intermediates are involved in both S_N1 and S_N2 processes, so both types of reaction can become difficult if the solvent polarity is very low.

11.6 Predicting S_N1 and S_N2 Reaction Mechanisms

To be able to predict the mechanism of a nucleophilic displacement, three characteristics of the reaction need to be examined. These are listed in order of importance:

1. structure of the electrophile
2. strength of the nucleophile
3. solvent used

Leaving groups are important in the rate-determining step of both S_N1 and S_N2 pathways, so it can be difficult to differentiate between these two mechanisms based on the leaving group.

11.6.1 Structure of the electrophile

By far the most important factor in determining a displacement mechanism is the nature and structure of the electrophile. If a carbocation can be sufficiently stabilized, the reaction will be S_N1; if not, the S_N2 pathway will prevail.

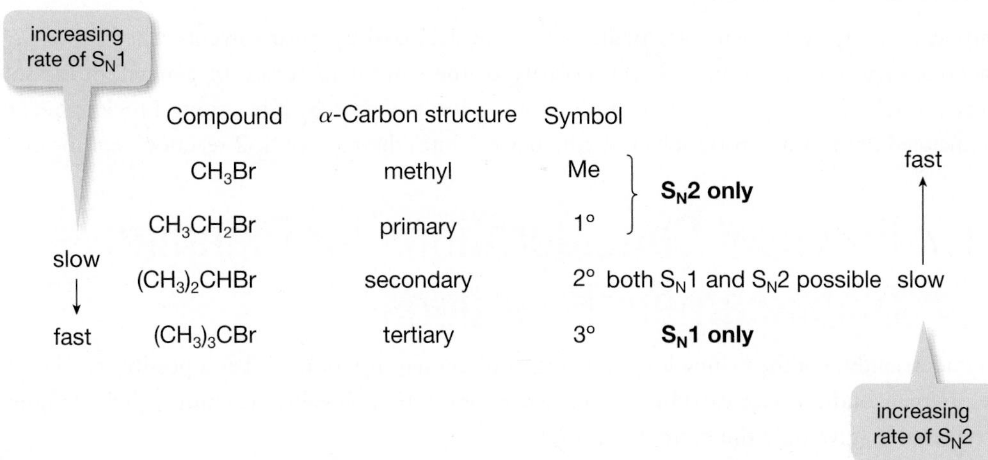

increasing rate of S_N1
Compound
α-Carbon structure
Symbol
CH3Br methyl Me
CH3CH2Br primary 1°
S_N2 only
(CH3)2CHBr secondary 2° both S_N1 and S_N2 possible
(CH3)3CBr tertiary 3° S_N1 only
slow
fast
fast
slow
increasing rate of S_N2

Methyl and primary substrates follow S_N2 pathways, whereas tertiary substrates displace by S_N1 pathways. Substrates that can form a stabilized carbocation tend to follow S_N1-type pathways.

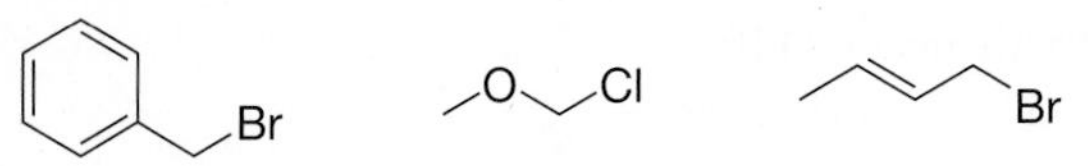

Br O Cl Br

Because reactions with secondary substrates can follow both paths depending on the reaction conditions, reaction analysis can be challenging with these substrates.

11.6.2 Strength of the nucleophile

Both the S_N1 and S_N2 pathways are possible in reactions involving secondary electrophiles and those that can form stabilized carbocations. In these cases, other factors determine whether one pathway is favoured over the other. The nucleophile is involved in the rate-determining step of the S_N2 reaction, and so good nucleophiles favour S_N2 reactions. The use of poor nucleophiles slows the S_N2 reaction, which allows S_N1 mechanisms to predominate (the nucleophile does not affect the rate of S_N1 reactions). The reaction pathway may also be influenced by the solvent used.

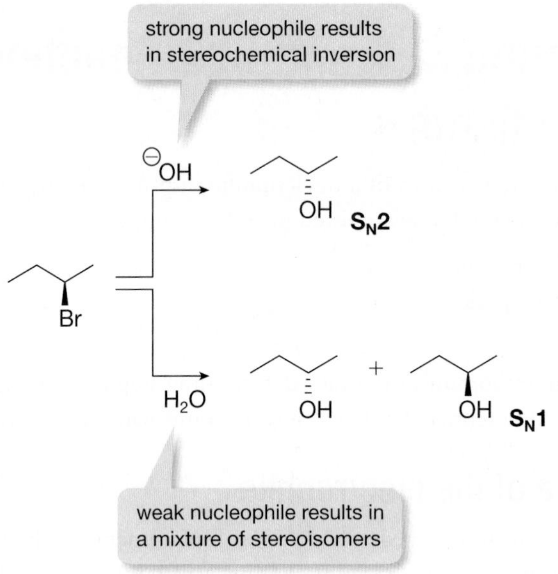

strong nucleophile results in stereochemical inversion

$^{\ominus}OH$ OH S_N2

Br

H_2O OH + OH S_N1

weak nucleophile results in a mixture of stereoisomers

11.6.3 Solvent used

Carbocations are very polar intermediates that are stabilized by polar solvents. The rate of S_N1 reactions increases significantly as the polarity of the solvent increases. In general, protic solvents strongly favour S_N1 reactions, and aprotic solvents favour S_N2 reactions. Nucleophilicity is enhanced in dipolar aprotic solvents, which accelerates the rates of S_N2 reactions significantly.

11.7 Practical Considerations for Planning Displacement Reactions

An understanding of the principles governing displacement reactions makes it possible to identify the optimal conditions for carrying out a reaction or synthesis. Ideally, a reaction will be designed to be reliable (give only the desired product).

For reliability, it is better to use S_N2 displacements instead of S_N1 displacements to avoid carbocation intermediates. Carbocations are very reactive and can participate in competing reactions, such as rearrangements that can reduce yields and complicate purification. As a result, S_N2 reactions often produce higher yields and cleaner products (fewer by-products in smaller amounts) than their S_N1 counterparts.

When there is no choice and S_N1 reactions must be used—for example, in the case of tertiary products—reaction conditions that strongly favour the S_N1 pathway (protic solvents, good nucleophiles) will result in fewer problems.

Planning a synthesis can sometimes involve the systematic disconnection of several bonds of the target molecule. The synthon approach is often very useful for this because mechanistic arrows can be used to dissect synthetic targets. The resulting fragments, termed *synthons*, can represent the *type* of reactivity required in the individual components (nucleophile or electrophile). For example, heteroatoms are potential sites for displacement reactions as many common nucleophiles are heteroatoms. Disconnection of the carbon-heteroatom bond in a way that generates a heteroatomic nucleophile and a carbon electrophile provides useful synthons for synthesis. Once suitable synthons are identified, appropriate substrates and reaction conditions can be devised to form these specific bonds.

11.8 Special Nucleophiles and Electrophiles Used in Displacement Reactions

11.8.1 Using acetate to make alcohols

Using hydroxide as a nucleophile to make secondary alcohols can be challenging because hydroxide is a strong enough base to promote other types of reactions, such as eliminations (Chapter 12). To encourage nucleophilic substitution reactions (also called *displacement reactions*) to take place, lower reaction temperatures and polar aprotic solvents are needed.

As an alternative procedure, a less basic nucleophile such as acetate can be used to make secondary alcohols in a two-step reaction sequence. Because of its reduced basicity, acetate is a nucleophile that can displace a leaving group, while minimizing the formation of other products.

the required C–O bond is formed

DMF

The resulting displacement product is an ester functional group (an acetate ester) that must be converted to the target OH group by a second reaction. The ester group can be cleaved by a hydrolysis reaction under basic conditions (Chapter 15) to produce the desired alcohol.

hydroxide cleaves the ester

NaOH, H_2O

Secondary electrophiles that can form charge-delocalized carbocations such as allylic or benzylic halides are reactive enough that displacement reactions with hydroxide proceed with relative ease.

allylic halide

benzylic halide

NaOH, acetone

11.8.2 Using alkoxides to make ethers

The **Williamson ether synthesis** is a way of making ethers from alkoxides and alkyl halides using an S_N2 displacement.

The **Williamson ether synthesis** is an S_N2 reaction between an alkoxide and an alkyl halide to form an ether.

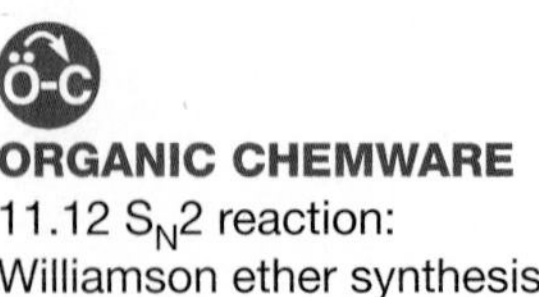

ORGANIC CHEMWARE 11.12 S_N2 reaction: Williamson ether synthesis

The alkoxide nucleophile is usually generated by deprotonating an alcohol with a non-nucleophilic base such as sodium hydride (NaH) in a polar aprotic solvent such as THF or DMF. Once the alkoxide has formed, adding the alkyl halide electrophile results in an S_N2 displacement of the halide leaving group to produce an ether.

1) NaH
2) ethyl bromide

The Williamson ether synthesis is particularly useful for producing non-symmetric ethers (R-O-R′). Two synthetic routes are possible depending on which group originates from the nucleophile and which group stems from the electrophile. When planning a synthesis utilizing this method, both disconnections need to be examined to identify which set of reagents will produce the most reliable reaction. For example, either an isopropoxide or benzyloxide nucleophile can be used in the synthesis of a benzylisopropyl ether.

benzylic electrophile (more reactive)

OTs + HO

OH + TsO

sterically hindered (less reactive)

Secondary electrophiles are sterically hindered and are not ideal substrates for S_N2 displacements. On the other hand, benzylic electrophiles react particularly well in nucleophilic displacement reactions. Of the two synthetic routes, the substitution of benzyl tosylate with isopropoxide is preferred over the displacement of isopropyl tosylate with benzyloxide because the first combination is most likely to follow an S_N2 reaction and be the most reliable method.

OH —NaH→ ONa —TsO (faster)→ O

HO —NaH→ NaO —OTs (slower)→ O

11.8.3 Using epoxide electrophiles in synthesis

Ethers are normally an unreactive functional group. An important exception is epoxides, which have a large amount of ring strain and consequently are easily opened. This ring opening proceeds via an S_N2 displacement; a strong nucleophile attacks an epoxide carbon to open the ring and form a new product.

The opening of an epoxide is regioselective. In basic conditions, the nucleophile reacts at the less substituted carbon because this provides easier access for the nucleophile.

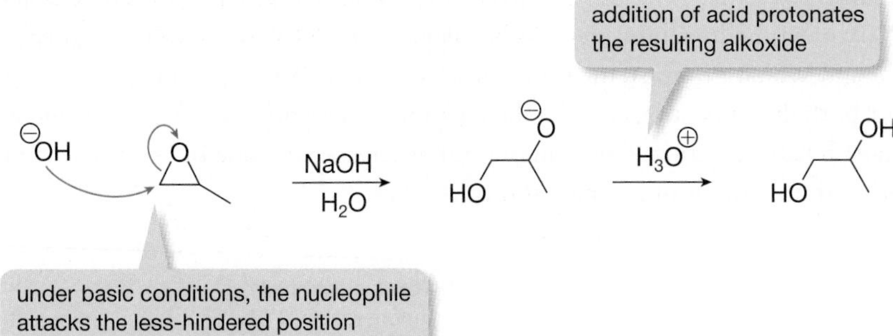

ORGANIC CHEMWARE
11.13 Epoxide opening (basic conditions)

In acidic conditions, the protonated epoxide undergoes nucleophilic displacement at the more substituted carbon—that is, at the position that can best stabilize a carbocation (Chapter 8). This reaction is a good example of an S_N2 reaction with some S_N1 character.

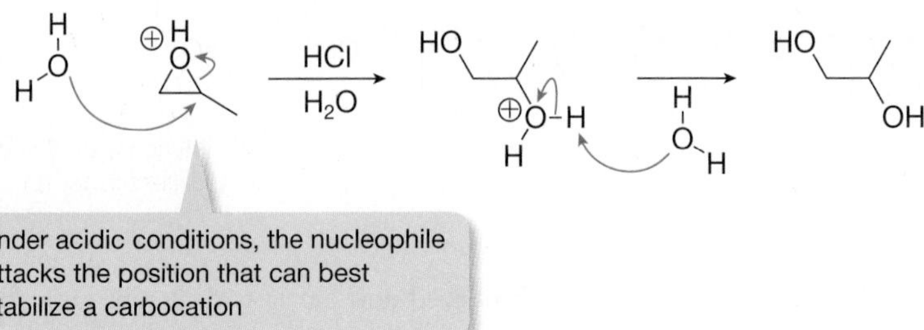

ORGANIC CHEMWARE
11.14 Epoxide opening (acidic conditions)

Whether the nucleophilic attack occurs at the more or the less substituted position, the opening of the epoxide ring follows the standard S_N2 displacement attack from the backside, inverting the configuration of any stereocentre at the α-site.

1) NaSEt
2) $H_3O^{\oplus}$
OH
EtS

HBr
HSEt
HO
SEt

under either acidic or basic conditions, inversion of configuration is observed

CHEMISTRY: EVERYTHING AND EVERYWHERE

The First Successful Cancer Drug

Mustard gas is a liquid compound that was used extensively as a chemical weapon in World War I. When sprayed on its victims, it causes blindness, lung damage, and extensive blistering. Mustard gas was so terrible a weapon that it was banned after the war, along with other chemical weapons.

Despite the ban, during World War II chemical weapons were manufactured by all armies. The liberty ship *John E. Harvey*, carrying tons of mustard gas, was sunk during an air raid in 1943, releasing its deadly cargo into an Italian harbour. Hundreds of people were killed by the toxic material. However, doctors who treated the survivors noticed an unusual toxicity on their patient's white blood cells, suggesting a potential benefit in treating leukemia. Experiments with tumours showed that mustard gas was an effective anti-cancer agent, but too toxic to use as a drug. Based on the mechanism by which the drug killed tumours, chemists designed a new drug with toxicity low enough for human use.

This new drug, mustine, became the first effective chemotherapy agent for cancer. Today, many of the drugs used to treat cancer employ the key functional group that was derived from the original chemical weapon.

Sulfur is an excellent nucleophile and readily displaces chloride to form the reactive intermediate.

very powerful electrophile undergoes S_N2 reaction with DNA

mustard gas

disrupted DNA function

Nitrogen is less nucleophilic than sulfur, so this compound is less likely to be converted to the active form. This increases its selectivity for cancer cells.

mustine

mustine melphalan cyclophosphamide

The reactivity of these drugs depends on the principle of anchimeric assistance: that is, one functional group participates in the reaction of another functional group within the same molecule. This assistance results in a reactive intermediate that facilitates the reaction. Many cases of anchimeric assistance result in highly stereoselective reactions.

11.8.4 Role of nucleophiles in carbon-carbon bond formation

Some carbon nucleophiles are so basic that they do not undergo controlled displacement reactions; rather, eliminations or other side reactions dominate (Chapter 12). Typically, to form carbon-carbon bonds using carbon nucleophiles, it is best to use nucleophiles in which the negatively charged nucleophilic carbon (carbanion) is stabilized by charge delocalization (as described by resonance) or by induction. Some of the carbon nucleophiles commonly used in displacements include acetylides, cyanide, and enolates. Acetylides and cyanide are discussed in the following section; enolates are described in Chapter 17.

11.8.4.1 Acetylides and cyanides

Sodium acetylide and cyanide are considerably less basic than many other carbanions. This can be explained by the valence bond model, in which the negative charge resides on an sp-hybridized carbon. The sp orbital has considerable s orbital character (50%!), holding the electrons of an sp-hybridized carbanion closer to the nucleus, and lowering the energy relative to sp^2- or sp^3-hybridized carbanions. Cyanides and acetylides are therefore strong nucleophiles that are able to undergo controlled S_N2 displacements to form carbon-carbon bonds. As with other carbon-based nucleophiles, displacement reactions work best with primary electrophiles.

KCN / acetonitrile

A nitrile product made using a cyanide nucleophile can be subsequently reduced to make a primary amine by hydrogenation over a metal catalyst or with a hydride reagent such as $LiAlH_4$. This procedure is a way to make primary amines that have one carbon more than the starting electrophile.

Br 1) KCN/acetonitrile 2) $LiAlH_4$ 3) NaOH H_3CO NH_2 H_3CO

Acetylides are produced from the deprotonation of acetylenes by using a very strong base such as $NaNH_2$, NaH, or BuLi (Chapter 7). Because acetylene itself has two hydrogens that can be removed, reactions can be done on both sides, sequentially.

HC≡CH 1) $NaNH_2$ 2) Br HC≡C 1) $NaNH_2$ 2) Br

As with nitriles, the π bonds of alkynes can be reduced by hydrogenation over a metal catalyst (for example, H_2 over palladium on carbon) to make alkanes or alkenes (Chapter 8). This is one of the most reliable ways to make carbon-carbon bonds in synthesis.

HC≡CH 1) $NaNH_2$ 2) Br HC≡C H_2 Pd/C

11.8.5 Primary amines

Amines are less basic than carbanions and are excellent nucleophiles for S_N2 displacement reactions. The difficulty in using amine nucleophiles for chemical synthesis is that adding electron-donating alkyl groups to an amine *increases* the nucleophilicity of the nitrogen, which makes controlling substitution reactions a challenge.

The substitution of propyl bromide with ammonia begins as expected, with ammonia displacing the bromide to form an alkylated ammonium ion that is then deprotonated by another molecule of ammonia.

Br NH_3 ⊕N H H_2 NH_3 NH_2 + NH_4Br

However, the propylamine product is more nucleophilic than ammonia, the original nucleophile. This typically results in over-alkylation, which can be difficult to suppress because the reaction conditions that promote the first substitution are more than sufficient to promote further alkylation.

Br NH_3 NH_2 Br NH Br N

Steric hindrance of a tertiary amine usually prevents accidental over-alkylation to produce the quaternary ammonium salt; however, a fourth alkyl group can be added if forcing conditions are used.

To form a primary amine without risk of over-alkylation, a masked amine equivalent is often used. Two common masked amines are phthalimide and azide.

11.8.5.1 Gabriel synthesis

The Gabriel synthesis uses a nucleophilic phthalimide salt to form a substitution product that can be converted to a primary amine. Potassium phthalimide reacts only once to form an *N*-alkylated product, a product that is less nucleophilic than the original phthalimide nucleophile.

The resulting substituted phthalimides can be converted to primary alkyl amines by heating with either hydroxide or hydrazine (NH_2NH_2; details in Chapter 15).

11.8.5.2 Azide

Sodium azide (NaN_3) can also be used to form a carbon-nitrogen bond, in the form of an alkyl azide. Reduction with hydrogen over a metal catalyst then converts the azide to a primary amine.

CHECKPOINT 11.8

You should now be able to design syntheses employing nucleophilic displacements.

SOLVED PROBLEM

Suggest a synthesis of the following molecule from acyclic molecules containing up to four carbons.

STEP 1: This molecule contains only one functional group, the nitrogen. To make this material requires reactions involving this functional group. The stereocentre in the molecule has to be controlled; this suggests an S_N2 reaction. Disconnecting the ring first gives a symmetrical intermediate and the primary amine for the S_N2 reaction.

STEP 2: Write the synthetic scheme including reagents.

NaN$_3$ / DMF; H$_2$ / Pd/C; Br(CH$_2$)$_4$Br / NaOH

PRACTICE PROBLEM

11.15 Propose methods to carry out the following transformations.

a) HO(CH$_2$)$_4$OH ⟶ N-methylpyrrolidine

b) OH ⟶ OH

c) Br ⟶ NH$_2$

INTEGRATE THE SKILL

11.16 Propose a synthesis of the following compound using materials containing four carbons or less.

11.9 Patterns in Nucleophilic Displacements on Saturated Carbons

Primary and methyl α-carbons that have leaving groups undergo S_N2 displacements in which the displacement step is rate-determining. In some cases, the mechanism involves extra steps, usually acid–base steps. Bases activate nucleophiles by converting them to their more reactive conjugate base form. Acids activate electrophiles by converting them to their conjugate acid form.

Sulfonate esters can be used to activate hydroxyl groups. All S_N2 displacements proceed with inversion of configuration. If the α-carbon is stereogenic, then this carbon will have the opposite configuration in the product.

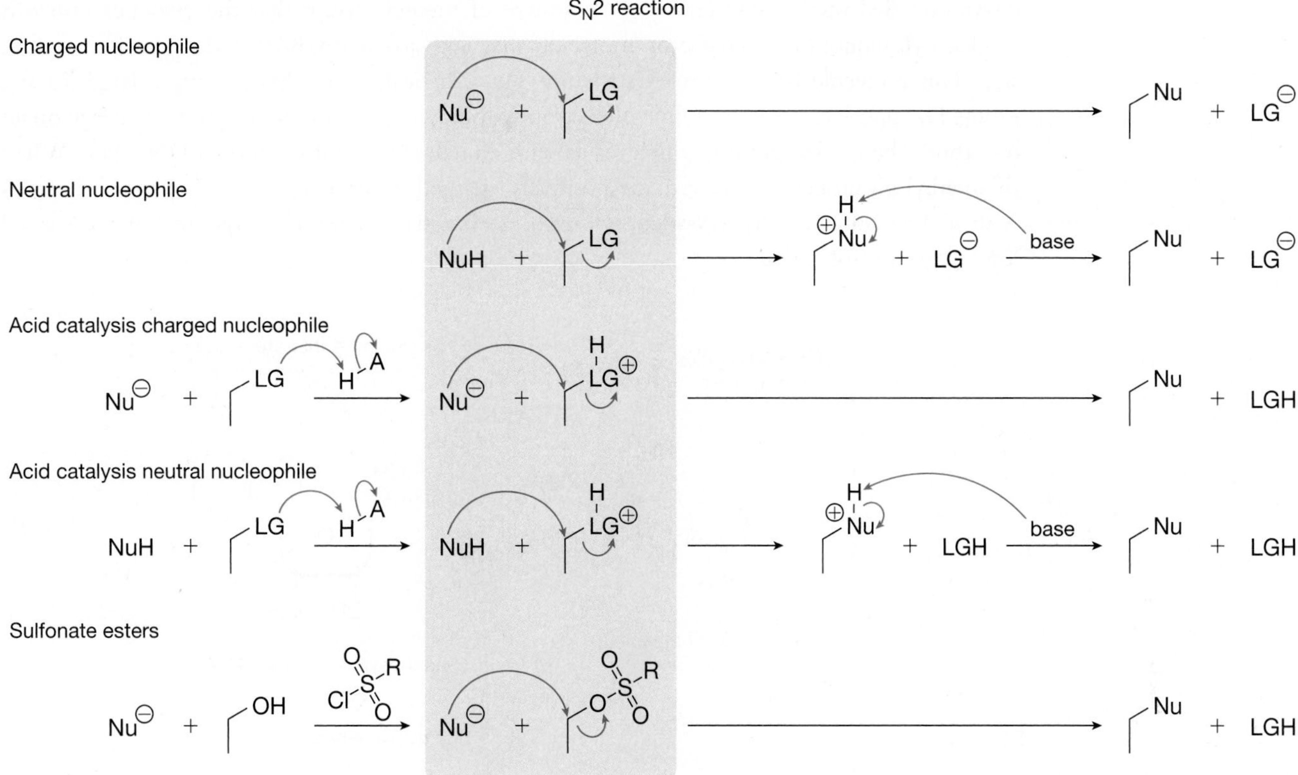

Tertiary substrates undergo S_N1 displacements. These reactions are dependent on the ability of the intermediate carbocation to be stabilized.

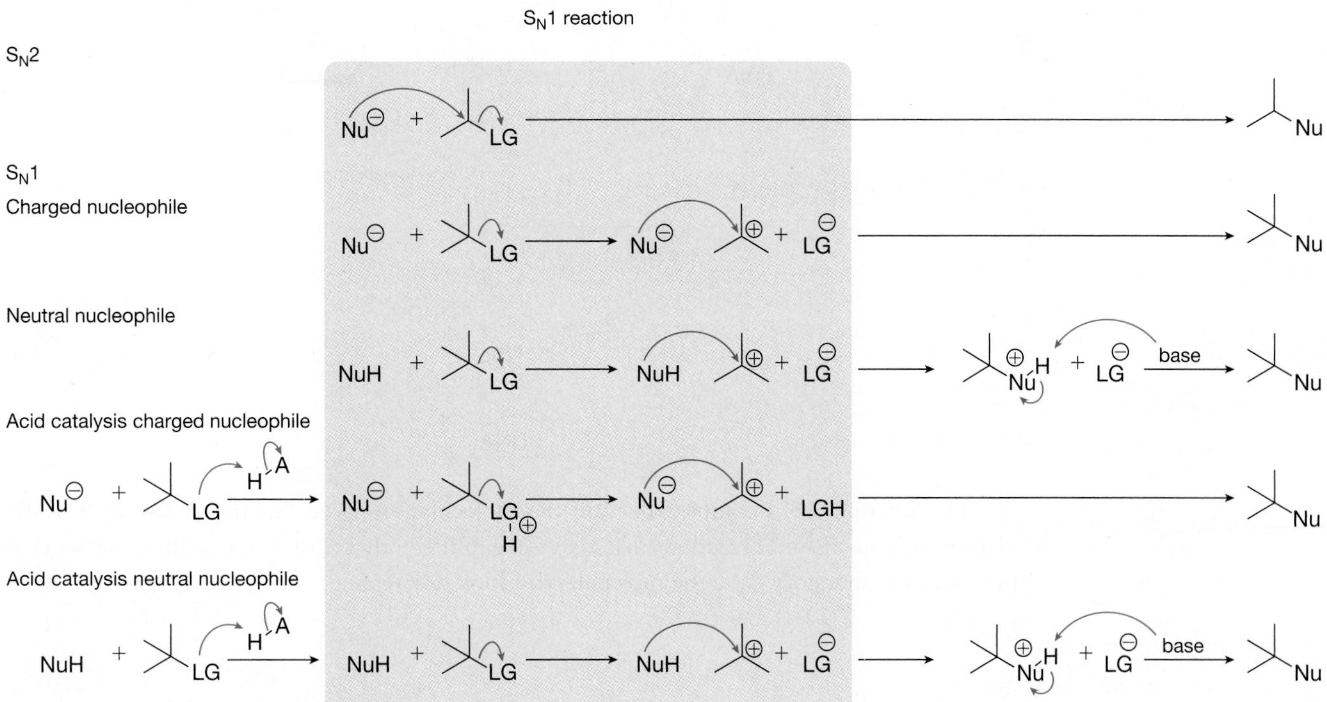

Bringing It Together

A common epigenetic change involves the attachment of methyl groups to various places in DNA. The body adds these methyl groups via S_N2 reactions carried out by specialized enzymes called methyl transferases. One source of methyl groups that the enzymes utilize is *S*-adenosylmethionine, a co-factor molecule that acts as electrophile in the enzymatic reaction. This molecule has a complex structure, but is basically a methyl bearing a large leaving group. The positively charged sulfur is electron withdrawing and induces a positive charge on an α-carbon (the methyl group), which reacts with a nucleophilic nitrogen on a DNA base. When the methyl group gets transferred, the positively charged sulfur leaves, taking a pair of electrons with it. The sulfur product, *S*-adenosylhomocysteine, has a neutral charge, making it a good leaving group (weak base).

The reaction looks complex because the molecules are large, but this is just an S_N2 displacement. The chemical reactions that take place in the body are the same as those we work to produce in a laboratory flask; the molecules just look different.

You Can Now

- Draw a mechanism for nucleophilic displacements on sp^3-hybridized electrophiles using the S_N2 mechanism.
- Identify relative nucleophilicity based on charge, electronegativity, polarizability, and charge delocalization (described by resonance).
- Identify the use of acid in catalyzing nucleophilic displacements of OH groups.
- Identify the use of sulfonate esters in nucleophilic displacements of OH groups.
- Predict the stereochemical outcome of S_N2 and S_N1 reactions.
- Draw a mechanism for nucleophilic displacements on sp^3-hybridized electrophiles using the S_N1 mechanism.
- Identify relative electrophilicity based on the degree of substitution or quality of the leaving group.
- Differentiate between reaction conditions favouring S_N1 or S_N2 mechanisms.
- Design syntheses using nucleophilic displacements.

A Mechanistic Re-View

S_N2 REACTION

Charged nucleophile

$$Nu^{\ominus} + R\text{–}LG \longrightarrow R\text{–}Nu + LG^{\ominus}$$

Neutral nucleophile

$$NuH + R\text{–}LG \longrightarrow R\text{–}\overset{\oplus}{Nu}H + LG^{\ominus} \xrightarrow{\text{base}} R\text{–}Nu + LG^{\ominus}$$

Acid catalysis charged nucleophile

$$Nu^{\ominus} + R\text{–}LG \xrightarrow{H\text{–}A} Nu^{\ominus} + R\text{–}\overset{\oplus}{LG}H \longrightarrow R\text{–}Nu + LGH$$

Acid catalysis neutral nucleophile

$$NuH + R\text{–}LG \xrightarrow{H\text{–}A} NuH + R\text{–}\overset{\oplus}{LG}H \longrightarrow R\text{–}\overset{\oplus}{Nu}H + LGH \xrightarrow{\text{base}} R\text{–}Nu + LGH$$

S_N1 REACTION

Charged nucleophile

$$Nu^{\ominus} + R_3C\text{–}LG \longrightarrow Nu^{\ominus} + R_3C^{\oplus} + LG^{\ominus} \longrightarrow R_3C\text{–}Nu$$

Neutral nucleophile

$$NuH + R_3C\text{–}LG \longrightarrow NuH + R_3C^{\oplus} + LG^{\ominus} \longrightarrow R_3C\text{–}\overset{\oplus}{Nu}H + LG^{\ominus} \xrightarrow{\text{base}} R_3C\text{–}Nu$$

Acid catalysis charged nucleophile

Nu⊖ + LG, H–A → Nu⊖ + LG(H)⊕ → Nu⊖ ⊕ + LGH → Nu

Acid catalysis neutral nucleophile

NuH + LG, H–A → NuH + LG(H)⊕ → NuH ⊕ + LGH → ⊕Nu H + LG⊖ H, base → Nu

TABLE 11.1 General Features of S_N1 and S_N2 Displacements

	S_N2	S_N1
Mechanism	One-step reaction (not including acid–base)	Two-step reaction (not including acid–base)
Rates	Rate depends on the concentration of both the nucleophile and electrophile rate = k[Nu][E⊕]	Rate depends only on the concentration of the electrophile rate = k[E⊕]
Electrophile structure and rate	Methyl > Primary > Secondary	Tertiary > Secondary
Nucleophile	Promoted by good nucleophiles	Promoted by weak nucleophiles
Stereochemistry	Inversion of configuration	Mixtures of stereoisomers
Leaving group	Good leaving groups give fast reactions	Good leaving groups give fast reactions
Solvent	Favoured in polar aprotic solvents	Favoured in protic solvents

Problems

11.17 Use arrow notation to show mechanisms and predict the products of the following reactions. Identify the nucleophile, the α-carbon of the electrophile, and the leaving group.

a) Br + ONa (cyclohexyl) → THF

b) NaCN + (phenyl, Br) → THF

c) Cl (cyclopentyl) → NaI, acetone

d) OH, OH → H_2SO_4

e) Br (cyclohexane) → NaSEt

f) OCH_3; H–C–CH_3; Br–C–H; CF_3 → NaCN

11.18 A solution of 0.25 M cyclohexylbromide and 0.36 M CH_3NH_2 is mixed. Measurements show that the reaction produces 3.6×10^{-7} mol·L^{-1} of product each second.

a) Calculate the rate constant for the reaction.

b) If the concentration of CH_3NH_2 is changed to 0.15 M, what is the initial rate of the reaction?

c) Predict the relative rate (faster or slower) of the reaction if $CH_3CH_2CH_2OH$ is used in place of CH_3NH_2, and justify your choice.

d) What could be done to the electrophile to increase the rate of the reaction? Briefly explain your reasoning.

11.19 Predict the effect on the reaction rate if $(CH_3)_3CONa$ (a strong base) is added to a mixture of CH_3CH_2I and CH_3CH_2SH. Write an equation to show what is happening and provide a brief explanation of the effect.

11.20 Write a mechanism for each of the following reactions. Use your mechanism to predict the products of the reaction.

a) 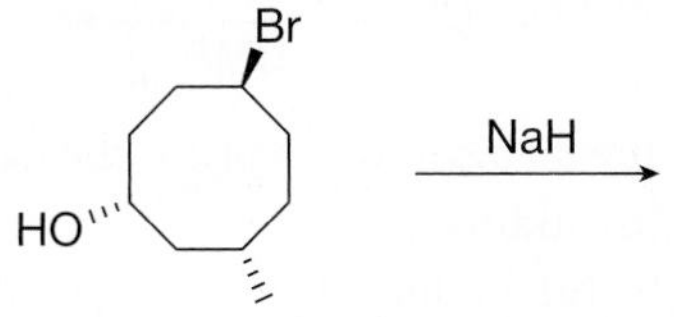 (cyclooctane with Br, HO) $\xrightarrow{NaH}$

b) 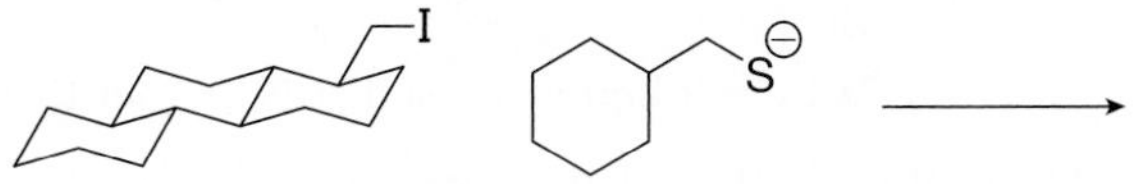 (steroid-like ring system with CH_2I) + (cyclohexylmethyl thiolate, $S^{\ominus}$) $\longrightarrow$

c) 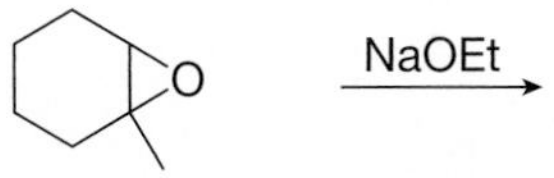 (epoxide) $\xrightarrow{NaOEt}$

d) 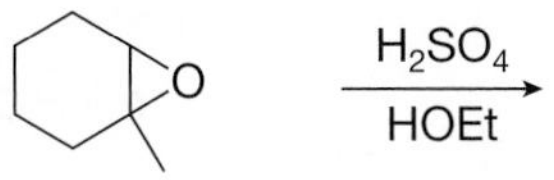 (epoxide) $\xrightarrow[HOEt]{H_2SO_4}$

11.21 Rank the following pairs of compounds in order of increasing S_N2 reactivity, and explain your choices.

a) 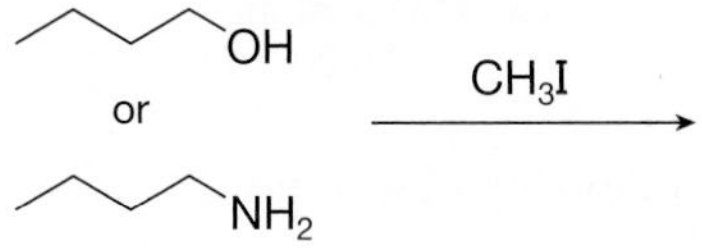 1-butanol (OH) or butylamine (NH_2) $\xrightarrow{CH_3I}$

b) 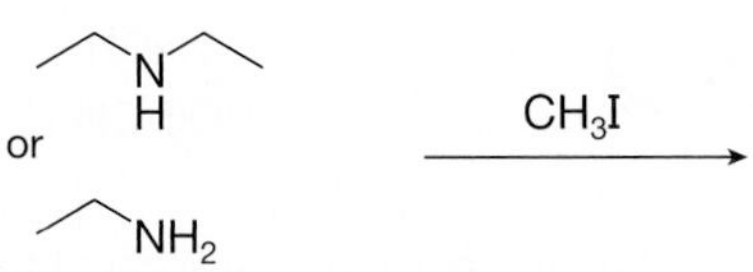 diethylamine (N–H) or ethylamine (NH_2) $\xrightarrow{CH_3I}$

c) 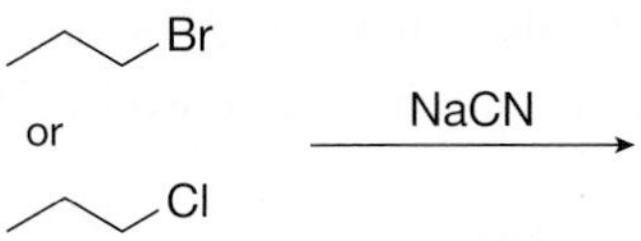 1-bromopropane (Br) or 1-chloropropane (Cl) $\xrightarrow{NaCN}$

d) 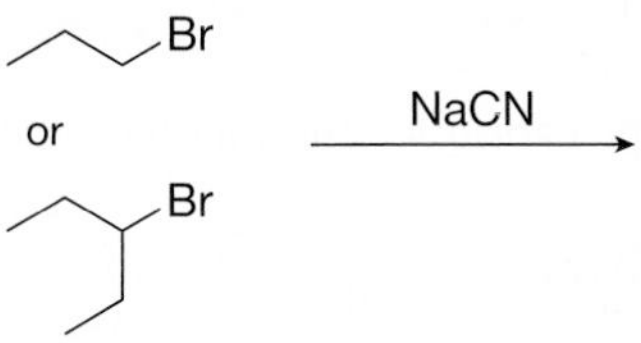 1-bromopropane (Br) or branched bromide (Br) $\xrightarrow{NaCN}$

e) 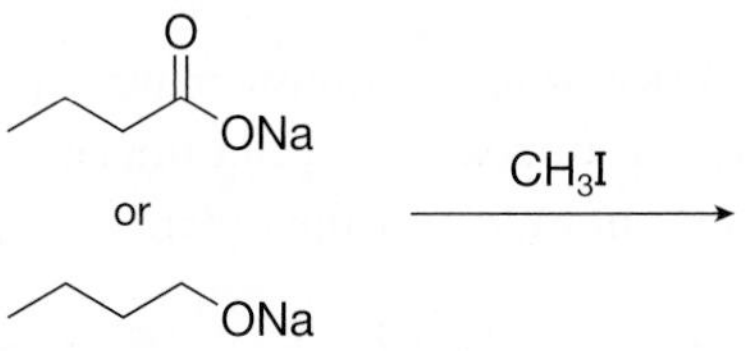 sodium carboxylate (O, ONa) or sodium butoxide (ONa) $\xrightarrow{CH_3I}$

11.22 Predict the products of the following reactions. If no reaction is possible, write NR in place of the products, and explain your reasoning. Write a mechanism for each possible transformation.

a) 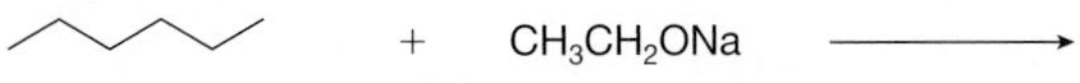 (alkane) + CH_3CH_2ONa $\longrightarrow$

b) 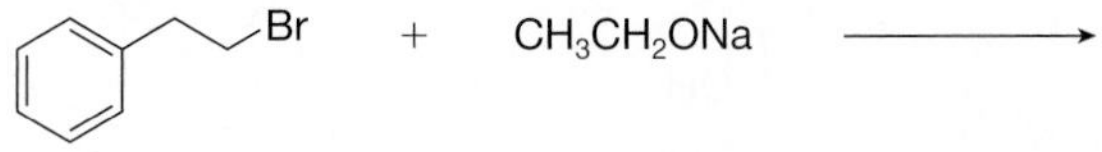 (2-phenylethyl bromide, Br) + CH_3CH_2ONa $\longrightarrow$

c)  (Newman projection: CH_3, H, H, H, Br, CH_3) + NaI $\longrightarrow$

d) 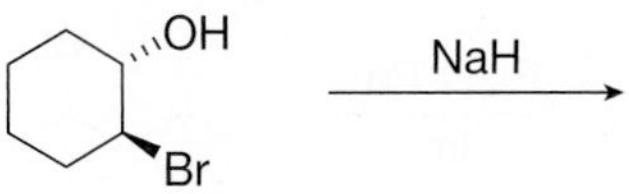 (cyclohexane with OH and Br) $\xrightarrow{NaH}$

e) 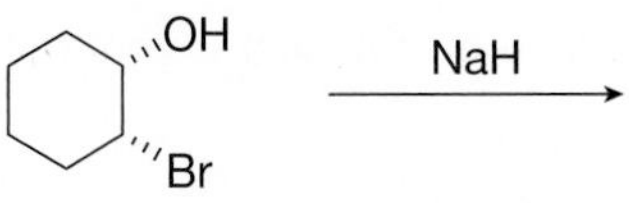 (cyclohexane with OH and Br) $\xrightarrow{NaH}$

f) 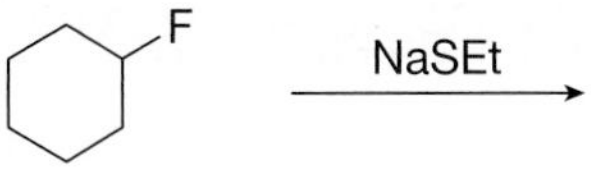 (cyclohexylmethyl fluoride, F) $\xrightarrow{NaSEt}$

11.23 a) Predict the product of the reaction of $(CH_3)_3CCl$ with 1-cyclohexylmethanol.

b) Write a mechanism for the reaction.

c) Give two ways in which the reaction could be accelerated.

d) If Ph_3CCl is used in place of $(CH_3)_3CCl$, what would be the effect on the reaction rate? Explain.

11.24 For the following, predict which set of reaction conditions will give the fastest reaction. Justify your choice.

a) 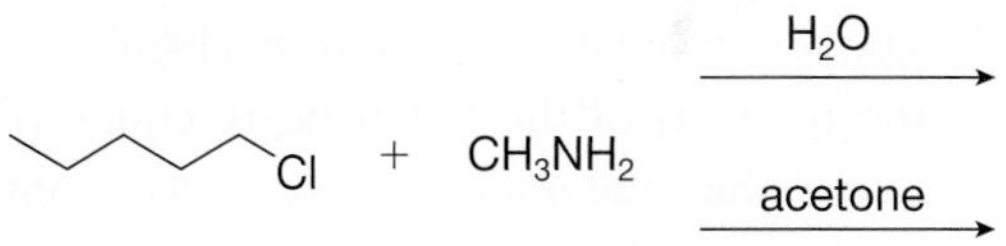 (1-chloropentane, Cl) + CH_3NH_2 $\xrightarrow{H_2O}$ or $\xrightarrow{acetone}$

b) 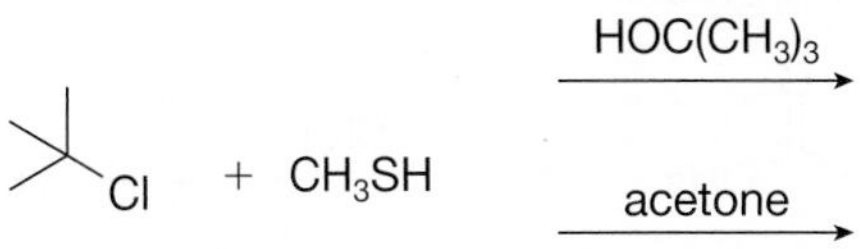 (tert-butyl chloride, Cl) + CH_3SH $\xrightarrow{HOC(CH_3)_3}$ or $\xrightarrow{acetone}$

c) 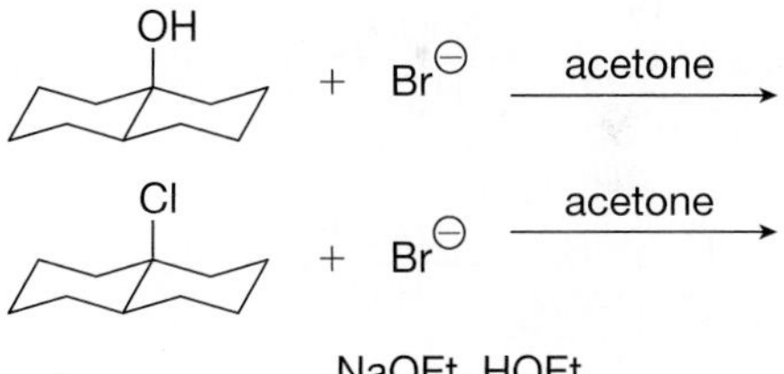 (decalin with OH) + $Br^{\ominus}$ $\xrightarrow{acetone}$

(decalin with Cl) + $Br^{\ominus}$ $\xrightarrow{acetone}$

d) 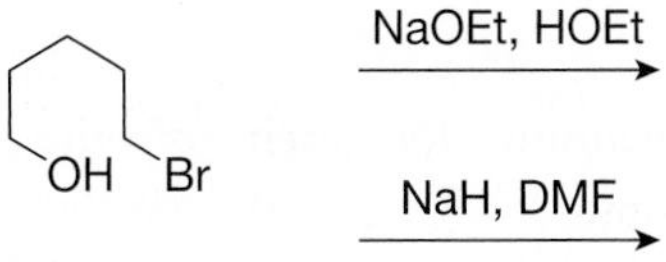 (OH, Br) $\xrightarrow{NaOEt, HOEt}$ or $\xrightarrow{NaH, DMF}$

11.25 Suggest reagents that could be used to make each of the following.

a) 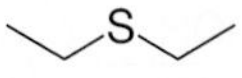 (diethyl sulfide, S)

b) 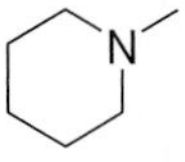 (N-methylpiperidine, N)

c) 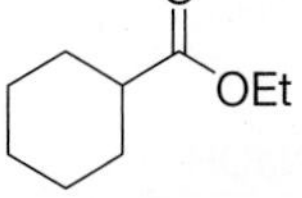 (ethyl cyclohexylacetate, O, OEt)

d) 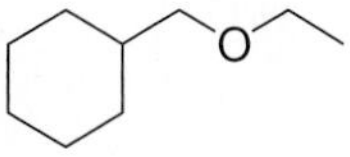 (cyclohexylmethyl ethyl ether, O)

11.26 The following reaction is an excellent way of converting alcohols into alkyl halides (Im = imidazole, an organic base).

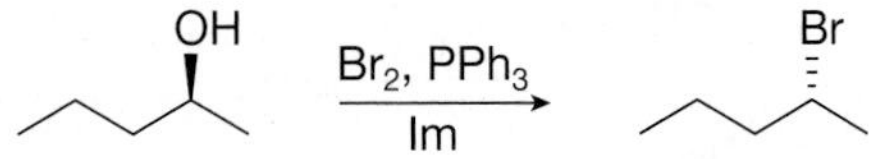

Br_2, PPh_3 / Im

a) Provide mechanisms for each step of the process.

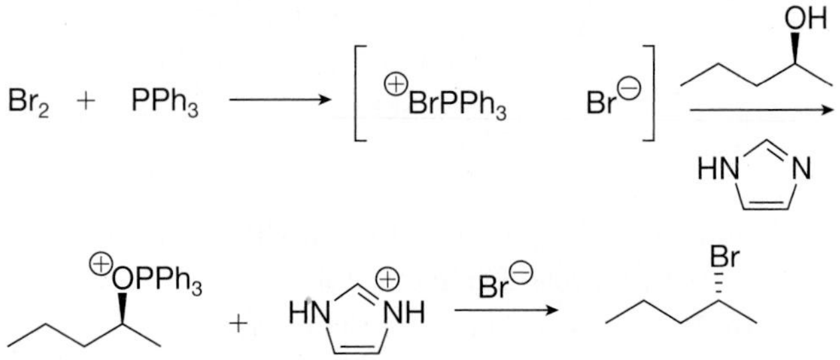

Br_2 + PPh_3 ⟶ [$^{\oplus}BrPPh_3$ $Br^{\ominus}$]

b) Predict the product of the following transformation by analogy with the reaction shown above.

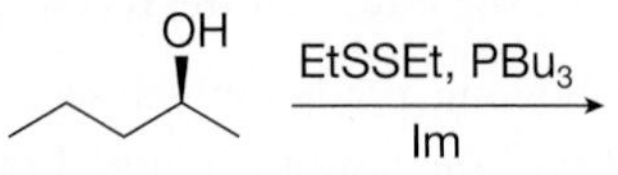

EtSSEt, PBu_3 / Im

11.27 Use arrow notation to show mechanisms and predict the products of the following reactions. Identify the nucleophile, the α-carbon of the electrophile, and the leaving group.

a)  ONa + Cl ⟶

b) 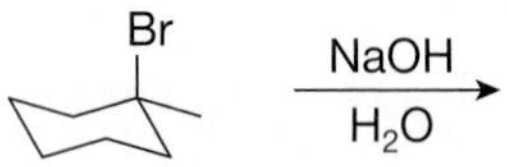 Br; NaOH / H_2O

c) 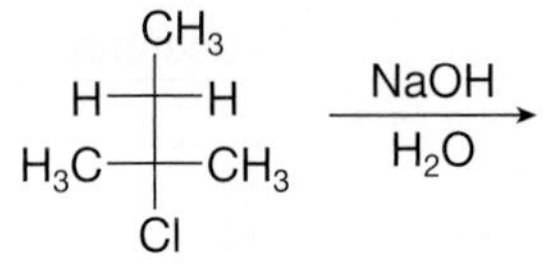 CH_3, H, H, H_3C, CH_3, Cl; NaOH / H_2O

11.28 Draw mechanisms for each of the following transformations.

a) 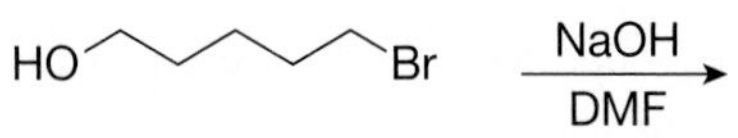 HO … Br; NaOH / DMF

b) 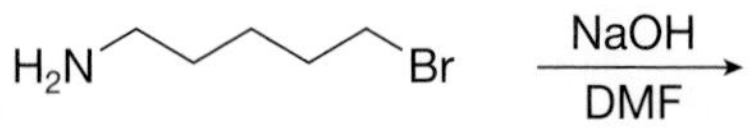 H_2N … Br; NaOH / DMF

c) 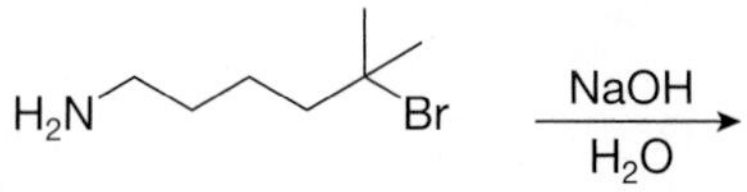 H_2N … Br; NaOH / H_2O

d) 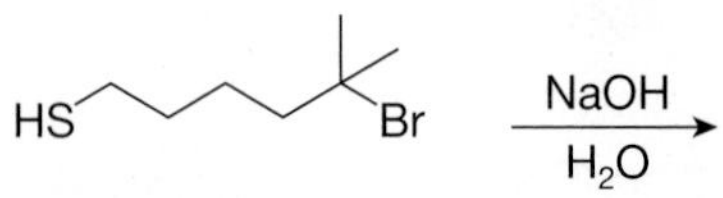 HS … Br; NaOH / H_2O

11.29 S_N2 reactions can be catalyzed by the addition of quaternary ammonium salts such as Bu_4NI.

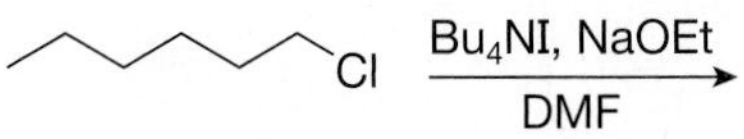

Cl; Bu_4NI, NaOEt / DMF

a) Write a mechanism to explain the accelerating effect of this additive.

b) Why is Bu_4NI preferred over other salts such as Bu_4NCl?

c) Why is it better to use Bu_4NI rather than NaI?

11.30 Predict the relative rates of the following reactions, and draw reaction coordinate diagrams to explain your predictions.

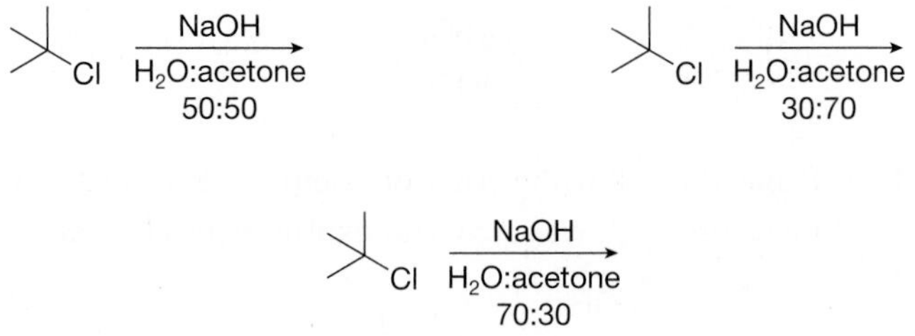

Cl; NaOH / H_2O:acetone 50:50

Cl; NaOH / H_2O:acetone 30:70

Cl; NaOH / H_2O:acetone 70:30

11.31 Explain the following observations.

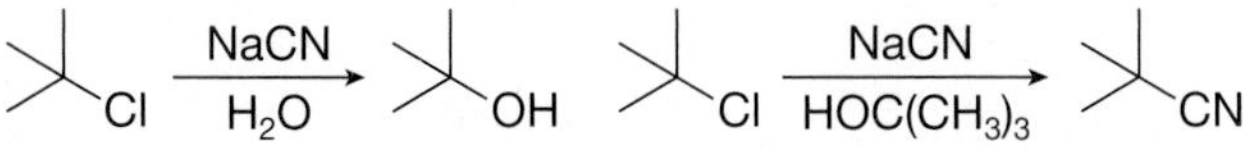

Cl; NaCN / H_2O ⟶ OH Cl; NaCN / $HOC(CH_3)_3$ ⟶ CN

11.32 When the following reactant is subjected to S_N2 reaction conditions, the unexpected product below is obtained. Provide a mechanism to explain this result.

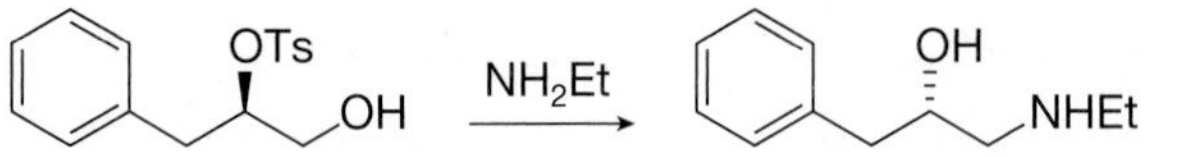

OTs, OH; NH_2Et ⟶ OH, NHEt

11.33 Suggest an explanation for the following observation.

Br; CH_3CH_2OH ⟶ OEt + OEt

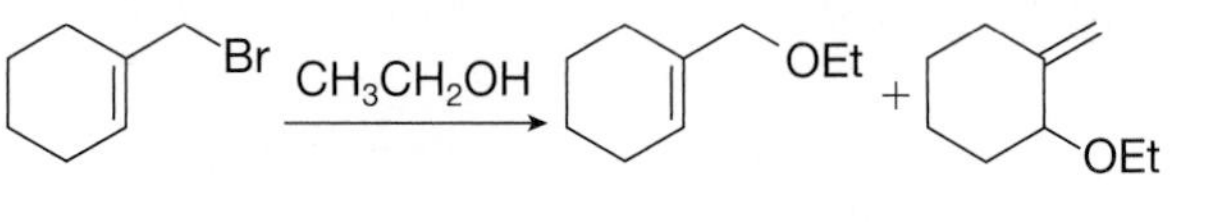

11.34 During the following chemical transformation, only one chloride is displaced. Explain the enhanced reactivity of one chloride over the other.

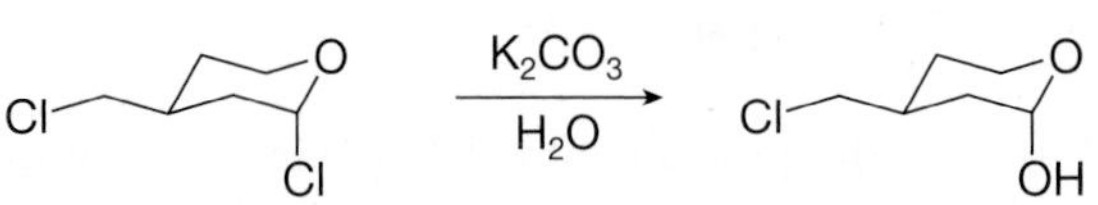

Cl, O, Cl; K_2CO_3 / H_2O ⟶ Cl, O, OH

11.35 Show how the following product is formed.

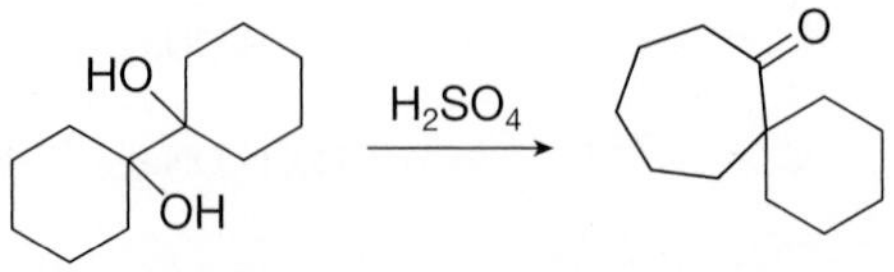

HO, OH; H_2SO_4 ⟶ O

MCAT STYLE PROBLEMS

11.36 Which of the following will most likely rearrange during an S_N1 reaction?

a) 2-bromobutane
b) 2-chloro-3-methylbutane
c) 2-chlorobutane
d) 2-bromo-2,3-dimethylbutane

11.37 Which of the following is the most likely product of the reaction shown?

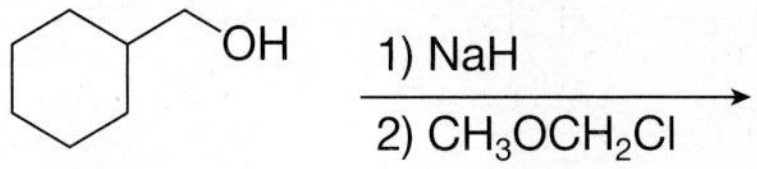

1) NaH
2) CH_3OCH_2Cl

a) 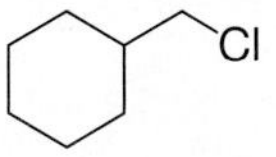 Cl
b) 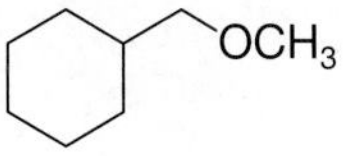 OCH_3
c) 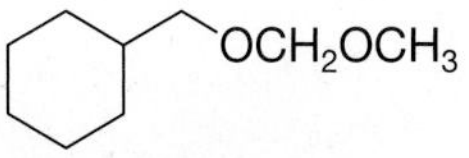 OCH_2OCH_3
d) 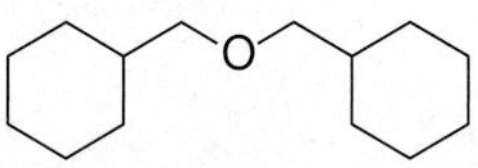 O

CHALLENGE PROBLEM

11.38 Chiral α-hydroxyacids can be prepared from many amino acids by treating them with sodium nitrite in aqueous acid. The reaction is very reliable and gives a product in which the configuration of the hydroxyl group is retained relative to that of the starting product. Provide a full mechanism to explain this observation.

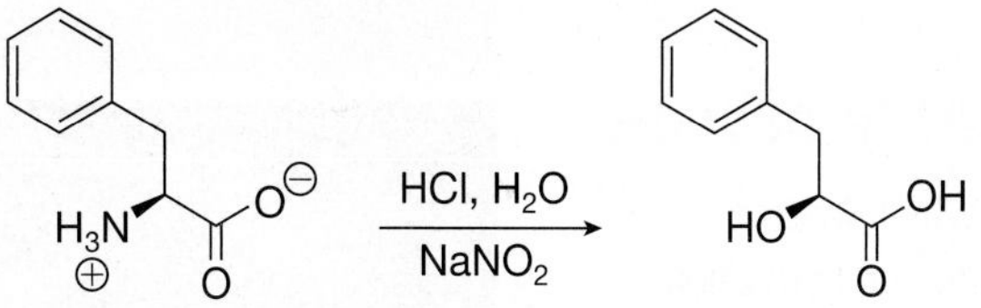

$H_3N^{\oplus}$ … $O^{\ominus}$ —— HCl, H_2O / $NaNO_2$ ——> HO … OH

12 Formation of π Bonds by Elimination Processes

ELIMINATION AND OXIDATION REACTIONS

CHAPTER OUTLINE

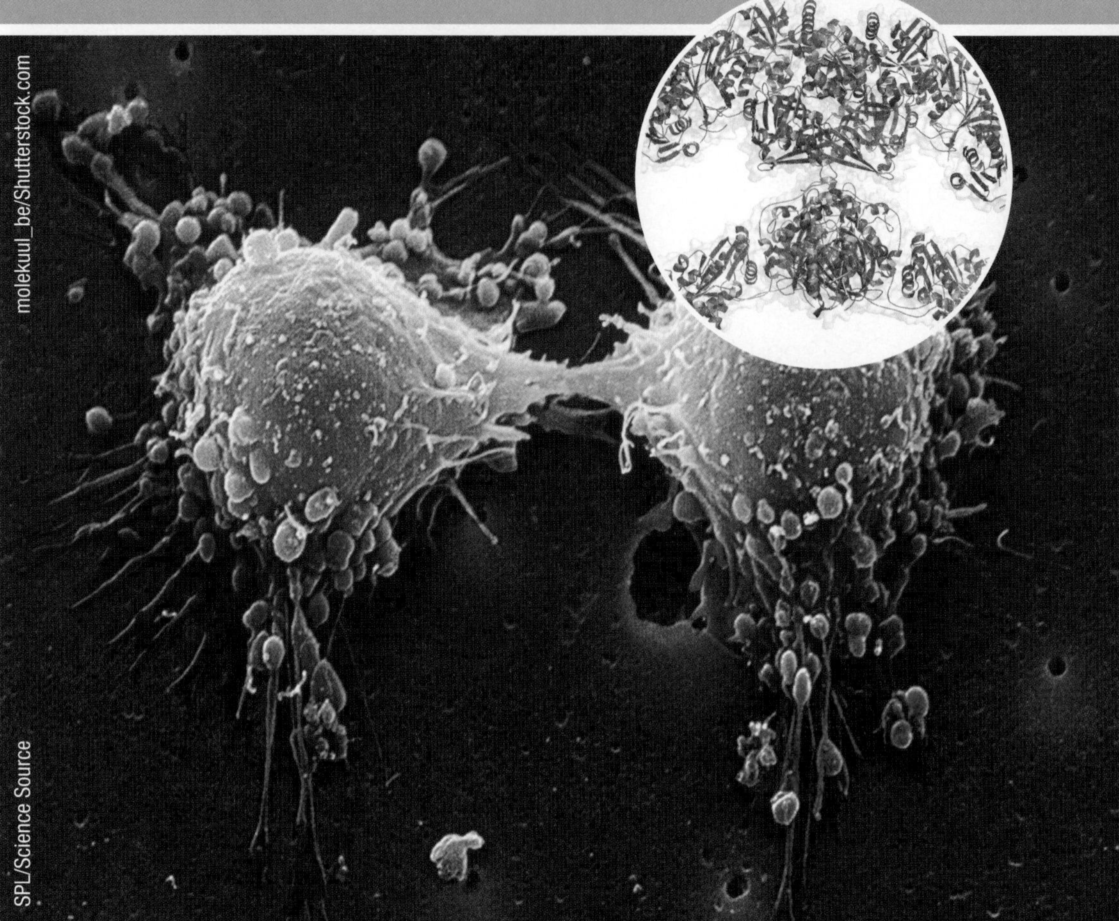

molekuul_be/Shutterstock.com

SPL/Science Source

Cancer cells, which undergo rapid division, contain elevated *fatty acid synthase* (inset) levels so that they can produce the amount of fatty acids required to sustain rapid growth. Fatty acid synthase is an emerging target for the development of new anti-cancer drugs.

12.1 Why It Matters

Fatty acids are key components of lipids, the type of molecules that our bodies use to store energy. They are also building blocks for cell membranes and other essential biomolecules. The biosynthesis of fatty acids plays a fundamental role in metabolism—the complex series of chemical processes that occur in living organisms to maintain life. Fatty acids are manufactured from a very simple two-carbon source, acetyl co-enzyme A (acetyl-SCoA), which is transformed in a stepwise series of reactions by a multi-enzyme complex called fatty acid synthase. One of the key steps involves a dehydration reaction, in which the elements of water (H_2O) are removed to form a π bond. The enzyme dehydratase catalyzes the reaction. The following simplified biosynthetic scheme highlights a dehydratase-catalyzed reaction that produces palmitate, a 16-carbon fatty acid.

acetyl-SCoA → → (dehydratase) → + H_2O → → palmitate

Dehydration is one of several reactions that belong to the general class of elimination reactions. This chapter describes different types of elimination and oxidation reactions used in the formation of π bonds.

12.2 Alkene Formation by E2 Elimination Reactions

Elimination reactions involve the removal of hydrogens and leaving groups from adjacent carbons to form π bonds.

H LG — elimination → + $H^{\oplus}$ + $:LG^{\ominus}$

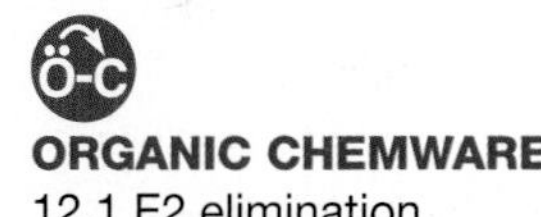

ORGANIC CHEMWARE
12.1 E2 elimination

Elimination reactions are classified based on the specific mechanism whereby the hydrogen and leaving group are removed, in particular by the order in which the removal takes place. The simplest mechanism is the **E2 elimination** (elimination bimolecular) reaction. In this reaction, a strong base removes a hydrogen from a carbon that is β to a leaving group and simultaneously ejects the leaving group to form a π bond between the α- and β-carbons.

E2 elimination involves the removal of a hydrogen and the concerted loss of a leaving group to form an alkene.

loss of leaving group

π bond forms between α- and β-carbons

$B:^{\ominus}$ — E2 → + BH + $:LG^{\ominus}$

base removes a hydrogen β to the leaving group

Kinetic measurements provide the following rate expression:

$$\text{rate} = k[\text{substrate}][\text{base}]$$

rate of the elimination depends on the concentration of the substrate and the base

This expression implies that both the substrate and the base are involved in the rate-determining step of the reaction, and this indicates that the elimination occurs in a single step (some eliminations involve acid–base reactions before the rate-determining step). As the base removes a proton from the β-carbon, the two electrons from that carbon-hydrogen bond flow between the α- and β-carbons to form the π bond and, simultaneously, expel the leaving group. All bonds are made and broken at the same time in a concerted fashion. Typically, E2 eliminations occur in reaction conditions with strong bases such as hydrides, hydroxides, and alkoxides.

The reaction energy diagram for an E2 elimination is shown in Figure 12.1. Because there is only one transition state, the diagram is similar to that of an S_N2 reaction in that all electron movements occur in a single step. In fact, E2 and S_N2 reactions often compete with one another (Section 12.5), and it is important to differentiate between them. However, the reaction conditions can be controlled in ways that favour the formation of either S_N2 or E2 products.

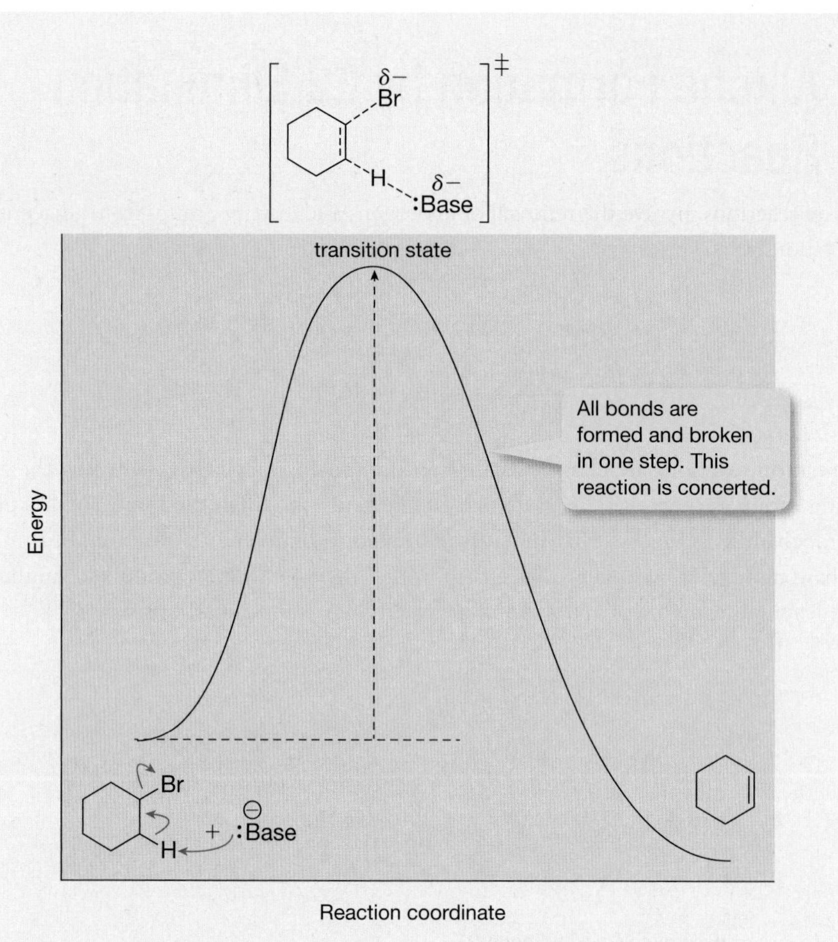

FIGURE 12.1 Reaction diagram for an E2 reaction.

12.2.1 Regioselectivity in E2 reactions

Mixtures of alkene products are often produced when the substrate is not symmetrical. In 1875, long before the mechanism was known, the Russian chemist Alexander Zaitsev described the results of elimination reactions involving unsymmetrical alkyl halides and found that one regioisomer tended to be formed more than the other.

Br
base
major + minor

This favoured product regioisomer is known as the Zaitsev product, or the product of a Zaitsev elimination in which the products can be predicted by **Zaitsev's rule**. The *modern interpretation* of Zaitsev's rule is that the elimination reactions favour the production of the most stable alkene. In the majority of reactions, the most stable alkene product is the one that is the most highly substituted.

The modern interpretation of **Zaitsev's rule** states that elimination reactions favour the most stable alkene product. This is often the most substituted alkene product.

Alkenes can be classified as mono-, di-, tri-, or tetrasubstituted, depending on the number of alkyl groups directly attached to the double bond. If the number of alkyl groups is one, then the alkene is monosubstituted; if the number is two, the alkene is disubstituted; and so on.

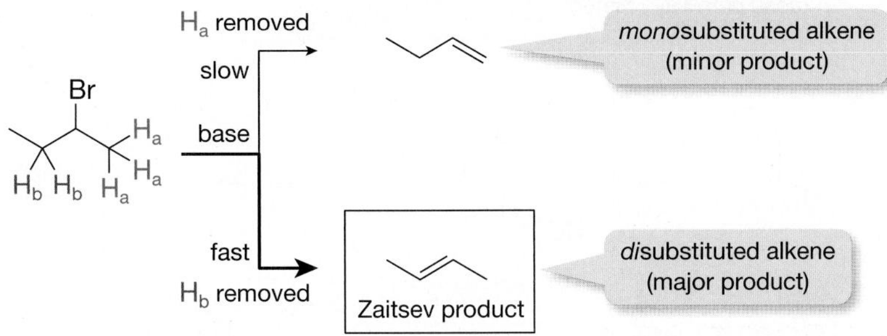

STUDENT TIP
To determine the major product of an elimination reaction, look for the most stable alkene.

STUDENT TIP
It is good practice to consider all of the possible isomers that can form during a reaction that can produce more than one product (such as an elimination).

Alkenes become stabilized in two general ways. First, they are stabilized by hyperconjugation with connected alkyl groups. When filled carbon-hydrogen or carbon-carbon σ molecular orbitals overlap with the empty $\pi^\star$ orbital of the alkene, a small sharing or delocalization of electrons occurs. This lowers the energy of the alkene and stabilizes it. This effect is similar to the stabilization of carbocations by hyperconjugation, in which the empty p orbital of the sp^2-hybridized carbocation accepts electron density. As the number of alkyl groups on a double bond increases, so does the extent of the possible hyperconjugation and therefore the stability of the alkene.

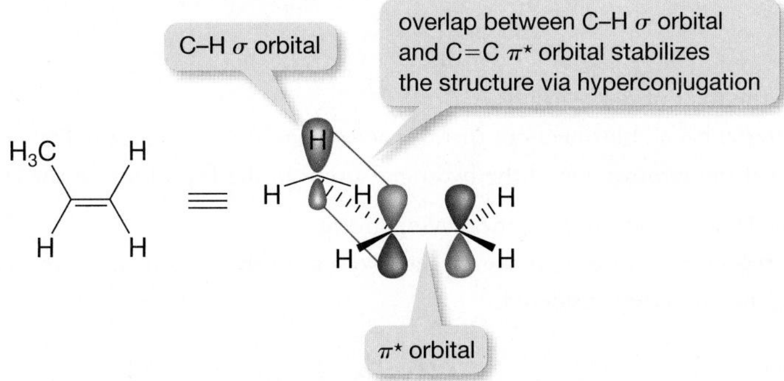

Alkenes are also stabilized by delocalization. As the extent of delocalization increases, so does the stability of the alkene. Elimination reactions that can form such stabilized products are very favourable.

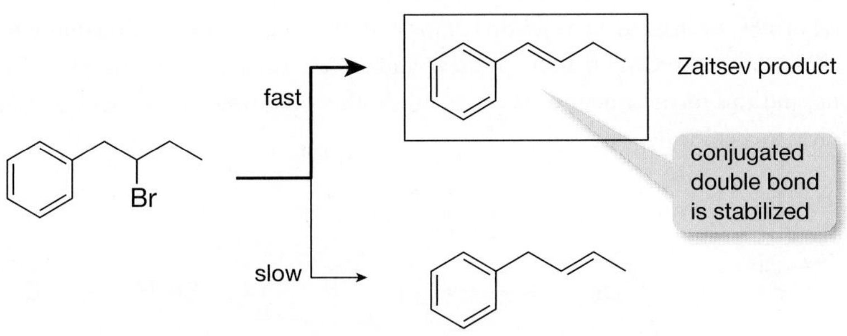

The stability of the double bonds in the products of an elimination affects the rate at which those products are formed. The transition state of the E2 reaction has considerable double-bond character, so factors that stabilize double bonds, such as increased substitution or delocalization, stabilize the transition states leading to those products and lower the activation energies for the formation of the most stable alkene substrates.

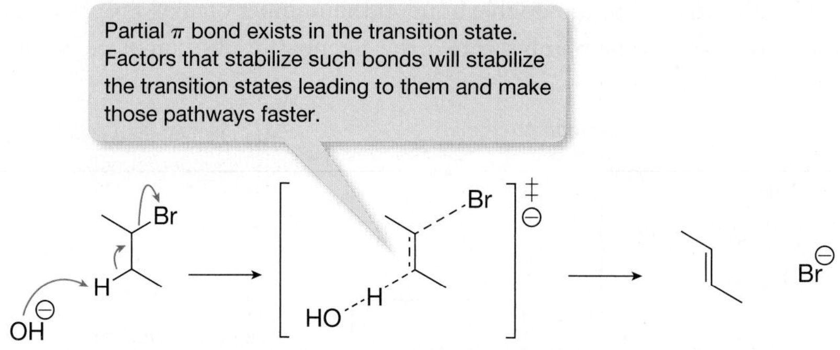

Partial π bond exists in the transition state. Factors that stabilize such bonds will stabilize the transition states leading to them and make those pathways faster.

Br H OH → [Br HO H]‡ → Br

CHECKPOINT 12.1

You should now be able to draw a mechanism and its corresponding product for an E2 elimination reaction and predict the major regioisomer formed.

SOLVED PROBLEM

Draw a mechanism showing the formation of both the major and minor elimination products for this reaction. Explain the origin of the expected regioselectivity.

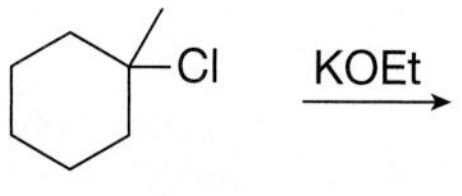

Cl KOEt →

STEP 1: Determine the roles of the reactants.

The starting material has a chlorine atom that can act as a good leaving group. The potassium ethoxide is a strong base, which can remove one of the hydrogen atoms in the β position to the leaving group.

STEP 2: Draw in the hydrogen atoms β to the leaving group.

There are three sets of β-hydrogens; however, the two sets on the cyclohexane ring are equivalent, and so only one set (H_b) needs to be considered.

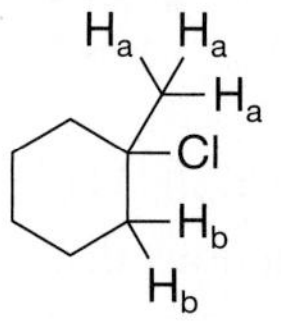

H_a H_a H_a Cl H_b H_b

STEP 3: Draw a mechanism for the reaction on one set of β-hydrogens.

In the removal of a H_a hydrogen, the reaction begins with the base sharing its electrons with the electrophilic hydrogen. This causes the C–H σ bond to break and the electrons to flow toward the carbon bearing the leaving group, and this forms a new C=C π bond. At the same time, the C–Cl bond breaks and the leaving group leaves as a chloride ion.

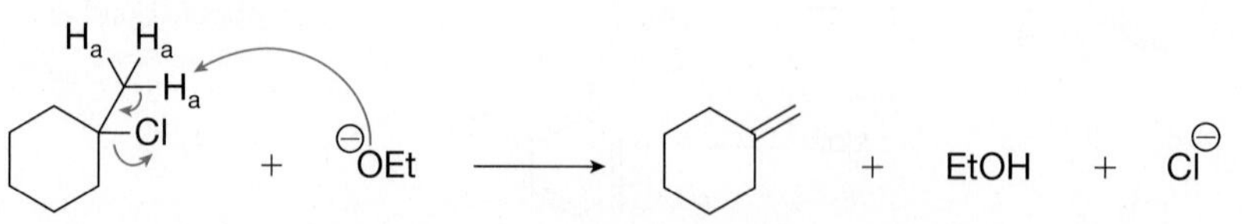

H_a H_a H_a Cl + OEt → + EtOH + Cl

STEP 4: Repeat Step 3 with the other set of β-hydrogens.

The mechanism for the removal of H_b proceeds by the same steps as above.

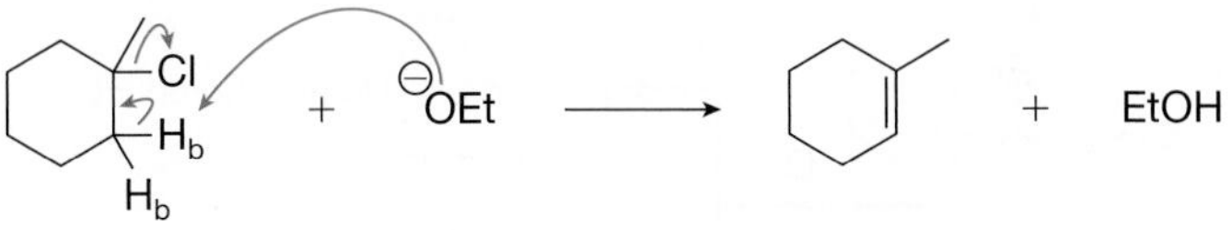

STEP 5: Compare the relative stability of the two alkene products and determine which is the major product.

When H_a is removed, the product is a disubstituted double bond; when H_b is removed, the resulting alkene is trisubstituted. Since alkene stability increases with increased substitution, the trisubstituted alkene product is the more stable product and therefore the major isomer formed.

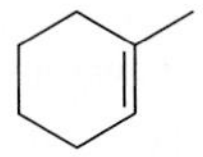

major isomer

PRACTICE PROBLEM

12.1 For each of the following reactions, draw a mechanism and show the major product.

a) 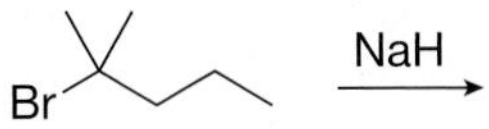 NaH

b) 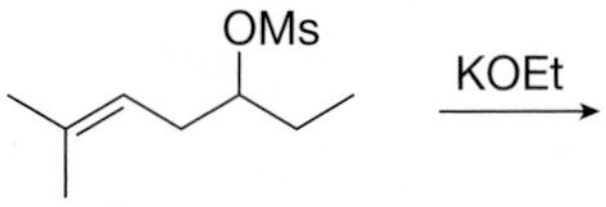 KOEt

c) 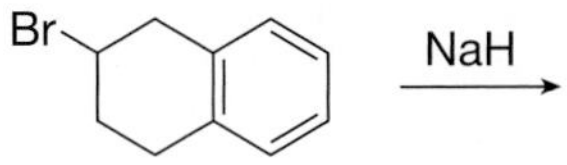 NaH

d) 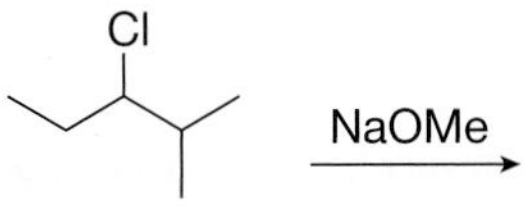 NaOMe

INTEGRATE THE SKILL

12.2 When 1-(1-bromoethyl)cyclohex-1-ene is heated in the presence of a strong base, two different dienes are formed. Draw the two diene products, along with a mechanism showing how each one forms.

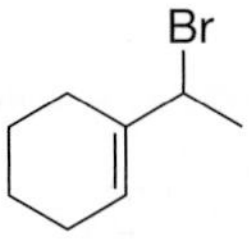

1-(1-bromoethyl)cyclohex-1-ene

12.2.2 Reversal of selectivity to give Hofmann products

Bulky bases are used to alter the regioselectivity of E2 reactions because they often favour the formation of the less stable (less substituted) alkene. Large, strong bases such as *tert*-butoxide (t-BuO$^-$) promote an E2 mechanism. However, because the base is so large, it cannot easily "reach" hydrogens that are in sterically hindered parts of the molecule. Instead, these bases tend to remove hydrogens from less crowded positions and produce less substituted alkenes. The less substituted (less stable) product obtained from an E2 elimination is called the **Hofmann product**.

The **Hofmann product** is the product of an elimination derived from the loss of the least hindered hydrogen.

large base (t-BuO$^{\ominus}$) can only reach the more accessible hydrogens at the less substituted position
Br
fast
Hofmann product
slow

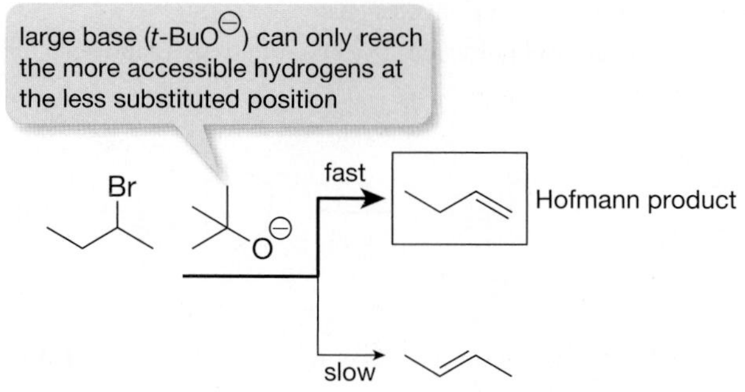

DID YOU KNOW?

Hofmann elimination gives the less substituted product instead of the more substituted (Zaitsev) product.

The **Hofmann elimination** is a special type of E2 reaction that involves a cyclic flow of electrons and gives the less substituted alkene as the major product. The reaction described here involves the use of a quaternary ammonium salt or a sulfonium ion as the leaving group. When such molecules are treated with a strong base—for example, hydroxide ion—the least substituted alkene is the major product.

base removes a hydrogen from one of the quaternary ammonium alkyl groups
base
CH_2
CH_3
elimination happens through an intramolecular removal of hydrogen

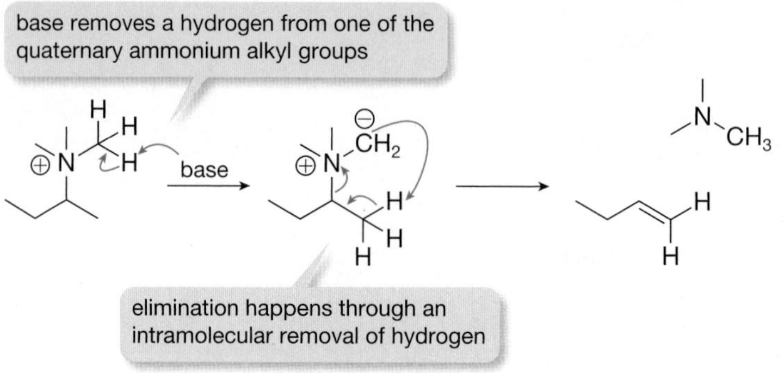

In this particular reaction, the leaving group is a tertiary amine, which is a very large functionality and plays a special role in the reaction. There is evidence to suggest that the first step in this process is the formation of an **ylide** (a group with negatively charged carbon adjacent to a positively charged heteroatom), which undergoes an elimination through a cyclic transition state. The major product of this reaction is normally the least substituted alkene.

An **ylide** is a group with adjacent positive and negative charges.

This reaction was originally studied by August Wilhelm von Hofmann, and today the reaction is named after him. The term *Hofmann product* is used today to describe the products of many elimination reactions, specifically those that give less substituted alkenes.

CHECKPOINT 12.2

You should now be able to predict whether the Zaitsev product or Hoffmann product will form in an E2 elimination reaction, depending on the reaction conditions.

SOLVED PROBLEM

Suggest reaction conditions for carrying out the following elimination reaction.

Br
?

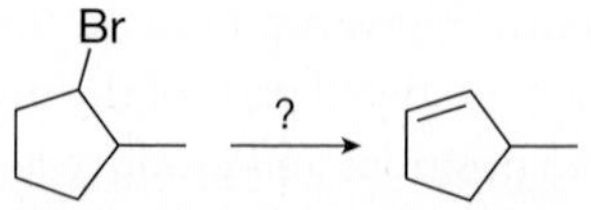

STEP 1: Redraw the starting material with the β-hydrogens included.

An elimination reaction involves the removal of a hydrogen β to a leaving group; therefore, we begin by drawing in the two sets of hydrogens β to the bromine atom.

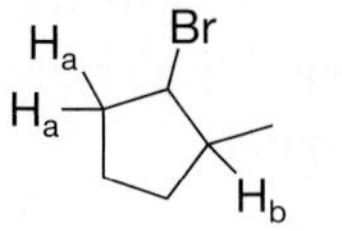

STEP 2: Determine which hydrogen was removed in forming the product shown.

The product is the result of removing one of the H_a hydrogens.

STEP 3: Compare the relative steric crowding of the two sets of β-hydrogens.

The H_a hydrogens are less crowded compared to the H_b hydrogen that is attached to a tertiary carbon.

STEP 4: Suggest reaction conditions appropriate for removal of the necessary β-hydrogen.

To favour the removal of a less sterically crowded H_a hydrogen, we use a very bulky base such as potassium *tert*-butoxide.

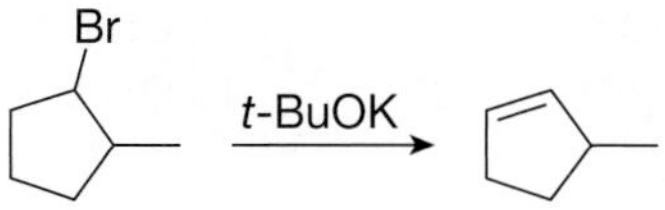

PRACTICE PROBLEM

12.3 Complete the following reactions by filling in the missing products or suggesting appropriate reaction conditions for carrying out the desired transformation.

a) 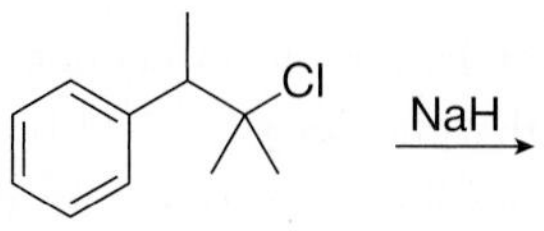 NaH →

b) 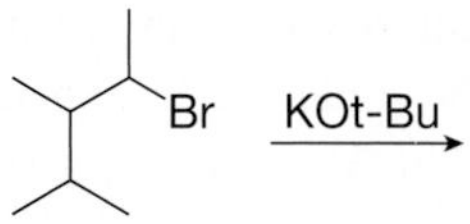 KOt-Bu →

c) 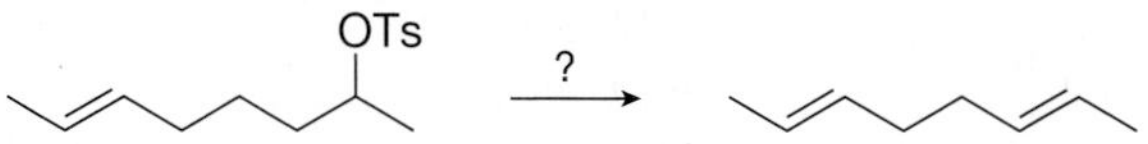 ? →

d) 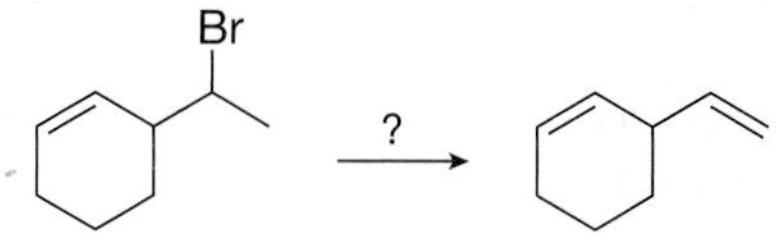 ? →

INTEGRATE THE SKILL

12.4 Draw an energy diagram for the reaction in Question 12.3a showing pathways to all products. How would the energy diagram look if the base was potassium tert-butoxide instead of sodium hydride?

12.2.3 Stereochemistry of E2 reactions

Stereoselectivity refers to the preferential formation of one stereoisomer over another in a chemical reaction. In elimination reactions, many alkenes can be formed in either the *Z* or *E* configuration (*cis* or *trans*). For example, in reactions involving elimination in acyclic molecules, the *E* product is often favoured. Several stereochemical factors affect this selectivity. In particular, relative bond orientation and the conformation of the molecule both play an important role in determining the stereochemistry of the product.

The first factor controlling stereochemistry is bond orientation. Four atoms are involved in a typical E2 reaction: the hydrogen, the two carbons, and the leaving group. All four must lie in the same plane, with the carbon-hydrogen bond and the carbon–leaving group bond *anti* to each other. This orientation allows for the best overlap of the orbitals involved; as the carbon-hydrogen bond breaks, the electrons from this bond can directly flow into the σ^* orbital of the carbon–leaving group bond, thus breaking that bond.

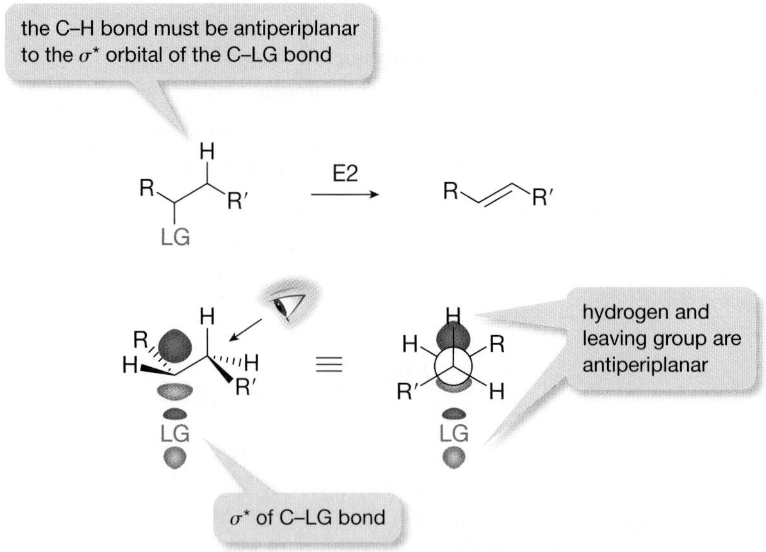

the C–H bond must be antiperiplanar to the σ^* orbital of the C–LG bond

hydrogen and leaving group are antiperiplanar

σ^* of C–LG bond

A **scissile** bond is a bond that is capable of being broken. Typically, the term is used to describe the bonds that are broken in a biological reaction.

In the following example, either one of two β-hydrogens can be removed to generate either the *E* or the *Z* product. To predict the major product, the rule that the **scissile** C–H bond must be antiperiplanar to the leaving group is imposed, and the two possible conformers for the reaction are examined. Newman projections, as shown in the following diagram, are a very useful tool for this kind of analysis. In the top and bottom pathways, the molecule has been rotated to place a different hydrogen antiperiplanar to the leaving group. This analysis shows that the *E* isomer is favoured because the transition state leading to its formation has less steric hindrance than the transition state that produces the *Z* isomer.

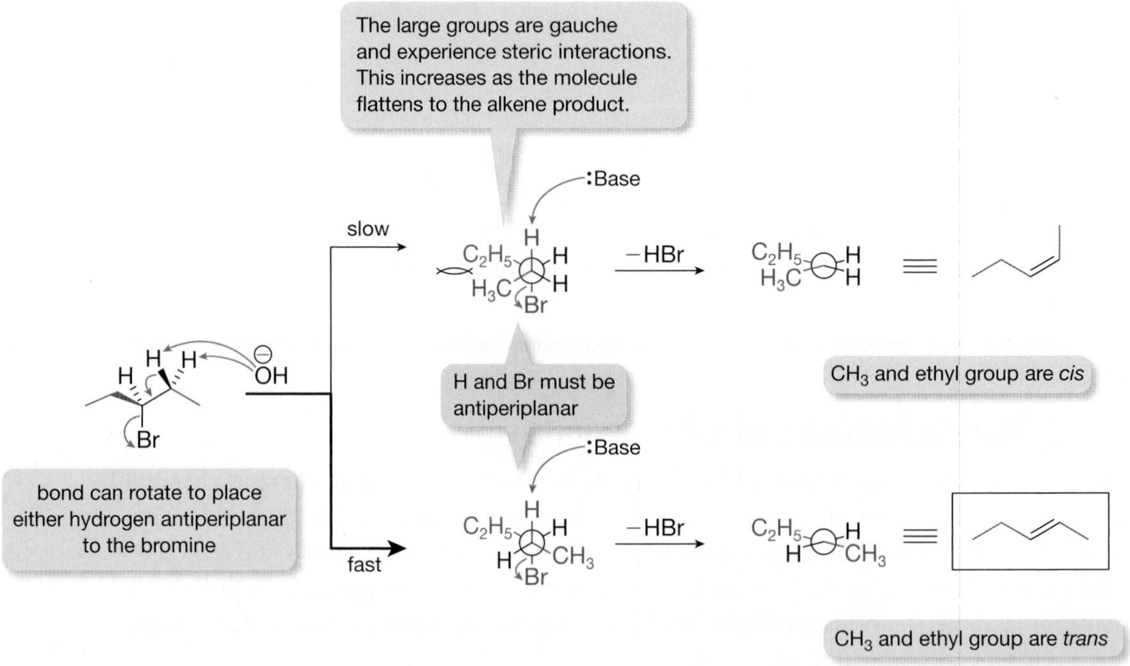

The large groups are gauche and experience steric interactions. This increases as the molecule flattens to the alkene product.

H and Br must be antiperiplanar

bond can rotate to place either hydrogen antiperiplanar to the bromine

CH_3 and ethyl group are *cis*

CH_3 and ethyl group are *trans*

The second factor that affects selectivity concerns the initial conformation of the molecule. When the hydrogen and leaving group are not antiperiplanar, the molecule must rotate, if possible, into a conformation that provides the proper alignment. Only then can the E2 elimination take place.

$EtO^{\ominus}$ — no reaction possible

H and Cl are *not* antiperiplanar

rotate around carbon-carbon bond

$EtO^{\ominus}$

H and Cl are antiperiplanar

E isomer formed selectively

STUDENT TIP

To analyze the stereochemical outcomes of E2 reactions, draw the substrate so that the carbon-hydrogen and carbon–leaving group bonds are in the plane of the paper. This helps in drawing the Newman projection.

These stereochemical requirements for E2 eliminations are commonly seen in substituted cyclohexane reactants. To react by an E2 pathway, the β-hydrogen and the leaving group in a cyclohexane must be *trans*-diaxial, as this is the only way they can be *anti* to one another. If they are not, the molecule must change conformations so that these two components are positioned in this orientation. Only then can an E2 elimination proceed.

In the following example, two diastereomers of a cyclohexane derivative bearing a tosyl leaving group (OTs) undergo an E2 elimination under basic conditions. Diastereomer A, in its more stable chair conformation, has two alkyl groups in equatorial positions and the OTs group in the axial position. There is an axial hydrogen at both of the β positions, so an E2 elimination can proceed from either of those two locations, resulting in rapid formation of two specific E2 elimination products.

Diastereomer B, in its more stable chair conformation, has all three ring substituents in the equatorial position. The OTs group is *not* axial and therefore has no antiperiplanar β-hydrogens. To undergo an E2 elimination, this compound must first flip to the higher energy chair in which the OTs group is axial—this is a slow and energetically unfavourable transition. In this conformation, only one of the two β-carbons bears an axial hydrogen that can participate in an elimination. The E2 elimination of diastereomer B forms only one elimination product, and in a *much* slower reaction compared to the elimination of diastereomer A.

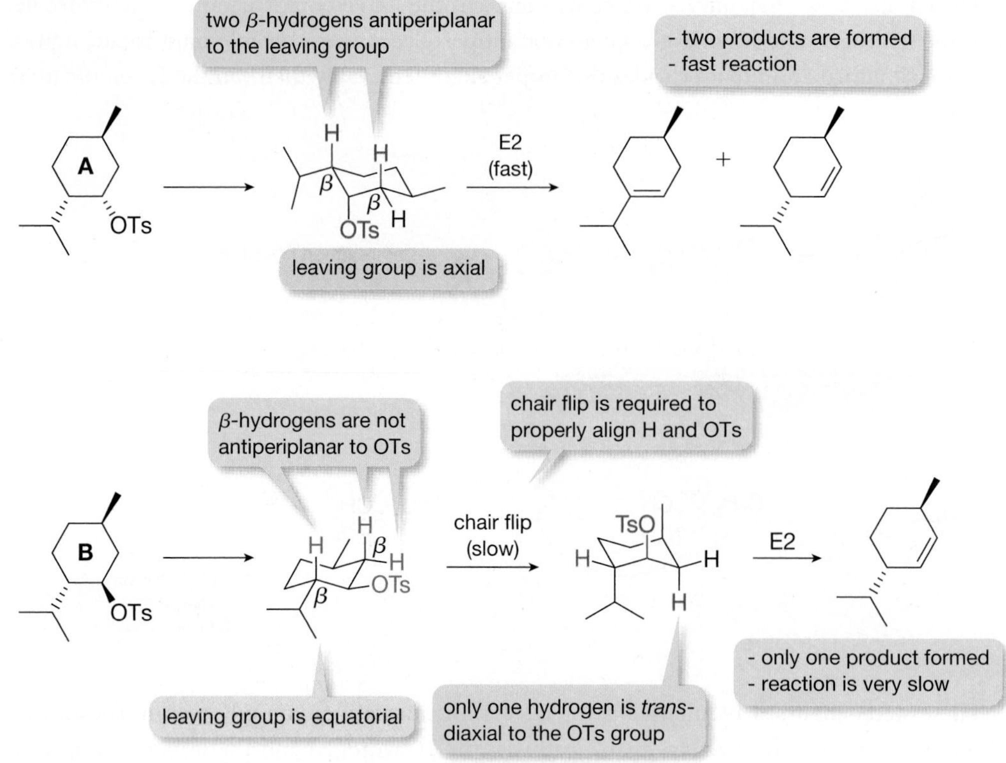

CHECKPOINT 12.3

You should now be able to predict the stereochemical outcome of an E2 elimination reaction.

SOLVED PROBLEM

Predict the major product of the following elimination reaction and explain the observed regio- and stereoselectivity.

Cl

NaOMe
MeOH

STEP 1: Determine the roles of the reactants.

The starting material has a chlorine atom that can act as a good leaving group. The sodium methoxide is a strong base, which can remove one of the hydrogen atoms β to the leaving group. Methanol is the solvent.

STEP 2: Draw in the hydrogen atoms β to the leaving group.

There are two sets of β-hydrogens, H_a and H_b.

Cl
H_a
H_a
H_a
H_b

STEP 3: Determine which set of β-hydrogens is more likely to be attacked by the base.

The base is small enough that it should not be influenced by steric crowding. Therefore, it will remove whichever hydrogen yields the more stable product: in this case, H_b, since its removal gives a trisubstituted double bond, compared to the removal of H_a, which gives a monosubstituted double bond.

STEP 4: Redraw the starting material in a conformation that places H_b so that it is antiperiplanar to the leaving group.

In the original conformation, both H_b and the Cl leaving group are pointing out of the plane of the page. To have these groups in an antiperiplanar arrangement, the C–C bond between them must be rotated until they are 180° apart.

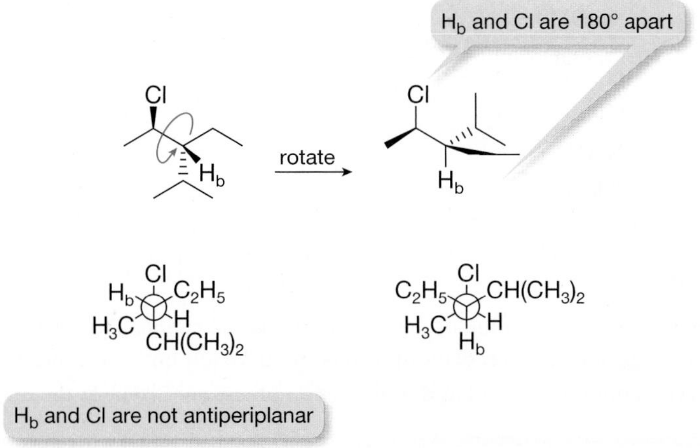

H_b and Cl are 180° apart

Cl, H_b — rotate → Cl, H_b

Cl, H_b, C_2H_5, H_3C, H, $CH(CH_3)_2$ — Cl, C_2H_5, $CH(CH_3)_2$, H_3C, H, H_b

H_b and Cl are not antiperiplanar

STEP 5: Draw a mechanism for the reaction. Be sure to include the stereochemistry of the final product as well. Using Newman projections is a good way to keep track of stereochemistry.

Now that the hydrogen and leaving group are antiperiplanar, the base attacks the hydrogen, causing the electrons from the C–H bond to flow toward the leaving group, creating a new C=C π bond. At the same time, the C–Cl σ bond breaks and the chloride ion leaves. The *E* geometry of the new π bond is set by the antiperiplanar conformation of the reactant.

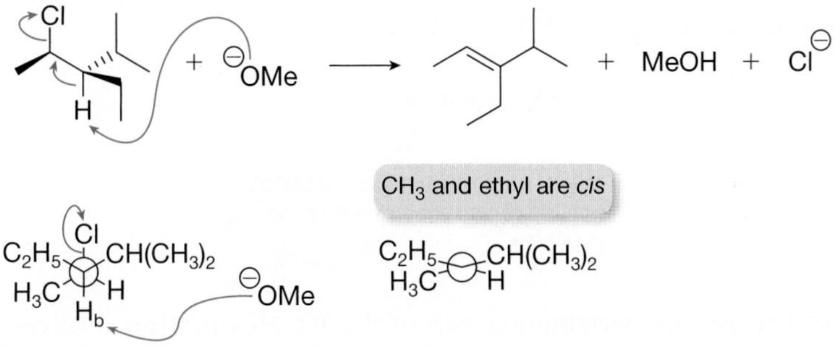

Cl, H + $^{\ominus}$OMe ⟶ + MeOH + $Cl^{\ominus}$

CH_3 and ethyl are *cis*

Cl, C_2H_5, $CH(CH_3)_2$, H_3C, H, H_b, $^{\ominus}$OMe — C_2H_5, $CH(CH_3)_2$, H_3C, H

PRACTICE PROBLEM

12.5 Predict the major product of the following reactions, and draw a mechanism showing their formation.

a) 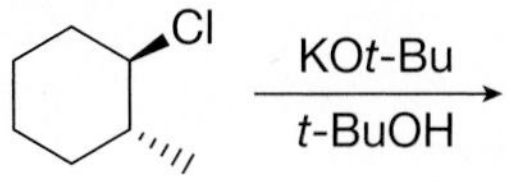 Cl — KO*t*-Bu / *t*-BuOH →

b) NaOMe / MeOH

c) KO*t*-Bu / *t*-BuOH

d) NaOMe / MeOH

e) Ph, Br; NaOMe / MeOH

f) Ph, Br; NaOMe / MeOH

INTEGRATE THE SKILL

12.6 Draw all possible alkenes that could be generated from an E2 elimination reaction of 3-bromo-3-methylhexene. Suggest conditions that can be used to make just one of the alkenes in a controlled manner (i.e., conditions that would lead to formation of just one alkene as the major product).

12.3 Alkene Formation by E1 Elimination Reactions

In addition to E2 reactions, alkenes are also formed by an E1 (unimolecular) elimination process. Kinetic experiments for the E1 type of reaction show that the rate of formation of the products depends only on the concentration of the substrate:

$$\text{rate} = k[\text{substrate}]$$

rate of an E1 elimination depends only on the concentration of the substrate

E1 elimination is a multi-step reaction in which the leaving group is lost in the first and rate-determining step and the adjacent proton is lost in the second step.

This means that the rate-determining step of the **E1 elimination** involves only the substrate and not the base. The only way this can happen is if the reaction occurs in two steps: one involving the substrate and one involving the base. To respect the octet rule, the leaving group must leave in the first step, which is then followed by a base-promoted loss of the β-hydrogen in a second step, resulting in an alkene product. The loss of the leaving group in the first step indicates the formation of a carbocation intermediate.

Step 1: loss of leaving group

Step 2: loss of β-hydrogen

H LG β α → :B⊖ H ⊕ → + BH

carbocation intermediate

With knowledge of this mechanism, a reaction diagram can be constructed showing that this two-step reaction pathway has many similarities to that of the S_N1 reaction (in fact, the S_N1 reaction often competes with the E1 elimination, giving mixtures of products that are common in both processes). The first step of the reaction has the highest activation energy, is rate determining, and leads to the formation of a carbocation. In the second step, the hydrogen is rapidly removed by the base to form the E1 elimination product.

ORGANIC CHEMWARE
12.2 Curved arrow notation: β-elimination

ORGANIC CHEMWARE
12.3 E1 elimination

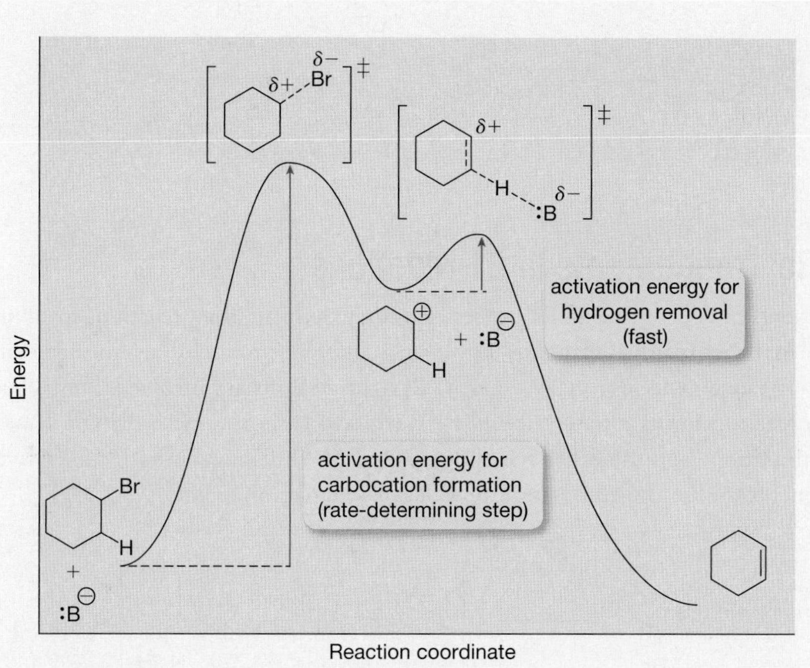

Substrates that can form stable carbocation intermediates (secondary, tertiary, allylic, or benzylic) are the ones that undergo elimination via the E1 pathway. Primary, vinyl, or aryl substrates do not undergo such E1 elimination reactions because they cannot form stable carbocations.

LG LG LG LG
good E1 substrates

LG LG LG
inappropriate E1 substrates

Typical leaving groups in these reactions include halides, water, tosylate, and mesylate (Chapter 11). As in the case of nucleophilic displacements, the OH group is not a good leaving group for E1 reactions and must be converted into one before it can leave. This can be achieved by the use of acid, or by converting it into a tosylate or mesylate.

In contrast to the E2 elimination, which requires strong bases, the E1 elimination can be carried out with weaker bases. This is because the hydrogen β to the carbocation is generally easier to remove (more acidic) than a typical carbon-bonded hydrogen. The nearby carbocation is very electron withdrawing and begins pulling electrons from the carbon–β-hydrogen bond. This weakens the bond, making it removable with weak bases.

12.3.1 Selectivity in E1 reactions

As in E2 reactions, E1 eliminations follow Zaitsev's rule of the preferential formation of the most stable alkene. Although the deprotonation step is not rate determining, its transition state has partial π-bond character, so factors that stabilize double bonds also stabilize this transition state. Consequently, the reactions tend to form the most stable alkenes.

E1 eliminations are usually less stereoselective than E2 reactions. The carbocation is flat and connected to the β-carbon by a single bond. This bond can rotate to give access to both possible stereoisomers of alkene. Because the second step is fast, the energy difference between the transition states leading to the *E* and *Z* isomers is small. This gives selectivity for the most stable isomer, but the ratio of products can be very close to 1.

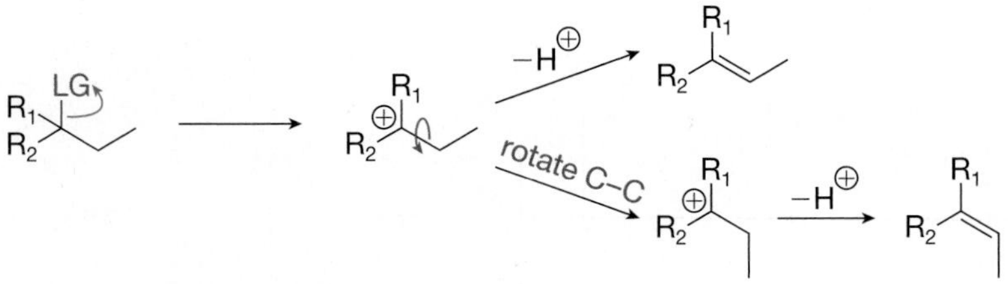

12.3.2 Rearrangements in E1 reactions

Rearrangements can occur during E1 processes, as they do in all other reactions involving carbocations. Unlike other types of reactions, however, a different product does not necessarily result from these rearrangements, especially when [1,2] hydride shifts are involved. This is because the elimination reaction forms a new π bond between two carbons: one bearing a leaving group (forms a carbocation), and the other, a hydrogen. Hydride shifts may just effectively switch these two components, and so the elimination product may remain the same.

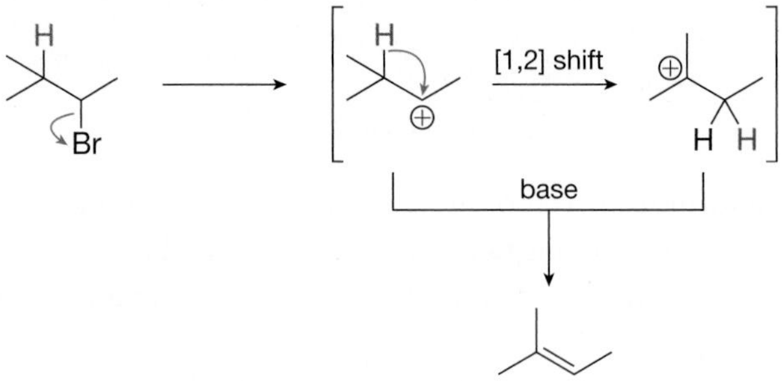

CHECKPOINT 12.4

You should now be able to draw a mechanism for an E1 elimination reaction and predict the major isomer formed, including its regio- and stereochemistry.

STUDENT TIP
Consider all of the possible carbocations that can form from rearrangements of the initial carbocation.

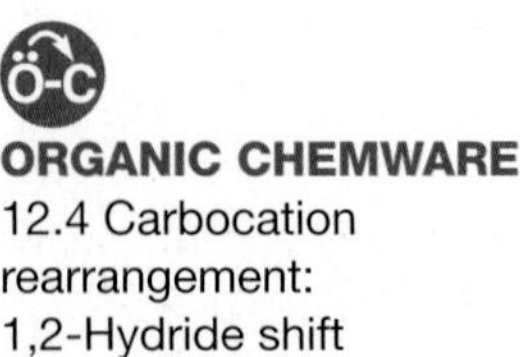

ORGANIC CHEMWARE
12.4 Carbocation rearrangement: 1,2-Hydride shift

SOLVED PROBLEM

Draw the expected elimination product for the following reaction and explain the observed selectivity using a reaction mechanism.

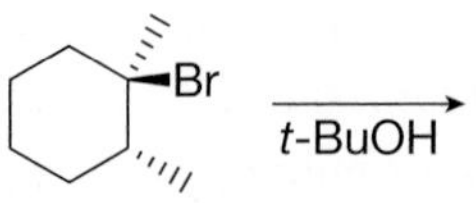

STEP 1: Determine the roles of the reactants.

The starting material has a bromine atom that can act as a good leaving group. The *tert*-butyl alcohol is a weak base that can remove a β-hydrogen from a carbocation if the elimination proceeds via an E1 mechanism. It is not, however, basic enough to remove a β-hydrogen directly from the alkyl bromide.

STEP 2: Draw a mechanism for the first step of the reaction.

The tertiary alkyl bromide readily loses its leaving group to form the tertiary carbocation intermediate, as shown here.

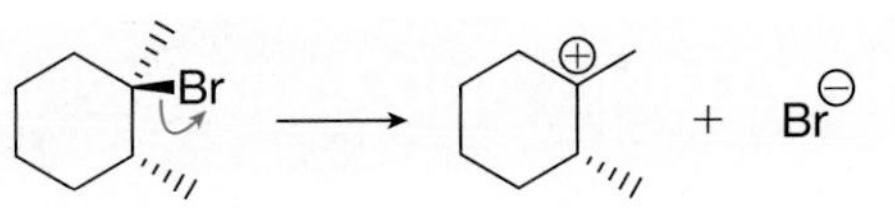

STEP 3: Draw in the hydrogen atoms β to the carbocation.

There are three sets of hydrogens that are β to the carbocation: H_a, H_b, and H_c.

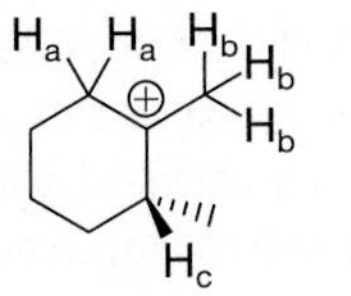

STEP 4: Determine which set of β-hydrogens is more likely to be attacked by the base.

The β-hydrogen removed will be the one that leads to the most stable alkene product—typically, the most substituted alkene, unless there is resonance available. For this substrate, the most stable product is a tetrasubstituted double bond that results from removal of H_c.

STEP 5: Draw a mechanism for the second step of the reaction.

A lone pair on the *tert*-butyl alcohol oxygen shares its electrons with the β-hydrogen, causing the C–H bond to break and a new C=C π bond to form.

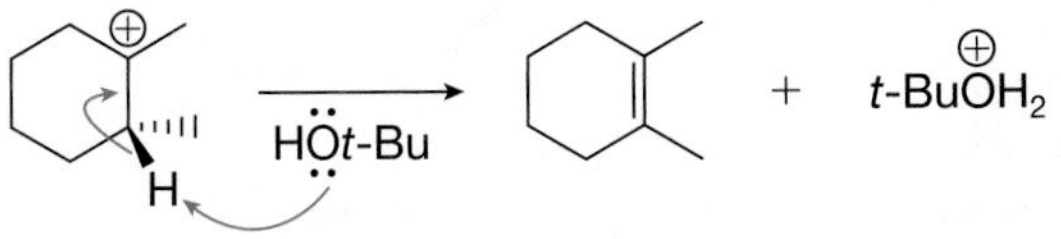

PRACTICE PROBLEM

12.7 Draw the expected elimination products for each of the following reactions. Where more than one isomer is possible, explain the expected selectivity.

a) 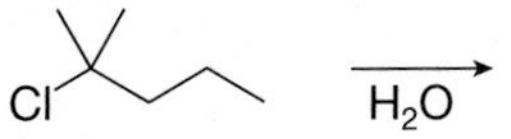 $\xrightarrow{H_2O}$

b) 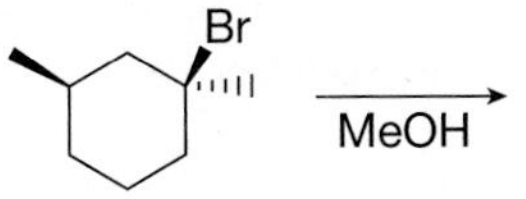 $\xrightarrow{MeOH}$

c) 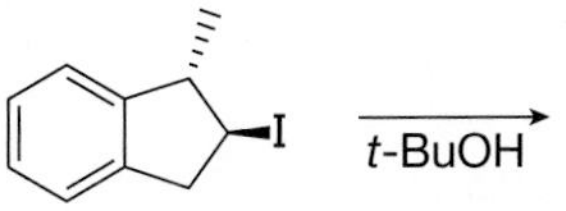 $\xrightarrow{t\text{-BuOH}}$

d)  $\xrightarrow{i\text{-PrOH}}$

INTEGRATE THE SKILL

12.8 Propose a mechanism for the following reaction.

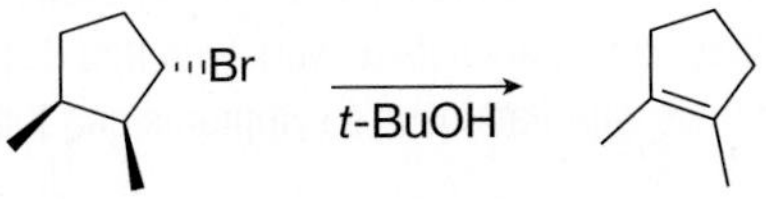

CHEMISTRY: EVERYTHING AND EVERYWHERE

Don't-Do-It-Yourself Drug Synthesis

Designer drugs are an attempt to create legal versions of illegal drugs. By making small changes to the chemical structure of a compound, underground chemists aim to produce a new substance that not only retains a particular drug's "recreational" effects but also remains technically legal because there is no current law against its usage.

In 1982, people with symptoms of advanced Parkinson's disease began arriving at emergency rooms in San Francisco. Many of these people were completely paralyzed, able to move only their eyes. Investigation quickly established that they were all illicit drug users who had experimented with a designer drug similar to Demerol, a narcotic painkiller.

Whoever made this designer drug had failed to consider side reactions that could occur during its preparation. The design involved switching the orientation of the ester group in the Demerol structure; their idea was to retain the narcotic effect of Demerol in an unregulated (legal) street drug.

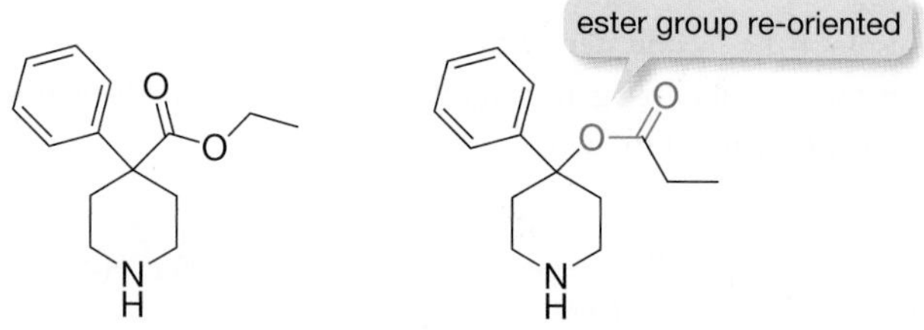

ester group re-oriented

Demerol

designer drug

In the last step of this designer drug synthesis, an esterification reaction generated HCl as a by-product. When the pH of the reaction was not properly controlled, the buildup of HCl initiated an E1 elimination reaction that formed a highly toxic chemical called MPTP. This material was a powerful neurotoxin that destroyed the *substantia nigra*, a part of the brain that controls movement.

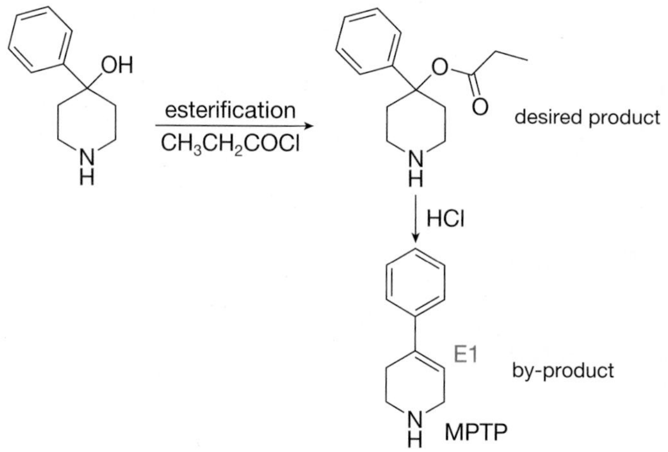

esterification CH_3CH_2COCl

desired product

HCl

E1

by-product

MPTP

This tragic episode nevertheless prompted research into a cure for Parkinson's disease. Until then, scientists could not do this research because they had no animals to work with that displayed symptoms of Parkinson's. By using MPTP, researchers created animals with Parkinson's symptoms and then studied them to develop modern treatments for this disease.

12.4 Dehydration and Dehydrohalogenation

Elimination reactions can be classified according to the type of leaving group. A **dehydration** reaction involves the elimination of hydroxyl leaving groups and the loss of water. **Dehydrohalogenation** involves the elimination of halides. Both dehydration and dehydrohalogenation can follow either an E1 or E2 pathway, depending on whether or not the substrate can form a carbocation. Although the leaving groups and the names change, the basic electron flows are the same.

Dehydration is the loss of water to produce an alkene from an alcohol.

Dehydrohalogenation is the loss of a hydrogen and a halogen to produce an alkene from an alkyl halide.

The major difference between the dehydration of alcohols and the dehydrohalogenation of alkyl halides concerns whether the reaction is promoted by an acid or a base. Most of the elimination mechanisms shown so far in this chapter are dehydrohalogenations done under basic conditions. E2 eliminations require a strong base, and E1 eliminations can use a weaker base.

Dehydrohalogenation

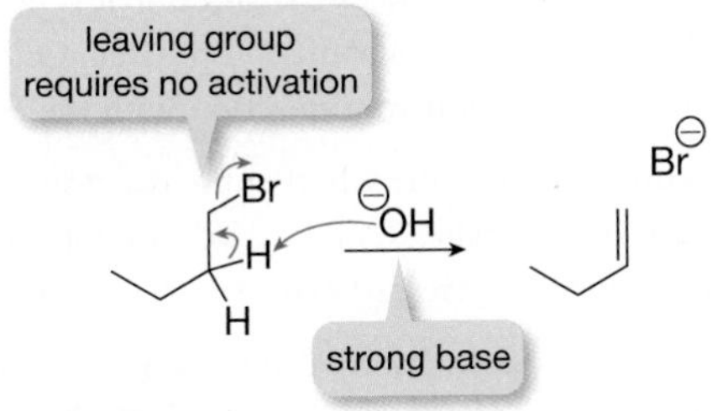

leaving group requires no activation

strong base

Dehydration

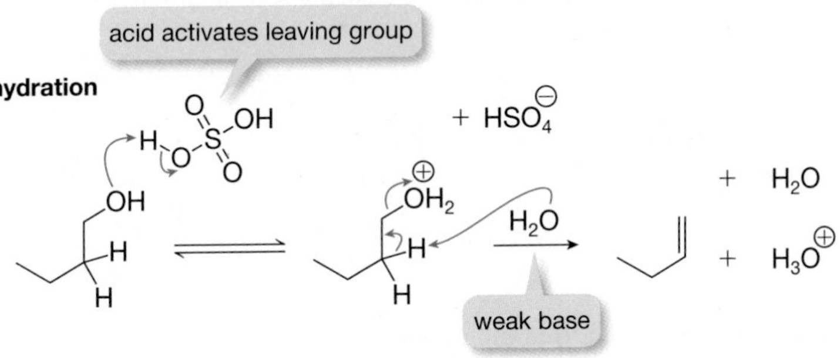

acid activates leaving group

weak base

By contrast, the elimination of alcohols (dehydration) generally requires the presence of a strong acid, such as H_2SO_4, to activate the hydroxyl leaving group. Under aqueous acidic conditions, the base that removes the hydrogen in the last step is likely H_2O, the strongest base available in the mixture.

ORGANIC CHEMWARE
12.5 Dehydrohalogenation

ORGANIC CHEMWARE
12.6 Alcohol dehydration

The mechanism of E1 dehydration is the reverse of the mechanism for the hydration of alkenes (Chapter 8). Le Chatelier's principle controls whether an alkene hydrates to form an alcohol or whether an alcohol eliminates to form an alkene. Specifically, concentrated strong acid favours elimination, whereas dilute strong acid favours hydration due to a large excess of water.

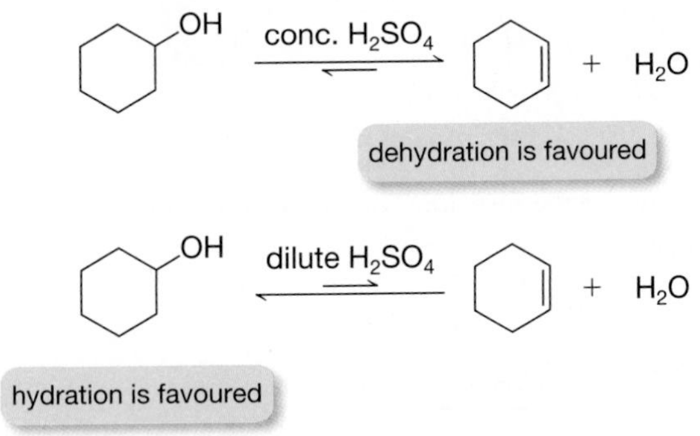

conc. H_2SO_4 — dehydration is favoured

dilute H_2SO_4 — hydration is favoured

CHECKPOINT 12.5

You should now be able to draw a mechanism and its corresponding product for both E1 and E2 dehydrations of an alcohol.

SOLVED PROBLEM

Draw the product of the following dehydration reaction and a mechanism showing its formation.

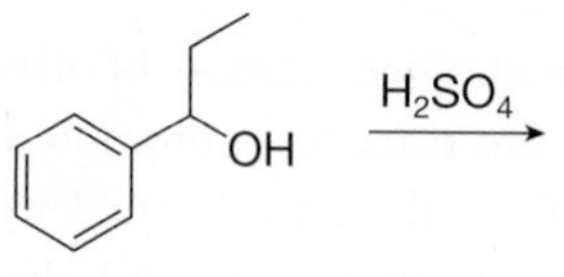

OH $\xrightarrow{H_2SO_4}$

STEP 1: Determine the roles of the reactants.

The starting material has an alcohol functional group that is not normally a good leaving group; however, in the presence of the strong acid, it becomes protonated. Once protonated, the OH becomes a good leaving group ready to undergo an elimination reaction with the loss of water.

STEP 2: Evaluate the structure to determine whether the reaction will proceed via an E1 or E2 mechanism.

The alcohol is secondary and benzylic, so the carbocation intermediate will be stabilized by both hyperconjugation and resonance. Therefore, we can expect an E1 mechanism for the dehydration.

STEP 3: Draw a mechanism for the formation of the carbocation intermediate.

Following protonation of the alcohol by sulfuric acid, the leaving group leaves, creating a carbocation intermediate and a molecule of water.

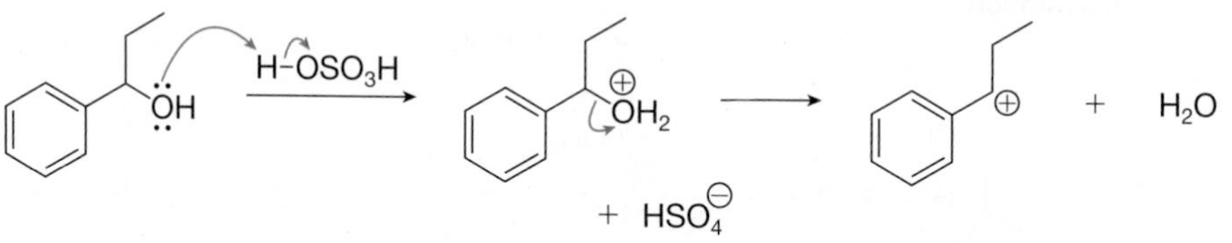

ÖH $\xrightarrow{H\text{–}OSO_3H}$ $\overset{\oplus}{O}H_2$ + $HSO_4^{\ominus}$ ⟶ ⊕ + H_2O

STEP 4: Draw in the hydrogen atoms β to the carbocation.

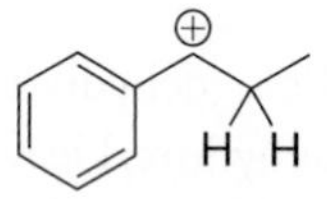

⊕ H H

STEP 5: Draw a mechanism for the second step of the reaction. Be sure to include the stereochemistry of the final product.

One of the β-hydrogens is removed by the weak conjugate base, and the electrons from the C–H bond form a new C=C π bond. The alkene formed is the more stable *trans* (*E*) alkene.

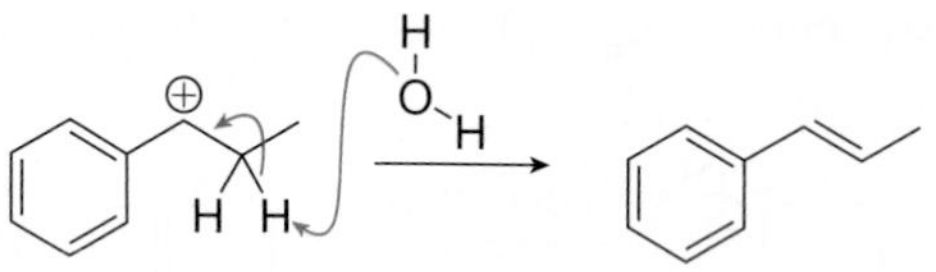

⊕ H H H–Ö–H ⟶

PRACTICE PROBLEM

12.9 Draw a mechanism showing the formation of the major product in each of the following reactions.

a) 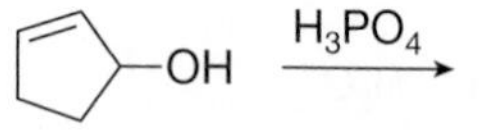 –OH $\xrightarrow{H_3PO_4}$

b) 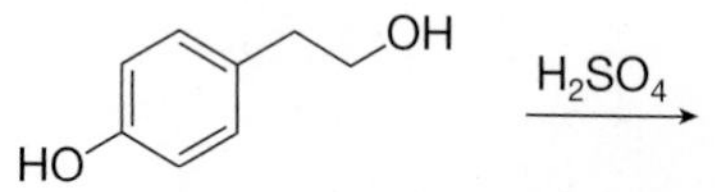 HO OH $\xrightarrow{H_2SO_4}$

c) 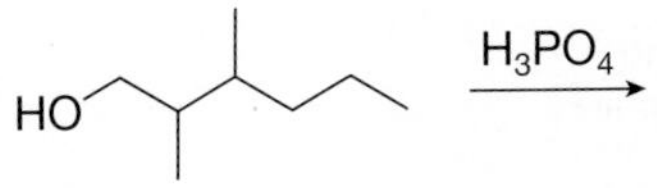 HO [structure] $\xrightarrow{H_3PO_4}$

d) 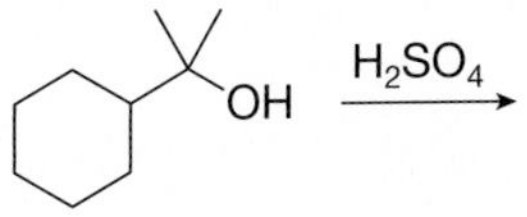 [structure] OH $\xrightarrow{H_2SO_4}$

INTEGRATE THE SKILL

12.10 Draw mechanisms for the E1 dehydration of cyclohexanol and the hydration of cyclohexene to show that these reactions are the reverse of each other.

12.5 Differentiation between Elimination Reactions and Nucleophilic Substitutions

One function of both nucleophiles and bases is the donation of a lone pair of electrons. This means that many bases act as nucleophiles, and most nucleophiles can act as bases. This creates a situation in which small amounts of nucleophilic displacement occur in elimination reactions, and some elimination occurs in a nucleophilic displacement. This can present problems when one is trying to achieve a specific transformation without creating side-products. A further complication arises because E1 and E2 elimination reactions are part of a reactivity continuum (as are S_N1 and S_N2 reactions) that displays characteristics of at least three reactions (the third reaction is described in Chapter 17). It is therefore important to identify factors that control the possible pathways.

First, the strength of the base is the most important factor that differentiates E1 reactions from E2 reactions. Strong bases (such as NaH or KO*t*-Bu) promote E2 reactions because they are involved in the rate-determining step. Conversely, weak bases are less likely to influence the rate of hydrogen abstraction, and so an E1 reaction is often the result.

Second, the strength of the nucleophile is the key factor that differentiates elimination reactions from substitutions. Good nucleophiles that are not very basic promote substitution, whereas weak nucleophiles tend to give elimination or follow first-order substitution pathways. For substitutions, tertiary substrates undergo S_N1 displacements only, primary substrates undergo S_N2 reactions only, and secondary substrates can follow either pathway.

Heat is also an important factor. In general, higher temperatures tend to promote eliminations, whereas nucleophilic displacements often predominate at lower temperatures.

When necessary, the solvent conditions can be considered. Protic solvents stabilize carbocations well and tend to promote first-order pathways. Dipolar aprotic solvents do not stabilize carbocations and are associated with bimolecular reactions.

To analyze such reactions, consider the following factors:

1. Determine if the reagent is a strong or weak (bulky) nucleophile and whether it is a strong or weak base.
   - Strong bases and nucleophiles favour second-order reactions. When this component is strong, the result will be either an S_N2 or E2 reaction.
   - Weak bases and nucleophiles favour first-order reactions. When this component is weak, the result will be either an S_N1 or E1 reaction.
2. Determine the level of substitution at the carbon that is carrying the leaving group.
   - Tertiary substrates undergo S_N1 reactions with a good nucleophile. When the nucleophile is more basic, E2 reactions are favoured. Note that S_N1 and E1 reactions can happen together, so product mixtures are common.

- Primary substrates undergo second-order reactions, either S_N2 or E2 reactions, depending on the nature of the nucleophile or base.

3. Consider the solvent and temperature when necessary.
   - For substrates that can undergo either unimolecular or bimolecular reactions (i.e., secondary, allylic, or benzylic halides), consider both the nucleophile and the solvent. Strong nucleophiles in polar aprotic solvents favour S_N2 reactions. Strong bases in polar aprotic solvents favour E2 reactions. The use of protic solvents stabilizes carbocations and tends to produce a mixture of S_N1 and E1 products. Higher temperatures tend to promote eliminations (both E1 and E2 reactions).

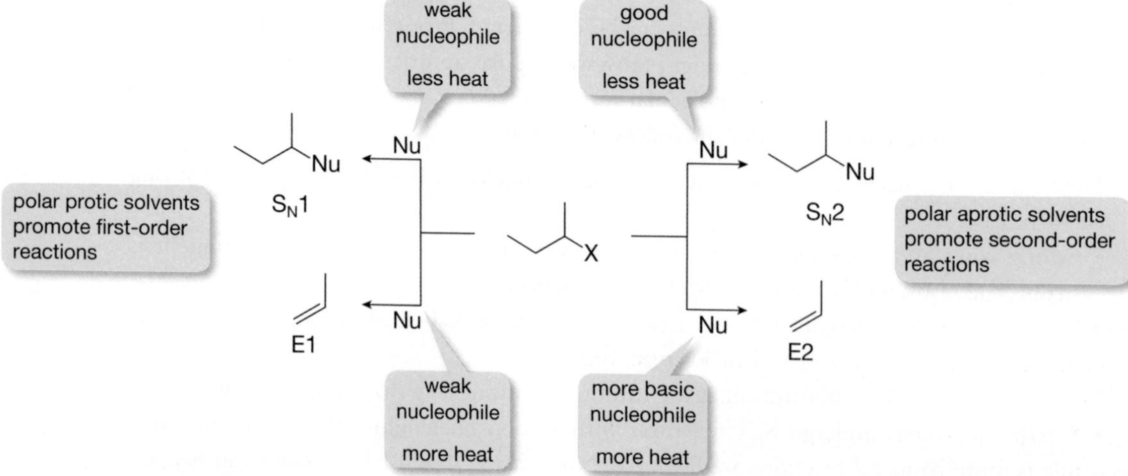

- Though weak nucleophiles or bases in protic solvents promote both S_N1 and E1 reactions, the activation energy for the conversion of carbocations into products is higher for E1 reactions than for S_N1 reactions. Therefore, the addition of heat (Δ) promotes more E1 reactions; lower temperatures tend to promote S_N1 reactions.

STUDENT TIP
Working systematically through the factors controlling eliminations and displacements is the most reliable way to analyze these reactions.

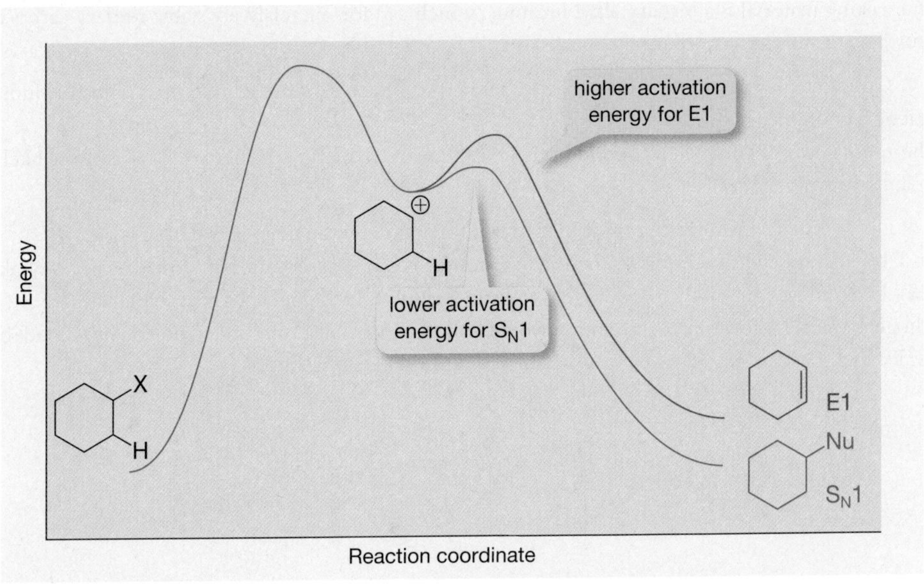

12.6 Designing Reactions for Selectivity

Ideally, organic reactions are designed to produce only a single product. Otherwise, the process is wasteful because side reactions occur and not all the reactant molecules become converted to the desired material.

Second-order reactions (E2/S_N2) are normally more reliable than first-order ones (E1/S_N1). Because carbocations are unstable entities, they may react in undesired and difficult-to-control ways. Since the use of strong, non-nucleophilic bases promotes second-order eliminations, elimination reactions are typically chosen that use strong non-nucleophilic bases such as NaH. For this reaction, the preferred solvents are polar aprotic solvents such as THF, DMF, or DMSO. Similarly, the use of strong nucleophiles and polar aprotic solvents can promote second-order substitution reactions. The vast majority of nucleophilic substitutions are done in these conditions. Conditions that favour first-order reactions are often avoided, unless strong controlling elements are in place.

CHECKPOINT 12.6

You should now be able to predict whether a given set of reactants and conditions is more likely to favour E1, E2, S_N1, or S_N2 reactivity. You should also be able to apply this information to the planning of short syntheses such that the reaction favours the formation of a single product.

SOLVED PROBLEM

Predict the major products of the following reactions.

a) 

b) 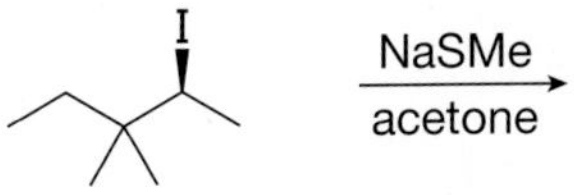

STEP 1: Beginning with the reaction in part (a), evaluate the reagents to determine whether there is a strong base or strong nucleophile.

The only base or nucleophile is propyl alcohol, which is neither very basic nor a particularly good nucleophile. Therefore, a second-order reaction (S_N2 or E2) will not likely take place.

STEP 2: Evaluate the starting material to determine whether it can easily form a carbocation.

The starting material is a tertiary alkyl bromide, which can form a relatively stable tertiary carbocation. This further suggests a first-order reaction (E1 or S_N1).

STEP 3: Consider the reaction conditions to see whether the solvent and the temperature can influence the reaction outcome.

The use of heat in the reaction conditions helps favour E1 over S_N1; therefore, E1 is most likely the dominant reaction taking place.

STEP 4: Draw a mechanism for the most likely reaction and draw the major product.

An E1 reaction begins with loss of the leaving group to form a carbocation intermediate. The weak base (PrOH) then removes one of the β-hydrogens to form a trisubstituted double bond. Note that there are other β-hydrogens available; however, their removal would lead to a less stable disubstituted double bond, and so they are less likely to react.

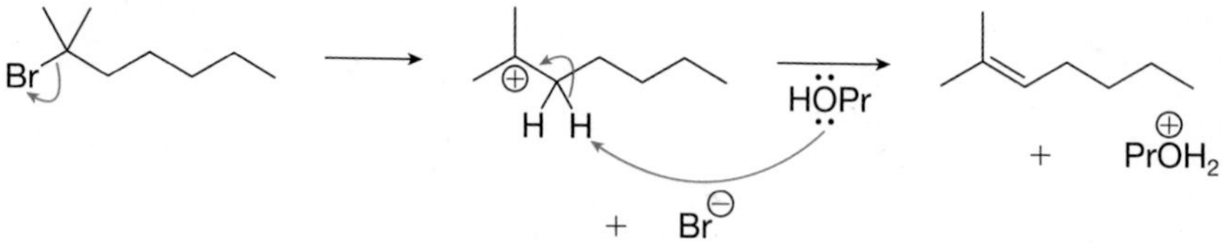

Br → ⊕ H H + $Br^{\ominus}$ —HÖPr→ + $Pr\overset{\oplus}{O}H_2$

STEP 5: Repeat Steps 1–4 for the reaction in part (b).

The starting material is a secondary alkyl iodide that can react by either a first-order or second-order reaction. However, the reagent (SMe^-) is an excellent nucleophile, and so a second-order reaction is favoured—in particular, an S_N2 reaction.

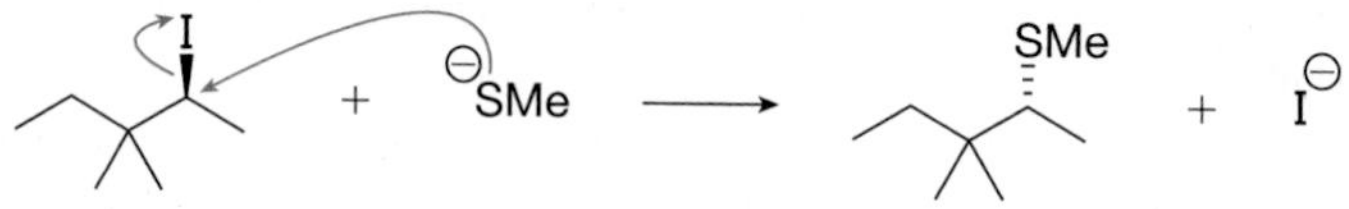

I + $^{\ominus}SMe$ → SMe + $I^{\ominus}$

PRACTICE PROBLEM

12.11 For each of the following reactions, predict the major products, and draw a mechanism that shows their formation.

a) 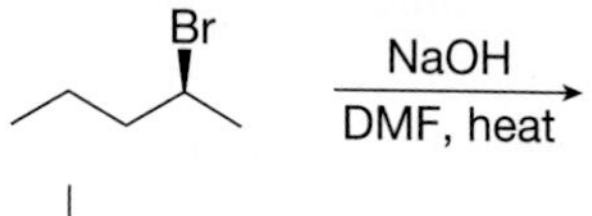 Br —NaOH / DMF, heat→

b) 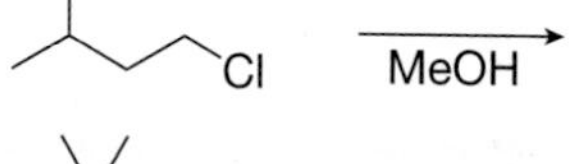 Cl —MeOH→

c) 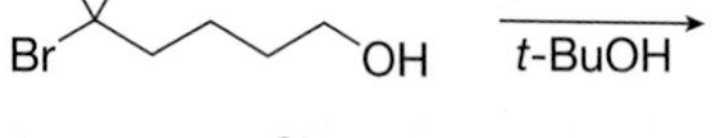 Br OH —t-BuOH→

d) 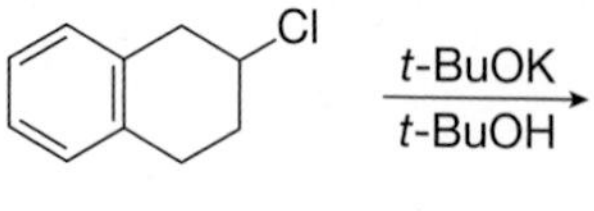 Cl —t-BuOK / t-BuOH→

INTEGRATE THE SKILL

12.12 a) For each of the following transformations, provide appropriate reaction conditions. Where necessary, include solvents or reaction temperatures (cold/warm).

i) Br ? →

ii) Cl ? →

b) The following reactions generate more than one product. Suggest changes that could make these reactions favour the formation of the indicated desired product rather than the other.

i) Cl NaSPh → + SPh

desired product

ii) OMs EtOH → + OEt

desired product

12.7 Oxidation of Alcohols: An Elimination Reaction

The oxidation of alcohols, which involves the removal of a hydrogen and a leaving group to form a π bond, is another type of elimination reaction. When the leaving group in an oxidation departs, it takes with it a pair of electrons. The substrate therefore loses electrons and by definition is oxidized. Typically, the term *elimination* is used when a new π bond forms between two carbon atoms. When the π bond involves a heteroatom, the reaction is called an oxidation. Whether the term *oxidation* or *elimination* is used, the flow of electrons in the two processes is essentially the same.

elimination removes a leaving group and a hydrogen to form a π bond
carbon-carbon π bonds are formed during eliminations
LG
elimination
:LG⊖
H
base
oxidation removes a leaving group and a hydrogen to form a π bond
carbon-heteroatom π bonds are formed during oxidations
OH
attach leaving group to the oxygen
O LG
H
base
oxidation
O
:LG⊖
leaving group removes electron pair, giving an oxidation

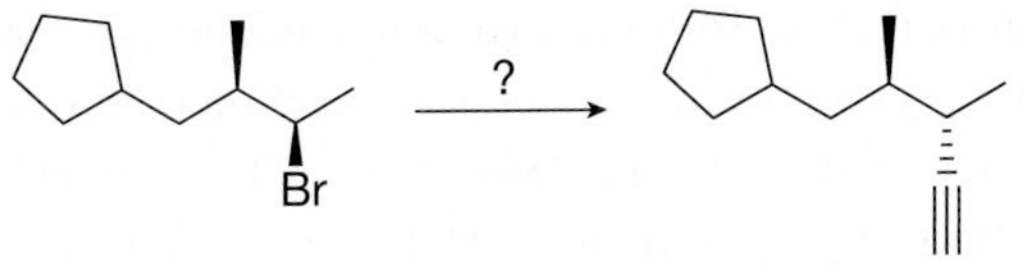

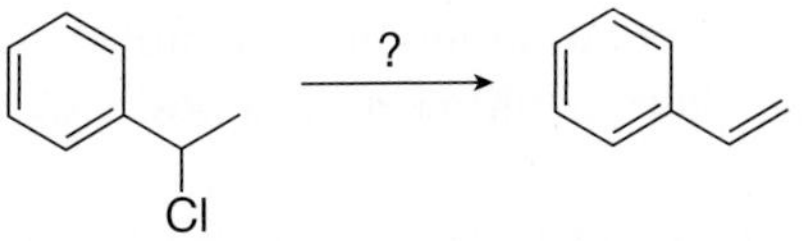

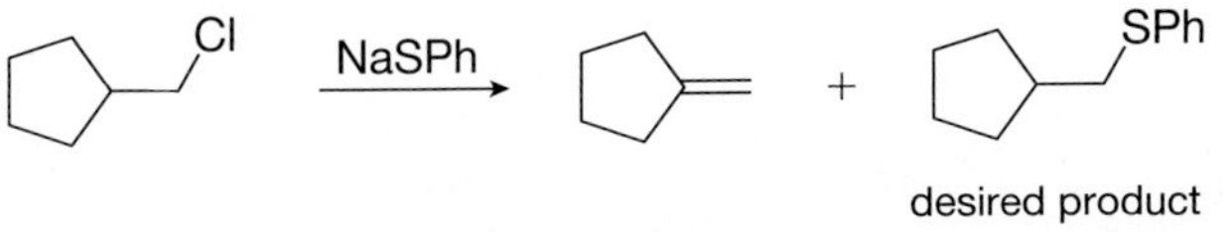

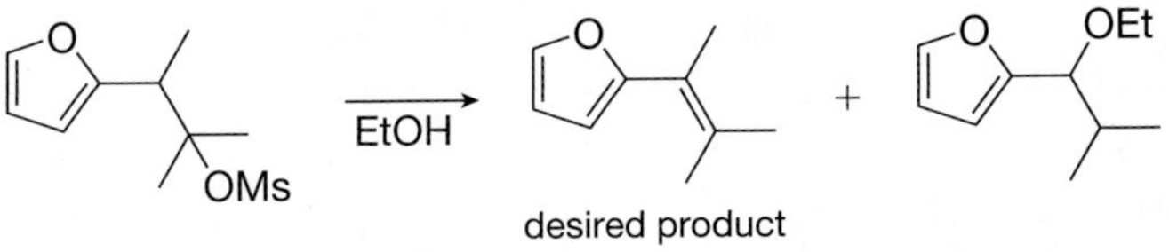

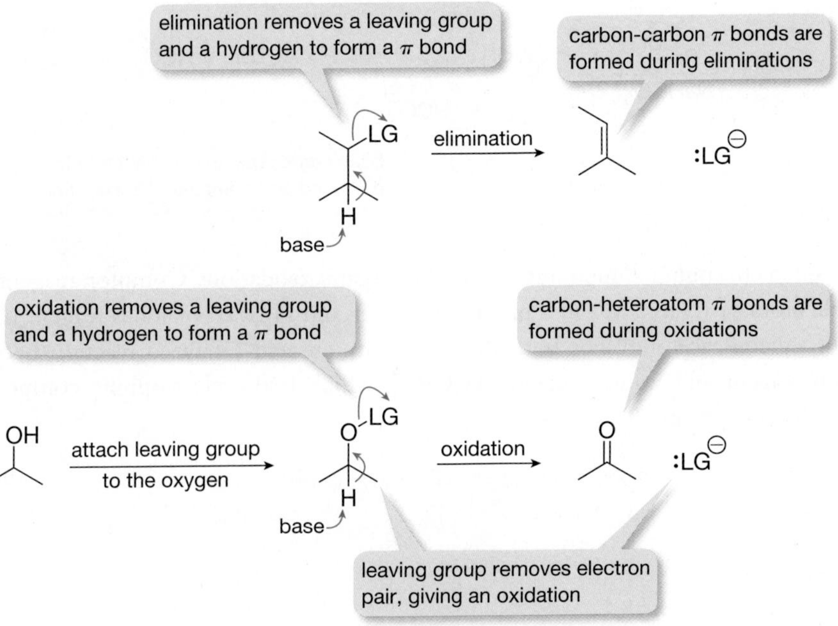

The general mechanism of oxidations first involves the attachment of a leaving group to one of the atoms that eventually forms the π bond; in most cases, the attachment is to the heteroatom. In the second step, an elimination reaction forms the π bond. The reagent that installs the leaving group is usually called an oxidizing agent. The strength of the oxidizing agent refers to its ability to carry out an oxidation, which corresponds to its ability to attract and withdraw electrons. Strong oxidizing agents exert a strong "pull" on electrons, whereas weak oxidants do not attract electrons as well.

Primary alcohols can be oxidized to form either aldehydes or carboxylic acids, depending on the conditions; secondary alcohols are oxidized to form ketones; and tertiary alcohols do not undergo oxidation reactions because they do not contain a hydrogen that can be removed to form a π bond.

OH, H, H — oxidation → O, H — oxidation → O, OH

primary alcohol — aldehyde — acid

OH, H — oxidation → O

secondary alcohol — ketone

OH — oxidation → no removable hydrogen, so no oxidation possible

tertiary alcohol

It is possible to oxidize primary and secondary alcohols by using halogens such as Br_2 as the oxidizing agent together with a base. Bromine is electrophilic and reacts with the nucleophilic hydroxyl group to form an intermediate that has a leaving group (Br) on oxygen; this intermediate then undergoes an E2-style elimination to form a new π bond.

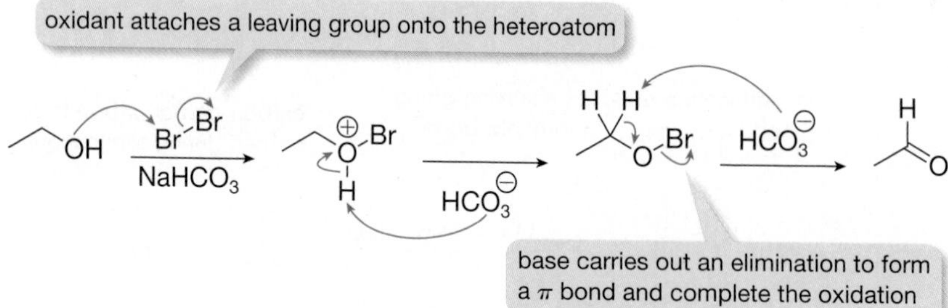

Very often chromium compounds are used in organic oxidations. Common reagents include sodium dichromate ($Na_2Cr_2O_7$ in a mixture of water and H_2SO_4) or Jones reagent (CrO_3 and H_2SO_4 in aqueous acetone). For both these reagents, an acid-catalyzed reaction between the chromium reagent and water generates H_2CrO_4, which is the electrophilic component that installs the leaving group.

$CrO_3 \xrightarrow[H_2O]{H_2SO_4} H_2CrO_4$

$Na_2Cr_2O_7 \xrightarrow[H_2O]{H_2SO_4} 2\ H_2CrO_4$

The chromium species H_2CrO_4 is in equilibrium with $HCrO_3^+$, which reacts with the OH group of the alcohol in a nucleophilic addition reaction; the nucleophilic OH adds to the chromium. The chromium group, now bonded to oxygen, acts as a leaving group, allowing the elimination step to proceed to form the π bond of the new carbonyl functional group.

$HCrO_3^+$ reacts with alcohol

leaving group on alcohol oxygen

elimination

equilibrium provides a small supply of $HCrO_3^+$, a powerful electrophile

When these reagents are used to oxidize primary alcohols, the resulting aldehyde is prone to further oxidation and the formation of a carboxylic acid. Aldehydes readily form hydrates (geminal diols) in the presence of water (Chapter 7). A second oxidation will occur if one of the OH groups of the hydrate reacts with the oxidizing agent. In this way, the intermediate aldehyde becomes further oxidized into the corresponding carboxylic acid.

ORGANIC CHEMWARE
12.7 Oxidation of alcohols

hydrate provides another OH and a removable hydrogen for further oxidation by the same mechanism as the first oxidation

To prevent the further oxidation of the aldehyde to the carboxylic acid, the oxidization of a primary alcohol needs to be done under anhydrous (no water) conditions. Such conditions inhibit the hydration of the aldehyde, and so the oxidation stops at the aldehyde stage. The required anhydrous conditions can be created with the use of an oxidizing reagent, such as pyridinium chlorochromate (PCC), which is soluble in non-polar organic solvents such as CH_2Cl_2.

Other reagents that can be used to create anhydrous conditions include pyridinium dichromate (PDC) and bromine with a non-nucleophilic base (Br_2 + $NaHCO_3$) in anhydrous methanol.

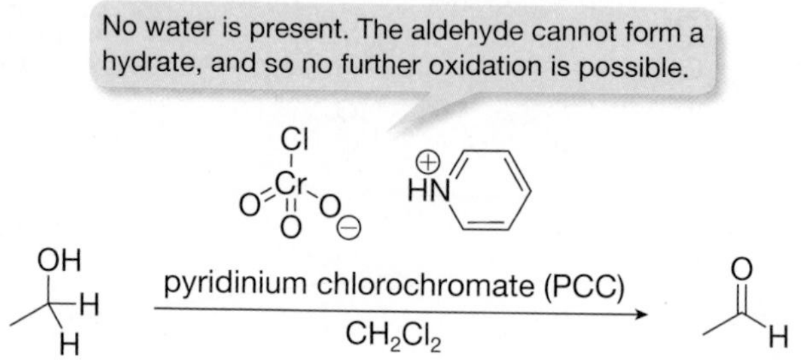

No water is present. The aldehyde cannot form a hydrate, and so no further oxidation is possible.

pyridinium chlorochromate (PCC)

CH_2Cl_2

CHECKPOINT 12.7

You should now be able to predict the products of alcohol oxidation and draw mechanisms for their reactions.

SOLVED PROBLEM

Predict the product of the following reaction, and draw a mechanism for its formation.

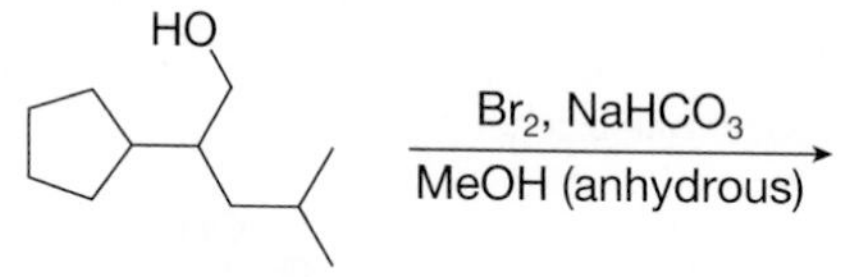

Br_2, $NaHCO_3$

MeOH (anhydrous)

STEP 1: Determine the roles of the reactants and the type of reaction taking place.

The reagents in this reaction (molecular bromine with a non-nucleophilic base) make a good oxidizing agent, and the starting material has a primary alcohol that can be oxidized to either an aldehyde or a carboxylic acid. Since no water is present (the solvent used is anhydrous), this suggests that the oxidation will stop at the aldehyde rather than continue to the carboxylic acid.

STEP 2: Draw a mechanism for the first step of the reaction (the installation of the leaving group).

One of the nucleophilic lone pairs on the OH oxygen shares its electrons with the electrophilic bromine, causing the Br–Br σ bond to break. A molecule of base then removes a proton from the oxonium ion.

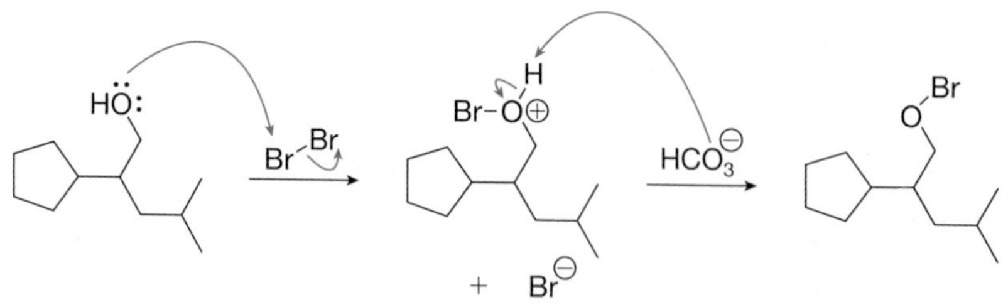

STEP 3: Draw a mechanism for the second step of the reaction (the oxidation).

The O–Br bond in the intermediate is weak and can readily break with bromine, creating a leaving group. A molecule of weak base therefore attacks one of the two hydrogens that are β to the Br leaving group, causing the C–H σ bond to break. The electrons from the C–H bond flow toward the oxygen, creating a new C=O π bond and causing the leaving group to leave as a bromide ion. The resulting product is an aldehyde, which in the absence of any water does not become further oxidized.

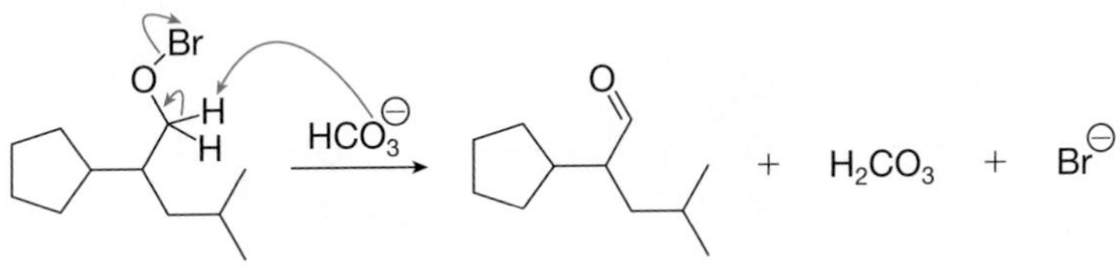

$$\xrightarrow{HCO_3^{\ominus}} \quad + \ H_2CO_3 \ + \ Br^{\ominus}$$

PRACTICE PROBLEM

12.13 a) Predict the products of the following reactions and draw mechanisms showing their formation.

i) 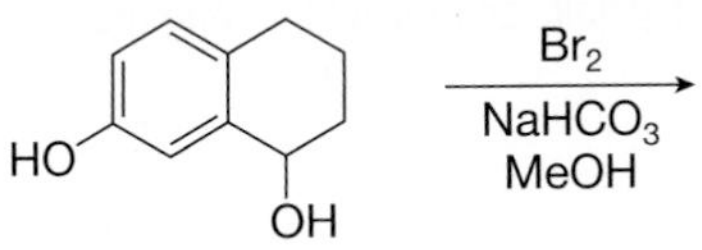 $\xrightarrow[\text{NaHCO}_3 \\ \text{MeOH}]{Br_2}$

ii) 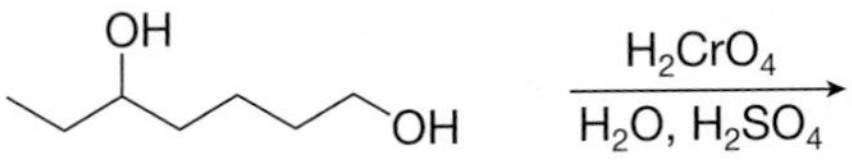 $\xrightarrow[H_2O,\ H_2SO_4]{H_2CrO_4}$

iii) 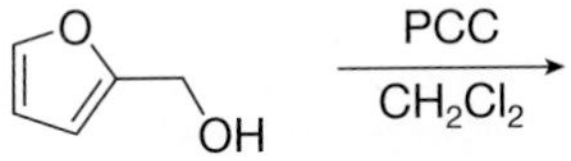 $\xrightarrow[CH_2Cl_2]{PCC}$

iv) 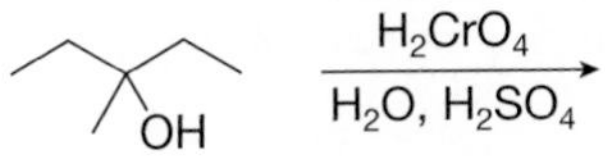 $\xrightarrow[H_2O,\ H_2SO_4]{H_2CrO_4}$

b) Suggest reagents and conditions to carry out the following transformations.

i)

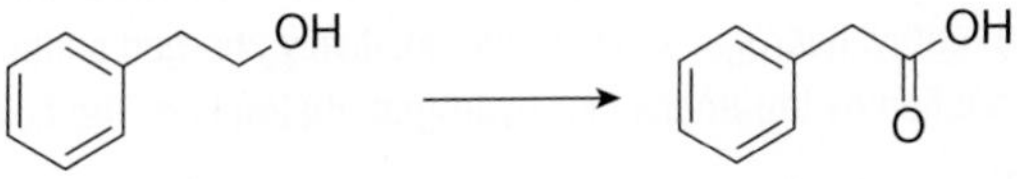

ii)

iii)

INTEGRATE THE SKILL

12.14 Propose a series of reagents that could be used to transform 4-methylpent-1-ene into the following compounds. More than one step is required.

a) 4-methylpentan-2-one
b) 4-methylpentanoic acid
c) 5-methylhexan-2-ol

DID YOU KNOW?

The oxidation of primary alcohols to form aldehydes is a common reaction in organic chemistry and requires methods that give very consistent results. One oxidation that is very reliable is the Swern oxidation, which consists of three separate reactions carried out in the same reaction flask! The first step in the sequence involves the mixing of oxalyl chloride and DMSO at a very low temperature (−78 °C). The two components react instantly, producing a chlorosulfonium ion (R_2SCl^+) and releasing a large quantity of carbon monoxide and carbon dioxide gases.

Continued

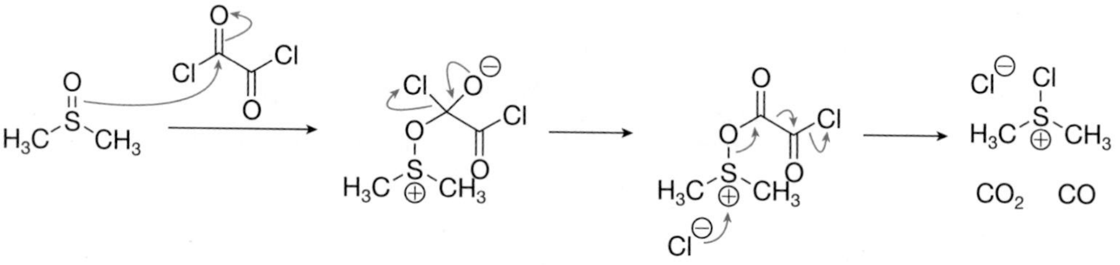

In the second step, a solution of the alcohol is added, which causes a nucleophilic displacement between the chlorosulfonium intermediate and the alcohol oxygen, which installs an excellent leaving group on this oxygen.

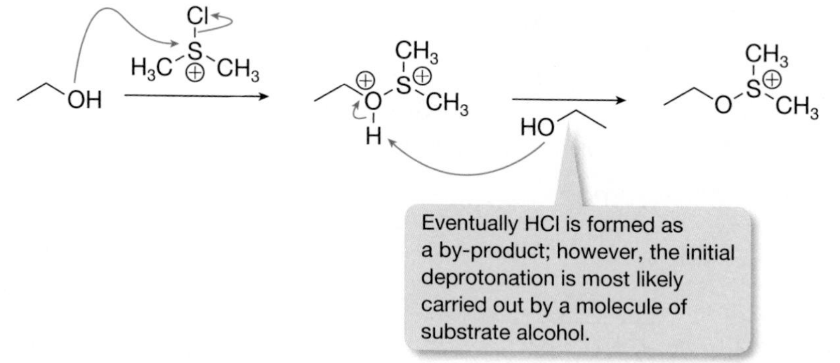

Eventually HCl is formed as a by-product; however, the initial deprotonation is most likely carried out by a molecule of substrate alcohol.

In the third step, a hindered base such as Et_3N is added to carry out the oxidation. At this point the mechanism involves the abstraction of one of the methyl hydrogens beside the positively charged sulfur to form an ylide, which then completes the oxidation reaction by intramolecular hydrogen abstraction. The flow of electrons is similar to that of the Hofmann elimination.

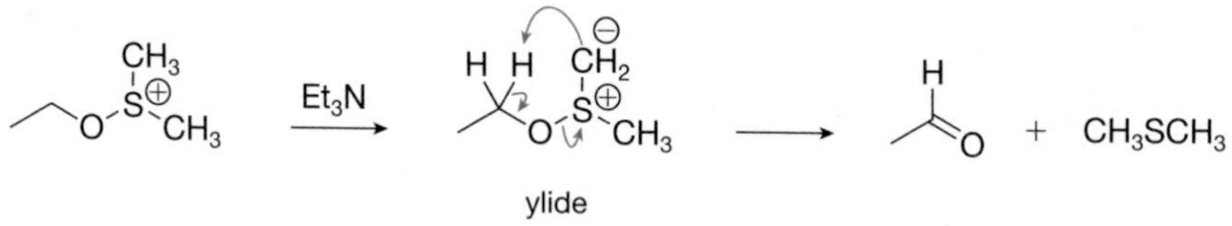

ylide

12.8 Patterns in Eliminations and Oxidations

Eliminations follow one of two general mechanisms, both involving the removal of a hydrogen and a leaving group to form a π bond. The E1 mechanism happens in two steps: the leaving group leaves to form a carbocation, which then forms a π bond as a base removes a proton in a separate step. Base catalyzed eliminations, such as the dehydrohalogenation reaction, proceed to form a carbocation, that then eliminates in the presence of a base. Dehydrations usually involve acid catalysis. This process adds a protonation pre-step, but the elimination itself follows the same electron movements as other E1 processes.

The E2 mechanism, involves a similar flow of electrons as the E1 reaction, but happens in a single concerted step. Strong base promotes reactions such as the dehydrohalogenation, which follows this general pattern. Dehydrations according to E2 kinetics usually require the installation of a leaving group such as a tosylate or meylate. The elimination itself follows an E2 pathway.

Similar reactions involving heteroatoms also form π bonds, and these are called oxidation reactions. The main difference between elimination reactions and oxidation reactions is the role of the oxidant, which attaches a leaving group to the heteroatom as part of the sequence. Chromium reagents are very common oxidants that can be used to perform a great many organic oxidations.

E1 elimination

base; H; LG; H; ⊕; + HBase + $LG^{\ominus}$

$HCO_3^{\ominus}$; H; Br; H; ⊕; + H_2CO_3 + $Br^{\ominus}$

H; OH; $H_3O^{\oplus}$; H_2O; H; $OH_2^{\oplus}$; H; ⊕; + $H_3O^{\oplus}$ + H_2O

E2 elimination

base; H; LG; + HBase + $LG^{\ominus}$

$CH_3O^{\ominus}$; H; Br; + CH_3OH + $Br^{\ominus}$

H; OH; Et_3N; Cl; S; Ar; O; O; $CH_3O^{\ominus}$; H; OTs; + CH_3OH + $TsO^{\ominus}$

Oxidation

H; H; O; LG_1; LG_2; base; H; O; LG_1; O; + HBase + $LG_1^{\ominus}$ $LG_2^{\ominus}$

H; H; O; O; Cr; O; $OH^{\oplus}$; H_2O; H; O; Cr; O; O; OH; O; + $H_3O^{\oplus}$ + $HCrO_3^{\ominus}$

H; OH; H; O; O; Cr; O; $OH^{\oplus}$; H_2O; H; OH; O; Cr; O; O; OH; OH; O; + $H_3O^{\oplus}$ + $HCrO_3^{\ominus}$

The chromium species becomes a leaving group which forms a π bond after an E2-type electron flow. Some functional groups can undergo more than one level of oxidation. Primary alcohols, for example, are first oxidized to form aldehydes. In the presence of water these groups are transformed into hydrates which can undergo a second oxidation step to form carboxylic acids. The movements of electrons in this second oxidation is consistent with the first oxidation.

Bringing It Together

Living organisms carry out oxidations and reductions using co-factors to transport the electrons that are either lost or gained in these processes. Two important co-factors are NAD^+ and NADP. Each of these molecules transports electron pairs as a formal unit of a hydride (H^-).

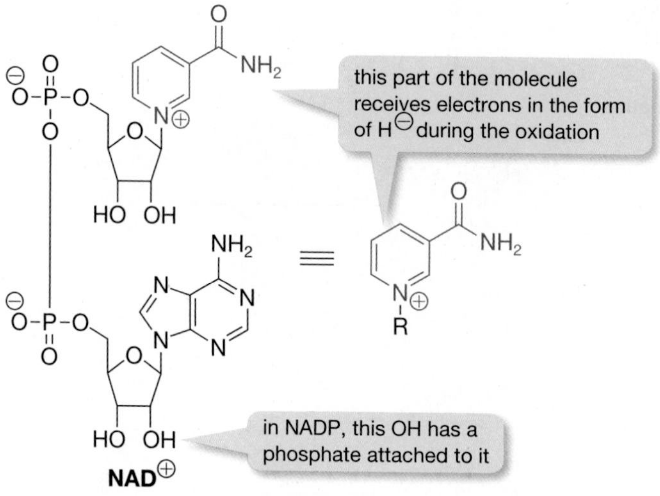

this part of the molecule receives electrons in the form of H⊖ during the oxidation

in NADP, this OH has a phosphate attached to it

NAD⊕

Living cells have enzymes called alcohol dehydrogenases that oxidize an alcohol into a carbonyl compound. These enzymes bind not only to the alcohol group but also to a molecule of the appropriate co-factor (NAD^+, for example). The enzyme also has a functionality that acts as a base and makes possible the elimination required to oxidize the alcohol. In biological oxidations, the electron flow involves an elimination in which the leaving group (hydride H^-) departs from the *carbon* rather than from the oxygen.

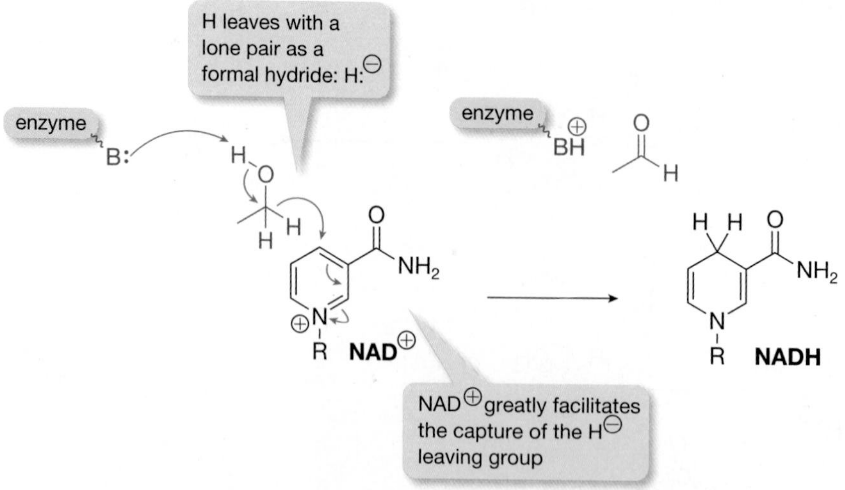

H leaves with a lone pair as a formal hydride: H:⊖

enzyme

enzyme

NAD⊕

NADH

NAD⊕ greatly facilitates the capture of the H⊖ leaving group

Because co-factor NAD^+ has a positively charged functional group, NAD^+ can accept electrons (in the form of a hydride) from the alcohol and carry out the oxidation. In the process, the co-factor NAD^+ becomes NADH (or NADPH), which living cells use in reductions on other molecules by donating electrons (in the form of hydride) and regenerating the aromaticity of their positively charged rings.

In the oxidation of alcohols, the electron flow is an elimination reaction, but the direction of the electron flow is reversed with respect to the carbon and heteroatom: the leaving group departs from the carbon instead of from the heteroatom. This occurs because the enzyme that carries out the reaction creates a special chemical environment that favours this. However, in essence, the reaction is an elimination/oxidation reaction.

You Can Now

- Draw mechanisms for the E1 and E2 elimination reactions of alkyl halides and alcohols (dehydration reactions).
- Distinguish between E1 and E2 reaction mechanisms using rate equations and reaction coordinate diagrams.
- Predict the major products formed from elimination reactions, including their regiochemistry and stereochemistry.
- Predict whether a given set of reactants and conditions is more likely to favour E1, E2, S_N1, or S_N2 reactivity.
- Incorporate substitutions and elimination reactions into the synthesis of small molecules, while controlling the reaction so that a single product is formed.
- Predict the products of alcohol oxidation and draw mechanisms for their formation.

A Mechanistic Re-View

ELIMINATIONS

E2 Elimination of halogens (dehydrohalogenation)

H H, Br → (CH_3ONa) → H

$CH_3O^{\ominus}$, H H, Br → H

Dehydration using E2 mechanism

H H, OH → (1) TsCl, py; 2) $NaOCH_3$) → H

H H, OH → (TsCl, py) → H H, OTs → (H_3C–$O^{\ominus}$) → H

E1 Elimination of halogens (dehydrohalogenation)

CH_3OH

CH_3OH

Dehydration using E1 mechanism

H_3O
Δ

OXIDATIONS

Oxidation of alcohols using Br_2

Br_2
$NaHCO_3$

$-HBr$

Oxidation of alcohols using chromium reagents

H_2CrO_4

Oxidation of alcohols (aldehydes) to acids

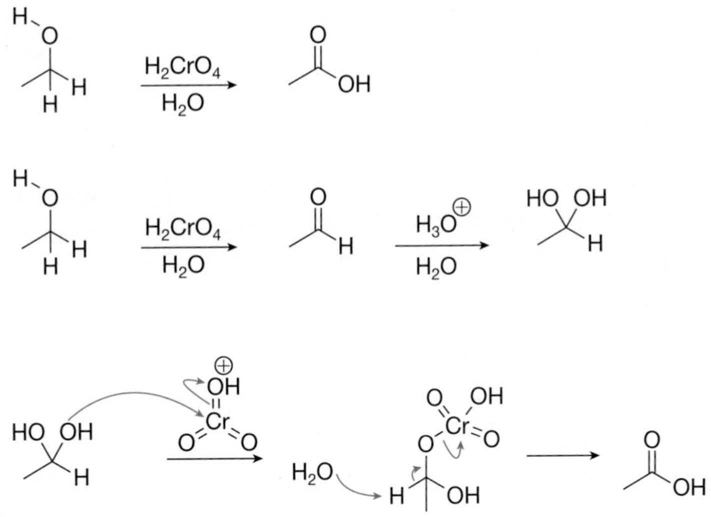

$\xrightarrow[H_2O]{H_2CrO_4}$

$\xrightarrow[H_2O]{H_2CrO_4}$ $\xrightarrow[H_2O]{H_3O^{\oplus}}$

Problems

12.15 Draw the products of the following elimination reactions. If more than one product is possible, indicate which is major and which is minor.

a) 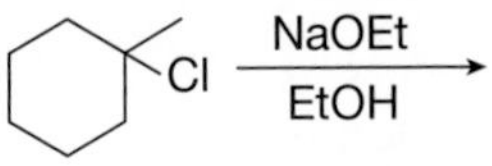 $\xrightarrow[EtOH]{NaOEt}$

b) 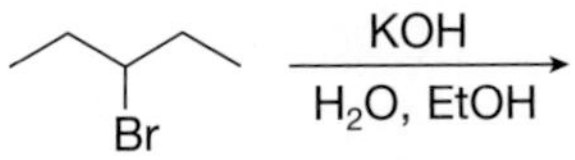 $\xrightarrow[H_2O, EtOH]{KOH}$

c) 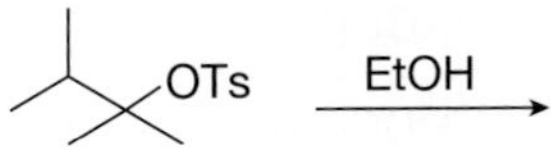 $\xrightarrow{EtOH}$

d) 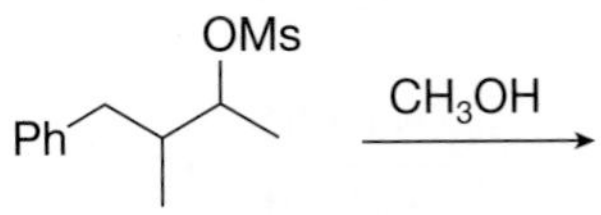 $\xrightarrow{CH_3OH}$

e) 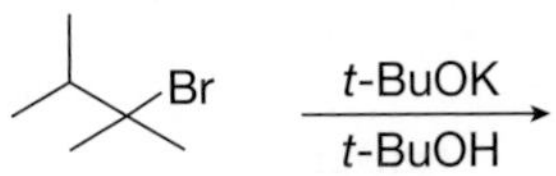 $\xrightarrow[t\text{-BuOH}]{t\text{-BuOK}}$

f) 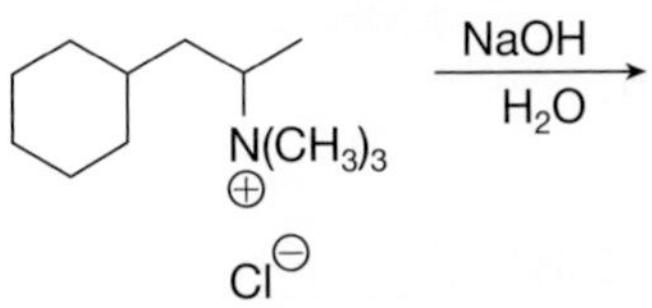 $\xrightarrow[H_2O]{NaOH}$

12.16 For the following reactions, draw the *expected major product*, including stereochemical information if relevant.

a) 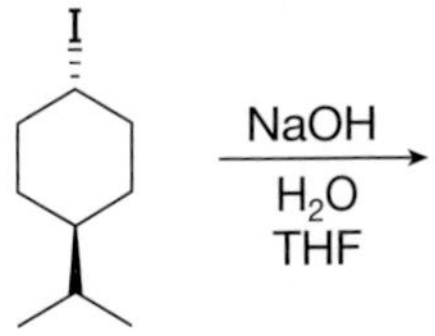 $\xrightarrow[H_2O, THF]{NaOH}$

b) 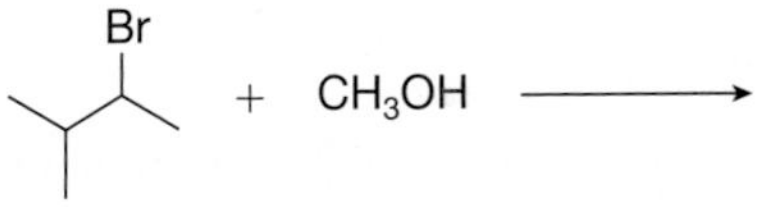 + CH_3OH ⟶

c) 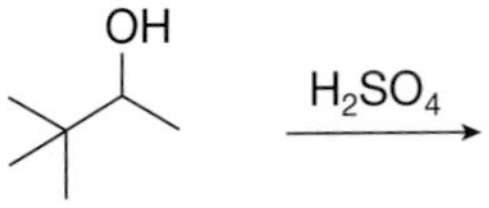 $\xrightarrow{H_2SO_4}$

d) 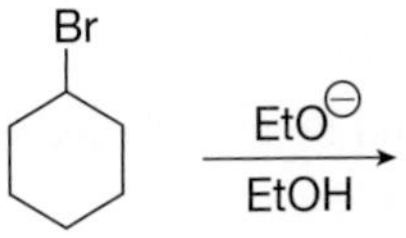 $\xrightarrow[EtOH]{EtO^{\ominus}}$

e) 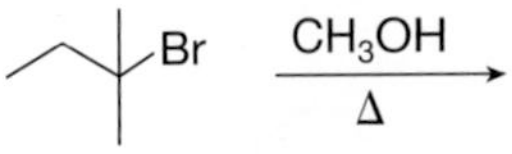 $\xrightarrow[\Delta]{CH_3OH}$

f) 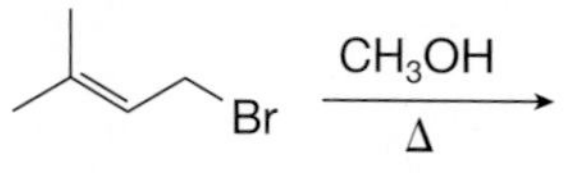 Br $\xrightarrow[\Delta]{CH_3OH}$

g) 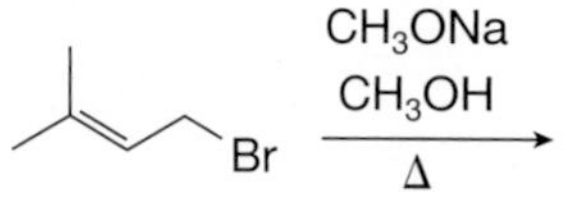 Br $\xrightarrow[\Delta]{CH_3ONa,\ CH_3OH}$

12.17 Provide a stepwise mechanism for the following reactions. Note that mechanisms include all intermediate species, all formal charges, and arrows showing the movement of electrons.

a) 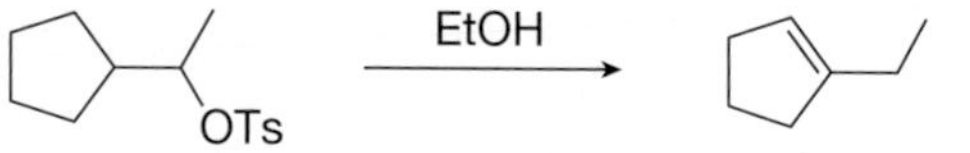 OTs $\xrightarrow{EtOH}$

b) 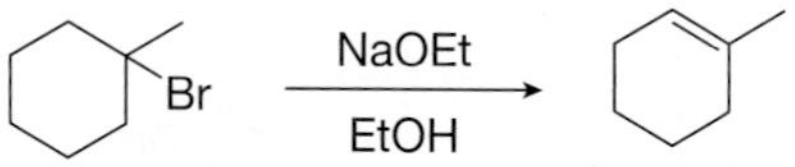 Br $\xrightarrow[EtOH]{NaOEt}$

12.18 If *t*-BuOK/*t*-BuOH were used in Question 12.17b instead of NaOEt/EtOH, would the product be the same or different? Explain.

12.19 Which of the following bases would give the most Hofmann product for the reaction?

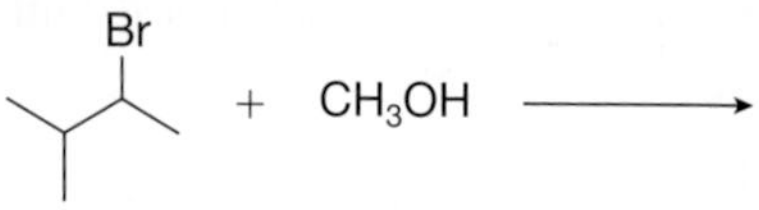

Br + CH_3OH ⟶

a) KOH
b) $NaOCH_3$
c) NaOEt
d) $NaOCH(CH_3)_2$
e) $KOC(CH_3)_3$

12.20 Draw all possible products (substitution and elimination) for the following reaction.

Cl $\xrightarrow[heat]{CH_3OH}$

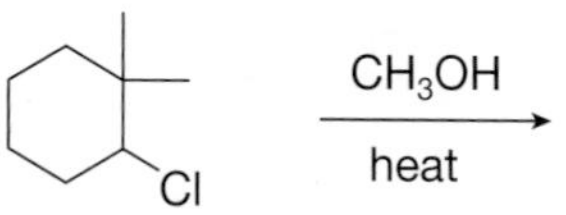

12.21 Suggest reagents that could be used to accomplish the following transformations.

a)  OH ⟶

b)  HO ⟶ O

c) 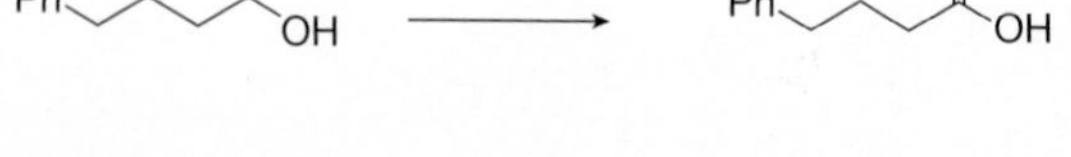 Ph OH ⟶ Ph O OH

d)  Br ⟶

e)  TsO ⟶

12.22 Show all the steps in the mechanism for this reaction, including the formation of *each* product.

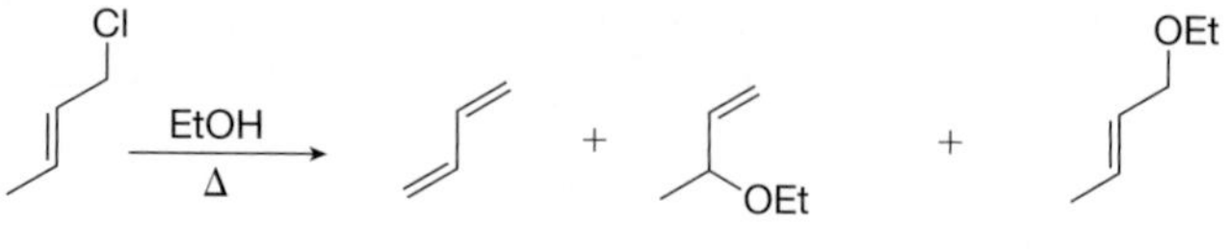

Cl $\xrightarrow[\Delta]{EtOH}$ + OEt + OEt

12.23 Show the product of the following reactions, including stereochemistry.

a) 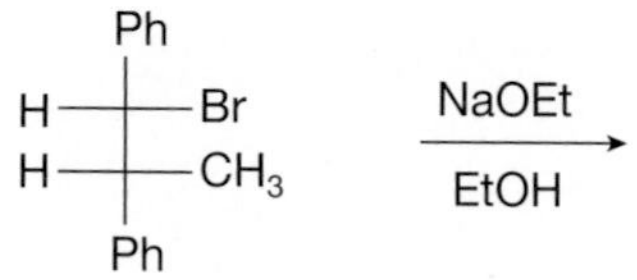 Ph, H, Br, H, CH_3, Ph $\xrightarrow[EtOH]{NaOEt}$

b) 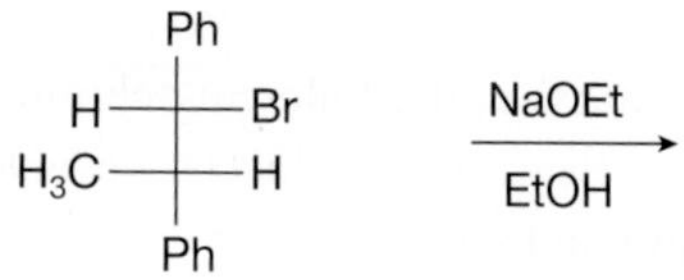 Ph, H, Br, H_3C, H, Ph $\xrightarrow[EtOH]{NaOEt}$

c) 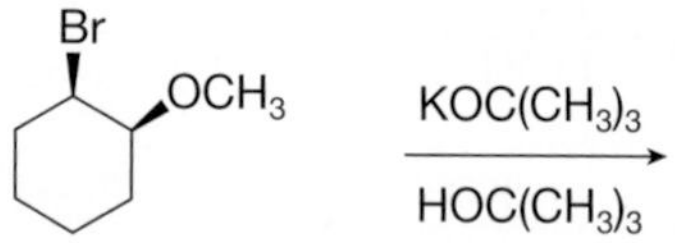 Br, OCH_3 $\xrightarrow[HOC(CH_3)_3]{KOC(CH_3)_3}$

d) 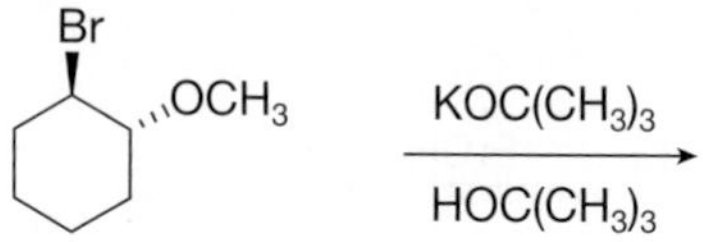 Br, OCH_3 $\xrightarrow[HOC(CH_3)_3]{KOC(CH_3)_3}$

e) 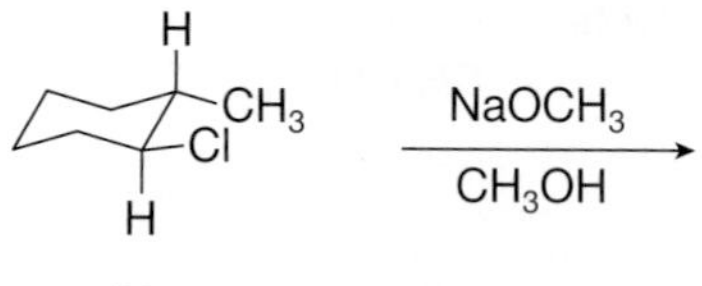 H, CH_3, Cl, H $\xrightarrow[CH_3OH]{NaOCH_3}$

f) 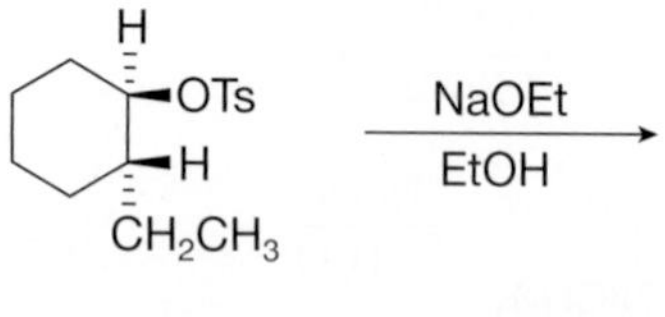 H, OTs, H, CH_2CH_3 $\xrightarrow[EtOH]{NaOEt}$

g) 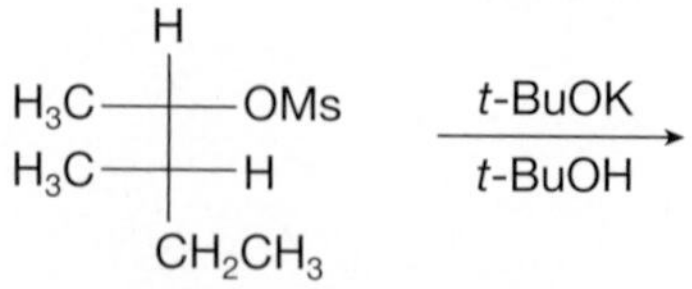 H, H_3C, OMs, H_3C, H, CH_2CH_3 $\xrightarrow[t\text{-BuOH}]{t\text{-BuOK}}$

h) CH_2CH_3, H_3C, H, H_3C, Br, Ph $\xrightarrow[EtOH]{NaOEt}$

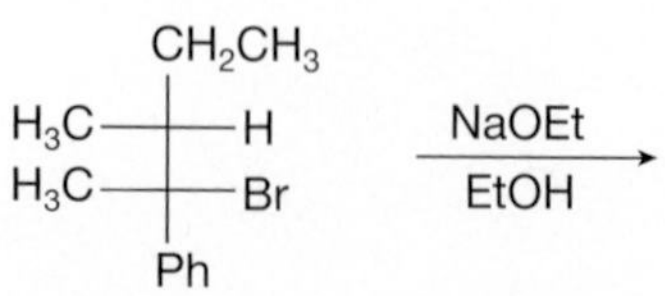

12.24 When heated in ethanol, this alkyl halide gives two substitution and two elimination products. Show the structures of these products and the mechanism for their formation.

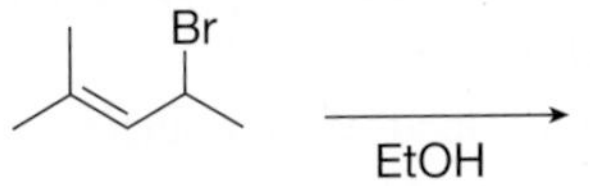

Br

EtOH

12.25 This elimination reaction gives a single product. Show its structure, and explain why it is the only product formed.

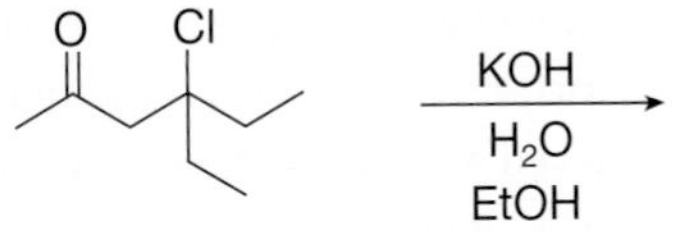

O Cl

KOH
H_2O
EtOH

12.26 Styrene can be converted to phenylacetylene in two steps, starting with bromination. Provide the structure of compound A and the reagents required to convert it to the desired product.

Br_2
CH_2Cl_2
?

compound A

12.27 For capital punishment in ancient Greece and Rome, hemlock was used as a poison; this is how Socrates was executed. Coniine, the active compound in hemlock, causes paralysis of the motor nerves and leads to death by asphyxiation. In 1881, to establish the structure of coniine, A.W. Hoffmann used the following reaction sequence as one of the degradation steps. Show the mechanism and the products formed in this reaction.

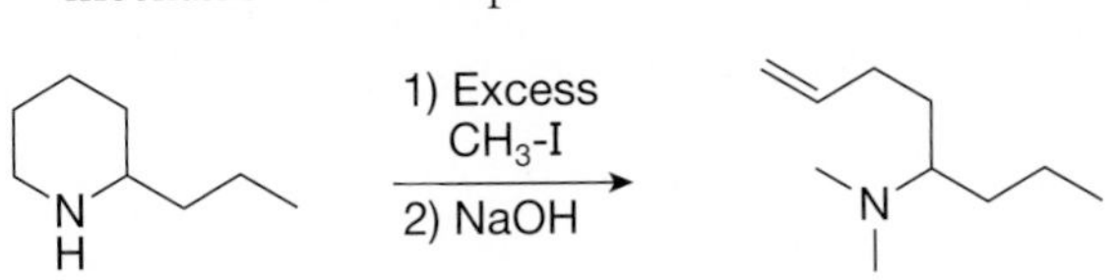

1) Excess
CH_3-I
2) NaOH

coniine

12.28 Predict the product of the following reactions.

a) 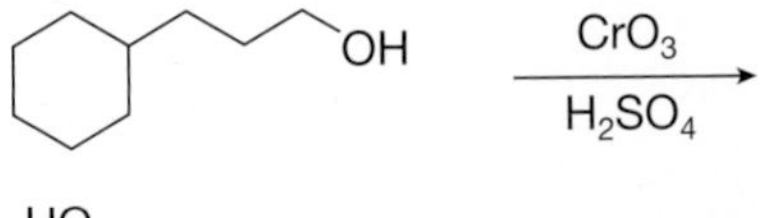 OH $\xrightarrow[H_2SO_4]{CrO_3}$

b) 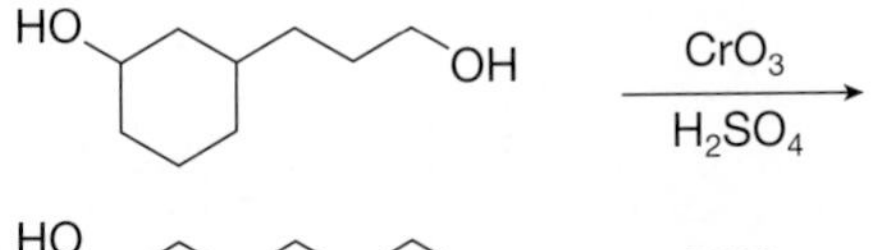 HO OH $\xrightarrow[H_2SO_4]{CrO_3}$

c) 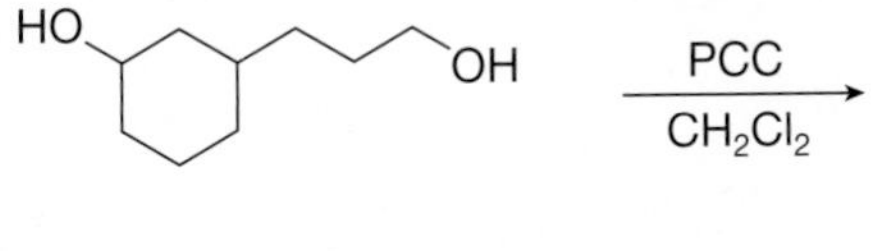 HO OH $\xrightarrow[CH_2Cl_2]{PCC}$

d) 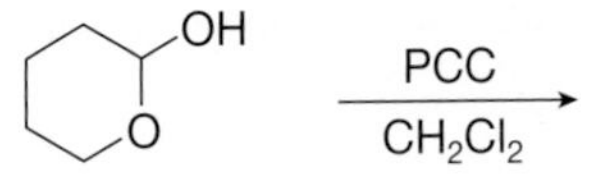 OH O $\xrightarrow[CH_2Cl_2]{PCC}$

e) 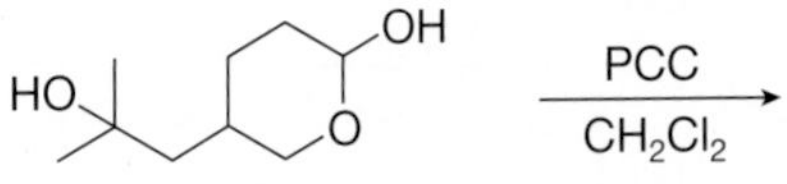 HO OH O $\xrightarrow[CH_2Cl_2]{PCC}$

12.29 Suggest an oxidizing agent that could be used to perform the following reactions.

a)  OH OH ⟶ O O

b) 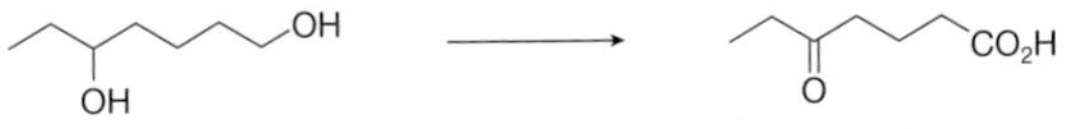 OH OH ⟶ O CO_2H

c) 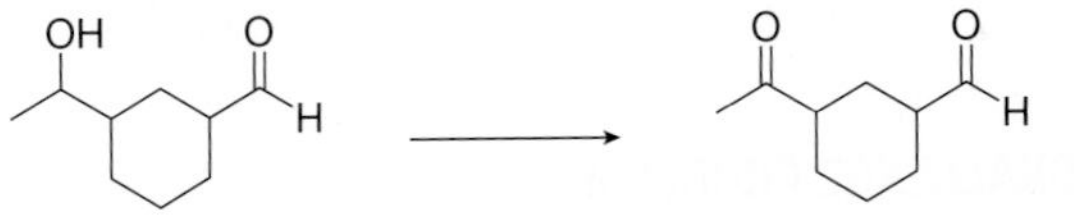 OH O H ⟶ O O H

d)  OH O H ⟶ O O OH

12.30 Use mechanisms to predict the product of the following oxidations.

a) 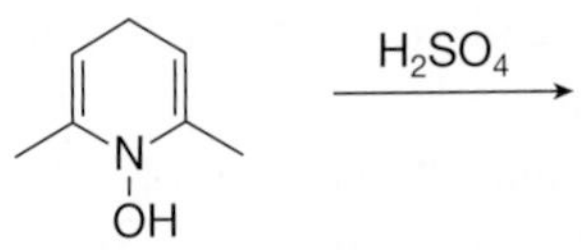 N OH $\xrightarrow{H_2SO_4}$

b) 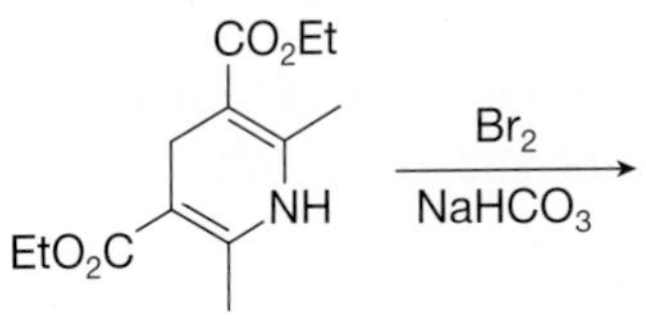 CO_2Et NH EtO_2C $\xrightarrow[NaHCO_3]{Br_2}$

12.31 Provide mechanisms to explain the following reaction.

HO OH $\xrightarrow[H_2SO_4]{Na_2Cr_2O_7}$ O O

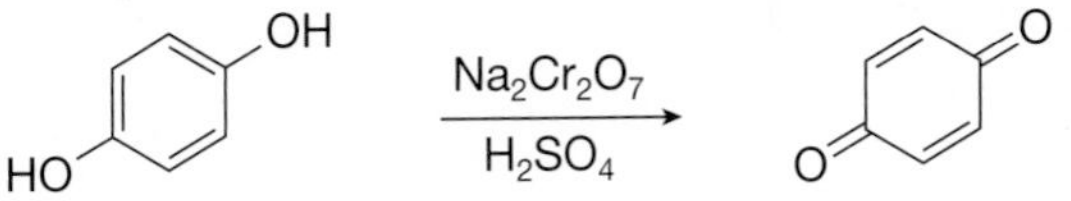

MCAT STYLE PROBLEMS

12.32 What kind of reaction is described by the following equation?

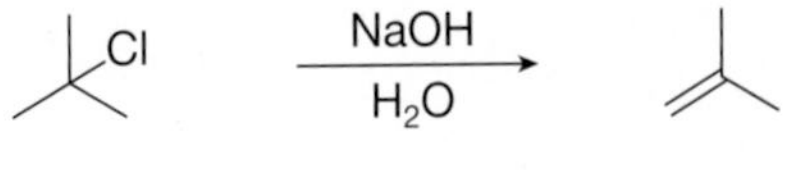

Cl $\xrightarrow[H_2O]{NaOH}$

a) S_N1
b) S_N2
c) E1
d) E2

12.33 Which of the following represents an oxidation reaction?

a) cyclohexanol (OH) $\xrightarrow{HBr}$ bromocyclohexane (Br)

b) cyclohexanol (OH) $\xrightarrow[K_2CO_3]{Br_2}$ cyclohexanone (O)

c) cyclohexanol (OH) $\xrightarrow{H_2SO_4}$ cyclohexene

d) cyclohexanol (OH) $\xrightarrow{NaH}$ NaO–cyclohexane

CHALLENGE PROBLEM

12.34 You need to make (*Z*)-(1-methyl-1-propenyl)benzene using H_2O and heat, from (*R*)-(1-bromo-1-methylpropyl)benzene. After the reaction is complete, you have three compounds present (two minor components and one major).

- Minor compound 1 has a molecular mass of 213.11 g/mol and is 56.36% C, 6.15% H; it is optically active.
- Minor compound 2 has a molecular mass of 132.20 g/mol and is 90.85% C, 9.15% H; it is optically inactive.
- Major compound has a molecular mass of 150.22 g/mol and is 79.96% C, 9.39% H; it is optically active.

Use the provided information to answer the following questions.

a) Draw (*R*)-(1-bromo-1-methylpropyl)benzene (the starting material).
b) Draw (*Z*)-(1-methyl-1-propenyl)benzene (the desired product).
c) What unwanted side-product(s) might you obtain, in competition with the desired elimination product?
d) What is the identity of minor product 1?
e) How would you alter conditions to eliminate minor product 1?
f) What is the identity of minor compound 2?
g) Draw a mechanism for the formation of minor compound 2 from (*R*)-4-(1-methylpropyl)toluene, using curved arrows to show electron movement.
h) What is the identity of the major product?
i) Draw a mechanism for the formation of major product from (*R*)-4-(1-bromo-1-methylpropyl)toluene, using curved arrows to show electron movement.
j) How would you alter the conditions or reagents to reverse the ratio between the major product and the minor compound 2? Be careful that in so doing you do not increase the amount of minor compound 1.

Structure Determination I

NUCLEAR MAGNETIC RESONANCE SPECTROSCOPY

CHAPTER OUTLINE

The phantasmal poison frog secretes alkaloids epiquinamide and epibatidine from its skin. Because these compounds have very different physiological effects, it is important to identify which structure is which.

13.1 Why It Matters

Many biological processes begin when a small signalling molecule binds to a protein through intermolecular forces and causes the protein to change shape. Nicotinic acetylcholine receptors (nAChRs) are proteins embedded in cell membranes that bind the biomolecule acetylcholine. When acetylcholine binds, nAChRs change their shape, opening to allow sodium or potassium ions to cross the membrane through an ion channel. An open nicotinic acetylcholine receptor is important for nerve signalling both in brain function and in other processes in the body, including muscle contraction. Molecules other than acetylcholine, such as nicotine, can also activate acetylcholine receptors. Nicotine binding to and opening nAChR ion channels in the brain leads to the calming and appetite-suppression effects of smoking and addiction. So the ability to better control these effects makes molecules that bind to nAChRs valuable as potential medications.

In 2003, a molecule called epiquinamide, isolated from the phantasmal poison frog, *Epipedobates tricolor*, was found to activate one subtype of nAChR. However, other researchers could not reproduce the original results, either with more epiquinamide isolated from frogs or with synthetic epiquinamide made in the lab.

Later, researchers discovered that even though the original batch of epiquinamide was greater than 99 percent pure, it was contaminated with 0.1 percent epibatidine, a molecule already known to activate the nAChRs.

epibatidine

epiquinamide

Even a very small amount of contamination in a medication can drastically change how it works. In this case, the contamination gave a false lead on a potential new drug. Depending on the contamination, however, an impurity could be dangerous, or even fatal. This is why the requirements to identify and remove impurities from medications are so strict.

Spectroscopy is the measurement of the interaction between a molecule and electromagnetic radiation, and reveals information about molecular properties, including structure.

Chemists use **spectroscopy** to determine the identity and amounts of compounds in a sample. When exposed to electromagnetic (EM) radiation, a molecule will absorb a radiation photon if the energy of the photon matches the energy gap between the ground state and an excited state of the molecule or nucleus within the molecule. Different molecules and nuclei have different and often distinctive energy gaps. Thus, the frequencies absorbed or emitted can yield information about the composition and structure of the compounds present.

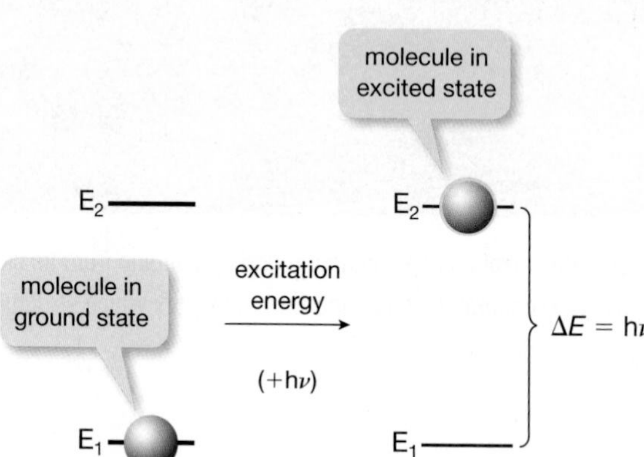

Every molecule has several types of ground and excited states, including vibrational, rotational, and electronic states. The energy gaps between these states correspond to different frequency ranges in the EM spectrum. This chapter describes nuclear magnetic resonance (NMR) spectroscopy, which uses radio frequency radiation. Spectroscopy that uses infrared radiation is discussed in Chapter 14. Ultraviolet and visible spectroscopy was discussed in Chapter 9.

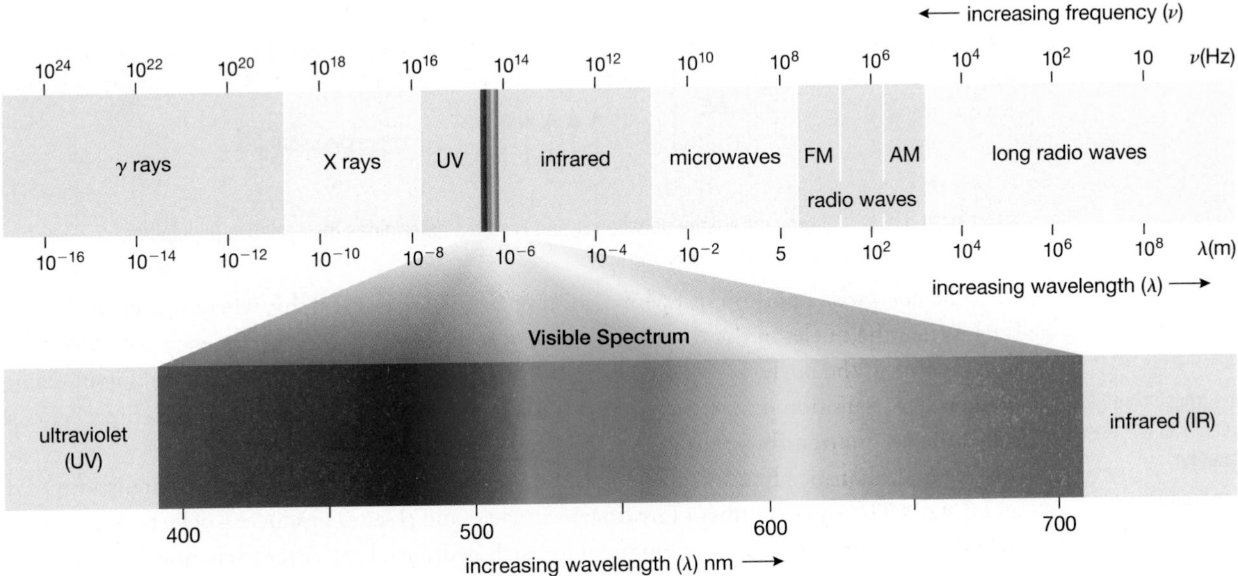

13.2 Magnetic Resonance in Organic Analyses

All atomic nuclei possess a quantum mechanical property called *spin*. Certain isotopes, including protons (^{1}H) and carbon-13 (^{13}C) have only two possible spin states: +½ or −½. The number of spin states that a nucleus has is determined by its atomic number and mass. Nuclei with even atomic numbers and masses have spins of zero. All other nuclei can occupy any of several spin states. Somewhat like the way a spinning charge generates a magnetic field, quantum spin produces a **magnetic moment** (μ).

The **magnetic moment** (μ) of a nucleus is a vector property related to the magnitude and direction of its spin.

How these nuclei respond to external magnetic fields can yield significant information about the structure of the molecule containing the nuclei. The magnetic moments of nuclei in an external magnetic field ($\mathbf{B}_0$) align either parallel to the external field (α, lower energy) or antiparallel to it (β, higher energy), as shown in Figure 13.1.

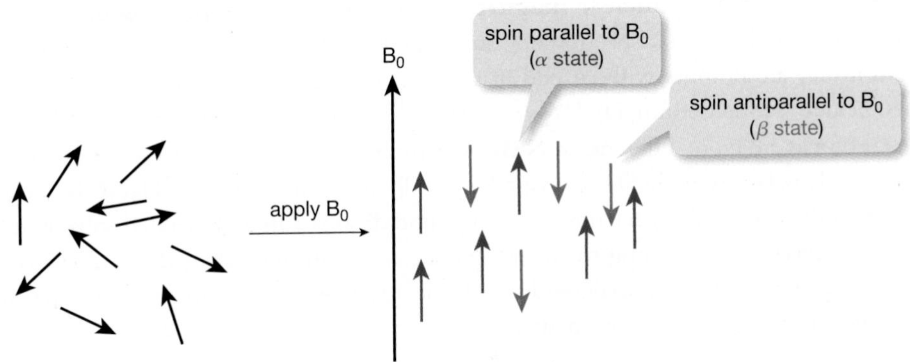

FIGURE 13.1 Effect of applying external magnetic field $\mathbf{B}_0$ to the magnetic moment of spin ±½ nuclei.

The energy difference (ΔE) between the α and β states is small, so the population of nuclei in the more stable α state at equilibrium is only slightly greater than the population in the β state (Figure 13.2). The magnitude of ΔE depends on the strength of the applied magnetic field, with a larger ΔE resulting from a stronger magnetic field.

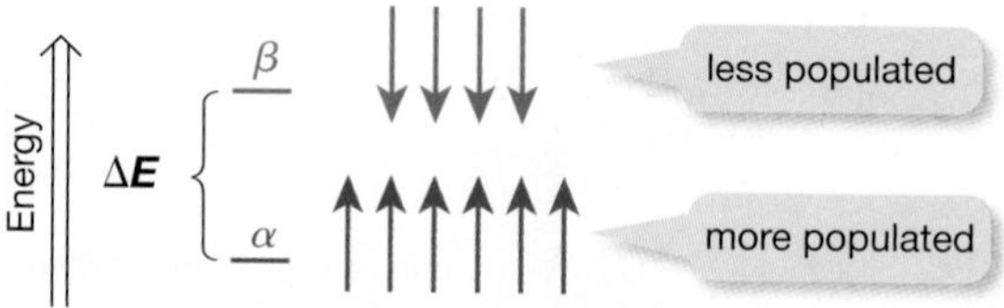

FIGURE 13.2 Energy difference between α and β states determines their relative populations.

Spin-flip is the transition of a nucleus between α and β states (low to high energy).

The **resonance frequency** of a signal is the frequency at which spin-flip occurs, and is proportional to the energy required.

The energy of a photon of EM radiation of frequency ν is $E = h\nu$, where h is Planck's constant. When the nuclei in the α state are exposed to EM radiation of a frequency corresponding to ΔE, some of the nuclei are excited from the α state to the higher-energy β state (Figure 13.3), while others transition downward from β to α. The frequency at which this **spin-flip** occurs is called the **resonance frequency**.

A rapid resonance frequency EM pulse applied to the sample equalizes the populations of α and β states. The equal number of hydrogens in the α and β states produces a detectable magnetic field that contains a frequency component for each hydrogen type. After excitation, this magnetic field weakens as the system relaxes back to equilibrium, a process called *free induction decay* (FID). During the relaxation stage, signal detection provides information about the magnetization experienced by each hydrogen type.

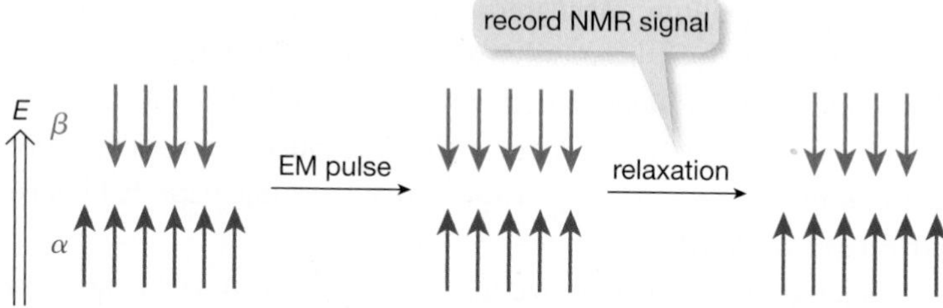

FIGURE 13.3 NMR pulse and signal generation.

Nuclear magnetic resonance (NMR) spectroscopy measures the energy differences between spinning nuclei in a magnetic field to determine the structures of molecules in solution.

The different chemical environments in a molecule cause variations in the magnetic field around the nuclei in the molecule. These variations cause the nuclei to absorb energy at different frequencies, measured using **nuclear magnetic resonance (NMR) spectroscopy**. The most common nuclei used in NMR spectroscopy are ^1H and ^{13}C. Other nuclei can provide an NMR spectrum if they are isotopes with an odd mass number, an odd atomic number, or both. Such isotopes include ^2H (deuterium, D), ^{14}N, ^{17}O, ^{19}F, and ^{31}P. Since the resonance frequency ranges for these isotopes are quite different, an NMR spectrometer can be tuned to selectively detect a particular nucleus. For example, ^2H resonances are well out of the range of a ^1H-NMR spectrum. Therefore, in the ^1H-NMR spectrum of a compound dissolved in a deuterated solvent, such as $CDCl_3$, the signals corresponding to the sample are much stronger than those of the solvent. NMR spectra always contain a small residual solvent signal, because it is impossible to replace all of the hydrogens in the solvent with deuterium.

DID YOU KNOW?

The EM frequencies used in NMR include the band used for FM radio broadcasts. Occasionally, an NMR instrument will consistently register an extra signal in ^{13}C-NMR spectra because of interference from a local radio station!

13.3 The NMR Instrument

A typical NMR instrument used for research has a large cylindrical superconducting magnet into which a sample tube can be inserted. The magnet must produce a strong, homogeneous magnetic field, which is most easily done when using superconducting magnets. To keep the magnet's conductors in a superconductive zero-resistance state, the magnet needs to be cooled with liquid helium. Coil antennas in the centre of the magnet serve as transmitters for exciting the sample and as detectors for recording the decay signals. NMR spectrometers that use permanent magnets instead of superconductors are smaller and much less costly, but have significantly lower resolution.

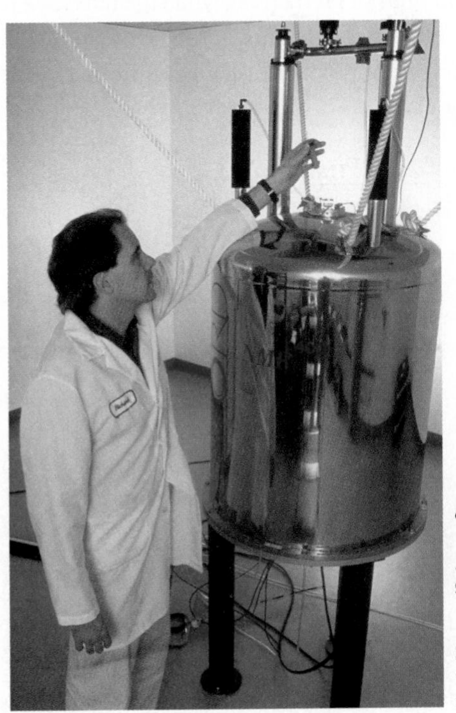

Hank Morgan/Science Source

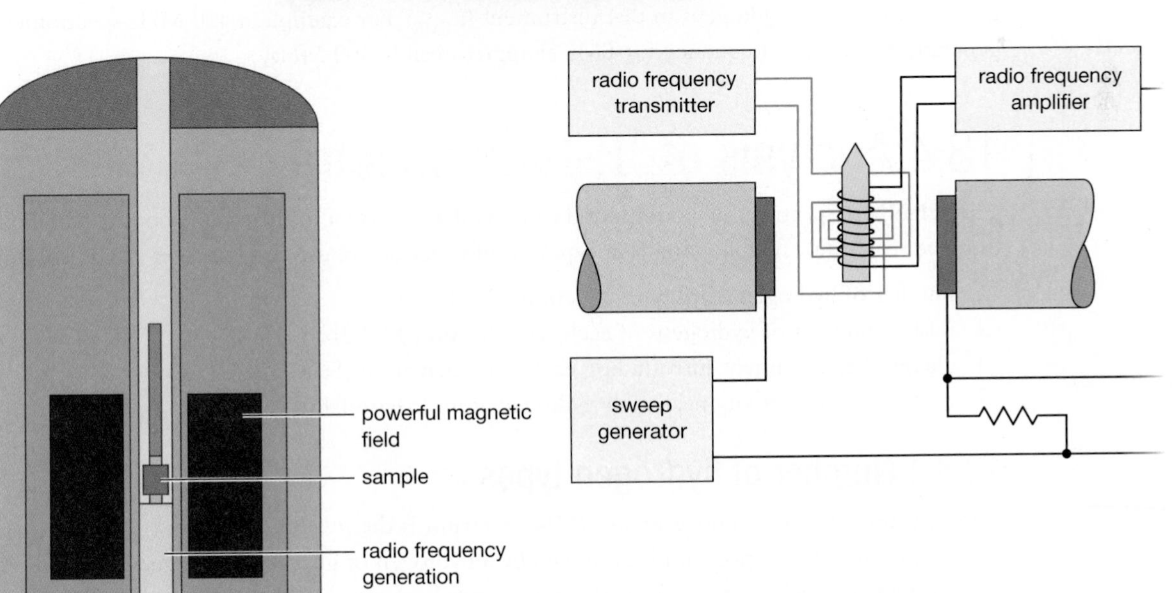

The NMR instrument detects signal strength as a function of the time. By using a Fourier transformation (a mathematical form of harmonic analysis), a computer can convert the time domain plot (FID) to a frequency domain plot (spectrum) which displays signal strength versus frequency. The frequencies at which the proton in each hydrogen atom resonates depend on the strength of the magnetic field used. It is impossible to manufacture magnets that produce identical frequencies; therefore, to compare data obtained on different instruments, there needs to be a relative frequency scale in which each resonance frequency is reported relative to a reference frequency. The difference between the sample frequency and the reference frequency is expressed in terms of the **chemical shift** (δ), which is the difference in the frequencies divided by the reference frequency. Typically, chemical shifts are expressed in parts-per-million (ppm) because the difference in resonance frequencies is in units of Hertz (Hz), while the reference resonates at MHz frequencies.

Chemical shift (δ) is the difference between the NMR resonance frequency of a sample and a reference frequency, tetramethylsilane.

Tetramethylsilane (TMS) is the compound used to provide the reference frequency for ^{1}H-NMR spectra.

For ^{1}H-NMR spectra, **tetramethylsilane** (TMS) is commonly used as a reference. Chemical shifts for ^{1}H-NMR spectra with TMS are calculated from equation 13.1, which produces a chemical shift of 0 ppm for TMS.

$$\frac{\nu_{obs} - \nu_{TMS}}{\nu_{TMS}} = \delta(\text{ppm}) \tag{13.1}$$

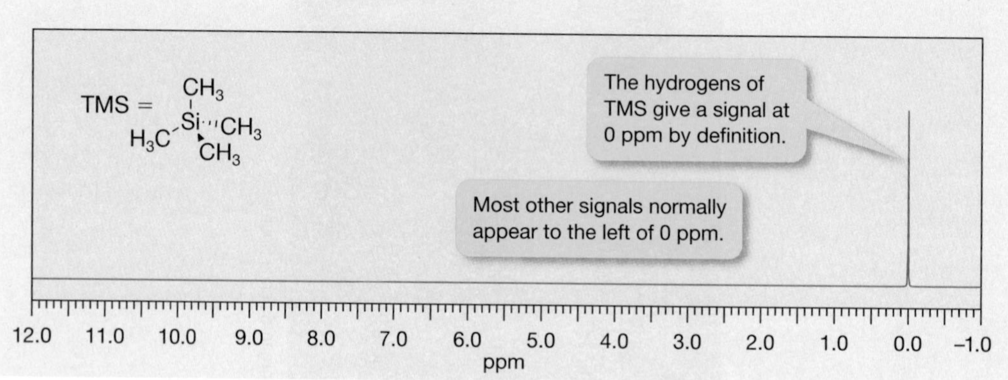

The strength of the magnets in NMR instruments is typically classified by the resonance frequency of the TMS hydrogens in that instrument (ν_{TMS}). For example, a 400 MHz spectrometer generates a resonance frequency for TMS at approximately 400 MHz.

13.4 Analysis of ^{1}H-NMR Spectra

The ^{1}H-NMR spectrum of a compound can reveal a variety of information about its composition and structure. The following four types of information are used to interpret NMR spectra:

1. number of hydrogen atom types (Section 13.4.1)
2. relative number of hydrogens of each type (Section 13.4.2)
3. chemical environment surrounding each hydrogen atom (Section 13.4.3)
4. molecular connectivity near each hydrogen atom (Section 13.4.4)

13.4.1 Number of hydrogen types

The first information to notice in an NMR spectrum is the *number* of different signals, or *resonances*. Each signal in a spectrum is produced by a hydrogen or group of hydrogens in a different chemical environment, which affects the energy gap between their α and β states. A signal may consist of single peak or a collection of peaks grouped closely together in a symmetric pattern.

For example, the following ^{1}H-NMR spectrum for bromoethane (CH_3CH_2Br) has two signals: one produced by the hydrogens in the CH_2 group, and the other from those in the CH_3 group. Both of these signals consist of a group of peaks.

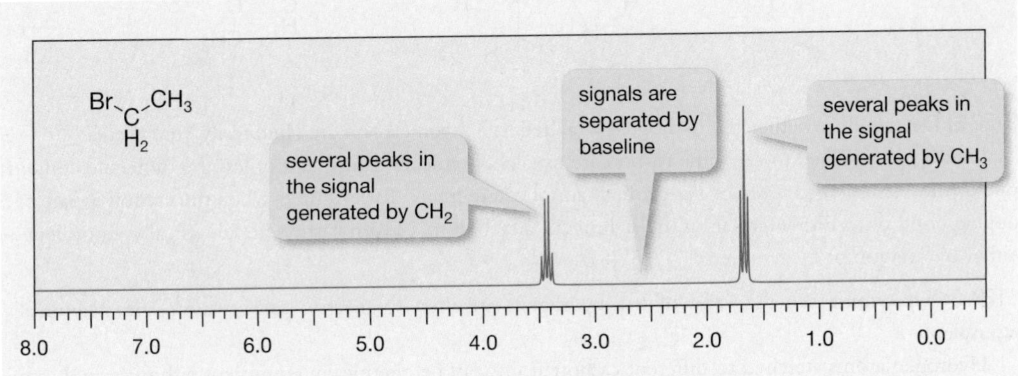

ORGANIC CHEMWARE
13.1 ^{1}H-NMR spectrum of bromoethane

The two hydrogens in the CH_2 group of CH_3CH_2Br are chemically equivalent by symmetry and have the same chemical shift, about 3.4 ppm. **Chemically equivalent hydrogens** are those in chemically identical environments and so have identical chemical shifts. Such hydrogen atoms are interchangeable by rotation about a single bond or by a plane of symmetry in a molecule.

The three hydrogens in the CH_3 group are chemically equivalent by rotation, and all have a chemical shift of about 1.7 ppm. The signals from the two hydrogen types in bromoethane are easy to distinguish since they are clearly separate; however, the hydrogen types in other compounds may produce overlapping signals.

Chemically equivalent hydrogens are in identical environments within a compound because they are interchangeable by the fast rotation of single bonds or by symmetry.

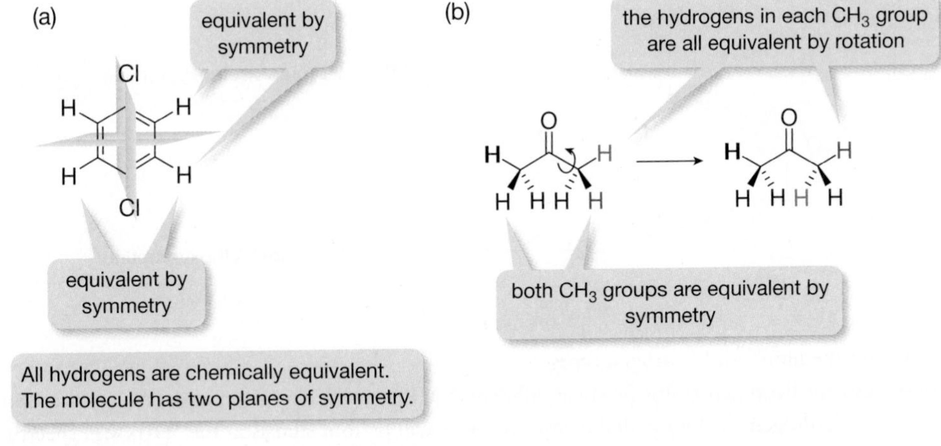

ORGANIC CHEMWARE
13.2 ^{1}H-NMR spectrum of 1,4-dichlorobenzene

CHECKPOINT 13.1

You should now be able to predict the number of different signals in a ^{1}H-NMR spectrum for a compound given its structure.

ORGANIC CHEMWARE
13.3 ^{1}H-NMR spectrum of 2-propanone

SOLVED PROBLEM

Determine the number of signals in the ^{1}H-NMR spectrum of the following compound.

ORGANIC CHEMWARE
13.4 ^{1}H-NMR spectrum of 2-butanone

STEP 1 (OPTIONAL): Redraw the structure of the compound to show all of the hydrogen atoms. You can partially expand either the Lewis structure or a line drawing.

STEP 2: Determine whether the hydrogens attached to a given carbon are chemically equivalent.

Hydrogens attached to the same carbon are usually chemically equivalent unless the molecule contains one or more chirality centres, or the groups cannot rotate freely (for example, when the carbon is part of a double bond or a ring). Here all of the hydrogens attached to a given carbon are chemically equivalent—either by rotation or by symmetry.

STEP 3: Determine whether any of the hydrogens attached to *different* carbon atoms are chemically equivalent.

Hydrogen atoms attached to different carbon atoms can be chemically equivalent either through symmetry or through rotation. This molecule has a plane of symmetry that makes the two methyl groups of the isopropyl group chemically equivalent. Expanding the Lewis structure at the isopropyl group and adding some perspective reveals a plane of symmetry in the plane of the page.

Another method for determining whether two protons are chemically equivalent is to replace each one in turn with different atom (e.g., a Br), and then name the resulting compounds. If the names are the same, the hydrogens are chemically equivalent. For example, replacing one of the hydrogens from either CH_3 on the isopropyl group yields 1-bromo-2-ethoxypropane. Therefore, the hydrogens in these methyl groups are chemically equivalent.

1-bromo-2-ethoxypropane

1-bromo-2-ethoxypropane

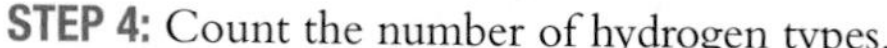

STUDENT TIP

The technique of replacing each hydrogen in turn with a different atom and then naming the resulting compounds works for hydrogens attached to the same carbon, as well as those attached to different carbons.

STEP 4: Count the number of hydrogen types.

Non-equivalent hydrogen atoms produce different NMR signals. The given structure has four sets of non-equivalent hydrogens (colour coded below), so there will be four signals in the ^{1}H-NMR spectrum.

PRACTICE PROBLEM

13.1 Predict the number of different signals in the ^{1}H-NMR spectrum for each of the following compounds.

a)

b) OH

c) O S

d)

e)

f) Br

INTEGRATE THE SKILL

13.2 The following ¹H-NMR spectrum is of a compound with the formula $C_5H_{10}O$. Which of the following structures is most likely to yield this spectrum? Explain your answer.

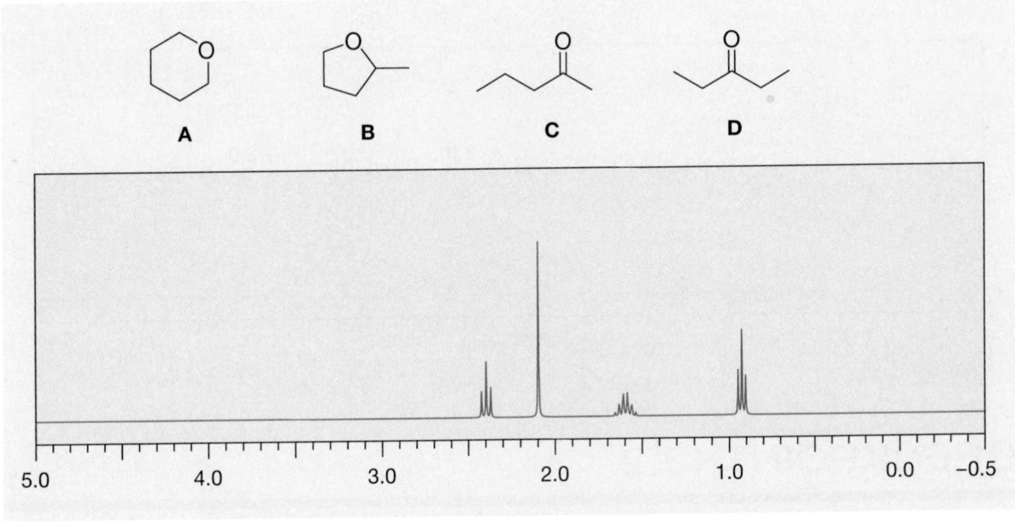

13.4.2 Relative number of hydrogens of each type

The intensity of each NMR signal corresponds to the *area* under the entire signal. The ratio of the areas under the signals indicates the relative amounts of each type of hydrogen atom producing that signal. NMR spectrometers **integrate** the signals to display the relative intensities. For example, integration of the spectrum of CH_3CH_2Br shows that the intensities of the two signals (areas under each signal) are in a 3:2 ratio; this corresponds to the relative number of hydrogens producing each signal. Diethyl ether ($CH_3CH_2OCH_2CH_3$) also produces two signals in a 3:2 integration ratio; however, the actual number of hydrogens producing each signal is six and four, respectively.

Integration refers to calculations of the area under a peak or set of peaks that determine the intensity of a signal in NMR spectrometry.

NMR spectrometers display the integrals either as a graph over each signal or as a numerical value, as shown in the following two spectra. To determine a relative intensity from an integration line, measure the vertical distance from the point where the graph line starts to curve up to the point where it levels out again (Figure 13.4).

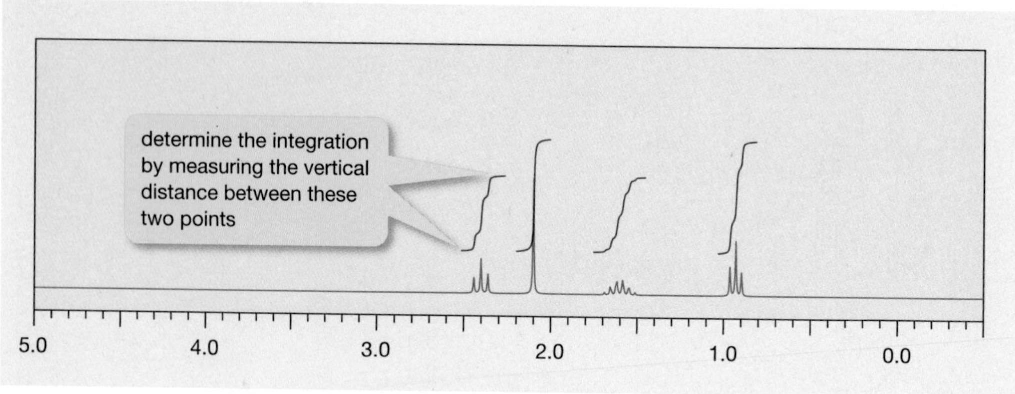

FIGURE 13.4 ^{1}H-NMR spectrum of 2-pentanone with integration graphs.

For example, integral graphs with heights of 18 mm and 12 mm indicate that the ratio of the corresponding types of hydrogen is 18:12, which simplifies to 3:2.

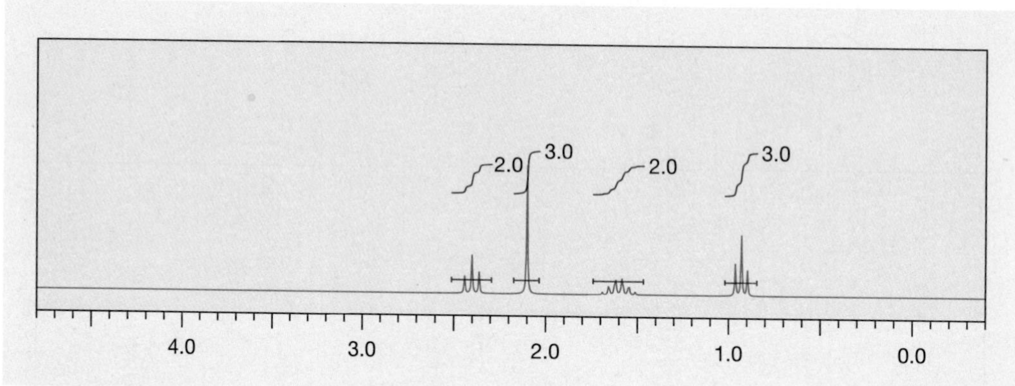

CHECKPOINT 13.2

ORGANIC CHEMWARE
13.5 ^{1}H-NMR spectrum of pentanone

You should now be able to determine the proportion of hydrogen atoms represented by a signal in a ^{1}H-NMR spectrum. You should also be able to predict the ratio of integration values for a given compound.

SOLVED PROBLEM

Determine the number and ratio of integrated signals in the ^{1}H-NMR spectrum of the following compound.

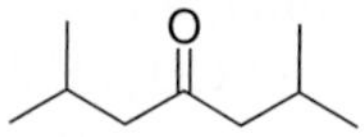

STEP 1 (OPTIONAL): Partially expand the Lewis structure to show the hydrogens.

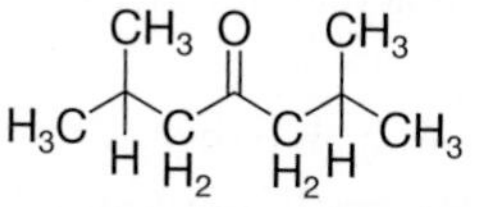

STEP 2: Determine which sets of hydrogens give rise to the three signals.

The plane of symmetry through the centre of the molecule means that the hydrogens on the left side of the molecule are chemically equivalent to the corresponding hydrogens on the right side. The hydrogens in methyl groups are all chemically equivalent to each other by symmetry and rotation.

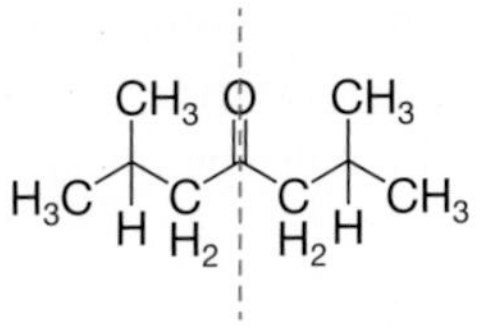

Therefore, there are three different types of hydrogens in the molecule (as shown here), and hence three ^{1}H-NMR signals.

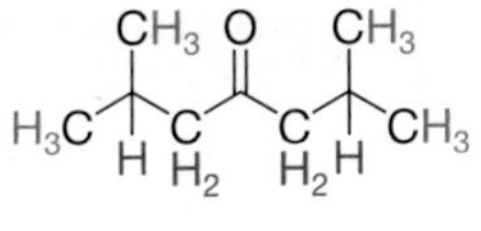

STEP 3: Determine the ratio of the numbers of hydrogens in each different chemical environment.

Twelve hydrogens contribute to the signal from the methyl groups (red); four hydrogens contribute to the CH_2 signal (green); and two hydrogens contribute to the CH signal (blue). Therefore, the ratio of the integrated signals will be 12:4:2, which simplifies to 6:2:1.

PRACTICE PROBLEM

13.3 Determine the number and ratio of integrated signals in the ^{1}H-NMR spectrum of each of the following compounds.

a)

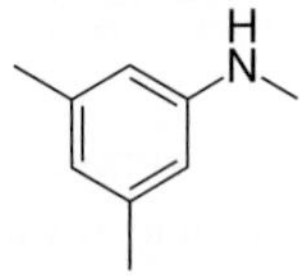

b)

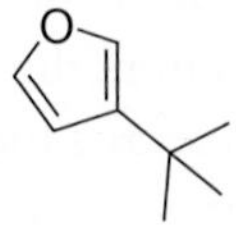

c)

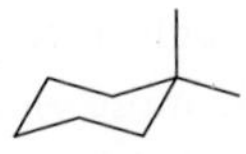

d)

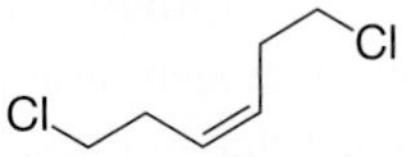

e)

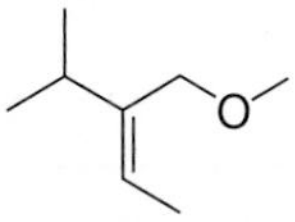

f)

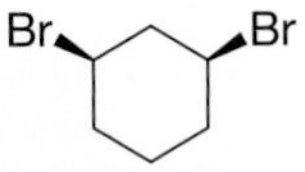

INTEGRATE THE SKILL

13.4 How could integration be us ed to distinguish between the constitutional isomers of bromo-n-pentane ($C_5H_{11}Br$)?

13.4.3 Chemical shift of the hydrogens

The ^{1}H-NMR spectrum of a compound provides information about the chemical environment surrounding each hydrogen atom, and therefore the chemical shift of the hydrogens. The energy gap between spin states for a particular nucleus depends on the net strength of the applied magnetic field and the small, induced magnetic fields of neighbouring *electrons*. When exposed to an external magnetic field $\mathbf{B}_0$, electrons in a compound circulate, creating small currents that generate local magnetic fields ($\sigma\mathbf{B}_0$) that *oppose* the external field. The principle that an induced current always generates a magnetic field opposing the applied field is known as Lenz's law.

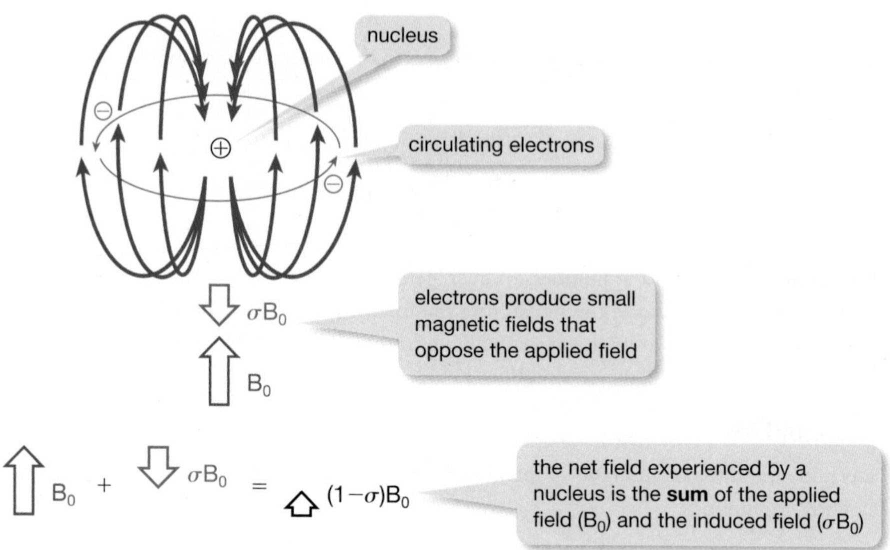

Consequently, the electrons near the nuclei of the hydrogen atoms in a sample influence the magnetic field experienced by each of these atoms. This local magnetic field is the sum of the applied field and the local induced fields of the nearby electrons. Because these induced fields oppose the applied field, electrons in a molecule partly shield the nuclei from the external field, locally weakening the effective magnetic field nearby. This **shielding** decreases the energy gap for spin-flip, which lowers the resonance frequency. By convention, ppm scales for NMR spectra display low frequencies (low δ) on the right and high frequencies (high δ) toward the left. So nuclei that are shielded by their environment—that is, surrounded by more electrons—produce signals toward the right side of the spectrum and are said to be **upfield**. Conversely, hydrogens that are less shielded (**deshielded**) by surrounding electrons (hydrogens near electron-withdrawing groups) produce signals that are **downfield** (toward the left side of the spectrum).

Shielding is the reduction in the magnetic field strength experienced by a nucleus due to the opposing magnetic fields of nearby electrons.

Upfield signals are toward the right of an NMR spectrum.

Deshielding is the reduction in shielding due to a lack of nearby electrons.

Downfield signals are toward the left side of an NMR spectrum.

downfield
deshielded
more electron poor

upfield
shielded
more electron rich

10.0 9.0 8.0 7.0 6.0 5.0 4.0 3.0 2.0 1.0 0.0

STUDENT TIP

The reference signal from TMS is usually the one farthest to the right in a spectrum. To help remember that shielding decreases the chemical shift (and moves the signal to the right), think of TMS as an acronym for Too Much Shielding.

Tetramethylsilane (TMS) is used as the reference compound in ^{1}H-NMR spectra because its 12 hydrogens are all chemically equivalent and highly shielded—largely due to the silicon atom, which is less electronegative than carbon. So TMS produces a signal that is far upfield, away from most other NMR resonances.

$$\begin{array}{c} CH_3 \\ | \\ H_3C{-}Si{-}CH_3 \\ | \\ CH_3 \end{array}$$

tetramethylsilane
(TMS)

Several important factors affect the chemical environment surrounding each hydrogen atom and therefore the chemical shift of the hydrogens. These include the effects of electronegativity, resonance, π-bond currents and anisotropy, and hydrogen bonding and exchange. These are discussed in the following sections.

13.4.3.1 Effect of electronegativity

Electronegative atoms deshield nearby hydrogen atoms and produce a downfield chemical shift. The greater the electronegativity of the atom, the greater is the downfield shift. Figure 13.5 compares the chemical shift of the hydrogens in methyl groups connected to atoms with increasing electronegativities.

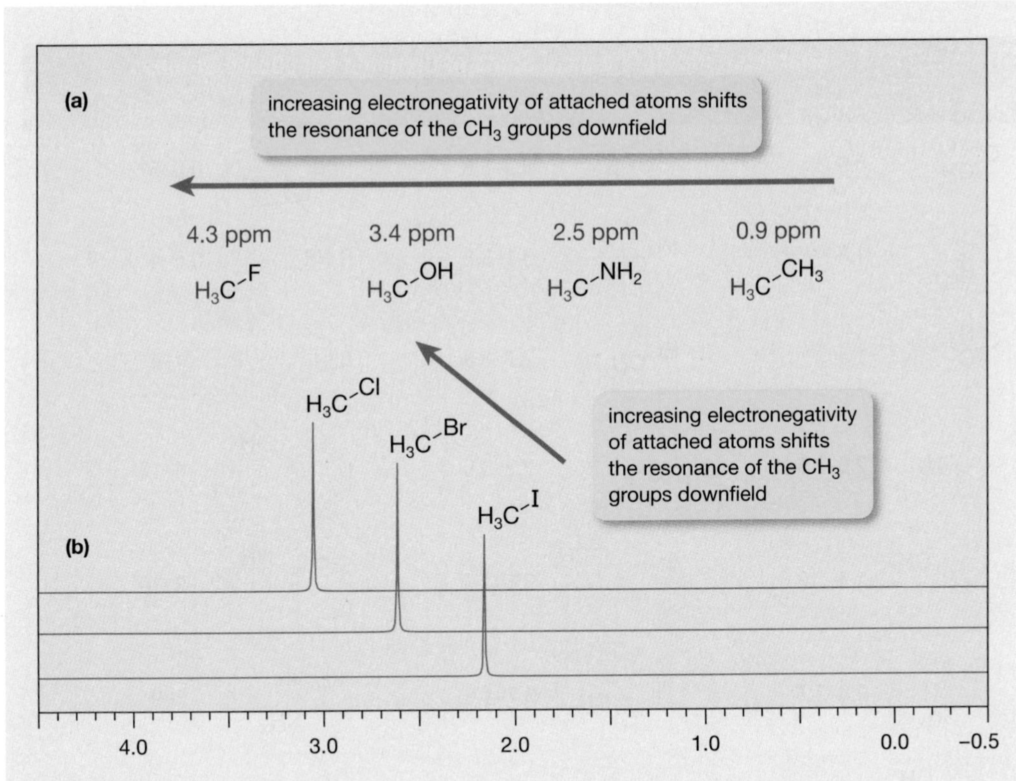

FIGURE 13.5 The effect of electronegativity on chemical shift of nearby hydrogens. a) Electronegativity effects of elements in the same row of the periodic table. b) Electronegativity effects of elements in the same column of the periodic table.

As the number of electronegative atoms attached to a carbon increases, the hydrogens on that carbon become increasingly deshielded, and their resonances shift further downfield (to the left) (Figure 13.6).

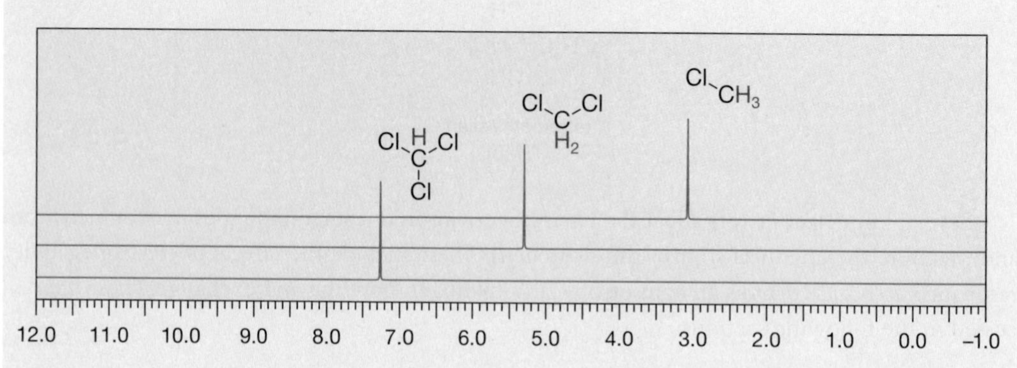

FIGURE 13.6 ^{1}H-NMR spectra of chloromethane, dichloromethane, and trichloromethane (chloroform). Each additional chlorine atom shifts the resonances of the hydrogens in the molecule downfield.

The effects of different functional groups on chemical shifts are consistent and predictable. Table 13.1 and Figure 13.7 show typical chemical shifts for hydrogens in various functional groups.

TABLE 13.1 Characteristic ^{1}H Chemical Shifts of Common Functional Groups

Structure	δ (ppm)	Structure	δ (ppm)	Structure	δ (ppm)
$C{-}\underline{CH_3}$	0.9	$-O{-}\underline{CH}_X$	3.3–3.6	$RO\underline{H}$	0.5–5*
$C{-}C\underline{H_2}{-}$	1.2	$Cl{-}\underline{CH}_X$	3.1–3.8	$R_2N\underline{H}$	0.5–4*
$C{-}C\underline{H}<$	1.5	$Br{-}\underline{CH}_X$	2.7–3.8	$RS\underline{H}$	1–4*
$C{=}C{-}\underline{CH}_X$	2.0–2.5	$I{-}\underline{CH}_X$	2.2–3.9	$C_6H_5O\underline{H}$	4–10*
$C{\equiv}C{-}\underline{CH}_X$	1.8–2.9	$C_6H_5O\underline{CH}_X$	3.8–4.7	$C_6H_5N\underline{H}_X$	3–10*
$C(=O){-}\underline{CH}_X$	2.0–2.7	$C(=O){-}O{-}\underline{CH}_X$	3.7–4.9	$C(=O){-}N\underline{H}_X$	5–9*
$C_6H_5\underline{CH}_X$	2.3–2.9	$C{=}C{-}\underline{H}$	4.5–6.5	$C(=O){-}O\underline{H}$	6–13*
$C{\equiv}C{-}\underline{H}$	2.8	$R{-}C_6H_4{-}\underline{H}$	6.6–8.2	$(CH_3)_4Si$	0
$>N{-}\underline{CH}_X$	2.3–3.0	$C(=O){-}\underline{H}$	9–10		

* Chemical shift of hydrogens on N, O, and S atoms vary, depending on temperature, concentration, and solvent polarity.

ORGANIC CHEMWARE
13.6 ¹H-NMR spectrum of propanone

ORGANIC CHEMWARE
13.7 ¹H-NMR spectrum of methyl ethanoate

ORGANIC CHEMWARE
13.8 ¹H-NMR spectrum of chloroethane

ORGANIC CHEMWARE
13.9 ¹H-NMR spectrum of propanal

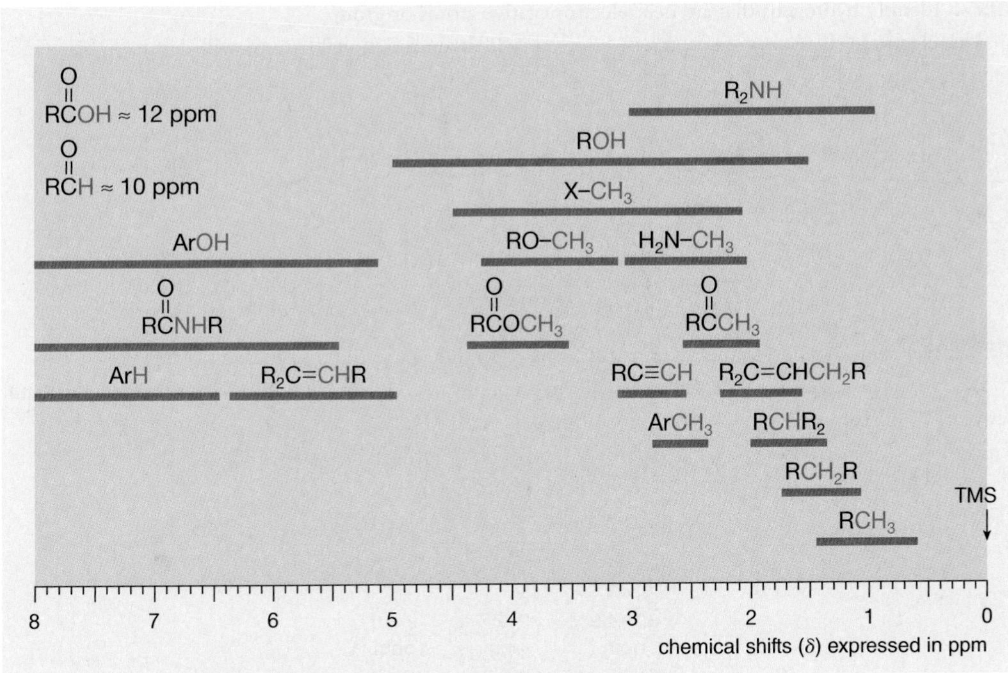

FIGURE 13.7 Hydrogen chemical shift ranges, according to hydrogen type.

Carbon is more electronegative than hydrogen, so replacing a hydrogen with a carbon in an alkyl group shifts the resonances of the remaining hydrogen atoms a bit downfield. For instance, the methyl hydrogens on ethane have a chemical shift of 0.9 ppm, whereas shifts for the hydrogens on the central carbon of propane and isobutene are 1.2 and 1.5 ppm, respectively (Figure 13.8).

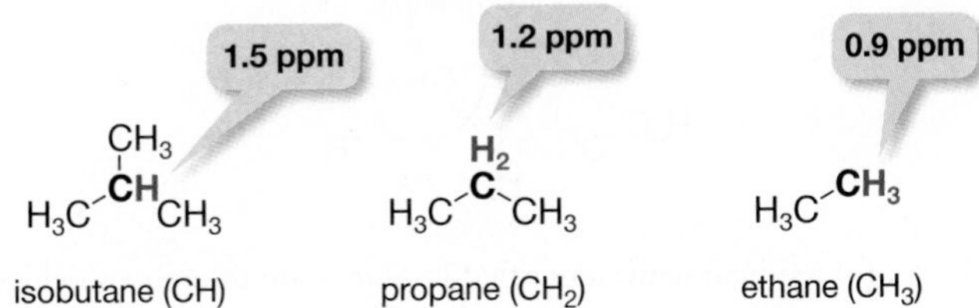

FIGURE 13.8 Increasing alkyl substitution increases the downfield shift.

CHECKPOINT 13.3

You should now be able to predict the approximate chemical shift of the ¹H-NMR signal of a hydrogen based on the neighbouring functional groups and the electronegativity of nearby atoms.

SOLVED PROBLEM

Determine which set of hydrogen atoms in the following compound has the highest chemical shift. Which set has the lowest chemical shift?

STEP 1 (OPTIONAL): Partially expand the Lewis structure to show the hydrogens.

STEP 2: Identify hydrogens that are near electronegative atoms or groups.

The electronegative oxygen atoms in ester and aldehyde functional groups will deshield the neighbouring hydrogens.

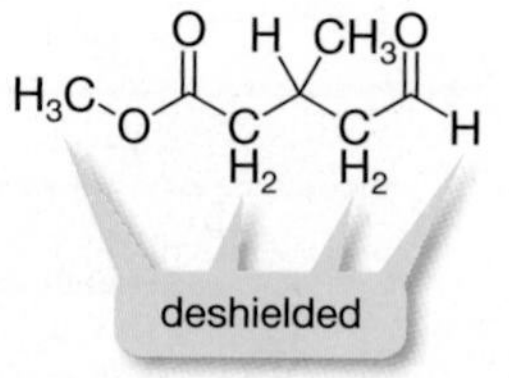

deshielded

STEP 3: Determine which of the hydrogens identified in Step 2 are the most deshielded.

Based on data in Table 13.1, the aldehyde hydrogens are the most deshielded, with a chemical shift between 9 and 10 ppm.

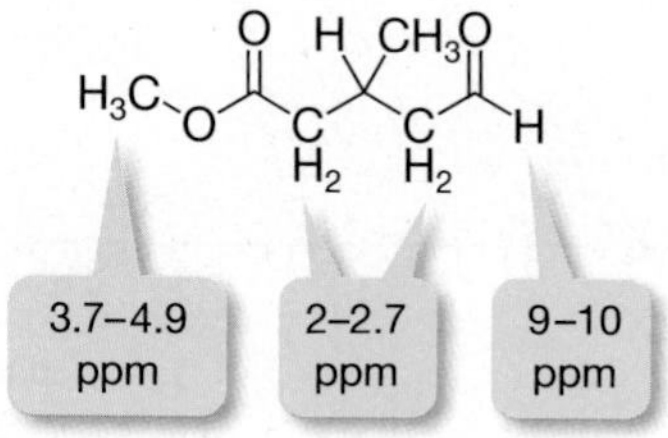

3.7–4.9 ppm
2–2.7 ppm
9–10 ppm

STEP 4: Identify any hydrogens that are far from electronegative atoms or groups, or are near electron-donating groups.

Only two sets of hydrogens are not directly adjacent to any heteroatoms or electron-withdrawing groups. Since these groups are furthest away from any electronegative atoms or groups, these hydrogens are likely the most shielded.

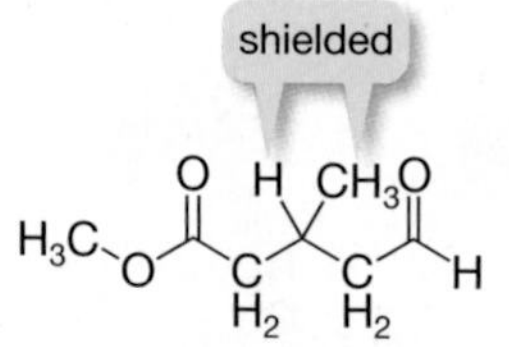

shielded

STEP 5: Determine which of three hydrogens identified in Step 4 are the most shielded.

Table 13.1 lists alkyl hydrogens as having chemical shifts between 0.9 ppm and 1.5 ppm. Since carbon is more electronegative than hydrogen, chemical shift increases with increasing alkyl substitution. Therefore, the hydrogen atoms of CH_3 are less shielded than the hydrogen of CH.

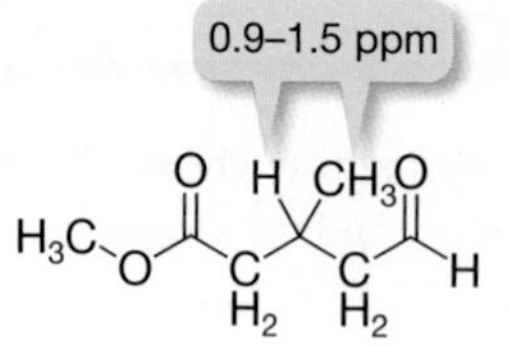

0.9–1.5 ppm

PRACTICE PROBLEM

13.5 a) Identify the most deshielded proton(s) in the following structures.

i)

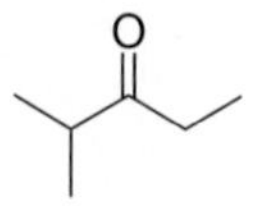

ii) 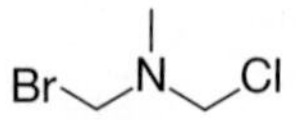 Br N Cl

iii)

b) For each pair of compounds, identify the one in which the methyl group(s) are more shielded.

i) CH_3OCH_3 or CH_3SCH_3

ii) CH_3F or CH_3Cl

iii) $CH_3C(CH_3)_3$ or $CH_3Si(CH_3)_3$

INTEGRATE THE SKILL

13.6 Identify the hydrogens that correspond to each signal in the following ^{1}H-NMR spectra.

a)

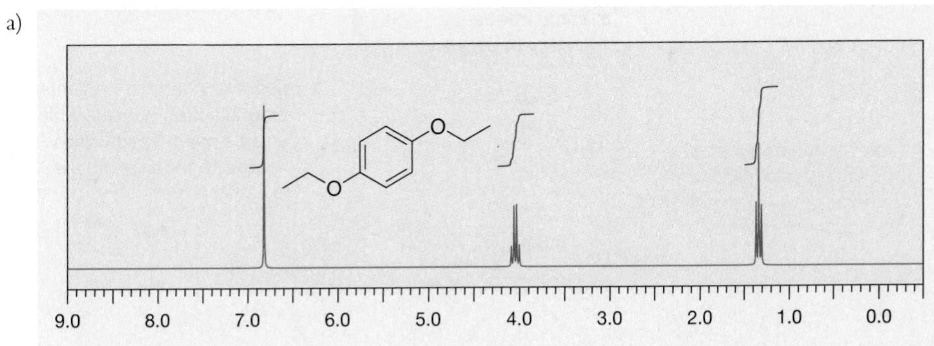

b)

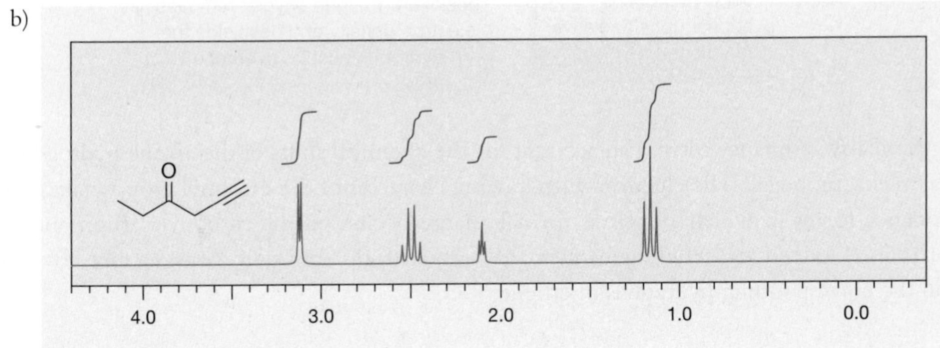

c)

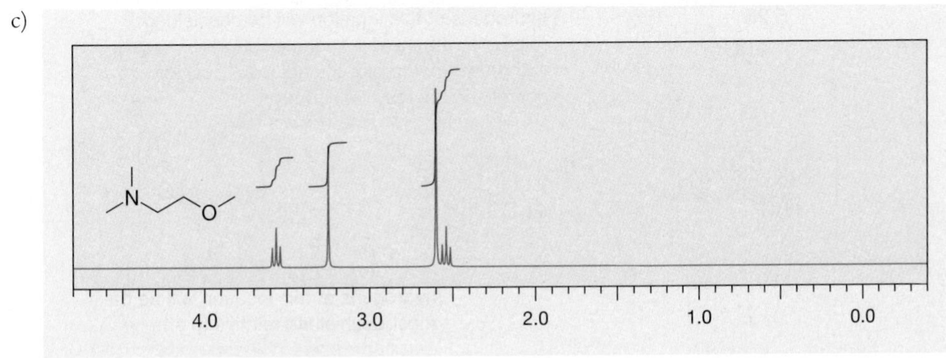

13.4.3.2 Effects of conjugation

Conjugation can have a powerful effect on chemical shift because conjugation involves the distribution of electrons in a molecule. The effects of conjugation on the chemical shift of the atoms in a functional group can be predicted by examining the resonance forms of that functional group.

STUDENT TIP

Resonance forms or resonance structures refer to the alternate structures that can represent a conjugated molecule. Hydrogens have resonance (or resonate) with specific electromagnetic frequencies, resulting in signals in a NMR spectrum.

For example, there is no conjugation in the double bond of an alkene, so the chemical shift of all the hydrogen atoms in such a group will be similar. In ethane, for example, all of the hydrogens resonate at 5.26 ppm. Substituting one of the hydrogens with an OCH_3 group produces methyl vinyl ether, which contains an enol ether group that experiences delocalization. One of the resonance contributors has a positive formal charge on the oxygen and a negative formal charge on a carbon. Negative charges represent areas of extra electron density and are therefore shielding. As a result, the chemical shift of the two CH_2 hydrogens in this molecule is shifted slightly upfield to 4.05 ppm. The positive charge on the oxygen is electron withdrawing. This formally charged oxygen will slightly deshield the CH hydrogen resonance to 6.61 ppm.

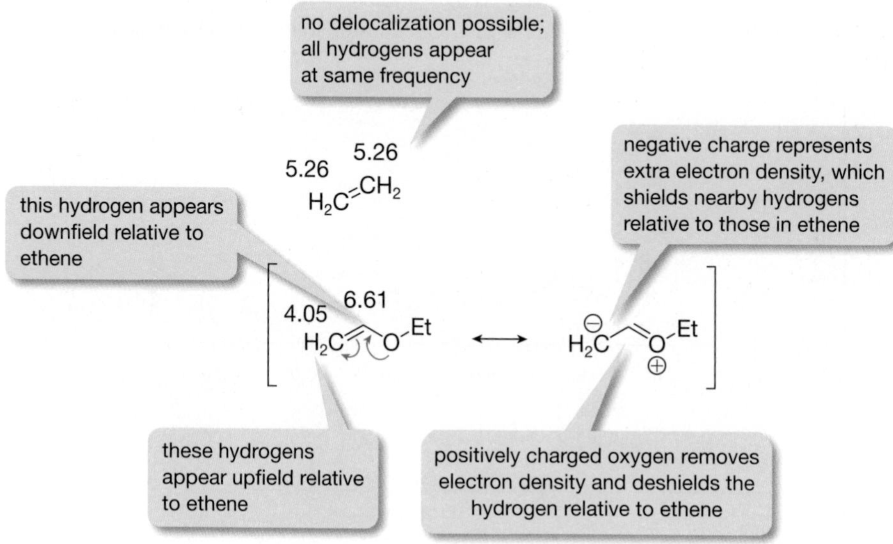

ORGANIC CHEMWARE
13.10 ^{1}H-NMR spectrum of ethyl vinyl ether

Similarly, resonance forms can account for the chemical shifts of the alkene hydrogen atoms in acrolein (propenal). The electron-withdrawing character of the carbonyl group makes possible resonance forms in which a positive formal charge occurs on the carbonyl carbon and on the β-carbon (Figure 13.9). Hydrogens near the carbocation centres are considerably less shielded than the corresponding hydrogens in ethene.

5.26 5.26 $H_2C{=}CH_2$

hydrogens at this location will be deshielded relative to ethene due to the electron-withdrawing nature of the adjacent positive charges in the resonance forms

6.43 6.31

hydrogens at this location will be the most deshielded relative to ethene

FIGURE 13.9 Resonance electron withdrawal within a π-system causes a downfield shift of the β-carbon.

ORGANIC CHEMWARE
13.11 ^{1}H-NMR spectrum of acrolein

Functional groups on aromatic rings can also dramatically affect chemical shifts. The ^{1}H-NMR spectrum of benzene (C_6H_6) is a singlet at 7.26 ppm. The electron-donating hydroxyl group in phenol allows conjugation effects that produce NMR signals upfield of 7.26 ppm (formal negative charges represent areas of extra electron density). The resonance structures for this molecule predict that the hydrogens at the *ortho* and *para* positions are more strongly shielded than those at the *meta* position.

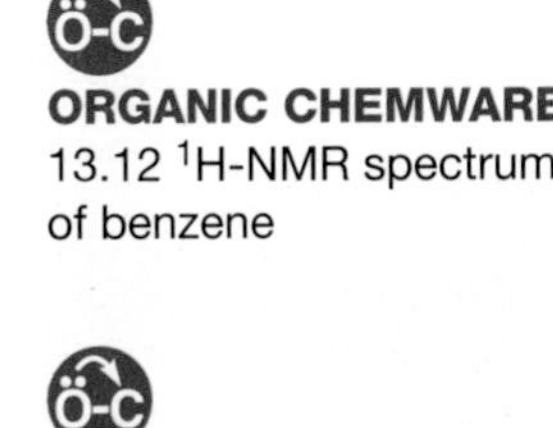

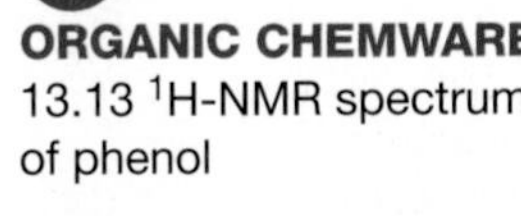

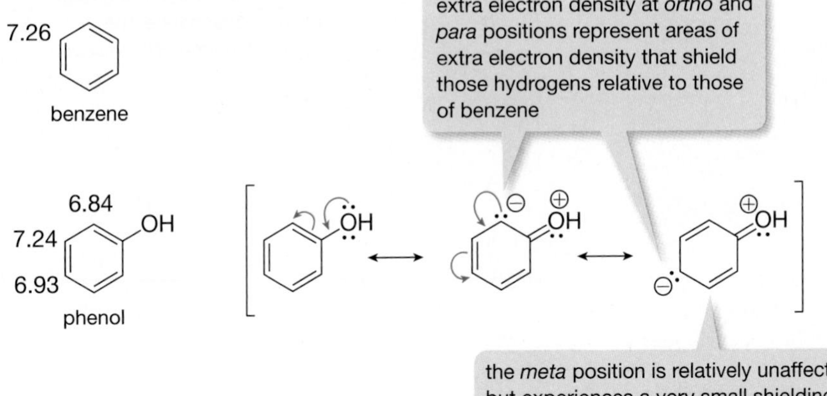

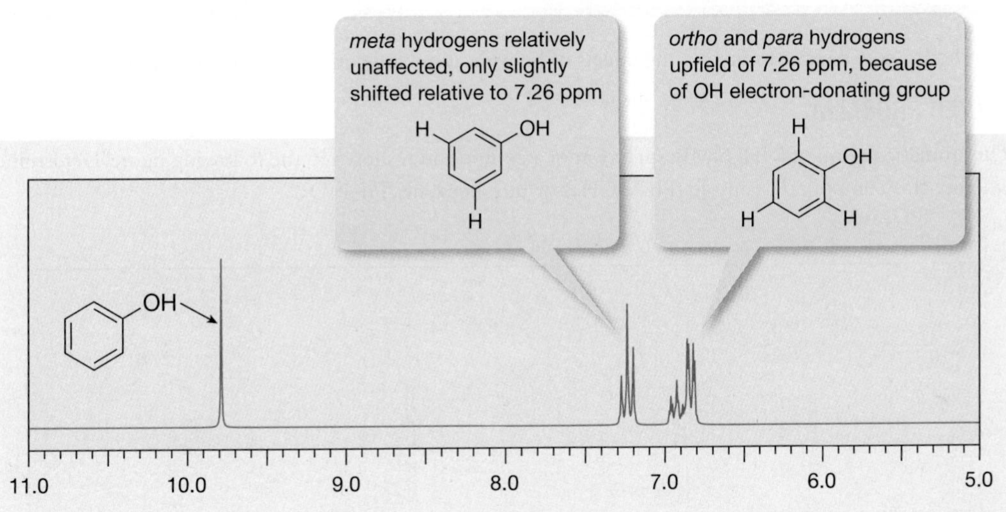

Similarly, the ^{1}H-NMR spectrum of benzaldehyde demonstrates the electron-withdrawing character of the aldehyde group as all five of the aryl hydrogens are deshielded relative to benzene.

7.26

benzene

7.88
7.53
7.63

benzaldehyde

The positive charges represent areas of less electron density. The hydrogens at these positions will be slightly deshielded relative to those of benzene.

The *meta* position is affected less strongly but is deshielded by the inductive effect of the nearby positive charges.

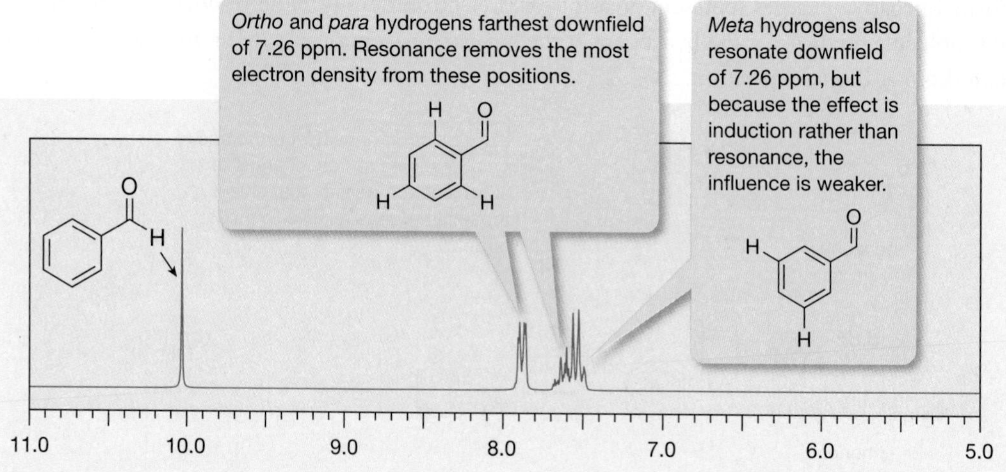

CHECKPOINT 13.4

You should now be able to predict the effects of hybridization and conjugation on chemical shift.

SOLVED PROBLEM

The aromatic region of a ^{1}H-NMR spectrum of a compound is shown in the following figure. Determine whether this compound is anisole (Ph–OCH_3) or nitrobenzene (Ph–NO_2).

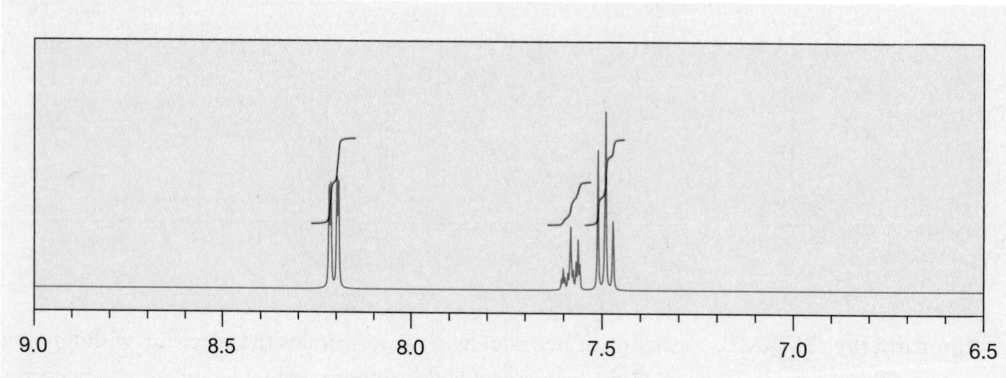

STEP 1: Determine whether the aromatic signals are above or below the chemical shift of benzene.

The chemical shifts are all above 7.26 ppm.

STEP 2: Determine whether the substituent on the compound is adding or removing electron density from the aromatic ring.

The higher chemical shifts indicate that the aromatic hydrogen atoms on the compound are less shielded than those in benzene. Therefore, some electron density is removed from the ring.

STEP 3: Use resonance structures to determine the electron distribution in the two possible compounds.

Nitro groups are electron withdrawing, and pull electron density out of the ring.

Methoxy groups are electron donating through resonance, and add electron density to the ring.

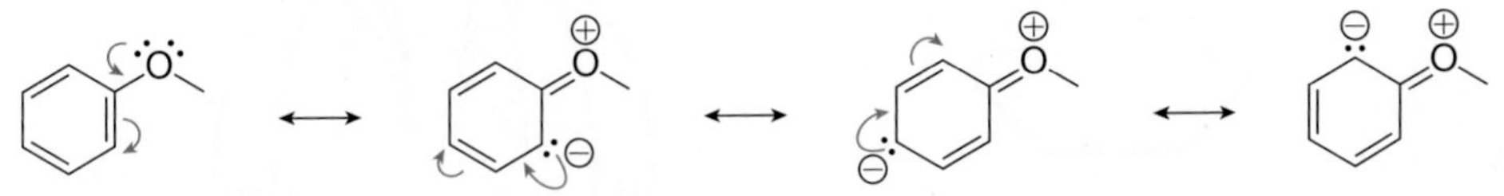

The resonance contributors combine to form the following resonance hybrids, which show low electron density in the ring for nitrobenzene and increased ring electron density for anisole.

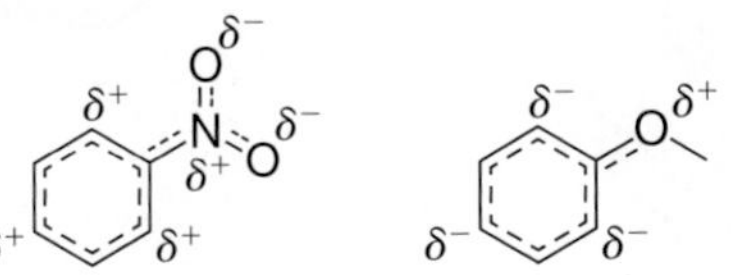

STEP 4: Determine the structure of the compound.

The aromatic hydrogens of the compound have been shifted downfield, indicating that they are electron poor relative to benzene. Therefore, the compound must be nitrobenzene, with its strongly electron-withdrawing nitro group.

PRACTICE PROBLEM

13.7 For each pair of compounds, explain the differences between the observed chemical shifts noted on their structures.

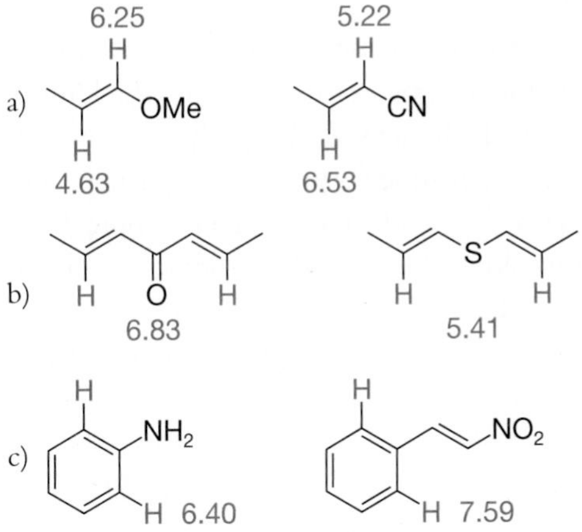

a) 6.25 H, OMe, H 4.63; 5.22 H, CN, H 6.53

b) H O H 6.83; S, H H 5.41

c) H, NH_2, H 6.40; H, NO_2, H 7.59

13.8 Describe the major differences you would expect to see in the ^{1}H-NMR spectra of the following two compounds.

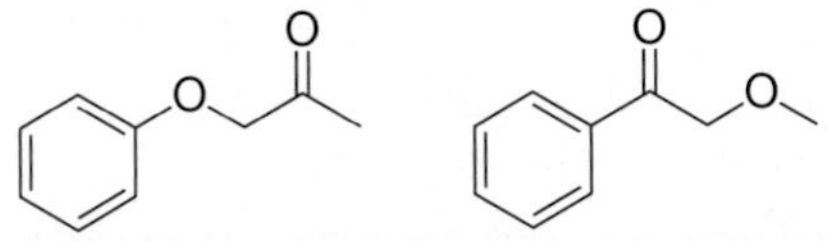

13.4.3.3 Effects of π-bond currents and anisotropy

In a magnetic field, the π electrons of aromatic rings circulate as ring currents. These ring currents produce a local magnetic field that strongly influences the chemical shifts of neighbouring hydrogens (Figure 13.10).

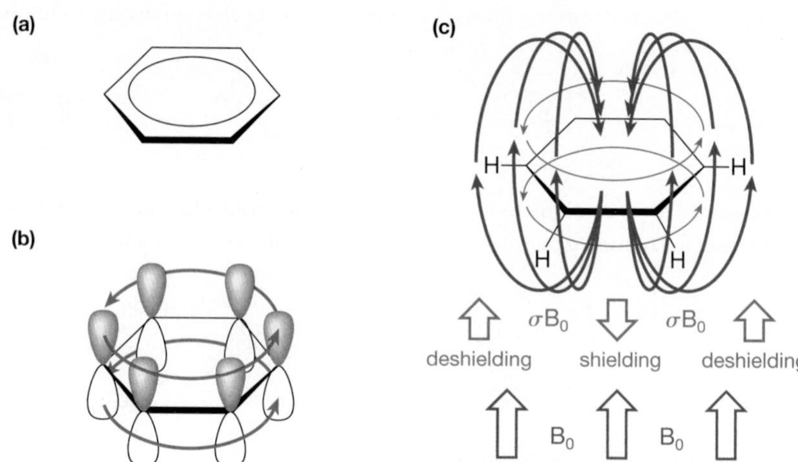

FIGURE 13.10 The ring current in an aromatic ring deshields hydrogen atoms bonded to the ring.

In the interior of benzene, the induced magnetic field opposes the applied field, creating an area of shielding over the two faces of the ring. However, outside of the aromatic ring, the induced field aligns *with* the applied field, creating a region in which the overall magnetic field is somewhat *stronger* than the applied field. Hydrogens in this region usually have chemical shifts greater than 7 ppm.

The π bonds in alkenes and carbonyls also have electron currents that affect chemical shifts. π bonds create a cone of shielding above and below the face of the π bond, shielding hydrogens that lie within these regions and deshielding hydrogens at the periphery. This effect is readily apparent in the spectra of aldehydes, where the hydrogens produce a signal far downfield (>9 ppm). A similar effect occurs in alkenes, which appear in the region of 5 to 7 ppm. In alkynes, the circulating currents form a shielding region along the axis of the bond and deshielding regions near the two ends of the bond, as shown in Figure 13.11.

Changes in chemical shift that depend on the orientation of a hydrogen are examples of **magnetic anisotropy**.

Magnetic anisotropy is the dependence of a magnetic property of a material on direction relative to an external magnetic field.

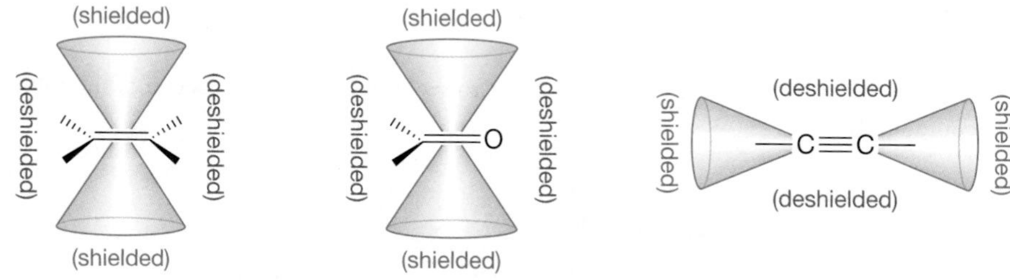

FIGURE 13.11 The electron currents in π bonds have shielding and deshielding effects.

ORGANIC CHEMWARE
13.15 ^{1}H-NMR spectrum of 2-methyl-1-butene

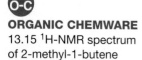

ORGANIC CHEMWARE
13.16 ^{1}H-NMR spectrum of 3-bromopropyne

13.4.3.4 Effects of hydrogen bonding and exchange

Hydrogen atoms bonded to heteroatoms such as oxygen, nitrogen, and sulfur are subject to hydrogen bonding, which is affected by temperature, concentration, and polarization of the solvent. Hydrogen bonding of a donor to an acceptor withdraws electrons from the donor and deshields the hydrogen.

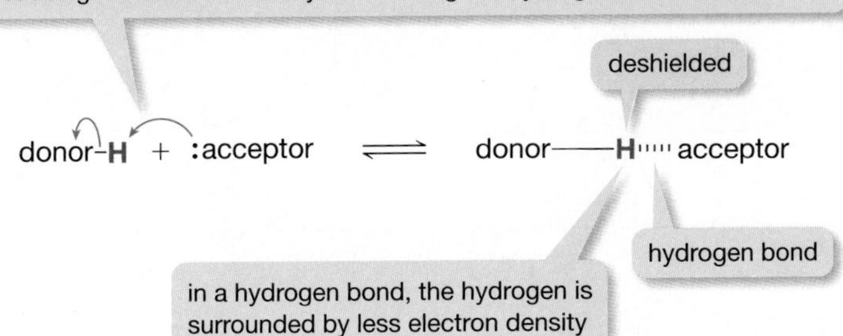

STUDENT TIP

The hydrogen bond donor/acceptor refers to the donation/acceptance of hydrogens, rather than electrons.

Because the extent of hydrogen bonding can vary from sample to sample, the chemical shift of the affected hydrogen also varies. For example, the hydroxyl hydrogens of a concentrated solution of an alcohol have more *inter*molecular hydrogen bonding than those of a dilute solution. Consequently, the signal for hydroxyl hydrogens in a concentrated sample appears further downfield than the corresponding signal for a dilute sample, as shown in Figure 13.12. The chemical shift range of hydrogen-bonded hydrogens can span several ppm!

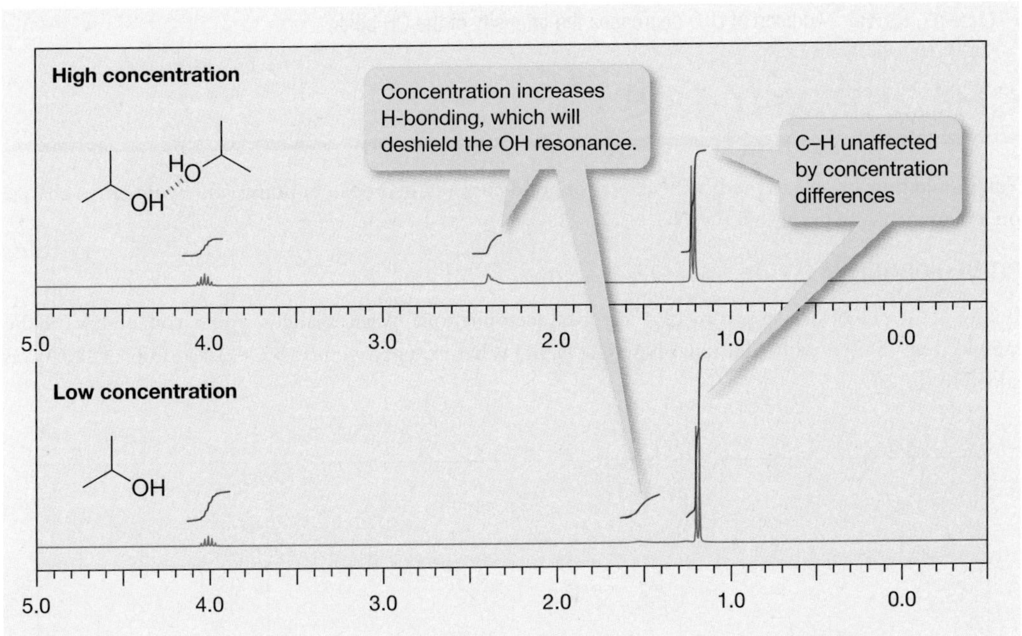

FIGURE 13.12 Changing the concentration of a sample can affect the chemical shift of hydrogen-bonded hydrogens.

ORGANIC CHEMWARE
13.17 ^{1}H-NMR spectrum of ethanol

Hydrogen atoms capable of hydrogen bonding may also exchange with other hydrogens through various acid–base reactions with other solutes, the solvent, or traces of water in the sample. A particular hydrogen could move from one chemical environment to another during the NMR experiment. Such exchanges broaden the resonance peaks, often so much as to be barely visible on a spectrum.

ORGANIC CHEMWARE
13.18 ^{1}H-NMR spectrum of 1-propanol

Signals from such exchangeable hydrogen atoms can be identified by recording a normal NMR spectrum and then adding a drop of D_2O and recording another spectrum (Figure 13.13). The exchange of any acidic hydrogens of the sample with deuterium causes their signal intensity to decrease or disappear completely from the spectrum since deuterium resonances are outside the range of a ^{1}H-NMR spectrum.

ORGANIC CHEMWARE
13.19 ^{1}H-NMR spectrum of 2-propanol

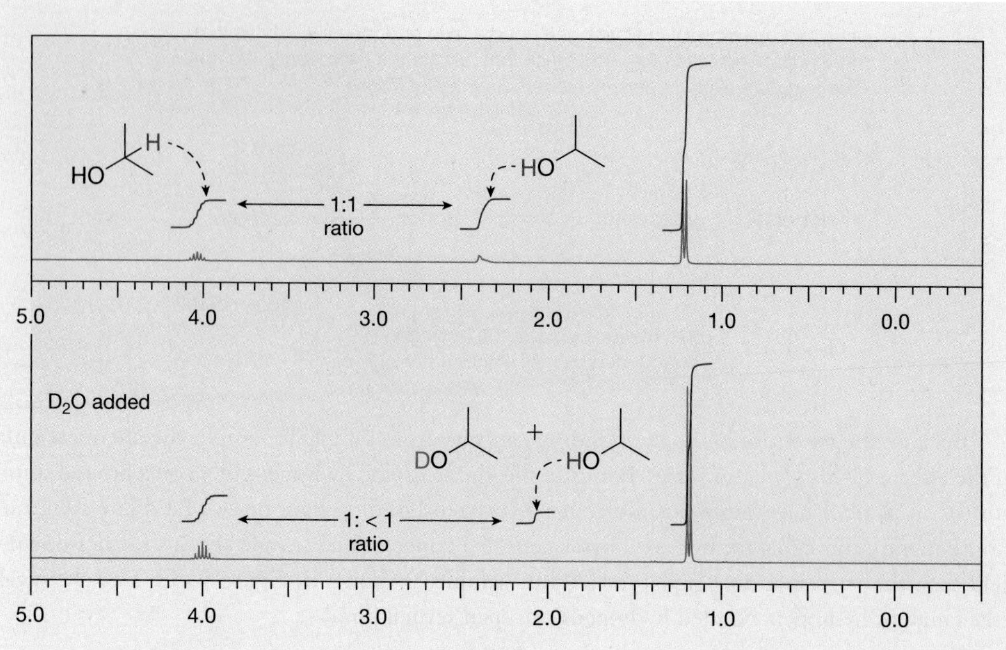

FIGURE 13.13 Addition of D_2O decreases the intensity of the OH peak.

CHECKPOINT 13.5

You should now be able to predict the effects of bond currents, hydrogen bonding, and hydrogen exchange on a hydrogen's chemical shift.

SOLVED PROBLEM

Below is the ¹H-NMR spectrum of 4′-hydroxyacetophenone. What changes would you observe if the sample were diluted with additional NMR solvent? What changes would you see if a drop of D_2O was added to the sample?

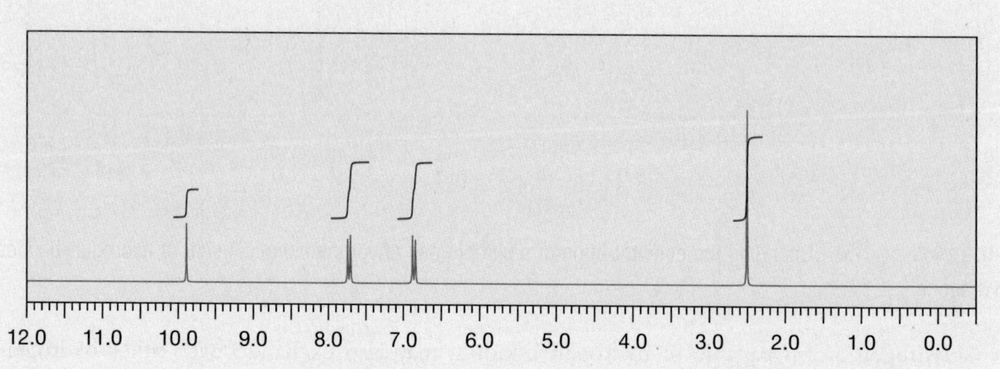

STEP 1: Determine whether any hydrogen atoms in the compound will be affected by diluting the sample or adding D_2O.

Only those hydrogens capable of hydrogen bonding have signals that are affected by sample concentration and the addition of D_2O. Typically, such hydrogens are attached to electronegative atoms, such as oxygen and nitrogen. A drawing of the structure shows that it has only one hydrogen capable of hydrogen bonding: the H on the OH group.

STEP 2: Determine the chemical shift range for the OH hydrogen.

According to Table 13.1, aromatic OH signals appear between 4 and 10 ppm. This broad range makes it difficult to identify the OH peak in the spectrum without first identifying the rest of the signals.

STEP 3: Identify the chemically non-equivalent hydrogens.

By symmetry, there are two pairs of chemically equivalent hydrogens on the ring. So the compound has four different types of hydrogen (marked as a, b, c, and d), which correspond to the four signals in the spectrum.

STEP 4: Look up the chemical shift ranges for the four types of hydrogens in the structure.

d 2.0–2.7 ppm

4–10 ppm

6.6–8.2 ppm

STEP 5: Identify the signals in the ¹H-NMR spectrum.

The electron-donating OH group shifts the signal for the b hydrogens upfield, while the electron-withdrawing ketone shifts the signal for the c hydrogens downfield. Therefore, the hydrogens producing the signals are as follows:

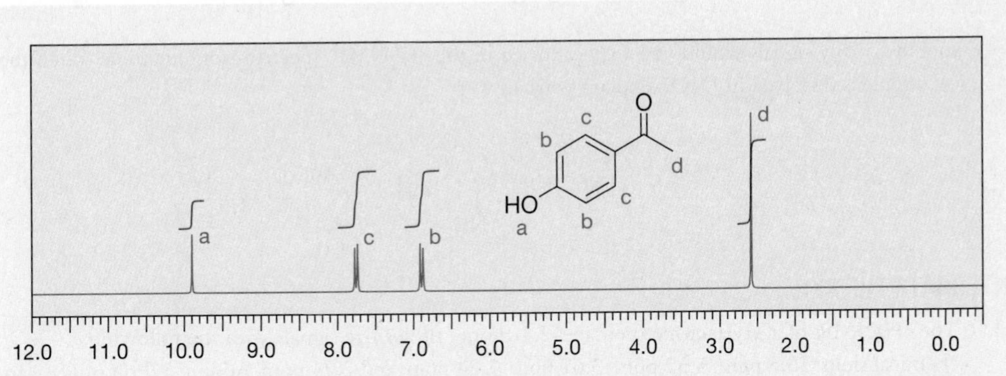

STEP 6: Predict the changes to the OH signal if the sample concentration is decreased.

If the sample is diluted, the amount of hydrogen bonding between molecules of 4′-hydroxyacetophenone decreases. Since hydrogen bonding has a deshielding effect, diluting the sample shifts the OH peak upfield.

STEP 7: Predict the changes to the OH signal if a drop of D_2O is added to the sample.

The hydrogen in the OH group will slowly exchange with deuterium from the D_2O. Since deuterium does not give a signal in a ¹H-NMR spectrum, the intensity of the OH signal will decrease, and might even disappear.

PRACTICE PROBLEM

13.9 a) Which hydrogens in the following structure are the most shielded, and which are the least shielded? Explain your answer.

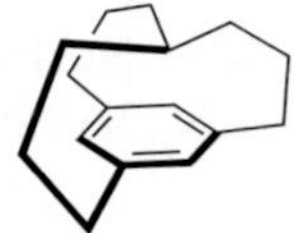

b) The ^{1}H-NMR spectrum of [18]-annulene (below) has two signals: one at 8.2 ppm and another at −1.9 ppm. Determine which hydrogen produces each signal, and explain the significant difference between the chemical shifts.

c) Which of the following compounds have a ^{1}H-NMR spectrum that changes when the concentration of the sample is doubled? Explain your choices, and describe what the changes would be.

A B C D E

d) How many signals would you expect to see in the ^{1}H-NMR spectrum of glutamine when the sample is dissolved in D_2O? Explain your answer.

INTEGRATE THE SKILL

13.10 The ^{1}H-NMR of acetylacetone (pentane-2,4-dione) shows five signals with the following chemical shifts: 15.5 ppm, 5.52 ppm, 3.61 ppm, 2.24 ppm, and 2.24 ppm. Adding a drop of D_2O to the sample and shaking it vigorously for several minutes causes the signals at 15.5 and 5.52 ppm to disappear. Determine which hydrogens produce the signals at 15.5 and 5.52 ppm, and explain why they disappears after the addition of D_2O. (Hint: Consider what other forms of acetylacetone may be present under equilibrium conditions. See Chapter 8 for assistance.)

13.4.4 Determining molecular connectivity near each hydrogen atom

The ^{1}H-NMR spectrum of a compound reveals information about molecular connectivity near each hydrogen atom. The chemical shift of a nucleus depends upon the magnetic fields of the electrons surrounding it (the chemical bonds). Minor, but important, contributors to the local magnetic field are the weak magnetic fields of nearby nuclei. They produce a phenomenon called **spin-spin coupling**, and this effect splits some NMR signals into several smaller peaks. The symmetric pattern of these small peaks can indicate how atoms are connected together in a given molecule.

Spin-spin coupling is an interaction between the magnetic fields of adjacent nuclei.

Consider two **adjacent hydrogens**, H_A and H_X, that have different chemical shifts. The spins of the H_X nuclei will align with or against $\mathbf{B}_0$ in nearly equal proportions. When aligned with $\mathbf{B}_0$, the spin of each H_X nucleus slightly increases the magnetic field experienced by the nearby H_A nuclei, and hence slightly increases their chemical shift. When aligned against $\mathbf{B}_0$, the spin of H_X slightly decreases the magnetic field experienced by the H_A nuclei and slightly decreases their chemical shift. The result is that half of the H_A nuclei experience a stronger magnetic field and half experience a weaker magnetic field. The signal for H_A therefore becomes split into two smaller resonances spaced equally about the chemical shift of H_A (Figure 13.14). Such a signal is called a *doublet*. Similarly, the signal for H_X is split into a doublet due to the effects of the spin of H_A.

Adjacent hydrogens are bonded to adjacent atoms, and are three bonds away from each other.

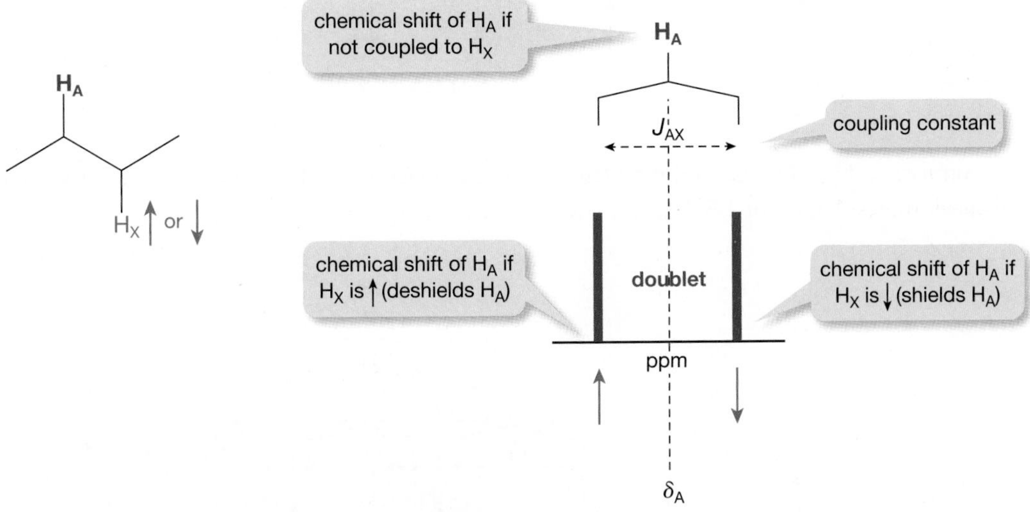

FIGURE 13.14 H_A splits into a doublet due to coupling with H_X.

The difference in frequency between the split peaks (measured in hertz) is called the **coupling constant** (J) and is independent of the strength of the applied magnetic field. The magnitude of J decreases as the number of bonds separating the atoms increases. Spin-spin coupling is observed primarily between hydrogen atoms that are no more than three bonds apart.

The **coupling constant** (J) is the difference in resonance frequency between peaks in a multiplet.

Because the nuclei of H_A and H_X interact magnetically, they are said to be coupled to each other. Each magnet (H_A and H_X) pulls equally on the other, which makes their coupling constants the same. Measurement of coupling constants can be used to determine which signals are coupled to each other in more complex NMR spectra, and this can be used to identify which nuclei are near others in a compound.

Now consider a H_A adjacent to two identical H_X. The two H_X protons have four possible spin combinations in which a nucleus can be aligned with (↑) or against (↓) the magnetic field. These four combinations are ↑↑, ↑↓, ↓↑, and ↓↓. There is an equal probability of each of these combinations, and so all four occur in equal amounts. The ↑↓ and ↓↑ spin combinations do not change the chemical shift of H_A, so a peak is observed at this base chemical shift with twice the intensity of the other peaks in the signal. The ↑↑ spin combination has a net deshielding effect on H_A, while the ↓↓ spin combination has a net shielding effect on H_A. As a result, the resonance of H_A appears as a triplet with the sub-peaks in an intensity ratio of 1:2:1, corresponding to ↑↑, (↑↓ + ↓↑), and ↓↓.

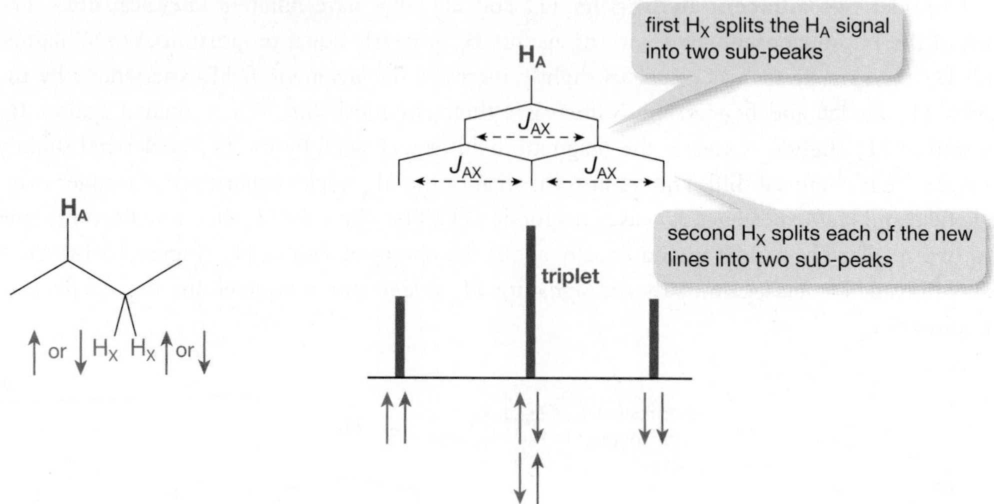

Similarly, a H_A adjacent to three identical H_X produces an NMR signal as a quartet with a sub-peak intensity ratio of 1:3:3:1.

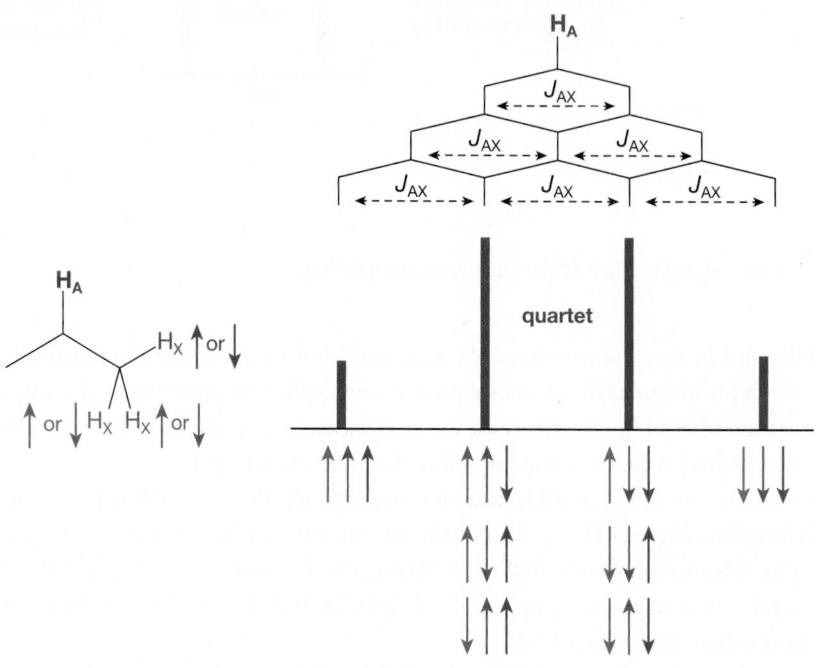

ORGANIC CHEMWARE
13.20 ^{1}H-NMR spectrum of 1,1-dichloroethane

ORGANIC CHEMWARE
13.21 ^{1}H-NMR spectrum of 1,1,2-trichloroethane

The ***n*+1 rule** states that the number of peaks in a multiplet is one more than *n*, the number of adjacent equivalent hydrogens.

The ***n*+1 rule** can be used to determine the multiplicity of a signal: the number of peaks is always one more than the number of equivalent adjacent hydrogens. A hydrogen with a single neighbour appears as a doublet, a hydrogen with two equivalent neighbours appears as a triplet, etc. The relative intensity of the individual peaks follows Pascal's triangle (Table 13.2).

TABLE 13.2 **Multiplet Patterns**

Multiplet	Abbreviation	# of adjacent H (n)	Intensity Ratio
Singlet	s	0	1
Doublet	d	1	1:1
Triplet	t	2	1:2:1
Quartet	q	3	1:3:3:1
Quintet	quint	4	1:4:6:4:1
Sextet	sext	5	1:5:10:10:5:1
Heptet	hept	7	1:6:15:20:15:6:1

STUDENT TIP

In the $n+1$ rule, n is the number of adjacent hydrogen atoms, *not* the number of peaks in a signal. Remember that a singlet has zero adjacent hydrogens. Having one adjacent hydrogen cannot give a signal with zero peaks.

13.4.4.1 Splitting trees

Splitting trees offer a visual reference on how the various neighbouring hydrogens split a signal into a multiplet. Consider the signal for a hydrogen, H_A. If H_A has no nearby hydrogens, its signal appears as a singlet at chemical shift δ_A. If H_A has a single hydrogen neighbour, H_X, the signal of H_A splits into a doublet ($n+1$) centred on δ_A and separated by the coupling constant J_{AX}. If H_A has two hydrogen neighbours, H_X, that are equivalent to each other, the signal of H_A splits first into a doublet ($n+1$) due to one of the H_X, then each peak splits into further doublets due to the other H_X (Figure 13.15). The result is a triplet with a 1:2:1 intensity ratio.

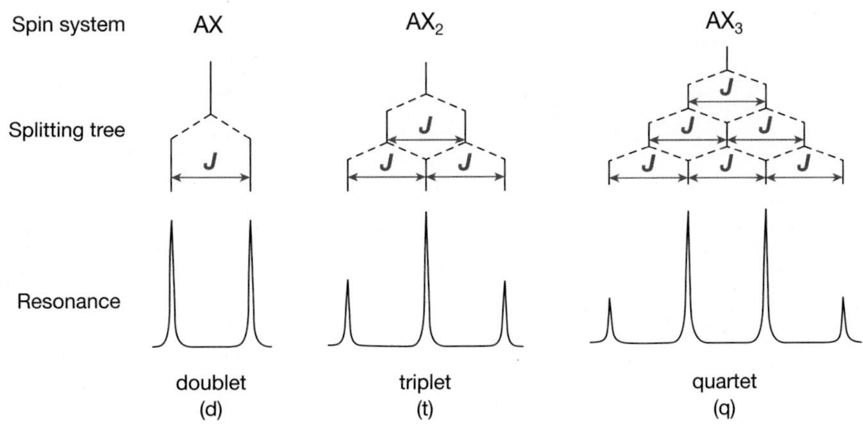

FIGURE 13.15 **Splitting trees predict the appearance of complex multiplets. Each signal corresponds to the resonance for nucleus A.**

ORGANIC CHEMWARE

13.22 ¹H-NMR spectrum of 1-bromopropane

Drawing a splitting tree for other systems follows the same logic. The splitting tree method can be used to identify coupling constants in complex splitting situations.

13.4.4.2 Diastereotopic hydrogens

Diastereotopic hydrogens are non-equivalent hydrogens on the same carbon.

Hydrogens with molecular asymmetry that makes them non-equivalent are called **diastereotopic hydrogens**. The signals produced by diastereotopic hydrogens are different from signals produced by molecules that follow the $n+1$ rule, as described in Section 13.4.4. Diastereotopic hydrogens, because they are non-equivalent, produce different signals and can have complex coupling that does not appear to follow the $n+1$ rule. Conformationally restricted molecules that have rings and/or double bonds commonly produce such patterns. As shown in Figure 13.16, the two terminal hydrogens of ethyl vinyl ether are diastereotopic. You can verify this property by mentally substituting one of the hydrogens with a deuterium, and comparing the resulting structure with that formed by replacing the other hydrogen: the two structures are diastereomers.

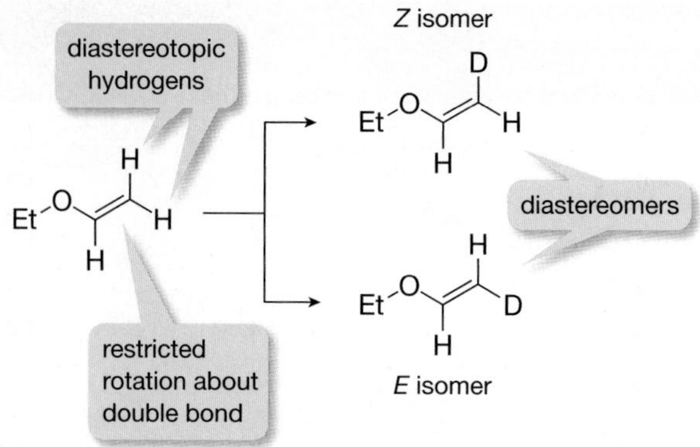

FIGURE 13.16 Diastereotopic hydrogens bonded to the same carbon on a terminal alkene.

Because the two terminal hydrogen atoms are in different chemical environments, they are non-equivalent and couple to each other. They also couple to the other CH group of the alkene, each with a different coupling constant (J). The resulting NMR signals can be quite complex, but can reveal much about the molecular structure. In the spectrum of ethyl vinyl ether, each alkene hydrogen signal is split into a doublet of doublets.

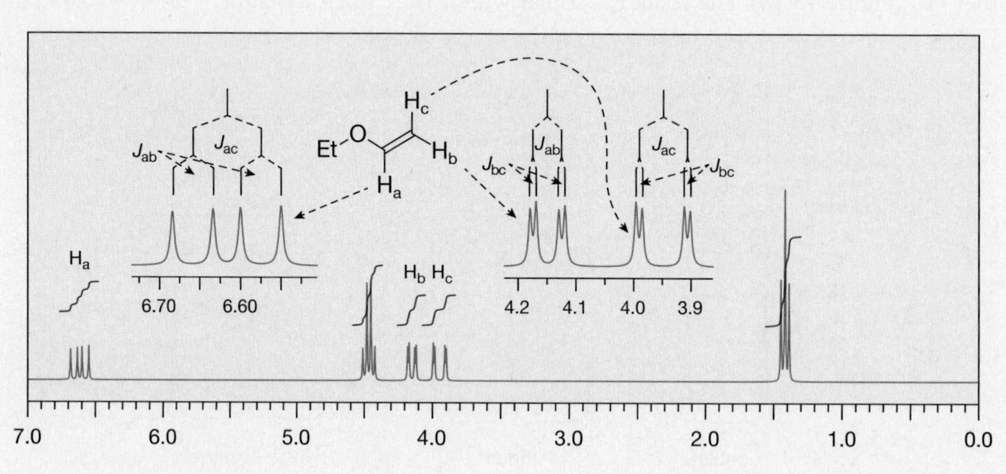

An alkene hydrogen adjacent to an allylic CH_2 and another alkene hydrogen is split differently by each, producing a doublet of triplets (dt). This multiplet may *appear* to violate the $n+1$ rule, since it is split into six peaks by the three adjacent hydrogens, not four peaks. However, the $n+1$ rule *is* obeyed, when n = the nearest neighbour of a given type. The signal is split once by the CH, and again by the CH_2.

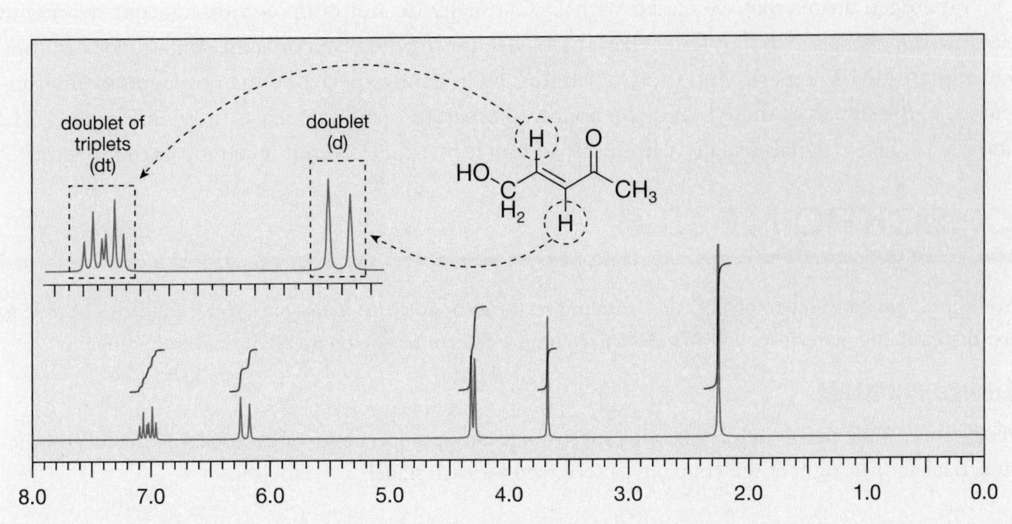

Hydrogen atoms on the same carbon but adjacent to a stereogenic centre are also diastereotopic. This configuration may produce a different signal for each hydrogen of a diastereotopic CH_2, such as the one in 2-butanol.

diastereotopic hydrogens

diastereomers

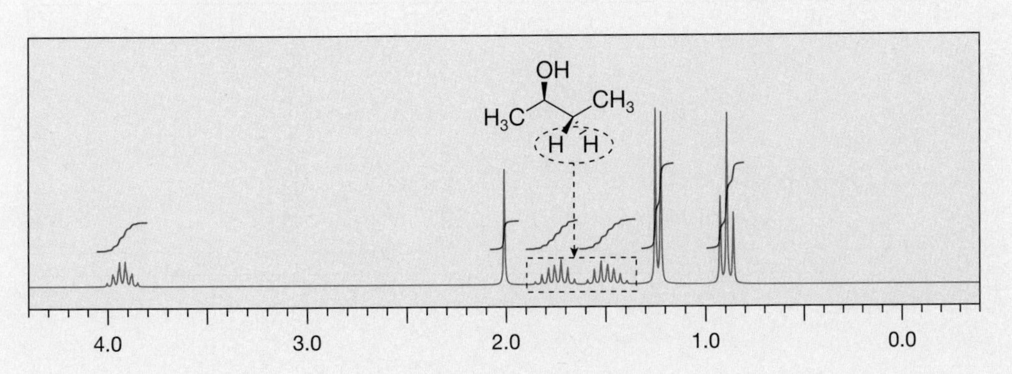

It is important to note that coupling between equivalent hydrogens is *not* observed. For example, 2-butene has two types of hydrogen. The alkene hydrogens are each next to a methyl group and the other alkene hydrogen. Because the alkene hydrogens are chemically equivalent (by symmetry), they do not couple to each other; they produce a quartet (due to coupling with the adjacent CH_3) only.

ORGANIC CHEMWARE
13.24 ^{1}H-NMR spectrum of 2-bromobutane

Hydrogen atoms that exchange with D_2O usually do not couple with adjacent hydrogens because the exchanges occur faster than the NMR spectrometer records the spin-spin transitions. During an NMR experiment, the exchanging hydrogens experience many magnetic environments, so the signals of these hydrogens appear as broad resonances, such as those in Figures 13.12 and 13.13. These signals usually flatten out so much that they are not apparent in the spectra.

CHECKPOINT 13.6

You should now be able to predict the coupling patterns for adjacent hydrogens. You should also be able to use the coupling patterns of a ^{1}H-NMR spectrum to determine the connectivity of the hydrogens.

SOLVED PROBLEM

Assign each of the signals in the following spectrum to the appropriate set of hydrogens in the given structure. Explain the origin of the coupling pattern seen for each signal.

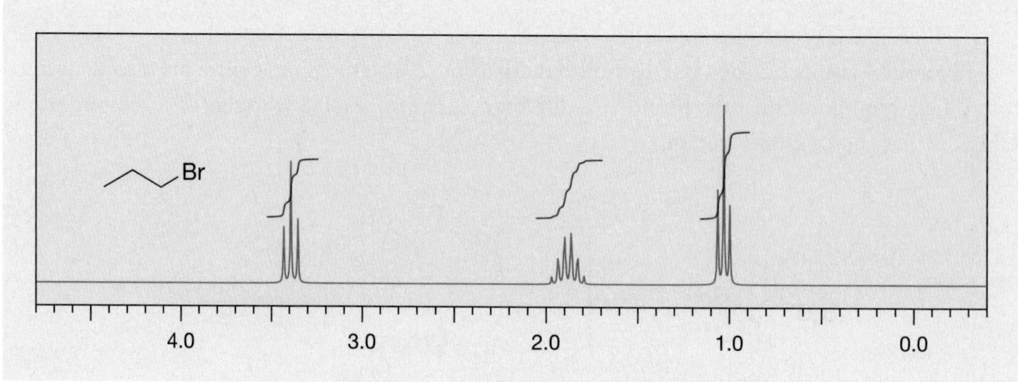

STEP 1: Label the peaks in the spectrum and identify their coupling patterns.

Signal a is a triplet, signal b is a sextet, and signal c is another triplet.

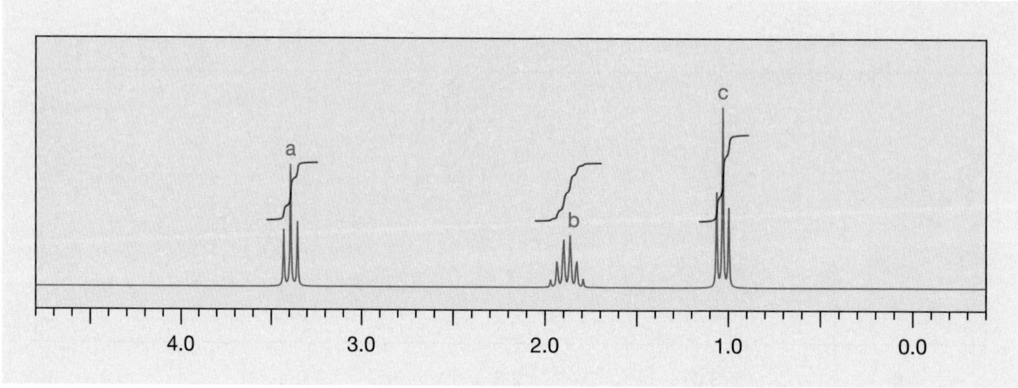

STEP 2: Determine the number of adjacent hydrogen atoms for each signal.

The $n+1$ rule indicates that a triplet results from a hydrogen that has two adjacent hydrogens ($n = 2$). Similarly, a sextet arises from a hydrogen with five adjacent hydrogens.

STEP 3: Determine the expected coupling pattern for each set of non-equivalent hydrogens.

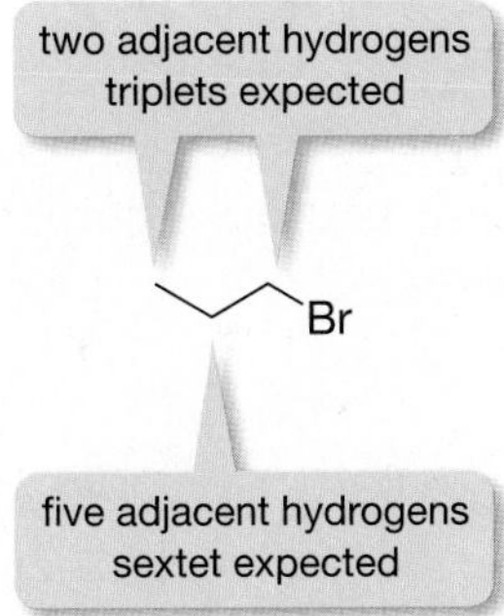

STEP 4: Use the coupling patterns, integration values, and chemical shifts to assign the signals in the spectrum.

Signal c is a triplet that integrates for three hydrogens, so it is produced by the methyl group, which is beside a CH_2. Signal a is a triplet that integrates for two hydrogens and can only be produced by the CH_2 next to the Br atom. These assignments are consistent with the chemical shifts: the CH_2 next to the electronegative Br atom is expected to be shifted downfield relative to the other peaks. The sextet signal b must therefore originate from the hydrogens on the middle carbon.

c a Br
b

PRACTICE PROBLEM

13.11 a) Predict the coupling pattern for each hydrogen in the following compounds. Include both the number of peaks in each signal and the relative intensities.

i) S

ii) O

iii) OH

iv) Br N

b) Match each compound with one of the following spectra. Explain your reasoning.

A B C

i)

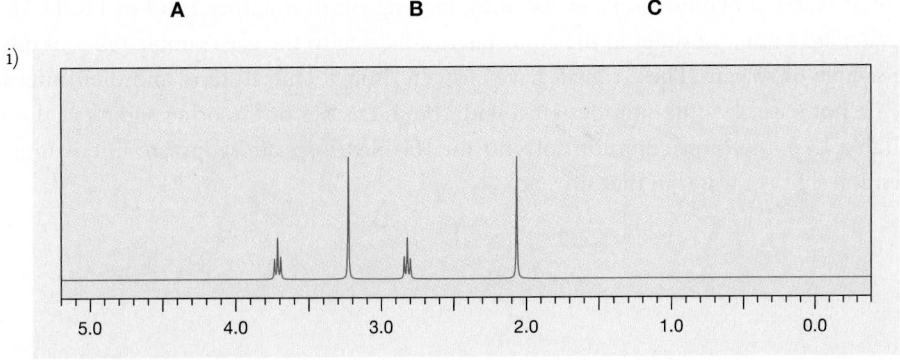

ii)

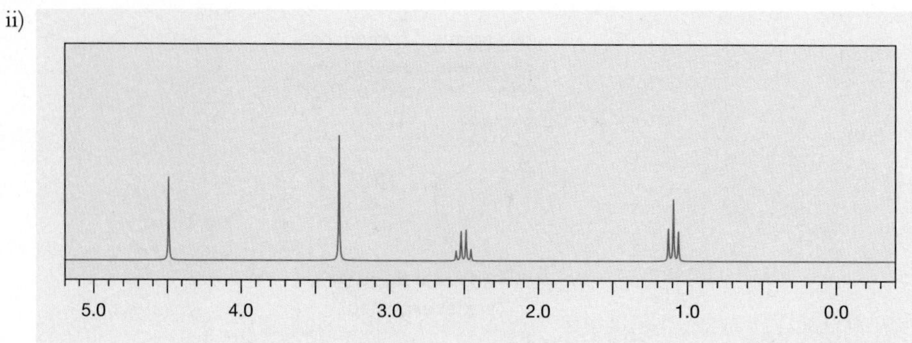

iii)

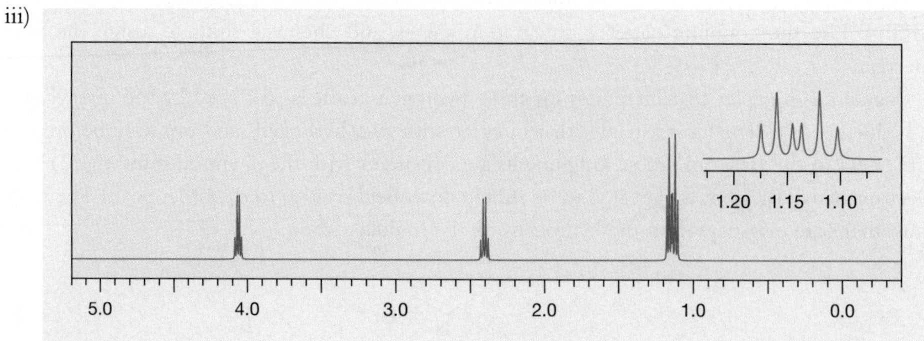

INTEGRATE THE SKILL

13.12 Explain how you could use ^{1}H-NMR spectra to distinguish between each pair of compounds.

a) Cl Cl

b) O O

c) NO_2 NO_2 NO_2 NO_2

13.4.5 Solvent signals in ^{1}H-NMR spectra

Deuterated solvents used for ^{1}H-NMR spectroscopy always contain a small residual amount of non-deuterated solvent, which creates an unwanted signal in the spectrum that needs to be ignored. Water contamination can also introduce signals from either H_2O or HOD. These traces of water are a result of water in the atmosphere adhering to the glass sample tubes or dissolving in the sample or solvent. These signals have characteristic chemical shifts, and their intensities usually do not have a whole-number ratio with the intensities of the other signals in the spectrum. In Table 13.3, the most common solvents for ^{1}H-NMR spectroscopy are listed along with the chemical shifts of water in that solvent.

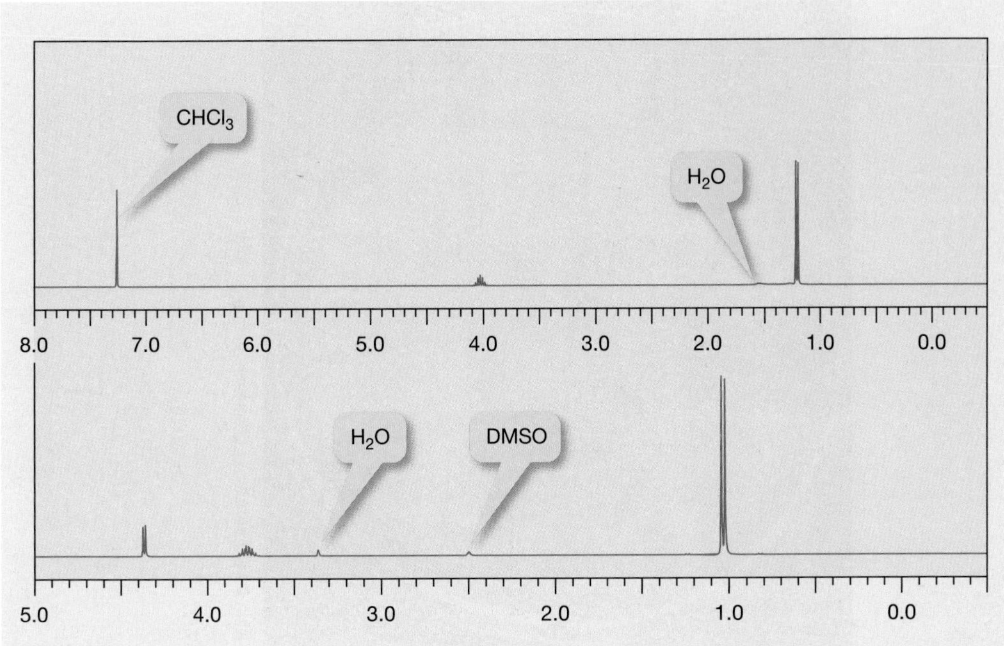

TABLE 13.3 ¹H-NMR Solvents

Solvent	Non-deuterated Signal	Water contamination
$CDCl_3$	Singlet, 7.26 ppm	1.55 ppm
DMSO-d_6	Heptet, 2.50 ppm	3.31 ppm
Benzene-d_6	Singlet, 7.16 ppm	0.4 ppm

It is important to be able to recognize solvent signals, because then they can be ignored. The correct structure can be identified only by using the signals that arise from that structure. Trying to interpret a spectrum without ignoring signals from solvent or other contaminants will never lead to the correct structure, only to frustration.

CHEMISTRY: EVERYTHING AND EVERYWHERE

Magnetic Resonance and Imaging

The same principles that produce nuclear magnetic resonance spectra of molecules can be applied to produce images of living tissue using a technique called magnetic resonance imaging (MRI). MRI produces a three-dimensional map of the different environments of the water molecules in living tissue. The two general types of magnetic resonance images are T1 and T2/T2* weighted. In a T1-weighted image, water molecules that relax more quickly from the β to the α state appear brighter. In T2/T2*, fast-relaxing water appears darker. In more dense tissue, water molecules relax more quickly, which leads to a contrast between different tissue types in a very high resolution. Figure 13.17 shows a 2-D horizontal (transverse) slice through the MRI of a human brain.

Continued

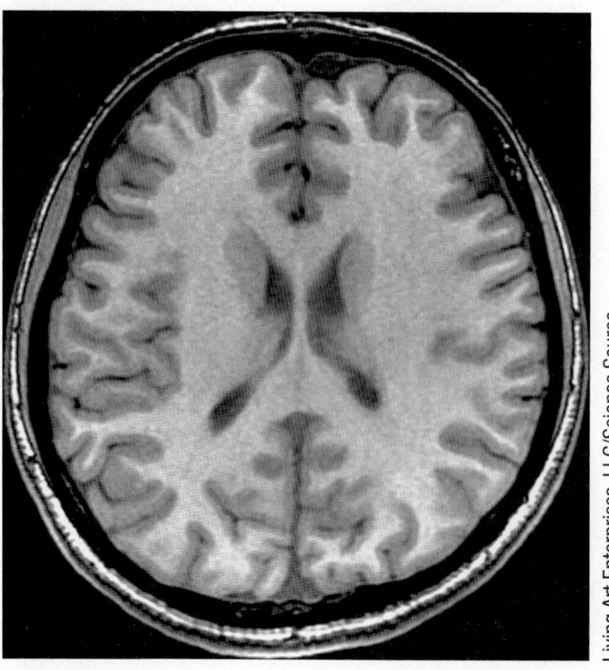

Living Art Enterprises, LLC/Science Source

FIGURE 13.17 Transverse plane T1-weighted MRI of a human brain.

The MRI instrument uses magnetic field gradients to determine XYZ coordinates for the various water signals in the region being scanned. Because the frequency required to excite a proton depends on the strength of the applied magnetic field, the resonance frequency of each water molecule indicates its location along the gradient. By controlling the gradients in all three dimensions, the location of the water hydrogen in 3-D space can be determined. Computer processing of successive closely spaced scans builds up a 3-D map of the density of the tissue, with denser areas appearing brighter.

A **contrast agent** is a compound that increases the contrast of an MRI image.

Greater difference between bright and dark areas gives the image greater resolution. Patients may be administered a **contrast agent**; this compound increases the relaxation rate of the hydrogens in water (Figure 13.18). Gadolinium compounds are commonly used as contrast agents because gadolinium speeds the relaxation process for water hydrogens. This property enables gadolinium to quickly accept excitation energy from any water to which the agent binds. As a result, tissue that absorbs the contrast agent appears brighter.

FIGURE 13.18 Gadovist, a paramagnetic contrast agent increases the relaxation rate of coordinated water, brightening T1-weighted images.

Blood pool MRI, sometimes called MR angiography, uses contrast agents within the circulatory system to give high definition images of blood vessels (Figure 13.19).

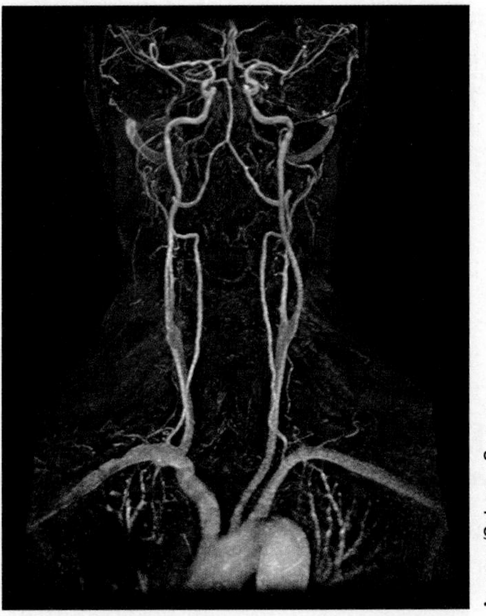

FIGURE 13.19 Gadolinium contrast agent MS-325 (Ablavar) binds to human serum albumin, highlighting the blood vessels in this image.

13.5 Determining Molecular Structures from NMR Spectra

NMR provides information about molecular structure that is not available from other types of spectroscopy. There is no one "right" way to deduce an unknown structure from NMR spectra. However, the following procedure is often useful:

1. If the molecular formula is known, calculate the molecule's degree of unsaturation.
2. Use integration to determine the number of hydrogen atoms that produce each signal.
3. Use splitting patterns to determine the number of adjacent hydrogens, and then determine the structure of as many fragments in the molecule as possible.
4. Use chemical shifts to identify nearby heteroatoms and functional groups.
5. Summarize what is known about the molecular structure.

The solved problems in Checkpoint 13.7 show how the information in Section 13.4 can be combined to solve the molecular structure using a ^{1}H-NMR spectrum.

13.5.1 Degree of unsaturation

Saturated molecules have the maximum number of σ bonds possible for the number of carbons they contain. Alkanes have a general molecular formula of $C_nH_{(2n+2)}$, where n is any positive integer. For example, hexane, with $n = 6$, has $2 \times 6 + 2$, or 14 hydrogens. An unsaturated molecule has fewer hydrogens than an alkane. For example, both hexene and cyclohexane have two fewer hydrogens, giving them both a molecular formula of C_6H_{12}.

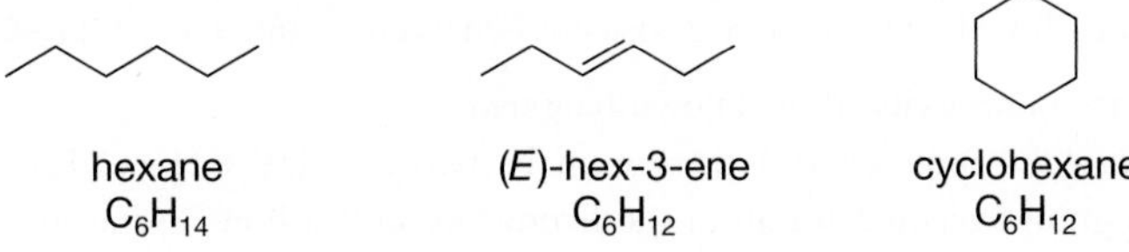

The **degree of unsaturation** (DoU) of a molecule is a measure of the number of rings and/or π bonds it has.

The absence of two hydrogens indicates that the molecule has either a π bond or a ring, both of which can be termed a **degree of unsaturation** (DoU). The simplest way to determine the degree of unsaturation is to use equation 13.2:

$$\text{DoU} = \frac{2\times C+2+N-X-H}{2} \tag{13.2}$$

C = # of carbons
N = # of nitrogens
X = # of halogens
H = # of hydrogens

CHECKPOINT 13.7

You should now be able to use a ^{1}H-NMR spectrum to determine the structure of an unknown compound, given its molecular formula.

SOLVED PROBLEM 1

Use the following spectrum to determine the structure of the compound, given that its molecular formula is $C_6H_{14}O$.

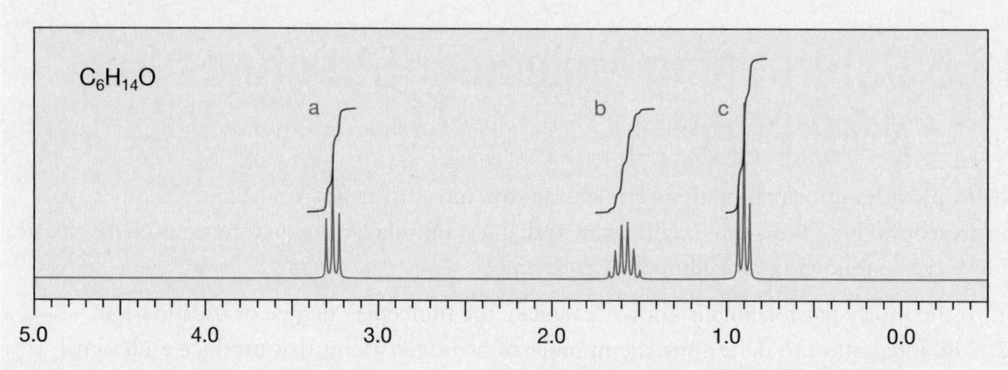

STEP 1: Determine the compound's degree of unsaturation using equation 13.2. The DoU is zero for this molecule (no rings or π bonds).

STEP 2: Determine the number of each type of hydrogen.

The ratio of the three integrated signals in the spectrum is 2:2:3. Because the molecule has 14 hydrogen atoms, the actual ratio must be 4:4:6. It is impossible for a carbon to hold six hydrogens; therefore, each resonance must result from more than one carbon, and so the structure must have some symmetry.

STEP 3: Use the coupling patterns to determine the structure of sections of the molecule.

Start with the simpler signals. Signal C is a triplet that integrates for six hydrogens, and so it must be produced by two identical CH_3 groups. The n+1 rule indicates that these methyl groups each have two adjacent hydrogens. Therefore, part of the molecule has the structure $C\underline{H}_3$–CH_2–.

The triplet signal A integrates for four hydrogens, and so it must be produced by two identical CH_2 groups. According to the n+1 rule, they each have two adjacent hydrogens and therefore must correspond to two identical $C\underline{H}_2$–CH_2 groups.

Signal B also integrates for four hydrogens, and must originate from two identical CH_2 groups. However, signal B is a sextet, so each CH_2 must have five adjacent hydrogens as in a CH_2–$C\underline{H}_2$–CH_3 group.

STEP 4: Determine the connectivity of the known fragments.

The integration of 4:4:6 indicates that there must be two identical CH_2–CH_2–CH_3 groups in this molecule. These two groups account for all of the hydrogens and carbons in the structure. There is also an unaccounted oxygen. The only way to assemble these fragments to form a symmetrical molecule is $CH_3CH_2CH_2OCH_2CH_2CH_3$.

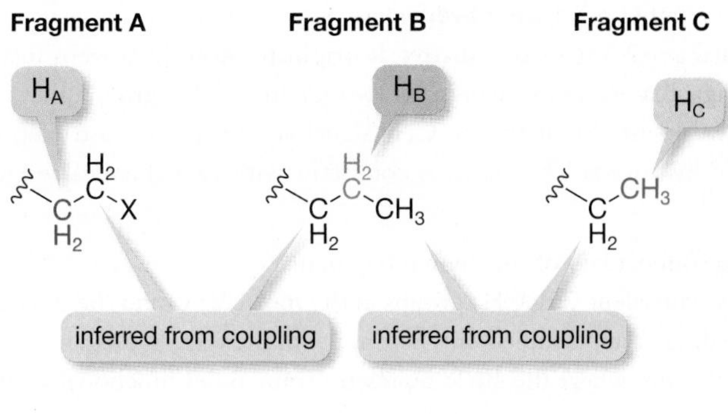

STEP 5: Use chemical shifts to determine the overall structure.

The chemical shift of the CH_2 triplet (3.25 ppm) indicates proximity to an electron-withdrawing atom or group, which must be the oxygen of the molecular formula. The resonance at 3.25 ppm is consistent with an ether. The chemical shifts of the resonances at 0.9 ppm and 1.6 ppm are consistent with the structure of dipropyl ether.

STUDENT TIP

Whenever an adjacent hydrogen is inferred to exist because of coupling, there must be a split signal elsewhere in the spectrum from that hydrogen.

SOLVED PROBLEM 2

Following is the ^{1}H-NMR spectrum for an isomer of $C_5H_{10}O$. Determine the structure of this isomer.

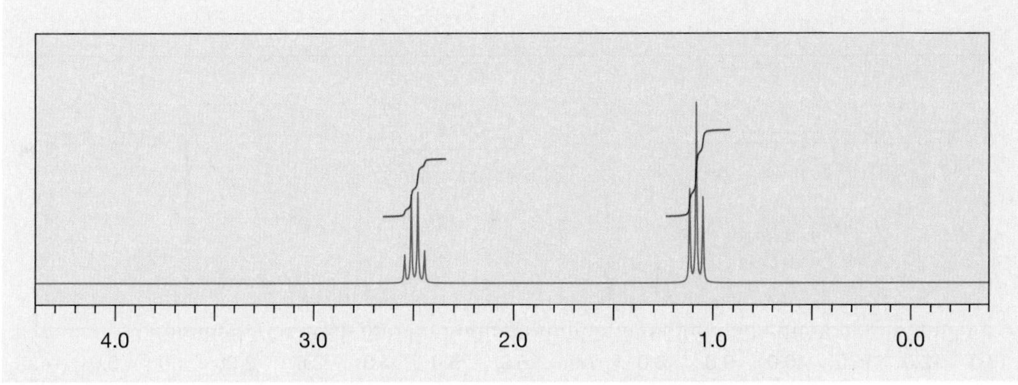

STUDENT TIP

Work with the simplest resonances first. From time to time summarize what you have deduced.

STEP 1: Determine the degree of unsaturation from the molecular formula.

The molecule contains five carbon atoms, ten hydrogens, and no nitrogens or halides. The oxygen atom does not affect the saturation. Substituting these values into the formula for degree of saturation gives the following:

$$\text{DoU} = \frac{2\times\text{C}+2+\text{N}-\text{X}-\text{H}}{2}$$

$$\text{DoU} = \frac{2\times5+2+0-0-10}{2}$$

$$\text{DoU} = 1$$

STEP 2: Determine the number of hydrogens associated with each signal.

The two signals have a 2:3 intensity ratio. Without the presence of symmetry, the integration accounts for only five hydrogens. The presence of ten hydrogens in the molecular formula establishes that the molecule must be symmetrical. Therefore, the signal at ~2.5 ppm must represent two chemically equivalent CH_2 groups, whereas the signal at ~1.0 ppm represents two chemically equivalent CH_3 groups.

STEP 3: Determine the number of adjacent hydrogens.

Since the CH_2 signal at ~2.5 ppm is a quartet, it originates from hydrogens that have three adjacent hydrogens. Since the signal at ~2.5 ppm represents two identical CH_2 groups, they must each be part of identical $-C\underline{H_2}-CH_3$ fragments. Similarly, the CH_3 signal at ~1.0 ppm is a triplet, indicating its hydrogens have two adjacent hydrogens. This signal is consistent with two chemically equivalent $-CH_2-C\underline{H_3}$ fragments.

STEP 4: Determine the connectivity of the known fragments.

The two chemically equivalent CH_2CH_3 groups in the molecule cannot be directly connected to each other since the CH_2 hydrogens are coupled only to one set of the CH_3 hydrogens. Therefore, the molecule must look something like this, where the circle represents some other functional group or heteroatom:

$$CH_3CH_2-(?)-CH_2CH_3$$

STEP 5: Use chemical shifts to determine the overall structure.

The two ethyl fragments leave one carbon and one oxygen unaccounted for in the molecular formula. You also know that the molecule has one degree of unsaturation. Therefore, the two ethyl fragments are probably linked by a carbon with a double bond to the oxygen. Alkyl groups next to a carbonyl have chemical shifts between 2.0 and 2.7 ppm, which is consistent with the quartet located at ~2.5 ppm. So the chemical shifts confirm that the structure of the molecule is as follows.

O

PRACTICE PROBLEM

13.13 Determine the structure of each unknown compound using the ^{1}H-NMR data provided along with the molecular formula.

a) $C_{10}H_{12}O_2$

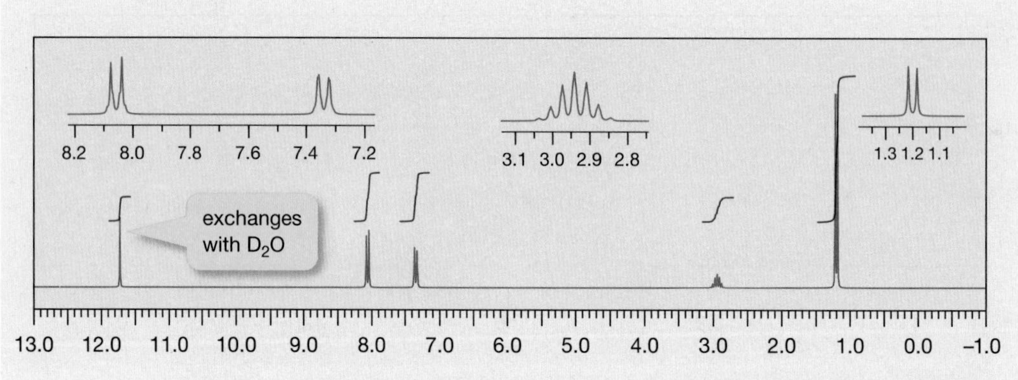

b) $C_6H_4Br_2$

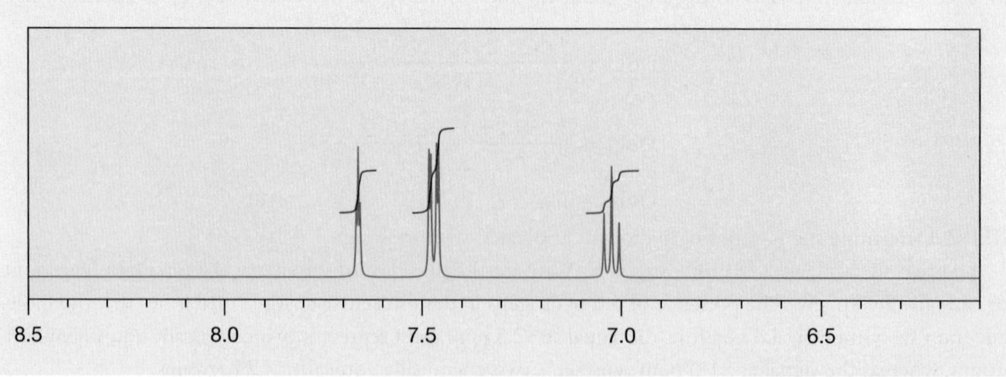

c) $C_7H_{14}O$

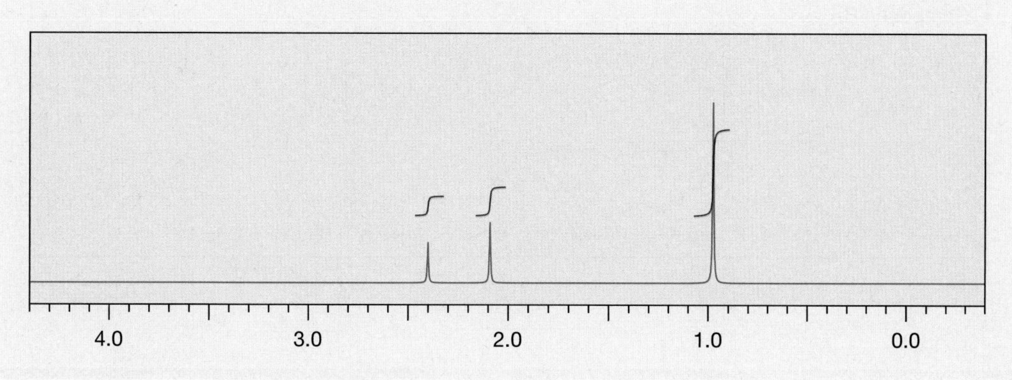

d) C_4H_7N

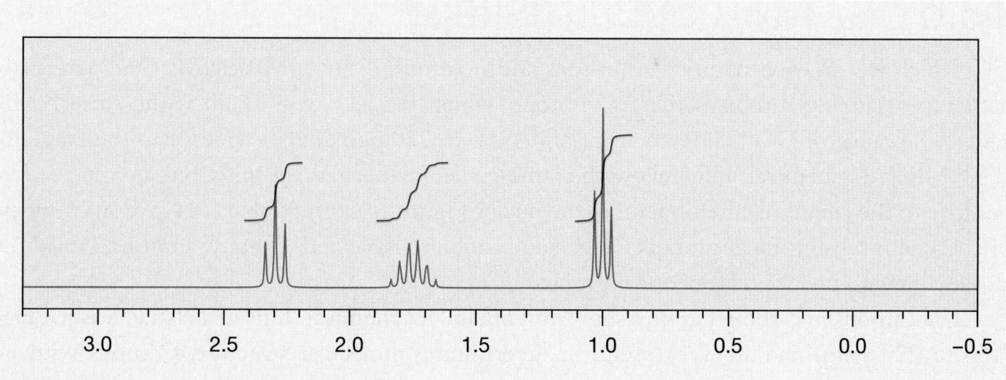

INTEGRATE THE SKILL

13.14 Compound A, with a molecular formula of C_6H_{12}, reacts with HCl to give compound B. The ^{1}H-NMR spectra of both compounds are shown below. Determine the structures of compounds A and B.

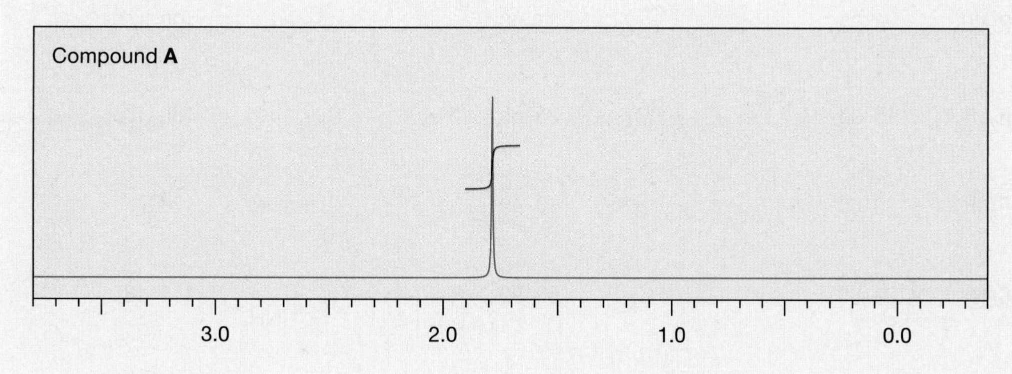

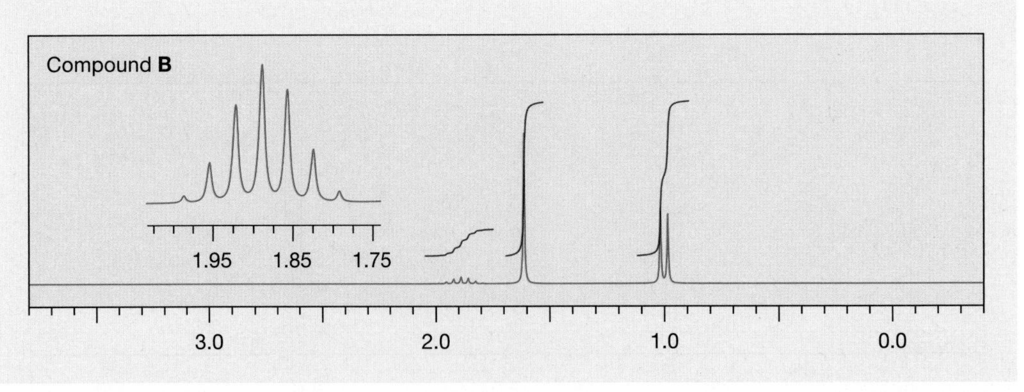

13.6 ^{13}C-NMR Spectroscopy

^{12}C, which has an even number of protons and neutrons, is not NMR active. One percent of naturally occurring carbon is the ^{13}C isotope, which has a +½ spin and is therefore NMR active. The range of ^{13}C-NMR chemical shifts is 0 to 220 ppm, much wider than the range for ^{1}H-NMR. ^{13}C chemical shifts follow the same general pattern as ^{1}H shifts, but are much more sensitive to the chemical environment of the nuclei. Figure 13.20 and Table 13.4 give an overview of ^{13}C chemical shift ranges for alkanes, alkenes, and aromatic and carbonyl carbons. Table 13.4 also lists ranges for specific functional groups.

The carbons in carbonyl groups show downfield ^{13}C chemical shifts in a characteristic range (160 to 220 ppm) and are very useful in determining molecular structure. Carbons with no attached hydrogens, called *quaternary carbons*, produce no ^{1}H-NMR signals, but are detectable in ^{13}C-NMR spectra.

TABLE 13.4 Characteristic ^{13}C Chemical Shifts of Common Functional Groups

Structure	δ (ppm)	Structure	δ (ppm)	Structure	δ (ppm)
R–**C**H$_3$	8–30	Cl–**C**	35–80	C=**C**	100–150
R$_2$**C**H$_2$	15–55	Br–**C**	25–65	(aromatic ring carbon, R-substituted benzene)	100–175
R$_3$**C**H	20–60	I–**C**	0–40	Benzene	128.5
C–N	30–65	C≡**C**	65–90	–**C**(=O)NH$_2$	155–185
C–O	40–80	N≡**C**	65–90	–**C**(=O)O(R,H)	155–185
–C(=O)–**C**	30–60	Ethene	123.3	–**C**(=O)(C,H)	185–220

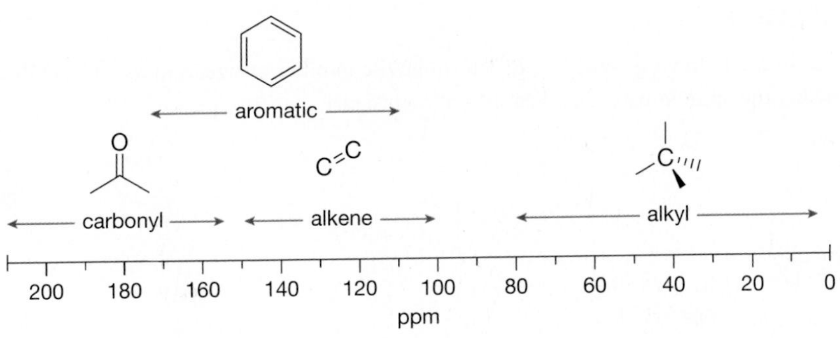

FIGURE 13.20 Carbon chemical shift ranges, according to carbon type.

STUDENT TIP

The chemical shifts of the carbons in a molecule are approximately 20 times the chemical shifts of the hydrogens attached to the carbons. For example, if the hydrogens in an alkyl chain resonate around 1.2 ppm, the carbons they are attached to resonate at 25 ppm. If the hydrogens of a CH_3–O group resonate at approximately 3.8 ppm, the corresponding carbon gives a signal at 75 ppm.

CHECKPOINT 13.8

You should now be able to predict the general features of the ^{13}C-NMR spectrum of a compound, including the number of different signals and their approximate chemical shifts.

SOLVED PROBLEM

How many signals appear in the ^{13}C-NMR spectrum of the following compound? Determine the approximate chemical shift of each signal.

STEP 1: Determine whether any of the carbons are chemically equivalent.

This molecule has no planes of symmetry, and none of the carbons are equivalent through rotation. So none of the carbons are chemically equivalent.

STEP 2: Determine the number of signals in the ^{13}C-NMR spectrum.

Since none of the nine carbon atoms in the molecule are chemically equivalent, the ^{13}C-NMR spectrum will have nine signals.

STEP 3: Determine the approximate chemical shift of each signal.

Label the carbons in the compound, and then use Table 13.4 to estimate the chemical shifts. These are listed in Table 13.5.

TABLE 13.5 ^{13}C-NMR Chemical Shift

Carbon Atom	Expected Chemical Shift Range (ppm)
a	8–30
b–g	100–175
h	15–55
i	35–80

PRACTICE PROBLEM

13.15 For each of the following compounds, determine the number of signals in its ^{13}C-NMR spectrum and predict the approximate chemical shift of each signal.

a) O O

b)

c) OH

d) NC NH_2

e) O

INTEGRATE THE SKILL

13.16 Below are the ^1H-NMR and ^{13}C-NMR spectra for a compound with the molecular formula $C_9H_{13}Cl$. Determine the structure of the compound and identify the atoms producing each peak in the spectra.

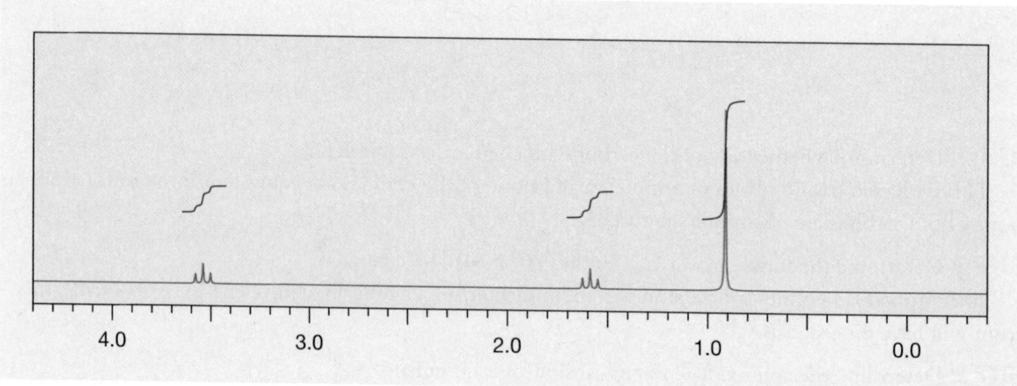

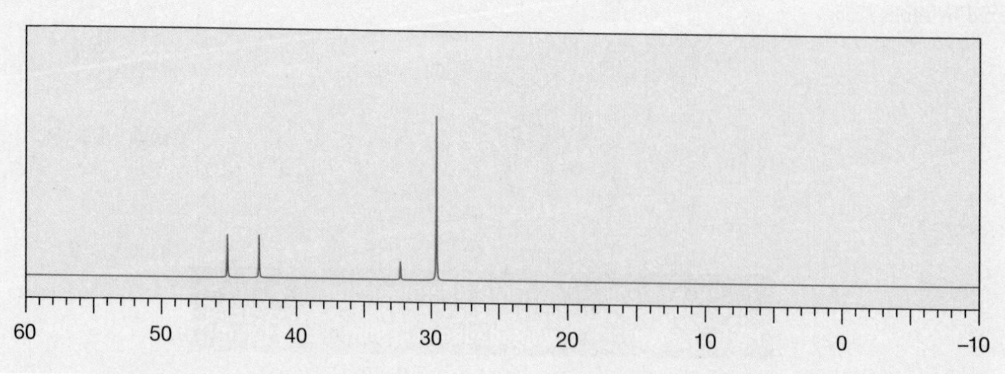

13.6.1 Solvent peaks in ^{13}C-NMR

It is difficult and costly to make solvents in which all the carbon atoms are ^{12}C, so ^{13}C spectra are recorded in solvents that have a natural abundance of the carbon isotopes (99% ^{12}C). The most common NMR solvents contain deuterium, which splits the solvent signal in characteristic ways, making it possible to distinguish between solvent and sample resonances. For example, the carbon resonance for $CDCl_3$ appears as a 1:1:1 triplet at 77.0 ppm; note that this solvent signal does not follow either the $n+1$ rule for the number of peaks or Pascal's triangle for their intensities.

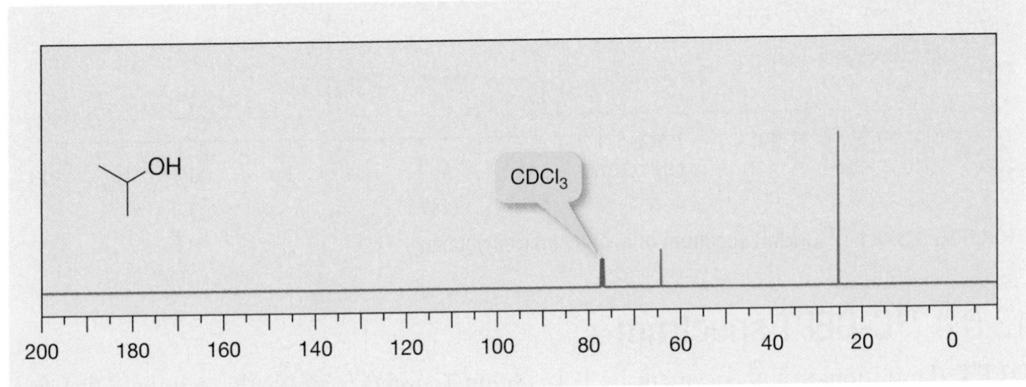

13.6.2 Proton decoupling and the nuclear Overhauser effect (nOe)

Since only 1.1 percent of natural carbon is ^{13}C, the likelihood of two adjacent carbons in a molecule both being ^{13}C is only 1/10 000. As a result, carbon-carbon coupling signals are too weak to be observed. However, most carbons in a molecule are bonded to hydrogens, a configuration that results in very strong C–H coupling and widely split ^{13}C-NMR signals. To simplify analysis and increase signal strength, the hydrogen resonances are normally **decoupled** from the carbons in ^{13}C spectroscopy. This is accomplished by transmitting a continuous signal of all the excitation frequencies of hydrogen while the instrument is recording in the carbon frequency range. This causes the hydrogens to flip very rapidly between the spin states, thus averaging out the coupling effects. In such decoupled spectra, each carbon signal appears as a singlet.

Decoupling a ^{13}C-NMR spectrum eliminates any splitting by attached hydrogens, making each carbon signal appear as a singlet.

Due to the continual flood of energy to the hydrogen atoms, some of that energy is transferred to the carbons of the compound. This **nuclear Overhauser effect (nOe)** causes the signals of carbons with directly connected hydrogens to appear stronger than signals from carbons with no hydrogens. This effect makes it impossible to integrate ^{13}C-NMR spectra.

The **nuclear Overhauser effect (nOe)** causes carbon signals in a decoupled spectrum to be taller if there are attached hydrogens. Carbons without attached hydrogens give very short signals.

13.6.3 Interpretation of ^{13}C-NMR

^{1}H-NMR gives information about the number of hydrogen types, chemical shift, integration, and coupling. ^{13}C-NMR gives information only about the number of carbon types and chemical shift.

ORGANIC CHEMWARE
13.25 ^{13}C-NMR spectrum of bromopropane

For example, the structure of 4-(2-hydroxyethyl)phenol, shown in Figure 13.21, has six sp^2-hybridized carbons, which appear between 100 ppm and 160 ppm. Because of symmetry, the six carbons produce only four signals. The two sp^3-hybridized carbons appear below 80 ppm. The two hydroxyl groups have no carbons and do not appear in a ^{13}C-NMR. However, their effect on the chemical shift is apparent. The downfield sp^3-hybridized carbon signal is from the alkyl carbon bonded to oxygen.

ORGANIC CHEMWARE
13.26 ^{13}C-NMR spectrum of 2-bromobutane

ORGANIC CHEMWARE
13.27 ^{13}C-NMR spectrum of 2-butanone

ORGANIC CHEMWARE
13.28 ^{13}C-NMR spectrum of isopropyl ethanoate

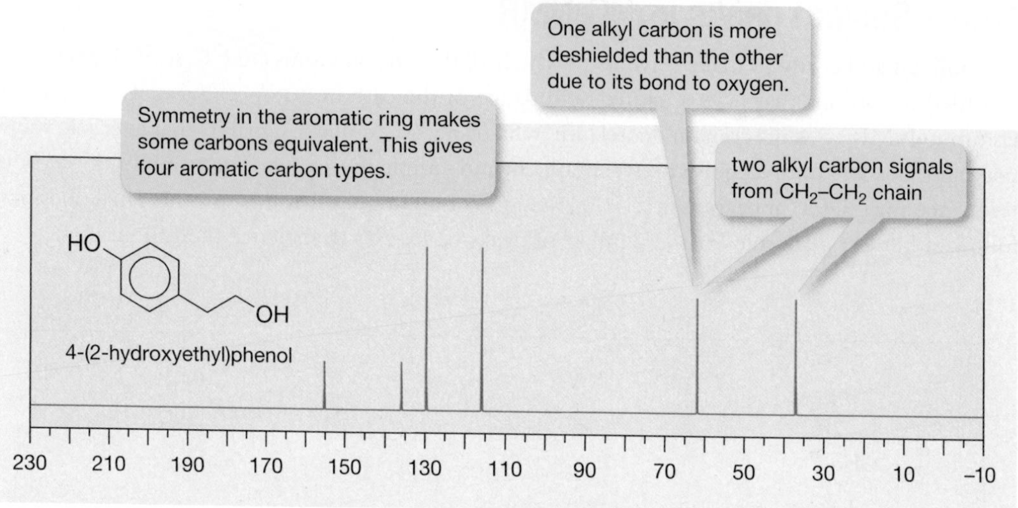

FIGURE 13.21 ^{13}C-NMR spectrum of 4-(2-hydroxyethyl)phenol.

13.6.4 ^{13}C-DEPT spectrum

A **DEPT** spectrum is a variation of a ^{13}C-NMR spectrum that distinguishes how many hydrogens are attached to each carbon.

DEPT (Distortionless Enhancement by Polarization Transfer) is an NMR technique that uses special radio frequency (RF) pulses to differentiate between carbons with different numbers of directly connected hydrogens. This technique changes the apparent phase of ^{13}C signals, making some signals disappear and moving others above or below the baseline of the spectrum.

In all three types of DEPT experiments, the signals of carbons without hydrogens (quaternary carbons) disappear. The quaternary carbons can be identified by observing which peaks are missing in a DEPT spectrum (Table 13.6). In DEPT-135, the most useful type, the CH_3 signals and CH signals both appear above the baseline (positive phase), whereas the CH_2 signals appear below the baseline (negative phase). It is usually possible to differentiate between CH and CH_3 carbons because the nOe makes the CH_3 carbon signals appear much stronger. The other two DEPT techniques are typically used only in special circumstances.

TABLE 13.6 **Summary of Information Conveyed by Phase in DEPT Spectra**

DEPT Type	CH_3	CH_2	CH	C
DEPT-45	Positive	Zero	Positive	Zero
DEPT-90	Zero	Zero	Positive	Zero
DEPT-135	Positive	Negative	Positive	Zero

The standard ^{13}C-NMR of 4-methyl-pentan-2-one has five carbon types, as shown in Figure 13.22. The carbonyl carbon signal at 207.5 ppm is absent from the DEPT-135 spectrum, indicating that this carbon is not connected to a hydrogen. The three methyl groups appear as two signals because of the symmetry of the isobutyl methyls. These signals and the CH resonances appear above the baseline in the DEPT spectrum, and the CH_2 signal appears below the baseline.

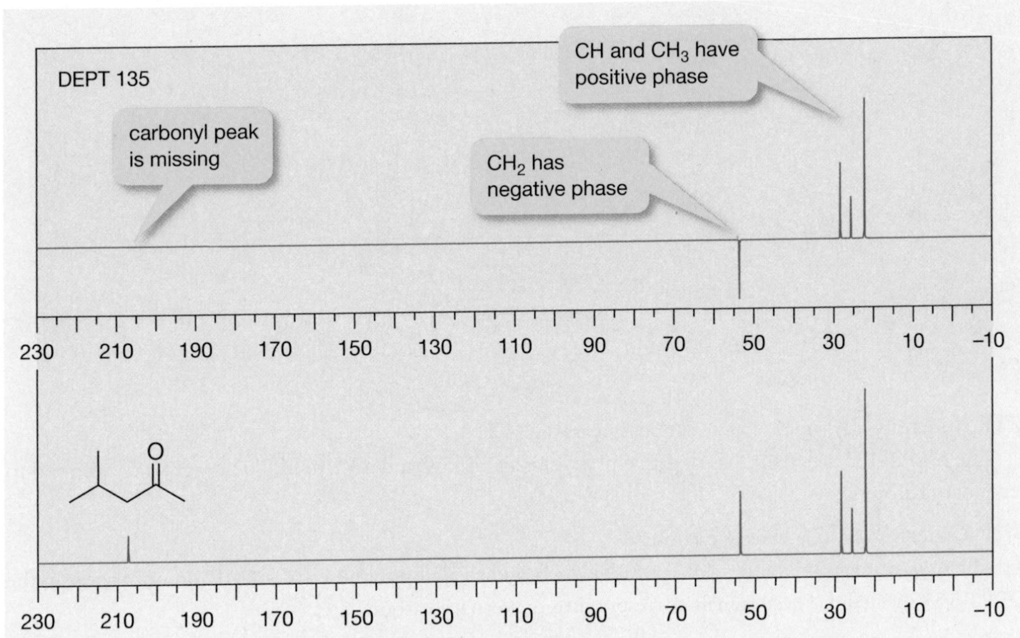

FIGURE 13.22 DEPT-135 and standard spectra of 4-methyl-2-pentanone.

CHECKPOINT 13.9

You should now be able to predict and interpret NMR data from a DEPT-135 experiment.

SOLVED PROBLEM

Below is the standard ^{13}C-NMR spectrum for a compound with the molecular formula $C_{10}H_{10}O_2$.

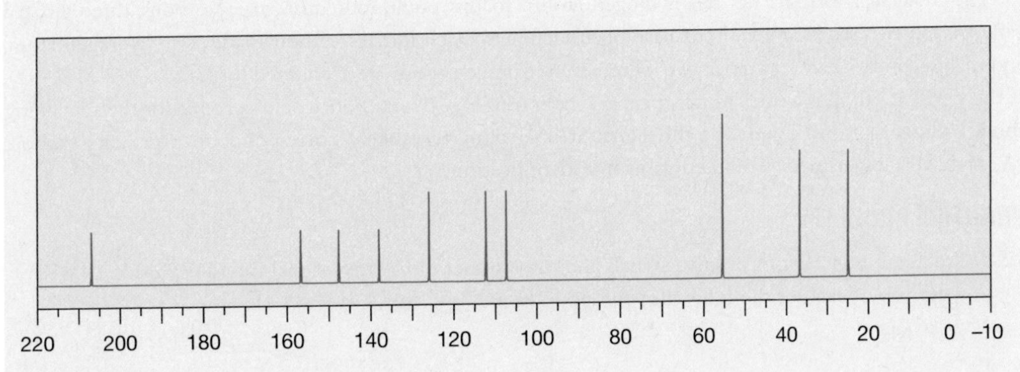

The DEPT-135 NMR experiment shows positive signals at 125.6, 114.3, 109.6, and 55.3 ppm, and negative signals at 35.9 and 28.1 ppm. Determine the number of quaternary, tertiary, secondary, and primary carbons present in the compound.

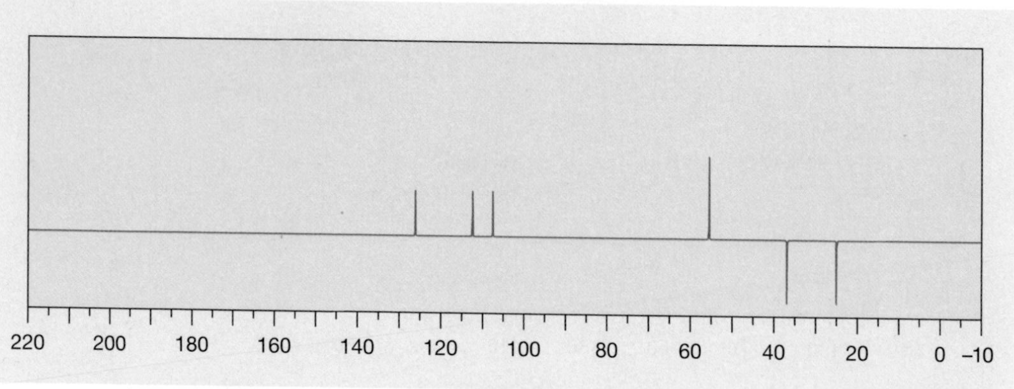

STEP 1: Identify any signals from secondary carbons (CH_2).

The DEPT-135 spectrum has negative phase signals at 35.9 and 28.1 ppm. These signals correspond to secondary carbons.

STEP 2: Identify the quaternary carbons.

The signals at approximately 207.3, 159.8, 154.3, and 133.5 ppm in the ^{13}C-NMR do not appear in the DEPT-135 spectrum. These signals correspond to quaternary carbons.

STEP 3: Compare the relative peak heights of the remaining signals.

The four remaining signals (at 125.6, 114.4, 109.6, and 55.3 ppm) all have positive phase in the DEPT-135 spectrum, indicating that they are either primary (CH_3) or tertiary (CH) carbons. Because of the nOe effect (Section 13.6.2), the relative heights of the ^{13}C signals may help distinguish between primary and tertiary carbons. Unfortunately, there is no obvious difference between the peak heights of these four signals.

STEP 4: Compare the chemical shifts of the remaining signals.

Three of the signals fall within the aromatic region of the spectrum, between 100 and 175 ppm. These signals can be identified as primary carbons because they are probably CHs that are part of an aromatic ring. The signal at 55.3 ppm could originate from either CH or CH_3.

STEP 5: Use the molecular formula to determine whether the signal at 55.3 ppm is a primary or tertiary carbon.

The molecular formula has ten hydrogen atoms. To this point, four quaternary carbons, three tertiary carbons, and two secondary carbons have been identified. Each tertiary carbon carries one hydrogen atom, and each secondary carbon carries two. Thus, seven hydrogen atoms are accounted for ($3 \times 1 + 2 \times 2 = 7$). If the signal at 56 ppm comes from a primary carbon (CH_3), this carbon would account for the remaining three hydrogen atoms. Therefore, the signal at 55.3 ppm very likely corresponds to a primary carbon. (A ^{1}H-NMR experiment would confirm this identification.)

PRACTICE PROBLEM

13.17 Predict the number of positive signals and the number of negative signals in a DEPT-135 NMR spectrum of each of the following compounds.

a) NH_2 / CN

b) Cl

c) O / O

d)

INTEGRATE THE SKILL

13.18 The ^{13}C-NMR spectrum for vinyl methacrylate is shown here. When a DEPT-135 NMR experiment was run, the signals at 19.1 and 144.8 ppm pointed up, and the signals at 126.3 and 97.4 ppm pointed down. Identify the carbon atoms producing each signal.

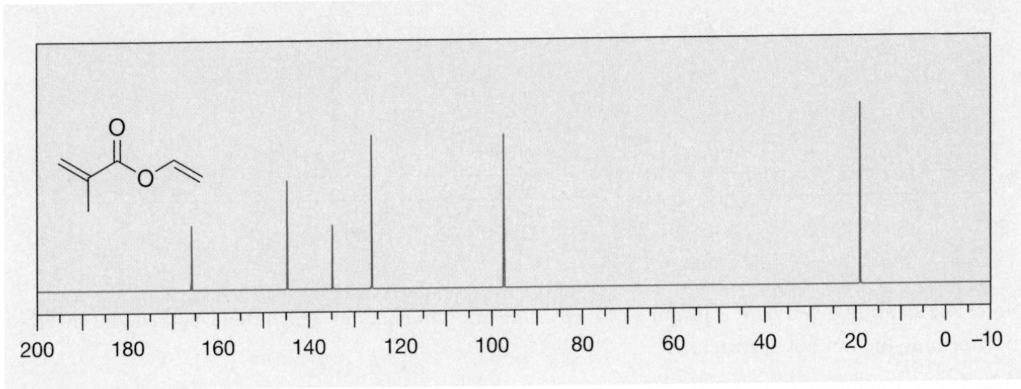

CHECKPOINT 13.10

You should now be able to combine ^{1}H-NMR and ^{13}C-NMR data to determine the structure of an unknown compound.

SOLVED PROBLEM

Below are the ^{1}H-NMR and ^{13}C-NMR spectra of an unknown compound. High-resolution data from the ^{1}H-NMR spectrum have been tabulated. Determine the structure of the compound.

^{1}H-NMR spectrum:

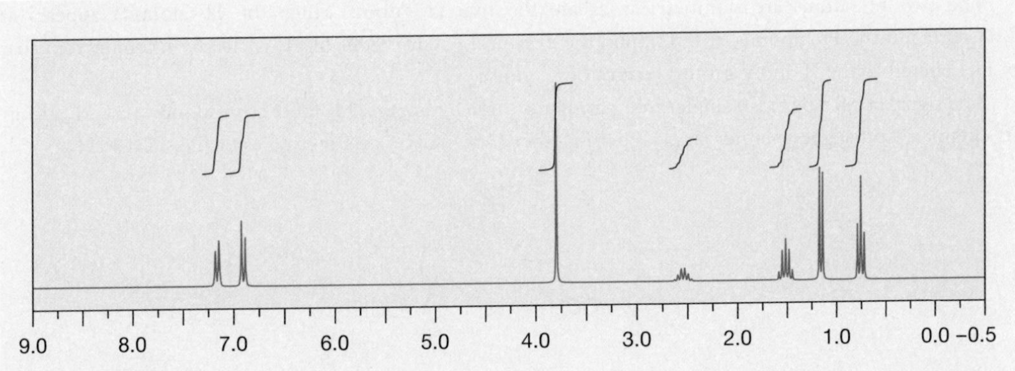

Signal	Chemical Shift, δ (ppm)	Integration	Coupling pattern
a	7.17	2	doublet
b	6.91	2	doublet
c	3.80	3	singlet
d	2.54	1	sextet
e	1.52	2	quintet
f	1.16	3	doublet
g	0.76	3	triplet

^{13}C-NMR spectrum:

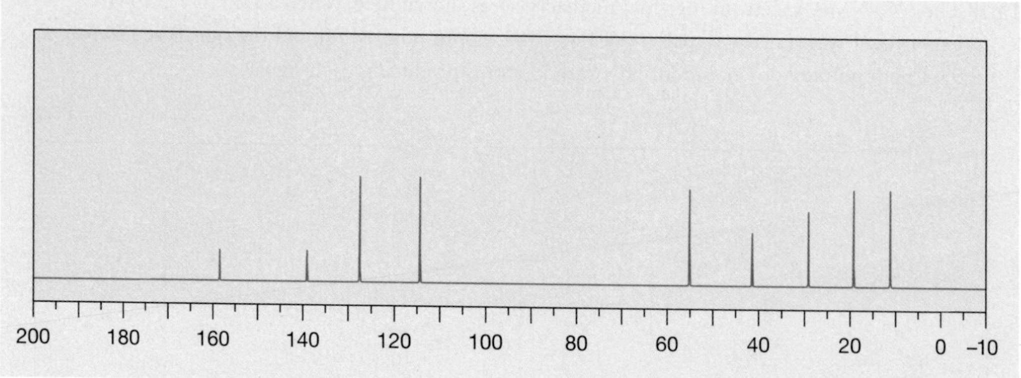

Note: The solution given here is just one of many different sequences in which the data could be analyzed to determine the unknown structure.

STEP 1: Use the integration and coupling patterns in the ^{1}H-NMR spectrum to determine the structure of fragments of the molecule.

It is often easiest to start with the simpler signals. Signals a and b in the ^{1}H-NMR spectrum are both doublets, which appear to couple to each other. This coupling pattern indicates one adjacent hydrogen; hence, the signals must come from side-by-side CH groups, that is, –CH–CH–. Since both signals integrate for two hydrogens, the molecule must have enough symmetry to contain two chemically equivalent –CH–CH– groups. The chemical shifts for the two signals are in the aromatic region, strongly suggesting a 1,4-disubstituted benzene ring.

The two H_a atoms are symmetrical, as are the two H_b atoms. Since the H_a doublet appears at 7.08 ppm and the H_b doublet at 6.82 ppm, the chemical environment of H_a is clearly different from that of H_b. Therefore the X and Y groups cannot be identical.

The signal at 3.75 ppm is a singlet integrating for three hydrogens. H_c must correspond to a CH_3 group. Its chemical shift suggests either an aryl methyl ether (ArO–CH_3) or a methyl ester (R–COOCH_3).

Signals d through g all have coupling patterns that can link them together. Signal g is a triplet integrating for three protons. This must be a CH_3 group coupled to a CH_2. Signal e is the only remaining signal in the spectrum integrating for two protons, so it must correspond to the CH_2 group that is coupled to signal g. Since signal e is a quintet, it corresponds to equivalent hydrogens with four adjacent hydrogens. The CH_3 group (signal g) accounts for three of the adjacent hydrogens, so the CH_2 group producing signal g must be attached to a CH group as well.

Signal f is a doublet integrating for three hydrogens and must be a $\underline{CH_3}$–CH group. Signal d is the only CH group, a sextet integrating for one hydrogen, which fits with both CH fragments shown above. Therefore, these fragments are connected by the CH. A sextet has five adjacent hydrogens, so the CH must also be bonded to a CH_3 group. This CH_3 group corresponds to signal f, which is a doublet integrating for three protons at 1.22 ppm. Adding this methyl group gives the following fragment.

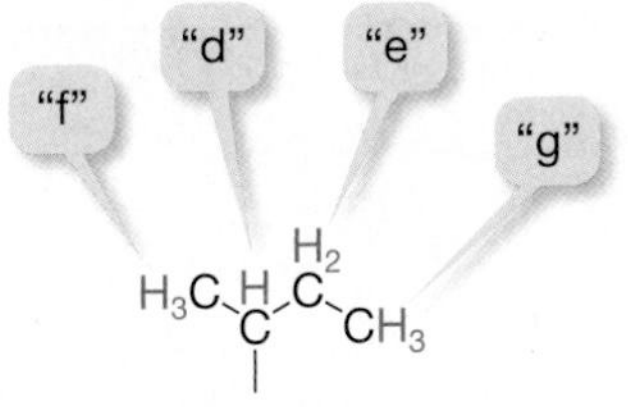

"f" "d" "e" "g"

H_3C–CH–CH_2–CH_3

The chemical shift of signal d is 2.53 ppm, suggesting that this fragment is attached to either an aromatic ring or a carbonyl.

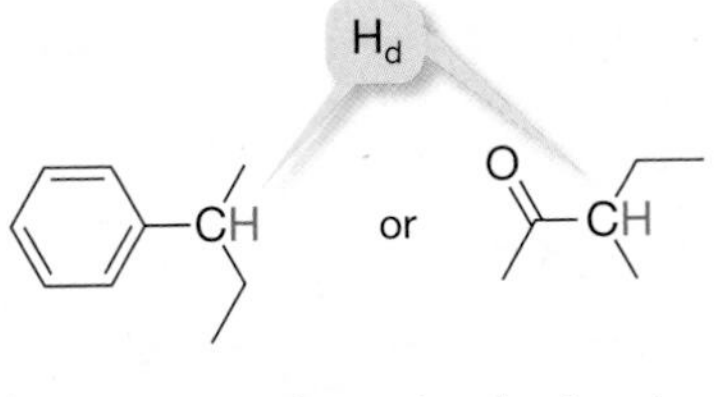

H_d

or

STEP 2: Use data from the ^{13}C-NMR spectrum to determine the functional groups connecting the fragments.

You now have three fragments to reconcile into a single structure:

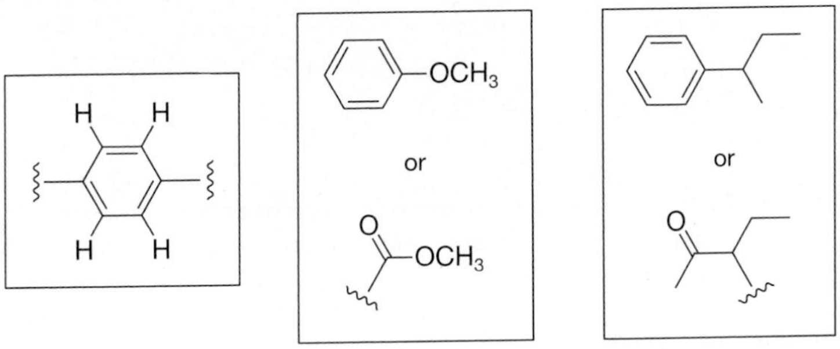

–OCH$_3$ or –OCH$_3$ | or

Carbonyl carbons appear between 155 and 220 ppm in ^{13}C-NMR spectra. There is a ^{13}C signal at 158.5 ppm, which could possibly come from an ester. However, we know the molecule contains a benzene ring that produces four signals in the aromatic region. Since the ^{13}C-NMR spectrum has only four such signals (including the signal at 158.5 ppm), none of these signals can be attributed to a carbonyl group. Therefore, the molecule is made up of the following three fragments:

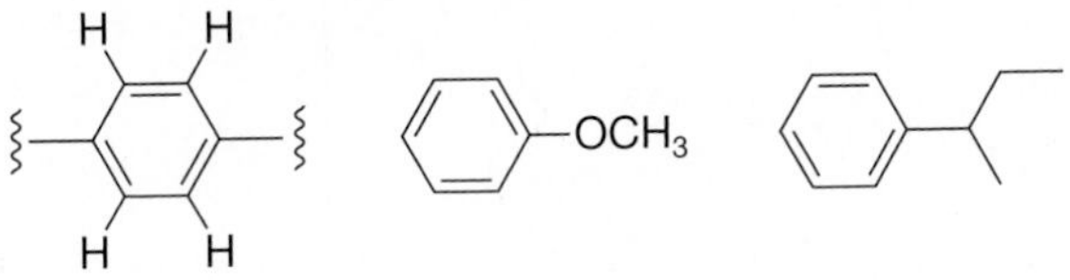

–OCH$_3$

There is only one way to combine these fragments:

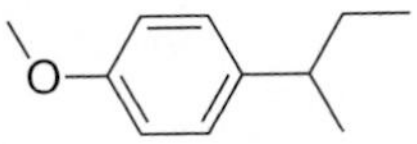

STEP 3: Confirm that the ^{13}C-NMR data fit with the proposed structure.

There are nine signals in the ^{13}C-NMR spectrum, which is the number expected for the proposed structure, given the symmetry in the benzene ring. As noted in Step 2, four of these signals are in the aromatic region between 100 ppm and 175 ppm, as expected for a benzene rig. The signal at 55.2 ppm fits with the methoxy carbon. The remaining four signals are in the aliphatic region. The signal at 41.4 ppm could correspond to the benzylic carbon connecting the *sec*-butyl group to the benzene ring.

PRACTICE PROBLEM

13.19 Determine the structures of the following unknown compounds using the data provided.

a) Molecular formula: $C_8H_{12}O$
^{1}H-NMR spectrum:

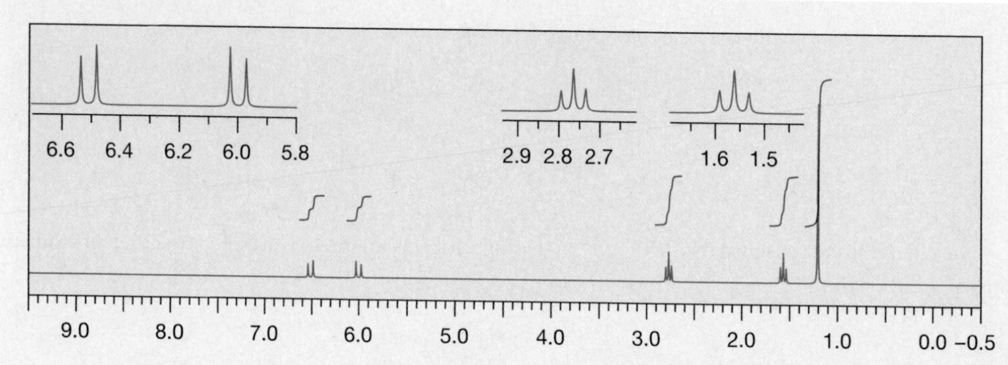

^{13}C-NMR spectrum:

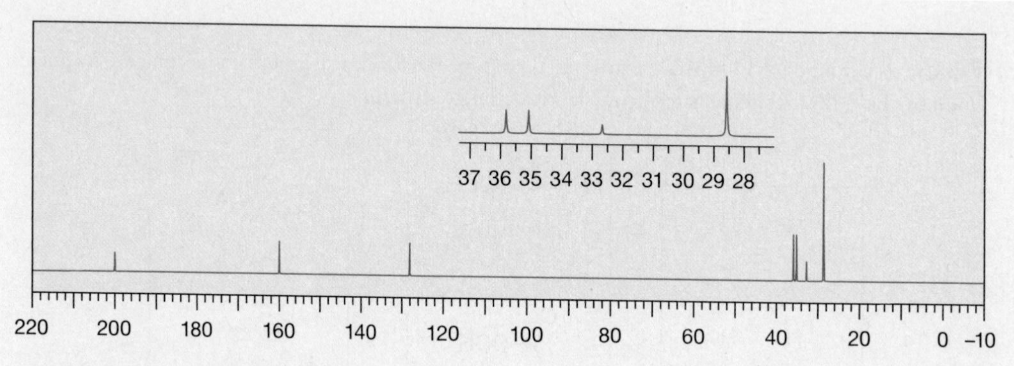

b) Molecular formula: $C_4H_8Br_2$
^{1}H-NMR spectrum:

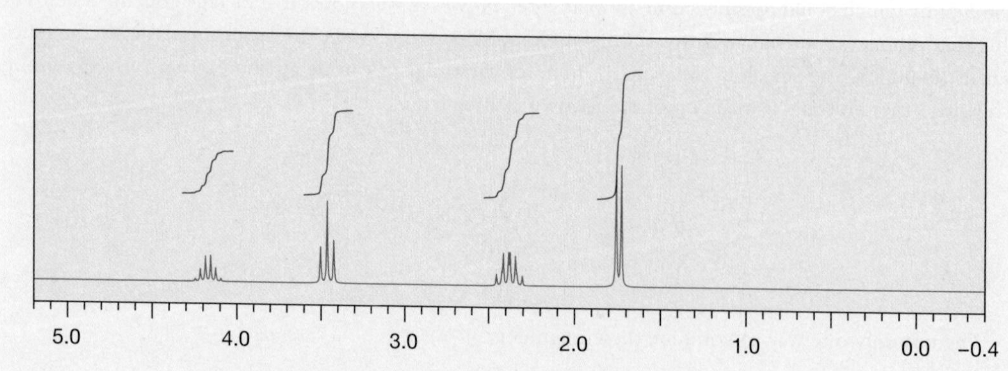

^{13}C-NMR spectrum:

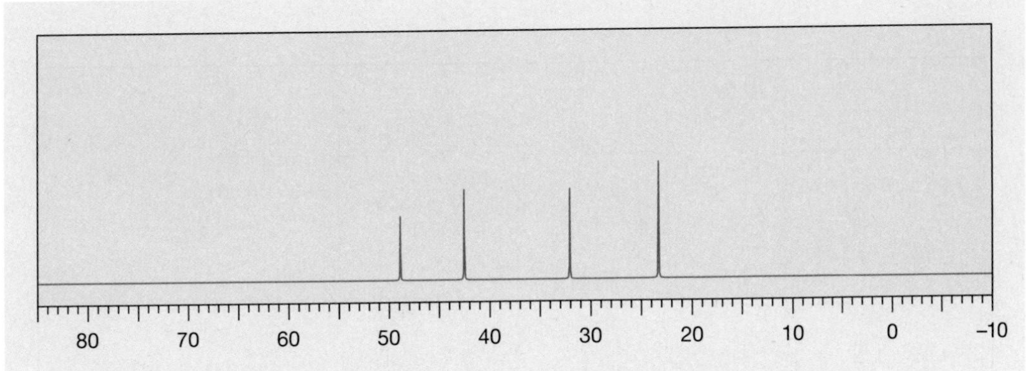

c) Molecular formula: $C_5H_9NO_2$
^1H-NMR spectrum:

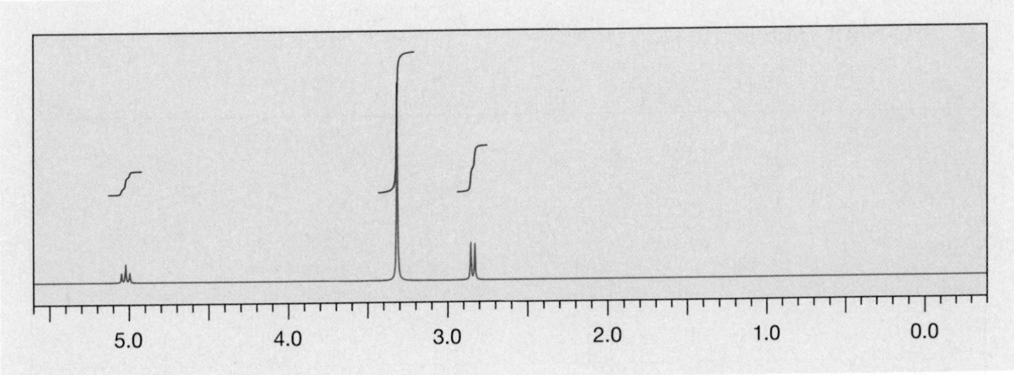

^{13}C-NMR spectrum:

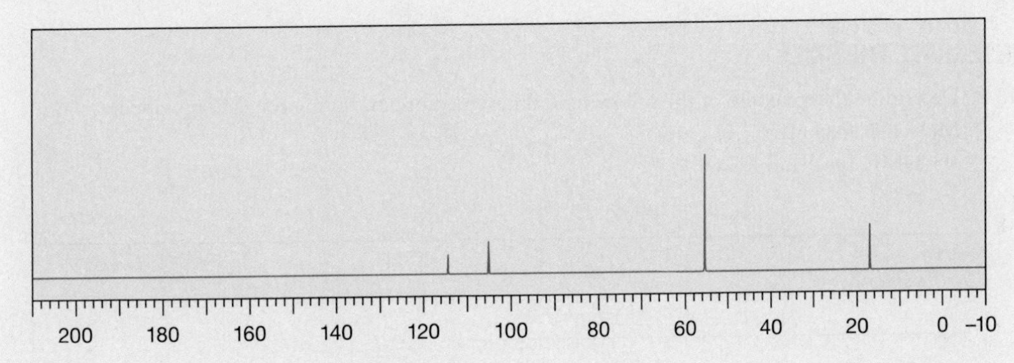

d) Molecular formula: $C_{10}H_{10}O_3$
^{1}H-NMR spectrum:

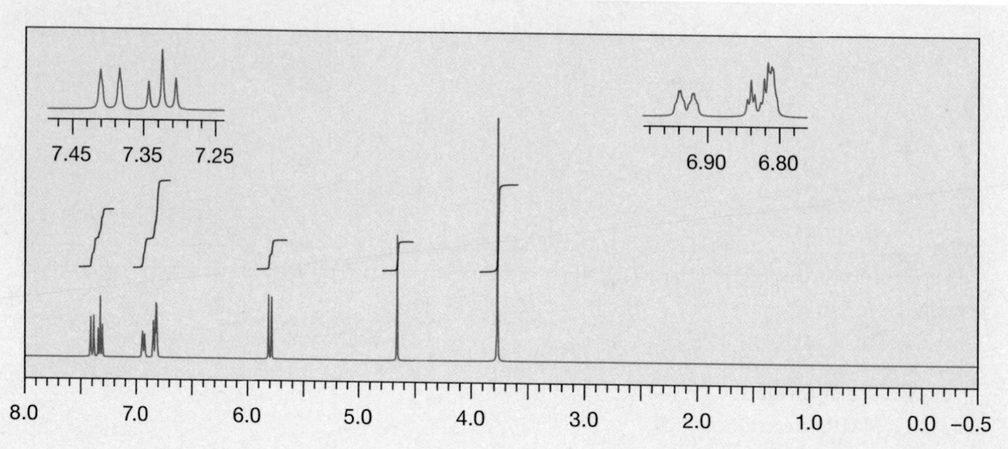

^{13}C-NMR spectrum:

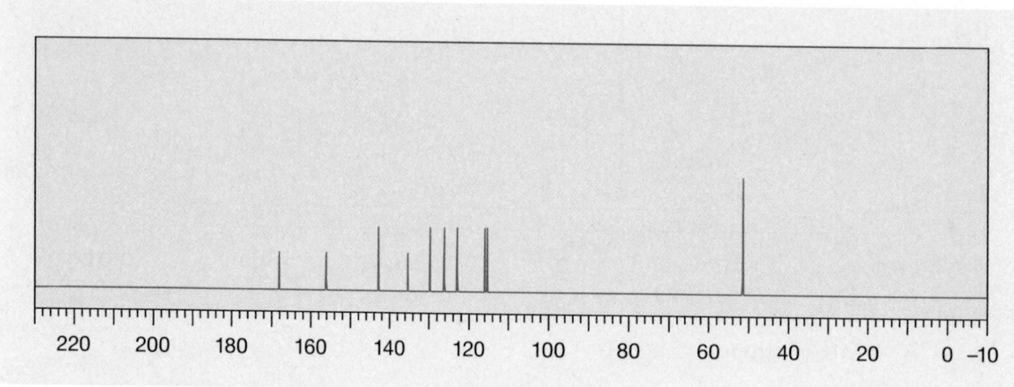

INTEGRATE THE SKILL

13.20 Determine the structure of the following unknown compound using the data provided.
Molecular formula: $C_{11}H_{13}NO$
^{1}H-NMR spectrum:

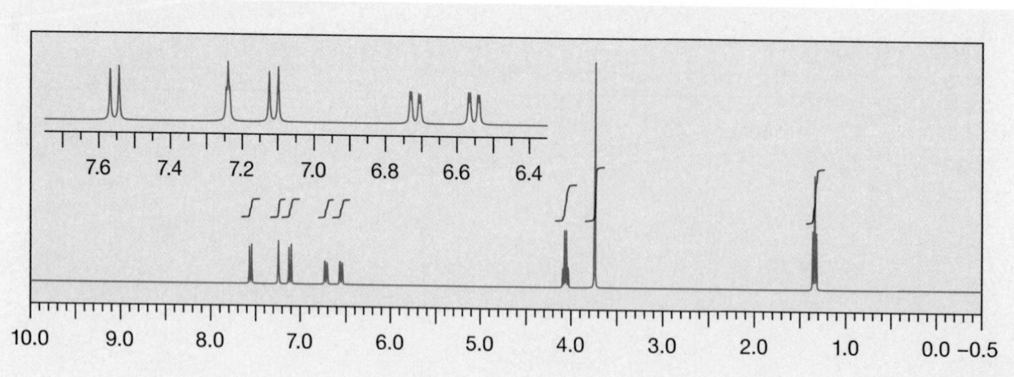

^{13}C-NMR spectrum:

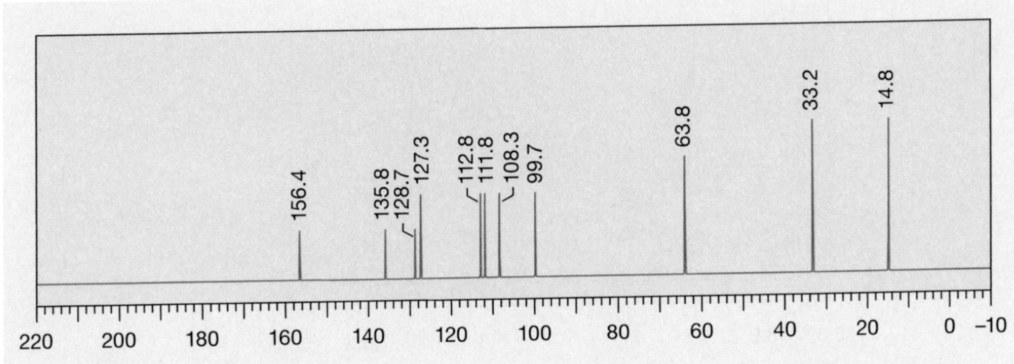

DEPT-135 spectrum:

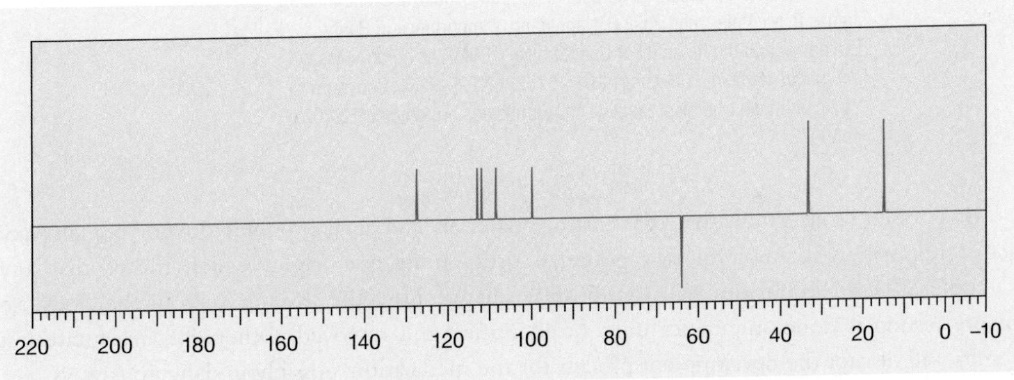

Bringing It Together

As discussed in the chapter opening, impurities in medications need to be identified and removed. The traditional ways to do this—using liquid or gas chromatography—generate significant waste or destroy the sample being tested. A new procedure currently under development is quantitative NMR (qNMR). By looking closely at an NMR spectrum for impurity peaks and comparing the integrations impurities and their concentrations can be identified.

For example, quercetin and analogs of quercetin have potential as drugs for osteoporosis and also for prostate, breast, and colon cancer. This makes quercetin a good starting material for potential drug synthesis. If a synthesis that starts with quercetin contains a significant amount of kaempferol, the end-product will likely have impurities that are difficult to remove. For example, the qNMR spectrum of a sample of commercially available quercetin revealed it was not 99 percent pure as advertised. Rather, the qNMR spectrum showed the sample actually contained 12.2 percent kaempferol.

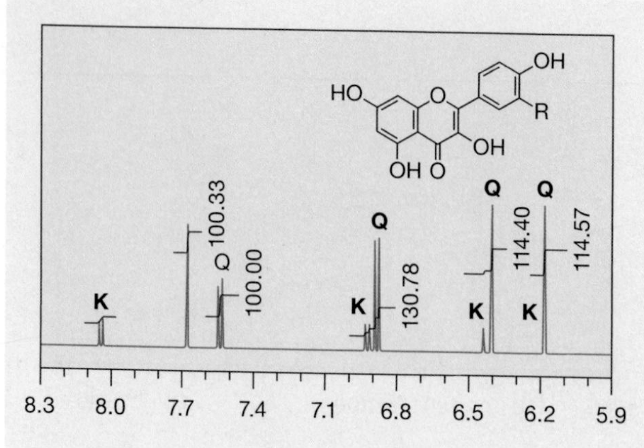

SOURCE: Re-drawn from Guido F. Pauli, Shao-Nong Chen, Charlotte Simmler, David C. Lankin, Tanja Gödecke, Birgit U. Jaki, J. Brent Friesen, James B. McAlpine, and José G. Napolitano, "Importance of Purity Evaluation and the Potential of Quantitative 1H NMR as a Purity Assay," *Journal of Medicinal Chemistry* 2014 57 (22), 9220-9231. Copyright © 2014 American Chemical Society. http://pubs.acs.org/doi/abs/10.1021/jm500734a

It is essential to know that the starting materials and final products do not contain any active impurities. Kaempferol is a potential drug in its own right, which means that any kaempferol-based impurities will significantly change how the mixture acts in the body. So before testing is done, impurities must be identified and removed. Otherwise, inaccurate test results will disrupt the development process for the medication, possibly in dangerous ways.

You Can Now

- Determine structural information from ^{1}H-NMR spectra, including the following:
  - number of hydrogen types and possible symmetry in the structure
  - number of hydrogens of each type
  - chemical shifts, which can indicate hybridization and nearby functional groups
  - number of hydrogens adjacent to each hydrogen type
- Use a ^{1}H-NMR spectrum to determine the structure of a fragment around each hydrogen type.
- Use chemical shifts and coupling patterns to assemble these fragments into a complete structure.
- Determine structural information from a ^{13}C-NMR spectrum, including the following:
  - number of carbon types and possible symmetry in the structure
  - chemical shifts, which can indicate hybridization and nearby functional groups
- Identify the number of hydrogens bonded to carbon using DETP-135.
- Determine a molecular structure using a ^{13}C-NMR spectrum in conjunction with a ^{1}H-NMR spectrum.

Problems

13.21 Which of the sp^3-hybridized hydrogens in each of the following compounds have the lowest electron density? Which have the highest density?

a) O

b) HS OH

c)

d) O

e)

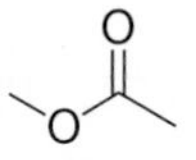

f)

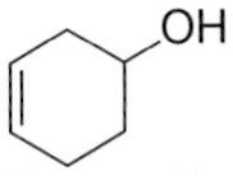

13.22 Which of the sp^2-hybridized hydrogens in the following compounds have the lowest electron density? Which have the highest?

a)

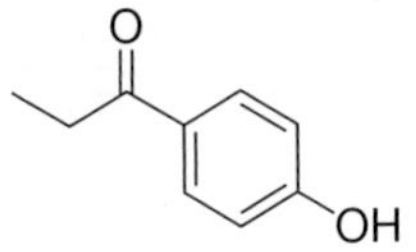

b)

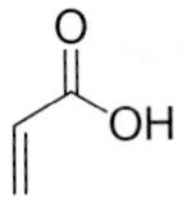

c)

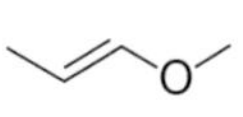

d)

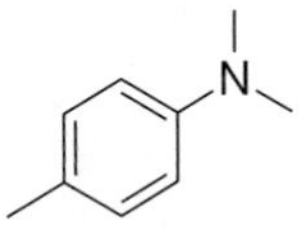

13.23 Will the signals for hydrogens with high electron density appear farther downfield than those for hydrogens with low electron density? Justify your answer.

13.24 Which of the following structures has a doublet in its ^{1}H-NMR spectrum? Why?

a)

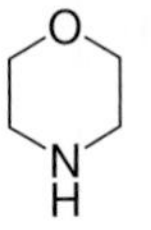

b)

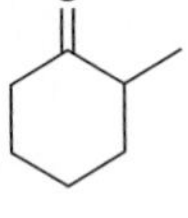

c)

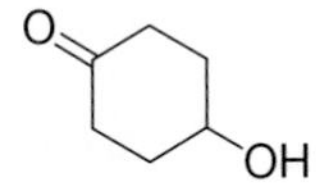

d)

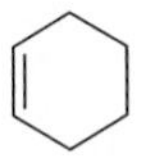

13.25 Predict the splitting pattern for each hydrogen type in the following compounds.

a)

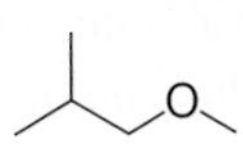

b)

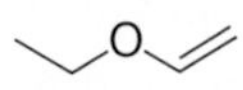

c)

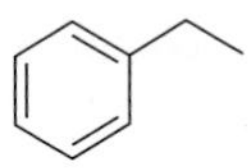

d)

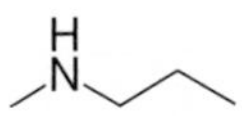

e)

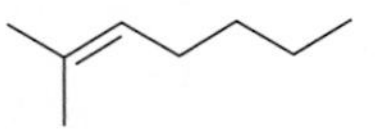

f)

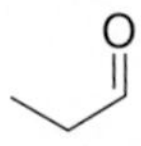

13.26 For each of the integration ratios given, calculate the lowest ratio of hydrogens of each type.

a) 0.52:0.99:0.98:1.57
b) 4.79:1.37:3.18:1.52:9.48
c) 9.64:4.12:4.16:6.21
d) 1.49:3.03:2.96:5.93:4.50

13.27 For each of the following ^{1}H-NMR signals, draw a structural fragment that would produce the observed chemical shift, integration ratio, and coupling.

a)

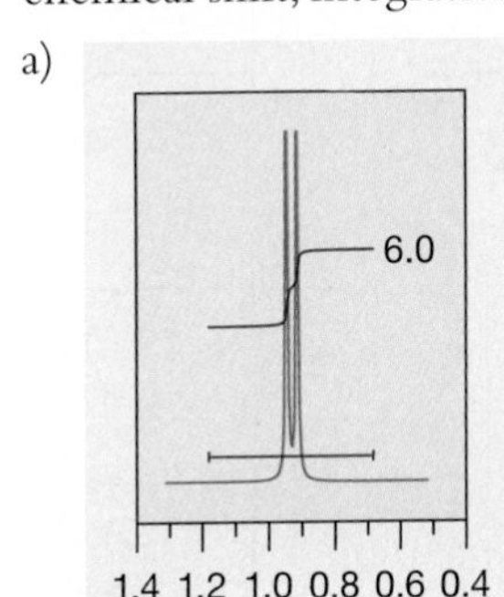

b)

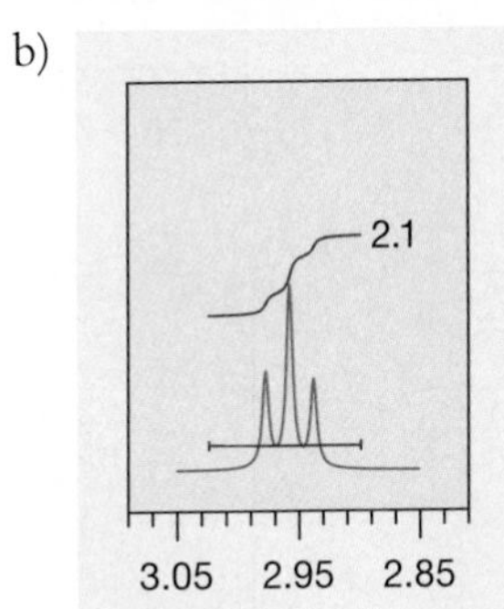

c)

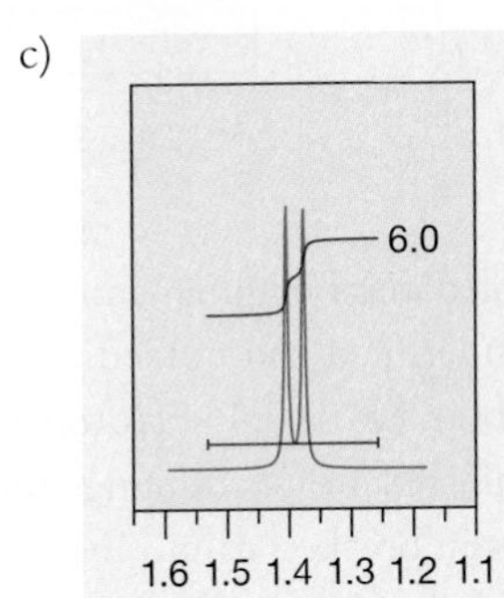

d)

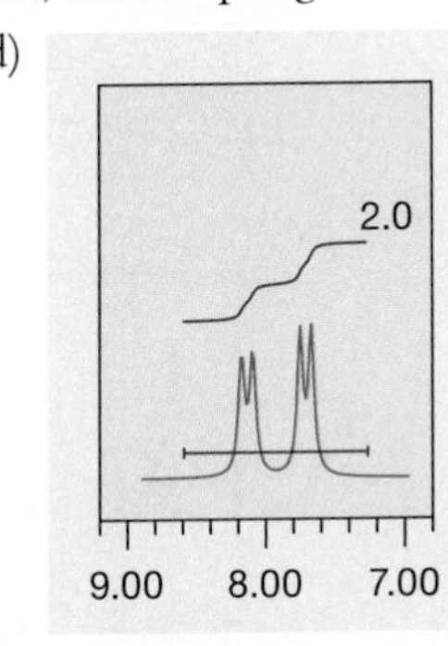

e)

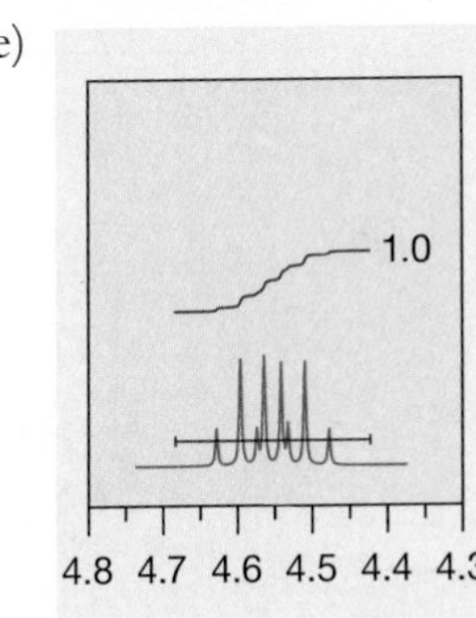

f)

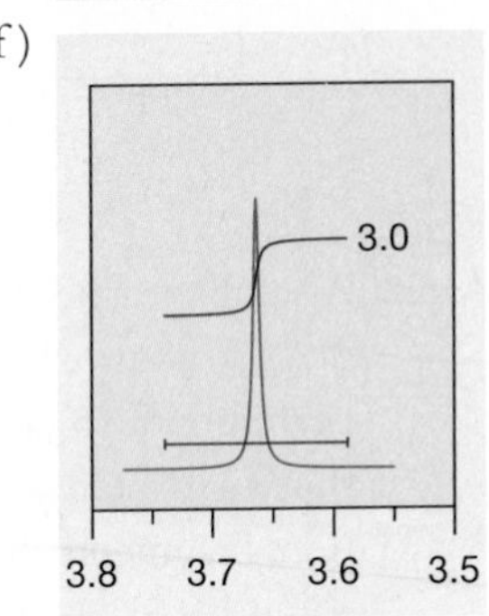

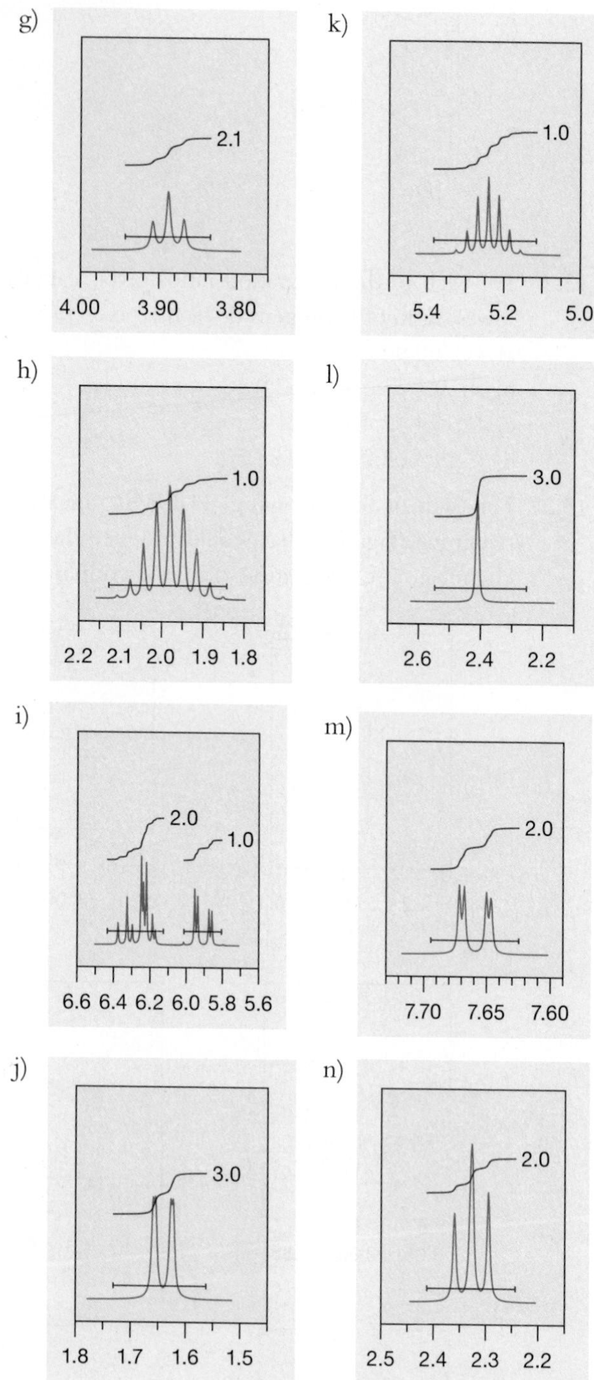

a) H_3C–CH_2– ; [H_3C–CH_2–C(=O)– or H_3C–CH_2–Ph] ; EWG-substituted benzene ; EDG-substituted benzene ; –OH

b) [–CH_2–CH_2–C(=O)– or –CH_2–CH_2–Ph] ; –O–CH_2–CH_2– ; [–C(=O)–O–CH_3 or Ph–O–CH_3] [–OH or –NH–]

c) –CH_2–C(=O)H ; –CH_2–CH_2–C(=O)H ; H_3C–CH_2–CH_2– ; H_3C–CH_2–

d) [–OH or –NH–] ; –CH=CH– ; –CH=CH–CH_2– ; [–CH–CH_2–O– or –CH–CH_2–Cl] ; phenyl

e) [–C(=O)–CH_3 or Ph–CH_3] ; monosubstituted benzene ; benzene ; EDG-substituted benzene

f) [–CH_2–CH_2–C(=O)– or –CH_2–CH_2–Ph] ; ?-substituted benzene ; [H_3C–C(=O)– or H_3C–Ph]

13.28 Assemble each set of structural fragments into a complete structure. The underlined and bolded hydrogens correspond to an observed signal. Hydrogens not shown in bold are inferred from coupling information. Fragments shown in brackets cannot be differentiated based on the spectrum, so you need to choose between the two options. Duplicate fragments may arise from different signals in the spectrum.

13.29 For each of the structures assembled for Question 13.28, predict the approximate chemical shift, integration, and coupling pattern of each hydrogen type.

13.30 Using the structural fragments from Questions 13.27, assemble the pieces that go together, based on coupling. You might not end up with a complete molecule.

13.31 Determine the structures that produce the following spectra. Explain your reasoning.

a)

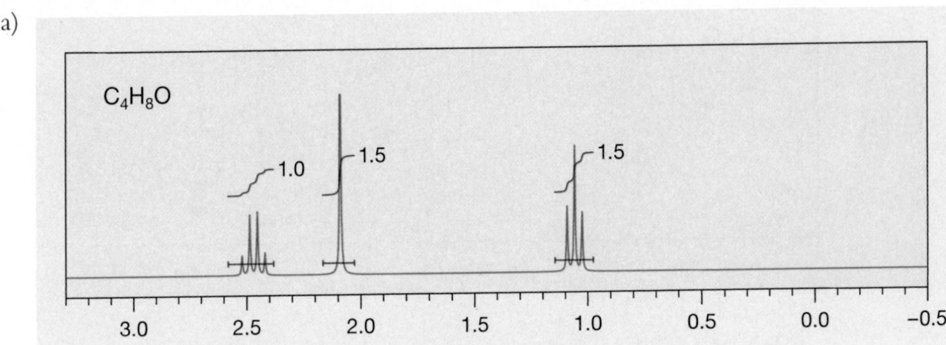

b)

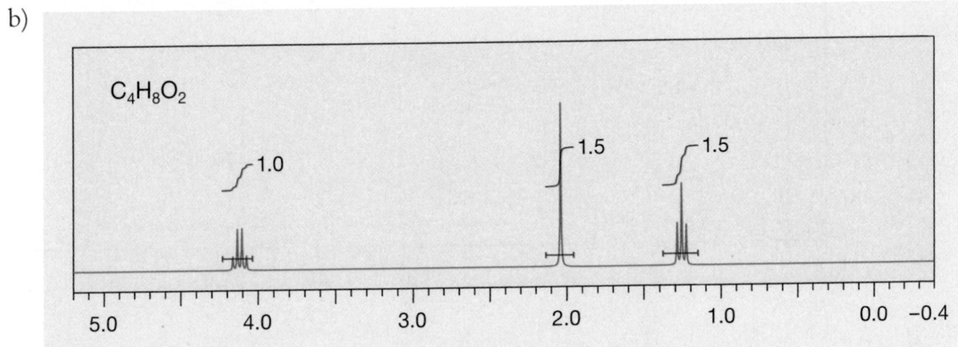

c)

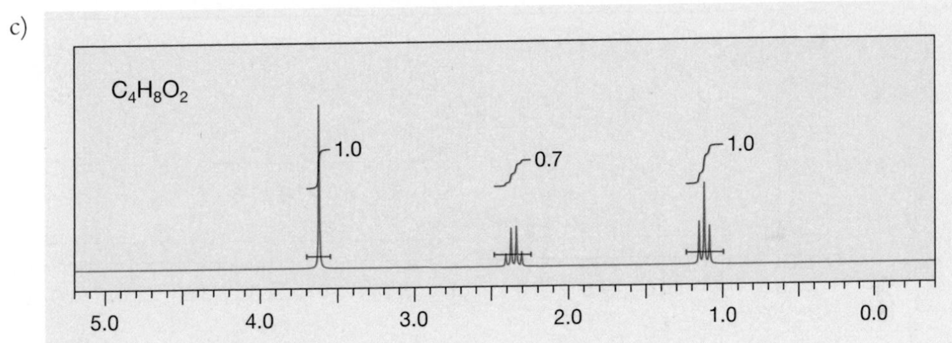

d)

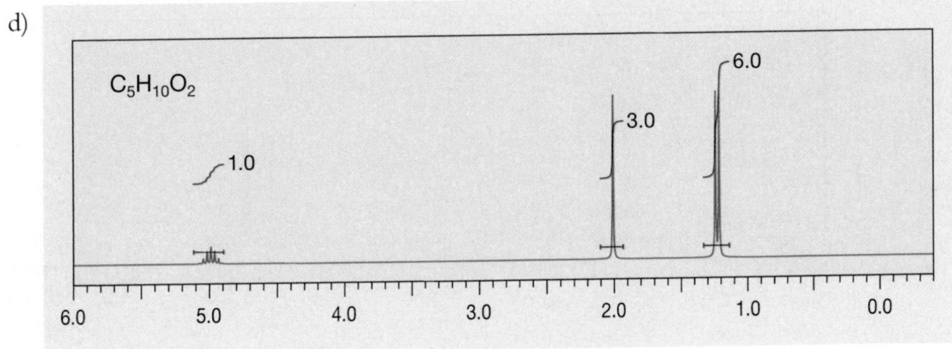

e)

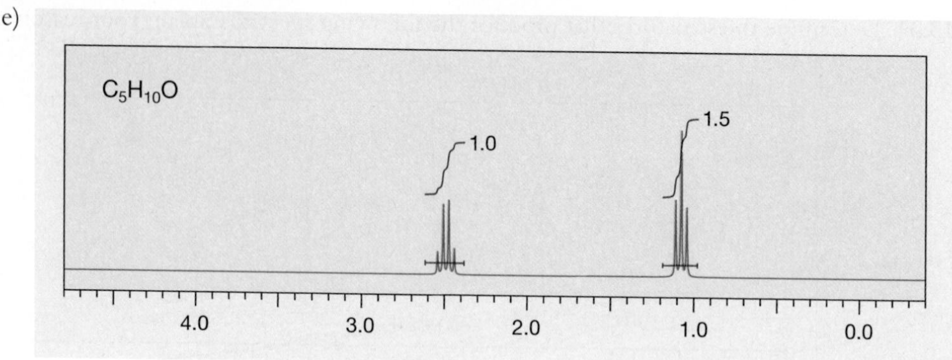
$C_5H_{10}O$
1.0
1.5
4.0
3.0
2.0
1.0
0.0

f)

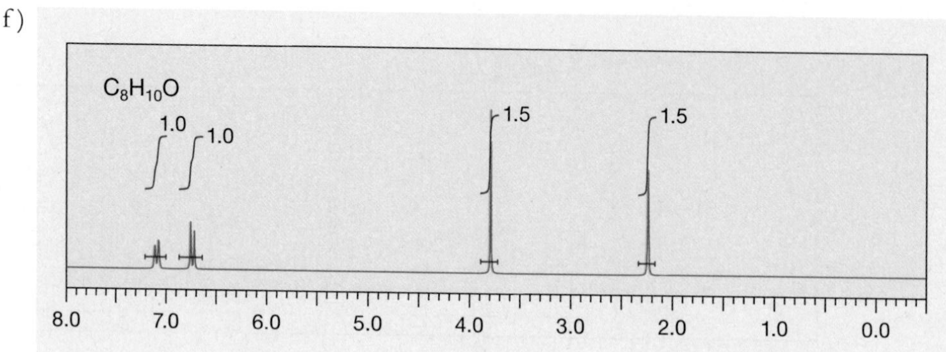
$C_8H_{10}O$
1.0
1.0
1.5
1.5
8.0
7.0
6.0
5.0
4.0
3.0
2.0
1.0
0.0

g)

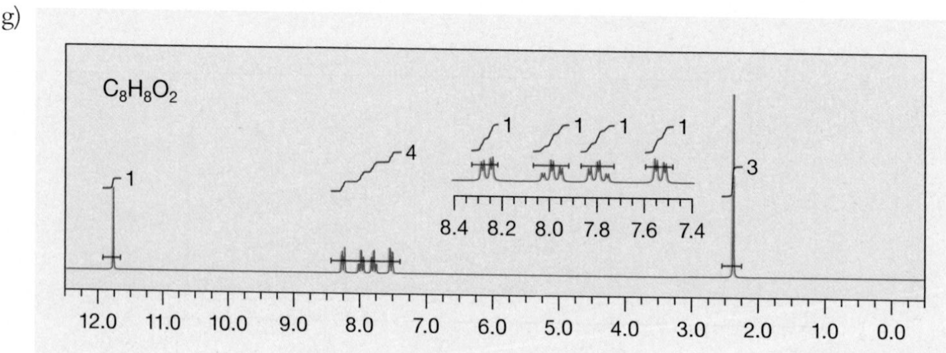
$C_8H_8O_2$
1
4
1
1
1
1
3
8.4 8.2 8.0 7.8 7.6 7.4
12.0 11.0 10.0 9.0 8.0 7.0 6.0 5.0 4.0 3.0 2.0 1.0 0.0

h)

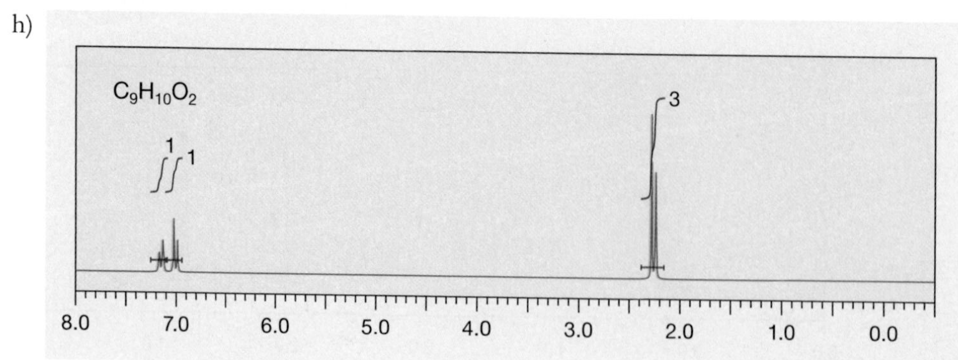
$C_9H_{10}O_2$
1
1
3
8.0
7.0
6.0
5.0
4.0
3.0
2.0
1.0
0.0

i)

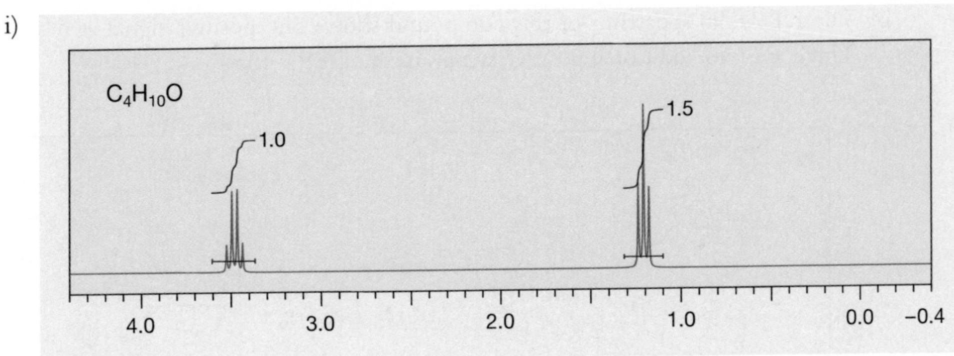

j)

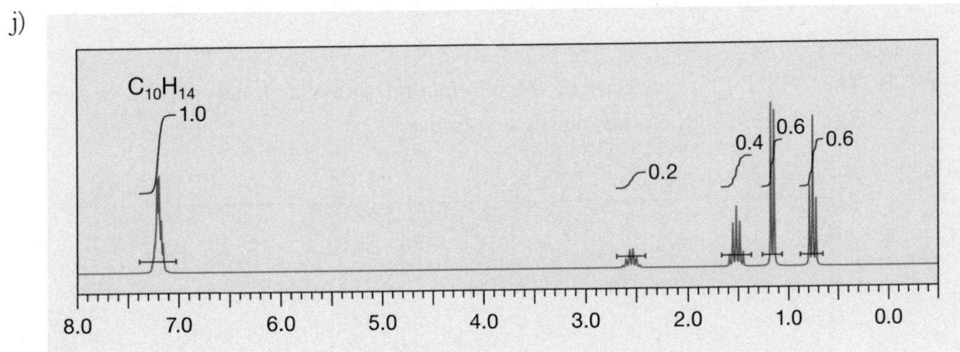

13.32 What structures correspond to the following ^{13}C-NMR spectra? All the structures contain at least one oxygen.

a)

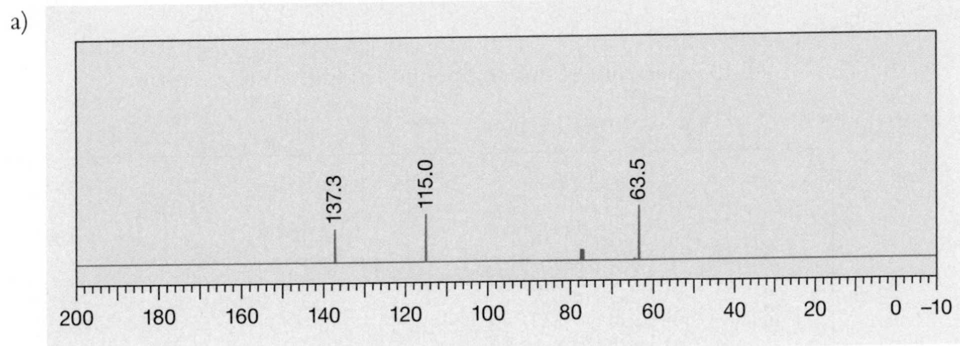

b) The DEPT-90 spectrum of this compound shows one positive signal at δ = 68 ppm. There are two additional positive signals in the DEPT-135.

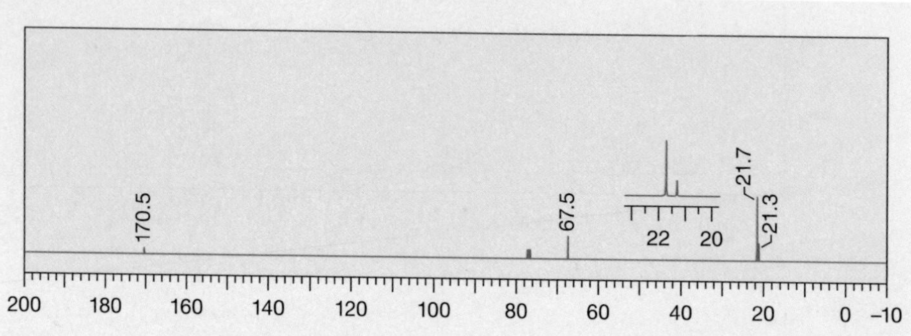

c) The DEPT-135 spectrum of this compound shows a positive and a negative signal at δ = 23 ppm. All the other signals are positive.

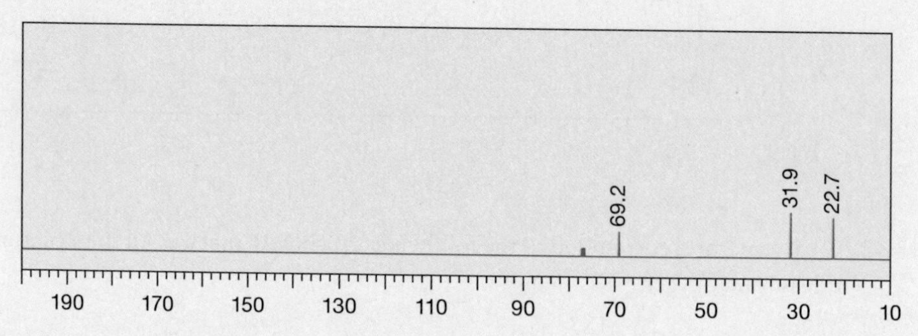

d) The DEPT-135 spectrum of this compound has four positive signals.

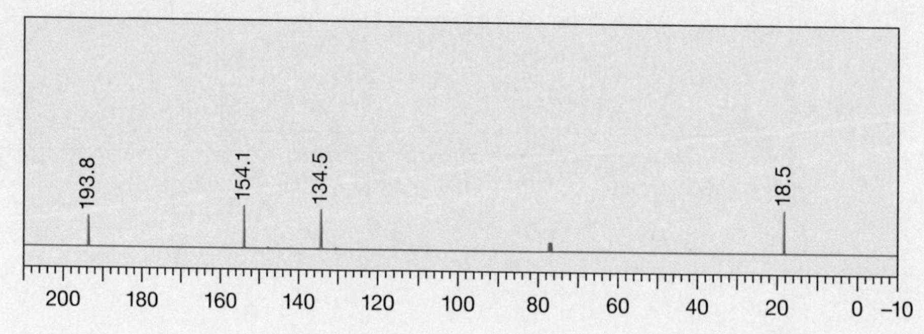

13.33 The following ^{1}H-NMR spectra correspond to two structural isomers of an alkyl bromide. What is the structure of each? Explain your reasoning.

a)

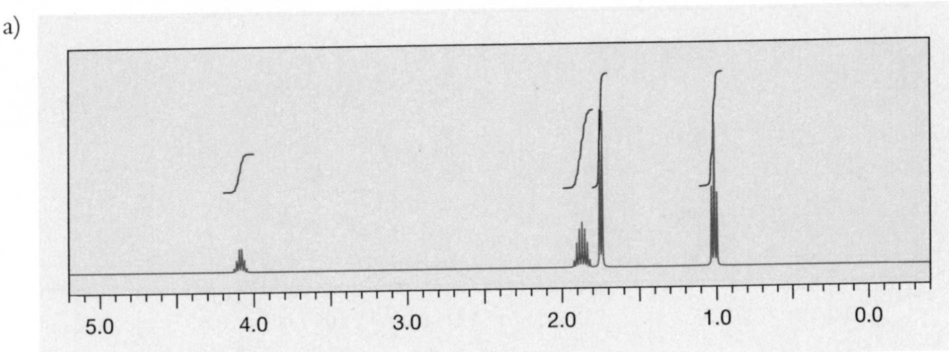

b)

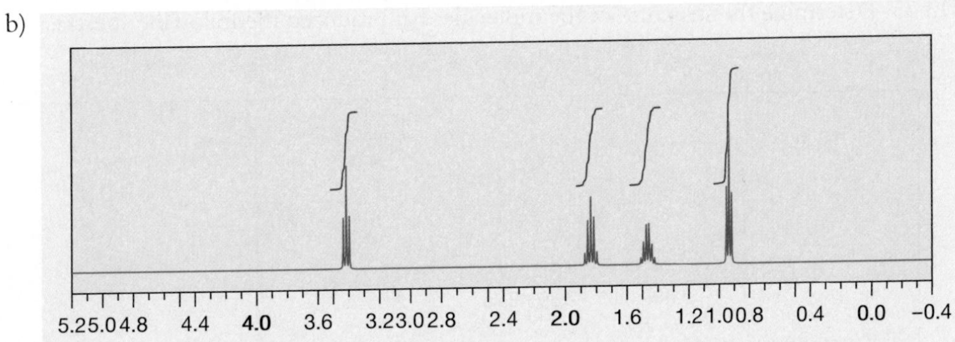

13.34 Spectrum A was produced by an alcohol containing six carbons, which was then oxidized to form the compound that produced spectrum B. Determine the structure of the alcohol and the oxidation product.

Spectrum A

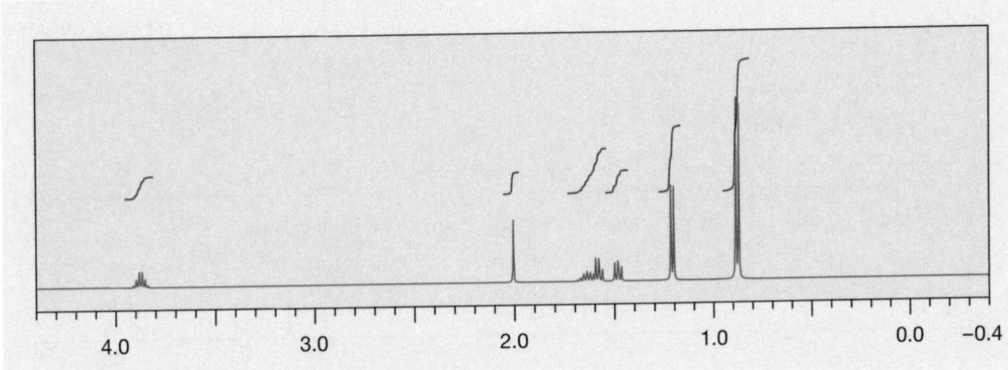

Spectrum B

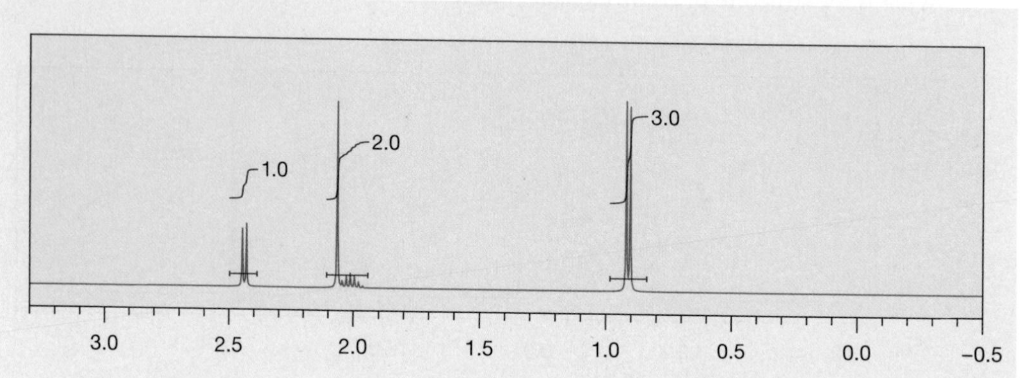

13.35 Determine the structure of the molecule that produced the following spectra.

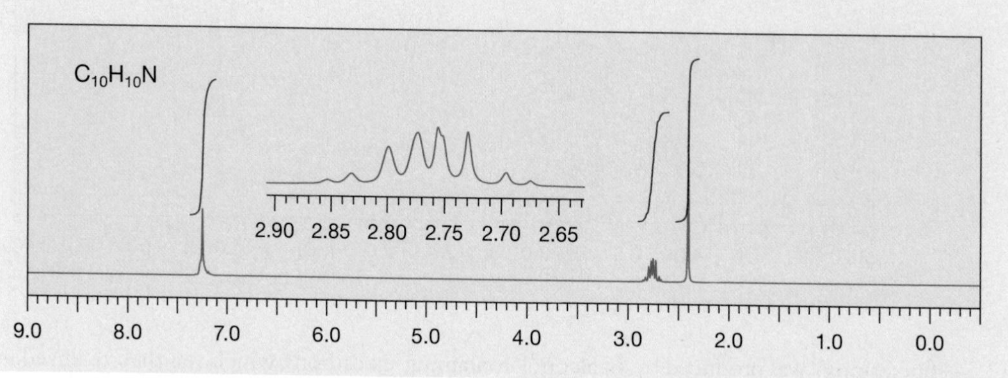

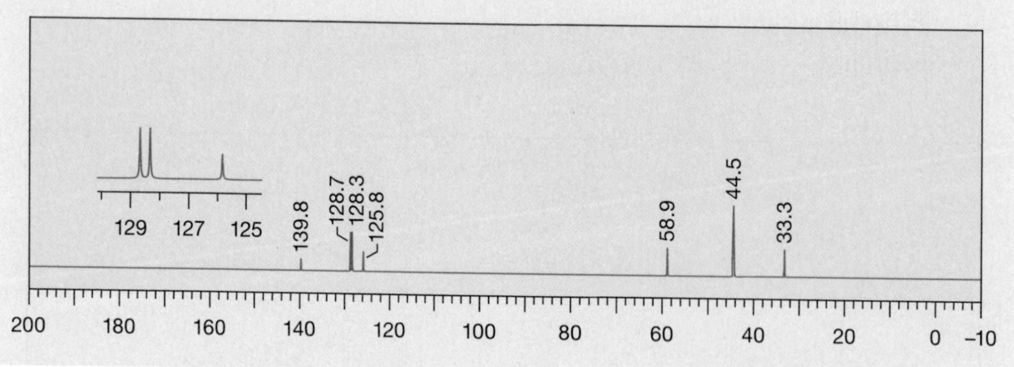

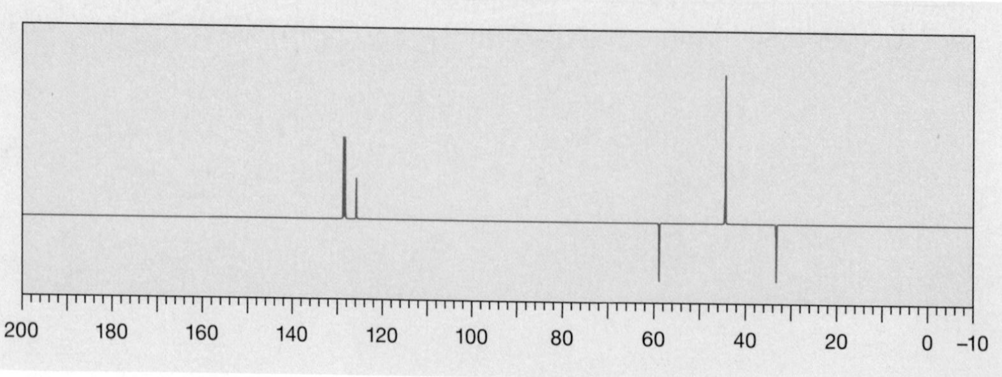

13.36 Determine the structure of the molecule that produced the following spectra.

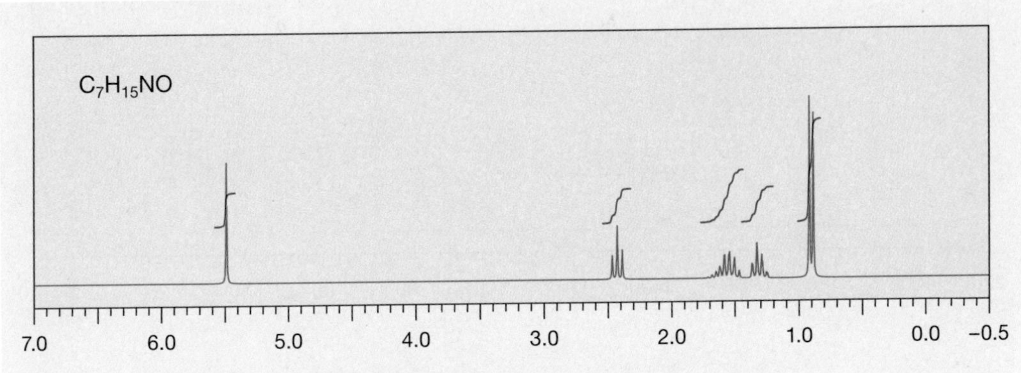

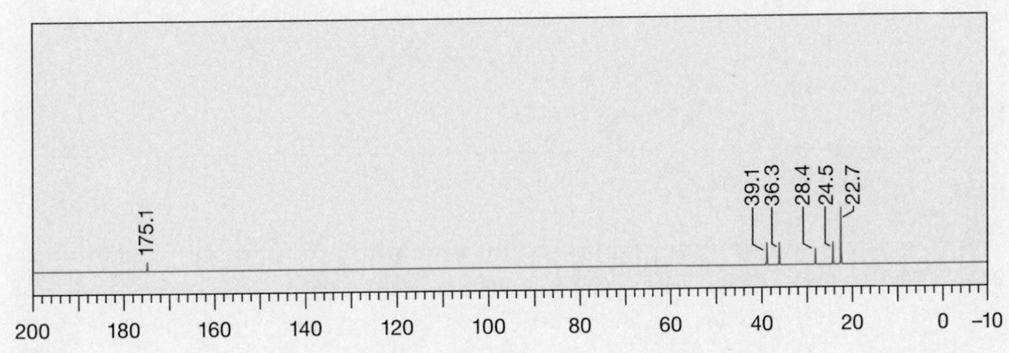

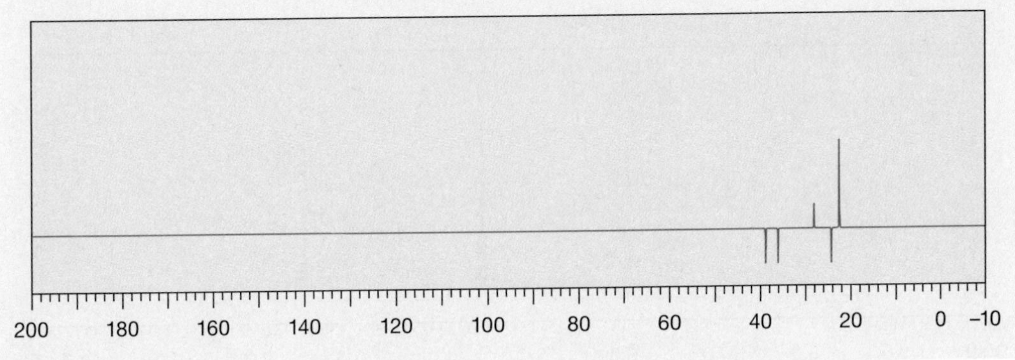

13.37 Determine the structure of the molecule that produced the following spectra.

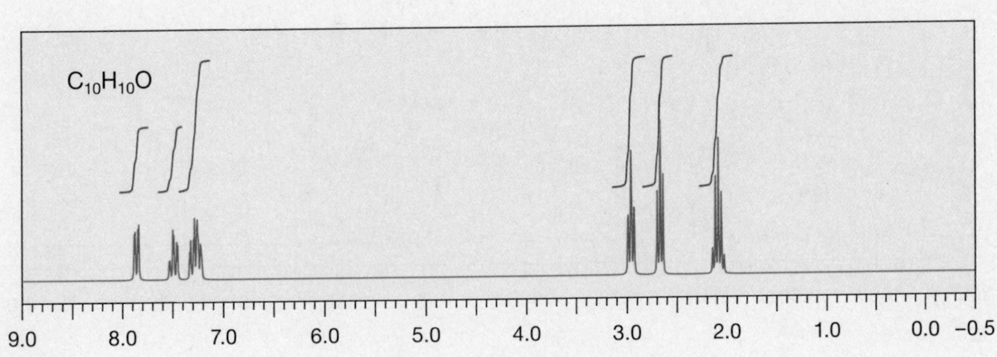

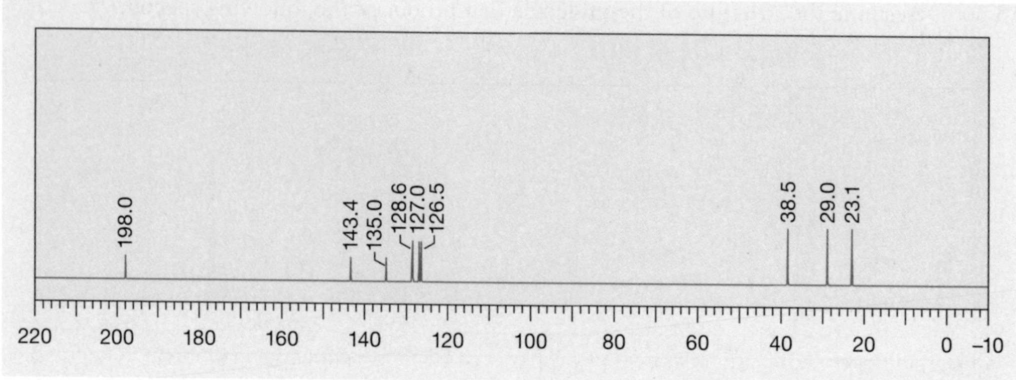

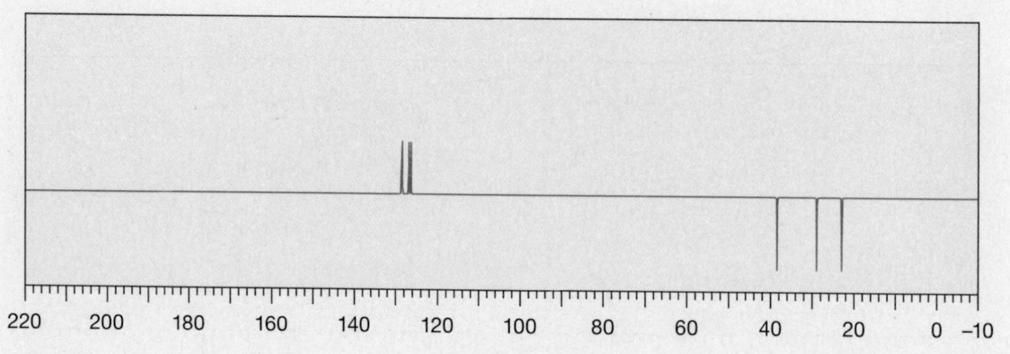

13.38 Determine the structure of the molecule that produced the following spectra.

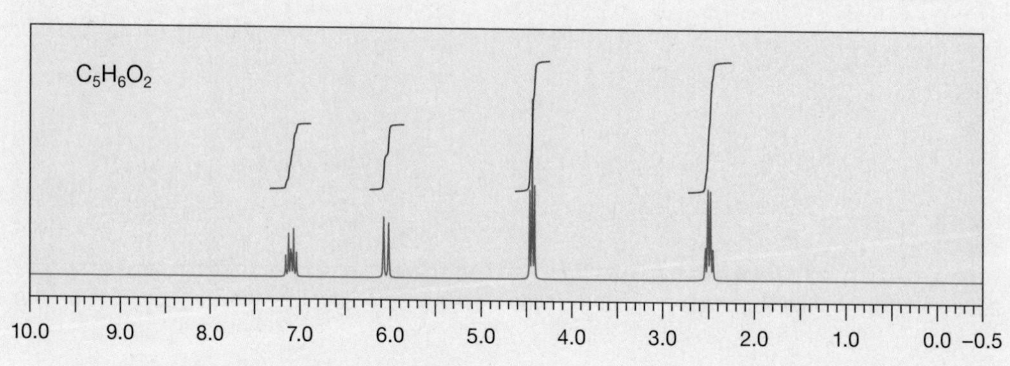

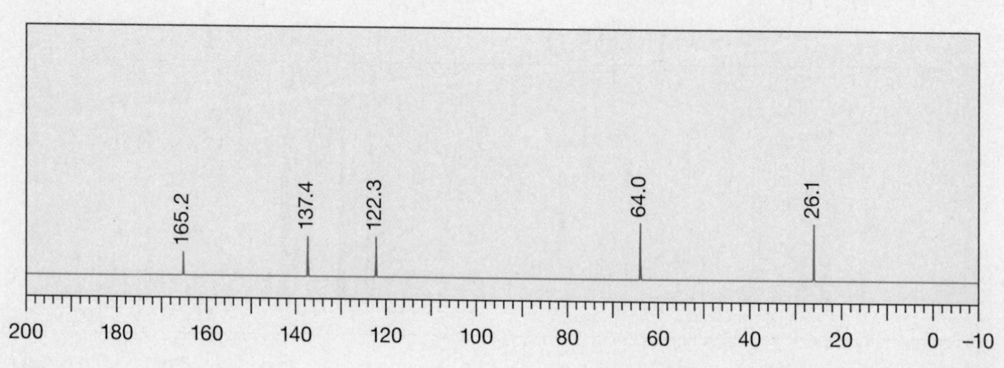

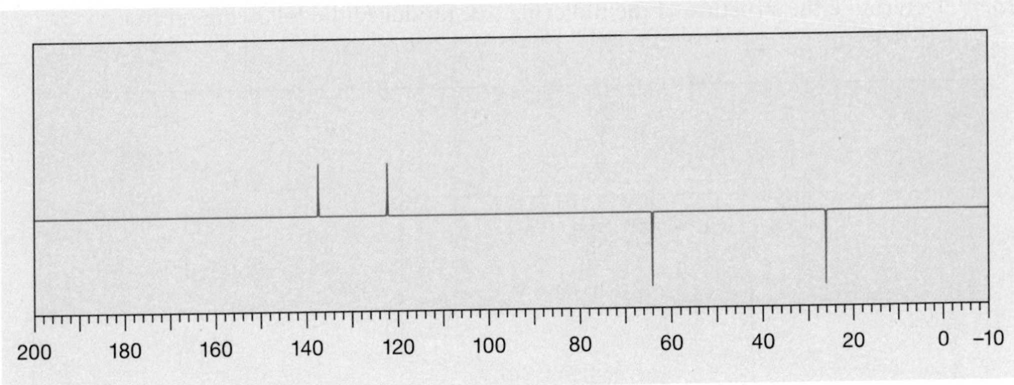

13.39 Determine the structure of the molecule that produced the following spectra.

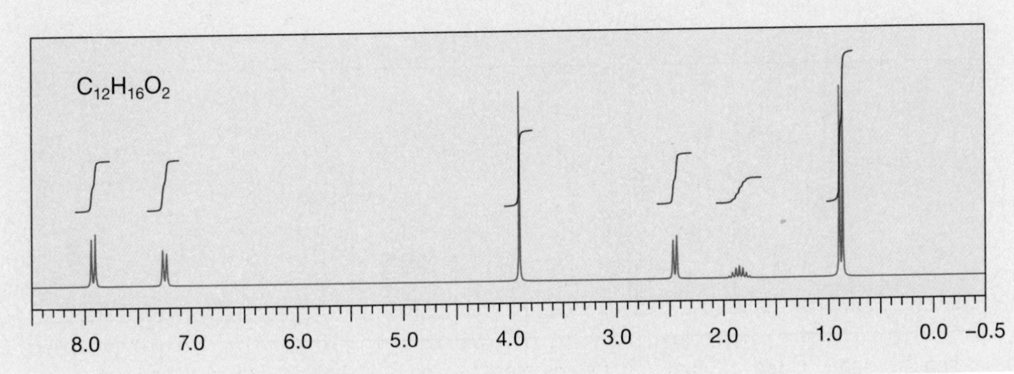

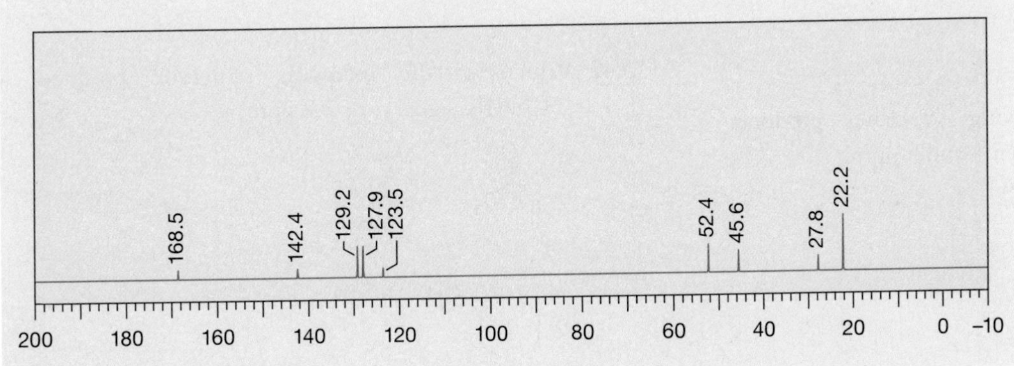

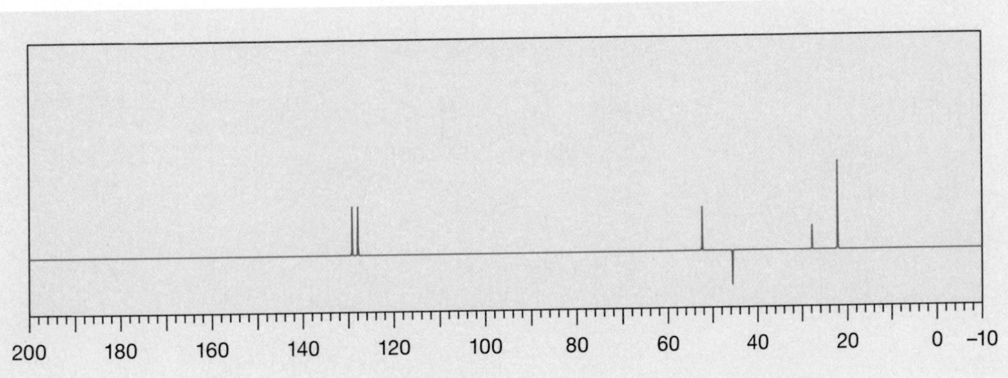

13.40 Determine the structure of the molecule that produced the following spectra.

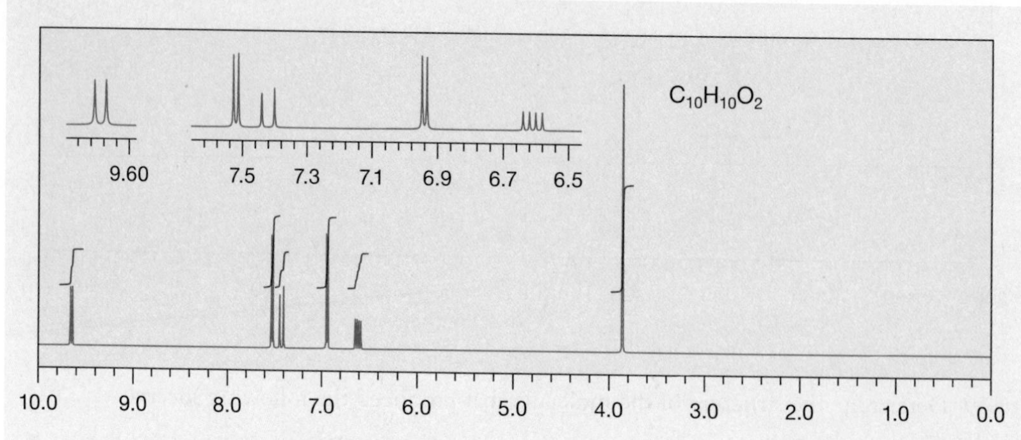

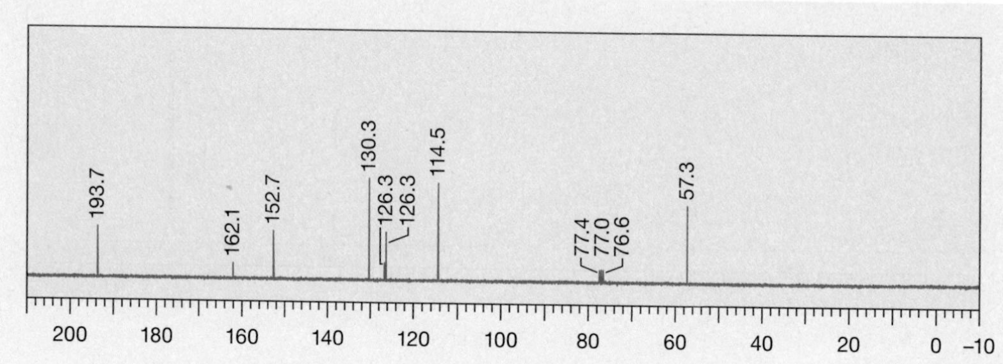

MCAT STYLE PROBLEMS

13.41 Which of the following structures produces a ^{1}H-NMR signal between 4 and 5 ppm?

a)

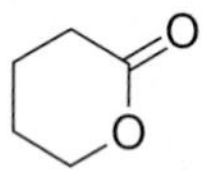

b)

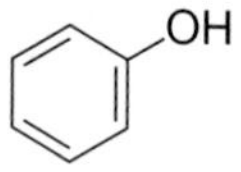

c)

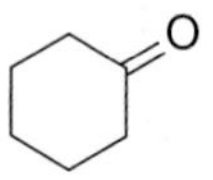

d)

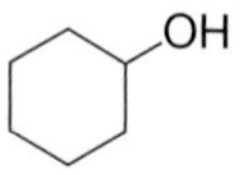

13.42 Which of the following structures produces a ^{1}H-NMR signal near 2.2 ppm?

a)

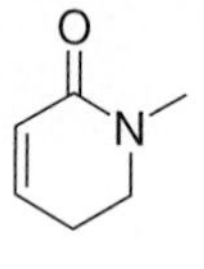

b)

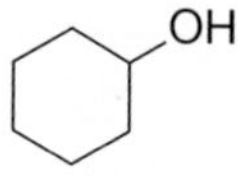

c)

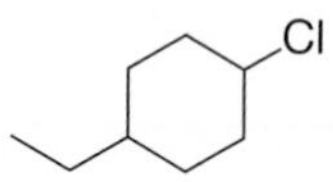

d)

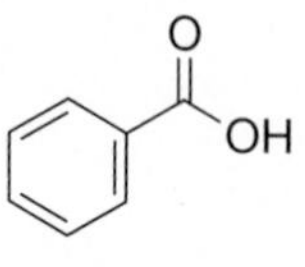

13.43 Which of the following is the spectrum for the amino acid valine?

a)

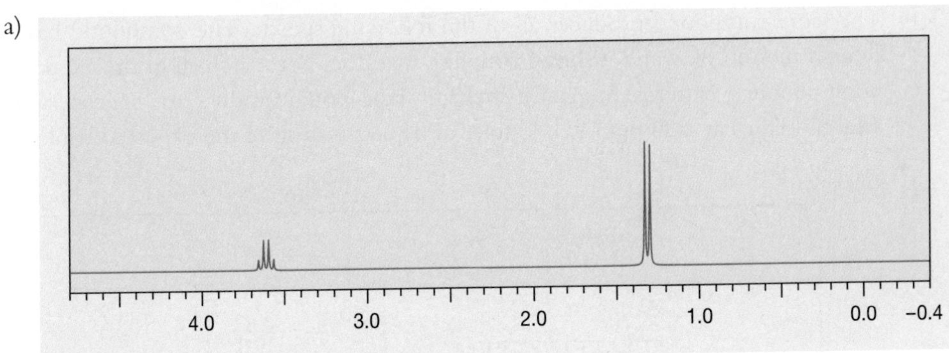

b)

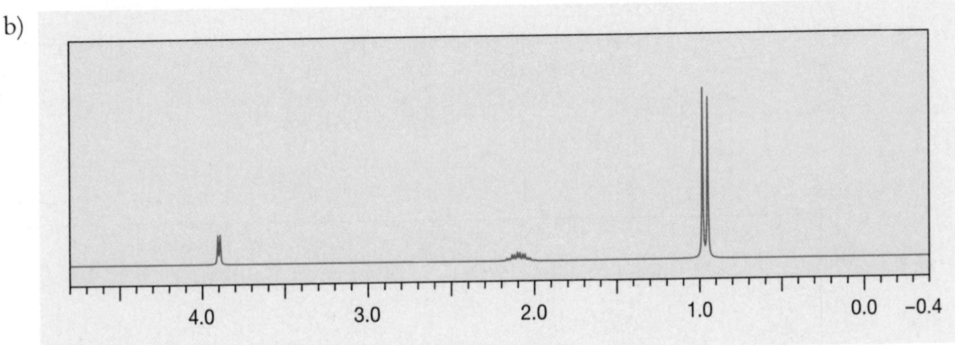

c)

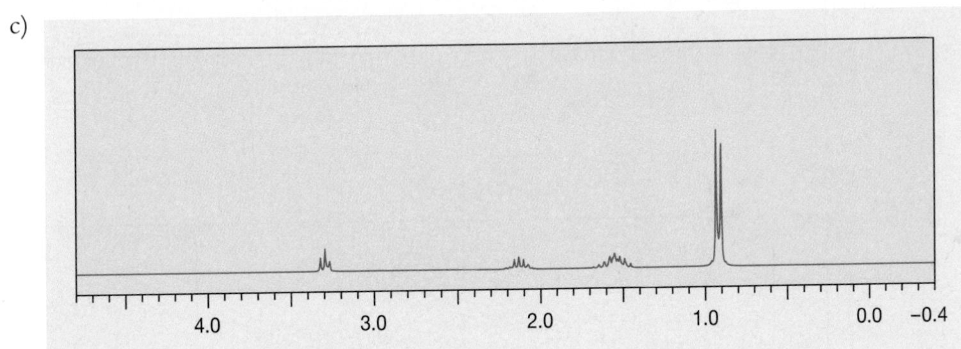

d)

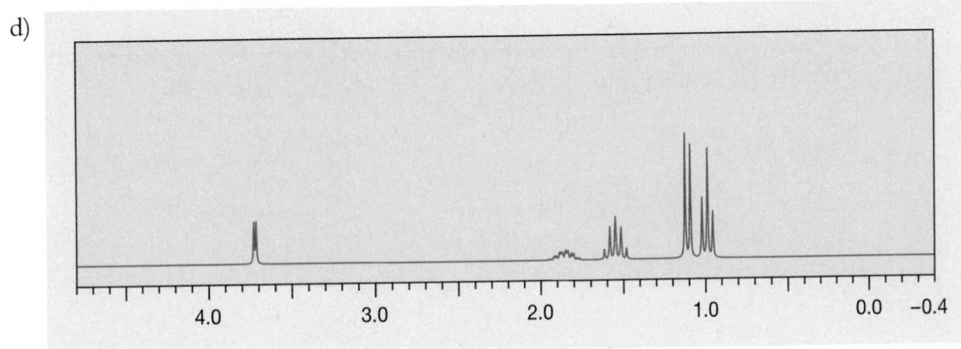

CHALLENGE PROBLEMS

13.44 The same compound produces all of the following spectra. The compound has a degree of unsaturation of 2, and it shows coupling that does not match all of the expected three-bond coupling patterns. Suggest a structure that would produce the spectra, and explain why the signal at 2.25 ppm is a doublet of triplets instead of the expected triplet.

a)

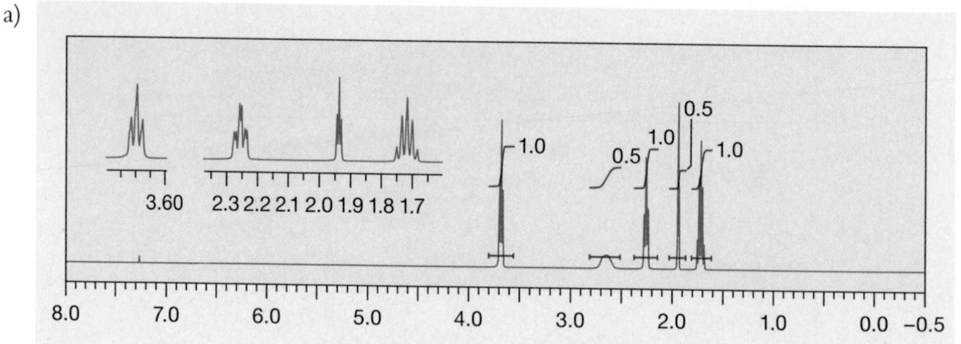

b)

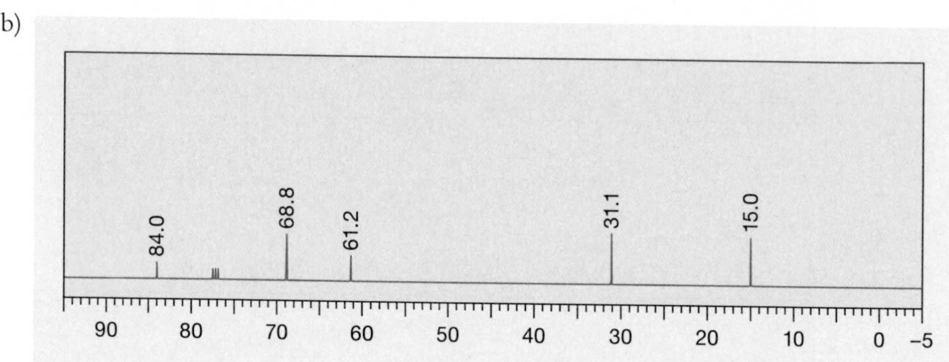

c)

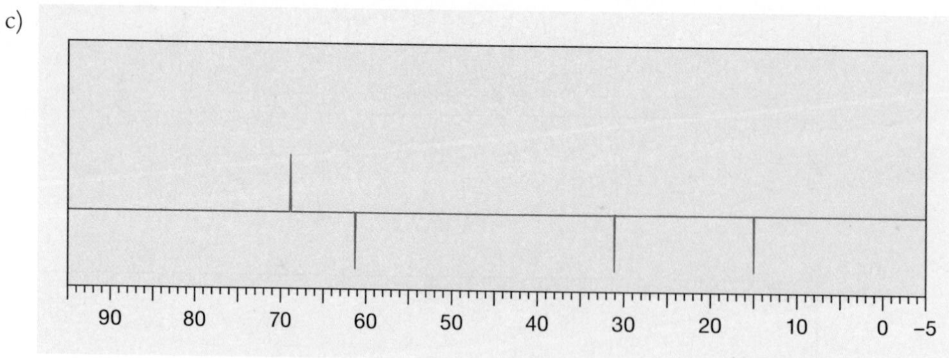

13.45 The following two ^{1}H-NMR spectra are produced by carveol and its oxidized analog, carvone.

OH O

carveol carvone

The alkyl region in each spectrum contains multiple, overlapping signals, and the integrations are complicated by the presence of different stereoisomers. Which spectrum corresponds to each structure? How do you know?

a)

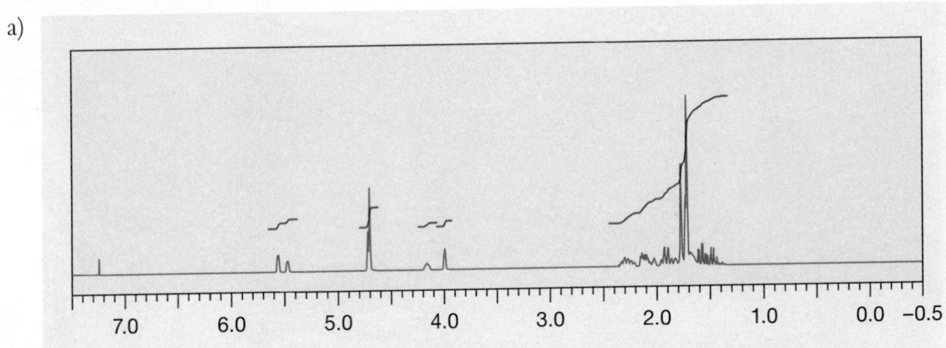

b)

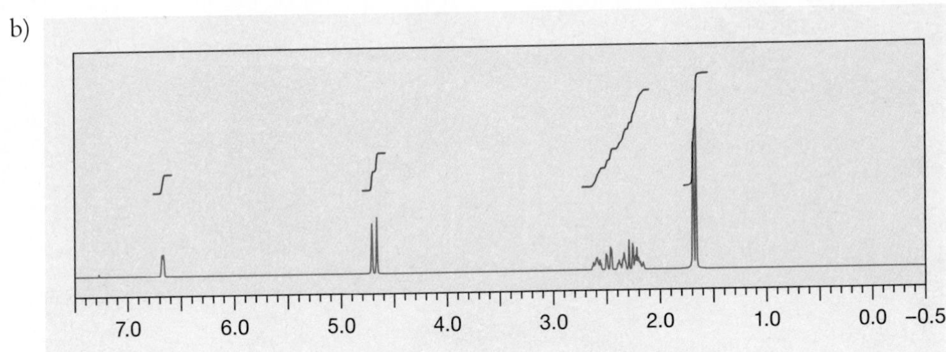

14 Structure Determination II

MASS SPECTROMETRY AND INFRARED SPECTROSCOPY

CHAPTER OUTLINE

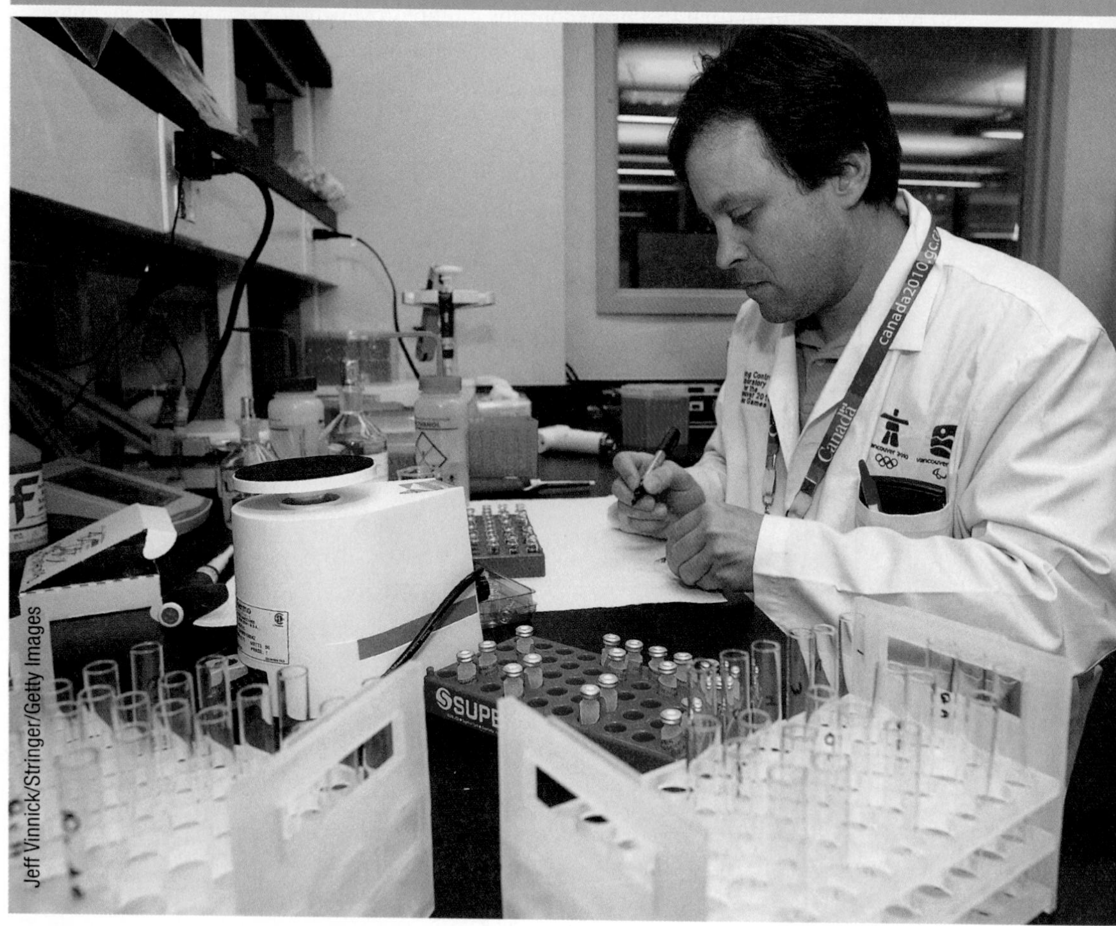

The anti-doping laboratory for the 2010 Winter Olympic Games in Vancouver analyzed over 2500 samples for prohibited substances that may have been used by athletes. Mass spectrometry is a technique used to identify the prohibited substances.

14.1 Why It Matters

The ability to determine the identity of an organic compound or a mixture of compounds is central to many scientific disciplines, including forensic science, nanotechnology, and new medicine discovery. As described in Chapter 13, nuclear magnetic resonance is commonly used to determine the structures of unknown substances. This procedure is especially valuable for a compound that is being made for the first time. However, for well-known compounds, other techniques—specifically, mass spectrometry (MS) and infrared (IR) spectroscopy—make it possible to identify the structure of organic compounds quickly. These approaches are used in situations where the structure of a compound can typically be determined by matching a sample of the compound with a known standard. They are also used in situations where speed is a more pressing concern than completeness.

Consider the case of a forensic scientist who receives from airport security officials a sample of a clear, colourless liquid taken from luggage left unattended at the baggage carousel. This substance is likely a known compound (the structure and properties of the compound have been previously measured), and may or may not be dangerous. It is important to identify it quickly. Can the identity of this substance be determined with certainty? Experience might lead the forensic scientist to suspect that the substance is 4-hydroxybutanoic acid (γ-hydroxybutanoic acid), commonly known as GHB or Liquid G. This compound and its sodium salt are illegal substances used at clubs and raves as an intoxicant known for its effects of euphoria and enhanced libido.

HO OH O

γ-hydroxybutanoic acid
(GBH)

Mass spectrometry (MS) and infrared (IR) spectroscopy are two of the most important and widely used methods for rapidly and reliably obtaining information about molecular structure. This chapter discusses these two methods in detail. At the end of this chapter we describe how these techniques can be used to determine the identity of this unknown sample picked up at the airport.

14.2 Mass Spectrometry

Mass spectrometry (MS) is an important technique for determining molecular structure. It measures the masses of individual molecules of a compound to provide their molecular weight. Mass spectrometry also breaks some molecules of the compound into fragments and measures their masses. The pattern of the fragments (their masses and relative abundances) can be used to identify the compound's structure.

Mass spectrometry (MS) is a method that gives information about the mass of a compound and the fragments from which it is formed.

The instrument used to measure the mass of molecules is a mass spectrometer. For a particular organic compound of interest, a small sample (much less than 1 mg) is introduced into the mass spectrometer (Figure 14.1), where a combination of heat and low pressure inside the instrument ($<10^{-9}$ atm) drives some molecules of the sample into the gas phase.

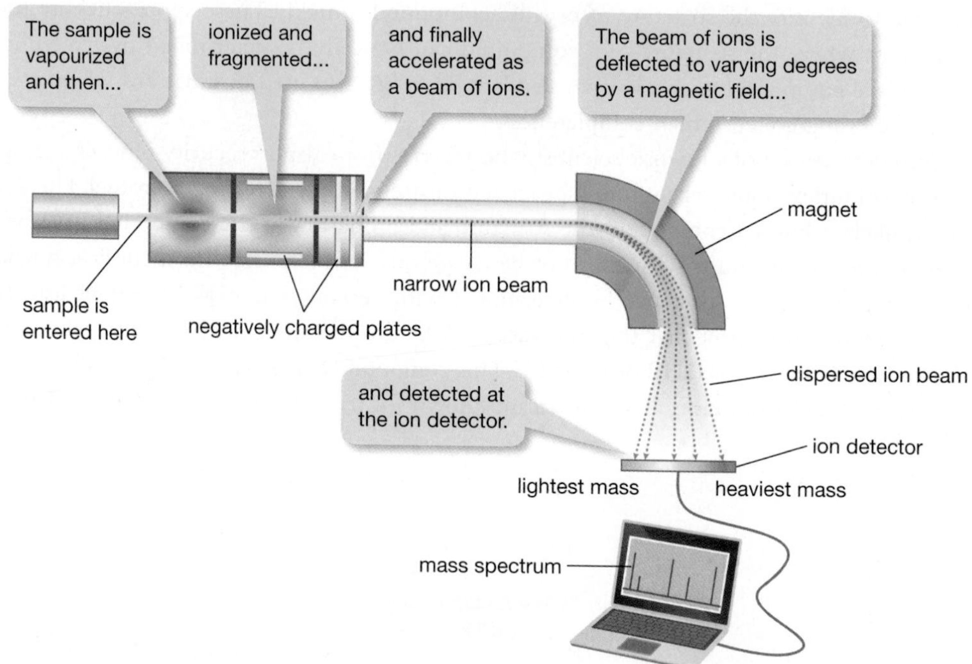

FIGURE 14.1 Schematic diagram of a mass spectrometer. Having greatest momentum, the heaviest ions—the molecular ions—are deflected the least by the magnet on their way to the detector.

The **molecular ion** ($M^{+\bullet}$ or more commonly M^+) is the cation formed by the loss of one electron from a molecule of the organic sample. The molecular ion is a radical.

A **radical** is a molecule with an unpaired electron.

The resulting vapour cloud passes through an ionization chamber where an electron is removed from a fraction of the vapour molecules, creating **molecular ions**, each with a +1 charge. The molecular ion is often written as $M^{+\bullet}$ (or just M^+). These organic cations are almost always **radicals**, because the loss of one electron from such a molecule leaves the resulting cation with one unpaired electron. (Radicals are described in Chapter 19.)

Ionization of the gaseous sample is accomplished in a number of ways. Among the most common is **electron impact**, a process in which the vapour of organic molecules passes through a beam of fast-moving electrons.

STUDENT TIP

Radicals are molecular species that have one unpaired electron—they are described in detail in Chapter 19.

$$\text{molecule} \xrightarrow[\text{beam}]{\text{electron}} [\text{molecular ion}]^{+\bullet} \quad (M^{+\bullet} \text{ or } M^+) \; + \; e^-$$

Electron impact is an ionization technique used in mass spectroscopy in which a sample vapour passes through a beam of electrons. Collisions between the electrons and the molecules in the beam produce positively charge ions.

The **mass-to-charge ratio** (m/z) is a ratio of an ion's mass divided by its charge. Since its charge is almost always +1 for organic compounds, m/z for an ion is effectively a measure of the mass of the ion.

The bombardment from these high-energy electrons ionizes a fraction of the organic molecules in the vapour. Molecular ions are inherently unstable, and many of them then split into smaller molecular fragments. A series of negatively charged plates (Figure 14.1) draws these positively charged molecular ions and fragment ions out of the ionization chamber, accelerates them, and focuses them into a narrow beam.

The beam of ions then passes between the poles of a strong magnet, where the magnetic field exerts a force on the particles because of their positive charge. This force changes the direction of flight of each ion in the beam by an amount that depends upon the ion's mass (m) and its charge (z). These two quantities are often expressed together as the **mass-to-charge ratio** (m/z) of the ion. For typical organic compounds, the majority of ions in the sample have a +1 charge, so any differences are due to changes in mass.

The heaviest ions of the beam have the greatest momentum, so their direction of flight is the least altered by the external magnetic force. Charged fragments with the smallest mass curve more than those with larger mass. Since all the ions have the same charge, the magnetic field effectively sorts the ions by their mass before they reach the detector. The detector counts the number of individual ions of each mass as they arrive and presents the data as a **mass spectrum**, such as the one shown in Figure 14.2.

A **mass spectrum** of an organic compound graphically presents the masses of its ions along the horizontal axis and their relative abundances along the vertical axis.

Five important characteristics of an organic compound can be learned by looking at a mass spectrum:

1. mass of the molecular ion (Section 14.3)
2. mass of the base peak (Section 14.3)
3. presence or absence of heavy isotopes (Section 14.3.1)
4. presence of an odd number of nitrogen atoms (Section 14.3.2)
5. fragments formed from the molecular ion (Section 14.4)

14.3 The Mass Spectrum

A mass spectrum presents graphically the masses of the ions found in the beam and their relative abundances. For example, Figure 14.2 shows the mass spectrum of 3-methylpentan-2-one.

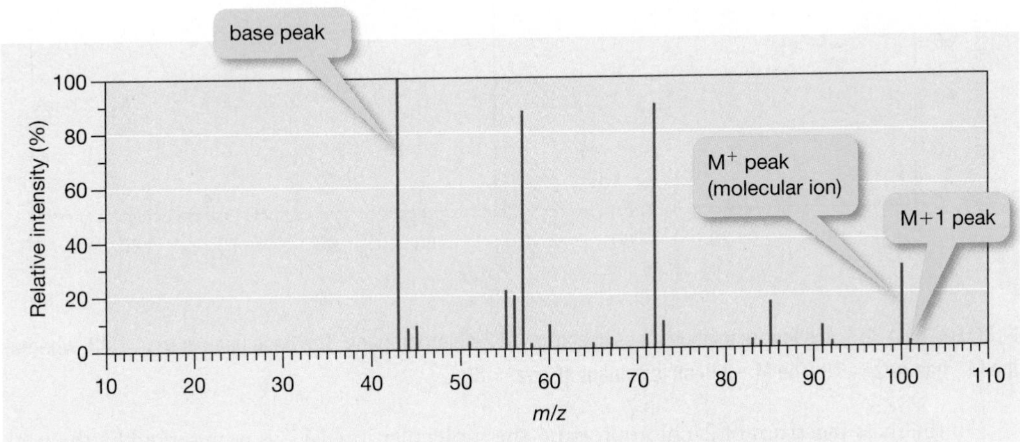

FIGURE 14.2 The low-resolution mass spectrum of 3-methylpentan-2-one, $C_6H_{12}O$.

Figure 14.2 is a **low-resolution mass spectrum** since it reports the detected ion masses—expressed m/z peaks along the horizontal axis—only to the nearest atomic mass unit (amu). This spectrum is a bar graph in which each vertical line (peak) corresponds to a detected ion. The location on the x-axis of each peak in the spectrum corresponds to the mass of the ions in that peak. The height of each peak (y-axis) represents the number (abundance) of ions of each mass. The tallest peak, called the **base peak** of the mass spectrum, is assigned an abundance of 100 percent. The height of all of the other peaks is reported as an abundance relative to that of the base peak. The base peak in Figure 14.2 shows an ion with a mass of 43 amu; the molecular ion peak (at 100 amu) shows an abundance of only about 30 percent of the base peak.

A **low-resolution mass spectrum** shows the mass of each peak to zero decimal places.

The **base peak** of a mass spectrum is the peak of highest intensity. It is assigned a height of 100 percent from which the relative intensities of all other peaks are measured.

In general, the peak corresponding to the heaviest mass appears farthest to the right in the spectrum and represents the mass of the molecular ion. This is known as the molecular ion peak. The identification of the molecular ion peak is important because it shows the mass of the whole molecule—that is, its molecular mass.

Because the mass peaks represent the detection of individual molecules, the presence of different isotopes results in peaks that represent different masses. The mass spectrum of Figure 14.2 has a molecular ion peak, M^+, at $m/z = 100$. It also has a very small M+1 (sometimes reported as $(M+1)^+$) peak at $m/z = 101$, which is called an isotope peak.

14.3.1 Isotope peaks

The height of an isotope peak reflects the percentage of the heavier isotopes in the compound. For example, the M+1 isotope peak due to ^{13}C is much smaller than the M^+ peak. Because elemental carbon is composed of 1.109% ^{13}C, the height of the M+1 peak relative to that of the M^+ peak is 1.1% times the number of carbons in the molecule. For example, 3-methylpentan-2-one, which has six carbons, has an M+1 peak roughly 6.6% the height of the M^+ peak (Figure 14.2).

When chlorine or bromine occur in a compound, the mass spectrum shows significant M+2 peaks; this is due to the high natural abundance of some of their isotopes. For example, the natural abundance of the ^{35}Cl and ^{37}Cl isotopes is 75.78% and 24.22%, respectively. Figure 14.3 shows the mass spectrum of 2-chloropropane.

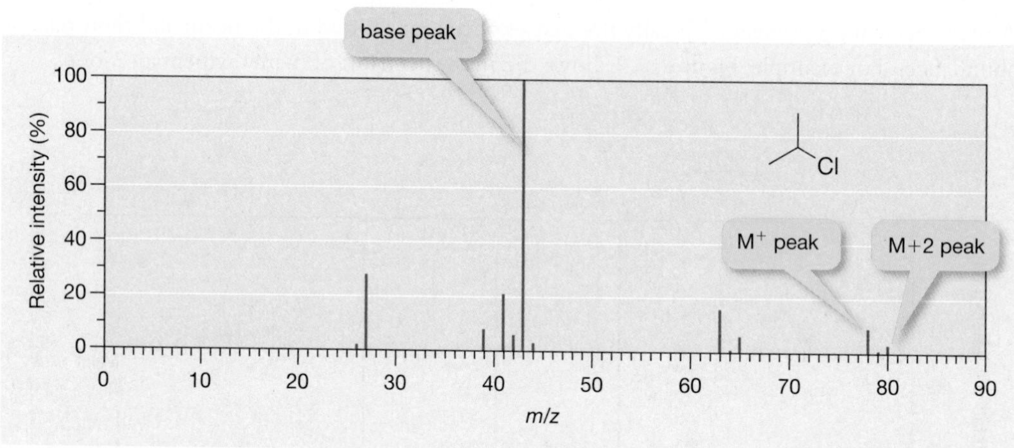

FIGURE 14.3 The low-resolution mass spectrum of 2-chloropropane. The base peak is $m/z = 43$, whereas the M^+ has $m/z = 78$. The M+2 peak is evident at $m/z = 80$.

In the mass spectrum of 2-chloropropane, the molecular ion, M^+, is represented by the peak at $m/z = 78$. This peak comes from the molecular ions in which all the Cl atoms are ^{35}Cl, and the formula is $C_3H_7{}^{35}Cl$. These molecular ions do indeed have a mass of 78 amu, as obtained from the total of their exact isotopic masses listed in the isotopic data in Table 14.1. However, the major isotopes of chlorine are ^{35}Cl and ^{37}Cl, in roughly a 3:1 ratio. Ions with a molecular formula of $C_3H_7{}^{37}Cl$ have an $m/z = 80$, and give a peak of about 33 percent of the height of the M^+ peak.

STUDENT TIP
Notice that these molecular masses are calculated by using the exact masses of particular isotopes of the elements, rather than by using atomic weights, which are the average atomic mass across all stable isotopes.

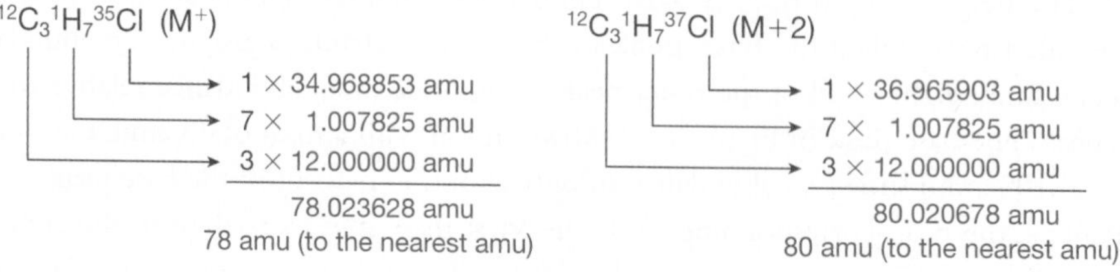

TABLE 14.1 Common Isotopes: Their Masses and Natural Abundances			
Element	Isotope	Mass (amu)	Natural Abundance (%)
Hydrogen	^{1}H	1.007825	99.9885
	^{2}H	2.014102	0.0115
Carbon	^{12}C	12.000000	98.93
	^{13}C	13.0033355	1.07
Nitrogen	^{14}N	14.003074	99.632
	^{15}N	15.000109	0.368
Oxygen	^{16}O	15.994915	99.575
	^{18}O	17.999160	0.205
Fluorine	^{19}F	18.998403	100.00
Chlorine	^{35}Cl	34.968853	75.78
	^{37}Cl	36.965903	24.22
Bromine	^{79}Br	78.918338	50.69
	^{81}Br	80.916291	49.31

In Figure 14.3, the two peaks at $m/z = 63$ and $m/z = 65$ also display the signature 3:1 ratio in their intensities, which indicates this fragment still contains a Cl atom. The base peak at $m/z = 43$ has no such partner at $m/z = 45$, which indicates this fragment does not contain a Cl atom.

The natural abundance of the ^{79}Br and ^{81}Br isotopes is 50.69% and 49.31%, respectively. As a result, the M^+ peak of a bromine-containing compound is always accompanied by an M+2 peak in a 1:1 ratio.

CHECKPOINT 14.1

You should now be able to identify the molecular ion peak (M^+) and any isotope peaks in a mass spectrum.

SOLVED PROBLEM

In the following mass spectrum, identify the molecular ion peak (M^+) and any isotope peaks (M+1 or M+2).

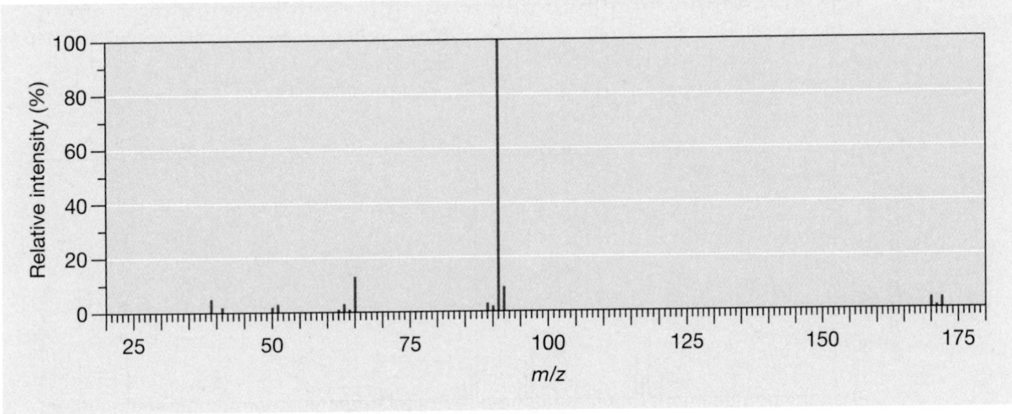

STEP 1: Find the peak or cluster of peaks appearing at the highest m/z.

In the spectrum shown, there is a group of peaks from 170 to 172 amu. In this spectrum, the peak at 91 is the base peak.

STEP 2: Take note of the spacing between peaks in the high *m*/*z* cluster. When there is a spacing of 2 *m*/*z* units, it may be because of an M+2 peak from a halide. When there is a spacing of only 1 *m*/*z* units, it may be because of ^{13}C.

In this spectrum, the peaks are 2 amu apart and have a typical height ratio of a bromide, suggesting there are is one Br atom.

STEP 3: Determine whether the peak with the highest mass has significant height, or if it is a small fraction of the height of the peak next to it.

In this spectrum, the peak representing the highest mass is the almost invisible peak at 173 amu, which is only a fraction of the height of the peak at 172. Another peak, at 171 is a fraction of the height of the peak at 170. This suggests that 170 is the M^+ peak; 171 is an M+1, due to the presence of ^{13}C. The peak at 172 amu must be an M+2 peak, due to the presence of ^{81}Br, and the tiny peak at 173 must be an M+3 peak corresponding to ^{81}Br and ^{13}C.

PRACTICE PROBLEM

14.1 In each of the following spectra, identify the M^+ peak and any isotope peaks.

a)

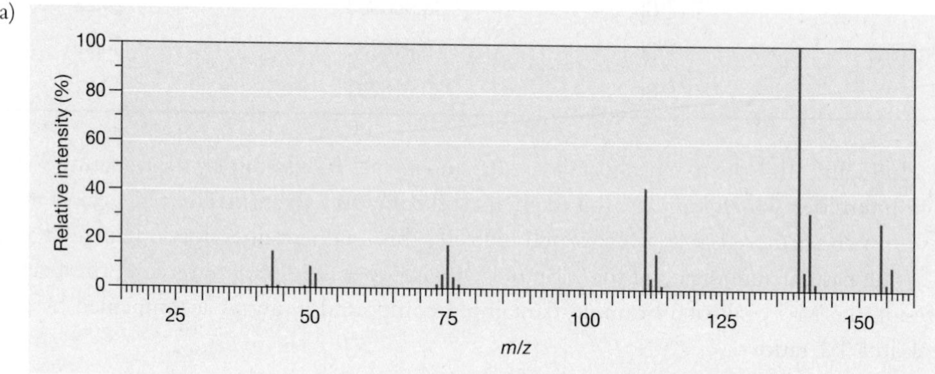

b)

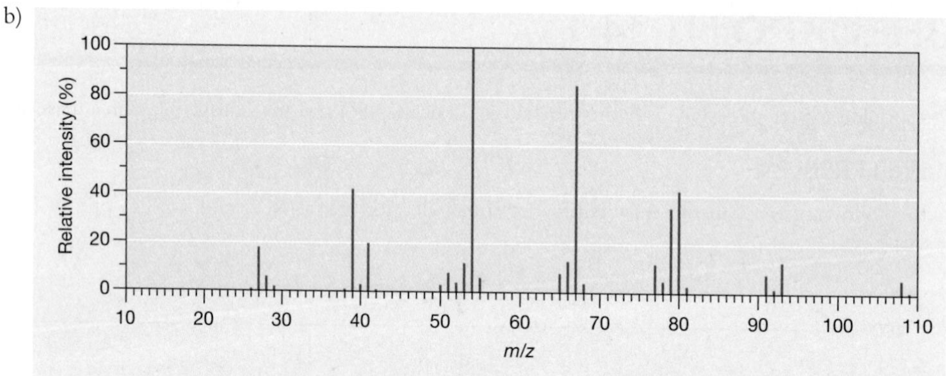

c)

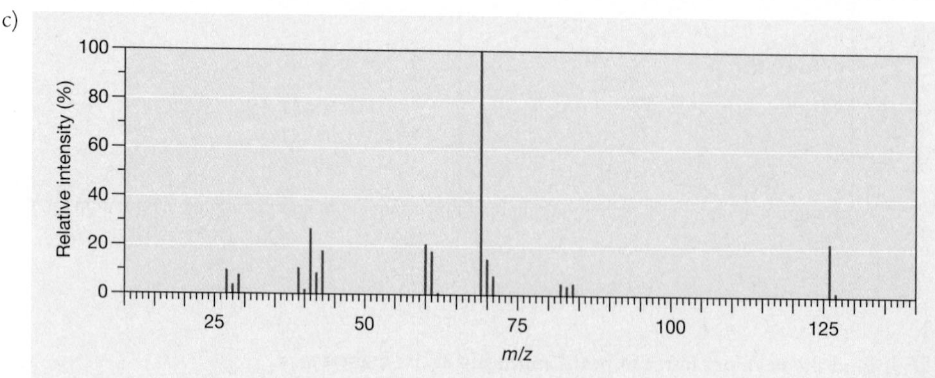

INTEGRATE THE SKILL

14.2 In the following spectrum, there is an M+2 peak *and* an M+4 peak, in addition to the M^+ peak. Label each of these peaks, and justify their height ratios.

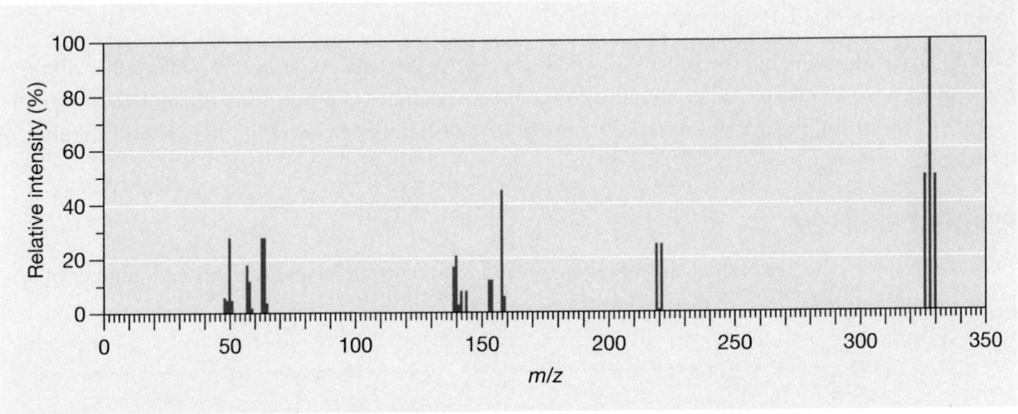

14.3.2 Number of nitrogens

Most organic molecules generate molecular ions that have an even mass (even-number mass value) because the major isotopes of atoms that form the backbone of organic molecules—carbon, nitrogen, oxygen, and sulfur—all have even masses. Atoms that attach to them, most often hydrogen and halides, have an odd-number mass. However, because carbon, oxygen, and sulfur form even numbers of bonds, there is an even overall mass. All molecules that contain only carbon, oxygen, sulfur, hydrogen, or halides always have an even mass.

Nitrogen forms an odd number of bonds, and so a molecule containing an odd number of nitrogens has an odd-number mass. When there is an even number of nitrogens in a molecule, the odd-number effect is cancelled (two odd numbers added together give an even number). So when a molecular ion peak has an odd m/z, this indicates the molecule has an odd number of nitrogen atoms.

CHECKPOINT 14.2

You should be now be able to identify when there is an odd number of nitrogen atoms in a molecular formula, based on the odd mass of the molecular ion peak.

SOLVED PROBLEM

Determine whether there is there an odd number of nitrogens in the following spectrum.

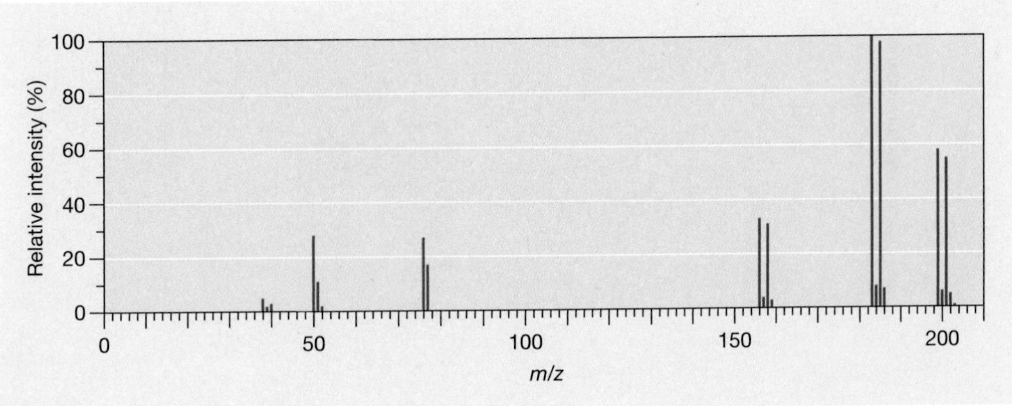

STEP 1: Follow the steps in Checkpoint 14.1 to identify the M^+ and isotope peaks.

In this spectrum, the peak cluster between 199 and 202 amu contains the M^+ peak at 199 amu and an M+1 peak at 200 amu. There is also an M+2 peak at 201 amu, with 98 percent the height of the M^+ peak. This peak arises due to the presence of bromine. There is also an M+3 peak at 202 amu, arising from molecules with both a ^{81}Br and ^{13}C.

STEP 2: Determine whether the M^+ peak is odd or even. An odd mass indicates an odd number of nitrogens. An even mass indicates either no nitrogens or an even number of nitrogens in the molecular formula.

In this spectrum, the M^+ peak at 199 amu is an odd-number mass, indicating an odd number of nitrogens.

PRACTICE PROBLEM

14.3 Determine whether each of the following spectra represents molecules with an odd number of nitrogen atoms.

a)

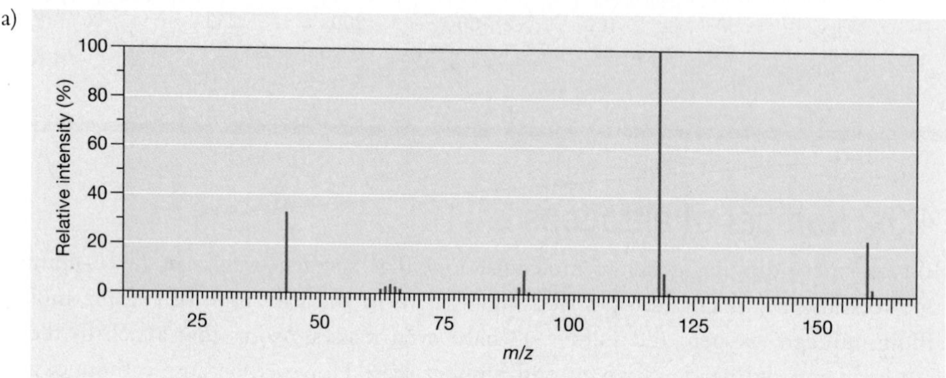

b)

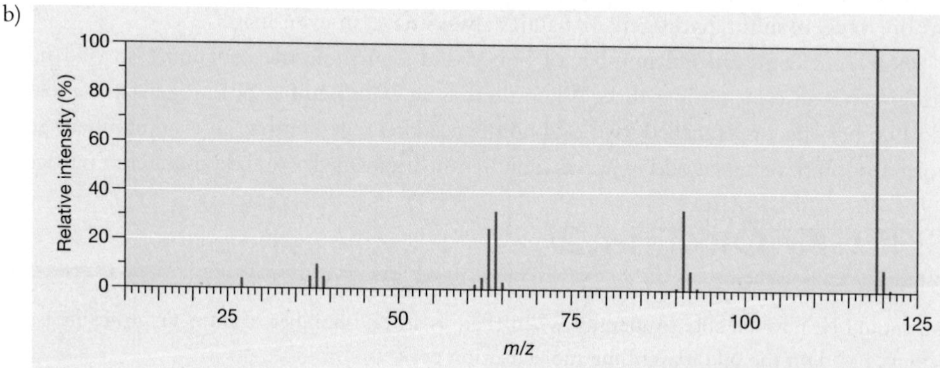

c)

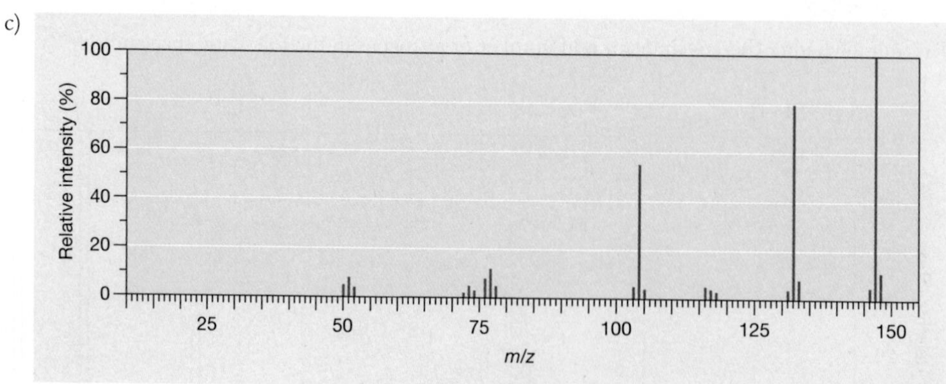

d)

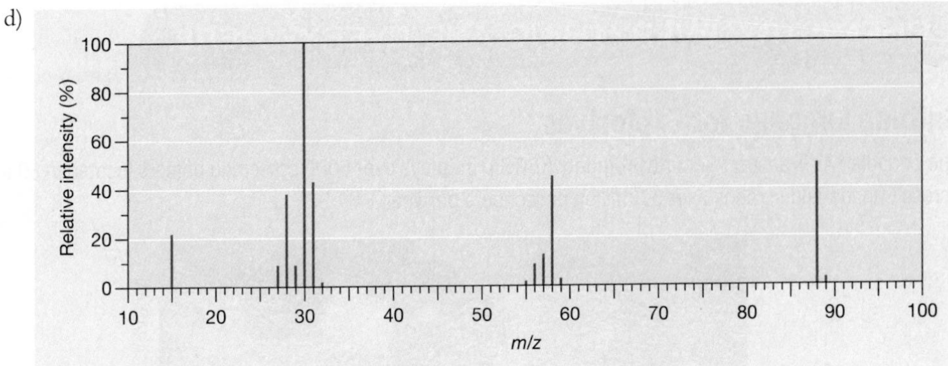

INTEGRATE THE SKILL

14.4 Knowing the mass of the molecular ion can simplify the identification of a compound's structure. Determine the structure of the compound represented by the mass and NMR spectra that follow.

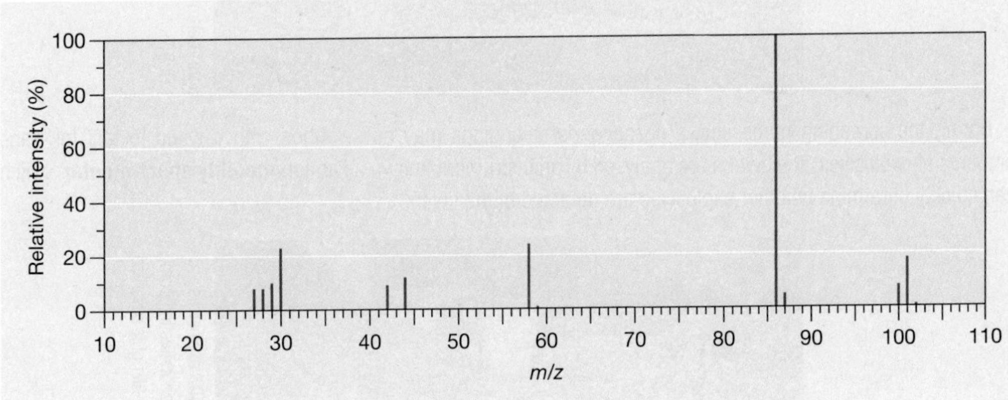

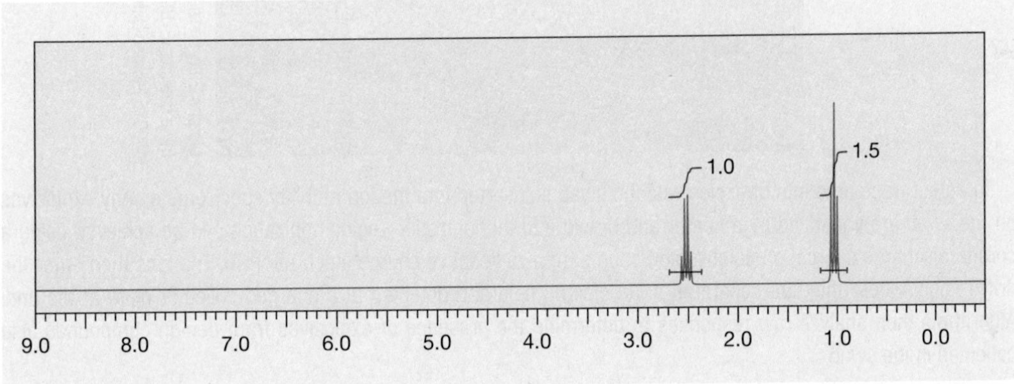

CHEMISTRY: EVERYTHING AND EVERYWHERE

Sniffing Luggage for Explosives

The Canadian Air Transport Security Authority (CATSA) employs over 6000 screening officers located in 89 airports across Canada and screens over 57 million passengers per year.

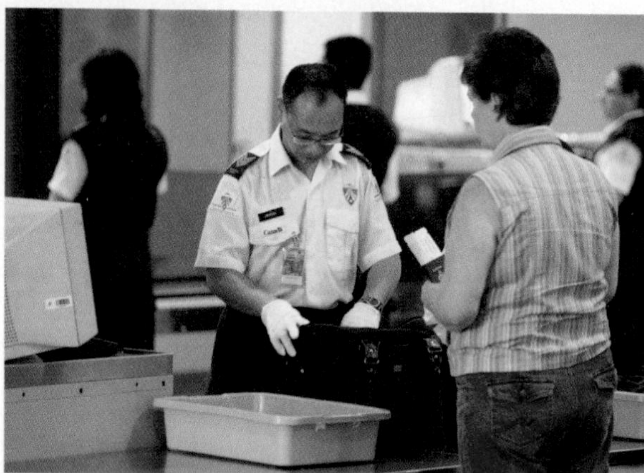

John Lehmann/Globe and Mail/THE CANADIAN PRESS

An **ion mobility spectrometer** separates ions in a magnetic field based on how fast they pass through a carrier gas that is flowing in the opposite direction.

During the screening processes, a passenger's belongings may be swabbed with a wand to test for trace amounts of explosives. The swabs are analyzed by an instrument known as an **ion mobility spectrometer**, which can quickly determine whether explosives are detected in the swab.

Morpho Detection's Itemiser® 4DX explosives trace detector (ETD). www.morphodetection.com. ©2015 Morpho Detection, LLC. All rights reserved.

To detect trace amounts of explosives, the swab is inserted into the ion mobility spectrometer. Any explosives on the swab are vaporized by a heater and drawn into the ionization region. Ionization can be achieved using a corona discharge source, photoionization source, or a radioactive source (such as ^{63}Ni). The ions then enter the drift region, where they are separated in an electric field and detected at a charged collector plate at the end. Algorithms then analyze raw responses to determine the presence of explosives from benign compounds also collected in the swab.

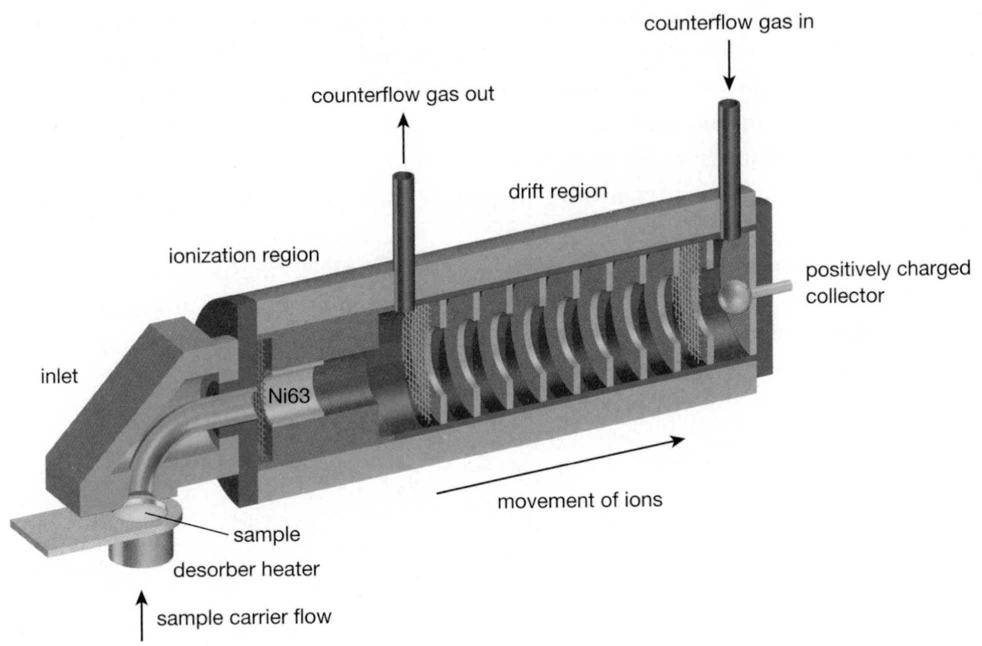

Different ions, and hence explosives, have different motilities against the counter-flowing gas and so require different amounts of time to reach the positively charged collector plate. By comparing the detection time against the times for known standards, the instrument can flag a package as potentially containing explosive materials. Explosives that can be identified by ion mobility spectrometry include RDX and PETN—these are found in the plastic explosive Semtex, which was used in the 1998 bombing of Pan Am flight 103 over Lockerbie, Scotland.

RDX PETN

Ion mobility instruments are very useful as screening techniques because they are portable, relatively easy to use, and work very quickly. However, a more sophisticated method, such as mass spectrometry or infrared spectroscopy, is needed to conclusively identify a substance.

14.4 Fragmentation of the Molecular Ion

In contrast to the molecular ion and isotope peaks resulting from detection of the complete molecule, most peaks in a mass spectrum have m/z values less than that of the molecular ion and result from the fragmentation of molecular ions. The set of fragment ion peaks and their relative abundances often serves as a "fingerprint" of the compound. However, not all fragments of the molecular ion are ions, and only charged fragments are detected by the ion detector of the mass spectrometer. Also, fragmentation is occasionally so extensive that the molecular ion peak can be weak or even absent.

molecular ion ($M^{+\bullet}$) ⟶ ionic fragments (detected) + neutral fragments (not detected)

Many (but rarely all) of the other peaks in a spectrum can be linked to specific fragmentations and may provide information about the structure. In Figure 14.2, the most prevalent ion peaks of 3-methylpentan-2-one are found at $m/z = 43$, 57, and 72 amu. Not surprisingly, the most abundant fragment ions arise either because they form easily by the cleavage of a weak bond or because their structure has some stability. Examining the pathways that transform the molecular ion into these ionic fragments may provide information about the structure of the molecule.

In the case of 3-methylpentan-2-one, two of the most common fragments are a butyl group or result from the loss of the butyl group. Bond cleavage next to the carbonyl gives fragments that are either stabilized by delocalization or secondary. These are more stable than the fragments that would be formed by breaking many other bonds.

$m/z = 57$

$m/z = 43$

14.5 High-Resolution Mass Spectrometry

An accurate determination of the molecular formula of the molecular ion is impossible using a low-resolution mass spectrum. That method reports ion masses only to the nearest amu, and many molecular formulae may create a molecular ion peak of a given mass. For example, following are some molecular formulae—using the most common isotopes of each element—that correspond to a molecular ion of 100 amu.

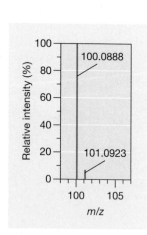

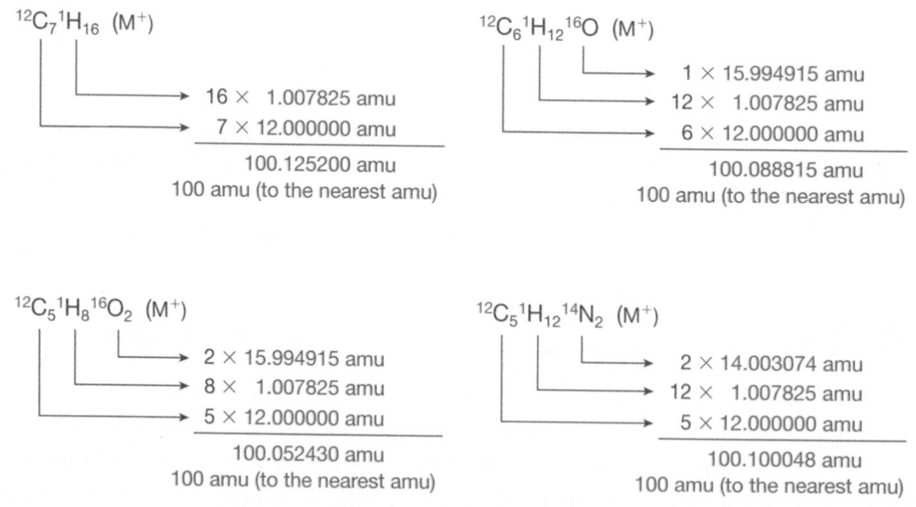

Relative intensity (%) — 100.0888, 101.0923 — m/z

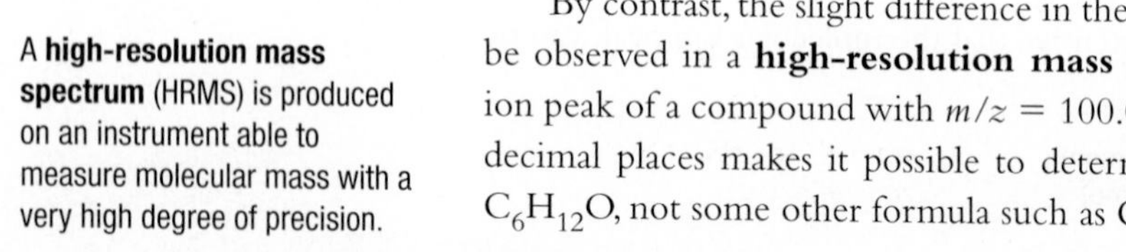

By contrast, the slight difference in the masses of ions with different molecular formulae can be observed in a **high-resolution mass spectrum** (HRMS). The high-resolution molecular ion peak of a compound with $m/z = 100.0888$ amu is shown in the margin. Resolution to four decimal places makes it possible to determine that the molecular ion must have the formula $C_6H_{12}O$, not some other formula such as C_7H_{16} or $C_5H_{12}N_2$.

A **high-resolution mass spectrum** (HRMS) is produced on an instrument able to measure molecular mass with a very high degree of precision.

High-resolution mass spectrometry is a good first step in determining the identity of an unknown compound—particularly one that does not even have an established molecular formula.

CHECKPOINT 14.3

You should now be able to interpret a high-resolution mass spectrum to determine the molecular formula.

SOLVED PROBLEM

Predict the exact mass of the following compound, as would be shown in a high-resolution mass spectrum.

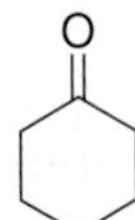

STEP 1: Determine the molecular formula by counting the atoms of each type in the structure.
Cyclohexanone is $C_6H_{10}O$.

STEP 2: Multiply the exact mass of the most common isotope for each atom type by the number of each atom type in the molecular formula. Exact masses are found in Table 14.1.

$$6 \times {}^{12}\text{C} = 6 \times 12.000000 = 72.000000$$
$$10 \times {}^{1}\text{H} = 10 \times 1.007825 = 10.07825$$
$$1 \times {}^{16}\text{O} = 1 \times 15.994915 = 15.994915$$

STEP 3: Add the masses of each atom type to give a combined exact mass for the entire formula.

$$72.000000 + 10.07825 + 15.994915 = 98.073165 \text{ amu}$$

PRACTICE PROBLEM

14.5 Predict the exact high-resolution mass of the following compounds.

a)

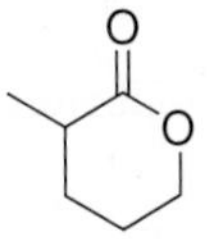

b)

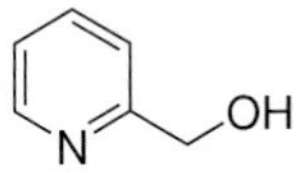

c)

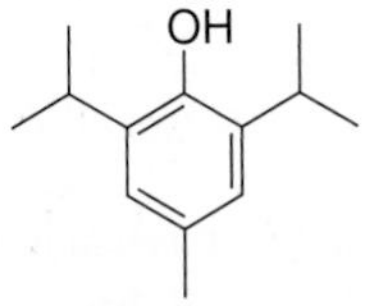

d)

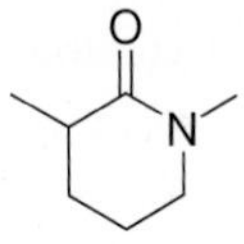

e)

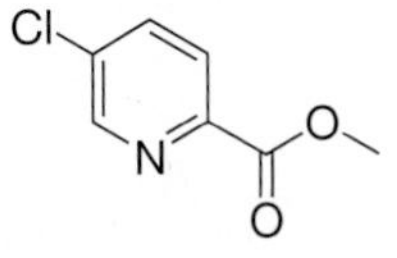

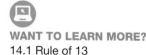

WANT TO LEARN MORE?
14.1 Rule of 13

INTEGRATE THE SKILL

14.6 A low-resolution mass spectrum gives an M^+ peak of 112. The molecular formula could be C_8H_{16}, $C_7H_{12}O$, or $C_6H_8O_2$. The high-resolution mass spectrum gives a mass of 112.0524 amu. What is the molecular formula?

14.6 Infrared Spectroscopy

The determination of molecular structure is simplified when the type of functional groups in a molecule are known. The most direct way to determine functional groups is by using infrared (IR) spectroscopy.

Light consists of photons—packets of oscillating electric and magnetic fields. Photons have an energy that depends on the frequency (ν) or the wavelength (λ) of that oscillation, as described by the following equation:

$$E = h\nu = \frac{hc}{\lambda}$$

Wavenumber ($1/\lambda$) is a convenient measure of light energy because it is directly proportional to a photon's energy. It is usually expressed in units of cm^{-1}.

This relation shows that the energy of a photon is proportional to $1/\lambda$. This quantity, usually expressed in the units of cm^{-1}, is called a **wavenumber**. As a convenient measure of light energy, this term is commonly used in infrared spectroscopy.

The bonds in a molecule absorb energy from infrared light, which causes a change in the bond vibrations. Because the bonds in different functional groups vibrate at characteristic energies, the wavenumber of the absorbed light reveals which functional groups are present in a molecule.

14.6.1 Measurement of bond vibrations by infrared spectroscopy

The bonds in organic molecules are constantly vibrating around a certain equilibrium distance (bond length) between them.

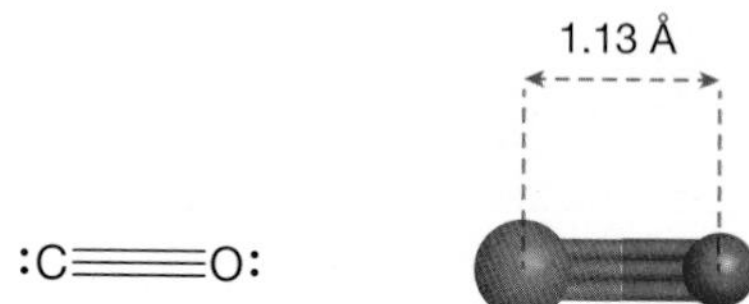

In the simplest terms, molecular vibrations are described as *stretching* or *bending* motions. Stretching modes of vibration either lengthen or shorten bonds to distances slightly longer (or shorter) than their most stable bond lengths. For carbon monoxide, this stretching vibration involves the regular lengthening and shortening of its C≡O bond relative to its equilibrium bond length (1.13 Å).

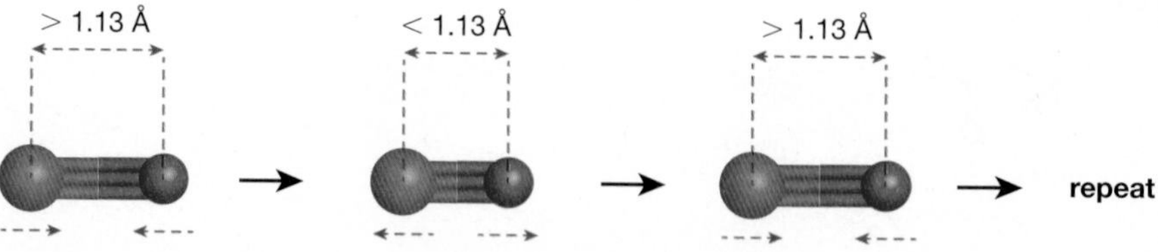

Bending modes of vibration increase (and decrease) the angles between atoms relative to the equilibrium bond angles in the molecule. Figure 14.4 shows some of the vibrations of formaldehyde. Molecular vibrations happen very rapidly, approximately 10^{13} times a second (s^{-1}); however, each different vibration occurs at a slightly different frequency.

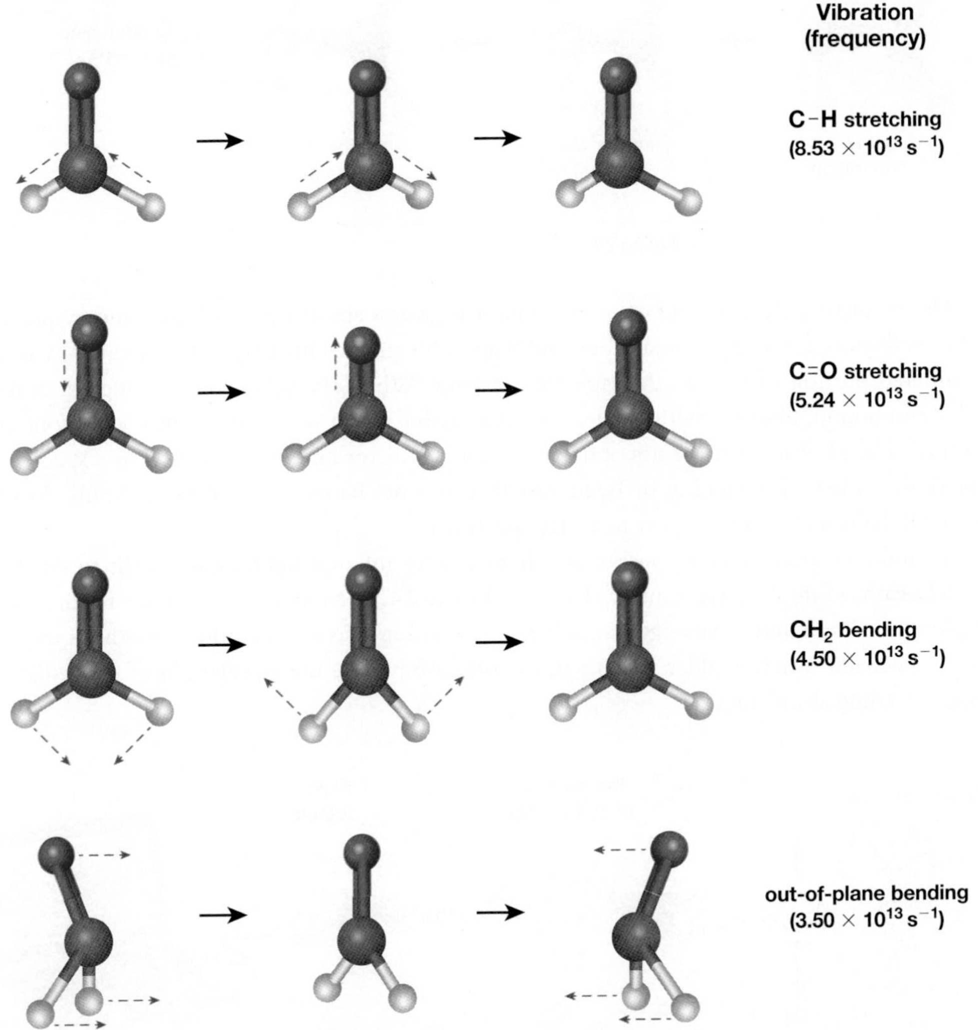

FIGURE 14.4 Four of the six modes of vibration for formaldehyde (amplitudes have been exaggerated to show them more clearly).

Molecules that have a greater number of atoms undergo more stretching and bending vibrations than molecules with fewer atoms. The number of these vibrations can be calculated using the formula $3n - 6$, where n is the number of atoms in the molecule. For example, formaldehyde has six different stretching and bending vibrations, all taking place simultaneously and continuously. Phenol, with more atoms, has 33 distinct vibrations.

formaldehyde	methanol	phenol
$3n-6 =$ 6 vibrations	$3n-6 =$ 12 vibrations	$3n-6 =$ 33 vibrations

Most vibrating bonds produce a change in bond dipoles. In formaldehyde, for example, the partial charges across the C=O bond become rapidly and repeatedly closer and further apart during the C=O bond stretch, creating a small, oscillating electric field.

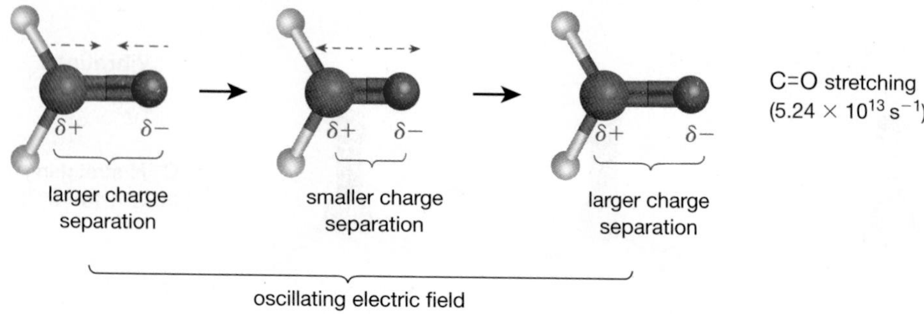

The oscillating electric field of most functional groups absorbs light of the same frequency as the oscillation: between wavenumbers 4000 and 600 cm^{-1}. This range of frequencies falls in the infrared portion of the electromagnetic spectrum. When IR light passes through a sample of the compound, photons with frequencies that match those of the molecular vibrations are absorbed. The photons that are not absorbed reach a detector and are converted to a spectrum. Vibrations—whether stretching or bending—that do not change the molecular dipole do not absorb IR light, and do not appear in an IR spectrum.

An infrared spectrometer consists of a high-quality infrared light source, a slit to create a parallel beam, a sample carrier, and a detector (Figure 14.5). Modern instruments irradiate the sample with all infrared wavelengths simultaneously and use Fourier transform methods to produce a spectrum, whereas older instruments continuously vary the wavelength of the radiation while recording absorbances.

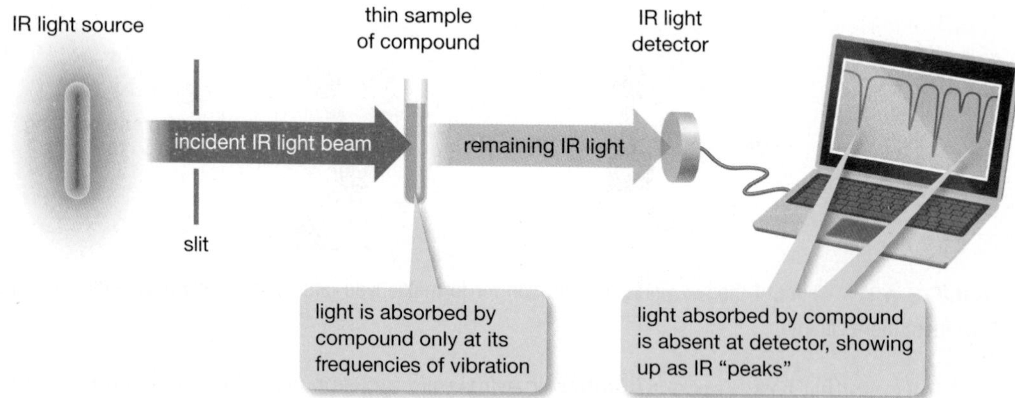

FIGURE 14.5 Light of wavenumbers between 4000 and 600 cm^{-1} projects through a narrow slit to the sample. Light of the same frequency as the frequency of the compound's molecular vibrations is absorbed by the compound.

14.7 Interpretation of Infrared Spectra

The simplest way to identify functional groups in an organic molecule is by examining the absorbances (absorption peaks) in an infrared spectrum. Because each functional group has characteristic bonds, functional groups have characteristic vibrations that can be distinguished in a spectrum (Figure 14.6). Like a mass spectrum, an infrared spectrum has many peaks. However, only some of the absorption peaks are needed to identify the functional groups in a molecule. The IR spectrum includes the following four key regions:

1. hydrogen region (3600–2700 cm^{-1})
2. triple bonds (2700–1900 cm^{-1})
3. double bonds (1900–1500 cm^{-1})
4. fingerprint region (1500–500 cm^{-1})

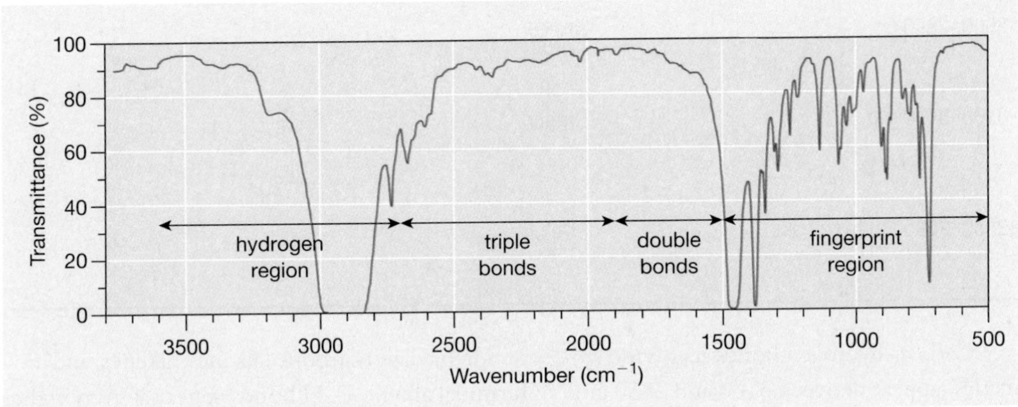

FIGURE 14.6 Infrared spectrum of 1-hexene showing key regions of the spectrum.

SOURCE: NIST, Office of Data and Informatics, Chemistry WebBook, http://webbook.nist.gov/

14.7.1 Hydrogen region (3600–2700 cm^{-1})

The hydrogen region of an IR spectrum, 3600–2700 cm^{-1}, gives information about the "types" of hydrogens, such as C–H bonds, as well as O–H and N–H bonds. Bonds involving hydrogen vibrate at the fastest frequencies due to the relative lightness of hydrogen. Several examples of C–H bond vibrations are shown in Figure 14.7 and Table 14.2. Because the location of some of these peaks overlap, both the location and the shape of the various peaks are important.

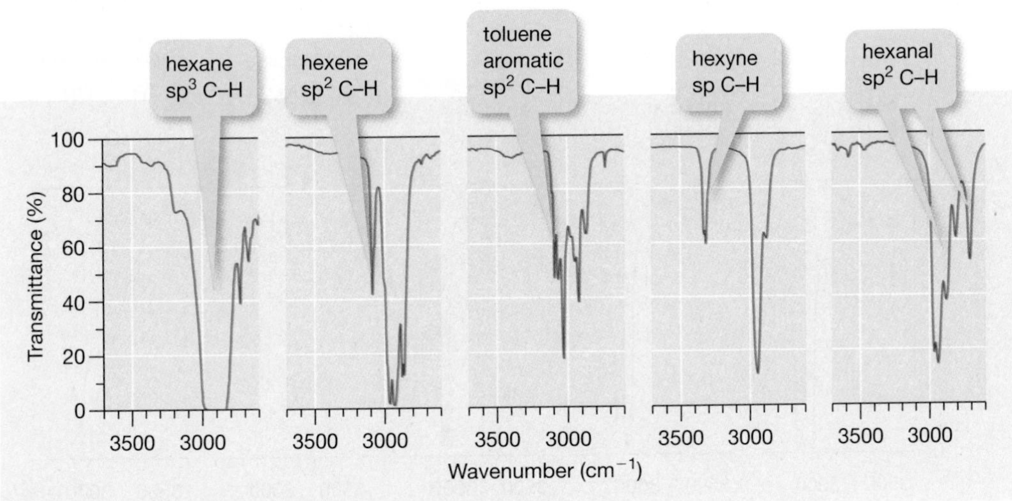

FIGURE 14.7 Carbon-hydrogen region of various compounds.

SOURCE: NIST, Office of Data and Informatics, Chemistry WebBook, http://webbook.nist.gov/

TABLE 14.2 IR Absorptions of C–H bonds

The intensity of signals in IR are reported as (v) = very strong, (s) = strong, (m) = medium, (w) = weak, (br) = broad

Wavenumber Range (cm^{-1})	Vibrational Mode	Atom Group
3350–3250 (s)	Stretch	C≡C–H
3095–3000 (w)	Stretch	C=C–H
3000–2800 (s)	Stretch	C–C–H
3100–3000 (m)	Stretch	(benzene ring) C–H
2830–2695 (m) two peaks	Stretch	O=C–H

Carbon-hydrogen bonds involving sp^3- or sp^2-hybridized carbons (alkanes, alkenes, and aromatics) appear between 3100 and 2850 cm^{-1}. Terminal alkyne C–H bonds appear at even higher wavenumbers: between 3350 and 3250 cm^{-1}. Farther to the right of the spectrum are aldehyde C–H bonds, which often appear as two medium intensity peaks near 2850 and 2700 cm^{-1}.

Several of the peaks in the IR spectrum shown in Figure 14.7 overlap with the peaks for O–H and N–H bonds, though the shapes of the absorbances can be quite different (Figure 14.8). In spite of their similar locations, alkyne C–H and alcohol O–H bonds can usually be distinguished, because alkyne peaks are narrower (sharper) than the broad O–H bond peaks.

In the IR spectrum, N–H bonds from amines or amides (NH_2) also overlap with alcohol O–H bonds (Table 14.3). N–H bonds are typically narrower than alcohol O–H bonds and broader than alkyne C–H bonds. Note that primary amines and amides have two distinct vibrational modes (symmetric and antisymmetric stretching) for NH_2 groups, giving them double N–H peaks. In contrast, secondary amines and amides produce only one N–H peak.

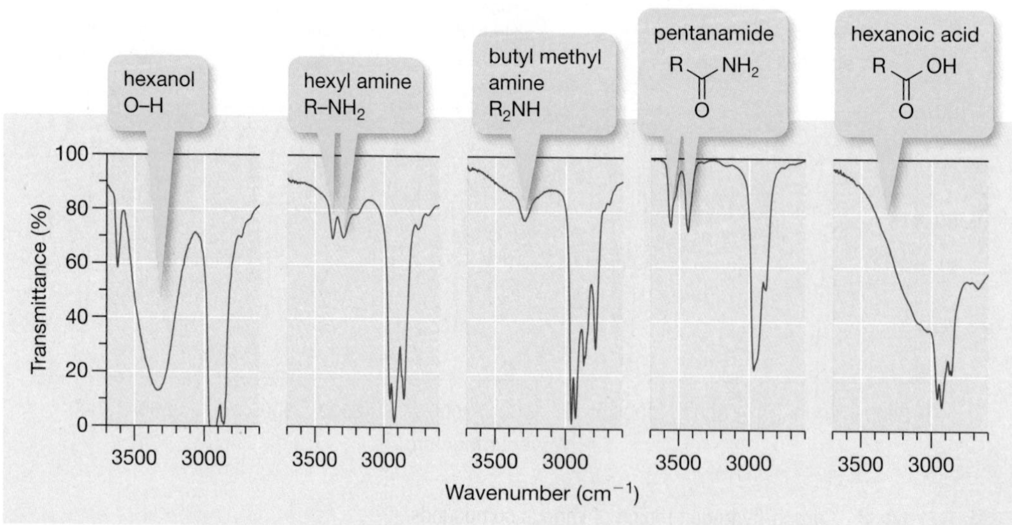

FIGURE 14.8 Related N–H and O–H bonds.

TABLE 14.3 IR Absorptions of O–H, N–H Bonds
The intensity of signals in IR are reported as (v) = very strong, (s) = strong, (m) = medium, (w) = weak, (br) = broad

Wavenumber Range (cm^{-1})	Vibrational Mode	Atom Group
3640–3200 (s)	stretch	Alcohol OH
3400–3250 (s)	stretch	Amide, amine NH
3300–2500 (br, s)	stretch	Carboxylic acid OH

Figure 14.8 also shows the O–H stretch of a carboxylic acid, which has a very different shape from that of an alcohol. Carboxylic acids are much broader (3300–2500 cm^{-1}) and more jagged than alcohol O–H peaks.

CHECKPOINT 14.4

You should now be able to identify the types of hydrogens in a compound, based on the hydrogen region of an infrared spectrum.

SOLVED PROBLEM

Based on the hydrogen region of the following spectrum, what types of hydrogens does the compound have?

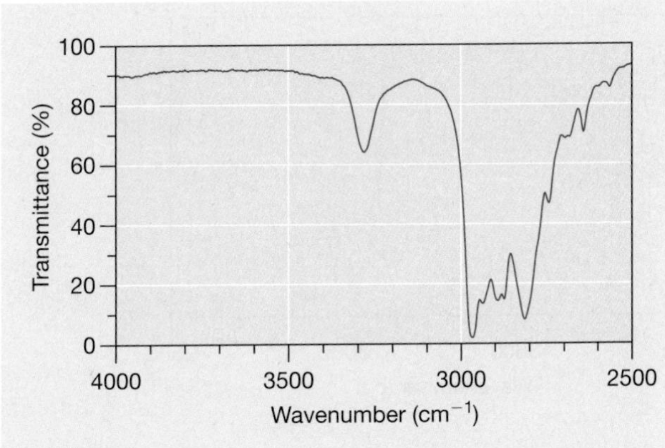

STEP 1: Consult Table 14.2 to identify the possible C–H bonds based on peak location.

The above spectrum has absorption peaks between 3000 and 2800 cm^{-1}, which may be alkyl C–H bonds. There is also a peak near 3000 cm^{-1}, which may be an sp C–H bond (alkyne).

STEP 2: Consult Figure 14.7 to identify what types of C–H bonds are present, based on peak appearance.

The absorption peaks between 3000 and 2800 cm^{-1} look like the alkyl C–H bond absorptions in Figure 14.7. The peak near 3300 cm^{-1} does not look like the example alkyne C–H signal. Instead, it is broader than would be expected from a signal arising from C–H bond. So this is likely a N–H or O–H bond.

STEP 3: Consult Table 14.3 to identify what O–H and N–H bonds are possible in the molecule, based on peak location.

The peak near 3300 cm^{-1}, which does not arise from an alkyne C–H bond because it is too broad, is in the correct location to be an alcohol O–H or N–H bond.

STEP 4: Consult Figure 14.8 to identify the types of O–H and N–H bonds are present, based on peak appearance.

The peak near 3300 cm^{-1} is too narrow to be an alcohol O–H bond. So this is likely a N–H bond, and because there is only one peak, it is a N–H, not a NH_2.

This compound has a secondary amine and alkyl C–H bonds.

PRACTICE PROBLEM

14.7 Identify the types of hydrogen in the compounds represented by the following spectra.

a)

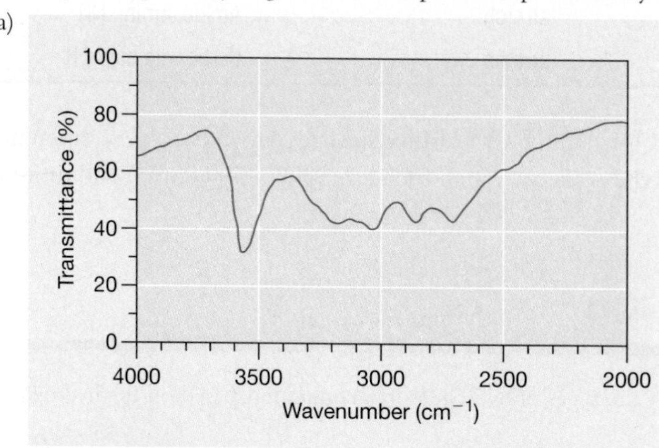

SOURCE: NIST, Office of Data and Informatics, Chemistry WebBook, http://webbook.nist.gov/

b)

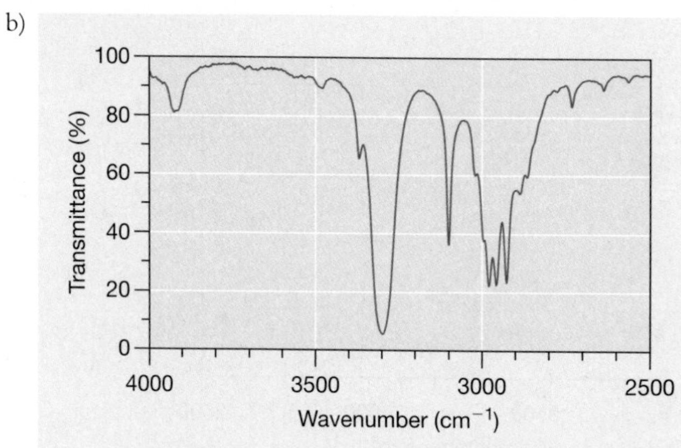

SOURCE: Copyright © 2016 Sigma-Aldrich Co. LLC. All Rights Reserved.

c)

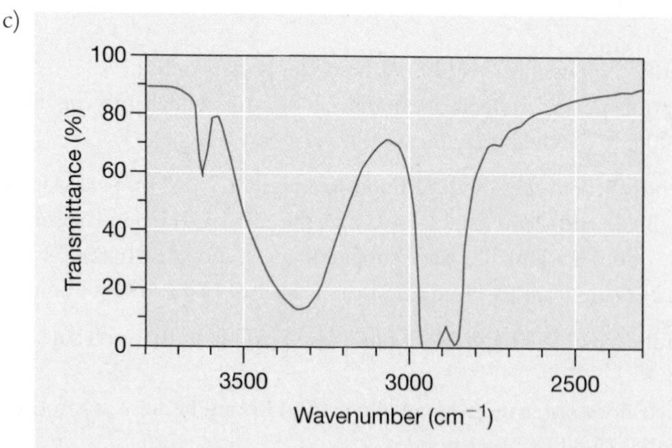

SOURCE: NIST, Office of Data and Informatics, Chemistry WebBook, http://webbook.nist.gov/

d)

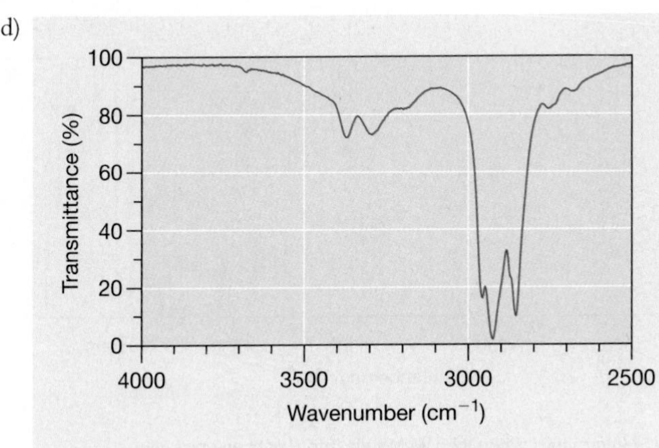

INTEGRATE THE SKILL

14.8 The ^{1}H-NMR spectra of methyl 2-hydroxybenzoate and 2-methoxybenzoic acid are predicted to be very similar, so these structures are difficult to distinguish using this method only.

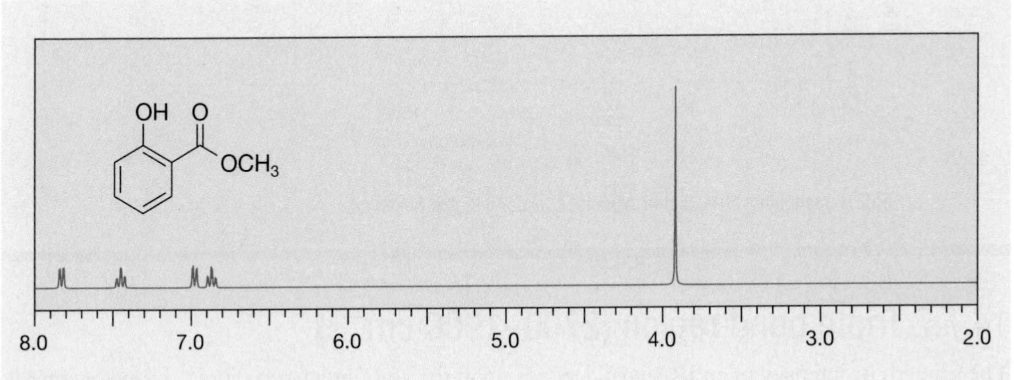

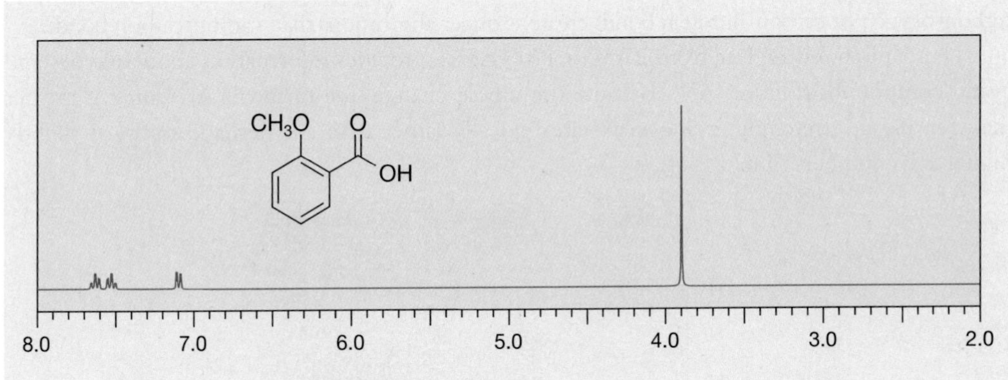

IR spectroscopy can more easily distinguish these two compounds. Which of the following spectra represent the ester, and which represents the carboxylic acid?

a)

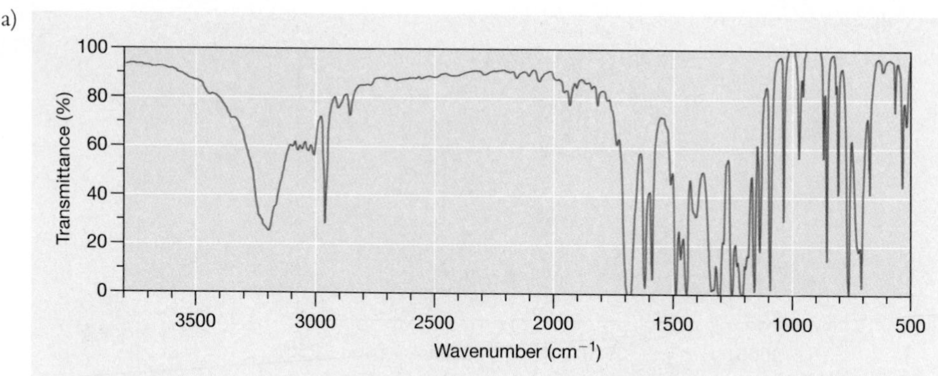

SOURCE: NIST, Office of Data and Informatics, Chemistry WebBook, http://webbook.nist.gov/

b)

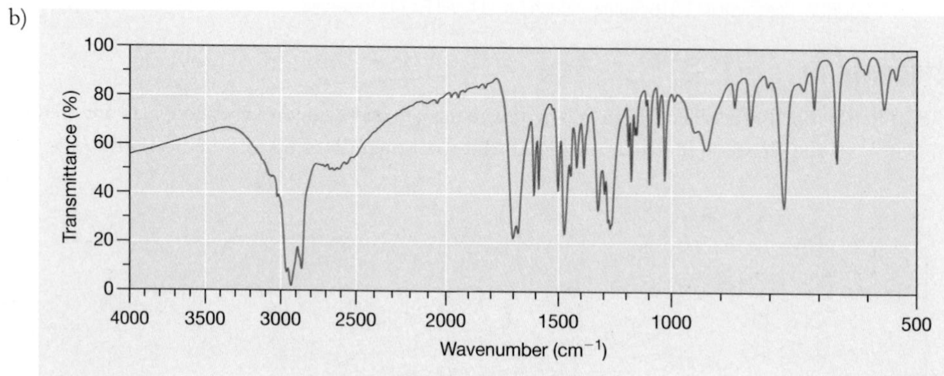

14.7.2 Triple bond region (2700–1900 cm^{-1})

The strength or intensity of an IR signal depends upon the amount a bond dipole changes as a result of molecular vibration. Specifically, a greater intensity of absorption of IR radiation is associated with greater change in the dipole. This is most noticeable in the double and triple bond regions because carbon-oxygen or carbon-nitrogen bonds create stronger absorptions than carbon-carbon bonds.

The triple bond region, from 2700 to 1900 cm^{-1}, provides information about alkynes and cyano compounds (Figure 14.9). Because the dipole change due to nitrile stretching is greater than for alkyne stretching, cyano peaks are typically larger than alkynes, and occur at slightly higher wavenumbers (Table 14.4).

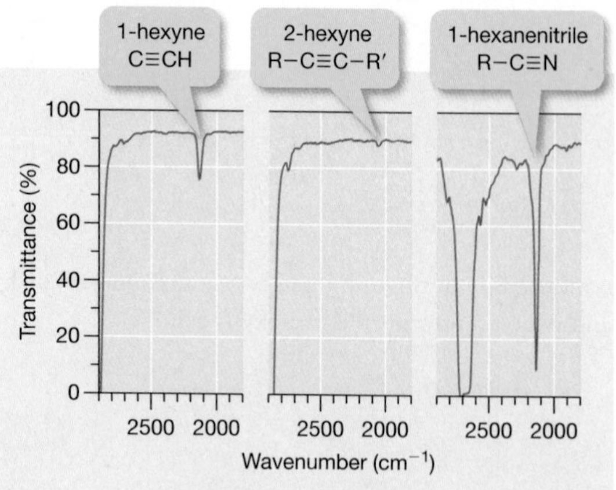

FIGURE 14.9 Terminal alkynes, internal alkynes, and nitriles.

SOURCE: NIST, Office of Data and Informatics, Chemistry WebBook, http://webbook.nist.gov/

TABLE 14.4 IR Absorptions of Triple Bonds
The intensity of signals in IR are reported as (v) = very strong, (s) = strong, (m) = medium, (w) = weak, (br) = broad

Wavenumber Range (cm^{-1})	Vibrational Mode	Atom Group
2260–2210 (v)	Stretch	R–C≡N
2260–2100 (w)	Stretch	R–C≡C–R
2250–2100 (m)	Stretch	R–C≡C–H

The stretching of the terminal alkyne, such as occurs in 1-hexyne, induces a greater change in the dipole, so the terminal alkyne produces a stronger signal than internal alkynes, such as 2-hexyne. Internal alkynes, which lack both an sp C–H bond and a strong dipole, are difficult to distinguish in an IR spectrum.

14.7.3 Double bond region (1900–1500 cm^{-1})

The double bond region, 1900–1500 cm^{-1}, provides information about C=O and C=C bonds. Because C=O bonds are more polar than C=C bonds, they produce larger peaks between 1850 and 1650 cm^{-1} (Table 14.5). C=C bonds appear as weak signals between 1700 and 1600 cm^{-1}. The C=O peaks are particularly important because carbonyls are common functional groups, and because the different types of carbonyl groups can often be differentiated based on the absorbances they produce at slightly different wavenumbers (Figure 14.10).

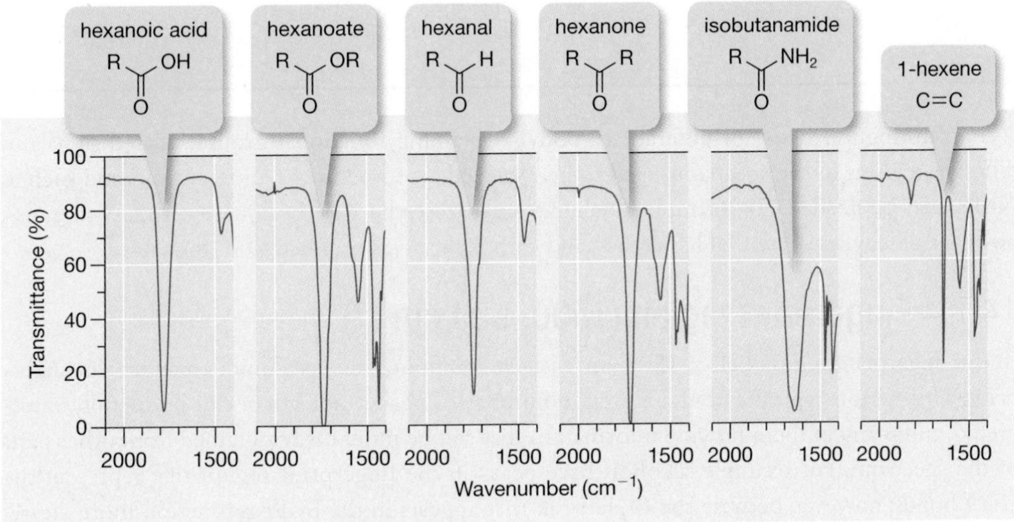

FIGURE 14.10 C=O and C=C bond stretches.

SOURCE: NIST, Office of Data and Informatics, Chemistry WebBook, http://webbook.nist.gov/

For example, acid chlorides have a C=O stretch as high as 1800 cm^{-1}, whereas amides appear between 1700 and 1630 cm^{-1}. This range of frequencies correlates with the strength of the double bond and the relative contributions of resonance forms in each group (see Chapter 15).

In practice, however, the range of absorption of one carbonyl functional group often overlaps with that of another, and so conclusively determining the type of carbonyl group present often requires some other kind of measurement (other IR absorbance or NMR).

TABLE 14.5 IR Absorptions of Double Bonds

The intensity of signals in IR are reported as (v) = very strong, (s) = strong, (m) = medium, (w) = weak, (br) = broad

Wavenumber Range (cm^{-1})	Vibrational Mode	Atom Group
1820–1760 (s)	Stretch	RC(=O)Cl
1800–1700 (s, br)	Stretch	RC(=O)OH
1750–1735 (s)	Stretch	RC(=O)OR
1740–1720 (s)	Stretch	RC(=O)H
1720–1708 (s)	Stretch	RC(=O)R
1700–1630 (s)	Stretch	RC(=O)NR_2
1680–1640 (m)	Stretch	C=C

To distinguish between different carbonyl-containing functional groups, related signals in different regions of the spectrum need to be analyzed, or sometimes a different method such as NMR may be needed. For example, aldehydes and carboxylic acids both have carbonyl stretches; however, aldehydes show C–H stretches, and carboxylic acids produce O–H signals.

14.7.4 Fingerprint region (1500–500 cm^{-1})

In the fingerprint region below 1500 cm^{-1}, each compound shows a unique pattern: its fingerprint. This region typically contains a large number of peaks, more than can be usefully interpreted, and many of them provide information that can be more easily obtained from other parts of the spectrum. For example, alcohols have peaks in the fingerprint region that represent the C–O bonds; however, because the O–H peak that appears in the hydrogen region more clearly identifies alcohols, it is not necessary to identify C–O bonds to identify alcohols.

Three functional groups have key absorbances that are observed in the fingerprint region: ester, ether, and nitro. Figure 14.11 shows the IR signals for ester, ether, and nitro.

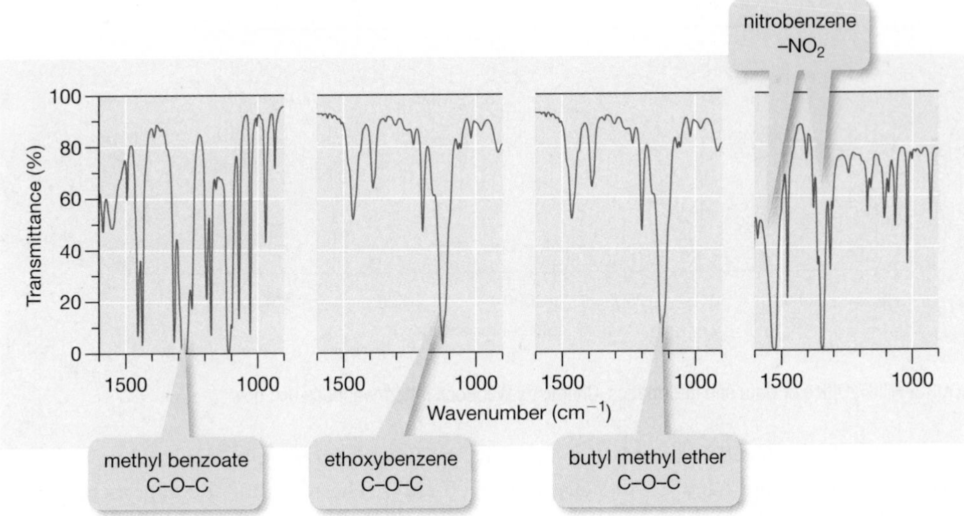

FIGURE 14.11 Important fingerprint signals: C–O–C and nitro peaks.

SOURCES: Copyright © 2016 Sigma-Aldrich Co. LLC. All Rights Reserved. NIST, Office of Data and Informatics, Chemistry WebBook, http://webbook.nist.gov/

The C–O–C peak of an ester or aryl ether appears as a strong signal between 1350 and 1200 cm^{-1} (Table 14.6). Alkyl ether peaks look similar, but occur at lower frequencies, between 1200 and 1100 cm^{-1}. Nitro groups produce two strong peaks near 1500 and 1350 cm^{-1}.

TABLE 14.6 Key IR Absorptions in the Fingerprint Region
The intensity of signals in IR are reported as (v) = very strong, (s) = strong, (m) = medium, (w) = weak, (br) = broad

Wavenumber Range (cm^{-1})	Vibrational Mode	Atom Group
1550–1475 (s) 1360–1290 (m)	Stretch	$R{-}N^{\oplus}(=O){-}O^{\ominus}$
1320–1000 (s)	Stretch	C–O–R

Most often, the fingerprint region is used for comparing the IR spectrum of an unknown compound to a library of known spectra to find the matching spectrum.

CHECKPOINT 14.5

You should now be able to identify the presence of double and triple bonds in a compound. You should also be able to identify specific carbonyl-containing functional groups, based on complementary peaks in other regions of the spectrum.

SOLVED PROBLEM

The following spectrum shows a C=O signal near 1700 cm^{-1}. Is this C=O part of a carboxylic acid, ester, amide, aldehyde, or ketone?

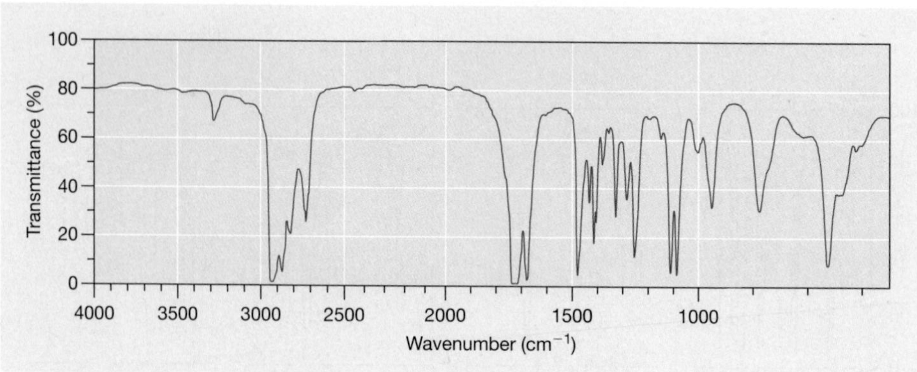

SOURCE: NIST, Office of Data and Informatics, Chemistry WebBook, http://webbook.nist.gov/

STEP 1: Consult Table 14.5 and Figure 14.10 to identify C=O signals based on location and appearance.

As stated, this compound has a C=O signal near 1700 cm^{-1}. The spectrum also appears to show a weaker C=C bond stretch at 1650 cm^{-1}.

STEP 2: Consult Table 14.2 and Figure 14.7 to identify C–H types based on location and appearance.

This spectrum shows some alkyl C–H signals near 3000 cm^{-1}. Also, there are signals representing an aldehyde C–H bond at 2850 and 2700 cm^{-1}.

STEP 3: Consult Table 14.3 and Figure 14.8 to identify O–H and N–H types based on location and appearance.

This spectrum shows no O–H or N–H types.

STEP 4: Consult Table 14.6 and Figure 14.11 to identify possible ester signals in the fingerprint region based on location and appearance.

There does not appear to be an ester C–O–C signal in the fingerprint region.

The presence of the aldehyde C–H and the absence of other ester, acid, and amide signals suggest that the C=O is part of an aldehyde.

PRACTICE PROBLEM

14.9 What functional groups are shown in the following spectra?

a)

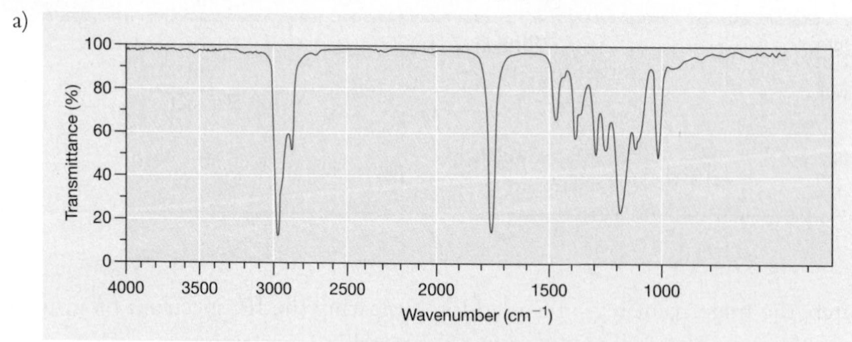

SOURCE: NIST, Office of Data and Informatics, Chemistry WebBook, http://webbook.nist.gov/

b)

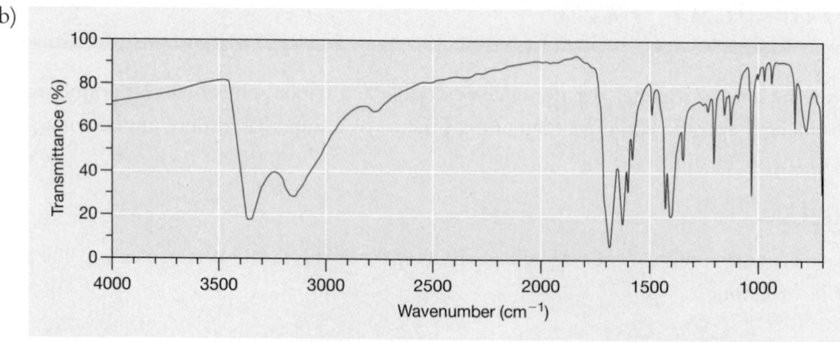

SOURCE: NIST, Office of Data and Informatics, Chemistry WebBook, http://webbook.nist.gov/

c)

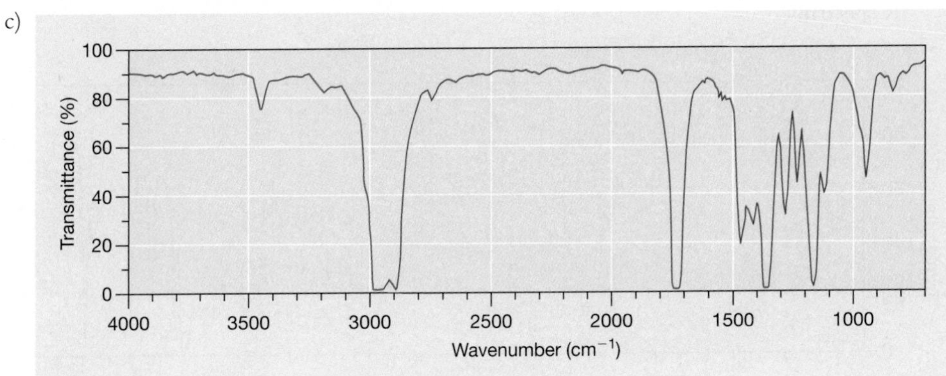

SOURCE: NIST, Office of Data and Informatics, Chemistry WebBook, http://webbook.nist.gov/

d)

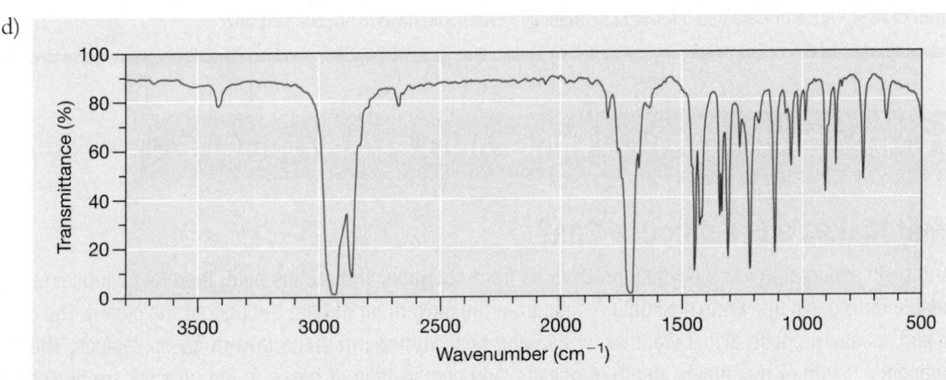

SOURCE: NIST, Office of Data and Informatics, Chemistry WebBook, http://webbook.nist.gov/

INTEGRATE THE SKILL

14.10 Neither ketones nor tertiary amides have complementary signals that help to distinguish them in an IR spectrum. Using information provided by the mass spectrum, suggest a viable structure for the following unknown compound.

Mass spectrum

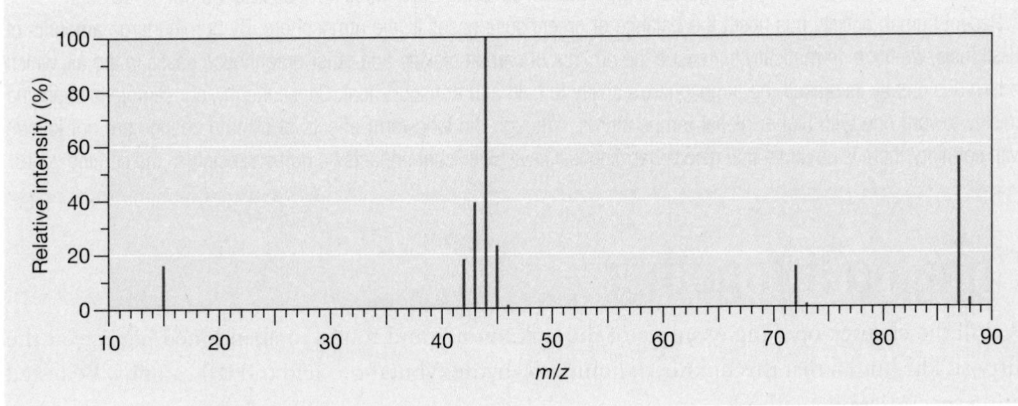

IR spectrum

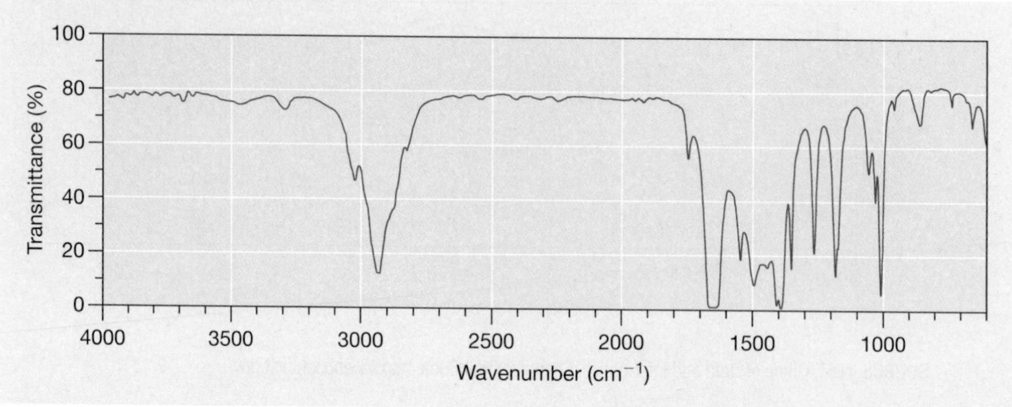

SOURCE: NIST, Office of Data and Informatics, Chemistry WebBook, http://webbook.nist.gov/

CHEMISTRY: EVERYTHING AND EVERYWHERE

What Makes a Greenhouse Gas?

Earth's atmosphere provides oxygen, protects us from radiation, and retains heat, thereby maintaining a climate suitable for organic life. Earth's climate results from the movement of heat throughout the planet. The circulating air and oceans produce global weather by carrying heat from warm areas toward cooler regions. The shape of continents, height of mountains, depth of oceans, and composition of gases in the atmosphere all contribute to Earth's climate. A significant influence comes from the gases in the atmosphere that carry the most heat.

Carbon dioxide is one such gas. This molecule is composed of two carbonyl groups, whose bonds are subject to four vibrational modes. These bond vibrations allow CO_2 to very strongly absorb infrared radiation, so CO_2 can "carry" large amounts of heat. This property makes CO_2 a "greenhouse gas" because it traps heat, which effectively insulates our atmosphere and maintains the temperature on the surface of the planet at a point above freezing.

A number of so-called greenhouse gases contribute to Earth's climate by holding and transporting heat throughout the atmosphere. These gases have molecular structures that enable the absorption of IR radiation. For example, methane's C–H bonds make it a good absorber of heat, as do the O–H bonds of water. A balance of greenhouse gases in the atmosphere helps to regulate global temperatures, ensuring Earth does not become too hot or too cold.

Recent human activity has upset the balance of greenhouse gases in the atmosphere. By burning large amounts of fossil fuels, we have dramatically increased the amount of carbon dioxide and other greenhouse gases in the air, which in turn has greatly increased the atmosphere's ability to hold and transport heat. Consequently, our climate is changing quickly toward one with higher global temperatures. Although the long-term effects of climate change are not known with certainty, there is no doubt that global warming will have significant impacts on human society in the coming years.

Bringing It Together

Recall the chapter-opening example of the unknown liquid found in abandoned luggage at the airport. The hunch that this unknown liquid is 4-hydroxybutanoic acid (GHB) can now be tested using spectroscopic methods.

HO O OH

γ-hydroxybutanoic acid
(GHB)

One tool often used at airports is low-resolution mass spectrometry. Mass spectrometry can provide a molecular mass, which can be used to calculate a molecular formula. However, many organic compounds can have the same molecular formula, so further information is needed to make a positive identification.

The IR spectrum of the unknown compound (Figure 14.12) can be compared to spectra in a reference library to find a matching fingerprint pattern.

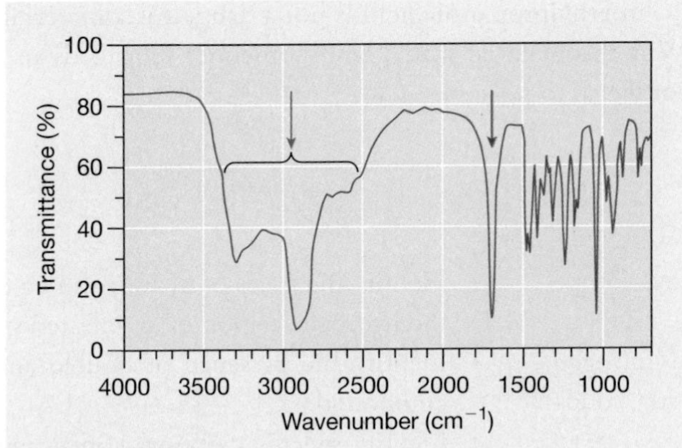

FIGURE 14.12 IR spectrum of the unknown sample from an airport.

SOURCE: NIST, Office of Data and Informatics, Chemistry WebBook, http://webbook.nist.gov/

After consulting the reference library, it was determined that the IR spectrum of the unknown compound does not match what is expected for GHB. The strong absorbance of a C=O stretch at 1700 cm^{-1} (green arrow) and the very broad band of overlapping alcohol and acid O–H stretches between 3500 and 2500 cm^{-1} (red arrow) suggest the molecule contains alcohol and carboxylic acid functional groups, which is consistent with the structure of GHB. However, the peaks in the fingerprint region of the spectrum do not match those of the GHB standard.

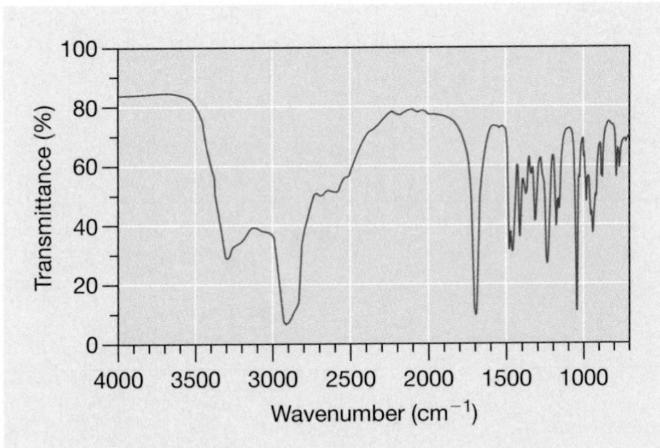

FIGURE 14.13 IR spectrum of 3-hydroxy-2,2-dimethylpropanoic acid from an IR reference library.

SOURCE: NIST, Office of Data and Informatics, Chemistry WebBook, http://webbook.nist.gov/

If not GHB, what *is* the compound? A comparison of the mass and IR spectra to standards provided in the reference library shows that the unknown compound is actually 3-hydroxy-2,2-dimethylpropanoic acid (Figures 14.13 and 14.14), not 4-hydroxybutanoic acid.

3-hydroxy-2,2-dimethylpropanoic acid

4-hydroxybutanoic acid (GHB)

FIGURE 14.14 Structures of 3-hydroxy-2,2-dimethylpropanoic acid and GHB.

3-Hydroxy-2,2-dimethylpropanoic acid is not a drug; it is a material used to make biodegradable plastics. Why this material would be in someone's luggage is an interesting question, but the material is not illegal to possess.

You Can Now

- Identify the molecular ion (M^+) peak and any isotope peaks in a mass spectrum.
- Identify the presence of an odd number of nitrogen atoms in the molecular formula, based on the odd-numbered mass of the molecular ion peak.
- Interpret a high-resolution mass spectrum to determine the molecular formula.
- Identify the types of hydrogens in a compound based on the hydrogen region of an infrared spectrum.
- Identify the presence of double and triple bonds in a compound.
- Identify specific carbonyl-containing functional groups, based on complementary peaks in other regions of the spectrum.

Problems

14.11 For each of the following mass spectra, identify the mass of the compound, the most common fragment, and any isotopes.

a)

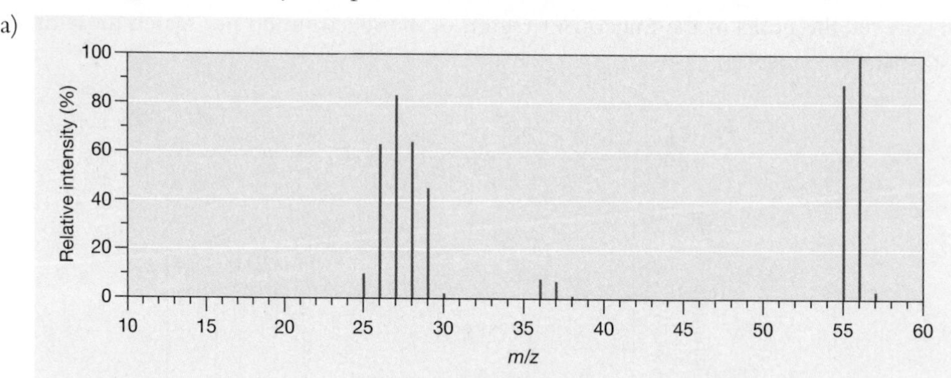

b)

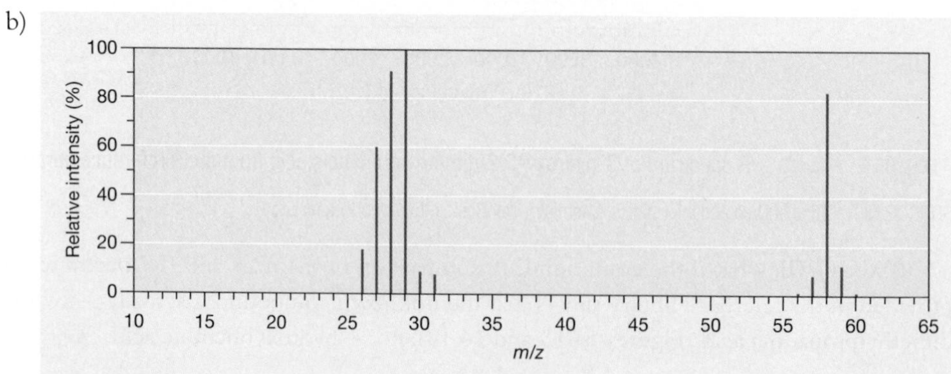

c)

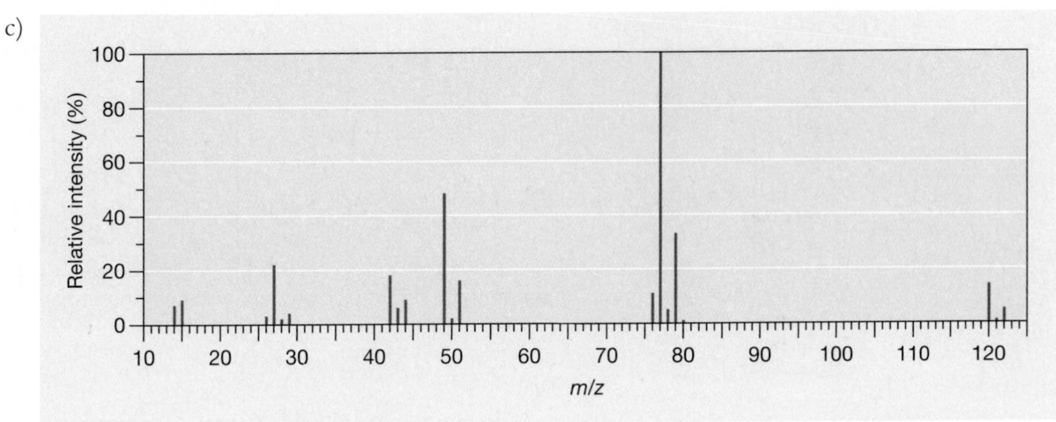

d)

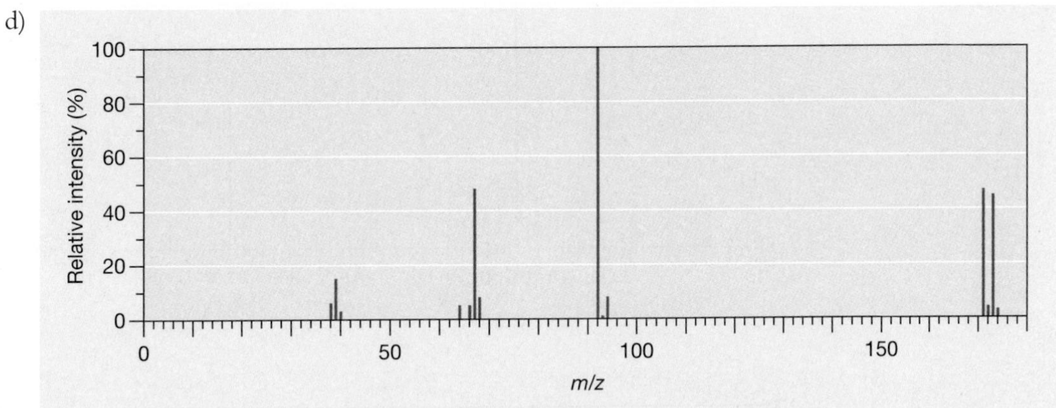

e)

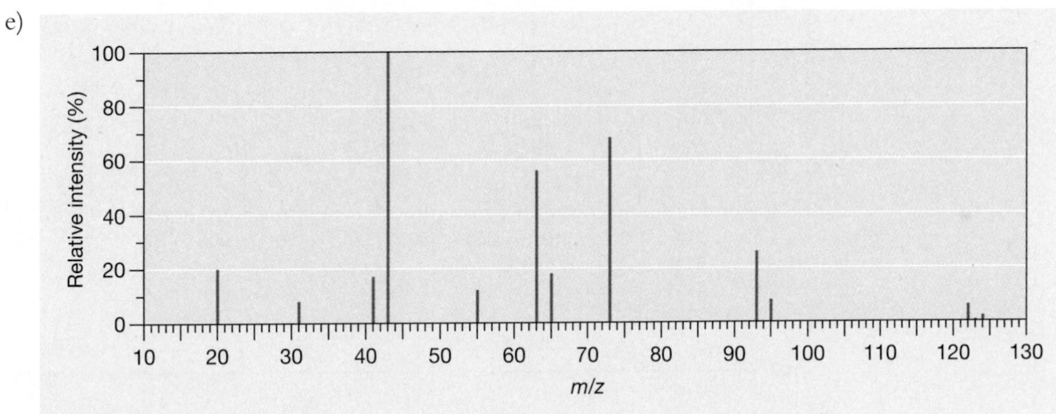

14.12 Match each compound (a–d) to its mass spectrum (1–4).

a)

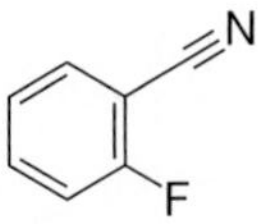

b)

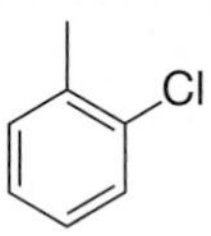

c)

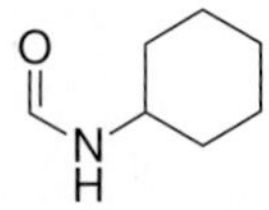

d)

1)

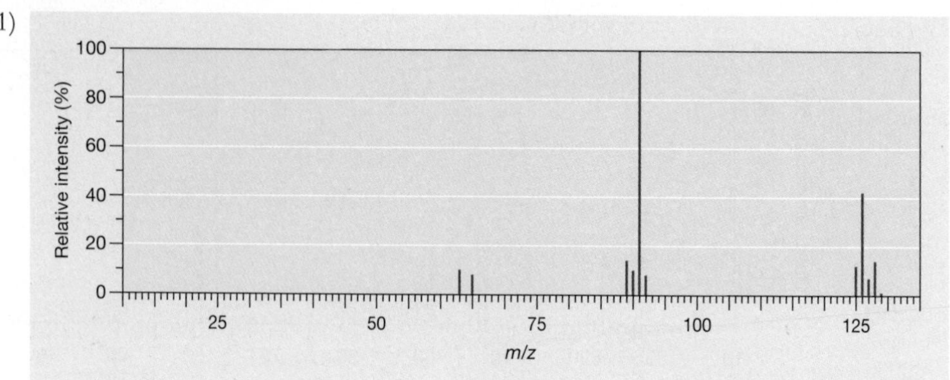

Relative intensity (%)
100
80
60
40
20
0
25
50
75
100
125
m/z

2)

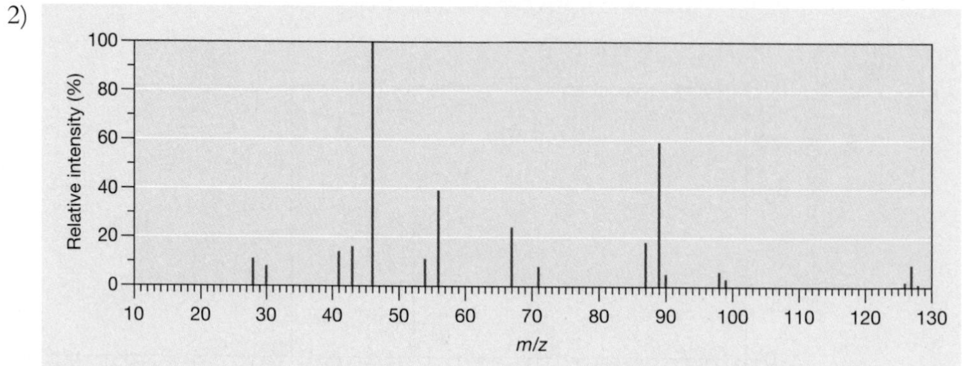

Relative intensity (%)
100
80
60
40
20
0
10
20
30
40
50
60
70
80
90
100
110
120
130
m/z

3)

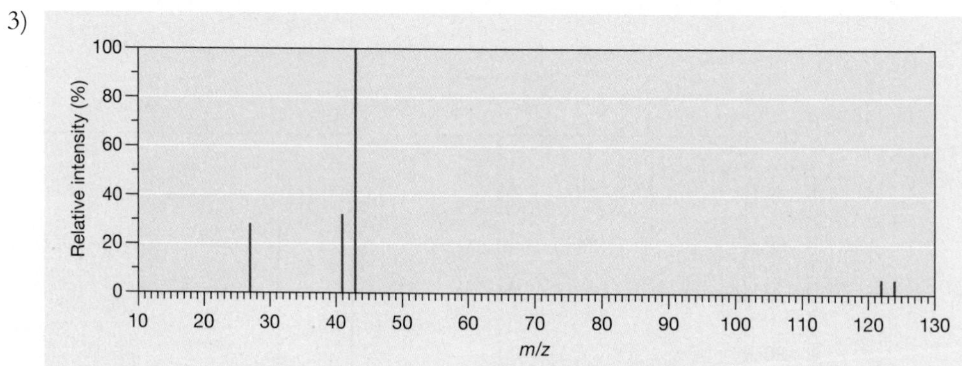

Relative intensity (%)
100
80
60
40
20
0
10
20
30
40
50
60
70
80
90
100
110
120
130
m/z

4)

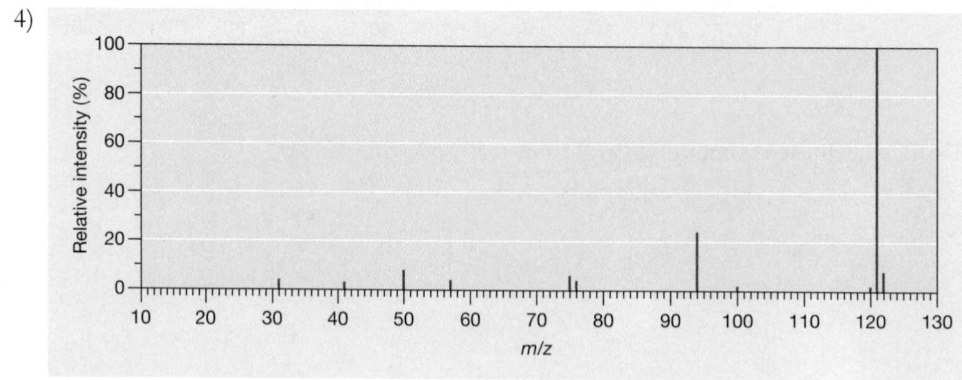

Relative intensity (%)
100
80
60
40
20
0
10
20
30
40
50
60
70
80
90
100
110
120
130
m/z

14.13 Match each compound (a–d) to its mass spectrum (1–4).

a) 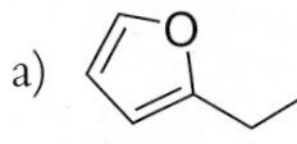

c)

b)

d)

1)

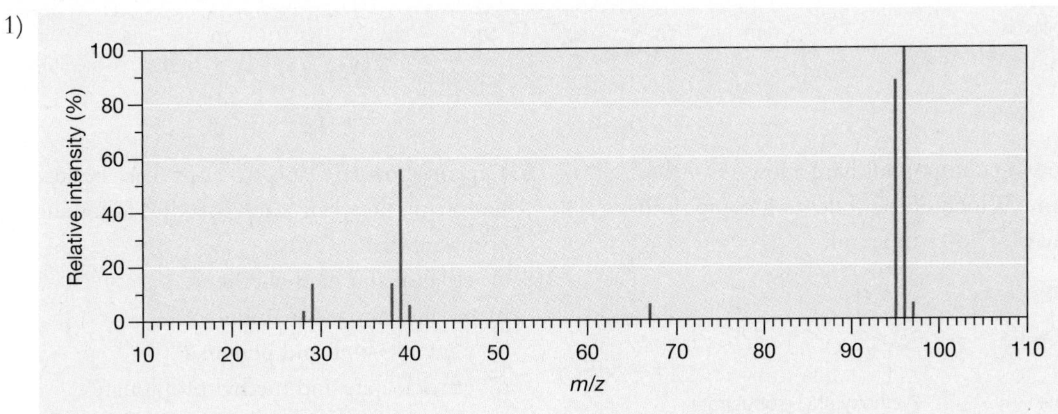

2)

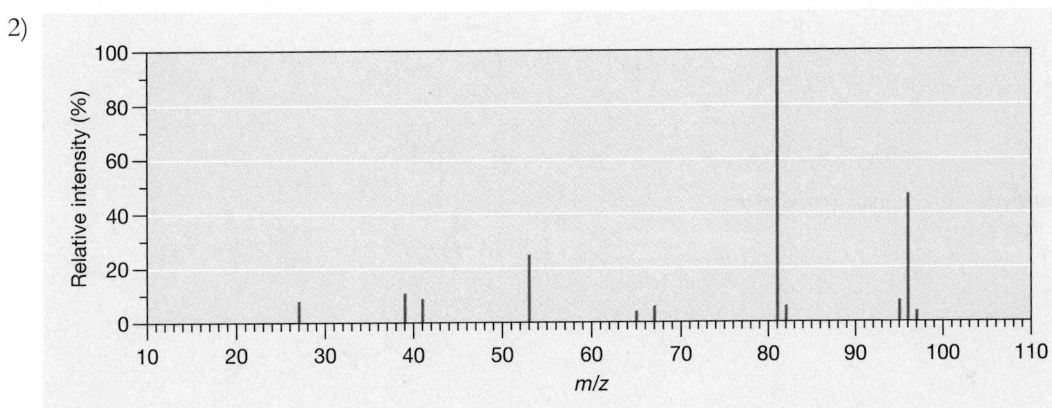

3)

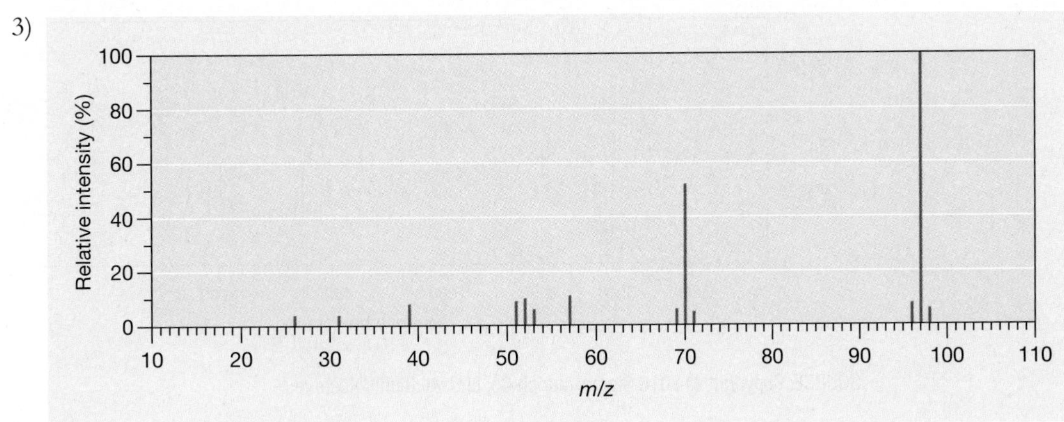

4)

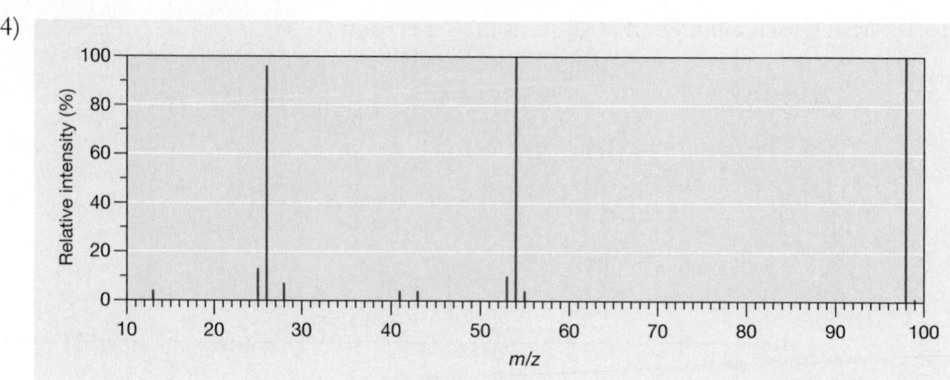

14.14 All of the following compounds have a low-resolution mass of 160 amu. Which of the following compounds has an exact mass of 160.0136 amu?

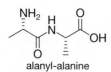

alanyl-alanine

2-ethoxyethyl isobutyrate

2,3,6-trifluorobenzaldehyde

diethyl malonate

14.15 Explain how IR spectroscopy can be used to differentiate between each of the following pairs of compounds.

a) ethanol and diethyl ether
b) ethanol and ethanamine
c) pentan-3-one and pentanal
d) ethyl acetate and methyl propionate

14.16 Match each of the compounds (a–d) to its IR spectrum (1–4).

a)

b)

c)

d)

1)

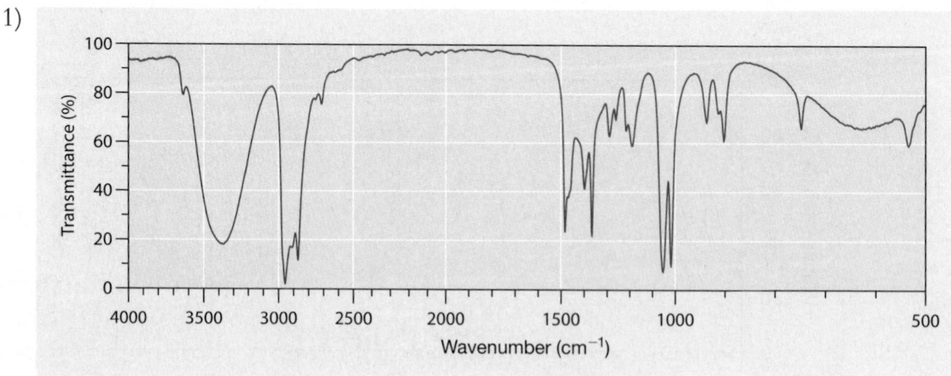

2)

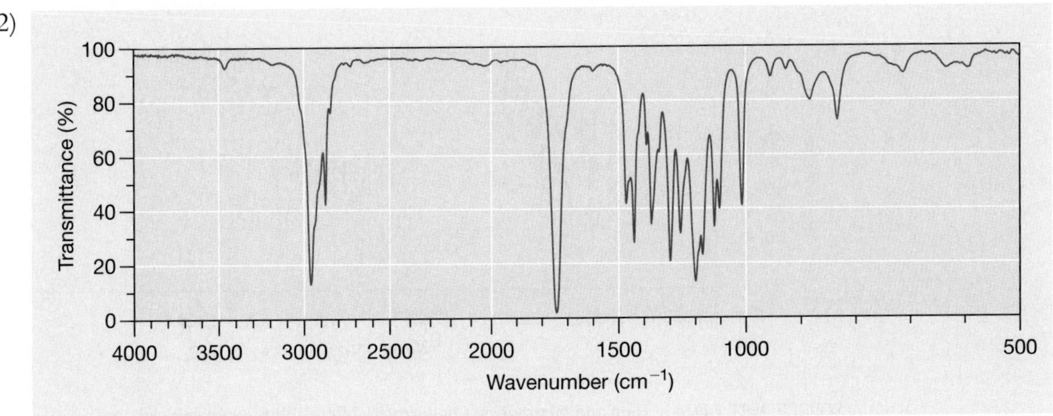

3)

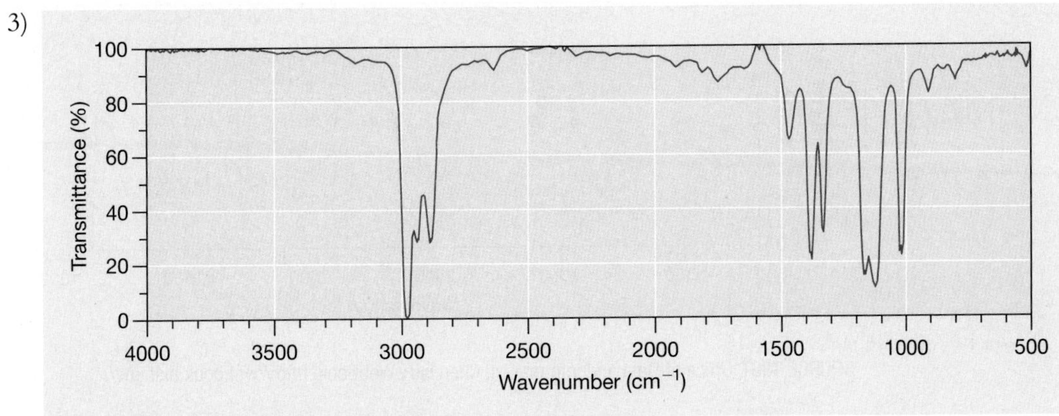

SOURCE: NIST, Office of Data and Informatics, Chemistry WebBook, http://webbook.nist.gov/

4)

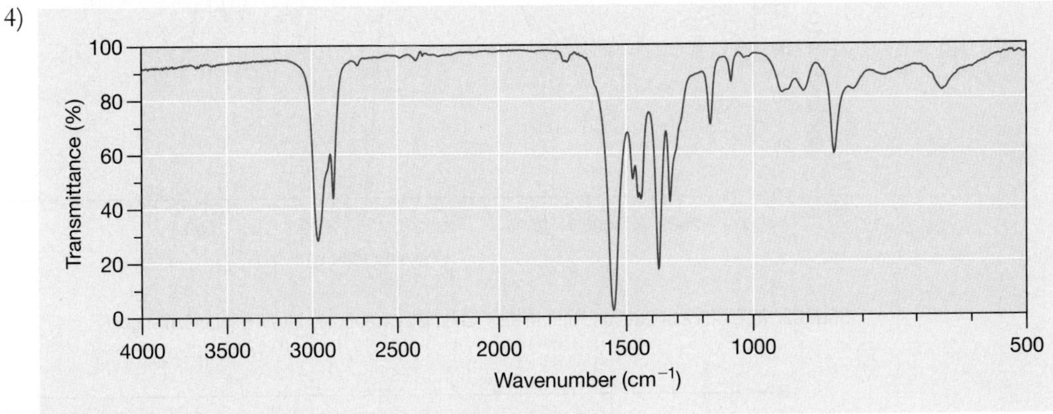

14.17 Match each compound (a–d) to its IR spectrum (1–4).

a) O, O

b) O, OH

c) O, HO

d) O, O

1)

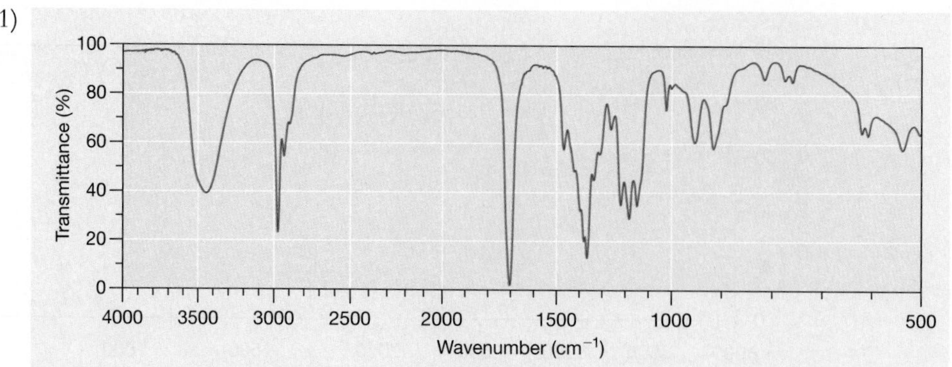

SOURCE: NIST, Office of Data and Informatics, Chemistry WebBook, http://webbook.nist.gov/

2)

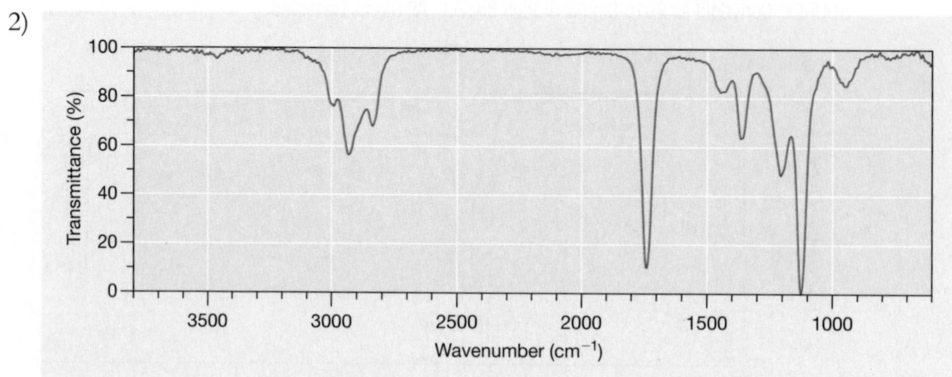

SOURCE: NIST, Office of Data and Informatics, Chemistry WebBook, http://webbook.nist.gov/

3)

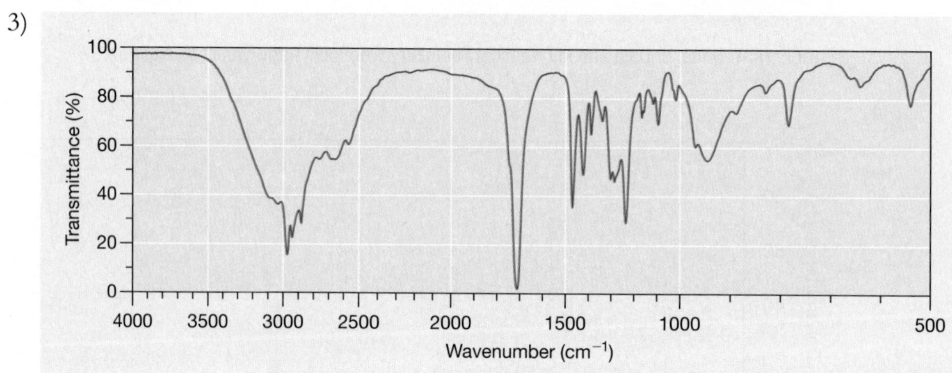

SOURCE: NIST, Office of Data and Informatics, Chemistry WebBook, http://webbook.nist.gov/

4)

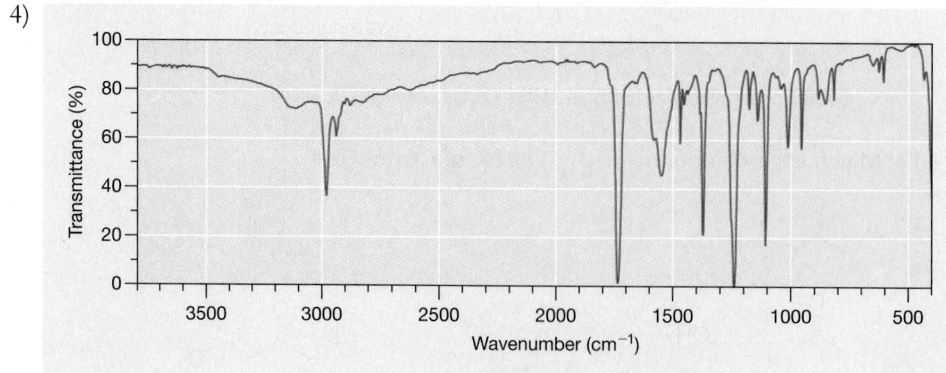

SOURCE: NIST, Office of Data and Informatics, Chemistry WebBook, http://webbook.nist.gov/

14.18 Match each compound (a–d) to its IR spectrum (1–4).

a) OCH_3, O

b) H, N

c) O, O

d) Cl

1)

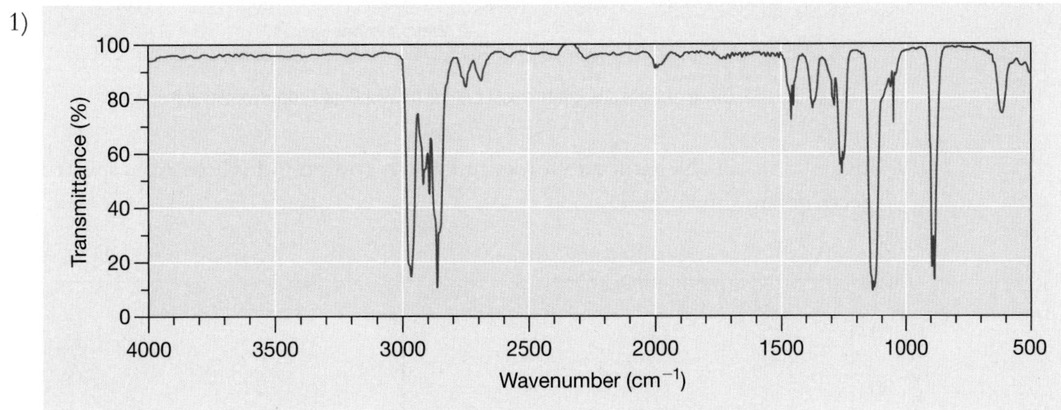

SOURCE: NIST, Office of Data and Informatics, Chemistry WebBook, http://webbook.nist.gov/

2)

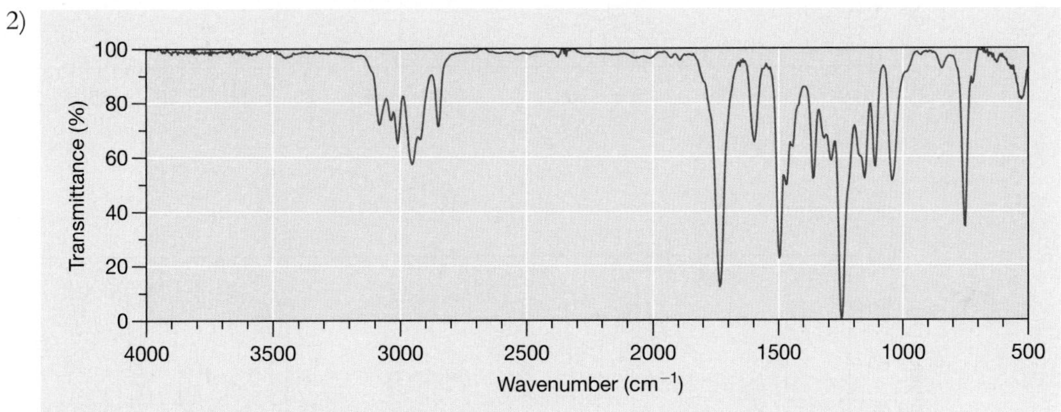

SOURCE: NIST, Office of Data and Informatics, Chemistry WebBook, http://webbook.nist.gov/

3)

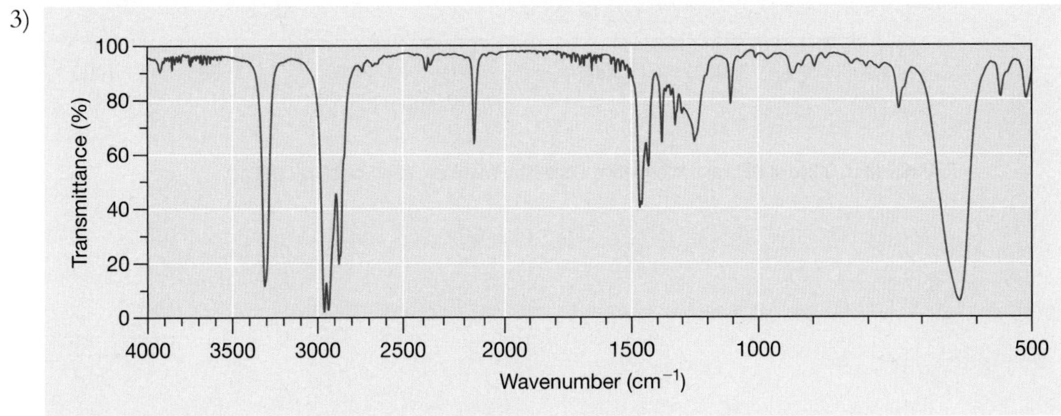

4) 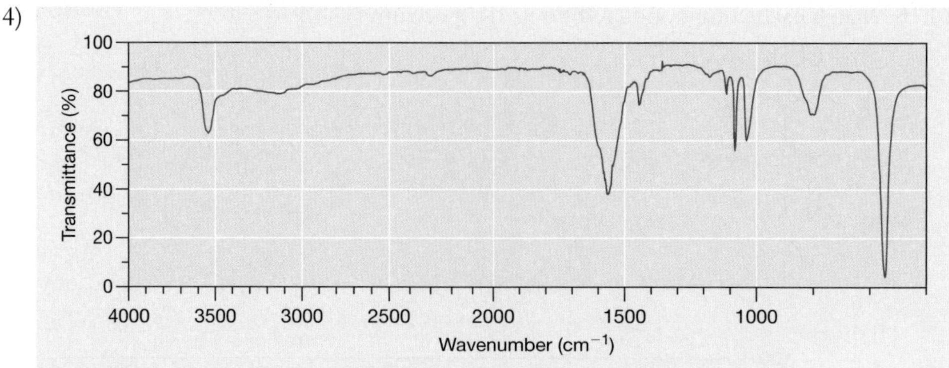

SOURCE: NIST, Office of Data and Informatics, Chemistry WebBook, http://webbook.nist.gov/

14.19 Suggest a reasonable structure for an unknown compound whose mass spectrum and IR spectrum are shown here.

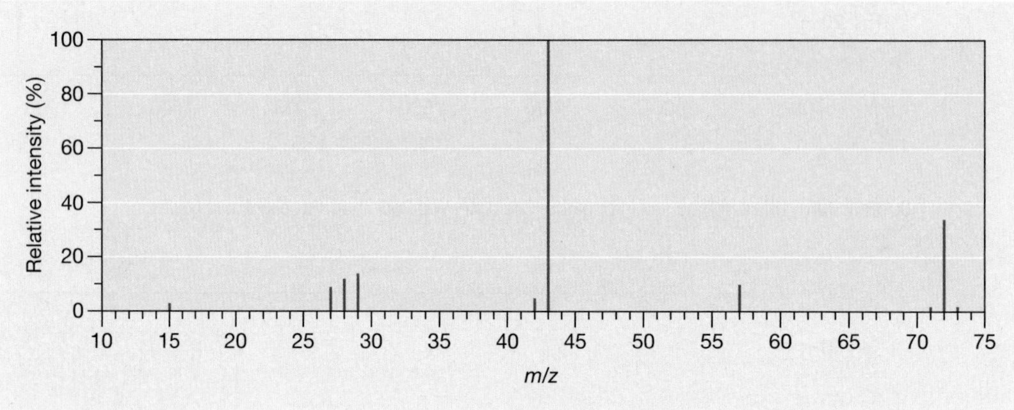

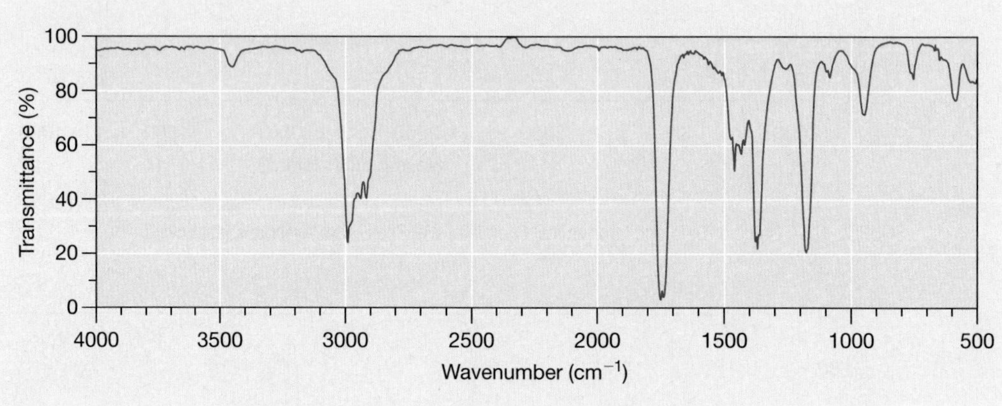

SOURCE: NIST, Office of Data and Informatics, Chemistry WebBook, http://webbook.nist.gov/

14.20 Suggest a reasonable structure for an unknown compound whose mass spectrum and IR spectrum are shown here.

Mass spectrum

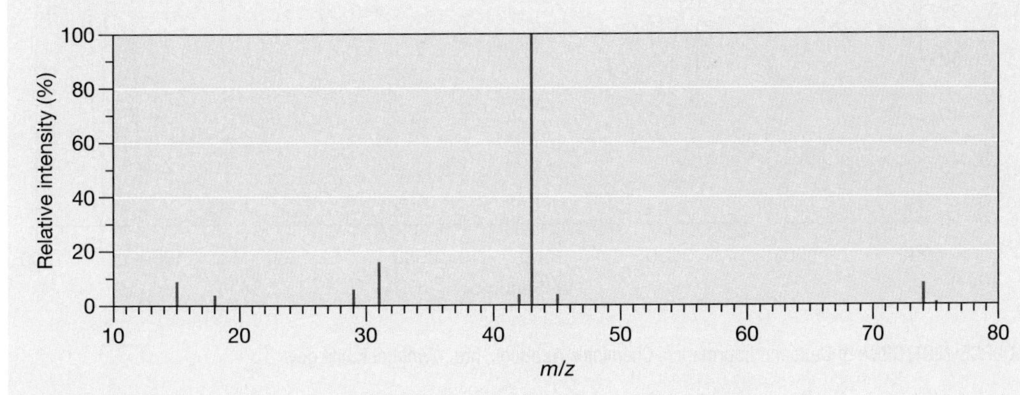

IR spectrum

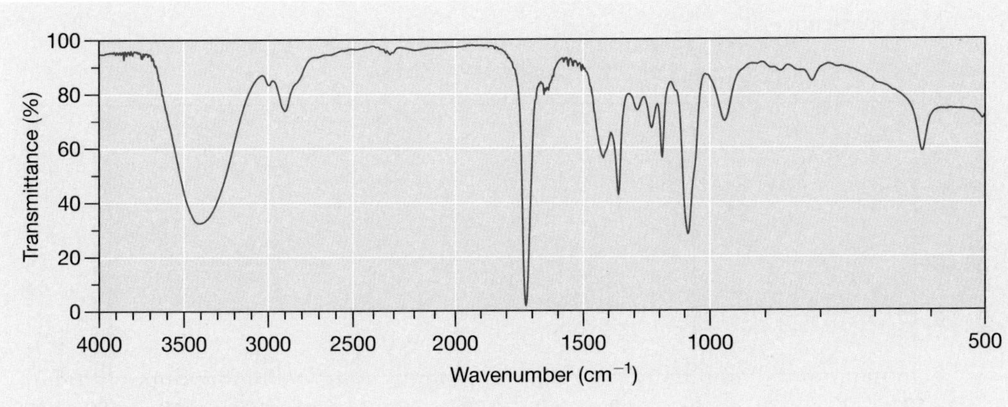

14.21 Suggest a reasonable structure for an unknown compound whose mass spectrum and IR spectrum are shown here.

Mass spectrum

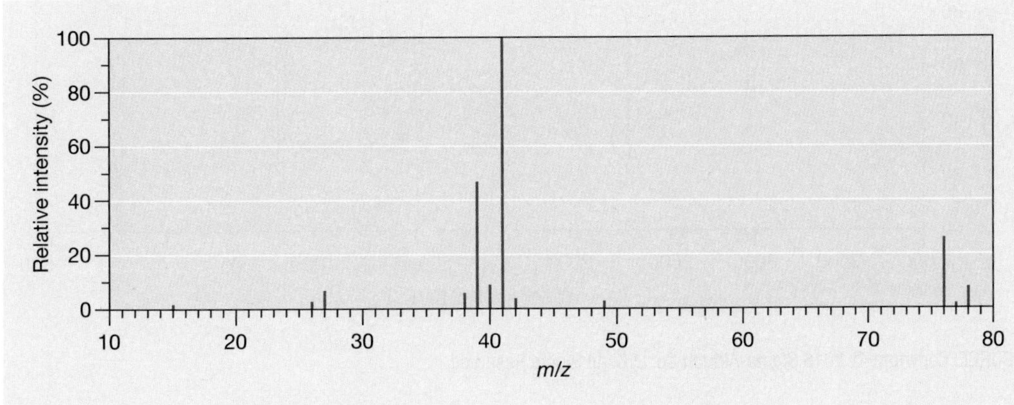

IR spectrum

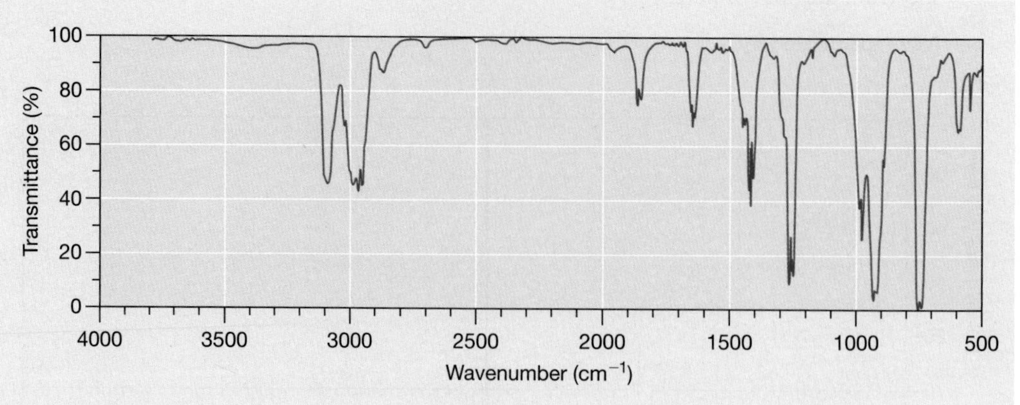

SOURCE: NIST, Office of Data and Informatics, Chemistry WebBook, http://webbook.nist.gov/

14.22 Suggest a reasonable structure for an unknown compound whose mass spectrum and IR spectrum are shown here.

Mass spectrum

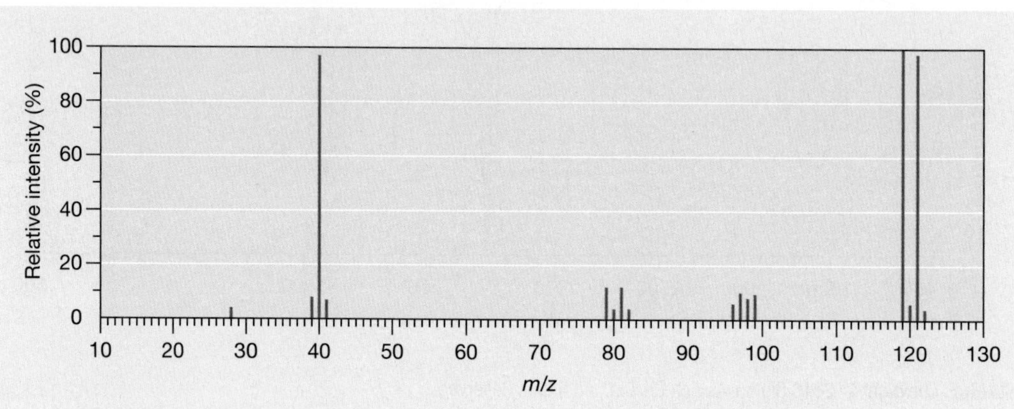

IR spectrum

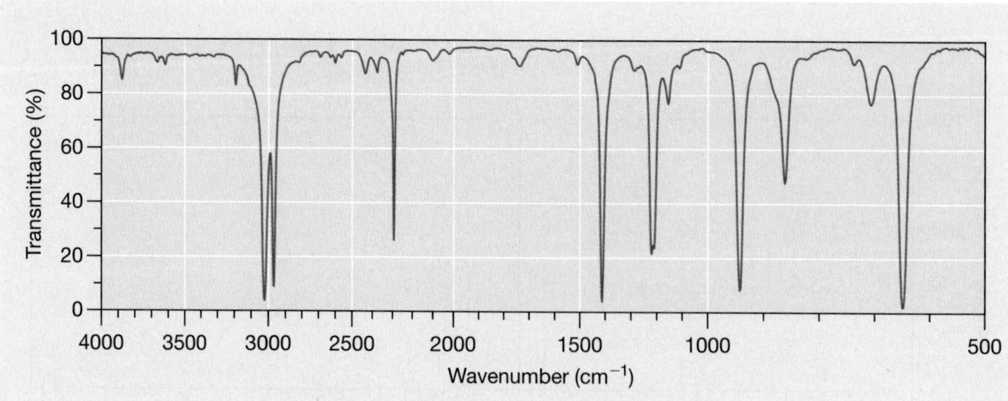

14.23 Suggest a reasonable structure for an unknown compound whose mass, IR, and ^{1}H-NMR spectra are shown here.

Mass spectrum

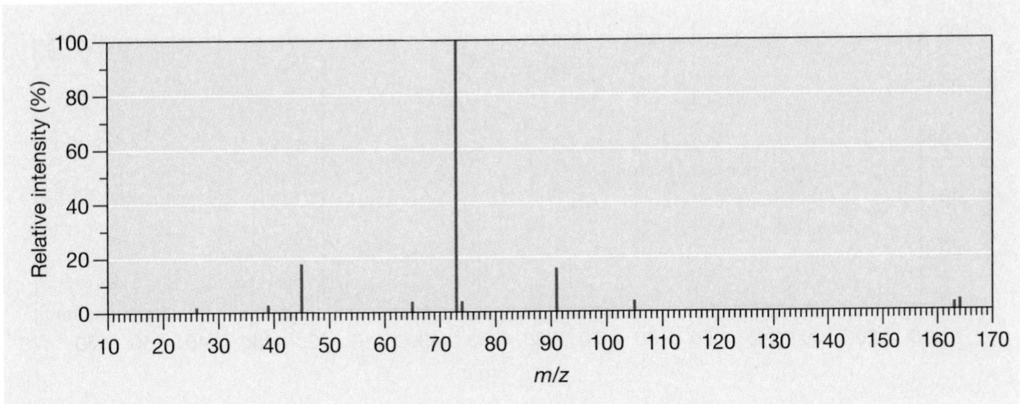

IR spectrum

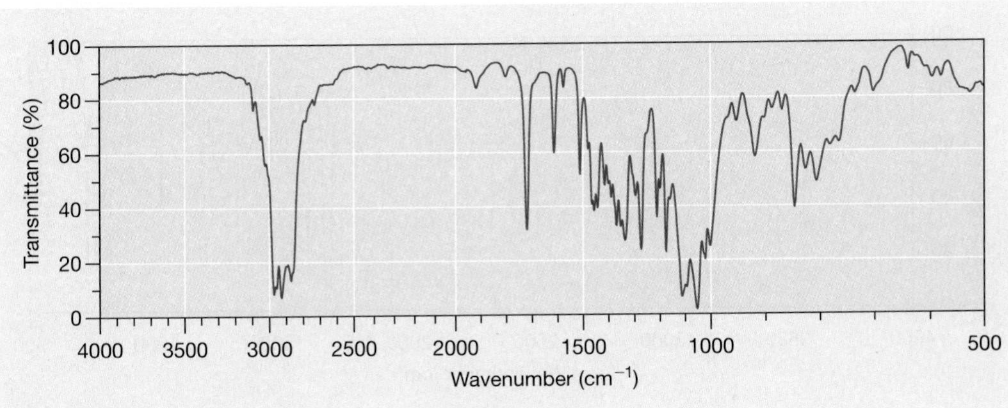

^{1}H-NMR spectrum

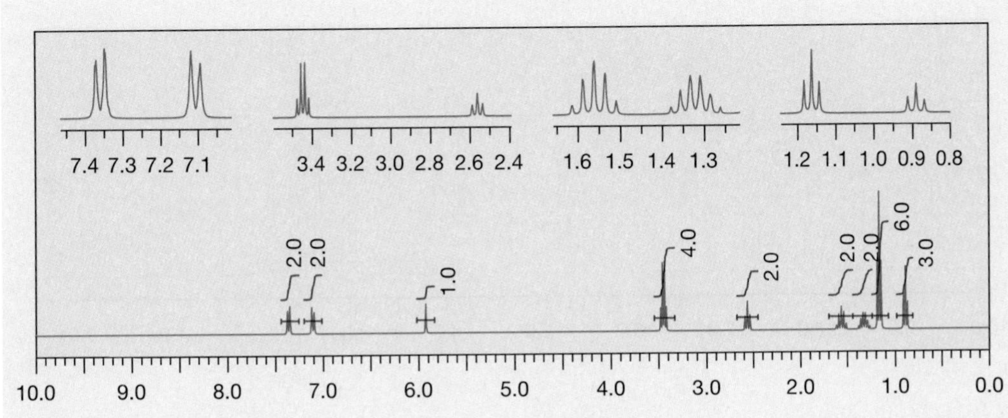

14.24 Suggest a reasonable structure for an unknown compound whose mass, IR, and ^{1}H-NMR, and ^{13}C-NMR spectra are shown here.

Mass spectrum

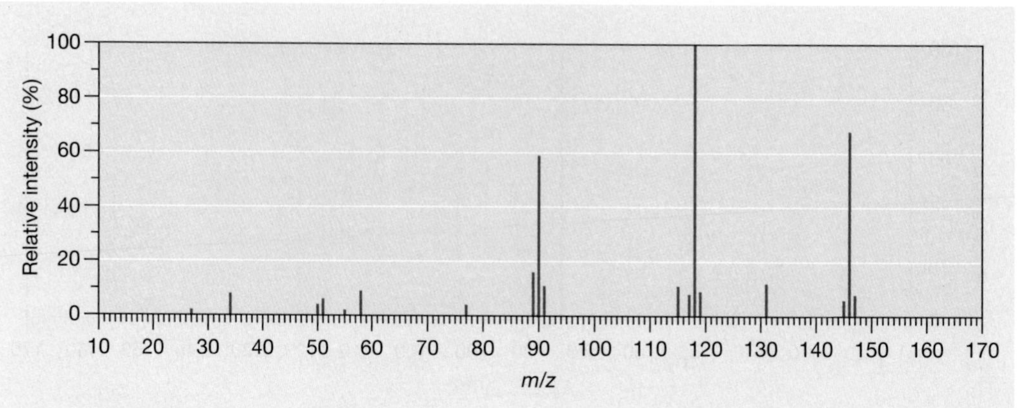

IR spectrum

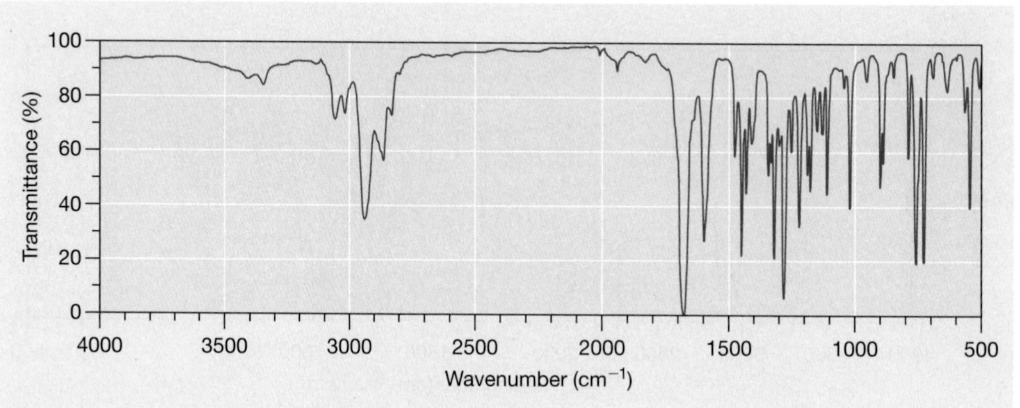

SOURCE: NIST, Office of Data and Informatics, Chemistry WebBook, http://webbook.nist.gov/

^{1}H-NMR spectrum

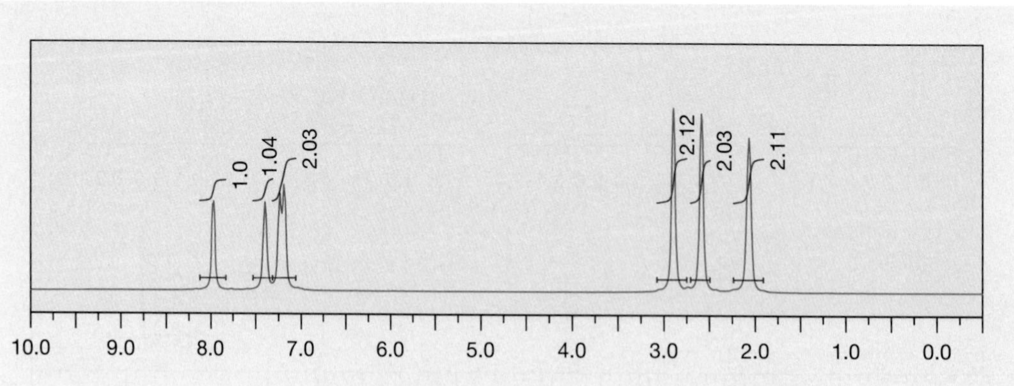

^{13}C-NMR spectrum

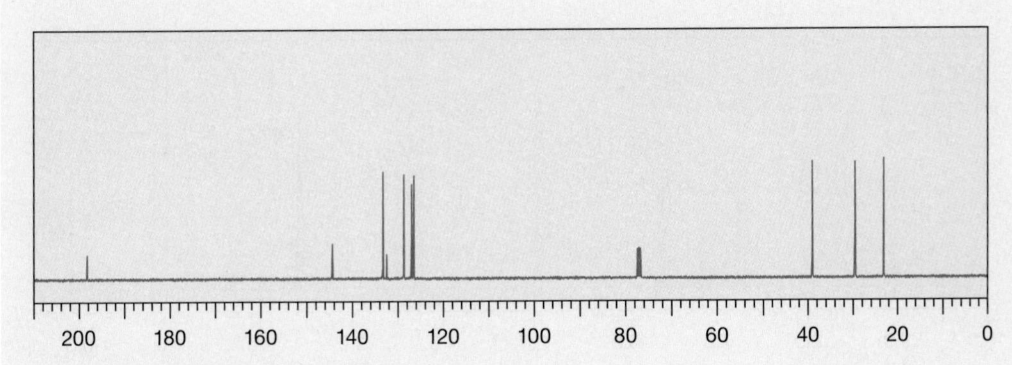

14.25 Suggest a reasonable structure for an unknown compound whose mass, IR, ^{1}H-NMR, and ^{13}C-NMR spectra are shown here.

Mass spectrum

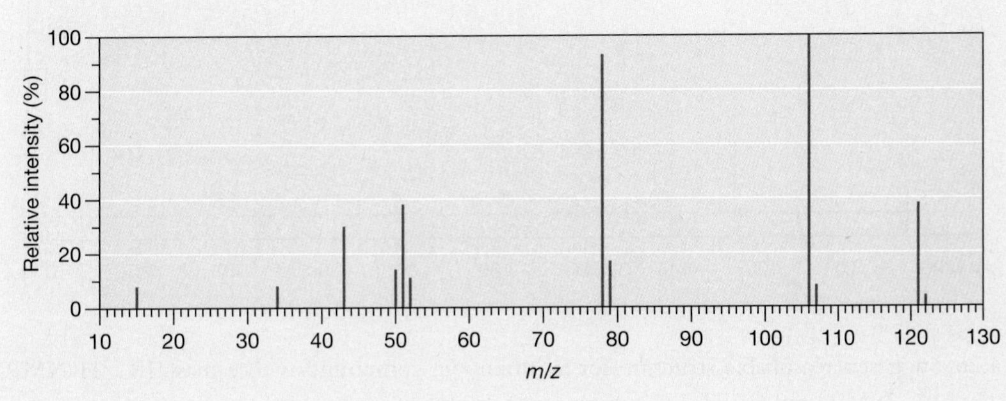

IR spectrum

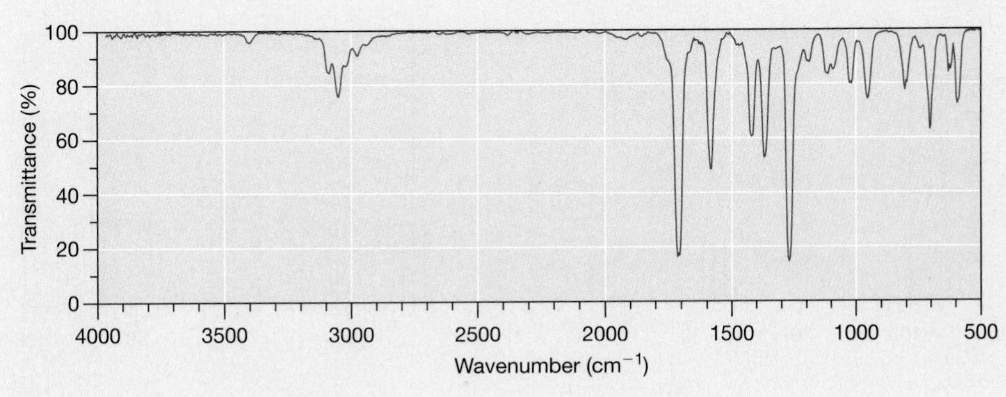

SOURCE: NIST, Office of Data and Informatics, Chemistry WebBook, http://webbook.nist.gov/

^{1}H-NMR spectrum

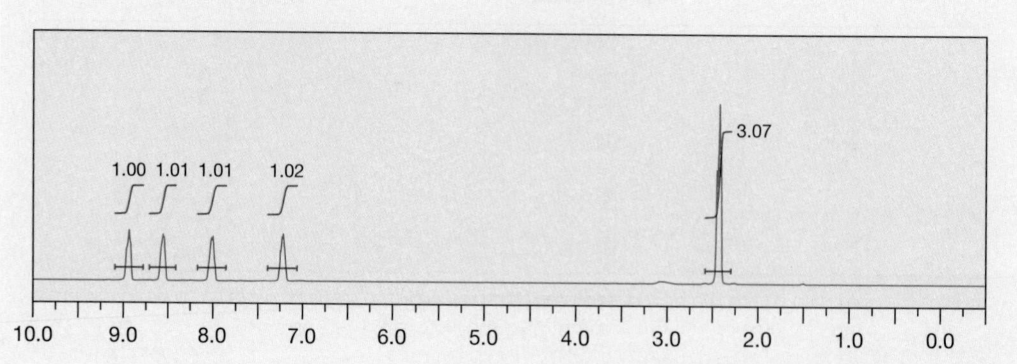

^{13}C-NMR spectrum

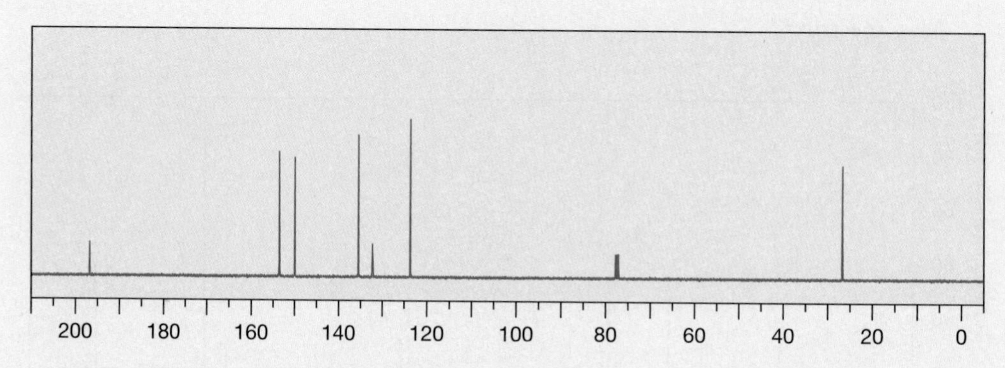

14.26 Suggest a reasonable structure for an unknown compound whose mass, IR, ^{1}H-NMR, ^{13}C-NMR, and DEPT-135 spectra are shown here.

Mass spectrum

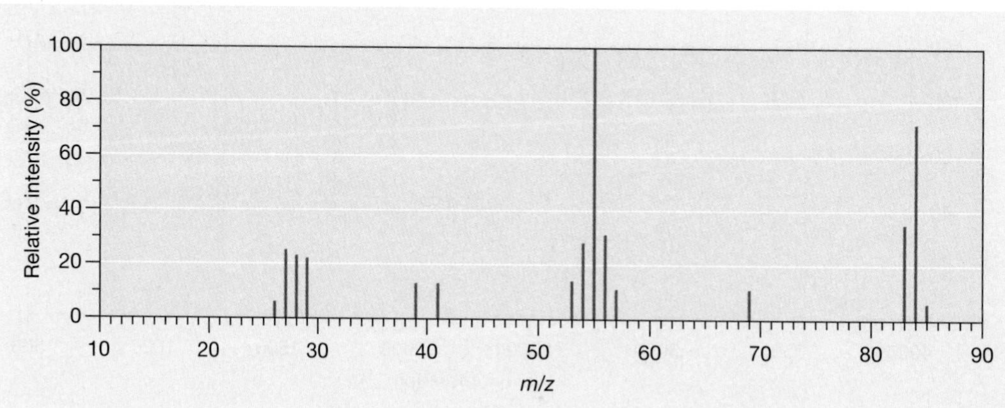

IR spectrum

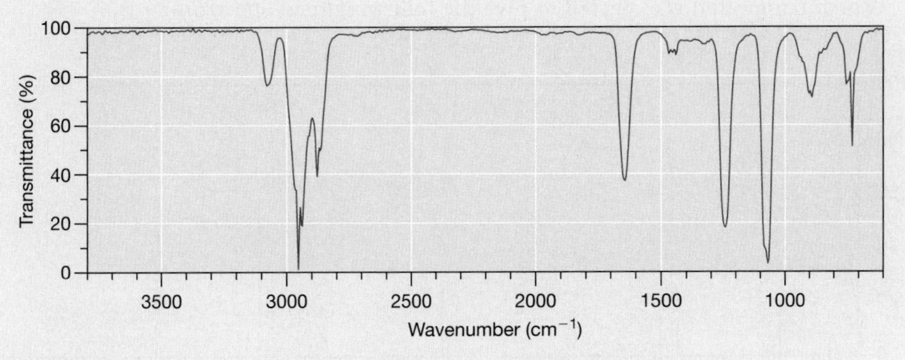

SOURCE: NIST, Office of Data and Informatics, Chemistry WebBook, http://webbook.nist.gov/

^{1}H-NMR spectrum

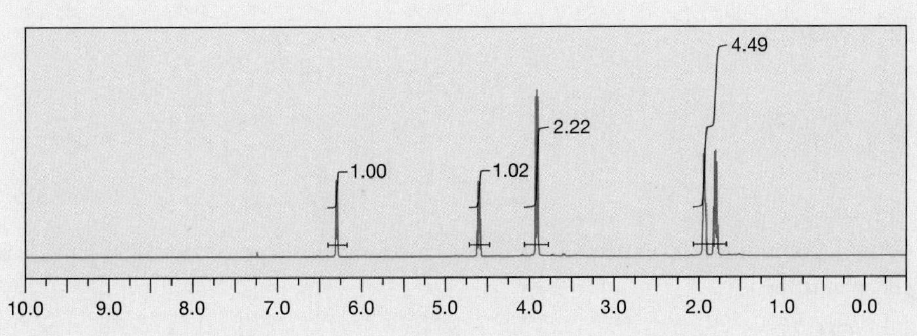

^{13}C-NMR spectrum

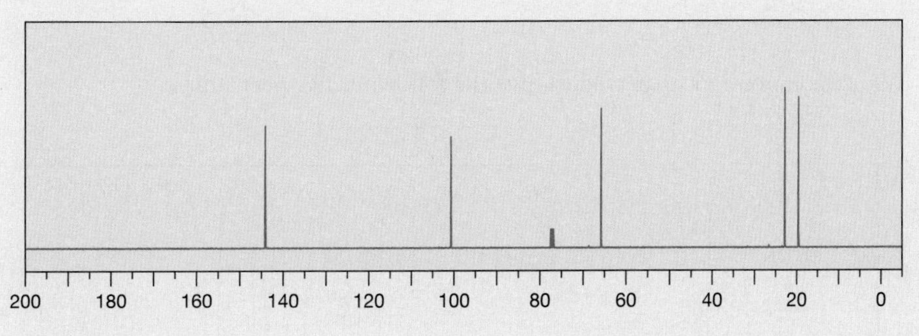

DEPT-135 spectrum

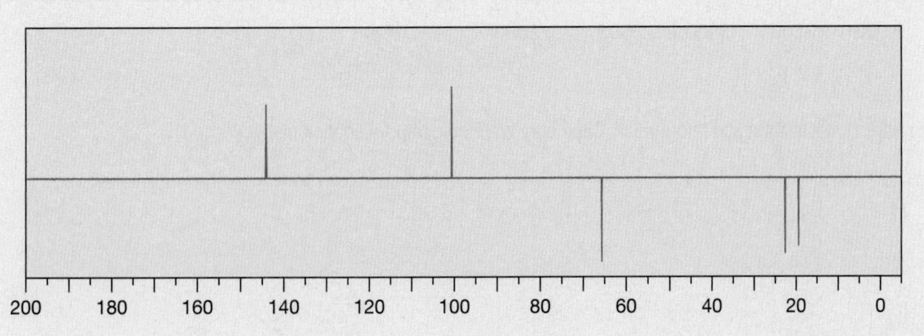

MCAT STYLE PROBLEMS

14.27 Which compound is expected to give the following mass spectrum?

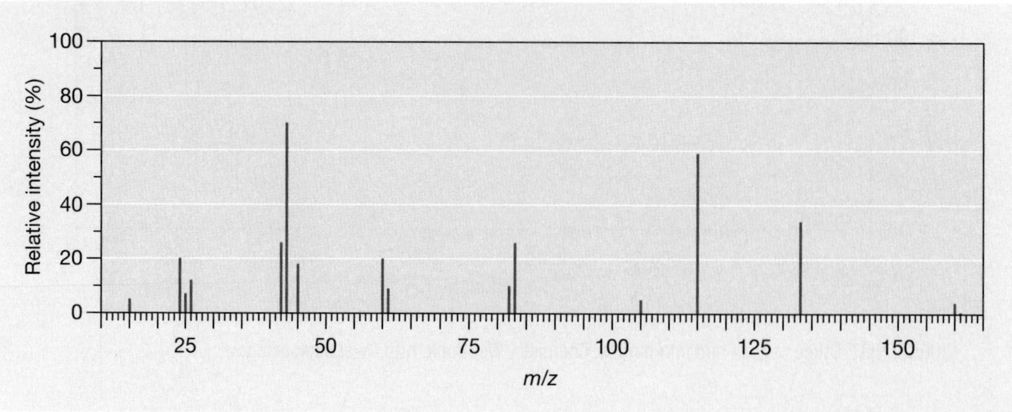

a) O O O O

b) NH_2 CF_3

c) Cl Cl

d) CF_3 F

14.28 Which compound is expected to show a strong absorption at 1750 cm^{-1} in the IR spectrum?

a) OH

b) O

c) O

d) O N O

14.29 Which compound is expected to give the following IR spectrum?

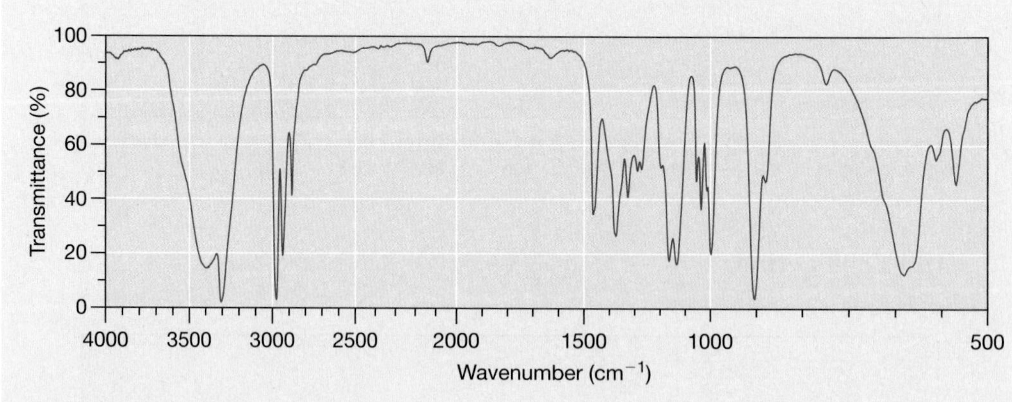

SOURCE: NIST, Office of Data and Informatics, Chemistry WebBook, http://webbook.nist.gov/

a) O OH

b) OH

c) O

d) OH

CHALLENGE PROBLEM

14.30 Compound A was reacted with Cl_2 and $AlCl_3$ to give Product B. Alternatively, Compound A was reacted with $NaBH_4$ in methanol to give Product C. Suggest structures for A, B, and C.

Compound A has a pair of doublets in the ^{1}H-NMR spectrum, at 7.82 and 7.52 ppm; a molecular ion peak at $m/z = 140$; and an M+2 peak at 142.

Compound A

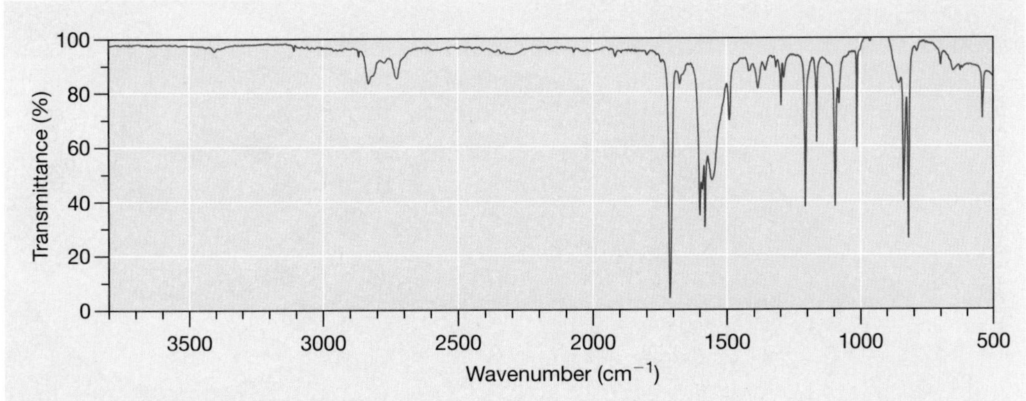

SOURCE: NIST, Office of Data and Informatics, Chemistry WebBook, http://webbook.nist.gov/

Product B

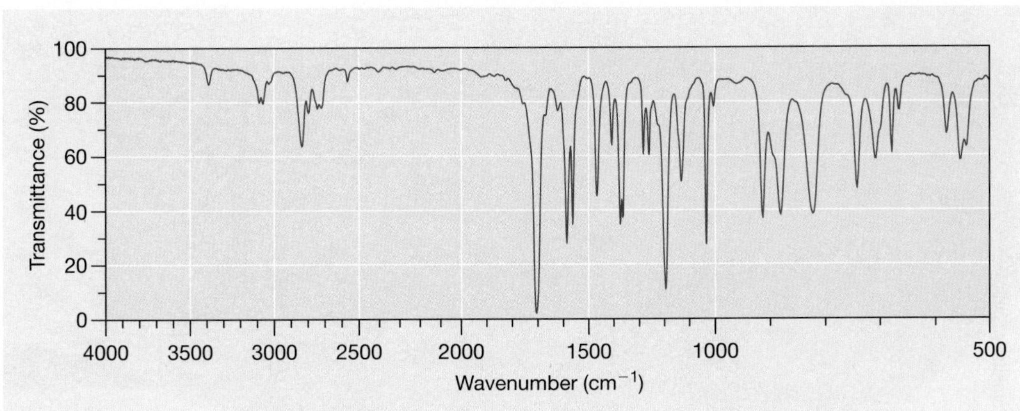

Product C

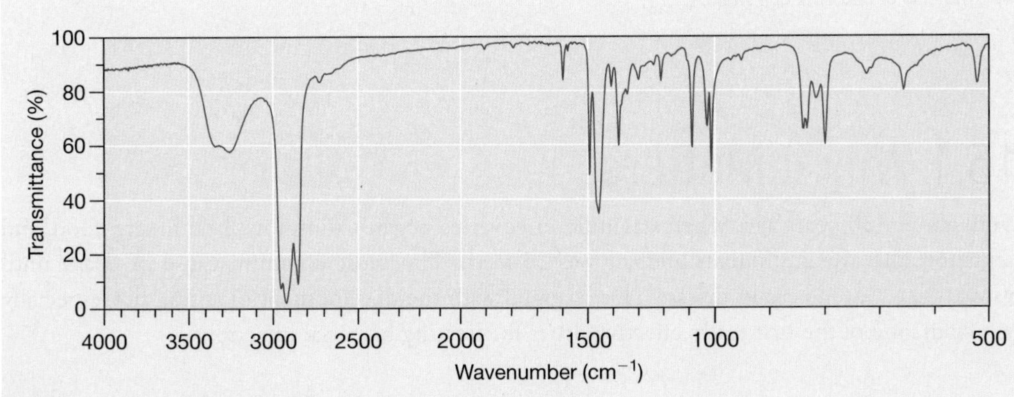

15 π Bond Electrophiles Connected to Leaving Groups

CARBOXYLIC ACID DERIVATIVES AND THEIR REACTIONS

CHAPTER OUTLINE

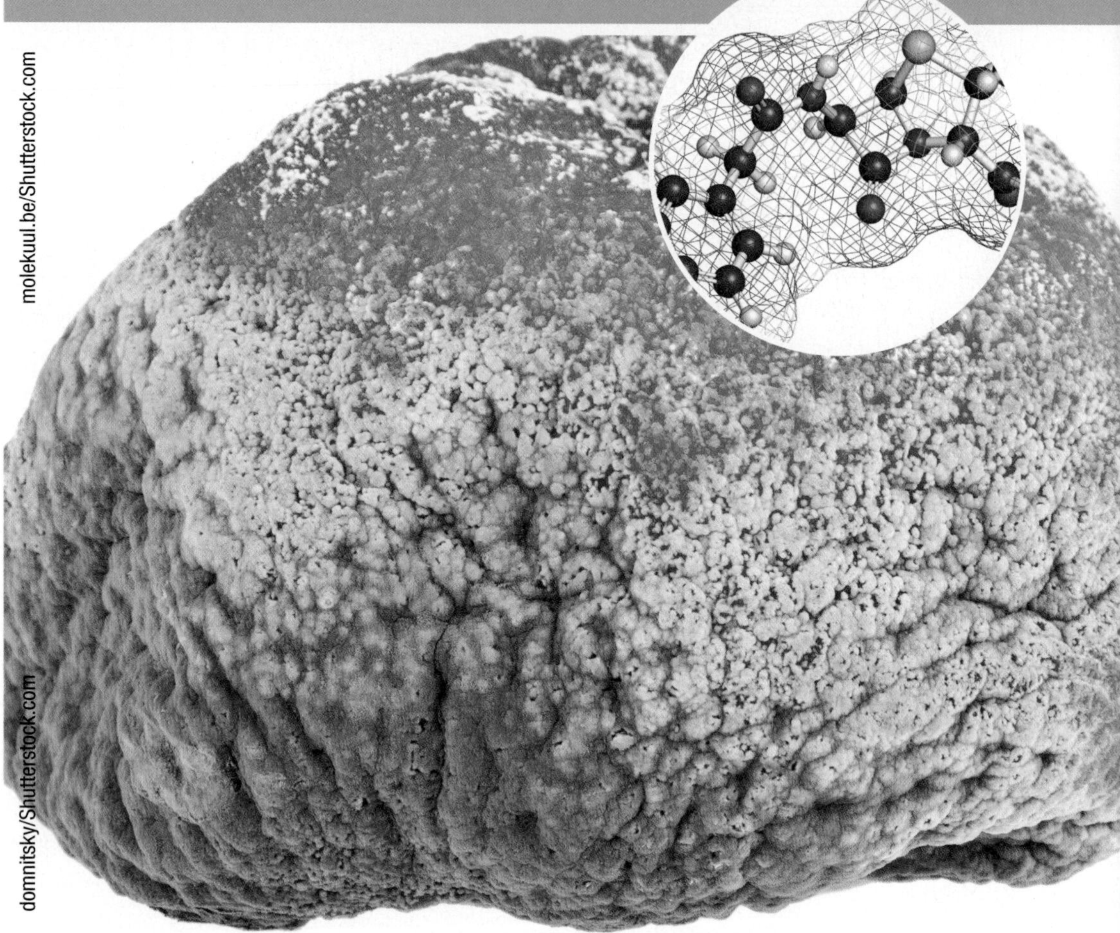

Penicillum chrysogenum mould is the source of penicillin, the world's first antibiotic. All antibiotics in the penicillin family contain an electrophilic carbonyl group that reacts with *transpeptidase*, an enzyme required for the synthesis of bacterial cell walls.

15.1 Why It Matters

Until about 150 years ago, when scientific discoveries began to improve health care, food, and sanitation, the average human lifespan was 35 years. The most common cause of death until 60 years ago was infectious disease. This changed with the development of antibiotics, especially penicillin, one of the first really effective drugs for treating bacterial infections.

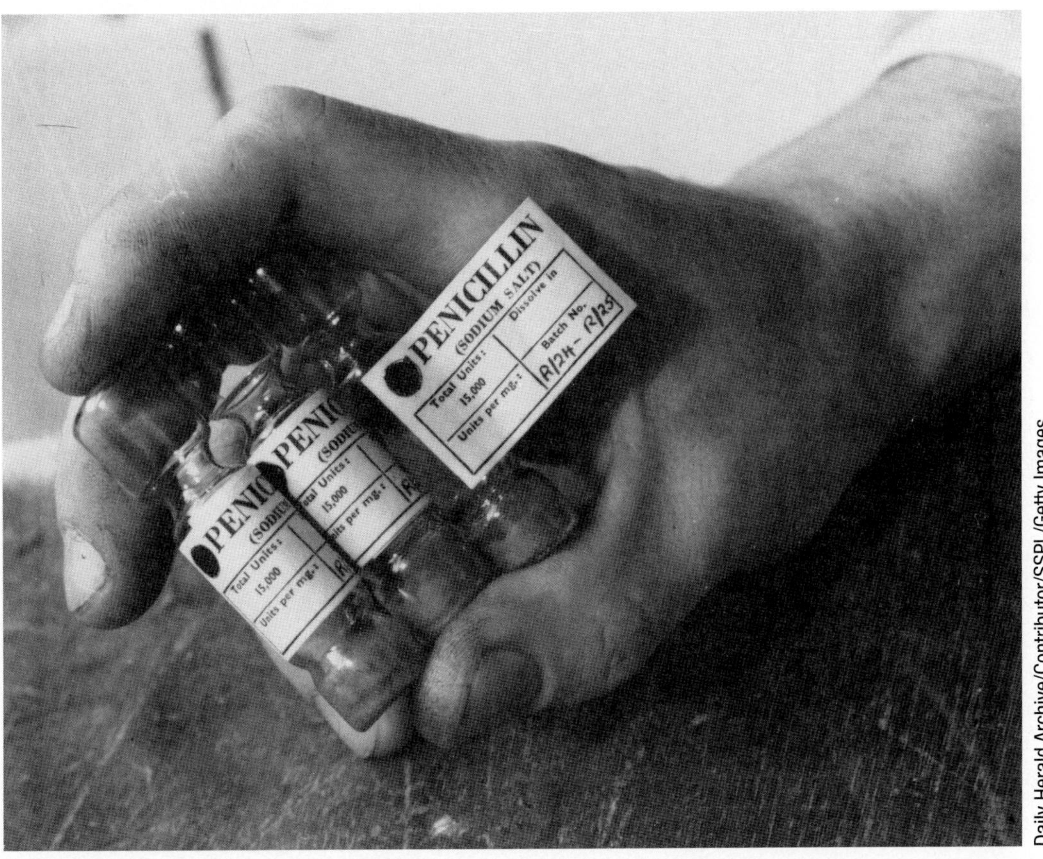

Daily Herald Archive/Contributor/SSPL/Getty Images

Penicillin disrupts the action of an enzyme—transpeptidase—that bacteria use to construct and repair their cell walls. As the following diagram shows, the drug has a carbonyl group, which is a very strong electrophile, coupled with an excellent leaving group. A nucleophilic hydroxyl group on the bacterial enzyme reacts with the carbonyl to open the small ring in the penicillin molecule and form a permanent bond to the bacterial enzyme. This bond blocks the catalyzing action of the enzyme and prevents bacteria from making cell walls. The bacteria die when their weakened cell walls burst. The commercially available penicillin used toward the end of World War II came from the fermentation of a mould found on a cantaloupe in a grocery store in Peoria, Illinois.

$-H^+$ $\longrightarrow$ $+H^+$ $\longrightarrow$

transpeptidase

the covalent bond is formed between penicillin and the enzyme

the modified enzyme is now inactive

Modern penicillins have been engineered to have more convenient and effective properties. Altering the structure can have significant effects on properties such as half-life and effectiveness against different bacterial strains. These structural alterations are possible because of the selective reactivity of certain carbonyl groups.

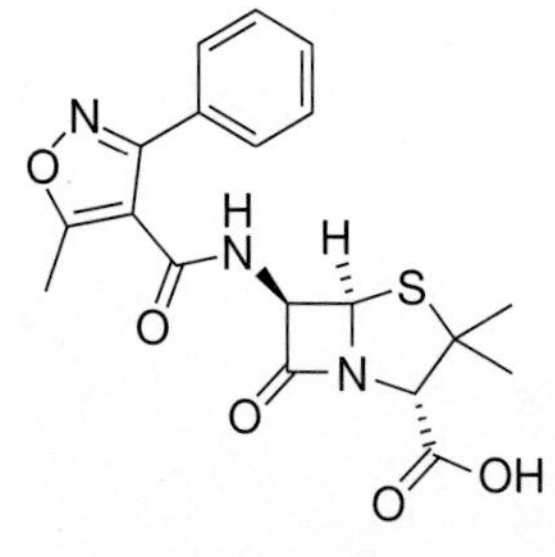

oxacillin 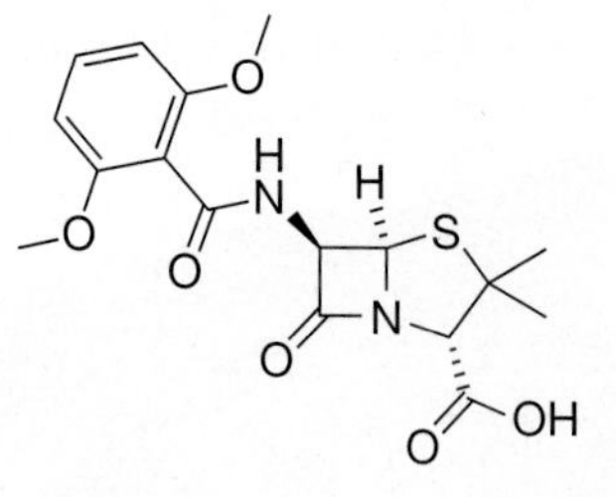 methicillin 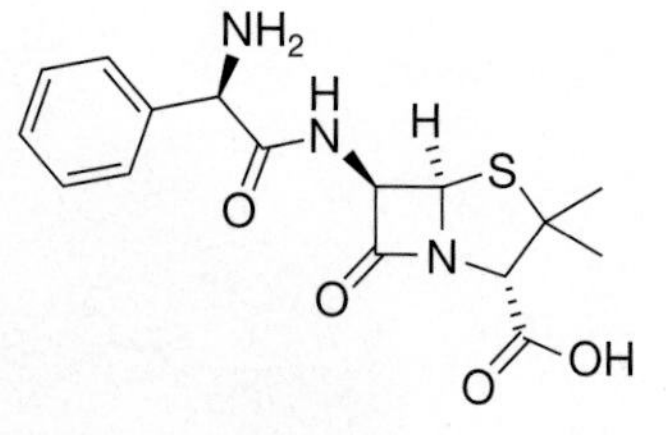 ampicillin

This chapter focuses on reactions with electrophilic π bonds that have a leaving group, such as the halogen-substituted nitrobenzene and carboxylic acid chloride shown here.

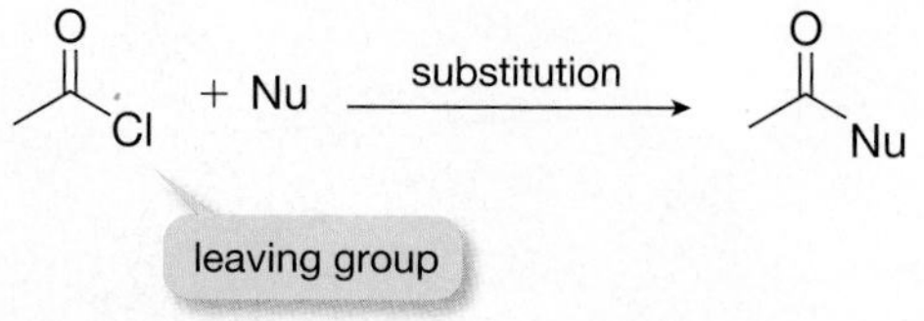

leaving group

substitution

leaving group

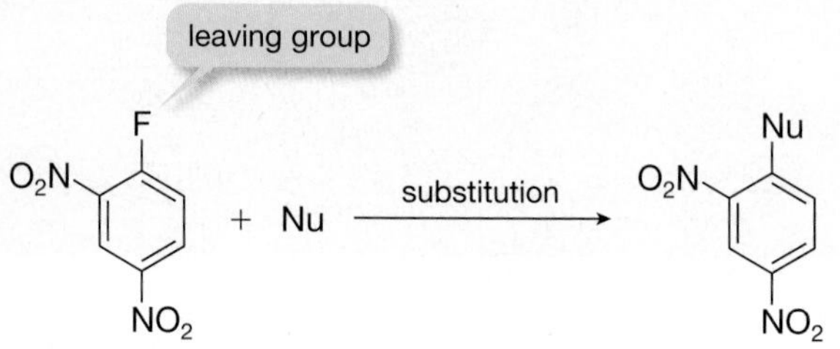

substitution

15.2 Substitution Reactions of Carboxylic Acid Derivatives

As described in Chapter 7, carbonyl groups are good electrophiles that react with nucleophiles to form bonds. When a nucleophile adds to a carbonyl, the addition of electrons to the carbonyl forces the weakest bond in the group—the C–O π bond—to break (to avoid exceeding an octet on carbon). This process creates a tetrahedral intermediate, in which the negatively charged oxygen can re-form the C–O π bond and expel the leaving group. This process is similar to a nucleophile displacing a leaving group in an S_N2 reaction; the difference is that the "nucleophile" is already part of the molecule. In functional groups that have a carbonyl directly connected to a heteroatom, the heteroatom can act as a leaving group for the tetrahedral intermediate. The resulting transformations are called substitution reactions because the product has the nucleophile substituted for the leaving group.

Addition-elimination is a two-step reaction that adds a nucleophile to a π bond, and then eliminates a leaving group to re-form the π bond.

The electron flow in these processes follows a pattern called **addition-elimination**. Addition of a nucleophile to a carbonyl breaks the weakest bond in the carbonyl (the π bond) and forms a tetrahedral intermediate, which then eliminates a leaving group to re-form the carbonyl bond. The first part of the mechanism for an addition-elimination reaction follows the same pattern as the addition of a nucleophile, such as a Grignard reagent, to a ketone or aldehyde (see Chapter 7).

Unlike ketones and aldehydes, an acid derivative can have a leaving group that allows the elimination step. These reactions are sometimes called **acylation** reactions because they add an acyl group to a nucleophile.

Acylation is the addition of an acyl (R–CO) group to a nucleophile.

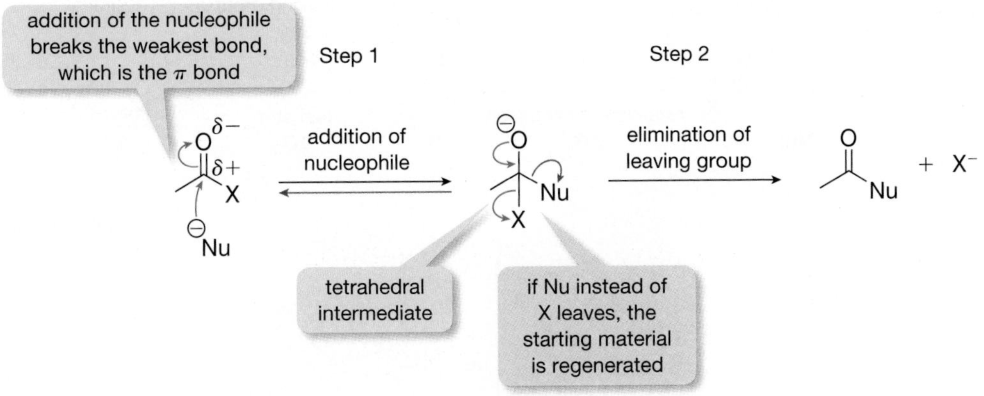

addition of the nucleophile breaks the weakest bond, which is the π bond

Step 1

Step 2

addition of nucleophile

elimination of leaving group

$+ X^-$

tetrahedral intermediate

if Nu instead of X leaves, the starting material is regenerated

15.3 Relative Reactivity in Nucleophilic Acyl Substitution Reactions

Carboxylic acid derivatives have the general formula RCOX, in which X is a group with a heteroatom bonded to the carbon atom of the carbonyl group.

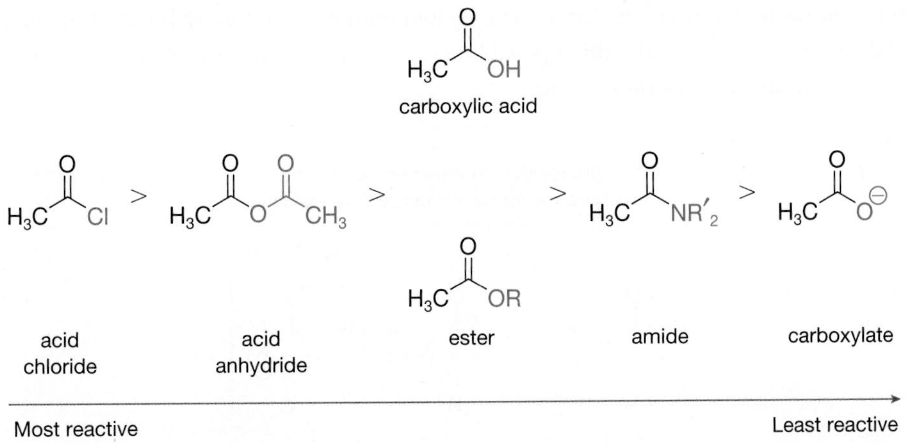

carboxylic acid

acid chloride > acid anhydride > ester > amide > carboxylate

Most reactive → Least reactive

These derivatives all react with nucleophiles through a mechanism in which a nucleophile donates electrons to the carbonyl carbon. Since the π bond of the carbonyl is the weakest bond in the structure, this bond breaks at the same time, forming a tetrahedral intermediate. This intermediate is a high-energy species. Consequently, its formation is the rate-determining step of the reaction.

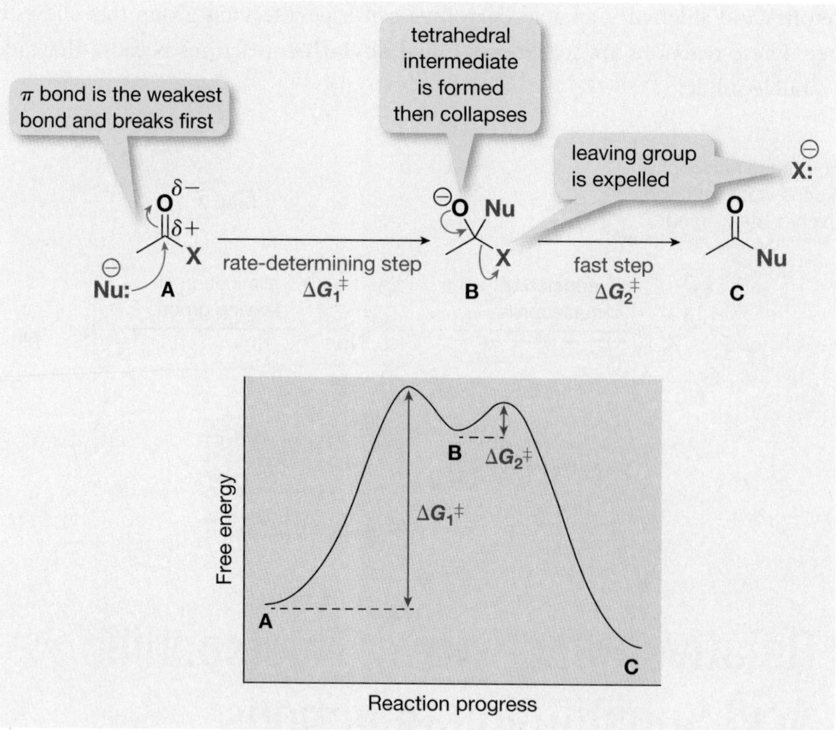

The rate at which this intermediate forms is used to determine whether a reaction will occur, and, if so, how fast or easily it will happen. This rate generally depends on the potential energy of the starting material, which is affected by electron delocalization.

Heteroatoms have lone pairs that are conjugated with the adjacent π bond of the carbonyl. The relative contributions of the three possible resonance forms determine the potential energy and the reactivity patterns of these groups.

the relative contributions of these resonance forms is a guide to reactivity

A B C

If the heteroatom is not a good electron donor, resonance form **B** will make a large contribution to the resonance hybrid. Because this form is the highest energy form, it raises the energy of the compound and makes the C=O group more reactive. The carbon of this form lacks an octet, making it reactive toward nucleophiles, which can donate electrons to this carbon.

If resonance form **C** makes a large contribution, the functional group will not be very reactive toward nucleophiles. Because no atom lacks an octet, this form lowers the energy and the reactivity of the compound. In this form, the X atom and the carbonyl carbon are strongly connected by a double bond, making them difficult to separate.

The groups in Figure 15.1 are ranked from most to least reactive based on the relative contribution of resonance form C. Acid chlorides are most reactive because halogens are poor electron-donating groups and form C makes a minimal contribution. In contrast, the nitrogen atom of an amide is an excellent electron donor. Resonance form C contributes a lot to the structure of amide carbonyls, making them much less reactive electrophiles than acid chlorides. Nitrogen is a better electron donor because its orbitals are similar in size to those of carbon, resulting in a greater overlap of p orbitals between nitrogen and carbon than between chlorine and carbon.

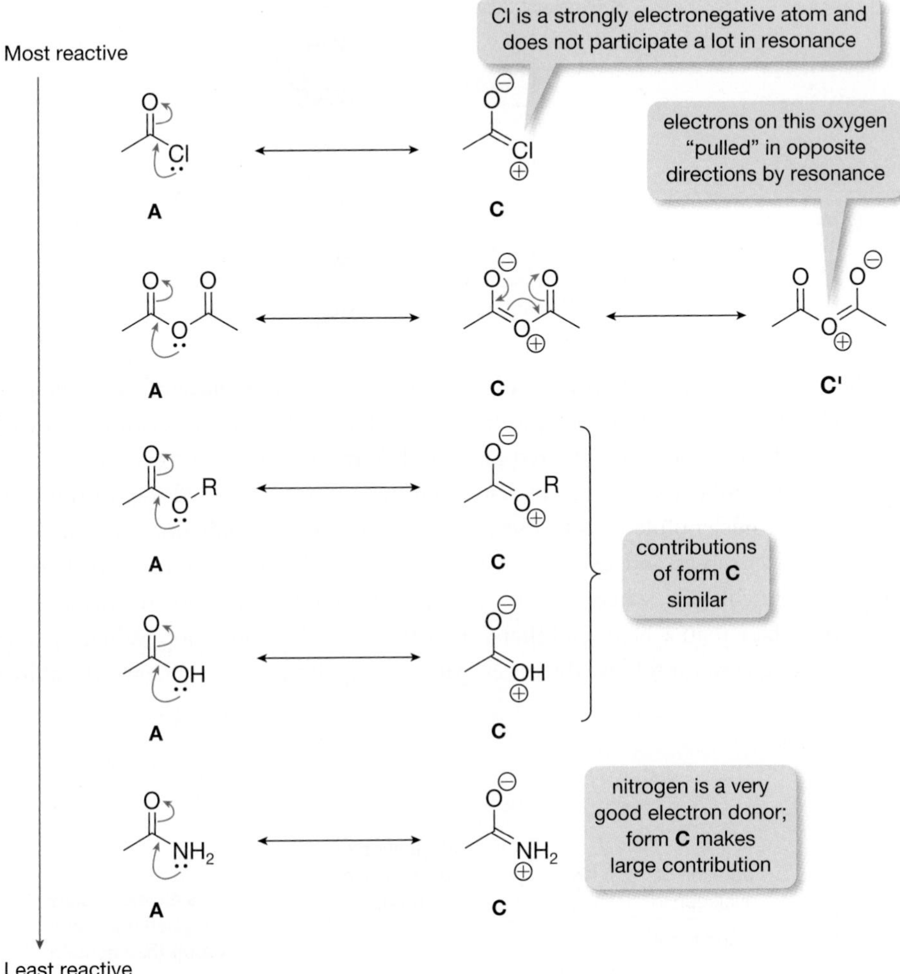

FIGURE 15.1 Relative reactivity of carbonyl-containing functional groups.

The properties of the possible leaving groups also affects whether or not a reaction is possible. The tetrahedral intermediate expels the best leaving group. If the X group is a better leaving group than the incoming nucleophile, a reaction is possible. Otherwise, the tetrahedral intermediate simply transforms back to the starting material.

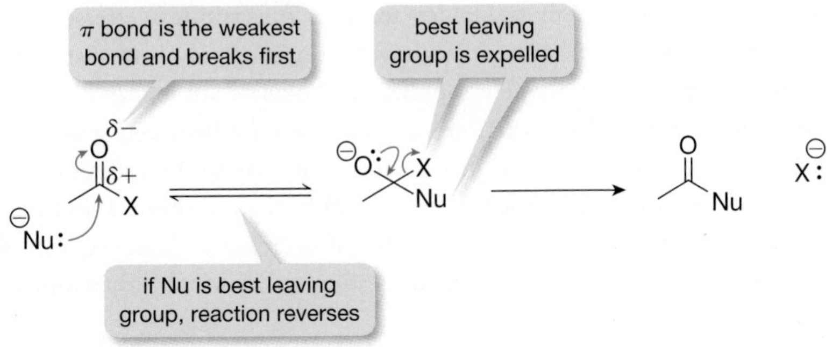

π bond is the weakest bond and breaks first

best leaving group is expelled

$\delta-$ $\delta+$ X Nu: Nu X:

if Nu is best leaving group, reaction reverses

As discussed in Chapters 7 and 11, weak bases are good leaving groups. The relative leaving group abilities of such X groups decrease as their relative base strengths increase.

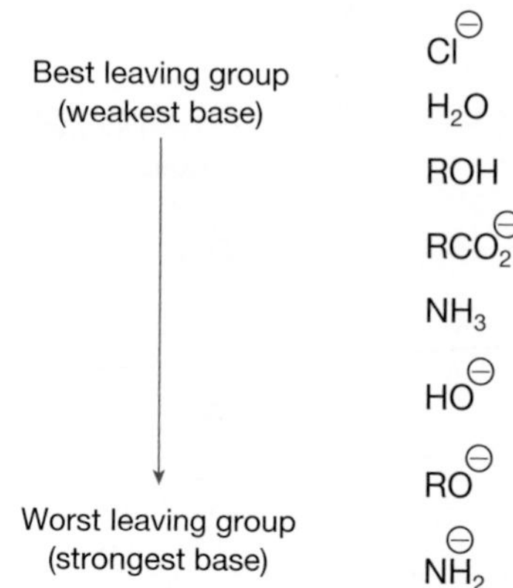

Together, the electrophilicity of the carbonyl carbon and the stability of the leaving group predict the relative reactivity of the acyl groups (alkanoyl in IUPAC nomenclature) and whether one particular acyl group can be converted into another. In the reaction shown in the following diagram, an acetate group reacts with an acid chloride to form an anhydride. In the first step of the reaction, the nucleophile adds to the carbonyl to form the tetrahedral intermediate. This intermediate then collapses and expels one of two possible leaving groups. Expelling acetate regenerates the acid chloride, whereas removal of the chloride produces an anhydride. Chloride is a much weaker base than acetate, and therefore is the better leaving group. When the tetrahedral intermediate collapses, it follows the easiest pathway, expelling Cl^- to generate the anhydride.

addition of nucleophile

π bond breaks

elimination of best leaving group

Cl is a weaker base and better leaving group than acetate

Figure 15.2 gives an overview of carbonyl reactivity. The more electrophilic carboxylic acid derivatives (on the left) can be converted to any of the less reactive forms (to the right) by the addition of the appropriate nucleophile. The reactions in this cycle can typically be done with either acid or base catalysts, though acid chlorides and anhydrides are reactive enough that their conversions usually do not need catalysts. The less reactive conversions of esters, acids, and amides do need catalysts. Acids and esters can be interconverted using alcohols or water as the nucleophile, as appropriate. The least reactive electrophile, a carboxylate anion, can be converted into a carboxylic acid through an acid–base reaction using any stronger Brønsted acid.

Carboxylic acid is the only functional group with a pathway to the groups on the left in the figure, and these transformations (discussed in Section 15.5) require very reactive reagents. When the carboxylic acid derivatives are arranged in order of decreasing reactivity, as in Figure 15.2, the carbonyl addition-elimination reactions form nesting cycles of transformations that all flow counter-clockwise.

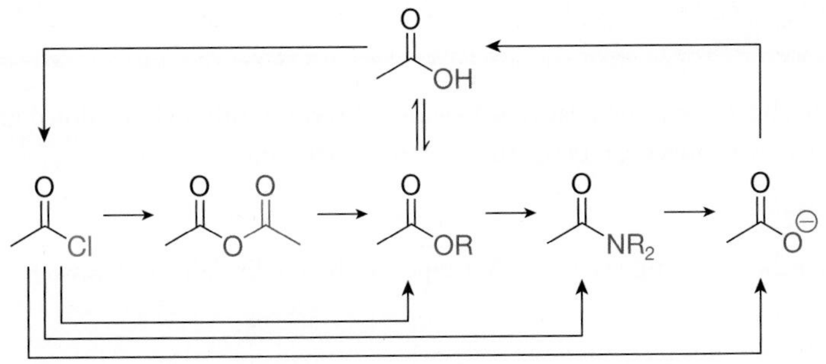

FIGURE 15.2 The pattern of interconversions of carboxylic acid derivatives.

15.3.1 Retrosynthetic analysis of nucleophilic acyl substitution reactions

Acyl substitution reactions produce one acyl compound from another by substituting a nucleophile for a leaving group on a carbonyl compound. From a synthetic point of view, the acyl product disconnects to an acyl cation synthon and a negatively charged nucleophile. The electrophilic synthon is derived from a carboxylic acid derivative with any leaving group, X, that stabilizes negative charge better than the nucleophile. The nucleophilic synthon arises from the deprotonation of a protonated nucleophile, as long as that nucleophile is a weaker leaving group than X.

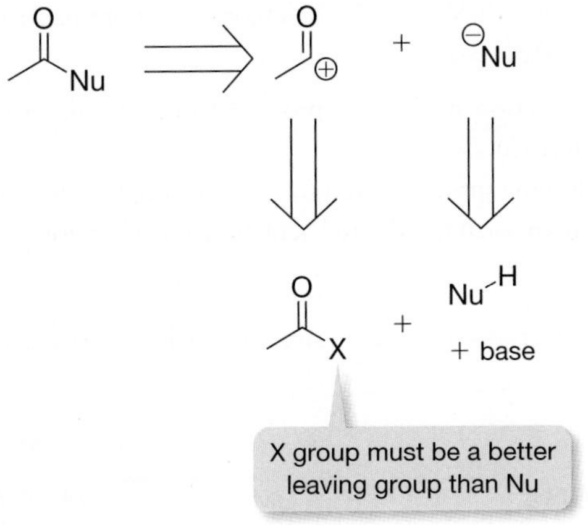

This pattern indicates that less reactive acid derivatives can be synthesized from other, more reactive electrophilic acid derivatives.

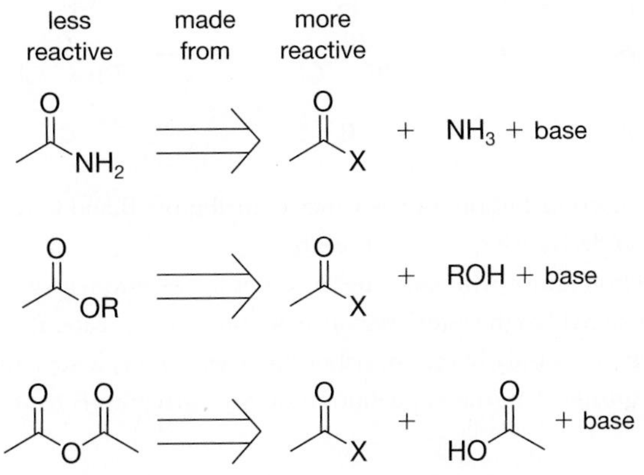

CHECKPOINT 15.1

You should now be able to rank the relative reactivities of various carbonyl functional groups and predict which carboxylic acid derivatives can be synthesized from each other.

SOLVED PROBLEM

Determine whether the following reaction will happen as shown. Explain your answer.

PhC(=O)OMe + NaCl ⟶ PhC(=O)Cl + NaOMe

METHOD A

STEP 1: Draw mechanism arrows for the proposed reaction.

The chloride nucleophile adds to the electrophilic carbonyl of the ester, giving a tetrahedral intermediate. The tetrahedral intermediate collapses, eliminating a methoxide leaving group.

PhC(=O)OMe + Cl⁻ ⟶ PhC(O⁻)(Cl)OMe ⟶ PhC(=O)Cl + ⁻OMe

STEP 2: Determine which of the two possible leaving groups on the tetrahedral intermediate is better.

The two possible leaving groups are the chloride ion and methoxide ion. Since chloride is the conjugate base of a very strong acid (HCl, $pK_a = -7$), the chloride is more likely to leave than methoxide, the conjugate base of a weak acid (MeOH, $pK_a = 15.5$).

STEP 3: Decide whether the reaction is likely to proceed based on which leaving group is likely to be expelled by the tetrahedral intermediate.

Since chloride is a better leaving group, the tetrahedral intermediate is most likely to collapse and give back the starting ester, rather than form the desired acid chloride. Therefore, the reaction will not happen.

METHOD B

STEP 1: Draw the resonance structures of the two carboxylic acid derivatives in the reaction.

[Ph–C(=O)–OMe (A) ⟷ Ph–C⁺(–O⁻)–OMe (B) ⟷ Ph–C(–O⁻)=O⁺Me (C)]

[Ph–C(=O)–Cl (A) ⟷ Ph–C⁺(–O⁻)–Cl (B) ⟷ Ph–C(–O⁻)=Cl⁺ (C)]

STEP 2: Evaluate the relative contributions of resonance contributors B and C to both structures to determine which carboxylic acid derivative is the most reactive.

Oxygen is a better electron donor than chlorine since it is less electronegative. Therefore, the contribution of ester resonance structure C to the ester resonance hybrid will be greater than the contribution of acid chloride resonance structure C. Similarly, the contribution of acid chloride structure B to the acid chloride resonance hybrid will be greater than the contribution of ester structure B to the ester resonance hybrid.

Thus, the acid chloride hybrid will have a greater partial positive charge on its carbonyl carbon, making the acid chloride more reactive than the ester.

STEP 3: Use the relative reactivities to determine whether the reaction will happen as shown.

Interconversion of carboxylic acid derivatives flows in the direction of decreasing reactivity. Therefore, making a more reactive acid chloride directly from a less reactive ester is not practical.

PRACTICE PROBLEM

15.1 Which of the following reactions is likely to proceed as written?

a) O, OMe + NHMe$_2$ ⟶ O, NMe$_2$ + MeOH

b) N, O + O, ONa ⟶ O, O, O + NaNMe$_2$

c) O, Cl + H$_2$O ⟶ O, OH + HCl

d) O, OEt + O, HO ⟶ O, O, O + EtOH

INTEGRATE THE SKILL

15.2 Select the feasible reactions among those given in Question 15.1. Then draw a mechanism and identify the addition step, the elimination step, and the tetrahedral intermediate for each of them.

15.3.2 Reactions with basic nucleophiles

A nucleophile in its conjugate base form can react directly with acid chlorides and anhydrides. The method is usually practical only when the nucleophile is oxygen based. Negatively charged nitrogen nucleophiles are too basic and can cause unwanted side reactions.

O, Cl + ⊖:O: ⟶ (EtOH) O, O + Cl⊖

A strong base such as NaH is normally employed to generate the required alkoxide from the alcohol.

ORGANIC CHEMWARE
15.1 Substitution at carbonyl: General mechanism (basic conditions)

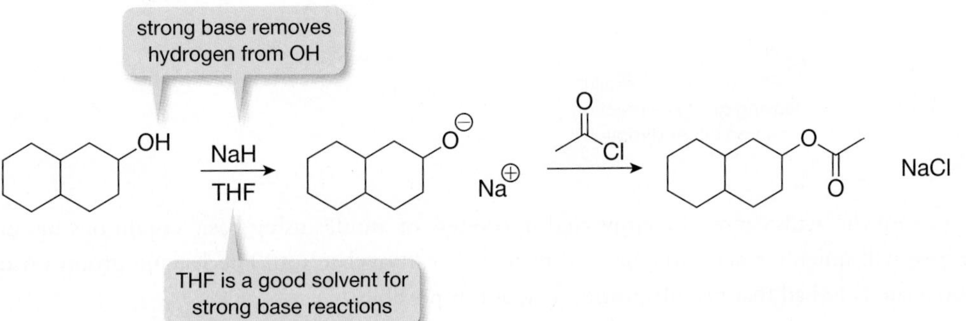

Acid anhydrides are less reactive than acid chlorides, but undergo the same reactions. When the anhydride is asymmetric (i.e., the acyl groups are different), reactions at the two carbonyl sites give different products. The result is a mixture of two products, unless one of the acyl groups is less reactive toward nucleophiles than the other.

symmetrical anhydride

either carbonyl electrophile will give the same product

asymmetrical anhydride

nucleophile attacks the less-hindered electrophilic carbonyl

product of nucleophilic attack at the more-hindered carbonyl is not observed

Ester hydrolysis is a hydrolysis reaction that converts an ester into a carboxylic acid.

Saponification is the conversion of an ester to a carboxylate under basic conditions.

Esters can be converted to acids using NaOH or KOH in water (basic conditions). This reaction is a type of **ester hydrolysis** called a **saponification**. Such reactions are irreversible because the product acid is immediately converted to the unreactive carboxylate. At the end of the reaction, an inorganic acid is added to convert the carboxylate to the carboxylic acid. Saponification of a triacyl glyceride (fat) produces soap.

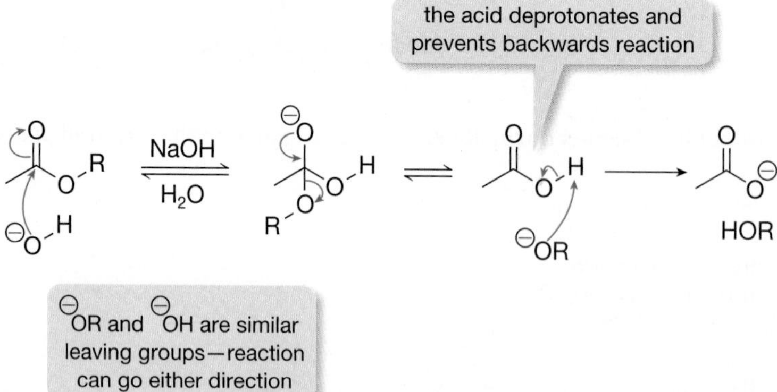
the acid deprotonates and prevents backwards reaction
NaOH
H_2O
HOR
⊖OR and ⊖OH are similar leaving groups—reaction can go either direction

STUDENT TIP
Basic hydrolysis follows the same mechanism, regardless of which carbonyl derivative is used.

ORGANIC CHEMWARE
15.2 Ester hydrolysis (basic conditions)

Carboxylic acids cannot be converted into esters or amides using basic conditions because the base will quickly react with the acid to form a carboxylate ion. The leaving group on the carboxylate is so bad that no substitution reaction is possible.

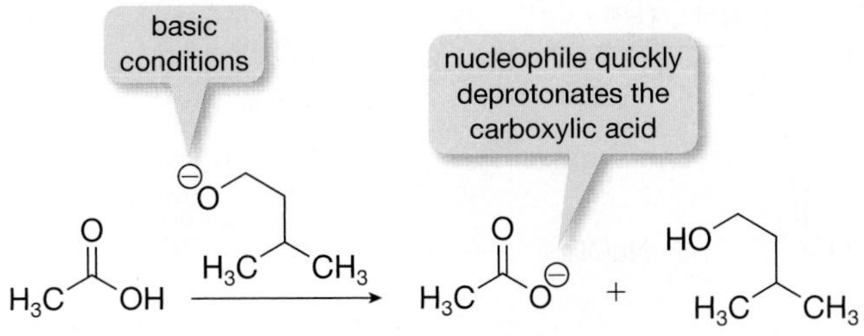

basic conditions

nucleophile quickly deprotonates the carboxylic acid

H_3C OH + $^{\ominus}$O (H$_3$C, CH$_3$) ⟶ H_3C O$^{\ominus}$ + HO (H$_3$C, CH$_3$)

CHECKPOINT 15.2

You should now be able to predict which carboxylic acid derivatives react directly with basic nucleophiles and draw the products and mechanisms of these reactions.

SOLVED PROBLEM

Draw a mechanism and predict the final products of the following reaction.

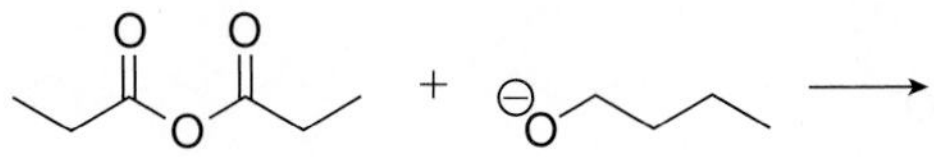

STEP 1: Determine the roles of the reactants.

The negatively charged alkoxide acts as the nucleophile, and one of the carbonyl groups of the anhydride acts as the electrophile. (Note: Since the anhydride is symmetrical, it does not matter which carbonyl group reacts.)

STEP 2: Draw the mechanism for the addition step of the reaction, and determine the intermediate product.

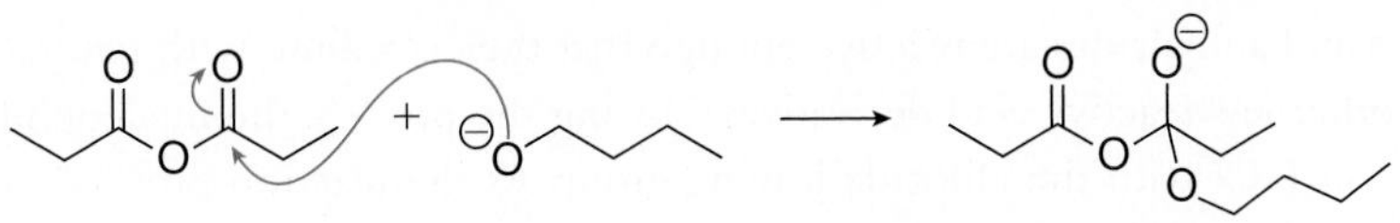

STEP 3: Draw the mechanism for the elimination step of the reaction, and determine the final products.

Since a carboxylate is a better leaving group than butoxide (pK_a of RCO_2H is about 5 versus about 15 for ROH), the final products are an ester and a carboxylate.

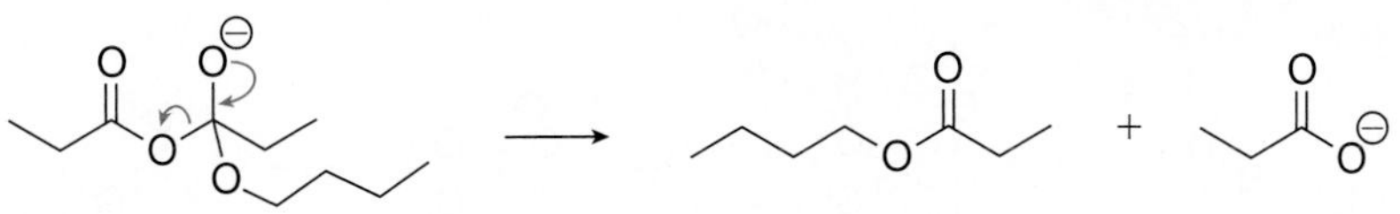

PRACTICE PROBLEM

15.3 Predict the products of the following reactions. Where appropriate, assume an aqueous work-up at the end of the reaction.

a) 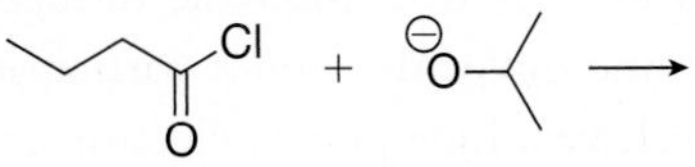 Cl + $^{\ominus}$O ⟶

b) 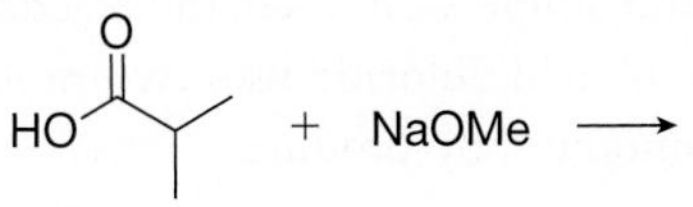 HO + NaOMe ⟶

c) 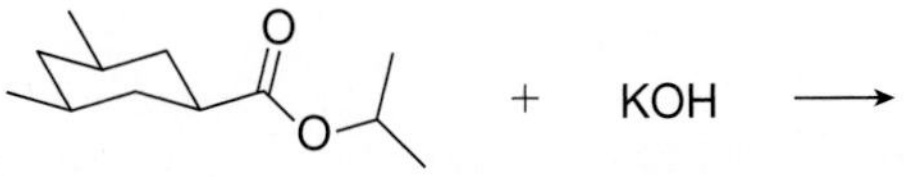 + KOH ⟶

d) 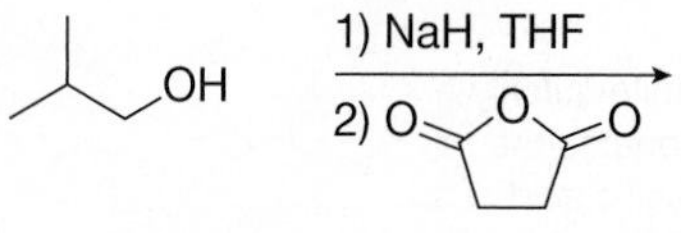 1) NaH, THF 2) [succinic anhydride]

e) 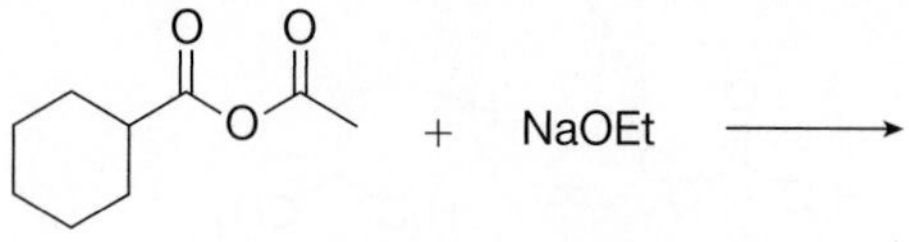 + NaOEt

INTEGRATE THE SKILL

15.4 The term *saponification* has its origins in soap making. Historically, soap was made by reacting vegetable and/or animal fats with lye (NaOH). Under the strongly basic conditions, the triglycerides in fats and oils hydrolyze to form glycerol and fatty acid salts. Following is the structure of a triglyceride common to beef fat. Draw the mechanism and products for its reaction with sodium hydroxide.

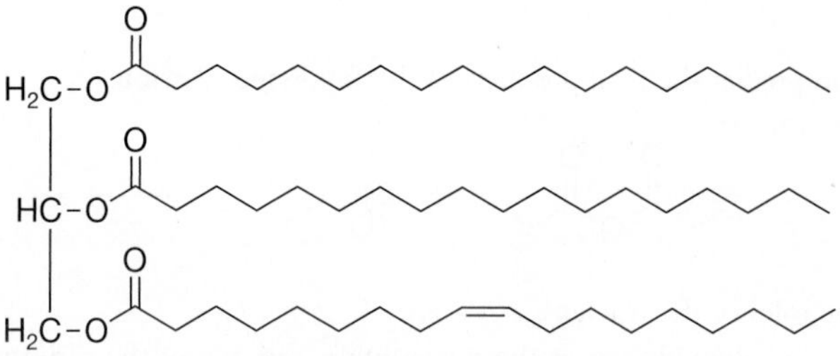

H_2C-O
HC-O
H_2C-O

15.3.3 Reactions with neutral nucleophiles

Acid chlorides and anhydrides are reactive enough that they combine with neutral nucleophiles to make any other less reactive acid derivatives. During the process, the nucleophilic atom loses a proton to form HCl with the chloride leaving group. As the reaction proceeds, the generated HCl can make the reaction mixture very acidic. Unless base is added, dangerous HCl gas may also be released from the reaction mixture.

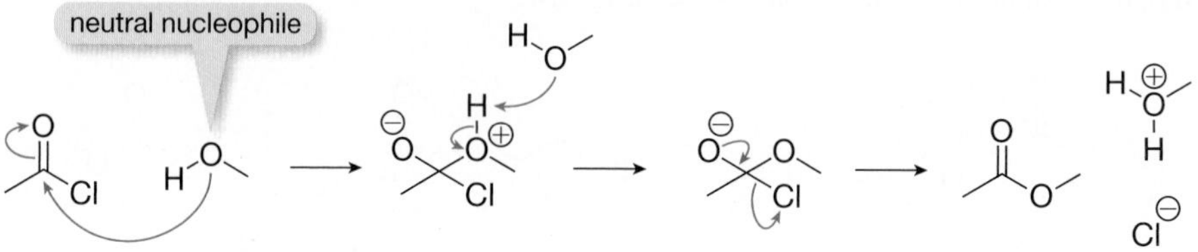

neutral nucleophile

ORGANIC CHEMWARE
15.3 Esterification (via acid chloride)

This excess acid can also destroy sensitive compounds or stop reactions. If nitrogen nucleophiles are used to make amides, the HCl by-product will react with the basic nitrogen of the amine nucleophile to form an ammonium salt. This amine molecule cannot participate in the desired reaction because an ammonium group does not have a lone pair of electrons, and is no longer nucleophilic. If a 1:1 mixture of acid chloride and amine were used, the reaction would stop at 50 percent completion because each molecule of acid chloride uses two molecules of amine: one to form the amide and one to form the ammonium by-product.

neutral nucleophile

50% of the desired product forms, releasing HCl

50% max

50% max

half of the nucleophile is protonated, making it non-nucleophilic

The reaction will transform all of the acid chloride if extra base is added to neutralize the HCl by-product. KOH is sometimes used for this purpose. Using excess amine is another way to supply extra base, but some amines can be expensive. A more common method is to use a tertiary amine base such as Et_3N or *i*-Pr_2NEt. These amines are very hindered and non-nucleophilic, and therefore function only as bases, not as nucleophiles.

nucleophile

non-nucleophilic base

non-nucleophilic base removes the HCl as it is produced

The mechanism of acylation reactions involving hindered amines employs neutral nucleophiles because the pK_a of the ammonium salt of these amines (10–12) is well below the pK_a of most nucleophiles (alcohols 16–18; amines 25–40).

base is not strong enough to deprotonate nucleophile

base is only strong enough to remove $H^{\oplus}$ from charged oxygen

Et_3NHCl

15.3.4 Acceleration of acylation substitution reactions using pyridine and DMAP

Pyridine is a weak base that is often used as a solvent or additive in acylation reactions because it can act as a reversible nucleophile. Pyridine reacts very quickly with an acid chloride to make a positively charged intermediate with a pyridinium leaving group. While this may look like an unreactive amide, the positive charge makes it much more reactive than the acid chloride. Consequently, nucleophiles usually undergo the acylation reaction much faster with this intermediate than with the original acid chloride. Since the pyridine is regenerated in the process, it acts as a catalyst. If an extra base (such as i-Pr_2NEt) is used in the reaction, only a small amount of pyridine is needed because of its constant regeneration in the reaction.

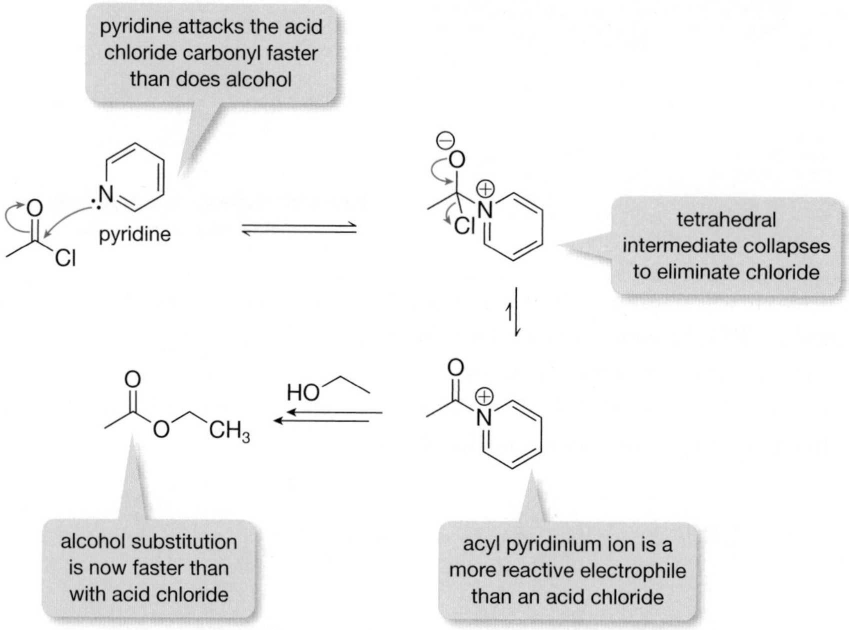

Knowledge of the pyridine mechanism has led to the development of an even better catalyst, dimethylaminopyridine (DMAP). The electron-donating p-dimethylamino group makes DMAP more nucleophilic than pyridine. Furthermore, as seen in the alternate resonance structures, the dimethylamino group stabilizes the acyl pyridinium ion by conjugation, thereby facilitating its formation.

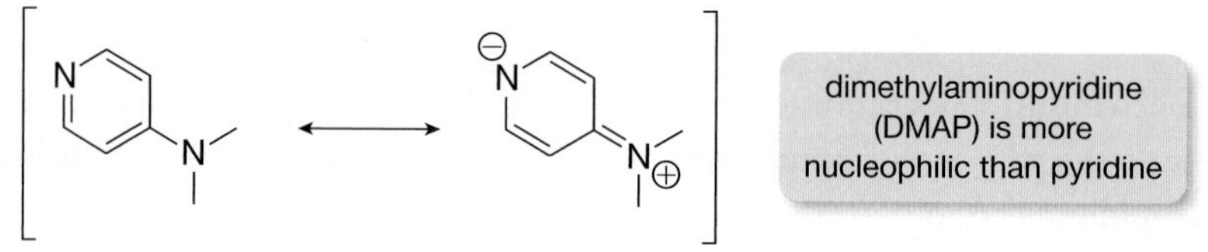

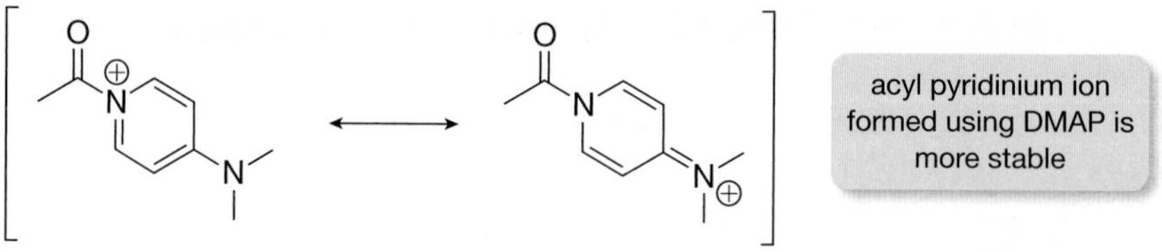

CHECKPOINT 15.3

You should now be able to predict which carboxylic acid derivatives react directly with neutral nucleophiles and draw mechanisms and products of these reactions.

SOLVED PROBLEM

Draw a mechanism and predict the final products of the following reaction.

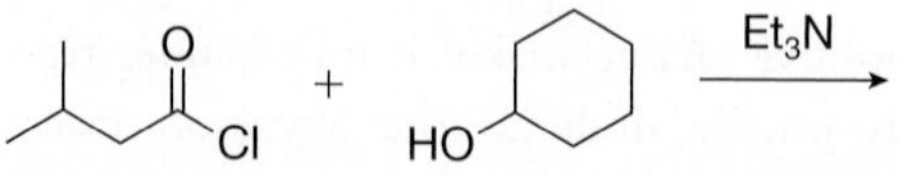

STEP 1: Determine the roles of the reactants.

The acid chloride is a good electrophile since it has a partial positive charge on the carbonyl carbon. Cyclohexanol has lone pairs available for sharing on its oxygen atom, making it a good nucleophile. The triethylamine has lone pairs, but it is a quite hindered tertiary amine, and hence unlikely to act as a nucleophile. However, the trimethylamine could act as a base.

STEP 2: Draw a mechanism for the addition step of the reaction, and use this mechanism to determine the intermediate product.

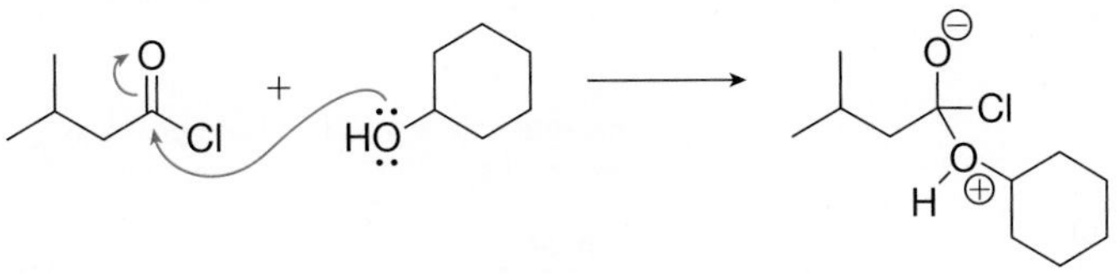

STEP 3: Draw a mechanism for the elimination step of the reaction to determine the final product.

The product of the first step has a positively charged oxonium ion. In the absence of the amine base, another molecule of cyclohexanol could deprotonate this intermediate. However, the amine is more basic than an alcohol and therefore serves as the base in this reaction. Following deprotonation, the tetrahedral intermediate collapses, eliminating the chloride leaving group.

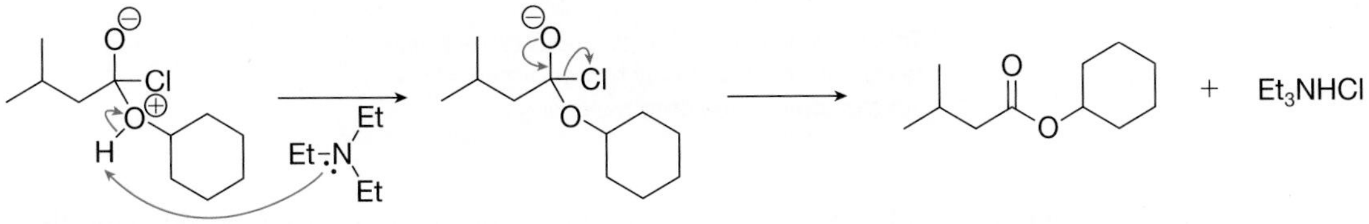

PRACTICE PROBLEM

15.5 Draw mechanisms for the following reactions, and predict the final products. Where appropriate, assume an aqueous work-up at the end of the reaction.

a) 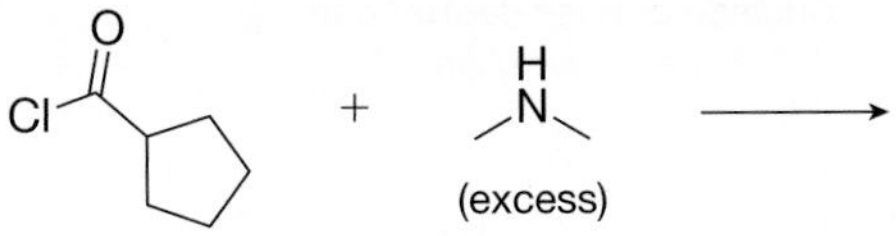 (excess)

b)

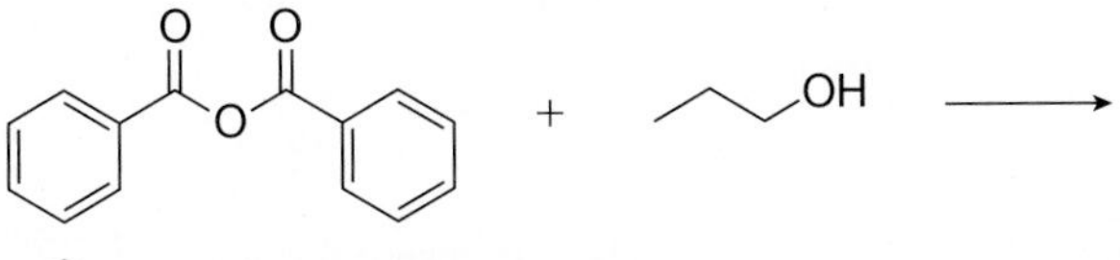

c) 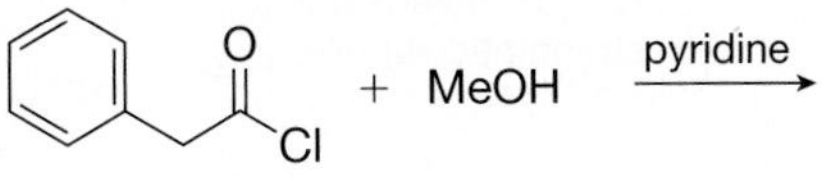 + MeOH, pyridine

d)

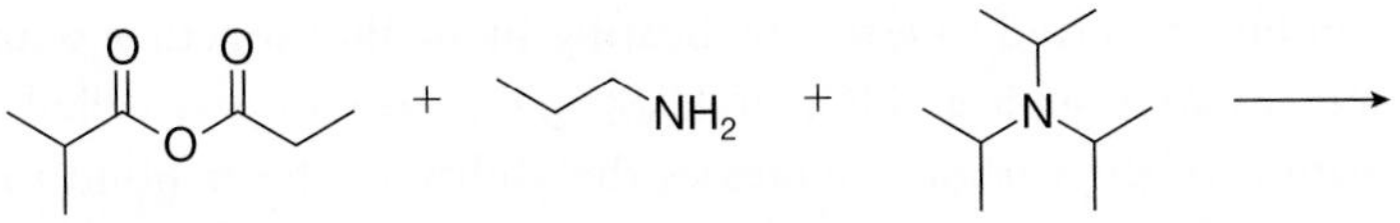

e) 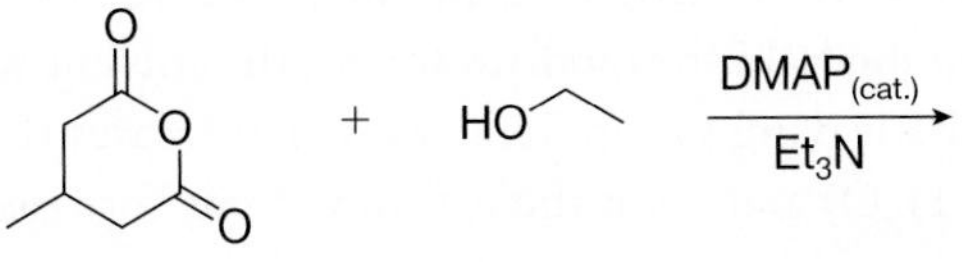 $\xrightarrow[\text{Et}_3\text{N}]{\text{DMAP}_{(\text{cat.})}}$

INTEGRATE THE SKILL

15.6 Draw an energy diagram for the reaction in Question 15.5a.

15.3.5 Acid catalysis in acyl substitution reactions

Esters can be converted to amides by a direct reaction with an amine, but these reactions sometimes require heating for a long time to complete. Reactions between acids and amines to make amides are possible, but they require such high temperatures that it is usually practical only for the simplest molecules. Most mixtures of an acid and an amine result in a fast acid–base reaction that generates both a deactivated electrophile (carboxylate) and a deactivated nucleophile (ammonium salt), instead of the amide product shown in the following diagram.

this reaction is possible but very slow

This reaction is possible but requires very high temperatures. It can only be performed with very simple compounds or to make rings.

Since acid catalysts greatly increase reactivities in transformations of esters and acids, these two reactions are practical. Acid accelerates reactions by protonating the electrophile, which *pulls* electrons, while base accelerates reactions by activating the nucleophile, which *pushes* electrons.

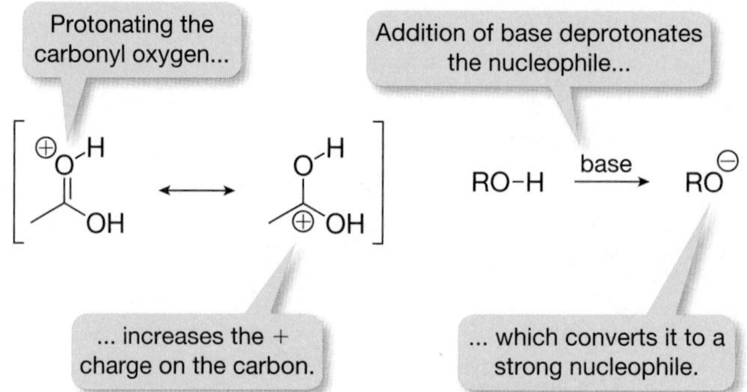

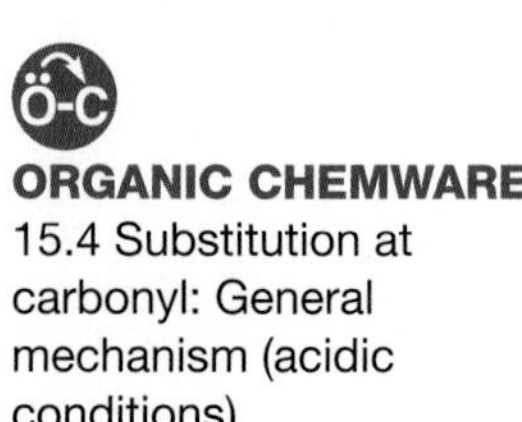

ORGANIC CHEMWARE
15.4 Substitution at carbonyl: General mechanism (acidic conditions)

Fischer esterification is the acid-catalyzed conversion of a carboxylic acid to an ester.

Carboxylic acids can be converted to esters by heating in alcohol together with a small amount of a strong acid catalyst such as HCl or H_2SO_4. This reaction is called **Fischer esterification**. Protonation of the carbonyl increases the ability of the π bond to act as a leaving group, which makes the carbonyl more electrophilic, and hence more likely to react with a neutral nucleophile. Once the tetrahedral intermediate forms, the solvent transfers a proton from the added nucleophile to the leaving group. This transfer increases the leaving ability of the OH group (turning it into H_2O) and helps the carbonyl to re-form as an ester.

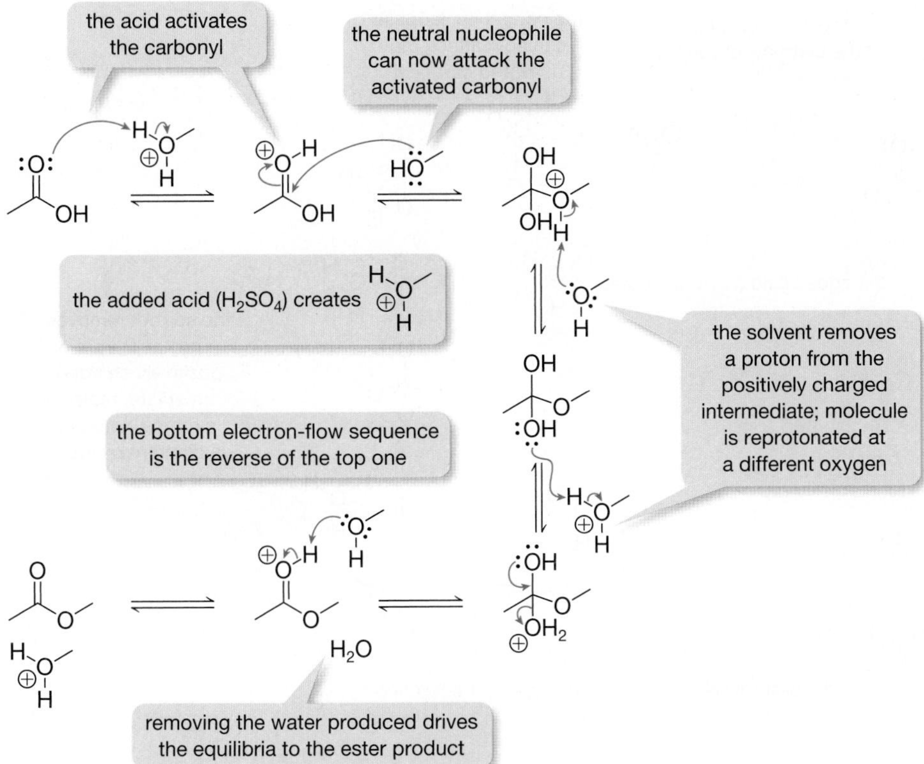

Esters can be transformed into other esters by a reaction called **transesterification**. The mechanism of this reaction is the same as Fischer esterification, but an ester appears in place of the carboxylic acid.

Transesterification is the conversion of one ester to another by heating it with an acid catalyst in an alcohol solvent.

Esters can be transformed into acids by heating them in water with an acid catalyst. Acid-catalyzed hydrolysis has the same mechanism as the Fischer esterification, but it uses water instead of alcohol as the nucleophile. Acids and esters are therefore in equilibrium in acidic conditions. Because there is little thermodynamic (energy) difference between carboxylic acids and esters, the reactions produce mixtures of products, unless the equilibrium is shifted according to Le Chatelier's principle. For example, Fischer esterification gives a good yield of the product ester if the water by-product is removed by distillation or drying agents.

H_2SO_4 + H_2O

Removing water (by distillation or using a drying agent) drives the reaction forward to make more ester. Adding water drives the reaction backwards.

Similarly, excess water can be added to an ester hydrolysis to favour the production of a carboxylic acid.

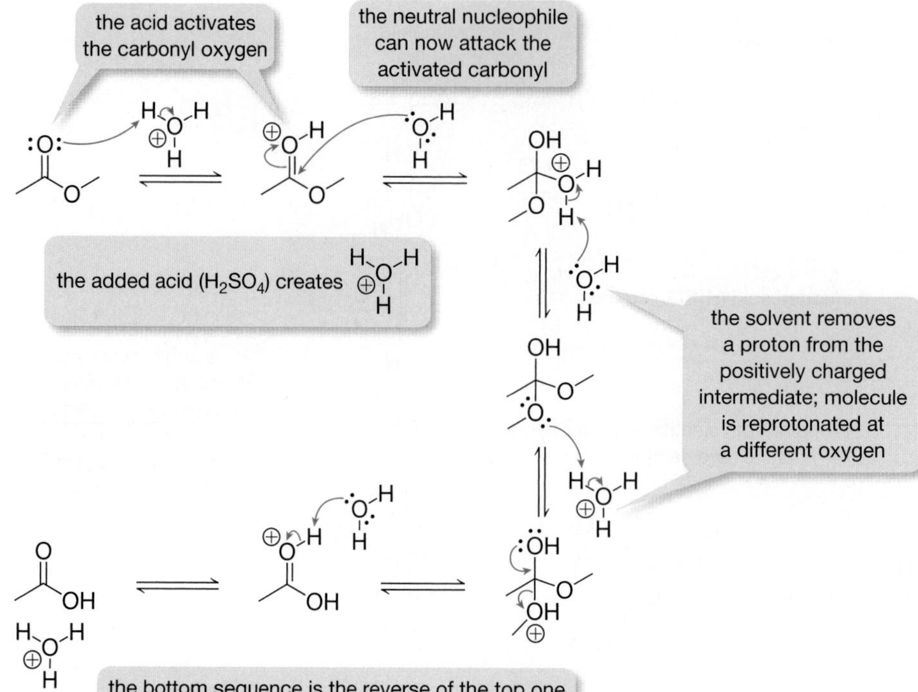

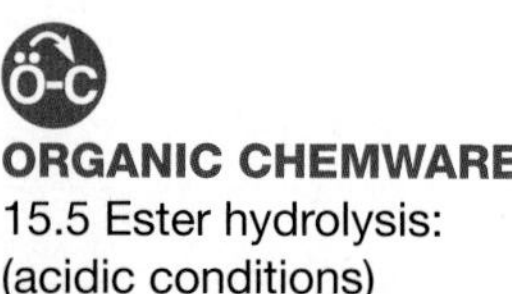

ORGANIC CHEMWARE
15.5 Ester hydrolysis: (acidic conditions)

ORGANIC CHEMWARE
15.6 Fischer esterification

CHECKPOINT 15.4

You should now be able to predict which carboxylic acid derivatives react with neutral nucleophiles in the presence of an acid catalyst and draw mechanisms and products of these reactions.

SOLVED PROBLEM

Draw a mechanism of the following reaction and predict the final products.

OMe — $H_2SO_{4\,(cat.)}$ →
OH

STEP 1: Determine the roles of the reactants.

The starting material contains an electrophilic ester and a nucleophilic alcohol in close enough proximity to each other to be able to react. Under neutral conditions, the ester does not react with the alcohol, but sulfuric acid catalyzes the reaction by protonating the ester, making it susceptible to nucleophilic attack.

STEP 2: Draw a mechanism for the catalyzing protonation and the products of this reaction.

OMe + $H-O-SO_2-OH$ → OMe + $HSO_4^{\ominus}$

STEP 3: Consider how the protonated ester could react with the other materials present.

Whenever a molecule has both a nucleophile (such as on OH) and an electrophile (a protonated ester), there is the potential for an intramolecular reaction. For an intramolecular reaction to happen, the reactive groups need to be next to each other. In this case, an intramolecular reaction can only happen from the minor conformation, with both substituents in the axial position.

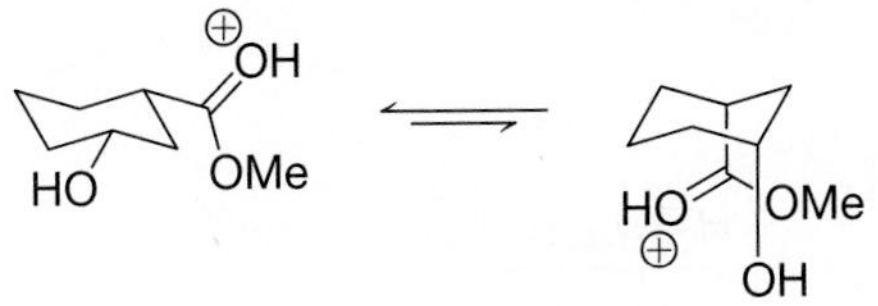

Because reactions depend on the concentration, the minor conformation leads to a slow reaction. However, an intramolecular reaction is much faster than an intermolecular reaction, which favours the internal reaction.

STEP 4: Complete the mechanism to predict the final product.

Once the ester is protonated, the alcohol functional group can attack it, forming a double-ring structure. A weak base (designated B: in the following figure) can protonate the methoxy group, making a good leaving group. The weak base could be the conjugate base of the acid catalyst, a molecule of the starting material (with its weakly basic alcohol group), or the methanol produced in the semi-final step of the reaction. Following a series of the proton transfers, the tetrahedral intermediate collapses, expelling a molecule of methanol.

Deprotonation then produces the final product, a cyclic ester.

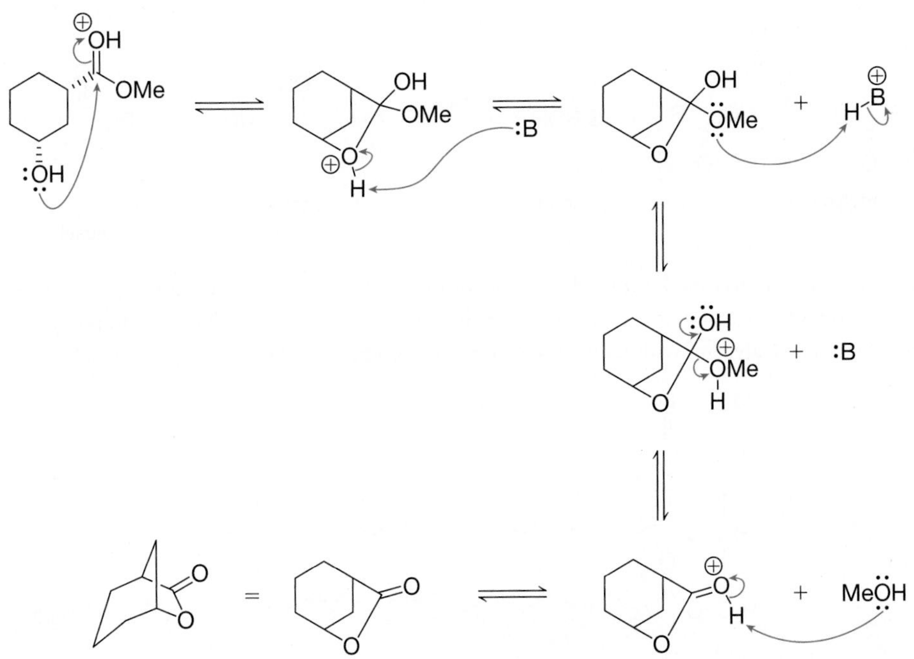

Note that in writing the mechanism there is some flexibility in terms of depicting what carries out the various proton transfer steps. Here, we used a generic base, B:, for some of the steps. This could have been the conjugate base of the acid catalyst, a molecule of the starting material (with its weakly basic alcohol group), or the methanol produced in the reaction.

PRACTICE PROBLEM

15.7 Draw mechanisms of the following reactions, and predict the final products. Where appropriate, assume an aqueous work-up at the end of the reaction.

a) 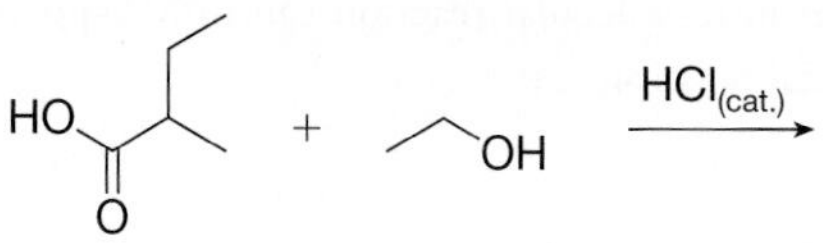 HO(C=O)CH(CH2CH3)CH3 + CH3CH2OH $\xrightarrow{HCl_{(cat.)}}$

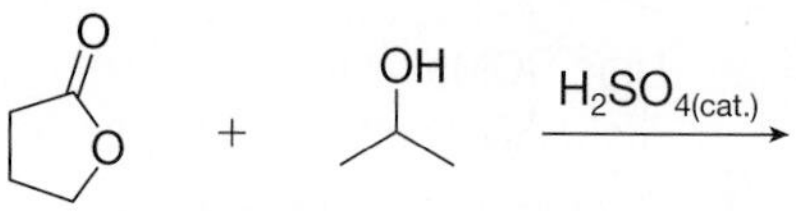

b) γ-butyrolactone + (CH3)2CHOH $\xrightarrow{H_2SO_{4(cat.)}}$

c) ester + CH3CH2NH2 $\xrightarrow{H_2SO_{4(cat.)}}$

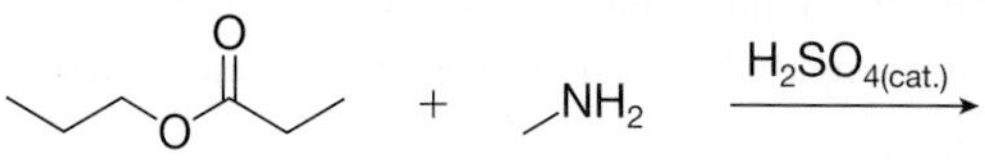

d) 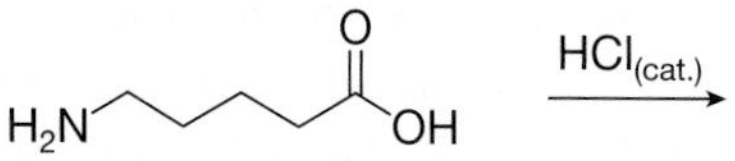 H_2N–(CH2)4–C(=O)OH $\xrightarrow{HCl_{(cat.)}}$

INTEGRATE THE SKILL

15.8 Biodiesel, a mixture of fatty acid esters, is made by transesterifying vegetable oils or animal fats. The general reaction proceeds as follows:

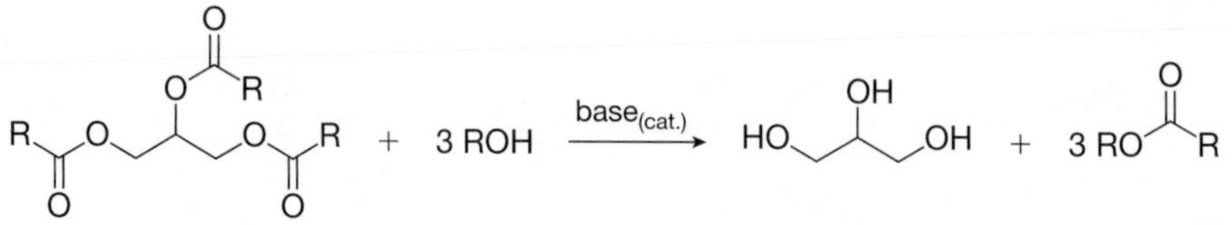

triglyceride + 3 ROH (alcohol) $\xrightarrow{base_{(cat.)}}$ HO–CH2–CH(OH)–CH2–OH (glycerol) + 3 RO(C=O)R (fatty acid esters (biodiesel))

The most common feedstock for biodiesel is soybean oil. The structure of trilinoleic glycerol, the most common triglyceride found in soybean oil, is shown here. Draw the mechanism and products of the reaction of trilinoleic glycerol with methanol. Assume the reaction is catalyzed by potassium hydroxide.

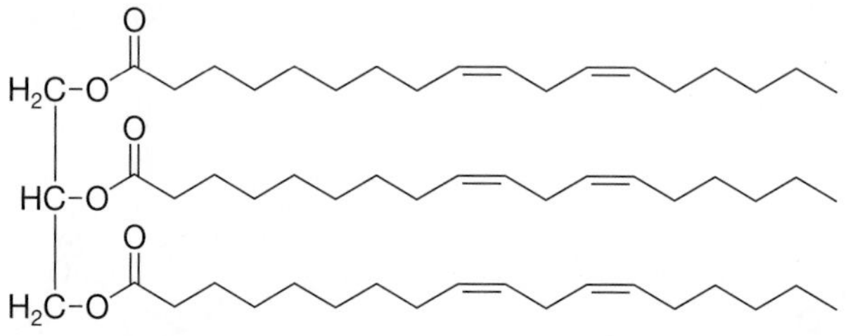

H_2C–O–C(=O)R
HC–O–C(=O)R
H_2C–O–C(=O)R

15.4 Reacting Poor Electrophiles Using Acids and Bases

Since amides have a strong electron-withdrawing substituent, they are the least reactive acyl compounds. They can be converted into other acyl groups, but only under forcing conditions or with highly reactive reagents. The only useful transformation of amides into another acyl group is their conversion to carboxylic acids, which requires high temperatures plus an acid or base reagent. Basic conditions are more difficult because the carbonyl of an amide is not very electrophilic (very small contribution from its electrophilic resonance form) and the leaving group NR_2^- is extremely poor.

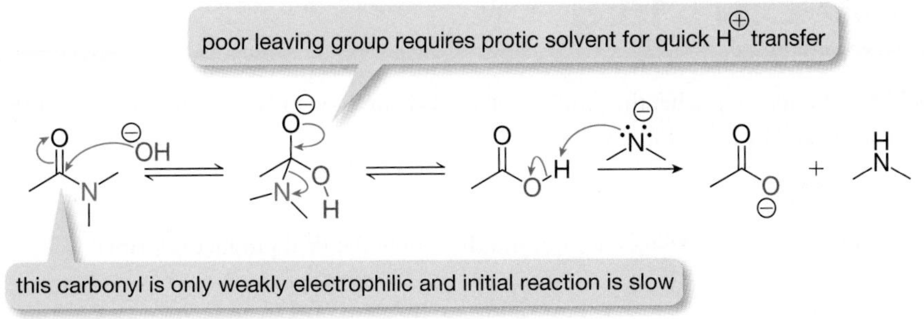

poor leaving group requires protic solvent for quick H⊕ transfer
this carbonyl is only weakly electrophilic and initial reaction is slow

ORGANIC CHEMWARE
15.7 Amide hydrolysis (basic conditions)

Acidic conditions usually provide easier reactions because the carbonyl oxygen is weakly basic (large contribution from resonance form C). Acids also convert the amine into a good leaving group. Since the nitrogen atom is a very good electron donor, the iminium ion is the lowest energy resonance form. Mechanisms are usually drawn using the lowest energy forms, so acidic hydrolysis of an amide is written using the iminium ion instead of an oxonium ion.

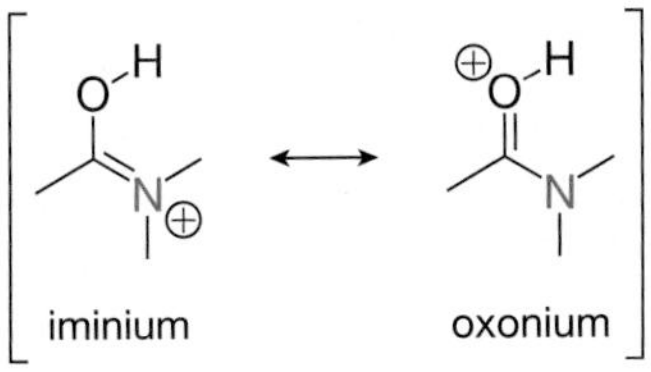

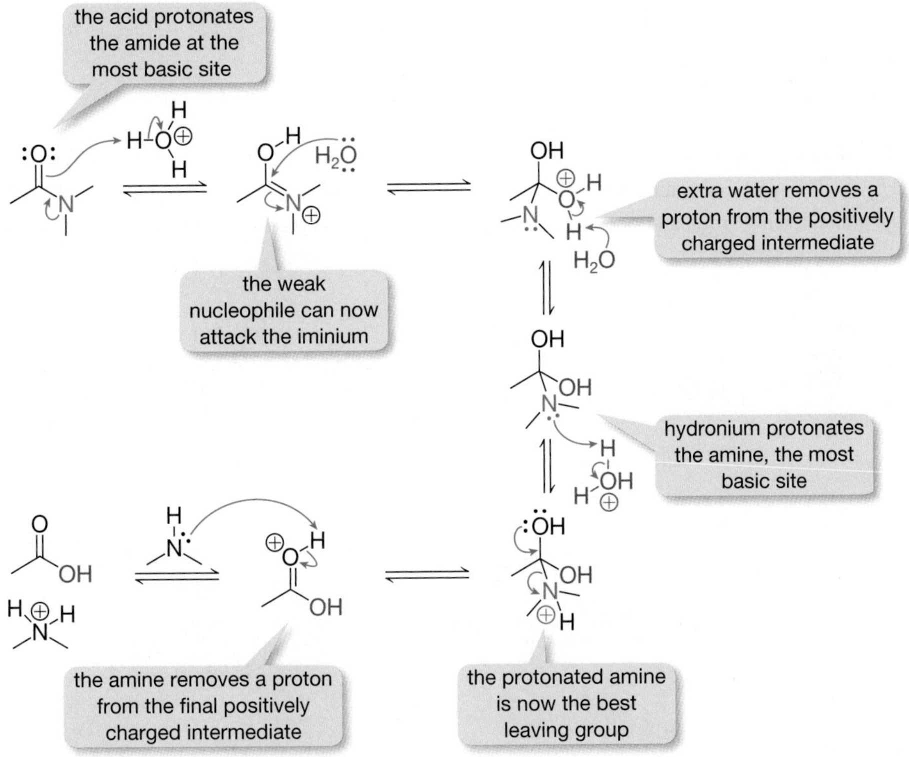

iminium
oxonium

the acid protonates the amide at the most basic site
the weak nucleophile can now attack the iminium
extra water removes a proton from the positively charged intermediate
hydronium protonates the amine, the most basic site
the amine removes a proton from the final positively charged intermediate
the protonated amine is now the best leaving group

ORGANIC CHEMWARE
15.8 Amide hydrolysis (acidic conditions)

In the presence of a strong acid or base, nitriles behave similarly to amides. The first addition of water creates an unsubstituted amide, which reacts as other amides do.

first equivalent of water makes an amide
second equivalent of water makes an acid

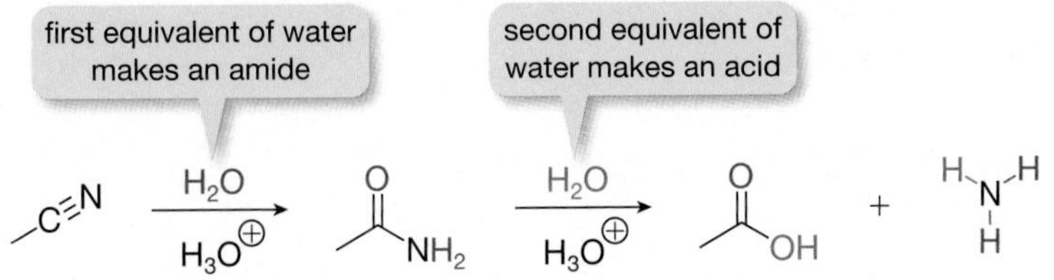

CHECKPOINT 15.5

You should now be able to predict the products of amide hydrolysis and draw a mechanism for the reaction under both acidic and basic conditions.

SOLVED PROBLEM

Draw a mechanism of the following reaction, and show how the final product is formed.

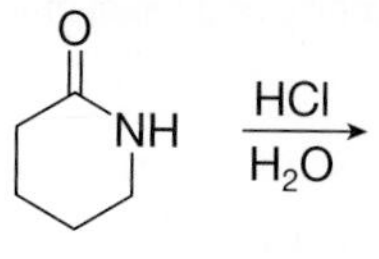

O NH HCl H_2O

STEP 1: Determine the roles of the reactants.

The amide, with its polarized carbonyl group, is a weak electrophile because the N atom is sufficiently electron donating to the carbonyl. The nucleophile present is water, which is not a good enough nucleophile to react with the amide under neutral conditions. However, the HCl acts as an acid catalyst by protonating the amide, making it susceptible to nucleophilic attack.

STEP 2: Draw the mechanism for the catalyzing protonation.

Recall that the iminium resonance structure is a stronger contributor to the resonance hybrid than the oxonium resonance structure, so the mechanism should proceed via the iminium ion as shown here.

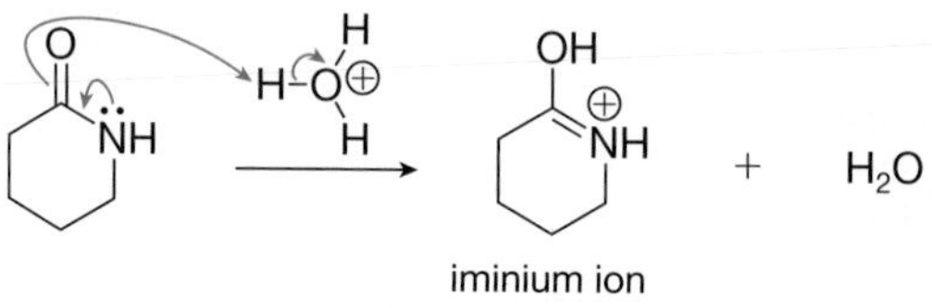

O NH H H_2O H OH NH + H_2O

iminium ion

STEP 3: Consider how the iminium ion can react with the other reagents. The electrophilic iminium ion is readily attacked by the lone pairs on water. After a series of proton transfers, the tetrahedral intermediate collapses, eliminating the amine leaving group and opening up the ring. Deprotonation then yields the neutral final product.

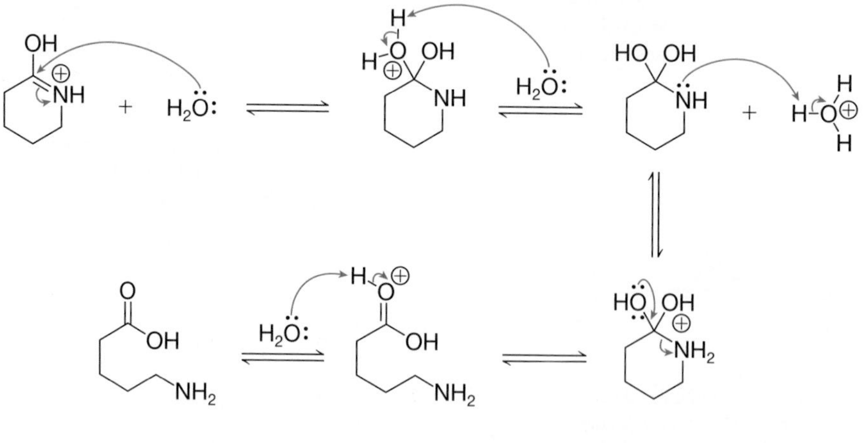

OH NH + $H_2\ddot{O}$: H H O OH NH $H_2\ddot{O}$: HO OH NH + H O H H

HO OH NH_2 H O OH NH_2 $H_2\ddot{O}$: O OH NH_2

PRACTICE PROBLEM

15.9 For each of the following reactions, predict the major products and draw mechanisms to show their formation.

a) 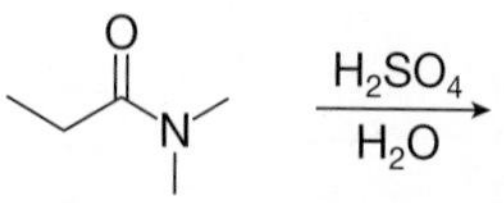 O N H_2SO_4 H_2O

b) 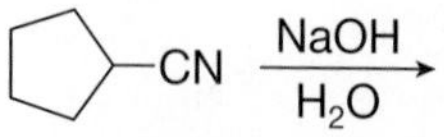 CN NaOH H_2O

c) KOH / H_2O

d) HCl / H_2O

INTEGRATE THE SKILL

15.10 Proteins are made up of amino acids linked together by peptide bonds, also called amide bonds. The following structure is a tripeptide made up of three amino acids. Draw the product(s) formed when this tripeptide reacts with a strong acid in water. Assign the absolute configuration to any chirality centres present in the products.

15.5 Carboxylic Acid Activation

Carboxylic acids can be converted to all of the other acyl groups by using a dehydrating agent, a reagent that absorbs water. Such reagents convert the OH of the acid into a leaving group that permits the desired transformation. One such reaction is the Fischer esterification described in Section 15.3.5. In this reaction, acid converts the OH group into water, which leaves readily.

Another common reaction converts carboxylic acids into acid chlorides—a process that involves the formation of a **mixed anhydride** by using thionyl chloride ($SOCl_2$) or oxalyl chloride (ClCOCOCl). These two reagents are more electrophilic than carboxylic acids are, so the acid acts as a nucleophile in the initial step of the process, generating a mixed anhydride structure in which the original carbonyl is now very electrophilic. The chloride generated in the first step then acts as a nucleophile to make the acid chloride, and the loss of SO_2 or CO_2 and CO gas drives the reaction sequence to completion.

A **mixed anhydride** is formed from two different types of acid, such as a carboxylic acid and a sulfonic acid. Typically, the carboxylic acid component is the more reactive electrophile.

the S tetrahedral intermediate collapses to eliminate Cl

eliminated Cl adds to activated carbonyl

carboxylic acid attacks S electrophile

mixed anhydride

HCl

new tetrahedral intermediate collapses, eliminating SO_2 and HCl

Acids can convert to anhydrides by reacting with acid chlorides. Symmetrical anhydrides are formed by treating a carboxylic acid with a strong dehydrating agent such as P_2O_5. With these reactions and those described in the preceding sections, all of the transformations in the nested cycles of reactivity that link all of the carboxylate derivatives are possible (Figure 15.3). Reactivity increases, moving left in the figure. Groups to the left can be converted to groups on the right. Moving left to right, transformations become more difficult and require acid or base catalysis. Transformations in the opposite direction have to pass through a carboxylic acid. All the chains of possible carbonyl addition-elimination reactions are loops that run counter-clockwise.

ORGANIC CHEMWARE
15.9 Transesterification (basic conditions)

ORGANIC CHEMWARE
15.10 Amide formation (via acid chloride)

ORGANIC CHEMWARE
15.11 Amide formation (via anhydride)

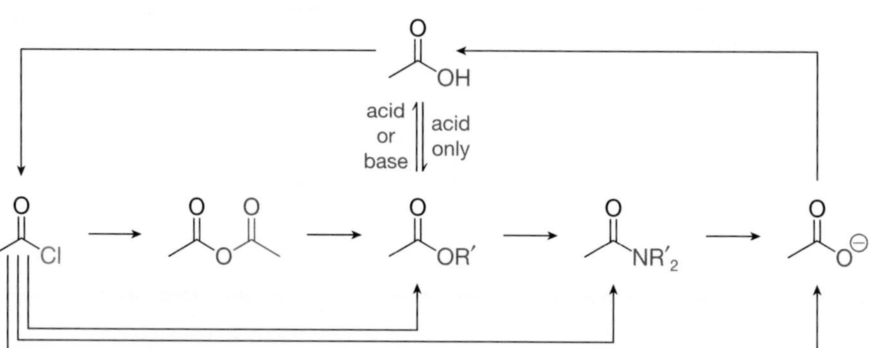

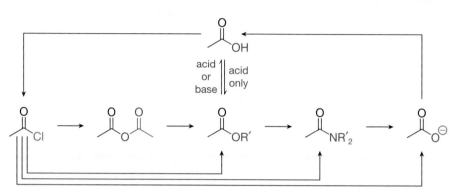

FIGURE 15.3 Interconversions of carboxylic acid derivatives.

15.5.1 Carboxylic acid activation to form esters and amides

Activating agents are reagents that convert a starting material to a more reactive intermediate in order to simplify its conversion to a desired product.

Carboxylic acids can be converted to esters and amides using a family of reagents called **activating agents.** These all remove water during the reaction between a carboxylic acid and a nucleophile, thus speeding the displacement of the OH group from the carboxylic acid. The key advantage of activating agents is that all of the reagents can be mixed together in a single reaction that runs at or below room temperature.

One of the most common applications of these agents is the chemical synthesis of proteins. Proteins are large molecules made by linking together amino acids, which in turn are small molecules containing an acid group and an amine group arranged such that the acid of one amino acid can form an amide with the amine of another. The properties of each protein depend on the particular amino acids present and on the sequence in which they are linked together. To make a specific protein, amino acids are joined one at a time in the proper order, using conditions that do not destroy the protein. The process can be repeated to make larger molecules.

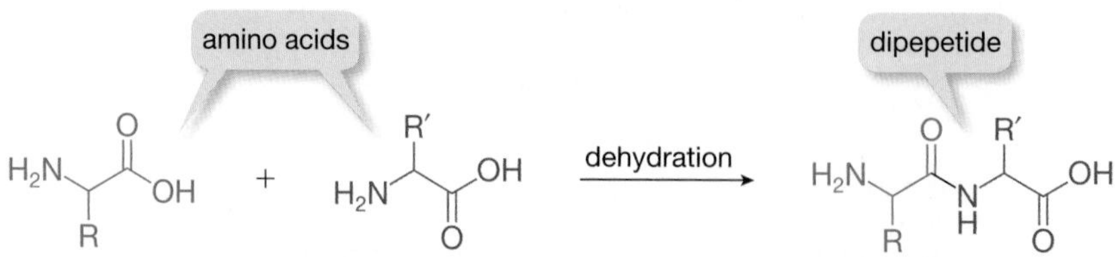

Acid chlorides cause technical problems if used to make amide bonds in proteins, and so special reagents have been designed to form amide bonds from a mixture of amine and acid. The oldest of these reagents is dicyclohexylcarbodiimide (DCC), which works as a dehydrating agent by capturing two hydrogens and an oxygen during amide bond formation to form dicyclohexylurea (DCU).

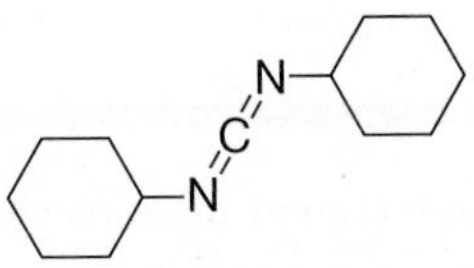

dicyclohexylcarbodiimide (DCC)

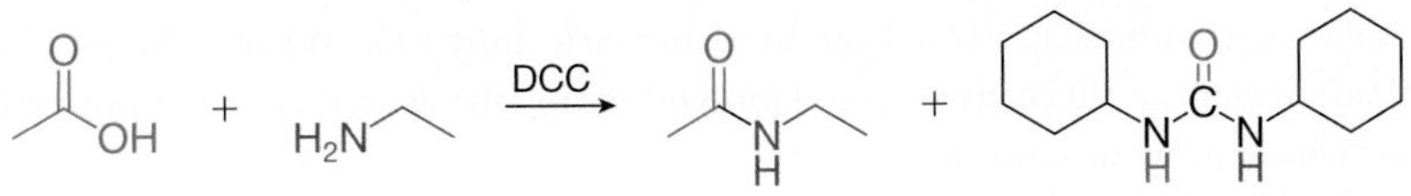

dicyclohexylurea (DCU)

The mechanism for carboxylic acid activation using DCC creates a reactive intermediate that resembles a mixed acid anhydride. Diimides make very good reagents for this because the protons can be easily transferred during the process.

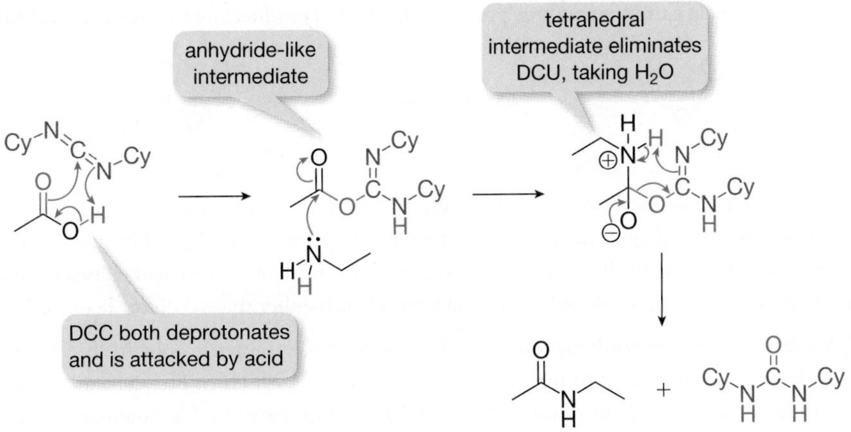

anhydride-like intermediate

tetrahedral intermediate eliminates DCU, taking H_2O

DCC both deprotonates and is attacked by acid

Other carbodiimides include diisopropylcarbodiimide (DIC), which has better solubility in organic solvents, and ethyl-dimethylaminopropylcarbodiimide hydrochloride (EDC), which produces a water-soluble by-product, making it easy to purify the final products by extraction.

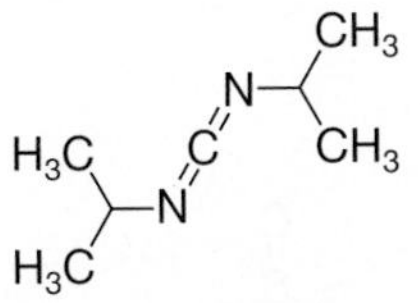

diisopropylcarbodiimide (DIC)

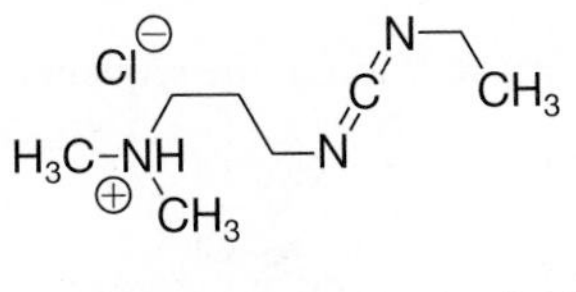

ethyl-dimethylaminopropylcarbodiimide hydrochloride (EDC)

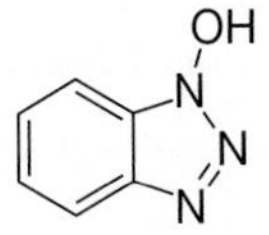

hydroxybenzotriazole (HOBt)

Hydroxybenzotriazole (HOBt) is sometimes added to speed up the amidation reaction. The nitrogen in the five-membered ring opposite the OH group acts as a nucleophile (similar to pyridine or DMAP) to make a more reactive intermediate.

CHECKPOINT 15.6

You should now be able to predict the products and draw mechanisms for dehydration reactions of carboxylic acids to carboxylic acid derivatives.

SOLVED PROBLEM

Consider the following synthesis proposed for an asymmetric anhydride. What is the problem with the proposed method? Suggest an alternative method for synthesizing the desired product from the acids shown and draw a mechanism for your synthesis.

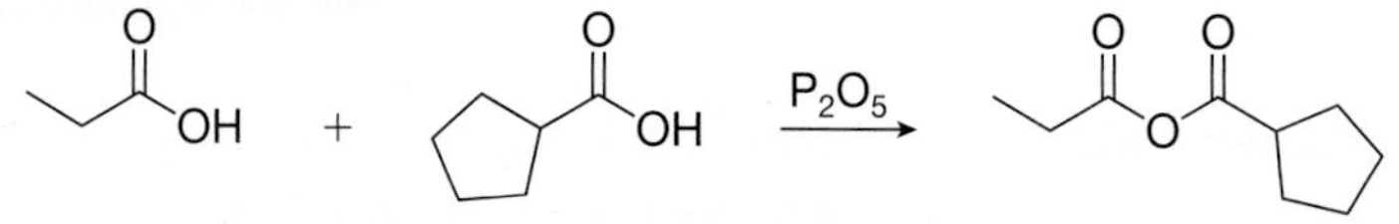

P_2O_5

STEP 1: Consider how the given reagents are likely to react.

P_2O_5 is a dehydrating agent used to make symmetrical anhydrides from a single carboxylic acid. The given reaction would yield a mixture of three products: the desired product and two symmetrical anhydrides.

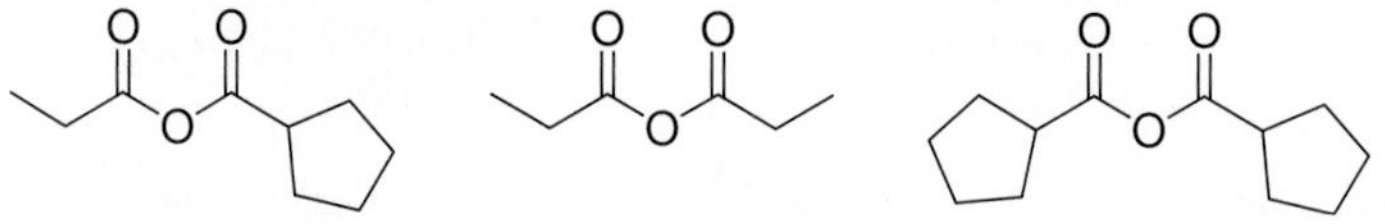

STEP 2: Consider alternative methods for synthesizing an asymmetric anhydride.

Carboxylic acid reacts with an acid chloride to make an asymmetric anhydride. The given starting materials are both carboxylic acids, so the first step is to convert one of them into an acid chloride. This conversion of an acid proceeds through a mixed anhydride formed with either thionyl chloride or oxalyl chloride.

STEP 3: Propose an alternative synthetic sequence for making the asymmetric anhydride.

Several different sequences are possible since it does not matter which carboxylic acid is converted to the acid chloride nor which reagent is used to carry out the transformation. The following is one possible sequence.

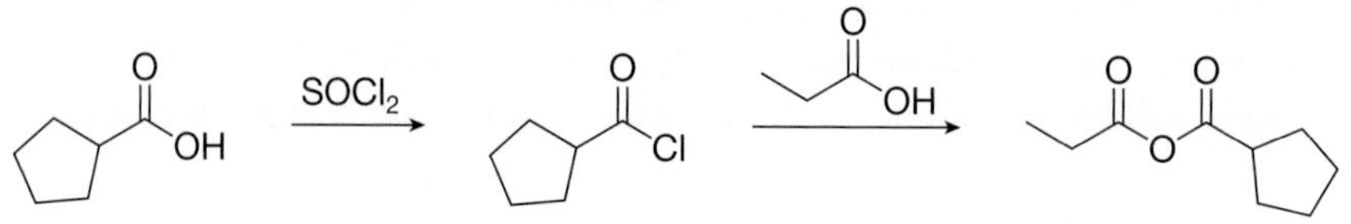

$SOCl_2$

STEP 4: Draw a mechanism for the first reaction in the sequence.

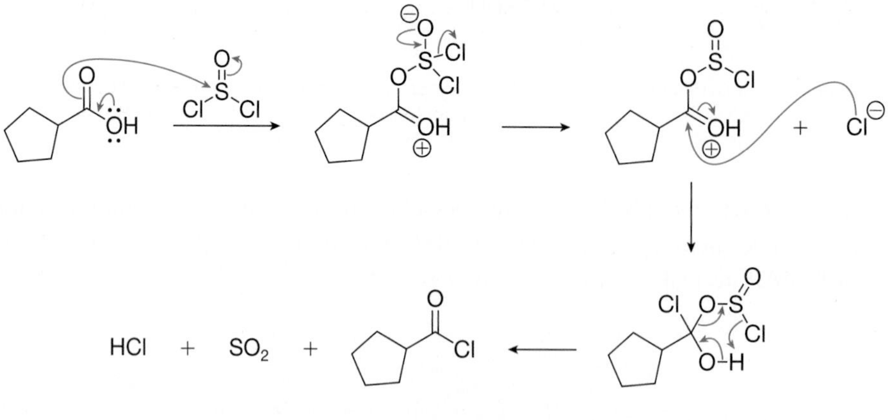

$HCl + SO_2 +$

STEP 5: Draw the mechanism for the second reaction.

The acid chloride is an excellent electrophile and the lone pairs on the carboxylic acid's OH group can function as a nucleophile. In the first step of the reaction, electrons flow from the acid to the acid chloride (the addition step). The tetrahedral intermediate quickly collapses, expelling a chloride leaving group (the elimination step). Deprotonating the anhydride gives the neutral final product.

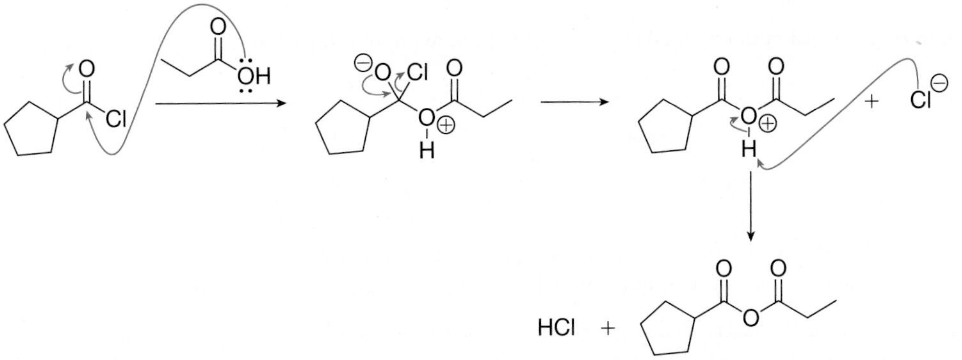

HCl +

PRACTICE PROBLEM

15.11 Draw the products of the following reactions. Draw mechanisms for the reactions in parts (b), (c), and (d).

a) 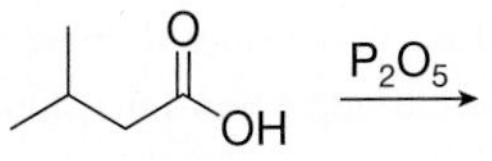 P_2O_5 (no mechanism required)

b) (*R*)-4-methylhexanoic acid $SOCl_2$

c) 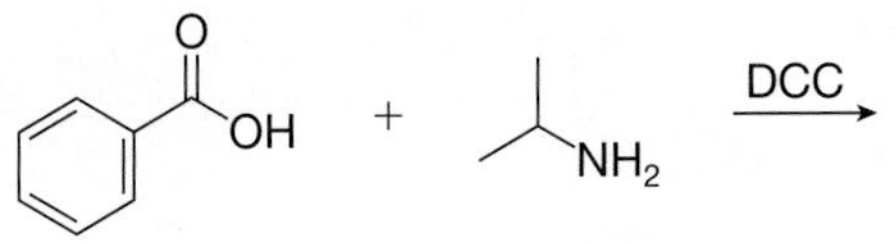 + DCC

d) 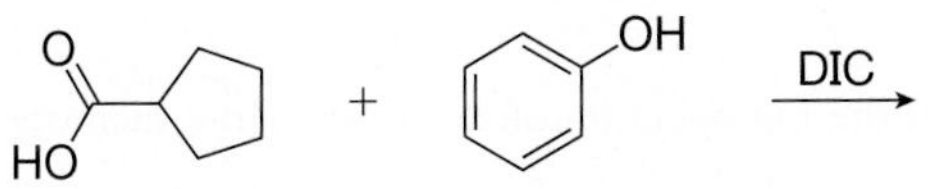 + DIC

INTEGRATE THE SKILL

15.12 a) Another reagent that can be used to catalyze carboxylic acids in the formation of amides and esters is carbonyldiimidazole (CDI). Use the mechanisms for acid chloride formation and DMAP catalysis to suggest a mechanism for the reaction of acetic acid and ethanol catalyzed by CDI. Note that the acid is mixed with CDI first, and the ethanol is added later.

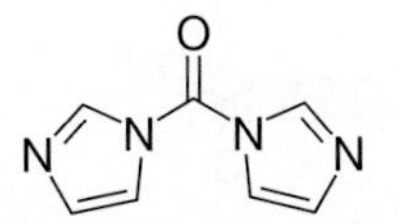

carbonyldiimidazole (CDI)

b) Propose a mechanism for the formation of an acid chloride from a carboxylic acid and oxalyl-chloride (ClCOCOCl). (Hint: The mechanism is similar to the one using thionyl chloride; the products of the reaction include CO_2, CO, and HCl.)

CHECKPOINT 15.7

You should now be able to suggest appropriate reagents and steps to convert any carboxylic acid derivative into any other carboxylic acid derivative.

SOLVED PROBLEM

Suggest reagents that could be used to carry out the following transformation.

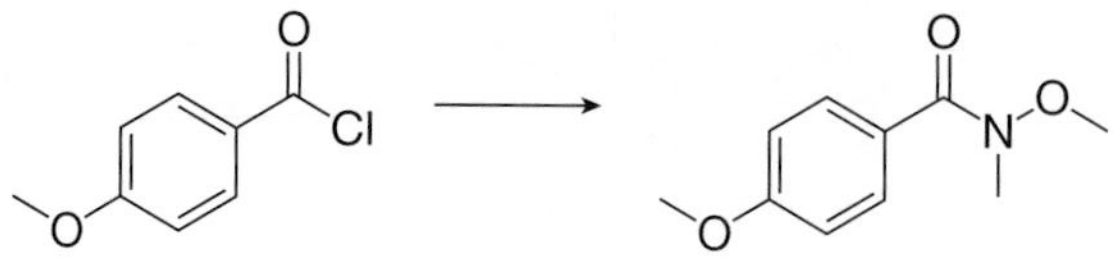

STEP 1: Identify the bonds that are broken and formed in the reaction.

A C–Cl bond has been broken and replaced with a C–N bond, making an amide from an acid chloride.

STEP 2: Determine the reactants needed for the transformation.

Since an acid chloride is a highly reactive electrophile, it does not require activation by an acid, nor does it require a negatively charged nucleophile. The chloride will react with a neutral nucleophile. The corresponding amine nucleophile for this reaction is $NH(CH_3)(OCH_3)$.

STEP 3: Consider whether a catalyst or any other reagents are needed to facilitate the transformation.

The acid chloride is reactive enough that no activation is needed for either the electrophile or nucleophile; however, the reaction between a neutral amine and acid chloride creates HCl as a by-product. To prevent this acid from reacting with the amine nucleophile and prematurely stopping the reaction, add a non-nucleophilic amine base such as pyridine or triethylamine. Therefore, the overall reaction is as follows:

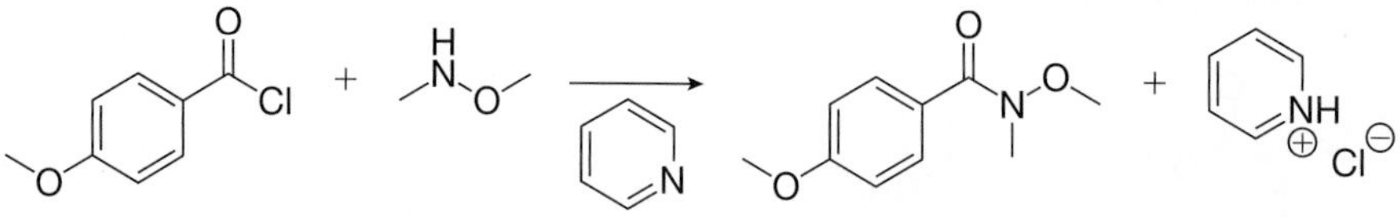

PRACTICE PROBLEM

15.13 Propose a method for carrying out each of the following transformations. More than one step may be required.

a)

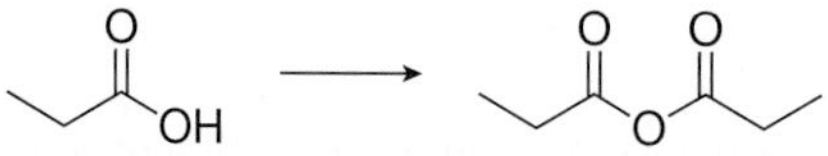

b)

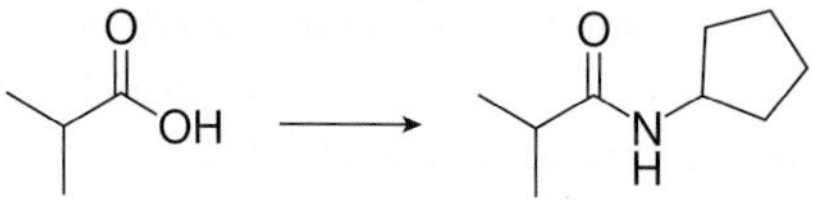

c)

d)

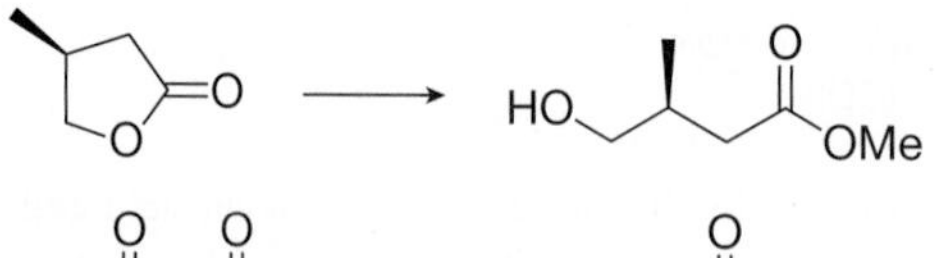

e)

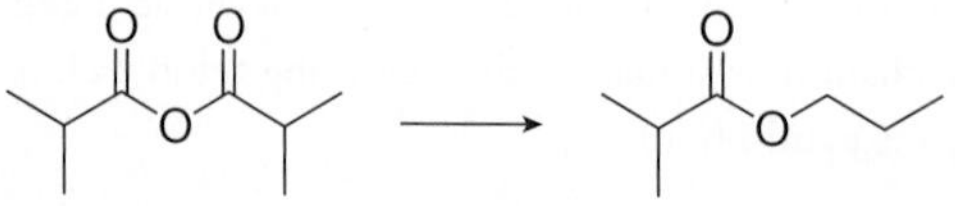

f)

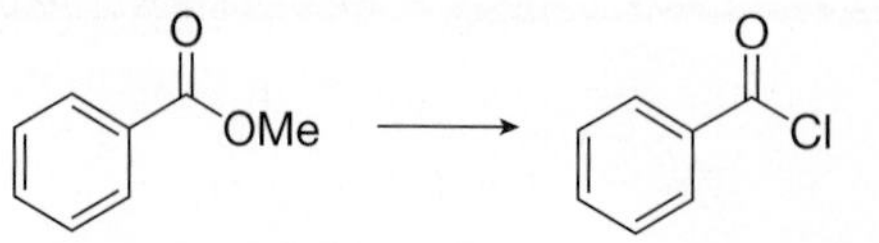

INTEGRATE THE SKILL

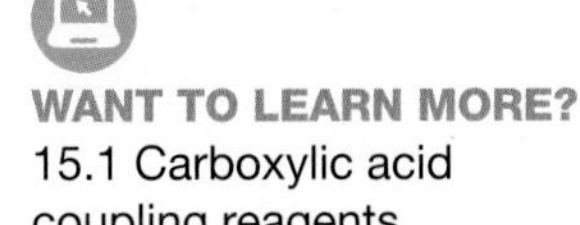

WANT TO LEARN MORE?
15.1 Carboxylic acid coupling reagents

15.14 Propose a series of reactions to transform ethynylbenzene into the ester shown here. You can use any reagents you wish, but all carbon atoms in the product must come from the ethynylbenzene.

?

ethynylbenzene

15.6 Reduction of Acid Derivatives with Nucleophilic Hydride Reagents

As described in Section 7.3.2, the nucleophilic hydride reagents lithium aluminum hydride ($LiAlH_4$, commonly abbreviated as LAH) and sodium borohydride ($NaBH_4$) reduce both aldehydes and ketones to alcohols. $LiAlH_4$ is much more reactive as a reducing agent than $NaBH_4$. This difference is not significant for reactions with aldehydes and ketones since these compounds have similar electrophilicity. However, the differing reactivities of the hydrides do affect reactions with other carbonyls, as shown in Figure 15.4. $NaBH_4$ will not react with a carbonyl that is less reactive than a ketone, but $LiAlH_4$ will reduce all the groups (amides usually require heating). $NaBH_3CN$ is typically used only to reduce iminium ions to amines in a reductive amination (Section 16.5.3).

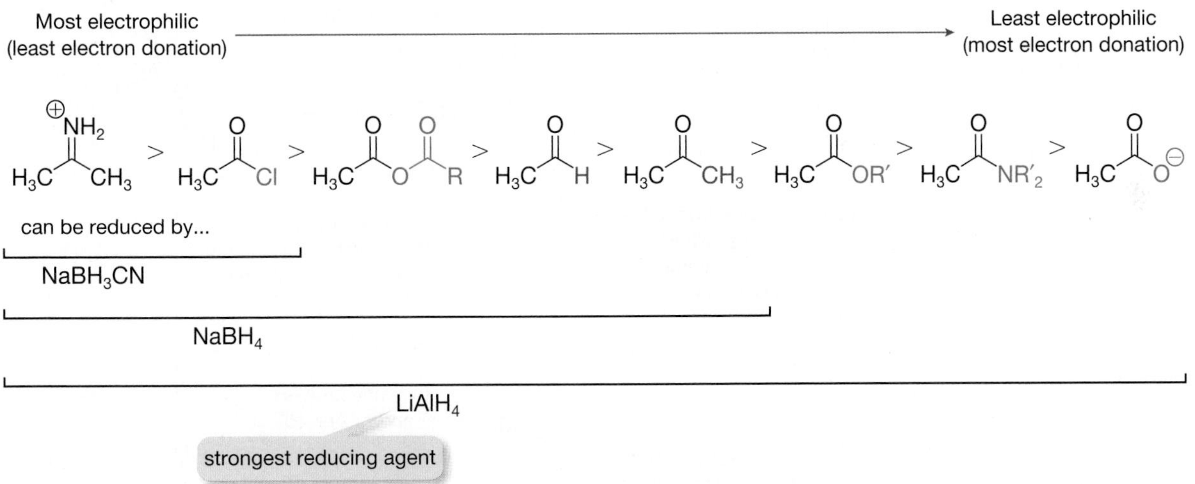

FIGURE 15.4 Order of electrophilicity of acid derivatives and related compound classes.

Nucleophilic hydrides add to carbonyl groups while breaking the π bond to form a tetrahedral intermediate. If the carbonyl has a leaving group, the addition-elimination mechanism then forms an intermediate aldehyde. The carbonyl in this intermediate reacts with the hydride reagent to form an alcohol. The overall result is the addition of two units of hydride, carrying out two reductions in one step. Such transformations normally involve a second reaction using water and acid or base to hydrolyze the oxygen-metal bond in the reduction product.

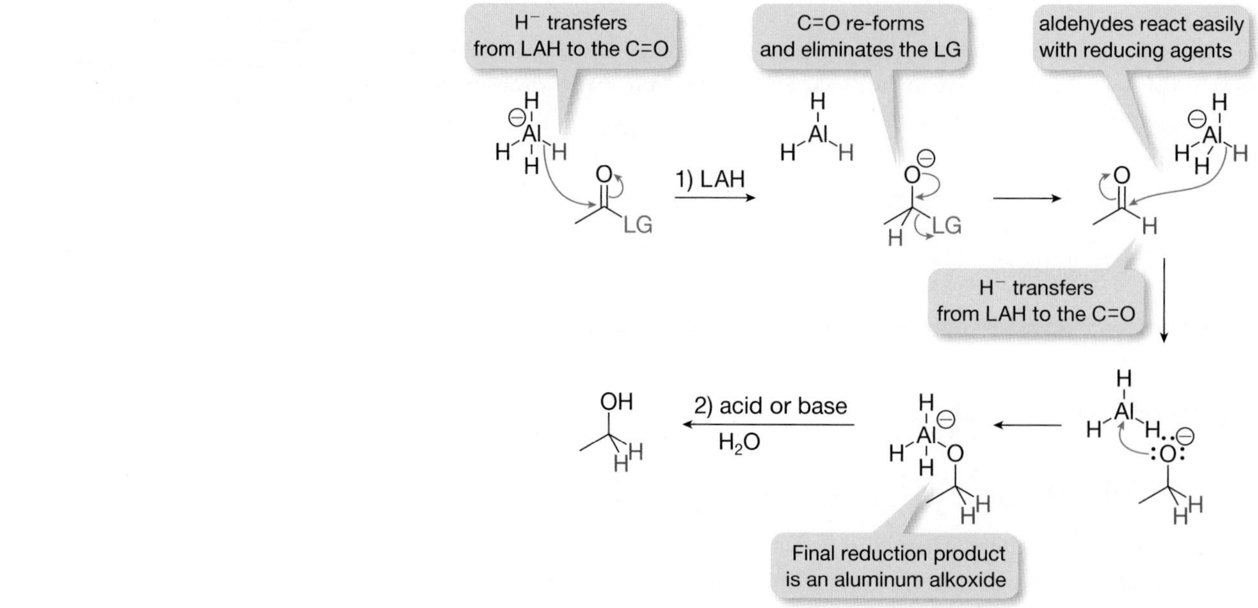

H⁻ transfers from LAH to the C=O

C=O re-forms and eliminates the LG

aldehydes react easily with reducing agents

1) LAH

H⁻ transfers from LAH to the C=O

2) acid or base

H_2O

Final reduction product is an aluminum alkoxide

The reduction of carboxylic acids with $LiAlH_4$ involves a modification of the leaving group. The hydrides in LAH are strongly basic and react quickly with the acidic hydrogen of a carboxylic acid, creating an aluminum-containing intermediate and giving off hydrogen gas. Since aluminum-oxygen bonds are very strong, the O–Al group is a good leaving group. When another hydride adds to the carbonyl, addition-elimination produces an aldehyde, which continues the overall reduction.

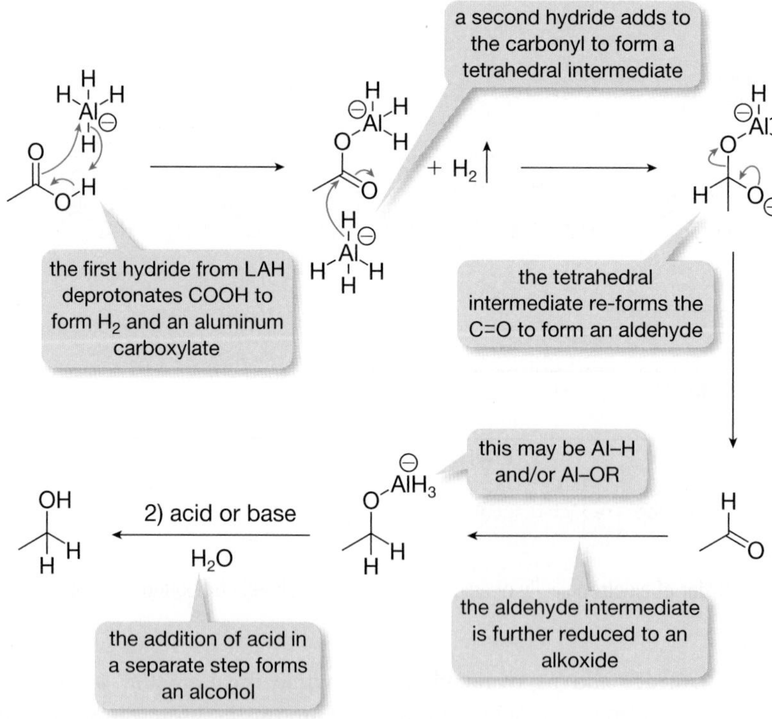

a second hydride adds to the carbonyl to form a tetrahedral intermediate

$+ H_2\uparrow$

the first hydride from LAH deprotonates COOH to form H_2 and an aluminum carboxylate

the tetrahedral intermediate re-forms the C=O to form an aldehyde

this may be Al–H and/or Al–OR

2) acid or base

H_2O

the addition of acid in a separate step forms an alcohol

the aldehyde intermediate is further reduced to an alkoxide

As the reduction reaction proceeds, Al–H bonds are replaced by Al–OR bonds. The addition of alkoxide makes aluminum hydride less reactive, but the reduction reaction continues as long as there are any remaining Al–H bonds.

In all examples so far, the carbonyl oxygen ends up as the alcohol oxygen of the product. In reactions with amides, the carbonyl oxygen is eliminated instead of the nitrogen, producing the corresponding amine. The tetrahedral intermediate formed after the first hydride addition cannot quickly collapse because the nitrogen-based leaving group is poor. Consequently, the oxygen has time to form a strong bond with the aluminum atom. Now the nitrogen is a more powerful electron donor than the oxygen, and the oxygen is a better leaving group than the nitrogen. Therefore, the tetrahedral intermediate collapses to form an iminium intermediate, which aluminum hydride rapidly reduces to an amine.

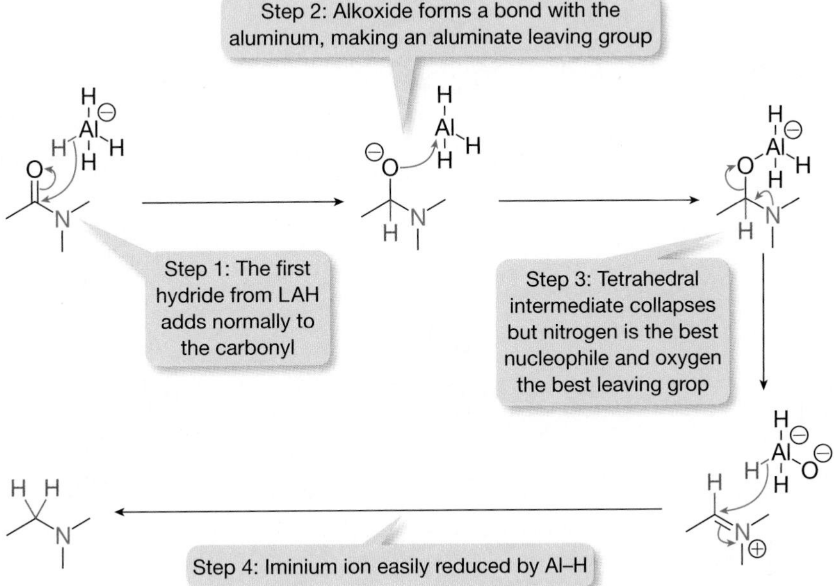

CHECKPOINT 15.8

You should now be able to determine which carboxylic acid derivatives react with nucleophilic reducing agents, such as $NaBH_4$ and $LiAlH_4$, and predict the products of these reactions. You should also be able to draw the mechanisms for these reductions.

SOLVED PROBLEM

Would you use $LiAlH_4$ or $NaBH_4$ as the reducing agent for the following reaction? Explain your selection, and draw a mechanism for the reaction.

?

STEP 1: Compare the reactivity of the functional group to the ranking in Figure 15.4.

$NaBH_4$ can react only with carbonyl groups that are at least as reactive as a ketone. Since esters are less reactive than ketones, this reaction could not be carried out with $NaBH_4$; $LiAlH_4$ is needed.

STEP 2: Determine the steps needed for the transformation.

Since $LiAlH_4$ is highly reactive, it needs to be used in an aprotic solvent such as THF. However, an acidic aqueous work-up is needed to hydrolyze the O–Al bonds at the end of the reaction. The overall reaction is therefore as follows:

1) $LiAlH_4$, THF
2) HCl, H_2O

STEP 3: Draw a mechanism for the reaction.

The nucleophilic hydride adds to the electrophilic ester, making a tetrahedral intermediate. The tetrahedral intermediate collapses, eliminating a methoxide leaving group and forming an aldehyde intermediate. Since aldehydes are even more reactive than esters, the aldehyde quickly reacts with a second equivalent of $LiAlH_4$, forming an alkoxide intermediate, which then attacks the AlH_3 by-product to form an aluminum alkoxide. The reaction then stops until the aqueous acid is added to hydrolyze the Al–O bond.

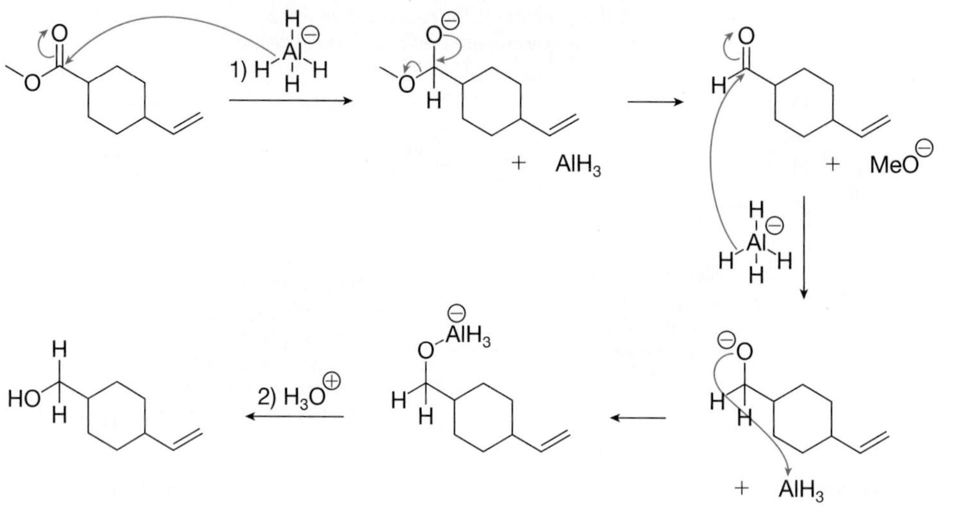

1) $H-AlH_3^{\ominus}$ + AlH_3 + $MeO^{\ominus}$ $H-AlH_3^{\ominus}$ + AlH_3 $^{\ominus}O-AlH_3$ 2) $H_3O^{\oplus}$ HO

PRACTICE PROBLEM

15.15 For each of the following reactions, predict the product and draw a mechanism for its formation.

a) 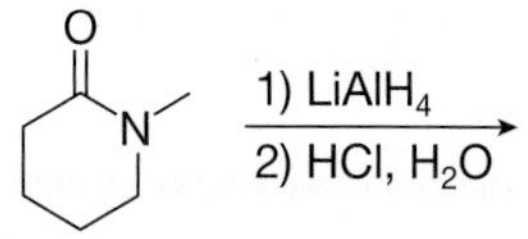 1) $LiAlH_4$ 2) HCl, H_2O

b) 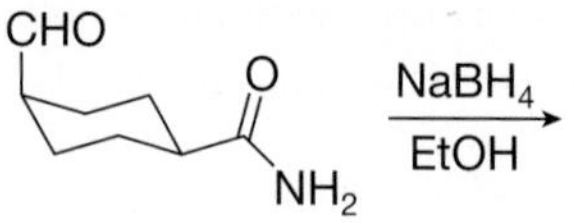 CHO, NH_2 — $NaBH_4$, EtOH

c) 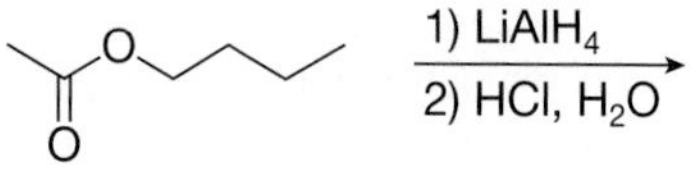 1) $LiAlH_4$ 2) HCl, H_2O

d) 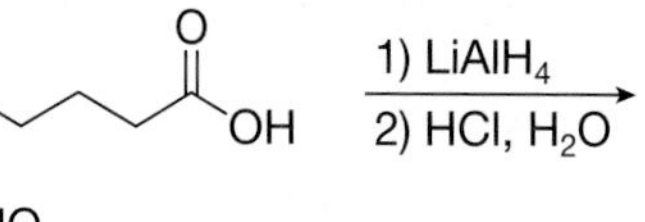 OH — 1) $LiAlH_4$ 2) HCl, H_2O

e) 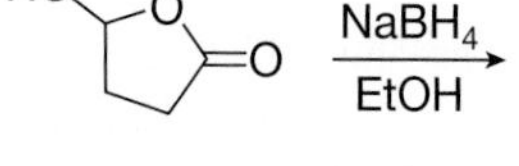 HO — $NaBH_4$, EtOH

INTEGRATE THE SKILL

15.16 Propose a series of reactions to carry out the following transformations.

a) 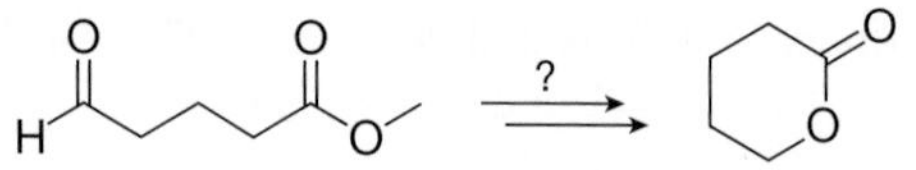 ?

b) 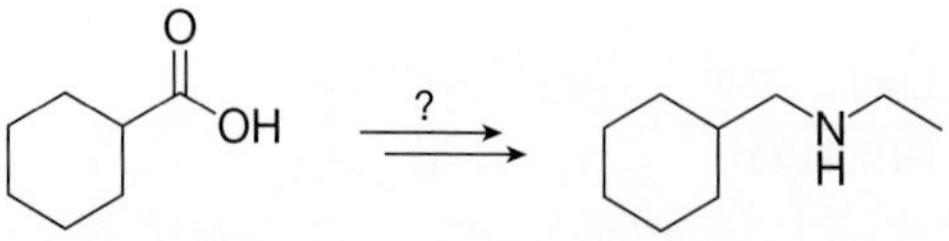 OH ?

15.7 Selectivity with Electrophilic Reducing Agents

Electrophilic reducing agents force acid derivatives to act as nucleophiles. The relative reactivities of acid derivatives in these reactions are the reverse of those for reactions with nucleophilic reducing reagents. The simplest electrophilic reducing reagent is borane (BH_3), which exists as a dimer (diborane) unless dissolved in a nucleophilic solvent, such as ether. Borane and other electrophilic hydride sources, such as diisobutylaluminum hydride (DIBAL) or another dialkyl boranes can selectively reduce some types of carbonyl compounds.

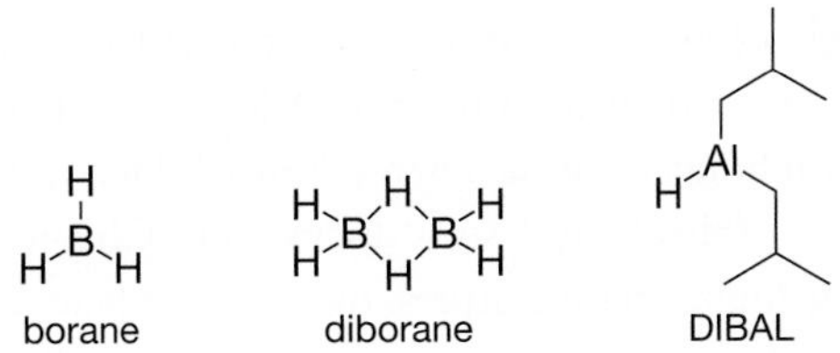

borane diborane DIBAL

Because boron has only six valence electrons, BH_3 is an electrophilic reagent that will react with nucleophiles. The carbonyl oxygen of acid derivatives has a significant negative charge due to the resonance forms discussed in Section 15.3. Functional groups that are highly stabilized by conjugation react quickly with reagents such as BH_3.

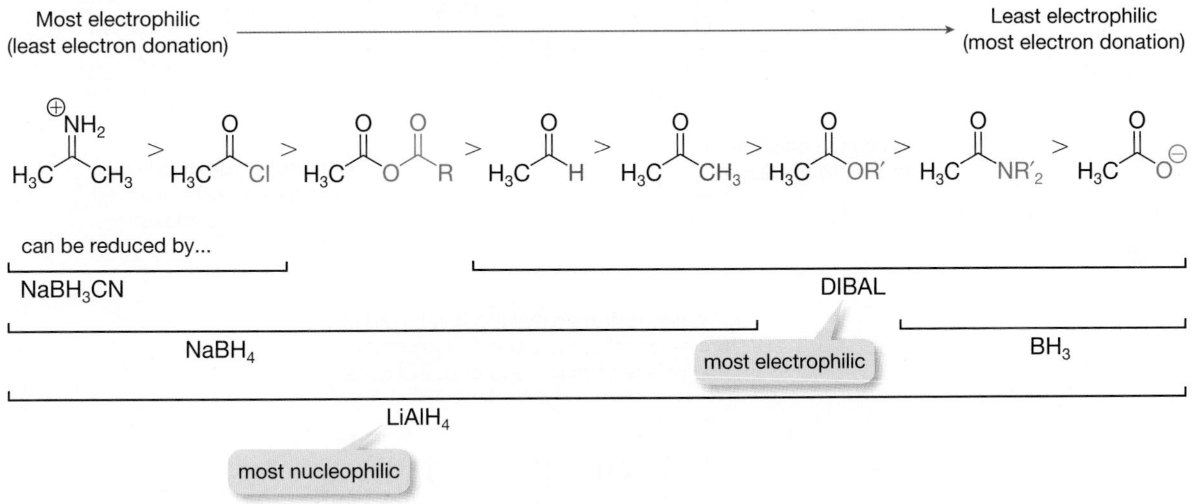

Most electrophilic (least electron donation) → Least electrophilic (most electron donation)

can be reduced by...

NaBH$_3$CN

DIBAL

NaBH$_4$

most electrophilic

BH$_3$

LiAlH$_4$

most nucleophilic

FIGURE 15.5 Reactivity of nucleophilic reducing agents and electrophilic reducing agents toward various carbonyl groups.

The opposite reactivities of nucleophilic and electrophilic reagents provide a way to selectively reduce some carbonyl groups in the presence of others. Carbonyls that are *less* reactive toward $LiAlH_4$ and $NaBH_4$ are *more* reactive toward BH_3. Similarly, carbonyls that are *more* reactive toward $LiAlH_4$ and $NaBH_4$ are *less* reactive toward BH_3.

1) $NaBH_4$
2) $H_3O^{\oplus}$
nucleophilic $NaBH_4$ reduces electrophilic carbonyls
1) B_2H_6, ether
2) $H_3O^{\oplus}$
electrophilic BH_3 reduces nucleophilic carbonyls

The nucleophilic oxygen of an amide carbonyl rapidly adds to borane, forming a charged nucleophilic species, which can reduce the nearby iminium ion into a tetrahedral intermediate. If the reaction is performed at low temperature, this intermediate is stable because the adjacent nitrogen donates electrons to form an iminium ion. Adding water to the reaction at this point hydrolyzes the oxygen-boron bond, creating a hemiaminal, which quickly hydrolyzes to produce an aldehyde. The mechanism of this hydrolysis is discussed in Chapter 16.

If the reaction is run at higher temperatures, the intermediate transforms into an iminium ion by expelling the oxygen leaving group. Another molecule of borane then reduces the iminium to an amine.

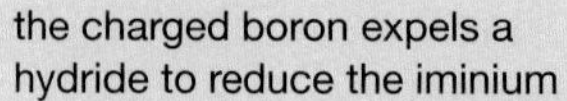
the charged boron expels a hydride to reduce the iminium

the carbonyl complexes to and activates the boron
the resulting boron complex is stable until either warmed or water is added

water destroys remaining borane and prevents further reduction as it hydrolyzes the borane complex to form an aldehyde
H_2O
hemiaminal
heat
heat results in iminium formation, which will lead to further reduction

Diisobutylaluminum hydride forms tetrahedral intermediates with amides and esters. At very low temperatures, the intermediate complex is stable and can be converted to the aldehyde by adding water. However, the structure of the compound being reduced can make the reaction very difficult to perform because the strained four-membered ring may open too easily.

water destroys remaining DIBAL and prevents further reduction as it hydrolyzes the aluminum complex to form an aldehyde

DIBAL

−78 °C

H_2O

hemiacetal

heat

the aluminum complex prevents further reduction until warmed or water is added

heat breaks the O–Al bond, forming a hemiacetal that collapses to an aldehyde without inactivating remaining DIBAL

Borane reacts vigorously with carboxylic acids to form hydrogen gas.

leaving group

Since the boron in the carbonyl product lacks an octet, it reduces the electron density of the oxygen to which it is connected. The change in electron distribution prevents this oxygen from participating in conjugation with the carbonyl, altering the reactivity of the group by creating an active leaving group. Thus the carbonyl is easily reduced by additional BH_3 to form the corresponding alcohol.

CHECKPOINT 15.9

You should now be able to determine which carboxylic acid derivatives react with electrophilic reducing agents, such as BH_3 and DIBAL, and predict the products of these reactions. You should also be able to draw the mechanisms for these reductions.

SOLVED PROBLEM

The reaction between borane and methyl cyclopentanecarboxylate yields different products depending on the reaction conditions. Predict the products that are formed from each set of conditions, and draw mechanisms that account for the different products.

1) DIBAL, −78 °C, then warm to 20 °C
2) HCl, H_2O

1) DIBAL, −78 °C
2) HCl, H_2O

STEP 1: Determine the type of reaction that is taking place.

DIBAL stands for diisobutylaluminum hydride, an electrophilic reducing agent. The reactant is an ester, which can be reduced by DIBAL to yield either an aldehyde or an alcohol, depending on the reaction conditions.

STEP 2: Predict the products of the reactions.

Under cold conditions, DIBAL reduces esters to their corresponding aldehydes. At room temperature, the reaction goes further, and the ester is reduced all the way to the corresponding alcohol. So the predicted products are as follows:

1) DIBAL, −78 °C, then warm to 20 °C
2) HCl, H_2O

1) DIBAL, −78 °C
2) HCl, H_2O

STEP 3: Draw mechanisms for the reactions.

The nucleophilic lone pairs on the ester attack the electron-deficient aluminum of DIBAL. The resulting intermediate has a negative charge on the aluminum; this nucleophilic hydride then attacks the electrophilic oxonium ion. The product of this hydride addition is stable at low temperatures owing to the neighbouring oxygen donating a lone pair to the aluminum atom.

AlH(*i*-Bu)$_2$

stable complex at cold temperatures

If the reaction is carried out at low temperatures, the reaction does not continue past the aluminum complex, as shown in the preceding diagram. When aqueous acid is added in the second step, any excess DIBAL is destroyed, and the Al–O bonds are hydrolyzed, producing the corresponding hemiacetal in equilibrium with its corresponding aldehyde (see Chapter 7).

H_3O^{+}

hemiacetal

aldehyde

+ MeOH

At higher temperatures, the cyclic intermediate is unstable and rapidly decomposes to give an aldehyde by expelling aluminum methoxide as a leaving group. Since there is still DIBAL present in solution, the aldehyde intermediate is quickly reduced by another molecule of DIBAL following the same mechanism as before. After the addition of aqueous acid, the Al–O bonds are hydrolyzed, producing the alcohol.

PRACTICE PROBLEM

15.17 Predict the products of the following reactions.

INTEGRATE THE SKILL

15.18 Suggest appropriate reagents and conditions to carry out the following transformations. Draw a mechanism for each reaction.

15.8 Multiple Addition of Organometallic Reagents to Acid Derivatives

Organometallic reagents such as Grignard reagents generally add twice to a carboxylate derivative because the initial tetrahedral intermediate expels the heteroatom leaving group to give a ketone, which is susceptible to Grignard addition. For example, the addition of excess phenylmagnesium bromide to ethyl acetate produces the tertiary alcohol as the product.

1) BrMg–Ph
2) $H_3O^{\oplus}$
OH
Ph
Ph

The intermediate in this process is the expected ketone. The presence of additional Grignard reagent results in a second addition of phenyl to form a tertiary alkoxide.

ketone undergoes second addition to make an alkoxide
1) excess BrMg–Ph
H_3C O Et
H_3C Ph
BrMg O
H_3C Ph Ph
2) $H_3O^{\oplus}$
OH
H_3C Ph Ph
addition-elimination with first equivalent produces a ketone
hydrolyzing magnesium alkoxide produces an alcohol

Because the ketone intermediate is more electrophilic than the ester starting material, it is not possible to stop the reaction at the ketone stage. If less than two equivalents of Grignard are used, the reaction will produce mainly the alcohol, leaving some of the ester unchanged.

Weinreb amide is an amide with a coordinating group that blocks a second addition of Grignard reagents; it is often used to synthesize ketones.

One way to prevent a second addition of Grignard reagent is to use a **Weinreb amide** (*N*-methyl-*N*-methoxy amide) in place of the starting ester. Once the first Grignard addition occurs, the methoxy oxygen of the amide is located in a perfect position to donate a pair of electrons to the magnesium, thereby forming a complex that does not allow the tetrahedral intermediate to collapse. Adding aqueous acid destroys any excess Grignard reagent and hydrolyzes the complex to a hemiaminal, which then falls apart into a ketone and amine.

O N O
PhMgBr
−78 °C
Br Mg O Ph N O
$H_3O^{\oplus}$
OH Ph N O H
$H_3O^{\oplus}$
Ph O
stable at low temperature
unstable hemiaminal

CHECKPOINT 15.10

You should now be able to predict the products formed when carboxylic acid derivatives are reacted with various organometallic reagents. You should also be able to draw the mechanisms for these reactions.

SOLVED PROBLEM

Predict the major product of the following reaction, and draw a mechanism for its formation.

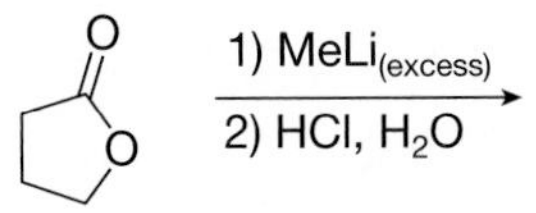

1) $MeLi_{(excess)}$
2) HCl, H_2O

STEP 1: Determine the roles of the reactants.

The highly electropositive lithium atom in methyllithium creates a polarized C–Li bond, making the carbon atom very nucleophilic. The ester, on the other hand, has a polarized carbonyl group with a partial positive charge on the carbon atom, making it a good electrophile.

STEP 2: Draw the mechanism for the addition of the first equivalent of methyllithium.

The methyllithium adds to the carbonyl, and electrons flow to the more electronegative O atom. The result of this addition is a tetrahedral intermediate, which quickly eliminates an alkoxide leaving group and re-forms the carbonyl group.

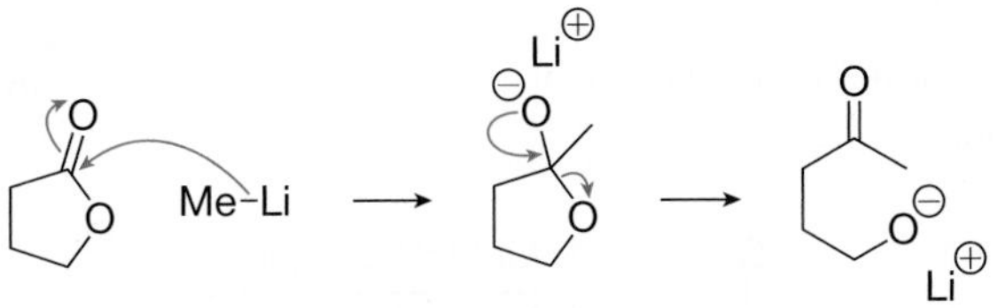

Me–Li

STEP 3: Determine the roles of the intermediates and remaining reactants.

The intermediate produced by the first addition-elimination sequence is a ketone. Since ketones are more electrophilic than esters, the intermediate quickly reacts with another equivalent of methyllithium.

STEP 4: Draw a mechanism for the addition of the second equivalent of methyllithium. Again, the methyllithium adds to the carbonyl, and electrons flow to the O atom. The resulting intermediate lacks a leaving group, so no further reaction takes place.

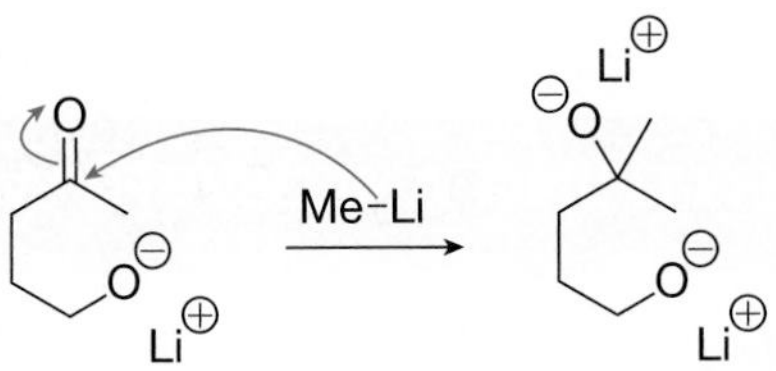

Me–Li

STEP 5: Determine the final product.

The HCl and water added in the second step combine to make hydronium ions, which protonate the two alkoxides.

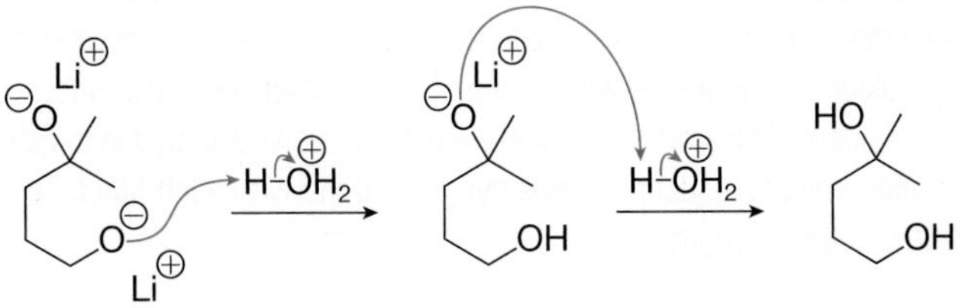

$H–OH_2^{\oplus}$ $H–OH_2^{\oplus}$

PRACTICE PROBLEM

15.19 Predict the products of the following reactions, and draw mechanisms for their formation.

a) 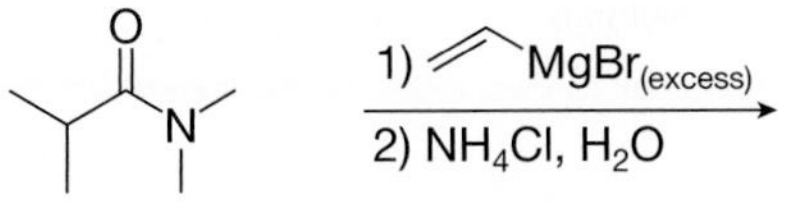 1) $MgBr_{(excess)}$
2) NH_4Cl, H_2O

b) 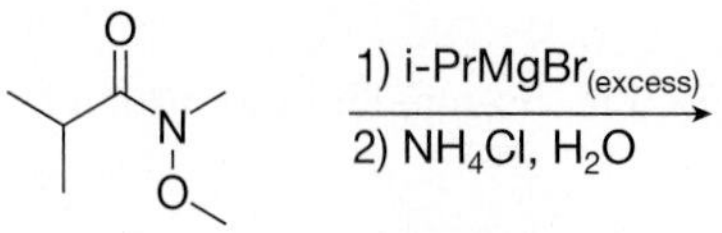 1) i-$PrMgBr_{(excess)}$ 2) NH_4Cl, H_2O

c) 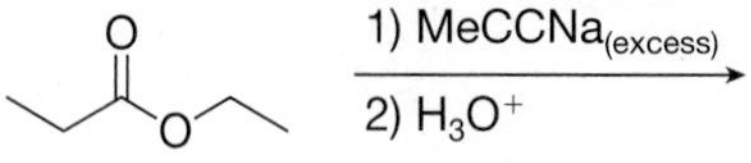 1) $MeCCNa_{(excess)}$ 2) H_3O^+

d) 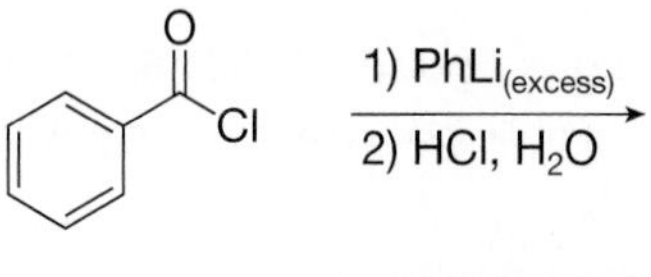 1) $PhLi_{(excess)}$ 2) HCl, H_2O

e) 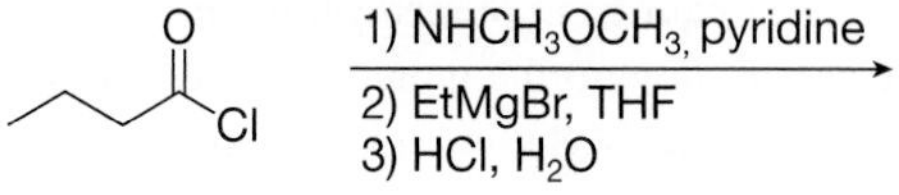 1) $NHCH_3OCH_3$, pyridine 2) EtMgBr, THF 3) HCl, H_2O

INTEGRATE THE SKILL

15.20 a) Propose a mechanism for the following transformation, and explain why the product shown is the major product.

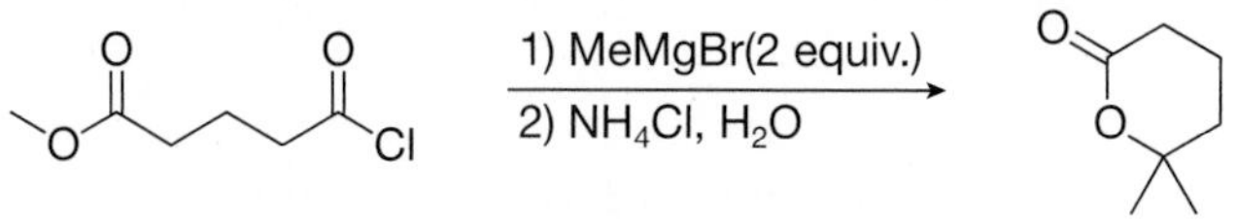

1) MeMgBr(2 equiv.) 2) NH_4Cl, H_2O

b) Would the reaction in part (a) work if the starting material contained two esters instead of an acid chloride and an ester? Why or why not?

c) What would the product of the reaction be in part (a) if four molar equivalents of the Grignard reagent were used?

CHEMISTRY: EVERYTHING AND EVERYWHERE

Nylon: Fabric from Amides

Nylons are a family of extremely strong materials that were engineered as synthetic versions of silk. Nylons are polymers, very large molecules formed from small repeating units joined together in a long chain. The size of the molecules gives these synthetic materials strength and flexibility. The amide bonds that hold the chains together are also a key determinant of the properties of nylons.

Synthetic carpet fibres are often made from nylon-6, which is based on a lactam called caprolactam. Mixing caprolactam with a small amount of acid and water causes a cascading series of reactions. The water opens one of the lactam molecules making a free amine, which then opens another molecule of lactam. This process forms chains of thousands of amide units. The amide groups in each chain are all oriented in the same direction, resulting in a structure in which amide units in adjacent chains hydrogen bond with each other. The vast number of these hydrogen bonds contributes to the strength of nylon.

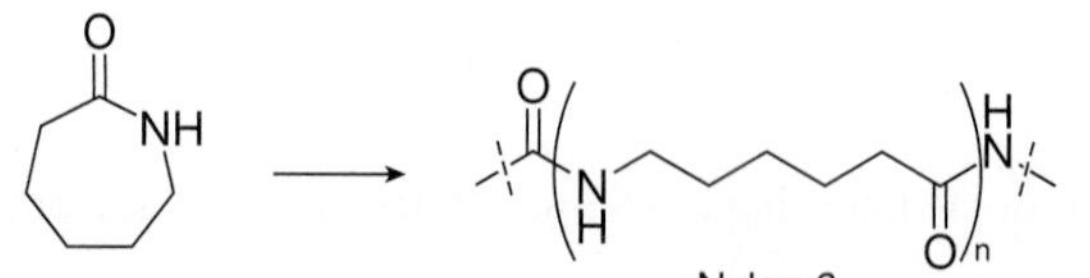

Nylon 6

15.9 The Aromatic Ring as an Electrophile

Electrophilic π bonds that have a leaving group can be found in many structures, including aromatic rings. Halogen-substituted aromatic rings that carry strong electron-withdrawing groups make possible the reactions known as **nucleophilic aromatic substitution** (S_NAr). These reactions proceed by the addition-elimination mechanism and have a tetrahedral intermediate that is similar to the one formed in carbonyl substitution reactions.

Nucleophilic aromatic substitution reactions exchange a nucleophile for a leaving group on an aromatic ring.

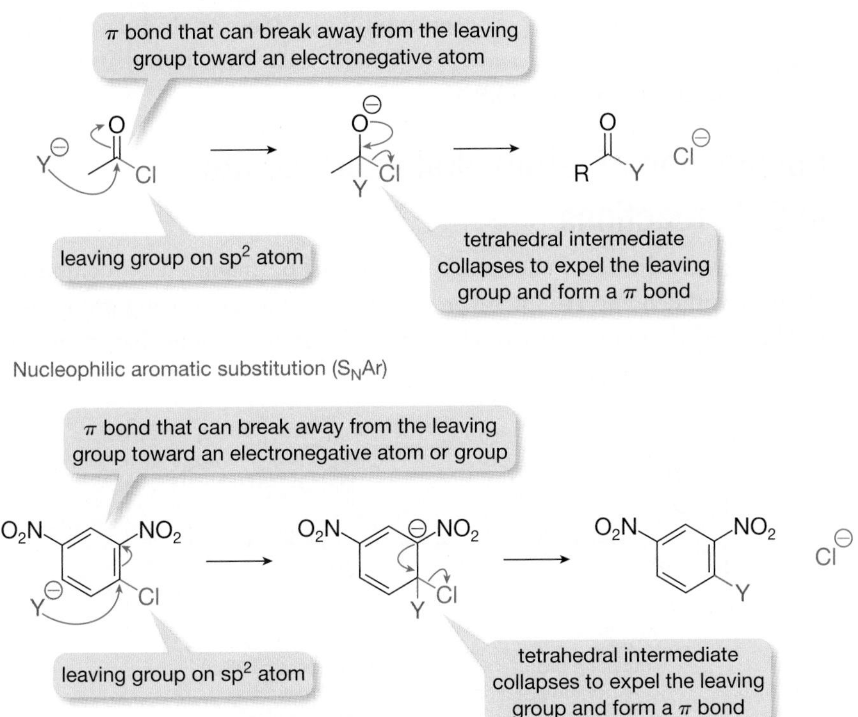

An S_NAr reaction requires both a strong electron-withdrawing activating group and a leaving group on the aromatic ring. Because a negative charge will reside *beside* the electron-withdrawing group in the tetrahedral intermediate, the group must pull very strongly to stabilize the charge. F, NO_2, CN, or carbonyl groups are the most common activating groups for S_NAr reactions.

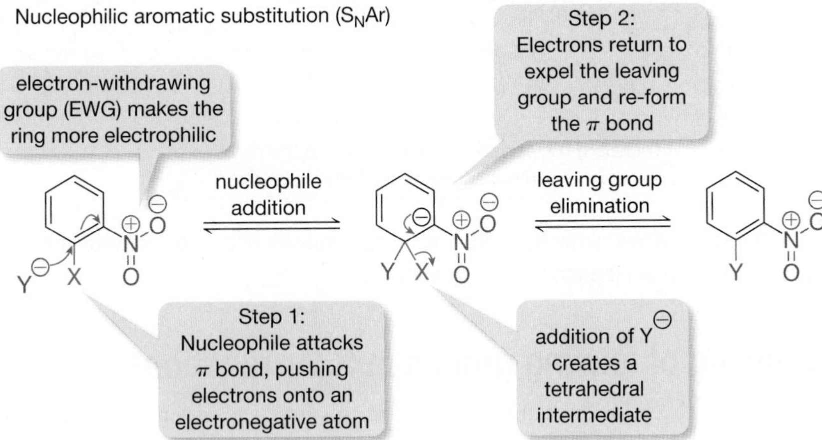

NO_2 and CN groups are excellent electron-withdrawing groups because they stabilize the negatively charged intermediate through electron delocalization.

stabilization of carbanion by nitro group

stabilization of carbanion by cyano group

Details of this general reaction are discussed in the following sections, including the effect of electron-withdrawing group conjugation on substitution (Section 15.9.1), the effect of different leaving groups (Sections 15.9.2 and 15.9.3), and a related reaction that does not require electron-withdrawing groups (Section 15.9.4).

15.9.1 Conjugation of tetrahedral intermediates in S_NAr reactions

The negative charge on the tetrahedral intermediate can be stabilized by conjugation if the activating electron-withdrawing groups are *ortho* and/or *para* to the leaving group, as shown in Figure 15.6. The strength and number of electron-withdrawing groups that are present determine how readily the reaction proceeds.

Activating group (NO_2) increases the electrophilicity of the reaction site through resonance.

Activating group cannot co-operate with the leaving group and the substrate reacts poorly.

There are no resonance forms in which the positive charge is directly beside the leaving group.

FIGURE 15.6 Electron-withdrawing groups with *para*-substitution (top) activate the reaction with nucleophiles. *Meta*-substitution (bottom) does not work well.

STUDENT TIP

To determine if an S_NAr reaction is possible, look for leaving groups *ortho* or *para* to a strong electron-withdrawing group.

15.9.2 Ranking of leaving groups in S_NAr reactions

For the displacement reactions encountered so far in this text, the order of leaving group ability is typically $I^- > Br^- > Cl^-$. Fluoride is the only strong conjugate base of the halogens and does not usually participate as a leaving group. This order of leaving group ability is *reversed* for S_NAr

reactions. For these reactions the *least* basic leaving group is the most effective because the order of events is reversed relative to other displacements such as S_N1 or S_N2.

In alkyl substitution reactions (S_N1 and S_N2), the rate-determining step is the one in which the carbon-leaving group bond is broken. This step happens faster with leaving groups that leave more easily. In nucleophilic aromatic substitution reactions (S_NAr), the rate-determining step is the one in which the nucleophile adds to the ring to form a tetrahedral intermediate. This addition disrupts the aromaticity in the ring, which makes this step difficult and rate-determining. The loss of the leaving group in the second step restores the aromatic bonds, causing a large decrease in energy, which makes the second step very fast.

Leaving-group ability is therefore not important in S_NAr reactions. What *is* important is the positive charge induced on the atom connected to the leaving group. Very electronegative atoms and strong electron-withdrawing groups induce significant positive charges, making the compound reactive with nucleophiles. The order of reactivity for halogen leaving groups in S_NAr reactions is therefore $F^- >> Cl^- > Br^- > I^-$. Since fluorine is the most electronegative halogen, it induces the largest positive charge on the reactive ring carbon, and this makes the rate of the nucleophile approach as much as 10 000 times greater than the rate with chlorine.

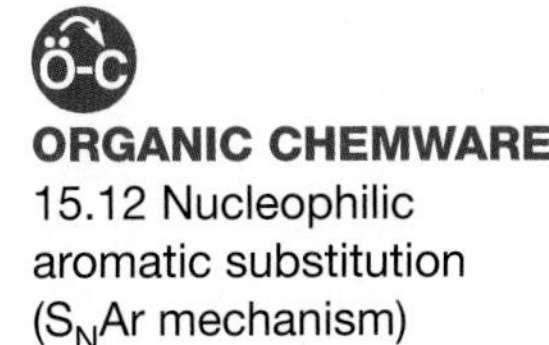

ORGANIC CHEMWARE
15.12 Nucleophilic aromatic substitution (S_NAr mechanism)

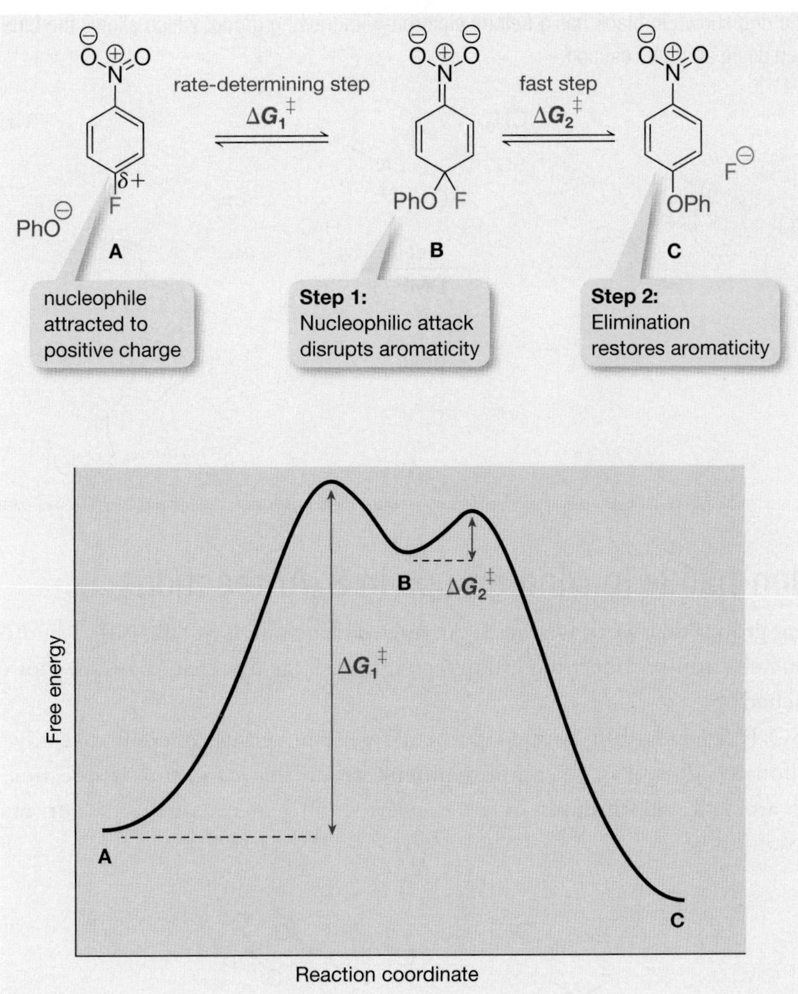

FIGURE 15.7 The reaction energy diagram shows the rate-determining step of an S_NAr reaction does not include the loss of the leaving group.

CHEMISTRY: EVERYTHING AND EVERYWHERE

Anti-estrogen Therapy

Raloxifene, marketed as Evista by Eli Lilly and Company, has been found to reduce the incidence of some types of breast cancer. The estrogen receptor, found in the nucleus of a cell, is a key part of the signalling that leads to cell division. In some types of tumour cells, there is a higher-than-normal concentration of estrogen receptor, which leads to an increased rate of cell division. The part of the raloxifine molecule that has two aromatic alcohols (shown here in red) binds to estrogen receptors, thereby preventing the proteins needed for cell division from binding. In this way, the drug interrupts the cell division cycle and slows tumour growth.

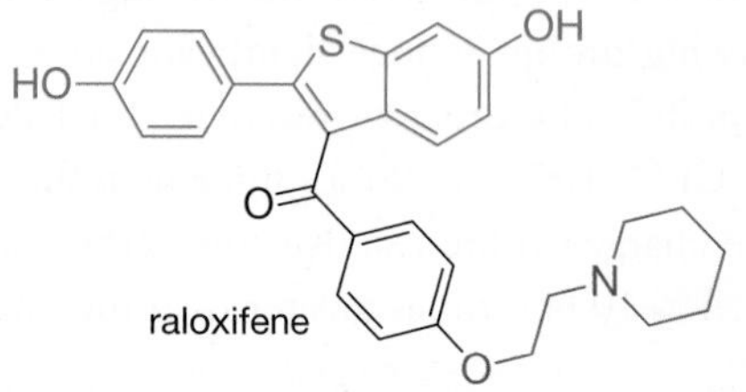

raloxifene

The aromatic ring shown in black has a ketone electron-withdrawing group, which allows the blue side-chain to be introduced using an S_NAr reaction.

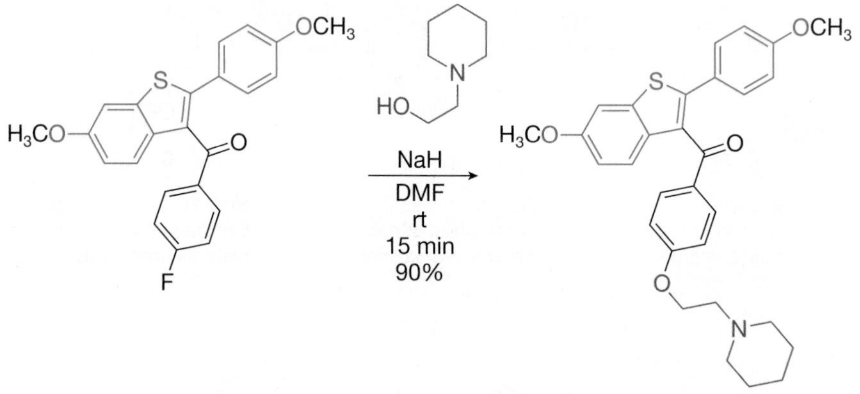

NaH, DMF, rt, 15 min, 90%

15.9.3 Non-halide leaving groups in S_NAr reactions

Other leaving groups that work well in S_NAr reactions include OSO_2R, NR_3^+ and NO_2^-. All of these strong electron-withdrawing groups can induce large $\delta+$ charges on the atom to which they are attached.

The OSO_2R leaving group family is particularly useful because phenols are easily converted into aryl sulfonates. Phenol is a versatile synthetic group that can act as a directing group in electrophilic aromatic substitutions or be converted into a leaving group for nucleophilic aromatic substitutions.

o, p-directing for S_EAr

leaving group for S_NAr

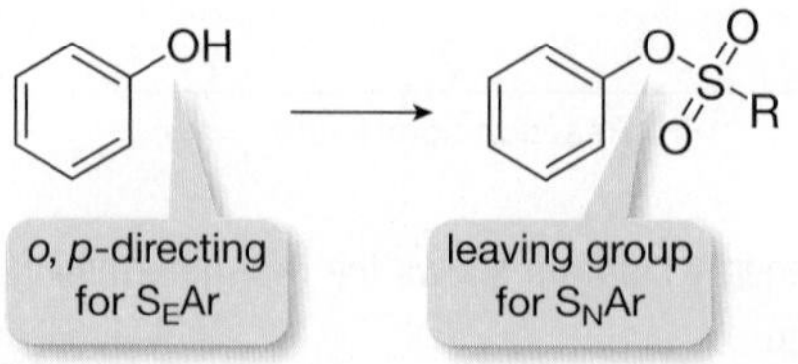

CHECKPOINT 15.11

You should now be able to recognize structures that are capable of undergoing nucleophilic aromatic substitution reactions and rank their reactivities. You should also be able to draw the products and mechanisms of these reactions.

SOLVED PROBLEM

Arrange the following chloronitrobenzenes in order of decreasing reactivity in S_NAr reactions. Draw a mechanism for the reaction of the most reactive chloronitrobenzene with sodium methoxide.

NO_2 NO_2 Cl **A** | NO_2 Cl **B** | NO_2 NO_2 Cl **C** | NO_2 Cl **D**

STEP 1: Sort the aromatics according to the number of activating groups they have.

Structures A and C have two strongly activating nitro groups, whereas structures B and D have only one.

STEP 2: Label the positions of the activating groups as *ortho*, *meta*, or *para* relative to the leaving group.

The leaving group in all four structures is a chlorine atom, so the positions of the nitro groups relative to the chlorine atom need to be classified.

Structure A: *ortho* and *meta*

Structure B: *meta*

Structure C: *ortho* and *para*

Structure D: *ortho*

STEP 3: Rank the structures based on the number and position of the nitro activating groups.

Since activating groups in the *meta* position are less activating than ones in the *ortho* and *para* positions, structures with *meta* nitro groups will be less reactive toward S_NAr. Structure C will be the most reactive because it has two activating groups: one *ortho* and the other *para*. Next is structure A, with its two activating groups, only one of which is in an *ortho* or *para* position. Structures B and D are less reactive because they each have only one activating group. Structure B is the least reactive since it has a *meta* nitro group, while structure D has an *ortho* activating group.

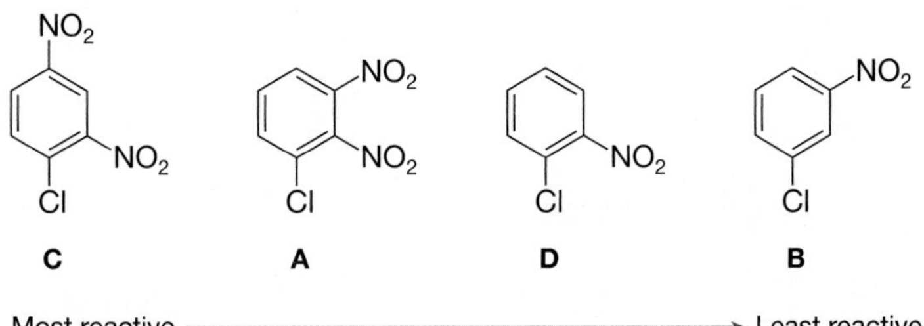

STEP 4: Write out the overall reaction equation for the reaction of structure C with sodium methoxide.

An S_NAr reaction replaces the leaving group on the aromatic ring with a nucleophile. Here, the leaving group is chlorine, and the nucleophile is methoxide. Therefore the overall reaction is as follows:

NO_2 NO_2 Cl —NaOMe→ NO_2 NO_2 O + NaCl

STEP 5: Draw the mechanism for the S_NAr reaction.

The methoxide nucleophile adds to the electrophilic carbon bearing the chlorine leaving group; this causes electrons from the π bond to flow toward the electron-withdrawing nitro group. In the second step, the negative charge moves back toward the carbon with the leaving group, breaking the C–Cl bond and eliminating a chloride ion.

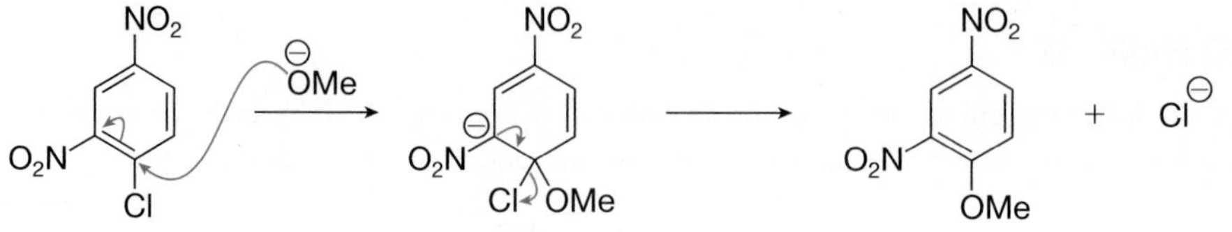

PRACTICE PROBLEM

15.21 a) For each pair of compounds, select the one that is the most reactive toward S_NAr. Justify your selection.

i)

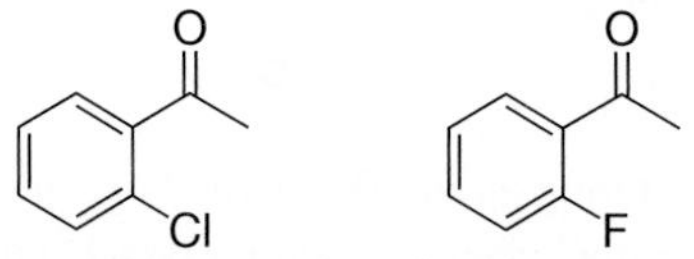

ii)

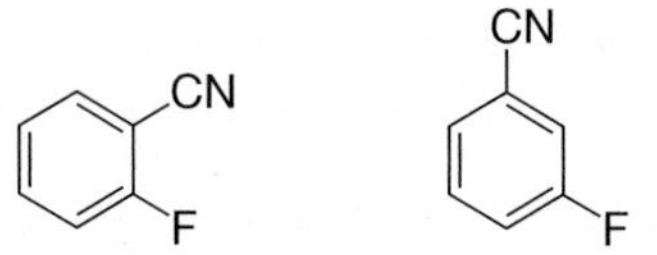

iii)

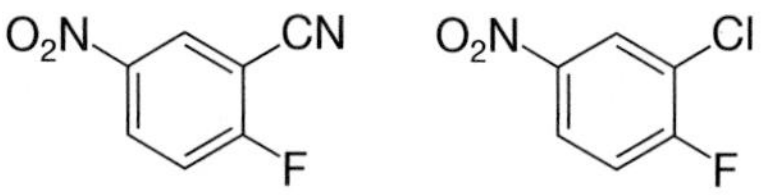

b) Predict the products of the following reactions.

i) NaOEt

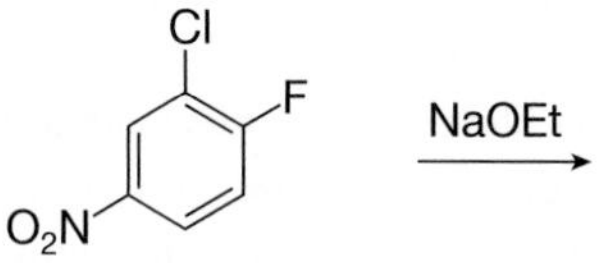

ii) H_2O

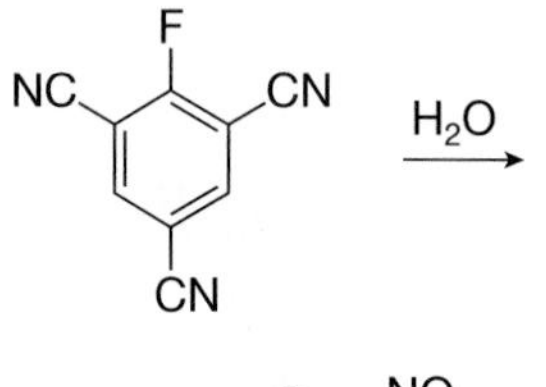

iii) $MeNH_2$

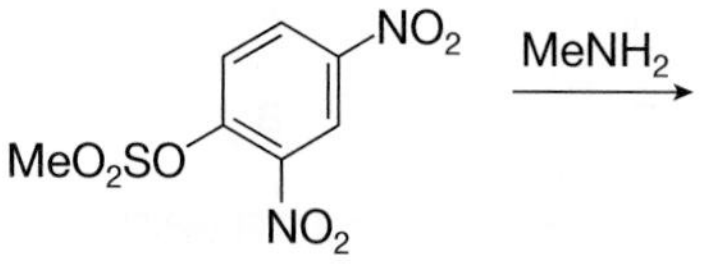

INTEGRATE THE SKILL

15.22 The substrate below has three potential leaving groups. Yet, when it is reacted with KOH, only a single S_NAr product is obtained. Predict the product formed, and use the reaction mechanism and resonance structures to explain the observed selectivity.

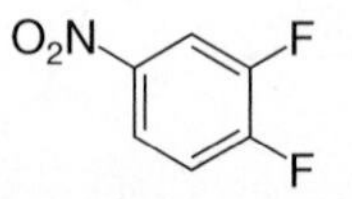

15.9.4 Diazonium salt substitutions

The addition-elimination mechanism for S_NAr described in Section 15.9 requires a strong electron-withdrawing group to activate the ring for nucleophilic attack. However, there is a set of reactions in which the leaving group is so good that substitution takes place at reasonable temperatures without an activating group.

Nitrosonium (NO^+) is a strong electrophile formed when nitrous acid or sodium nitrite is mixed with hydrochloric acid. Nitrosonium can be used to make aromatic diazonium ions (ArN_2^+) from aryl amines. The N_2^+ group on an aryl diazonium salt forms nitrogen gas when displaced, making it an extremely good leaving group. Diazonium ions are also strongly electron withdrawing and activate aromatic rings for S_NAr substitution. The nucleophilic substitution of diazonium salts usually requires either an acid or metal cation activator to proceed.

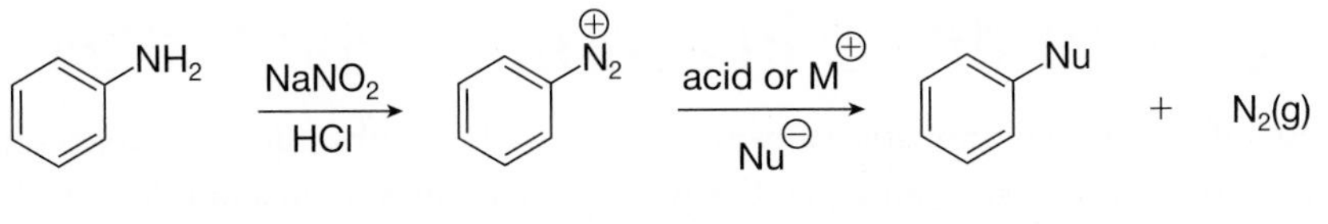

Specific diazonium salt reactions are shown in Figure 15.8. Several of these reactions give products that cannot be synthesized using any of the other reactions described thus far. For example, there are no S_EAr (electrophilic aromatic substitution) reactions that are commonly used to form iodo- or fluorobenzene products. Iodobenzene can be made from an aryl diazonium salt and potassium iodide, and the **Schiemann reaction** gives a fluorobenzene using tetrafluoroboric acid. Substitution using the copper (I) salts of chlorine, bromine, or cyanide is known as the **Sandmeyer reaction**. Reacting an aryl diazonium salt with aqueous acid is a simple method for synthesizing phenols.

The **Schiemann reaction** uses diazonium salts and tetrafluoroboric acid to convert anilines into aryl fluorides.

The **Sandmeyer reaction** uses copper salts to substitute Cl, I, or CN for an aryl diazonium salt.

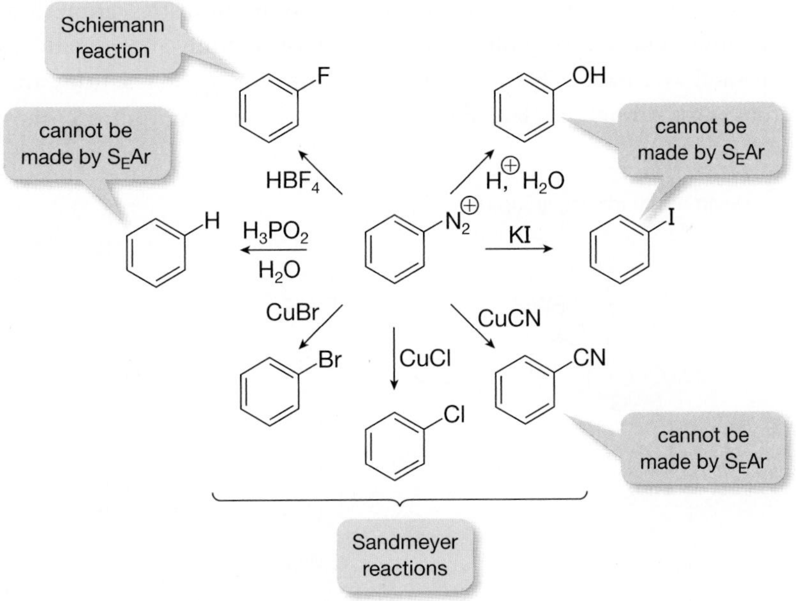

FIGURE 15.8 Aryl diazonium salts give a variety of products.

DID YOU KNOW?

A diazonium ion can be reduced by substitution with the hydride obtained from the conversion of hypophosphorous or phosphinic acid (H_3PO_2) to phosphorous acid (H_3PO_3).

15.10 Substitutions in Aromatic Synthesis

The synthesis of substituted aromatic rings using nucleophilic substitution is conceptually the opposite of electrophilic substitutions (see Chapter 10). This allows nucleophilic substitution to be used as a complementary method to broaden the scope of possible aromatic rings that can be synthesized.

The general retrosynthesis of S_NAr reactions is outlined in Section 15.10.1. The use of S_NAr and diazonium salt substitutions, along with S_EAr reactions to form more complicated substituted aromatic rings, is described in Section 15.10.2.

15.10.1 Retrosynthetic analysis in S_NAr reactions

Nucleophilic aromatic substitution reactions add a nucleophile to an aromatic ring that has at least one strong electron-withdrawing substituent. From a synthetic point of view, a nucleophile-substituted aromatic ring disconnects to an aromatic cation and a reactive anionic nucleophile. The electrophilic synthon is derived from an aromatic ring with a leaving group such as a halide. The best leaving groups for S_NAr reactions are those that attract electrons most strongly, such as F, Cl, OR, OSO_2R, and even nitro groups.

In S_NAr reactions, the most reactive aromatic rings are those with electron-withdrawing groups *ortho* and/or *para* to the leaving group. Electron-withdrawing groups *meta* to the leaving group are unable to stabilize the negative charge in the reaction intermediate.

STUDENT TIP

The best leaving groups for S_NAr reactions are the worst leaving groups for S_N1 or S_N2 reactions.

15.10.2 Aromatic substitutions using diazonium salts

The substitution of diazonium salts provides a way to form products that are difficult or impossible to make using other aromatic substitutions. For example, the synthesis of *meta*-bromochlorobenzene presents a challenge because both halides are *ortho/para* directors for electrophilic reactions. Both *ortho* and *para* products are formed regardless of the order of addition.

whichever is installed first prevents the other from adding to the *meta* position

both halides are *ortho/para*-directing

halogens are *ortho/para* directors

a mixture of *ortho/para* products is formed

Disconnecting either one of the halogens of a diazonium is a reasonable synthetic approach. Diazonium ions are made from the nitrosation of anilines. The nitro group is a *meta* director, which can be used to direct the halogen to the desired location.

The two synthetic routes to *meta*-bromochlorobenzene shown here use both S_NAr and S_EAr reactions. Once the *meta*-directing nitro group is put in place with a nitration reaction (H_2SO_4 + HNO_3), electrophilic halogenations will selectively produce the *meta* isomer. Reduction of the nitro group to an amine sets up the formation of the diazonium salt by nitrosation. The Sandmeyer reaction substitutes the remaining halogen on the carbon that previously held the nitro group.

The N_2 of the diazonium salt can be removed entirely using phosphinic acid, instead of being replaced with another substituent that remains on the ring. With this method, a *meta*-directing group is converted to an *ortho*- or *para*-directing group, and then removed in the last step.

CHECKPOINT 15.12

You should now be able to predict the products formed when aryl diazonium salts react with a variety of nucleophiles. You should also be able to incorporate these reactions into multi-step syntheses of substituted aromatic compounds.

SOLVED PROBLEM

Predict the final product in the following reaction sequence.

STEP 1: Determine the roles of the reactants in the first step.

Sodium nitrite and hydrochloric acid form nitrosonium ions when combined. Nitrosonium ions can convert aryl amines into aryl diazonium salts.

STEP 2: Draw the product of the first step of the reaction.

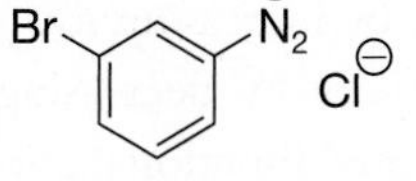

Br $N_2^{\oplus}$ $Cl^{\ominus}$

STEP 3: Determine the roles of the reactants in the second step.

The aryl diazonium salt has an excellent leaving group (N_2 gas) with a positive charge, making the salt the electrophile in the reaction. HBF_4 is a source of fluoride ions, which serve as the nucleophile.

STEP 4: Draw the product of the second step of the reaction.

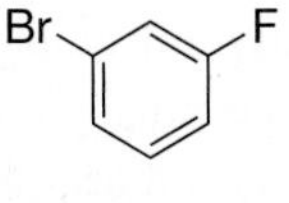

Br F

PRACTICE PROBLEM

15.23 Predict the products of the following reactions.

a) 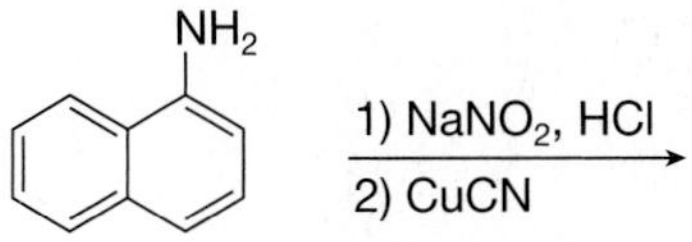 NH_2 — 1) $NaNO_2$, HCl; 2) CuCN →

b) 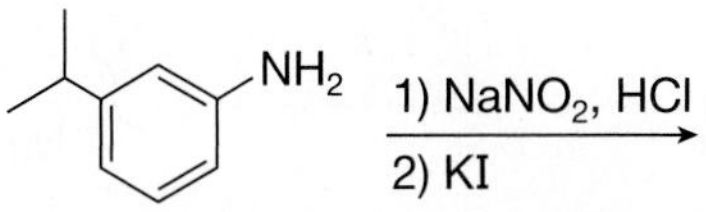 NH_2 — 1) $NaNO_2$, HCl; 2) KI →

c) 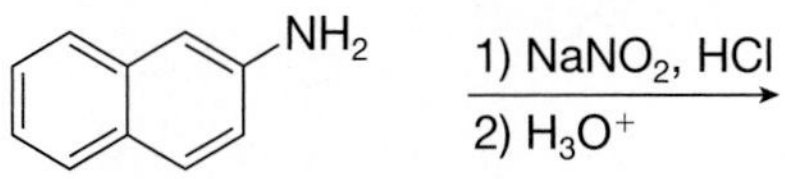 NH_2 — 1) $NaNO_2$, HCl; 2) H_3O^+ →

d) 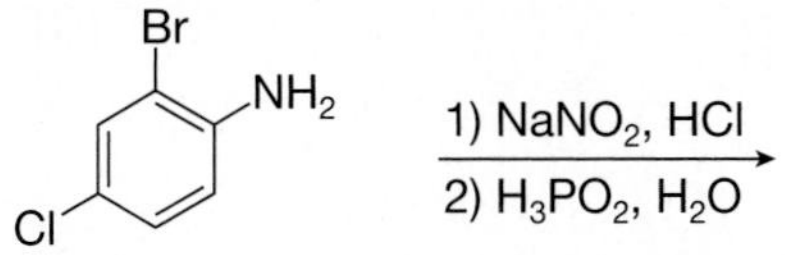 Br, NH_2, Cl — 1) $NaNO_2$, HCl; 2) H_3PO_2, H_2O →

INTEGRATE THE SKILL

15.24 Propose a series of transformations that could be used to prepare the following compounds from benzene.

a) 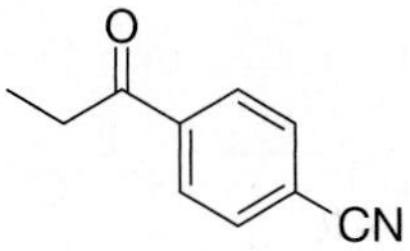 O, CN

b)

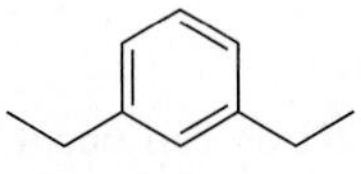

c) 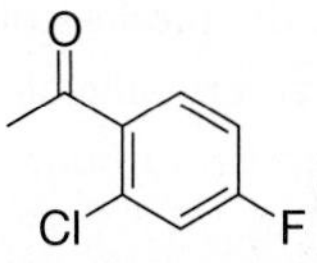 O, Cl, F

15.11 Patterns in Addition-Elimination Reactions

The reactivity of a carbonyl is controlled by the groups that are attached to it. Electron-withdrawing groups increase reactivity by increasing the positive charge on the carbonyl carbon. Electron-donating groups lower reactivity by decreasing the positive charge on the carbonyl carbon. Nucleophilic additions to carbonyl functional groups are controlled by the leaving group that is present. More reactive carbonyl groups can be converted into less reactive forms, but the only way to convert an unreactive group into a reactive one is to hydrolyze the unreactive group into a carboxylic acid and then convert the acid.

A nucleophile adds to a carbonyl at the same time that the π bond (the weakest bond of the group) breaks, creating a tetrahedral intermediate. This intermediate then collapses to re-form a π bond and expel a leaving group. If neutral nucleophiles are added to acid chlorides or anhydrides, the hydrogen of the nucleophile is eventually removed, forming a molecule of acid. Basic conditions will prevent this acid formation and also increase the reactivity of the nucleophile. Acid catalysts can also be used to accelerate the addition of nucleophiles to carbonyl groups. The acid reacts to make the carbonyl more electrophilic but is regenerated at the end of the reaction.

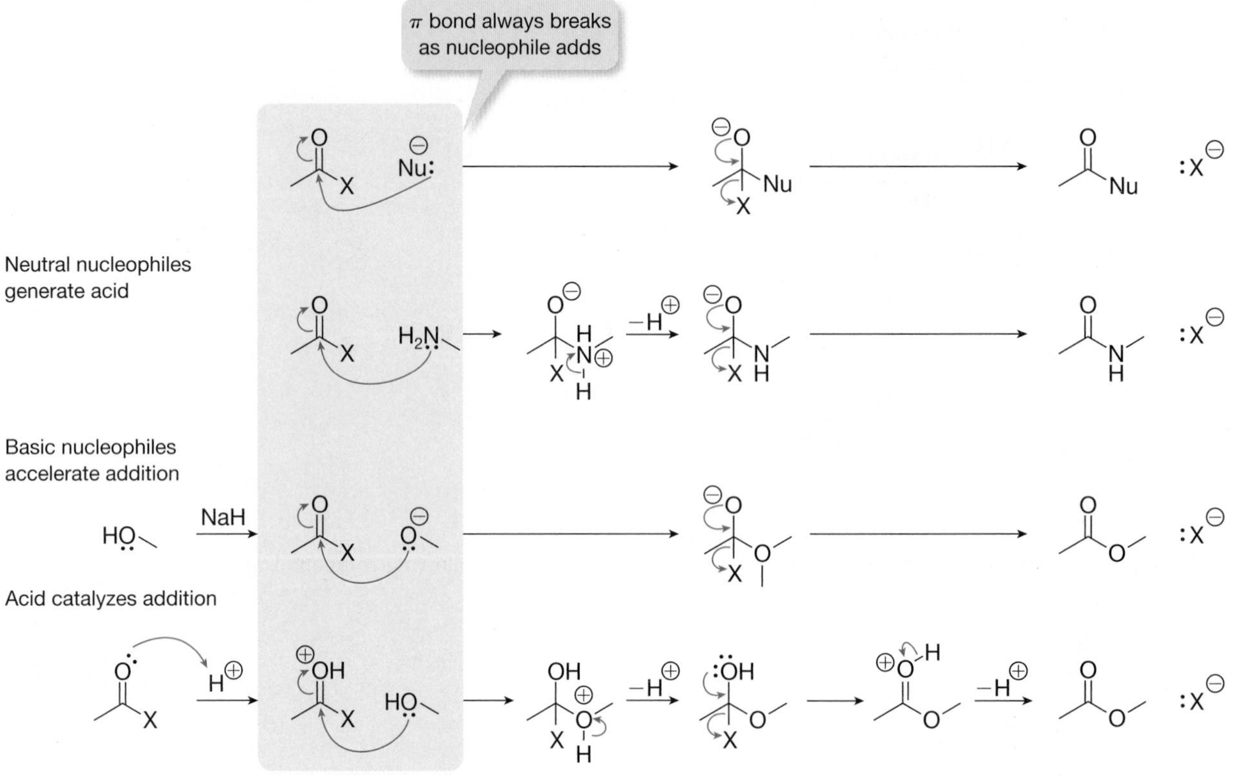

π bond always breaks as nucleophile adds

Neutral nucleophiles generate acid

Basic nucleophiles accelerate addition

Acid catalyzes addition

If the product of an addition is reactive enough, a second addition reaction can occur, forming an alcohol. The second addition follows the same pattern of reactivity since the product of the first addition is a carbonyl compound. Because the resulting carbon-carbon bond is very difficult to break, these reactions normally produce tertiary alcohols when the nucleophilic atom is a carbon.

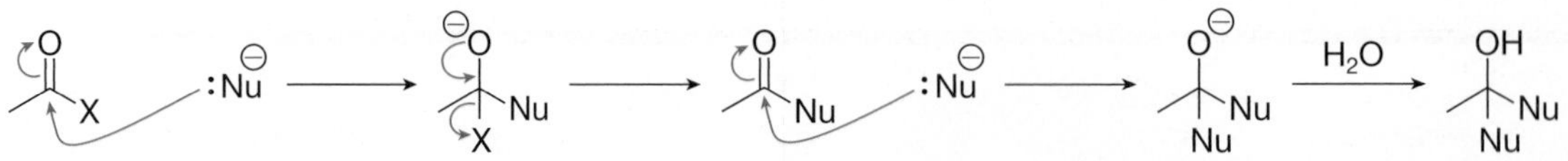

Other functional groups in which an sp^2-hybridized atom is connected to a leaving group undergo similar reactions in which a nucleophile adds to the sp^2 atom, breaking a π bond and forming a tetrahedral intermediate. This intermediate then collapses to expel the leaving group and generate the product. When the electrophile is an aromatic ring, the initial intermediate is stabilized by the presence of electron-withdrawing groups *ortho* or *para* to the leaving group.

X :Nu⁻ → Nu X → Nu :X⁻

EWG EWG X :Nu⁻ → EWG EWG Nu X → EWG EWG Nu :X⁻

The mechanism of addition-elimination for both the acyl substitution and S_NAr reactions are initiated by a nucleophile bonding to a carbon with a leaving group. A retrosynthesis for the product of either reaction uses a synthon with a positive charge on the carbon that will be bonded to the nucleophile. In acyl substitution, this positive charge is on the carbonyl carbon. For S_NAr, the charge is on the aromatic carbon *ortho* and/or *para* to the electron-withdrawing groups.

Nu ⟹ :Nu⁻ ⟹ X :Nu–H

$O-CH_3$ ⟹ $:OCH_3^-$ ⟹ Cl $HOCH_3$

EWG EWG Nu ⟹ EWG EWG :Nu⁻ ⟹ EWG EWG X :Nu–H

EWG EWG SPh ⟹ EWG EWG :SPh⁻ ⟹ EWG EWG F HSPh

Bringing It Together

Morphine is an effective pain killer and sedative that has been used medicinally for at least 5000 years, and it is still one of the best drugs available for managing severe pain. Unfortunately, morphine can also produce euphoric effects and be addictive.

In 1897, the Bayer company was trying to manufacture a safer, less addictive substitute for morphine. The drug they produced, heroin, was made by using acetic anhydride and pyridine to replace the hydroxyl groups on morphine with esters. Heroin is much more lipophilic (soluble in fats) than morphine, which allows heroin to diffuse across cell membranes and reach the brain very quickly. In the brain, enzymes called esterases catalyze an addition-elimination hydrolysis that removes the acetyl groups and regenerates morphine. This process results in a much higher concentration of morphine in the brain than is possible by taking morphine itself.

morphine

heroin

Although heroin turned out to be more potent and more addictive than morphine, Bayer marketed Heroin (originally a trademarked name) as a non-addictive cough suppressant, pain reliever, and cure for morphine addiction. For many years, heroin was sold over the counter in most drug stores. Concerned by mounting evidence that heroin is dangerously addictive, the United States restricted heroin to prescribed medical uses in 1914, and banned the drug completely in 1924. Except for strictly controlled use in medical treatment and research, heroin is now illegal in almost all countries, including Canada.

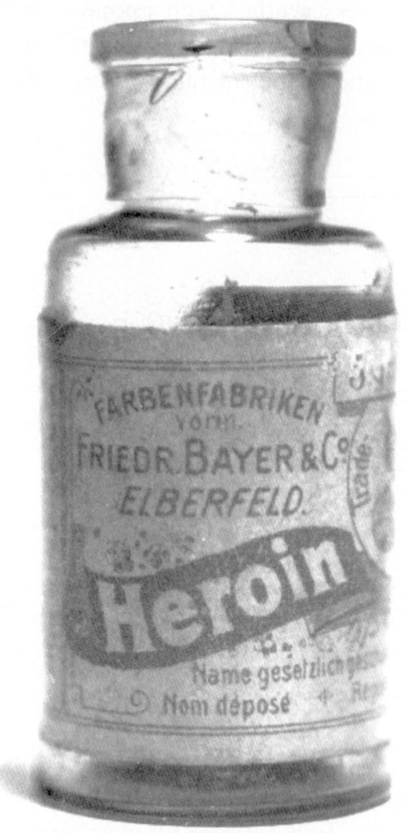

Mpv_51/Public Domain

You Can Now

- Determine the relative reactivity of various carboxylic acid derivatives based on the relative contributions of their resonance forms.
- Use the relative reactivity of carboxylic acid derivatives to determine which derivatives can be directly prepared from each other.
- Predict the products formed when carboxylic acid derivatives are reacted with a variety of nucleophiles under basic, neutral, and acidic conditions.
- Draw mechanisms for addition-elimination reactions of carboxylic acid derivatives with various nucleophiles under basic, neutral, and acidic conditions.
- Explain the need for added base when amines are reacted with acid chlorides and anhydrides.
- Show how the addition of catalytic pyridine or DMAP can accelerate acylation reactions.
- Show how carboxylic acids are activated via reagents like $SOCl_2$, P_2O_5, and DCC.
- Explain the differences in selectivity observed when reacting nucleophilic reducing agents (e.g., $NaBH_4$ and $LiAlH_4$) and electrophilic reducing agents (BH_3 and DIBAL) with various carboxylic acid derivatives, and draw mechanisms for these reactions.
- Explain how reaction temperature is used to control the selectivity of reductions carried out with BH_3 and DIBAL.
- Explain why organometallic reagents typically add twice when reacted with carboxylic acid derivatives, and describe procedures for limiting the addition to a single equivalent.
- Recognize structures that are capable of undergoing nucleophilic aromatic substitution reactions, and rank their reactivities.
- Predict the products of nucleophilic aromatic substitution and draw mechanisms for these reactions.
- Predict the products formed when aryl diazonium salts react with a variety of nucleophiles.

A Mechanistic Re-View

NUCLEOPHILIC ACYL SUBSTITUTIONS

With basic nucleophile

$H_3C-C(=O)-X \xrightarrow{^{\ominus}Nu} H_3C-C(=O)-Nu$

With neutral nucleophile

$CH_3C(=O)Cl \xrightarrow{CH_3OH} CH_3C(=O)OCH_3 + CH_3OH_2^{\oplus} + Cl^{\ominus}$

Pyridine activation

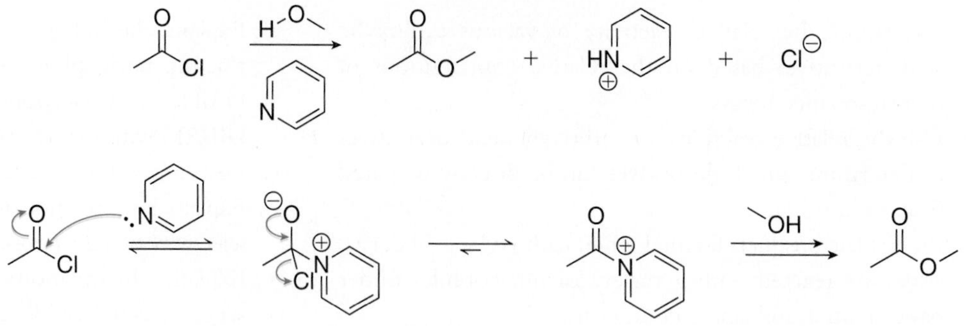

Acid catalysis—Fischer esterification

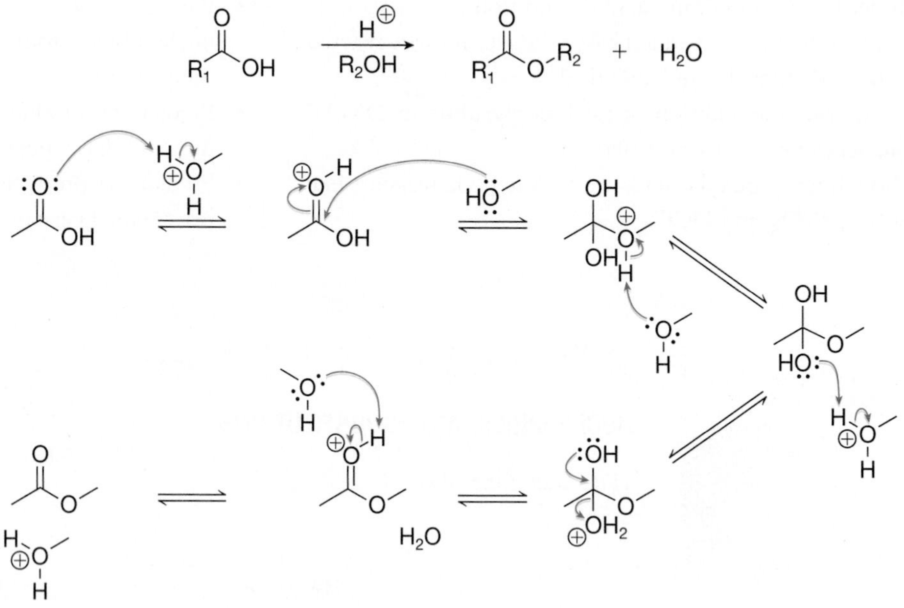

Acid catalysis—ester hydrolysis

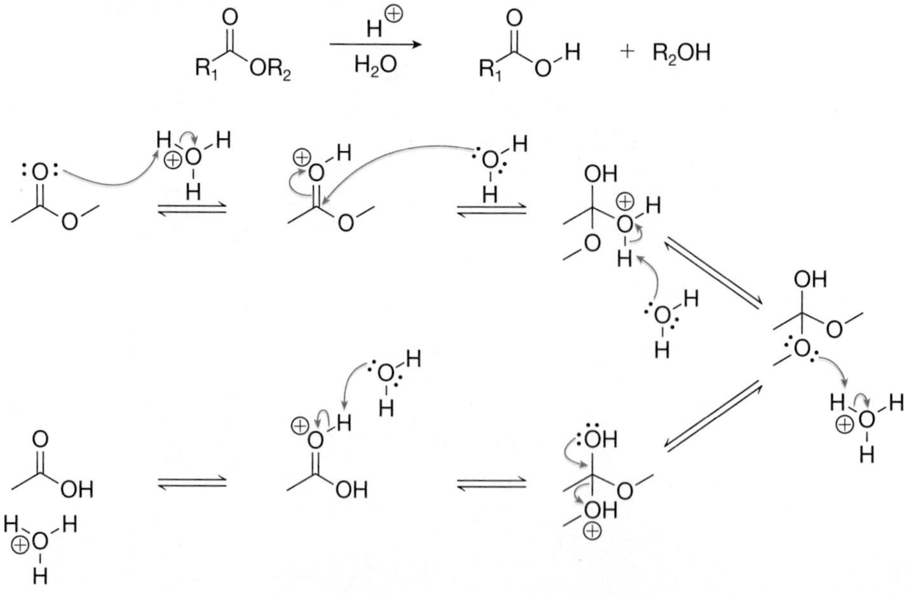

Amide hydrolysis—basic conditions

Amide hydrolysis—acidic conditions

Carboxylic acid activation—thionyl chloride

Carboxylic acid activation—DCC

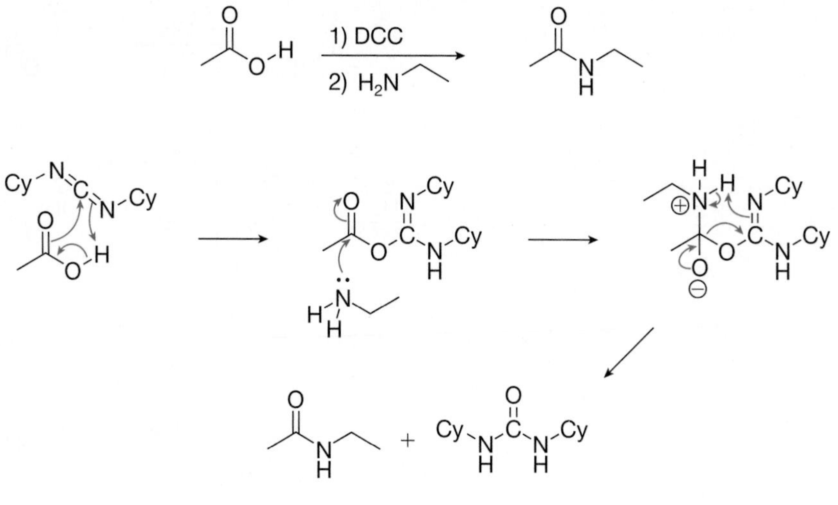

1) DCC
2) H_2N

Cy-N=C=N-Cy
O
O-H

N-Cy
O-C-N-Cy
H

H-N-H
H

H
⊕N-H
N-Cy
C
O-N-Cy
H
O
⊖

O
N
H
+
Cy-N-C-N-Cy
H H

General reduction—$LiAlH_4$

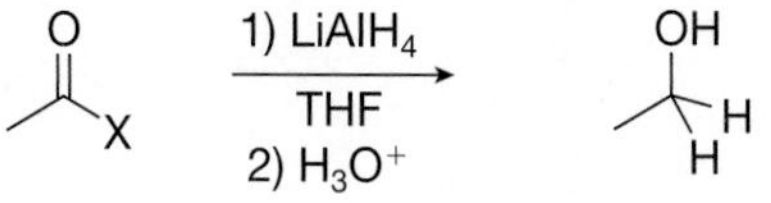

O
X
1) $LiAlH_4$
THF
2) H_3O^+
OH
H
H

X = Cl^-, RCO_2^-, RO^-, HO^-

H
⊖Al
H H
H
O
X
1) LAH
H
H-Al
H ⊖
O
X
H
O
H
H
⊖Al
H H
H

2) acid
or base
H_2O
OH
H
H
H
⊖
Al
H
O
H
H
H
H-Al
H
⊖
:O
H
H

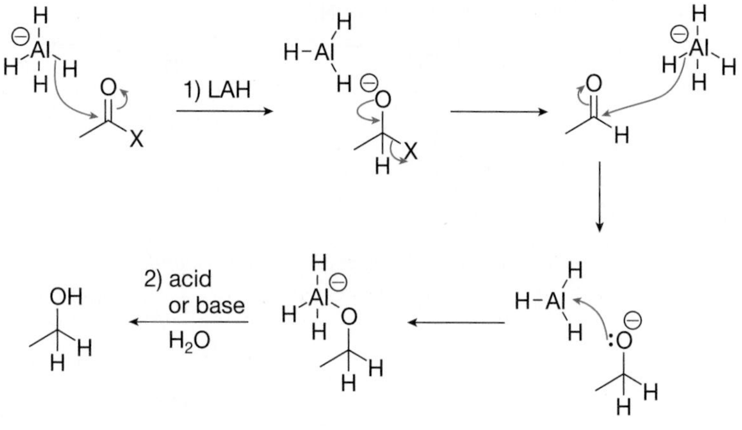

Carboxylic acid reduction—$LiAlH_4$

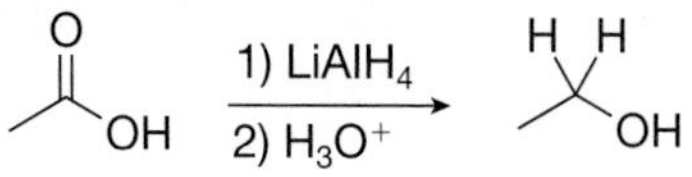

O
OH
1) $LiAlH_4$
2) H_3O^+
H H
OH

Amide reduction—$LiAlH_4$

1) $LiAlH_4$

2) H_3O^+

Amide reduction—BH_3

1) B_2H_6, ether

2) $H_3O^{\oplus}$

H_2O

heat

Ester reduction—DIBAL

1) 1 equiv. DIBAL −78 °C
2) H_3O^+

−78 °C

H_2O

hemiacetal

heat

ORGANOMETALLIC ADDITION

Grignard addition

1) BrMg
2) H_3O^+

S_NAR REACTIONS

X includes halides (F >> Cl > Br > I), nitro, OR; Y includes anionic O, N, S compounds

X includes halides (F >> Cl > Br > I), nitro, OR; Y includes anionic O, N, S compounds

Problems

15.25 Which of the following transformations can be carried out directly? Explain your answer.

a) amide → carboxylic acid
b) acid chloride → ester
c) ester → anhydride
d) ester → carboxylic acid
e) anhydride → amide
f) carboxylic acid → amide

15.26 Predict the products of the following reactions.

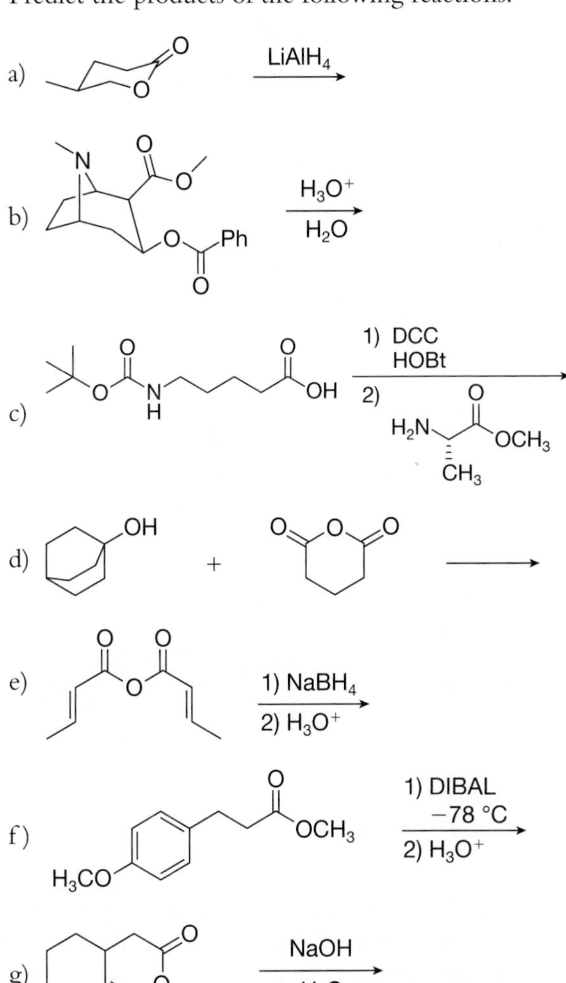

h) EtO — 1) 1 equiv. PhMgBr 2) H$_3$O$^+$, heat

i) O–CH$_3$ — 1) CH$_3$MgI$_{(excess)}$ 2) H$_3$O$^+$, heat

j) OH + — HN

k) OH — 1) SOCl$_2$ 2) MeNH$_2$, pyridine

l) CN — 1) NaOH, H$_2$O, heat 2) H$_3$O$^+$ 3) P$_2$O$_5$

m) N — HCl, H$_2$O heat

15.27 Predict the products of the following reactions.

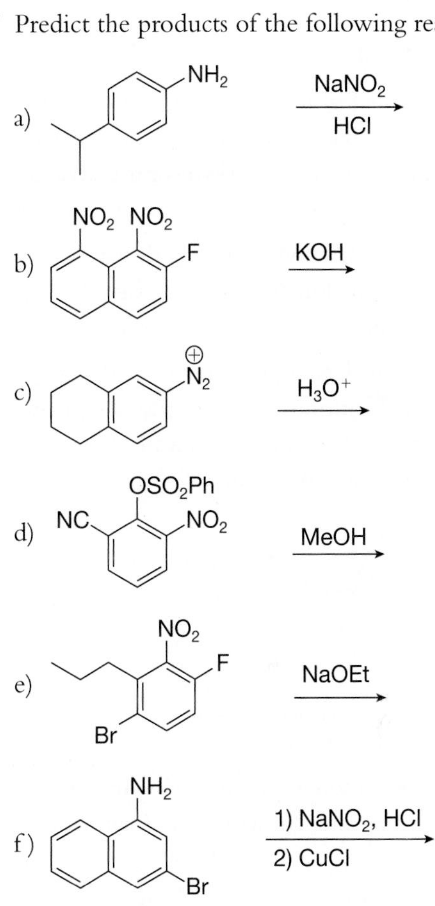

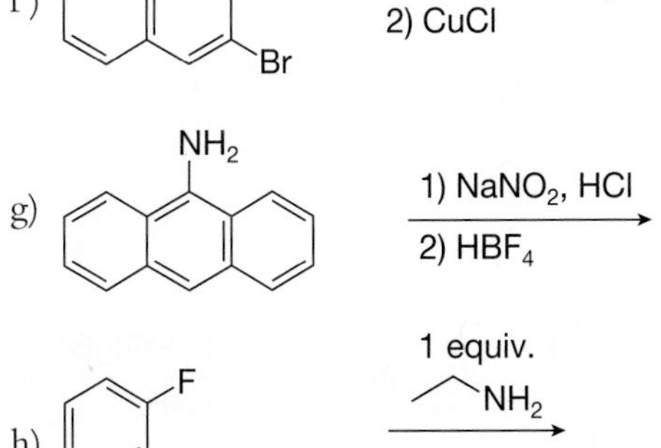

15.28 Rank the following sets of aromatic compounds in order of increasing reactivity in S_NAr reactions.

a) 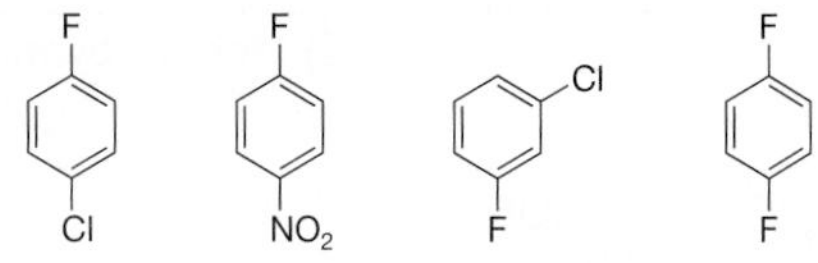 A B C D

b) 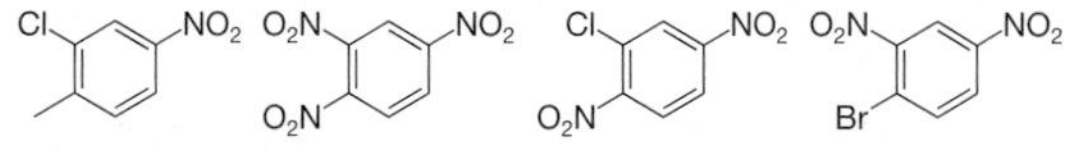 A B C D

c) 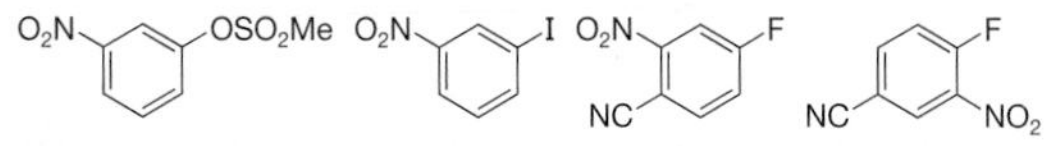 A B C D

15.29 Why is it essential to use at least two equivalents of an amine nucleophile or to add one equivalent of a non-nucleophilic base such as pyridine or triethyl amine when making an amide from an acid chloride or anhydride?

15.30 Pyridine and triethyl amine both have a lone pair of electrons to donate. Why can they be said to be non-nucleophilic?

15.31 Treatment of desacetylmatricarin with aqueous acid, followed by removal of water, generates a mixture of the starting material and a second lactone. What is the structure of the new product?

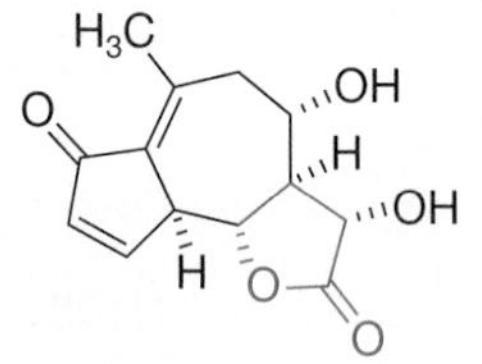

desacetylmatricarin

15.32 Predict the products of the following reactions and draw mechanisms showing the formation of these products.

a) 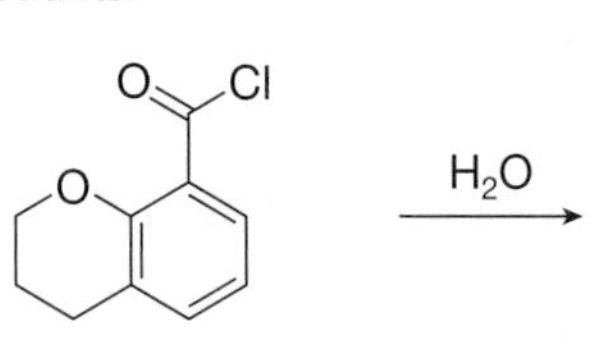 H_2O

b) 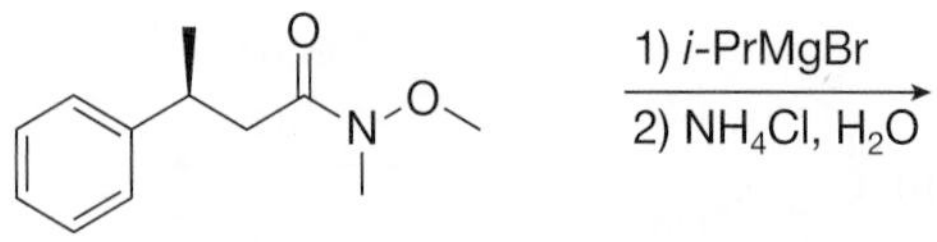 1) *i*-PrMgBr 2) NH_4Cl, H_2O

c) 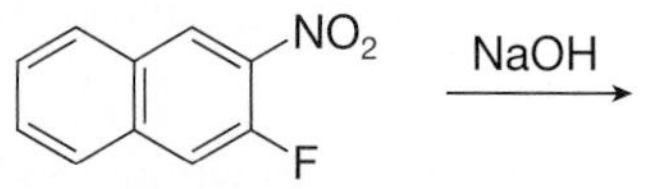 NaOH

d) 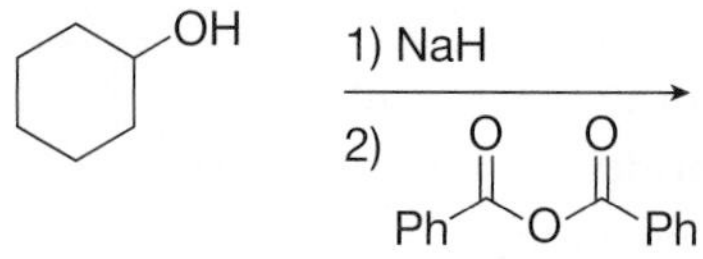 1) NaH 2) (PhCO)$_2$O

e) 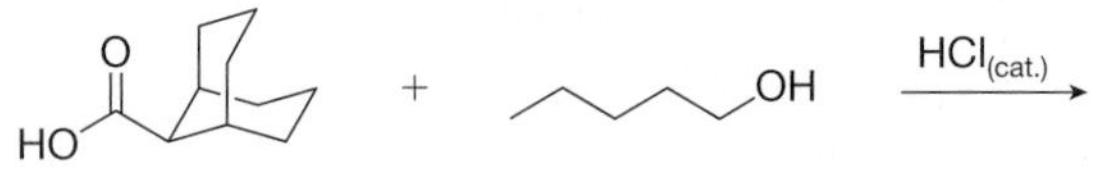 + $HCl_{(cat.)}$

f) 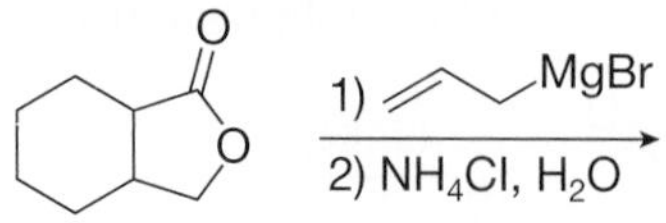 1) allyl-MgBr 2) NH_4Cl, H_2O

g) 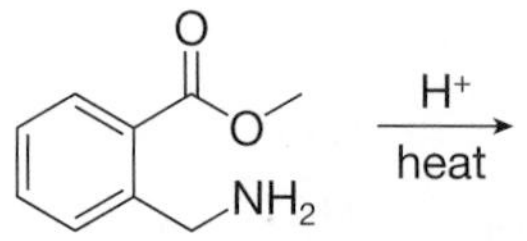 H^+ heat

h) + DMAP Et_3N

i) 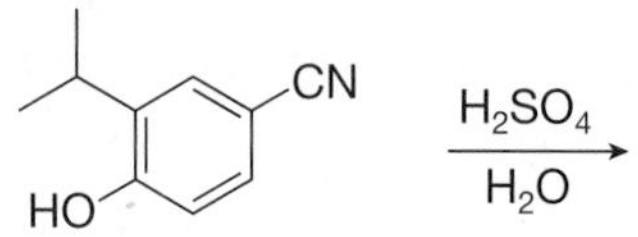 H_2SO_4 H_2O

j) 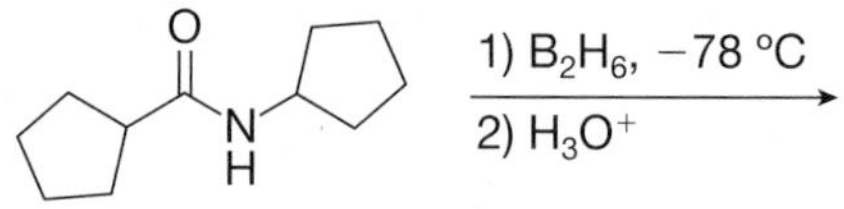 1) B_2H_6, −78 °C 2) H_3O^+

k) 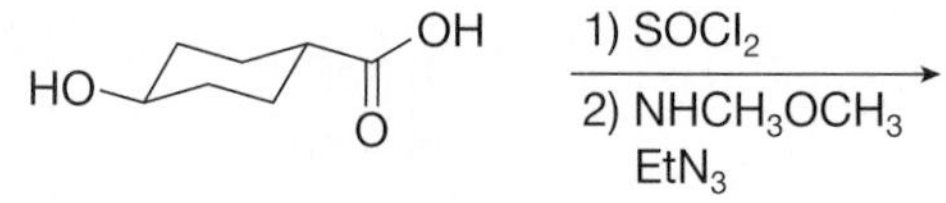 1) $SOCl_2$ 2) $NHCH_3OCH_3$ EtN_3

l) 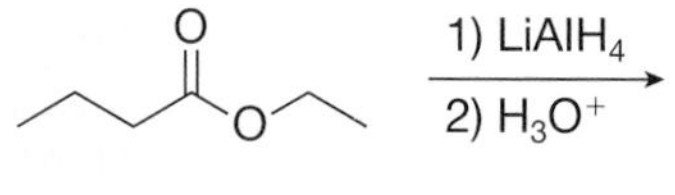 1) $LiAlH_4$ 2) H_3O^+

15.33 The *haloform reaction*, shown here, is a way to oxidize a methyl ketone. Propose a reasonable mechanism for the second step, the loss of the CBr_3 group. (Note: The mechanism of the first step, bromination, is described in Chapter 16.)

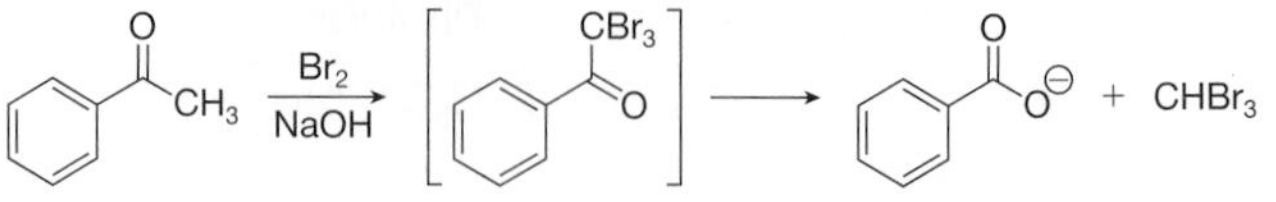

Br_2 NaOH → [ ] → + $CHBr_3$

15.34 Isobutyl chloroformate (IBCF) is often used to make activated carbonyl groups for the conversion of acids into amides and esters. The acid is dissolved in THF with a hindered base such as Et_3N, and then IBCF is added to make an anhydride. Shortly afterward, an alcohol or amine is introduced, and the product forms quickly. An example of this process is shown here.

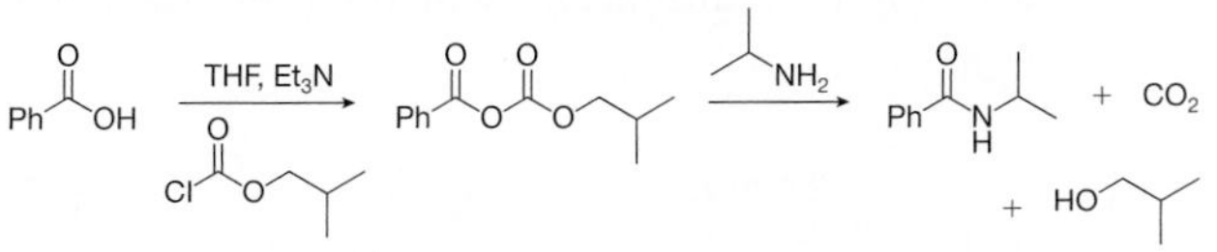

THF, Et_3N NH_2 + CO_2 + HO

a) Draw a mechanism for the formation of the anhydride.
b) Explain why only one of the carbonyl groups reacts with nucleophiles.
c) Draw a mechanism for the formation of the final product.

15.35 The chemical name for Aspirin is acetylsalicylic acid. How can Aspirin be synthesized from salicylic acid, shown here?

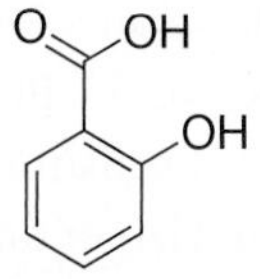

salicylic acid

15.36 List the reagents necessary for the following transformations.

a) 
b) 
c) 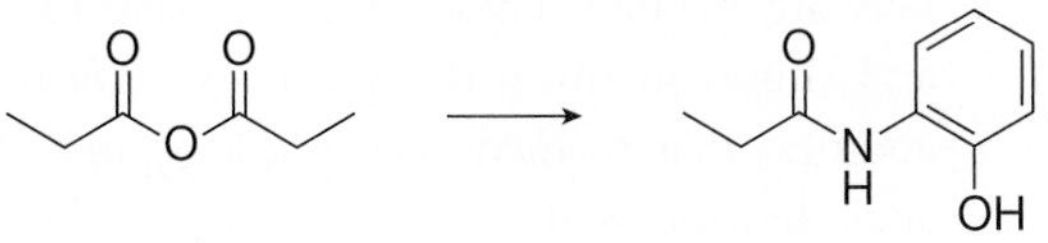
d) 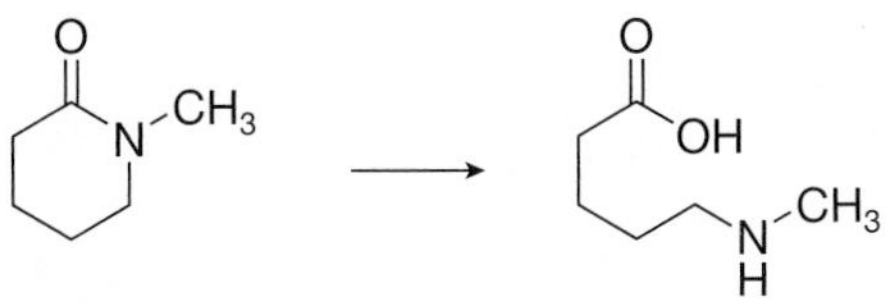
e) 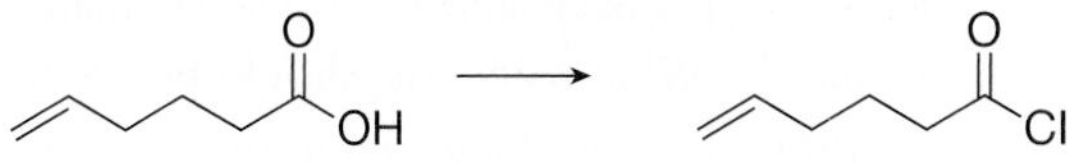
f) 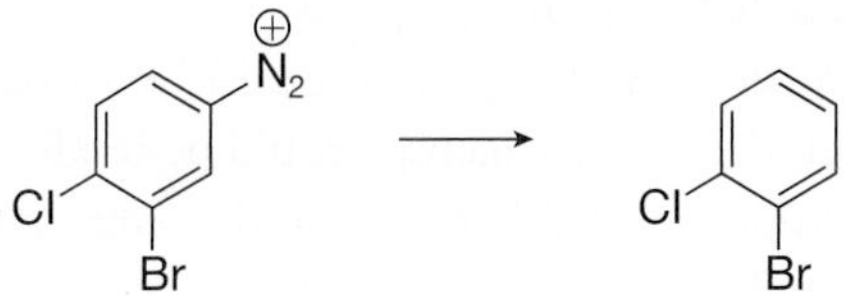
g) 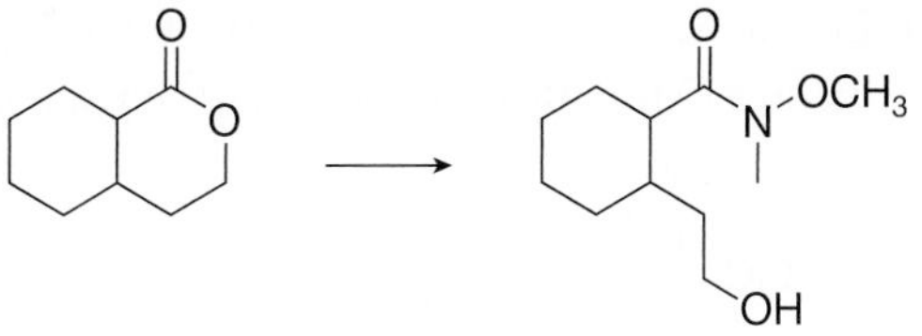
h) 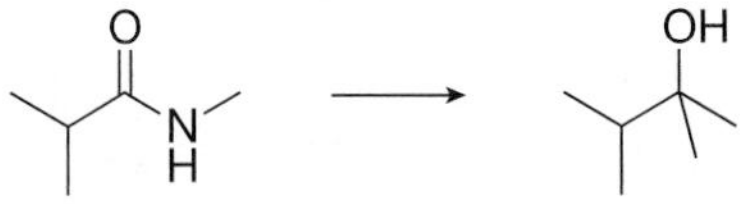
i) 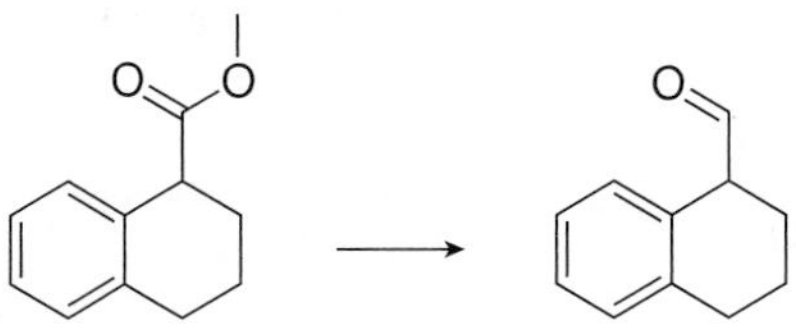
j)

15.37 Which of the nitro groups in the following structure would be displaced by a reaction with a nucleophile? Explain why.

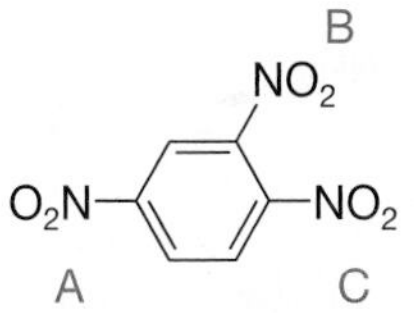

A B C

15.38 The Ritter reaction can be used to make sterically hindered primary or secondary amines that are difficult to make in other ways.

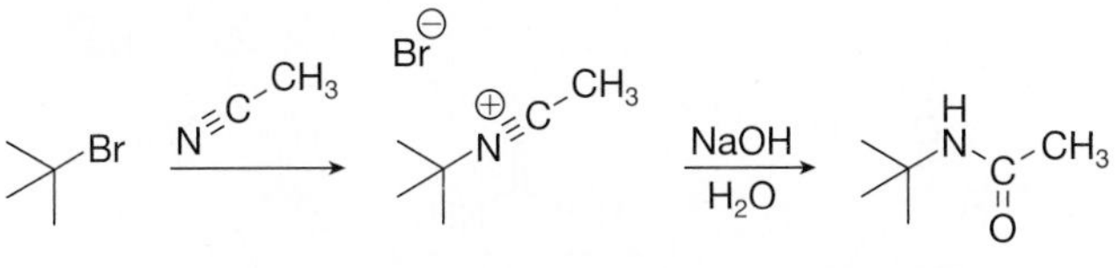

NaOH, H_2O

a) Propose a mechanism for the conversion of a tertiary bromide to an amide.

b) What primary amine could be made from this amide? What reagent(s) should be used?
c) Propose a mechanism for the conversion of the amide to a primary amine.
d) What secondary amine could be made from this amide? What reagent(s) could be used?
e) Propose a mechanism for the conversion of the amide to a secondary amine.

15.39 List the reagents necessary for the following transformations. More than one step may be required.

a) 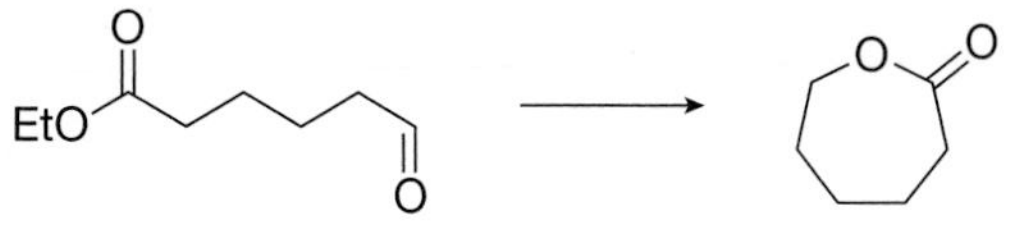 EtO … O → (lactone)

b) 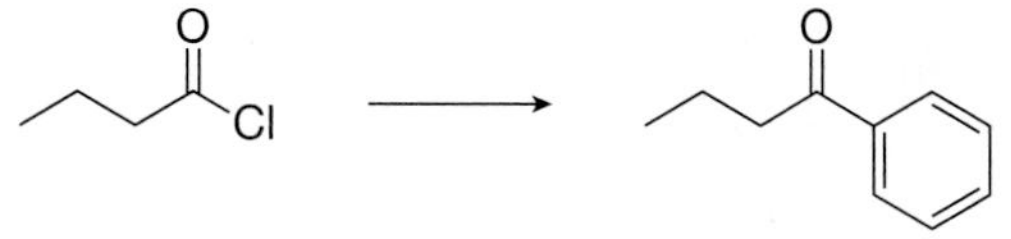 Cl … → Ph ketone

c) 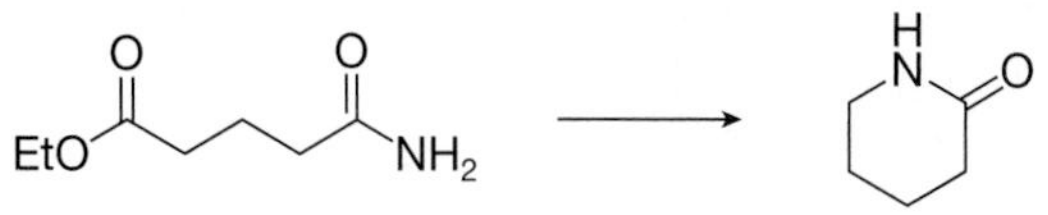 EtO … NH_2 → (lactam, H N O)

d) 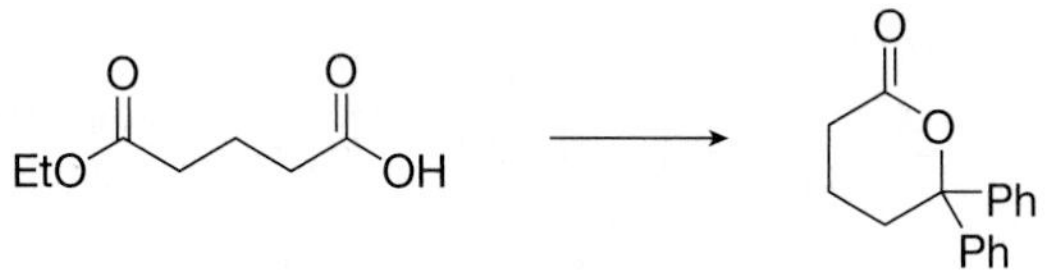 EtO … OH → (lactone, Ph, Ph)

e) 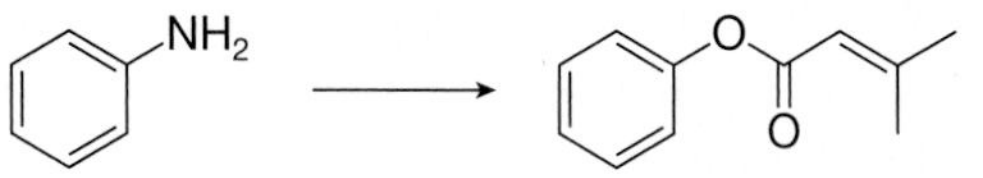 NH_2 → (ester)

f)  NH_2 → (ester)

15.40 The final step in the synthesis of sodium amytol, a barbiturate prescribed for insomnia or epilepsy (and sometimes used as a "truth serum"), is shown here. Propose a reasonable mechanism for this conversion.

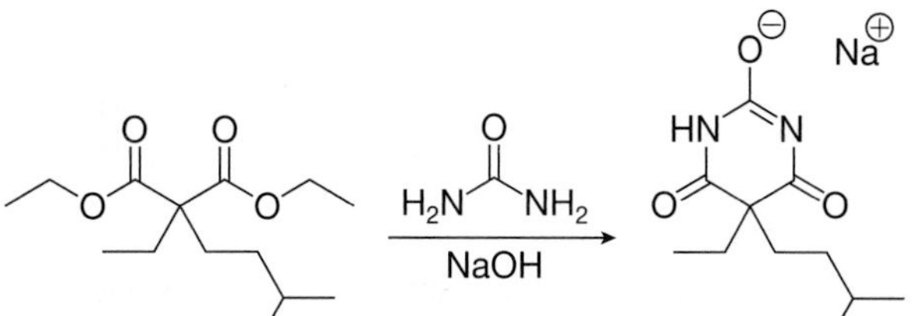

H_2N–C(O)–NH_2, NaOH → HN, N, $O^{\ominus}$ $Na^{\oplus}$

15.41 Chemoluminescence is the generation of light from a chemical reaction. The reaction of bis-(2,4,6-trichlorophenyl)oxalate (TCPO) and hydrogen peroxide leads to 1,2-dioxetanedione, as the following reaction sequence shows. Dioxetanedione falls apart to form carbon dioxide and energy. The energy is transferred to a compound known as a fluorophore, such as 9,10-diphenylanthracene, which releases the energy as light.

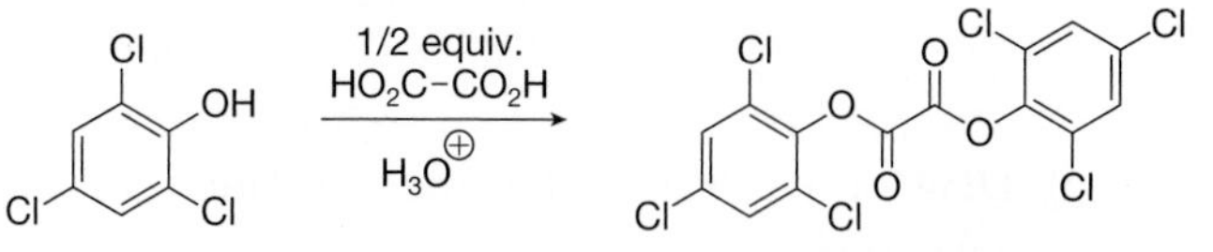

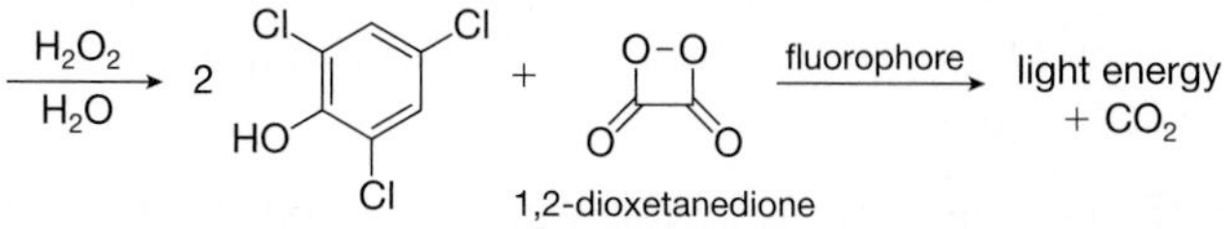

1/2 equiv. $HO_2C{-}CO_2H$, $H_3O^{\oplus}$ → TCPO

H_2O_2 / H_2O → 2 (2,4,6-trichlorophenol) + 1,2-dioxetanedione —fluorophore→ light energy + CO_2

a) Propose a reasonable mechanism for the formation of TCPO and its reaction with hydrogen peroxide to form 1,2-dioxetanedione.
b) Build a model of TCPO and predict the likely low-energy conformation. (Remember that chlorine is four times larger than hydrogen even though they are the same size in a model kit.)
c) The conversion of 1,2-dioxetanedione to carbon dioxide results in the release of light, with the help of a fluorophore. Where does the energy for this light come from?

15.42 Nitrobenzene was subjected to four reactions. These were reaction with Fe/HCl, treatment with $NaNO_2$/HCl followed by heating in aqueous acid, reaction with acid chloride/pyridine, and reaction with Cl_2/$AlCl_3$. Depending on the order in which the two reactions are carried out, two different final products are formed. These products have identical mass and almost identical IR spectra. Specify the order of addition that formed each product, and identify the acid chloride used.

Product A: ^{1}H-NMR

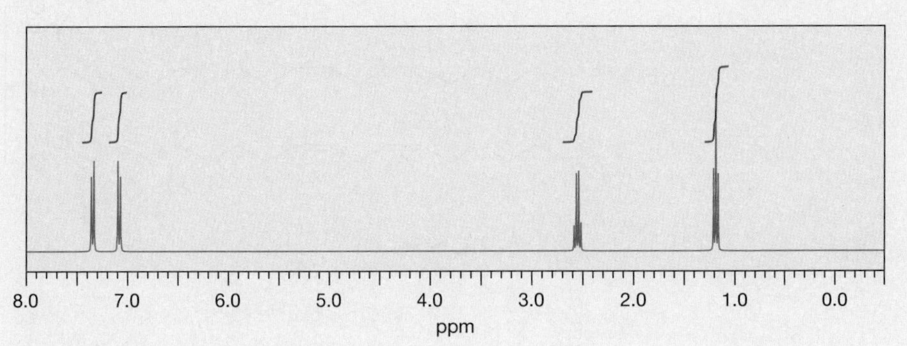

Product A: ^{13}C-NMR

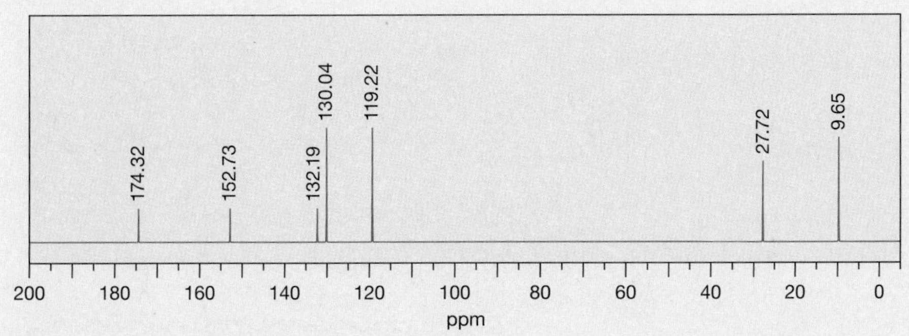

Product B: ^{1}H-NMR

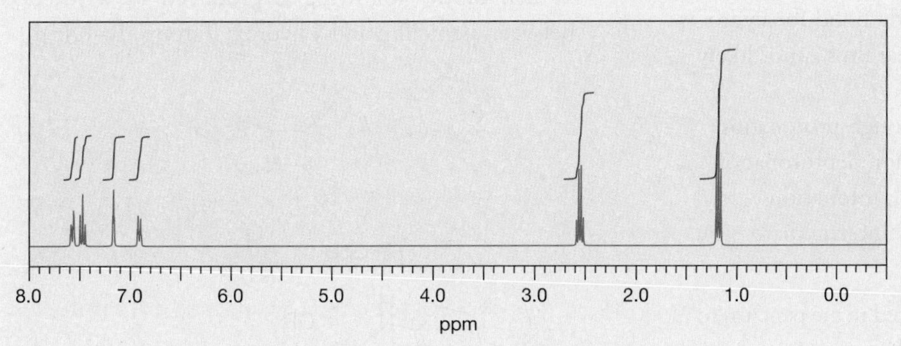

Product B: ^{13}C-NMR

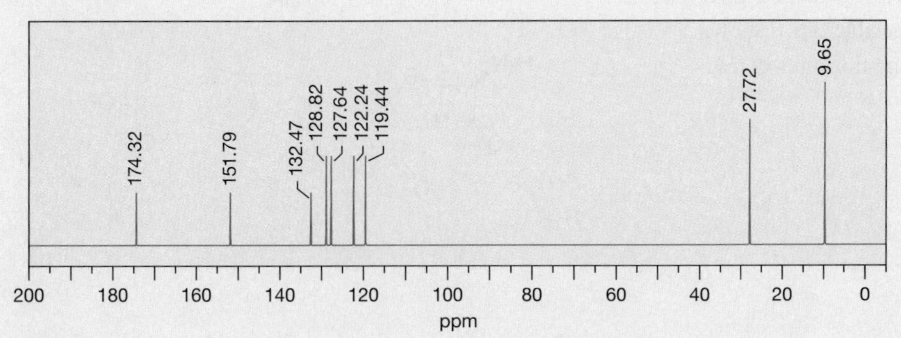

15.43 Synthesize the following compounds, starting with benzene and any other compound that has fewer than four carbons.

a) OH OH

b) O Cl

c) O O N H

d) O O O_2N

MCAT STYLE PROBLEMS

15.44 Proteases are natural enzymes that hydrolyze the amide bonds found in proteins. Various enzymes use different mechanisms for activating the amide bond for hydrolysis. Which of the following mechanisms is not likely to be used by a protease enzyme?

a) activating the leaving group through protonation
b) activating the nucleophile through deprotonation
c) activating the carbonyl through protonation
d) activating the tetrahedral intermediate by nucleophilic addition

15.45 Pancreatic lipase is an enzyme produced in the pancreas to aid in the digestion of fats. The overall reaction carried out by this enzyme is akin to ester hydrolysis. Specifically, the reaction catalyzes the partial hydrolysis of a triglyceride to a monoglyceride, releasing two molecules of fat. A triglyceride commonly found in many vegetable oils is shown here. Select the correct products of digestion after its reaction with pancreatic lipase.

O O O O O O

a) O O OH OH + 2 OH

b) O O OH OH + 2 OH OH

c) O O O OH OH O + 2 O $\ominus$ O

d) O O O O O OH O + OH

15.46 Penicillin works by reacting with hydroxyl groups in an enzyme that stitches bacterial cell walls together. Which of the following is produced by a reaction between penicillin and a hydroxyl group (R-OH)?

H N H S O N O O OH

penicillin

a) H N OR SH O N O O OH

b) H_2N H S O N O OH + OR O

c)

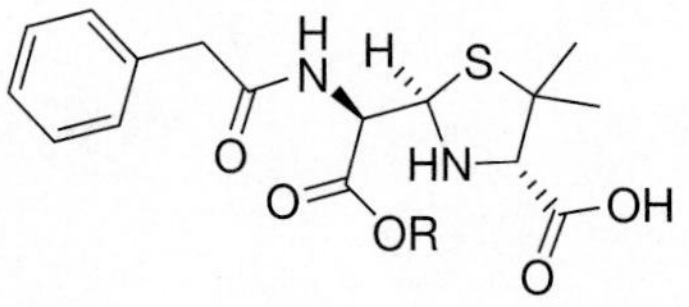

d)

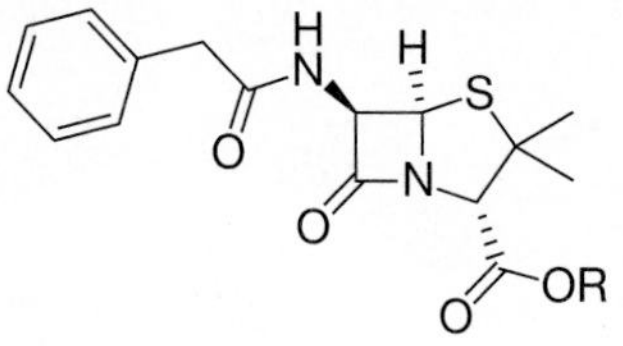

CHALLENGE PROBLEM

15.47 Weinreb amides prevent a second equivalent of a Grignard reagent adding to a carboxylic acid derivative. Theses amides form a five-membered cyclic structure, which is relatively stable under the reaction conditions. Many carboxylic acid derivatives can form similar stable five- and six-membered cyclic intermediates, provided the right neighbouring substituents are present. Examine the following structures, and determine which ones could, given the right nucleophile, undergo a single addition to form a ketone instead of a tertiary alcohol. Explain your selection with appropriate diagrams.

a) 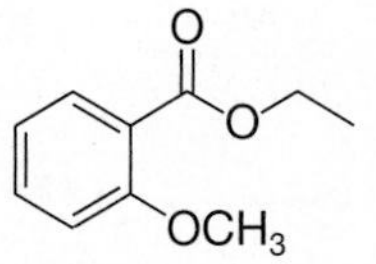 OCH_3

b) 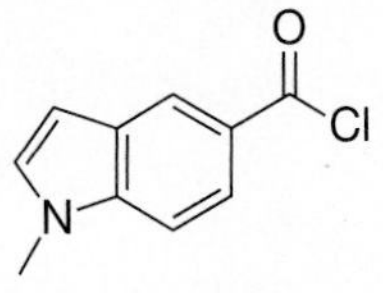 Cl N

c) 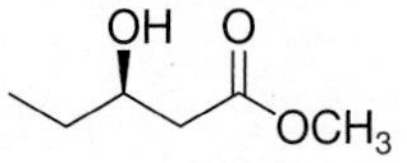 OH O OCH_3

d)

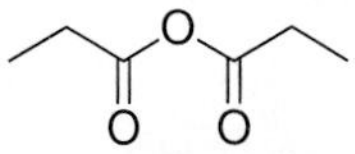

e) 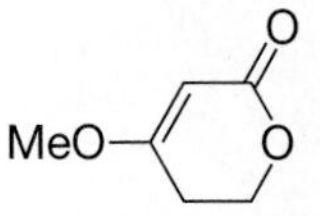 MeO

16 π Bonds with Hidden Leaving Groups

REACTIONS OF ACETALS AND RELATED COMPOUNDS

CHAPTER OUTLINE

The brown colour and mouth-watering aroma of freshly baked bread arises from the chemical reactions of hidden leaving groups. Inset: Ball-and-stick model of glucose, which is joined together as a polymer to make the starches present in bread.

16.1 Why It Matters

When food is cooked, chemical reactions take place that change the flavour, appearance, and smell of the food. One of the most important is the Maillard reaction, which involves the reaction between sugars and amino acids to form new molecules. Through a complex series of events, the Maillard reaction produces the substances that give food flavour, and because many of these molecules are coloured, they produce the browning of food as it is cooked. The coloured compounds, called melanoid pigments, have structures related to pigments in skin (melanin).

The Maillard reaction slowly combines food molecules through reactions that involve carbonyl groups and produce water as a by-product. The water is the hidden leaving group in the starting carbonyls. As you cook the food, some of the steam comes from the water that is given off as the sugars and amino acids react together.

released as steam

carbohydrate

amino acid

heat

imine

H_2O

further reactions

heterocycles

This chapter describes conditions and mechanisms of acetal formation and hydrolysis, as well as the classification of acetals and hemiacetals in carbohydrates as either α- or β-anomers. The chapter explains mechanisms for imine and enamine formation and hydrolysis and also the synthesis of five- and six-membered heterocycles.

16.2 Formation and Reactivity of Acetals

Carbonyl groups are good electrophiles. They react with nucleophiles to form products in which the carbonyl carbon is tetrahedral. An example of this kind of reaction is the formation of a hydrate, which is catalyzed by an acid or a base (see Chapter 7).

acid activates the carbonyl group

ketone or aldehyde

oxonium ion

hydrate

Hydrates are functional groups in which two OH groups are connected to an sp^3-hybridized carbon. The formation of hydrates is a reversible process; the flow of electrons in the opposite direction can regenerate the starting carbonyl compound.

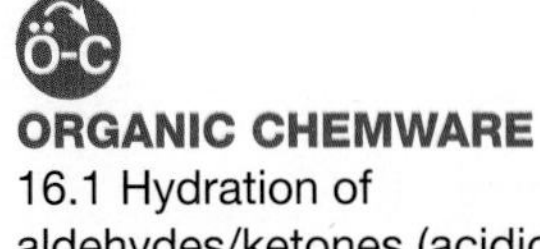

ORGANIC CHEMWARE
16.1 Hydration of aldehydes/ketones (acidic conditions)

ORGANIC CHEMWARE
16.2 Hydration of aldehydes/ketones (basic conditions)

acid protonates the OH to make it a better leaving group

hydrate

oxonium ion

ketone or aldehyde

These two processes—the addition of a nucleophile to a carbonyl and the removal of a leaving group from an sp^3-hybridized carbon to form a carbonyl—can be combined into a multi-step reaction. When this happens, the oxygen of the carbonyl becomes a hidden leaving group in the form of H_2O. The removal or replacement of this hidden leaving group opens up a very large body of organic chemistry. Many reactions and functional groups can be accessed using this hidden leaving group principle, and these processes share common patterns of electron flow.

Hemiacetals are compounds with an OH and an OR group bonded to an sp^3-hybridized carbon.

Acetals are compounds with two OR groups bonded to an sp^3-hybridized carbon.

When an aldehyde or a ketone is reacted with an alcohol in the presence of an acid, a **hemiacetal** is formed as a product by the same mechanism as the addition of water to form a hydrate (see the discussion of hemiacetal formation in Chapter 7). The only difference between hemiacetal formation and hydrate formation is the nucleophile (water or alcohol) involved in the reactions. Hemiacetals are usually not very stable. In the presence of excess alcohol and a trace of acid, they can be converted to **acetals**.

STUDENT TIP
The term *acetal* is used regardless of whether the starting carbonyl is an aldehyde or a ketone. Historically, the term *ketal* was used exclusively to describe acetals that were made from ketones.

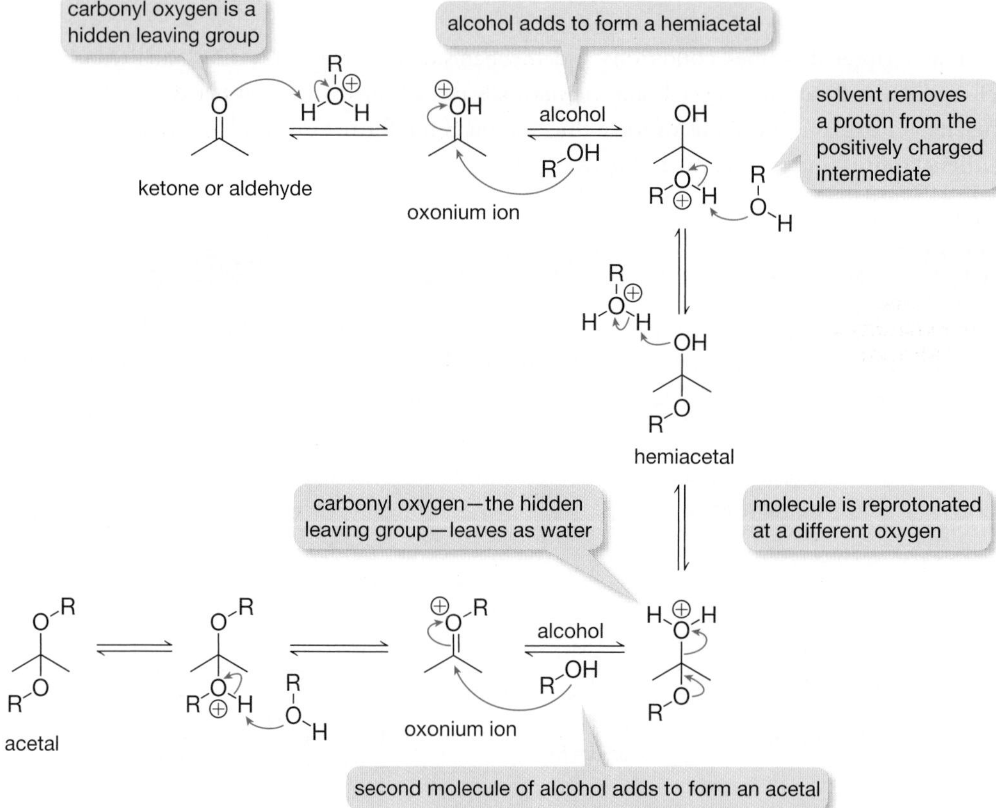
carbonyl oxygen is a hidden leaving group
alcohol adds to form a hemiacetal
solvent removes a proton from the positively charged intermediate
ketone or aldehyde
alcohol
oxonium ion
hemiacetal
carbonyl oxygen—the hidden leaving group—leaves as water
molecule is reprotonated at a different oxygen
alcohol
acetal
oxonium ion
second molecule of alcohol adds to form an acetal

ORGANIC CHEMWARE
16.3 Acetal and hemiacetal formation (acidic conditions)

During acetal formation, the hemiacetal is first converted to an oxonium by protonation and the removal of water (hidden leaving group). Then a molecule of alcohol reacts with the oxonium ion and, after losing a proton, forms an acetal. Proton transfers like the one in hemiacetal formation involve two different protons and are not direct reactions. A surrounding solvent molecule removes the first proton to form the hemiacetal, and a *different* solvent molecule supplies the proton to the oxygen to form the oxonium.

Acetals consist of two alkoxy groups (OR groups) connected to an sp^3-hybridized carbon. They are formed from carbonyl groups only in acidic conditions because the acid is required to convert the OH group of the hemiacetal into a good leaving group.

STUDENT TIP

The oxygen of the carbonyl group (or other heteroatoms of related functionalities such as imines) can serve as a hidden leaving group during the formation of acetals and many other functional groups.

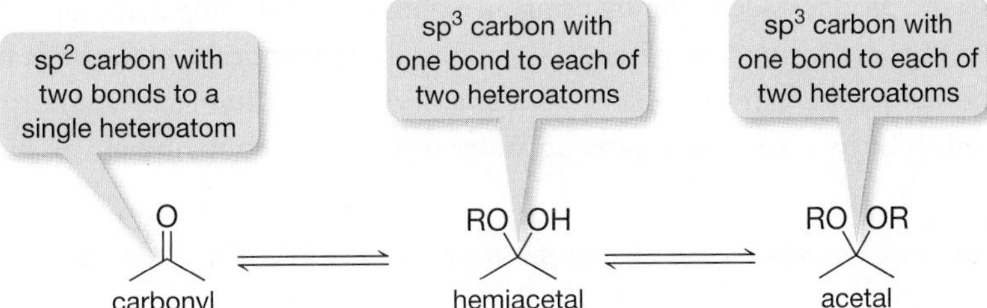

16.2.1 Reversibility of acetal formation

Acetal formation is an equilibrium process because all steps in the sequence can run in both directions. The patterns of electron flow repeat through the reaction sequences; what changes are the nucleophile and leaving groups. When an acetal is converted into the corresponding carbonyl form, the term *hydrolysis* is used. This is because water (hydro) is added to break (lyse) the acetal. The overall mechanism of acetal hydrolysis is simply the reverse of acetal formation.

STUDENT TIP

A common error when analyzing acetal mechanisms is to protonate the "wrong" oxygen. If you are stuck, try protonating a different heteroatom on the functional group. For acetal formation, H_2O is a hidden leaving group, and OH groups should be protonated. For hydrolysis, the OR groups need to be protonated.

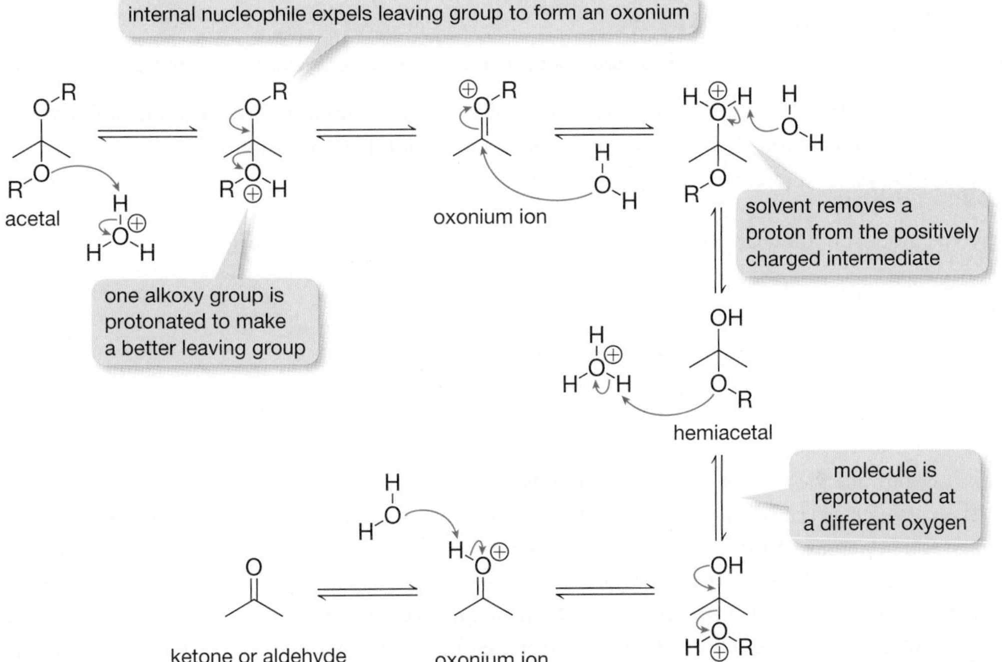

Therefore, the overall equation for acetal formation is as follows:

STUDENT TIP

H^+ is a shorthand notation for acid. It is often used as a reminder that catalytic acid is required in certain reactions.

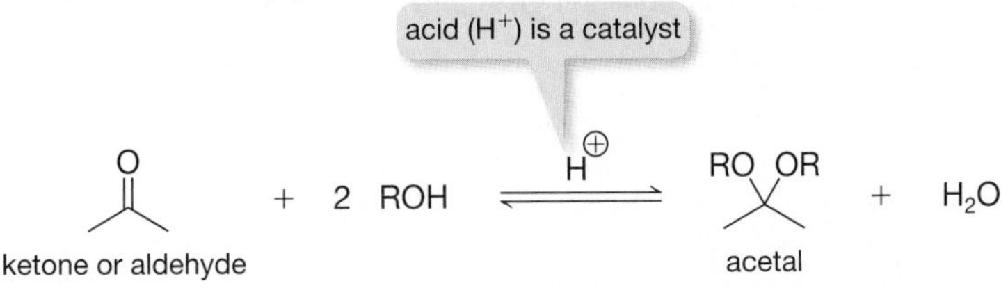

Compared to irreversible reactions, equilibrium reactions produce mixtures that result in a lower yield of the desired product. Since Le Chatelier's principle controls equilibria, chemical synthesis reactions can be shifted toward one side or the other to ensure that mixtures are minimized. For example, if the synthesis of an acetal from a carbonyl is desired, the equilibrium can be made to favour the product side either by using an excess of one starting material or by removing the products as they are formed. In practice, it is often more practical to remove water than one of the other products. This can be done by adding a drying agent (typically an inorganic reagent that absorbs water) or by physically removing water from the reaction, usually by distillation.

using an excess of either reagent can shift the equilibrium toward the products

ketone or aldehyde + 2 ROH ⇌ ($H^{\oplus}$) acetal (RO OR) + H_2O

the reaction is an equilibrium

the removal of water shifts the equilibrium to the product side

CHECKPOINT 16.1

You should now be able to predict the conditions under which an acetal is formed and hydrolyzed. You should be able to draw the products of these reactions along with their mechanisms.

SOLVED PROBLEM

Draw the product of the following reaction and a mechanism showing its formation.

OH $\xrightarrow[\text{MeOH}]{\text{HCl}_{(\text{trace})}}$

STEP 1: Determine the roles of the reactants and the type of reaction taking place.

The starting material contains an electrophilic carbonyl group and a nucleophilic alcohol, placed close enough (six atoms apart) to allow a reaction. In the presence of the acid catalyst (HCl), these will react to form a hemiacetal. The fact that only trace (small) amounts of acid are required is a good clue that this reagent is a catalyst. Note that this intramolecular reaction occurs faster than the intermolecular reaction that is possible between the solvent (MeOH) and the starting material. In the presence of excess MeOH, the hemiacetal can be expected to further react under the acidic conditions to yield an acetal.

STEP 2: Draw a mechanism for the formation of the hemiacetal (if needed, see the mechanisms in Chapter 7).

The acid catalyst first activates the carbonyl group, making it susceptible to nucleophilic attack by the OH group on the same molecule. Deprotonation gives the bicyclic hemiacetal as an intermediate.

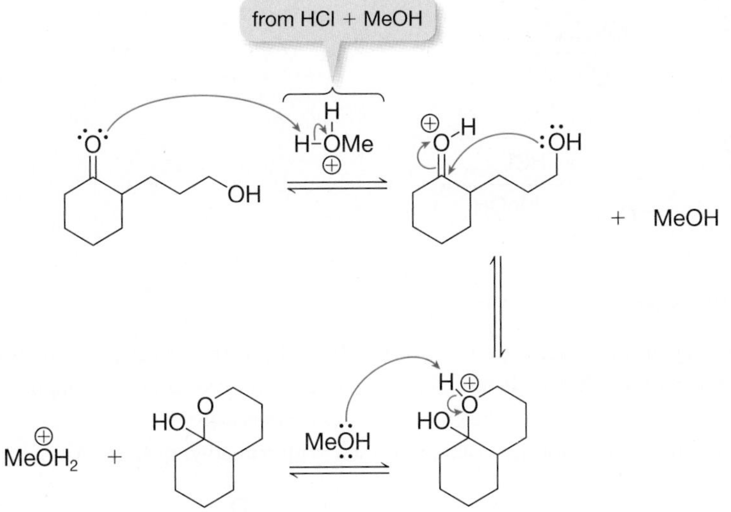

from HCl + MeOH

H–OMe (H, ⊕) + MeOH; MeOH; MeOH₂⊕ +; HO, O

STEP 3: Draw a mechanism for the conversion of the hemiacetal into an acetal.

The OH of the hemiacetal is a hidden leaving group. Once it is protonated, it becomes H_2O, which is readily pushed out of the molecule by the nucleophilic lone pairs on the adjacent oxygen atom. The resulting oxonium ion is positively charged and highly electrophilic. A molecule of solvent (MeOH) donates electrons to the oxonium to give a protonated acetal. Deprotonation then gives the final product.

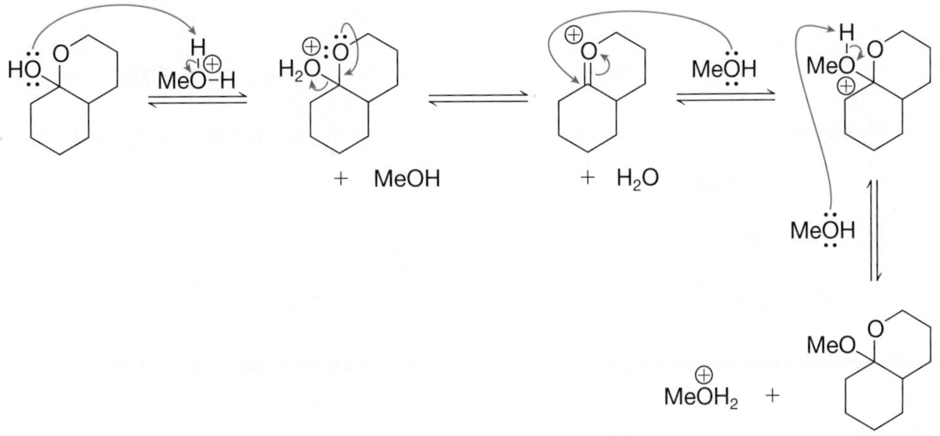

MeO–H (H⊕); + MeOH; + H_2O; MeOH; MeOH; MeO; MeOH₂⊕ +

PRACTICE PROBLEM

16.1 Draw the products of the following reactions and a mechanism for their formation.

a) 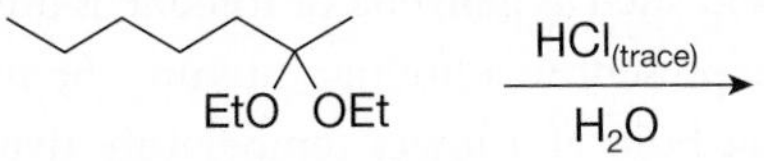 EtO OEt $\xrightarrow[H_2O]{HCl_{(trace)}}$

b) 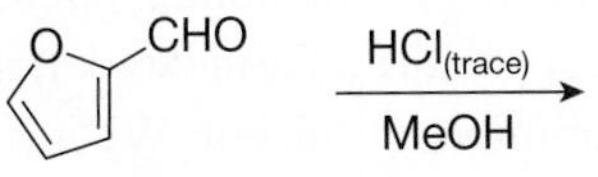 CHO $\xrightarrow[MeOH]{HCl_{(trace)}}$

c) 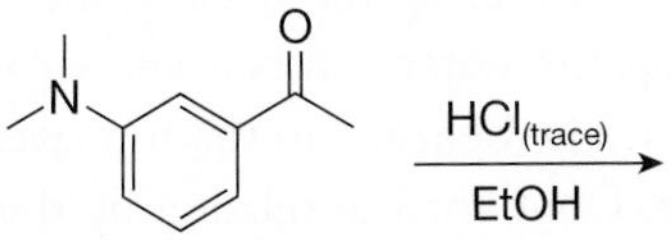 N, O $\xrightarrow[EtOH]{HCl_{(trace)}}$

d) [cyclic acetal: 2-methoxytetrahydrofuran] $\xrightarrow[H_2O]{HCl_{(trace)}}$

e) [3-oxocyclopentanecarbaldehyde] $\xrightarrow[MeOH]{HCl_{(trace)}}$

STUDENT TIP
Orthoesters such as trimethyl orthoformate usually require special conditions to form. They are not commonly produced under the conditions used for acetal formation.

INTEGRATE THE SKILL

16.2 Chemical reagents that react with water, such as trimethyl orthoformate, are sometimes used to remove water for acetal production. The reagent also converts carbonyl compounds to their dimethyl acetals, which can then be converted into other acetals.

a) Propose a mechanism for the reaction of water with trimethyl orthoformate.

$HC(OMe)_3$ + H_2O $\xrightarrow{H^{\oplus}}$ HC(=O)OMe + 2 MeOH

trimethyl orthoformate

b) Why is the reaction in part (a) not reversible?

c) Propose a mechanism for the following reaction.

$HC(OMe)_3$ + cyclohexanone $\xrightarrow{H^{\oplus}}$ 1,1-dimethoxycyclohexane (MeO OMe) + HC(=O)OMe

d) Show, using a mechanism, how the following can occur. What is the by-product of the reaction?

[2-methoxy-1,3-dioxolane with MeO_2C and CO_2Me substituents] $\xrightarrow[TsOH(hydrate)]{PhCHO}$ [2-phenyl-1,3-dioxolane with MeO_2C and CO_2Me substituents]

16.2.2 Removal of water during acetal formation

In acetal formation, H_2O is a hidden leaving group, and water needs to be removed. Experimentally, the water that is formed from a reaction can be removed by an azeotropic distillation as the reaction proceeds. In an azeotropic distillation, a solvent such as benzene or toluene is used to dissolve the mixture. When both water and benzene are present in a boiling mixture, the molecules interact to form an azeotrope: a vapour mixture that boils at a lower temperature than that of the individual components. Because the azeotropic mixture has the lowest boiling point, it leaves the reaction flask before the solvent does. A special glassware apparatus, a Dean–Stark trap, can be used to separate the water from the azeotrope during the continuous distillation. When a reaction mixture boils in a Dean–Stark trap, the azeotrope vapours travel up the neck of the apparatus and into a condenser. The cooled liquid drops into a trap, the water and benzene separate into distinct layers, and water is collected on the bottom layer. The benzene in the top layer returns to the reaction flask and recycles back into the reaction. Continued distillation by this process removes all the water present.

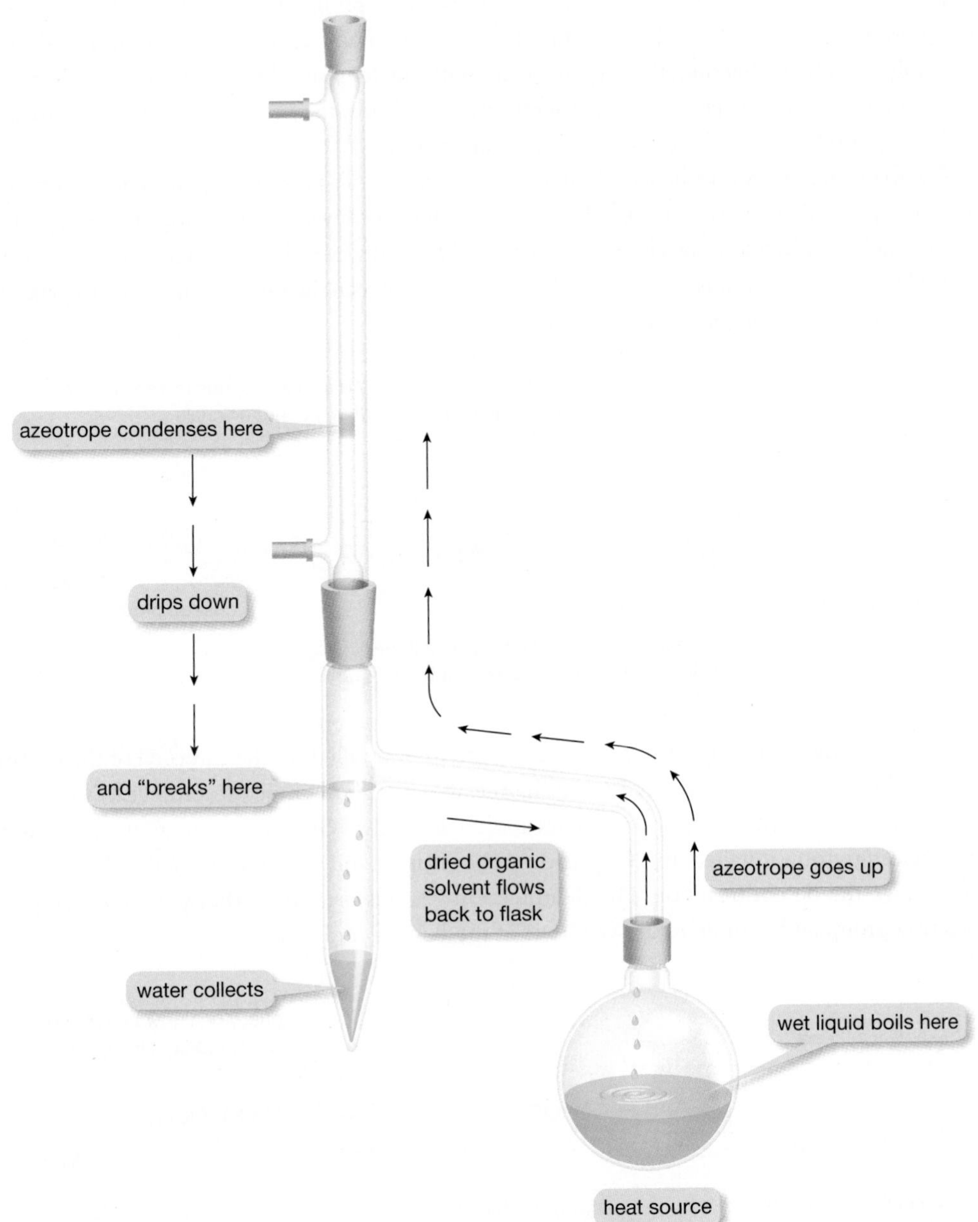

16.2.3 Acetals used as protecting groups

Synthesizing molecules from simpler ones is a major goal of organic chemistry. Most molecules contain more than one functional group. This attribute can create problems for chemical synthesis during some reactions. If more than one functional group is capable of reacting under a set of conditions, different products may be formed. To ensure that a reaction occurs at a particular functional group, the other functional groups present in the molecule may have to be temporarily inactivated. This is done by the use of a **protecting group**. A protecting group changes a functional group into a related functional group that will not react. Once the desired reaction is complete, the protecting group can be removed during a "deprotection" step, thereby restoring the original functionality. The characteristics of an effective protecting group include the ease with which it can be installed and removed and its inability to participate in the reactions for which the functional group is being protected.

A **protecting group** is a temporary change installed on a functional group to reduce the chemical reactivity of that functional group.

The reactivity pattern of acetals makes them useful as protecting groups for aldehydes and ketones, as well as for alcohols. First, acetals do not contain any acidic hydrogens and so do not react easily with bases. Second, they do not react with nucleophiles because they have no bonds that can be easily broken (π bonds or leaving groups). Third, they are easily installed onto carbonyl compounds and can be removed using acid catalysis.

The following reaction sequences illustrate the importance of a protecting group. The first reaction sequence depicts a desired reaction whereby an alkyl bromide is converted into a Grignard reagent, then reacted with benzaldehyde, to produce a secondary alcohol. However, this desired reaction sequence is impossible to carry out because a cross-reaction can occur between the Grignard intermediate and the carbonyl group already present in the molecule.

desired reaction is Grignard addition of one molecule with the carbonyl group of a second molecule

O, Br; Mg, Et_2O; O, MgBr; 1) PhCHO, 2) NH_4Cl; O, Ph, OH

desired reaction

desired product

Grignard formation is impractical if the molecule already contains a carbonyl

To achieve the desired reaction, a protecting group is installed on the carbonyl of the starting compound to prevent the undesired reaction. Acetals are a good choice to protect this carbonyl because they are generally inert toward bases and nucleophiles (Grignard reagents are both). With the protecting group installed, the Grignard can be formed and allowed to react with benzaldehyde to give the desired reaction. The deprotection step restores the carbonyl by removing the protecting group and completes the synthesis of the desired molecule.

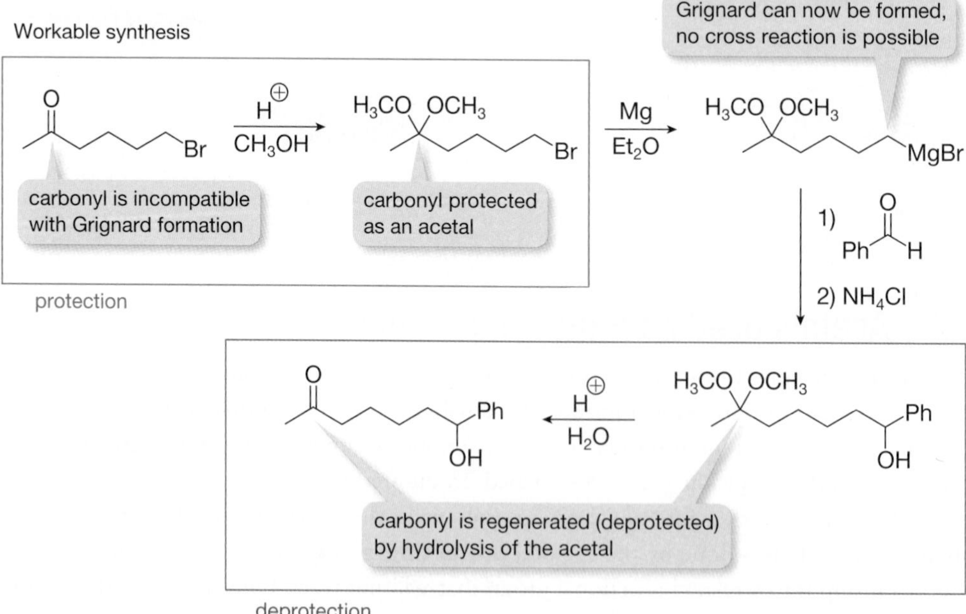

Protection and deprotection introduce additional steps into a synthetic sequence. However, they are necessary when synthesis involves molecules that contain functional groups that are incompatible with a desired chemical reaction.

During chemical synthesis, it is common to use diols to form protecting groups for carbonyls. When diols are used, the reaction forms a cyclic acetal in which both oxygens are embedded in a ring, and this increases the stability of the acetal. The stability of the ring structure means it is more difficult to hydrolyze a cyclic acetal than an acyclic one, and this is because intramolecular reactions normally occur faster than intermolecular ones. In the first step of acetal hydrolysis, an oxonium ion is formed after the removal of one alcohol unit. An acyclic acetal produces a separate molecule of alcohol during this process. The reverse reaction therefore requires that two molecules come together in the right way to react, and this process happens relatively slowly. In a cyclic acetal, the alcohol unit remains attached to the oxonium by a tether that holds the OH group close to the oxonium; this makes the ring closure in the reverse reaction much faster. Consequently, the cyclic acetal is more difficult to "remove" than the acyclic one. This extra stability leads them to be used commonly as protecting groups.

Acetals can also be used to protect OH groups. In carbohydrate chemistry, this is done to differentiate various OH groups in sugar molecules. For this purpose, a carbonyl-containing compound can be used to install the acetal protecting group.

In the preceding example, an acetal was used to protect a diol. However, acetals can also be used to protect lone hydroxyl groups. A common reagent for this purpose is called dihydropyran (DHP). In the presence of acid, DHP reacts with an alcohol to form a tetrahydropyranyl (THP) protecting group. In the reaction, DHP is protonated by the acid to give an oxonium ion intermediate, which is then attacked by the alcohol to form the THP acetal. The reagent contains a functional group—called an enol ether—that reacts very quickly with acid to form an oxonium (stabilized carbocation).

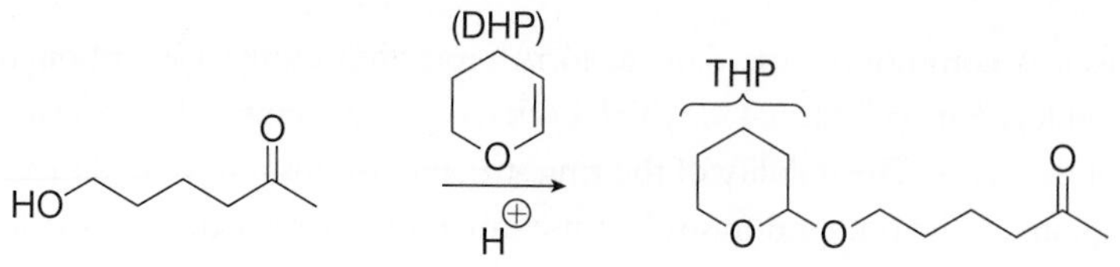

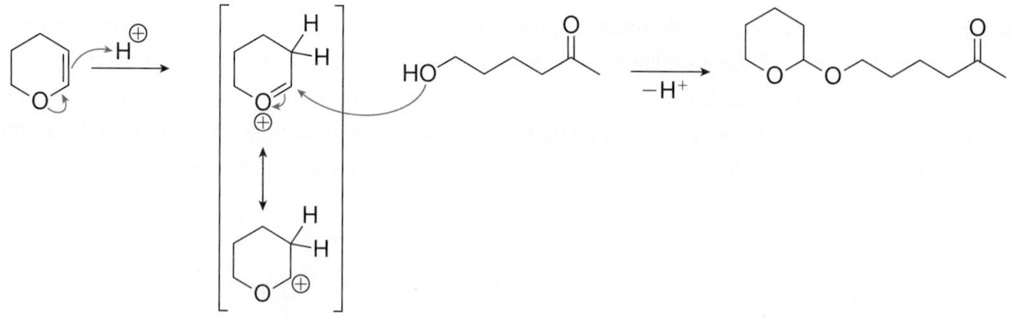

(DHP)

THP

HO O H⊕ O O O O

H⊕ H H HO O $-H^+$ O O O

To remove the THP group, the molecule is treated with water and acid.

16.2.4 Retrosynthetic analysis of hidden leaving groups

Acetals, hemiacetals, and hydrates all disconnect to the corresponding carbonyl form. Similarly, carbonyls can be produced from the corresponding acetals.

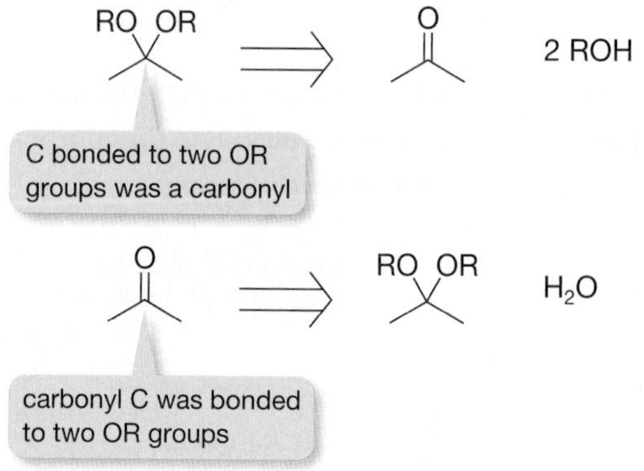

RO OR ⟹ O 2 ROH

C bonded to two OR groups was a carbonyl

O ⟹ RO OR H_2O

carbonyl C was bonded to two OR groups

The interchange between these groups is called *functional group interconversion* because the two groups can be interchanged directly.

CHECKPOINT 16.2

You should now be able to plan short syntheses involving the use of an acetal protecting group. You should also be able to predict when such a protecting group is required and provide the conditions for its formation and removal.

SOLVED PROBLEM

Propose a method of carrying out the following transformation.

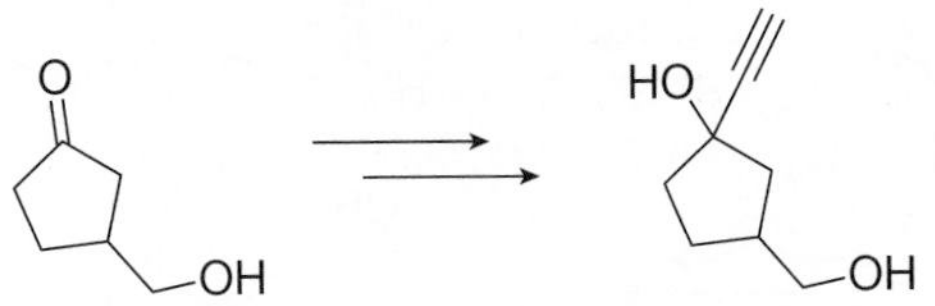

STEP 1: Count the number of carbon atoms in the starting material and the final product.

The starting material has six carbon atoms while the final product has eight.

STEP 2: Determine which bonds have been broken or formed in the synthesis.

The C=O π bond has been broken, and new C–H and C–C σ bonds formed.

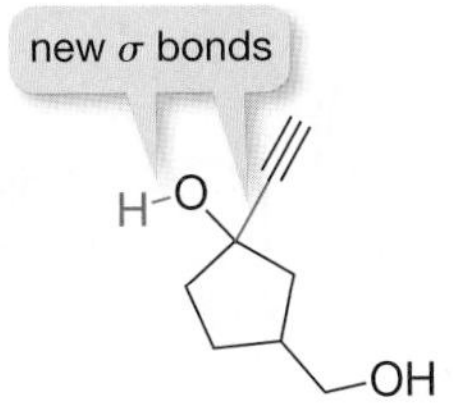

STEP 3: Disconnect the new bonds in the product to obtain the appropriate starting materials and reagents.

The new C–C σ bond could be generated from a ketone and an acetylide ion in a nucleophilic addition reaction.

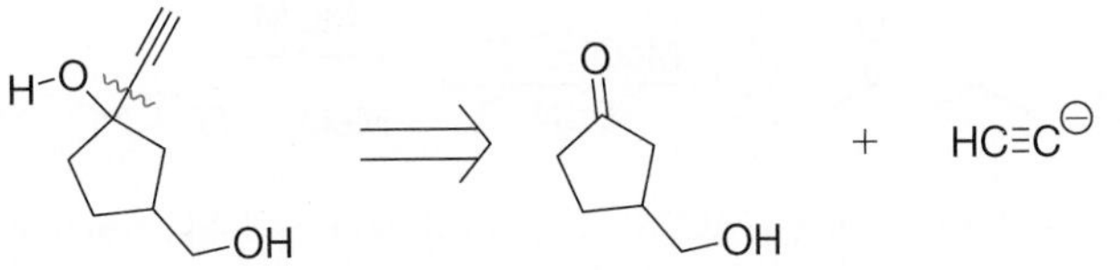

STEP 4: Evaluate the feasibility of the forward synthesis.

Acetylide ions are highly basic, and the starting material contains an acidic OH group. Rather than the desired nucleophilic attack on the carbonyl, these two reactants would undergo an acid–base reaction when mixed.

STEP 5: Propose a modification to the above route that would allow the nucleophilic addition reaction to take place instead of the acid–base reaction.

To prevent the acid–base reaction from occurring, the alcohol can be protected as a THP acetal. Since acetals are stable in basic conditions, the acetal will not interfere with the nucleophilic addition of the acetylide ion. After the addition reaction is complete, the THP group can be removed by hydrolysis. In fact, this step could be combined with the acidic aqueous workup required at the end of the acetylide ion addition.

STEP 6: Write out the overall reaction sequence for the synthesis, including the protection and deprotection of the alcohol.

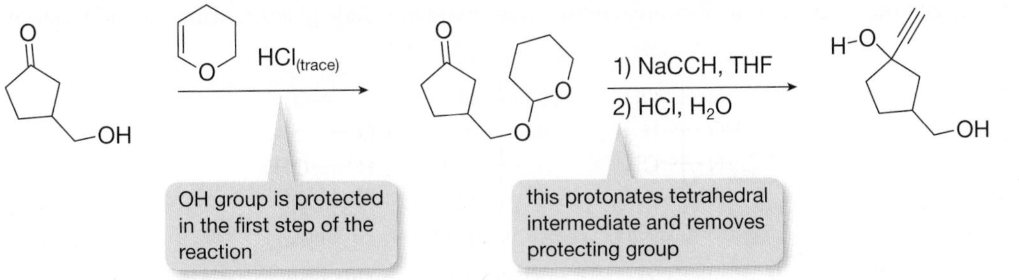

PRACTICE PROBLEM

16.3 Suggest a series of reagents that could be used to carry out the following transformations.

a) 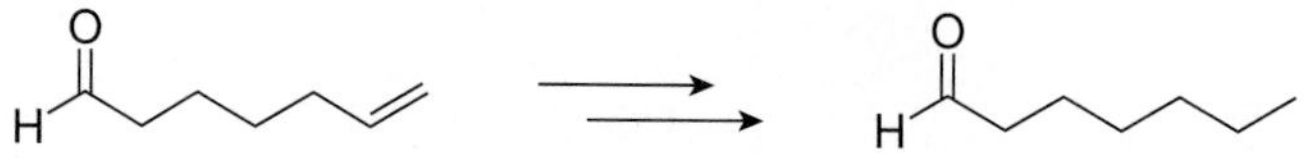 (aldehyde with terminal alkene) ⟶⟶ (saturated aldehyde)

b) 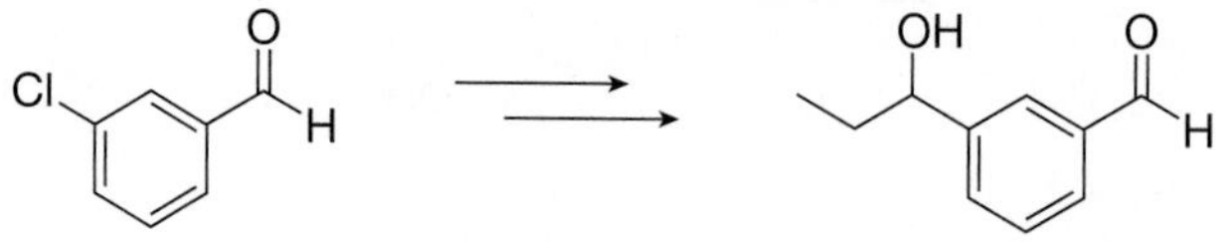 (3-chlorobenzaldehyde) ⟶⟶ (3-(1-hydroxypropyl)benzaldehyde)

c) 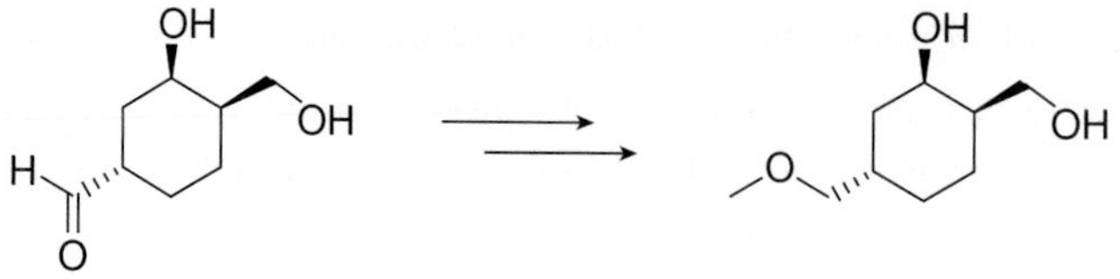 (cyclohexane bearing OH, CH₂OH and CHO) ⟶⟶ (cyclohexane bearing OH, CH₂OH and CH₂OCH₃)

d) 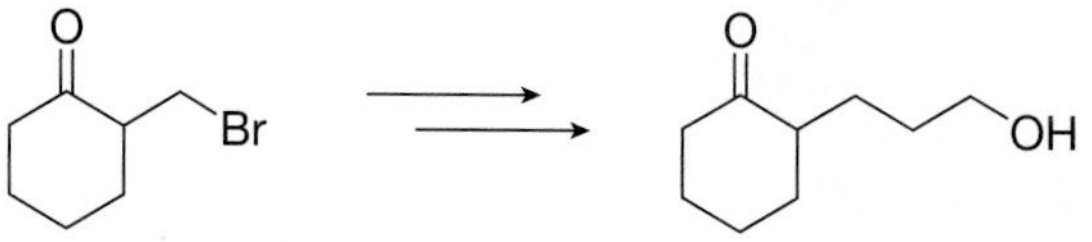 (2-(bromomethyl)cyclohexanone) ⟶⟶ (2-(3-hydroxypropyl)cyclohexanone)

INTEGRATE THE SKILL

16.4 Methoxymethyl (MOM) groups are commonly used as protecting groups for hydroxyls (OH). The reagent that is used to install this group is called MOM-Cl.

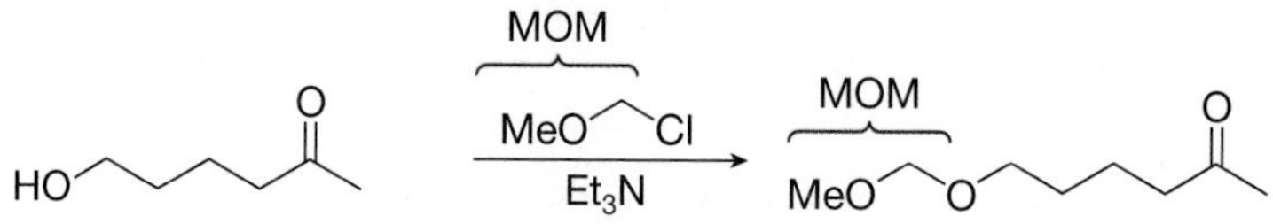

a) The mechanism for installing a MOM group involves an S_N1-like reaction (oxonium). Give a mechanism for its reaction with the compound shown.

b) MOM groups are removed using H_2O and acid. Give a mechanism for this process.

16.3 Acetals in Sugars and Carbohydrates

Acetals and hemiacetals are very important in carbohydrate chemistry. Carbohydrates (sugars) are used by living things for a variety of purposes: as energy sources, structural elements, and communication devices. Carbohydrates are composed of carbon chains to which are connected oxygen-containing functional groups (hydrates of carbon) such as hydroxyls, aldehydes, and carboxylic acids. Some carbohydrates also incorporate nitrogen or other heteroatoms.

Monosaccharides consist of a single continuous carbon chain bearing a variety of oxygen-based functionalities. These molecules are used as building blocks to create larger molecules that are important for metabolism and function. Examples of monosaccharides include glucose, mannose, and fructose.

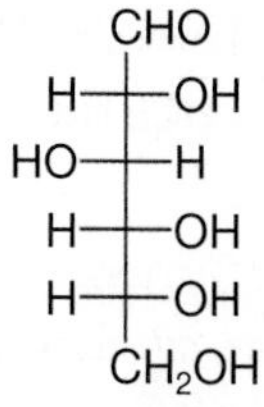

glucose	mannose	fructose
CHO	CHO	CH_2OH
H—OH	HO—H	=O
HO—H	HO—H	HO—H
H—OH	H—OH	H—OH
H—OH	H—OH	H—OH
CH_2OH	CH_2OH	CH_2OH

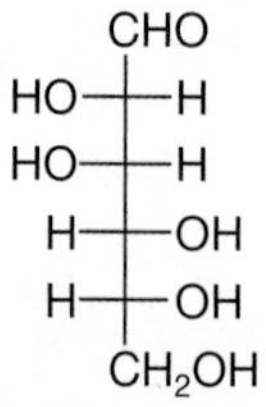

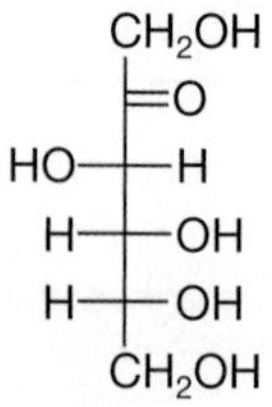

Monosaccharides can be connected to make larger molecules called polysaccharides. The connections between the monosaccharide subunits are made by forming acetals, whereby an OH group from one sugar and a carbonyl group from the other sugar make the link. Polysaccharides can contain as few as two subunit sugars or can be composed of thousands.

Acyclic hemiacetals are usually not very stable. The OH of the hemiacetal is easily removed, and this makes the oxygen of the OH a fairly good nucleophile that can "expel" the other oxygen and convert the hemiacetal into the corresponding carbonyl compound and alcohol. If the carbonyl and the hydroxyl group are located on different molecules, hemiacetal hydrolysis will happen faster than hemiacetal formation, which makes the equilibrium favour the carbonyl form. If the carbonyl and the hydroxyl group are located on the same molecule, hemiacetal formation may be faster than hemiacetal hydrolysis, in which case the reaction will favour hemiacetal formation. The presence of carbonyls and hydroxyl groups on the same carbon chain in carbohydrates results in these groups interacting to form cyclic acetals and hemiacetals.

alcohol and carbonyl are favoured

alcohol and carbonyl on separate molecules

alcohol and carbonyl on same molecule

new bond

hemiacetal form is favoured due to intramolecular reaction

Most carbohydrates with four or more carbons predominately exist as cyclic hemiacetals rather than as open-chain compounds because the ring form is favoured over the open-chain form. In the acyclic form, the alcohol group and the carbonyl are located on the same molecule, and due to their close proximity they are extremely likely to react. Therefore, the equilibrium shifts toward the cyclic hemiacetal product. This process is highly favoured when five- and six-membered rings are formed from the acetal linkage.

new bond

close proximity of OH and carbonyl group makes it easy for them to react

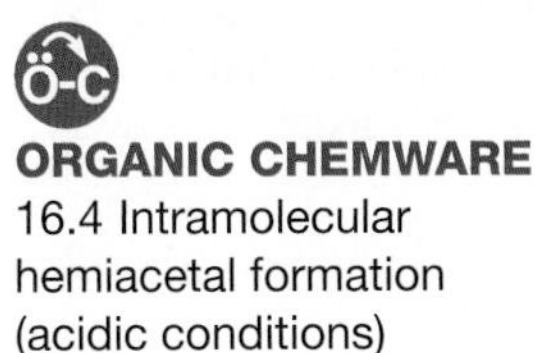

ORGANIC CHEMWARE
16.4 Intramolecular hemiacetal formation (acidic conditions)

Glucose exists as an equilibrium mixture of an open-chain form and several cyclic hemiacetal forms. Both five- and six-membered cyclic forms are possible, but the six-membered rings are heavily favoured because they can adopt chair conformations, which reduce ring strain.

The formation of a cyclic hemiacetal generates an sp^3 centre at the site of the acetal carbon that has four different substituents. This gives rise to two stereoisomers called **anomers**—stereoisomers that arise by a reaction at the carbonyl carbon of a carbohydrate.

An **anomer** is one of the stereoisomers formed when the carbonyl carbon of a sugar is converted to a cyclic acetal or hemiacetal.

The **anomeric carbon** is the carbon that gives rise to the two anomers.

The acetal carbon is called the **anomeric carbon** because it gives rise to anomers. Anomers exist because the mechanism of acetal formation converts the sp^2-hybridized carbon of a carbonyl into the stereogenic sp^3-hybridized carbon of an acetal or hemiacetal.

anomers
anomeric carbon
cyclic products with six-membered ring
glucose
acyclic form in trace amounts
H_2O
38.9%
60.9%
anomeric carbon
cyclic products with five-membered ring
0.1%
0.1%
anomers

16.3.1 Naming anomers

A special nomenclature exists to describe the configuration of the anomeric carbon relative to the ring carbon of the sugar that is furthest from the anomeric position. The symbols α and β describe the relative configuration between this furthest-ring carbon and the anomeric carbon of a cyclic sugar. If the substituents at the anomeric carbon and the ring-carbon that is furthest from the anomeric carbon are on opposite sides of the ring (*trans*), the anomeric carbon has an α configuration, and the molecule is called the α anomer. If the substituents at the furthest-ring carbon and at the anomeric carbon are on the same face of the ring (*cis*), a β anomer is produced.

STUDENT TIP
A shortcut to remember α and β that works for most carbohydrates is that the hemiacetal OH of the α-anomer (**a**lpha) is normally **a**xial.

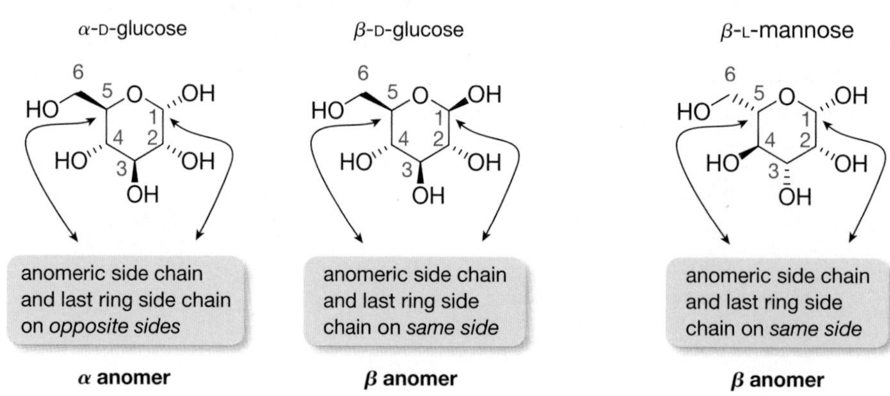

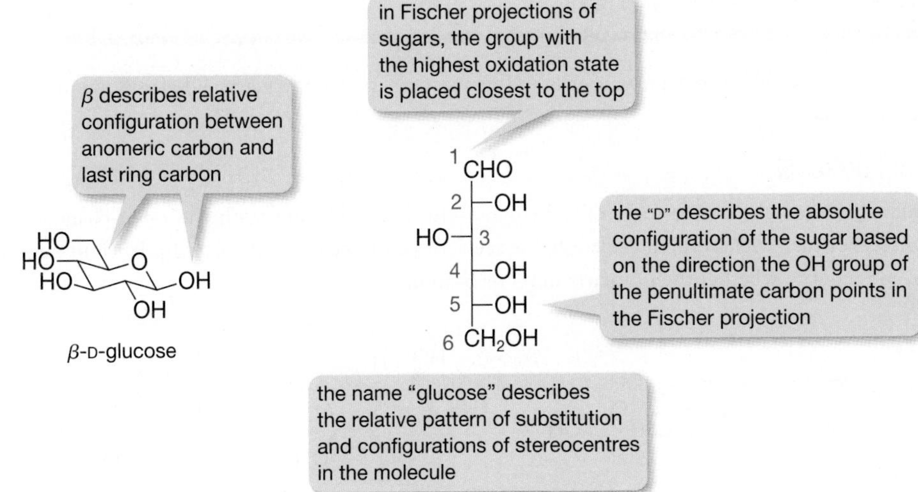

in Fischer projections of sugars, the group with the highest oxidation state is placed closest to the top
β describes relative configuration between anomeric carbon and last ring carbon
the "D" describes the absolute configuration of the sugar based on the direction the OH group of the penultimate carbon points in the Fischer projection
the name "glucose" describes the relative pattern of substitution and configurations of stereocentres in the molecule
CHO
OH
HO
OH
OH
CH₂OH
β-D-glucose

When sugar units are connected together by means of acetals, the bond at the anomeric carbon is called a glycosidic bond (glycosidic linkage), and the structures that are formed are called glycosides. Glycosidic bonds can adopt one of two configurations, α or β, depending on the relative configuration between the anomeric carbon and ring carbon furthest from the anomeric position.

The configuration of glycosidic bonds affects the properties of the compound formed. For example, glucose forms two polysaccharides that differ only in the configuration of the anomeric position. In starch, an important energy-storage molecule, all the anomeric carbons have an α configuration. When the body needs energy, enzymes digest the starch into component glucose units that are then used as fuel. In the human digestive system, enzymes in saliva, stomach, and gastrointestinal tract break down starch very quickly.

starch

cellulose

Cellulose, the most abundant organic chemical on this planet, consists of large numbers of glucose units connected by β glycosidic bonds. Primarily a structural material, cellulose provides the strength and toughness of materials such as wood, paper, and cotton. Most mammals cannot digest cellulose (except ruminants, which carry special bacteria in their stomachs that digest cellulose). The cellulose portion of food therefore passes through the digestive system as fibre. Fibre helps to maintain the healthy functioning of the intestines and remove harmful substances from food. Many of these substances are non-polar and become adsorbed (stick) to the fibre, rather than remain dissolved in the water present in the intestine. This reduces the amounts of such substances taken in to the body, because the body can usually only absorb materials from the intestine that are in solution. The fibre passes out of the digestive system, taking the non-polar materials with it.

CHECKPOINT 16.3

You should now be able to recognize acetals and hemiacetals in carbohydrates and classify them as either α- or β-anomers.

SOLVED PROBLEM

The following four cyclic structures of D-mannose are in equilibrium with the open-chain form of D-mannose. For each cyclic structure, identify the acetal or hemiacetal functional group, label the anomeric carbon, and indicate whether the isomer is an α- or β-anomer.

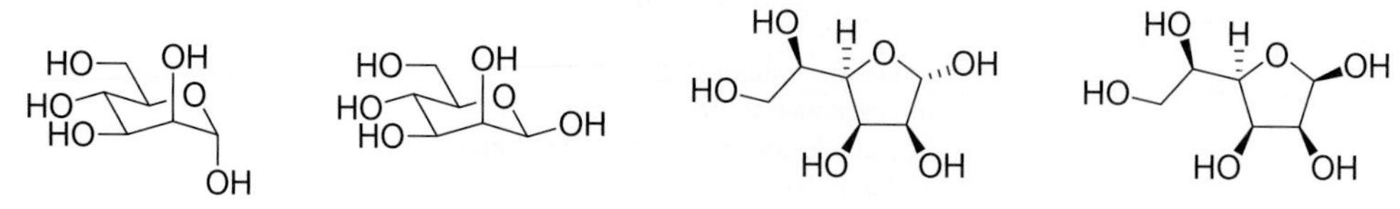

A B C D

STEP 1: Identify any acetal functional groups in the structures.

An acetal consists of two alkoxide groups (OR) attached to an sp^3-hybridized carbon atom. There are no acetals in any of the structures.

STEP 2: Identify any hemiacetal functional groups in the structures.

A hemiacetal has a hydroxyl group (OH) and an alkoxide group (OR) attached to the same sp^3-hybridized carbon atom. Structures A–D all contain one hemiacetal, identified here in red.

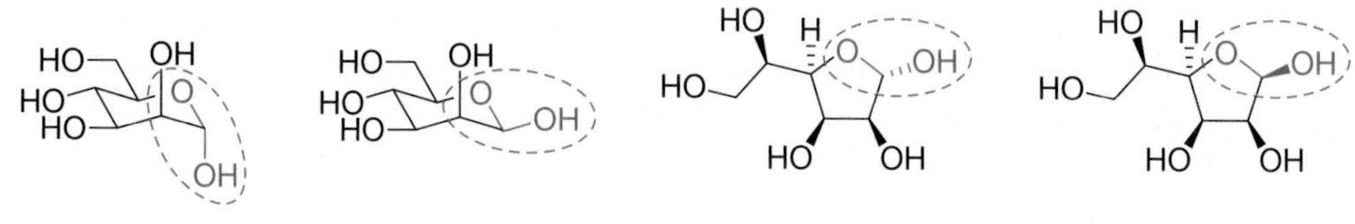

A B C D

STEP 3: Locate the sp^3-hybridized carbon in the centre of each hemiacetal, and label it as the anomeric carbon.

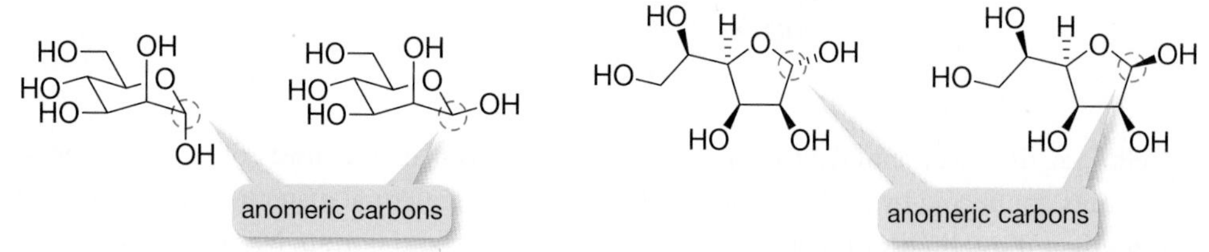

anomeric carbons

anomeric carbons

A B C D

STEP 4: Number the carbon atoms in each structure, starting with the anomeric carbon as number 1.

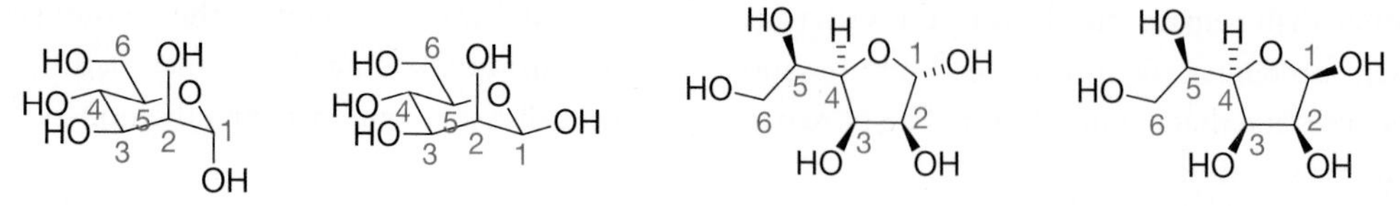

A B C D

STEP 5: Compare the relative stereochemistry at the penultimate carbon (C-5) and the anomeric carbon (C-1).

In structures A and C, the OH groups at carbons 1 and the side chains on the last ring carbon (C-5 in A and C-4 in C) are pointing in the opposite direction, so these hemiacetals are α-anomers. In structures B and D, the OH groups at carbons 1 and the side chains on the last ring carbon (C-5 in B and C-4 in C) are pointing in the same direction, so these hemiacetals are β-anomers.

PRACTICE PROBLEM

16.5 For each of the following carbohydrates, identify the acetal and hemiacetal functional groups, label the anomeric carbon(s), and indicate whether the α- or β-anomer is present.

a) Lactose, a disaccharide composed of galactose and glucose, is a naturally occurring sugar found in milk.

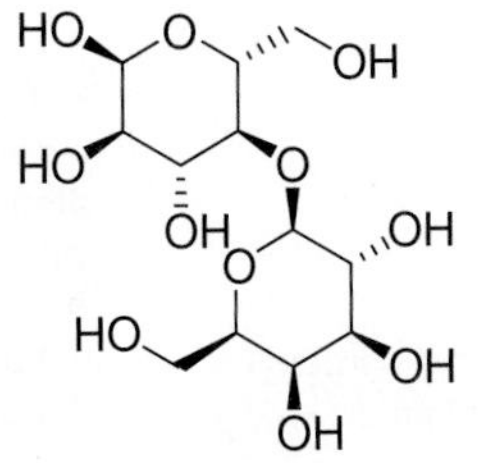

lactose

b) Sucrose (table sugar) is a disaccharide composed of glucose and fructose.

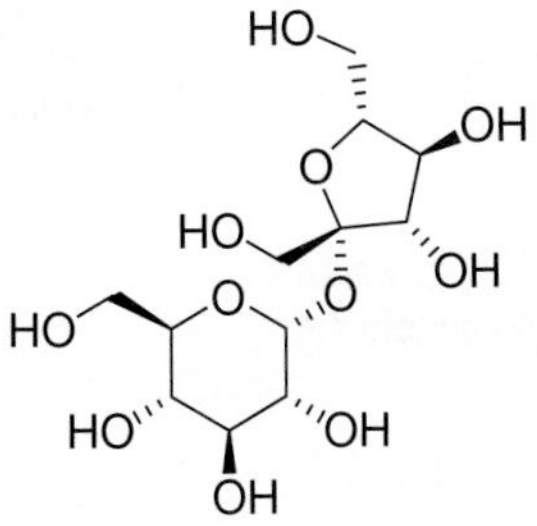

c) Raffinose, a trisaccharide composed of fructose, glucose, and galactose, is found in many vegetables, legumes, and whole grains.

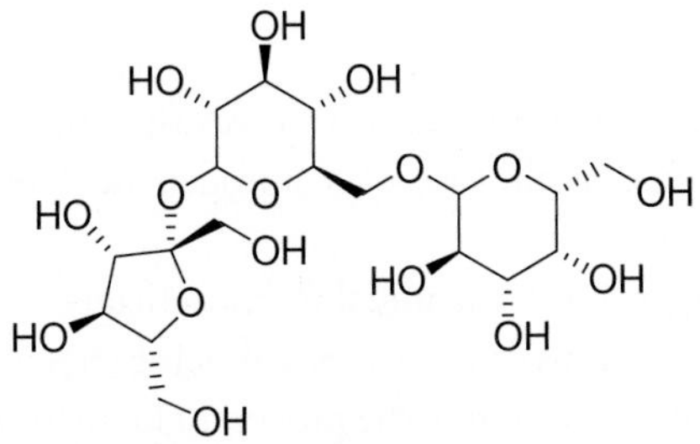

raffinose

INTEGRATE THE SKILL

16.6 a) Draw the four possible hemiacetals of D-galactose, and label each as either the α- or β-anomer.

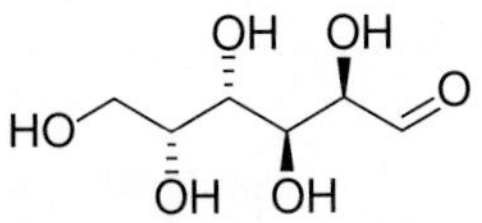

D-galactose

b) What are the relationships between the cyclic structures drawn in part (a)? Identify any enantiomers, diastereomers, and constitutional isomers.

16.4 Aminals and Imines

Nitrogen can form multiple bonds to carbon, and nitrogen-containing functional groups participate in the same types of chemical reactions as the analogous oxygen-based groups do.

An **imine** consists of a carbon connected to a nitrogen by a double bond.

When carbonyl groups are reacted with amines under the proper conditions, the result is the formation of an **imine** (an older term for a substituted imine is Schiff base). The mechanism of imine formation involves two steps: (1) nucleophilic attack of an amine onto a carbonyl carbon to form a hemiaminal; and (2) the removal of water from the hemiaminal to produce an iminium ion (analogous to an oxonium ion), which gives an imine. These two steps have different requirements with respect to the pH of the surrounding environment. The steps of the reaction are shown in the following diagram.

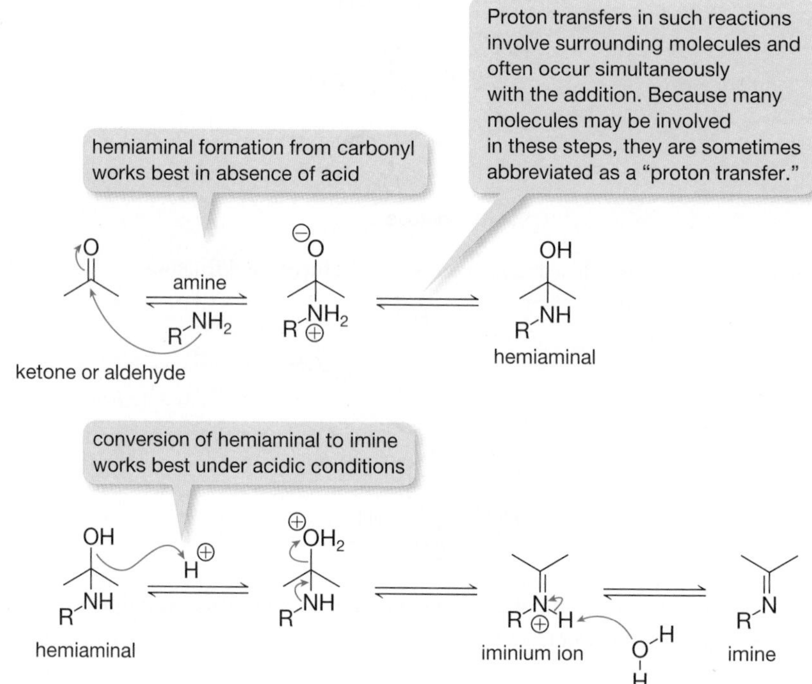

ORGANIC CHEMWARE
16.5 Imine formation

The reaction sequence of imine formation involves the same mechanistic steps that occur in acetal formation. What differs is the use of a nitrogen nucleophile rather than an oxygen counterpart in the initial reaction.

The first step in imine formation is the initial reaction that uses a nitrogen nucleophile in the formation of the hemiaminal. Because nitrogen is a much better nucleophile than oxygen, the carbonyl group does not require activation in the process of hemiaminal formation. Consequently, an acid is not needed as a catalyst and is, in fact, detrimental to the reaction.

The initial nucleophilic reaction between an amine and a carbonyl forms a charged intermediate. The next reaction involves the rapid removal of H^+ from the nitrogen atom and the addition of H^+ to the oxygen atom to prepare for the rapid formation of the hemiaminal. These steps are extremely rapid and may occur simultaneously. In fact, the reaction is more properly shown with simultaneous proton transfers with solvent.

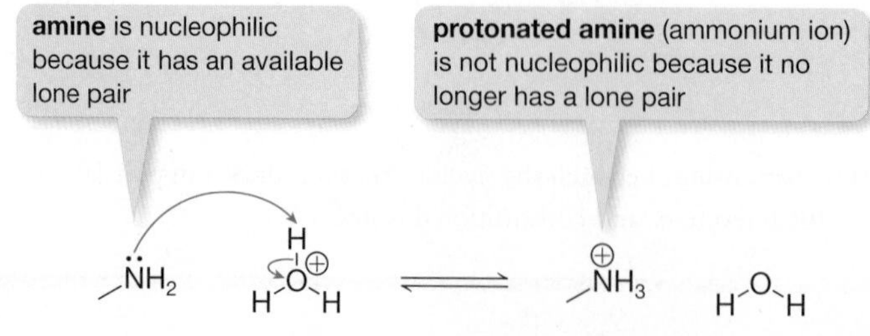

DID YOU KNOW?

Proton transfers like the one in hemiaminal formation involve two different protons and are not direct reactions. A surrounding solvent molecule removes the proton from the nitrogen, while a *different* solvent molecule supplies the proton to the negative oxygen. Further hydrogen transfers between the solvent balance the charges. These transfers normally happen at the same time as the addition or elimination reactions.

hemiaminal

In the first step of imine formation, a strong acid will slow the process because amines are basic and react quickly with acid, forming an ammonium salt. The nitrogen atom of an ammonium salt does not have a free lone pair and cannot be a nucleophile. So the first step is inhibited by acid. The second step of imine formation is accelerated by acid. In fact, it requires an acid to convert the hydroxyl group into a better leaving group (OH_2^+). Therefore, the requirements for acid catalysis in the two steps are opposite to each other. This means that imine formation can occur only in a very narrow pH range, where an acid and a base can co-exist in reasonable amounts: a pH range between 4 and 6. Acetic acid (abbreviated as AcOH or HOAc) is commonly used as a solvent for this process because the pK_a of acetic acid (HOAc) is 4.76, and its use sets the pH of the reaction near the optimal point for imine formation.

STUDENT TIP

A common error in showing the mechanism of a proton transfer is to indicate a proton transfer via a four-membered transition state. It is important to remember that the two protons involved in a proton transfer are normally not the same and are provided by different molecules.

this step works best above pH 4

this step works best below pH 6

ketone or aldehyde — amine — hemiaminal — iminium ion — imine

Imines are usually unstable and are rapidly hydrolyzed to the corresponding carbonyl forms. The process of imine hydrolysis readily occurs when an acid is present. This is because the acid reacts with the amine that is formed, which shifts the equilibrium toward products according to Le Chatelier's principle.

imine — amine — ammonium salt

The mechanism of imine hydrolysis follows the same pathway as the imine formation reaction, but in the reverse direction (the electrons flow in the opposite direction). Because of their instability, acyclic imines are not often isolated. To form acyclic imines, water must be removed and the reaction driven toward the imine.

Some acyclic imines are stable, specifically the ones in which the imine group is conjugated, especially with an aromatic system. Cyclic imines are normally more stable than their open-chain counterparts, and so they can be isolated. The reason for this is the same reason that cyclic hemiacetals are more stable than their corresponding acyclic counterparts; the two components that come together to form the hemiaminal (amine and carbonyl) are located on the same molecule and can react very quickly with each other.

16.4.1 Conversion to enamines

Iminium ions form when carbonyl compounds react with secondary amines. These ions are unable to convert to the corresponding imine because there is no H^+ on the nitrogen that can be removed. Instead, these molecules tend to lose a H^+ from the carbon next to the iminium ion and thereby form enamines. The process is analogous to enol formation.

aldehyde or ketone + secondary amine ⇌ iminium ion

nitrogen does not have a hydrogen that can be deprotonated

hydrogen α to iminium ⇌ enamine

Enamines, the nitrogen equivalent of enols, carry out the same chemistry that enols do. They are important as nucleophiles and therefore are frequently used in aldol-type processes (Chapter 17). Unlike enols, secondary enamines are often stable enough to be isolated. This is because there is no group (H) on the nitrogen atom that can be easily removed.

ORGANIC CHEMWARE
16.6 Tautomerization of imines

CHECKPOINT 16.4

You should now be able to predict the conditions under which imines and enamines are formed and hydrolyzed. You should be able to draw the mechanisms and products of these reactions.

SOLVED PROBLEM

Predict the product of the following reaction, and draw a mechanism for its formation.

AcOH →

STEP 1: Determine the roles of the reactants and the type of reaction taking place.

The starting material contains both an electrophilic ketone and a nucleophilic amine. In the presence of the weakly acidic solvent (acetic acid), these two functional groups react to form an imine via a hemiaminal intermediate.

STEP 2: Draw a mechanism for the formation of the hemiaminal.

The nucleophilic amine shares its electrons with the electrophilic ketone, creating a new C–N σ bond and breaking the C=O π bond. The tetrahedral intermediate is quickly protonated by a molecule of acetic acid, followed by a deprotonation of the nitrogen atom to give the neutral hemiaminal intermediate.

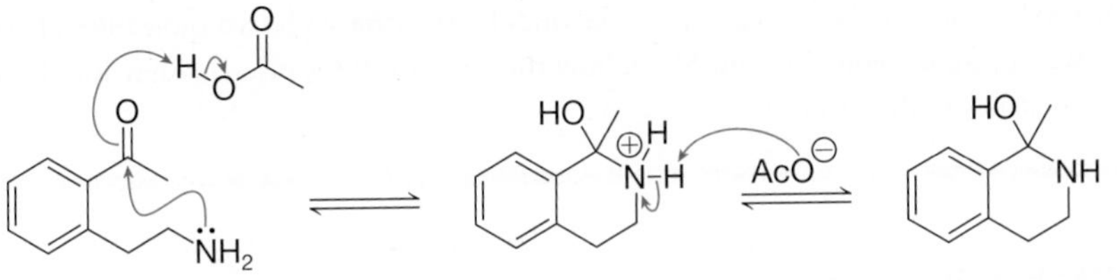

STEP 3: Draw a mechanism for the conversion of the hemiaminal into the imine.

The OH of the hemiaminal is a hidden leaving group. Following protonation by acetic acid, the H_2O leaving group is readily pushed out of the molecule by the nucleophilic lone pairs on the nitrogen atom. Deprotonation of the nitrogen gives the final imine product.

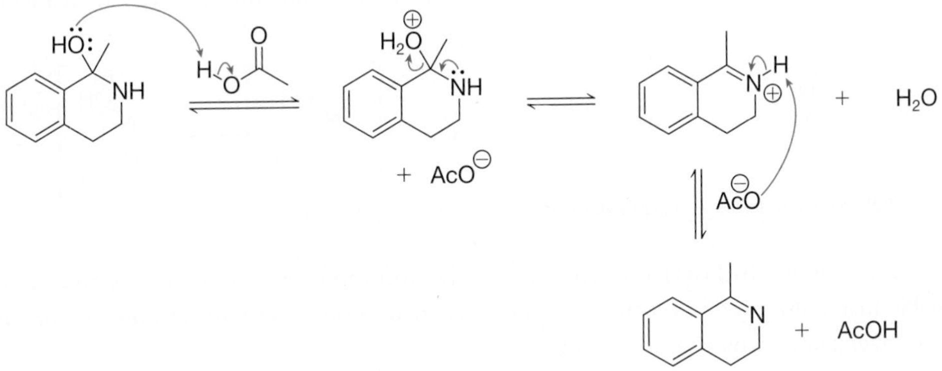

PRACTICE PROBLEM

16.7 For each of the following reactions, predict the major products and draw a mechanism for their formation.

a) 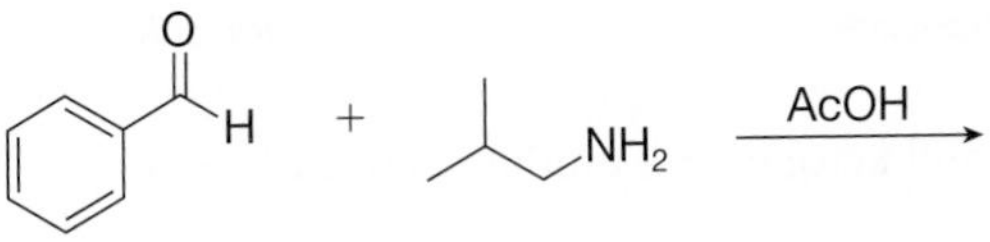 + AcOH →

b) 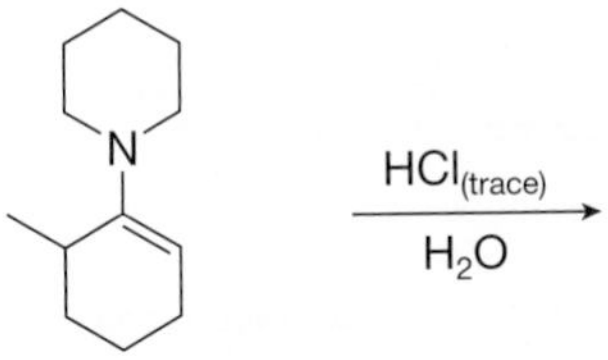 $HCl_{(trace)}$, H_2O →

c) 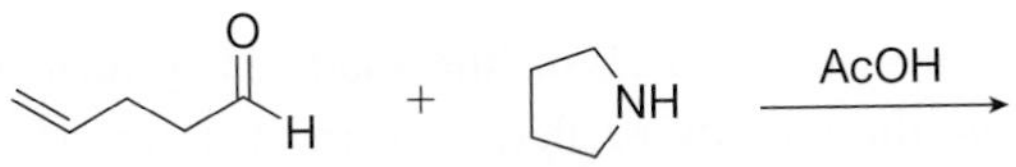 + AcOH →

d) 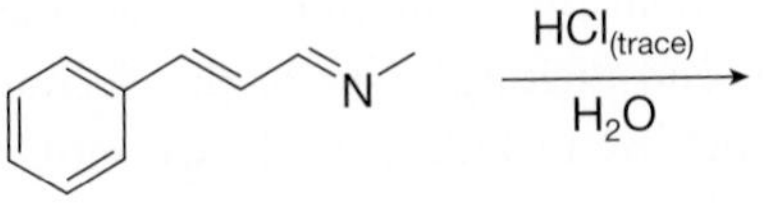 $HCl_{(trace)}$, H_2O →

INTEGRATE THE SKILL

16.8 a) Can pyridine (C_6H_5N) be used to form an imine? An enamine? Why or why not?

b) When a Grignard reagent reacts with a nitrile, a ketone is formed upon aqueous acid workup. Show how this reaction takes place by drawing the key intermediates in the reaction.

c) When a Grignard reagent reacts with a Weinreb amide, a ketone is formed upon aqueous workup. Show how this reaction takes place by drawing the key intermediates in the reaction.

d) Carboxylic acids can be converted into ketones by reacting with two molecules of alkyl lithium, followed by an aqueous workup. Show how this reaction takes place by drawing the key intermediates in the reaction.

16.4.2 Stable imines

Some imines are especially stable. Stable imines are formed when strongly nucleophilic amines are used in a reaction—in particular, when heteroatoms are directly attached to nitrogen. The nitrogen atoms of such nucleophiles react very rapidly with carbonyl groups. The imine products thus formed are less reactive than other imines because the non-bonded electrons on the directly attached heteroatom can delocalize with the π system of the imine (described by resonance).

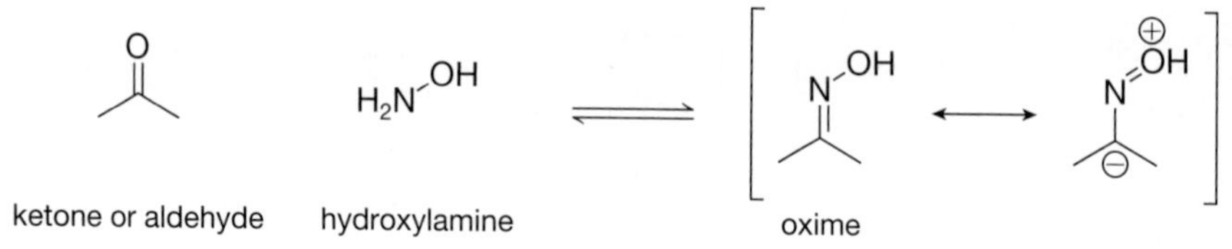

ketone or aldehyde　hydroxylamine　oxime

One such amine, hydroxylamine, reacts with aldehydes and ketones to form oximes. Oximes are stable materials, and often form solids. Likewise, hydrazines react with carbonyl compounds to form hydrazones, which are also stable.

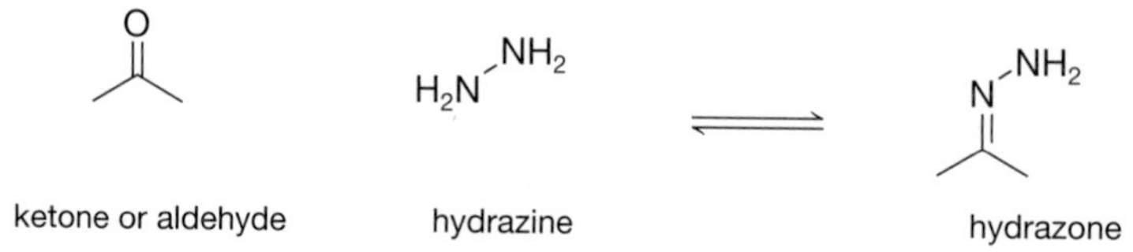

ketone or aldehyde　hydrazine　hydrazone

Semicarbazides react with aldehydes and ketones and produce semicarbazones.

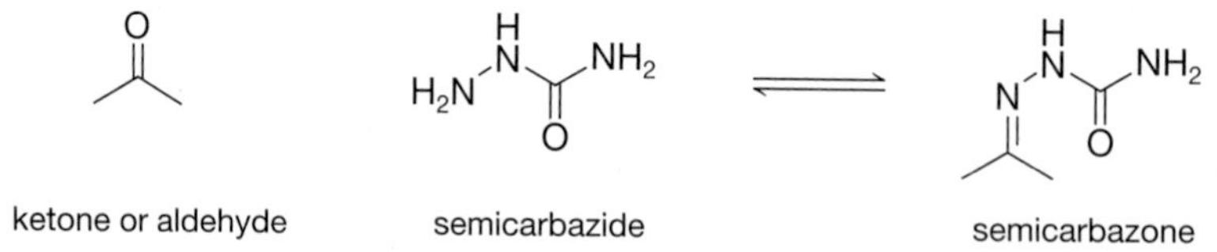

ketone or aldehyde　semicarbazide　semicarbazone

Oximes, hydrazones, and semicarbazones are often crystalline solids. They form the basis of a series of chemical tests used to prove the identity of aldehydes and ketones. Historically, unsubstituted semicarbazones and hydrazones were particularly useful for this purpose, and entire books were written listing the characteristic colour and melting point of the derivatives that could be formed from individual aldehydes or ketones. By making several such molecules and comparing their physical properties, scientists could identify the original compound. Modern instrumentation has made these tests largely obsolete, but for over 100 years these molecules represented one of the best ways to identify aldehydes and ketones.

16.4.3 Reductive amination

Amines can be made by nucleophilic displacement reactions such as S_N1 or S_N2 reactions. However, this process is often complicated by the nucleophilic nature of the product amines. For example, it is sometimes difficult to obtain a mono-alkylated product from an S_N2 reaction when amine nucleophiles are involved because the product is nucleophilic and can continue to react. Alkyl groups are electron donating (induction), and so each alkylation produces a product that is more reactive than the last. This results in mixtures in which the multi-alkylated products are present in the largest amounts. The result is a mixture of the desired compound and several by-products, unless a control is in place to prevent over-reaction.

Ph⌒NH₂ Br⌒ → Ph⌒N⊕H(Et)H ⊖OH → Ph⌒N(Et)H → (Br⌒) Ph⌒N⊕(Et)₂H ⊖OH → Ph⌒N(Et)₂

desired product

by-product (dialkylated)

This problem can be overcome by a strategy called *reductive amination*: a reaction sequence that involves imine formation followed immediately by a reduction. This precisely controls the number of groups that are added to the nitrogen. Because the nitrogen of an iminium ion is not nucleophilic, only one addition takes place.

Ph⌒NH₂ + O=⌒ —H⊕→ [Ph⌒N⊕(H)=⌒] —$NaBH_3CN$→ Ph⌒N(H)⌒

desired product

the imine or iminium is formed as an intermediate in the process

The two reactions are normally carried out in tandem; that is, they happen one after the other in the same reaction mixture. Because imine formation requires some acid, and because imines must be activated by acid, the method employs hydride-containing molecules that are stable in acidic solutions. Such hydride reagents are not strongly nucleophilic and usually react only with activated electrophiles such as iminium ions. $NaBH_4$ cannot be used because it not only reacts with the acid that activates the imine (Chapter 7) but also reacts very quickly with carbonyls. $NaBH_3CN$ is a commonly used reagent because the CN group is an electron-withdrawing group, and it reduces the reactivity of the reagent enough for it to resist exposure to acid. The slightly deactivated nature of the reagent also means that it will selectively reduce iminiums faster than carbonyls.

CHEMISTRY: EVERYTHING AND EVERYWHERE

Amination in Living Systems

Amination reactions are key to many metabolic pathways. These reactions occur frequently when amino groups become part of compounds such as amino acids. In living systems, enzymes called transmutases carry out trans-aminations (transfer of nitrogen from one molecule to another) using pyridoxal phosphate (from vitamin B6) as a nitrogen source in the production of amino acids.

pyridoxal phosphate

imine is formed between amine and carbonyl

pyruvic acid

pyruvate transmutase

all reactions are carried out by an enzyme

conjugation with the aromatic ring and iminium lowers the pK_a of the benzylic hydrogens enough to allow them to be removed by the enzyme

enzyme

iminium

electron flow "flips" the iminium to a new location

enzyme

protonation gives a new iminium ion

alanine

hydrolysis of the new iminium releases the amino acid together with vitamin B6

The key to this process is the formation of an iminium ion (which is stabilized by the nearby phenoxide) between the pyridoxal phosphate and a ketone, such as that in pyruvic acid. The C=N π bond in this intermediate is conjugated with the carbonyl group and the aromatic ring. The resulting delocalization makes possible the removal of a proton from the benzylic position of the ring. This sets up an electron flow that ultimately results in a proton being put back on the molecule, but at the position α to the carboxylate at the bottom. Hydrolysis of the newly formed iminium releases the new amino acid product together with vitamin B6 as a by-product. This reaction illustrates how amino groups are introduced into amino acids within the chemical environment of living things (the mechanism for H^+ transfer involves enolization; see Chapter 17).

16.4.4 Wolff–Kishner reduction

The Wolff–Kishner reduction provides a method to remove carbonyl groups by reducing a hydrazone with a strong base. In the process, the two nitrogen atoms of the hydrazine are converted into N_2 gas. Since gases make excellent leaving groups that diffuse out of solution, the reaction is irreversible, despite the fact that all the preceding steps are equilibria.

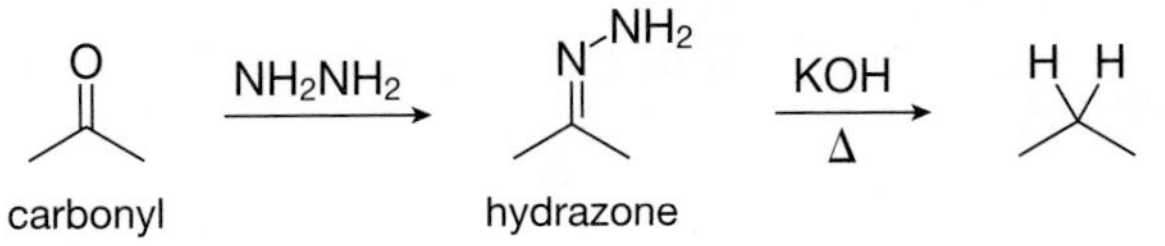

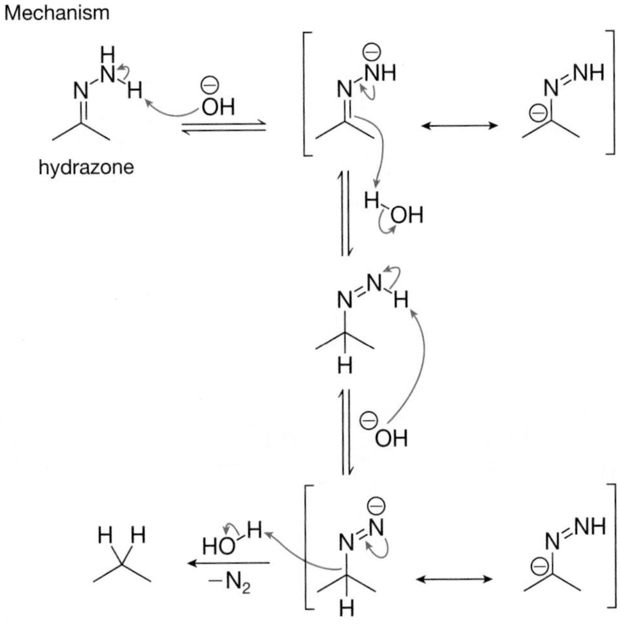

carbonyl —NH_2NH_2→ hydrazone —KOH, Δ→

Mechanism

hydrazone

CHECKPOINT 16.5

You should now be able to predict the products of reduction reactions involving imine intermediates. This includes reductive aminations and Wolff–Kishner reductions. You should also be able to draw the mechanisms for these reactions.

SOLVED PROBLEM

Draw a mechanism for both steps of the following reaction and predict the final product.

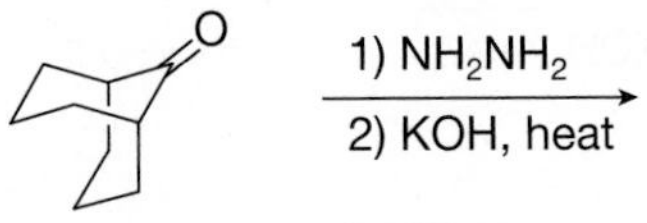

1) NH_2NH_2
2) KOH, heat

STEP 1: Determine the roles of the reactants and the type of reaction taking place.

The starting material contains an electrophilic carbonyl group that can react with the nucleophilic hydrazine to form a hydrazone. In the second step, a strong base is present with heat. These conditions should be recognized as part of the Wolff–Kishner reaction that reduces hydrazones to alkanes.

STEP 2: Draw a mechanism for the formation of the hydrazone following the steps for imine formation from Checkpoint 16.4.

The mechanism for hydrazone formation is the same as for imine formation. However, since no solvent was specified in the reaction, it is not clear what species is facilitating the various proton transfers in the reaction. It could be a molecule of solvent, the reactants themselves, or an intermediate generated in the reaction. For simplicity, we have used H–A here to signify a generic weak acid.

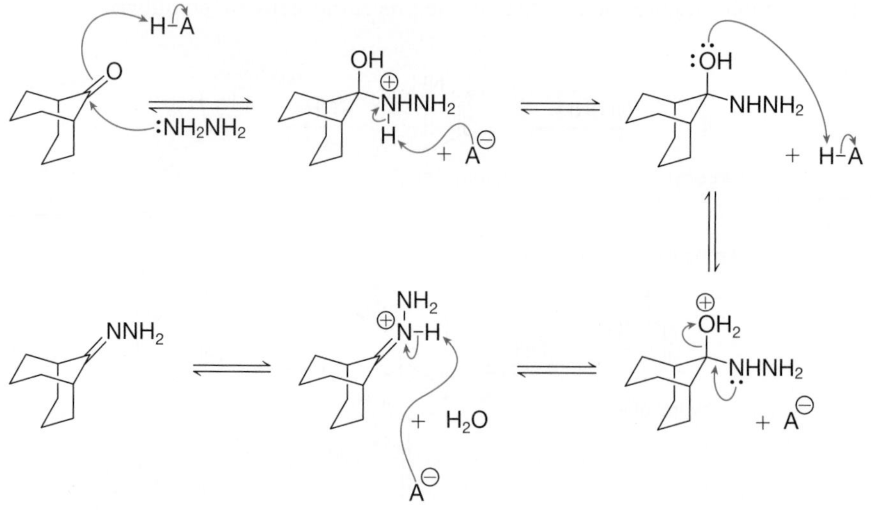

STEP 3: Draw a mechanism for the reduction of the hydrozone.

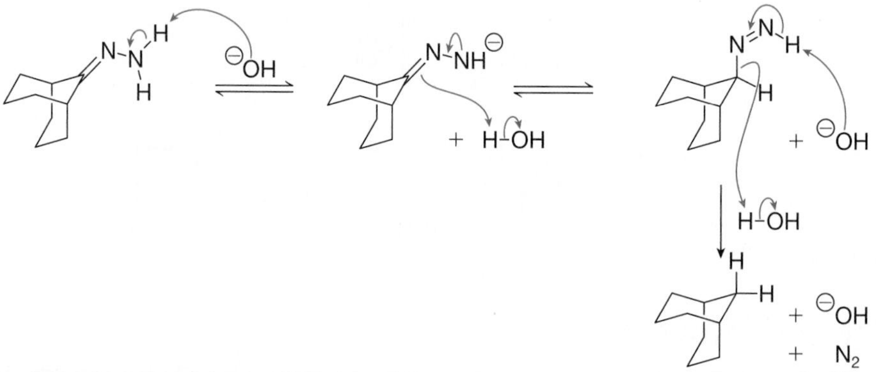

Note that the final step of the reaction is irreversible due to the evolution of nitrogen gas. All other steps in the reaction are reversible.

PRACTICE PROBLEM

16.9 Predict the products of the following reactions.

a) 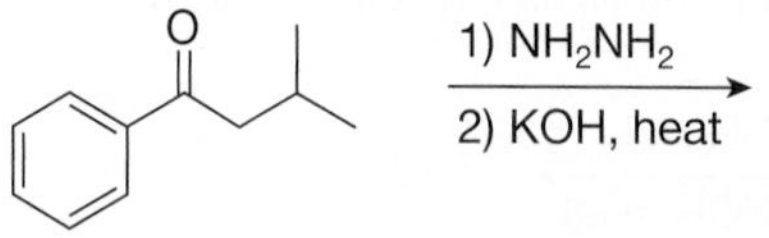 1) NH_2NH_2 2) KOH, heat

b) 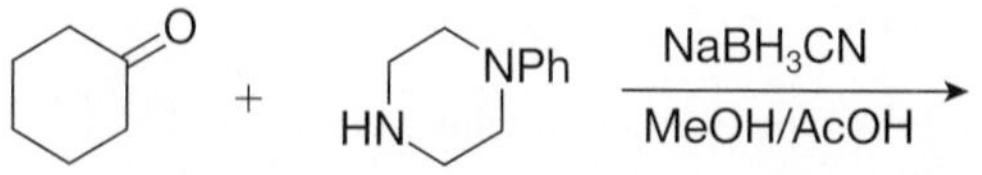 + ; $NaBH_3CN$, MeOH/AcOH

c) 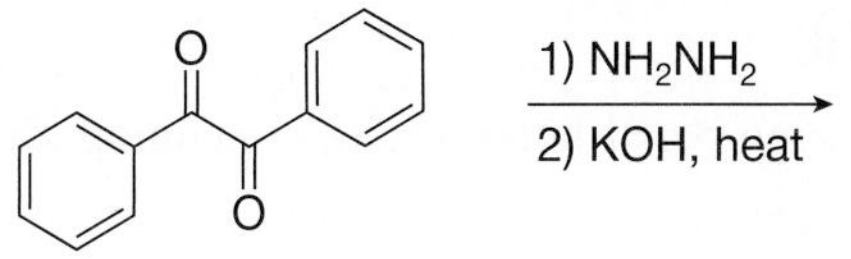 1) NH_2NH_2
2) KOH, heat

d) 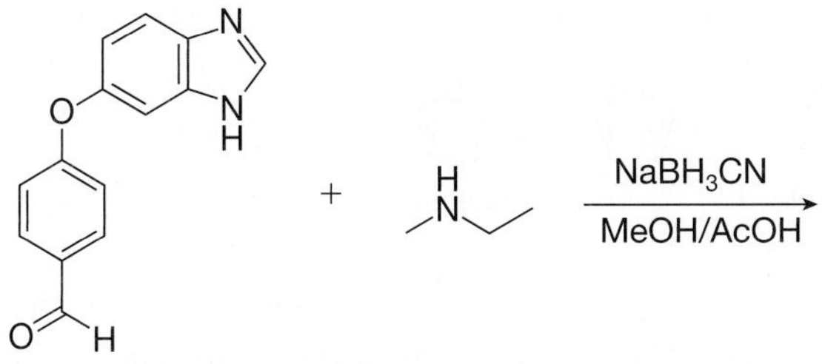 + NaBH$_3$CN
MeOH/AcOH

INTEGRATE THE SKILL

16.10 a) Draw a mechanism for the following reductive amination reaction.

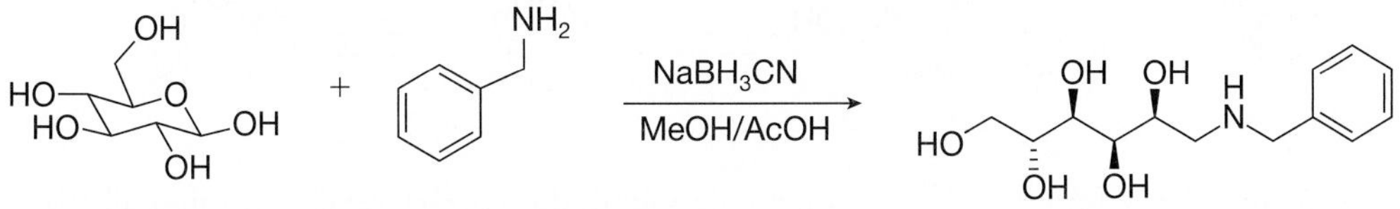

NaBH$_3$CN
MeOH/AcOH

b) The Wharton reaction is mechanistically very similar to the Wolff–Kishner reduction. Suggest a mechanism for the Wharton reaction shown here.

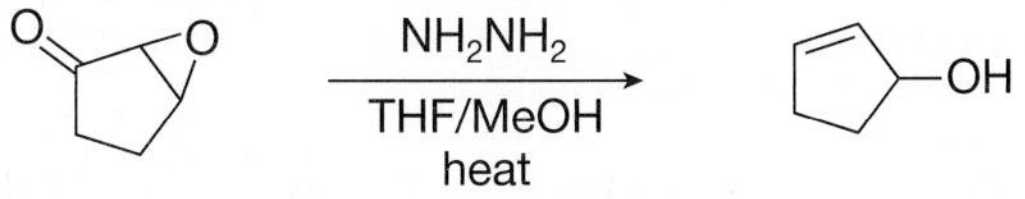

NH_2NH_2
THF/MeOH
heat

16.5 Heterocycle Formation Using Hidden Leaving Groups

The presence of rings imparts rigidity and contributes to the three-dimensional structures of organic molecules; both are key features of the proper functioning of living things. A ring structure formed entirely from carbon atoms is called a **carbocycle**. If one or more of the atoms that constitute the ring is a heteroatom, the ring is a called a **heterocycle**.

A **carbocycle** is a ring composed entirely of carbon atoms.

A **heterocycle** is a ring of atoms in which one or more of the atoms are heteroatoms. The term *heterocycle* is also used to describe molecules that contain such rings.

The formation of many heterocycles, especially aromatic heterocycles, includes reactions involving hidden leaving groups, enol chemistry, and hydride eliminations (oxidations). The order of the reactions varies, but these processes can be described as a sequence of smaller functional group transformations: (1) the formation of oxoniums or iminiums from carbonyl compounds; (2) enol or enol ether formation from carbonyls or oxoniums; (3) hydrolysis of enol ethers, oxoniums, or iminiums; and (4) the generation of double bonds by means of oxidizing agents (leaving groups). By combining these operations, a great many heterocycles can be produced from simpler materials.

The formation of aromatic products is an important part of these reaction sequences. Because the sub-reactions involved are mostly equilibria, there must be a means of shifting the equilibria toward the products. The thermodynamic stability of the aromatic rings tends to drive these equilibria toward the ultimate heterocyclic products.

16.5.1 Making pyrroles and furans

Pyrroles, furans, and thiophenes are simple, five-membered heterocycles containing a nitrogen, oxygen, or sulfur, respectively.

pyrrole furan thiophene

For each of these rings, the heterocycle is attached to a double bond, giving an enol ether–like substructure (enol ethers react with acid via Markovnikov-type mechanisms; see Chapter 7). The double bond is an obvious point of bond disconnect in the retrosynthesis of such heterocycles.

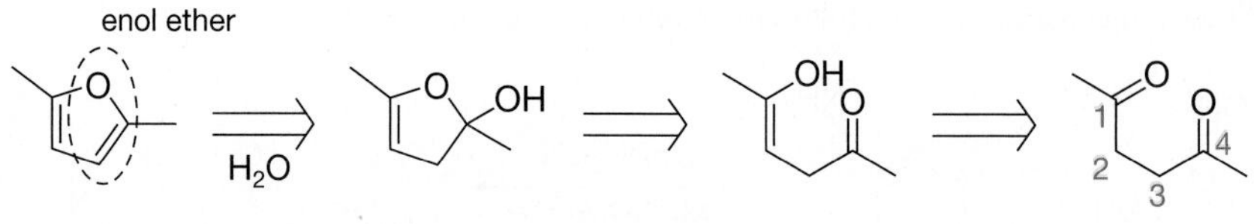

enol ether H_2O

To form a furan from a 1,4-dicarbonyl compound, the correct catalyst is often all that is needed. The driving force for the reaction is the formation of an aromatic compound through the loss of a molecule of water. All reactions are equilibria and may require the removal of water to shift things toward the products. Powdered P_2O_5 is a dehydrating agent that is also a source of acid to catalyze the transformation.

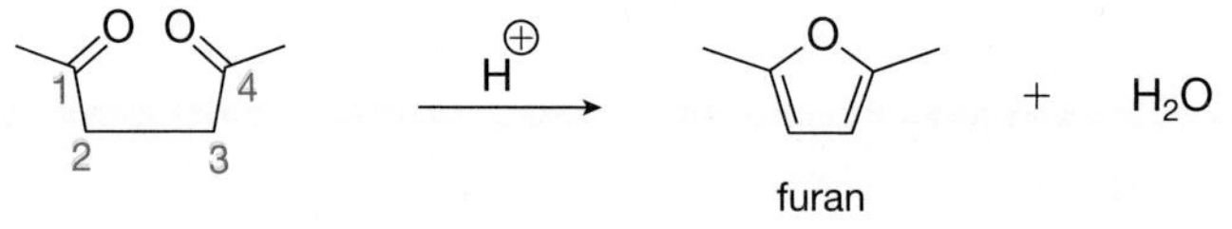

$H^\oplus$ furan + H_2O

The mechanism of the process involves enol formation and acetal chemistry. Notice how the acid activates the carbonyl groups, making the α-hydrogen more acidic and the carbonyl carbon more electrophilic.

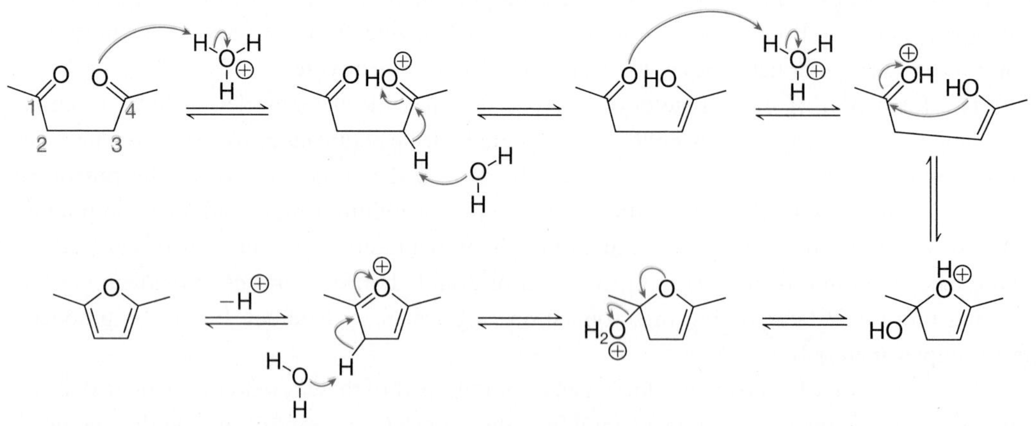

P_2O_5 is a good dehydrating agent for this purpose because it reacts quickly with water, generating phosphoric acid (H_3PO_4) in the process.

$$P_4O_{10} + 6\,H_2O \longrightarrow 4\,H_3PO_4$$

STUDENT TIP

P_4O_{10} is a strong dehydrating agent; it removes water from other materials. The empirical formula of this substance is P_2O_5, and so the material is usually referred to as phosphorus pentoxide or P_2O_5.

The formation of pyrroles or thiophenes follows a similar sequence. The difference here is that an initial exchange must happen between one of the carbonyl groups and the added heteroatom to form an imine or thiocarbonyl. These exchanges follow acetal chemistry pathways (imine formation or thioacetal formation).

NH_3

thiophene

Related heterocycles can be formed by mixing the appropriate dicarbonyl compound and reaction partner. 1,3-Dicarbonyl compounds react with hydroxylamine or hydrazine to form isoxazoles and pyrazoles, respectively. If an aminal is used as the heteroatom donor, a pyrimidine results.

isoxazole

pyrazole

pyrimidine

16.5.2 Preparation of benzimidazoles and benzoxazoles

Benzimidazoles and benzoxazoles are aromatic heterocycles that have fused ring systems. They are formed by the reaction of *ortho*-disubstituted benzene rings with acid chlorides or esters. Heat or dehydrating agents are sometimes required to "persuade" the rings to close after the initial reaction. The sequence involves an acylation followed by imine formation. A molecule of water is released in this second reaction.

benzimidazole

benzoxazole

16.5.3 Making heterocycles using oxidizing agents

In some cases, an oxidation step is necessary to form an aromatic heterocycle. For example, pyridines contain one nitrogen in a six-membered ring. The structure can be formed from 1,5-dicarbonyl starting materials and ammonia. After imine formation and ring closure, the molecule must be oxidized (lose a pair of electrons). Two hydrogens are removed as part of this process.

There are two general methods to oxidize the molecule: (1) use of a leaving group; or (2) elimination of a hydride equivalent. Both methods remove a pair of electrons from the ring, thus oxidizing it. The first strategy requires the attachment of an oxidizing agent (leaving group) to the lone pair of electrons on the nitrogen. Nitric acid, which breaks down to the strong electrophile NO_2^+, is often used for this purpose. (Recall that it is also the electrophile used for the nitration of benzene; see Chapter 10.)

The leaving group will form a bond with the electrons on the nitrogen. These electrons will be removed when the leaving group leaves, thus oxidizing the ring.

Another leaving group involves the use of hydroxylamine for the initial heterocycle formation. This creates an exocyclic OH group that can be converted to a leaving group. When the OH is removed, it takes a pair of electrons with it; this results in the oxidation of the ring to the aromatic pyridine.

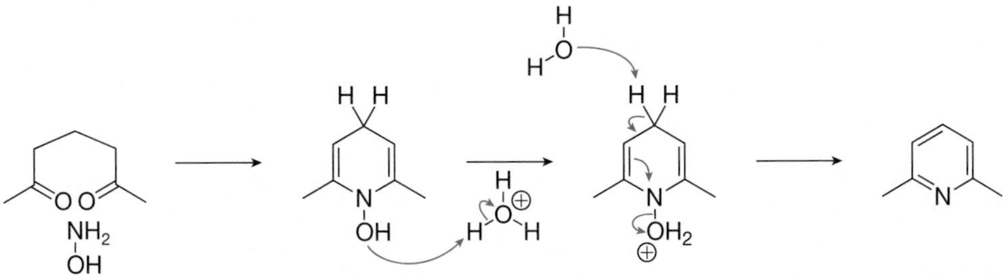

The second method to aromatize the molecule involves specialized oxidants such as 2,3-dichloro-5,6-dicyano-1,4-benzoquinone (DDQ). This molecule accepts a pair of electrons in the form of hydride (H^-) from the dihydropyridine, thus oxidizing the ring. Donation of a pair of electrons from the nitrogen facilitates this process because the initial product is aromatic. Final removal of a proton gives the pyridine.

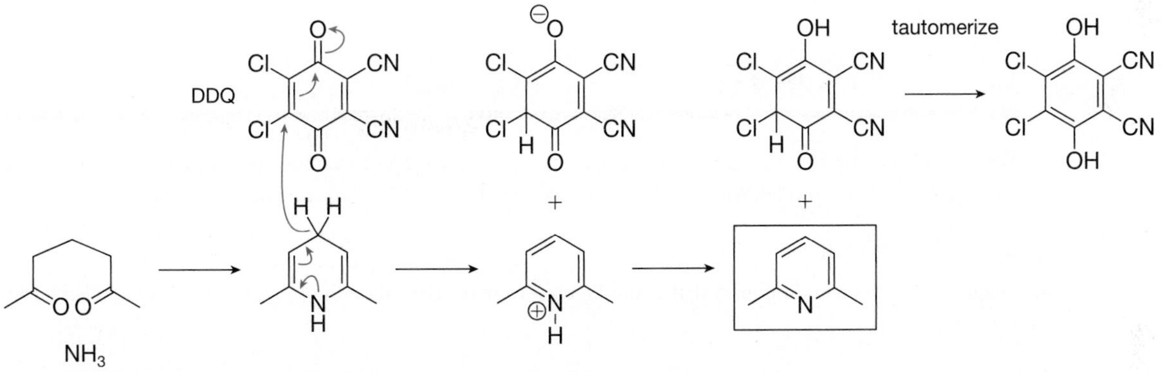

CHEMISTRY: EVERYTHING AND EVERYWHERE

Heterocycle Oxidation and Reduction in Living Systems

Pyridine rings are important in many biochemical processes. For example, many enzymes use co-factors to carry electrons that are added or removed as part of redox reactions. At the centre of co-factor NADH, a molecule with a complex structure, is a pyridine. This functional group carries electrons in the form of hydride and becomes oxidized or reduced as necessary. The oxidized form is called NAD^+, and the reduced form is NADH. If an enzyme carries out an oxidation reaction, it uses NAD^+ as a co-factor. When the oxidation is carried out, the electrons are removed from the substrate molecule in the form of a hydride and transferred to NAD^+, which becomes NADH. During a reduction reaction, the process is reversed (Chapter 12).

Continued

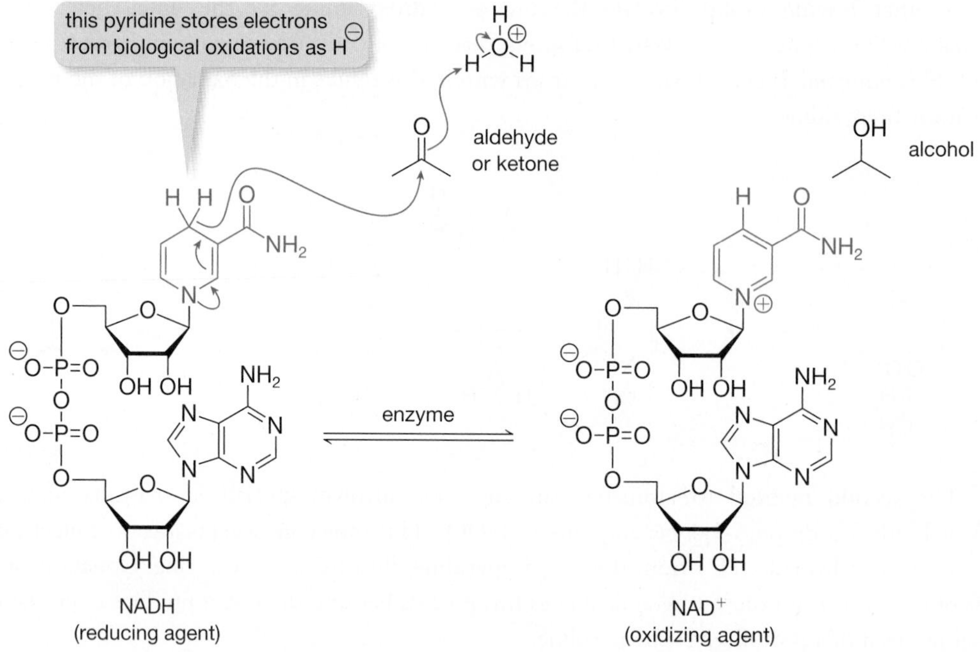

this pyridine stores electrons from biological oxidations as $H^{\ominus}$

aldehyde or ketone

alcohol

enzyme

NADH
(reducing agent)

NAD^{+}
(oxidizing agent)

CHECKPOINT 16.6

You should now be able to suggest appropriate starting materials and reagents for the synthesis of five- and six-membered heterocycles. You should also be able to draw mechanisms for these reactions.

SOLVED PROBLEM

Suggest a dicarbonyl compound that could be used to make the following heterocycle. Draw a mechanism for the reaction.

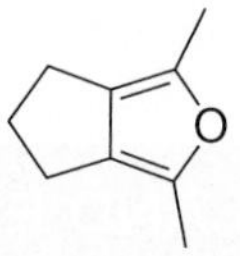

STEP 1: Using the retrosynthetic scheme in Section 16.2.4 as a guide, propose a series of disconnections to generate the appropriate dicarbonyl starting material.

Disconnect one of the double bonds next to the heteroatom to create a hemiacetal intermediate. The hemiacetal disconnects to an enol ether and a carbonyl by breaking the C–O bond. Finally, the enol ether can be disconnected to a second carbonyl to give a 1,4-dicarbonyl starting material.

$-H_2O$

hemiacetal

enol

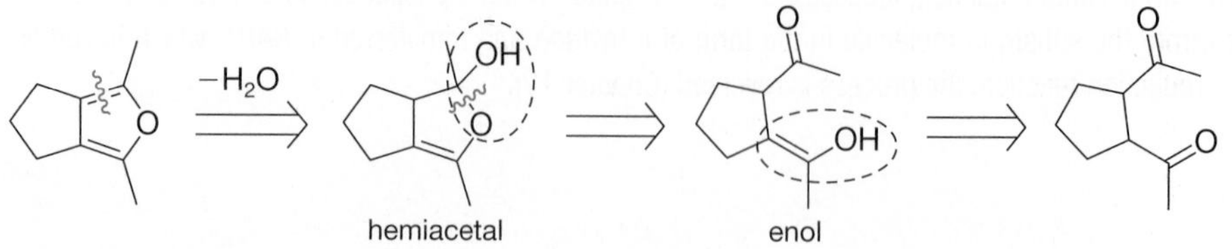

STEP 2: Write an overall reaction equation for the proposed synthesis.

The reaction involves enol formation, hemiacetal formation, and dehydration of a hemiacetal. These steps can all be catalyzed by acid, so the reaction conditions will require an acid catalyst. The overall reaction can be written as follows:

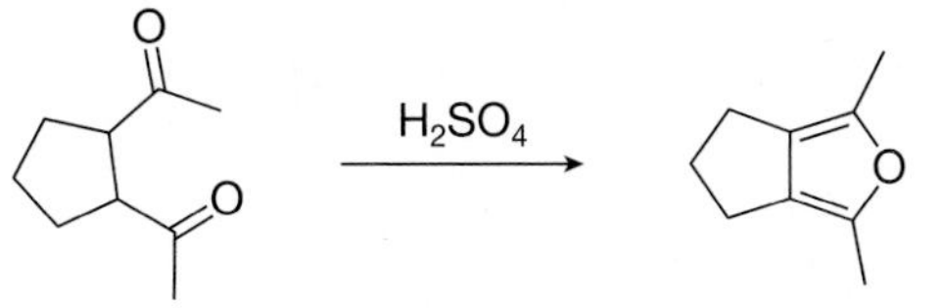

H_2SO_4

STEP 3: Draw a mechanism for the formation of the enol ether and its cyclization to form the hemiacetal.

Formation of the enol ether begins by protonating one of the carbonyl groups. This increases the acidity of the α-hydrogens, making them easily removed by even a weak base such as the conjugate base of the sulfuric acid. The enol ether is nucleophilic at both its π bond and the oxygen atom; however, attack of the π bond on the carbonyl would give a highly strained three-membered ring. Therefore, the reaction proceeds through nucleophilic attack of the oxygen's lone pairs on the adjacent carbonyl group. Deprotonation of the oxygen atom yields a stable, cyclic hemiacetal. The reaction is catalyzed by $H_3O^{\oplus}$ formed by the reaction between H_2SO_4 and the H_2O produced in the reaction.

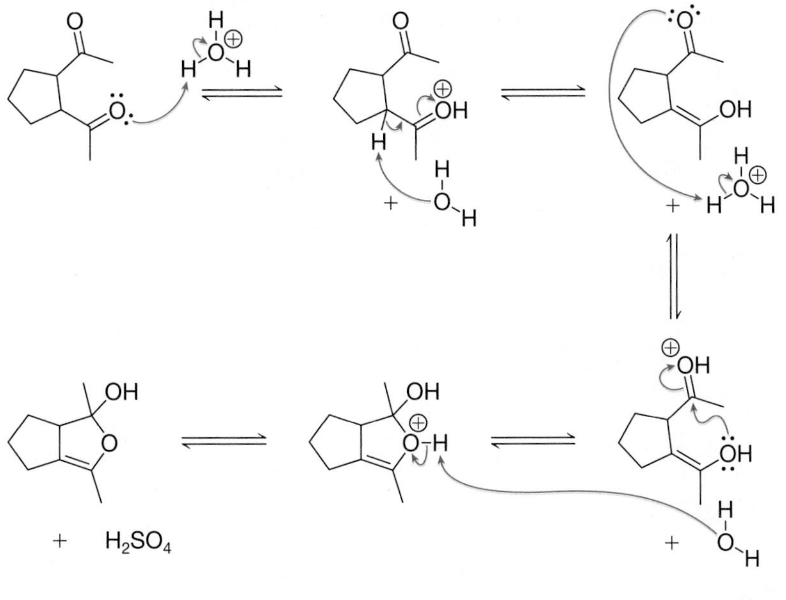

+ H_2SO_4

STEP 4: Draw a mechanism for the dehydration of the hemiacetal.

The hidden leaving group of the hemiacetal is revealed after protonating the OH group. This allows the lone pairs on the enol ether oxygen to push out H_2O as a leaving group, creating an oxonium ion. Deprotonation of the oxonium ion at the tertiary hydrogen generates a highly stable aromatic furan ring, thereby making the reaction irreversible.

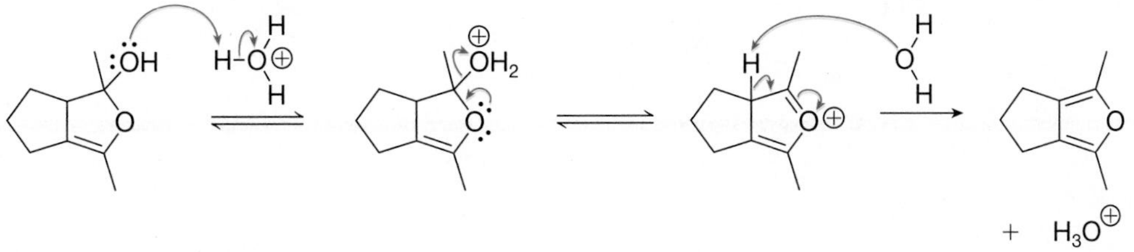

+ $H_3O^{\oplus}$

PRACTICE PROBLEM

16.11 a) Propose a mechanism for each of the following reactions.

i) 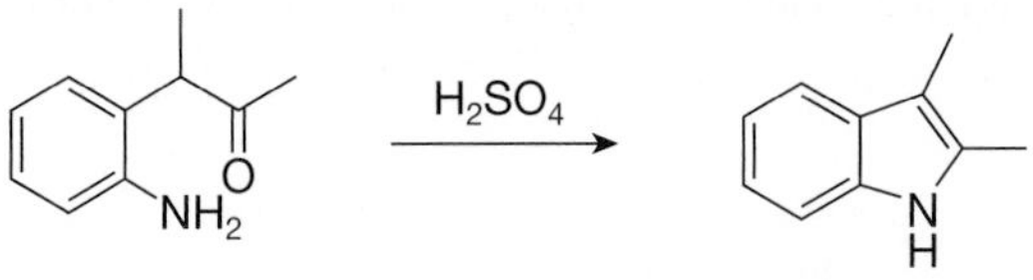 H_2SO_4

ii) 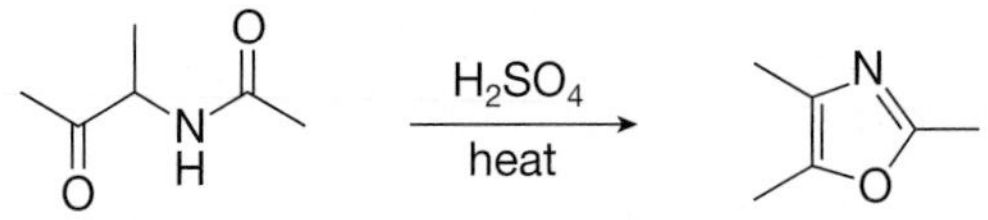 H_2SO_4, heat

iii) 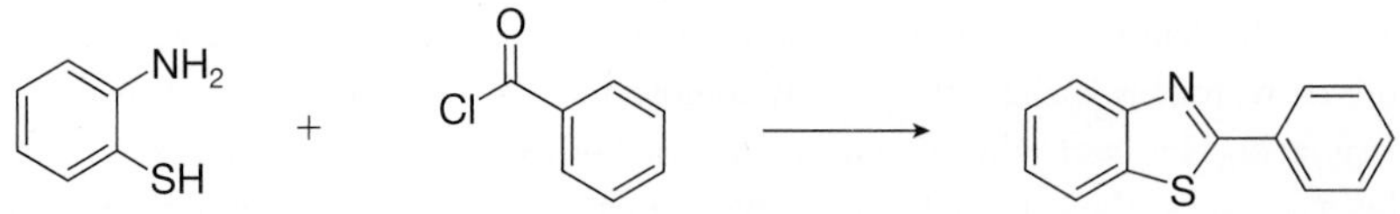 +

iv) 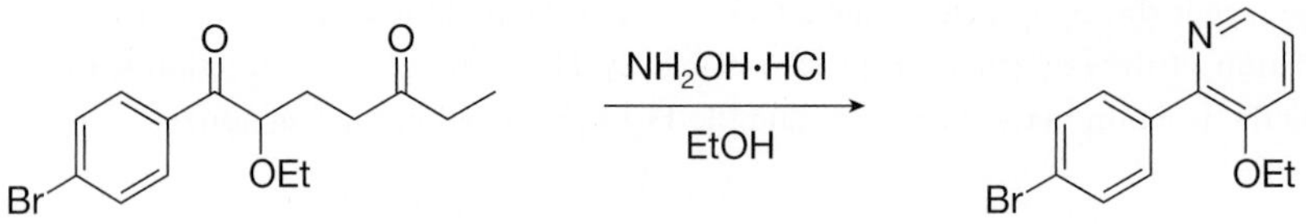 $NH_2OH{\cdot}HCl$, EtOH

b) Suggest a method of synthesizing the following heterocycles.

i)

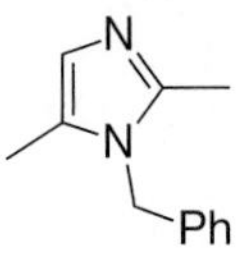

ii)

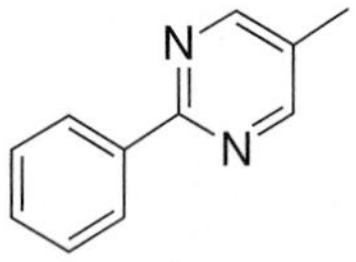

iii)

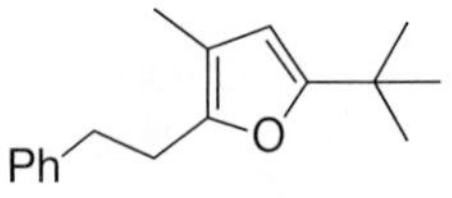

iv)

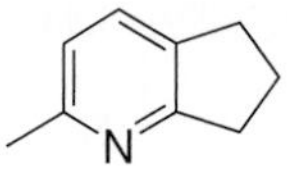

INTEGRATE THE SKILL

16.12 An alternative method for generating a pyrrole ring involves reacting an α-aminoketone with a second carbonyl compound via the Knorr pyrrole synthesis. Propose a mechanism for this reaction. Hint: The reaction involves an enamine intermediate.

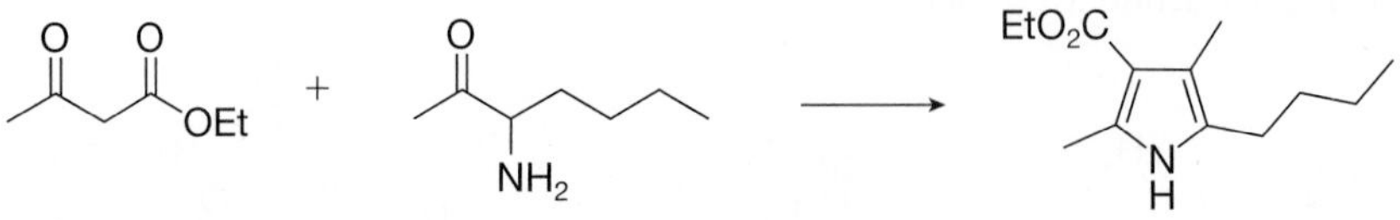

+

16.6 Patterns in Hidden Leaving Groups

Many functional groups that contain carbon heteroatom π bonds (such as carbonyls) can be converted to functional groups containing at least two heteroatoms joined to a central carbon by σ bonds.

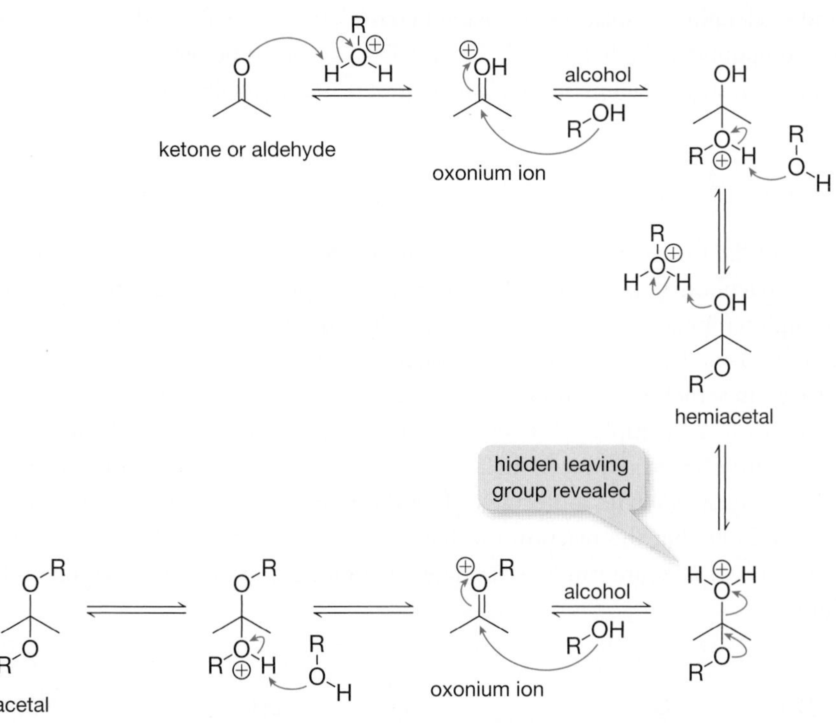

Most of the transformations are acid catalyzed because the acid is needed to convert the heteroatom in the starting carbonyl into a hidden leaving group. All of the reactions are equilibria, and the pattern of electron flow is similar in the forward and reverse directions.

Carbonyl exchanges follow a similar pattern of electron flow.

Many organic transformations incorporate aspects of this chemistry into their reactions. Any heteroatom with a π bond will behave according to these transformations, as will functional groups that consist of an sp^3-hybridized carbon connected to two heteroatoms by σ bonds (reverse reaction).

Heterocycle formation involves combinations of these types of reactions together with enolization and nucleophilic displacement. Many heterocycles can be formed from the appropriate dicarbonyl compound and a heterocycle source. Six-membered heterocycles sometimes require an extra oxidation step (attachment of a leaving group and then elimination).

Bringing It Together

Bakelite, one of the first synthetic plastics, was used extensively to make many consumer items such as toys, telephones, and radios. It is a thermoset polymer, a gigantic molecule that is formed at high temperature. Heating Bakelite resin (liquid) caused the formation of polymer molecules that formed a large three-dimensional network. This made a tough solid that would keep the shape in which it was moulded. Such plastics transformed manufacturing because it was very easy to make complex parts by injecting liquid resin into a mould and heating it to solidify the resin. The formation of Bakelite employs a hidden leaving group—water—that is released from formaldehyde molecules when the resin is heated. The reaction forms a new carbon-carbon bond, but the reaction mechanism is very much like that of acetal formation. So the formation of Bakelite can be considered a combination of enol acylation and addition to oxoniums.

2 phenol + formaldehyde $\xrightarrow{\text{H-Cl}}$ ... $\xrightarrow{\text{HCHO/phenol, repeat}}$ Bakelite

You Can Now

- Predict the conditions under which acetals and hemiacetals are formed and hydrolyzed.
- Draw mechanisms for the formation and hydrolysis of acetals.
- Identify the circumstances under which an acetal protecting group would be needed in a synthesis, and incorporate its use into a synthetic plan.
- Identify the anomeric carbon in cyclic carbohydrates, and use its stereochemistry to classify the molecule as either the α- or β-anomer.
- Predict the conditions under which imines and enamines are formed and hydrolyzed.
- Draw mechanisms for the formation and hydrolysis of imines and enamines.
- Predict the products of reduction reactions involving imine intermediates, such as reductive amination reactions and Wolff–Kishner reductions.
- Draw mechanisms for reductive aminations and Wolff–Kishner reductions.
- Suggest appropriate starting materials and reagents for the synthesis of five- and six- membered heterocycles.
- Draw mechanisms for the formation of five- and six-membered heterocycles formed via enol ether, acetal, and related chemistry.

A Mechanistic Re-View

Acetal formation

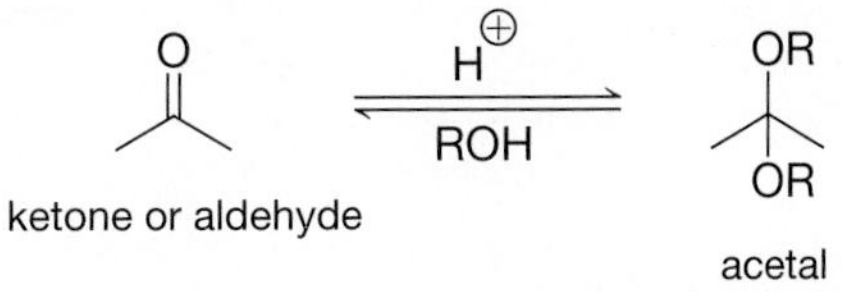

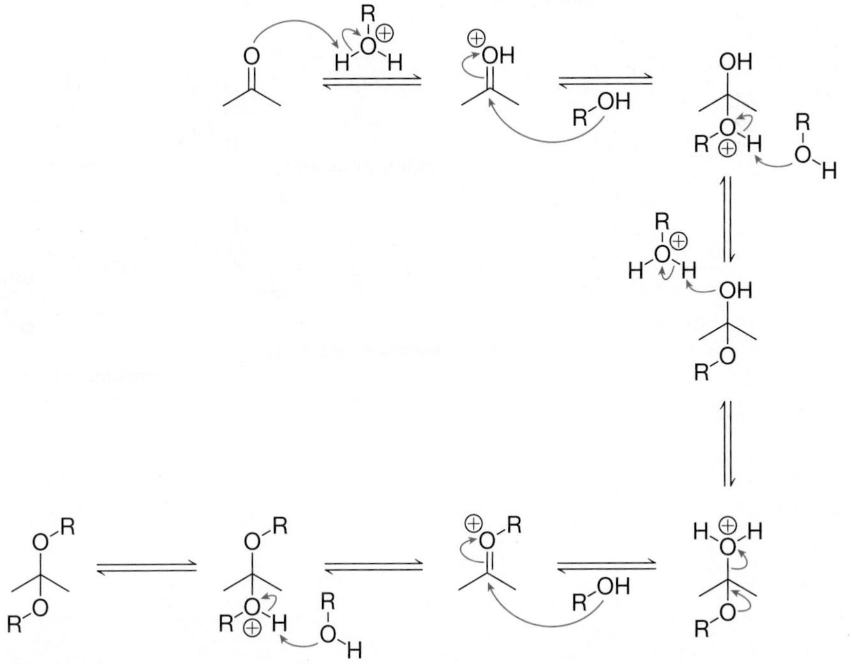

ketone or aldehyde

$H^{\oplus}$

ROH

acetal

Acetal hydrolysis

acetal

ketone or aldehyde

Carbonyl exchange

HOAc

ketone or aldehyde

imine

ketone or aldehyde

amine

oxonium ion

hemiaminal

hidden leaving group revealed

imine

iminium ion

Enamine formation

secondary amine

aldehyde or ketone

enamine

Oxime formation

ketone or aldehyde

hydroxylamine

oxime

Hydrazone formation

ketone or aldehyde

hydrazine

hydrazone

Semicarbazone formation

ketone or aldehyde

semicarbazide

semicarbazone

Reductive amination

$NaBH_3CN$

Wolff–Kishner reduction

Heterocycle formation

Heterocycle formation with oxidation

Problems

16.13 Predict the product of the following transformations.

a) 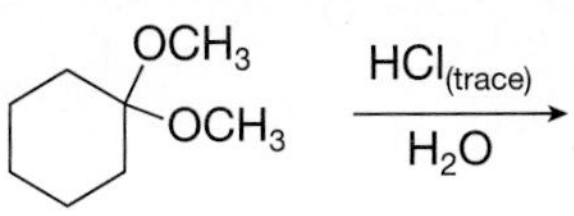 $HCl_{(trace)}$, H_2O

b) 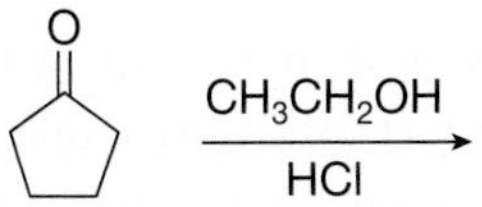 CH_3CH_2OH, HCl

c) 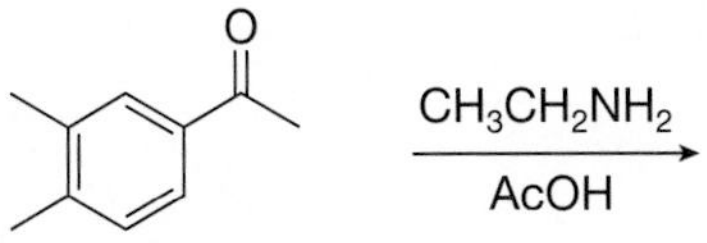 $CH_3CH_2NH_2$, AcOH

d) 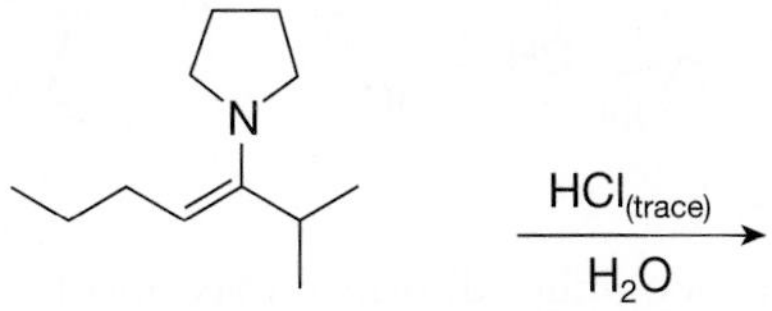 $HCl_{(trace)}$, H_2O

e) 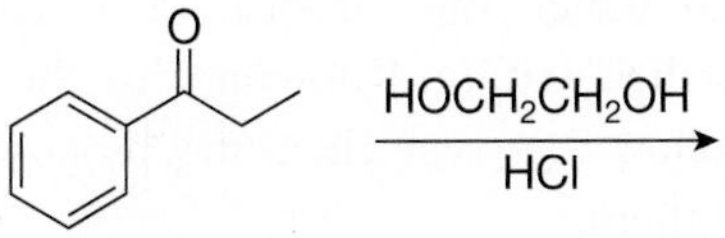 $HOCH_2CH_2OH$, HCl

f) 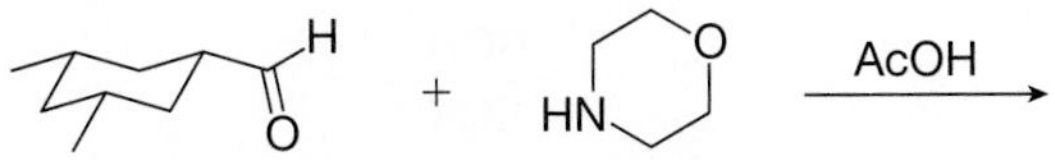 + HN(morpholine), AcOH

g) 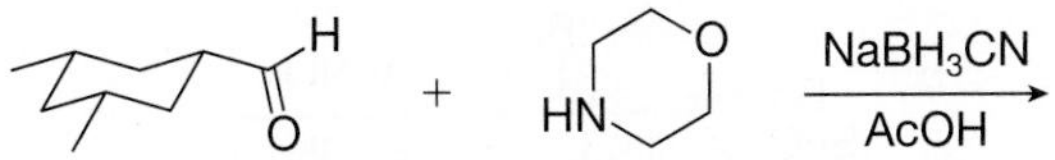 + HN(morpholine), $NaBH_3CN$, AcOH

h) CH_3CH_2OH, HCl

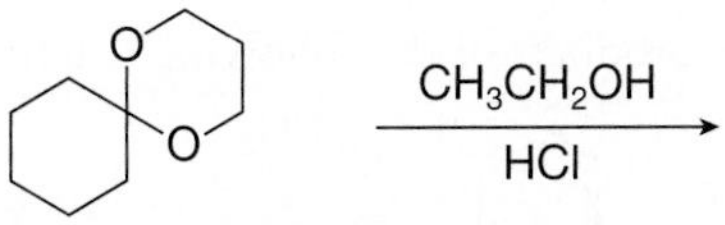

16.14 Predict the major products that result when cyclohexanone is reacted with each of the following.

a) excess CH_3OH and trace HCl
b) NH_2NH_2, then K*t*-BuO and heat
c) NH_2OH
d) $HOCH_2CH_2CH_2OH$ and trace HCl
e) $CH_3CH_2CH_2NH_2$ and AcOH
f) $HSCH_2CH_2SH$ and $TiCl_4$
g) pyrrolidine and acetic acid
h) $PhCH_2CH(OH)CO_2H$ and trace HCl
i) $HSCH_2CHCH_3OH$ and trace H_2SO_4
j) catechol and H_2SO_4
k) $PhNH_2$, AcOH, and $NaBH_3CN$

16.15 Predict the products of the following transformations.

a) 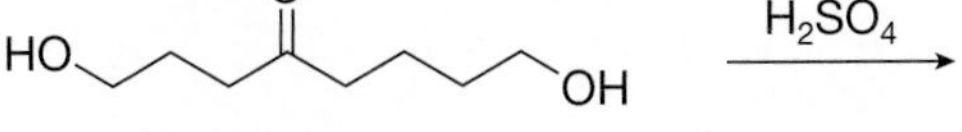 HO ... OH, H_2SO_4

b) 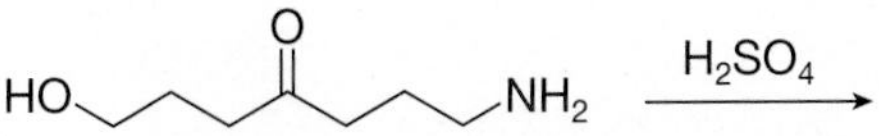 HO ... NH_2, H_2SO_4

c) 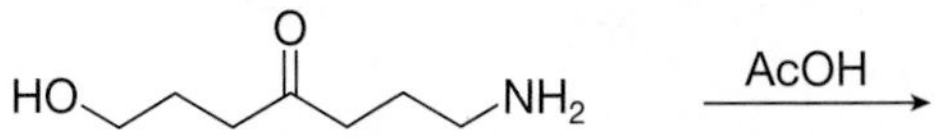 HO ... NH_2, AcOH

d) 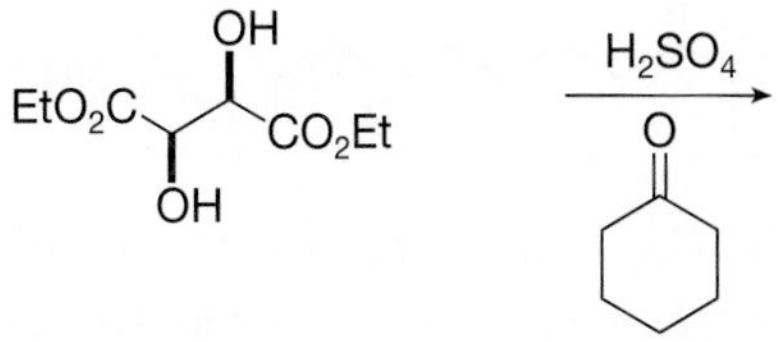 EtO_2C, OH, OH, CO_2Et; H_2SO_4, cyclohexanone

e) 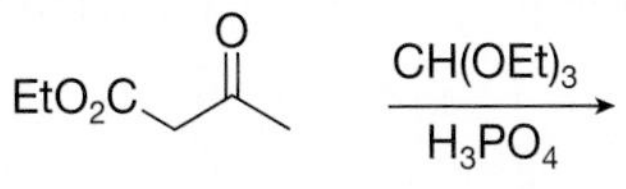 EtO_2C; $CH(OEt)_3$, H_3PO_4

16.16 Predict the products of the following reactions, and draw a mechanism showing their formation.

a) 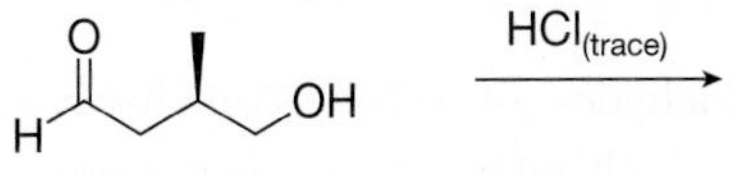 H, O, OH; $HCl_{(trace)}$

b) 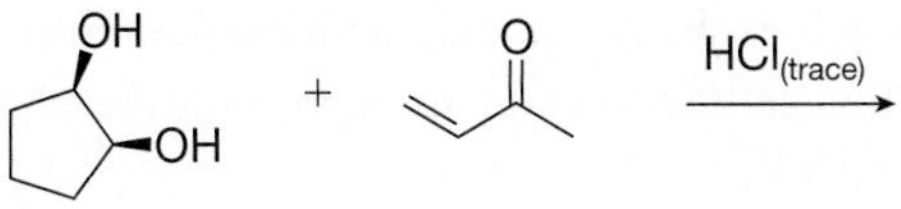 OH, OH + O; $HCl_{(trace)}$

c) 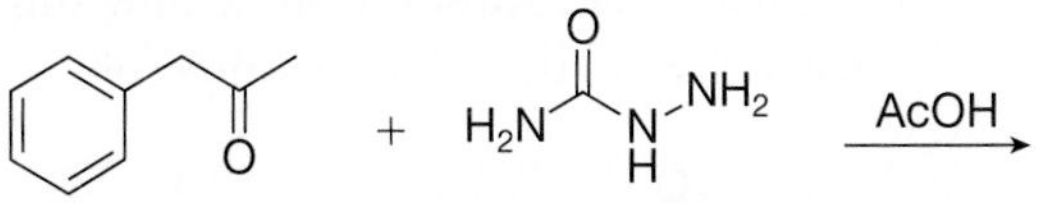 O + H_2N–C(O)–NH–NH_2; AcOH

d) 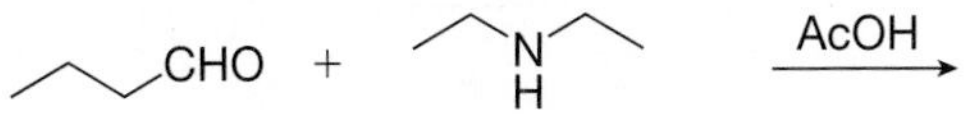 CHO + diethylamine (N–H); AcOH

e) 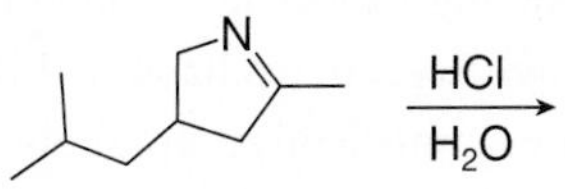 N; HCl, H_2O

f) 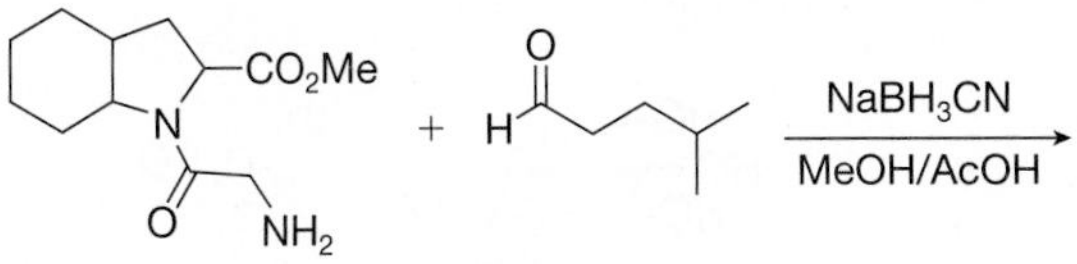 CO_2Me, N, O, NH_2 + H, O; $NaBH_3CN$, MeOH/AcOH

g) 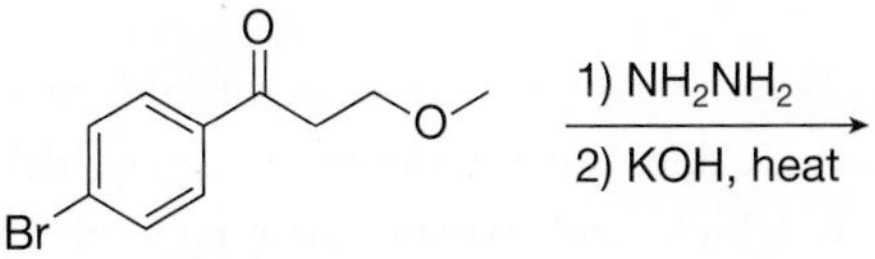 Br, O, O; 1) NH_2NH_2 2) KOH, heat

16.17 Predict the products of the following reactions, and draw a mechanism showing their formation.

a) 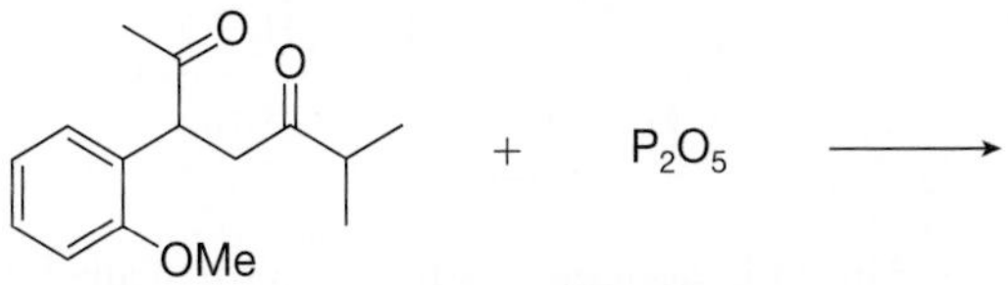 O, O, OMe + P_2O_5 ⟶

b) 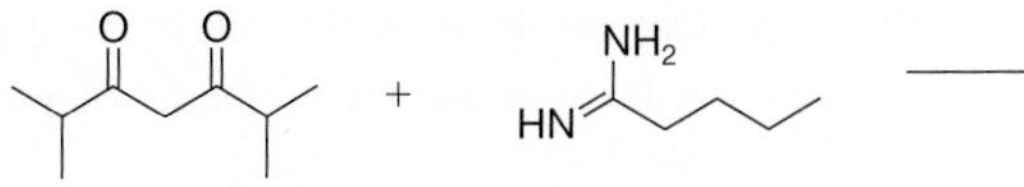 O, O + NH_2, HN ⟶

c) 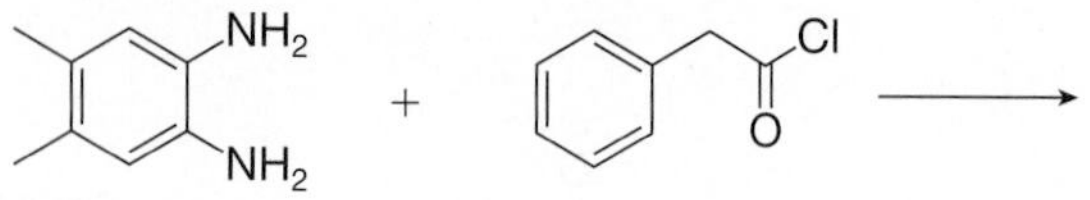 + $\longrightarrow$

d) 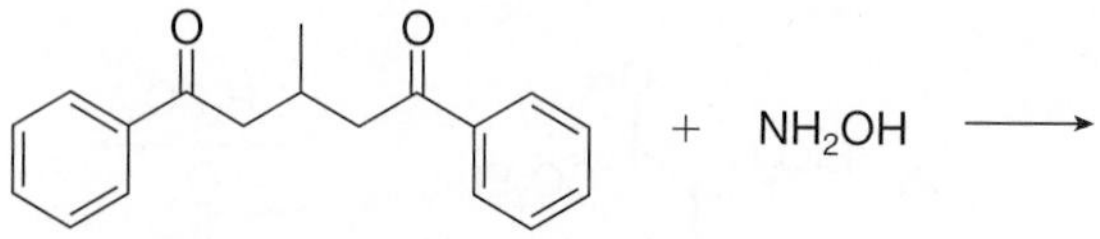 + NH_2OH $\longrightarrow$

16.18 The following scheme shows the removal of a special protecting group for acids and esters. Use a mechanism to show how this protecting group is removed to produce the acid.

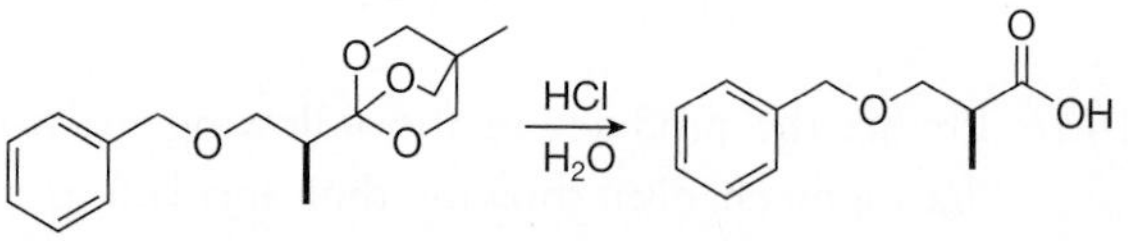

HCl / H_2O

16.19 When aldehyde- or ketone-containing compounds are mixed with water, the oxygen atoms of the carbonyls exchange quickly with oxygens from the water. This can be shown by using isotopically labelled water ($H_2{}^{18}O$).

a) Provide a mechanism to show how this exchange takes place in the following molecule.

$H_2{}^{18}O$

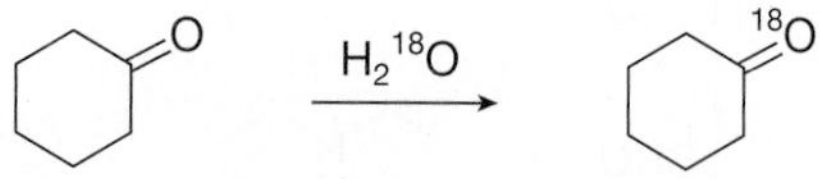

b) When the following molecule is added to labelled water, only one oxygen exchanges. Identify the oxygen and use a mechanism to show why it is exchanged.

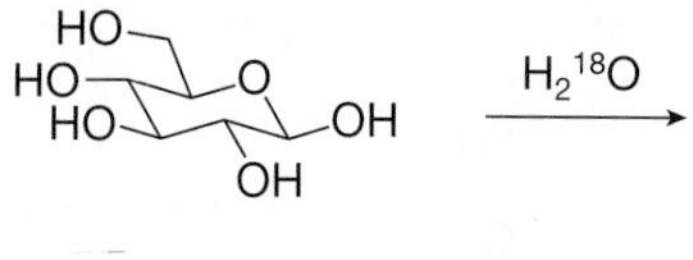

$H_2{}^{18}O$

c) Two oxygens are exchanged in this molecule. One exchanges quickly, the other slowly. Identify which is which, and explain their rates of exchange by drawing a mechanism.

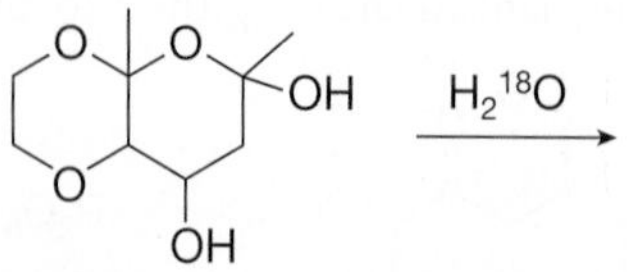

$H_2{}^{18}O$

16.20 Almonds contain a strong poison called amygdalin, which releases cyanide after an enzymatic reaction. Show mechanistically how this can occur (the enzyme can be considered as a giant molecule of acid).

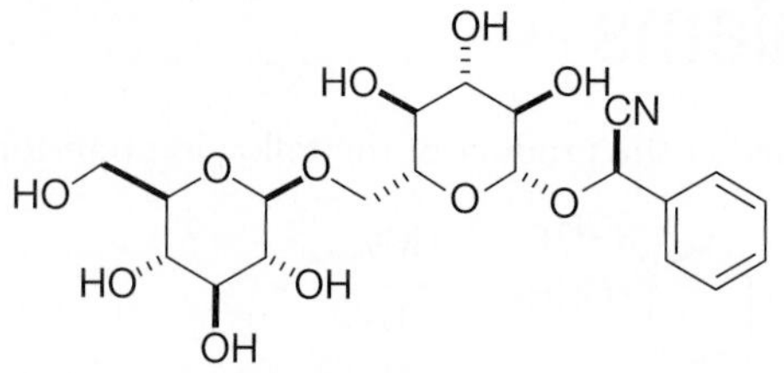

16.21 When glucose is dissolved in acetone with a small amount of acid, a product called diacetone glucose is produced. Use a mechanism to explain the formation of this product.

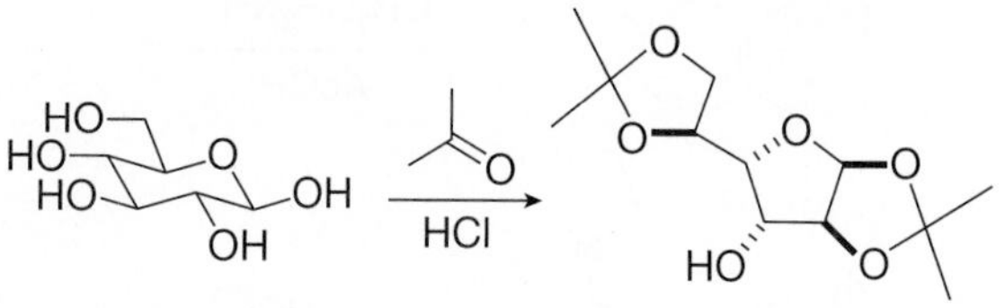

HCl

16.22 Many nucleophilic displacements involve S_N1-like pathways in which the carbocation is stabilized by an adjacent heteroatom. Knowing that the following reactions follow S_N1-type chemistry, propose products for each of them.

a) 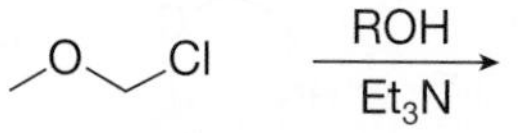 ROH / Et_3N

b) 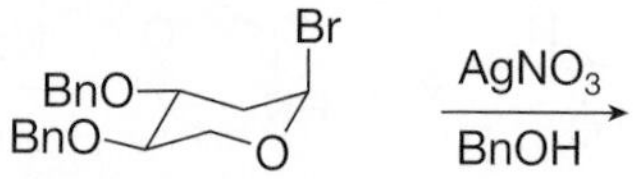 $AgNO_3$ / BnOH

16.23 Predict the products of the following reactions.

a)

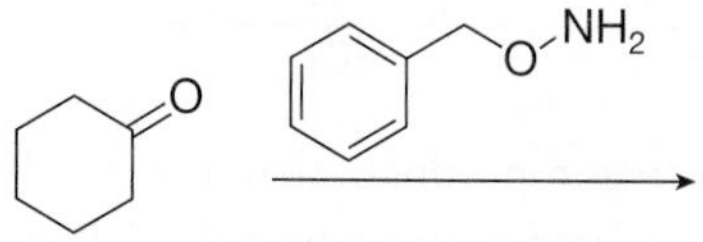

b)

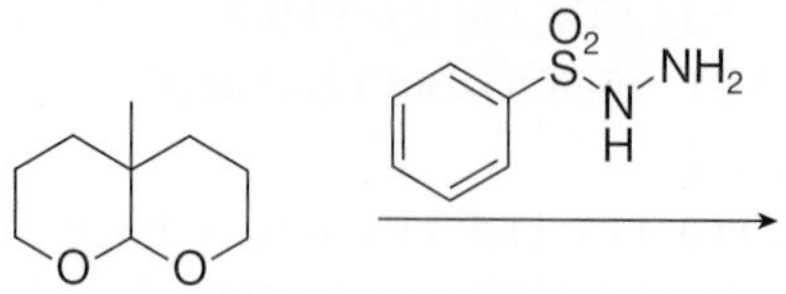

c) 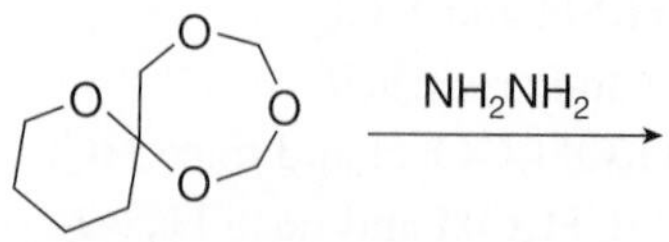 NH_2NH_2

16.24 Predict the product of the following transformations.

a) 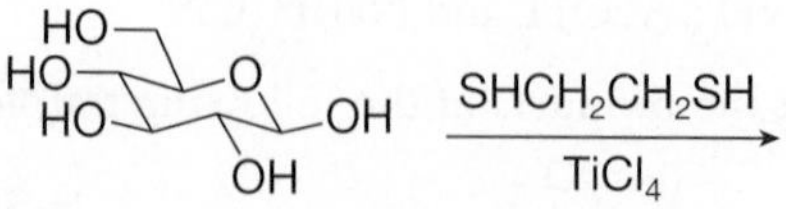 $SHCH_2CH_2SH$ / $TiCl_4$

b) 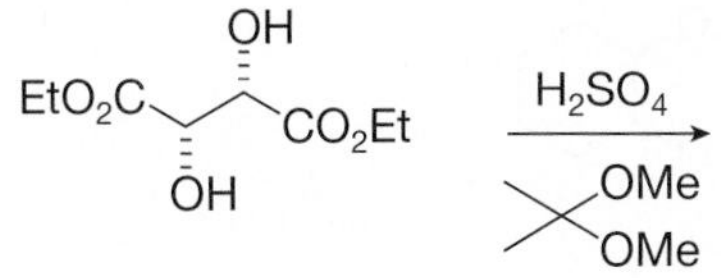 H_2SO_4, 2,2-dimethoxypropane

c) 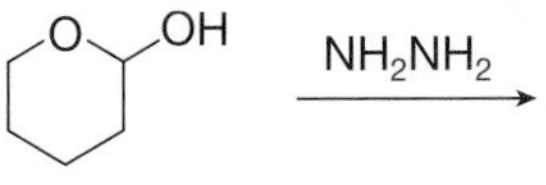 NH_2NH_2

16.25 Thioacetals are very stable functional groups that require special installation reactions using reagents that form very strong bonds to oxygen. Based on this information, identify the structures of the products and propose mechanisms for their formation.

a) 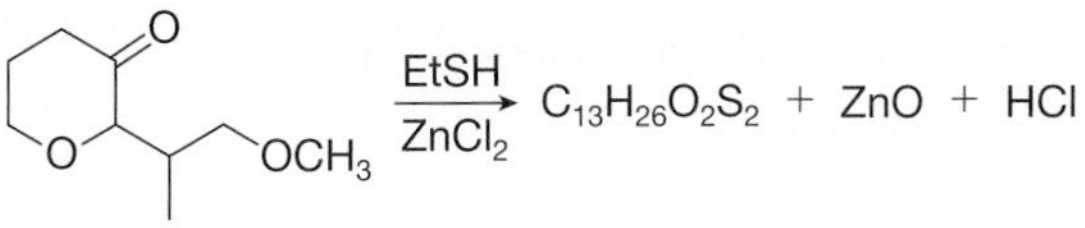 EtSH, $ZnCl_2$ → $C_{13}H_{26}O_2S_2$ + ZnO + HCl

b) 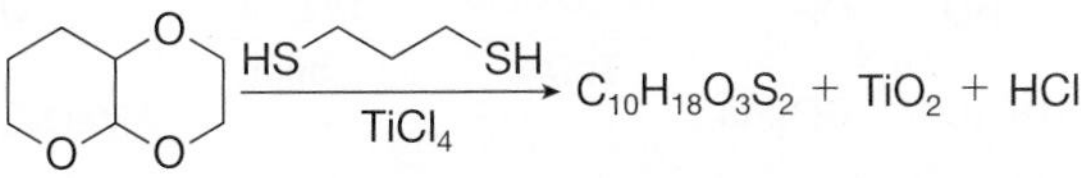 HS–SH, $TiCl_4$ → $C_{10}H_{18}O_3S_2$ + TiO_2 + HCl

16.26 Thioacetals are very stable functional groups that require special removal reactions using reagents that form very strong bonds to sulfur. Based on this information, propose mechanisms for the following observations.

a) 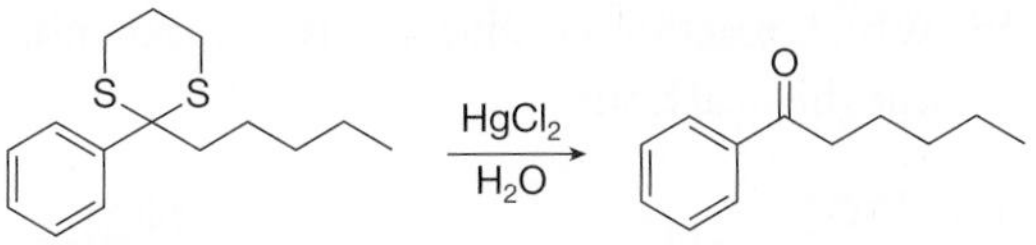 $HgCl_2$, H_2O

b) 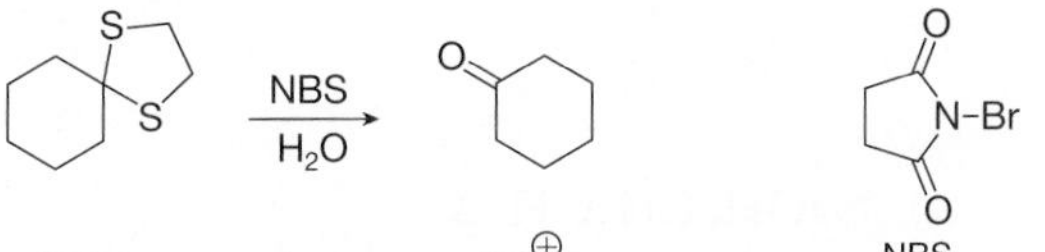 NBS, H_2O

(NBS is an excellent source of $Br^{\oplus}$)

NBS

16.27 Write mechanisms to account for the following observations.

a) 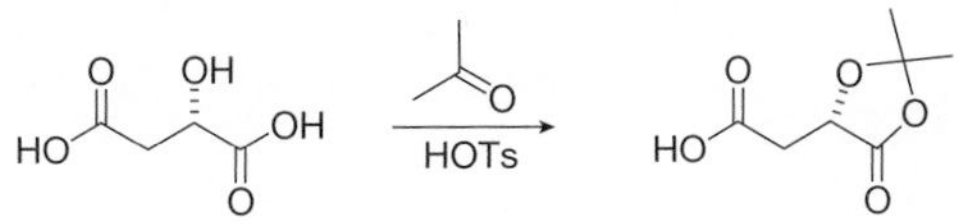 HOTs

b) 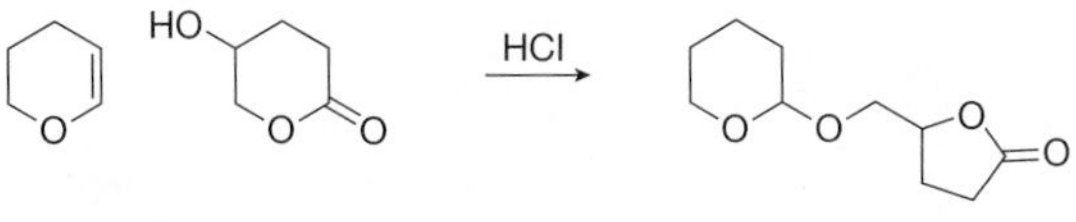 HCl

c) 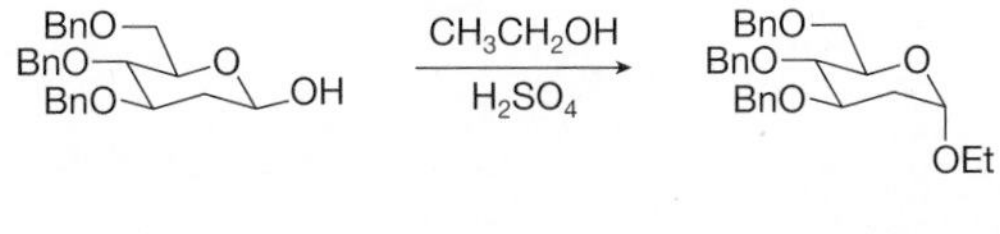 CH_3CH_2OH, H_2SO_4

d) 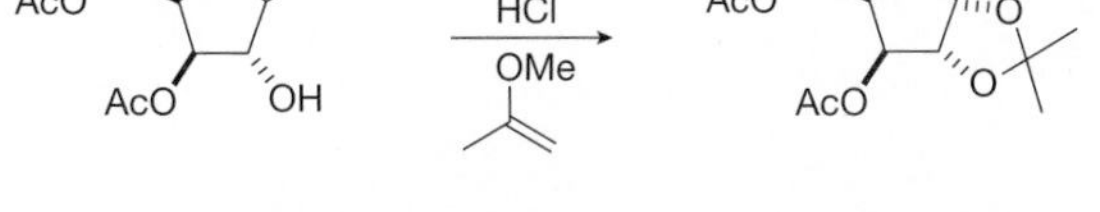 HCl, OMe

e) 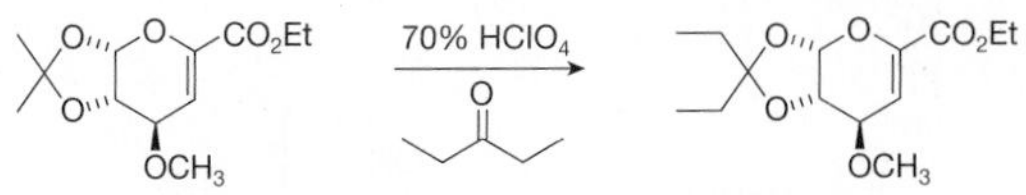 70% $HClO_4$

f) 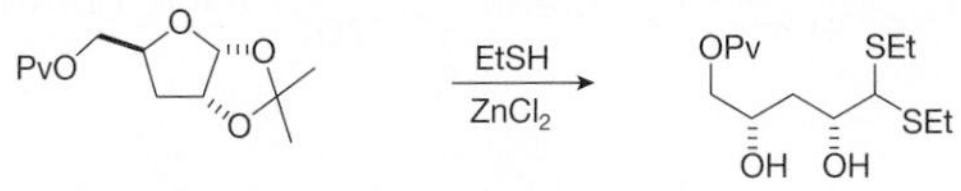 EtSH, $ZnCl_2$

16.28 The synthesis of 3-TC, a drug used to treat AIDS, begins with the following reaction. Provide a mechanism to show how this works.

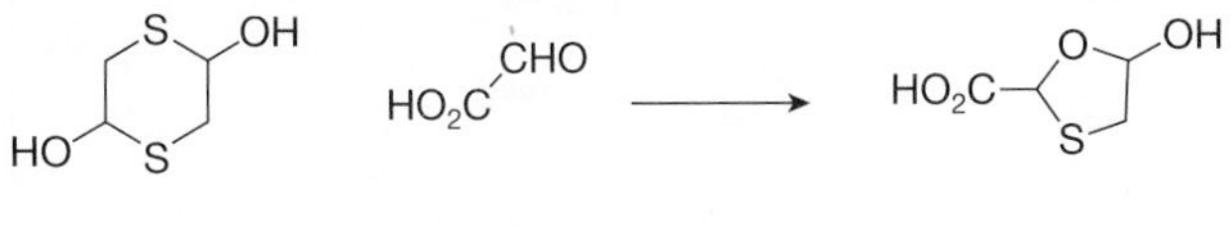

16.29 Provide mechanisms to explain the formation of each product.

a) 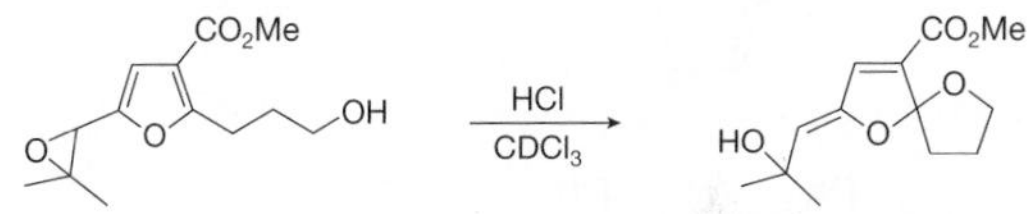 HCl, $CDCl_3$

b) 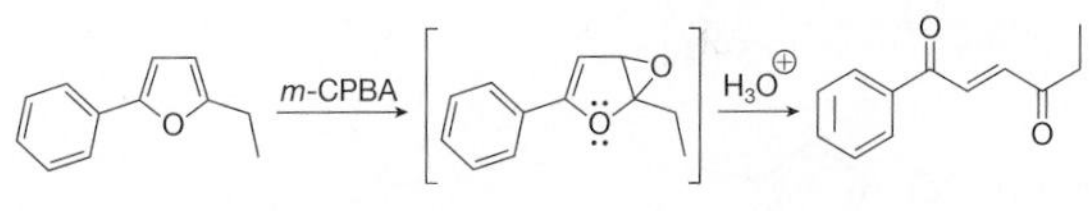 *m*-CPBA, $H_3O^{\oplus}$

c) 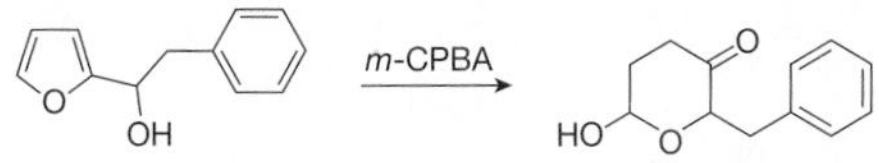 *m*-CPBA

d) 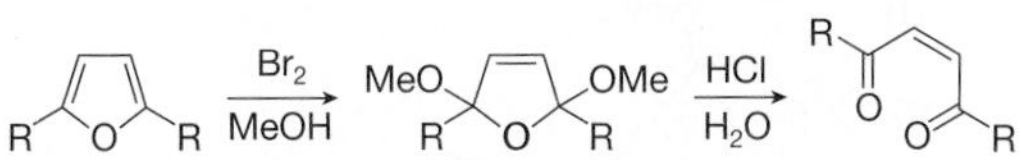 Br_2, MeOH; HCl, H_2O

16.30 Provide a mechanism to explain the following reactions.

a) 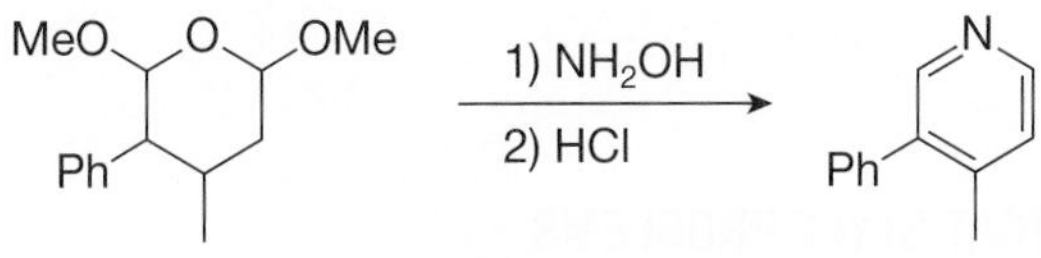 1) NH_2OH 2) HCl

b) 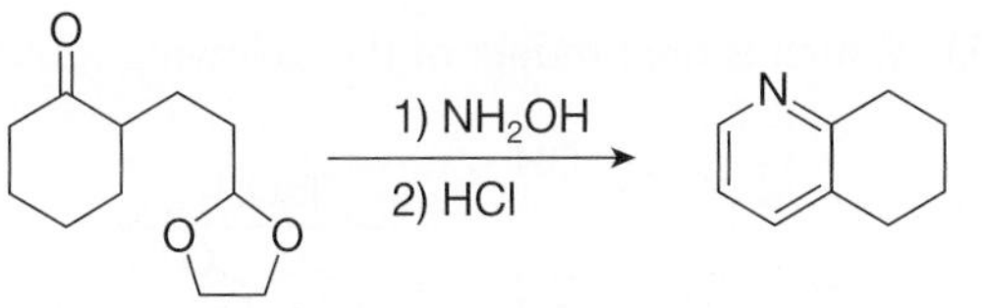 1) NH_2OH 2) HCl

16.31 Show how each of the following reactions occur.

a)

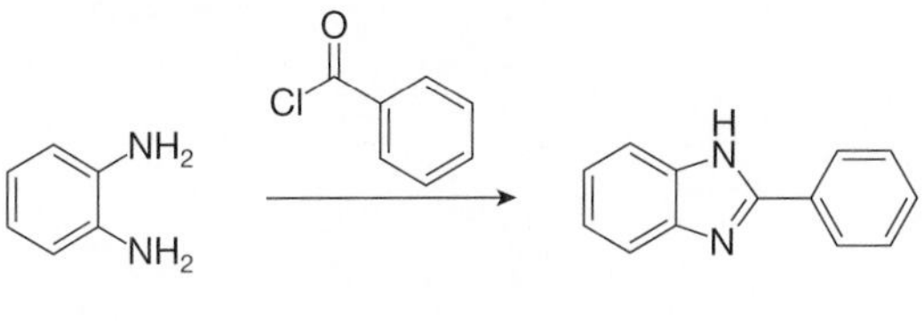

b)

CbzHN, H, O; NaHSO$_3$; CbzHN, OH, $SO_3^{\ominus}$; NaCN; CbzHN, OH, CN; HCl, EtOH; CbzHN, OH, OEt, $NH_2^{\oplus}$; HS, H_2N, heat; CbzHN, OH, S, N

16.32 Show how each of the following heterocycles could be synthesized.

a) Ph, O

b) Ph, O, Ph, N

c) S

d) N, NH

e) N, Ph

MCAT STYLE PROBLEMS

16.33 Which is the product of the following reaction?

OH, OH; TsOH; O

a) O, O

b) O, O

c) O, O

d) O, O

16.34 What are the major by-products of the following reaction?

HO, OH, MeO$_2$C, CO$_2$Me; CH(OMe)$_3$, TsOH; PhCHO, heat; Ph, O, O, MeO$_2$C, CO$_2$Me

a) H_2O and $HC(O)OCH_3$
b) CH_3OH and H_2O
c) $HC(O)OCH_3$ and H_2O
d) CH_3OH and $HC(O)OCH_3$

16.35 Which reaction conditions are a good choice to carry out this reaction?

Ph, O, O, O, Ph; HO, HO, O, Ph

a) NaOH, CH_3OH, Δ
b) $HgCl_2$, H_2O
c) HCl, CH_3OH, Δ
d) $NaBH_4$, H_2O

CHALLENGE PROBLEM

16.36 Viagra can be made according to the following sequence of reactions. Show how each of these reactions works by drawing a mechanism for each one.

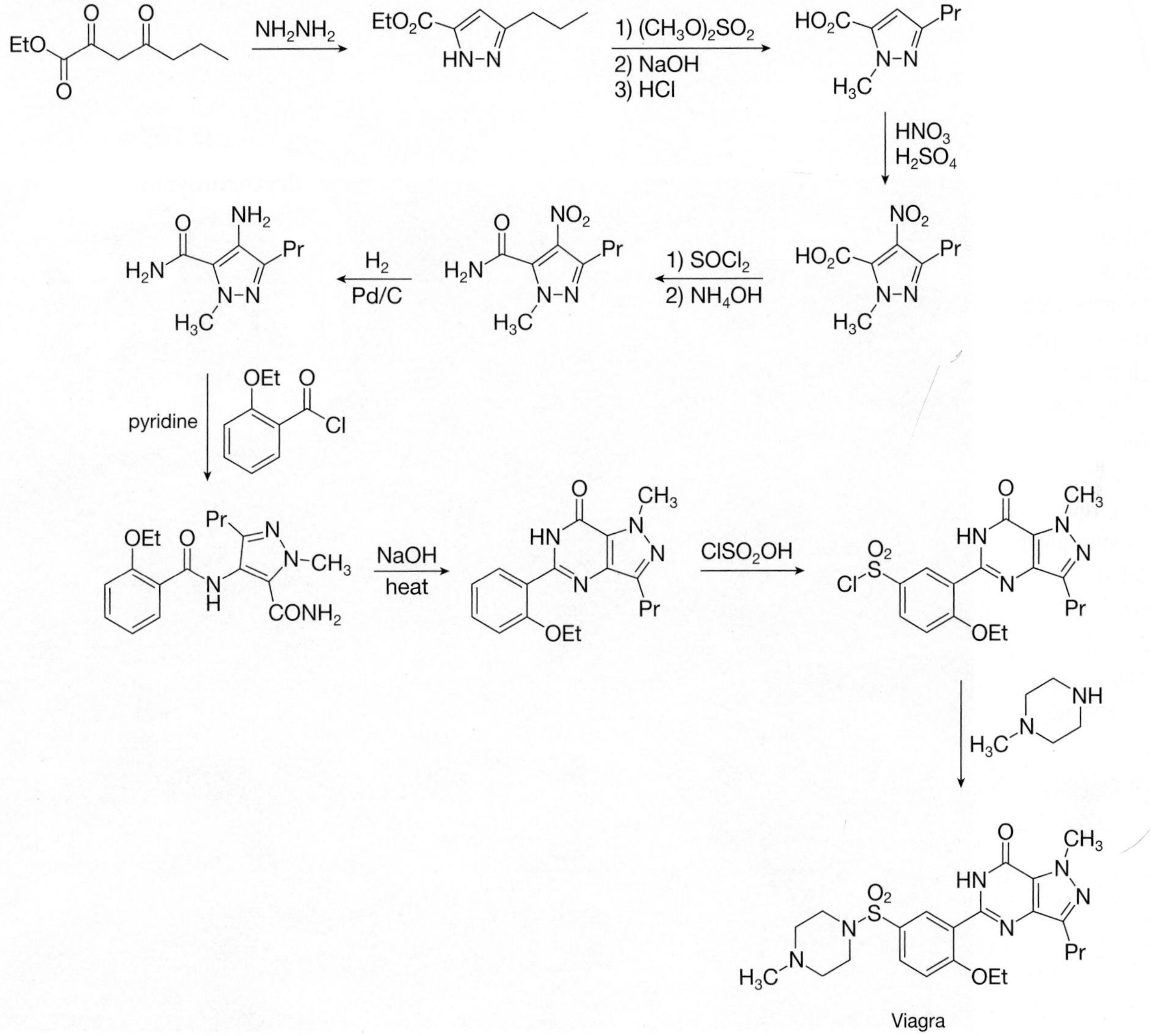

NH_2NH_2

1) $(CH_3O)_2SO_2$
2) NaOH
3) HCl

HNO_3
H_2SO_4

1) $SOCl_2$
2) NH_4OH

H_2
Pd/C

pyridine

NaOH
heat

$ClSO_2OH$

Viagra

17

Carbonyl-Based Nucleophiles

ALDOL, CLAISEN, WITTIG, AND RELATED ENOLATE REACTIONS

CHAPTER OUTLINE

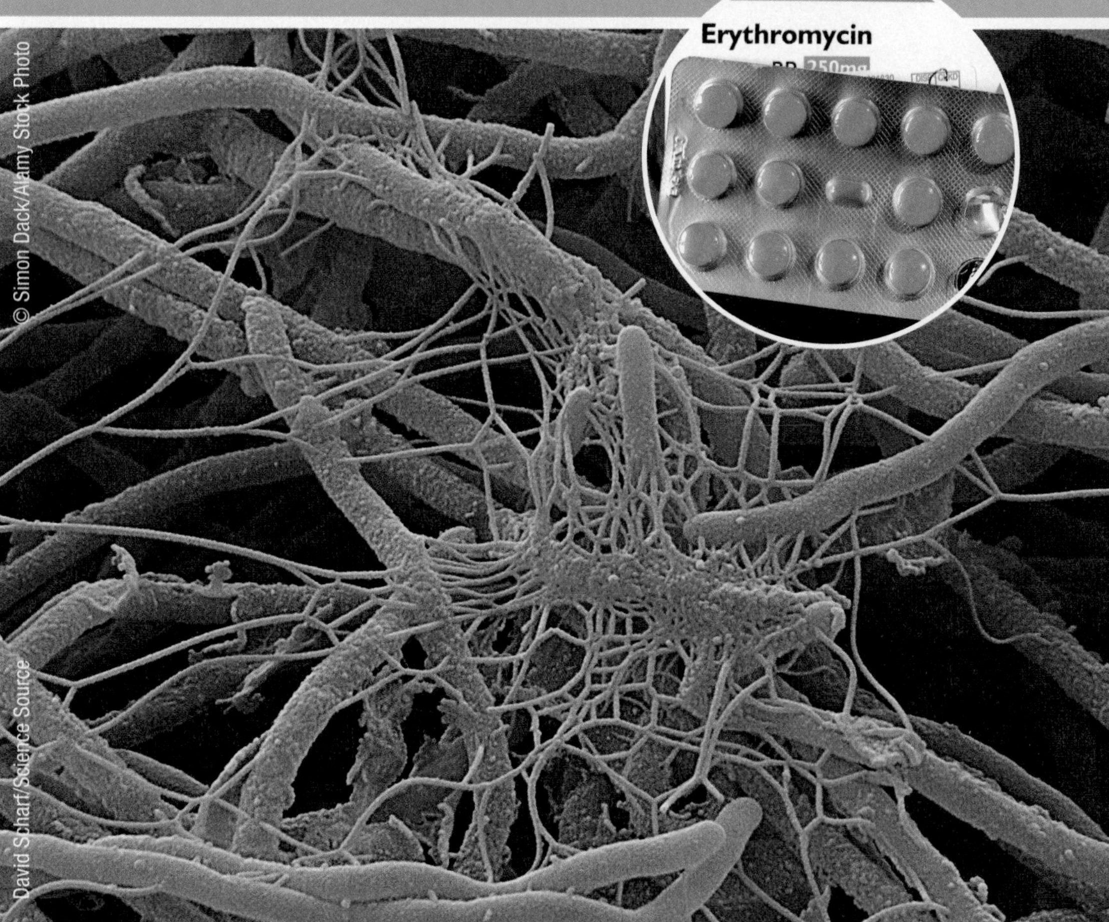

Macrolide antibiotics such as erythromycin are produced by *Streptomyces erythreus* and are often the antibiotic of choice for patients allergic to penicillin. Unlike penicillin, which inhibits the synthesis of bacterial cell walls, macrolide antibiotics function by inhibiting bacterial protein synthesis.

17.1 Why It Matters

Produced by *Streptomyces* soil bacteria, erythromycin is an antibiotic often used to treat patients who are allergic to penicillin. Erythromycin binds to the centre of one subunit of the bacterial ribosome that is responsible for protein synthesis (Figure 17.1), preventing protein synthesis and thereby inhibiting bacterial growth.

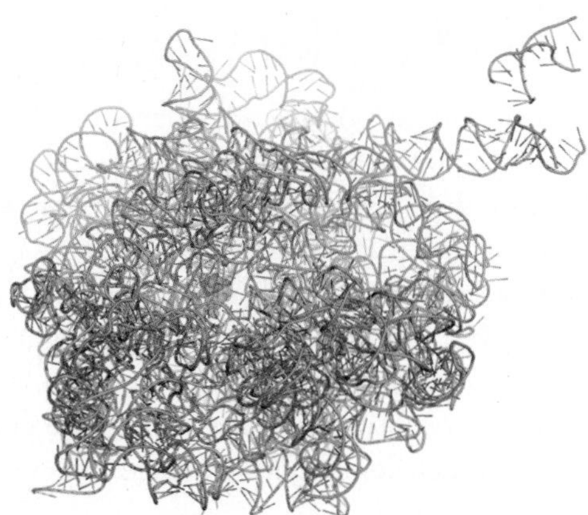

FIGURE 17.1 X-ray crystal structure of erythromycin (green) bound to the 50S subunit of the ribosome of *E. coli* bacteria (PDB 1JZY).

Figure 17.2 shows the structures of erythromycin, lovastatin, and rapamycin—three natural compounds that have useful medicinal properties. Only lovastatin, the simplest of these three drugs, is currently manufactured by organic synthesis. Erythromycin and rapamycin are obtained from natural sources; synthesis of commercial quantities of these complex structures is not yet feasible.

erythromycin (antibacterial)

lovastatin (cholesterol-lowering)

rapamycin (immunosuppressant)

FIGURE 17.2 Three drugs isolated from natural sources.

How do soil bacteria form the bonds needed to produce erythromycin? Such bacteria cannot carry out reactions that require highly reactive substances, high temperatures, or anhydrous conditions. For example, making carbon-carbon bonds in a living organism requires a nucleophile with enough reactivity to attack moderate electrophiles but enough stability to exist in an aqueous environment. **Enolates**, alkenes that carry a negatively charged oxygen atom, have these balanced properties.

An **enolate** is an alkene with a negatively charged oxygen atom as one of its substituents. It is the conjugate base of an enol.

enol enolate

This chapter describes the mechanisms of enolate reactions with a variety of electrophiles, including additions to carbonyls, S_N2 reactions, and halogenation reactions. It is through a series of addition to carbonyl reactions that the bacteria form the C–C bonds that make up erythromycin. Similar reactions can be used in the laboratory to synthesize similar molecules in order to fine-tune the desired molecular properties.

17.2 The Acidity of Carbonyl Compounds

An **α-carbon** is a carbon directly connected to a carbonyl carbon or other functional group.

Enolizable compounds can be converted to enolates.

As described in earlier chapters, carbonyl groups act as electrophiles in a variety of reactions. However, these groups can also behave as nucleophiles. The key to the nucleophilic behaviour of carbonyl groups is how easily they form carbanions. The hydrogens on the **α-carbon** of a carbonyl-containing functional group are relatively easy to remove because the resulting carbanion (conjugate base) is stabilized by electron conjugation with the carbonyl oxygen (electronegative atom). Such enolate carbanions are usually represented by the oxyanion resonance form because this form contributes the most to the actual structure. Carbonyl compounds that can be deprotonated to form an enolate are called **enolizable compounds**. By extension, the α-carbons and α-hydrogens in such compounds are sometimes referred to as enolizable.

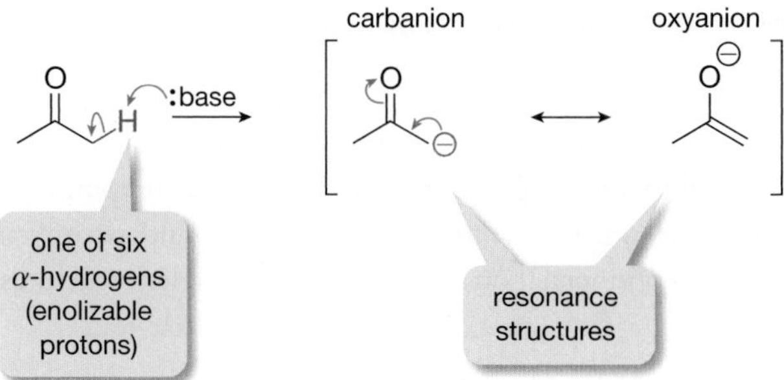

STUDENT TIP

The ease with which protons are removed from organic molecules is sometimes called the acidity of the protons. These molecules are not acidic, but behave like acids according to the Brønsted definition. Higher pK_a values indicate that the protons are more difficult to remove.

The nature of the carbonyl group influences the acidity of the α-hydrogen. The presence of a group on the carbonyl that is electron donating destabilizes the negatively charged conjugate base—the more readily the substituent group donates electrons, the less the acidity of the α-hydrogen. An α-hydrogen is removed as a proton.

Aldehydes have the poorest electron-donating substituent (H), and hence the most acidic α-hydrogen (lowest pK_a). As shown in Figure 17.3, ketones, esters, and amides have increasingly stronger electron-donating substituents (by conjugation) and correspondingly decreased acidity of the α-hydrogen.

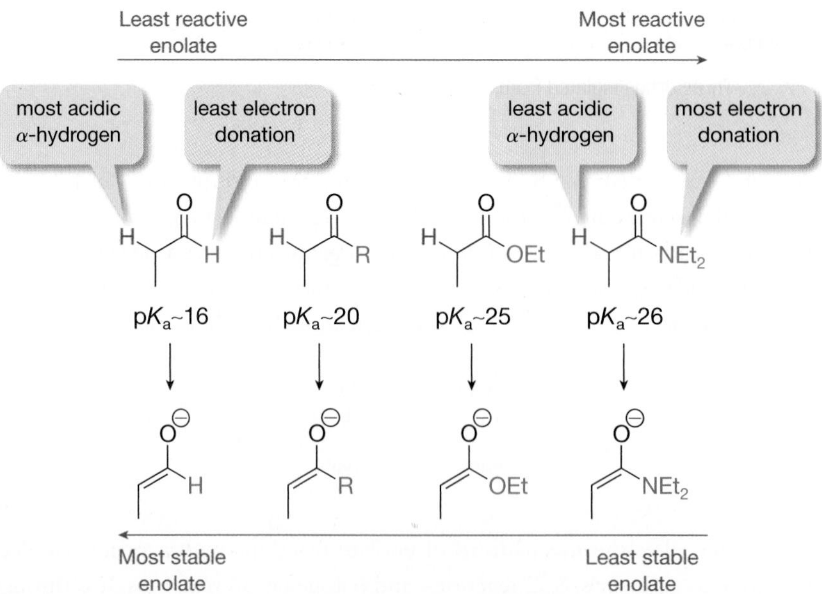

FIGURE 17.3 Aldehydes, ketones, esters, and amides are all capable of forming enolates, although with different stabilities.

17.2.1 Deprotonation using weak bases

Since all the pK_a values in Figure 17.3 are higher than the pK_a of water (15.7), deprotonation of a carbonyl compound is energetically unfavourable when hydroxide is used as the base. Nonetheless, a small amount of enolate is formed in the presence of OH^-, and this enolate can participate in chemical reactions. Because only a small amount of enolate is available to react in any instance, the use of weak bases to form enolates may create a mixture of products. This method must be used carefully.

equilibrium lies to side with weaker acid

weaker acid + weaker base ⇌ stronger base + stronger acid

STUDENT TIP

The effective strength of a base can depend on the other reagents in the reaction. Hydroxide, though generally considered a strong base, is a weak base in reactions with enols because hydroxide ion is not reactive enough to deprotonate every molecule of a carbonyl compound to form enolates.

17.2.2 Keto-enol equilibria

The unfavourable equilibrium that produces enolates using weak bases also results in the formation of a neutral nucleophile called an enol. As discussed in Chapter 8, enols rapidly undergo **tautomerization**, or tautomerizes to the corresponding carbonyl form because of the nucleophilicity of the alkene, which is enhanced by the presence of an electron-donating hydroxyl group. One hydrogen atom changes location to produce the ketone form. The reverse reaction, the formation of an enol, can occur under both acidic and basic conditions. Most carbonyl-containing compounds exist in equilibrium with small amounts of the enol form (less than 1% enol).

Tautomerization is the conversion of a compound into a structural isomer that differs by the location of one atom.

Tautomerization in acid

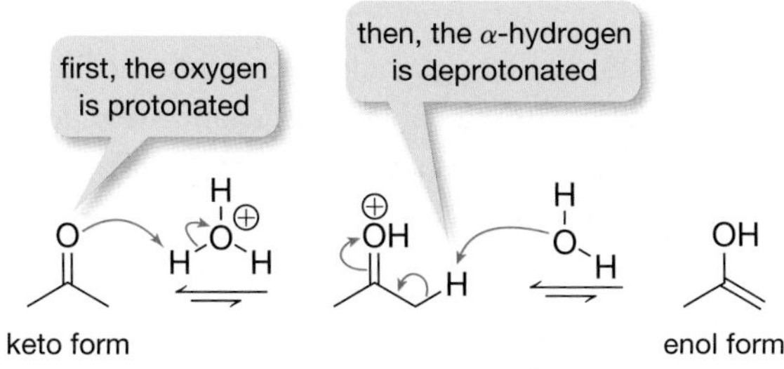

Tautomerization in base

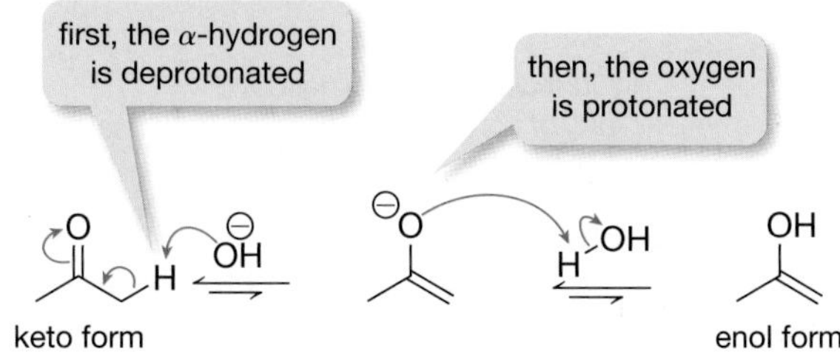

ORGANIC CHEMWARE

17.1 Tautomerization of ketones (basic conditions)

ORGANIC CHEMWARE

17.2 Tautomerization of ketones (acidic conditions)

Tautomerization is responsible for the slow racemization of carbonyl-containing compounds that form enols by deprotonation at a stereogenic centre. When such enols transform back into the carbonyl form, the hydrogen can add to either face of the π bond, resulting in a racemic mixture of products. This process can be catalyzed by an acid or a weak base.

STUDENT TIP
Keto-enol tautomerization occurs in the presence of an acid or weak base catalyst, but not in the presence of strong base.

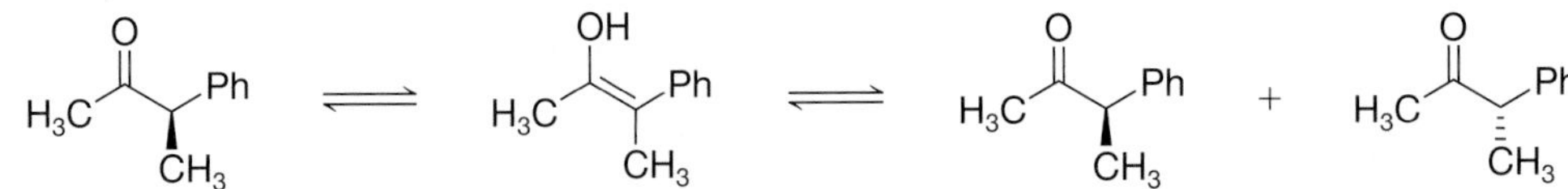
O
OH
Ph
H3C
CH3
+

17.2.3 Deprotonation with strong bases in carbonyl chemistry

Quantitative deprotonation is an irreversible deprotonation that converts all of the carbonyl-containing molecules in the starting material to enolates.

Nitrogen bases are much stronger than oxygen bases, and can irreversibly form enolates from aldehydes, ketones, esters, and amides. The conversion of all carbonyl-containing molecules in a solution to enolates is referred to as a **quantitative deprotonation**. Since nitrogen bases are potential nucleophiles, sterically hindered amine bases are used to generate enolates. Because the nitrogen atoms in these bases are very crowded, they are unable to function as nucleophiles.

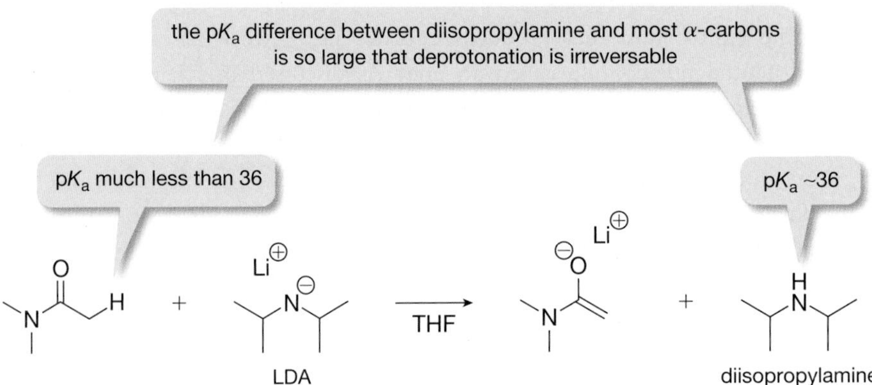
the pKa difference between diisopropylamine and most α-carbons is so large that deprotonation is irreversable
pKa much less than 36
pKa ~36
THF
LDA
diisopropylamine

ORGANIC CHEMWARE
17.3 Enolate alkylation

Lithium diisopropylamide (LDA) is a strong amide base with steric hindrance. This base is commonly used to make enolates since it is strong enough to quantitatively deprotonate the α-position of carbonyl compounds without adding to the carbonyl. The pK_a difference between the conjugate acid of these amide and most α-hydrogens is so large that the equilibrium between the carbonyl compound and the enolate is completely (quantitatively) shifted toward the enolate. In general, a pK_a difference of 10 or more results in quantitative deprotonation.

STUDENT TIP
An amide is a functional group with a nitrogen adjacent to a carbonyl. Amide is also the term for the conjugate base of an amine. A negatively charged amine is an amide, in the same way that a negatively charged chlorine is a chloride.

lithium diisopropylamide (LDA)
pK_a of conjugate acid = 36

Amide bases are much stronger bases than hydroxide. Since they react violently with water, they must be used in anhydrous aprotic solvents such as THF. All these bases are normally used at low temperatures to prevent side reactions. The most common temperature is −78 °C because this temperature is easy to maintain using dry ice. LDA is commercially available; however, it is best prepared immediately before use by reacting diisopropylamine with butyl lithium (BuLi), a very strong base. A reaction using a strong base must be done in separate steps; the base generates the enolate before an electrophile is added.

diisopropylamine ($pK_a \sim 36$) + butyl lithium (BuLi) $\xrightarrow[-78\ °C]{THF}$ LDA + butane ($pK_a \sim 60$)

CHECKPOINT 17.1

STUDENT TIP

Enolate reactions that use a weak base do not convert all of the starting material to form the enolate. Reactions using these conditions are usually done in one step. Other enolate reactions require quantitative deprotonation using a strong base. Reactions that occur in these conditions involve sequential reactions. Take care to note whether a strong base (nitrogen anion) or weak base (oxygen anion) is used in each reaction.

You should now be able to identify the enolizable protons in carbonyl compounds and rank their relative acidities. You should also be able to use pK_a data to predict whether the reactions of carbonyl compounds with various bases will be partial or quantitative.

SOLVED PROBLEM

Identify the enolizable proton(s) in each of the following carbonyl compounds. Draw two acid–base reactions for each compound, one in which the carbonyl compound is only partially deprotonated, and one in which it is quantitatively deprotonated.

STEP 1: Expand the Lewis structure at each α-carbon to show the α-hydrogens. The α-carbons are the ones directly attached to the carbonyl carbon. The amide has one α-carbon, whereas the ketone has two. The enolizable protons are the α-hydrogens attached to the α-carbons.

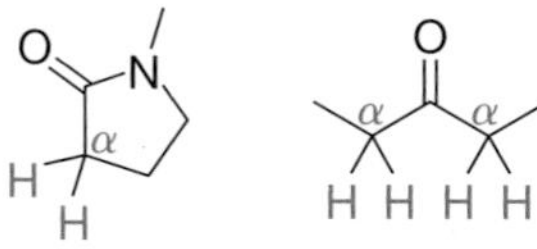

STEP 2: Draw the conjugate base for each structure.

Since the ketone has two sets of enolizable protons, consider whether two different enolates are possible. Here the ketone is symmetrical, so both sets of protons are equivalent and only one enolate is possible. Each conjugate base has two possible resonance structures. Draw the enolate using the most stable resonance contributor.

STEP 3: Look up the approximate pK_a value for each set of enolizable protons.

The enolizable protons on an amide have a pK_a of approximately 26, whereas the enolizable protons on a ketone have a pK_a of approximately 20.

STEP 4: For each compound, write a reaction equation for incomplete deprotonation using a base with a conjugate acid that has a pK_a near or somewhat below that of the enolizable proton. Many different bases could be used, but alkoxide bases are the most commonly used for partial deprotonation of α-hydrogens. The amide is less acidic than the ketone, so a stronger base such as *tert*-butoxide could be used (pK_a of *tert*-butanol = 18); methoxide would be a good choice for enolizing the ketone (pK_a of methanol = 15.5).

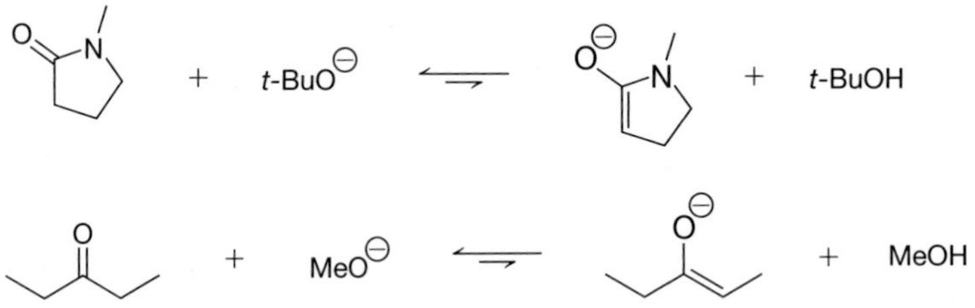

+ t-BuO⁻ ⇌ + t-BuOH

+ MeO⁻ ⇌ + MeOH

STEP 5: For each compound, write a reaction equation for quantitative deprotonation using a non-nucleophilic base with a conjugate acid having a pK_a at least 10 higher than that of the enolizable proton.

Quantitative deprotonation of an enolizable proton requires a very strong base. Using LDA ensures that both reactions will be complete (in deprotonation reactions, the counter ion is often omitted).

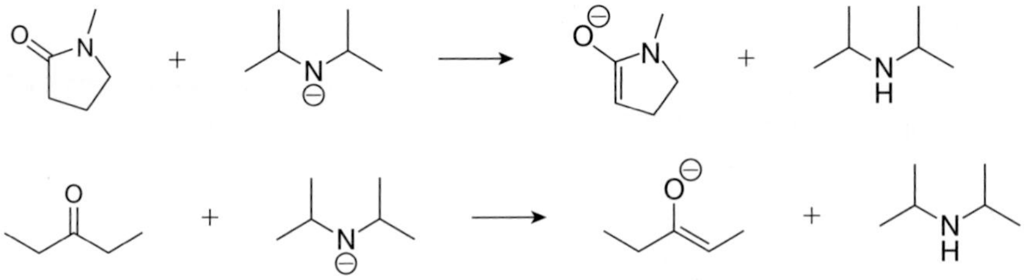

PRACTICE PROBLEM

17.1 For each of the following pairs of reactions, draw the products of the reactions, and predict which products will be favoured. Indicate whether quantitative deprotonation will occur.

a) LDA

LDA

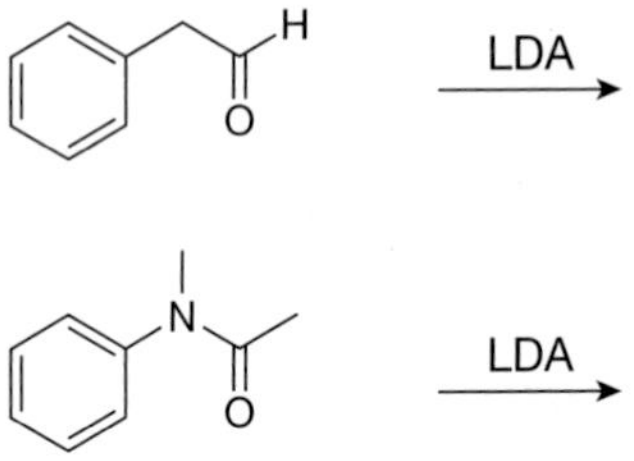

b) NaOMe

NaOMe

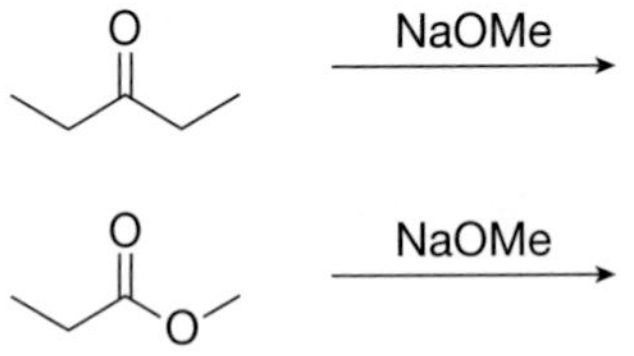

c) LDA

NaOEt

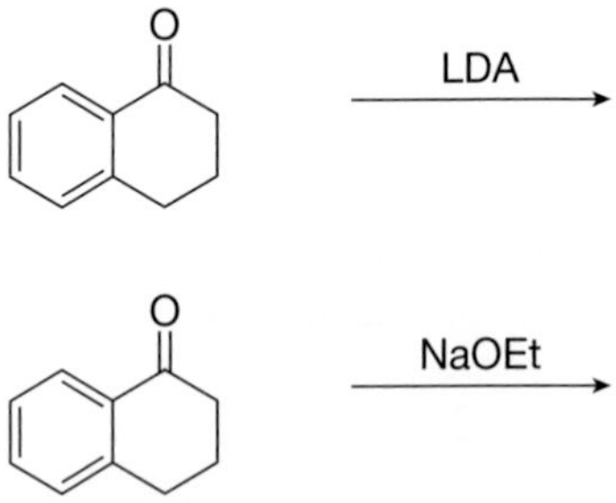

d)

KO*t*-Bu

NMe

KO*t*-Bu

INTEGRATE THE SKILL

17.2 The equilibrium constant, K, for the deprotonation of acetone by hydroxide is 6.31×10^{-20}.

a) Write the equation for the equilibrium constant, and use it to calculate how much acetone remains when one mole of hydroxide is added to one mole of acetone.

b) If acetaldehyde is used instead of acetone, how does the proportion of unreacted starting material change?

17.3 Reactions of Enolates with Electrophiles

Enols and enolates are nucleophilic functional groups that add electrophiles at the α-carbon. By manipulating the reaction conditions, a variety of electrophiles can be added with a high degree of control. These reactions can be catalyzed by an acid or a base, although base catalysis is *much* more common. The pattern of these processes depends on the catalyst: acid catalysts activate electrophiles, and base catalysts activate nucleophiles.

Acid catalysis

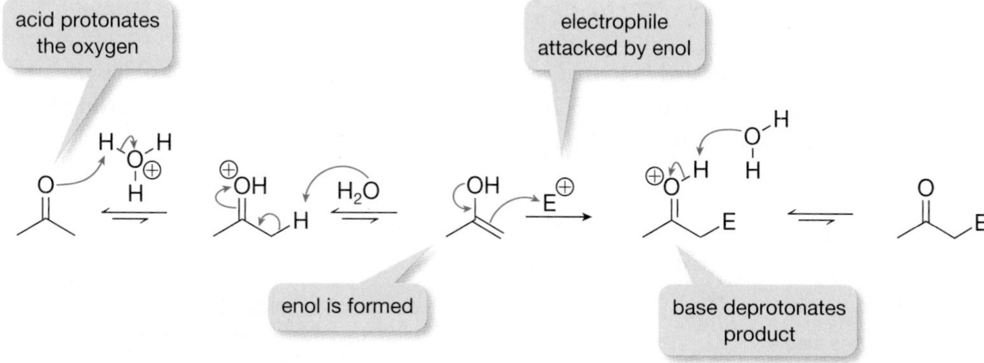

Base catalysis

base deprotonates the α-hydrogen

electrophile attacked by enolate

Enols and enolates can react with the same electrophiles as described for reactions with nucleophiles in previous chapters. As described in the following sections, they react with halogens and undergo S_N2 reactions with alkyl halides. Enolates also add to carbonyls in a way that is similar to how other nucleophiles add to carbonyls, as discussed in earlier chapters. The additions of enols to carbonyls, similar to the reactions in Chapter 7, are discussed in Section 17.4. Addition-elimination reactions, parallel to those in Chapter 15, are described in Section 17.5.

17.3.1 α-Halogenation of carbonyl compounds

Most carbonyl groups can be halogenated by replacing α-hydrogens with halogens. Although this reaction occurs in the presence of acid or base, acid catalysis is common because the reaction generates acid (HBr) as a by-product.

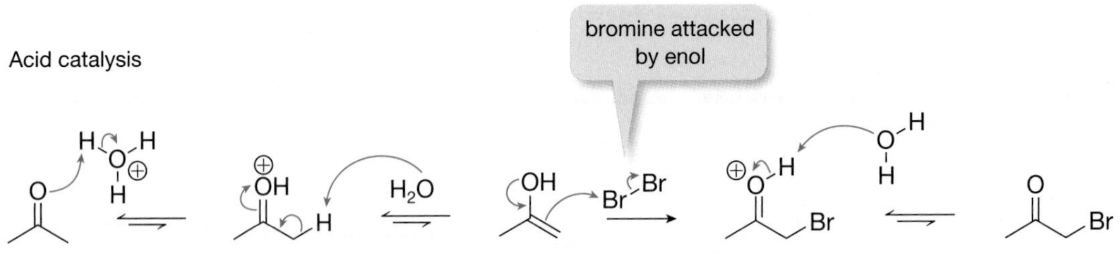

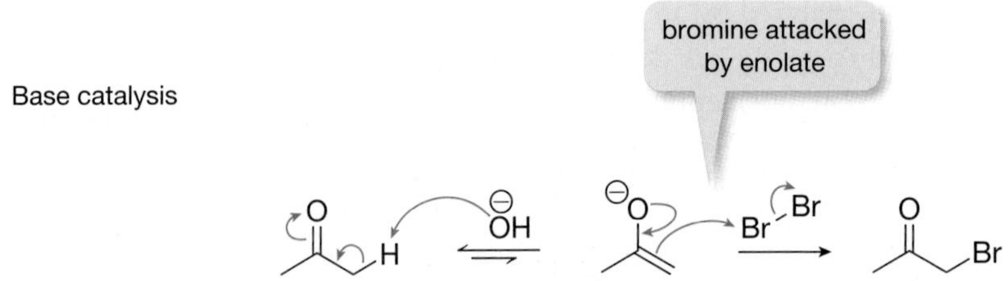

Acid catalysis

bromine attacked by enol

H_2O

Base catalysis

bromine attacked by enolate

For example, when a methyl ketone is mixed with bromine in the presence of base, a further reaction occurs. The brominated enolate is more stable than the enolate of the starting ketone due to the inductive effect of the electronegative halogen.

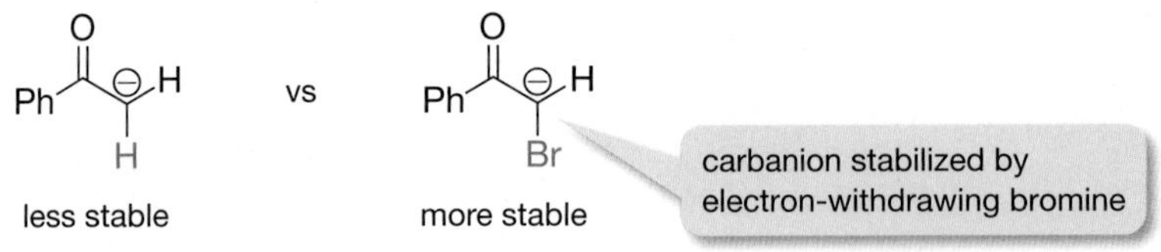

less stable vs more stable

carbanion stabilized by electron-withdrawing bromine

As a result, the brominated product, instead of the remaining starting ketone, is deprotonated. Bromine then replaces a second α-hydrogen on the enolate, further increasing its stability. The dibrominated enolate is preferentially deprotonated again, and quickly undergoes a third bromination. The result is a trihalogenated ketone, with all three α-hydrogens replaced by halogen atoms.

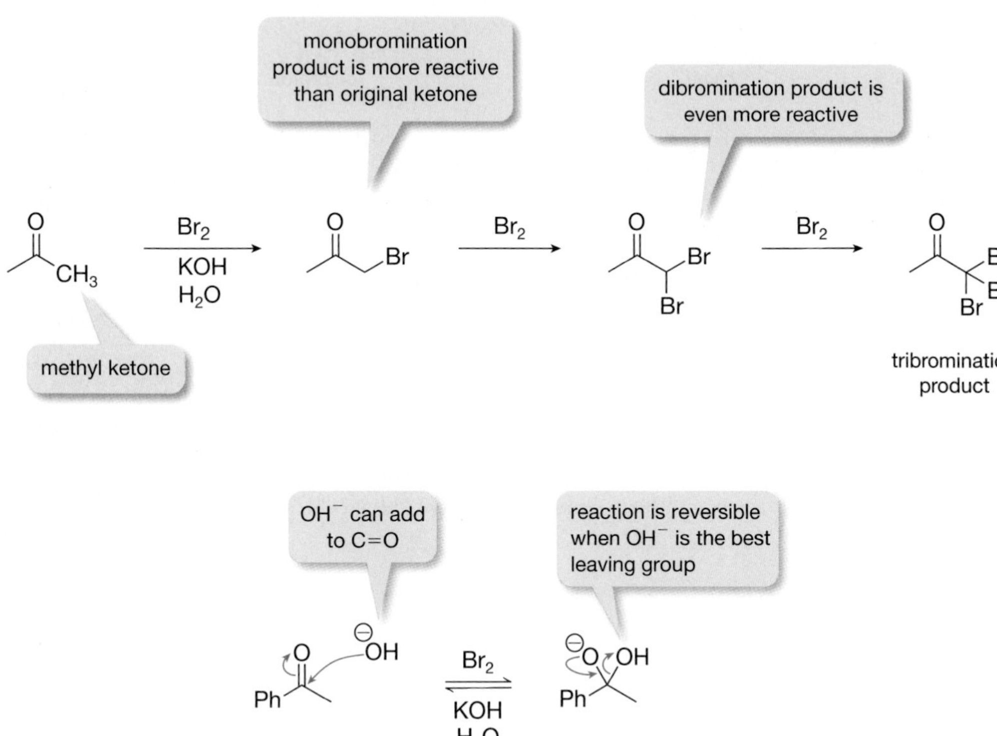

STUDENT TIP

In α-halogenation, the intended role of hydroxide is to act as a base. However, it can also act as a nucleophile, and both processes occur. When hydroxide acts as a nucleophile, the reaction is reversible if the hydroxide is the best leaving group in the reverse reaction.

When a tribrominated ketone is formed in the presence of hydroxide, the nucleophilic role of hydroxide becomes important. If OH^- adds to the carbonyl of a tribrominated ketone, the tetrahedral intermediate has an excellent leaving group, the CBr_3^- group (the negative charge is stabilized by the three electronegative bromines). The tribrominated ketone undergoes an addition-elimination reaction to form a carboxylic acid. This acid is immediately deprotonated by the CBr_3^- leaving group, producing a carboxylate and a molecule of bromoform. Similar reactions occur with other halogens.

CBr₃⁻ is the best leaving group
CBr₃⁻ deprotonates carboxyl group
C=O attacked by hydroxide
CBr₃⁻ eliminated to form carboxylic acid
low reactivity of carboxylate prevents reaction from reversing
$HCBr_3$ bromoform

This reaction is an example of a **haloform reaction**. Using Cl_2 forms chloroform ($CHCl_3$), Br_2 forms bromoform ($CHBr_3$), and I_2 forms iodoform (CHI_3). The iodoform reaction was once used as a chemical test for the methyl ketone functional group. Iodoform is a yellow solid, which is easy to detect; its formation in this reaction indicates the presence of a methyl ketone. Today, the test is rarely used, but the haloform reaction remains a useful method for making carboxylic acids from methyl ketones.

A **haloform reaction** converts a methyl ketone to a carboxylic acid and a haloform.

STUDENT TIP

The haloform reaction is done with a weak base. The hydroxide acts as a base to form small amounts of enolate, and also acts as a nucleophile. A strong base could not accomplish both roles.

CHECKPOINT 17.2

You should now be able to predict the products formed by reactions of enolizable carbonyl compounds with halogens under both acidic and basic conditions. You should also be able to draw mechanisms for these reactions.

SOLVED PROBLEM

Predict the product of the following reaction, and draw a mechanism showing its formation.

STEP 1: Determine the roles of the reactants.

The carbonyl group itself is electrophilic; however, in the presence of acid the carbonyl is in equilibrium with its enol form, which can act as a nucleophile. The nucleophilic enol can then react with the electrophilic bromine.

STEP 2: Draw a mechanism for the formation of the enol.

First, the acid protonates a lone pair of electrons on the carbonyl. The resulting positive charge on the carbonyl oxygen increases the acidity of the α-hydrogens, allowing one of them to be removed by a weakly basic water molecule, thereby generating the enol.

STEP 3: Draw a mechanism for the reaction of the enol with bromine.

The nucleophilic enol attacks the electrophilic bromine while simultaneously breaking the Br–Br bond. Then a molecule of water removes the proton from the carbonyl to give the final brominated product.

PRACTICE PROBLEM

17.3 For each of the following reactions, predict the major product, and draw a mechanism showing its formation.

a) 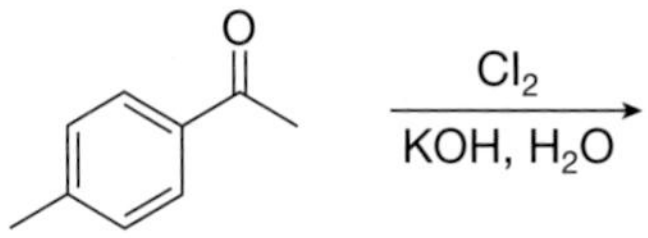 $\xrightarrow[\text{KOH, } H_2O]{Cl_2}$

b) 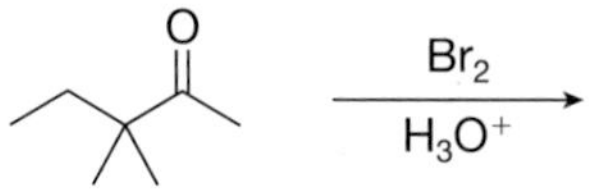 $\xrightarrow[H_3O^+]{Br_2}$

c) 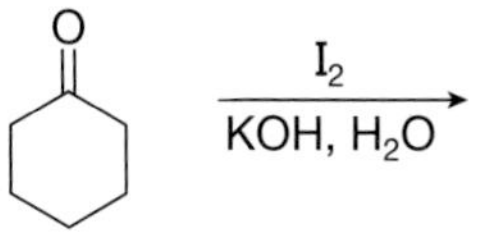 $\xrightarrow[\text{KOH, } H_2O]{I_2}$

d) 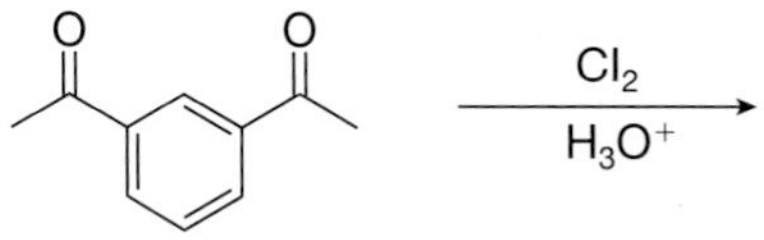 $\xrightarrow[H_3O^+]{Cl_2}$

INTEGRATE THE SKILL

17.4 The Hell–Volhardt–Zelinsky reaction adds an α-halogen, such as a bromine, to a carboxylic acid, as shown in this example.

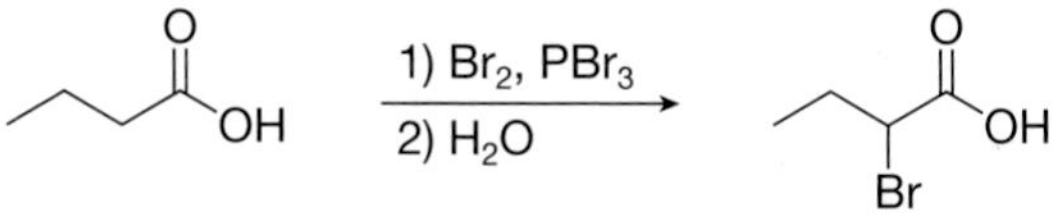

OH $\xrightarrow[\text{2) } H_2O]{\text{1) } Br_2\text{, } PBr_3}$ OH Br

In the first step, the PBr_3 converts the carboxylic acid to an acid bromide, forming HBr as a by-product. The acid bromide is in equilibrium with its enol tautomer, which is then brominated. Draw the mechanisms for the tautomerization of the acid bromide to the enol form and the α-bromination that follows.

17.3.2 Alkylation of enolates

Once an enolate has been prepared using LDA or a similar strong base, an electrophile can be added to form a new product. If the electrophile is an alkyl halide, the result is an alkylation reaction. Such alkylations are S_N2 reactions, so the alkyl halide must be able to react by an S_N2 pathway. Primary alkyl halides react readily with enolates, secondary halides react slowly, and tertiary halides do not react. Most alkylations use bromine or iodine as the leaving group. Chloride is not a strong enough leaving group for most enolate alkylation reactions.

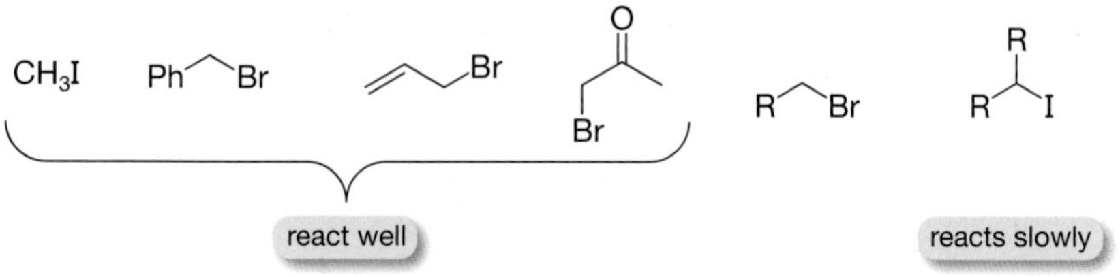

CH_3I Ph Br Br Br R Br R I R

react well

reacts slowly

Most ester and amide enolates undergo alkylation reactions readily. Since the carbonyl of these compounds can be electrophilic, care must be taken to ensure that the enolate forms quickly and completely. The normal procedure is to add a solution of the carbonyl compound to a solution of LDA at −78 °C.

ORGANIC CHEMWARE
17.4 Enolate alkylation

17.3.2.1 Alkylation of ketones

Ketones can be alkylated in similar conditions, but such reactions can produce regioisomers. While it is possible to choose reaction conditions that form an enolate on the more substituted side (a thermodynamic enolate), it is far more common for the enolate to form on the less substituted side (a kinetic enolate).

enolates can form on either side of the ketone carbonyl

LDA

OLi

thermodynamic enolate

CH_3I

OLi

kinetic enolate

CH_3I

regioisomers

Generally, conditions are chosen such that the enolate formation is irreversible. Irreversible deprotonation favours the kinetic product, normally the *less* substituted enolate. Using an excess of a very strong base at a very low temperature prevents the enolate from becoming protonated, and thus ensures an irreversible reaction.

LDA
THF, −78 °C

OLi

kinetic enolates (less substituted double bond) form the fastest and usually at the less substituted position

Thermodynamic enolates and the conditions required to form them are discussed in further detail in Chapter 18.

CHECKPOINT 17.3

You should now be able to predict the products of reactions of carbonyl compounds with alkyl halides in the presence of a strong base. You should also be able to draw mechanisms for these reactions.

STUDENT TIP
Alkylation requires a strong base to ensure that all of the starting carbonyl compound is enolized. Any remaining carbonyl material could lead to undesired side reactions.

SOLVED PROBLEM

Predict the major product of the following reaction, and draw a mechanism for its formation.

1) LDA, THF, −78 °C
2) MeBr

STEP 1 (OPTIONAL): Draw out the structure of any abbreviated reactants.

LDA stands for lithium diisopropylamide.

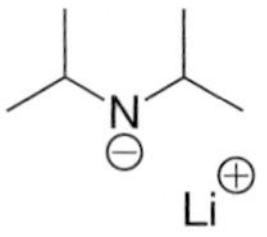

STEP 2: Determine the roles of the reactants in each step.

LDA is a very strong base capable of quantitatively deprotonating the enolizable protons of a carbonyl group. Therefore, the first step is an acid–base reaction. The deprotonated carbonyl compound is in its enolate form, which acts as a nucleophile, reacting with the electrophilic alkyl halide, MeBr.

STEP 3: Draw mechanism for the formation of the enolate with LDA.

Although either set of α-hydrogens could react with LDA, the kinetic enolate is typically favoured. So the less substituted enolate is the major product.

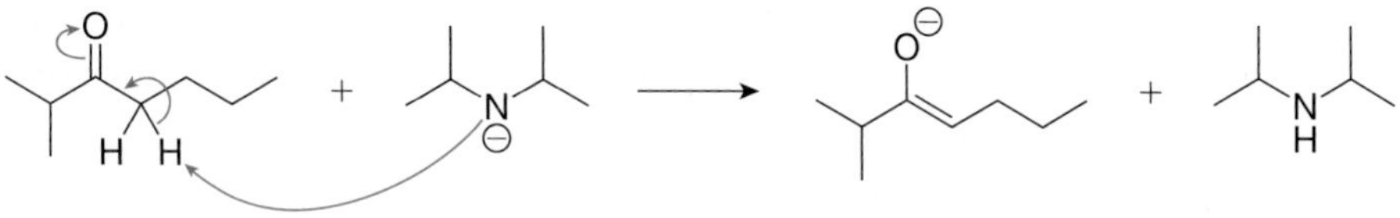

STEP 4: Draw a mechanism for the reaction of the enolate with the alkyl halide.

The nucleophilic enolate shares its electrons with the electrophilic carbon on methyl bromide. This flow of electrons breaks the C–C π bond, while simultaneously re-forming the C=O π bond of the enolate and breaking the C–Br bond on the methyl bromide.

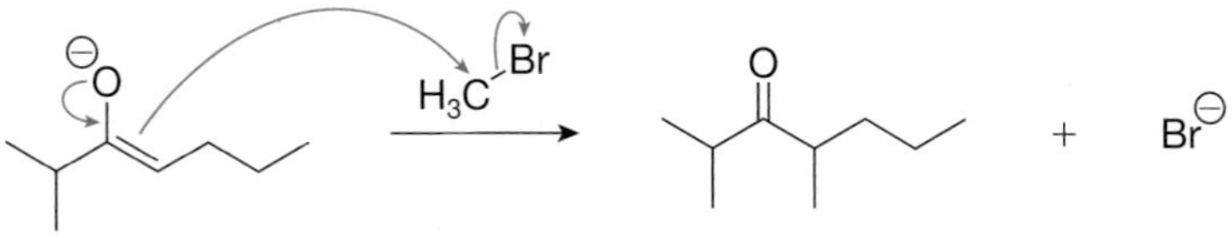

PRACTICE PROBLEM

17.5 For each of the following reactions, predict the major product, and draw a mechanism for its formation.

a) 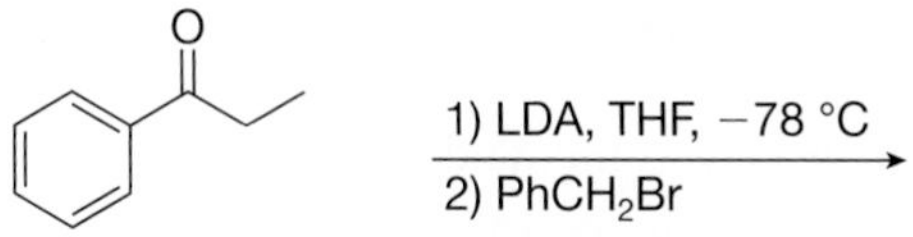 1) LDA, THF, −78 °C; 2) $PhCH_2Br$

b) 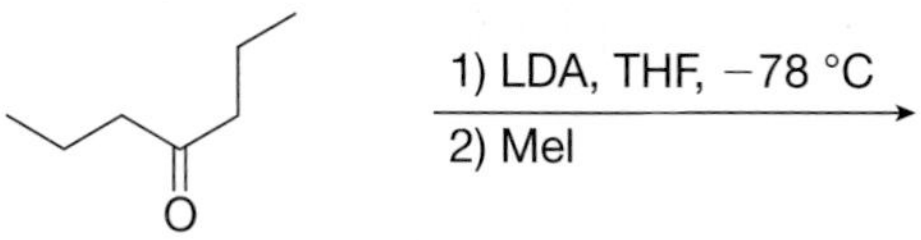 1) LDA, THF, −78 °C; 2) MeI

c) OHC 1) LDA, THF, −78 °C; 2) Br

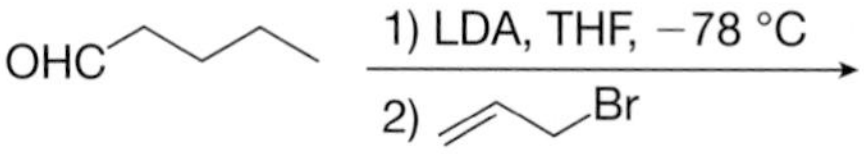

d) 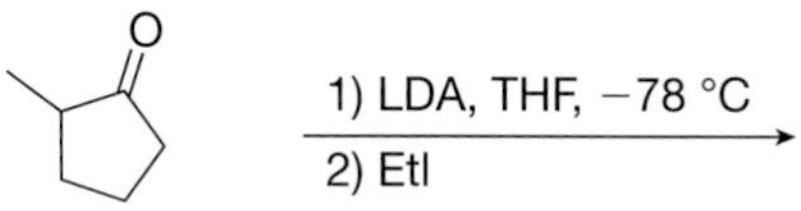 1) LDA, THF, −78 °C; 2) EtI

e) 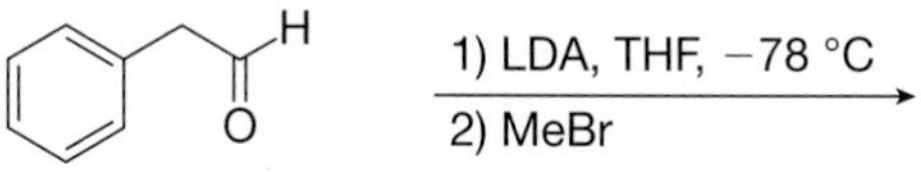 H 1) LDA, THF, −78 °C; 2) MeBr

INTEGRATE THE SKILL

17.6 The following ketone has two sets of enolizable protons. Treating this ketone with a strong base, such as LDA, under typical kinetic conditions produces only the more substituted enolate.

a) Draw the enolate formed, and explain why it is favoured.

b) Draw the product that results from reacting the favoured enolate with (*R*)-6-iodo-4-methylhex-1-ene.

17.3.2.2 Alkylation of enamines

Imines and enamines have the same electronic structure as ketones and enols. So imines and enamines can be used as equivalents of enols and enolates. However, the C=N bond of an imine is less electron withdrawing than the carbonyl bond because nitrogen is less electronegative than oxygen. As a result, imine α-hydrogens are less acidic than those of carbonyl groups, and aza-enolates are more reactive than carbonyl enolates.

Enamines are a commonly used family of nitrogen-based nucleophiles. Since nitrogen is a better electron donor than oxygen, enamines are slightly more reactive and more controllable equivalents for enols in alkylation reactions. The alkylation of enamines usually requires very reactive electrophiles, such as methyl, allylic, and benzylic alkyl halides, and does not proceed well with other halides, even primary ones.

STUDENT TIP

Though they are reactive enough to be alkylated, enamines react only with the most reactive alkylating agents: methyl, allylic, and benzylic halides. Imine anions require strong base for their formation.

Secondary amines cannot form imines with carbonyl compounds because these amines cannot lose a final hydrogen from the nitrogen to form a neutral imine. Instead, the charged iminium ions lose hydrogen from the α-carbon to form an enamine, which often can be isolated and purified.

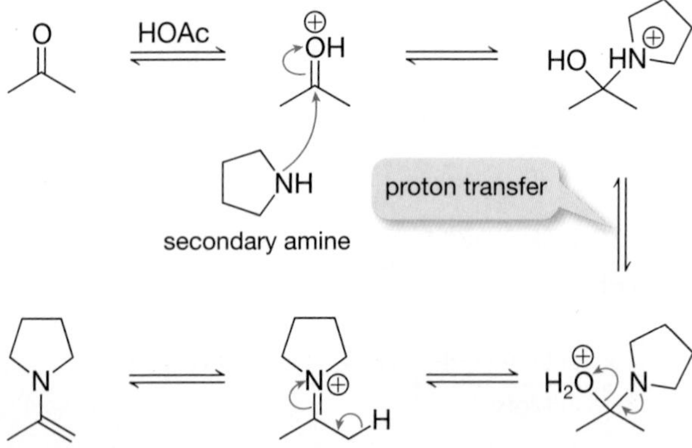

Because enamines are formed under equilibrium conditions, the thermodynamic product is favoured. This product is the *less* substituted enamine because of steric interference between the more substituted double bond and the secondary amine. This interference results from conjugation involving the lone pair on the nitrogen, which gives this atom substantial sp^2 character. This character makes all of the bonds connected to the enamine coplanar, increasing the steric effects.

more substituted double bond generates steric interaction with pendant groups

favoured product

enamine has two resonance structures and nitrogen is essentially sp^2 hybridized, restricting rotation around C–N bond

The electron-donating ability of the nitrogen makes the enamine a good nucleophile, and this group adds electrophiles selectively. Only one addition is possible because the product of that alkylation is an iminium ion, which is not nucleophilic. After the reaction with the electrophile is complete, the addition of water hydrolyzes the iminium ion to produce the corresponding carbonyl compound. Enamines are useful intermediates for alkylations of carbonyl compounds.

after first reaction is complete, water and acid hydrolyzes the iminium and produces a carbonyl

$PhCH_2Br$ → H_3O^+ →

Ph Ph

reaction stops after one addition

STUDENT TIP

Enamines are neutral nucleophiles, analogous to enols. They do not require either a weak or strong base. Because they are neutral instead of anionic, they are less reactive than enolates and are only alkylated by very reactive electrophiles.

17.3.2.3 Nucleophilic carbonyl groups in synthesis

Electrophilic addition to a carbonyl group provides a powerful way to make complex molecules by attaching a halogen, an alkyl group, or other electrophile at the α-position of the carbonyl group. Disconnecting at the α-carbon of the carbonyl group (between atoms 2 and 3) gives a synthon in which the carbonyl functional group acts as a nucleophile. This synthon is obtained from a carbonyl compound containing an α-hydrogen that can be removed to form an enolate. The electrophile (E^+) is derived from a halogen or from an alkyl group containing the appropriate leaving group.

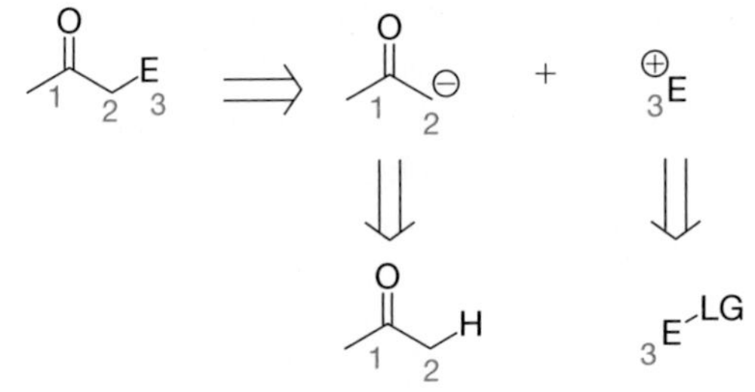

E
1 2 3
1 2
3E
H
1 2
3 E LG

The haloform reaction converts *methyl ketones* into carboxylic acids. Therefore, a carboxylic acid can be disconnected back to the appropriate methyl ketone.

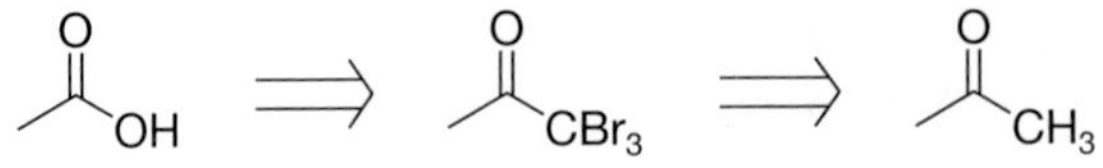

OH CBr$_3$ CH$_3$

CHECKPOINT 17.4

You should now be able to propose syntheses involving the electrophilic addition of halogens and alkyl groups to carbonyl compounds.

SOLVED PROBLEM

Propose a method for carrying out the following transformation. More than one step may be required.

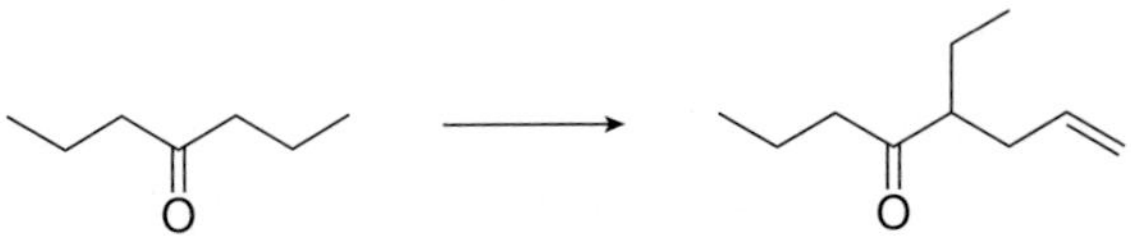

O O

STEP 1: Identify which bonds have been broken and/or formed in the synthesis.

A new C–C bond has formed alpha to the carbonyl (at C-2).

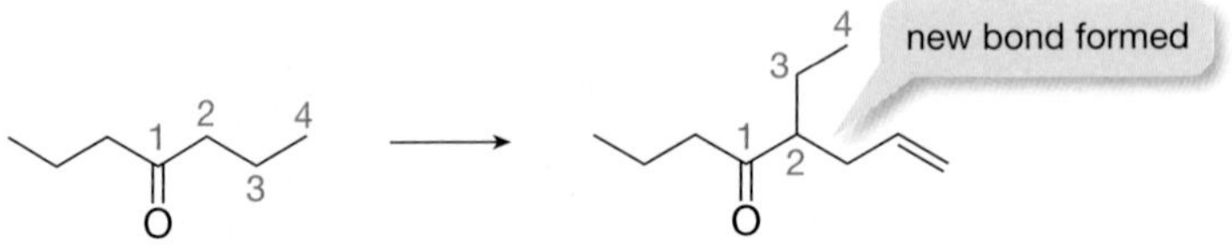

1 2 3 4 O
new bond formed
1 2 3 4 O

STEP 2: Disconnect the new bond to determine the appropriate synthons.

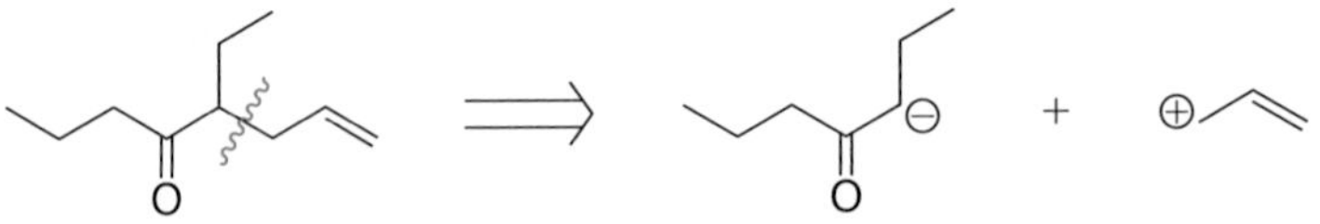

O O

STEP 3: Determine which reagents could be used as synthon equivalents.

The enolate synthon can be obtained from the starting ketone using a strong base such as LDA. The allyl cation synthon is equivalent to an electrophilic allyl halide such as allyl bromide.

STEP 4: Write reaction equation for the synthesis.

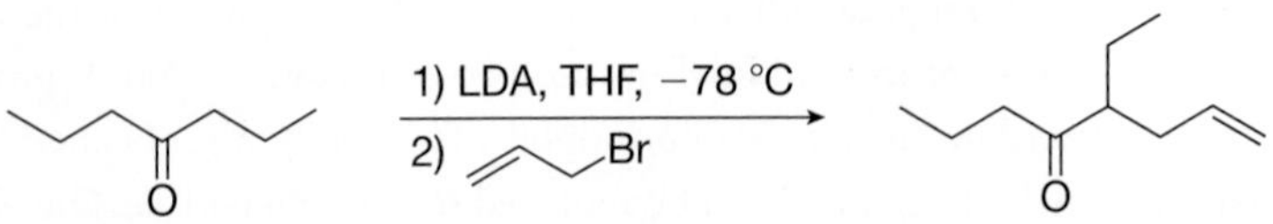

1) LDA, THF, −78 °C
2) Br

O O

PRACTICE PROBLEM

17.7 Suggest appropriate reagents for the following transformations. More than one step may be necessary.

a)

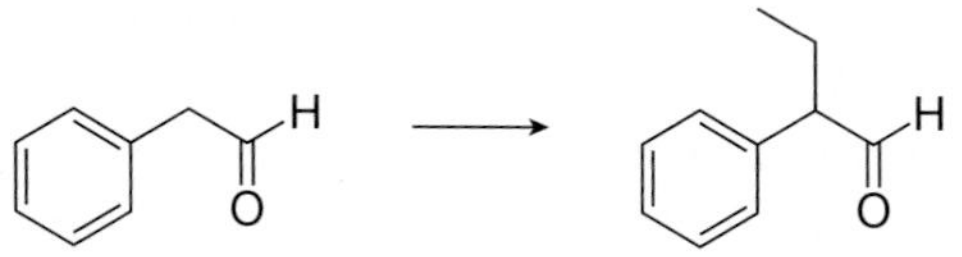

b)

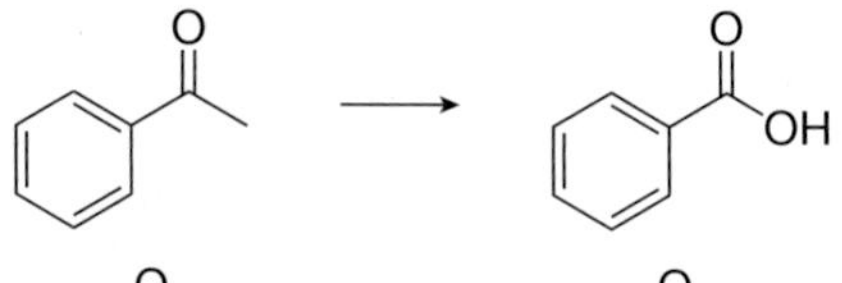

c)

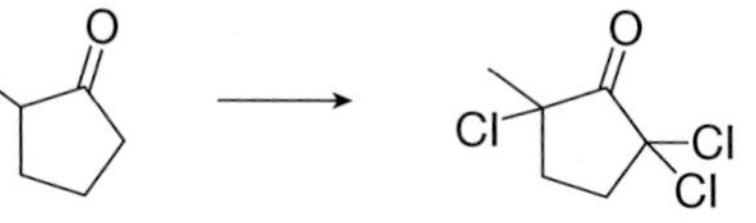

d)

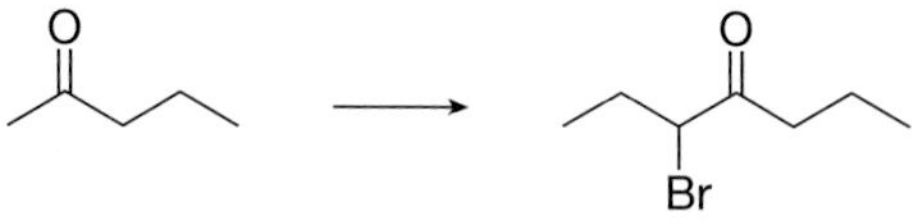

e)

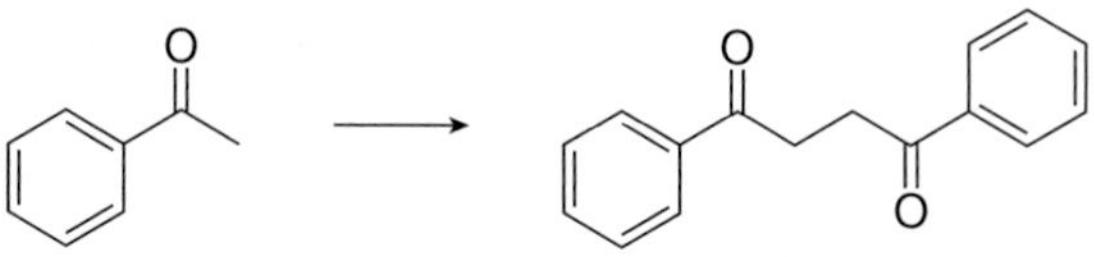

INTEGRATE THE SKILL

17.8 Suggest appropriate reagents for the following transformations. More than one step will be required.

a)

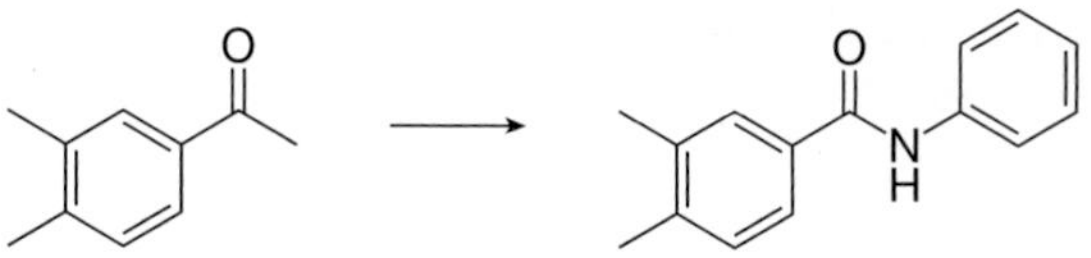

b)

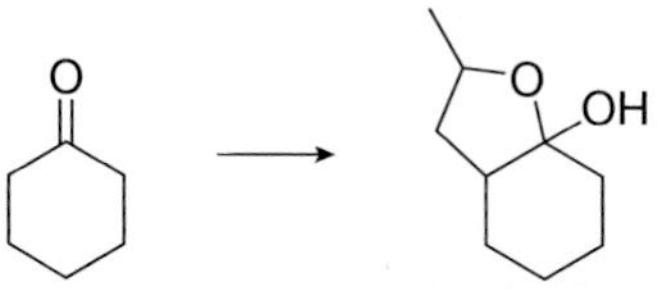

17.4 The Aldol Reaction

An aldehyde stirred with a base such as KOH produces a compound that has double the molar mass of the original aldehyde. This product is an **aldol**, and its formation is the result of an **aldol reaction**.

An **aldol** is a compound with an OH group bonded to a carbon atom in the β-position beside a carbonyl group.

An **aldol reaction** joins two carbonyl compounds to form a β-hydroxy carbonyl compound.

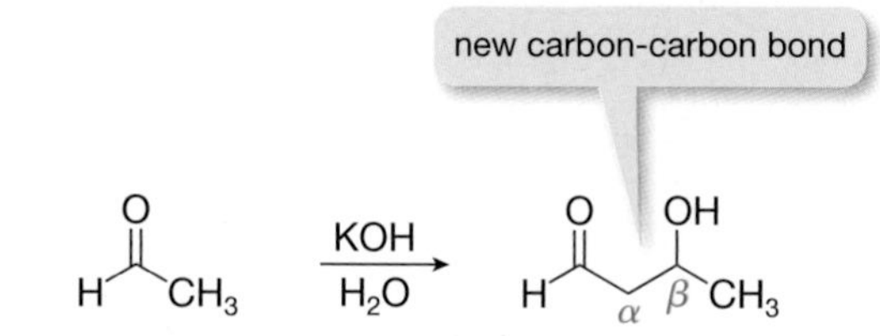

MW = 44 g/mol

aldol MW = 88 g/mol

In this reaction, the aldehyde acts both as a nucleophile and an electrophile. Treating an aldehyde with an alkoxide base establishes an equilibrium between the aldehyde and the corresponding enolate. Only a small amount of enolate is present, but this is enough to cause the aldol reaction to occur. The enolate is a nucleophile, and the aldehyde is an electrophile, so the two react to produce an alkoxide. The solvent quickly protonates this product, forming the aldol (which is a dimer of the aldehyde). As the enolate is consumed, more of the starting aldehyde transforms into the enolate (in accordance with Le Chatelier's principle), so the aldol reaction continues.

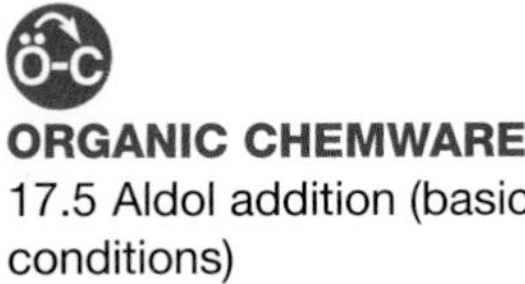

ORGANIC CHEMWARE
17.5 Aldol addition (basic conditions)

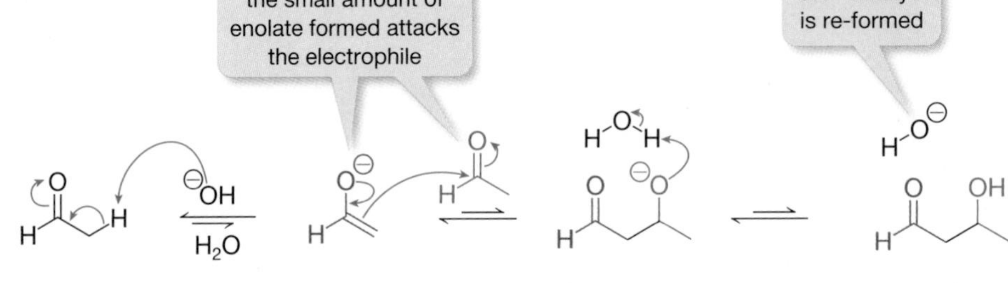
the small amount of enolate formed attacks the electrophile
OH⁻ catalyst is re-formed
OH
H₂O

A self-addition reaction involves the combination of two identical molecules—the addition of the nucleophilic form (enolate) to the electrophilic form (carbonyl).

In a retro-aldol reaction, the reverse of the aldol reaction, an aldol fragments into an enolate and an aldehyde.

In such **self-addition reactions**, the enolate nucleophile is formed from the same aldehyde that acts as the electrophile, which limits the range of products that can be formed. The aldol reaction can also be done using ketones. However, in this case, the addition step tends to reverse, and the product undergoes a **retro-aldol reaction** that regenerates the starting ketone. The products of ketone-based aldol reactions are tertiary alcohols, which are sterically hindered (very crowded). Consequently, the equilibrium in these reactions usually favours the starting material.

equilibrium lies to the left because product has high steric strain

ketone

aldol made from ketone

The aldol reaction can also be catalyzed by an acid. The mechanism involves a few more steps, but the key events follow the same pattern as the base-catalyzed process. As in other acid catalyses, an acid activates the electrophile. Acid catalysis also facilitates β-elimination (see Chapters 12 and 18). The acid can catalyze the formation of an enol in the product and, at the same time, protonate the hydroxyl group to make an excellent leaving group. The result is an elimination reaction that forms an α,β-unsaturated carbonyl compound. The formation of enones is discussed in more detail in Section 17.4.3.

equilibrium lies to the left because product has high steric strain
elimination favours enone product and pulls equilibrium forward

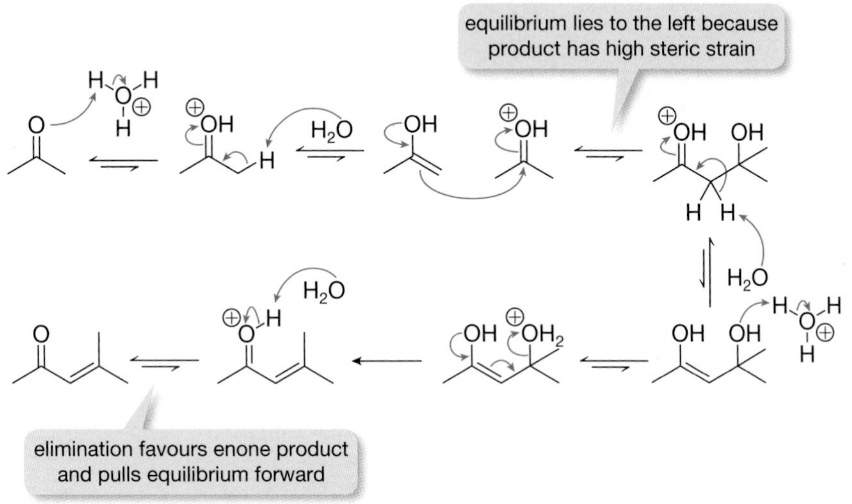

The following sections discuss variations of the aldol reaction, including ways to avoid self-condensation, the aldol condensation, and ring-forming aldol reactions.

CHECKPOINT 17.5

You should now be able to predict the products formed in an aldol reaction and draw a mechanism for their formation under both acidic and basic conditions.

SOLVED PROBLEM

Predict the product of the following reaction, and draw a mechanism for its formation.

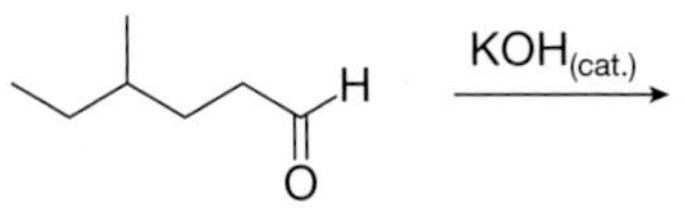

$\xrightarrow{KOH_{(cat.)}}$

STEP 1: Determine the roles of the reactants.

The carbonyl of an aldehyde is a good electrophile; however, given that the hydroxide ions are present only in catalytic amounts, there is no obvious nucleophile. Carbonyl compounds with enolizable protons can act as nucleophiles in the presence of acids and bases. The nucleophile in this reaction is an enolate formed from a molecule of the aldehyde.

STEP 2: Draw mechanism for the formation of the enolate.

The basic hydroxide ion removes one of the weakly acidic α-hydrogens to yield an enolate and a molecule of water. Note that the reaction is in equilibrium because the base is not strong enough to fully deprotonate the carbonyl compound.

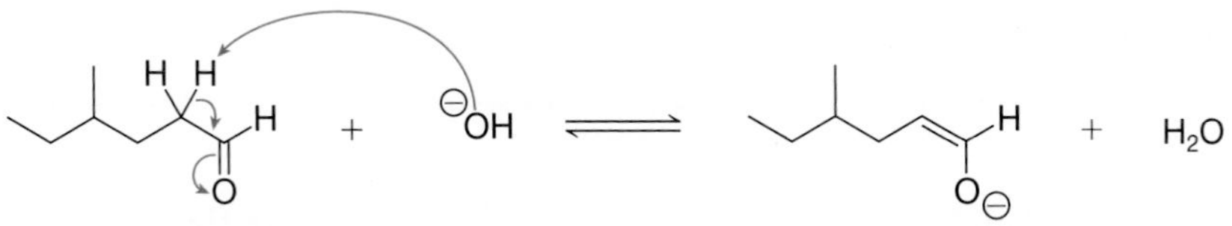

STEP 3: Draw a mechanism for the reaction of the enolate with a molecule of the aldehyde. The nucleophilic enolate adds to a molecule of unreacted aldehyde, thereby forming a new C–C single bond and breaking the C=O π bond of the aldehyde. At the same time, the carbonyl of the enolate is re-formed.

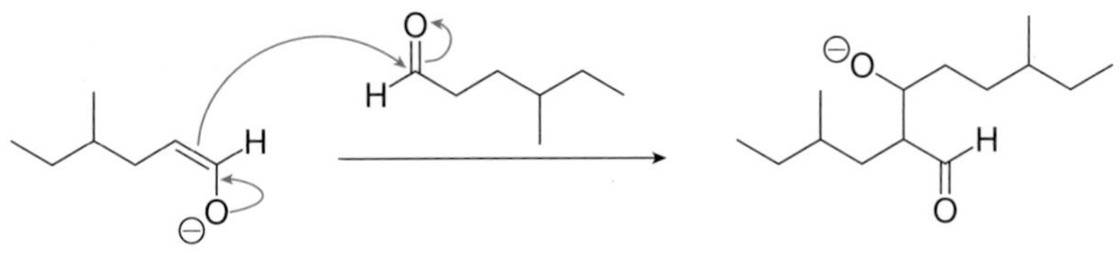

STEP 4: Use the conjugate acid of the base catalyst to protonate the alkoxide intermediate and generate the neutral final product. Note that this step regenerates the hydroxide catalyst.

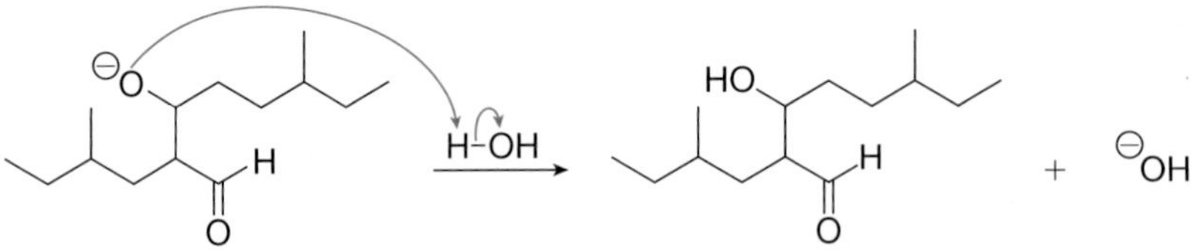

PRACTICE PROBLEM

17.9 For each of the following reactions, predict the major product, and draw a mechanism for its formation.

a) 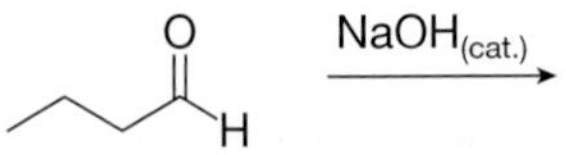 $NaOH_{(cat.)}$

b) 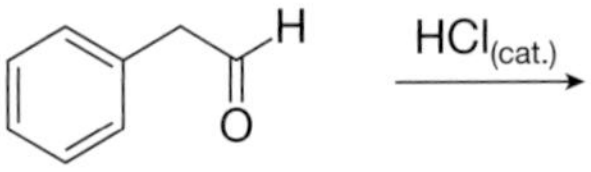 $HCl_{(cat.)}$

c) 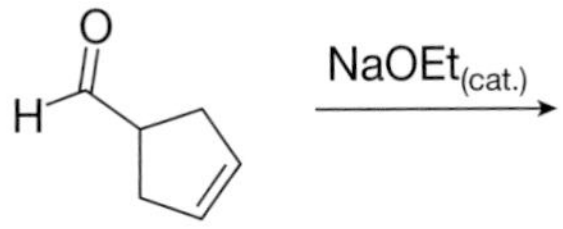 $NaOEt_{(cat.)}$

d) 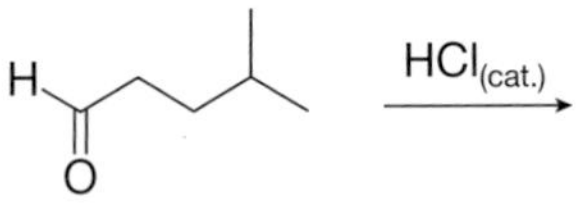 $HCl_{(cat.)}$

INTEGRATE THE SKILL

17.10 a) Which of the following compounds cannot undergo an aldol reaction? Explain your selections.

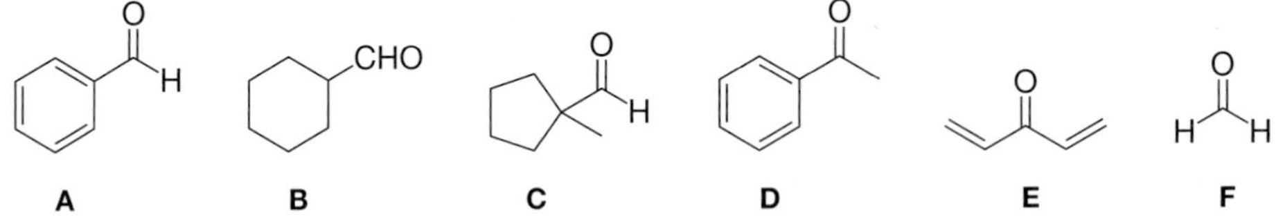

b) If the aldehyde in the Solved Problem were enantiomerically pure, how many stereoisomers of the final product could be formed?

17.4.1 Crossed aldols: The Claisen–Schmidt reaction

Combining two different aldehydes with a hydroxide base produces a mixture of four products because each aldehyde can act as a nucleophile and an electrophile.

Crossed aldols are aldols formed from two different aldehydes.

Two of the aldol products are dimers of the two aldehydes. The other two products are **crossed aldols**, formed by reactions between the two different aldehydes. This mixture of dimer and crossed aldols can be very difficult to purify.

KOH, EtOH

crossed aldol dimer dimer crossed aldol

Crossed aldol reactions can be useful if the reagents are chosen such that only one aldol product is possible. The **Claisen–Schmidt reaction** uses an enolizable ketone nucleophile with an aldehyde electrophile that *cannot* form an enolate because it has no α-hydrogens.

The **Claisen–Schmidt reaction** is an aldol reaction that uses a ketone enolate and a non-enolizable aldehyde.

examples of compounds without enolizable hydrogens

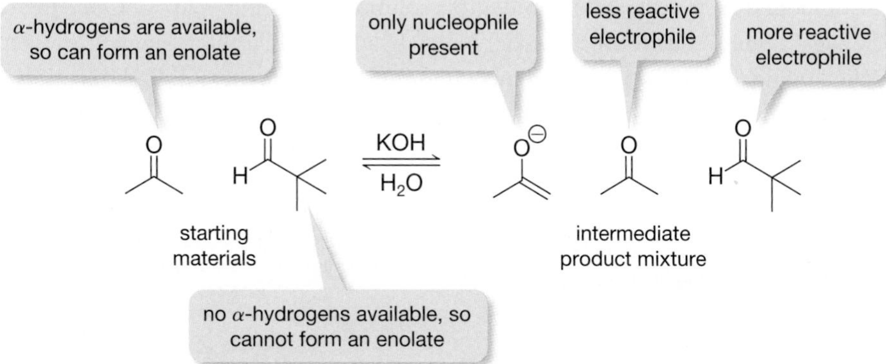

Because the aldehyde component has no α-hydrogens, it cannot be a nucleophile and can function only as an electrophile. The ketone can self-condense, but the product undergoes a retro-aldol reaction and reverts to the starting ketone. This means that the ketone cannot be an electrophile in this reaction and can function only as a nucleophile. The reaction of the ketone with the aldehyde is also not easily reversible. Le Chatelier's principle applies, and the only product is the crossed aldol.

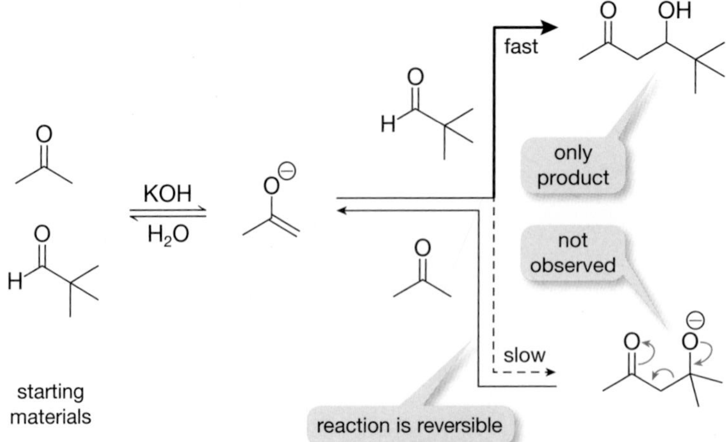

STUDENT TIP
Self-addition reactions and Claisen–Schmidt crossed aldol reactions are done with weak base.

This reaction is performed by mixing the electrophile aldehyde with base and then slowly adding the nucleophile ketone. This procedure speeds formation of the crossed aldol by reducing self-addition of the ketone. The overall reaction can be summarized as follows:

$$\text{acetone} + \text{pivaldehyde} \xrightarrow[H_2O]{KOH} \text{aldol product}$$

17.4.2 Crossed aldols using strong bases

STUDENT TIP
Crossed aldol reactions in which both the nucleophile and electrophile are enolizable must be done using a strong base. Using a weak base would give a complex mixture of products.

Most crossed aldol reactions use strong bases to control enolate formation. This method can be used to bond almost any aldehyde to another carbonyl compound (aldehydes are excellent electrophiles but usually poor nucleophiles). The nucleophile is generated by adding a solution of carbonyl compound (ketone, ester, or amide) to a solution of LDA at −78 °C. Once the enolate forms, the electrophile (usually an aldehyde) is slowly added while the temperature is kept low. Because the leaving group (OLi^-) is extremely poor in the aprotic solvent, β-elimination cannot occur. Instead, the electrophile reacts with the enolate to form an alkoxide. Once this reaction is complete, a weak acid is added in a separate reaction to generate the aldol.

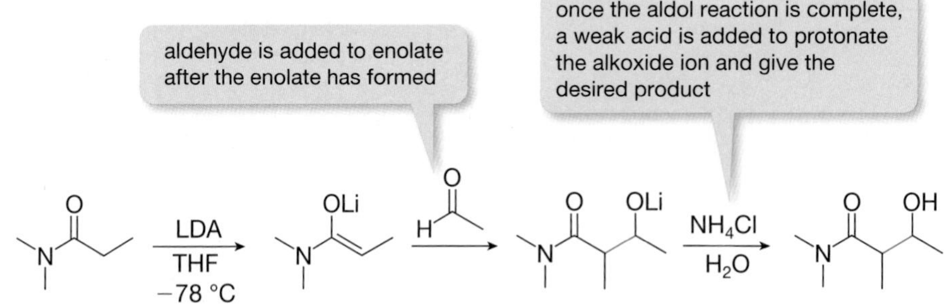

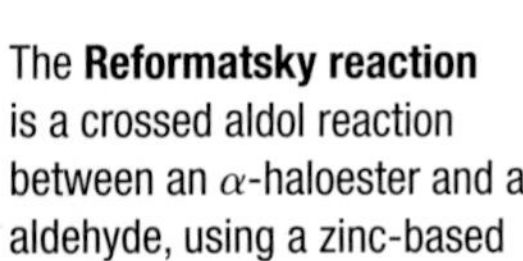

The **Reformatsky reaction** is a crossed aldol reaction between an α-haloester and an aldehyde, using a zinc-based nucleophile.

The **Reformatsky reaction** synthesizes a crossed aldol by combining an α-haloester with an aldehyde and zinc dust.

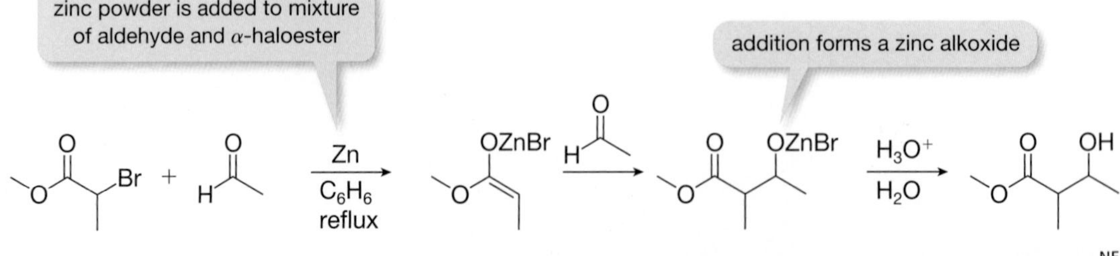

The zinc forms a zinc enolate, which has some similarities to Grignard reagents. Organozinc reagents have lower reactivities than Grignard reagents, and they add to aldehydes but not to esters.

The mechanism by which the zinc enolates form is special; it involves the zinc donating electrons to the carbonyl oxygen.

Zn: Br Zn C_6H_6 reflux Zn Br ZnBr H OZnBr

The first step of the mechanism is similar to the Clemmensen reduction of carbonyls (Chapter 19), in which zinc (Zn) or tin (Sn) donate electrons to the C=O bond. The available α-halide that can be eliminated redirects the mechanism to form an enolate instead of an alcohol.

Clemmensen reduction

Zn: Ph Zn HCl Zn Ph HCl OH Ph H

Reformatsky reaction

Zn: Br Zn C_6H_6 reflux Br Zn ZnBr

DID YOU KNOW?

Some crossed aldols are made by using a Lewis acid instead of a Brønsted acid. The formation and reactivity of the enol can be controlled by the size and nature of the Lewis acid.

aldehyde cannot enolize

ketone has two possible sites for enolization

Lewis acid

H

Bu B Bu OTf

iPr N Et iPr −78 °C

Bu_2B

enol forms on less substituted side of ketone

C–C bond formed on the less substituted side

Continued

The mechanism of this reaction follows the same pattern as the bromination of enols—the Lewis acid acts like a large proton. The Lewis acid forms a boron-enolate that reacts with the aldehyde to produce a compound that has a strong oxygen-boron bond, which stops further reaction. When the mixture is worked up, the aldol product is obtained.

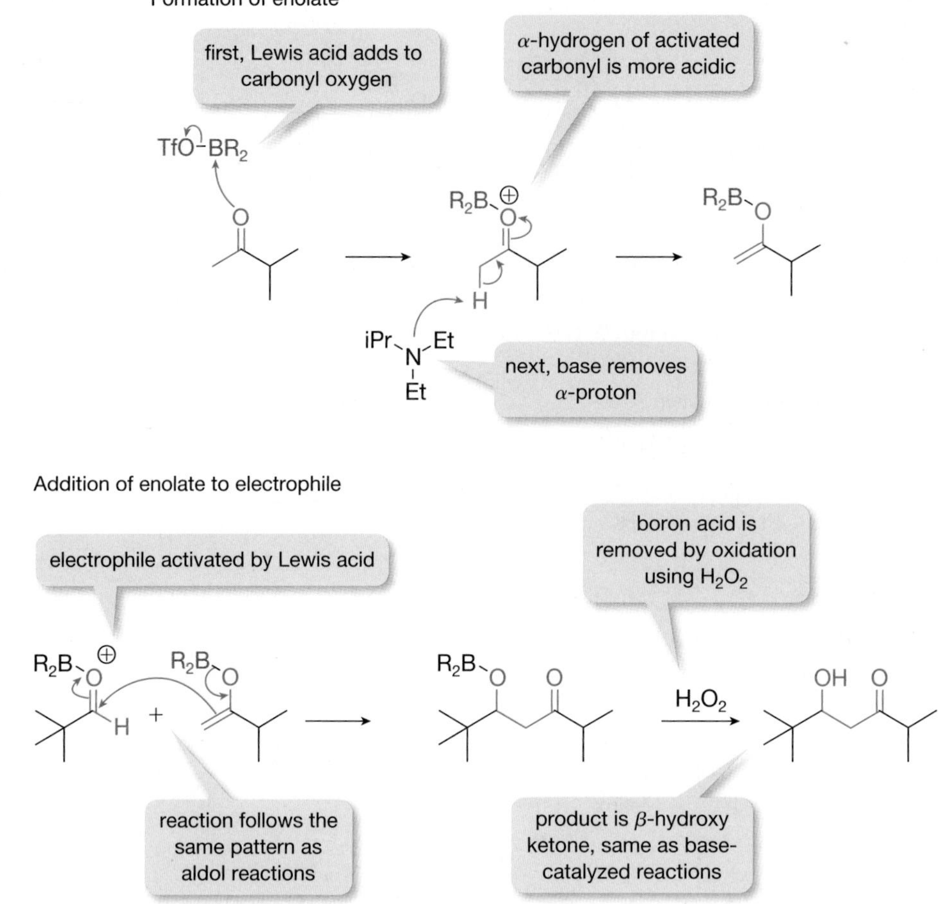

CHECKPOINT 17.6

You should now be able to predict the products formed in crossed aldol reactions, including the Claisen–Schmidt and Reformatsky variations. You should also be able to draw mechanisms for these reactions.

SOLVED PROBLEM

The following reaction can produce four different products.

A + **B** $\xrightarrow{KOH_{(cat.)}}$

a) Draw the four possible products.
b) Suggest a way of carrying out this reaction such that only one aldol product forms—the one that results from compound A acting as a nucleophile and compound B acting as an electrophile.

STEP 1: Determine the roles of the reactants.

Both of the available carbonyl compounds could act as electrophiles. In addition, both have enolizable protons, which could be removed by the catalytic base to form nucleophilic enolates. Since each aldehyde can act as either the electrophile or the nucleophile (enolate), the reaction yields a mixture of products.

STEP 2: Draw the four possible electrophile-nucleophile pairs that can result from the two aldehydes.

The structure of an aldehyde does not change when acting as an electrophile. However, when acting as a nucleophile under basic conditions, an aldehyde needs to be drawn in its enolate form. The four possible combinations are as follows:

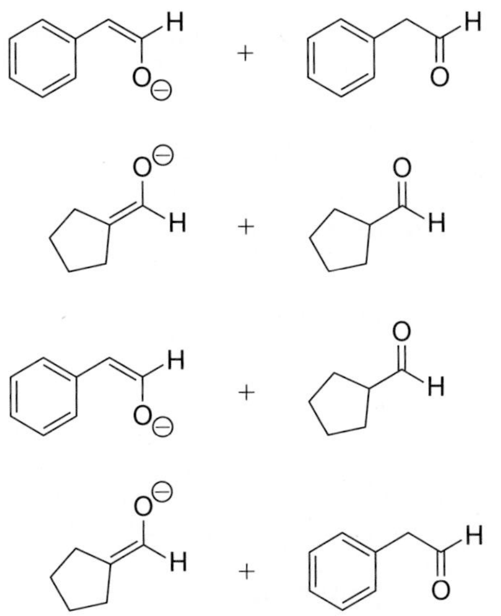

STEP 3: Draw the aldol product for each pairing shown in Step 2.

You may find it helpful to draw the mechanism arrows for the addition step when you draw the products.

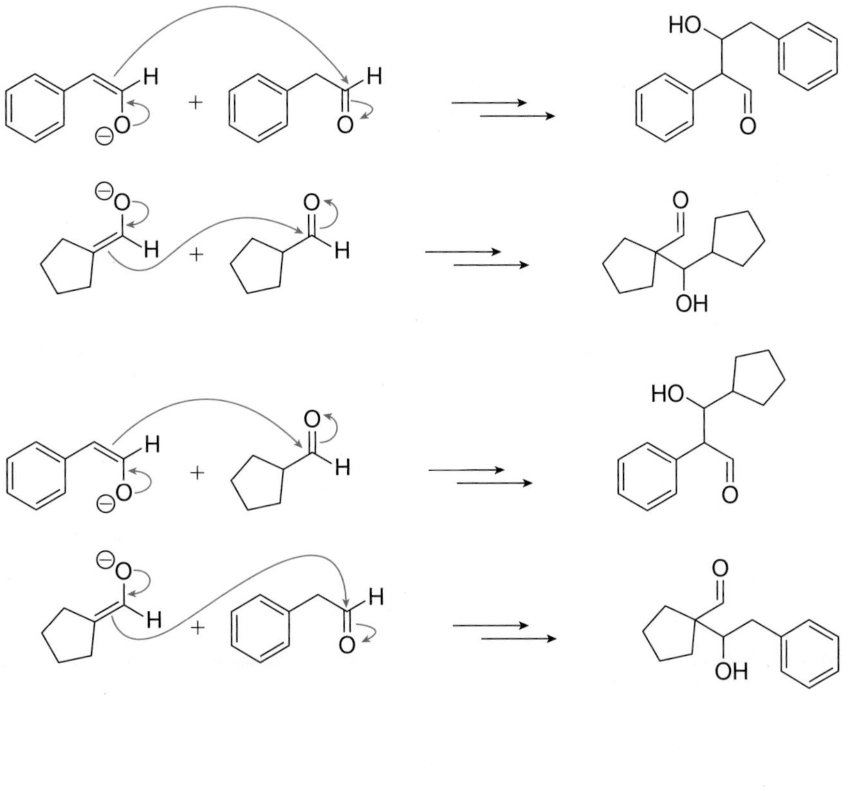

STEP 4: Identify the product that forms when compound A acts as a nucleophile and compound B acts as an electrophile.

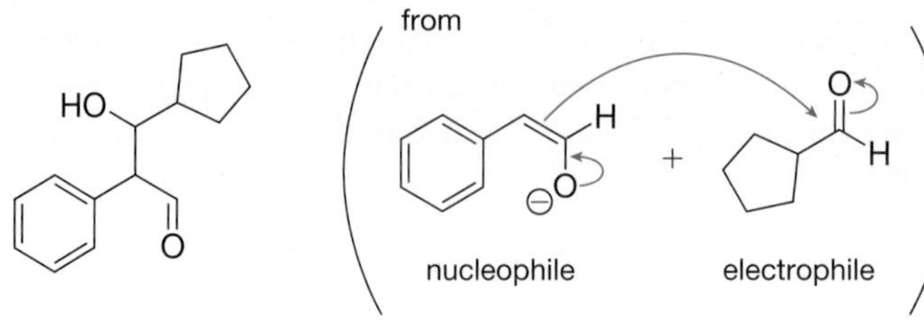
HO
O
from
H
O
+
O
H
nucleophile
electrophile

STEP 5: Determine which methods for controlling a crossed aldol reaction would be appropriate.

You can ensure that compound A acts as the nucleophile by pretreating it with a strong base such as LDA to completely convert it into its enolate form. Once the enolate of A forms, compound B can be added to the solution of enolate.

STEP 6: Write the overall reaction equation for the controlled crossed aldol reaction.

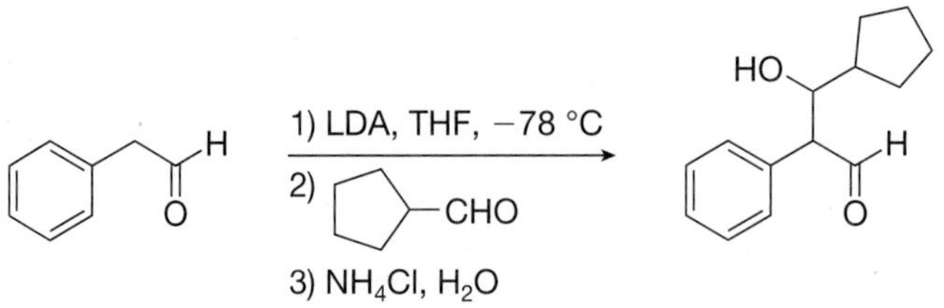
H
O
1) LDA, THF, −78 °C
2) CHO
3) NH4Cl, H2O
HO
H
O

PRACTICE PROBLEM

17.11 Predict the products of the following reactions, and draw mechanisms for their formation.

a) 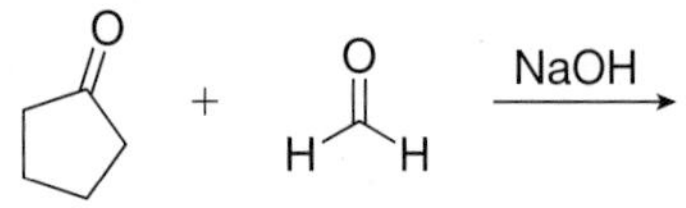
O
+
O
H H
NaOH

b) 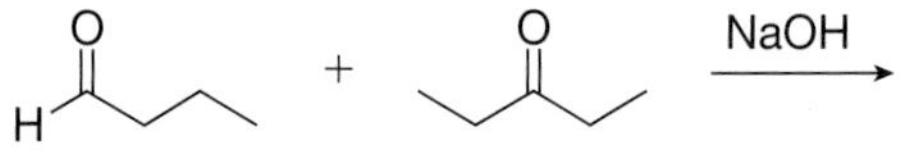
O
H
+
O
NaOH

c) 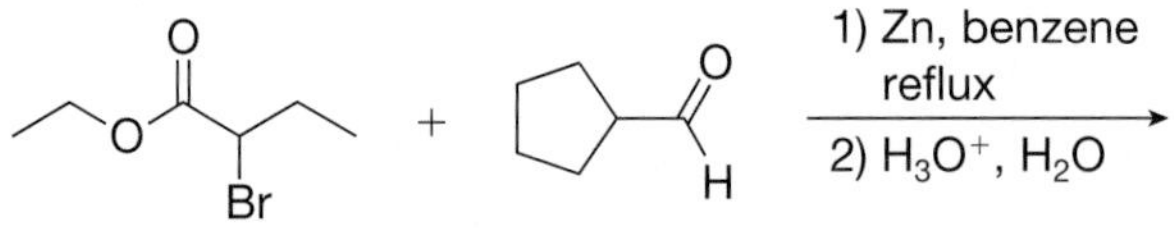
O
O
Br
+
O
H
1) Zn, benzene
reflux
2) H3O+, H2O

d) 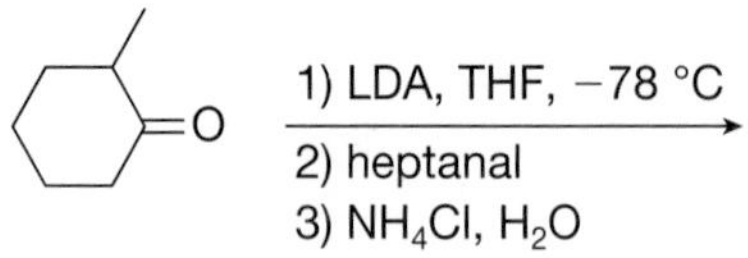
O
1) LDA, THF, −78 °C
2) heptanal
3) NH4Cl, H2O

INTEGRATE THE SKILL

17.12 In Question 17.11a, a second product is possible from the self-condensation of the ketone. Draw this product, and explain why its formation does not reduce the overall yield of the desired crossed aldol product. Draw a mechanism to help explain your answer.

17.4.3 Elimination and the aldol condensation

The aldol reaction produces a product that has a hydroxyl group in a 1,3 relationship with a carbonyl. This hydroxyl group, sometimes called a β-hydroxy group, is capable of acting as a leaving group under certain conditions. Aldol products are therefore susceptible to a β-elimination that produces an α,β-unsaturated carbonyl. When the aldol addition is followed by the elimination of water, the reaction is referred to as an **aldol condensation**. Two factors contribute to the tendency to undergo a β-elimination: the acidity of the α-hydrogen and the leaving group ability of the hydroxyl group.

An **aldol condensation** combines two aldehydes or ketones to produces enols or enones, with the loss of water.

STUDENT TIP
A condensation is a reaction in which two molecules combine, with the loss of water or alcohol.

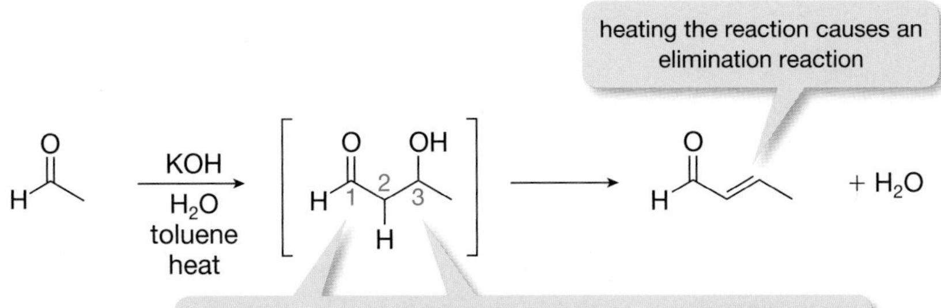

An aldol reaction performed in a protic solvent at low temperature usually produces the aldol containing a β-hydroxyl group. Since this product does not enolize readily and the OH^- group is a poor leaving group, the reaction stops at this point.

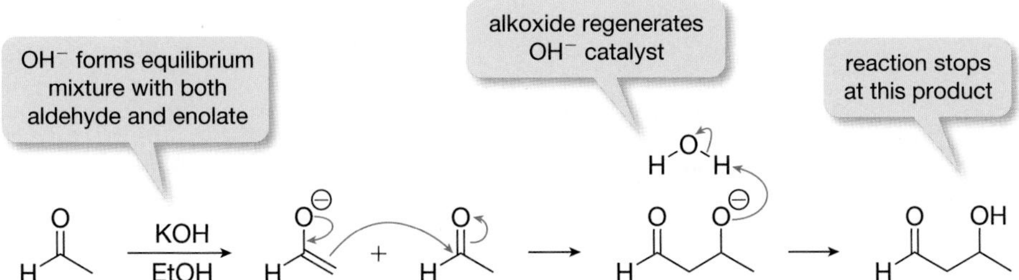

Heating the reagents makes the formation of a second enolate possible. The carbon of this enolate (carbanion) is located beside the OH group (potential leaving group), so an elimination reaction is possible. Since the OH^- group is a very poor leaving group, the elimination reaction is possible only in a protic solvent. As the OH^- group leaves, it can receive a proton from a nearby solvent molecule; this converts the hydroxyl group into H_2O, which is a good leaving group. Both heat (Δ) and a protic solvent are necessary for this β-elimination.

protic solvent supplies proton to allow poor leaving group (OH^-) to leave

KOH

H_2O

toluene, heat

α,β-unsaturated carbonyl

heat provides energy to form an enolate

base deprotonates α-hydrogen

aldol condensation product

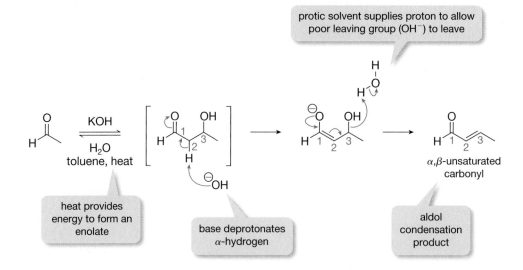

An **E1$_{cb}$ reaction** is a unimolecular elimination reaction in which an acidic proton is removed to make the conjugate base *before* the leaving group leaves.

This elimination is an **E1$_{cb}$ reaction**, a unimolecular elimination in which the rate-determining step is the formation of the conjugate base. The formation of water prevents the retro-aldol reaction and drives the process forward to the unsaturated product. Carbanion (enolate) formation is rate limiting in this process. The loss of the hydroxyl to form the double bond happens *after* this rate-determining step. Consequently, the reaction shows first-order kinetics.

Aldol reactions of ketones that form the α,β-unsaturated product are productive because the β-elimination is an irreversible reaction, and the retro-aldol process is not an issue. Le Chatelier's principle applies, and the reaction runs to completion, forming the unsaturated product.

STUDENT TIP

E1$_{cb}$ reactions are different from E$_2$ reactions. The formation of α,β-unsaturated carbonyls involves enolate intermediates, which should be shown in mechanisms.

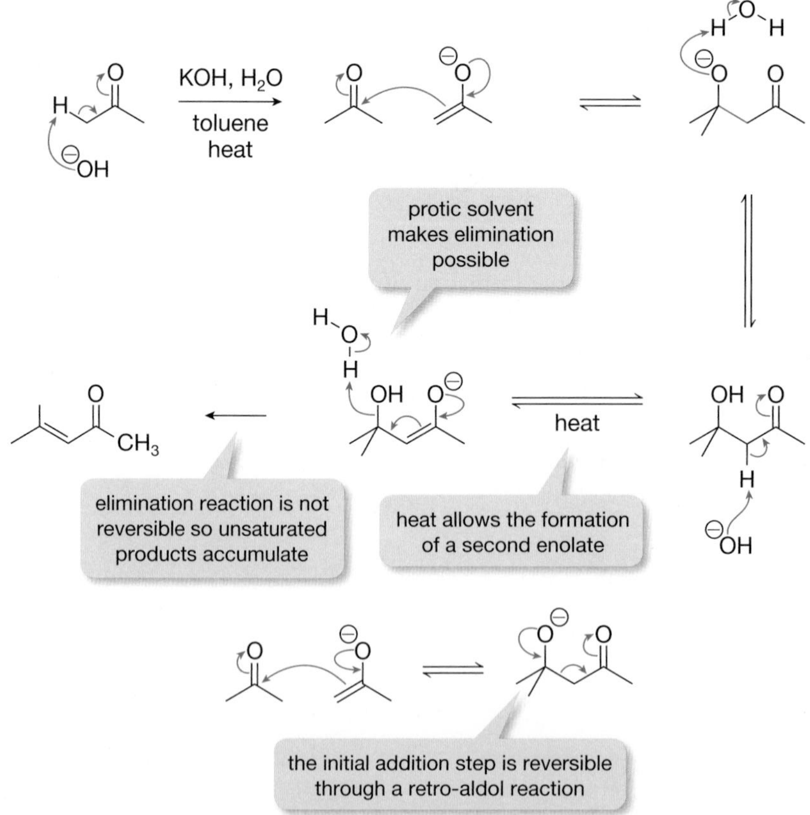

Acid-catalyzed aldols are prone to β-eliminations because good leaving groups are generated, which facilitate the elimination process and result in the formation of unsaturated products.

When strong bases are used in aldol reactions, β-eliminations usually do not occur. The strong bases require aprotic solvents, so the aldol products do not have good-enough leaving groups for β-elimination.

CHECKPOINT 17.7

You should now be able to draw a mechanism for the dehydration of an aldol product under both acidic and basic conditions. You should also be able to predict under what circumstances the aldol condensation product will be favoured over the β-hydroxy product.

SOLVED PROBLEM

Consider the following aldol condensation. What reaction conditions would ensure that dehydration takes place? Draw a mechanism for the reaction using the selected conditions.

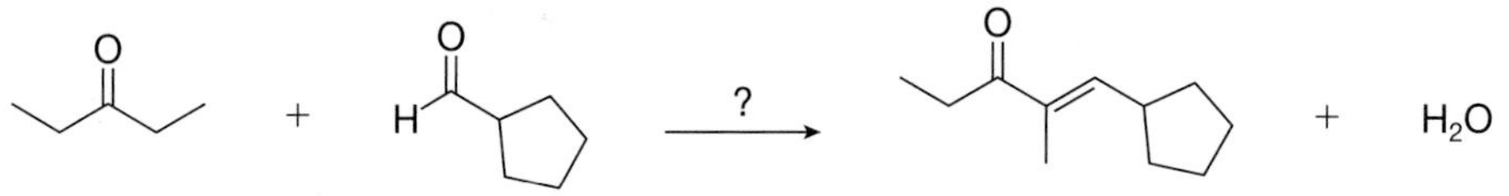

+ H ? + H_2O

STEP 1: Examine the type of aldol reaction that takes place (ignoring, for now, the dehydration step), and determine what conditions would be appropriate for this reaction.

The desired reaction is a crossed aldol reaction where multiple products are possible. In order to achieve control in this type of reaction, the ketone should be pretreated with a strong base such as LDA, followed by the addition of the aldehyde (see Checkpoint 17.6).

STEP 2: Determine whether the conditions identified in Step 1 are likely to dehydrate the aldol product into the desired enone.

Under strongly basic low-temperature conditions, dehydration does not normally occur because the required solvents are aprotic. These conditions prevent the aldol product from generating a good-enough leaving group for β-elimination to occur.

STEP 3: Suggest modifications to the reaction conditions to ensure dehydration of the aldol product into the enone.

Suitable modifications to the aldol reaction step are not possible, but the aqueous workup at the end of the reaction can be modified to induce dehydration. Normally, a weak acid (such as NH_4Cl in H_2O) would be added to protonate the final aldol product; however, if a stronger acid was used (such as HCl in H_2O) and the mixture was heated, the aldol product would likely undergo an acid-catalyzed elimination of water to form the enone.

STEP 4: Write an overall reaction equation, and include the necessary conditions for the desired aldol condensation reaction.

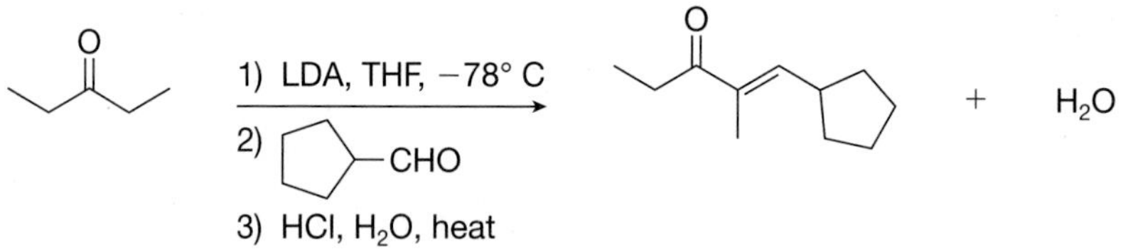

1) LDA, THF, −78° C
2) cyclopentyl–CHO
3) HCl, H_2O, heat
+ H_2O

STEP 5: Draw a mechanism for the formation of the aldol product. (You can use the steps in Checkpoint 17.5 as a guide.)

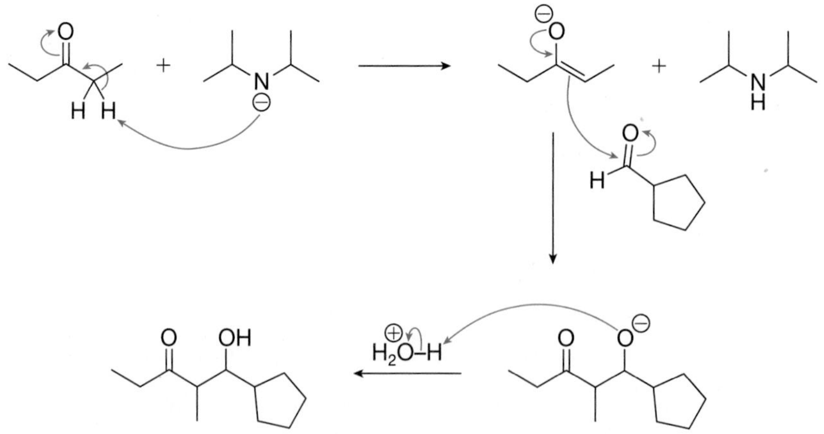

STEP 6: Draw a mechanism for the acid-catalyzed dehydration of the aldol product.

Following protonation of the carbonyl group, the α-hydrogen becomes acidic enough that a molecule of water is able deprotonate the intermediate product, thereby generating the corresponding enol. An intramolecular elimination of water then irreversibly forms the desired enone.

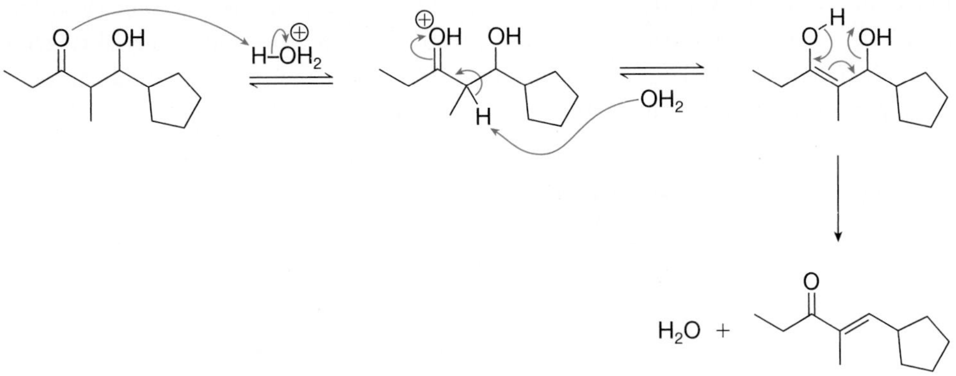

H_2O +

PRACTICE PROBLEM

17.13 Predict the major products of the following reactions, and draw mechanisms for their formation.

a) 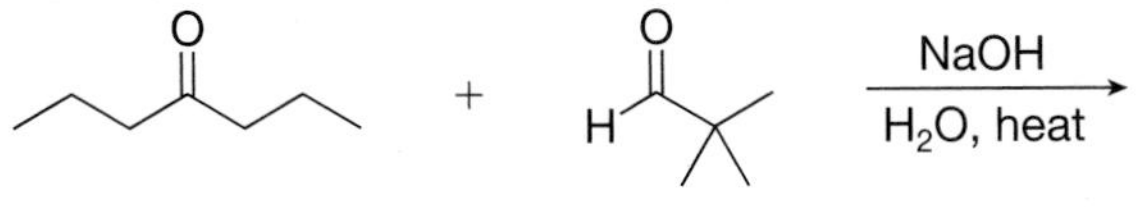 NaOH / H_2O, heat

b) 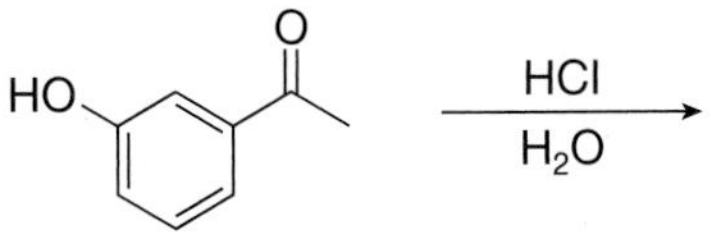 HCl / H_2O

c) 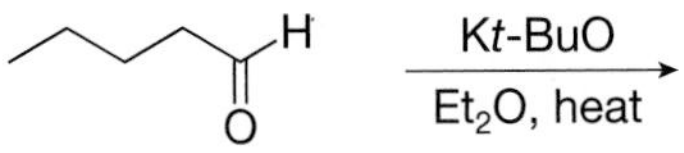 K*t*-BuO / Et_2O, heat

d) 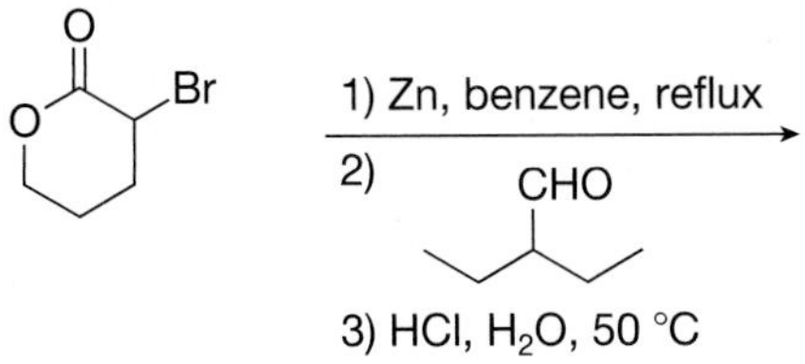 1) Zn, benzene, reflux
2) CHO
3) HCl, H_2O, 50 °C

e) 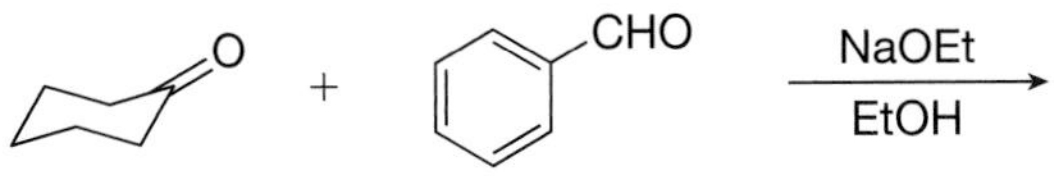 NaOEt / EtOH

INTEGRATE THE SKILL

17.14 Draw a reaction coordinate diagram for the $E1_{cb}$ elimination of water from an aldol product.

17.4.4 Intramolecular aldol reactions

Alkylation and aldol reactions can occur intramolecularly when the nucleophile and electrophile are part of the same molecule. The products of such transformations are cyclic molecules. In aldol reactions and related processes, especially those involving ketone electrophiles, the addition step is potentially reversible. The starting materials and addition products are in equilibrium, so the stability of the final product is an important factor in determining yield and selectivity for rings of less than six atoms. If two or more products are possible, the reaction will favour the ring with the least strain since it has the lowest energy. In general, the strains in small rings cause the stabilities of the rings to have this pattern: $S_3 < S_4 << S_5 < S_6$, where S_n is the relative stability of a ring of n atoms.

Rings larger than six atoms tend to form very slowly because the distance between the reacting sites is large. The bonds between the reacting sites tend to adopt staggered conformers that hold the reacting sites apart. The probability that the reacting sites will become close enough to react is very low for sites that are many more than six atoms apart, and so special reagents are normally required to make large rings. The effects of stability and speed of closure account for the high percentage of five- and six-membered rings in organic materials.

the ring with the least amount of strain will be formed

the smaller ring is formed preferentially because it forms much faster

STUDENT TIP
Intramolecular aldol reactions are done using a weak base.

CHECKPOINT 17.8

You should now be able to predict the products of an intramolecular aldol reaction, including the expected ring size where more than one size is possible.

SOLVED PROBLEM

Draw all possible products for the following reaction, and predict the major product(s).

NaOH, H_2O

STEP 1: Identify the roles of the reactants and reagents.

Both of the carbonyl groups on the starting material could act as an electrophile. In addition, both carbonyl groups have at least one set of enolizable protons, which could be removed in the presence of the base (NaOH) to form nucleophilic enolates. Water is the solvent in the reaction, as indicated by H_2O under the arrow.

STEP 2: Identify all possible enolizable protons, and draw the corresponding enolates.

There are three sets of enolizable protons and hence three possible enolates that can form in the presence of base.

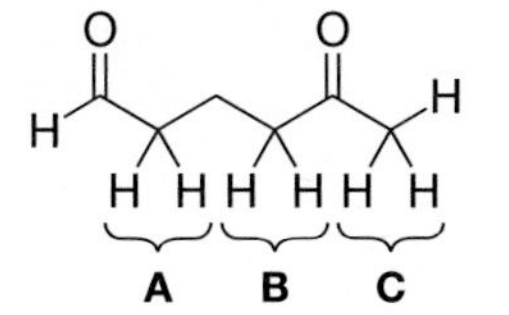

H H H H H H
A B C

Three possible enolates

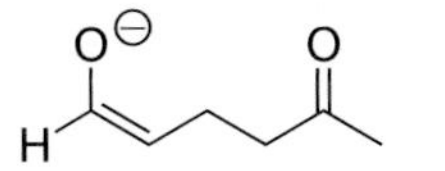

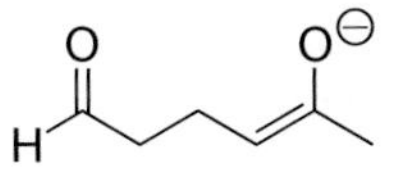

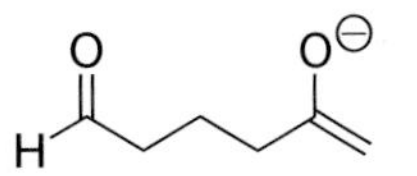

A B C

STEP 3: For each possible enolate, draw the corresponding aldol addition product.

You may find it helpful to number the carbons and draw the mechanism for the initial addition step.

A

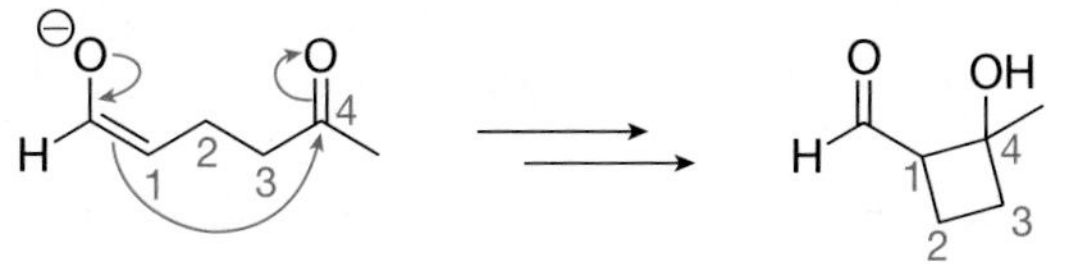

B

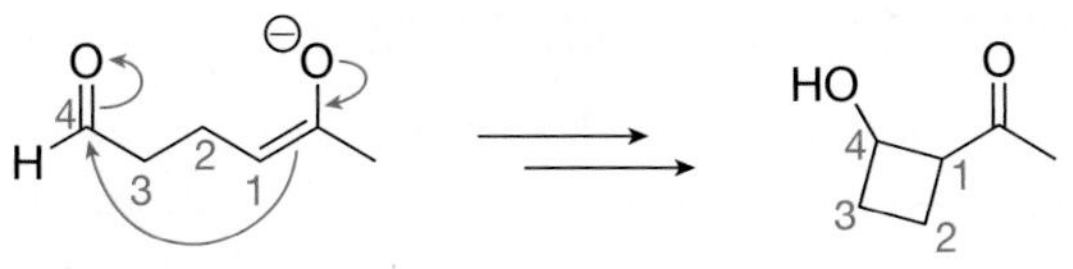

C

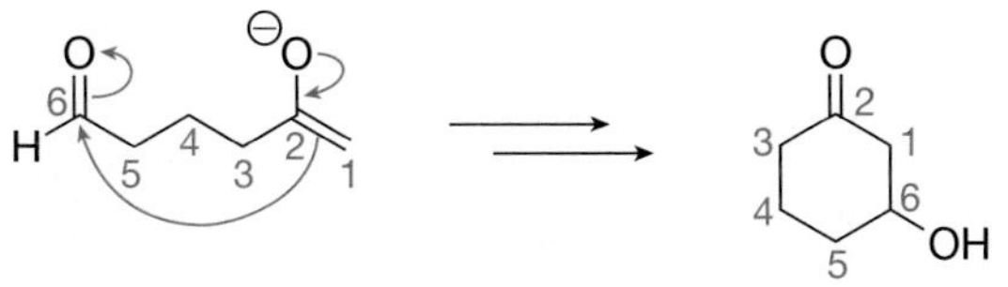

STEP 4: Determine which product is the most stable.

Two of the products have a highly strained four-membered ring and are unlikely to form in any appreciable amount. Therefore, the major product is the one with the stable six-membered ring.

STEP 5: Determine whether the aldol product is likely to dehydrate under the reaction conditions.

Since the reaction conditions do not include heat, the $E1_{cb}$ dehydration step is unlikely to occur, and the aldol product, rather than the enone, is the most likely product. Therefore, the overall reaction is as follows:

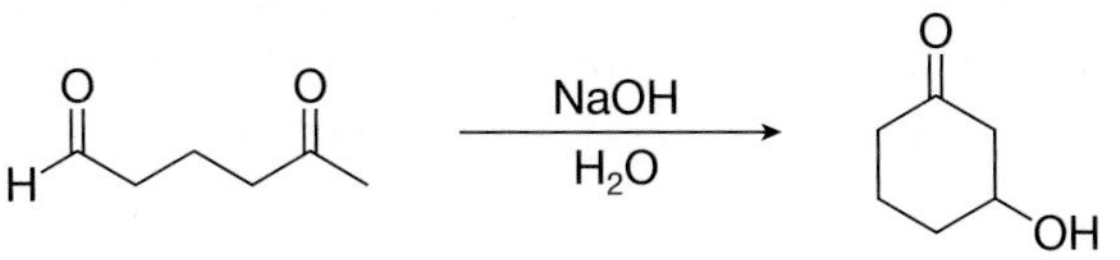

NaOH
H_2O

PRACTICE PROBLEM

17.15 Predict the major products for each of the following reactions.

a) 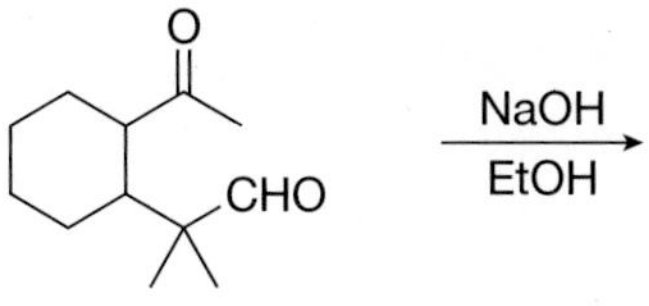 CHO NaOH EtOH

b) 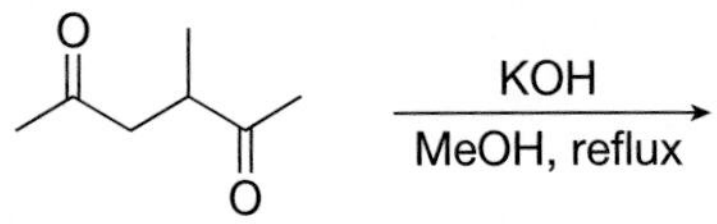 KOH / MeOH, reflux

c) 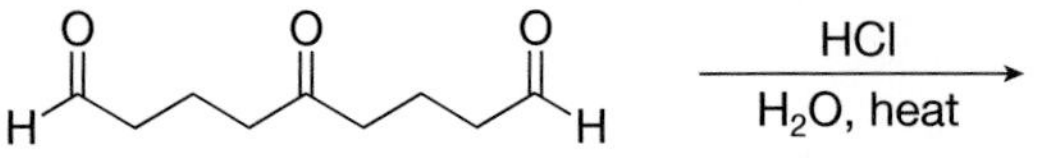 HCl / H_2O, heat

d) 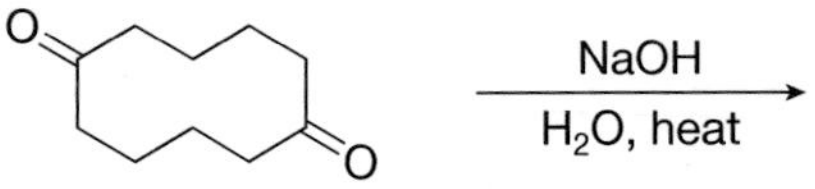 NaOH / H_2O, heat

INTEGRATE THE SKILL

17.16 Propose a mechanism for the following transformation.

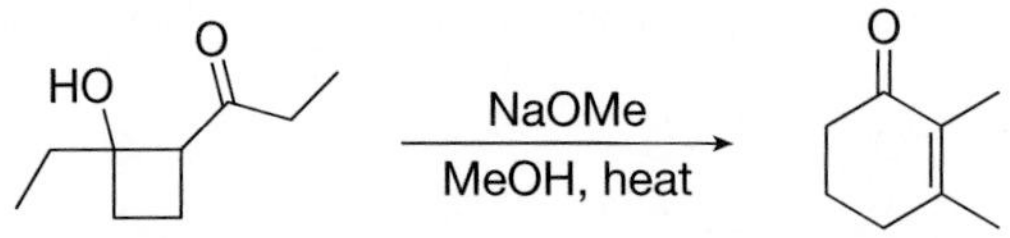

NaOMe / MeOH, heat

17.4.5 Retrosynthetic analysis of aldols

Aldol reactions produce molecules in which a hydroxy group is located β to a carbonyl or related functional group. From a synthetic point of view, this 1,3-pattern of oxygen atoms disconnects to an enolate synthon and an electrophilic carbonyl. The nucleophilic synthon is derived from the appropriate electron-withdrawing group by removal of the α-hydrogen. The electrophilic synthon is, of course, a carbonyl.

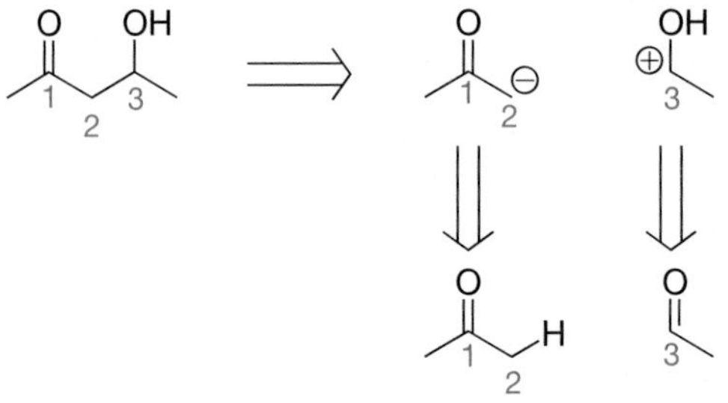

This pattern also applies for other 1,3-dihydroxy compounds (which can be derived from corresponding carbonyl products). Similarly, the pattern applies for α,β-unsaturated carbonyls in which the alkene carbon furthest from the carbonyl is derived from a hydroxyl group in an aldol. In each case, the bond forms between atoms 2 and 3 (the α- and β-carbons).

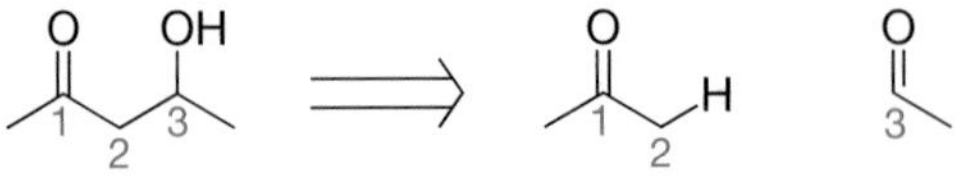

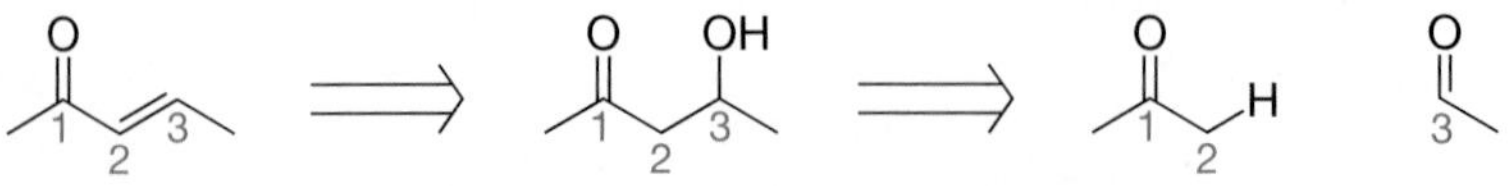

CHECKPOINT 17.9

You should now be able to propose syntheses of molecules involving aldol addition and condensation reactions.

SOLVED PROBLEM

Propose a method for synthesizing the following compound from an aldol reaction of a single dicarbonyl compound.

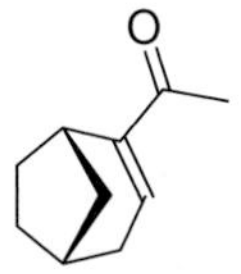

STEP 1: Examine the structure to find any β-hydroxy carbonyls or α,β-unsaturated carbonyls (enones). The structure contains an α,β-unsaturated carbonyl (enone), shown in red.

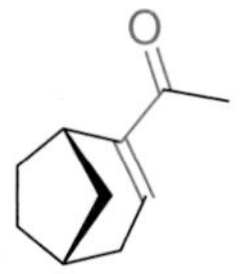

STEP 2: Number the atoms of the enone starting with 1 at the carbonyl carbon.

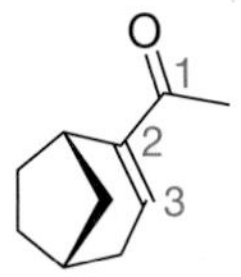

STEP 3: Disconnect the π bond between atoms 2 and 3 to show the β-hydroxy aldol product.

Enones are readily prepared from the dehydration of an aldol product. When the π bond is disconnected, the hydroxyl group bonds to carbon 3. This location is β to the carbonyl group.

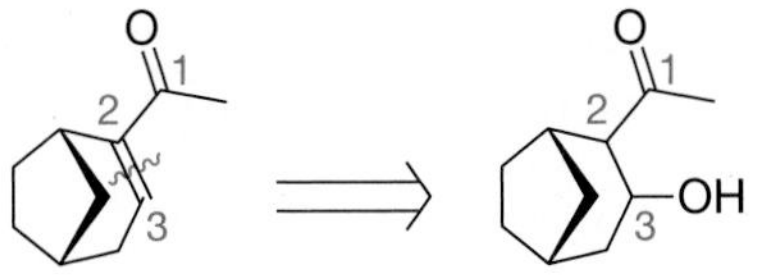

STEP 4: Disconnect the σ bond between atoms 2 and 3 to give a dicarbonyl compound.

β-Hydroxy carbonyls are readily prepared from an aldol reaction. When the σ bond is disconnected, the hydroxyl group become a carbonyl.

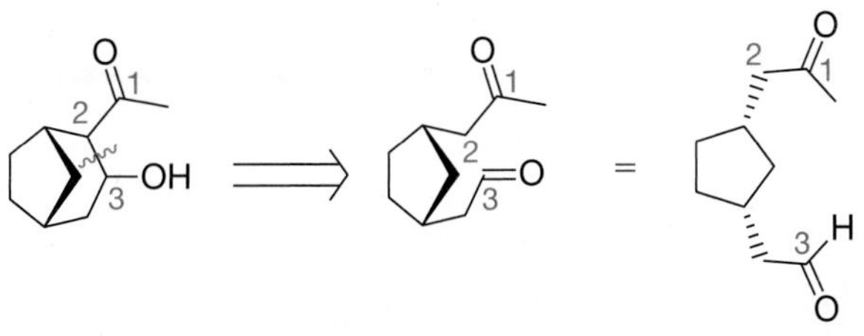

STEP 5: Determine the appropriate reaction conditions to produce the desired compound from the bicarbonyl in Step 4.

For an intramolecular aldol reaction, a weak base such as hydroxide is generally used to ensure that deprotonation is incomplete. The resulting equilibrium favours the most stable product. Since the desired product is the enone rather than the aldol, the reaction should be carried out in a protic solvent with heat. The overall reaction is as follows:

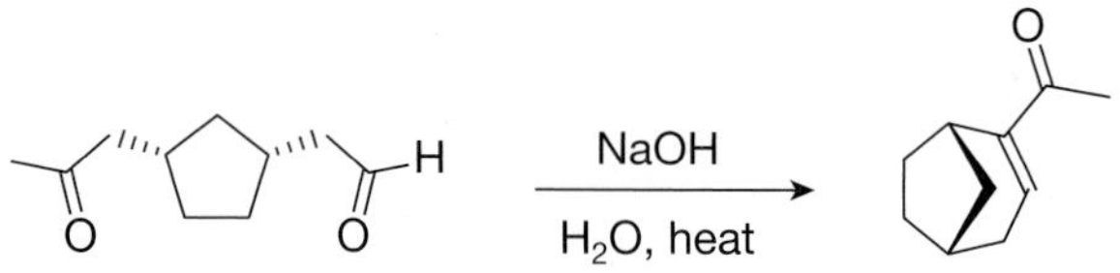

NaOH

H_2O, heat

PRACTICE PROBLEM

17.17 Determine appropriate starting materials and reaction conditions to synthesize each of the following compounds using aldol reactions and aldol condensations.

a)

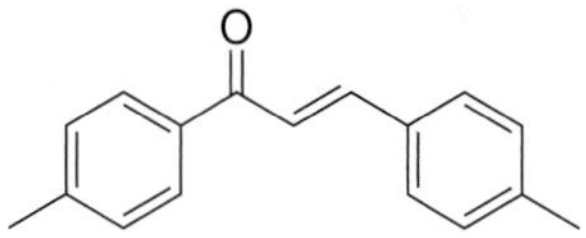

b)

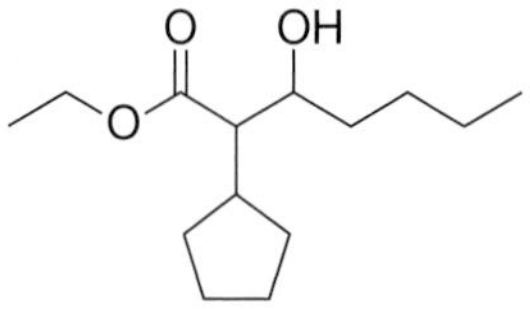

c)

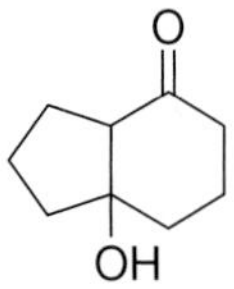

d)

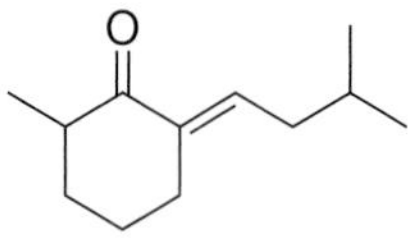

INTEGRATE THE SKILL

17.18 Propose a series of reactions that could be used to carry out the following transformations. You can use any reagents that do not have more than three carbon atoms.

a)

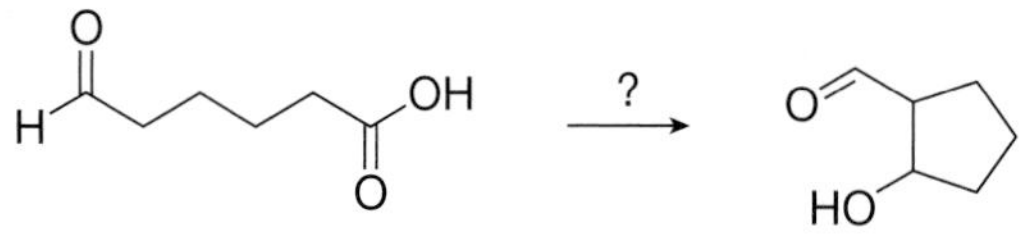

b)

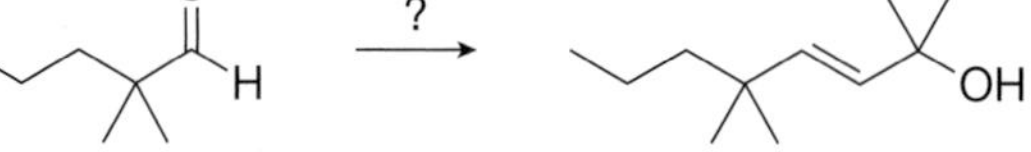

17.5 Preparation of Dicarbonyl Compounds: The Claisen Condensation

A **Claisen condensation** is the formation of a β-ketoester from esters; it is similar to an aldol addition and has an ester as the electrophile.

In a **Claisen condensation**, esters exposed to alkoxide bases in protic solvents form enolates that undergo self-addition similar to that in aldol reactions, followed by the elimination of alcohol. Because esters undergo addition-elimination reactions, the product is a 1,3-dicarbonyl called a β-ketoester. The alkoxide used in a Claisen condensation must match the alkoxy portion of the ester. Otherwise, a base-catalyzed transesterification will create a mixture of products.

Aldol addition

KOEt / EtOH

Claisen condensation

KOEt / EtOH

STUDENT TIP

An aldol condensation combines aldehydes or ketones with the loss of water, forming a C=C double bond. A Claisen condensation combines esters, with the loss of alcohol and the formation of a C=O double bond.

The mechanism for the first two stages of a Claisen condensation is similar to that for the aldol reaction. The alkoxide base forms a small amount of nucleophilic enolate that attacks the electrophilic ester to form a tetrahedral intermediate. This intermediate then expels the alkoxide leaving group to form the β-ketoester.

Aldol addition

weak base forms enolate in the presence of reactive electrophile

addition step forms an alkoxide

alkoxide is protonated to end the addition reaction

KOH / EtOH

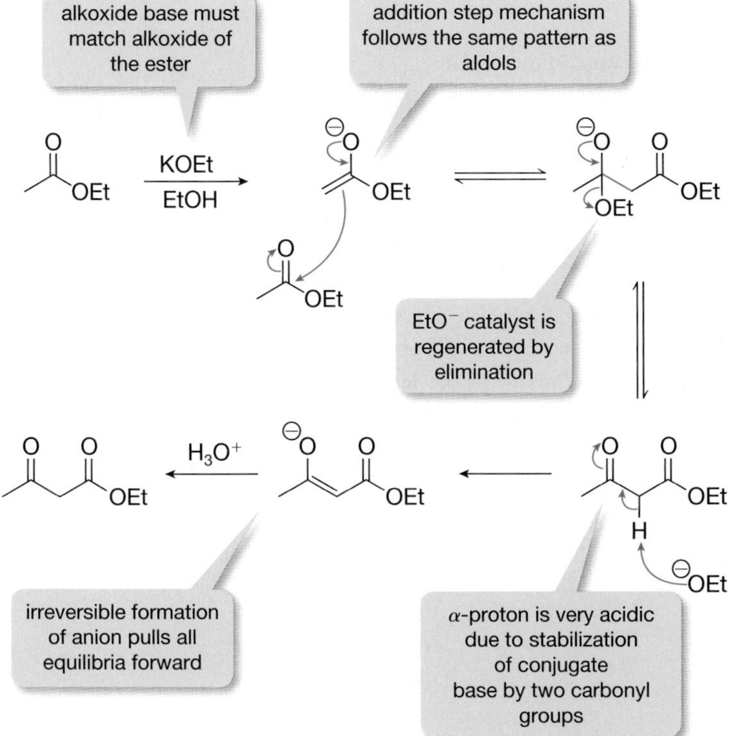
Claisen condensation
alkoxide base must match alkoxide of the ester
addition step mechanism follows the same pattern as aldols
KOEt
EtOH
EtO^- catalyst is regenerated by elimination
H_3O^+
irreversible formation of anion pulls all equilibria forward
α-proton is very acidic due to stabilization of conjugate base by two carbonyl groups

ORGANIC CHEMWARE
17.6 Claisen ester condensation

All steps in the formation of a β-ketoester involve equilibria, but the last step in the condensation is essentially irreversible. In the β-ketoester, the two electron-withdrawing carbonyls on either side of the α-carbon increase the acidity of the α-hydrogen significantly. The pK_a of the α-hydrogen of a β-ketoester is between 9 and 11, so this hydrogen is *much* more acidic than alcohol (pK_a of 16 to 18). The difference is large enough that the deprotonation of the β-ketoester in the reaction mixture is essentially irreversible. This deprotonation shifts the equilibria of the preceding steps such that all of the starting ester transforms into an enolate of the β-ketoester. When that reaction is complete, the addition of an acid generates the β-ketoester product.

The intramolecular version of the Claisen condensation produces a ring and is called a **Dieckmann cyclization**.

A **Dieckmann cyclization** is an intramolecular Claisen condensation that forms a ring.

CHECKPOINT 17.10

You should now be able to predict the products formed in a Claisen condensation and draw mechanisms for their formation.

SOLVED PROBLEM

Predict the major product of the following reaction.

O
OMe
1) NaOMe, MeOH
2) NH_4Cl, H_2O

STEP 1: Identify the roles of the reagents and reactants in the first step.

The starting material contains an ester with an electrophilic carbonyl group, as well as acidic α-hydrogens. In the presence of the NaOMe base, the ester is in equilibrium with its enolate form, which can then react with another molecule of ester in a Claisen condensation. Note that NaOMe can also act as a nucleophile, which causes a transesterification reaction. However, you can ignore this reaction since it simply regenerates the starting material.

STEP 2: Draw a mechanism for the formation of the enolate.

O
OMe
H H
+
$\ominus$
OMe
O$^\ominus$
OMe
+
MeOH

STEP 3: Draw a mechanism for the addition of the enolate to a second molecule of ester.

The nucleophilic enolate adds to the electrophilic carbonyl of the ester. The resulting tetrahedral intermediate readily collapses, expelling methoxide as the leaving group and regenerating the C=O π bond. The remaining α-hydrogen is now very acidic due to the two neighbouring carbonyl groups, so the methoxide generated in the previous step quickly deprotonates the dicarbonyl compound, thus re-forming the enolate.

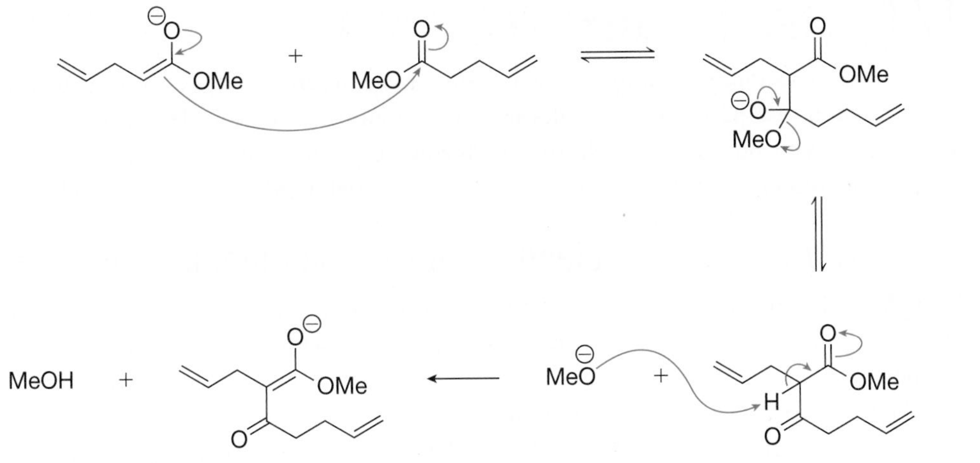

OMe + MeO ⇌ OMe MeO

MeOH + OMe ⟵ MeO + H OMe

STEP 4: Use the aqueous acid added in the second step of the reaction to protonate the enolate and give the major product: a 1,3-dicarbonyl compound.

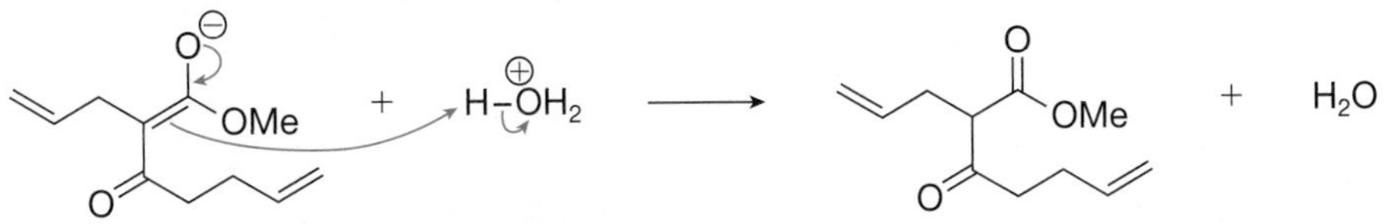

OMe + $H-OH_2^+$ ⟶ OMe + H_2O

PRACTICE PROBLEM

17.19 Predict the major products of the following reactions.

a) 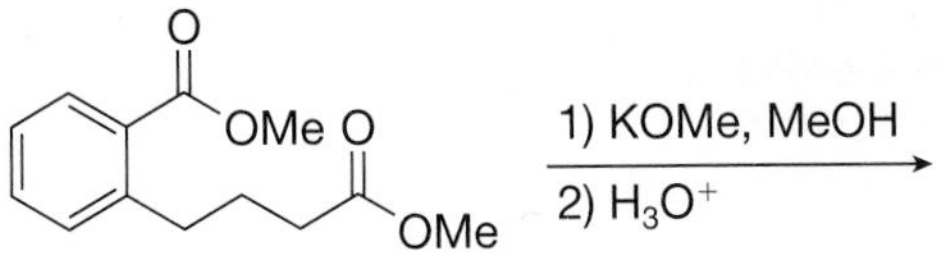 OMe O OMe — 1) KOMe, MeOH 2) H_3O^+

b) 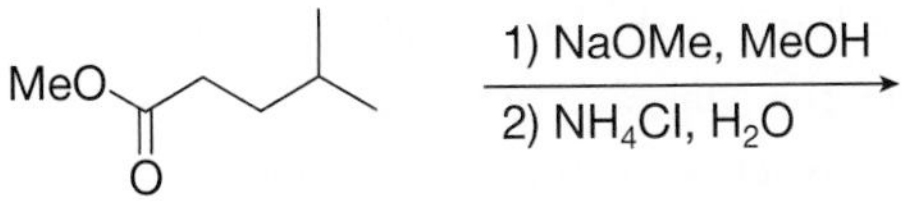 MeO — 1) NaOMe, MeOH 2) NH_4Cl, H_2O

c) 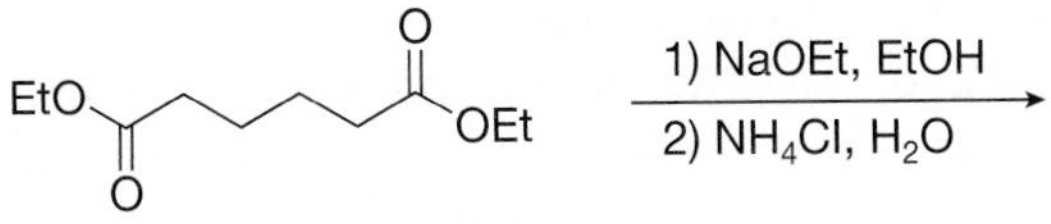 EtO OEt — 1) NaOEt, EtOH 2) NH_4Cl, H_2O

d) 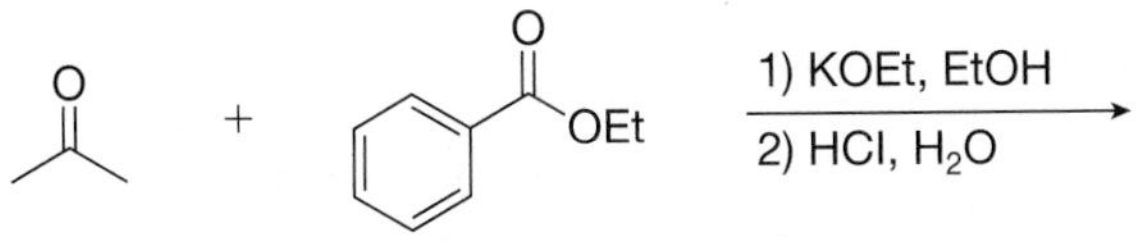 + OEt — 1) KOEt, EtOH 2) HCl, H_2O

INTEGRATE THE SKILL

17.20 a) Mechanistically, the aldol reaction is very similar to the Claisen condensation. However, the aldol reaction requires only a catalytic amount of base, whereas a Claisen condensation generally needs at least one equivalent of base. Explain this difference using a reaction mechanism and any substrates of your choice.

b) The Claisen condensation does not work well on esters that have only one enolizable proton. For example, the self-condensation of the following ester fails to give a significant amount of the Claisen product. Use a mechanism to explain this poor yield.

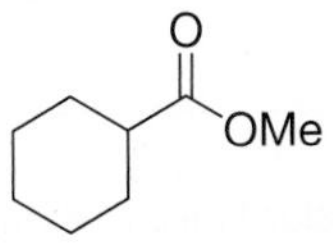

OMe

17.6 Aldol-Related Reactions

A number of reactions related to both aldol and Claisen reactions follow a similar mechanism. These reactions use related electrophiles such as iminium ions and nitriles, or they use related nucleophiles with non-carbonyl electron-withdrawing groups—such as nitriles, nitro groups, imines and enamines, and phosphonium salts—that form stabilized anions that react like enolates.

17.6.1 Nitrogen-based electrophiles in the Mannich reaction

The **Mannich reaction** is an aldol-type reaction that produces a β-aminoketone instead of a β-hydroxyketone.

In the **Mannich reaction**, an aldehyde and a ketone combine with an amine to form a β-amino carbonyl compound. The reaction is most commonly run with formaldehyde as the aldehyde component, but other aldehydes also work. The reaction uses a low concentration of acid to catalyze the formation of an iminium ion from the aldehyde and amine. This iminium intermediate acts as an electrophile in an aldol-type addition. The actual nucleophile is an enol rather than an enolate because acid is used as a catalyst (a strong base will not exist in acidic solution).

An iminium ion is very electrophilic; the π bond in the ion is readily attacked by an enolate. The ion reacts with the ketone enol to produce the Mannich product.

Carbonyl-containing groups are able to react as carbanion nucleophiles because the electron-withdrawing carbonyl stabilizes the negative charge on the α-carbon. Many other functional groups can stabilize carbanions in the same way, readily forming carbanions analogous to enolate nucleophiles. Compounds containing these groups can undergo self-condensations and crossed condensations with a variety of electrophiles. Such reactions have mechanistic similarities to aldol reactions and Claisen condensations.

17.6.2 The Thorpe reaction

Since the hydrogens adjacent to a nitrile have a pK_a similar to that of esters, nitriles can undergo a self-condensation—the **Thorpe reaction**—which is similar to an aldol condensation. The deprotonation of a nitrile makes an anion with two resonance forms; the form with the charge on the nitrogen makes the greatest contribution. Unless very strong base is used, this anion is generated in small amounts, resulting in a self-condensation that forms a β-iminonitrile, which tautomerizes to a conjugated cyanoenamine.

The **Thorpe reaction** is a reaction that uses nitriles instead of aldehydes or ketones.

cyanoenamine

β-iminonitrile

The cyanoenamine product can be hydrolyzed to a β-ketonitrile in equilibrium with the corresponding conjugated enol. Performing the hydrolysis at a higher temperature can further hydrolyze the nitrile to produce a carboxylic acid. The carboxylic acid, being a β-keto acid, spontaneously undergoes decarboxylation to a simple ketone (Section 17.7.2). Thus, the Thorpe reaction can be used to produce a variety of compounds.

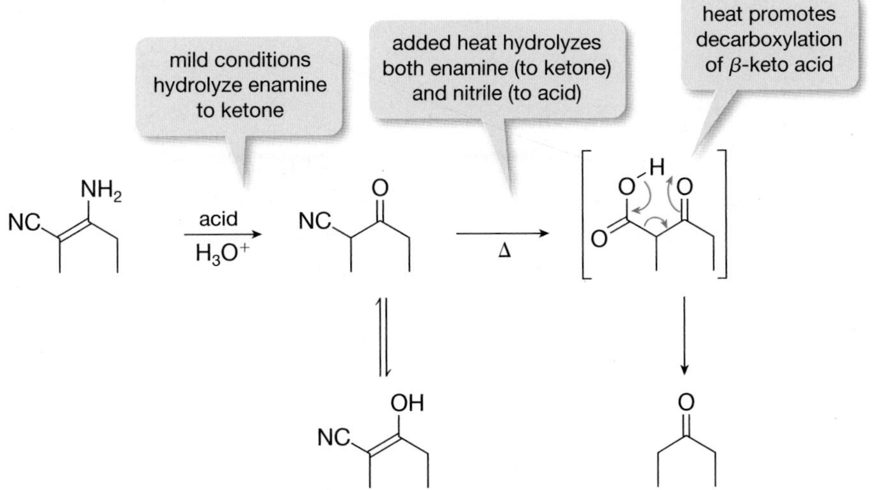
mild conditions hydrolyze enamine to ketone
added heat hydrolyzes both enamine (to ketone) and nitrile (to acid)
heat promotes decarboxylation of β-keto acid
acid
H3O+
Δ

The **Thorpe–Ziegler reaction** is an intramolecular Thorpe condensation of nitriles.

STUDENT TIP
Like aldol reactions, the Thorpe reaction can be done with a weak base for self-addition or with a strong base if crossed reactions are desired.

An intramolecular Thorpe reaction produces cyclic ketones, and is known as the **Thorpe–Ziegler reaction**.

17.6.3 The Henry reaction

The **Henry reaction**, also known as the **nitroaldol reaction**, produces β-hydroxy nitro compounds. This reaction works well as a crossed condensation because nitroalkanes are more acidic than aldehydes and ketones and cannot be electrophiles. In a mixture of nitroalkane and aldehyde, a weak base selectively deprotonates the more acidic nitroalkane, forming a nucleophile similar to an enolate.

The **Henry** or **nitroaldol reaction** uses the anion of a nitroalkane as the nucleophile in an aldol-type reaction.

This nucleophile adds to the aldehyde or ketone electrophile to produce the β-hydroxy nitro product. Note the 1,3-relationship between the hydroxyl group and the nitrogen of the nitro group.

aldehyde electrophile + nitro α-carbon nucleophile $\xrightarrow{\text{base}}$ β-hydroxy nitro product

The Henry reaction is a key step in the synthesis of L-acosamine, a subunit of the anthracycline family of anti-cancer drugs. In the following reaction sequence, the fluoride from TBAF (tetrabutylammonium fluoride) acts as a weak base to deprotonate a nitroalkane, which then forms a bond to the aldehyde.

TBAF, MTBE; HCl, MeOH; H_2, Pd/C

TBAF

STUDENT TIP

The Henry reaction can be done with a weak base because nitro groups are not electrophilic.

CHECKPOINT 17.11

You should now be able to predict the products of aldol-like reactions, including the Mannich, Thorpe, and Henry reactions. You should also be able to draw mechanisms for these reactions.

SOLVED PROBLEM

Depending on the workup conditions used, treating the following compound with a catalytic amount of alkoxide base can yield a variety of cyclic compounds with different functional groups. Draw the possible cyclic structures that could be made, and list the necessary reaction conditions for each.

STEP 1: Determine the roles for the reactants and the type of reaction taking place.

The starting material has two electrophilic nitrile groups. The alkoxide base (such as NaOEt) is present only in a catalytic amount, so this base is unlikely to function as a nucleophile. The base could react with the weakly acidic α-hydrogens from one of the nitriles to form an enolate equivalent. This reaction would yield a molecule with a nucleophilic enolate and an electrophilic nitrile in close proximity. The enolate could then add intramolecularly to the nitrile in a Thorpe–Ziegler reaction.

STEP 2: Draw the possible anions formed when the starting compound reacts with the base.

Since the molecule is symmetrical, only one anion is possible.

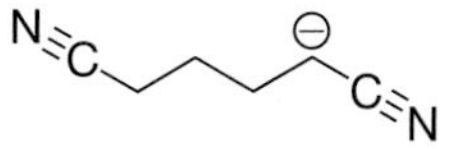

STEP 3: Determine the initial cyclic product formed when the anion adds to the nitrile.

You may find it helpful to number the carbon atoms when drawing the mechanism arrows for the nucleophilic attack. The initial addition of the anion onto the nitrile gives a five-membered ring with a nitrile and an imine. The imine readily tautomerizes to give the more stable conjugated enamine.

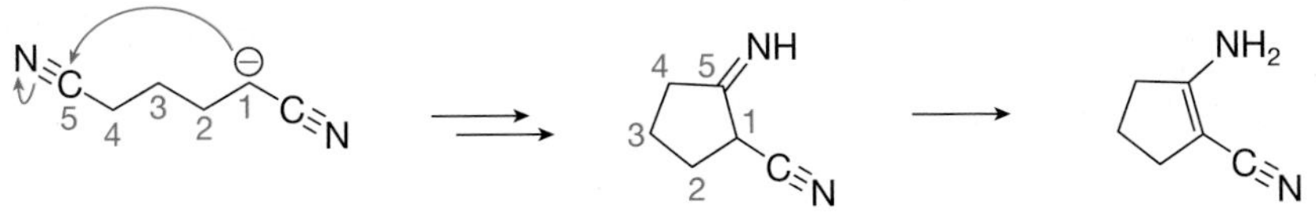

STEP 4: Determine what other structures could be generated from the enamine.

As discussed in Section 16.4, hydrolyzing the imine with acid and water yields a ketone. Alternatively, if the acidic solution is heated, the nitrile can hydrolyze to a β-keto acid, which then readily decarboxylates to give a simple ketone.

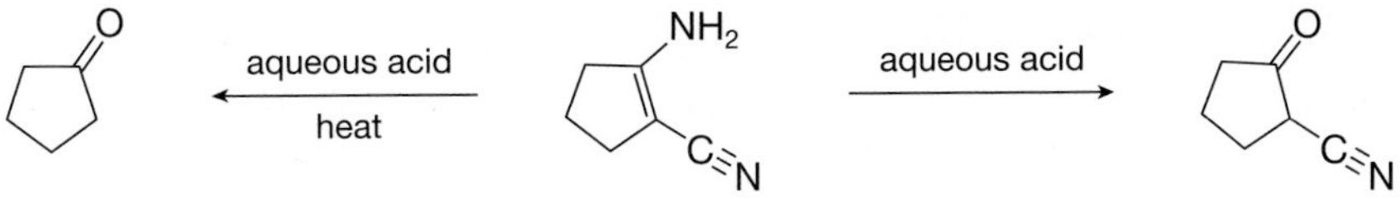

aqueous acid, heat ← → aqueous acid

STEP 5: Write an overall reaction equation for the formation of each cyclic structure.

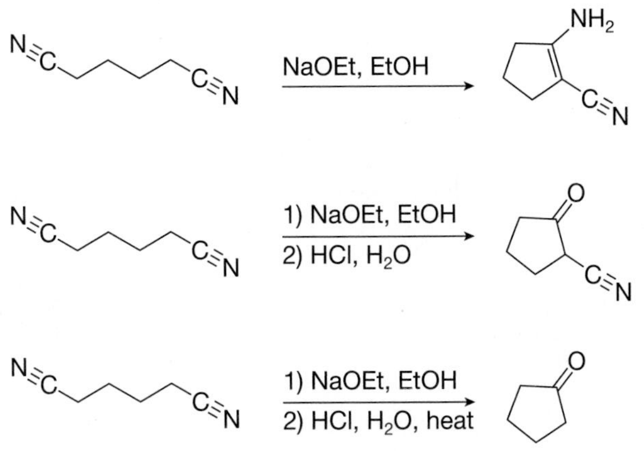

NaOEt, EtOH

1) NaOEt, EtOH
2) HCl, H_2O

1) NaOEt, EtOH
2) HCl, H_2O, heat

PRACTICE PROBLEM

17.21 For each of the following reactions, predict the major product, and draw a mechanism for its formation.

a) 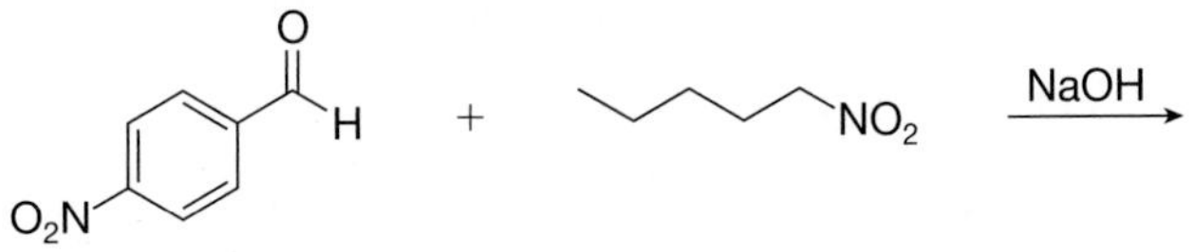 O_2N–C$_6$H$_4$–CHO + NO_2 (nitroalkane) → NaOH

b) 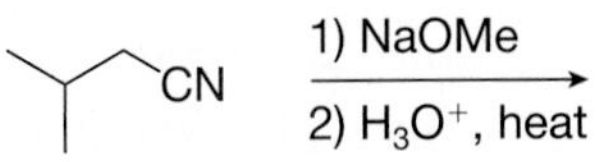 CN → 1) NaOMe 2) H_3O^+, heat

c) NH_2 + (ketone) → CH_2O, $HCl_{(cat.)}$

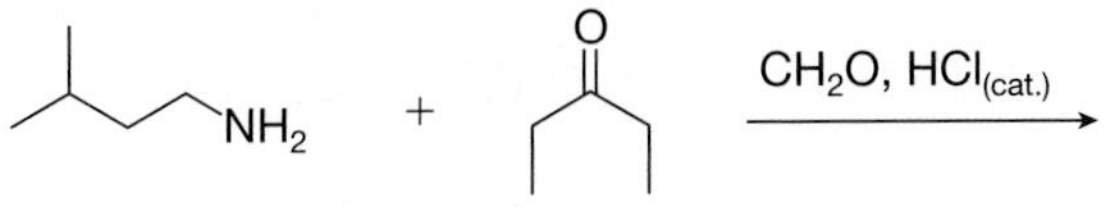

d) 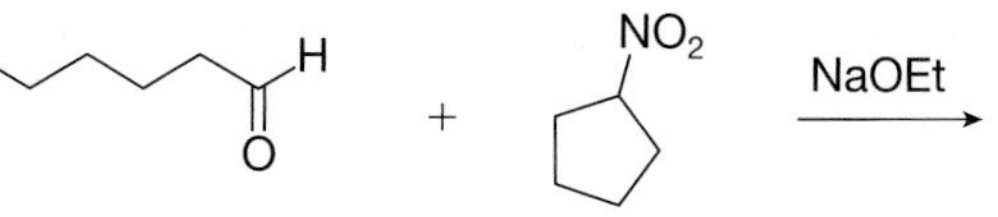 [hexanal + nitrocyclopentane, NaOEt]

e) 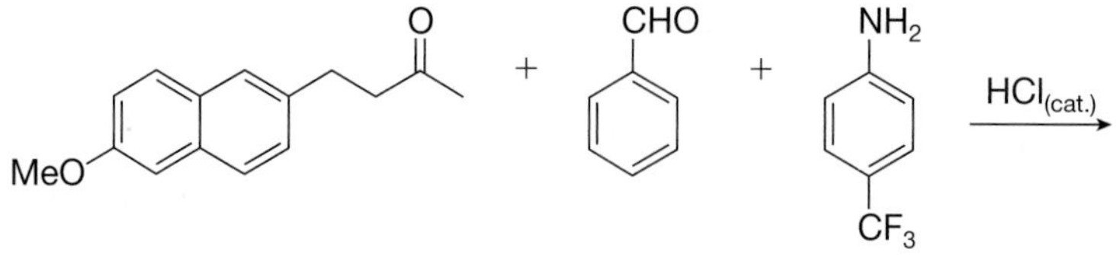 [MeO-naphthyl ketone + benzaldehyde (CHO) + 4-(trifluoromethyl)aniline (NH_2, CF_3), $HCl_{(cat.)}$]

INTEGRATE THE SKILL

17.22 The mechanism of a *benzoin condensation* is similar to that of the Thorpe reaction and uses cyanide as a catalyst. Suggest a mechanism for the following benzoin condensation.

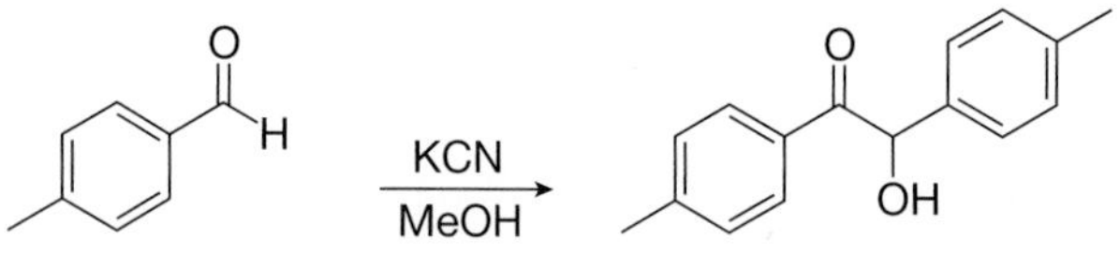

[4-methylbenzaldehyde, KCN / MeOH → 4,4′-dimethylbenzoin (O, OH)]

17.6.4 Imine anions and enamines

As discussed in Section 17.3.2.2, enamines and imine anions can be considered enolate equivalents.

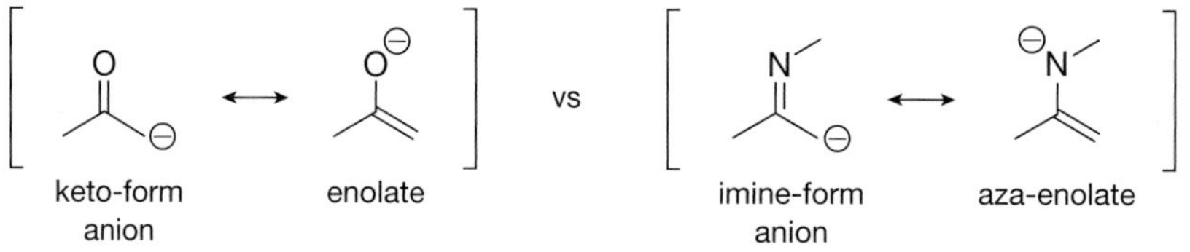

keto-form anion ⟷ enolate vs imine-form anion ⟷ aza-enolate

Enamines are common in biological systems. For example, a key step in the natural production of fructose 1,6-bisphosphate is a reaction of dihydroxyacetone phosphate (DHAP) with glyceraldehyde-3-phosphate, catalyzed by the enzyme aldolase A. An enamine is a critical intermediate in this reaction.

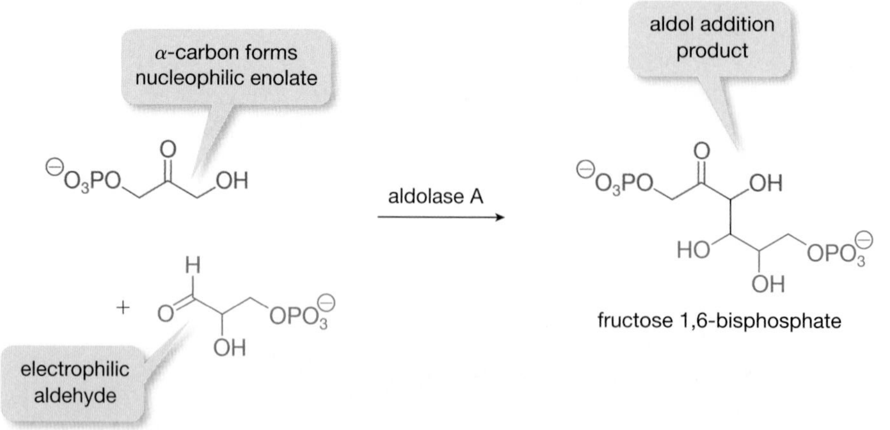

The active site of aldolase A contains an amine (side-chain of lysine) and a base (side-chain of tyrosine). When DHAP enters the active site, it condenses with the lysine NH_2 group to form an iminium ion. The positive charge on the iminium ion increases the acidity of the α-hydrogen of DHAP, allowing it to be removed by the basic oxygen of the tyrosine. This step forms an enamine, which then attacks the carbonyl of glyceraldehyde to form a six-carbon compound with a new C–C bond. Hydrolysis of the iminium ion releases the aldol product and regenerates the enzyme.

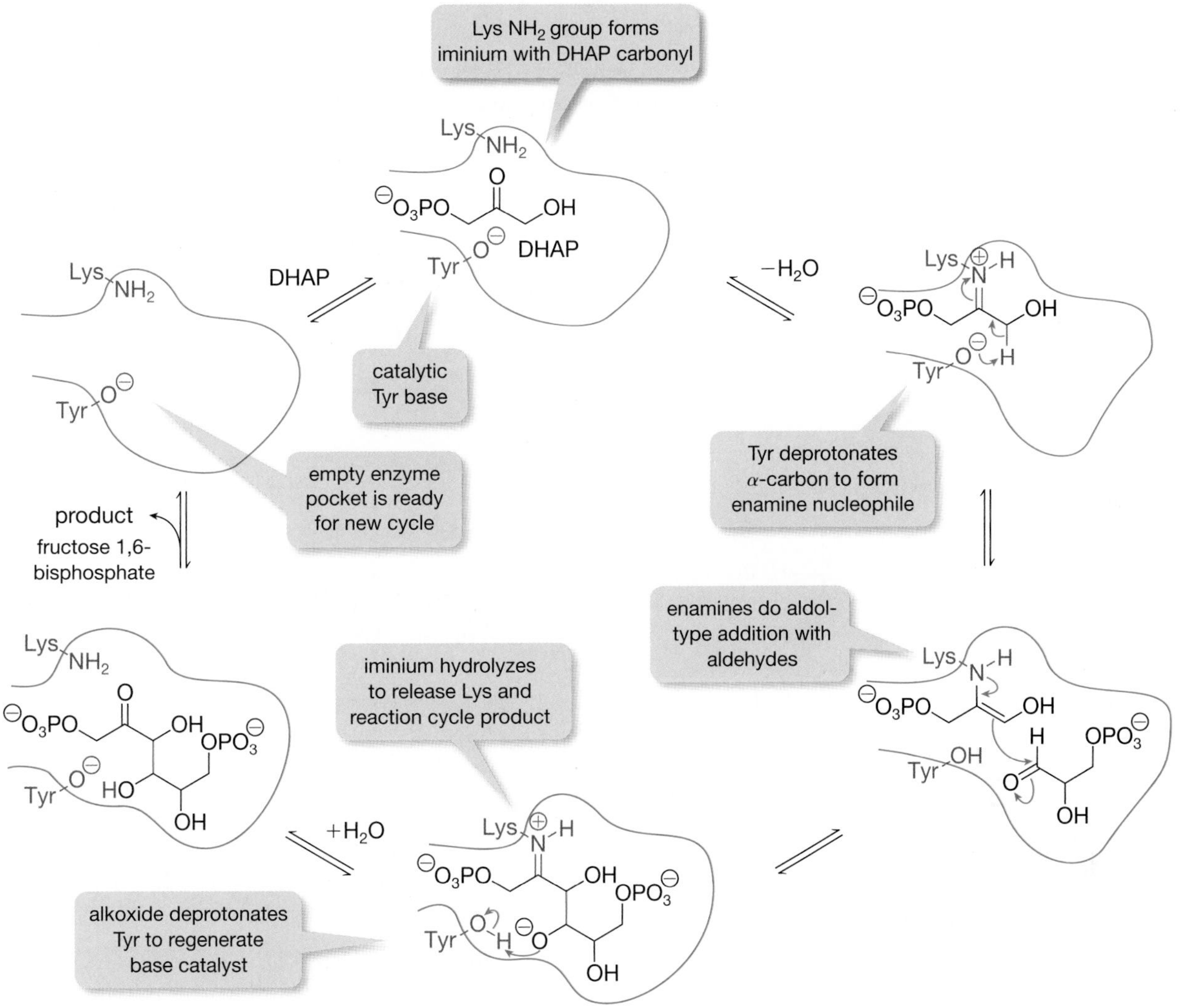

The use of aza-enolates is uncommon, primarily because the imines they are formed from are difficult to work with. One exception is crossed aldol reactions involving aldehydes. Aldehydes are so reactive that it is difficult to suppress self-condensation, even when forming the enolates with LDA. Because the imines are less electrophilic, making the corresponding aza-enolate provides a way to produce crossed aldols from aldehydes. Hydrolyzing the imine product at the end of the reaction produces the carbonyl compounds.

CHECKPOINT 17.12

You should now be able to predict the products formed when imine anions and enamines react with electrophiles such as alkyl halides and carbonyl compounds. You should also be able to draw mechanisms for these reactions.

SOLVED PROBLEM

Provide the missing products from the following stepwise reaction. For each step, draw a mechanism that shows how the product is formed.

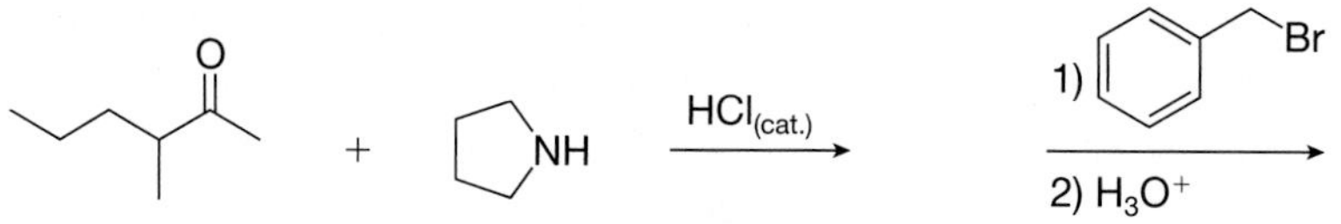

STEP 1: Determine the roles of the reagents in the first step of the reaction.

The carbonyl of the ketone is a good electrophile, and the amine is a good nucleophile. In the presence of an acid catalyst, these two reactants readily form an enamine.

STEP 2: Draw a mechanism for the formation of the enamine. (For guidance, refer to Chapter 16.)

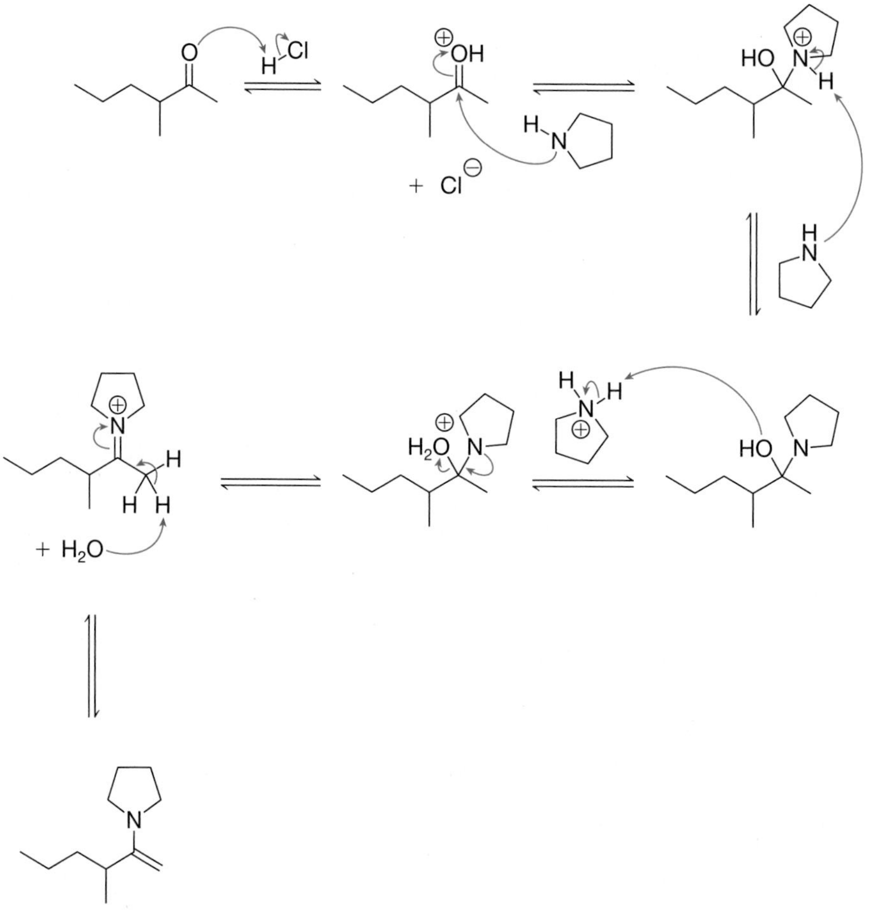

STEP 3: Determine the roles of the reagents and reactants in the second step of the reaction.

The enamine is nucleophilic at the π bond and can react with the electrophilic alkyl bromide.

STEP 4: Draw a mechanism for the enamine alkylation.

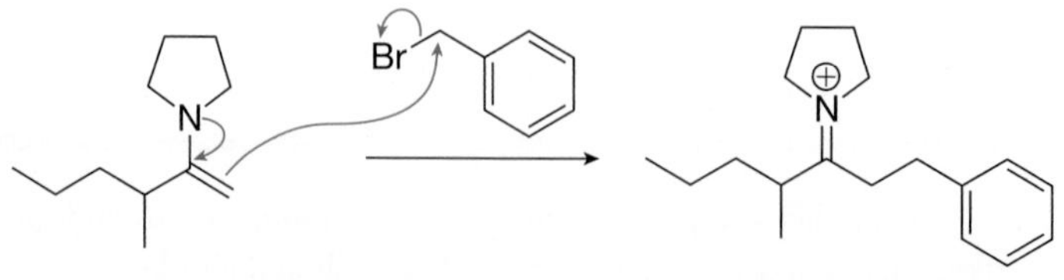

STEP 5: Determine the roles of the reagents and reactants in the third step of the reaction.

The imine is electrophilic and, in the presence of acid, readily reacts with nucleophilic water to hydrolyze the functional group.

STEP 6: Draw a mechanism for the imine hydrolysis step. (For guidance, refer to Chapter 16).

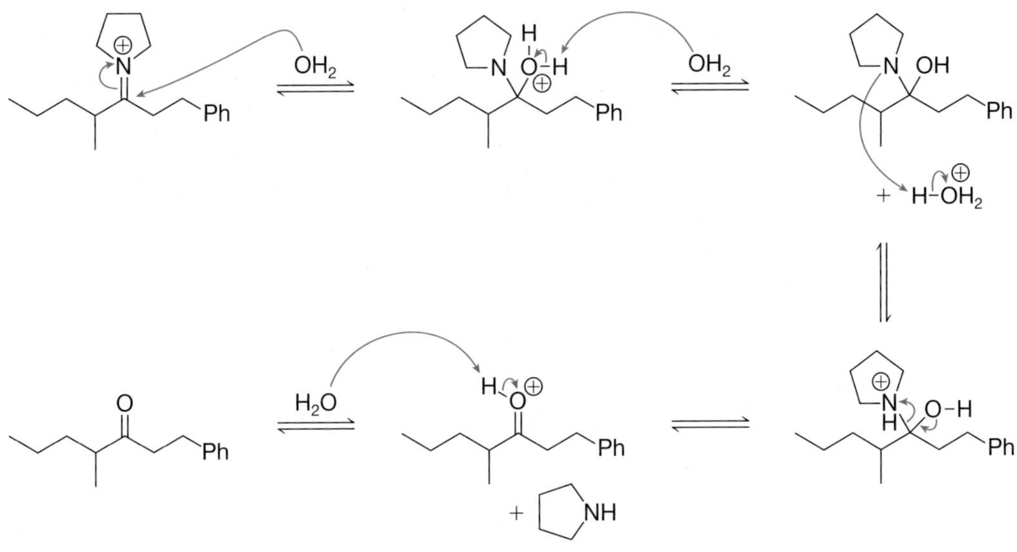

PRACTICE PROBLEM

17.23 Predict the products of the following reactions, and draw mechanisms for their formation.

a) 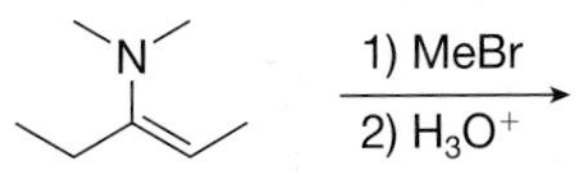 1) MeBr 2) H_3O^+

b) 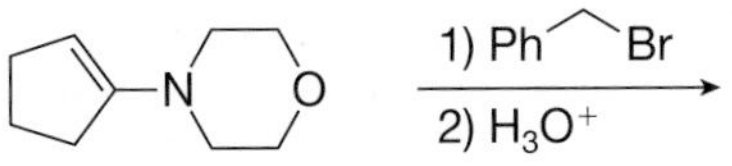 1) Ph Br 2) H_3O^+

c) 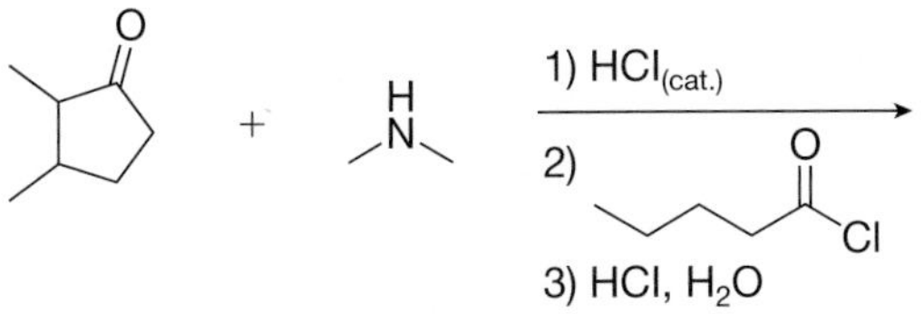 + 1) $HCl_{(cat.)}$ 2) Cl 3) HCl, H_2O

d) 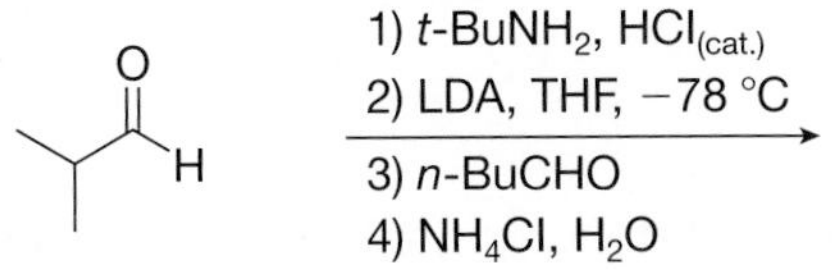 H 1) t-$BuNH_2$, $HCl_{(cat.)}$ 2) LDA, THF, −78 °C 3) n-BuCHO 4) NH_4Cl, H_2O

INTEGRATE THE SKILL

17.24 Provide two methods for carrying out the following transformations. One of these methods should involve an enamine.

a) 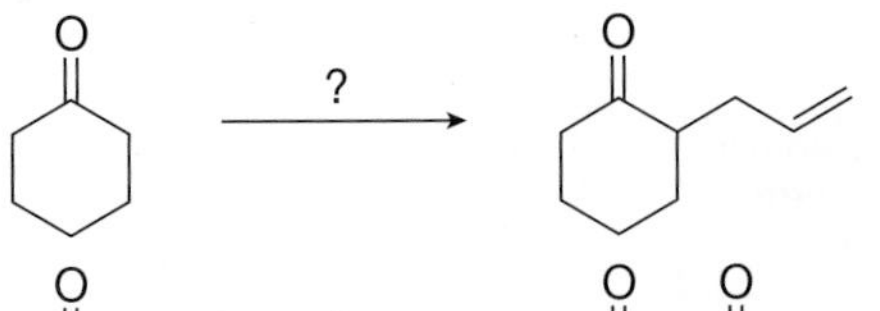 ?

b) 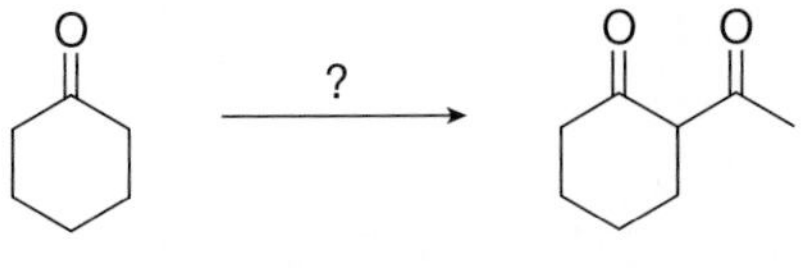 ?

17.6.5 The Wittig reaction

The **Wittig reaction** makes alkenes out of aldehydes with predominantly *cis* selectivity, using a phosphonium ylide.

The **Wittig reaction** features a phosphorus-stabilized carbanion that reacts with the carbonyl of an aldehyde or ketone. The initial product of this reaction is unstable and reacts further to form an alkene. The process is an excellent way to make carbon-carbon bonds. With aldehydes, *cis* and *trans* isomers of the product are possible. Normally, the reaction favours the *cis* product.

cis alkene (major product)

trans alkene (minor product)

$Ph_3P^+CH_2CH_3$, *n*-BuLi

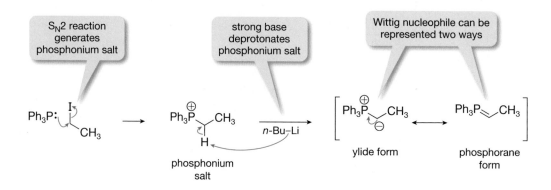

The nucleophile in the Wittig reaction arises from the deprotonation of a phosphonium salt, which is produced by an S_N2 reaction between triphenylphosphine (Ph_3P) and the appropriate akyl halide. This nucleophile is often represented as a neutral species with formal charges, known as an **ylide**; it may also be shown as a **phosphorane**, which has a C=P double bond. The phosphorane form is possible because phosphorus, which has 3d orbitals, can have more than eight valence electrons.

An **ylide** is a neutral compound with positive and negative charges on adjacent atoms.

A **phosphorane** is a Wittig ylide resonance form with a C=P double bond.

STUDENT TIP

Formation of a Wittig ylide typically requires a strong base.

S_N2 reaction generates phosphonium salt

strong base deprotonates phosphonium salt

Wittig nucleophile can be represented two ways

n-Bu–Li

phosphonium salt

ylide form

phosphorane form

The ylide adds to the carbonyl of an aldehyde or ketone (but not to esters and amides) to form a four-membered **oxaphosphetane** ring. The strain in this ring coupled with the high affinity of phosphorus for oxygen causes the ring to break apart into triphenylphosphine oxide and the product alkene. When *E* and *Z* isomers of the alkene are possible, the Wittig reaction generally favours the *Z* isomer.

An **oxaphosphetane** is a four-membered ring containing an oxygen atom and a phosphorus atom.

four-membered ring breaks to form alkene

ylide

oxaphosphetane

Z isomer is favoured

A simplified way to represent the addition of the ylide to the carbonyl is to treat the reaction as a nucleophilic addition to a carbonyl. As the carbonyl π bond breaks, the electrons from this bond form a new bond with the positively charged phosphorus, generating an oxaphosphetane that has two alkyl groups *cis* to each other. This stereochemistry is maintained when the ring fragments to make the alkene. The actual reaction between the aldehyde and ylide is more complex; it is described in detail in Chapters 18 and 20.

ORGANIC CHEMWARE
17.7 Wittig reaction

CHECKPOINT 17.13

You should now be able to predict the products formed in a Wittig reaction and draw mechanisms for their formation.

SOLVED PROBLEM

Predict the major products of the following reaction, and draw a curved arrow mechanism showing their formation.

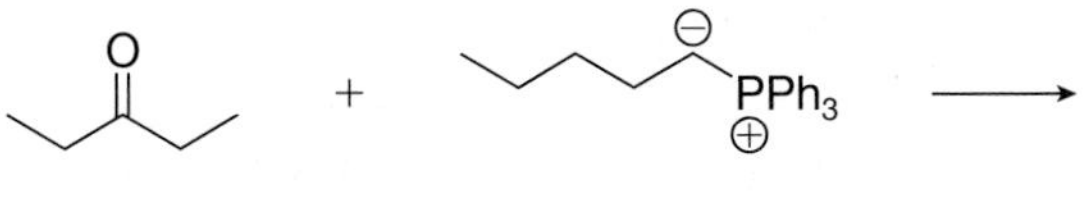

STEP 1: Determine the roles of the reactants.

The phosphorous-containing compound is an ylide, which readily reacts with electrophilic carbonyls such as the ketone reagent.

STEP 2: Align the substrates so the partial charges on the carbonyl face opposite charges on the ylide.

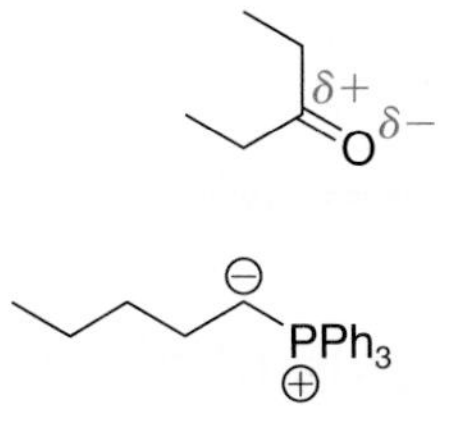

STEP 3: Draw a mechanism for the reaction.

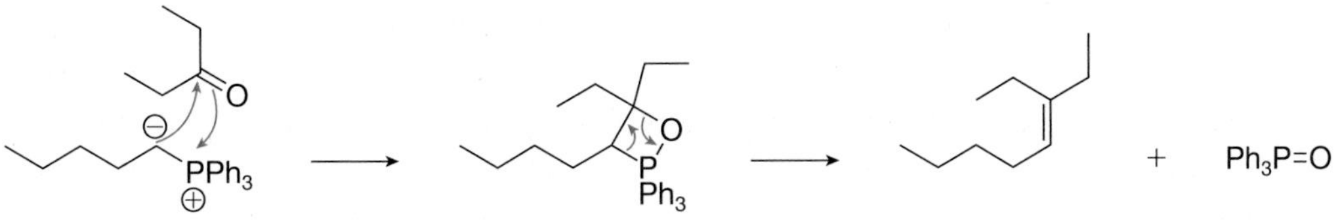

STEP 4: Check whether *E*/*Z* isomers of the final product are possible, and, if so, draw the *Z* isomer n as the favoured isomer.

In this reaction, no *E*/*Z* isomers of the final product are possible.

PRACTICE PROBLEM

17.25 Predict the major products of the following reactions, and draw mechanisms for their formation.

a)

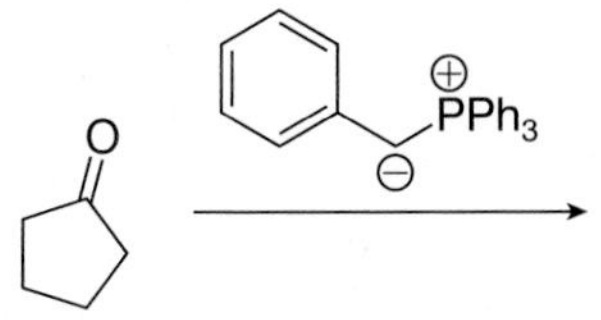

b)

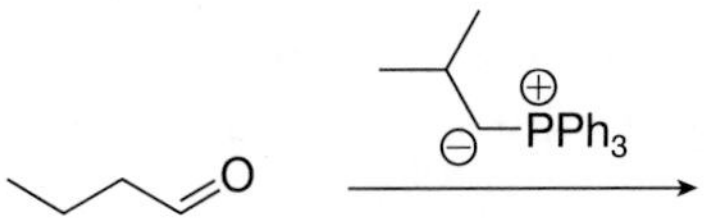

c)

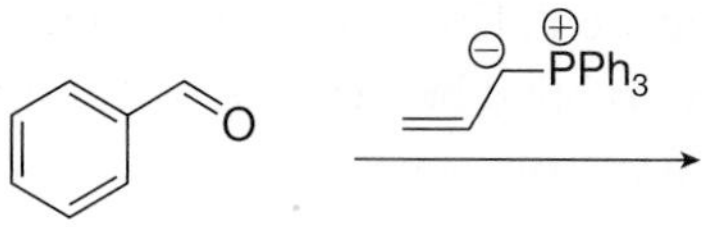

PPh_3

1) Ph_3P
2) *n*-BuLi
3)

d) 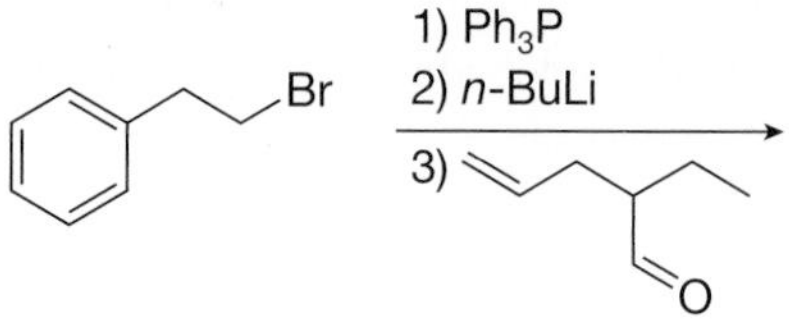 Br

O

INTEGRATE THE SKILL

17.26 a) Explain why the Wittig reaction is a better method than dehydration for making the following alkene.

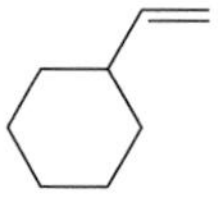

b) The Wittig reaction forms an alkene from a carbonyl compound and a phosphonium ylide. The alkene shown here could potentially be formed from two different carbonyl compounds. Draw the two possible starting materials, and determine which one is the better choice. Explain your answer.

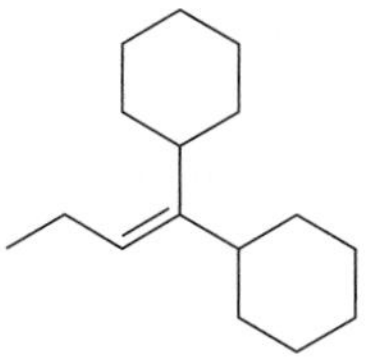

17.6.6 Retrosynthetic analysis of aldol-related reactions

Aldol reactions produce molecules in which a hydroxy group is located β to a carbonyl or related functional group. If the hydroxyl group eliminates, an unsaturated carbonyl compound forms. Claisen reactions produce a β-keto carbonyl instead of a β-hydroxyl.

Aldol

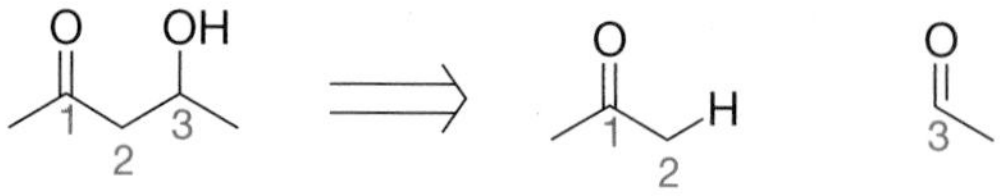

O OH 1 2 3 ⟹ O 1 2 H O 3

Aldol condensation

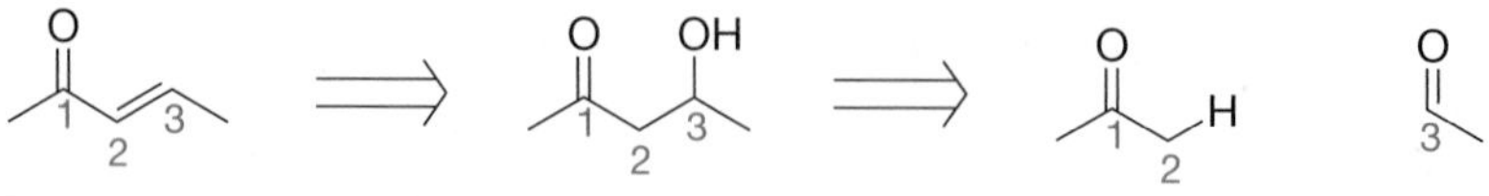

O 1 2 3 ⟹ O OH 1 2 3 ⟹ O 1 2 H O 3

Claisen

O O RO 1 2 3 ⟹ O RO 1 2 H O 3 OR

From a synthetic point of view, aldol-related reactions such as the Mannich or Thorpe reactions parallel the 1,3 pattern of oxygen atoms with heteroatoms, such as a β-amino carbonyl or a β-hydroxynitrile products. Instead of oxygens bonded to the first and third carbon, the Mannich and Thorpe reactions form a product that has a nitrogen atom. The Henry reaction uses a nitro group in place of a carbonyl, which replaces the first carbon of the 1,3 system with a nitrogen.

Mannich

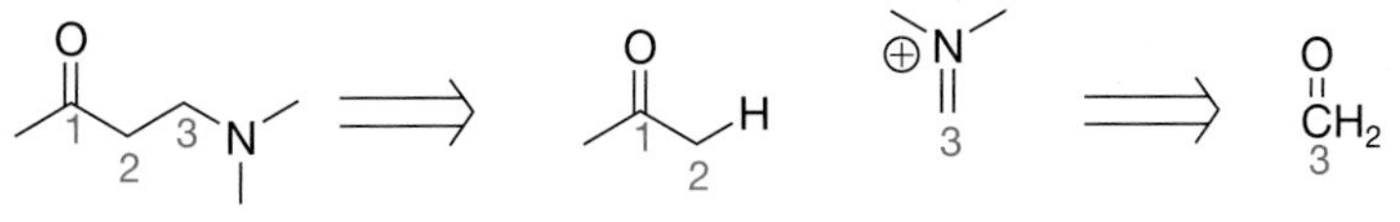

Thorpe

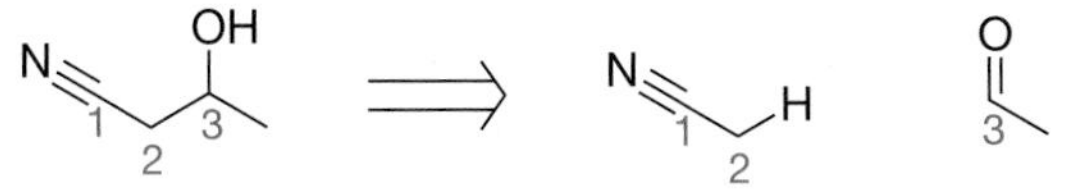

Henry

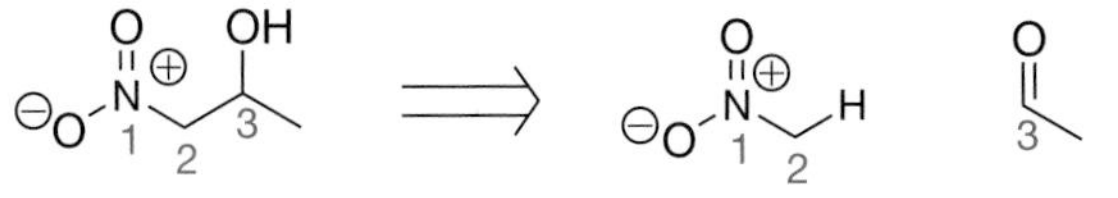

Enamine

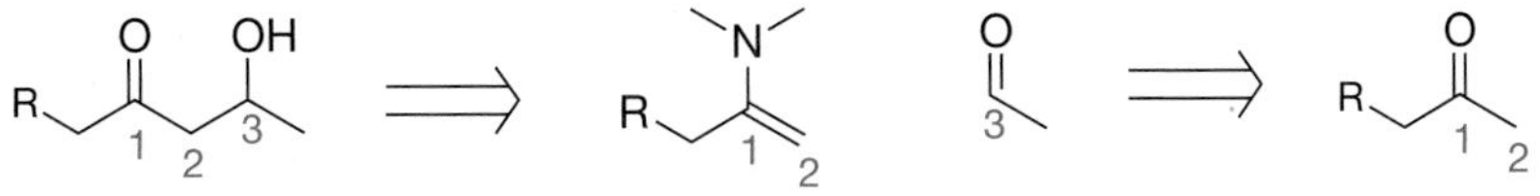

The Wittig reaction uses a similar mechanism to make *cis*-alkenes.

Wittig

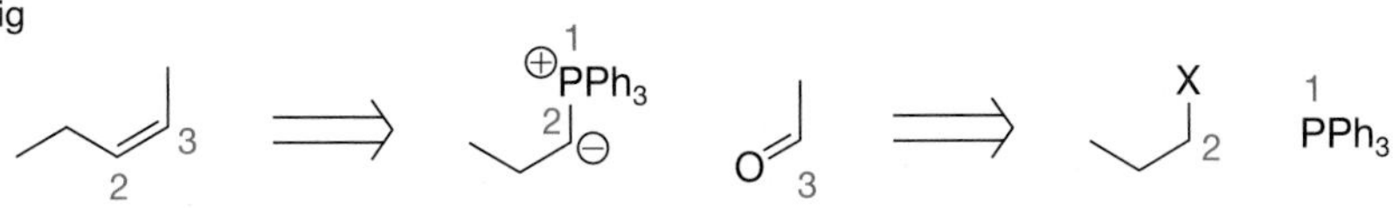

CHECKPOINT 17.14

You should now be able to propose syntheses of molecules involving aldol-like reactions, such as the Claisen condensation, the Mannich reaction, the Thorpe reaction, the Henry reaction, the reactions of enamines and imines, and the Wittig reaction.

SOLVED PROBLEM

Propose a synthesis for the following compound starting with any carbonyl compounds containing no more than seven carbon atoms.

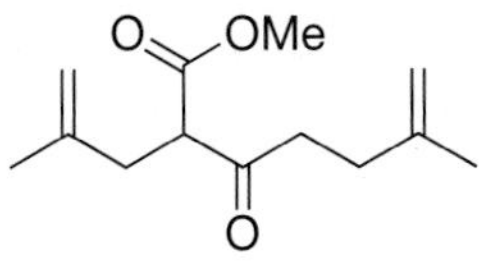

STEP 1: Determine where any major disconnections may be needed. There are 13 carbon atoms in the target molecule, nearly double the number of carbons allowed in the starting material. So the molecule likely needs to be disconnected into two roughly equal-sized synthons.

STEP 2: Identify what functional groups are present, as well as any significant patterns between them. (Refer to Section 17.6.6 for help identifying important relationships.) The desired molecule contains an ester, a ketone, and two alkenes. The 1,3 relationship between the ester and ketone suggests that a Claisen condensation could be used.

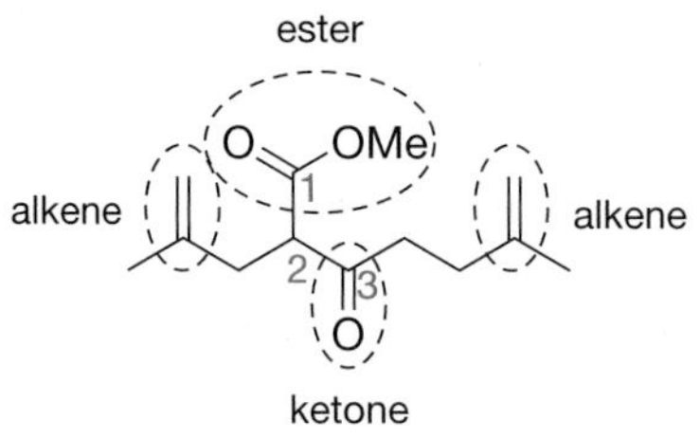

ester

alkene O OMe alkene

1 2 3

O

ketone

STEP 3: Disconnect the molecule between carbons 2 and 3 of the dicarbonyl functional group, and draw the corresponding synthons.

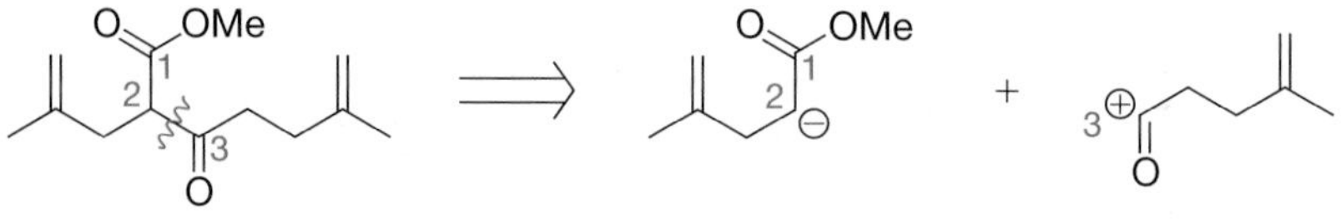

O OMe 1 2 3 O ⟹ O OMe 1 2 ⊖ + 3⊕ O

STEP 4: Identify molecules that correspond to the synthons, and confirm that neither has more than seven carbon atoms.

The nucleophilic synthon is the enolate of the corresponding ester. The electrophilic synthon is equivalent to the same ester that has OMe as the leaving group. Therefore, the reaction is a self-condensation between two molecules of the same ester.

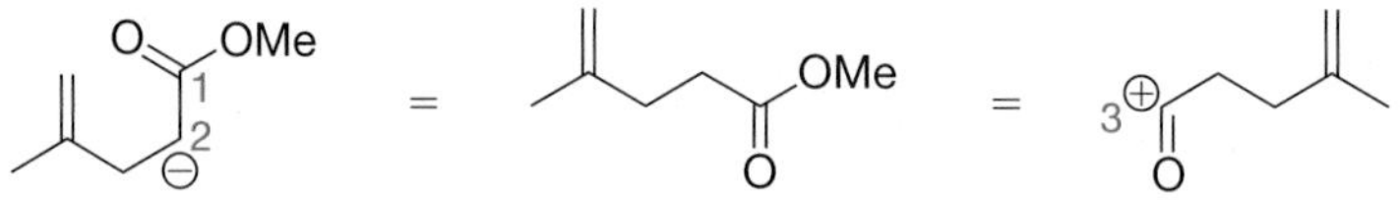

O OMe 1 2 ⊖ = OMe O = 3⊕ O

STEP 5: Write an overall reaction equation for the synthesis.

A Claisen condensation requires a stoichiometric amount of base. To avoid a transesterification reactions, select an alkoxide base that matches the leaving group of the ester: in this case, OMe. To obtain a neutral product, add aqueous acid in a second work-up step.

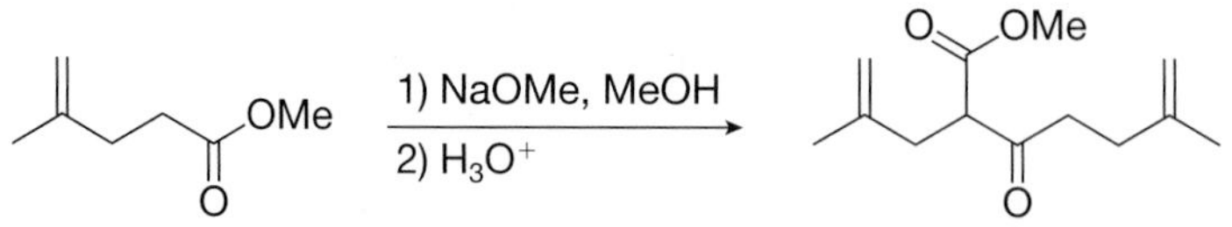

OMe O → 1) NaOMe, MeOH; 2) H_3O^+ → O OMe O

PRACTICE PROBLEM

17.27 a) Suggest reagents that could be used to carry out the following syntheses:

i) 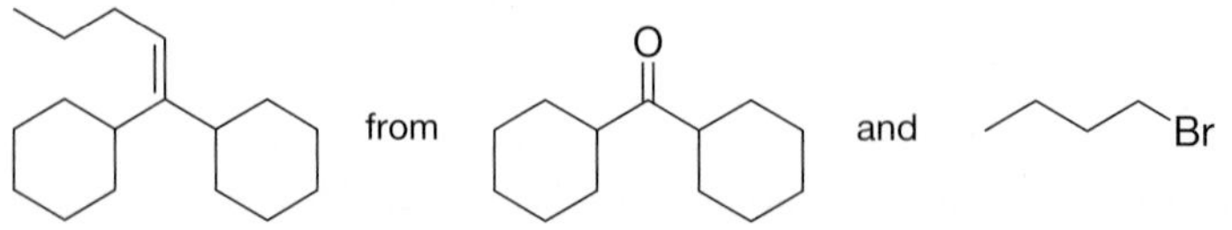 from O and Br

ii) 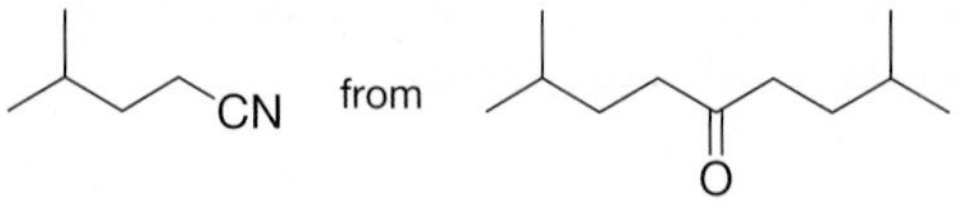 CN from O

iii) 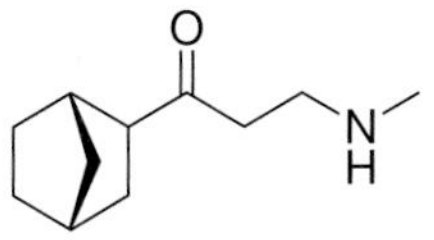 N H from

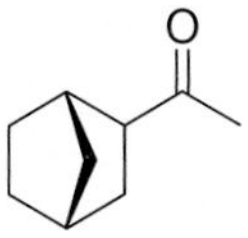

b) Propose a method of synthesizing the following compounds, starting with any carbonyl compound with no more than seven carbon atoms.

i) 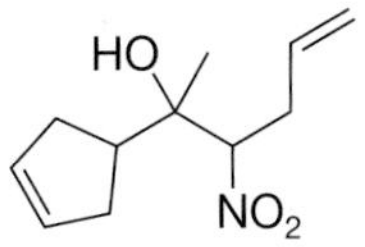 HO NO_2

ii) 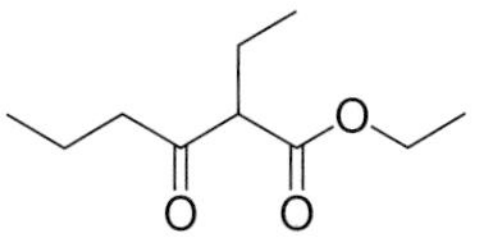 O O O

iii) 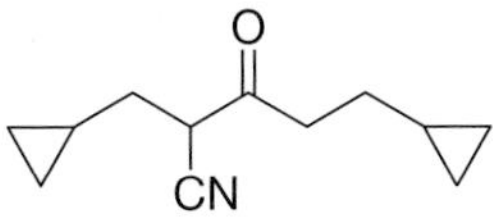 O CN

INTEGRATE THE SKILL

17.28 Propose a method of synthesizing the following compounds, starting with any compounds with no more than seven carbon atoms. More than one step will be required.

a) 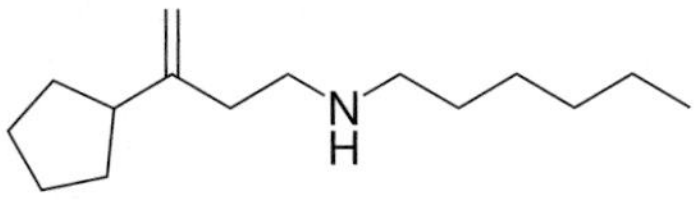 N H

b) 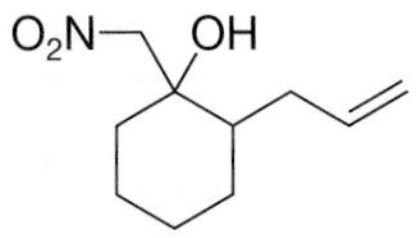 O_2N OH

17.7 1,3-Dicarbonyl Compounds

Dicarbonyl compounds provide an excellent way to control reactivity in aldol and alkylation reactions because the deprotonation of 1,3-dicarbonyls occurs only in one place on the nucleophilic component.

The large family of nucleophiles with electron-withdrawing groups connected in a 1,3 relationship to each other includes combinations of ketones, esters, amides, nitriles, phosphonium salts, and nitro compounds. The two electron-withdrawing groups lower the pK_a of the α-hydrogen between them to such an extent that enolate formation is possible in alcohols or in aqueous conditions using hydroxide, alkoxide, or amine bases.

anion is resonance-stabilized by two carbonyl groups

charge resides mainly on the oxygens

carbon is most nucleophilic site

Three resonance forms are possible for 1,3-dicarbonyl compounds. It is most common to show one of the enolate forms because these forms tend to contribute more to the structure of the anion. Other electron-withdrawing groups—such as nitrile and nitro groups—stabilize adjacent carbanions and form nucleophiles similar to enolates.

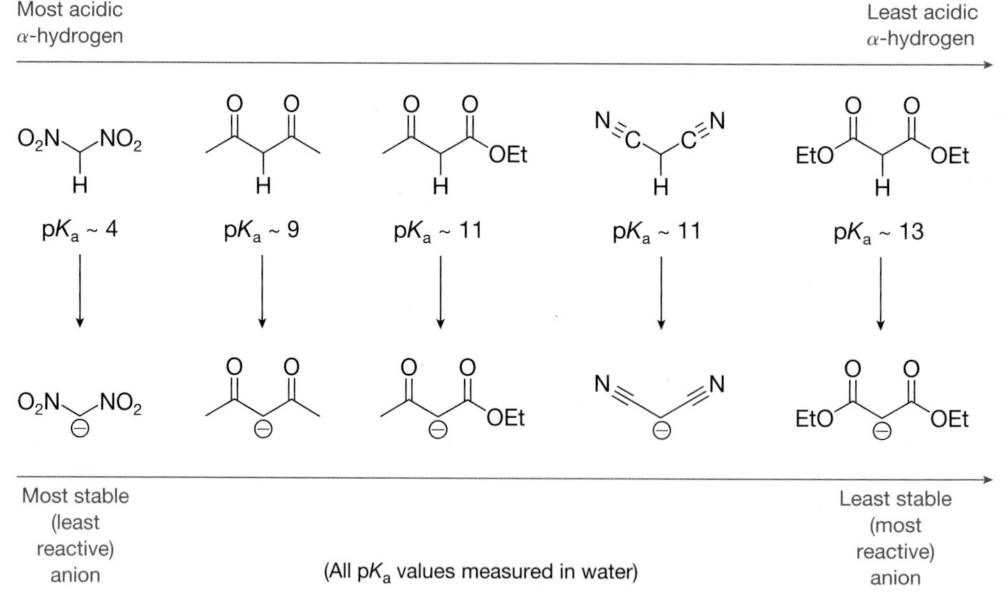

STUDENT TIP

Treating an ester with hydroxide leads to ester hydrolysis that forms the carboxylate. To avoid hydrolysis or transesterification when working with esters, it is important to use an alkoxy base that matches the ester.

17.7.1 Selective alkylations with 1,3-dicarbonyl compounds

1,3-Dicarbonyl enolates act as nucleophiles in S_N2 reactions, and the principles that control S_N2 selectivity in electrophiles apply. Reactive electrophiles such as allylic, benzylic, and primary alkyl halides work best. Secondary halides also react, although competition with E2 reactions sometimes reduces the yields.

It is possible to perform two alkylations at the carbon between the carbonyls in 1,3-dicarbonyl nucleophiles. The second alkylation is usually more difficult than the first, and it often requires a stronger base because the alkyl group from the first addition is electron donating and also hinders the α-carbon.

The resulting change in reactivity is useful because it reduces the chance of multiple addition of the first alkyl group. When two different groups need to be added, it is usually wise to add the smaller group first because this will reduce the steric hindrance during the second addition.

first alkylation

second alkylation uses a stronger base

1) KOEt, EtOH
2) Et–Br

1) KOtBu
2) Bu–Br

CHECKPOINT 17.15

You should now be able to rationalize the increased acidity of 1,3-dicarbonyl compounds using resonance structures and rank the relative acidities of various 1,3-dicarbonyls. You should also be able to predict the products of their reactions with base and alkyl halides.

SOLVED PROBLEM

Predict the product of the following reaction, and draw a mechanism for its formation.

1) NaOMe
2) Br

STEP 1: Identify the roles of the reactants in the first step of the reaction.

The NaOMe could either act as a base or a nucleophile. The starting material contains two ester groups, either of which could act as an electrophile. However, because the esters are in a 1,3 relationship, the α-hydrogens between them are very acidic. Acid–base reactions are generally very fast, so the first step of this reaction is the deprotonation of one of the α-hydrogens by methoxide to give an enolate ion.

STEP 2: Draw a mechanism for the first step of the reaction.

STEP 3: Repeat Steps 1 and 2 for the second step of the reaction.

The enolate ion formed in the first step is a good nucleophile and reacts with the electrophilic alkyl bromide in an alkylation.

PRACTICE PROBLEM

17.29 a) Predict the products of the following reactions.

i) OEt
1) KO*t*-Bu
2) Br

STUDENT TIP

Any reaction with a dicarbonyl compound or equivalent can be done using a weak base. The acetoacetic ester and the malonic ester syntheses that follow were developed before strong organic bases became common. These reactions are useful when there is reason to avoid strongly basic conditions.

ii) CN; O — 1) KO*t*-Bu 2) EtBr

iii) O; O — 1) NaOEt 2) MeBr 3) KO*t*-Bu 4) *n*-PrBr

b) Rank the following structures in order of increasing acidity.

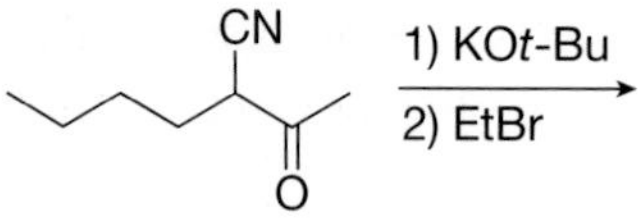

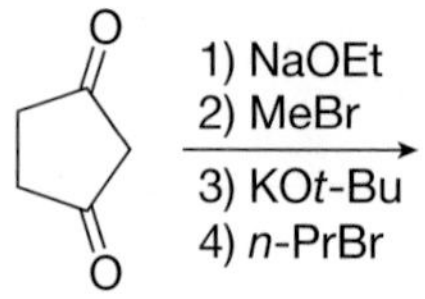

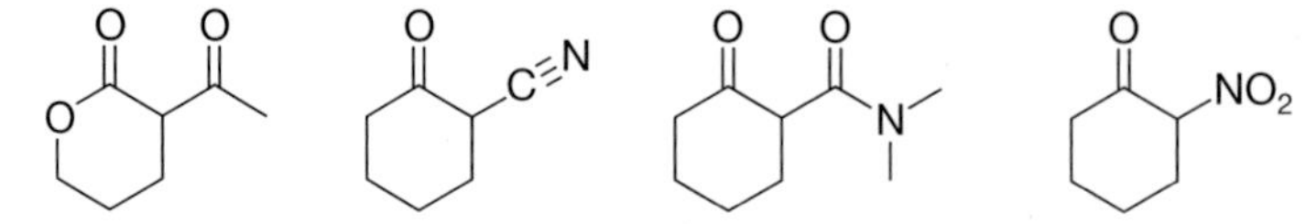

O O O; O C≡N; O O N; O NO_2

INTEGRATE THE SKILL

17.30 a) Predict the product of the following reaction, and draw a mechanism for its formation.

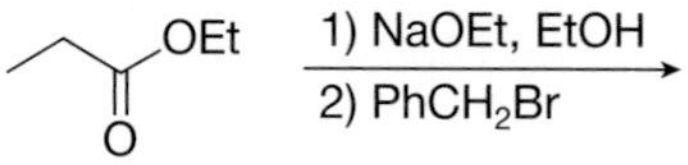

OEt O — 1) NaOEt, EtOH 2) $PhCH_2Br$

b) Consider the following transformation, which does not work satisfactorily. Use a mechanism to explain why the expected product does not form, and suggest an alternative method for its synthesis, starting from a diester.

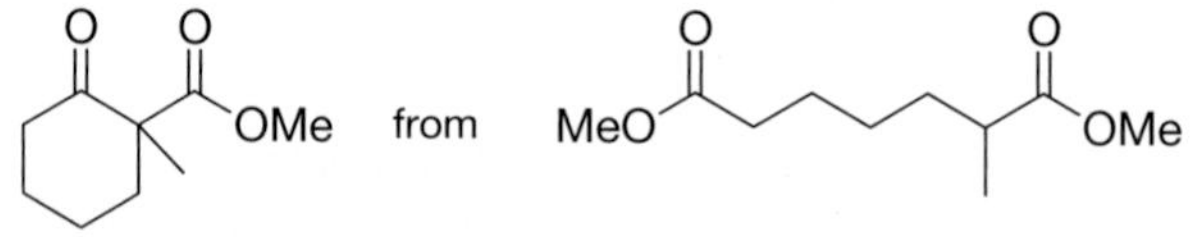

O O OMe from MeO O O OMe

17.7.2 Decarboxylation of dicarbonyl materials

A **decarboxylation** reaction removes carbon dioxide from a β-keto acid.

1,3-Dicarbonyls provide a mild and highly controllable way to alkylate simple carbonyl compounds. Certain dicarbonyl products can be converted into monocarbonyl compounds; hence, these dicarbonyls are *synthetic equivalents* to the monocarbonyls. The advantages of using dicarbonyls include selectivity (only one α-position) and mild reaction conditions (no need for a strong base or ultra-cold temperatures). This method requires an extra step, **decarboxylation**, after the alkylation. The 1,3-dicarbonyl substrate must have at least one group that can be converted to a carboxylic acid.

the ester group provides control for alkylation and easy proton removal
once alkylation is done, the ester is decarboxylated
O O OEt
1) NaOEt
2) Br
O O OEt
1) KOH
2) HCl, heat
O

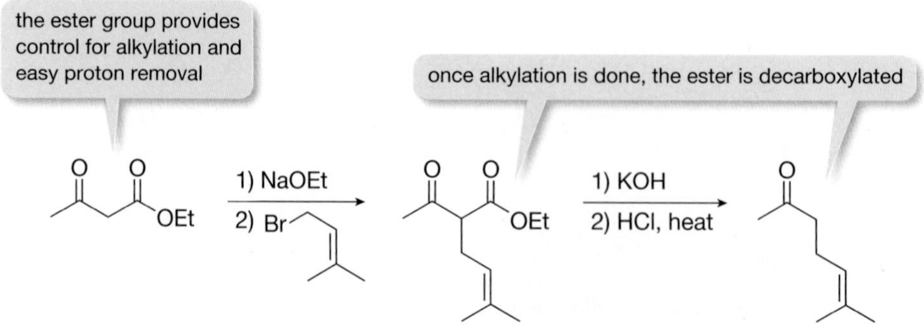

Hydrolysis of the ester group of a β-ketoester produces a β-keto acid. As discussed in Chapter 15, this reaction can be carried out with either an acid or base catalyst (saponification). Gently warming this β-keto acid causes it to lose CO_2 and form the monoketone product. The key to this reaction is the electron-withdrawing ability of the second carbonyl group, which accepts electrons from the CO_2 when it leaves. The reaction is irreversible because the CO_2 gas is lost from the reaction mixture. The resulting enol intermediate quickly tautomerizes to the carbonyl form.

KOH

HCl

H+ from carboxyl group is transferred to ketone

enol tautomerizes to ketone

ketone is converted to enol form

CO_2 gas from carboxyl is released

The most stable conformation of β-keto acids is one in which the hydrogen of the carboxylic acid is hydrogen bonded to the oxygen of the ketone. This conformation allows a transition state with a six-membered ring that eases the transfer of the hydrogen from the acid to the ketone as the CO_2 is lost. The ketone converts to an enol by accepting electrons from the departing carboxylate bond. This enol then quickly tautomerizes into the product ketone.

The combination of alkylation and decarboxylation provides an excellent way to synthesize mono-, di-, and tri-alkylated ketones, nitriles, esters, and nitro compounds. When performed with an acetoacetic ester, the reaction is often called the **acetoacetic ester synthesis**. This synthesis produces substituted acetone derivatives using a weak base. The retrosynthesis of 4-phenyl-2-butanone shows that this compound can be made using an acetone enolate synthon. Starting with acetone would require the use of a strong base, such as LDA. Alternatively, the synthesis can be done using a weaker alkoxide base and a dicarbonyl enolate.

The **acetoacetic ester synthesis** is a traditional synthesis of substituted acetone from acetoacetic ester, using a removable ester as an activating group.

product looks like it was made by S_N2 reaction using acetone enolate nucleophile

synthons

or

+ NaOEt

+ LDA

synthetic equivalents

product may have come from acetoacetic ester and weak base

Direct alkylation of acetone does not work using a weak base. However, the product can be formed by the alkylation, hydrolysis, and decarboxylation of a β-ketoester.

Actual forward synthesis

alkoxide base must match ester alkoxide

base hydrolyzes ester to CO_2H

1) NaOEt 2) $PhCH_2Br$

1) KOH 2) H_3O^+ heat

dicarbonyl increases acidity, prevents equilibrium

enolate is nucleophile in S_N2 reaction

heat and acid causes decarboxylation of β-keto acid

The malonic ester synthesis produces substituted acetic acids or acetate esters from malonic ester.

The **malonic ester synthesis** is a similar reaction that converts malonic esters to mono- and disubstituted acetates. All the steps of either of these ester syntheses can be done in the same reaction flask, so the reaction sequence can be written with a single arrow.

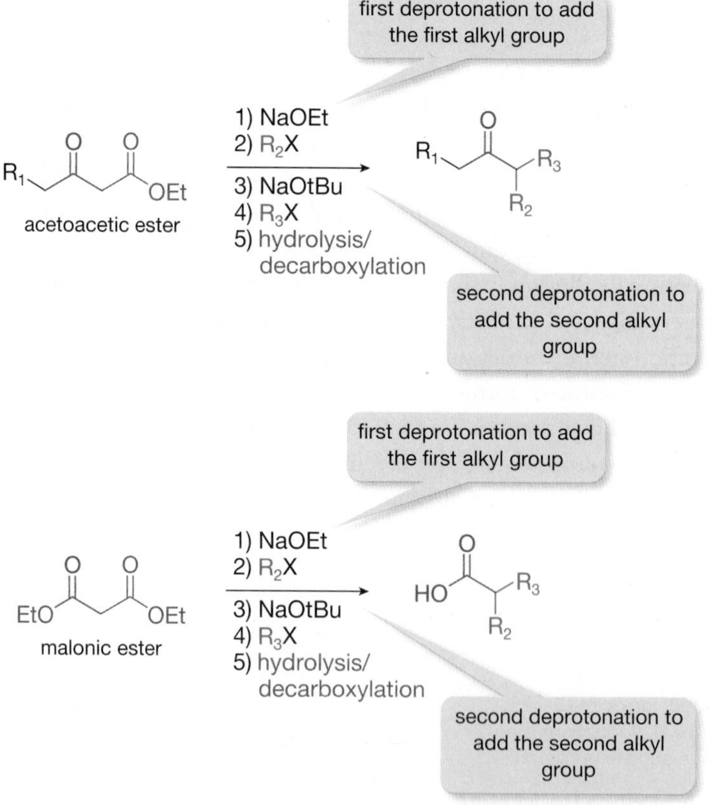

Similar sequences can be used to make nitriles or nitro compounds by employing nitrile or nitro electron-withdrawing groups positioned β to an ester group that can be removed through decarboxylation.

CHECKPOINT 17.16

You should now be able to predict the conditions under which β-ketoesters will decarboxylate, and draw the products and mechanisms of these reactions. You should also be able to combine decarboxylation methods with the reactions of 1,3-dicarbonyls in syntheses such as the acetoacetic ester synthesis and the malonic ester synthesis.

WANT TO LEARN MORE?
17.1 Krapcho cleavage

SOLVED PROBLEM

Predict the products formed after each step in the following multi-step reaction. Draw a mechanism for the final step in the reaction.

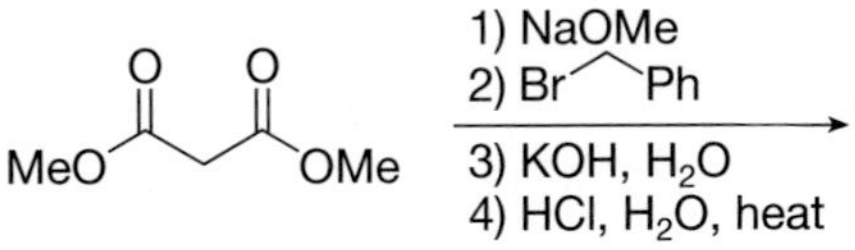

MeO OMe O O
1) NaOMe
2) Br Ph
3) KOH, H_2O
4) HCl, H_2O, heat

STEP 1: Identify the roles of the reactants in the first step of the reaction.

The NaOMe could either act as a base or a nucleophile. The starting material contains two ester groups, either of which could act as an electrophile. However, the esters are in a 1,3 relationship, making the α-hydrogens between them very acidic. Since acid–base reactions are generally very fast, the first step of this reaction is the deprotonation of one of the α-hydrogens to form an enolate ion.

STEP 2: Draw the product of the first step in the reaction.

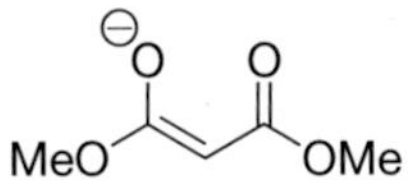

⊖O O
MeO OMe

STEP 3: Identify the roles of the reactants in the second step of the reaction.

The enolate ion is a good nucleophile, which can react with the electrophilic alkyl bromide added in the second step.

STEP 4: Draw the product of the second step in the reaction.

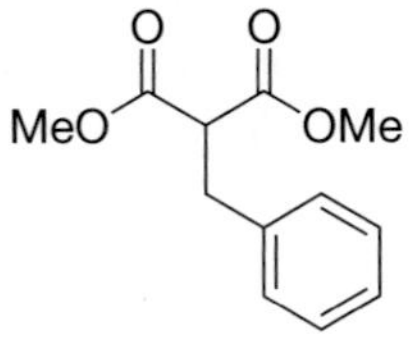

O O
MeO OMe

STEP 5: Identify the roles of the reactants in the third step of the reaction.

The aqueous basic conditions readily hydrolyze the two esters to their corresponding carboxylates.

STEP 6: Draw the product of the third step in the reaction.

Under the basic conditions, the carboxylic acids remain deprotonated.

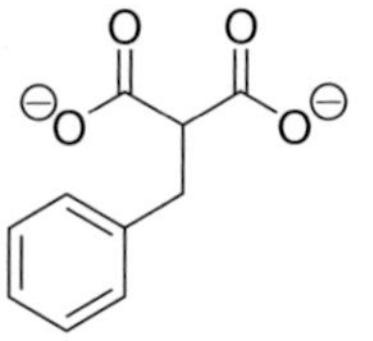

O O
⊖O O⊖

STEP 7: Identify the roles of the reactants in the fourth step of the reaction.

The aqueous acid protonates the two carboxylate groups, and the heat encourages decarboxylation of the 1,3-diacid.

STEP 8: Draw a mechanism for the fourth step of the reaction.

Under the acidic conditions, the carboxylate groups are both protonated.

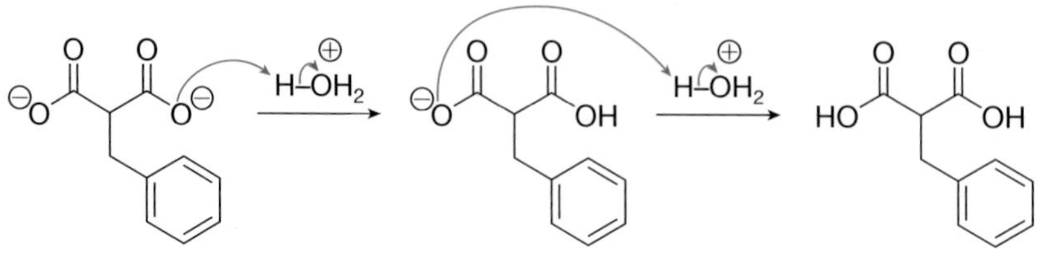

H–OH$_2$ ⊕ → H–OH$_2$ ⊕ →

OH HO OH

Following protonation, the diacid decarboxylates, thereby expelling carbon dioxide gas and forming an enol.

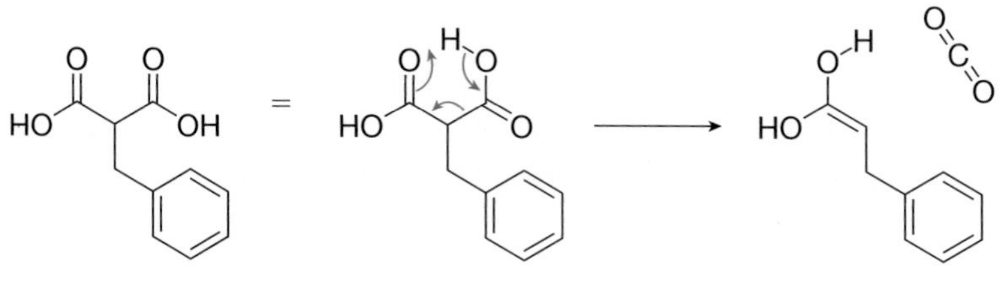

HO OH = HO → HO O–H O=C=O

The enol then tautomerizes to the carboxylic acid to give the final product.

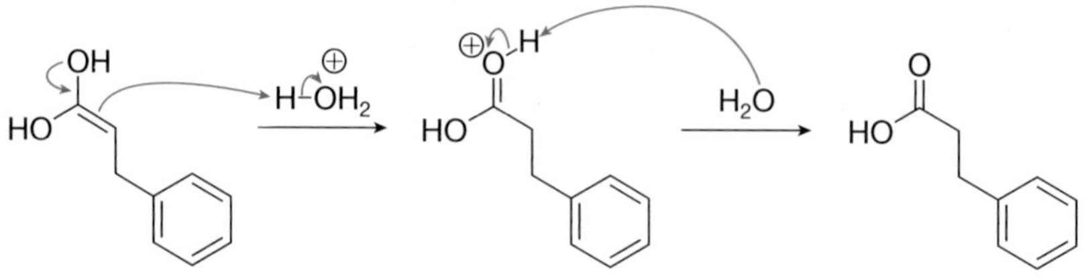

OH HO H–OH$_2$ ⊕ → HO H_2O → HO

The overall reaction is therefore as follows:

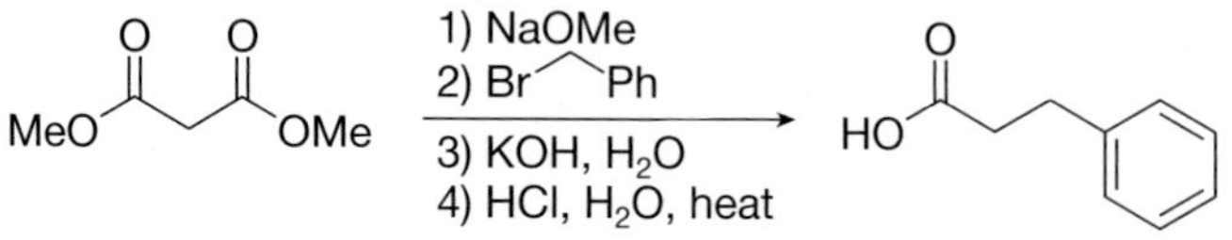

MeO OMe 1) NaOMe 2) Br Ph 3) KOH, H_2O 4) HCl, H_2O, heat → HO

PRACTICE PROBLEM

17.31 Draw the final product in the following reaction sequences. Draw a mechanism for each step of the reactions of parts (a), (b), and (c).

a) 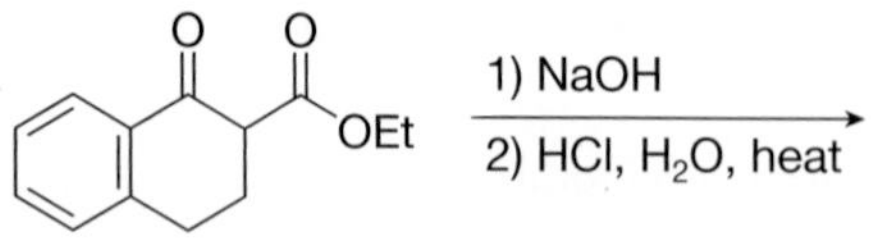 OEt 1) NaOH 2) HCl, H_2O, heat →

b) 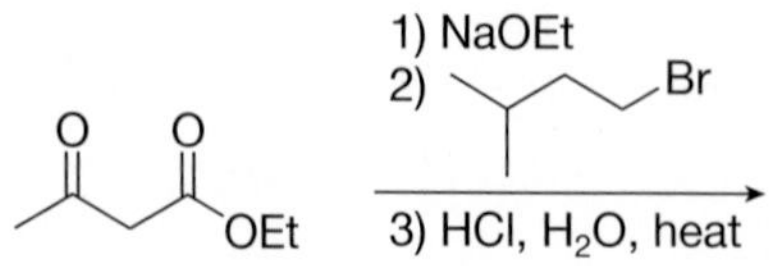 OEt 1) NaOEt 2) Br 3) HCl, H_2O, heat →

c) 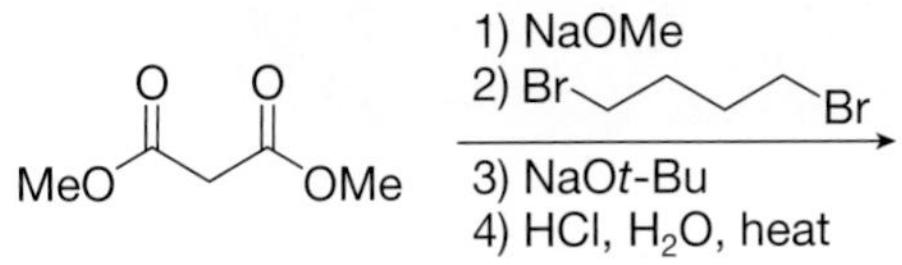 MeO–C(=O)–CH₂–C(=O)–OMe

1) NaOMe
2) Br–(CH₂)₄–Br
3) NaO*t*-Bu
4) HCl, H_2O, heat

d) 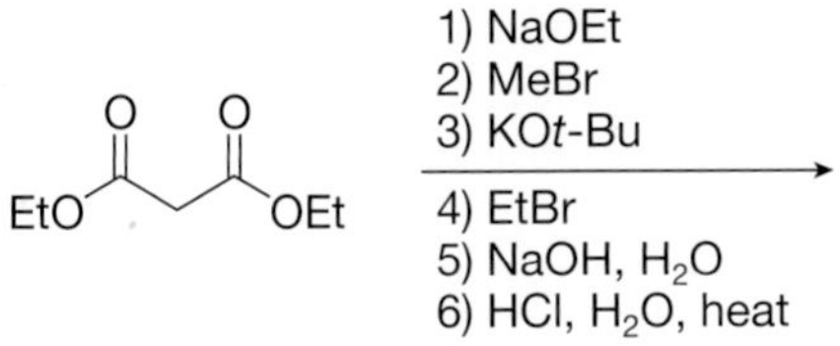 EtO–C(=O)–CH₂–C(=O)–OEt

1) NaOEt
2) MeBr
3) KO*t*-Bu
4) EtBr
5) NaOH, H_2O
6) HCl, H_2O, heat

INTEGRATE THE SKILL

17.32 a) What is the error in the following mechanism?

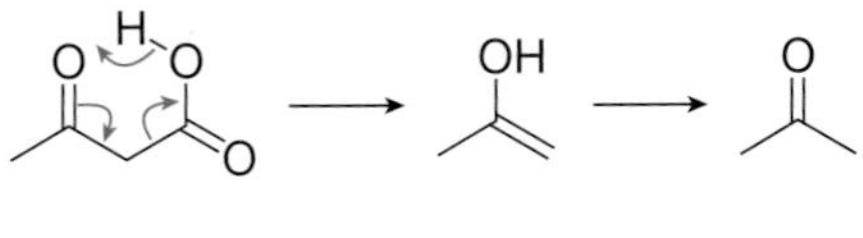

b) Only one of the two following compounds can be prepared using malonic ester synthesis. Which compound cannot be prepared by this method? Explain why.

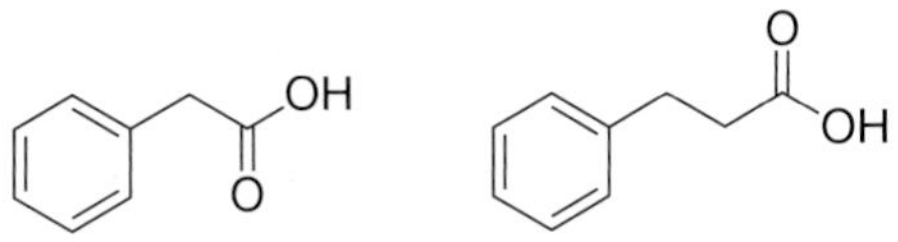

A **B**

17.7.3 Condensation of dicarbonyls with aldehydes: The Knoevenagel condensation

Dicarbonyl enolates react very quickly with aldehydes under mildly basic conditions, a process known as the **Knoevenagel condensation**. Because the α-hydrogens of dicarbonyl compounds are so acidic, weak bases such as pyrrolidine (pK_a of 11) are often used for this reaction. Knoevenagel condensations normally produce α,β-unsaturated products directly due to the acidity of the α-hydrogen that makes the E1$_{cb}$ elimination extremely fast. It is unusual to be able to isolate aldol products from these reactions.

The **Knoevenagel condensation** transforms a dicarbonyl compound to an α,β-unsaturated compound by means of addition to an aldehyde or ketone, followed by dehydration.

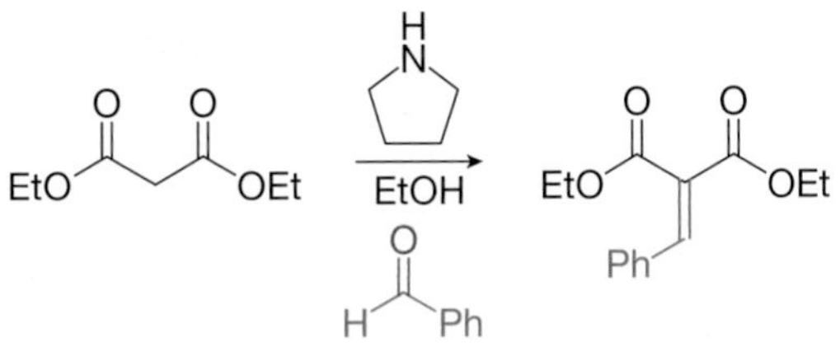

EtO–C(=O)–CH₂–C(=O)–OEt → (pyrrolidine, EtOH, PhCHO) → EtO–C(=O)–C(=CHPh)–C(=O)–OEt

The α,β-unsaturated products can be decarboxylated to produce the corresponding mono-carbonyl compounds.

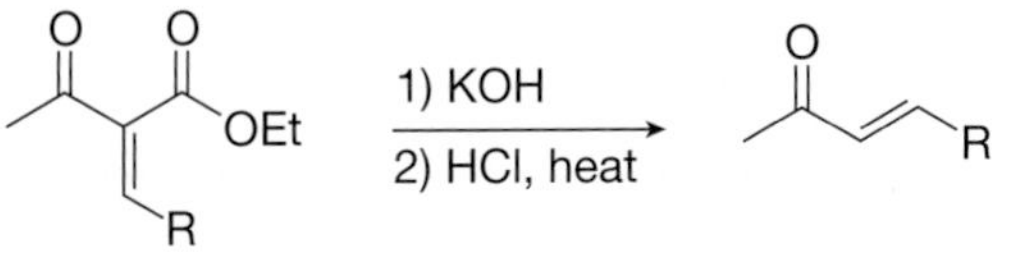

1) KOH
2) HCl, heat

CHEMISTRY: EVERYTHING AND EVERYWHERE

Sweet Grass and Blood Clots

The intramolecular Knoevenagel condensation that forms coumarin is an example of a condensation that forms a ring. Under basic conditions, salicylaldehyde undergoes transesterification with Meldrum's acid to form an enol acid. This acid condenses intramolecularly with the aldehyde and then decarboxylates to form coumarin.

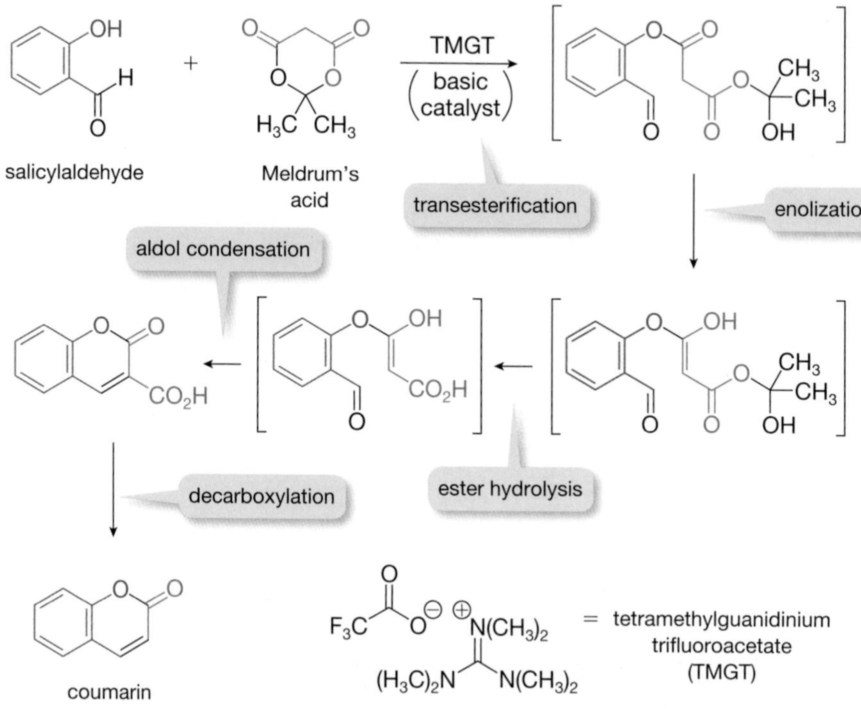

salicylaldehyde + Meldrum's acid → TMGT (basic catalyst) [transesterification] → enolization → ester hydrolysis → aldol condensation → decarboxylation → coumarin

F_3C–CO_2^- $^+$$N(CH_3)_2$=C(N(CH_3)_2)_2 = tetramethylguanidinium trifluoroacetate (TMGT)

Coumarin is an important natural structure found in many plants. It gives sweet grass and sweet clover a pleasant fragrance, and is used in perfumes. Coumarins that have a bulky substituent at carbon 3 and a hydroxy group at carbon 4, such as dicoumarol or warfarin, inhibit blood clotting. Warfarin is prescribed in carefully controlled doses to prevent strokes, heart attacks, and blood clots. It is also used as a rat poison because higher doses can cause internal bleeding.

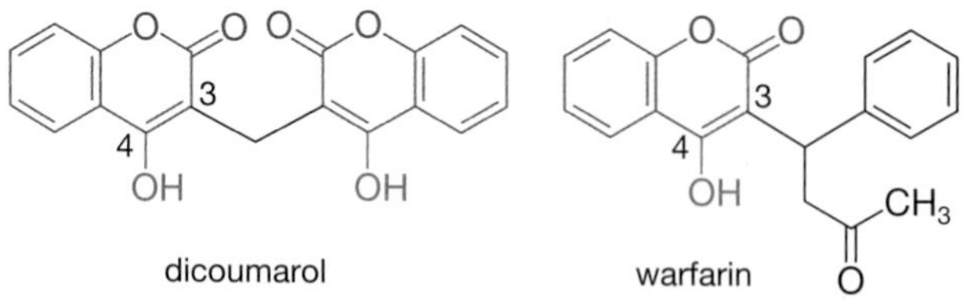

dicoumarol warfarin

CHECKPOINT 17.17

You should now be able to predict the products of a Knoevenagel reaction, and draw mechanisms for their formation.

SOLVED PROBLEM

Predict the product of the following reaction, and draw a mechanism showing its formation.

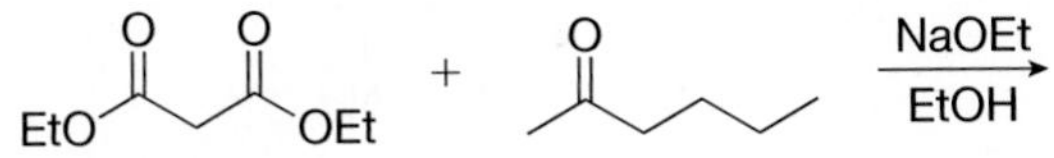

EtO(C=O)CH₂(C=O)OEt + 2-hexanone $\xrightarrow[\text{EtOH}]{\text{NaOEt}}$

STEP 1: Identify the roles of the reactants.

The starting materials are electrophiles with electron-poor carbonyl groups; however, diethyl malonate is also very acidic at its α-hydrogens. In the presence of the ethoxide base, an acid–base reaction happens first, creating a nucleophilic enolate, which can then react with the electrophilic ketone. Ethanol is the solvent.

STEP 2: Draw a mechanism for the reaction. The mechanism begins with a rapid deprotonation of the malonate by ethoxide to give an enolate. The nucleophilic enolate then adds to the electrophilic carbon of the ketone, thereby forming a new C–C bond while simultaneously breaking the C–O π bond.

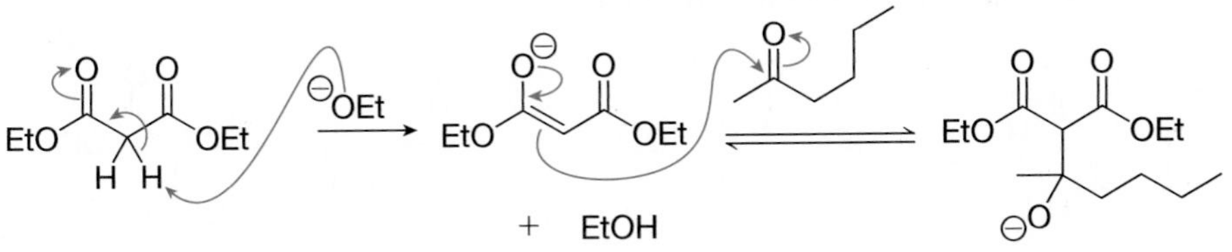

EtO OEt ⊖OEt → EtO OEt + EtOH ⇌ EtO OEt ⊖O

A series of proton transfers neutralizes the alkoxide ion and re-forms the enolate ion. In the protic solvent, this enolate ion readily eliminates a molecule of water in an $E1_{cb}$ reaction to give the final condensation product.

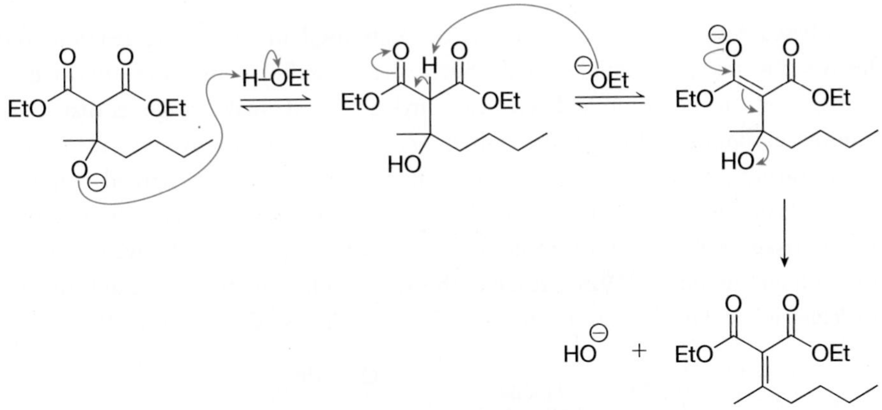

EtO OEt ⊖O ⇌ H–OEt ⇌ EtO OEt HO ⇌ ⊖OEt ⇌ EtO OEt HO → HO⊖ + EtO OEt

The overall reaction is as follows:

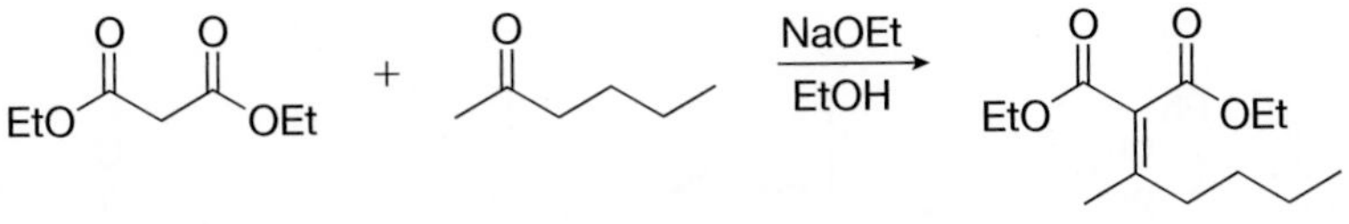

EtO OEt + 2-hexanone $\xrightarrow[\text{EtOH}]{\text{NaOEt}}$ EtO OEt

PRACTICE PROBLEM

17.33 Predict the products of the following reactions, and draw mechanisms for their formation.

a) 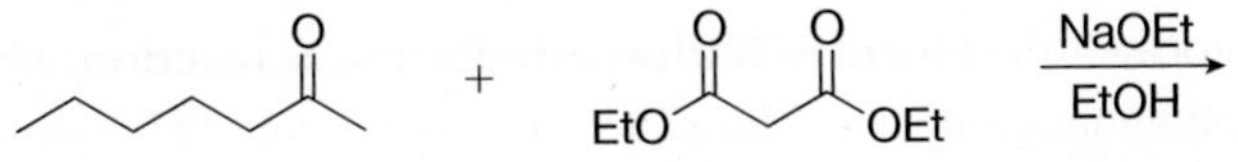 2-heptanone + EtO(C=O)CH₂(C=O)OEt $\xrightarrow[\text{EtOH}]{\text{NaOEt}}$

b) 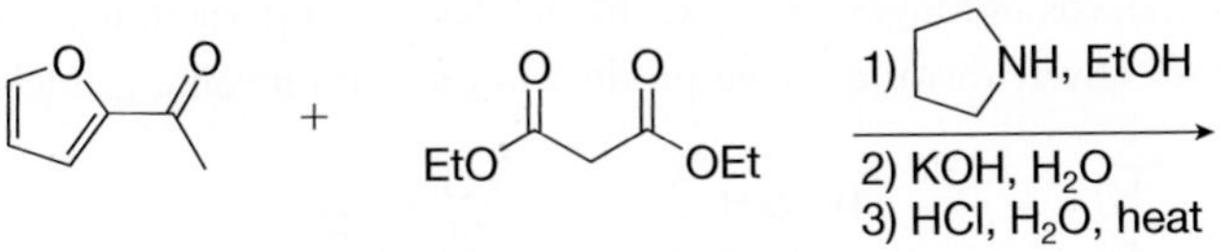 2-acetylfuran + EtO(C=O)CH₂(C=O)OEt $\xrightarrow{}$ 1) pyrrolidine (NH), EtOH; 2) KOH, H_2O; 3) HCl, H_2O, heat

c) 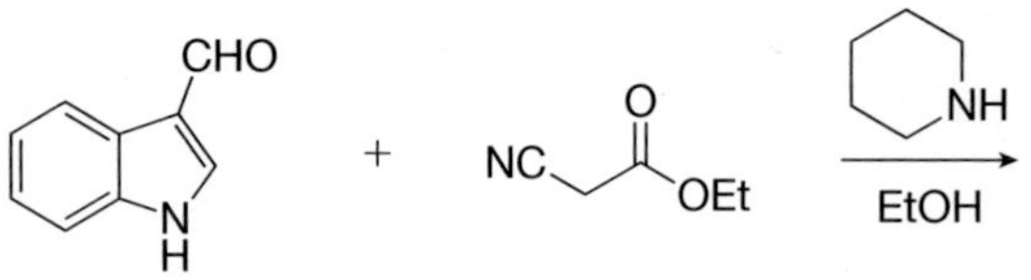 CHO, N H + NC OEt O → (piperidine NH) EtOH

d) 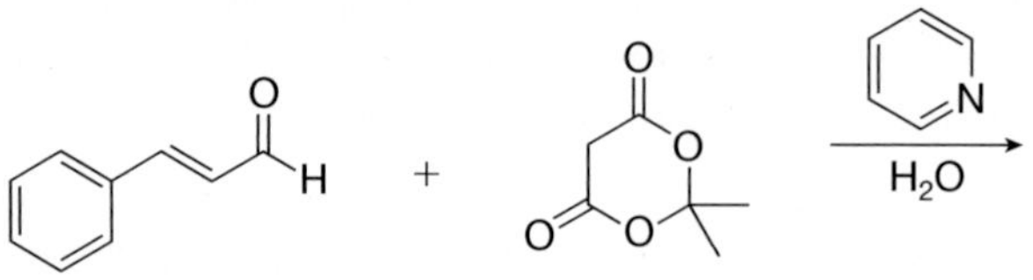 O H + O O O O → (pyridine N) H_2O

INTEGRATE THE SKILL

17.34 Propose a mechanism for the following reaction.

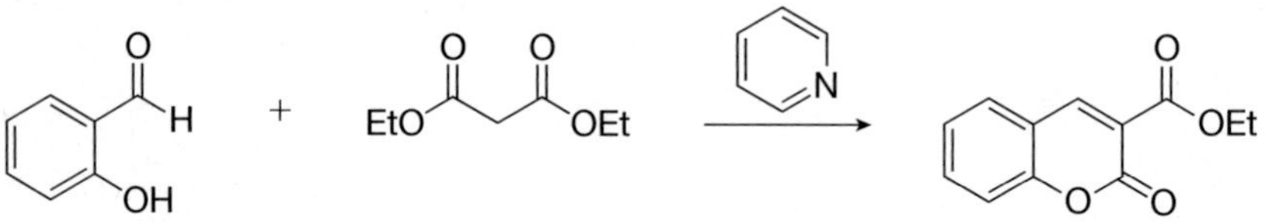

O H OH + EtO O O OEt → (pyridine N) O OEt O O

17.7.4 Stabilized ylides and the Horner–Wadsworth–Emmons reaction

Phosphorus-based functional groups such as the salts used in the Wittig reaction provide a very effective way to make double bonds. Since the phosphorus in these groups is electron withdrawing, they can be combined with carbonyl groups to make reagents that behave like 1,3-dicarbonyl compounds. In fact, the α-hydrogens of phosphonium salts are so acidic that the salts are usually deprotonated, and the reagent is stored as the uncharged conjugate base. Many such reagents are commercially available. When reacted with an aldehyde or ketone, these reagents form oxaphosphetane in the same way as in the Wittig reaction. However, in contrast to a reaction with an unstabilized Wittig reagent, this reaction favours the *anti*-oxaphosphetane and leads to a *trans*-alkene. The reasons for this selectivity are discussed in Chapter 18.

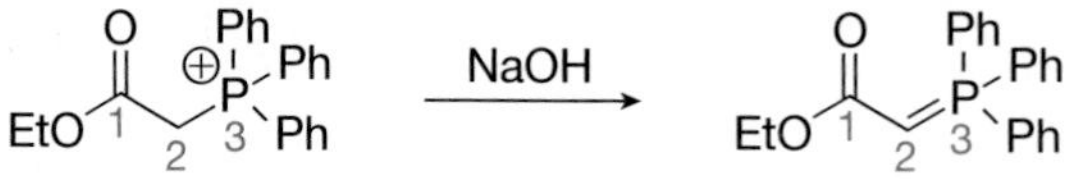

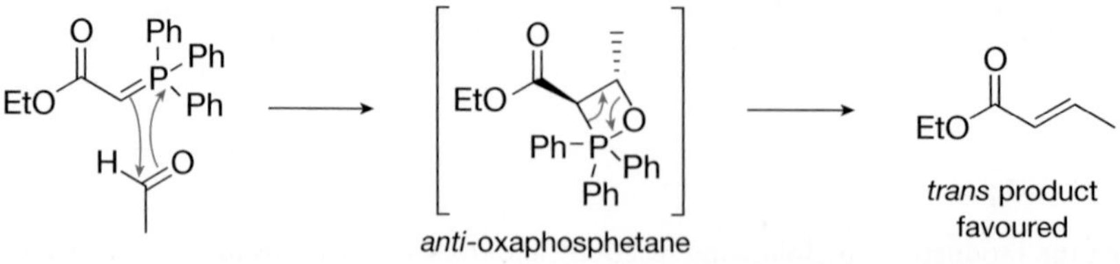

EtO 1 2 3 O ⊕P Ph Ph Ph —NaOH→ EtO 1 2 3 O P Ph Ph Ph

EtO O P Ph Ph Ph + H O → [EtO O O Ph–P Ph Ph] → EtO O

anti-oxaphosphetane

trans product favoured

The **Horner–Wadsworth–Emmons reaction** forms alkenes from aldehydes with *trans* selectivity by using a stabilized enolate equivalent.

A popular variation of this process is the **Horner–Wadsworth–Emmons reaction**, which uses a phosphite in place of the phosphonium group. The phosphite is not as electron withdrawing as the phosphonium salt, so this reaction needs a stronger base. Like the stabilized Wittig reaction, the Horner–Wadsworth–Emmons reaction selectively forms the *trans* product in reactions involving aldehydes.

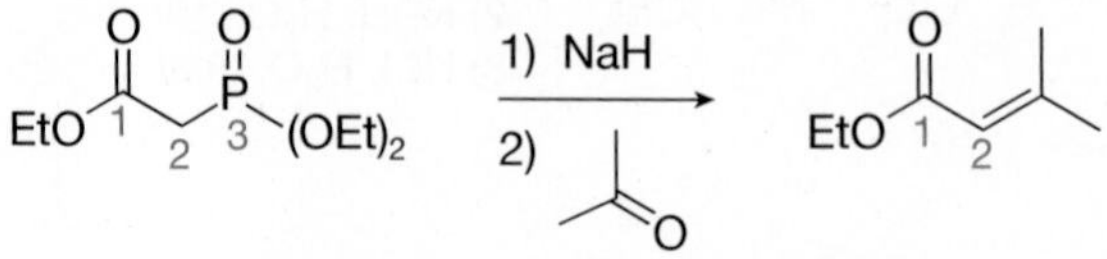

EtO 1 2 3 O O P $(OEt)_2$ —1) NaH 2) O→ EtO 1 2 O

CHECKPOINT 17.18

You should now be able to predict the products of a Wittig reaction using stabilized ylides, including the Horner–Wadsworth–Emmons reaction. You should also be able to draw mechanisms for these reactions.

SOLVED PROBLEM

Predict the major product of the following reaction, and draw a mechanism for its formation.

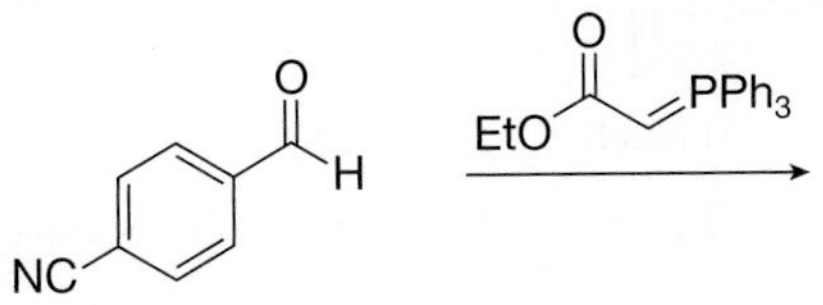

STEP 1: Determine the roles of the reactants.

The reagent is a stabilized ylide, which reacts with electrophilic carbonyls, such as the aldehyde found in the starting material.

STEP 2: Draw the resonance structure of the stabilized ylide, where the C=P π bond is broken and the formal charges are visible.

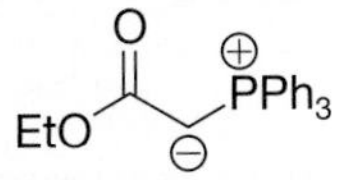

STEP 3: Align the substrates so that the partial charges on the carbonyl face opposite charges on the ylide.

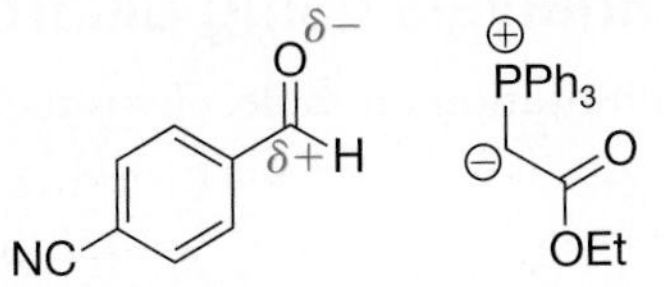

STEP 4: Draw a mechanism for the reaction.

Since stabilized ylides favour the formation of *E* isomers, draw the oxaphosphetane intermediate and the final product with the largest groups *trans* to each other.

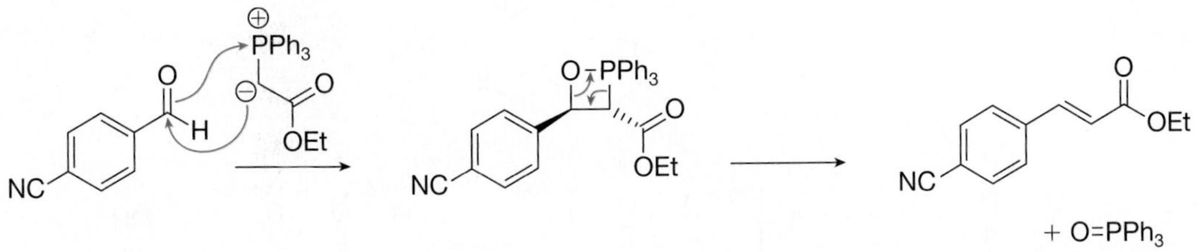

+ O=PPh$_3$

PRACTICE PROBLEM

17.35 Predict the major product in the following reactions, and draw a mechanism for its formation.

a)

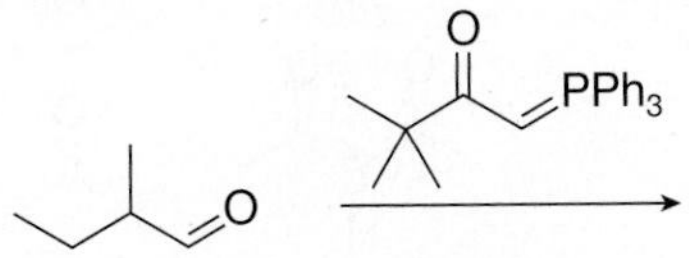

b)

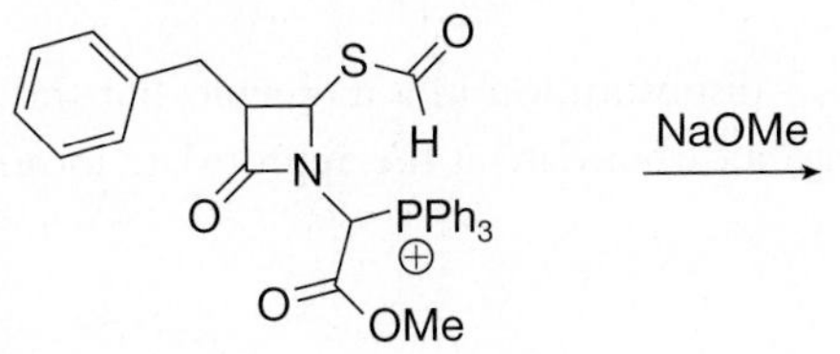

NaOMe

c) 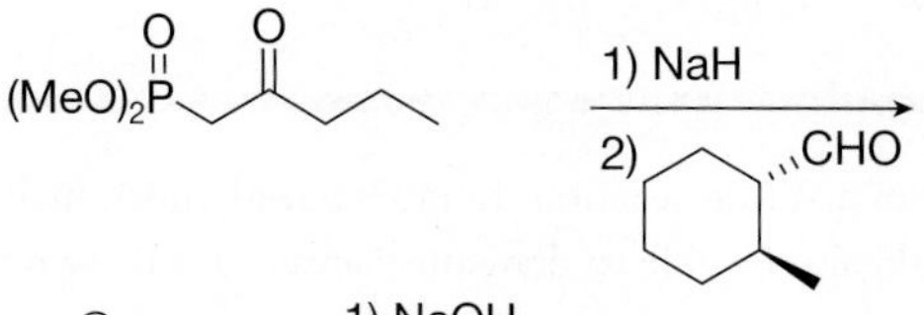 $(MeO)_2P$ ... 1) NaH 2) CHO

d) 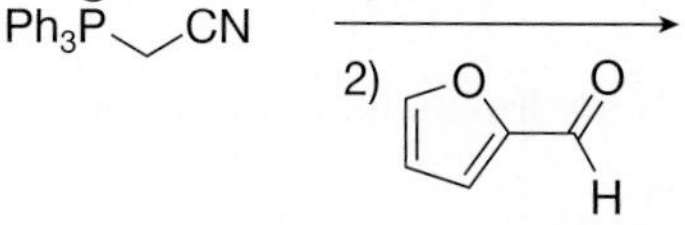 $Ph_3\overset{\oplus}{P}$ CN 1) NaOH 2) O, O, H

e) 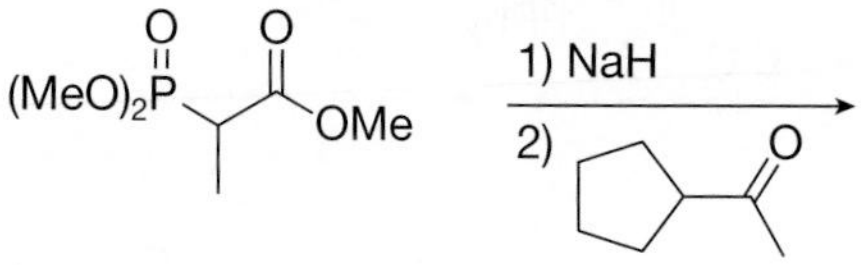 $(MeO)_2P$ ... OMe 1) NaH 2) O

INTEGRATE THE SKILL

17.36 Suggest three different methods for carrying out the following transformation using reactions from this chapter. More than one step may be necessary.

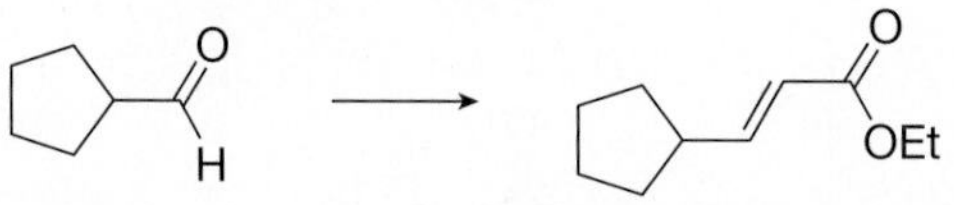

O, H ⟶ O, OEt

17.7.5 Retrosynthetic analysis using dicarbonyl compounds

The 1,3 pattern of heteroatom substitution on molecules is the key identifier for 1,3-dicarbonyl disconnections.

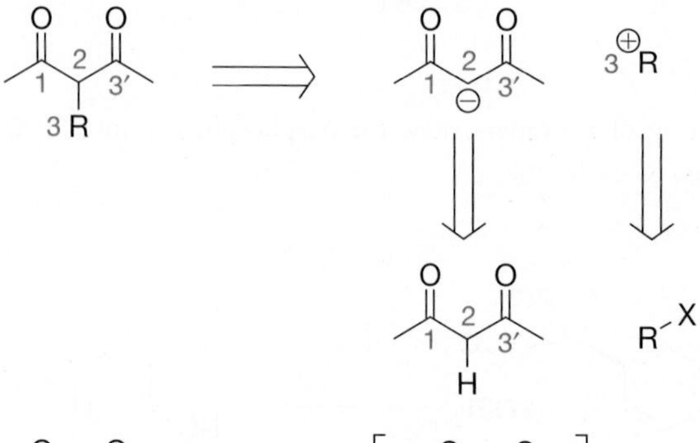

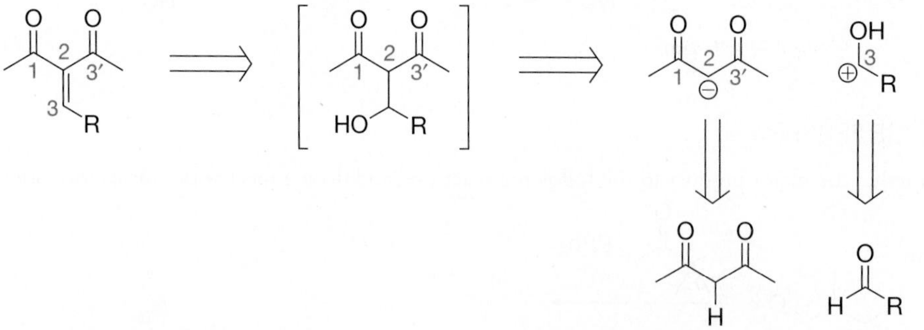

O O 2 1 3′ 3 R ⟹ O O 2 1 ⊖ 3′ $3\overset{\oplus}{R}$

O O 2 1 3′ H R^{X}

O O 2 1 3′ 3 R ⟹ [O O 2 1 3′ HO R] ⟹ O O 2 1 ⊖ 3′ OH 3 ⊕ R

O O H O H R

The chemistry can be applied to existing 1,3-disubstitution in a molecule, but the decarboxylation reaction provides selectivity by "adding" a carboxylate at the appropriate location.

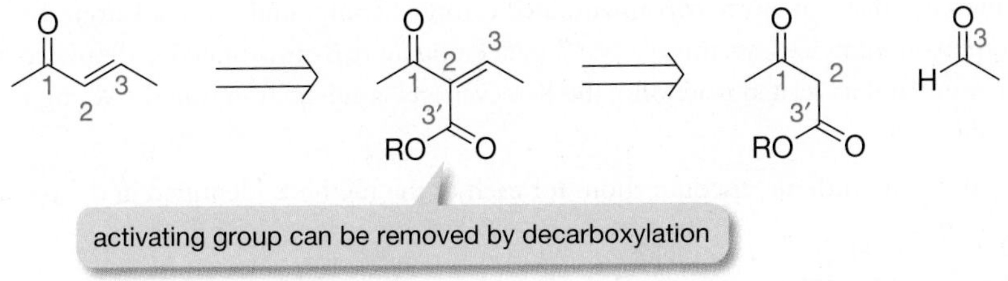

activating group can be removed by decarboxylation

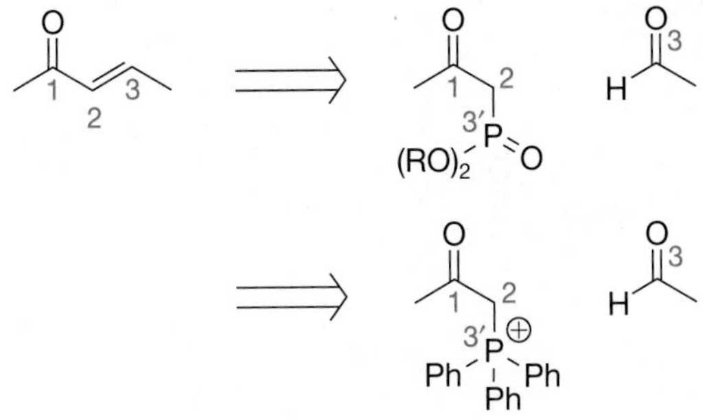

Decarboxylation makes the retrosynthesis appear more complex, but once you identify the correct synthon for each disconnection, you can choose an appropriate dicarbonyl synthetic equivalent.

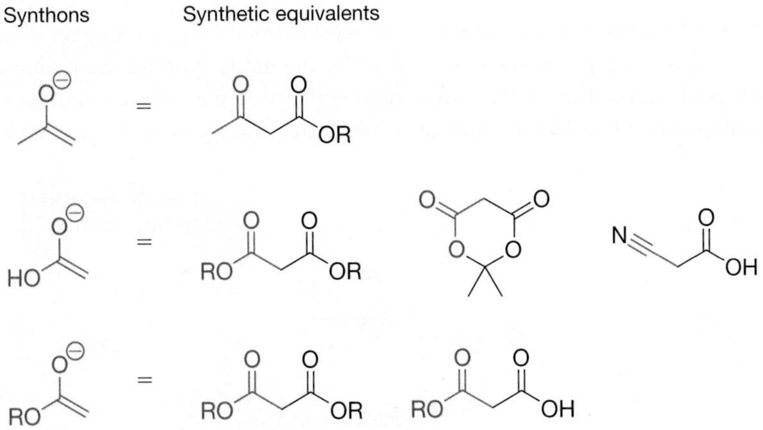

Synthons Synthetic equivalents

CHECKPOINT 17.19

You should now be able to propose syntheses of molecules involving the reactions of 1,3-dicarbonyls and related compounds.

SOLVED PROBLEM

Propose a method for the following synthesis using as few steps as possible.

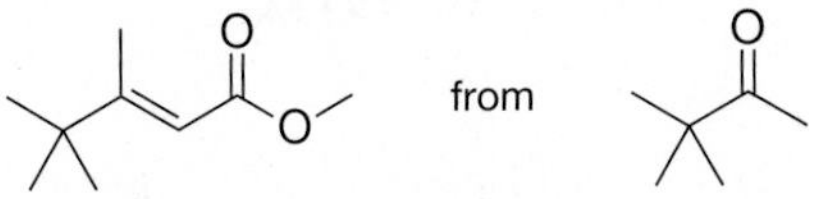

from

STEP 1: Identify the available functional groups and any significant relationships among them.

The product has an ester and an alkene in a 1,3-relationship. Together, they make up an α,β-unsaturated carbonyl.

STEP 2: List methods that can create α,β-unsaturated carbonyl compounds from a ketone.

At this point, you know at least three ways of generating an α,β-unsaturated carbonyl compound: the aldol condensation (and its related reactions), the Knoevenagel condensation, and the Wittig reaction using a stabilized ylide.

STEP 3: Draw the retrosynthetic disconnections for each of the methods identified in the previous step.

$-H_2O$

aldol-like

$-CO_2$

Wittig

Ph_3P

Knoevenagel

STEP 4: Determine which disconnection would be the most efficient for generating the desired product.

The aldol-like disconnection would require generating an enolate from the ester component using a strong base, such as LDA, and then reacting the enolate with a very hindered ketone. Since this addition reaction is reversible (via a retro-aldol reaction), it is unlikely to yield much of the desired product.

sterically hindered tertiary alkoxide

OMe

reversible reaction

The Knoevenagel method also involves an enolate reacting with a hindered ketone, but the reaction is in equilibrium and is driven to completion by the irreversible $E1_{cb}$ elimination of water in the last step. Therefore, this reaction would work reasonably well; however, additional steps are needed to generate the final product via decarboxylation. This process is complicated by the presence of the C=C bond.

CO_2Me

irreversible elimination

?

The third option is a Wittig reaction using a stabilized ylide. Stabilized ylides are selective for the *E* isomer, so the reaction would give the desired product as the major isomer. Since the reaction has just one step, it is an efficient method of generating the target compound.

STEP 5: Draw the overall synthesis of the target compound.

Ph$_3$P

PRACTICE PROBLEM

17.37 Suggest reagents that could be used to carry out the following transformations in as few steps as possible.

a) EtO, O, CN ? → CN

b) O, H ? → O, OMe

c) O O OMe ? → O, OMe

d) O O OEt ? → O

INTEGRATE THE SKILL

17.38 Propose two different syntheses for each of the following structures, using any starting materials with no more than six carbon atoms. Use reactions described in this chapter for at least one synthesis for each compound.

a) 4-methylheptanoic acid

b) O, OEt

c) O, O

17.8 Patterns in Enolate Chemistry

The α-hydrogen of a carbonyl group is relatively acidic and can be removed with a strong base because conjugation enables the carbonyl to act as an electron-withdrawing group. The oxygen of the carbonyl is electronegative, and negative charges tend to reside on this atom; this makes the enolate the predominant resonance form. Removal of the α-hydrogen forms an enolate, which makes the α-hydrogen of the carbonyl group nucleophilic. Nitriles, nitro compounds, and some heterocycles can serve as electron-withdrawing groups in much the same way.

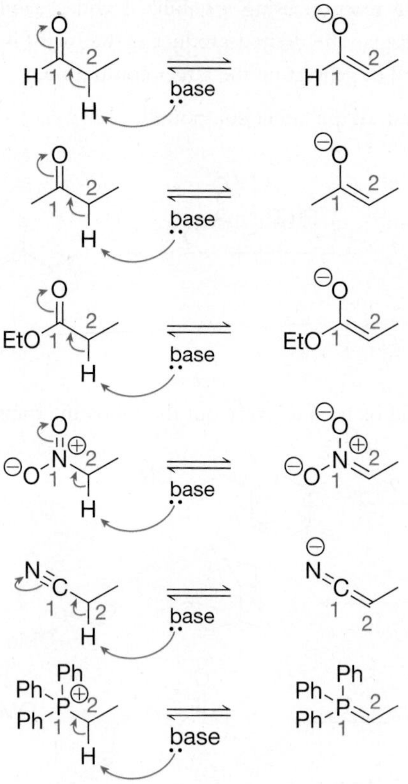

Enolization is much easier if two electron-withdrawing groups are arranged in a 1,3 pattern, which greatly enhances the acidity of the α-hydrogen between them.

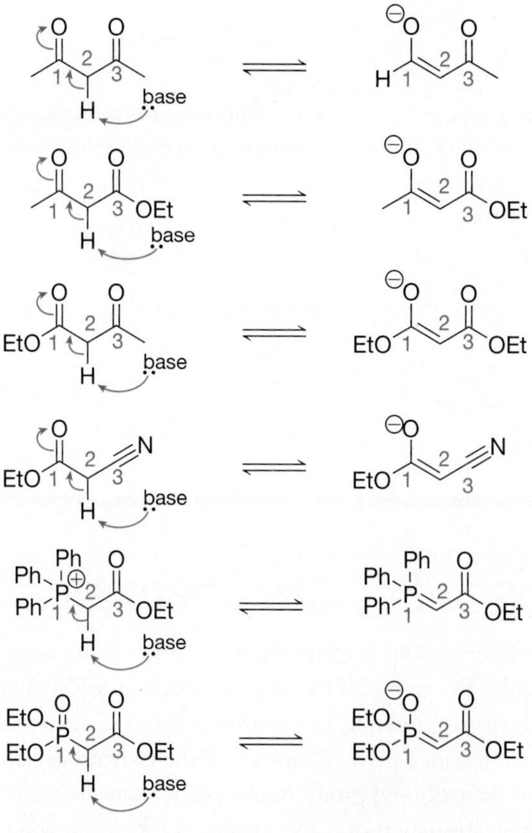

Enolates react with a variety of electrophiles. With alkyl halides they form new carbon-carbon bonds, whereas reactions with sources of electrophilic halogen form carbon-halogen bonds. The reactions can be performed with weak bases, but usually work better with strong bases that completely convert the carbonyl to an enolate. For enolate formation, oxygen bases such as hydroxide or alkoxide are weak because they cannot fully deprotonate monocarbonyl compounds. Compared to enolate formation using oxygen bases, the acidity of the α-hydrogen in 1,3-dicarbonyl compounds is much greater. This property facilitates most enolate chemistry.

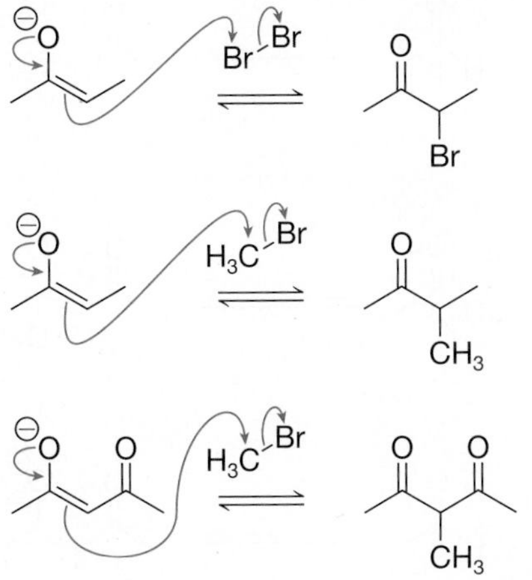

The use of acid in these reactions adds some mechanistic steps, but the fundamental flow of electrons is the same. As in other reactions, bases activate nucleophiles, and acids activate electrophiles.

Base

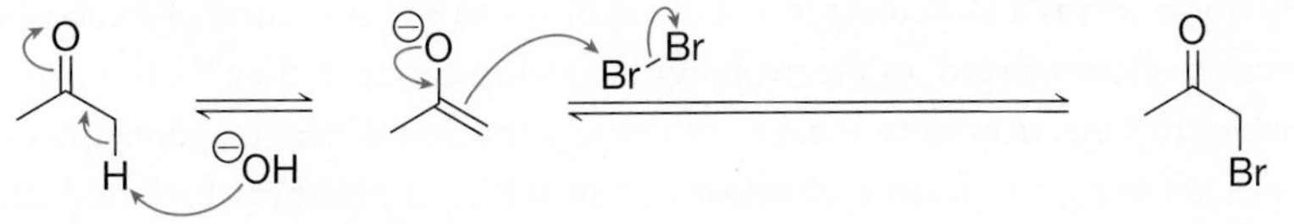

Acid

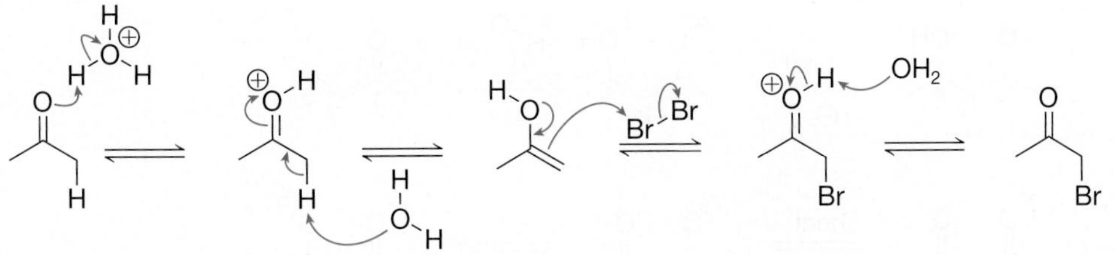

Aldol reactions use carbonyls as electrophiles. The products of these reactions have a 1,3 relationship between the original carbonyl group (carbon 1) and the electrophilic carbon (carbon 3) of the electrophile.

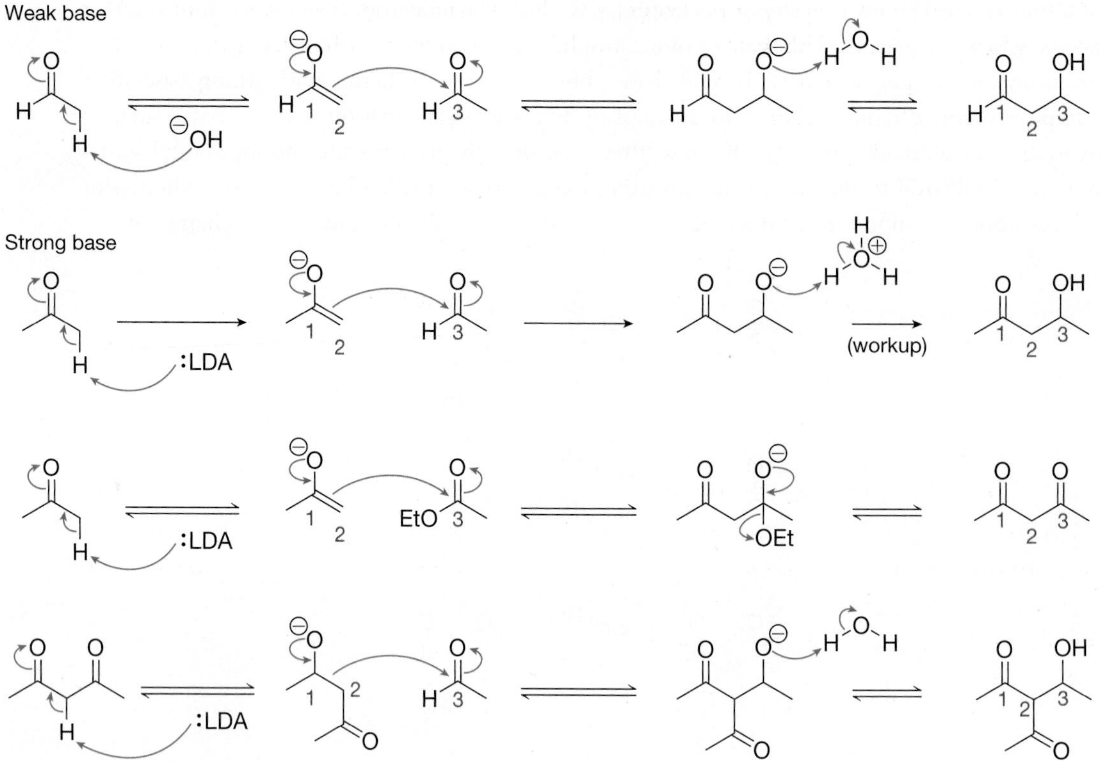

Weak base

Strong base

(workup)

:LDA

EtO

OEt

β-elimination forms an α,β-unsaturated carbonyl product when the aldol has an easily removable α-hydrogen and a suitable leaving group. The Wittig reaction has a similar pattern of electron movement, but the elimination removes a group rather than an H^+. The Claisen condensation forms a β-ketoester, which enolizes in the presence of excess base. In this enolization, electrons flow toward an electron-withdrawing group rather than a leaving group; conjugation enables this electron flow. Reactions using stabilized Wittig reagents or the Horner–Wadsworth–Emmons sequence feature eliminations in which a phosphorus-based group takes the role of a proton. Decarboxylation also involves electron flow toward an electron-withdrawing group.

heat

heat

Δ

CO_2

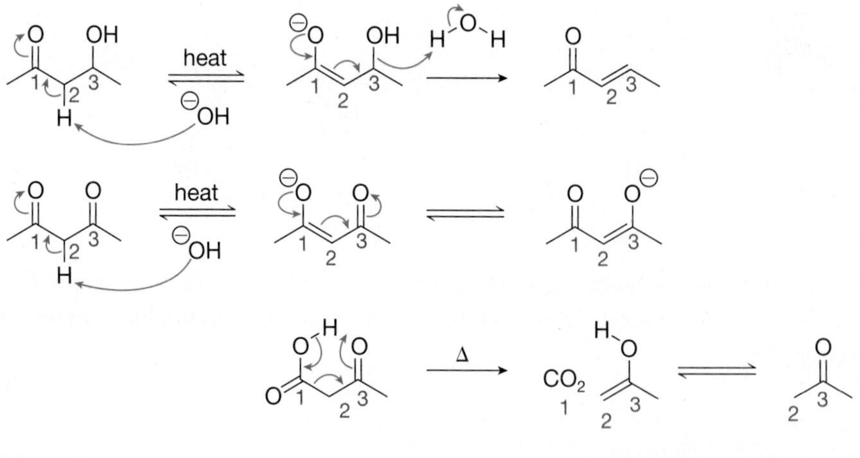

A large family of reactions involves a 1,3 arrangement of functional groups in a molecule. This 1,3 pattern is a key point of disconnection for retrosynthetic analysis. An α,β-unsaturated compound can be disconnected to the corresponding 3-hydroxy carbonyl compound, which in turn derives from an aldol process. The 1,3 pattern exists in dicarbonyl compounds and also applies to monocarbonyl compounds because the *addition* of a carboxylate gives a synthon that can be more easily controlled than a monocarbonyl.

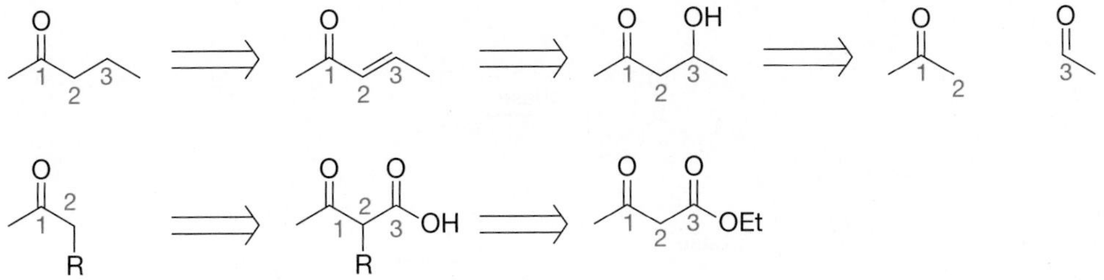

Bringing It Together

As mentioned in Section 17.1, structurally complex drugs, including erythromycin, are difficult to synthesize. A close look at the structure of erythromycin (Figure 17.4) reveals many sections with oxygen atoms bonded to carbons that have a 1,3 relationship; three such areas are shown in red. There is no doubt that the *Streptomyces* bacteria uses aldol processes to make erythromycin.

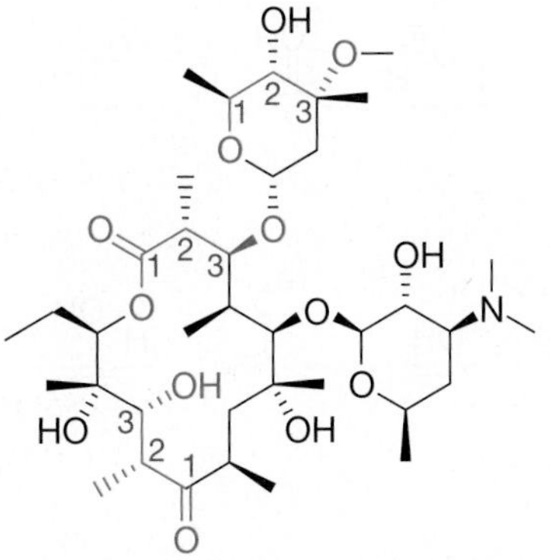

FIGURE 17.4 The 1,3 pattern appears throughout the structure of erythromycin.

The reactions described in this chapter show how enolates and related nucleophiles are crucial to the production of compounds for drugs, plants, foodstuffs, and even perfumes.

You Can Now

- Identify the enolizable protons on carbonyl compounds and related structures, and rank their relative acidities according to number and types of adjacent functional groups.
- Predict which bases will quantitatively deprotonate an α-hydrogen and which bases will form an enolate in equilibrium with its parent compound.
- Draw the mechanisms and products for the reactions of enolizable carbonyl compounds and related structures with various electrophiles, including halogens, alkyl halides, aldehydes, ketones, esters, and ylides.
- Determine the major products of above reactions by applying key concepts from this chapter, including
  - the different mechanisms and reactivities observed under acidic and basic conditions
  - the effect of base strength on whether a reaction is in equilibrium
  - the effect of heating the reagents
  - the relative stabilities of different ring sizes
  - the stereoselectivity of reagents such as ylides and stabilized ylides
- Use the reactions in this chapter to synthesize various molecules.

A Mechanistic Re-View

Enolization

Tautomerization

- In acid

- In base

Addition of electrophile

- In acid

- In base

α-Halogenation

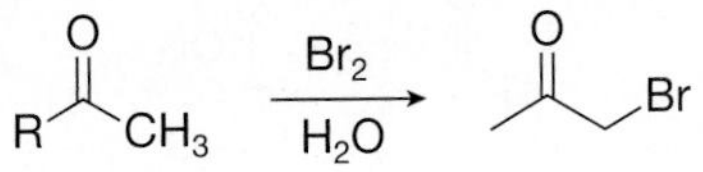

• In acid

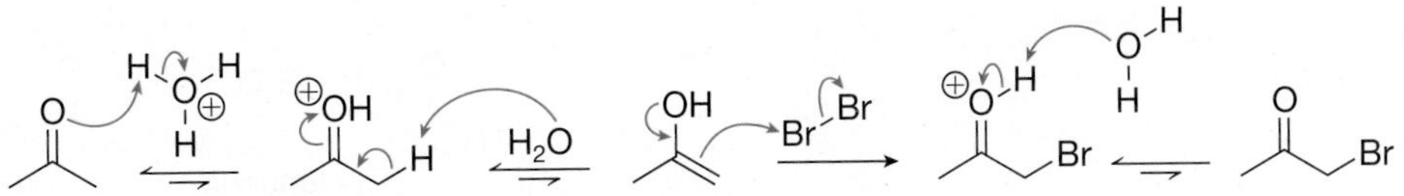

• In base

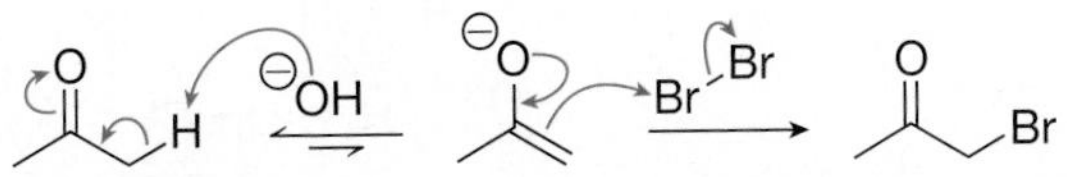

Haloform reaction

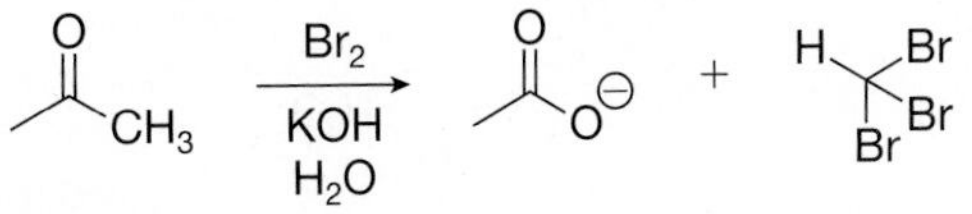

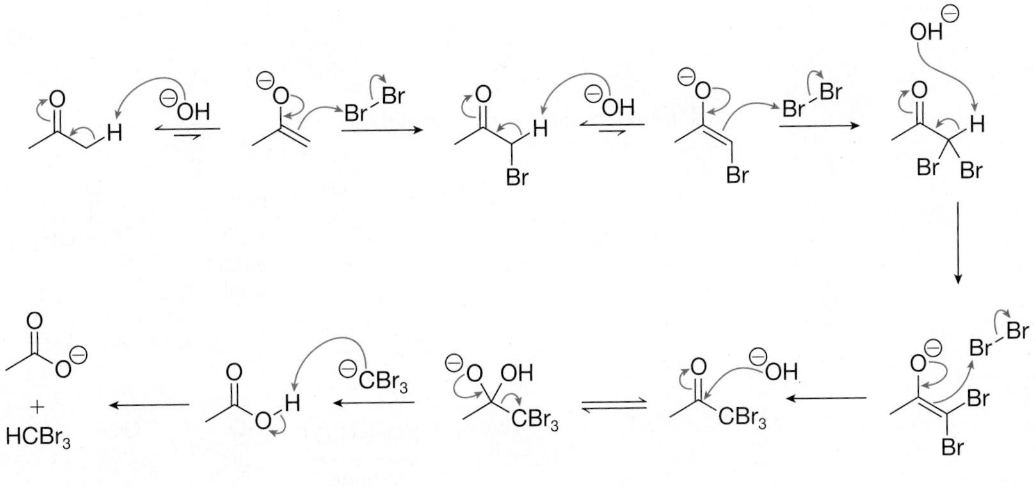

Ketone alkylation

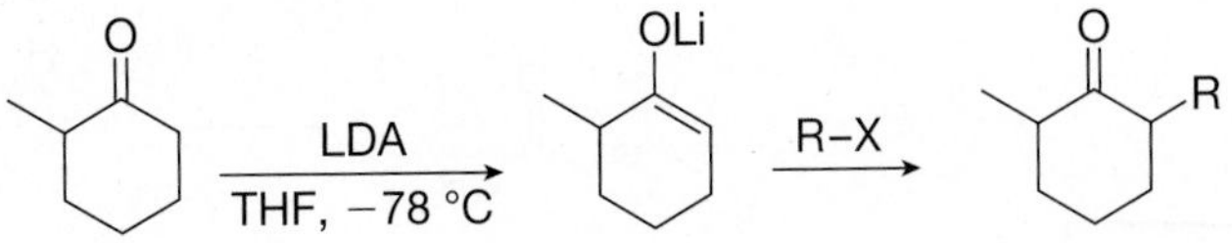

Aldol addition

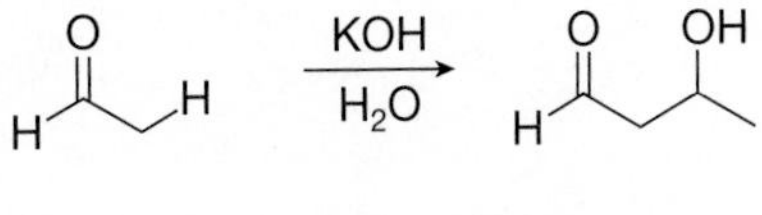

KOH

H_2O

Aldol condensation—Acid catalysis

H_3O^+

H_2O
toluene
heat

HCl

H_2O
heat

H_2O

Aldol condensation—Base catalysis

KOH

H_2O
toluene
heat

CH_3

KOH, H_2O

toluene
heat

heat

CH_3

Claisen condensation

KOEt
EtOH

OEt

H_3O^+

Dicarbonyl alkylation

1) KOH
2) R–Br

R

Decarboxylation of β-keto acids

Acetoacetic ester synthesis

1) $NaOR_1$
2) R_2X
3) $NaOR_1$
4) R_3X
5) hydrolysis/decarboxylation

Malonic ester synthesis

1) $NaOR_1$
2) R_2X
3) $NaOR_1$
4) R_3X
5) Krapcho cleavage (for ester) or hydrolysis/decarboxylation (for acid)

or

ALDOL-RELATED REACTIONS

Mannich reaction

$H^{\oplus}$

Thorpe reaction

base

acid/water
heat

strong
acid/water
heat

strong
acid/water
high heat

:base

base

Henry reaction

$$R_1CHO + R_2CH_2NO_2 \xrightarrow{\text{base}} R_1CH(OH)CH(NO_2)R_2$$

Enamine alkylation

$$CH_3COCH_3 \xrightarrow[\text{Cat. } H_2SO_4]{\text{pyrrolidine (N–H)}} \text{enamine } (=CH_2) \xrightarrow[2)\ H_3O^+]{1)\ R–X} CH_3COCH_2R$$

Unstabilized Wittig reaction

$$CH_3CHO \xrightarrow[n\text{-BuLi}]{Ph_3P^+CH_2CH_3} \text{(Z)-2-butene} + Ph_3P{=}O$$

$$Ph_3P{:} + CH_3CH_2I \longrightarrow Ph_3P^{\oplus}CH_2CH_3 \xrightarrow{n\text{-Bu–Li}} \left[Ph_3P^{\oplus}–\overset{\ominus}{C}HCH_3 \longleftrightarrow Ph_3P{=}CHCH_3 \right]$$

$$Ph_3P^{\oplus}–\overset{\ominus}{C}HCH_3 + CH_3CHO \longrightarrow [\text{oxaphosphetane}] \longrightarrow Ph_3P{=}O + \text{(Z)-2-butene}$$

Stabilized Wittig reaction

$$CH_3CHO \xrightarrow[n\text{-BuLi}]{Ph_3P^+CH_2CO_2Et} CH_3CH{=}CHCO_2Et\ (E) + Ph_3P{=}O$$

$$EtO_2CCH_2P^{\oplus}Ph_3 \xrightarrow[^{\ominus}OH]{NaOH} EtO_2CCH{=}PPh_3$$

$$EtO_2CCH{=}PPh_3 + CH_3CHO \longrightarrow [\text{oxaphosphetane}] \longrightarrow EtO_2CCH{=}CHCH_3$$

Problems

17.39 Each of the following compounds can be made using an enolate-type nucleophile. For each compound, identify the carbon that was the nucleophile and the new bond that was formed.

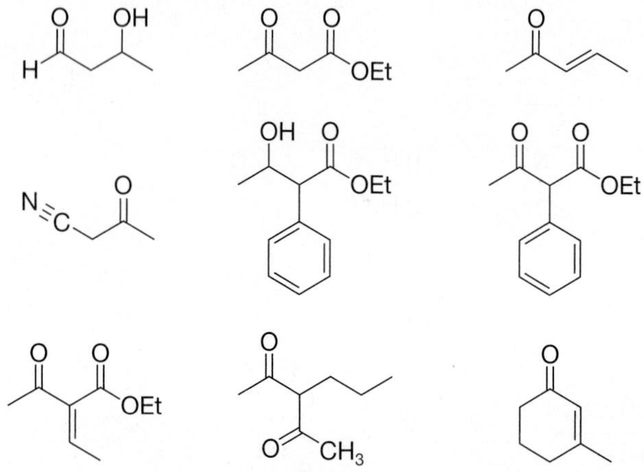

17.40 List suitable starting materials for the synthesis of each compound in Question 17.39.

17.41 Predict the products of the following reactions.

a) 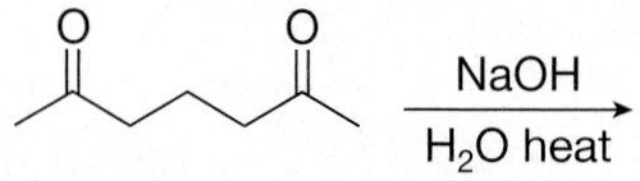 NaOH / H_2O heat

b) 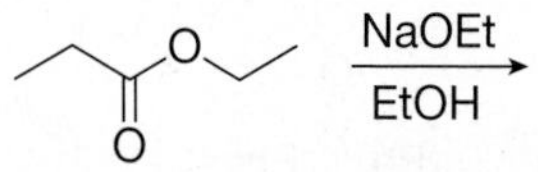 NaOEt / EtOH

c) 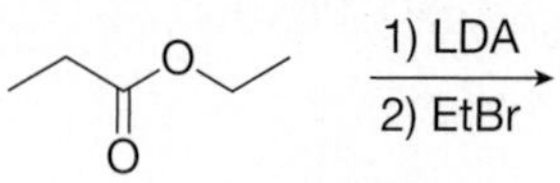 1) LDA 2) EtBr

d) 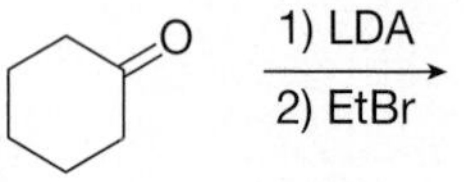 1) LDA 2) EtBr

e) 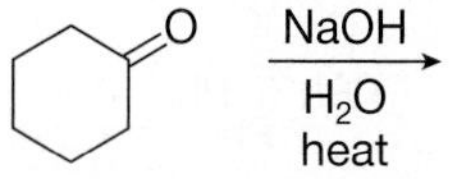 NaOH / H_2O heat

f) 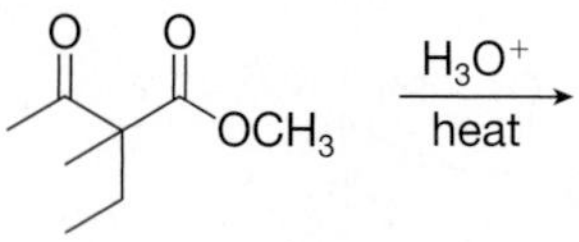 H_3O^+ / heat

g) 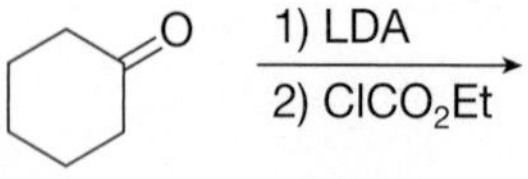 1) LDA 2) $ClCO_2Et$

h) Br_2 / NaOH / H_2O

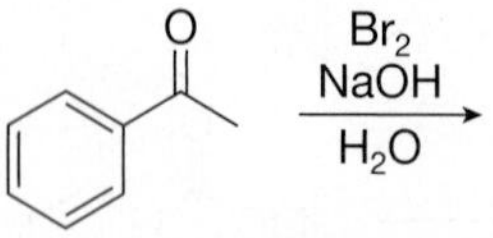

i) 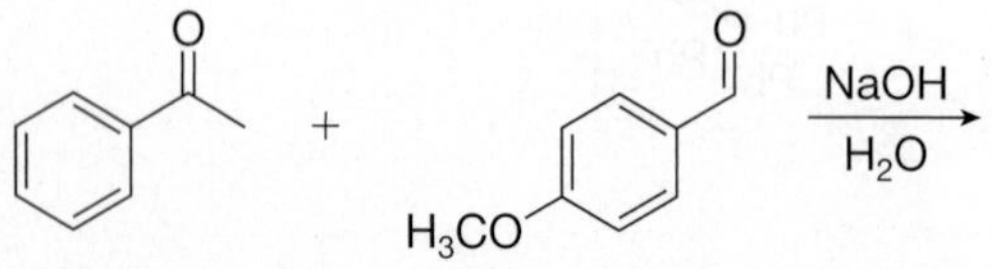 + NaOH / H_2O

j) 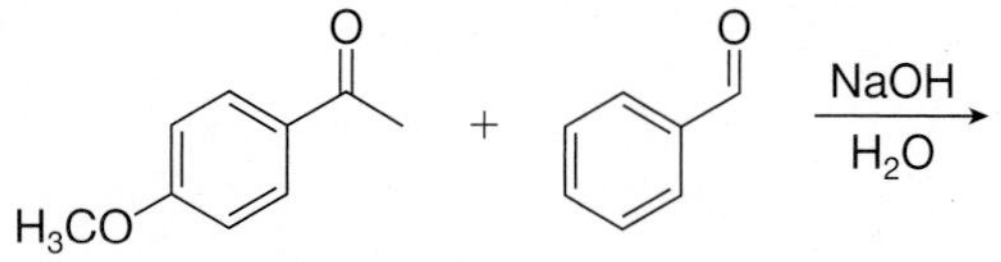 + NaOH / H_2O

k) 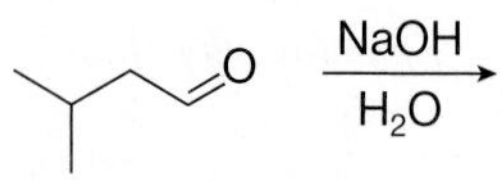 NaOH / H_2O

l) 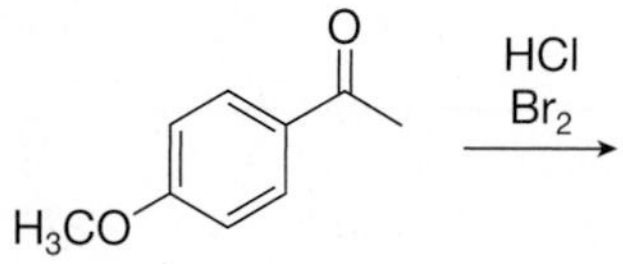 HCl / Br_2

m) 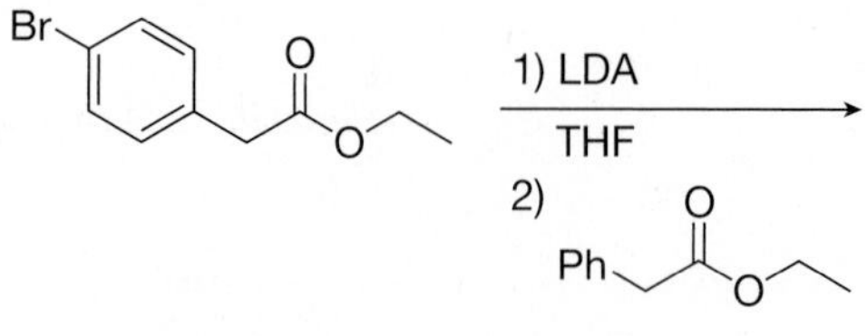 1) LDA / THF 2)

17.42 Predict the products of the following reactions.

a) 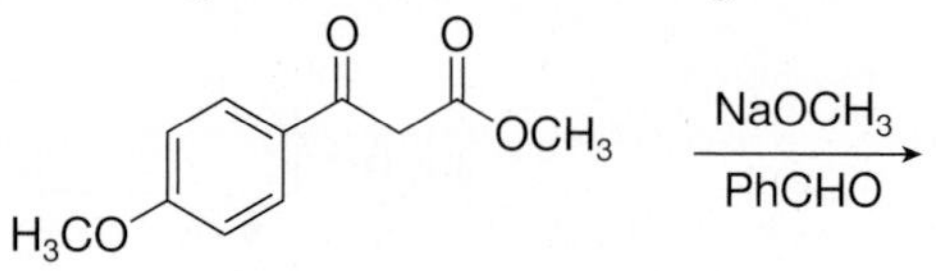 NaOCH$_3$ / PhCHO

b) 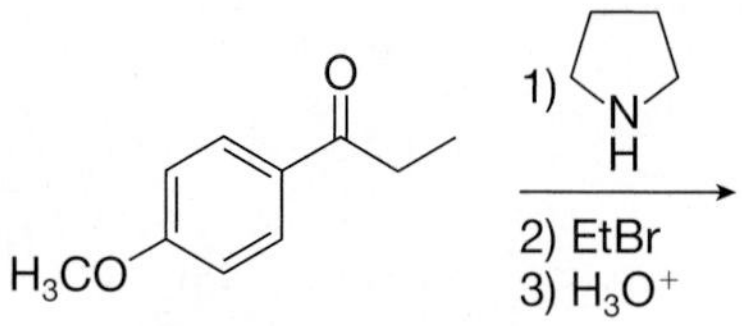 1) pyrrolidine 2) EtBr 3) H_3O^+

c)

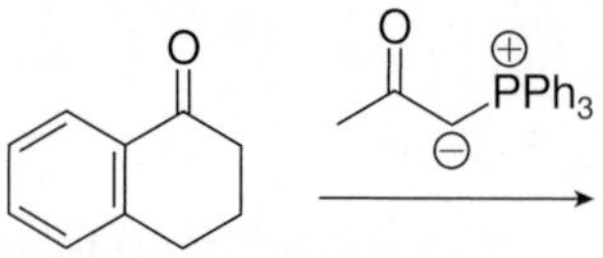

d) 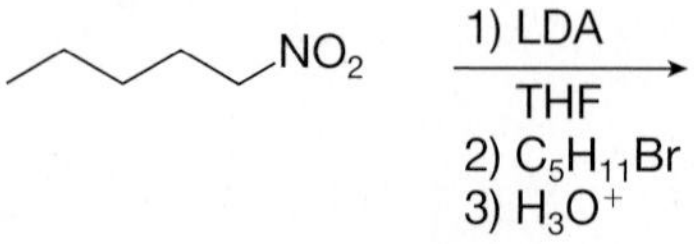 1) LDA / THF 2) $C_5H_{11}Br$ 3) H_3O^+

e) 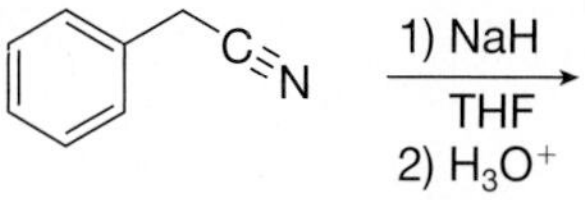 1) NaH / THF 2) H_3O^+

f)

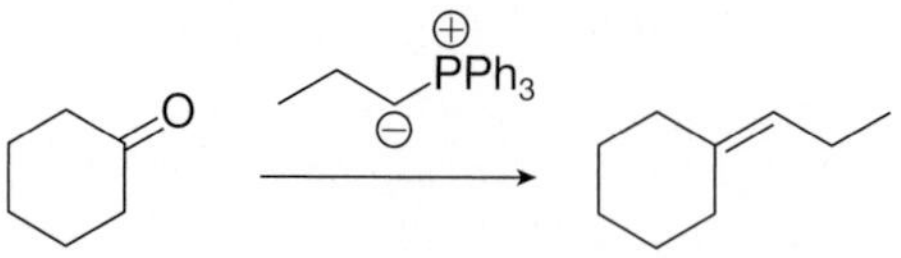

17.43 What reagents are required for the following transformations?

a)

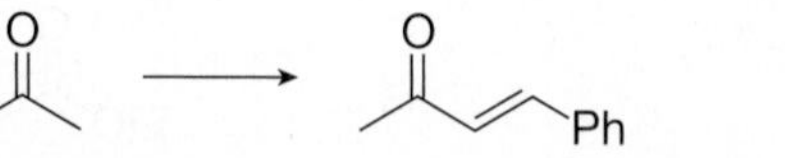

b)

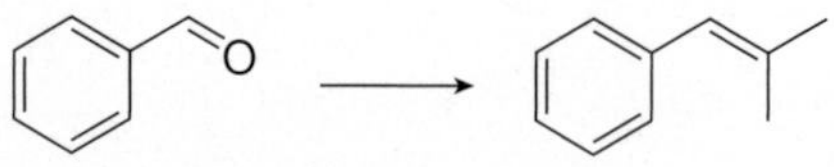

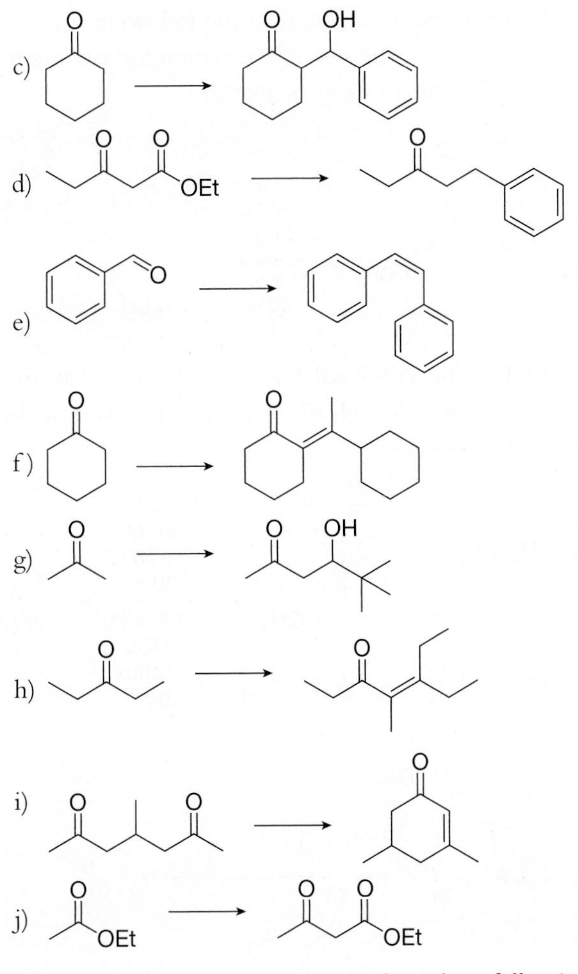

g) Br, O → O, O

h) Br, O → O, OH, Br

i) O → O, Br

j) O → O

k) $^{\ominus}$O–N$^{\oplus}$(=O), O, H → $^{\ominus}$O–N$^{\oplus}$(=O), HO

l) O → O

17.44 What reagents are required for the following transformations?

a) EtO, O, O, OEt → EtO, O

b) EtO, O, O, OEt → HO, O

c) EtO, O → EtO, O

d) O → O, O

e) O, O, OEt → O

f) O, O, OEt → O, O, OEt

17.45 What reagents are required for the following transformations? More than one step will be needed.

a) O, O → O

b) Ph, O, O → Ph, Ph, H, N, O, O

c) O, O → Ph, O, O

d) O → O, Br

e) O → O

f) H_3CO, O, O, OCH_3 → HO, O

17.46 Show the intermediates and final product of the following synthesis.

Benzene → (CH_3COCl, $AlCl_3$) → **A** C_8H_8O → (1) $LiAlH_4$, THF; 2) H_3O^+) → **B** $C_8H_{10}O$ → ($H_2SO_{4(conc.)}$) → **C** C_8H_8 → (1) B_2H_6, THF; 2) NaOH, H_2O_2) → **D** $C_8H_{10}O$ → ($NaCr_2O_7$, H_2SO_4) → **E** $C_8H_8O_2$ → ($H_2SO_{4(cat.)}$, ethanol) → **F** $C_{10}H_{12}O_2$ → (1) LDA, THF, −78 °C; 2) ClC(O)OEt) → **G** $C_{13}H_{16}O_4$ → (1) NaOEt; 2) EtBr) → **H** $C_{15}H_{20}O_4$ → ($H_2NC(O)NH_2$, heat) → **I** $C_{12}H_{12}N_2O_3$

17.47 There is a shorter synthetic route to intermediate C in Question 17.46. What is intermediate J, and what reagents are required for the conversion to product C?

Benzene → ($ClCH_2CHO$, $AlCl_3$) → **J** C_7H_6O → **C** (styrene)

17.48 The Mannich reaction combines an amine, an aldehyde, and a ketone to form a β-aminoketone. The less substituted α-carbon of the ketone and the amine are both nucleophiles that attack the aldehyde carbon.

2-Butanone + 4-chloroaniline (H_2N–C₆H₄–Cl) + PhCHO → ($HCl_{(cat.)}$, EtOH, 90%) → β-aminoketone (HN–C₆H₄–Cl, Ph)

a) Which nucleophile attacks the aldehyde first, and what intermediate is formed?

b) Draw the mechanism for the addition of the ketone to the intermediate formed in part (a).

17.49 Draw the mechanism of the following reaction. Why does only one side of the symmetric starting material dehydrate to form an alkene?

1,2-bis(4-methoxyphenyl)ethane-1,2-dione (H_3CO, OCH_3) → (KOH, EtOH, acetone, 2 h, 86%) → cyclopentenone product with OH and two 4-methoxyphenyl groups (H_3CO, OCH_3)

17.50 For the following reaction scheme, what are intermediates A and B? Draw the mechanism for their formation.

Methyl (4-methoxyphenyl)acetate (H_3CO, OCH_3) → (1) LDA, ether, −78 °C, 20 h; 2) AcOH, HCl, reflux, 5 h) → **A** 40%

A → (1) LDA, THF, −78 °C, 1 h; 2) 2-bromo-1-(4-methoxyphenyl)ethanone (Br, OCH_3), 1.5 h) → **B** 61%

B → (KOH, CH_3OH, RT, 20 min.) → **C** 54% (cyclopentenone bearing three 4-methoxyphenyl groups; H_3CO, OCH_3, H_3CO)

17.51 The intramolecular aldol condensation of 2,7-octanedione shown here yields a cyclic compound. Draw mechanisms for the formation of a five-membered and a seven-membered ring. Explain why only the five-membered ring is formed.

2,7-Octanedione → (NaOH, H_2O) → 1-acetyl-2-methylcyclopentene, not the seven-membered ring product

17.52 What starting materials could be used to make the following compounds using the Thorpe reaction?

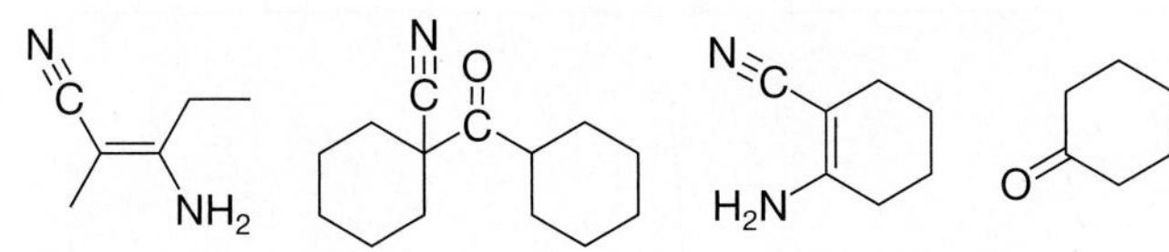

17.53 What is the mechanism for the formation of the coumarin shown here, and what reagents are necessary for the transformation?

17.54 A cyclic diester, known as Meldrum's acid, can be used to make a variety of substituted acid derivatives. Propose a mechanism for the following transformations.

a) 1) LDA, −78 °C, THF; 2) BnBr; Δ

b) H_2O; H_2NR; HOR

17.55 Compound A, with a molecular weight of 88 g/mol, was treated with sodium methoxide in methanol and stirred at room temperature overnight. After adding sufficient $HCl_{(aq)}$ to make the solution acidic, compound B was isolated. Compound B has a molecular weight of 144 g/mol. Draw the structure of each compound.

^{1}H-NMR of Compound A

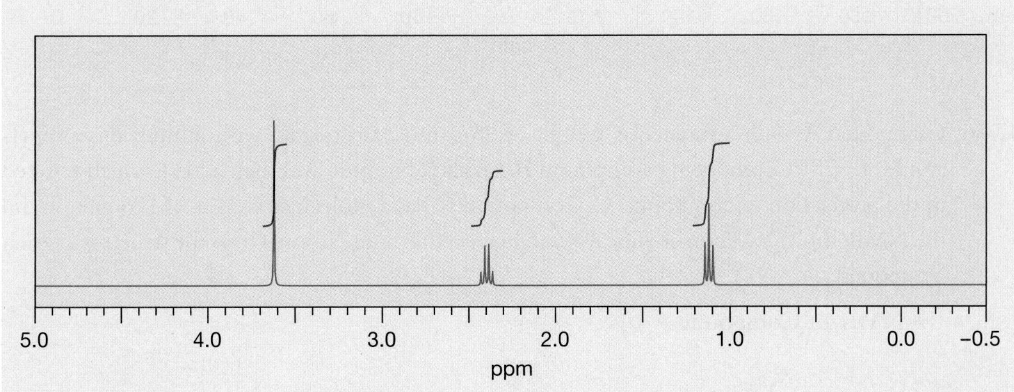

^{13}C-NMR of Compound A

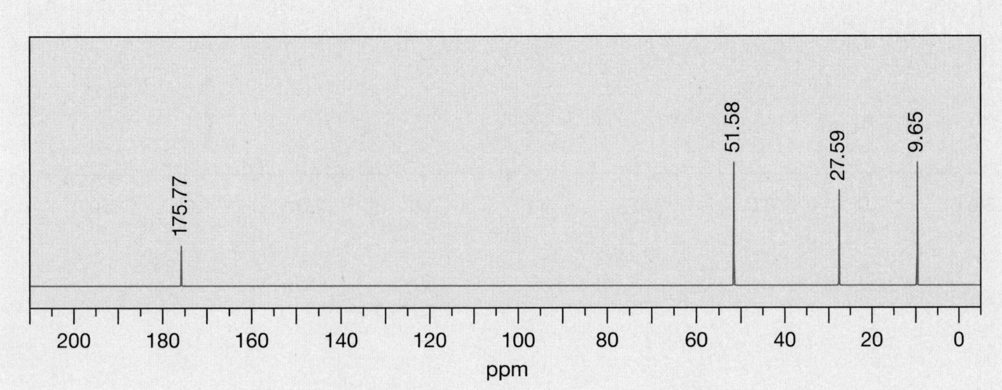

^{1}H-NMR of Compound B

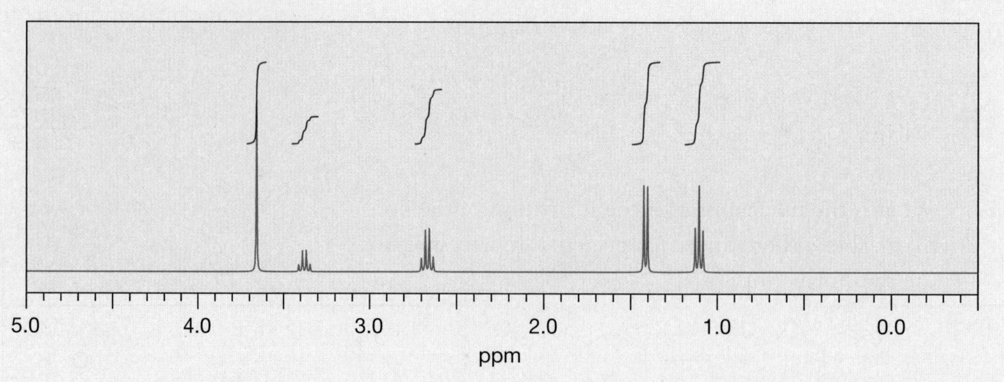

^{13}C-NMR of Compound B

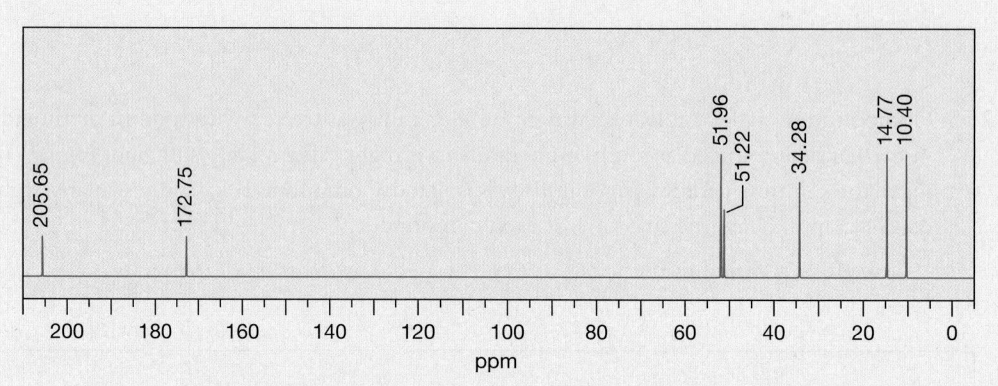

17.56 Compound A, with a molecular weight of 86 g/mol, was treated with lithium diisopropylamide at −78 °C, producing compound B. An alkyl bromide was then added, which resulted in the production of compound C. Compound C has a molecular weight of 176 g/mol, and the NMR shows two non-identical hydrogens on the same carbon. Draw the structure of each compound.

^{1}H-NMR of Compound A

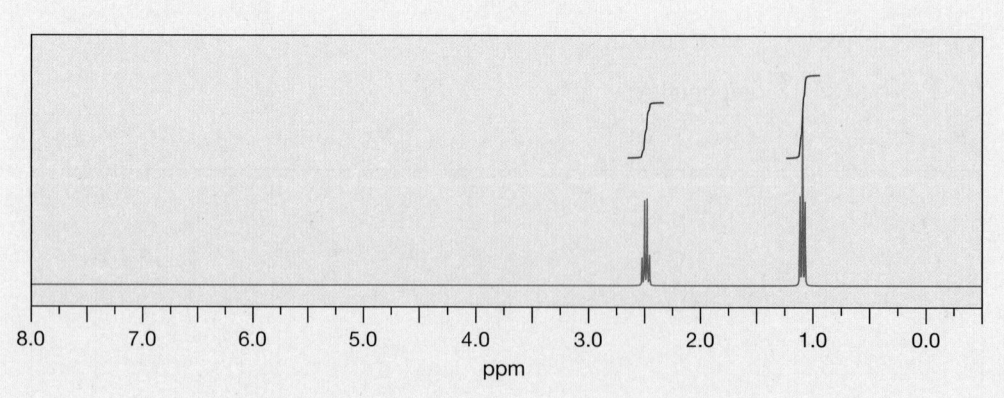

^{13}C-NMR of Compound A

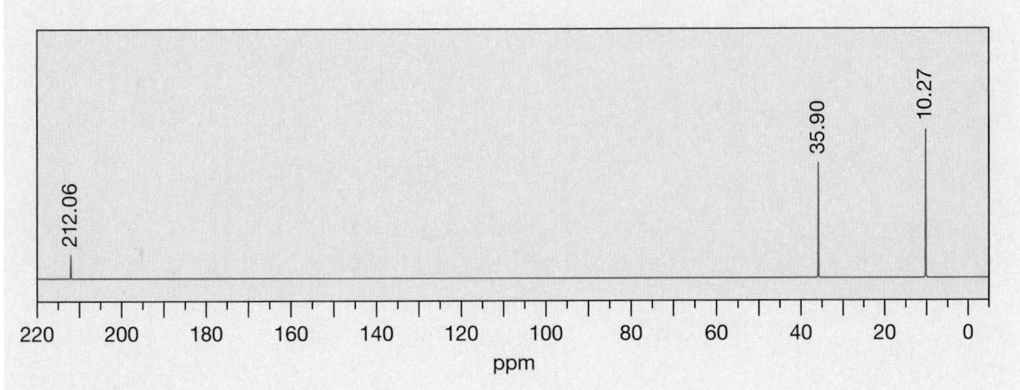

^{1}H-NMR of Compound B

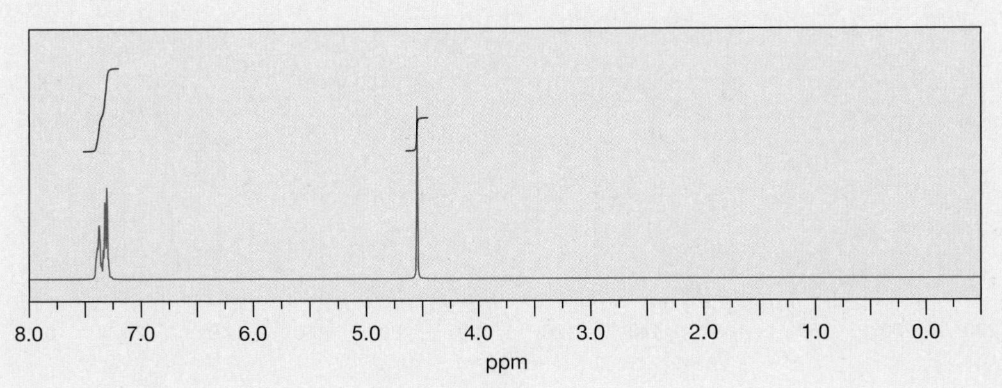

^{13}C-NMR of Compound B

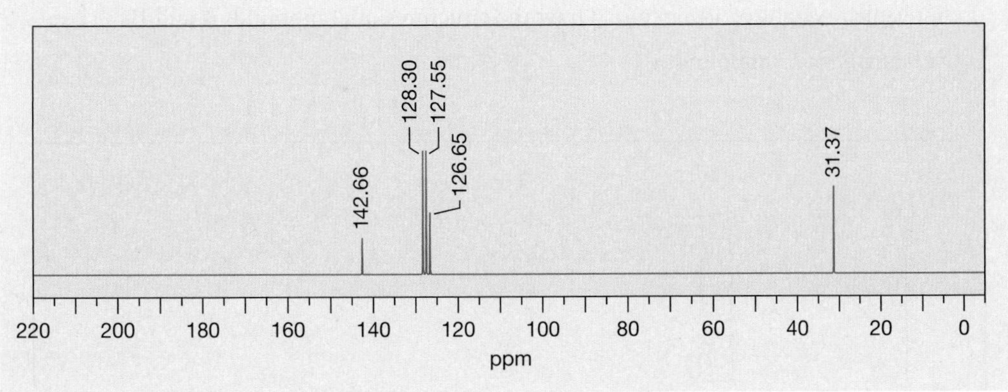

^{1}H-NMR of Compound C

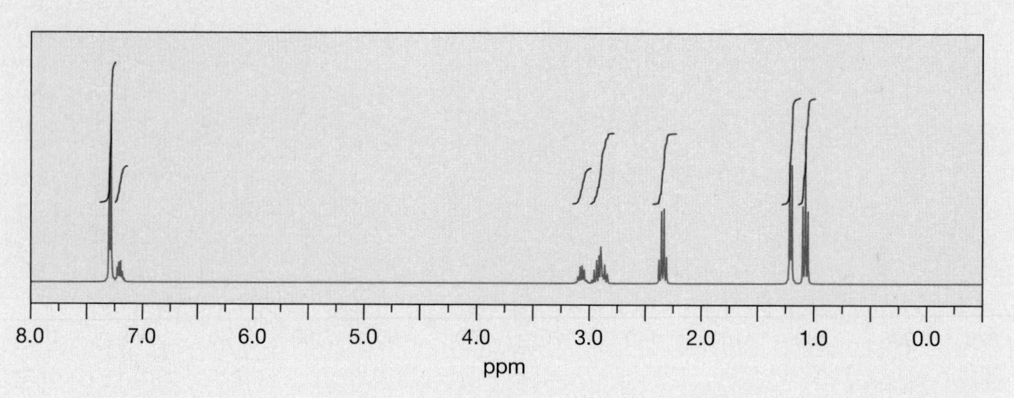

^{13}C-NMR of Compound C

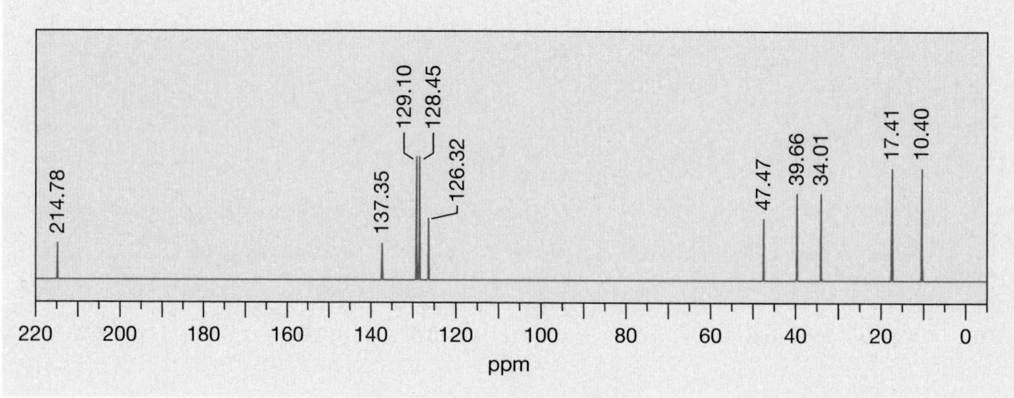

17.57 Compound A, with a molecular weight of 106 g/mol, was treated with an alkyltriphenylphosphonium salt and strong base. After completion, compound B was isolated: with a molecular weight of 146 g/mol. Draw the structures of compounds A and B.

^{1}H-NMR of Compound A

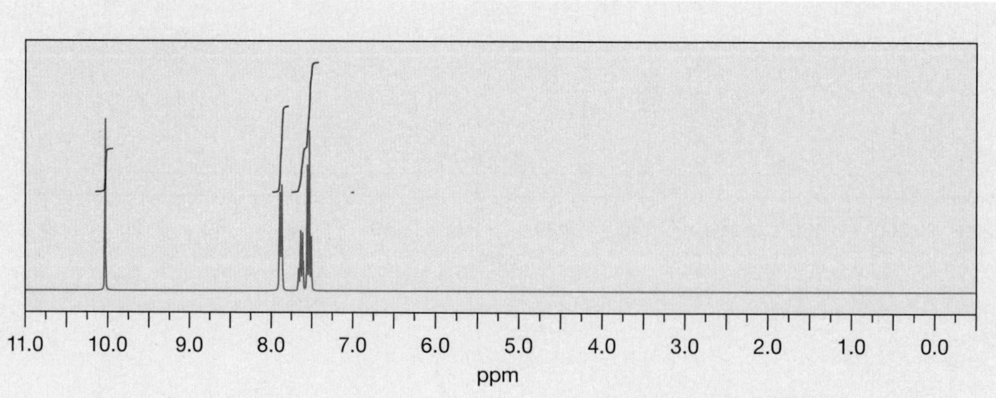

^{13}C-NMR of Compound A

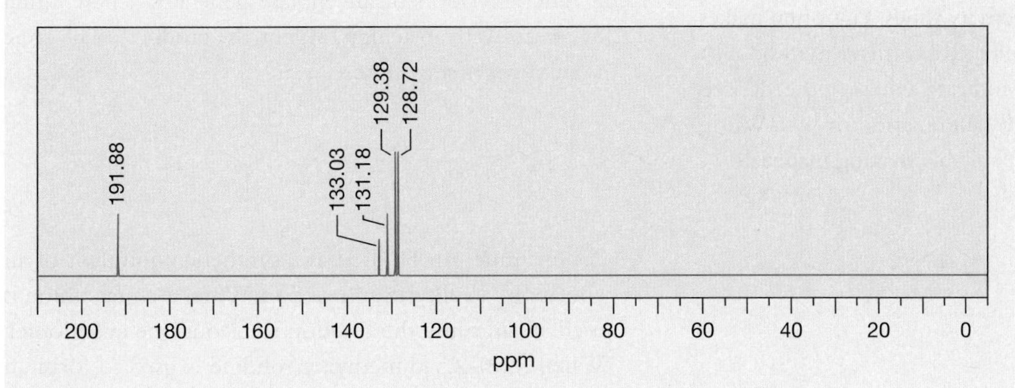

^{1}H-NMR of Compound B

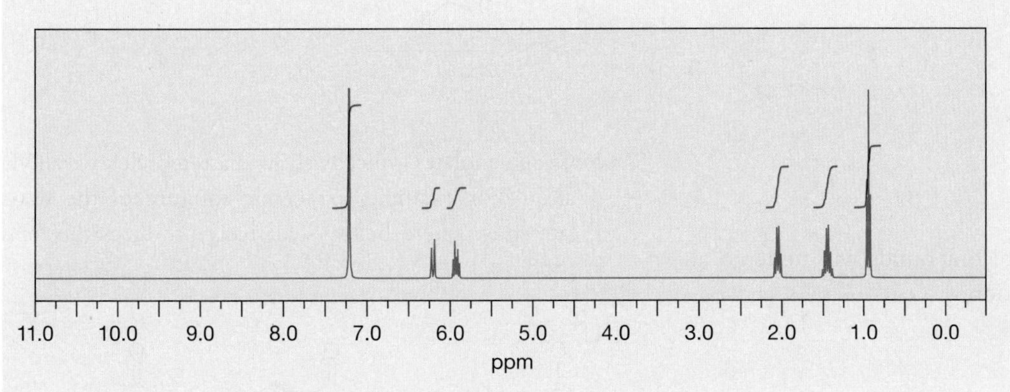

^{13}C-NMR of Compound B

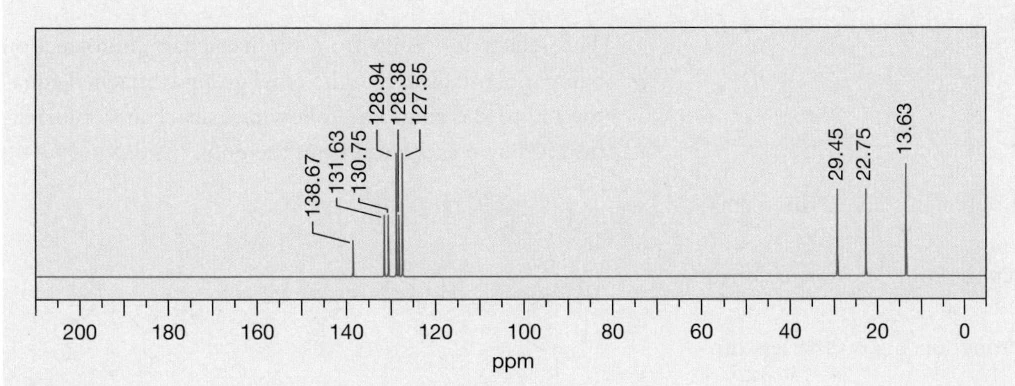

17.58 E1 and E2 elimination reactions can be used to make alkenes, but these reactions are complicated by competing reactions and selectivity issues. This often makes a Wittig reaction the preferred synthetic method. For the following products, indicate whether the desired product can be formed by elimination or by a Wittig reaction. In each case list suitable starting materials.

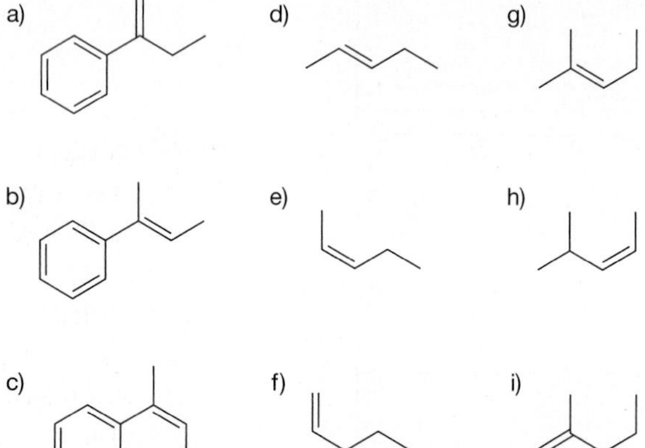

MCAT STYLE PROBLEMS

17.59 Which of the following compounds will undergo a successful haloform reaction?

a) $CH_3CH_2COCH_2CH_3$
b) $CH_3CH_2CO_2CH_3$
c) C_6H_5CHO
d) $CH_3CH_2COCH_3$

17.60 Which of the following are produced when an ester reacts with alkoxide base?

a) a β-hydroxy ester
b) a β-ketoester
c) a carboxylic acid
d) a ketone

17.61 Which of the following statements about the formation of enolates are true?

I. Hydroxide base results in complete formation of enolate.
II. Deprotonation of ketones occurs on the less hindered α-carbon.
III. β-ketoesters can be extensively deprotonated using weak base.

a) I, II, III
b) I, II
c) I, III
d) II, III

CHALLENGE PROBLEMS

17.62 The alkylation of an enolate generates a new chiral centre. With an achiral reagent, the product is a racemic mixture, as shown here.

1) LDA
2) CH_3I

An enamine can be used as a synthetic equivalent of an enolate in an alkylation reaction. When a chiral amine is used, it can cause the reaction to favour one enantiomer. When (*S*,*S*)-2,5-dimethylpyrrolidine is used to form an enamine, what is the absolute configuration of the major product, *R* or *S*?

1) (*S*,*S*)-2,5-dimethylpyrrolidine
2) CH_3I
3) H_3O^+

17.63 Cyclic enolates typically show diastereoselective alkylation. For example, a racemic mixture of the enolate shown here below selectively produces the *trans* product.

This selectivity results from the half-chair conformation being more stable when the ethyl group is attached equatorially to the ring. The following half-chair conformation is drawn as viewed from the enolate side.

Provide a reasonable justification for the observed stereoselectivity in the product, based on the stereochemistry of the half-chair conformation of the enolate.

Selectivity and Reactivity in Enolate Reactions

CONTROL OF STEREOSELECTIVITY AND REGIOSELECTIVITY

18

CHAPTER OUTLINE

Dmitry Kalinovsky/Shutterstock.com

Chemical syntheses performed at an industrial level are performed on a large scale. To maximize the yield of the desired product, the reactions performed need to be stereoselective.

18.1 Why It Matters

Stereochemistry plays a very important role in how molecules behave in a biological system.

The ability to control both stereoselectivity and regioselectivity is essential to our ability to copy and modify molecules from nature and synthesize useful products. In the case of erythromycin (see Chapter 17), changing the configuration of even a single stereogenic centre can dramatically alter the overall conformation of the molecule and disrupt the antibiotic activity. The 1,3-dioxy segments of erythromycin are important features that control the conformation of the ring core. Consider the erythromycin isomer in which the β-hydroxyl group in the 1,3-dioxy segment has the *anti*-relative configuration (see Figure 18.1). In one of these segments, highlighted in blue, the α-methyl and β-hydroxyl groups are on the same side of the molecule (i.e., a *syn*-aldol relative configuration) and point away from the viewer.

the CH_3 and OH groups are *syn*

the *anti*-isomer does not have antibacterial activity

FIGURE 18.1 Erythromycin has *syn*-aldol segments. The *anti*-aldol version does not have any antibacterial activity.

This single change in configuration alters the conformation of the entire molecule. As shown in Figure 18.2, the preferred conformations of the two erythromycin isomers are different, so we cannot assume that the isomer retains any antibiotic activity. Ultimately, to synthesize a drug such as erythromycin, the stereochemical configuration of all the ring substituents needs to be controlled.

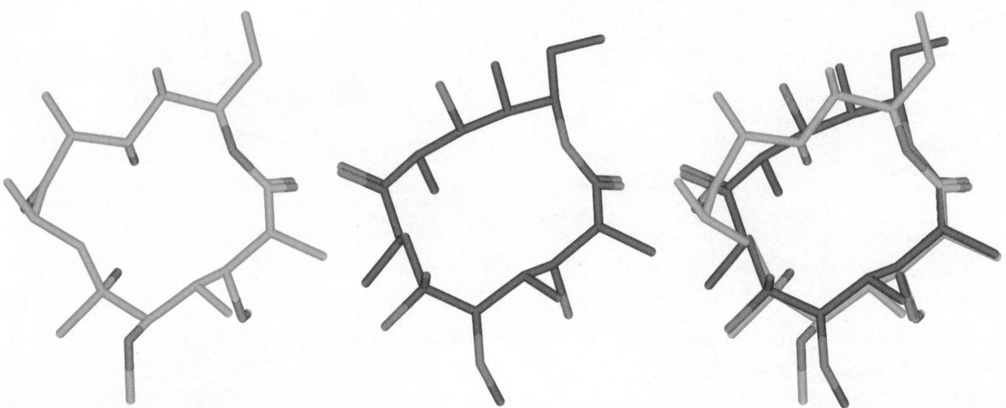

FIGURE 18.2 Changing the configuration of only one hydroxyl group can change the preferred conformation of the whole molecule (erythromycin shown in green, *anti*-isomer in blue, oxygen atoms in red). Hydrogens and attached sugars are omitted for clarity.

The absolute configuration of chiral molecules can also affect their biological activity. For example, the enantiomers of propoxyphene have either anti-tussive (1*R*,2*S*) or analgesic (1*S*,2*R*) activity.

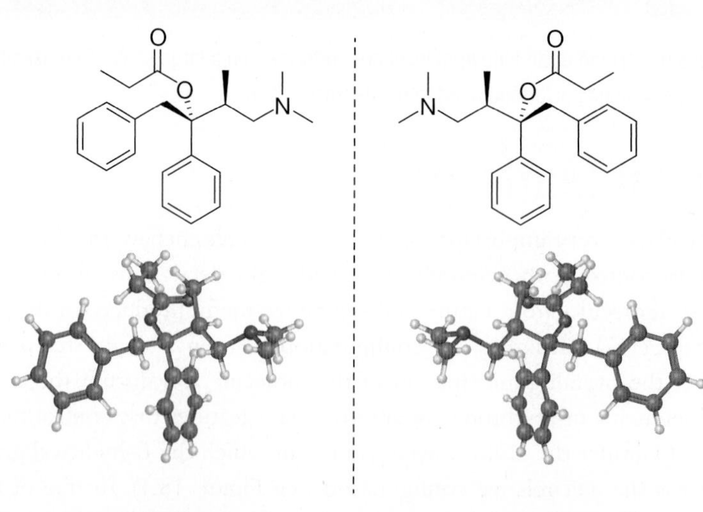

(1*R*,2*S*)-propoxyphene
(Novrad)
anti-tussive (cough suppressant)

(1*S*,2*R*)-propoxyphene
(Darvon)
analgesic (pain reliever)

Even compounds with a single stereogenic centre can have vastly different biological activities depending on their absolute configuration. Consider the anti-depressant drug, citalopram, a selective serotonin reuptake inhibitor (SSRI). The *S*-enantiomer has anti-depressant properties, whereas the *R*-enantiomer is inactive. Originally sold as a racemic mixture called Celexa, it is now also available as a single *S*-enantiomer called Lexapro (Cipralex). These examples illustrate that stereochemical control is a critical aspect of organic synthesis.

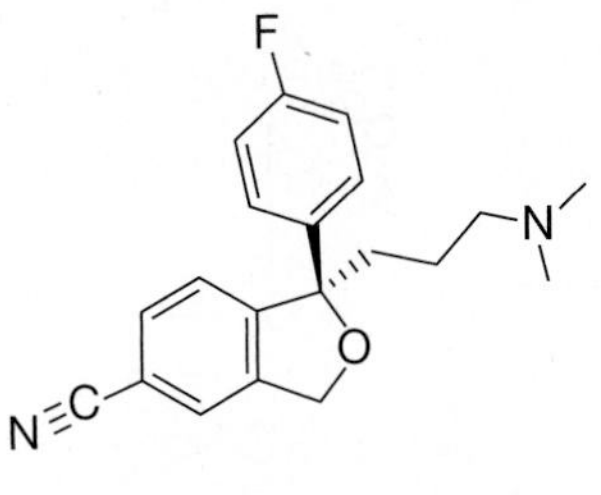

(*R*)-citalopram 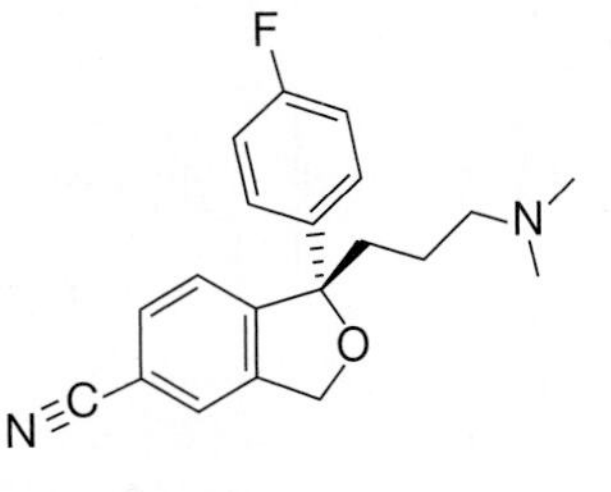 (*S*)-citalopram

This chapter builds on the chemistry of carbanions introduced in Chapter 17 and elaborates on how to control both the regio- and stereoselectivity of several organic reactions.

18.2 Regioselectivity in α,β-Unsaturated Electrophiles

When a chemical reaction, such as a nucleophilic addition, can follow more than one pathway, it raises the question of selectivity. When the different reaction pathways add the nucleophile to different parts of the electrophile, the selectivity question is one of regioselectivity.

18.2.1 Direct and conjugate addition

Aldol condensation can be used to make α,β-unsaturated carbonyls (see Chapter 17). These functional groups have two electrophilic carbons, so they are known as **ambident electrophiles**. Each of the electrophilic carbons is a potential site for the addition of a nucleophile. **Direct addition (1,2-addition)** refers to the process whereby a nucleophile is added directly to the electrophilic carbon of the carbonyl group. **Conjugate addition (1,4-addition)** refers to the addition of a nucleophile to the electrophilic β-carbon of an α,β-unsaturated carbonyl.

Ambident electrophile is a functional group with two electrophilic sites.

Direct addition (1,2-addition) is the addition of a nucleophile to the carbonyl carbon of an α,β-unsaturated carbonyl group.

Conjugate addition (1,4-addition) is the addition of a nucleophile to the β-carbon of an α,β-unsaturated carbonyl group.

1,2-addition (direct)

1,4-addition (conjugate)

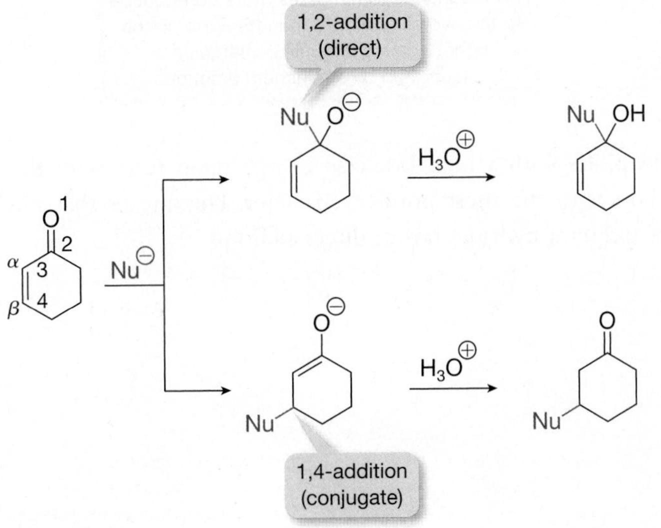

The double bond and carbonyl of an α,β-unsaturated system are conjugated and act like a single functional group that combines the reactivity of both groups. The electron-withdrawing nature of the carbonyl has a strong influence on this conjugated system and dominates the chemistry of the system. Unlike isolated (unconjugated) alkenes, the double bond of an α,β-unsaturated carbonyl usually reacts like an electrophile. Analysis of resonance forms reveals this electrophilic character of the β-carbon (third structure). The carbon-carbon double bond is still capable of reacting like a nucleophile (alkene), but only with very strong electrophiles.

Resonance contributors show positive charges on carbon. These carbons lack octets and are therefore electrophilic.

ORGANIC CHEMWARE
18.1 Conjugate addition

18.2.2 The basis of regioselectivity

The two electrophilic carbons in an α,β-unsaturated carbonyl have sufficiently different reactivity that nucleophiles can differentiate between them, and selectivity between direct and conjugate addition is possible. Although there is no pattern in selectivity that applies to all nucleophiles, some generalizations can be made (Table 18.1).

First, the two charged resonance forms of an α,β-unsaturated carbonyl do not make equal contributions to structure. The middle form makes a slightly larger contribution because the two unlike charges are separated by a minimum amount. This creates a situation in which the carbonyl carbon has more positive character than the β-carbon does.

This resonance form makes more contribution to the overall structure than the form on the right because the unlike charges are separated by a minimum amount.

Second, nucleophiles with a high localized charge often react with the carbonyl carbon, which is the carbon with the most positive character. This means that nucleophiles such as Grignard reagents and most hydrides favour direct addition.

Third, neutral nucleophiles favour conjugate addition. These include weak nucleophiles, such as alcohols and water, as well as stronger, more polarizable nucleophiles such as thiols and phosphines. Due to the neutral nucleophile's lack of charge, its rate of reaction with the carbonyl carbon decreases, and its rate of reaction with the more polarizable β-carbon increases.

Fourth, certain nucleophiles will undergo either conjugate or direct addition depending on the reaction conditions. When conditions allow the nucleophile to act as a leaving group in a reversible reaction, the major product is determined by the stability of the product or a key intermediate, instead of by the speed of the addition step. Such reactions are discussed in the following section.

18.2.2.1 Thermodynamic versus kinetic control

The principle of thermodynamic versus kinetic control sometimes governs regioselectivity and stereoselectivity. This principle describes how two important energy parameters—kinetics and thermodynamics—operate in chemical reactions. Kinetics are governed by the activation energy, designated as $\Delta G^{\ddagger}$, and so the speed of reactions becomes important. Kinetics dominate when reactions are irreversible. Thermodynamics are governed by the free energy change between reactants and products, designated as ΔG. Thermodynamics dominate when reactions are reversible.

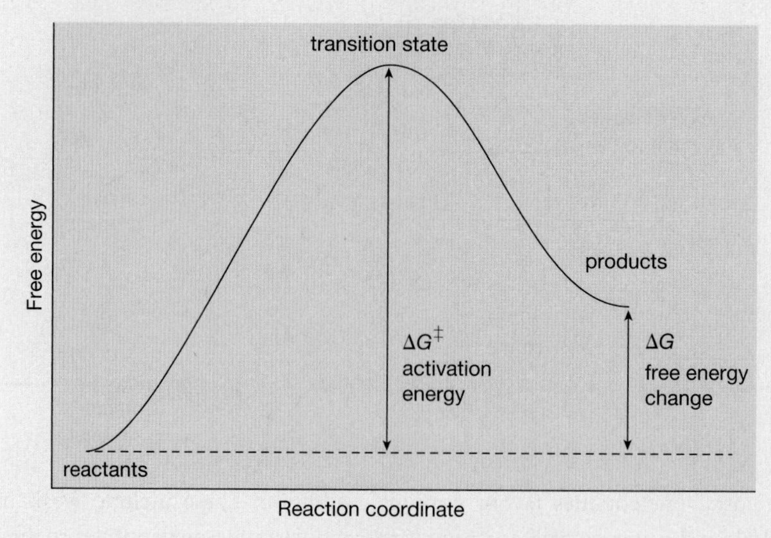

The activation energy controls how quickly a reaction proceeds. In general, molecules tend to react by the quickest (lowest energy) pathway, and so most molecules in a sample react by the pathway that has the smallest activation energy ($\Delta G^{\ddagger}$). When the reaction follows a mechanism that is irreversible, the distribution of products depends on the rate at which each product forms. Products that form quickly are produced in large amounts; products that form slowly accumulate small amounts. The major product in this kind of reaction results from the lowest energy pathway and is called the kinetic product. Such reactions are said to be under **kinetic control**. The ratio of products in such reactions depends on the difference in the activation energies for each pathway.

A reaction under **kinetic control** uses reaction conditions, such as reduced reaction temperature, that favour the most quickly formed product.

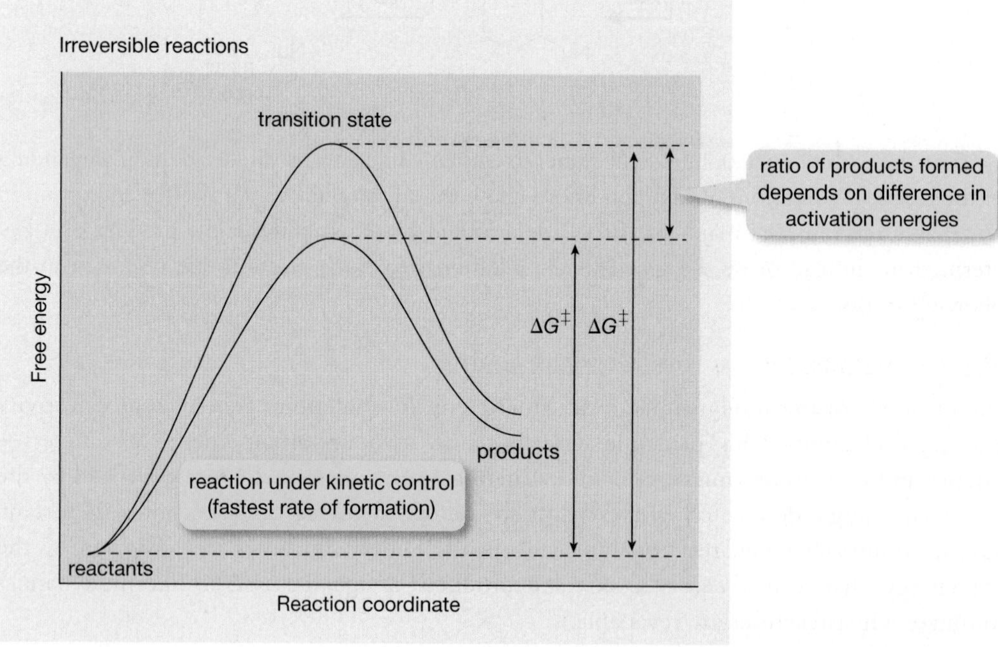

In contrast, when a reaction is reversible (an equilibrium reaction), the products that are formed depend on their relative stability or energy content. Unlike irreversible reactions, equilibria favour the most stable (lowest energy) products. The major product that is formed under these conditions is called the thermodynamic product, and the ratio of products depends on the difference in energy between the possible products formed. This type of reaction is said to be under **thermodynamic control**.

A reaction under **thermodynamic control** uses reaction conditions, such as increased temperature, that promote an equilibrium that leads to the most stable product.

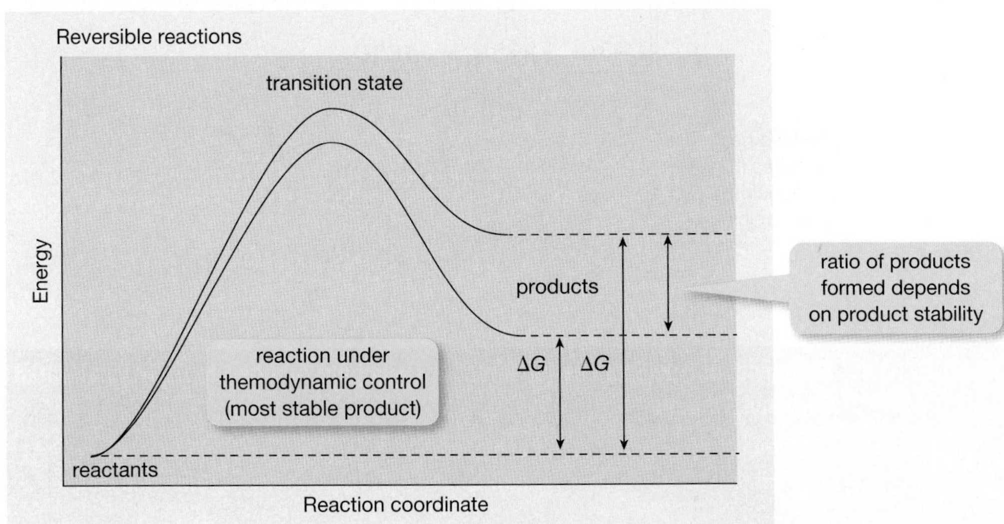

Some nucleophiles add preferentially to the carbonyl carbon when the reaction occurs at lower temperatures. This is because the fast addition at the most highly charged site is irreversible unless heat is applied. When the temperature is raised, the addition step becomes a reversible equilibrium. This favours addition to the β-carbon, which produces a lower energy intermediate. Some examples include amines, unstabilized enolates, and cyanide, all of which form the kinetic product when cooled, but the thermodynamic product when heated.

cooler conditions give the faster forming product
NaCN, HCN 5–10 °C
NC OH
kinetic product
NaCN, HCN 80 °C
CN
thermodynamic product
warmer conditions give the product resulting from the more stable intermediate

The cyanide does not act as a leaving group unless sufficient heat is added. Regardless of the temperature, direct addition is faster than conjugate addition. Therefore, in direct addition, the kinetic product is favoured under cooler temperatures. When heat is applied, the intermediates formed by either direct or conjugate addition become reversible. Conjugate addition forms a more stable intermediate, which is less likely to reverse and leads to the thermodynamic product.

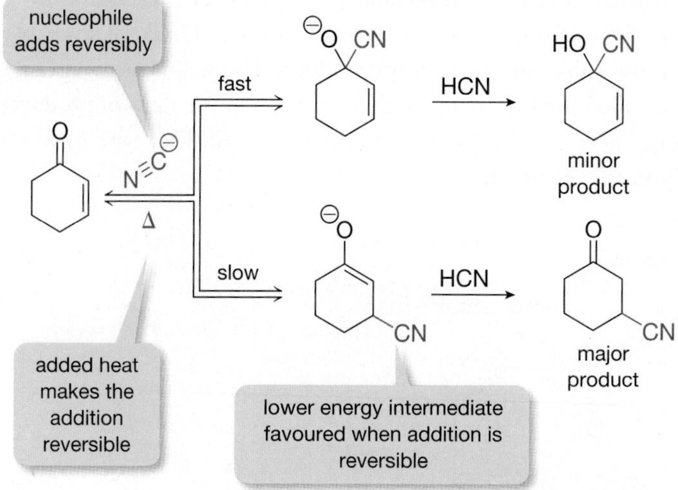

WANT TO LEARN MORE?
18.1 Klopman–Salem equation

TABLE 18.1 Selectivity of Nucleophiles for Direct or Conjugate Addition to α,β-Unsaturated Carbonyl Systems

	Direct (1,2) Addition	Conjugate (1,4) Addition	Kinetic or Thermodynamic Control
Nucleophile	RMgX, RLi, most hydrides (H^-)	RCu, R_2CuLi, stabilized enolates, H_2O, ROH, RSH, RPH_2, R_2PH	RNH_2, R_2NH, simple enolates, CN^-

18.2.3 Cuprate and Grignard reagents

Grignard reagents and cuprates are two types of organometallic reagents. They are both nucleophilic and can be used to make carbon-carbon bonds with organic electrophiles, but they interact with those electrophiles in different ways. Grignard reagents add to carbonyl groups and, when used with α,β-unsaturated carbonyls, tend to form 1,2-addition products (direct addition). This occurs because the carbon of a Grignard reagent carries a strong negative character, which tends to favour the addition of the nucleophile to the electrophilic carbon closest to the electronegative oxygen (this carbon has the most positive character). Most hydride reagents, such as $LiAlH_4$, also favour direct addition for the same reason.

Cuprates are a type of organometallic nucleophile, with a negatively charged copper. They favour conjugate addition to α,β-unsaturated carbonyl compounds.

Orthogonal reactivity describes reagent pairs that have opposite and complementary reactivities.

Cuprates can be of the general form RCu (a lower order cuprate) or R_2CuLi (a higher order cuprate). Cuprates have what is referred to as **orthogonal reactivity**—compared to Grignard reagents, they react in opposite ways. Whereas Grignard reagents generally react in direct fashion (1,2-addition) with α,β-unsaturated carbonyl groups, cuprates react in a conjugate fashion (1,4-addition).

1) CH_3MgBr
2) H_3O^+
H_3C OH
Grignard reagents undergo 1,2-addition (direct)
1) $(CH_3)_2CuLi$
2) H_3O^+
CH_3
cuprate reagents undergo 1,4-addition (conjugate)

Cuprates are made by mixing alkyl lithium reagents with copper salts (usually CuI or CuCN) at low temperature. The resulting compounds are not very stable, so typically they are made immediately before use.

$$2\ H_3C{-}Li + CuI \longrightarrow (H_3C)_2Cu^{\ominus}\ Li^{\oplus} + LiI$$

Cuprates react differently because, unlike alkyl lithium and Grignard reagents, cuprates carry a negatively charged metal, which changes the mechanism by which they react. In these reagents, the *copper*, not the alkyl group, first acts as the nucleophile in the unsaturated system, and the copper undergoes conjugate addition instead of direct addition. In the next step, the copper transfers *one* of its alkyl groups to the β-carbon and gains a lone pair of electrons from the copper-carbon bond. This process, called **reductive elimination**, is the opposite of the insertion of magnesium into a carbon-halide bond to form a Grignard reagent. In the case of a cuprate, the reductive elimination forms the new C–C bond and generates an enolate. Once all of the cuprate molecules have reacted, an electrophile such as H^+ is added to make a ketone.

Reductive elimination is a reaction involving the removal of a metal atom from between two alkyl groups and the return of a pair of electrons to the metal.

reductive elimination transfers methyl from copper to β-carbon

1) $\ominus$ Li $\oplus$, $H_3C{-}Cu{-}CH_3$; 2) $E^{\oplus}$; CH_3; CH_3Cu; E

nucleophilic copper participates in reversible conjugate addition to enone

generated enolate attacks an electrophile added in a second step

electron pair is transferred to the copper (reduction)

Alternatively, the enolate could be alkylated or be used in an aldol-type reaction, creating an extra carbon-carbon or even a carbon-heteroatom bond. A cuprate conjugate addition, such as an enolate alkylation reaction sequence, is an example of a **tandem reaction**, because two transformations occur one after another in a single reaction sequence.

A **tandem reaction** is a set of reactions, performed in sequence. The product of one reaction acts as the starting material in the next.

CHECKPOINT 18.1

You should now be able to predict the products formed when organometallic reagents, such as cuprates and Grignard reagents, are reacted with α,β-unsaturated carbonyl compounds. You should also be able to draw mechanisms for these reactions.

SOLVED PROBLEM

Predict the major product of the following reaction, and draw a mechanism for its formation.

1) Me_2CuLi
2) MeI

STEP 1: Determine the roles of the reactants and the type of reaction taking place in the first step.

The starting material is an α,β-unsaturated carbonyl with two electrophilic carbons: the carbonyl carbon and the β-carbon. The reactant from the first step is a nucleophilic cuprate. Nucleophiles can potentially react at either electrophilic position on an α,β-unsaturated carbonyl; however, copper-based nucleophiles tend to react at the β-carbon. Therefore, this reaction is a conjugate addition reaction.

STEP 2: Draw a mechanism for the first step of the reaction.

The nucleophilic cuprate first adds to the electrophilic β-carbon of the α,β-unsaturated carbonyl, generating an enolate intermediate. A reductive elimination then occurs to form a new C–C bond and release methyl copper (MeCu).

+ MeCu

STEP 3: Repeat Steps 1 and 2 until the final product is obtained.

The intermediate after the first step of the reaction is a nucleophilic enolate. The reagent added in the second step is electrophilic methyl iodide. The enolate shares its electrons with the methyl iodide, expelling an iodide ion and reforming the C=O π bond to make the final product shown.

H_3C–I

PRACTICE PROBLEM

18.1 Predict the products of the following reactions, and draw mechanisms showing their formation.

a) 1) PhMgBr; 2) NH_4Cl, H_2O

b) 1) Et_2CuLi; 2) HCl, H_2O

c) 1) MeMgBr; 2) HCl, H_2O

d) 1) Me_2CuLi; 2) HCl, H_2O

e) 1) (vinyl)$_2$CuLi; 2) Br

f) 1) (isobutyl)$_2$CuLi; 2) Cl

INTEGRATE THE SKILL

18.2 Suggest reagents that could be used to carry out the following transformations. More than one step may be required.

a) b) c)

OEt HO OEt O OEt

18.2.4 Michael reaction

The **Michael reaction**, or Michael addition, is a specific type of conjugate addition. Under basic conditions, a β-dicarbonyl compound such as a 1,3-diketone, a β-ketoester, or a malonate ester can add a nucleophile to the β-carbon of an α,β-unsaturated carbonyl compound. The properties of the enolate favours conjugate (1,4) instead of direct (1,2) addition. This is because the charge is spread over several atoms in the nucleophile, making the addition more easily reversible than it is with simple enolates. This puts the reaction under thermodynamic control.

The **Michael reaction** is the conjugate addition of a 1,3-dicarbonyl compound to an α,β-unsaturated carbonyl compound. Note that the α,β-unsaturated component is not limited to carbonyls. Other electron-withdrawing groups may be used.

enone has two electrophilic sites

NaOH
H_2O

1,3-diketone, β-ketoester or malonate ester can add to the β-carbon of unsaturated electrophile

The mechanism of the Michael reaction begins with the removal of one α-hydrogen from a 1,3-dicarbonyl compound by a base to form a stabilized enolate. Conjugate addition of the enolate to the β-carbon of the unsaturated electrophile yields an initial addition product: an enolate intermediate originating from addition to the electrophilic component. Protonation of the enolate intermediate by an acid (H_2O) gives the expected Michael addition product, which regenerates the hydroxide.

A highly acidic α-hydrogen remains in this product, which can further react with NaOH, driving the reaction equilibrium forward and generating the Michael product as an anion. Note that a stoichiometric amount of base must be used in this reaction, and an acid needs to be added at the end of the reaction to neutralize and isolate the Michael product.

ORGANIC CHEMWARE
18.2 Michael reaction

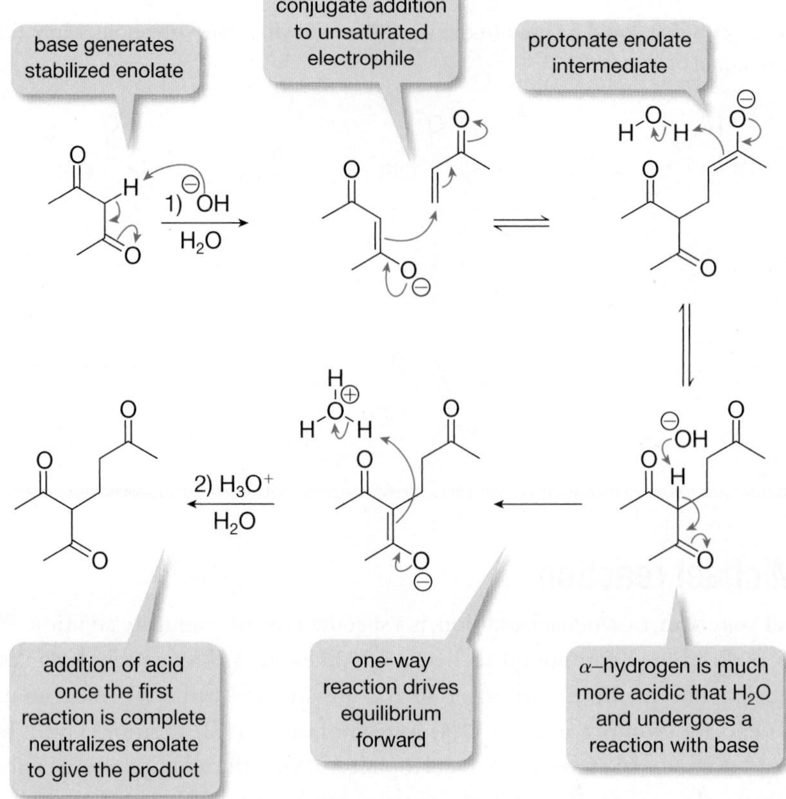
base generates stabilized enolate
conjugate addition to unsaturated electrophile
protonate enolate intermediate
1) $^{\ominus}$OH
H_2O
2) H_3O^+
H_2O
addition of acid once the first reaction is complete neutralizes enolate to give the product
one-way reaction drives equilibrium forward
α–hydrogen is much more acidic that H_2O and undergoes a reaction with base

A **Michael acceptor** is any α,β-unsaturated carbonyl compound capable of undergoing a Michael reaction.

A **Michael donor** is any nucleophile that favours conjugate addition to an α,β-unsaturated carbonyl compound.

α,β-Unsaturated carbonyl compounds are often referred to as **Michael acceptors** due to their ability to accept the dicarbonyl enolate nucleophile. The enolate is referred to as the **Michael donor**.

Reactions similar to a Michael reaction can involve different nucleophiles, and they proceed by 1,4-addition to an α,β-unsaturated system. All 1,4-additions, Michael reactions or not, are referred to as conjugate additions. However, the term *Michael reaction* is reserved for conjugate addition reactions involving the addition of 1,3-dicarbonyls.

HS⌒Ph, Tf_2NH (0.1 equiv.), CH_2Cl_2

HO⌒Ph, $HBF_4{\cdot}OMe_2$, CH_3CN, −20 °C

H_2NCO_2Bn, Tf_2NH (0.1 equiv.), CH_3CN, −20 °C

STUDENT TIP

Although a conjugate addition using a nucleophile other than a β-dicarbonyl enolate is occasionally referred to as a Michael or Michael-type addition, this is incorrect. A conjugate addition is a Michael addition only when a β-dicarbonyl enolate is the nucleophile.

CHECKPOINT 18.2

You should now be able to draw the products and curved arrow mechanisms for a Michael addition and other similar conjugate addition reactions.

SOLVED PROBLEM

Propose a mechanism for the synthesis of the blood thinner, warfarin, shown here:

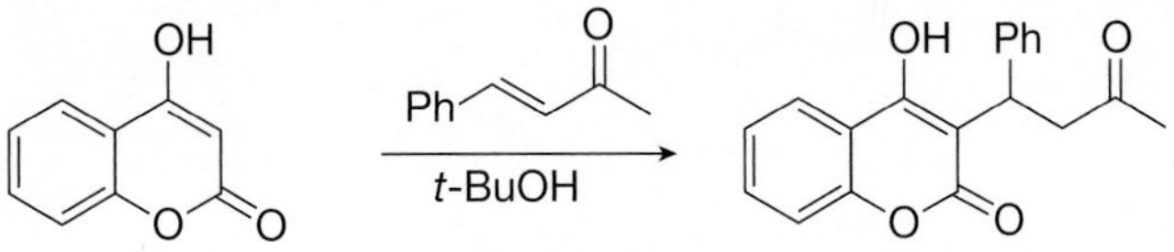

STEP 1: Determine the roles of the reactants and the type of reaction taking place.

The starting material contains a nucleophilic enol ether; the reactant is an α,β-unsaturated carbonyl with two electrophilic sites. Looking at the position of the new bond in the product, we can deduce that the enol ether added to the β-carbon of the unsaturated carbonyl in a conjugate addition reaction.

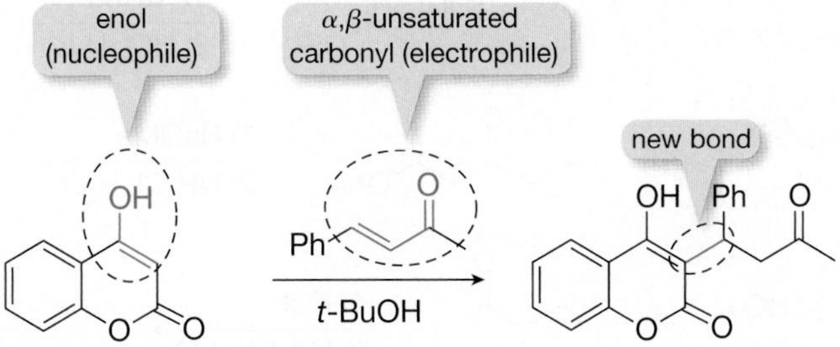

STEP 2: Draw a curved arrow mechanism for the conjugate addition step using the principle that electrons flow from nucleophile to electrophile to guide you.

Electrons flow from the oxygen of the enol ether to the π bond, which then shares its electrons with the β-carbon of the unsaturated carbonyl.

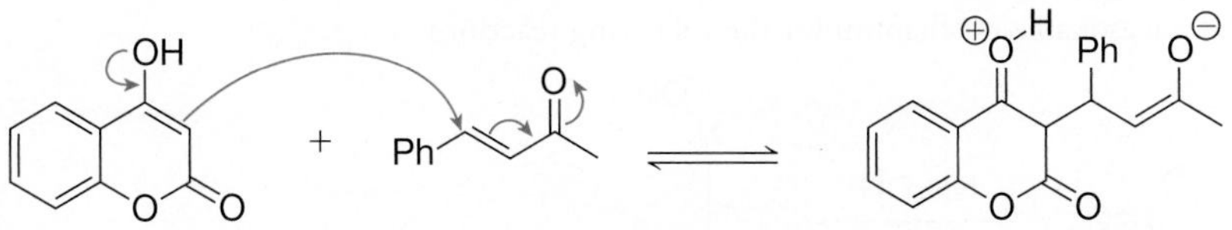

STEP 3: Use molecules of solvent to carry out the necessary proton-transfer steps to neutralize the intermediate and provide the final product.

The intermediate formed immediately after the conjugate addition has both a positive charge on the protonated carbonyl and a negative charge on the enolate oxygen. The α-hydrogen between the two carbonyl groups is made very acidic by the protonated carbonyl, so a molecule of *tert*-butanol is basic enough to remove this proton and re-form the enol ether. The conjugate acid of *tert*-butanol can then act as an acid to protonate the enolate to give the final neutral product.

+ HO-*t*-Bu ⟶ + H-O⁺(H)-*t*-Bu

+ HO*t*-Bu

PRACTICE PROBLEM

18.3 Predict the products of the following reactions, and draw curved arrow mechanisms showing their formation.

a) + 1) NaOMe 2) HCl, H_2O

b) + 1) NaOEt 2) HCl, H_2O

c) + 1) NaOMe 2) NH_4Cl, H_2O

d) + 1) KOH, 2) NH_4Cl, H_2O

e) + PhSH ⟶

INTEGRATE THE SKILL

18.4 Suggest a reasonable mechanism for the following reactions:

a) NH_2OH ⟶

b) MeOH/H_2O ⟶

CHECKPOINT 18.3

You should now be able to predict whether a reagent will react with an α,β-unsaturated carbonyl via direct or conjugate addition based on the nucleophile's structure and the reaction conditions.

SOLVED PROBLEM

Predict the major product of the following reaction.

1) $NaNH_2$, THF
2) NH_4Cl, H_2O

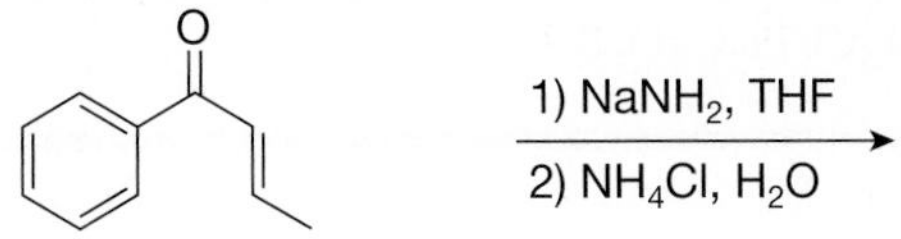

STEP 1: Determine the roles of the reactants and the type of reaction taking place.

The starting material contains an α,β-unsaturated carbonyl with two electrophilic sites. The reactant in the first step of the reaction is a negatively charged amide ion (NH_2^-). Highly charged nucleophiles are more likely to react via direct addition than conjugate addition; therefore, the first step of this reaction is likely to involve a nucleophilic attack of the amide onto the carbonyl carbon.

STEP 2: Draw a mechanism for the reaction.

The nucleophilic nitrogen adds to the carbonyl carbon, breaking the C=O π bond and forming a tetrahedral intermediate. The tetrahedral intermediate is then protonated in the second step by the hydronium ions formed from NH_4Cl and water.

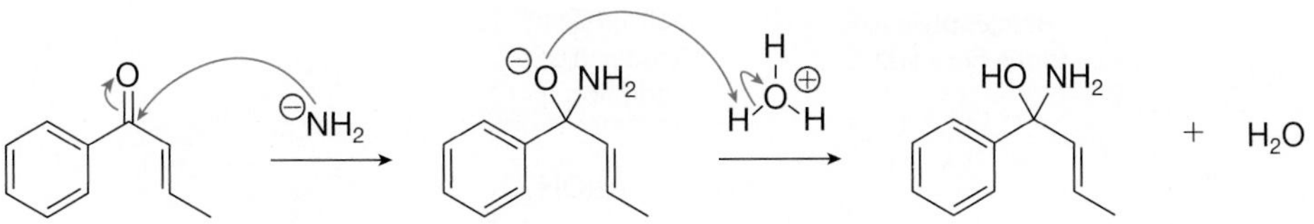

$^{\ominus}NH_2$ → $^{\ominus}O$ NH_2 — $H_3O^{\oplus}$ → HO NH_2 + H_2O

PRACTICE PROBLEM

18.5 Predict the products of the following reactions, and draw mechanisms showing their formation.

a) 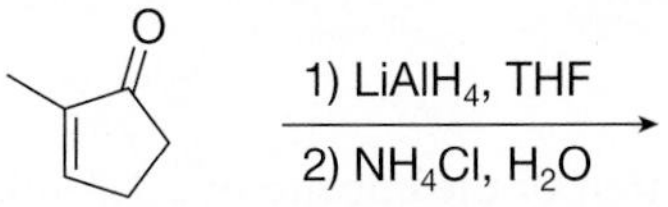 1) $LiAlH_4$, THF; 2) NH_4Cl, H_2O

b) 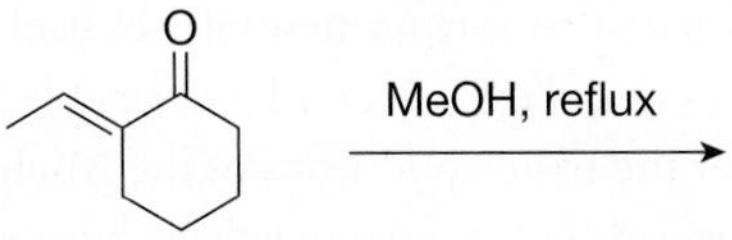 MeOH, reflux

c) 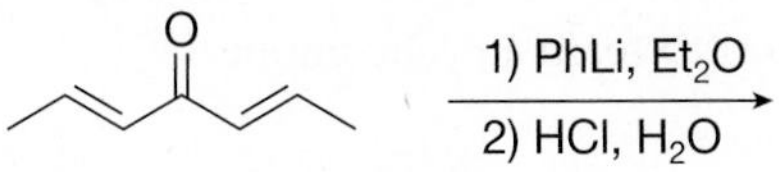 1) PhLi, Et_2O; 2) HCl, H_2O

d) 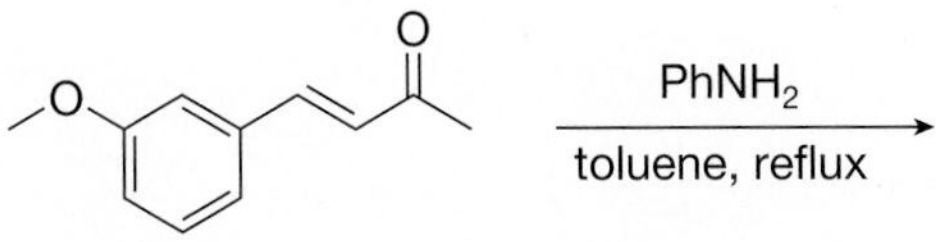 $PhNH_2$; toluene, reflux

e) 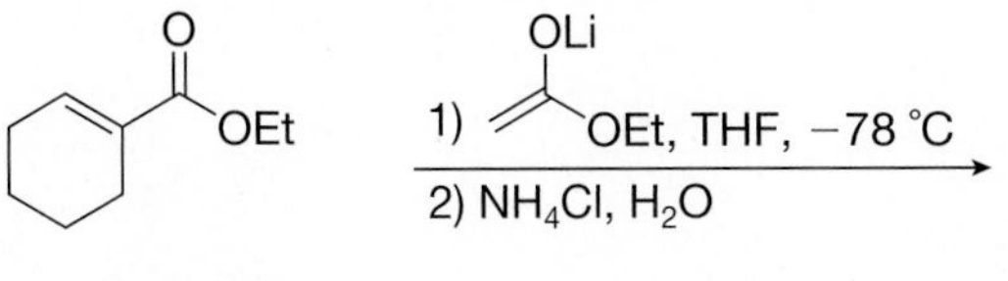 1) OLi / OEt, THF, −78 °C; 2) NH_4Cl, H_2O

INTEGRATE THE SKILL

18.6 The following compound has both an α,β-unsaturated carbonyl and an isolated double bond. Predict where each of the following reagents is most likely to react on the molecule.

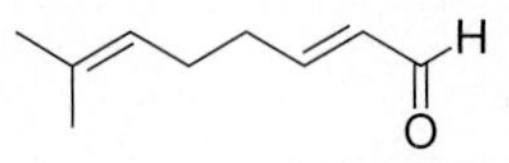

Cl_2, NaOH, $LiAlH_4$, OsO_4, CH_3MgI, HOO^-, CH_3OH, O_3, Ph_2PH, HC≡CLi, NaI, $PhCH_2OH$, EtLi, CH_3SH, $PhCH_2MgCl$, Pr_2CuLi, ICl

18.3 Using Michael Additions to Generate Complex Organic Molecules

18.3.1 The Robinson annulation stands the test of time

A **Robinson annulation** is a ring-forming process that involves a Michael addition, followed by a ring-closing aldol condensation. Often a Robinson annulation forms a cyclohexenone.

The Michael addition is part of a method for the synthesis of cyclohexenones. A Michael addition followed by an aldol condensation to form a new ring is known as a **Robinson annulation**. The reaction was developed by Robert Robinson in 1935 and has been used in the synthesis of many natural products, including steroids.

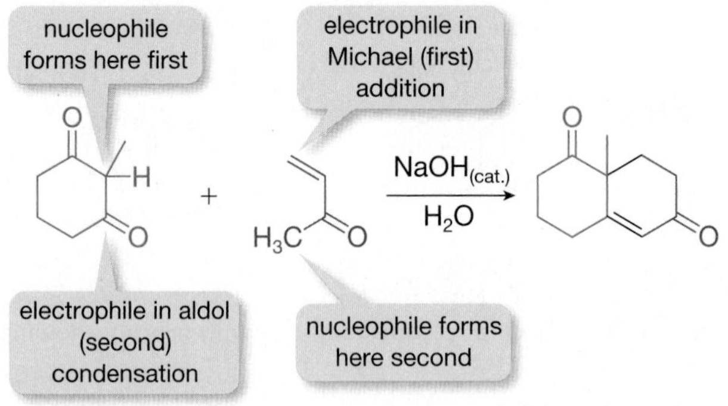

A Robinson annulation begins with conjugate addition of the enolate of a 1,3-diketone to an enone. The enolate that is generated is then protonated to form a neutral Michael addition product. In the example shown, the methyl group does not allow a second irreversible deprotonation from the α-carbon of the Michael donor. Under the basic conditions of the Michael addition, a small amount of terminal enolate is formed instead. This enolate cyclizes by undergoing an intramolecular aldol condensation and elimination to form a cyclic enone.

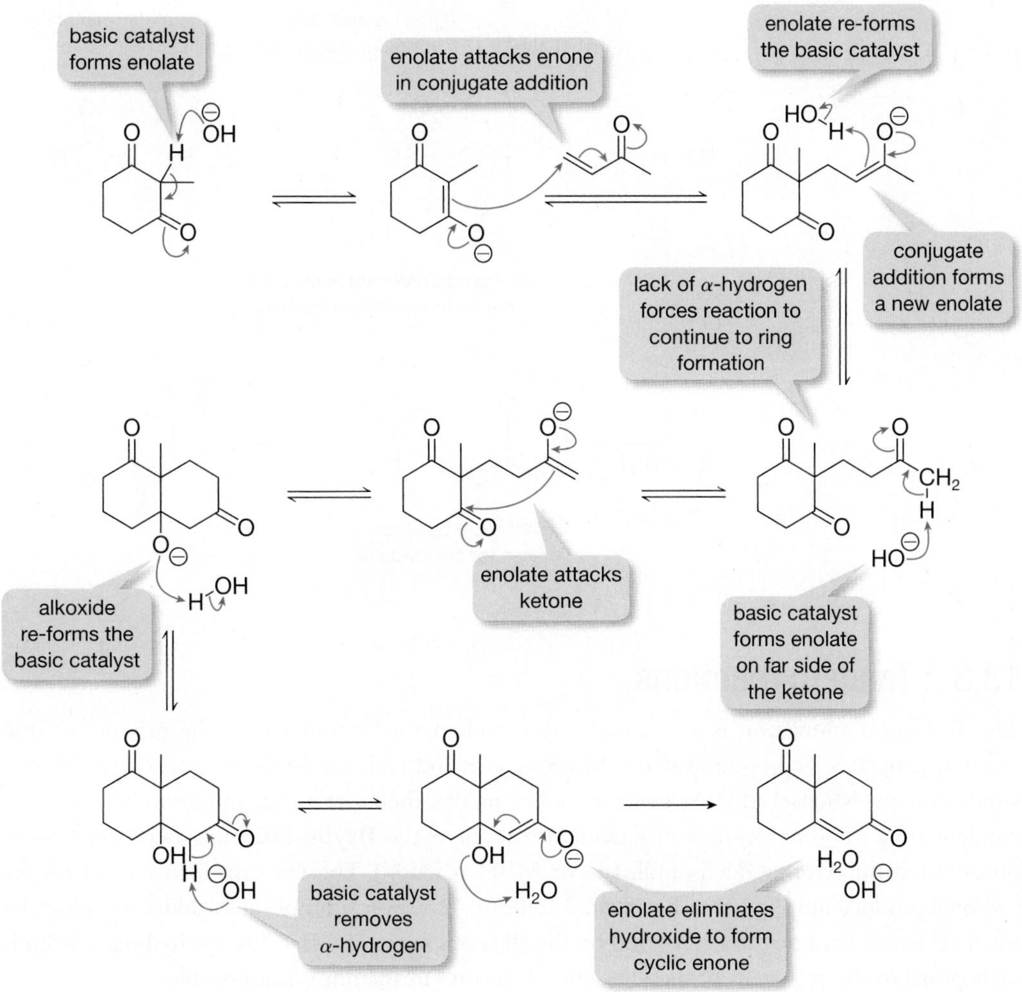

Any Michael addition can become a Robinson annulation if the Michael acceptor can also enolize. However, a second enolizable hydrogen on the Michael donor is more acidic and will be preferentially removed by the base, preventing the formation of the necessary terminal enolate. The enol formed from the 1,3-dicarbonyl cannot cyclize because the ring that would result is too strained (four-membered ring) and is too difficult to form. Other conjugate additions using acceptors without enolizable α-hydrogens, such as α,β-unsaturated esters or phenyl ketones, cannot undergo the cyclization of the Robinson process.

When an enolizable group is present on the Michael acceptor and a second enolizable hydrogen is absent on the Michael donor, a Michael addition can be followed by the intramolecular aldol condensation that is required for a Robinson annulation.

18.3.2 Tandem reactions

The Baylis–Hillman reaction temporarily converts an α,β-unsaturated carbonyl compound into an enolate, which adds to an aldehyde electrophile.

The Robinson annulation is an example of a tandem reaction, in which the product of one reaction generates a compound that undergoes a second reaction. In the case of the Robinson annulation, the Michael addition generates a ketone that then carries out an intramolecular aldol condensation. Another example of a tandem reaction is the **Baylis–Hillman reaction** (sometimes called the Morita–Baylis–Hillman, or MBH reaction). The reaction forms a β-hydroxy carbonyl product and appears to be an aldol addition. However, a simple aldol addition cannot be the mechanism because the C–H bond on the alkene is not acidic. Because the hydrogen atom is orthogonal to the π-system of the C=O bond, enolate formation is not possible.

The mechanism of the Baylis–Hillman reaction involves a tertiary amine catalyst in a conjugate addition. Just as in the addition of cuprates, this generates an enolate that can undergo further reactions. After an aldol-type addition to an aldehyde occurs, the catalyst is eliminated by an $E1_{cb}$ reaction that regenerates the alkene.

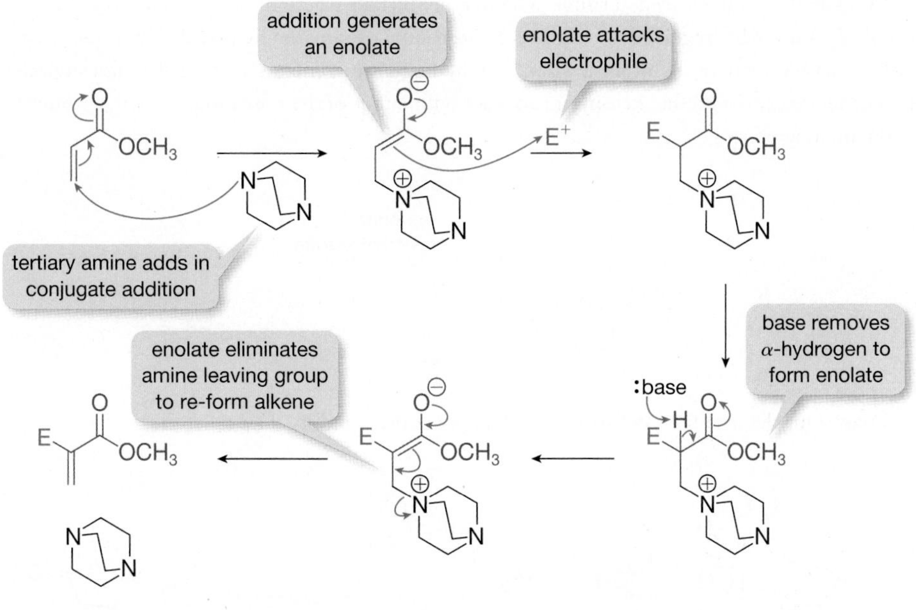

addition generates an enolate

enolate attacks electrophile

tertiary amine adds in conjugate addition

base removes α-hydrogen to form enolate

enolate eliminates amine leaving group to re-form alkene

CHECKPOINT 18.4

You should now be able to predict the product of tandem reactions involving conjugate addition reactions on unsaturated carbonyls—in particular, the Robinson annulation and the Baylis–Hillman reaction. You should also be able to draw mechanisms for these reactions.

SOLVED PROBLEM

Predict the major product of the following reaction, and draw a mechanism showing its formation.

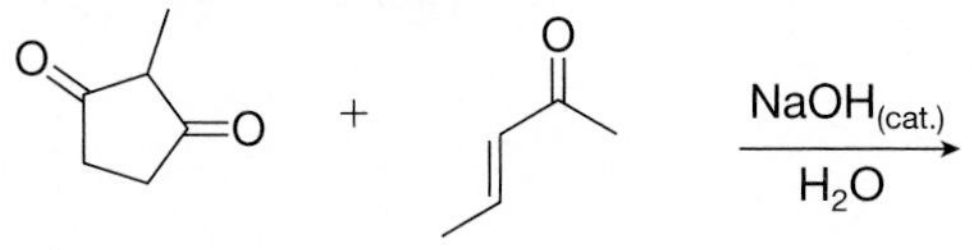

$NaOH_{(cat.)}$, H_2O

STEP 1: Determine the roles of the reactants and the type of reaction taking place.

Both of the starting materials are electrophilic; however, the 1,3-dicarbonyl also has an acidic α-hydrogen that is readily deprotonated in the presence of the hydroxide base. This creates a nucleophilic enolate to react with an electrophilic α,β-unsaturated carbonyl. Since the enolate is stabilized by two carbonyl groups, we can expect this nucleophile to add in a conjugate addition reaction (Michael addition).

STEP 2: Draw a mechanism for the formation of the Michael addition product.

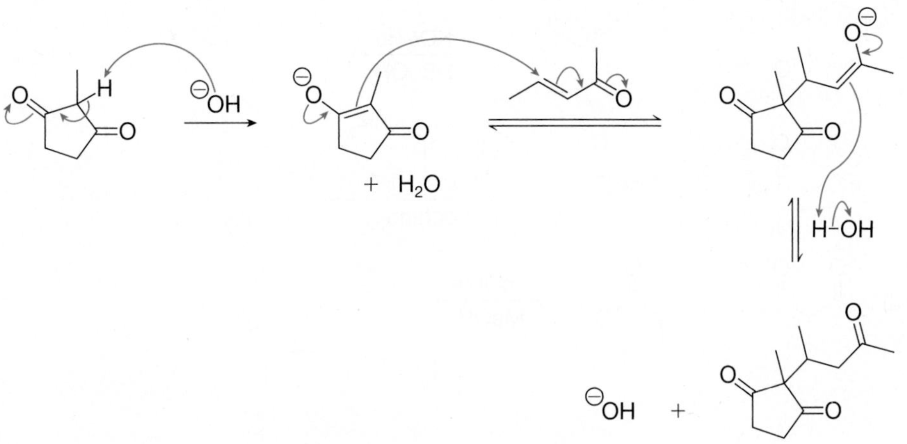

STEP 3: Evaluate the products to determine whether any further reactions are possible from the reaction.

Since there is no α-hydrogen between the 1,3-dicarbonyl groups, the hydroxide left in the reaction can potentially react with any of the other less acidic α-hydrogens to generate new enolate nucleophiles. If an enolate is generated at the methyl ketone, it can react with either of the remaining carbonyl groups to form a stable six-membered ring.

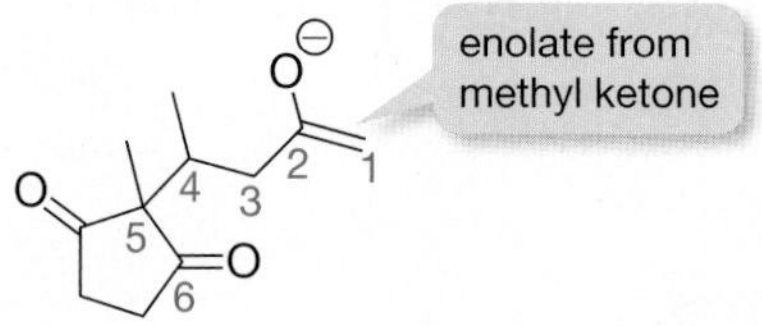

enolate from methyl ketone

STEP 4: Draw a mechanism for the formation of a new enolate and its aldol reaction.

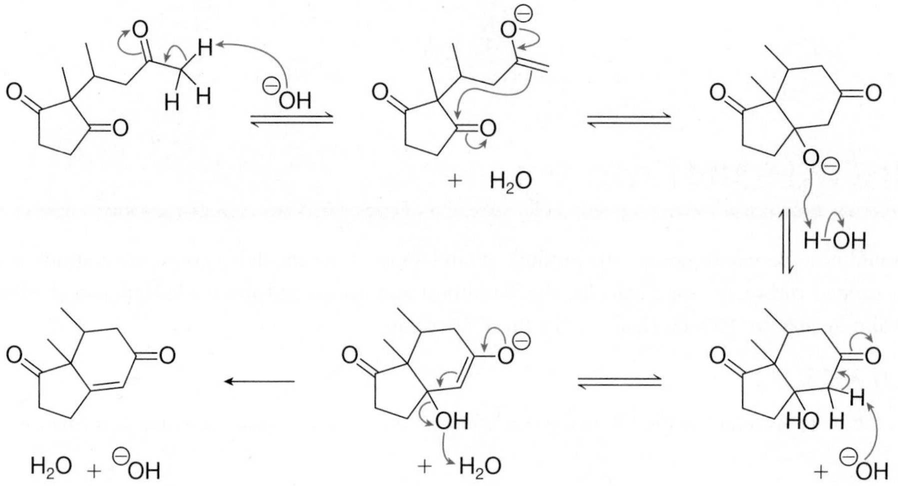

PRACTICE PROBLEM

18.7 Predict the products of the following reactions:

a) 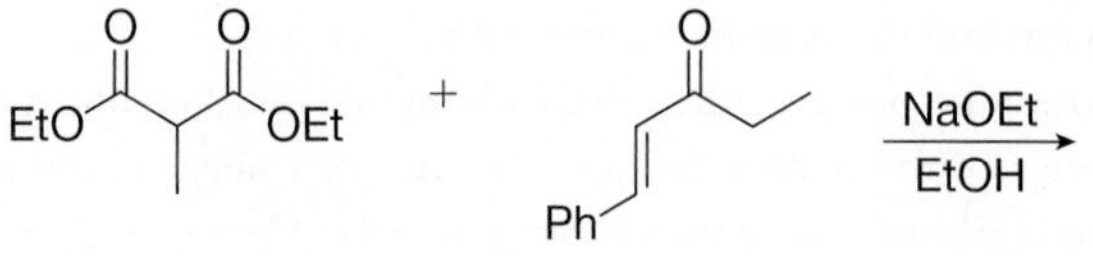 EtO OEt + Ph; NaOEt, EtOH

b) 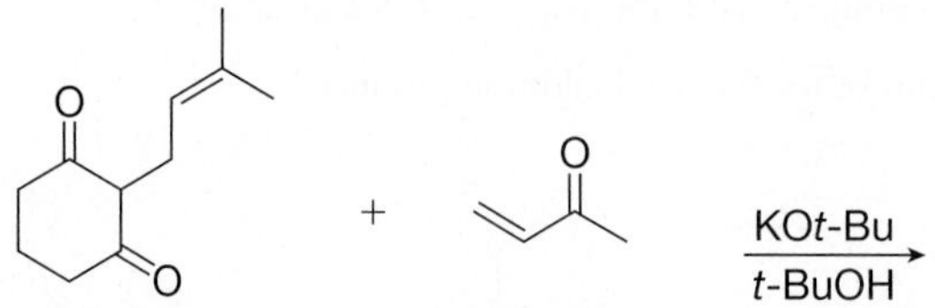 + ; KOt-Bu, t-BuOH

c) 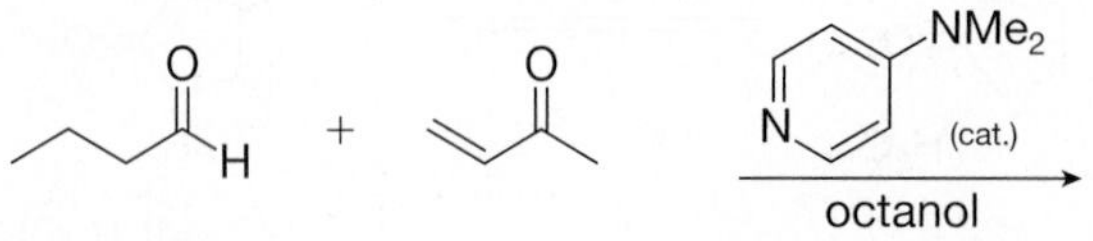 H + ; NMe_2 (cat.), octanol

d) 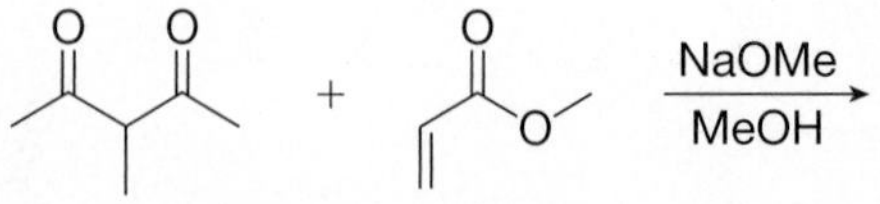 + ; NaOMe, MeOH

INTEGRATE THE SKILL

18.8 Propose a mechanism for the following tandem reaction.

18.4 Regioselectivity in Ketone Nucleophiles

18.4.1 Ketone enolates

When ketones are used to make enolates, regioselectivity becomes a consideration. Unsymmetrical ketones can be deprotonated at either α-carbon, which leads to different enolates and ultimately different products. If the unsymmetrical substitution is very close to the carbonyl, it is usually possible to control which enolate forms.

Such regioselectivity is not an issue with other enolizable groups such as esters or amides, because only one side of the carbonyl can be deprotonated in those functional groups.

18.4.2 Kinetic versus thermodynamic enolates

Enolates that are made with a strong hindered base (LDA) typically form irreversibly. Because the difference in pK_a between the base (~35) and the enolizable hydrogen (~22) is very large, once the enolate forms it cannot reprotonate. Under such kinetic control, less substituted enolates are formed selectively. The **kinetic enolate** forms faster because less substituted α-carbons are more accessible to a hindered base, and so the activation energy for their formation is lower.

Enolates that are allowed to equilibrate before reacting with an electrophile tend to form the more substituted enolate. This **thermodynamic enolate** forms because it contains a more substituted alkene and is therefore slightly more stable than the other enolate.

A **kinetic enolate** forms under conditions of kinetic control (irreversible), and the hydrogen is removed from the less substituted α-carbon.

A **thermodynamic enolate** forms under conditions of thermodynamic control (reversible), and the hydrogen is removed from the more substituted α-carbon.

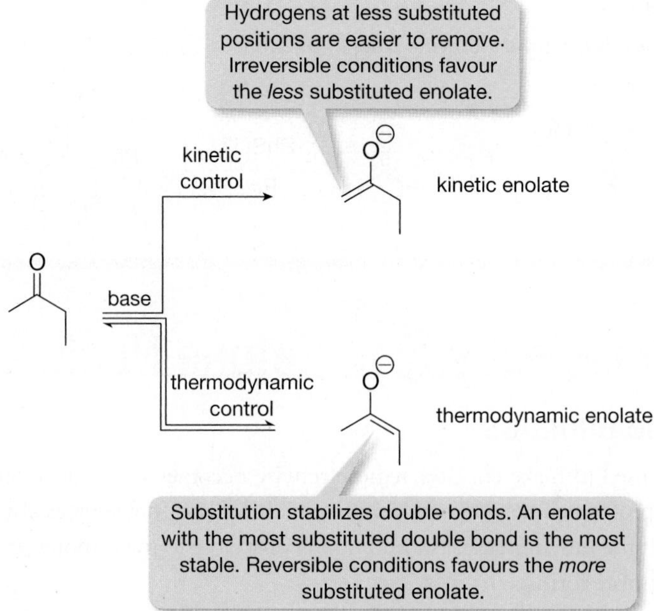

When a base is added to a flask containing a ketone, an excess of ketone is always present in the solution as the base is added. When a limiting amount of base is added to a slight excess of ketone and the mixture is allowed to warm, the excess ketone molecules provide a small supply of protons. This allows the enolates to equilibrate until the most stable enolate accumulates as the major product. The resulting thermodynamic enolate is the major product in a mixture of enolates that can be used as is, or trapped as a mixture of silyl enol ethers: stable products that can be purified to remove the minor component.

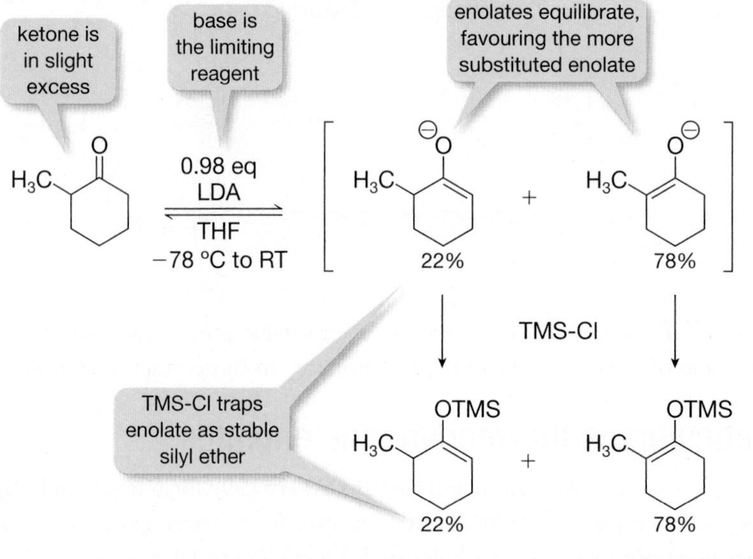

Inverse addition is the addition of an enolizable carbonyl to a solution of base, used to form a kinetic enolate.

Alternatively, the ketone can be added to a slight excess of base at low temperature. This method is called **inverse addition**, and it ensures that there is always a large excess of base relative to ketone. This prevents equilibration because there are never any hydrogens available that could protonate the enolate molecules. Under these conditions, the enolate that forms is the one with the smallest activation energy. The resulting kinetic enolate can be used directly or trapped as the silyl enol ether.

Silyl enol ethers, whether formed from the kinetic or thermodynamic enolate, can be converted back to enolates after purification. With the addition of methyl lithium, the lithium enolate forms.

18.4.3 Mukaiyama aldol reaction

Silyl enol ethers can be used directly in aldol reactions in a process called a **Mukaiyama aldol reaction**. The Mukaiyama aldol reaction is an excellent way to obtain only one product from a crossed aldol reaction without using strong base (see Section 17.4.2). It also provides good control of regioselectivity with ketone enolates.

A **Mukaiyama aldol reaction** is a Lewis acid–catalyzed aldol addition, using a silyl enol ether as a trapped enolate equivalent.

In a Mukaiyama aldol reaction, a Lewis acid such as $TiCl_4$ coordinates to the carbonyl of the aldehyde electrophile. In the process, the titanium loses one chloride ion, which then attacks the TMS group of the enol ether to release the trapped enolate that in turn attacks the activated aldehyde. Because silyl enol ethers can be purified and the undesired regioisomer can be discarded, the desired products can be formed in high purity.

$TiCl_4$ Lewis acid activates the electrophile

chlorine ion frees the trapped enolate by removing TMS

the enolate adds to the activated aldehyde in an aldol addition

once the reaction is complete, workup with water gives the aldol product

CHECKPOINT 18.5

You should now be able to predict whether a ketone will form the kinetic or thermodynamic enolate, given the ketone's structure and the reaction conditions. You should also be able to draw the products formed when these enolates react with other electrophiles—either directly or as their silyl enol ethers, as in the Mukaiyama aldol reaction.

SOLVED PROBLEM

Predict the major product of the following reaction. Suggest modifications that could be made to make the reaction favour the minor product.

1) 0.95 equiv. LDA THF, −78 °C to rt
2) CHO
3) HCl, H_2O

STEP 1: Determine the roles of the reactants and the type of reaction taking place.

The starting material has an electrophilic carbonyl group with acidic protons at both α-carbons. In the first step of the reaction, a strong, non-nucleophilic base (LDA) has been added. Therefore, an acid–base reaction in the first step will likely yield an enolate. The enolate can then react with the electrophilic aldehyde added in the second step of the reaction in a crossed aldol reaction.

STEP 2: Determine whether the reaction conditions will favour the kinetic or the thermodynamic enolate.

Since less than 1 molar equivalent of the base has been added, there will be some leftover ketone in the reaction mixture with the enolate. This will lead to an equilibrium that favours the formation of the thermodynamic enolate. Note that the increase in temperature from −78 °C to room temperature (rt) further encourages the establishment of an equilibrium.

STEP 3: Draw the thermodynamic enolate.

The thermodynamic enolate is the more stable enolate—in this case, the one with the more substituted double bond.

STEP 4: Draw the product formed between the thermodynamic enolate and the aldehyde electrophile.

Although the question does not ask for a mechanism, it is often helpful to draw in a partial mechanism, showing where new bonds are formed and old bonds are broken.

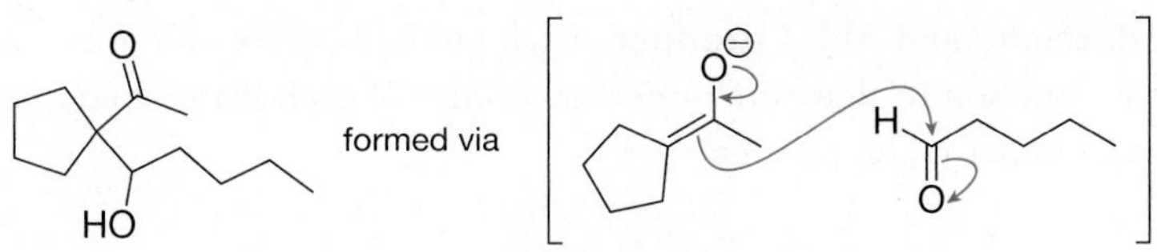

formed via

STEP 5: Propose conditions that would favour the formation of the kinetic enolate.

To form the kinetic enolate, there needs to be an excess of base at all times during the deprotonation. This is achieved by using a small excess of base (e.g., 1.05 equivalents) and adding the ketone to the base, rather than the other way around.

PRACTICE PROBLEM

18.9 Predict the products of the following reactions, and draw mechanisms to show their formation.

a) 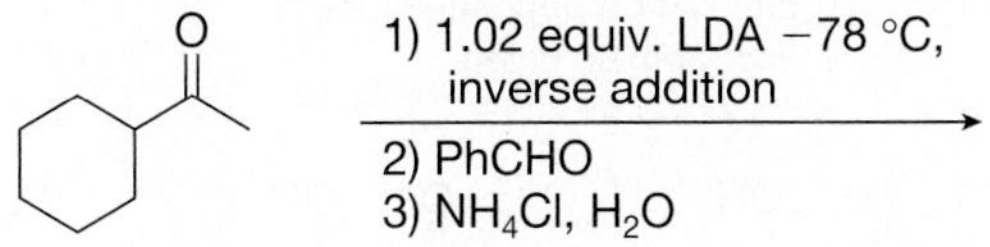 1) 1.02 equiv. LDA −78 °C, inverse addition; 2) PhCHO; 3) NH_4Cl, H_2O

b) 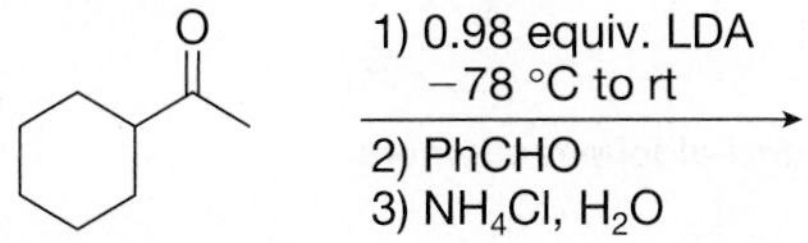 1) 0.98 equiv. LDA −78 °C to rt; 2) PhCHO; 3) NH_4Cl, H_2O

c) OTMS + (aldehyde) 1) $TiCl_4$, CH_2Cl_2; 2) H_2O

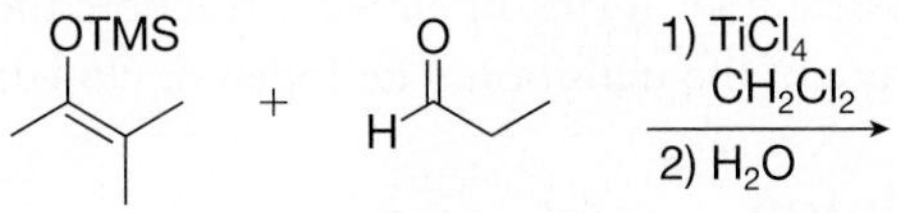

d) 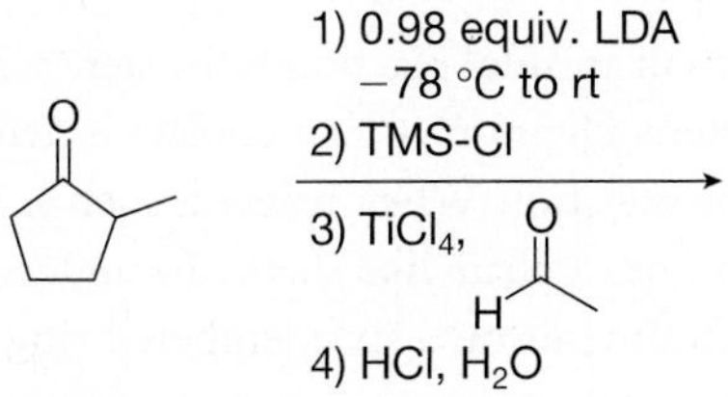 1) 0.98 equiv. LDA −78 °C to rt; 2) TMS-Cl; 3) $TiCl_4$, (aldehyde); 4) HCl, H_2O

INTEGRATE THE SKILL

18.10 a) Select appropriate reagents and conditions to complete the following transformations.

i)

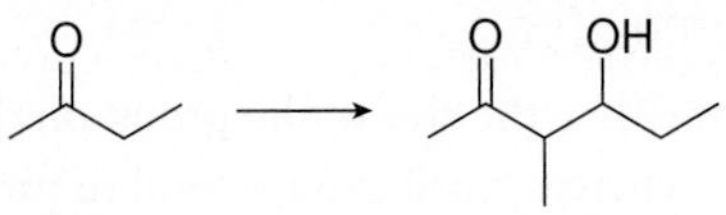

ii)

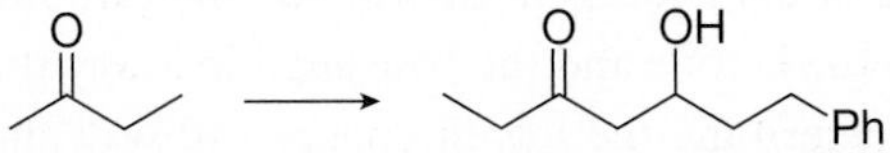

b) Draw an energy diagram comparing the formation of the kinetic and thermodynamic enolates of 2-methylcyclohexanone.

18.5 Stereoselectivity in Aldol Processes

A ***syn*-aldol product** is an aldol in which the substituent on the α-carbon and the hydroxy at the β-position are *syn* with respect to the extended carbon chain.

An ***anti*-aldol product** is an aldol in which the substituent on the α-carbon and the hydroxy at the β-position are *anti* with respect to the extended carbon chain.

The product of an aldol reaction contains two new stereocentres. These centres can form either with the α-alkyl group oriented in the same direction as the hydroxyl (***syn*-aldol product**) or in the opposite direction (***anti*-aldol product**). Each product (*syn* or *anti*) forms as a mixture of enantiomers. It is common to draw only one enantiomer of each diastereomer as it is assumed that the other enantiomer is also present.

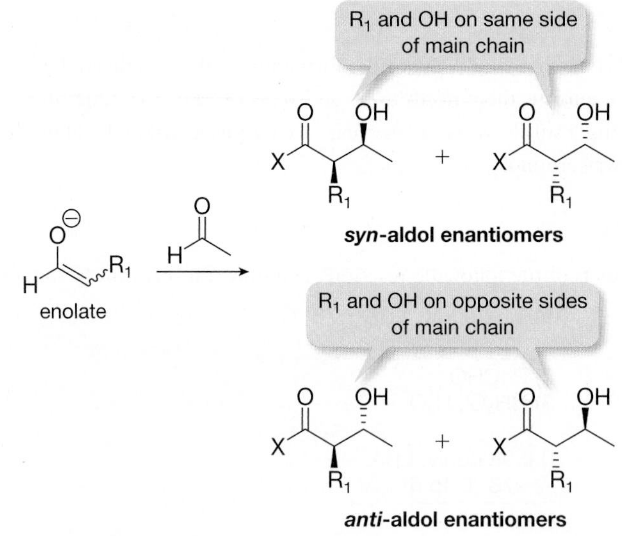

Two important factors control which diastereomer forms in an aldol reaction: the stereochemistry of the enolate (*E* or *Z*) and the nature of the transition state (open or closed).

18.5.1 *E* enolates versus *Z* enolates

The first factor that controls which diastereomer forms in an aldol reaction is the stereochemistry of the enolate. Normally under kinetic control, the stereochemistry of an enolate is determined by the relative energies of the transition states for deprotonation. When reagents such as LDA are used as the base, the transition state for deprotonation has a chair-like shape. To understand the interactions in these transition states, consider that substituents on a six-membered ring "prefer" equatorial-like (pseudo-equatorial) locations. Three general situations arise concerning the relation between transition state and the geometry of the enolate; these situations correspond to whether the enolate is formed from a ketone, an ester, or an amide.

18.5.1.1 Controlling the geometry of ketone enolates

The *E* or *Z* geometry of a ketone enolate is determined by the size of the group on the *other* side of the carbonyl. Large groups favour the *Z* enolate, whereas small groups tend to produce the *E* enolate. The transition state for the removal of the hydrogen from beside the carbonyl involves a flow of electrons between six atoms, and so the ketone and the base align in a way that resembles a chair structure. Interactions in the chair determine the lowest energy pathway, and hence the geometry of the enolate formed.

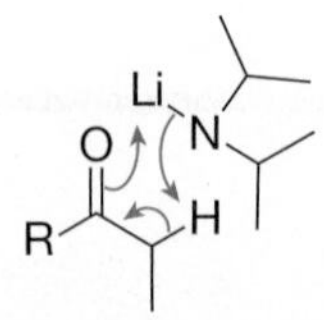

Because there is a great deal of sp^2 character in the atoms making up the chair, interactions between adjacent groups in the structure become important. If there is a large R group on the ketone component, interactions between this group and the group at the α-position determine the geometry of the enolate formed. These interactions are strongest when the α-substituent is in the pseudo-equatorial position.

weaker 1,3-interaction with *i*-Pr group
fast
slow
Z enolate
E enolate
R group is large—strong interactions with adjacent α-group

The geometry of a ketone enolate can be more carefully controlled by using a counter ion that contains boron. These enolates are formed by adding the appropriate boron halide (which is very electrophilic due to the electron-deficient boron atom) to the ketone together with a hindered amine base. The boron reagent is a Lewis acid, which forms an oxonium and lowers the pK_a of the α-hydrogen enough to allow the amine to form the enolate. The size of the groups on the boron and the base can be used to determine the geometry of the enolate that forms.

favoured
disfavoured
E boron enolate formed selectively
Z boron enolate
The large Cy (cyclohexyl) groups and α-group interfere. The α-group tends to align *anti* to the oxygen.

Hünig's base is ethyl diisopropyl amine (sometimes called diisopropylethylamine), a sterically hindered tertiary amine base.

Large groups on boron favour *E* enolates. The *Z* enolate can be produced by using smaller groups on the boron atom together with a bulky amine base such as ethyl diisopropyl amine, also known as **Hünig's base**. The smaller groups on the boron reduce the interactions with the substituent on the ketone. A bulky base approaches the ketone from a direction that is angled slightly away from the boron, which tends to enhance the preference for the ketone substituent to be *syn* to the oxygen.

The linear alkyl chains are relatively small and allow the CH_3 to orient *syn* to the oxygen.

slow

(*E*)-(O) boron enolate

i-Pr_2NEt

fast

(*Z*)-(O) boron enolate formed selectively

The larger amine base must approach from an angle away from the boron.

After undergoing an aldol addition, boron enolates require a workup that includes H_2O_2 to remove the boron, similar to the hydroboration-oxidation reactions of alkenes.

The use of an unsymmetrical ketone introduces problems with the regioselectivity of enolate formation (kinetic or thermodynamic). Another approach to address this problem is to use an ester or tertiary amide enolate and convert the product to a ketone after the aldol reaction is complete. Amide enolates are particularly convenient to work with, and using the Weinreb amide provides a convenient way to convert to the ketone.

1) LDA
2) $E^{\oplus}$

1) RMgBr, Me_3Al
2) $H_3O^{\oplus}$

18.5.1.2 Controlling the geometry of ester and amide enolates

When esters are deprotonated using LDA, the removal reaction involves a six-membered transition state that resembles a chair, and the groups on the ester tend to "prefer" pseudo-equatorial positions. Because the oxygen of the OR group on the ester is much smaller than the alkyl group of a ketone, interactions between the OR group and the group on the α-position are very small. This gives an opposite selectivity for esters than observed for ketones. The strongest interaction happens between the pseudo-axial α-substituent and one of the groups on the amine. This leads to a lower energy transition state for the *E* enolate, which forms selectively.

stronger 1,3-interaction with *i*-Pr group

slow

fast

Z enolate

E enolate favoured

Oxygen is small. Interactions with this group are very weak.

STUDENT TIP

In situations where sp^2-hybridized atoms are part of the six-membered ring, chairs are often used to represent the structure. However, the sp^2 atoms do not allow the structure to be a true chair; rather, it becomes chair-like. In this chair-like structure, the substituents cannot be exactly equatorial or axial. Instead, they are pseudo-equatorial or pseudo-axial (almost but not quite).

Enolates can be formed only from tertiary amides. This is because any hydrogens bonded to the nitrogen of a primary amide are more easily removed than the hydrogens bonded to carbons.

LDA

this hydrogen is most easily removed

The Hammond postulate can be used to facilitate the analysis of the stereoselectivity of enolate formation in amides. Because the transition state that leads to an enolate occurs late in the reaction, this transition state resembles the enolate, and the interactions in this enolate predict differences in the energies of the transitions states leading to this intermediate.

Conjugation is very strong in an amide, and the nitrogen in this group has considerable sp^2 character, making the C–N bond almost a full double bond. The atoms making up the amide functional group (N–C–O) are almost fully sp^2-hybridized and coplanar as a result of this.

resonance contribution is very strong—the atoms of the amide group are almost coplanar

When enolates are formed from amides, the flat geometry of the amide group creates very strong interactions between the groups on the nitrogen and the group on the enolate. These interactions strongly disfavour the *E* form. When amides are deprotonated, the *Z* isomer is usually the only isomer produced.

CHECKPOINT 18.6

You should now be able to predict the geometry of an enolate formed under a variety of conditions and explain the selectivity using appropriate drawings.

SOLVED PROBLEM

Select appropriate conditions to make the following enolates, and explain the origin of selectivity in each case.

STEP 1: Assign each enolate as either *E* or *Z*.

The enolate on the left is the *E* enolate; the enolate on the right is the *Z* enolate.

STEP 2: Determine conditions that can be used to favour the formation of the *E* enolate from a ketone.

The *E* enolate of a ketone can be favoured by using a bulky boron counter ion together with a hindered amine base, such as Cy_2BCl with Et_3N.

STEP 3: Draw a mechanism for the formation of the *E* enolate, showing the steric interactions (or lack of interactions) that lead it to be the favoured enolate.

A lone pair on oxygen attacks the electrophilic boron, causing the chloride leaving group to leave. The boron now acts like a Lewis acid, making the α-protons of the carbonyl acidic enough that the amine base can remove them to form the enolate.

During the deprotonation step, the alkyl group on the α-carbon orients itself away from the carbonyl group to avoid steric interactions between it and the bulky cyclohexyl groups on the boron. This results in the formation of the *E* enolate rather than the *Z* enolate.

STEP 4: Repeat Steps 2 and 3 for the *Z* enolate.

Ketones favour the *Z* enolate when treated with LDA, provided that the R group on the non-enolizing side of the ketone is large. Since the R group here is a bulky *tert*-butyl group, the treatment of this ketone with LDA will likely strongly favour the *Z* enolate. Deprotonation by LDA proceeds via a cyclic six-membered transition state. To minimize steric interactions with the bulky *tert*-butyl group, the α-methyl group of the ketone is oriented in the pseudo-axial position.

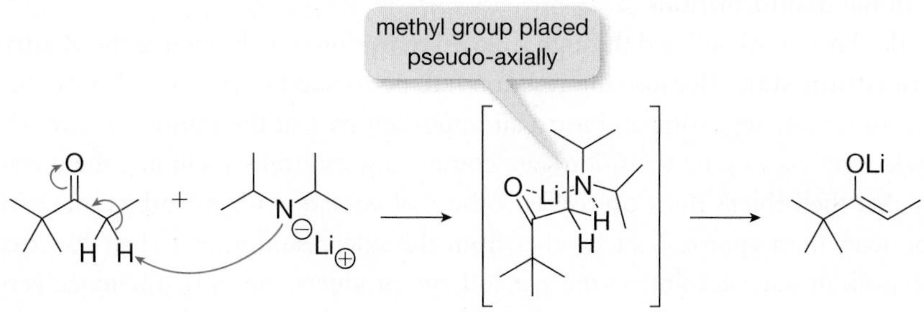

methyl group placed pseudo-axially

PRACTICE PROBLEM

18.11 Draw the expected enolate for each of the following reactions, and use a drawing to explain the predicted selectivity:

a) 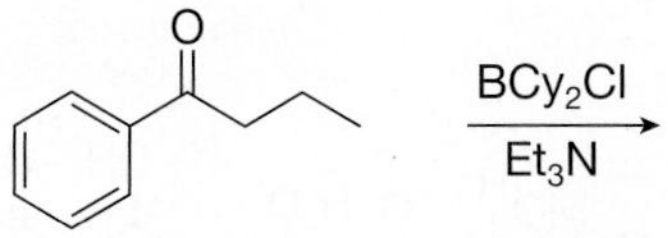 BCy_2Cl, Et_3N

b) 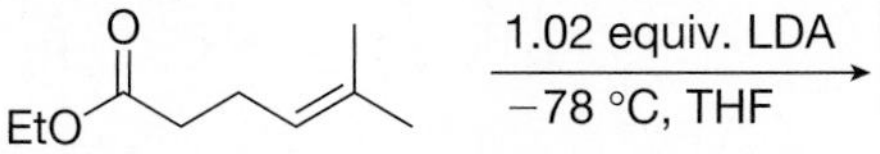 1.02 equiv. LDA, −78 °C, THF

c) BBu_2Cl, i-Pr_2NEt

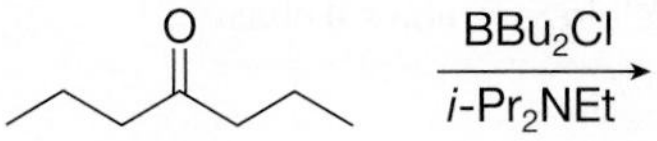

d) 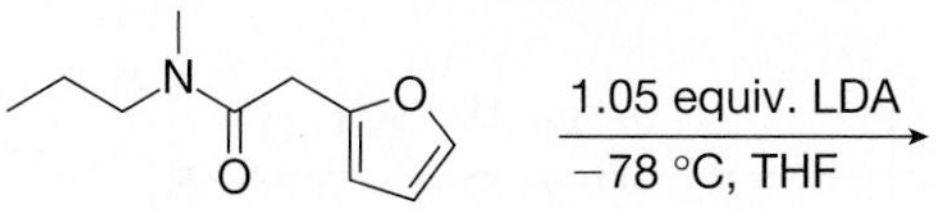 1.05 equiv. LDA, −78 °C, THF

e) 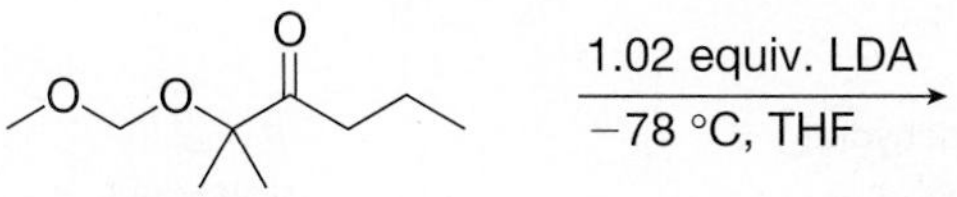 1.02 equiv. LDA, −78 °C, THF

INTEGRATE THE SKILL

18.12 Predict the regio- and stereochemistry of the enolates formed in the following reactions.

a) 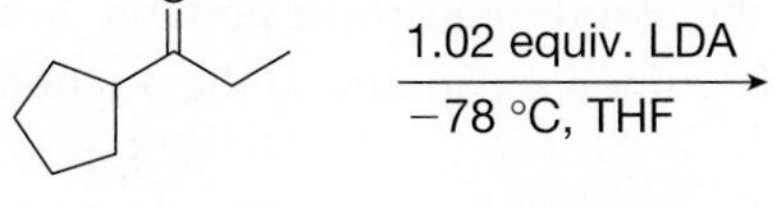 1.02 equiv. LDA, −78 °C, THF

b) 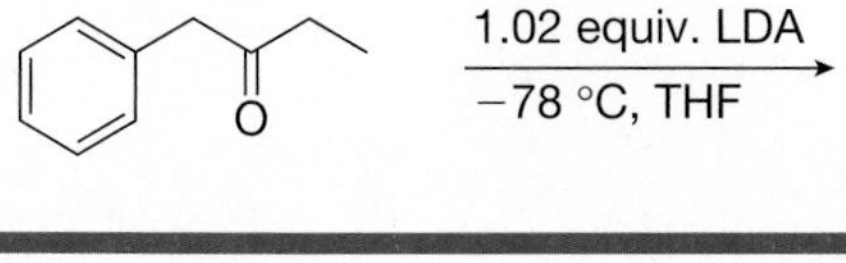 1.02 equiv. LDA, −78 °C, THF

18.5.2 The Zimmerman–Traxler transition state model

An **open transition state** is a transition state in which the nucleophile approaches the electrophile from the least sterically hindered direction.

A **closed transition state** is a transition state in which the nucleophile and electrophile are complexed and form a ring. Closed transition states offer better control of stereochemistry than open transition states.

The **Zimmerman–Traxler transition state** is the closed transition state for aldol additions that have a coordinating counter ion. The six-membered ring chair explains the observed stereochemistry of the aldol product.

The second factor that controls diastereoselectivity in aldol reactions is the nature of the transition state. Two types of transition states are possible for aldol reactions. In an **open transition state**, the electrons flow in one direction from one component to the other. In a **closed transition state**, the electrons flow in a ring from one component to the other and back.

The best control of stereochemistry happens with closed transition states because all of the components are close together and interact. In these transition states, the components can align only in certain ways, and this restriction gives higher selectivity to the reactions. Open transition states allow the molecules to approach each other in several ways, which leads to less stereoselectivity in the transformations.

Many aldol reactions proceed through a closed transition state known as the **Zimmerman–Traxler transition state**. Because the reactions are performed in aprotic solvents, the bonds to the metal counter ion are strong and form an important part of the transition state. The enolate and aldehyde react via a cyclic transition state comprising six atoms, including the metal. Because of the way that the orbitals must overlap (π orbital of enolate with π^* orbital of carbonyl), the two components must approach each other from the side, resulting in a chair-like structure. In the chair transition states leading to the *syn* and *anti* products, the only difference is the orientation of the groups on the aldehyde (R_3 and H). When the R_3 of this group is pseudo-axial, interactions in the chair create sterics that disfavour this transition state. When an aldehyde reacts with an *E* enolate, the reaction tends to produce an *anti*-aldol product.

When a *Z* enolate is the nucleophilic partner, the reaction also favours the pathway in which the substituent on the aldehyde component adopts the pseudo-equatorial position. In this case, however, the groups on the enolate are exchanged (R_2 is now axial), and so the reaction favours the *syn* product.

transition state with R_3 pseudo-equatorial has lowest energy

Z enolate favours *syn*-aldols

Z enolate

strong diaxial-like interaction between R_1 and R_3

Z enolate disfavours *anti*-aldols

The Mukaiyama aldol reaction proceeds by an open transition state, and so the nucleophile is free to approach the electrophile from either side. In this situation, there is little control over the stereochemistry of the reaction, and the products tend to form as mixtures of isomers.

One of the stereocentres in the macrocycle of erythromycin has a *syn*-aldol geometry. To get the *syn*-aldol product, a *Z* enolate is needed, which requires the use of the larger boron group with the smaller base.

1) $B(Cy)_2Cl$, NEt_3
2) R_1CHO
3) H_2O_2 workup

CHECKPOINT 18.7

You should now be able to use the open and closed transition state models to predict whether an aldol reaction will favour the *syn* or the *anti* product.

SOLVED PROBLEM

Determine the reaction conditions necessary to obtain the *syn*-aldol product from the *Z* enolate for the following reactants. Draw the transition state for the reaction.

STEP 1: Select conditions appropriate for the formation of the *Z* enolate and draw its structure.

The *Z* enolate of a ketone is favoured by reacting the ketone with LDA, provided the R group on the non-enolizing side is sufficiently large. Since this ketone has a sterically bulky phenyl group in this position, the LDA will likely favour the *Z* enolate. This means the counter ion of the enolate will be lithium, as shown here:

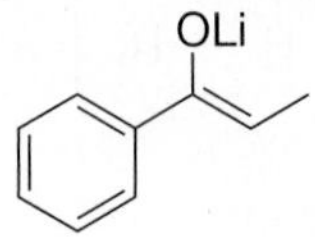

OLi

STEP 2: Draw the Zimmerman–Traxler transition state using the enolate drawn in the previous step and the provided aldehyde.

Use the transition states shown above as a template by replacing the appropriate R groups ($R_1 = R_3 =$ Ph, $R_2 =$ Me). This places the bulky phenyl group of the aldehyde in a pseudo-equatorial position, thereby avoiding 1,3-diaxial interactions across the ring. It is a good habit to look at both transition states to decide which is better.

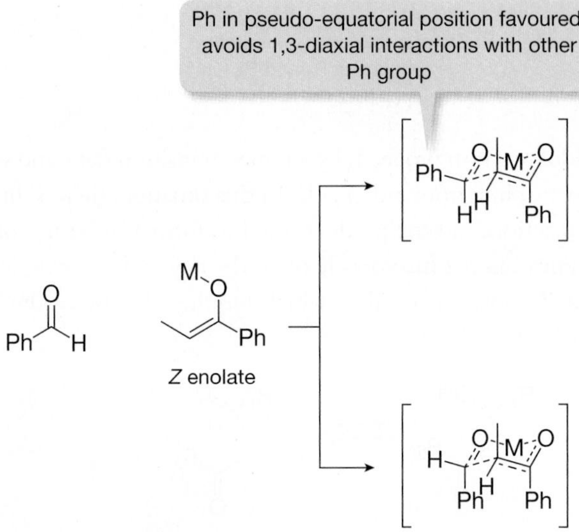

Ph in pseudo-equatorial position favoured; avoids 1,3-diaxial interactions with other Ph group

Z enolate

STEP 3: Draw the product of the reaction with appropriate stereochemistry using the transition states as a guide.

Keeping all the atoms in the same place as they were in the transition state, replace the lithium ion with a hydrogen, and redraw the bonds to correspond to the aldol product. Use a line bond drawing to unfold the structure and show the final product.

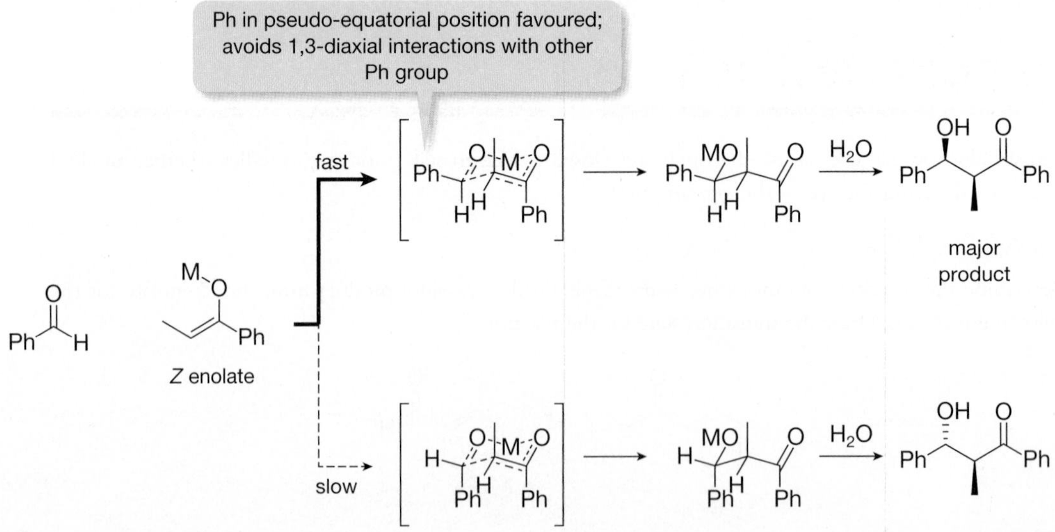

Ph in pseudo-equatorial position favoured; avoids 1,3-diaxial interactions with other Ph group

fast

slow

Z enolate

H_2O

major product

PRACTICE PROBLEM

18.13 Specify the reaction conditions necessary for the following transformations.

a) 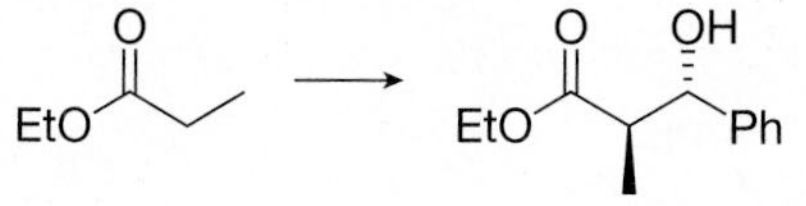

b) 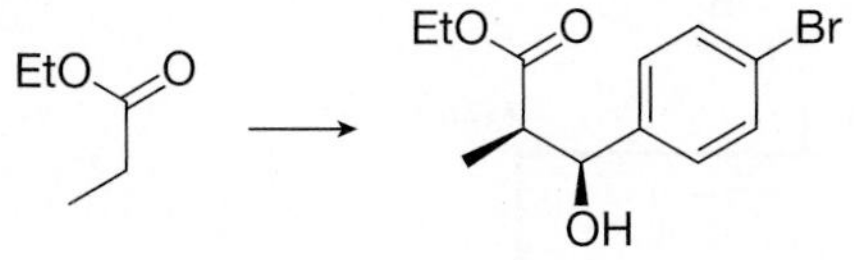

c) 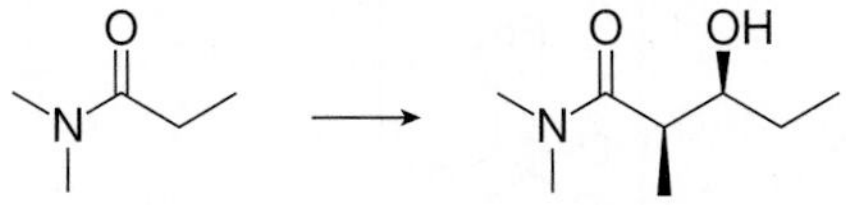

d) 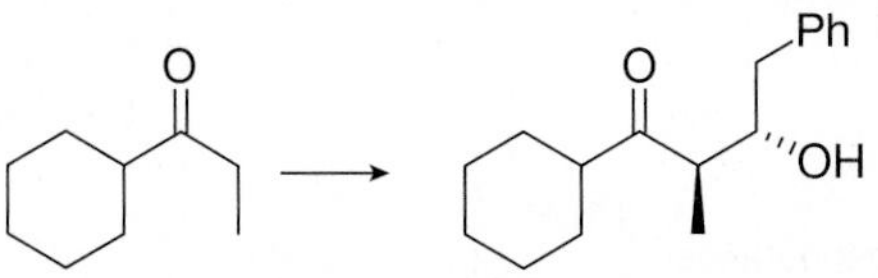

INTEGRATE THE SKILL

18.14 Below is the structure of zincophorin, a naturally occurring antibiotic. Identify two segments in the molecule that could be synthesized using an aldol reaction, and show how the reaction could be used to control their relative stereochemistry.

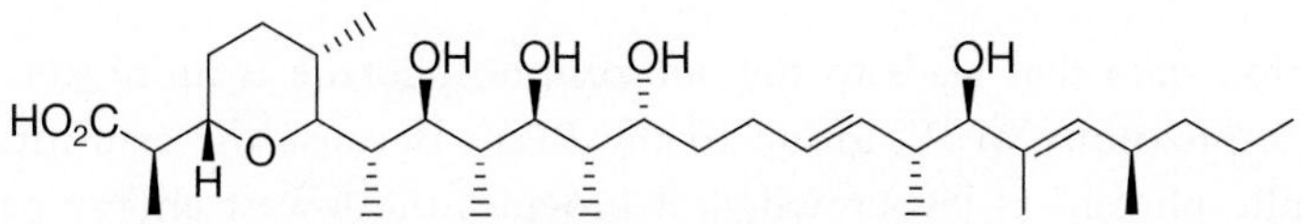

18.6 Stereoselectivity in Alkene-Forming Processes

18.6.1 Basis for Wittig reaction selectivity

The Wittig reaction is an efficient way to make alkenes and is also one of the best ways to make carbon-carbon bonds. Using alkyl-substituted ylides in the Wittig reaction typically produces Z alkenes.

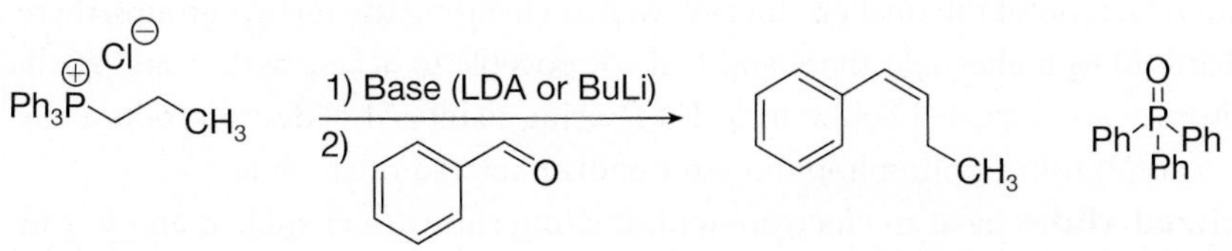

The Wittig reaction combines an aldehyde with an ylide through a transition state that is usually closed. The ylide is a nucleophile that adds to the carbonyl carbon of an aldehyde. At the same time, the π bond of the aldehyde breaks, and the electrons released form a bond to the phosphorus of the ylide. The result is an oxaphosphetane, which contains a four-membered ring.

Because of the way the orbitals must interact, the ylide and carbonyl approach each other at right angles in the transition state. Interactions between the R_2 group and PPh_3 control the energies of the two possible transition states.

The transition state that leads to the *anti*-oxaphosphetane is more sterically hindered, due to the close proximity of R_2 group to the PPh_3. Because the transition state leading to the *syn*-oxaphosphetane is less crowded, it provides the lowest energy pathway, and the *syn*-oxaphosphetane forms faster. Once each oxaphosphetane is formed, it quickly collapses to form the phosphine oxide and alkene. The final ratio between the *cis*- and *trans*-alkenes will be the same as the ratio between *syn*- and *anti*-oxaphosphetane rings, because the stereochemistry in the oxaphosphetane is preserved in the alkenes.

18.6.2 Variations on the Wittig reaction provide alternate stereochemistry

The stereochemistry of the Wittig product is a result of how the size (sterics) and charge (electronics) of the reactants work together in the rate-determining transition state. Altering the phenyl rings on the triphenylphosphine, for example, changes both the charge stability and the steric considerations. This changes the relative rate of formation of the *syn*- and *anti*-oxaphosphetanes, which changes the E/Z ratio of the final product. As well as changing the phenyl groups, there are other changes that lead to higher selectivity and make it possible to access both E and Z alkenes. Two of these alternatives, discussed below, include (1) using stabilized ylides to produce the E alkene selectively; and (2) using a phosphonium ester enolate instead of an ylide.

Stabilized ylides are ylides with an electron-withdrawing group adjacent to the negative charge, such as a carbonyl.

Stabilized ylides have an electron-withdrawing substituent such as an ester or a ketone located in such a way (1,3) as to stabilize the resulting ylide (Chapter 17). Unlike the alkyl-substituted ylides, stabilized ylides produce the E alkene selectively. In the transition state, there is a partial negative charge on the aldehyde oxygen and a partial negative charge on the adjacent electron-withdrawing group. The amount of repulsion between those negative charges in the transition state determines which oxaphosphetane forms fastest.

stabilized ylide

$\delta-$ charges are close together; electron repulsion raises the energy of this transition state

slow

fast

oxaphosphatanes

disfavoured

syn

anti

When the oxygen and electron-withdrawing group are close together, the partial negative charges repel each other and raise the energy of that transition state relative to the transition state in which the partial negative charges are separated. The lower energy transition state leads irreversibly to the *anti*-oxaphosphetane. Once formed, the oxaphosphetanes quickly collapse to form the corresponding alkene.

A reaction related to the Wittig reaction with stabilized ylides—known as the **Horner–Wadsworth–Emmons (HWE) reaction**—uses a phosphonate ester instead of a phosphonium salt. Like reactions with stabilized ylides, the HWE reaction is commonly used to make *E* alkenes with an electron-withdrawing group.

The **Horner–Wadsworth–Emmons (HWE) reaction** is a Wittig-type reaction that uses a phosphonium ester enolate instead of an ylide. This typically results in the formation of *E* alkenes.

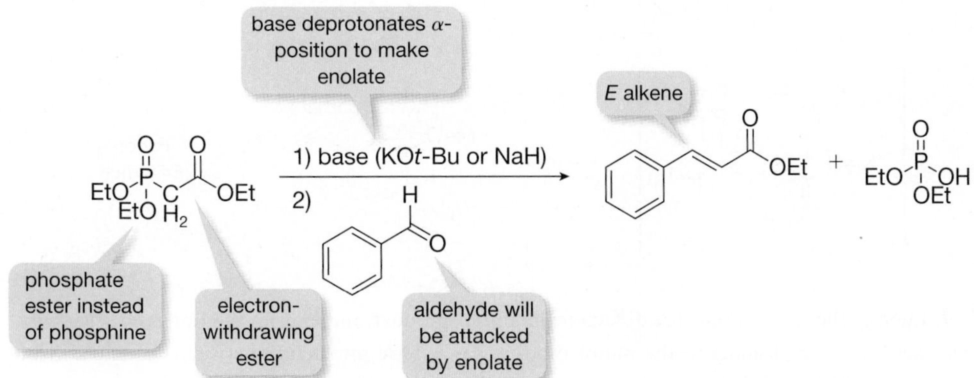

STUDENT TIP

Because there is no phosphonium salt, the nucleophile in a HWE reaction has a negative charge, making it a stabilized enolate, not an ylide.

CHECKPOINT 18.8

You should now be able to rationalize the stereochemical outcome of the Wittig reaction and its variations using appropriate drawings for the possible transition states.

SOLVED PROBLEM

Draw the transition states leading to both the major and minor products of the following reaction. Identify the interactions that lead to the major product being favoured.

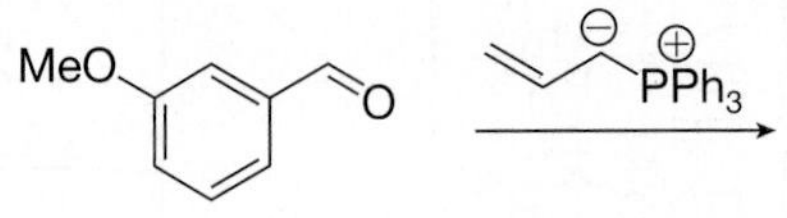

MeO PPh3

STEP 1: Determine the roles of the reactants and type of reaction taking place.

The presence of the ylide reagent tells us that this is a Wittig reaction between the aldehyde and the unstabilized ylide.

STEP 2: Draw the major and minor products of the reaction, including stereochemistry.

Unstabilized ylides favour the formation of *Z* alkenes, so the major and minor products will be as follows:

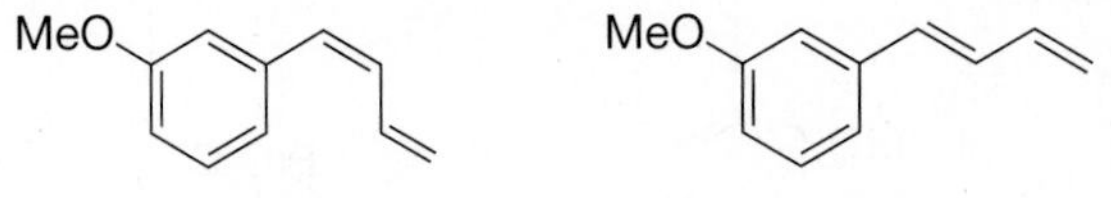

MeO major (*Z* isomer)

MeO minor (*E* isomer)

STEP 3: Draw the transition states leading to the two products.

Use the transition states for the HEW and the Wittig reaction using stabilized ylides shown in Section 18.6.2 as templates to draw the transition states for this reaction, and fill in the appropriate R groups ($R_1 = C_4H_4OCH_3$, $R_2 = CHCH_2$). It is always a good idea to show both transition states.

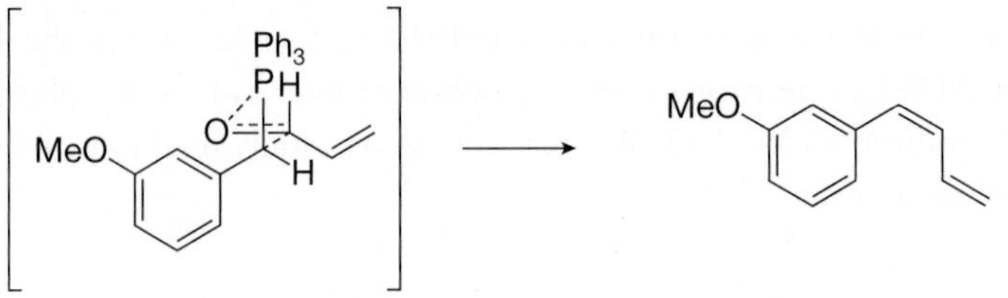

Ph3 P H O H MeO → MeO major (*Z* isomer)

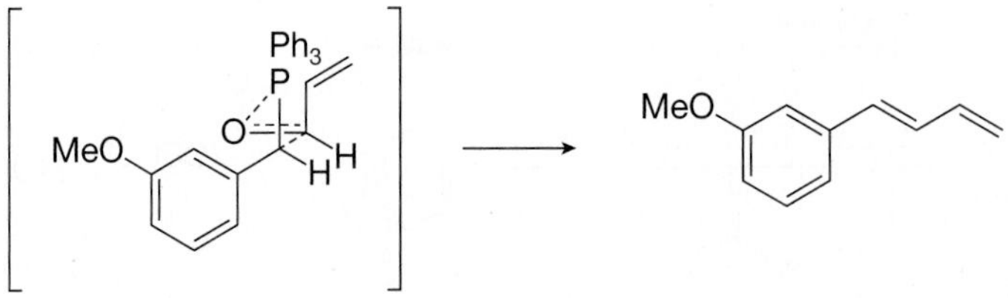

Ph3 P O H H MeO → MeO minor (*E* isomer)

STEP 4: Identify the interactions that disfavour the minor product, and explain the observed selectivity.

The transition state leading to the minor product has a steric interaction between the triphenylphosphine group and the vinyl group. This makes it higher in energy than the alternative transition state. The other pathway is faster and produces the major product of the reaction.

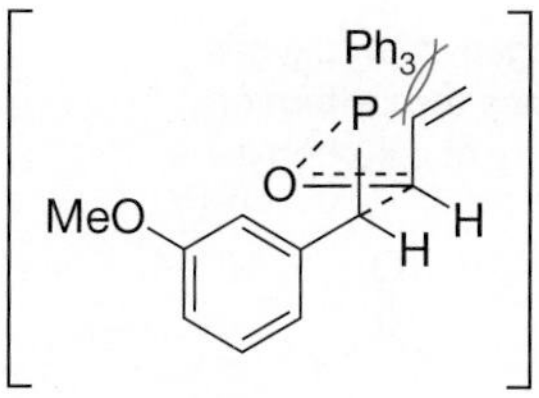

PRACTICE PROBLEM

18.15 Draw the major product for each of the following reactions, and explain the selectivity using appropriate transition state diagrams.

a)

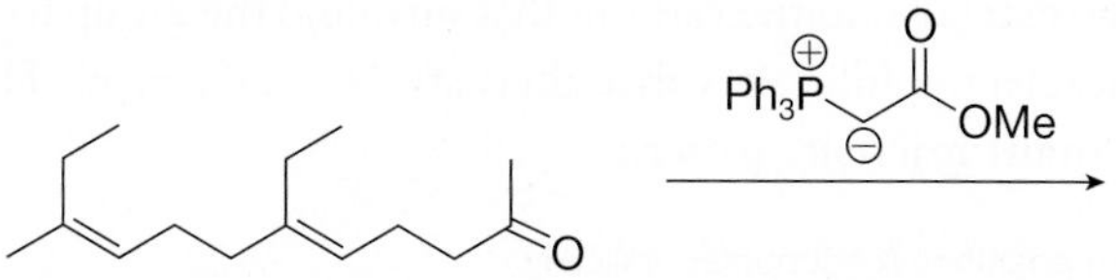

b)

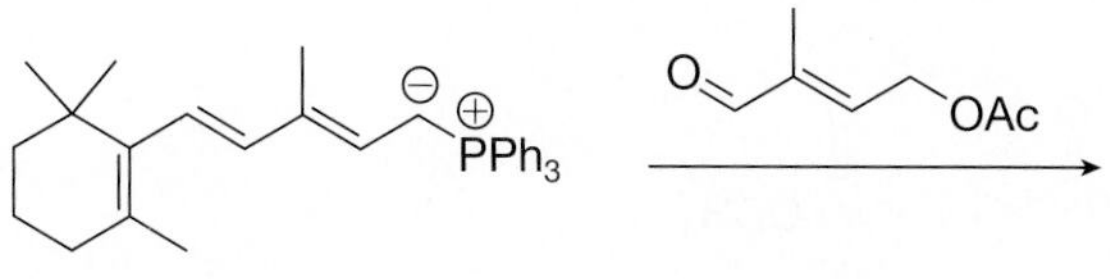

c)

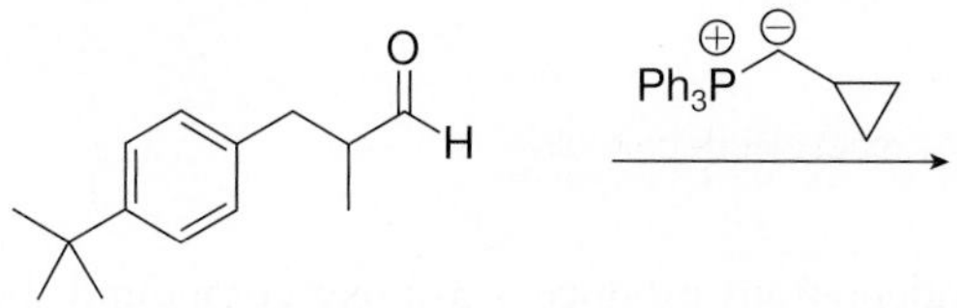

d) 1) LDA 2)

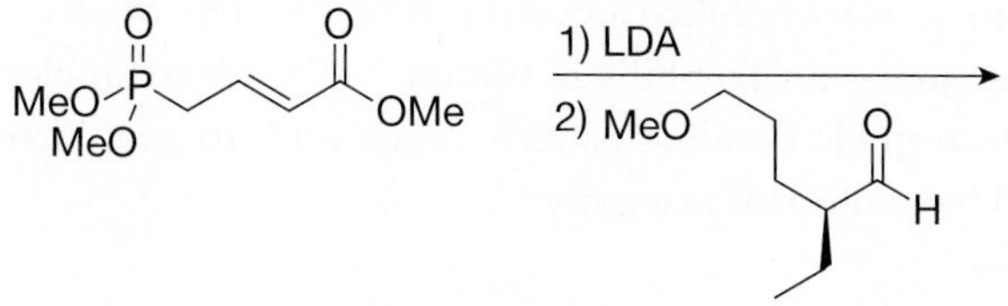

INTEGRATE THE SKILL

18.16 Draw a reaction coordinate diagram for the reaction in Question 18.15a, showing the differences in energies of the two possible pathways.

18.7 Umpolung Reactions

Umpolung reactions involve changes in the reactivity of an atom. In these reactions, atoms that are normally nucleophilic become electrophiles, and atoms that are usually electrophiles become nucleophilic. An example of an umpolung reaction that should be familiar is the conversion of an alkyl halide electrophile to a Grignard reagent nucleophile.

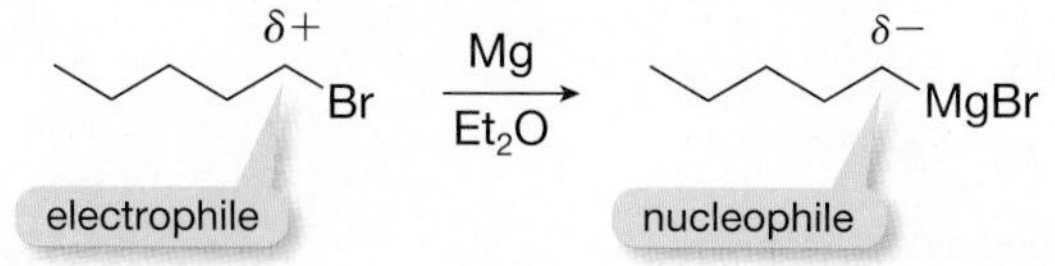

An **umpolung reaction** involves changing the reactivity of an atom from nucleophilic to electrophilic, or vice versa. *Umpolung* is German for "reversal of polarity."

$\delta+$ Br $\xrightarrow[Et_2O]{Mg}$ $\delta-$ MgBr

electrophile nucleophile

The functional groups in each molecule, and the carbon atoms that surround them, have a reactivity pattern based on the electronegativity of the atoms in the functional group. The oxygen in a carbonyl is the most electronegative atom and sets up a pattern of reactivity in the carbonyl in which the carbonyl carbon acts as an electrophile (+) and the α-carbon acts as a nucleophile.

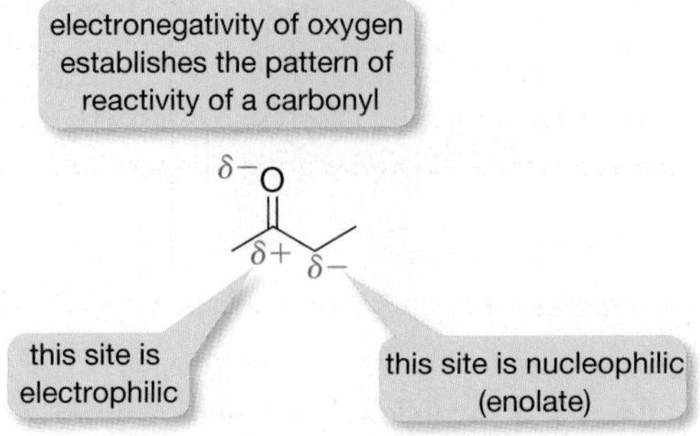

This pattern of reactivity can be extended to the carbons that surround the group to establish a reactivity pattern of nucleophilic/electrophilic sites that alternate (+, −, +, −, +). This alternating reactivity pattern is a **consonant** reactivity pattern.

Consonant molecules have an alternating pattern of nucleophilic-electrophilic reactivity potential in the atoms along a chain (+ − + − + ···) between functional groups.

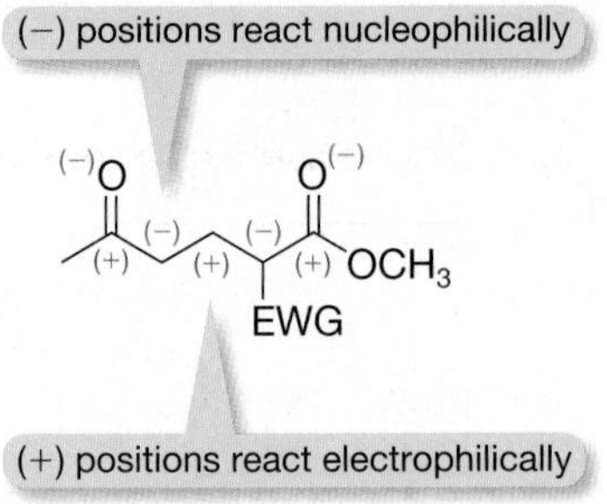

Both aldol additions and Claisen condensations produce 1,3-dioxy compounds because a nucleophilic α-carbon (−) bonds to an electrophilic carbonyl carbon (+). The Michael addition produces 1,5-dioxy compounds because a nucleophilic α-carbon (−) bonds to an electrophilic β-carbon (+). This nucleophile-electrophile bonding pattern always leads to an odd-numbered dioxy substitution and is produced by consonant reactivity.

Claisen condensation

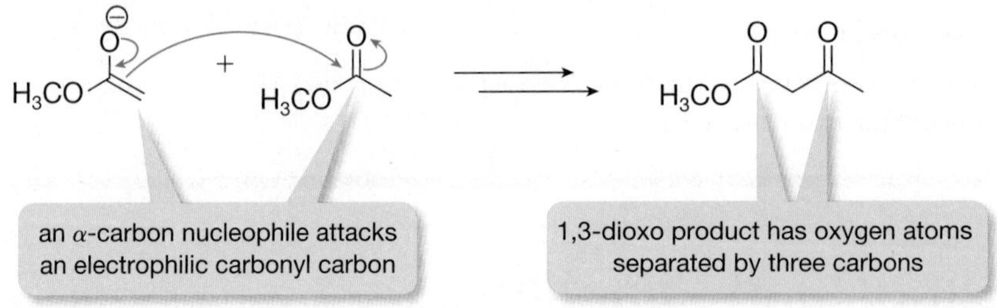

Michael addition

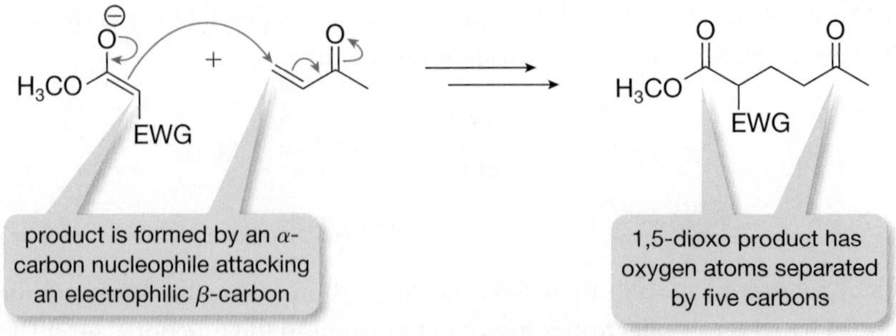

Molecules with oxy substituents separated by an even number of atoms display **dissonant** reactivity. In this case, the alternating pattern of nucleophilic and electrophilic atoms between the functional groups conflicts, and the normal pattern of reactivity alternation is not possible.

Dissonant molecules have a nucleophilic-electrophilic reactivity potential pattern with matching reactivity on adjacent carbons along a chain (+ − − + or − + + −) between functional groups.

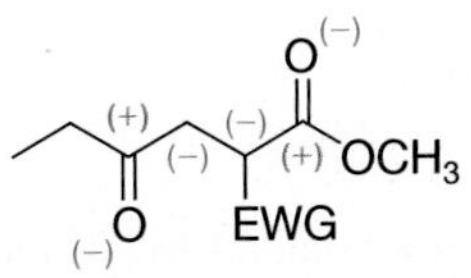

To construct molecules that display dissonant reactivity requires special methods. This is because these molecules do not have a reactivity pattern that "matches" the natural nucleophilicity and electrophilicity (established by electronegativity) of the components used to assemble them.

One method of making such molecules is to employ an umpolung reagent, which reverses the normal nucleophile and electrophile reactivity pattern of a functional group. Umpolung reagents can be nucleophiles or electrophiles. In an umpolung nucleophile, a group is modified to transform an electrophilic atom into a nucleophilic one. This is most commonly done by attaching heteroatoms that can stabilize carbanions in some way. Umpolung electrophiles are often made by activating leaving groups with adjacent electron-withdrawing groups.

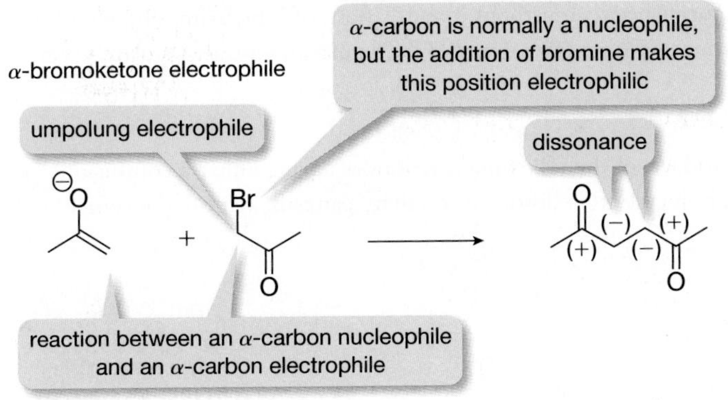

α-bromoketone electrophile

umpolung electrophile

α-carbon is normally a nucleophile, but the addition of bromine makes this position electrophilic

dissonance

reaction between an α-carbon nucleophile and an α-carbon electrophile

CHECKPOINT 18.9

You should now be able to distinguish between consonant or dissonant reactivity patterns and be able to recognize when an umpolung reagent is used or needed in a synthesis.

SOLVED PROBLEM

Determine whether the following reaction involves umpolung reactivity. Explain your answer. Note that you do not need to know the details of the reaction or its mechanism to answer this question.

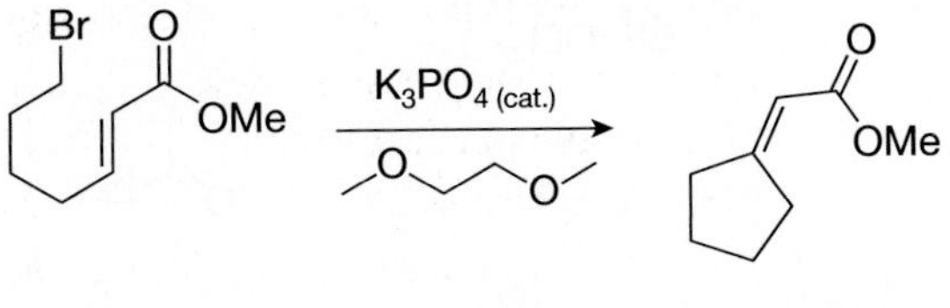

K_3PO_4 (cat.)

STEP 1: Determine which new bonds are formed in the reaction.

Numbering the carbons in the reactant and product makes it easier to see that a new bond has formed between carbons 3 and 7.

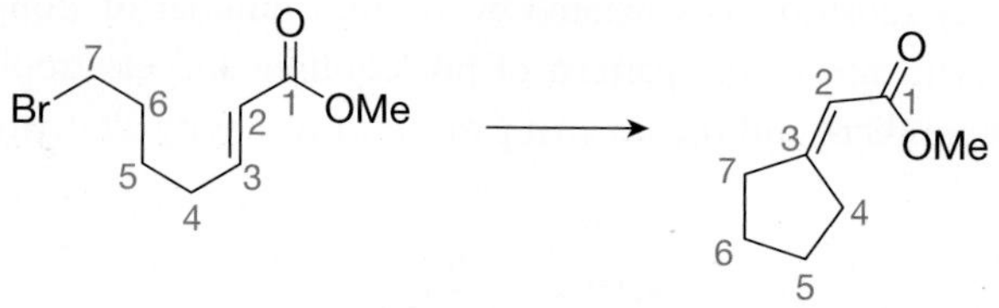
Br
OMe
O
OMe
O

STEP 2: Assign the atoms near where the new bond forms as either nucleophilic (+) or electrophilic (–).

It is easiest to start with atoms next to heteroatoms or other functional groups. Since bromine is very electrophilic, the carbon next to it is expected to react as an electrophile. On the other end of the molecule, the carbonyl carbon and β-carbon are electrophilic, and the α-carbon is nucleophilic (recall the + and – signs alternate).

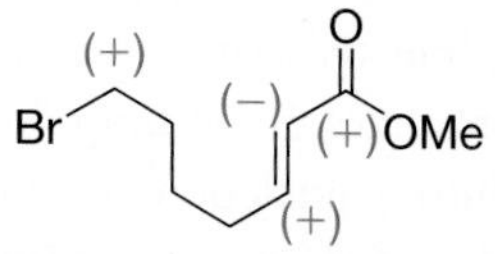
(+)
Br
(–)
(+)OMe
O
(+)

STEP 3: Compare the signs on the atoms involved in the formation of the new bond to determine whether their polarity matches and whether umpolung reactivity is involved.

The new bond forms between carbons 3 and 7, both of which are inherently electrophilic. For these two atoms to be connected, a reversal of one of their polarities must take place via an umpolung reaction.

PRACTICE PROBLEM

18.17 Classify the products of the following reactions as having either a consonant or dissonant reactivity pattern. For products with a dissonant reactivity patterns, identify the source of the umpolung in the reaction.

a)

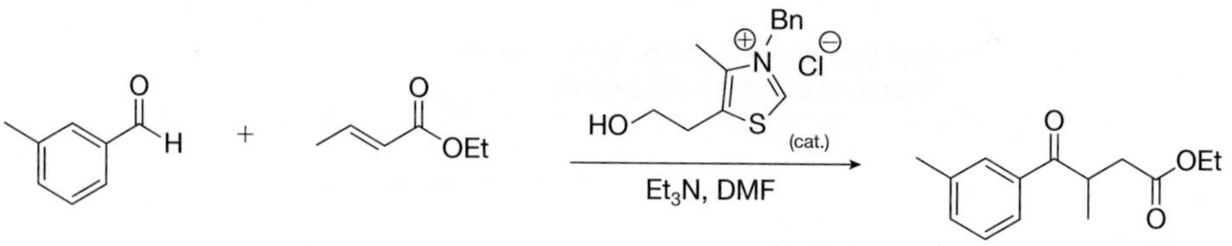
H
OEt
HO
Bn
N
Cl
S
(cat.)
Et3N, DMF
OEt

b)

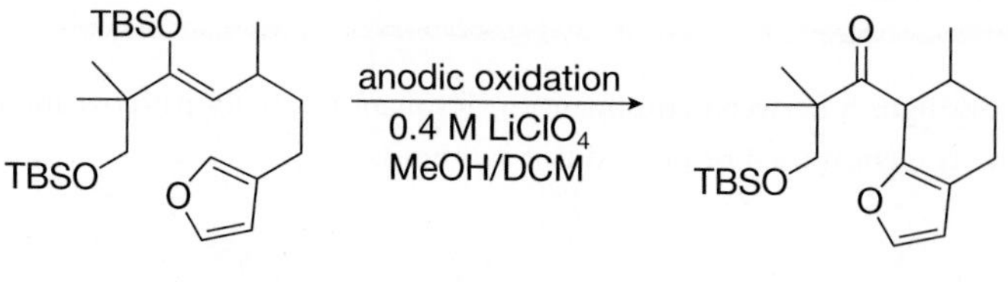
TBSO
TBSO
anodic oxidation
0.4 M LiClO4
MeOH/DCM
TBSO

c)

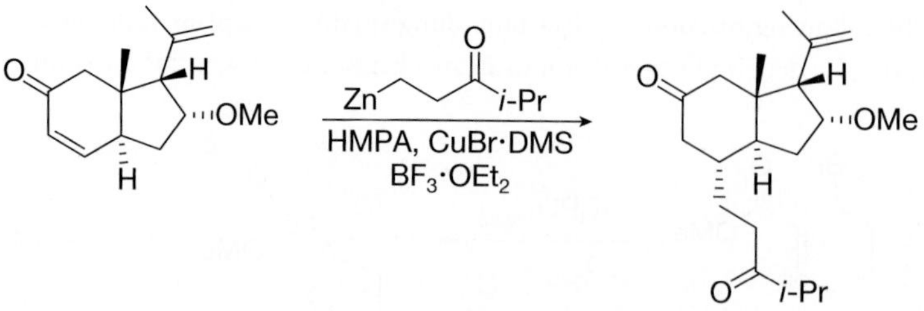
H
OMe
H
Zn
i-Pr
HMPA, CuBr·DMS
BF3·OEt2
H
OMe
H
i-Pr

d)

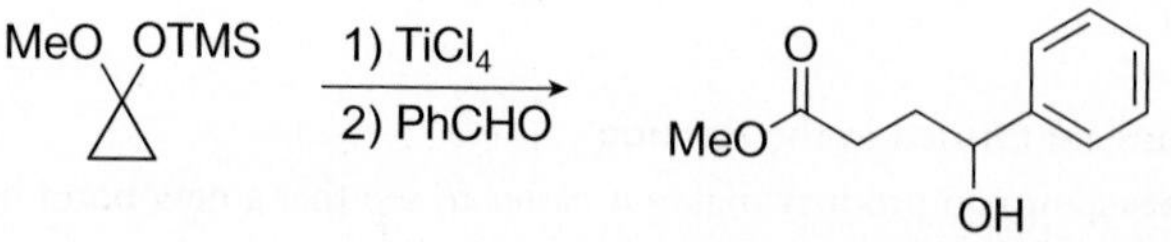
MeO OTMS
1) TiCl4
2) PhCHO
MeO
OH

INTEGRATE THE SKILL

18.18 Not all examples of umpolung reactivity result in structures with dissonant reactivity patterns. The following example of an amide-forming reaction involves umpolung chemistry. Use the partial mechanism provided to explain why this reaction is considered an example of umpolung chemistry. Contrast this reactivity with one of the more traditional methods of forming an amide bond, as described in Chapter 15.

Ph, Br, NO_2 + t-BuNH$_2$ → NIS, H_2O, K_2CO_3 / THF, 0 °C → Ph, H, N, t-Bu, O

via

Ph, N, O, O, Br, I, N, t-Bu, H → Ph, H, N, t-Bu, O_2N, Br

18.7.1 Umpolung nucleophiles

18.7.1.1 Dithianes

The hydrogen of a thioacetal can be removed by a strong base such as BuLi to make a nucleophilic anion. The addition of an electrophile to this anion gives a substituted thioacetal, which can be hydrolized to form a carbonyl using mercuric oxide (HgO) in water.

electrophile added to carbonyl; carbonyl appears to behave like a nucleophile

R, H, O → R, E, O

$BF_3 \cdot OEt_2$, SH SH

$HgCl_2$, HgO, H_2O, CH_3OH

S, S, R, H (dithiane) → BuLi → S, S, C, R (acyl anion equivalent) → $E^{\oplus}$ → S, S, R, E

Typically, this sequence is performed using 1,3-propanedithiol to form the thioacetal from an aldehyde. The resulting six-membered ring is called a dithiane. Overall, this thioacetal functions as a carbonyl anion equivalent. The normal reactivity of a carbonyl carbon is as an electrophile (electron acceptor). When converted to a dithiane, the carbon acts as a nucleophile (electron donor). Dithiane anions can react with a variety of carbonyl electrophiles as well as alkyl halides, and this makes it possible to create a large variety of products. After hydrolysis, the carbonyl is regenerated, restoring the carbonyl functionality to its original electrophilic state.

Depending on the type of electrophile chosen, an aldehyde group can be used to create a variety of structures. If another aldehyde is used as the electrophile, the result is an acyloin or α-hydroxyketone. Alternatively, if an acid chloride, carbon dioxide, or chloroformate is used as the electrophile, the sequence produces a diketone, α-keto acid, or α-ketoester, respectively. Alkylation with an alkyl halide converts an aldehyde into a ketone.

18.7.1.2 Cyanohydrins

Chapter 7 describes the synthesis of cyanohydrins, and Chapter 17 includes the Thorpe reaction, in which a nitrile acts as the stabilizing group of an enol-like nucleophile. A cyanohydrin can also be used to activate an aldehyde hydrogen for deprotonation.

To form a nucleophilic carbanion, the OH hydrogen of a cyanohydrin must be protected from deprotonation, usually as an acetal or silyl group. For example, an enol ether may be used to protect the alcohol of a cyanohydrin because an acetal is impervious to a base. Treating the protected cyanohydrin with a strong base removes the hydrogen α to the nitrile, making the former carbonyl carbon nucleophilic. This acyl anion equivalent can then act as a nucleophile in an S_N2 alkylation. After the new alkyl group is in place, the addition of aqueous acid hydrolyzes the acetal protecting group, and the resulting cyanohydrin collapses to form a ketone.

DID YOU KNOW?

Special acetal protecting groups such as $-OCH_2OCH_3$ (methoxy methyl or MOM) are useful because they are unreactive toward base or nucleophiles. The MOM acetal can be installed on an OH group using a special reagent called MOM chloride (CH_3OCH_2Cl).

It is often useful to have protecting groups that can be removed under selective reaction conditions, and silyl groups form a large family of such reagents. These groups are usually added by mixing the alcohol with a silyl chloride (R_3SiCl) and a weak hindered base.

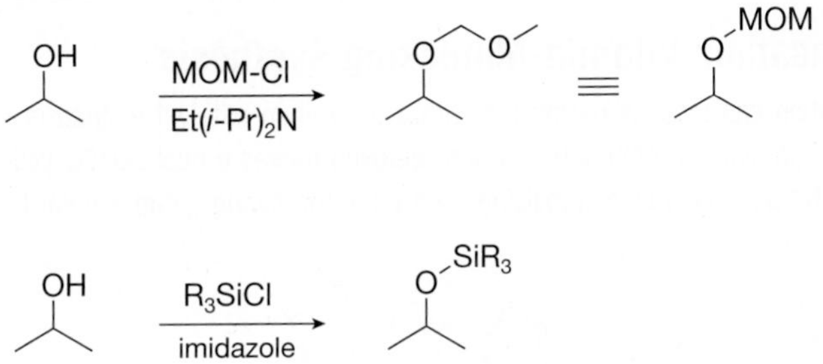

OH — MOM-Cl, Et(*i*-Pr)$_2$N → O–CH$_2$–O– ≡ MOM

OH — R_3SiCl, imidazole → O–SiR_3

Silyl ethers, such as trimethylsilyl (TMS) and *tert*-butyldimethylsilyl (TBDMS) are stable to basic conditions. Both are selectively removed using a fluoride ion, which does not usually react with other organic functional groups. This gives an extremely selective way to remove silyl groups. A common reagent for silyl group removal is tetrabutylammonium fluoride (TBAF), which serves as an organic-soluble source of fluoride.

–OTMS
trimethylsilyl ether

–OTBDMS
tert-butyldimethylsilyl ether

18.7.1.3 Benzoin condensation

Cyanohydrin nucleophiles can be used in weakly basic conditions if a very strong electrophile, such as an aldehyde, is used. A small amount of cyanide (catalytic amount) is used as a catalyst in the **benzoin condensation** to convert a small amount of aldehyde into an acyl anion equivalent, which quickly adds to another molecule of aldehyde. Once this has occurred, the cyanohydrin reverts to a carbonyl.

The **benzoin condensation** is the self-condensation of aldehydes to make α-hydroxy ketones. For this reaction to happen, an electrophilic aldehyde carbonyl must become nucleophilic.

CHEMISTRY: EVERYTHING AND EVERYWHERE

Benzoin Condensation: Vitamin-Mimicking Synthesis

The benzoin condensation can also be performed with a thiazolium catalyst instead of cyanide. The positively charged thiazolium has an acidic proton, which when removed makes a nucleophilic ylide. Once this ylide adds to an aldehyde, the $C{=}N^+$ bond acts as a stabilizing electron-withdrawing group, similar to the $C{\equiv}N$ of a nitrile.

thiazole thiazolium

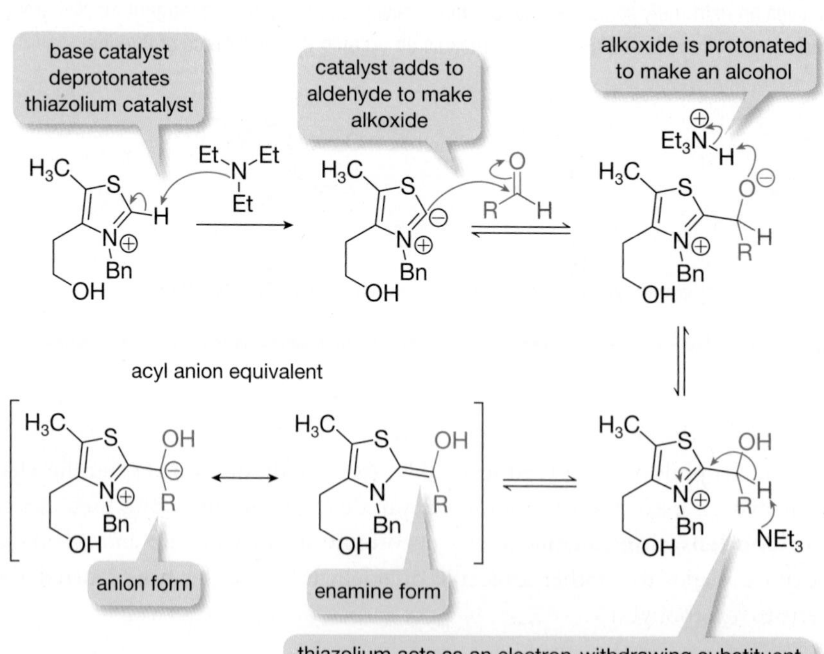
base catalyst deprotonates thiazolium catalyst
catalyst adds to aldehyde to make alkoxide
alkoxide is protonated to make an alcohol
acyl anion equivalent
anion form
enamine form
thiazolium acts as an electron-withdrawing substituent to allow deprotonation and form a nucleophile

The thiazolium salt catalyst was designed to mimic the biological activity of thiamine (vitamin B1).

Vitamin B1 is used as a nucleophilic catalyst co-factor by certain enzymes in the body. One enzyme that uses thiamine as a benzoin-type catalyst is transketolase. The thiamine co-factor in transketolase adds to the ketone of fructose-6-phosphate (shown in red). A retro-aldol-type reaction breaks off four carbons to create a two-carbon acyl anion equivalent. This then adds in a benzoin-type condensation to glyceraldehyde-3-phosphate (shown in blue) to make xylulose-5-phosphate.

18.7.2 Umpolung electrophiles

The principle of umpolung chemistry can also be applied to electrophiles, although these are less common than umpolung nucleophiles. α-Bromocarbonyls are excellent electrophiles. The adjacent carbonyl enhances the leaving group ability of the bromide such that these functions will react with many nucleophiles.

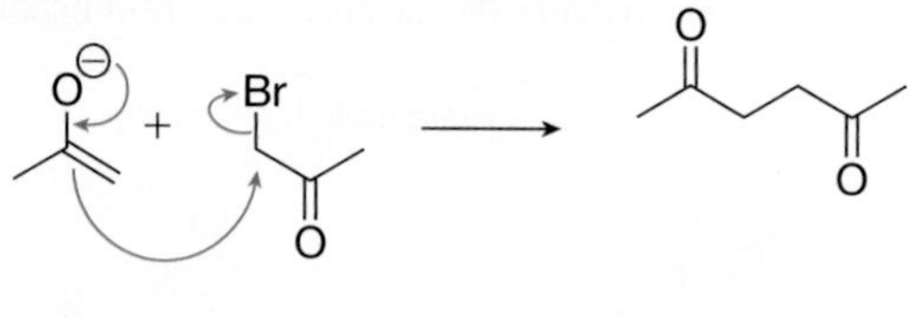

Occasionally, this reaction produces an epoxide instead of the desired 1,4-diketone. Usually this happens if the temperature of the reaction is not carefully controlled. This reaction, known as **Darzen's condensation**, occurs when the enolate removes an α-hydrogen from the bromoketone instead of performing the desired substitution; this switches the nucleophile and electrophile pattern. The resulting alkoxide oxygen then displaces the bromide to form an epoxide.

Darzen's condensation produces an epoxy ketone from an aldol addition with an α-halide.

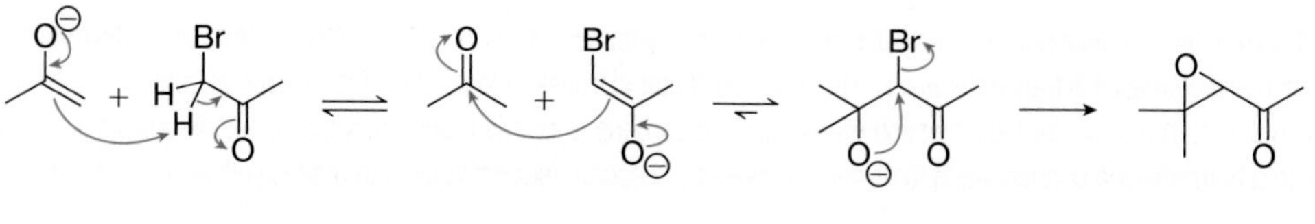

CHECKPOINT 18.10

You should now be able to draw the products formed from various umpolung reactions and incorporate these reactions into the synthesis of molecules that have dissonant reactivity patterns. You should also be able to draw mechanisms for these reactions.

SOLVED PROBLEM

Draw the intermediates formed after each step of the following reaction, as well as the final product. Identify the umpolung nucleophile and/or electrophile, and explain how its polarity is reversed.

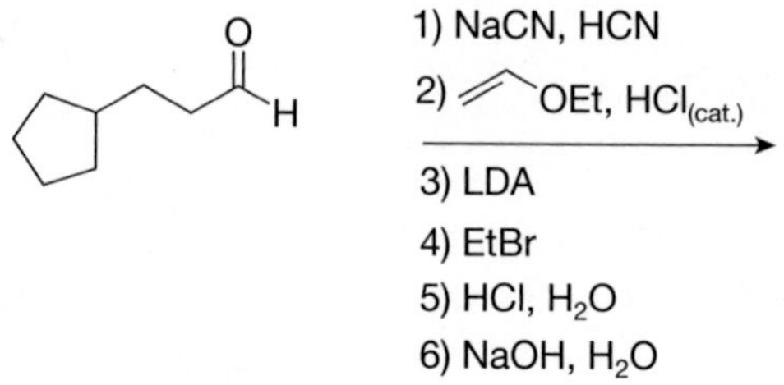

1) NaCN, HCN
2) OEt, $HCl_{(cat.)}$
3) LDA
4) EtBr
5) HCl, H_2O
6) NaOH, H_2O

STEP 1: Determine the roles of the reactants in the first step and the type of reaction taking place.

The starting material has an electrophilic aldehyde that can react with the nucleophilic cyanide ions to form a cyanohydrin.

STEP 2: Draw the product of the first reaction.

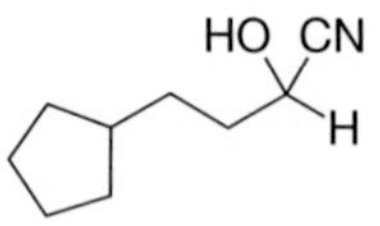

HO CN
H

STEP 3: Repeat Steps 1 and 2 for the remaining steps.

The ethyl vinyl ether added in the second step is a common protecting group for alcohols. Under the acidic conditions, the alcohol of the cyanohydrin is converted to an acetal. In the third step, LDA, a strong base, is added. The most acidic proton in the molecule is the one alpha to the nitrile group. Deprotonation of this hydrogen creates an excellent nucleophile, which reacts with the electrophilic ethyl bromide added in the fourth step of the reaction. The aqueous acid in the fifth step hydrolyzes the acetal, and finally, the aqueous base in the last step reverses the cyanohydrin formation and restores the carbonyl group.

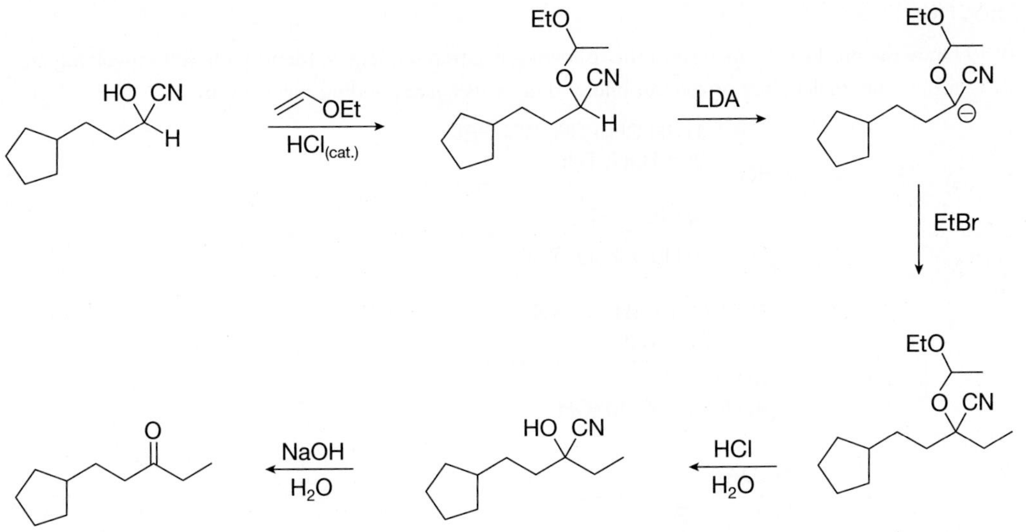

STEP 4: Examine the overall reaction and determine which new bonds are formed in the reaction.

The net effect of the reaction sequence is to convert an aldehyde into a ketone by creating a new C–C bond.

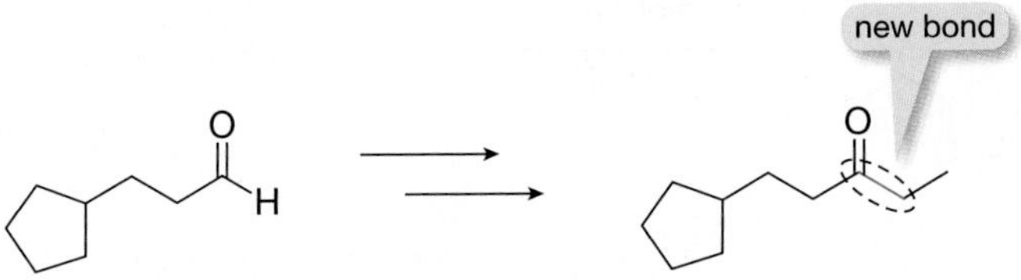

STEP 5: Assign the atoms in the reactants involved in the formation of the new bond as either nucleophilic (+) or electrophilic (–).

Both the carbonyl carbon and the carbon with the bromine attached are electrophilic in nature.

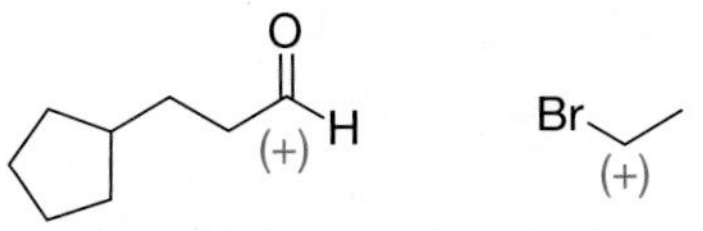

STEP 6: Compare the signs on the atoms involved in the formation of the new bond to determine whether their polarity matches and whether umpolung reactivity is involved.

The carbon atoms that are bonded together in this reaction are both electrophilic. Therefore, umpolung must be involved in the reaction in order to couple these two atoms of mismatched polarity.

STEP 7: Identify the step in the reaction where the polarity was reversed, and identify the umpolung reagent.

Although the carbonyl carbon is normally electrophilic, in this reaction it reacts as a nucleophile after being converted into a cyanohydrin and then deprotonated. The cyanohydrin can be considered an umpolung nucleophile.

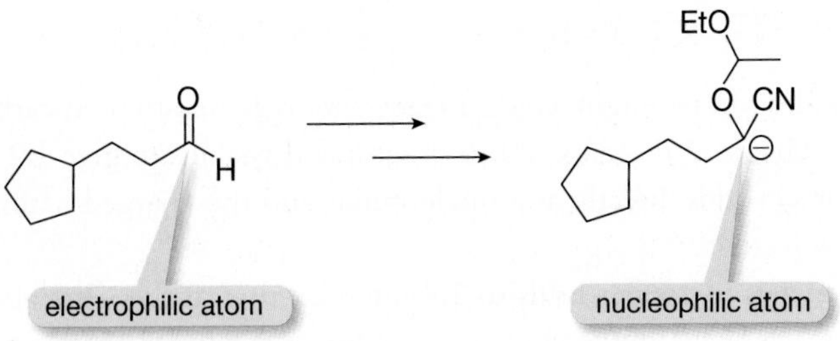

PRACTICE PROBLEM

18.19 Draw the products formed from the following reaction sequences. Identify the steps involving an umpolung nucleophile or electrophile, and draw its corresponding mechanism.

a) CHO
1) $SH(CH_2)_3SH$, $BF_3{\cdot}OEt_2$
2) *n*-BuLi, THF
3) Ph (epoxide, O)
4) HgO, H_2O, THF

b) O, H
1) $SH(CH_2)_3SH$, $BF_3{\cdot}OEt_2$
2) *n*-BuLi, THF
3) EtCOCl
4) HgO, H_2O, MeOH

c) O
1) LDA
2) Br, OEt, O

d) O, H, O
$NaCN_{(cat.)}$
DMF

INTEGRATE THE SKILL

18.20 Propose a reaction or series of reactions that could be used to carry out the following syntheses in as few steps as possible.

a) O, H → O, HO

b) O, H → O, HO

c) O → O, O

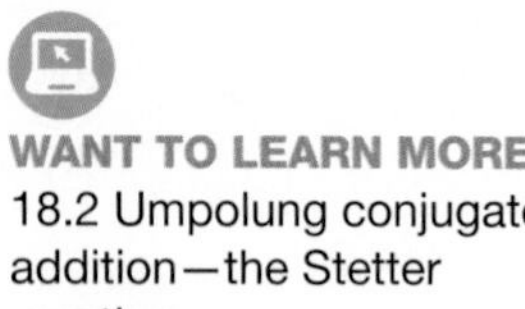

WANT TO LEARN MORE?
18.2 Umpolung conjugate addition—the Stetter reaction

18.8 Patterns in Enolate Reactions

Carbanion nucleophiles can be engineered to react with α,β-unsaturated carbonyls with a high degree of selectivity. Grignard reagents add to unsaturated systems to give 1,2 products. In these nucleophiles, the carbon adds directly as a nucleophile, and the charged nature of the Grignard reagent favours 1,2 selectivity.

Switching the metals to copper leads to 1,4-addition due to the generation of a negatively charged metal that reacts as the initial nucleophile. Once the copper of the cuprate has added to the 4-position, reductive elimination transfers an alkyl group, resulting in a 1,4-addition product.

Stabilized carbanions such as 1,3-dicarbonyls also tend to add in a conjugate or 1,4-pattern, giving Michael products.

The Robinson annulation is an example of a tandem process in which a Michael addition is immediately followed by an aldol condensation to produce a cyclic structure.

Many types of addition/elimination reactions exist, which create the appearance of direct nucleophilic displacements. Any carbonyl with a β-heteroatom can function as an enhanced electrophile. The mechanism is not S_N2, but instead involves an $E1_{cb}$ elimination followed by a 1,4-addition. This reaction mode facilitates reactivity relative to the corresponding S_N2 displacement.

Grignard

RMgBr

BrMgO R

$H_3O^{\oplus}$

HO R

Cuprate

R_2CuLi

OLi

$H_3O^{\oplus}$

Michael

$H_3O^{\oplus}$

Robinson

Conjugate amine addition

R_2NH

OMe

R_2NH

$R_2NH_2^{\oplus}$

Bayliss-Hillman

Regioselectivity can be controlled by using an enolate, thereby adding an electron-withdrawing group to provide extra stabilization for the carbanion. A convenient group for this purpose is the ester, which can later be removed by decarboxylation.

Regioselectivity can also be controlled by choosing the appropriate reaction conditions. Kinetic control is the most commonly used method and usually gives the best selectivity. Kinetic conditions produce the enolate that forms the fastest.

Thermodynamic conditions may be used to produce the most stable enolate.

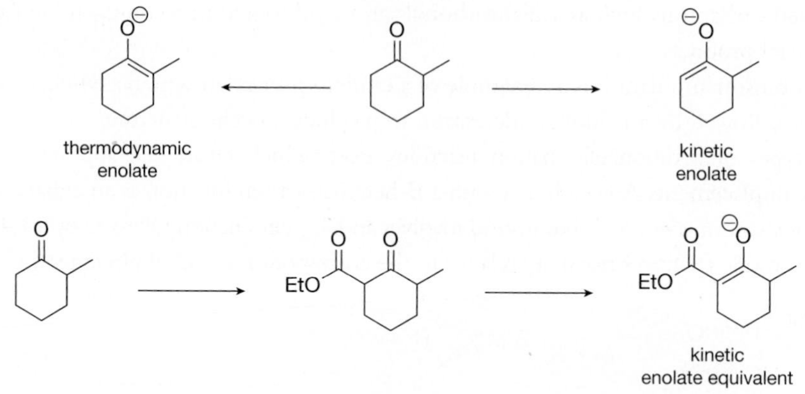

thermodynamic enolate

kinetic enolate

EtO

kinetic enolate equivalent

Acyclic enolates can form with an *E* or *Z* geometry. This has consequences for subsequent reactions because this geometry controls the stereochemistry of the product that is formed. Ketone enolates generally favour an *E* geometry because in the transition state for deprotonation there is steric hindrance with the group on the other side of the carbonyl.

Stereoselectivity can be manipulated by using boron-based counter ions. In these reactions, steric interference with the groups on the boron produces selectivity for the *E* or *Z* enolate, depending on the boron reagent chosen.

The alkoxy group on the far side on the carbonyl in an ester is small, and so this group does not contribute much interference in the deprotonation transition state. Esters tend to form Z enolates preferentially.

Secondary amides are very selective for the *E* enolate because the steric interactions from the amide nitrogen groups are very strong.

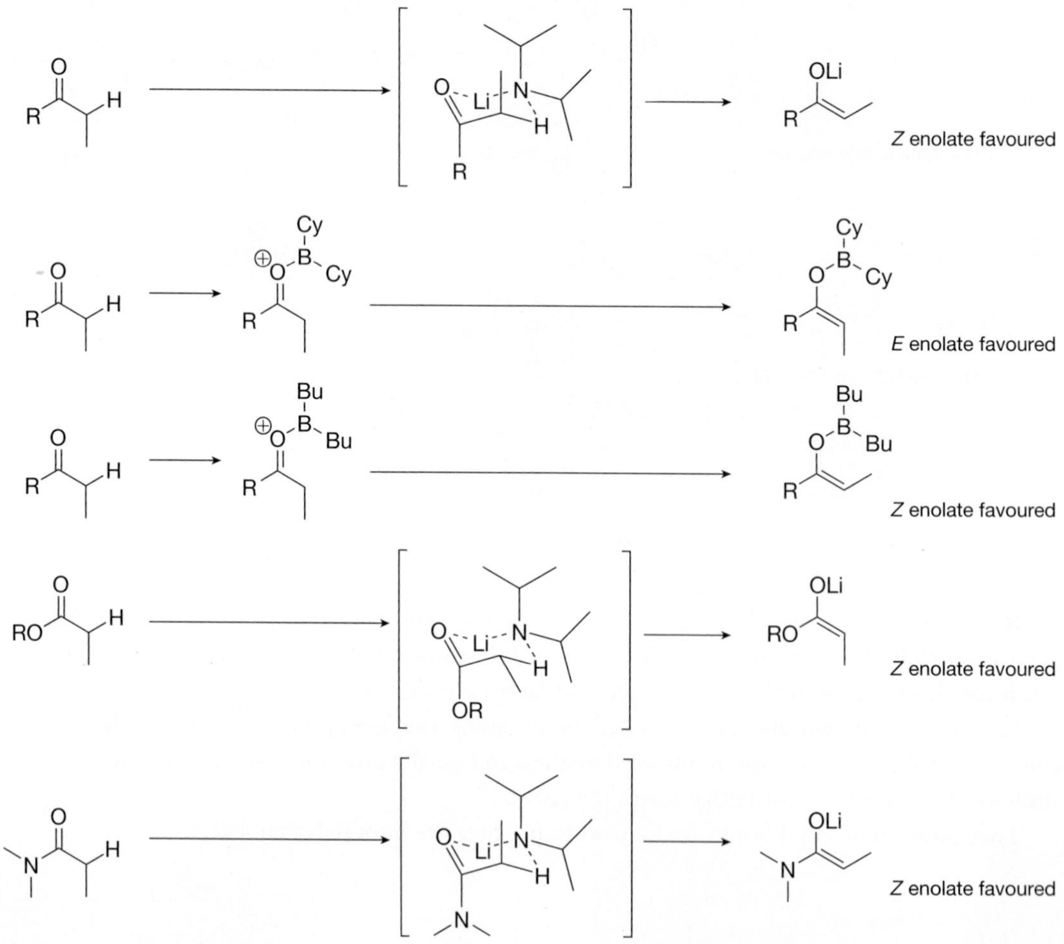

Z enolate favoured

E enolate favoured

Z enolate favoured

Z enolate favoured

Z enolate favoured

Patterns exist in alkene-forming reactions involving Wittig reagents. Phosphorus ylides add to carbonyls to form intermediate oxaphospheniums that collapse to form the corresponding alkene. Interactions in the cyclic transition states leading to the oxaphospheniums determine the favoured stereochemistry of the oxaphospheniums, and this stereochemistry is preserved in the alkene final product.

Alkyl phosphorus ylides (non-stabilized Wittig reagents) are controlled by steric interactions in the transition state. These steric interactions control which face of the aldehyde will be approached by the Wittig reagent. Since these interactions favour *syn* stereochemistry in the oxaphosphenium, the Wittig reaction produces the *cis*-alkene. Stabilized Wittig reagents have small negative charges on the oxygen and EWG components in the transition states, and these properties control which face of the ylide will approach the aldehyde. Electron-electron repulsion dictates that these negative charges be spread as far apart as possible, and this interaction favours the *anti*-geometry in the oxaphosphenium, resulting in a preference for the *trans*-alkene.

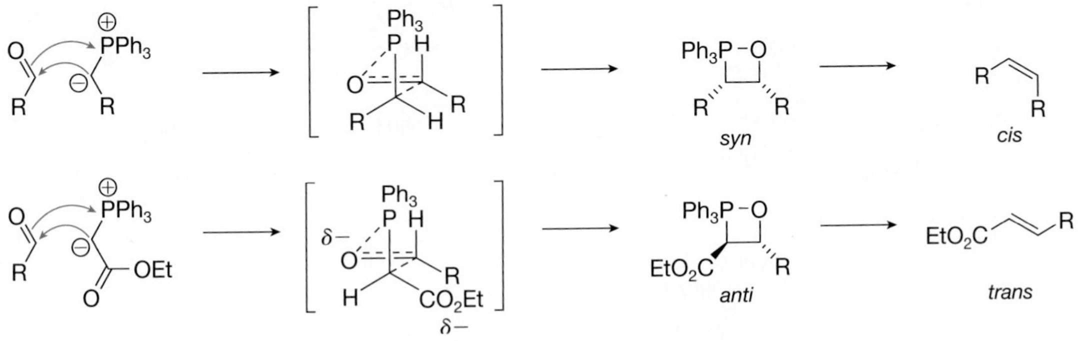

syn cis

anti trans

Umpolung reagents reverse the normal reactivity pattern of functional groups to allow for the construction of molecules that have dissonant reactivity patterns. Two popular carbonyl umpolungs are dithianes and cyanohydrins. The attachment of certain electron-withdrawing groups directly to the carbon of a carbonyl group makes it possible to reverse the reactivity of that carbon from electrophilic to nucleophilic. Once the umpolung has reacted with the appropriate electrophile, the carbonyl can be "unmasked" by reversing the conditions under which the electron-withdrawing group was installed.

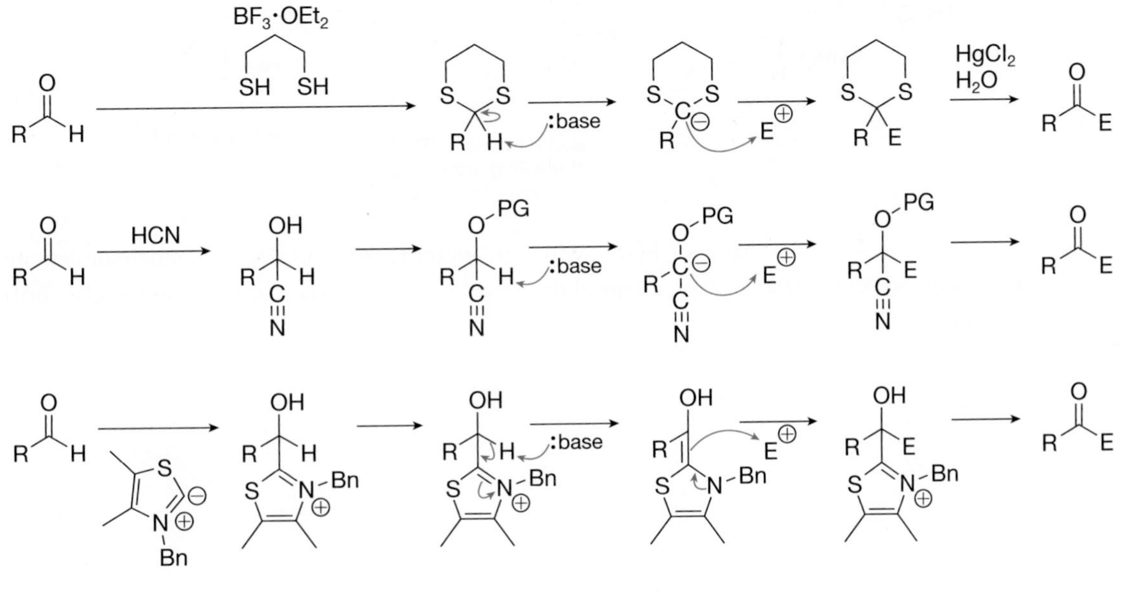

$BF_3 \cdot OEt_2$ SH SH :base $HgCl_2$ H_2O

HCN :base

:base

Bringing It Together

Many molecules made in nature by enzymes are produced as single stereoisomers with a very high degree of control. To produce similar compounds in a flask requires the ability to control both regio- and stereoselectivity using small molecules. For example, prostaglandin GE_2 (PGE_2), synthesized in 1969 by a team led by E.J. Corey, has two alkenes: one *cis,* one *trans.* Starting from an advanced intermediate, now known as Corey's lactone, a Horner–Wadsworth–Emmons reaction adds a side-chain with a *trans* alkene as part of an enone.

A number of steps lead to the preparation of a second aldehyde that undergoes a Wittig reaction to form a *cis*-alkene. Removing the THP protecting groups reveals PGE_2.

The core and side-chains of PGE_2, with the appropriate *cis* and *trans* stereochemistry, are part of the structure of travoprost, a topical drug used to reduce intraocular pressure in glaucoma patients.

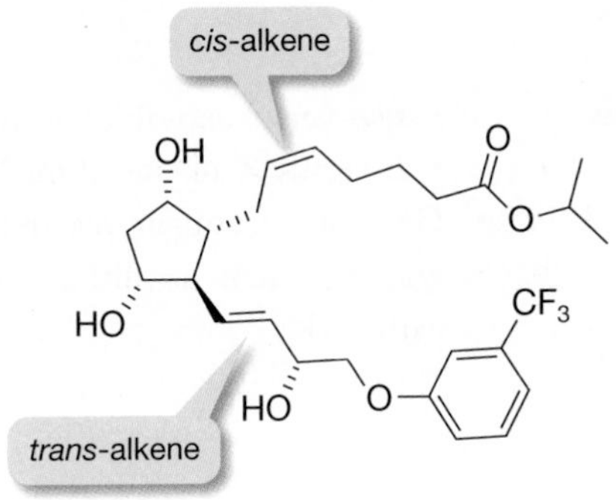

cis-alkene

trans-alkene

travoprost

Estrone, a naturally occurring estrogen, is used to synthesize ethynyl estradiol, a contraceptive. Estrone was synthesized in 1977 by K.P.C. Vollhardt, using conjugate addition and a thermodynamic enolate, among other steps. The synthesis starts with the attachment of two side-chains to the cyclopentanone D-ring.

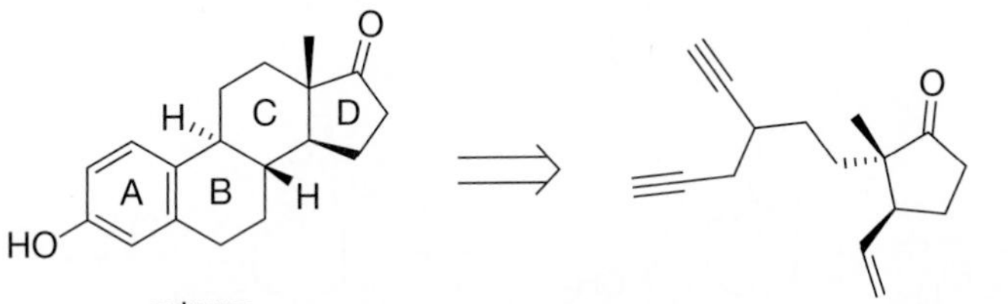

estrone

The first step is a conjugate addition using a Grignard reagent with added copper. Vinyl Grignard reagent would typically undergo direct addition to an enone. However, the included copper exchanges with the magnesium to create a low-order cuprate that does conjugate addition. The resulting enolate intermediate was trapped by the addition of TMSCl.

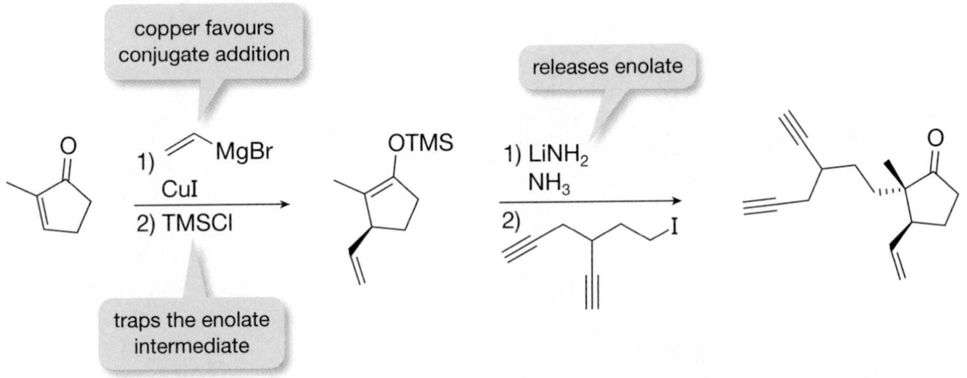

copper favours conjugate addition

1) MgBr CuI

2) TMSCl

traps the enolate intermediate

OTMS

releases enolate

1) $LiNH_2$ NH_3

2) I

Trapping the enolate preserves the ability to add the alkyl chain to the more substituted α-carbon. The thermodynamic enolate formed on the more substituted side by the conjugate addition makes the alkylation relatively easy, as compared to trying to avoid kinetic deprotonation.

The ability to control both regio- and stereochemistry is essential to our use of synthesis to copy and modify molecules from nature and provide us with useful products.

You Can Now

- Predict the major product of regioselective reactions, including
  - direct versus conjugate addition of α,β-unsaturated carbonyls
  - additions using kinetic versus thermodynamic enolates
- Predict the stereochemistry of stereoselective reactions, including
  - *syn*- versus *anti*-aldol products using E versus Z enolates
  - E versus Z alkenes using Wittig and related reactions
- Generate dissonant reactivity patterns using umpolung reagents such as dithianes, cyanohydrins, and α-halo carbonyls.

A Mechanistic Re-View

CONJUGATE ADDITION

Cuprate addition

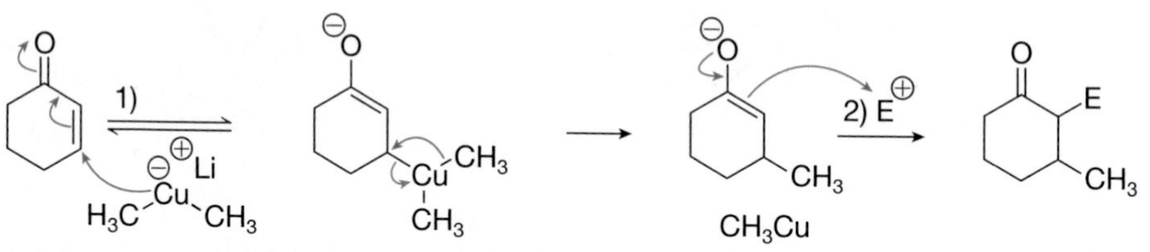

Michael addition

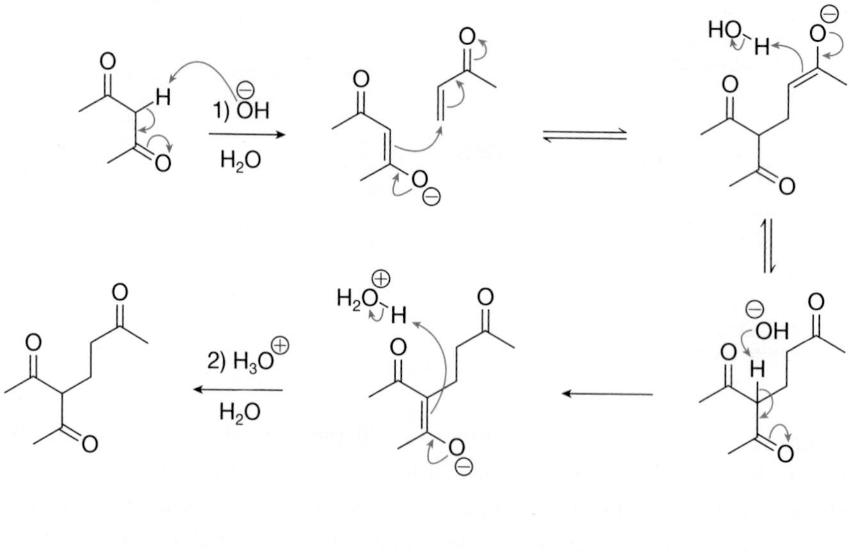

Robinson annulation

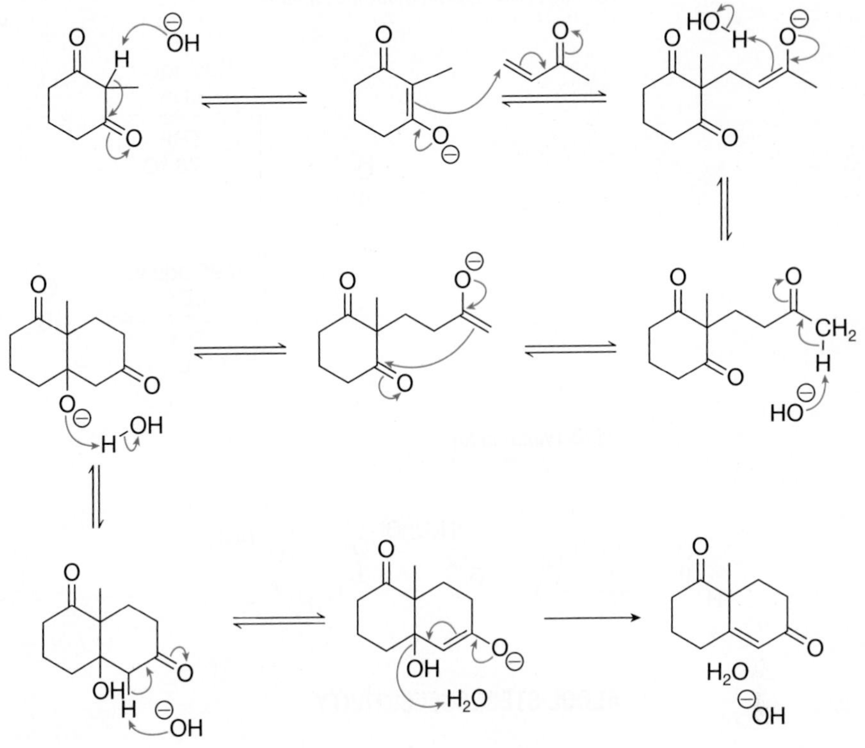

Baylis–Hillman reaction

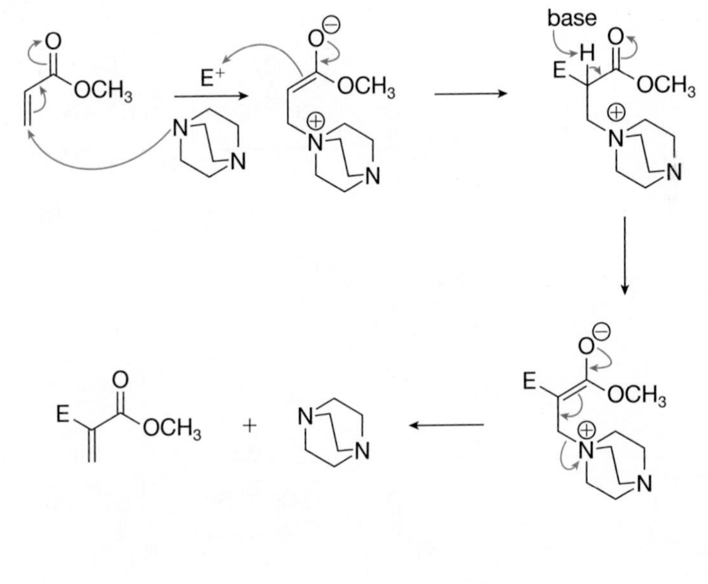

ALDOL REGIOSELECTIVITY

Kinetic versus thermodynamic enolate

1.05 equiv. LDA

THF −78 °C

kinetic enolate

0.98 equiv. LDA

THF −78 °C to rt

thermodynamic enolate

Mukaiyama aldol

$(H_3C)_3Si$

$TiCl_3$

H_3O^+

H_2O

ALDOL STEREOSELECTIVITY

E versus Z enolates

$B(Cy)_2$

fast

slow

NEt_3

(*E*)-(O) boron enolate formed selectively

$B(Bu)_2$

slow

fast

i-Pr_2NEt

(*Z*)-(O) boron enolate formed selectively

Zimmerman–Traxler model

fast
slow
E enolate
fast
slow
Z enolate

WITTIG VARIATIONS

slow
fast
oxaphosphetanes
anti
syn

slow

fast

oxaphosphetanes

syn

anti

UMPOLUNG REACTIONS

Dithianes

BF_3-OEt_2

SH SH

$HgCl_2$
HgO
H_2O
CH_3OH

base

E^+

Cyanohydrins (benzoin condensation)

$NaCN_{(cat.)}$

DMF

Bromoketone

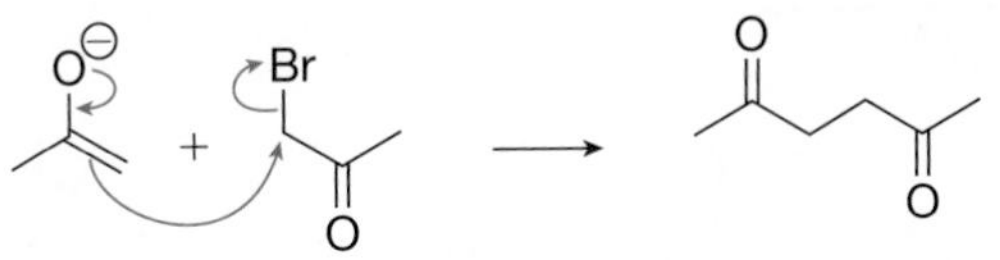

Darzen's condensation

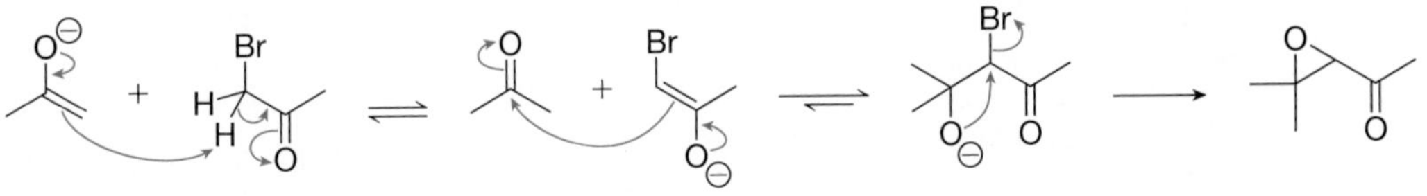

Problems

18.21 Select the reagent(s) needed to do the following conversions.

a)

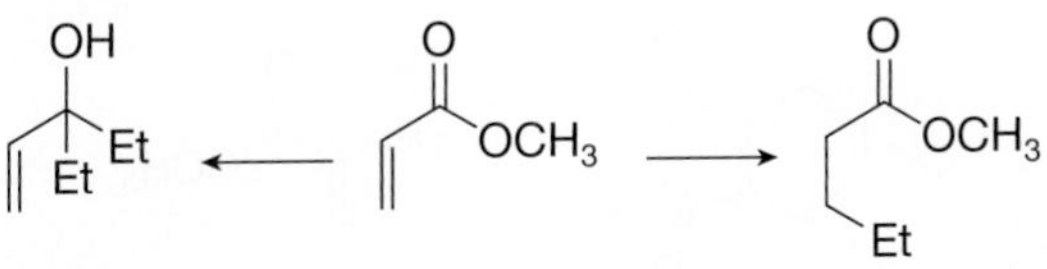

b)

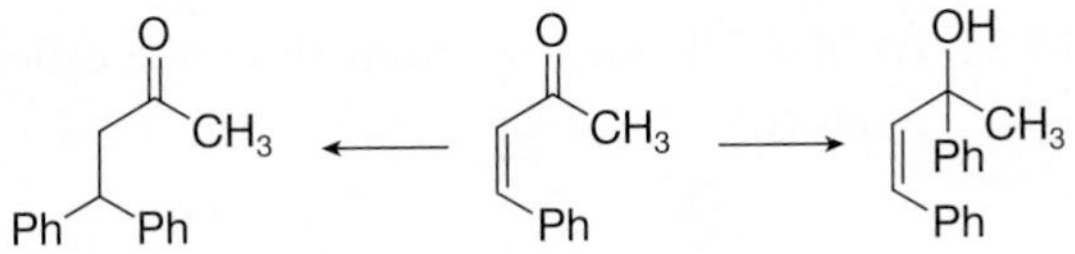

c)

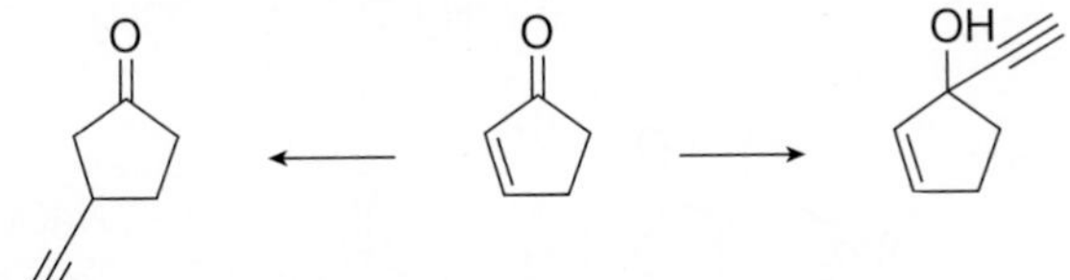

18.22 Select the appropriate conditions for the following transformations.

a)

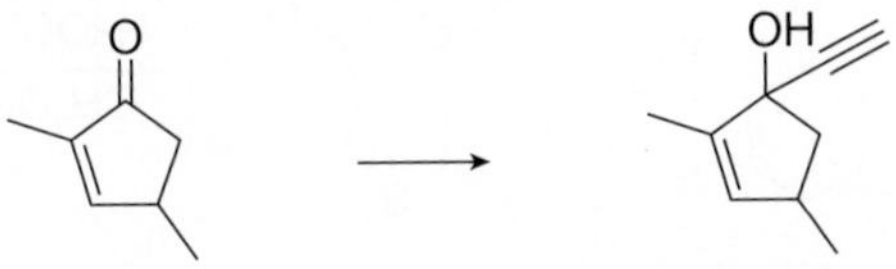

b)

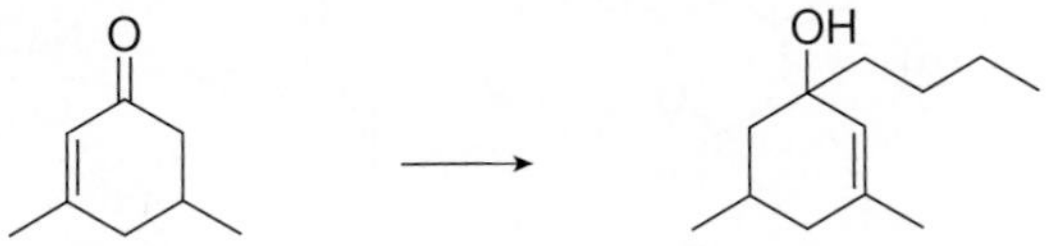

c)

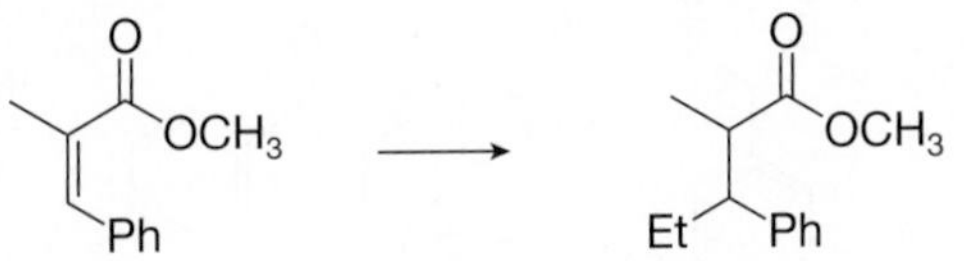

d)

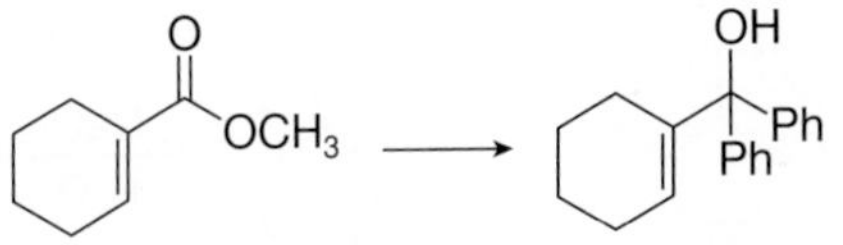

e)

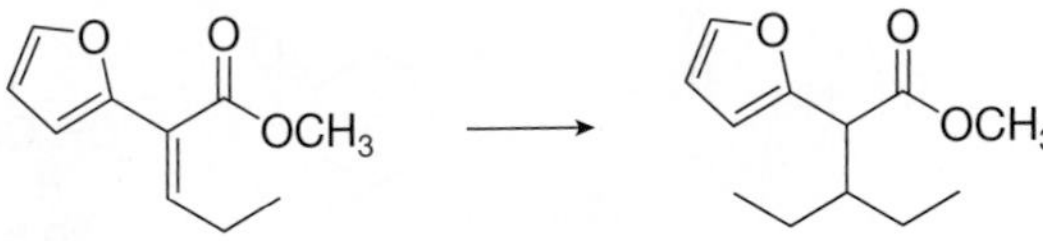

f)

18.23 For the following reactions, show the expected major product.

a) 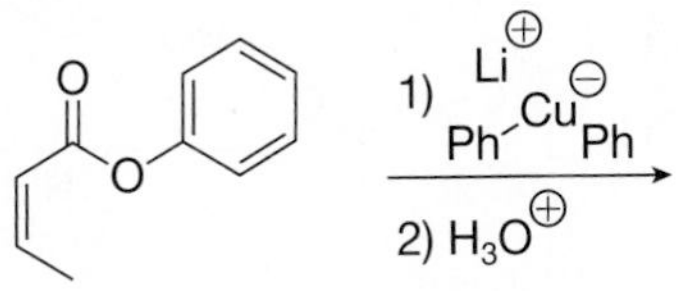 1) $Li^{\oplus}$ $Ph_2Cu^{\ominus}$ 2) $H_3O^{\oplus}$

b) 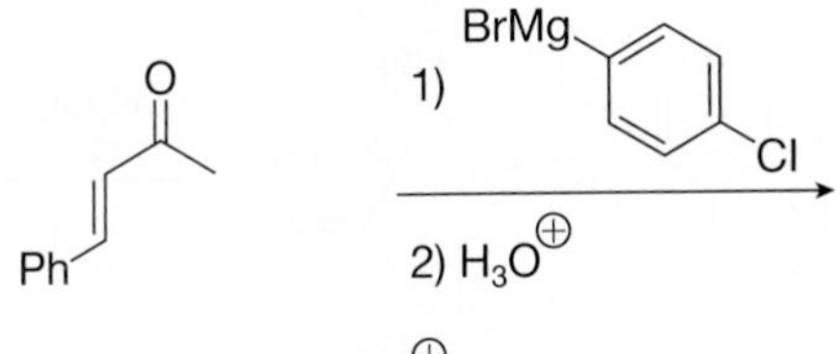 1) BrMg–C_6H_4–Cl 2) $H_3O^{\oplus}$

c) 1) $Li^{\oplus}$ $Ph_2Cu^{\ominus}$ 2) $H_3O^{\oplus}$

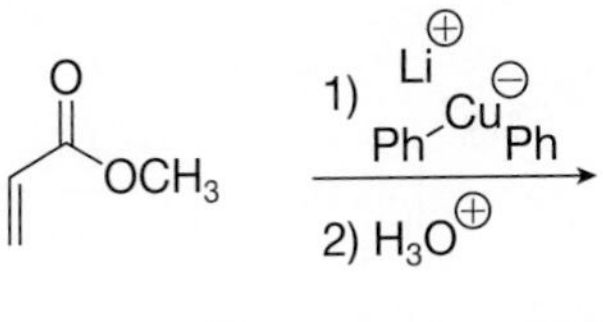

d) 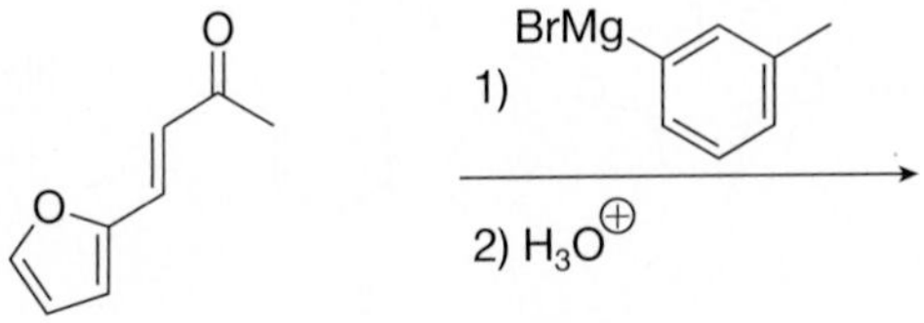 1) BrMg–(3-methylphenyl) 2) $H_3O^{\oplus}$

e) 1) $(HCC)_2CuLi$ 2) $H_3O^{\oplus}$

f) Ph + NaOH / H_2O

g) 1) Et_2CuLi 2) $H_3O^{\oplus}$

h) + HN, HS toluene

i) + NaOH / H_2O

j) NH + CH_2Cl_2 rt 20 h

k) + NaOH / H_2O

l) H + HO, S, N⊕, $Bn_{(cat.)}$ / NEt_3

m) + HO CH_2Cl_2 rt

n) + HS toluene

o) + NaOH / H_2O

p) OCH_3 + NaOCH$_3$ / CH_3OH

q) OH CH_2Cl_2 rt

r) + OCH_3 NaOCH$_3$ / CH_3OH

s) NH + H base catalyst CH_2Cl_2 rt 20 h

t) + Br OCH_3 NaOCH$_3$ / CH_3OH

18.24 For the following reactions, show the expected major product.

a) Br + H_3CO CH_3 NaOCH$_3$ / CH_3OH

b) + OCH_3 NaOCH$_3$ / CH_3OH

c) + S NaOCH$_3$ / CH_3OH

d) + NaOCH$_3$ / CH_3OH

e) + NaOCH$_3$ / CH_3OH

18.25 The epoxidation of an alkene using peracids such as *m*-CPBA results in a poor yield when the alkene is conjugated to a carbonyl or other electron-withdrawing group. The use of peroxide and base will result in a much higher yield of epoxide with electron-poor alkenes. Propose a reasonable mechanism for this epoxidation, and explain why it is more efficient than using a peracid.

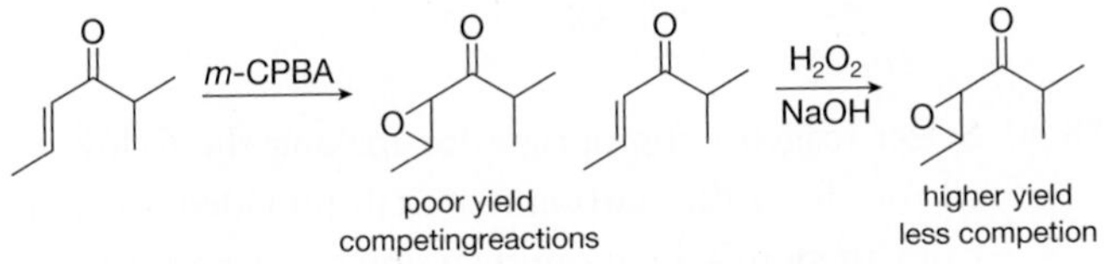

m-CPBA — poor yield competingreactions; H_2O_2, NaOH — higher yield less competion

18.26 Propose a reasonable mechanism for the following reaction.

CHO, OH + Ph, NO_2 — base, CH_2Cl_2, rt → CHO, NO_2, O, Ph

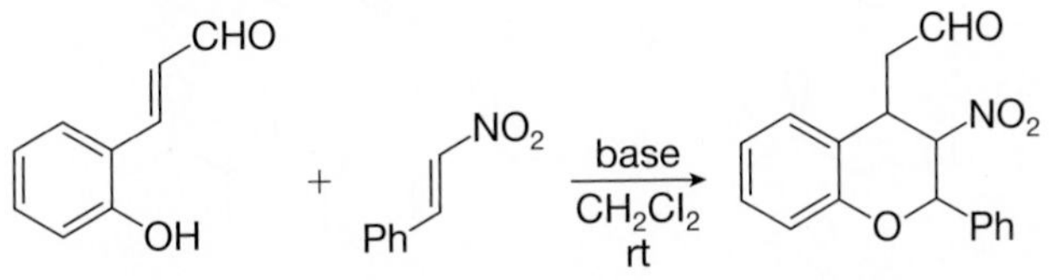

18.27 Select appropriate conditions for the following transformations.

a) 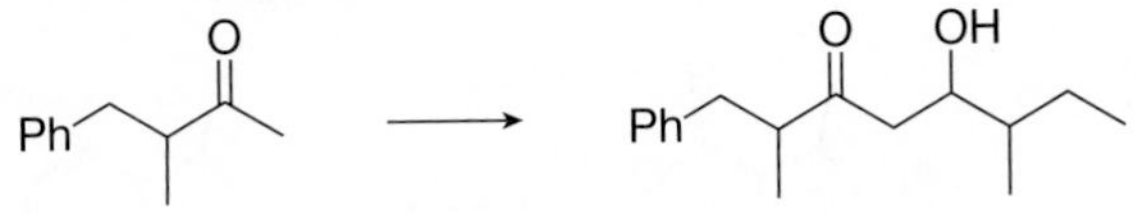 Ph → Ph, O, OH

b) 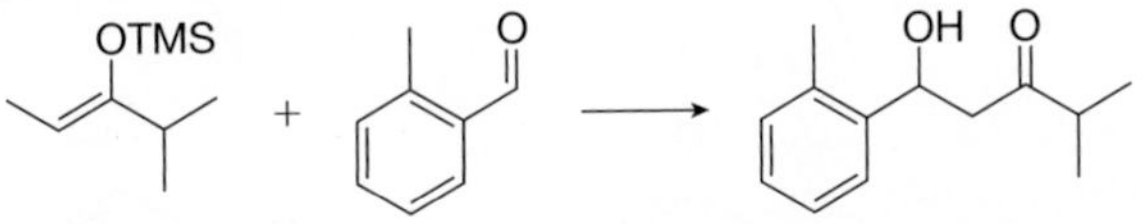 OTMS + O → OH O

c) 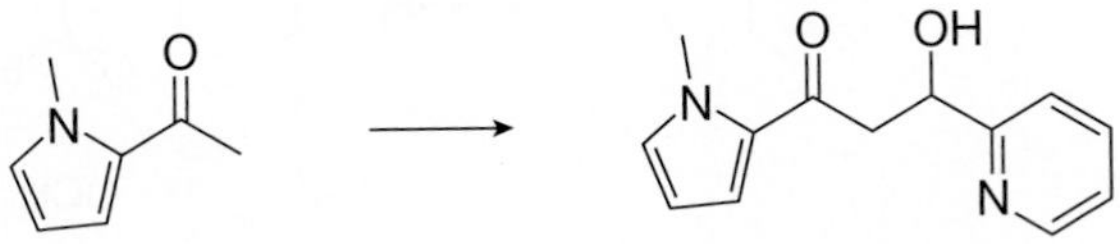 N, O → N, O, OH, N

d) 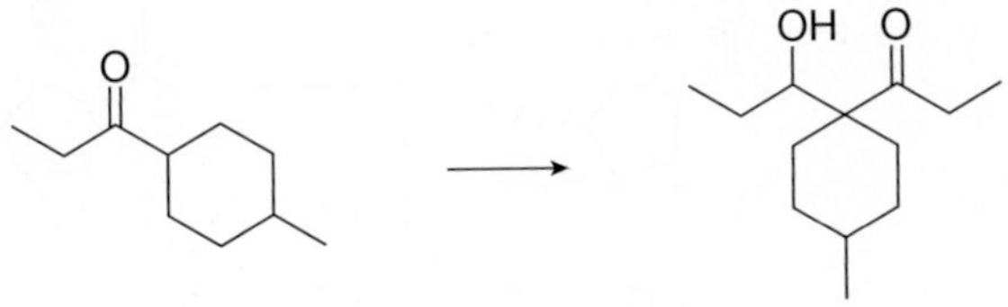 O → OH O

e) 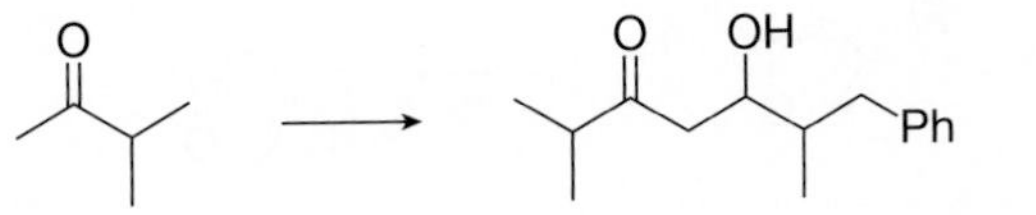 O → O OH, Ph

f) 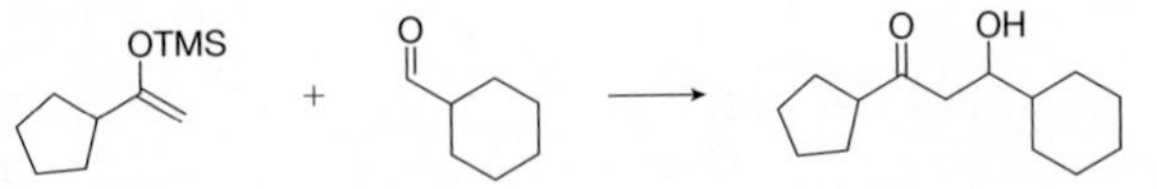 OTMS + O → O OH

g) 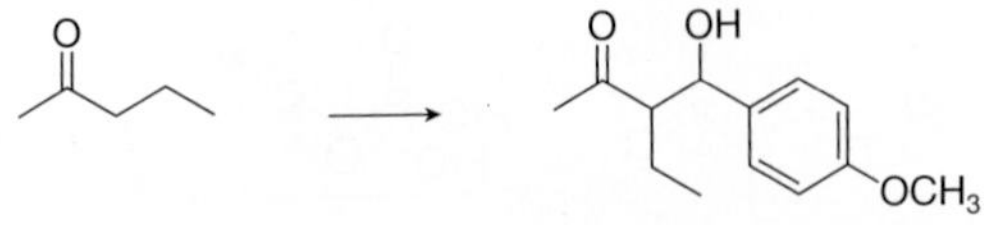 O → O OH, OCH_3

h) 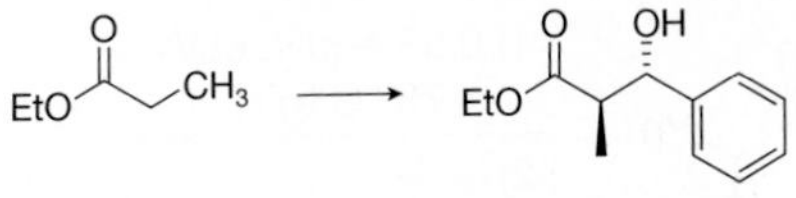 EtO, O, CH_3 → EtO, O, OH + enantiomer

i) 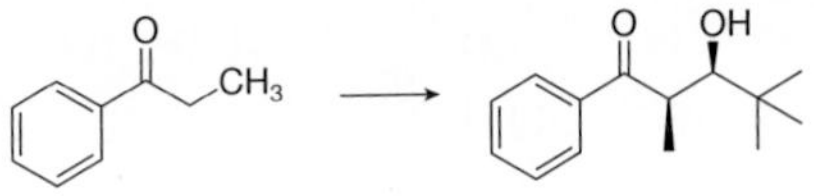 O, CH_3 → O OH + enantiomer

j) 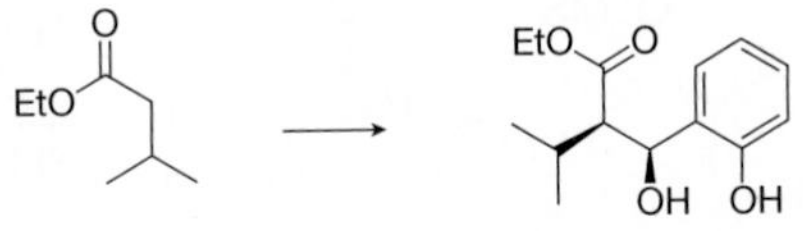 EtO, O → EtO, O, OH OH + enantiomer

k) 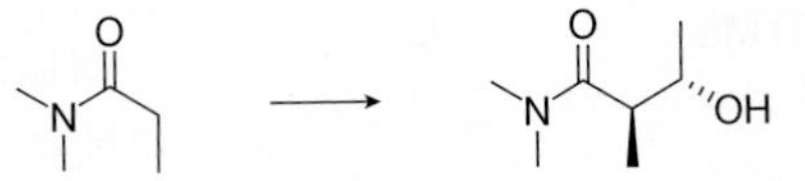 N, O → N, O, OH + enantiomer

18.28 Predict the major product of the following reactions.

a) 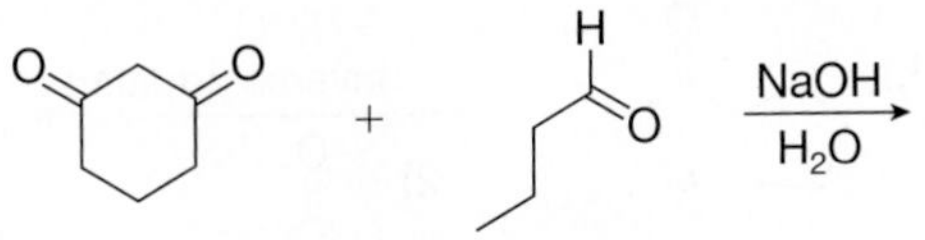 O, O + H, O — NaOH, H_2O →

b) 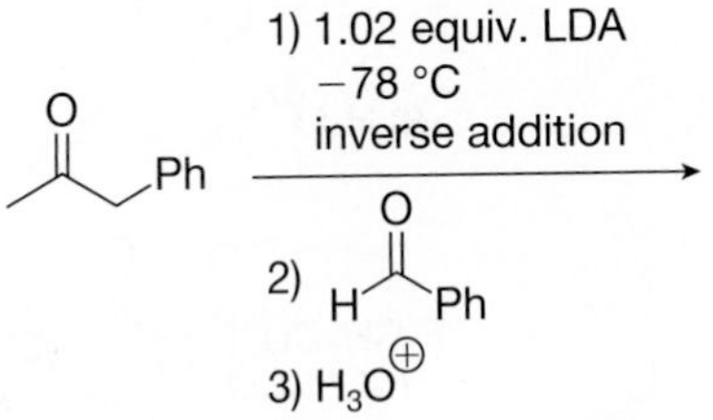 O, Ph — 1) 1.02 equiv. LDA −78 °C inverse addition; 2) H, O, Ph; 3) $H_3O^{\oplus}$ →

c) OTMS + Ph, O, H — 1) $TiCl_4$ CH_2Cl_2; 2) H_2O →

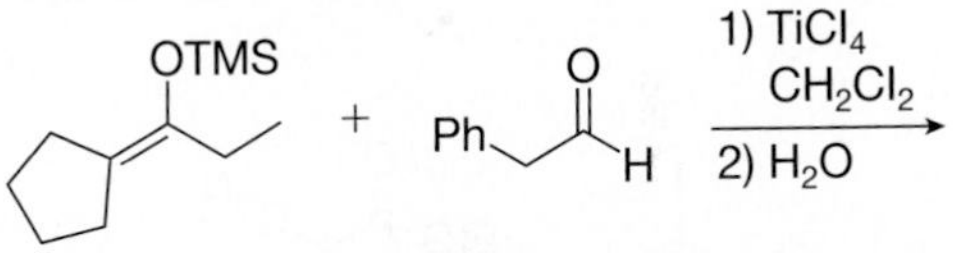

d) 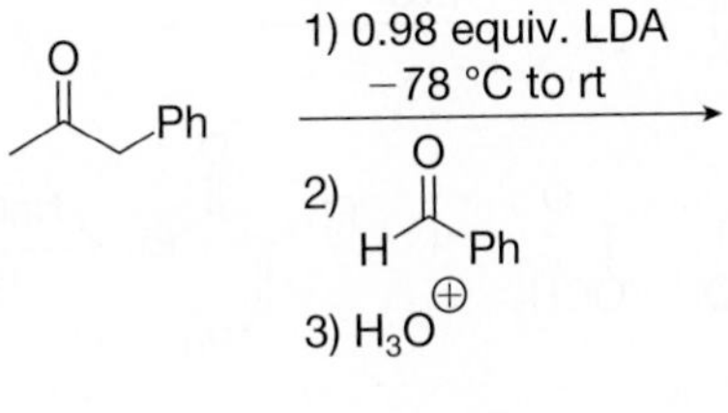 O, Ph — 1) 0.98 equiv. LDA −78 °C to rt; 2) H, O, Ph; 3) $H_3O^{\oplus}$ →

e) 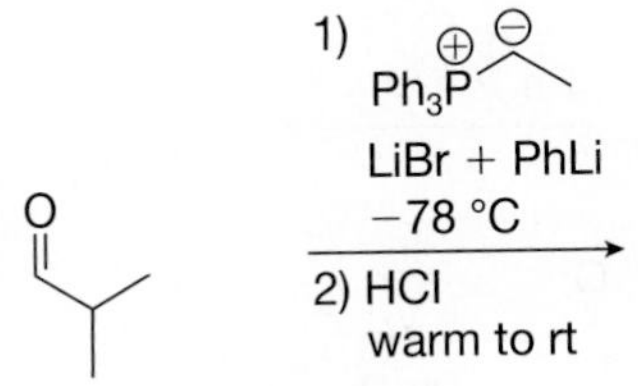 O — 1) $Ph_3P^{\oplus}$ $^{\ominus}$, LiBr + PhLi −78 °C; 2) HCl warm to rt →

f) 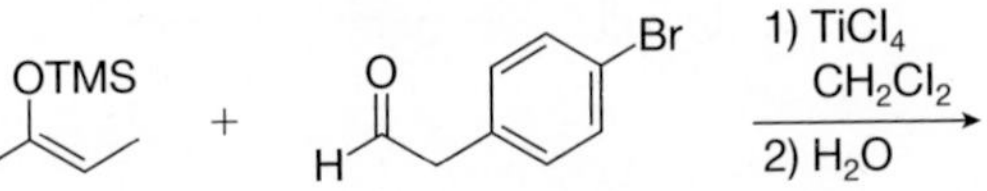 OTMS + H, O, Br — 1) $TiCl_4$ CH_2Cl_2; 2) H_2O →

g) 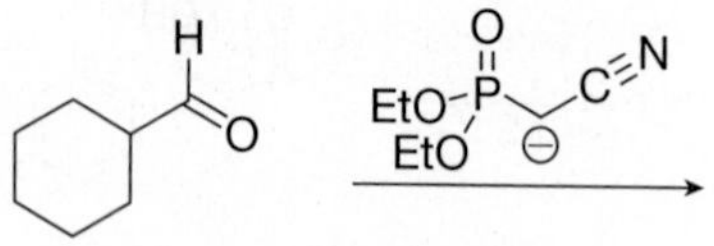 EtO–P(=O)(EtO)–CH^-–C≡N

h) 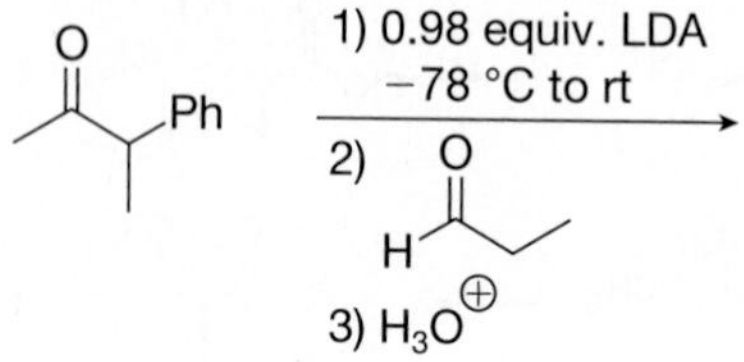 1) 0.98 equiv. LDA, −78 °C to rt; 2) propanal; 3) H_3O^+

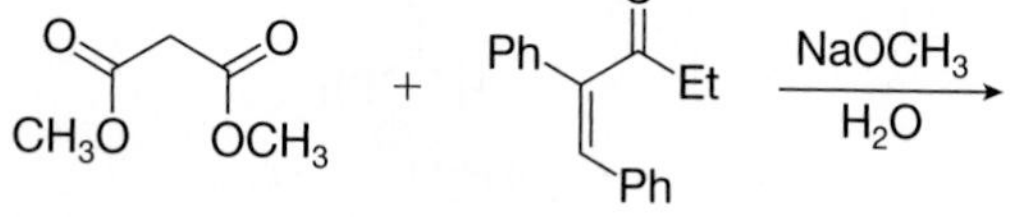

i) 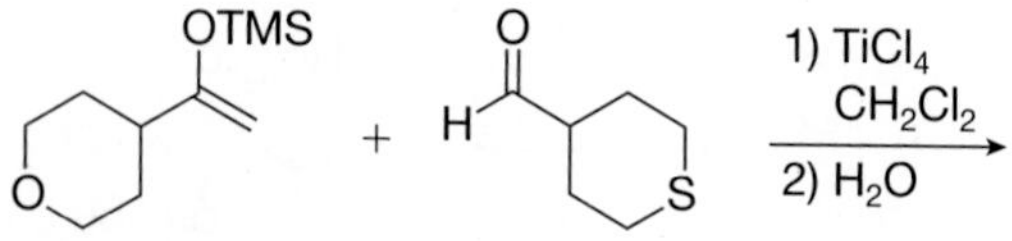 OTMS + H(C=O)– ; 1) $TiCl_4$, CH_2Cl_2; 2) H_2O

j) 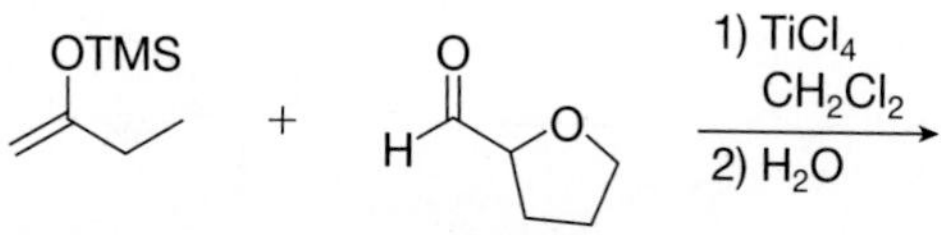 OTMS + ; 1) $TiCl_4$, CH_2Cl_2; 2) H_2O

k) 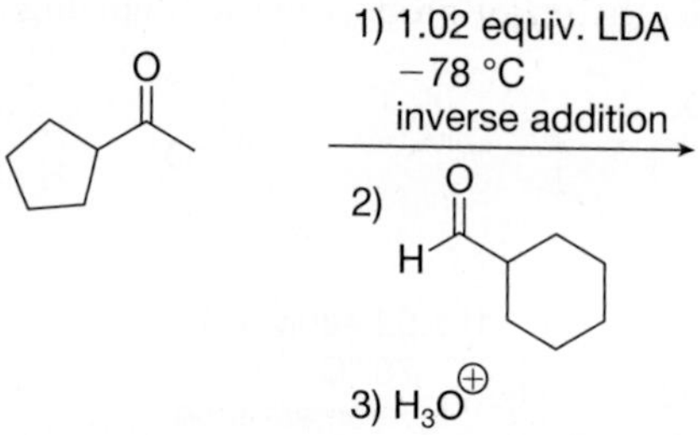 1) 1.02 equiv. LDA, −78 °C, inverse addition; 2) cyclohexanecarbaldehyde; 3) H_3O^+

l) 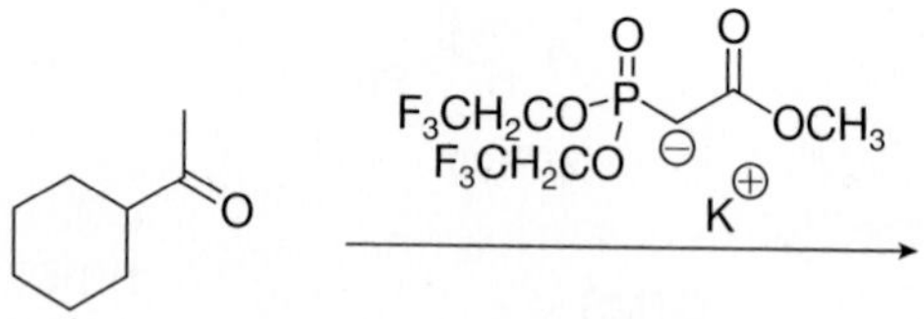 F_3CH_2CO–P(=O)(F_3CH_2CO)–CH^-–C(=O)OCH_3 K^+

m) 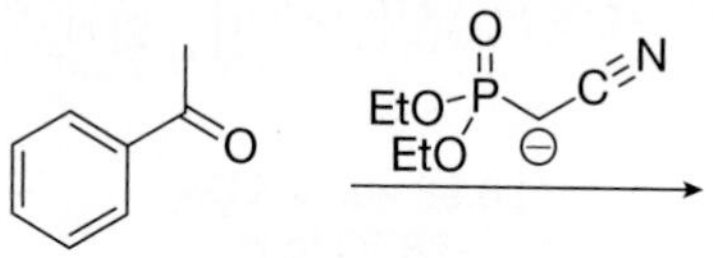 EtO–P(=O)(EtO)–CH^-–C≡N

n) CH_3O, OCH_3 + Ph, Et, Ph; NaOCH$_3$, H_2O

o) 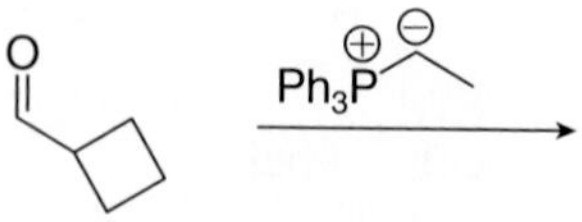 Ph_3P^+–CH^-–

p) Ph_3P^+–CH^-–

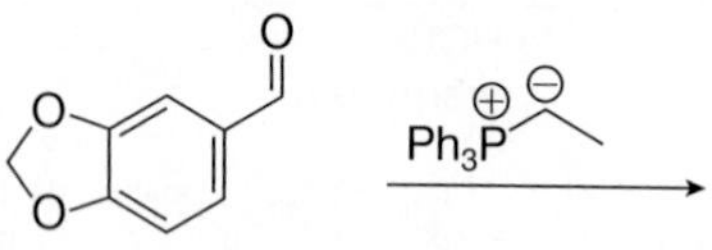

18.29 What conditions should be used to close the ring of erythromycin by forming the C–C bond, shown in red, with the proper stereochemistry?

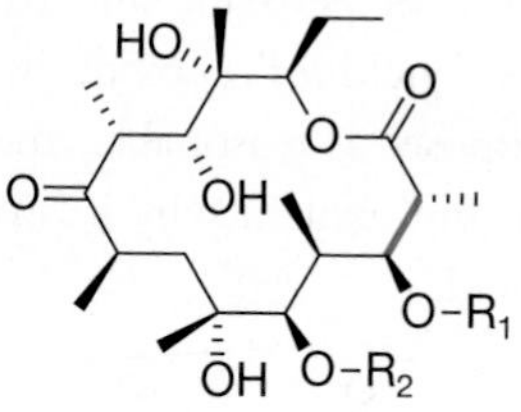

HO, O, O, O, OH, O–R_1, OH, O–R_2

18.30 Select reagents appropriate for making the following alkenes from the starting materials provided. In some cases, there may be more than one correct answer.

a) 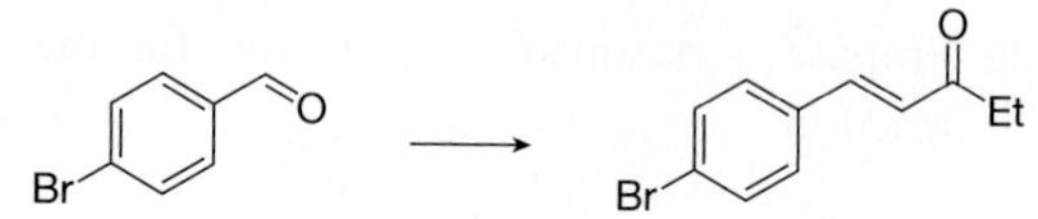 Br → Br, Et

b) 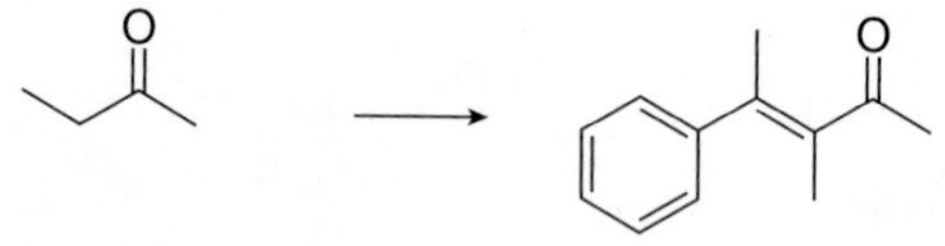 →

c) 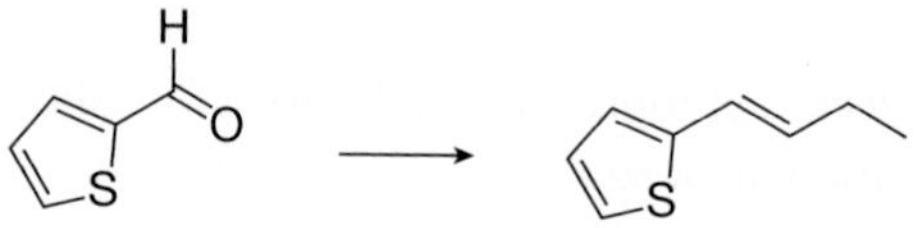 H, O, S → S

d) H, O → Ph

e) H, O → O, Et

f) O, OCH_3 → OCH_3, O, OCH_3

g) O, H, O → O

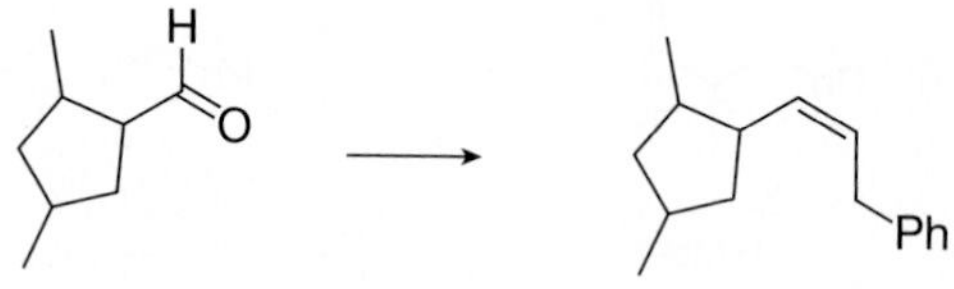

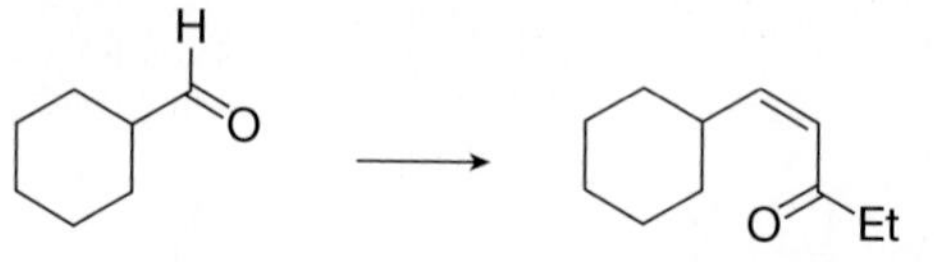

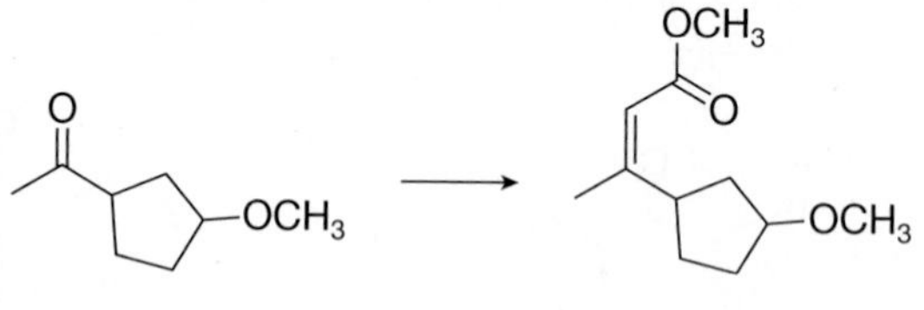

h) O, O → O

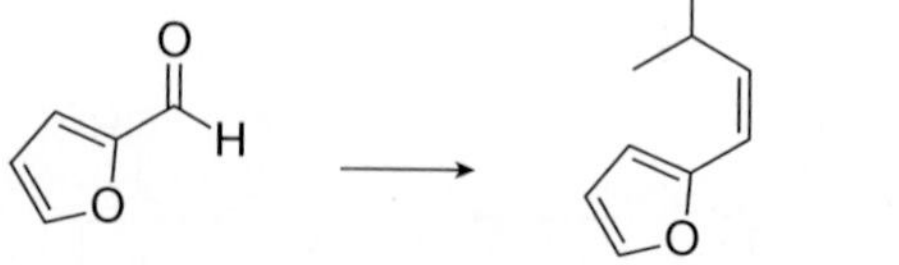

18.31 Select the reagent(s) needed to do the following conversions.

a)

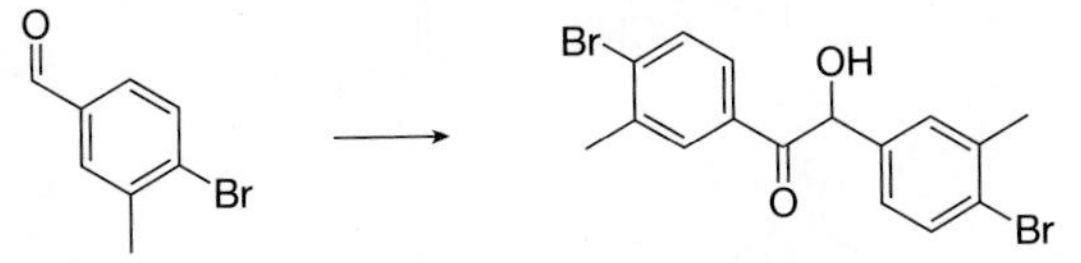

b)

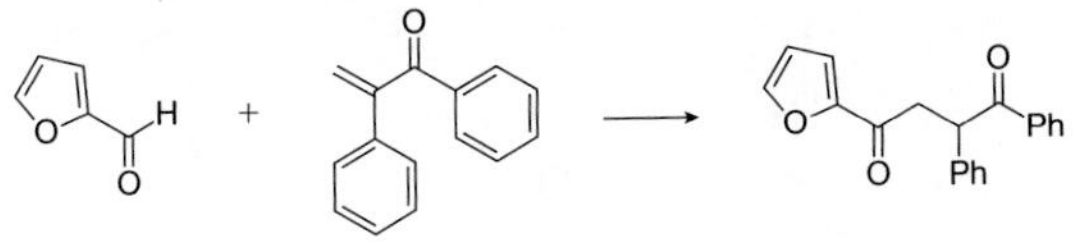

c)

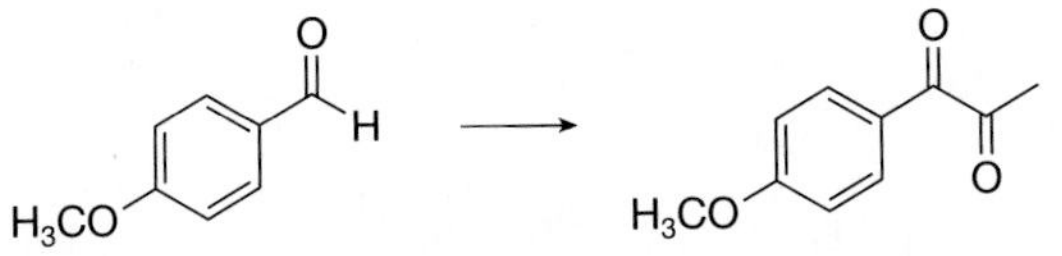

d)

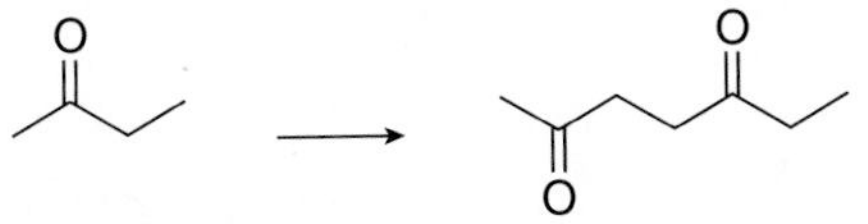

e)

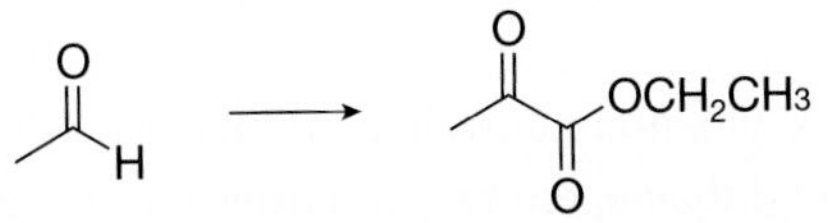

f) 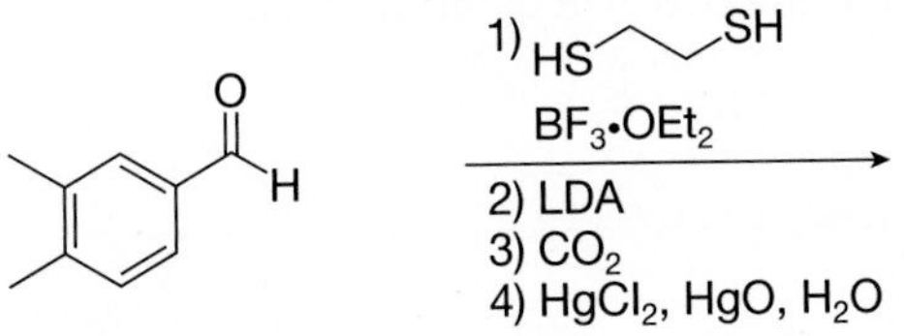 1) HS~~SH, $BF_3{\bullet}OEt_2$; 2) LDA; 3) CO_2; 4) $HgCl_2$, HgO, H_2O

g) 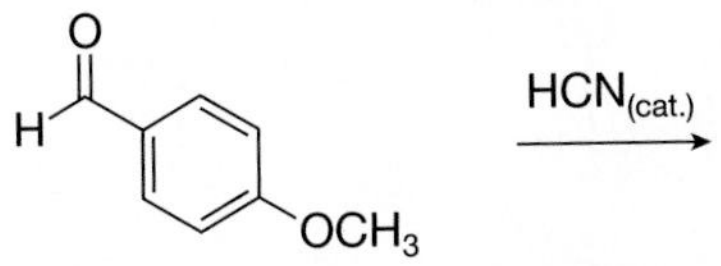 $HCN_{(cat.)}$

18.32 The combination of a Henry reaction with a Nef reaction produces an acyl anion equivalent that is selective for conjugate addition. Propose a reasonable mechanism for the following reaction sequence.

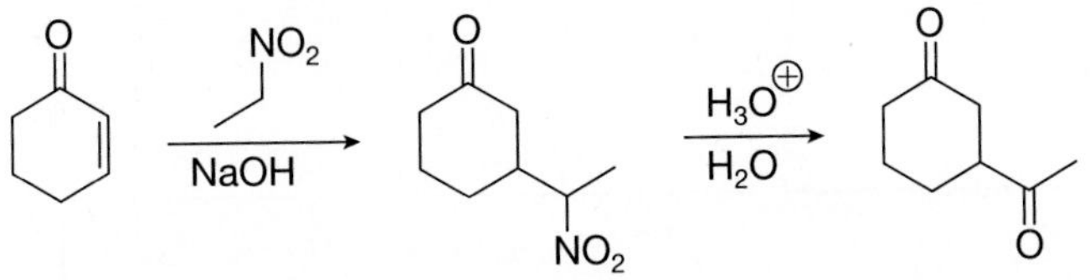

NaOH; $H_3O^{\oplus}$, H_2O

18.33 One synthesis of ketones discussed in this chapter includes an S_N2 displacement using a deprotonated dithiane. Unlike oxygen-based acetals, thioacetals are both formed and hydrolyzed using Lewis acids. Propose a reasonable mechanism for the hydrolysis of a substituted dithiane, using Hg^{2+} salts and water.

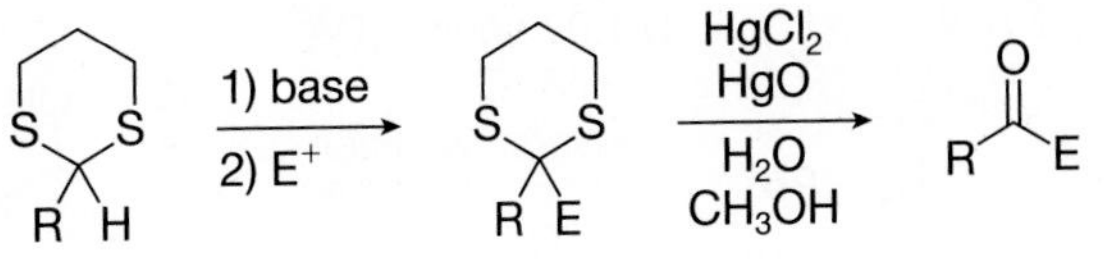

1) base, 2) E^+; $HgCl_2$, HgO, H_2O, CH_3OH

18.34 What are the intermediates and the final product of the following reaction scheme?

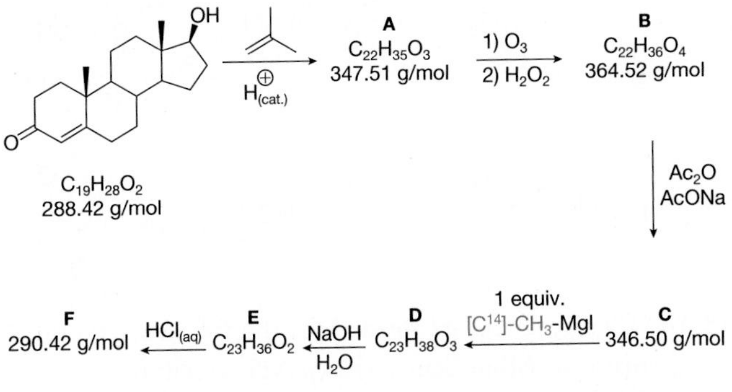

$C_{19}H_{28}O_2$ 288.42 g/mol → ($H^{\oplus}_{(cat.)}$) A $C_{22}H_{35}O_3$ 347.51 g/mol → (1) O_3 2) H_2O_2) B $C_{22}H_{36}O_4$ 364.52 g/mol → (Ac_2O, AcONa) C 346.50 g/mol → (1 equiv. [C^{14}]-CH_3-MgI) D $C_{23}H_{38}O_3$ → (NaOH, H_2O) E $C_{23}H_{36}O_2$ → ($HCl_{(aq)}$) F 290.42 g/mol

18.35 Using 2-phenylethanonitrile, formaldehyde, ethyl bromide, and methyl acetate as your only sources of carbon, synthesize glutethimide, a sedative and hypnotic drug.

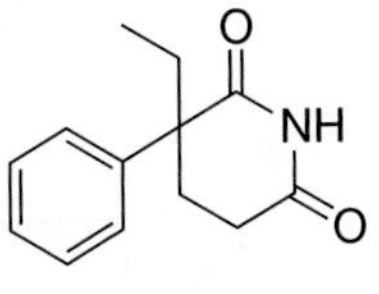

glutethimide

18.36 What is the problem with each of the following reactions?

a) 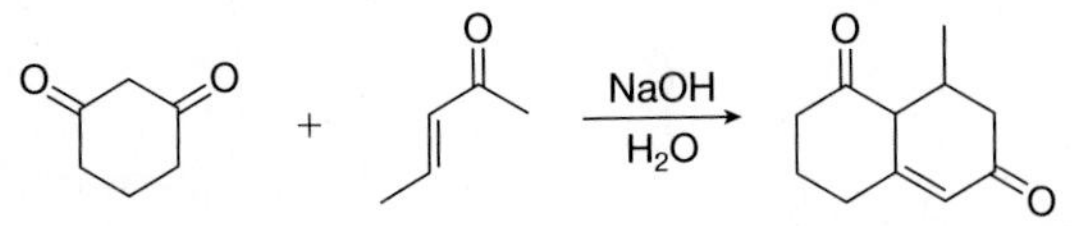 NaOH, H_2O

b) 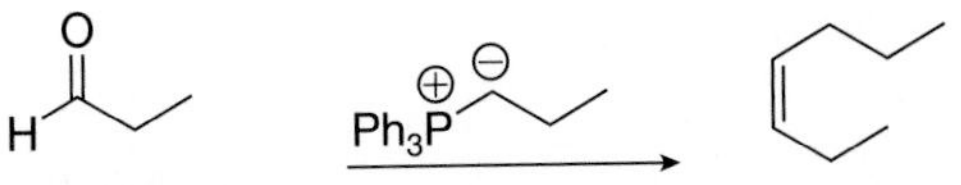 $Ph_3P^{\oplus}$

c) 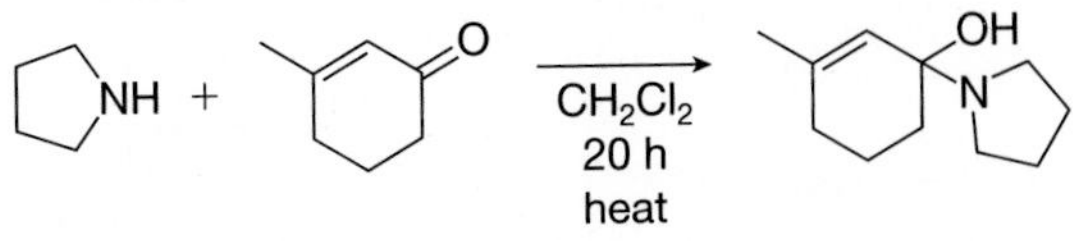 CH_2Cl_2, 20 h, heat

d) 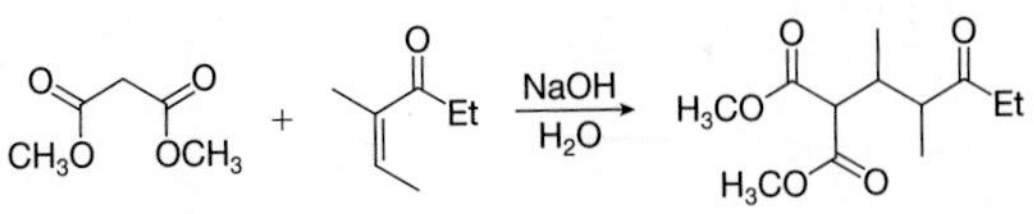 NaOH, H_2O

e) 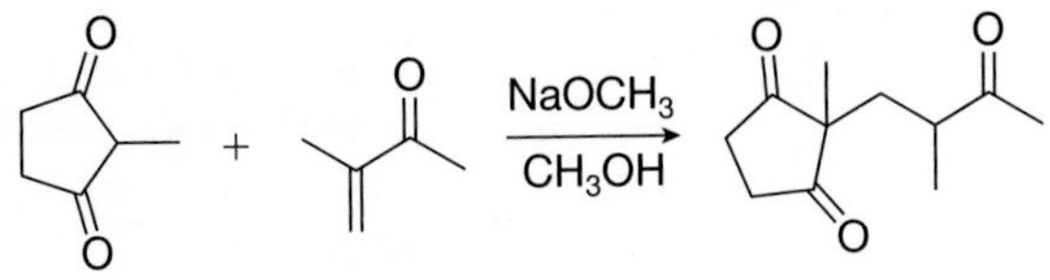 $NaOCH_3$, CH_3OH

f) 1) 1.02 equiv. LDA, −78 °C, inverse addition; 2) PhCHO; 3) $H_3O^{\oplus}$

g) + PhCH$_2$CHO; 1) $TiCl_4$, CH_2Cl_2; 2) H_2O

h) PhCH$_2$CHO + CH$_3$CHO; $HCN_{(cat.)}$

18.37 What reagents are needed for the following transformations? More than one step will be required.

a)

b)

c)

d)

e) + enantiomer

18.38 Aldehyde A was converted to a dithiane acyl anion equivalent and reacted with two different acid chlorides. After removal of the dithiane, the two reactions produced B and C. What are the structures of A, B, C, and the two acid chlorides used?

^{1}H-NMR of Compound A

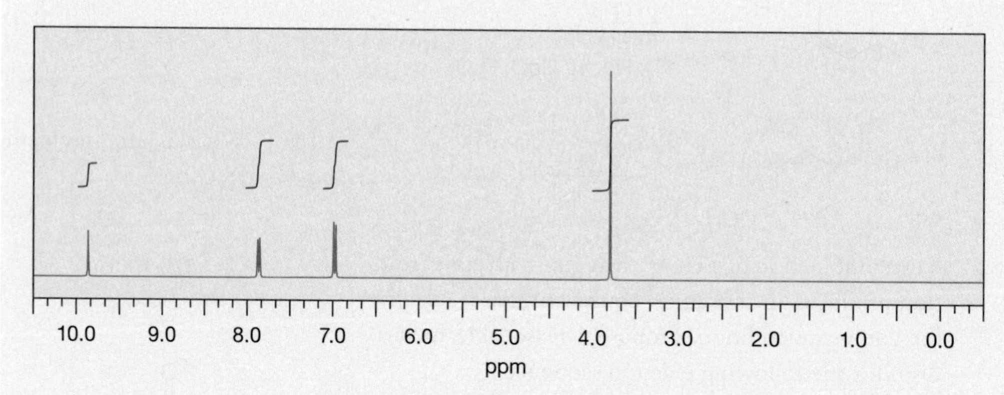

^{13}C-NMR of Compound A

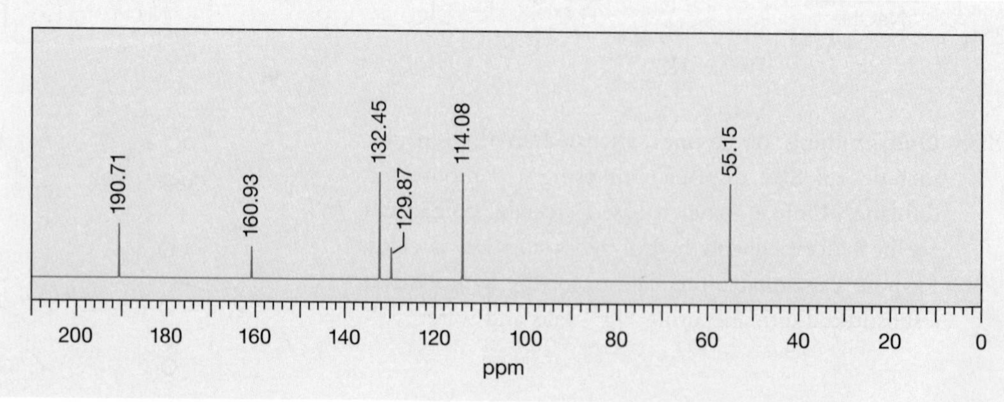

¹H-NMR of Compound B

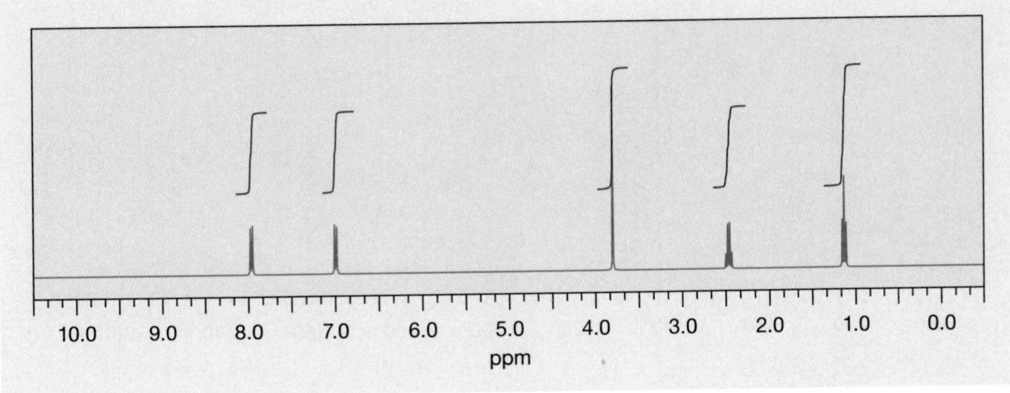

¹³C-NMR of Compound B

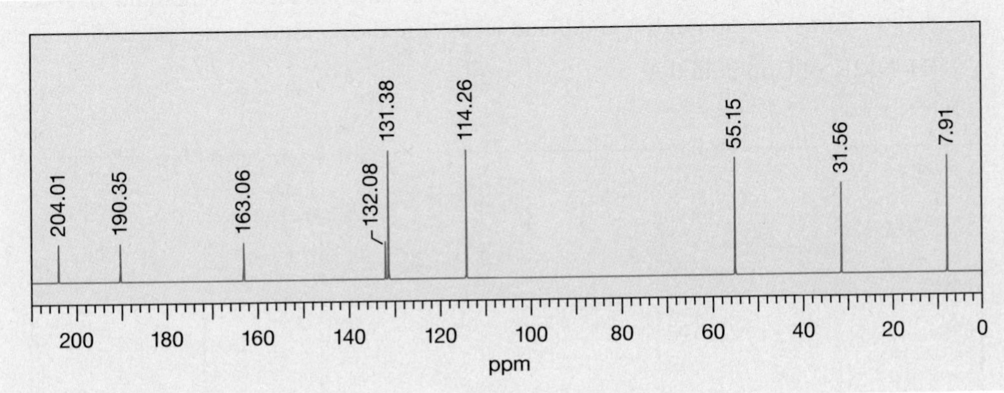

¹H-NMR of Compound C

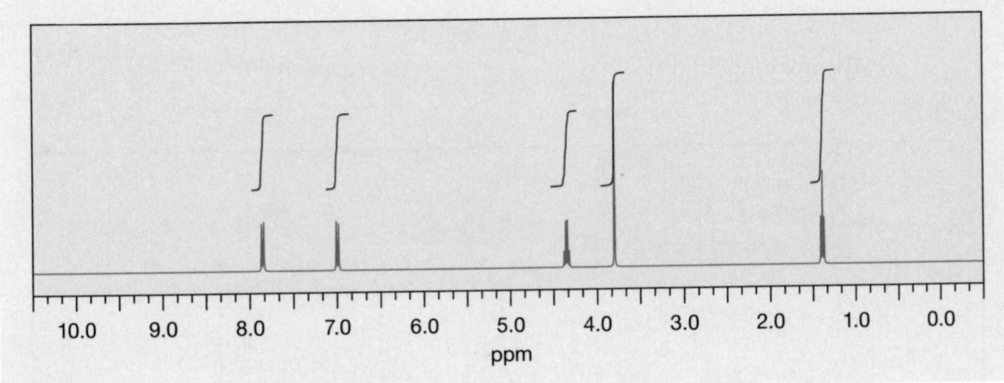

^{13}C-NMR of Compound C

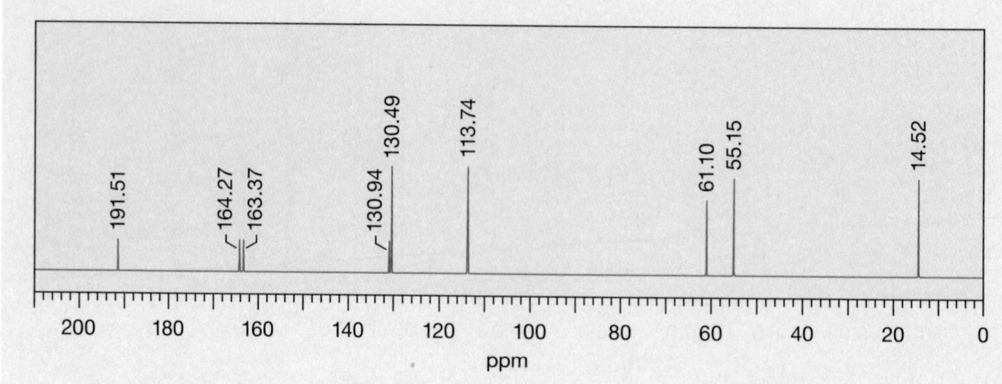

18.39 Butanone was reacted with a strong base and benzaldehyde to produce compound A. Using different reaction conditions, the same reactants produced compound B. What is the structure of compounds A and B and what were the conditions used to produce each?

^{1}H-NMR of Compound A

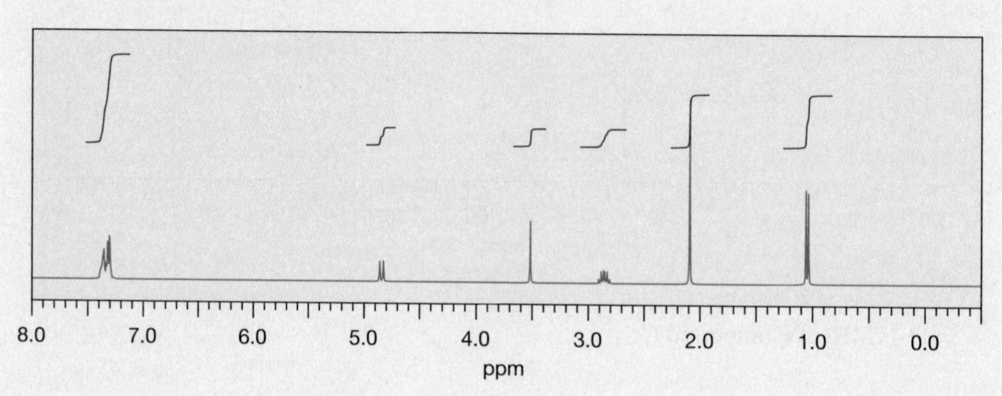

^{1}H-NMR of Compound B

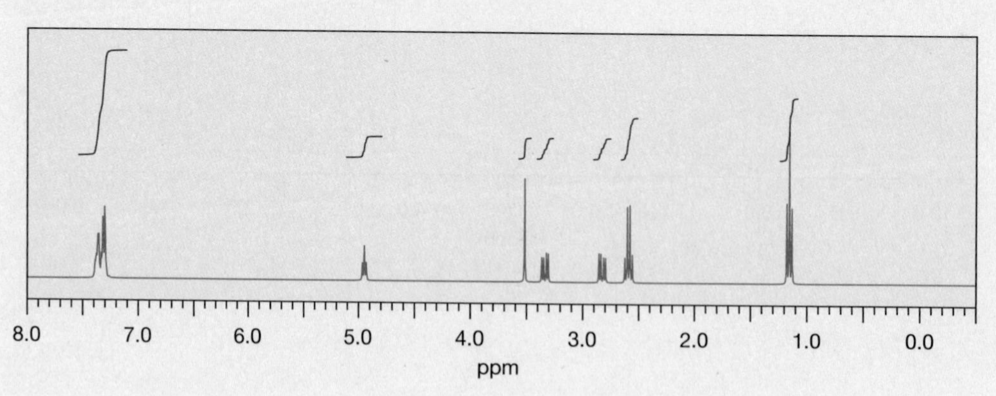

18.40 Compound A and compound B are both symmetrical cyclic diones. A was reacted with C in the presence of base to produce D. B was also reacted with C under catalytic base to form E. What is the structure of each compound?

^{1}H-NMR of Compound A

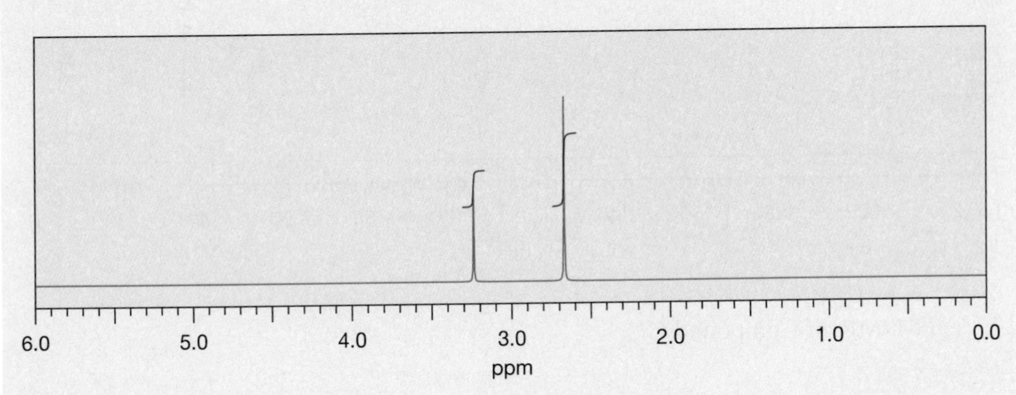

^{13}C-NMR of Compound A

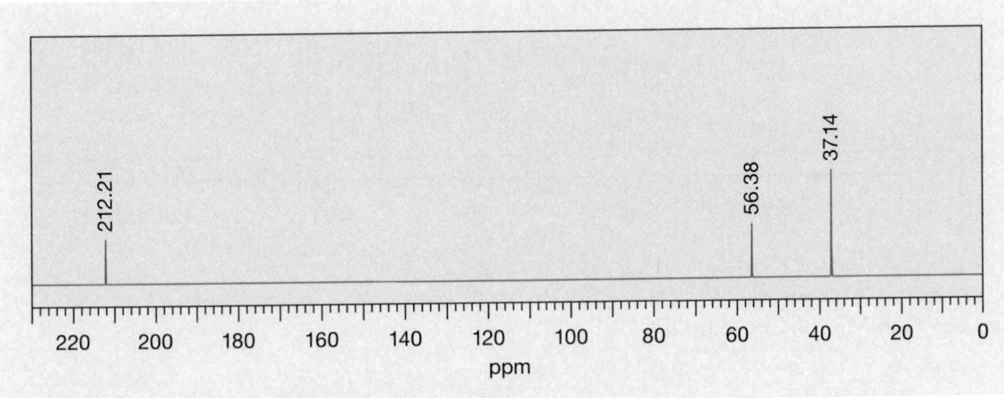

^{1}H-NMR of Compound B

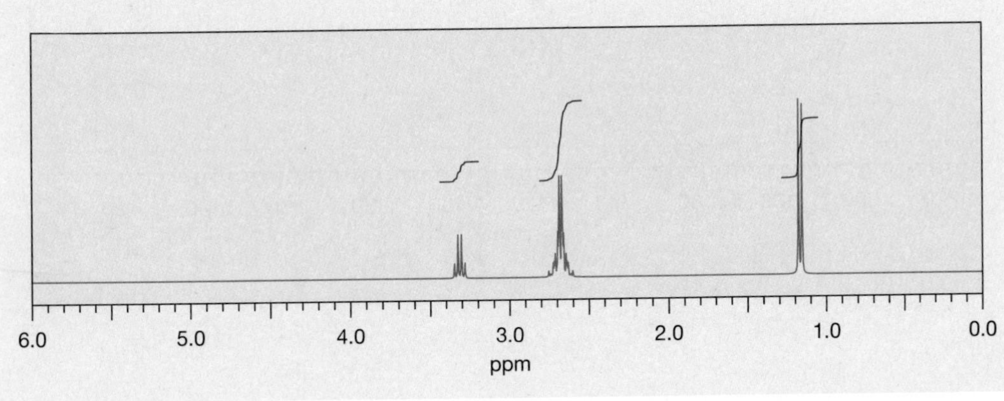

^{13}C-NMR of Compound B

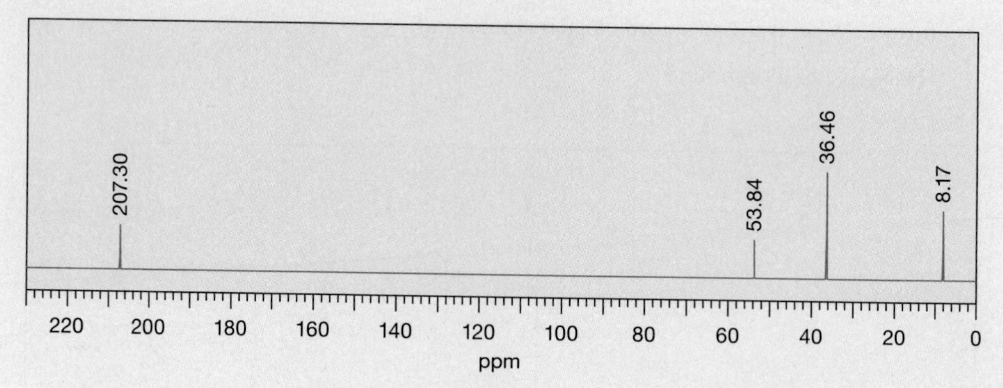

^{1}H-NMR of Compound C

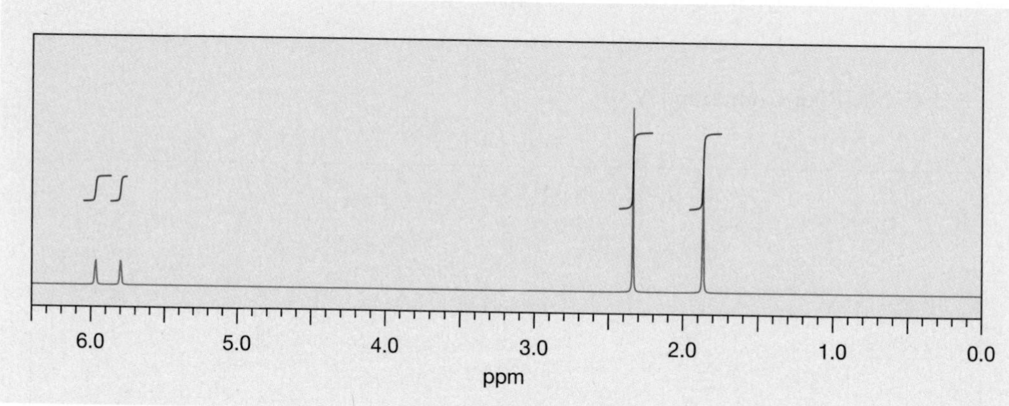

^{13}C-NMR of Compound C

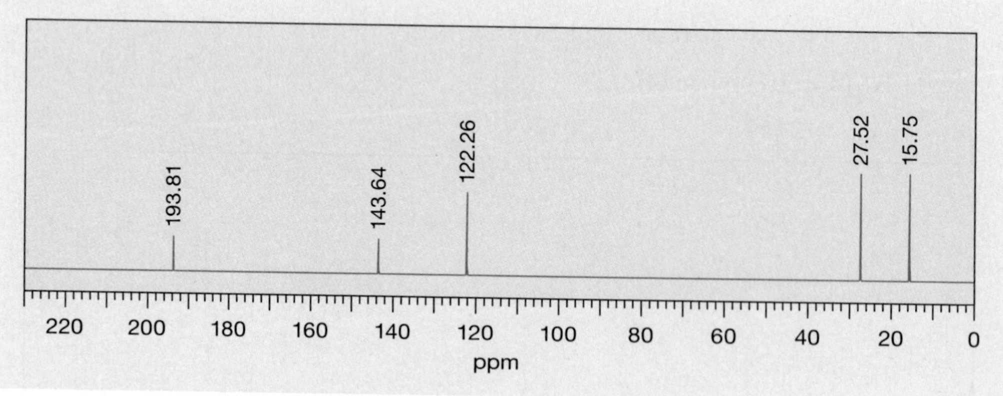

^{1}H-NMR of Compound D

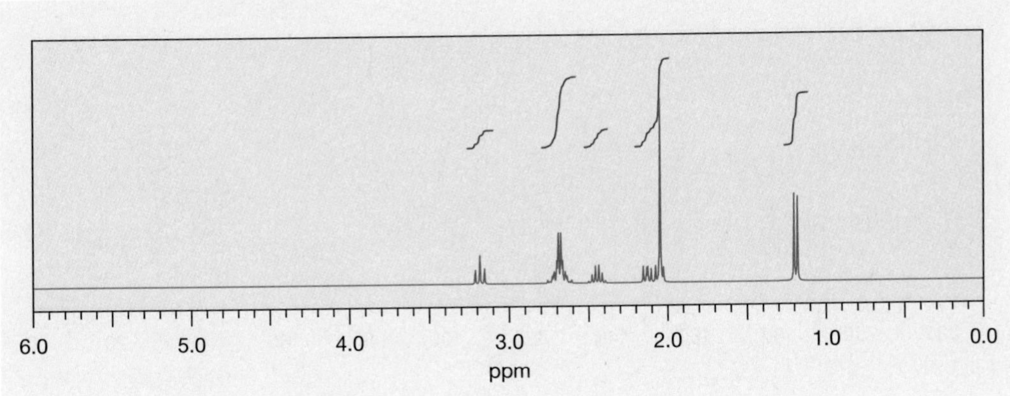

^{13}C-NMR of Compound D

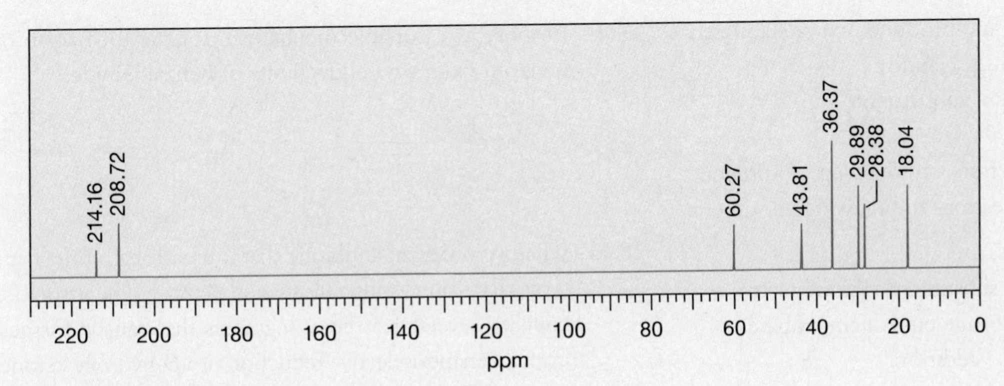

^{1}H-NMR of Compound E

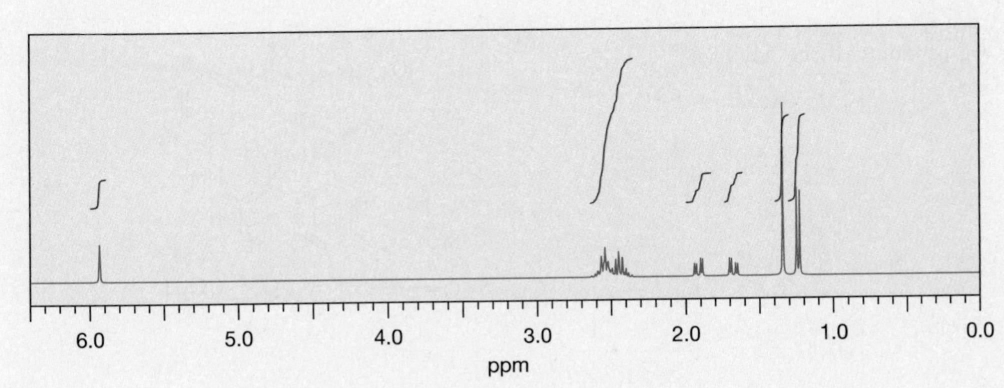

^{13}C-NMR of Compound E

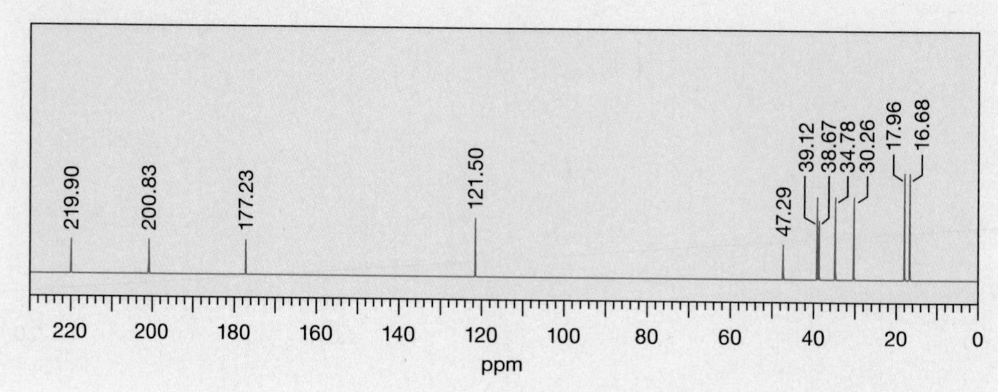

MCAT STYLE PROBLEMS

18.41 Which of the following are products when stabilized ylides are used in the Wittig reaction?

a) *cis*-alkenes with an alkyl substituent

b) *trans*-alkenes with an alkyl substituent

c) *cis*-alkenes with an electron-withdrawing substituent

d) *trans*-alkenes with an electron-withdrawing substituent

18.42 Thermodynamic enolates are

I. formed on the most substituted side of a ketone.

II. formed with less than one equivalent of base.

III. the product of inverse addition.

a) I, II, III

b) I, II

c) I, III

d) II, III

18.43 Which of the following are products of the Michael reaction?

a) β-hydroxy ketones

b) 1,4-diketones

c) 1,5-diketones

d) cyclohexenones

CHALLENGE PROBLEMS

18.44 Propose a reasonable mechanism for the formation of benzoin from two equivalents of benzaldehyde.

18.45 Rifamycin B is an antibiotic that is useful against the bacteria that cause tuberculosis and leprosy. The structure contains several 1,3-dioxy segments that can be formed from an enantioselective reduction of a β-hydroxy ketone. In the following structure, identify two 1,3-dioxy segments and show how their β-hydroxy ketone precursors could be formed using an enolate.

Radicals

HALOGENATION, POLYMERIZATION, AND REDUCTION REACTIONS

19

Flaxseed oil is high in polyunsaturated fatty acids and is touted for its nutritional benefits. However, because of its high percentage of polyunsaturated fatty acids, it becomes rancid more quickly upon exposure to air than do oils that have lower percentages of polyunsaturated fatty acids.

CHAPTER OUTLINE

19.1 Why It Matters

Naturally occurring oils and fats are compounds known as triglycerides (triacylglycerides). Triglycerides consist of three fatty acids that are long-chain carboxylic acids esterified to a single molecule of glycerol. Fatty acids that have no carbon-carbon double bonds are called saturated fatty acids. Those that contain one double bond are monounsaturated, and those that contain more than one double bond are polyunsaturated. Oils are liquids at room temperature and contain triglycerides that typically have several double bonds. Fats are solids at room temperature and contain fewer carbon-carbon double bonds than do oils.

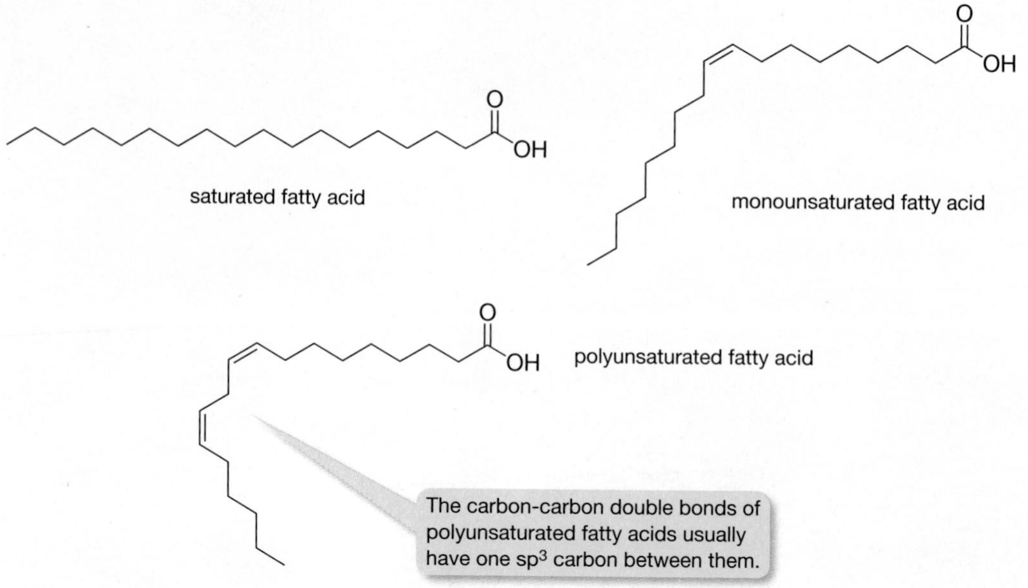

saturated fatty acid

monounsaturated fatty acid

polyunsaturated fatty acid

The carbon-carbon double bonds of polyunsaturated fatty acids usually have one sp^3 carbon between them.

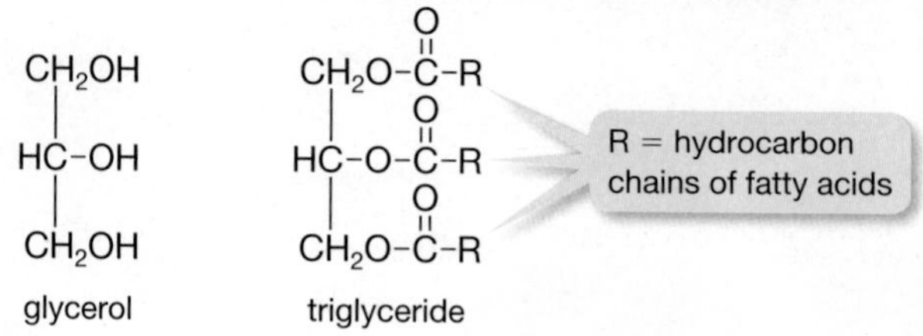

glycerol

triglyceride

R = hydrocarbon chains of fatty acids

Oxidative cleavage reactions are oxidation reactions that lead to bond breakage and the fragmentation of a molecule.

All polyunsaturated fatty acids—and the triglycerides that contain them—are susceptible to **oxidative cleavage reactions** that occur in the presence of atmospheric oxygen. These reactions lead to rancid fats. This occurs because the hydrocarbon chain of the fatty acid breaks down into various odoriferous aldehydes and carboxylic acids that are characteristic of the smell of rancidity.

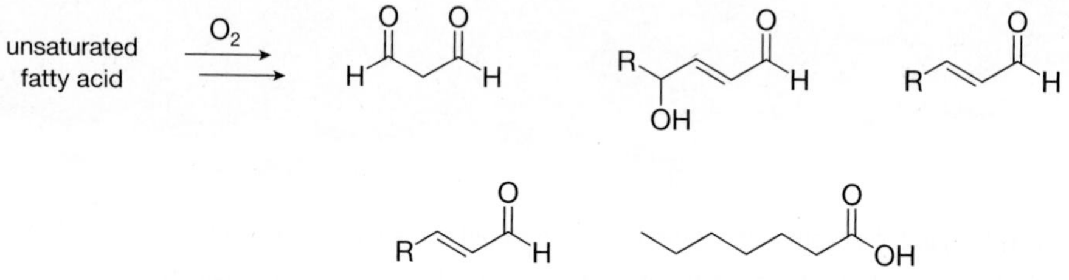

unsaturated fatty acid $\xrightarrow{O_2}$

examples of decomposition products

Radicals are chemical species containing one or more unpaired electrons.

The process of oxidative cleavage involves **radicals**, also known as free radicals. These are chemical species that contain one or more unpaired electrons. Because radicals are electron deficient due to their lack of a full octet, they are highly reactive electrophiles. This chapter describes the common reactivity patterns of radicals, specifically the variety of reagents and conditions that can be used to initiate radical chain reactions, including free-radical halogenation of alkanes and alkenes, benzylic halogenation, dehalogenation of an alkyl halide, and dissolving metal reactions.

19.2 Bond Breakage and Formation

The chemical reactions and processes discussed in previous chapters typically involve the movement of electrons *as a pair* when bonds are broken or formed. This movement of a pair of electrons is denoted by a regular curved arrow.

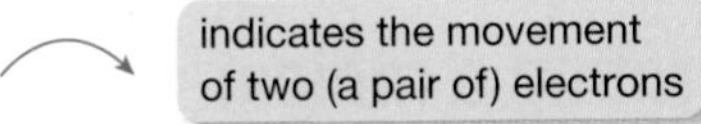
indicates the movement
of two (a pair of) electrons

This regular, curved, double-headed arrow indicates a type of bond breakage known as **heterolytic cleavage**. In this case, the two electrons move to one, and only one, of the two bonding atoms.

Heterolytic cleavage is a bond-breaking reaction where both bonding electrons migrate to one of the two bonded atoms.

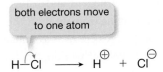

both electrons move to one atom

H—Cl ⟶ $H^{\oplus}$ + $Cl^{\ominus}$

In the case of radicals, which contain unpaired electrons, the reaction processes that make or use radicals involve the movement of a *single* electron. This type of electron movement is denoted by a single-headed curved arrow, also called a fishhook arrow.

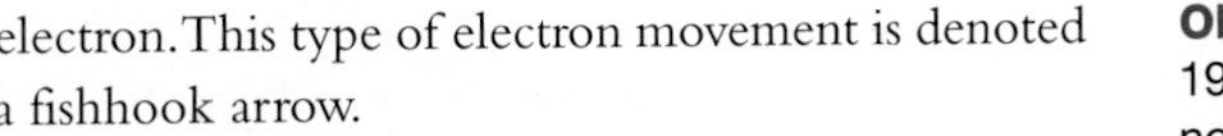

ORGANIC CHEMWARE
19.1 Curved arrow notation: Break/form X–Y σ bond

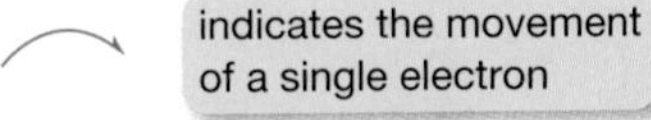
indicates the movement
of a single electron

Radicals form by a type of bond breakage known as **homolytic cleavage**. In this case, the two electrons in a bond are equally divided between the two constituent atoms. One electron moves to one atom, and the other electron moves to the other atom. For example, molecular chlorine can break into two chlorine radicals (atoms):

Homolytic cleavage is a bond-breaking process in which the two bonding electrons become split between the two originally bonded atoms.

ORGANIC CHEMWARE
19.2 Curved arrow notation: Break/form X–Y σ bond (radical mechanism)

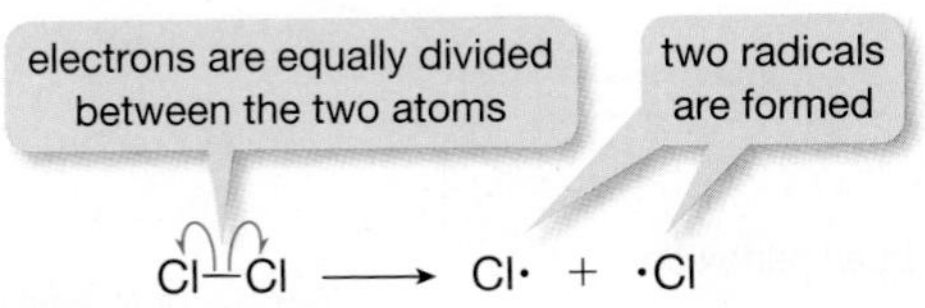
electrons are equally divided
between the two atoms
two radicals
are formed
Cl—Cl ⟶ Cl• + •Cl

In the reverse process, when two radicals combine to form a bond, each radical supplies one electron. Thus, two fishhook arrows show the two electrons coming together to form a bond between two radicals or between a radical and an atom. For radical reactions, it is common to show bond formation using fishhook arrows that point to the space between the atoms being bonded.

STUDENT TIP
In processes involving regular curved arrows, the charges on the atoms involved in bond breakage or formation often change. Reactions involving fish-hook arrows may or may not involve changes in the charges. If you are not sure, use the formal charge method to check for changes in charge.

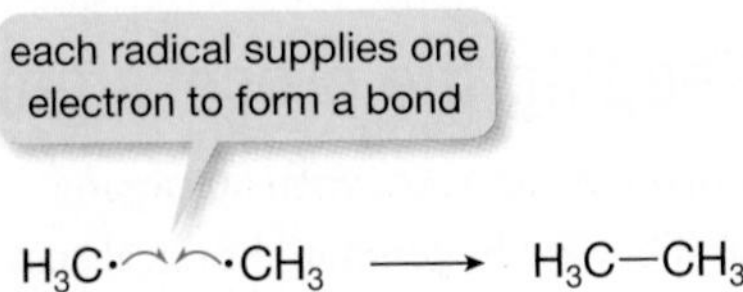
each radical supplies one
electron to form a bond
H3C• •CH3 ⟶ H3C—CH3

This chapter describes three primary ways that generate carbon-centred radicals: (1) the removal of an atom (such as a hydrogen or a halogen atom), (2) the addition of another radical to a π bond, and (3) the addition of a single electron to a π bond.

19.3 Radical Chain Reactions

A **radical chain** reaction is a three-phase reaction that characterizes many radical processes.

Initiation refers to the first phase (beginning) of the radical chain reaction, which involves the formation of one or more radicals.

Propagation refers to the second phase of the radical chain reaction, in which repeated reactions or steps form new radicals.

Termination refers to the third and final phase of the radical chain reaction, which involves the combination of radicals and leads to the destruction of radicals.

Perhaps the most interesting aspect of radical chemistry is that reactions involving radicals often take place as a **radical chain**—that is, the reaction of one radical leads to the generation of another radical. Regardless of the organic starting material and the type of radical involved, all radical reactions follow a general mechanism that can be divided into three phases: **initiation**, **propagation**, and **termination**.

19.3.1 Initiation: Generating a radical

Initiation refers to a process that generates one or more free radicals from a non-radical source, typically by the homolytic cleavage of a single bond. In the laboratory, initiation can be accomplished by heating (indicated by the symbol Δ) an appropriate reagent or by irradiating it with light of a suitable wavelength (indicated by the symbol $h\nu$). Reagents that generate radicals in this way are called initiators and usually contain a bond that is very weak (low bond enthalpy), so homolytic cleavage occurs readily. Such compounds include halogens and halogen-containing compounds, organic peroxides, and azo compounds. The following diagram shows common examples of organic radical initiators. When peroxides or azo compounds are used as initiators, a second reaction with one of the other components is necessary to make the "working" radical involved in the chain.

Halogens and halogen-containing compounds

$$Cl{-}Cl \xrightarrow{h\nu} Cl\cdot + \cdot Cl$$

$$Br{-}Br \xrightarrow{h\nu} Br\cdot + \cdot Br$$

$$\text{NBS} \xrightarrow{h\nu} \text{N}\cdot + \cdot Br$$

N-bromosuccinamide (NBS)

Organic peroxides

$$\text{di-}tert\text{-butyl perioxide} \xrightarrow{\Delta} (CH_3)_3CO\cdot + \cdot OC(CH_3)_3$$

di-*tert*-butyl perioxide

Azo compounds

release of N_2 gas drives the reaction forward

$$\text{AIBN} \xrightarrow{\Delta} NC\text{–}\dot{C}(CH_3)_2 + N{\equiv}N + \cdot C(CH_3)_2CN$$

azobisisobutyronitrile (AIBN)

19.3.2 Propagation: Keeping the process going

Once a small amount of radical forms, it can react with an organic compound to produce a product and also generate another radical. The newly generated radical may be of the same type as the one originally made during initiation, so the process can repeat over and over in a chain reaction.

For example, a bromine radical can abstract (remove) a hydrogen atom from a compound such as methane to generate a methyl radical. Abstracting a hydrogen atom in this way involves both the nucleus (i.e., a proton) and one electron; it is not the same as removing H^+ (a proton) or H^- (a hydride ion).

STUDENT TIP

A proton does not contain any electrons. A hydrogen atom comprises a proton and one electron. A hydride ion contains a proton and two electrons.

homolytic cleavage

two single electrons combine to form a new bond

$H_3C{-}H \; \cdot Br \longrightarrow H_3C\cdot + H{-}Br$

alkyl radical

bromine radical is consumed

The methyl radical can then react with a molecule of Br_2 to produce bromomethane and another bromine radical, which can go on to react with another alkane molecule.

new bromine radical is generated

$H_3C\cdot \; Br{-}Br \longrightarrow H_3C{-}Br + \cdot Br$

methyl radical — bromomethane

Because the mechanism of this bromination involves a radical, the mechanism is aptly named **free-radical halogenation**. The bromine radical that forms can undergo the same reaction with another molecule of methane, hence the term *chain reaction*.

Free-radical halogenation is a type of halogenation reaction. In this reaction, a halogen radical replaces a hydrogen atom on an alkane.

19.3.3 Termination: Destroying the radical

A chain reaction continues until some kind of competing reaction combines two radicals to form a new bond, effectively terminating the process. In the case of the free-radical bromination of methane, possible competing reactions include the combining of two bromine radicals, one bromine radical and one methyl radical, and even two methyl radicals. Any of these reactions reduce the overall number of radicals in the reaction.

$Br\cdot \; \cdot Br \longrightarrow Br{-}Br$

$H_3C\cdot \; \cdot Br \longrightarrow H_3C{-}Br$

$H_3C\cdot \; \cdot CH_3 \longrightarrow H_3C{-}CH_3$

The overall chain reaction therefore contains three phases. For example, the bromination of methane can be summarized as follows:

Initiation

$Br{-}Br \xrightarrow{h\nu} Br\cdot + \cdot Br$

Propagation

$H_3C{-}H \; \cdot Br \longrightarrow H_3C\cdot + H{-}Br$

$H_3C\cdot \; Br{-}Br \longrightarrow H_3C{-}Br + \cdot Br$

Termination

$Br\cdot \; \cdot Br \longrightarrow Br{-}Br$

$H_3C\cdot \; \cdot Br \longrightarrow H_3C{-}Br$

$H_3C\cdot \; \cdot CH_3 \longrightarrow H_3C{-}CH_3$

CHECKPOINT 19.1

You should now be able to identify radical reactions and use the appropriate curved arrows that depict the movement of single electrons.

SOLVED PROBLEM

Use curved arrows to show the mechanism of the following reaction.

phenol + •OH ⟶ phenoxyl radical + H_2O

phenol hydroxyl radical phenoxyl radical

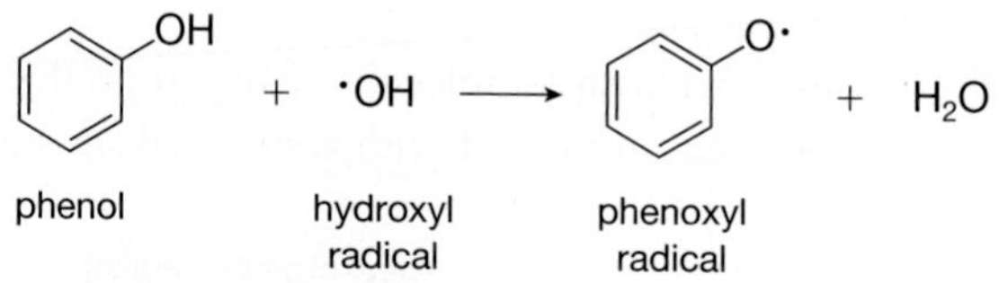

STEP 1: Recognize that the reaction involves radicals and homolytic cleavage. Phenol has been turned into a radical, and the hydroxyl radical has been turned into water. In other words, a hydrogen atom has been transferred from phenol to the hydroxyl radical.

STEP 2: Identify the bonds that are broken and those that are formed.

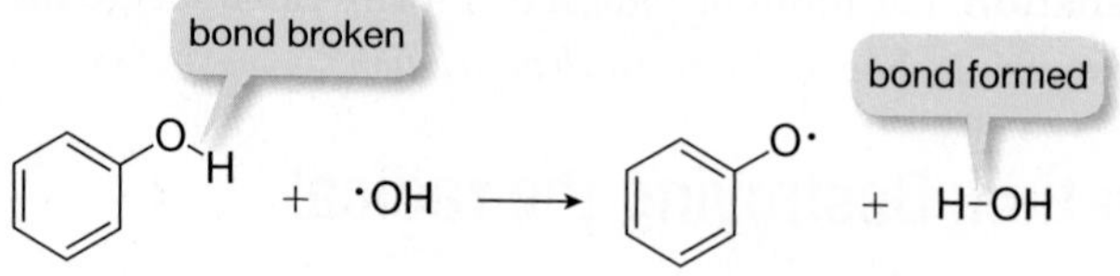

bond broken

bond formed

+ •OH ⟶ + H–OH

STEP 3: Use fishhook arrows to indicate the movement of single electrons.

+ •OH ⟶ + H–OH

PRACTICE PROBLEM

19.1 Use curved arrows to show mechanisms of the following reactions.

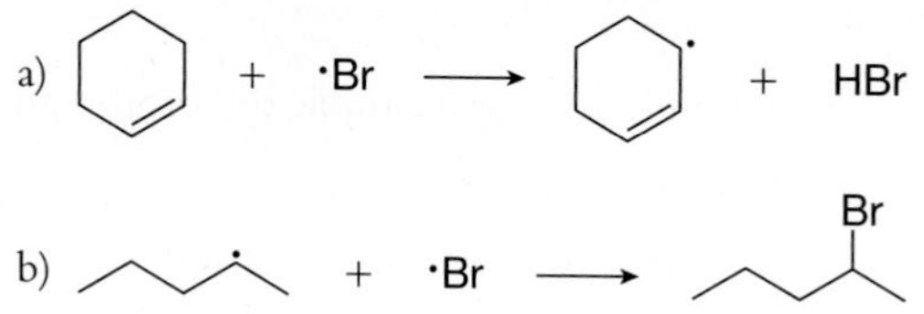

a) + •Br ⟶ + HBr

b) + •Br ⟶ Br

INTEGRATE THE SKILL

19.2 Use the proper arrows to show the heterolytic and homolytic cleavages of the OH bond in phenol to form a hydrogen radical and a proton, respectively. For each of the products, use mechanistic arrows to derive the contributing resonance forms.

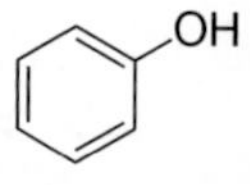

OH

19.3.4 Depicting radical chain reactions

Radical chain reactions may be thought of as cyclical reactions. As the reaction components are fed into the cycle, products are released. The cyclic part of a radical reaction involves the conversion of one radical into another. In the free-radical bromination of methane, bromine radicals generated by the homolytic cleavage of Br_2 during the initiation step begin the reaction cycle. The following notation shows the intermediates involved in the cycle of propagation reactions: when the starting materials methane and Br_2 are fed into the cycle, the products (CH_3Br and HBr) are released.

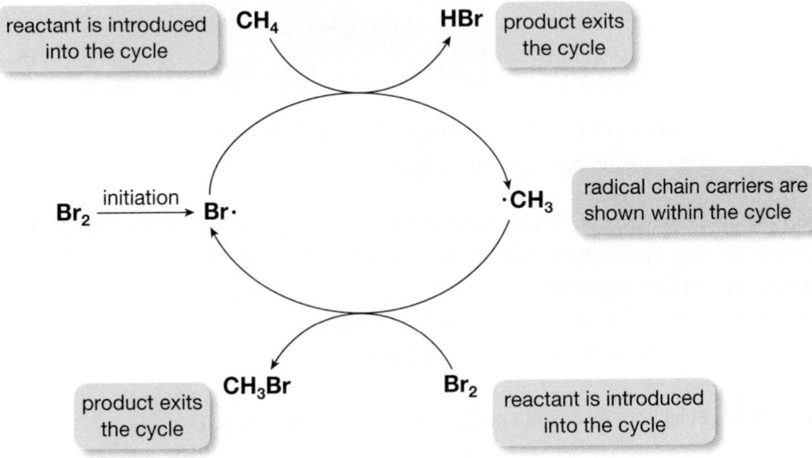

This compact notation represents all steps in the chain reaction. It is important to remember when using this notation that each turn of the cycle involves a *different* molecule. The Br· formed during the initiation phase is not the same Br· that forms after the reaction of Br_2. Similarly, each $\cdot CH_3$ produced is made from a different molecule of CH_4 each time.

CHEMISTRY: EVERYTHING AND EVERYWHERE

Radical Chemistry: Protecting Us from Ultraviolet Radiation

The ozone layer, tens of kilometres above the ground, shields life on our planet against some of the sun's ultraviolet rays. Contrary to popular belief, the ozone layer is not primarily composed of ozone. Its composition is approximately 78 percent N_2, 21 percent O_2, and 1 percent Ar, which is not much different from that of the atmosphere at ground level. The concentration of O_3 is highest in the "ozone layer," but the absolute amount of O_3 is actually very small: only about 4 ppm, on average! If an area of the layer has a lower-than-normal ozone concentration, the layer is said to have an *ozone hole*; this is not a true hole (an empty space), but rather simply an area of abnormally low ozone concentration.

Continued

studio023/iStock/Thinkstock

The blue tint of the ozone layer can be seen from orbiting spacecraft or from high-flying planes.

The O_2 and O_3 in the ozone layer protect us from the harmful effects of ultraviolet light under 242 nm (O_2) and between 242–310 nm (O_3) by absorbing light at those wavelengths. Ironically, when O_2 and O_3 perform their intended functions, they are destroyed because the absorption of light causes a photochemical reaction to occur. Because the products formed can recombine to regenerate O_2 and O_3, the two species are constantly being broken down and recreated. The overall process is called the Chapman cycle.

1. O_2 absorbs ultraviolet light under 242 nm and breaks down into two oxygen radicals.

atmospheric oxygen is actually a biradical

$$\cdot\ddot{O}-\ddot{O}\cdot \xrightarrow{< 242\text{ nm}} \cdot\ddot{O}\cdot + \cdot\ddot{O}\cdot$$

2. An oxygen radical reacts with O_2 to form ozone.

$$\cdot\ddot{O}\cdot + \cdot\ddot{O}-\ddot{O}\cdot \longrightarrow \left[O_3 \longleftrightarrow \ominus O-O^{\oplus}=O \longleftrightarrow O=O^{\oplus}-O\ominus \right]$$

3. Ozone absorbs ultraviolet light in the range of 242–310 nm.

$$O_3 \xrightarrow{242\text{–}310\text{ nm}} \cdot\ddot{O}\cdot + \cdot\ddot{O}-\ddot{O}\cdot$$

4. Ozone can also combine with an oxygen radical to regenerate O_2.

$$\cdot\ddot{O}\cdot + O_3 \longrightarrow 2\ \cdot\ddot{O}-\ddot{O}\cdot$$

Due to the delicately balanced kinetic parameters of the four steps, the absolute amounts of O_2 and O_3 in the ozone layer have remained fairly constant over time. The concentration of O_3, however, can become reduced if additional processes cause the degradation of O_3. Compounds known as *ozone-depleting substances* are responsible for many of the processes that lead to the degradation of ozone.

A well-known family of ozone-depleting substances is the chlorofluorocarbons (CFCs). CFCs were commonly used as refrigerants in air conditioners and refrigerators, and as propellants in aerosol cans. Thanks to the Montreal Protocol, that was signed in 1987, their manufacture and usage have gradually been phased out.

CFCs are relatively inert in the lower atmosphere and do not easily degrade. This allows them to persist for long periods, and, over time, they make their way up into the ozone layer, where an ozone-degrading chain reaction occurs. The chlorine radical not only causes ozone to decompose but also consumes oxygen radicals, which prevents the propagation phase—the ozone-generating step—of the Chapman cycle from occurring. It is estimated that a single chlorine radical can destroy several million ozone molecules before it is consumed by a termination reaction.

Initiation

the C–Cl bond is much weaker than the C–F bond

$$CCl_2F_2 \xrightarrow{< 240\ nm} \cdot CClF_2 + \cdot Cl$$

Propagation

$$O_3 + \cdot Cl \longrightarrow O_2 + ClO\cdot$$

$$ClO\cdot + \cdot\ddot{O}\cdot \longrightarrow O_2 + \cdot Cl$$

any available oxygen radical is used, so it cannot combine with O_2 to form O_3

Nowadays, hydrofluorocarbons (HFCs) such as 1,1,1,2-tetrafluoroethane have largely replaced CFCs as refrigerants and propellants in new products. HFCs are not considered ozone-depleting substances for two reasons. First, because HFCs have a carbon-hydrogen bond, they are degraded relatively quickly by oxidation in the lower atmosphere, making their travel up into the ozone layer much less likely. Second, they do not have a weak carbon-chlorine bond that can be homolytically cleaved by UV light.

$F_3C{-}CH_2F$

1,1,1,2-tetrafluoroethane
(R134a or HFC-134a)

Unfortunately, even though HFCs are not ozone-depleting, they are greenhouse gases with a substantial global warming potential (GWP) rating. In October of 2016, the parties to the Montreal Protocol agreed to amend the protocol and phase out the use of HFCs. Modern refrigerants called hydrofluoroolefins (HFOs) have been developed to replace HFCs. HFOs have a lower GWP and an even shorter atmospheric lifetime, and one of these, 2,3,3,3-tetrafluoropropene, is being phased in as a refrigerant for air conditioning systems in vehicles.

$H_2C{=}CF{-}CF_3$

2,3,3,3-tetrafluoropropene
(R1234yf or HFO-1234yf)

19.4 Stability of Carbon Radicals

In reactions that involve simple organic compounds such as methane or ethane, the same type of radical forms regardless of where hydrogen abstraction takes place. However, most organic compounds contain non-equivalent carbon atoms, so hydrogen abstraction can form more than one type of radical.

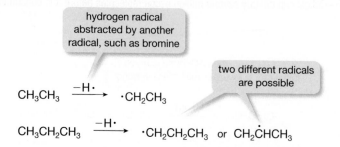

hydrogen radical abstracted by another radical, such as bromine

two different radicals are possible

$$CH_3CH_3 \xrightarrow{-H\cdot} \cdot CH_2CH_3$$

$$CH_3CH_2CH_3 \xrightarrow{-H\cdot} \cdot CH_2CH_2CH_3 \text{ or } CH_2\dot{C}HCH_3$$

When the same reactant forms more than one kind of radical, the rate of formation of each radical determines how much of each radical is produced. Carbon-centred radicals have only seven valence electrons, instead of a full octet, so they are electron deficient. Carbocations are also electron deficient (they have only six valence electrons). Significantly, the structural features that stabilize carbocations also stabilize radicals: the degree of substitution and the amount of delocalization.

Radicals that are more substituted (bonded to more alkyl groups) than other radicals are more stable as a result of the electron-donating character of alkyl groups through hyperconjugation. The trend for alkyl radical stability parallels that for alkyl carbocation stability. Note, however, that the carbon atoms of radicals are sp^3-hybridized, while carbocations are sp^2-hybridized.

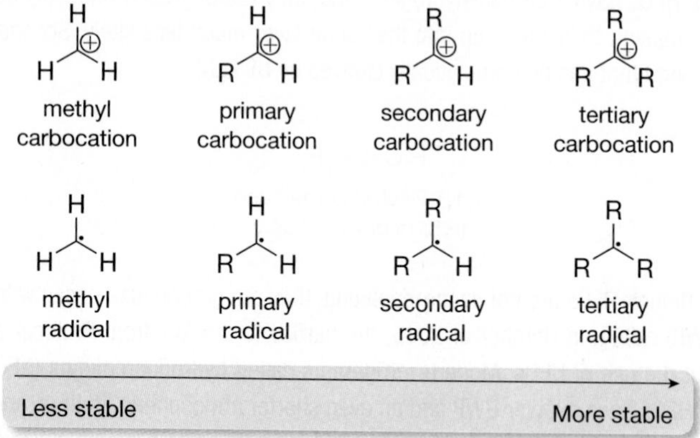

Radicals that are delocalized are usually more stable than radicals that are not delocalized. This effect is normally much stronger than the stabilization by hyperconjugation, and delocalized radicals form much more readily than alkyl radicals do. As with delocalized anions or cations, the delocalization of radicals commonly involves an adjacent π bond. Due to p orbital involvement, the carbon atoms of radicals that are delocalized are sp^2-hybridized.

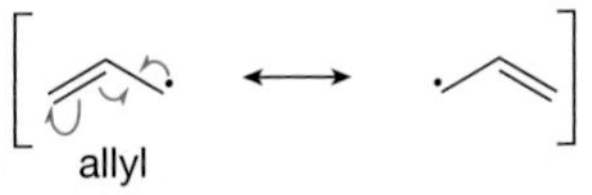

allyl radical

α-carbon radical

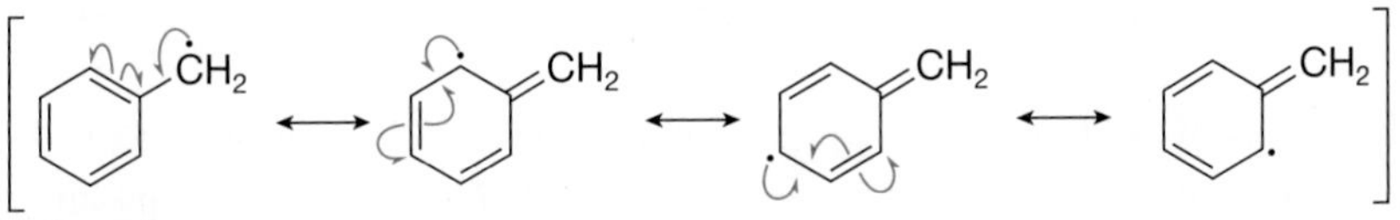

benzyl radical

However, in contrast to what occurs with carbocations, radicals are not stabilized by an adjacent atom with one or more lone pairs (such as a nitrogen or oxygen atom). This is because double-bonded resonance forms exceed the octet on the heteroatom and are therefore insignificant. For example, an alkoxy carbocation can be stabilized by an adjacent oxygen, but an alkoxy radical cannot be stabilized by an adjacent oxygen because the formation of a double bond would exceed the octet rule at the oxygen atom.

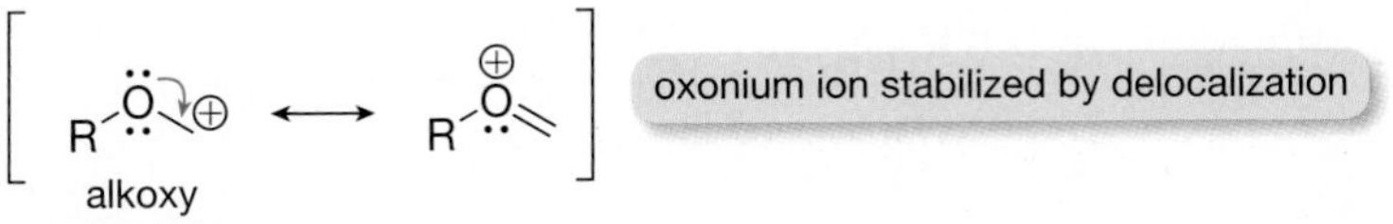

alkoxy carbocation

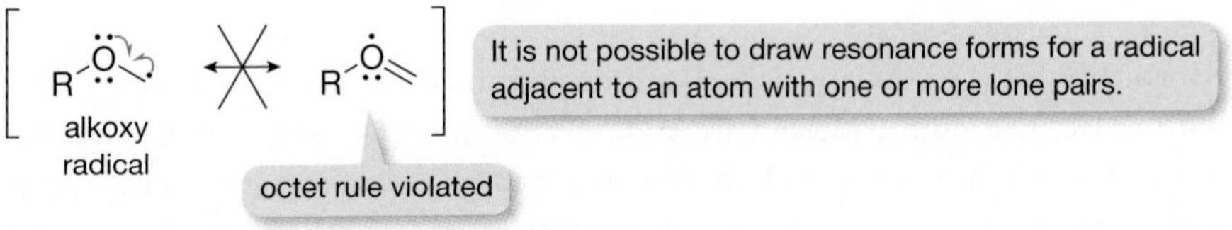

oxonium ion stabilized by delocalization

alkoxy radical

It is not possible to draw resonance forms for a radical adjacent to an atom with one or more lone pairs.

octet rule violated

19.5 Free-Radical Halogenation

Free-radical halogenation involves the replacement of a hydrogen atom of an organic compound with a halogen atom. This is an example of a reaction that forms a carbon radical by atom abstraction (removal), the first of the three common ways for generating carbon-centred radicals. The following sections describe free-radical halogenation of alkanes, alkenes, and compounds containing a benzylic carbon atom. These processes are all synthetically useful radical reactions.

19.5.1 Halogenation of alkanes

The radical halogenation of alkanes generally produces mixtures. The product distribution can be controlled in some cases, depending on the structure of the alkane and the type of halogen used. Halogenation processes that use bromine—bromination reactions—typically provide the highest selectivity, whereas the use of chlorine or fluorine gives little or no selectivity.

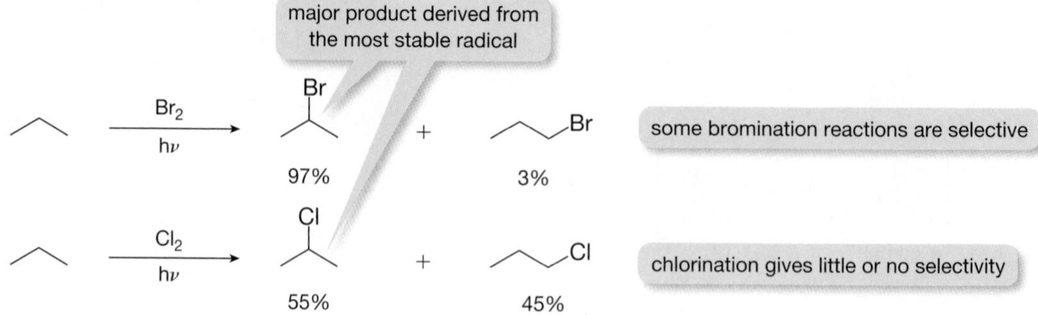
major product derived from the most stable radical
Br_2
hν
Br
Br
97%
3%
some bromination reactions are selective
Cl_2
hν
Cl
Cl
55%
45%
chlorination gives little or no selectivity

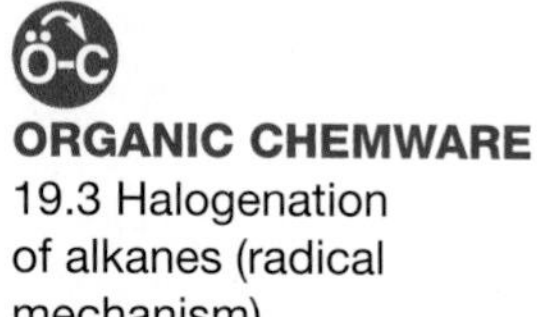
ö-c

ORGANIC CHEMWARE
19.3 Halogenation of alkanes (radical mechanism)

The first mechanistic step of the propagation phase involves the removal of a hydrogen atom to form a carbon-centred radical and a molecule of acid. The overall enthalpy change for this process can be calculated by using the bond-dissociation energies (BDE) for both the hydrogen-halogen bond that is formed and the hydrogen-carbon bond that is broken.

bond dissociation energy (BDE) of a secondary C–H bond is 98.5 kcal/mol

BDE of a Br–H bond is 87 kcal/mol

secondary + ·Br ⟶ + H–Br ΔH = BDE *(bonds broken)* – BDE *(bonds formed)* = +11 kcal/mol

primary + ·Br ⟶ + H–Br $\Delta H = +14$ kcal/mol

BDE of a primary C–H bond is 101 kcal/mol

The formation of a secondary radical is more favoured than the formation of a primary radical, but the fact that both pathways are endothermic (positive ΔH) is more significant. The Hammond postulate indicates that endothermic reactions have late transition states—that is, they more closely resemble the product. When transition states are late, the energy difference between the two possible transition states is relatively large. As a result, endothermic bromination significantly favours the pathway involving the most stable radical.

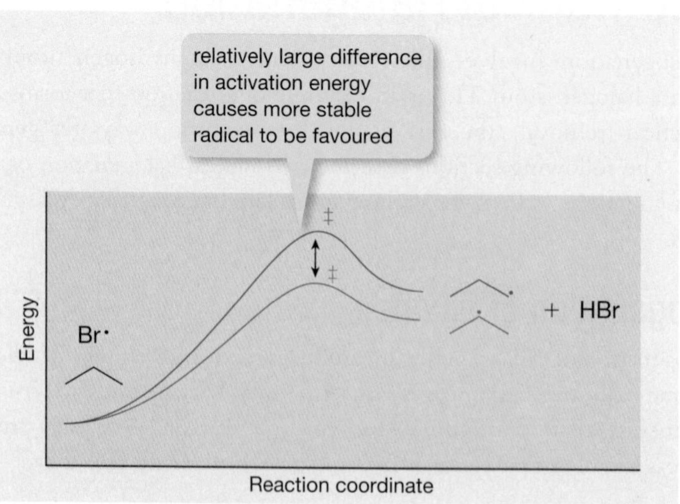
relatively large difference in activation energy causes more stable radical to be favoured
‡
‡
Energy
Br·
+ HBr
Reaction coordinate

In the case of chlorination, the high strength of the hydrogen-chlorine bond results in exothermic pathways (negative ΔH) that form both primary and secondary carbon radicals. Applying the Hammond postulate in this case shows that the two transition states will be early. When this happens, the energy differences between the two transitions become relatively small, so the selectivity for the formation of the carbon-centred radicals is low. The overall result is a low selectivity for the production of the secondary product.

BDE of a secondary C–H bond is 98.5 kcal/mol

BDE of a Cl–H bond is 101 kcal/mol

secondary + ·Cl ⟶ radical + H–Cl ΔH = BDE *(bonds broken)* – BDE *(bonds formed)* = –5 kcal/mol

primary + ·Cl ⟶ radical + H–Cl ΔH = –2 kcal/mol

BDE of a primary C–H bond is 101 kcal/mol

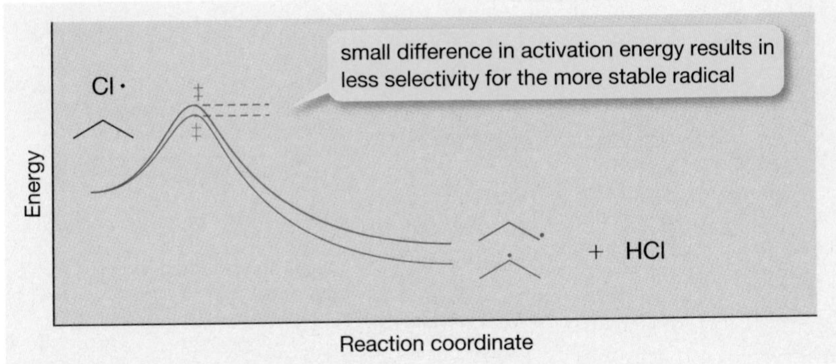

The free-radical halogenation reaction of alkanes is generally useful only if the energetics of the reaction provide high selectivity or if efficient methods exist to separate the product mixtures that are formed. Radical halogenations (and other types of radical processes) are much more efficient when some kind of controlling element is in place to ensure that product mixtures are minimized.

19.5.2 Allylic halogenation

Compared to the free-radical halogenation reactions of alkanes, those of alkenes are much more selective. The π bond of an alkene provides the possibility of conjugation and delocalization, which increases the stability of the radical formed and makes these reactions much more controllable. Hydrogen abstraction preferentially takes place *beside* the alkene, at what is called the allylic position. Because it is delocalized, the resulting allyl radical is formed almost exclusively, so the main halogenation product formed is the allyl halide.

Br2
hν
Br
allyl halide products
the allylic position is the one next to an alkene
Br
major product
allyl radical is resonance-stabilized
Of the two possible products, the one with the more substituted double bond predominates.

allyl radical

Allylic halogenation reactions are highly selective, but there is an important consideration: when a halogen such as a Br_2 is used, free-radical halogenation is in competition with the electrophilic addition of the halogen to the alkene (Chapter 8). This can give rise to two products, which are formed by two very different mechanisms.

Br2
hν
Br
allylic halogenation product (radical reaction)
Br2
(dark)
Br
Br
electrophilic addition product

Electrophilic addition is difficult to avoid because the alkene reacts directly with Br_2. Allylic halogenation, being a free-radical halogenation, first requires a bromine radical to be generated by homolytic cleavage of Br_2. The formation of the bromine radical in the presence of light is actually an equilibrium reaction that favours the starting material, so the problem of competition by electrophilic addition is exacerbated by low light or by a high concentration of Br_2 in the reaction mixture.

$$Br_2 \underset{}{\overset{h\nu}{\rightleftharpoons}} 2\,Br\cdot$$

reacts directly with alkene
required for allylic halogenation to occur

To minimize the amount of electrophilic addition product made in the reaction, a radical source other than Br_2 is typically used. This alternate source, *N*-bromosuccinamide (NBS), virtually eliminates the electrophilic addition of Br_2 to an alkene. In the presence of light or a peroxide initiator, NBS forms a bromine radical. This radical can subsequently abstract a hydrogen atom from the allylic position of an alkene to make an allyl radical.

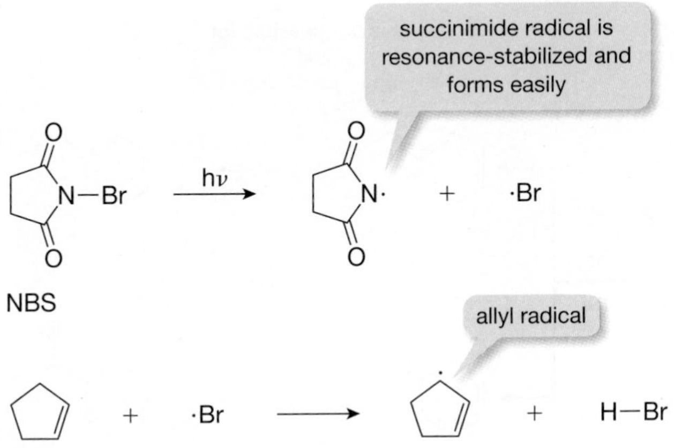

succinimide radical is resonance-stabilized and forms easily

NBS

allyl radical

Even though NBS is used as the source of bromine radical, the actual species that reacts with the allyl radical is thought to be molecular bromine (Br_2). Molecular bromine is formed *in situ*, but only in very small amounts (thereby minimizing competition from electrophilic addition), when HBr reacts with a second molecule of NBS.

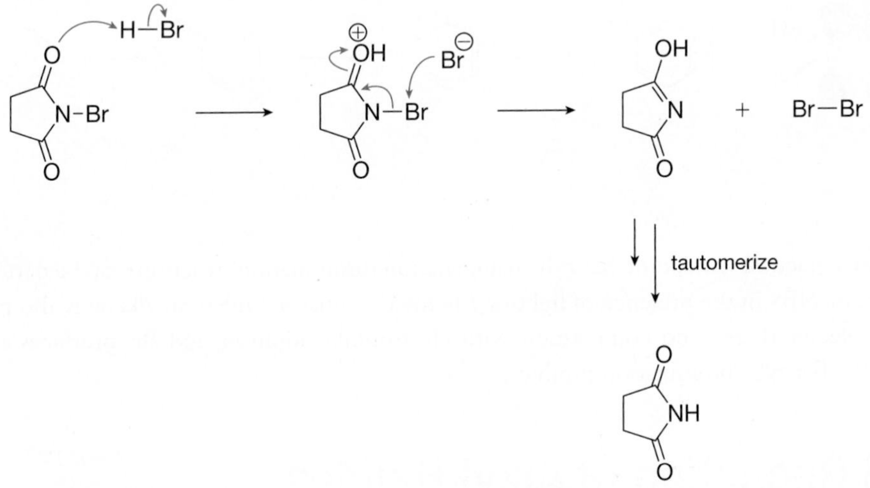

tautomerize

The resulting molecular bromine reacts with the allyl radical, forming the bromination product and another bromine radical that continues the chain reaction.

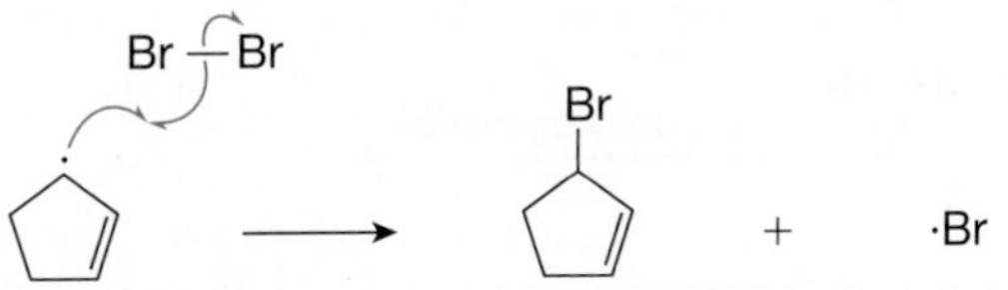

19.5.3 Benzylic halogenation

Like allylic halogenation reactions, benzylic halogenation reactions are also very selective because the benzyl radical is stabilized by delocalization. Because of this stabilization by delocalization, the major product is almost exclusively the one with the halogen bonded to the benzylic carbon.

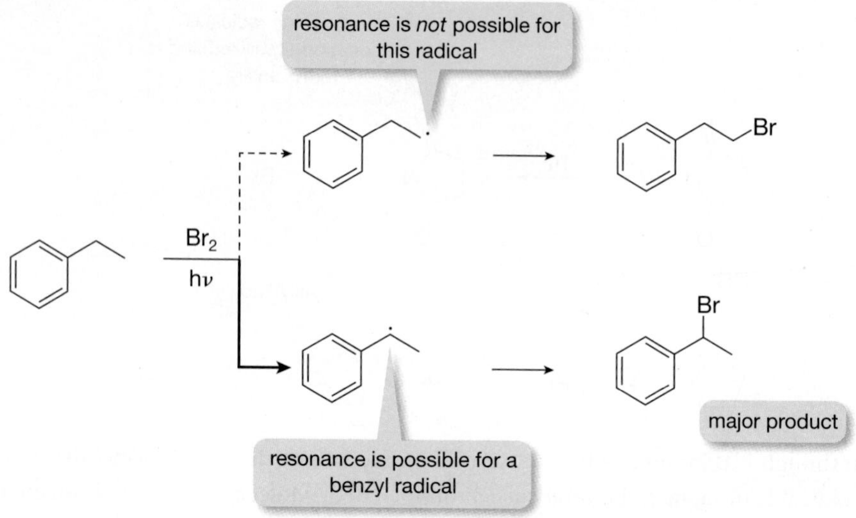

resonance is *not* possible for this radical

Br_2

hν

Br

Br

major product

resonance is possible for a benzyl radical

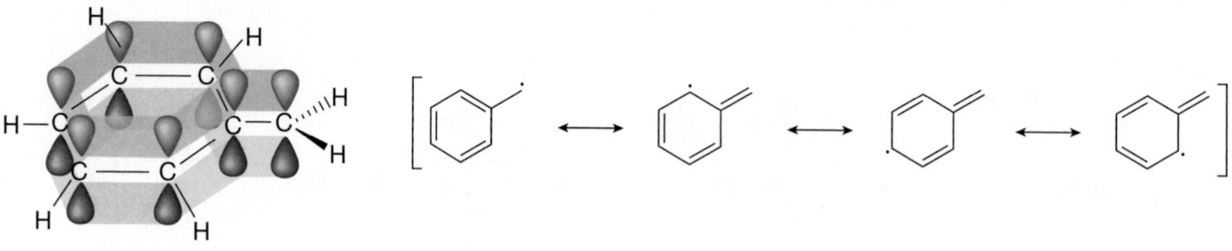

benzyl radical

From a practical standpoint, benzylic halogenation (bromination) reactions can be performed using Br_2 or NBS in the presence of light or a peroxide initiator. Unless an alkene is also present in the molecule there is no competition with electrophilic addition, and Br_2 produces a good yield of the benzylic bromination product.

19.6 Reduction of Alkyl Halides

Dehalogenation is a reduction reaction that replaces the halogen atom of an alkyl halide with a hydrogen atom.

Another example of the generation of a carbon-centred radical by the removal of an atom from a molecule involves the reduction of an alkyl halide to form an alkane. Also known as **dehalogenation**, this reaction removes a halogen atom and replaces it with a hydrogen atom.

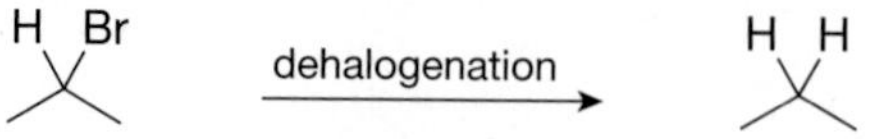

H Br —dehalogenation→ H H

To carry out dehalogenation, organotin compounds such as tributyltin hydride (Bu_3SnH) are used along with an initiator such as azobisisobutyronitrile (AIBN). The organotin compounds found in many types of radical transformations can stabilize radicals very effectively. Dehalogenation begins when a radical is initiated, and Bu_3SnH donates a hydrogen *atom* to the alkyl radical formed by the decomposition of the initiator.

NC–C(CH$_3$)$_2$–N=N–C(CH$_3$)$_2$–CN (AIBN) $\xrightarrow{\Delta}$ 2 NC–Ċ(CH$_3$)$_2$ + N_2

AIBN

NC–Ċ(CH$_3$)$_2$ + H–SnBu$_3$ ⟶ NC–C(CH$_3$)$_2$–H + ·SnBu$_3$

butyl groups provide solubility in organic solvents

Sn radical formed

In the propagation phase, the tin radical formed in the initiation phase abstracts a halogen atom from the alkyl halide and forms another alkyl radical. The alkyl radical thus formed abstracts a hydrogen atom from another molecule of Bu_3SnH, which gives the desired alkane and regenerates another tin radical to continue the chain reaction.

Br–R + ·SnBu$_3$ ⟶ R· + Br–SnBu$_3$

R· + H–SnBu$_3$ ⟶ R–H + ·SnBu$_3$

19.7 Anti-Markovnikov Addition of Hydrogen Bromide

As seen in Chapter 8, the Markovnikov addition of hydrogen bromide to an alkene proceeds through a carbocation intermediate, and the major regioisomer formed (the Markovnikov product) is the one derived from the most stable carbocation.

primary carbocation is less stable

HBr

Br

minor product

Br

major product (Markovnikov product)

secondary carbocation is more stable

It is possible to obtain the *anti*-Markovnikov product as the *major* product, but the reaction conditions, and hence the mechanism for this reaction, are very different. Typically, the formation of the anti-Markovnikov product is achieved by combining the alkene with hydrogen bromide and a peroxide that functions as a radical initiator.

HBr / ROOR → Br (major product (anti-Markovnikov)) + Br (minor product)

The route by which this product forms involves a free-radical reaction chain. First, the peroxide undergoes homolytic cleavage to form two alkoxy radicals in the presence of heat (or even light).

R–O–O–R —Δ→ R–O· + ·O–R

Next, one alkoxy radical abstracts a hydrogen atom from hydrogen bromide, forming a bromine radical.

R–O· H–Br ⟶ R–OH + ·Br

Because the bromine radical is electron deficient, and therefore electrophilic, it reacts with the nucleophilic π bond and adds to one of the alkene carbons. The bromine radical adds preferentially to one of the two alkene carbons such that it results in the formation of the more stable (more substituted) alkyl radical. This reaction is an example of the addition of a radical to a π bond: the second of the three primary ways that generate carbon-centred radicals.

most stable radical (more favourable)

less stable radical (less favourable)

The alkyl radical now abstracts a hydrogen atom from hydrogen bromide, generating another bromine radical that can continue the cycle.

H—Br ⟶ Br + ·Br

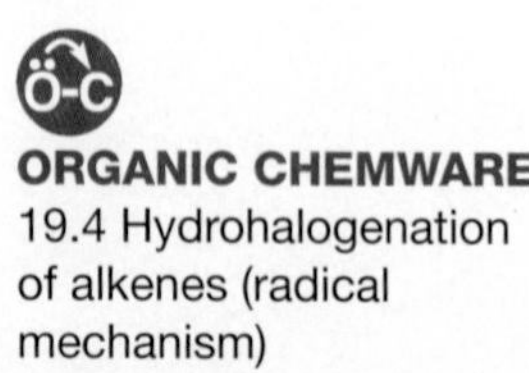

ORGANIC CHEMWARE
19.4 Hydrohalogenation of alkenes (radical mechanism)

The free-radical addition and the electrophilic addition to an alkene have similar controlling principles in that each involves the formation of an intermediate that is electron poor (has an incomplete octet). In electrophilic addition, the nucleophilic alkene adds the electrophilic H^+ such that the most stable *carbocation* intermediate is produced. The electrophilic carbocation is then attacked by the nucleophilic Br^- ion to give the Markovnikov product.

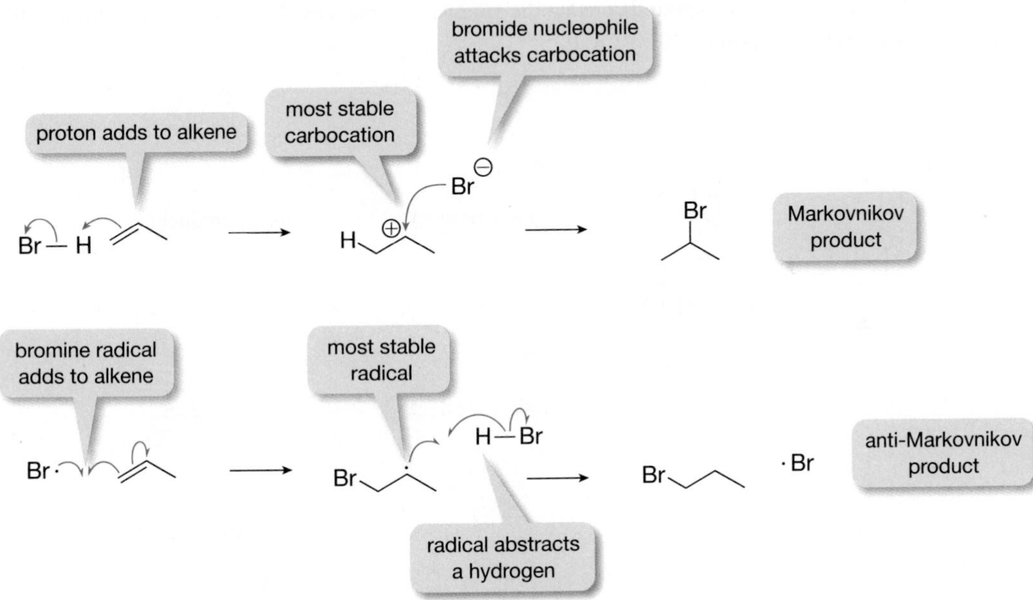

In free-radical addition, the nucleophilic alkene adds the electrophilic Br radical in a way that results in the most stable *radical* being formed. The same factors that stabilize a carbocation stabilize a radical (alkyl substitution and delocalization), and so the most stable carbocation and most stable radical are at the same position. The carbon radical then abstracts a hydrogen atom from another molecule of hydrogen bromide to give the anti-Markovnikov product.

Both reactions proceed through an electron-poor intermediate and follow pathways in which this intermediate is best stabilized. In each case, the electron-poor atom (radical or carbocation) is at the same (most substituted) position because this is the position best able to stabilize the intermediates. Because the order in which the hydrogen and bromine add is different in each sequence, the major products are different.

CHECKPOINT 19.2

You should now be able to identify the alkene needed to make an alkyl bromide, either by electrophilic addition or by the free-radical addition of hydrogen bromide.

SOLVED PROBLEM

What alkene should be used to make the following product in the highest yield, either by the electrophilic addition or the free-radical addition of hydrogen bromide?

Br carbon bearing Br

STEP 1: Locate the carbon that bears the bromine atom. This carbon atom must have been one of the carbon atoms of the alkene.

STEP 2: Draw the possible starting alkenes based on the location of the bromine atom.

A B

STEP 3: For each of the two compounds, predict the product(s) formed by the electrophilic addition (HBr) and by the free-radical addition (HBr/ROOR).

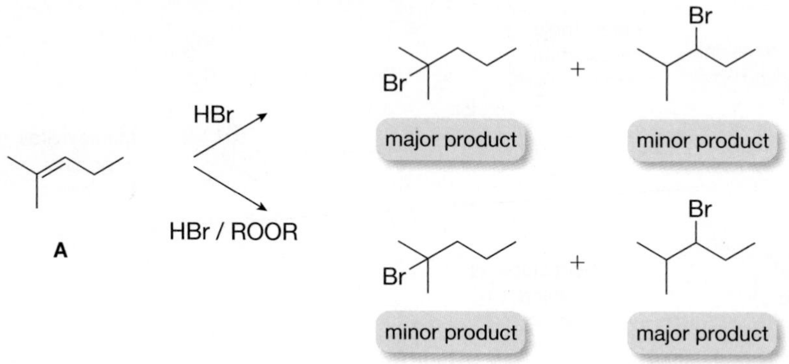

HBr
HBr / ROOR
A
Br + Br
major product
minor product
Br + Br
minor product
major product

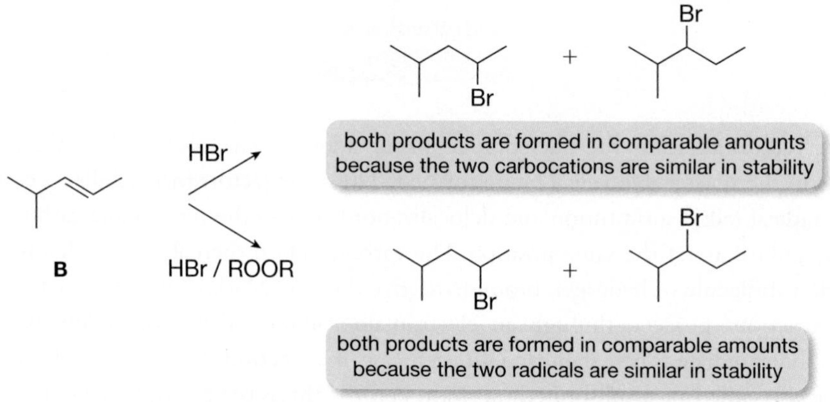

HBr
HBr / ROOR
B
Br + Br
both products are formed in comparable amounts because the two carbocations are similar in stability
Br + Br
both products are formed in comparable amounts because the two radicals are similar in stability

STEP 4: Identify the alkene and the reaction that forms the desired product in the greatest yield. In this case, alkene A reacting by the free-radical addition of hydrogen bromide would be the best choice.

PRACTICE PROBLEM

19.3 For each of the following compounds, identify the alkene that could be used to make the compound, in the highest yield, by electrophilic addition or free-radical addition of hydrogen bromide.

a) 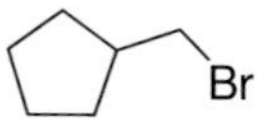 Br

b) 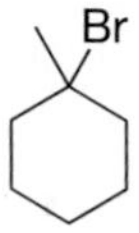 Br

c) 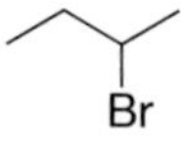 Br

INTEGRATE THE SKILL

19.4 What alkene should be used to make the following product in the highest yield, either by the electrophilic addition or the free-radical addition of hydrogen bromide?

Br

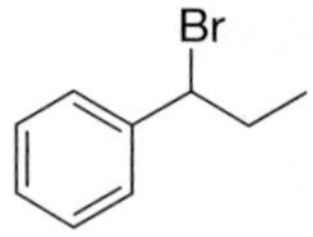

19.8 Polymerization of Alkenes

A great number of materials, such as plastics, are classified as **polymers**. Polymers are composed of repeating molecules known as **monomers**. In polymers, many monomers connect together to form very large chains or networks. The process of joining the monomers together is called **polymerization**. One common method of polymerization involves a type of radical reaction in which one monomer at a time adds to a growing polymer chain without the loss of any other atoms or molecules. Each time a monomer is added, addition of a radical to a π bond forms a new carbon-centred radical that continues the chain. Polymers formed by this method are known as **addition polymers**. The mechanism by which addition polymers form involves the same three phases previously seen: initiation, propagation, and termination.

Polymers are large molecules composed of many repeating units.

Monomers are the small building blocks and the repeating units found in polymers.

Polymerization is the process of reacting monomers together to form a polymer.

Addition polymers are polymers made by combining monomers one by one without losing any atom or molecule.

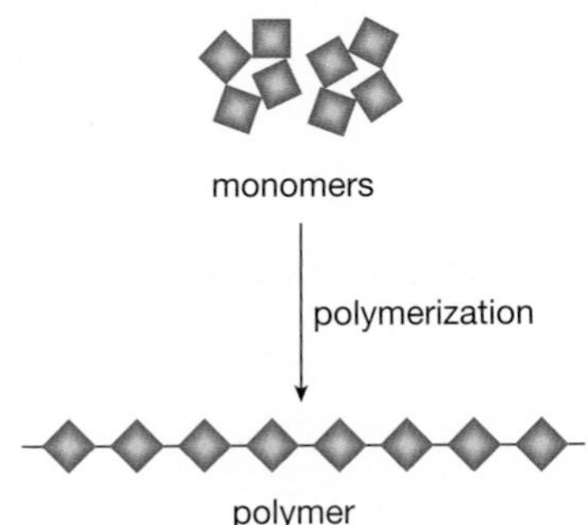

The synthesis of polyethylene, an addition polymer, involves a process that connects together many alkene monomers. In this process, the carbon-carbon π bond of each monomer is broken, and the electrons are used to form a σ bond to another monomer. The number of monomers that become connected to make a single molecule can be extremely large (thousands). The gigantic size of the molecules formed gives polymers their useful properties. In fact, objects made of certain types of plastic are essentially one gigantic molecule. Some common addition polymers are shown in Table 19.1.

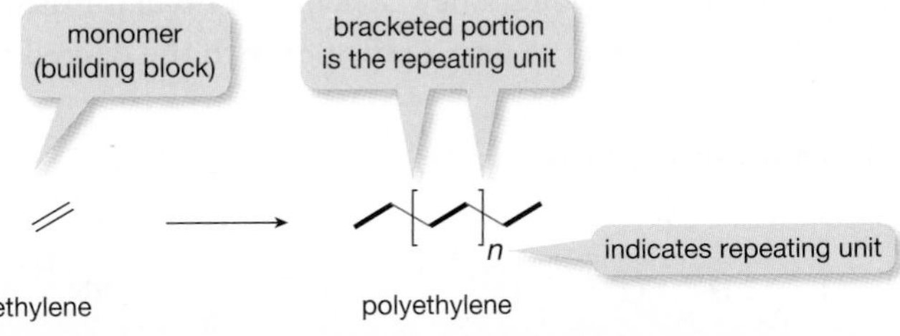

TABLE 19.1 Different Types of Addition Polymers and Typical Applications

Monomer	Polymer	Common Uses	
ethylene (H, H, H, H)	polyethylene (n)	Bags, bottles, water pipes, food containers	Hurst Photo/Shutterstock.com
propylene	polypropylene (n)	Bags, bottles, banknotes, storage boxes, clothing, furniture, food containers	PorChonlawit/ Shutterstock.com

Continued

TABLE 19.1 Different Types of Addition Polymers and Typical Applications (Continued)

Monomer	Polymer	Common Uses	
 tetrafluoroethylene	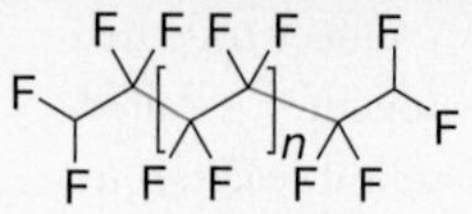 polytetrafluoroethylene (PTFE)	Non-stick coatings, chemical resistant materials	 WitR/iStock/Thinkstock
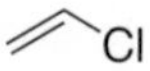 vinyl chloride	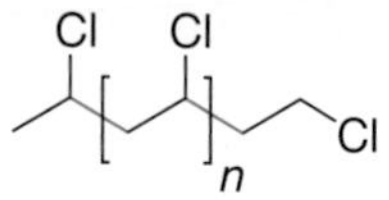 polyvinyl chloride (PVC)	Inflatable toys, water pipes, electrical tape, signage	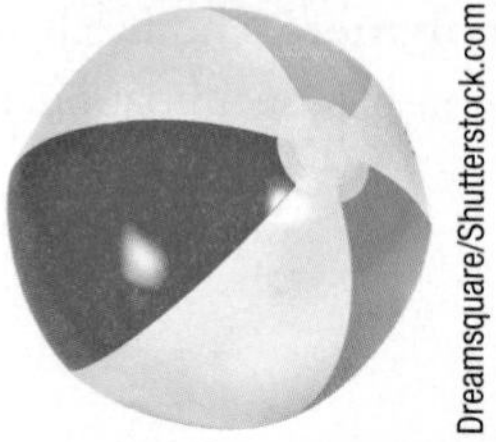 Dreamsquare/Shutterstock.com
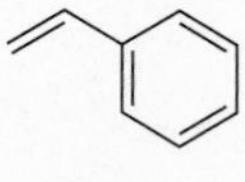 styrene	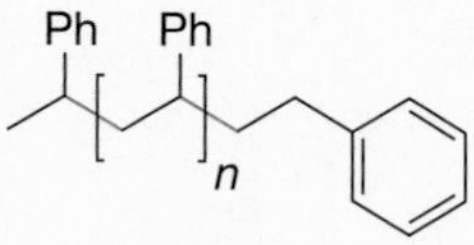 polystyrene	Insulation, food containers, packaging materials	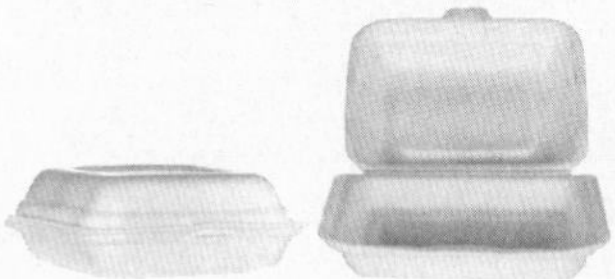 Boonchuay1970/Shutterstock.com

One common industrial method for making addition polymers involves the use of an organic peroxide initiator, which undergoes homolytic cleavage to form two alkoxyl radicals. The reaction of an alkoxyl radical with an alkene leads to the formation of an alkyl radical. The newly formed alkyl radical is electrophilic and can react with another molecule of alkene, which is the nucleophile. Through this reaction, propagation occurs. Each addition of alkene puts extra atoms on the polymer chain, while providing a new attachment point for the next molecule. This process typically repeats 700–1800 times to give a polymer with the desired properties.

Initiation (formation of alkyl radical)

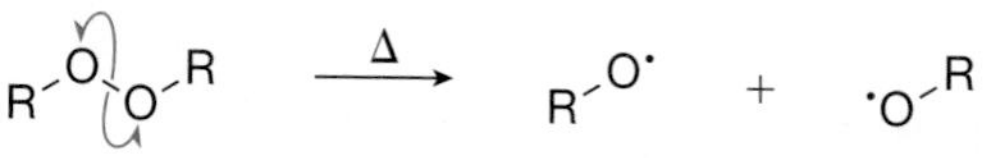

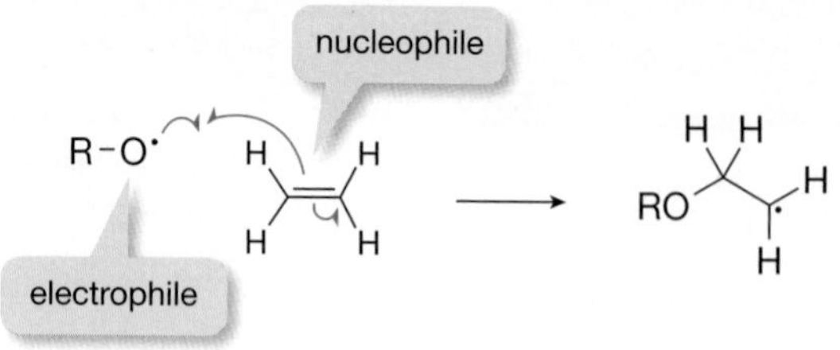

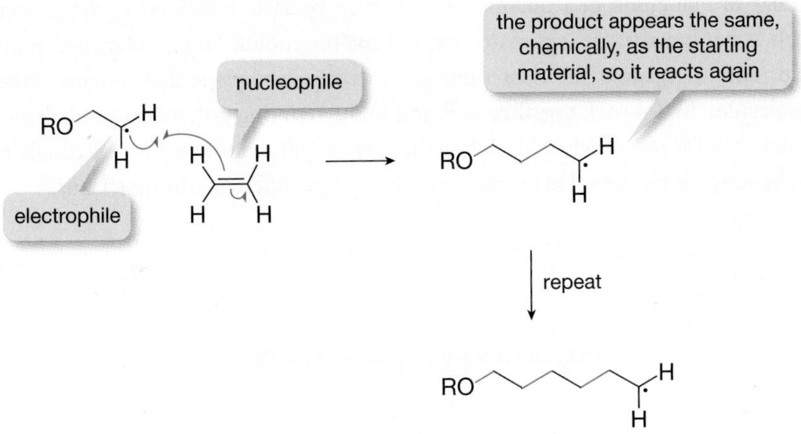

Termination takes place when any two radicals combine, for example, when two alkyl chains join together.

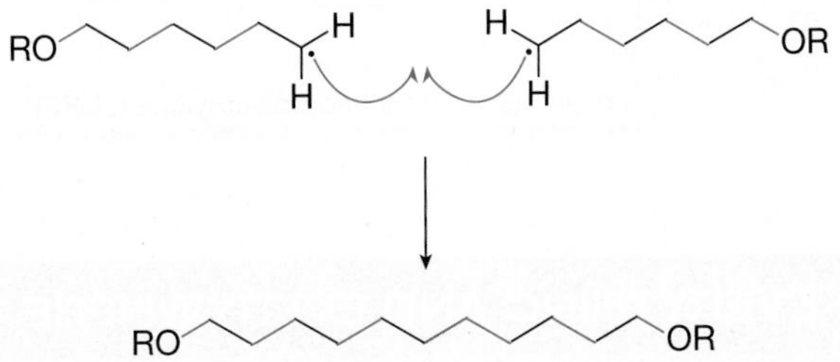

One interesting aspect of polyethylene synthesis is that the final product is usually branched, as opposed to linear. Branching occurs because a hydrogen atom in the product can be abstracted by an alkyl radical. As a result, the new radical can undergo propagation with another alkene.

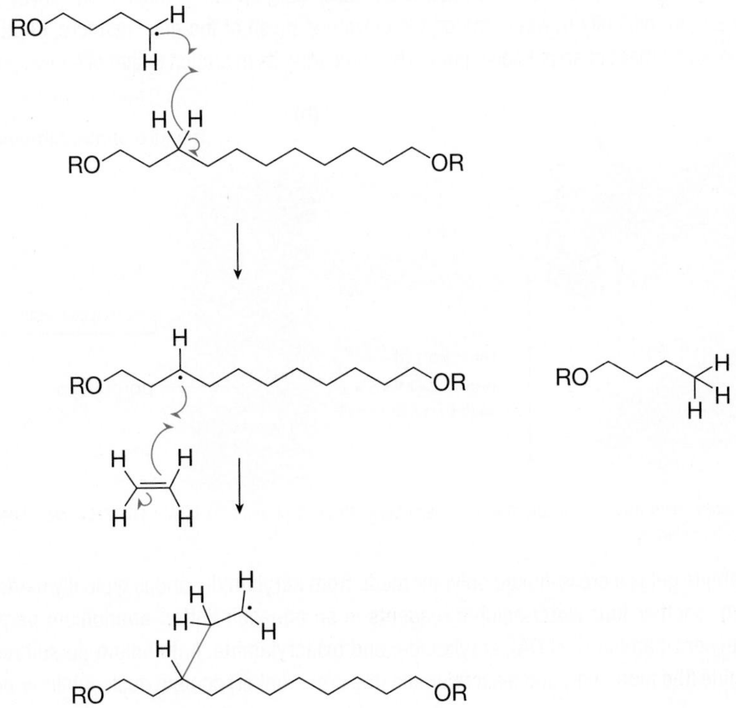

Zonda/Shutterstock.com

ifong/Shutterstock.com

The extent of branching in a polymer is important because it influences the properties of the polymer, such as hardness, transparency, density, and melting point. Highly branched polyethylene is more flexible, more transparent, less dense, and softer than polyethylene that contains fewer branches. Branched molecules do not pack together well, and so the resulting polymer is a soft, flexible material. The molecules of linear polyethylene pack together very tightly, producing a very tough, rigid, dense, and chemically resistant product. These two forms have very different commercial uses.

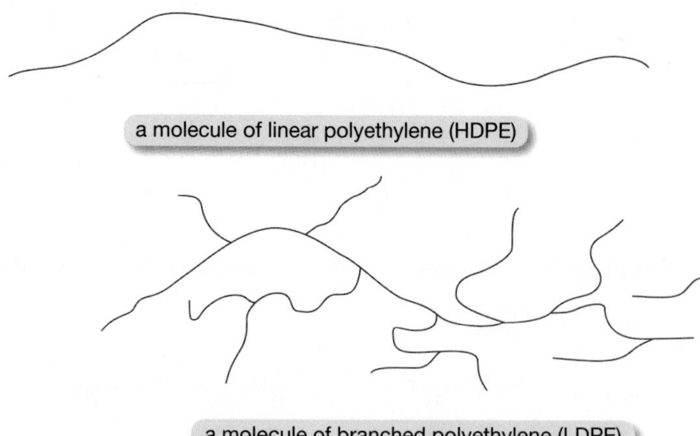

CHEMISTRY: EVERYTHING AND EVERYWHERE

Polyacrylamide Gel Electrophoresis

Polyacrylamide gel electrophoresis (PAGE) is one of the most commonly used techniques in biochemistry to separate proteins. In this technique, a sample containing protein is "loaded" onto the top of a polymeric gel containing a detergent used to denature (unfold) the protein. When voltage is applied, the proteins move from the top to the bottom of the polyacrylamide gel according to their size (molecular mass). Large proteins move more slowly than small ones because they have more difficulty moving through the polymeric mesh of the gel. Therefore, PAGE can provide an estimate of the molecular mass of an unknown protein by comparing its migration to that of known protein standards.

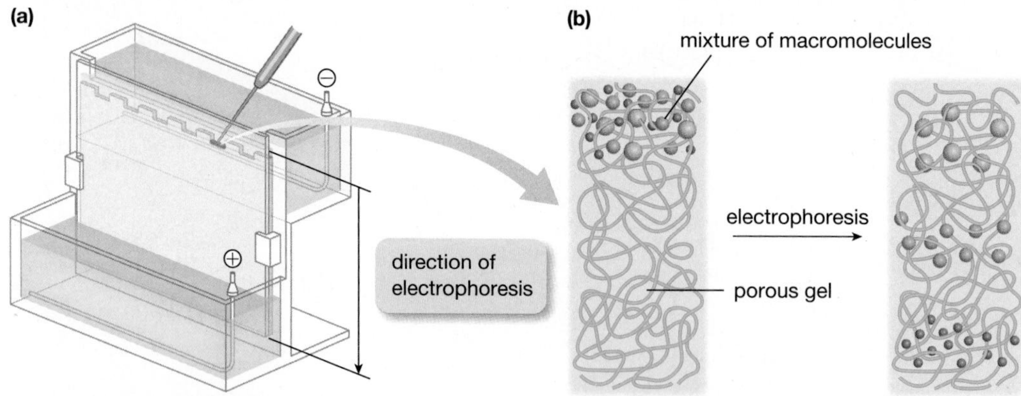

BIOCHEMISTRY 8e, by Jeremy Berg, John L. Tymoczko, Gregory J. Gatto, Lubert Stryer, Copyright 2015 by W.H. FREEEMAN AND COMPANY. Used with permission of the publisher.

The polyacrylamide gel is a cross-linked polymer made from acrylamide, and is typically made shortly before use by combining together four water-soluble reagents in an aqueous buffer: ammonium persulfate, *N*,*N*,*N*′,*N*′-tetramethylethylenediamine (TMEDA), acrylamide, and bisacrylamide. Ammonium persulfate is a peroxide, and both acrylamide (the monomer) and bisacrylamide (the cross-linker) contain nucleophilic π bonds (alkenes).

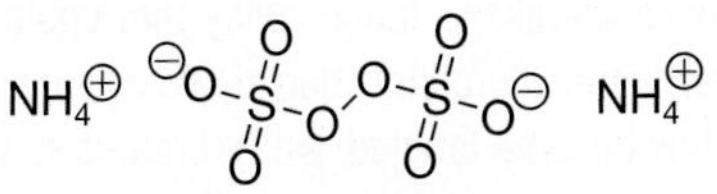

ammonium persulfate

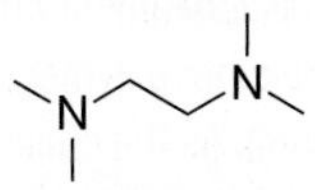

N, *N*, *N*′, *N*′-tetramethylethylenediamine (TMEDA)

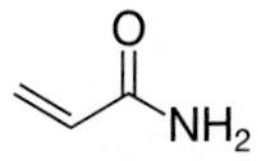

acrylamide

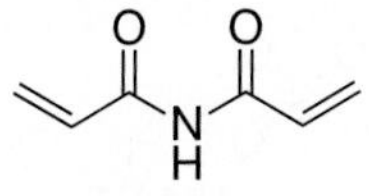

bisacrylamide

The polymerization is initiated when the sulfate radical—formed by the reaction of ammonium persulfate with TMEDA—adds to acrylamide. The newly formed radical allows propagation to occur.

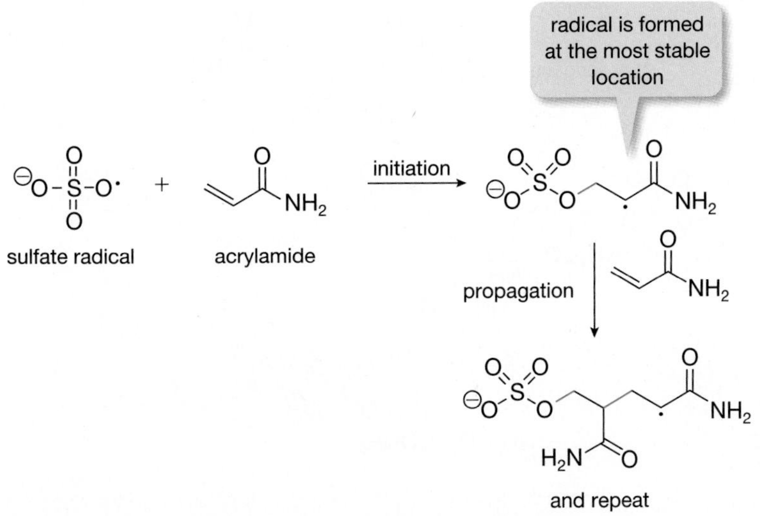

radical is formed at the most stable location

sulfate radical + acrylamide —initiation→

propagation

and repeat

Because the reaction mixture contains a small amount of bisacrylamide, the propagation can take place on a random basis on bisacrylamide instead of acrylamide. After bisacrylamide has been incorporated, propagation using acrylamide can continue. This creates a branch on the growing polymer chain.

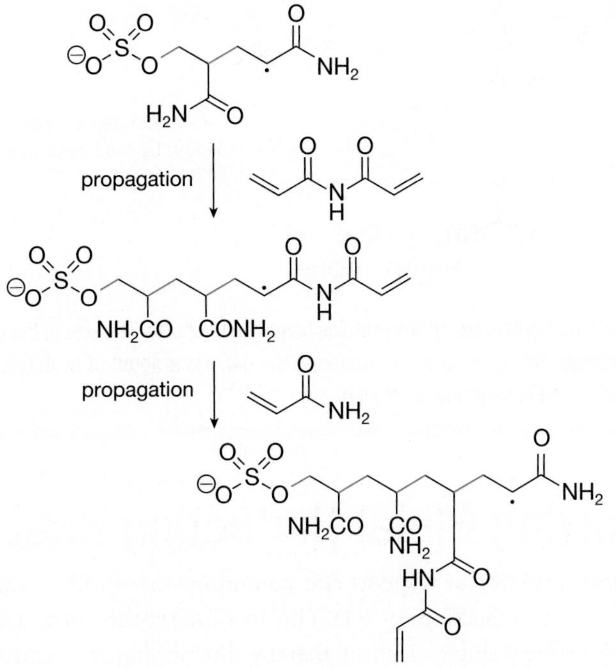

propagation

propagation

Continued

The incorporation of bisacrylamide into the polymer introduces an alkene functionality that could react with another radical, such as another polymer radical. This would result in the interconnection of two polymer molecules, also known as cross-linking. As the cross-linked chains grow, they become tangled, which creates spaces (pores) in the bulk material. The size of the pores can be controlled by the amount of branching that is allowed (by the ratio of bisacrylamide to acrylamide).

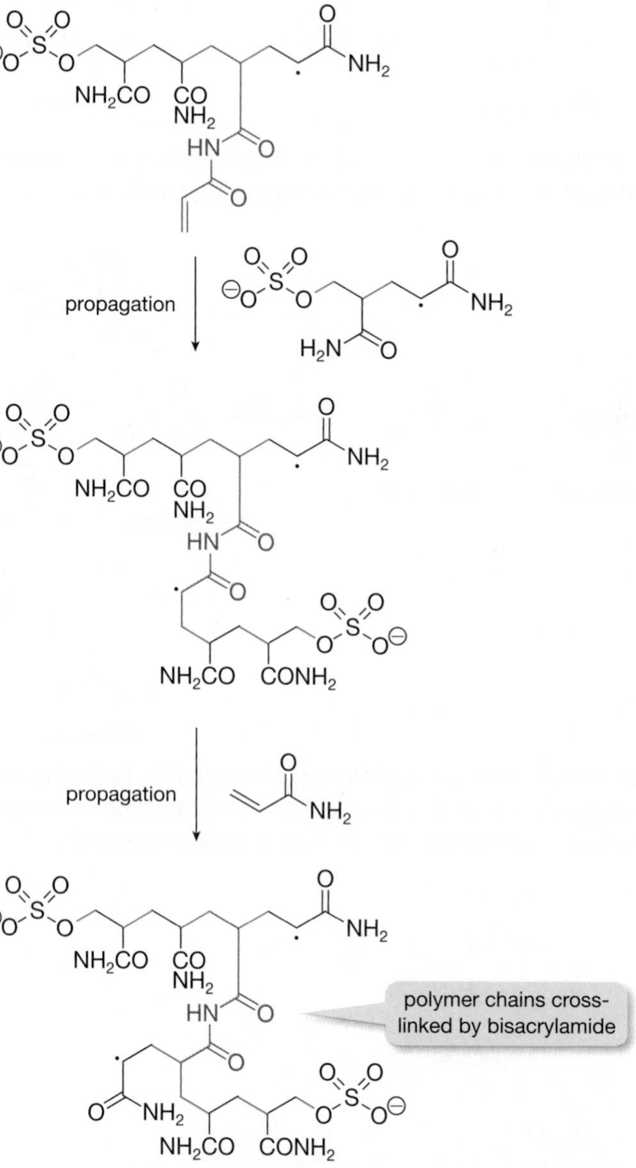

propagation

propagation

polymer chains cross-linked by bisacrylamide

Polyacrylamide gel electrophoresis separates proteins based on their size. The size of the gel pores is determined by the frequency of the cross-linkages, so when preparing the gel, the amount of bisacrylamide added relative to the amount of acrylamide used is carefully controlled.

19.9 Dissolving Metal Reduction Reactions

This chapter describes a variety of reagents and conditions that can be used to initiate radical chain reactions (in particular, Section 19.3.1). This section explains how a metal can initiate a radical reaction by donating a single electron, thereby functioning as a reducing agent. For such reactions, the metal is typically dissolved in a suitable solvent—one that does not react with, but may solvate, the electron released by the metal. One commonly used metal is sodium, which can be dissolved in liquid ammonia to give a solution containing **solvated electrons**. Electrons that are solvated by ammonia have a characteristic blue colour. Because various functional groups on the organic compounds in these solutions accept these electrons, these compounds become reduced. The following sections describe dissolving metal reactions that reduce π bonds. Common types of such reactions include the reduction of alkynes to alkenes, aromatics to 1,4-cyclohexadienes, and carbonyl groups to alcohols. These reactions all involve the addition of a single electron to a π bond, and represent the third method by which a carbon-centred radical may be generated.

Solvated electrons are individual electrons that are surrounded by a solvent.

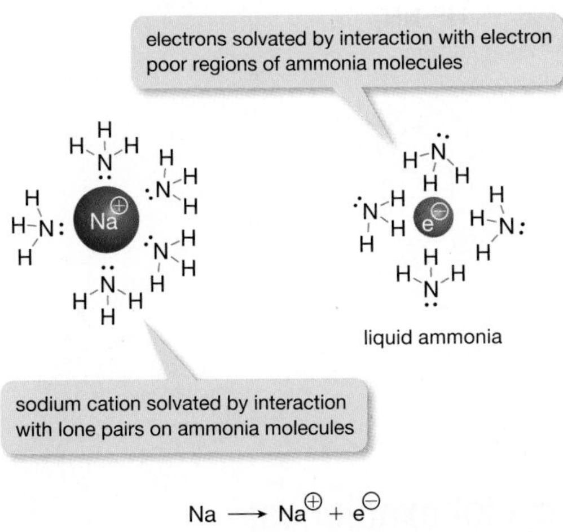
electrons solvated by interaction with electron poor regions of ammonia molecules
liquid ammonia
sodium cation solvated by interaction with lone pairs on ammonia molecules
Na ⟶ Na⊕ + e⊖

14/20
14/20

© 1994 Richard Megna - Fundamental Photographs

19.9.1 Reduction of alkynes to alkenes

When a metal is dissolved in liquid ammonia, the resulting solution can stereoselectively reduce an alkyne to a *trans* alkene. The *trans* alkene is preferentially formed because it is less sterically hindered than the *cis* alkene. The reduction stops at this point and does not continue to form the alkane.

$$\text{alkyne} \xrightarrow[\text{NH}_3(\text{l})]{\text{Na}} \text{trans alkene}$$

Mechanistically, the reaction involves an intermediate known as a **radical anion**, formed by the addition of an electron to a π bond of the alkyne. This ion can form by the homolytic cleavage of one of the alkyne π bonds as the extra electron is added. Alternatively, the formation of this radical anion can be considered as a kind of nucleophilic addition, in which an electron nucleophile adds to one of the alkyne carbons, breaking a π bond by a heterolytic process. The negatively charged carbon atom of a radical anion is highly basic, and removes a proton from the ammonia solvent to form a new carbon-hydrogen bond.

A **radical anion** is a chemical species that is negatively charged and has an unpaired electron.

This protonation reaction creates a carbon radical that now accepts a second electron to form a carbanion, which is rapidly protonated by the solvent. Overall, two electrons along with two protons are transferred to the triple bond, increasing the electron count on the group and thereby reducing it.

sodium has one valence electron

$Na\cdot \xrightarrow{NH_3} Na^{\oplus} + e^{\ominus}$

radical anion

anion

trans alkene

alkyl groups adopt a *trans* arrangement to minimize steric hindrance

nucleophilic addition of an electron to one of the alkyne carbons can also generate the radical anion

19.9.2 Reduction of aromatics to cyclohexadienes: The Birch reduction

The **Birch reduction** is a reaction that reduces aromatic rings to 1,4-cyclohexadienes.

Related to the process that reduces alkynes to alkenes is the dissolving metal reaction called the **Birch reduction**, in which an aromatic ring is reduced to a non-conjugated 1,4-cyclohexadiene. This reaction also uses $Na/NH_3(l)$, but a small amount of an alcohol is often introduced to the reaction mixture to act as a proton source.

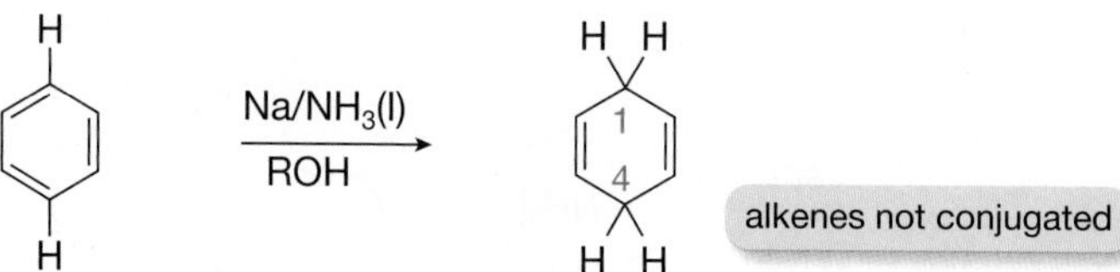

STUDENT TIP

Reduction of an alkyne using $Na/NH_3(l)$ forms a *trans* alkene, whereas reduction using H_2 and Lindlar's catalyst forms a *cis* alkene.

The net result of this process is the addition of two hydrogen atoms to opposite carbons (C-1 and C-4) of the benzene ring. As a result, the Birch reduction is a 1,4-reduction.

radical anion is resonance-stabilized

electron repulsion favours the resonance forms in which the non-bonding electrons are as far apart as possible

The addition of an electron can be considered as a nucleophilic addition to the π system. Note that the π bonds break heterolytically.

The Birch reduction reaction commences with the transfer of an electron to the aromatic ring. The reaction can be considered a kind of nucleophilic addition (electron nucleophile) that results in heterolytic cleavage of the corresponding π bonds. This generates a radical anion, a delocalized intermediate that is best described by contributing resonance structures that hold the non-bonded electrons at the maximum separation from each other. Protonation from the solvent or the alcohol additive then generates a radical that accepts a second electron (from another sodium atom); this makes a second anion that, in turn, abstracts a proton to form the product. This second anion is also stabilized by delocalization and tends to protonate so as to form the non-conjugated 1,4-cyclohexadiene. Such reactions are regioselective and form different isomers depending on the substituents attached to the aromatic ring. If the ring is connected to an electron-withdrawing group, the product will have one of the new tetrahedral carbons beside this group. This occurs because the intermediate anions (and radicals) are stabilized by the electron-withdrawing group when the negative charge is located adjacent to it. In the following mechanisms (and some other mechanisms involving metals), note the mix of single and paired electron movements.

STUDENT TIP
When working with metals, it may be helpful to explicitly draw the valence electrons of the metal. This helps to keep track of the oxidation states of the metals and to balance the electrons in the reaction.

electron-withdrawing group stabilizes adjacent anions

Electron-donating groups give the opposite selectivity because the intermediates are destabilized when the negatively charged atom is adjacent to the electron donor.

electron-donating group destabilizes the resonance forms that have a negative charge adjacent to it

19.9.3 Reduction of carbonyl groups to alcohols: The Bouveault–Blanc reduction

Esters can be reduced to alcohols by using an aluminum hydride reagent such as $LiAlH_4$ or diisobutylaluminum hydride (DIBAL). These reagents work very well, but can be very expensive, and their high reactivity creates issues for industrial use.

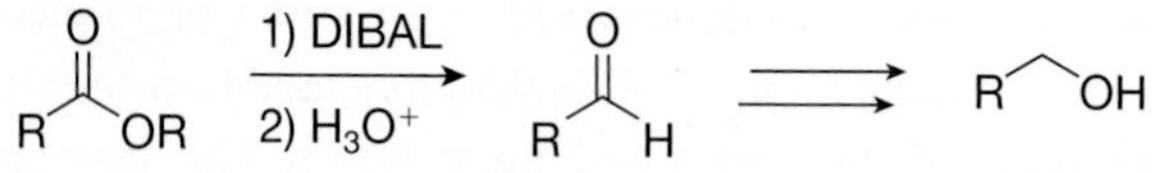

The **Bouveault–Blanc reduction** uses sodium in ethanol to reduce an ester to an alcohol.

The **Bouveault–Blanc reduction** is an industrially viable method for the synthesis of alcohols from esters. In a way that is similar to the Birch reduction, sodium metal functions as a single-electron donor. However, the solvent used in this case is generally an inexpensive alcohol such as ethanol. As in other reactions that involve dissolving metals, the mechanism of the Bouveault–Blanc reduction reaction involves radical anions as intermediates.

As shown previously, electrons can react in a nucleophile-like manner, adding to π bonds to form radical anions. In doing so, the electron adds to the most electrophilic atom in a π-bonded functional group, breaking the π bond so as to place a lone pair on the most electronegative atom. This pattern is similar to the addition of negatively charged nucleophiles to carbonyls. Proton transfer results in a new radical that at this point accepts another electron to form a hemiacetal intermediate.

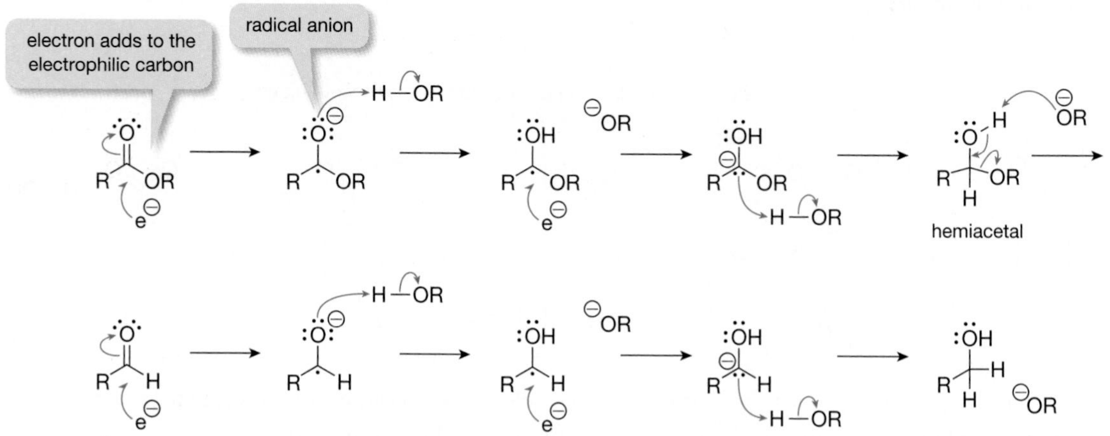

electron adds to the electrophilic carbon

radical anion

hemiacetal

In the existing basic conditions, this hemiacetal is quickly converted to the aldehyde, which undergoes further reduction in the same way: addition of an electron in a nucleophile-like manner to form a radical anion. Not surprisingly, the same reaction conditions can be used to reduce other carbonyl functionalities such as aldehydes and ketones.

19.9.4 Reduction of aldehydes and ketones to alkanes: The Clemmensen reduction

The **Clemmensen reduction** is a reaction that reduces an aldehyde or ketone to an alkane.

The **Clemmensen reduction** uses a zinc-mercury alloy in concentrated aqueous hydrogen chloride to reduce a carbonyl group (aldehyde or ketone) to an alkane.

Zn/Hg

$HCl_{(conc.)}$

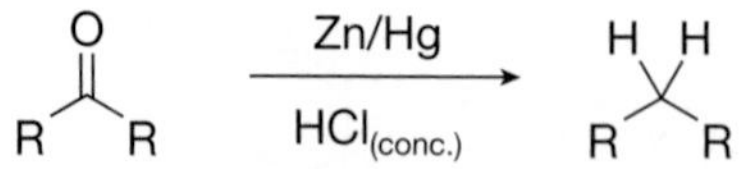

The mechanism by which this reaction proceeds is not entirely known. However, it has recently been proposed that a ZnCl radical, formed by the reaction of the alloy with hydrogen chloride, reacts with the aldehyde or ketone. In the hydrogen chloride solution, the zinc metal and the mercury in the alloy react to form a zinc radical. This radical adds to the carbonyl group, forming a carbon radical that reacts with water to produce a hydroxyl radical. The overall process adds a pair of electrons to the carbon chain, thus reducing it. Then homolytic cleavage of the carbon-oxygen bond makes a carbon radical that captures a hydrogen atom from water to form the reduced product.

The reaction works particularly well with carbonyl groups that are bonded to an aromatic ring. This is because the carbon radical that forms is stabilized by delocalization.

Zn: Cl HgCl ⟶ ·ZnCl ·HgCl
·ZnCl
carbon radical abstracts hydrogen atom from H_2O

The oxygen radicals that form combine to make various zinc species that in the presence of hydrogen chloride and water are hydrolyzed to zinc chloride.

$$\cdot\ddot{O}H \quad \cdot ZnCl \longrightarrow HOZnCl \xrightarrow{HCl} ZnCl_2 \quad H_2O$$

$$\cdot\ddot{O}{-}ZnCl \quad \cdot ZnCl \longrightarrow ClZn{-}O{-}ZnCl \xrightarrow{HCl} 2\ ZnCl_2 \quad H_2O$$

CHEMISTRY: EVERYTHING AND EVERYWHERE

Antioxidants and Artificial Preservatives

How can we prevent unsaturated fatty acids from being oxidized, and in this way to prevent a fat or an oil from going rancid? One method is to add a **radical scavenger**. Radical scavengers are compounds that "capture" highly reactive free radicals, thereby terminating radical chain reactions. One naturally occurring radical scavenger is vitamin E (alpha-tocopherol), which can scavenge radicals, such as hydroxyl radicals, by reacting with them.

Radical scavengers are chemical compounds that react with, or scavenge, other radicals.

highly reactive hydroxyl radical
phenolic O–H bond is weak and easily cleaved homolytically
H_2O
phenoxyl radical is resonance-stabilized and is sterically hindered by adjacent methyl groups

When vitamin E reacts with a radical, it also becomes a radical. However, the radical of vitamin E is stabilized by delocalization and surrounded by bulky methyl groups, so it is much less reactive and therefore less damaging to biological materials than is the hydroxyl radical. In other words, vitamin E prevents the biological material from being oxidized by the hydroxyl radical, which is why vitamin E is classified as an **antioxidant** vitamin.

Antioxidants are compounds that prevent *other* compounds (such as DNA, lipids, etc.) from being oxidized. This is accomplished by allowing the antioxidants themselves to be oxidized.

Continued

By having a thorough understanding of the chemical mechanism of vitamin E's antioxidant action, structural analogs that mimic the function of vitamin E were invented. These compounds, such as butylated hydroxyanisole (BHA) and butylated hydroxytoluene (BHT), are artificial preservatives that have the similar chemical moieties found in vitamin E.

butylated hydroxyanisole (BHA)

butylated hydroxytoluene (BHT)

19.10 Patterns in Radical Reactions

The formation of carbon-centred radicals primarily occurs by three methods: the abstraction of an atom, (Section 19.5), the addition of a radical to a π bond (Section 19.7), and the addition of a single electron to a π bond (Section 19.9).

In the case of radical formation by the abstraction of a hydrogen atom, a carbon-centred radical with a neutral charge is formed. Because the carbon atom is electron deficient, the radical that is preferentially formed can be predicted by applying the same factors that stabilize carbocations: delocalization and the presence of electron-donating alkyl groups. Atom abstraction preferentially takes place at a location that leads to the formation of the most stable radical.

Radicals can also be formed by adding a radical to a π bond creating a carbon-centred radical with an overall neutral charge. The radical that is preferentially formed can be predicted by considering the same factors that stabilize carbocations.

A radical can form when a single electron adds to a π bond, as happens in dissolving metal reactions. This type of reaction generates a radical anion, and the interaction of this ion with different groups introduces selectivity to this reaction type. If an electron adds to a carbonyl group, the reaction is regioselective in that the electron adds to the electrophilic carbon instead of the oxygen. The negative charge resides on the oxygen atom, which is the most electronegative atom in the carbonyl functional group. This reaction follows the pattern of nucleophilic addition (the electron can be considered as a special nucleophile). Note the combination of single- and double-headed arrows in this mechanism.

If an electron adds to an aromatic ring, the radical anion that forms is the one that separates the non-bonded electrons by the greatest amount (1,4-relationship). Once again, both arrow types are used to show electron movements. In the radical anion formed, the preferential position of the negative charge depends on the presence or absence of electron-withdrawing and electron-donating groups. To determine the preferential position, the relative position of the negatively charged atom and any of these groups need to be examined because electron-withdrawing groups *stabilize* anions, and electron-donating groups *destabilize* them.

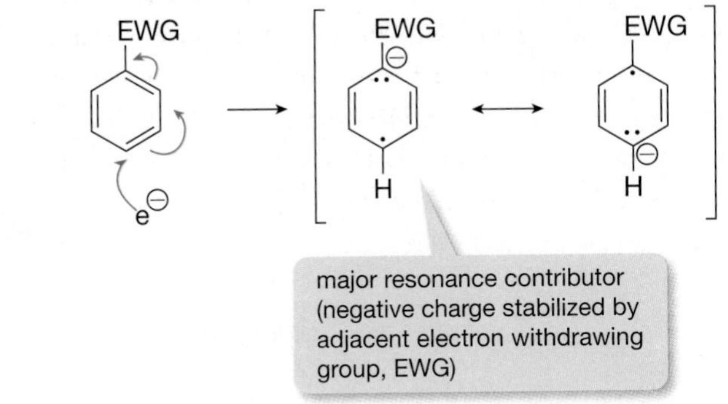

EWG EWG EWG H H e

major resonance contributor (negative charge stabilized by adjacent electron withdrawing group, EWG)

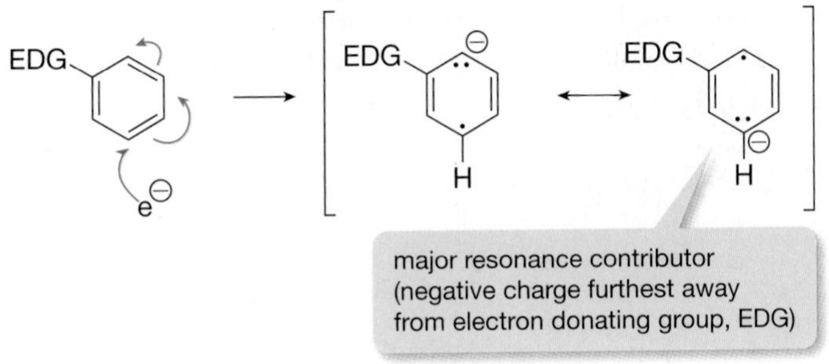

EDG EDG EDG H H e

major resonance contributor (negative charge furthest away from electron donating group, EDG)

Bringing It Together

Saturated fatty acids do not contain any carbon-carbon double bonds, monounsaturated fatty acids contain one carbon-carbon double bond, and polyunsaturated fatty acids contain two or more carbon-carbon double bonds. In most polyunsaturated fatty acids, the double bonds are each separated by a single methylene (CH_2) group.

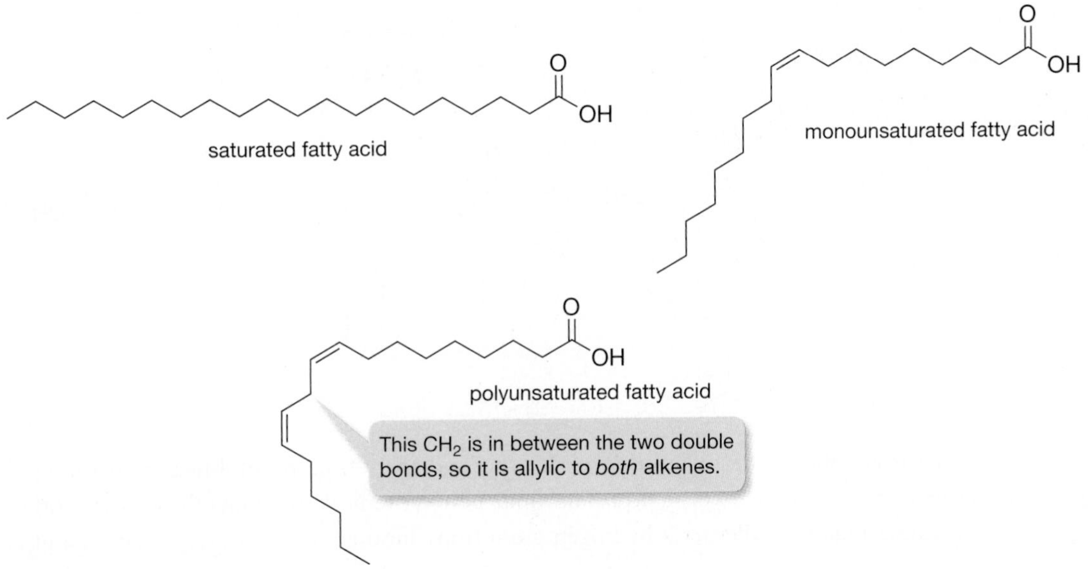

O OH saturated fatty acid

O OH monounsaturated fatty acid

O OH polyunsaturated fatty acid

This CH_2 is in between the two double bonds, so it is allylic to *both* alkenes.

The fact that a polyunsaturated fatty acid has methylene groups that are allylic to more than one alkene has an impact on its reactivity. Specifically, it is easier to abstract a hydrogen atom from a polyunsaturated fatty acid than from a monounsaturated or saturated fatty acid because the resulting radical is stabilized by delocalization involving more than one alkene, thus increasing the stability of the radical. When such molecules are exposed to air, hydrogen abstraction initially takes place by reaction with atmospheric oxygen.

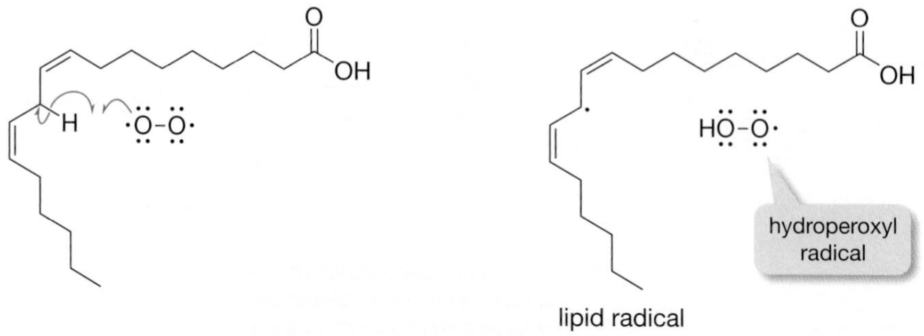

hydroperoxyl radical

lipid radical

In the process, the fatty acid loses an electron and is effectively oxidized. The hydroperoxyl radical that forms is a short-lived but highly reactive species that decomposes into products such as hydroxyl radicals. These extremely reactive hydroxyl radicals are even better at abstracting hydrogen atoms from a fatty acid.

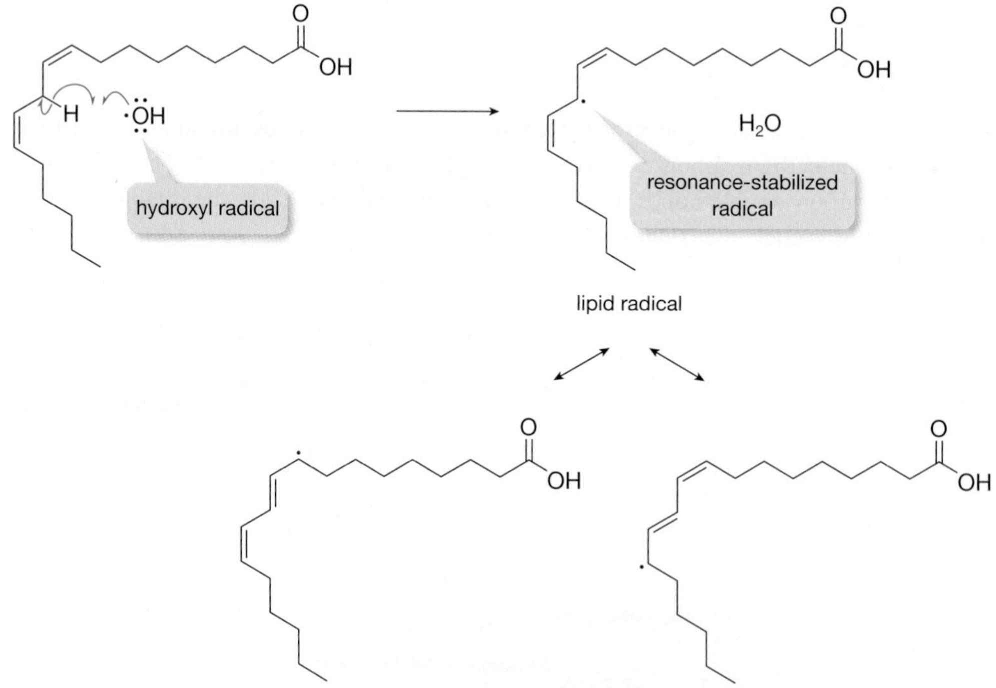

hydroperoxyl radical

hydroxyl radical

hydroxyl radical

H_2O

resonance-stabilized radical

lipid radical

Radicals are also prone to reaction with other radicals. Oxygen combines with the lipid radical to form a lipid peroxyl radical. The lipid peroxyl radical subsequently uses the other oxygen atom's radical to abstract a hydrogen atom from another fatty acid, regenerating a lipid radical and forming a lipid hydroperoxide. The lipid hydroperoxide is unstable and undergoes a variety of decomposition reactions to give small, odoriferous aldehydes, ketones, and carboxylic acids that allow us to detect when a fat or oil has become rancid.

lipid radical

lipid peroxyl radical

lipid hydroperoxide

examples of odoriferous decomposition products

You Can Now

- Explain, identify, and use arrow notation to show homolytic cleavage, heterolytic cleavage, and bond-formation reactions.
- Describe the general steps in radical chain reactions.
- Draw radical chain reactions using a circular cycle.
- Predict the relative stabilities of carbon radicals, and identify the factors that increase their stabilities.
- Draw a mechanism for, and predict the major product of, the free-radical halogenation of an alkane.
- Draw a mechanism for, and predict the major product of, allylic and benzylic halogenation reactions.
- Draw a mechanism for the dehalogenation of an alkyl halide.
- Draw a mechanism for, and predict the major product of, the radical addition of hydrogen bromide to an alkene.
- Identify the repeating units of monomers in addition polymers.
- Explain the general mechanistic patterns in dissolving metal reactions.

A Mechanistic Re-View

RADICAL CHAIN REACTIONS

Initiation

$$Cl{-}Cl \xrightarrow{h\nu} Cl\cdot + \cdot Cl$$

N-Br (N-bromosuccinimide) $\xrightarrow{h\nu}$ N· (succinimidyl radical) + ·Br

tBuO–OtBu $\xrightarrow{h\nu}$ tBuO· + ·OtBu

Propagation

$$H_3C{-}H \;\; \cdot Br \longrightarrow H_3C\cdot + H{-}Br$$

$$H_3C\cdot \;\; Br{-}Br \longrightarrow H_3C{-}Br + \cdot Br$$

Br· + propene ⟶ Br–CH₂–ĊH–CH₃ — most stable radical (more favourable)

Br–CH₂–ĊH–CH₃ + H–Br ⟶ Br–CH₂CH₂CH₃ + ·Br

Termination

$$Br\cdot \;\; \cdot Br \longrightarrow Br{-}Br$$

$$H_3C\cdot \;\; \cdot Br \longrightarrow H_3C{-}Br$$

$$H_3C\cdot \;\; \cdot CH_3 \longrightarrow H_3C{-}CH_3$$

RO~~~~~ĊH₂ + H₂Ċ~~~~~OR ⟶ RO~~~~~~~~~~OR

ANTI-MARKOVNIKOV (FREE-RADICAL) ADDITION

Addition of HBr to alkene

Br· → Br⌄·⌄ H—Br → Br⌄⌄ + ·Br

DISSOLVING METAL REACTIONS

Stereoselective reduction of alkyne to alkene

$H_3C-C\equiv C-CH_3$ → $e^{\ominus}$ → H–NH_2 → CH_3 → $NH_2^{\ominus}$

Birch reduction of benzene

$e^{\ominus}$ → H–OR → $^{\ominus}OR$

Bouveault–Blanc reduction of esters

R–C(=O)–OR → $e^{\ominus}$ → H–OR → $^{\ominus}OR$ → hemiacetal

R–C(=O)–H → $e^{\ominus}$ → H–OR → $^{\ominus}OR$

Clemmensen reduction of aldehydes and ketones

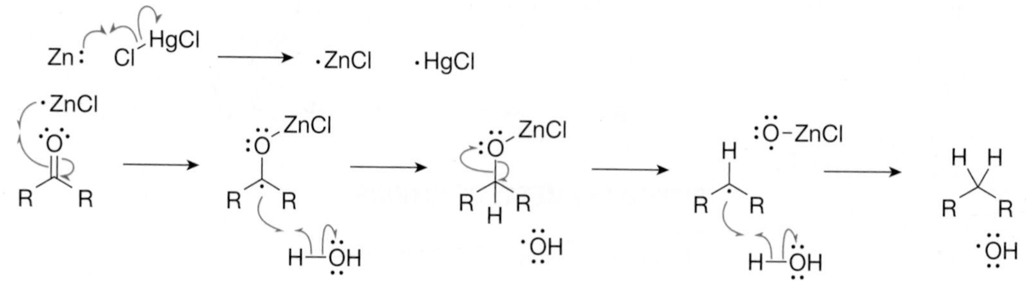

Zn: Cl–HgCl ⟶ ·ZnCl ·HgCl

·ZnCl ... ZnCl ... H–ÖH ... ZnCl ... ·ÖH ... :Ö–ZnCl ... H–ÖH ... ·ÖH

Problems

19.5 Complete the Lewis structures for each of the following compounds by adding all of the implied hydrogen atoms and electron pairs.

a)

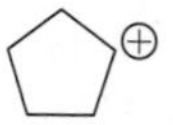

b)

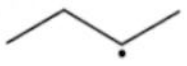

c)

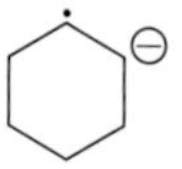

d)

19.6 For each of the following compounds, use the proper arrow notation to show the homolytic cleavage of the indicated bond. Draw the products formed.

a)

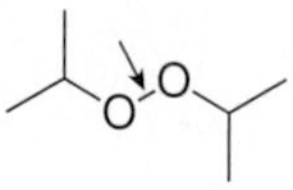

b) 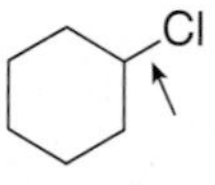 Cl

c) 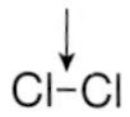 Cl–Cl

19.7 For each of the following pairs of radicals, use the proper arrow notation to show how they combine to form a bond. Draw the product formed.

a) 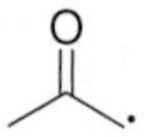 and ·Br:

b) 2 equivalents of ·ÖH

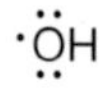

c)  and

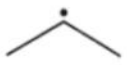

19.8 Which radical in each pair is more stable, if any? Justify your answer.

a)

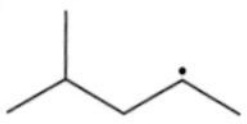

b)

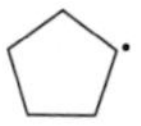

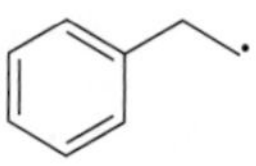

c)

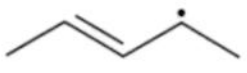

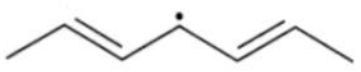

d)

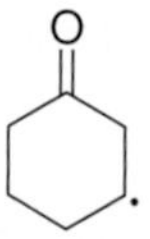

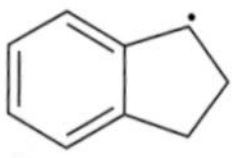

19.9 Which of the following fatty acids produces the most stable radical upon abstraction of a hydrogen from one of its carbons?

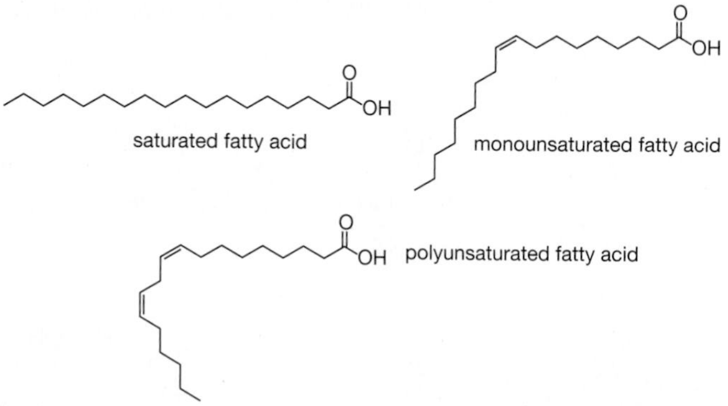

saturated fatty acid

monounsaturated fatty acid

polyunsaturated fatty acid

19.10 Examine the following radicals. For those that are delocalized, draw resonance contributors.

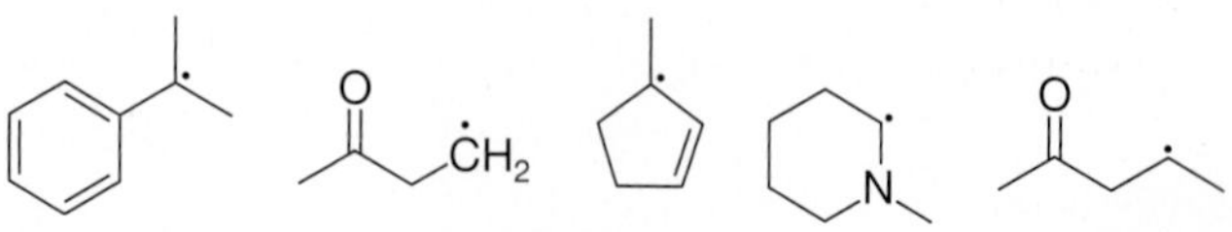

19.11 Draw the major product(s) formed by free-radical monobromination of each of the following compounds. Hint: Do not forget about stereochemistry.

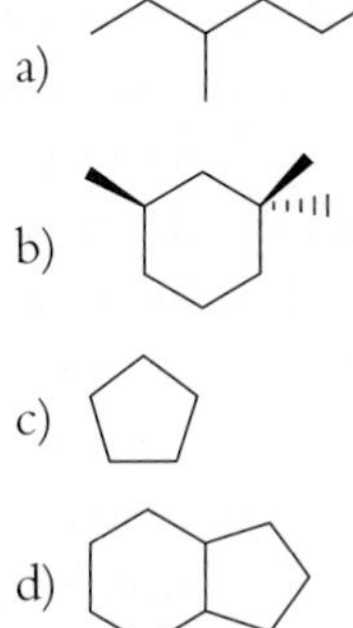

a)

b)

c)

d)

19.12 When 1-butene undergoes free-radical bromination, two products are formed. Write a mechanism that shows how the two products are formed.

NBS, $h\nu$ → Br + Br

19.13 What is the major free-radical halogenation product formed when the following compound is treated with NBS in the presence of light?

19.14 The increased selectivity of the free-radical bromination of an alkane over that of a free-radical chlorination can be explained by the Hammond postulate. Why?

19.15 Explain how the following two products could be made from 1-hexene and hydrogen bromide. Write mechanisms to show their formation.

Br Br

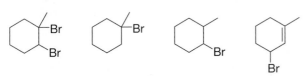

19.16 Explain how the following four products could be made from 1-methylcyclohexene and an appropriate reagent. Write mechanisms to show their formation.

Br Br Br Br Br

19.17 The following compound can undergo a radical polymerization reaction to form an addition polymer that is an important sealant. Draw a short segment of the polymer and indicate the repeating unit.

19.18 In general, how does branching affect the physical properties of a polymer?

19.19 Draw a mechanism for the polymerization of tetrafluoroethylene using a peroxide initiator. Is it possible to form branched polytetrafluoroethylene?

19.20 What is the major product formed in each of the following sequential reactions?

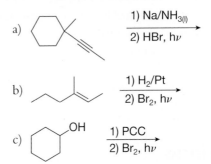

a) 1) $Na/NH_{3(l)}$ 2) HBr, $h\nu$

b) 1) H_2/Pt 2) Br_2, $h\nu$

c) OH 1) PCC 2) Br_2, $h\nu$

19.21 Draw resonance structures for the radical anions shown here. In each set of resonance structures, identify the resonance contributor that is most stable.

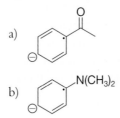

a) O

b) $N(CH_3)_2$

19.22 Prepare a list of all the different ways by which an aldehyde could be reduced to an alcohol. Which of these methods involve the nucleophilic attack of the aldehyde carbon?

19.23 Which one of the following compounds is most likely to donate a hydrogen atom to a radical? Justify your answer.

a)

b) OH

c) OH

19.24 Propose a reaction scheme for each of the following transformations. Note that there may be more than one possible reaction scheme.

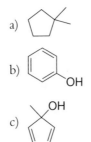

a) OH ⇒

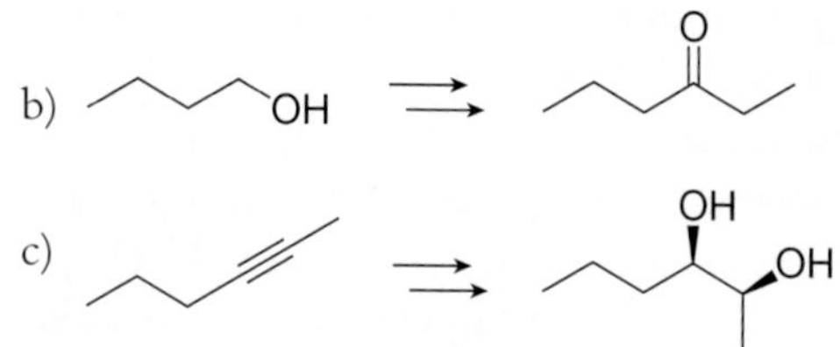

19.25 Species known as *radical cations* are commonly formed in mass spectrometry and as highly reactive intermediates in chemical reactions. Draw resonance structures for each of the following radical cations.

MCAT STYLE PROBLEMS

19.26 Which one of the following best describes the sequence of events in a radical chain reaction?

a) protonation, nucleophilic attack, deprotonation
b) electrophilic attack, protonation, termination
c) initiation, propagation, termination
d) deprotonation, electrophilic attack, termination

19.27 The factors that affect radical stability are most similar to factors that affect which one of the following?

a) carbanion stability
b) conjugate base stability
c) covalent bond strength
d) carbocation stability

19.28 The following compound is an addition polymer that is used as a sealant. Identify the alkene building block used to synthesize this polymer.

A B C D

CHALLENGE PROBLEM

19.29 Drugs are metabolized in the body using special enzymes called cytochromes, one of which is cytochrome P450 (abbreviated CYP_{450}). These enzymes use a heme group, which carries a reactive iron atom that is the actual oxidizer. During drug metabolism, this iron atom removes electrons from the drug (along with any attached atoms). Several pathways are possible, including a radical-based one. Because oxidation involves the loss of electrons, drugs are susceptible to oxidation at electron-rich locations. Some of the more common reactions are shown here. Propose a radical-based mechanism for each of these metabolic reactions. Note: The structure of the CYP_{450} is abbreviated showing the active iron only.

a) alkyl group of drug

b) methoxy group of drug

c) aromatic group of drug

d) aromatic group of drug

Reactions Controlled by Orbital Interactions

RING CLOSURES, CYCLOADDITIONS, AND REARRANGEMENTS

20

CHAPTER OUTLINE

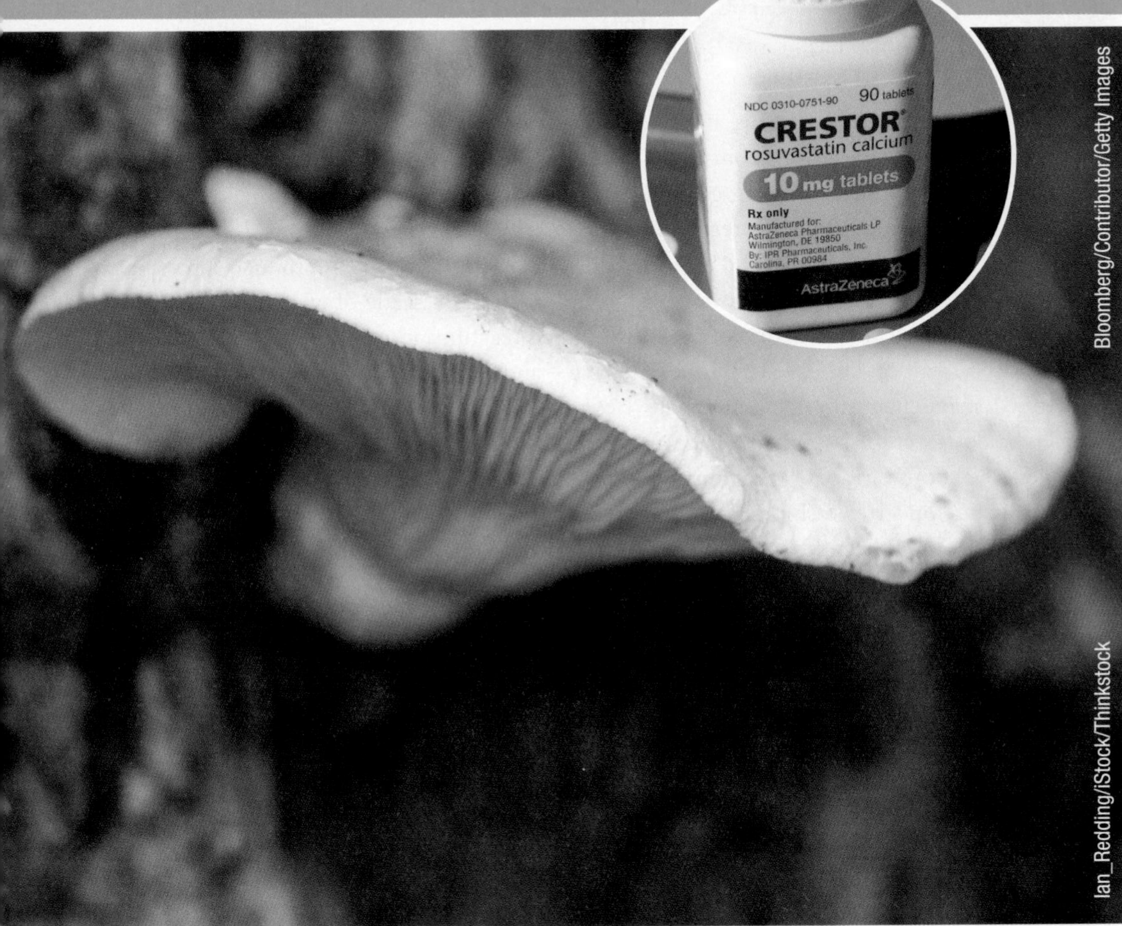

Bloomberg/Contributor/Getty Images

Ian_Redding/iStock/Thinkstock

Oyster mushrooms, a natural source of cholesterol drug lovastatin (Crestor).

20.1 Why It Matters

Lovastatin is a naturally occurring compound produced by oyster mushrooms and by the mould responsible for the colour of red yeast rice. Because it inhibits the production of cholesterol in the body, lovastatin has become an important drug to treat cardiovascular disease, sold under the tradename of Crestor by the pharmaceutical company AstraZeneca. The biosynthesis of lovastatin by these organisms includes a chemical step, known as a Diels–Alder reaction, whereby its two core rings form simultaneously.

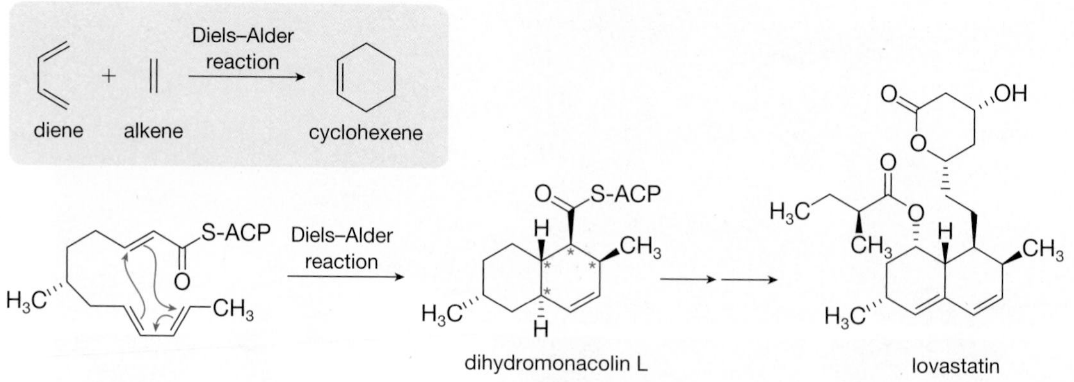

The Diels–Alder reaction forms a new cyclohexene ring by re-organizing six π electrons in cyclic fashion over six carbon atoms. The Diels–Alder reaction that forms lovastatin is *intra*-molecular because both participating groups, an alkene and a diene, are part of the same molecule. Under enzymatic control, this reaction forms a cyclohexene ring, a second six-membered ring, and four stereogenic centres. The intermediate, dihydromonacolin L, is produced as a single stereoisomer and then converted to lovastatin through a further cascade of enzymatic reactions.

The Diels–Alder reaction is one of the most studied reactions in organic chemistry. It forms two new carbon-carbon bonds—a cyclohexene ring—and can generate several chiral centres of well-defined stereochemistry. Its applications to organic synthesis are wide-ranging. However, the Diels–Alder reaction is perhaps more interesting from the mechanistic standpoint as it is a pericyclic reaction: a reaction that involves a cyclic transition state that re-organizes several electron pairs in concerted fashion. For such reactions, the distinction between nucleophile and electrophile becomes blurred, yet they proceed in a precise and predictable fashion. As it turns out, when explaining the Diels–Alder reaction only the π orbitals of the interacting alkene and diene need to be considered. Specifically, the molecular orbitals (MOs) constructed from their π orbital components are analyzed.

The majority of this chapter deals with three types of pericyclic reactions: cycloadditions, sigmatropic rearrangements, and electrocyclic-ring-opening reactions. Some non-pericyclic rearrangements (non-pericyclic because they do not have a cyclic transition state) are also described.

20.2 π Molecular Orbitals

A **pericyclic reaction** is one in which the moving electrons flow in a ring during the transition state.

A **pericyclic reaction** is one in which the electrons move in a cyclic transition state, such as occurs in the decarboxylation of β-keto acids (Chapter 17). To understand the mechanism of a pericyclic reaction and how it leads to the stereochemistry of its product, it is essential to first understand the molecular orbitals involved.

Pericyclic reactions can occur only if the symmetry of each of the interacting molecular orbitals matches the others. To determine the symmetry of the interacting orbitals, it is necessary to determine (1) which orbitals are involved in a particular pericyclic reaction, and (2) the symmetry of these orbitals based on the phases of their lobes at each atomic location. Before these can be considered, a model system of the conjugated π molecular orbitals must be developed.

The linear combination of atomic orbitals (LCAOs) provides a convenient approximation of the structure of molecular orbitals. A review of the following key points about molecular orbitals (MOs) can facilitate working with LCAOs.

- The number of MOs is always equal to the number of atomic orbitals (AOs) being combined.
- Any orbital can hold a maximum of two electrons.

- MOs are populated with valence electrons according to the Aufbau principle (lowest energy orbitals first).
- Combining (i.e., overlapping) two orbitals that have the same phase produces a bonding MO, lower in energy than the original AOs.
- Combining two orbitals that have opposing phases produces an anti-bonding MO, higher in energy that the original AOs.
- Sigma bonding (σ) MOs are lower energy than pi (π) anti-bonding MOs.
- Sigma anti-bonding ($\sigma^\star$) MOs are higher energy than pi ($\pi^\star$) MOs.
- π orbital diagrams are drawn with the most stable MO (Ψ_1) at the bottom and the higher energy MOs above, as the number of nodes increases.

According to the LCAO, the π and $\pi^\star$ molecular orbitals in a double bond can be represented by the 2p atomic orbitals on each carbon that makes up the bond. The in-phase overlap of two atomic orbitals represents a π molecular orbital. This orbital has no node between atoms and is lower in energy than either of the two atomic orbitals. It is a bonding orbital because, when populated with two electrons, it holds together the two contributing atoms: there is a continuous region of electron density between both nuclei. The different orbital phases (positive and negative) are represented by different colours.

Out-of-phase overlap of the same two atomic orbitals produces a $\pi^\star$ molecular orbital. This orbital has one node between atoms, is higher in energy than both of the atomic orbitals contributing to it, and is anti-bonding (⋆) because, when populated with two electrons, it forces the two atoms apart: the node "exposes" charge repulsion between the positively charged nuclei. The representations of the π and $\pi^\star$ orbitals shown on the right in the next diagram are based on quantum mechanical calculations.

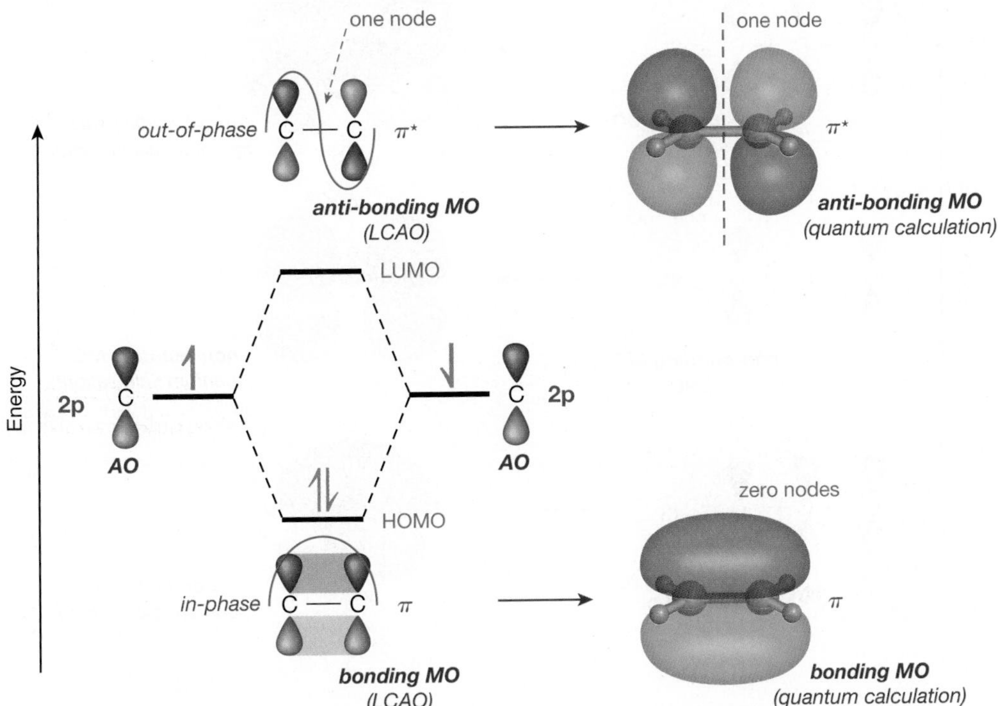

In the ground state of an alkene, the two valence electrons forming the π bond are paired and occupy the lowest energy π molecular orbital; the $\pi^\star$ orbital is unoccupied. The π orbital is the highest occupied molecular orbital (HOMO) of the functional group, and is the orbital equivalent of a nucleophile. The $\pi^\star$ is the lowest unoccupied molecular orbital (LUMO) of the

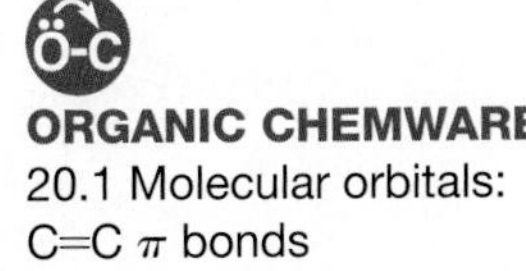

ORGANIC CHEMWARE
20.1 Molecular orbitals: C=C π bonds

In the electronic configuration of a molecule, the **frontier molecular orbitals** are the highest energy occupied (HOMO) and lowest energy unoccupied (LUMO) orbitals.

functional group; it can be viewed as the orbital equivalent of an electrophile. The HOMO and LUMO are called the **frontier molecular orbitals**; they are at the frontier of reactivity. Many organic reactions are interpreted based on the interaction between the HOMO of one component and the LUMO of another.

When π bonds are conjugated in organic molecules, the adjacent π bonds actually mix to form a single orbital system. In a line structure, the system is drawn as alternating single and double bonds, but the actual molecule consists of one continuous system of π electrons across all the conjugated atoms. These systems are often described by the LCAO system representing the molecular orbital pattern. The allyl system consists of three molecular orbitals spanning the three atoms making up the system. Conjugated molecular orbitals are typically designated as $\Psi_1, \Psi_2 \ldots \Psi_n$, starting from the lowest energy (most stable) orbital at the bottom of the diagram. Ψ_1 has no nodes and has the lowest energy. Ψ_2 has higher energy and has one node, which is located on the central carbon (i.e., the probability of finding electrons at that point is zero). Ψ_3 has the highest energy and has two nodes. The relative phase of the lobes of the orbitals changes at each node, and the nodes are distributed in a symmetrical way.

Within the context of conjugated π-systems, **symmetric** MOs have the same phase at their two terminal lobes.

Within the context of conjugated π-systems, **antisymmetric** MOs have the opposite phase at their two terminal lobes.

The molecular orbitals represented in the following diagram derive from the LCAO system and also calculations based on quantum mechanical methods. The Ψ_1 and Ψ_3 orbitals are **symmetric** (in terms of a vertical plane of symmetry) because the orbital phases at the two terminal carbons are the same. Ψ_2 is called **antisymmetric** because the orbital phases at the two terminal carbons are opposite each other. In such orbital diagrams, the lowest energy MO is always symmetric and has no nodes. The other molecular orbitals alternate between antisymmetric and symmetric as their energy increases.

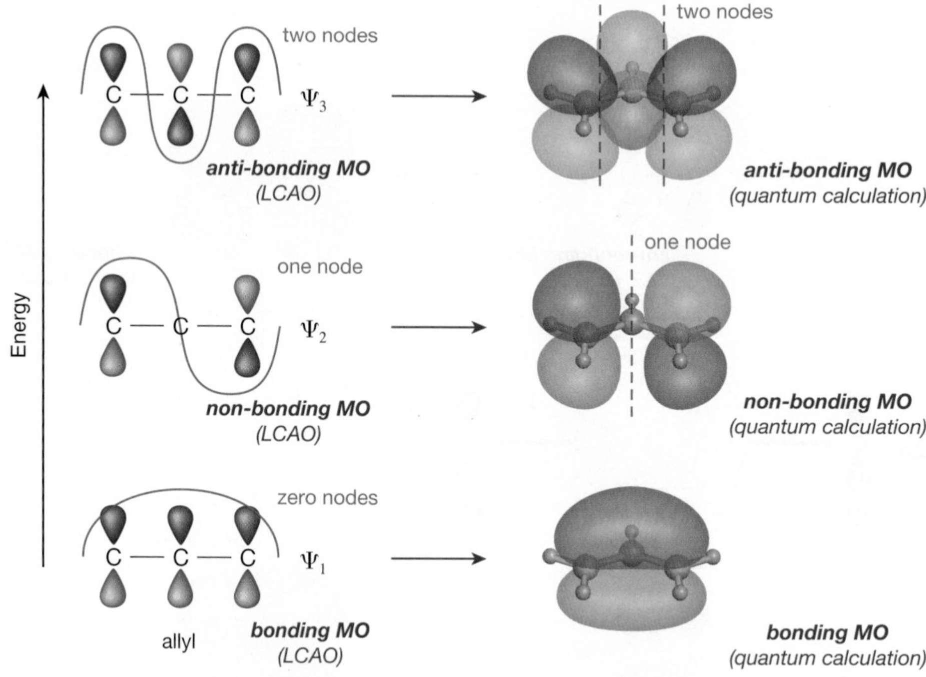

The molecular orbitals for a four-carbon conjugated system can be determined in similar fashion. Once again, the relative energy of the orbitals increases as the number of nodes increases. As before, Ψ_1 is symmetric, Ψ_2 is antisymmetric, Ψ_3 is symmetric, and Ψ_4 is antisymmetric in this stack of orbitals. Molecular orbitals for longer linear conjugated systems can also be constructed by following the same principles.

STUDENT TIP

Orbital nodes that fall directly on atoms can be represented by a large dot. The dot indicates zero probability of finding an electron at that location.

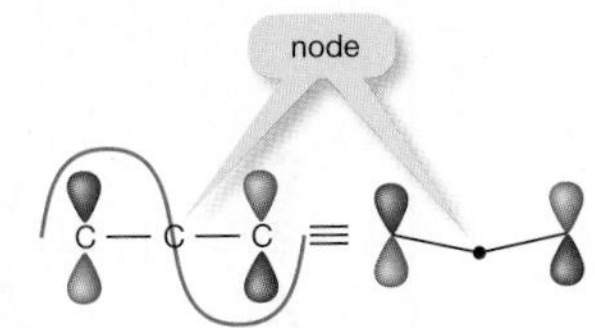

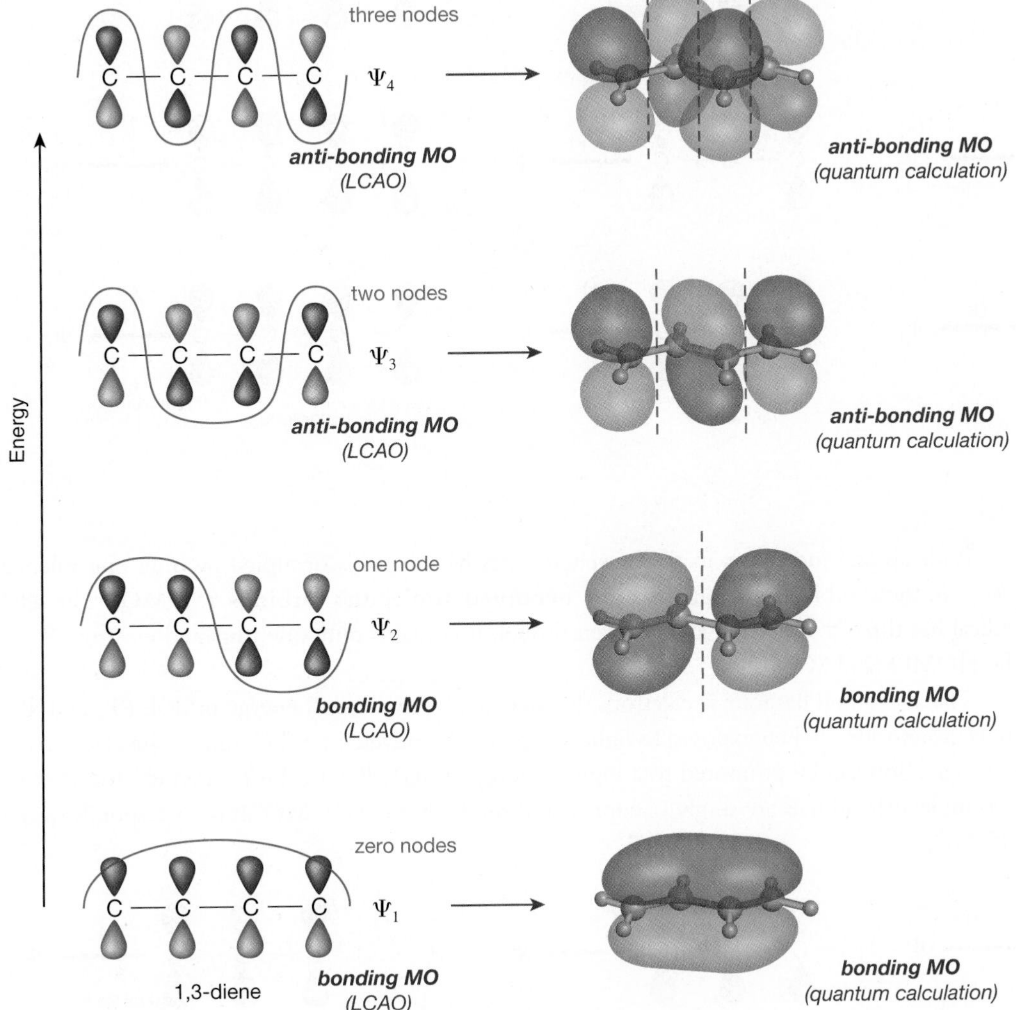

Which orbital is the HOMO and which is the LUMO depends on the number of π electrons in the conjugated system and whether the system is in the ground state or some excited state. In the ground (lowest energy) state, the allyl cation has only two π electrons, so Ψ_1 is the HOMO and Ψ_2 is the LUMO. The allyl anion contains four π electrons, so Ψ_2 is now the HOMO and Ψ_3 is the LUMO. In the ground state, a conjugated diene has four π electrons, so its Ψ_2 is its HOMO, and its Ψ_3 is its LUMO.

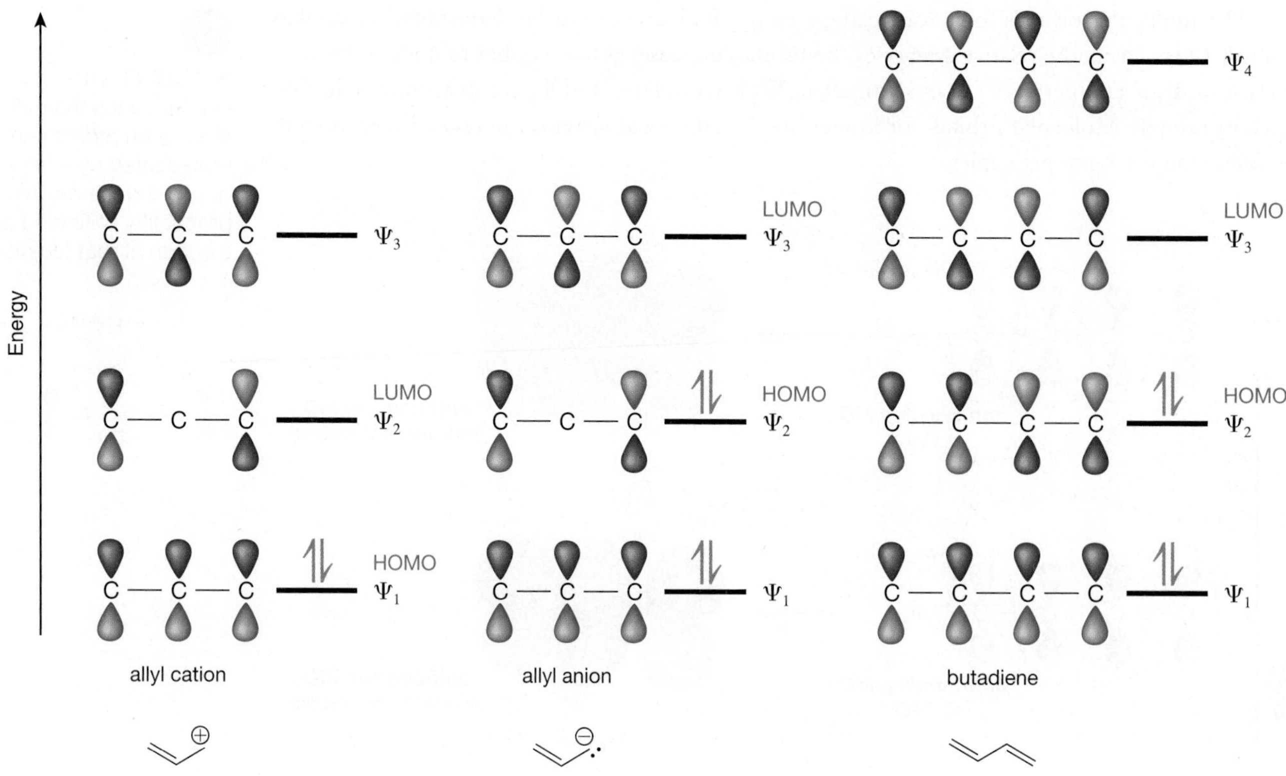

A singly occupied molecular orbital (SOMO) is any MO that is occupied by a single electron.

Radicals and molecules in their excited states have orbitals occupied by only one valence electron; these orbitals are called **singly occupied molecular orbitals (SOMO).** The allyl radical has three π electrons, and so, even though it contains only one unpaired electron, Ψ_2 is the HOMO (and SOMO).

The allyl anion has four π electrons that occupy the two lowest energy orbitals (Ψ_1 and Ψ_2) in its ground state. When exposed to light of a particular wavelength ($h\nu$), one of the electrons of the allyl anion can be promoted to a higher energy orbital (Ψ_3). In this excited electronic state, two molecular orbitals are singly occupied, and the higher energy MO thus corresponds to the HOMO.

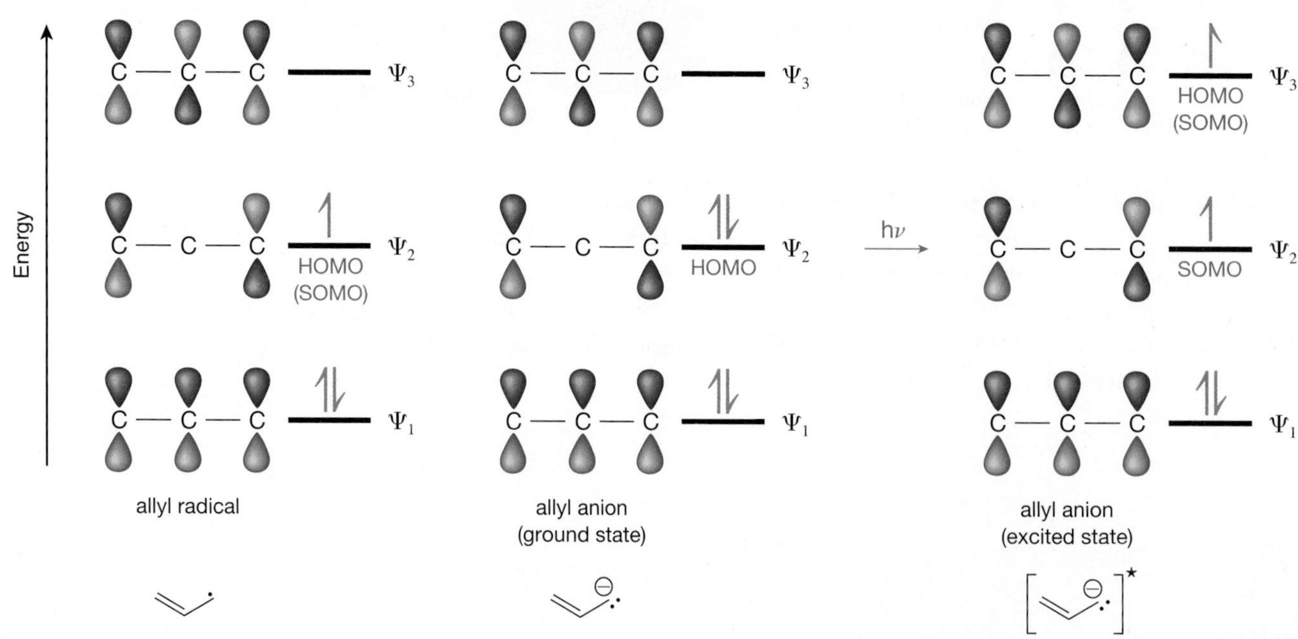

CHECKPOINT 20.1

You should now be able to predict the relative energy levels, with phases and nodes, of π molecular orbitals, including odd-numbered systems.

SOLVED PROBLEM

Draw the MOs in an orbital diagram for a five-atom conjugated system, such as methoxybutadiene. Label the HOMO and the LUMO as well as the bonding and anti-bonding orbitals.

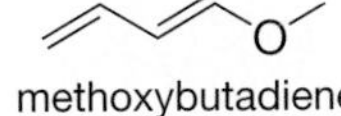

methoxybutadiene

STEP 1: Determine the number of orbitals contributing to the π molecular orbital. Note there is one p orbital for each atom in the conjugated system.

In the case of methoxybutadiene, there are five conjugated atoms, each with providing one p orbital to the π-system.

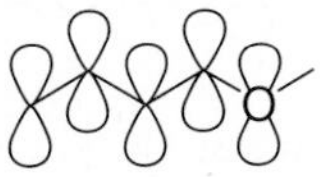

STEP 2: Combine the atomic orbitals to create an equal number of molecular orbitals. In the case of linear conjugated systems, each π molecular orbital has a different, equally spaced, energy level.

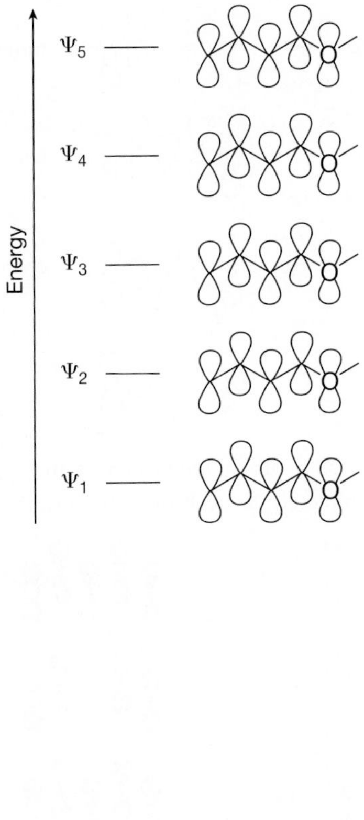

STEP 3: Identify the number of nodes for each MO. The lowest energy MO has no nodes. Each MO, in ascending order of energy, has one more node.

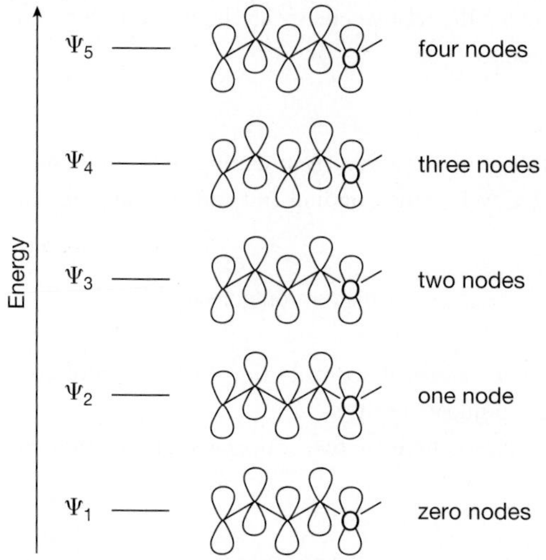

Energy

Ψ_5 — four nodes

Ψ_4 — three nodes

Ψ_3 — two nodes

Ψ_2 — one node

Ψ_1 — zero nodes

STEP 4: Identify the location of nodes. In each case, the placement of nodes must be symmetrical.

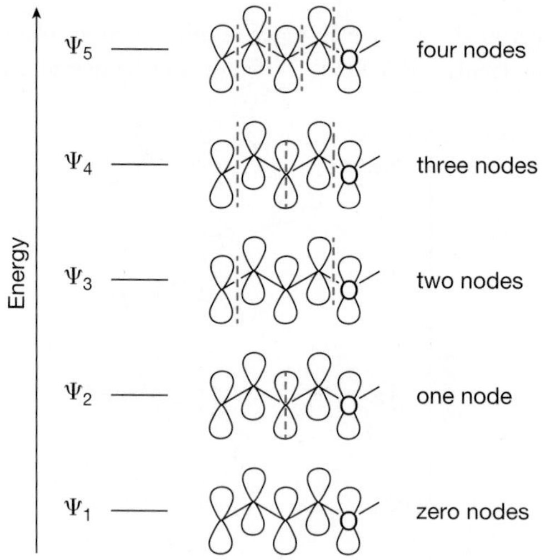

Energy

Ψ_5 — four nodes

Ψ_4 — three nodes

Ψ_3 — two nodes

Ψ_2 — one node

Ψ_1 — zero nodes

STEP 5: Add appropriate phases to each p-orbital lobe to create the appropriate nodes. In cases where the node falls directly on an atom, it completely "cancels out" that atom's p orbital.

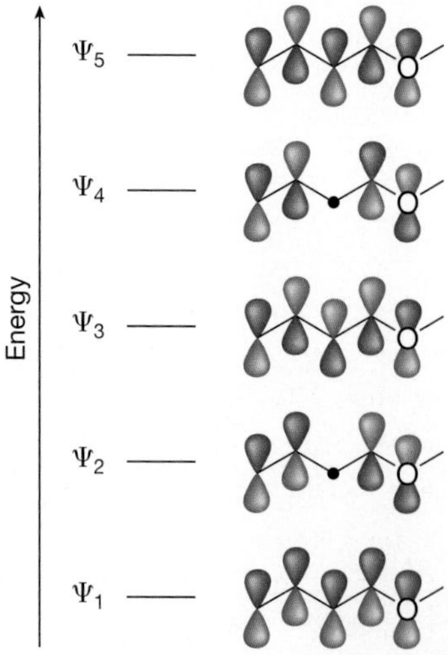

Energy

Ψ_5 —

Ψ_4 —

Ψ_3 —

Ψ_2 —

Ψ_1 —

STEP 6: Add the appropriate number of π electrons, filling the various MOs, starting from the lowest energy level.

To calculate this you may wish to consult the line drawing, adding any lone pairs as necessary. In this case, there are two lone pairs on oxygen, but only one can participate in conjugation. For conjugation to occur, the oxygen must be sp^2-hybridized, which places one of the lone pairs in an sp^2 orbital and unavailable to resonate.

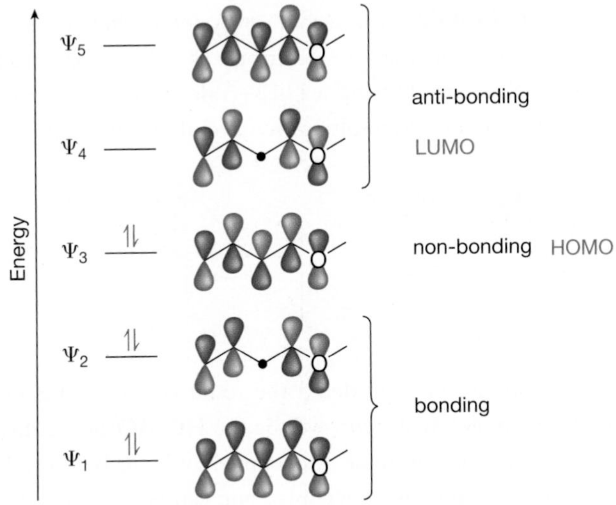

Energy

Ψ_5 — anti-bonding

Ψ_4 — LUMO

Ψ_3 ⇅ non-bonding HOMO

Ψ_2 ⇅ bonding

Ψ_1 ⇅

PRACTICE PROBLEM

20.1 Show molecular orbital energy diagrams for the following molecules:

a)

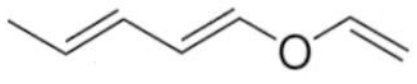

b)

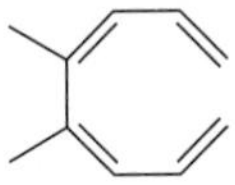

c)

INTEGRATE THE SKILL

20.2 Radical allylic bromination of 1,2-dimethylcyclohexene does not result in bromination at the allylic position. Explain this observation based on the molecular orbitals of the radical intermediate.

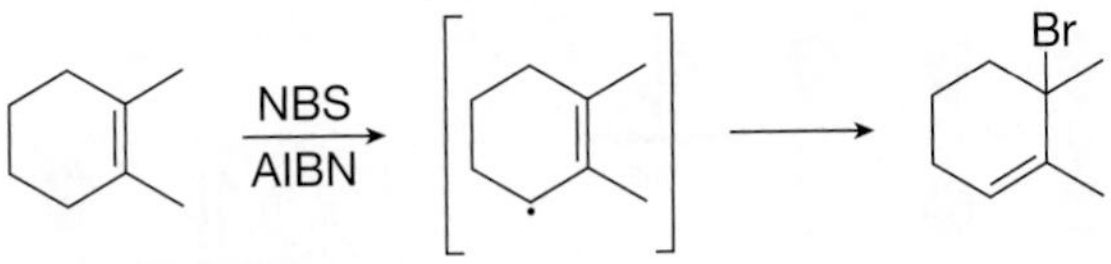

NBS, AIBN → Br

20.3 Cycloadditions

Cycloadditions are a class of pericyclic reactions in which two molecular components, usually present on two separate molecules, add to each other to form a ring, sometimes called a **cycloadduct**. Cycloadditions involve interactions between π-systems. To account for the interactions between these systems, their frontier molecular orbitals must be analyzed. Depending on their symmetry requirements, some cycloadditions are promoted by heat, including the Diels–Alder reaction (Sections 20.3.1 and 20.3.2) and dipolar cycloadditions (Section 20.3.5). Others are promoted photochemically, such as many [2+2] cycloadditions (Section 20.3.3).

A **cycloadduct** is the product of a cycloaddition.

20.3.1 The Diels–Alder reaction: A thermal cycloaddition

20.3.1.1 Mechanism

A **dienophile** refers to a π-bonded functional group that reacts with a diene in a Diels–Alder reaction.

The Diels–Alder reaction is one of the most widely studied cycloadditions in organic chemistry and is an example of a [4+2] cycloaddition. The numbers in brackets refer to the atoms in each reacting functional group that forms the ring. This reaction involves a diene (four-carbon π-system) and a **dienophile** (two-carbon π-system)—hence, the designation [4+2]. Occasionally, a Diels–Alder reaction may be described as a [4π+2π] reaction, based on the number of participating π electrons. During a Diels–Alder reaction, two of the three participating π bonds form two new σ bonds, resulting in a cyclohexene product.

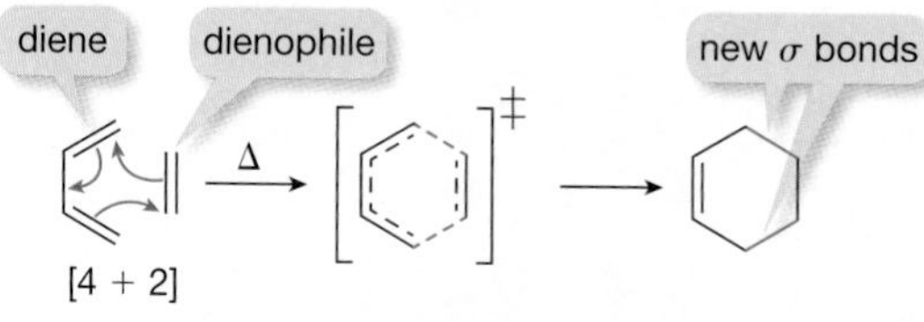

ORGANIC CHEMWARE
20.2 Diels–Alder reaction: General mechanism

ORGANIC CHEMWARE
20.3 Diels–Alder reaction: Cyclopentadiene + ethene

The Diels–Alder reaction can be considered the result of an overlap between the frontier orbitals of the two components. Normally, this will be the HOMO of the diene and the LUMO of the dienophile. The HOMO of the diene is occupied and can be considered a nucleophile, whereas the LUMO of the dienophile is unoccupied and can be considered an electrophile. The smaller the energy gap between the HOMO and the LUMO, the better the overlap between these orbitals and the lower the activation energy for the reaction. In most cases, the dienophile carries an electron-withdrawing group that lowers the energy of the LUMO and improves this overlap.

STUDENT TIP
The dienophile acts as the electrophile, and the addition of electron-withdrawing groups enhances this ability. The diene acts as a nucleophile, and the addition of electron-donating groups enhances this ability.

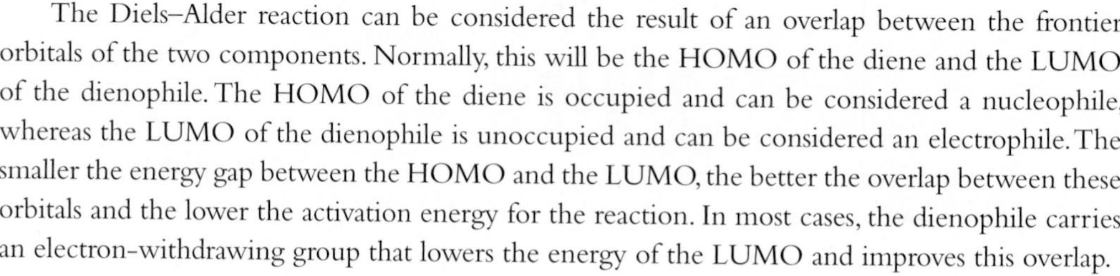

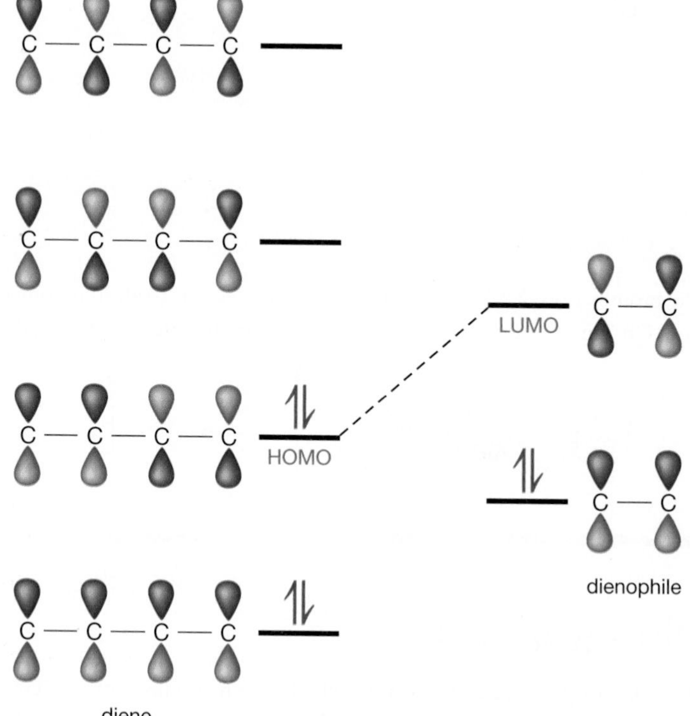

s-cis and ***s-trans*** refer to *cis/trans* isomerism around a single bond where rotation is restricted by conjugation. The *s* stands for sigma bond (σ bond).

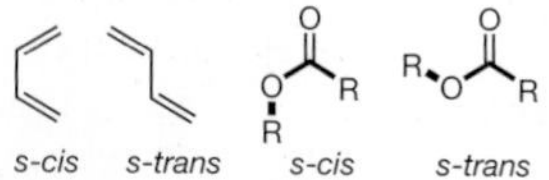

For the Diels–Alder reaction to take place, the diene must be in its higher energy ***s-cis*** conformation, which is typically in equilibrium with the ***s-trans*** form for acyclic dienes. Also, the symmetry of the two interacting molecular orbitals must match. Normally, both the HOMO and the LUMO are antisymmetric, so the interacting lobes at the ends of the two π-systems can overlap constructively to form two new σ bonds. This forms a new cyclohexene ring in a boat-like conformation.

HOMO
phase match
phase match
LUMO
HOMO
phase match
phase match
LUMO
σ
σ

The addition of an electron-donating group to the diene raises the energy of all of the π molecular orbitals, including the HOMO. Likewise, the addition of an electron-withdrawing group to the dienophile lowers the energy of its LUMO. Either or both of these factors can narrow the HOMO–LUMO energy gap and increase the rate of the Diels–Alder reaction. Most intermolecular Diels–Alder reactions employ one or more electron-withdrawing groups on the dienophile, and this has more influence on the activation energy of the process than does the addition of an electron-donating group to the diene.

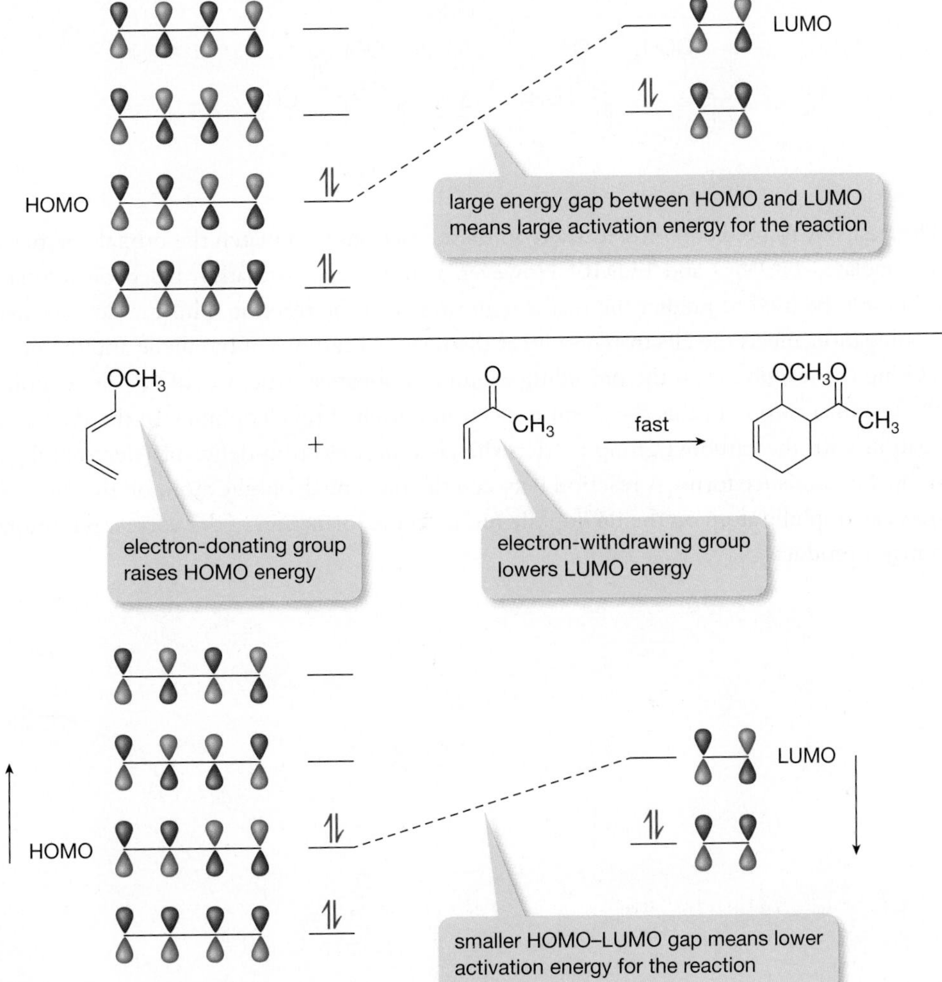
LUMO
HOMO
large energy gap between HOMO and LUMO means large activation energy for the reaction
OCH3
O
CH3
fast
OCH3O
CH3
electron-donating group raises HOMO energy
electron-withdrawing group lowers LUMO energy
LUMO
HOMO
smaller HOMO–LUMO gap means lower activation energy for the reaction

STUDENT TIP

Diels–Alder reactions generally work best when one component carries electron-donating groups and the other carries electron-withdrawing groups. The most common Diels–Alder reactions employ electron-withdrawing groups on the dienophile.

20.3.1.2 Regioselectivity

If the diene and the dienophile are both unsymmetrical, the Diels–Alder reaction can, in principle, give rise to two regioisomers based on the relative position of the substituents. Depending on how the two components approach each other, this can give either "*ortho*" or "*meta*"-substituted cyclohexenes.

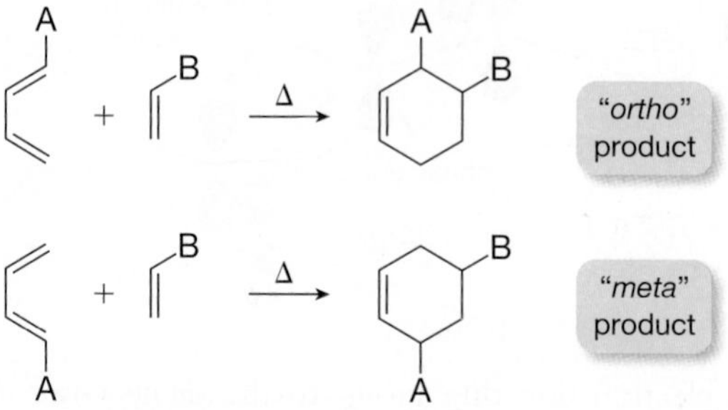

A + B $\xrightarrow{\Delta}$ A, B "*ortho*" product

A + B $\xrightarrow{\Delta}$ B, A "*meta*" product

STUDENT TIP

Products in a Diels–Alder reaction are sometimes described using the terms *ortho/meta/para*, which are otherwise reserved for disubstituted aromatic rings. This terminology is based on the cyclic aromatic transition state and, though common, is not technically correct when applied to the product.

The regioselectivity of the reaction can often be controlled by the strategic placement of electron-donating and/or electron-withdrawing groups on the diene and dienophile. These groups influence the electron density at specific carbon positions on both diene and dienophile such that electron-rich carbons on one reactant favour a reaction with electron-poor carbons on the other reactant. For example, the following reaction gives the "*ortho*" regioisomer as the major product.

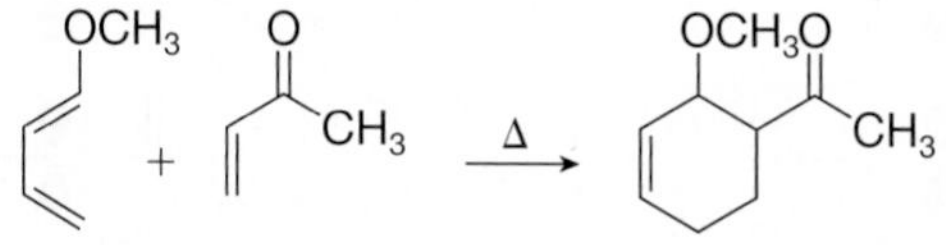

OCH_3 + O, CH_3 $\xrightarrow{\Delta}$ OCH_3 O, CH_3 "*ortho*"

The best way to predict regioselectivity in these reactions is to match the orbital coefficients of the calculated HOMO and LUMO. However, a simple approximation based on resonance forms can also be used to predict the major regioisomer of the reaction. This method compares how conjugation affects the electron density at the terminal carbons of the diene and the dienophile. Using the reaction from the preceding example, resonance structures of the diene indicate that the terminal carbon of this π-system is electron enriched (nucleophilic). In the dienophile, conjugation with the carbonyl group renders the β-carbon electron deficient (electrophilic), as shown by the resonance forms. A reaction between the most nucleophilic atom on the diene and the most electrophilic atom on the dienophile results in the formation of the correct regioisomer as the major product.

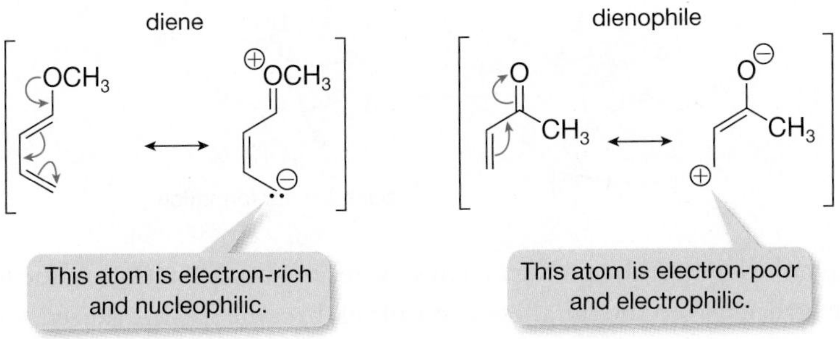

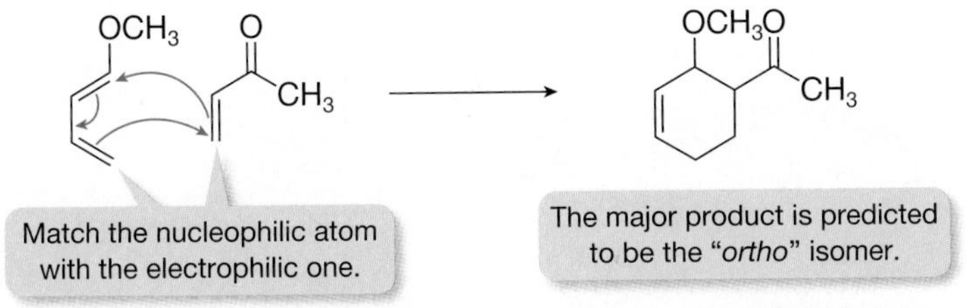

diene

dienophile

This atom is electron-rich and nucleophilic.

This atom is electron-poor and electrophilic.

Match the nucleophilic atom with the electrophilic one.

The major product is predicted to be the "*ortho*" isomer.

The same method can be used to predict the regioselectivity of other Diels–Alder reactions. With 2-methoxybutadiene, resonance forms show that the methoxy group enriches the electron density of C-1, making this atom nucleophilic. In this example, a reaction between the nucleophilic atom of the diene and the electrophilic atom of the dienophile results in the "*para*" product.

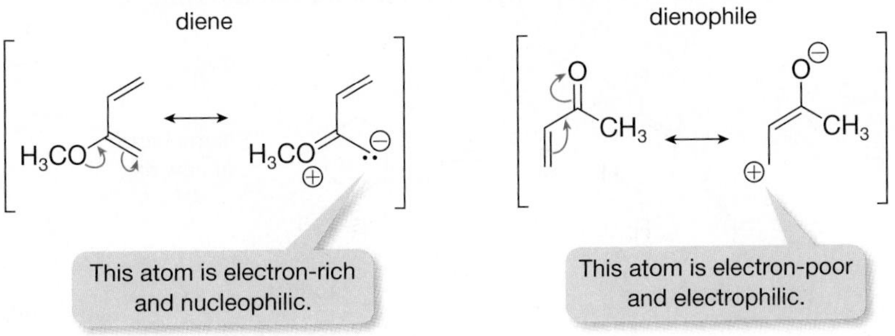

diene

dienophile

This atom is electron-rich and nucleophilic.

This atom is electron-poor and electrophilic.

Match the nucleophilic atom with the electrophilic one.

The major product is predicted to be the "*para*" isomer.

WANT TO LEARN MORE?
20.1 Diels–Alder regioselectivity

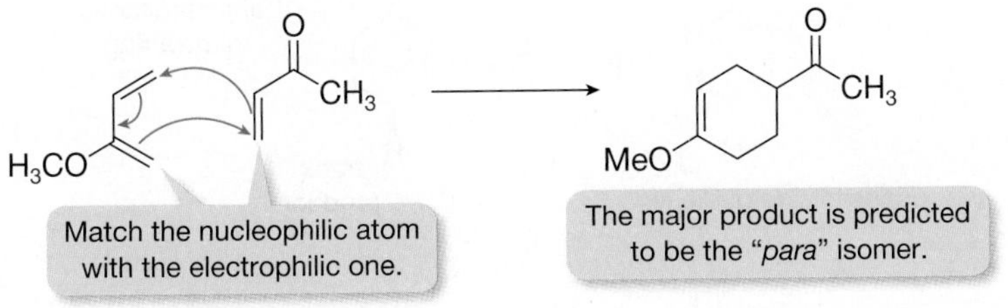

Neither substitution pattern will result in a "*meta*"-substituted product as the major product for rings with two substituents.

20.3.1.3 Stereochemistry

Three important factors govern the stereochemistry of the Diels–Alder reaction. These include the relative position of substituents on the diene and the dienophile (*cis*/*trans*), the orientation of the dienophile as it approaches the diene (*endo*/*exo*), and the direction of approach (top/bottom).

First, the stereochemical outcome can be very specific because the reaction product maintains the relative geometry of any substituents on the diene or on the dienophile. Because the Diels–Alder reaction is concerted, the two σ bonds form simultaneously, preventing any alteration of the substituent orientations along the reaction pathway.

STUDENT TIP

Unless otherwise noted, an approach from either the top or the bottom is equally likely, and both approaches result in a racemic mixture that contains both enantiomers. For clarity, the figures in this chapter show the dienophile approaching the diene only from below, producing one enantiomer. Do not forget that approach from the top to form the other enantiomer is also happening.

boat-like conformation

For example, the Diels–Alder product retains the relative configurations of the four groups on the dienophile (g, h, i, j) even though these two carbons become sp^3-hybridized. Consequently, a dienophile with two *cis* substituents gives a cyclohexene product in which these two substituents remain *cis* to each other.

substituents on the double bond are *cis* to each other

same substituents on the sp^3 carbons are *cis* to each other

For the diene, consider only the substituents at the two terminal carbons. With the diene in its *s-cis* conformation, the relative positions of those four substituents (*cis*/*trans*) on these carbons map directly onto the corresponding sp^3-hybridized carbons in the ring.

same side of diene

same faces of new ring

cis product

opposite side of diene

opposite faces of new ring

trans product

A substituent is ***endo*** if it points toward the alkene of the cyclohexene product of a Diels–Alder reaction.

A substituent is ***exo*** if it points away from the alkene of the cyclohexene product of a Diels–Alder reaction.

Second, the geometry of the approaching dienophile relative to that of the diene influences the stereochemistry of the Diels–Alder reaction. A non-symmetric dienophile can approach the diene in one of two ways, and these two approaches give rise to two possible products. In the ***endo*** product, the electron-withdrawing substituent on the dienophile is oriented toward the alkene of the cycloadduct, whereas in the ***exo*** product, this substituent is oriented away from the alkene. Bicyclic structures (those with two fused rings) can best illustrate the structures of these products.

dienophile approaches with the electron-withdrawing group underneath the diene

endo

electron-withdrawing group is oriented toward the alkene

dienophile approaches with the electron-withdrawing group away from the diene

exo

electron-withdrawing group is oriented away form the alkene

ORGANIC CHEMWARE
20.4 Diels–Alder reaction: Cyclopentadiene + maleic anhydride

Most Diels–Alder reactions favour the formation of the *endo* rather than the *exo* cycloadduct. This *endo* selectivity has been attributed to **secondary orbital overlap**: $\pi^\star$ orbitals, when they occur on the dienophile substituent, can overlap π orbitals of the diene (filled orbitals of one component overlap with empty orbitals of the other). This overlap happens only when the components approach each other in an *endo* fashion and the overlap lowers the energy of the transition state (a form of delocalization), leading to the *endo* product.

Secondary orbital overlap is a favourable interaction between a molecular orbital lobe of a dienophile substituent and an orbital lobe of the diene that stabilizes the *endo* transition state without forming a bond. One of the involved orbitals is filled and the other is empty.

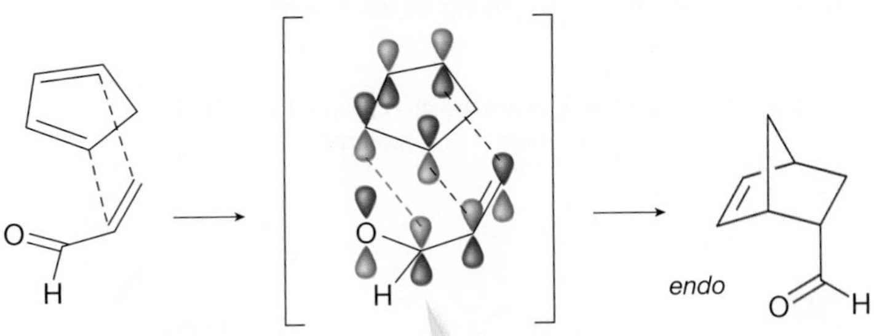
π* orbitals of the electron-withdrawing group lie under the π orbitals of the diene
O
H
O
H
endo
O
H
overlap between a π* lobe of EWG and lobe of the diene lowers the reaction activation energy

In recent years, the explanation for *endo* selectivity based on the concept of secondary orbital overlap has been questioned because it only partially accounts for the observed *endo* selectivity, and then only for dienophiles that have a substituent with $\pi^\star$ orbitals. Many alkyl-substituted dienophiles, in which there is no possibility of secondary orbital overlap, also show *endo* selectivity. This suggests that sterics or perhaps favourable bond angles in the transition state may be responsible for *endo* selectivity. Regardless of the exact reason for *endo* selectivity, the Diels–Alder reaction is selective for the *endo* product, and this selectivity appears to result from kinetic effects (effects in the transition state).

To draw the correct Diels–Alder product based on the *endo* geometry, carefully draw each component in a way that mimics their approach.

1. Draw the dienophile so that the substituent that will be *endo* (usually the electron-withdrawing group) points toward the diene component.

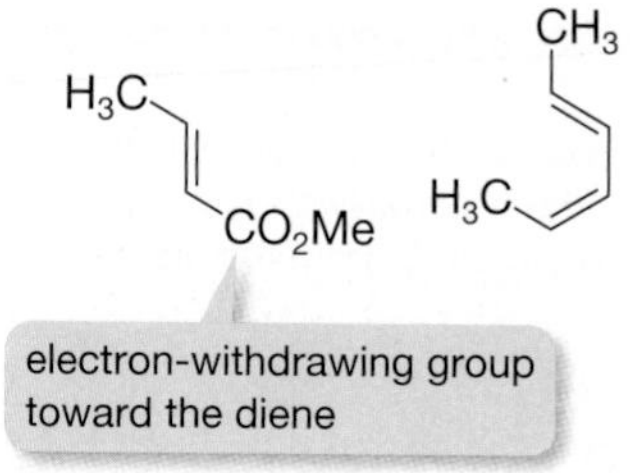

2. On the terminal positions of the diene and dienophile, add any implied hydrogens. At each location one substituent will lie to the right, and the other will lie toward the left.

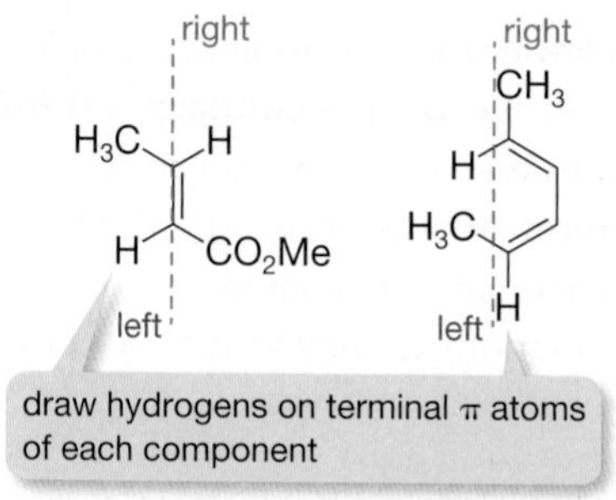

3. In the product, the substituents that were to one side at each location (all left or all right) will be on the same face of the molecule (all up or all down).

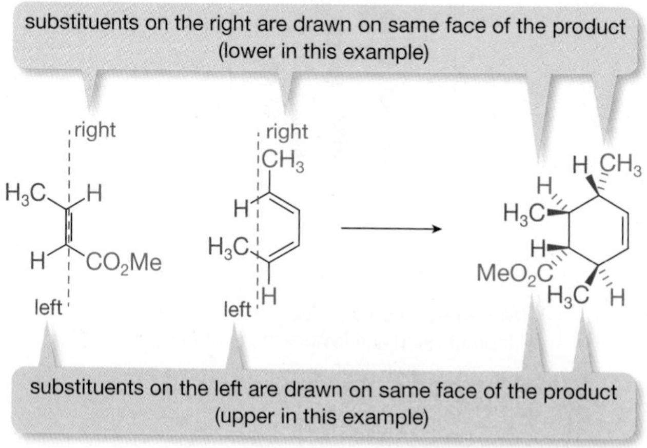

4. Redraw the final molecule as a line structure (optional) or in a three-dimensional representation, with the ring in a boat-like conformation.

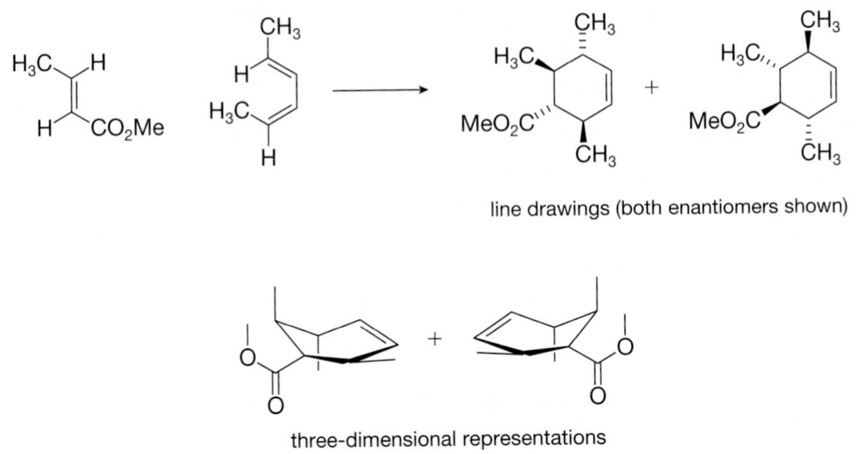

H_3C H H CO_2Me CH_3 H H_3C H → CH_3 H_3C MeO_2C CH_3 + CH_3 H_3C MeO_2C CH_3

line drawings (both enantiomers shown)

three-dimensional representations

Third, the direction from which the dienophile approaches the diene controls the formation of Diels–Alder enantiomers. All examples illustrated so far show the dienophile reacting from the bottom face of the diene. However, both the diene and the dienophile have simple, planar structures, so the dienophile can approach from either above or below the plane of the diene. Unless one or both of the components contains stereogenic centres near their respective π-system, both top and bottom approaches are equally likely and both enantiomers will form in equal proportions (a racemic mixture). In practice only one enantiomer is drawn with the understanding that a racemate is represented.

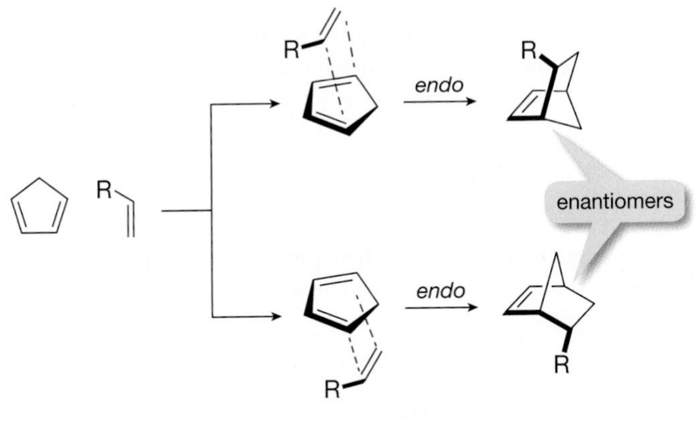

R *endo* R enantiomers *endo* R R

CHECKPOINT 20.2

You should now be able to predict the products of [4+2] cycloaddition between a diene and a dienophile, including the correct regio- and stereoselectivity.

SOLVED PROBLEM

What is the expected product of the following reaction?

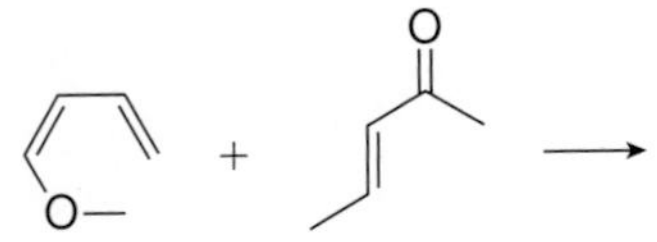

O + O →

STEP 1: Determine resonance structures of the diene and dienophile that show the relative partial charges.

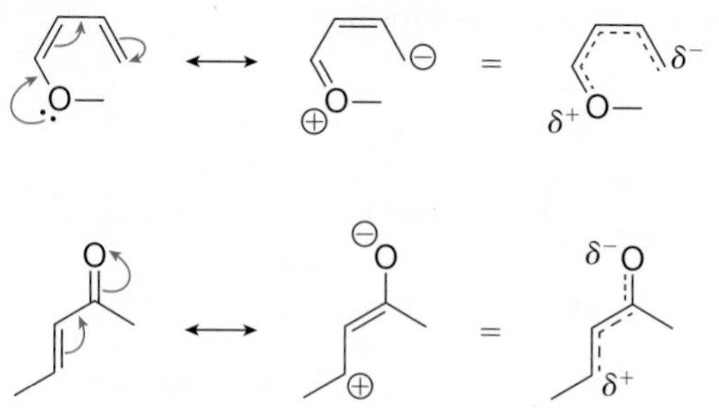

STEP 2: Position the diene and dienophile in the correct orientation for a Diels–Alder cycloaddition, such that the partial negative on the diene aligns with the partial positive on the dienophile.

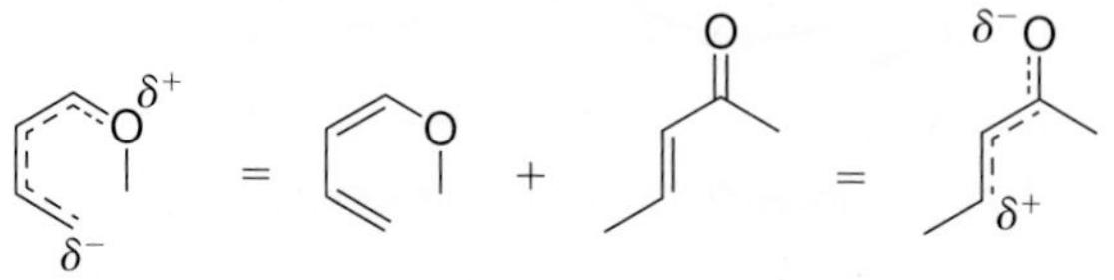

STEP 3: With the nucleophilic and electrophilic components identified, draw the reagents such that the nucleophilic atom of one aligns with the electrophilic atom of the other. This will illustrate the correct regiochemistry in the product.

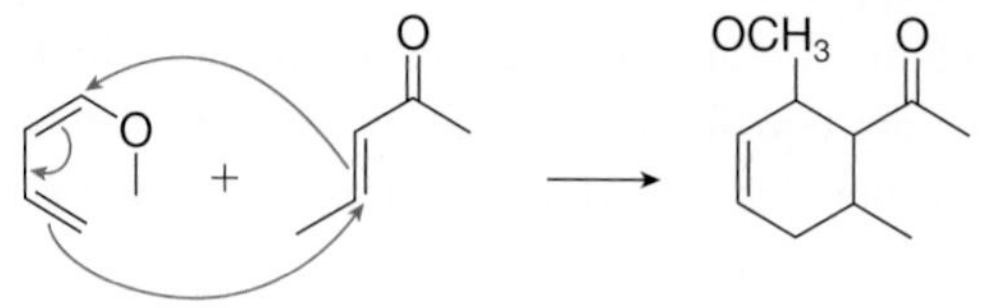

STEP 4: To determine the relative stereochemistry or the product, re-draw the dienophile so that the substituent that will be *endo* (usually the electron-withdrawing group) points toward the diene component.

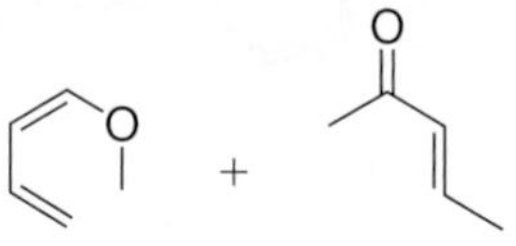

STEP 5: On the terminal positions of the diene and dienophile, add any implied hydrogens. At each location one substituent will lie to the right, and the other will lie toward the left.

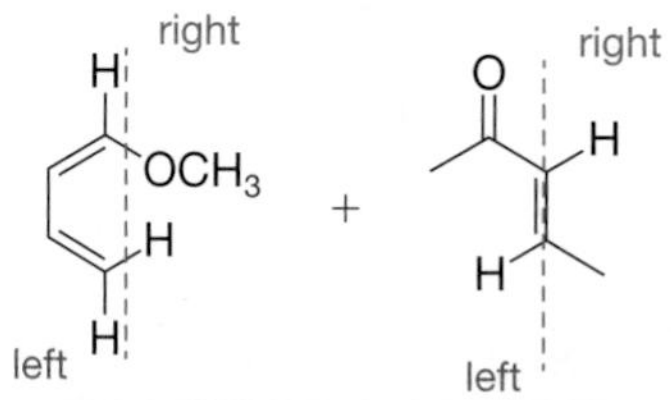

STEP 6: In the product, the substituents that were to one side at each location (all left or all right) will be on the same face of the molecule (all up or all down).

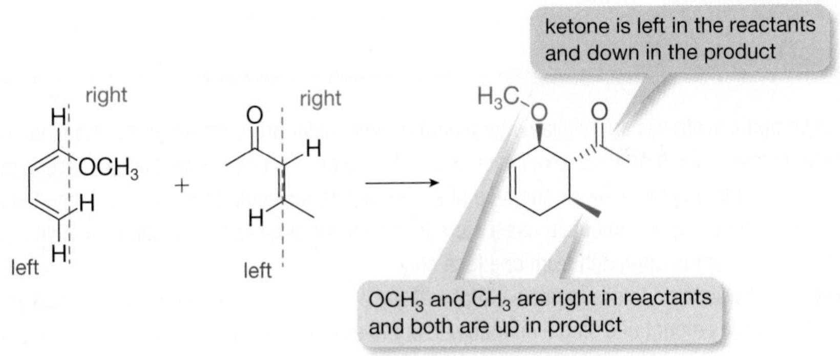

ketone is left in the reactants and down in the product

right H OCH$_3$ H H left + O right H H left → H$_3$C O O

OCH$_3$ and CH$_3$ are right in reactants and both are up in product

PRACTICE PROBLEM

20.3 Predict the product of a cycloaddition reaction for the following molecules. Justify the stereochemistry, using the appropriate molecular orbitals.

a) 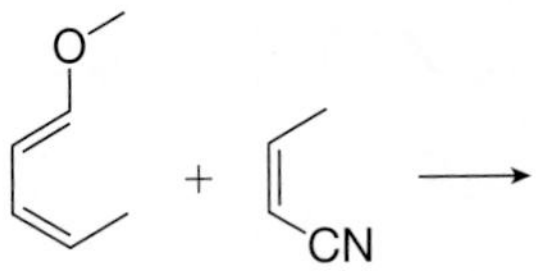 O + CN →

b) 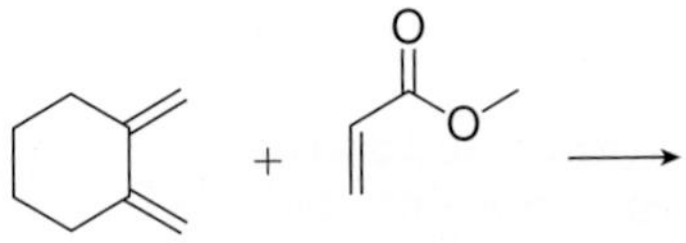 + O O →

c) 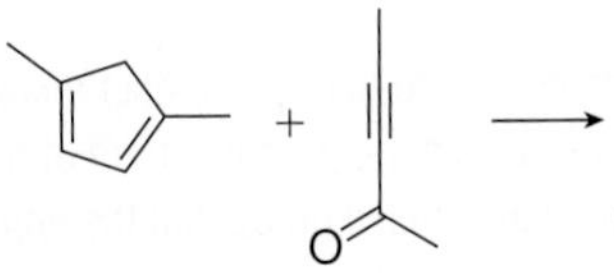 + O →

d) 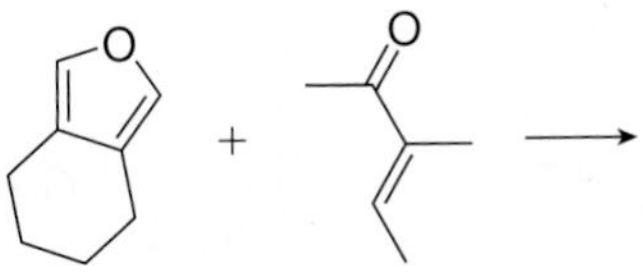 O + O →

e) 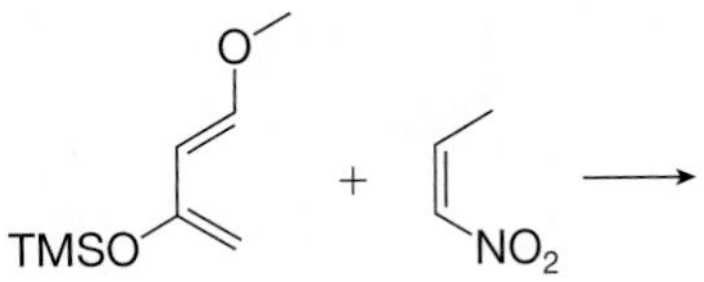 O TMSO + NO$_2$ →

f) 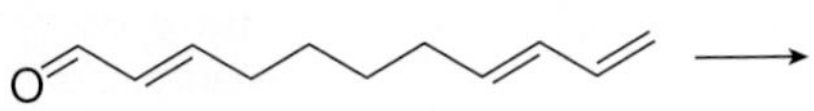 O →

INTEGRATE THE SKILL

20.4 The Diels–Alder reaction is a reliable way to form six-membered rings. With the proper substituents, cyclohexenes can be converted to aromatic rings. For the following reaction, show the intermediate formed from the Diels–Alder reaction and the reagents required to convert it to a benzene ring.

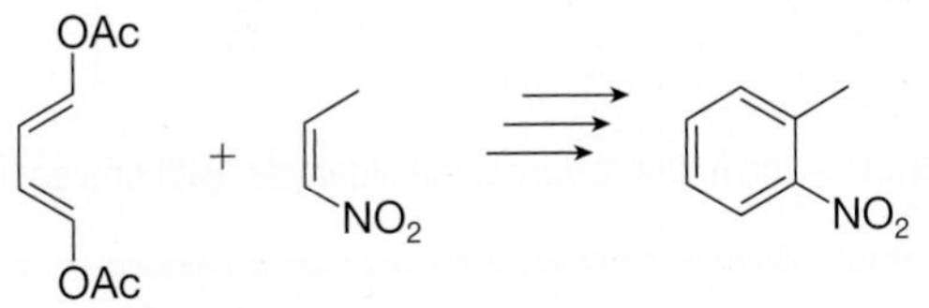

OAc OAc + NO$_2$ ⟶⟶⟶ NO$_2$

DID YOU KNOW?

Methods of asymmetric synthesis are available to generate Diels–Alder products as single enantiomers starting from non-chiral reactants. Each procedure relies on the participation of a single enantiomer reaction component. The function of this component is to block one face of each reacting molecule to favour the formation of just one enantiomer. If either the diene or dienophile has a stereogenic centre close to the reaction site, this centre can cause the other component to approach from one face only.

The most efficient way to do this is to use a chiral catalyst because only a small amount of catalyst is required to make a large amount of a single enantiomer. In the example shown here, a chiral organic amine forms an iminium ion intermediate with the aldehyde dienophile. The positively charged nitrogen of this complex is very electron withdrawing and lowers the LUMO energy of the dienophile, thus catalyzing the reaction.

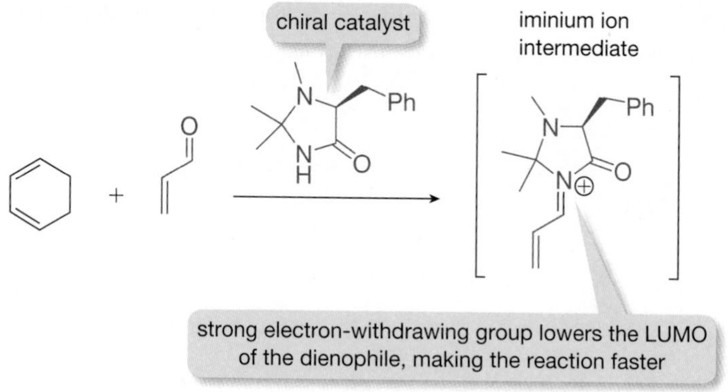

Because the catalyst has a stereogenic centre, the transition states leading to the two product enantiomers are different (they are diastereomers), have different activation energies, and so the products form at different rates. The approach of the diene from above the dienophile is blocked by the phenyl group, but the approach from below is unhindered.

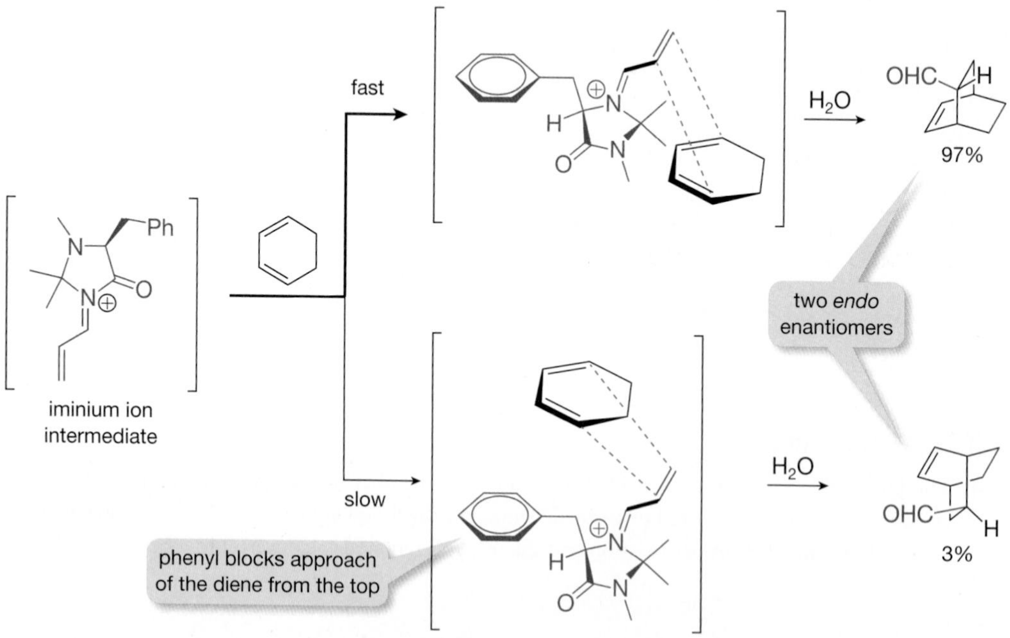

After the ring forms, hydrolysis of the imine re-forms the aldehyde, with one enantiomer strongly favoured.

20.3.2 Hetero-Diels–Alder reaction

In what is known as a **hetero-Diels–Alder reaction**, the heteroatoms in either the diene or the dienophile can be incorporated into the ring of the cycloadduct by a [4+2] reaction. In the first example, a nitrogen atom becomes integrated into a diene, and the intramolecular hetero-Diels–Alder reaction produces cycloadducts that have a nitrogen-containing ring.

A **hetero-Diels–Alder reaction** contains a non-carbon atom in either the diene or the dienophile and forms a heterocycle instead of a cyclohexene.

4:1
98%

In the second example, the ketone carbonyl of an α-ketoester serves as a dienophile for a hetero-Diels–Alder reaction. The ester group is electron withdrawing, which lowers the LUMO of the adjacent carbonyl dienophile. A Lewis acid further increases the reactivity of the dienophile. The two electron-donating groups activate the diene, thereby raising the HOMO. The initial cycloadduct is then treated with acid to hydrolyze the silyl enol ether and eliminate the OMe group.

electron-rich diene
electron-poor dienophile
hydrolyze TMS enol ether and eliminate OMe
Lewis acid Δ
H_3O^+
hetero-Diels–Alder adduct

20.3.3 Photochemical [2+2] cycloadditions

A cycloaddition that forms a four-membered ring is a [2+2] reaction: for example, the addition of two alkenes. Due to the symmetry of the reacting orbitals, such [2+2] cycloadditions using four π electrons require photochemical conditions (they are photochemically allowed), but cannot happen with heat (they are thermally forbidden).

$h\nu$ [2+2] cyclization

This reaction cannot occur thermally because the phases of the HOMO and the LUMO do not permit constructive overlap in a concerted way. Instead, light energy excites an electron from the HOMO of one alkene into the next highest energy orbital (the new HOMO), which has the required phase to combine with the LUMO of the other alkene.

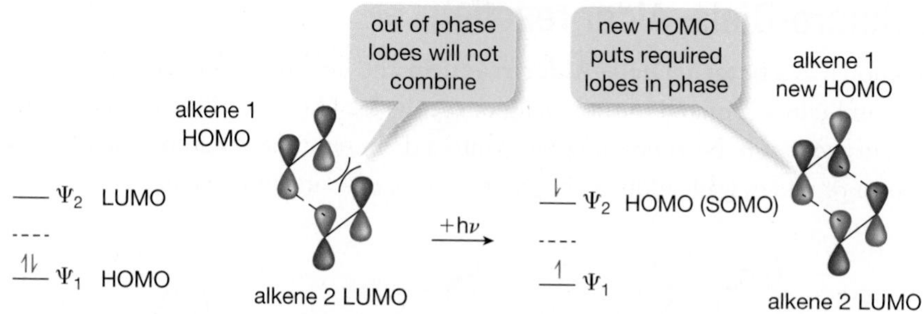

CHEMISTRY: EVERYTHING AND EVERYWHERE

Photochemistry and Cancer

As a result of exposure to sunlight, molecular scars called thymine dimers form on DNA. The reaction, initiated by UV radiation from the sun, involves the photoexcitation of thymine and then the [2+2] cycloaddition of adjacent bases to form a dimer.

This kind of chemical change alters the structure of the DNA and leads to potential permanent damage. In extreme cases, this leads to cancer; so be sure to use sunscreen to protect your skin!

20.3.4 Thermal [2+2] cycloadditions

Thermal [2+2] cycloadditions appear to violate the required orbital symmetry rules used in thermal [4+2] and photochemical [2+2] reactions. Thermal [2+2] reactions are forbidden, and only occur in special cases. Some of these processes are stepwise, instead of concerted. Examples such as the Wittig reaction and [2+2] reactions involving ketenes and isocyanates involve p or π orbitals that are orthogonal to each other. Either of these circumstances can allow a thermal [2+2] reaction to proceed provided the reagents approach each other in the correct way.

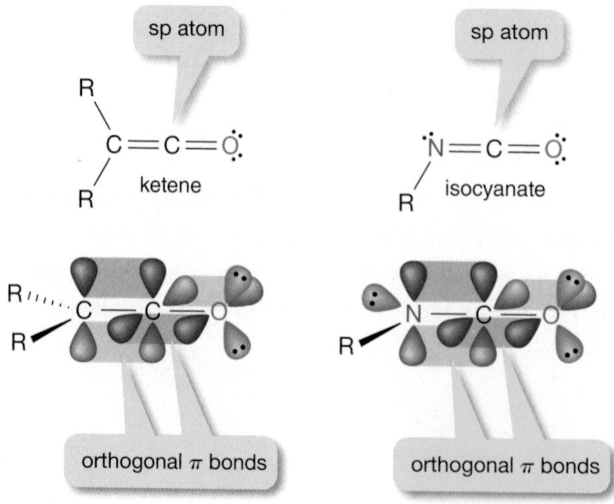

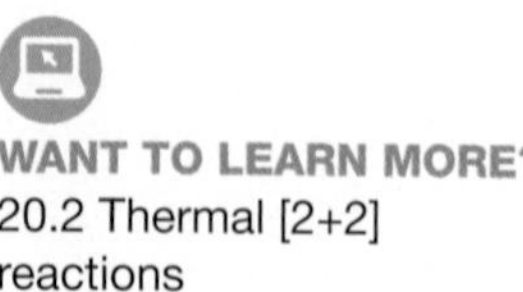

WANT TO LEARN MORE?
20.2 Thermal [2+2] reactions

CHECKPOINT 20.3

You should be able to recognize when a [2+2] cyclization is an allowed pericyclic reaction that can proceed through photochemical excitation. You should also be able to recognize when a [2+2] cyclization must occur thermally and involve orthogonal orbitals.

SOLVED PROBLEM

Can the following reaction proceed under thermal or photochemical conditions? What is the expected product of the following reaction?

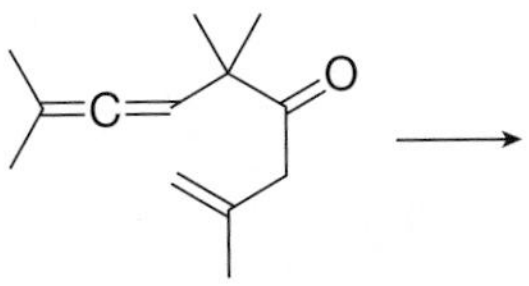

STEP 1: Determine whether there is an atom with orthogonal p orbitals, such as a sp-hybridized atom. Orthogonal π orbitals allow a thermal reaction. Otherwise, cyclization must happen photochemically.

This reactant can undergo thermal cyclization, because of the allene.

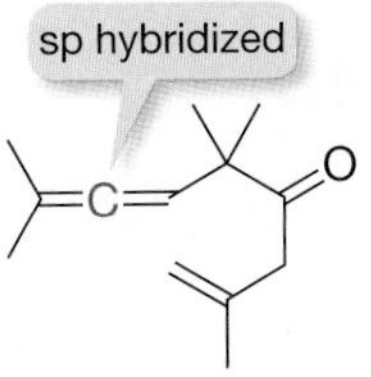

sp hybridized

STEP 2: Draw mechanism arrows for the possible [2+2] cyclizations. Consider each pathway to determine whether each is viable.

There are two possible cyclizations that could occur, depending on which side of the allene reacts.

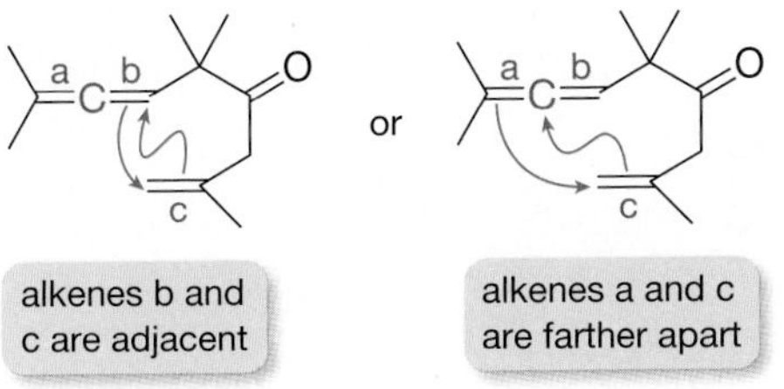

alkenes b and c are adjacent

or

alkenes a and c are farther apart

STEP 3: Draw the products formed from the possible mechanisms, and consider whether each one is a viable product.

The five-membered ring has less strain from the remaining alkene than the six-membered ring product.

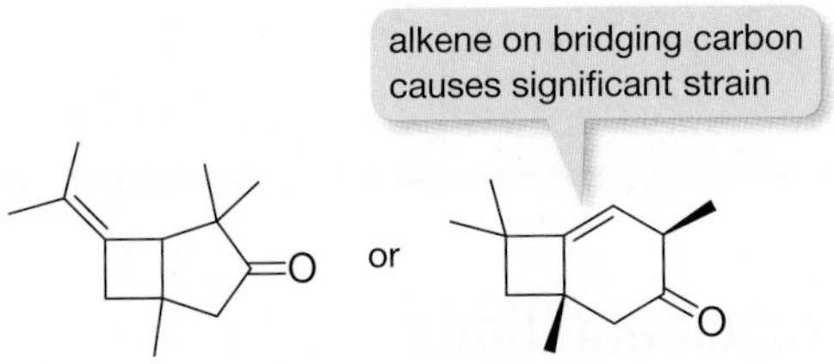

alkene on bridging carbon causes significant strain

The expected final product is the bicyclic [3.1.0] product.

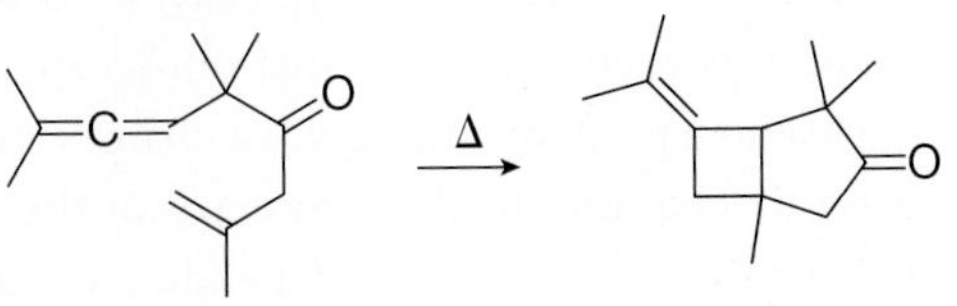

Δ

PRACTICE PROBLEM

20.5 a) Predict whether heat or light is required for each of the following reactions.

i) Cl, O, NEt_3, C=O

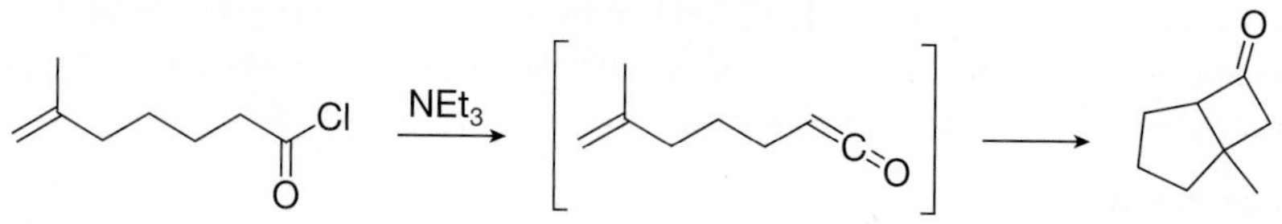

ii) O, O, O, O

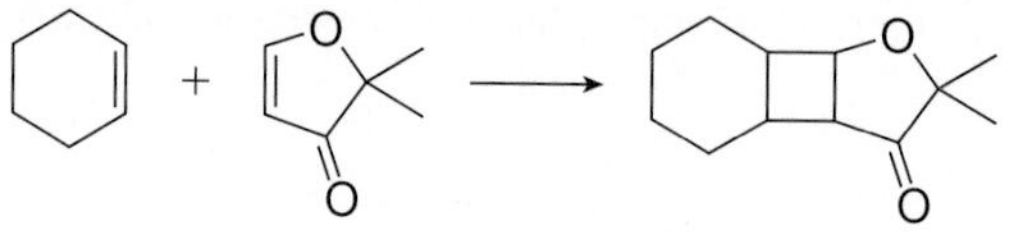

iii) O, O

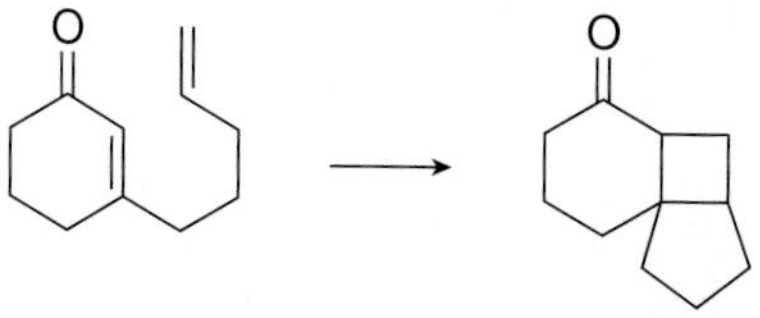

b) What is the expected product of each of the following reactions?

i) C, O, Δ

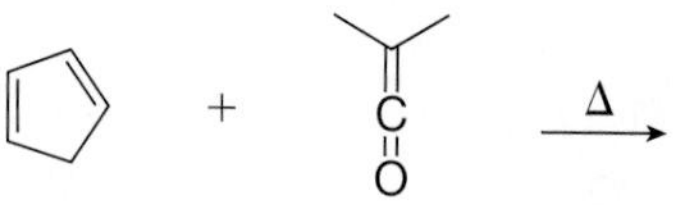

ii) Si, O, O, C, Cl, Cl, Δ

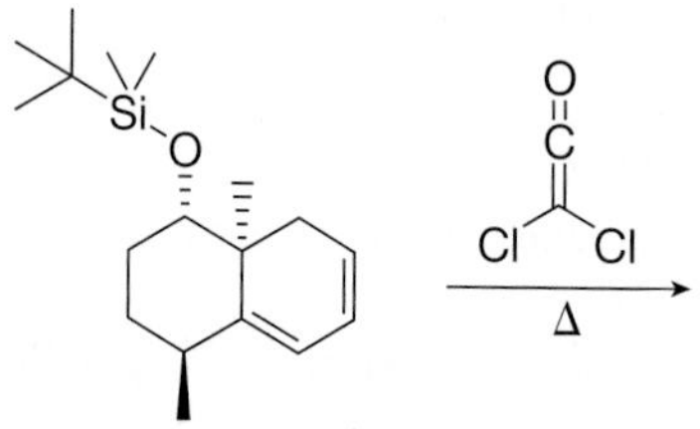

iii) O, hν

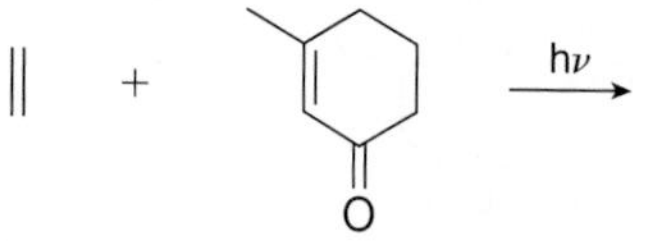

INTEGRATE THE SKILL

20.6 The following reactants give a different product when reacted under thermal or photochemical conditions. Predict the correct product of each reaction.

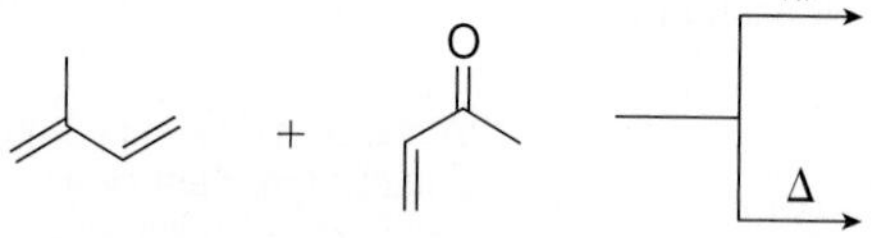

O, hν, Δ

A **1,3-dipole** is a functional group with a three-atom sequence that has a nucleophilic site and an electrophilic site at positions 1 and 3, respectively. Dipoles can undergo [3+2] cycloadditions with alkenes or alkynes (i.e., dipolarophiles).

A **dipolarophile** is a compound that can react with a 1,3-dipole in a [3+2] cycloaddition reaction. It contains either an alkene or alkyne functional group.

A **zwitterion** is a neutral molecule that contains balancing, opposite charges.

20.3.5 1,3-Dipolar cycloadditions

In a [3+2] cycloaddition, a **1,3-dipole** reacts with a **dipolarophile**. A 1,3-dipole is a three-atom functional group that can react with an alkene. Typically, a 1,3-dipole carries a negatively charged atom at one end and a positively charged atom that can accept electrons from an adjacent π bond. Molecules that carry balanced positive and negative charges are called **zwitterions**. The [3+2] reaction can be understood according to the symmetry of the frontier orbitals involved. In these reactions, the HOMO or the LUMO varies, depending on the dipole. The 1,3-dipole normally has three π orbitals, whereas the dipolarophile has two.

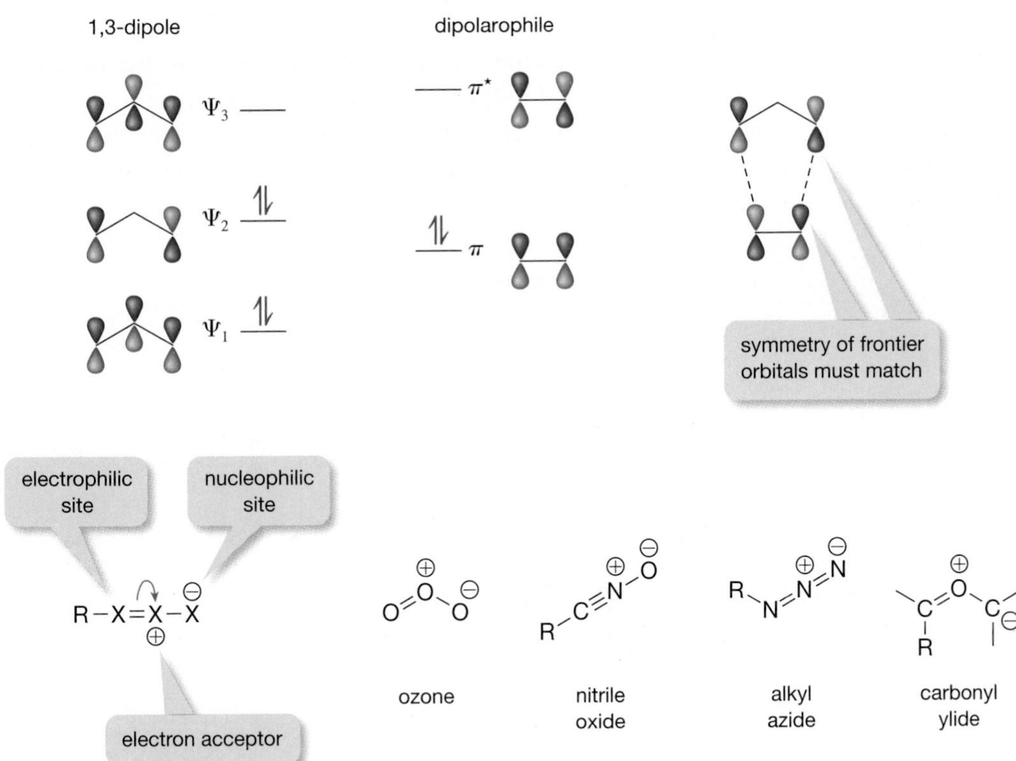

Various 1,3-dipoles can be used to form five-membered rings from alkenes or alkynes, although certain combinations are more common than others. For example, the reaction of an alkyl azide (R-N_3) with an alkyne is known as the Huisgen cycloaddition (Section 20.3.5.1).

If an alkyne is used, the resulting ring will contain an extra carbon-carbon double bond. Many types of products are possible from these reactions, which gives a wide variety of heterocycles. Regioselectivity in these reactions can be predicted using orbital coefficients obtained from quantum calculations (as in the Diels–Alder reaction).

20.3.5.1 Huisgen cycloaddition

The [3+2] cycloaddition of azides with alkynes produces triazoles in a reaction known as the Huisgen cycloaddition. This reaction is high yielding, works with a wide range of substrates, and can be highly regio- and stereospecific. For these reasons, the term *click reaction* was coined, referring to how molecular units can be readily joined together like LEGO™ blocks.

[3+2]
Δ

The reaction works well by simple heating. It works faster in the presence of a copper(I) catalyst in a polar aprotic solvent, even at room temperature; however, the reaction changes to a stepwise mechanism, and is no longer a true pericyclic reaction.

20.3.6 Other cycloadditions

There are several reactions of alkenes that involve mechanistic steps that are cycloadditions. Ozonolysis (Section 20.3.6.1) results in the cleavage of an alkene to form a carbonyl. Osmium tetroxide and potassium permanganate (Sections 20.3.6.2 and 20.3.6.3) both convert alkenes into 1,2-diols, with *syn*-stereochemistry.

20.3.6.1 Ozonolysis

A common cycloaddition is the initial reaction of ozone with an alkene in the ozonolysis reaction. The first step of this process is a [3+2] cycloaddition between ozone (a 1,3-dipole) and an alkene. The second step is a retro [3+2] reaction (ring opening) in which the electron flow starts at a different atom. One of the open components then flips over, and a second [3+2] reaction forms an ozonide.

Ozonolysis reaction: Initial formation of an ozonide

molozonide

this flips over

nucleophilic alkene adds to O_3

ring opens by a slightly different reverse electron flow

ozonide

dipole adds to carbonyl

Ozonides are explosive and usually not isolated. Instead, they are immediately reacted with a reducing agent such as Me_2S, PPh_3, or Zn in acetic acid to give the corresponding dicarbonyl product. This final reduction reaction is a type of retro [3+2] reaction.

Me_2S or PPh_3 or Zn/HOAc

ozonide

Me_2SO or $OPPh_3$ or ZnO

When hydrogen peroxide is used instead of a reducing agent for the second step, carboxylic acids are formed instead of aldehydes.

20.3.6.2 Osmium tetroxide dihydroxylation

Alkenes react with various reagents to make diols by the apparent addition of hydrogen peroxide (HOOH) across the double bond; this is a dihydroxylation reaction. These additions follow the general concept of the alkene acting as a nucleophile while accepting a pair of electrons from the electrophile to form a ring. Osmium tetroxide (OsO_4) oxidizes double bonds via cycloaddition to form osmate esters, which then hydrolyze to form 1,2-diols (also known as vicinal diols). The two new OH bonds are always *cis* to each other with respect to the starting alkene because the initial reaction is a concerted cycloaddition.

cycloaddition
osmate ester
H_2O
hydrolysis
cis 1,2-diol

OsO_4 is a highly toxic and expensive reagent, so the osmate esters are rarely isolated. Instead, the dihydroxylation is normally carried out with a catalytic amount of OsO_4. During the reaction the osmium(VI) by-product is oxidized to OsO_4(VIII) using a stoichiometric amount of inexpensive co-oxidant. During this re-oxidation, the osmium ester is hydrolyzed by water present in the mixture to form the 1,2-diol. A typical co-oxidant is NMO (*N*-methylmorpholine-*N*-oxide) as it can selectively re-oxidize the osmium species back to OsO_4—and not oxidize the alkene substrate—during the reaction.

STUDENT TIP
Many mechanisms involving metals are easier to understand if you draw the electron pairs on the metal.

co-oxidant
Os(VIII)
NMO
Os(VI)
osmate ester
Os(VIII)
H_2O
hydrolysis
H_2O
OsO_4 regenerated

20.3.6.3 Permanganate dihydroxylation

Potassium permanganate ($KMnO_4$) reacts with double bonds under basic conditions to form diols directly. The initial addition is similar to that of OsO_4, and the intermediate is never isolated. Since the $KMnO_4$ does not act as a catalyst, equimolar quantities of alkene and $KMnO_4$ must be used.

KOH
H_2O

CHECKPOINT 20.4

You should now be able to predict the products of [3+2] cycloaddition between a dipole and a dipolarophile.

SOLVED PROBLEM

What is the expected product of the following reaction?

STEP 1: Identify the nucleophilic and electrophilic ends of the 1,3-dipole. Typically, the nucleophilic atom has a negative charge; the electrophilic atom has a partial positive charge due to conjugation.

nucleophilic atom

electrophilic atom

STEP 2: Identify the nucleophilic and electrophilic atoms of the dipolarophile. The nucleophilic π bond attacks the dipole, resulting in the partial formation of a cation on the more substituted carbon (electrophilic site).

transition state

partial cation character on the more substituted carbon

STEP 3: Align the dipole and dipolarophile such that the nucleophilic sites on one component align with the electrophilic sites on the other component. Draw the cycloaddition mechanism arrows to form the product.

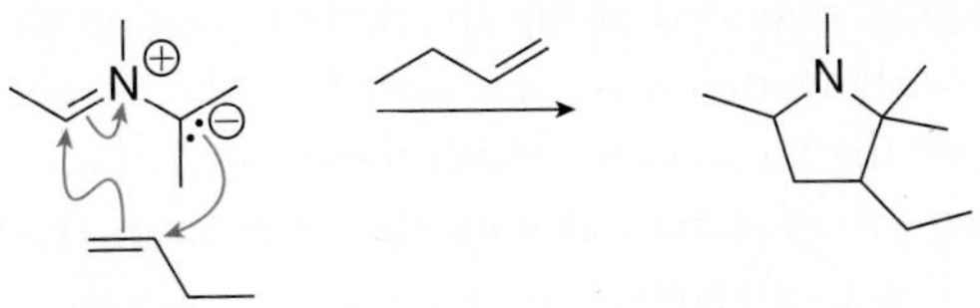

PRACTICE PROBLEM

20.7 Predict the product of electrocyclic ring closure under photochemical conditions for the following molecules. Justify the stereochemistry using the appropriate molecular orbitals.

a)

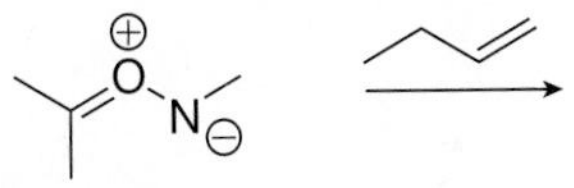

b) $\ominus$—N≡N $\oplus$

c)  —C≡N–O ($\oplus$ $\ominus$)

d)

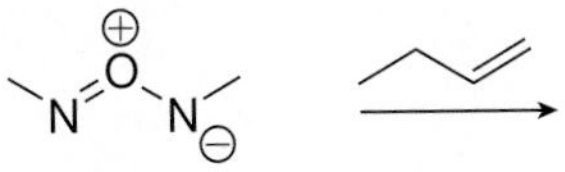

e)

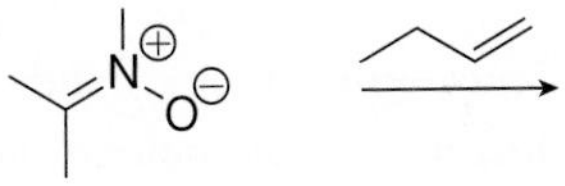

f) 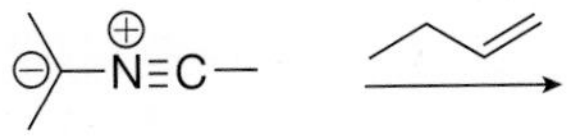 $\ominus$—N≡C— ($\oplus$)

INTEGRATE THE SKILL

20.8 Ozonolysis typically results in the cleavage of an alkene and the formation of two aldehydes, such as the ozonolysis of cyclohexene.

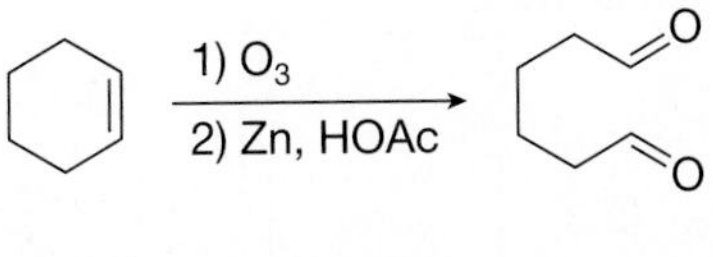

1) O_3
2) Zn, HOAc

During chemical synthesis, sometimes it is useful to differentiate the two oxidized carbons. For the following reaction, propose a mechanism that results in the formation of one aldehyde and one aldehyde protected as an acetal.

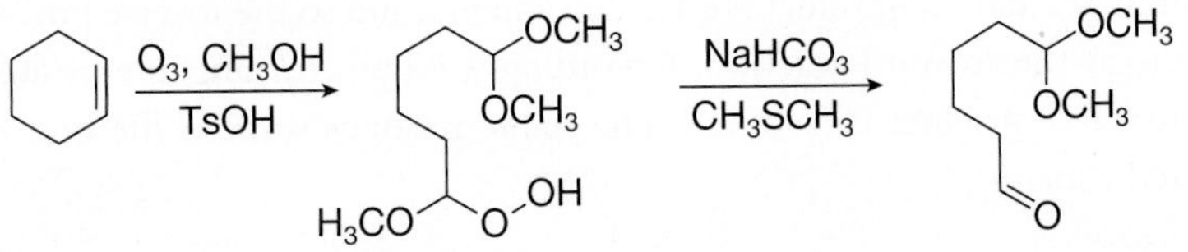

O_3, CH_3OH / TsOH; OCH_3, OCH_3, OH, H_3CO, O; $NaHCO_3$ / CH_3SCH_3; OCH_3, OCH_3, O

20.4 Sigmatropic Rearrangements

A **sigmatropic rearrangement** is one in which one σ bond is replaced by another, keeping the total number of σ bonds the same.

The second class of pericyclic reactions is **sigmatropic rearrangement**. These reactions involve the movement of a σ- (sigma-) bonded group to another position along a π-system. The classification of sigmatropic rearrangements is represented by the numbers given to the atoms in the migrating group and by the location to which these atoms migrate. For example, in a [1,3] rearrangement, a migrating group reattaches with the same atom [1,*x*] to a new position three atoms away [*x*,3]. In a [3,3] rearrangement, the migrating group, shown here in red, re-forms a σ bond at the third atom from the original σ bond. The migrating group migrates to the third atom from the original σ bond, making this a [3,3] rearrangement.

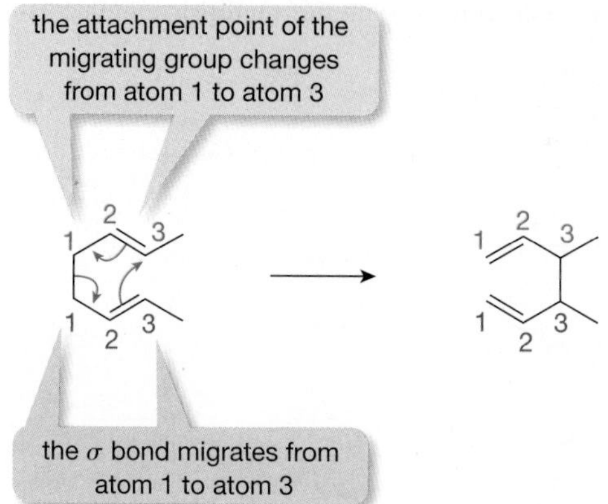

The following sections describe [3,3] rearrangements—the Cope and Claisen rearrangements—and [1,*n*] rearrangements, specifically, hydrogen rearrangements and alkyl rearrangements.

20.4.1 [3,3] rearrangements

20.4.1.1 Cope rearrangement

The **Cope rearrangement** is the [3,3]-sigmatropic rearrangement of a 1,5-hexadiene.

The **Cope rearrangement** is a [3,3] sigmatropic reaction of the six carbons of a 1,5-hexadiene. The reaction involves the circular flow of six valence electrons (two π bonds and one σ bond) that can be represented by three curved arrows.

Both the substrate and the product are 1,5-hexadienes, and so the reverse process can happen at the same time as the forward reaction. Accordingly, Cope rearrangements are equilibria in which the more stable product is favoured. The stable product is often the one with the most highly substituted alkenes.

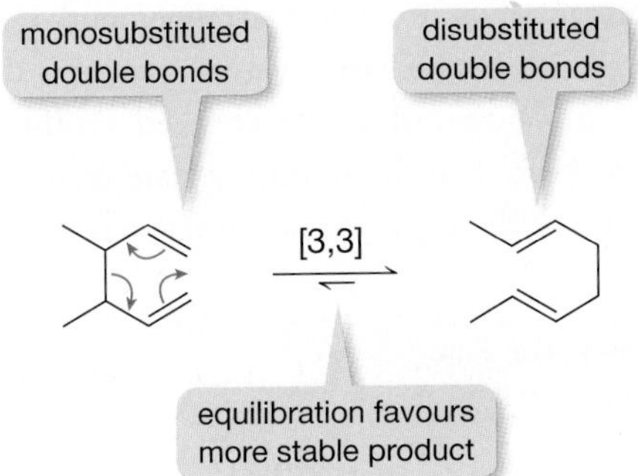

The nature of the substituents on the double bonds, as well as steric or ring strain, contribute to the stability of the reactant or product and also control which product the reaction favours. For example, alkenes with substituents carrying non-bonded electrons that participate in conjugation with the π bond (such as enol ethers) can influence the product distribution.

Non-bonded electrons on the oxygen are conjugated with the adjacent π bond. This material is stabilized and this reaction favours the products.

favoured product

Like other concerted pericyclic reactions, Cope rearrangements are stereospecific. Therefore, the configuration of any substituents on the sp^3-hybridized carbons controls the stereochemistry of the products.

The σ bond breaks, with both sp^3-hybridized atoms rotating in the same direction: either clockwise or counter-clockwise. As the atoms rehybridize, the orbital phases align with the matching adjacent p orbitals to form π bonds. Likewise, the terminal sp^2-hybridized carbons rotate in the same direction as they rehybridize to form the new σ bond.

σ bond breaks as atoms rotate in the same direction

orbitals aligned

σ bond forms as atoms rotate in the same direction

The Cope rearrangement involves a cyclic rearrangement of six bonded carbons, so the reaction favours a chair-like transition state. For any given reaction, the atoms can rotate in either direction, and two chair structures are possible. The steric effects usually favour the one in which the most groups are equatorial. Note that the transition state involves double bonds, so substituents are not exactly equatorial, and the term *pseudo-equatorial* is used.

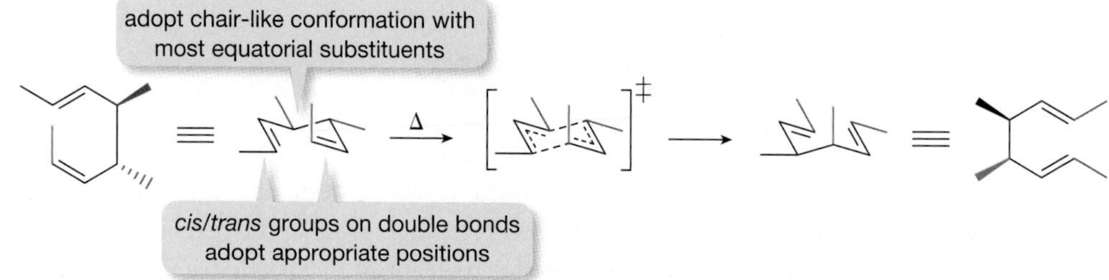

ORGANIC CHEMWARE
20.5 Cope rearrangement

20.4.1.2 Claisen rearrangement

A **Claisen rearrangement** is the [3,3]-sigmatropic rearrangement of an allylic vinyl ether.

The **Claisen rearrangement** is a close analog of the Cope rearrangement. It is the [3,3]-sigmatropic rearrangement of an allylic vinyl ether. During the reaction, a carbonyl group forms, which tends to inhibit the reverse reaction from occurring.

O ‡ O O

In the Claisen rearrangement, the product stereochemistry follows the same pattern as occurs in the Cope rearrangement; it is based on a favoured chair-like transition state.

O H_3C CH_3 ≡ O H_3C CH_3 → O H_3C CH_3 ‡ chair-like → O H_3C CH_3 ≡ O H_3C CH_3 95%

The original discovery of the Claisen rearrangement came from heating aryl allyl ethers, which produce 2-allyl substituted phenols. To form these compounds, the initial rearrangement product is a cyclohexadiene in which the aromaticity has been disrupted. Then, a very fast keto-enol tautomerization takes place, which re-aromatizes the ring to produce a phenol as the final product.

O [3,3] O H keto-enol tautomerization OH aromaticity restored

CHECKPOINT 20.5

ORGANIC CHEMWARE
20.6 Claisen rearrangement

You should now be able to recognize compounds that are able to undergo [3,3] sigmatropic rearrangements. You should also be able to predict the product of a [3,3] sigmatropic rearrangement.

SOLVED PROBLEM

Predict the expected product of the following rearrangement and indicate whether the product or starting material would be favoured at equilibrium.

$\xrightarrow{\Delta}$

N
H

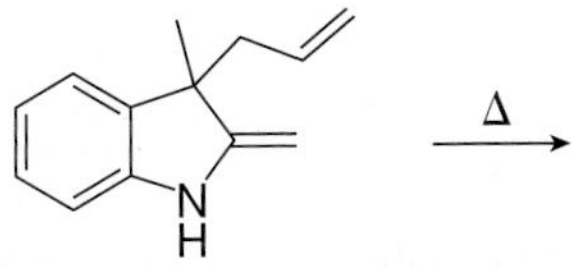

STEP 1: Identify any hexa-1,5-diene systems in the substrate—that is, two alkenes with two atoms or three σ bonds separating them.

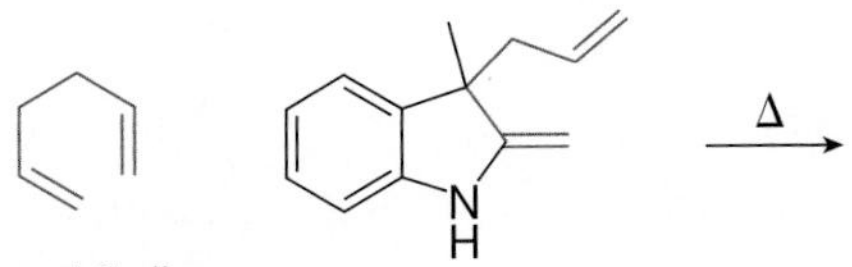

hexa-1,5-diene

STEP 2 (OPTIONAL): Redraw the compound in a conformation that shows more clearly the potential for a six-membered-ring transition state.

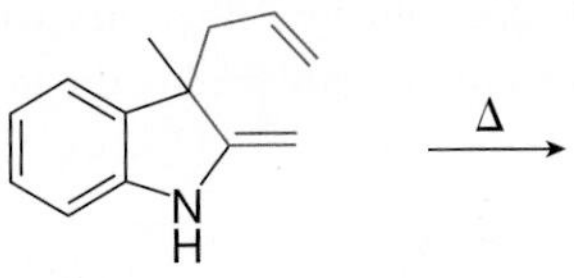

STEP 3: Draw the mechanism arrows for the rearrangement, moving the electrons through a cyclic transition state. Redraw the compound with the rearranged bonds.

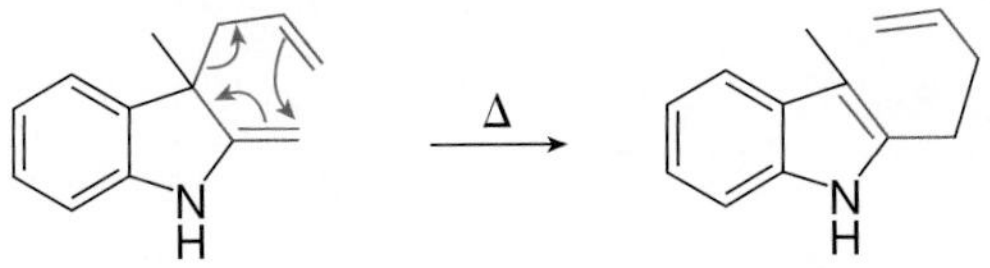

STEP 4 (IF NECESSARY): Redraw the product with appropriate bond angles.

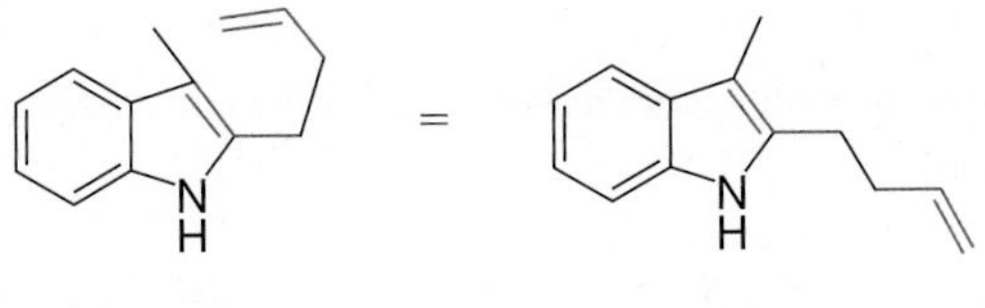

STEP 5: Compare the starting material with the rearranged product to determine which is more thermodynamically stable.

Both the starting material and the rearranged product have a monosubstituted alkene. The rearranged product has an aromatic ring that replaces a three-atom conjugated system in the starting material. In this case, the rearranged product is favoured at equilibrium.

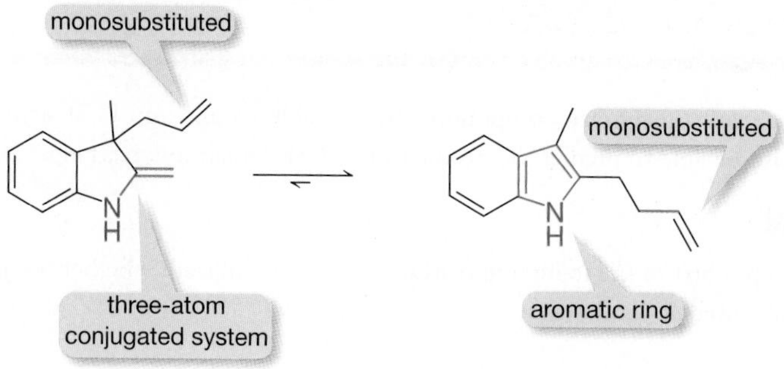

monosubstituted

monosubstituted

three-atom conjugated system

aromatic ring

PRACTICE PROBLEM

20.9 a) Identify which compound is more stable in each of the following pairs of compounds.

i)

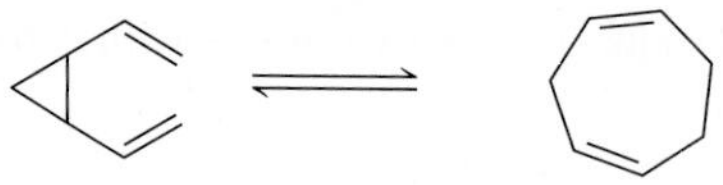

ii)

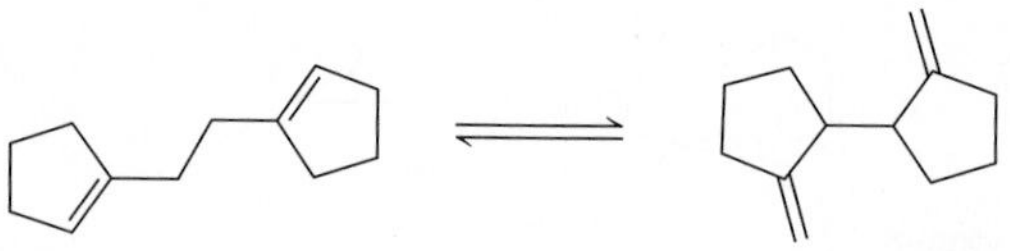

iii)

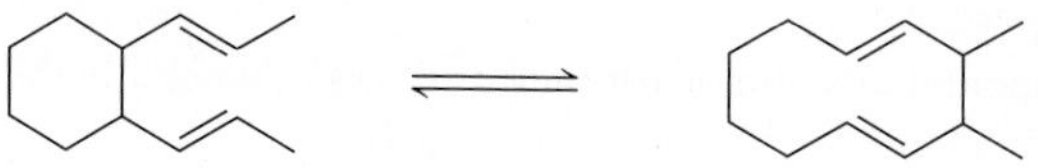

b) For the following compounds, determine whether they will undergo a [3,3] sigmatropic rearrangement. For appropriate compounds, predict the rearranged product.

i) 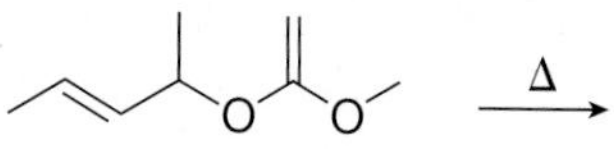 $\xrightarrow{\Delta}$

ii) 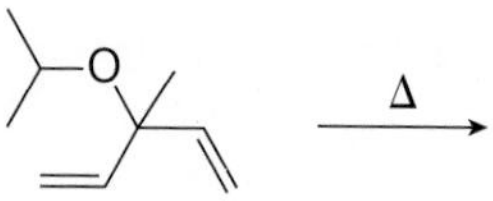 $\xrightarrow{\Delta}$

iii) 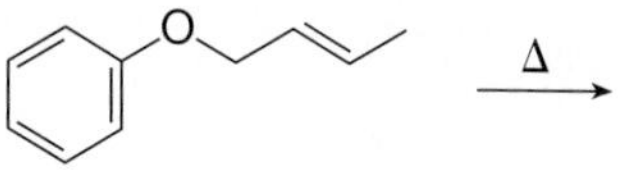 $\xrightarrow{\Delta}$

iv) 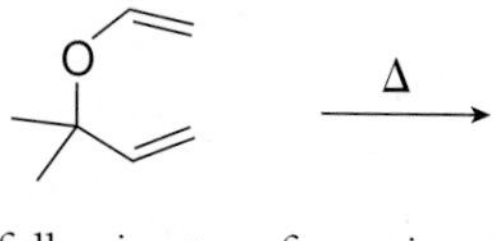 $\xrightarrow{\Delta}$

c) The following transformation occurs in two steps. Show a mechanism for each step and the required intermediate.

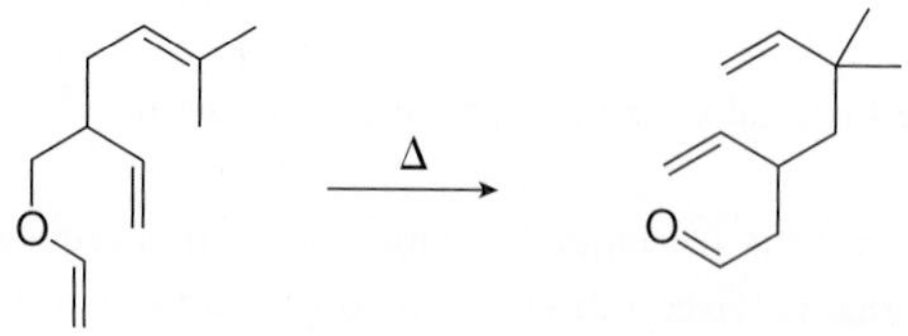

$\xrightarrow{\Delta}$

INTEGRATE THE SKILL

20.10 a) The Carroll rearrangement is a sigmatropic rearrangement that requires basic conditions to proceed and results in a release of one equivalent of carbon dioxide. Show the intermediate that forms when the following reagent is treated with base, and predict the product formed.

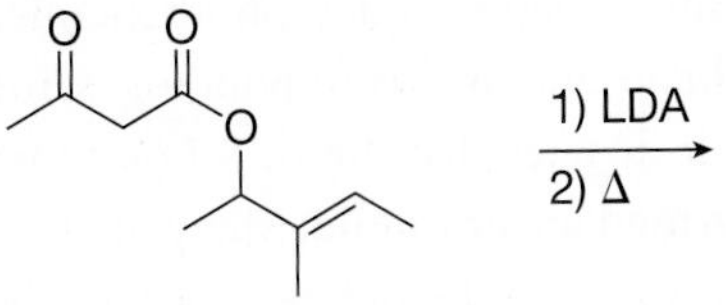

1) LDA
2) Δ

b) The Nazarov cyclization is a related rearrangement of dienones that proceeds under acidic conditions to form a cyclopentenone. It begins with the protonation of the dienone carbonyl. Propose a reasonable mechanism for the rearrangement that follows. More than one step will be required.

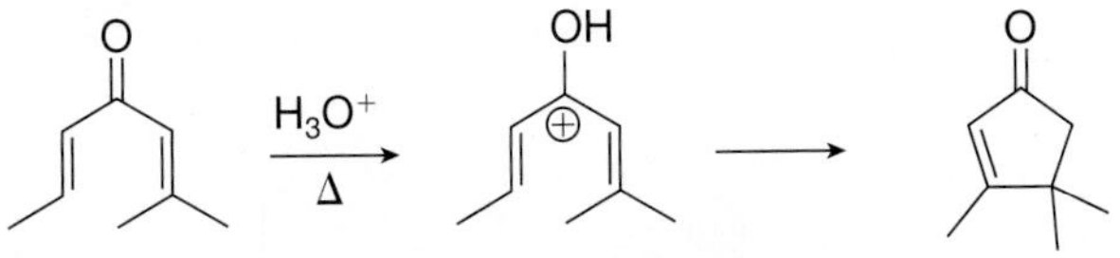

H_3O^+
Δ

20.4.2 [1,*n*] rearrangements

20.4.2.1 Hydrogen rearrangements

Under certain conditions, conjugated alkene systems rearrange by "shifting" a hydrogen from one end of the π-system to the other in a concerted manner. This pericyclic reaction has no intermediates; however, using a diradical model, this reaction is often *imagined* as a migration of a hydrogen radical. Interpreting the stereochemistry of these sigmatropic rearrangements involves considering the process as a homolytic cleavage of the C–H bond and analyzing the orbital symmetry of the resulting conjugated radical. Note that these are *not* radical reactions, but the diradical model helps explain the reactivity and stereoselectivity of sigmatropic rearrangements.

H radical

resonance structure of carbon radical

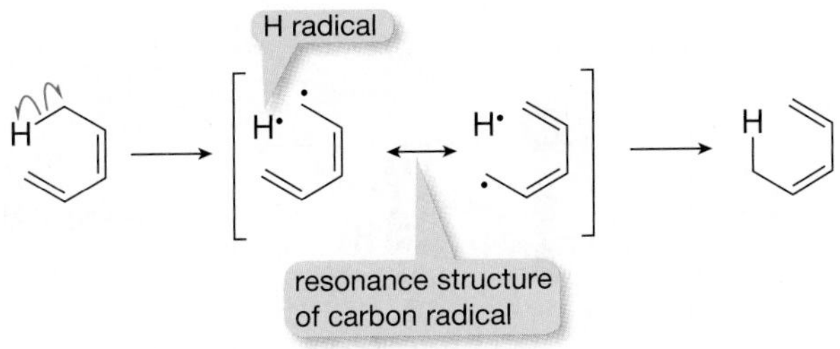

Under thermal conditions, 1,3-dienes can undergo [1,5] hydrogen shifts. The HOMO of the pentadienyl radical is symmetric (Ψ_3), so a hydrogen must migrate along the same face of the π-system as the original C–H bond. This type of migration is called a **suprafacial** rearrangement.

A rearrangement is **suprafacial** when the migrating group stays on the same face of the molecule as it moves.

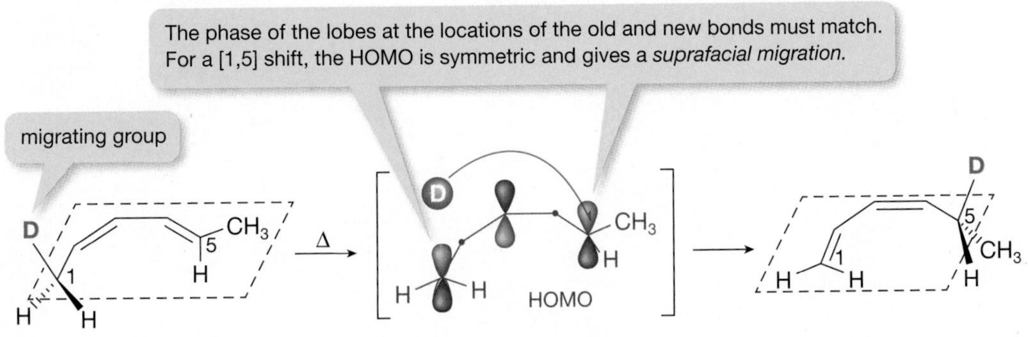

The phase of the lobes at the locations of the old and new bonds must match. For a [1,5] shift, the HOMO is symmetric and gives a *suprafacial migration.*

migrating group

HOMO

A rearrangement is **antarafacial** when the migrating group moves to the opposite face of a molecule.

When the [1,5] hydride shift occurs under photochemical conditions, photoexcitation changes the HOMO of the diene and its corresponding symmetry, so the resulting radical HOMO becomes Ψ_4. This result dictates that the new bond forms on the opposite face of the π-system. Such a migration is termed **antarafacial**. Many open-chain dienes are flexible enough to allow the migrating hydrogen to reach the opposite face of the molecule in a [1,5] hydrogen shift. Antarafacial rearrangements are not possible in small ring structures.

Photoexcitation changes the HOMO of the pentadienyl radical to Ψ_4, which is antisymmetric. To match phases, the migrating group must form a new bond on the opposite side of the π system, an *antarafacial migration.*

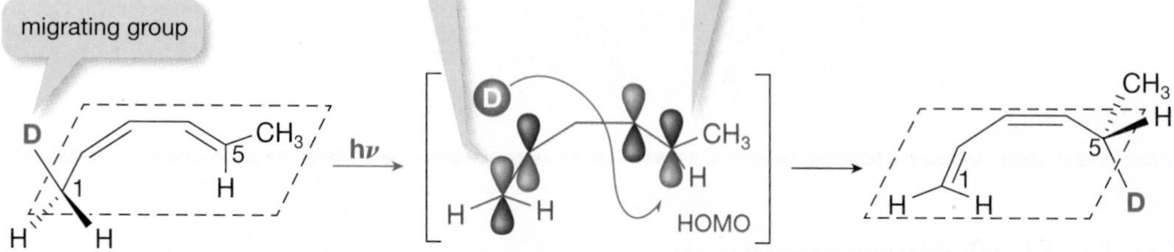

Under thermal conditions, allylic hydrogens cannot undergo a [1,3] hydrogen shift. The HOMO of an allyl radical is the Ψ_2 orbital, which is antisymmetric and allows only antarafacial migration. Because the chain is short, it is rigid and cannot twist to allow the hydrogen to form a bond on one face at the same time as it breaks a bond on the other. Consequently, thermal [1,3] hydrogen shifts do not occur and are said to be "symmetry forbidden."

In its ground state, the HOMO of an allyl radical (Ψ_2) is antisymmetric, allowing only for antarafacial migration to occur. However, the conjugated system is too short and rigid for the rearrangement to take place.

migrating group

Δ

HOMO

Although [1,3] hydrogen shifts cannot take place under thermal conditions, they do occur under conditions of photochemical excitation. In this case, the HOMO of the allyl radical is Ψ_3, which is a symmetric orbital that allows suprafacial migration to occur.

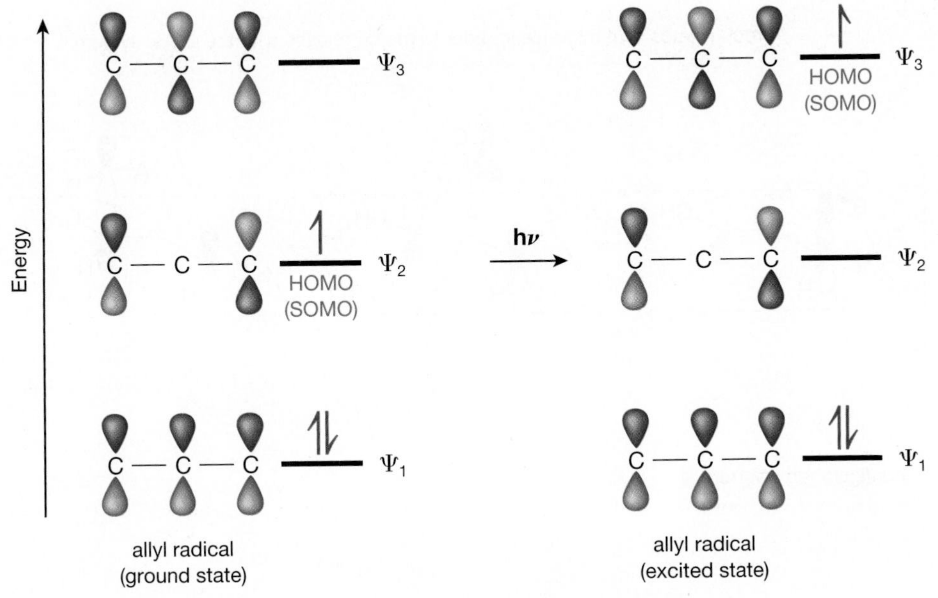

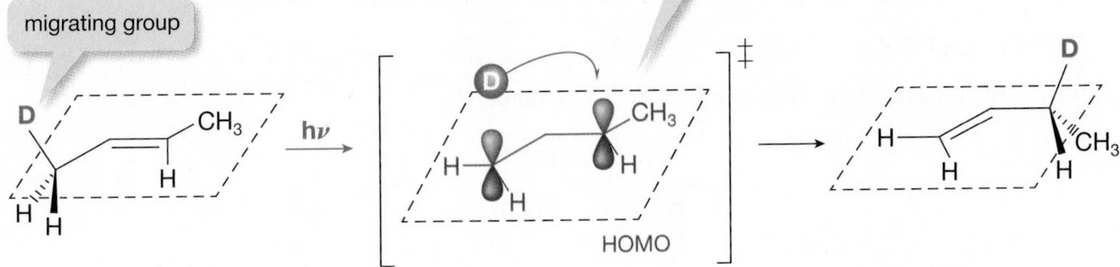

20.4.2.2 Alkyl rearrangements

In contrast to [1,3] hydrogen rearrangements that cannot occur under thermal conditions, alkyl groups can undergo sigmatropic thermal rearrangement reactions, and these reactions have interesting stereochemical outcomes. Under thermal conditions, the HOMO of the allyl radical is antisymmetric, so the two terminal lobes have opposite phases. Conceptually, homolytic cleavage of the alkyl-allyl bond produces a carbon radical that is sp^2-hybridized. As that carbon migrates suprafacially on the carbon chain, it can rotate 180° to have its orbital lobe remain in phase with the matching lobe on the allyl radical. If the migrating sp^3-hybridized carbon is stereogenic, the result is a net inversion of its stereochemical configuration.

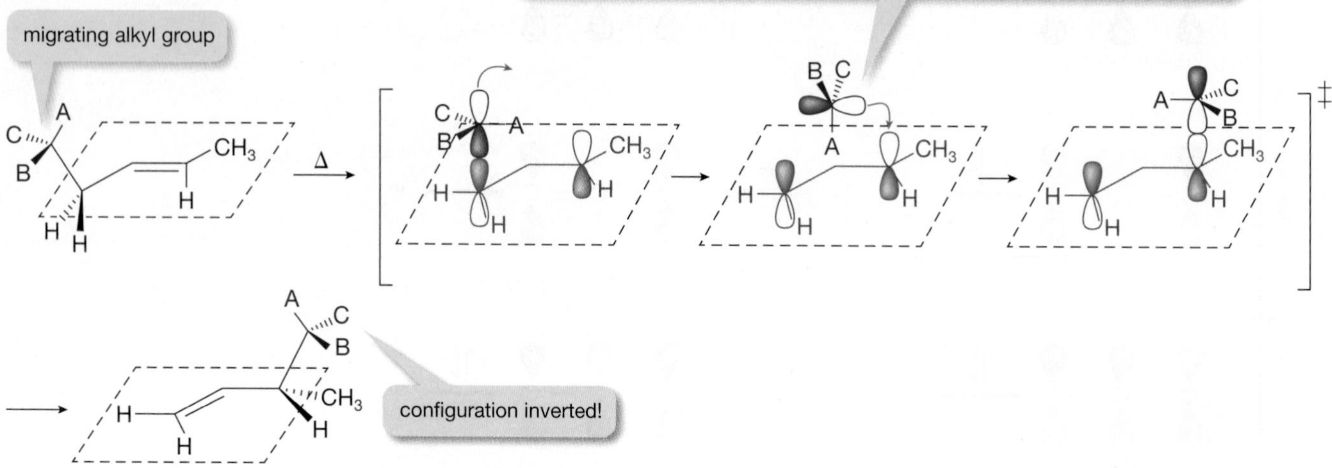
migrating alkyl group
group rotates as it migrates in order to match phases with the allyl π system
configuration inverted!

CHECKPOINT 20.6

You should now be able to predict the product of a [1,2] sigmatropic hydrogen rearrangement, with correct stereochemistry. You should also be able to predict the product of an alkyl rearrangement.

SOLVED PROBLEM

What is the product of the following [1,5] hydrogen shift?

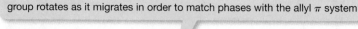

STEP 1: Identify the hydrogen that will migrate and break the C–H bond to form two radicals. The migrating hydrogen will be bonded to the sp^3-hybridized carbon adjacent to the conjugated alkene system.

STEP 2: Rotate the σ bonds to bring the ends of the π-system together.

STEP 3: Draw the HOMO (SOMO) of the conjugated radical.

The five-atom, five-electron HOMO that results from a thermal reaction has two nodes. The p-orbital lobes at each end of the HOMO are symmetric, with matching phase.

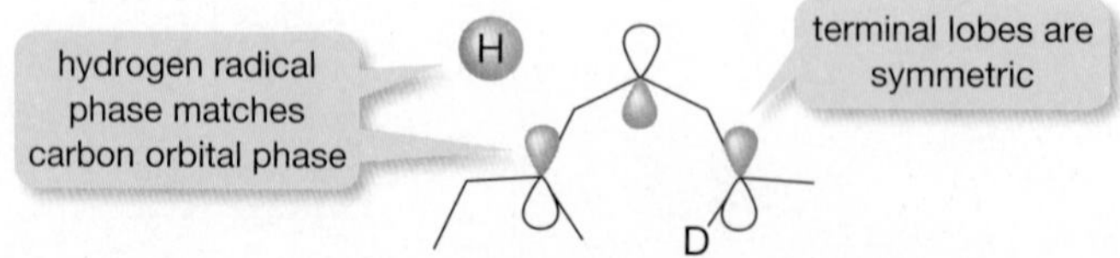
hydrogen radical phase matches carbon orbital phase
terminal lobes are symmetric

STEP 4: Move the hydrogen to the other end of the π-system, on the side of the molecule with the matching phase of p orbital.

Because the HOMO is symmetric, the migration must be suprafacial. The migrating hydrogen starts and ends the migration above the plane of the diene.

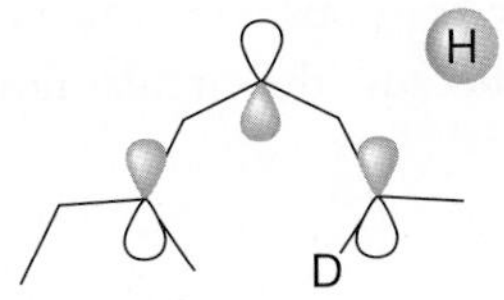

STEP 5: Draw the mechanism arrows for the migration of the hydrogen, showing the new location of the π bonds.

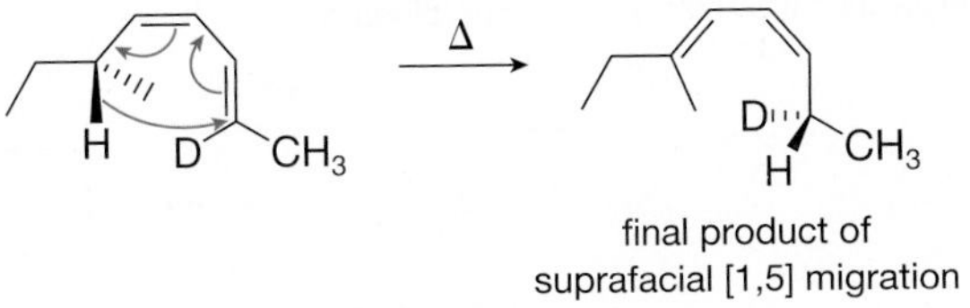

final product of suprafacial [1,5] migration

PRACTICE PROBLEM

20.11 For the following reactions, predict the major product of hydrogen migration.

a)

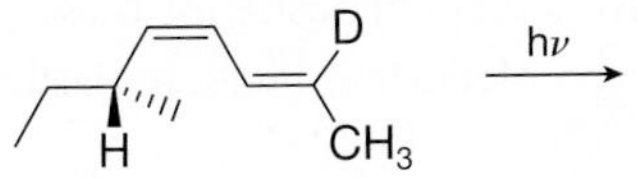

b)

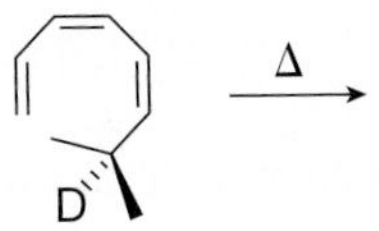

c)

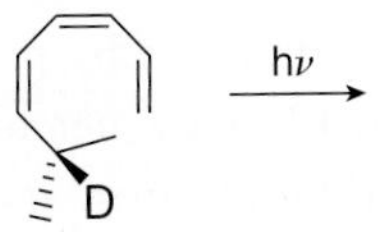

INTEGRATE THE SKILL

20.12 Show the mechanism for the rearrangements shown here.

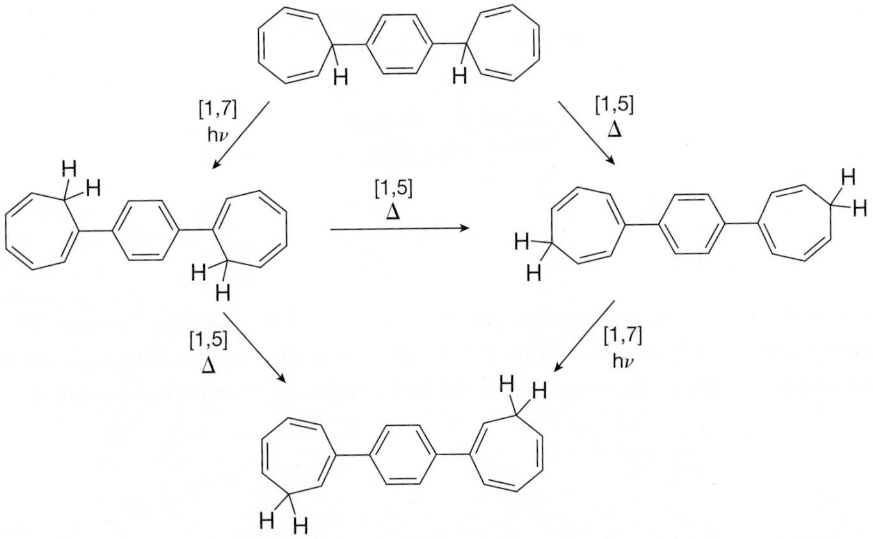

20.5 Electrocyclic Reactions

The third class of pericyclic reactions includes electrocyclic reactions, which involve ring formation or ring opening by a concerted motion of electrons. Such a reaction transforms two sp^2-hybridized carbons into sp^3-hybridized carbons, which are potential stereocentres. Cycloadditions are similar to electrocyclic reactions because the circular flow of electrons in both types of reactions is essentially the same.

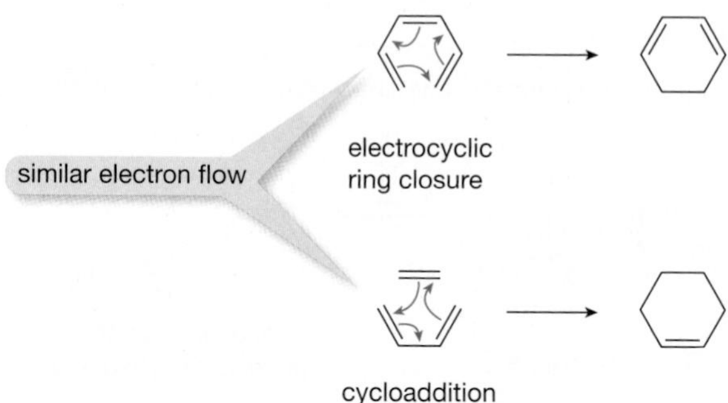

The four important types of electrocyclic reactions include two electrocyclic ring closures—thermal (Section 20.5.1) and photochemical (Section 20.5.2)—and two corresponding electrocyclic ring-opening reactions (Section 20.5.3). The relative stereochemistry of these reactions can be predicted by analyzing the symmetry of the relevant molecular orbitals, in particular, the symmetry of the open-chain HOMO.

20.5.1 Thermal electrocyclic ring closure

Thermal electrocyclic ring closure refers to a process whereby conjugated π-systems can cyclize to form a ring upon heating. In these reactions, a pair of π electrons from one end of a conjugated π-system forms a new σ bond with a carbon at the opposite end, thereby producing a ring. While this happens, the other π electrons become rearranged among neighbouring positions. The curved arrow notation makes it possible to track which bonds form and which ones are broken during thermal electrocyclic ring closure. The direction of electron movement (clockwise or counter-clockwise) is unimportant; the reactions are concerted, so all the electron reorganizations occur at the same time.

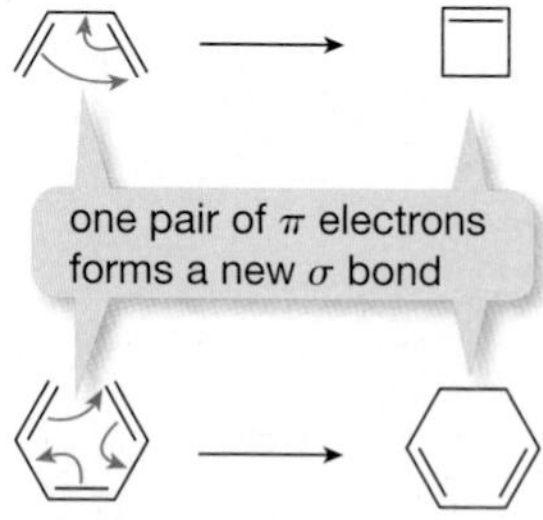

Electrocyclic ring closure reactions are stereospecific. For example, heating (2*E*,4*E*)-2,4-hexadiene produces only *trans*-2,3-dimethylcyclobutane as the product. Although the reaction mechanism does not change, heating the (2*E*,4*Z*) isomer produces only the *cis* product:

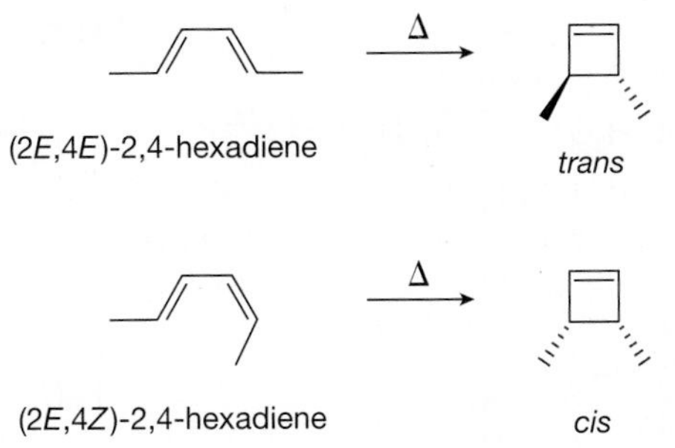

(2*E*,4*E*)-2,4-hexadiene $\xrightarrow{\Delta}$ *trans*

(2*E*,4*Z*)-2,4-hexadiene $\xrightarrow{\Delta}$ *cis*

To explain cyclization reactions, the symmetry of their relevant MOs must be analyzed. In principle, all the molecular orbitals involved in the reaction should be considered. However, it is much simpler to focus solely on the frontier orbitals (HOMO and LUMO) as these interact to form the key bonds in the product. The HOMO is especially informative because it mixes to form the σ bond in the ring.

The HOMO of a four-atom π-system in the ground state is the Ψ_2 orbital, the terminal lobes of which must overlap to form the σ bond.

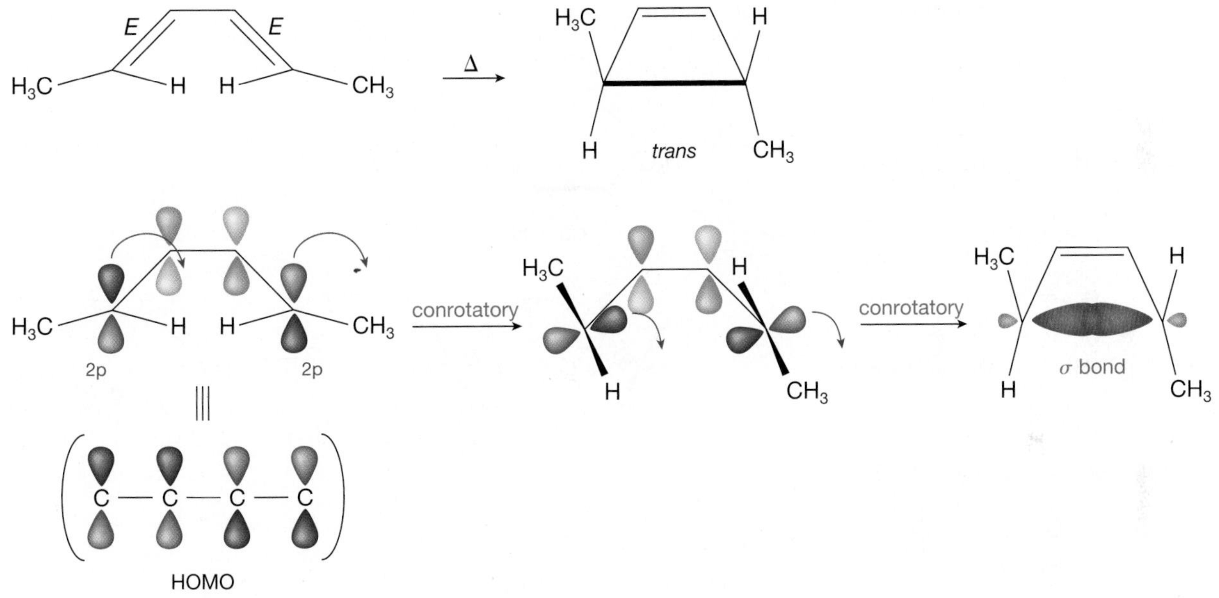

E E H_3C H H CH_3 $\xrightarrow{\Delta}$ H_3C H H *trans* CH_3

H_3C H H CH_3 2p 2p $\xrightarrow{\text{conrotatory}}$ H_3C H H CH_3 $\xrightarrow{\text{conrotatory}}$ H_3C H σ bond H CH_3

C — C — C — C

HOMO

To overlap and form a new σ bond, the terminal carbons of the π-system must rotate in such a way that the phases of the lobes on those atoms match to form a σ bond. In a four-atom system, these atoms must simultaneously rotate in the same direction. This motion is called **conrotatory**. In the preceding figure, both carbons are shown to rotate clockwise to form a *trans* isomer. It is important to remember that the cyclization can also proceed counter-clockwise, producing the product enantiomer.

A reaction is **conrotatory** if the end carbons rotate in the same direction as they rehybridize to form a C–C bond.

The relative stereochemistry of the product resulting from cyclization of the (2*E*,4*Z*) diene isomer can be accounted for in the same way.

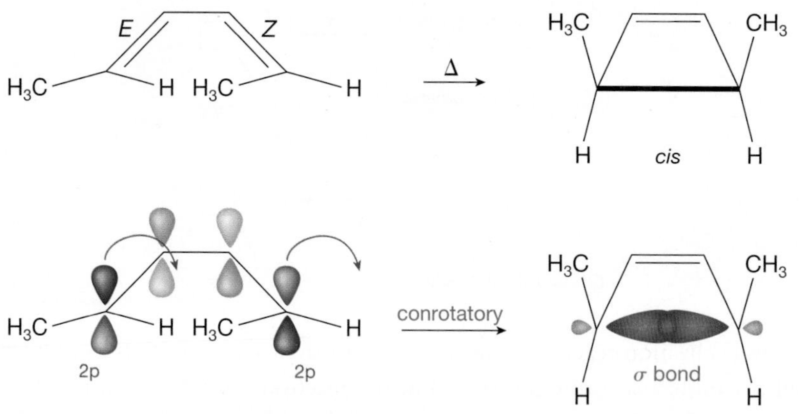

The conrotatory closure is a result of the orbital symmetry of the HOMO, which arises from the fact that it is a four-atom π-system. This rotates the two methyl groups to the same side of the cyclobutane ring, resulting in the production of a *cis*-product stereoisomer.

Cyclization of a conjugated triene can also be analyzed using the HOMO symmetry. The triene is a six-atom π-system, in which the HOMO is the Ψ_3 orbital. Heating generates *cis*-5,6-dimethyl-1,3-cyclohexadiene.

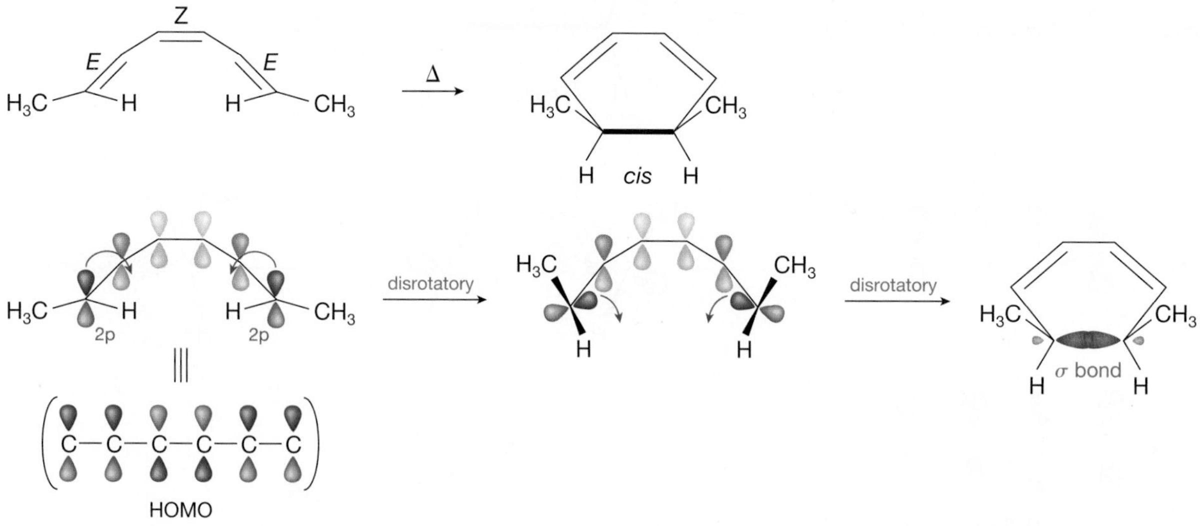

A reaction is **disrotatory** if the end carbons rotate in opposite directions as they rehybridize to form a C–C bond.

In this reaction, the phases of the two terminal lobes of the HOMO are symmetric. To match phases so that cyclization can take place, the two participating carbons must simultaneously rotate in opposite directions to form the new σ bond. This motion is called **disrotatory**.

CHECKPOINT 20.7

You should now be able to predict the products of electrocyclic ring closing reactions under thermal conditions, including the correct stereochemistry resulting from conrotatory or disrotatory motion.

SOLVED PROBLEM

Predict the product of the thermal electrocyclic ring closure of (2*Z*,4*Z*,6*E*)-octa-2,4,6-triene, including the correct stereochemistry.

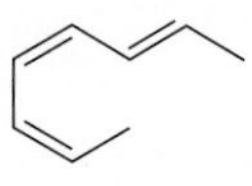

STEP 1: Draw the HOMO for the conjugated system.

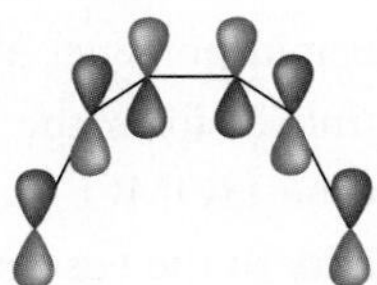

STEP 2: Add the remaining substituents in the proper orientation.

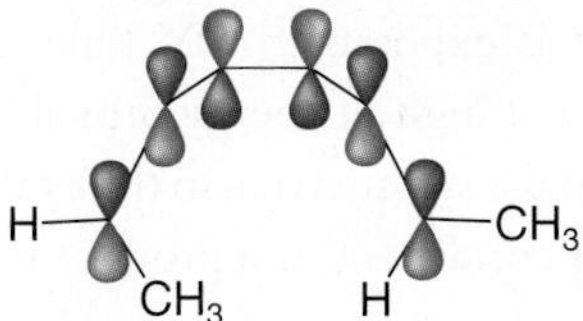

STEP 3: Identify the required rotation of the terminal carbons that will result in an in-phase overlap to form a σ bond. The rotation determines the final orientation of the substituents in the product.

In this reaction the orbital is symmetric, which requires conrotatory motion.

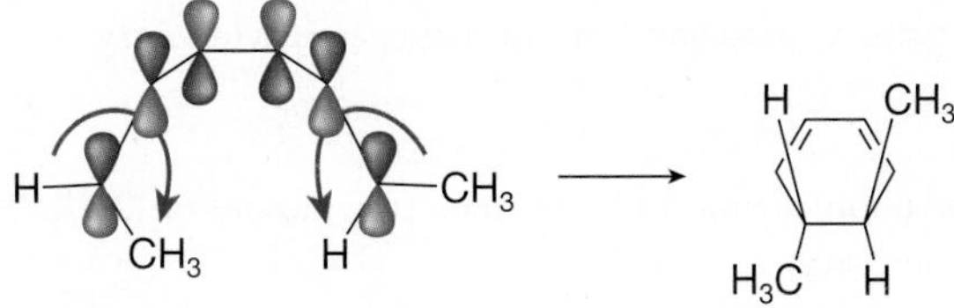

STEP 4: Draw the correct mechanism arrows to predict the correct locations of the new double and single bonds.

H— —CH_3 ⟶

CH_3 H

PRACTICE PROBLEM

20.13 Predict the product of electrocyclic ring closure under thermal conditions for the following molecules. Justify the stereochemistry using the appropriate molecular orbitals.

a) $\xrightarrow{\Delta}$

b) $\xrightarrow{\Delta}$

c) $\xrightarrow{\Delta}$

INTEGRATE THE SKILL

20.14 Ring closure of an eight-π-electron system results in a six-π-electron conjugated ring that undergoes a second ring closure to form a bicyclic product. Predict the intermediate and product of successive electrocyclic ring closures under thermal conditions for the following molecule. Justify the stereochemistry using the appropriate molecular orbitals.

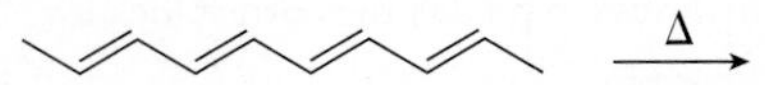 $\xrightarrow{\Delta}$

20.5.2 Photochemical electrocyclic ring closure

As discussed in Chapter 9, when conjugated polyenes absorb energy in the form of UV light, a single electron becomes excited and moves from the HOMO into the next highest energy molecular orbital, which then becomes the HOMO. The symmetry of the new HOMO continues to control the overall stereoselectivity of the ensuing electrocyclic reaction. In a four-atom π-system, the HOMO for ground-state reactions corresponds to Ψ_2, an antisymmetric orbital, whereas in photochemical reactions the HOMO is Ψ_3, a symmetric orbital.

When (2*E*,4*E*)-2,4-hexadiene is exposed to UV light, an electron is promoted from the ground-state HOMO (Ψ_2) to the next highest energy orbital (Ψ_3). The two terminal lobes of this orbital have the same phase (the orbital is symmetric), so the cyclization must proceed by disrotatory ring closure. Under photochemical conditions, the product is *cis*-3,4-dimethylcyclobutane.

CHECKPOINT 20.8

You should now be able to predict the products of electrocyclic ring closing reactions under photochemical conditions, including the correct conrotatory or disrotatory stereochemistry.

SOLVED PROBLEM

Predict the product of the photochemical electrocyclic ring closure of (2*Z*,4*Z*,6*E*)-octa-2,4,6-triene, as well as the correct stereochemistry.

STEP 1: Draw the SOMO for the conjugated system.

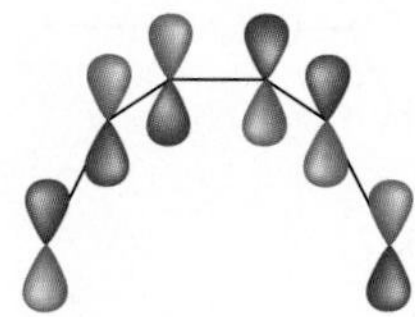

STEP 2: Add the remaining substituents in the proper orientation.

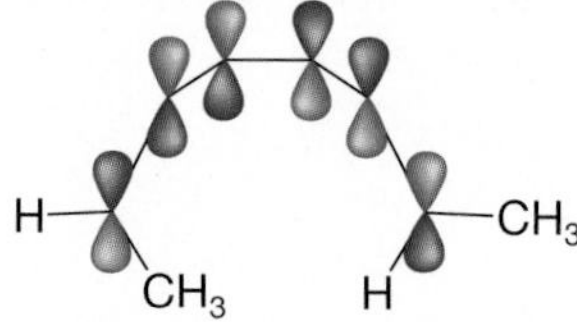

STEP 3: Identify the required rotation of the terminal carbons that will result in an in-phase overlap to form a σ bond. The rotation determines the final orientation of the substituents in the product.

In this reaction the orbital is symmetric, which requires conrotatory motion.

H CH_3 CH_3 H → H H H_3C CH_3

STEP 4: Draw the correct mechanism arrows to predict the correct locations of the new double and single bonds.

H CH_3 CH_3 H →

PRACTICE PROBLEM

20.15 Predict the product of electrocyclic ring closure under photochemical conditions for the following molecules. Justify the stereochemistry using the appropriate molecular orbitals.

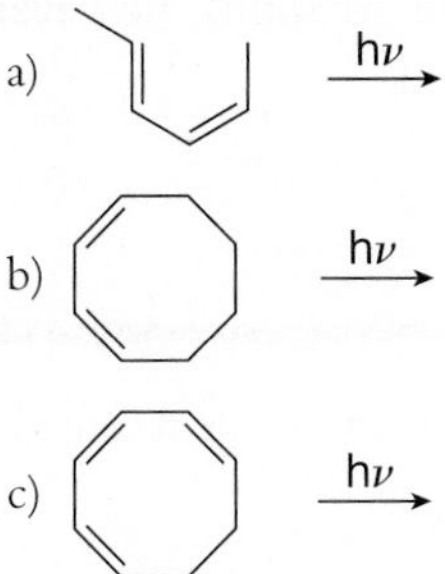

c) $\xrightarrow{h\nu}$

INTEGRATE THE SKILL

20.16 The ring closure of [14]-annulene is shown without stereochemical information. Based on the molecular orbitals involved, suggest the stereochemistry of the product when the reaction occurs under photochemical or thermal conditions.

20.5.3 Electrocyclic ring opening

Electrocyclic reactions are reversible. During ring closures, one σ bond is formed, and the product has one fewer π bond than the starting material. This situation usually favours the formation of rings based on the relative energies of these bond types—a low-energy bond (σ) replaces a high-energy bond (π)—typically lowering the energy of the molecule. However, in some cases, cyclizations introduce considerable ring strain in the product. Because electrocyclic ring closures are equilibria and the position of each equilibrium depends on the relative energies of the reactant and product, a particular electrocyclic reaction may favour ring closure or ring opening. In general, low-strain rings tend to favour ring closure, and high-strain rings tend to favour ring opening.

six-membered ring
low ring strain

ring closure
favoured

four-membered ring
high ring strain

ring opening
favoured

Electrocyclic ring-opening reactions are the exact reverse of the corresponding ring-closure reactions. The orbital symmetry rules that account for the electrocyclic ring closures also apply to the ring-opening reactions. Accordingly, a ring opens in a manner that matches the orbital configuration of the product's HOMO.

Upon heating, *trans*-3,4-dimethylcyclobutene undergoes electrocyclic ring opening. Only conrotatory ring opening results in the phases of the two terminal lobes of the butadiene HOMO (Ψ_2) being antisymmetric. Consequently, only the (*E*, *E*) product is observed. Conrotatory ring opening could also occur in the other direction (counter-clockwise rotation); however, this would produce the very strained (*Z*, *Z*) isomer, and this is not observed.

CHECKPOINT 20.9

You should now be able to determine whether ring-closing or ring-opening reactions are favoured under equilibrium conditions.

SOLVED PROBLEM

Predict the direction of equilibrium in the thermal electrocyclic ring closure of (2*Z*,4*Z*,6*E*)-octa-2,4,6-triene to determine which species predominates.

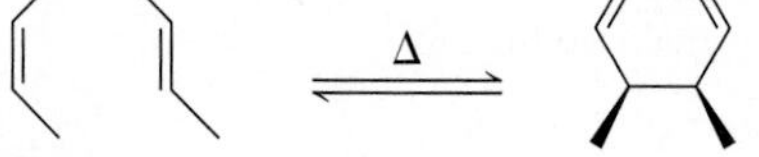

Δ

STEP 1: Examine the closed-ring form for ring strain.

In the case of six-membered rings, ring closure does not introduce extra strain, in comparison to the open form of the reactant.

six-membered ring has little significant strain

STEP 2: Examine both the open and closed forms to identify significant steric strain.

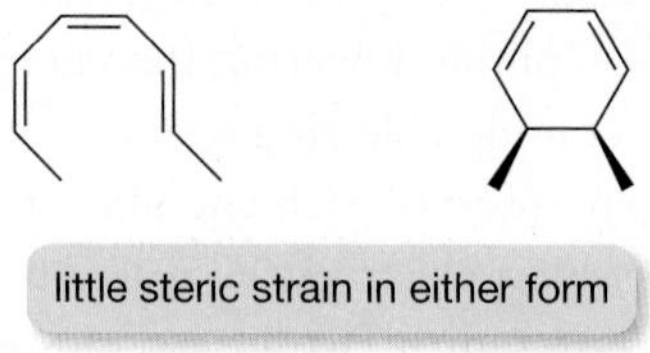

little steric strain in either form

STEP 3: In the absence of ring and steric strain, ring closure is favoured due to the extra stability of σ bonds, compared to the stability π bonds.

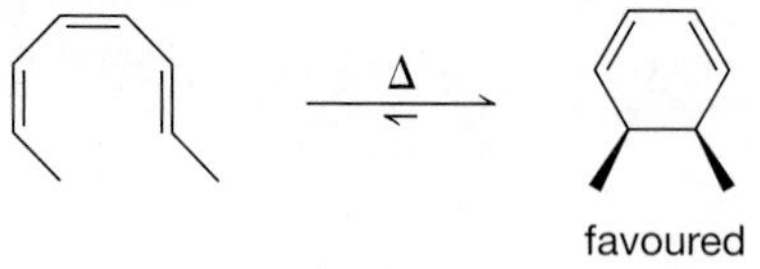

Δ

favoured

PRACTICE PROBLEM

20.17 Predict the product favoured at equilibrium for the following reactions.

a) hν

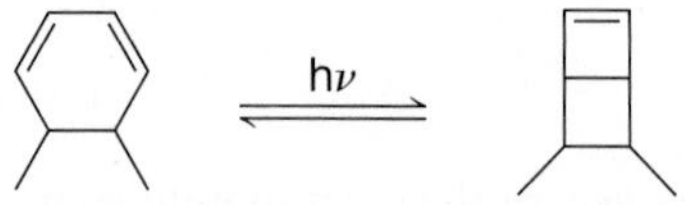

b) Δ

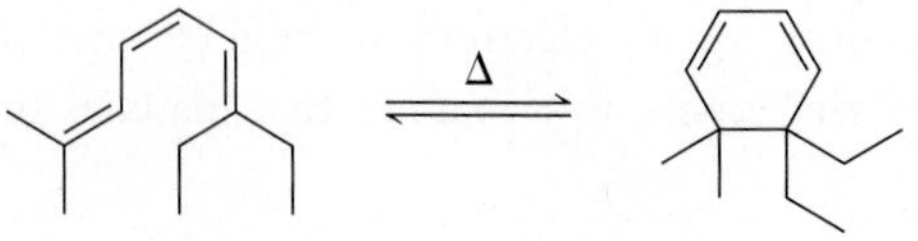

c) 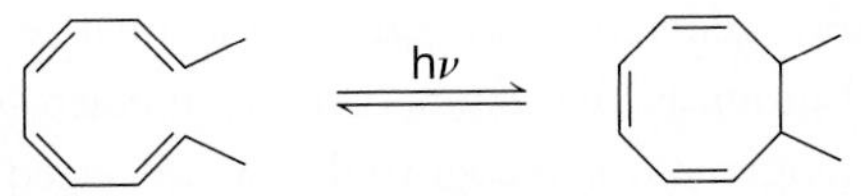 $\xrightleftharpoons{h\nu}$

INTEGRATE THE SKILL

20.18 *cis*-Dimethylcyclobutene undergoes ring opening through both clockwise and counter-clockwise conrotation. *trans*-Dimethylcyclobutene follows only one pathway. Is this pathway clockwise or counter-clockwise? Why does it not follow both pathways?

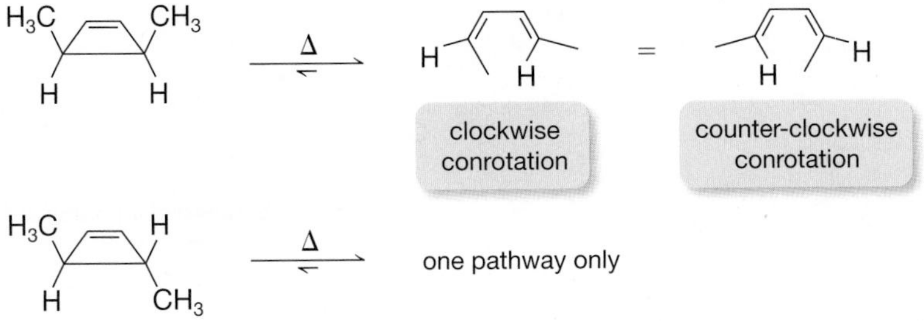

clockwise conrotation

counter-clockwise conrotation

one pathway only

20.6 Non-sigmatropic Rearrangements

Sigmatropic rearrangements are examples of pericyclic reactions in which orbital symmetry governs the stereochemistry and regiochemistry. Other types of rearrangements, which are not sigmatropic, occur in non-pericyclic reactions. The orientation of the molecular orbitals involved also control the course of these reactions, although orbital symmetry is not a key factor. In non-pericyclic reactions, all rearrangements involve unstable intermediates at some point in the process. In the cases discussed in this section, a migrating group displaces a leaving group. These rearrangement reactions include the Baeyer–Villiger oxidation, which converts a ketone to an ester, and the Hofmann rearrangement, which converts a primary amide to an amine.

20.6.1 Baeyer–Villiger oxidation

The **Baeyer–Villiger oxidation** involves a 1,2 migration that converts a ketone to an ester by inserting an oxygen into one of the carbon-carbon bonds of the ester. To carry out this reaction peracids are used, and the most common reagents are *m*-CPBA and MMPP (Chapter 8).

The **Baeyer–Villiger oxidation** involves the use of a peracid reagent to convert a ketone to an ester through a rearrangement.

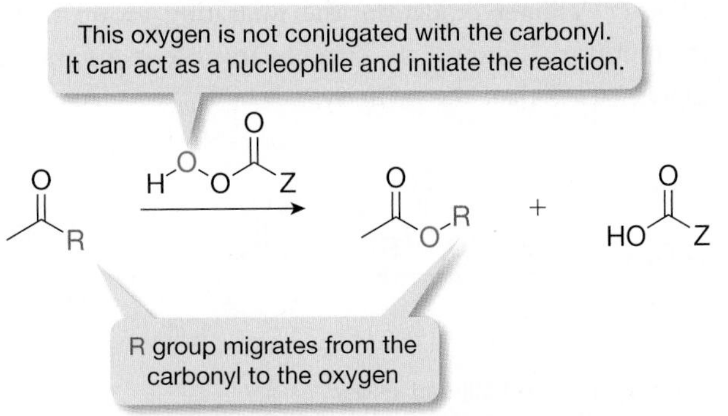

This oxygen is not conjugated with the carbonyl. It can act as a nucleophile and initiate the reaction.

R group migrates from the carbonyl to the oxygen

The mechanism of the Baeyer–Villiger oxidation is essentially a two-step reaction. The first step is the nucleophilic addition of the peracid reagent to the ketone carbonyl group (protonated in the presence of a trace amount of carboxylic acid) to form a tetrahedral intermediate. In the second step, the tetrahedral intermediate collapses, and of one of the ketone alkyl groups migrates to the adjacent peracid oxygen (a 1,2-migration), generating an ester as the final rearrangement product. Note how a lone pair on the OH group in the tetrahedral intermediate assists in "pushing out" the migrating R group. The rearrangement is recognized as a concerted process where the alkyl group migrates simultaneously with the loss of the leaving group.

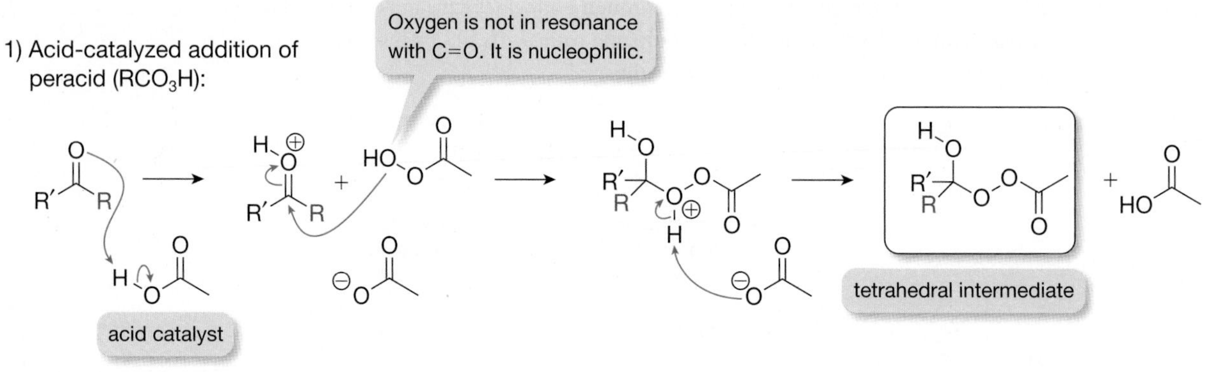

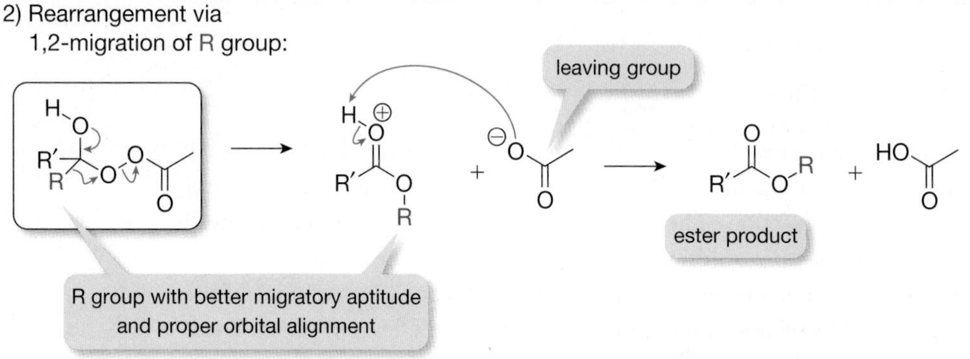

1) Acid-catalyzed addition of peracid (RCO_3H):

Oxygen is not in resonance with C=O. It is nucleophilic.

acid catalyst

tetrahedral intermediate

2) Rearrangement via 1,2-migration of R group:

leaving group

ester product

R group with better migratory aptitude and proper orbital alignment

Migratory aptitude describes the relative potential of a group to migrate during a rearrangement reaction.

One of the two ketone R groups rearranges—which one depends on their migratory aptitude, that is, their relative potential to migrate during the reaction, and also their alignment with the O–O bond that is broken. The **migratory aptitude** of ketone substituents matches the ability of the migrating group to stabilize a positive charge. In general, those groups that can best stabilize cations will have the highest migratory aptitude.

$$H > 3° \text{ alkyl} > 2° \text{ alkyl} \approx Ph > Ph\text{–}CH_2 > 1° \text{ alkyl} > CH_3$$

In the following Baeyer–Villiger oxidation, the migrating group that can best stabilize a carbocation (best electron donor) "follows" the two electrons of its migrating σ bond. Thus, the secondary carbon migrates, rather than the primary.

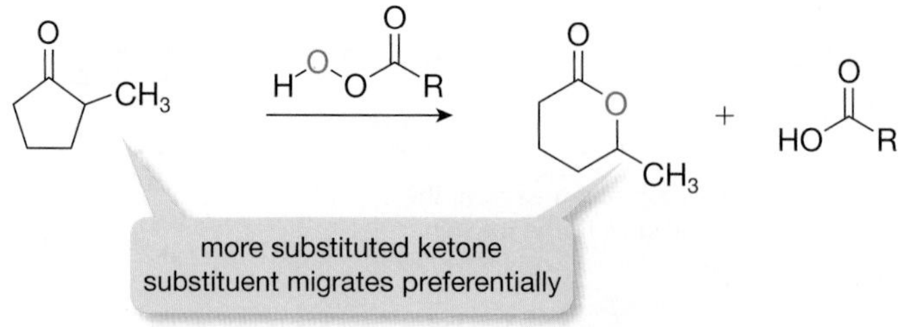

more substituted ketone substituent migrates preferentially

For rearrangement to occur, the migrating R group favours the *anti* approach to the O–O bond. This arrangement allows for optimal overlap of the occupied σ orbital of the migrating R group with the receiving σ^* orbital of the O–O bond. The migration of the R group transfers electrons into the σ^* orbital to displace the carboxylate leaving group as the O–O bond breaks. For this to happen in concerted fashion, the OH lone pair must also be *anti* to the migrating group.

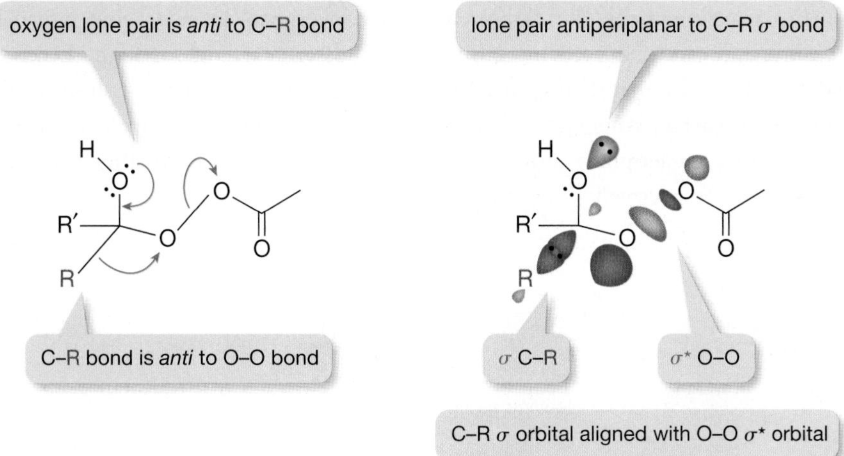
oxygen lone pair is *anti* to C–R bond
lone pair antiperiplanar to C–R σ bond
C–R bond is *anti* to O–O bond
σ C–R
σ* O–O
C–R σ orbital aligned with O–O σ* orbital

20.6.2 Hofmann rearrangement and related reactions

The **Hofmann rearrangement** involves a similar 1,2 non-sigmatropic migration that converts a primary amide to an amine by loss of the carbonyl group.

The **Hofmann rearrangement** converts an amide to an amine through a rearrangement that releases carbon dioxide.

Br_2, NaOH, H_2O; NH_2 + $O=C=O$

The first part of this mechanism is similar to the base-catalyzed bromination of a ketone, which produces an *N*-bromoamide. Deprotonation of the bromoamide by a base induces a 1,2-migration of the alkyl group over to the nitrogen as the bromine leaving group departs to form an isocyanate. In the presence of water, an isocyanate hydrolyzes readily to form a carbamic acid, which then decarboxylates to produce the corresponding amine and carbon dioxide.

same mechanism as bromination of ketones

amide deprotonated under basic conditions

1,2-migration of alkyl group

carbamic acid collapses to amine and CO_2

proton transfer

carbamic acid

isocyanate

The **Curtius rearrangement** converts a carboxylic acid to an isocyanate that is similar to the intermediate of the Hofmann rearrangement.

The **Lossen rearrangement** converts a hydroxamic acid to an isocyanate.

Two related rearrangements, the **Curtius rearrangement** and the **Lossen rearrangement**, form isocyanates through conversion of carboxylic acids and hydroxamic acids, respectively. Both reactions have a nitrogen atom that bears a leaving group (as in the Hofmann rearrangement). In the Curtius rearrangement, the carboxylic acid converts to an acyl azide intermediate in which the nitrogen bears a N_2^+ leaving group. In the Lossen rearrangement, the hydroxyl group of the hydroxamic acid converts to a tosylate leaving group. In either case, heating the intermediate induces a rearrangement via 1,2-migration of their alkyl group to form an isocyanate. In the presence of a nucleophilic amine or an alcohol, the isocyanate further reacts to form a carbamate or a urea, respectively.

Curtius rearrangement

NaN_3 ... leaving group ... $+ N_2$

Lossen rearrangement

TsCl, base ... base, Δ ... leaving group ... + TsOH

CHEMISTRY: EVERYTHING AND EVERYWHERE

A Spicy Rearrangement

Myrosinase, also known as thioglucoside glucohydrolase, is an enzyme found in a variety of plants, including wasabi, mustard, cabbage, broccoli, and kale. These plants store the enzyme in tiny vesicles, along with molecules called glucosinolates. When the plant is cut or chewed, the vesicles break, releasing and mixing these two components. Myrosinase breaks down the glucosinolate into D-glucose and products that then undergo a Lossen-like rearrangement to generate isothiocyanates.

glucosinolate — myrosinase → D-glucose + ... → isothiocyanate + $HOSO_3^-$

The strong, bitter taste of isothiocyanates deter animals from eating plants that produce these chemicals. However, when prepared properly and in the right combination of flavours, these isocyanates produce a "spicy" taste and effect that people enjoy. Wasabi is a type of horseradish that contains large amounts of sinigrin (an allyl glucosinolate) and myrosinase. When the tissues of the plant roots are damaged, the components mix, resulting in the production of allyl isothiocyanate. This volatile and unstable chemical produces the strong "kick" of wasabi, which is often felt in the nose (volatility) and does not last very long (instability). Due to the instability of this compound, high-quality restaurants prepare wasabi directly from the root immediately before it is used. This preserves the myrosinase and gives maximum amounts of the allyl isothiocyanate. Powdered wasabi is less potent and must be mixed fresh with water to activate the myrosinase. The strongest taste is produced a few minutes after mixing.

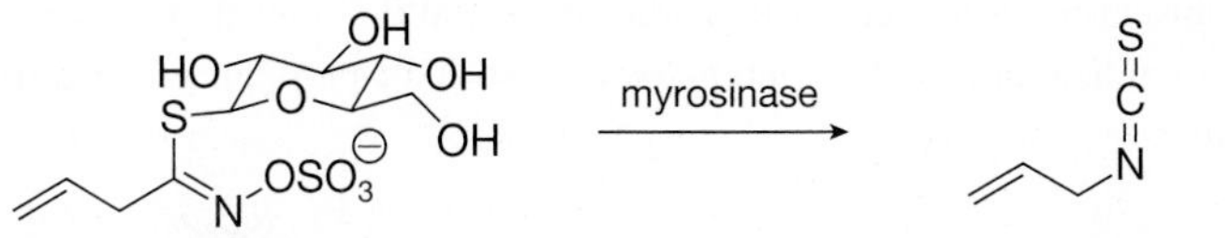

sinigrin

myrosinase

allyl isothiocyanate

CHECKPOINT 20.10

You should now be able to predict the product of rearrangements that include a [1,2] migration. You should also be able to rank substituents according to migratory aptitude.

SOLVED PROBLEM

Predict the product of the following reaction.

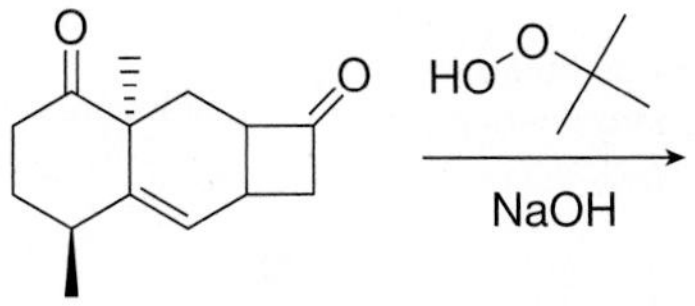

NaOH

STEP 1: Determine the nature of the reagents used to identify the type of reaction.

Peroxide is deprotonated by a base to form a nucleophilic oxidizing agent.

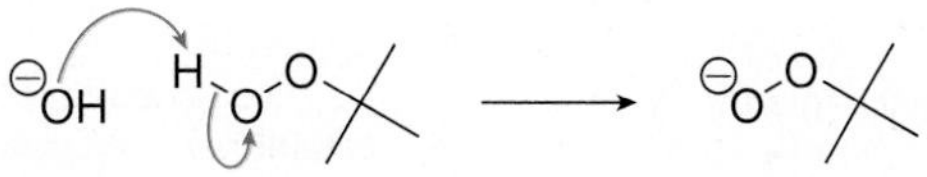

nucleophilic
oxidizing agent

STEP 2: Identify compatible reactive groups in the substrate.

The nucleophilic oxidizing agent requires an electrophilic reactive partner. This substrate has two electrophilic carbonyl groups.

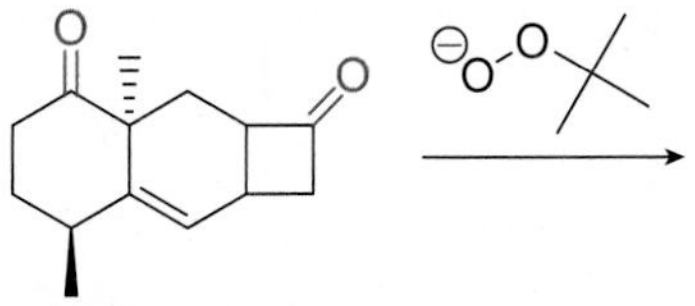

STEP 3: Identify the most reactive reaction site.

The rehybridization to sp^3 that results from nucleophilic attack reduces the ring strain of a cyclic ketone. Comparing the reactivity of cyclobutanone and cyclohexanone, the more highly strained cyclobutanone is more reactive because it undergoes greater reduction in ring strain.

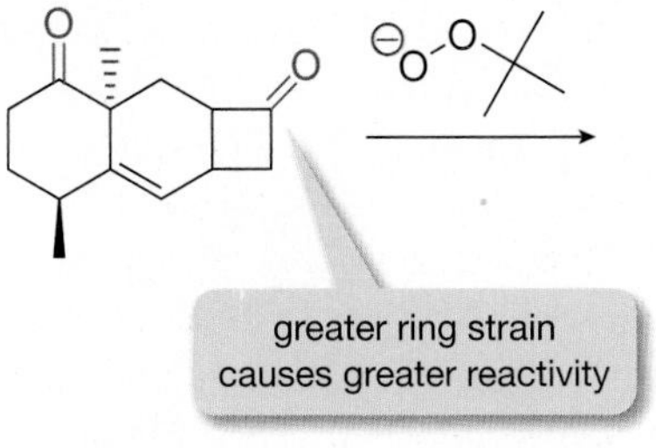

greater ring strain
causes greater reactivity

STEP 4: Draw mechanism arrows for the nucleophilic attack and the intermediate formed.

The resulting intermediate has an electrophilic oxygen with a *t*-butoxide leaving group that is suitable for an alkyl migration.

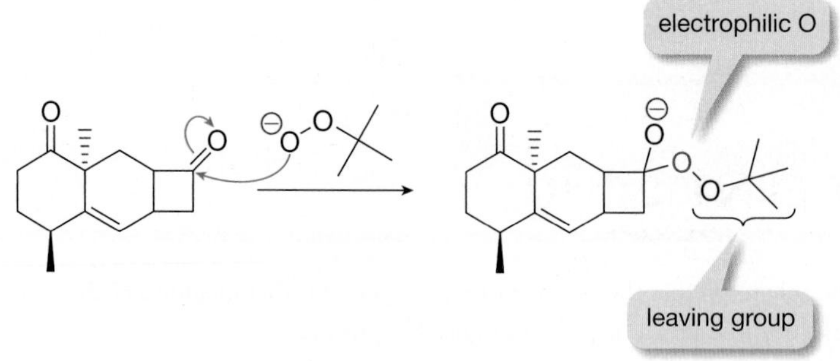

electrophilic O

leaving group

STEP 5: Draw the mechanism for the re-formation of the carbonyl and the loss of the leaving group. Consider each of the possible migrating groups to identify the substituent with the greatest migratory aptitude.

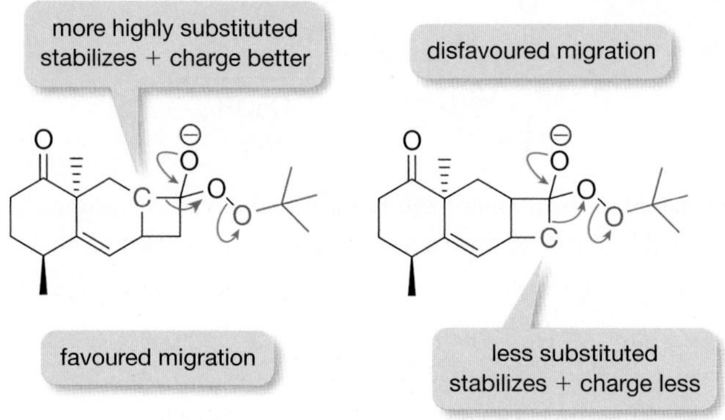

more highly substituted stabilizes + charge better

disfavoured migration

favoured migration

less substituted stabilizes + charge less

STEP 6: Draw the product resulting from the favoured migration.

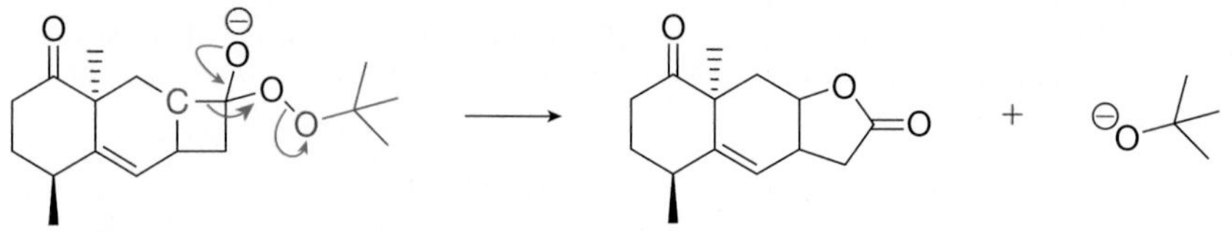

PRACTICE PROBLEM

20.19 a) For the following carbonyl compounds, identify the substituent with the greater migratory aptitude in a Baeyer–Villiger reaction.

i)

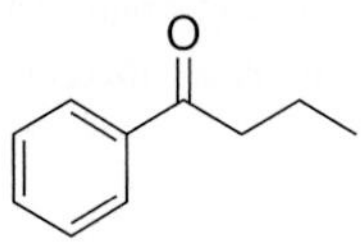

ii)

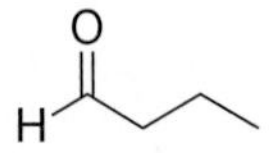

iii)

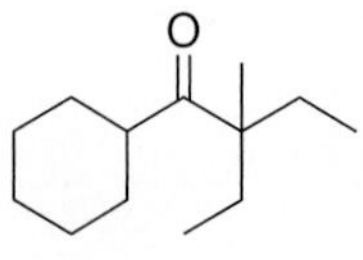

iv)

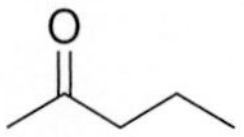

b) Predict the product of the following reactions.

i) [heptanoic acid] $\xrightarrow[2)\ NaN_3,\ \Delta]{1)\ SOCl_2}$

ii) [N,N-dimethyl 3-methylcyclohexanecarboxamide] $\xrightarrow[NaOCH_3]{Br_2}$

iii) [chiral 2-methyl-2-phenylbutanoic acid] $\xrightarrow[2)\ NaN_3,\ \Delta]{1)\ SOCl_2}$

iv) [1-phenylcyclopentane carboxylic acid, HO_2C] $\xrightarrow[2)\ NaN_3,\ \Delta]{1)\ SOCl_2}$

v) [2-hydroxy-3-phenylpropanamide, OH, NH_2] $\xrightarrow[NaOH]{Br_2}$

INTEGRATE THE SKILL

20.20 a) During the last step of a hydroboration reaction, hydrogen peroxide and hydroxide are added to transform the intermediate borane into a hydroxyl group. This reaction is very similar to a Bayer–Villiger reaction. Propose a mechanism to explain this.

$$\text{CH}_3\text{CH}_2\text{CH}_2\text{BR}_2 \xrightarrow[\text{NaOH, } H_2O]{H_2O_2} \text{CH}_3\text{CH}_2\text{CH}_2\text{OH}$$

b) The Favorskii reaction results in the rearrangement of an α-haloketone, when treated with an alkoxide base. Propose a reasonable mechanism for this transformation.

[2-chlorocyclohexanone] $\xrightarrow[\Delta]{^{\ominus}OCH_2CH_3}$ [ethyl cyclopentanecarboxylate, OEt]

20.7 Patterns in Pericyclic Reactions and Other Rearrangements

A circular flow of π electrons characterizes pericyclic reactions, including cycloadditions such as the Diels–Alder and electrocyclization reactions. These reactions are carried out under thermal or photochemical conditions, and the reaction outcomes are predicted by interpreting their molecular orbitals. The symmetry of the relevant HOMOs and LUMOs determines whether a particular reaction is or is not allowed.

electrocyclization

HOMO

Diels–Alder

HOMO

LUMO

The circular flow of six π electrons is the same for Diels–Alder and [3+2] dipolar cycloadditions; certain electrocyclic reactions; and [3,3] sigmatropic rearrangements, including the Cope and Claisen rearrangements. [3,3] rearrangements proceed through chair-like transition states, which determine the stereochemistry of the products.

Diels–Alder

[3+2] dipolar addition

electrocyclic reaction

Cope rearrangement

Claisen rearrangement

In electrocyclic reactions—either ring opening or closing—the terminal carbons rotate as they form a new σ bond, and this rotation is either conrotatory or disrotatory, depending on the relative phase of the interacting π orbitals at both ends. This depends on both the number of π electrons and whether the reaction is done under thermal or photochemical conditions.

Δ

HOMO

disrotatory

[1,*n*] hydrogen shifts can be either antarafacial or suprafacial, depending on the number of atoms involved and whether the reactions take place under thermal or photochemical conditions. The outcome of these reactions can be readily explained by considering the reactions as diradicals and examining the relevant HOMOs and, specifically, their symmetry.

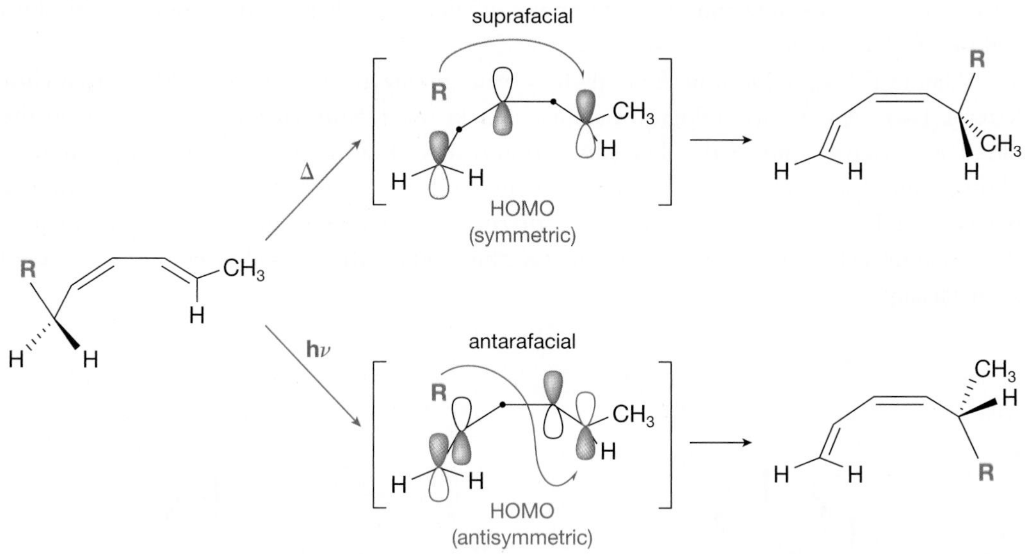

The 1,2-rearrangement pathway is characteristic of the Baeyer–Villiger oxidation and also the Hofmann, Curtius, and Lossen rearrangements. In this pathway a group migrates to an adjacent heteroatom with the assistance of a lone pair on oxygen and the displacement of a leaving group. Proper orbital alignments are important for the reactions to take place.

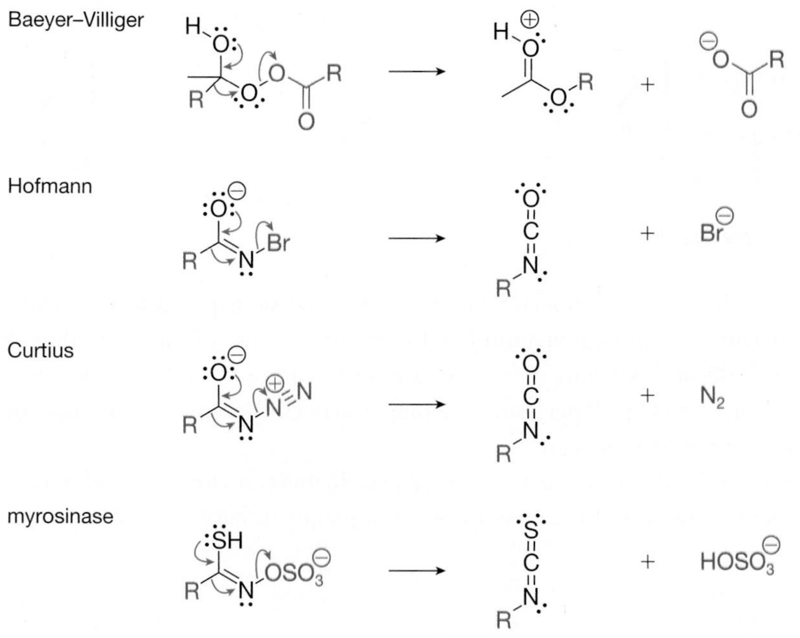

Bringing It Together

Vitamin D3 (cholecalciferol) is a human hormone essential for intestinal calcium absorption. Cholecalciferol binds to the vitamin D receptor, which is located in cellular nuclei, and initiates the transcription of calcium transport proteins. A cholecalciferol deficiency reduces the body's capacity to absorb calcium and can lead to bone softening, which is called osteomalacia in adults and rickets in children.

Vitamin D3 is produced in the body by a series of enzymatic reactions starting from cholesterol. Two steps of this biological manufacture do not require enzymes; they occur in the blood as it circulates under the skin. Absorption of UV-B radiation from sunlight promotes a photochemical electrocyclic ring opening (conrotatory) of 7-dehydrocholesterol to produce pre-vitamin D3, which then undergoes a thermal [1,7]-hydride shift (antarafacial) to produce vitamin D3. Although it is not evident in the product, this second reaction does proceed antarafacially!

HO H
7-dehydrocholesterol
hν
(conrotatory)
HO H
pre-vitamin D3

H H H C
HO H
pre-vitamin D3
[1,7]-hydride shift
(antarafacial)
HO H
vitamin D3

People who live in the northern hemisphere may not be exposed to sufficient sunlight for their bodies to produce enough vitamin D3. In summer, the use of sunscreen that blocks UV-B radiation (270–300 nm) prevents the production of vitamin D3 (SPF 15 sunscreen diminishes production of vitamin D by 98 percent). In winter, warm clothes and winter coats cover the skin, so vitamin D3 cannot be produced.

To counteract this deficiency, many foods, especially milk, are fortified with vitamin D, which provides sufficient amounts of the vitamin for most people to have healthy bones and teeth.

You Can Now

- Predict the relative energy levels, with phases and nodes, of π molecular orbitals, including odd-numbered systems.
- Predict the products of electrocyclic ring-closing reactions under both thermal and photochemical conditions, including the correct conrotatory or disrotatory stereochemistry.
- Determine whether ring-closing or ring-opening reactions are favoured under equilibrium conditions.
- Predict the products of [4+2] cycloaddition between a diene and a dienophile, including the correct regio- and stereoselectivity.
- Predict the products of [3+2] cycloaddition between a dipole and a dipolarophile.
- Recognize when a [2+2] cyclization is an allowed pericyclic reaction that can proceed through photochemical excitation.
- Recognize when a [2+2] cyclization must happen thermally, using orthogonal orbitals.
- Recognize compounds that are capable of undergoing [3,3] sigmatropic rearrangements.
- Predict the product of a [3,3] sigmatropic rearrangement.
- Predict the product of [1,n] sigmatropic hydrogen rearrangements, with correct stereochemistry.
- Predict the product of an alkyl rearrangement.
- Rank substituents according to migratory aptitude.
- Predict the product of rearrangements that include a [1,2] migration.

A Mechanistic Re-View

CYCLOADDITIONS

Diels–Alder reaction

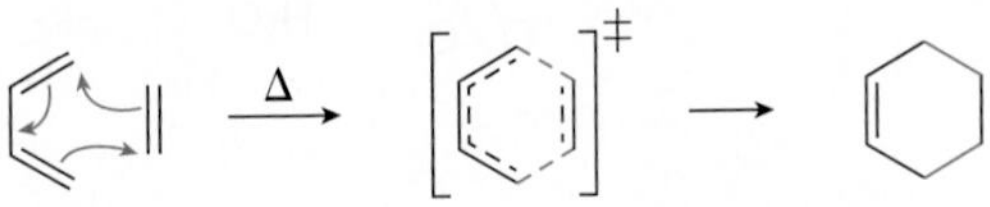

1,3-DIPOLAR CYCLOADDITIONS

Huisgen cycloaddition

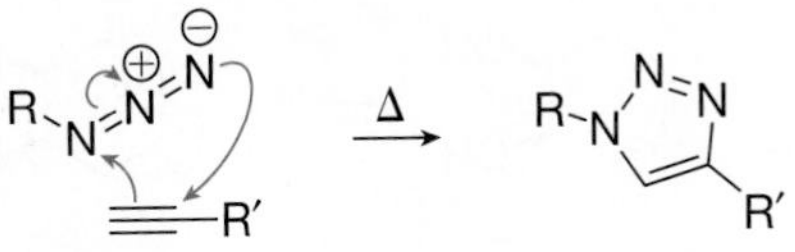

Ozonolysis

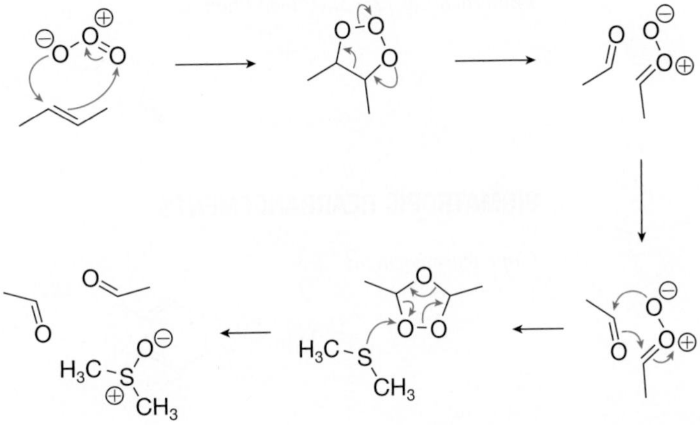

Osmium tetroxide dihydroxylation (catalytic OsO_4)

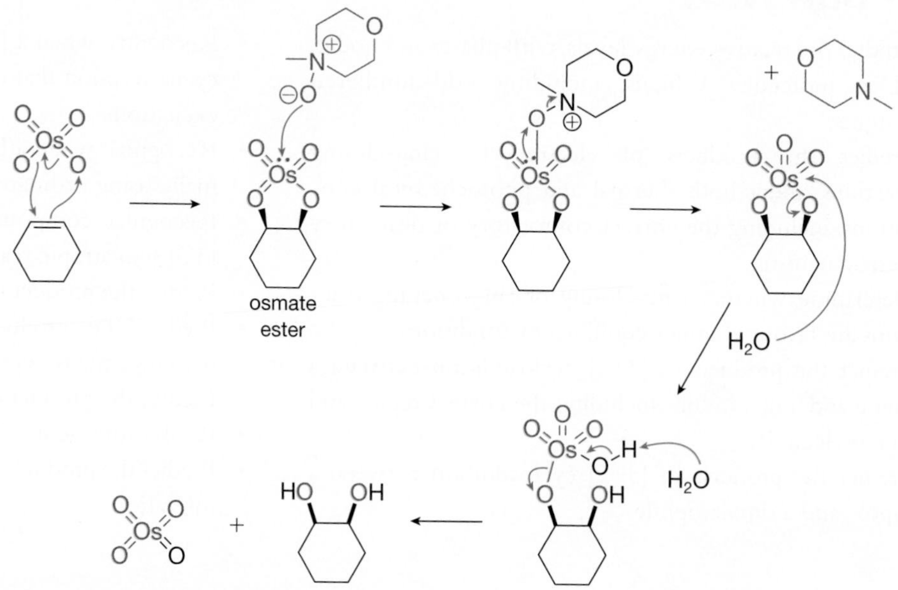

osmate ester

H_2O

Permanganate dihydroxylation

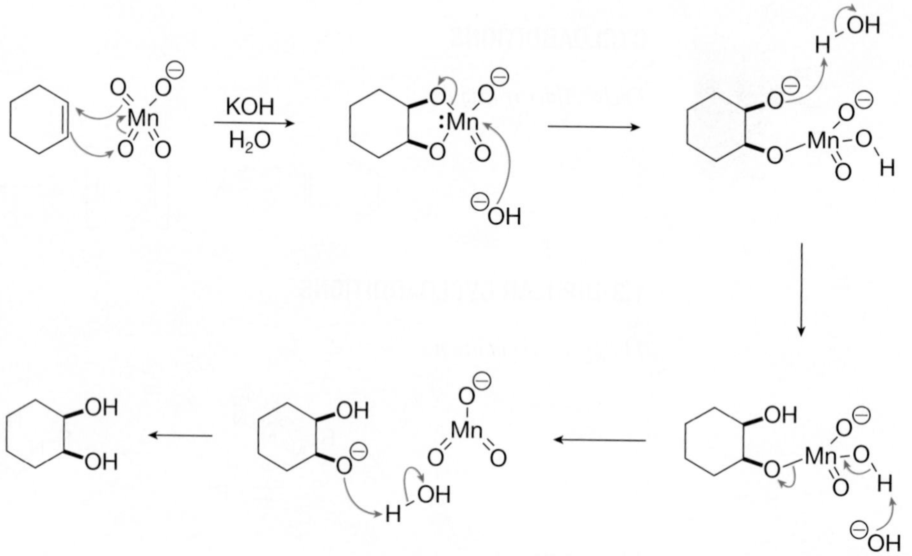

KOH
H_2O

Photochemical [2+2] cycloaddition

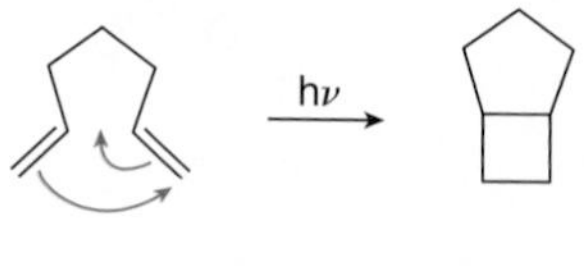

hν

SIGMATROPIC REARRANGEMENTS

Cope rearrangement

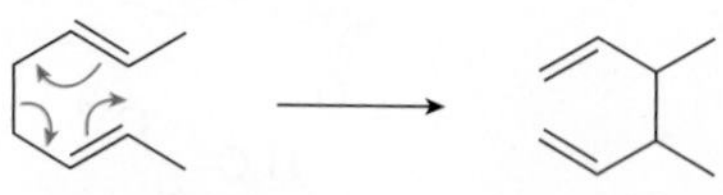

Claisen rearrangement

[1,5] hydrogen rearrangement

Suprafacial

Antarafacial

[1,3] hydrogen rearrangement

[1,3] alkyl shift

ELECTROCYCLIC REACTIONS

Ring closure

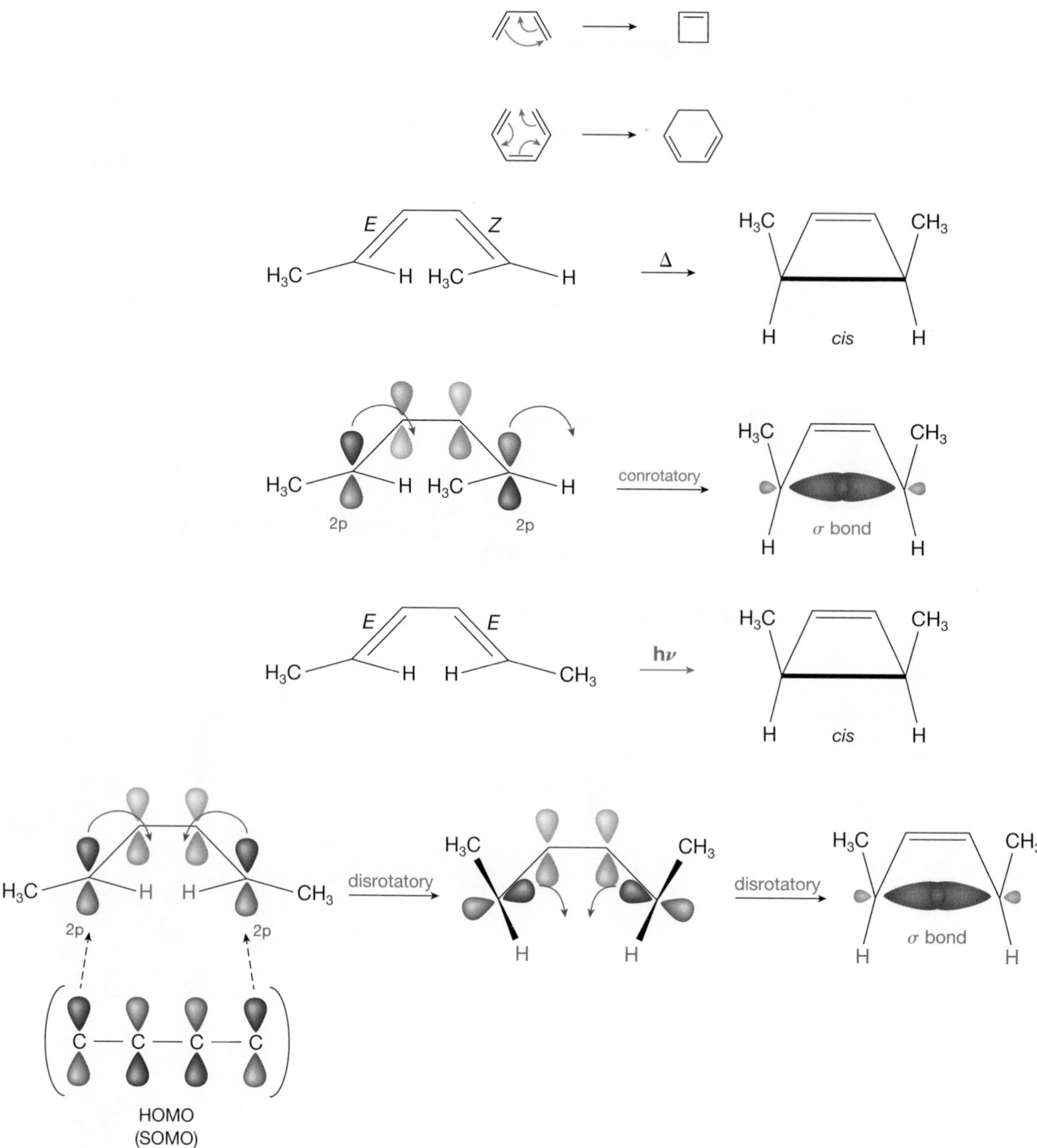

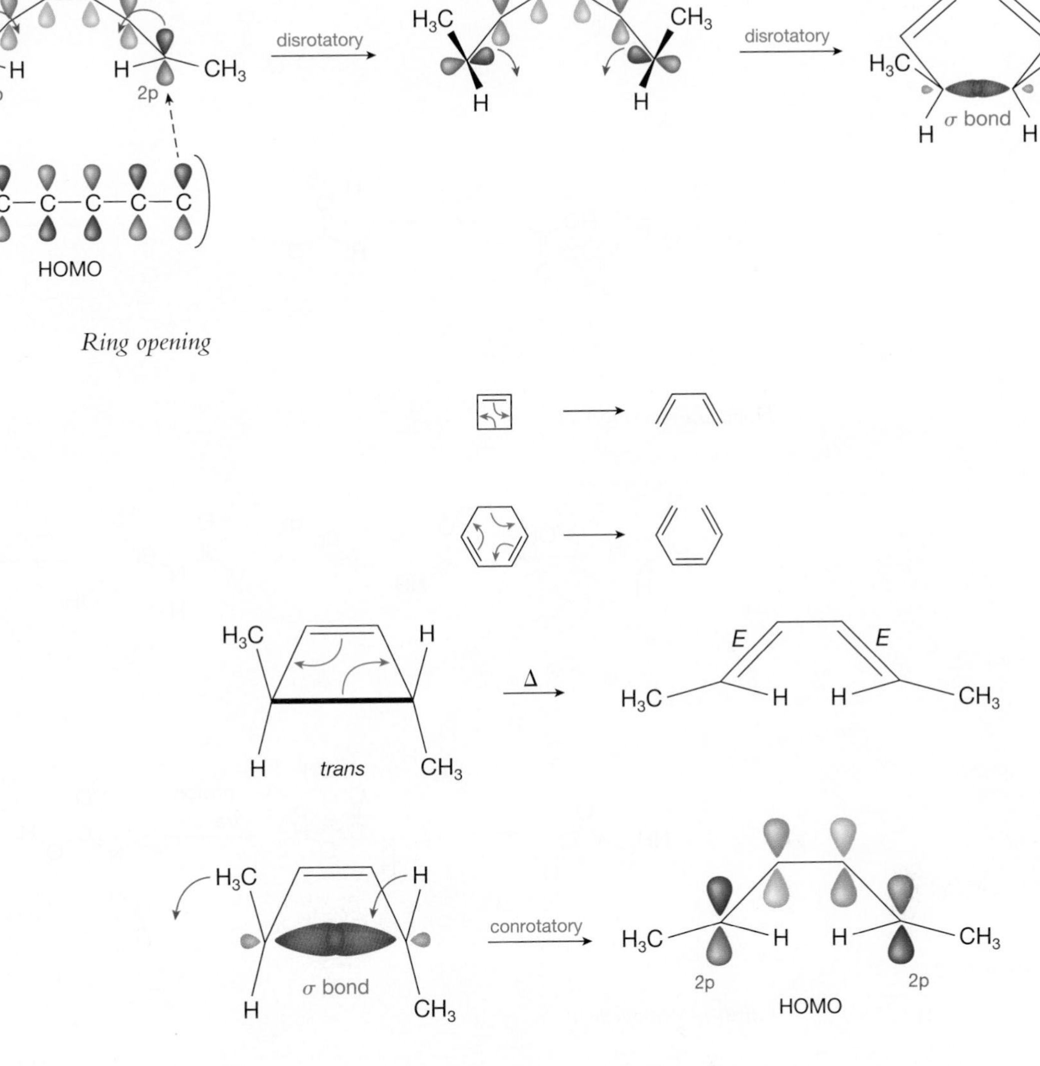

Ring opening

NON-SIGMATROPIC REARRANGEMENTS

Baeyer–Villiger rearrangement

Hofmann rearrangement

proton transfer

Curtius rearrangement

NaN_3

Lossen rearrangement

TsCl

base

base

Δ

+ TsOH

Problems

20.21 Show the missing products in the following sequence of reactions. Indicate whether the reactions are conrotatory or disrotatory.

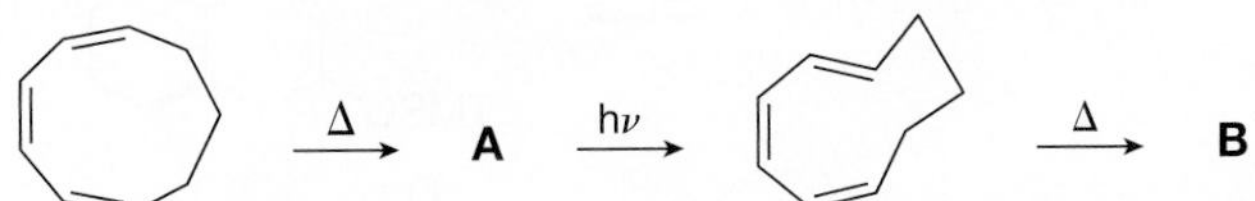

20.22 Explain what is happening in the following reactions. Are the processes conrotatory or disrotatory? Why are different products formed? Hint: Build models of the products.

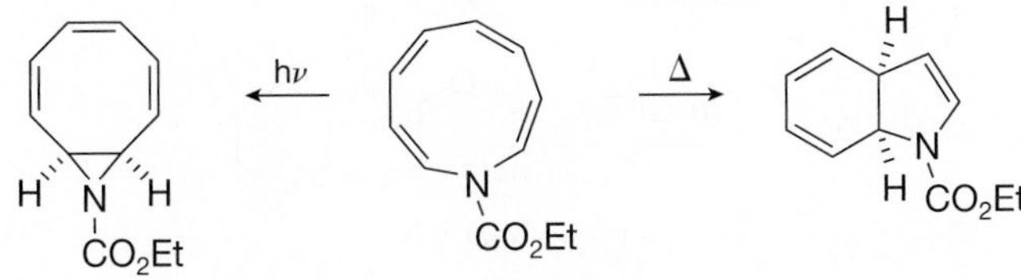

20.23 Identify the mode (conrotatory or disrotatory) of ring closure or opening in the following. Are the hydrogens in the products *syn* or *anti*?

a) hν →

b) H, Me, H, Me Δ →

c) H, H hν → Δ →

20.24 Show the products of the following reactions.

a) Ph, Me Δ →

b) Δ →

c) OMe, H Δ →

20.25 Show the products of the following reactions, considering both regiochemistry and stereochemistry.

a) OMe + CO_2Me →

b) HO + O →

c) EtO + EtO_2C, S →

20.26 Show the product of the following reactions, and give a mechanism for the transformations.

a) O, OH; 1) $SOCl_2$ 2) NaN_3 3) Δ, EtOH →

b) O, NH_2, Cl, N, Cl; Br_2, NaOH, H_2O, Δ →

c) O, H; 1) NH_2OH 2) H_3PO_4 →

d) O, H; 1) *m*-CPBA 2) $LiAlH_4$ →

20.27 Explain the formation of A and B in the following reaction. Identify the direction of rotation, and give a name for each reaction. Identify the stereochemistry at the ring fusion of compounds A and C.

A Δ → B Δ → C

20.28 Indicate whether the following reactions are thermal, photochemical, or whether a chemical reagent is required.

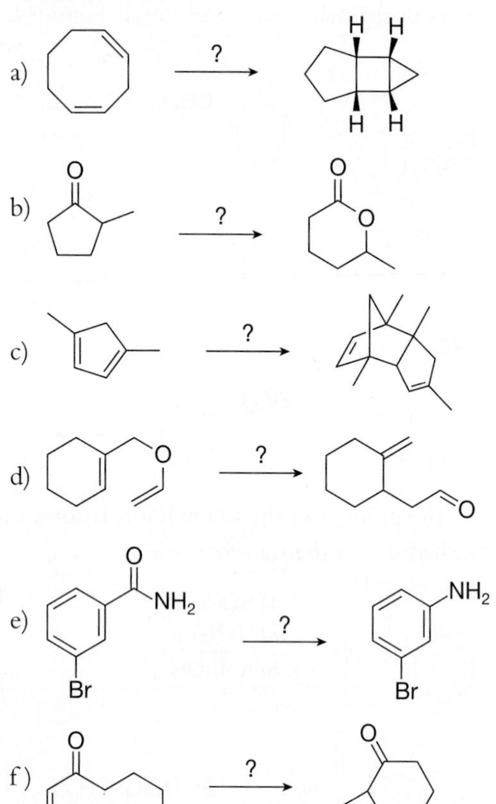

20.29 (*R*,2*E*,4*Z*)-6-Methylocta-2,4-diene undergoes [1,5] hydrogen migration instead of electrocyclic ring closure. Suggest a rationale for this observation.

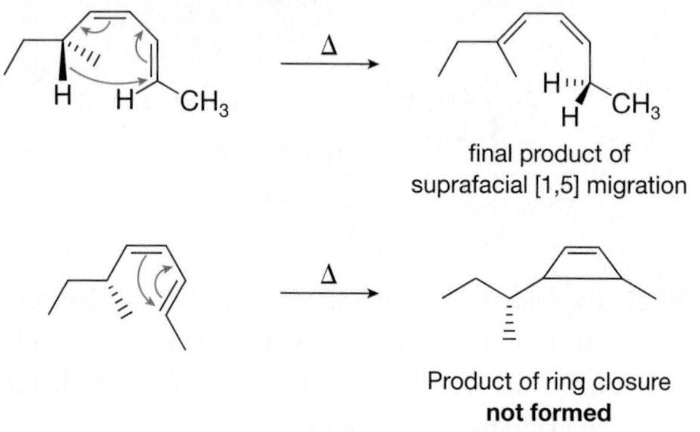

H H CH$_3$ — Δ → H CH$_3$ H

final product of suprafacial [1,5] migration

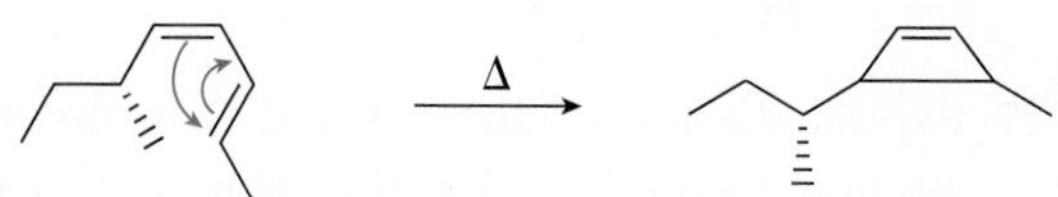

Product of ring closure **not formed**

20.30 The synthesis of a monosubstituted cyclopentadiene results in a mixture of three isomers. Explain why this happens.

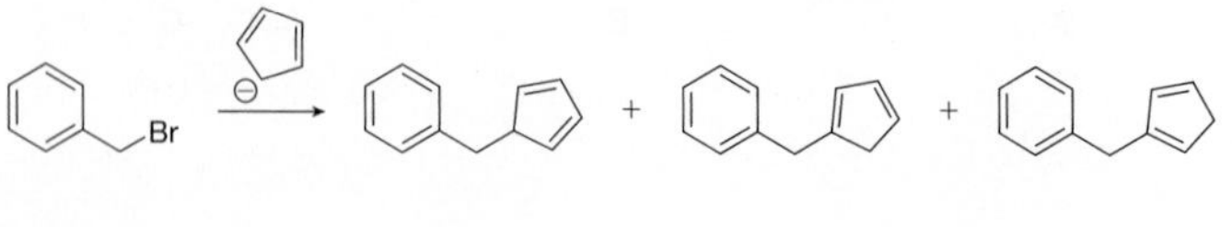

Br ⊖ → + +

20.31 What are the conditions or starting materials required to make the following products? Use reactions from this chapter.

a) 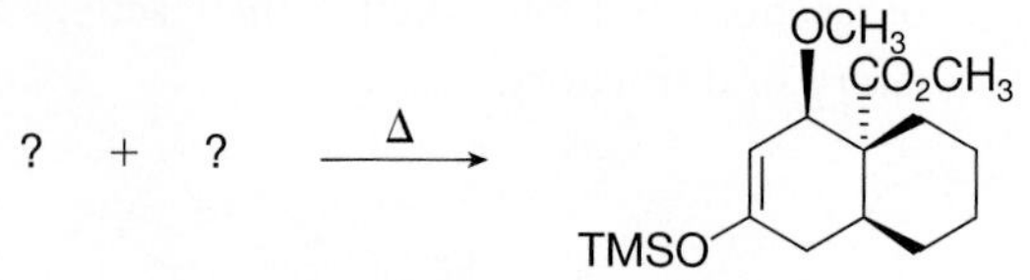 ? + ? Δ→ OCH$_3$ CO$_2$CH$_3$ TMSO

b) 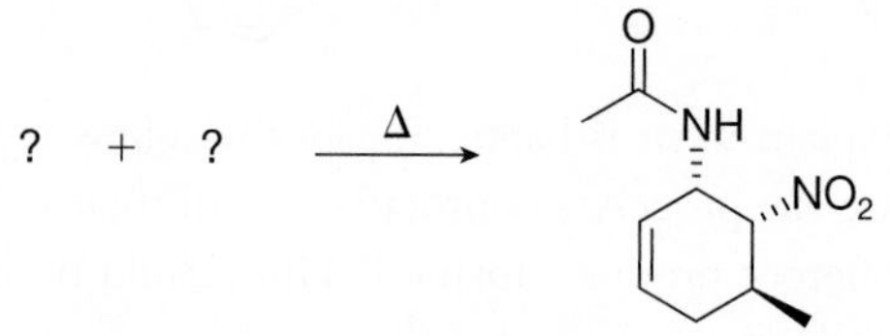 ? + ? Δ→ O NH NO$_2$

c) 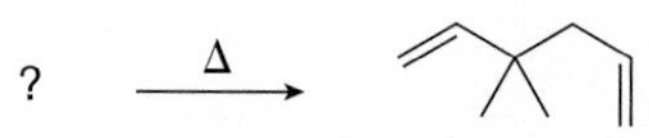 ? Δ→

d) 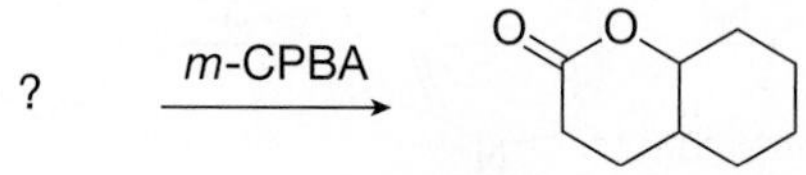 ? *m*-CPBA→ O O

e) 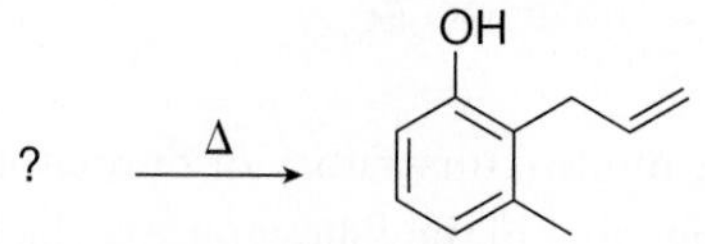 ? Δ→ OH

f) 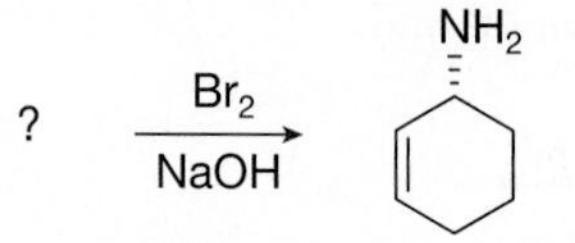 ? Br$_2$ NaOH→ NH$_2$

20.32 For the following reactions, show the missing product, and indicate what kind of reaction is occurring. Show the stereochemistry of products.

a) 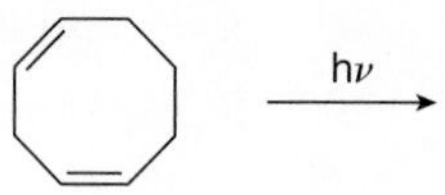 hν→

b) 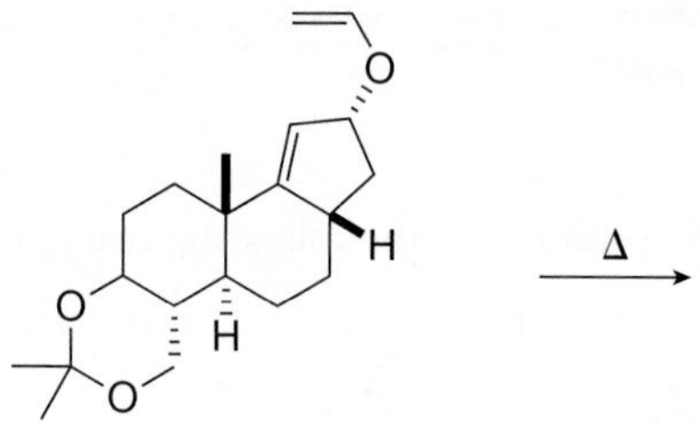 O H O H O Δ→

c) 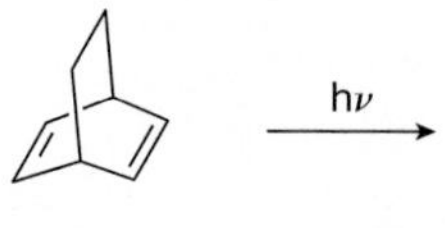 hν→

d) 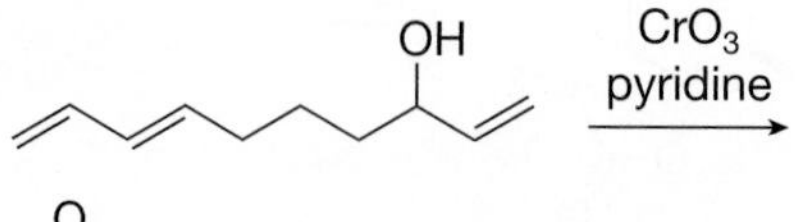 OH CrO$_3$ pyridine→

e) 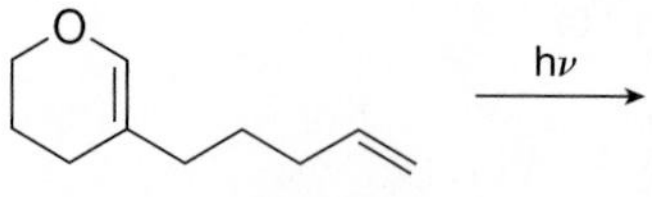 O hν→

f) 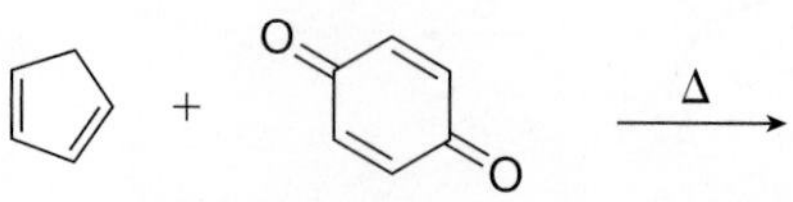 + O O Δ→

g) 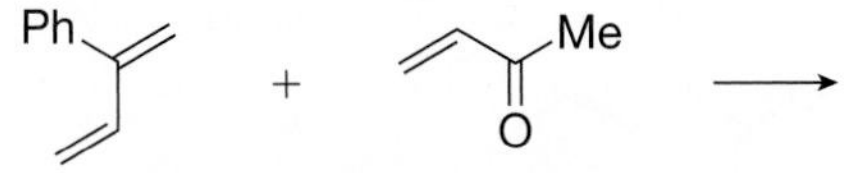 Ph + Me O ⟶

h) 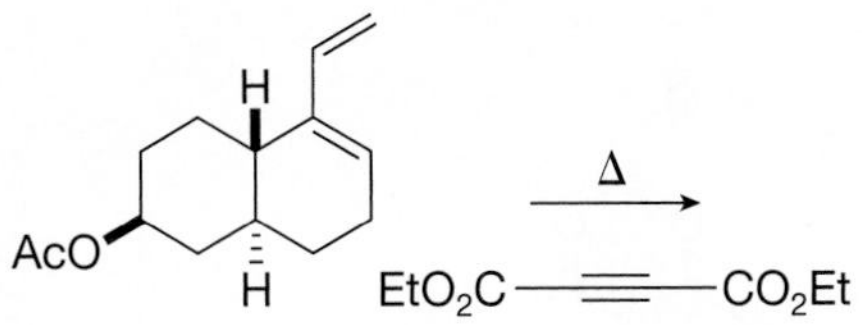 AcO H H + EtO_2C—≡—CO_2Et $\xrightarrow{\Delta}$

i) 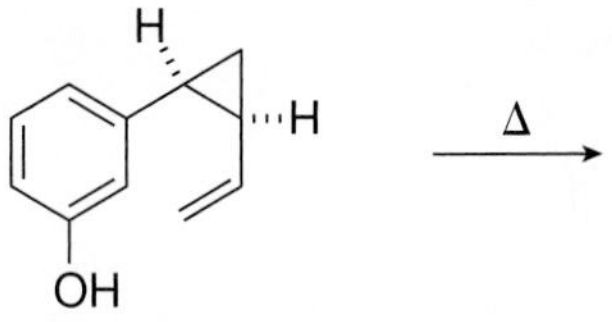 H H OH $\xrightarrow{\Delta}$

j) 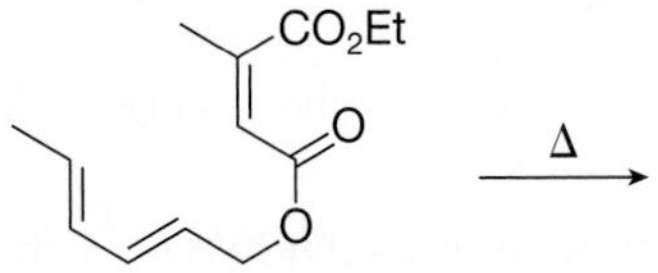 CO_2Et O O $\xrightarrow{\Delta}$

k) 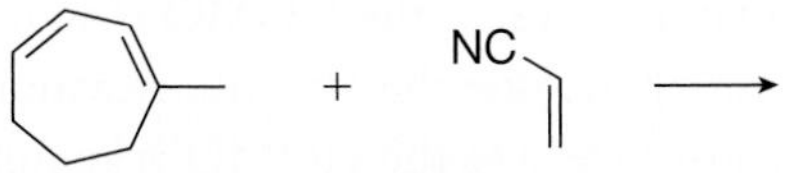 + NC ⟶

l) 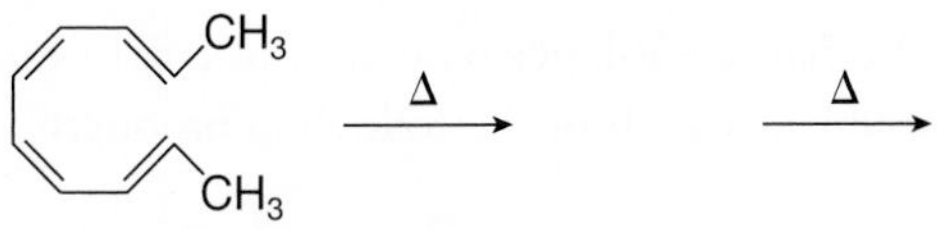 CH_3 CH_3 $\xrightarrow{\Delta}$ $\xrightarrow{\Delta}$

m) 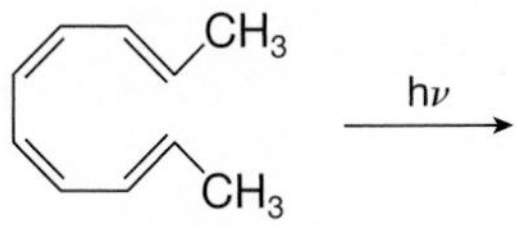 CH_3 CH_3 $\xrightarrow{h\nu}$

n) 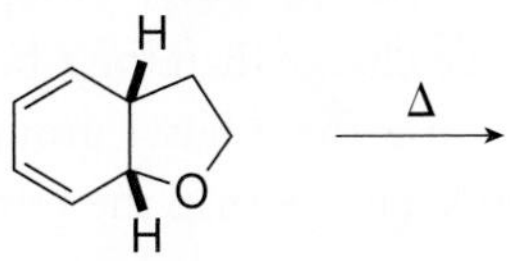 H O H $\xrightarrow{\Delta}$

o) CH_3 CH_3 $\xrightarrow{\Delta}$

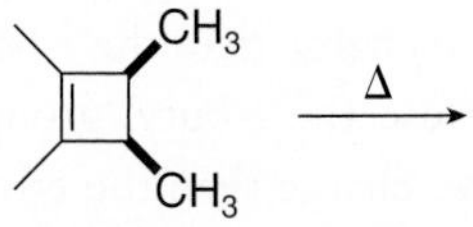

p) 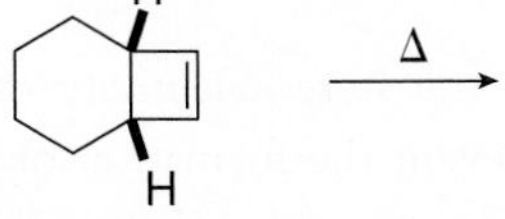 H H $\xrightarrow{\Delta}$

20.33 Show the product of the following reaction. Heating gives a 4π electrocyclic ring opening followed by a cycloaddition reaction. Account for the stereochemistry of the product.

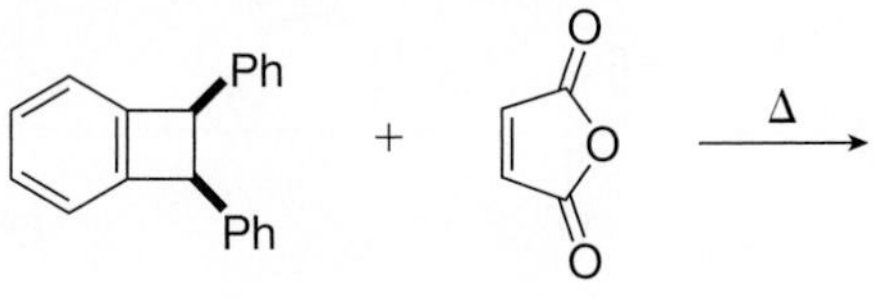

Ph Ph + O O O $\xrightarrow{\Delta}$

20.34 Propose a synthesis of the following products using the indicated starting materials.

a) 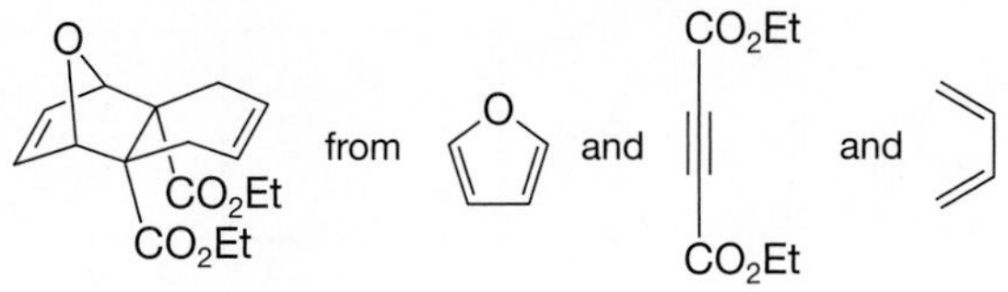 O CO_2Et CO_2Et from O and CO_2Et ≡ CO_2Et and

b) 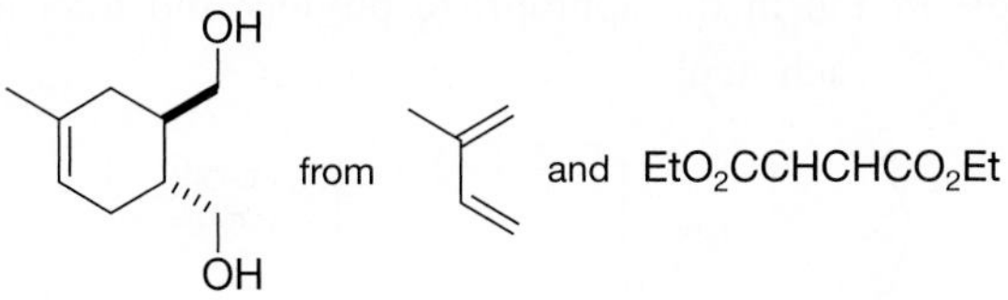 OH OH from and $EtO_2CCHCHCO_2Et$

c) 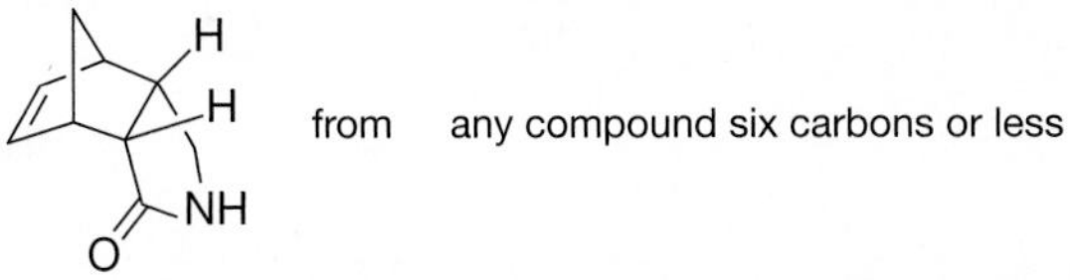 H H NH O from any compound six carbons or less

20.35 The nucleophilic reactions of cyclopropanes involve S_N1-type displacements leading to the formation of open-chain products. This is because the cyclopropyl carbocation formed when the chloride leaving group departs undergoes an electrocyclic ring-opening reaction.

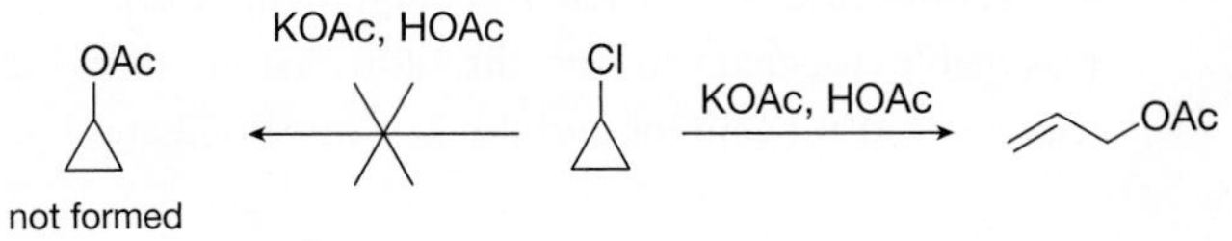

OAc (not formed) ⟵✕— KOAc, HOAc — Cl $\xrightarrow{\text{KOAc, HOAc}}$ OAc

a) Explain the above reaction by drawing a mechanism.
b) Is the opening in this reaction conrotatory or disrotatory?

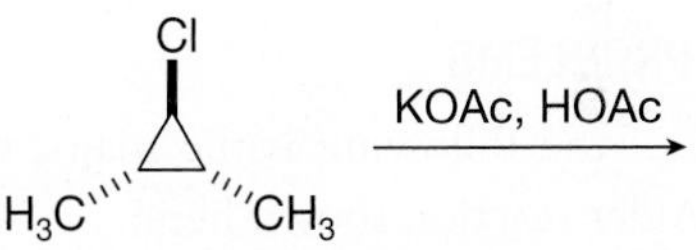

Cl H_3C CH_3 $\xrightarrow{\text{KOAc, HOAc}}$

20.36 The endriandric acids are a family of complex natural products isolated from an Australian plant. These substances are highly unusual because they occur in racemic form, suggesting that the chemical reactions the plant uses to form them do not involve enzymes. Three of the endriandric acids (D, E, and A) arise from a common precursor. Endriandric acids D and E form by means of a thermal 8π electrocyclization followed by a thermal 6π ring closure. The A isomer is produced when endriandric acid E undergoes an intramolecular Diels–Alder reaction. Use mechanisms to derive the structures of all three isomers (D, E, and A).

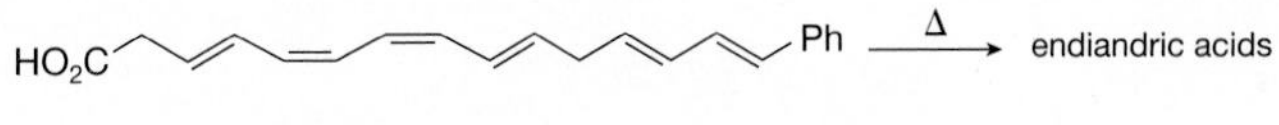

HO_2C Ph $\xrightarrow{\Delta}$ endriandric acids

20.37 Provide the missing structure, and explain the conversion to the final product.

CO_2Et CO_2Et Δ Δ CO_2Et CO_2Et $+CO_2$

20.38 Fill in the appropriate product and intermediate for each step.

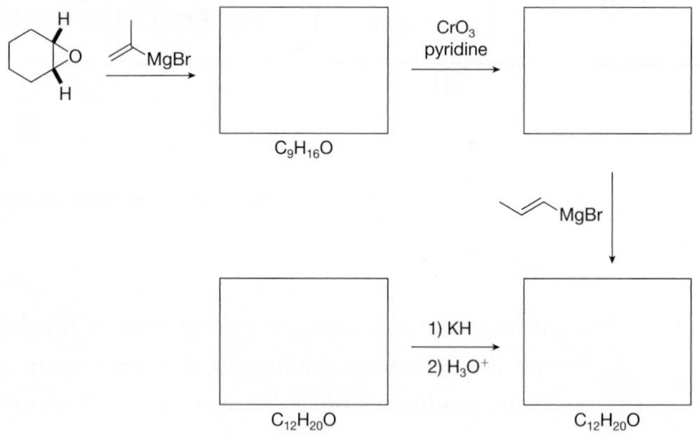

H O H MgBr $C_9H_{16}O$ CrO_3 pyridine MgBr 1) KH 2) H_3O^+ $C_{12}H_{20}O$ $C_{12}H_{20}O$

20.39 The following synthesis involves a [2+2] cycloaddition, followed by an organometallic addition to a carbonyl and a thermal rearrangement. Propose a reasonable mechanism for this conversion, showing the electron movement and the key intermediates.

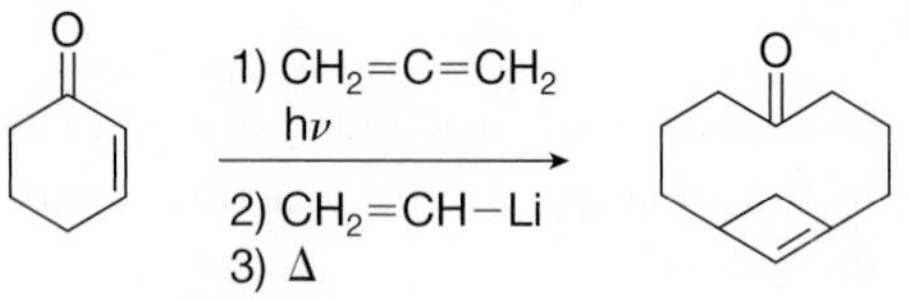

1) $CH_2{=}C{=}CH_2$ $h\nu$
2) $CH_2{=}CH{-}Li$
3) Δ

MCAT STYLE PROBLEMS

20.40 Which of the following is the major product of the Diels–Alder reaction shown here?

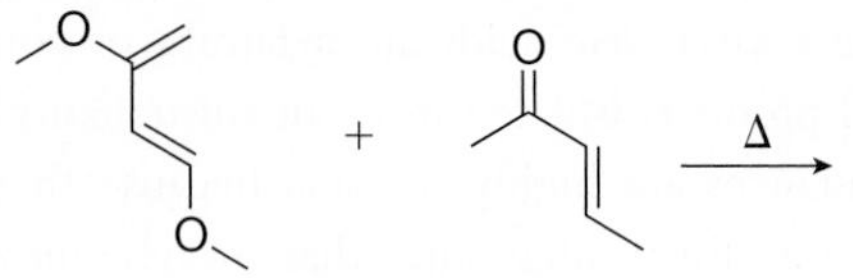

a)

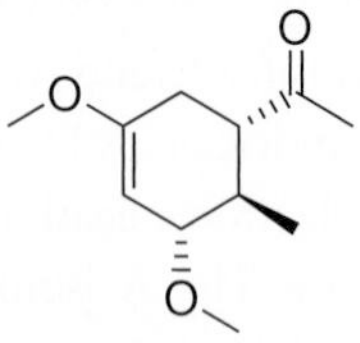

b)

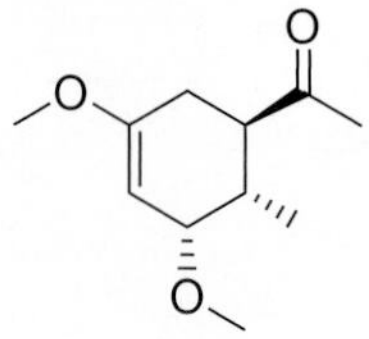

c)

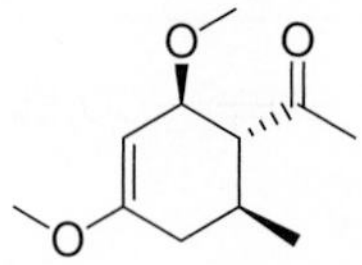

d)

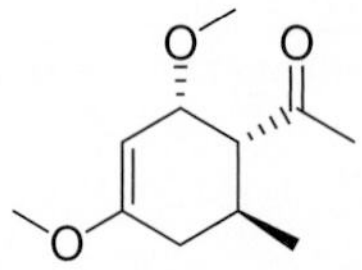

20.41 The thermal electrocyclic ring closure of 1,3,5-hexatriene is

a) conrotatory, because the HOMO is symmetric.
b) conrotatory, because the HOMO is asymmetric.
c) disrotatory, because the HOMO is symmetric.
d) disrotatory, because the HOMO is asymmetric.

20.42 The Baeyer–Villiger oxidation of ethyl *t*-butyl ketone results in which of the following products?

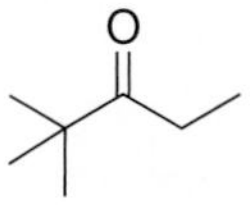

a) an ethyl ester, because the ethyl group is more stable with a positive charge than the *t*-butyl group
b) an ethyl ester, because the *t*-butyl group is more stable with a positive charge than the ethyl group
c) a *t*-butyl ester, because the ethyl group is more stable with a positive charge than the *t*-butyl group
d) a *t*-butyl ester, because the *t*-butyl group is more stable with a positive charge than the ethyl group

CHALLENGE PROBLEM

20.43 One way to control the stereoselectivity of a Diels–Alder reaction, to prevent the formation of a racemic mixture, is to add the amino acid proline to the reaction of aldehyde dienophiles. How does the addition of proline change the reaction to prevent the formation of one of the possible enantiomers?

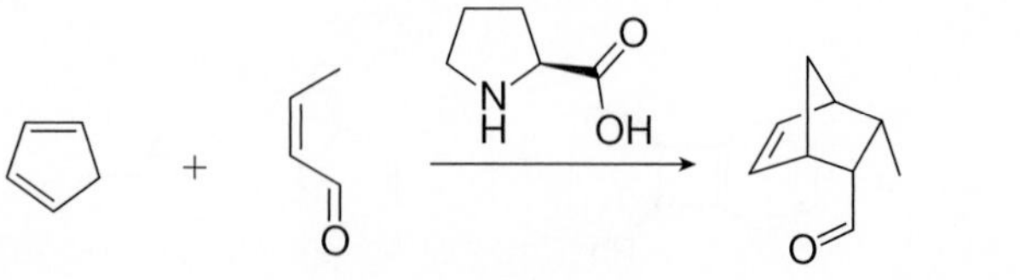

APPENDIX A

Answers to Checkpoint Problems

CHAPTER 1

PRACTICE PROBLEM 1.1

a) S — $1s^22s^22p^63s^23p^4$

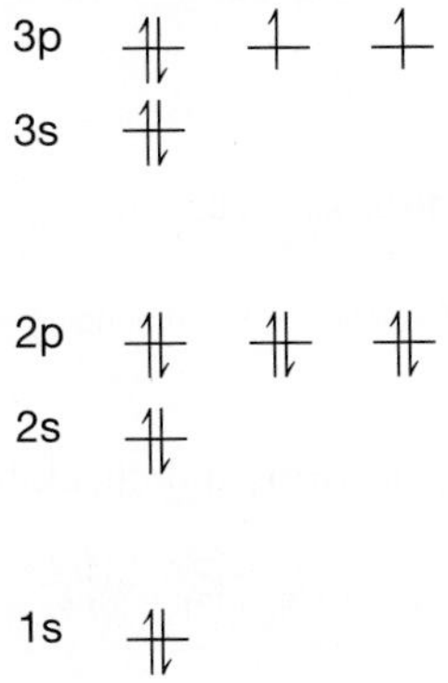

3p ⥮ ↿ ↿
3s ⥮

2p ⥮ ⥮ ⥮
2s ⥮

1s ⥮

b) Cl — $1s^22s^22p^63s^23p^5$

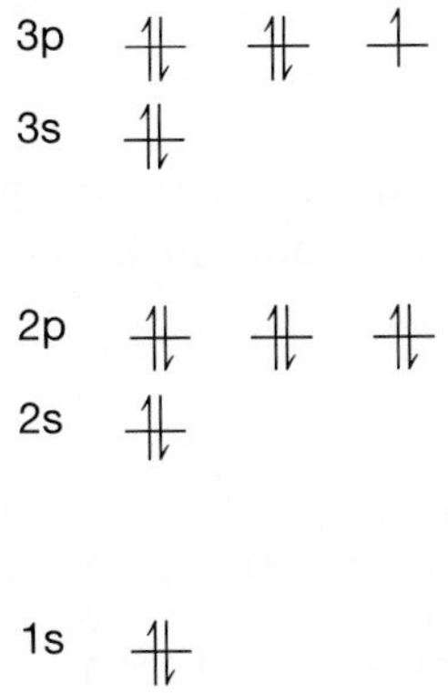

3p ⥮ ⥮ ↿
3s ⥮

2p ⥮ ⥮ ⥮
2s ⥮

1s ⥮

c) Na^+ — $1s^22s^22p^6$

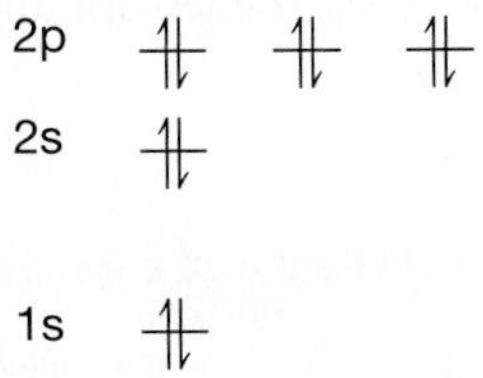

2p ⥮ ⥮ ⥮
2s ⥮

1s ⥮

PRACTICE PROBLEM 1.2

a) Count valence electrons.

$CH_3CH_2NH_3$

2 carbons (group 4)	2 × 4 = 8
7 hydrogens (group 1)	7 × 1 = 7
1 nitrogen (group 5)	1 × 5 = 5
	20 valence electrons

Build a basic bonding framework and account for electrons used.

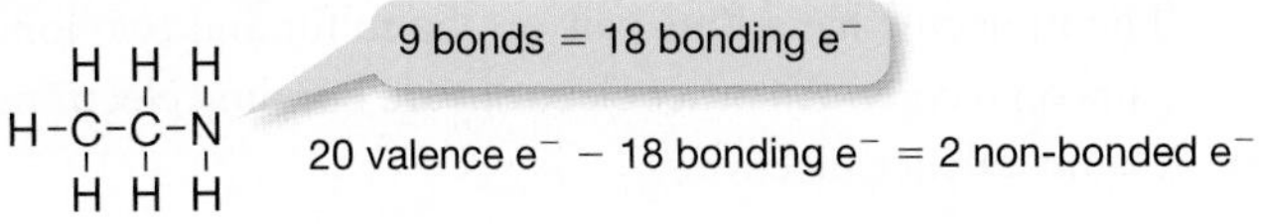

9 bonds = 18 bonding e^-

20 valence e^- − 18 bonding e^- = 2 non-bonded e^-

Add remaining electrons and check for formal charges.

FC = (5) − (3) − (2) = 0

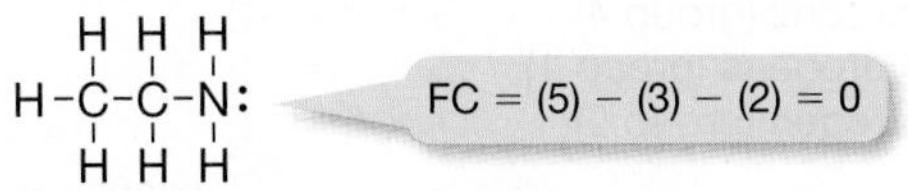

The molecule has a lone pair on the nitrogen. All other electrons are bonding electrons.

b) Count valence electrons.

$CH_3S(O)CH_3$ (oxygen is connected to sulfur only)

2 carbons (group 4)	2 × 4 = 6
6 hydrogens (group 1)	6 × 1 = 6
1 oxygen (group 6)	1 × 6 = 6
1 sulfur (group 6)	1 × 6 = 6
	26 valence e^-

Build a basic bonding framework and account for electrons used.

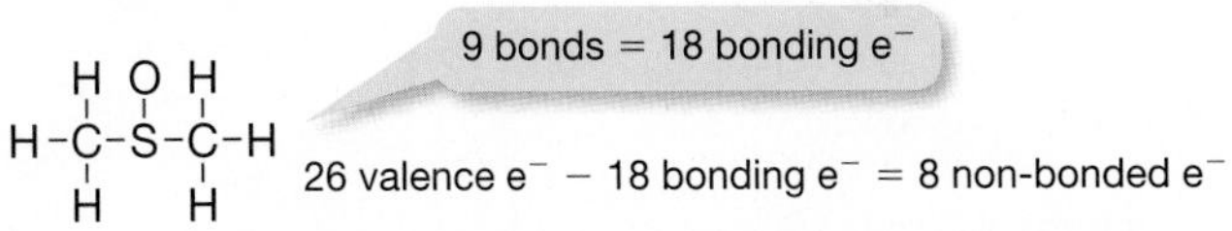

9 bonds = 18 bonding e^-

26 valence e^- − 18 bonding e^- = 8 non-bonded e^-

Add remaining electrons and check for formal charges.

FC = (6) − (1) − (6) = −1

FC = (6) − (3) − (2) = +1

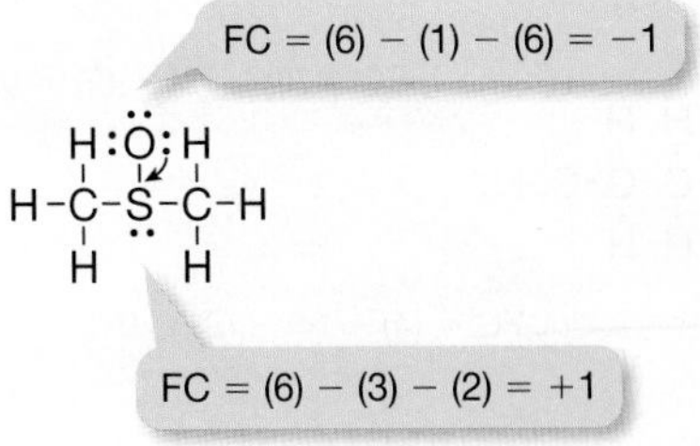

There are formal charges on the S and O atoms. They can be removed by making an additional bond between O and S.

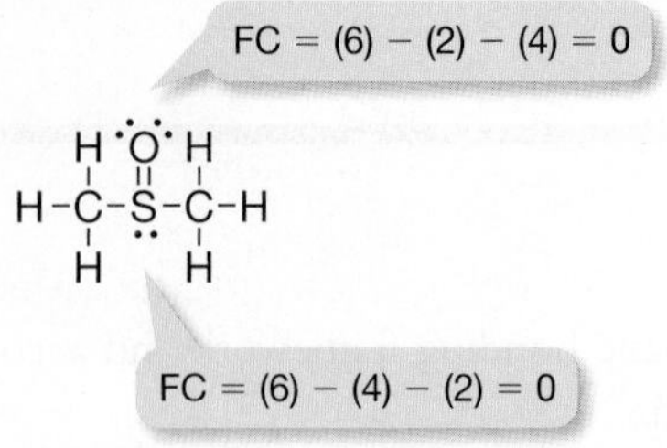

FC = (6) − (2) − (4) = 0

FC = (6) − (4) − (2) = 0

The molecule has a lone pair on the sulfur and two lone pairs on oxygen. All other electrons are bonding electrons.

c) Count valence electrons.

CH_3CH_2CN (nitrogen is connected to one carbon only)

3 carbons (group 4)	3 × 4 = 12
5 hydrogens (group 1)	5 × 1 = 5
1 nitrogen (group 5)	1 × 5 = 5
	22 valence e^-

Build a basic bonding framework and account for electrons used.

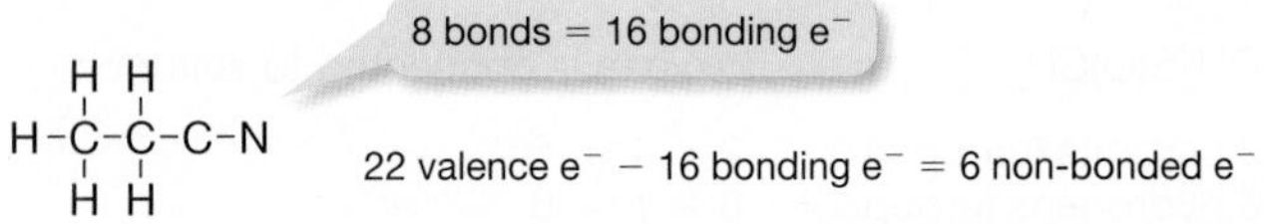

8 bonds = 16 bonding e^-

22 valence e^- − 16 bonding e^- = 6 non-bonded e^-

Add remaining electrons and check for formal charges.

FC = (5) − (1) − (6) = −2

FC = (4) − (2) − (0) = +2

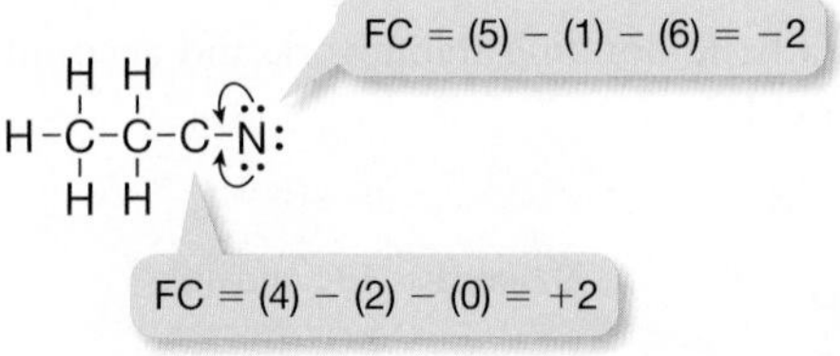

The formal charges on C and N show the carbon needs more electrons and the N has too many. Forming two more bonds between C and N alleviates this problem.

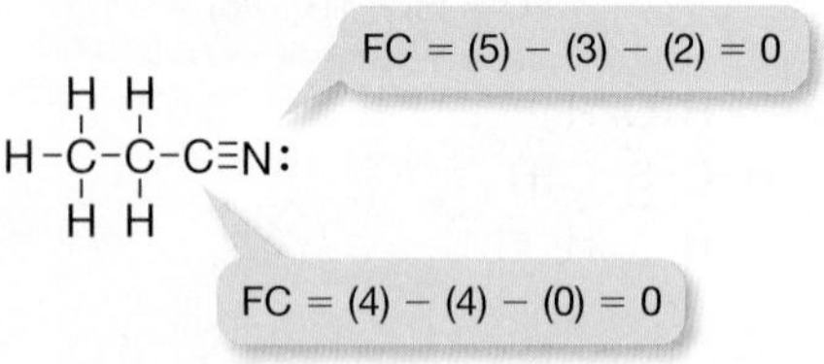

FC = (5) − (3) − (2) = 0

FC = (4) − (4) − (0) = 0

The molecule has a lone pair on the nitrogen. All other electrons are bonding electrons.

d) Count valence electrons.

$(CH_3)_2CHO^-$

3 carbons (group 4)	3 × 4 = 12
7 hydrogens (group 1)	7 × 1 = 7
1 oxygen (group 6)	1 × 6 = 6
1 negative charge	1 × 1 = 1
	26 valence e^-

Build a basic bonding framework and account for electrons used.

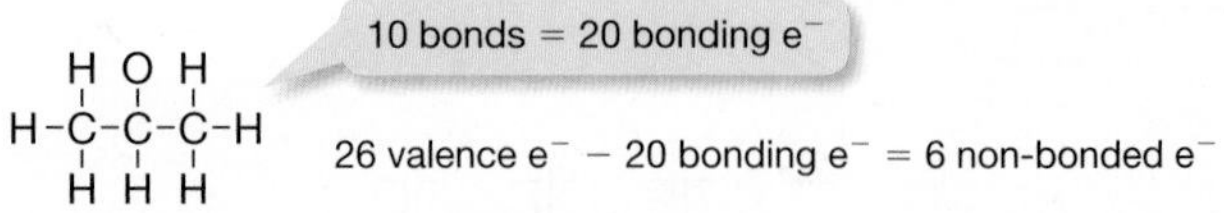

10 bonds = 20 bonding e^-

26 valence e^- − 20 bonding e^- = 6 non-bonded e^-

Add remaining electrons and check for formal charges.

FC = (6) − (1) − (6) = −1

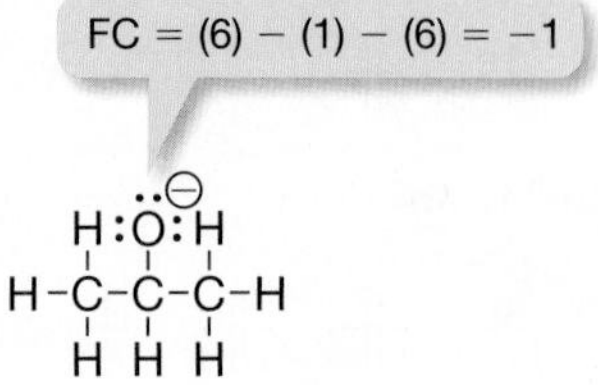

The oxygen atom has three lone pairs and a positive charge. All other electrons are bonding electrons.

e) Count valence electrons.

$(CH_3)_4N^{\oplus}$

4 carbons (group 4)	4 × 4 = 16
12 hydrogens (group 1)	12 × 1 = 12
1 nitrogen (group 5)	1 × 5 = 5
1 positive charge	1 × −1 = −1
	32 valence e^-

Build a basic bonding framework and account for electrons used.

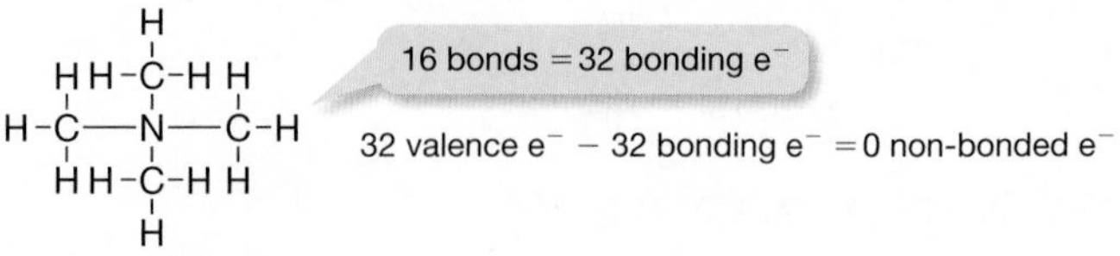

16 bonds = 32 bonding e^-

32 valence e^- − 32 bonding e^- = 0 non-bonded e^-

Add remaining electrons and check for formal charges.

FC = (5) − (4) − (0) = +1

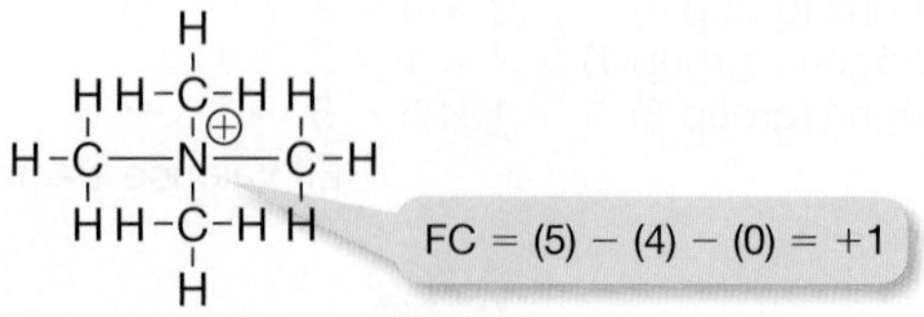

The ion has a formal positive charge on the nitrogen. All electrons are bonding electrons.

f) Count valence electrons.

$HSO_3^{\ominus}$ (hydrogen is connected to oxygen only)

1 hydrogen (group 1)	1 × 1 =	1
3 oxygens (group 6)	3 × 6 =	18
1 sulfur (group 6)	1 × 6 =	6
1 negative charge	1 × 1 =	1
		26 valence e^-

Build a basic bonding framework and account for electrons used.

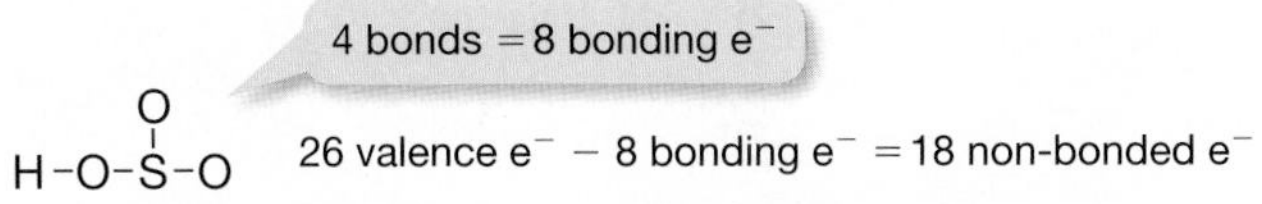

4 bonds = 8 bonding e^-

26 valence e^- − 8 bonding e^- = 18 non-bonded e^-

Add remaining electrons and check for formal charges.

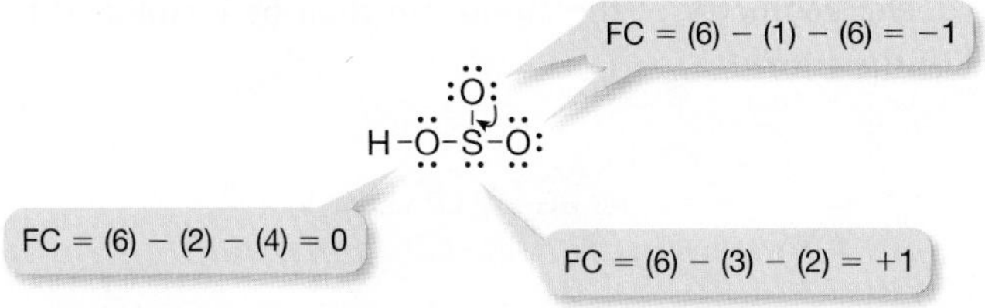

FC = (6) − (1) − (6) = −1

FC = (6) − (2) − (4) = 0

FC = (6) − (3) − (2) = +1

There are formal charges on the sulfur (+1) and two of the oxygens (−1). These can be reduced by forming a double bond between sulfur and either of the oxygen atoms carrying a formal charge.

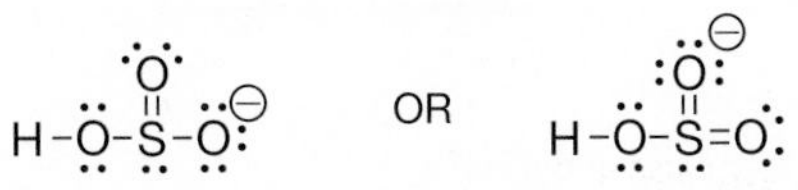

OR

The ion has seven lone pairs on oxygen atoms and one lone pair on sulfur. All other electrons are bonding electrons. One oxygen has a formal negative charge.

g) Count valence electrons.

$HSO_3^{\oplus}$ (hydrogen is connected to oxygen only)

1 hydrogen (group 1)	1 × 1 =	1
3 oxygens (group 6)	3 × 6 =	18
1 sulfur (group 6)	1 × 6 =	6
1 positive charge	1 × −1 =	−1
		24 valence e^-

Build a basic bonding framework and account for electrons used.

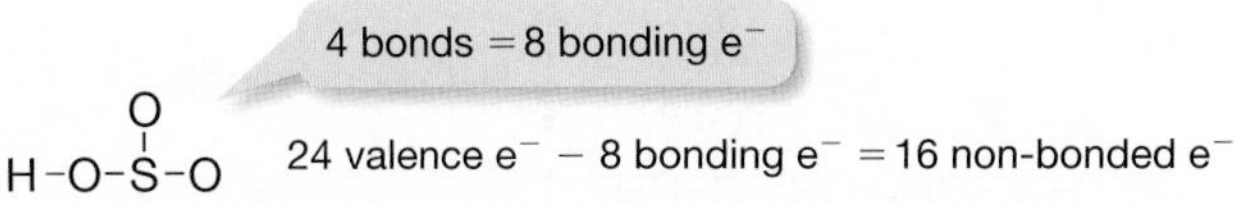

4 bonds = 8 bonding e^-

24 valence e^- − 8 bonding e^- = 16 non-bonded e^-

Add remaining electrons (on oxygen first) and check for formal charges.

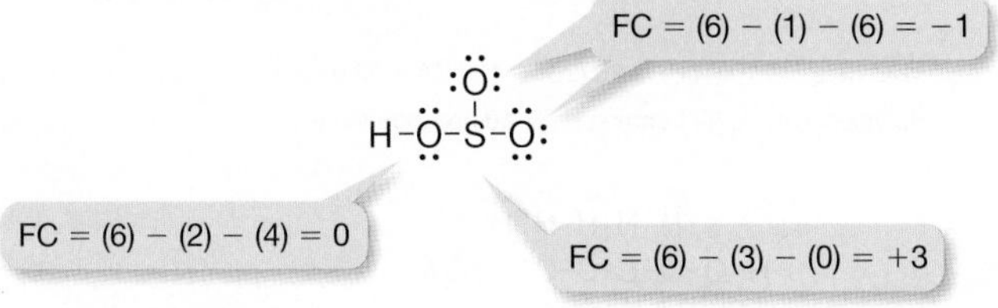

FC = (6) − (1) − (6) = −1

FC = (6) − (2) − (4) = 0

FC = (6) − (3) − (0) = +3

There are formal charges on the sulfur (+3) and two of the oxygens (−1). These can be reduced by forming double bonds between sulfur and both charged oxygen atoms. This expands the octet of the sulfur but, since it is a third-row element, this is allowed.

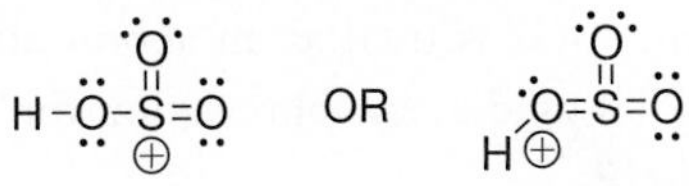

OR

The final ion has six lone pairs on oxygen atoms. The sulfur has a formal positive charge. All other electrons are bonding electrons.

PRACTICE PROBLEM 1.3

a) valence e^- = 9 (9H) + 16 (4C) + 7 (1Cl) = 32 e^-

32 valence e^- − 26 bonding e^- = 6 non-bonded e^-

Lewis Structure

dipole

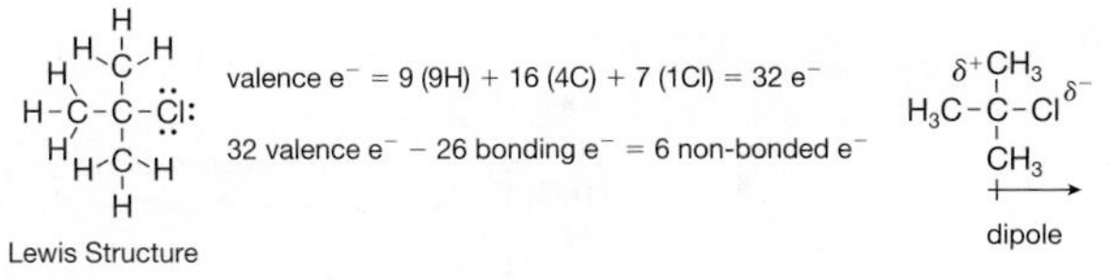

b) Lewis Structure

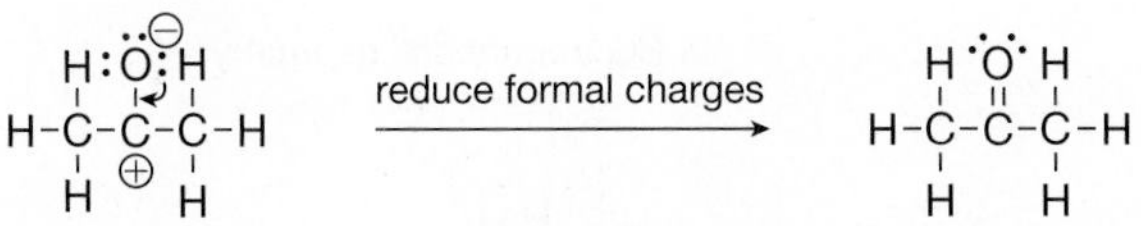

reduce formal charges

valence e^- = 9 (9H) + 12 (3C) + 7 (1Cl) = 28 e^-

28 valence e^- − 18 bonding e^- = 6 non-bonded e^-

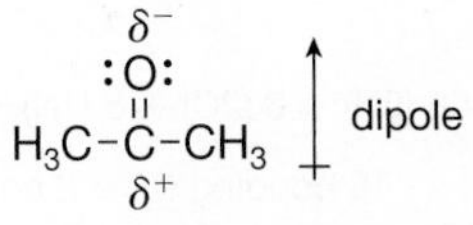

dipole

c) valence e^- = 12 (12H) + 20 (5C) + 6 (1O) = 38 e^-

38 valence e^- − 34 bonding e^- = 4 non-bonded e^-

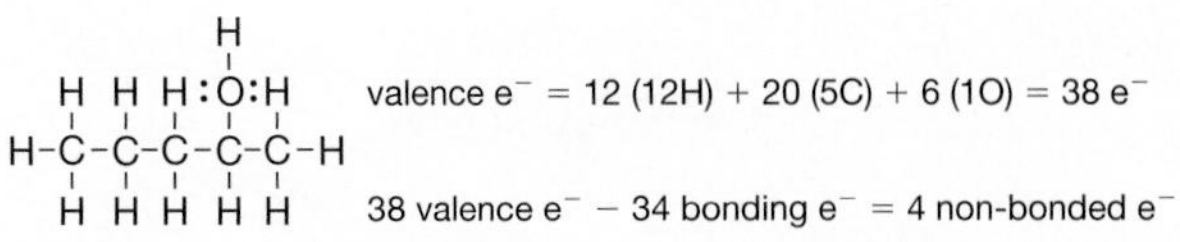

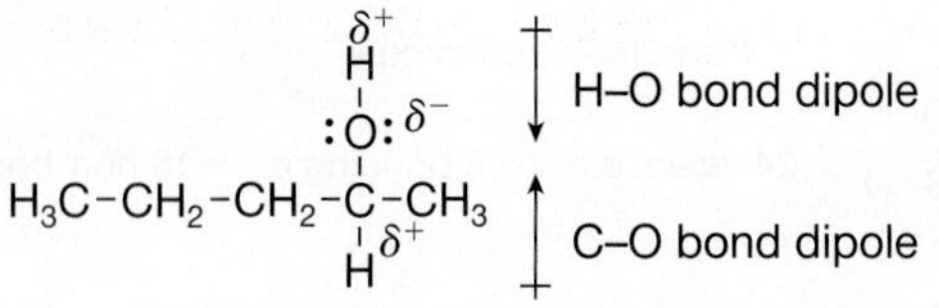

H–O bond dipole

C–O bond dipole

d)

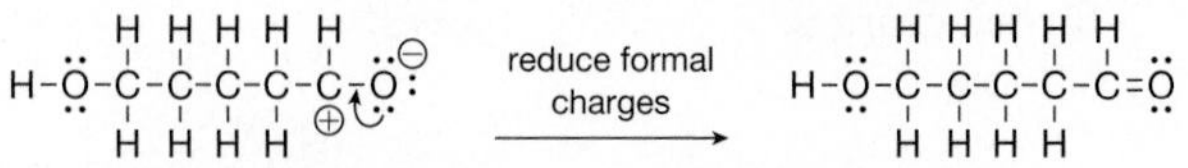

reduce formal charges

valence e$^-$ = 10 (10H) + 20 (5C) + 12 (2O) = 42 e$^-$

42 valence e$^-$ − 32 bonding e$^-$ = 10 non-bonded e$^-$

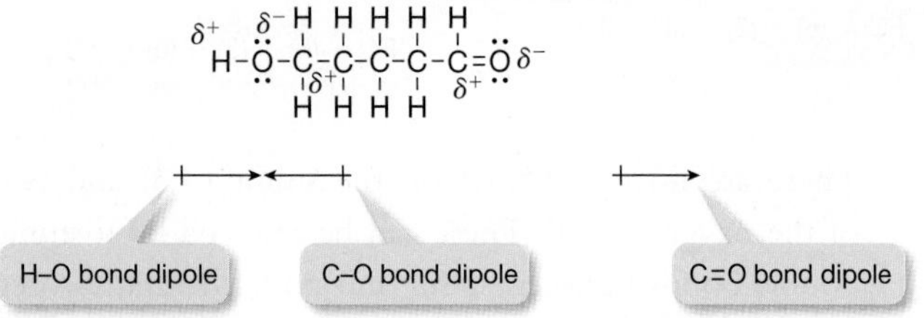

H–O bond dipole

C–O bond dipole

C=O bond dipole

PRACTICE PROBLEM 1.4

In the solutions, "BG" is used as an abbreviation for "bond group" and "LP" is used as an abbreviation for "lone pair."

a)

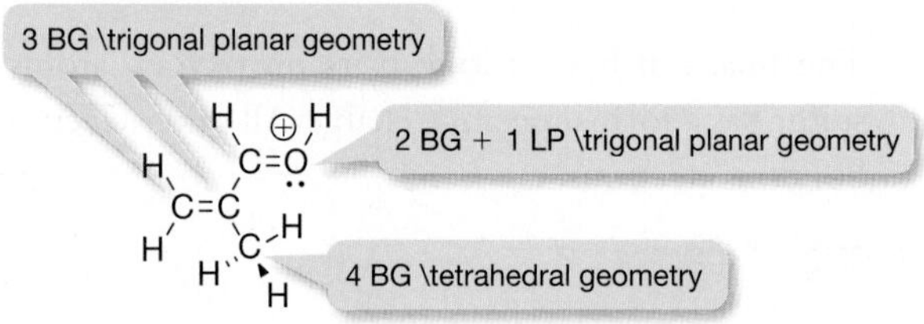

3 BG \trigonal planar geometry

2 BG + 1 LP \trigonal planar geometry

4 BG \tetrahedral geometry

b)

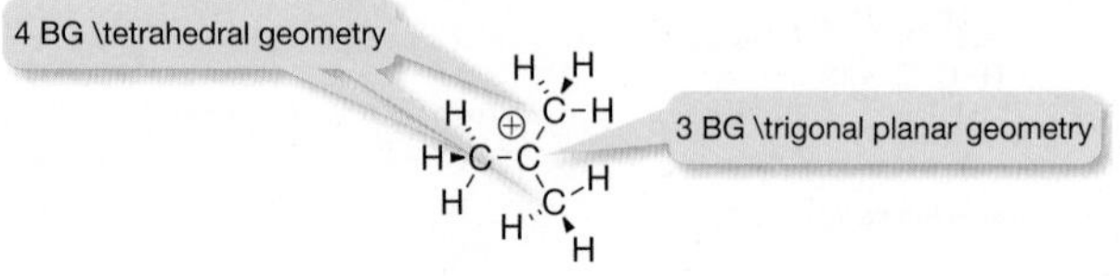

4 BG \tetrahedral geometry

3 BG \trigonal planar geometry

c)

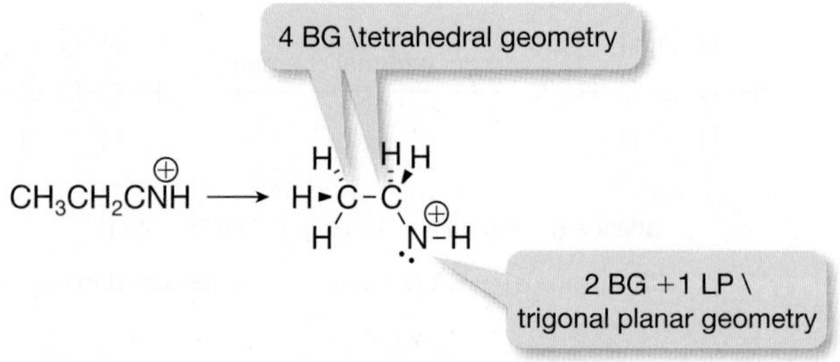

4 BG \tetrahedral geometry

CH_3CH_2CNH ⟶

2 BG +1 LP \ trigonal planar geometry

valence e$^-$ = 6 (6H) + 8 (2C) + 5 (1N)−1 (+ charge) = 18 e$^-$

18 valence e$^-$ − 16 bonding e$^-$ = 2 non-bonded e$^-$

INTEGRATE THE SKILL 1.5

All carbons have four bonds and so will not have lone pairs. Lone pairs are added to the nitrogen atoms according to the formal charges indicated. "BP" refers to shared pairs of electrons in bonds between atoms where each atom formally has one of the electrons.

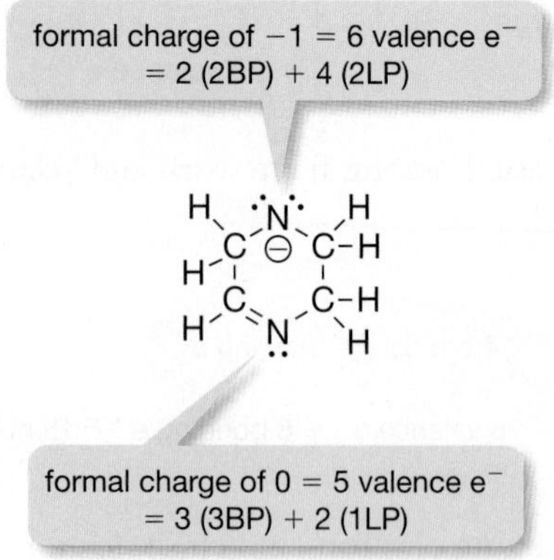

formal charge of −1 = 6 valence e$^-$ = 2 (2BP) + 4 (2LP)

formal charge of 0 = 5 valence e$^-$ = 3 (3BP) + 2 (1LP)

The geometry of the atoms can then be established for all of the atoms.

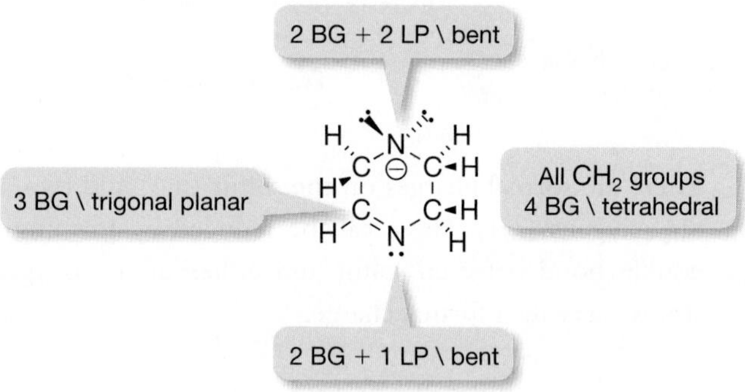

2 BG + 2 LP \ bent

3 BG \ trigonal planar

All CH_2 groups 4 BG \ tetrahedral

2 BG + 1 LP \ bent

PRACTICE PROBLEM 1.6

a)

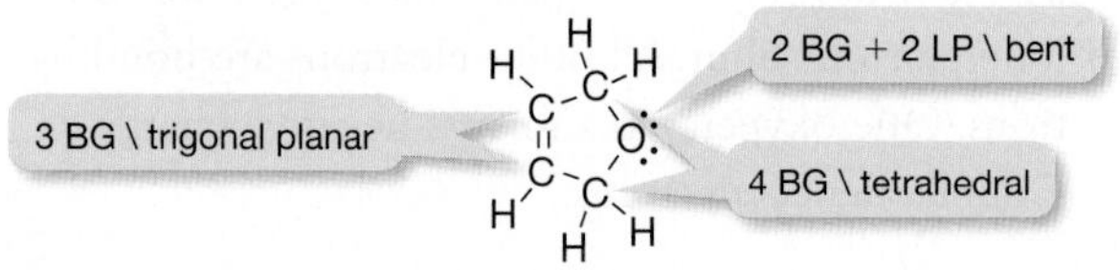

2 BG + 2 LP \ bent

3 BG \ trigonal planar

4 BG \ tetrahedral

b)

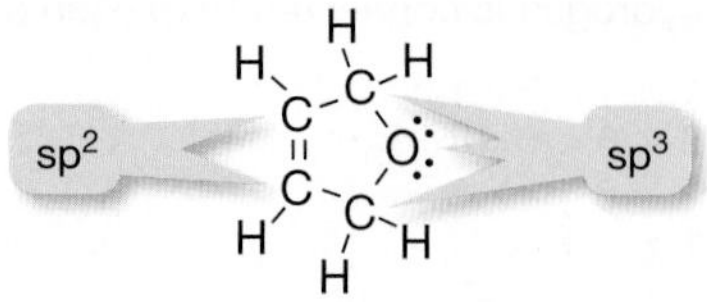

sp^2

sp^3

INTEGRATE THE SKILL 1.7

a) Accounting for all of the electrons leaves four non-bonded electrons to add as lone pairs. These are added to the oxygen first (most electronegative), leaving the carbon with a formal positive charge.

valence e^- = 5 (5H) + 8 (2C) + 6 (1O) −1 (+charge) = 18 e^-

18 valence e^- − 14 bonding e^- = 4 non-bonded e^-

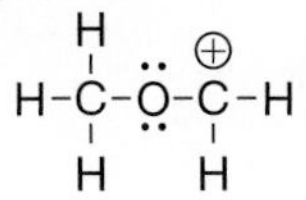

b) Electron geometry:

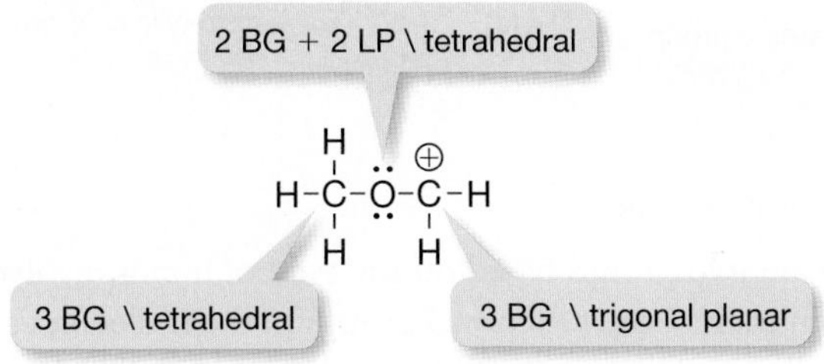

2 BG + 2 LP \ tetrahedral

3 BG \ tetrahedral

3 BG \ trigonal planar

c) Hybridization:

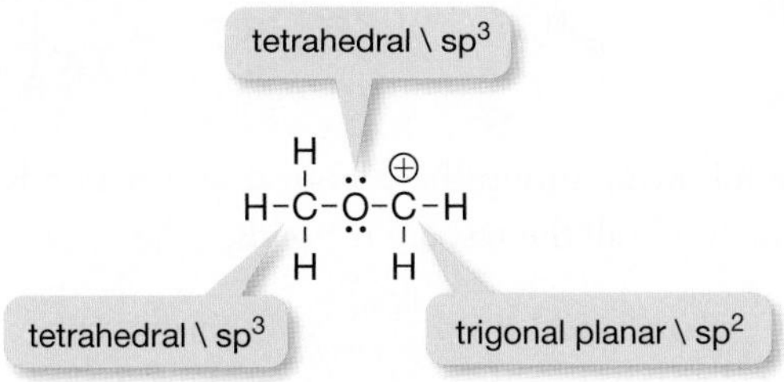

tetrahedral \ sp^3

tetrahedral \ sp^3

trigonal planar \ sp^2

d) The charged carbon is not saturated, so a second bond to the oxygen can be formed. This moves the formal charge to the oxygen atom.

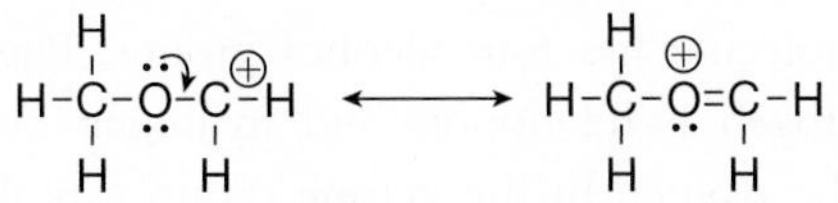

This leads to the following geometries for the new structure.

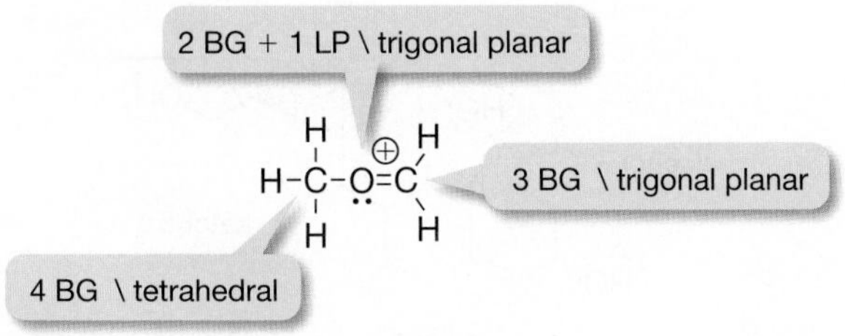

2 BG + 1 LP \ trigonal planar

3 BG \ trigonal planar

4 BG \ tetrahedral

The corresponding hybridizations would then be

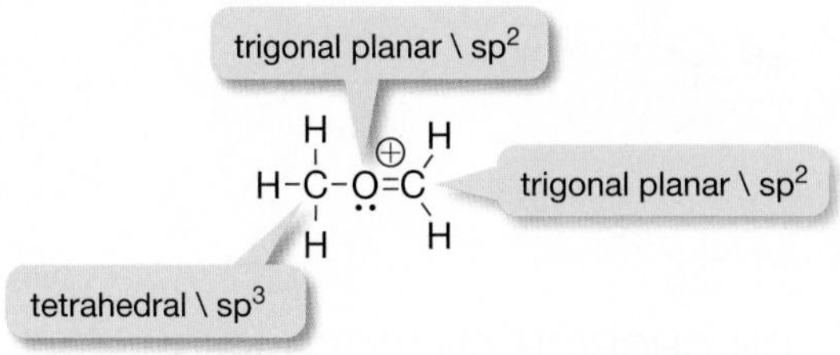

trigonal planar \ sp^2

trigonal planar \ sp^2

tetrahedral \ sp^3

PRACTICE PROBLEM 1.8

a)

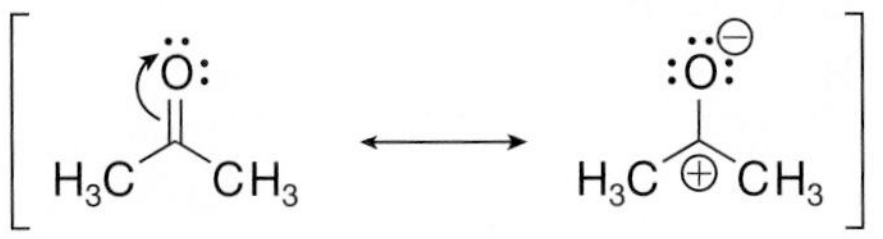

b)

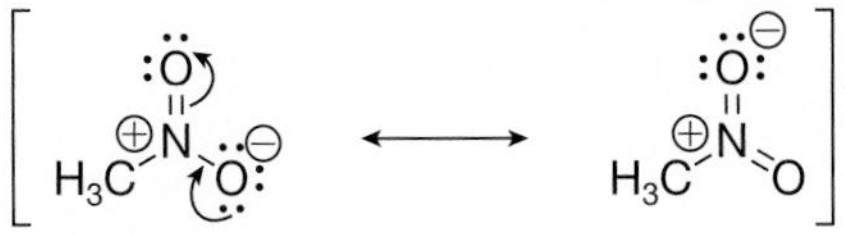

c)

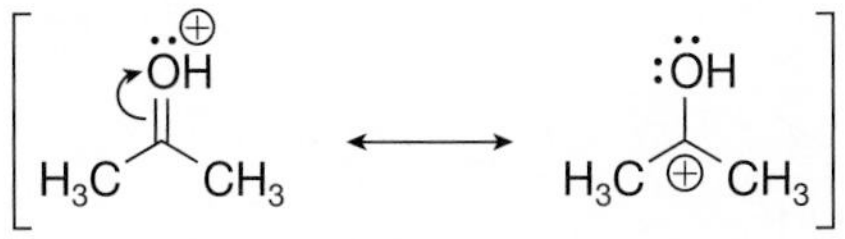

d)

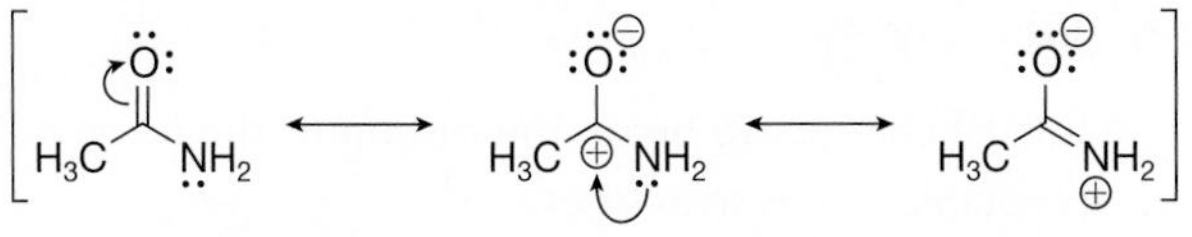

INTEGRATE THE SKILL 1.9

Drawing the basic structure leaves a formal positive charge on the carbon atom and lone pairs on each nitrogen atom.

valence e^- = 6(6H) + 4 (1C) + 15 (3N) −1 (+charge) = 24 e^-

24 valence e^- − 18 bonding e^- = 6 non-bonded e^-

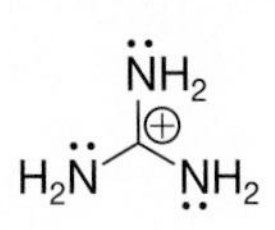

Three more resonance forms can be produced by forming a double bond between each of the nitrogen atoms and the central carbon atom. This leaves the formal charge on a nitrogen atom for each of these new forms.

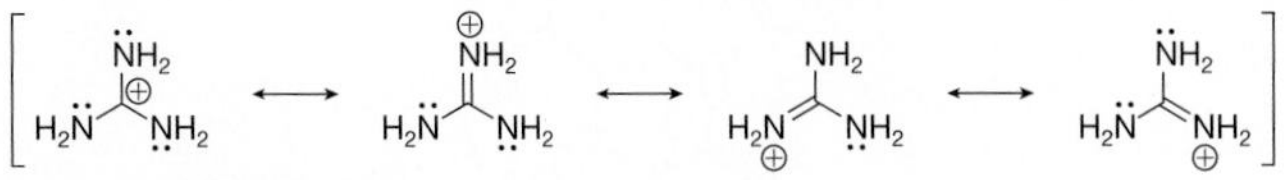

PRACTICE PROBLEM 1.10

a) i)

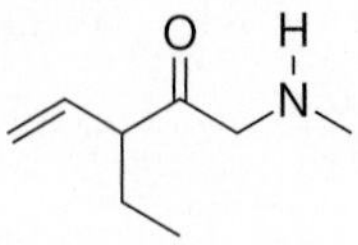

ii)

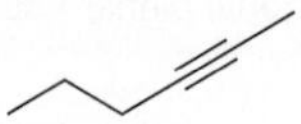

b) i) $CH_2CHCH(CH_2CH_3)COCH_2NHCH_3$

ii) $CH_3CH_2CH_2CCCH_3$

c) i)

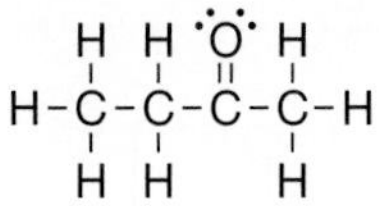

ii)

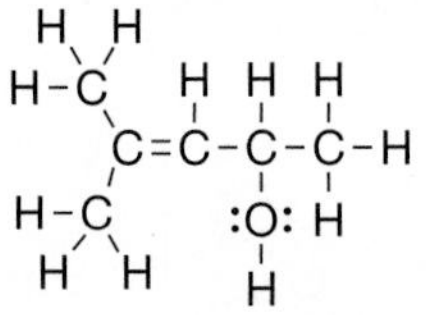

d) i)

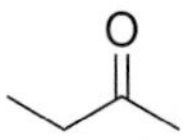

ii)

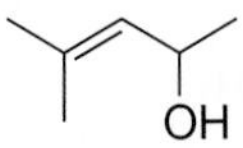

INTEGRATE THE SKILL 1.11

The formal charge can be on the carbon or the oxygen. Both are acceptable Lewis structures.

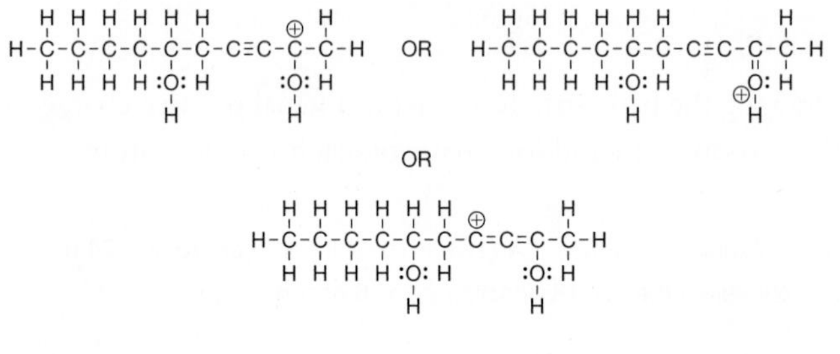

OR

OR

CHAPTER 2

PRACTICE PROBLEM 2.1

Functional groups are shown in the following molecules.

a)

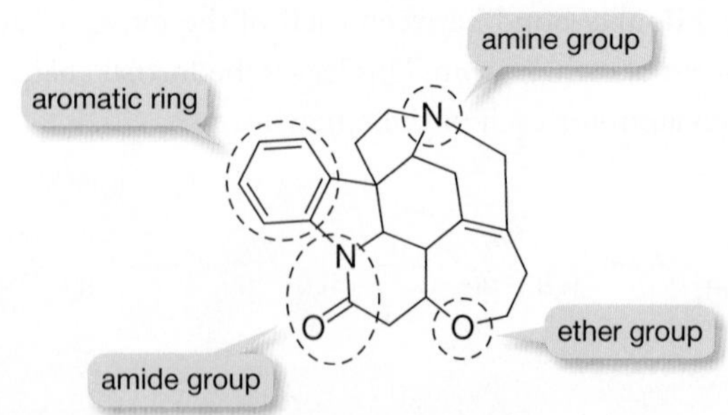

b)

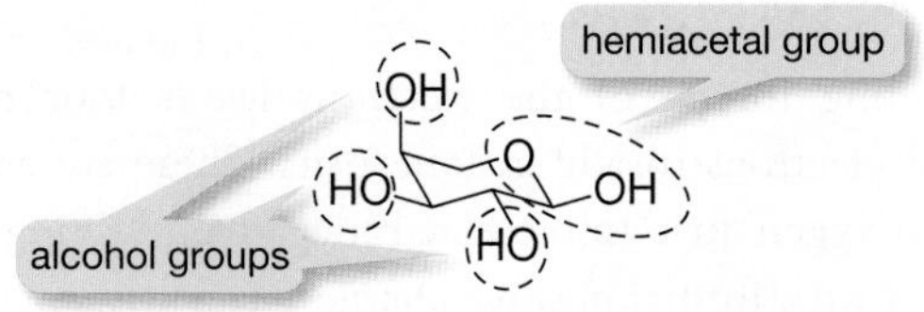

hemiacetal group
alcohol groups

c)

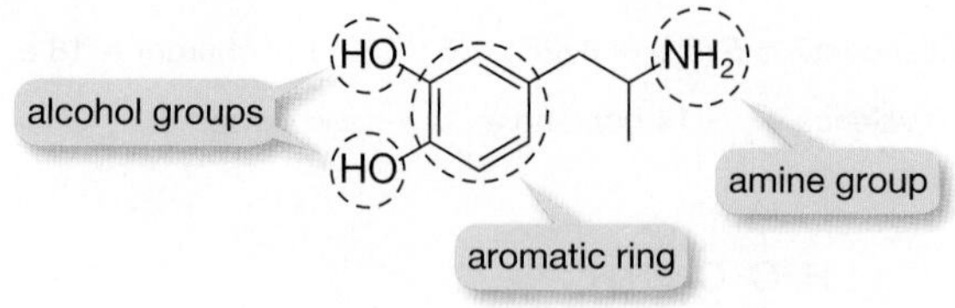

alcohol groups
amine group
aromatic ring

d)

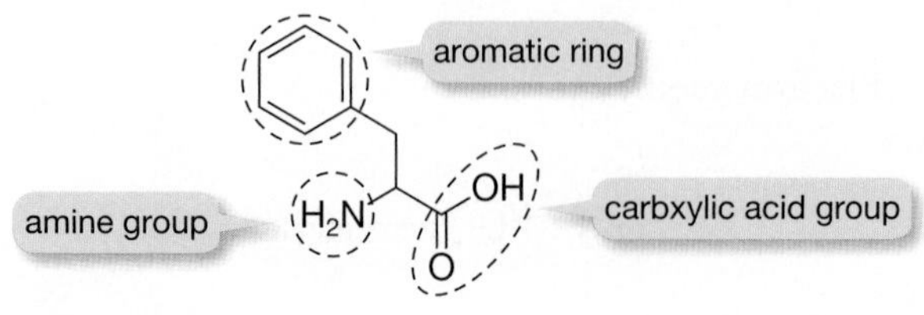

aromatic ring
amine group
carbxylic acid group

INTEGRATE THE SKILL 2.2

There are two groups based on the type of bonds involved (α or π) in each of the functional groups in the molecules above. The following groups are all formed using only α bonds:

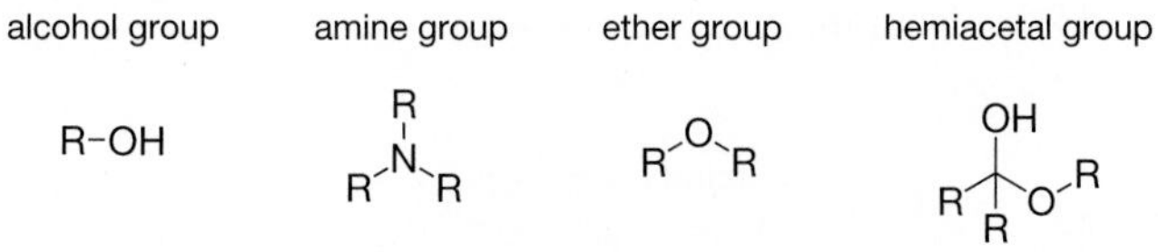

alcohol group	amine group	ether group	hemiacetal group
R–OH			

The following groups have both α and π bonds. The π bonds are noted; all the rest are α bonds.

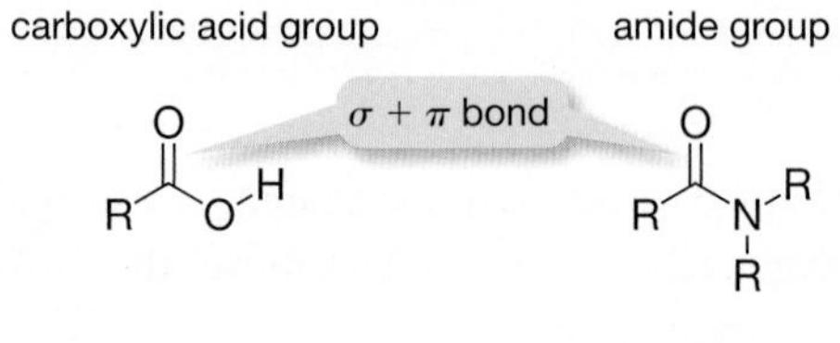

carboxylic acid group

amide group

$\sigma + \pi$ bond

PRACTICE PROBLEM 2.3

a) The molecule has four alcohol groups. These can act as hydrogen bond donors and hydrogen bond acceptors. The oxygen in the ketone group can also act as a hydrogen bond acceptor.

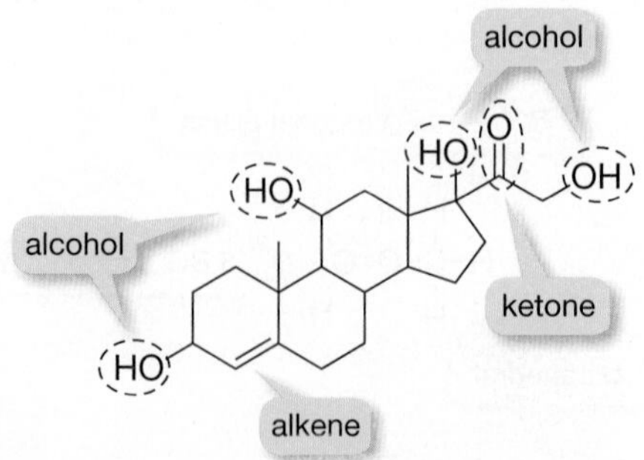

b) The molecule has three types of functional groups. The amine groups with no attached hydrogen atoms can act as hydrogen bond acceptors. The alcohol groups will be hydrogen bond donors and hydrogen bonds acceptors. The amide groups will act as hydrogen bond donors and hydrogen bond acceptors.

amide groups

amine groups

alcohol groups

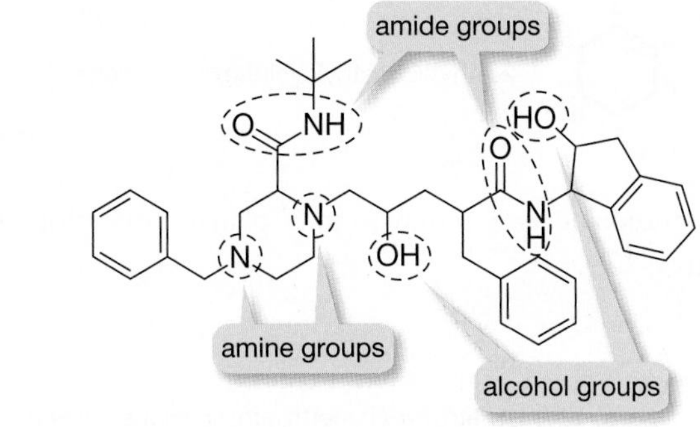

INTEGRATE THE SKILL 2.4

Functional groups and hybridizations for atoms in each group are shown.

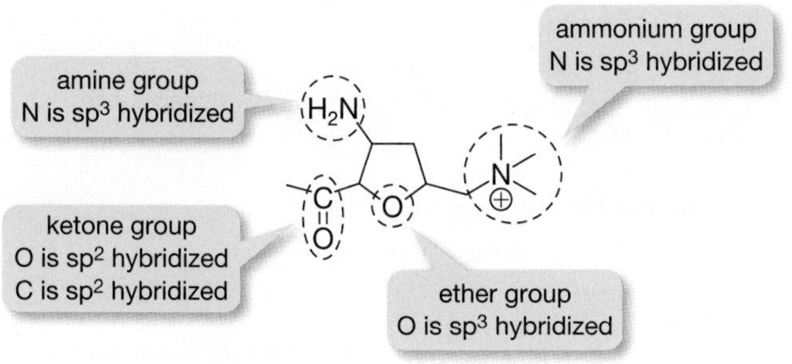

amine group
N is sp^3 hybridized

ammonium group
N is sp^3 hybridized

ketone group
O is sp^2 hybridized
C is sp^2 hybridized

ether group
O is sp^3 hybridized

PRACTICE PROBLEM 2.5

The two molecules are drawn below, with hydrophobic and hydrophilic regions indicated.

hydrophilic groups

hydrophobic region

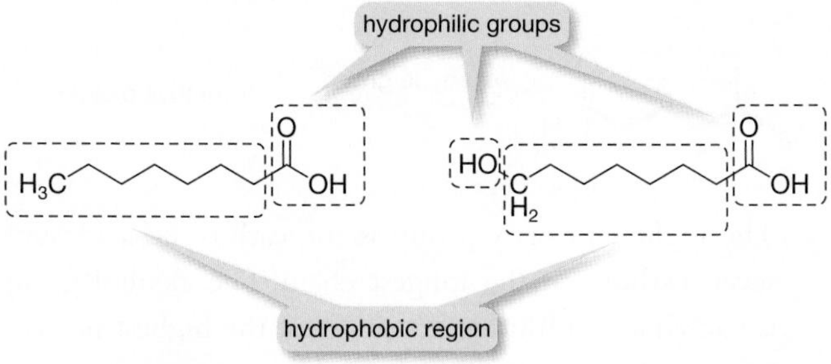

The alcohol-containing molecule would be more soluble in water, since it has an extra hydrogen bond site (donor and acceptor). The long hydrocarbon chain in the other molecule would interact favourably with long-chain hydrocarbon solvents like hexanes and make it more soluble in hexanes.

more soluble in hexanes

more soluble in water

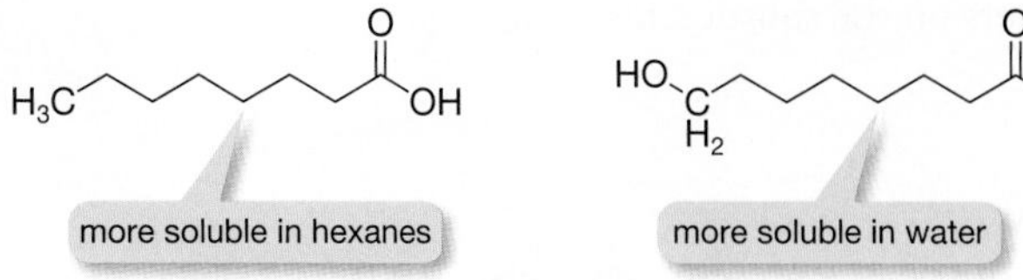

INTEGRATE THE SKILL 2.6

Lewis structure of ethyl acetate—$CH_3CO_2CH_2CH_3$—with bond dipoles and hydrophobic and hydrophilic regions indicated:

hydrophilic

hydrophobic

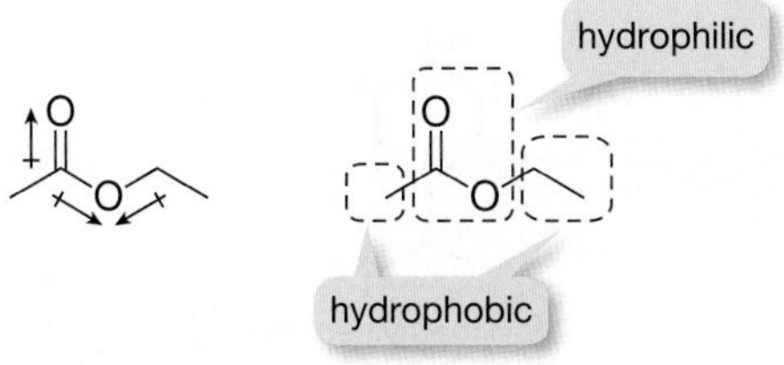

PRACTICE PROBLEM 2.7

a) Five-carbon main chain:

$CH_3CH_2CH_2CH_2CH_3$

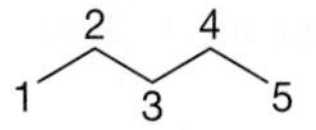

pentane

b) Seven-carbon main chain:

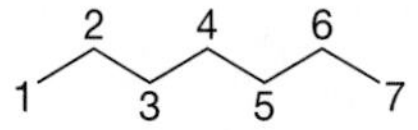

heptane

c) Ten-carbon main chain:

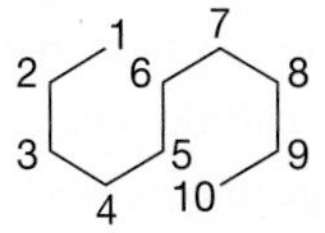

decane

d) The largest group of carbons are in the six-membered ring (in bold) with a four-carbon substituent.

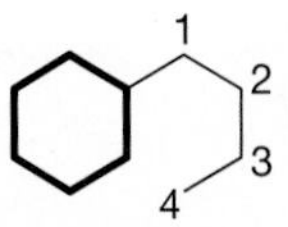

butylcyclohexane

e) The main chain has nine carbons (in bold) with three sidechains. The sidechains are given in alphabetical order and the main chain is numbered to minimize the numbers on the substituents.

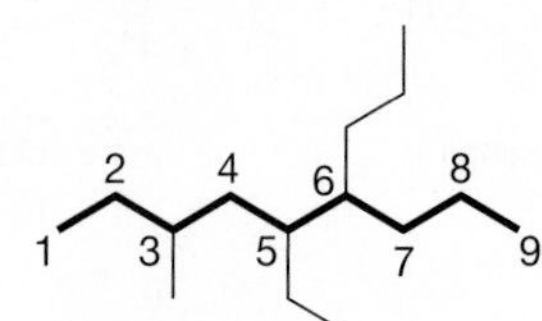

5-ethyl-3-methyl-6-propylnonane

f) The main chain has seven carbons (in bold) with two methyl sidechains.

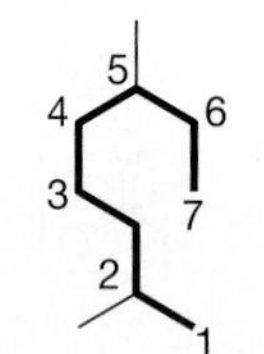

2,5-dimethylheptane

g)

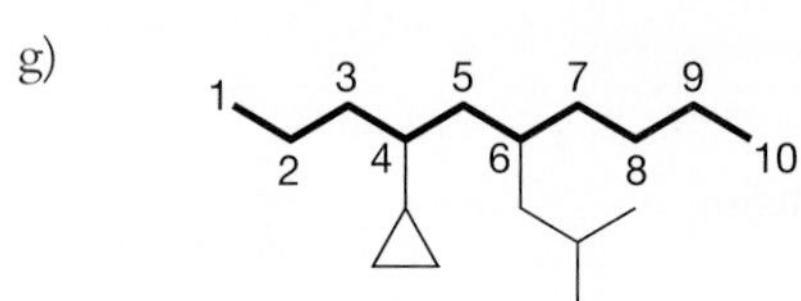

4-cyclopropyl-6-(2-methylprop-1-yl)decane

INTEGRATE THE SKILL 2.8

a) The IUPAC name for this compound is *isooctane*.

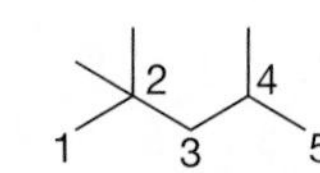

2,2,4-trimethylpentane

b) Octane is a straight-chain molecule. It can efficiently pack with other octane molecules to establish dispersion attractions. Isooctane is a branched molecule and, therefore, has a shorter carbon chain to effectively interact with other isooctane molecules. The branches will also inhibit effective packing of the molecule. So, octane will have greater dispersion attractions and a higher boiling point.

PRACTICE PROBLEM 2.9

a)

2-heptene

b)

1-hexyne

c) The double bonds are in the ring, so it is named as a *cyclohexadiene*, shown in bold.

2-ethyl-6-methylcyclohexa-1,3-diene

d) The molecule is a redrawn in a planar projection for clarity.

1-methyl-4-(1-methylethyl)cyclohex-1-ene

INTEGRATE THE SKILL 2.10

First, draw the main chain, numbering the atoms in the chain. This helps with substituents.

dodec-3-ene

Now, add the substituents.

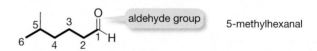

4-methyl-5,5-dipropyldodec-3-ene

PRACTICE PROBLEM 2.11

a) The only functional group is an aldehyde, so the suffix is *-al*. Numbering is from the functional group end.

aldehyde group

5-methylhexanal

b) The highest priority group is the carboxylic acid, with seven carbons in the longest chain. The alcohol group is named as a substituent, as it is not the highest priority functional group.

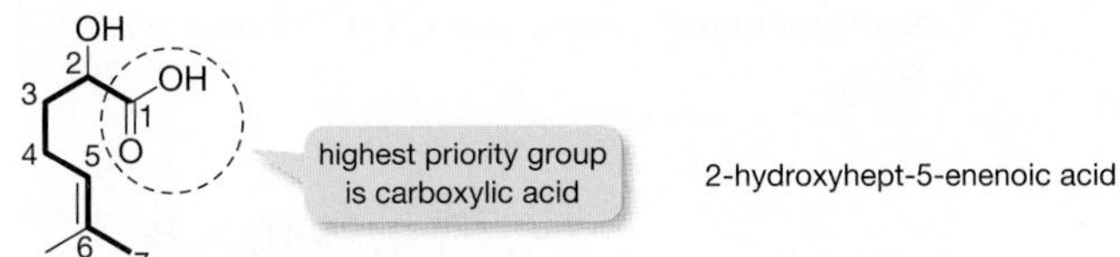
highest priority group
is carboxylic acid
2-hydroxyhept-5-enenoic acid

c) The ketone is the highest priority group. The substituents are named alphabetically.

4-chloro-2-methylcyclohexanone

d) The highest priority group is the alcohol. The locations of the double bond and the alcohol must be specified.

hept-1-en-4-ol

INTEGRATE THE SKILL 2.12

The systematic name of the molecule is based on an eight-carbon main chain (in bold). The highest priority functional group is the alcohol. Numbering starts at the alcohol end and leads to the following name.

3,7-dimethyloct-6-en-1-ol

There are many possible results from searches. Wikipedia is one location with a large amount of chemical information that is generally properly referenced. Searching for "3,7-dimethyloct-6-en-1-ol" leads to the following page: https://en.wikipedia.org/wiki/Citronellol. The common name of this molecule is *citronellol*, reminiscent of *citronella oil*, of which citronellol is a component. It is often used to ward off biting insects like mosquitos and blackflies.

PRACTICE PROBLEM 2.13

The root is cyclohexa-1,3-diene, so the framework upon which to build the molecule is

1-(4-methylheptan-2-yl) is a seven-carbon group attached by the second carbon in the chain. This group also has a methyl attached on the fourth carbon. This is connected to carbon 1 in the ring.

"5,5-difluoro" refers to two fluorine atoms, which are both attached to carbon 5 in the ring.

5,5-difluoro-1-(1,4-dimethylhexanyl)cyclohexa-1,3-diene

CHAPTER 3

PRACTICE PROBLEM 3.1

a) This is an eclipsed conformation:

back C

front C

b) This is a staggered conformation:

c) This is an eclipsed conformation:

d) This is a staggered conformation:

INTEGRATE THE SKILL 3.2

a) Newman projections of staggered and eclipsed 2,2,2-trifluoroethan-1-ol:

b) Line bond drawing of a Newman projection:

PRACTICE PROBLEM 3.3

a) Newman projections for the staggered and eclipsed conformations of butane:

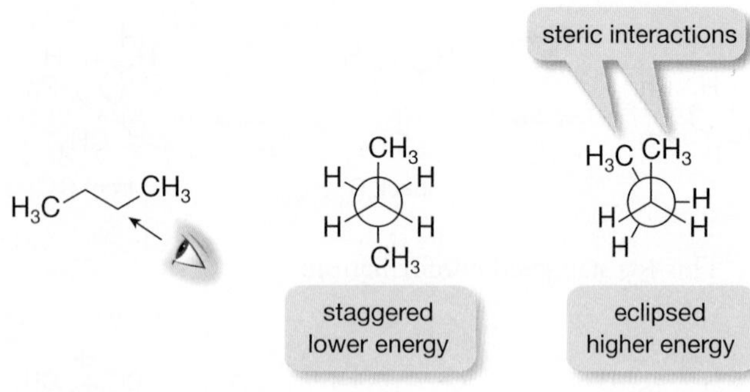

b) i) 2-methylpentane (along the C2–C3 bond with C2 in front):

ii) 1-bromopentane (along the C1–C2 bond with C1 in front):

iii) N,N,O-trimethylhydroxylamine (along the O-N bond with the O in front):

INTEGRATE THE SKILL 3.4

a) Use the energy diagram in Figure 3.3.

i) Energy barrier for gauche → anti is (5 kcal/mol – 0.6 kcal/mol) = 4.4 kcal/mol.

ii) Energy barrier for anti → gauche is (5 kcal/mol – 0 kcal/mol) = 5 kcal/mol.

iii) The highest energy barrier for gauche (60°) to gauche (300°) is anti → gauche (300°).
The energy barrier for this is 5 kcal/mol.

b) The only thing that would change are the x-axis labels. The relation between the H–H torsion angle and the Cl–Cl torsion angle is shown below.

This would give the following graph:

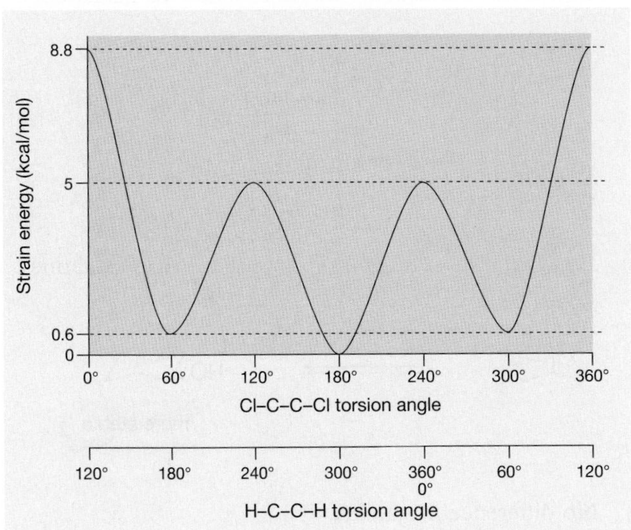

PRACTICE PROBLEM 3.5

a) The three-membered ring would have more angle strain.

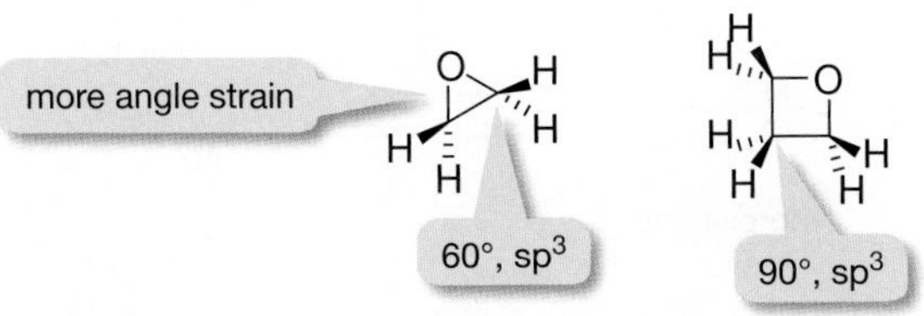

b) The biggest difference would be the amide having more angle strain than the alcohol. This is a result of the carbonyl carbon being sp^2-hybridized (120° angle preferred), while the carbon with the alcohol group is sp^3-hybridized (109° angle preferred).

sp^2, 90° more strain

sp^3, 90° less strain

c) The cyclopropene would have more angle strain due to hybridization.

sp^3, 60° less strain

sp^2, 60° more strain

d) The alkene would have more angle strain due to hybridization.

sp^2, 60° more strain

sp^3, 60° less strain

INTEGRATE THE SKILL 3.6

a) If cyclohexane were planar, there would be 12 eclipsed H–H interactions.

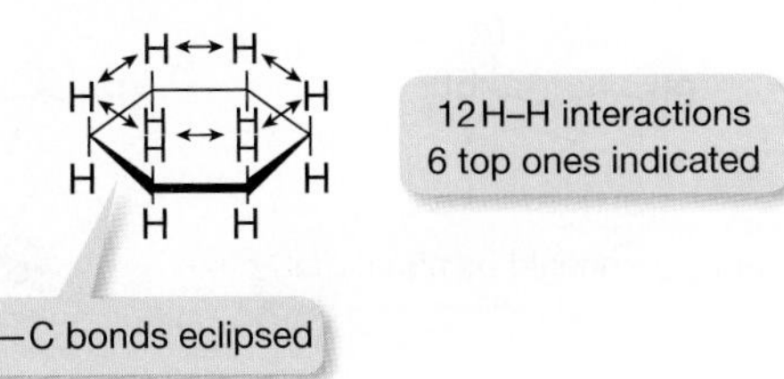

b) Planar cyclohexane would have angle strain, since planar cyclohexane would require 120° bond angles, while the carbon atoms would prefer tetrahedral (109°) angles.

planar, 120° angles
sp^3, 109° preferred

PRACTICE PROBLEM 3.7

a) i) 1,1,3,3-tetramethylcyclohexane

ii) 4-bromo-1,1-dimethylcyclohexane (Br in an equatorial position)

iii)

b) i) The C–F bond should have a break in it to indicate that it's behind the ring C–C bond.

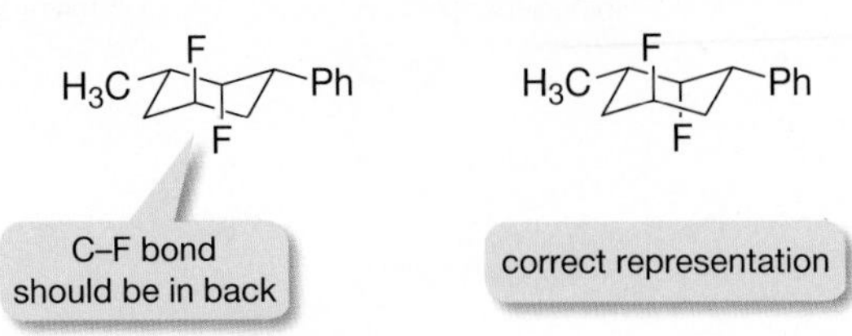

ii) Adjacent axial positions need to point in opposite directions (one up, one down). Assuming the ethyl groups are *cis*, the correct structure is shown.

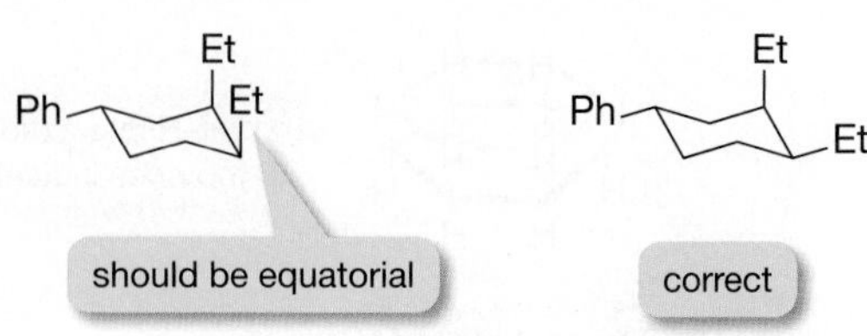

iii) The main bonds in the cyclohexane ring should not be horizontal.

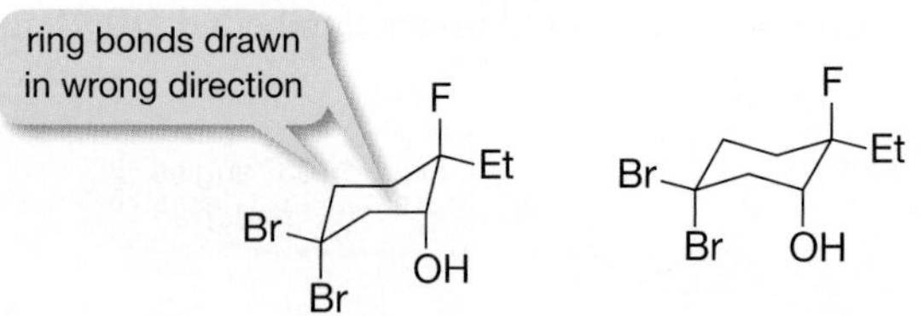

INTEGRATE THE SKILL 3.8

a) Newman projection of a chair structure with hydrogen atoms:

b) Two differences between an envelope conformation of cyclopentane and a chair conformation of cyclohexane:

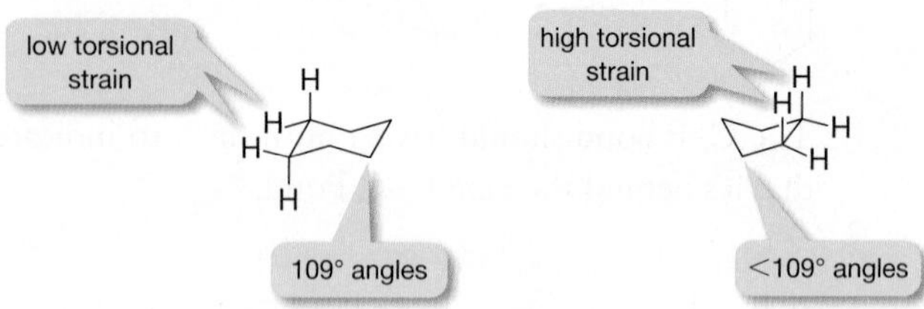

PRACTICE PROBLEM 3.9

a) Equatorial substituents are more stable than axial ones.

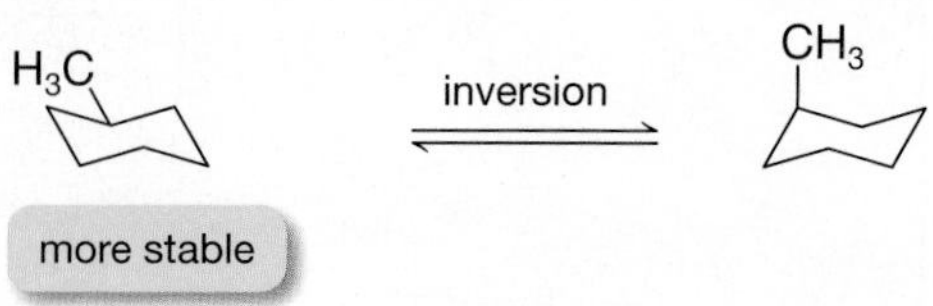

b) Equatorial substituents are more stable than axial ones.

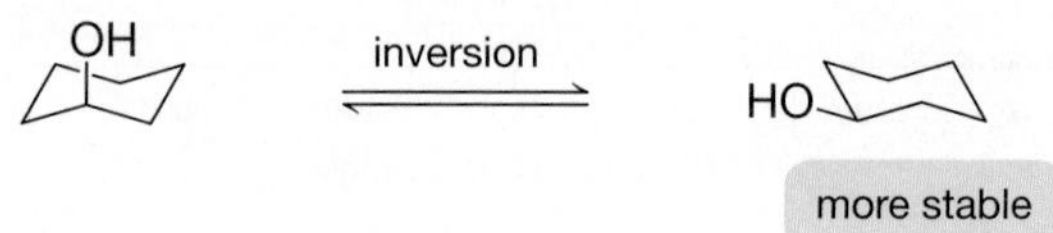

c) No difference in stability.

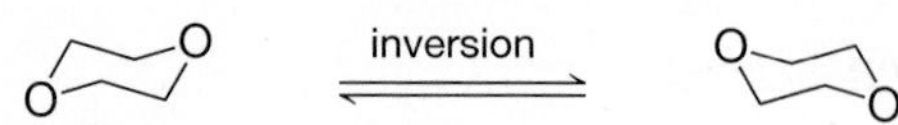

d) Equatorial substituents are more stable than axial ones.

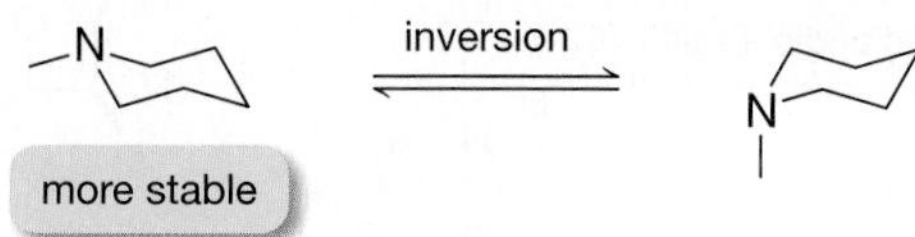

INTEGRATE THE SKILL 3.10

a) Newman projections for the two possible chair conformations of methylcyclohexane:

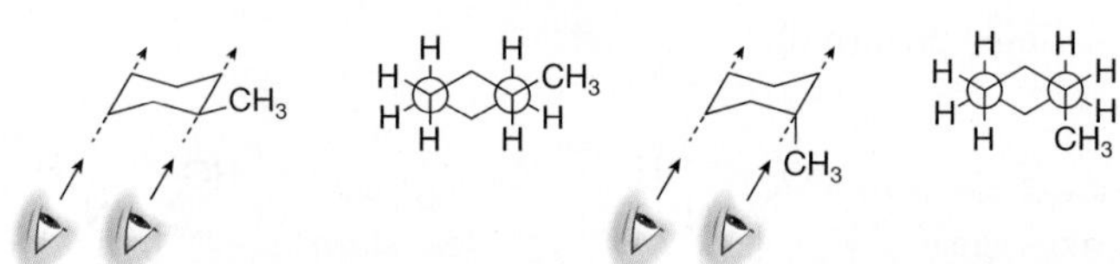

b) The ethyl group is bulkier than the ethynyl group at the carbon attached to the ring. So, torsional interactions would be larger for the ethyl group. The energy difference between axial and equatorial positions would be larger for ethylhexane, making equatorial preferred. The fact that ethynylcyclohexane has almost equal amounts of axial and equatorial conformers means there is very little difference in energy between them.

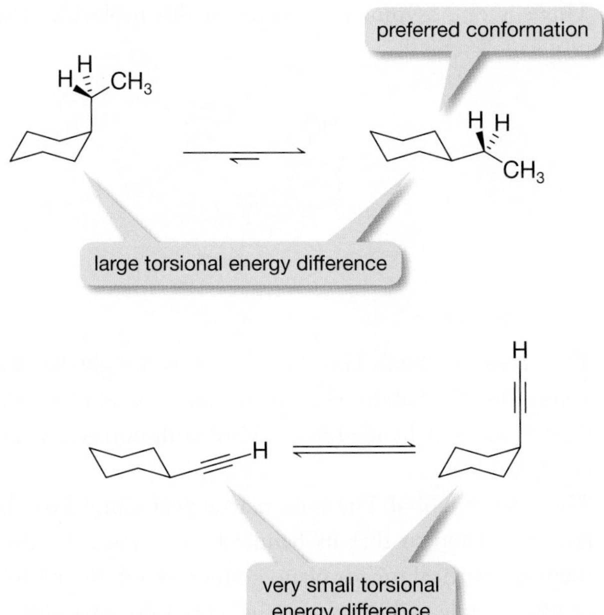

PRACTICE PROBLEM 3.11

a) Only one substituent can be equatorial. The largest group is the bromine atom, so the preferred conformation will have it in an equatorial position.

inversion

favoured largest group equatorial

b) Only one substituent can be equatorial. The largest group is the chlorine atom, so the preferred conformation will have it in an equatorial position.

inversion

favoured largest group equatorial

c) The favoured conformation has all but one group in an equatorial position, which should be the lowest energy.

inversion

favoured, most substituents are equatorial

d) The favoured conformation has the two largest substituents in equatorial positions.

inversion

favoured, largest groups equatorial

INTEGRATE THE SKILL 3.12

In both of the chair conformers of *cis*-1,4-di-*tert*-butylcyclohexane, at least one of the very bulky tertiary butyl groups is in an axial position. The methyl groups would be in very close proximity to the adjacent axial hydrogens.

inversion

1,3-diaxial interactions

A twist-boat conformation allows the axial t-butyl group to move out of a pure axial position and, thus, can reduce its steric interactions with axial hydrogen atoms.

CHAPTER 4

PRACTICE PROBLEM 4.1

a) There are no chirality centres in this molecule, so it cannot be a chiral molecule.

b) There is one chirality centre.

mirror image

Rotating the mirror image molecule so the main chain is lined up like the original shows that this mirror image is not superimposable.

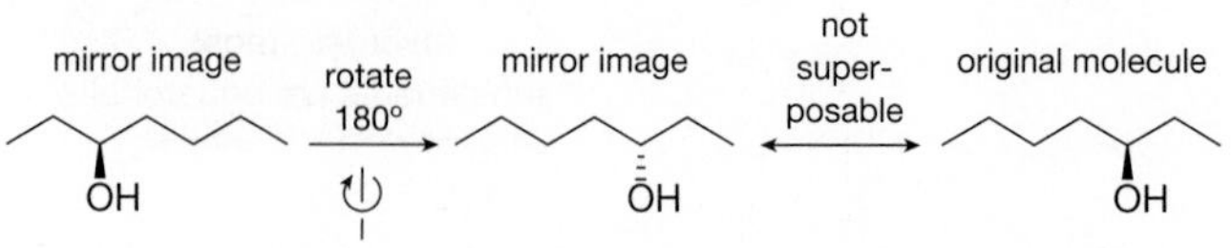

mirror image rotate 180° mirror image not superposable original molecule

This is a chiral molecule.

c) This molecule has two chirality centres. Looking at the mirror image, if it is rotated 180° on the plane of the page, the original orientation of the molecule results.

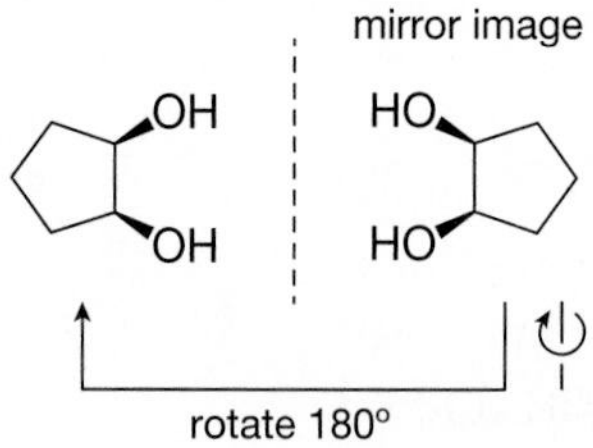

mirror image

rotate 180°

d) This molecule is chiral. First, drawing the mirror image and then rotating it to put it in the same orientation as the original molecule shows it to be non-superposable. In the final orientation of the mirror image, it is clear that each of the chiral centres has the opposite configuration of the original molecule.

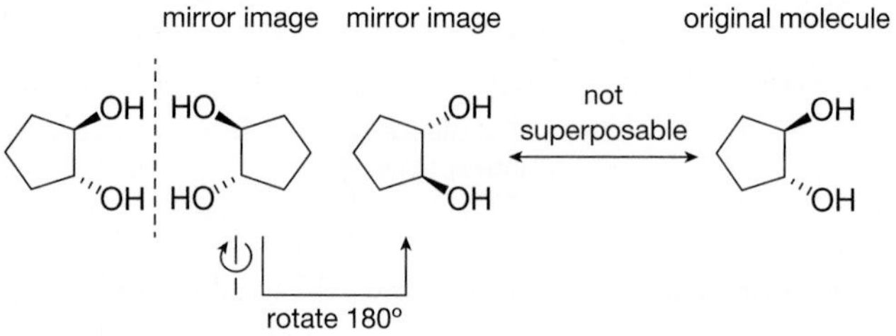

mirror image mirror image not superposable original molecule

rotate 180°

INTEGRATE THE SKILL 4.2

a) The first step is to draw the molecule and determine whether there are chirality centres. Looking at the drawing below, there are no asymmetric carbon centres in this molecule and it is, therefore, not chiral.

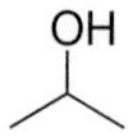

b) This molecule has one chirality centre. The molecule is, therefore, chiral.

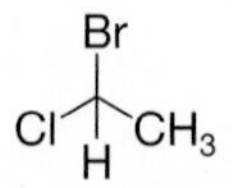

c) There is one asymmetric centre in this molecule. This molecule is chiral.

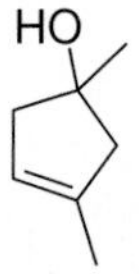

PRACTICE PROBLEM 4.3

a) The screw is chiral. The threads follow a right-handed helix along the length. The mirror image would have the threads in a left-handed helix. Most common screws are right-handed.

b) The mug is achiral. The mug may appear chiral, but, the mug is rotated so that its handle faces toward the user, there is a vertical mirror plane going through the middle of the handle, which splits the cup into two mirror images. Since it has this internal mirror plane, it cannot be chiral.

c) The doghouse is achiral. A vertical mirror plane running along the peak of the roof splits it into a left half and a right half, which are mirror images. The presence of this mirror plane means that the doghouse is not chiral.
Aside: If the name of the dog were on the front, in most cases the doghouse would be chiral. Try to come up with a name where this would *not* be true.

d) If the molecule is achiral, the two chiral carbons will be mirror images of each other. Rotating about a C–C bond so the presence or absence of symmetry is easier to assess, it is clear that there is **no** mirror plane in the molecule that reflects the left half into the right half and the right half into the left half. This molecule is chiral.

not a plane of symmetry

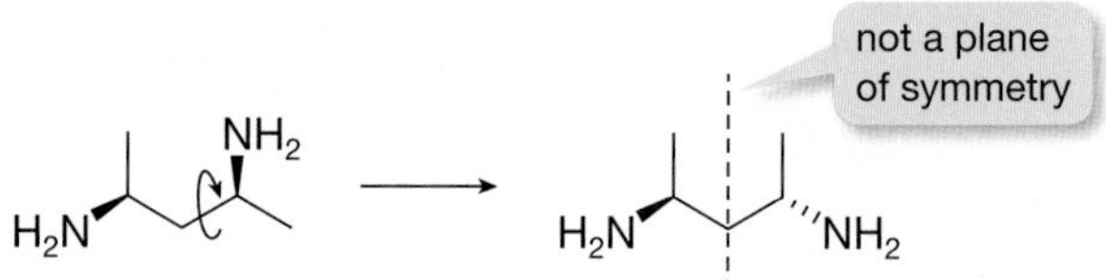

e) There are two chirality centres, so an achiral compound is possible. Redrawing the molecule reveals the internal mirror plane of symmetry that reflects the top half into the bottom half and vice versa. This molecule is achiral (meso).

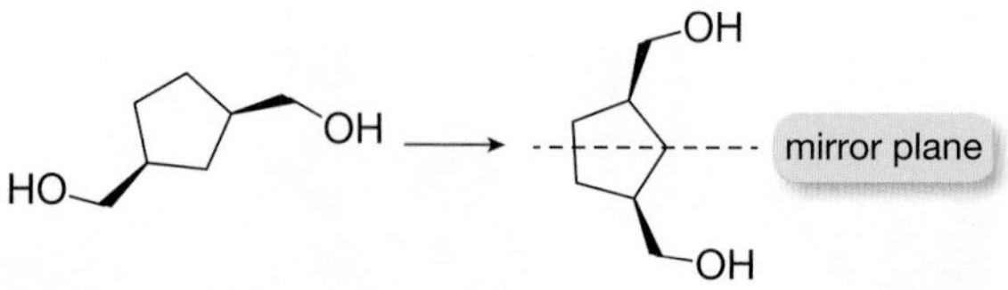

mirror plane

f) There are no chirality centres in this molecule. There is a mirror plane running through the middle of the molecule, bisecting the bromine and chlorine atoms. This molecule is achiral.

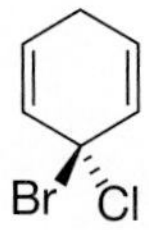

Br Cl

INTEGRATE THE SKILL 4.4

a) To see the symmetry, a boat form is necessary.

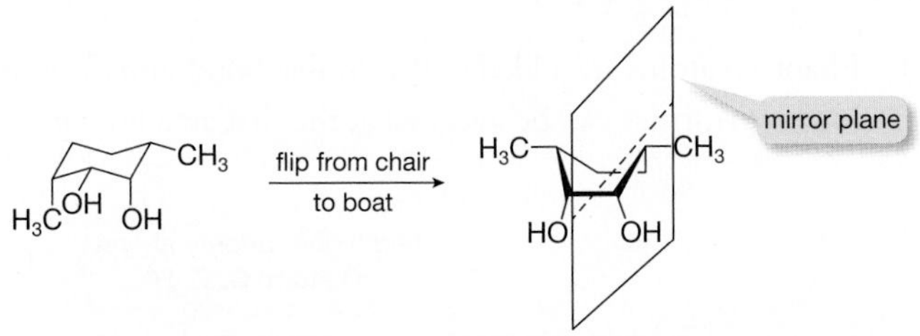

CH_3 flip from chair to boat H_3C CH_3 mirror plane
H_3C OH OH HO OH

b) Rotation about the C–C bond is required until the same groups on the front carbon and back carbon are eclipsed. The symmetry is most easily seen by then rotating the molecule by 90°, as shown.

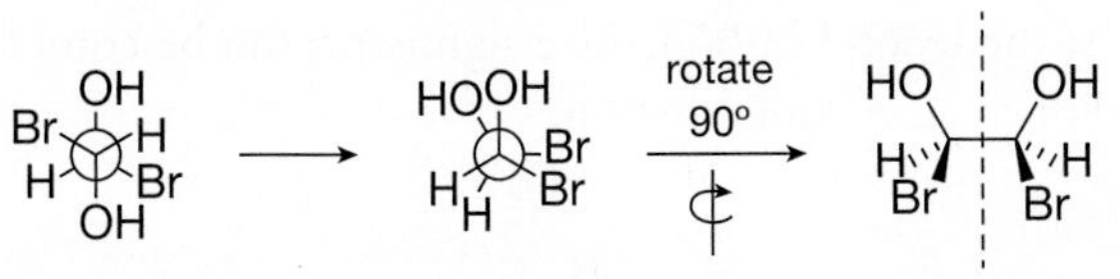

rotate 90°

c) Rotating around the centre C–C bond by 180° to eclipse equivalent groups shows the symmetry in this molecule.

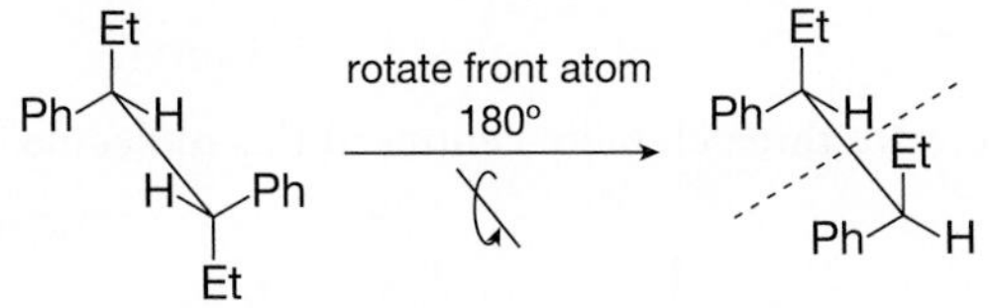

rotate front atom 180°

PRACTICE PROBLEM 4.5

a) There are two sp^3 carbons with four different substituents.

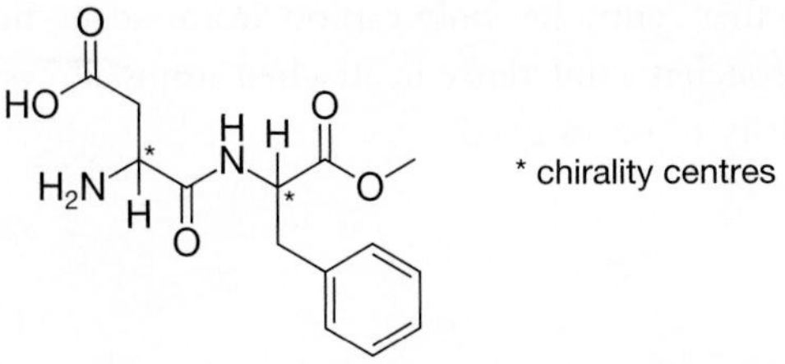

* chirality centres

b) There are two chiral sp^3 carbon atoms, as shown.

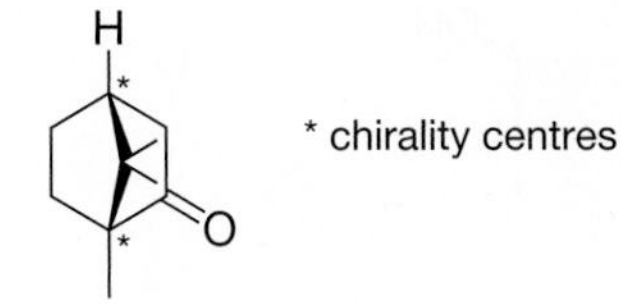

* chirality centres

c) The five carbon atoms indicated are chiral sp^3 centres.

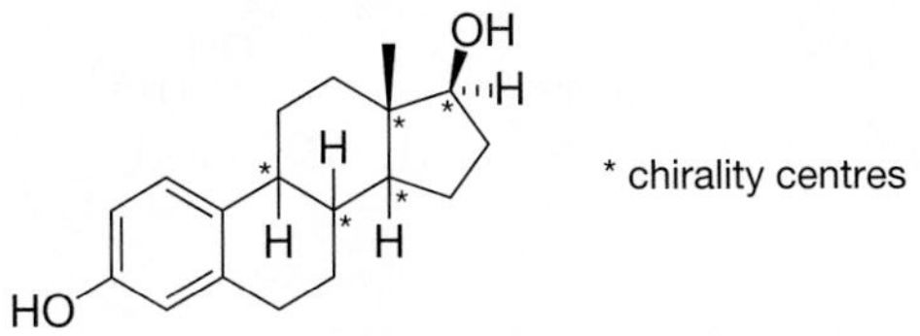

* chirality centres

d) There are two chirality centres: one an sp^3 carbon and the other a sulfoxide group.

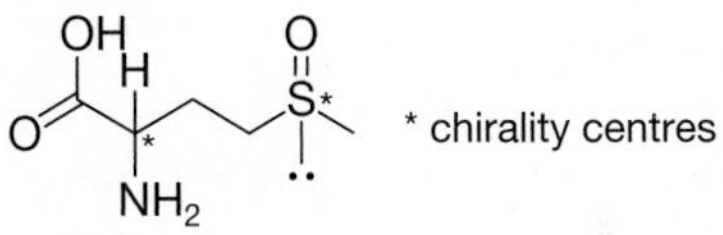

* chirality centres

INTEGRATE THE SKILL 4.6

The constitutional isomers contain no rings or double bonds, since the molecular formula has *n* C atoms and 2*n* + 2 H atoms (see textbook Section 2.2; *n* = 5 in this case), which is the ratio found in an alkane. Formation of a ring or double bond would require removing two hydrogen atoms. So, the molecules will be ethers or alcohols and contain no rings. Drawing each basic carbon framework then adding in the oxygen in all possible unique sites gives the 14 isomers. Chirality centres are indicated where they exist.

Pentane Backbone

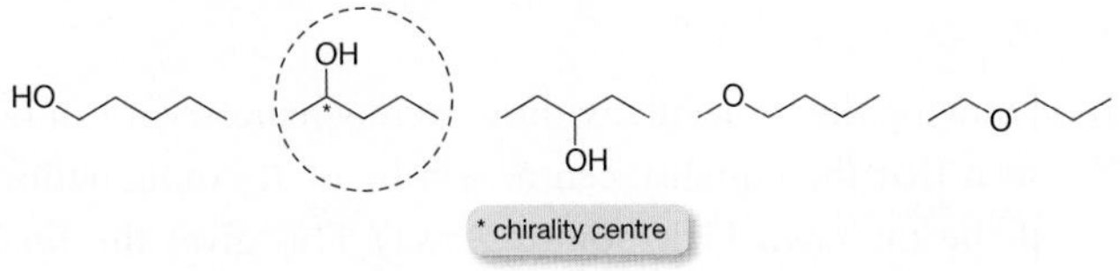

* chirality centre

Methylbutane Backbone

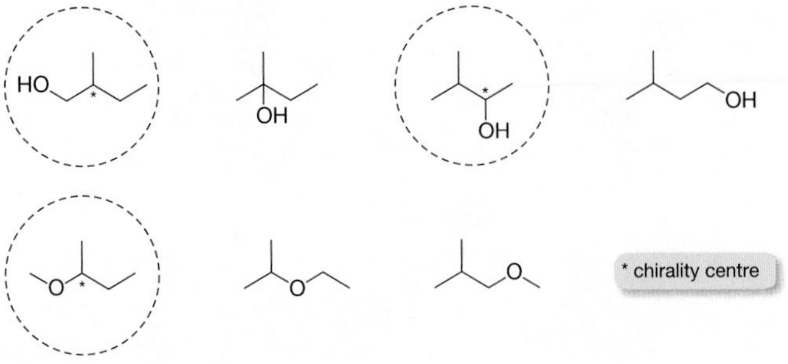

* chirality centre

Dimethylpropane Backbone

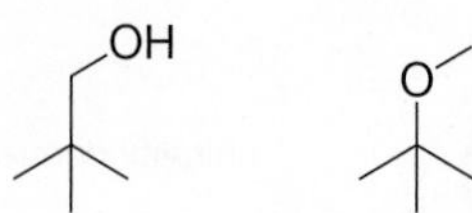

PRACTICE PROBLEM 4.7

a)

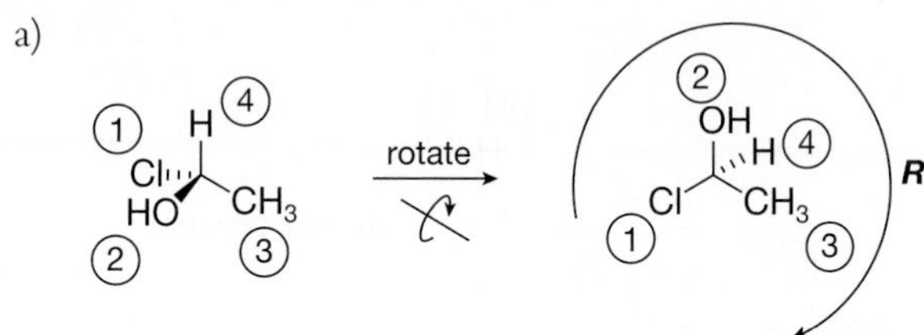

b)

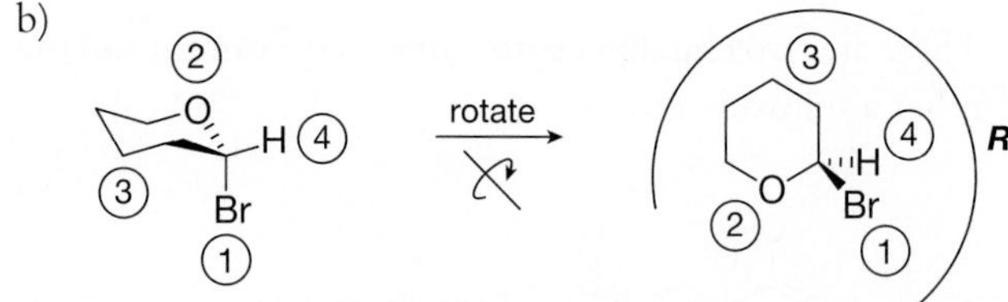

c)

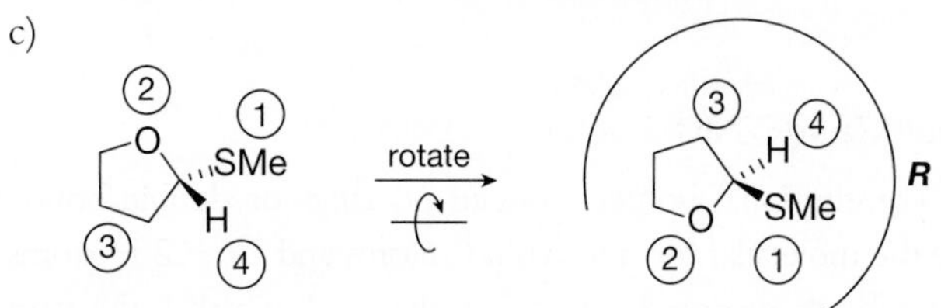

d)

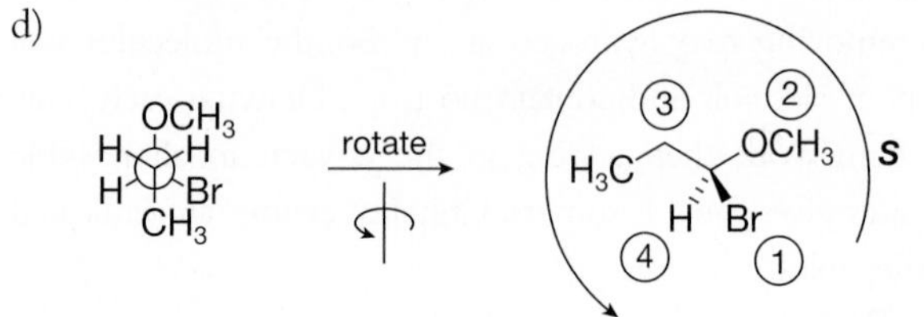

INTEGRATE THE SKILL 4.8

a) Drawing the molecule without stereochemistry, it can be seen that the chirality centre will be of *R* configuration if the undrawn H is pointing away. This gives the final structure with the stereochemistry indicated.

b) After assigning the priorities on the chiral carbon, the lowest priority group is the methyl group, so it should be pointing away to assign the stereochemistry. If the determined stereochemistry was incorrect, switching any two groups would give rise to the correct *R* stereochemistry.

PRACTICE PROBLEM 4.9

a) Phantom atoms are added to the double bond, and the two lowest priorities can be assigned at the first attachment.

At the second carbon, the assignments can be completed. This is an *S* configuration.

b) There are three chirality centres in this molecule.

The first centre has only carbon atoms at the first comparison, but a difference in attached atoms allows the first priority to be assigned.

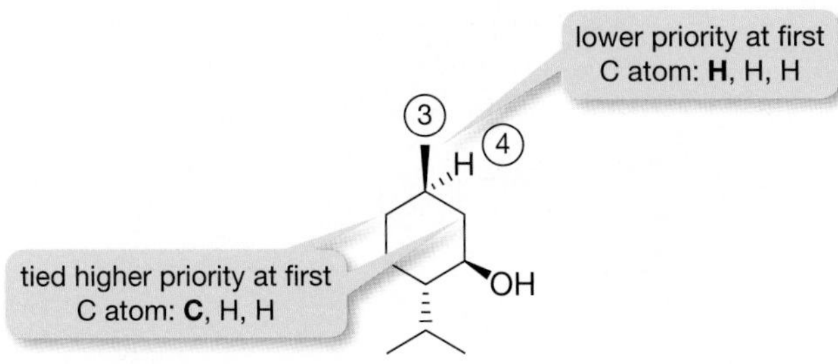

At the second pair of C atoms to be compared, there is a difference and the groups can be fully assigned priorities. This centre has an *R* configuration.

The second centre is simpler to assign. The O atom is top priority and the attached carbon atoms are second and third as indicated. This is an *R* configuration.

For the third centre, it is useful to rotate the lowest priority H atom to the back. This is an *S* configuration.

c) There are four chirality centres in this sugar molecule, as shown.

Centre 1: *S* Configuration

Centre 2: *S* Configuration

Centre 3: *S* Configuration

Centre 4: *R* Configuration

d) There is one chirality centre in this molecule. It is of *S* configuration.

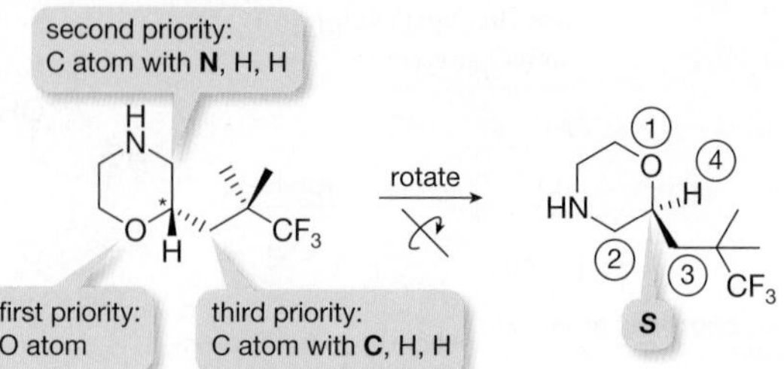

INTEGRATE THE SKILL 4.10

a)

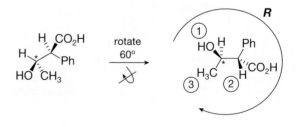

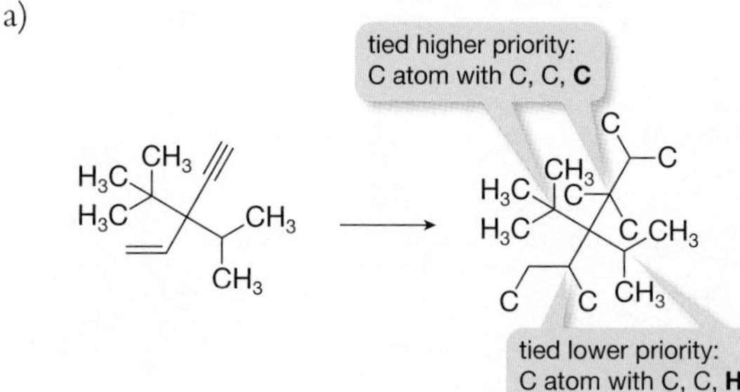

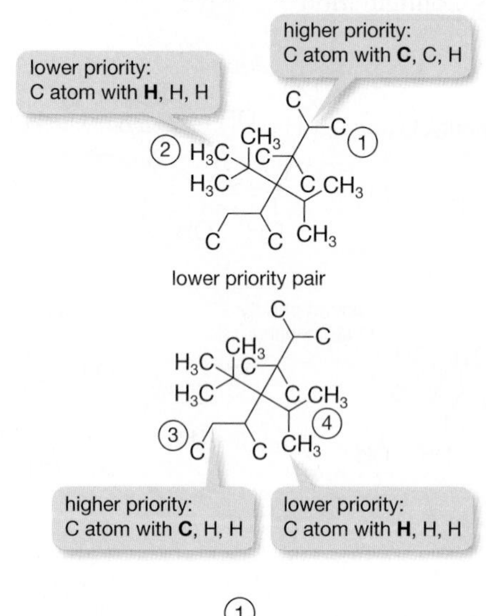

t-Bu, i-Pr, **S**

b)

third priority:
C atom with **C**, **C**, H

second priority:
C atom with **O**, O, **H**

first priority:
C atom with **O**, O, **O**

fourth priority:
C atom with **C**, **H**, H

S

PRACTICE PROBLEM 4.11

a)

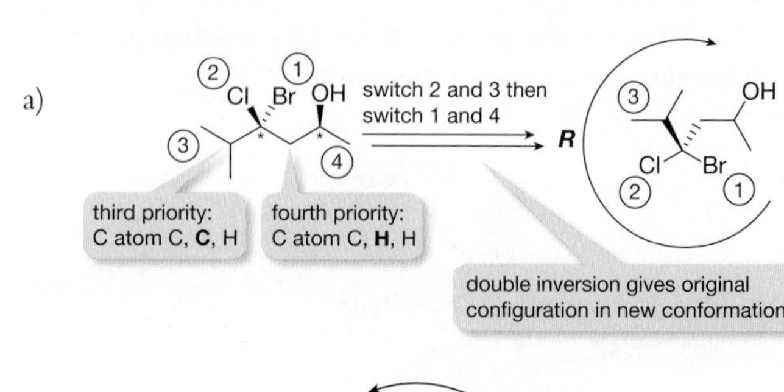

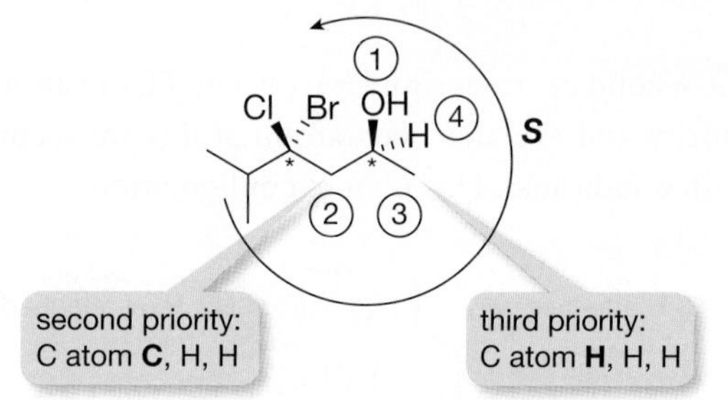

b)

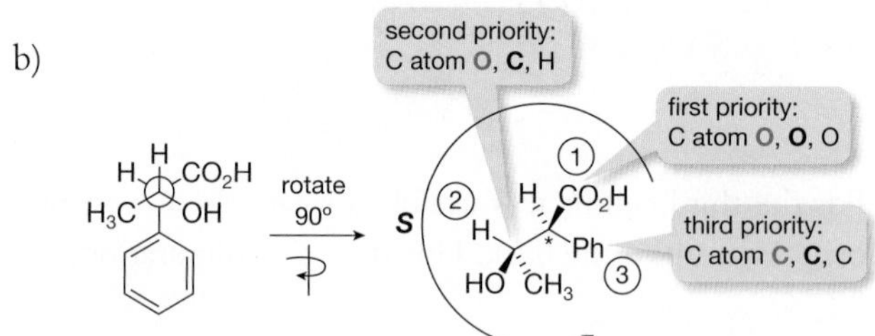

rotate 60°

R

c)

fourth priority:
C atom with **C**, C, C

third priority:
C atom with **O**, O, O

second priority:
C atom with **F**, F, F

rotate 60°

R

d)

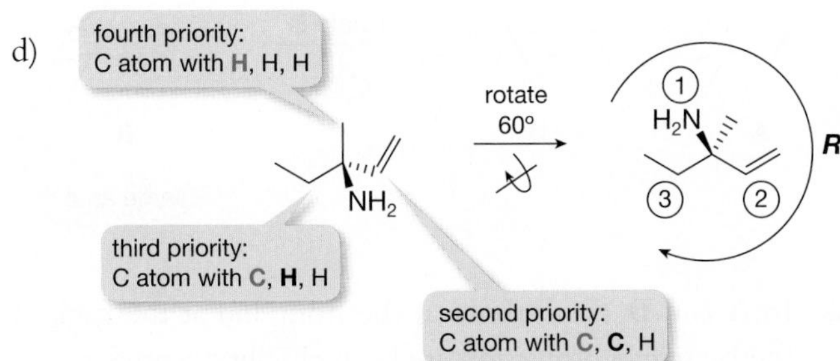

INTEGRATE THE SKILL 4.12

Enantiomers of compounds in Question 4.11.

a) (2*S*, 4*R*)-4-bromo-4-chloro-5-methyl-hexan-2-ol

(2*R*, 4*S*)-4-bromo-4-chloro-5-methyl-hexan-2-ol

enantiomer

b) (2*S*, 3*R*)-3-hydroxy-2-phenylbutanoic acid

(2*S*, 3*R*)

(2*R*, 3*S*)-3-hydroxy-2-phenylbutanoic acid

enantiomer

c) (*R*)-3,3,3-trifluoro-2-methoxy-2-phenylpropanoic acid

(*R*)

(*S*)-3,3,3-trifluoro-2-methoxy-2-phenylpropanoic acid

enantiomer

d) (*R*)-3-amino-3-methyl-1-pentene

(*R*)

(*S*)-3-amino-3-methyl-1-pentene

enantiomer

PRACTICE PROBLEM 4.13

a) The pair must be enantiomers, since 2-butanol has only one chirality centre. To confirm the enantiomeric relationship, the quickest approach is to assign the stereochemistry of the structure and compare it to (*R*)-2-butanol.

rotate 180°

(*S*)-2-butanol

b) The first step would be to redraw the Newman projection and assign configurations.

rotate 90° — rotate H to back

Next, assign configurations to the other molecule.

rotate H to back

Since the configuration at only one chirality centre has changed, these must be diastereomers.

c) There are many centres to consider here, so redrawing the chair into the first form will allow a direct comparison of each chiral centre without assigning an absolute configuration to each. First, add missing H atoms to the ring, then assign up (solid wedge) or down (dashed wedge) to each position for the non-hydrogen substituent.

Now, compare the two molecules directly. The configuration differs in only one position, so they must be diastereomers.

configuration differs

d) These are best compared to see whether they are mirror images of each other.

First, reorient **B** by rotation.

rotate 90°

Comparing the two molecules side-by-side, it is clear that they are mirror images.

mirror plane

One more rotation of **B** and comparison with **A** shows that these mirror images are not superposable and are, therefore, enantiomers:

rotate 180°

superposable here

not superposable here

INTEGRATE THE SKILL 4.14

a) These molecules are identical even though they are drawn as mirror images.

rotate **B**

same as **A**

b) In **A** and **B**, the carbons at the front and at the back of the Newman projection are both chirality centres.

To compare, first reorient **B** into a mirror image of **A**.

rotate **B** 180°

rotate carbon at back around C–C bond

Putting both molecules side by side with a mirror plane shows that the groups on the back carbon are reflected and are therefore of opposite configuration. The groups on the front carbon are not reflected and, therefore, are of the same configuration. **A** and **B** are diastereomers.

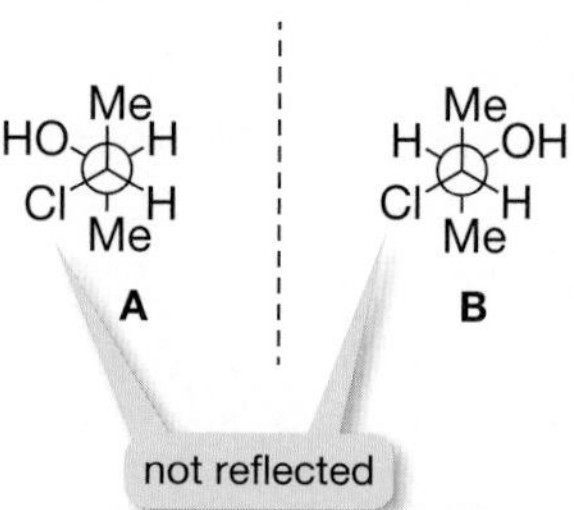

c) These are constitutional isomers, since their –OH groups are positioned at different carbons around the ring. One is 1,3,5 substituted and the other is 1,2,4.

1,3,5 substituted

1,2,4 substituted

PRACTICE PROBLEM 4.15

a) Draw 2,3,5,6-tetrahydroxy-2-methylheptanoic acid and determine the chirality centres. There are 4 here for 16 possible stereoisomers.

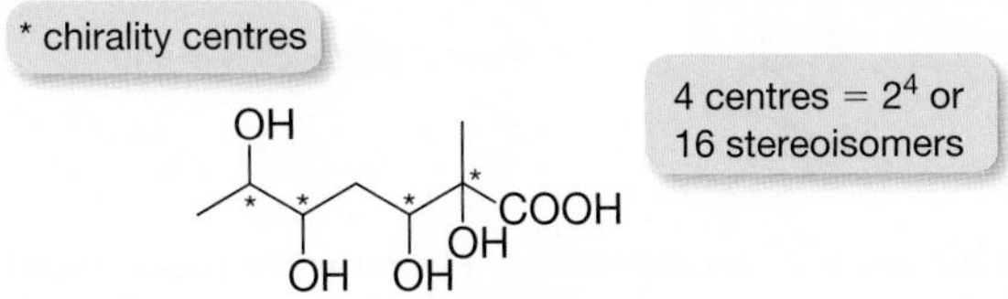

b) Draw 1,3,5-trimethylcyclopent-1-ene and determine the chirality centres. There are two here for four possible stereoisomers.

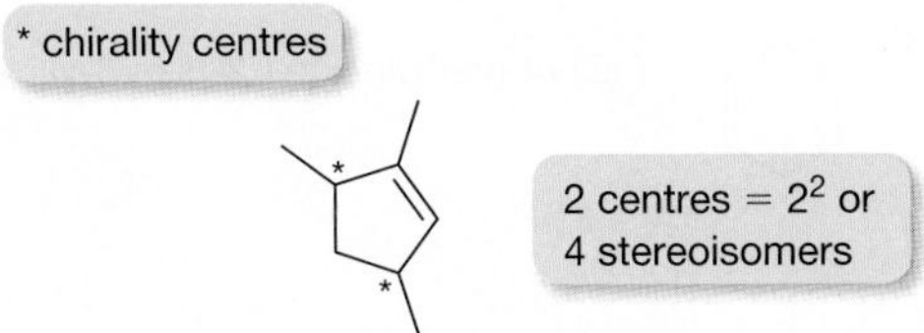

c) There are eight possible stereoisomers for this molecule.

d) There are 7 chiral centres and 128 possible stereoisomers.

INTEGRATE THE SKILL 4.16

R configurations for molecules in Practice Problem 4.15.

PRACTICE PROBLEM 4.17

a) This molecule is achiral. There are no carbon atoms with four different substituents.

b) The molecule has a mirror plane of symmetry that reflects the two centres into each other, so this is a meso compound (achiral).

c) This molecule has a mirror plane, so this is a meso compound (achiral).

d) There are two chirality centres. Redrawing the ring shows there is no plane of symmetry between the two chiral centres, so this is not a meso compound. This is a chiral molecule.

e) There are two chirality centres in this molecule. Redrawing the molecule in a different representation shows there is a mirror plane, so this is a meso compound (achiral).

rotate 90° — rotate centre bond 30° — mirror plane

f) There are two chirality centres. As there is no mirror plane, this is not a meso compound. It is a chiral molecule.

H_2N NH_2

INTEGRATE THE SKILL 4.18

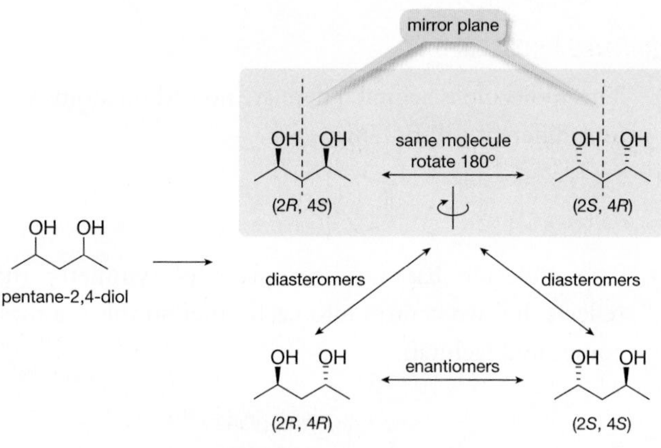

Upon first inspection, the (2*R*,4*S*) and (2*S*,4*R*) stereoisomers of pentane-2,4-diol appear to be enantiomers. However, due to the symmetry in the molecule, these two are actually the same molecule. Therefore, there are only three stereoisomers of pentane-2,4-diol.

PRACTICE PROBLEM 4.19

a) The ring structure means there can only be one isomer. So, the correct name is 4-propylcyclohex-1-ene.

(*E*)-4-propylcyclohexene

b) The first double bond is symmetric at one end, so no *E*/*Z* can be assigned. The correct name is (4*E*,6*E*)-2-methylocta-2,4,6-triene.

(2*E*,4*E*,6*E*)-2-methylocta-2,4,6-triene

c) *E*/*Z* should be used instead of *cis*/*trans* in naming. This is (*E*)-3-methylhex-3-ene.

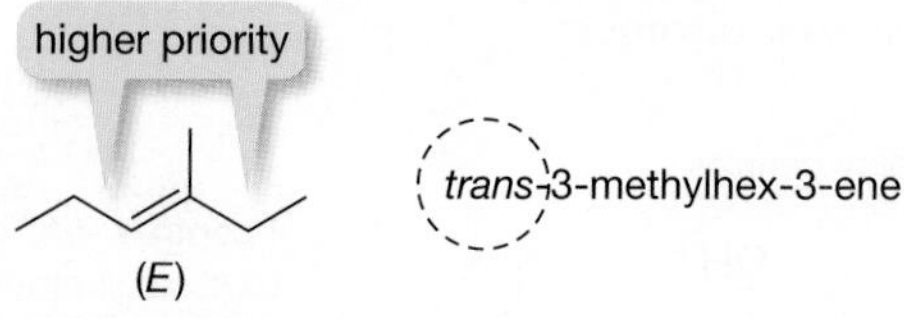

d) This is an *E* arrangement. The correct name would be (*E*)-(4-methylhexa-1,3-dien-3-yl)benzene.

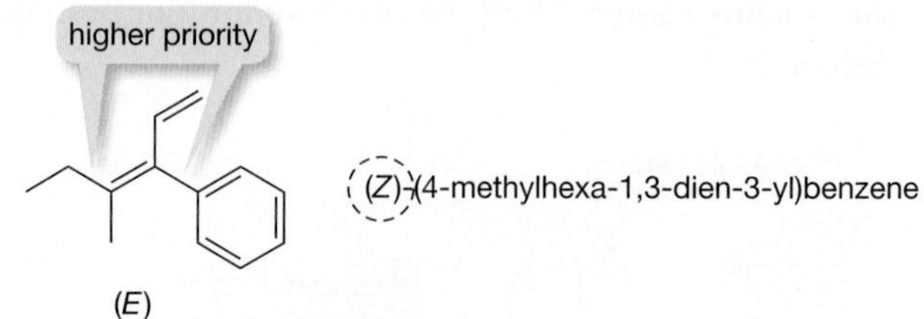

INTEGRATE THE SKILL 4.20

a) There are three chirality centres, so 2^3, or eight, stereoisomers are possible.

OH — * chirality centres

b) There is one asymmetric double bond, so there would be two stereoisomers of this compound.

O — H — asymmetric double bond

c) There are seven stereochemical sites (five chirality centres, two asymmetric double bonds), so there are 2^7, or 128, possible stereoisomers.

asymmetric double bond — HO — CO_2H — HO — OH — * chirality centres

PRACTICE PROBLEM 4.21

a) First, calculate the concentration of the compound in g/mL.

$$\textit{concentration} = \frac{100.0\text{ mg}}{50.0\text{ mL}} \times \frac{1\text{ g}}{1000\text{ g}}$$

$$\textit{concentration} = \frac{0.0020\text{ g}}{\text{mL}}$$

To use the specific rotation formula, we need the concentration in g per 100 mL.

$$\frac{0.0020\text{ g}}{\text{mL}} = \frac{x}{100\text{ mL}} \quad x = 0.20\text{ g} \quad c = \frac{0.20\text{ g}}{100\text{ mL}}$$

Now, calculate the rotation using the specific rotation equation.

$$[\alpha]_D = \frac{100 \cdot \alpha}{c \cdot L} = \frac{100 \cdot -0.004°}{\frac{0.20\text{ g}}{100\text{ mL}} \times 1.0\text{ dm}} = -2.0°$$

This would be reported as

$$[\alpha]_D = -2.0° \ (c\ 0.20, H_2O)$$

b) The enantiomer should have the same absolute rotation in the opposite direction, so it should be +2.0° (c 0.20, H_2O).

c) This rotation is 20× what was measured in part (a). To increase the measured rotation 20×, the concentration must be increased 20×. The original concentration was 0.20 g/100 mL, so a twentyfold increase would give a concentration of 4.0 g/100 mL.

d) The concentration of this solution is

$$\textit{concentration} = \frac{30.0\text{ mg}}{10.0\text{ mL}} = 3.00\frac{\text{mg}}{\text{mL}}$$

$$= 0.00300\frac{\text{g}}{\text{mL}} = 0.300\frac{\text{g}}{100\text{ mL}}$$

This is 1.5× the concentration in part (a), so the rotation would increase one and one half times to −0.006°.

INTEGRATE THE SKILL 4.22

There are two groups of stereoisomers to consider. The first are the meso compounds. There are two of them and, as meso compounds, neither is optically active.

In drawing all other stereoisomers, it turns out there are only two distinct chiral molecules and these are enantiomers of each other. Therefore, determining the specific rotation of one would give the specific rotation of the other (as the opposite sign). As these are the only chiral stereoisomers, they are the only ones that would have an optical rotation to measure, so one measurement is sufficient.

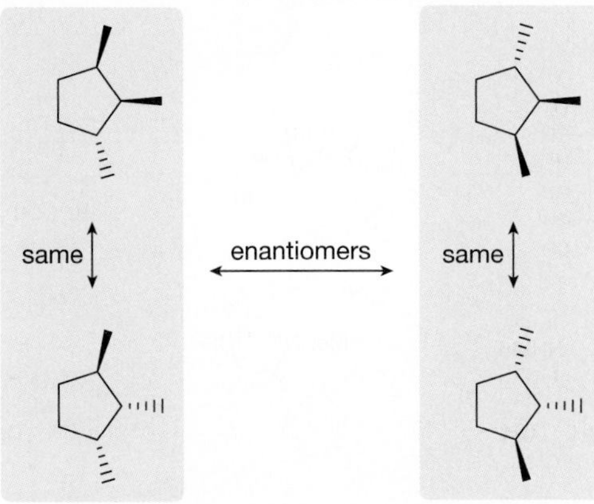

PRACTICE PROBLEM 4.23

a) $$\text{optical purity} = \%\text{ee}\ \frac{|\text{observed}[\alpha]|}{|[\alpha]\text{ of pure enantiomer}|} \times 100\%$$

$$= \frac{6.7}{13.5} \times 100\% = 50\%$$

The optical purity is 50%.

b) There is an excess of the (*S*) isomer, since the observed rotation is positive. Therefore, 50% is pure (*S*) and 50% is the racemate (50/50 mixture of (*R*)/(*S*)). Overall, there is 50% plus 25%, or 75% (*S*), and 25% (*R*).

c) If the ratio of enantiomers was 9:1, there would be 8 parts of excess (*S*) and a 1 part:1 part racemate of (*R*)/(*S*). So, it would be 8/10 pure, or 80% ee.

d) For 99% purity, the racemate must be 1% of the total mixture. So, (*S*) would be 99% plus 0.5% from the racemate. The ratio would be 99.5% (*S*):0.5% (*R*).

INTEGRATE THE SKILL 4.24

First, calculate the specific rotation of the original sample.

$$c = \frac{0.4\text{ g}}{1\text{ mL}} = \frac{40\text{ g}}{100\text{ mL}}$$

$$[\alpha]_D = \frac{100 \cdot \alpha}{c \cdot L} = \frac{100 \cdot +20.0°}{\frac{40\text{ g}}{100\text{ mL}} \cdot 0.5\text{ dm}} = +100.0°$$

This rotation arises from the 50% excess. A pure sample of the compound would be twice as optically active, so it would have specific rotation of +200.0° (c 80, H_2O).

PRACTICE PROBLEM 4.25

a)

b)

c)

d)

e)

f)

INTEGRATE THE SKILL 4.26

a) R R R R

enantiomer

b) S R

enantiomer

c) S

enantiomer

PRACTICE PROBLEM 4.27

There are three chirality centres in fructose, so there will be a maximum of eight (2^3) stereoisomers. There is one enantiomer of (D)-fructose, (L)-fructose. All the other stereoisomers are diastereomers of (D)-fructose.

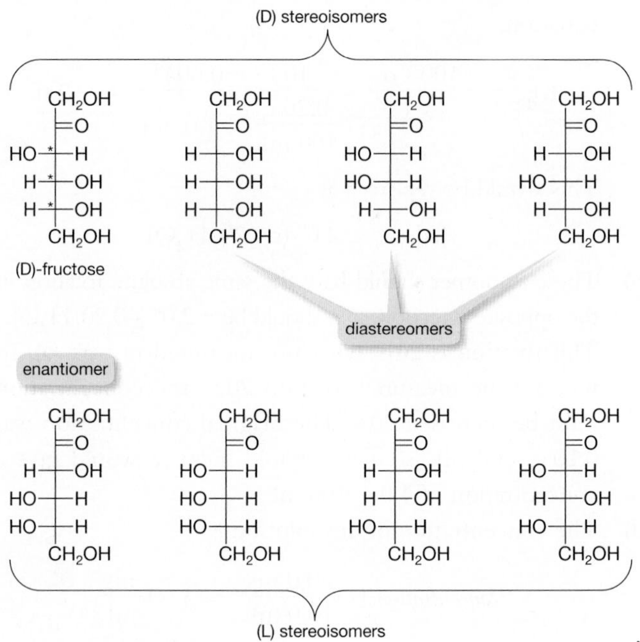

INTEGRATE THE SKILL 4.28

a) rotate 90° rotate 90°

(L)-aspartic acid

b) rotate 90° rotate 90°

(D)-cysteine

c) rotate 90° rotate 90°

(L)-histidine

CHAPTER 5

PRACTICE PROBLEM 5.1

a)

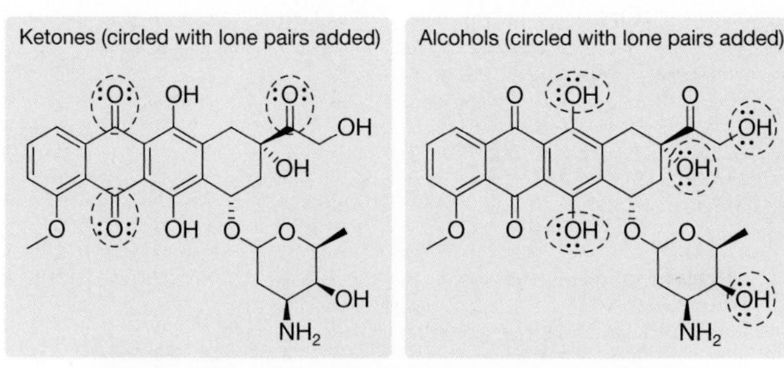

Other groups (circled and labelled with lone pairs added)

acetal

ether

amine

b) Note: the carbamate functional group is not included in Table 2.1

Ketones (circled with lone pairs added)

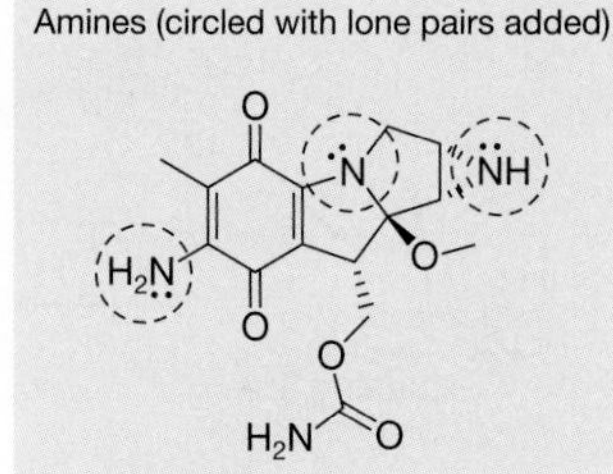

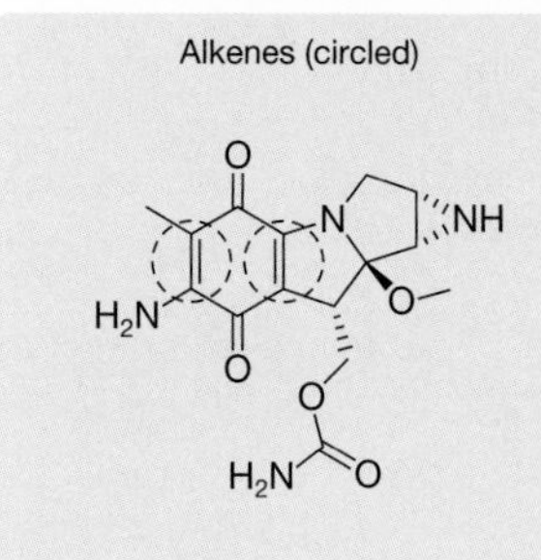

Other groups (labelled and circled with lone pairs added)

carbamate (see note)

ether

c)

Esters (circled and lone pairs added)

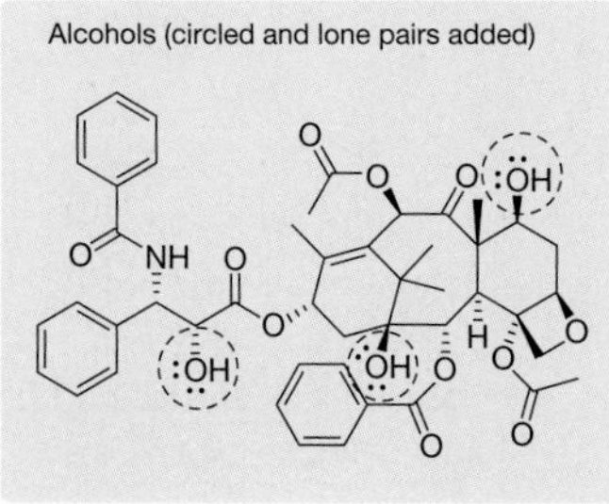

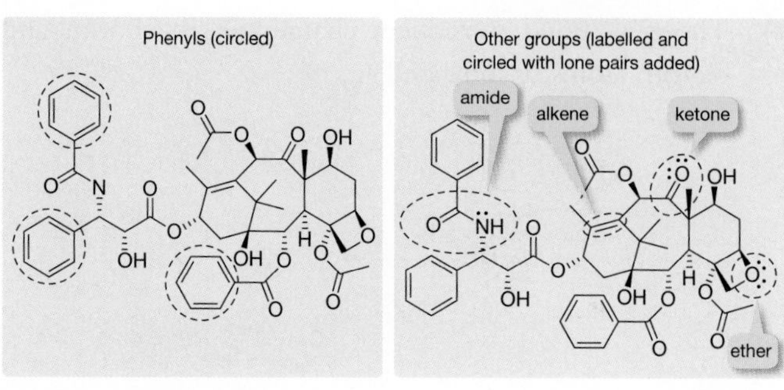

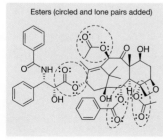

INTEGRATE THE SKILL 5.2

There are many possible answers to the question. One is shown below.

one possible solution

PRACTICE PROBLEM 5.3

a) Here, both of the bonding electrons go to nitrogen. The bond was formally one electron from N and one from C. Since both are now on N, the nitrogen has gained an electron and the carbon has lost an electron, leading to the final charge location on C.

shared bond electrons now both on N

carbon has lost its shared bond electron

b) The hydroxide ion contributes both electrons in forming the new C–O bond and so has formally lost a valence electron. The carbon still has four valence electrons and so remains neutral. The carbonyl oxygen gained a valence electron, as it is no longer sharing the bond electron with carbon, and is negatively charged.

FC = 6 − 1 − 6 = −1

FC = 4 − 4 − 0 = 0

FC = 6 − 2 − 4 = 0

c) The new bond introduces charges on the boron and oxygen atoms, as indicated.

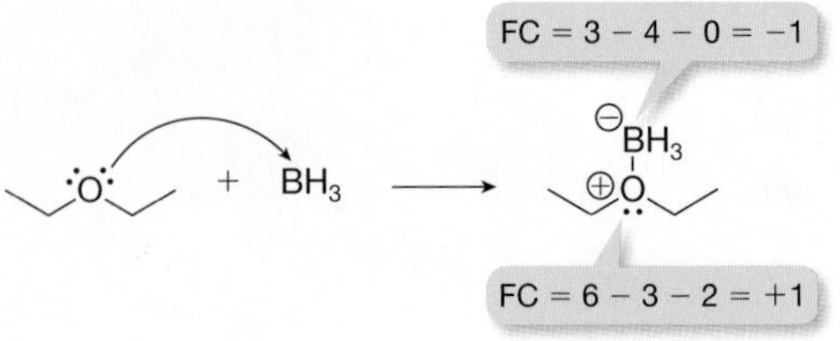

FC = 3 − 4 − 0 = −1

FC = 6 − 3 − 2 = +1

d) A new C–H bond forms from the $CH_3^{\ominus}$ lone pair electrons and the broken C–H bond electrons end up on the alkyne carbon.

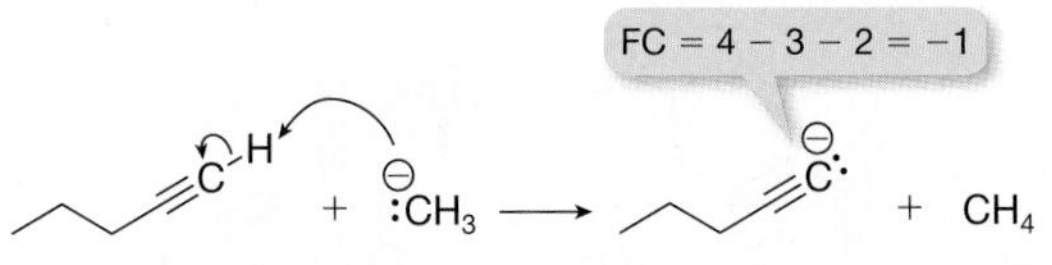

FC = 4 − 3 − 2 = −1

e) The $Cl^{\ominus}$ removes an H to make HCl. The C–H bond electrons make a new π bond.

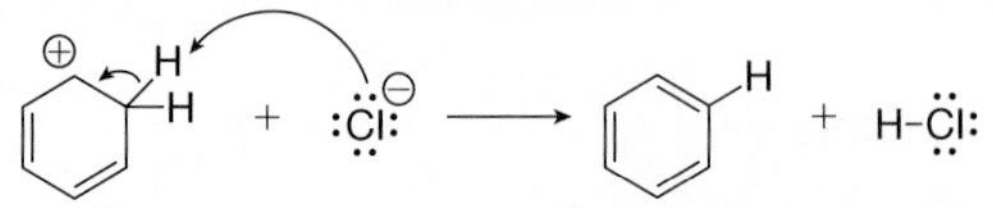

f) A new C–H bond forms by formal addition of $H^{\ominus}$ to the C=O carbon.

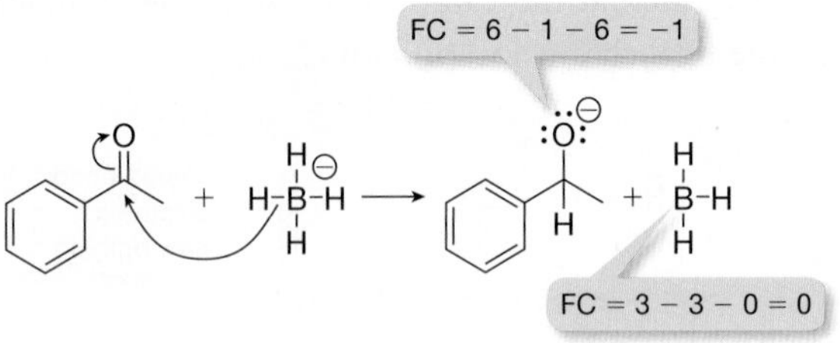

FC = 6 − 1 − 6 = −1

FC = 3 − 3 − 0 = 0

INTEGRATE THE SKILL 5.4

a) The product is incorrect. A benzene ring and HBr would form, as in part (e) of Practice Problem 5.3.

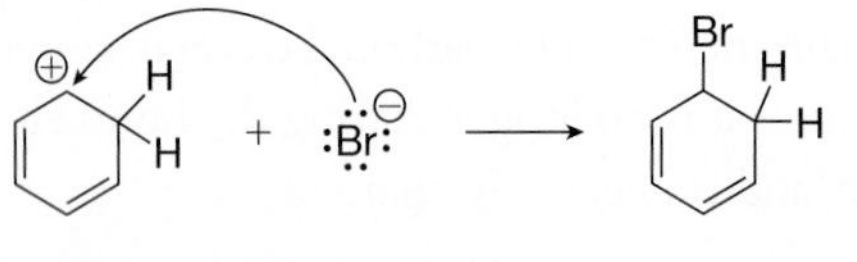

b) The mechanism was missing the arrow for movement of the C=N double bond electrons to the N atom.

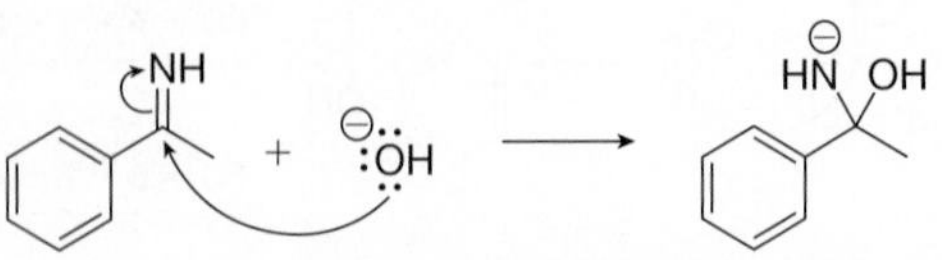

c) The lone pair electrons on O make the bond to H. The arrow was pointing in the wrong direction. As well, adding lone pair electrons on the O is recommended when one is just getting started working with mechanisms.

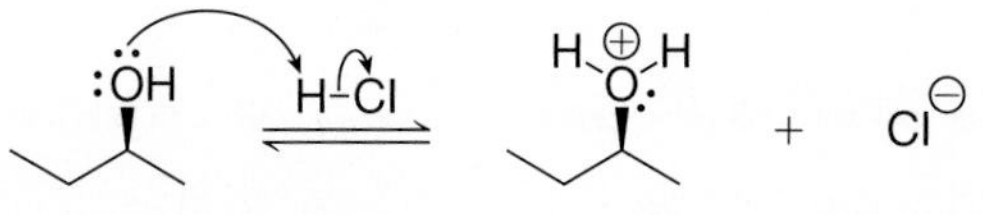

PRACTICE PROBLEMS 5.5

a)

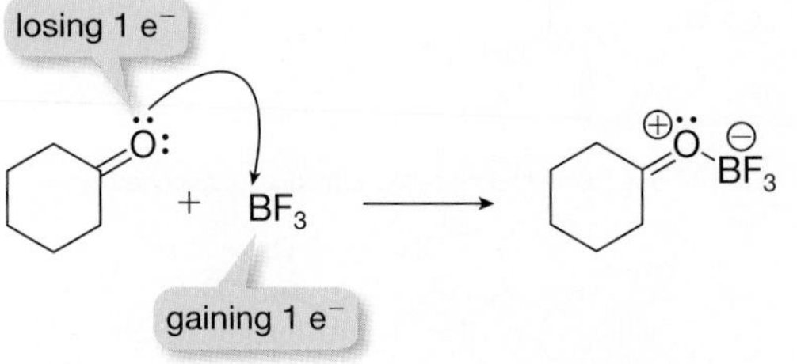

losing 1 e^-

gaining 1 e^-

b)

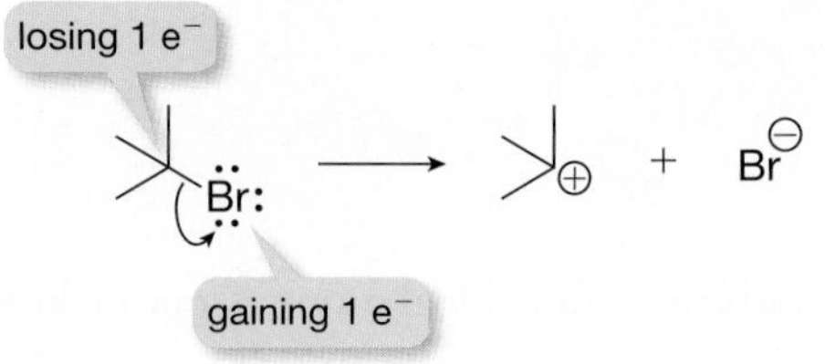

losing 1 e^-

gaining 1 e^-

c)

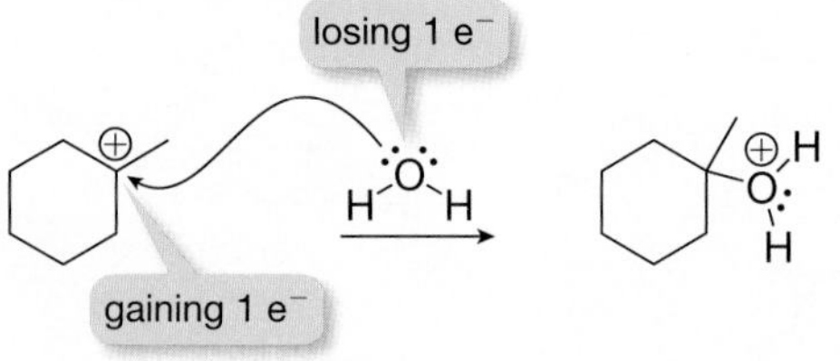

losing 1 e^-

gaining 1 e^-

d)

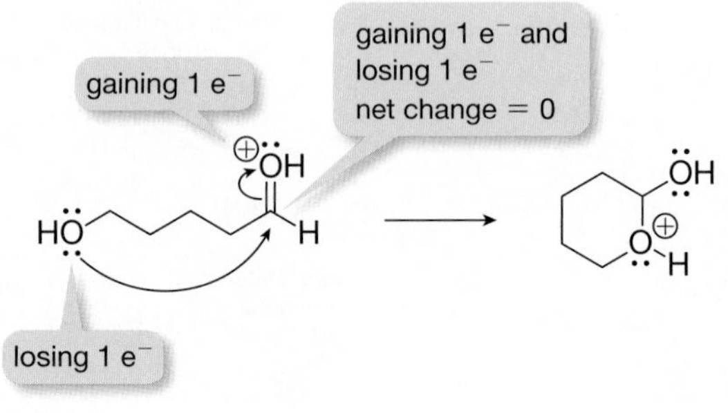

gaining 1 e^-

gaining 1 e^- and losing 1 e^- net change = 0

losing 1 e^-

e)

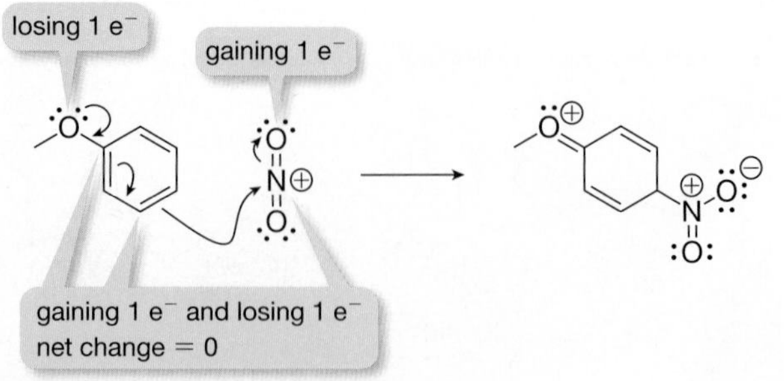

losing 1 e^-

gaining 1 e^-

gaining 1 e^- and losing 1 e^- net change = 0

INTEGRATE THE SKILL 5.6

a)

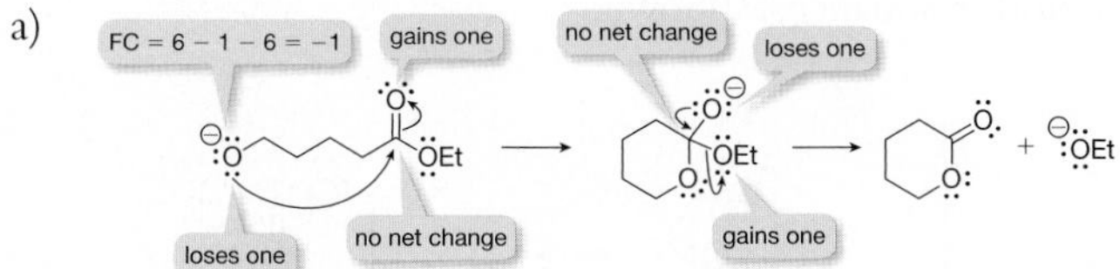

b)

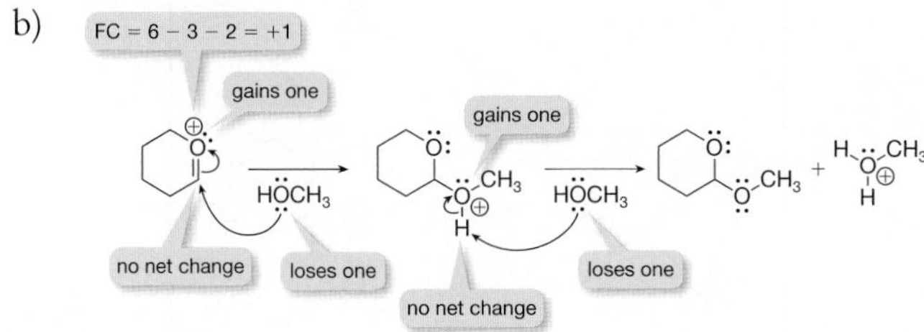

PRACTICE PROBLEM 5.7

a) Intermolecular reaction

b) Intramolecular reaction

c) Intramolecular reaction

d) Intermolecular reaction

INTEGRATE THE SKILL 5.8

a) This reaction is intramolecular. Two chirality centres are created in the product.

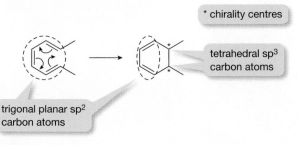

* chirality centres

tetrahedral sp3 carbon atoms

trigonal planar sp2 carbon atoms

b) This is an intermolecular reaction. Two chirality centres are produced in the product.

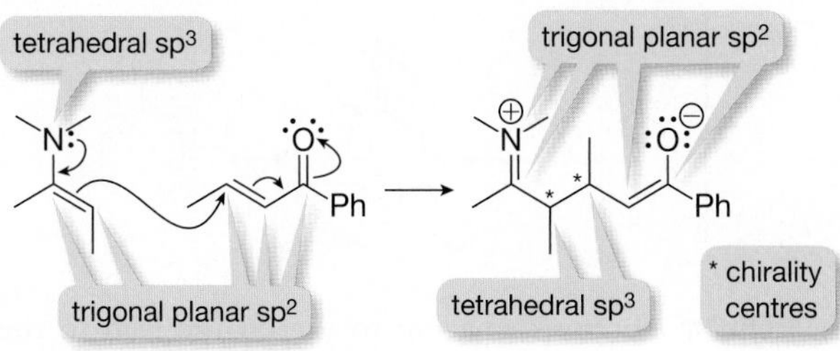

c) This is an intramolecular reaction. There are five chirality centres formed in the product.

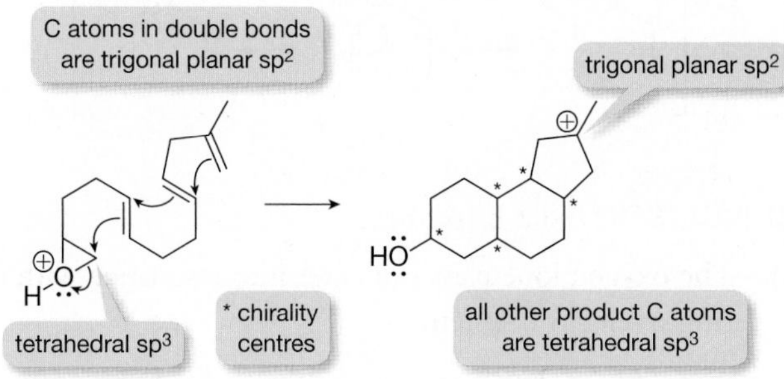

PRACTICE PROBLEM 5.9

a) This molecule has a lone pair beside a carbocation, leading to the resonance forms indicated below.

b) No resonance is possible because the charge is isolated from the double bond by a saturated CH_2 group.

c) No resonance is possible because the charge is isolated from the double bond by a saturated CH_2 group.

d) No resonance is possible because the lone pair is isolated from the double bond by a saturated carbon atom.

e) The charge is adjacent to two π systems, resulting in the following possible resonance forms.

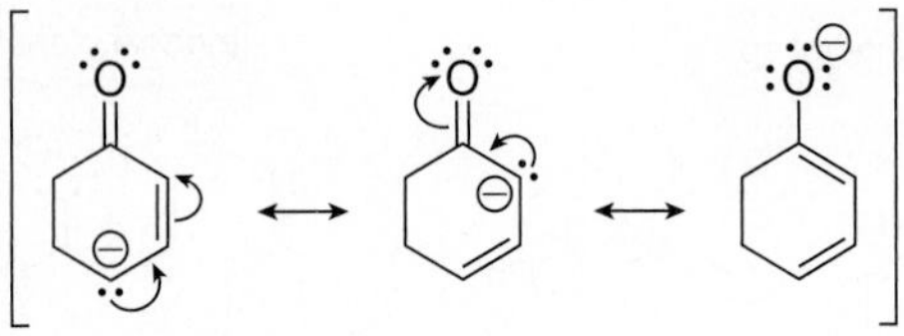

f) The π electrons adjacent to the charge lead to the following resonance forms.

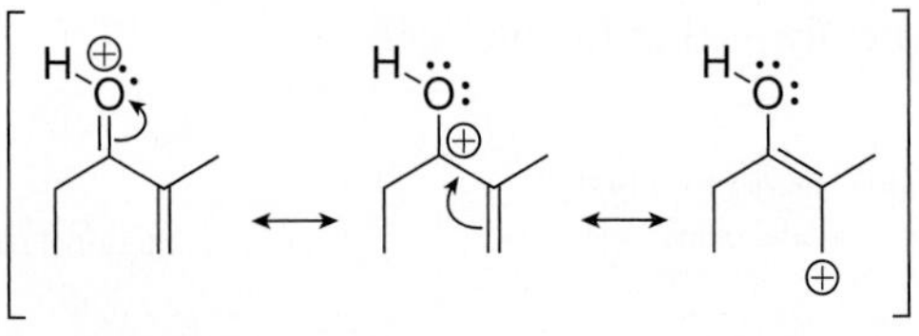

INTEGRATE THE SKILL 5.10

a) The oxygen lone pairs can enter into resonance with the ring π-electron system.

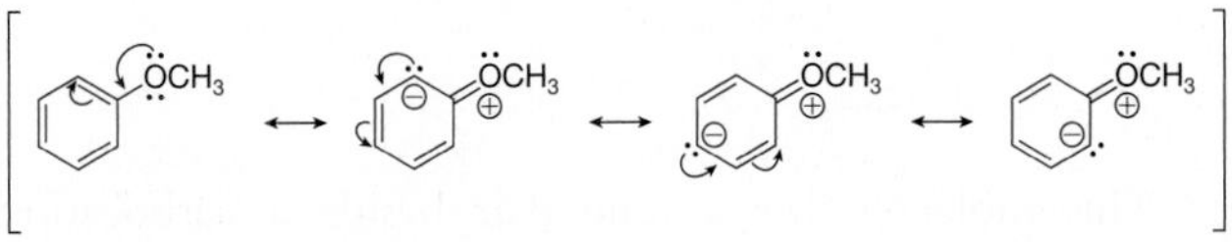

The result of this resonance is to give some negative character to the *ortho* and *para* positions in the ring. These will be attracted to positive charges. As well, the lone pairs on the oxygen atom will be attracted to positive charges.

b) The molecule has possible resonance due to the lone pair being next to a π bond, leading to the following resonance forms.

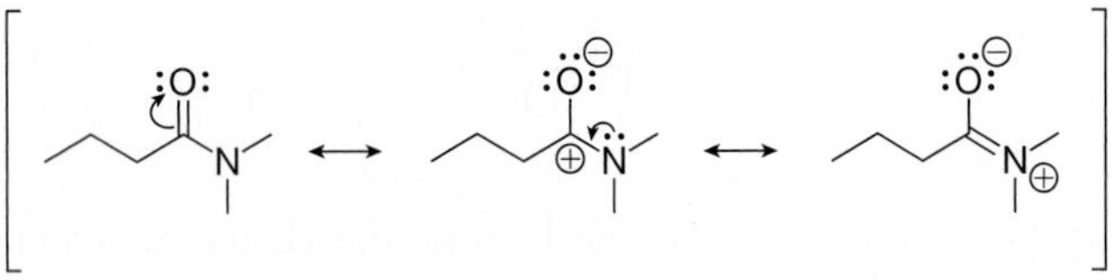

The form on the right has a formal negative charge on the oxygen, making it most attractive to positive charges.

c) The oxygen lone pair next to a double bond can enter into resonance, as shown.

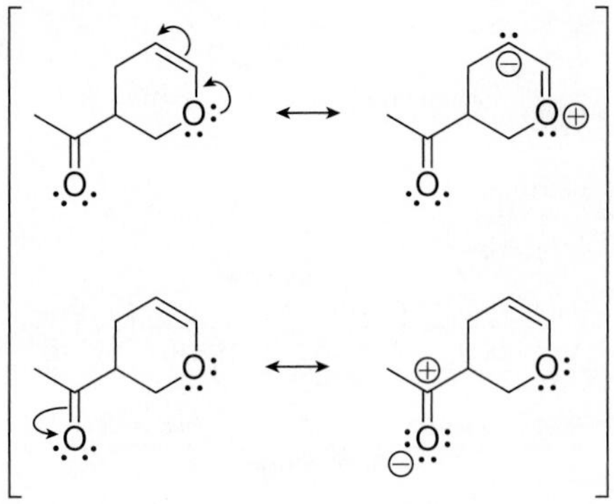

This results in a formal negative charge on the carbon indicated, which would be attracted to positive charges. The lone pairs will be attracted to positive charges as well.

PRACTICE PROBLEM 5.11

a) The major difference in these forms is the location of the negative charge. It will be more stable on the most electronegative atom, which is N in this case.

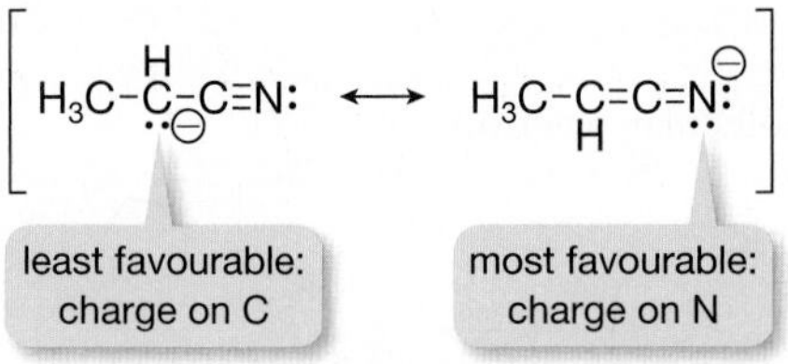

least favourable: charge on C

most favourable: charge on N

b) The presence and separation of charges is the major factor distinguishing these forms.

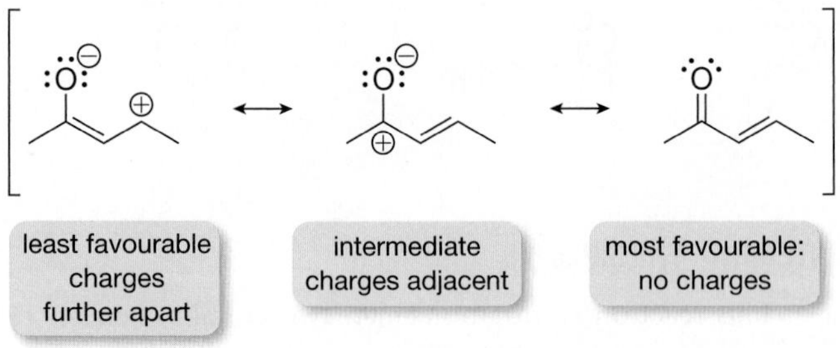

least favourable charges further apart

intermediate charges adjacent

most favourable: no charges

c) The least favourable form has the negative charge on the least electronegative atom, C. The other two structures have similar π systems, and so the charge on O would be equal in quality.

least favourable: charge on C

similar: charge on O

d) Because the bonding in each is the same, the two resonance forms on the right are of equal importance in describing the bonding in the molecule. All atoms in these two resonance forms have filled valence orbitals. The resonance structure on the left, with an empty valence orbital on the carbon atom, will be of less importance than the other two.

least important

most important and equivalent

e) The first two resonance forms each have a carbon atom with an empty valence orbital; so the third resonance form, with each atom having all orbitals filled, will be the most important contributor to the bonding. The carbon atom having the positive charge in the first resonance form is positioned further from the electronegative O atom. This will be a more important contributor to the bonding than the middle resonance form in which the carbon atom bearing the positive charge is adjacent to the O atom.

intermediate: positive charge further from electronegative O

least important: positive charge closer to electronegative O

most important: all atoms full octets

INTEGRATE THE SKILL 5.12

The ring electrons react with the bromine molecule, producing a positively charged intermediate and a bromide ion.

The two main determinants of the relative contribution of each resonance form are whether the atoms have filled or empty valence orbitals and the location of the positive charge in the intermediate.

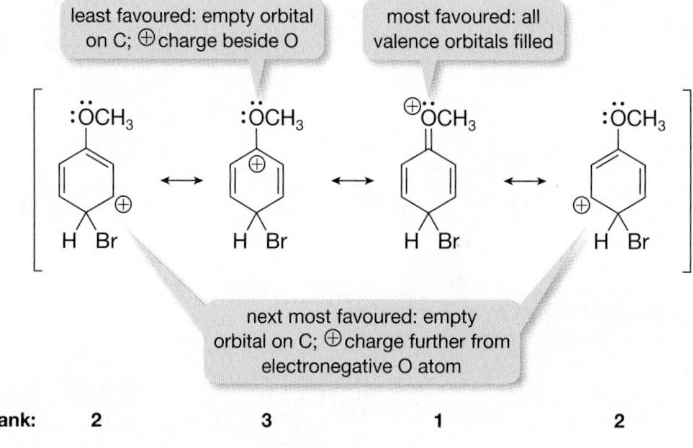

PRACTICE PROBLEM 5.13

a) There are three apparent groupings, shown below.

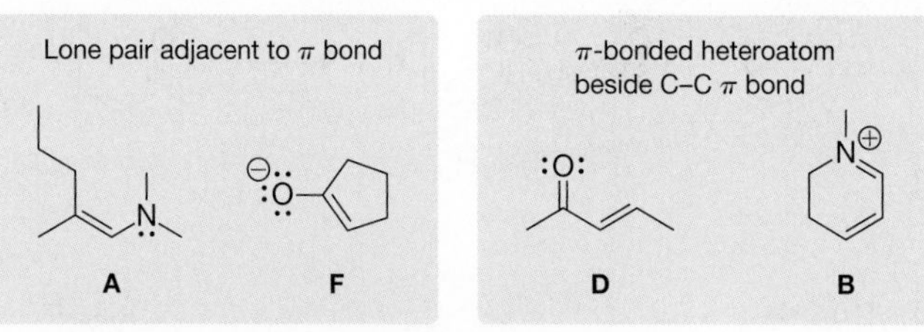

Lone pair adjacent to 2 π bonds

C E G

b) Lone pair adjacent to π bond

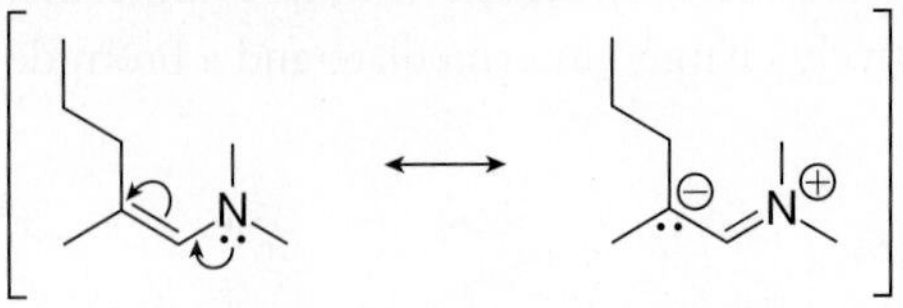

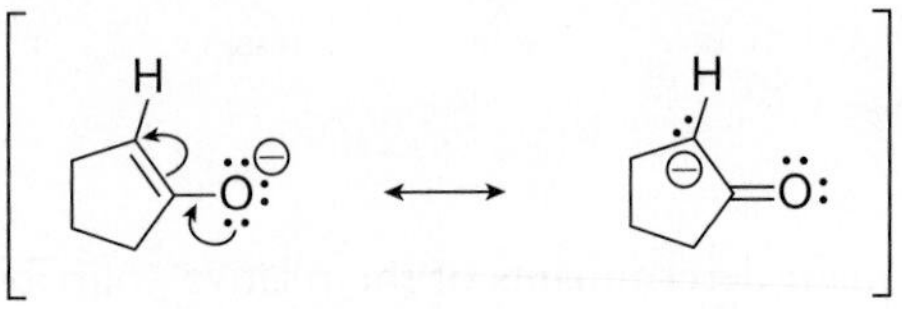

π-bonded heteroatom beside C–C π bond

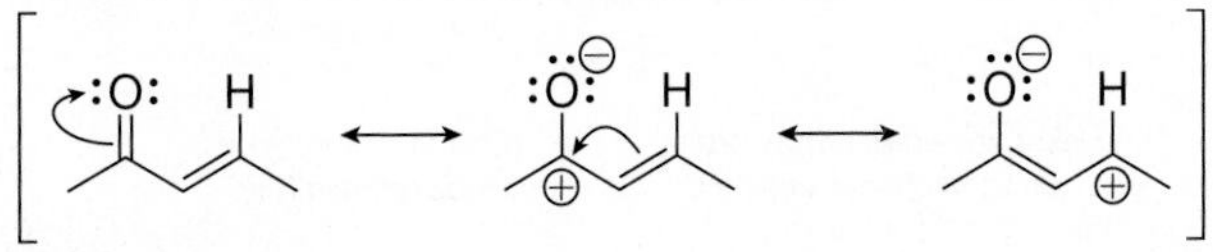

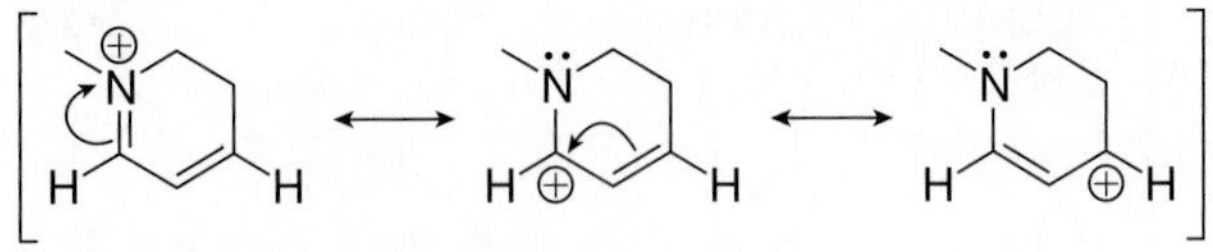

Lone pair adjacent to 2 π bonds

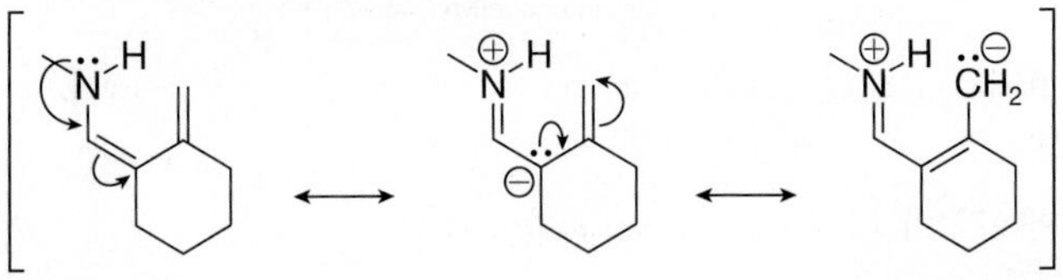

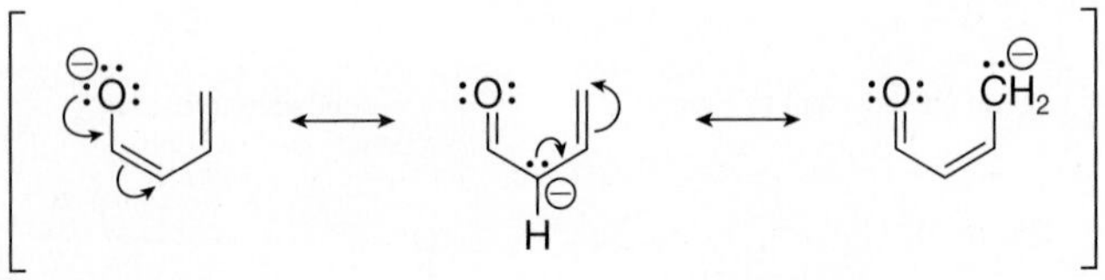

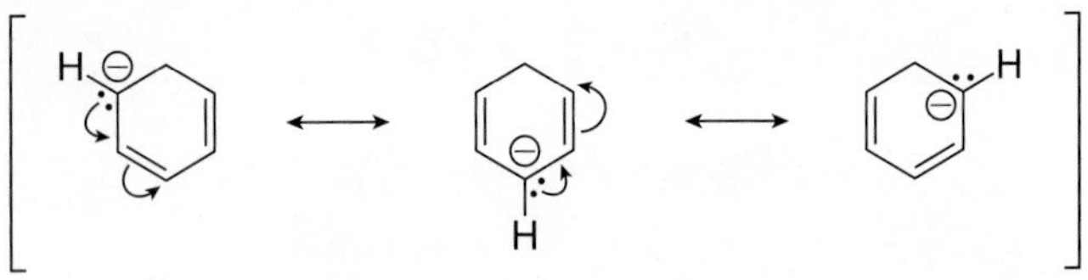

INTEGRATE THE SKILL 5.14

a) The basic form is a double bond beside a heteroatom with lone pairs.

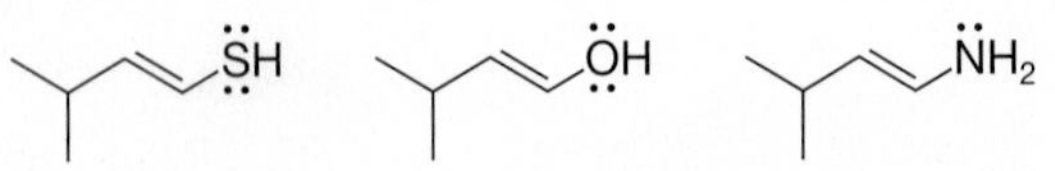

b) These each have a double-bonded hetereoatom with an adjacent C–C double bond.

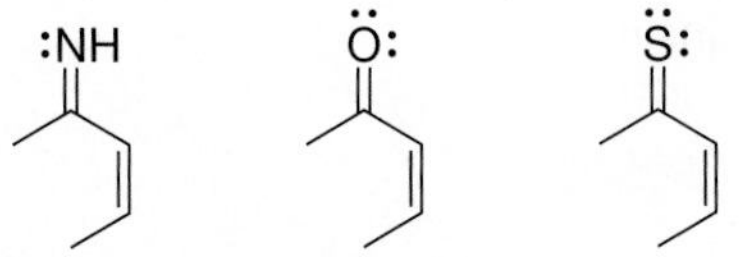

CHAPTER 6

PRACTICE PROBLEM 6.1

a)

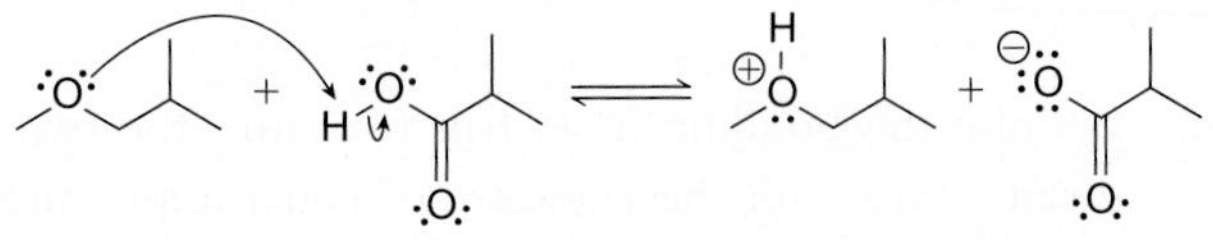

b)

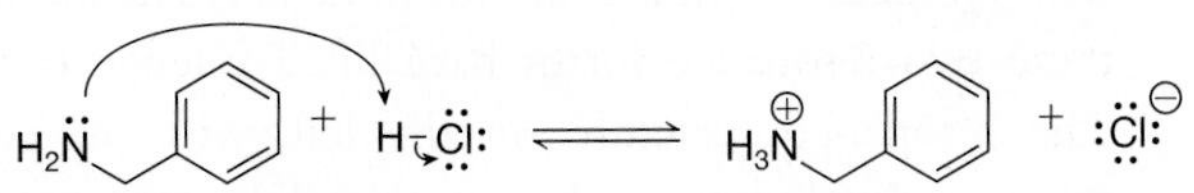

c)

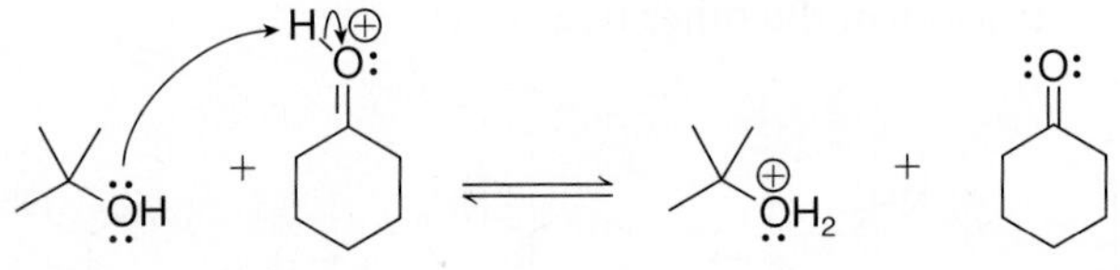

d)

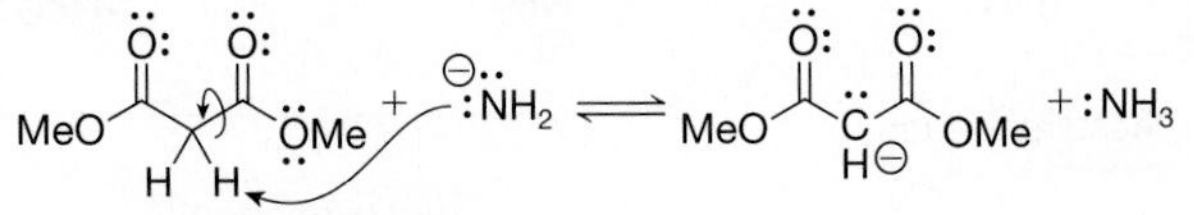

INTEGRATE THE SKILL 6.2

a)

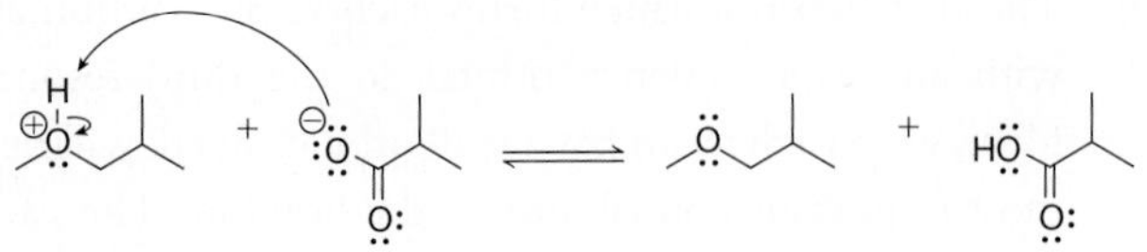

b)

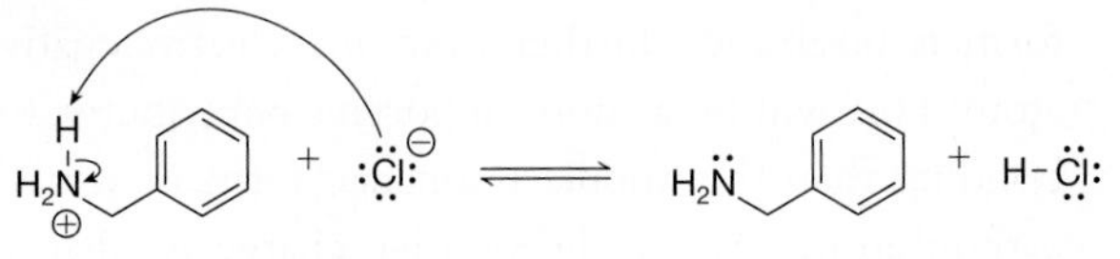

c)

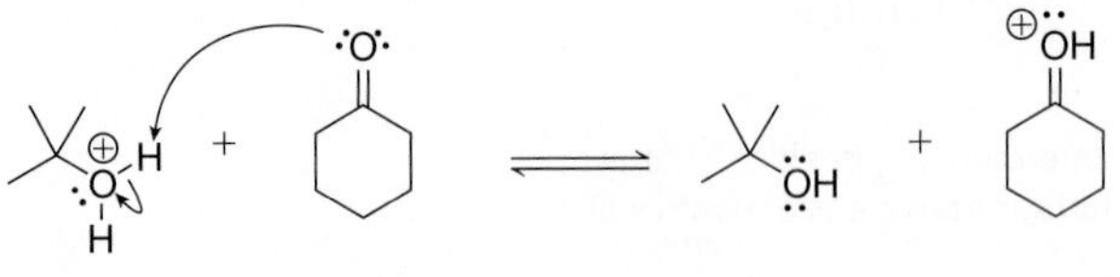

d)

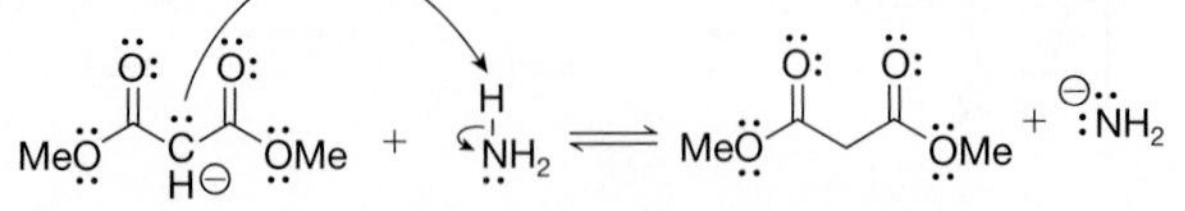

PRACTICE PROBLEM 6.3

a) Since the equilibrium favours the reactants, they must be lower in energy than the products.

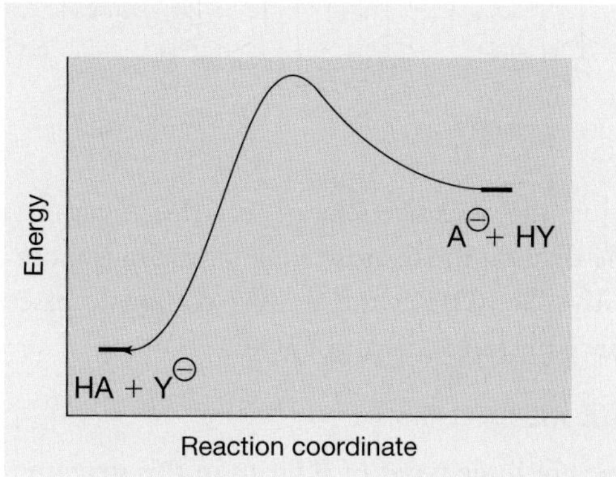

b) Since neither reactants nor products are favoured, they must be equal in energy.

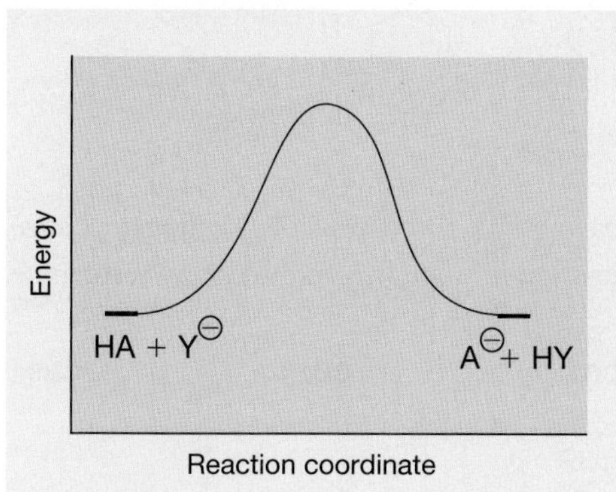

c) Since the conjugate base A^- is less stable than the base Y^-, the reactants should be lower in energy and be favoured.

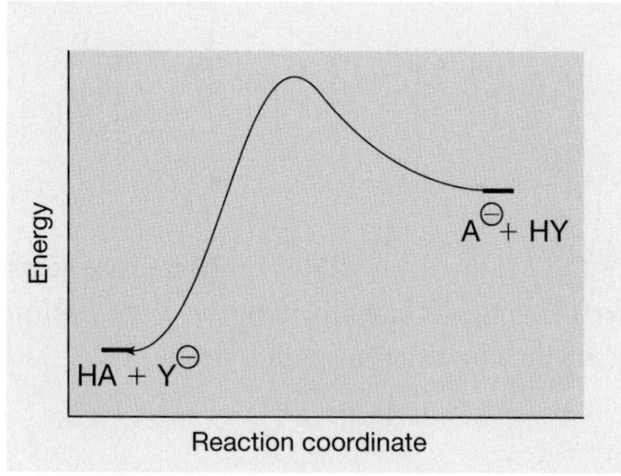

d) Since the conjugate base A^- is more stable than the base Y^-, the products should be lower in energy and be favoured.

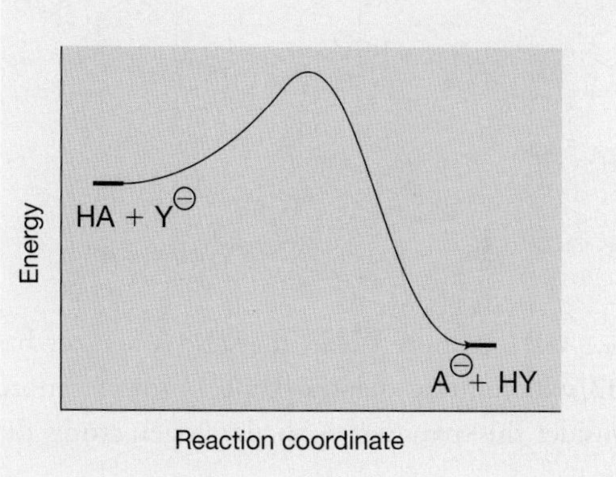

INTEGRATE THE SKILL 6.4

Hydrofluoric acid (HF) is a weak acid that only partially dissociates in water. In contrast, hydrobromic acid (HBr) is a strong acid that completely dissociates in water.

a) Water is the base in both reactions.

$$HF + H_2O \rightleftharpoons F^{\ominus} + H_3O^{\oplus}$$

$$HBr + H_2O \rightleftharpoons Br^{\ominus} + H_3O^{\oplus}$$

b)

$$:\ddot{F}-H + H_2\ddot{O}: \rightleftharpoons :\ddot{F}:^{\ominus} + H_3O:^{\oplus}$$

$$:\ddot{Br}-H + H_2\ddot{O}: \rightleftharpoons :\ddot{Br}:^{\ominus} + H_3O:^{\oplus}$$

c) Since HBr fully dissociates, its equilibrium heavily favours the products. Therefore, this energy diagram, with the products much lower in energy than the reactants, is a better representation of the dissociation of HBr.

PRACTICE PROBLEMS 6.5

a) There are two types of explicitly drawn protons in the structure, so there are two conjugate bases to consider.

Type 1 HÖ– ... –ÖH Type 2

CB1 CB2

CB1 is more stable than **CB2**, since it is resonance stabilized.

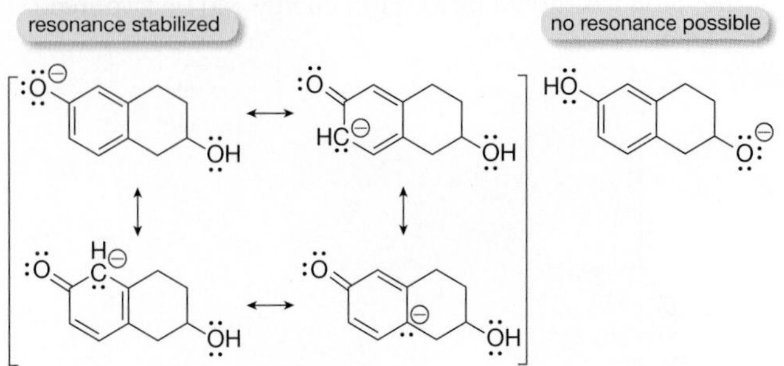

Since **CB1** is more stable, it is also a weaker base than **CB2**, which means that the **Type 1** proton is more acidic.

b) Consider the two most acidic hydrogen atoms shown:

Type 1 HO OH Type 2

There are two conjugate bases to be considered: **CB1** and **CB2**.

CB1 CB2

The charge on **CB2** is resonance stabilized, while the charge on **CB1** is not. **CB2** is the more stable and weaker base so the **Type 2** proton is the most acidic.

c) Type 1 HO SH Type 2

The two conjugate bases to be considered for the indicated protons are **CB1** and **CB2**.

CB1 CB2

The negative charge on the S atom is more dispersed (it has one extra electron over a larger atomic volume) than on the smaller O atom and, therefore, **CB2** is more stable and the weaker base. So, the **Type 2** proton is the most acidic.

d) The two most acidic protons are indicated below.

Type 1 Type 2

Removing either a **Type 1** or **Type 2** proton forms conjugate bases, which are resonance stabilized.

CB1 CB2

Having the negative charge on the N atom is more stable than on the C atom since nitrogen is more electronegative. So, **CB1** is the weaker conjugate base and the **Type 1** proton is the most acidic.

INTEGRATE THE SKILL 6.6

a) There are three types of protons in this structure.

Type 1 Type 2 Type 3

Comparing the three possible conjugate bases, **CB2** is the most stable since it is stabilized by resonance.

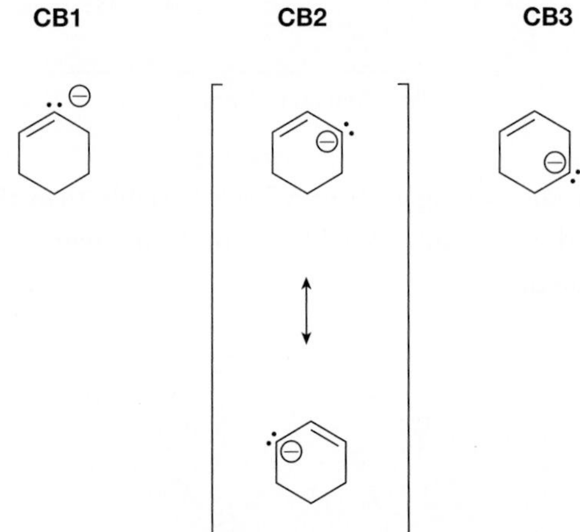

Since **CB2** is the most stable conjugate base, it is also the weakest conjugate base, making the Type 2 protons the most acidic. The reaction with a strong base is as follows.

H + :B ⇌ + HB

b) The three most acidic protons are shown in the diagram below.

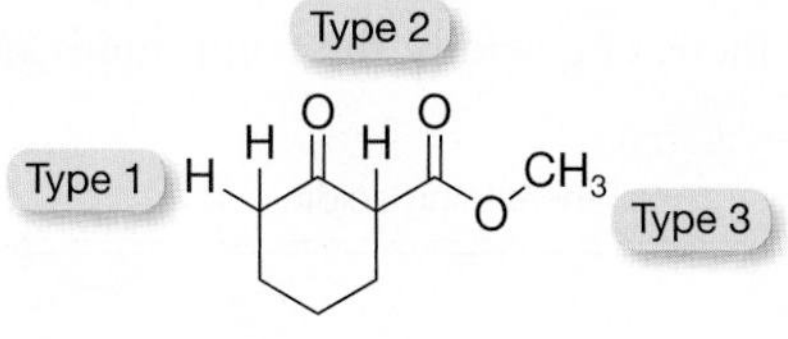

Type 2

Type 1

Type 3

The resulting conjugate bases are:

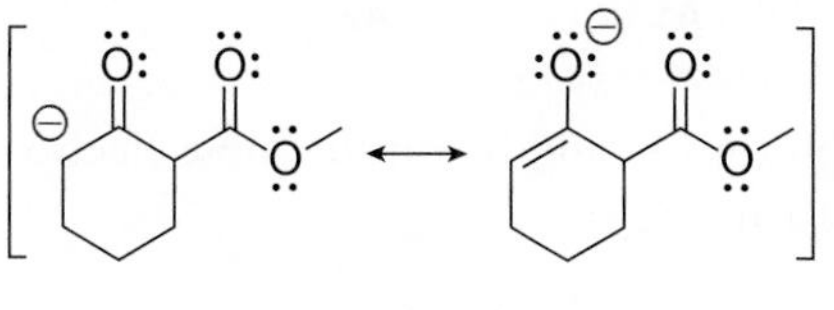

CB1

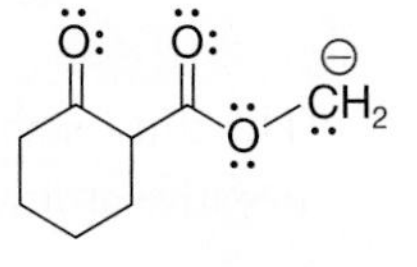

CB3

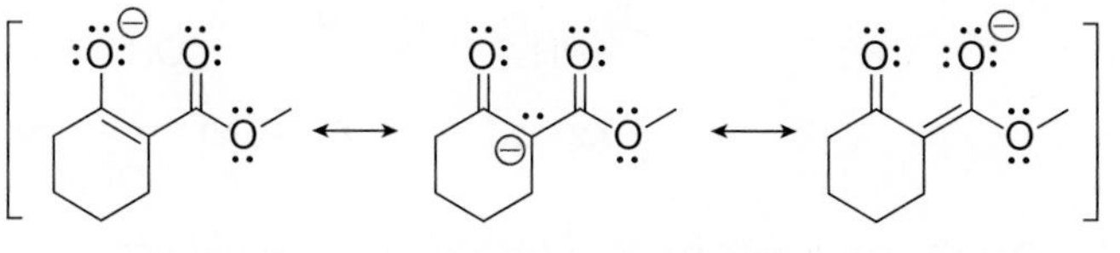

CB2

The least acidic are **Type 3** since **CB3** has a localized charge on a carbon atom.

CB1 and **CB2** are both resonance stabilized, but **CB2** distributes the charge onto two carbonyl oxygens, leading to a more stable conjugate base. Therefore, the **Type 2** proton is the most acidic. The reaction with a strong base is as follows.

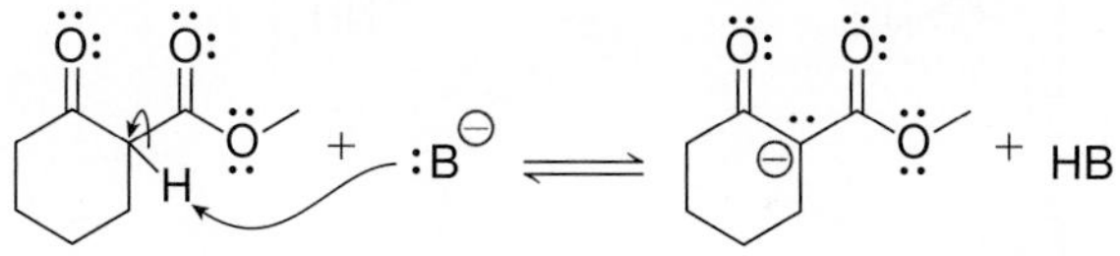

c) There are three general types of protons in this structure.

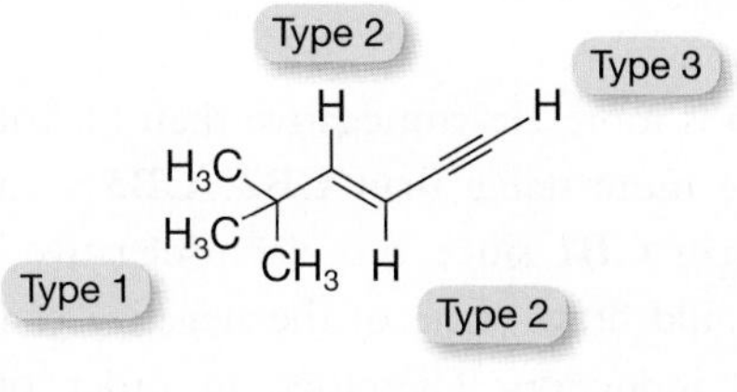

Type 2

Type 3

Type 1

Type 2

The major difference between them is the hybridization of the carbon atom to which they are attached. Removing a **Type 1** proton would leave the negative charge in an sp^3 hybrid orbital, removing a **Type 2** would leave the negative charge in an sp^2 orbital, and removing a **Type 3** would leave the negative charge in an sp orbital. None of the three types is stabilized by resonance. The negative charge is most stable in the sp orbital because it has the highest s orbital character. (An sp orbital has 50% s character/50% p character, while an sp^2 orbital is only 33% s character.) This high s orbital character gives the highest probability of finding the negative charge close to the nucleus, resulting in the electronegativity order of $sp > sp^2 > sp^3$. So, **Type 3** is the most acidic proton and the reaction with a strong base is as follows.

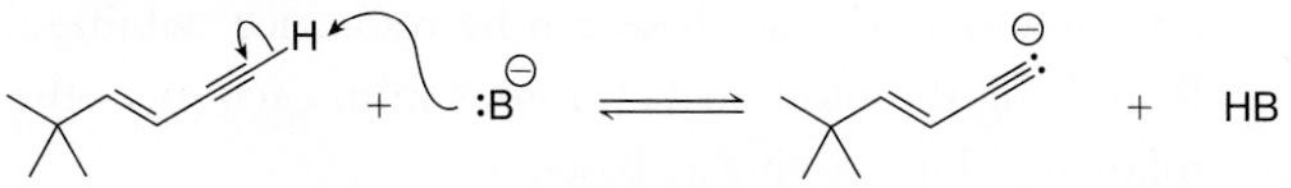

PRACTICE PROBLEM 6.7

a)

A1 A2 A3 A4

Since the acids are all neutral, the stability of their conjugate bases should be compared. Removing the most acidic proton in each, the four conjugate bases are:

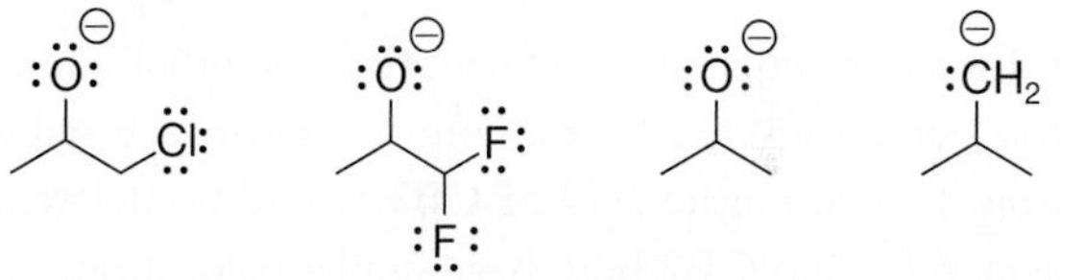

CB1 CB2 CB3 CB4

CB4 is the least stable since its negative charge is on the least electronegative C atom and there are no other features to help stabilize it.

The other three all have a negative charge on O; however, they differ in the number and type of nearby electron-withdrawing groups. **CB2** will be the most stable since it has two very electronegative F atoms nearby, which stabilize the charge through induction. **CB1** also has some stabilization through induction; however, there is only one Cl atom and it is less electronegative than F. **CB3** has no nearby electronegative atoms.

Therefore, in order of increasing stability, the conjugate bases are ranked **CB4, CB3, CB1, CB2.**

The more stable the conjugate base, the weaker a base it is, and the more acidic its conjugate acid. Therefore, in order of increasing acidity, the compounds are ranked as:

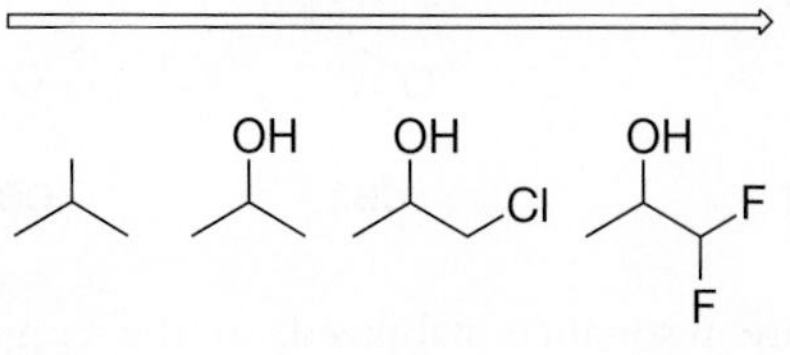

increasing acidity

A4 A3 A1 A2

b)

A1 A2 A3

The most acidic protons in each of these would be the ones whose conjugate base can be resonance stabilized. Removing the most acidic protons from each gives the following three conjugate bases.

CB1

CB2

CB3

CB3 has fewer resonance forms than the other conjugate bases and, thus, the least amount of resonance stabilization. The conjugate acid of **CB3** would be the weakest acid. **CB1** and **CB2** both have similar delocalization, but **CB1** has the negative charge on a more electronegative O atom in one of its resonance forms. This would make **CB1** more stable than **CB2**. Therefore, in order of increasing acidity, the compounds are ranked as:

increasing acidity

A3 A2 A1

c)

A1 A2 A3

Removing the most acidic proton in each gives the following conjugate bases.

CB1 CB2 CB3

None are resonance stabilized, so the conjugate bases differ mainly by having their charges in sp^3, sp^2, or sp hybrid orbitals. **CB2** (sp) would be more stable than **CB1** (sp^2), which is more stable than **CB3** (sp^3). Therefore, in order of increasing acidity, the compounds are ranked as:

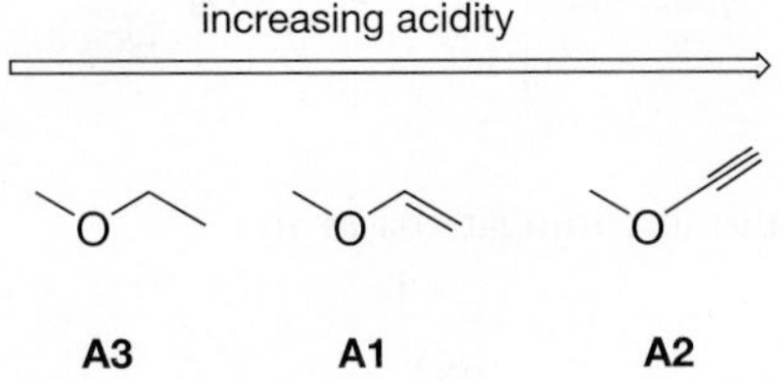

A3 A1 A2

d) The most acidic protons would be on the most electronegative atoms (N, O).

A1 A2 A3

So, the conjugate bases of each of these would be:

CB1

CB2

CB3

Since O is more electronegative than N, both **CB1** and **CB3** are more stable than **CB2. CB3** would be more stable than **CB1** since the electronegative Br atom in **CB3** would draw some of the negative charge to itself through induction. Therefore, in order of increasing acidity, the compounds are ranked as:

increasing acidity

A2 A1 A3

INTEGRATE THE SKILL 6.8

From the reaction coordinate diagram:

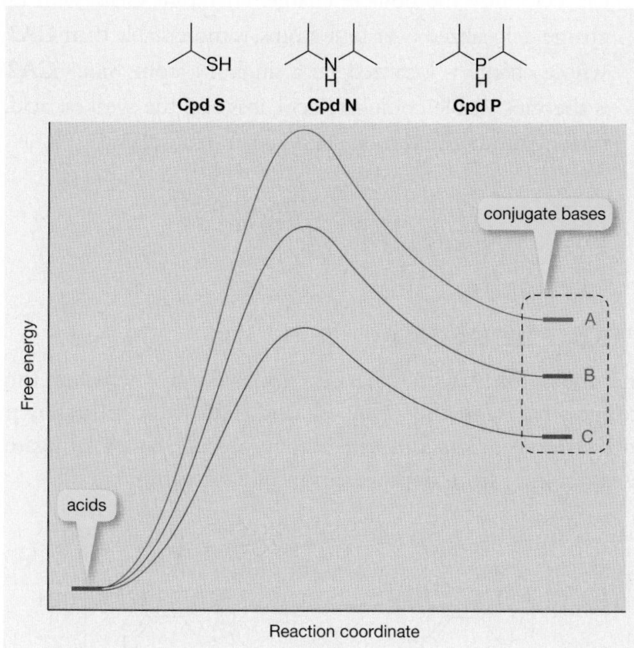

The conjugate bases (products) need to be compared in terms of their relative stabilities to identify acids A, B, and C. There are two pairwise comparisons to be made, keeping all other factors as constant as possible.

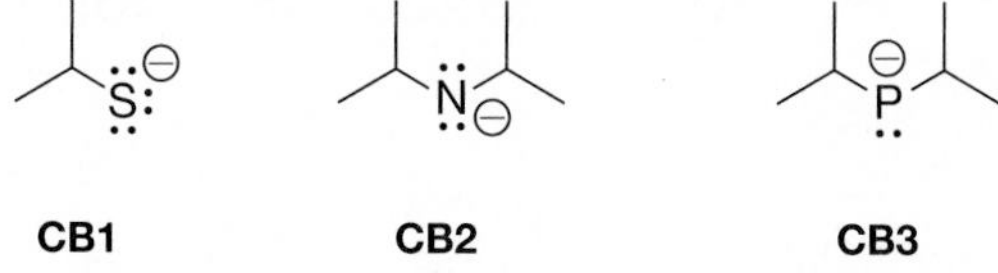

- S is more electronegative than P and should better stabilize the charge on the conjugate base. Therefore, **CB1** should be more stable than **CB3**.
- P is a larger atom than N and is a better stabilizer of negative charge because of its larger size. Therefore, **CB3** should be more stable than **CB2**.

Based on these comparisons:

acid A (N–H) acid B (P–H) acid C (SH)

PRACTICE PROBLEM 6.9

a) Since the acids are charged, their stability—and, therefore, their acidity—can be compared directly.

H_2N^{+} vs HN NH^{+}

Both acids have resonance stabilization; however, only the right-hand structure has another resonance contributor with a complete octet on all atoms.

Therefore, the right-hand structure is more stable and less acidic, making the left-hand structure the stronger acid.

b) Since the bases all have negative charges, they can be compared directly.

A B C

Compound **C** is the most stable since it has adjacent F atoms, which are very electronegative and help stabilize the negative charge by induction. Compound **B** is the least stable since it has alkyl groups adjacent to the negative charge, which are slightly electron-donating, making the anion less stable. Compound **A** has a stability in between the others. Since the more stable a base is, the less basic it is, the compounds are ranked in order of increasing basicity as: **C**, **A**, **B**.

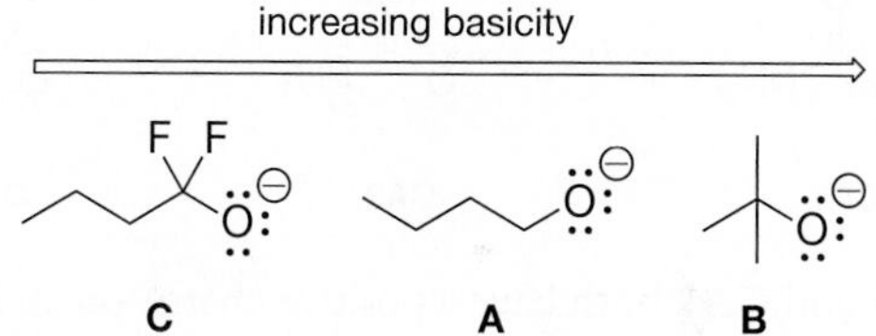

c) The compound has three carboxylic acid functional groups, which are all acidic.

However, they differ in the atoms located nearby. To determine which is more acidic, consider the three conjugate bases and their relative stability.

CB1 CB2 CB3

CB1 and **CB2** have nearby Br and Cl atoms, which are electronegative and can help stabilize the negative charge through induction. Since Cl is more electronegative than Br, **CB2** is more stable, making its corresponding carboxylic acid the most acidic.

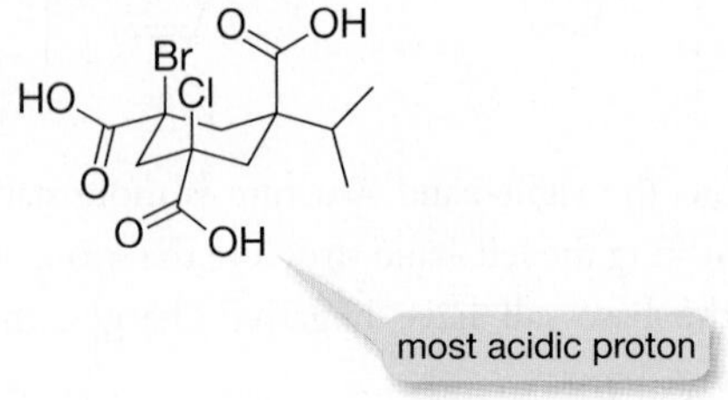

most acidic proton

d) There are three basic atoms in this structure, **A**, **B**, and **C**, each with lone pairs available for sharing.

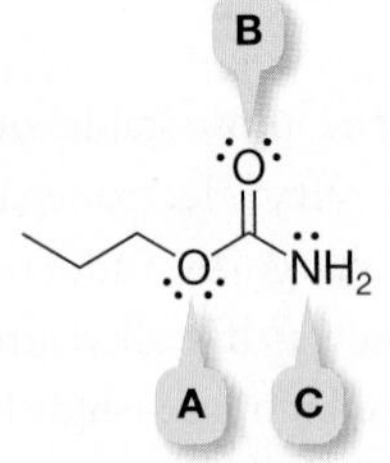

To determine the most basic site, consider their conjugate acids and compare their relative stability:

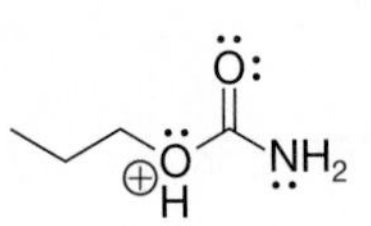

CA1 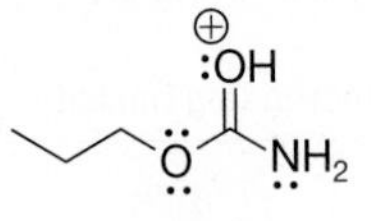 CA2 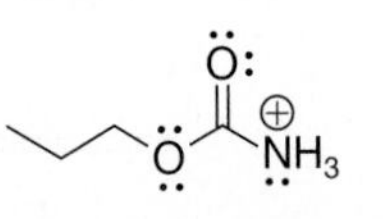 CA3

CA1 and **CA2** both have a positive charge on an O atom; however, **CA2** benefits from resonance stabilization of the charge while **CA1** does not. (Note, resonance structures can be drawn for **CA1**; however, they do not involve the protonated oxygen atom and, therefore, do not delocalize the positive charge and stabilize the cation.)

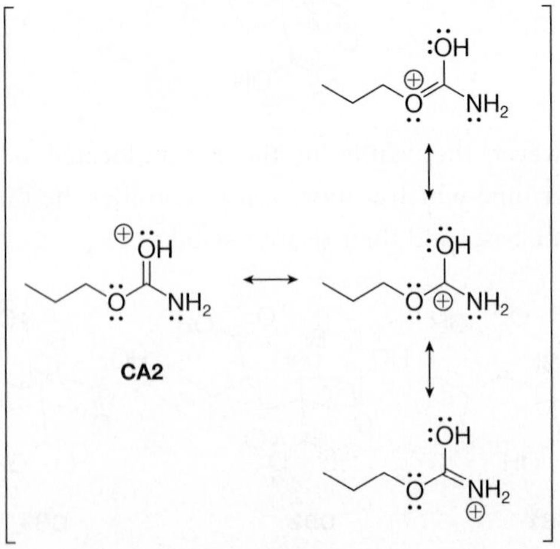

CA2

CA3 has its charge on a less electronegative N atom; however, like **CA1**, it does not have any resonance structures that can delocalize the positive charge. Therefore, **CA2**, with its charge delocalized over four atoms, is more stable than **CA3** whose charge is localized on a single N atom. Since **CA2** is the most stable conjugate acid, it is also the weakest acid, making site **B**, the carbonyl oxygen, the most basic.

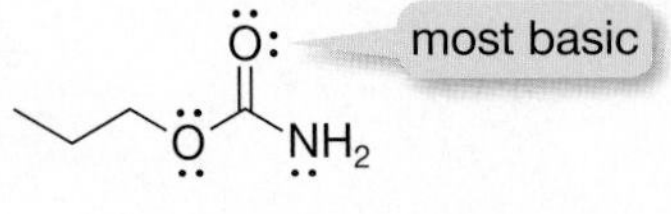

most basic

INTEGRATE THE SKILL 6.10

The most likely acid base reaction will be between the strongest base and the strongest acid. The hydrogens shown are the more acidic ones in each molecule. The conjugate bases resulting from deprotonation of each group are:

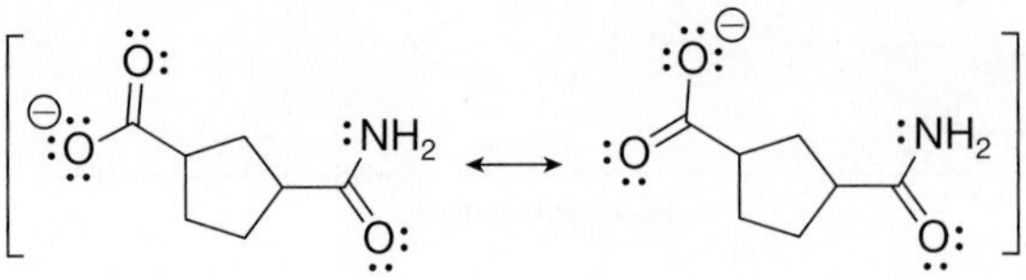

CB1

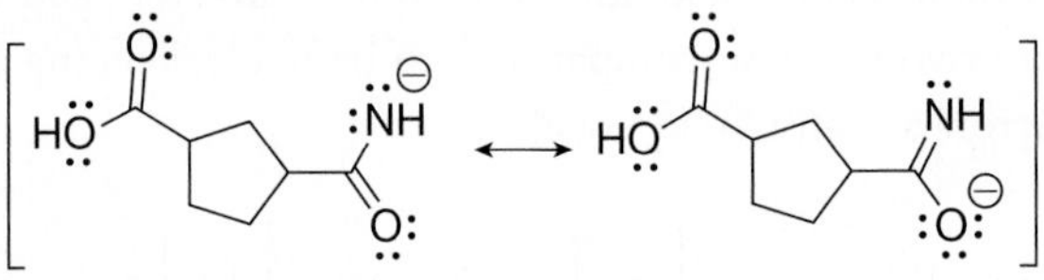

CB2

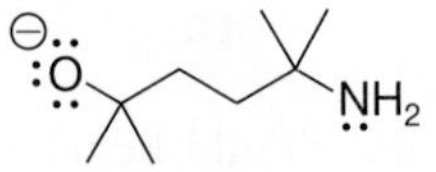

CB3 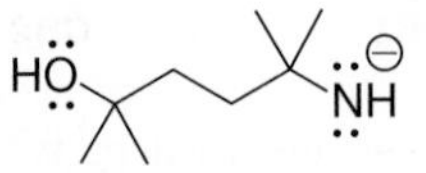 CB4

The carboxylic acid group should be the most acidic since **CB1** is the most stable conjugate base. It is both resonance stabilized and the negative charge is carried by O atoms rather than the less electronegative N atom. **CB3** and **CB4** have no resonance stabilization; additionally, **CB2** has the negative charge on N. Since the carboxylic acid group on the first molecule is the acid, the base must reside in the second molecule.

To compare the bases, their conjugate acids should be considered.

CA1 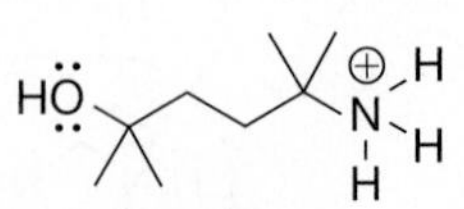 CA2

Since N is more electropositive than O, **CA2** should be more stable than **CA1**. This makes the amino group the most basic in this molecule.

This results in the following being the most likely acid–base reaction:

PRACTICE PROBLEM 6.11

a) The pK_a data indicates that the conjugate acid (**CA**) is weaker than the **acid**.

base acid CA CB

pK_a = 7.8 pK_a ~ 11

Therefore, the products are favoured by the reaction.

b) The **acid** is stronger than the **CA**, so the equilibrium favours the products.

acid base CB CA

pK_a ~ 9.0 pK_a = 16.5

c) The **CA** is stronger than the **acid**, so the equilibrium favours the reactants.

acid base CB CA

pK_a ~ −2.2 pK_a = −4.7

d) The **CA** is stronger than the **acid**, so the equilibrium favours the reactants.

base acid CA CB

pK_a = 40 pK_a ~ 24

INTEGRATE THE SKILL 6.12

a) The pK_a values for the indicated groups are given below.

A: pK_a ~ 38
B: pK_a ~ 16
C: pK_a ~ 4
D: pK_a ~ 25

Using this data, we would rank the groups as: **A** (least acidic), **D**, **B**, **C** (most acidic).

b) To determine the relative basicity of these lone pairs, we compare the pK_a values for their conjugate acids.

AH$^+$: pK_a ~ 10 **BH$^+$:** pK_a ~ −5 **CH$^+$:** pK_a ~ −2

The strongest acid among the protonated forms would correspond to the weakest base among the neutral forms. Therefore, the groups would be ranked **B** (least basic), **C**, **A** (most basic).

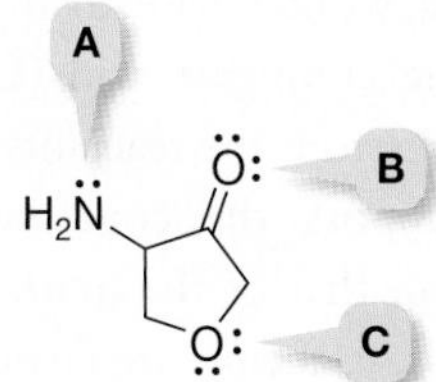

PRACTICE PROBLEM 6.13

a)

less stable due to electron donation; stronger base

acid base CB CA

pK_a ~ 16 pK_a = 18

The **base** is less stable than the conjugate base (**CB**) due to the nearby alkyl groups, which are weakly electron donating. Since the **base** is less stable, it is a stronger base than **CB**, so the conjugate acid (**CA**) is a weaker acid than the **acid**. Therefore, the products will be favoured. The pK_a data also supports this conclusion: the pK_a of the **CA** is higher than that of the **acid**, so **CA** is a weaker acid. Therefore, the products are favoured.

b)

base + acid ⇌ CA + CB

pK_a = 1.0 pK_a = 5.0

The two charged species are the **acid** and **CA**. The nitro group ($-NO_2$) is destabilizing the charge in the acid since it is an electron-withdrawing group. The methoxy group (–OMe) in the **CA** is an electron-donating group and is a stabilizing influence on the charge. This is reflected in the pK_a values. So, the **acid** is the stronger acid and the **base** would be a stronger base for similar reasons. Therefore, the products should be favoured.

c)

base + acid ⇌ CA + CB

pK_a ~ > 51 pK_a = 41

The **base** is more stable than the **CB** due to resonance stabilization of the negative charge by the aromatic ring. Therefore, it is a weaker base than the conjugate base (**CB**), making the conjugate acid (**CA**) a stronger acid than the **acid**. Therefore, the reactants should be favoured. The pK_a data supports this conclusion: the pK_a of the **CA** is higher than that of the **acid**, so **CA** is a weaker acid. Therefore, the reactants are favoured.

d)

acid + base ⇌ CB + CA

pK_a ~ 11 pK_a ~ –7

The positive charge on the **acid** will be more stable than the positive charge on the **CA**. In the **acid**, the positive charge resides on a more electropositive N atom rather than the more electronegative O atom in the **CA**.

The **CA** is further destabilized by the nearby O atom in the ester group. So, the **CA** should be the least stable and strongest acid, which is reflected in the pK_a values shown. It follows that the **CB** is the stronger base (i.e., the more electron-rich lone pair) and, therefore, the reaction should favour the reactants.

INTEGRATE THE SKILL 6.14

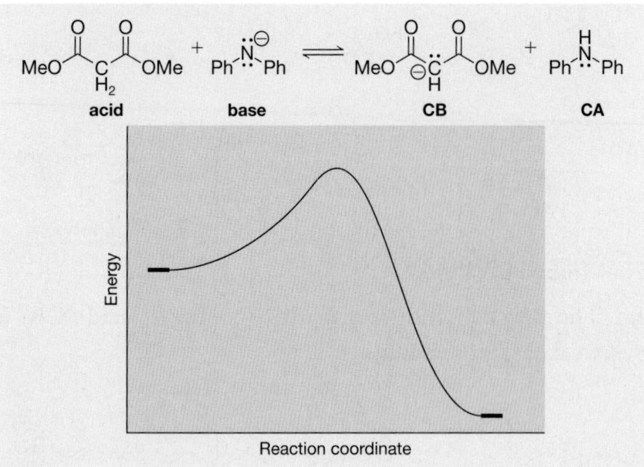

The energy diagram indicates that the products are favoured. Therefore, the strongest acid and strongest base should be the reactants. This means the lower pK_a (i.e., the stronger acid) should be for dimethylmalonate (**acid**) versus diphenylamine (**CA**).

CHAPTER 7

PRACTICE PROBLEM 7.1

Water is assumed to be the solvent where necessary.

a)

electrophile nucleophile

b)

electrophile nucleophile

c)

electrophile nucleophile

d)

electrophile nucleophile

INTEGRATE THE SKILL 7.2

a) Electrophile A

i) Removing the added bromide leaves a carbocation fragment, which can form a ketone.

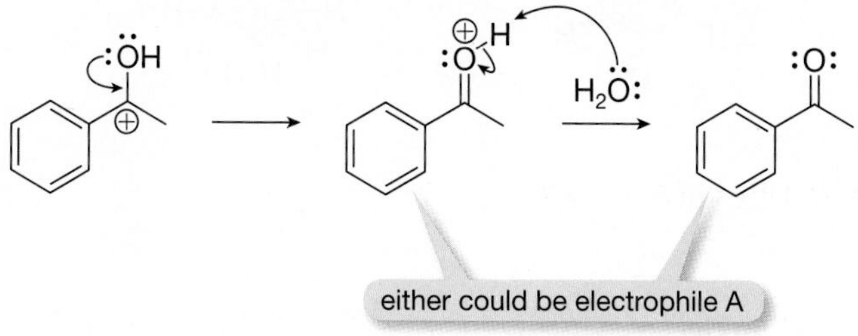

either could be electrophile A

ii) Reaction mechanism

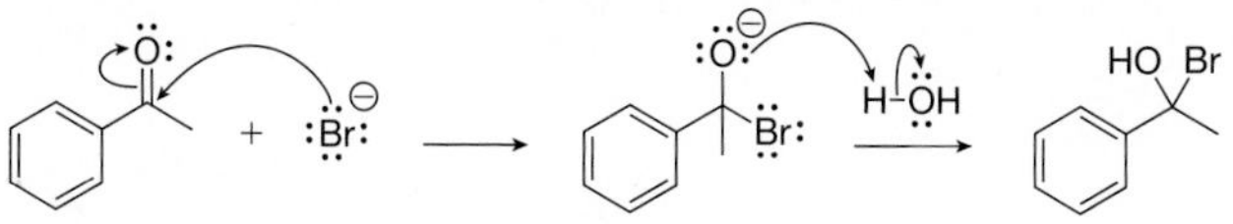

iii) Reverse reaction mechanism

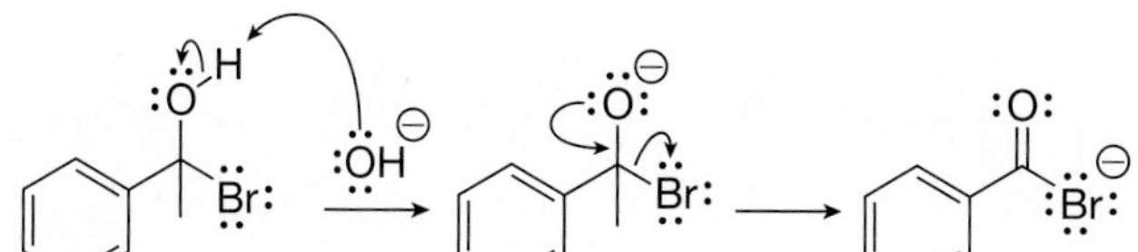

b) Electrophile B

i) Removing the added hydroxyl gives the following carbocation, which can form an imine.

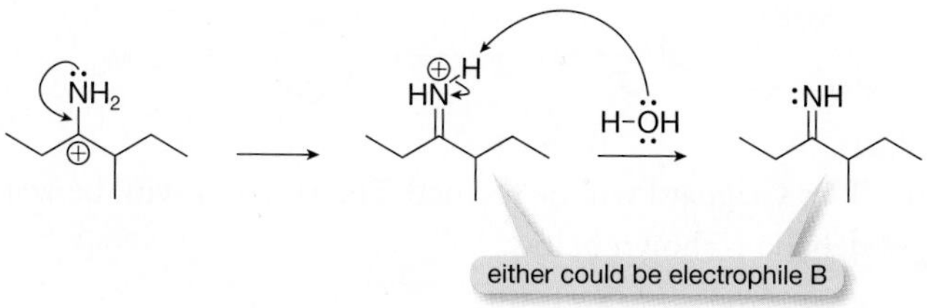

either could be electrophile B

ii) Reaction mechanism

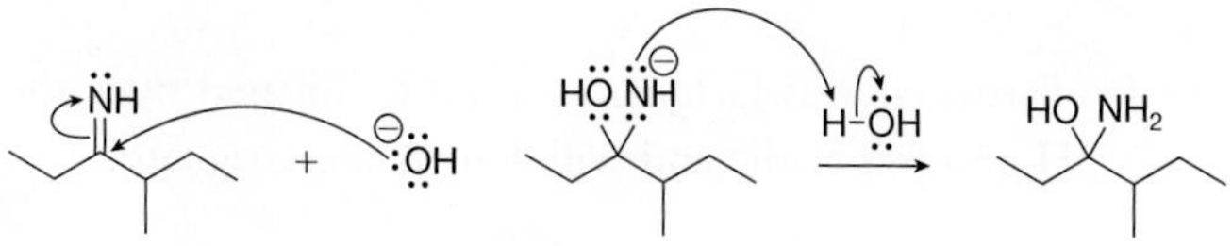

iii) Reverse reaction mechanism

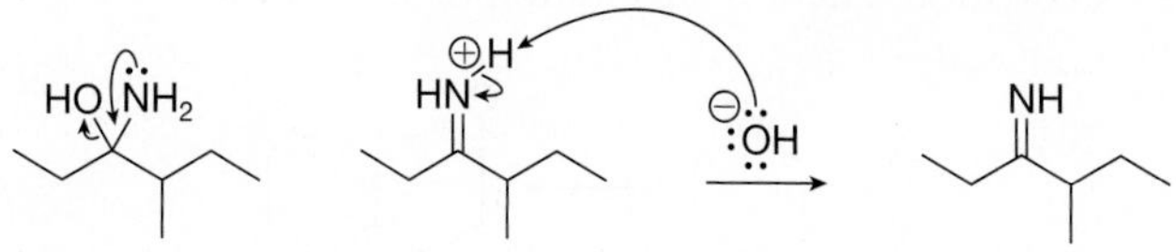

PRACTICE PROBLEM 7.3

a) EtOH solvent

i)

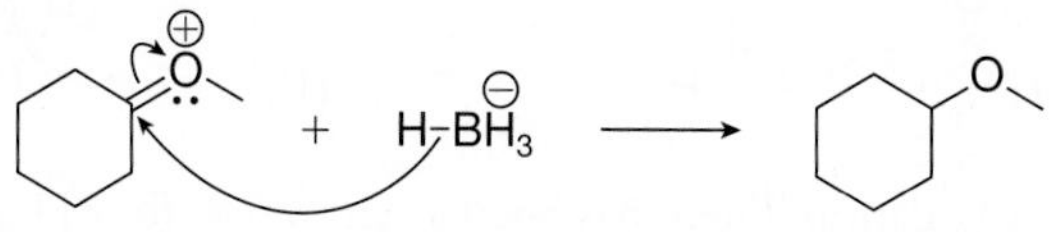

ii)

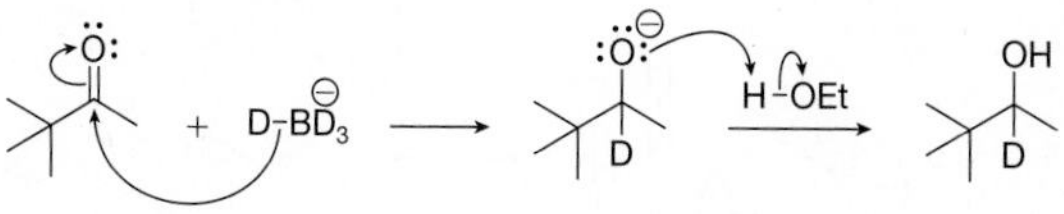

b) Et_2O solvent, aqueous acid workup in second step

i)

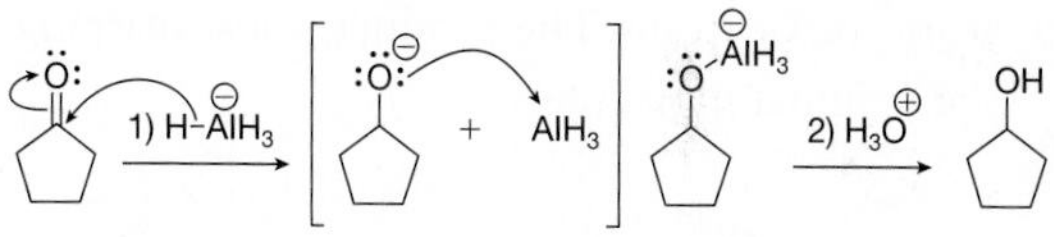

ii)

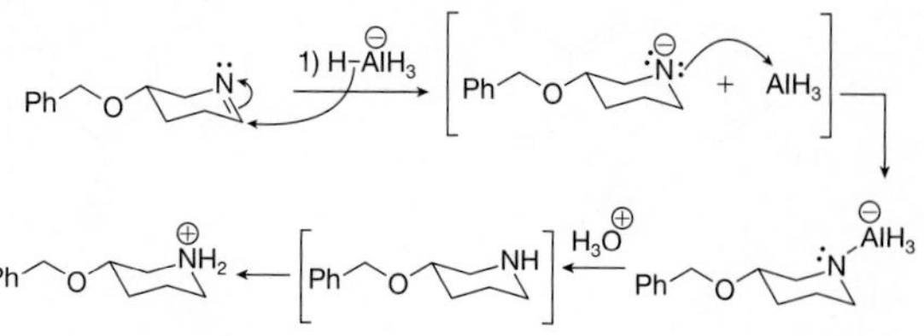

INTEGRATE THE SKILL 7.4

a) The initial imine formed will be reduced again. There are two hydride additions.

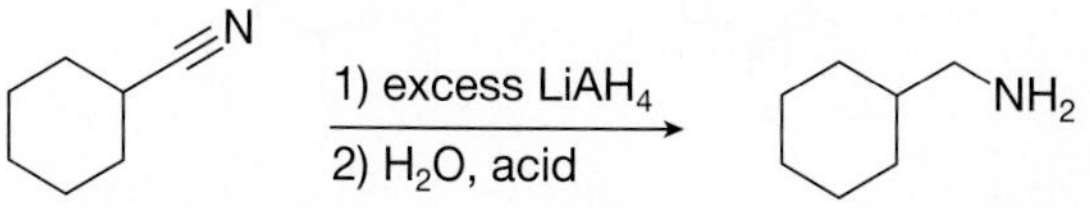

1) excess $LiAH_4$
2) H_2O, acid

b) Both carbonyl groups will be reduced since there is an excess of $LiAlH_4$.

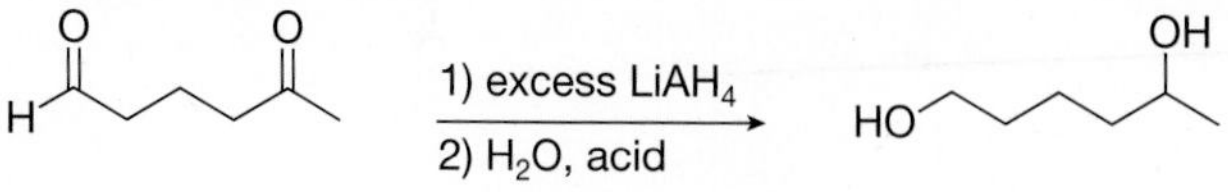

1) excess $LiAH_4$
2) H_2O, acid

PRACTICE PROBLEM 7.5

Relevant hydrogens have been added for clarity.

a) Oxidation. There is an addition of an O atom.

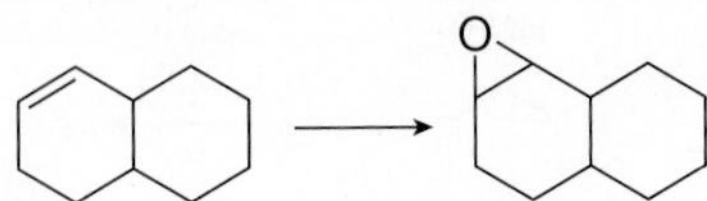

b) Reduction. One of the aldehydes has had two H atoms added.

c) Oxidation. There has been a removal of four H atoms, so two oxidations have occurred.

d) Neither. There is no net change in the number of H atoms or O atoms. The product is just rearrangements of the original molecule.

INTEGRATE THE SKILL 7.6

Relevant hydrogens have been added for clarity.

a) Reduction. There has been a loss of an O atom.

b) Oxidation. There has been an addition of an O atom.

c) Reduction. Two H atoms have been added.

d) Reduction. Six H atoms have been added to phenol. This means three reductions have occurred.

PRACTICE PROBLEM 7.7

a)

b)

c)

d)

INTEGRATE THE SKILL 7.8

a) The Grignard reagent cannot be prepared in the presence of a protic solvent like ethanol. Any Grignard reagent formed will be destroyed by deprotonating the alcohol.

b) This Grignard will be formed. The reaction with benzaldehyde is shown below.

c) Similar to (a), this Grignard cannot be formed since the –OH group is acidic and will destroy the Grignard.

PRACTICE PROBLEM 7.9

a)

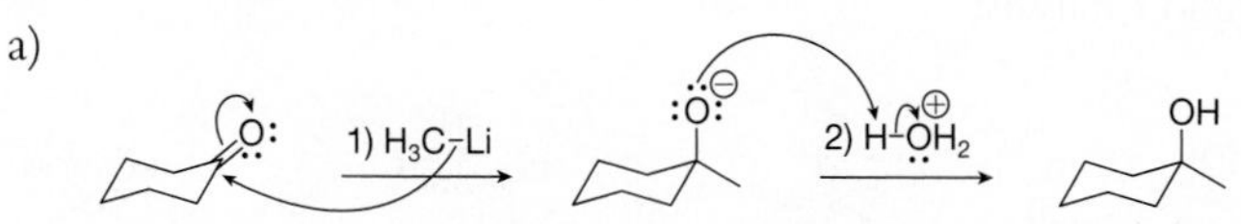

b)

1) Ph MgBr — 2) $H-OH_2$ — Ph

c)

1) ≡—Ph — 2) $H-OH_2$ — Ph, OH

d)

1) Ph-Li — 2) H-OH — Ph

INTEGRATE THE SKILL 7.10

The carbonyl carbon is more electrophilic than the imine carbon. This is because the oxygen is more electronegative than the nitrogen atom and will induce a greater partial positive charge on the carbon.

1) MeLi (1 equivalent), THF
2) NH_4Cl, H_2O

electrophilic site

PRACTICE PROBLEM 7.11

a) i) CN, HCl — H-Cl

ii) CN — H-Cl

b) i) CN

ii) AcO, OAc, CN

INTEGRATE THE SKILL 7.12

The product of the first step is an alcohol. Since the hydroxyl proton is acidic, it would react with the $LiAlH_4$ and produce an oxide, which would revert to the starting ketone.

acidic proton

1) KCN, HCl (slow addition) — 2) $LiAlH_4$

PRACTICE PROBLEM 7.13

a) i) OH, OH_2 — OH + H_2O:

ii) NH, OSO_2Me — NH + O-S-Me

iii) :O: :Cl: — :O H + :Cl:

b) i) NC O-H, OH — + CN

ii) OH — + H_2O: — + OH

INTEGRATE THE SKILL 7.14

OH — CN — NC, OH +

PRACTICE PROBLEM 7.15

a)

$+ :\ddot{Cl}:^{\ominus}$

$+$ HCl

b)

$+ {}^{\ominus}OH$

c)

$\times 2$

$+ 2\ :\ddot{Cl}:^{\ominus}$

$+ 2$ HCl

d)

$+ {}^{\ominus}OH$

INTEGRATE THE SKILL 7.16

Acid catalysis:

Base catalysis:

PRACTICE PROBLEM 7.17

a)

$+ {}^{\ominus}\ddot{O}{-}CH_3$

b)

$+$ HCl:

c)

d)

INTEGRATE THE SKILL 7.18

The stereochemistry of the double bond does not allow the −OH group to get near enough to the C=O group to react. In order for this to happen, the double bond would have to have a Z configuration. Therefore, this will react like a simple ketone, as shown below.

PRACTICE PROBLEM 7.19

a) There is a favoured product since the stereocentre makes the two faces inequivalent.

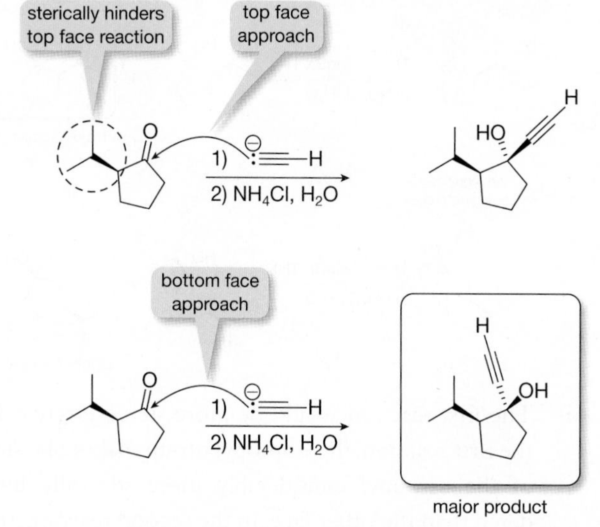

b) There is no preference and the product will be a racemic mixture.

c) The reaction does not create a stereocentre, so there is only one product.

d) Only one product is formed.

e) Stereocentres are present, so there are two products. The alkyl groups near the imide make addition to the face furthest from them less sterically hindered.

INTEGRATE THE SKILL 7.20

a) i)

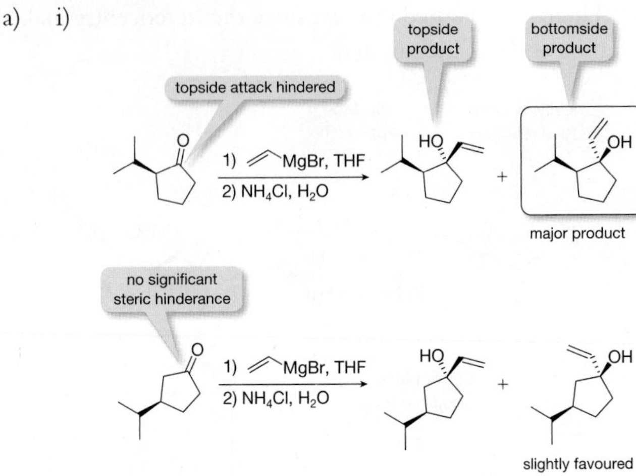

ii) The first reaction would be more stereoselective. In the first reaction, the isopropyl group makes one side of the carbonyl considerably more sterically hindered than the other face. In the second reaction, the isopropyl group is further from the reaction centre and, therefore, sterically influences the reaction outcome to a lesser degree.

b) i) Both molecules have a sterically bulky group that restricts access to one face of the carbonyl. Each reaction will produce a greater amount of product resulting from the addition of CN^- to the less crowded face of the carbonyl group.

ii) The second reaction will be more stereoselective since the larger t-butyl group would be more effective at blocking the top face of the carbonyl than the i-propyl group.

c) i) Both reactions will form a major product from hydride addition to the face away from the bulky groups.

ii) The second reaction will be more stereoselective, as it has more steric hinderance on one face of the carbonyl.

CHAPTER 8

PRACTICE PROBLEM 8.1

a)

First, add the H atoms around the double bond and identify the roles of the reagents.

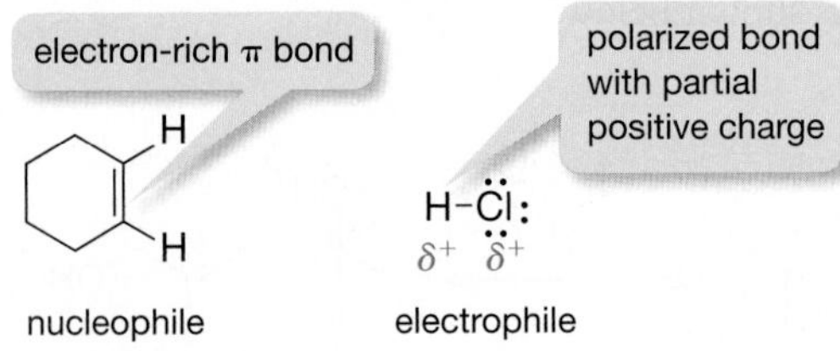

Now, draw mechanistic arrows to show the flow of electrons from the nucleophile to the electrophile. Be sure to direct the arrow toward the δ^+ part of the electrophile and continue to push the electron pairs to the ultimate electron sink (Cl^-).

Next, draw the products of this step, using the arrows to determine the products.

There is no preference as to which carbon the H adds to, since cyclohexene is symmetrical.

Now, repeat the process for the new set of intermediate reactants.

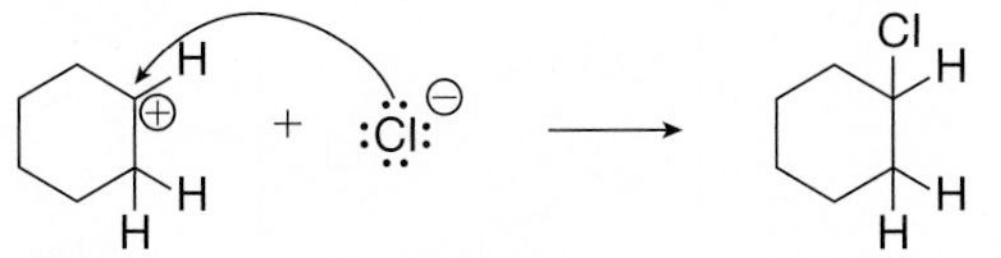

Redraw the steps to show the total mechanism, leaving out added lone pairs and hydrogens.

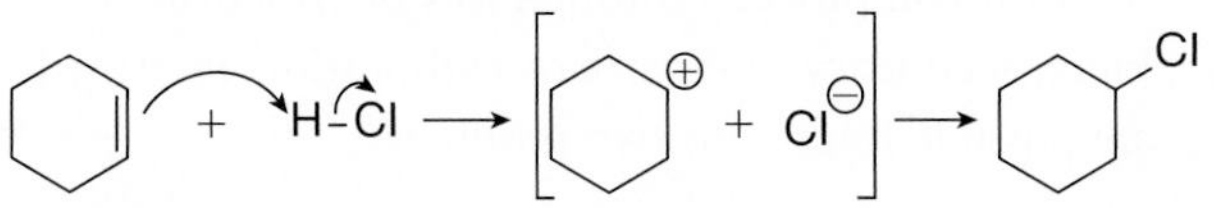

b)

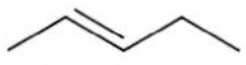

First, identify the reagents adding H atoms to the alkene and add mechanistic arrows to show electron flow from the nucleophilic π-bond to the electrophile.

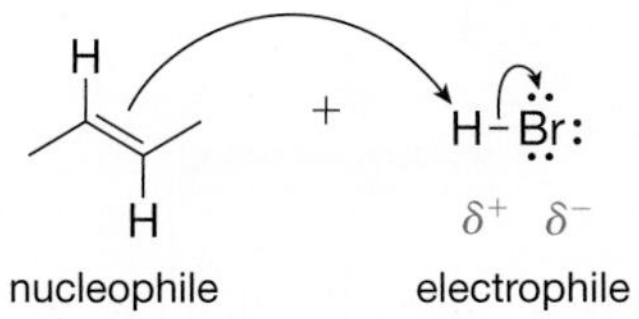

The resulting products would be

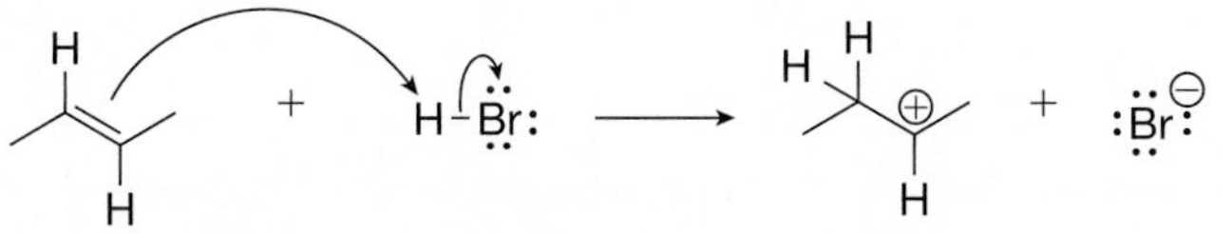

Reacting the intermediate nucleophile **(Nu)** and electrophile (E) gives the final product.

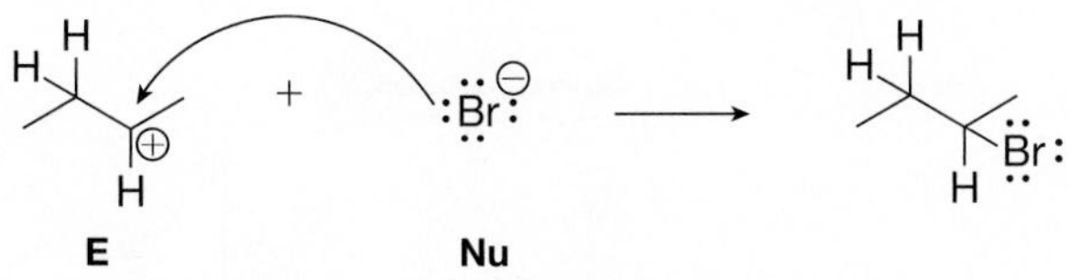

The overall reaction mechanism would be

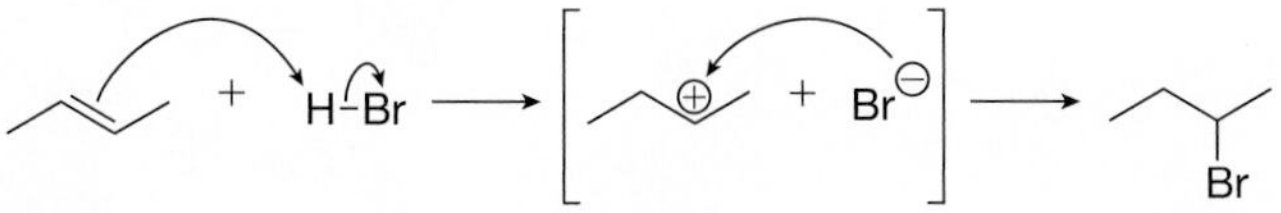

c)

Leaving out the lone pairs this time to avoid clutter, the two reagents are the nucleophilic alkene **Nu** and the electrophilic hydrogen on **E**. Donating the π-electrons to the H gives

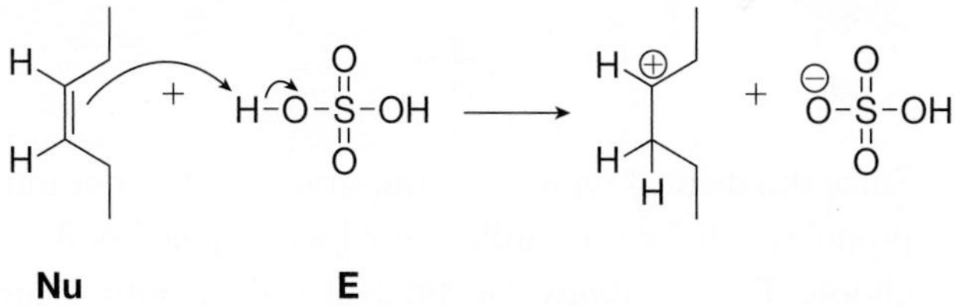

The second step in the reaction is the nucleophilic, charged oxygen on the hydrogen sulfate ion donating electrons to the charged carbon in the electrophilic carbocation.

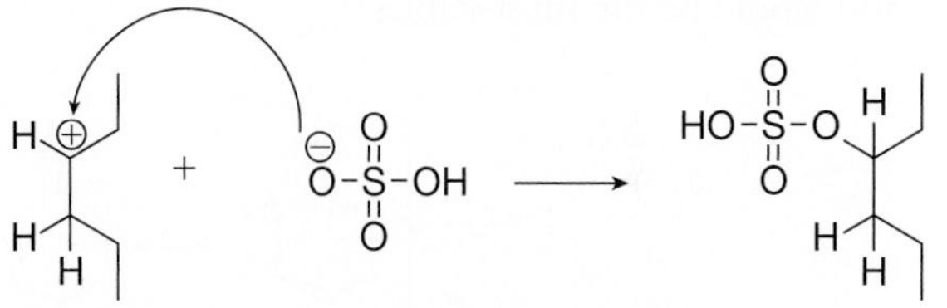

The overall mechanism would be

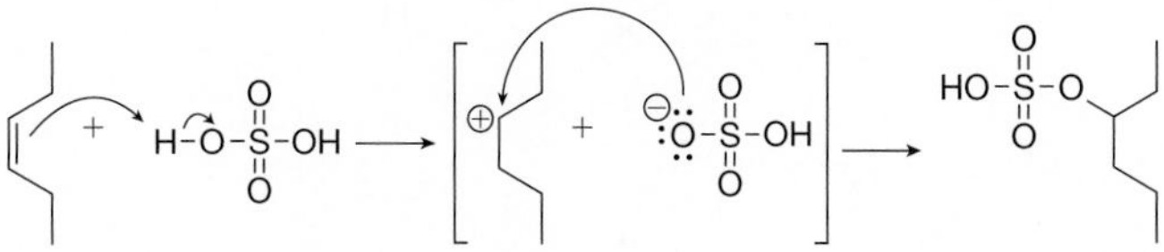

d)

Add H atoms to the alkene and expand the structure of the carboxylic acid. Then, react the alkene π electrons with the electrophilic H atom in the acid.

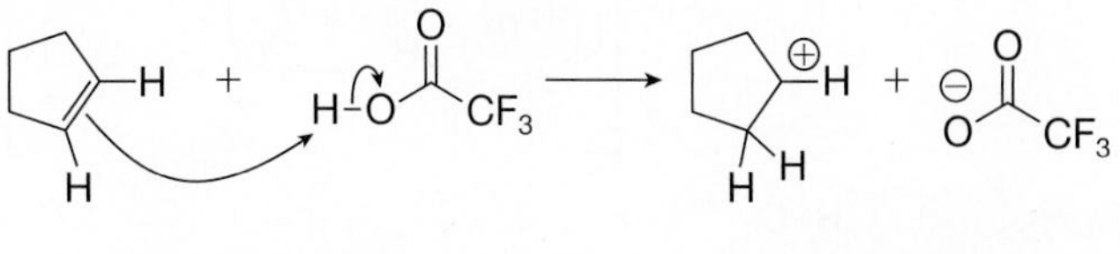

The next step is the carboxylate anion reacting with the positive carbocation.

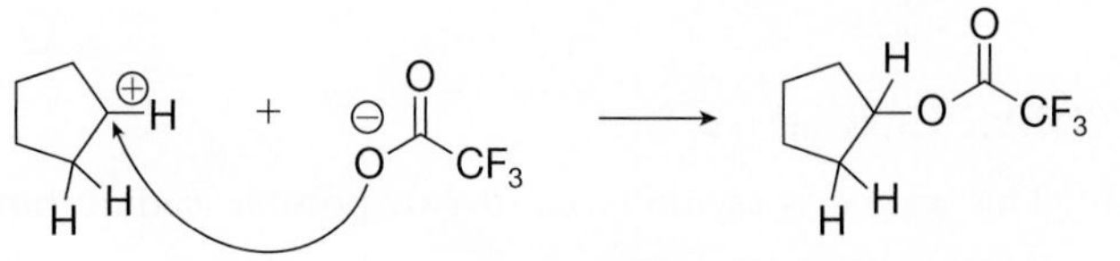

Putting it all together would give the overall mechanism

INTEGRATE THE SKILL 8.2

a) + HCl ⟶

Since this diene is symmetric, the same initial carbocation products will form regardless of which double bond you choose. The possibility for different carbocations comes from reacting either end of the alkene with the acidic proton. Picking the top double bond arbitrarily gives two products shown below, with the added H atom shown for clarity. The top one has the resonance forms indicated and would be the most stable.

b) + HCl ⟶

As in part (a), the symmetry here means reacting either double bond will result in the same set of products.

PRACTICE PROBLEM 8.3

a) This alkene is asymmetric, so two possible carbocation intermediates can form.

primary carbocation

minor

major

tertiary carbocation

The tertiary carbocation is more stable and most of the reaction will proceed through this intermediate.

b) Here, secondary and tertiary carbocation intermediates are possible leading to two products.

tertiary carbocation

major

minor

secondary carbocation

The major product goes through the more stable tertiary intermediate.

c) This alkene is symmetric, so only one possible intermediate forms, leading to one product.

d) The asymmetric alkene has two possible intermediate carbocations.

secondary carbocation

minor

major

tertiary carbocation

The major product forms through the more stable tertiary carbocation intermediate.

e) The asymmetric alkene will form two different carbocation intermediates, as shown below.

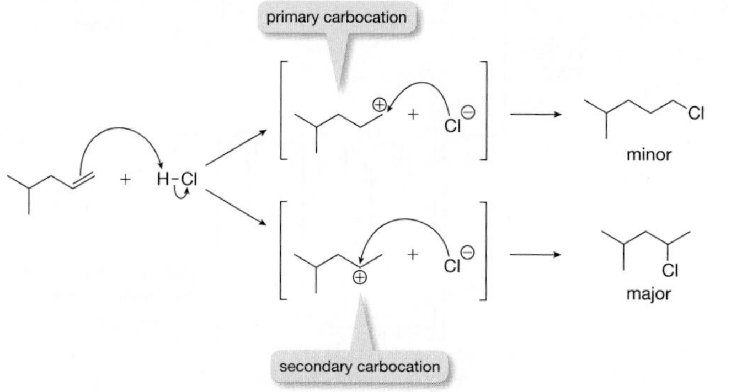

The secondary carbocation will be preferred and lead the major product indicated.

INTEGRATE THE SKILL 8.4

a) This is a symmetric diene, so reaction of one or the other double bond would lead to the same products. Considering reacting just one double bond leads to the following possible reaction pathways. Both reactions generate secondary carbocations as intermediates.

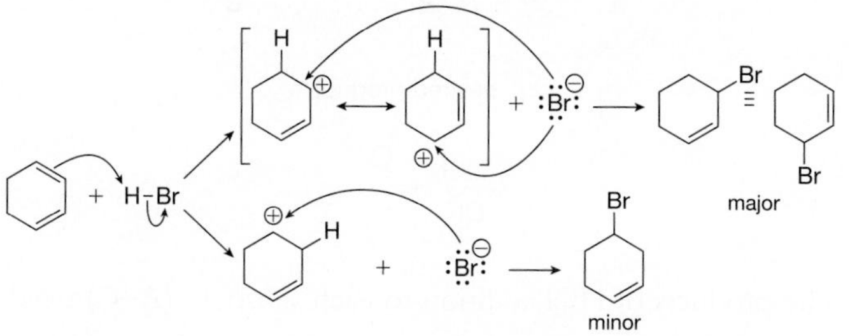

The top reaction proceeds through a resonance-stabilized intermediate, but reaction of either resonance form leads to the same final product. This is the favoured pathway, as this would be the most stable carbocation and therefore the one that will be formed most rapidly.

The bottom reaction leads to the product shown, which would be the minor product.

b) This is a symmetric diene, so reacting either double bond would lead to the same products. Both pathways proceed through secondary carbocations.

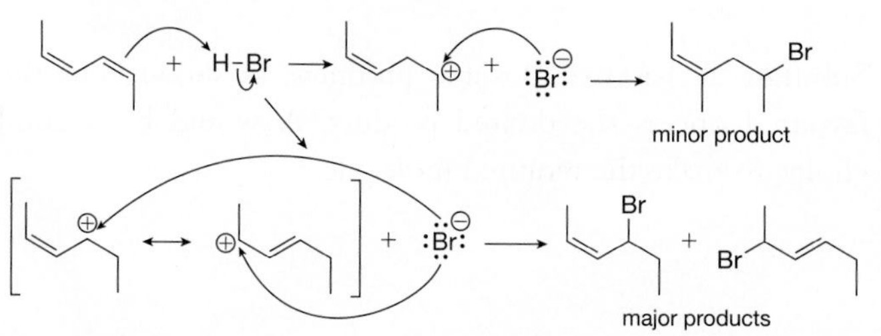

The bottom pathway will be favoured due to the more stable resonance-stabilized carbocation, which will be formed more rapidly than the secondary carbocation of the top pathway. Two products result from this pathway, since the carbocation sites in the resonance forms are not symmetric. A third, distinct, minor product is formed via the bottom pathway, as indicated.

c) To simplify the diagrams, consider the reaction of each double bond separately. The first bond would proceed through two secondary carbocation intermediates and lead to the two products shown.

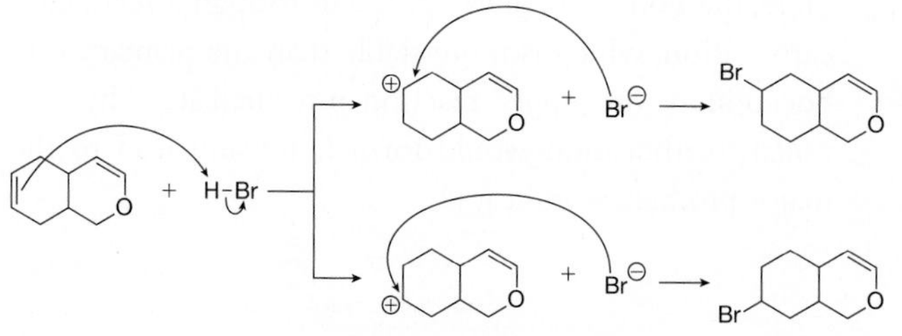

Reaction of the other double bond would also proceed through two pathways each, with a secondary carbocation intermediate, as shown.

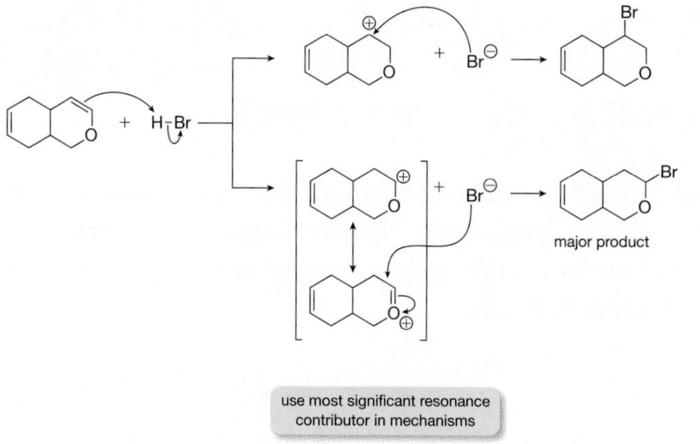

The choice of major product is due to the resonance stabilization of the carbocation in the bottommost reaction pathway. This is the most stable carbocation and would lead to the formation of the major product.

PRACTICE PROBLEM 8.5

a) The reaction proceeds through two carbocations of very different stability. Since primary carbocations are less stable than tertiary carbocations, the bottom pathway would be preferred and result in the major product.

b) Here, the bottom pathway proceeds though a secondary carbocation, which is more stable than the primary carbocation in the upper reaction intermediate. The secondary carbocation would form faster and lead to the major product, as indicated.

c) The two carbocation intermediates are secondary and tertiary respectively. Tertiary being more stable. the bottom reaction will be favoured and lead to the major product shown.

d) The possible reaction pathways for this substrate are shown below.

The most favourable pathway is the bottom one, with the more stable resonance-stabilized secondary cation intermediate. This leads to the major product indicated.

INTEGRATE THE SKILL 8.6

The products of HCl addition to each substrate (**A–C**) need to be considered separately.

Substrate **A** has two possible products, as shown, but the favoured one is the desired product. **A** would be a good choice to make the required molecule.

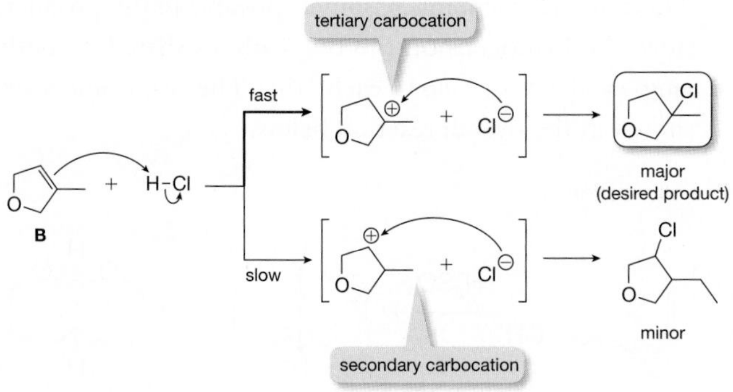

tertiary carbocation

fast

slow

major
(desired product)

minor

secondary carbocation

Substrate **B** also produces the desired molecule as the major product.

The reaction of substrate **C** is shown below. Although the desired product could form,

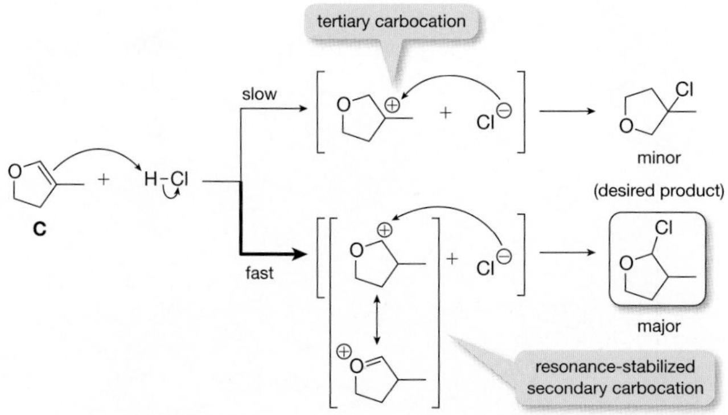

tertiary carbocation

slow

fast

minor
(desired product)

major

resonance-stabilized
secondary carbocation

it would not be the favoured reaction pathway. The bottom reaction pathway is preferred due to the resonance stabilization of the secondary carbocation, as shown.

So, compounds **A** and **B** would be the best molecules to use.

PRACTICE PROBLEM 8.7

a) Sulphuric acid is strong, so it will be completely dissociated in aqueous solution.

The hydronium ion will be the electrophile reacting with the alkene.

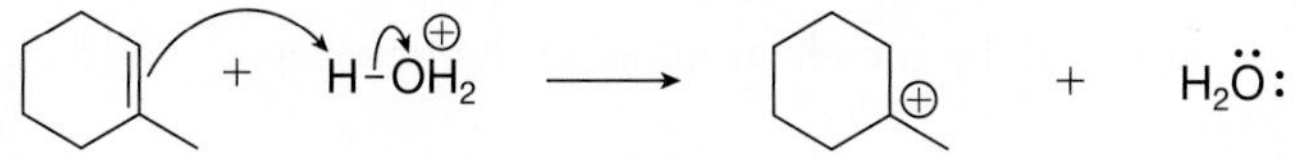

Water will then be the nucleophile reacting with the carbocation.

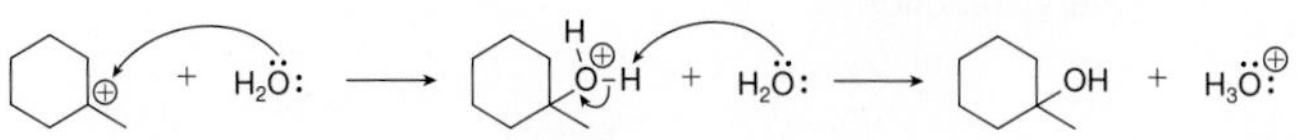

The protonated alcohol is deprotonated by water, regenerating the hydronium ion.

Overall reaction

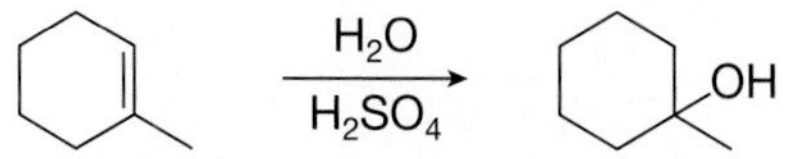

$\xrightarrow[H_2SO_4]{H_2O}$

b) Phosphoric acid is a strong acid and will protonate the basic alcohol group to some extent.

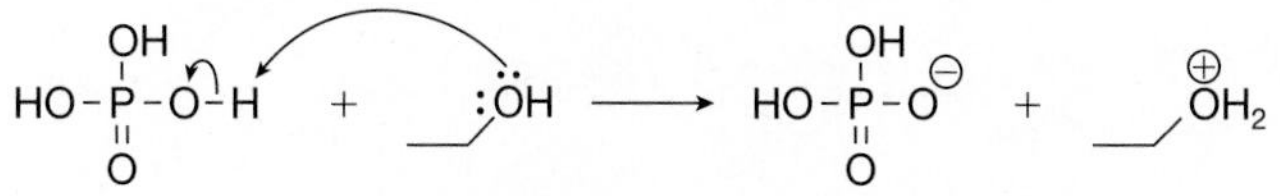

The protonated alcohol group will be the electrophile reacting with the alkene.

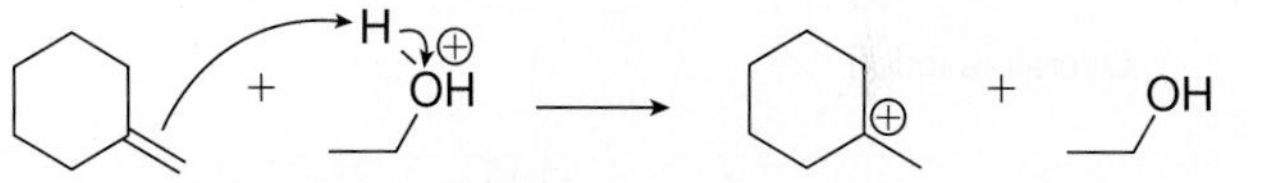

Ethanol will then be the nucleophile reacting with the carbocation.

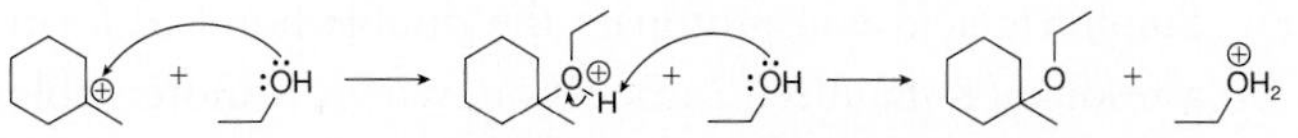

The protonated ether is deprotonated by alcohol regenerating the protonated alcohol.

Overall reaction

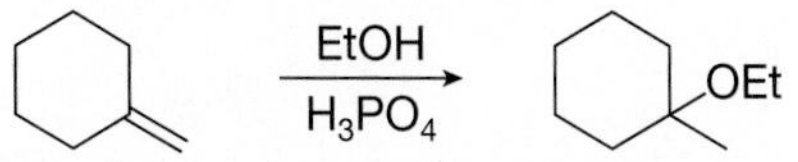

$\xrightarrow[H_3PO_4]{EtOH}$

c) Sulfuric acid will react with the double bond to generate a tertiary carbocation.

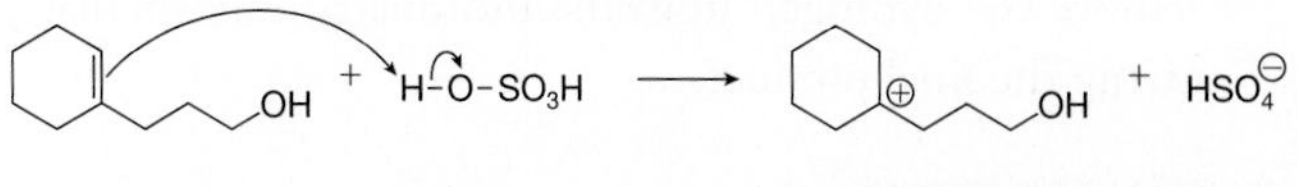

The alcohol group in the molecule is the best nucleophile available and will react with the carbocation.

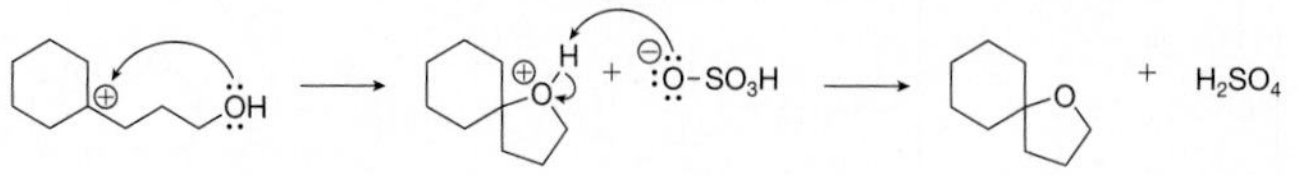

The slightly basic hydrogen sulphate ion will deprotonate the ether product regenerating sulphuric acid.

Overall reaction

H_2SO_4

d) Phosphoric acid will react with the double bond to generate a tertiary carbocation.

The alcohol group in the molecule is the best nucleophile available and will react with the carbocation.

The slightly basic dihydrogen phosphate ion will deprotonate the ether product, regenerating phosphoric acid.

Overall reaction

H_3PO_4

e) Sulphuric acid will protonate the double bond to form a resonance-stabilized carbocation, which is more stable than the tertiary carbocation that could form.

The alcohol group will be the nucleophile reacting with the carbocation. Then, the hydrogen sulphate will remove the hydrogen from the protonated intermediate, giving the final product.

use most significant resonance form in mechanisms

There are stereochemical isomers possible in this product, since the hydrogens on the two carbons shared by both rings can be *cis* or *trans* to each other. The two isomers are shown in the overall reaction below.

Overall reaction

H_2SO_4

cis trans

INTEGRATE THE SKILL 8.8

The reaction proceeds from reactant **A** through intermediates **B** and **C** to final product **D**, as shown in the figure below.

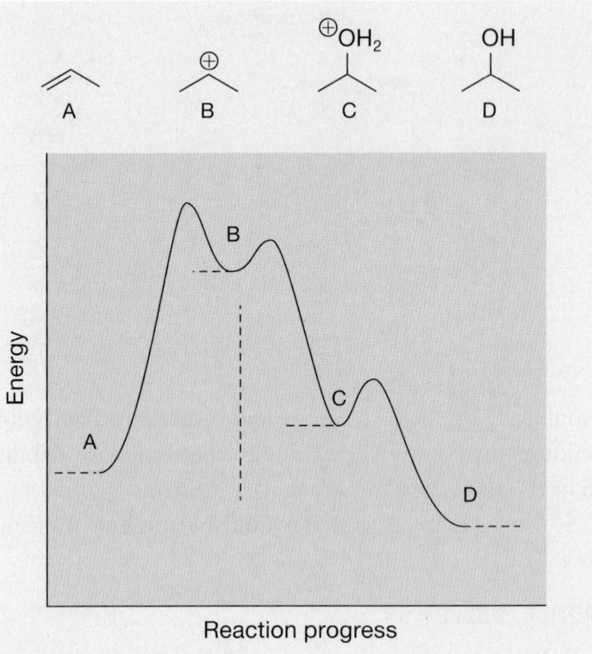

The product **D** is more stable than the reactant **A**, so it is represented with the lower energy. Both intermediates are less stable than either **A** or **D**. Comparing **B** to **C**, **B** is shown at the highest energy (least stable), since the only bonds are broken going from **A** → **B**, but an new C-O bond is made going from **B** → **C**. This would make **C** significantly lower in energy than **B**.

PRACTICE PROBLEM 8.9

a) The π bond will be the nucleophile and the mercury atom will be the electrophile in the first step.

The second step will be the nucleophilic methanol reacting with the electrophilic carbon in the mercurium ion. Either side leads to the same regioisomer (but not stereoisomers), so only one pathway needs to be considered.

HgOAc Me–OH → HgOAc H O Me OAc → HgOAc O Me + HOAc

The only regioisomer will be

+ MeOH $\xrightarrow{Hg(OAc)_2}$ HgOAc Me–O

b) Mercuric acetate reacts to form a symmetric intermediate.

OAc HgOAc → HgOAc + OAc

Only one regioisomer will form, since attack at either side of the mercurium ion results in the same regioisomer.

HO HgOAc → H O OAc HgOAc → O HgOAc + HOAc

The acetate ion will deprotonate the intermediate to form the final product.

+ HO $\xrightarrow{Hg(OAc)_2}$ O HgOAc

c) The initial product is asymmetric, so regioisomers need to be considered.

OAc HgOAc → HgOAc + OAc

There are two pathways from this intermediate, each leading to a different regioisomer.

Path A

HO HgOAc → H O OAc HgOAc → O HgOAc + HOAc

Path B

HO HgOAc → H O OAc HgOAc → O HgOAc + HOAc

Path A will be the favoured one, with the incoming nucleophile attacking the tertiary carbon, which would have the greater carbocation character.

+ HO $\xrightarrow{Hg(OAc)_2}$ fast → O HgOAc (major product); slow → O HgOAc

d) The initial product is asymmetric.

OAc HgOAc → HgOAc + OAc

The two pathways the reaction can proceed through are outlined below as Path A and Path B.

Path A

HgOAc Me–OH → OAc H Me O HgOAc → Me O HgOAc + HOAc

Path B

HgOAc Me–OH → HgOAc Me O H OAc → HgOAc Me O + HOAc

In Path A, the methanol reacts with a primary carbon, while in Path B, the methanol reacts at the secondary site. There will be more positive charge on the secondary site, so Path B will be the preferred pathway.

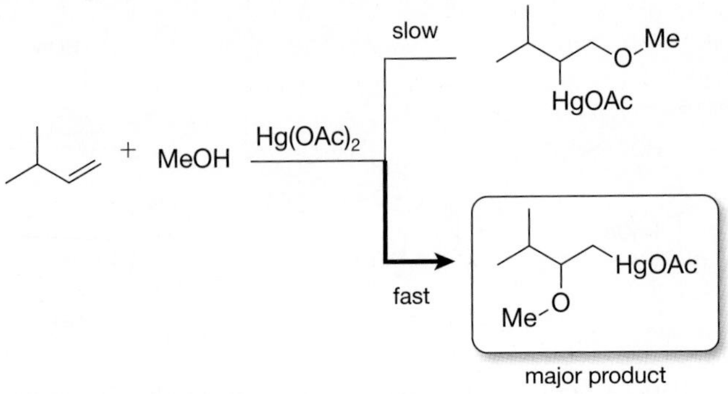

slow
HgOAc
Me
$Hg(OAc)_2$
+ MeOH
HgOAc
fast
Me
major product

INTEGRATE THE SKILL 8.10

The two products that result are stereoisomers. They are produced from the mercuric acetate approaching from either side of the double bond. These are labelled Path A and Path B below. The arrows are not meant to denote electron movement; rather, the approach of the mercuric acetate to the double bond. Products A and B are mirror images of each other and are, therefore, **enantiomers** of each other.

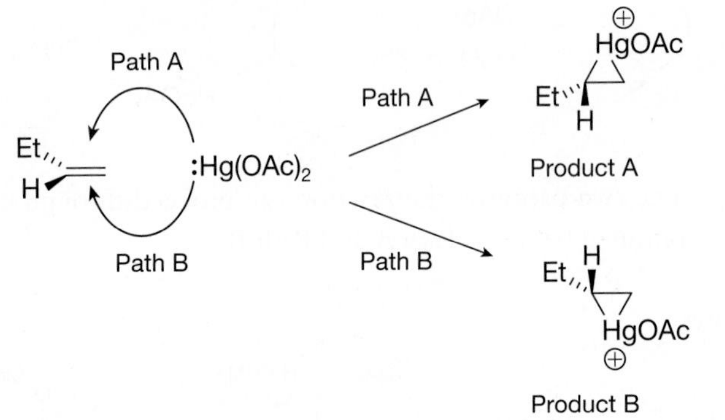

≡ Et

Path A
Path A
Et
H
$:Hg(OAc)_2$
HgOAc
Product A
Path B
Path B
Product B

PRACTICE PROBLEM 8.11

a) The bromonium ion formed has a tertiary carbon, which is the preferred site for the incoming nucleophile. The nucleophile adds *trans* to the bromine leading to a *trans* product.

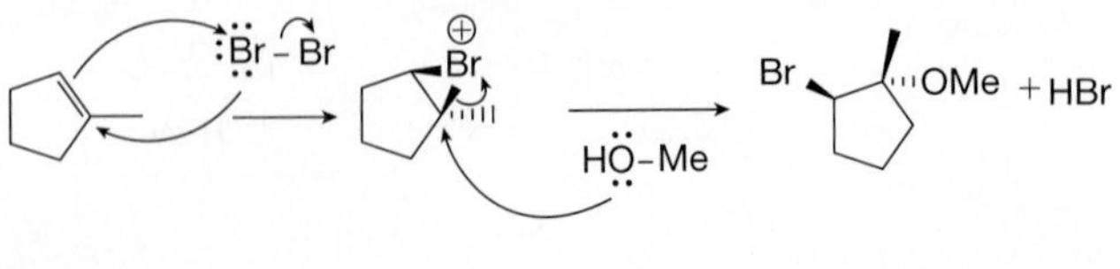

Br – Br
HÖ–Me
Br OMe + HBr

b) The initial ion has primary and secondary carbons at the reactive site. The secondary site will react fastest and lead to the indicated product.

Br–Br:
HÖ–Et
Br OEt

c)

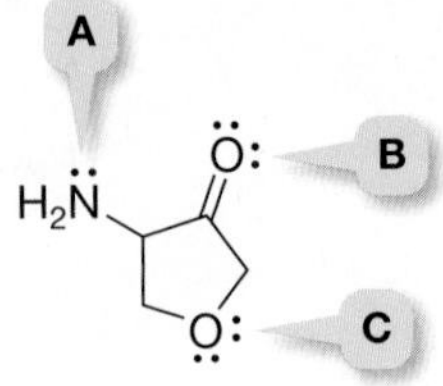

A
B
C
H_2N

d)

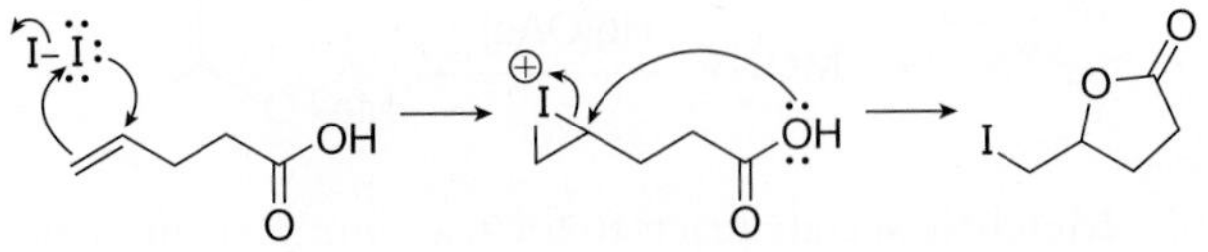

I–I:
OH
I

e)

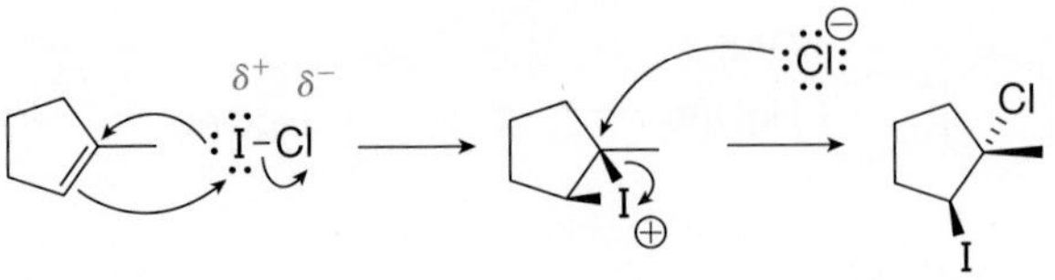

δ+ δ−
:I–Cl
:Cl:
Cl
I

f)

Cl_2
$H_2O:$
OH
Cl

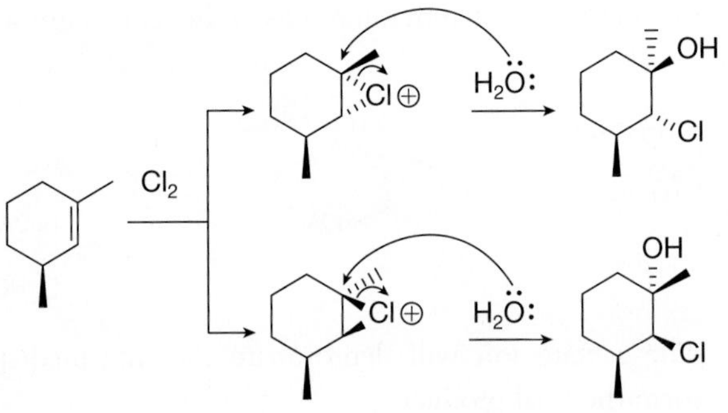

INTEGRATE THE SKILL 8.12

Reactions reversed to move from products to reactants.

a)

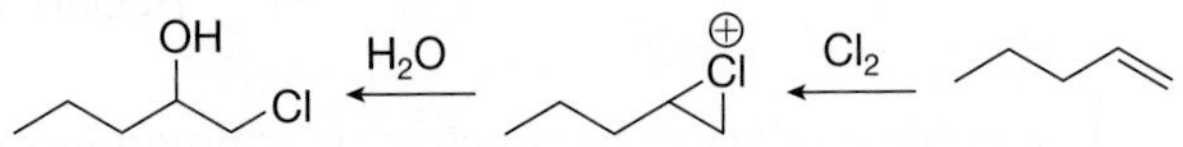

OH
Cl
H_2O
Cl
Cl_2

b)

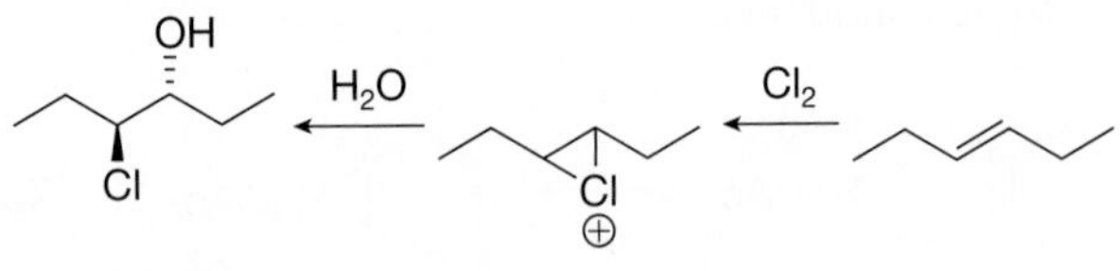

OH
H_2O
Cl_2
Cl
Cl

c)

OH
Cl
rotation around single bond
H_2O
Cl_2
OH
Cl
Cl

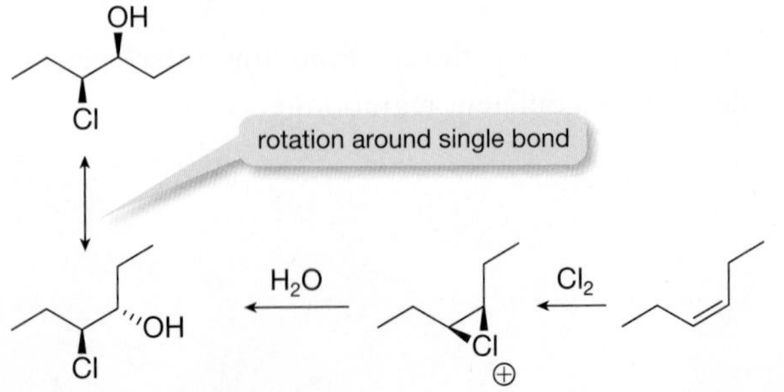

d) HO ... Cl_2 ... HO

PRACTICE PROBLEM 8.13

a) $H_2O + H_2SO_4 \rightleftharpoons H_3O^{\oplus} + HSO_4^{\ominus}$

b) $H_2O + HBr \rightleftharpoons H_3O^{\oplus} + Br^{\ominus}$

c) 1) NaOEt
2) NH_4Cl, H_2O

d)

INTEGRATE THE SKILL 8.14

This transformation could potentially be accomplished by using Cl_2 and water. The regiospecificity of the addition would give the desired product.

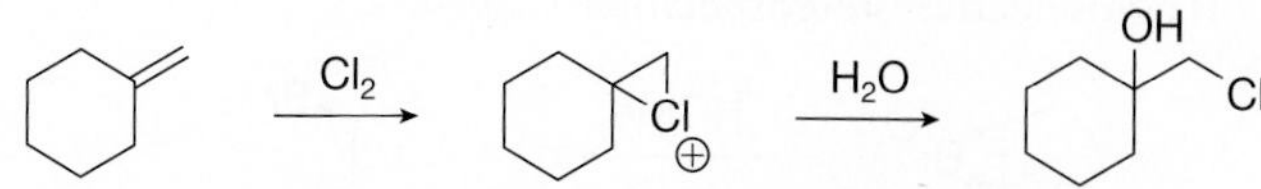

Epoxidation and ring opening with chloride could give the desired product. In order to get the regiochemistry correct, the ring opening would have to be done in neutral conditions. This would bring about preferred chloride attack at the primary carbon.

1) mCPBA

2) NaCl

3) NH_4Cl, H_2O

PRACTICE PROBLEM 8.15

Any chiral molecules will be produced as racemic mixtures.

a) BH_2 → H_2O_2, NaOH → OH

b) BH_2 → H_2O_2, NaOH → OH

c) BH_2 → H_2O_2, NaOH → OH

d) BH_2 → H_2O_2, NaOH → OH

e) BH_2 → H_2O_2, NaOH → OH

INTEGRATE THE SKILL 8.16

a) Regioselective, stereospecific

BH_3 → BH_2

b) Regioselective, stereospecific

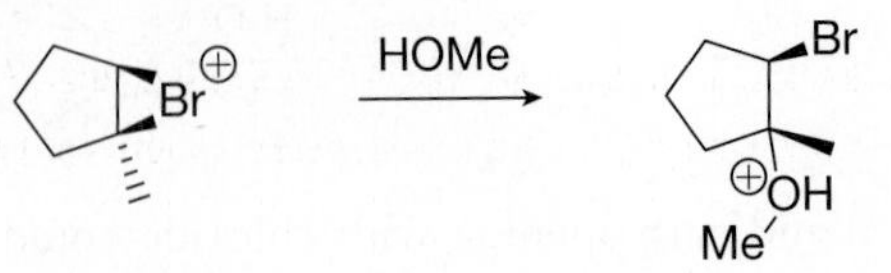

HOMe

Br

OH

Me

c) Stereospecific

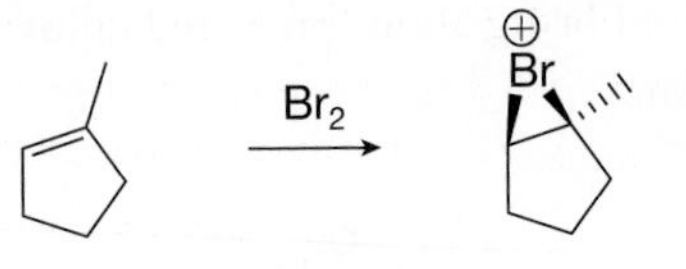

Br_2

Br

d) Stereoselective

OH

OH

H

O

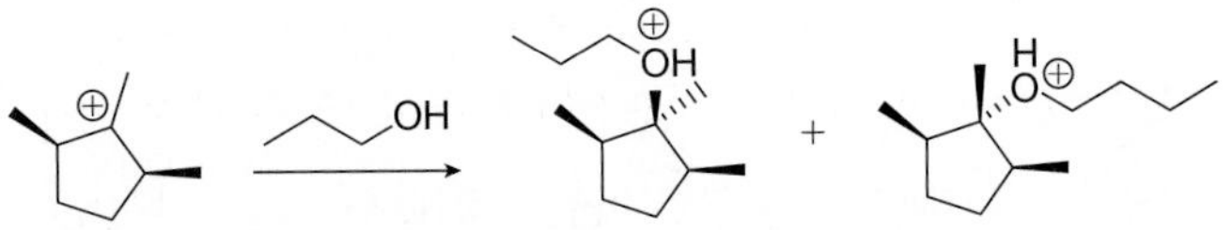

PRACTICE PROBLEM 8.17

a) Not conjugated, same as alkene reaction

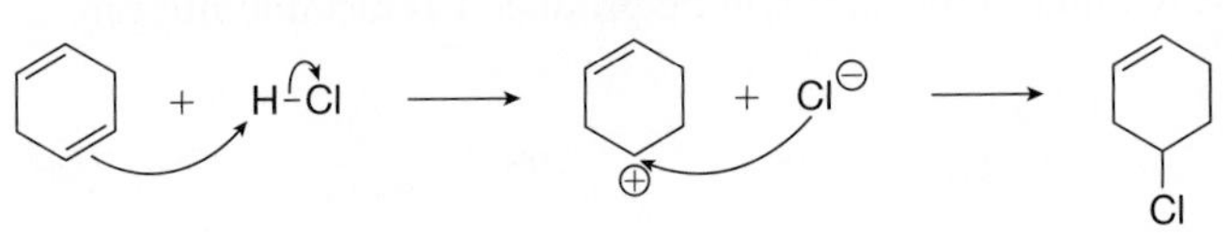

H–Cl

$Cl^{\ominus}$

Cl

b)

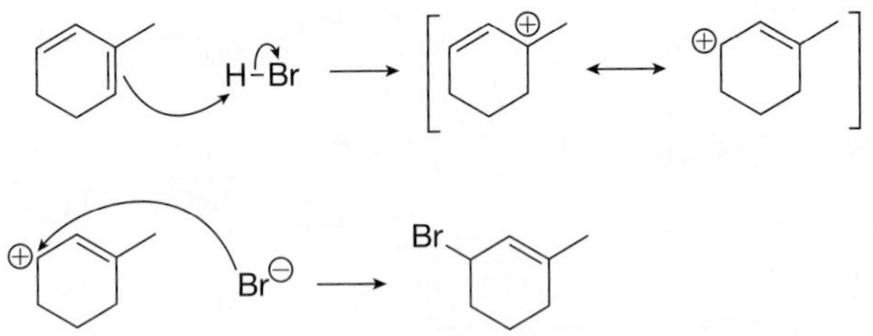

H–Br

Br

$Br^{\ominus}$

c)

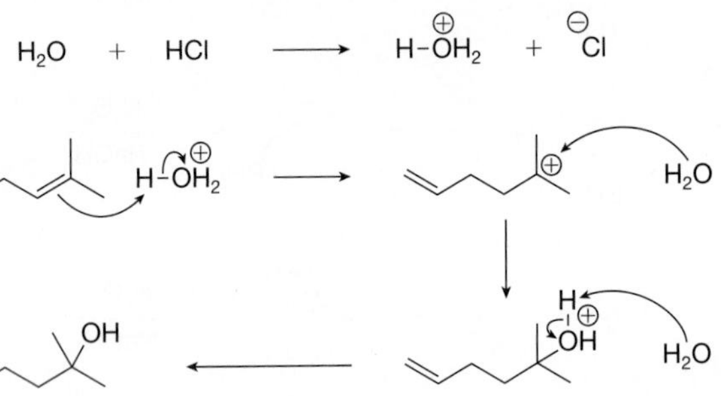

H_2O + HCl ⟶ $H{-}OH_2^{\oplus}$ + $Cl^{\ominus}$

$H{-}OH_2$

H_2O

OH

H

OH

H_2O

d)

H–Br

$Br^{\ominus}$

Br

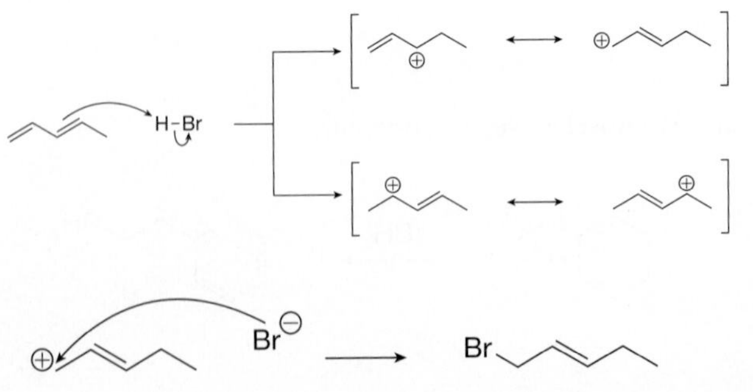

INTEGRATE THE SKILL 8.18

a)

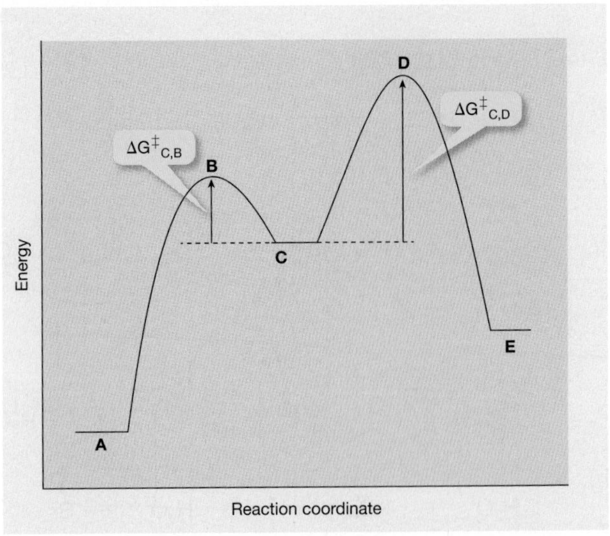

$\Delta G^{\ddagger}_{C,B}$

$\Delta G^{\ddagger}_{C,D}$

A

B

C

D

E

Energy

Reaction coordinate

b) The kinetic product is formed fastest with the smallest energy barrier. The smallest barrier is **C** → **B** ($\Delta G^{\ddagger}_{C,B}$), so **A** would be the kinetic product.

c) The thermodynamic product would be **A**, as it is the most stable.

d) **B** and **D** would be the transition states in this diagram.

PRACTICE PROBLEM 8.19

a)

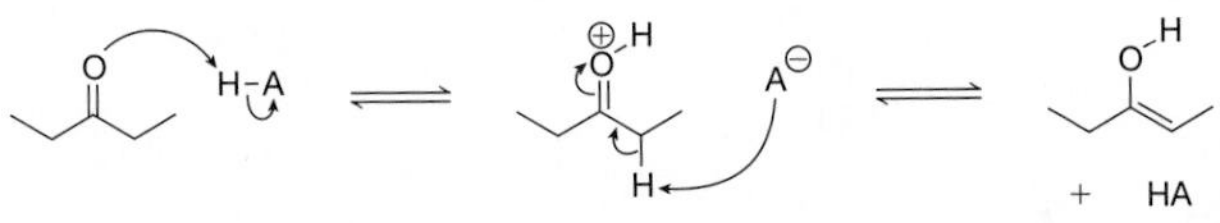

H–A

$A^{\ominus}$

+ HA

b)

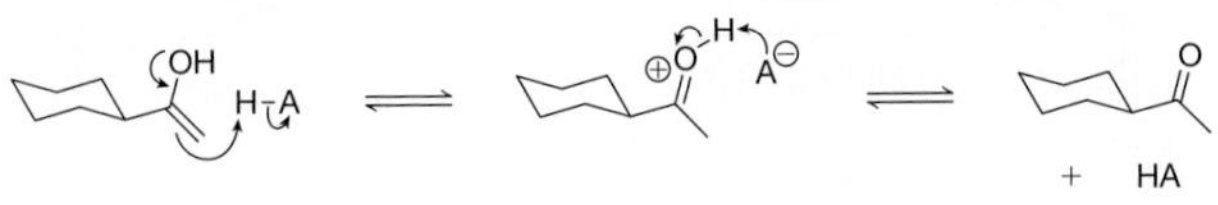

OH

H–A

$A^{\ominus}$

+ HA

c)

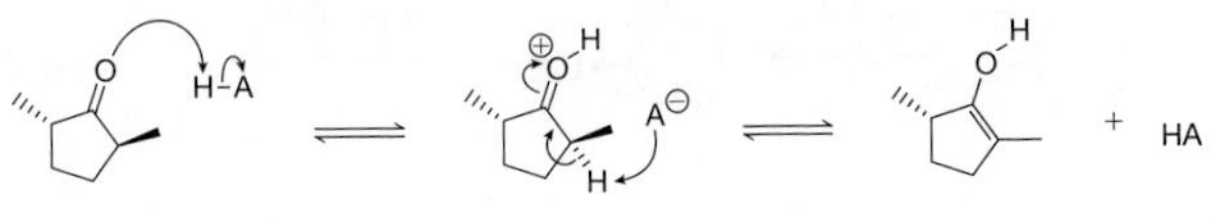

H–A

$A^{\ominus}$

+ HA

d)

OH

H–A

$A^{\ominus}$

+ HA

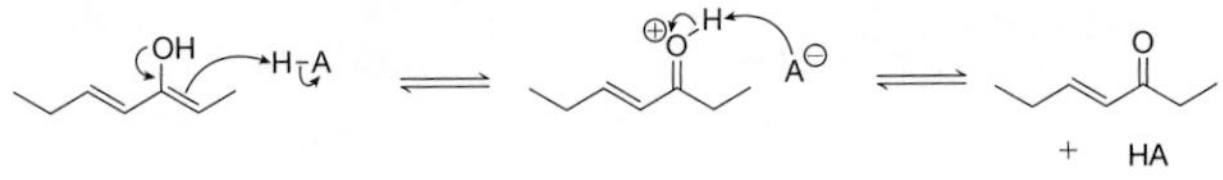

INTEGRATE THE SKILL 8.20

a)

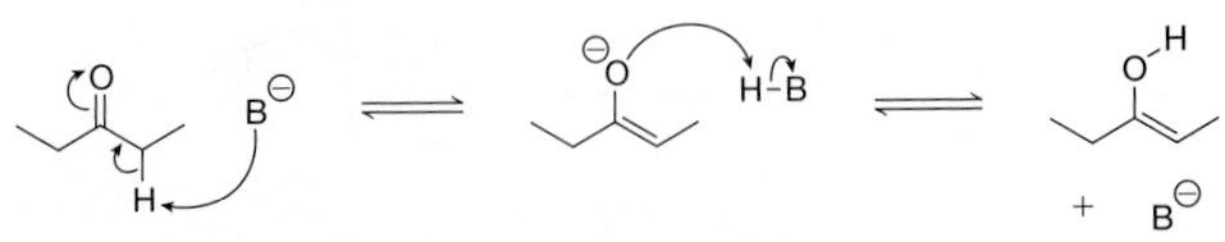

$B^{\ominus}$

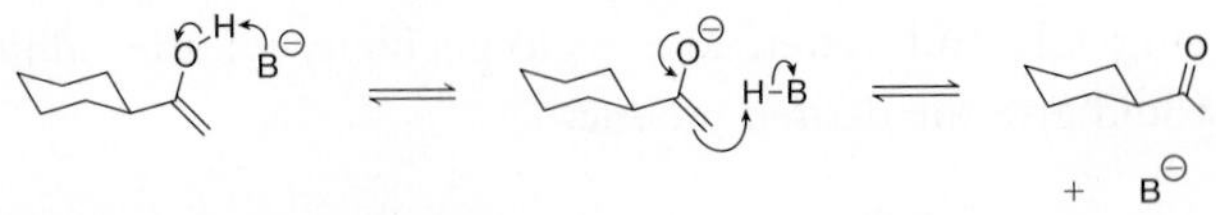

H–B

+ $B^{\ominus}$

b)

$B^{\ominus}$

H–B

+ $B^{\ominus}$

CHAPTER 9

PRACTICE PROBLEM 9.1

a) i) = 9 freely rotatable bonds
= 4 rotate at elevated temperature

rosuvastatin

ii) = 2 freely rotatable bonds
= 6 rotate at elevated temperature

imitinib

iii) = 5 freely rotatable bonds
= 2 rotate at elevated temperature

minocycline

b) i) shortest C–C bond

ii) shortest C–C bond

iii) shortest C–C bond

c) The lowest heat of hydrogenation is the most stable double bond.

i) NH group is in conjugation with the CO in the amide. This stabilizes the CO and makes it less reactive.

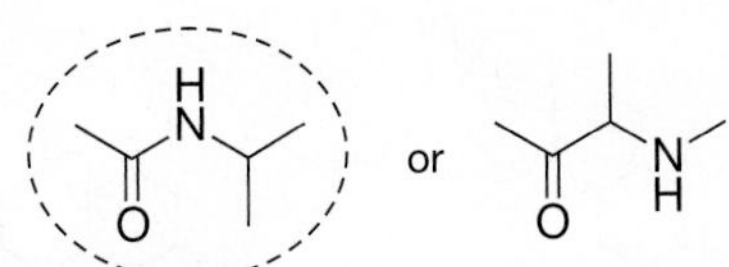

ii) The double bond is in conjugation with the oxygen in the molecule on the right, making it more stable.

or

iii) The ether group extends the conjugation in the molecule on the left, making it the most stable.

or HO

INTEGRATE THE SKILL 9.2

Bond **b** would be the shortest, as it has the most double bonds of the three resonance forms.

a = s, s, d
b = d, d, s
c = s, s, d

a : single
b : double
c : single

a : single
b : double
c : single

a : double
b : single
c : double

PRACTICE PROBLEM 9.3

a) The molecule with the longer λ_{max} is circled, below.
- The molecules with the longest conjugated π systems will have the longer λ_{max}.
- Longer π systems have smaller ΔE's from a smaller HOMO–LUMO gap.

- Longer wavelengths correspond to a smaller energy gap, remembering $\Delta E = hc/\lambda$.

Note: Conjugated systems are highlighted in bold.

i)

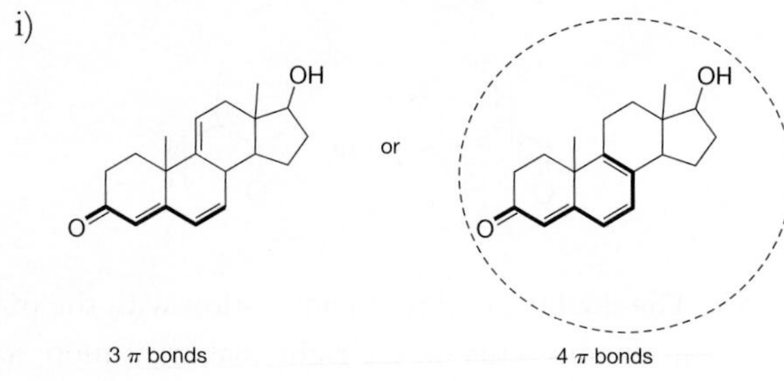

ii)

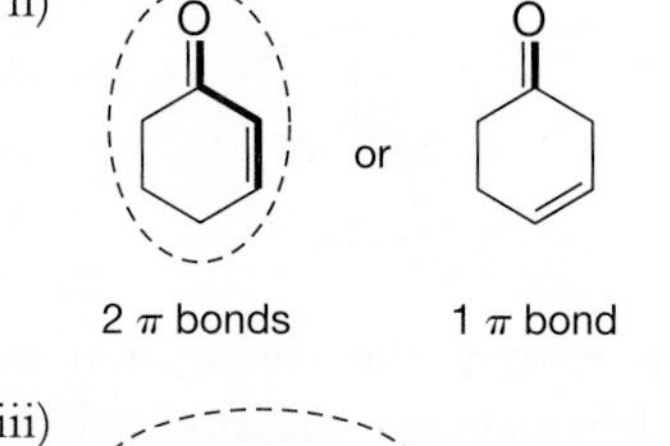

iii)

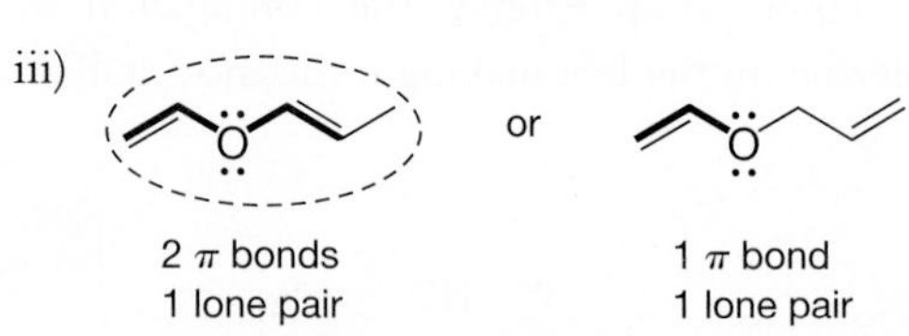

b) i) The molecule is absorbing in the blue-green range of the visual spectrum, so its colour will be the compliment of that: red.

ii) This is absorbing blue-violet and will appear yellow-orange in colour.

iii) This is absorbing blue-green and will appear red in colour.

INTEGRATE THE SKILL 9.4

The molecule is absorbing photons in the UV, so all visible light would be reflected. So, it would appear white.

The excited molecule would release photons of slightly longer wavelength (lower energy), which would be in the violet range of the visible spectrum. Since these are emitted photons, the colour corresponds to their energy. So, the fluorescence would be purple in colour.

PRACTICE PROBLEM 9.5

a) There are 12 π electrons, or $4n$ ($n = 3$). This is anti-aromatic.

b) There are 10 π electrons, or $4n + 2$ ($n = 2$). This is aromatic.

c) This is not a cyclic system, so it cannot be aromatic.

d) There are 8 π electrons, or $4n$ ($n = 2$). This is anti-aromatic.

e)

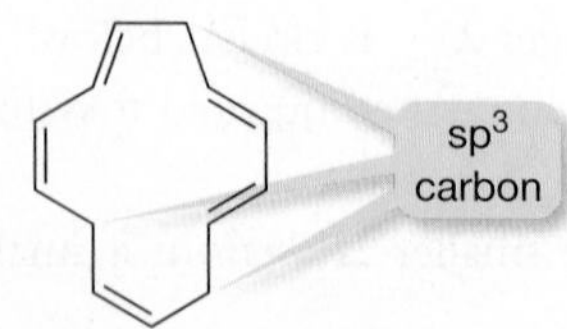

The system is not conjugated since the saturated carbon atoms break up the π bond system.

INTEGRATE THE SKILL 9.6

Aromatic rings are shown in blue, below.

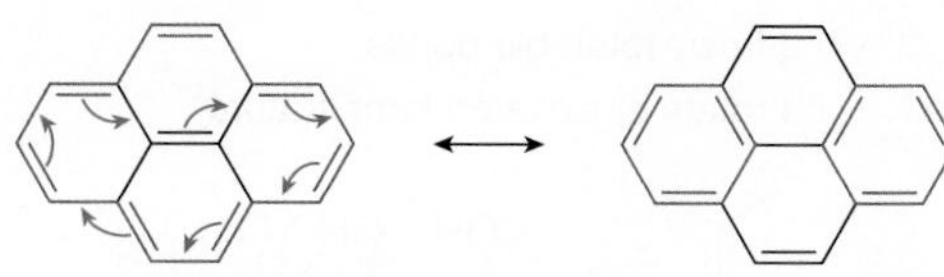

PRACTICE PROBLEM 9.7

a)

Aromatic: 2 π electrons (1 double bond), $4n + 2$ ($n = 0$).

b)

Anti-aromatic: 8 π electrons (2 double bonds, 2 lone pairs), $4n$ ($n = 2$).

c)

Aromatic: 6 π electrons (2 double bonds, 1 lone pair), $4n + 2$ ($n = 1$).

d)

Aromatic: 10 π electrons (4 double bonds, 1 lone pair), $4n + 2$ ($n = 2$).

e)

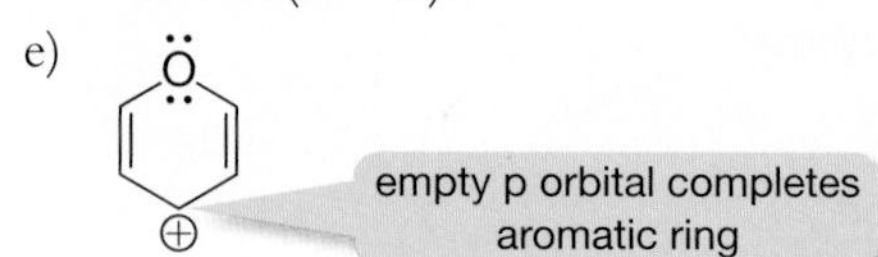

Aromatic: 6 π electrons (2 double bonds, 1 lone pair), $4n + 2$ ($n = 2$).

f)

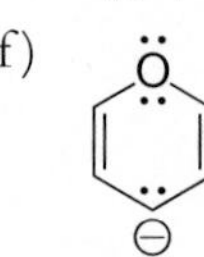

Anti-aromatic: 8 π electrons (2 double bonds, 2 lone pairs), $4n$ ($n = 2$).

INTEGRATE THE SKILL 9.8

aromatic ($4n + 2$, $n = 2$)
2 double bonds + 1 lone pair

aromatic ($4n + 2$, $n = 2$)
3 double bonds

caffeine

PRACTICE PROBLEM 9.9

a) Two π electrons and three p orbitals

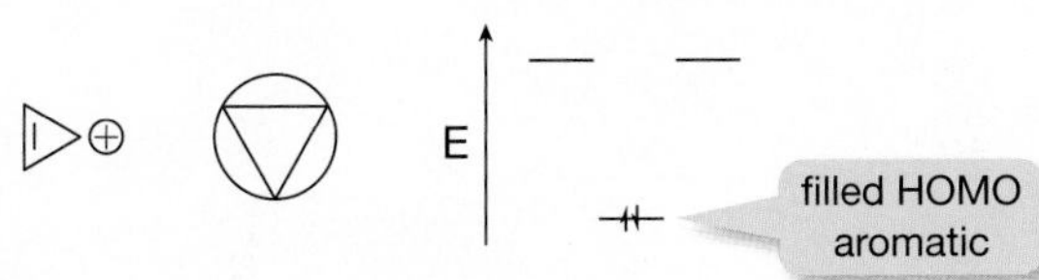

b) Six π electrons and seven p orbitals

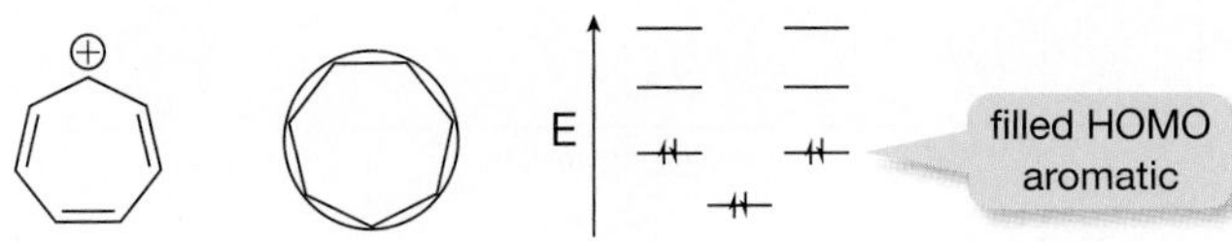

INTEGRATE THE SKILL 9.10

The only difference between the three systems is the number of π electrons in the MO's.

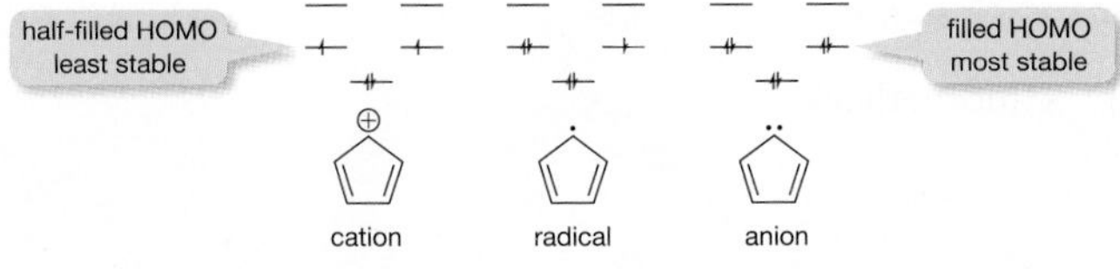

PRACTICE PROBLEM 9.11

a) i)

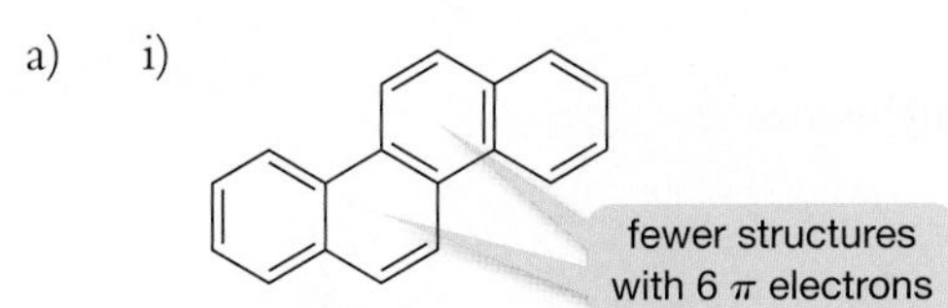

ii)

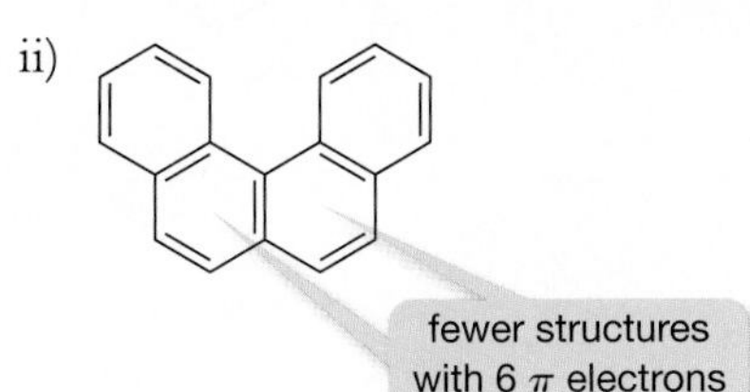

iii)

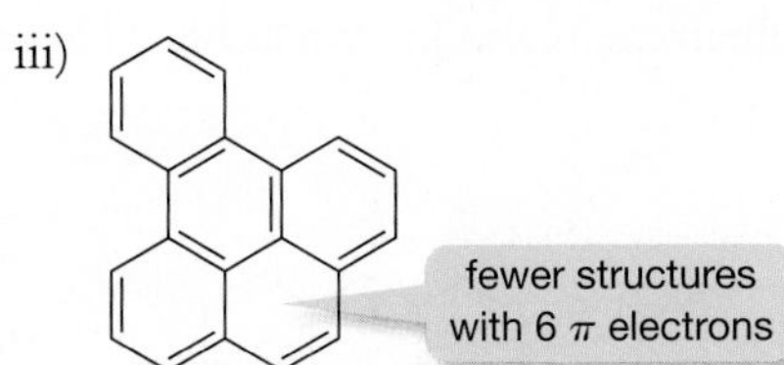

b) i) [18]-annulene should be aromatic ($4n + 2$, $n = 4$). There are no steric interactions that would preclude the molecule from being planar.

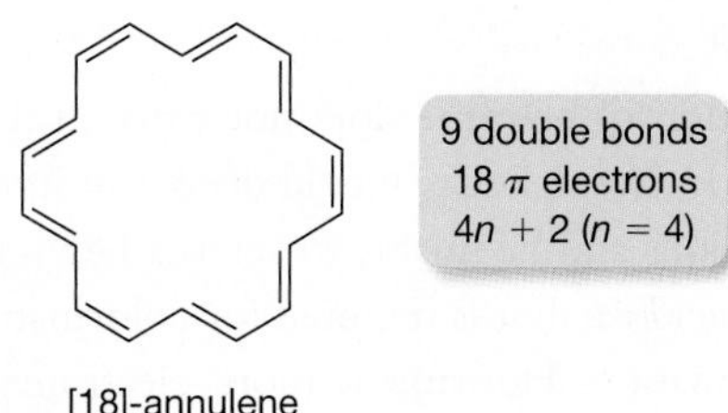

ii) [22]-annulene should be aromatic ($4n + 2$, $n = 5$). There are no steric interactions that would preclude the molecule from being planar.

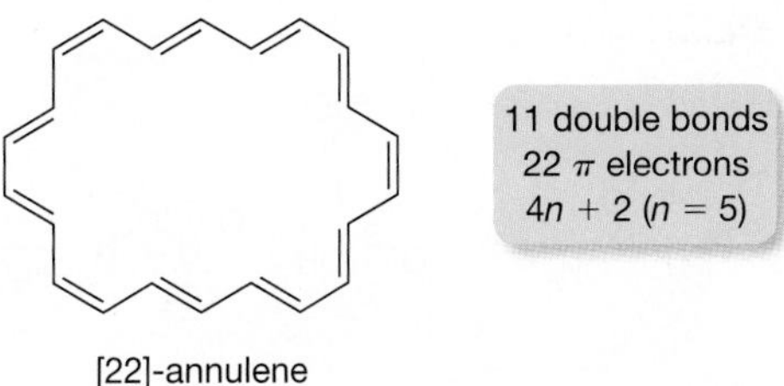

INTEGRATE THE SKILL 9.12

Aromaticity in coronene. Two aromatic rings shown in blue and red in the resonance form.

CHAPTER 10

PRACTICE PROBLEM 10.1

a) i)

ii)

b) i)

ii)

INTEGRATE THE SKILL 10.2

a) Bromination of benzene does not work in the presence of water. In water, $FeBr_3$ would dissociate into Fe^{3+} and bromide ions and be unable to act as a Lewis acid.

b) A Lewis acid catalyst is required for chlorination but not for fluorination. Fluorine is more electronegative than chlorine, so it does not need a Lewis acid catalyst to enhance its electrophilicity. In practice, fluorine is actually too reactive, both as an electrophile and as an oxidant. Many modern sources of F^+ have been developed.

PRACTICE PROBLEM 10.3

a) i)

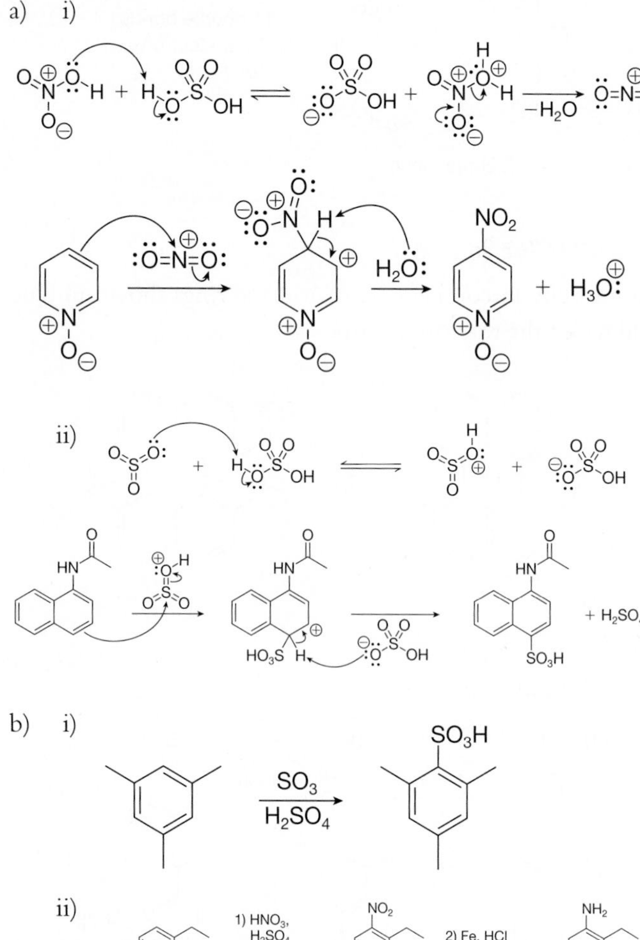

INTEGRATE THE SKILL 10.4

The reaction coordinate diagrams for the sulfonation of benzene and the desulfonation of benzenesulfonic acid are shown below.

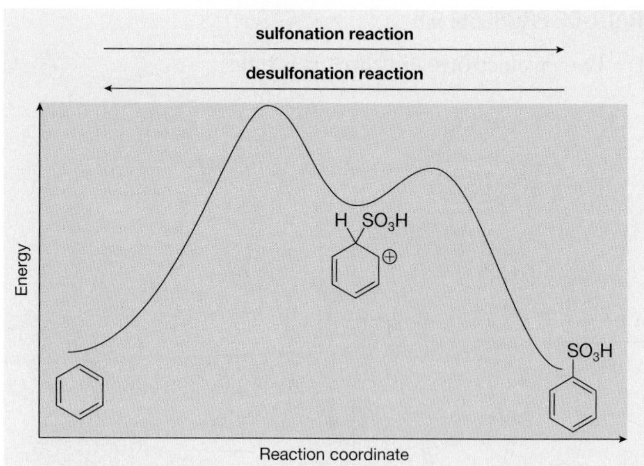

The reaction diagrams for both transformations will be the same, just reversed. The reactions have the same intermediates and reactions, just reversed, depending on whether sulfonation or desulfonation is being described. This is known as the Principle of Microscopic Reversibility.

PRACTICE PROBLEM 10.5

a) Chloroethane:

+ Cl $\xrightarrow{AlCl_3}$ + HCl

b) (1-bromoethyl)benzene:

Br + $\xrightarrow{AlCl_3}$ + HBr

c) 2-chloroheptane:

+ Cl $\xrightarrow{AlCl_3}$ + HCl

d) (Bromomethyl)benzene is also known as *benzyl bromide.*

Br + $\xrightarrow{AlCl_3}$ + HBr

INTEGRATE THE SKILL 10.6

a) i) The primary carbocation will rearrange to a benzylic one, giving a different product than the one shown.

Carbocation formation and rearrangement:

$AlCl_3$ + $Cl-AlCl_3$

Alkylation:

$Cl-AlCl_3$ + HCl + $AlCl_3$

ii) The initial 1° carbocation will rearrange into a more stable 3° carbocation before alkylating the ring.

$AlCl_3$ $AlCl_4$ $Cl-AlCl_3$ + HCl + $AlCl_3$

b) The alkene will react with the HCl to form a carbocation and $AlCl_3$ reacts with the chloride ion formed.

H–Cl $AlCl_3$ $AlCl_4$

The carbocation then reacts with the aromatic ring to form the final product.

$Cl-AlCl_3$ + HCl + $AlCl_3$

PRACTICE PROBLEM 10.7

a) i) $AlCl_3$ $AlCl_3$ + $AlCl_4$

MeO $Cl-AlCl_3$ + HCl + $AlCl_3$

ii) $AlCl_3$ $AlCl_3$ + $AlCl_4$

Cl_3Al-Cl + HCl + $AlCl_3$

b) i) $AlCl_3$ OH

ii) + CO $\xrightarrow{HCl}$ H

INTEGRATE THE SKILL 10.8

In both failed reactions from Question 10.6, the initial carbocation formed by the alkyl halide rearranged, leading to products different from the ones desired. Carbocation rearrangement is avoided by using an acylation reaction followed by reduction of the ketone to an alkane.

i) Acylation:

Cl + $\xrightarrow{AlCl_3}$ + HCl

Reduction:

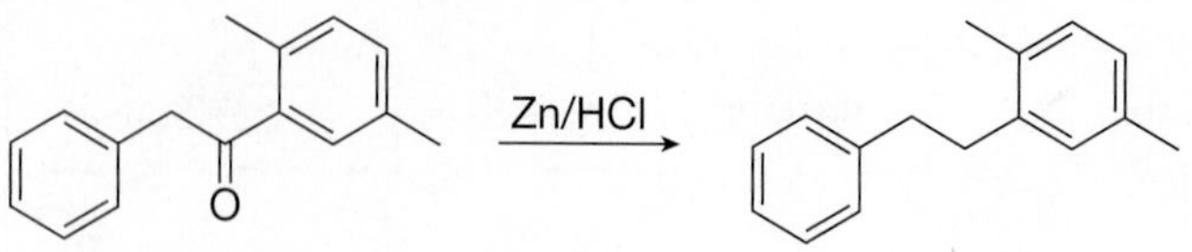
Zn/HCl
O

ii) Acylation:

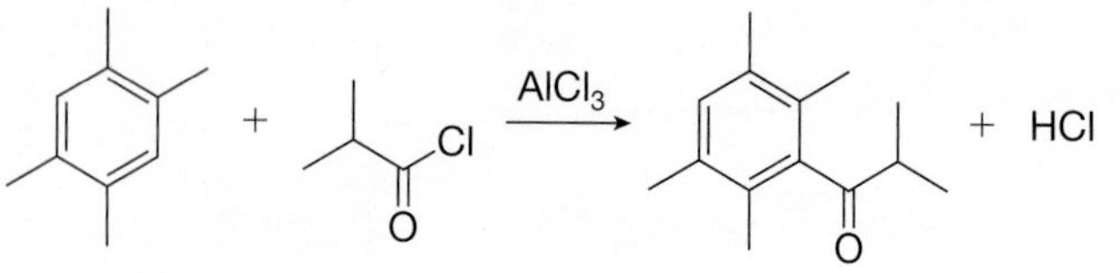
+
Cl
O
AlCl3
+ HCl
O

Reduction:

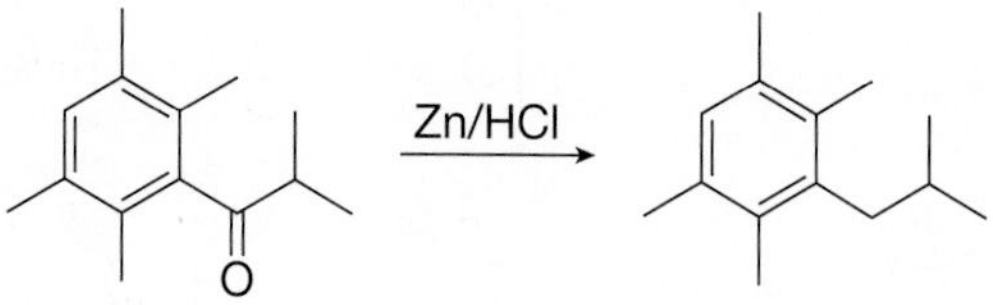
Zn/HCl
O

PRACTICE PROBLEM 10.9

a) The methyl group is *ortho*/*para* directing and the *para* position is less sterically hindered, so *para* substitution is preferred.

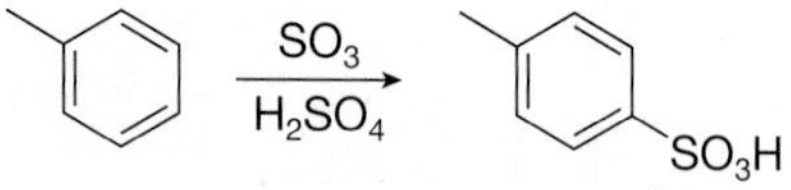
SO3
H2SO4
SO3H

b) The nitro group is *meta* directing. Step 2 reduces the nitro groups to amines.

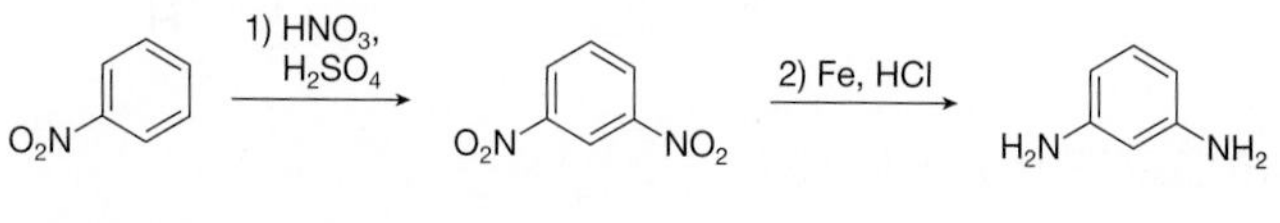
O2N
1) HNO3,
H2SO4
O2N
NO2
2) Fe, HCl
H2N
NH2

c) The sidechain is *meta* directing.

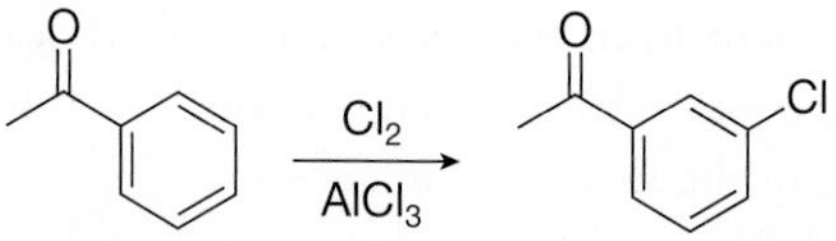
O
Cl2
AlCl3
O
Cl

d) The sidechain is *ortho*/*para* directing. The major product will be *para* substituted due to steric hindrance around the *ortho* site.

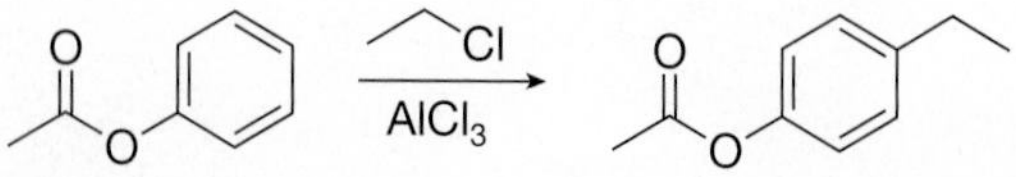
O
O
Cl
AlCl3
O
O

e) The bromine is *meta* directing.

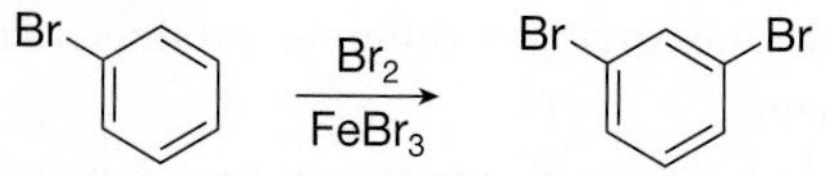
Br
Br2
FeBr3
Br
Br

f) The OH group is *ortho*/*para* directing.

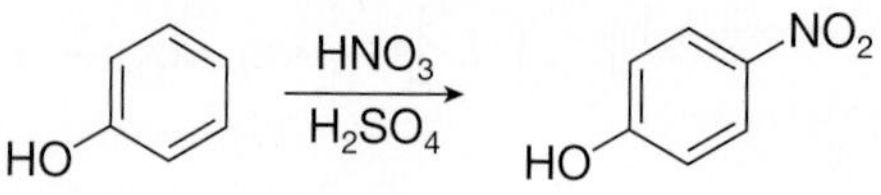
HO
HNO3
H2SO4
HO
NO2

INTEGRATE THE SKILL 10.10

The major product from alkylation hydroxylated biphenyl with a *t*-butyl group will be *ortho* substitution on the hydroxylated ring.

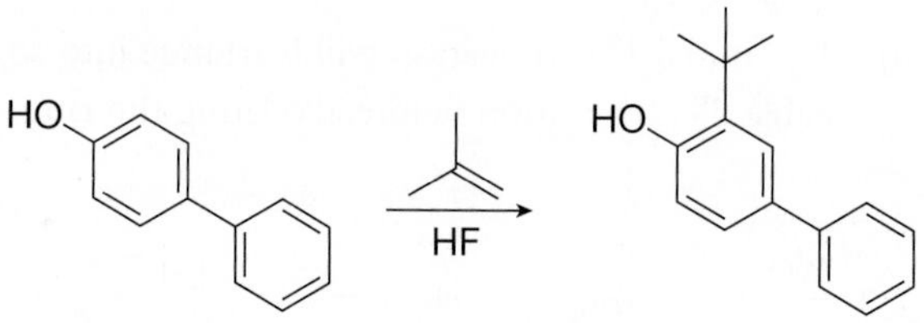
HO
HF
HO

The hydroxyl group is *ortho*/*para* directing and activates the ring. *Para* substitution is impossible since the *para* position is already occupied by the other phenyl ring.

The hydroxyl group helps to stabilize the carbocation intermediate for the *ortho* addition.

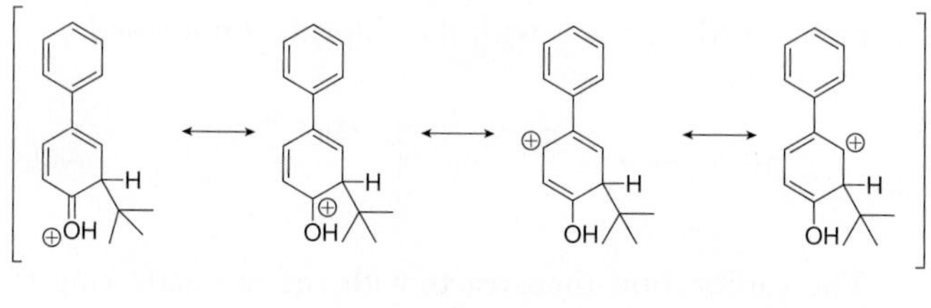
H
OH
H
OH
H
OH
H
OH

Substitution *meta* to the hydroxyl group does not allow for delocalization of the positive charge by the hydroxyl group.

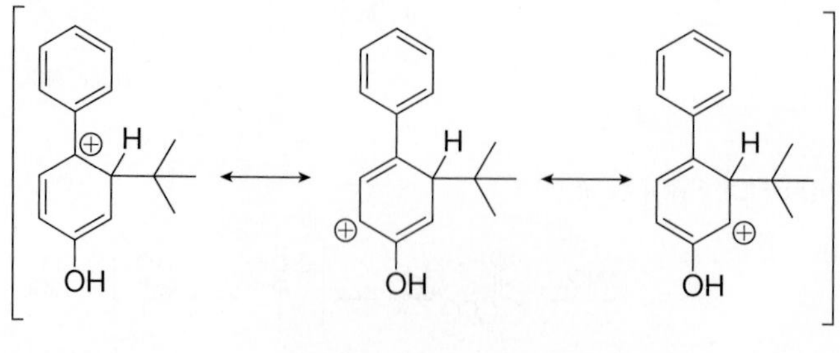
H
OH
H
OH
H
OH

PRACTICE PROBLEM 10.11

a) The basic amino group needs to be protected before the sulfonation reaction. It would protonate and become a *meta*-directing ammonium ion, which would give the wrong substitution product.

NH_2 → (Cl, O; pyridine) → HN, O → (SO_3, H_2SO_4) → HN, O; SO_3H → (NaOH) → NH_2; SO_3H

b) Bromination produces HBr as a by-product. This would react with the amino group to form a *meta*-directing ammonium group. In addition, the amine is activating, so multiple bromination could be possible. Protecting the amine as an amide prevents these complications.

NH_2 → (Cl, O; pyridine) → HN, O → (Br_2, $FeBr_3$) → HN, O; Br → (NaOH) → NH_2; Br

c) The ethoxy group and the added alkyl group are both activating, so multiple alkylation could be a problem. Acylation followed by reduction will prevent this possible problem.

O → (Cl, O; $FeCl_3$) → O, O → (Zn/HCl) → O

d) Chlorination followed by hydrolysis of the ester will give the required product.

O, O → (Cl_2, $FeCl_3$) → O, O; Cl → (NaOH, H_2O) → HO; Cl

INTEGRATE THE SKILL 10.12

HO → (O, Cl; pyridine) → O, O (activating; deactivating) → (HNO_3, H_2SO_4) → O, O; NO_2

The nitration step is selective because of the different effects of the ester group. The reactive ring is connected via the oxygen, which has lone pairs and is activating. The other ring is bonded to the carbonyl, which is electronegative and destabilizing to arenium ion formation.

PRACTICE PROBLEM 10.13

a) The methyl groups are *ortho/para* directing. *Para* substitution to one methyl is *ortho* to the other methyl so they reinforce each other. The *ortho/ortho* site is unlikely due to steric considerations.

→ (SO_3, H_2SO_4) → SO_3H

b) The NO_2 group is meta directing and the OH group is *ortho/para* directing. Both of these reinforce substitution at the site *para* to the OH group.

O_2N, OH → (HNO_3, H_2SO_4) → NO_2; O_2N, OH

c) The acyl group (meta directing) and the methoxy group (*ortho/para* directing) are opposing each other. The methoxy is a stronger activator so should control the regiochemistry. The site *para* to the acyl is least hindered and most likely to be substituted.

O; O → (Cl_2, $AlCl_3$) → O; Cl; O

d) Both substituents are *ortho/para* directing. Since there are only *ortho* sites the stronger activator will determine the regiochemistry. The ester group is the stronger of the two activators.

O, O → (Cl, $AlCl_3$) → O, O

e) The both groups are *ortho/para* directing and are reinforcing each other. The site *para* to the phenyl group is the least sterically crowded and most accessible to the reactants.

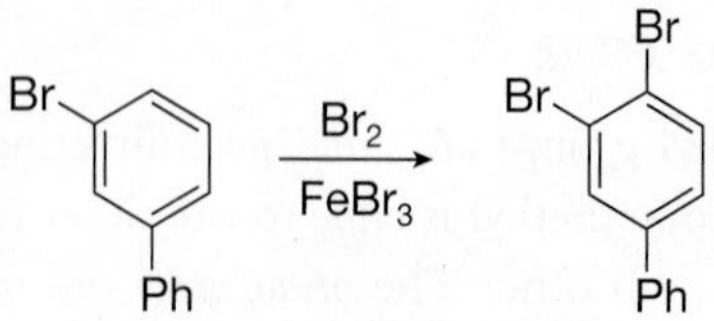

Br Ph Br2 FeBr3 Br Br Ph

f) All three groups are reinforcing substitution *ortho* to the OH group. The side *para* to the ethyl group is most sterically accessible.

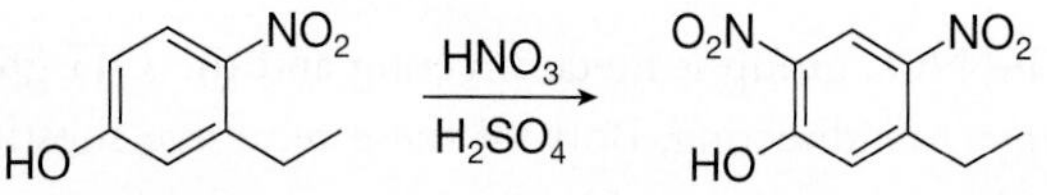

NO_2 HO HNO_3 H_2SO_4 O_2N NO_2 HO

g) The OH and Cl groups are both *ortho* directing and the acyl is *meta* directing. This puts all three in competition for directing the substitution. The OH group is the strongest director and should control the regiochemistry. The least sterically hindered position *ortho* to the OH should be substituted.

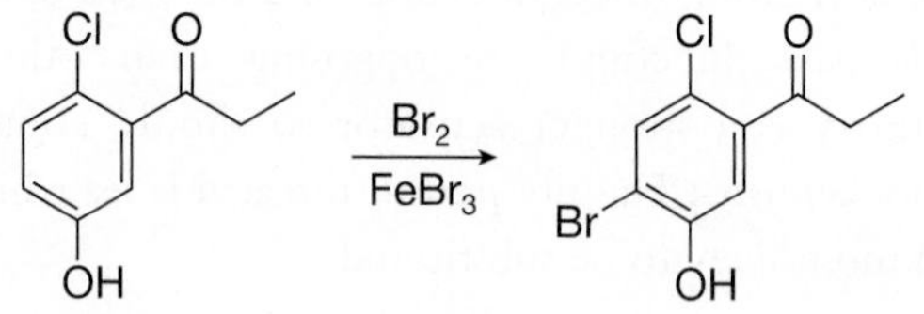

Cl O OH Br2 FeBr3 Cl O Br OH

INTEGRATE THE SKILL 10.14

The OH substituted ring should be more reactive since the OH is strongly activating. The acyl group is deactivating.

Ortho substitution is favoured due to the extra stabilization of the arenium intermediate by the oxygen lone pairs. Both of these factors would lead to the major product shown below.

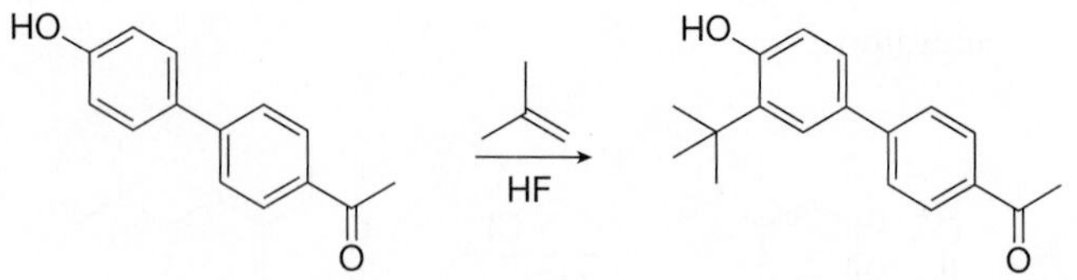

HO O HF HO O

Ortho arenium stabilization:

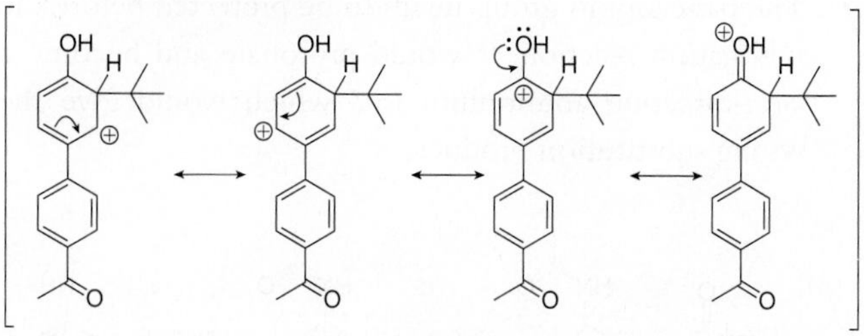

The other three possible intermediates have less-delocalized positive charges in the arenium ion.

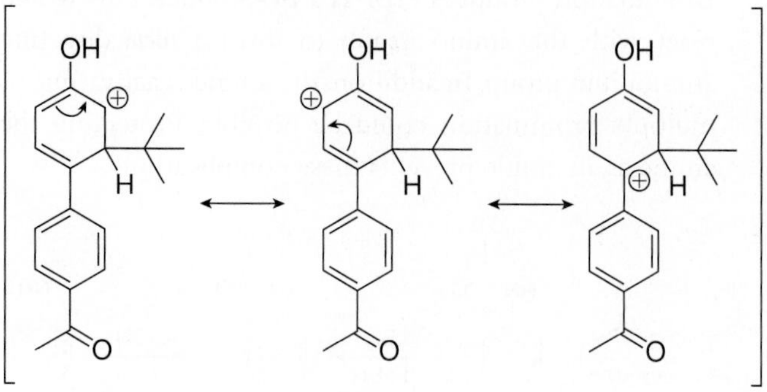

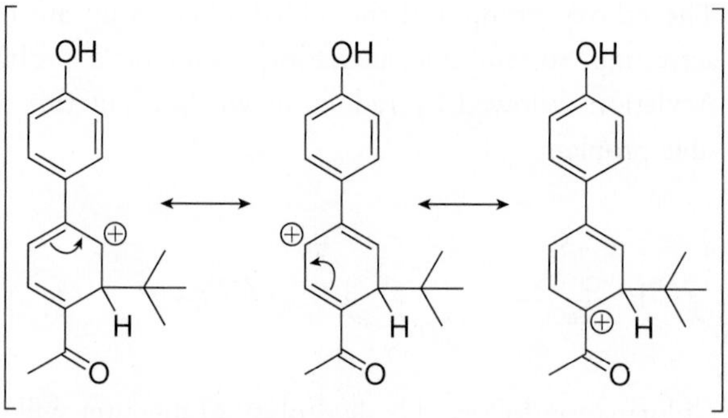

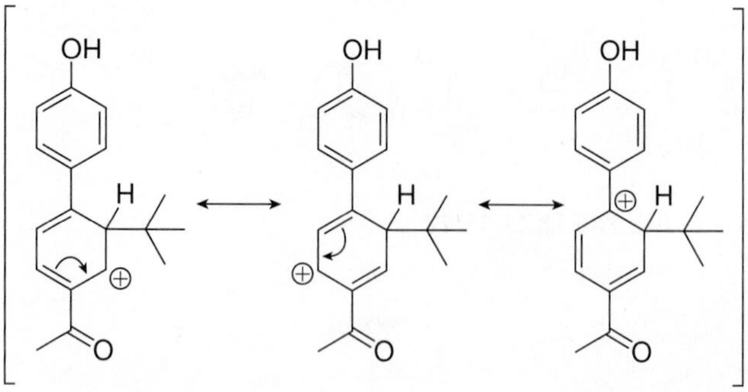

PRACTICE PROBLEM 10.15

a) Substitution *ortho* to the oxygen is preferred in furan reactions.

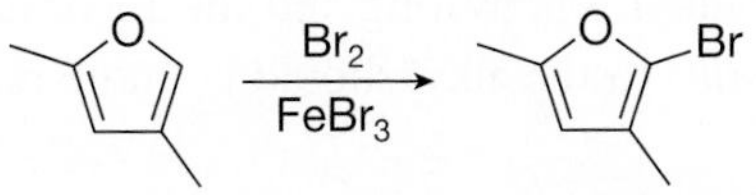

Br_2 / $FeBr_3$

b) The heterocyclic ring will substitute preferentially. The preferred site of reactivity is controlled by the formation of the most stable arenium ion intermediate.

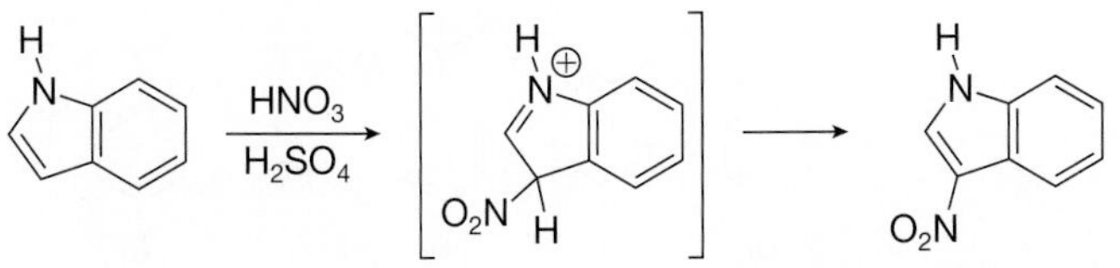

HNO_3 / H_2SO_4 ; O_2N

c) This ring is similar to a furan and should have the same regiochemistry.

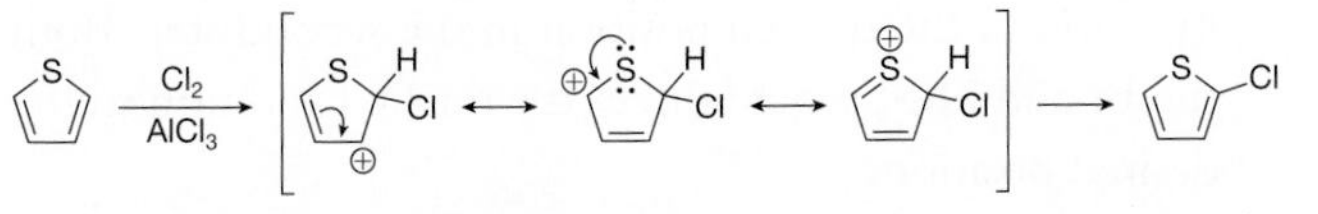

Cl_2 / $AlCl_3$

d) Substitution is at the 3 position since it forms the most stable intermediate carbocation.

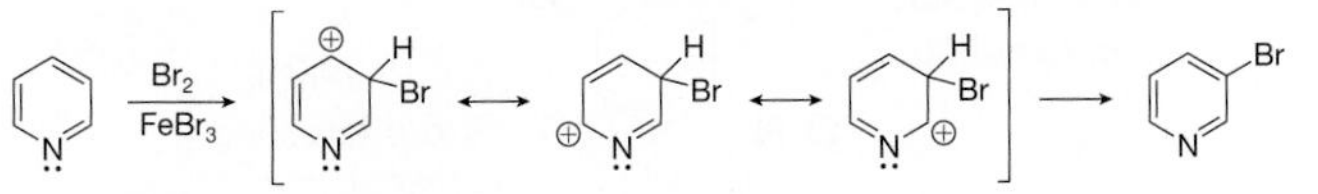

Br_2 / $FeBr_3$

Bromine addition at the 2- or 4 positions leads to intermediates with a positive charge on the electronegative nitrogen atom.

substitution at 2

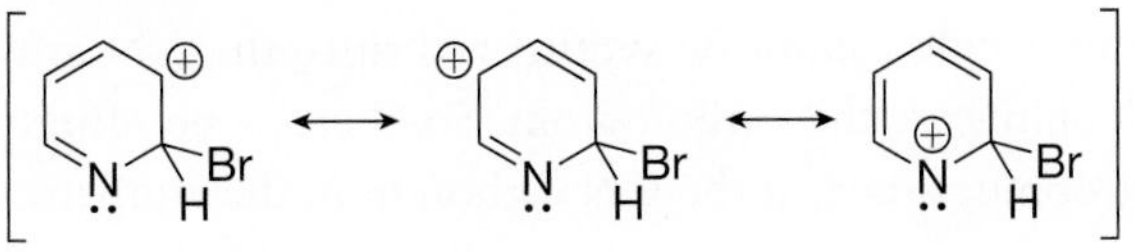

substitution at 4

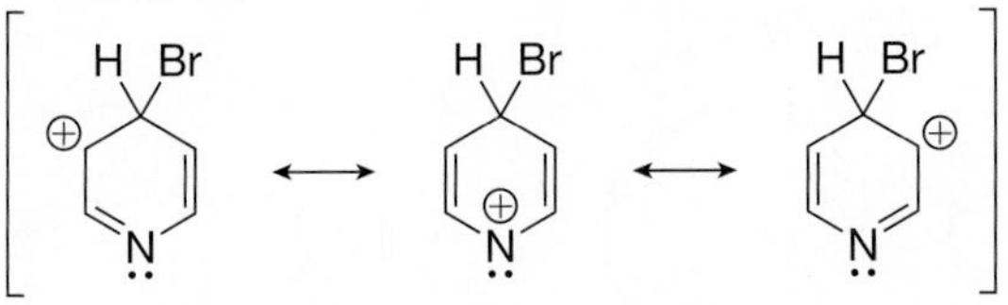

INTEGRATE THE SKILL 10.16

The acylation of pyridine should proceed as in part (d) of Question 10.15.

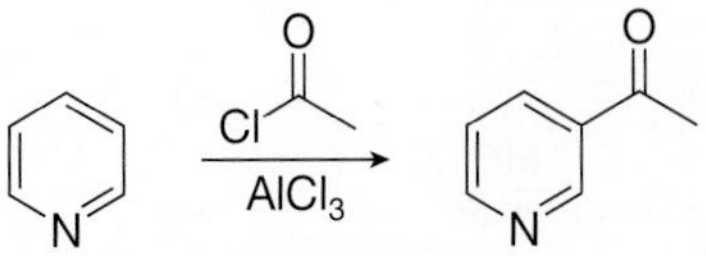

$AlCl_3$

Drawing resonance structures for the pyridine N-oxide, the electron density has been altered, with formal negative charges at the 2- and 4 positions.

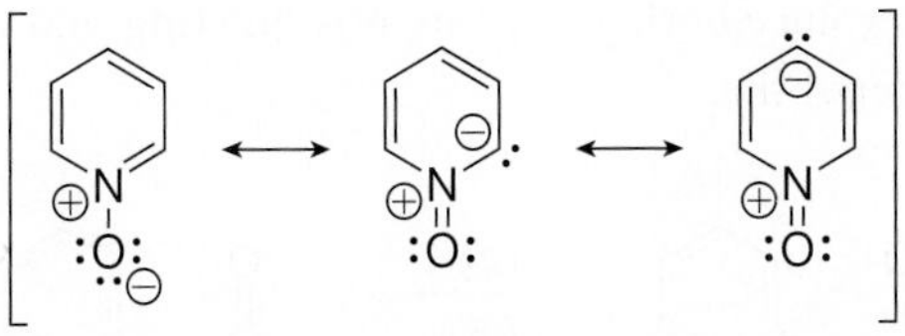

This makes the 2- and 4 positions more nucleophilic and these will be the preferred sites of substitution. The product of the acylation would therefore be

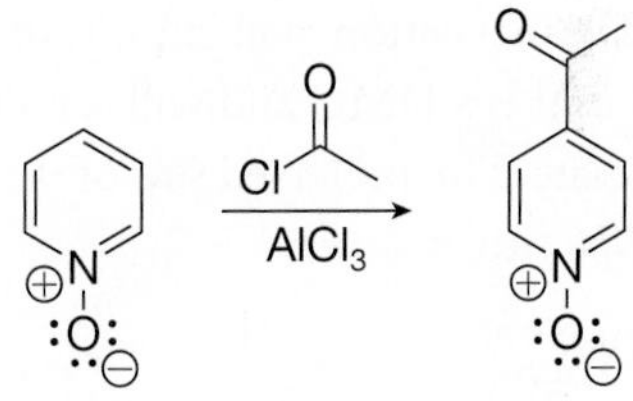

$AlCl_3$

PRACTICE PROBLEM 10.17

a) i)

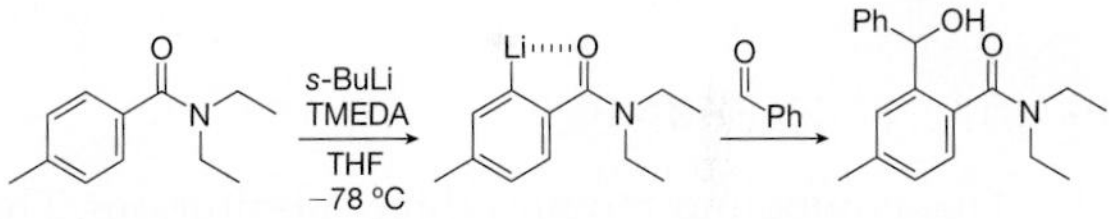

s-BuLi, TMEDA, THF, −78 °C

ii) There are two *ortho* sites near the DMG. The most accessible one will react more quickly.

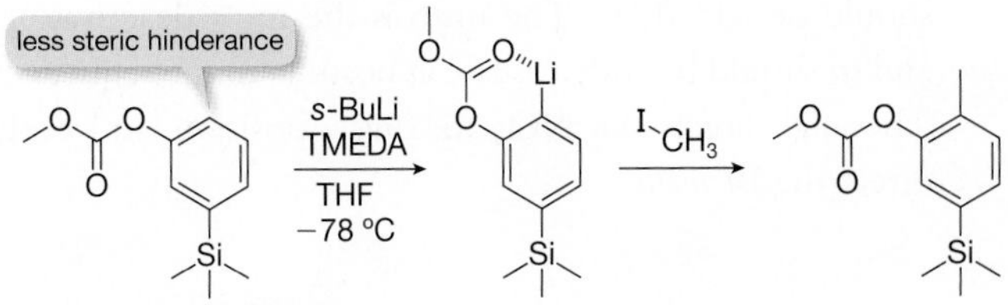

less steric hinderance

s-BuLi, TMEDA, THF, −78 °C ; $I{-}CH_3$

iii)

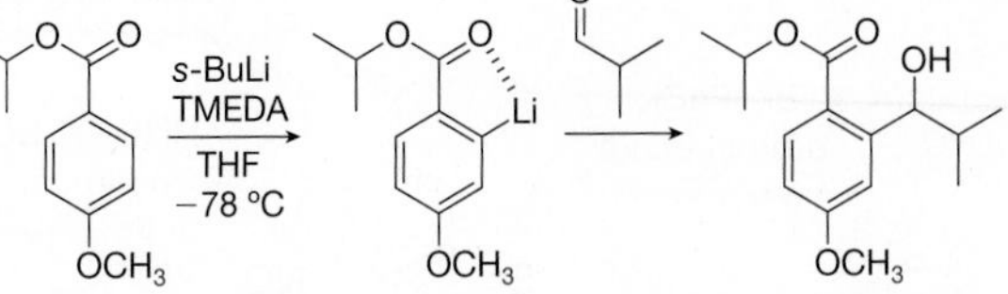

s-BuLi, TMEDA, THF, −78 °C ; OCH_3

b) i)

s-BuLi, TMEDA, THF, −78 °C

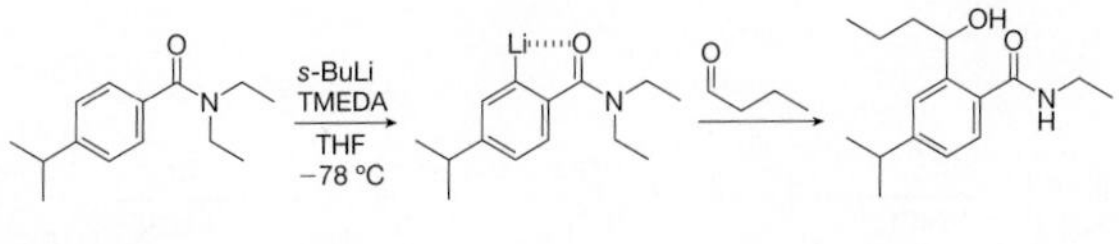

ii) 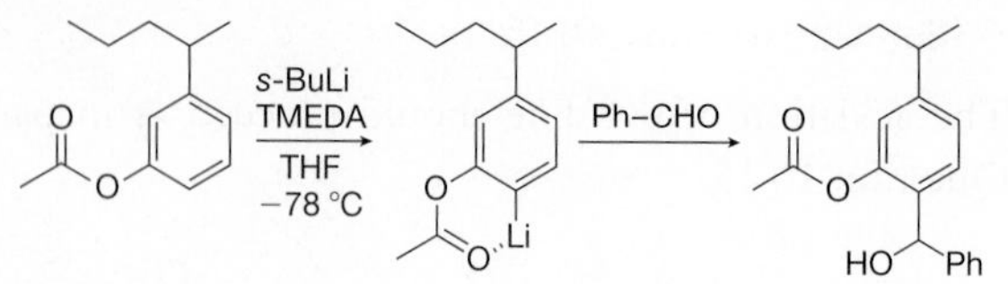 s-BuLi, TMEDA, THF, −78 °C; Ph–CHO

iii) 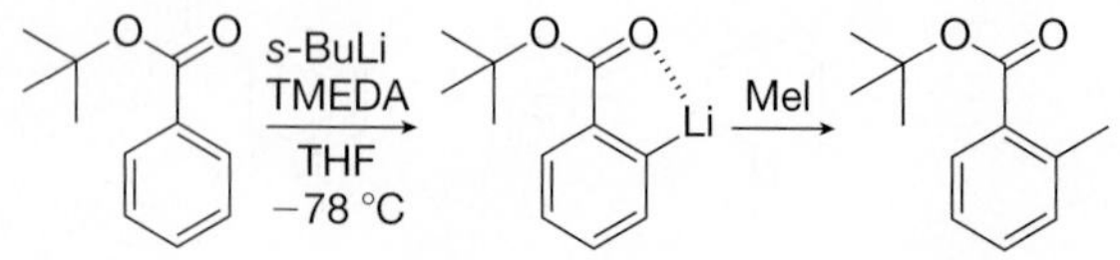 s-BuLi, TMEDA, THF, −78 °C; MeI

INTEGRATE THE SKILL 10.18

Friedel–Crafts methylation should substitute the position *para* to the ester since both groups are *o/p* directing, and that is the least hindered site.

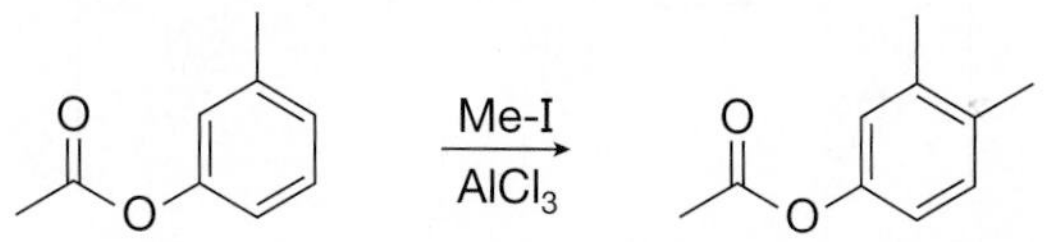

Me-I, $AlCl_3$

Formylation using Gatterman–Koch conditions followed by reduction might be preferred to avoid multiple methylation since all the substituents on the ring are activating.

Directed *ortho* metalation will substitute the *ortho* position to the ester, as it is a DMG and will stabilize the organolithium intermediate. The preferred site of methylation is the least hindered *ortho* site.

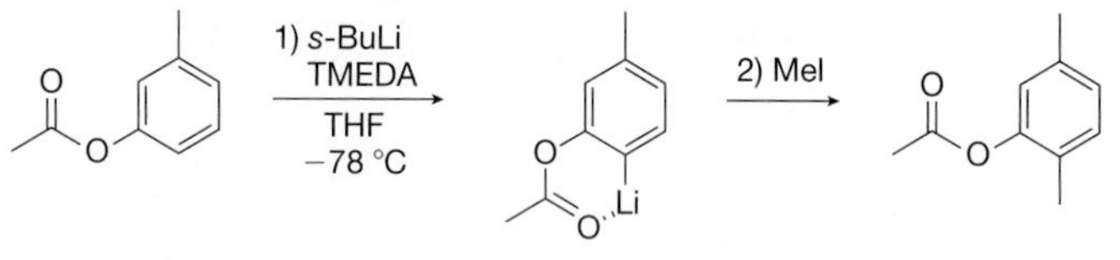

1) s-BuLi, TMEDA, THF, −78 °C; 2) MeI

PRACTICE PROBLEM 10.19

a) This compound requires three substitutions. The groups are all *meta* to each other, so adding *meta*-directing groups first is optional. This means the bromine (*o/p* directing) should be added last. The nitro is the most deactivating and so should be added as late as possible in the sequence. The acyl directs the NO_2 to a *meta* position, and both direct the Br *meta*.

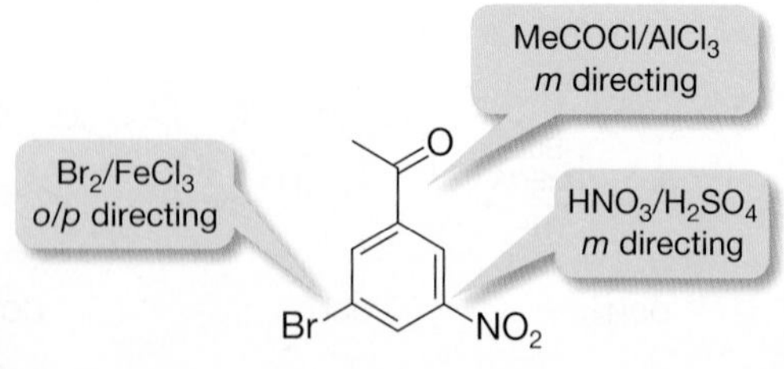

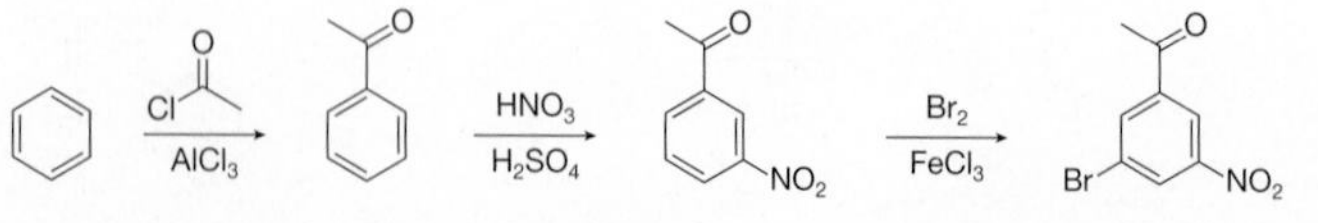

MeCOCl/$AlCl_3$ *m* directing; $Br_2/FeCl_3$ *o/p* directing; HNO_3/H_2SO_4 *m* directing

$AlCl_3$; HNO_3, H_2SO_4; Br_2, $FeCl_3$

b) All groups are *ortho/para* directing. Adding Br first directs the alkyl addition *para* without activating the ring and causing multiple substitutions. Both groups activate the remaining positions for the final addition of the Cl group. Since the alkyl is activating and the Br deactivating, the position *ortho* to the alkyl should be more reactive.

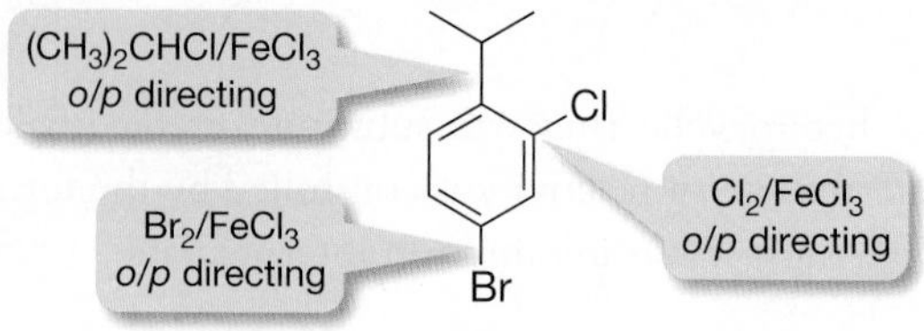

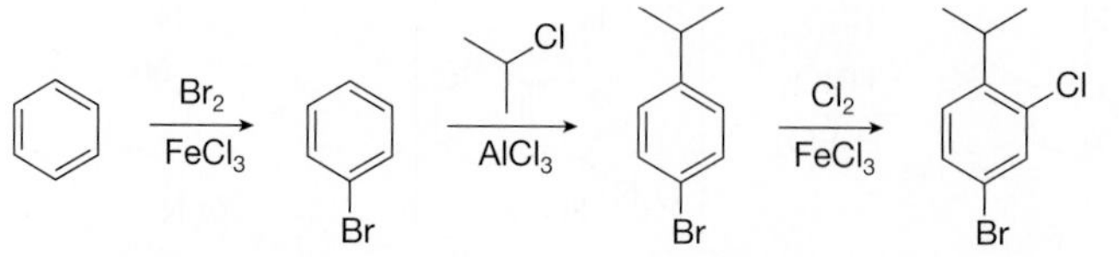

$(CH_3)_2CHCl/FeCl_3$ *o/p* directing; $Br_2/FeCl_3$ *o/p* directing; $Cl_2/FeCl_3$ *o/p* directing

Br_2, $FeCl_3$; $AlCl_3$; Cl_2, $FeCl_3$

c) The Cl would direct addition *para* to itself, a site where no substitution is required. So, it should be added last. The Br is a *para* director, so adding it first would promote nitration in the desired position in the second step. Both the Br and NO_2 would direct the final Cl addition to the desired position.

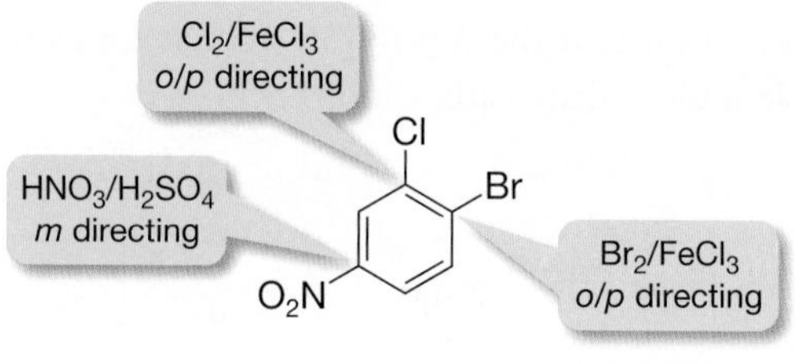

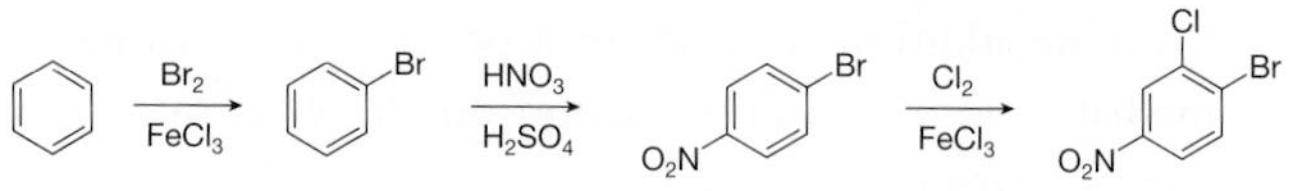

$Cl_2/FeCl_3$ *o/p* directing; HNO_3/H_2SO_4 *m* directing; $Br_2/FeCl_3$ *o/p* directing

Br_2, $FeCl_3$; HNO_3, H_2SO_4; Cl_2, $FeCl_3$

INTEGRATE THE SKILL 10.20

The molecule cannot be synthesized since the CN group is nucleophilic at the carbon atom. So, there is no affinity for electrophilic attack at the CN carbon from the aromatic ring.

CHAPTER 11

PRACTICE PROBLEM 11.1

a)

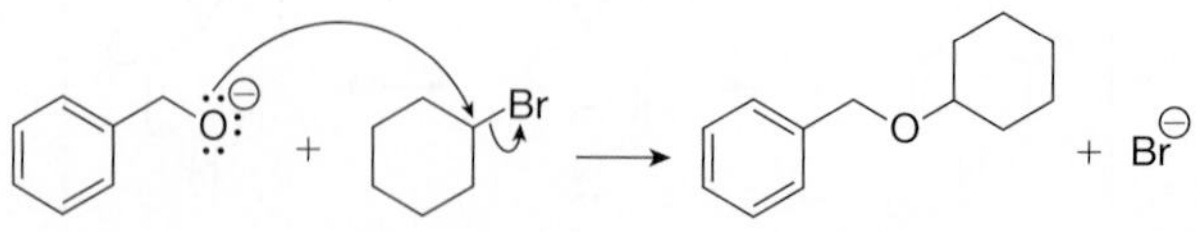

+ Br⁻

b)

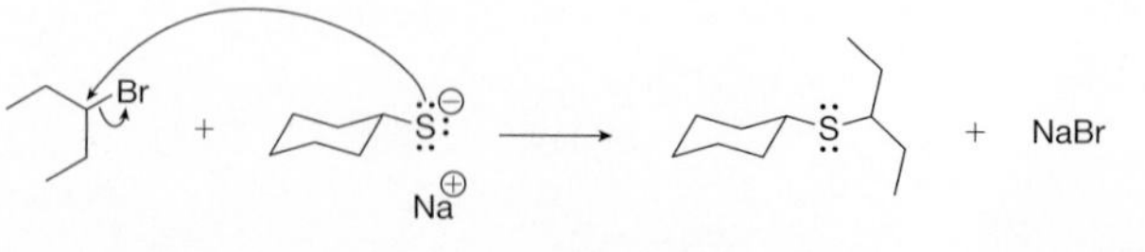

+ NaBr

c)

INTEGRATE THE SKILL 11.2

The first step in the reaction would be a bromination of the alkene by the Br_2. This produces a dibromide, so there are two possible sites for substitution. The alcohol will act as the nucleophile.

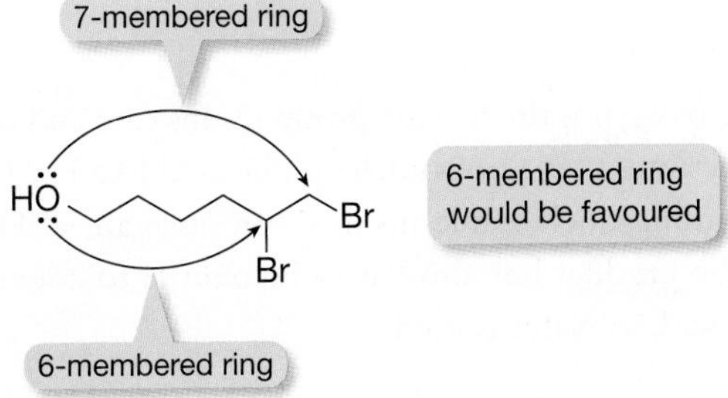

The major product would be the six-membered ring, and ideally the reaction would be done in dilute solution to avoid intermolecular reactions. The bicarbonate serves as base to deprotonate the initial substitution product to give a neutral final product.

PRACTICE PROBLEM 11.3

a) To invert the stereochemistry in a regular structural drawing, change wedge bonds to dashed wedges or vice versa.

inverted stereochemistry

b) To invert the stereochemistry in Fischer projections, switch the side of attachment.

inverted stereochemistry

c) To invert stereochemistry in Newman projections, swap positions of any two groups on a carbon.

inverted stereochemistry

INTEGRATE THE SKILL 11.4

a) Expected enantiomerically pure product:

b) The *ee* of this product is only 20 percent. The product is an alkyl iodide. Since iodine is a good leaving group, the product formed is actually a substrate for substitution. So, after some initial product is formed, there is a competing reaction that inverts the stereochemistry but does not consume the iodide nucleophile.

Inversion reaction

This inverted product will also invert to form the initial product.

Reinversion reaction

c) The expected product should be in excess. It is the only product formed in the initial reaction of the alkyl chloride starting material. If the reaction is allowed to proceed for long enough, the product would be a racemic mixture. The only way to account for an observed *ee* of 20 percent is an excess of the product from the initial substitution of the alkyl chloride.

PRACTICE PROBLEM 11.5

The better nucleophile is circled.

a) OH or NH_2

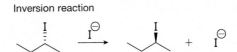

Nitrogen is less electronegative and more effective at sharing electrons than oxygen.

b)

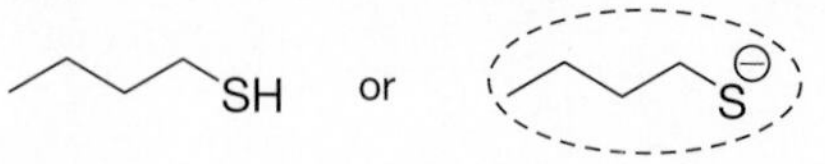

SH or S⊖

A charged nucleophile is more nucleophilic than its conjugate acid.

c)

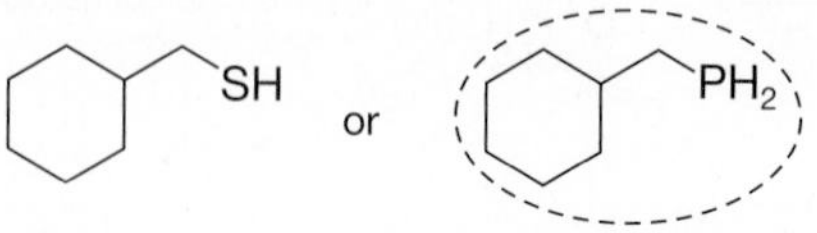

SH or PH$_2$

Phosphorus is less electronegative than sulphur and so will be a better electron donor.

d)

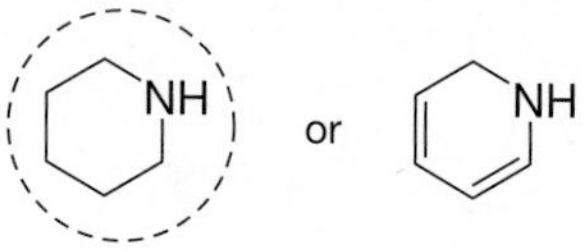

NH or NH

The saturated amine will be more nucleophilic since the lone pair in the other molecule is part of a resonance delocalization and less available for donation.

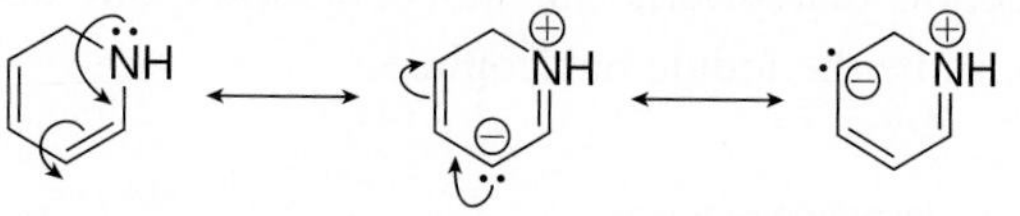

NH ⟷ NH ⟷ NH

INTEGRATE THE SKILL 11.6

a) The base in this reaction is not nucleophilic and serves to deprotonate the initial product to give a neutral product. The amine is a stronger nucleophile than the alcohol.

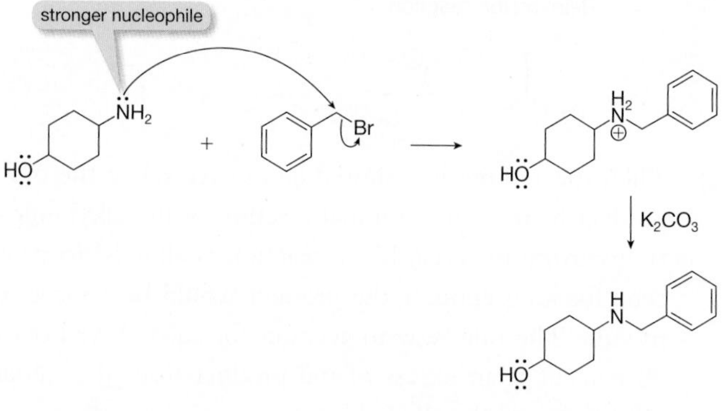

stronger nucleophile

HO, NH$_2$ + Br → HO, H_2N⊕ ; K_2CO_3 ↓ HO, HN

b) Sodium hydride is a very strong base and will react with acidic protons on amines and alcohols. Since alcohols are more acidic, the initial product would be an alkoxide ion.

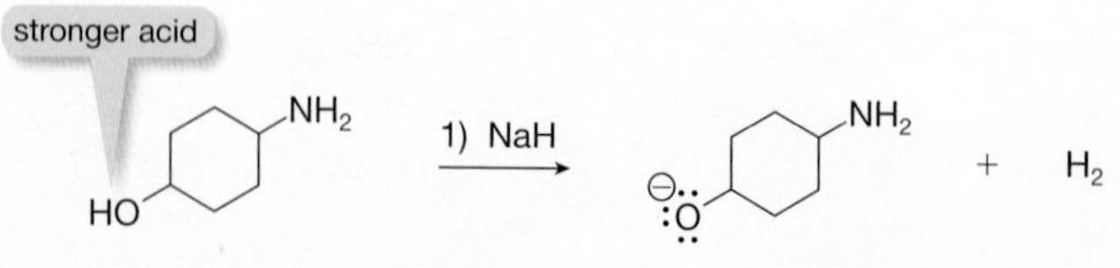

stronger acid

HO, NH$_2$ —1) NaH→ ⊖O, NH$_2$ + H_2

In the second step, the alkoxide substitutes the bromine to give the final product.

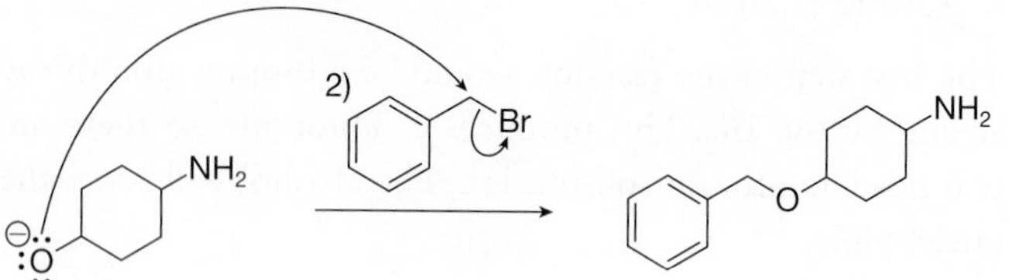

⊖O, NH$_2$ —2) Br→ O, NH$_2$

PRACTICE PROBLEM 11.7

a) The oxygen is the leaving group in this case and needs to be activated by a non-nucleophilic acid like H_2SO_4. The incoming nucleophile needs to produce an –OH group in the product but also has to be neutral to coexist with the acid, so water is used.

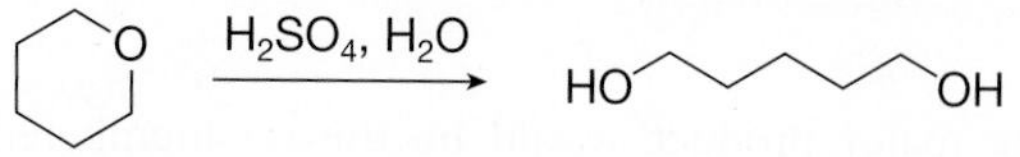

O —H_2SO_4, H_2O→ HO OH

b) The leaving groups are alcohols and so need some activation. The incoming nucleophile is an amine and is basic, so acid catalysis cannot be used. The alcohols need to be activated by forming sulfonate esters and then reacted with a primary amine.

HO OH —(× 2) Ts–Cl, pyridine→ Ts O O Ts

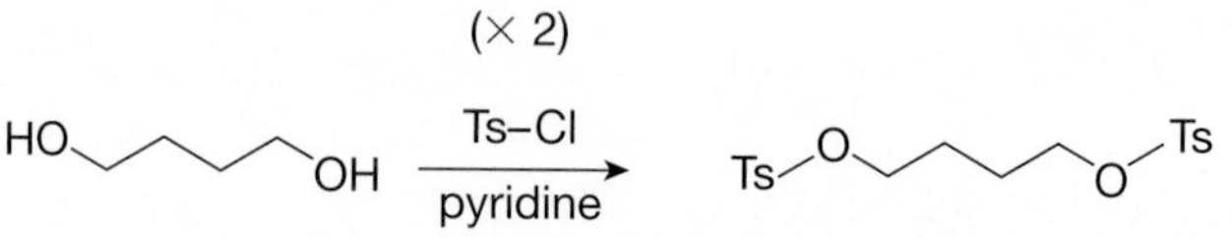

The activated product is then reacted with a primary amine. The protonated intermediates will be neutralized by the amine. Note that the amine would be added to the reaction slowly to avoid reacting both ends with different amines before the intramolecular reaction occurs with the second substitution.

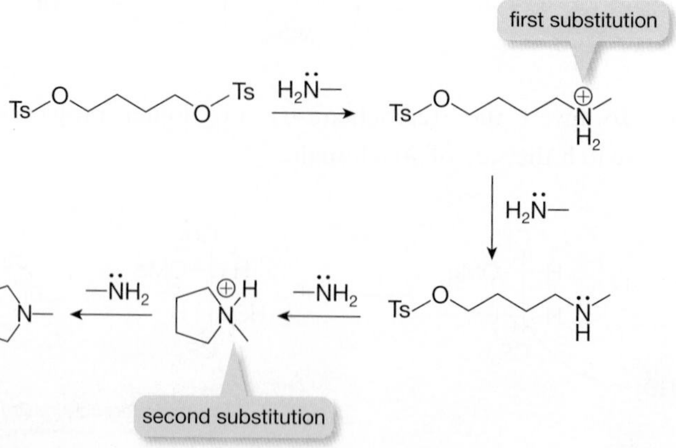

first substitution

Ts O O Ts —$H_2\ddot{N}$—→ Ts O ⊕N H_2 ; $H_2\ddot{N}$— ↓ Ts O N H ; ←—$\ddot{N}H_2$— ⊕H N ←—$\ddot{N}H_2$— N—

second substitution

c) The HBr acts as both activator (acid) and nucleophile ($Br^{\ominus}$).

HBr

OH Br Br

INTEGRATE THE SKILL 11.8

The aldehyde could be converted into an alcohol by reduction.

$NaBH_4$ ethanol

Then, the alcohol could be a nucleophile and a leaving group by first activating it with a sulfonate ester and reacting it with more alcohol.

1) Ts–Cl pyridine 2) NaH

It could also be coupled by using an acid catalyst. In this case, removing the water as it is produced would be advantageous to avoid competing reactions.

$H_2SO_{4(cat)}$ + H_2O

PRACTICE PROBLEM 11.9

a) This is a tertiary alkyl halide and so would proceed via an S_N1 mechanism.

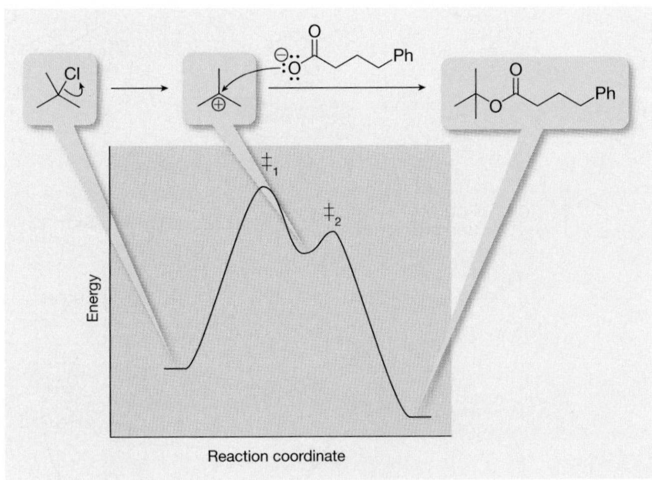

b) This is a tertiary alkyl halide and so would proceed via an S_N1 intramolecular mechanism.

HO Br K_2CO_3 O

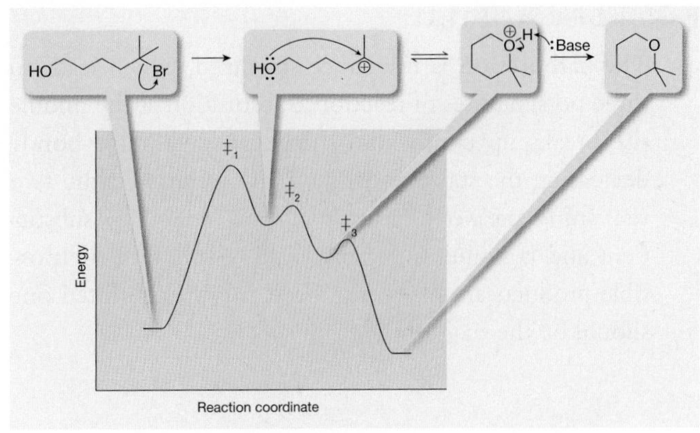

c) This is a tertiary alkyl halide and so would proceed via an S_N1 two-step mechanism.

Br CH_3 H CH_3 = Br NaHS HS

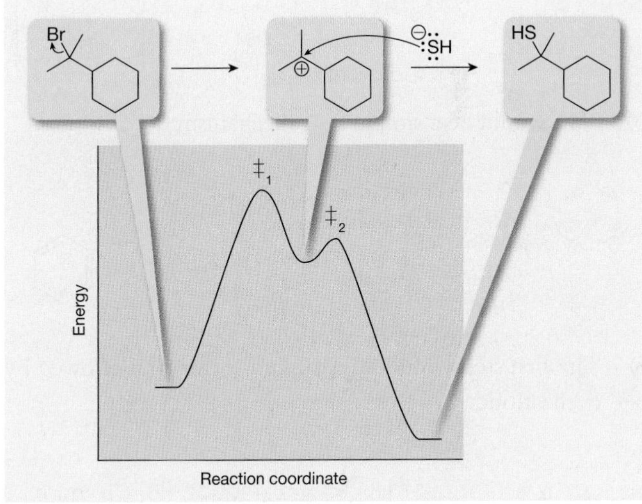

INTEGRATE THE SKILL 11.10

Inductive electron donation by the alkyl groups to the negatively charged carbon make it less stable and, therefore, a stronger base. The more substituted the carbanion, the more basic it will be. The bases would be ranked as shown below.

Li Li Li

most basic least basic

Hyperconjugation cannot occur since the carbanion does not have any unfilled orbitals into which electron density can be donated.

PRACTICE PROBLEM 11.11

a) The carbocation is resonance stabilized, which leads to three possible sites of reaction. Substitution at the middle site breaks up conjugation between the double bonds, decreasing the stability of the final product. Of the two remaining sites, one is further from the methyl substituent and is slightly less sterically hindered. All the possible products are likely produced, but the indicated one should be the major one produced.

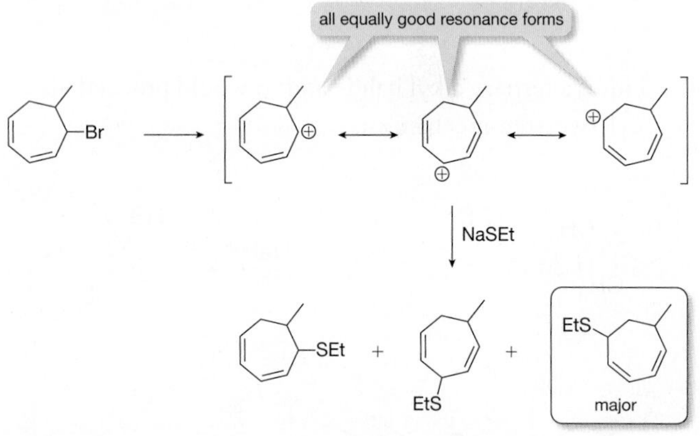

all equally good resonance forms

NaSEt

SEt + EtS + EtS major

b) This would be a simple S_N2 mechanism.

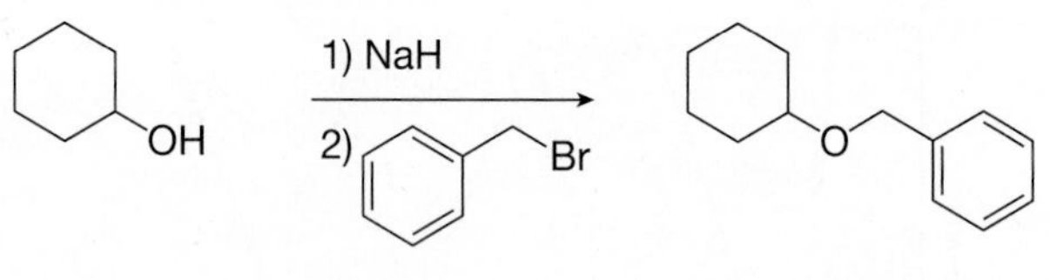

1) NaH
2) Br

c) The first step would be oxonium formation followed by cyclization.

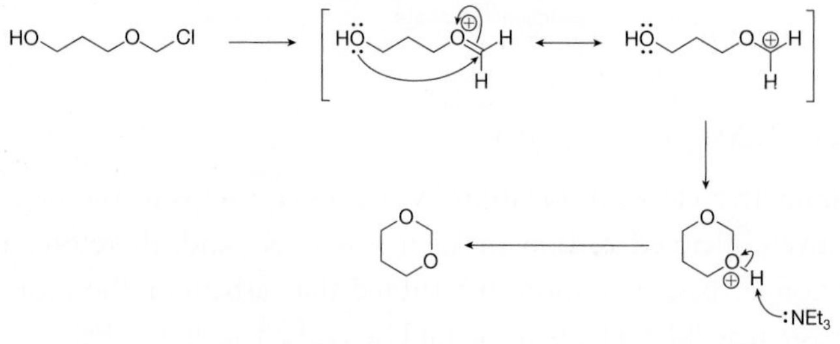

HO O Cl

:NEt$_3$

INTEGRATE THE SKILL 11.12

The molecule is an acetal. Under acidic conditions with water present, it will hydrolyse to a carbonyl group and two alcohol groups.

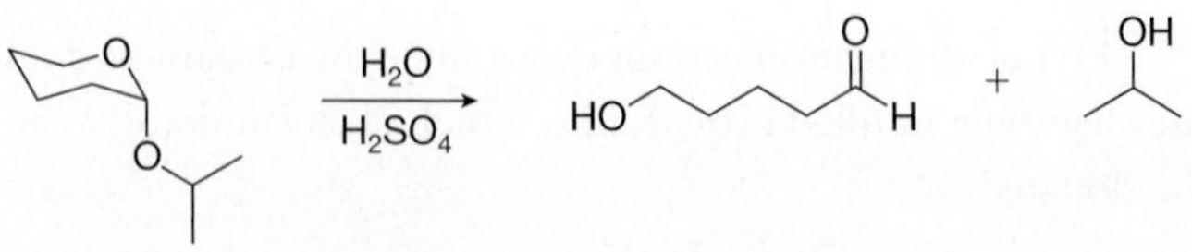

H_2O
H_2SO_4

The mechanism involves multiple activation and substitution steps, as shown below.

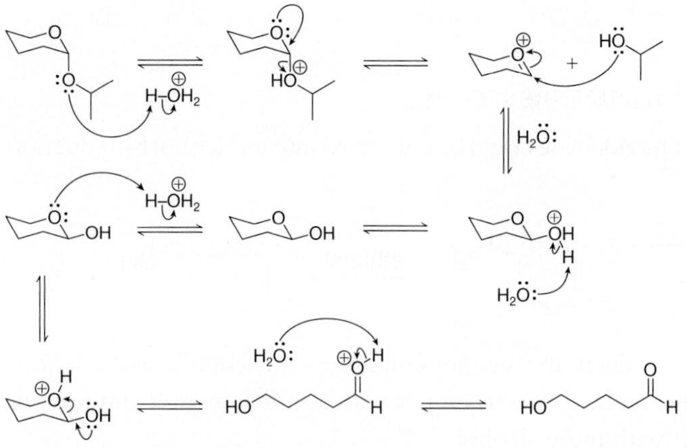

PRACTICE PROBLEM 11.13

a)

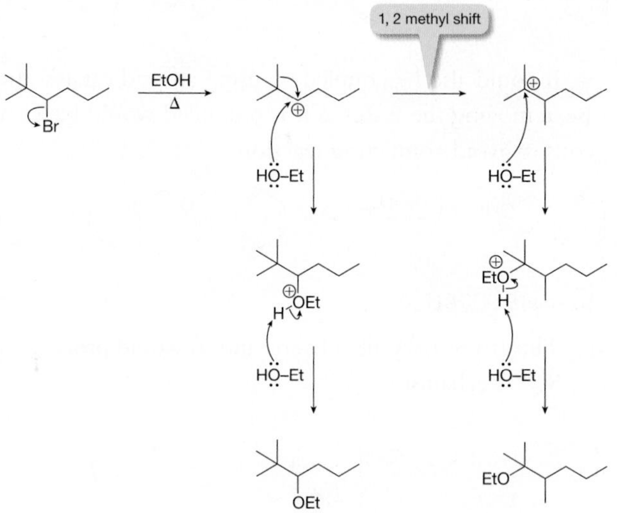

1, 2 methyl shift

EtOH
Δ

HÖ–Et

OEt

EtO

b)

1, 2 alkyl shift

CH_3OH
Δ

$CH_3\ddot{O}H$

OCH_3

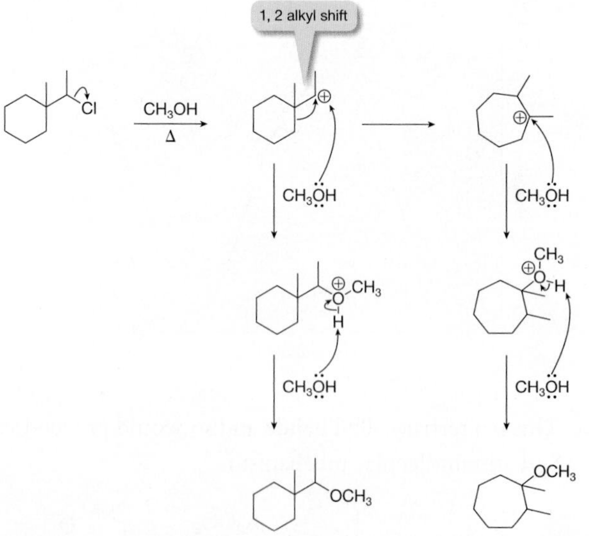

c)

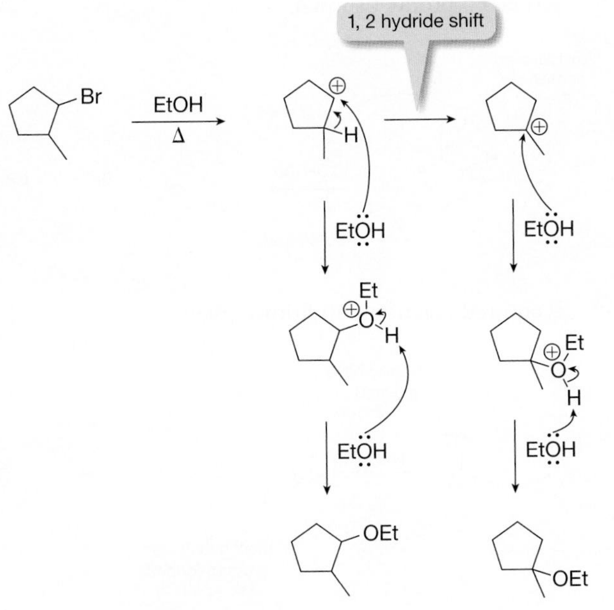

1, 2 hydride shift

EtOH

Δ

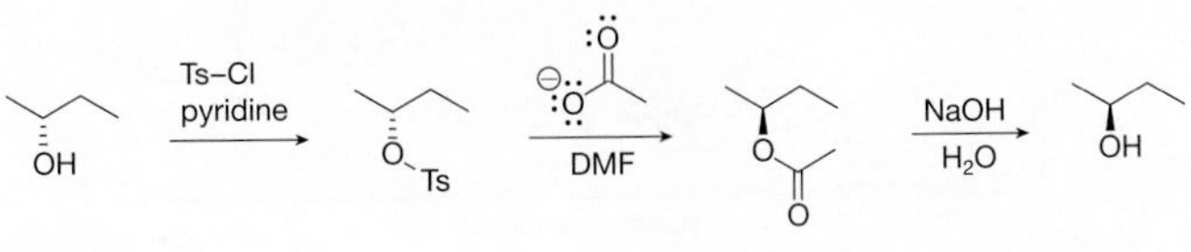

c) Two S_N2 substitutions are necessary to give a product with the same configuration as the initial alkyl bromide. Reaction with acetate in DMF followed by hydrolysis of the ester formed would give the inverted alcohol. This can then be activated by tosyl chloride. Azide addition followed by reduction to amine is done to ensure a primary amine is formed.

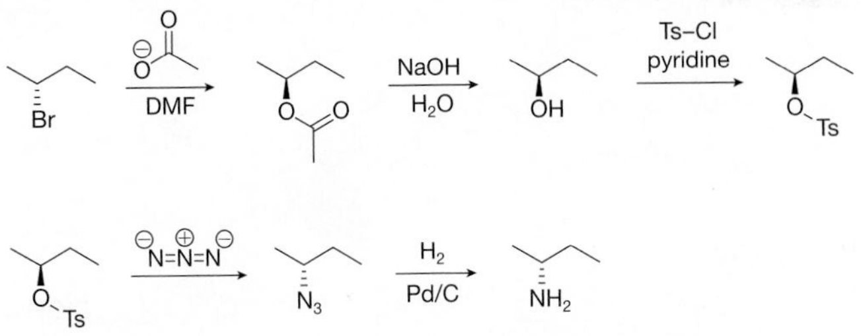

DMF NaOH H_2O Ts–Cl pyridine

H_2 Pd/C

INTEGRATE THE SKILL 11.14

Pinacol rearrangement. The acid is a concentrated solution (3–5 M, usually)

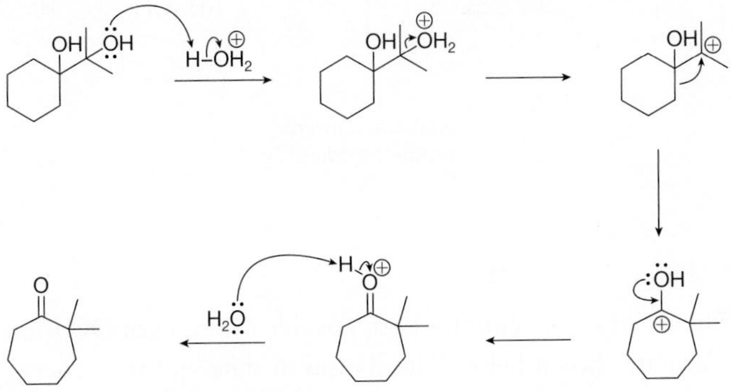

PRACTICE PROBLEM 11.15

a) Neutral reaction conditions are needed for the basic amine nucleophile. So activation of the alcohol groups in advance is necessary.

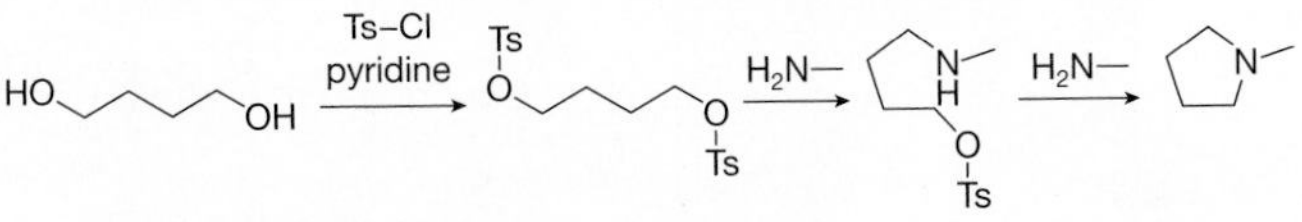

Ts–Cl pyridine

b) Acetate addition/hydrolysis is used to avoid elimination reactions possible if hydroxide was used as the nucleophile.

Ts–Cl pyridine DMF NaOH H_2O

INTEGRATE THE SKILL 11.16

The molecule is easily broken into three four-carbon fragments.

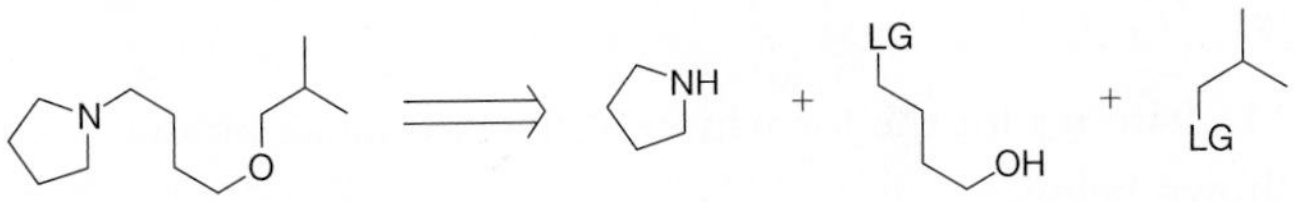

Since nitrogen is a better nucleophile than oxygen, the first reaction will be to alkylate the amine.

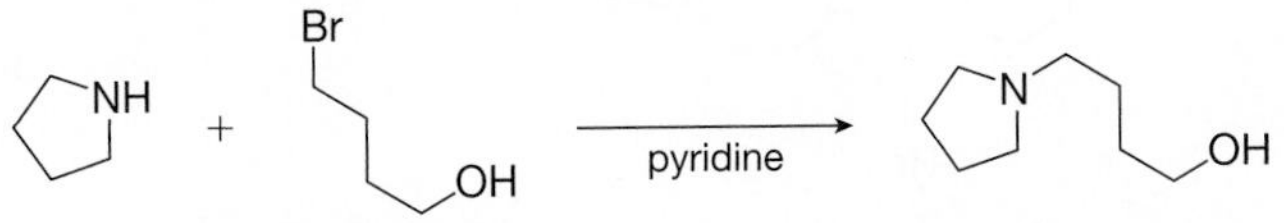

pyridine

The second reaction to complete the molecule would be O-alkylation by substitution.

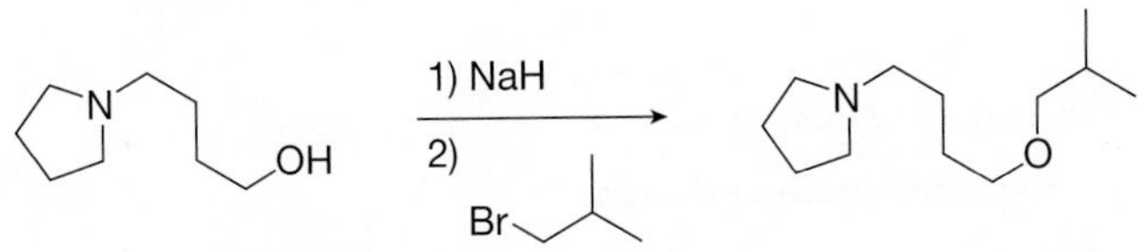

1) NaH

2)

CHAPTER 12

PRACTICE PROBLEM 12.1

The acidic hydrogens to be considered are shown as H_a and H_b.

a)

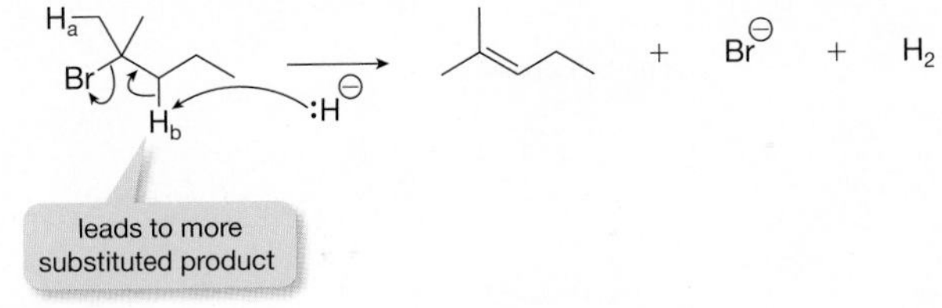

leads to more substituted product

b) OMs, H_a, H_b, :ÖEt → + $\overset{\ominus}{\text{OMs}}$ + EtOH

gives conjugated product; more stable

E isomer favoured

c) gives conjugated product; more stable

Br, H_a, H_b, $\ddot{H}^{\ominus}$ → + $Br^{\ominus}$ + H_2

d) Cl, H_a, H_b, :ÖEt → + $Cl^{\ominus}$ + EtOH

leads to more substituted product

INTEGRATE THE SKILL 12.2

Mechanisms for the formation of the two diene products are shown below.

Br, H, :base →

Br, H, :base →

PRACTICE PROBLEM 12.3

a) Expected product shown:

leads to more substituted product

H_a, Cl, H_b, $\ddot{H}^{\ominus}$ → + $Cl^{\ominus}$ + H_2

non-bulky base

b) Expected product shown:

least hindered reaction site

H_a, H_b, Br, $^{\ominus}$:Ö–*t*-Bu → + $Br^{\ominus}$ + *t*-BuOH

bulky base

c) Required reaction conditions given:

non-bulky base required

OTs, NaH → + H_2 + $\overset{\ominus}{\text{OTs}}$

most substituted product formed

d) Required reaction conditions given:

bulky base required

Br, KO*t*-Bu → + *t*-BuOH + $Br^{\ominus}$

least substituted product formed

INTEGRATE THE SKILL 12.4

The reaction diagrams for both possible reactions in Question 12.3a are shown below. The transition state and final energy for the bulky base are both higher since the molecules being formed are less stable than those formed via the other pathway.

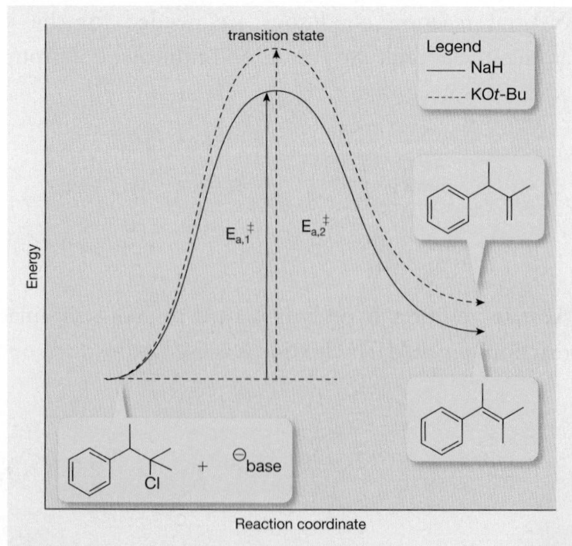

PRACTICE PROBLEM 12.5

Major products and mechanisms are shown.

a)

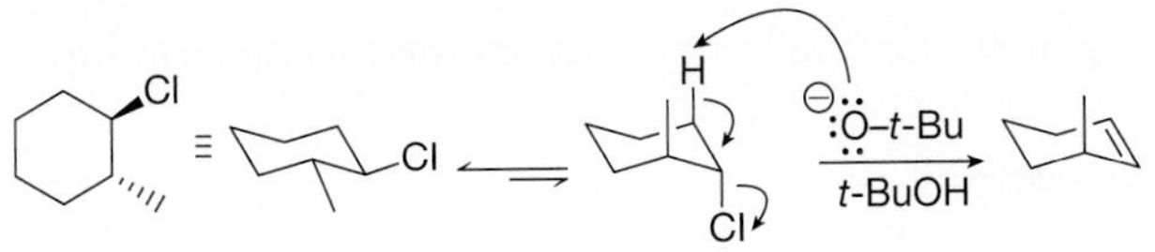

:Ö–t-Bu
t-BuOH

b)

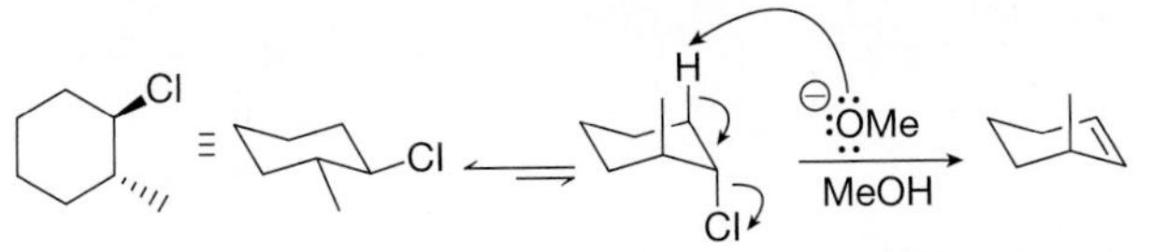

:ÖMe
MeOH

c)

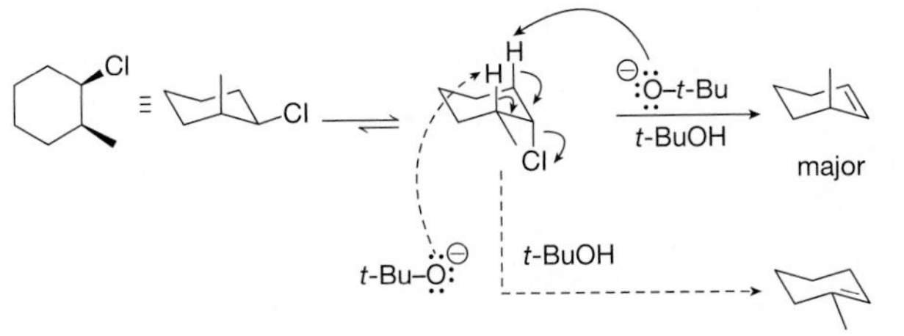

:Ö–t-Bu
t-BuOH
major
t-Bu–Ö:
t-BuOH

d)

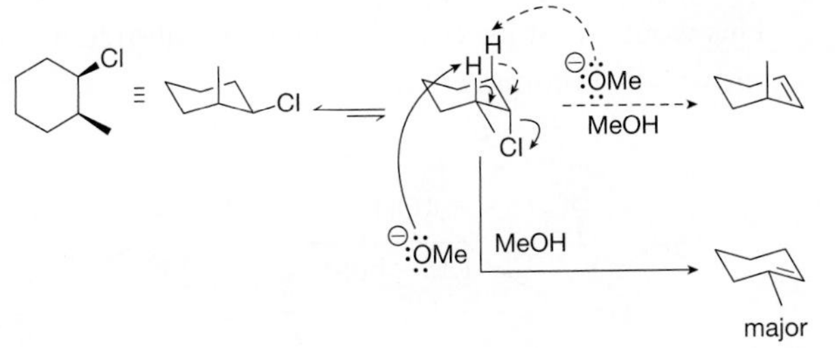

:ÖMe
MeOH
:ÖMe MeOH
major

e)

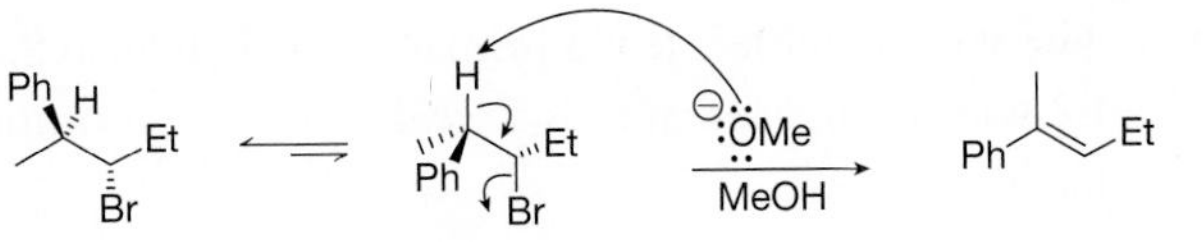

:ÖMe
MeOH

f)

:ÖMe
MeOH

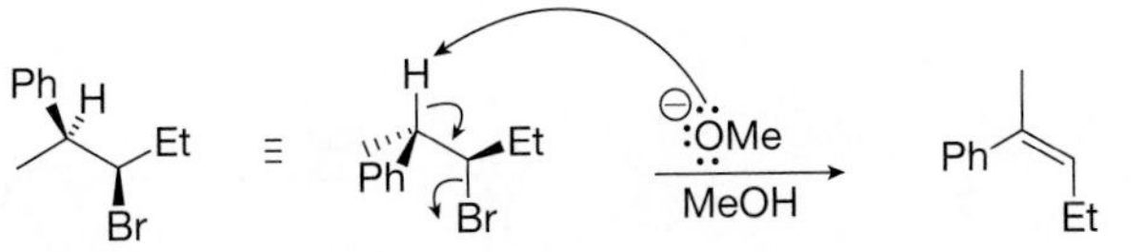

INTEGRATE THE SKILL 12.6

The three different types of hydrogens (H_a, H_b, and H_c) that could participate in an E2 reaction are shown with their corresponding E2 products.

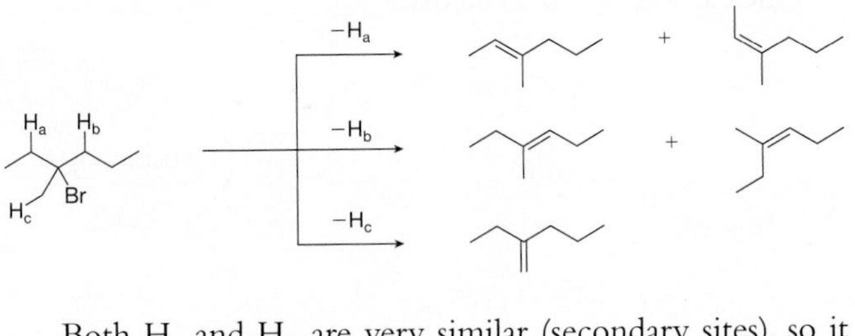

$-H_a$
$-H_b$
$-H_c$

Both H_a and H_b are very similar (secondary sites), so it would be very difficult to differentiate them with a reagent. However, H_c is primary and less sterically hindered. So, using a very large base might favour one major product.

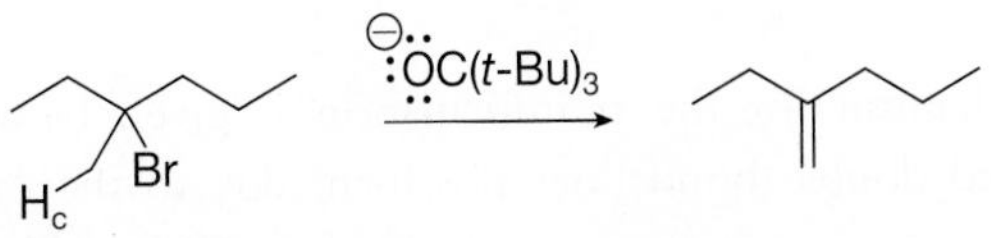

:ÖC(t-Bu)$_3$

PRACTICE PROBLEM 12.7

a) Two products are possible by removing either H_a or H_b, but the Zaitsev product from removing H_b will be favoured.

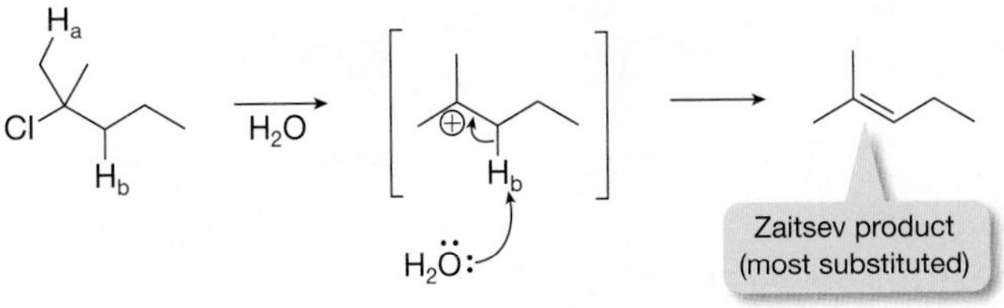

H_2O

Zaitsev product (most substituted)

b) The indicated alkene may be slightly favoured since the H_b is in a slightly less hindered position and might be more easily removed. Both alkenes are equally substituted.

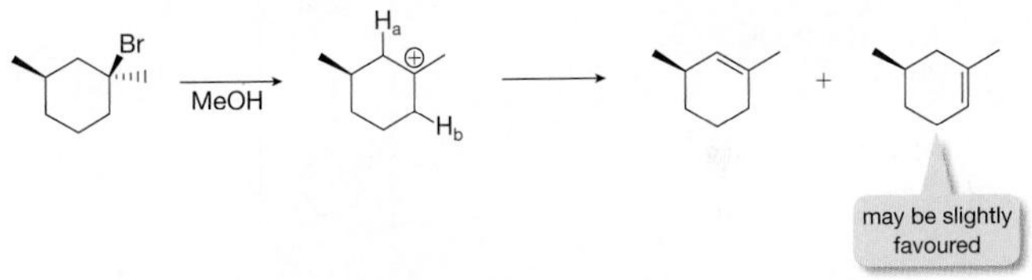

MeOH

may be slightly favoured

c) Since the solvent is large and bulky, the least hindered hydrogen will be removed preferentially.

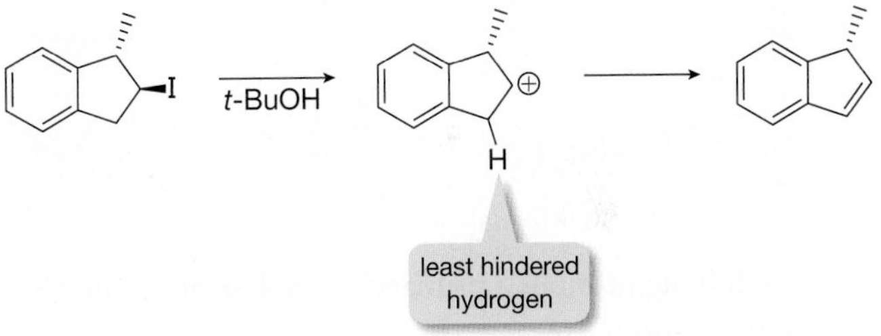

t-BuOH

least hindered hydrogen

d) Two products are possible, but the conjugated diene formed by removing H_a will be more stable than the product with two isolated double bonds formed by removing H_b. The double bond formed will be E to minimize steric interactions.

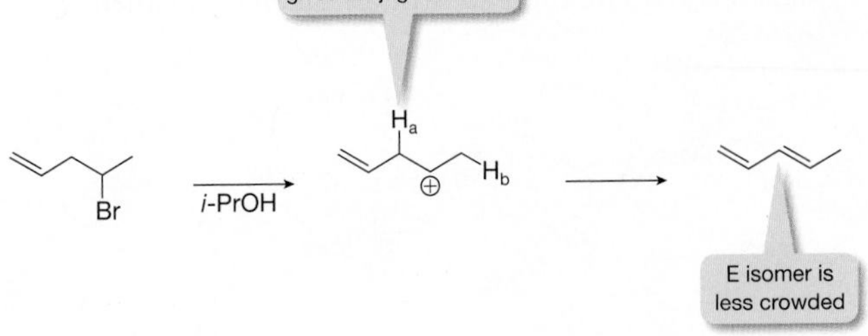

gives conjugated diene

i-PrOH

E isomer is less crowded

INTEGRATE THE SKILL 12.8

A mechanism for the transformation is given below. The expected double bond does not form due to the carbocation rearranging to a more stable 3° form. The double subsequently formed is the most substituted one.

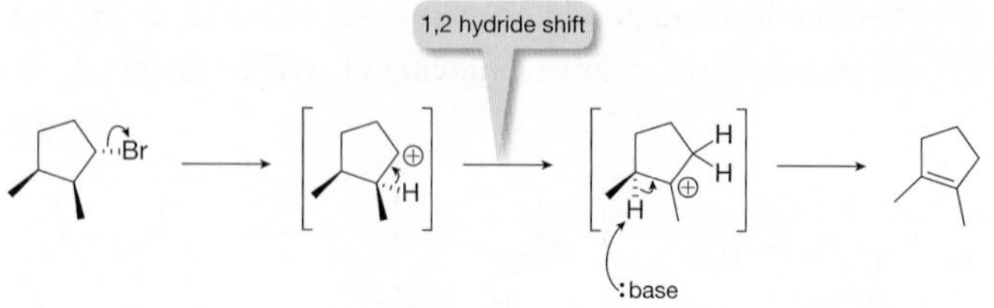

PRACTICE PROBLEM 12.9

a) A resonance-stabilized carbocation suggests this is an E1 mechanism.

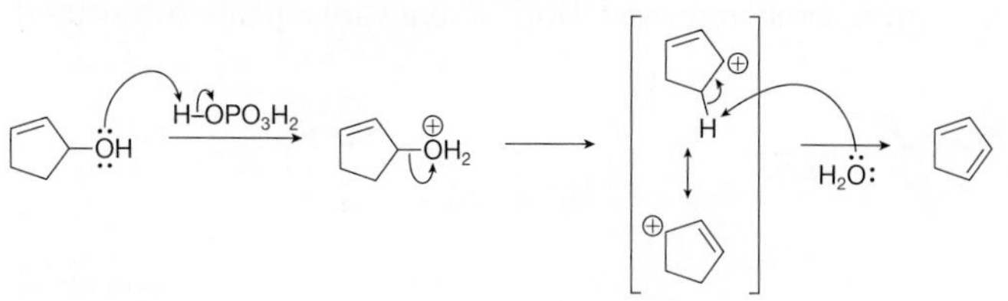

b) The aromatic –OH group is in resonance and will not eliminate. The leaving group is a primary alcohol, so the mechanism should be E2.

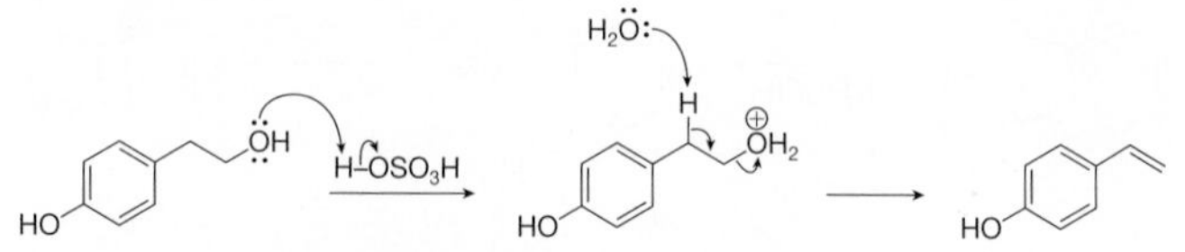

c) The leaving group is a primary alcohol, so an E2 mechanism is expected.

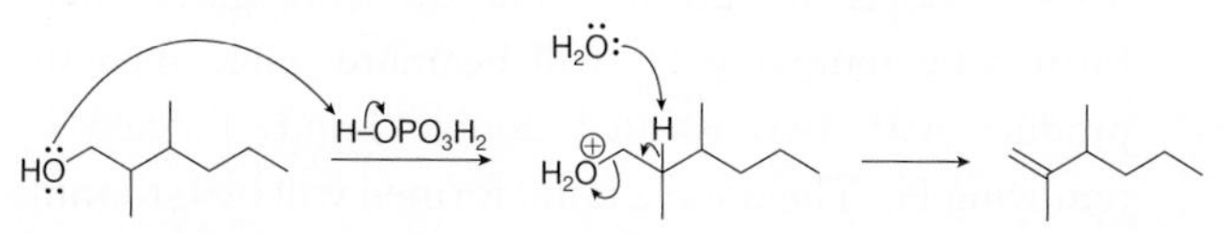

d) The leaving group is on the tertiary carbon, so the mechanism is E1. The most substituted alkene is formed.

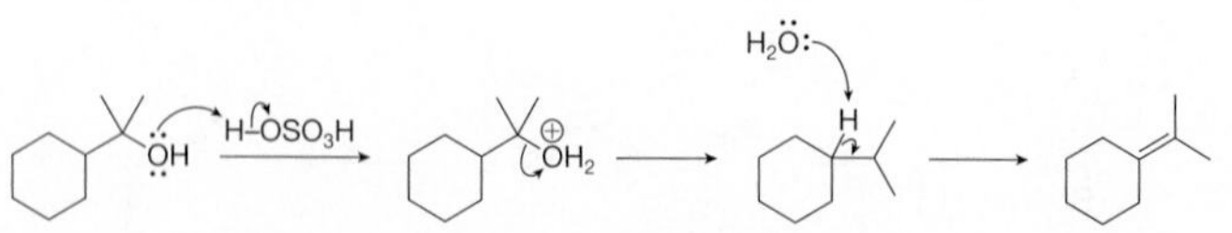

INTEGRATE THE SKILL 12.10

E1 dehydration of cyclohexanol and hydration of cyclohexene are shown below with the hydration drawn in reverse to highlight the similarities between the two mechanisms.

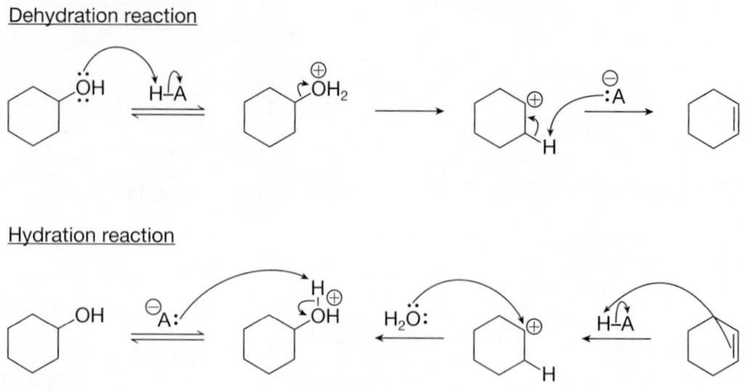

PRACTICE PROBLEM 12.11

a) A strong base and heat promote an elimination reaction. This would most likely be E2 since the substrate is a secondary alkyl bromide.

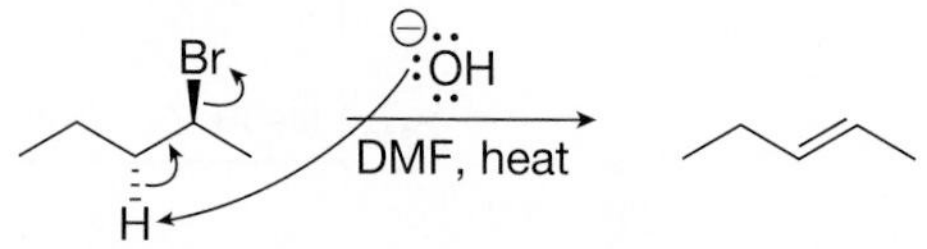

b) The starting molecule is a primary alkyl halide, so S_N2 or E2 will occur. Methanol is a weak base, so substitution is likely.

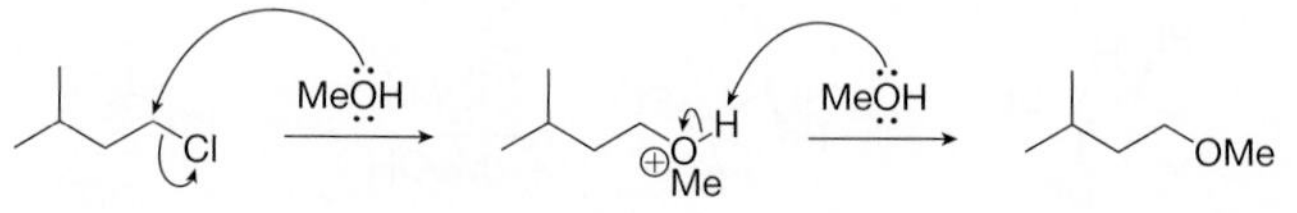

c) The best leaving group (Br) is on a tertiary carbon, so S_N1 or E1 are likely. The lack of heat suggests S_N1 will be the preferred pathway. Since the bulky solvent would be a poor nucleophile, the alcohol group must react with the carbocation intermediate.

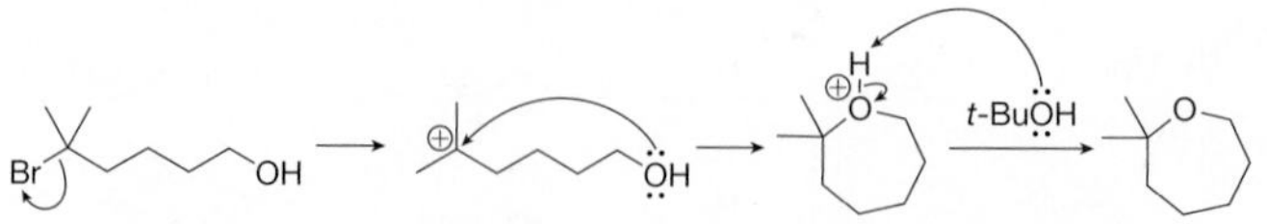

d) The base is bulky and protic. Therefore, an elimination reaction is expected. The most stable alkene that can be formed is shown. It is in conjugation with the aromatic ring.

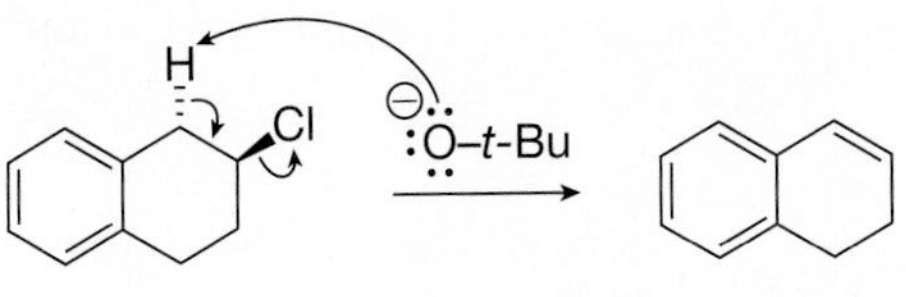

:O–t-Bu

INTEGRATE THE SKILL 12.12

a) i) Cold, aprotic conditions favour the S_N2 reaction required.

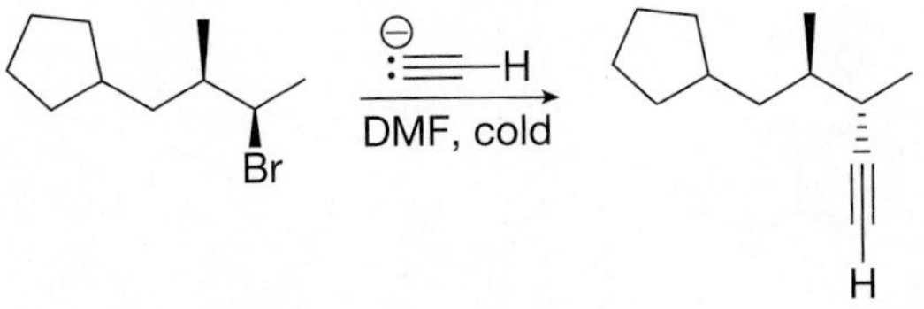

DMF, cold

Br

H

ii) Using a bulky base and high temperature will favour elimination.

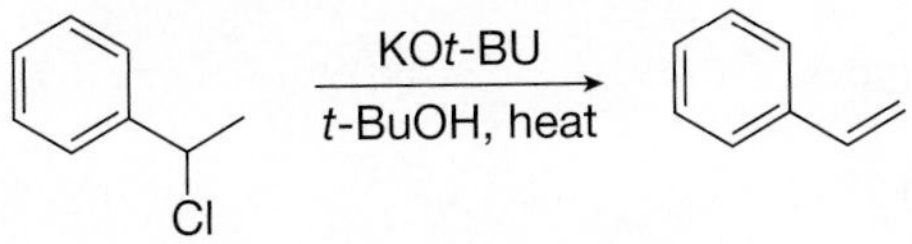

KOt-BU

t-BuOH, heat

Cl

b) i) Keeping the reaction cool will reduce the elimination reaction.

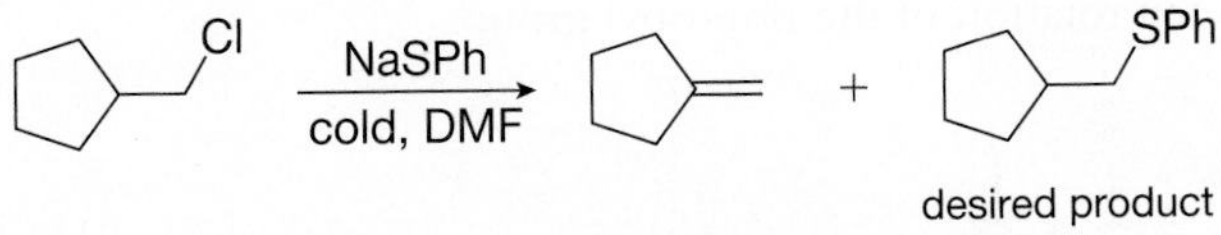

Cl NaSPh cold, DMF + SPh

desired product

ii) Substitution can be supressed by using a bulkier protic solvent. Adding a strong base, which is a weak nucleophile, will also enhance elimination.

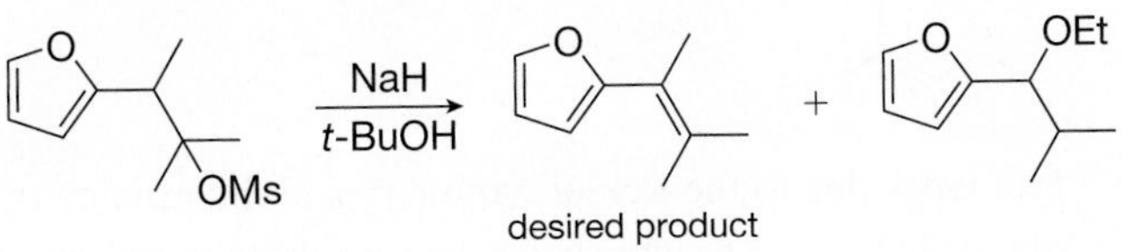

NaH t-BuOH + OEt

OMs

desired product

PRACTICE PROBLEM 12.13

a) i)

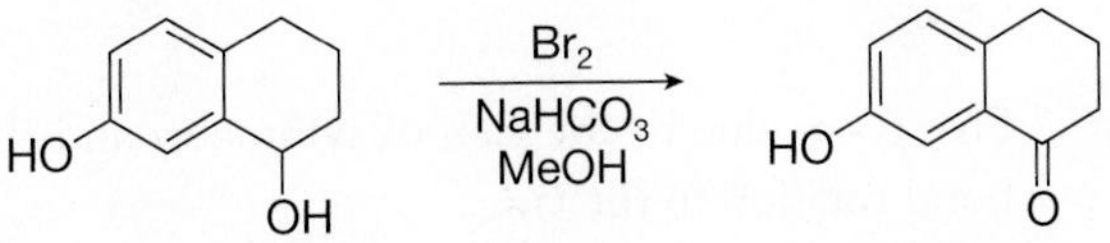

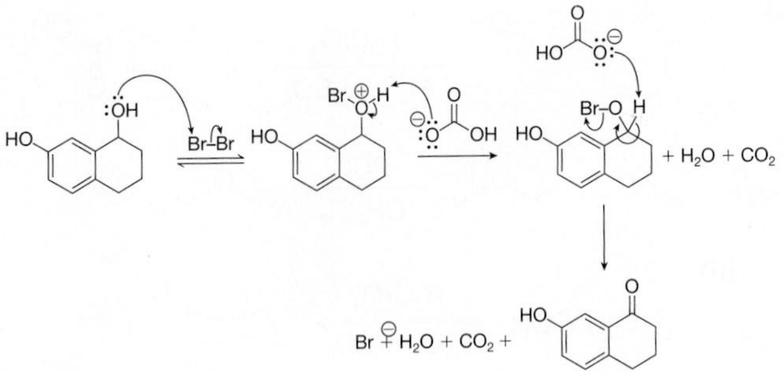

HO OH Br_2 $NaHCO_3$ MeOH HO O

HO :OH Br–Br HO Br H O OH HO Br–O H + H_2O + CO_2

Br + H_2O + CO_2 + HO O

ii) OH OH H_2CrO_4 H_2O, H_2SO_4 O OH O

OH Cr O HO: OH OH Cr OH Cr O H H_2O: H_2O HO Cr O OH H_2O: H HO Cr O H_2O:

O OH H H_2O HO Cr O O OH OH H_2O O O OH Cr O

O H OH H_2O HO Cr O O OH O

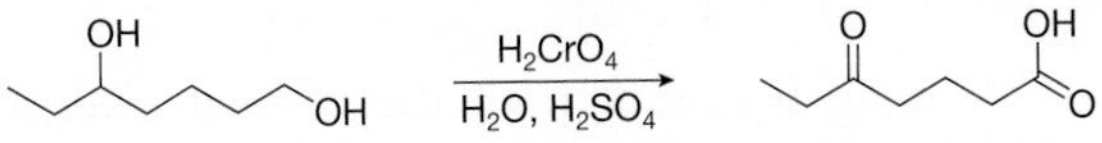

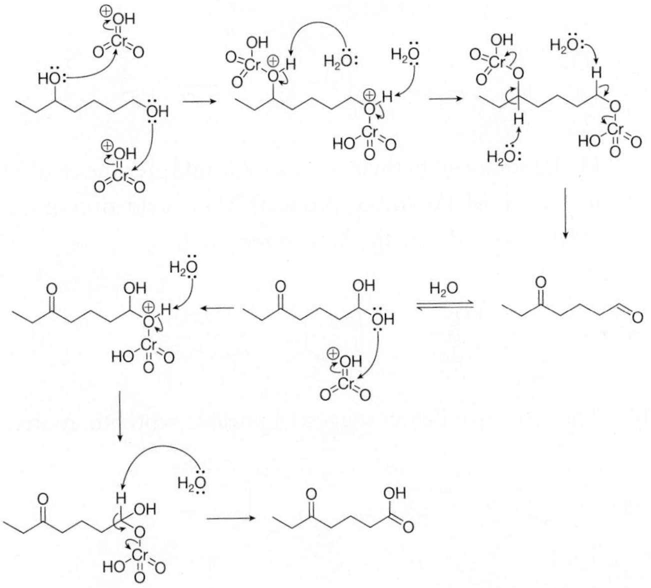

iii) O OH PCC CH_2Cl_2 O O

Cl–Cr–O O OH O H O=Cr–O Cl O O O O=Cr–OH Cl O

O O O H H_2O: O O=Cr–OH O

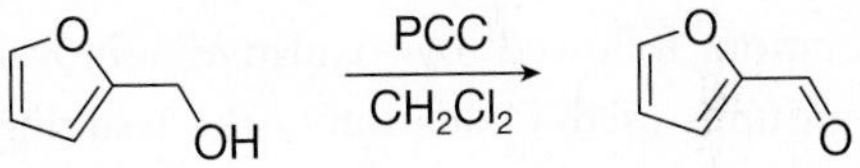

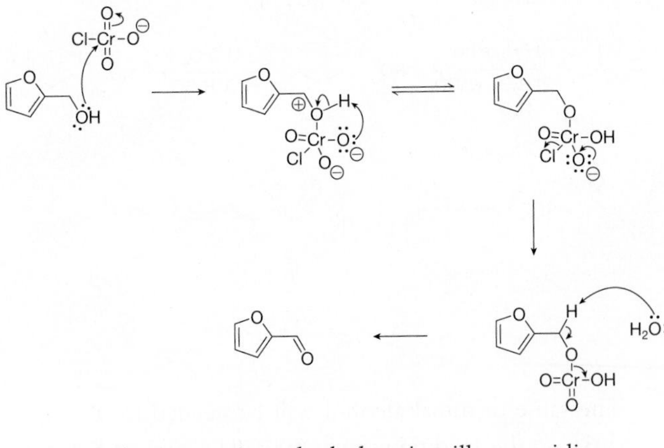

iv) This is a tertiary alcohol, so it will not oxidize.

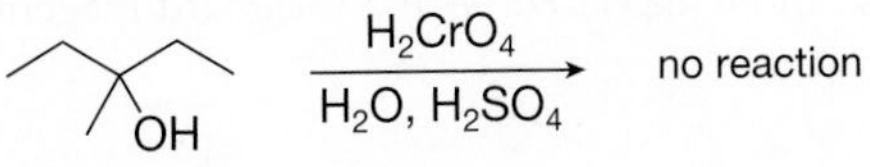

OH H_2CrO_4 H_2O, H_2SO_4 no reaction

b) i) [reaction scheme: H_2CrO_4 / H_2O, H_2SO_4]

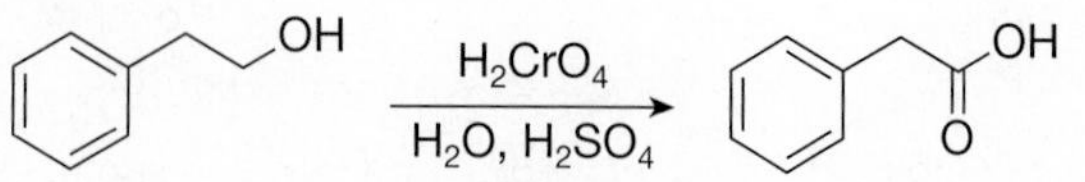

ii) [reaction scheme: PCC / CH_2Cl_2]

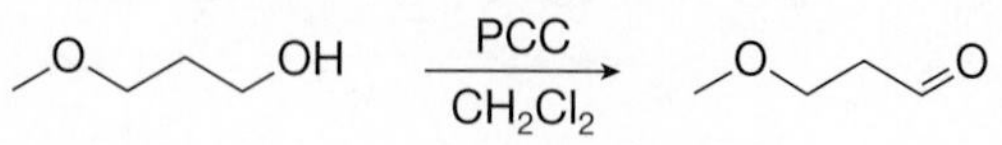

iii) [reaction scheme: H_2CrO_4 / H_2O, H_2SO_4]

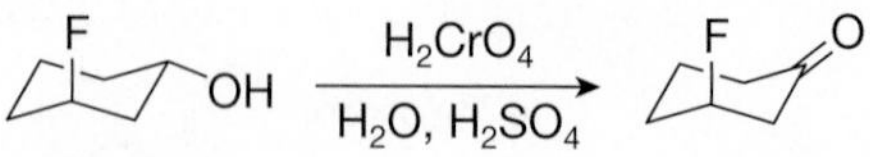

INTEGRATE THE SKILL 12.14

a) Working backward from the product gives a possible pathway.

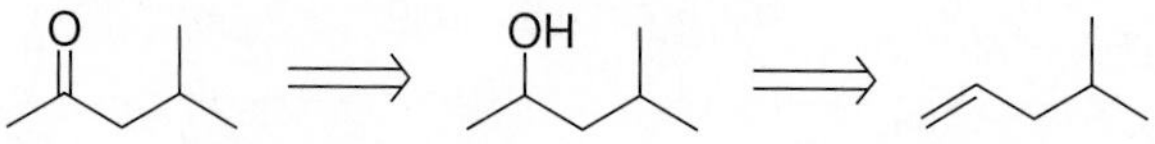

Hydration of 4-methylpent-1-ene would give the alcohol indicated (Markovnikov product). Then, oxidation of the alcohol would give the ketone required.

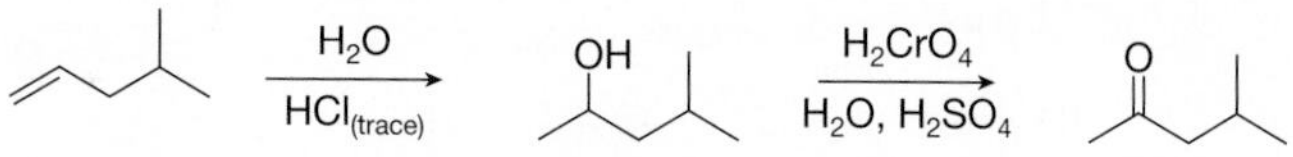

[reaction scheme: H_2O / $HCl_{(trace)}$; H_2CrO_4 / H_2O, H_2SO_4]

b) The sequence below suggests a possible synthetic route.

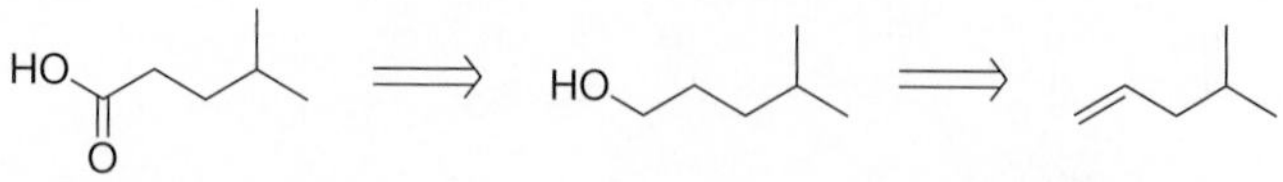

The first hydration needs to be anti-Markovnikov, so hydroboration followed by oxidative removal of the borane group is used. Oxidation of the resulting alcohol gives the desired product.

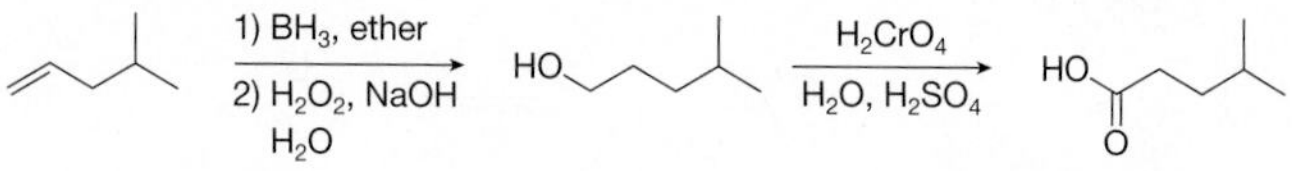

[reaction scheme: 1) BH_3, ether; 2) H_2O_2, NaOH, H_2O; H_2CrO_4 / H_2O, H_2SO_4]

c) [retrosynthesis scheme]

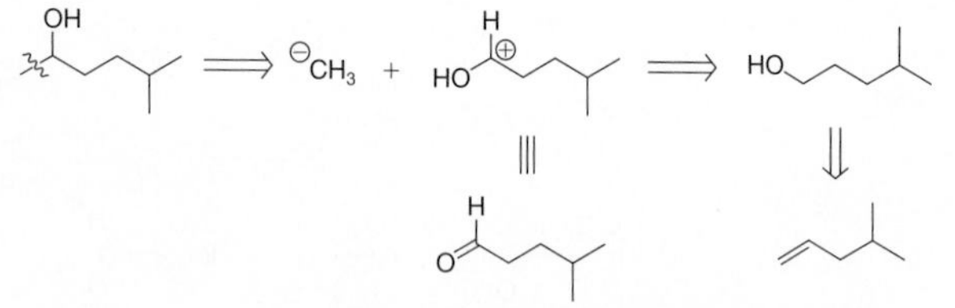

The same terminal alcohol will be needed for this synthesis, so the method used in (b) will be applied. Selective oxidation of the alcohol with PCC gives the aldehyde, which can be alkylated with a Grignard reagent.

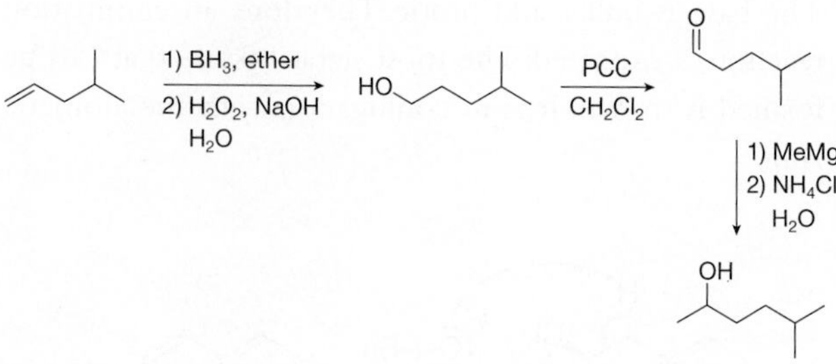

[reaction scheme: 1) BH_3, ether; 2) H_2O_2, NaOH, H_2O; PCC / CH_2Cl_2; 1) MeMgBr; 2) NH_4Cl, H_2O]

CHAPTER 13

PRACTICE PROBLEM 13.1

Each type of equivalent hydrogen giving a signal in the ^{1}H-NMR spectrum is indicated by letters in the diagrams. For implied hydrogen atoms, the carbon atom they are attached to is labelled.

a)

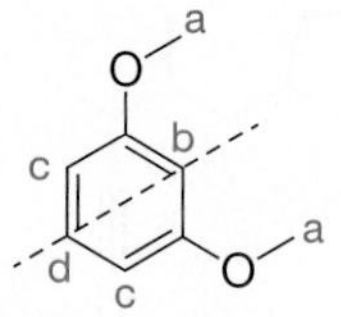

Four types, due to the symmetry of the molecule.

b)

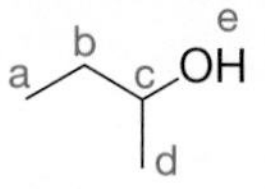

Five types.

c)

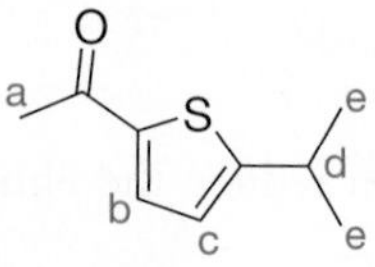

Five types, due to the interchange of the methyl groups by rotation of the isopropyl group.

d)

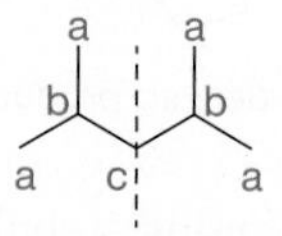

Three types, due to the symmetry of the molecule and the interchange of the methyl groups by rotation.

e)

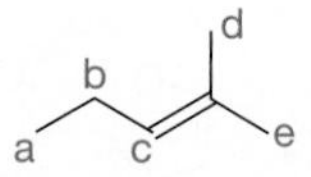

Five types, due to the lack of symmetry and the inability of the methyl groups to interchange by rotation of the alkene.

f)

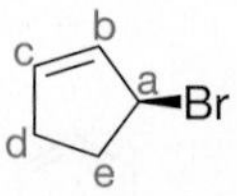

Five types, due to the lack of symmetry and the lack of bond rotation in the ring.

INTEGRATE THE SKILL 13.2

The spectrum shows four signals, corresponding to four hydrogen types. Compound **C** is the only structure that has four hydrogen types.

A **B** **C** **D**

PRACTICE PROBLEM 13.3

The attached carbon atom is labelled for implied hydrogen atoms.

a) Five types, a:b:c:d:e = 6:1:2:1:3.

b) Four types, a:b:c:d = 1:1:1:9.

c) Four types, a:b:c:d = 1:2:2:3.

d) Three types, a:b:c = 2:2:1.

e) Five types, a:b:c:d:e:f = 6:1:1:3:2:3.

f) The bromines would prefer an equatorial conformation. This causes most of the protons to be inequivalent. There are seven types: a, b, c, d, e, f, g in a 1:1:2:2:2:1:1 ratio.

Four types, a:b:c:d = 1:2:1:1.

INTEGRATE THE SKILL 13.4

Both 2-bromopentane and 3-bromopentane have a signal that integrates for one hydrogen. The 1-bromopentane isomer will not have this signal.

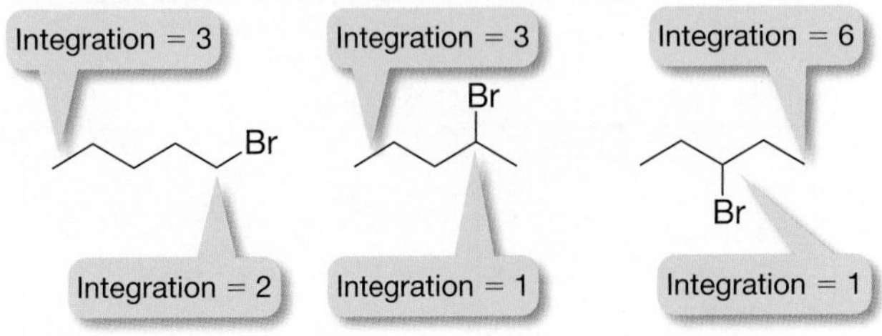

The 2-bromopentane isomer has two different methyl groups that will each integrate for three hydrogens. The methyl groups of 3-bromopentane are identical, and will be represented by one signal that integrates for six hydrogens.

PRACTICE PROBLEM 13.5

a) The locations of the most deshielded protons are indicated by blue arrows.

i)

ii)

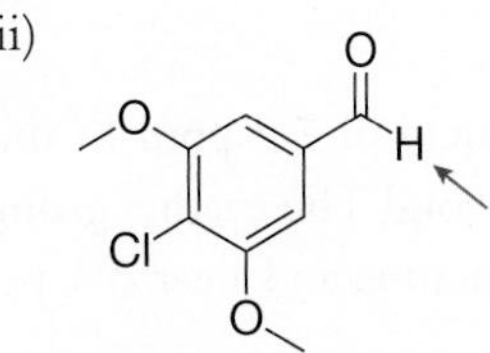

iii)

b) More-shielded methyl groups are shown in bold.

i) $\mathbf{CH_3OCH_3}$ or CH_3SCH_3

ii) $\mathbf{CH_3}$F or CH_3Cl

iii) $\mathbf{CH_3}$C($\mathbf{CH_3}$)$_3$ or $CH_3Si(CH_3)_3$

INTEGRATE THE SKILL 13.6

Structures are labelled with letters for corresponding ^{1}H-NMR signals.

a)

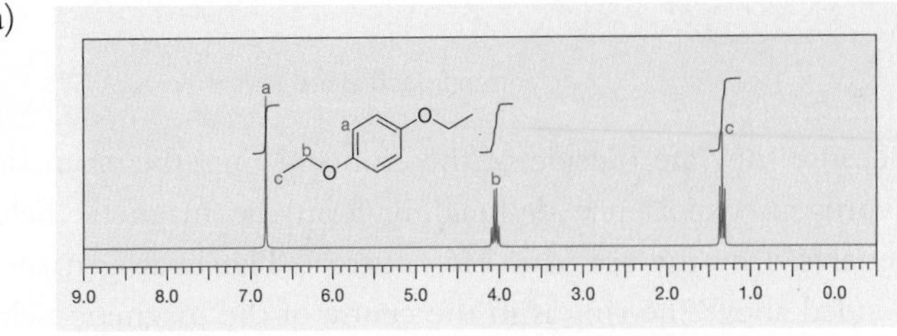

b)

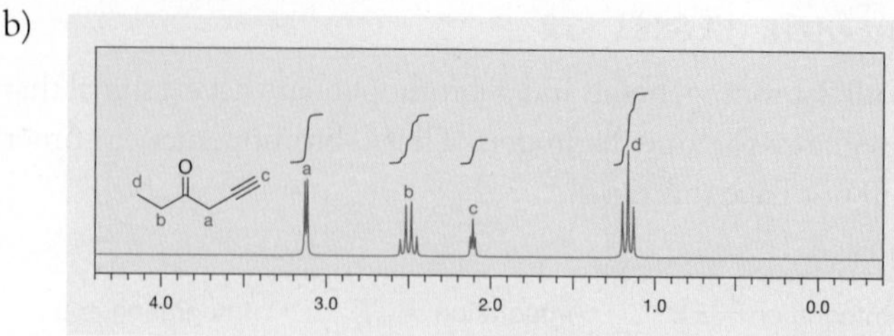

c)

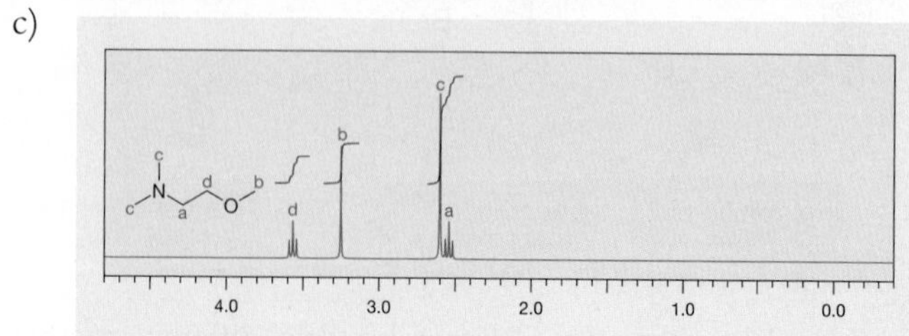

PRACTICE PROBLEM 13.7

a) The methoxy group is electron-donating to the β carbon by resonance, while the nitrile is electron-withdrawing. Both substituents are electron-withdrawing by induction from the α carbon. Because oxygen is so electronegative, the methoxy is more electron-withdrawing than the nitrile.

b) The sulfur is electron-donating to the β carbon by resonance, while the ketone is electron-withdrawing.

c) The amino group is electron-donating to the *ortho* carbon by resonance, while the nitro group is electron-withdrawing.

INTEGRATE THE SKILL 13.8

The aromatic signals will be upfield of 7.2 ppm in the first structure, but downfield in the second. The methyl group will be near 2.1 ppm in the first structure, and near 3.4 ppm in the second.

PRACTICE PROBLEM 13.9

a)

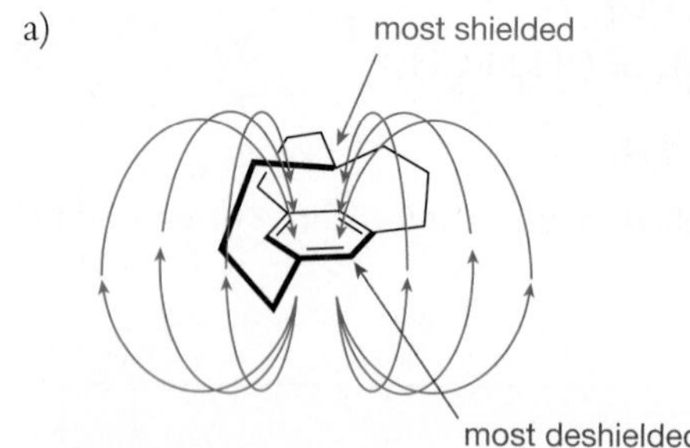

Because they are outside of the aromatic ring, the aromatic hydrogens experience deshielding from the magnetic field generated by the aromatic ring current. The hydrogen suspended above the ring is in the centre of the magnetic field and experiences a shielding effect.

b)

H −1.9 ppm
H 8.2 ppm

[18]-annulene has $4n + 2$ p electrons and shows some aromatic character. The downfield signal represents hydrogens outside of the ring. The hydrogens in the centre of the ring experience shielding from the ring current.

c)

A B C D E

Compounds **A**, **C**, and **D** all have hydrogens that will be involved in hydrogen bonds (shown in blue). Doubling the concentration will increase the hydrogen bonding and decrease the electron density on those hydrogens. This will move the signals downfield at higher concentrations.

d)

$\xrightarrow{D_2O}$

In D_2O, the N–H and O–H groups exchange H for D, which does not appear in an ^{1}H-NMR spectrum. This will leave glutamine with only three signals.

INTEGRATE THE SKILL 13.10

Acetylacetone is in equilibrium with an enol form. The signals that disappear are the protons (deuterons in D_2O), which take part in the enol-keto tautomerization process.

δ 15.5
δ 3.61
δ 5.52

Deuterium exchange

The signals from the hydroxyl (δ 15.5) and the alpha hydrogen (δ 5.52) of the enol form and the methylene (δ 3.61) of the keto form will disappear in D_2O.

PRACTICE PROBLEM 13.11

a) Multiplets and relative intensities

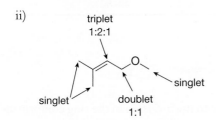

i) heptet 1:6:15:20:15:6:1; doublet, 1:1

ii) triplet 1:2:1; singlet; doublet 1:1; singlet

iii) doublet 1:1; singlet; triplet 1:2:1; quartet 1:3:3:1; doublet 1:1

iv) singlet; singlet; singlet; doublet 1:1; triplet 1:2:1

b) i) Compound **C**. This spectrum has one singlet near 2.1 ppm, a position typical of methyl hydrogens adjacent to a carbonyl. Compound **C** is the only structure that has such a methyl group.

ii) Compound **A**. With two downfield singlets, this spectrum represents a compound with two types of hydrogen adjacent to an oxygen atom, but with no adjacent hydrogens. This describes compound **A**.

iii) Compound **B**. This is the only spectrum without any singlet signals. Compound **B** is the only structure without any methyl groups with no adjacent hydrogens.

INTEGRATE THE SKILL 13.12

a) The most downfield signal in each spectrum will be the hydrogens adjacent to the chlorine. Whether the integration is 1 or 2, relative to the other signals, will distinguish the two compounds.

b) The most downfield alkyl hydrogens will be adjacent to the oxygen. Whether each is a quartet or a singlet will distinguish the two compounds.

c) Due to symmetry differences, one compound will have two aromatic hydrogen types while the other will have three types.

PRACTICE PROBLEM 13.13

a) $C_{10}H_{12}O_2$

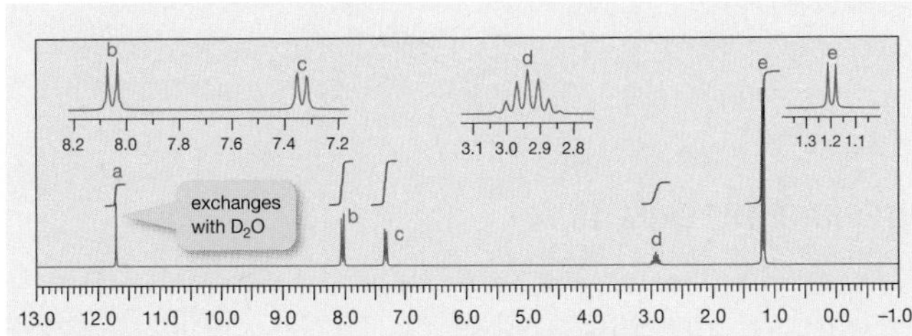

b) $C_6H_4Br_2$

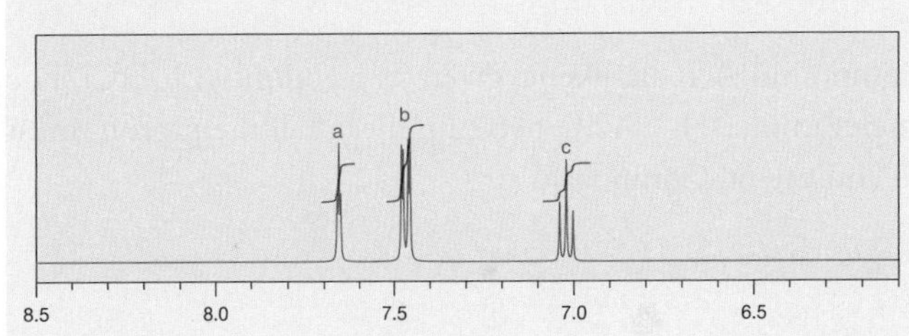

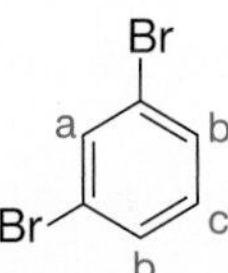

c) $C_7H_{14}O$

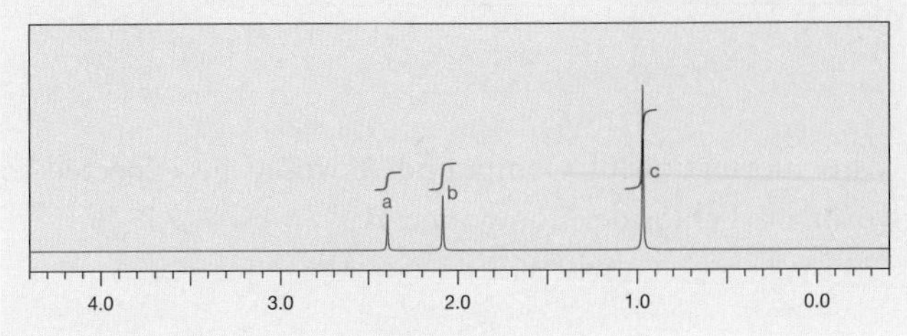

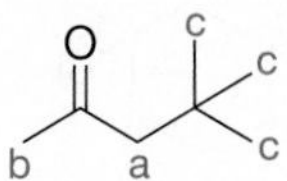

d) C_4H_7N

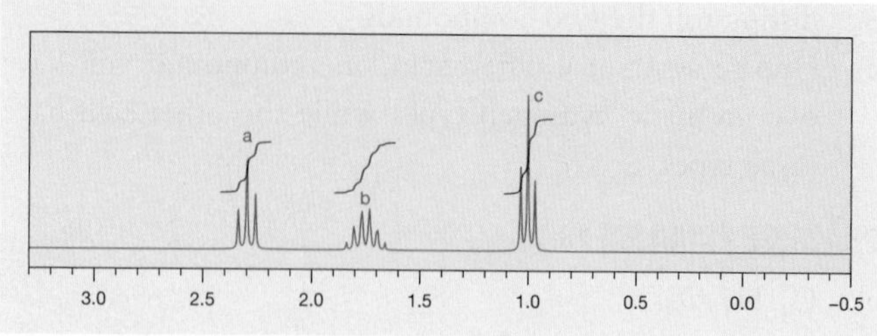

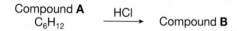

INTEGRATE THE SKILL 13.14

Compound **A** C_6H_{12} $\xrightarrow{HCl}$ Compound **B**

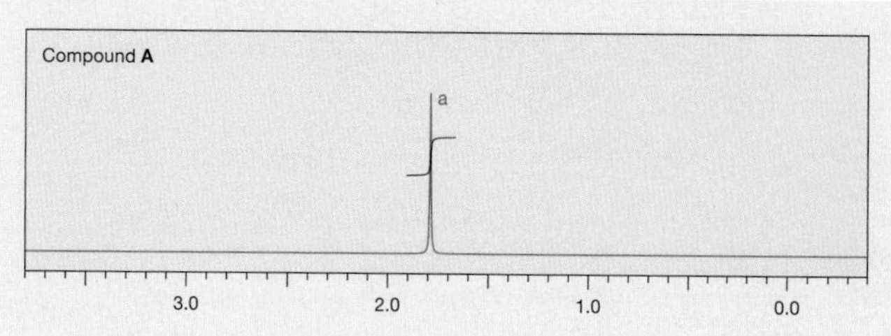

Compound **A** is an alkene, given its reaction with HCl. The singlet in the ^{1}H-NMR spectrum means all the protons must be equivalent. Compound **A**:

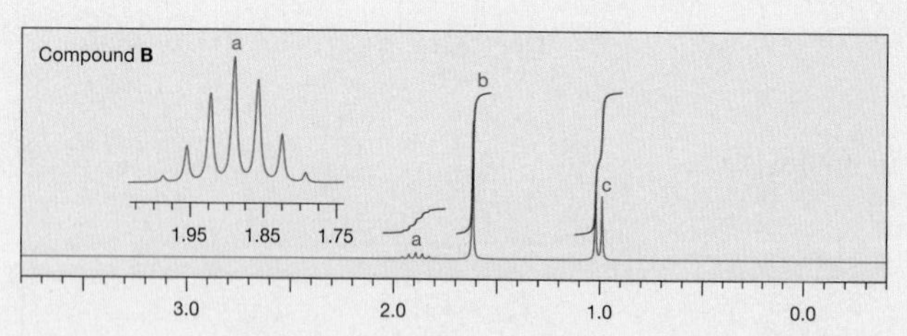

Hydrochlorination of Compound **A** would be expected to give an alkyl chloride. Compound **B**:

PRACTICE PROBLEM 13.15

a)

Four types

Carbon atom	Expected chemical shift range (ppm)
a, b	100–150
c	155–185
d	40–80

b)

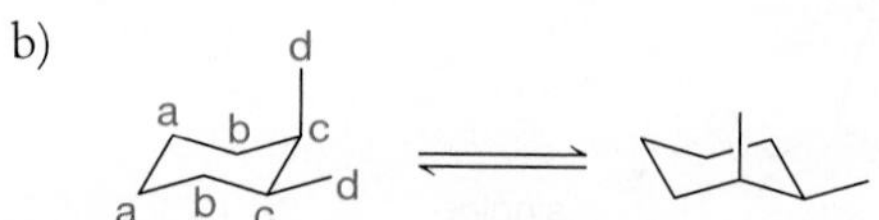

Four types, due to the inversion of the ring conformation.

Carbon atom	Expected chemical shift range (ppm)
a, b	15–55
c	20–60
d	8–30

c)

Five types

Carbon atom	Expected chemical shift range (ppm)
a	8–30
b, c	100–150
d	15–55
e	40–80

d)

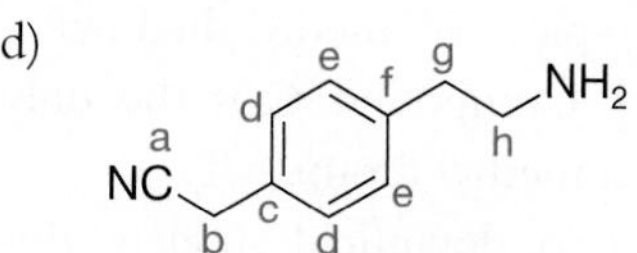

Eight types, due to the symmetry of the aromatic ring.

Carbon atom	Expected chemical shift range (ppm)
a	65–90
b, g, h	15–55
c, d, e, f	100–175

e)

Five types, due to the interconversion of methyl groups by bond rotation.

Carbon atom	Expected chemical shift range (ppm)
a, e	8–30
b, c	40–80
d	15–55

INTEGRATE THE SKILL 13.16

$C_6H_{13}Cl$

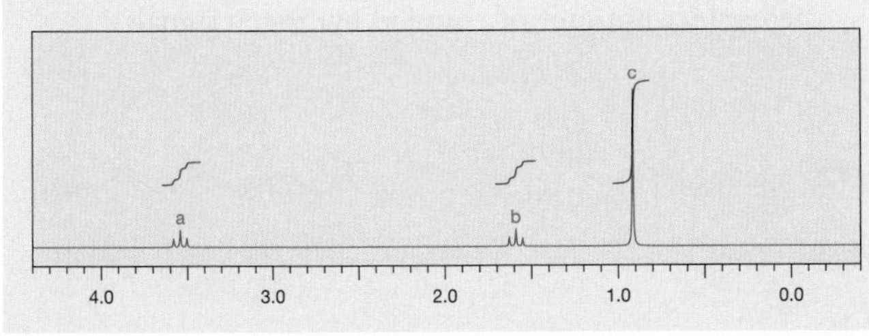

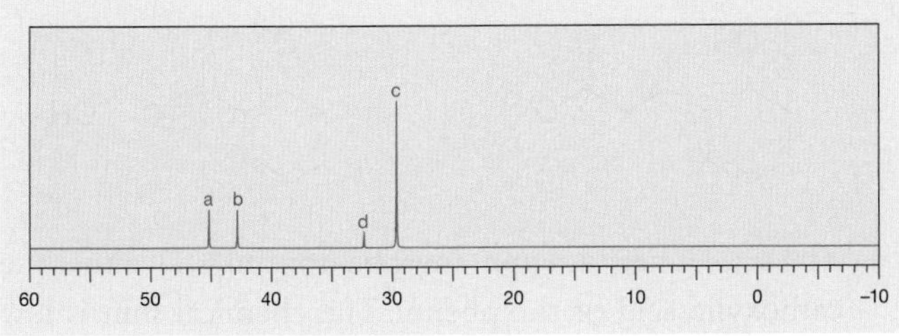

PRACTICE PROBLEM 13.17

Carbon atoms that will give a signal are labelled (+) and (–). Unlabelled carbon atoms will be missing in the DEPT-135 spectra.

a)

Four positive signals, one negative signal, and four signals missing.

b)

Three positive signals, four negative signals, and one signal missing.

c)

Five positive signals, two negative signals, and one signal missing.

d)

Eight positive signals, one negative signal, and two signals missing.

INTEGRATE THE SKILL 13.18

The ^{13}C -NMR spectrum for vinyl methacrylate is shown below. DEPT-135 NMR: 19.1 ppm (+), 144.8 ppm (+), 126.3 ppm (–), 97.4 ppm (–).

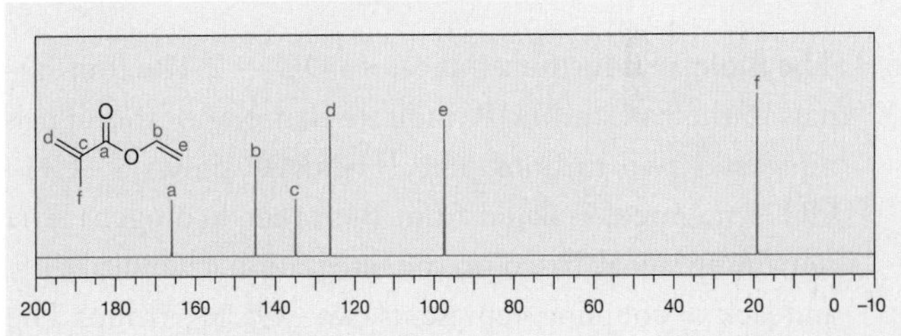

PRACTICE PROBLEM 13.19

a) The molecular formula indicates DU = 3. The carbon NMR shows the presence of one ketone and one alkene, suggesting the presence of a ring for the third unsaturation. The hydrogen NMR indicates –CH=CH– and $-CH_2-CH_2-$ fragments and two equivalent CH_3 groups. There is one remaining carbon with no hydrogens.

Assembling these fragments so that the –CH=CH– and $-CH_2-CH_2-$ pieces are not adjacent provides only one structure.

b) The molecular formula indicates DU = 0. The carbon NMR has four signals, indicating that there is no symmetry. This must be a straight chain butane with two bromines attached. The ^{1}H-NMR has signals with integration of 1, 2, 2, and 3. The two possible structures consistent with these integrations are:

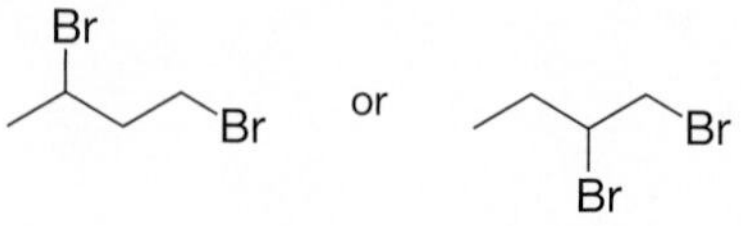

Br Br or Br Br

Analyzing the coupling differentiates the structures, however. The methyl group is a doublet, indicating that the bromine must be attached to the carbon adjacent to methyl. The CH signal is a sextet, indicating that it is between a CH_3 and a CH_2. One structure is consistent with this information.

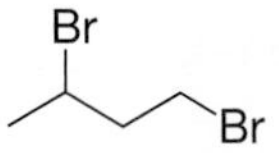

Br Br

c) The molecular formula indicates a DU = 2. The four signals in the carbon NMR indicate that one of the signals represents two carbons. The ^{1}H-NMR shows a –CH–CH_2– fragment isolated from the other hydrogens. The remaining signal, based on the integration, chemical shift, and lack of coupling, represents two –OCH_3 groups. The remaining atoms are one C and one N, which can combine to be a nitrile –C≡N. This is consistent with the DU of 2.

The nitrile can attach to the CH_2 side of the –CH–CH_2– fragment and the equivalent methoxy groups can attach to the CH side to give the following compound. The unusual downfield shift of the alkyl –CH–CH_2– fragment is the result of having three electron-withdrawing groups.

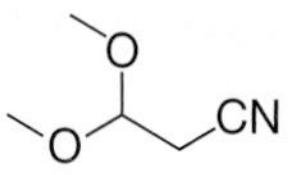

O O CN

d) The molecular formula indicates DU = 6. A DU greater than 4 suggests an aromatic ring, which is consistent with the chemicals shifts in both the ^{1}H-NMR and ^{13}C-NMR. There are nine carbon signals above 100 ppm, suggesting that the aromatic ring is not symmetrical and that there is an alkene and a carbonyl. The carbon signal near 50 ppm and the hydrogen signal near 3.8 indicate a methyl group attached to either an aryl ether or an ester.

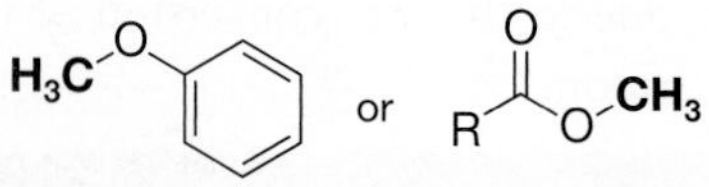

H_3C O or O R O CH_3

The carbonyl that must be present is below 185 ppm, indicating that it is a carboxylic acid or an ester, not an aldehyde or ketone.

The remaining signals have seven hydrogens and eight aromatic or alkene carbons. The molecular formula has three oxygens, two of which are part of the ester or acid. There are four possible ways to assemble the fragments that result in an asymmetrical aromatic ring and the correct number of coupled hydrogen signals.

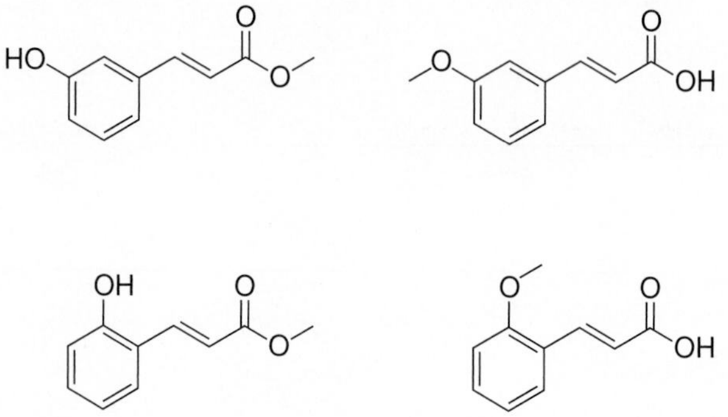

HO O O O O OH

OH O O O O OH

The singlet near 4.6 ppm must be from an OH, either the carboxylic acid or the phenol. The chemical shift is low for a carboxylic acid and toward the low end of the range for phenols, suggesting that the compound is a methyl ester, either *ortho-* or *meta-*substituted.

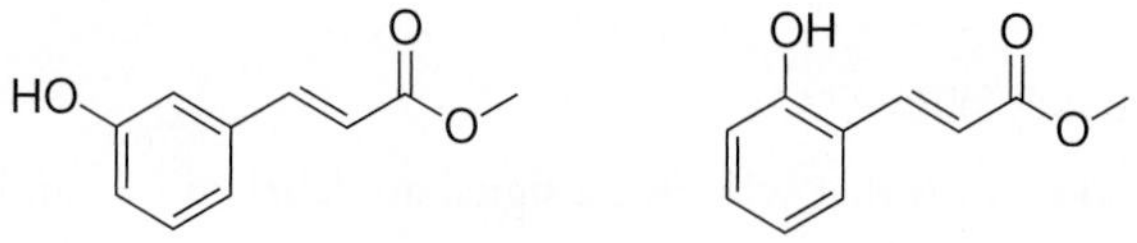

HO O O OH O O

There is some overlap of aromatic hydrogen signals, which makes it difficult to distinguish between the two possibilities. There is, however, only one triplet. The others appear to be doublets or overlapping doublets. The *meta*-substituted ring is expected to have one triplet and a series of doublets. The *ortho*-substituted ring would be expected to have two triplets, suggesting that the actual compound is *meta*-substituted.

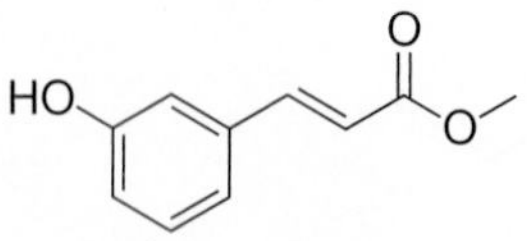

HO O O

INTEGRATE THE SKILL 13.20

The molecular formula indicates DU = 6, which is suggestive of an aromatic ring with two additional unsaturations. These may be a ring and a π bond, two rings, or two π bonds. All the NMR spectra indicate the presence of an aromatic ring, based on the chemical shift of the signals at 6.5–8.0 ppm in the ^{1}H-NMR and 95–160 ppm in the ^{13}C-NMR.

The ^{1}H-NMR shows five distinct hydrogens in the aromatic region. This seems to suggest a monosubstituted benzene ring; however, this cannot be the case. A monosubstituted ring can have a maximum of three signals, due to symmetry. This molecule cannot have symmetry and must include at least one additional π bond.

Summarizing the ^{1}H-NMR data into a chart results in the following:

H Type	d (ppm)	Integration	Coupling
a	7.6	1	d (n=1)
b	7.2	1	s (n=0)
c	7.1	1	d (n=1)
d	6.7	1	d (n=1)
e	6.6	1	d (n=1)
f	4.1	2	q (n=3)
g	3.7	3	s (n=1)
h	1.3	3	t (n=2)

Comparing the ^{13}C-NMR to the DEPT-135 shows that there are three sp^2 carbon signals missing in the DEPT spectrum. This suggests that there may be a tri-substituted aromatic ring. All the other aromatic signals are CH signals.

The singlet for H_b, along with the other aromatic doublets, suggests that the aromatic ring has a 1,2,4-substitution pattern. The other four signals for H_a, H_c, H_d, and H_e are the other two hydrogens on the aromatic ring and two hydrogens on the other π bond.

The remaining ^{1}H-signals are a three-hydrogen singlet (H_g), a two-hydrogen quartet (H_f) with a three-hydrogen triplet (H_h) that combine as an ethyl group. The downfield chemical shift of H_f suggests that it is an ethoxy group. H_g is also far downfield and may be bonded to the nitrogen atom. The chemical shift is farther downfield than expected for a methyl group bonded to a nitrogen; this may be the result of an aromatic ring.

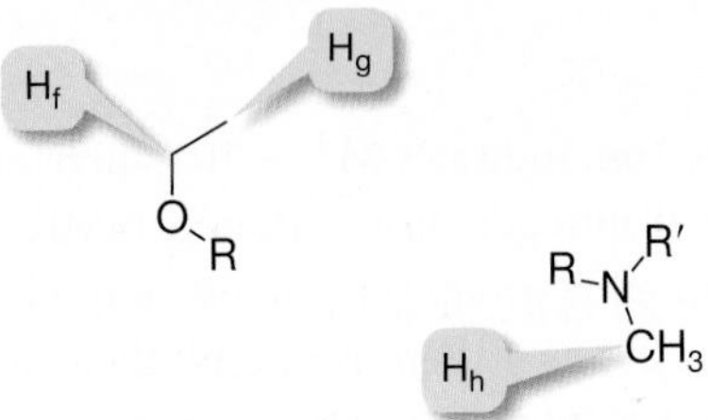

There are no other fragments arising from any of the spectra. The remaining unsaturation must be a ring, from the two adjacent substitutions on the aromatic ring. The nitrogen, which has two open bonding sites, and the alkene can form such a ring. The ethoxy can occupy the other substituted position on the ring.

This gives two possible structures:

These two structures are not easily distinguished. However, if the first structure is the correct structure, the singlet on the aromatic ring would represent a hydrogen between two electron-donating substituents: an O and an N. If the second structure is correct, the singlet is from a hydrogen that is adjacent to only one electron-donating substituent. The chemical shift of H_b, 7.2 ppm, is perhaps too far downfield for a hydrogen between two electron-donating groups, making the second structure more likely.

CHAPTER 14

PRACTICE PROBLEM 14.1

a) $M^+ = 154$
$M+2 = 156$
3:1 ratio indicates Cl

b) $M^+ = 108$
$M+1 = 109$ from ^{13}C

c) $M^+ = 126$
$M+1 = 127$ from ^{13}C

INTEGRATE THE SKILL 14.2

$M^+ = 326$
$M+2 = 328$
$M+4 = 330$
1:2:1 indicates 2 × Br

PRACTICE PROBLEM 14.3

a) $M^+ = 160$, even
b) $M^+ = 119$, odd
c) $M^+ = 147$, odd
d) $M^+ = 88$, even

INTEGRATE THE SKILL 14.4

The mass spectrum indicates $M^+ = 101$; therefore, there is an odd number of nitrogen atoms. Proton NMR indicates the presence of an ethyl group (or groups) only. The mass of the ethyl group is 29, so if there is one N, there are three ethyl groups ($14 + 3 \times 29 = 101$). This suggests trimethylamine. If there are three nitrogens (odd mass), then $101 - 42 = 59$. This is two ethyl groups *and another proton* not seen in the NMR, so we can rule this out. There is insufficient mass present to have an ethyl group if there are five nitrogens ($101 - 5 \times 14 = 11$). Therefore, the compound is triethyl amine.

PRACTICE PROBLEM 14.5

a) $C_6H_{10}O_2$:114.0681
b) C_6H_7NO:109.0528
c) $C_{13}H_{20}O$:192.1514
d) $C_7H_{13}NO$:127.0997
e) $C_7H_6NO_2Cl$:171.0087

INTEGRATE THE SKILL 14.6

The answer is $C_6H_8O_2$. The mass of C_6H_{18} is 112.1252 and the mass of $C_7H_{12}O$ is 113.0888.

PRACTICE PROBLEM 14.7

a)

p-hydroxybenzoic acid

There are O–H stretch signals (alcohol and carboxylic acid hydrogen) and C–H stretch signals (aromatic hydrogens).

b)

6-methyl-5-hepten-1-yne

C–H stretches are present for alkane, alkene, and alkyne hydrogens.

c)

1-octanol

There are O–H stretch signals (alcohol) and C–H stretch signals (alkane hydrogen).

d)

3-propoxypropylamine

There are C–H stretch signals (alkane) and N–H stretch signals (primary amine hydrogen).

INTEGRATE THE SKILL 14.8

Spectrum (a) is methyl 2-hydroxybenzoate, as determined by the presence of the alcohol O–H stretching signal at 3300 cm^{-1}.

Spectrum (b) is 2-methoxybenzoic acid, as determined by the presence of the carboxylic acid O–H stretch signal from 2500 to 3300 cm^{-1}.

PRACTICE PROBLEM 14.9

a) The triple bond stretch at 2125 cm^{-1} indicates this could be an alkyne (di-substituted, since there is no C–H stretch near 3300) or a nitrile.
b) The carbonyl stretch at 1720 cm^{-1} indicates this could be an amide or a ketone.
c) The carbonyl stretch (signal at ~1740 cm^{-1}) and the C–O–C stretch at 1170 cm^{-1} would make this an ester.
d) The carbonyl stretch at 1710 cm^{-1} would be for a ketone.

INTEGRATE THE SKILL 14.10

The mass spectrum suggests 87 is the M^+; therefore, there is an odd number of N. The infrared spectrum shows a band at 1660, suggesting amide, which is consistent with MS. No NH stretches are Apparent above 3000, so the amide must be di-substituted on N. The base peak at 44 in MS suggests a $(CH_3)_2N^+$ ion. The fragment at 72 confirms a methyl group, since $87 - 72 = 15$. Carbonlyl CO plus $(CH_3)_2N$ leaves CH_3. Therefore, the compound is

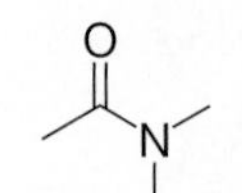

dimethylacetamide

CHAPTER 15

PRACTICE PROBLEM 15.1

a) This reaction will proceed since the products should be favoured. The resonance forms for the amide are more stable than the ester, with a positive charge on a nitrogen versus an oxygen.

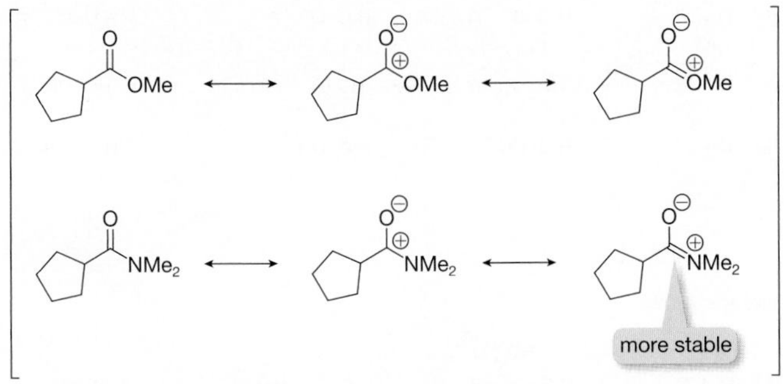

more stable

b) This will not proceed, as it produces a stronger base (amide) than the reacting acetate ion.

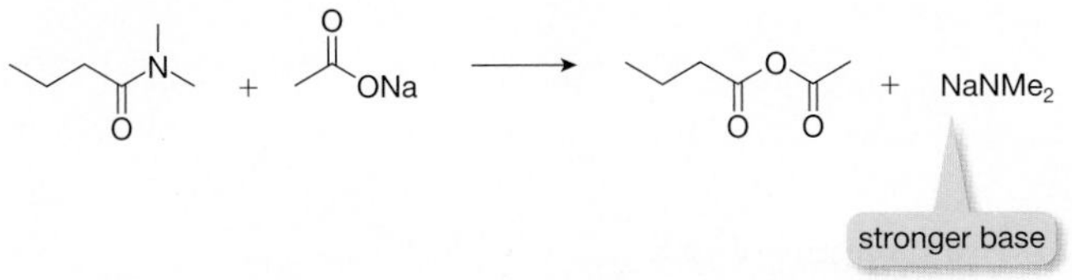

+ ONa ⟶ + $NaNMe_2$

stronger base

c) This reaction will proceed, as the acid chloride is less stable (more reactive) than the product carboxylic acid.

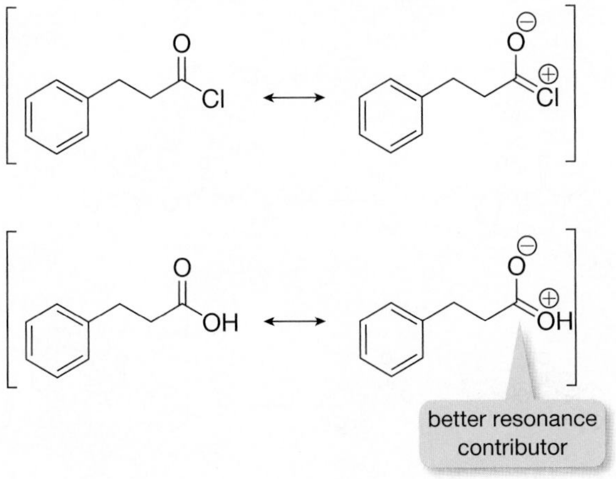

better resonance contributor

d) This reaction will not proceed. The product acid anhydride is more reactive than the starting ester.

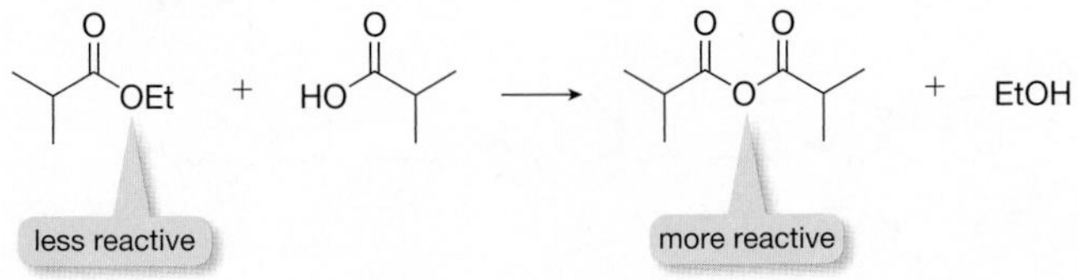

OEt + HO ⟶ + EtOH

less reactive

more reactive

INTEGRATE THE SKILL 15.2

Reaction from part (a):

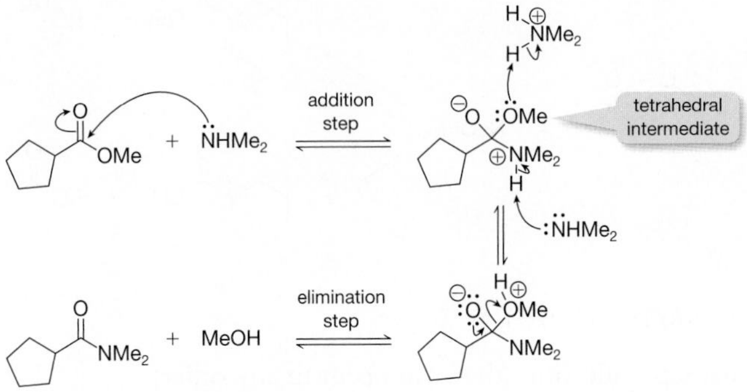

addition step

tetrahedral intermediate

elimination step

+ MeOH

Reaction from part (c):

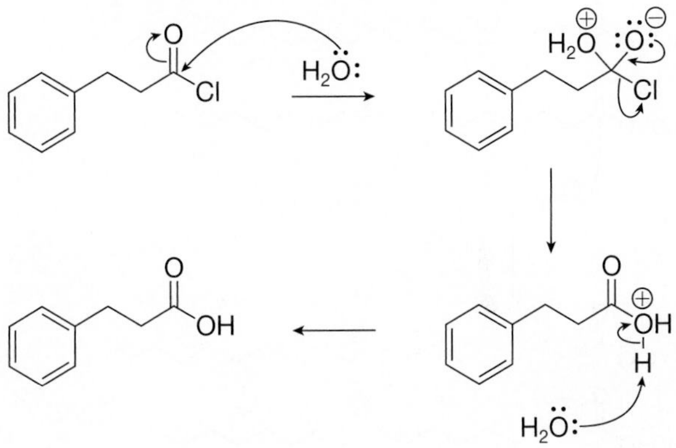

PRACTICE PROBLEM 15.3

a)

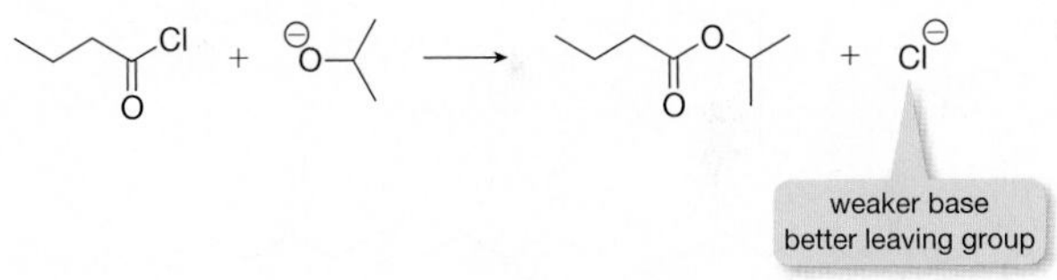

weaker base
better leaving group

b) The methoxide base would be neutralized by the acid. The carboxylate would then be neutralized back to its acid form in the workup.

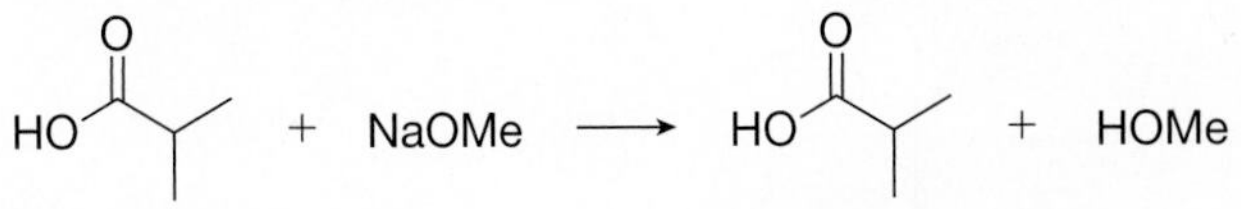

+ NaOMe ⟶ + HOMe

c) + KOH ⟶ +

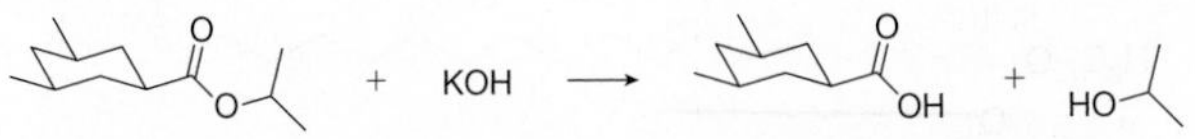

d) Only one of the carbonyls will react since the product of the first addition is a carboxylic acid, which will not react under these conditions.

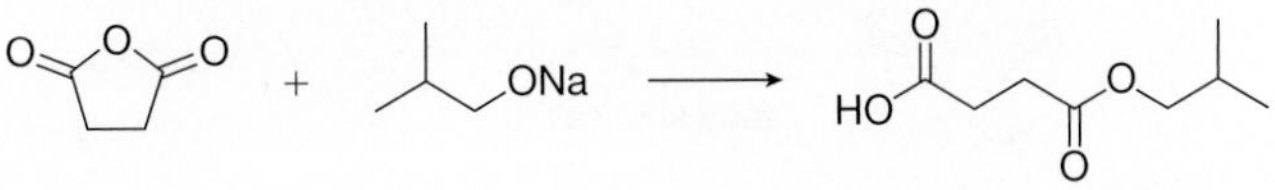

+ ONa ⟶

e) Either carbonyl can react with the ethoxide, so two sets of products can be formed.

INTEGRATE THE SKILL 15.4

First saponification (they can occur in any order)

fatty acid salt

For the rest of the saponification, the alkyl chains have been simplified in the figures.

Second saponification

Third saponification

PRACTICE PROBLEM 15.5

a) The amine acts as a nucleophile and as a base in this reaction.

b)

c)

d) The carbonyl group that reacts is less sterically hindered than the other.

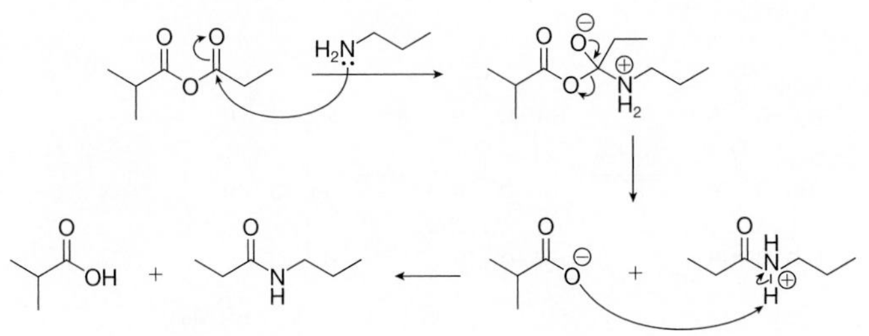

However, the other carbonyl group would react to some extent and form the other possible amide.

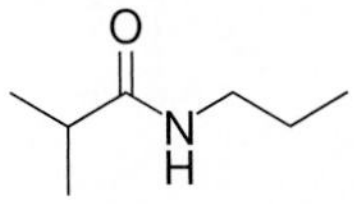

e) The tertiary amine will not react, as it is too bulky. The more basic trimethylamine removes the proton from the intermediate, leaving the catalytic DMAP available for further reaction with starting material.

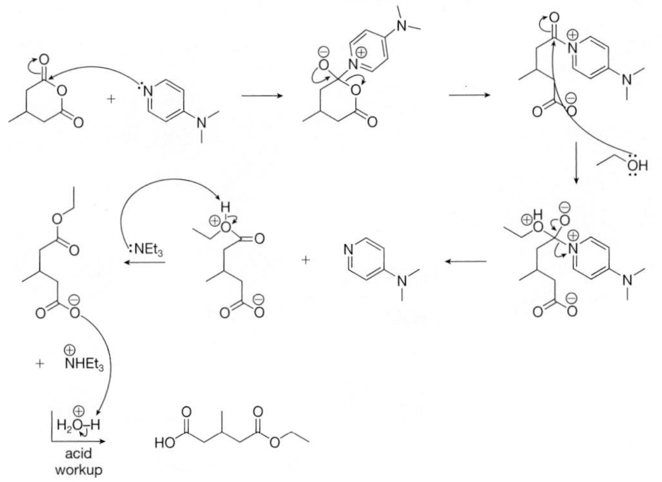

INTEGRATE THE SKILL 15.6

Energy diagram for part (a), above.

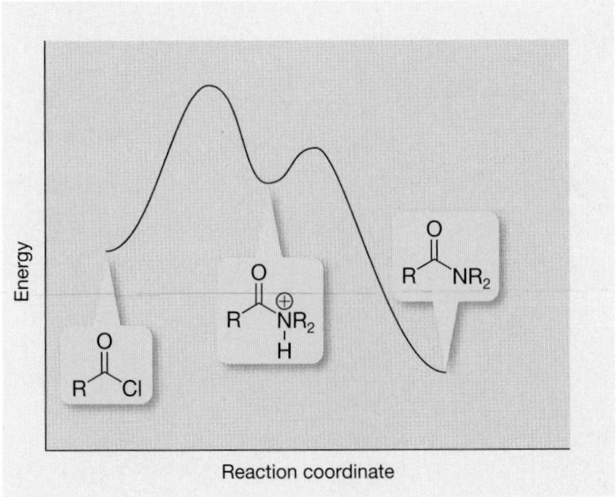

PRACTICE PROBLEM 15.7

a)

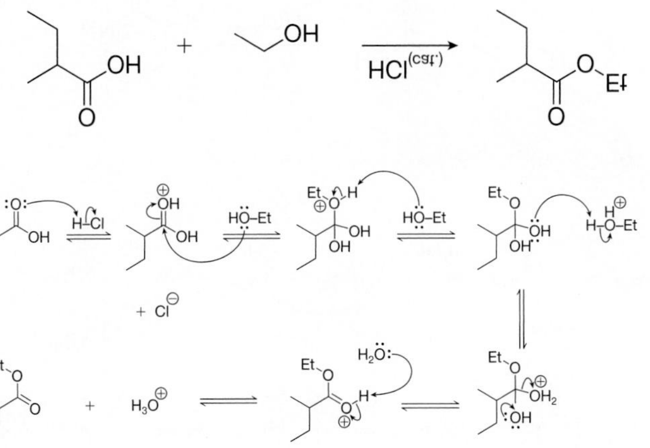

b)

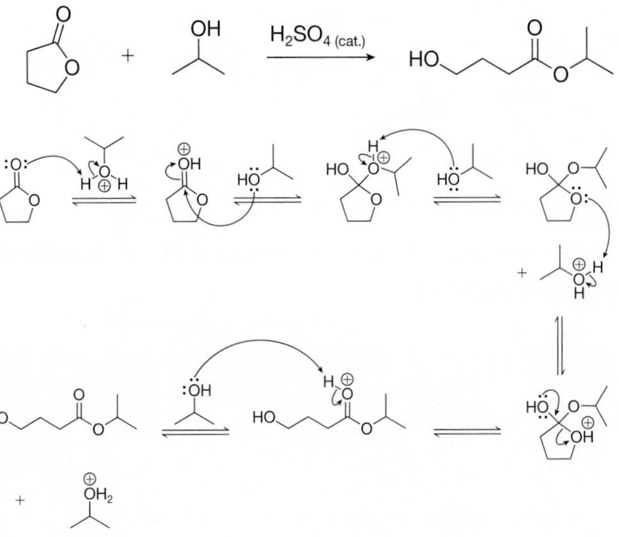

c)

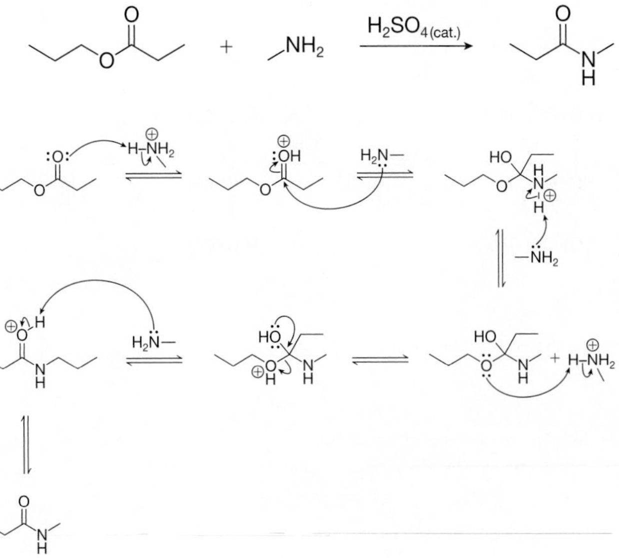

d)

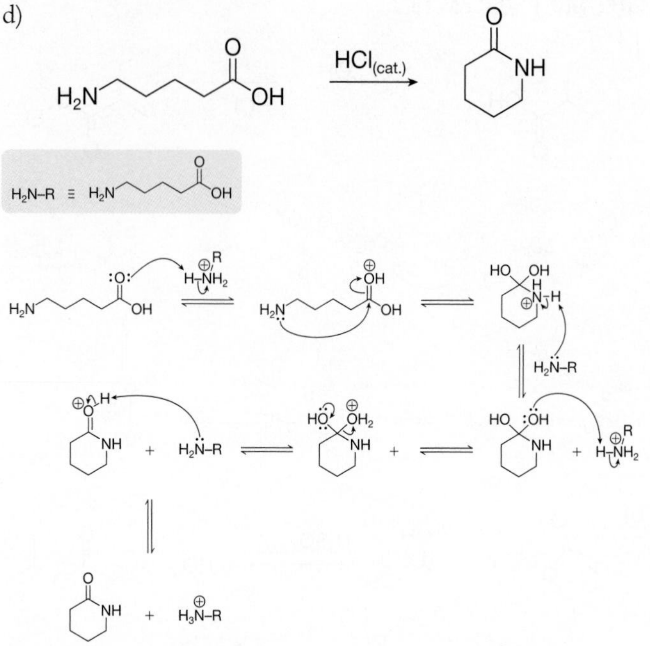

H_2N–(CH$_2$)$_4$–COOH $\xrightarrow{HCl_{(cat.)}}$ (cyclic lactam, NH)

H_2N–R ≡ H_2N–(CH$_2$)$_4$–COOH

INTEGRATE THE SKILL 15.8

The carbon chains have been replaced with –R to make the mechanisms clearer.

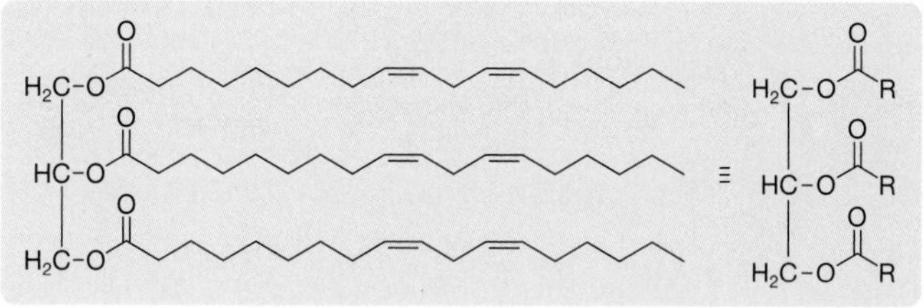

Overall reaction

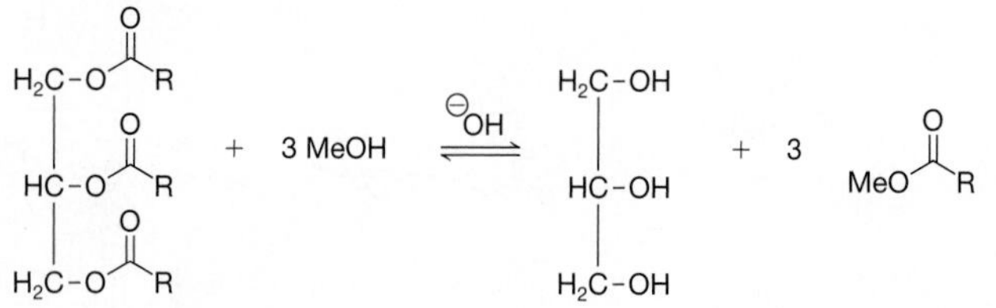

Triglyceride + 3 MeOH $\overset{\ominus OH}{\rightleftharpoons}$ glycerol + 3 MeO–C(=O)–R

Mechanism

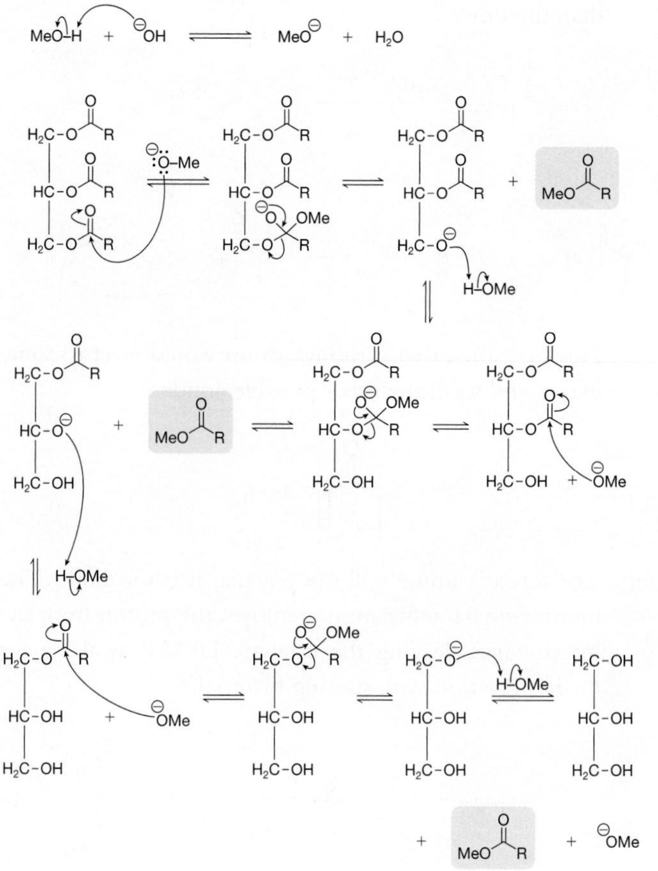

MeO–H + $^{\ominus}$OH ⇌ MeO$^{\ominus}$ + H_2O

PRACTICE PROBLEM 15.9

a) $\xrightarrow[H_2O]{H_2SO_4}$

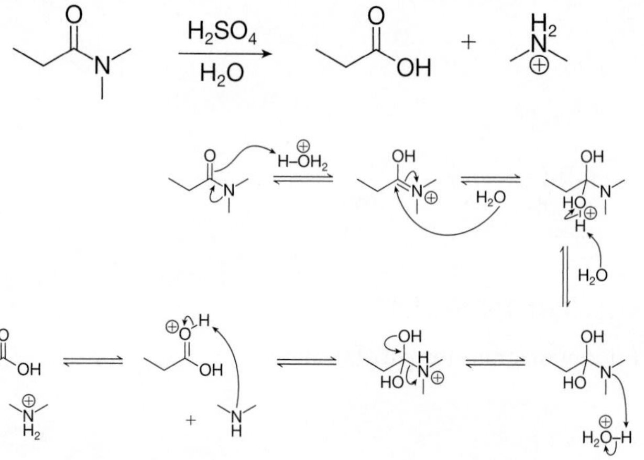

b) cyclopentyl–CN $\xrightarrow[H_2O]{NaOH}$ cyclopentyl–$CO_2^{\ominus}$ + NH_3

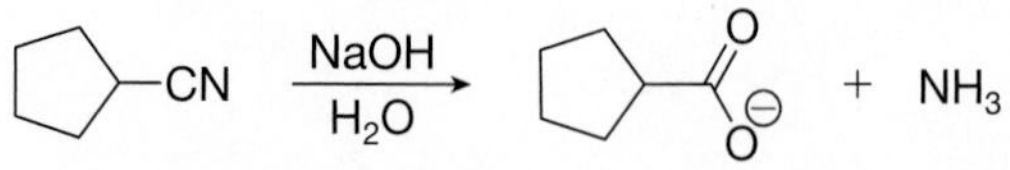

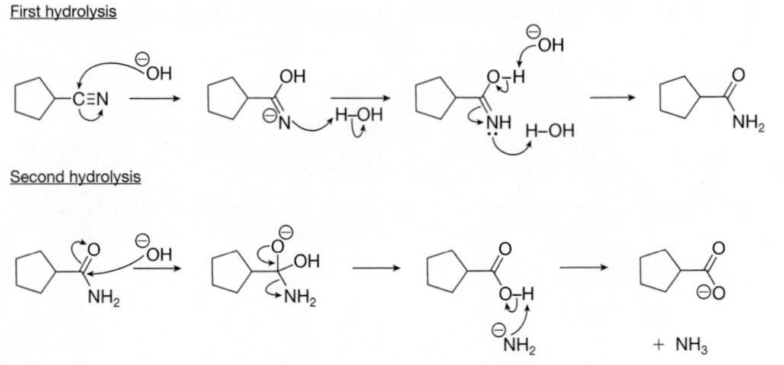

First hydrolysis

Second hydrolysis

c)

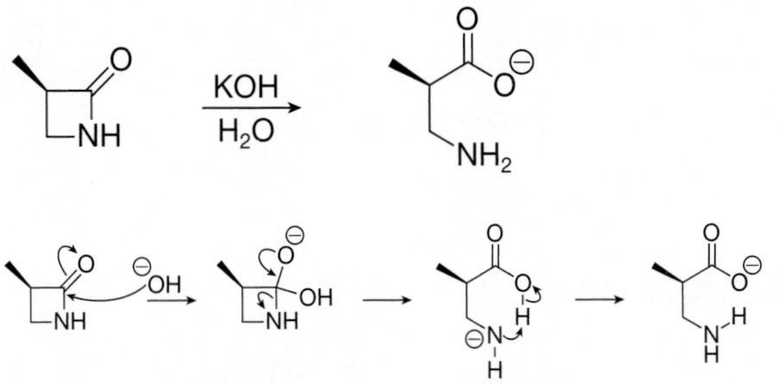

KOH / H_2O

d)

HCl / H_2O

First hydrolysis

Second hydrolysis

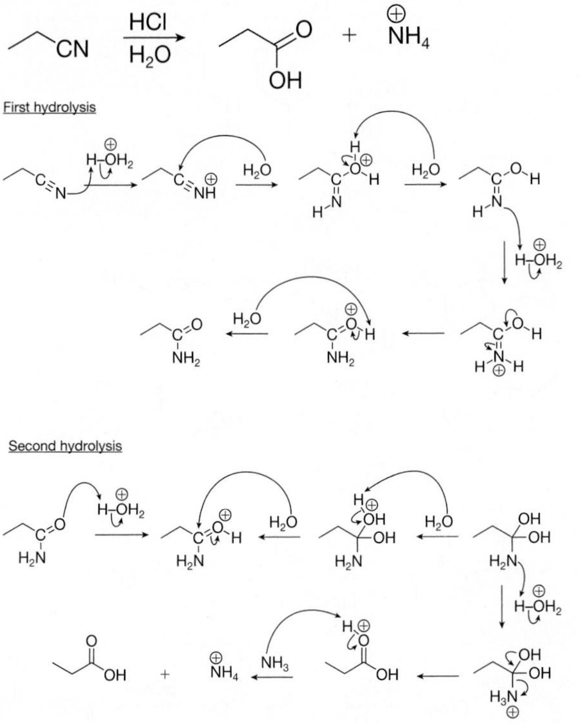

INTEGRATE THE SKILL 15.10

The three amino acids are connected by two peptide bonds, as shown below.

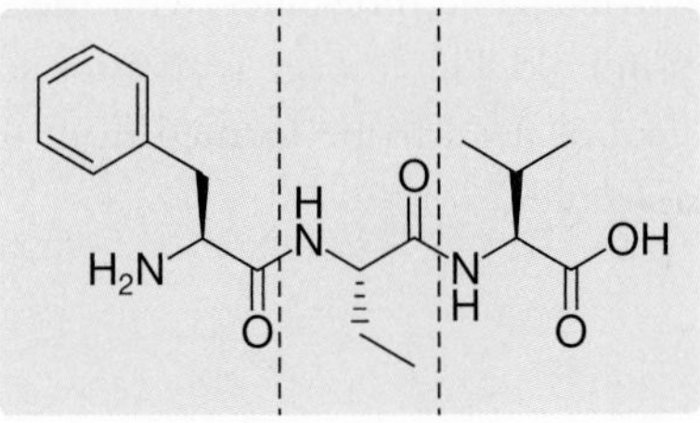

Acid hydrolysis would give the three amino acids below. Since none of the chiral centres are involved in the reaction, their configurations should be retained and the amines would be protonated under acid conditions.

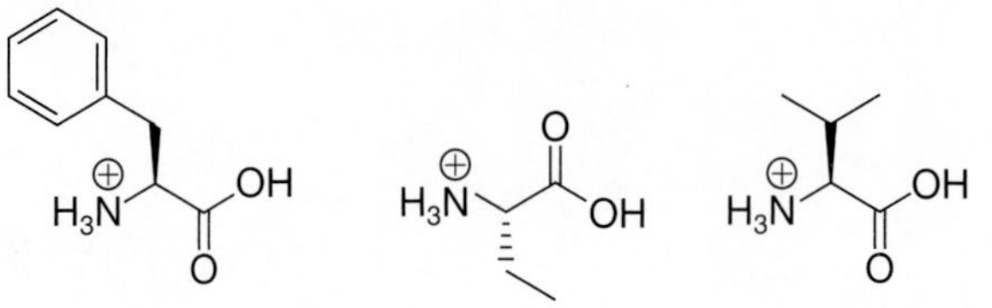

PRACTICE PROBLEM 15.11

a)

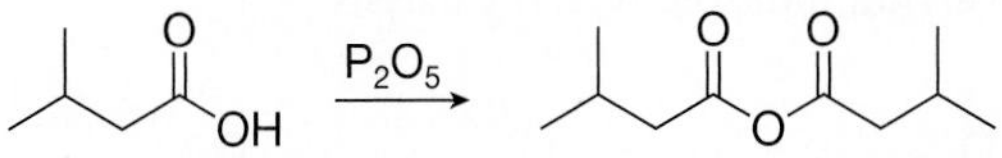

P_2O_5

b)

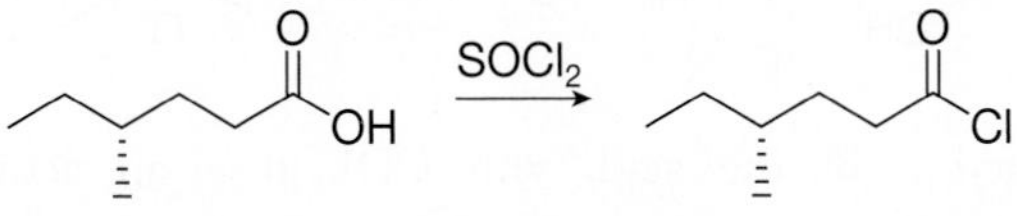

$SOCl_2$

(*R*)-4-methylhexanoic acid

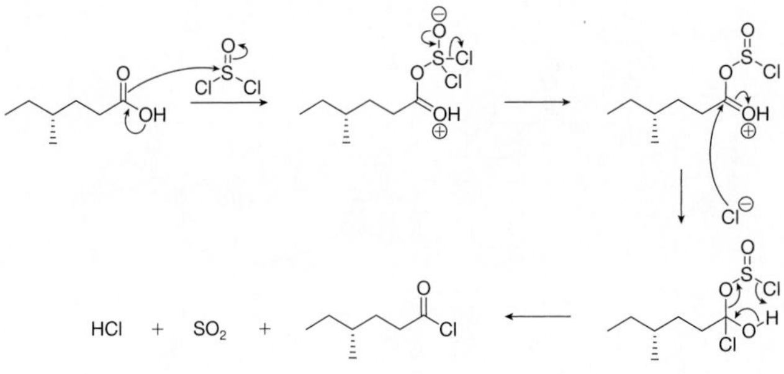

HCl + SO_2 +

c)

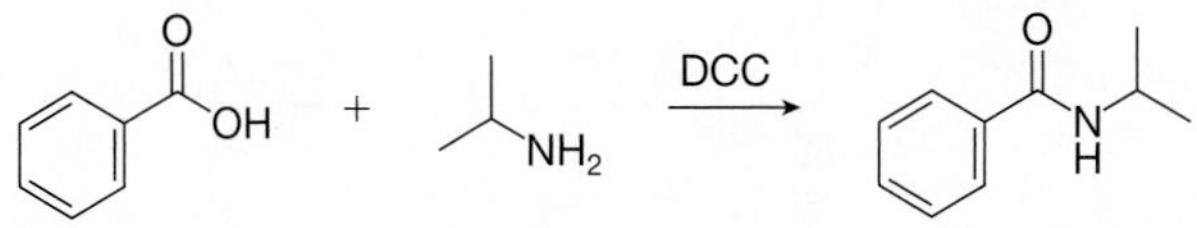

DCC

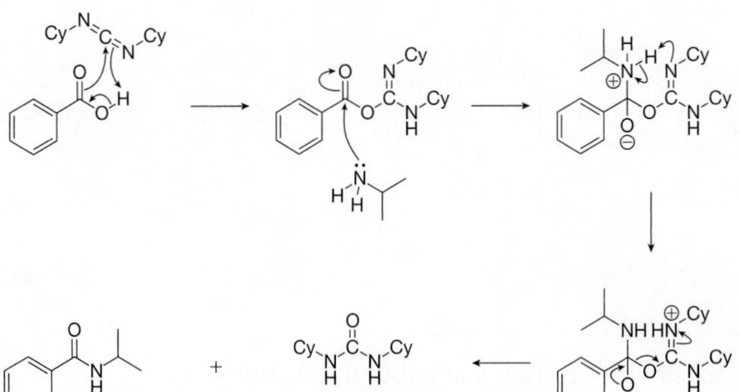

d)

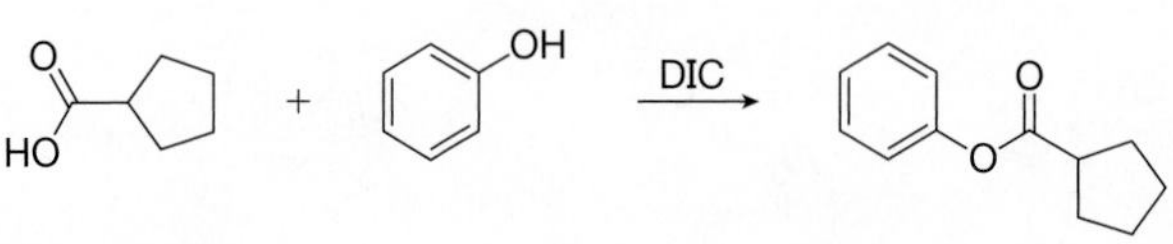

DIC

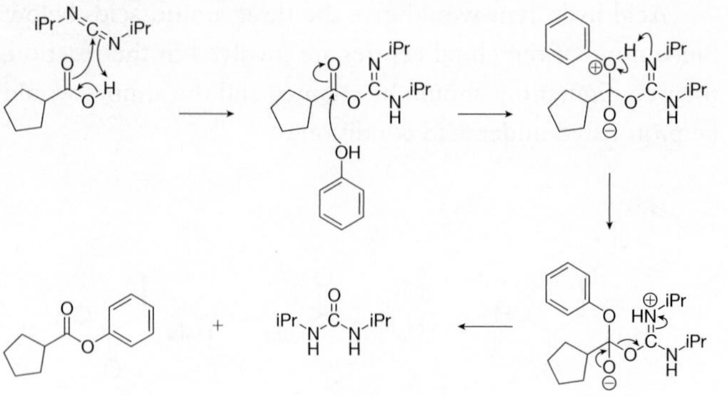

INTEGRATE THE SKILL 15.12

a) Carbonyldiimidazole (CDI) catalysis:

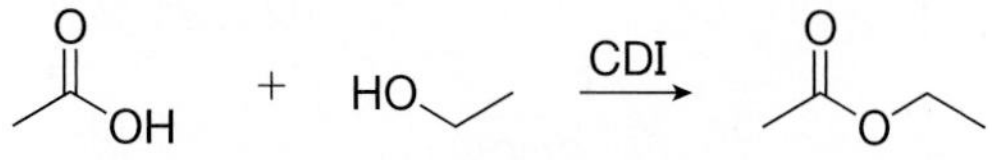

Reaction of the acid with CDI gives an activated carboxylic acid.

Activation

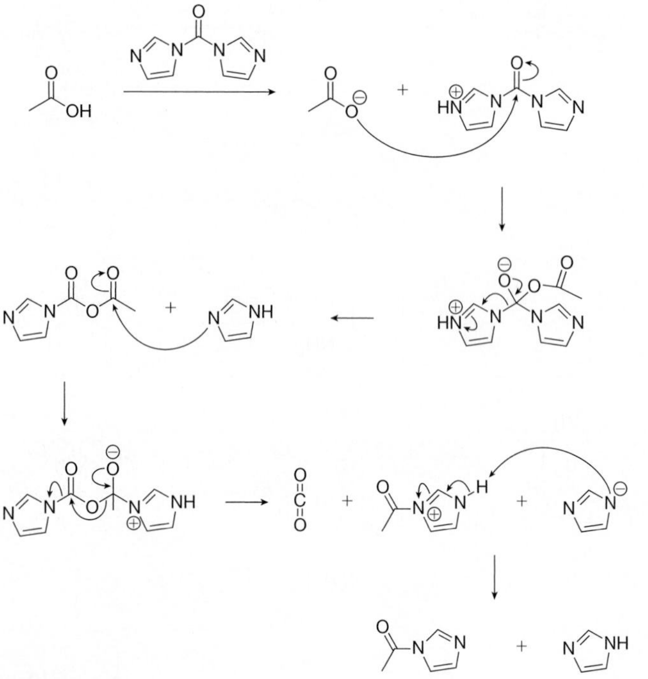

This product is then reacted with ethanol.

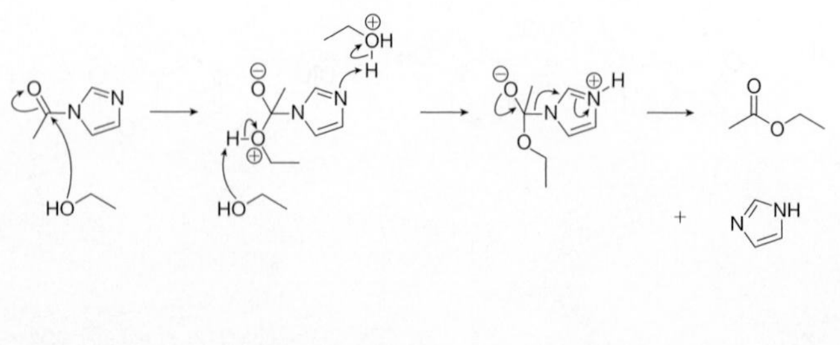

b) Acid chloride formation, using oxalyl chloride:

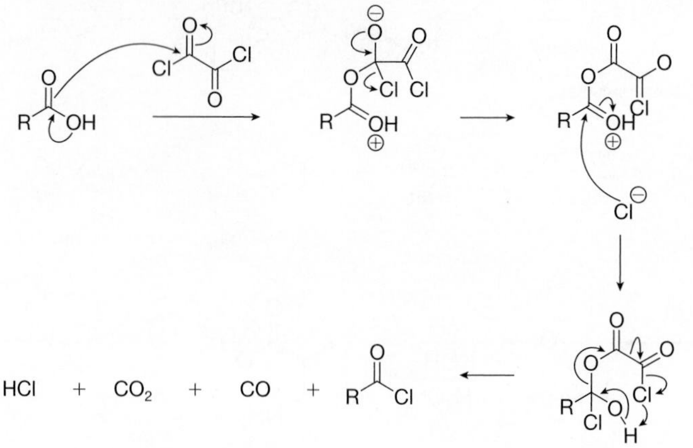

$HCl + CO_2 + CO + $ RCOCl

PRACTICE PROBLEM 15.13

a) P_2O_5

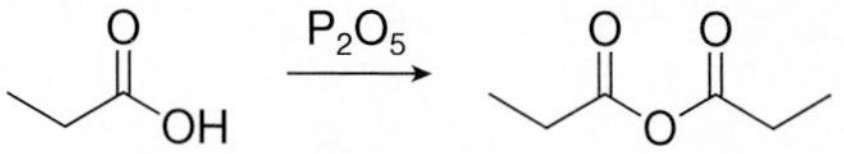

b) DCC, NH_2

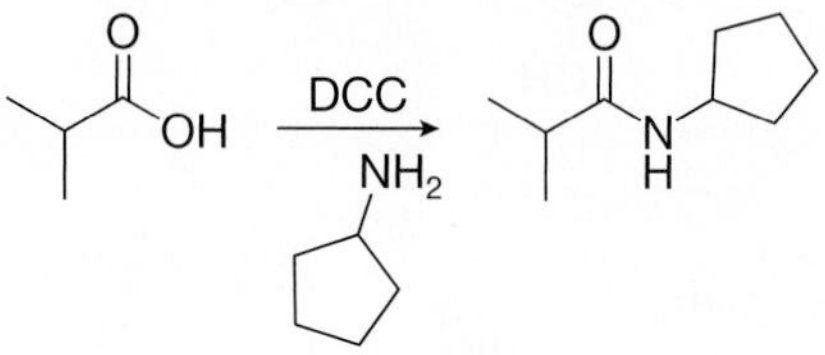

c) H_2SO_4, H_2O

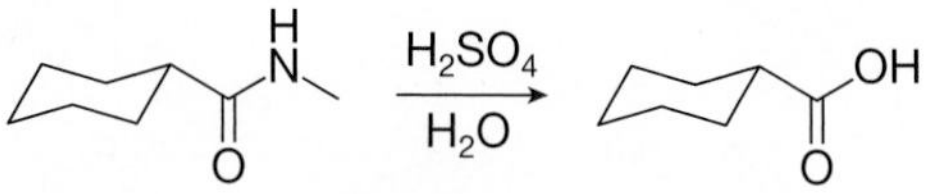

d) $H_2SO_{4(cat)}$, MeOH

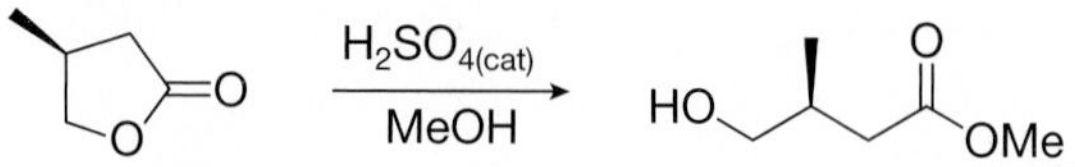

e) pyridine, HO

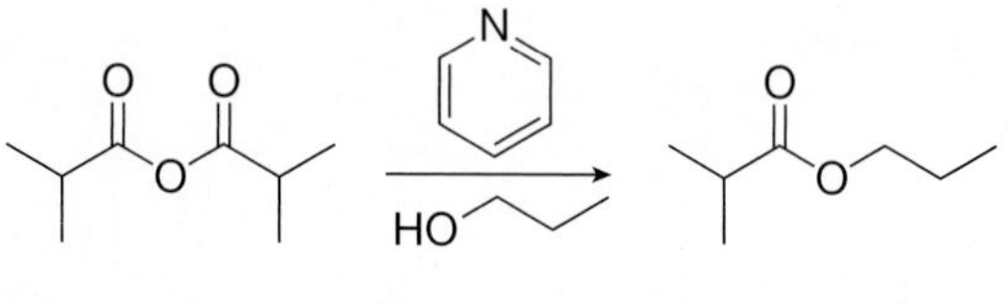

f) H_2SO_4, H_2O; $SOCl_2$

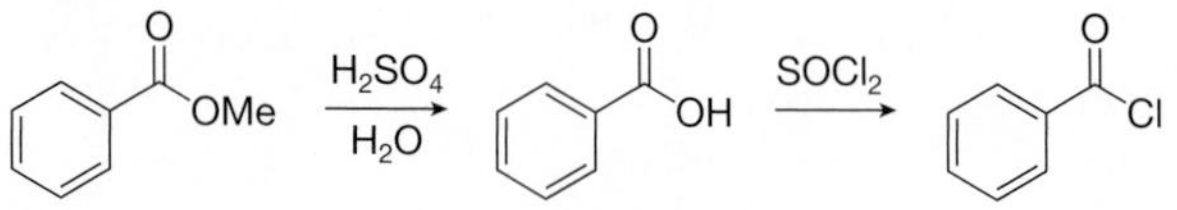

INTEGRATE THE SKILL 15.14

Looking at the carbon framework, the product is formed from two ethynylbenzene molecules, with the eight carbons of each shown in bold. The linkage is an ester, so a carboxylic acid and an alcohol need to be formed from separate ethynylbenzene molecules.

The alcohol could come from Markovnikov hydration of a terminal alkene, which could be made from the ethynylbenzene by hydrogenation.

Therefore, the steps for the alcohol synthesis would be:

Lindler's catalyst, H_2; H_2O, $HCl_{(trace)}$

The carboxylic acid could be formed from a primary alcohol, which could come from anti-Markovnikov hydration of the same terminal alkene used to make the alcohol.

The reactions to synthesize the acid from the alkene would be:

1) BH_3, ether 2) H_2O_2, H_2O, NaOH; $HCrO_4$, H_2O

The final esterification will be enhanced by activating the acid with thionyl chloride.

1) $SOCl_2$ 2) HO–CH(CH₃)Ph, pyridine

PRACTICE PROBLEM 15.15

a)

1) $LiAlH_4$ 2) HCl, H_2O

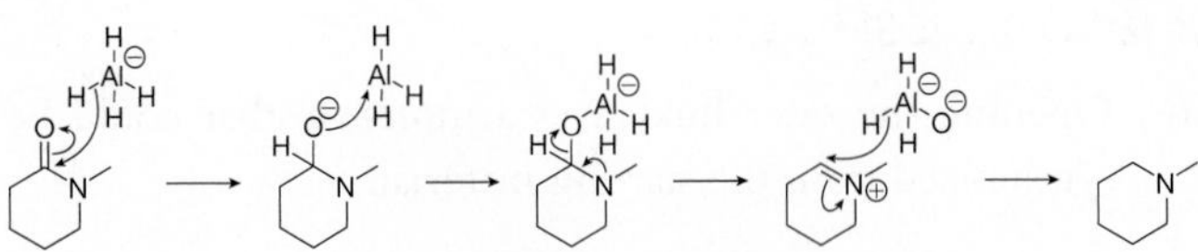

b) Sodium borohydride will not reduce an amide.

$NaBH_4$, EtOH

c)

1) $LiAlH_4$ 2) HCl, H_2O

d)

1) $LiAlH_4$ 2) HCl, H_2O

e) The molecule contains an ester and a hemiacetal, a functional group that is in equilibrium with the corresponding aldehyde. Sodium borohydride will not reduce ester carbonyls but will reduce aldehydes.

$NaBH_4$, EtOH

INTEGRATE THE SKILL 15.16

a) Opening the ester link gives a molecule that could be synthesized from the starting material.

If the aldehyde were reduced, then a transesterification would give the final product.

b) The amine could be produced by reduction of an amide. The reaction scheme would be

PRACTICE PROBLEM 15.17

a) Low temperature causes formation of the aldehyde.

b)

c) Warming the reaction mixture results in full reduction of the carbonyl group.

d)

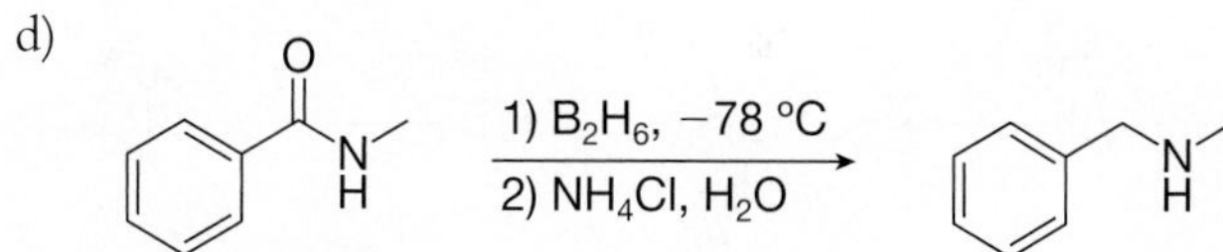

INTEGRATE THE SKILL 15.18

a) Borohydride is selective for aldehydes over amides.

b) BH_3 will reduce acids and not esters.

c) A strong reducing agent is needed. $LiAlH_4$ would be a possible choice.

PRACTICE PROBLEM 15.19

a)

b)

c)

d)

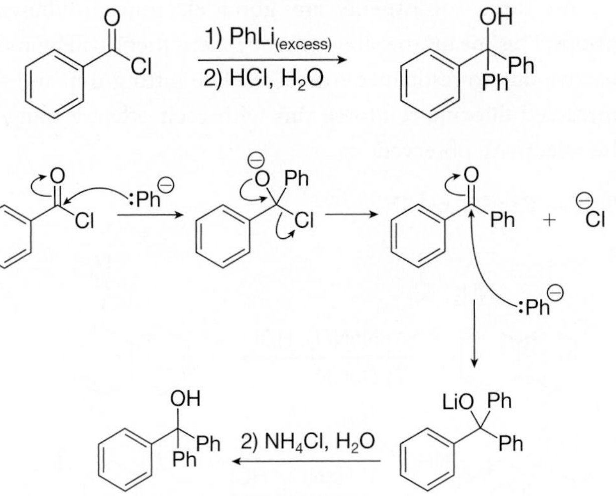

1) $PhLi_{(excess)}$
2) HCl, H_2O

2) NH_4Cl, H_2O

e)

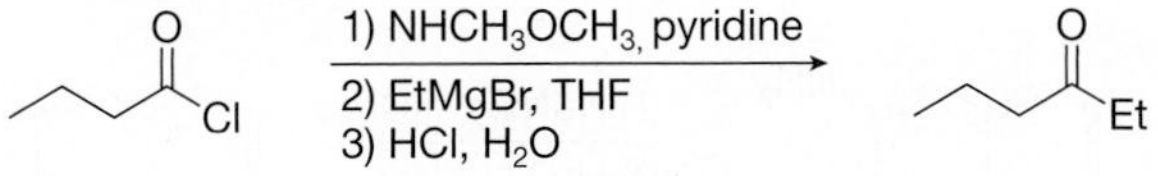

1) $NHCH_3OCH_3$, pyridine
2) EtMgBr, THF
3) HCl, H_2O

Step 1:

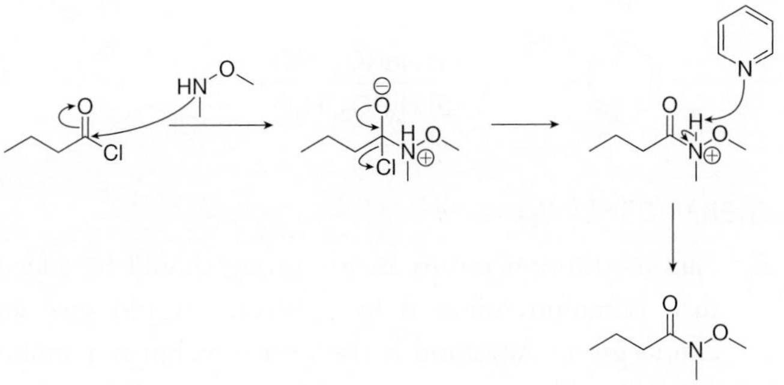

Steps 2 and 3:

−78 °C

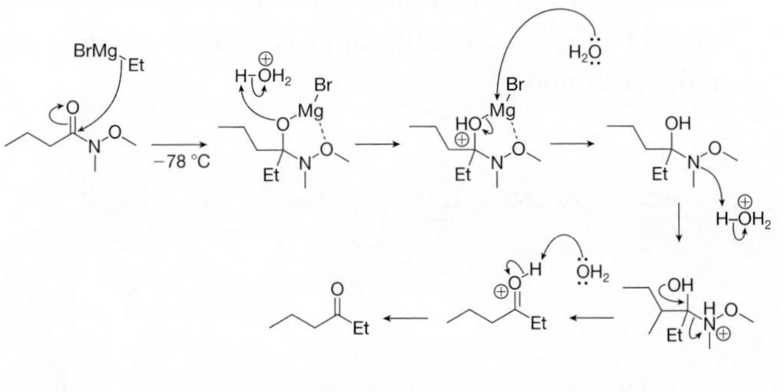

INTEGRATE THE SKILL 15.20

a) The first step would be to dialkylate the acid chloride. The selectivity would be due to the higher reactivity of the acid chloride and the subsequent ketone intermediate. Since there are only two equivalents of Grignard, the most reactive sites are the only ones transformed.

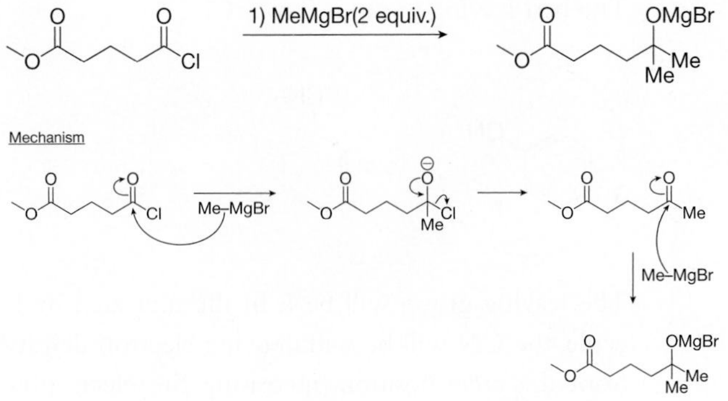

1) MeMgBr(2 equiv.)

Mechanism

The acidic aqueous workup would also be the conditions for transesterification. The alkoxide would be neutralized, then normal esterification would proceed. This is the reaction that leads to the observed product.

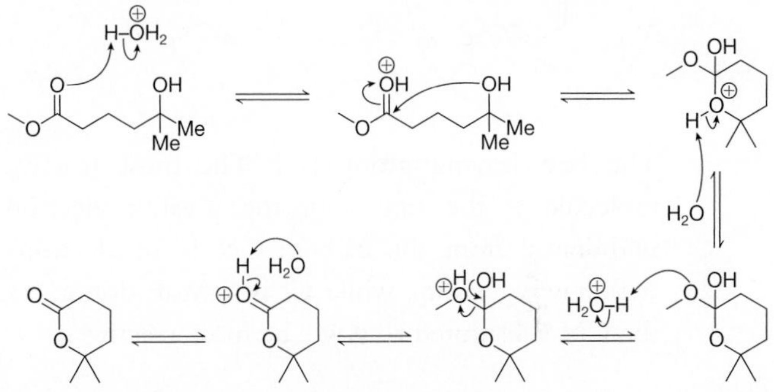

b) The observed reaction hinges on the selectivity of the alkylation steps. It would work with a diester as well since the second alkylation step is faster (ketone reactivity > ester reactivity) and would still give an ester and alcohol as the functional groups. The subsequent transesterification means the ester that was previously present is irrelevant, so even asymmetric diesters would work.

c) With four molar equivalents of the Grignard reagent, both carbonyls would be fully alkylated.

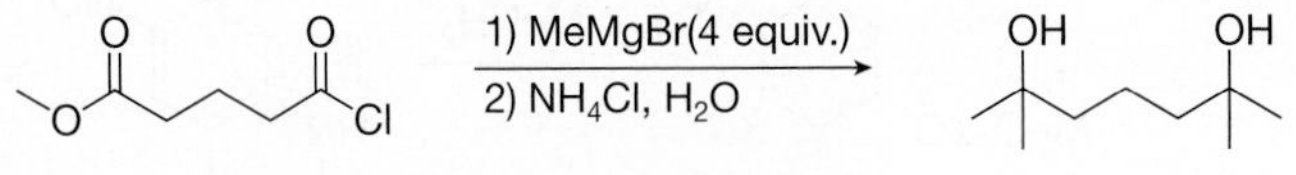

1) MeMgBr(4 equiv.)
2) NH_4Cl, H_2O

PRACTICE PROBLEM 15.21

a) The compound that is most reactive toward S_NAr is circled.

i)

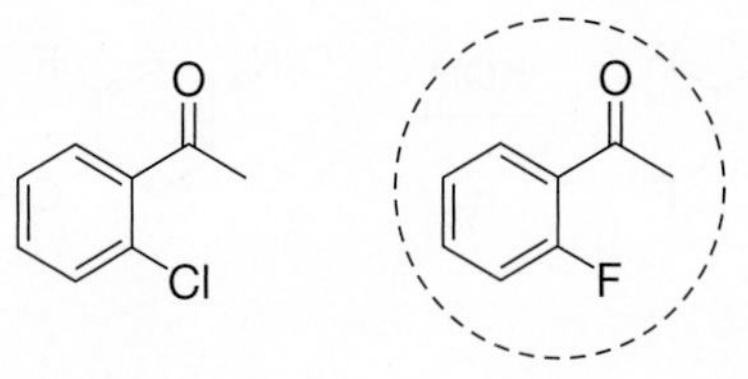

The best leaving group is F over Cl.

ii)

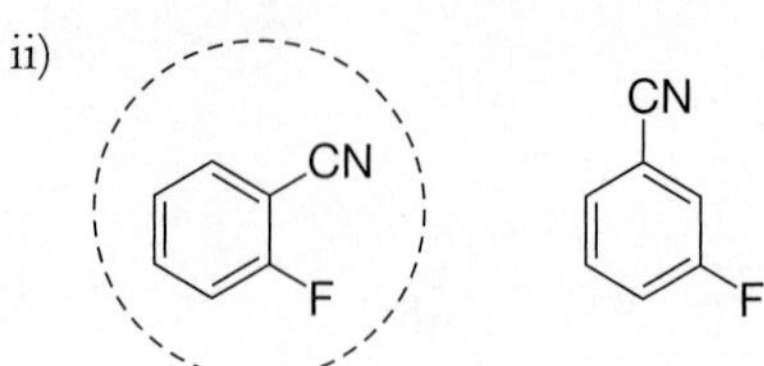

The leaving group will be F. In the indicated molecule, the CN will be withdrawing electron density from the *ortho* position, increasing the electrophilicity of that carbon.

iii)

O_2N CN F O_2N Cl F

The best leaving group is F. The most reactive molecule is the one with the greatest electron withdrawal from the carbon. CN is an electron-withdrawing group, while Cl is a weak donor. So, the CN-substituted ring will be more reactive.

b) i)

Cl F O_2N —NaOEt→ Cl OEt O_2N

ii)

F NC CN CN —H_2O→ OH NC CN CN

iii)

NO_2 MeO_2SO NO_2 —$MeNH_2$→ NO_2 HN NO_2

INTEGRATE THE SKILL 15.22

A fluoride will be the better leaving group.

O_2N F F —KOH→ O_2N F OH

All three substituents are good electron-withdrawing groups. This means the sites *ortho* or *para* to them will be more reactive due to resonance structures. The nitro group and the unreacted fluorine reinforce this with each other, leading to the selectivity observed.

PRACTICE PROBLEM 15.23

a)

NH_2 —1) $NaNO_2$, HCl; 2) CuCN→ N≡C

b)

NH_2 —1) $NaNO_2$, HCl; 2) KI→ I

c)

NH_2 —1) $NaNO_2$, HCl; 2) H_3O^+→ OH

d)

Br NH_2 Cl —1) $NaNO_2$, HCl; 2) H_3PO_2, H_2O→ Br Cl

INTEGRATE THE SKILL 15.24

a) *Para* substitution means an *o*/*p* group should be added first. Nitration followed by reduction would give an amino group. Acylation is the next step, but the amino group would react under Friedel-Crafts conditions, so it is protected as an amide. The amide is still an *o*/*p* director, so acylation gives the *para* product. The cyanide is introduced by deprotecting the amine, diazotization and cyanide substitution.

—HNO_3, H_2SO_4→ NO_2 —Fe, HCl→ NH_2 —Cl(C=O)CH₃, pyridine→ NHC(O)CH₃ —ClC(O)CH₂CH₃, $AlCl_3$→ (H N O, O) —$H_2SO_{4(cat)}$, H_2O→ NH_2 (O) —1) $NaNO_2$, HCl; 2) CuCN→ CN (O)

b) Akylation would give *o*/*p* substitution, and over-alkylation would be a problem. The regioisomer desired could be formed by acylation first (*meta* director/deactivating) followed by alkylation. Reduction of the ketone would give the required product.

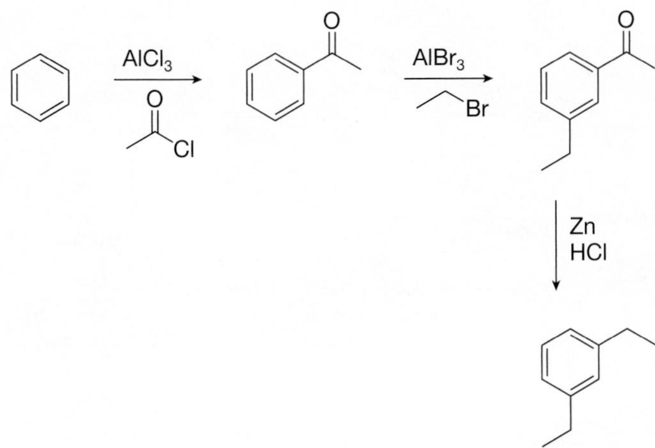

c) To control the substitution patterns, the substituents need to be manipulated. The directing properties of the substituted rings below give the desired regiochemistry.
The first addition is a nitro group. This will direct the chlorination to a *meta* position. The nitro group is now directing to the wrong position for the acylation. Reduction to the amine at this stage changes the directing properties to the *para* position, which is desired.

meta directing

NO_2 — HNO_3 / H_2SO_4; Cl_2 / $FeCl_3$; Fe / HCl; NH_2, Cl

Since the amino group will react under acylation conditions, it is first protected as an amide. The reaction sequence is completed by acylation followed by diazotization and fluorination.

pyridine; $AlCl_3$; $H_2SO_{4(cat)}$ / H_2O; 1) $NaNO_2$, HCl 2) HBF_4

CHAPTER 16

PRACTICE PROBLEM 16.1

a) EtO OEt — $HCl_{(trace)}$ / H_2O

b) $HCl_{(trace)}$ / MeOH — OMe, OMe

c)

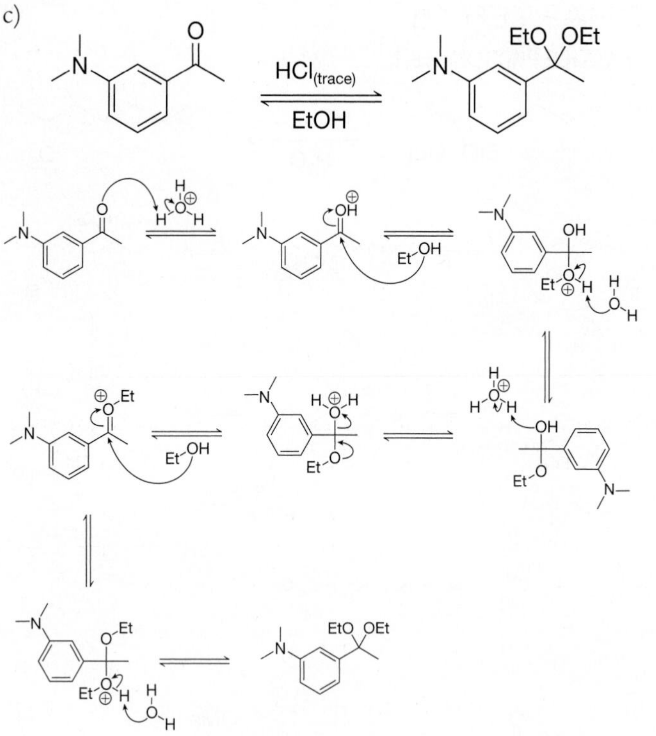

$$\text{HCl}_{(trace)} \;/\; \text{EtOH}$$

d)

$$\text{HCl}_{(trace)} \;/\; H_2O$$

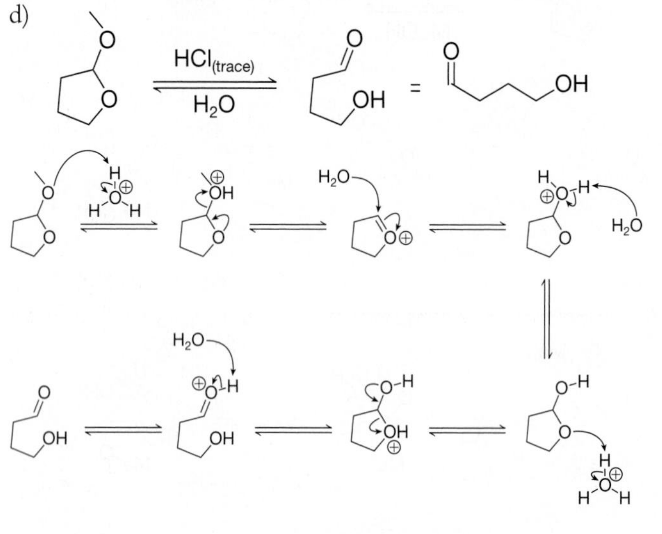

e) Similar mechanistic steps are shown simultaneously to keep the diagram compact. In reality, this would not be the case.

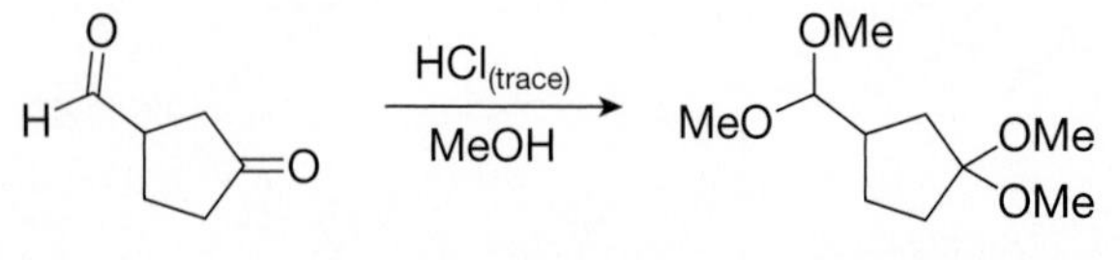

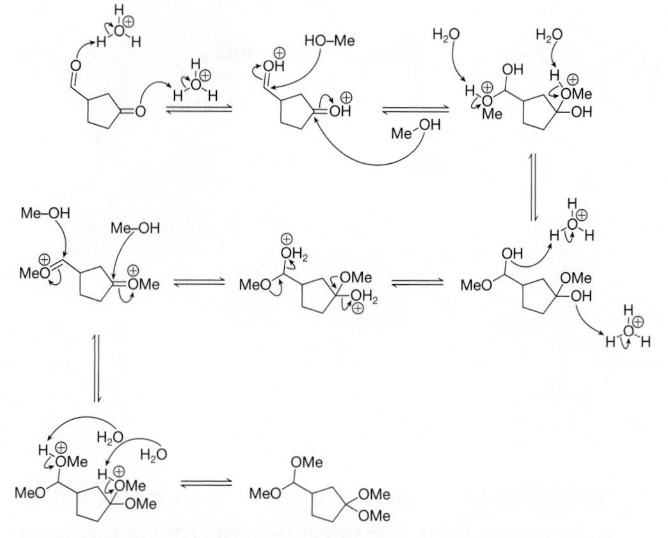

$$\text{HCl}_{(trace)} \;/\; \text{MeOH}$$

INTEGRATE THE SKILL 16.2

a) Orthoformate hydrolysis:

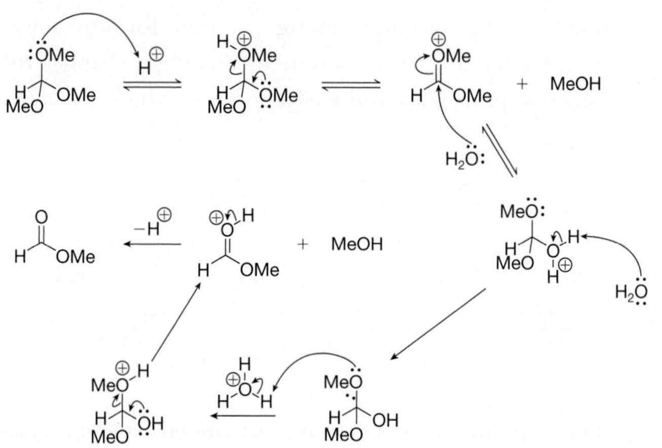

b) The reaction is not reversible since the final product is not a ketone or aldehyde.

c)

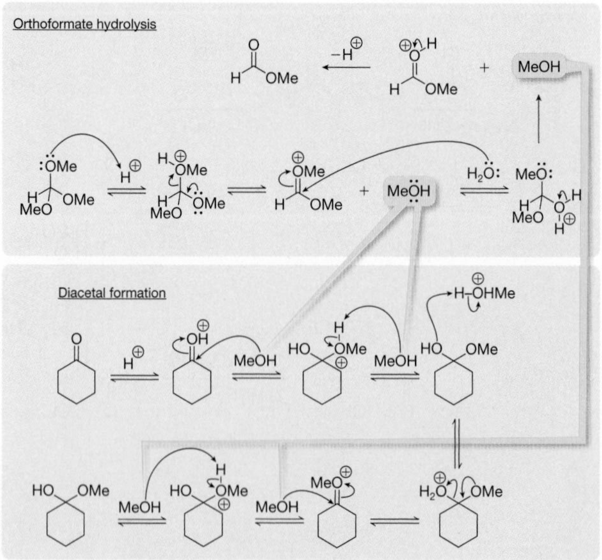

Orthoformate hydrolysis

Diacetal formation

d) The by-product would be methyl formate (HCOOMe).

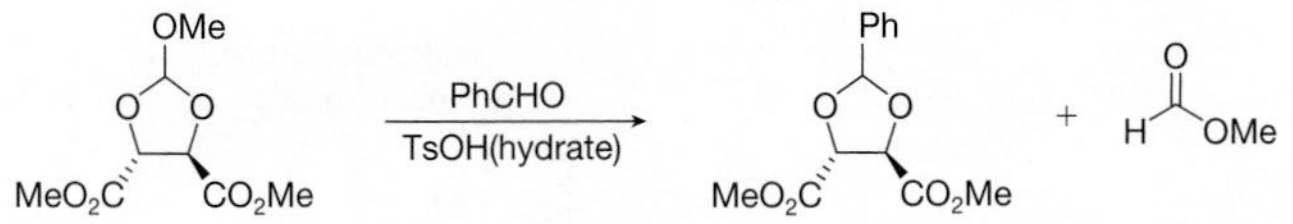

Orthoformate hydrolysis:

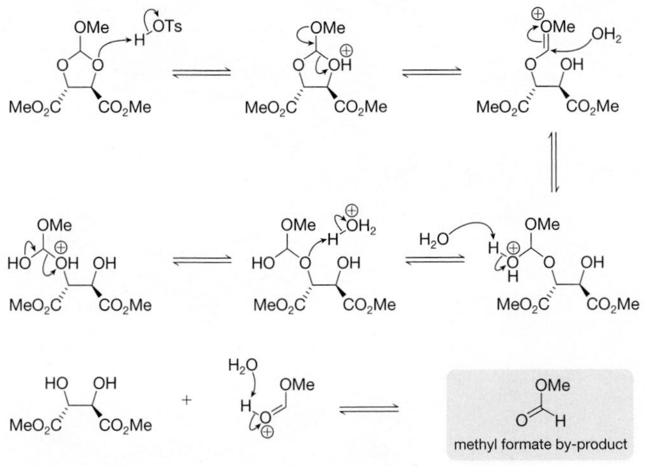

Acetal formation:

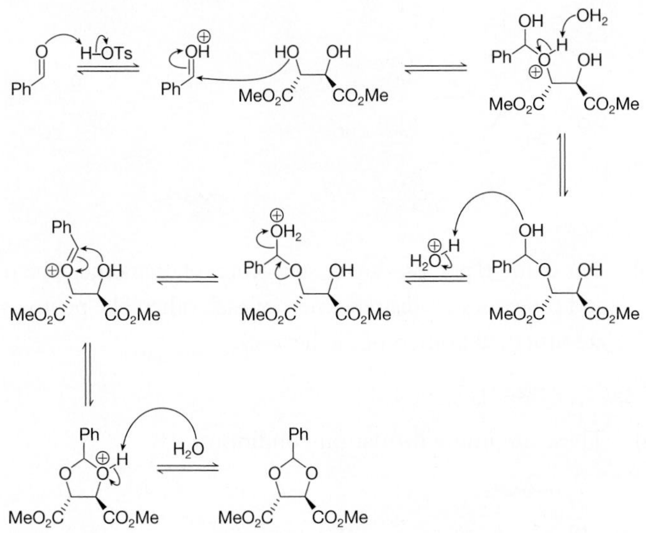

PRACTICE PROBLEM 16.3

a) No protection needed, as aldehydes will not reduce under these conditions.

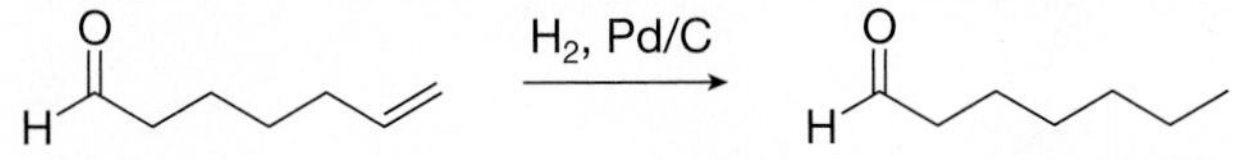

b) Making a Grignard reagent and then reacting it with an aldehyde would produce the desired alcohol group. However, the existing carbonyl would react. So, first the aldehyde is protected and then the Grignard can be prepared.

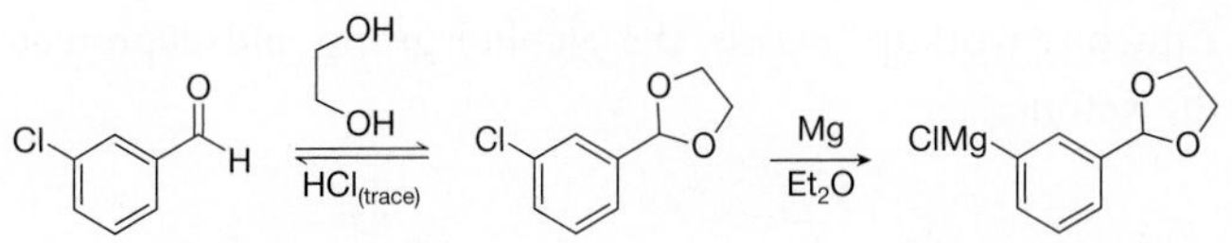

Now, the Grignard is reacted with an aldehyde to form a magnesium salt intermediate. Aqueous workup produces the alcohol and also reforms the original aldehyde.

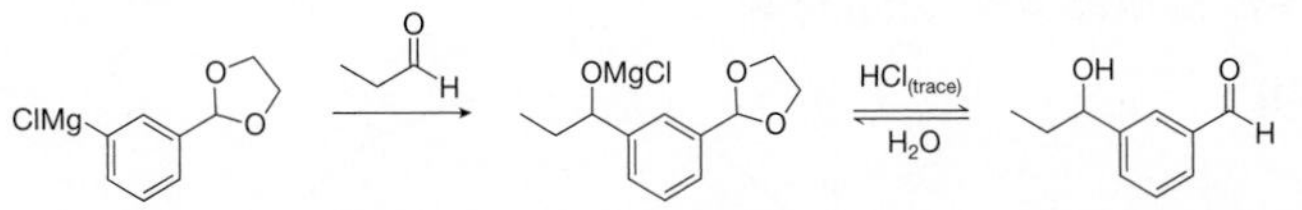

c) The aldehyde group could form the desired ether by nucleophilic substitution if it were reduced to an alcohol. The existing alcohol groups need to be protected before these reactions are carried out.

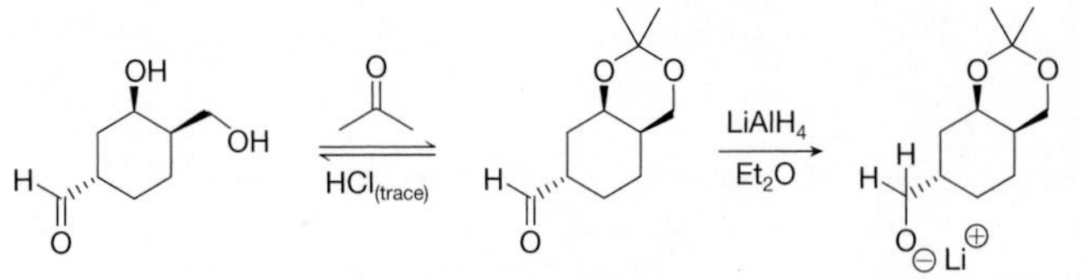

The alkoxide can be used directly as a nucleophile to form the ether. Aqueous workup will then release the original alcohol groups.

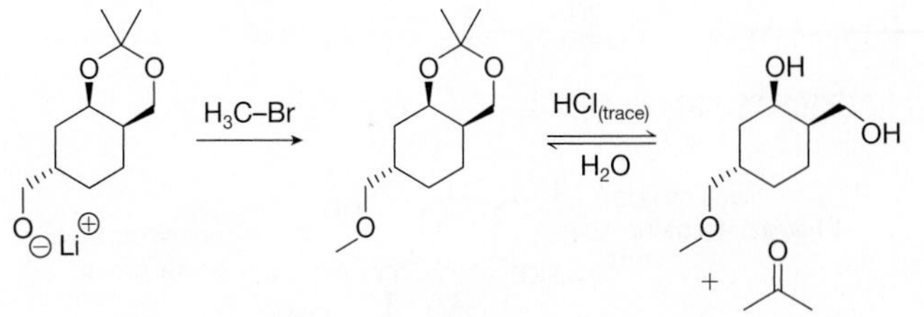

d) The product requires adding two carbon atoms and an alcohol group. Nucleophilic attack on a two-carbon epoxide would accomplish this. The starting material would be the nucleophile by forming a Grignard at the alkyl bromide site. The ketone would need to be protected first to avoid reaction with the Grignard.

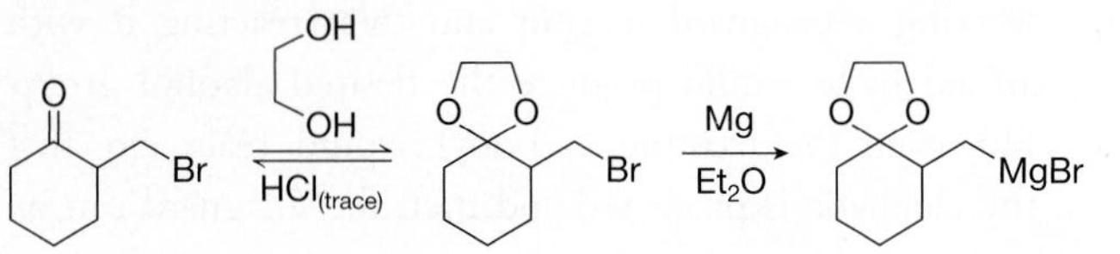

Then the Grignard can be reacted with the epoxide. Aqueous workup releases the alcohol group and deprotects the ketone.

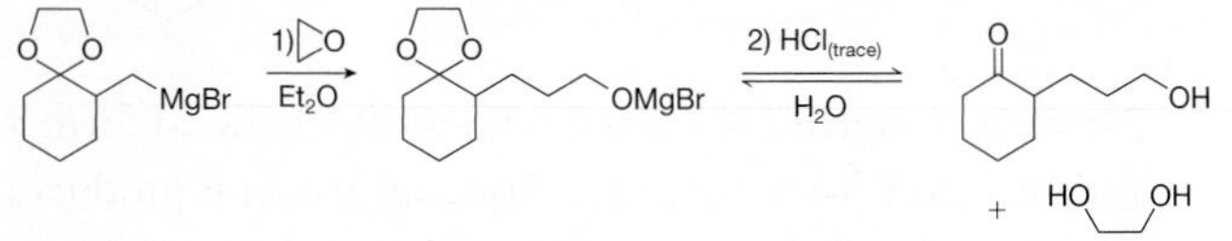

INTEGRATE THE SKILL 16.4

a)

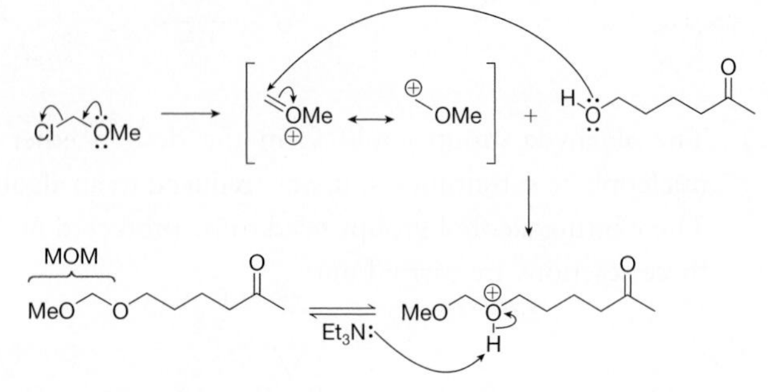

b)

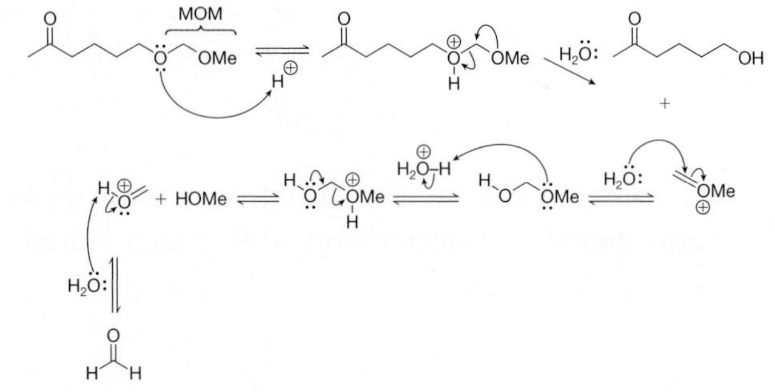

PRACTICE PROBLEM 16.5

a) α-anomeric carbon hemiacetal group; β-anomeric carbon acetal group

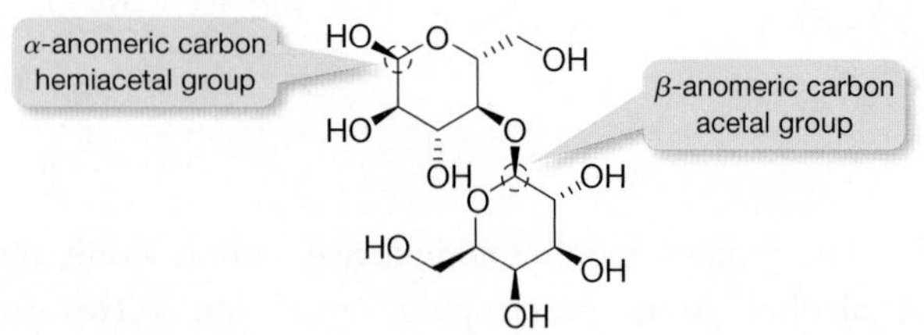

b) β-anomeric carbon acetal group; α-anomeric carbon acetal group

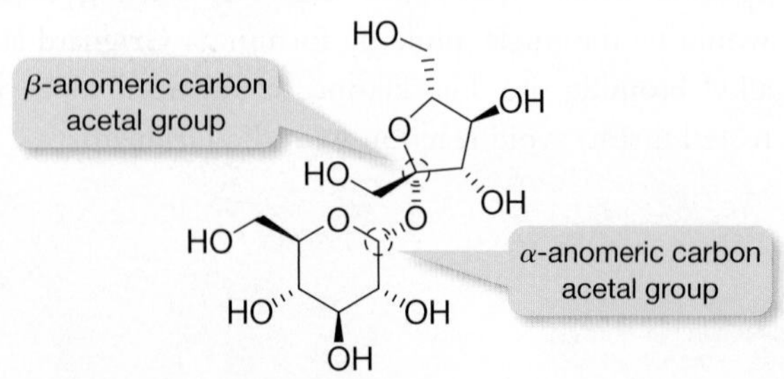

c) Stereochemistry is not shown for two of the anomeric carbons, so a/b designations cannot be assigned.

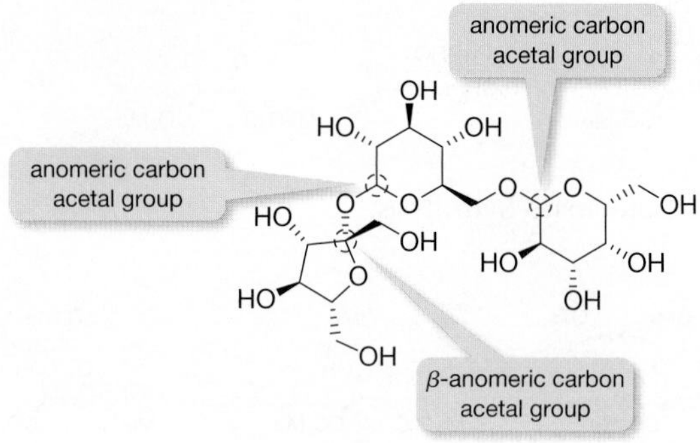

anomeric carbon acetal group; anomeric carbon acetal group; β-anomeric carbon acetal group

INTEGRATE THE SKILL 16.6

a)

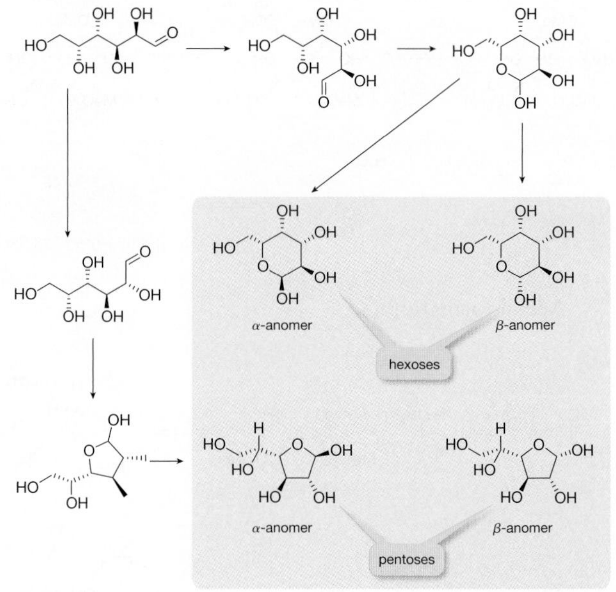

α-anomer; β-anomer; hexoses

α-anomer; β-anomer; pentoses

b) The α and β pentoses are diastereomers of each other. The α and β hexoses are diastereomers of each other. The pentoses are structural isomers of the hexoses.

PRACTICE PROBLEM 16.7

a) These are imine formation conditions.

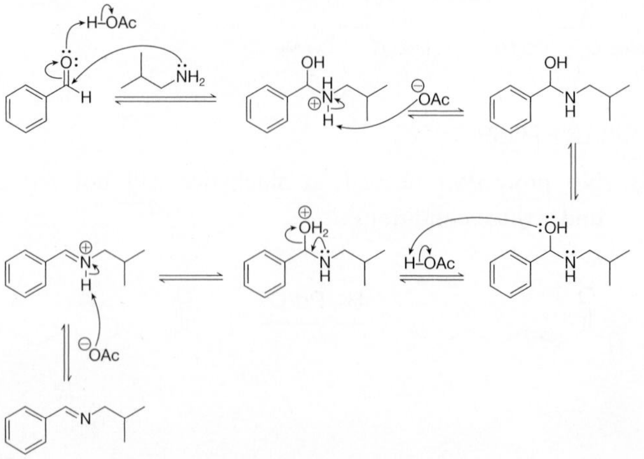

b) These are hydrolysis conditions.

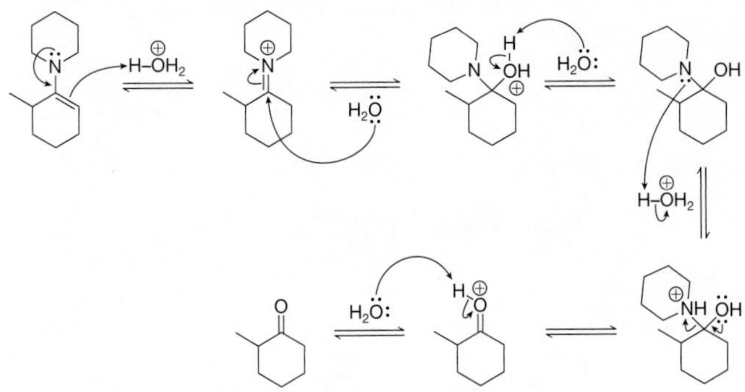

c) These are enamine formation conditions.

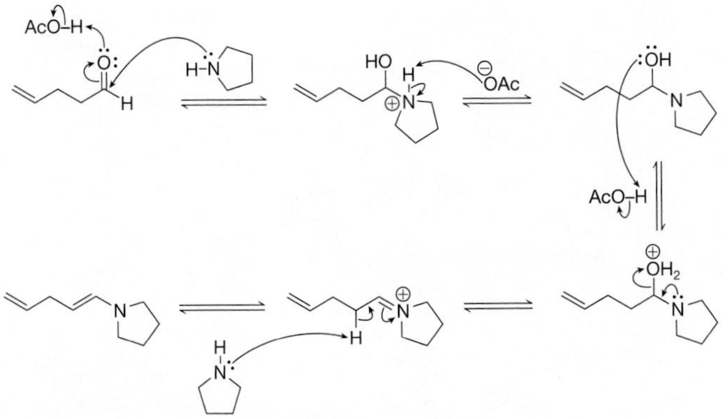

d)

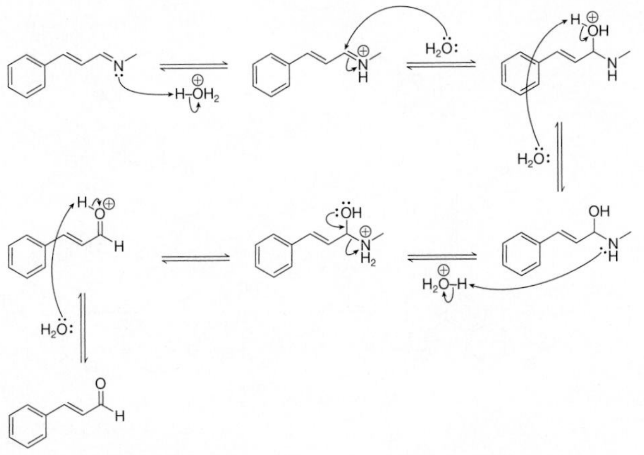

INTEGRATE THE SKILL 16.8

a) Pyridine cannot form an imine because the nitrogen cannot form a double bond, since its π electrons are tied up in the aromatic ring.

b)

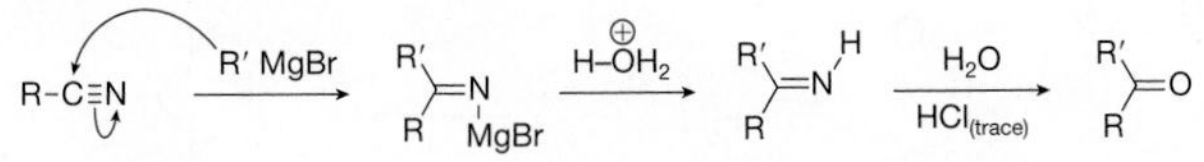

c)

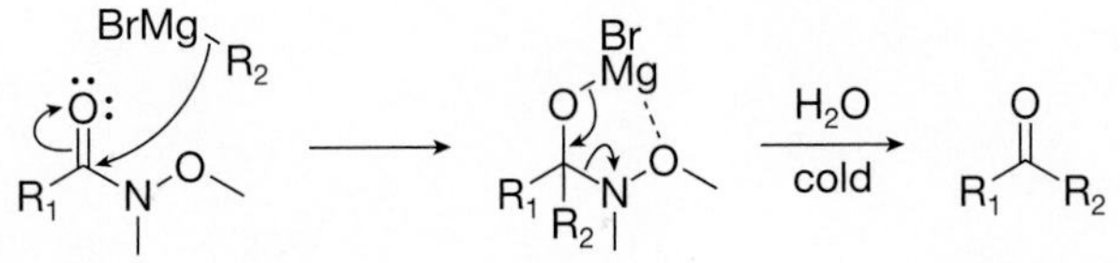

d) Step 1

Step 2

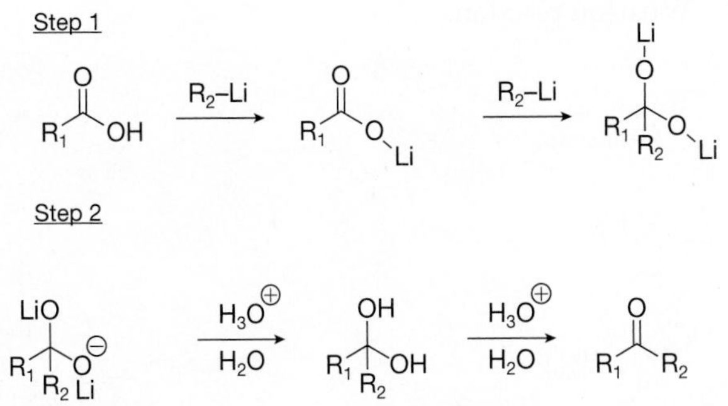

PRACTICE PROBLEM 16.9

a) These are Wolff–Kischner reaction conditions.

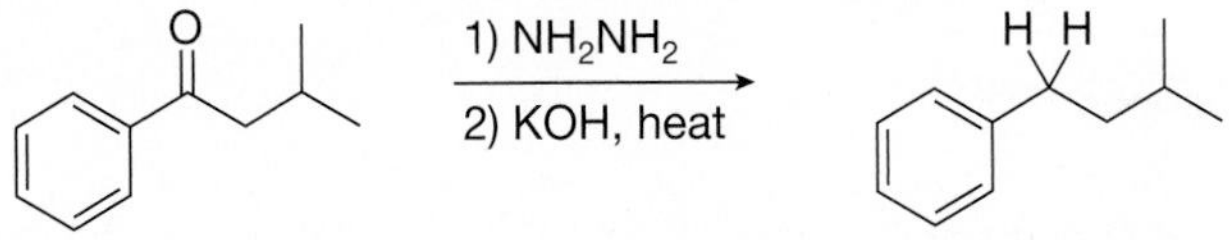

b) Reductive amination:

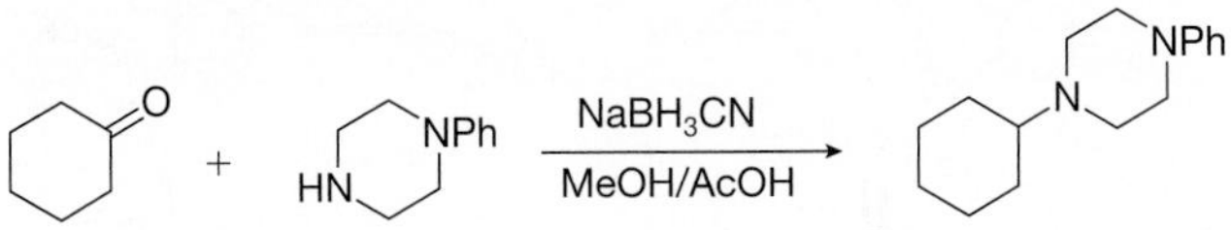

c)

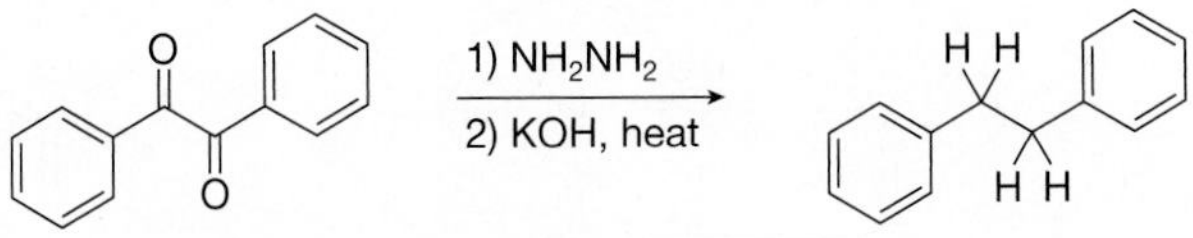

d)

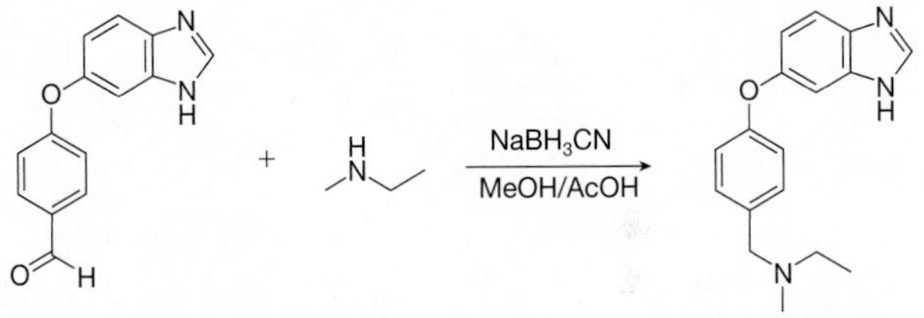

INTEGRATE THE SKILL 16.10

a) The sugar reacts in its linear form (aldehyde group) to form an imine, which is then reduced by the hydride reagent.

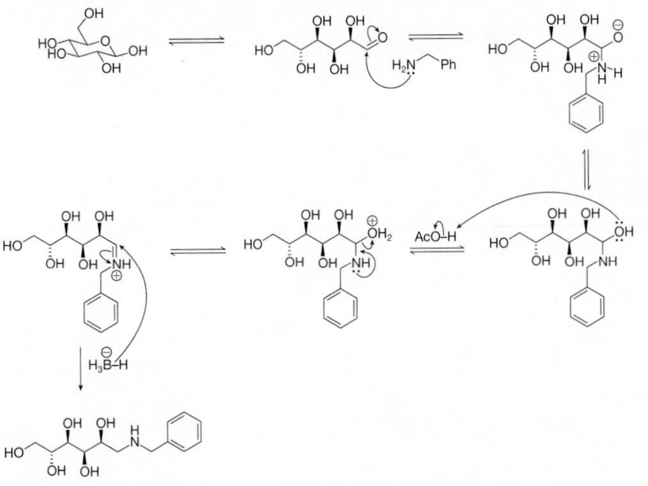

b) Wharton reaction:

Hydrazone formation

$H_2\ddot{N}$–NH_2 H–OMe MeOH MeOH MeOH MeO–H MeOH MeOH H–OMe HO

PRACTICE PROBLEM 16.11

a) i) $H–\overset{\oplus}{O}H_2$ $H_2\overset{\oplus}{O}$–H H_2O H_2O

ii) $H–\overset{\oplus}{O}H_2$ H_2O $H_2\overset{\oplus}{O}$–H $H_3O^{\oplus}$ H_2O

iii) Cl–H H_2O

iv) H_2N–OH EtOH EtOH EtOH EtOH

b) i)

ii)

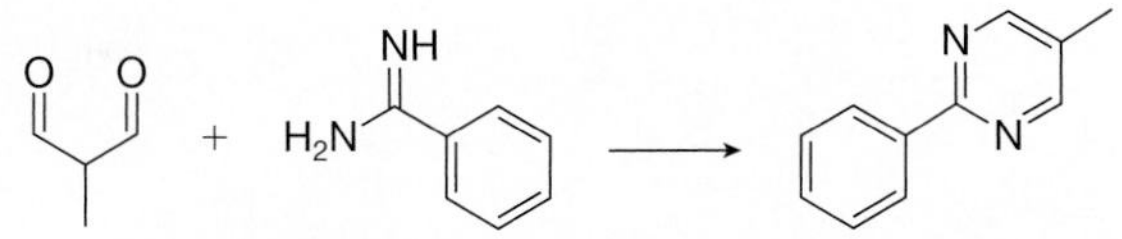

iii)

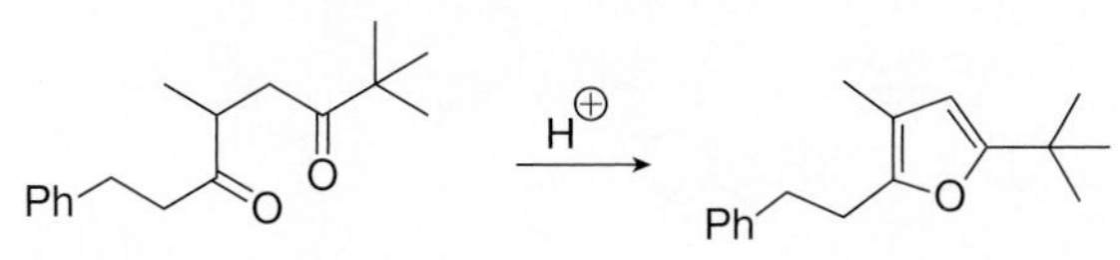

iv)

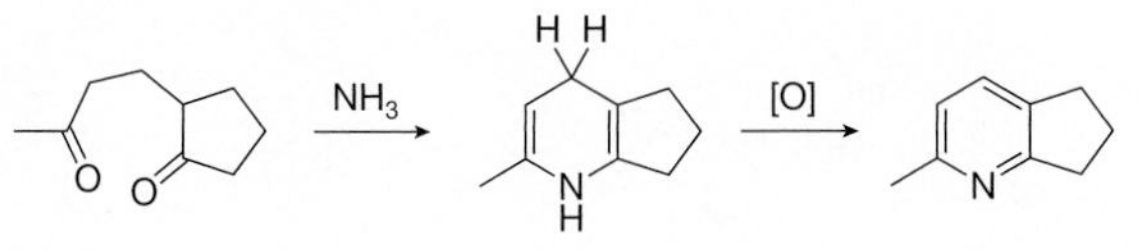

INTEGRATE THE SKILL 16.12

Enamine formation

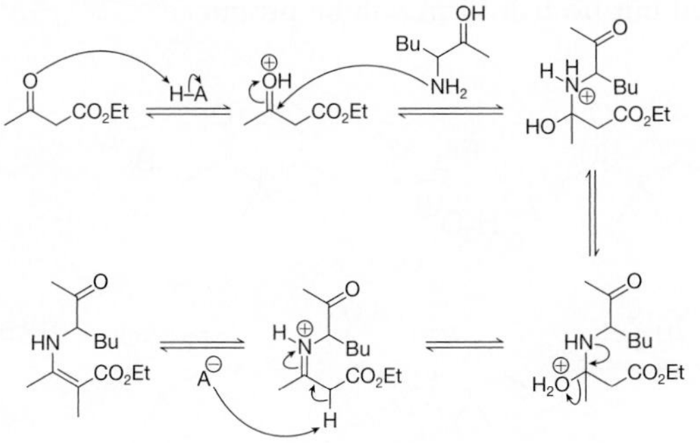

Ring closure

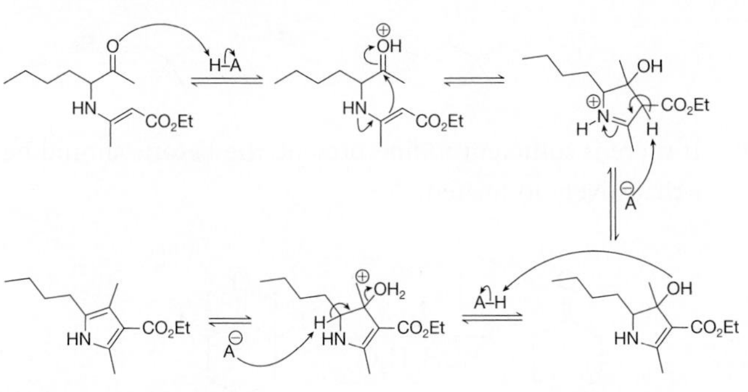

CHAPTER 17

PRACTICE PROBLEM 17.1

a) The pK_a difference is great enough to completely deprotonate the aldehyde. This will be the favoured product.

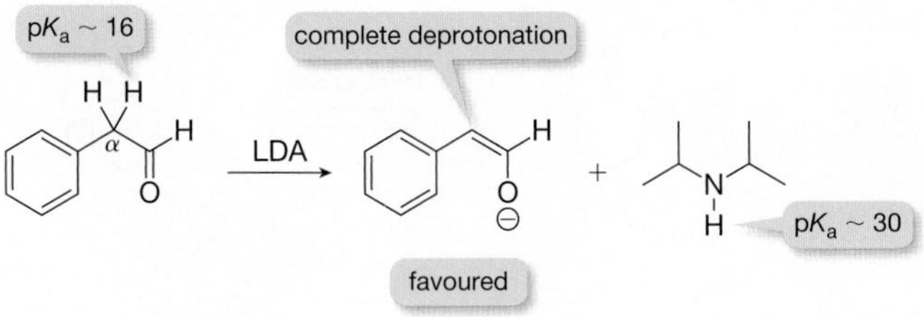

pK_a ~ 16 complete deprotonation LDA pK_a ~ 30 favoured

The pK_a of the amide is much lower than the conjugate acid of LDA, so this reaction will completely form the enolate.

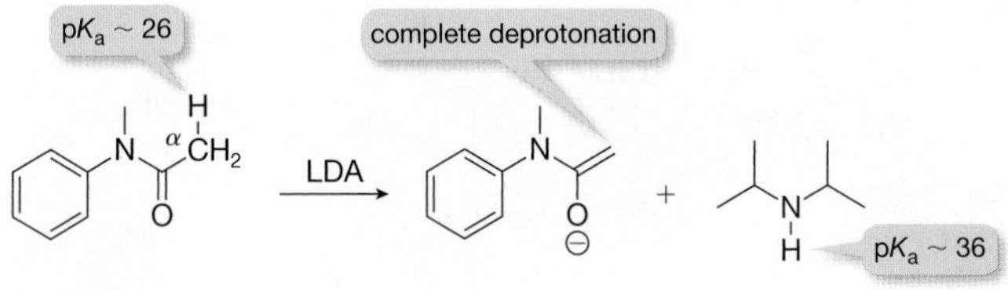

pK_a ~ 26 complete deprotonation LDA pK_a ~ 36

b) Both enolizable protons are weaker acids than the conjugate base formed (methanol). Both reactions will result in only partial deprotonation of the starting material.

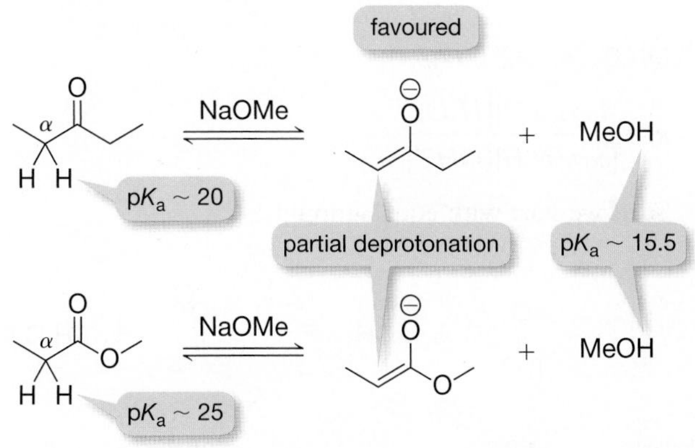

favoured NaOMe MeOH pK_a ~ 20 partial deprotonation pK_a ~ 15.5 NaOMe MeOH pK_a ~ 25

c) The stronger base LDA will produce the favoured reaction with complete deprotonation. Sodium ethoxide will only partially deprotonate the starting ketone.

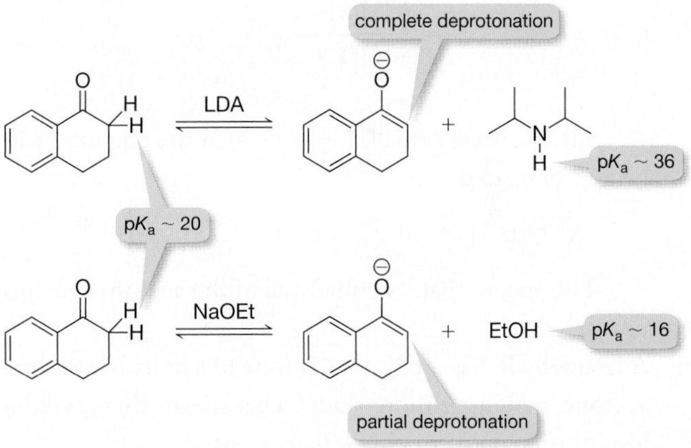

complete deprotonation LDA pK_a ~ 36 pK_a ~ 20 NaOEt EtOH pK_a ~ 16 partial deprotonation

d) Neither starting material will be completely deprotonated. The ester group has slightly more acidic a protons and will be slightly favoured.

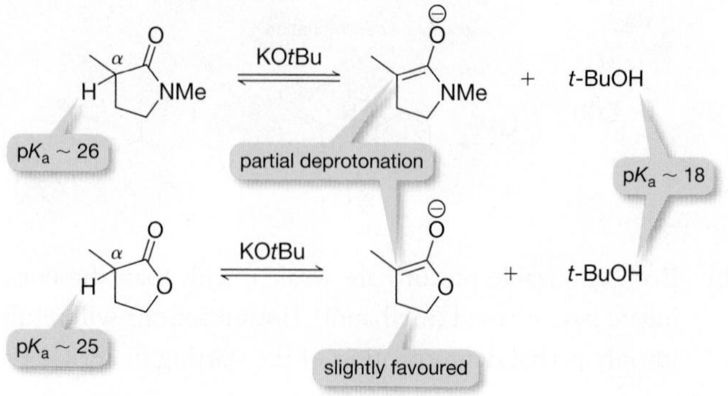

KOtBu
α
H NMe
NMe
+ t-BuOH
$pK_a \sim 26$
partial deprotonation
$pK_a \sim 18$
KOtBu
+ t-BuOH
$pK_a \sim 25$
slightly favoured

INTEGRATE THE SKILL 17.2

a) $K = \dfrac{[acetone^{\ominus}][H_2O]}{[acetone\ H][OH^{\ominus}]}$

So, if we start with equal amounts:

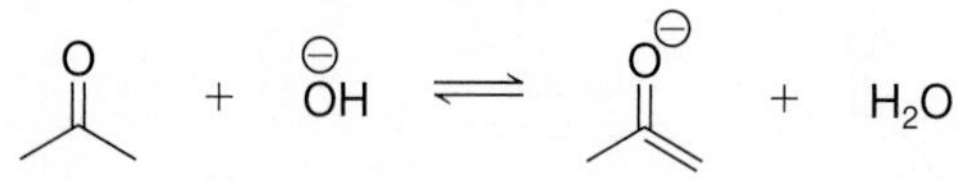

	(acetone) +	$OH^{\ominus}$ ⇌	(enolate) +	H_2O
Initial:	1.0	1.0	0	0
Change:	− x	− x	+ x	+ x
Final:	1.0 − x	1.0 − x	x	x

$$6.31 \times 10^{-20} = \frac{[x][x]}{[1 - x][1 - x]}$$

Since x is very small, $1 - x \approx 1$, so the equation can be approximated as

$$x^2 = 6.31 \times 10^{-20} \quad \text{or} \quad x = 2.51 \times 10^{-10}$$

This means that, essentially, all of the acetone remains unreacted.

b) Acetaldehyde has a pK_a that is four pH units lower than acetone, making it more acidic. That means there will be less unreacted acetaldehyde than acetone.

PRACTICE PROBLEM 17.3

a) Under basic conditions, the methyl ketone should undergo a haloform reaction to give a carboxylic acid.

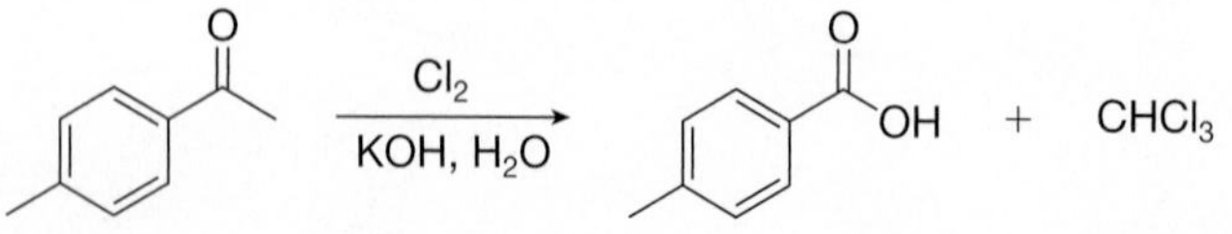

Cl_2 / KOH, H_2O → OH + $CHCl_3$

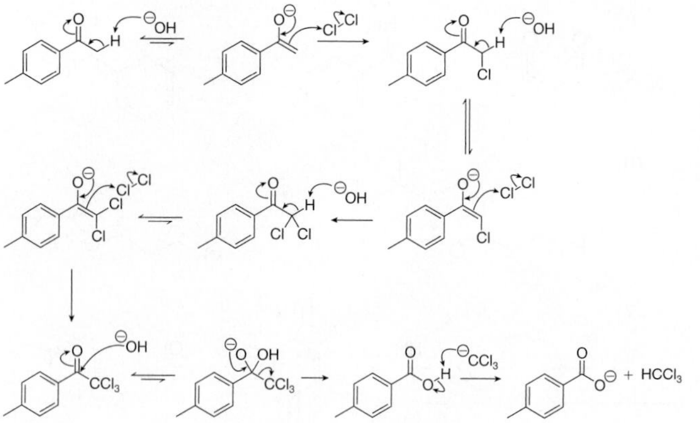

Cl Cl
Cl
Cl Cl
+ $HCCl_3$

b) Under acidic conditions, the methyl ketone will be brominated but no haloform will be produced.

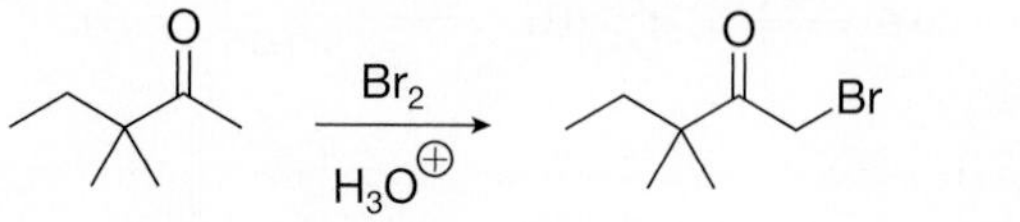

Br_2 / $H_3O^{\oplus}$ → Br

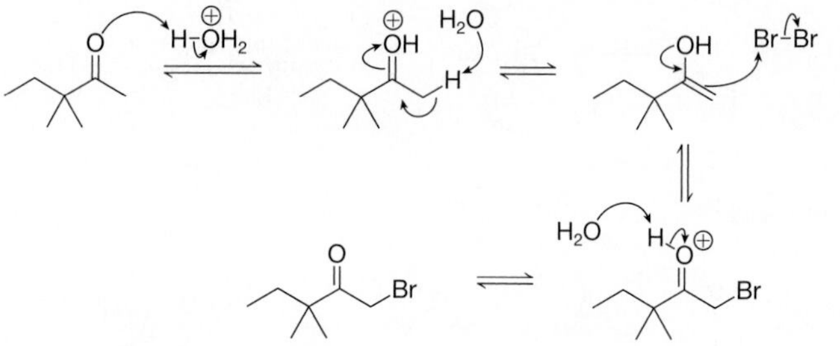

H_2O
H
OH Br–Br
H_2O
Br
Br

c) If there is sufficient iodine present, the ketone should be exhaustively iodinated.

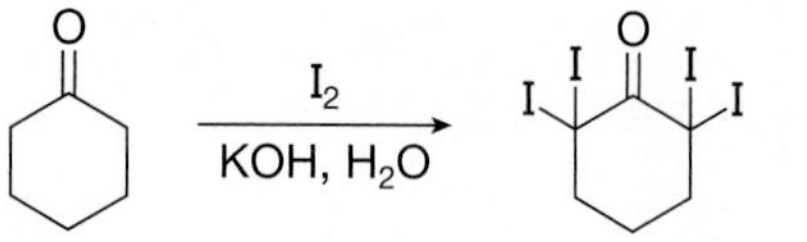

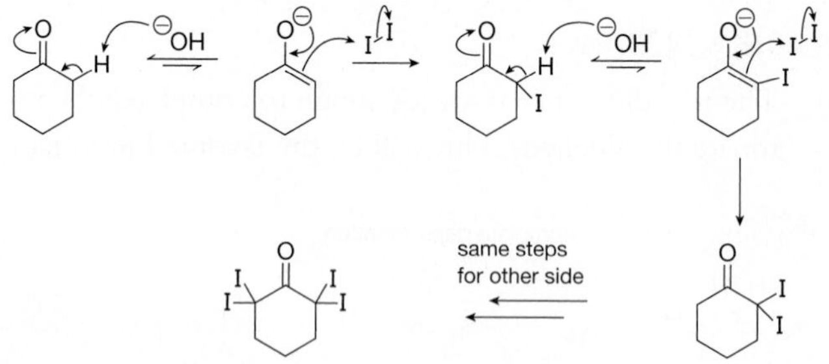

I_2 / KOH, H_2O → I I I I

I–I
I
I–I
I
same steps for other side

d) Both ketones are equally reactive, so both should be chlorinated under acidic conditions.

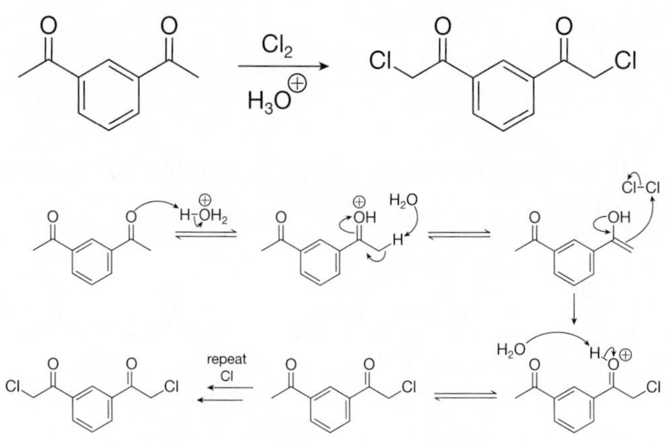

INTEGRATE THE SKILL 17.4

In the first step, the PBr_3 converts the carboxylic acid to an acid bromide, forming HBr as a by-product. The acid bromide is in equilibrium with its enol tautomer, which is then brominated. Draw the mechanisms for the tautomerization of the acid bromide to the enol form and the α-bromination that follows.

Enol formation

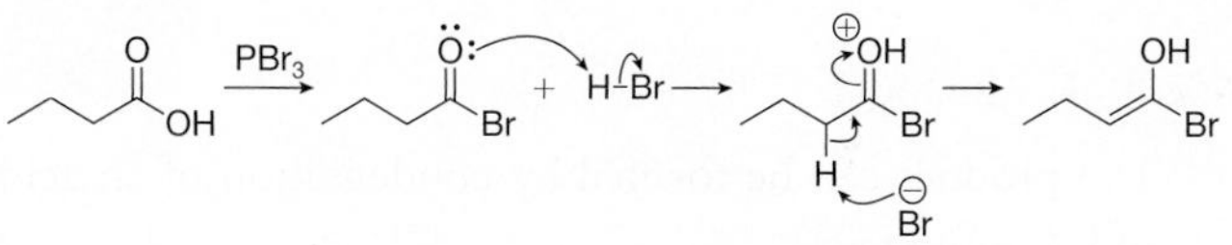

PBr_3

Bromination

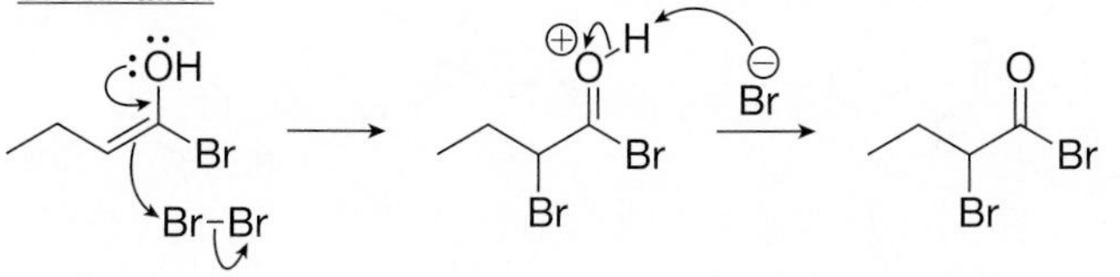

The final carboxylic acid is formed by reacting the acid bromide with water.

PRACTICE PROBLEM 17.5

a)

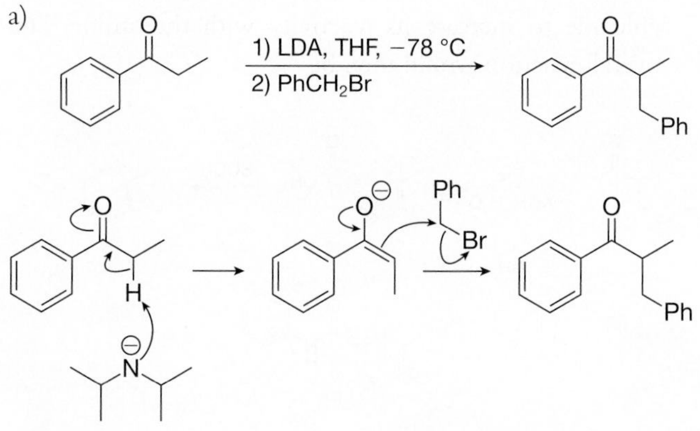

1) LDA, THF, −78 °C
2) $PhCH_2Br$

b) This is a symmetric ketone, so no regioisomers need to be considered.

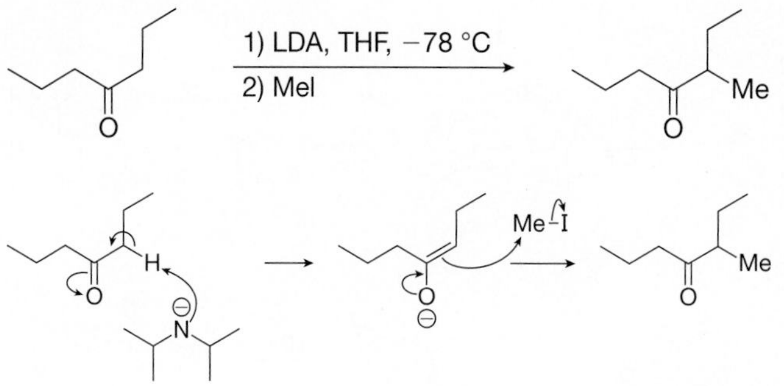

1) LDA, THF, −78 °C
2) MeI

c)

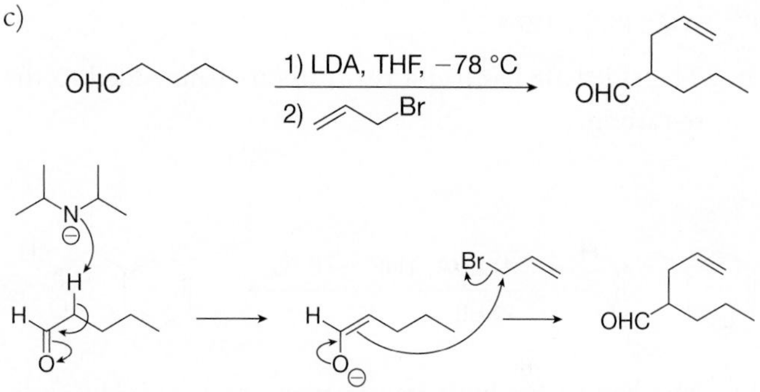

1) LDA, THF, −78 °C
2) allyl bromide

d) There are two different α-hydrogen sites that could give different regioisomers. Reaction of the kinetically favoured site will produce the major product.

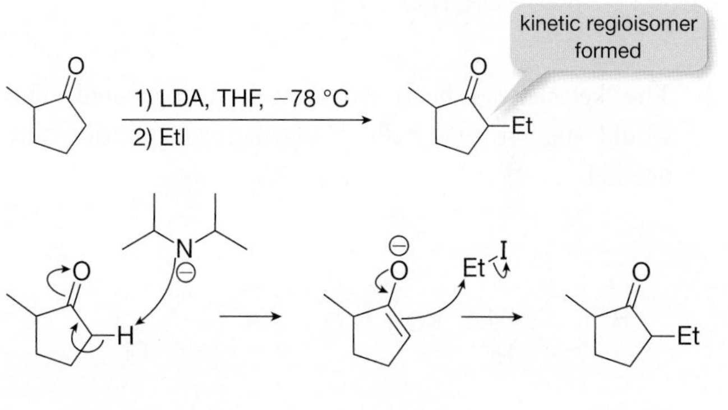

1) LDA, THF, −78 °C
2) EtI

kinetic regioisomer formed

e)

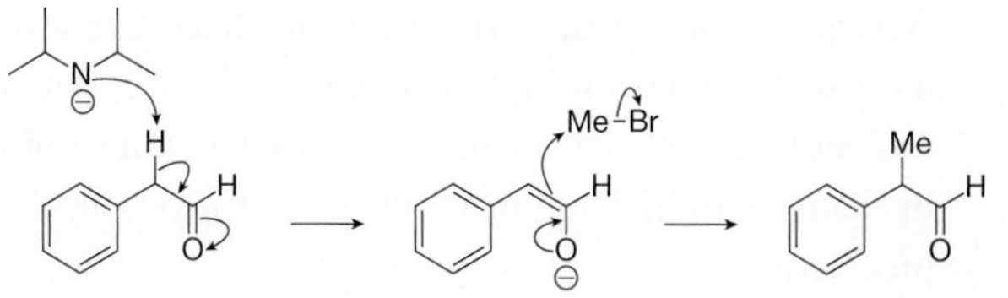

INTEGRATE THE SKILL 17.6

a)

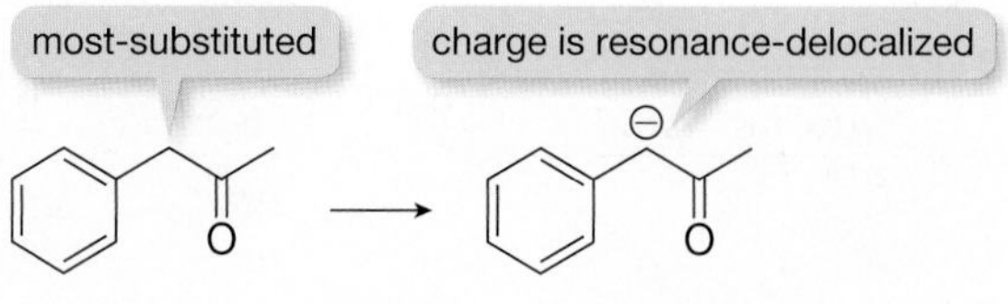

most-substituted

charge is resonance-delocalized

The more-substituted enolate is formed since the carbanion is stabilized by delocalizing the charge through the aromatic ring.

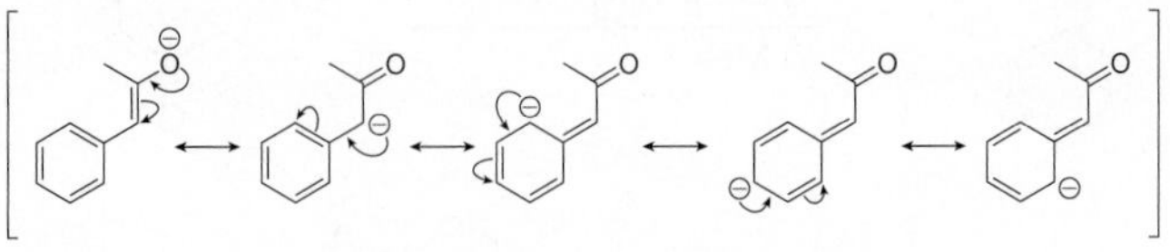

b)

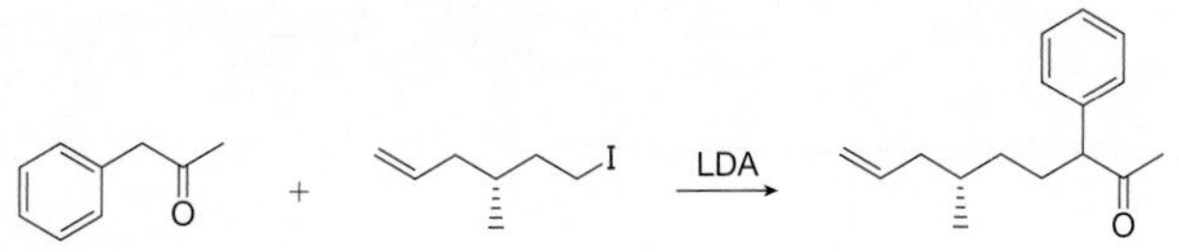

LDA

PRACTICE PROBLEM 17.7

a) The aldehyde has had a two carbon chain added to the α-carbon.

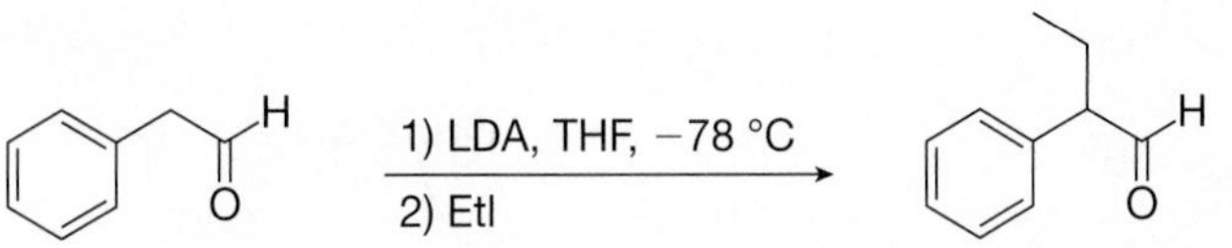

1) LDA, THF, −78 °C
2) EtI

b) The ketone has been transformed to an acid. This indicates a haloform reaction has occurred.

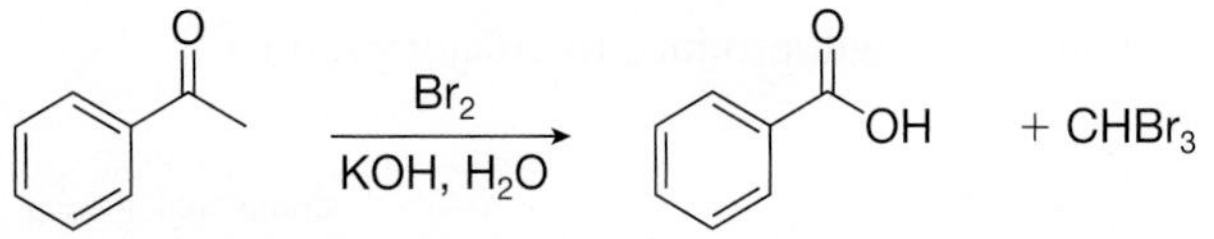

Br_2 / KOH, H_2O ; OH + $CHBr_3$

c) The ketone has been exhaustively halogenated. This would suggest that basic halogenation conditions are needed.

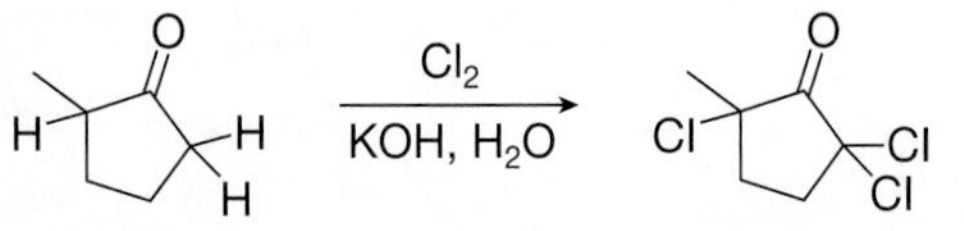

Cl_2 / KOH, H_2O

d) A two-carbon chain and a bromine have been added. This will require multiple reactions. Breaking down the molecule, an alkylation of the methyl ketone followed by halogenation with bromine would give the desired product.

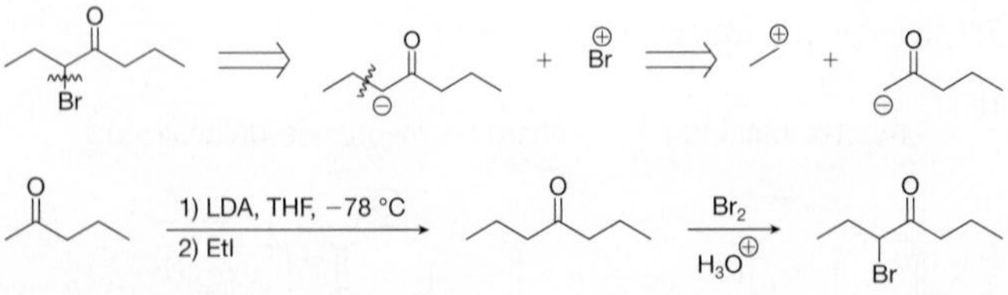

1) LDA, THF, −78 °C
2) EtI

Br_2 / $H_3O^{\oplus}$

e) Breaking the molecule apart, the product can be formed by an alkylation of an enolate.

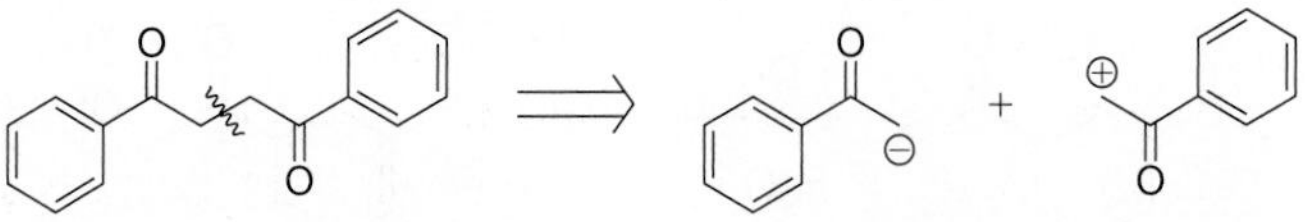

If the required bromo ketone is available the reaction would be

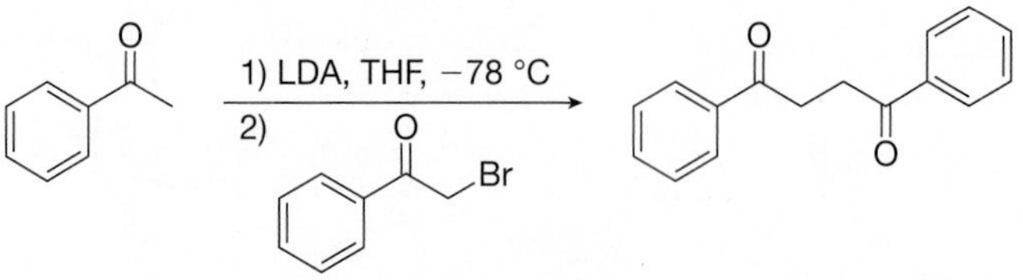

1) LDA, THF, −78 °C
2)

If only the starting ketone is used, the alkyl bromide needed can be generated by bromination under acidic conditions.

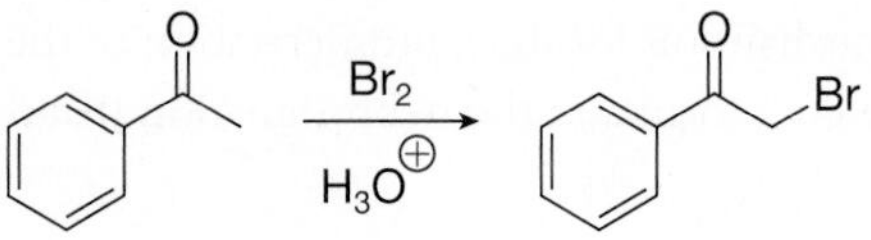

Br_2 / $H_3O^{\oplus}$

INTEGRATE THE SKILL 17.8

a) The product can be formed by condensation of an acid and an amine.

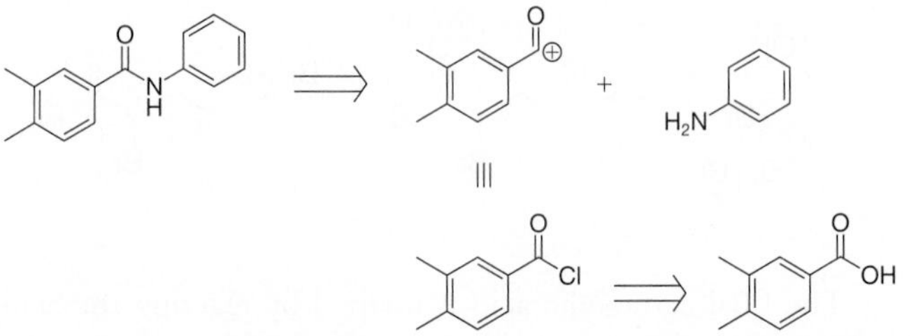

The acid can be produced by using haloform conditions on the initial ketone. The acid is converted to an acid chloride to increase its reactivity with the amine. The overall reaction would then be

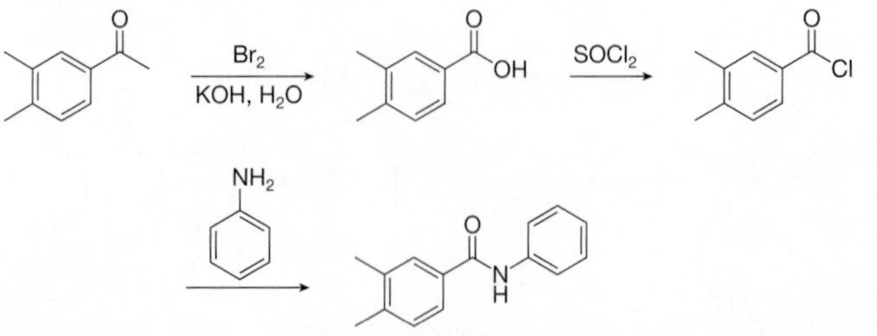

Br_2 / KOH, H_2O ; $SOCl_2$; NH_2

b) The final product is a hemiacetal, which requires a sidechain with an alcohol to be added to the cyclohexanone. The sidechain will be added by alkylation of an enolate with a carbon electrophile (alkyl halide) with an alcohol functional group on it. Since an alcohol would interfere with the alkylation, it must be masked in some way for this step. One possibility is to use an ester, which can be hydrolyzed to produce an alcohol.

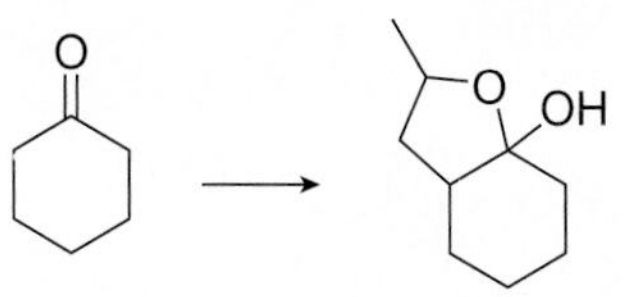

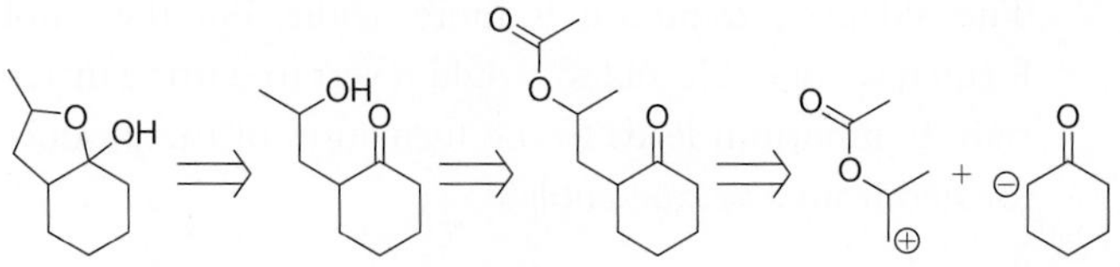

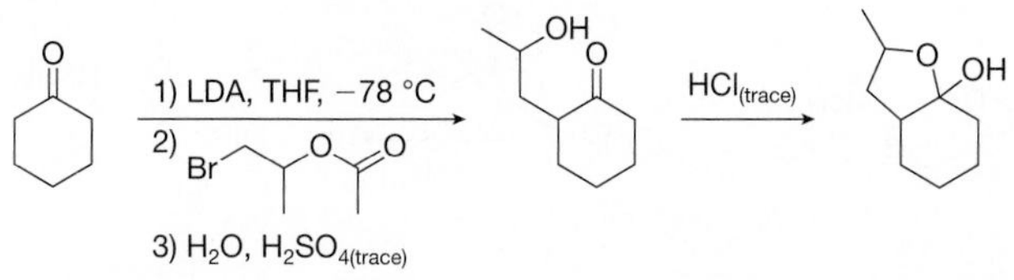

1) LDA, THF, −78 °C
2) Br
3) H_2O, $H_2SO_{4(trace)}$

$HCl_{(trace)}$

Another option is to use an epoxide as the electrophile, which would produce an alcohol functional group on workup, which can then form the hemiacetal. Under basic conditions, the least-substituted carbon of the epoxide is attacked by the nucleophile, giving the correct regioisomer for the synthesis.

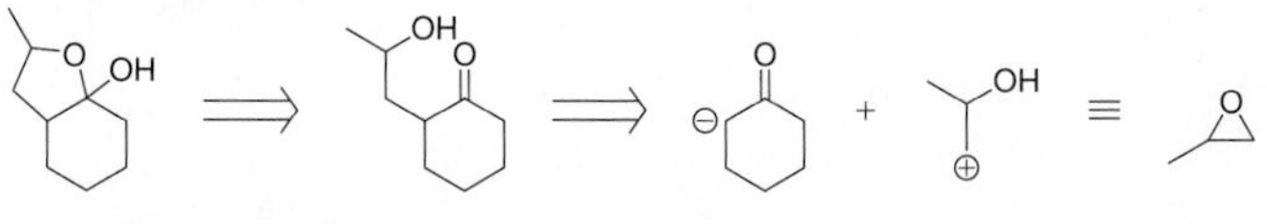

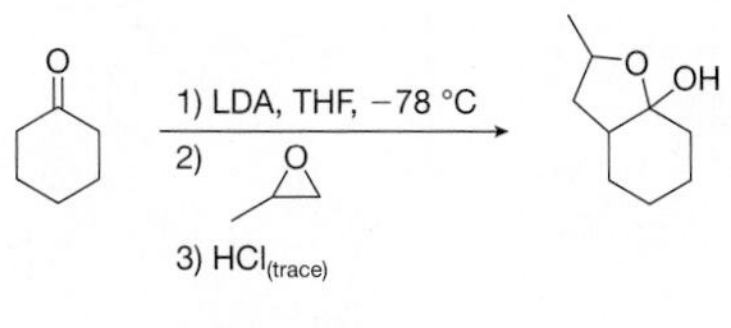

1) LDA, THF, −78 °C
2)
3) $HCl_{(trace)}$

PRACTICE PROBLEM 17.9

a)

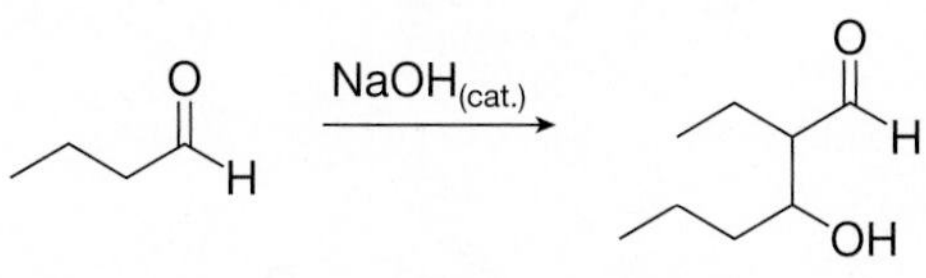

$NaOH_{(cat.)}$

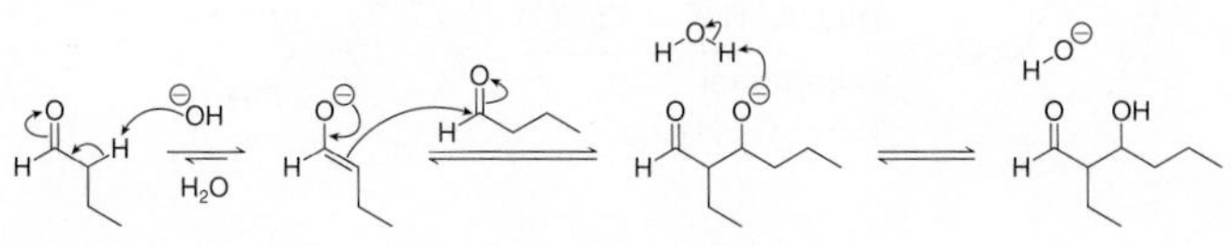

b)

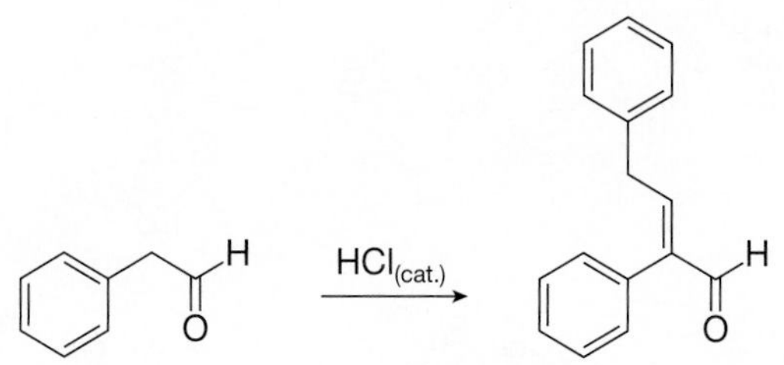

$HCl_{(cat.)}$

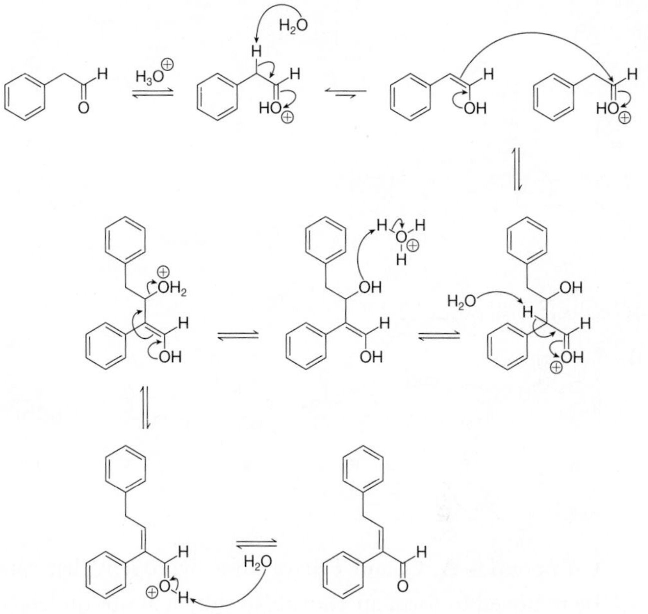

c)

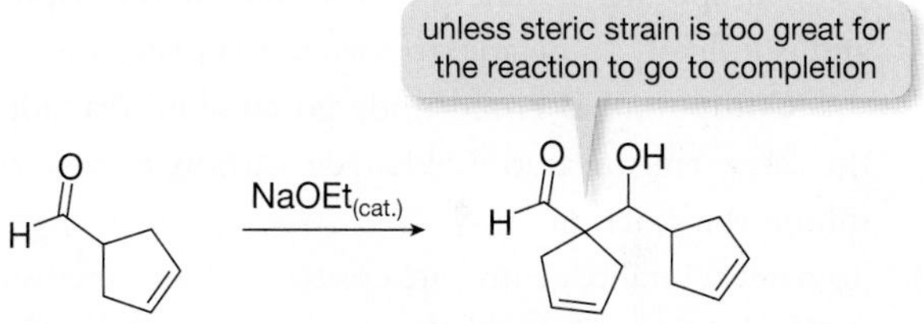

$NaOEt_{(cat.)}$

unless steric strain is too great for the reaction to go to completion

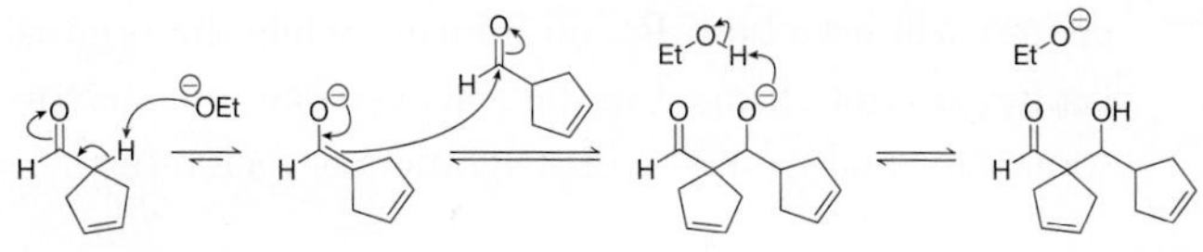

d)

$HCl_{(cat.)}$

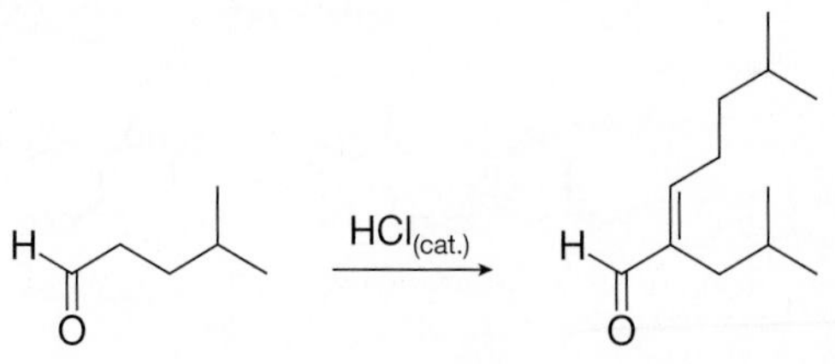

INTEGRATE THE SKILL 17.10

a)

A B C D E F

difficult

Compounds **A**, **C**, and **F** have no a-hydrogens that can be removed to form an enolate, so they will not undergo an aldol reaction.

Compound **E** has double bonds at the a-position and is unlikely to be able to form an enolate.

Compound **B** could undergo an aldol reaction, but the large ring near the aldehyde carbon may sterically inhibit the reaction.

b) Two new chirality centres are created, so four stereoisomers would be formed. Only the two newly formed chirality centres will have both R and S forms, while the original centres are not changed in the reaction. The four stereoisomers formed would be diastereomers of each other.

new chirality centre

$KOH_{(cat.)}$

new chirality centre

PRACTICE PROBLEM 17.11

a)

NaOH

b)

NaOH

The aldehyde α-proton is more acidic, but the aldol formed is not stable and so would revert to starting materials. Equilibrium leads to the formation of the product formed from a ketone enolate

c)

1) Zn, benzene reflux
2) H_3O^+, H_2O

$H_3O^{\oplus}$
H_2O

d)

1) LDA, THF, −78 °C
2) heptanal
3) NH_4Cl, H_2O

H–OH₂ O C₆H₁₃ H O OH

b)

HCl / H_2O

enol formation is assisted by acid

another molecule of starting material is protonated and is attacked by enol

HCl / H_2O heat

intramolecular rearrangement

H_2O

c)

Kt-BuO / Et_2O, heat

INTEGRATE THE SKILL 17.12

The self-condensation product from part (a) in the preceding question would be

NaOH

The reason that this reaction would not reduce the yield is that this reaction is reversible, and the equilibrium lies on the side of the initial ketone due to steric hindrance in the addition product.

all steps are reversible

very hindered carbon

favoured direction of equilibria

Although the crossed-aldol product forms reversibly, the aldol product formation is moderately exothermic, so eventually all of the ketone will be converted.

PRACTICE PROBLEM 17.13

a)

NaOH / H_2O, heat

d)

1) Zn, benzene, reflux
2) CHO
3) HCl, H_2O, 50 °C

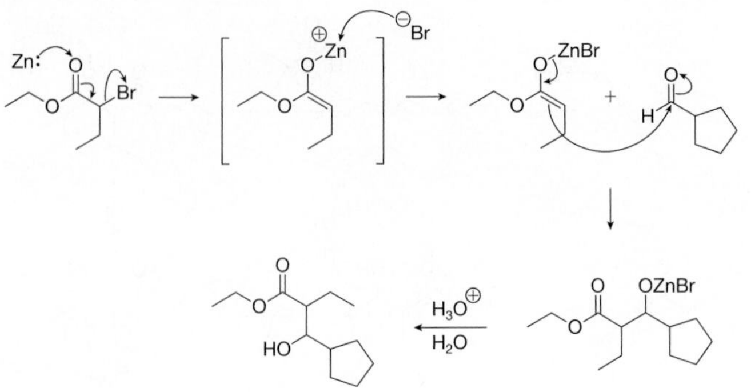

e)

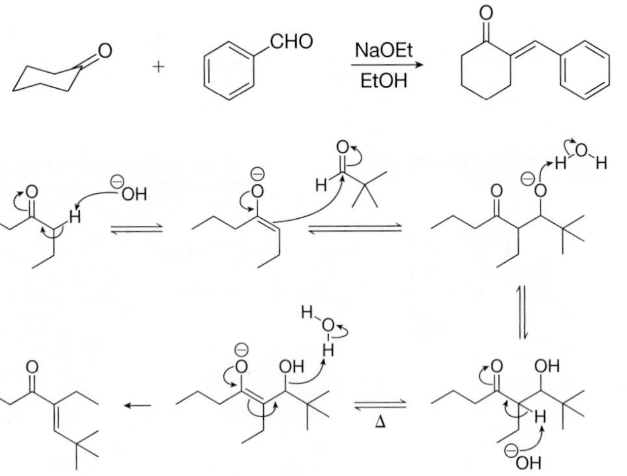

INTEGRATE THE SKILL 17.14

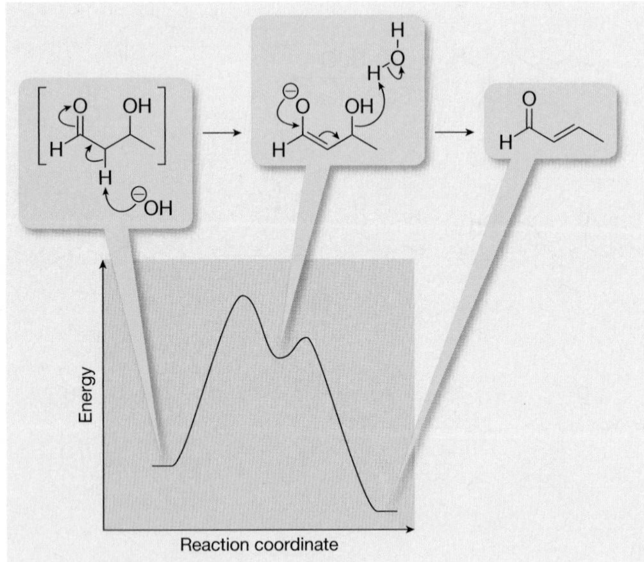

PRACTICE PROBLEM 17.15

a) There are two α-hydrogens that could be removed to form enols that would react with the aldehyde to form rings. The enol formed at C4 would make a four-membered ring, while the enol at C6 would form a six-membered ring. The latter is much more stable and would be preferred. The conditions would favour the aldol product.

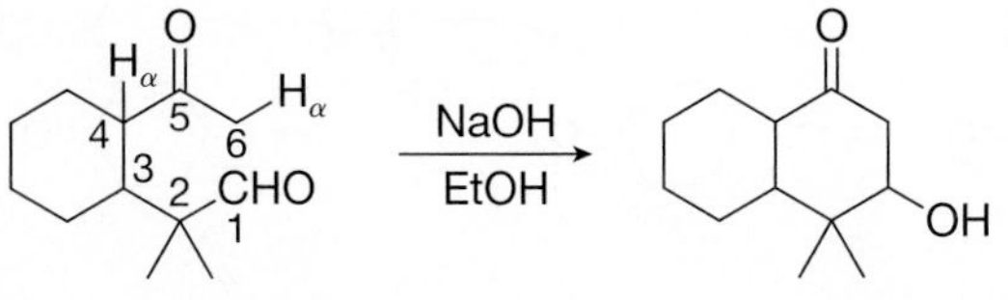

b) There are four potential sites for enol formation. C3 and C4 would lead to three-membered rings and can be ruled out. Enolization at C1 and C6 leads to similar enolates. However, C2 is less sterically hindered and will be easier for the enolate to attack. The conditions would produce an enone product by dehydration.

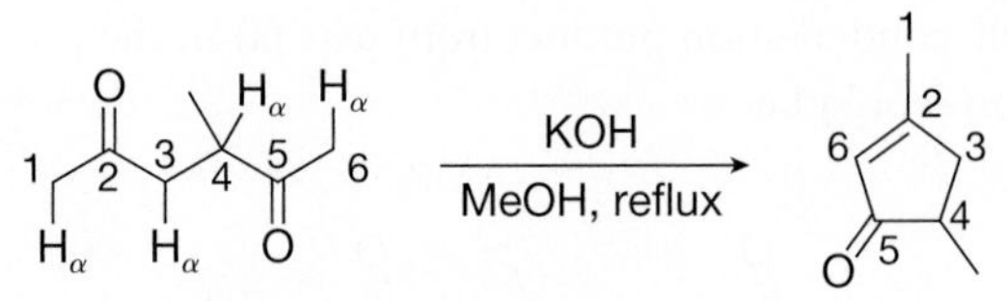

c) There are four possible sites for enol formation. Since an eight-membered ring is unlikely, reaction between a C8 enol and the C1 aldehyde is not expected. The aldehyde is more likely to be the electrophile, so enol formation at C4 or C6 is most reasonable. Reacting the C6 enol with the C1 aldehyde gives a six-membered ring, which is most probable. The reaction conditions would dehydrate the initial product into an enone. The reaction could be quite complicated, as the final product could still react intermolecularly to form aldol condensation products.

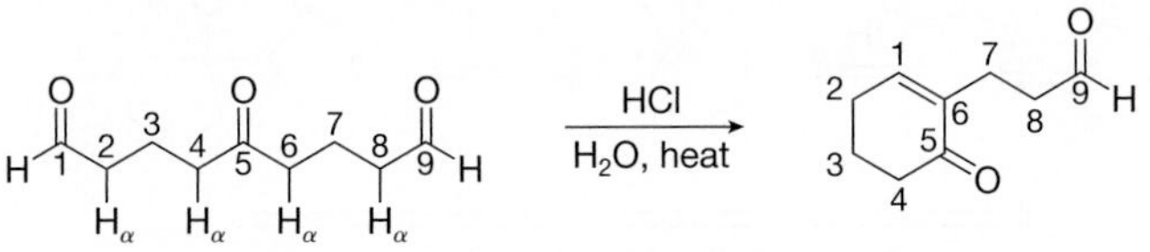

d) The all the α-hydrogen sites would give the same enol due to the symmetry of the molecule. Forming a bond from C2 to C6 gives a five-membered ring fused to a seven-membered ring. The reaction conditions would dehydrate the initial aldol into the enone shown.

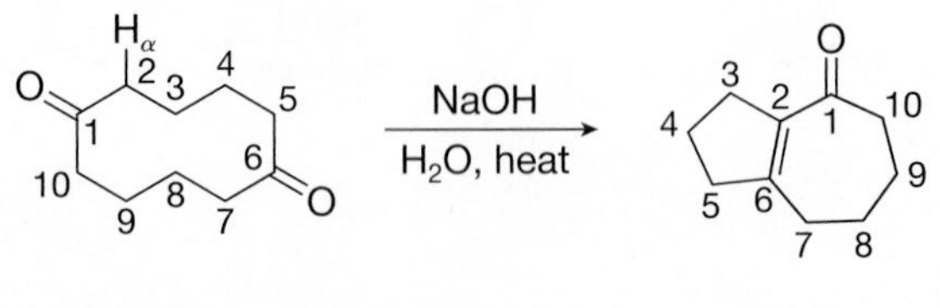

INTEGRATE THE SKILL 17.16

The strained four-membered ring can be viewed as an aldol product. Due to the steric strain, it could undergo a retroaldol to form the parent diketone, compound **A**, shown.

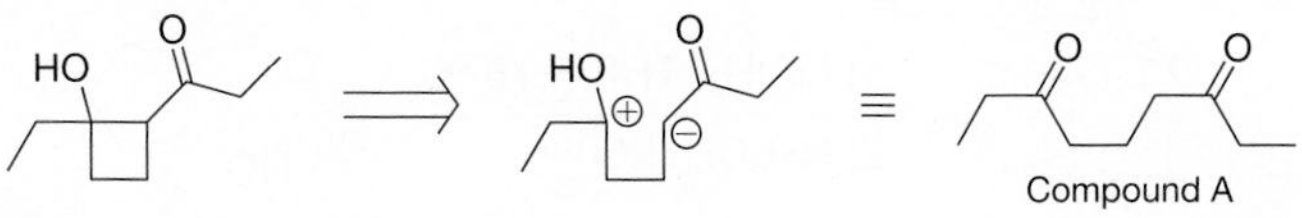

Compound A

The enone product formed can be shown to be derived from an aldol condensation of compound **B**.

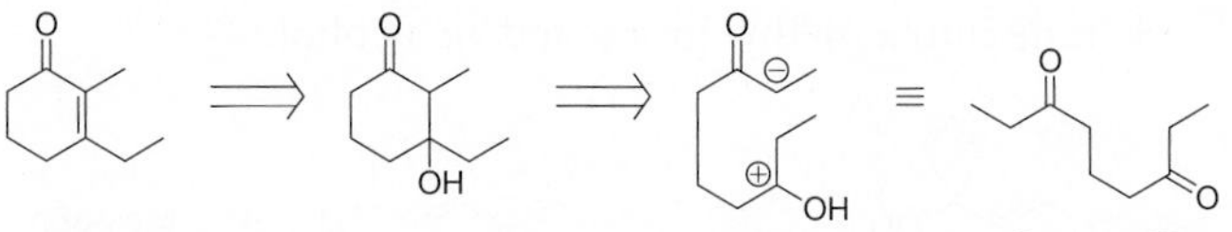

Compound B

Compound **A** and compound **B** are the same molecule. The overall reaction would first be a retroaldol to form the diketone.

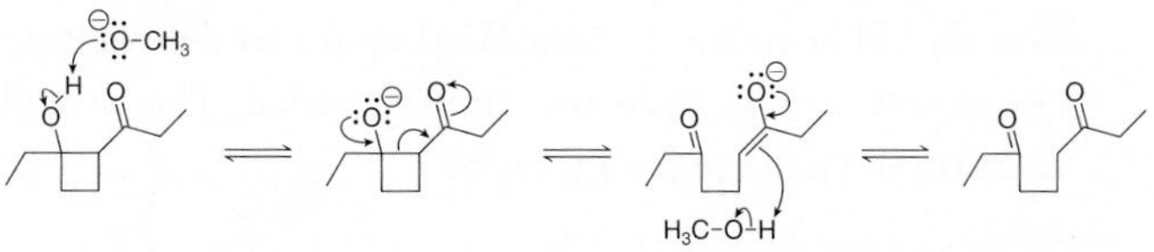

This is followed by aldol condensation and dehydration.

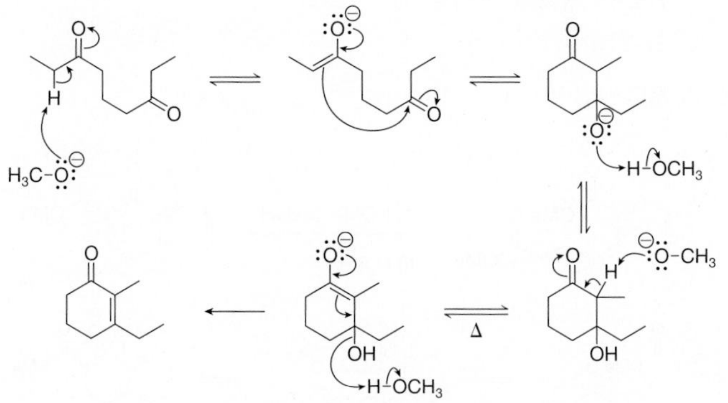

PRACTICE PROBLEM 17.17

a) Breaking down the product, it is the enone from dehydration of an aldol condensate.

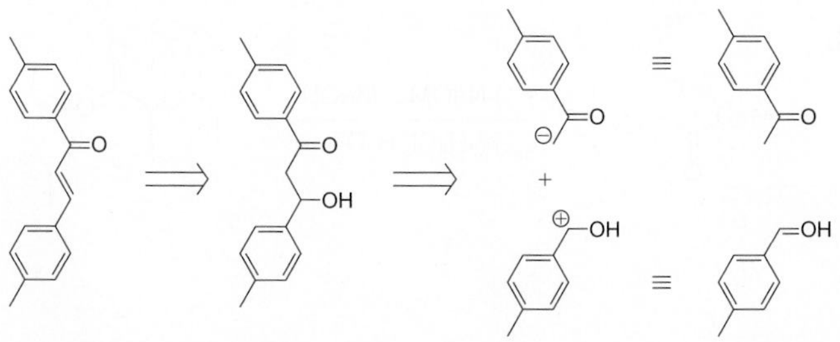

So, conditions that favour enone formation should be used.

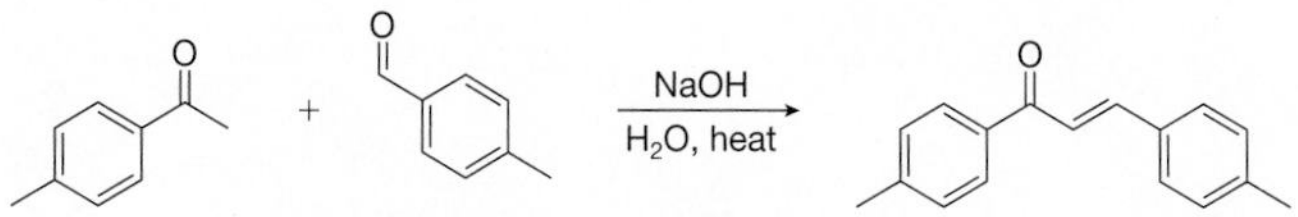

NaOH
H_2O, heat

b) Reaction between the enolate shown and an aldehyde will give the product indicated.

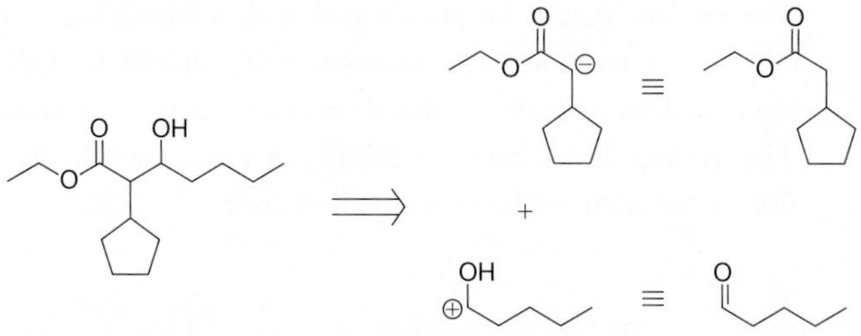

The conditions will have to be mild to avoid dehydration of the aldol product to an enone. To avoid competing reactions, the enolate should be formed first with a strong base followed by aldehyde addition. Workup should be mild hydrolysis conditions.

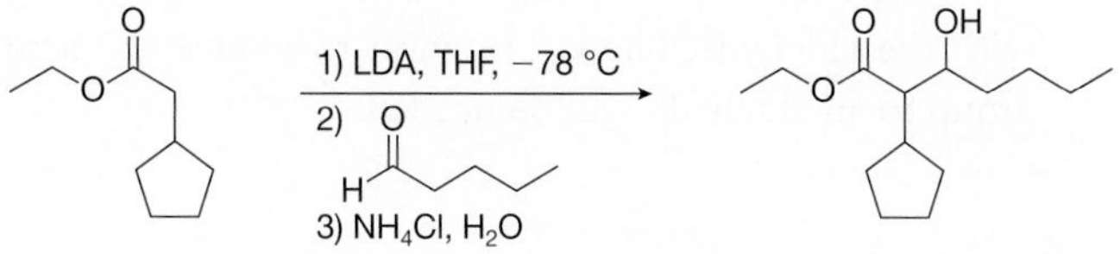

1) LDA, THF, −78 °C
2) (aldehyde)
3) NH_4Cl, H_2O

c) There are two types of reactive hydrogens. Removing H_A to form an enolate, which then reacts with the ketone, forms the desired product. The enolate from removing H_B would have to form a four-membered ring to react, which is unfavourable.

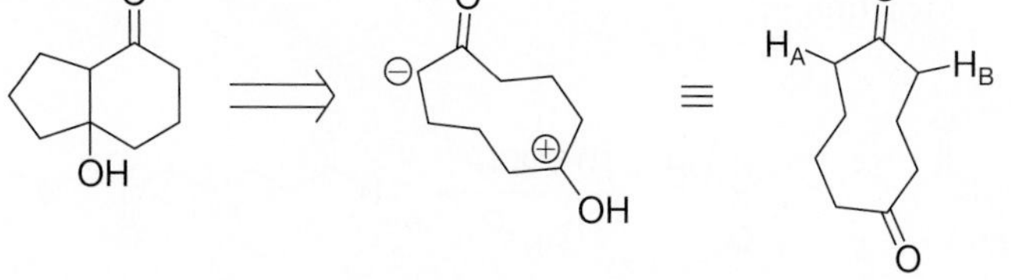

Non-dehydrating conditions are used to isolate the aldol product.

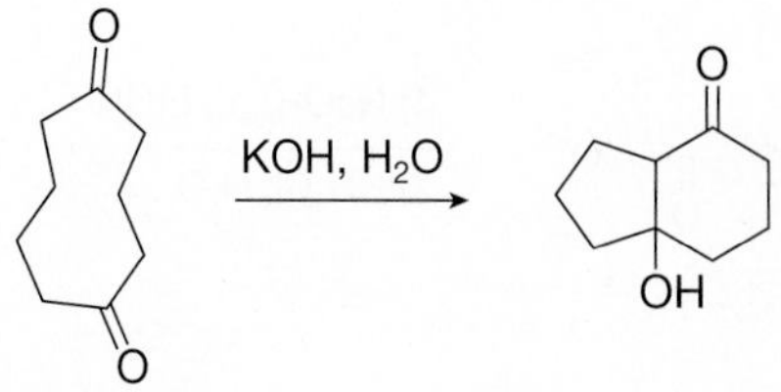

KOH, H_2O

d) The aldol can be formed from the ketone and aldehyde shown.

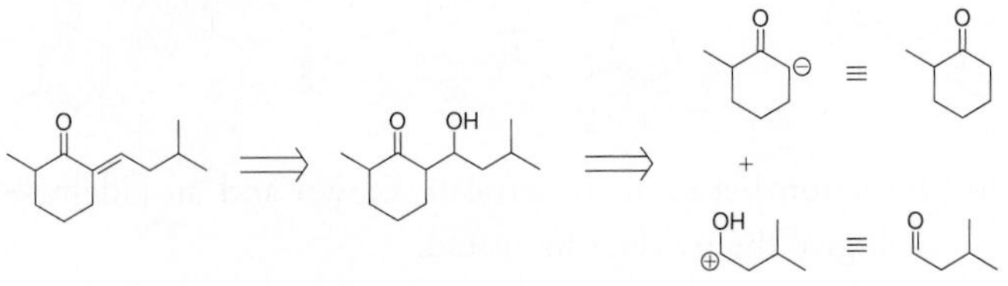

The enolate should be pre-formed with a strong base to avoid cross reactions. The regioselectivity should be OK since the least sterically hindered enolate is the one required. The workup should have conditions that will cause dehydration of the aldol product into the final enone.

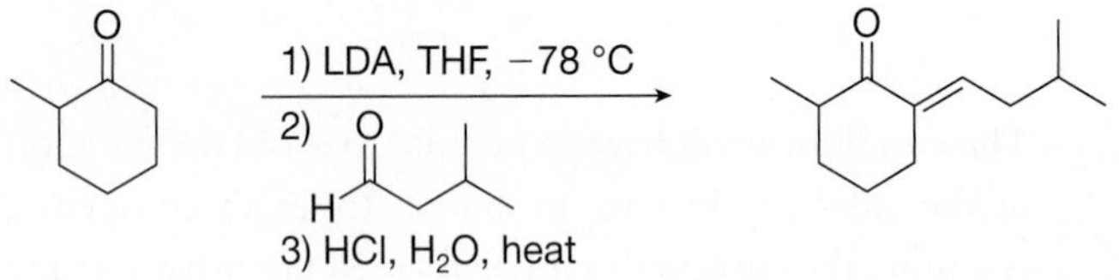

1) LDA, THF, −78 °C
2) aldehyde
3) HCl, H_2O, heat

INTEGRATE THE SKILL 17.18

a) The ring can be formed from a carbanion at C5 reacting with the aldehyde. Then, a reaction to reduce the acid group to an aldehyde will be needed.

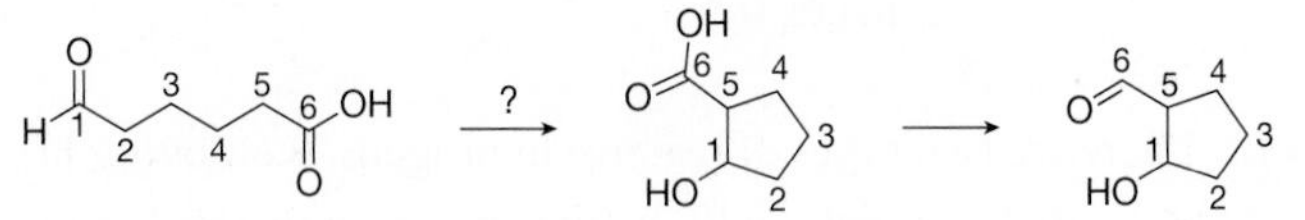

The aldol reaction will first require masking the carboxylic acid group. Selective reduction of the carboxylic acid group could be done by reduction of a Weinreb amide, which could also act as the protecting group for the aldol reaction.

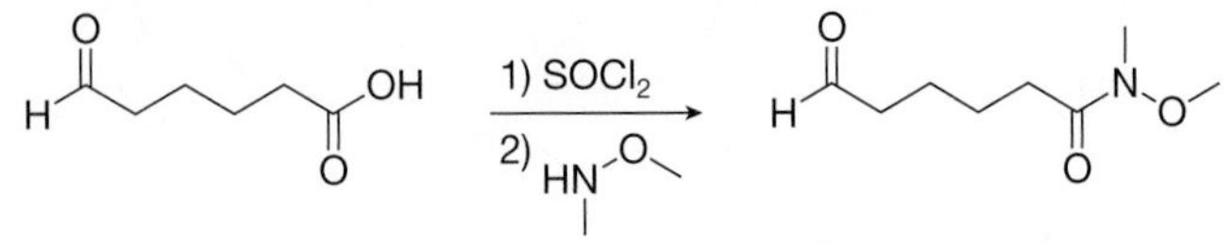

1) $SOCl_2$
2) HN(OMe)Me

Now, catalytic aldol addition using a base will form the ring. A mild workup retains the alcohol group.

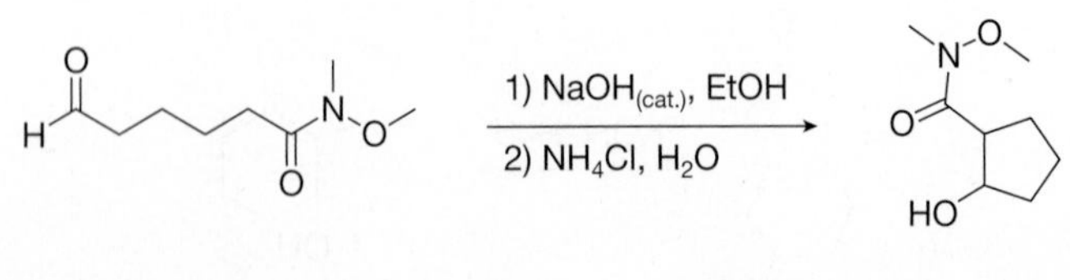

1) $NaOH_{(cat.)}$, EtOH
2) NH_4Cl, H_2O

Reduction of the Weinreb amide with $LiAlH_4$ will give the final aldehyde. A mild acid workup will be needed to release the alcohol.

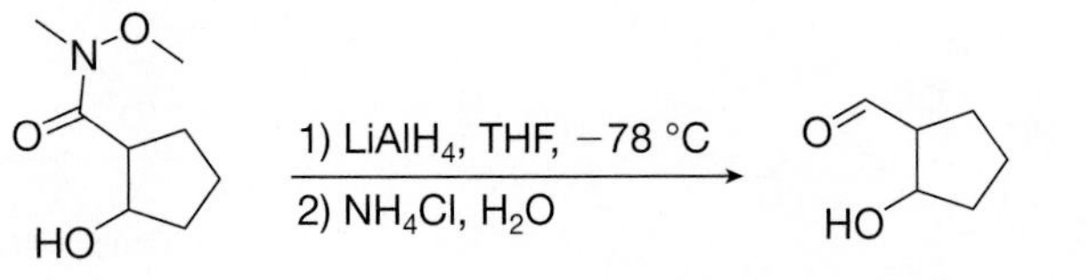

1) $LiAlH_4$, THF, −78 °C
2) NH_4Cl, H_2O

b) The product has a similar structure to an enone, if the alcohol (and one methyl group) is replaced by a ketone. A methyl Grignard addition to a ketone would convert a ketone into a methyl group and an alcohol.

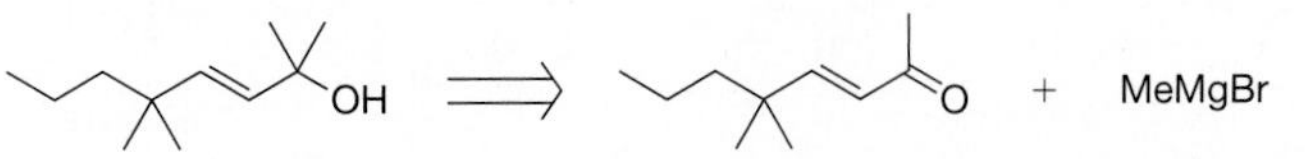

+ MeMgBr

The enone needed is then just a reaction of the starting aldehyde with an acetone enolate. Since acetone can self-condense, pre-forming the enolate before aldehyde addition should be more efficient. Workup under dehydrating conditions will provide the enone needed. The overall reaction sequence would then be

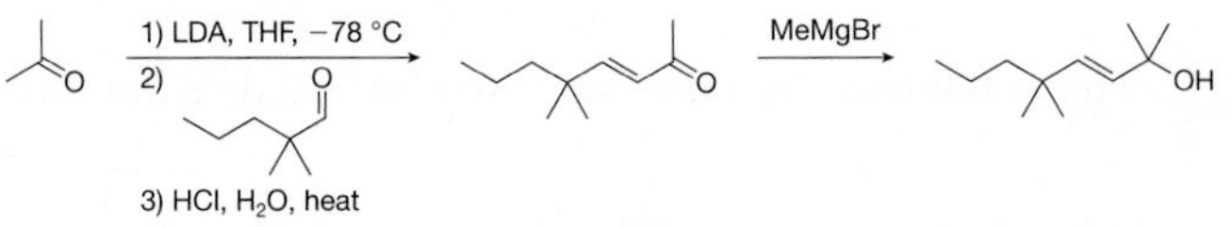

1) LDA, THF, −78 °C
2) aldehyde
3) HCl, H_2O, heat

MeMgBr

PRACTICE PROBLEM 17.19

a)

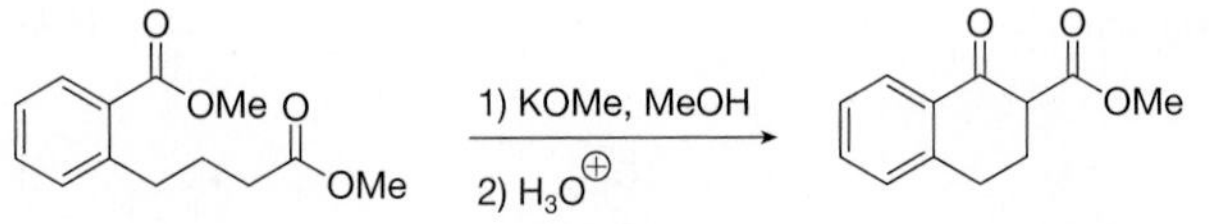

1) KOMe, MeOH
2) H_3O^+

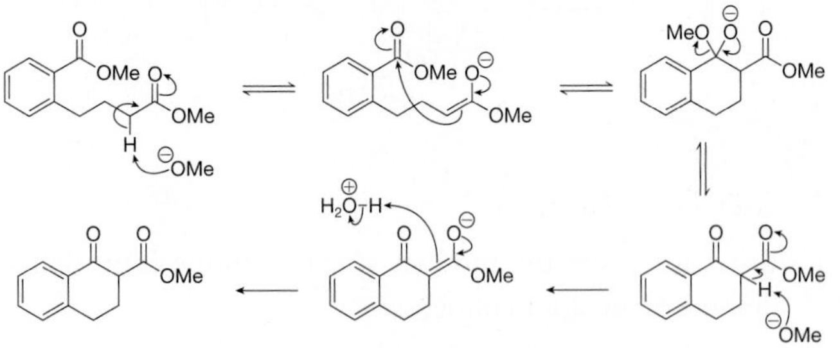

b)

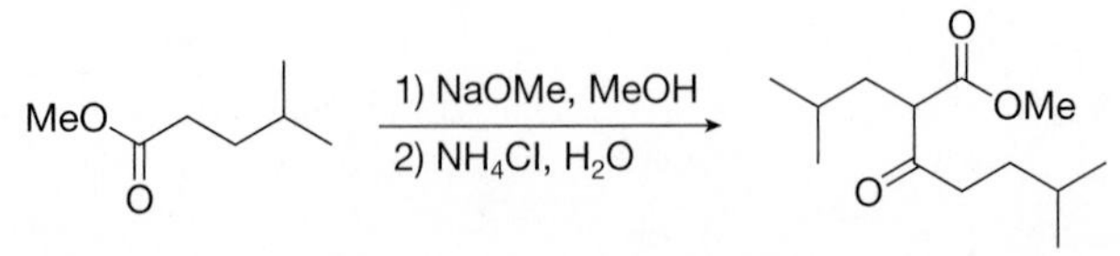

1) NaOMe, MeOH
2) NH_4Cl, H_2O

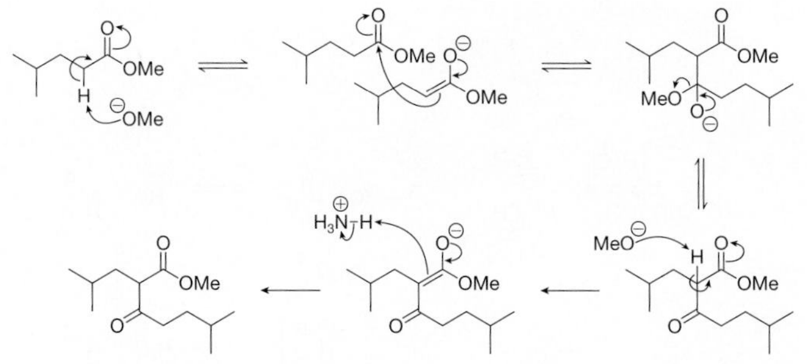

c)

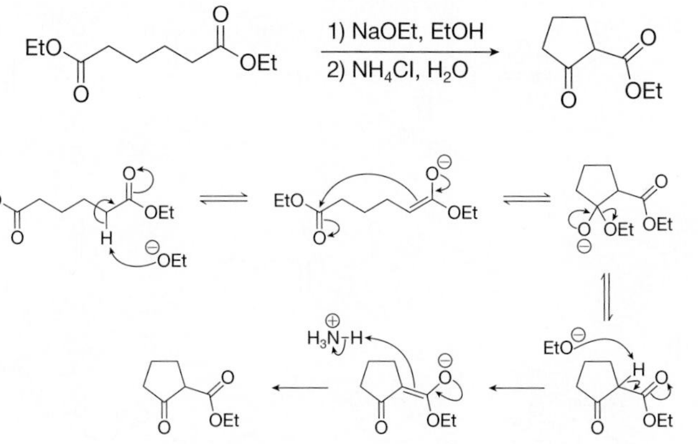

1) NaOEt, EtOH
2) NH_4Cl, H_2O

d)

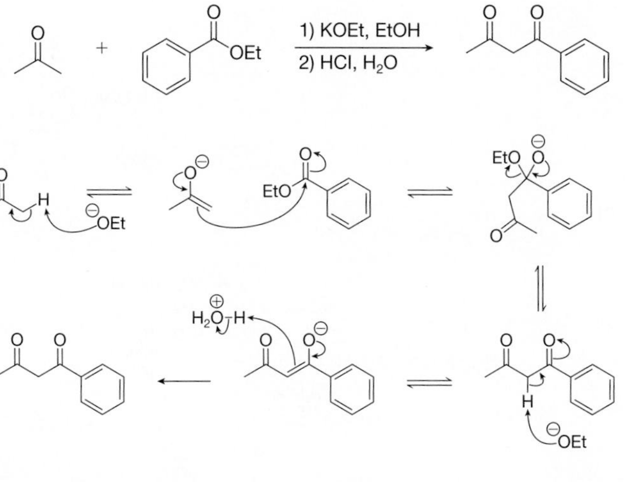

1) KOEt, EtOH
2) HCl, H_2O

INTEGRATE THE SKILL 17.20

a) The aldol condensation is driven forward by the formation of the aldol product and the base is regenerated.

Aldol

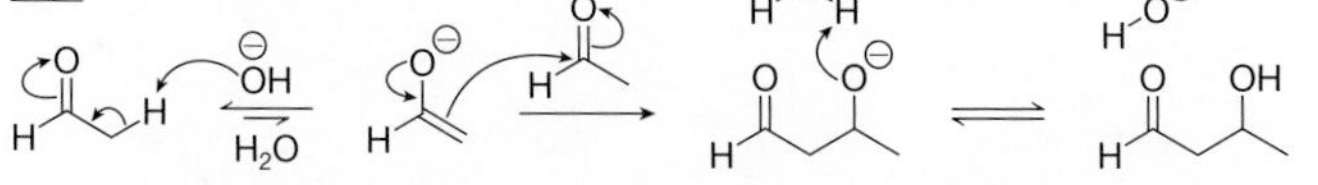

In the Claisen condensation, the driving force is the irreversible formation of the final anion. This step consumes one equivalent of base.

Claisen condensation

1) KOEt
EtOH

2) H_3O^+

irreversible formation of anion pulls all equilibria forward

base consumed

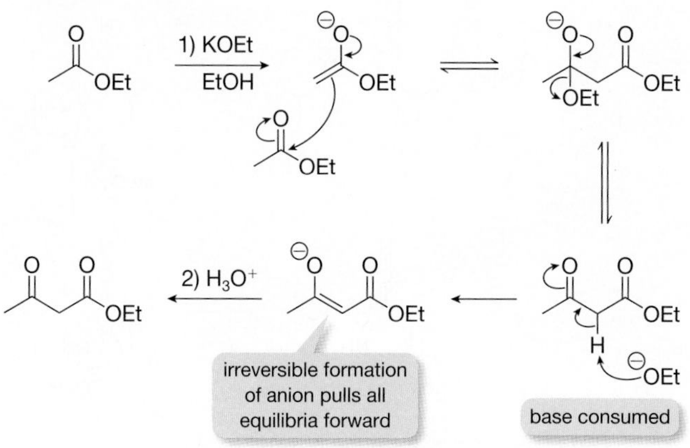

b) With only a single α-proton, there is no way drive the reaction by formation of the anion by hydrogen abstraction.

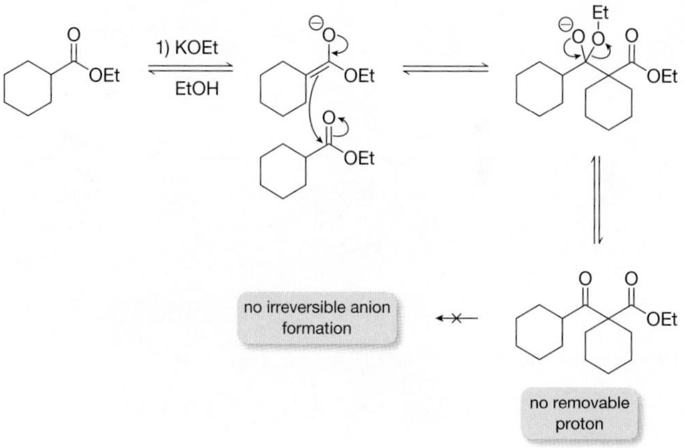

1) KOEt
EtOH

no irreversible anion formation

no removable proton

PRACTICE PROBLEM 17.21

a)

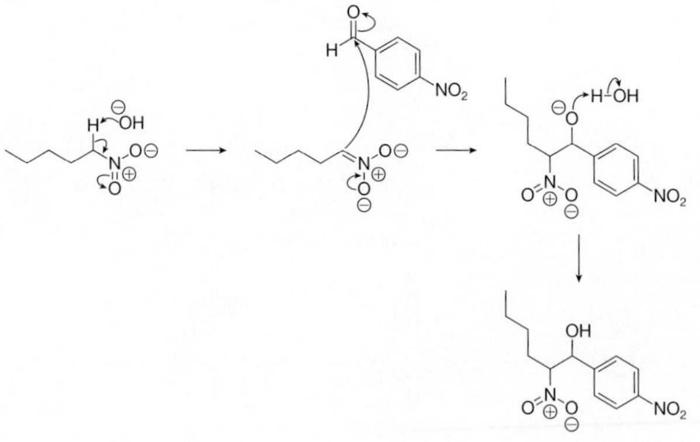

b) Step 1 (NaOMe)

MeO–H

OMe

Step 2 ($H_3O^{\oplus}$, heat)

H_2O

H_2O

Nitrile hydrolysis (Ch. 15)

$\equiv$ R–C≡N

c) $\equiv$ R-NH_2

H-Cl

H_2N–R

$\equiv$

d) OEt

e) F_3C–C$_6$H$_4$–NH_2 $\equiv$ Ar–NH_2

Ph

Ar–NH_2

MeO

$\equiv$

INTEGRATE THE SKILL 17.22

PRACTICE PROBLEM 17.23

a)

Step 1

Step 2

b)

Step 1

Step 2

c)

d)

Step 1

Step 3

Step 2

Acid workup:

Step 4

INTEGRATE THE SKILL 17.24

a)

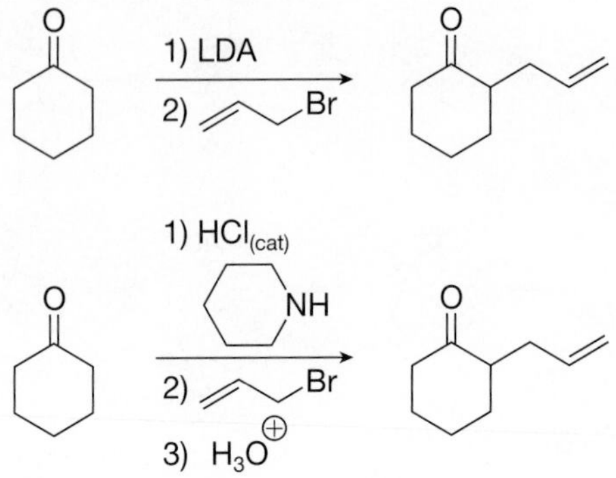

1) LDA
2) allyl bromide (Br)

1) $HCl_{(cat)}$, piperidine (NH)
2) allyl bromide (Br)
3) H_3O^+

b)

1) LDA
2) acetyl chloride (Cl)

1) $HCl_{(cat)}$, pyrrolidine (NH)
2) acetyl chloride (Cl)
3) H_3O^+

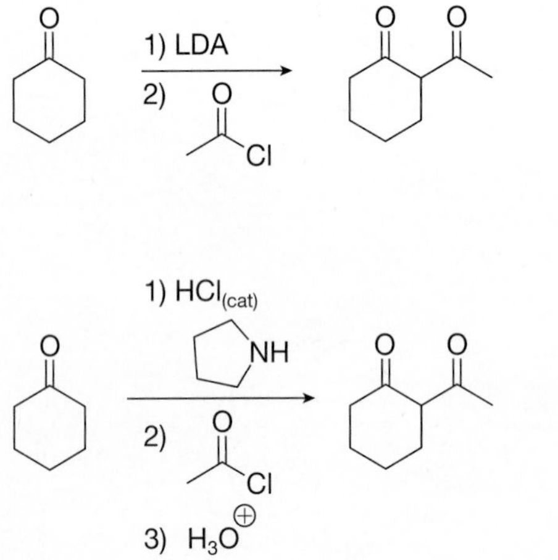

PRACTICE PROBLEM 17.25

a) PPh$_3$ … + $Ph_3P{=}O$

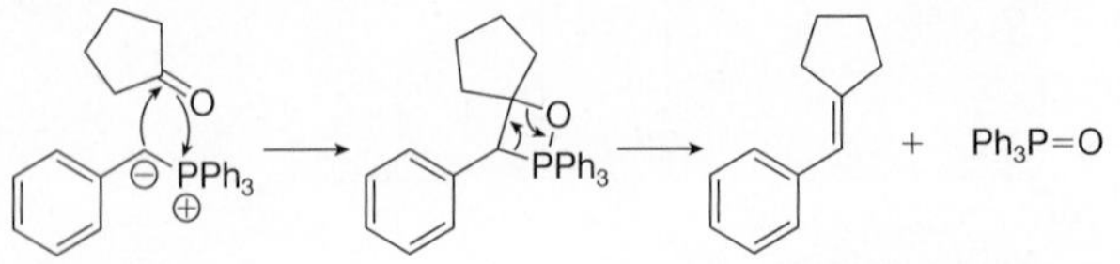

b) PPh$_3$ … + $Ph_3P{=}O$

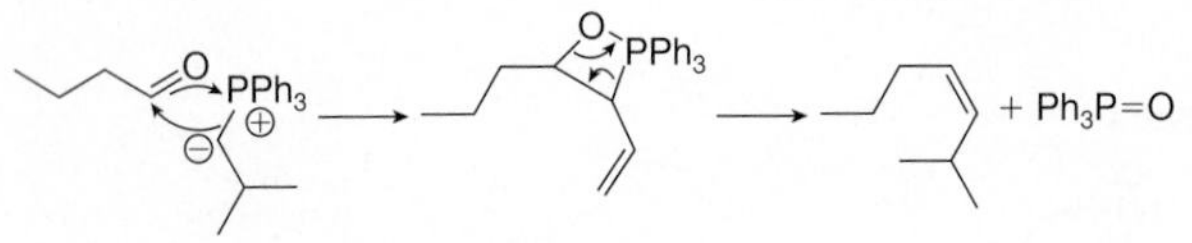

c) PPh$_3$ … + $Ph_3P{=}O$

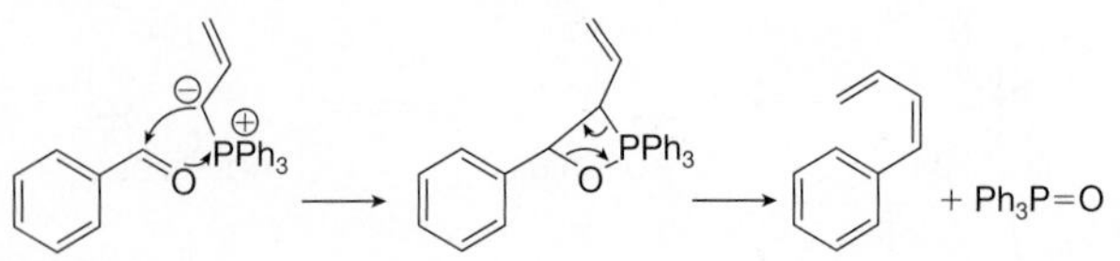

d)

Steps 1 and 2

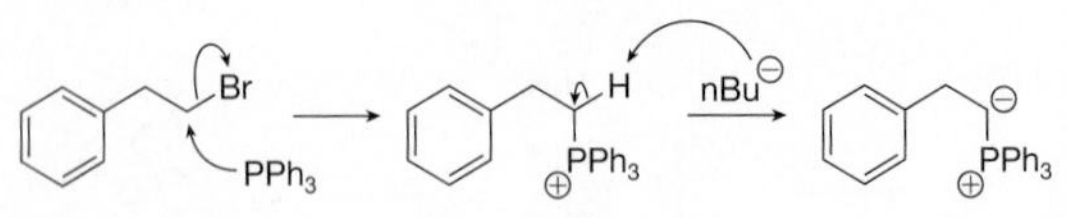

Br, PPh$_3$, H, nBu$^-$, PPh$_3$

Step 3

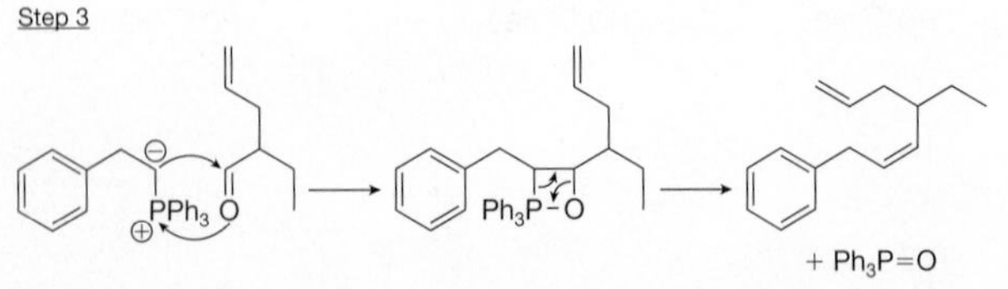

PPh$_3$ O, Ph_3P–O, + $Ph_3P{=}O$

INTEGRATE THE SKILL 17.26

a) This is a terminal alkene. The carbocation that would be needed to form it by dehydration would rearrange and the product would most likely be

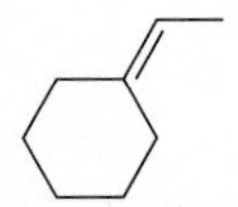

b) Forming the four-membered intermediate shows which would be the better ketone. One intermediate is highly sterically crowded while the other has the large groups separated.

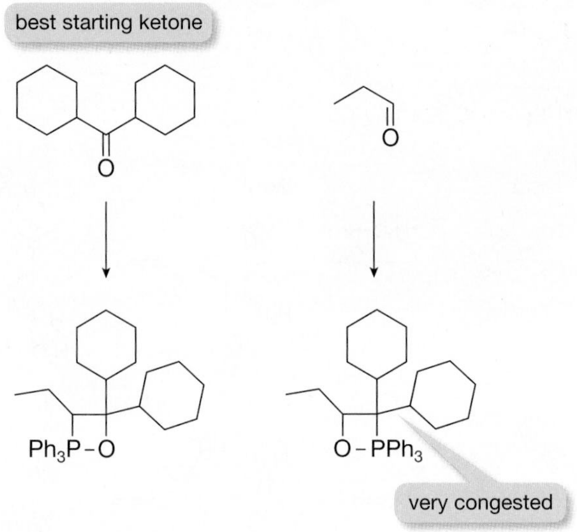

best starting ketone

Ph_3P–O

O–PPh_3

very congested

PRACTICE PROBLEM 17.27

a) i) Wittig reaction conditions

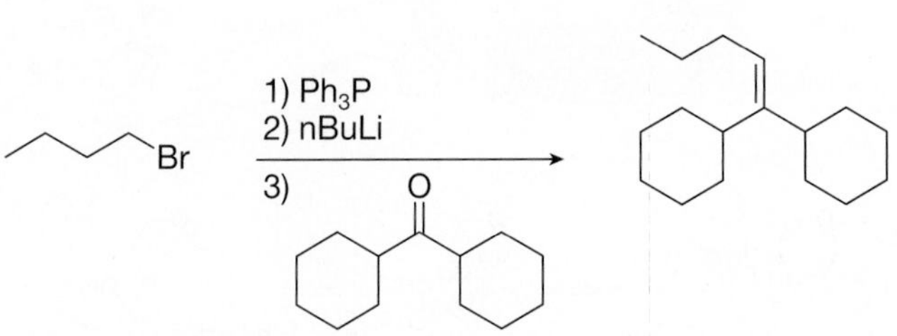

1) Ph_3P
2) nBuLi
3) dicyclohexyl ketone (O)

Br

ii)

CN

1) NaOEt
EtOH
2) H_3O^+
heat

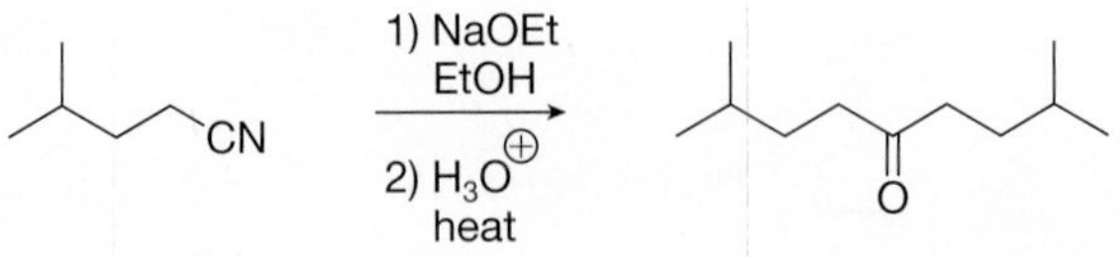

iii) H_2N / H, H, =O / $HCl_{(cat)}$

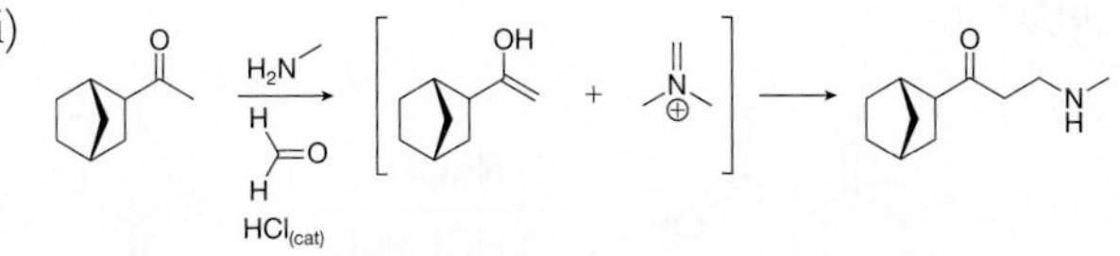

b) i) Analysis:

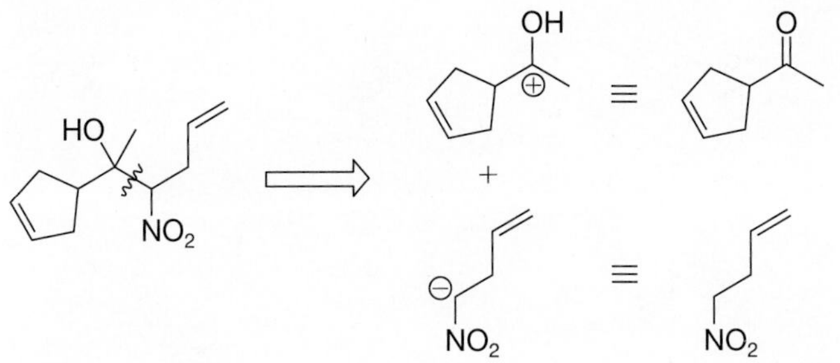

NO_2 / NO_2 / NO_2

Synthesis:

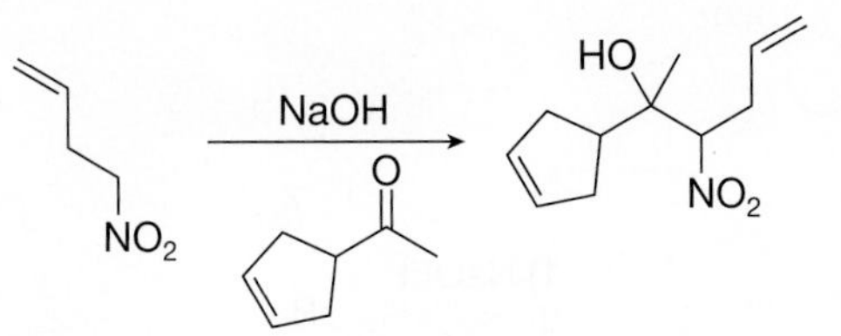

NaOH

NO_2 / NO_2

ii) Analysis:

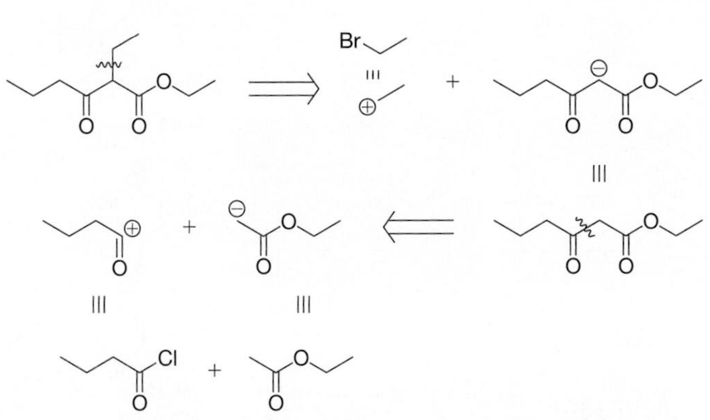

Br / Cl

Synthesis:

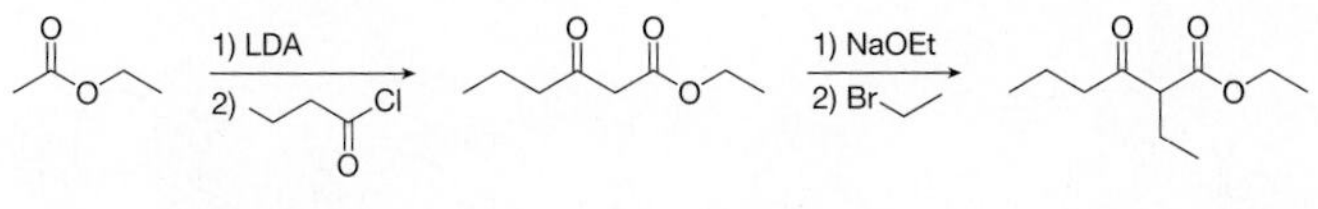

1) LDA 2) Cl

1) NaOEt 2) Br

iii) Analysis: Both of the final synthons have the same equivalent reagent.

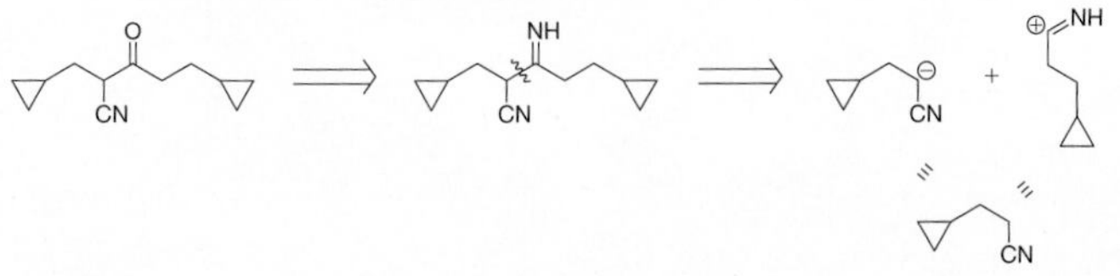

NH / CN / CN / CN / NH / CN

Synthesis:

NaOH / $H_3O^{\oplus}$

CN / NH / CN / CN

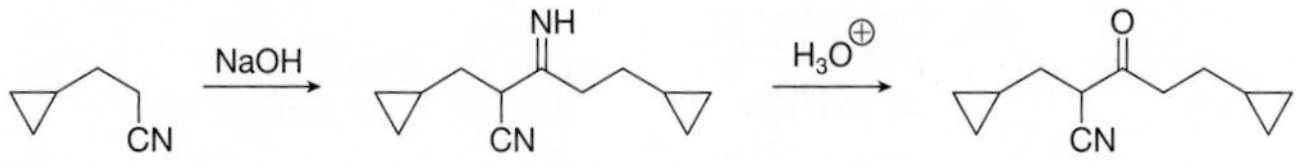

INTEGRATE THE SKILL 17.28

a) Synthesis:

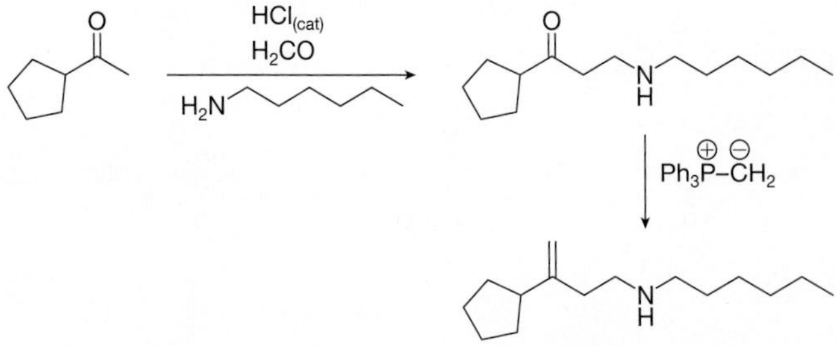

$HCl_{(cat)}$ / H_2CO / H_2N

$Ph_3\overset{\oplus}{P}-\overset{\ominus}{C}H_2$

b) Analysis:

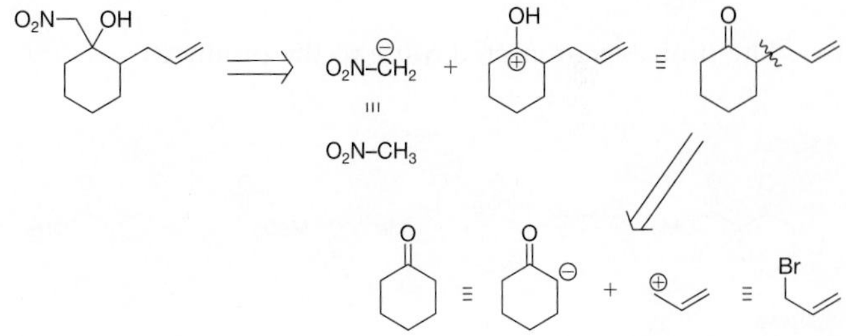

O_2N / OH / $O_2N-\overset{\ominus}{C}H_2$ / O_2N-CH_3 / Br

Synthesis:

O_2N-CH_3 / LDA

1) LDA 2) Br

1) $O_2N-\overset{\ominus}{C}H_2$ 2) $H_3O^{\oplus}$

O_2N / OH

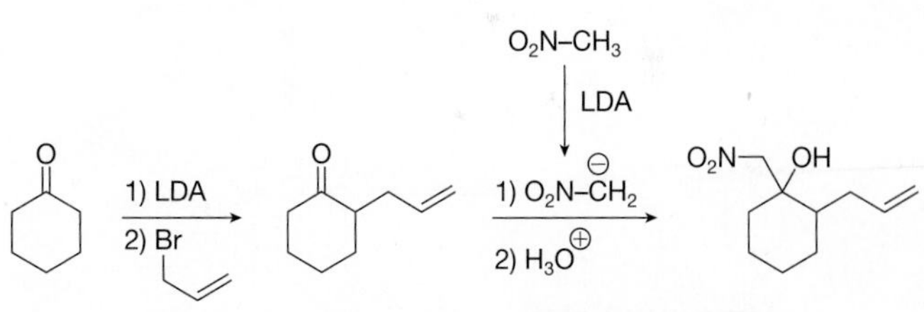

PRACTICE PROBLEM 17.29

a) i) OEt / 1) *t*-BuOK 2) Br / OEt

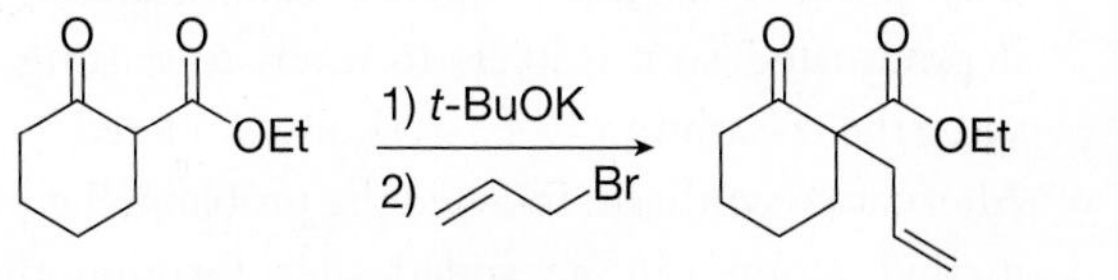

ii) CN / 1) *t*-BuOK 2) EtBr / CN

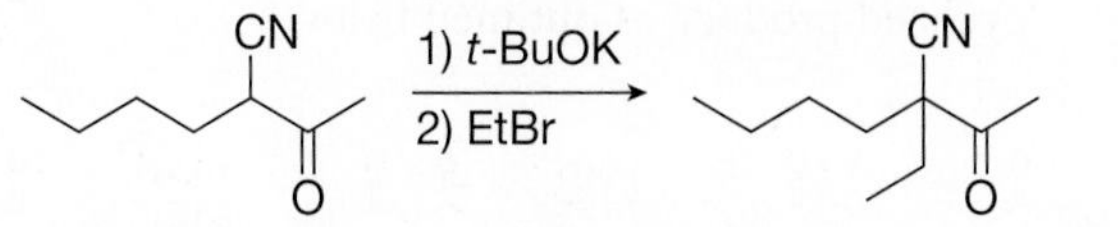

iii) 1) NaOEt 2) MeBr 3) *t*-BuOK 4) *n*-PrBr

b) least acidic → most acidic

INTEGRATE THE SKILL 17.30

a)

b) The probable attempted route to the product:

Analysis

Synthesis

can't deprotonate at α-carbon so all reactions are reversible

The product of this sequence cannot be irreversibly deprotonated, so it is likely to revert to starting material since the α-carbon is very sterically hindered.

Alternative synthesis: To avoid the problem, the offending methyl group can be added after forming the initial cyclized product, as outlined below.

1) $MeO^{\ominus}$ 2) $H_3O^{\oplus}$ → 1) $MeO^{\ominus}$ 2) MeI

PRACTICE PROBLEM 17.31

a) 1) NaOH 2) HCl, H_2O heat

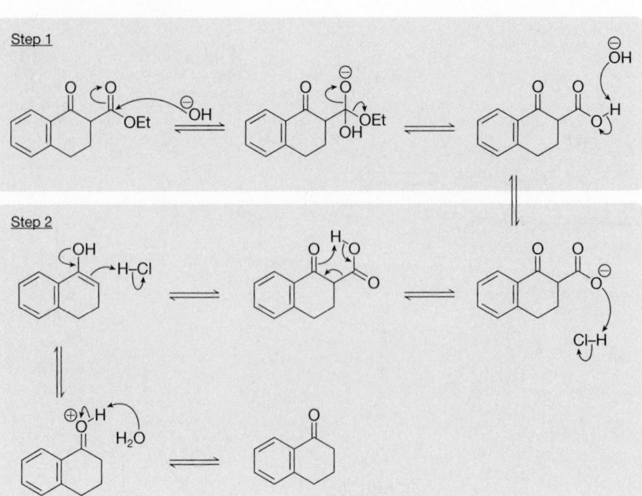

b) 1) NaOEt 2) [alkyl bromide] 3) HCl, H_2O, heat

Step 1

Step 2

Step 3

c) 1) NaOMe 2) Br(CH2)4Br 3) *t*-BuONa 4) HCl, H_2O, heat

Step 1

Step 2

Step 3

Step 4

same steps convert other ester to acid

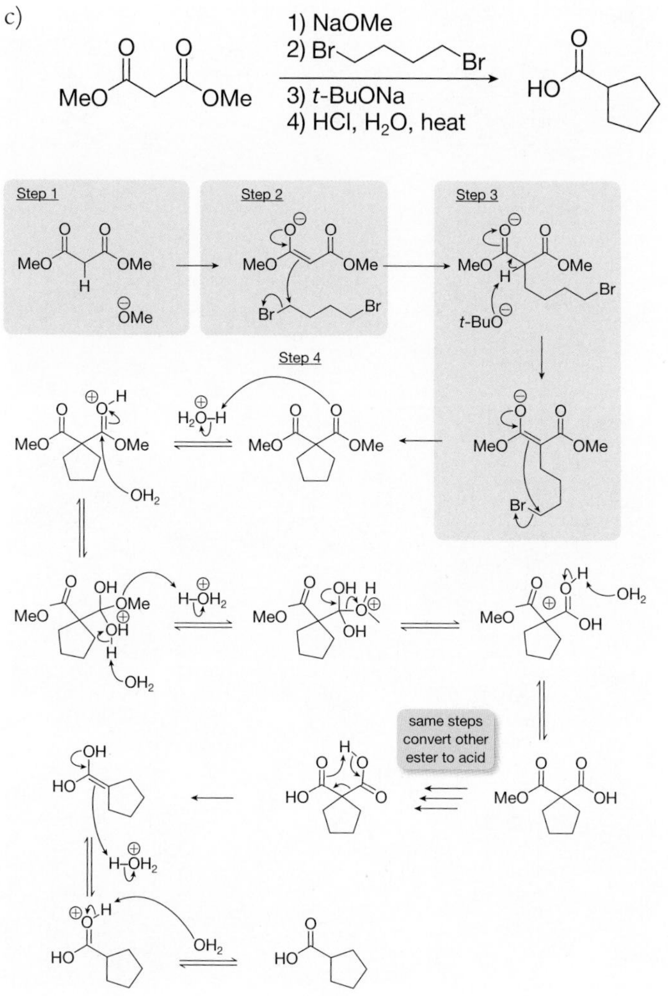

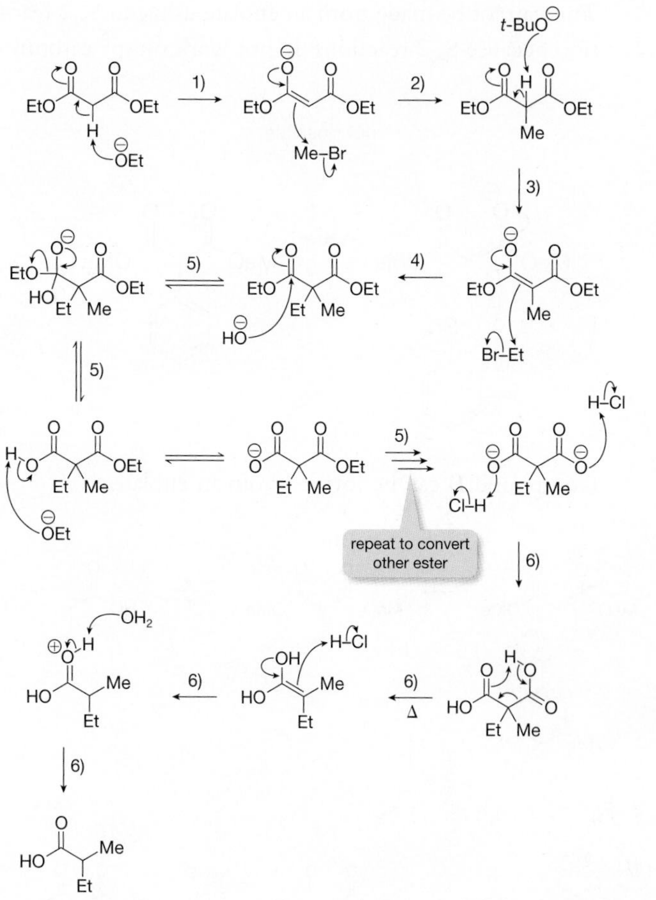

d) 1) NaOEt 2) MeBr 3) *t*-BuOK 4) EtBr 5) NaOH, H_2O 6) HCl, H_2O, heat

repeat to convert other ester

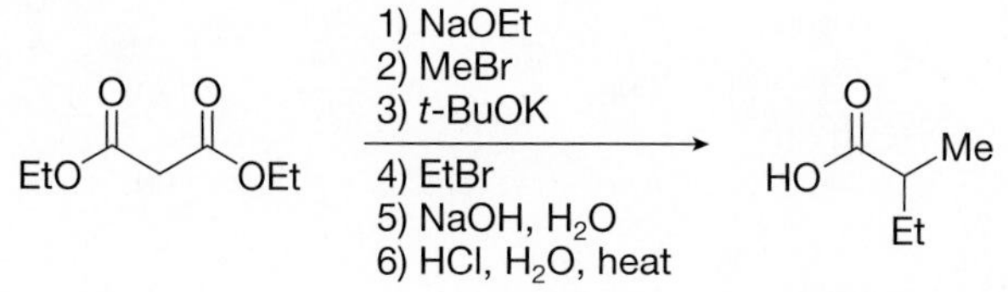

INTEGRATE THE SKILL 17.32

a) The electron movement arrows are going the wrong way. As drawn, they have a hydride moving to an electronegative oxygen, which is not probable. The better figure (shown below) has nucleophilic pairs moving to electrophilic sites.

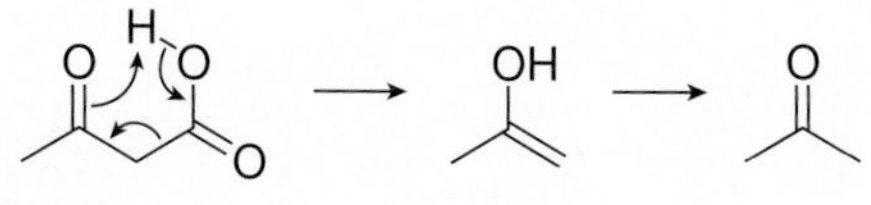

b) Compound **A** must come from

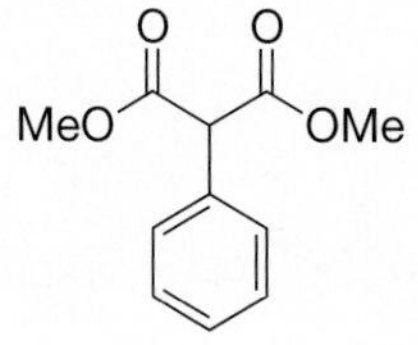

This cannot be made from an enolate using an S_N2 reaction because S_N2 reactions do not work on sp^2 carbons.

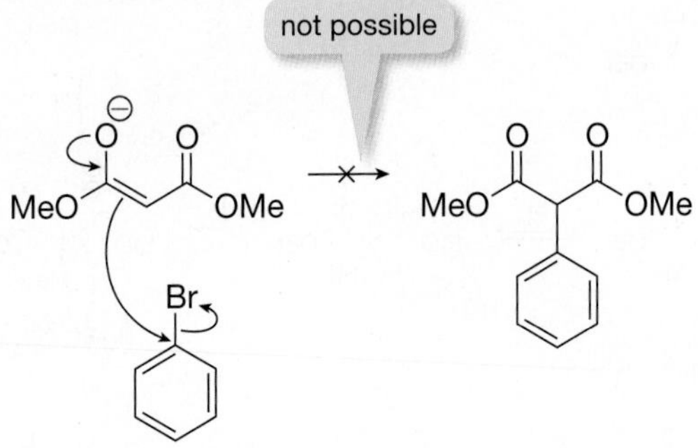
not possible
MeO
OMe
Br
MeO
OMe

Compound **B** can be formed from an enolate.

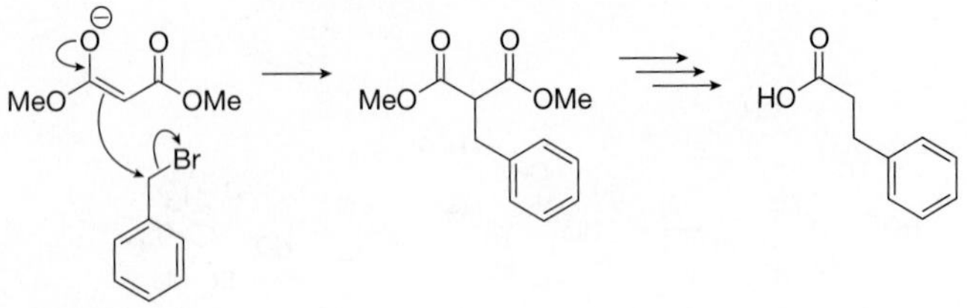
MeO
OMe
Br
MeO
OMe
HO

PRACTICE PROBLEM 17.33

a)

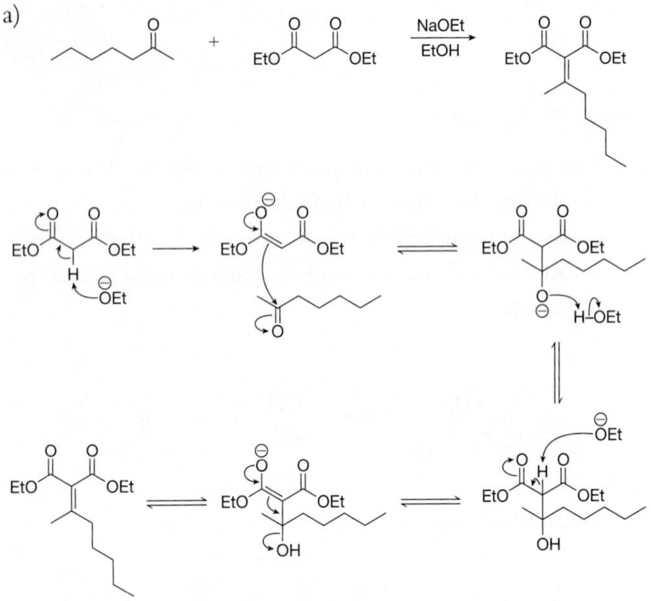
EtO
OEt
NaOEt
EtOH
EtO
OEt
H–OEt
OH

b)

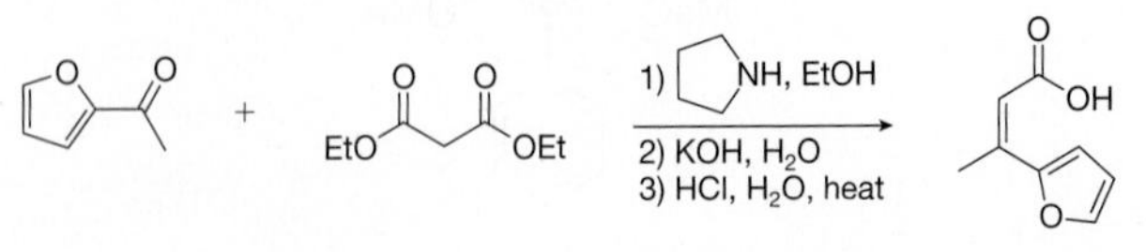
EtO
OEt
1) NH, EtOH
2) KOH, H2O
3) HCl, H2O, heat
OH

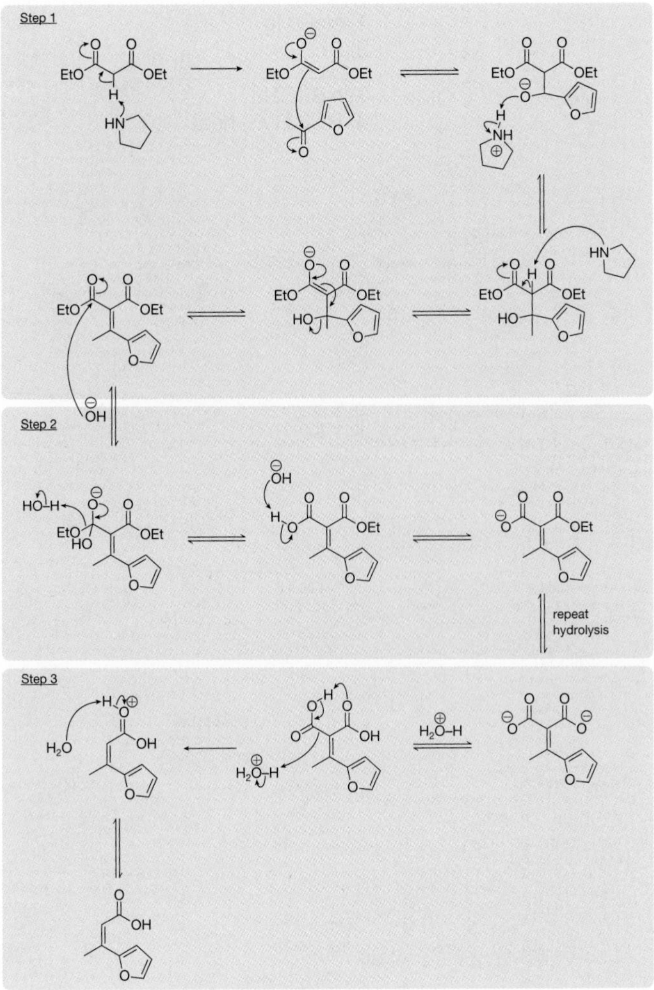
Step 1
EtO
OEt
HN
NH
Step 2
HO–H
OH
repeat
hydrolysis
Step 3
H2O
OH
H2O–H

c)

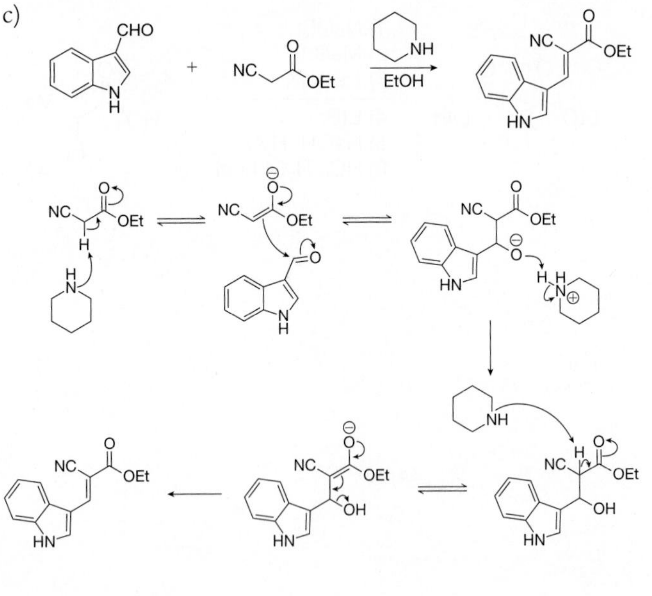
CHO
NC
OEt
NH
EtOH
NC
OEt
HN
OH

d)

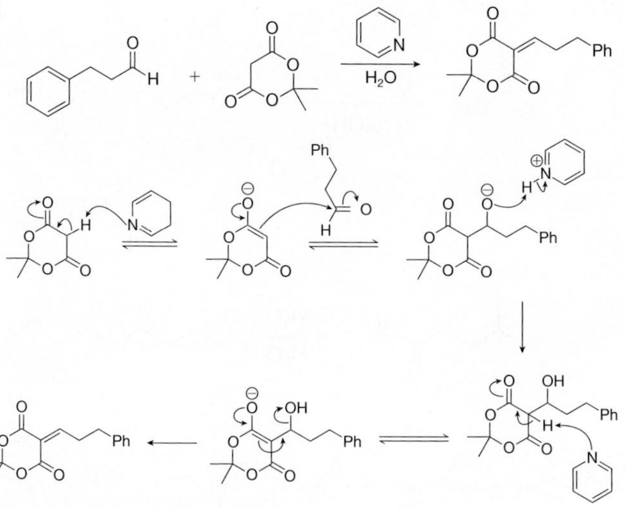

INTEGRATE THE SKILL 17.34

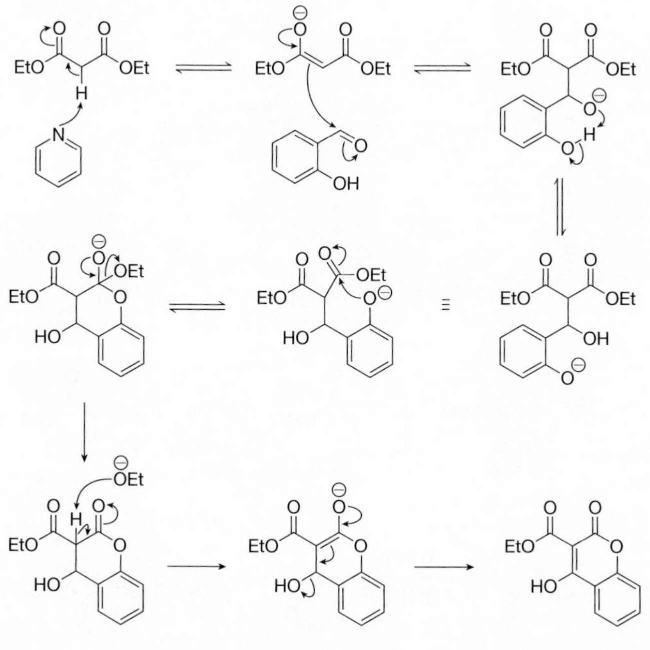

PRACTICE PROBLEM 17.35

a) *Trans*, due to the carbonyl group. Details in Chapter 18.

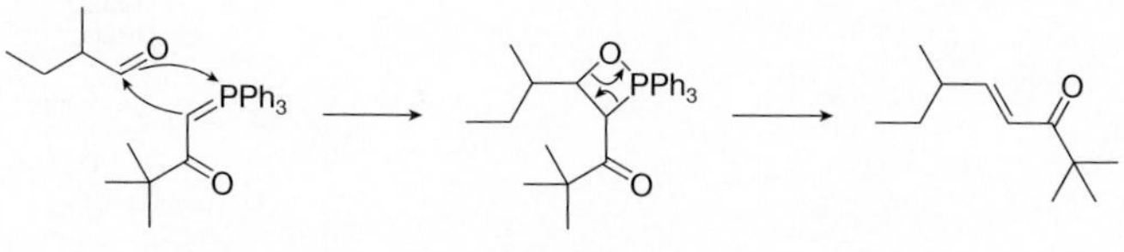

b)

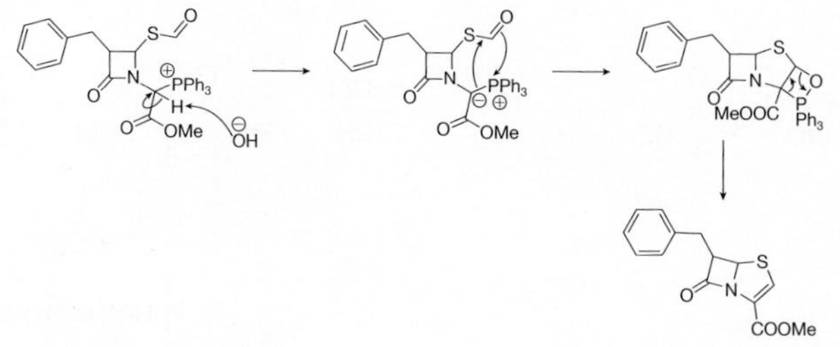

c)

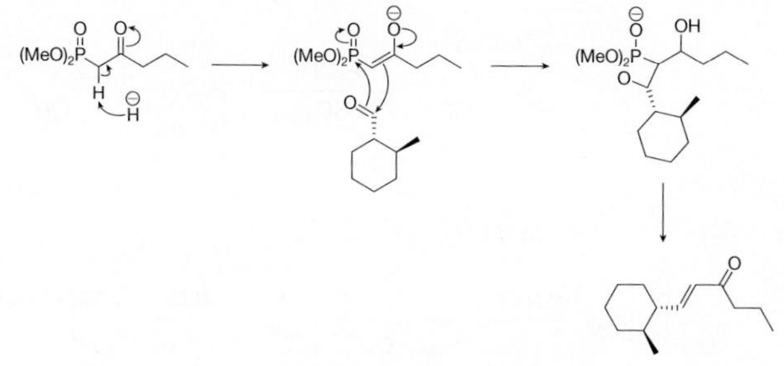

d)

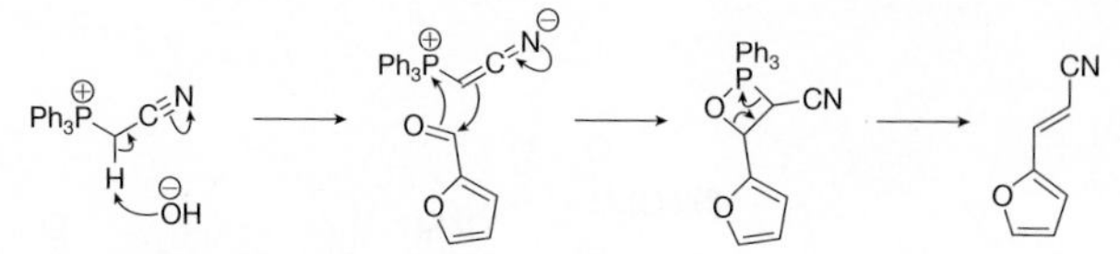

e)

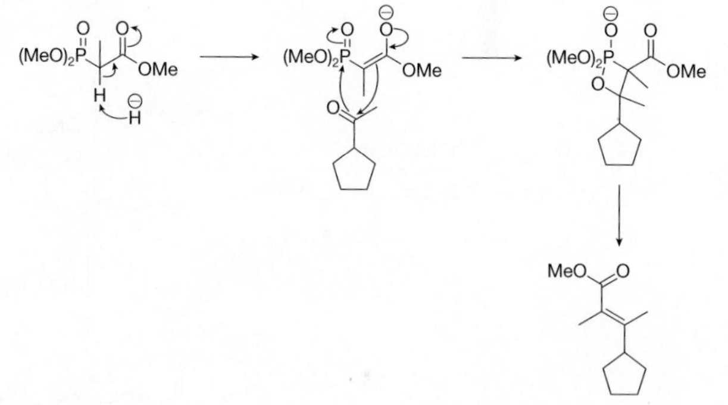

INTEGRATE THE SKILL 17.36

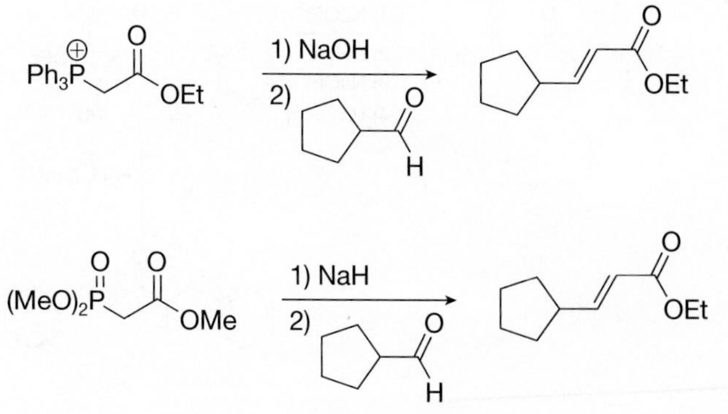

NaOEt, EtOH

HCl, H_2O, heat

EtOH, $H_2SO_{4(cat)}$

PRACTICE PROBLEM 17.37

a) 1) NaH 2) [benzyl bromide]; H_3O, Δ

b) $(MeO)_2P(O)CH_2CO_2Me$, $NaOCH_3$

c) 1) $NaOCH_3$ 2) [3-methoxybenzaldehyde]; H_3O^+, heat

d) 1) NaOEt 2) CH_3–I 3) NaOEt 4) Ph–CH_2–Br; H_3O^+, heat

INTEGRATE THE SKILL 17.38

a) 1) PPh_3 2) NaOH 3) [2-methylpentanal]; H_2/Pt; H_3O^+, H_2O

H_2SO_4, EtOH; 1) NaOEt 2) [1-bromo-2-methylpentane]; H_3O^+, heat

b) H_2SO_4, EtOH; 1) NaOEt 2) Br–$(CH_2)_5$–Br 3) NaO*t*Bu; H_3O^+, Δ; H_2SO_4, EtOH

1) Mg, Et_2O 2) CO_2 3) H_3O; 1) $SOCl_2$ 2) pyridine, EtOH

c) 1) PPh_3 2) NaOH 3) [pent-4-enal]

H_2SO_4, EtOH; 1) NaOEt 2) [pent-4-enal] 3) H_3O^+; LiCl, DMSO, heat

CHAPTER 18

PRACTICE PROBLEM 18.1

a) BrMg–Ph; O–MgBr; Ph; NH$_4$Cl; OH; Ph

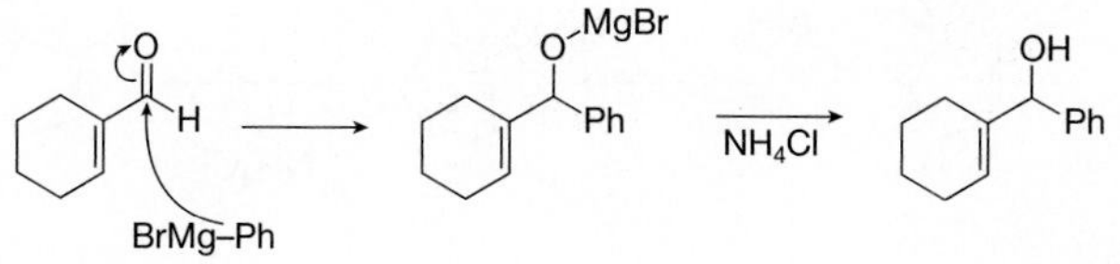

b) 1) $\oplus$Li; $\ominus$Cu, Et, Et; OLi; Cu, Et, Et; OLi; Et; EtCu; O; Et

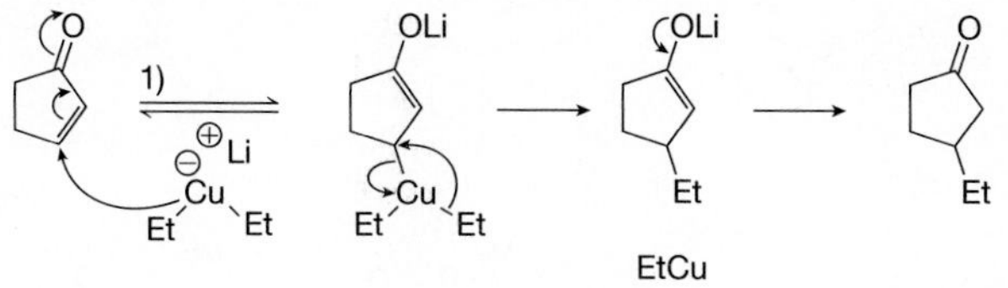

c) OEt; BrMg–Me; OEt; Me; Me; BrMg–Me; O–MgBr; Me; Me; H$_3$O$^\oplus$; OH

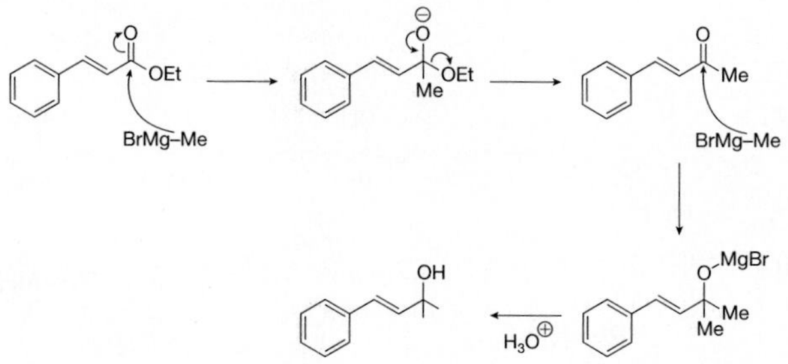

d) OEt; $\ominus$Cu, Me, Me; Me, Me, Cu, OLi, OEt; OLi, OEt; H$_3$O$^\oplus$; O, OEt

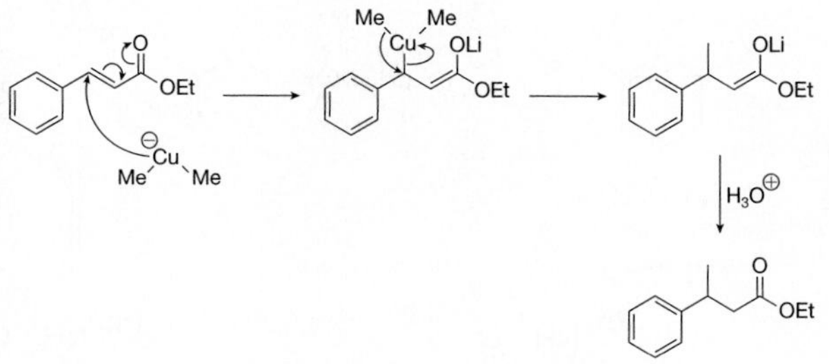

e) ( )$_2$CuLi; Cu; Br

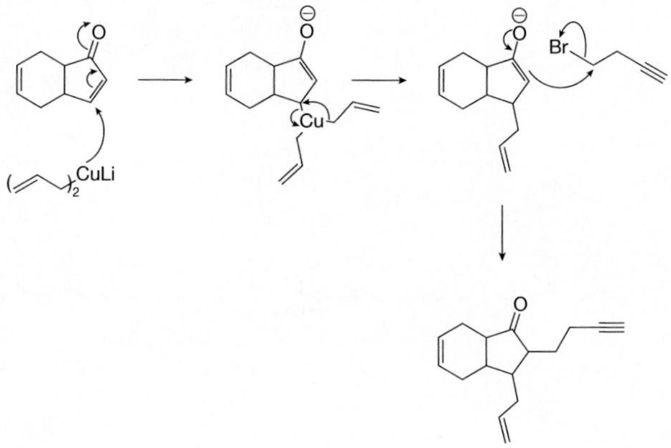

f) LiCu( )$_2$; C≡N; Cu; C=N$^\ominus$; Cl; C≡N

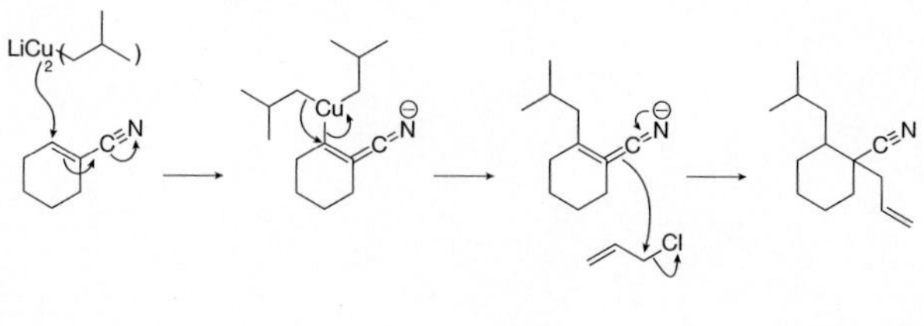

INTEGRATE THE SKILL 18.2

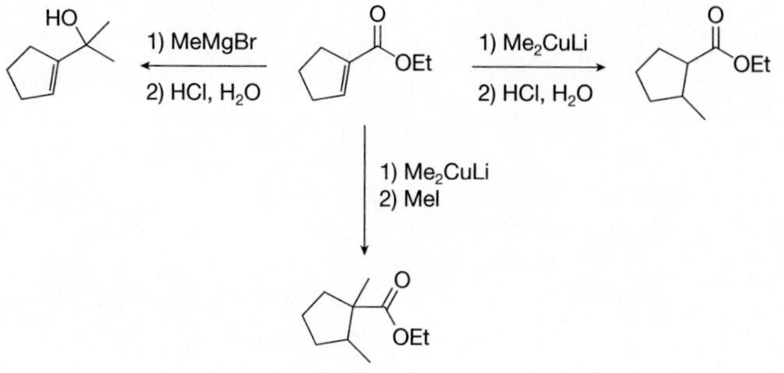

HO; 1) MeMgBr 2) HCl, H$_2$O; O, OEt; 1) Me$_2$CuLi 2) HCl, H$_2$O; O, OEt

1) Me$_2$CuLi 2) MeI; O, OEt

PRACTICE PROBLEM 18.3

a) H; $\ominus$OMe; OMe; H–OMe; OMe; H; $\ominus$OMe; OMe; H$_2$O–H$^\oplus$; OMe; OMe

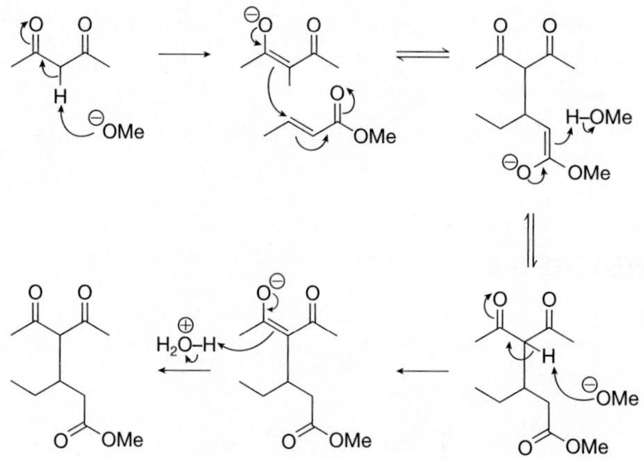

b) EtO, OEt, H, $\ominus$OEt; EtO, OEt; EtO, OEt, H–OEt; EtO, OEt, H, $\ominus$OEt; H$_2$O–H$^\oplus$; EtO, OEt; EtO, OEt

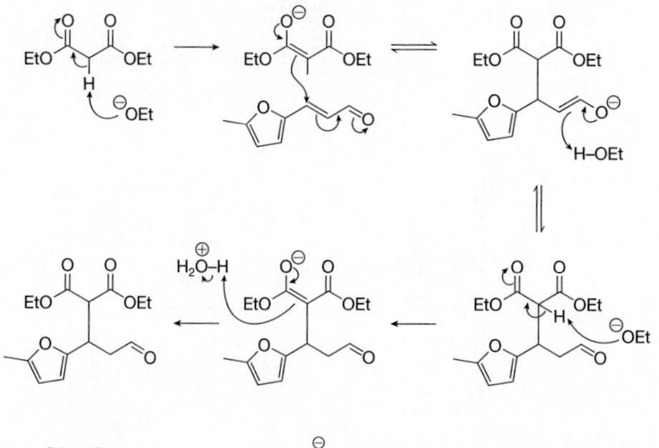

c) OMe, H, $\ominus$OMe; OMe, C≡N; OMe, C=N$^\ominus$, H–OMe; OMe, H, C≡N, $\ominus$OMe; H$_2$O–H$^\oplus$; OMe, C≡N; OMe, C≡N

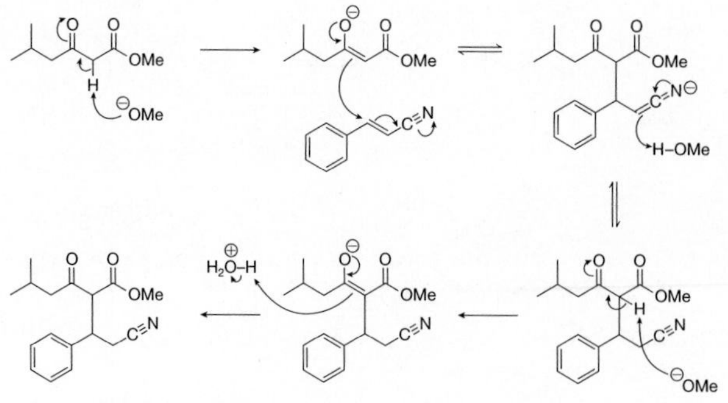

d)

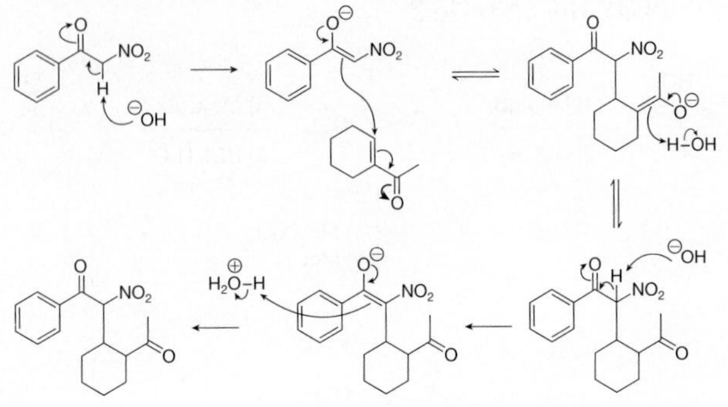

e)

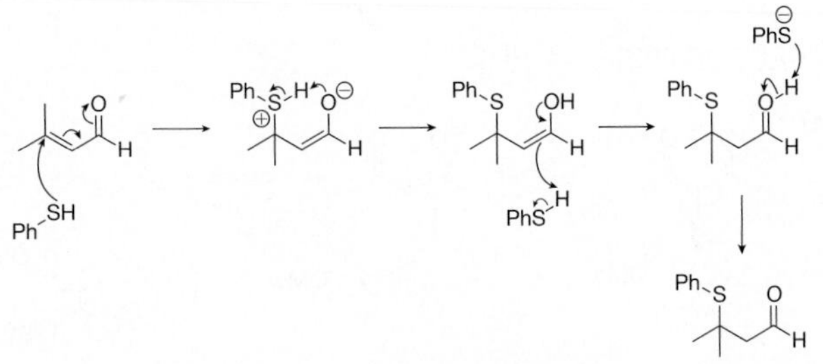

INTEGRATE THE SKILL 18.4

a)

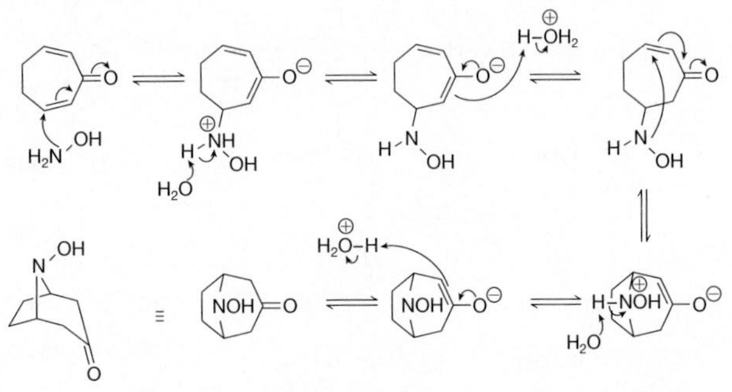

b)

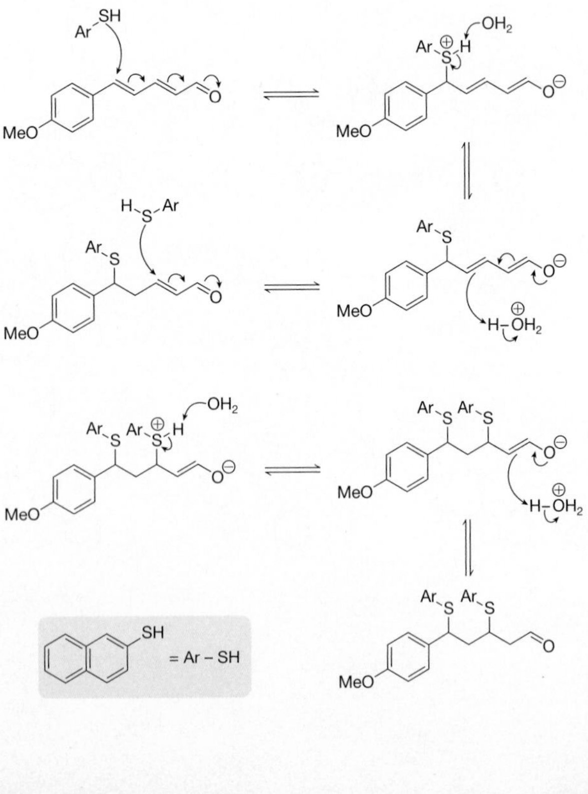

= Ar – SH

PRACTICE PROBLEM 18.5

a)

H_2O
NH_4Cl

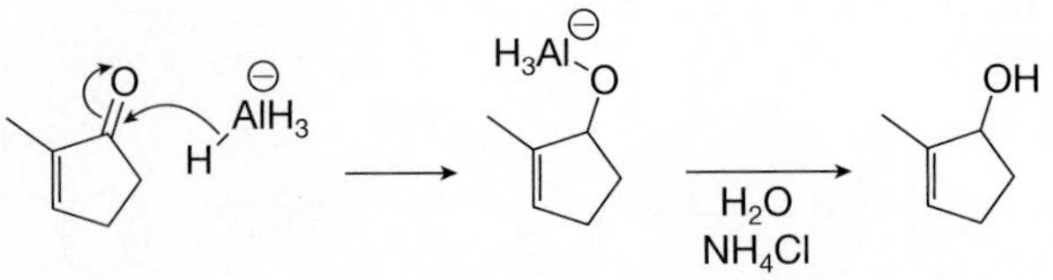

b)

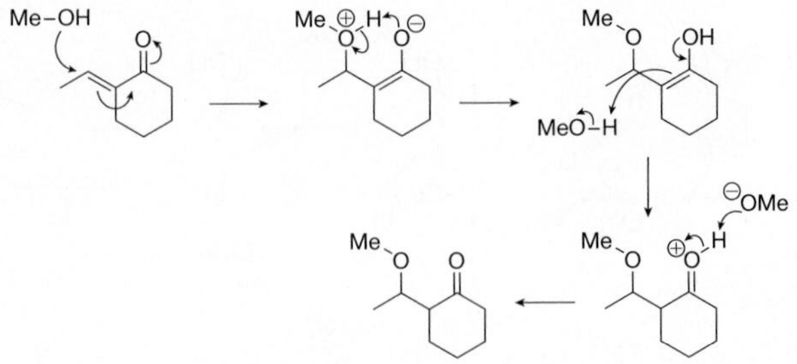

c)

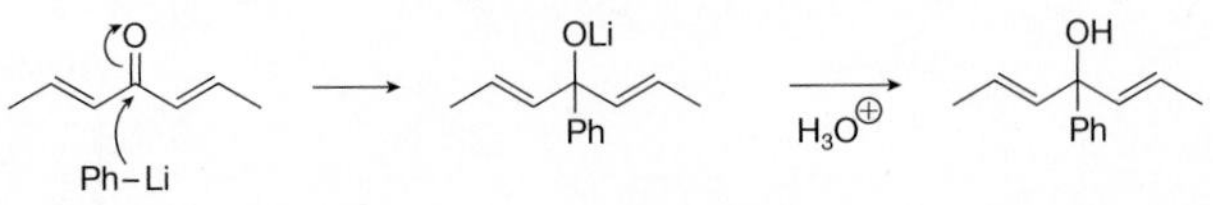

d)

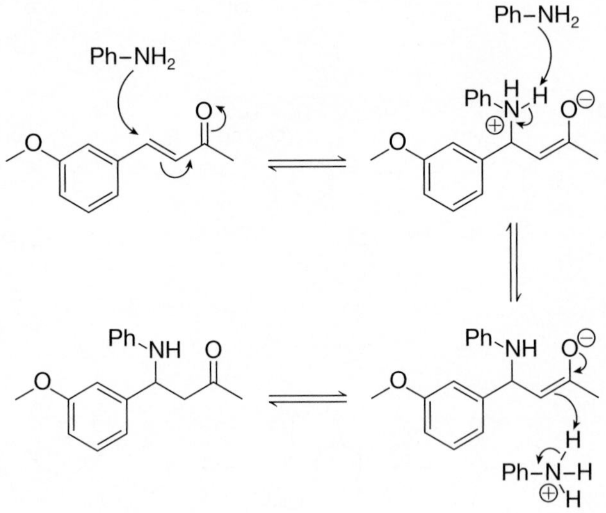

e)

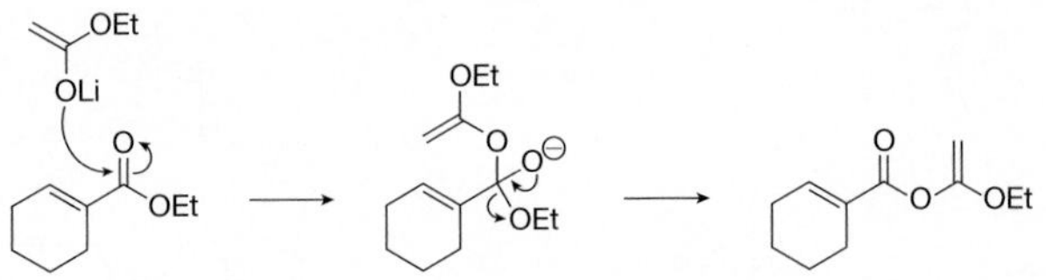

INTEGRATE THE SKILL 18.6

The sites of reaction for the reagents are outlined in the diagram below for each of the unsaturations in the molecule.

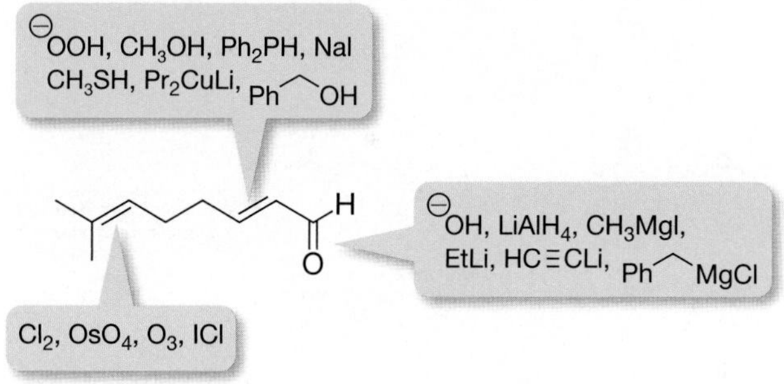

$^{\ominus}$OOH, CH_3OH, Ph_2PH, NaI, CH_3SH, Pr_2CuLi, Ph⌒OH

$^{\ominus}$OH, $LiAlH_4$, CH_3MgI, EtLi, HC≡CLi, Ph⌒MgCl

Cl_2, OsO_4, O_3, ICl

PRACTICE PROBLEM 18.7

a) NaOEt / EtOH

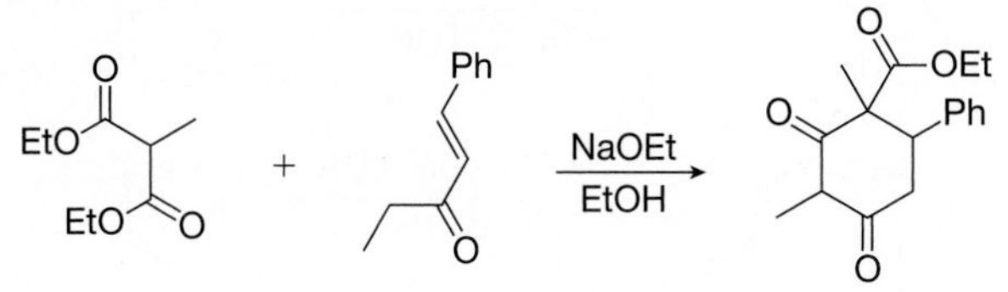

b) KO*t*-Bu / *t*-BuOH

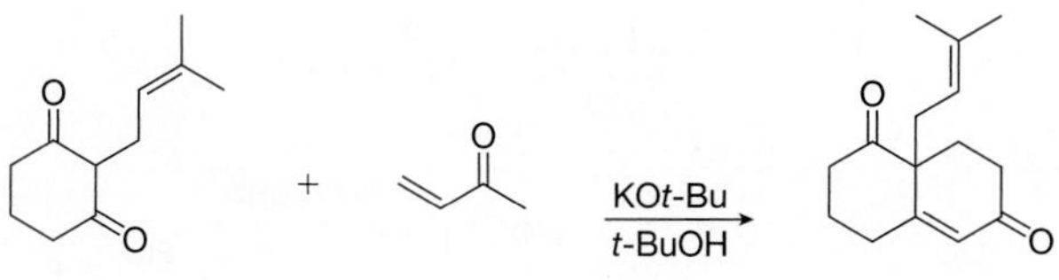

c) (cat.) / octanol

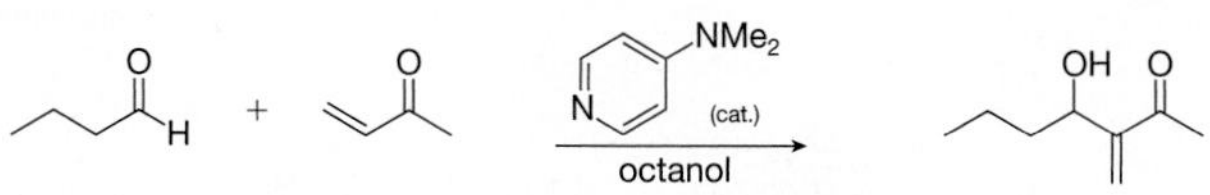

d) NaOMe / MeOH

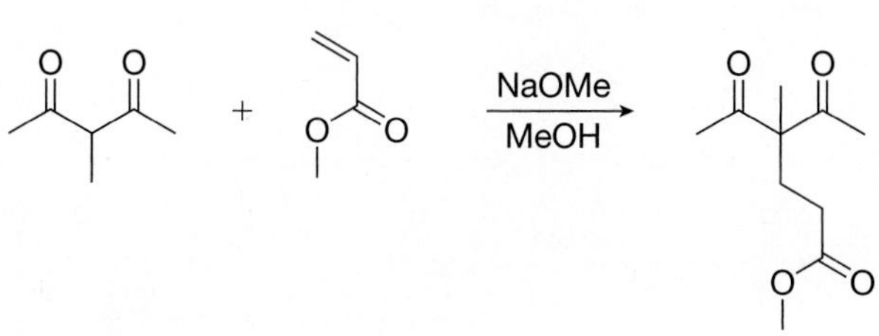

INTEGRATE THE SKILL 18.8

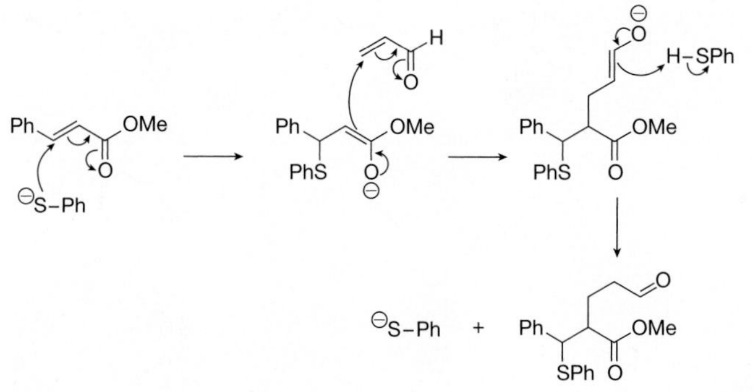

PRACTICE PROBLEM 18.9

a) Kinetic enolate is formed in the first step.

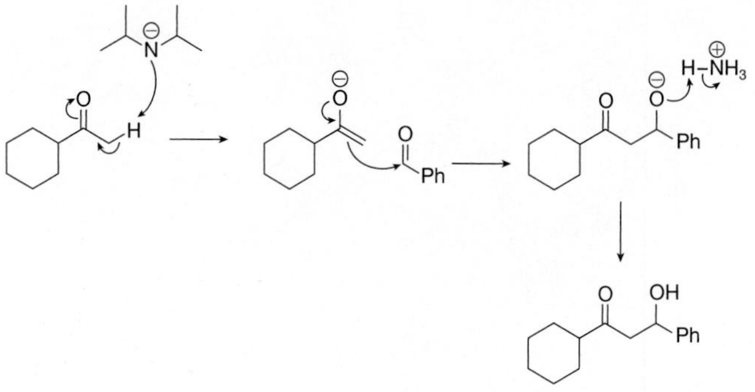

b)

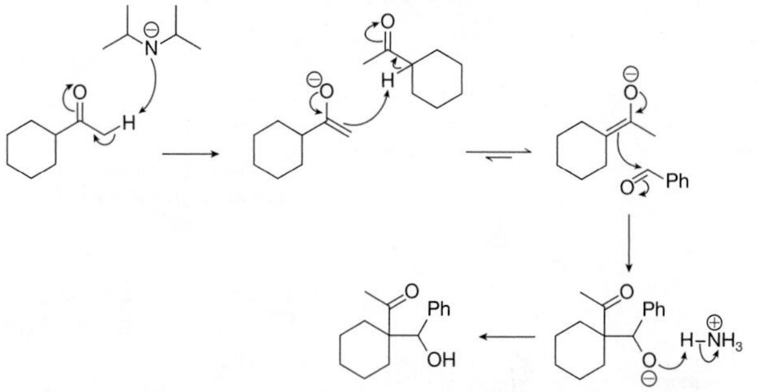

c) H_2O

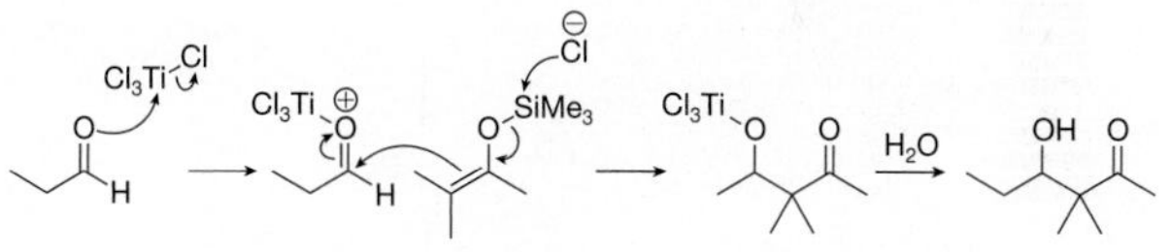

d) H_2O

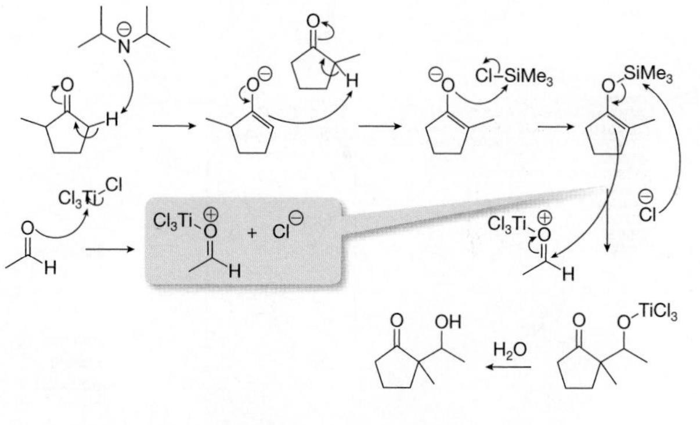

INTEGRATE THE SKILL 18.10

a) i) 1) LDA (0.98 eq.) −78° → rt
2) EtCHO
3) NH_4Cl, H_2O

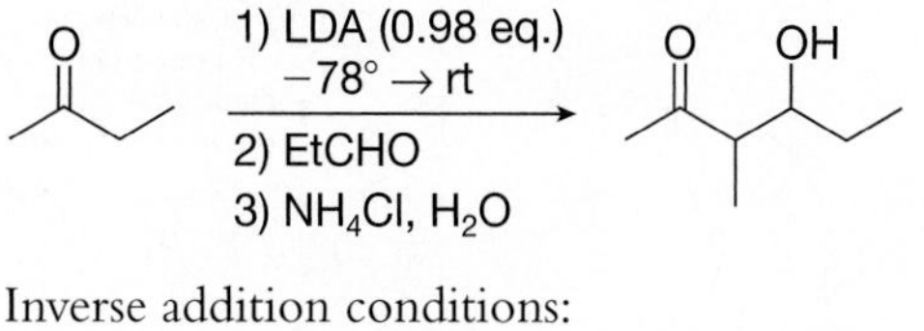

ii) Inverse addition conditions:

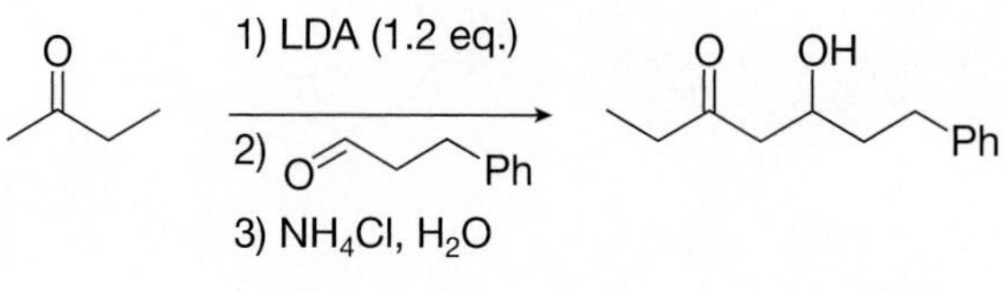

1) LDA (1.2 eq.)
2) O⌒⌒Ph
3) NH_4Cl, H_2O

b)

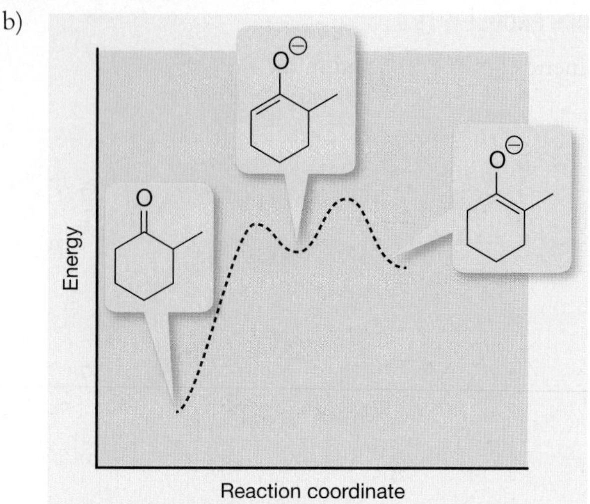

PRACTICE PROBLEM 18.11

a)

large BCy_2 group pushes alkyl chain away

small NEt_3 base fits into this space

b)

chair-like transition state

minimal steric repulsion

c)

smaller unbranched R-groups don't force chain away

larger base requires approach from side opposite to B

d)

steric repulsion between groups prevents formation of other isomer

e)

steric interactions determine alkene stereochemistry

only 1 type of α-hydrogen controls regiochemistry

INTEGRATE THE SKILL 18.12

a) 1.02 equiv. LDA, −78 °C, THF

b) 1.02 equiv. LDA, −78 °C, THF

PRACTICE PROBLEM 18.13

a) 1) LDA (1.02 eq.); 2) Ph–CHO

+ enantiomer

b) 1) LDA (1.02 eq.); 2)

+ enantiomer

c) 1) LDA (1.02 eq.); 2)

+ enantiomer

d) 1) LDA (1.02 eq.) or 1) Cl–BBu_2 2) i-Pr_2NEt

INTEGRATE THE SKILL 18.14

Site 2 · Site 1

Site 1:

anti

1) Cy_2BCl

Site 2:

needs LDA

syn requires *Z* enolate; esters favour *E*

PRACTICE PROBLEM 18.15

a)

E

b) There are two transition states: **A** and **B**. The reaction will proceed through **B** since the aldehyde chain (R_2) is further from the large PPh_3 group. This reduces the energy of the transition state.

R_1 = ylide chain
R_2 = aldehyde chain

A slow · B slow

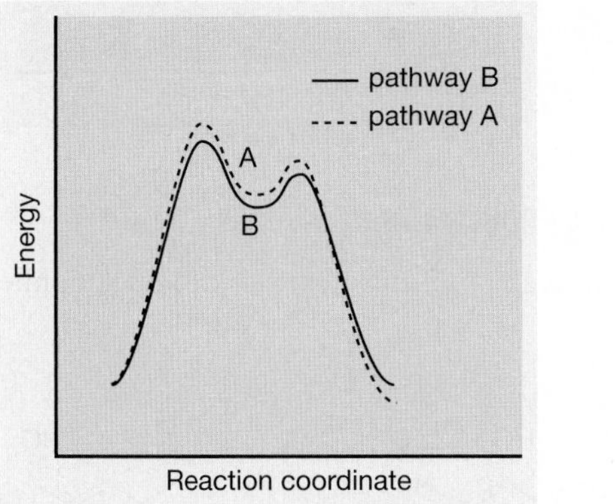

c) The selectivity is controlled in a fashion similar to part (b).

d)

1) LDA
2) MeO

INTEGRATE THE SKILL 18.16

Reaction coordinate diagram for the reaction in Question 18.15, part (a):

R = aldehyde chain

A slow · B fast

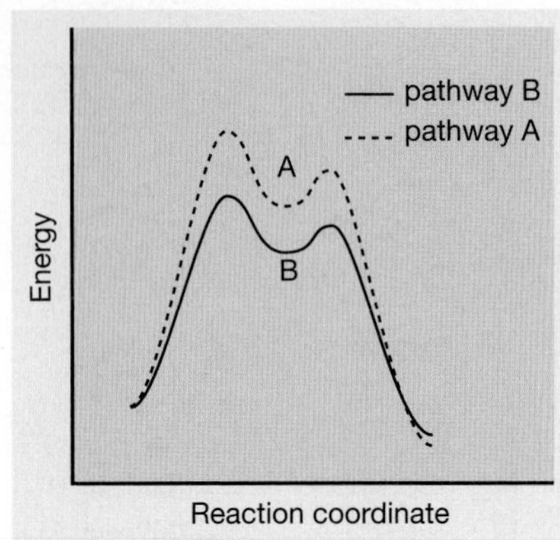

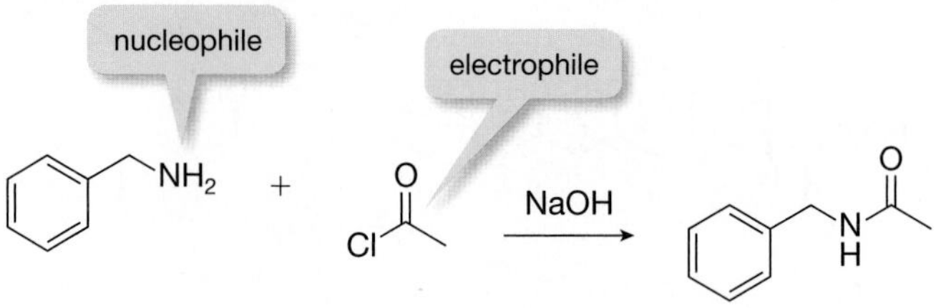

PRACTICE PROBLEM 18.17

a) This reaction is dissonant via the nucleophile:

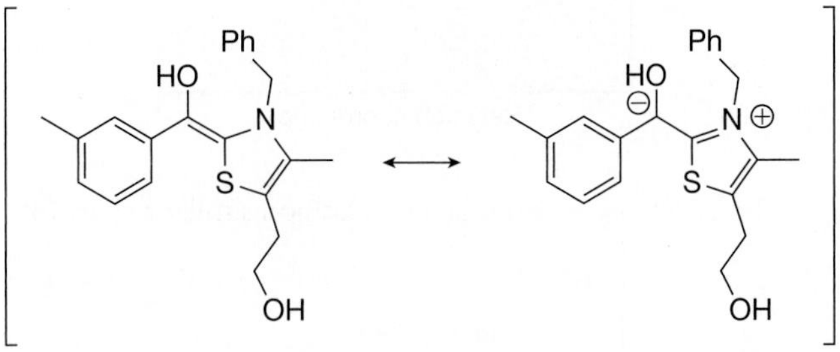

b) The reaction is consonant.

c) This reaction is dissonant via the reagent:

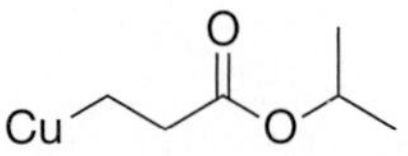

d) The reaction is dissonant via the reagent:

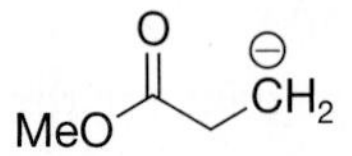

INTEGRATE THE SKILL 18.18

Reaction of NIS with *t*-butylamine gives the amine reagent:

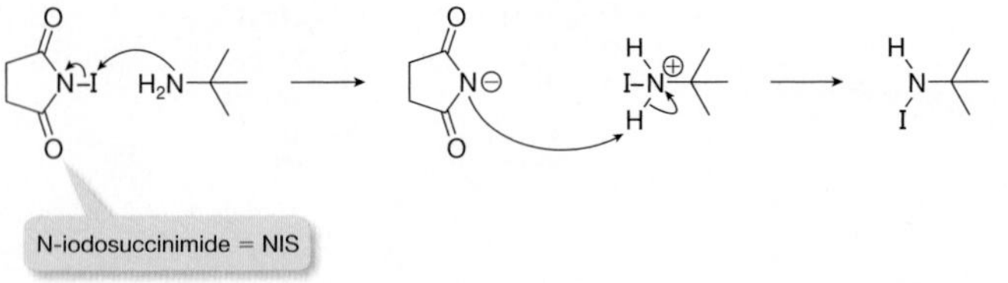

N-iodosuccinimide = NIS

Amines are usually nucleophiles, but the reaction with NIS puts an electronegative iodine on the nitrogen. This makes it electrophilic with an iodine leaving group attached.

A more traditional method would use nitrogen as the nucleophile; for example:

nucleophile

electrophile

NaOH

PRACTICE PROBLEM 18.19

a) 1) $BF_3 \cdot OEt_2$; 2) *n*-BuLi, THF; 4) HgO, H_2O, THF

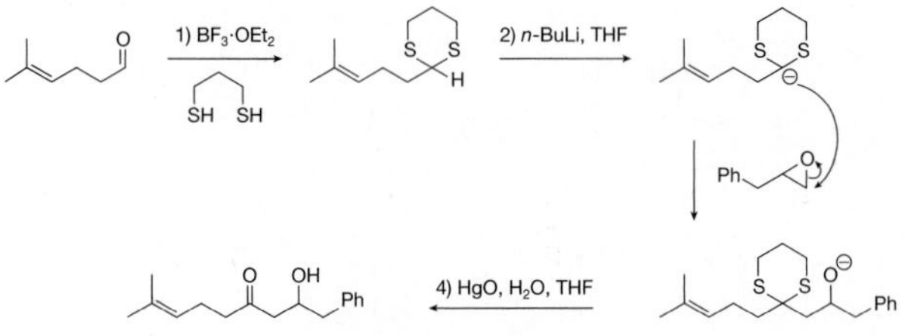

b) 1) $BF_3 \cdot OEt_2$; 2) *n*-BuLi, THF; 3); 4) HgO, H_2O, MeOH

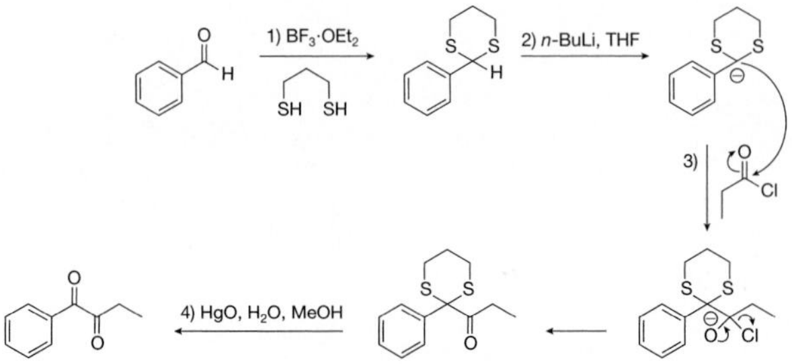

c) 1) LDA; 2)

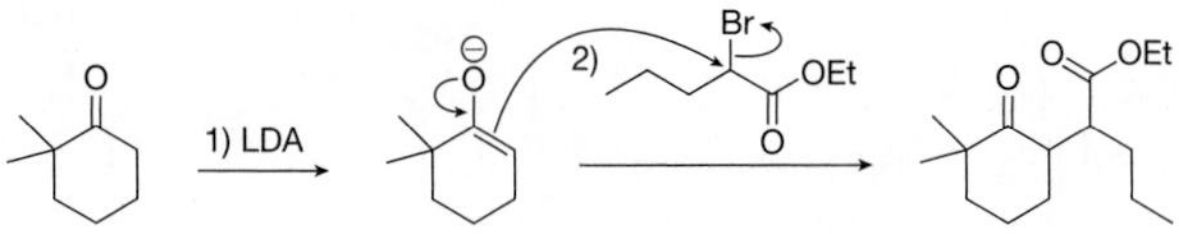

d)

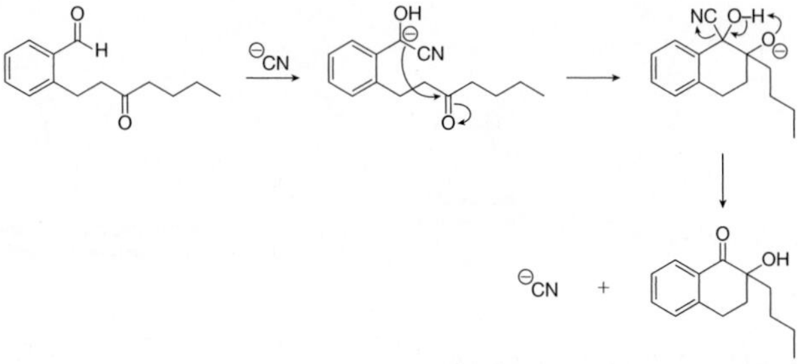

INTEGRATE THE SKILL 18.20

a) $^{\ominus}CN_{(cat)}$

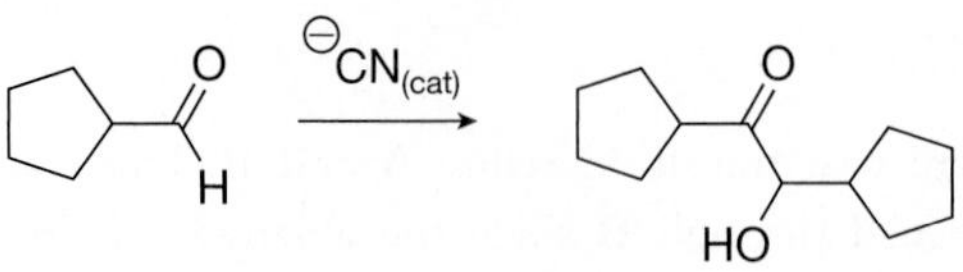

b) 1) HCN 2) acid OEt; 1) LDA 2) cyclohexanecarbaldehyde 3) $H_3O^{\oplus}$

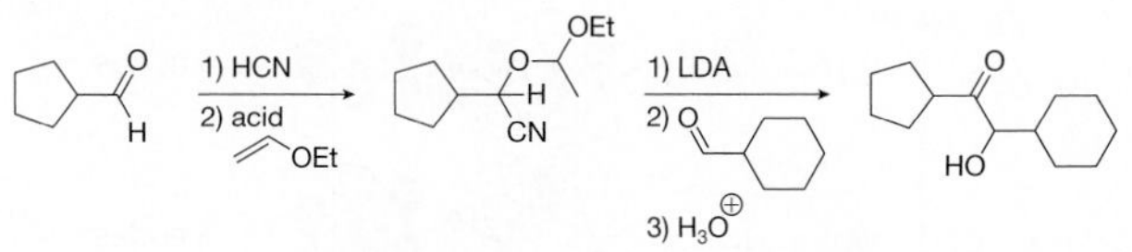

c) 1) 1.05 equiv. LDA THF, −78 °C 2) Br–CH₂C(O)Ph

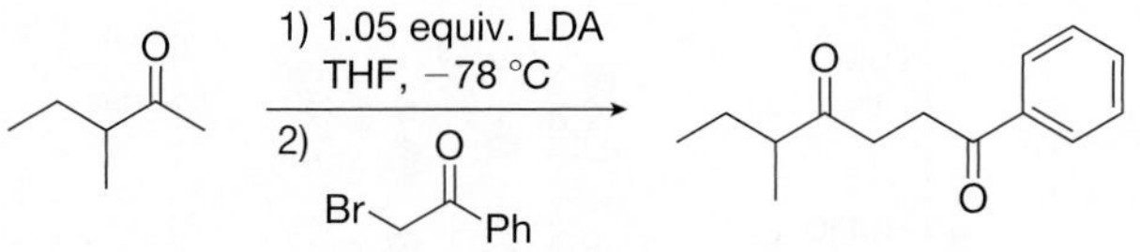

CHAPTER 19

PRACTICE PROBLEM 19.1

a) bond breaking; bond forming

+ ·Br ⟶ + H–Br

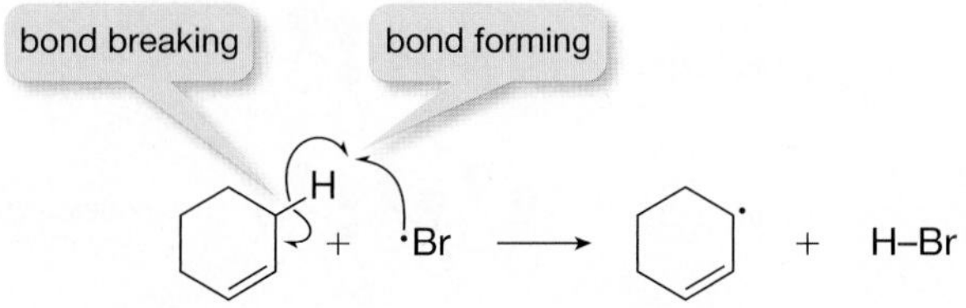

b) + Br· ⟶

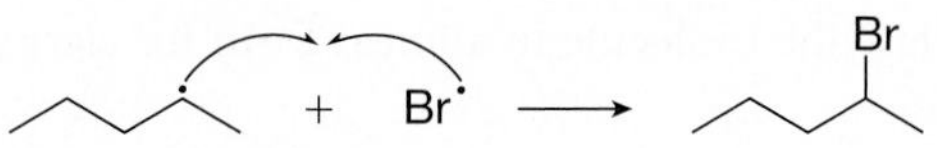

INTEGRATE THE SKILL 19.2

Homolytic O–H bond breaking (loss of hydrogen radical):

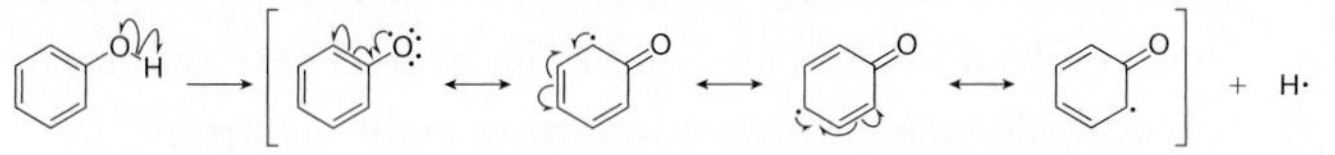

+ H·

Heterolytic O–H bond breaking (loss of proton):

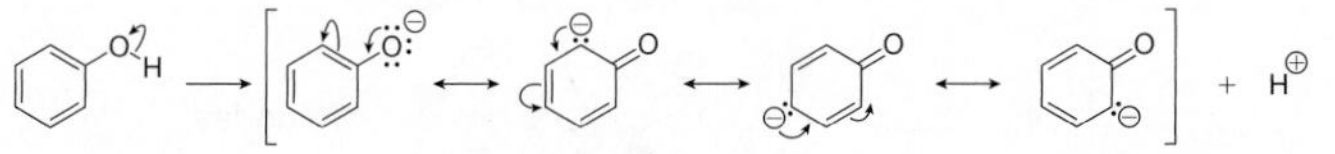

+ $H^{\oplus}$

PRACTICE PROBLEM 19.3

a) There is only one possible alkene that could be reacted to form the product indicated. Electrophilic addition would give the Markovnikov product, while free-radical hydrobromination gives the anti-Markovnikov product. Free-radical addition would be the preferred route.

HBr

HBr ROOR

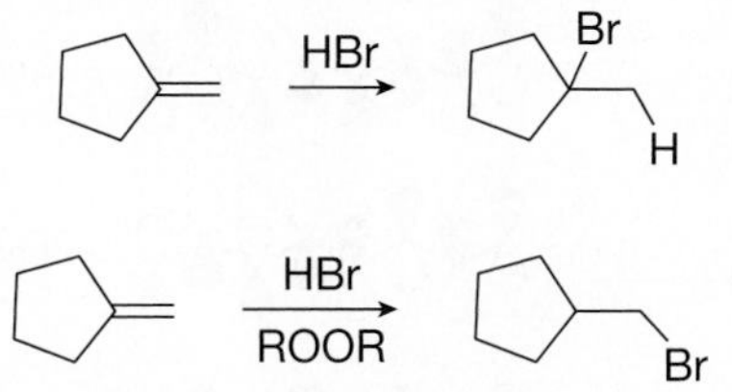

b) There are two possible alkene precursors to this molecule. For the first, hydrobromination would give the desired product, while free-radical addition would give the wrong product.

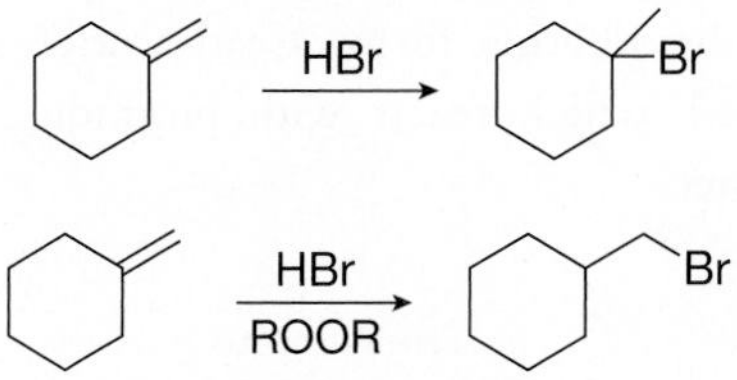

HBr

HBr ROOR

For the second, hydrobromination would also give desired molecule as the major product.

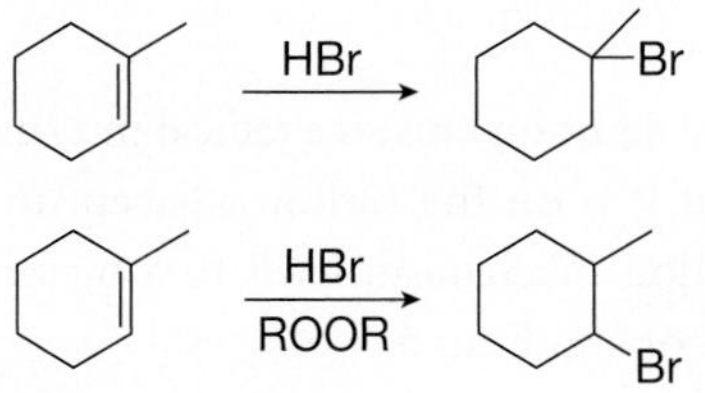

HBr

HBr ROOR

Hydrobromination of either alkene would be the preferred route.

c) There are two possible precursor alkenes to consider. The internal alkene gives the indicated product via hydrobromination or radical addition.

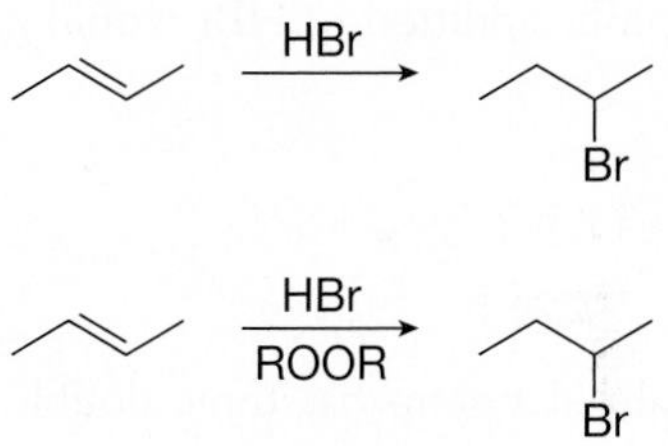

HBr

HBr ROOR

The terminal alkene would only give the preferred product via hydrobromination.

HBr

HBr ROOR

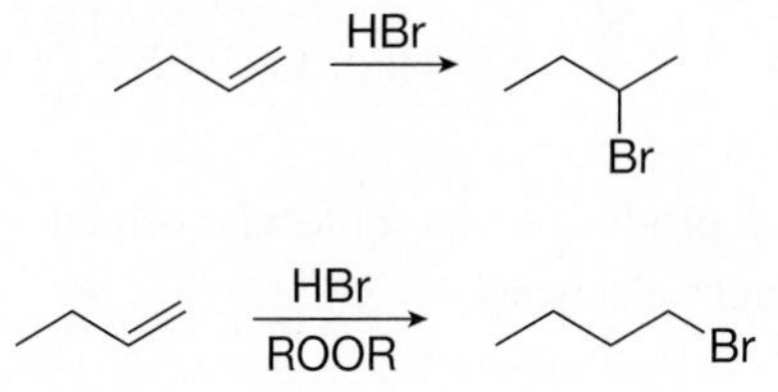

Hydrobromination of the internal alkene would be preferred since there are no competing products.

INTEGRATE THE SKILL 19.4

Due to the aromatic ring, there is only one alkene that can be considered with an unsaturation adjacent to the ring. Both reactions produce an unstable intermediate, which would be most stabilized by being adjacent to the aromatic ring.

Electrophilic addition forms a carbocation that is resonance stabilized, which reacts with bromide to form the required product.

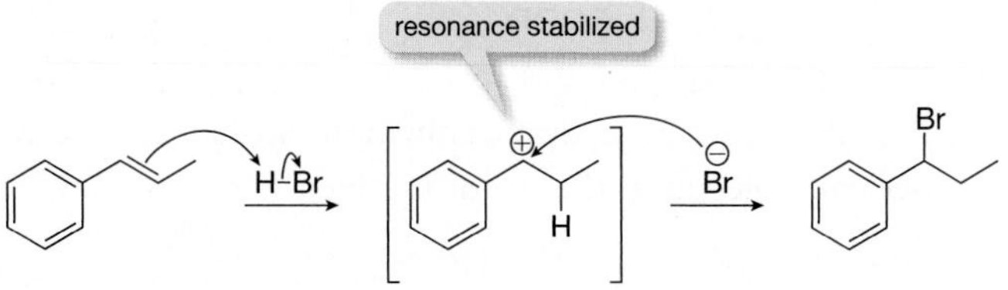

resonance stabilized

H–Br → Br⁻ →

Free radical addition creates a radical intermediate, which is most stable if it is on the carbon adjacent to the benzene ring. This reaction mechanism leads to a major product different than the one indicated.

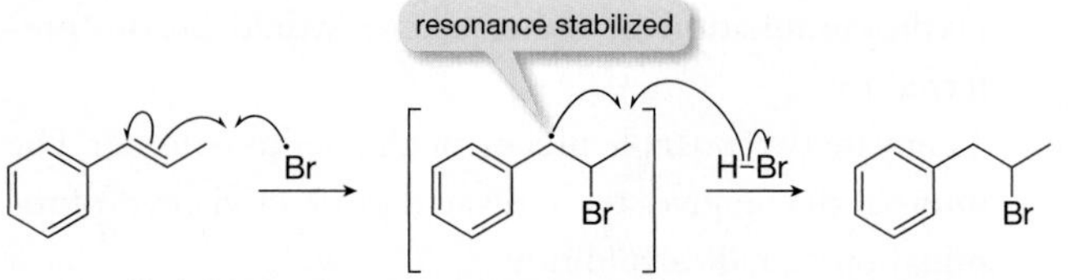

resonance stabilized

Br → H–Br →

So, electrophilic addition of HBr would give the correct major product.

CHAPTER 20

PRACTICE PROBLEM 20.1

a) The delocalized system has three double bonds and an oxygen lone pair containing seven p orbitals and eight electrons.

These will produce seven molecular orbitals containing a total of eight electrons.

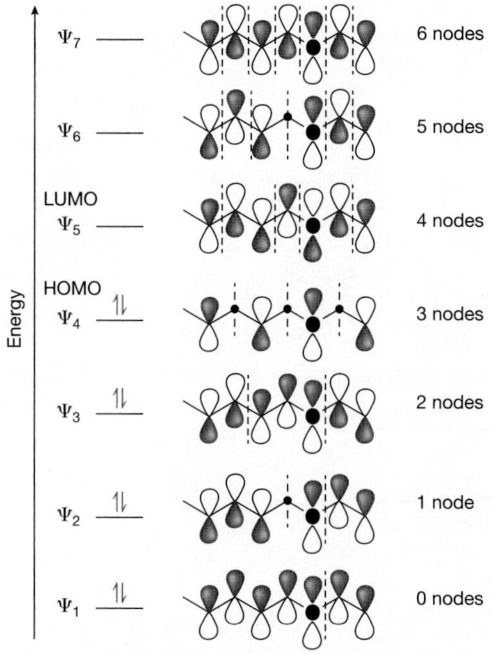

Energy

Ψ_7	—	6 nodes
Ψ_6	—	5 nodes
LUMO Ψ_5	—	4 nodes
HOMO Ψ_4	⇅	3 nodes
Ψ_3	⇅	2 nodes
Ψ_2	⇅	1 node
Ψ_1	⇅	0 nodes

b) First, redraw the molecule in a linear form for clarity.

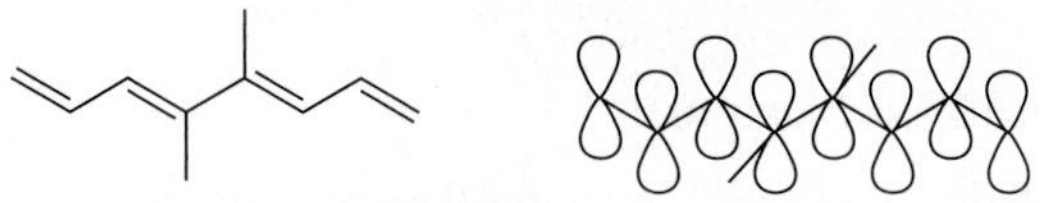

There are four double bonds in the delocalized system, with eight p orbitals and eight electrons. These will form eight molecular orbitals containing eight electrons.

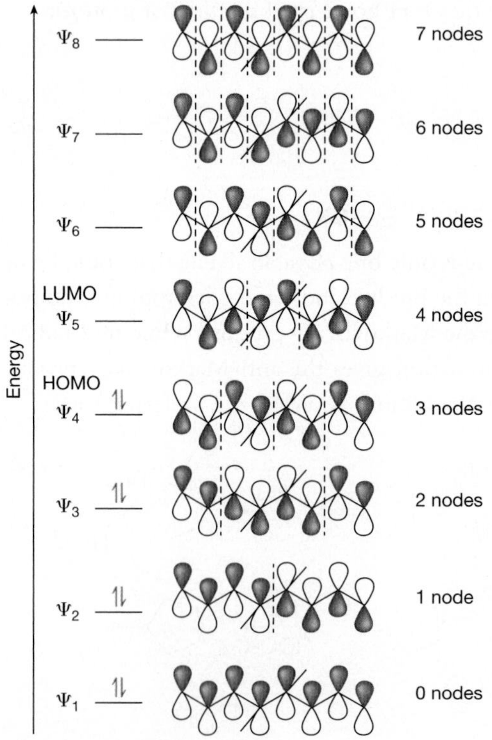

Energy

Ψ_8	—	7 nodes
Ψ_7	—	6 nodes
Ψ_6	—	5 nodes
LUMO Ψ_5	—	4 nodes
HOMO Ψ_4	⇅	3 nodes
Ψ_3	⇅	2 nodes
Ψ_2	⇅	1 node
Ψ_1	⇅	0 nodes

c) The three double bonds have six p orbitals and six electrons.

These form six molecular orbitals containing six electrons.

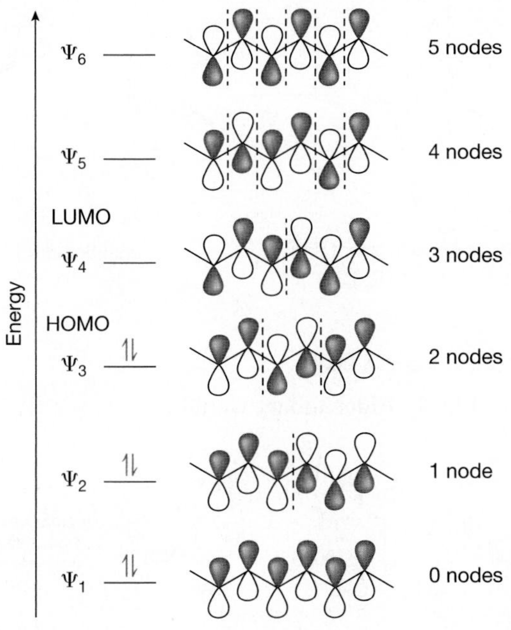

INTEGRATE THE SKILL 20.2

The bromination step has the unpaired electron reacting with the incoming bromine radical.

Ψ3

Ψ2

Ψ1

electron in HOMO reacts with incoming Br radical

preferred site of Br• addition

Br• can only add to sites that have a lobe. It cannot add at a node location.

From the molecular orbital, the radical electron is located on two carbon atoms. Based on the reactivity observed, the bromine radical prefers to add to the more substituted carbon. This could be because the radical is stabilized by the extra substitution and more electron density resides on that carbon.

PRACTICE PROBLEM 20.3

a) Using resonance forms, the ether end of the diene and the nitrile end of the dienophile should be aligned.

The correct stereochemistry can be assigned using the left/right procedure outlined in the chapter. Since the reaction is not controlled by any external chiral molecules, the product will be a racemic mixture.

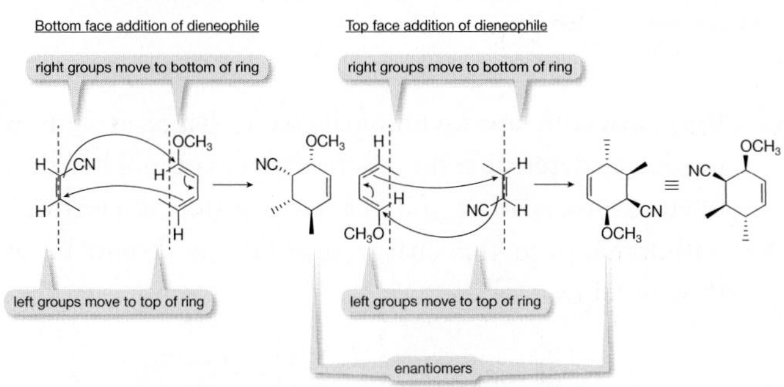

b) The diene is symmetric, so partial charges need not be determined. The dienophile should be aligned in an *endo* orientation. There is only one chiral carbon formed, so the stereochemistry produced will just be the racemic mixture.

enantiomers

c) The diene is symmetric, so only the *endo* orientation of the dienophile needs to be accounted for to determine the stereochemistry. The only chiral carbons produced are from the alkyne approaching from the top or the bottom of the diene.

enantiomers

d) The stereochemistry of the product is complex, so the details of the reaction geometry need to be accounted for. Orienting the dieneophile *endo* and using the left/right procedure gives the correct stereochemistry for the product.

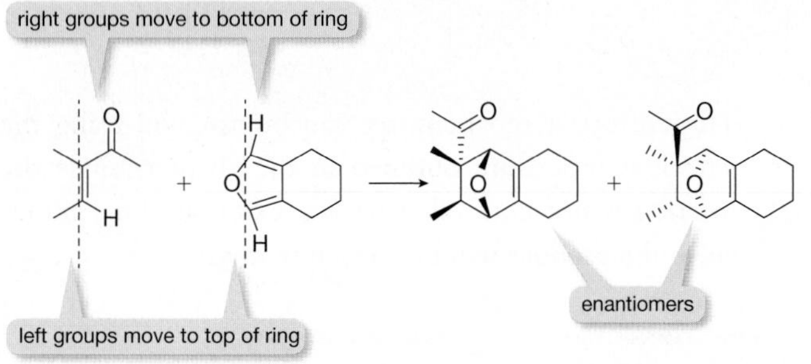
right groups move to bottom of ring
left groups move to top of ring
enantiomers

e) Both reactants are asymmetric, so a charge analysis is needed to determine the reaction orientation. The ether is electron-donating and the nitro group is electron-withdrawing, so the charge distribution should be as shown below.

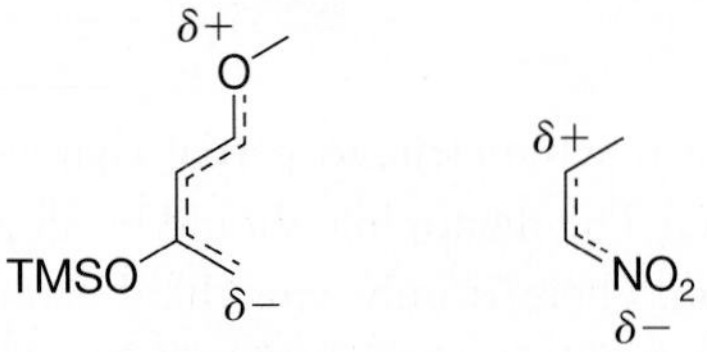
δ+
TMSO
δ−
δ+
NO2
δ−

Therefore, the dienophile should align with the NO_2 group *endo* to the diene and nearest the ether group. Using the left/right protocol, the correct stereochemistry of the product can be determined.

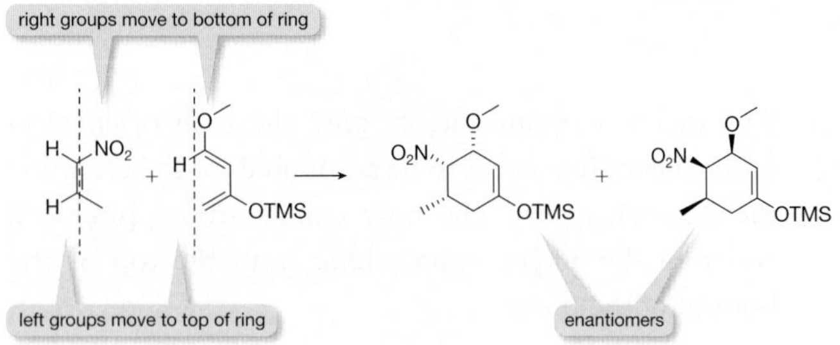
right groups move to bottom of ring
OTMS
OTMS
OTMS
left groups move to top of ring
enantiomers

f) The diene and dienophile are in the same molecule. The first step is to redraw the molecule in the proper geometry for the reaction to occur.

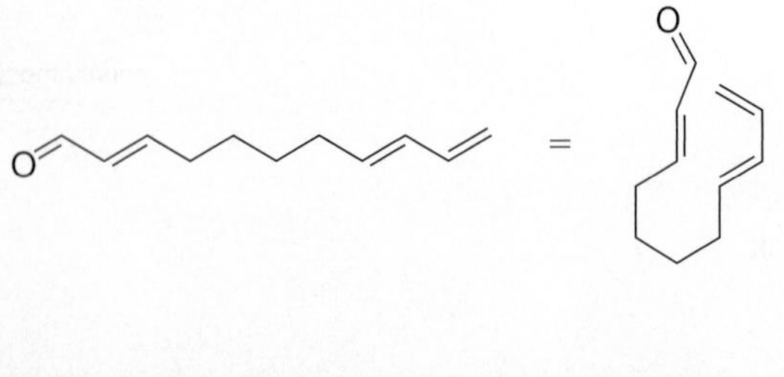

Then, the stereochemistry can be resolved orienting the electron-withdrawing group *endo* and assigning the relative stereochemistry for the substituents that produce chiral carbon atoms.

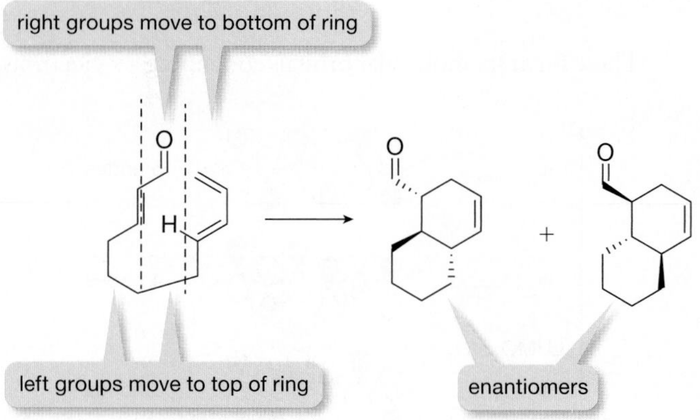
right groups move to bottom of ring
left groups move to top of ring
enantiomers

INTEGRATE THE SKILL 20.4

The initial Diels–Alder adduct would be

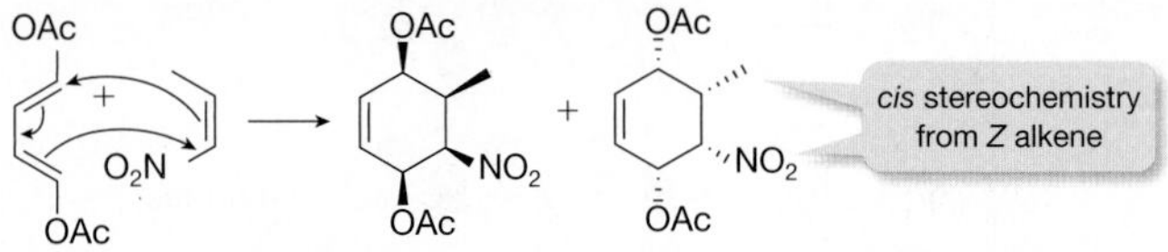
OAc
OAc
OAc
OAc
OAc
OAc
cis stereochemistry from Z alkene

The acetate groups need to be lost to form the final product. This is most easily done by first hydrolysing the ester to give a diol, then dehydrating the diol to form the double bonds.

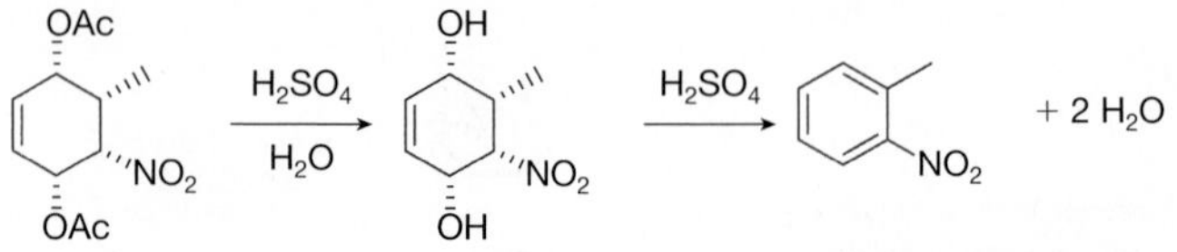
H2SO4
H2O
H2SO4
+ 2 H2O

PRACTICE PROBLEM 20.5

a) i) The intermediate has orthogonal π orbitals, so a thermal reaction is possible.

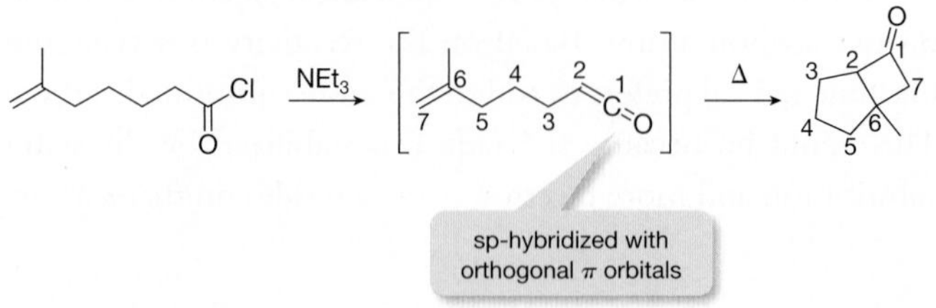
NEt3
Δ
sp-hybridized with orthogonal π orbitals

ii) There are no orthogonal sp-hybridized systems, so this reaction must occur under photochemical conditions.

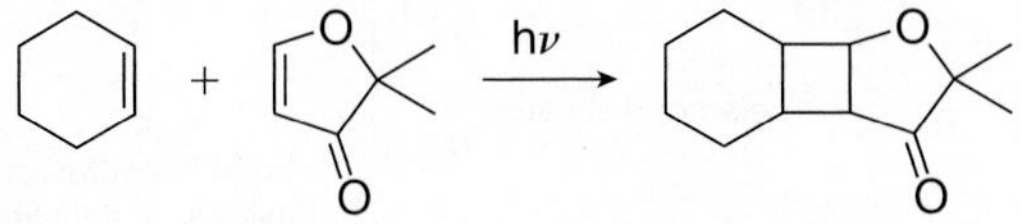

iii) There are no orthogonal sp-hybridized systems, so this reaction must occur under photochemical conditions.

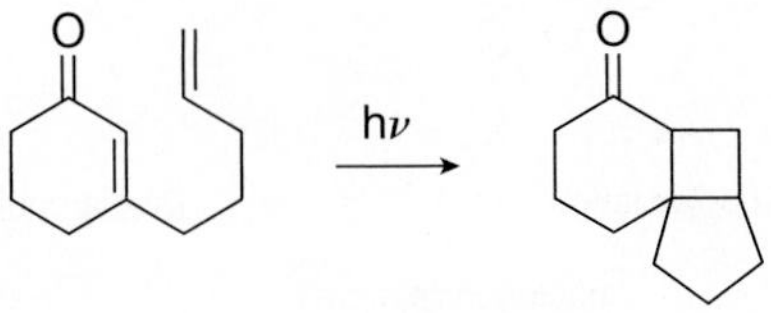

b) i) [2+2]:

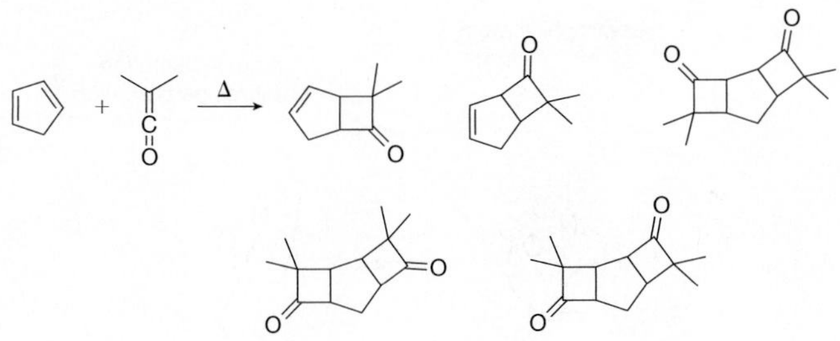

Diels–Alder:

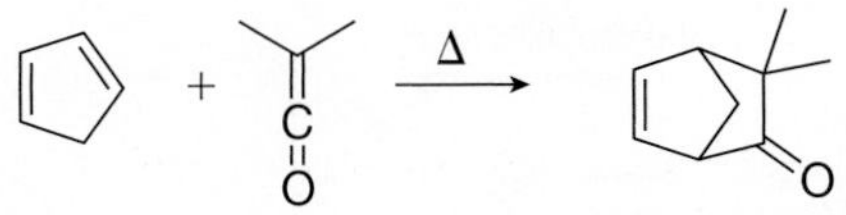

ii) [2+2]: The products for reaction with the other double bond is not shown since steric hindrance would inhibit reaction at this position.

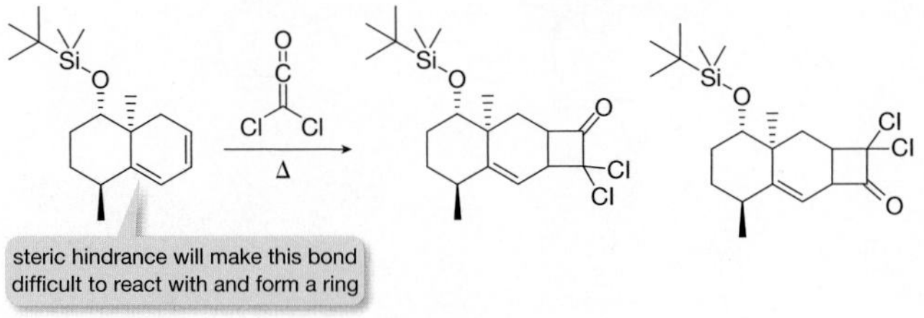

steric hindrance will make this bond difficult to react with and form a ring

Diels–Alder: Preparation of a model shows the bottom face of the diene is blocked by the CH_3 group, so attack must occur from the top face. Electronic *and* steric effects work in tandem to govern the regiochemistry of the incoming dienophile.

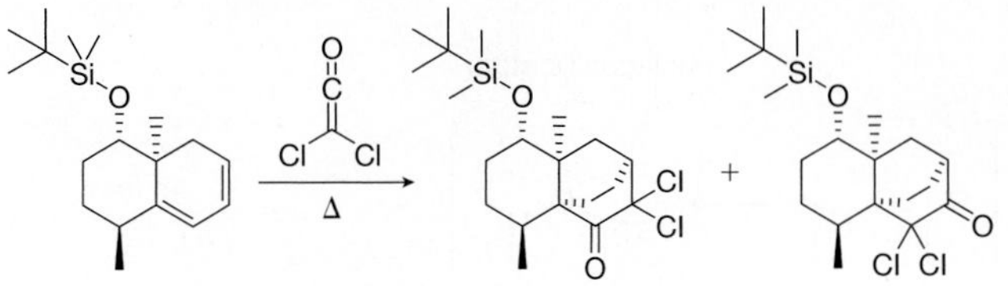

iii)

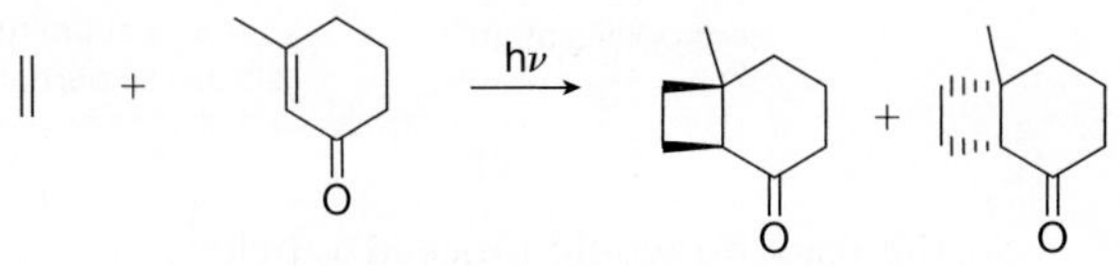

INTEGRATE THE SKILL 20.6

The thermal reaction will be a [4+2] Diels–Alder type of reaction since a diene and a dienophile are present.

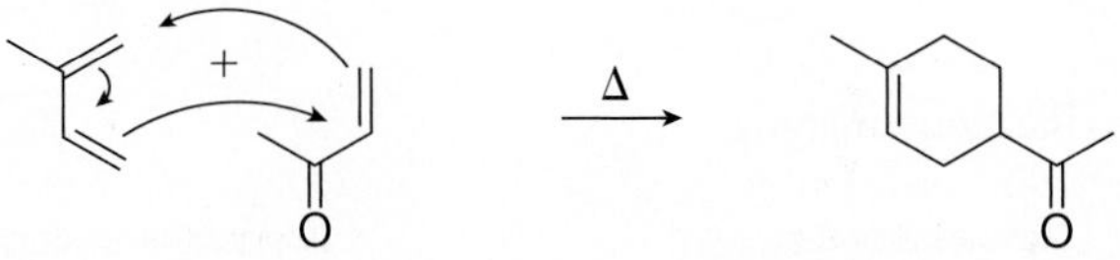

Under photochemical conditions, the molecular orbital symmetry will produce a [2+2] cycloaddition.

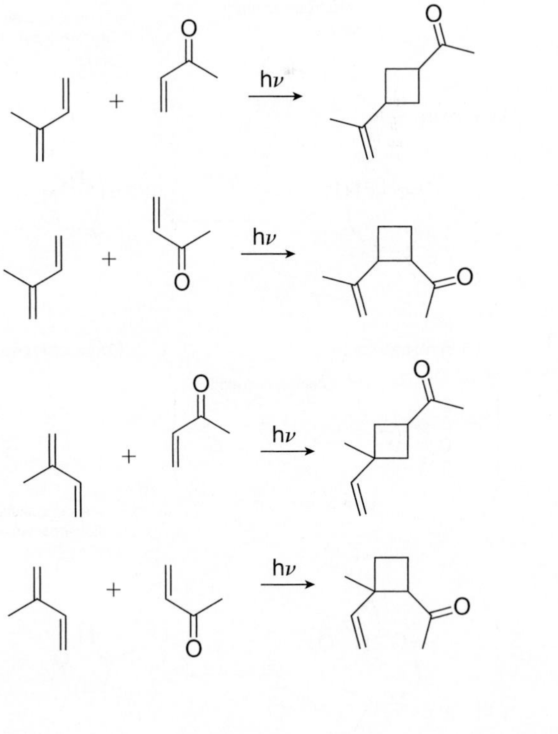

PRACTICE PROBLEM 20.7

a) The reactants are first examined to determine the polarization of the molecules.

1, 3 -dipole polarization

Dipolarophile polarization

nucleophilic atom

electrophilic atom

more substitution stabilizes partial cation

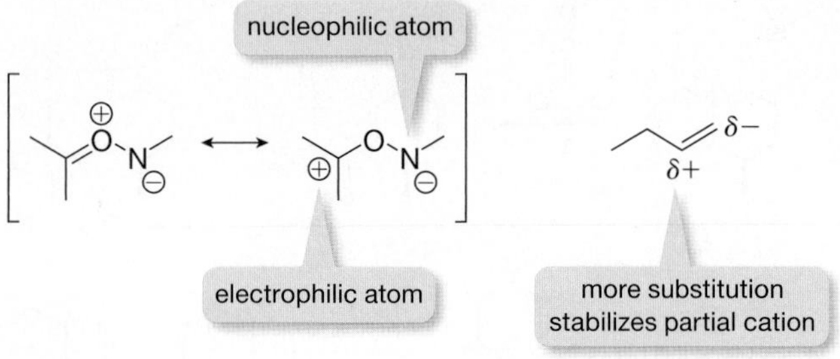

So, the reaction would proceed as below.

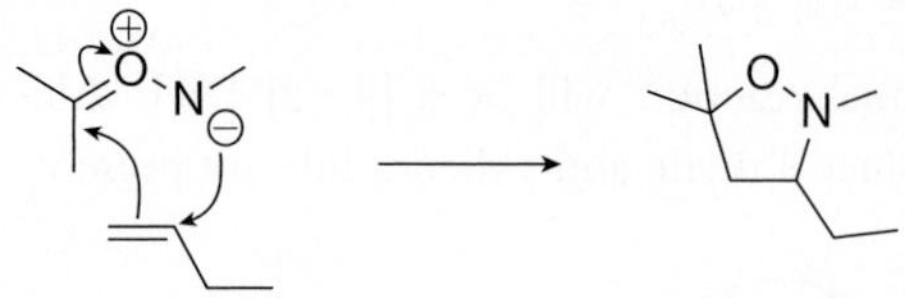

b) Reagent analysis:

1,3-dipole polarization

Dipolarophile polarization

nucleophilic atom

electrophilic atom

more substitution stabilizes partial cation

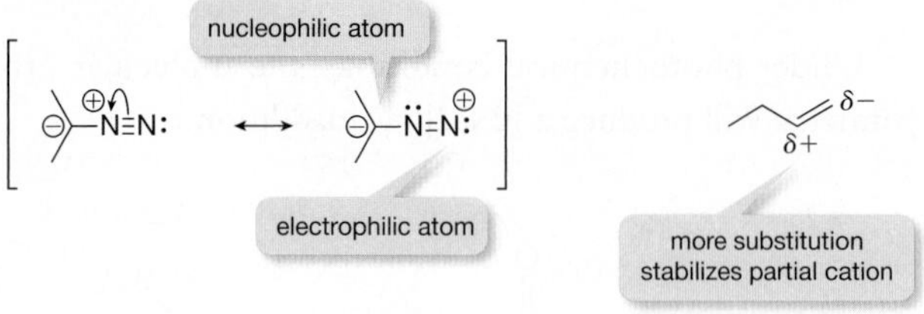

Reaction:

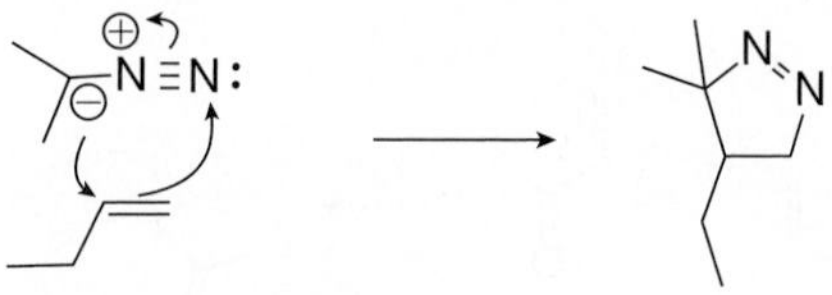

c) 1,3-dipole polarization

Dipolarophile polarization

nucleophilic atom

electrophilic atom

more substitution stabilizes partial cation

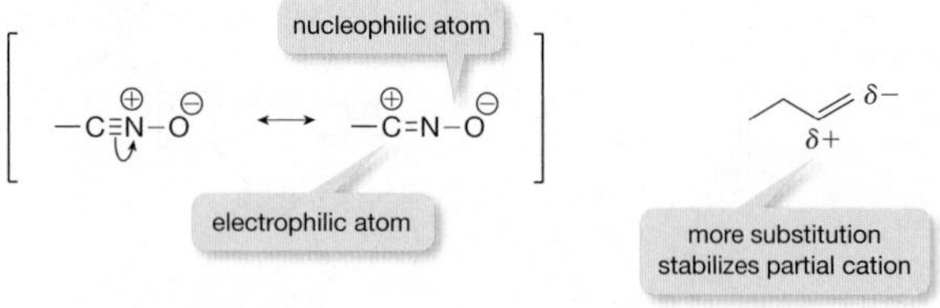

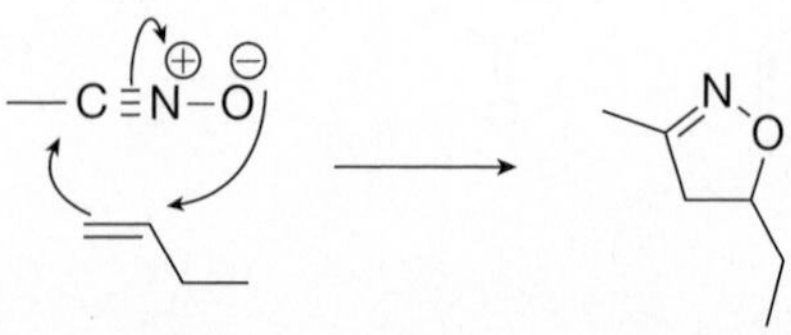

d) 1,3 -dipole polarization

Dipolarophile polarization

nucleophilic atom

electrophilic atom

more substitution stabilizes partial cation

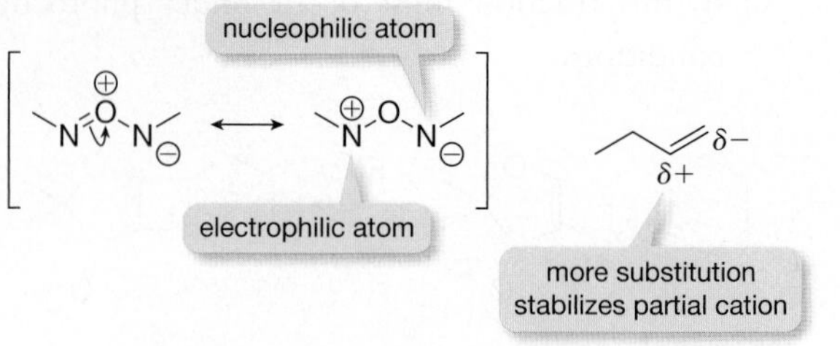

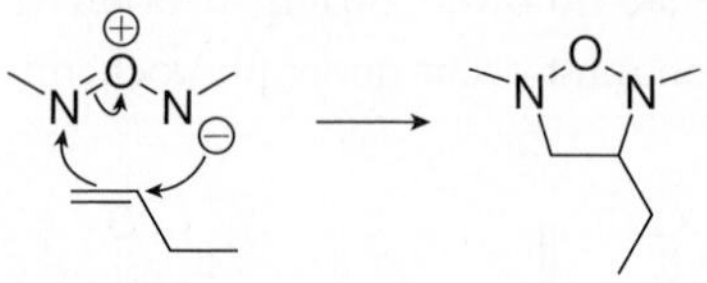

e) 1,3 -dipole polarization

Dipolarophile polarization

nucleophilic atom

electrophilic atom

more substitution stabilizes partial cation

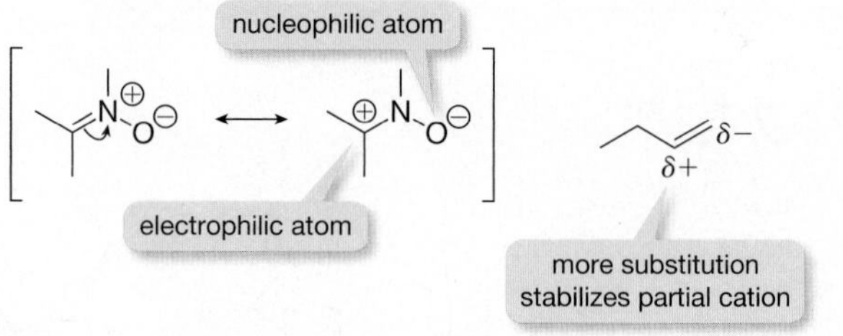

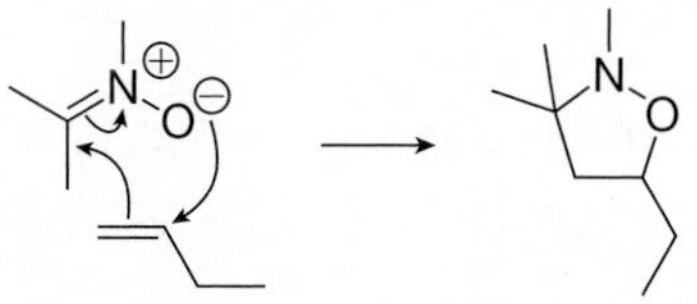

f) 1,3 -dipole polarization

Dipolarophile polarization

nucleophilic atom

electrophilic atom

more substitution stabilizes partial cation

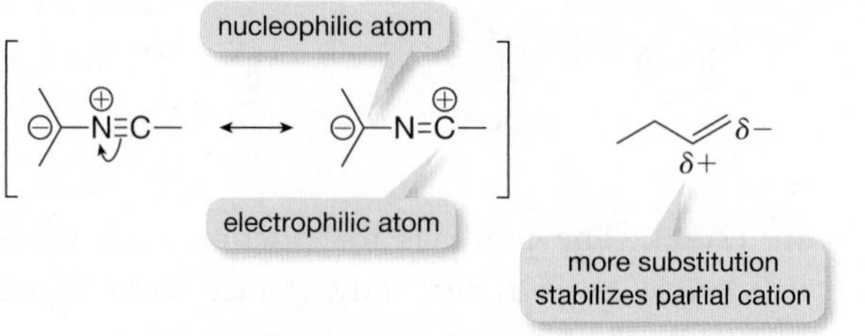

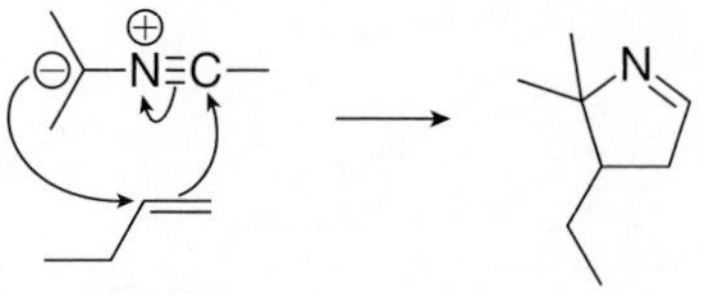

INTEGRATE THE SKILL 20.8

Mechanism:

Step 1

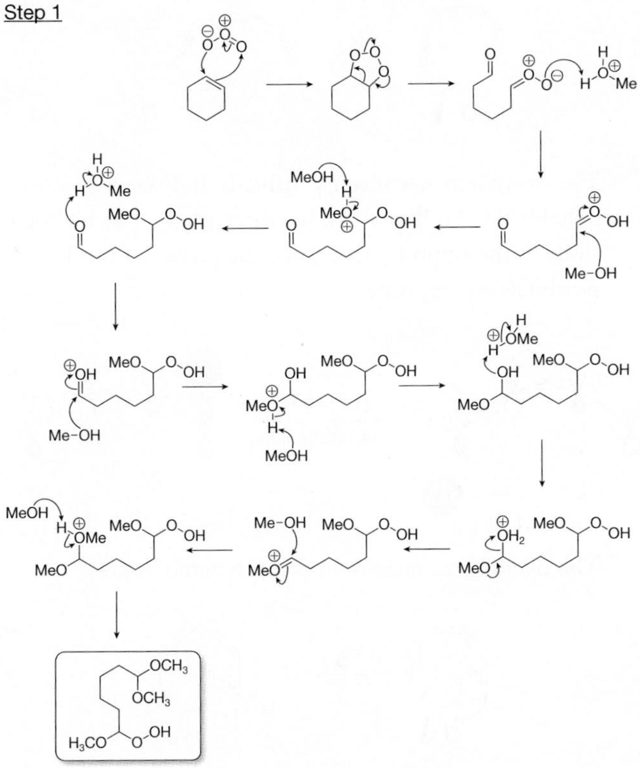

Step 2

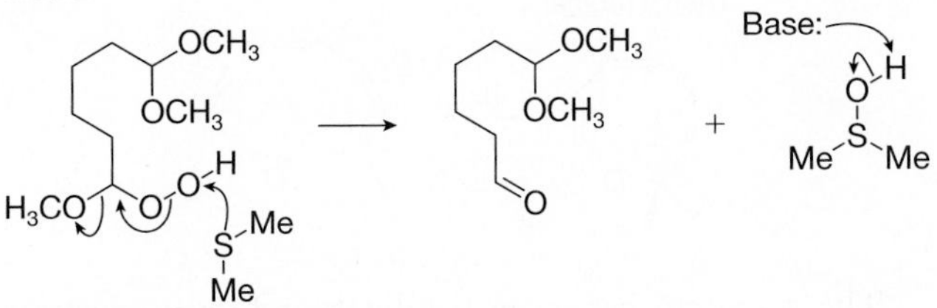

PRACTICE PROBLEM 20.9

a) The more stable molecule is indicated:

i) highly strained cyclopropane; more stable

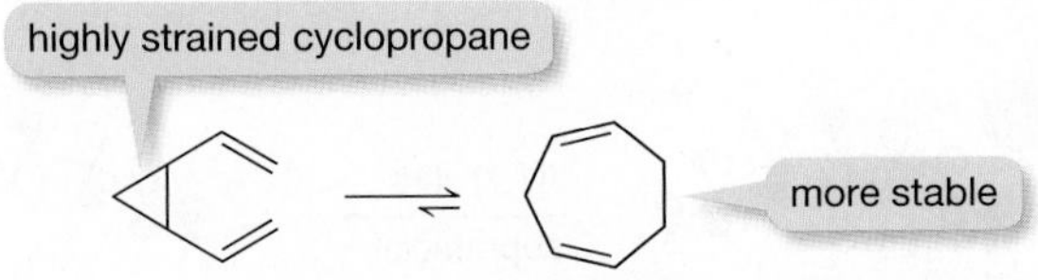

ii) more highly substituted alkenes are more stable

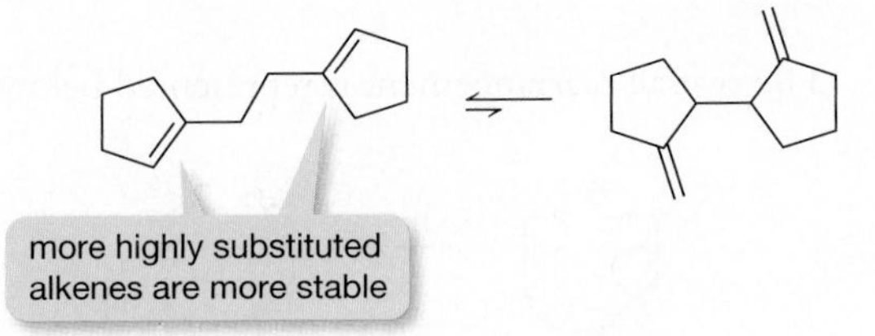

iii) more stable; H atoms on double bonds are forced together, causing strain

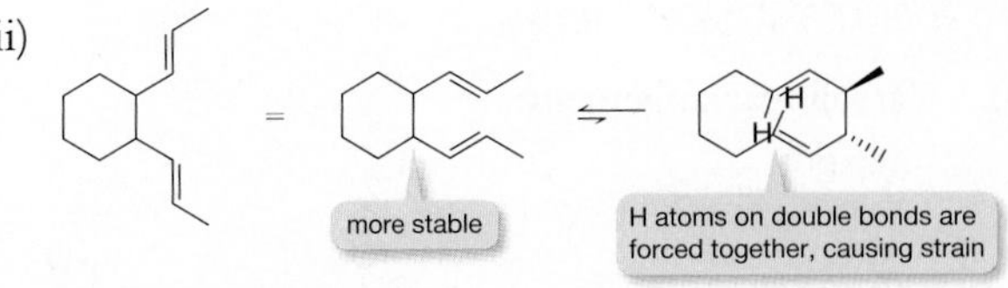

b) [3,3] sigmatropic rearrangements are shown where possible.

i)

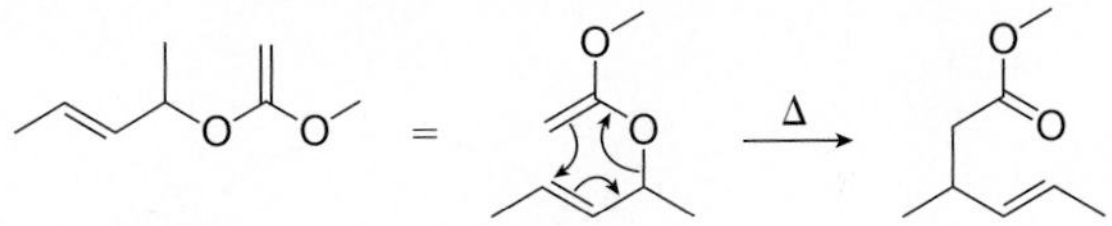

ii) not the correct configuration

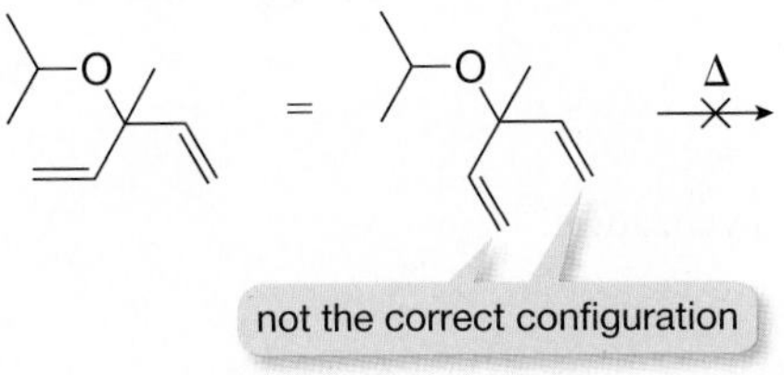

iii) The initial keto product of the rearrangement will undergo keto-enol tautomerization to restore the aromatic ring.

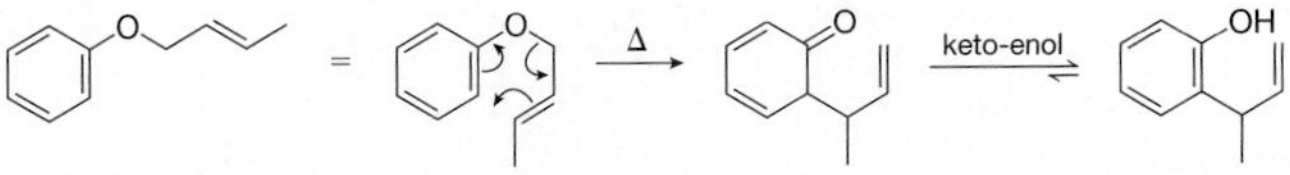

iv)

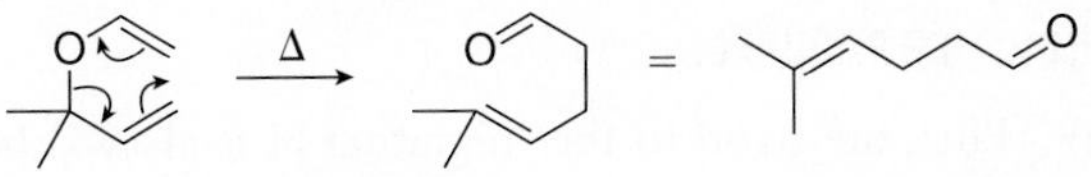

c) There is only one group that has the proper configuration for a [3,3] Cope rearrangement. Redrawing the 1,5-hexadiene system and rearranging the bonds gives the first product.

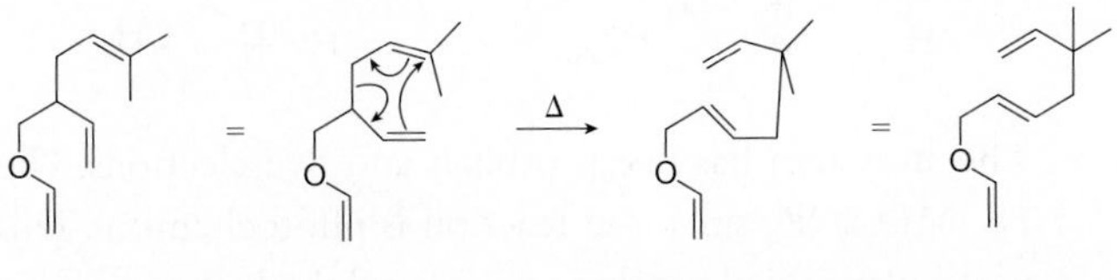

This rearranged system now has an allylic vinyl ether that can undergo a Claisen rearrangement to produce the ketone in the final product.

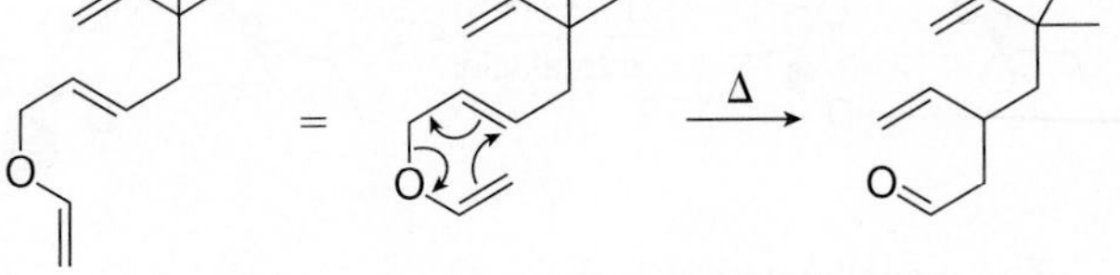

INTEGRATE THE SKILL 20.10

a) Carroll rearrangement:

b) Nazarov cyclization:

PRACTICE PROBLEM 20.11

a) First, the bond to the migrating H is cleaved homolytically and the ends of the π-system are brought together.

The π-system has five p orbitals and five electrons. The HOMO is Ψ_4 since the reaction is photochemical. This molecular orbital has three nodes, and the hydrogen shifts to the opposite face since the phases of the 1 and 5 p orbitals are different.

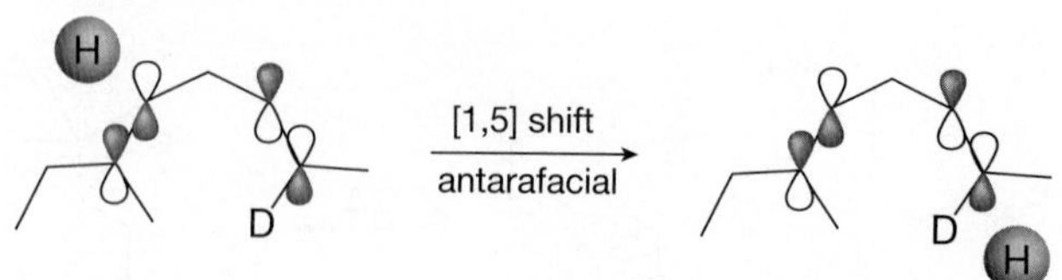

The overall reaction is then

b) The migrating group in this case is the deuterium atom.

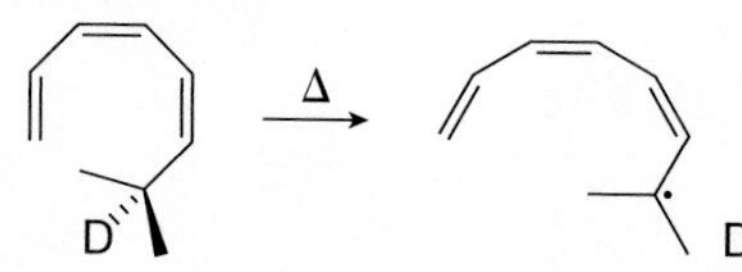

The π-system has seven p orbitals and seven electrons. The HOMO is Ψ_4, which has three nodes. The hydrogen shifts to the opposite face since the phases of the 1 and 7 p orbitals are opposite.

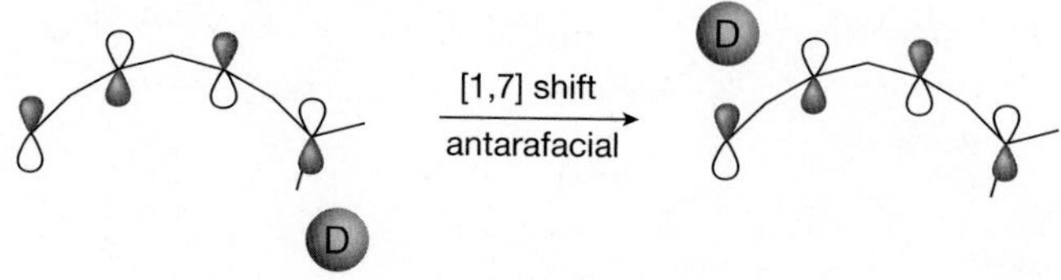

The overall rearrangement is represented below.

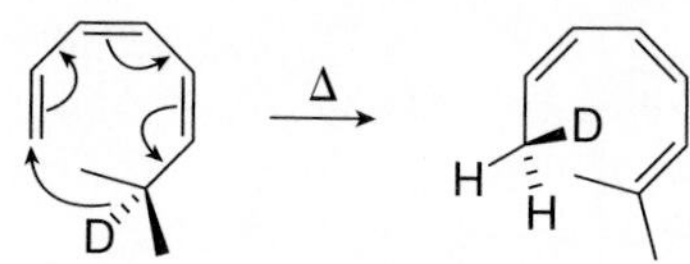

c) The migrating group in this case is the deuterium atom.

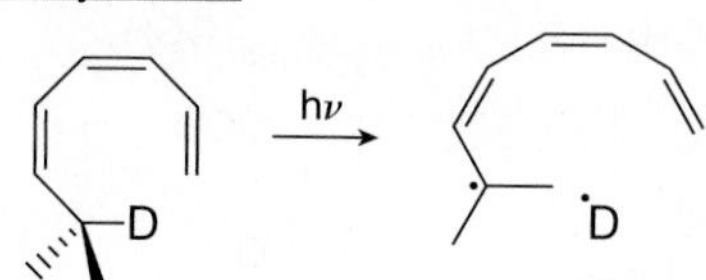

The π-system has seven p orbitals and seven electrons. Due to the reaction being under photochemical conditions, the reacting molecular orbital is Ψ_5, which has four nodes. The hydrogen movement is suprafacial since the HOMO is symmetric.

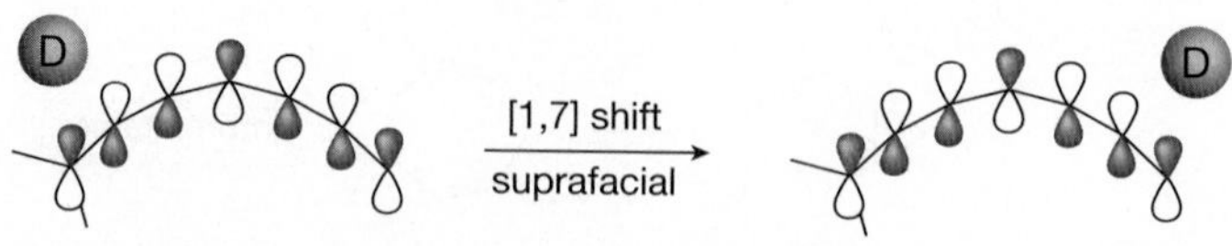

The overall rearrangement is represented below.

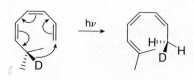

INTEGRATE THE SKILL 20.12

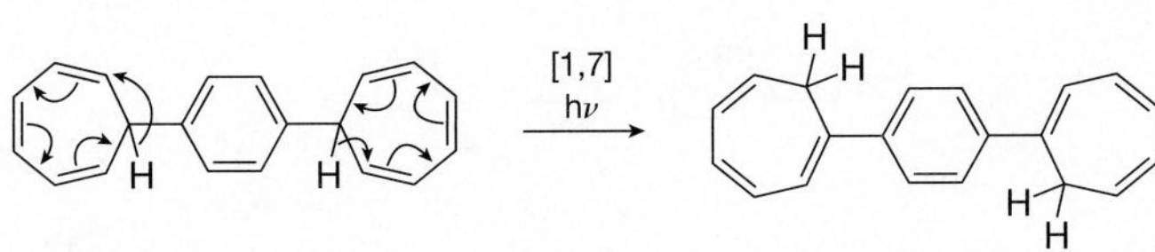

[1,5] Δ

[1,5] Δ

[1,5] Δ

[1,7] hν

PRACTICE PROBLEM 20.13

a) There are four electrons in the conjugated system. The HOMO will be Ψ_2, which has one node and is antisymmetric.

conrotatory motion needed

Ψ_2

opposite phases

Conrotatory ring closure leads to *cis* methyl groups.

H_3C H_3C H_3C CH_3

The overall reaction would therefore be represented as

Δ

b) There are four electrons in the conjugated system. The HOMO will be Ψ_2, which has one node and is antisymmetric.

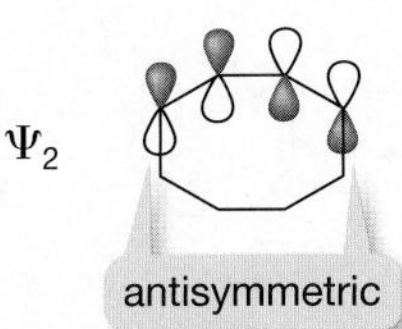

Conrotatory ring closure is required and will produce *trans* links to the hexane ring.

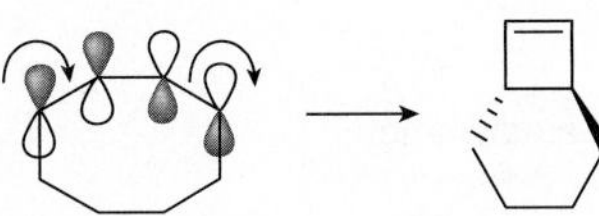

The overall reaction would therefore be represented as

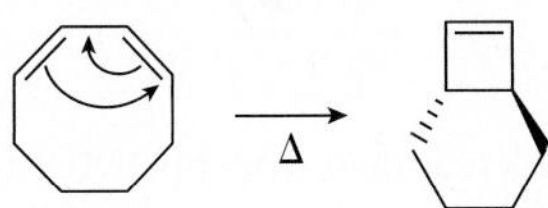

c) The three π bonds will form six molecular orbitals with six electrons in them. The HOMO will be Ψ_3, which has two nodes and is symmetric.

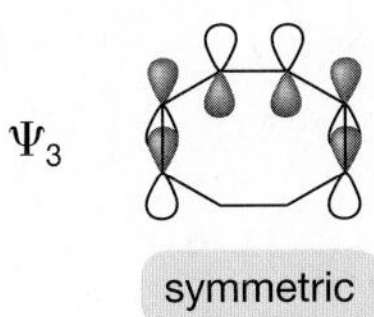

The ring closure will be disrotatory and produce *cis* linkages to the cyclobutane ring.

Therefore, the overall mechanism would be

Δ

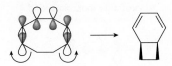

INTEGRATE THE SKILL 20.14

The overall series of reactions described would be (no stereochemistry is shown)

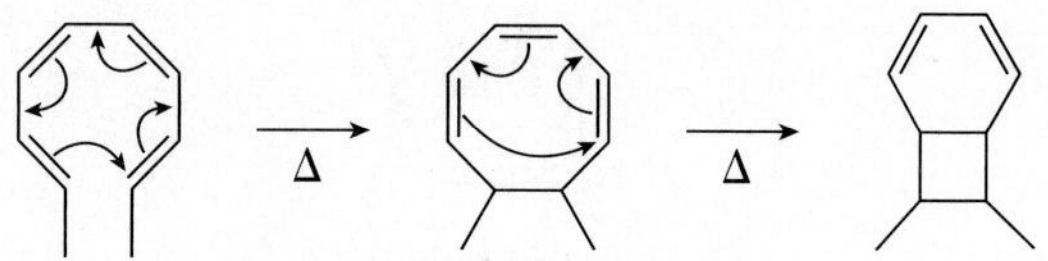

For the first reaction, the HOMO for the starting tetraene will be Ψ_4, which has three nodes and is antisymmetric. Therefore, the ring closure will be conrotatory and give the stereochemistry indicated.

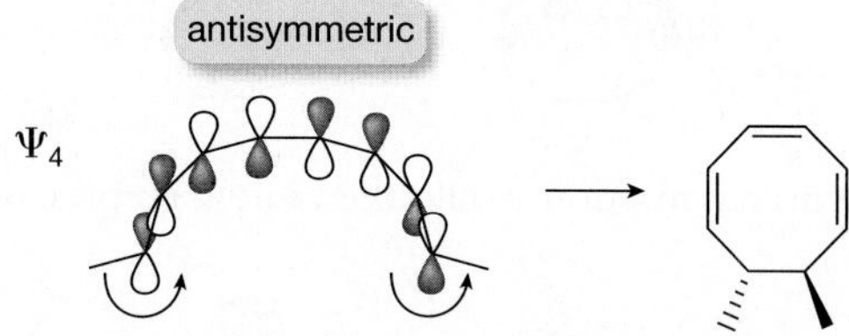

For the second reaction, the HOMO for the triene will be Ψ_3 (three nodes, symmetric). Ring closure will be disrotatory and produce the stereochemistry indicated below.

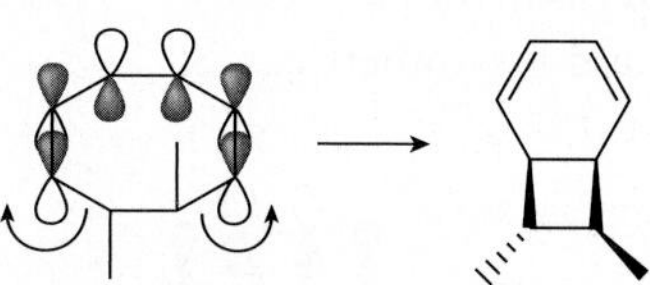

PRACTICE PROBLEM 20.15

a) There are four p orbitals and four electrons in the conjugated system. The SOMO will be Ψ_3 (two nodes, symmetric) after photochemical excitation.

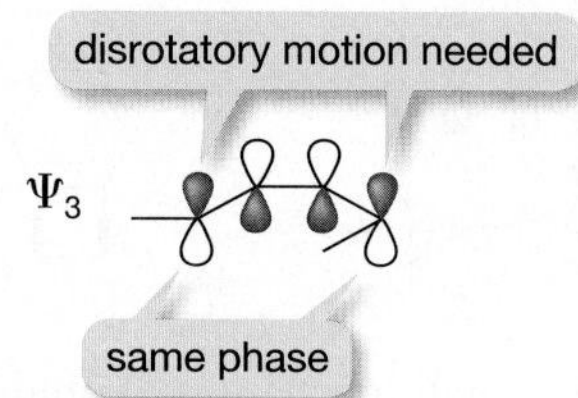

Disrotatory ring closure leads to *trans* methyl groups.

H_3C ... H_3C → H_3C ... CH_3

The overall reaction would therefore be represented as

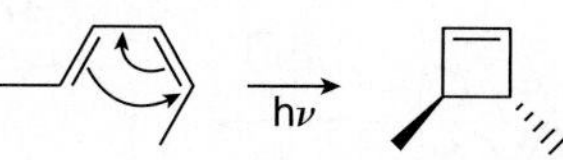

b) There are four electrons in the conjugated system. The SOMO will be Ψ_3 (two nodes, symmetric).

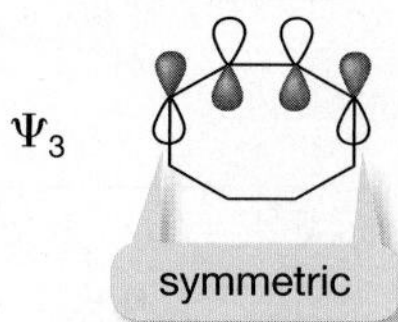

Conrotatory ring closure is required and will produce *cis* links to the hexane ring.

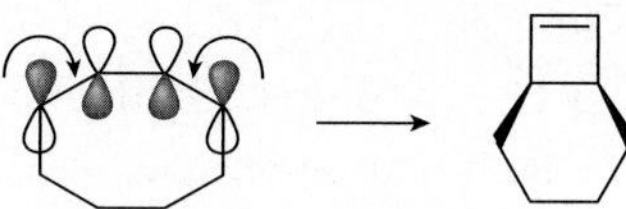

The overall reaction would therefore be represented as

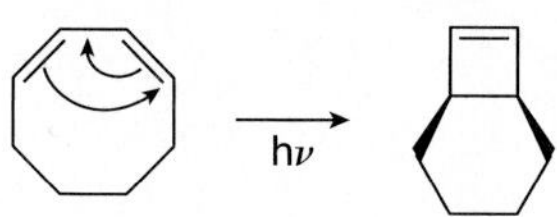

c) The three π bonds will form six molecular orbitals with six electrons in them. The SOMO will be Ψ_4 (three nodes, antisymmetric).

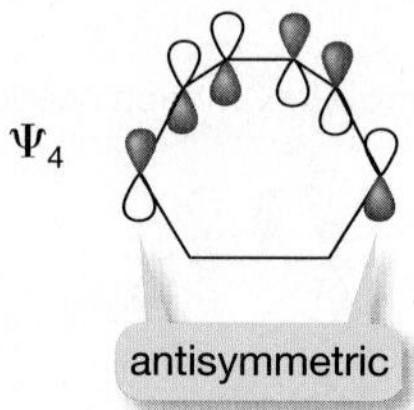

The ring closure will be conrotatory and produce trans linkages to the cyclobutane ring

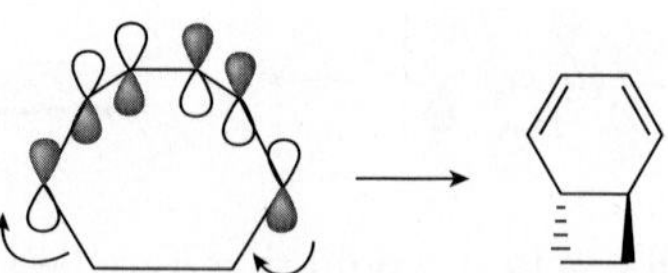

Therefore, the overall mechanism would be

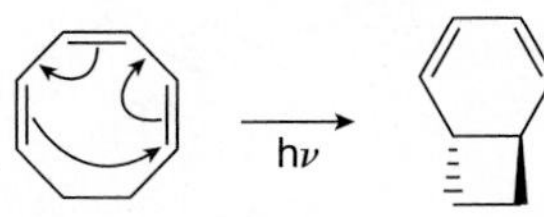

INTEGRATE THE SKILL 20.16

The electron movements required for this cyclization are shown below.

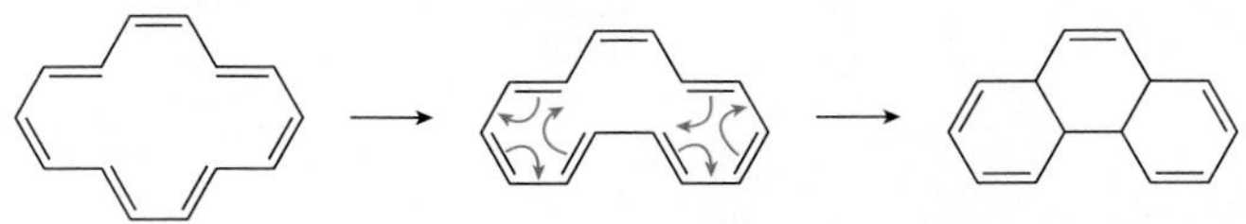

These are six electron ring closures. Thermally, this reaction must occur from the Ψ_3, which is symmetric. The ring closure of a symmetric orbital proceeds in a disrotatory fashion, giving *cis* stereochemistry. Photochemically, the reaction uses the Ψ_4, which is antisymmetric and must happen in a conrotatory fashion.

PRACTICE PROBLEM 20.17

a) The product has highly strained four-membered rings, so the reactant will be more stable and the favoured component in the equilibrium.

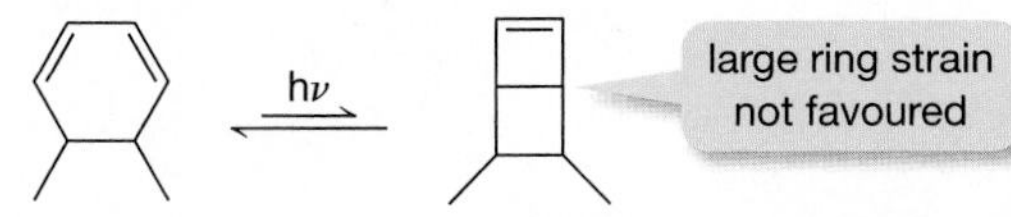

b) Neither product has any significant bond angle strain, but the product has more σ bonds and is more stable. The equilibrium will favour product formation.

Δ

c) Neither product has any significant bond angle strain, but the product has more σ bonds and is more stable. The equilibrium will favour product formation.

INTEGRATE THE SKILL 20.18

The two possible products for ring opening *trans*-dimethylcyclobutene are shown below. Only one product is observed since the clockwise pathway is much higher in energy. The two methyl groups would be forced close to each other, creating a large amount of strain.

counter-clockwise conrotation

clockwise conrotation

observed product

steric strain

PRACTICE PROBLEM 20.19

a) Products are shown for the substituent with the greater migratory aptitude in a Baeyer–Villiger reaction.

i) Phenyl has a greater migratory aptitude than a primary carbon

possible migrating substituents

higher migratory ability

ii) Hydrogen has a higher migratory aptitude than an alkyl group.

highest migrating ability

iii) Tertiary sites have a higher migratory aptitude than secondary sites.

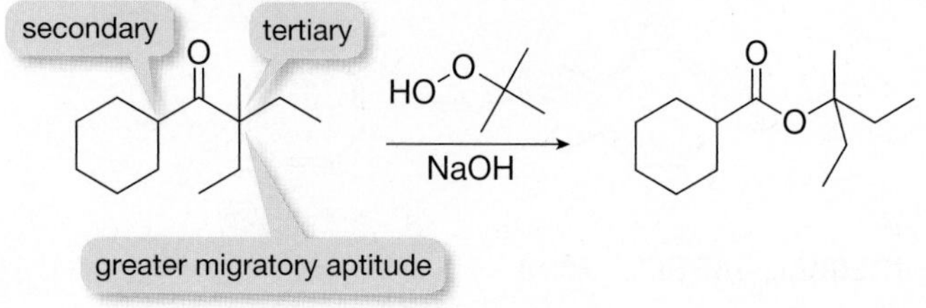

secondary

tertiary

HO–O–C(CH3)3 / NaOH

greater migratory aptitude

iv) Methyl groups have the least ability to migrate in a Baeyer–Villiger reaction.

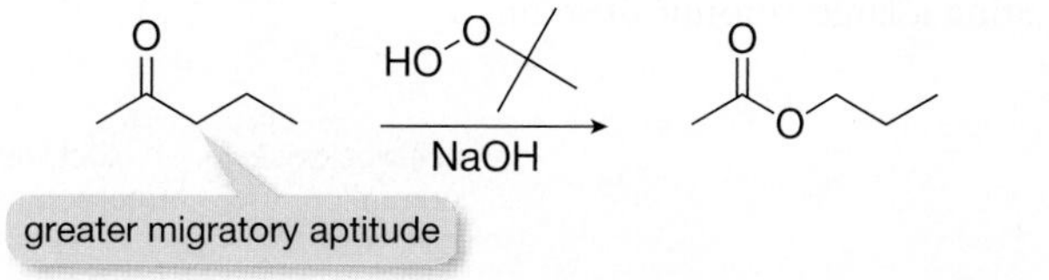

HO–O–C(CH3)3 / NaOH

greater migratory aptitude

b) i)

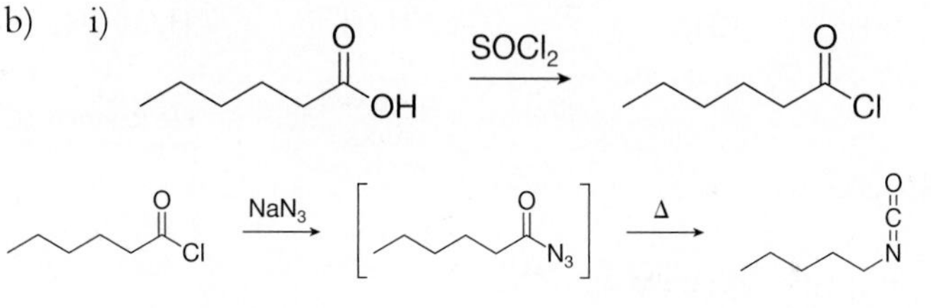

$SOCl_2$

NaN_3

Δ

ii)

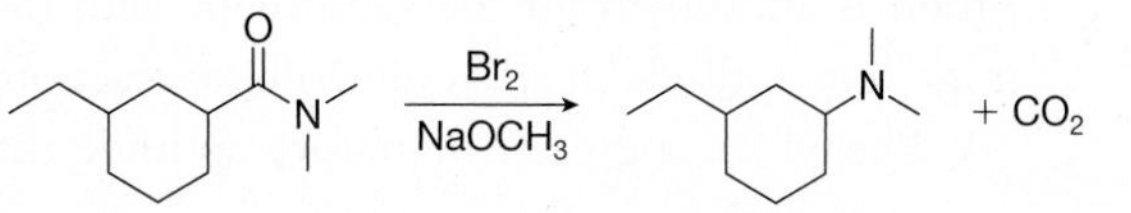

Br_2 / $NaOCH_3$ + CO_2

iii)

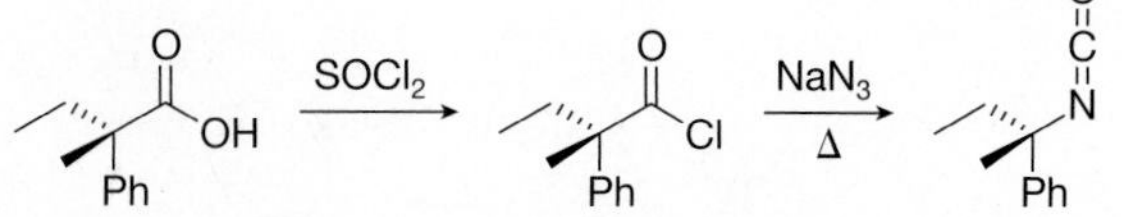

$SOCl_2$ NaN_3 / Δ

Ph Ph Ph

iv)

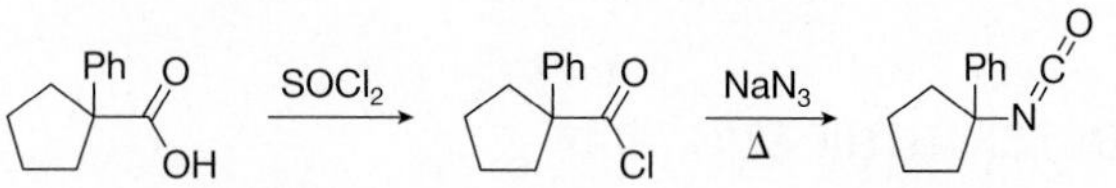

Ph $SOCl_2$ Ph NaN_3 / Δ Ph

v)

OH NH_2 Br_2 / NaOH OH NH_2

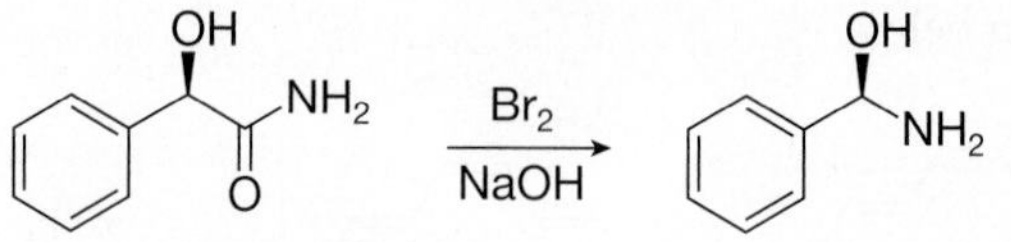

INTEGRATE THE SKILL 20.20

a)

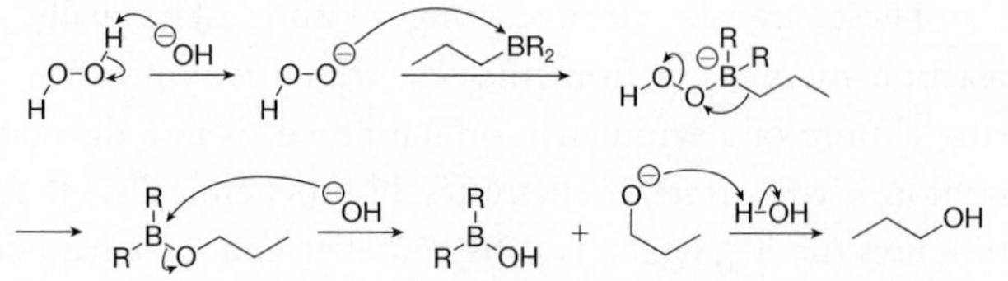

OH BR_2 OH OH

b)

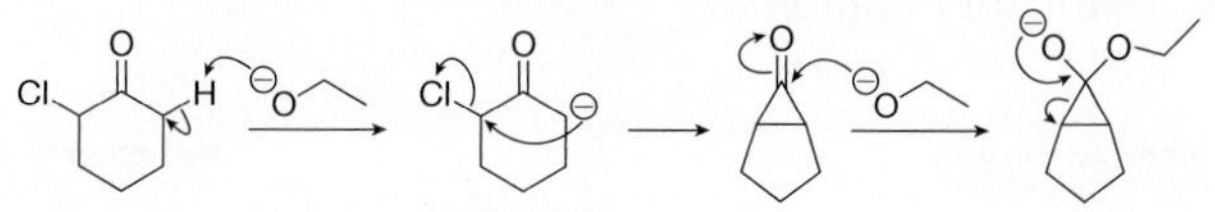

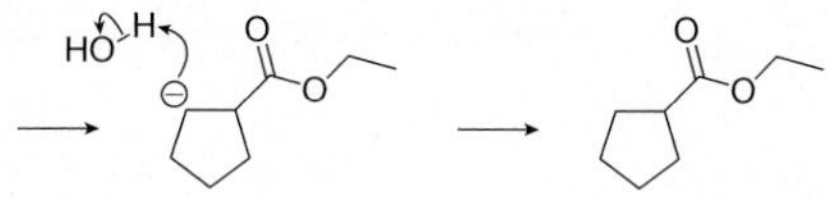

APPENDIX B

Common Errors in Organic Structures and Mechanisms

1. POORLY DRAWN CHAIRS

It only takes a few seconds to draw a chair properly. Here are some common mistakes to avoid:

a) There should be no horizontal or vertical lines in the ring structure.

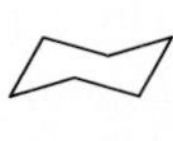

incorrect incorrect correct

The problem with the structure on the left is that axial bonds cannot be depicted by vertical lines. The structure in the middle has another common error: it does not distinguish between the two chair forms. The structure on the right provides a clear way to draw axial and equatorial bonds.

Avoid this problem by drawing the first "V" such that the two end tips are at the same horizontal level.

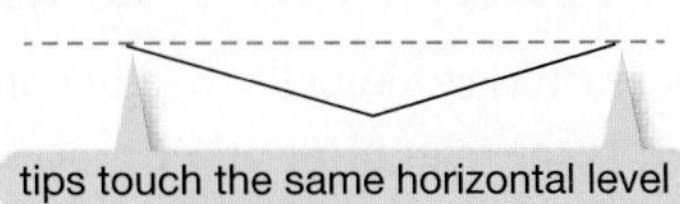

b) The axial bonds should point in the same direction as the "V" to which they are connected.

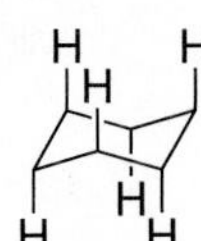

incorrect correct

Avoid this problem by using each "V" as a guide. Start at one clear location and alternate around the ring.

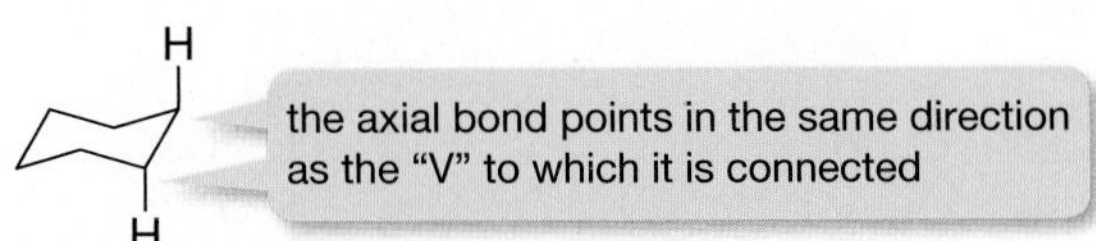

c) The equatorial bonds should be parallel to ring bonds.

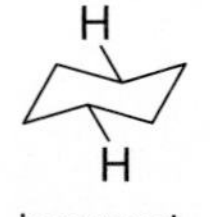

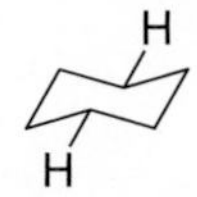

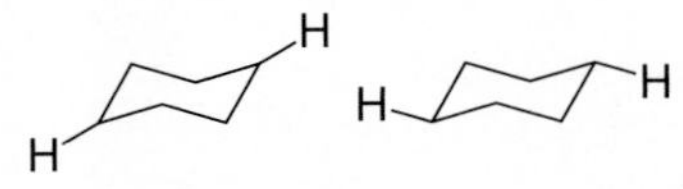

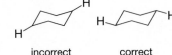

incorrect correct incorrect correct

Avoid this problem by making sure each equatorial bond is parallel to a ring bond.

d) The bottom three carbons are assumed to lie toward the viewer.

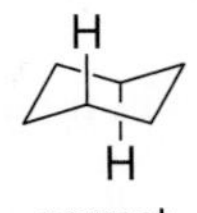

incorrect correct

2. USING ARROWS TO SHOW THE MOVEMENT OF ATOMS

This is very common with hydrogens.

using arrows to show the movement of atoms is a common error (especially with hydrogens)

Remember that curved arrows show the movement of electrons, not atoms. Errors like this can be prevented by drawing all components. If you commonly make the error, adding lone pairs helps.

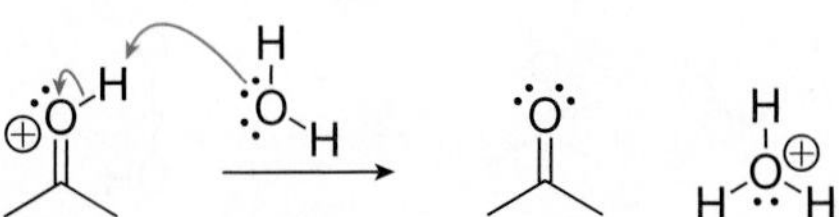

3. NON-STOP RESONANCE

This is very common in ring structures, but occurs with all types of resonance analysis.

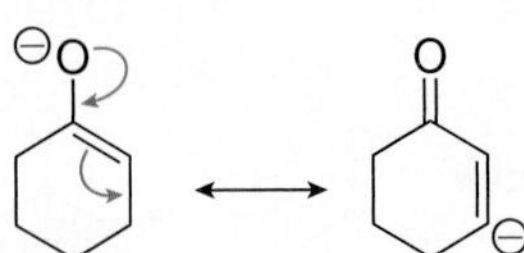

Remember that resonance involves π electrons only. σ bonds do not normally participate. One solution is to make sure to add hydrogens and lone pairs before starting your analysis.

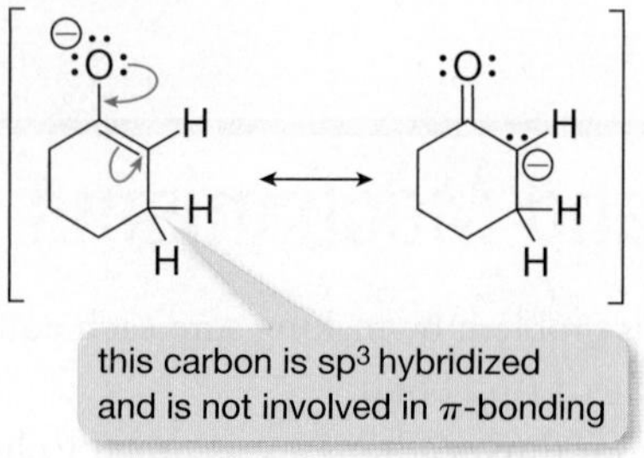

this carbon is sp³ hybridized and is not involved in π-bonding

4. ADDING OR SUBTRACTING CARBONS

This is a common mistake when working with carbocations.

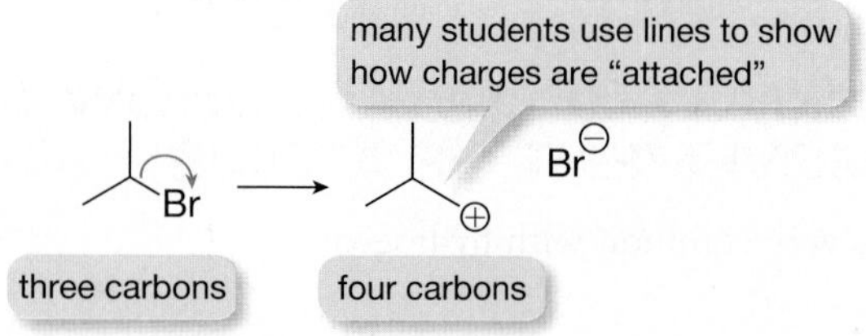

many students use lines to show how charges are "attached"

Remember that the charge is not "connected" to an atom but is actually on the atom. Avoid this error by counting carbons or adding labels.

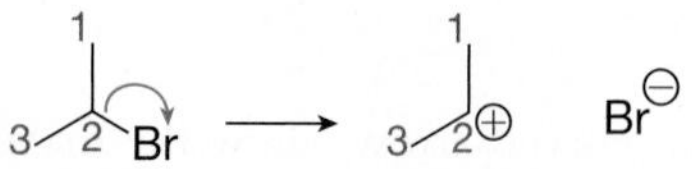

5. MISPLACING CHARGES AND LONE PAIRS

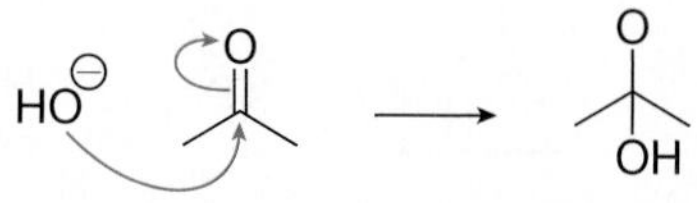

Avoid this by counting charges to make sure the total charge on the reactants and products is the same. Using curved arrow notation provides a second way to avoid this. Use the direction of the arrow and modify the charge integer accordingly.

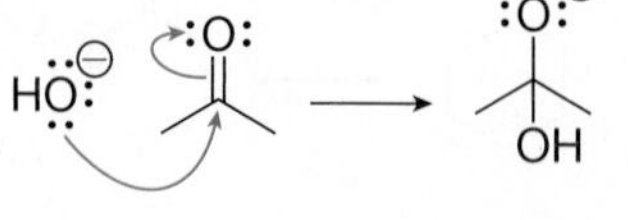

6. HYPERVALENT ATOMS

This is a common mistake when drawing line structures, especially when working with positively charged atoms.

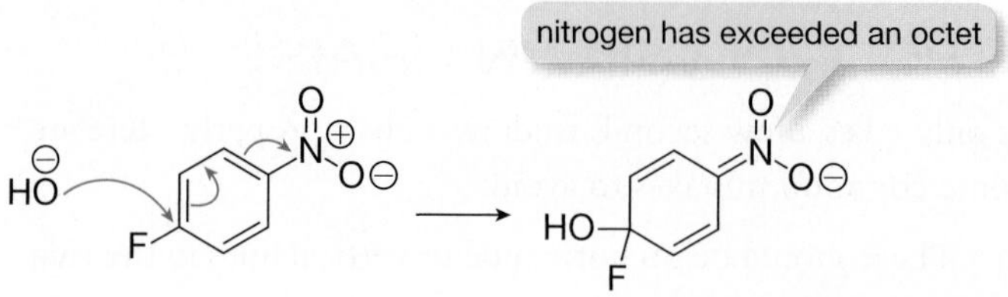

nitrogen has exceeded an octet

Avoid this by explicitly drawing lone pairs and hydrogens on functional groups and neighbouring carbons. Take the time to determine whether or not an atom can accept electrons.

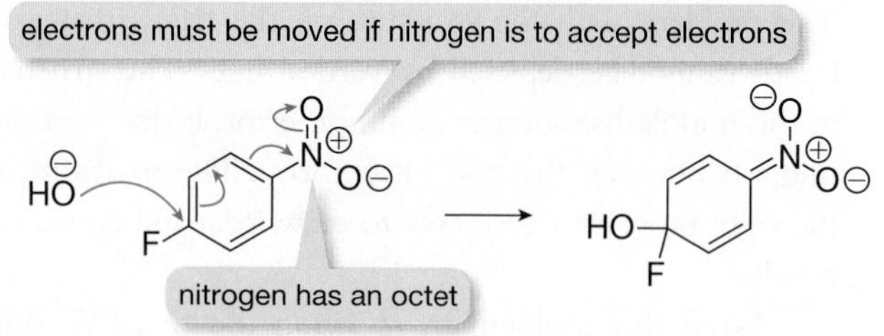

electrons must be moved if nitrogen is to accept electrons
nitrogen has an octet

7. MIXING ACIDS AND BASES

Acids and bases react very quickly together, and such reactions will happen faster than most other processes. Be careful of reacting positive and negative functional groups. Such salt-like interactions are not common in organic mechanisms.

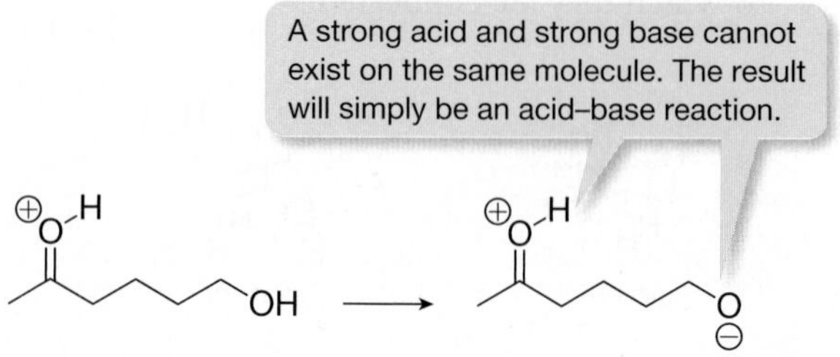

A strong acid and strong base cannot exist on the same molecule. The result will simply be an acid–base reaction.

In general, mechanisms will involve neutral molecules or only one charge type (positive or negative).

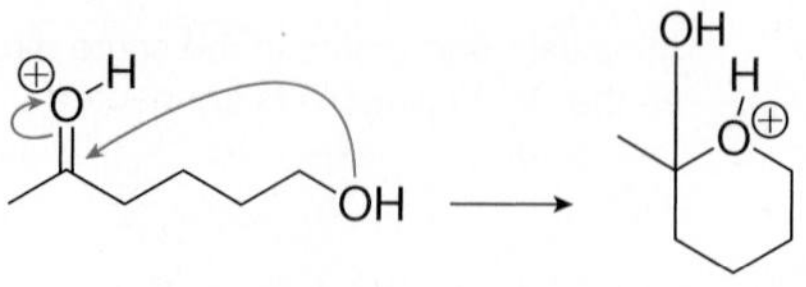

8. USING RESONANCE STRUCTURES

Many Internet help sites suggest using resonance forms in mechanisms. This can be a good guide to reactivity, but be careful of using them too extensively.

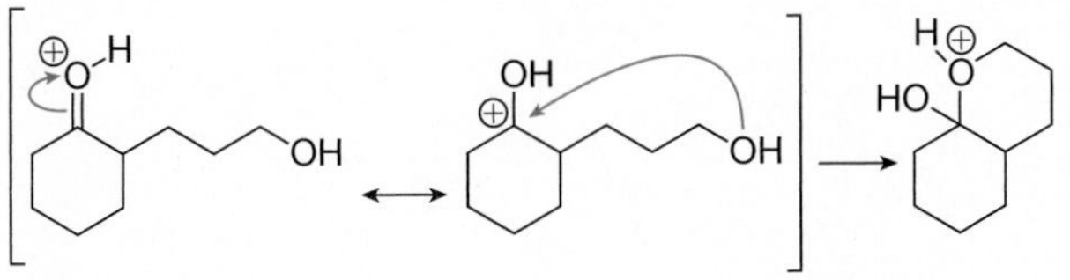

It is better practice to use the best resonance contributors when drawing mechanisms as this will usually result in an electron flow that best reflects reality. Other resonance forms can be analyzed as a guide to reactivity.

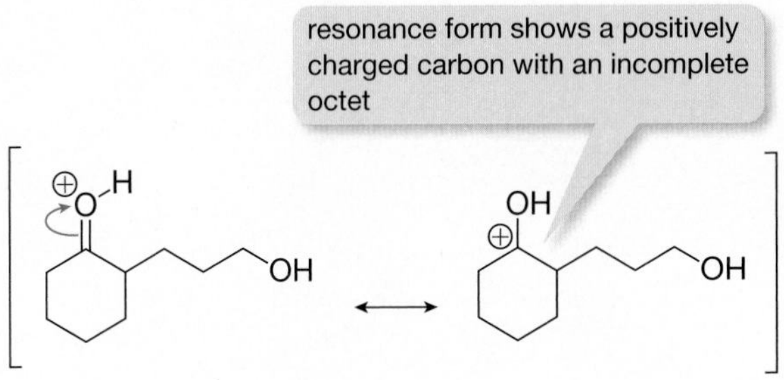

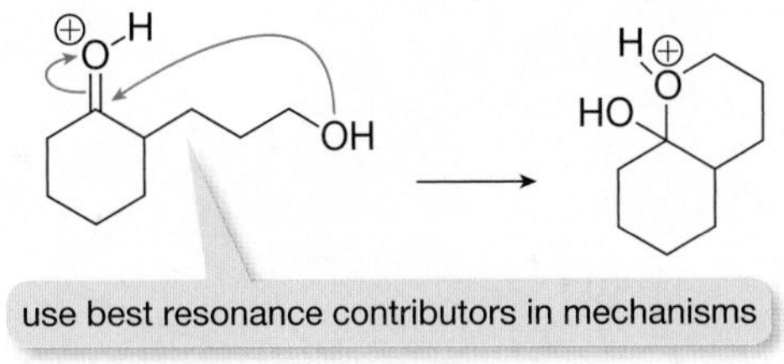

resonance form shows a positively charged carbon with an incomplete octet

use best resonance contributors in mechanisms

APPENDIX C

pK_a Values[1] of Selected Organic Compounds

Compound	pK_a	Compound	pK_a	Compound	pK_a
HBr	–9	$C_6H_5CO_2H$	4.2	$CH_2(C{\equiv}N)_2$	11
$C_6H_5C(\overset{\oplus}{O}H)OH$	–8	$C_6H_5NH_3^{\oplus}$	4.6	H_2O_2	11.6
$C_6H_5C(\overset{\oplus}{O}H)CH_3$	–6	CH_3CO_2H	4.8	CCl_3CH_2OH	12.2
HCl	–7	$C_5H_5\overset{\oplus}{N}H$	5.2	$CH_3OC(O)CH_2C(O)OCH_3$	13
$H_3C\overset{\oplus}{O}(H)CH_3$	–3.8	H_2CO_3	6.4	$CHCl_2CH_2OH$	12.9
H_2SO_4	–3	imidazolium (HN–$\overset{\oplus}{N}H$)	6.9	CH_3CHO	13.6
$H_3C\overset{\oplus}{O}(H)H$	–2.2	H_2S	7.0	CH_2ClCH_2OH	14.3
CH_3SO_3H	–2.6	C_6H_5SH	7.8	C_5H_6 (cyclopentadiene)	15
H_3O^+	–1.7	$CH_3C(O)OOH$	8.2	CH_3OH	15.5
HNO_3	–1.3	$CH_3C(O)CH_2C(O)CH_3$	9	H_2O	15.7
CF_3CO_2H	–0.2	NH_4^+	9.2	CH_3CONH_2	15.1
CCl_3CO_2H	0.6	H–C≡N	9.4	$(CH_3)_3COH$	17
$CHCl_2CO_2H$	1.3	$^-CO_2CH_2NH_3^+$	9.7	CH_3COCH_3	20
$CH_2(NO_2)CO_2H$	1.6	$(CH_3)_3NH^+$	9.8	HC≡CH	24
H_3PO_4	2.1	C_6H_5OH	10.0	N≡C–CH_3	25
CH_2ClCO_2H	2.9	CH_3CH_2SH	10.5	$(C_6H_5)_2CH_2$	34
HF	3.2	CH_3NO_2	10.3	H_2	36
O_2N–C_6H_4–CO_2H	3.4	$CH_3C(O)CH_2C(O)OCH_3$	11	NH_3	38
$CH_3OCH_2CO_2H$	3.6			$C_6H_5CH_3$	41
HCO_2H	3.8			C_6H_6	43
				CH_2=CH–CH_3	43
				CH_2=CH_2	44
				CH_3CH_3	48

[1] pK_a values obtained in aqueous solutions at 25 °C. pK_a values less than 0 and greater than 15.7 are corrected values from measurements in other solvents.

APPENDIX D

NMR and IR Spectroscopic Data

1H-NMR CORRELATION CHART

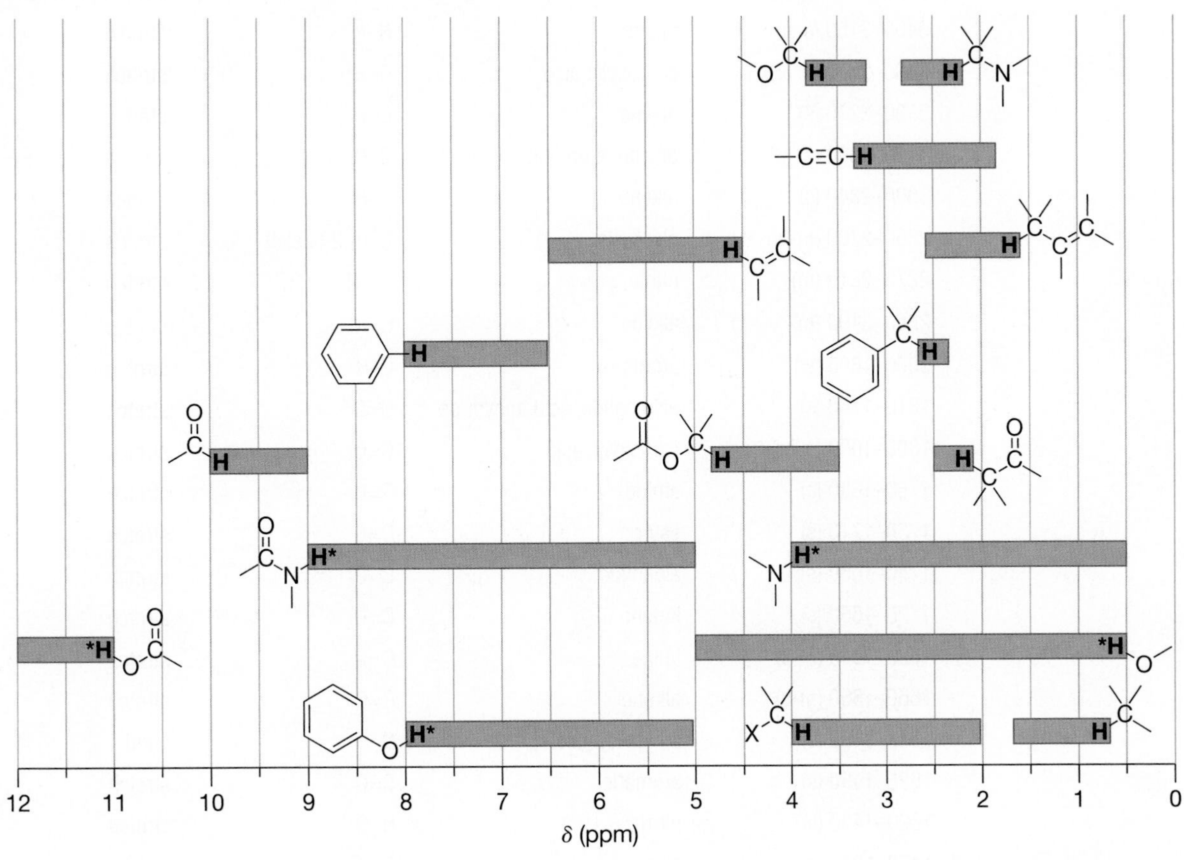

* Peaks may appear broad or absent; chemical shifts will vary with concentration and solvent.

13C-NMR CORRELATION CHART

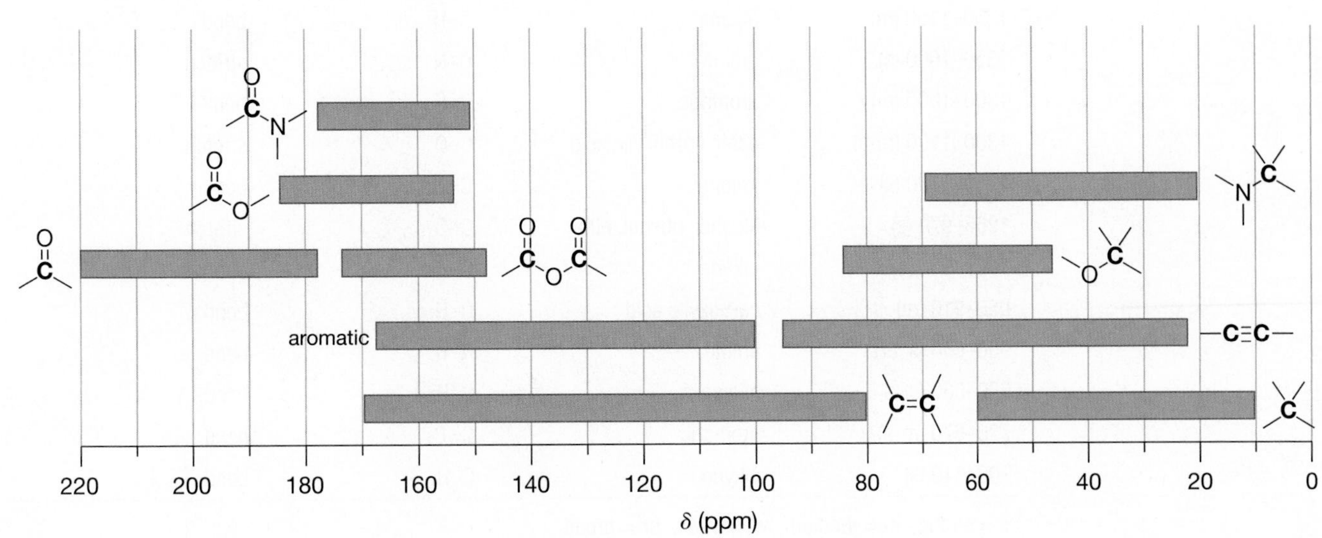

Table of Characteristic IR Absorptions			
Wavenumber in cm^{-1} (Intensity[1])	Functional Group(s)	Bond Type	
3650–3600 (s)	alcohol	O–H	stretch (free)
3500–3300 (m)	amine	N–H	stretch
3400–3300 (s, br)	alcohol	O–H	stretch (H-bonded)
3400–3150 (s, br)	amide	N–H	stretch
3400–2500 (s, br)	carboxylic acid	O–H	stretch
3330–3270 (s)	alkyne	C–H	stretch
3100–3000 (s)	alkene, aromatic	C–H	stretch
3000–2840 (s)	alkane	C–H	stretch
2860–2700 (m-w)	aldehyde	C–H (2 peaks)	stretch
2270–2210 (m)	nitrile	C≡N	stretch
2260–2100 (m)	alkyne	C≡C	stretch
2000–1660 (w)	aromatic	C–H	stretch
1810–1760 (s)	acid halide, acid anhydride	C=O	stretch
1800–1680 (s, br)	carboxylic acid	C=O	stretch
1760–1630 (s)	amide	C=O	stretch
1750–1715 (s)	ester	C=O	stretch
1740–1660 (s)	aldehyde	C=O	stretch
1720–1660 (s)	ketone	C=O	stretch
1690–1640 (m-s)	imine	C=N	stretch
1660–1560 (m-s)	alkene	C=C	stretch
1640–1490 (s)	amides, amine	N–H	bend
1625–1450 (s)	aromatic	C=C	stretch
1600–1490 (s)	nitro	N–O	stretch
1470–1350 (m)	alkane	C–H	bend
1450–1375 (m)	alkane	C–H (CH_3) 2 peaks	bend
1440–1320 (m-w)	alcohol	O–H	bend
1390–1300 (m)	nitro	N–O	stretch
1350–1150 (m)	alkane	C–H	bend
1350–1000 (s)	amine	C–N	stretch
1300–1000 (m)	aromatic	C–H	bend
1300–1100 (m-s)	ester, carboxylic acid	C–O	stretch
1300–1100 (s)	ketone	C=O	bend
1265–970 (s)	alcohol, phenol, ether	C–O	stretch
1000–650 (s)	alkene	C–H	bend
950–910 (m)	carboxylic acid	O–H	bend
900–680 (s, br)	amine	N–H	bend
900–660 (s)	alkene	C–H	bend
720–670 (m-s)	aromatic	C=C	bend
700–610 (s)	alkyne	C–H	bend

[1] s = strong, m = medium, w = weak, br = broad

APPENDIX E

Periodic Table of the Elements

Key:

28 - Atomic number
Ni - Symbol
Nickel - Name
58.69 - Average atomic mass

METAL
METALLOID
NON-METAL
PROPERTIES UNKNOWN

1	2	3	4	5	6	7	8	9	10	11	12	13	14	15	16	17	18
1 H Hydrogen 1.008																	2 He Helium 4.003
3 Li Lithium 6.941	4 Be Beryllium 9.012											5 B Boron 10.81	6 C Carbon 12.01	7 N Nitrogen 14.01	8 O Oxygen 16.00	9 F Fluorine 19.00	10 Ne Neon 20.18
11 Na Sodium 22.99	12 Mg Magnesium 24.305											13 Al Aluminum 26.98	14 Si Silicon 28.09	15 P Phosphorus 30.97	16 S Sulfur 32.06	17 Cl Chlorine 35.45	18 Ar Argon 39.95
19 K Potassium 39.10	20 Ca Calcium 40.08	21 Sc Scandium 44.96	22 Ti Titanium 47.87	23 V Vanadium 50.94	24 Cr Chromium 52.00	25 Mn Manganese 54.94	26 Fe Iron 55.85	27 Co Cobalt 58.93	28 Ni Nickel 58.69	29 Cu Copper 63.38	30 Zn Zinc 65.38	31 Ga Gallium 69.72	32 Ge Germanium 72.63	33 As Arsenic 74.92	34 Se Selenium 78.97	35 Br Bromine 79.90	36 Kr Krypton 83.80
37 Rb Rubidium 85.47	38 Sr Strontium 87.62	39 Y Yttrium 88.91	40 Zr Zirconium 91.22	41 Nb Niobium 92.91	42 Mo Molybdenum 95.95	43 Tc Technetium [97.91]	44 Ru Ruthenium 101.1	45 Rh Rhodium 102.9	46 Pd Palladium 106.4	47 Ag Silver 107.9	48 Cd Cadmium 112.4	49 In Indium 114.8	50 Sn Tin 118.7	51 Sb Antimony 121.8	52 Te Tellurium 127.6	53 I Iodine 126.9	54 Xe Xenon 131.3
55 Cs Cesium 132.9	56 Ba Barium 137.3	57–71 La Lanthanoids	72 Hf Hafnium 178.5	73 Ta Tantalum 180.9	74 W Tungsten 183.8	75 Re Rhenium 186.2	76 Os Osmium 190.2	77 Ir Iridium 192.2	78 Pt Platinum 195.1	79 Au Gold 197.0	80 Hg Mercury 200.6	81 Tl Thallium 204.4	82 Pb Lead 207.2	83 Bi Bismuth 209.0	84 Po Polonium 209.0	85 At Astatine [210.0]	86 Rn Radon [222.0]
87 Fr Francium [223.0]	88 Ra Radium [226.0]	89–103 Ac Actinoids	104 Rf Rutherfordium	105 Db Dubnium	106 Sg Seaborgium	107 Bh Bohrium	108 Hs Hassium	109 Mt Meitnerium	110 Ds Darmstadtium	111 Rg Roentgenium	112 Cn Copernicium	113 Uut Ununtrium	114 Fl Flerovium	115 Uup Ununpentium	116 Lv Livermorium	117 Uus Ununseptium	118 Uuo Ununoctium

57 La Lanthanum 138.9	58 Ce Cerium 140.0	59 Pr Praseodymium 140.1	60 Nd Neodymium 144.2	61 Pm Promethium [144.9]	62 Sm Samarium 150.4	63 Eu Europium 152.0	64 Gd Gadolinium 157.3	65 Tb Terbium 158.9	66 Dy Dysprosium 162.5	67 Ho Holmium 164.9	68 Er Erbium 167.3	69 Tm Thulium 168.9	70 Yb Ytterbium 173.0	71 Lu Lutetium 175.0
89 Ac Actinium [227.0]	90 Th Thorium 232.0	91 Pa Protactinium 231.0	92 U Uranium 238.0	93 Np Neptunium [237.0]	94 Pu Plutonium [244.1]	95 Am Americium [243.1]	96 Cm Curium [247.1]	97 Bk Berkelium [247.1]	98 Cf Californium [251.1]	99 Es Einsteinium [252.1]	100 Fm Fermium [257.1]	101 Md Mendelevium [258.1]	102 No Nobelium [259.1]	103 Lr Lawrencium [266.1]

IUPAC Periodic Table of the Elements

Key: atomic number / **Symbol** / name / standard atomic weight

1	2	3	4	5	6	7	8	9	10	11	12	13	14	15	16	17	18
1 **H** hydrogen [1.007, 1.009]																	2 **He** helium 4.003
3 **Li** lithium [6.938, 6.997]	4 **Be** beryllium 9.012											5 **B** boron [10.80, 10.83]	6 **C** carbon [12.00, 12.02]	7 **N** nitrogen [14.00, 14.01]	8 **O** oxygen [15.99, 16.00]	9 **F** fluorine 19.00	10 **Ne** neon 20.18
11 **Na** sodium 22.99	12 **Mg** magnesium [24.30, 24.31]											13 **Al** aluminium 26.98	14 **Si** silicon [28.08, 28.09]	15 **P** phosphorus 30.97	16 **S** sulfur [32.05, 32.08]	17 **Cl** chlorine [35.44, 35.46]	18 **Ar** argon 39.95
19 **K** potassium 39.10	20 **Ca** calcium 40.08	21 **Sc** scandium 44.96	22 **Ti** titanium 47.87	23 **V** vanadium 50.94	24 **Cr** chromium 52.00	25 **Mn** manganese 54.94	26 **Fe** iron 55.85	27 **Co** cobalt 58.93	28 **Ni** nickel 58.69	29 **Cu** copper 63.55	30 **Zn** zinc 65.38(2)	31 **Ga** gallium 69.72	32 **Ge** germanium 72.63	33 **As** arsenic 74.92	34 **Se** selenium 78.97	35 **Br** bromine [79.90, 79.91]	36 **Kr** krypton 83.80
37 **Rb** rubidium 85.47	38 **Sr** strontium 87.62	39 **Y** yttrium 88.91	40 **Zr** zirconium 91.22	41 **Nb** niobium 92.91	42 **Mo** molybdenum 95.95	43 **Tc** technetium	44 **Ru** ruthenium 101.1	45 **Rh** rhodium 102.9	46 **Pd** palladium 106.4	47 **Ag** silver 107.9	48 **Cd** cadmium 112.4	49 **In** indium 114.8	50 **Sn** tin 118.7	51 **Sb** antimony 121.8	52 **Te** tellurium 127.6	53 **I** iodine 126.9	54 **Xe** xenon 131.3
55 **Cs** caesium 132.9	56 **Ba** barium 137.3	57-71 lanthanoids	72 **Hf** hafnium 178.5	73 **Ta** tantalum 180.9	74 **W** tungsten 183.8	75 **Re** rhenium 186.2	76 **Os** osmium 190.2	77 **Ir** iridium 192.2	78 **Pt** platinum 195.1	79 **Au** gold 197.0	80 **Hg** mercury 200.6	81 **Tl** thallium [204.3, 204.4]	82 **Pb** lead 207.2	83 **Bi** bismuth 209.0	84 **Po** polonium	85 **At** astatine	86 **Rn** radon
87 **Fr** francium	88 **Ra** radium	89-103 actinoids	104 **Rf** rutherfordium	105 **Db** dubnium	106 **Sg** seaborgium	107 **Bh** bohrium	108 **Hs** hassium	109 **Mt** meitnerium	110 **Ds** darmstadtium	111 **Rg** roentgenium	112 **Cn** copernicium	113 **Uut** ununtrium	114 **Fl** flerovium	115 **Uup** ununpentium	116 **Lv** livermorium	117 **Uus** ununseptium	118 **Uuo** ununoctium

57 **La** lanthanum 138.9	58 **Ce** cerium 140.1	59 **Pr** praseodymium 140.9	60 **Nd** neodymium 144.2	61 **Pm** promethium	62 **Sm** samarium 150.4	63 **Eu** europium 152.0	64 **Gd** gadolinium 157.3	65 **Tb** terbium 158.9	66 **Dy** dysprosium 162.5	67 **Ho** holmium 164.9	68 **Er** erbium 167.3	69 **Tm** thulium 168.9	70 **Yb** ytterbium 173.0	71 **Lu** lutetium 175.0
89 **Ac** actinium	90 **Th** thorium 232.0	91 **Pa** protactinium 231.0	92 **U** uranium 238.0	93 **Np** neptunium	94 **Pu** plutonium	95 **Am** americium	96 **Cm** curium	97 **Bk** berkelium	98 **Cf** californium	99 **Es** einsteinium	100 **Fm** fermium	101 **Md** mendelevium	102 **No** nobelium	103 **Lr** lawrencium

INTERNATIONAL UNION OF
PURE AND APPLIED CHEMISTRY

For notes and updates to this table, see www.iupac.org. This version is dated 8 January 2016.

GLOSSARY

A

α-carbon (Sect. 11.2): An α-carbon is the site of a chemical reaction during a nucleophilic displacement. (p. 496); (Sect. 17.2): An α-carbon is a carbon directly connected to a carbonyl carbon or other functional group. (p. 812)

absolute configuration (Sect. 4.4): The absolute configuration of a chirality centre describes the exact three-dimensional arrangement of the atoms around a chirality centre. (p. 140)

acetals (Sect. 16.2): Acetals are compounds with two OR groups bonded to an sp^3-hybridized carbon. (p. 766)

acetoacetic ester synthesis (Sect. 17.7): The acetoacetic ester synthesis is a traditional synthesis of substituted acetone from acetoacetic ester, using a removable ester as an activating group. (p. 867)

achiral (Sect. 4.2): Achiral objects are superposable on their mirror images. (p. 129)

activating (Sect. 10.6): A substituent is activating if it increases the rate of reaction. In the context of S_EAr reactions, electron-donating groups are generally activating. (p. 450)

activating agents (Sect. 15.5): Activating agents are reagents that convert a starting material to a more reactive intermediate in order to simplify its conversion to a desired product. (p. 720)

acylation (Sect. 15.2): Acylation is the addition of an acyl (R–CO) group to a nucleophile. (p. 699)

acyl group (Sect. 10.4): An acyl group consists of a carbonyl connected to an alkyl group. (p. 446)

addition polymers (Sect. 19.8): Addition polymers are polymers made by combining monomers one by one without losing any atom or molecule. (p. 991)

addition-elimination (Sect. 15.2): Addition-elimination is a two-step reaction that adds a nucleophile to a π bond, and then eliminates a leaving group to re-form the π bond. (p. 698)

adjacent hydrogens (Sect. 13.4): Adjacent hydrogens are bonded to adjacent atoms, and are three bonds away from each other. (p. 603)

aldehyde (Sect. 7.2): An aldehyde is a functional group that consists of a carbonyl connected to at least one hydrogen. (p. 274)

aldol (Sect. 17.4): An aldol is a compound with an OH group bonded to a carbon atom in the β-position beside a carbonyl group. (p. 827)

aldol condensation (Sect. 17.4): An aldol condensation combines two aldehydes or ketones to produces enols or enones, with the loss of water. (p. 837)

aldol reaction (Sect. 17.4): An aldol reaction joins two carbonyl compounds to form a β-hydroxy carbonyl compound. (p. 827)

alkanes (Sect. 2.2): Alkanes are hydrocarbons that have only single bonds between carbon atoms. All the carbons are sp^3 hybridized. (p. 49)

ambident electrophile (Sect. 18.2): Ambident electrophile is a functional group with two electrophilic sites. (p. 901)

amide (Sect. 6.4): The term amide can refer to more than one functional group: a negatively charged nitrogen or a more complex structure (see Chapter 15). (p. 243)

angle strain (Sect. 3.4): Angle strain is a strain that arises from bond angles that do not permit maximum orbital overlap between the atoms of a molecule. (p. 98)

annulene (Sect. 9.5): Annulene is a general term for any cyclic, fully conjugated system. (p. 422)

anomer (Sect. 16.3): An anomer is one of the stereoisomers formed when the carbonyl carbon of a sugar is converted to a cyclic acetal or hemiacetal. (p. 777)

anomeric carbon (Sect. 16.3): The anomeric carbon is the carbon that gives rise to the two anomers. (p. 778)

antarafacial (Sect. 20.4): A rearrangement is antarafacial when the migrating group moves to the opposite face of a molecule. (p. 1046)

anti-addition (Sect. 8.6): In anti-addition, the two new atoms or groups are added to opposite faces of a double bond. (p. 360)

***anti*-aldol product** (Sect. 18.5): An *anti*-aldol product is an aldol in which the substituent on the α-carbon and the hydroxy at the β-position are *anti* with respect to the extended carbon chain. (p. 924)

anti-aromaticity (Sect. 9.3): Anti-aromaticity is the severe instability incurred from having $4n$ π electrons in a delocalized ring. Most often, anti-aromaticity prevents products from having a planar conformation or from forming. (p. 412)

anti-bonding molecular orbital (Sect. 1.9): An anti-bonding molecular orbital ($\sigma^\star$) has nodes between the adjacent atoms of the molecule. Electrons in anti-bonding orbitals destabilize the molecule and force atoms apart. (p. 33)

anti-conformation or antiperiplanar (Sect. 3.3): The antiperiplanar or anti-conformation of a molecule CH_2X–CH_2Y has a torsion angle of 180° between substituents X and Y. (p. 94)

anti-Markovnikov product (Sect. 8.7): An anti-Markovnikov product places the nucleophilic atom on the carbon of the alkene that is the site of the less stable carbocation. These products arise from Markovnikov mechanisms in which a $\delta+$ charge is found on the site that best stabilizes it. (p. 367)

anti-Markovnikov reactions (Sect. 8.7): Anti-Markovnikov reactions are those that produce anti-Markovnikov products. (p. 367)

antioxidants (Sect. 19.9): Antioxidants are compounds that prevent *other* compounds (such as DNA, lipids, etc.) from being oxidized. This is accomplished by allowing the antioxidants themselves to be oxidized. (p. 1001)

antisymmetric (Sect. 20.2): Within the context of conjugated π-systems, antisymmetric MOs have the opposite phase at their two terminal lobes. (p. 1014)

arenium ion (Sect. 10.3): An arenium ion is a 1,3-cyclohexadienyl cation, the common intermediate in electrophilic aromatic substitution reactions. (p. 434)

aromaticity (Sect. 9.3): Aromaticity is a property of certain fully conjugated rings that confers special stability and other chemical properties. (p. 410)

atomic orbital (Sect. 1.2): An atomic orbital (AO) maps out, point by point in the volume of space surrounding the nucleus, the likelihood (probability) of finding its electron at each point. It is a map of probability. Atomic orbitals are often represented as a surface within which the electron(s) may be found 95 percent of the time. (p. 2)

atropisomers (Sect. 4.3): Atropisomers are stereoisomers that result from hindered rotation around a single bond. (p. 139)

axial (Sect. 3.5): Axial substituents of a six-membered ring are displaced in directions perpendicular to the equatorial plane of the ring. (p. 102)

B

Baeyer–Villiger oxidation (Sect. 20.6): The Baeyer–Villiger oxidation involves the use of a peracid reagent to convert a ketone to an ester through a rearrangement. (p. 1057)

base peak (Sect. 14.3): The base peak of a mass spectrum is the peak of highest intensity. It is assigned a height of 100 percent from which the relative intensities of all other peaks are measured. (p. 651)

Baylis–Hillman reaction (Sect. 18.3): The Baylis–Hillman reaction temporarily converts an α,β-unsaturated carbonyl compound into an enolate, which adds to an aldehyde electrophile. (p. 916)

benzoin condensation (Sect. 18.7): The benzoin condensation is the self-condensation of aldehydes to make α-hydroxy ketones. For this reaction to happen, an electrophilic aldehyde carbonyl must become nucleophilic. (p. 943)

bimolecular (Sect. 11.3): The rate of a bimolecular reaction depends on the concentration of two reactants because two molecules are involved in the rate-determining step. (p. 497)

Birch reduction (Sect. 19.9): The Birch reduction is a reaction that reduces aromatic rings to 1,4-cyclohexadienes. (p. 998)

bond dipole (Sect. 1.5): A bond dipole is a dipole created across a chemical bond. It is the result of differences in electronegativity between the nuclei involved. (p. 12)

bonding molecular orbital (Sect. 1.9): A bonding molecular orbital is a molecular orbital with a high electron density in the volume of space between the atoms of the molecule. Electrons in bonding molecular orbitals stabilize a molecule. (p. 32)

Bouveault–Blanc reduction (Sect. 19.9): The Bouveault–Blanc reduction uses sodium in ethanol to reduce an ester to an alcohol. (p. 1000)

C

carbanion (Sect. 7.5): A carbanion is a negatively charged carbon. (p. 289)

carbocation (Sect. 8.2): A carbocation is a species containing a carbon atom with a +1 formal charge. (p. 332)

carbocation rearrangement (Sect. 8.5): A carbocation rearrangement changes the structure of the carbocation, often through a hydride or alkyl shift, to provide a more stable intermediate. (p. 357)

carbocycle (Sect. 16.5): A carbocycle is a ring composed entirely of carbon atoms. (p. 791)

carbonyl (Sect. 7.2): A carbonyl is a functional group that consists of a carbon and oxygen connected by a double bond. (p. 273)

chair conformation (Sect. 3.5): The chair conformation of a six-membered ring has a zig-zag configuration with the ring atoms alternately above and beneath the equatorial plane. (p. 102)

chemical shift (Sect. 13.3): Chemical shift (δ) is the difference between the NMR resonance frequency of a sample and a reference frequency, tetramethylsilane. (p. 582)

chemically equivalent hydrogens (Sect. 13.4): Chemically equivalent hydrogens are in identical environments within a compound because they are interchangeable by the fast rotation of single bonds or by symmetry. (p. 583)

chiral (Sect. 4.2): Chiral objects are not superposable on their mirror images. (p. 128)

chirality (Sect. 4.2): Chirality is the ability of objects to exist as non-superposable mirror images of each other. (p. 128)

chirality centres (Sect. 4.3): Chirality centres are atoms that are connected to four different groups. These atoms are also called **asymmetric centres**, **stereogenic centres**, or **stereocentres**. (p. 135)

***cis* isomer** (Sect. 3.7): The *cis* isomer of a cycloalkane has its two substituents on the same face of the ring. (p. 112)

Claisen condensation (Sect. 17.5): A Claisen condensation is the formation of a β-ketoester from esters; it is similar to an aldol addition and has an ester as the electrophile. (p. 846)

Claisen rearrangement (Sect. 20.4): A Claisen rearrangement is the [3,3]-sigmatropic rearrangement of an allylic vinyl ether. (p. 1042)

Claisen–Schmidt reaction (Sect. 17.4): The Claisen–Schmidt reaction is an aldol reaction that uses a ketone enolate and a non-enolizable aldehyde. (p. 831)

Clemmensen reduction (Sect. 19.9): The Clemmensen reduction is a reaction that reduces an aldehyde or ketone to an alkane. (p. 1000)

closed transition state (Sect. 18.5): A closed transition state is a transition state in which the nucleophile and electrophile are complexed and form a ring. Closed transition states offer better control of stereochemistry than open transition states. (p. 930)

concerted (Sect. 8.6): In a concerted reaction, all bond-forming and bond-breaking events happen in the same step. (p. 359)

concerted reaction (Sect. 11.3): A concerted reaction is one in which multiple steps or processes occur together and essentially at the same time. (p. 497)

configuration (Sect. 4.2): The configuration of a molecule is the three-dimensional arrangement of the bonds that connect the atoms. The configuration of a molecule is permanent. (p. 127)

conformation (Sect. 3.2): A conformation is a particular arrangement of atoms in a molecule resulting from rotation about the single bonds of the molecule. Molecules that exist in different conformations are called *conformers* or *rotamers*. (p. 88)

conformational analysis (Sect. 3.2): Conformational analysis is the study of the geometries and resulting energies of the conformations of a molecule. (p. 92)

conjugate addition (1,4-addition) (Sect. 18.2): Conjugate addition (1,4-addition) is the addition of a nucleophile to the β-carbon of an α,β-unsaturated carbonyl group. (p. 901)

conjugated dienes (Sect. 8.7): In conjugated dienes, the two alkenes are separated by exactly one single bond. (p. 374)

conjugated system (Sect. 8.7): A conjugated system is a region of a molecule with interacting π orbitals. The electrons in a conjugated system are delocalized over all of the atoms involved. (p. 375)

conrotatory (Sect. 20.5): A reaction is conrotatory if the end carbons rotate in the same direction as they rehybridize to form a C–C bond. (p. 1051)

consonant (Sect. 18.7): Consonant molecules have an alternating pattern of nucleophilic-electrophilic reactivity potential in the atoms along a chain (+ − + − + ...) between functional groups. (p. 938)

constitutional isomers (Sect. 4.2): Constitutional isomers are molecules that have the same chemical formula, but their atoms are connected in different ways. (p. 127)

contrast agent (Sect. 13.4): A contrast agent is a compound that increases the contrast of an MRI image. (p. 612)

Cope rearrangement (Sect. 20.4): The Cope rearrangement is the [3,3]-sigmatropic rearrangement of a 1,5-hexadiene. (p. 1040)

coupling constant (Sect. 13.4): The coupling constant (J) is the difference in resonance frequency between peaks in a multiplet. (p. 603)

covalent bond (Sect. 1.3): A covalent bond is the energetically favourable sharing of two electrons; this holds atoms in close proximity to each other. (p. 6)

crossed aldols (Sect. 17.4): Crossed aldols are aldols formed from two different aldehydes. (p. 830)

cuprates (Sect. 18.2): Cuprates are a type of organometallic nucleophile, with a negatively charged copper. They favour conjugate addition to α,β-unsaturated carbonyl compounds. (p. 906)

Curtius rearrangement (Sect. 20.6): The Curtius rearrangement converts a carboxylic acid to an isocyanate that is similar to the intermediate of the Hofmann rearrangement. (p. 1060)

cyanohydrin (Sect. 7.7): A cyanohydrin is an sp^3-hybridized carbon connected to a hydroxyl group and a nitrile. (p. 300)

cyclic (Sect. 2.2): A cyclic molecule has at least one ring of atoms. Acyclic molecules have no rings. (p. 50)

cycloadduct (Sect. 20.3): A cycloadduct is the product of a cycloaddition. (p. 1019)

D

Darzen's condensation (Sect. 18.7): Darzen's condensation produces an epoxy ketone from an aldol addition with an α-halide. (p. 946)

deactivating (Sect. 10.6): A substituent is deactivating if it decreases the rate of reaction. In the context of S_EAr reactions, electron-withdrawing groups are deactivating. (p. 450)

decarboxylation (Sect. 17.7): A decarboxylation reaction removes carbon dioxide from a β-keto acid. (p. 866)

decoupling (Sect. 13.6): Decoupling a ^{13}C-NMR spectrum eliminates any splitting by attached hydrogens, making each carbon signal appear as a singlet. (p. 621)

degenerate (Sect. 9.4): Degenerate describes multiple orbitals that have the same energy and the same number of nodes. (p. 418)

degenerate atomic orbitals (Sect. 1.2): Degenerate atomic orbitals are any set of orbitals that have the same energy value. (p. 4)

degree of unsaturation (Sect. 13.5): The degree of unsaturation (DoU) of a molecule is a measure of the number of rings and/or π bonds it has. (p. 614)

dehalogenation (Sect. 19.6): Dehalogenation is a reduction reaction that replaces the halogen atom of an alkyl halide with a hydrogen atom. (p. 986)

dehydration (Sect. 12.4): Dehydration is the loss of water to produce an alkene from an alcohol. (p. 557)

dehydrohalogenation (Sect. 12.4): Dehydrohalogenation is the loss of a hydrogen and a halogen to produce an alkene from an alkyl halide. (p. 557)

delocalization (Sect. 1.8): Delocalization occurs when π electrons are shared by more than two atoms. (p. 27)

delocalized electrons (Sect. 1.8): Delocalized electrons are not associated with a single atom or bond. Instead, they are shared among several atoms; they are delocalized. (p. 27)

demercuration (Sect. 8.4): Demercuration is a reaction that replaces the mercury in a C−Hg bond with a hydrogen atom. (p. 354)

DEPT (Sect. 13.6): A DEPT spectrum is a variation of a ^{13}C-NMR spectrum that distinguishes how many hydrogens are attached to each carbon. (p. 622)

deshielding (Sect. 13.4): Deshielding is the reduction in shielding due to a lack of nearby electrons. (p. 588)

diastereomers (Sect. 4.6): Diastereomers are stereoisomers that are not enantiomers. (p. 152)

diastereotopic hydrogens (Sect. 13.4): Diastereotopic hydrogens are non-equivalent hydrogens on the same carbon. (p. 605)

1,3-diaxial interactions (Sect. 3.7): 1,3-diaxial interactions are steric strains that arise from interpenetration of the electron clouds of ring substituents separated by two carbon atoms. (p. 109)

Dieckmann cyclization (Sect. 17.5): A Dieckmann cyclization is an intramolecular Claisen condensation that forms a ring. (p. 848)

dienes (Sect. 8.7): Dienes are compounds containing two alkenes. (p. 374)

dienophile (Sect. 20.3): A dienophile refers to a π-bonded functional group that reacts with a diene in a Diels–Alder reaction. (p. 1020)

dihedral angle (Sect. 3.2): The torsion or dihedral angle is the angle between the bonds to a substituent on each of the atoms of the single bond that constitutes the axis of rotation. (p. 89)

dipolarophile (Sect. 20.3): A dipolarophile is a compound that can react with a 1,3-dipole in a [3+2] cycloaddition reaction. It contains either an alkene or alkyne functional group. (p. 1034)

1,3-dipole (Sect. 20.3): A 1,3-dipole is a functional group with a three-atom sequence that has a nucleophilic site and an electrophilic site at positions 1 and 3, respectively. Dipoles can undergo [3+2] cycloadditions with alkenes or alkynes (i.e., dipolarophiles). (p. 1034)

dipole-dipole interaction (Sect. 2.3): A dipole-dipole interaction is an attractive force between the negative end of a permanent dipole in a molecule and the positive end of a permanent dipole in a neighbouring molecule (or vice versa). (p. 55)

direct addition (1,2-addition) (Sect. 18.2): Direct addition (1,2-addition) is the addition of a nucleophile to the carbonyl carbon of an α,β-unsaturated carbonyl group. (p. 901)

directed metalation group (Sect. 10.8): Directed metalation group (DMG) is a substituent that favours deprotonation at the adjacent *ortho* position for a directed *ortho* metalation reaction. (p. 473)

directed *ortho* metalation (Sect. 10.8): Directed *ortho* metalation (DOM) is an aromatic substitution reaction that first deprotonates the position *ortho* to a directed metalation group and then reacts with an electrophile. (p. 472)

disconnection (Sect. 10.9): A disconnection is a retrosynthetic step, an imaginary "reverse" reaction. (p. 477)

displacement reaction (Sect. 11.2): In a displacement reaction, one component of a molecule is replaced by another. These reactions are sometimes called replacements or substitutions. When the new component is provided by a nucleophile, the term *nucleophilic displacement* is used. (p. 495)

disrotatory (Sect. 20.5): A reaction is disrotatory if the end carbons rotate in opposite directions as they rehybridize to form a C–C bond. (p. 1052)

dissonant (Sect. 18.7): Dissonant molecules have a nucleophilic-electrophilic reactivity potential pattern with matching reactivity on adjacent carbons along a chain (+ − − + or − + + −) between functional groups. (p. 939)

downfield (Sect. 13.4): Downfield signals are toward the left side of an NMR spectrum. (p. 588)

E

E1 elimination (Sect. 12.3): E1 elimination is a multi-step reaction in which the leaving group is lost in the first and rate-determining step and the adjacent proton is lost in the second step. (p. 552)

$E1_{cb}$ reaction (Sect. 17.4): An $E1_{cb}$ reaction is a unimolecular elimination reaction in which an acidic proton is removed to make the conjugate base *before* the leaving group leaves. (p. 838)

E2 elimination (Sect. 12.2): E2 elimination involves the removal of a hydrogen and the concerted loss of a leaving group to form an alkene. (p. 541)

early transition state (Sect. 8.3): In an early transition state, the structure of the transition state most resembles the structure of the starting materials. (p. 343)

eclipsed conformation (Sect. 3.2): An eclipsed conformation has the substituents on the front atom aligned with those on the back atom. (p. 89)

electron-donating groups (Sect. 8.3): Electron-donating groups are atoms or groups that donate electron density. (p. 339)

electron impact (Sect. 14.2): Electron impact is an ionization technique used in mass spectroscopy in which a sample vapour passes through a beam of electrons. Collisions between the electrons and the molecules in the beam produce positively charge ions. (p. 650)

electronegativity (Sect. 1.5): The electronegativity of an atom is its ability to pull electrons toward itself from the surrounding atoms to which it is bonded. The greater the electronegativity, the greater is the ability of the atom to draw electrons from its neighbours. (p. 12)

electrophile (Sect. 7.2): An electrophile is a group or atom that accepts electrons. (p. 274)

electrophilic aromatic substitution (Sect. 10.2): Electrophilic aromatic substitution is a reaction where hydrogen on an aromatic ring is replaced by an electrophile. (p. 433)

electrostatic attraction (Sect. 1.3): Electrostatic attraction is the attraction of opposite charges to each other. (p. 6)

electrostatic potential map (Sect. 2.3): An electrostatic potential map is a plot of the forces on a point charge measured at a fixed distance from a molecule. It is often interpreted as regions of negative and positive charge on a molecule. (p. 54)

electrostatics (Sect. 2.3): Electrostatics are attractive forces that result from a full formal charge. (p. 55)

enantiomeric excess (Sect. 4.11): The enantiomeric excess (ee) of a sample is the ratio of the amount of one isomer to the total amounts of all isomers in the sample. (p. 168)

enantiomers (Sect. 4.2): Enantiomers are non-superposable mirror images. (p. 128)

enantioselective (Sect. 7.11): An enantioselective reaction is a reaction that produces more of one enantiomer than the other. (p. 320)

endo (Sect. 20.3): A substituent is *endo* if it points toward the alkene of the cyclohexene product of a Diels–Alder reaction. (p. 1024)

enolate (Sect. 17.1): An enolate is an alkene with a negatively charged oxygen atom as one of its substituents. It is the conjugate base of an enol. (p. 811)

enolizable compounds (Sect. 17.2): Enolizable compounds can be converted to enolates. (p. 812)

envelope conformation (Sect. 3.4): The envelope conformation of a five-membered ring has one ring atom that is displaced from the plane of the other four ring atoms. (p. 101)

epoxide (Sect. 8.7): An epoxide is a cyclic ether with a ring consisting of one oxygen and two carbon atoms. (p. 364)

equatorial (Sect. 3.5): Equatorial substituents are displaced in directions nearly parallel to the equatorial plane. (p. 102)

ester hydrolysis (Sect. 15.3): Ester hydrolysis is a hydrolysis reaction that converts an ester into a carboxylic acid. (p. 706)

ethereal solution (Sect. 7.5): An ethereal solution is one in which the solvent is an ether such as Et_2O or THF. (p. 289)

ethers (Sect. 8.4): Ethers are chemical compounds with two alkyl groups bonded to one oxygen atom. (p. 349)

exo (Sect. 20.3): A substituent is *exo* if it points away from the alkene of the cyclohexene product of a Diels–Alder reaction. (p. 1024)

F

Fischer esterification (Sect. 15.3): Fischer esterification is the acid-catalyzed conversion of a carboxylic acid to an ester. (p. 712)

formal charge (Sect. 1.4): The formal charge (FC) of an atom describes a deficit or excess of electrons based on formally comparing the number of electrons an atom shares in a Lewis structure with the number of electrons it should have to be electrically neutral. A Lewis structure is not complete without formal charges. (p. 7)

free-radical halogenation (Sect. 19.3): Free-radical halogenation is a type of halogenation reaction. In this reaction, a halogen radical replaces a hydrogen atom on an alkane. (p. 975)

Friedel–Crafts acylation (Sect. 10.4): Friedel–Crafts acylation, a variation of the Friedel–Crafts alkylation, substitutes an acyl group in the place of a hydrogen atom on an aromatic ring. (p. 446)

Friedel–Crafts alkylation (Sect. 10.4): Friedel–Crafts alkylation is a reaction that substitutes an alkyl group in place of a hydrogen atom on an aromatic ring using an alkyl halide and a Lewis acid. (p. 441)

frontier molecular orbitals (Sect. 20.2): In the electronic configuration of a molecule, the frontier molecular orbitals are the highest energy occupied (HOMO) and lowest energy unoccupied (LUMO) orbitals. (p. 1014)

Frost circles (Sect. 9.4): Frost circles are a mnemonic device for predicting the relative energy of the molecular orbitals of an aromatic system. (p. 420)

functional group (Sect. 2.2): A functional group is an atom or group of atoms that, because of its structure, exhibits its own distinct pattern of chemical reactivity. (p. 49)

G

gauche conformations (Sect. 3.3): The gauche conformations of a molecule $CH_2X–CH_2Y$ have a torsion angle of either 60° or 300° between substituents X and Y. (p. 94)

geminal (Sect. 8.7): Geminal describes two groups located on the same carbon. (p. 381)

Grignard reagent (Sect. 7.5): A Grignard reagent contains a carbon bonded to a magnesium atom. (p. 289)

ground-state electron configuration (Sect. 1.2): The ground-state electron configuration is the one of lowest energy: that is, the most stable one. All other arrangements, which are necessarily of higher energy, are called *excited states*. (p. 4)

H

half-chair (Sect. 3.6): The half-chair is a conformation with maximum energy on the pathway between chair and twist-boat conformations of a six-membered ring. (p. 108)

half-life (Sect. 5.1): Half-life is the time required for half of a given amount of substance to be consumed in a reaction. (p. 187)

haloform reaction (Sect. 17.3): A haloform reaction converts a methyl ketone to a carboxylic acid and a haloform. (p. 819)

halohydrins (Sect. 8.6): Halohydrins contain a halogen bonded to one carbon atom and a hydroxyl group bonded to an adjacent carbon atom. (p. 360)

halonium ion (Sect. 8.6): A halonium ion is a three-membered ring that contains a halogen with a +1 formal charge. If the halogen is chlorine, a chloronium is formed. Bromine produces a bromonium, and iodine gives an iodonium. (p. 359)

Hammond postulate (Sect. 8.3): The Hammond postulate states that the structure of a transition state resembles the species nearest to it in free energy. (p. 343)

heat of hydrogenation (Sect. 9.2): The heat of hydrogenation is the amount of heat released when an alkene (or alkyne) is completely saturated by a catalytic hydrogenation reaction. It can be used as an indicator of molecular stability. (p. 403)

hemiacetal (Sect. 7.9): A hemiacetal is an sp^3-hybridized carbon connected to an OH group and to an OR group. (p. 311); (Sect. 16.2): Hemiacetals are compounds with an OH and an OR group bonded to an sp^3-hybridized carbon. (p. 766)

Henry or nitroaldol reaction (Sect. 17.6): The Henry or nitroaldol reaction uses the anion of a nitroalkane as the nucleophile in an aldol-type reaction. (p. 851)

heterocycle (Sect. 16.5): A heterocycle is a ring of atoms in which one or more of the atoms are heteroatoms. The term *heterocycle* is also used to describe molecules that contain such rings. (p. 791)

hetero-Diels–Alder reaction (Sect. 20.3): A hetero-Diels–Alder reaction contains a non-carbon atom in either the diene or the dienophile and forms a heterocycle instead of a cyclohexene. (p. 1031)

heterolytic cleavage (Sect. 19.2): Heterolytic cleavage is a bond-breaking reaction where both bonding electrons migrate to one of the two bonded atoms. (p. 973)

high-resolution mass spectrum (Sect. 14.5): A high-resolution mass spectrum (HRMS) is produced on an instrument able to measure molecular mass with a very high degree of precision. (p. 660)

highest occupied molecular orbital (HOMO) (Sect. 9.2): The highest occupied molecular orbital (HOMO) is the highest energy orbital that contains electrons. (p. 401)

Hofmann elimination (Sect. 12.2): Hofmann elimination gives the less substituted product instead of the more substituted (Zaitsev) product. (p. 546)

Hofmann product (Sect. 12.2): The Hofmann product is the product of an elimination derived from the loss of the least hindered hydrogen. (p. 545)

Hofmann rearrangement (Sect. 20.6): The Hofmann rearrangement converts an amide to an amine through a rearrangement that releases carbon dioxide. (p. 1059)

HOMO–LUMO gap (Sect. 9.2): The HOMO–LUMO gap is the energy difference between the highest occupied molecular orbital (HOMO) and the lowest unoccupied molecular orbital (LUMO). (p. 406)

homolytic cleavage (Sect. 19.2): Homolytic cleavage is a bond-breaking process in which the two bonding electrons become split between the two originally bonded atoms. (p. 973)

Horner–Wadsworth–Emmons (HWE) reaction (Sect. 17.7): The Horner–Wadsworth–Emmons reaction forms alkenes from aldehydes with *trans* selectivity by using a stabilized enolate equivalent. (p. 874); (Sect. 18.6): The HWE reaction is a Wittig-type reaction that uses a phosphonium ester enolate instead of an ylide. This typically results in the formation of *E* alkenes. (p. 935)

Hückel's rule (Sect. 9.3): Hückel's rule states that the number of π electrons that participate in an aromatic ring structure must be a multiple of $(4n+2)$ where $n = 0, 1, 2, 3\ldots$ (p. 412)

Hünig's base (Sect. 18.5): Hünig's base is ethyl diisopropyl amine (sometimes called diisopropylethylamine), a sterically hindered tertiary amine base. (p. 926)

hybrid orbitals (Sect. 1.7): Hybrid orbitals are atomic orbitals that form a bonding geometry by a suitable mixing of the 2s and 2p orbitals of the atom. (p. 20)

hydrate (Sect. 7.9): A hydrate is a functional group formed by the addition of water to a carbonyl group. (p. 306)

hydrocarbon (Sect. 2.2): A hydrocarbon is a compound consisting of only carbon and hydrogen atoms. (p. 48)

hydrogen bond acceptors (Sect. 2.3): Hydrogen bond acceptors are functional groups in which oxygen or nitrogen atoms have lone electron pairs. (p. 57)

hydrogen bond donors (Sect. 2.3): Hydrogen bond donors are functional groups in which oxygen or nitrogen atoms are connected to hydrogens and can participate in hydrogen bonding. (p. 57)

hydrogen bonding (Sect. 2.3): Hydrogen bonding (H-bonding) is an attractive force between a lone pair of a nitrogen or oxygen atom in a group and a hydrogen atom that is covalently bonded to a N or O atom in a neighbouring group. (p. 56)

hydrogenation (Sect. 8.7): Hydrogenation is the addition of H_2 across a π bond. (p. 372)

hydrolysis (Sect. 7.5): Hydrolysis is a reaction with water that decomposes a functional group into other components. (p. 291)

hydrophilic (Sect. 2.4): A hydrophilic ("water-loving") molecule or group of atoms is polar enough to form favourable intermolecular interactions, including hydrogen bonding, with water. (p. 64)

hydrophobic (Sect. 2.4): A hydrophobic ("water-fearing") compound establishes only weak London forces with surrounding water molecules; therefore, water molecules maintain hydrogen bonds with each other, rather than forming new ones with the solute. (p. 63)

hyperconjugation (Sect. 8.3): Hyperconjugation is the interaction of the empty p orbital of the carbocation with filled σ bonds on adjacent carbon atoms. (p. 339)

I

imine (Sect. 7.2, Sect. 16.4): An imine is a functional group that consists of a carbon connected to a nitrogen by a double bond. (pp. 275, 782)

iminium (Sect. 7.2): An iminium ion contains a positively charged nitrogen connected by a double bond to a carbon. (p. 276)

induction (Sect. 6.4): Induction is the removal of electron density from an atom by a strongly electronegative atom nearby. (p. 244)

initiation (Sect. 19.3): Initiation refers to the first phase (beginning) of the radical chain reaction, which involves the formation of one or more radicals. (p. 974)

integration (Sect. 13.4): Integration refers to calculations of the area under a peak or set of peaks that determine the intensity of a signal in NMR spectrometry. (p. 585)

intermolecular forces (Sect. 2.3): Intermolecular forces are different kinds of weak attractive forces that molecules exert on each other when they are in close proximity. (p. 54)

intermolecular reactions (Sect. 5.4): Intermolecular reactions occur between two or more molecules. (p. 203)

intramolecular reactions (Sect. 5.4): Intramolecular reactions occur within a single molecule. (p. 204)

inverse addition (Sect. 18.4): Inverse addition is the addition of an enolizable carbonyl to a solution of base, used to form a kinetic enolate. (p. 920)

ion mobility spectrometer (Sect. 14.3): An ion mobility spectrometer separates ions in a magnetic field based on how fast they pass through a carrier gas that is flowing in the opposite direction. (p. 658)

ionic bonds (Sect. 1.3): Ionic bonds result from the transfer of electrons from one atom to another, which creates opposite charges that are attracted to each other. (p. 6)

isolated dienes (Sect. 8.7): In isolated dienes, the two alkenes are separated by more than one single bond. (p. 374)

isomers (Sect. 4.2): Isomers are molecules that have the same atoms, but their atoms are connected in different ways. (p. 127)

K

ketone (Sect. 7.2): A ketone is a functional group that consists of a carbonyl connected to two carbon groups. (p. 275)

kinetic control (Sect. 18.2): A reaction under kinetic control uses reaction conditions, such as reduced reaction temperature, that favour the most quickly formed product. (p. 904)

kinetic enolate (Sect. 18.4): A kinetic enolate forms under conditions of kinetic control (irreversible), and the hydrogen is removed from the less substituted α-carbon. (p. 919)

kinetic product (Sect. 8.7): The kinetic product is the product that is formed the fastest. (p. 376)

Knoevenagel condensation (Sect. 17.7): The Knoevenagel condensation transforms a dicarbonyl compound to an α, β-unsaturated compound by means of addition to an aldehyde or ketone, followed by dehydration. (p. 871)

L

late transition state (Sect. 8.3): In a late transition state, the structure of the transition state most resembles the structure of the products of that step. (p. 343)

leaving groups (Sect. 7.8): Leaving groups remove electron pairs from functional groups. (p. 303)

Lewis acid (Sect. 6.8): A Lewis acid is an electron pair acceptor. A Lewis base is an electron pair donor. (p. 265)

Lewis structure (Sect. 1.4): The Lewis structure depicts the bonding in a molecule. A line between the participating atoms represents the two shared valence electrons of each covalent bond. Non-bonded electrons are represented by dots. (p. 6)

Lindlar catalyst (Sect. 8.7): A Lindlar catalyst is a palladium catalyst, deactivated with lead acetate and quinoline. Because it is less reactive, it only partially reduces alkynes, stopping at the *cis*-alkene product. (p. 385)

localized bond (Sect. 1.7): A localized bond involves the sharing of two electrons by means of the overlap of two atomic orbitals on two adjacent atoms. It is confined to the region between the two atoms. (p. 20)

London forces (Sect. 2.3): London forces (dispersion forces) are attractive interactions that exist between *all* molecules in close proximity to each other, regardless of whether or not they engage in other intermolecular interactions. They are the result of small temporary dipoles induced in each molecule by the other. (p. 57)

Lossen rearrangement (Sect. 20.6): The Lossen rearrangement converts a hydroxamic acid to an isocyanate. (p. 1060)

low-resolution mass spectrum (Sect. 14.3): A low-resolution mass spectrum shows the mass of each peak to zero decimal places. (p. 651)

lowest unoccupied molecular orbital (LUMO) (Sect. 9.2): The lowest unoccupied molecular orbital (LUMO) is the lowest energy orbital that does not contain any electrons. (p. 401)

M

magnetic anisotropy (Sect. 13.4): Magnetic anisotropy is the dependence of a magnetic property of a material on direction relative to an external magnetic field. (p. 598)

magnetic moment (Sect. 13.2): The magnetic moment (μ) of a nucleus is a vector property related to the magnitude and direction of its spin. (p. 579)

malonic ester synthesis (Sect. 17.7): The malonic ester synthesis produces substituted acetic acids or acetate esters from malonic ester. (p. 868)

Mannich reaction (Sect. 17.6): The Mannich reaction is an aldol-type reaction that produces a β-aminoketone instead of a β-hydroxyketone. (p. 850)

Mass spectrometry (Sect. 14.2): Mass spectrometry (MS) is a method that gives information about the mass of a compound and the fragments from which it is formed. (p. 649)

mass spectrum (Sect. 14.2): A mass spectrum of an organic compound graphically presents the masses of its ions along the horizontal axis and their relative abundances along the vertical axis. (p. 651)

mass-to-charge ratio (Sect. 14.2): The mass-to-charge ratio (m/z) is a ratio of an ion's mass divided by its charge. Since its charge is almost always $+1$ for organic compounds, m/z for an ion is effectively a measure of the mass of the ion. (p. 650)

mercurinium (Sect. 8.4): A mercurinium ion is a three-membered ring containing two carbon atoms and a positively charged mercury ion. (p. 352)

meso compounds (Sect. 4.7): Meso compounds contain more than one stereocentre and have superposable mirror images. (p. 158)

***meta* directing groups** (Sect. 10.6): *Meta* directing groups are substituents that favour substitution at the *meta* position, relative to the directing group. (p. 450)

methyl carbocations (Sect. 8.3): Methyl carbocations are not bonded to any alkyl groups. **Primary**, **secondary**, and **tertiary carbocations** are respectively bonded to one, two, and three alkyl groups. (p. 338)

Michael acceptor (Sect. 18.2): A Michael acceptor is any α,β-unsaturated carbonyl compound capable of undergoing a Michael reaction. (p. 910)

Michael donor (Sect. 18.2): A Michael donor is any nucleophile that favours conjugate addition to an α,β-unsaturated carbonyl compound. (p. 910)

Michael reaction (Sect. 18.2): The Michael reaction is the conjugate addition of a 1,3-dicarbonyl compound to an α,β-unsaturated carbonyl compound. Note that the α,β-unsaturated component is not limited to carbonyls. Other electron-withdrawing groups may be used. (p. 909)

migratory aptitude (Sect. 20.6): Migratory aptitude describes the relative potential of a group to migrate during a rearrangement reaction. (p. 1058)

mixed anhydride (Sect. 15.5): A mixed anhydride is formed from two different types of acid, such as a carboxylic acid and a sulfonic acid. Typically, the carboxylic acid component is the more reactive electrophile. (p. 719)

molecular ion (Sect. 14.2): The molecular ion ($M^{+\bullet}$ or more commonly M^+) is the cation formed by the loss of one electron from a molecule of the organic sample. The molecular ion is a radical. (p. 650)

molecular orbital (Sect. 1.9): A molecular orbital (MO) maps out, point by point in space, the probabilities of finding electrons in the volume of space around the nuclei of a molecule. Molecular orbitals are often represented as a shape that depicts the probability of finding an electron within that volume of space 95 percent of the time. (p. 32)

molecule in an excited state (Sect. 9.2): A molecule in an excited state has absorbed energy. In the case of an electronic excited state, one of the lower energy electrons has been promoted to a higher energy molecular orbital. (p. 406)

molecule in the ground state (Sect. 9.2): A molecule in the ground state is at its lowest energy. (p. 406)

monomers (Sect. 19.8): Monomers are the small building blocks and the repeating units found in polymers. (p. 991)

Mukaiyama aldol reaction (Sect. 18.4): A Mukaiyama aldol reaction is a Lewis acid–catalyzed aldol addition, using a silyl enol ether as a trapped enolate equivalent. (p. 921)

N

***n*+1 rule** (Sect. 13.4): The $n+1$ rule states that the number of peaks in a multiplet is one more than n, the number of adjacent equivalent hydrogens. (p. 604)

nitration (Sect. 10.4): Nitration is the substitution of a NO_2 group in place of a hydrogen atom on an aromatic ring. (p. 437)

nitrile (Sect. 7.2): A nitrile is a functional group that consists of a carbon connected to a nitrogen by a triple bond. (p. 276)

non-bonded (Sect. 1.4): Non-bonded (lone pair) electrons reside on one atom, occupying space around that atom. (p. 6)

non-polar (Sect. 2.3): A non-polar molecule has no overall dipole, or a very small one. (p. 57)

nuclear magnetic resonance (NMR) spectroscopy (Sect. 13.2): Nuclear magnetic resonance (NMR) spectroscopy measures the energy differences between spinning nuclei in a magnetic field to determine the structures of molecules in solution. (p. 580)

nuclear Overhauser effect (nOe) (Sect. 13.6): The nuclear Overhauser effect (nOe) causes carbon signals in a decoupled spectrum to be taller if there are attached hydrogens. Carbons without attached hydrogens give very short signals. (p. 621)

nucleophile (Sect. 7.2): A nucleophile is a group or atom that donates or shares electrons. (p. 274)

nucleophilic aromatic substitution (Sect. 15.9): Nucleophilic aromatic substitution reactions exchange a nucleophile for a leaving group on an aromatic ring. (p. 737)

nucleophilicity (Sect. 11.3): Nucleophilicity is an expression of how fast (how easily) a nucleophile is able to react (donate electrons). (p. 502)

O

open transition state (Sect. 18.5): An open transition state is a transition state in which the nucleophile approaches the electrophile from the least sterically hindered direction. (p. 930)

optical purity (Sect. 4.11): The optical purity of a sample is the ratio of the specific rotation of a sample to the specific rotation of a pure enantiomer of the same compound. (p. 168)

optical rotation (Sect. 4.10): Optical rotation is rotation of plane-polarized light by a substance or mixture. (p. 164)

optical rotation (α) (Sect. 4.10): The optical rotation (α) is the angle that a sample rotates plane-polarized light. (p. 165)

optically active (Sect. 4.10): Optically active molecules rotate plane polarized light. (p. 165)

optically pure (Sect. 4.11): In optically pure samples, all the molecules are the same substance with the same absolute configuration. (p. 168)

orthogonal reactivity (Sect. 18.2): Orthogonal reactivity describes reagent pairs that have opposite and complementary reactivities. (p. 906)

***ortho/para* directing groups** (Sect. 10.6): *Ortho/para* directing groups are substituents that favour substitution at the *ortho* and *para* positions, relative to the directing group. (p. 450)

oxaphosphetane (Sect. 17.6): An oxaphosphetane is a four-membered ring containing an oxygen atom and a phosphorus atom. (p. 858)

oxidative cleavage reactions (Sect. 19.1): Oxidative cleavage reactions are oxidation reactions that lead to bond breakage and the fragmentation of a molecule. (p. 972)

oxocarbenium (Sect. 7.2): An oxocarbenium ion contains a positively charged oxygen connected by a double bond to a carbon. (p. 276)

oxymercuration-demercuration (Sect. 8.4): Oxymercuration-demercuration is a Markovnikov reaction that adds water to an alkene using a mercury compound. (p. 351)

P

π bond (Sect. 1.5): A π bond (pronounced *pi* bond) is a covalent bond in which the highest probability of finding the shared electrons occurs equally above and below the line through the nuclei. (p. 11)

pericyclic reaction (Sect. 20.2): A pericyclic reaction is one in which the moving electrons flow in a ring during the transition state. (p. 1012)

phosphorane (Sect. 17.6): A phosphorane is a Wittig ylide resonance form with a C=P double bond. (p. 858)

polar (Sect. 2.3): A polar molecule has a net overall dipole. (p. 55)

polar covalent bond (Sect. 1.5): A polar covalent bond is a covalent bond in which there is a significant difference in electronegativity between the atoms involved. (p. 13)

polar protic solvents (Sect. 2.3): Polar protic solvents are those solvents capable of acting as hydrogen bond donors. (p. 57)

polarizable (Sect. 11.3): Polarizable atoms easily form dipoles. These atoms have large, diffuse electron clouds, which are easily disturbed by nearby charges. (p. 503)

polymerization (Sect. 19.8): Polymerization is the process of reacting monomers together to form a polymer. (p. 991)

polymers (Sect. 19.8): Polymers are large molecules composed of many repeating units. (p. 991)

propagation (Sect. 19.3): Propagation refers to the second phase of the radical chain reaction, in which repeated reactions or steps form new radicals. (p. 974)

protecting group (Sect. 16.2): A protecting group is a temporary change installed on a functional group to reduce the chemical reactivity of that functional group. (p. 771)

Q

quantitative deprotonation (Sect. 17.2): Quantitative deprotonation is an irreversible deprotonation that converts all of the carbonyl-containing molecules in the starting material to enolates. (p. 814)

quantized (Sect. 1.2): Quantized refers to the particular fixed value of the energy of an atomic orbital. (p. 4)

R

racemic mixture (Sect. 4.11): A racemic mixture (racemate) is a mixture of equal amounts of both enantiomers of a compound. (p. 168)

radical (Sect. 14.2): A radical is a molecule with an unpaired electron. (p. 650)

radical anion (Sect. 19.9): A radical anion is a chemical species that is negatively charged and has an unpaired electron. (p. 997)

radical chain (Sect. 19.3): A radical chain reaction is a three-phase reaction that characterizes many radical processes. (p. 974)

radical scavengers (Sect. 19.9): Radical scavengers are chemical compounds that react with, or scavenge, other radicals. (p. 1001)

radicals (Sect. 19.1): Radicals are chemical species containing one or more unpaired electrons. (p. 972)

rate-determining step (Sect. 8.3): The rate-determining step of a multi-step reaction mechanism is the step with the highest activation energy. Because it is the slowest step, it controls the overall rate of the reaction. (p. 335)

reaction intermediate (Sect. 8.2): A reaction intermediate is a chemical species formed in one step of a chemical reaction but consumed in a subsequent step. (p. 332)

reductive elimination (Sect. 18.2): Reductive elimination is a reaction involving the removal of a metal atom from between two alkyl groups and the return of a pair of electrons to the metal. (p. 907)

Reformatsky reaction (Sect. 17.4): The Reformatsky reaction is a crossed aldol reaction between an α-haloester and an aldehyde, using a zinc-based nucleophile. (p. 832)

regioisomers (Sect. 8.3): Regioisomers are constitutional isomers that are formed from a chemical reaction. (p. 336)

regioselective (Sect. 8.3): Regioselective reactions favour the formation of one regioisomer over another. (p. 337)

relative stereochemistry (Sect. 8.6): Relative stereochemistry refers to spatial positions of two substituents *relative* to each other. The terms *cis* and *trans* refer to the relative stereochemistry of two substituents. (p. 361)

resonance (Sect. 1.8): Resonance is a *tool* used to describe the delocalization of π electrons in a molecule. (p. 26); (Sect. 5.5): Resonance is a method of describing the structure and properties of a molecule that cannot be drawn as a single Lewis structure. (p. 208)

resonance form (Sect. 1.8): A resonance form is one of a set of related Lewis structures used to describe bonding situations in which π electrons are not confined to individual atoms or bonds. Each resonance form provides a bonding description based upon a different location of the molecule's π electrons. (p. 27)

resonance frequency (Sect. 13.2): The resonance frequency of a signal is the frequency at which spin-flip occurs, and is proportional to the energy required. (p. 580)

retro-aldol reaction (Sect. 17.4): In a retro-aldol reaction, the reverse of the aldol reaction, an aldol fragments into an enolate and an aldehyde. (p. 828)

retrosynthesis (Sect. 10.9): Retrosynthesis is a technique of planning chemical synthesis in which a target molecule is analyzed in terms of what it can be made from. (p. 477)

ring inversion (Sect. 3.4): Ring inversion is the conversion of a cyclic molecule from one conformation to another by rotation about the single bonds of the ring. (p. 99)

Robinson annulation (Sect. 18.3): A Robinson annulation is a ring-forming process that involves a Michael addition, followed by a ring-closing aldol condensation. Often a Robinson annulation forms a cyclohexenone. (p. 914)

S

σ bond (Sect. 1.5): A σ bond (pronounced *sigma* bond) is a covalent bond in which the direct line through the nuclei presents the highest probability of finding the shared electrons. (p. 11)

σ molecular orbital (Sect. 1.9): A σ molecular orbital (σ MO) has its amplitude concentrated along the axis between the nuclei of the molecule. (p. 32)

s-cis **and** ***s-trans*** (Sect. 20.3): *s-cis* and *s-trans* refer to *cis/trans* isomerism around a single bond where rotation is restricted by conjugation. The *s* stands for sigma bond (σ bond). (p. 1020)

Sandmeyer reaction (Sect. 15.9): The Sandmeyer reaction uses copper salts to substitute Cl, I, or CN for an aryl diazonium salt. (p. 743)

saponification (Sect. 15.3): Saponification is the conversion of an ester to a carboxylate under basic conditions. (p. 706)

saturated (Sect. 2.2): A saturated molecule has no π bonds or rings and therefore has the maximum amount of hydrogen that its atoms can accommodate. (p. 49)

Schiemann reaction (Sect. 15.9): The Schiemann reaction uses diazonium salts and tetrafluoroboric acid to convert anilines into aryl fluorides. (p. 743)

scissile (Sect. 12.2): A scissile bond is a bond that is capable of being broken. Typically, the term is used to describe the bonds that are broken in a biological reaction. (p. 548)

secondary orbital overlap (Sect. 20.3): Secondary orbital overlap is a favourable interaction between a molecular orbital lobe of a dienophile substituent and an orbital lobe of the diene that stabilizes the *endo* transition state without forming a bond. One of the involved orbitals is filled and the other is empty. (p. 1025)

self-addition reaction (Sect. 17.4): A self-addition reaction involves the combination of two identical molecules—the addition of the nucleophilic form (enolate) to the electrophilic form (carbonyl). (p. 828)

shielding (Sect. 13.4): Shielding is the reduction in the magnetic field strength experienced by a nucleus due to the opposing magnetic fields of nearby electrons. (p. 588)

sigmatropic rearrangement (Sect. 20.4): A sigmatropic rearrangement is one in which one σ bond is replaced by another, keeping the total number of σ bonds the same. (p. 1040)

singly occupied molecular orbital (SOMO) (Sect. 20.2): A singly occupied molecular orbital (SOMO) is any MO that is occupied by a single electron. (p. 1016)

solute (Sect. 2.4): The solute is the material that is dissolved in the solvent. (p. 62)

solvated electrons (Sect. 19.9): Solvated electrons are individual electrons that are surrounded by a solvent. (p. 997)

solvent (Sect. 2.4): A solvent is a liquid medium in which compounds may be dissolved. It may be water or an organic substance. (p. 62)

specific rotation $[\alpha]$ (Sect. 4.10): The specific rotation ($[\alpha]$) is the angle that a sample rotates plane-polarized light after correction for sample size and concentration. (p. 165)

spectroscopy (Sect. 13.1): Spectroscopy is the measurement of the interaction between a molecule and electromagnetic radiation, and reveals information about molecular properties, including structure. (p. 578)

spin-flip (Sect. 13.2): Spin-flip is the transition of a nucleus between α and β states (low to high energy). (p. 580)

spin-spin coupling (Sect. 13.4): Spin-spin coupling is an interaction between the magnetic fields of adjacent nuclei. (p. 602)

stabilized ylides (Sect. 18.6): Stabilized ylides are ylides with an electron-withdrawing group adjacent to the negative charge, such as a carbonyl. (p. 934)

staggered conformation (Sect. 3.2): A staggered conformation has substituents on adjacent atoms as far away from each other as possible. (p. 90)

stereoisomers (Sect. 4.2): Stereoisomers have the same atoms connected in the same sequence, but they differ in the three-dimensional arrangement of those atoms. (p. 127)

stereoselective (Sect. 8.4): Stereoselective reactions are those that *favour* the formation of one stereoisomer, though the other stereoisomer may also be produced as a minor product. These reactions tend to involve stepwise mechanisms. (p. 347)

steric strain (Sect. 3.3): Steric strain is a repulsive force within a molecule that arises from the interpenetration of the electron clouds of atoms that are close to each other but not directly bonded. (p. 94)

substituent (Sect. 2.2): A substituent is a particular atom or group of atoms that replace a hydrogen atom in an organic molecule. (p. 49)

substrate (Sect. 11.2): A substrate is the main organic molecule of interest in a chemical reaction. (p. 496)

sulfonation (Sect. 10.4): Sulfonation is the substitution of an SO_3H group in place of a hydrogen atom on an aromatic ring. (p. 438)

suprafacial (Sect. 20.4): A rearrangement is suprafacial when the migrating group stays on the same face of the molecule as it moves. (p. 1045)

symmetric (Sect. 20.2): Within the context of conjugated π-systems, symmetric MOs have the same phase at their two terminal lobes. (p. 1014)

***syn* addition** (Sect. 8.7): In *syn* addition, the two new atoms or groups are added to the same face of an alkene. (p. 368)

***syn*-aldol product** (Sect. 18.5): A *syn*-aldol product is an aldol in which the substituent on the α-carbon and the hydroxy at the β-position are *syn* with respect to the extended carbon chain. (p. 924)

synthesis (Sect. 10.9): Synthesis is the process of making complex molecules from simpler ones. (p. 476)

synthon (Sect. 10.9): A synthon is a fragment resulting from a disconnection that shows the general reactivity (nucleophile/electrophile) of the fragment. (p. 478)

T

tandem reaction (Sect. 18.2): A tandem reaction is a set of reactions, performed in sequence. The product of one reaction acts as the starting material in the next. (p. 907)

tautomerization (Sect. 17.2): Tautomerization is the conversion of a compound into a structural isomer that differs by the location of one atom. (p. 813)

termination (Sect. 19.3): Termination refers to the third and final phase of the radical chain reaction, which involves the combination of radicals and leads to the destruction of radicals. (p. 974)

tetramethylsilane (TMS) (Sect. 13.3): Tetramethylsilane (TMS) is the compound used to provide the reference frequency for ^{1}H-NMR spectra. (p. 582)

thermodynamic control (Sect. 18.2): A reaction under thermodynamic control uses reaction conditions, such as increased temperature, that promote an equilibrium that leads to the most stable product. (p. 905)

thermodynamic enolate (Sect. 18.4): A thermodynamic enolate forms under conditions of thermodynamic control (reversible), and the hydrogen is removed from the more substituted α-carbon. (p. 919)

thermodynamic product (Sect. 8.7): The thermodynamic product is the product that is the most stable. (p. 376)

thiocarbonyl (Sect. 7.2): A thiocarbonyl is a functional group that consists of a carbon connected to a sulfur by a double bond. (p. 276)

Thorpe reaction (Sect. 17.6): The Thorpe reaction is a reaction that uses nitriles instead of aldehydes or ketones. (p. 851)

Thorpe–Ziegler reaction (Sect. 17.6): The Thorpe–Ziegler reaction is an intramolecular Thorpe condensation of nitriles. (p. 851)

torsion (Sect. 3.2): The torsion or dihedral angle is the angle between the bonds to a substituent on each of the atoms of the single bond that constitutes the axis of rotation. (p. 89)

torsional strain (Sect. 3.2): Torsional strain is a strain within a molecule that arises from repulsions between electrons in bonds on adjacent atoms. (p. 89)

***trans* isomer** (Sect. 3.7): The *trans* isomer of a cycloalkane has one substituent bonded to each face of the ring. (p. 112)

transesterification (Sect. 15.3): Transesterification is the conversion of one ester to another by heating it with an acid catalyst in an alcohol solvent. (p. 713)

U

umpolung reaction (Sect. 18.7): An umpolung reaction involves changing the reactivity of an atom from nucleophilic to electrophilic, or vice versa. *Umpolung* is German for "reversal of polarity." (p. 937)

unsaturated (Sect. 2.2): An unsaturated molecule has at least one π bond or a ring to which hydrogen atoms may be added. (p. 50)

upfield (Sect. 13.4): Upfield signals are toward the right of an NMR spectrum. (p. 588)

V

valence electrons (Sect. 1.2): Valence electrons occupy valence orbitals. (p. 5)

valence orbitals (Sect. 1.2): An atom's valence orbitals are the occupied orbitals of highest energy (and any accompanying empty orbitals of similar high energy). (p. 5)

vinyl halides (Sect. 8.7): Vinyl halides are compounds that contain a halogen bonded directly to the carbon atom of a double bond. (p. 380)

W

wavefunction (Sect. 1.2): A wavefunction is a mathematical description of a particular quantum state of an electron or other particle. (p. 3)

wavenumber (Sect. 14.6): Wavenumber ($1/\lambda$) is a convenient measure of light energy because it is directly proportional to a photon's energy. It is usually expressed in units of cm^{-1}. (p. 662)

Weinreb amide (Sect. 15.8): Weinreb amide is an amide with a coordinating group that blocks a second addition of Grignard reagents; it is often used to synthesize ketones. (p. 734)

Williamson ether synthesis (Sect. 11.8): The Williamson ether synthesis is a way of making ethers from alkoxides and alkyl halides using an S_N2 displacement. (p. 526)

Wittig reaction (Sect. 17.6): The Wittig reaction makes alkenes out of aldehydes with predominantly *cis* selectivity, using a phosphonium ylide. (p. 858)

Y

ylide (Sect. 12.2): An ylide is a group with adjacent positive and negative charges. (p. 546); (Sect. 17.6): An ylide is a neutral compound with positive and negative charges on adjacent atoms. (p. 858)

Z

Zaitsev's rule (Sect. 12.2): The modern interpretation of Zaitsev's rule states that elimination reactions favour the most stable alkene product. This is often the most substituted alkene product. (p. 543)

Zimmerman–Traxler transition state (Sect. 18.5): The Zimmerman–Traxler transition state is the closed transition state for aldol additions that have a coordinating counter ion. The six-membered ring chair explains the observed stereochemistry of the aldol product. (p. 930)

zwitterion (Sect. 20.3): A zwitterion is a neutral molecule that contains balancing, opposite charges. (p. 1034)

INDEX

Page numbers followed by t *or* f *refer to tables and figures on respective pages.*

B

C

D

E

F

G

N

O

P

Q

R

S

T

U